# IJCAI-97

Proceedings of the Fifteenth International
Joint Conference on Artificial Intelligence

**Nagoya, Japan**
**August 23–29, 1997**

Volume 1

Sponsored by the
International Joint Conferences on Artificial Intelligence, Inc. (IJCAII)
Japanese Society for Artificial Intelligence

Copyright © 1997 International Joint Conferences on Artificial Intelligence, Inc.
All rights reserved.

Edited by Martha E. Pollack

Distributed by
Morgan Kaufmann Publishers, Inc.
340 Pine Street, 6th Floor
San Francisco, CA 94104
http://www.mkp.com

Printed in Korea.

Design, composition, production, and manufacturing management by
Professional Book Center, Denver, Colorado.

# Brief Contents

## Volume 1

Ordering Information . . . . . . . . . . . . . . . . . . . . . . . . . . . . . . . . . . . . . . . . . . . . . . . . . . iv
Foreword . . . . . . . . . . . . . . . . . . . . . . . . . . . . . . . . . . . . . . . . . . . . . . . . . . . . . . . . . . . v
IJCAI-97 Conference Organization . . . . . . . . . . . . . . . . . . . . . . . . . . . . . . . . . . . . . vi
Local Arrangements Committee . . . . . . . . . . . . . . . . . . . . . . . . . . . . . . . . . . . . . . . vi
Program Committee . . . . . . . . . . . . . . . . . . . . . . . . . . . . . . . . . . . . . . . . . . . . . . . vii
IJCAI Organization. . . . . . . . . . . . . . . . . . . . . . . . . . . . . . . . . . . . . . . . . . . . . . . . vii
Cooperating Societies and Organizations . . . . . . . . . . . . . . . . . . . . . . . . . . . . . . viii
Japanese National Committee Executive Board. . . . . . . . . . . . . . . . . . . . . . . . . . viii
JSAI Organization. . . . . . . . . . . . . . . . . . . . . . . . . . . . . . . . . . . . . . . . . . . . . . . . . ix
Corporate Sponsors . . . . . . . . . . . . . . . . . . . . . . . . . . . . . . . . . . . . . . . . . . . . . . . ix
IJCAI-97 Awards. . . . . . . . . . . . . . . . . . . . . . . . . . . . . . . . . . . . . . . . . . . . . . . . . . x
Panel . . . . . . . . . . . . . . . . . . . . . . . . . . . . . . . . . . . . . . . . . . . . . . . . . . . . . . . . . . . x
Invited Speakers . . . . . . . . . . . . . . . . . . . . . . . . . . . . . . . . . . . . . . . . . . . . . . . . . . xi
IJCAI-97 Reviewers . . . . . . . . . . . . . . . . . . . . . . . . . . . . . . . . . . . . . . . . . . . . . . . xii
Contents . . . . . . . . . . . . . . . . . . . . . . . . . . . . . . . . . . . . . . . . . . . . . . . . . . . . . . . xv
    AI Challenges . . . . . . . . . . . . . . . . . . . . . . . . . . . . . . . . . . . . . . . . . . . . . . . 1
    Automated Reasoning . . . . . . . . . . . . . . . . . . . . . . . . . . . . . . . . . . . . . . . . 59
    Case-Based Reasoning . . . . . . . . . . . . . . . . . . . . . . . . . . . . . . . . . . . . . . 223
    Cognitive Modeling. . . . . . . . . . . . . . . . . . . . . . . . . . . . . . . . . . . . . . . . . 265
    Computer-Aided Education . . . . . . . . . . . . . . . . . . . . . . . . . . . . . . . . . . 327
    Constraint Satisfaction . . . . . . . . . . . . . . . . . . . . . . . . . . . . . . . . . . . . . . 349
    Diagnosis and Qualitative Reasoning. . . . . . . . . . . . . . . . . . . . . . . . . . . . 431
    Distributed Artificial Intelligence. . . . . . . . . . . . . . . . . . . . . . . . . . . . . . . 575
    Expert Systems. . . . . . . . . . . . . . . . . . . . . . . . . . . . . . . . . . . . . . . . . . . . 653
    Game Playing . . . . . . . . . . . . . . . . . . . . . . . . . . . . . . . . . . . . . . . . . . . . 675
    Information Retrieval . . . . . . . . . . . . . . . . . . . . . . . . . . . . . . . . . . . . . . . 705
Author Index . . . . . . . . . . . . . . . . . . . . . . . . . . . . . . . . . . . . . . . . . . . . . . . . . . . 792

## Volume 2

Ordering Information . . . . . . . . . . . . . . . . . . . . . . . . . . . . . . . . . . . . . . . . . . . . . . . . . . iv
IJCAI-97 Conference Organization . . . . . . . . . . . . . . . . . . . . . . . . . . . . . . . . . . . . . v
Local Arrangements Committee . . . . . . . . . . . . . . . . . . . . . . . . . . . . . . . . . . . . . . . v
Program Committee . . . . . . . . . . . . . . . . . . . . . . . . . . . . . . . . . . . . . . . . . . . . . . . vi
IJCAI Organization. . . . . . . . . . . . . . . . . . . . . . . . . . . . . . . . . . . . . . . . . . . . . . . . vi
Contents . . . . . . . . . . . . . . . . . . . . . . . . . . . . . . . . . . . . . . . . . . . . . . . . . . . . . . . vii
    Learning . . . . . . . . . . . . . . . . . . . . . . . . . . . . . . . . . . . . . . . . . . . . . . . . 795
    Natural-language Processing and Graphical Presentation. . . . . . . . . . . . . . 949
    Neural Networks. . . . . . . . . . . . . . . . . . . . . . . . . . . . . . . . . . . . . . . . . 1063
    Planning and Scheduling . . . . . . . . . . . . . . . . . . . . . . . . . . . . . . . . . . . 1153
    Probabilistic Reasoning . . . . . . . . . . . . . . . . . . . . . . . . . . . . . . . . . . . . 1273
    Robotics. . . . . . . . . . . . . . . . . . . . . . . . . . . . . . . . . . . . . . . . . . . . . . . 1337
    Search . . . . . . . . . . . . . . . . . . . . . . . . . . . . . . . . . . . . . . . . . . . . . . . . 1379
    Temporal Reasoning . . . . . . . . . . . . . . . . . . . . . . . . . . . . . . . . . . . . . . 1409
    Vision . . . . . . . . . . . . . . . . . . . . . . . . . . . . . . . . . . . . . . . . . . . . . . . . 1473
    Panel. . . . . . . . . . . . . . . . . . . . . . . . . . . . . . . . . . . . . . . . . . . . . . . . . 1509
    Videos. . . . . . . . . . . . . . . . . . . . . . . . . . . . . . . . . . . . . . . . . . . . . . . . 1523
    Doctoral Consortium Abstracts . . . . . . . . . . . . . . . . . . . . . . . . . . . . . . . 1537
    Invited Speakers. . . . . . . . . . . . . . . . . . . . . . . . . . . . . . . . . . . . . . . . . 1547
    Awards . . . . . . . . . . . . . . . . . . . . . . . . . . . . . . . . . . . . . . . . . . . . . . . 1649
Author Index . . . . . . . . . . . . . . . . . . . . . . . . . . . . . . . . . . . . . . . . . . . . . . . . . . 1653

# Ordering Information

The following is a list of proceedings of IJCAI conferences available from Morgan Kaufmann Publishers. To place an order or receive information regarding these and other Morgan Kaufmann publications, please use the following address:

**Morgan Kaufmann Publishers, Inc., 340 Pine Street, 6th Floor, San Francisco, California 94104 USA**
**Telephone: 1 800 745-7323; FAX: 1 415 982-2665; email: orders@mkp.com; http://www.mkp.com**

To order, please send a US$ check or money order, or provide credit card debit authorization for Visa, MasterCard, or American Express (include name on card, card number, expiry date, etc.) Shipping and Handling: US—for surface post, include $4.50 for the first volume and $2.50 for each additional volume; for international surface post to locations outside the US, include $6.50 for the first volume and $4.50 for each additional volume (allow 6–8 weeks for delivery); for international airmail: $30/volume (allow 1–2 weeks for delivery). California residents, add appropriate tax. Be sure to provide your full shipping address.

*Lower price listed below available to IJCAI conference registrants and members of national/regional AI societies only.*

IJCAI-97
Nagoya, Japan
2 volumes; ISBN 1-55860-480-4
$75/$56.25

IJCAI-95
Montréal, Québec
2 volumes; ISBN 1-55860-363-8
$75/$56.25

IJCAI-93
Chambéry, France
2 volumes, ISBN 1-55860-300-X
$75/$56.25

IJCAI-91
Sydney, Australia
2 volumes, ISBN 1-55860-160-0
$75/$56.25

IJCAI-89
Detroit, Michigan
2 volumes, ISBN 1-55860-094-9
$75/$56.25

IJCAI-87
Milan, Italy
2 volumes; ISBN 0-934613-43-5
$65/$48.75

IJCAI-85
Los Angeles, California
2 volumes; ISBN 0-934613-02-8
$65/$48.75

IJCAI-83
Karlsruhe, West Germany
2 volumes; ISBN 0-934613-043-4
$65/$48.75

IJCAI-81
Vancouver, British Columbia
2 volumes; ISBN 0-934613-044-2
$65/$48.75

IJCAI-79
Tokyo, Japan
2 volumes; ISBN 0-934613-47-8
$65/$48.75

IJCAI-77
Cambridge, Massachusetts
2 volumes; ISBN 0-934613-48-6
$65/$48.75

IJCAI-75
Tbilisi, Georgia USSR
ISBN 0-934613-20-6
$65/$48.75

IJCAI-73
Stanford, California
ISBN 0-934613-58-3
$65/$48.75

IJCAI-71
London, England
ISBN 0-934613-34-6
$65/$48.75

IJCAI 14-Year Set
1971–1997
24 volumes; ISBN 1-55860-481-2
$885/$641.25

**IJCAI-89, 91, -93, and -95 VIDEOTAPE PROGRAMS:** Short video programs selected by the video committee give an excellent overview of current applications involving AI technology in robotics, vision, language, planning, modeling, design, control, computer-aided instruction, expert systems, and diagnosis. The tapes are approximately 2 hours in running time (1995 videa: approx. 4 hours). NTSC-VHS or PAL-VHS, 1/2-inch formats available. Shipping in US/Canada: $3.50/first tape, $1.00/each additional tape; Canadian and international customers, please inquire.

**1995 Video:** ISBN 1-55860-388-3; $66.50/$49.88        **1993 Video:** ISBN 1-55860-309-3; $66.50/$49.88
**1991 Video:** ISBN 1-55860-183-X; $66.50/$49.88        **1989 Video:** ISBN 1-55860-047-3; $66.50/$49.88

European and Australian/New Zealand customers preferring to order locally may do so. Contact the following Morgan Kaufmann offices. (Prices may vary from above; please inquire as to pricing, availability, and shipping requirements). **UK/EUROPE:** Morgan Kaufmann-UK, Afterhurst, Ltd., 27 Church Road, Hove, East Sussex BN3 2FA, UK; Tel.: 44 1 273 748427; FAX: 44 1 273 722180; email: mkp.europe@erlbaum.co.uk. **AUSTRALASIA:** Morgan Kaufmann-Australia: Thomas Nelson Australia, 102 Dodds Street, South Melbourne, Victoria 3205, Australia; Tel.: 61 3 685411; Toll Free: 1 800 654831; FAX: 1 800 641 823; email: customerservice@nelson.com.au. For other international representatives, contact Morgan Kaufmann USA.

# Foreword

An International Joint Conference on Artificial Intelligence has been held every other year since 1969. The papers in this proceedings demonstrate that in 1997, AI remains a vital and important field of inquiry, and IJCAI remains the premiere conference for presenting the highest quality work in AI.

In 1997, 882 papers from 45 countries were submitted. Each paper received at least three reviews, overseen and coordinated by the members of the Program Committee. All reviewing was anonymous: reviewers were not told the identities of the authors of the papers they read. In many cases, after writing initial assessments, reviewers were put in touch with one another to discuss a paper further by e-mail. Additional face-to-face discussions took place among the Program Committee members at a meeting in Pittsburgh, PA, USA, on March 22–23. In the end, 216 papers were accepted for full presentation and inclusion in this proceedings: an acceptance rate of just under 25%. We also selected three Distinguished Papers, which will be presented at a special session at IJCAI-97.

Two innovations in the technical program are worth mentioning here:

- The program this year includes a set of nine "Challenge Papers," which issue specific and highly focused technical challenges relating to artificial intelligence. Each challenge paper is expected to instigate a narrowly focused program of research culminating in papers submitted to IJCAI-99 that address the technical challenge and describe progress made on it.

- We are experimenting this year with poster sessions that will supplement the traditional IJCAI talks. The Program Committee identified a number of papers that presented exciting and promising work but that we could not include in these proceedings, and we invited the authors of those papers to present their work at IJCAI-97 poster sessions. Short abstracts of work presented in the poster sessions will be distributed to all conference participants.

There will be a wealth of intellectual activities at the conference. In addition to the contributed papers and poster sessions, there will be invited talks by leading AI researchers from around the world, conference reports, video presentations, a panel, tutorials, workshops, exhibits, a robot-soccer competition, a computer-Go championship, and evening lectures by the IJCAI award winners. There will also be a consortium for doctoral students, whose participants have been invited to include brief abstracts of their dissertations in these proceedings.

Organizing the program for a conference of IJCAI's scope would be unthinkable without the help of many people. First and foremost I would like to thank the members of the Program Committee—I could not have asked for a better group of people to work with. All the members were thorough, responsive, and responsible in fulfilling their duties. Similar dedication was shown the program subchairs: Mark Drummond, workshop chair; Bernhard Nebel, tutorial chair; and Toyoaki Nishida, video chair. They were ably assisted by their Local Arrangements Committee liaisons: Hiroaki Kitano (workshops) and Ryohei Nakano (tutorials). Tom Dean oversaw the Challenge Paper track and Loren Terveen organized the Doctoral Consortium. Chris Welty and Hiroshi Okuno, the IJCAI-97 "webmeisters," were always ready and willing to reply to my seemingly endless requests to add information to the IJCAI home page. Carol Hamilton and her staff at the AAAI office—Rick Skalsky, Annette Eldredge, and Annette Boken—provided invaluable administrative assistance. The allocation of papers to reviewers was supported by software developed by Ramesh Patil, and I thank him greatly for making his programs available to us. Finally, I would like to thank my secretary, Keena Walker, for handling many, many tasks, without letting any drop, and three University of Pittsburgh graduate students—Colleen McCarthy, Atif Memon, and Ioannis Tsamardinos—for giving up a lovely March weekend to assist at the Program Committee meeting.

*Martha E. Pollack*
*Pittsburgh, PA, USA*

# IJCAI-97 Conference Organization

## Conference Committee

**President, Board of Trustees**
C. Raymond Perrault
SRI International
Artificial Intelligence Center
333 Ravenswood Avenue, EJ 247
Menlo Park, CA 94025 USA

**Conference Chair**
Michael P. Georgeff
Australian AI Institute
171 Latrobe Street, Level 6
Melbourne, Victoria 3000 Australia

**Secretary-Treasurer**
Ronald J. Brachman
AT&T Labs–Research
180 Park Avenue
P.O. Box 971
Florham Park, NJ 07932-0971 USA

**National Committee Chair**
Koichi Furukawa
Keio University
Faculty of Environmental Information
5322 Endo, Fujisawa-shi
Kanagawa, 252 Japan

**Program Chair**
Martha E. Pollack
University of Pittsburgh
Department of Computer Science
Pittsburgh, PA 15260 USA

## Advisory Committee

Mike Brady, Oxford University (England)
Randall Davis, MIT AI Lab (USA)
Rina Dechter, University of California, Irvine (USA)
Danail Dochev, Bulgarian Acadamy of Sciences (Bulgaria)
R. G. (Randy) Goebel, University of Alberta (Canada)
Carl Gustav Jansson, Royal Institute of Technology/University of Stockholm (Sweden)
Sarit Kraus, Bar-Ilan University (Israel)
Tom Mitchell, Carnegie Mellon University (USA)
Stuart Russell, University of California, Berkeley (USA)
Zhongzhi Shi, Chinese Academy of Sciences (China)
Mario Tokoro, Sony Computer Science Laboratory, Inc. (Japan)

# Local Arrangements Committee

**Chair:** Koichi Furukawa, Keio University
**Vice-Chair:** Mitsuru Ishizuka, The University of Tokyo
**Vice-Chair:** Hiroaki Kitano, Sony Computer Science Laboratory, Inc.
**Finance Chair:** Masato Yamazaki, Yokkaichi University
**Exhibition Chair:** Motoi Suwa, Electrotechnical Laboratory
**Publicity Chair:** Hiroshi Okuno, NTT Basic Research Laboratories
**Site Management Chair:** Hidenori Itoh, Nagoya Institute of Technology
**Workshop Liaison:** Hiroaki Kitano, Sony Computer Science Laboratory, Inc.
**Tutorial Liaison:** Ryohei Nakano, NTT Communication Science Laboratories

# Program Committee

**Program Commitee Chair:** Martha E. Pollack, University of Pittsburgh (USA)
Richard Alterman, Brandeis University (USA); Rachel Ben-Eliyahu, Ben Gurion University (Israel);
Christian Bessière, LIRM Montpellier (France); Susanne Biundo, German Research Center for AI [DFKI] (Germany);
Robert Bolles, SRI International (USA); Craig Boutilier, University of British Columbia (Canada);
Robert Dale, Microsoft Research Institute and Macquarie University (Australia); Tom Dean, Brown University (USA);
Ed Durfee, University of Michigan (USA); Boi Faltings, Swiss Federal Institute of Technology (Switzerland);
Eva Hajicova, Charles University (Czech Republic); Toru Ishida; Kyoto University (Japan);
Yumi Iwasaki, Stanford University (USA); Richard Korf, University of California, Los Angeles (USA);
Yasuo Kuniyoshi, Electrotechnical Laboratory (Japan); Maurizio Lenzerini, University of Rome "La Sapienza" (Italy);
Vladimir Lifschitz, University of Texas (USA); Sridhar Mahadevan, University of South Florida (USA);
Katharina Morik, University of Dortmund (Germany); Juzar Motiwalla, National University of Singapore (Singapore);
Michael Mozer, University of Colorado (USA); Louise Pryor, Harlequin, Ltd. (UK);
Daniela Rus, Dartmouth College (USA); Lorenza Saitta, University of Torino (Italy);
Tetsuo Tomiyama, The University of Tokyo (Japan); Frank van Harmelen, Vrije University (The Netherlands)
**Tutorials Chair:** Bernhard Nebel, Albert-Ludwigs-Universität Freiburg (Germany)
**Workshops Chair:** Mark Drummond, Electric Time Software (USA)
**Video Track Chair:** Toyoaki Nishida, Nara Institute of Science and Technology (Japan)

# IJCAI Organization

## Trustees

**Chair:** C. Raymond Perrault, SRI International (USA)
Michael P. Georgeff, Australian AI Institute (Australia)
Luigia Carlucci Aiello, University of Rome "La Sapienza" (Italy)
Wolfgang Wahlster, German Research Center for AI [DFKI] (Germany)
Barbara J. Grosz, Harvard University (USA)
Martha E. Pollack, University of Pittsburgh (USA)
Thomas Dean, Brown University (USA)

## Secretariat

**Secretary-Treasurer:** Ronald J. Brachman, AT&T Bell Labs–Research (USA)
Priscilla Rasmussen, Academic and Research Conference Services (USA)
Chris Welty, Electronic Information Officer, Vassar College (USA)

## Former Trustees

Wolfgang Bibel, Technical University Darmstadt (Germany)
Alan Bundy, University of Edinburgh (Scotland)
Alan Mackworth, University of British Columbia (Canada)
Saul Amarel, Rutgers University (USA)
Patrick J. Hayes, University of Illinois (USA)
D. Raj Reddy, Carnegie Mellon University (USA)
Erik Sandewall, Linköping University (Sweden)
Alistair D. C. Holden, University of Washington (USA)
Max B. Clowes (deceased), formerly University of Sussex (England)
Donald E. Walker (deceased), formerly Bellcore (USA)
Woodrow W. Bledsoe (deceased), formerly University of Texas, Austin (USA)

# Cooperating Societies and Organizations

IJCAI-97 acknowledges with thanks the cooperation of the following societies and organizations:

The Association for Natural Language Processing (NLP)
The Information Processing Society of Japan (IPSJ)
The Institute of Electronics, Information and Communication Engineers (IEICE)
The Institute of Electrical Engineers of Japan (IEEJ)
The Institute of Image Information and Television Engineers (ITE)
The Japan Information Processing Development Center (JIPDEC)
The Japan Society for Fuzzy Theory and Systems (JSFTS)
The Japan Society for Precision Engineering (JSPE)
The Japan Society for Software Science and Technology (JSSST)
The Japan Society of Mechanical Engineers (JSME)
The Japanese Cognitive Science Society (JCSS)
The Japanese Neural Network Society (JNNS)
Robotics Society of Japan (RSJ)
The Society of Instrument and Control Engineers (SICE)

# Japanese National Committee Executive Board

**Japanese National Chair:** Setsuo Ohsuga, Waseda University
**Japanese National Vice-Chair:** Teruo Fukumura, Chukyo University
**Japanese National Vice-Chair:** Saburo Tsuji, Wakayama University
**Japanese National Vice-Chair:** Koichi Furukawa, Keio University
**Finance Chair:** Masato Yamazaki, Yokkaichi University
**Exhibition Chair:** Motoi Suwa, Electrotechnical Laboratory
**Fundraising:** Shigeru Sato, Fujitsu Laboratories, Ltd.
**Local Arrangements Vice-Chair:** Mitsuru Ishizuka, The University of Tokyo
**Local Arrangements Vice-Chair:** Hiroaki Kitano, Sony Computer Science Laboratory, Inc.
**Industrial Liaison:** Yoshiaki Shirai, Osaka University
**General Affairs:** Mario Tokoro, Keio University
**Publicity Chair:** Hiroshi Okuno, NTT Basic Research Laboratories
**Video Track Chair:** Toyoaki Nishida, Nara Institute of Science and Technology
**JSAI Chair:** Hidehiko Tanaka, The University of Tokyo
**JSAI Vice Chair:** Ken-ichi Mori, Toshiba Corporation
**JSAI Past Chair:** Shuji Doshita, Kyoto University
**JSAI Past Chair:** Masamichi Shimura, Science University of Tokyo
**Local Arrangements:** Ken Satoh, Hokkaido University
**Accounting/PCO Management:** Kenji Kimura, Central Japan Industrial Association
**Observer, Liaison to JSAI:** Yuichi Ohta, Tsukuba University

# JSAI Organization

**Chair:** Hidehiko Tanaka, The University of Tokyo
**Vice-Chair:** Ken-ichi Mori, Toshiba Corporation
**Vice-Chair:** Katsuhiro Shirai, Waseda University

# Corporate Sponsors

IJCAI-97 gratefully acknowledges the generous contribution of the following corporations and organizations:

AdIn Research, Inc.
Canon Inc.
FANUC Ltd.
FUJITSU FACOM INFORMATION PROCESSING CORPORATION
FUJITSU LIMITED
Fuji Xerox Co., Ltd.
HEWLETT.PACKARD LABORATORIES JAPAN, INC.
Hitachi Ltd.
IBM Japan
JUSTSYSTEM Corporation
Kokusai Denshin Denwa Co., Ltd.
Matsushita Electric Industrial Co., Ltd.
Mitsubishi Electric Corporation
Mitsubishi Research Institute, Inc.
NEC Corporation
Nihon Cisco Systems K.K.
Nihon Sun Microsystems K.K.
Nihon Unisys, Ltd.
Nippon Telegraph and Telephone Corporation
NTT Advanced Technology Corporation
NTT Intelligent Technology Co., Ltd.
NTT Software Corporation
NTT DATA CORPORATION
Ohmsha Ltd.
Oki Electric Industry Co., Ltd.
Sharp Corporation
Sony Corporation
Tokyo Gas Co., Ltd.
TOSHIBA CORPORATION
TOYOTA CENTRAL R&D LABS., INC.
TOYOTA MOTOR CORPORATION
Yokogawa Electric Corporation

The Inamori Foundation
The Telecommunications Advancement Foundation
Support Center for Advanced Telecommunications Technology Research

*This publication was partly supported by the "Grant-in-Aid for Publication of Scientific Research Result" of The Ministry of Education, Science, Sports and Culture, Japan.*

# IJCAI-97 Awards

## The IJCAI-97 Computers and Thought Award

Leslie P. Kaelbling, Brown University

Computers and Thought Lecture: *Why Robbie Can't Learn: The Difficulty of Learning in Autonomous Agents.* In recent years, machine-learning methods have enjoyed great success in a variety of applications. Unfortunately, online learning in autonomous agents has not generally been one of them. Reinforcement-learning methods that were developed to address problems of learning agents have been most successful in offline applications. In this talk, I will briefly review the basic methods of reinforcement learning, point out some of their shortcomings, argue that we are expecting too much from such methods, and speculate about how to build complex, adaptive autonomous agents. (Not submitted for publication in this proceedings.)

## The IJCAI-97 Award for Research Excellence

Aravind K. Joshi, University of Pennsylvania

Research Excellence Lecture: *Relationship Between Natural Language Processing and AI.* The use of constrained formal/computational systems just adequate for modeling various aspects of language—syntax, semantics, pragmatics and discourse, among others—has proved to be not only an effective research strategy but has led to deeper understanding of these aspects, with implications to both machine processing as well as human processing. This approach enables one to distinguish between universal and stipulative constraints. The other approach is to start with the most general and most powerful formal/computational system and use it to model these phenomena, thus making all constraints stipulative, in a sense. The tension between these approaches, together with the increasing use of empirical methods combining structural and statistical information, has made the relationship between natural language processing and AI quite stormy. I will review some of the past and current research following the first approach and also suggest how the stormy relationship could be improved.

## IJCAI-97 Distinguished Papers

Fangzhen Lin, The Hong Kong University of Science and Technology
*Applications of the Situation Calculus To Formalizing Control and Strategic Information: The Prolog Cut Operator*

Jaime G. Carbonell, Yiming Yang, Robert E. Frederking, Ralf D. Brown, Yibing Geng, and Danny Lee,
Carnegie Mellon University
*Translingual Information Retrieval: A Comparative Evaluation*

Timothy Huang and Stuart Russell, University of California, Berkeley
*Object Identification in a Bayesian Context*

# Panel

*The Next Big Thing*
**Chair:** Munindar P. Singh, North Carolina State University
**Panelists:** Daniel G. Bobrow, Xerox Palo Alto Research Center
Michael N. Huhns, University of South Carolina
Margaret King, ISSCO, University of Geneva
Hiroaki Kitano, Sony Computer Science Laboratory, Inc.
Ray Reiter, University of Toronto

# Invited Speakers*

Wolfgang Bibel, Technical University Darmstadt (Germany)
*Let's plan it deductively!*

Margaret A. Boden, University of Sussex (UK)
*Creativity and Artificial Intelligence*

Cristiano Castelfranchi, National Research Council and University of Siena (Italy)
*Modeling Social Action for AI Agents*

Ernst D. Dickmanns, University of the German Army, Munich (Germany)
*Vehicles Capable of Dynamic Vision*

Masayuki Inaba, The University of Tokyo (Japan)
*Remote-Brained Robots*

Kathleen R. McKeown, Columbia University (USA)
*Generating Multimedia Briefings: Language Generation in a Coordinated Multimedia Environment*

Leora Morgenstern, IBM T. J. Watson Research Center (USA)
*Inheritance Comes of Age: Applying Nonmonotonic Techniques to Problems in Industry*

Hiroshi Motoda, Osaka University (Japan) and Kenichi Yoshida, Hitachi, Ltd. (Japan)
*Machine Learning Techniques to Make Computers Easier to Use*

Nicola Muscettola, P. Pandurang Nayak, and Brian C. Williams,
Recom Technologies, NASA Ames Research Center (USA)
and Barney Pell, Caelum Research Corporation, NASA Ames Research Center (USA)
*The New Millennium Remote Agent: To Boldly Go Where No AI System Has Gone Before*

Luc Steels, Sony Computer Science Laboratory (France) and VUB AI Laboratory (Belgium)
*The Origins of Syntax in Visually Grounded Robotic Agents*

Richard S. Sutton, University of Massachusetts, Amherst (USA)
*Reinforcement Learning: Lessons for AI*

Pascal Van Hentenryck, Brown University (USA)
*Numerica: a Modeling Language for Global Optimization*

---

★ Not all invited speakers elected to submit a paper for publication in the conference proceedings.

# IJCAI-97 Reviewers

Agnar Aamodt
Naoki Abe
Norihiro Abe
Beth Adelson
Pieter Adriaans
David Aha
Hans Akkermans
Rachid Alami
Dean Allemang
Klaus Dieter Althoff
Elisabeth André
P. Annandan
Ronald Arkin
Alessandro Artale
Minoru Asada
Hajime Asama
Kevin Ashley
Paolo Avesani
Nikolaos Avouris
Ruth Aylett
Franz Baader
Fahiem Bacchus
Christer Bäckström
Amitava Bagchi
Michael Bain
Leemon Baird
Ruzena Bajcsy
Michael Ballantyne
Shumeet Baluja
Chitta Baral
J. E. Barrett
John Bateman
Mathias Bauer
Peter Baumgartner
Roberto Bayardo
Michael Beetz
Siegfried Bell
Massimo Benerecetti
Salem Benferhat
Belaid Benhamov
Brandon Bennett
Francesco Bergadano
Ralph Bergmann
Luc Berthouze
Leopoldo Bertossi
Philippe Besnard
Wolfgang Bibel
Larry Birnbaum
Gautam Biswas
Jim Blythe
Daniel Bobrow

Igor Boguslavskij
Karl Bohringer
Christian Boitet
Kalina Bontcheva
Alex Borgida
Terrance Boult
Richard Boulton
Paolo Bouquet
Justin Boyan
Xavier Boyen
Bob Boyer
Ronen Brafman
Giorgio Brajnik
Bert Bredeweg
Chris Brew
Gerhard Brewka
Amy Briggs
Eric Brill
Peter Brockhausen
Jo Brook
Rodney Brooks
Chris Brown
Alan Bundy
Sasa Buvac
Thomas By
Marco Cadoli
Rui Camacho
Bob Carpenter
Claudio Carpineto
David Carter
Rich Caruana
Cristiano Castelfranchi
Thierry Castell
Amedeo Cesta
P. P. Chakrabarti
Rama Chellappa
Steve Chien
Assef Chmeiss
Howie Choset
Berthe Choueiry
Alessandro Cimatti
Alex Coddington
William Cohen
David Cohn
Tony Cohn
Bob Collier
Mike Collins
Chris I. Connolly
Luca Console
Greg Cooper
Ernesto Costa

Gary Cottrell
Cregg Cowan
Matt Crocker
Padraig Cunningham
Roger Dannenberg
Adnan Darwiche
Peter Dayan
Giuseppe DeGiacomo
Luc De Raedt
Virginia de Sa
Rina Dechter
Keith Decker
Dennis DeCoste
Kenneth DeJong
Alvaro del Val
Louise Dennis
Barbara Di Eugenio
Thomas Dietterich
John Dillenburg
Patrick Doherty
Pedro Domingos
Bruce Donald
Francesco Donini
Marco Dorigo
Juergen Dorn
Jon Doyle
Richard Doyle
Brian Drabble
Thomas Drakengren
Denise Draper
Didier Dubois
Daniel Dvorak
Saso Džeroski
Thomas Eiter
Renee Elio
Charles Elkan
Werner Emde
Elisabet Engdahl
J. Engelfriet
Joeri Engelfriet
Eithan Ephrati
Mike Erdmann
Kutluhan Erol
Kave Eshghi
Rino Falcone
Luis Fariñas del Cerro
Art Farley
Usama Fayyad
Dieter Fensel
Jacques Ferber
George Ferguson

Innes A. Ferguson
Jim Firby
Martin A. Fischler
Peter Flach
Nick Flann
Ken Forbus
Maria Fox
Enrico Franconi
Nir Friedman
Alan M. Frisch
Daniel Frost
Pascal Fua
Alex Fukunaga
Ulrich Furbach
Claire Gardent
Olivier Gascuel
Les Gasser
Ralph Gasser
Erann Gat
Hector Geffner
Michael Gelfond
Rochard Genisson
Ian Gent
Konstantinos Georgatos
Abigail Gertner
Zoubin Ghahramani
Malik Ghallab
Chiara Ghidini
Enrico Giunchiglia
Fausto Giunchiglia
Robert Givan
Natalie Glance
Piotr Gmytrasiewicz
Randy Goebel
Ashok Goel
Robert Goldman
Moises Goldszmidt
Ravi Goplanan
Diana Gordon
Geoff Gordon
Georg Gottlob
Nick Gotts
Jonathan Gratch
Russell Greiner
Benjamin Grosof
Adam Grove
Peter Grunwald
Rod Grupen
Hans Werner Guesgen
Vineet Gupta
Vu Ha

Peter Haddawy
Greg Hager
Walther von Hahn
Petr Hajek
Jan Hajic
Larry Hall
Babak Hamidzaheh
Steve Hanks
Stephen J. Hanson
Fumio Hara
Alois Haselböck
Fumio Hattori
Dale Hawkins
Satoru Hayamizu
Peter Hellwig
Roger Hersch
Dave Hershberger
Andreas Herzig
Tetsuya Higuchi
Larry Hines
Erhard Hinrichs
Thomas Hinrichs
Shigeoki Hirai
Kazuo Hiraki
Janet Hitzeman
Sepp Hochreiter
Wolfgang Höppner
Achim Hoffman
Rob Holte
Se June Hong
Masahiro Hori
Ian Horswill
Tsutomu Hoshino
Eduard Hovy
David Hsu
Michael Huhns
Takashi Ikegami
Masayuki Inaba
Nitin Indurkhya
Felix Ingrand
Katsumi Inoue
Amar Isli
Liliana Ironi
Yoshiteru Ishida
Hiroshi Ishiguro
Masatoshi Ishikawa
Takashi Ishikawa
Mitsuru Ishizuka
Mateja Jamnik
Michael Jampel
Carl Gustaf Jansson
Ray Jarvis
Philippe Jegou
Fredrick H. Jelinek
Nick Jennings

Thorsten Joachims
Peter Jonsson
Aravind Joshi
Leo Joskowicz
Froduald Kabanza
Hermann Kaindl
Sing Bing Kang
Ravi Kapadia
Ronald Kaplan
Deepak Kapur
Jussi Karlgren
Lars Karlsson
Simon Kasif
Lydia Kavraki
Mark Keane
Gerard A. M. Kempen
Didier Keymeulen
Joerg-Uwe Kietz
Margaret King
John Kingston
Takashi Kiriyama
David Kirsh
Masaharu Kitamura
Yoshinobu Kitamura
Mark Klein
Volker Klingspor
Jana Koehler
Sven König
John Kolen
Daphne Koller
Nobo Komagata
Kurt G. Konolige
Igor Kononenko
Daniel Korn
Yoshiyuki Koseki
Keith Kotay
Teruo Koyama
Alex Kozlov
Vladik Kreinovich
Amruth Kumar
Kenneth Kunen
Yasuo Kuniyoshi
Yoshinori Kuno
Kazuhiro Kuwabara
Philippe Laborie
Simon Lacroix
Peter Ladkin
Christopher Lain
Gerhard Lakemeyer
Kevin Lang
Edward Large
Javier Larrosa
Alex Lascarides
Tung Leng Lau
Steve LaValle

Nada Lavrac
Claude Le Pape
David Leake
Hing Yan Lee
Shie-Jue Lee
Hector Levesque
Alon Levy
Fangzhen Lin
Shieu-Hong Lin
Guido Lindner
Michael Littman
Yunhui Liu
Jorge Lobo
Luca Locchi
Uve Loerch
Tony Loeser
John D. Lowrance
Peter Lucas
Witold Lukaszewicz
Monika Lundell
Quang-Tuan Luong
Kevin Lynch
Chris Manning
Giovanni Manzini
Victor Marek
Shaul Markovitch
James Martin
Fabio Massacci
Maja Mataric
Hitoshi Matsubara
Toshihiro Matsui
Takashi Matsuyama
Stan Matwin
David McAllester
Norman McCain
Andrew McCallum
James McDowell
Sheila McIlraith
Gerard Medioni
Chris Meek
Pedro Meseguer
Ryszard Michalski
Francois Michaud
Risto Miikkulainen
Andrei Mikheev
Silvia Miksch
Alfredo Milani
Rob Miller
Robert Milne
Grigori Mints
Manavendra Misra
Jun Miura
Kazuo Miyashita
Riichiro Mizoguchi
Perry Moerland

Raymond Mooney
Andrew Moore
Johanna Moore
Leora Morgenstern
Yoram Moses
Pieter Mosterman
Hiroshi Motoda
Joerg Müller
Prasanna Mulgaonkar
Jean-Pierre Muller
Paul Munro
Robin Murphy
David Musliner
Makoto Nagao
Yuichi Nakamura
Hideyuki Nakashima
Gholemreza Nakhaeizadeh
Daniele Nardi
Bernhard Nebel
Wolfgang Nejdl
Peter Nelson
Randal Nelson
Filippo Neri
Eric Neufeld
Ilkka Niemela
Steven Norton
Gordon Novak
Werner Nutt
Daniel O'Leary
Angelo Oddi
Hans Juergen Ohlbach
Yuichi Ohta
Seishi Okamoto
Eugenio Oliveira
Patrick Olivier
Massimo Omologo
Ei-Ichi Osawa
Joseph O'Sullivan
Jens Otten
Stefanos Padelis
David Page
Dinesh Pai
Cyril Pain-Barre
Luigi Palopoli
Michael Papazoglou
Young-Tack Park
Lynne Parker
Ron Parr
Peter Patel-Schneider
Dan Patterson
Barney Pell
Jeff Pelletier
Joseph Pemberton
Anna Perini
Wim Peters

Avi Pfeffer
Rolf Pfeifer
Peter Pirolli
Fiora Pirri
Aske Plaat
Enric Plaza
Enric Plaza i Cervera
Polly Pook
David Poole
Henri Prade
Alun Preece
Chris Price
Armand Prieditis
Patrick Prosser
Halina Przymusinska
Teodor Przymusinski
Pearl Pu
Jean-Francois Puget
Arcot Rajasekar
T. Rajkumar
Ashwin Ram
Anand Rao
Christopher Rasmussen
Thomas Rath
Gopalan (Ravi)
   Ravichandran
Jean-Charles Règin
Alexander Reinefeld
Luis Paulo Reis
Ehud Reiter
Ray Reiter
Francesco Ricci
Barry Richards
Jeff Rickel
Anke Rieger
Vincent Risch
Irina Rish
Robert Rist
M. Risto
Graeme Ritchie
Robert Rodosek
Nico Roos
Riccardo Rosati
Jeff Rosenschein
Francesca Rossi
William Rounds
Marie-Christine Rousset
Henry Rowley
Eytan Ruppin
Bikash Sabata
Philip Sabes
Anna Sagvall-Hein

Lakhdar Saïs
Yasuki Saito
Shigeyuki Sakane
Rafal Salvstowicz
Eduardo Sanchez
Thomas Sandholm
Giulio Sandini
Abdul Sattar
I. Saxena
Jonathan Schaeffer
Andrea Schaerf
Marco Schaerf
Torsten Schaub
Thomas Schiex
Juergen Schmidhuber
Stephan Schmitt
Arno Schönegge
Nia Schraudolf
Juergen Schröter
Eddie Schwalb
Camilla Schwind
Michèle Sebag
Roberto Sebastiani
Alberto Maria Segre
Bart Selman
Gianna Semeraro
Sandip Sen
Petr Sgall
Paul Shaw
Takanori Shibata
Hideo Shimazu
Solomon Eyal Shimony
Yoshiaki Shirai
Masahiko Shizawa
Jeff Shrager
Jaime Simao Sichman
Carles Sierra
Hans Ulrich Simon
Reid Simmons
Munindar Singh
John Slaney
Barbara Smith
Barry Smyth
Lucca Spalazzi
Richard Sproat
Ashwin Srinivasan
Jamie Stark
Mark Steedman
Sam Steel
Lynn Andrea Stein
Donald Steiner
Werner Stephan

Mark Stevenson
Oliviero Stock
Peter Stone
Peter Struss
Devika Subramanian
Shigeki Sugano
Toshiharu Sugawara
Kurt Sundermeyer
Rich Sutton
Hiroyuki Suzuki
Prasad Tadepalli
Gentaro Taga
Tomoichi Takahashi
Hideaki Takeda
Hirokazu Taki
Milind Tambe
Ah Hwee Tan
Yao-Hua Tan
Hiroshi Tanaka
Jun Tani
Austin Tate
Ahmed Tawfik
Annette ten Teije
Joshua Tenenbaum
Moshe Tennenholtz
Sylvie Thiebaux
Michael Thielscher
Henry Thompson
Henry Tirri
Hans Tompits
Jan Top
Pietro Torasso
Carme Torras
Paolo Traverso
Brigitte Trousse
Mirek Truszczynski
Edward Tsang
Takashi Tsubouchi
Jun-ichi Tsujii
Katsuhiko Tsujino
Hudson Turner
Phil Turner
Yasusi Umeda
Tanja Urbancic
Raul Valdes-Perez
L. W. N. Van Der Torre
Peter van Beek
Thierry van de Merckt
Wiebe van der Hoek
Leon van der Torre
Manuela Veloso
Floor Verdenius

Gerard Verfaillie
Egon Verharen
Lluis Vila
Marie-Catherine Vilarem
Adolfo Villafiorita
Theo Vosse
Benjamin Wah
Marilyn Walker
Richard Wallace
Toby Walsh
Weihsin Wang
Takashi Washio
Stephen Watkinson
Geoff Webb
Hans Weigard
Gerhard Weiss
Michael Wellman
Dietrich Wettscherek
Jon Whittle
Marco Wiering
Markus Wiese
Janet Wiles
Yorick Wilks
Randy Wilson
Mats Wiren
Cees Witteveen
David Wolpert
Sharon Wood
Michael Wooldridge
Stefan Wrobel
Dekai Wu
Yasushi Yagi
Masanobu Yamamoto
David Yarowsky
Albert Yeap
Dit-Yan Yeung
Kenneth Yip
Makoto Yokoo
Naokazu Yokoya
Massimo Zancanaro
Chengqi Zhang
Nevin L. Zhang
Weixiong Zhang
Feng Zhao
Jieyu Zhao
Qi Zhao
Roland Zito-Wolf
Tatjana Zrimec
Jean-Daniel Zucker

# Contents

## VOLUME 1

## AI CHALLENGES

### CHALLENGE 1

The Predictive Toxicology Evaluation Challenge
*A. Srinivasan, R. D. King, S. H. Muggleton, and M. J. E. Sternberg* ... 4

Challenge: What is the Impact of Bayesian Networks on Learning?
*Nir Friedman, Moises Goldszmidt, David Heckerman, and Stuart Russell* ... 10

Adaptive Web Sites: an AI Challenge
*Mike Perkowitz and Oren Etzioni* ... 16

### CHALLENGE 2

The RoboCup Synthetic Agent Challenge 97
*Hiroaki Kitano, Milind Tambe, Peter Stone, Manuela Veloso, Silvia Coradeschi, Eiichi Osawa, Hitoshi Matsubara, Itsuki Noda, and Minoru Asada* ... 24

Understanding Three Simultaneous Speeches
*Hiroshi G. Okuno, Tomohiro Nakatani, and Takeshi Kawabata* ... 30

Distributed Vision System: A Perceptual Information Infrastructure for Robot Navigation
*Hiroshi Ishiguro* ... 36

### CHALLENGE 3

Challenges in bridging plan synthesis paradigms
*Subbarao Kambhampati* ... 44

Ten Challenges in Propositional Reasoning and Search
*Bart Selman, Henry Kautz, and David McAllester* ... 50

Challenge: How IJCAI 1999 can Prove the Value of AI by Using AI
*James Geller* ... 55

## AUTOMATED REASONING

### AUTOMATED REASONING 1: BELIEF REVISION

Qualitative Relevance and Independence: A Roadmap
*Didier Dubois, Luis Fariñas del Cerro, Andreas Herzig, and Henri Prade* ... 62

The Complexity of Belief Update
*Paolo Liberatore* ... 68

Anytime Belief Revision
*Mary-Anne Williams* ... 74

### AUTOMATED REASONING 2: BELIEF REVISION

Towards Generalized Rule-based Updates
*Yan Zhang and Norman Y. Foo* ... 82

Representation Theorems for Multiple Belief Changes
*Dongmo Zhang, Shifu Chen, Wujia Zhu, and Zhaoqian Chen* ... 89

Nonmonotonic Reasoning and Multiple Belief Revision
*Dongmo Zhang, Shifu Chen, Wujia Zhu, and Hongbing Li* ... 95

### AUTOMATED REASONING 3: THEOREM PROVING

High Performance ATP Systems by Combining Several AI Methods
*Jörg Denzinger, Marc Fuchs, and Matthias Fuchs* ... 102

Equational Reasoning using AC Constraints
*David A. Plaisted and Yunshan Zhu* ... 108

Strategies in Rigid-Variable Methods
*Andrei Voronkov* ... 114

### AUTOMATED REASONING 4: PROPOSITIONAL KBS

Tractable Cover Compilations
*Yacine Boufkhad, Éric Grégoire, Pierre Marquis, Bertrand Mazure, and Lakhdar Saïs* ... 122

A Four-Valued Fuzzy Propositional Logic
*Umberto Straccia* ... 128

### AUTOMATED REASONING 5: DESCRIPTION LOGIC

Autoepistemic Description Logics
*Francesco M. Donini, Daniele Nardi, and Riccardo Rosati* ... 136

Reifying Concepts in Description Logics
*Liviu Badea* ... 142

### AUTOMATED REASONING 6: NONMONOTONISM

Circumscribing Inconsistency
*Philippe Besnard and Torsten H. Schaub* ... 150

A default interpretation of defeasible network
*Xianchang Wang, Jia-Huai You, and Li Yan Yuan* ... 156

A Cumulative-Model Semantics for Dynamic Preferences on Assumptions
*Ulrich Junker* ... 162

Compiling reasoning with and about preferences into default logic
*James P. Delgrande and Torsten H. Schaub* ... 168

### AUTOMATED REASONING 7: NONMONOTONISM FOR LOGIC PROGRAMMING

Learning Extended Logic Programs
*Katsumi Inoue and Yoshimitsu Kudoh* .......... 176

Compiling Prioritized Circumscription into Extended Logic Programs
*Toshiko Wakaki and Ken Satoh* .............. 182

### AUTOMATED REASONING 8: MODAL LOGIC

Prefixed Tableaux Systems for Modal Logics with Enriched Languages
*Philippe Balbiani and Stéphane Demri* .......... 190

A Set-Theoretic Approach to Automated Deduction in Graded Modal Logics
*A. Montanari and A. Policriti* ................ 196

On evaluating decision procedures for modal logic
*Ullrich Hustadt and Renate A. Schmidt* ......... 202

### AUTOMATED REASONING 9: ANALOGY

Preduction: A Common Form of Induction and Analogy
*Jun Arima* ................................ 210

Analogy and Abduction in Automated Deduction
*Gilles Défourneaux and Nicolas Peltier* ......... 216

## CASE-BASED REASONING

### CASE-BASED REASONING 1

How Similar is VERY YOUNG to 43 Years of Age? On the Representation and Comparison of Polymorphic Properties
*Werner Dubitzky, Alfons Schuster, John G. Hughes, David A. Bell, and Kenneth Adamson* .......... 226

The Competence of Sub-Optimal Theories of Structure Mapping on Hard Analogies
*Tony Veale and Mark Keane* ................ 232

An Average-Case Analysis of the *k*-Nearest Neighbor Classifier for Noisy Domains
*Seishi Okamoto and Nobuhiro Yugami* ......... 238

### CASE-BASED REASONING 2

Learning to Integrate Multiple Knowledge Sources for Case-Based Reasoning
*David B. Leake, Andrew Kinley, and David Wilson* ... 246

Aggregating Features and Matching Cases on Vague Linguistic Expressions
*Alfons Schuster, Werner Dubitzky, Philippe Lopes, Kenneth Adamson, David A. Bell, John G. Hughes, and John A. White* ........................ 252

Using Case-Based Reasoning in Interpreting Unsupervised Inductive Learning Results
*Tu Bao Ho and Chi Mai Luong* .............. 258

## COGNITIVE MODELING

### COGNITIVE MODELING 1

Acquisition of Human Feelings in Music Arrangement
*Masayuki Numao, Masashi Kobayashi, and Katsuyuki Sakaniwa* .................. 268

Using Data and Theory in Multistrategy (Mis)Concept(ion) Discovery
*Raymond Sison, Masayuki Numao, and Masamichi Shimura* .................. 274

An Aggregation Procedure for Building Episodic Memory
*Olivier Ferret and Brigitte Grau* .............. 280

### COGNITIVE MODELING 2

An Achievement Test for Knowledge-Based Systems: QUEM
*Caroline Clarke Hayes and Michael I. Parzen* ..... 288

A Functional Theory of Design Patterns
*Sambasiva R. Bhatta and Ashok K. Goel* ........ 294

Mental Tracking: A Computational Model of Spatial Development
*Kazuo Hiraki, Akio Sashima, and Steven Phillips* ... 301

### COGNITIVE MODELING 3

In the Quest of the Missing Link
*Guilherme Bittencourt* ...................... 310

Implementing BDI-like Systems by Direct Execution
*Michael Fisher* ............................ 316

Managing decision resources in plan execution
*Michael Freed and Roger Remington* .......... 322

## COMPUTER-AIDED EDUCATION

Use of Abstraction and Complexity Levels in Intelligent Educational Systems Design
*Ruddy Lelouche and Jean-François Morin* ....... 329

Reasoning Symbolically About Partially Matched Cases
*Kevin D. Ashley and Vincent Aleven* ........... 335

Task Ontology Makes It Easier To Use Authoring Tools
*Mitsuru Ikeda, Kazuhisa Seta, and Riichiro Mizoguchi* .................... 342

## CONSTRAINT SATISFACTION

### CONSTRAINT SATISFACTION 1: CONSTRAINT PROGRAMMING

Semiring-based Constraint Logic Programming
*Stefano Bistarelli, Ugo Montanari, and Francesca Rossi* ...................... 352

Computational Complexity of Multi-way, Dataflow Constraint Problems
*Gilles Trombettoni and Bertrand Neveu* ......... 358

## CONSTRAINT SATISFACTION 2: SAT

Heuristics Based on Unit Propagation for
Satisfiability Problems
  *Chu Min Li and Anbulagan* . . . . . . . . . . . . . 366

Hidden Gold in Random Generation of SAT
Satisfiable Instances
  *Thierry Castell and Michel Cayrol* . . . . . . . . . . . . 372

Discrete Lagrangian-Based Search for Solving
MAX-SAT Problems
  *Benjamin W. Wah and Yi Shang* . . . . . . . . . . . . . . 378

Learning Short-Term Weights for GSAT
  *Jeremy Frank* . . . . . . . . . . . . . . . . . . . . . . . 384

## CONSTRAINT SATISFACTION 3: LOCAL CONSISTENCY

Local consistency for ternary numeric constraints
  *Boi Faltings and Esther Gelle* . . . . . . . . . . . . . . 392

Arc consistency for general constraint networks:
preliminary results
  *Christian Bessière and Jean-Charles Régin* . . . . . . . . 398

Constraint Satisfaction over Connected Row
Convex Constraints
  *Yves Deville, Olivier Barette,
  and Pascal Van Hentenryck* . . . . . . . . . . . . . . . . 405

## CONSTRAINT SATISFACTION 4

Some Practicable Filtering Techniques for the
Constraint Satisfaction Problem
  *Romuald Debruyne and Christian Bessière* . . . . . . . . 412

Structuring Techniques for Constraint
Satisfaction Problems
  *Rainer Weigel and Boi V. Faltings* . . . . . . . . . . . . 418

Merging constraint satisfaction subproblems to avoid
redundant search
  *Javier Larrosa* . . . . . . . . . . . . . . . . . . . . . . 424

# DIAGNOSIS AND QUALITATIVE REASONING

## DIAGNOSIS 1

Locating Faults in Tree-Structured Networks
  *Christopher Leckie and Michael Dale* . . . . . . . . . . . 434

Diagnosing Tree Structured Systems
  *Markus Stumptner and Franz Wotawa* . . . . . . . . . . . . 440

Event-Based Reasoning for Short Circuit Diagnosis in
Power Transmission Networks
  *Gianfranco Lamperti and Paolo Pogliano* . . . . . . . . . 446

## DIAGNOSIS 2

Exploiting domain knowledge for
approximate diagnosis
  *Annette ten Teije and Frank van Harmelen* . . . . . . . . 454

Semantically Guided Theorem Proving for
Diagnosis Applications
  *Peter Baumgartner, Peter Fröhlich, Ulrich Furbach,
  and Wolfgang Nejdl* . . . . . . . . . . . . . . . . . . . . 460

A Static Model-Based Engine for
Model-Based Reasoning
  *Peter Fröhlich and Wolfgang Nejdl* . . . . . . . . . . . . 466

## DIAGNOSIS 3

Polynomial Temporal Band Sequences for
Analog Diagnosis
  *Etienne Loiez and Patrick Taillibert* . . . . . . . . . . 474

Fundamentals of Model-Based Diagnosis of
Dynamic Systems
  *Peter Struss* . . . . . . . . . . . . . . . . . . . . . . 480

Comparative Analysis of Structurally Different
Dynamical Systems
  *H. de Jong and F. van Raalte* . . . . . . . . . . . . . . 486

## QUALITATIVE REASONING 1: MODELING SUPPORT

A Web-Based Compositional Modeling System for
Sharing of Physical Knowledge
  *Yumi Iwasaki, Adam Farquhar, Richard Fikes,
  and James Rice* . . . . . . . . . . . . . . . . . . . . . . 494

A Causal Time Ontology for Qualitative Reasoning
  *Yoshinobu Kitamura, Mitsuru Ikeda,
  and Riichiro Mizoguchi* . . . . . . . . . . . . . . . . . . 501

## QUALITATIVE REASONING 2: PERCEPTION AND BELIEF

Qualitative Reasoning about Perception and Belief
  *Alvaro del Val, Pedrito Maynard-Reid II,
  and Yoav Shoham* . . . . . . . . . . . . . . . . . . . . . 508

Rule-based Contact Monitoring Using Examples
Obtained by Task Demonstration
  *Pavan Sikka and Brenan J. McCarragher* . . . . . . . . . 514

## QUALITATIVE REASONING 3: GEOMETRIC AND SPATIAL REASONING

On the Complexity of Qualitative Spatial Reasoning:
A Maximal Tractable Fragment of the Region
Connection Calculus
  *Jochen Renz and Bernhard Nebel* . . . . . . . . . . . . . 522

Automation of Diagrammatic Reasoning
  *Mateja Jamnik, Alan Bundy, and Ian Green* . . . . . . . . 528

Structural Inferences from Massive Datasets
  *Kenneth Yip* . . . . . . . . . . . . . . . . . . . . . . . 534

## QUALITATIVE REASONING 4: CAUSALITY

Qualitative Analysis of Causal Graphs with
Equilibrium Type-Transition
  *Koichi Kurumatani and Mari Nakamura* . . . . . . . . . . 542

Action Localness, Genericity and Invariants in STRIPS
*Norman Y. Foo, Abhaya Nayak, Maurice Pagnucco, Pavlos Peppas, and Yan Zhang* ............... 549

Causality, Constraints and the Indirect Effects of Actions
*Hector Geffner* .......................... 555

## QUALITATIVE REASONING 5

Redesigning a Problem-Solver's Operators to Improve Solution Quality
*Eleni Stroulia and Ashok K. Goel* .............. 562

Formal Specifications for Hybrid Dynamical Systems
*Pieter J. Mosterman and Gautam Biswas* .......... 568

# DISTRIBUTED ARTIFICIAL INTELLIGENCE

## DISTRIBUTED AI 1: INTERAGENT COMMUNICATION

Middle-Agents for the Internet
*Keith Decker, Katia Sycara, and Mike Williamson* .... 578

Semantics and Conversations for an Agent Communication Language
*Yannis Labrou and Tim Finin* ................. 584

Persuasion among Agents: An Approach to Implementing a Group Decision Support System Based on Multi-Agent Negotiation
*Takayuki Ito and Toramatsu Shintani* ........... 592

## DISTRIBUTED AI 2: COORDINATION AND COOPERATION

Cooperation Structures
*Mark d'Inverno, Michael Luck, and Michael Wooldridge* .................... 600

Exploration and Adaptation in Multiagent Systems: A Model-based Approach
*David Carmel and Shaul Markovitch* ............ 606

The Effects of Runtime Coordination Strategies Within Static Organizations
*Edmund H. Durfee and Young-pa So* ............ 612

## DISTRIBUTED AI 3: MULTIAGENT ALGORITHMS

Dynamic Prioritization of Complex Agents in Distributed Constraint Satisfaction Problems
*Aaron Armstrong and Edmund Durfee* .......... 620

A dynamic theory of incentives in multi-agent systems
*Yoav Shoham and Katsumi Tanaka* ............ 626

On the Gains and Losses of Speculation in Equilibrium Markets
*Tuomas Sandholm and Fredrik Ygge* ........... 632

## DISTRIBUTED AI 4: MULTIAGENT ALGORITHMS

The Use of Meta-level Information in Learning Situation-Specific Coordination
*M. V. Nagendra Prasad and Victor R. Lesser* ........ 640

Analysis of Inheritance Mechanisms in Agent-Oriented Programming
*Lobel Crnogorac, Anand S. Rao, and Kotagiri Ramamohanarao* ................. 647

# EXPERT SYSTEMS

An Expert System Using Nonmonotonic Techniques for Benefits Inquiry in the Insurance Industry
*Leora Morgenstern and Moninder Singh* .......... 655

Can We Benefit from Metrics in KBS Development?
*Stefan Kramer, Hermann Kaindl, and Stefan Schlee* ... 662

Multi-Perspective Modelling of the Air Campaign Planning process
*John Kingston, Anna Griffith, and Terri Lydiard* ..... 668

# GAME PLAYING

## GAME PLAYING 1: GO

A Model of Strategy for the Game of Go Using Abstraction Mechanisms
*Patrick Ricaud* .......................... 678

An Evolutionary Algorithm Extended by Ecological Analogy and its Application to the Game of Go
*Takuya Kojima, Kazuhiro Ueda, and Saburo Nagano* ..................... 684

## GAME PLAYING 2

Search Versus Knowledge in Game-Playing Programs Revisited
*Andreas Junghanns and Jonathan Schaeffer* ........ 692

Learning Strategies in Games by Anticipation
*Christophe Meyer, Jean-Gabriel Ganascia, and Jean-Daniel Zucker* .................. 698

# INFORMATION RETRIEVAL

## INFORMATION RETRIEVAL DISTINGUISHED PAPER

Translingual Information Retrieval: A Comparative Evaluation
*Jaime G. Carbonell, Yiming Yang, Robert E. Frederking, Ralf D. Brown, Yibing Geng, and Danny Lee* ....... 708

## INFORMATION RETRIEVAL 1

Adaptive Personal Information Filtering System that Organizes Personal Profiles Automatically
*Toshiki Kindo, Hideyuki Yoshida, Tetsuro Morimoto, and Taisuke Watanabe* ................... 716

An Index Navigator for Understanding and
Expressing User's Coherent Interest
   *Yukio Ohsawa and Masahiko Yachida*............ 722

Wrapper Induction for Information Extraction
   *Nicholas Kushmerick, Daniel S. Weld,
and Robert Doorenbos* .............. 729

## INFORMATION RETRIEVAL 2

COSPEX: A System for Constructing Private
Digital Libraries
   *Masanori Sugimoto, Norio Katayama,
and Atsuhiro Takasu*...................... 738

Using a Bayesian Network Induction Approach for
Text Categorization
   *Wai Lam, Kon Fan Low, and Chao Yang Ho* ...... 745

Toward Structured Retrieval in Semi-structured
Information Spaces
   *Scott B. Huffman and Catherine Baudin* ......... 751

## INFORMATION RETRIEVAL 3

The Self-Organizing Desk
   *Daniela Rus and Peter de Santis* .............. 758

A Learning System for Selective Dissemination
of Information
   *Gianni Amati, Fabio Crestani,
and Flavio Ubaldini*...................... 764

WebWatcher: A Tour Guide for the World Wide Web
   *Thorsten Joachims, Dayne Freitag,
and Tom Mitchell*........................ 770

## INFORMATION RETRIEVAL 4

Recursive Plans for Information Gathering
   *Olivier M. Duschka and Alon Y. Levy* ........... 778

Efficiently Executing Information-Gathering Plans
   *Marc Friedman and Daniel S. Weld*............ 785

## AUTHOR INDEX ..................... 792

# VOLUME 2

# LEARNING

## LEARNING 1

Unbiased Assessment of Learning Algorithms
   *Tobias Scheffer and Ralf Herbrich* .............. 798

Is Nonparametric Learning Practical in Very High
Dimensional Spaces?
   *Gregory Z. Grudic and Peter D. Lawrence* ....... 804

Discovering Admissible Models of Complex Systems
Based on Scale-Types and Identity Constraints
   *Takashi Washio and Hiroshi Motoda* .......... 810

## LEARNING 2: REINFORCEMENT LEARNING

An Adaptive Architecture for Modular Q-Learning
   *Takayuki Kohri, Kei Matsubayashi,
and Mario Tokoro*........................ 820

A convergent Reinforcement Learning algorithm
in the continuous case based on a
Finite Difference method
   *Rémi Munos*............................ 826

Ants and Reinforcement Learning: A Case Study in
Routing in Dynamic Networks
   *Devika Subramanian, Peter Druschel,
and Johnny Chen*......................... 832

## LEARNING 3: DECISION TREES

Integrating Models of Discrimination and
Characterization for Learning from Examples in
Open Domains
   *Paul Davidsson* ......................... 840

Decision Tree Grafting
   *Geoffrey I. Webb* ........................ 846

Noise-Tolerant Windowing
   *Johannes Fürnkranz* ..................... 852

## LEARNING 4: CLASSIFICATION

Ensembles as a Sequence of Classifiers
   *Lars Asker and Richard Maclin*............... 860

Stacked Generalization: when does it work?
   *Kai Ming Ting and Ian H. Witten*.............. 866

## LEARNING 5: APPLICATIONS

Alignment Algorithms for Learning to Read Aloud
   *Charles X. Ling and Handong Wang* ........... 874

Socially Embedded Learning of the Office-Conversant
Mobile Robot *Jijo-2*
   *Hideki Asoh, Satoru Hayamizu, Isao Hara, Yoichi
Motomura, Shotaro Akaho, and Toshihiro Matsui* ..... 880

## LEARNING 6: LOGIC AND ILP

Tractable Induction and Classification in First Order
Logic Via Stochastic Matching
   *Michèle Sebag and Céline Rouveirol*............ 888

RHB+: A Type-Oriented ILP System Learning from
Positive Data
   *Yutaka Sasaki and Masahiko Haruno*............ 894

Integrating Explanatory and Descriptive Learning
in ILP
   *Yannis Dimopoulos, Saso Džeroski,
and Antonis Kakas*....................... 900

## LEARNING 7: DYNAMIC ENVIRONMENTS

Combining Knowledge Acquisition and Machine
Learning to Control Dynamic Systems
   *G. M. Shiraz and C. Sammut* ................ 908

Skill reconstruction as induction of LQ controllers with subgoals
*Dorian Šuc and Ivan Bratko* . . . . . . . . . . . . . . . . . 914

Learning Topological Maps with Weak Local Odometric Information
*Hagit Shatkay and Leslie Pack Kaelbling* . . . . . . . . 920

**LEARNING 8**

Discovering Interesting Holes in Data
*Bing Liu, Liang-Ping Ku, and Wynne Hsu* . . . . . . . . 930

An Analysis on Crossovers for Real Number Chromosomes in an Infinite Population Size
*Tatsuya Nomura* . . . . . . . . . . . . . . . . . . . . . . . . . 936

Minimum Splits Based Discretization for Continuous Features
*Ke Wang and Han Chong Goh* . . . . . . . . . . . . . . . 942

# NATURAL-LANGUAGE PROCESSING AND GRAPHICAL PRESENTATION

**NATURAL-LANGUAGE PROCESSING 1: GENERATION**

Dynamically Improving Explanations: A Revision-Based Approach to Explanation Generation
*Charles B. Callaway and James C. Lester* . . . . . . . . 952

Exploiting the Addressee's Inferential Capabilities in Presenting Mathematical Proofs
*Detlef Fehrer and Helmut Horacek* . . . . . . . . . . . . 959

Proof Verbalization as an Application of NLG
*Xiaorong Huang and Armin Fiedler* . . . . . . . . . . . . 965

**NATURAL-LANGUAGE PROCESSING 2: MACHINE TRANSLATION**

Corpus-Based Chinese-Korean Abstracting Translation System
*Jun-Jie Li and Key-Sun Choi* . . . . . . . . . . . . . . . . 972

A Hybrid Approach to Interactive Machine Translation—Integrating Rule-based, Corpus-based, and Example-based Method
*Kiyoshi Yamabana, Shin-ichiro Kamei, Kazunori Muraki, Shinichi Doi, Shinko Tamura, and Kenji Satoh* . . . . . . 977

Improving Performance of Transfer-Driven Machine Translation with Extra-Linguistic Information from Context, Situation and Environment
*Hideki Mima, Osamu Furuse, and Hitoshi Iida* . . . . . 983

**NATURAL-LANGUAGE PROCESSING 3: DIALOGUE AND DISCOURSE**

Dynamic, User-Centered Resolution in Interactive Stories
*Nikitas M. Sgouros* . . . . . . . . . . . . . . . . . . . . . . . 990

"Tall", "Good", "High" — Compared to What?
*Steffen Staab and Udo Hahn* . . . . . . . . . . . . . . . . 996

Charts, interaction-free grammars, and the compact representation of ambiguity
*Marc Dymetman* . . . . . . . . . . . . . . . . . . . . . . . . 1002

**NATURAL-LANGUAGE PROCESSING 4: DIALOGUE AND DISCOURSE**

On the Interaction of Metonymies and Anaphora
*Katja Markert and Udo Hahn* . . . . . . . . . . . . . . . 1010

Computing Parallelism in Discourse
*Claire Gardent and Michael Kohlhase* . . . . . . . . . 1016

Content Ordering in the Generation of Persuasive Discourse
*Chris Reed and Derek Long* . . . . . . . . . . . . . . . . 1022

**NATURAL-LANGUAGE PROCESSING 5: DIALOGUE AND DISCOURSE**

ARTIMIS: Natural Dialogue Meets Rational Agency
*M. D. Sadek, P. Bretier, and F. Panaget* . . . . . . . . . 1030

An Information-based Approach for Guiding Multi-Modal Human-Computer-Interaction
*Matthias Denecke* . . . . . . . . . . . . . . . . . . . . . . . 1036

Interactive Disambiguation of Natural Language Input: a Methodology and Two Implementations for French and English
*Hervé Blanchon* . . . . . . . . . . . . . . . . . . . . . . . . 1042

**GRAPHICS**

A Method of Generating Calligraphy of Japanese Character using Deformable Contours
*Lisong Wang, Tsuyoshi Nakamura, Minkai Wang, Hirohisa Seki, and Hidenori Itoh* . . . . . . . . . . . . . 1050

The Representation and Use of a Visual Lexicon for Automated Graphics Generation
*Michelle X. Zhou and Steven K. Feiner* . . . . . . . . 1056

# NEURAL NETWORKS

**NEURAL NETS 1: RULE EXTRACTION**

On the Efficient Classification of Data Structures by Neural Networks
*Paolo Frasconi, Marco Gori, and Alessandro Sperduti* . . . . . . . . . . . . . . . . . . . 1066

On the Role of Hierarchy for Neural Network Interpretation
*Jürgen Rahmel, Christian Blum, and Peter Hahn* . . . 1072

Law Discovery using Neural Networks
*Kazumi Saito and Ryohei Nakano* . . . . . . . . . . . . 1078

Active Diagnosis by Self-Organization: An Approach by The Immune Network Metaphor
*Yoshiteru Ishida* . . . . . . . . . . . . . . . . . . . . . . . . 1084

## NEURAL NETS 2: LANGUAGE AND STRUCTURE PROCESSING

Meaning and the Mental Lexicon
  Will Lowe . . . . . . . . . . . . . . . . . . . . . . . . . . . 1092

Extracting Propositions from Trained Neural Networks
  Hiroshi Tsukimoto . . . . . . . . . . . . . . . . . . . . . 1098

Convergence time characteristics of an associative memory for natural language processing
  Nigel Collier . . . . . . . . . . . . . . . . . . . . . . . . . 1106

## NEURAL NETS 3: NEUROBIOLOGICALLY INSPIRED COMPUTATION

Combining Probabilistic Population Codes
  Richard S. Zemel and Peter Dayan . . . . . . . . . . . . 1114

Self-Organization and Segmentation with Laterally Connected Spiking Neurons
  Yoonsuck Choe and Risto Miikkulainen . . . . . . . . 1120

A Music Stream Segregation System Based on Adaptive Multi-Agents
  Kunio Kashino and Hiroshi Murase . . . . . . . . . . 1126

## NEURAL NETS 4: LEARNING ALGORITHMS AND ARCHITECTURES

An effective learning method for max-min neural networks
  Loo-Nin Teow and Kia-Fock Loe . . . . . . . . . . . . 1134

Avoiding Overfitting with BP-SOM
  Ton Weijters, H. Jaap van den Herik, Antal van den Bosch, and Eric Postma . . . . . . . . . 1140

Evolvable Hardware for Generalized Neural Networks
  Masahiro Murakawa, Shuji Yoshizawa, Isamu Kajitani, and Tetsuya Higuchi . . . . . . . . . . . . . . . . . . . . . 1146

# PLANNING AND SCHEDULING

## PLANNING 1: RELATIONS AMONG TECHNIQUES

Prioritized Goal Decomposition of Markov Decision Processes: Toward a Synthesis of Classical and Decision Theoretic Planning
  Craig Boutilier, Ronen I. Brafman, and Christopher Geib . . . . . . . . . . . . . . . . . . . . . 1156

Model Minimization, Regression, and Propositional STRIPS Planning
  Robert Givan and Thomas Dean . . . . . . . . . . . . 1163

Automatic SAT-Compilation of Planning Problems
  Michael D. Ernst, Todd D. Millstein, and Daniel S. Weld . . . . . . . . . . . . . . . . . . . . . . 1169

## PLANNING 2: REACTIVE PLANNING

A Reactive Planner for a Model-based Executive
  Brian C. Williams and P. Pandurang Nayak . . . . . . 1178

Modeling Command Entities
  Michael D. Howard . . . . . . . . . . . . . . . . . . . . 1186

## PLANNING 3: PLANNING UNDER UNCERTAINTY

Vision-Motion Planning of a Mobile Robot considering Vision Uncertainty and Planning Cost
  Jun Miura and Yoshiaki Shirai . . . . . . . . . . . . . . 1194

Handling Duration Uncertainty in Meta-Level Control of Progressive Processing
  Abdel-Illah Mouaddib and Shlomo Zilberstein . . . . 1201

## PLANNING 4: REASONING ABOUT PLANS

Adaptive goal recognition
  Neal Lesh . . . . . . . . . . . . . . . . . . . . . . . . . . . 1208

Reasoning about Plans
  Witold Lukaszewicz and Ewa Madalińska-Bugaj . . . 1215

Reasoning about concurrent execution, prioritized interrupts, and exogenous actions in the situation calculus
  Guiseppe De Giacomo, Yves Lespérance, and Hector J. Levesque . . . . . . . . . . . . . . . . . . . 1221

Learning to Improve both Efficiency and Quality of Planning
  Tara A. Estlin and Raymond J. Mooney . . . . . . . . 1227

## PLANNING 5: APPLICATIONS AND SUPPORT

Robust Periodic Planning and Execution for Autonomous Spacecraft
  Barney Pell, Erann Gat, Ron Keesing, Nicola Muscettola, and Ben Smith . . . . . . . . . . . . 1234

System Assistance in Structured Domain Model Development
  Susanne Biundo and Werner Stephan . . . . . . . . . 1240

Par-KAP: a Knowledge Acquisition Tool for Building Practical Planning Systems
  Leliane Nunes de Barros, James Hendler, and V. Richard Benjamins . . . . . . . . . . . . . . . . . 1246

## SCHEDULING

Combining Local Search and Look-Ahead for Scheduling and Constraint Satisfaction Problems
  Andrea Schaerf . . . . . . . . . . . . . . . . . . . . . . . 1254

Automatic Generation of Heuristics for Scheduling
  Robert A. Morris, John L. Bresina, and Stuart M. Rodgers . . . . . . . . . . . . . . . . . . . 1260

Development of Iterative Real-time Scheduler to Planner Feedback
*Charles B. McVey, Ella M. Atkins, Edmund H. Durfee, and Kang G. Shin* ......... *1267*

# PROBABILISTIC REASONING

## PROBABILISTIC REASONING DISTINGUISHED PAPER

Object Identification in a Bayesian Context
*Timothy Huang and Stuart Russell* ........... *1276*

## PROBABILISTIC REASONING 1: EFFICIENCY

Probabilistic Partial Evaluation: Exploiting rule structure in probabilistic inference
*David Poole* ........................ *1284*

Space-efficient inference in dynamic probabilistic networks
*John Binder, Kevin Murphy, and Stuart Russell* .... *1292*

Mini-Buckets: A General Scheme for Generating Approximations in Automated Reasoning
*Rina Dechter* ....................... *1297*

## PROBABILISTIC REASONING 2: CAUSAL DISCOVERY

A Study of Causal Discovery With Weak Links and Small Samples
*Honghua Dai, Kevin Korb, Chris Wallace, and Xindong Wu* .................. *1304*

ILP with Noise and Fixed Example Size: A Bayesian Approach
*Eric McCreath and Arun Sharma* ........... *1310*

Learning probabilities for noisy first-order rules
*Daphne Koller and Avi Pfeffer* ............. *1316*

## PROBABILISTIC REASONING 3

A Symmetric View of Utilities and Probabilities
*Yoav Shoham* ....................... *1324*

PRISM: A Language for Symbolic-Statistical Modeling
*Taisuke Sato and Yoshitaka Kameya* ........... *1330*

# ROBOTICS

## ROBOTICS 1

Multi-Robot Exploration of an Unknown Environment, Efficiently Reducing the Odometry Error
*Ioannis M. Rekleitis, Gregory Dudek, and Evangelos E. Milios* ............. *1340*

Active Mobile Robot Localization
*Wolfram Burgard, Dieter Fox, and Sebastian Thrun* .. *1346*

Reactive Combination of Belief Over Time Using Direct Perception
*Robin R. Murphy, Dale K. Hawkins, and Marcel J. Schoppers* ................ *1353*

## ROBOTICS 2

Scaling the Dynamic Approach to Autonomous Path Planning: Planning Horizon Dynamics
*Edward W. Large, Heneik I. Christensen, and Ruzena Bajcsy* ................. *1360*

Learning to Coordinate Controllers—Reinforcement Learning on a Control Basis
*Manfred Huber and Roderic A. Grupen* ........ *1366*

Situated Actions and Cognition
*Jacques Penders and Peter J. Braspenning* ........ *1372*

# SEARCH

## SEARCH 1: DEPTH-FIRST SEARCH

Interleaved Depth-First Search
*Pedro Meseguer* ..................... *1382*

Depth-bounded Discrepancy Search
*Toby Walsh*. ....................... *1388*

## SEARCH 2: BIN PACKING

From Approximate to Optimal Solutions: Constructing Pruning and Propagation Rules
*Ian P. Gent and Toby Walsh* ............... *1396*

An Approximate 0-1 Edge-Labeling Algorithm for Constrained Bin-Packing Problem
*Ho Soo Lee and Mark Trumbo* ............. *1402*

# TEMPORAL REASONING

## TEMPORAL REASONING DISTINGUISHED PAPER

Applications of the Situation Calculus To Formalizing Control and Strategic Information: The Prolog Cut Operator
*Fangzhen Lin*. ...................... *1412*

## TEMPORAL REASONING 1

Reasoning by Regression: Pre- and Postdiction Procedures for Logics of Action and Change with Nondeterminism
*Marcus Bjäreland and Lars Karlsson* .......... *1420*

Change, Change, Change: three approaches
*Tom Costello* ....................... *1426*

## TEMPORAL REASONING 2

Reasoning with Incomplete Initial Information and Nondeterminism in Situation Calculus
*Lars Karlsson* ....................... *1434*

Defeasible Specifications in Action Theories
  *Chitta Baral and Jorge Lobo* . . . . . . . . . . . . . . . *1441*

Reasoning about Action in Polynomial Time
  *Thomas Drakengren and Marcus Bjäreland* . . . . . . . *1447*

## TEMPORAL REASONING 3

Qualitative Temporal Reasoning with Points and Durations
  *Isabel Navarrete and Roque Marin* . . . . . . . . . . . *1454*

On Finding a Solution in Temporal Constraint Satisfaction Problems
  *Alfonso Gerevini and Matteo Cristani* . . . . . . . . . *1460*

Towards a Complete Classification of Tractability in Allen's Algebra
  *Thomas Drakengren and Peter Jonsson* . . . . . . . . . *1466*

# VISION

### VISION 1

Comparing Random Starts Local Search with Key Feature Matching
  *J. Ross Beveridge, Christopher R. Graves, and Jim Steinborn* . . . . . . . . . . . . . . . . . . *1476*

Chain of Circles for Matching and Recognition of Planar Shapes
  *Jae-Moon Chung and Noboru Ohnishi.* . . . . . . . . *1482*

Name-It: Naming and Detecting Faces in Video by the Integration of Image and Natural Language Processing
  *Shin'ichi Satoh, Yuichi Nakamura, and Takeo Kanade* . . . . . . . . . . . . . . . . . . . *1488*

### VISION 2: STEREO VISION

Neural network based photometric stereo using illumination planning
  *Yuji Iwahori, Wataru Kato, Md. Shoaib Bhuiyan, Robert J. Woodham, and Naohiro Ishii* . . . . . . . . *1496*

A General Expression of the Fundamental Matrix for Both Perspective and Affine Cameras
  *Zhengyou Zhang and Gang Xu* . . . . . . . . . . . . . *1502*

# PANEL

The Next Big Thing: Position Statements
  *Munindar P. Singh, Daniel G. Bobrow, Michael N. Huhns, Margaret King, Hiroaki Kitano, and Ray Reiter* . . . . . . . . . . . . . . . . . . . . . *1511*

# VIDEOS

Robust Real-Time Face Tracking and Gesture Recognition
  *J. Heizmann and A. Zelinksy* . . . . . . . . . . . . . . *1525*

PAC—Personality and Cognition: an interactive system for modelling agent scenarios
  *Lin Padgham and Guy Taylor* . . . . . . . . . . . . . . *1531*

# DOCTORAL CONSORTIUM ABSTRACTS

Modularity in Computer Assisted Reasoning Systems
  *Alessandro Agostini* . . . . . . . . . . . . . . . . . . . *1539*

Describing Time-Varying Data
  *Sarah Boyd* . . . . . . . . . . . . . . . . . . . . . . . . *1540*

Automation of Diagrammatic Proofs in Mathematics
  *Mateja Jamnik* . . . . . . . . . . . . . . . . . . . . . . *1541*

Toward the Automatic Discovery of Misconceptions
  *Raymund C. Sison* . . . . . . . . . . . . . . . . . . . . *1542*

Control Structures for Software Agents
  *Hongjun Song* . . . . . . . . . . . . . . . . . . . . . . *1543*

Algorithm Evolution for Signal Understanding
  *Astro Teller* . . . . . . . . . . . . . . . . . . . . . . . . *1544*

The use of neural network approach in financial asset management
  *Francesco Virili* . . . . . . . . . . . . . . . . . . . . . *1545*

Constrained Object Hierarchy—An Architecture for Intelligent Systems
  *Hongxue Wang* . . . . . . . . . . . . . . . . . . . . . *1546*

# INVITED SPEAKERS

Let's plan it deductively!
  *W. Bibel* . . . . . . . . . . . . . . . . . . . . . . . . . *1549*

Creativity and Artificial Intelligence
  *Margaret A. Boden* . . . . . . . . . . . . . . . . . . . *1563*

Modeling Social Action for AI Agents
  *Cristiano Castelfranchi* . . . . . . . . . . . . . . . . . *1567*

Vehicles Capable of Dynamic Vision
  *Ernest D. Dickmanns* . . . . . . . . . . . . . . . . . . *1577*

Remote-Brained Robots
  *Masayuki Inaba* . . . . . . . . . . . . . . . . . . . . . *1593*

Generating Multimedia Briefings: Language Generation in a Coordinated Multimedia Environment
  *Kathleen R. McKeown* . . . . . . . . . . . . . . . . . *1607*

Inheritance Comes of Age: Applying Nonmonotonic
Techniques to Problems in Industry
    *Leora Morgenstern* ................... 1613

Machine Learning Techniques to Make Computers
Easier to Use
    *Hiroshi Motoda and Kenichi Yoshida* ........... 1622

The Origins of Syntax in Visually Grounded
Robotic Agents
    *Luc Steels* ...................... 1632

Numerica: a Modeling Language for
Global Optimization
    *Pascal Van Hentenryck* ................ 1642

# AWARDS

Research Excellence Award: Relationship Between
Natural Language Processing and AI
    *Aravind K. Joshi* .................... 1651

**AUTHOR INDEX** .................. **1653**

# AI CHALLENGES

# AI CHALLENGES

## Challenge 1

# The Predictive Toxicology Evaluation Challenge

A. Srinivasan
S.H. Muggleton
Oxford University Computing Laboratory
Wolfson Building Parks Road, Oxford
U.K.

R.D. King*
M.J.E. Sternberg
Biomolecular Modelling Laboratory
Imperial Cancer Research Fund
44 Lincoln's Inn Fields, London
U.K.

## Abstract

Can an AI program contribute to scientific discovery? An area where this gauntlet has been thrown is that of understanding the mechanisms of chemical carcinogenesis. One approach is to obtain Structure-Activity Relationships (SARs) relating molecular structure to cancerous activity. Vital to this are the rodent carcinogenicity tests conducted within the US National Toxicology Program by the National Institute of Environmental Health Sciences (NIEHS). This has resulted in a large database of compounds classified as carcinogens or otherwise. The Predictive-Toxicology Evaluation project of the NIEHS provides the opportunity to compare carcinogenicity predictions on previously untested chemicals. This presents a formidable challenge for programs concerned with knowledge discovery. Desirable features of this problem are: (1) involvement in genuine scientific discovery; (2) availability of a large database with expert-certified classifications; (3) strong competition from methods used by chemists; and (4) participation in true blind trials, with results available by next IJCAI. We describe the materials and methods constituting this challenge, and provide some initial benchmarks. These show the Inductive Logic Programming tool Progol to be competitive with current state-of-the-art. The challenge described here is aimed at encouraging AI programs to avail themselves the opportunity of contributing to an enterprise with immediate scientific value.

## 1 Introduction

Programs developed under the umbrella of Machine Learning are increasingly being used for "knowledge dis-

---
*Now at:Department of Computer Science, The University of Wales Aberystwyth

covery" tasks. Early specialised programs (for example, [Feigenbaum et al., 1971; Langley et al., 1983]) have given way to more general-purpose ones (for example, [Muggleton, 1995; Muggleton and Feng, 1990]) which have been applied with some success in areas of biochemistry ([King et al., 1996; 1992; Muggleton et al., 1992]). While the experimental studies reported are preliminary, they have at least one commendable feature, namely, they constitute examples of AI programs participating in true scientific discovery tasks. By "true" here, we mean problems where existing scientific knowledge is incomplete, the descriptions found automatically were unknown to experts in the field, and have been acknowledged by publication in peer-reviewed journals in the field. Given the promise shown by machine learning programs in biology and chemistry, this paper describes a challenging test-bed with the following desirable features: (1) a widespread scientific interest in any new results; (2) the availability of a large database of chemicals with classifications certified by experts; (3) strong competition from methods developed by expert chemists; and (4) the opportunity to participate in true blind trials.

The problem concerns obtaining a better understanding of the molecular mechanisms of chemical carcinogenesis. This is central to the prevention of many environmentally induced cancers. One approach is to form Structure Activity Relationships (SARs) that empirically relate molecular structure with ability to cause cancer. This work has been greatly advanced by the long term carcinogenicity tests of compounds in rodents (utilising both genders of one rat and mouse strain) by the US National Toxicology Program (NTP) of the National Institute of Environmental Health Sciences (NIEHS: [Huff and Haseman, 1991]). So far, the NTP tests have resulted in a database of more than 300 compounds that have been shown to be carcinogens or otherwise. The NIEHS Predictive Toxicology Evaluation (or PTE) project ([Bristol et al., 1996]) is closely associated with the NTP. The PTE project identifies a group of assays that are scheduled or ongoing in the NTP. These chemicals form the "test" set for researchers. Predictions for presence or absence of carcinogenicity activity are compared against true activity as observed in the rodents. The first such blind trial, PTE-1, is now complete. The second, PTE-2, is ongoing, and true activity levels will be available by June 1998. It is the prediction of, and reasons for, carcinogenic activity in chemicals constituting PTE-2 that we commend as a challenge for AI programs concerned with knowledge discovery from databases.

This paper is organised as follows. Section 2 presents the statement of the challenge. Section 3 summarises the data available in the NTP database, and the chemicals in PTE-1 and PTE-2. Section 4 sets out the evaluation criteria. Section 5 describes the results obtained using the Inductive Logic Programming (ILP) system Progol [Muggleton, 1995]. These results are intended to provide initial benchmarks for future entries. Section 6 concludes this paper including information on submitting entries to the challenge.

## 2 The PTE Challenge

The aim is to obtain a theory for predicting the carcinogenicity of 30 compounds currently undergoing rodent bioassays in the NTP (called PTE-2: see below). The performance of the theory is to be evaluated according to the criteria described in Section 4.

## 3 Materials

### 3.1 The NTP database

A compilation of 330 chemicals is available directly from the database of the National Cancer Institute and NTP ([Huff and Haseman, 1991], and via the Internet at *http://ntp-server.niehs.nih.gov/*. These compounds represent all the organic chemicals that have completed NTP reports at the time of writing this paper. Of the 330 compounds, 182 (55%) are classified carcinogenic, and the remaining 148 non-carcinogenic. Carcinogenicity is determined by analysis of long term rodent bioassays. For the purposes of this challenge, compounds classified by the NTP as equivocal are considered non-carcinogenic, as this allows direct comparison with other SAR predictive methods. No analysis is made of differences in incidence between rat and mouse cancer, or the role of sex, or particular organ sites. 39 of the 330 compounds in the NTP database formed the first of the blind trials (PTE-1) conducted by the PTE project. Results from the bioassays for these chemicals are now available, and show 22 (56%) to be carcinogenic, and the remaining 17 to be non-carcinogenic. Further details of these compounds are available in [Bahler and Bristol, 1993]. The 330 chemicals make this database very large and diverse, making it a great challenge to learn in.

### The PTE-2 compounds

The second round of blind trials (PTE-2) consists of 30 compounds (of which 5 are inorganic). These are fully described in [Bristol et al., 1996], where the schedule of events suggest that all bioassay results will be available by July 1998.

### 3.2 Other information available

The NTP has recently made available a number of structural attributes (features) describing a large section of their database. These descriptions are available at *http://ntp-server.niehs.nih.gov/Main_Pages*. The other information available is in the form of the atom and bond connectivity of the compounds (including those in PTE-1,2). This is described further in Section 5.

## 4 Evaluation

In [Bristol et al., 1996], the goal of predicitive toxicology (PT) is summarised as "...the ultimate value and most important goal of PT research may lie in the development of its potential to identify, characterise, and understand the various mechanisms or modes of action that determine the type and level of response observed when biological systems are exposed to chemicals...". Given this emphasis on understandability of models, we follow [Muggleton et al., 1996] in using the following definition for comparing the performance of rival theories.

**Definition 1 Performance comparison.** *If the predictive accuracies of two theories are statistically equivalent then the theory with better explanatory power has better performance. Otherwise the one with higher accuracy has better performance.*

We now elaborate further on the methods for evaluating predictive accuracy and "explanatory power".

### 4.1 Predictive accuracy

Predictive accuracy is taken to be the proportion of compounds in PTE-2 whose predicted classification (carcinogenic or non-carcinogenic) agrees with that rodent bioassays. Significant differences in predictive accuracy are best assessed by McNemar test for changes [Bland, 1989]. This test exploits the fact that the different prediction methods are applied to the same data and is based on counting the examples where the methods disagree about predictions. We suggest that differences be judged to be significant at least at $P = 0.10$.

### 4.2 Explanatory power

In the absence of an expert chemist to act as adjudicator, we propose that the simple criterion that theories are judged to have "explanatory power" – a boolean property – if some or all of it can be represented diagrammatically as chemical structures. The intuition underlying this is that such structural alerts form the preferred mode of discourse amongst chemists.

## 5 An experiment with the ILP system Progol

### 5.1 Progol

We refer the reader to [Muggleton, 1995] for complete descriptions of the ILP system Progol. In the current context, Progol is provided with a set of carcinogenic ("positive") and non-carcinogenic ("negative") examples from the NTP database together with background knowledge $B$ about these compounds (see Section 5.2). The aim is to generate a theory (expressed as a set of rules) which explains all the carcinogens in terms of the background knowledge whilst remaining consistent with the non-carcinogens. To achieve this Progol 1) randomly selects a positive example $e_i$; 2) uses inverse entailment to construct the most specific hypothesis $\perp(B, e_i)$ which explains $e_i$ in terms of $B$; 3) finds a rule $D_i$ which generalises $\perp(B, e_i)$ and which maximally compresses a set of entailed example $E_i$; and 4) adds $D_i$ to the theory $H$ and repeats from 1) with examples not covered so far until no more compression is possible. Compression is here defined as the difference, in numbers of descriptors, between $E_i$ and $D_i$.

## 5.2 Background knowledge

The generic atom/bond representation used in an earlier study [King et al., 1996; Srinivasan et al., 1996] is used. This consists of two basic relations to represent structure: *atom* and *bond*. For example, the fact *atom(127,127_1,c,ar_c_6_ring,-0.133)* states that in compound 127, atom no. 1 is of element carbon, and of type aromatic carbon in a 6 membered ring, and has a partial charge of -0.133. The type of the atom and its partial charge were taken from the molecular modelling package QUANTA, although any similar modelling package would have been suitable. Equivalently, *bond(127,127_1,127_2, ar)* states that in compound 127, atom no. 1 and atom no. 2 are connected by an aromatic bond. In QUANTA, a partial charge assignment is based on a specific molecular neighbourhood. This has the effect that a specific molecular sub- structure can be identified by an atom type and partial charge. This relational representation is completely general for chemical compounds and no special attributes need to be invented. The structural information of these compounds was represented by $\approx$ 18,300 facts of background knowledge.

Information was also given about the results of Salmonella mutagenicity tests for each compound. The mutagenic compounds were represented by the relation Ames, e.g. *ames(127)* states that compound 127 is mutagenic. The Progol algorithm allows for the inclusion of complex background knowledge, either in the form of facts, or in the form of arbitrary Prolog programs. In this study we included the background knowledge of chemical groups defined in [Srinivasan et al., 1996], along with the structural alerts in [Ashby et al., 1989]. All the information used is available in Prolog form at a prescribed Internet site: *http://www.comlab.ox.ac.uk/oucl/groups/machlearn*.

## 5.3 Results and discussion

The 39 compounds comprising PTE-1 were excluded and rules for carcinogenicity obtained using Progol. The resulting theory consists of 18 rules. Figure 1 tabulates a comparative evaluation on PTE-1. More details on the rules obtained are available in [King and Srinivasan, 1996]. Figure 2 tabulates the predictions made by the Progol theory for compounds in PTE-2. The first three entries in Figure 1 have been marked out for special attention because they had access to additional information in the form of short-term rodent (*in-vivo*) tests. The first two entries also require a degree of expert evaluation. The Ashby structural alerts are based on electrophilic attack on DNA, which makes them statistically dependent on the Ames test. It is also worth noting that the TIPT and Benigni methods rely on structural alerts derived by the Ashby method for their explanatory component. CASE, TIPT and Progol are the only data-driven inductive methods, and Progol is the only automated method capable of identifying new structural alerts. With these comments in place, the results in Figure 1 offer significant encouragement for machine learning programs on the following counts. First, we point out that one PTE-1, the results of Progol, TIPT and CASE demonstrate performance that is competitive with the current state-of-the-art – Progol has marginally the highest accuracy of all methods that do not use rodent tests. Second, the relatively low accuracies of all methods is primarily due to the diversity of compounds involved. It does however leave the door open for significant improvement. Progol, for example, achieves its performance with a very low-level atom/bond representation of compounds. Enriching this background information with the new structural descriptors available from the NTP could significantly improve its accuracy. These comments are reinforced by early results of using the Progol theory to classify compounds in PTE-2. Figure 2 shows that tentative classifications are available from the NTP for 13 chemicals. Progol's theory has correctly classified 7 of these. This should be seen in context of the performance of other theories listed in [Bristol et al., 1996] which shows that most chemists and automated methods have not been able to better this count. This should provide further impetus for participation by other AI programs.

## 6 Conclusions

The field of toxicology is a rich source of difficult scientific problems, and there is a pressing need for analysis methods that can advance our understanding of the issues involved. We believe that the Predictive Toxicology Evaluation trials being conducted by the US National Institute for Environmental Health Sciences afford AI programs a unique opportunity to participate in obtaining an improved understanding of the molecular mechanisms underlying chemical carcinogenesis. Should they be successful, it would also constitute a noteworthy example of a realistic application of AI techniques.

### Entering the PTE Challenge

Entries to the PTE Challenge can be submitted via *http://www.comlab.ox.ac.uk/oucl/groups/machlearn/PTE*.

| Method | Type | Accuracy | Explanation | Performance ranking |
|---|---|---|---|---|
| Ashby [Tennant et al., 1990] | Chemist | 0.77 | yes | 1 |
| RASH [Jones and Easterly, 1991] | Biological potency analysis | 0.72 | yes | 2 |
| TIPT [Bahler and Bristol, 1993] | Propositional machine learning | 0.67 | yes | 2 |
| Progol [Muggleton, 1995] | Inductive logic programming | 0.64 | yes | 2 |
| Benigni [Benigni, 1995] | Expert-guided regression | 0.62 | yes | 2 |
| DEREK [Sanderson and Earnshaw, 1991] | Expert system | 0.57 | yes | 2 |
| Bakale [Bakale and McCreary, 1992] | Chemical reactivity analysis | 0.63 | no | 3 |
| TOPKAT [Enslein et al., 1990] | Statistical discrimination | 0.54 | yes | 4 |
| CASE [Rosenkranz and Klopman, 1990] | Statistical correlation analysis | 0.54 | yes | 4 |
| COMPACT [Lewis et al., 1990] | Molecular modelling | 0.54 | yes | 4 |

Figure 1: Benchmarks on PTE-1. Methods above the central horizontal line had access to short-term rodent tests, which were unavailable to others. Further, the Ashby and RASH methods require a degree of subjective evaluation, making them semi-automatic. The performance ranking is obtained using the combined accuracy-explanation criterion described earlier.

Entries are accepted here for compounds in PTE-2. A submission requires the following: (a) name of entry; (b) predictions for the compounds in PTE-2; (c) whether or not the theory has explanatory power; and (d) a short description of the technique used for prediction. This is sufficient to compute automatically the accuracy and performance ranking of each entry.

**Acknowledgements**

This research was supported partly by the Esprit Basic Research Action Project ILP II, the SERC project project 'Experimental Application and Development of ILP' and an SERC Advanced Research Fellowship held by Stephen Muggleton. Stephen Muggleton is a Research Fellow of Wolfson College Oxford. R.D. King was at Imperial Cancer Research Fund during the course of much of the early work on this problem. We would also like to thank Professor Donald Michie and David Page for interesting and useful discussions concerning the use of ILP for predicting biological activity.

**References**

[Ashby et al., 1989] J. Ashby, R.W. Tennant, E. Zeiger, and S. Stasiewicz. Classification according to chemical structure, mutagenicity to salmonella and level of carcinogenicity of a further 42 chemicals tested for carcinogenicity by the U.S. National Toxicology Program. *Mutation Research*, 223:73–103, 1989.

[Bahler and Bristol, 1993] D. Bahler and D. Bristol. The induction of rules for predicting chemical carcinogenesis. In *Proceedings of the 26th Hawaii International Conference on System Sciences*, Los Alamitos, 1993. IEEE Computer Society Press.

[Bakale and McCreary, 1992] G. Bakale and R.D. McCreary. Prospective ke screening of potential carcinogens being tested in rodent bioassays by the US National Toxicology Program. *Mutagenesis*, 7:91–94, 1992.

[Benigni, 1995] R. Benigni. Predicting chemical carcinogenesis in rodents: the state of the art in the light of a comparative exercise. *Mutation Research*, 334:103–113, 1995.

[Bland, 1989] M. Bland. *An Introduction to Medical Statistics*. Oxford University Press, Oxford, 1989.

[Bristol et al., 1996] D.W. Bristol, J.T. Wachsman, and A. Greenwell. The NIEHS Predictive-Toxicology Evaluation Project. *Environmental Health Perspectives*, pages 1001–1010, 1996. Supplement 3.

[Enslein et al., 1990] K. Enslein, B.W. Blake, and H.H. Borgstedt. Prediction of probability of carcinogenecity for a set of ntp bioassays. *Mutagenesis*, 5:305–306, 1990.

[Feigenbaum et al., 1971] E.A. Feigenbaum, B.G. Buchanan, and J. Lederberg. On generality and problem solving: a case study using the DENDRAL program. In D. Michie, editor, *Machine Intelligence 6*. Edinburgh University Press, Edinburgh, 1971.

[Huff and Haseman, 1991] J. Huff and J. Haseman. Long-term chemical carcinogenesis experiments for identifying potential human cancer hazards. *Environmental Health Perspectives*, 96(3):23–31, 1991.

[Jones and Easterly, 1991] T.D. Jones and C.E. Easterly. On the rodent bioassays currently being conducted on 44 chemicals: a RASH analysis to predict test results from the National Toxicology Program. *Mutagenesis*, 6:507–514, 1991.

[King and Srinivasan, 1996] R.D. King and A. Srinivasan. Prediction of rodent carcinogenicity bioassays from molecular structure using inductive logic programming. *Environmental Health Perspectives*, 104(5):1031–1040, 1996.

[King et al., 1992] R.D. King, S.H. Muggleton, and M.J.E. Sternberg. Drug design by machine learning: The use of inductive logic programming to model the

| Compound Id. | Name | Actual | Progol prediction |
|---|---|---|---|
| 6533-68-2 | Scopolamine hydrobroamide | - | + |
| 147-47-7 | 1,2-Dihydro-2,2,4-trimethyquinoline | + | + |
| 8003-22-3 | D&C Yellow No. 11 | + | + |
| 78-84-2 | Isobutyraldehyde | - | + |
| 125-33-7 | Primaclone | + | + |
| 84-65-1 | Anthraquinone | T.B.A. | + |
| 518-82-1 | Emodin | T.B.A. | + |
| 5392-40-5 | Citral | T.B.A. | + |
| 104-55-2 | Cinnamaldehyde | T.B.A. | + |
| 76-57-3 | Codeine | - | - |
| 75-52-8 | Nitromethane | - | - |
| 109-99-9 | Tetrahydrofuran | + | - |
| 1948-33-0 | t-Butylhydroquinone | - | - |
| 100-41-4 | Ethylbenzene | + | - |
| 126-99-8 | Chloroprene | + | - |
| 127-00-4 | 1-Chloro-2-Propanol | T.B.A. | - |
| 11-42-2 | Diethanolamine | T.B.A. | - |
| 77-09-8 | Phenolphthalein | + | - |
| 110-86-1 | Pyridine | T.B.A. | - |
| 1300-72-7 | Xylenesulfonic acid, Na | - | - |
| 98-00-0 | Furfuryl alcohol | T.B.A. | - |
| 111-76-2 | Ethylene glycol monobutyl ether | T.B.A. | - |
| 115-11-7 | Isobutene | T.B.A. | - |
| 93-15-2 | Methyleugenol | T.B.A. | - |
| 434-07-1 | Oxymetholone | T.B.A. | - |
| 10026-24-1 | Cobalt sulfate heptahydrate | T.B.A. | not predicted |
| 1313-27-5 | Molybdenum trioxide | T.B.A. | not predicted |
| 1303-00-0 | Gallium arsenide | T.B.A. | not predicted |
| 7632-00-0 | Sodium nitrite | T.B.A. | not predicted |
| 1314-62-1 | Vanadium pentozide | T.B.A. | not predicted |

Figure 2: Progol predictions for PTE-2. The first column are the compound identifiers in the NTP database. The column headed "Actual" are tentative classifications from the NTP. Here the entry T.B.A. means "to be announced" – confirmed classifications will be available by July, 1998. An entry "+" means carcinogenic, and "-" means non-carcinogenic. The 5 compounds not predicted are inorganic compounds – Progol's rules are applicable to organic compounds only.

structure-activity relationships of trimethoprim analogues binding to dihydrofolate reductase. *Proc. of the National Academy of Sciences*, 89(23):11322–11326, 1992.

[King et al., 1996] R.D. King, S.H. Muggleton, A. Srinivasan, and M.J.E. Sternberg. Structure-activity relationships derived by machine learning: The use of atoms and their bond connectivities to predict mutagenicity by inductive logic programming. *Proc. of the National Academy of Sciences*, 93:438–442, 1996.

[Langley et al., 1983] P. Langley, G.L Bradshaw, and H. Simon. Rediscovering chemistry with the Bacon system. In R. Michalski, J. Carbonnel, and T. Mitchell, editors, *Machine Learning: An Artificial Intelligence Approach*, pages 307–330. Tioga, Palo Alto, CA, 1983.

[Lewis et al., 1990] D.F.V. Lewis, C. Ionnides, and D.V. Parke. A prospective toxicity evaluation (COMPACT) on 40 chemicals currently being tested by the National Toxicology Program. *Mutagenesis*, 5:433–436, 1990.

[Muggleton and Feng, 1990] S.H. Muggleton and C. Feng. Efficient induction of logic programs. In *Proceedings of the First Conference on Algorithmic Learning Theory*, Tokyo, 1990. Ohmsha.

[Muggleton et al., 1992] S. Muggleton, R. King, and M. Sternberg. Predicting protein secondary structure using inductive logic programming. *Protein Engineering*, 5:647–657, 1992.

[Muggleton et al., 1996] S.H. Muggleton, A. Srinivasan, R.D. King, and M.J.E. Sternberg. Biochemical knowledge discovery using Inductive Logic Programming. In R. Michalski, M. Kubat, and I. Bratko, editors, *Methods and Applications of Machine Learning, Data Mining and Knowledge Discovery*. John Wiley, 1996.

[Muggleton, 1995] S. Muggleton. Inverse Entailment and Progol. *New Gen. Comput.*, 13:245–286, 1995.

[Rosenkranz and Klopman, 1990] H.S. Rosenkranz and G. Klopman. Predicition of the carcinogenecity in rodents of chemicals currently being tested by the US

National Toxicology Program. *Mutagenesis*, 5:425–432, 1990.

[Sanderson and Earnshaw, 1991] D.M. Sanderson and C.G. Earnshaw. Computer prediction of possible toxic action from chemical structure. *Human Exp Toxicol*, 10:261–273, 1991.

[Srinivasan et al., 1996] A. Srinivasan, S.H. Muggleton, R.D. King, and M.J.E. Sternberg. Theories for mutagenicity: a study of first-order and feature based induction. *Artificial Intelligence*, 85:277–299, 1996.

[Tennant et al., 1990] R.W. Tennant, J. Spalding, S. Stasiewicz, and J. Ashby. Prediction of the outcome of rodent carcinogenicity bioassays currently being conducted on 44 chemicals by the National Toxicology Program. *Mutagenesis*, 5:3–14, 1990.

# Challenge:
# What is the Impact of Bayesian Networks on Learning?

| Nir Friedman | Moises Goldszmidt | David Heckerman | Stuart Russell |
|---|---|---|---|
| Computer Science Div. | SRI International | Microsoft Research | Computer Science Div. |
| University of California | 333 Ravenswood Ave. | One Microsoft Way | University of California |
| Berkeley, CA 94720 | Menlo Park, CA 94025 | Redmond, WA 98052 | Berkeley, CA 94720 |

## Abstract

In recent years, there has been much interest in *learning* Bayesian networks from data. Learning such models is desirable simply because there is a wide array of off-the-shelf tools that can apply the learned models as expert systems, diagnosis engines, and decision support systems. Practitioners also claim that adaptive Bayesian networks have advantages in their own right as a non-parametric method for density estimation, data analysis, pattern classification, and modeling. Among the reasons cited we find: their semantic clarity and understandability by humans, the ease of acquisition and incorporation of prior knowledge, the ease of integration with optimal decision-making methods, the possibility of causal interpretation of learned models, and the automatic handling of noisy and missing data.

In spite of these claims, and the initial success reported recently, methods that learn Bayesian networks have yet to make the impact that other techniques such as neural networks and hidden Markov models have made in applications such as pattern and speech recognition. In this paper, we challenge the research community to identify and characterize domains where induction of Bayesian networks makes the critical difference, and to quantify the factors that are responsible for that difference. In addition to formalizing the challenge, we identify research problems whose solution is, in our view, crucial for meeting this challenge.

## 1 Introduction

A Bayesian network is a graphical representation of the joint probability distribution for a set of variables. The representation was originally designed to encode the uncertain knowledge of an expert [Wright, 1921; Howard and Matheson, 1981; Pearl, 1988], and indeed today, they play a crucial role in modern expert systems, diagnosis engines, and decision support systems [Heckerman et al., 1995]. They also have become the representation of choice among researchers interested in uncertainty in AI. One often-cited merit of Bayesian networks is that they have formal probabilistic semantics and yet can serve as a natural mirror of knowledge structures in the human mind [Spirtes et al., 1993; Heckerman et al., 1995; Pearl, 1995]. This facilitates the encoding and interpretation of knowledge in terms of a probability distribution, enabling inference and optimal decision making.

A Bayesian network consists of two components. The first is a directed acyclic graph in which each vertex corresponds to a random variable. This graph represents a set of conditional independence properties of the represented distribution: each variable is probabilistically independent of its non-descendants in the graph given the state of its parents. This graph captures the qualitative *structure* of the probability distribution, and is exploited for efficient inference and decision making. Thus, while Bayesian networks can represent arbitrary probability distributions, they provide computational advantage for those distributions that can be represented with a simple structure. The second component is a collection of *local interaction models* that describe the conditional probability $p(X_i|\mathbf{Pa}_i)$ of each variable $X_i$ given its parents $\mathbf{Pa}_i$ (see Figure 1). Together, these two components represent a unique joint probability distribution over the complete set of variables $\mathbf{X}$ [Pearl, 1988]. The joint distribution is given by the following equation:

$$p(\mathbf{X}) = \prod_{i=1}^{n} p(X_i|\mathbf{Pa}_i) \quad (1)$$

It can be shown that this equation implies the conditional independence semantics of the graphical structure given earlier.

Equation 1 shows that the joint distribution specified by a Bayesian network has a factored representation as the product of individual local interaction models. Sparse Bayesian networks therefore correspond to concise representations of joint distributions. If the number of parents of any variable is bounded by a constant $k$, then (for most reasonable representations of the local interaction models, including all discrete models) the Bayesian network requires a number of parameters that is *linear* in the number of variables, instead of exponential for an unstructured representation. This observation is, of course, directly relevant to the learning problem, since concise parameterizations lead to statistically efficient learning—*provided* that the problem domain admits of a sparse structure of conditional dependencies. The latter assumption is of course directly related to the usefulness of Bayesian networks as models of human knowledge structures.

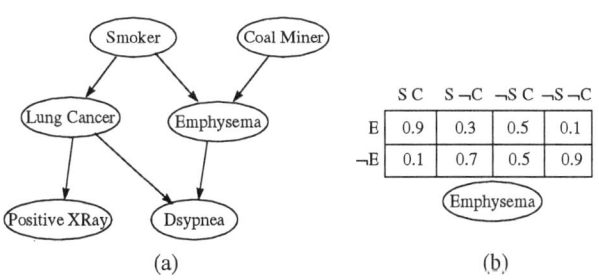

Figure 1: (a) A simple probabilistic network showing a proposed causal model. (b) A node with associated conditional probability table. The table gives the conditional probability of each possible value of the variable *Emphysema*, given each possible combination of values of the parent nodes *Smoker* and *CoalMiner*.

The characterization given by Equation 1 is a strictly mathematical characterization in terms of probabilities and conditional independence statements. An informal connection can be made between this characterization and the intuitive notion of *direct causal influence*. It has been noted that if the edges in the network structure correspond to causal relationships, where a variable's parents represent the direct causal influences on that variable, then resulting networks are often very concise and accurate descriptions of the domain. Thus it appears that in many practical situations, a Bayesian network provides a natural way to encode causal information. We can state this more precisely as the Causal Markov Assumption (CMA): if a network is constructed simply by connecting variables to other variables that they directly causally influence, then the resulting network *interpreted according to Equation 1* will correctly reflect the conditional independence statements that hold in the domain.

The naturalness of using causal information directly in constructing formally characterizable knowledge structures has made it possible to encode the knowledge of many experts. As a result, Bayesian networks have been incorporated into many expert systems, diagnosis engines, and decision-support systems [Heckerman et al., 1995]. Nonetheless, it is often difficult and time-consuming to construct Bayesian networks from expert knowledge alone.

This observation, together with the fact that data is becoming increasingly available and cheaper to acquire has led to a growing interest in using data to learn both the structure and probabilities of a Bayesian network. Several groups have worked on learning structure from scratch [Spirtes et al., 1993; Pearl, 1995; Friedman et al., 1997] or with weak constraints such as variable ordering [Cooper and Herskovits, 1992, for example], while others have worked on learning structure by refining an initial model [Heckerman et al., 1994]. Learning probabilities, which is non-trivial when the network contains hidden variables or the data set has missing values, can be done by a variety of methods including EM [Lauritzen, 1991; Lauritzen, 1995; Spiegelhalter et al., 1993; Olesen et al., 1992; Spiegelhalter and Cowell, 1992; Heckerman, 1996] gradient-based methods [Laskey, 1990; Golmard and Mallet, 1991; Neal, 1992; Russell et al., 1995], and Monte-Carlo techniques [Neal, 1993].

These researchers have cited several benefits of using the Bayesian-network representation, with its causal interpretation, as a tool for learning:

1. *Incorporation of prior knowledge.* Bayesian networks facilitate the translation of human knowledge into probabilistic form, making it suitable for refinement by data.

2. *Validation and insight.* In many cases, a learned Bayesian network can be given a causal interpretation. Consequently, a Bayesian network is more easily understood than "black box" representations such as neural networks. As an immediate byproduct, people will more readily accept the recommendations of a Bayesian network than those of a model justified only by its raw predictive performance. In addition, users are more likely to gain insights from Bayesian networks.

3. *Learning causal interactions.* Unlike purely probabilistic relationships, causal relationships allow us to make predictions given direct interventions or manipulations of the world. Therefore, by learning with Bayesian networks, there is a hope that we can make better predictions in the face of intervention. Learning causal relationships is crucial in scientific discovery, where interventional studies are often expensive or impossible. Similarly, the ability to learn

causal relationships is crucial for intelligent agents that must act in their environment on the basis of acquired knowledge.

Other benefits of using Bayesian networks for learning are derived from their probabilistic semantics. Because sophisticated yet efficient methods have been developed for using a Bayesian network to answer probabilistic queries, they can be used both for predictive inference and diagnostic (or abductive) inference. This is in contrast to standard regression and classification methods (e.g., feed forward neural networks and decision trees) that encode only the probability distribution of a target variable given several input variables. Whereas the Bayesian-network representation can describe the casual ordering in the domain, there are no restrictions as to the directions of the queries. Thus, there is no inherent notion of inputs and outputs of the network. This property also allows Bayesian networks to reason efficiently with missing values, by computing the marginal probability of the query given the observed values. One other cited benefit of the Bayesian-network representation, which derives from its probabilistic nature, is that it can be used to determine optimal decisions.

Even though these claims are compelling, it is yet to be demonstrated that these issues provide tangible advantages in applications. The purpose of this paper is therefore to challenge researchers to characterize and quantify these claims, including the specification of domains where they made a difference in the efficiency of learning (e.g., through the use of prior knowledge), in the quality of the resulting model (e.g., a new causal theory that is accepted by the experts), or in the deployment the system (e.g., through combination with utility estimation).

We hope that this challenge will focus the research community on a high-impact research agenda. We believe that in order to meet the challenge, at least three kinds of activities will take place:

1. We believe that experience in applications provide valuable lessons. Thus, we are interested in "success stories," that is papers that describe applications where learning Bayesian networks has led to significant advantages over other methods. These papers should attempt to distill the characteristics of the problem that made Bayesian networks the preferred solution.

2. We propose a series of "bake-offs" to experimentally evaluate how Bayesian networks and alternative approaches can exploit prior knowledge, deal with missing data, and learn causal models. These bake-offs will allow for controlled study and evaluation of the impact of the various alternatives. Section 2 describes our proposal for organizing these bake-offs and evaluating the results.

3. We identify specific technical research problems whose solution is, in our view, crucial for meeting the challenge. In Section 3, we outline these problems.

For a comprehensive assessment of the state of the art of the field, we refer the reader to Heckerman [1996] as well as the papers cited earlier.

## 2 Experimental Bake-Offs

As mentioned in the introduction, we propose a series of bake-off competitions with the objective to evaluate the extent to which the special features of Bayesian networks benefit the learning task and the resulting models. To this end, we will maintain a web site, where data sets, background information about them, and evaluation criterion will be made available.

We are currently assembling several collections of data sets, both syntactic and real, for the bake-offs described below. In order to preserve the validity of these experiments, some of them will be done using "blind" evaluation. That is, participants in the bake-off will have access to a portion of the training data and will have to register the learned models by a certain date. The learned models will be tested on unseen data by the central web server.[1]

Our hope is that these data sets will provide appropriate test beds for testing theories and new algorithms. We encourage practitioners and researchers interested in other induction methods to participate in these bake-offs and to use these datasets.[2]

The exact evaluation criteria will be decided based on inputs from participants and the discussions that will follow the presentation of this challenge. These criteria will include various error measures such as log-loss, cross-entropy or KL distance, and classification accuracy, and will depend on the different learning strategies (e.g., batch learning and incremental learning).

We propose to focus these bake-offs on three issues: incorporation of background knowledge, handling of missing data, and learning causal interactions.

**Background knowledge.** The main problem with experiments testing the influence of background knowledge in the learning process is to make the expertise readily available to all participants in a way that does not provide advantages to any particular learning method.

---

[1] This stricture is intended to get around the irresistible tendency, noted during the Statlog project, for researchers to "peek" at test data and report "best" results selected from runs with different knob settings.

[2] Toward this end, we plan to submit these data sets to both the UCI machine learning repository and the XXX repository at Toronto.

(A similar problem has arisen with experimental studies of inductive logic programming methods; we expect to compare notes.) We are considering two strategies. The first one is to provide summary of background expert knowledge in the form of free-form text and tables. The second strategy is to provide data about a domain familiar enough that anybody can be regarded as an expert, and define a prediction task in that domain. One such domain is that of TV shows. Data could be provided about shows, viewers characteristics etc., and the task would be to predict the shows that new subjects will like based on other shows they like.

**Missing data.** The basic problem of coping with missing values and hidden variables in the data set is addressed very simply in Bayesian networks, because likelihoods can be computed no matter what subset of variables are available as evidence. The tricky problem arises when the data is missing due to specific values that other variables take. In this case, the *failure to observe a variable* may in itself be informative about the true state of the world [Rubin, 1978]. In principle, a successful induction algorithm would be able to take advantage of a good model about the relationship between the state of the world and what variables are missing.

For this challenge we will provide both synthetic data and real life data. The former allows controlled experiments that account for the number of missing values and the dependence of omissions on the true state of the world. We also plan to provide data sets where the target task involves a large amount of incomplete information.

**Causal interactions.** In this study, we will attempt to learn cause and effect from observational studies. Ideally, we will also have interventional data to verify the real causal structure of the domain. We are currently investigating data sets in social sciences and epidemiology; the University of Michigan survey data archive contains thousands of data sets, some running into the gigabytes, that might be very suitable. We will also try to provide synthetic data as follows. We will contact experts that will provide us with causal models for their domain (e.g., epidemiology), from which we will create synthetic data. Since prior knowledge plays a significant role in the induction of causal theories, these experts would also provide summary of the prior knowledge they consider reasonable for the domain they created (e.g., known temporal ordering relations and possible latent causes).

We plan to evaluate the learned causal models as follows. First, we will measure how well they predict the effects of interventions (using standard statistical measures). Second, we will measure how many causal interactions were correctly and incorrectly identified.

# 3 Technical Challenges

Many researchers are now concentrating on learning in more expressive probabilistic models, including hybrid (discrete and continuous) models [Lauritzen and Wermuth, 1989], mixed (undirected and directed) models [Buntine, 1994; Cooper, 1995; Spirtes et al., 1995], dynamic Bayesian network models representing stochastic processes [Russell et al., 1995], and stochastic grammars [Stolcke and Omohundro, 1993]. Another important problem is the specification of prior distributions over parameters—most current work makes strong assumptions such as parameter independence and likelihood equivalence. MacKay [1992] and others are working on hierarchical models that relax the assumption of parameter independence. A third area of active research is the development of efficient approximation algorithms for probabilistic inference—a key component of learning—including Monte-Carlo [Thomas et al., 1992] and variational methods [Saul et al., 1996].

There are two technical challenges that we believe are critical to the success of Bayesian networks and for which much work needs to be done. One challenge is the efficient handling of incomplete data. One important subcomponent of the first task is the creation of search methods for Bayesian networks with hidden variables. Clever search strategies are needed to constrain the infinite search space. In addition, as mentioned in the previous section, learning with incomplete data is particularly difficult when the mere failure to observe some variable is informative about the true state of the world. For example, the fact that a patient drops out of a drug study may suggest that the he or she could not tolerate the effects of the drug. Several researchers have developed basic principles and methods for dealing with such situations, including Rubin [1978], Robins [1986], Cooper [1995], Spirtes et al. [1995], and Chickering and Pearl [1996], but more work needs to be done to connect these basic principles with graphical models and to make these methods more efficient.

A second challenge is the creation of simple but expressive probability distributions for the local interaction models in a Bayesian network. Most work on learning with Bayesian networks concentrates on discrete variables where each variable is associated with a set of multinomial distributions, one distribution for each configuration of its parents. Thiesson [1995] discusses a class of local likelihoods for discrete variables that use fewer parameters. Geiger and Heckerman [1994] and Buntine [1994] discuss simple linear local likelihoods for continuous variables that have continuous and discrete variables. Buntine [1994] also discusses a general class of local likelihoods from the exponential family. Nonetheless, alternative likelihoods for discrete and continuous variables are desired. Local likelihoods with fewer parameters might allow for the selection of correct models with less data as demonstrated empirically by Friedman and Goldszmidt [1996]. In addition, local likelihoods that express more accurately the data generating process would allow for easier interpretation of the resulting models.

## 4 Concluding Remarks

There are of course a number of other promising research avenues, challenges, and needs that directly concern learning Bayesian networks and their impact on real world applications for data analysis, pattern recognition and classification, knowledge discovery, and others. There is for example the belief that Bayesian networks will play a significant role in issues of feature engineering, selection and abstraction, as well as in the formulation of the learning problem and the various tradeoff evaluations of knowledge representation versus quality of prediction of the induced models. Our particular choice of problems and challenges were based not only on immediate relevance, but more important, on the fact that progress can be accomplished in the next two years.

Needless to say, our hope is that researchers and practitioners of all induction methods will participate in our challenge so that a fair comparison can be made and true scientific knowledge can be distilled.

## Acknowledgments

The participation of Nir Friedman and Stuart Russell was funded by ARO through MURI DAAH04-96-1-0341, and by NSF through FD96-34215.

## References

[Buntine, 1994] Buntine, W. (1994). Operations for learning with graphical models. *Journal of Artificial Intelligence Research*, 2:159–225.

[Chickering and Pearl, 1996] Chickering, D. and Pearl, J. (1996). A clinician's tool for analyzing non-compliance. In *Proceedings of the Thirteenth National Conference on Artificial Intelligence (AAAI-96)*, Portland, OR, volume 2, pages 1269–1276.

[Cooper, 1995] Cooper, G. (1995). Causal discovery from data in the presence of selection bias. In *Proceedings of the Fifth International Workshop on Artificial Intelligence and Statistics*, pages 140–150, Fort Lauderdale, FL.

[Cooper and Herskovits, 1992] Cooper, G. and Herskovits, E. (1992). A Bayesian method for the induction of probabilistic networks from data. *Machine Learning*, 9:309–347.

[Friedman et al., 1997] Friedman, N., Geiger, D., and Goldszmidt, M. (1997). Bayesian network classifiers. *Machine Learning*, to appear.

[Friedman and Goldszmidt, 1996] Friedman, N. and Goldszmidt, M. (1996) Learning Bayesian networks with local structure. In *Proceedings of Twelfth Conference on Uncertainty in Artificial Intelligence,* Portlan, OR. Morgan Kaufmann.

[Geiger and Heckerman, 1994] Geiger, D. and Heckerman, D. (1994). Learning Gaussian networks. In *Proceedings of Tenth Conference on Uncertainty in Artificial Intelligence,* Seattle, WA, pages 235–243. Morgan Kaufmann.

[Golmard and Mallet, 1991] Golmard, J.-L. and Mallet, A. (1991). Learning probabilities in causal trees from incomplete databases. *Revue d'Intelligence Artificielle*, 5:93–106.

[Heckerman, 1995] Heckerman, D. (1995). A Bayesian approach for learning causal networks. In *Proceedings of Eleventh Conference on Uncertainty in Artificial Intelligence,* Montreal, QU, pages 285–295. Morgan Kaufmann.

[Heckerman, 1996] Heckerman, D. (1996). A Tutorial on learning with Bayesian networks. Microsoft Research Technical Report MSR-TR-95-06. Updated Nov. 1996.

[Heckerman et al., 1994] Heckerman, D., Geiger, D., and Chickering, M. (1994). Learning Bayesian networks: The combination of knowledge and statistical data. Technical Report MSR-TR-94-09, Microsoft Research, Redmond, Washington.

[Heckerman et al., 1995] Heckerman, D., Mamdani, A., and Wellman, M. (1995). Real-world applications of Bayesian networks. *Communications of the ACM*, 38.

[Howard and Matheson, 1981] Howard, R. and Matheson, J. (1981). Influence diagrams. In Howard, R. and Matheson, J., editors, *Readings on the Principles and Applications of Decision Analysis*, volume II, pages 721–762. Strategic Decisions Group, Menlo Park, CA.

[Laskey, 1990] Laskey, K. B. (1990). Adapting connectionist learning to Bayes networks. *International Journal of Approximate Reasoning*, 4:261–282.

[Lauritzen, 1991] Lauritzen, S. L. (1991). The EM algorithm for graphical association models with missing data. Technical Report TR-91-05, Department of Statistics, Aalborg University.

[Lauritzen, 1995] Lauritzen, S. L. (1995). The EM algorithm for graphical association models with missing data. *Computational Statistics and Data Analysis*, 19:191–201.

[Lauritzen and Wermuth, 1989] Lauritzen, S. and Wermuth, N. (1989). Graphical models for associations between variables, some of which are qualitative and some quantitative. *Annals of Statistics*, 17:31–57.

[MacKay, 1992] MacKay, D. (1992). A practical Bayesian framework for backpropagation networks. *Neural Computation*, 4:448–472.

[Neal, 1992] Neal, R. M. (1992). Connectionist learning of belief networks. *Artificial Intelligence*, 56:71–113.

[Neal, 1993] Neal, R. (1993). Probabilistic inference using Markov chain Monte Carlo methods. Technical Report CRG-TR-93-1, Department of Computer Science, University of Toronto.

[Olesen et al., 1992] Olesen, K. G., Lauritzen, S. L., and Jensen, F. V. (1992). aHUGIN: A system for creating adaptive causal probabilistic networks. In *Proceedings of the Eighth Conference on Uncertainty in Artificial Intelligence (UAI-92)*, Stanford, California. Morgan Kaufmann.

[Pearl, 1988] Pearl, J. (1988). *Probabilistic Reasoning in Intelligent Systems: Networks of Plausible Inference*. Morgan Kaufmann, San Mateo, CA.

[Pearl, 1995] Pearl, J. (1995). Causal diagrams for empirical research. *Biometrika*, 82:669–710.

[Robins, 1986] Robins, J. (1986). A new approach to causal inference in mortality studies with sustained exposure results. *Mathematical Modelling*, 7:1393–1512.

[Rubin, 1978] Rubin, D. (1978). Bayesian inference for causal effects: The role of randomization. *Annals of Statistics*, 6:34–58.

[Russell et al., 1995] Russell, S., Binder, J., Koller, D., and Kanazawa, K. (1995). Local learning in probabilistic networks with hidden variables. In *Proceedings of the Fourteenth International Joint Conference on Artificial Intelligence (IJCAI-95)*, pages 1146–52, Montreal, Canada. Morgan Kaufmann.

[Saul et al., 1996] Saul, L., Jaakkola, T., and Jordan, M. (1996). Mean field theory for sigmoid belief networks. *Journal of Artificial Intelligence Research*, 4:61–76.

[Spiegelhalter et al., 1993] Spiegelhalter, D., Dawid, P., Lauritzen, S., and Cowell, R. (1993). Bayesian analysis in expert systems. *Statistical Science*, 8:219–282.

[Spiegelhalter and Cowell, 1992] Spiegelhalter, D. J. and Cowell, R. G. (1992). Learning in probabilistic expert systems. In Bernardo, J. M., Berger, J. O., Dawid, A. P., and Smith, A. F. M., editors, *Bayesian Statistics 4*, Oxford. Oxford University Press.

[Spirtes et al., 1993] Spirtes, P., Glymour, C., and Scheines, R. (1993). *Causation, Prediction, and Search*. Springer-Verlag, New York.

[Spirtes et al., 1995] Spirtes, P., Meek, C., and Richardson, T. (1995). Causal inference in the presence of latent variables and selection bias. In *Proceedings of Eleventh Conference on Uncertainty in Artificial Intelligence*, Montreal, QU, pages 499–506. Morgan Kaufmann.

[Stolcke and Omohundro, 1993] Stolcke, A. and Omohundro, S. (1993). Hidden Markov model induction by Bayesian model merging. In *Advances in Neural Information Processing Systems 5*, volume 5, pages 11–18, San Mateo, CA. Morgan Kaufmann.

[Thiesson, 1995] Thiesson, B. (1995). Score and information for recursive exponential models with incomplete data. Technical report, Institute of Electronic Systems, Aalborg University, Aalborg, Denmark.

[Thomas et al., 1992] Thomas, A., Spiegelhalter, D., and Gilks, W. (1992). Bugs: A program to perform Bayesian inference using Gibbs sampling. In Bernardo, J., Berger, J., Dawid, A., and Smith, A., editors, *Bayesian Statistics 4*, pages 837–842. Oxford University Press.

[Wright, 1921] Wright, S. (1921). Correlation and causation. *Journal of Agricultural Research*, 20:557–585.

# Adaptive Web Sites: an AI Challenge

Mike Perkowitz*   Oren Etzioni
Department of Computer Science and Engineering, Box 352350
University of Washington, Seattle, WA 98195
{map, etzioni}@cs.washington.edu
(206) 616-1845   Fax: (206) 543-2969

## Abstract

The creation of a complex web site is a thorny problem in user interface design. First, different visitors have distinct goals. Second, even a single visitor may have different needs at different times. Much of the information at the site may also be dynamic or time-dependent. Third, as the site grows and evolves, its original design may no longer be appropriate. Finally, a site may be designed for a particular purpose but used in unexpected ways.

Web servers record data about user interactions and accumulate this data over time. We believe that AI techniques can be used to examine user access logs in order to automatically improve the site. We challenge the AI community to create **adaptive web sites**: sites that automatically improve their organization and presentation based on user access data.

Several unrelated research projects in plan recognition, machine learning, knowledge representation, and user modeling have begun to explore aspects of this problem. We hope that posing this challenge explicitly will bring these projects together and stimulate fundamental AI research. Success would have a broad and highly visible impact on the web and the AI community.

## 1 Introduction

The World Wide Web is becoming a key medium for information dissemination, entertainment, and communication. Examples include personal home pages, on-line malls, university course information, and much more. Many web sites quickly sprout intricate collections of pages and hyperlinks as they begin to mirror the complexity of the information they convey.

Designing a rich web site so that it readily yields its information can be tricky. Unlike the oyster that contains a single pearl, a web site often contains myriad facts, images, and hyperlinks. Many different visitors approach a popular web site — each with his or her own goals and concerns. Consider, for example, the web site for a typical computer science department. The site contains an amalgam of research project descriptions, course information, lists of graduating students, pointers to industrial affiliates, and much more. Each nugget of information is of value to someone who would like to access it readily. One might think that a well organized hierarchy would solve this problem, but we've all had the experience of banging our heads against a web site and crying out "it's got to be here *somewhere*...".

The problem of good web design is compounded by several factors beyond the fact that different visitors have distinct goals. First, the same visitor may seek different information at different times. Second, many sites outgrow their original design, accumulating links and pages in unlikely places. Third, a site may be designed for a particular kind of use, but be used in many different ways in practice; the designer's *a priori* expectations may be violated. Too often web site designs are fossils cast in HTML, while web navigation is dynamic, time-dependent, and idiosyncratic. We challenge the AI community to address this problem by creating **adaptive web sites:** *web sites that automatically improve their organization and presentation by learning from user access patterns.*

In essence, web design is a problem in user interface design. However, in contrast with vendors of shrink-wrapped software, few web site designers can afford to subject their web sites to formal usability testing in special labs. Fortunately, web users interact directly with a server maintained by the inventors of the service or authors of the content being served. As a result, data on their behavior is recorded in web server logs (see Figure 1). Because this raw data is overwhelming for an overworked webmaster to process regularly, web server logs are ripe targets for automated analysis.

Our challenge then is this: how can we build a web site which improves itself over time in response to user interactions with the site? This challenge poses a number of

---

*This research was funded in part by Office of Naval Research grant 92-J-1946, by ARPA / Rome Labs grant F30602-95-1-0024, by a gift from Rockwell International Palo Alto Research, and by National Science Foundation grant IRI-9357772.

```
24hrlab-214.sfsu.edu - - [21/Nov/1996:00:01:05 -0800] "GET /home/jones/collectors.html HTTP/1.0" 200 13119
24hrlab-214.sfsu.edu - - [21/Nov/1996:00:01:06 -0800] "GET /home/jones/madewithmac.gif HTTP/1.0" 200 855
24hrlab-214.sfsu.edu - - [21/Nov/1996:00:01:06 -0800] "GET /home/jones/gustop2.gif HTTP/1.0" 200 25460
x67-122.ejack.umn.edu - - [21/Nov/1996:00:01:08 -0800] "GET /home/rich/aircrafts.html HTTP/1.0" 404 617
x67-122.ejack.umn.edu - - [21/Nov/1996:00:01:08 -0800] "GET /general/info.gif HTTP/1.0" 200 331
203.147.0.10 - - [21/Nov/1996:00:01:09 -0800] "GET /home/smith/kitty.html HTTP/1.0" 200 5160
24hrlab-214.sfsu.edu - - [21/Nov/1996:00:01:10 -0800] "GET /home/jones/thumbnails/awing-bo.gif HTTP/1.0" 200 5117
```

Figure 1: Typical user access logs, these from a computer science web site. Each entry corresponds to a single request to the server and includes originating machine, time, and URL requested. Note the series of accesses from each of two users (one from SFSU, one from UMN).

difficult, but not impossible, questions:

- **What kinds of generalizations can we draw from user access patterns and what kinds of changes could we make?** Suppose we maintain a web site containing information about various automobiles, organized by manufacturer. We observe that visitors who look at the *Ford Windstar* minivan page also tend to look at the *Dodge Caravan* and *Mazda MPV* minivan pages. We might therefore create a new page for minivans, which cuts across the existing manufacturer-based organization and provides a new *view* of the site.

- **How do we design a site for adaptivity?** We might specifically design parts of the site to be changeable. For example, we might present our users with a "tour guide" (as in [Armstrong *et al.*, 1995]) and have changes to the site be presented as the agent's suggestions. Alternatively, we might annotate our HTML with directives stating where and how changes can be made. Or we may provide semantic information about the entire site, allowing the agent to reason about the relationships between everything, perhaps representing the entire site as a database (see [Fernandez *et al.*, 1997]).

- **How do we effectively collaborate with a human webmaster to suggest and justify potential adaptations?** Suppose the human webmaster is still responsible for the final product. Instead of changing web pages directly, our system might accumulate observations and suggested changes and present them to the webmaster, clearly explaining its observations and justifying the changes it recommends.

- **How do we move beyond one-shot learning algorithms to web sites that continually improve with experience?** Over time, our adaptive web site will accumulate a great deal of data about its users and should be able to use its rich history to continually evolve and improve.

Our department maintains a web site for its introductory computer science course. This site contains schedules, announcements, assignments, and other information important to the hundreds of students who take the course every quarter. Enough information is available that important documents can be hard to find or entirely lost in the clutter. Imagine, however, if the site were able to determine what was important and make that information easiest to find. Important pages would be available from the site's front page. Important links would appear at the top of the page or be highlighted. Timely information would be emphasized, and obsolete information would be quietly moved out of the way.

There are several factors that make this challenge both appropriate and timely for the AI community. First, the growing popularity and complexity of the web underscores the importance of the challenge. Second, virtually all existing web sites are *not* adaptive, yet data to support the learning process is readily available in web server logs. Clearly, here is an opportunity for AI! Finally, a number of disconnected projects in machine learning [Armstrong *et al.*, 1995], data mining, knowledge representation, plan recognition [Kautz, 1987; Pollack, 1990], and user modeling [Fink *et al.*, 1996] have begun to explore aspects of the problem. Framing the problem explicitly in this paper could help bring these disparate approaches together.

We pose our challenge as a particular task to be accomplished by any means available. Many advances in artificial intelligence, both practical and theoretical, have come about in response to such task-oriented approaches. The quest to build a better chess-playing computer, for example, has led to many advances in search techniques (e.g., [Anantharaman *et al.*, 1990]). The autonomous land vehicle project at CMU [Thorpe, 1990] has resulted in not only a highway-cruising vehicle but also breakthroughs in vision, robotics, and neural networks. The quest to build autonomous software agents has similarly led to both practical and theoretical advances. For example, the Internet Softbot project has yielded both deployed softbots and advances in planning, knowledge representation, and machine learning [Etzioni, 1996].

We believe that the goal of creating self-improving web sites is a similar task: one whose accomplishment will require breakthroughs in different areas of AI. In this paper we discuss possible approaches to this task and how to evaluate the community's progress. In section 2, we present two basic approaches to creating an adaptive web site. We illustrate both with ongoing research and examples. In section 3, we discuss how to evaluate research on this challenge, discussing practical alternatives as well as open questions. Throughout, we pose **Challenge** questions intended to suggest research directions and illustrate where the open questions lie.

## 2 Approaches to adaptive web sites

Sites may be adaptive in two basic ways. First, the site may focus on *customization*: modifying web pages in real time to suit the needs of individual users. Second, the site may focus on *optimization*: altering the site itself to make navigation easier for all. We illustrate these two basic approaches with examples drawn from current AI research. Whether we modify our web pages online or offline, we must use information about user access patterns and the structure of our site. Much of this information is available in access logs and in the site's HTML, but this may not be sufficient; we also discuss how to support adaptivity with *meta-information* — information about page content. Finally, we examine other issues that arise in designing adaptive web sites.

### 2.1 Customization

*Customization* is adjusting the site's presentation for an individual user. Customization allows fine-grained improvement, since the interface may be completely tailored to each individual user. One way for a site to respond to particular visitors is to allow *manual customization*: allowing users to specify display options that are remembered during the entire visit and from one visit to the next. The Microsoft Network (at http://www.msn.com), for example, allows users to create home pages with customized news and information displays. Every time an individual visits her MSN home page, she sees the latest pickings from the site presented according to her customizations.

Path prediction, on the other hand, customizes *automatically* by attempting to guess where the user wants to go and taking her there more quickly. A path prediction system must answer at least the following questions.

- **What are we predicting?** We may try to predict the user's next step. For example, if we can predict what link on a page a particular user will follow, we might highlight the link or bring it to the top of the page. Alternatively, we may try to predict the user's eventual goal; if we can determine what page at the site a visitor is looking for, we can present it to her immediately.

- **On what basis do we make predictions?** We might use only a particular individual's actions to predict where she will go next. On the other hand, we might generalize from multiple users to gather data more quickly.

- **What kinds of modifications do we make on the basis of our predictions?** We may do as little as highlighting selected links (by making them bold or putting graphics around them, for example) or as much as synthesizing a brand new page that we *think* the user wants to see.

The WebWatcher [Armstrong *et al.*, 1995] (see http://www.cs.cmu.edu/ ~webwatcher/) learns to predict what links users will follow on a particular page as a function of their specified interests. WebWatcher observes many users over time and attempts to learn, given a user's current page and stated interests, where she will go next. A link that WebWatcher believes you are likely to follow will be highlighted graphically and duplicated at the top of the page. Visitors to a site are asked, in broad terms, what they are looking for. Before they depart, they are asked if they found what they wanted. WebWatcher uses the paths of people who indicated success as examples of successful navigations. If, for example, many people who were looking for "personal home pages" follow the "people" link, then WebWatcher will tend to highlight that link for future visitors with the same goal.

Instead of predicting a user's next action based on the actions of many, we might try to predict the user's ultimate goal based on what she has done so far. *Goal recognition* [Kautz, 1987; Pollack, 1990] is the problem of identifying, from a series of actions, what an agent is trying to accomplish. Lesh and Etzioni [Lesh and Etzioni, 1995] pose this problem in a domain-independent framework and investigate it empirically in the Unix domain: by watching over a user's shoulder, can we figure out what she is trying to accomplish (and offer to accomplish it for her)? They model user actions as planning operators. Assuming users behave somewhat rationally, they use these actions' precondition/postcondition representation to reason from what a user has done to what she must be trying to do. In the web domain, we observe a visitor's navigation through our site and try to determine what page she is seeking. If we can do this quickly and accurately, we can then offer the desired page immediately.

> **Challenge:** Can we formalize user navigation of the web as a planning process that is amenable to goal recognition? Do user actions on the web carry enough evidence of their purpose?

The AVANTI Project [Fink *et al.*, 1996] (see http://zeus.gmd.de/ projects/avanti.html) focuses on dynamic customization based on users' needs and tastes. As with the WebWatcher, AVANTI relies partly on users providing information about themselves when they enter the site. Based on what it knows about the user, AVANTI attempts to predict both the user's eventual goal and her likely next step. AVANTI will prominently present links leading directly to pages it thinks a user will want to see. Additionally, AVANTI will highlight links that accord with the user's interests. AVANTI is illustrated on a hypothetical Louvre Museum web site. For example, when a disabled tourist comes to the site, links regarding disabled access and tourist information are emphasized. AVANTI relies on users providing some information about themselves in an initial dialogue; the site then uses this information to guide its customization throughout the user's exploration of the site. AVANTI also attempts to guess where the user might go based on what she has looked at so far. For example, if our disabled tourist looks at a number of paintings at the site, AVANTI will emphasize paintings links as it continues to serve pages. As with the WebWatcher, we might ask if we can avoid AVANTI's requirement that users explicitly provide information.

## 2.2 Optimization

Whereas customization focuses on individuals, *optimization* tries to improve the site as a whole. Instead of making changes for each user, the site learns from all users to make the site easier to use. This approach allows even new users, about whom we know nothing, to benefit from the improvements.

We may view a web site's design as a particular point in the vast space of possible designs. Improving the site, then, corresponds to searching in this space for a "better" design. Assuming we have a way of measuring "better", we may view this as a classical AI search problem. One possible quality metric would be to measure the amount of *effort* a visitor needs to exert on average in order to find what she is looking for at our site. Effort is defined as a function of the number of links traversed and the difficulty of finding those links. For example, a site whose most popular local page is buried five links away from the front page could be improved by making that page accessible from a readily obvious link on the front page. We can navigate through this space by performing transformations on the site — adding or removing links, rearranging links, creating new web pages, etc. If we guarantee that each transformation improves the quality of the site, we are performing a hillclimbing search.

> **Challenge:** How large is this search space and what is an appropriate search strategy? Can we restructure the space to avoid searching large portions of it?

In [Perkowitz and Etzioni, 1997] we sketch the design of a system with a repertoire of transformations that aim to improve a site's organization; transformations include rearranging and highlighting links as well as synthesizing new pages. Our system learns from common patterns in the user access logs and decides how to transform the site to exploit those patterns and make the site easier to navigate. For example, the web site for our department's introductory computer science course contains a web page for each homework assignment given during the course. After each assignment's due date, a solution set for that assignment is made available. Our system would observe that after an assignment's due date many visitors look at the solution set; in fact, the most recent solution set is one of the most popular pages at the site. This observation would lead the system to *promote* the solution set by giving it a prominent link on the front page. Promotion — making the link to a page more prominent — is a simple but effective transformation. We have implemented a form of promotion on an existing web site and have found that approximately 10% of our 10,000-15,000 daily page accesses are through automatically generated links; roughly 25% of all visitors click through at least one such link. Of course, we note that promoting a link may be a self-fulfilling prophecy — making a page more prominent may increase its popularity, artificially inflating the site's apparent success at adaptation.

A more ambitious transformation is *clustering* — synthesizing a brand new web page that contains links to a set of related objects. From available data, the system must infer that a set of pages at the site are related and group them together. This inference might be based on content (e.g., when a number of pages cover the same topic) or on user navigation patterns (e.g., when visitors to one page are particularly likely to visit certain others). As final exams approach, students tend to look at multiple solutions sets on each visit. Even though the solution pages are not linked together directly, visitors navigate from one to another (via intervening pages) on their own. This pattern suggests that the solution sets form a meaningful group in our visitors' heads, which does not appear on our web site — solution sets are only linked to from their respective assignment pages. Our system would create a new page with a link to each solution set and make this new page available to visitors to the site. We are currently implementing clustering transformations based on user navigation data.

## 2.3 Meta-information

A web site's ability to adapt can be hampered by the limited knowledge about its content and structure provided by HTML. For example, suppose that a page contains a list of links. Is it appropriate to add a new link at the top of the list? The answer depends on the contents of the list — an adaptive site should not add a link to a course's home page to a list of links to faculty home pages; furthermore, if the list is in alphabetical order then a new item can only be added at the appropriate point. Clearly, a site's ability to adapt could be enhanced by providing it with *meta-information*: information about its content, structure, and organization. In this section, we discuss means of providing an adaptive site with this sort of information.

One way to provide meta-information is to represent the site's content in a formal framework with precisely defined semantics such as a database or a semantic network. This approach is pioneered by the STRUDEL web-site management system [Fernandez et al., 1997] which attempts to separate the information available at a web site from its graphical presentation. Instead of manipulating web sites at the level of pages and links, web sites may be specified using STRUDEL's view-definition language. In addition, web sites may be created and updated by issuing STRUDEL queries. For example, a corporation might create home pages for its employees by merging data from its "manager" and "employee" databases. A page would be created for every person in either database. Furthermore, each manager's page would have links to her employees, and vice-versa.

This approach would facilitate adaptivity because STRUDEL would enable a site to reason about its logical description and detect cases where adaptations would violate the existing logic. Furthermore, an adaptive site could easily transform itself by issuing STRUDEL queries; STRUDEL provides the mechanisms to automatically update the site appropriately. The drawback of the STRUDEL approach is that it requires the site's entire content to be encoded in a set of databases or in *wrappers* that map web pages and other information sources into STRUDEL. The cost of constructing such wrappers for existing web sites, and particularly for relatively unstructured sites, appears to be high.

A lighter-weight approach is to annotate an existing web site with meta-content tags. In this approach, a formal description of the content coexists with HTML documents. We may choose how much of the site to annotate and how complex our annotations will be. Yet, meta-content annotation still facilitates reasoning about the connections between parts of the site and still provides guidance as to where and how to make changes.

One approach of this type is Apple's Meta-Content Format (see http://mcf.research.apple.com). MCF is an attempt to establish a standard for meta-content annotation for the web. When a user visits an MCF-enhanced site with an MCF-enabled browser, she can choose to navigate the site in a three-dimensional representation of the site's structure, as determined from the site's MCF annotation. SHOE [Luke et al., 1997] (at http://www.cs.umd.edu/projects/plus/SHOE/), takes a different tack. SHOE is a language for adding simple ontologies to web pages. SHOE adds basic ontological declarations to HTML; a page can refer to a particular ontology and declare classifications for itself and relations to other pages. In their example, a man's home page is annotated with information about him, such as the fact that he is a person, his name, his occupation, and his wife's identity (she has her own home page). SHOE is designed to facilitate the exploration of agents and the workings of search tools, but ontological annotation could also support adaptation.

While lighter-weight than STRUDEL, meta-content tagging also has clear disadvantages. First, because the meta-content annotation is separated from the actual content, it has to be updated manually as the content changes. Second, since the meta-content is attached to existing HTML, it provides no direct support for automatic adaptation; Any adaptation must still modify the original HTML.

Each of the approaches described so far require a fair amount of effort to build and maintain the content descriptions. If we wish only to facilitate adaptation, this effort may be overkill. An alternative that we are actively investigating is to use an extremely lightweight annotation system designed specifically for adaptivity. These annotations would be in the form of directives to the adaptive system telling it where it may (or may not) make changes and what kinds of changes it might make. For example, we might add a *list* tag to HTML to allow us to describe the elements in a list and how they are ordered. A list might be declared as `<list order="unordered">`, which tells the system it may reorder the list in any way it chooses. Or a list might be declared `<list order="popularity">`, in which case the system will draw upon data from access logs to determine how to present the list. A list declared `<list order="alphabetical">` or `<list order="chronological">` can be modified by additions or deletions so long as its original ordering constraint is preserved.

We present tags of this sort as part of an "Adaptive HTML" language called A-HTML in [Perkowitz and Etzioni, 1997]. Our intention is to extend HTML to a higher level of abstraction, allowing a web designer to describe objects in terms of their time-relevance, organization, and interrelationships. Note that this approach does not require the global establishment of an A-HTML standard; the adaptive site uses a server capable of interpreting A-HTML and translating it into standard HTML at runtime. Only the resulting HTML is served in response to page requests.

## 2.4 Open Questions

The quest for a self-improving web site raises a number of related questions. An adaptive site will be active twenty-four hours a day, seven days a week. The site will constantly be ingesting and analyzing data, adjusting its concepts and models, and updating its own structure and presentation. Over time, this constant cycle will reflect many hours of experience and refinement. In the past, AI research has focused on single trials and short-lived entities: systems that run their experiments and shut down, to start again the next day with a blank slate. Although such an approach may be applied to the adaptive site challenge, the most intelligent site will surely be one that continually accumulates knowledge about pages, users, content, and itself.

User interface design is difficult enough for human beings to perform well. Yet an adaptive web site will have to take into account all the artistry of good design in its self-improvements. We can limit the scope of the system's ability to change itself, thus ensuring that it cannot do too much harm, but this means we also limit its scope for improvement. On the other hand, giving the system free rein for radical transformation might mean giving it free rein for radical screwup.

> **Challenge:** How do we formalize the concept of good design? How do we limit the potential for harm without overly limiting the potential for good?

We might instead put the AI system in the role of advisor to a human master. Instead of making changes under cover of night, our AI system must now intelligently present suggestions to a human being, complete with explanation and justification. Such a solution frees us from the problem of changing details without changing design but presents us with a new interface challenge.

> **Challenge:** How does our adaptive web site communicate its suggestions to a webmaster?

## 3 Evaluation

Although the problem of measuring the quality of a web site design is thorny, we have identified several preliminary approaches. Progress on the design of adaptive web sites will include more sophisticated methods of evaluating a site's usability. We propose a basic metric for how usable a site is: how much effort must a user exert on average in order to find what she wants? As discussed in section 2.2, effort can be defined as a function of the number of links traversed and the difficulty of finding the links on their pages. The standard daily access log may be used to approximately measure user effort.

However, standard log data is not sufficient to know everything about visitor navigation. For example, standard logs do not distinguish between individuals connecting from the same location or record which link a user followed. However, software is available to provide more complete information. WebThreads, for example (see http://www.webthreads.com), allows a site to track an individual user's progress, including both pages visited and links followed. Along with analysis of our site's structure, data from a system like WebThreads is sufficient for us to measure user effort.

Analysis of our user logs provides much information about how users interact with the site. In addition, we may use controlled tests with subjects. Such tests have the advantage of allowing us to observe users as they interact with the site – we get much more information than is encoded in user access logs. As subjects perform tasks such as finding information, downloading software, or locating documents, we may gather data such as:

- Whether the subject succeeded at the task (or realized it was not solvable).
- How long the subject took to solve the goal.
- How much exploration was required.

Careful observation of test subjects would complement the limited access data we get on all of the site's regular visitors. Of course, we can also rely on intermediate measures such as encouraging users to fill out feedback forms and send e-mail messages.

## 4 Conclusion

This paper posed the challenge of using AI techniques to radically transform web sites from today's inert collections of HTML pages and hyperlinks to intelligent, evolving entities. Adaptive web sites can make popular pages more accessible, highlight interesting links, connect related pages, and cluster similar documents together. An adaptive web site can perform these self-improvements autonomously or advise a site's webmaster, summarizing access information and making suggestions. The improvements can happen in real-time as a visitor is navigating the site, or offline based on observations culled from many visitors.

This paper juxtaposed a number of disconnected projects from knowledge representation, machine learning, and user modeling that are investigating aspects of the problem. We believe that posing the challenge explicitly, in this paper, will help to cross-fertilize existing efforts and alert new researchers to the problem. Success in the next two years will have a broad and highly visible impact on the web and the AI community.

## References

[Anantharaman et al., 1990] T. Anantharaman, M. Campbell, and F. Hsu. Singular extensions: adding selectivity to brute-force searching. *Artificial Intelligence*, 43(1):99–109, 1990.

[Armstrong et al., 1995] R. Armstrong, D. Freitag, T. Joachims, and T. Mitchell. Webwatcher: A learning apprentice for the world wide web. In *Working Notes of the AAAI Spring Symposium: Information Gathering from Heterogeneous, Distributed Environments*, pp. 6–12, Stanford University, 1995. AAAI Press. To order a copy, contact sss@aaai.org.

[Etzioni, 1996] O. Etzioni. Moving up the information food chain: softbots as information carnivores. In *Proc. 14th Nat. Conf. on AI*, 1996.

[Fernandez et al., 1997] Mary Fernandez, Daniela Florescu, Jaewoo Kang, Alon Levy, and Dan Suciu. System demonstration - strudel: A web-site management system. In *ACM SIGMOD Conference on Management of Data*, 1997.

[Fink et al., 1996] J. Fink, A; Kobsa, and A. Nill. User-oriented adaptivity and adaptability in the avanti project. In *Designing for the Web: Empirical Studies*, Microsoft Usability Group, Redmond (WA)., 1996.

[Kautz, 1987] H. Kautz. *A Formal Theory Of Plan Recognition*. PhD thesis, University of Rochester, 1987.

[Lesh and Etzioni, 1995] Neal Lesh and Oren Etzioni. A sound and fast goal recognizer. In *Proc. 14th Int. Joint Conf. on AI*, pp. 1704–1710, 1995.

[Luke et al., 1997] S. Luke, L. Spector, D. Rager, and J. Hendler. Ontology-based web agents. In *Proceedings of the First International Conference on Autonomous Agents*, 1997.

[Perkowitz and Etzioni, 1997] M. Perkowitz and O. Etzioni. Adaptive sites: Automatically learning from user access patterns. Technical Report UW-CSE-97-03-01, University of Washington, Department of Computer Science and Engineering, March 1997.

[Pollack, 1990] M. Pollack. Plans as complex mental attitudes. In P. Cohen, J. Morgan, and M. Pollack, eds., *Intentions in Communication*, pp. 77–101. MIT Press, Cambridge, MA, 1990.

[Thorpe, 1990] C. Thorpe, ed. *Vision and Navigation: the Carnegie Mellon Navlab*. Kluwer Academic Publishing, Boston, MA, 1990.

# AI CHALLENGES

## Challenge 2

# The RoboCup Synthetic Agent Challenge 97

**Hiroaki Kitano***
Sony Computer Science Laboratory
kitano@csl.sony.co.jp

**Milind Tambe**
ISI/USC
tambe@isi.edu

**Peter Stone**
Carnegie Mellon University
pstone@cs.cmu.edu

**Manuela Veloso**
Carnegie Mellon University
mmv@cs.cmu.edu

**Silvia Coradeschi**
Linkoeping University
silco@ida.liu.se

**Eiichi Osawa**
Sony Computer Science Laboratory
osawa@csl.sony.co.jp

**Hitoshi Matsubara**
ElectroTechnical Laboratory
matsubar@etl.go.jp

**Itsuki Noda**
ElectroTechnical Laboratory
noda@etl.go.jp

**Minoru Asada**
Osaka University
asada@mech.eng.osaka-u.ac.jp

## Abstract

RoboCup Challenge offers a set of challenges for intelligent agent researchers using a friendly competition in a dynamic, real-time, multi-agent domain. While RoboCup in general envisions longer range challenges over the next few decades, RoboCup Challenge presents three specific challenges for the next two years: (i) learning of individual agents and teams; (ii) multi-agent team planning and plan-execution in service of teamwork; and (iii) opponent modeling. RoboCup Challenge provides a novel opportunity for machine learning, planning, and multi-agent researchers — it not only supplies a concrete domain to evalute their techniques, but also challenges researchers to evolve these techniques to face key constraints fundamental to this domain: real-time, uncertainty, and teamwork.

## 1 Introduction

RoboCup (The World Cup Robot Soccer) is an attempt to promote AI and robotics research by providing a common task, Soccer, for evaluation of various theories, algorithms, and agent architectures [Kitano, et al., 1995; Kitano et al., 1997a; 1997b]. Defining a standard problem in which various approaches can be compared and progress can be measured provides fertile grounds for engineering research. Computer chess has been a symbolic example of the standard challenge problems. A salient feature of computer chess is that progress can be measured via actual games against human players.

For an agent (a physical robot or a synthetic agent) to play soccer reasonably well, a wide range of technologies need to be integrated and a number of technical breakthroughs must be made. The range of technologies spans both AI and robotics research, such as design principles of autonomous agents, multi-agent collaboration, strategy acquisition, real-time reasoning and planning, intelligent robotics, sensor-fusion, and so forth. RoboCup consists of three competition tracks:

**Real Robot League:** Using physical robots to play soccer games.

**Software Agent League:** Using software or synthetic agents to play soccer games on an official soccer server over the network.

**Expert Skill Competition:** Competition of robots which have special skills, but are not able to play a game.

RoboCup offers a software platform that forms the basis of the software or synthetic agent league. The goal is to enable a wider range of research in synthetic (or "virtual reality") environments, that are today proving to be critical in training, entertainment, and education[Tambe et al., 1995]. The software agent league also promotes research on network-based multi-agent interactions, computer graphics, and physically realistic animation — a set of technologies which potentially promotes advanced use of internet.

## 2 Technical Challenges in RoboCup

RoboCup offers significant long term challenges, which will take a few decades to meet. However, due to the clarity of the final target, several subgoals can be derived, which define mid term and short term challenges. One of the major reasons why RoboCup is attractive to so many researchers is that it requires the integration of a broad range of technologies into a team of complete agents, as opposed to a task-specific functional module. The long term research issues are too broad to compile as a list of specific items. Nevertheless, the challenges involve a broad range of technological issues ranging from the development of physical components, such as high performance batteries and motors, to highly intelligent real time perception and control software.

The mid term technical challenges, which are the target for the next 10 years, can be made more concrete, and a partial list of specific topics can be compiled. Following is a partial list of research areas involved in RoboCup, mainly targeted for the mid term time span: (1) agent architecture in general, (2) combining reactive approach and modeling/planning approach, (3) real-time recognition, planning, and reasoning, (4) reasoning and action

---

*Corresponding Author: Hiroaki Kitano, Sony Computer Science Laboratory, 3-14-13 Higashi-Gotanda, Shinagawa, Tokyo 141, Japan. kitano@csl.sony.co.jp RoboCup Home Page: http://www.robocup.org/RoboCup. RoboCup Mailing List: robocup@csl.sony.co.jp

in dynamics environment, (5) sensor fusion, (6) multi-agent systems in general, (7) behavior learning for complex tasks, (8) strategy acquisition, and (9) cognitive modeling in general.

In addition to these technologies, providing a network-based soccer server with high quality 3D graphics capabilities requires advancement of technologies for the real time animation of simulated soccer players and network-based interactive multi-user server system. These are key technologies for network-based services in the coming years.

The RoboCup Challenge shall be understood in the context of larger and longer range challenges, rather than as a one-shot challenge. Thus, we wish to provide a series of short term challenges, which naturally leads to the accomplishment of the mid term and long term challenges. RoboCup challenge is organized into three major classes; (1) Synthetic Agent Challenge, (2) Physical Agent Challenge, and (3) Infrastructure Challenge. The RoboCup Synthetic Agent Challenge deal with technologies which can be developed using software simulator, which is described in this paper. The RoboCup Physical Agent Challenge intends to promote research using real robot, and thus requires longer-time frame for each challenge to be accomplished. Details of this challange is described in [Asada et al., 1997], and carried out together with the RoboCup Synthetic Agent Challenge but in more moderate timeframe. The Infrstructure Challenge will be presented to facilitate research to establish infrastructure aspect of RoboCup, AI, and robotics in general. Such challenge includes education programs, common robot platforms and components standard, automatic commentary systems and intelligent studio systems for RoboCup games.

## 3 Overview of The RoboCup Synthetic Agent Challenge

For the RoboCup Synthetic Agent Challenge 97, we offer three specific targets, critical not only for RoboCup but also for general AI research. These challenges will specifically deal with the software agent league, rather than the real robot league.

The fundamental issue for researchers who wish to build a team for RoboCup is to design a multiagent system that behaves in real-time, performing reasonable goal-directed behaviors. Goals and situations change dynamically and in real-time. Because the state-space of the soccer game is prohibitively large for anyone to hand-code all possible situations and agent behaviors, it is essential that agents learn to play the game strategically. The research issues in this aspect of the challenge involve:

(1) machine learning in a multiagent, collaborative and adversarial environment, (2) multiagent architectures, enabling real-time multiagent planning and plan execution in service of teamwork, and (3) opponent modelling.

Therefore, we propose the following three challenges as areas of concentration for the RoboCup Synthetic Agent Challenge 97:

- Learning challenge
- Teamwork challenge
- Opponent modeling challenge

Evaluating how well competing teams meet these challenges in RoboCup is clearly difficult. If the task is to provide the fastest optimization algorithm for a certain problem, or to prove a certain theorem, the criteria are evident. However, in RoboCup, while there may be a simple test set to examine basic skills, it is not generally possible to evaluate the goodness of a team until it actually plays a game. Therefore, a standard, highly skilled team of opponents is useful to set an absolute basis for such evaluation. We hope to use hand-coded teams, possibly with highly domain-specific coordination, to provide such a team of opponents. Indeed, in a series of preliminary competitions such as PreRoboCup-96 held at the IROS-96 conference, and several other local competitions, teams with well-designed hand-coded behaviors, but without learning and planning capabilities, have performed better than teams with learning and planning schemes. Of course, these hand-coded teams enjoyed the advantage of very low game complexities in initial stages of RoboCup — increasingly complex team behaviors, tactics and strategies will necessitate agents to face up to the challenges of learning, teamwork and opponent modeling.

Therefore, responses to this challenge will be evaluated based on (1) their performance against some standard hand-coded teams as well as other teams submitted as part of the competition; (2) behaviors where task specific constraints are imposed, such as probabilistic occurance of unexpected events, (3) a set of task specific sequences, and (4) novelty and technical soundess of the apporach.

## 4 The RoboCup Learning Challenge

### 4.1 Objectives

The objectives of the RoboCup Learning Challenge is to solicit comprehensive learning schemes applicable to the learning of multiagent systems which need to adapt to the situation, and to evaluate the merits and demerits of proposed approaches using standard tasks.

Learning is an essential aspect of intelligent systems. In the RoboCup learning challenge, the task is to create a learning and training method for a group of agents. The learning opportunities in this domain can be broken down into several types:

1. Off-line skill learning by individual agents;
2. Off-line collaborative learning by teams of agents;
3. On-line skill and collaborative learning;
4. On-line adversarial learning.

The distinction between off-line and on-line learning is particularly important in this domain since games last for only 20 minutes. Thus on-line techniques, particularly if they are to learn concepts that are specific to an individual game, must generalize very quickly. For example, if a team is to learn to alter its behavior against an individual opponent, the team had better be able to improve its performance before the game is over and a new opponent appears. Such distinctions in learning can be applied to a broad range of multi-agent systems which involve learning capabilities.

## 4.2 Technical Issues

Technical issues anticipated in meeting this challenge are the development of novel learning schemes which can effectively train individual agents and teams of agents in both off-line and on-line methods. One example of possible learning scheme for meeting this challenge is as follows:

**Off-line skill learning by individual agents:** learning to intercept the ball or learning to kick the ball with the appropriate power when passing.

Since such skills are challenging to hand-code, learning can be useful during a skill development phase. However, since the skills are invariant from game to game, there is no need to relearn them at the beginning of each new game [Stone and Veloso, 1997].

**Off-line collaborative learning by teams of agents:** learning to pass and receive the ball.

This type of skill is qualitatively different from the individual skills in that the behaviors of multiple agents must be coordinated. A "good" pass is only good if it is appropriate for the receivers receiving action, and vice versa. For example, if the passer passes the ball to the receiver's left, then the receiver must at the same time move to the left in order to successfully complete a pass. As above, such coordination can carry over from game to game, thus allowing off-line learning techniques to be used [Stone and Veloso, 1997].

**On-line skill and collaborative learning:** learning to play positions.

Although off-line learning methods can be useful in the above cases, there may also be advantages to learning incrementally as well. For example, particular aspects of an opposing teams' behavior may render a fixed passing or shooting behavior inefective. In that case, the ability to adaptively change collaborative or individual behaviors during the course of a game, could contribute to a team's success.

At a higher level, team issues such as role (position) playing on the field might be best handled with adaptive techniques. Against one opponent it might be best to use 3 defenders and 8 forwards; whereas another opponent might warrant a different configuration of players on the field. The best teams should have the ability to change configurations in response to events that occur during the course of a game.

**On-line adversarial learning:** learning to react to predicted opponent actions.

If a player can identify patterns in the opponents' behaviors, it should be able to proactively counteract them. For example, if the opponent's player number 4 always passes to its teammate number 6, then player 6 should always be guarded when player 4 gets the ball.

## 4.3 Evaluation

For challenge responses that address the machine learning issue (particularly the on-line learning issue), evaluation should be both against the publicly available teams and against at least one previously unseen team.

First, teams will play games against other teams and publicly available teams under normal circumstances. This evaluates the team's general performance. This involves both AI-based and non-AI based teams.

Next, teams will play a set of defined benchmarks. For example, after fixing their programs, challengers must play a part of the game, starting from the defined player positions, with the movement of the opponents pre-defined, but not disclosed to the challengers. After several sequences of the game, the performance will be evaluated to see if it was able to improve with experience. The movement of the opponents are not coded using absolute coordinate positions, but as a set of algorithms which generates motion sequences. The opponent algorithms will be provided by the organizers of the challenge by withholding at least one successful team from being publicly accessible.

Other benckmarks which will clearly evaluate learning performance will be announced after discussion with challenge participants.

## 5 The RoboCup Teamwork Challenge

### 5.1 Objectives

The RoboCup Teamwork Challenge addresses issues of real-time planning, re-plannig, and execution of multi-agent teamwork in a dynamic adversarial environment. Major issues of interest in this specific challenge for the 97-99 period are architectures for real-time planning and plan execution in a team context (essential for teamwork in RoboCup). In addition, generality of the architecture for non-RoboCup applications will be an important factor.

Teamwork in complex, dynamic multi-agent domains such as Soccer mandates highly flexible coordination and communication to surmount the uncertainities, e.g., dynamic changes in team's goals, team members' unexpected inability to fulfil responsibilities, or unexpected discovery of opportunities. Unfortunately, implemented multi-agent systems often rely on preplanned, domain-specific coordination that fails to provide such flexibility. First, it is difficult to anticipate and preplan for all possible coordination failures; particularly in scaling up to complex situations. Thus, it is not robust enough for dynamic tasks, such as soccer games. Second, given domain specificity, reusability suffers. Furthermore, planning coordination on the fly is difficult, particularly, in domains with so many possible actions and such large state spaces. Indeed, typical planners need significantly longer to find even a single valid plan. The dynamics of the domain caused by the unpredictable opponent actions make the situation considerably more difficult.

A fundamental reason for these teamwork limitations is the current agent architectures. Architectures such as Soar [Newell, 1990], RAP [Firby, 1987], IRMA [Pollack,

1991], and BB1 [Hayes-Roth et al., 1995] facilitate an individual agent's flexible behaviors via mechanisms such as commitments and reactive plans. However, teamwork is more than a simple union of such flexible individual behaviors, even if coordinated. A now well-known example (originally from [Cohen and Levesque, 1991]) is ordinary traffic, which even though simultaneous and coordinated by traffic signs, is not teamwork. Indeed, theories of teamwork point to novel mental constructs as underlying teamwork, such as team goals, team plans, mutual beliefs, and joint commitments [Grosz, 1996; Cohen and Levesque, 1991], lacking in current agent architectures. In particular, team goals, team plans or mutual beliefs are not explicitly represented; furthermore, concepts of team commitments are absent. Thus, agents cannot explicitly represent and reason about their team goals and plans; nor flexibly communicate/coordinate when unanticipated events occur. For instance, an agent cannot itself reason about its coordination responsibilities when it privately realizes that the team's current plan is unachievable — e.g., that in the best interest of the team, it should inform its teammates. Instead, agents must rely on domain-specific coordination plans that address such contigencies on a case-by-case basis.

The basic architectural issue in the teamwork challenge is then to construct architectures that can support planning of team activities, and more importantly execution of generated team plans. Such planning and plan execution may be accomplished via a two tiered architecture, but the entire system must operate in real-time. In RoboCup Soccer Server, sensing will be done in every 300 to 500 milli-seconds, and action command can be dispatched every 100 milli-second. Situation changes at milli-second order, thus planning, re-planning, and execution of plans must be done in real-time.

## 5.2 Technical Issues

We present a key set of issues that arise assuming our particular two tiered planning and plan-execution approach to teamwork. Of course, those who approach the problem from different perspective may have different issues, and the issues may change depending on the type of architecture employed.

The following is the envisioned teamwork challenge in this domain: (i) a team deliberatively accumulates a series of plans to apply to games with different adversarial teams; (ii) game plans are defined at an abstract level that needs to be refined for real execution; (iii) real-time execution in a team-plan execution framework/architecture that is capable of addressing key contigencies. Such an architecture also alleviates the planning concerns by providing some "commonsense" teamwork behaviors — not all of the coordination actions are required to be planned in detail as a result. The key research tasks here are:

**Contingency planning for multiagent adversarial game playing:** Before a game starts, one would expect the team to generate a strategic plan for the game that includes contingency plan segments that are to be recognized and eventually slightly adapted in real-time. Two main challenges can be identified in this task:

- Definition of strategic task actions with probabilistic applicability conditions and effects. Uncertainty in the action specification is directly related to the identification of possible probabilistic disruptive or favorable external events.

- Definition of objectives to achieve. In this domain, the goal of winning and scoring should be decomposed in a variety of more concrete goals that serve the ultimate final scoring goal. Examples are actions and goals to achieve specific attacking or defending positioning.

**Plan decomposition and merge:** A correspondence between team actions and goals and individual actions and goals must be set. The team plan decomposition may create individual goals that are not necessarily known to all the team players. Furthermore, within the contingency team plan, it is expected that there may be a variety of adversary-independent and adversary-dependent goals. The decomposition, coordination, and appropriate merge of individual plans to the service of the main team plan remain open challenging research tasks. RoboCup provides an excellent framework to study these issues.

**Executing Team Plans:** Team plan execution during the game is the determining factor in the performance of the team. It addresses the coordination contigencies that arise during the execution, without the need for detailed, domain-specific coordination plans. Execution also monitors the contingency conditions that are part of the global contingency team plan. Selection of the appropriate course of action is driven by the state information gathered by execution.

## 5.3 Evaluations

The Teamwork Challenge scenario described above has been idealized by several AI researchers, at least in the planning and multiagent communities. RoboCup, both in its simulated and real leagues, provides a synergistic framework to develop and/or test dynamic planning multiagent algorithms.

Specifically, we are planning to evaluate the architecture and teams in the following evaluation scheme:

**Basic Performance:** The team must be able to play reasonably well against both the best hand-coded teams, which has no planning, and against other planning-based systems. Relative performance of the team can be measured by actually playing a series of games against other unknown teams. Thus, basic performance will be measured by:

- Performance against hand-coded teams.
- Performance against other teams.

**Robustness:** The robustness in teamwork means that the team, as a whole, can continue to carry out the mission even if unexpected changes, such as accidental removal of the players in the team, sudden change of team conposition, or changes in operation environment. For example, if one of players in the team was disabled, the team should be able to cope with such accidents, by taking over the role of disabled players, or reformulating their team strategy. Thus, this evalution represents a set of unexpected incidents during the game, such as:

- Some players will be disabled, or their capability will be significantly undermined by these accidents. Also, some disabled players may be enabled later in the game.
- Opponent switch their strategy, and the team must cope with their new strategy in real time.
- Some of opponent's players will be disabled, or their performance will be significantly undermined. These disabled players may come back to the game later.
- Teammate changes during the game.
- Weather factor changes.

The RoboCup Teamwork Challenge therefore is to define a general set of teamwork capabilities to be integrated with agent architectures to facilitate flexible, reusable teamwork. The following then establish the general evaluation criteria:

**General Performace:** General performance of the team, thus the underlying algorithms, can be measured by a series of games against various teams. This can be divided into two classes (1) normal compeitions where no accidental factors involved, and (2) contigency evaluaiton where accidental factors are introduced.

**Real-Time Operations:** The real-time execution, monotoring, and replanning of the contingency plan is an important factor of the evaluaiton. For any team to be successful in the RoboCup server, it must be able to react in real time: sensory information arrives between 2 and 8 times a second and agents can act up to 10 times a second.

**Generality:** Reuse of architecture in other applications: Illustrate the reuse of teamwork capabilities in other applications, including applications for information integration on the internet, entertainment, training, etc.

**Conformity with Learning:** Finally, given the premises above and the complexity of the issues, we argue and challenge that a real-time multiagent planning system needs to have the ability to be well integrated with a learning approach, i.e., it needs to refine and dynamically adapt and refine its complete behavior (individual and team) based on its past experience.

Other issues such as reuse of teamwork architecture within the RoboCup community, and planning for team players that are not yet active in order to increase their probability of being useful in future moves, such as role playing and positioning of the team players that *do not* have the ball, will be considered, too.

## 6 RoboCup Opponent Modeling Challenge

Agent modeling – modeling and reasoning about other agent's goals, plans, knowledge, capabilities, or emotions — is a key issue in multi-agent interaction. The RoboCup opponent modeling challenge calls for research on modeling a team of opponents in a dynamic, multi-agent domain. The modeling issues in RoboCup can be broken down into three parts:

**On-line tracking:** Involves individual players' real-time, dynamic tracking of opponents' goals and intentions based on observations of actions. A player may use such tracking to predict the opponents' play and react appropriately. Thus if a player predicts that player-5 is going to pass a ball to player-4, then it may try to cover player-4. Such on-line tracking may also be used in service of deception. The challenges here are (i) real-time tracking despite the presence of ambiguity; (ii) addressing the dynamism in the world; (iii) tracking teams rather than only individuals – this requires an understanding of concepts involved in teamwork.

On-line tracking may feed input to the on-line planner or the on-line learning alogrithm.

**On-line strategy recognition:** "Coach" agents for teams may observe a game from the sidelines, and understand the high-level strategies employed by the opposing team. This contrasts with on-line tracking because the coach can perform a much higher-level, abstract analysis, and in the absence of real-time pressures, its analysis can be more detailed.

The coach agents may then provide input to its players to change the team formations, or play strategy.

**Off-line review:** "Expert" agents may observe the teams playing in an after-action review, to recognize the strenghts and weaknesses of the teams, and provide an expert commentary. These experts may be trained on databases of human soccer play.

These issues pose some fundamental challenges that will significantly advance the state of the art in agent modeling. In particular, previous work has mostly focused on plan recognition in static, single-agent domains, without real-time constraints. Only recently has attention shifted to dynamic, real-time environments, and modeling of multi-agent teamwork [Tambe, 1996b].

A realistic challenge for IJCAI-99 will be to aim for on-line tracking. Optimistically, we expect some progress towards on-line strategy recognition; off-line review will likely require further research beyond IJCAI-99.

For evaluation, we propose, at least, following evaluation to be carried out to measure the progress:

**Game Playing:** A team of agents plays against two types of teams:

- One or two unseen RoboCup team from IJCAI-97, shielded from public view.
- The same unseen RoboCup teams from IJCAI-97 as above, but modified with some new behaviors. These teams will now deliberately try out new adventurous strategies, or new defensive strategies.

**Disabled Tracking:** Tracking functionality of the agents will be turned off, and compared with normal performance.

**Deceptive Sequences:** Fake teams will be created which generates deceptive moves. The challenger's agent must be able to recognize the opponent's deceptive moves to beat this team.

For each type of team, we will study the performance of the agent-modelers. Of particular interest is variations seen in agent-modelers behaviors given the modification in the opponents' behaviors. For each type of team, we will also study the advise offered by the coach agent, and the reviews offered by the expert agents, and the changes in them given the changes in the opponents' behaviors.

## 7 Managing Challenges

In order to facilitate technical progress based on the RoboCup challenge, we offer basic resources and opportunities.

**The RoboCup Challenge Committee:** The RoboCup Challenge Committee will be formed to execute the challenge initiative. The commitee will include members of the international executive committee for RoboCup and distinguished researchers not directly involved in RoboCup. The committee will create specific tasks and criteria for evaluation, as well as providing technical advice for the challengers.

**Resources:** In the RoboCup home page, basic software resources and technical information can be obtained. (http://www.robocup.org/RoboCup) Software includes the Soccer Server system, which is a server system for the simulation track, and various sample teams. In addition, sample test sequences will be provided. The home page also provides a set of papers and technical documents related to RoboCup.

**Competitions:** A series of RoboCup competitions are planned to provide opportunities to test ideas. As international events, we are planning to have RoboCup-98 in Paris (The Official Event of the World Cup), RoboCup-98 Victoria (as a part of IROS-98 conference), and RoboCup-98 Singapore (as a part of PRICAI-98 Conference). Several local competitions will be organized by local committee in each region. The final evaluation and exhibit of the results will be made at IJCAI-99.

**Workshops:** Workshops will be organized at major international conferences, as well as at local workshops, in order to faciliate exchange of information, to have technical discussions, and to get feedback on the status of the challengers in relation to the overall framework of the challenge.

## 8 Conclusion

The RoboCup Challenge-97 offers a set of three fundamental challenges, focused on learning, real-time planning, and opponent modeling. Learning and real-time planning of multi-agent systems were chosen as the first set of challenges because they are essential technical issues for RoboCup, as well as for general AI systems using a multi-agent approach. Together with the physical agent challenge, these challenges will be be a basis for the RoboCup Challenge-99, and for longer research enterprises.

## References

[Asada et al., 1997] Asada, M., Kuniyoshi, M., Drogoul, A., Asama, H., Mataric, M., Duhaut, D., Stone, P., and Kitano, H., "The RoboCup Physical Agent Challenge: Phase-I," To appear in *Applied Artificial Intelligence (AAI) Journal*, 1997.

[Cohen and Levesque, 1991] Cohen, P. R. and Levesque, H. J., "Confirmation and Joint Action", *Proceedings of International Joint Conf. on Artificial Intelligence*, 1991.

[Firby, 1987] Firby, J., "An investigation into reactive planning in complex domains", *Proceedings of National Conf. on Artificial Intelligence*, 1987.

[Grosz, 1996] Grosz, B., "Collaborating Systems", *AI magazine*, 17, 1996.

[Hayes-Roth et al., 1995] Hayes-Roth, B. and Brownston, L. and Gen, R. V., "Multiagent collaobration in directed improvisation", *Proceedings of International Conf. on Multi-Agent Systems*, 1995.

[Jennings, 1995] Jennings, N., "Controlling cooperative problem solving in industrial multi-agent systems using joint intentions", *Artificial Intelligence*, 75, ,195-240, 1995.

[Kitano et al., 1997a] Kitano, H., Asada, M., Osawa, E., Noda, I., Kuniyoshi, Y., Matsubara, H., "RoboCup: The Robot World Cup Initiative", *Proc. of the First International Conference on Autonomous Agent (Agent-97)*, 1997.

[Kitano et al., 1997b] Kitano, H., Asada, M., Osawa, E., Noda, I., Kuniyoshi, Y., Matsubara, H., "RoboCup: A Challenge Problem for AI", *AI Magazine*, Vol. 18, No. 1, 1997.

[Kitano, et al., 1995] Kitano, H. and Asada, M. and Kuniyoshi, Y. and Noda, I. and Osawa, E., "RoboCup: The Robot World Cup Initiative", *IJCAI-95 Workshop on Entertainment and AI/Alife*, 1995.

[Newell, 1990] Newell, A., *Unified Theories of Cognition*, Harvard Univ. Press, Cambridge, Mass., 1990.

[Pollack, 1991] Pollack, M., "The uses of plans", *Artificial Intelligence*, 57, 1992.

[Stone and Veloso, 1997] Stone, P. and Veloso, M., "A layered approach to learning client behaviors in the robocup soccer server," *To appear in Applied Artificial Intelligence (AAI) Journal*, 1997.

[Tambe, 1996a] Tambe, M., "Tracking dynamic team activity", *Proceedings of National Conf. on Artificial Intelligence*, 1996.

[Tambe, 1996b] Tambe, M., "Teamwork in real-world, dynamic environments", *Proc. International Conf. on Multiagent Systems*, 1996.

[Tambe et al., 1995] Tambe, M. and Johnson, W. and Jones, R. and Koss, F. and Laird, J. and Rosenbloom, P. and Schwamb, K., "Intelligent agents for interactive simulation environments", *AI Magazine*, 16, 1995.

# Understanding Three Simultaneous Speeches

Hiroshi G. Okuno, Tomohiro Nakatani, and Takeshi Kawabata
NTT Basic Research Laboratories
3-1 Morinosato-Wakamiya, Atsugi, Kanagawa 243-01, Japan
okuno@nue.org, nakatani@horn.brl.ntt.co.jp, kaw@idea.brl.ntt.co.jp

## Abstract

Understanding three simultaneous speeches is proposed as a challenge problem to foster artificial intelligence, speech and sound understanding or recognition, and computational auditory scene analysis research. Automatic speech recognition under noisy environments is attacked by speech enhancement techniques such as noise reduction and speaker adaptation. However, the signal-to-noise ratio of speech in two simultaneous speeches is too poor to apply these techniques. Therefore, novel techniques need to be developed. One candidate is to use speech stream segregation as a front-end of automatic speech recognition systems. Preliminary experiments on understanding two simultaneous speeches show that the proposed challenge problem will be feasible with speech stream segregation. The detailed plan of the research on and benchmark sounds for the proposed challenge problem is also presented.

## 1 Introduction

Recently emerges a new research on understanding arbitrary sound mixtures including non-speech sounds and music. Their understanding represents a challenging and little-studied area of artificial intelligence, automatic speech recognition/understanding, and signal processing. This interdisciplinary research area is called *computational auditory scene analysis* (hereafter, CASA).

At a crowded party, one can attend one conversation and then switch to another one. This phenomenon is known as the *cocktail party effect* [Cherry, 1953]. As seen in the cocktail-party effect, humans have the ability to selectively attend to sound from a particular source, even when it is mixed with other sounds. Current automatic speech recognition systems can understand *clean* speech well in relatively noiseless laboratory environments, but break down in more realistic, noisier environments.

Computers also need to be able to decide which parts of a mixed acoustic signal are relevant to a particular purpose — which part should be interpreted as speech, for example, and which should be interpreted as a door closing, an air conditioner humming, or another person interrupting. CASA focuses on the computer modeling and implementation for the understanding of acoustic events

The research topics concerning CASA include modeling, signal processing, sound representational, control and system architecture, and applications as well as sensor integration. Some of these topics were discussed at the IJCAI-95 workshop on Computational Auditory Scene Analysis [Rosenthal and Okuno, 1997].

At the AAAI-96, the panel entitled "Challenge Problems for Artificial Intelligence", Brooks proposed two problems concerning sounds [Selman et al., 1996]:

- Challenge 1: Speech understanding systems that are based on different principles other than hidden Markov models.
- Challenge 2: Noise understanding systems.

Although CASA shares the above interests, its ultimate goals go further; understanding general acoustic signals such as voiced speech, music and/or other sounds from real-world environments.

We propose the problem of *Understanding Three Simultaneous Speeches*[1] (hereafter, *the challenge*) as a challenge problem for artificial intelligence, in particular, for CASA. A computer capable of listening to several things simultaneously is called *Prince Shotoku Computer* after the Japanese legendary that Prince Shotoku (A.D. 574–622) could listen to ten people's petitions simultaneously [Okuno et al., 1995]. Since psychoacoustic studies have recently showed that humans cannot listen to more than two things simultaneously [Kashino and Hirahara, 1996], CASA research would make computer

---

[1] The selection of the word "*simultaneous*" or "*concurrent*" is controversial. The former carries more physical senses, while the latter carries more mental senses; e.g., "separation of simultaneous talkers", "simultaneous voices separation", and "separation of concurrent sentences" make sense. We adopt "simultaneous" because the proposed challenge problem won't pursuit understanding what each speaker talks about. Understanding what a speaker says without speech recognition, for example, is beyond our problem.

audition more powerful than human audition, similar to the relationship of an airplane's ability to that of a bird.

The rest of this paper is organized as follows: Section 2 explains the research issues for the challenge, in particular, its relevance and significance to AI. Section 3 presents its feasibility by showing the result of preliminary experiments on understanding two simultaneous speeches with/without interfering sounds. Section 4 discusses the detailed plan of our challenge problem. Concluding remarks are given in Section 5.

## 2 Research Issues for Understanding Three Simultaneous Speeches

In this section, we explain the reasons why we focus on three simultaneous speeches, not two simultaneous speeches and how significant the challenge is to AI researches. We also discuss several research issues involved in realizing understanding three simultaneous speeches. The main research areas related to the challenge are automatic speech recognition (ASR), signal processing, speech understanding, computational auditory scene analysis, and psychoacoustics.

### 2.1 Automatic Speech Recognition (ASR)

At present, one of the hottest topics of ASR research is how to make ASR systems more robust so that they can perform well outside *laboratory conditions* [Hansen *et al.*, 1994]. Conventional approaches for robust ASR are speech enhancement and many techniques for speech enhancement such as noise reduction and speaker adaptation have been developed [Hansen *et al.*, 1994; Minami and Furui, 1995].

One possible approach is to enhance a speech by employing noise reduction techniques. Once a speech is enhanced, it can be subtracted from a mixture of sounds in waveform. By repeating this procedure to the residue (remaining sounds), it seems possible to extract most speeches from a mixture of sounds.

This approach, however, works only up to two simultaneous speeches. The reason is as follows; Most conventional noise reduction techniques assume that the signal-to-noise ratio (SNR) of speech is 0 dB or better. The SNR of speech in a mixture of two simultaneous speeches is approximately 0 dB and thus noise reduction techniques can be applied to two simultaneous speeches. However, new techniques need to be developed for understanding three simultaneous speeches.

### 2.2 Signal Processing

Speech separation is more aggressive approach than noise reduction. Adaptive filters are used for speech separation [Ramalingam, 1994]. Spatial information on the sound source plays an important role in separating a speech from a mixture of sounds. This mechanism is called *localization*, which is performed by using a dummy head microphone (called *binaural sounds*) [Blauert, 1983; Bodden, 1993] or by using microphone arrays [Hansen *et al.*, 1994; Stadler and Rabinowitz, 1993]. For a pair of microphones, localization can be obtained better from binaural sounds than from stereo sounds.

Adaptive window technique for localizing two simultaneous voices by using two microphones is also developed for speech enhancement in real-time [Banks, 1993]. Procedures for enhancing the intelligibility of a target speaker (talker) in the presence of a simultaneous talker is developed by using harmonic selection and cepstral filtering [Stubbs and Summerfield, 1991]. Classification tasks within an automated two-speech separation system are performed by neural net [Roger *et al.*, 1989].

Most of these systems can separate a speech from a mixture of two simultaneous speeches. A speech separation system is developed by using harmonic structure and directional information and can extract one speech from a mixture of more than two overlapping speeches [Luo and Denbigh, 1994].

Since the spectrum of speech separated by speech separation techniques developed so far is distorted, they cannot be applied continuously to the remaining signals. We need to develop new techniques for the challenge. In addition, a segregated speech cannot be used as a input to automatic speech recognition systems due to spectral distortion. We also need to develop an interfacing technique between speech separation and ASR.

### 2.3 Computational Auditory Scene Analysis (CASA)

Speech enhancement technologies developed so far focus on only one speech and treat other speeches or sounds as noise. CASA takes an opposite approach. First, it deals with the problems of handling mixture of sounds to develop methods and technologies. Then it applies these to develop ASR systems that work in a real-world environment. The main research topic of CASA is *sound stream segregation*, a process that segregates sound streams that have consistent acoustic attributes from a mixture of sounds.

In extracting acoustic attributes, some systems assume the humans auditory model of primary processing and simulate the processing of cocklear mechanism [Brown, 1992; Slaney *et al.*, 1994]. Brown and Cooke designed and implemented the system that builds various auditory maps for input sounds and integrates them to segregate speech from input sounds [Brown, 1992; Brown and Cooke, 1992]. An auditory map represents acoustic attributes such as onset, offset, AM and FM modulations, and formants. Since the integration process becomes complicated when treating a mixture of sounds under the real-world environments, the blackboard architecture is used to simplify this integration process [Cooke *et al.*, 1993].

To design a more flexible and expandable system, control mechanisms are needed. IPUS (*Integrated Processing and Understanding Signals*) [Lesser *et al.*, 1993] integrates signal processing and signal interpretation into the blackboard system. IPUS has various interpretation knowledge sources which understand actual sounds such as hair driers, footsteps, telephone rings, fire alarms, and waterfalls [Nawab, and Lesser, 1992].

Nakatani *et al.* took a multi-agent approach to sound stream segregation which extracts individual sound stream from a mixture of sounds by agents each of which traces harmonic structure with directional information [Nakatani *et al.*, 1994]. They use the Fourier transformation instead of the auditory model because the former is easy to implement and its properties are well analyzed.

## 2.4 Psychoacoustics

Psychoacoustic people have studied the human auditory mechanism extensively as auditory scene analysis [Bregman, 1990], but computer modeling has not been exploited yet. Emerging computational auditory scene analysis research focuses on computer modeling and has prompted interdisciplinary studies with psychoacoustic and AI and signal processing communities. In addition, our challenge problem has fostered psychoacoustic studies on how many simultaneous speeches human can listen to. Kashino *et al* claimed that human could listen to at most two things simultaneously by performing various experiments [Kashino and Hirahara, 1996]. If this is true, the challenge will attempt to make computer audition superior to human's capability of listening.

## 3 Preliminary Experiments

In this section, we demonstrate the feasibility of the challenge by describing the preliminary experiments on understanding two simultaneous speeches (up-to-date information of our AAAI-96 paper [Okuno *et al.*, 1996]). This problem is attacked by speech stream segregation, one of the main research topics of computational auditory scene analysis. The whole system consists of two components, speech stream segregation and speech recognition, as is shown in Figure 1.

First speech streams are extracted from a mixture of speeches, and then each speech stream is recognized by conventional automatic speech recognition system.

### 3.1 Speech Stream Segregation

Human voice consists of harmonic sounds such as vowel and voiced consonants, and non-harmonic sounds such as unvoiced consonants. By assuming the structure of "Vowel (V) + Consonant (C) + Vowel (V)" of speech, speech stream segregation is realized by the following two subprocesses:

(1) extracting and grouping harmonic stream fragments (*harmonic structure extraction*), and

(2) restoring non-harmonic parts by residue (*residue substitution*).

Rough flow of the computation is depicted in Figure 2.

Harmonic structures are extracted from a binaural input by the Bi-HBSS (Binaural Harmonics-Based Stream Segregation) system [Nakatani *et al.*, 1995; Nakatani *et al.*, 1996]. Bi-HBSS uses a harmonic structure and the

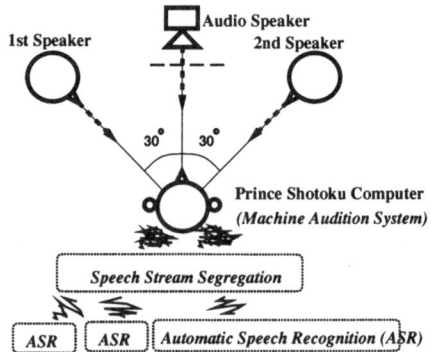

Figure 1: System architecture and sound sources for experiments on understanding two simultaneous speeches.

direction of sound source as cues of segregation. Bi-HBSS adopts a pair of HBSSes [Nakatani *et al.*, 1994] for the right and left channel to extract harmonic stream fragments. It determines the fundamental frequency ($F_0$) of a harmonic stream fragment by coordinating the pair of HBSSes. The direction of sound source is identified by calculating the interaural time difference (ITD) and interaural intensity difference (IID) of a pair of harmonic stream fragments of the same $F_0$ extracted by the pair of HBSS. Harmonic stream fragments are grouped by the direction of the sound source.

The residue obtained by subtracting harmonic structures from an input sound is substituted for non-harmonic parts of a group. If a group ends with non-harmonic parts, the residue is substituted for 150 msec. The idea of residue substitution is similar to the psychophysical observation known as *auditory induction* [Green *et al.*, 1995; Warren, 1970]. It is a phenomena that human listeners can perceptually restore a missing sound component if it is very brief and masked by appropriate sounds.

### 3.2 Automatic Speech Recognition

The automatic speech recognition system, HMM-LR [Kita *et al.*, 1990], is used to recognize speech streams. HMM-LR is based on hidden Markov model of each phonetic transition, in spite of Rodney Brooks' challenge problem. The parameters of HMM-LR are trained by a set of 5,240 words uttered by five speakers.

Since the spectrum of speech streams segregated by the speech stream segregation is distorted due to binaural input, binauralized training data is used to recover from the degradation of the performance of recognition [Okuno *et al.*, 1996].

### 3.3 Performance Evaluation

The performance of automatic speech recognition is usually measured by the *cumulative accuracy up to the 10th candidate* (or simply *cumulative accuracy*) of word recognition, since ASR returns the first about 10 candidates of each word. Such candidates are further selected by successive speech understanding systems. Therefore,

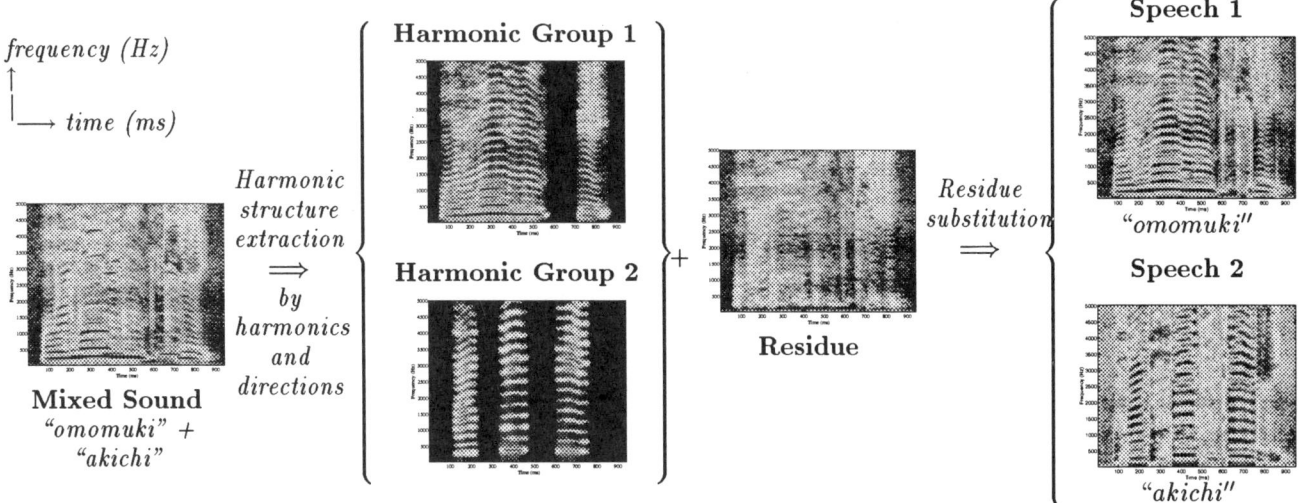

Figure 2: **Speech stream segregation**: From binaural sounds consisting of two Japanese words *"omomuki"* and *"akichi"*, two harmonic groups are extracted by tracing harmonic structures and the direction of sound sources. The residue is also generated by subtracting harmonic groups from input. Then, the residue is substituted for missing non-harmonic parts (black parts in the spectrogram) of each group, and thus speech streams are generated.

Table 1: Three sets of benchmark sounds

| No. | Speaker 1 | Speaker 2 | Audio Speaker |
|---|---|---|---|
| Double | Woman 1 | Woman 2 | — |
| Triple | Woman 1 | Woman 2 | Sound 1 |
| Triple' | Woman 1 | Woman 2 | Sound 2 |

Table 2: Error rates in the word recognition caused by an interfering speaker without/with third sound.

| Benchmark | Speaker 1 | Speaker 2 |
|---|---|---|
| Double | 76.19% | 95.50% |
| Triple | 94.99% | 95.70% |
| Triple' | 94.99% | 95.90% |

we adopted the same measurement with open tests. By open tests, we mean that the training and benchmark (testing) data are disjoint.

We used three sets of 500 benchmark sounds; one set of 500 two-sound mixtures and two sets of 500 three-sound mixtures (Table 1). The first sound is uttered by the first speaker at 30° to the left from the center, and the second sound is uttered after 150 msec by the second speaker at 30° to the right from the center (Figure 1). To recognize the first speech in the mixed sound directly by HMM-LR, the utterance of the second speaker is delayed by 150 msec.

The third sound with $F_0$ of 250 Hz is an intermittent harmonic sound from the center. It starts before the first speaker and repeats to last for 1 sec with 50 msec of pause. The average power ratios of the first and second sounds to the third sound in benchmarks **Triple** and **Triple'** are 1.7 dB and -1.3 dB, respectively.

The *error rate caused by interfering sounds* is defined as follows. Let the cumulative accuracy of recognition of original data up to the 10th candidate be $\mathcal{CA}_{org}$, and let the cumulative accuracy of recognition of (non-binaural) mixed sounds up to the 10th candidate be $\mathcal{CA}_{mix}$. The error rate caused by interfering sounds, $\mathcal{E}$, is calculated as $\mathcal{E} = \mathcal{CA}_{org} - \mathcal{CA}_{mix}$.

To evaluate the performance of speech stream segregation, *error reduction rate* is defined. Let the cumulative accuracy of recognition up to the 10th candidate be $\mathcal{CA}_{seg}$. The error reduction rate, $\mathcal{R}_{seg}$, is calculated as follows:

$$\mathcal{R}_{seg} = \frac{\mathcal{CA}_{seg} - \mathcal{CA}_{mix}}{\mathcal{CA}_{org} - \mathcal{CA}_{mix}} \times 100 = \frac{\mathcal{CA}_{seg} - \mathcal{CA}_{mix}}{\mathcal{E}} \times 100.$$

The original cumulative accuracies of word recognition uttered by single speakers, Woman 1, and Woman 2, are 94.99%, and 96.10%, respectively. The error rate by interfering sounds is shown in Table 2.

Error reduction rates by speech stream segregation for the three benchmark sets are shown in Figure 3.3. The **Ideal** shows the upper limits of error reduction, which are calculated for the case in which the utterances of a single speaker are recognized after speech stream segregation. For **Double**, 77% of errors caused by an interfering speaker were reduced by speech stream segregation. By additional noise, the SNR of each speech is decreased further (by about 1 dB and 2 dB for **Triple** and **Triple'**, respectively), but, 55% and 49% of errors are reduced respectively.

Since this performance was attained without using any features specific to human voices, we believe that understanding three simultaneous speeches is a short-term research problem.

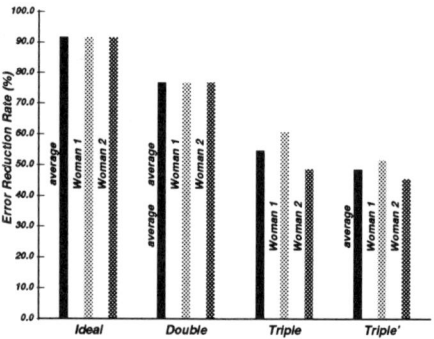

Figure 3: Error reduction rates for two speeches

## 4 Detailed Plan for the Challenge

The research issues depend on an approach taken by a challenger. Some possible approaches are listed below:

- Either speech separation system or speech stream segregation may be exploited.
- Speech separation or speech stream segregation may run either incrementally or in batch.
- Multiple speech enhancement or separation systems may run concurrently to extract all speeches or one system may extract all speeches.
- Speech segregation/separation system may be either used as a front-end to ASR or integrated with ASR.
- Top-down or hybrid approaches needed for continuous speech recognition or understanding may be employed, although word recognition is requested by the challenge.

We only give a general guideline on the benchmarks and evaluation criteria in this paper. Further information will be made available at the URL of http://www.nue.org/CASA97/.

### 4.1 Benchmark Sounds

The common platform for the challenge is quite important in order to share and transfer the methodology and technology developed by each challenger. Monaural data of speech used for the challenge should should be widely available. The current candidates are as follows:

- The DARPA TIMIT Acoustic-Phonetic Continuous Speech Corpus for English speeches. It contains a total of 3,600 sentences by 360 speakers uttering the same 10 sentences. (http://www.ldc.upenn.edu/)
- The continuous speech corpus developed by the Acoustic Society of Japanese for Japanese speeches. It contains a total of 9,600 sentences by 64 speakers uttering some of 503 sentences, which were recorded by ATR.

Since these corpus are copyrighted, only the combination of word utterances will be made available. One benchmark set contains a total of 200 combination of words; three speakers, arbitrary combination of men and women, utter a different word simultaneously.

The acoustic field is made simple enough to produce benchmark sounds easily. A sound source (speaker) should be placed from 1.4 meters to 2 meters from the microphone on the floor in a free-field without reverberation. Several combinations of speaker positions selected from 0°, 30°, 45°, 60°, 90°, 120°, 135°, 150°, and 180° may be strongly recommended. The challenge does not assume that any speaker move during speaking. However, challengers may attack the problem of moving speakers.

A mixture of sounds may be either recorded or generated artificially. The number of microphones should be less than or equal to 3. If a challenger wants to use a binaural input, it may be generated artificially by using Head-Related Transfer Function (HRTF), which specifies the spectral transformation of a binaural sound. The data of HRTF for the KEMAR dummy head microphone is available from MIT [Gardner and Martin, 1994]. For this HRTF, a sound source should be placed 1.4 meters from the dummy head microphone. In our preliminary experiments, all sound sources were placed at the distance of 2 meters from the dummy head microphone.

### 4.2 Measurement of Evaluation

The measurement of performance evaluation is error reduction rate as well as cumulative accuracy of up to 10th candidate, both of which are defined in the previous section. The first measurement may be important because it is rather independent for automatic speech recognition systems used.

The first stage of the challenge investigates the performance of word recognition.

### 4.3 Tentative Schedule

- The challenge problem will be presented at IJCAI-97 as well as IJCAI-97 workshop on Computational Auditory Scene Analysis.
- The guideline on the benchmarks will be made available by the end of 1997.
- Intermediate progress reports will be presented at AAAI-98 or an appropriate conference in 1998.
- Final progress reports will be submitted in Jan., 1999 and will be presented at IJCAI-99.

## 5 Conclusions

In this paper, we proposed *understanding three simultaneous speeches* as a new challenge and standard AI problem. It provides rich research issues for a wide range of AI including automatic speech recognition, speech understanding, and CASA, as well as psychoacoustics. We expect that research on the challenge would play an important role in realizing the "Prince Shotoku Computer" or powerful computer audition systems.

# References

[Banks, 1993] D. Banks. Localization and separation of simultaneous voices with two microphones. In *IEE Proceedings I*, Vol.140, No.4, pp.229-34, 1993.

[Blauert, 1983] J. Blauert. *Spatial Hearing: the Psychophysics of Human Sound Localization*. MIT Press, 1983.

[Bodden, 1993] M. Bodden. Modeling human sound-source localization and the cocktail-party-effect. *Acta Acustica* 1:43-55, 1993.

[Bregman, 1990] A.S. Bregman. *Auditory Scene Analysis - the Perceptual Organization of Sound*. MIT Press, 1990.

[Brown, 1992] G.J. Brown. Computational auditory scene analysis: A representational approach. Ph.D diss., Dept. of Computer Science, University of Sheffield, 1992.

[Brown and Cooke, 1992] G.J. Brown, and M.P. Cooke. A computational model of auditory scene analysis. In *Proc. of Intern'l Conf. on Spoken Language Processing*, 523-526.

[Cherry, 1953] E.C. Cherry. Some experiments on the recognition of speech, with one and with two ears. *J. of Acoustic Society of America* 25:975-979, 1953.

[Cooke et al., 1993] M.P. Cooke, G.J. Brown, M. Crawford, and P. Green. Computational Auditory Scene Analysis: listening to several things at once. *Endeavour*, 17(4):186-190, 1993.

[Gardner and Martin, 1994] B. Gardner, and K. Martin. HRTF Measurements of a KEMAR Dummy-Head Microphone. *MIT Media Lab Perceptual Computing - Technical Report*, #280, May 1994. http://sound.media.mit.edu/KEMAR.html

[Green et al., 1995] P.D. Green, M.P. Cooke, and M.D. Crawford. Auditory Scene Analysis and Hidden Markov Model Recognition of Speech in Noise. In *Proc. of 1995 International Conference on Acoustics, Speech and Signal Processing*, vol.1:401-404, IEEE, 1995.

[Hansen et al., 1994] J.H.L. Hansen, R.J. Mammone, and S. Young. Editorial for the special issue on robust speech processing". *IEEE Transactions on Speech and Audio Processing* 2(4):549-550, 1994.

[Kashino and Hirahara, 1996] M. Kashino, and T. Hirahara. One, two, many – Judging the number of concurrent talkers. *J. of Acoustical Society of America*, 99 (4) Pt.2, 2596.

[Kita et al., 1990] K. Kita, T. Kawabata, and K. Shikano. HMM continuous speech recognition using generalized LR parsing. *Transactions of Information Processing Society of Japan*, 31(3):472-480, 1990.

[Lesser et al., 1993] V. Lesser, S.H. Nawab, I. Gallastegi, and F. Klassner. IPUS: An Architecture for Integrated Signal Processing and Signal Interpretation in Complex Environments. In *Proc. of Eleventh National Conference on Artificial Intelligence*, 249-255, AAAI, 1993.

[Luo and Denbigh, 1994] H.Y. Luo, and P.N. Denbigh. A speech separation system that is robust to reverberation, In *Proc. of International Conference on Speech, Image Processing and Neural Networks*, vol.1:339-42, IEEE, 1994.

[Minami and Furui, 1995] Y. Minami, and S. Furui. A Maximum Likelihood Procedure for A Universal Adaptation Method based on HMM Composition. In *Proc. of 1995 International Conference on Acoustics, Speech and Signal Processing*, vol.1:129-132, IEEE, 1995.

[Nakatani et al., 1994] T. Nakatani, H.G. Okuno, and T. Kawabata. Auditory Stream Segregation in Auditory Scene Analysis with a Multi-Agent System. In *Proc. of 12th National Conference on Artificial Intelligence*, 100-107, AAAI, 1994.

[Nakatani et al., 1995] T. Nakatani, H.G. Okuno, and T. Kawabata. Residue-driven architecture for Computational Auditory Scene Analysis. In *Proc. of 14th International Joint Conference on Artificial Intelligence*, vol.1:165-172, IJCAI, 1995.

[Nakatani et al., 1996] T. Nakatani, M. Goto, and H.G. Okuno. Localization by harmonic structure and its application to harmonic sound stream segregation. In *Proc. of 1996 International Conference on Acoustics, Speech and Signal Processing*, IEEE, 1996.

[Nawab, and Lesser, 1992] S.H. Nawab and V. Lesser. Integrated Processing and Understanding of Signals. In Oppenheim, A.V. and Nawab, S.H. (Eds.) *Symbolic and Knowledge-Based Signal Processing*, 251-285. Prentice-Hall, 1992.

[Nawab et al, 1995] S.H. Nawab, C.Y. Espy-Wilson, R. Mani, and N.N. Bitar. Knowledge-Based analysis of speech mixed with sporadic environmental sounds. In [Rosenthal and Okuno, 1997].

[Okuno et al., 1995] H.G. Okuno, T. Nakatani, and T. Kawabata. Cocktail-Party Effect with Computational Auditory Scene Analysis — Preliminary Report —. In *Symbiosis of Human and Artifact* vol.2:503-508, Elsevier, 1995.

[Okuno et al., 1996] H.G. Okuno, T. Nakatani, and T. Kawabata. Interfacing Sound Stream Segregation to Speech Recognition Systems — Preliminary Results of Listening to Several Things at the Same Time. In *Proc. of 13th National Conference on Artificial Intelligence*, pp.1082-1089, 1996.

[Ramalingam, 1994] C.S. Ramalingam and R. Kumaresan. Voiced-speech analysis based on the residual interfering signal canceler (RISC) algorithm. In *Proc. of 1994 International Conference on Acoustics, Speech, and Signal Processing*, pp.473-476, IEEE, 1994.

[Roger et al., 1989] C. Rogers, D. Chien, M. Featherston, and K. Min. Neural network enhancement for a two speaker separation system. In *Proc. of 1989 International Conference on Acoustics, Speech and Signal Processing*, pp.357-60, IEEE, 1989.

[Rosenthal and Okuno, 1997] D. Rosenthal and H.G. Okuno (Eds.). *Computational Auditory Scene Analysis*, Lawrence Erlbaum Associates. Forthcoming.

[Selman et al., 1996] B. Selman, R.A. Brooks, T. Dean, E. Horovitz, T.M. Mitchell, and N.J. Nilsson. Challenge Problems for Artificial Intelligence, In *Proc. of 13th National Conference on Artificial Intelligence*, pp.1340-1345, 1996.

[Slaney et al., 1994] M. Slaney, D. Naar, and R.F. Lyon. Auditory Model Inversion For Sound Separation. In *Proc. of 1994 International Conference on Acoustics, Speech, and Signal Processing*, vol.2:77-80, IEEE, 1994.

[Stadler and Rabinowitz, 1993] R.W. Stadler and W.M. Rabinowitz. On the potential of fixed arrays for hearing aids. *J. of Acoustic Society of America* 94(3) Pt.1:1332-1342, 1993.

[Stubbs and Summerfield, 1991] R.J. Stubbs and Q. Summerfield. Effects of signal-to-noise ratio, signal periodicity, and degree of hearing impairment on the performance of voice-separation algorithms, *J. of Acoustical Society of America*, 89(3):1383-93, 1991.

[Warren, 1970] R.W. Warren. Perceptual restoration of missing speech sounds. *Science*, 167:392-393, 1970.

# Distributed Vision System: A Perceptual Information Infrastructure for Robot Navigation

### Hiroshi Ishiguro
Department of Information Science, Kyoto University
Sakyo-ku, Kyoto 606-01, Japan
E-mail: ishiguro@kuis.kyoto-u.ac.jp

## Abstract

This paper proposes a *Distributed Vision System* as a *Perceptual Information Infrastructure* for robot navigation in a dynamically changing world. The distributed vision system, consisting of vision agents connected with a computer network, monitors the environment, maintains the environment models, and actively provides various information for the robots by organizing communication between the vision agents. In addition to conceptual discussions and fundamental issues, this paper provides a prototype of the distributed vision system for navigating mobile robots.

## 1 Introduction

Many researchers are tackling to develop autonomous intelligent mobile robots which behave in a real world in robotics and artificial intelligence. For limited environments such as offices and factories, several types of mobile robots have been developed. However, it is still hard to realize autonomous robots behaving in dynamically changing real worlds such as an outdoor environment. To develop such robots which can adapt to the dynamic worlds is the original purpose of robotics and artificial intelligence.

**Attention control**
As discussed in technical papers on *Active Vision* [Ballard 89], the main reason lies in attention control to select viewing points according to various events relating to the robot. Two kinds of the attention control exist; one is called *Temporal Attention Control* and the other is *Spatial Attention Control*. If the robot has a single vision, it needs to change its gazing direction in a time slicing manner to simultaneously execute several vision tasks. The control of gazing direction is *Temporal Attention Control*. For example, the robot has to detect free regions even while gazing at the targets. We, human, solve this complex temporal attention control with its sophisticated mechanisms of memory and prediction. Further, the vision fixed on the robot body sometimes cannot provide proper information for the vision tasks. For example, when a robot estimates collisions with a moving obstacle, the side view in which both the robot itself and the obstacle are observed may be more proper than the view from the robot. This view point selection is called *Spatial Attention Control*.

**Difficulties in autonomous robots**
To realize the attention control is difficult with current technologies for autonomous robots. The following reasons can be considered.

- Active vision systems need a flexible body for acquiring proper visual information like a human. However, vision systems of previous mobile robots are fixed on the mobile platforms and it is generally difficult to build mobile robots which can acquire visual information from arbitrary viewing points in a 3D space.

- An ideal robot builds environment models by itself and uses them for executing commands from humans operators. However, to build a consistent model for a wide dynamic environment and maintain it is basically difficult for a single robot. We, humans, sometimes need helps of other persons to acquire information on the environment.

One of the promising research directions to solve the above-mentioned problems is to develop an infrastructure which provides sufficient information for the robots. This paper discusses such an infrastructure. The infrastructure in this paper differs from the infrastructure for mobile robots which move in factories. Our purpose is not to develop systems which support individual functions of the robots such as guide lines and landmarks for locomotion, but to develop a *Perceptual Information Infrastructure* (PI$^2$) which actively provides various information for real world agents, such as robots and humans. That is,

> The PI$^2$ monitors the environment, maintains the dynamic environment models, and provides information for the real world agents

As the PI$^2$, this paper proposes a *Distributed Vision System* (DVS). The DVS consists of multiple cameras which have own computing resource and communication links with others. The camera, called *Vision Agent* (VA), provides visual information required by the vision-guided mobile robots. The VAs at various locations provide sufficient visual information for attention control of the robots, and they maintain dynamic environment models by organizing communication between them. Of course, we can use other sensors in the infrastructure. The camera, however, is the most compact and low cost passive sensor to acquire various kinds of information and many interesting vision research issues are still remained. Another purpose of this research approach is to deal with the issues and develop real applications of computer vision through the DVS.

In addition to conceptual discussions and research issues, a prototype of the DVS and experimental results using it are shown. The author has confirmed the DVS can robustly navigate the mobile robot in a complex real world.

**Related works**

Recently novel research approaches using distributed sensors and robots have been proposed in robotics. For example, the *Robotic Room* proposed by Mizoguchi and others [Mizoguchi 96] support human activities with sensors and robots embedded in a room. Their interests are to design mechanisms and develop sensor system for executing well-defined local tasks. On the other hand, the purpose of this paper is to propose a flexible sensor system utilized by various kinds of robotics systems as an information infrastructure.

Several vision systems which utilize multiple cameras has been reported, especially, in multimedia. Moezzi and others proposed the concept of *Immersive Video* and developed a vision system using precisely calibrated cameras for building a precise geometrical model of an outdoor environment. Pinhanez and Bobick [Pinhanez 96] developed a system which dynamically selects cameras providing proper views for broadcasting a TV show. We, however, consider a demerit of the systems is to use calibrated cameras and geometrical models of the world. Geometrical representations of environments obtained by the calibrated cameras lack robustness and flexibility of the systems. In order to solve the problems, this paper proposes an alternative approach for modeling dynamic environments, which dynamically and locally estimates the camera parameters and directly represents robot tasks.

In distributed artificial intelligence, several fundamental works dealing with systems using multiple sensors have been reported. Lesser [Lesser 83] proposed a *Distributed Vehicle Monitoring Testbet* (DVMT) as an example of distributed sensing problems, and Durfee [Durfee 91] proposed *Partial Global Planning* which is a planning method for globally analyzing signals provided by multiple signal processing agents. The DVS, basically, can be considered as a kind of the distributed sensing systems, such as the DVMT, but deals with vision sensors and communicates with robots. And further, the purpose in the DVS is not to globally analyze the signals, but to navigate mobile robots with local information by representing the navigation tasks in the VA network.

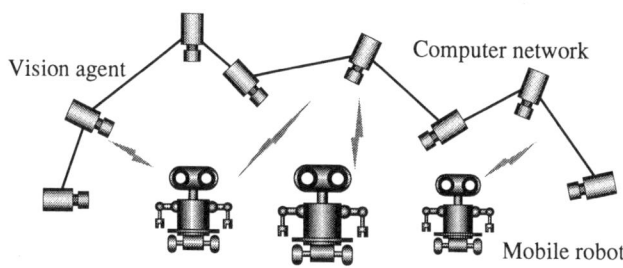

Figure 1: Distributed Vision System

## 2 Distributed Vision System

### 2.1 Concept of distributed vision

In order to simultaneously execute the vision tasks, an autonomous robot needs to change its visual attention. The robot, generally, has a single vision sensor and a single body, therefore the robot needs to make complex plans to execute the vision tasks with the single vision sensor. *Active Vision* proposed by Ballard [Ballard 89] is a research direction to solve the complex planning problem with active camera motions bring proper visual information and to enable real-time and robust information processing. That is, the active vision system needs a flexible body to acquire the proper visual information like a human. However, the vision system of previous mobile robots is fixed on the mobile base and it is generally difficult to build autonomous robots which can acquire visual information from arbitrary viewing points in a 3D space.

Our idea to solve the problem is to use many VAs embedded in the environment and connected them with a computer network (See Fig. 1). Each VA independently observes events in the local environment and communicates with other VAs through the computer network. Since the VAs do not have any constraints in the mechanism like autonomous robots, we can install a sufficient number of VAs according to tasks, and the robots can acquire necessary visual information from various viewing points. As a new concept to generalize the idea, the author proposes *Distributed Vision* that multiple vision agents embedded in an environment recognize dynamic events by communicating each other. In the distributed vision, the attention control problems are dealt as dynamic organization problems of communication between the vision agents.

The DVS is not a standard computer network. It is an extended computer network which bridges between physical worlds and virtual worlds building in the computer network. Current computer networks transmit only data, such as images and characters. However, as the services and functions of the computer networks are extended, more efficient and intelligent communication between computers are required. The author calls such

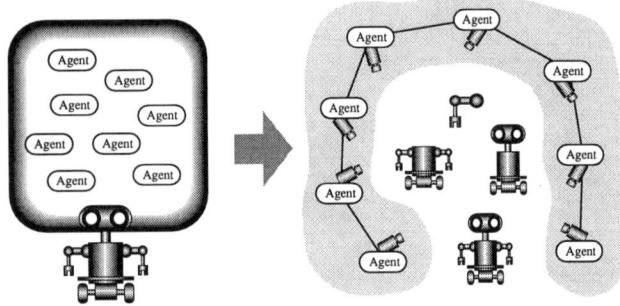

Figure 2: From an autonomous robot to robots integrated with environments

a future computer network *Perceptual Information Infrastructure* (PI$^2$). The PI$^2$ observes physical worlds, maintains dynamic models in the computer network and supports robots and humans. The DVS is one example of the PI$^2$.

In robotics, the PI$^2$ enables robust and flexible robotic systems by offering necessary information and develops a new research area. As shown in Fig. 2, a previous autonomous robot consists of a mechanical body and software agents. That is, the intelligence produced by the software agents. On the other hand, the intelligent information processing of robots supported by the PI$^2$ is done by agents embedded in the environment. Development of *Robots Integrated with Environments* is an important research direction for realizing useful robotic systems.

## 2.2 Design policies for the DVS

The VAs are designed based on the following idea:

**Tasks of robots are closely related to local environments.**

For example, when a mobile robot executes a task of approaching a target, the task is closely related to a local area where the target locates. This idea allows to give VAs specific knowledge for recognizing the local environment, therefore each VA has a simple but robust information processing capability.

More concretely, the VAs can easily detect dynamic events since they are fixed in the environment. A vision-guided mobile robot of which camera is fixed on the body has to move for exploring the environment, therefore there exist a difficult problem to recognize the environment through the moving camera. On the other hand, the VA in the DVS easily analyzes the image data and detects moving objects by constructing the background image for the fixed viewing point.

All of the VAs, basically, have the following common visual functions:

- Detecting moving obstacles by constructing the background image and comparing with it.
- Tracking detected obstacles by a template matching method.
- Identifying mobile robots based on given models.
- Finding relations between moving objects and static objects in the images.

The DVS, which does not keep the precise camera positions for robustness and flexibility, autonomously and locally calibrates the camera parameters with local coordinate systems according to demand (the detail is discussed in Section 4.3). That is, the VAs iterate to establish representation frames for communicating with other agents.

The VAs identifies objects with the motions observed in the images in addition to the visual features since they can provide reliable motion information from the fixed viewing points. The author considers the DVS can solve the correspondence problem more robustly and flexibility than the previous vision systems.

The DVS organizes communication between VAs in order to execute given tasks. The design policy that a VA executes particular subtasks in the local environment allows to solve the organization problem in a hierarchical manner. That is, global tasks given to the DVS, generally, can be decomposed into the subtasks and the VAs execute them. However, the subtasks often need to be simultaneously executed and the combinations often change according to various situations. Therefore, the VAs should be globally and locally organized to execute the global tasks. The organization of VAs is the most important research issue of the DVS.

## 3 Fundamental issues

### 3.1 Communication between VAs

Remarkable difference of the DVS with previous computer systems is the DVS has two kinds of communication. In addition to communication with a computer network, VAs in the DVS communicate by observing common events. When two VAs which have own local internal representations simultaneously observe a robot from different viewing points, they may synchronously update their local internal representations. The VAs share symbolic and non-symbolic information through the computer network and the observations, respectively. It is an important research issue how to establish sophisticated and flexible communication links through the two types of communication. The author is especially interested in the non-symbolic communication which is difficult to deal with in previous frame works.

### 3.2 Dynamic environment model

The robust detection of dynamic events enables to hierarchically represent the environment. We, basically, consider static environment models should be generated from dynamic environment models representing the dynamic events. The dynamic environment models give meanings to static objects represented in the static models. For example, a gray region in the images, which is a road in the outdoor environment, is defined as a region where the robot can move. The DVS which can easily detect the dynamic events is a promising system for realizing the hierarchical environment models.

## 3.3 Organization of VAs

The DVS needs to organize the VAs for acquiring the dynamic environment models. Let us imagine a DVS navigating a mobile robot. In order to avoid moving obstacles and detect free regions, the mobile robot needs visual information provided by VAs locating around it, and in order to go toward a destination, it also needs information about subgoals from VAs locating along the robot path. That is, the VAs should be locally and globally organized in order to provide proper information for the robot navigation. In the organization process, the DVS represents given tasks by organizing the VAs. For realizing the organization, it is necessary to develop new methods which deal with the total process including image understanding by the VAs, task understanding, and task execution through the symbolic and non-symbolic communication links.

## 3.4 Distributed model

The dynamic environment models are not shared by all VAs, but distributed over VAs. Robots access to the dynamic environment models through dynamic organizations of the VAs. For example, when a robot avoids a moving obstacle, the DVS continuously organizes the VAs located around it and navigates it. Further, if troubles occur in a VA, other VAs take the place of the VA. To distribute the models in the VA network is important for realizing flexibility and robustness of the DVS.

# 4 A prototype of the DVS

This section discusses a developed prototype of the DVS [Tanaka 97]. The prototype system briefly deals with the fundamental issues discussed in Section 3, but it does not completely solve them. The issues are carefully dealt with in the feature works.

## 4.1 Mobile robot navigation

The outline for mobile robot navigation by the DVS is as follows. First, a human operator teaches tasks by manually controlling a robot. The human operator does not directly give task models or behavior models of the robot, but gives examples to the DVS. While the robot moves, each VA tracks it within the visual field with simple image processing functions discussed in section 2.2. Then the DVS decomposes the given example paths into several components which can be maintained by each VA and memorizes then by organizing the VAs. After organizing the VAs, the DVS autonomously navigates the mobile robot while the VAs communicate each other. All of the VAs monitor the robot motions and send messages to other VAs according to the memorized organization patterns for global and local tasks.

## 4.2 The architecture

A VA consists basic modules and memory modules as shown in Fig. 3. For the basic modules, the

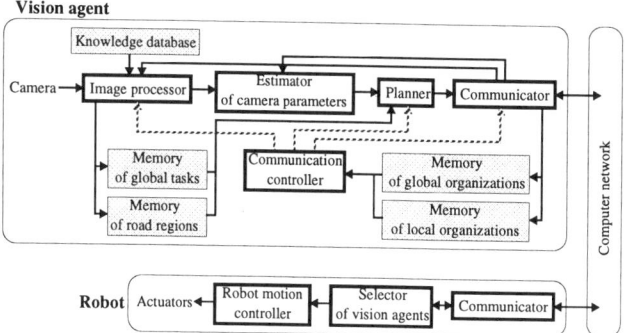

Figure 3: The architecture of the DVS

VA has *Image processor*, *Estimator*(Estimator of camera parameters), *Planner*, *Communicator*, and *Controller*(Communication controller). For the memory modules, it has a knowledge database for image processing, memories to memorize global and local tasks, and memories to maintain relations with other VAs for executing the global and local tasks. In this experimentation, the global task is to navigate toward goals and the local task is to avoid obstacles.

*Image processor* detects moving robots and tracks them by referring to the knowledge database which stores visual features of robots. *Estimator* receives the results and estimates camera parameters for establishing representation frames for sharing robot motion plans with other VAs. *Planner* plans robot actions based on the estimated camera parameter and sends them to the robot through *Communicator*. The robot corrects the plans, selects proper plans, and executes the plans. The selected plans are sent back to the VAs and memorized. The memorized plans are directly applied in the same situations of the VAs and the robot by *Controller*.

## 4.3 Global and local organization

For selecting and integrating the robot motion plans from VAs, all plans at a time should be represented with a common representation frame. In the case of the DVS, the robot motion is represented with a $X - Y$ robot path-centered coordinate system. The coordinate transformation from the camera frame of a VA to the robot path-centered coordinate system is represented with two parameters $\alpha$ and $\beta$. Since the vision data is very noisy and the obtained plans are simple, the DVS assumes orthographic projection for the obtained image and represents the coordinate transform with only two parameters of camera rotations. As a more sophisticated method, it is possible to use an automatic calibration method proposed by Hosoda and others [Hosoda 94].

Fig. 4 shows data flows between functions for globally and locally organizing VAs. *Estimator* computes $\alpha$, $\beta$ and their error estimates $\Delta\alpha$ and $\Delta\beta$ for the coordinate transformation (Estimating and updating camera parameters). *Planner* plans robot motions with the obtained coordinate system (Planning a robot motion).

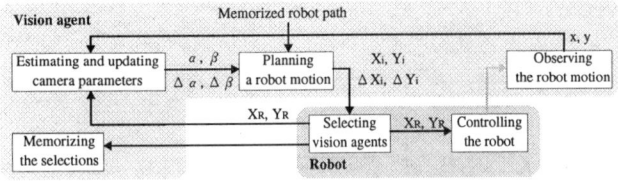

Figure 4: Global and local organization

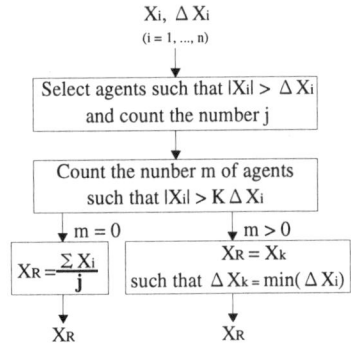

Figure 5: Plan selection for the global task

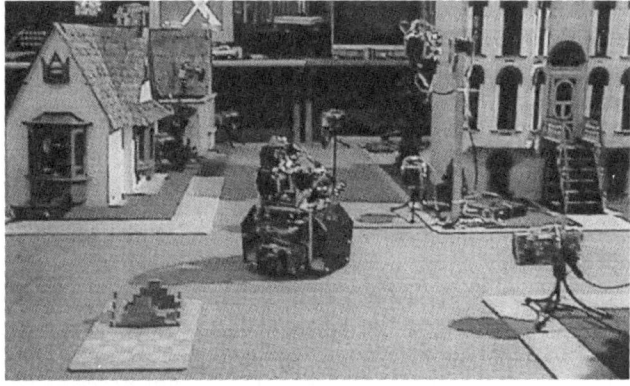

Figure 6: Model town

Figure 7: An example path and a robot trajectory navigated by the DVS

The plans $X_i$ and $Y_i$ and their error estimates $\Delta X_i$ and $\Delta Y_i$ by agent $i$ are send to the robot and the robot selects proper plans based on the error estimations (Selecting VAs). While the robot executes the integrated plans $X_R$ and $Y_R$, it sends the plans to the VAs. The VAs memorize the selection results as organization patterns.

The robot selects the proper plans for global as shown in Fig. 5 by referring to the error estimates. For the plans such that $|X_i| > \Delta X_i$, the robot computes the mean value if the error estimate is large ($|X_i| > K\Delta X_i$, K=2), otherwise, it selects a plan which has the minimum error estimate. The algorithm means that if $|X_i| > \Delta X_i$ the plan is incorrect and if $|X_i| > K\Delta X_i$ the plan is roughly correct. For the local task, avoiding obstacles, the robot selects vision agents such that $|X_i| > \Delta X_i$. The VAs generate negative values of $X_i$ and $Y_i$ as the plans to avoid obstacles.

### 4.4 Experimental setup

Fig. 6 shows a model town and a mobile robot used in the experimentation. The model town, of which reduced scale is 1/12, has been made for representing enough realities of an outdoor environment, such as shadows, textures of trees, lawns and houses. Sixteen VAs have been established in the model town and used for navigating the mobile robot.

Images taken by the VAs are sent to an image encoder which integrates sixteen images into one image (the size of each image is reduced from $512 \times 512 pixels$ to $128 \times 128 pixels$). Then, the image is sent to a color frame grabber and a motion estimation processor which can detect optical flows in real time. The main computer, Sun Spark Station 10, executes the vision functions by using data from the color frame grabber and the motion estimation processor. This system, unfortunately, cannot compute in parallel. The author is currently developing a new parallel computing system using twenty four C40 DSPs.

### 4.5 Experimental results

Fig. 7(a) shows an example path taught by a human operator in the teaching phase. Fig. 7(b) shows a robot trajectory autonomously navigated by the DVS. Because of simplicity of the image processing, the DVS could robustly navigate the mobile robot in a complex environment.

Fig. 8 shows images taken by VAs in the autonomous navigation phase. The vertical axis and the horizontal axis indicate the time stamp and the ID numbers of the VAs, respectively. The white boxes and the black boxes indicate selected VAs for the global and local tasks, respectively. Here, all VAs which simultaneously observe the robot are locally organized. As shown in Fig. 8, the DVS have dynamically organized the VAs for executing the global and local tasks.

The experimentation shows important aspects of the DVS. The DVS can memorizes the tasks for navigating the robot along a path by organizing the VAs and iterate to select proper VAs for robustly executing the tasks in

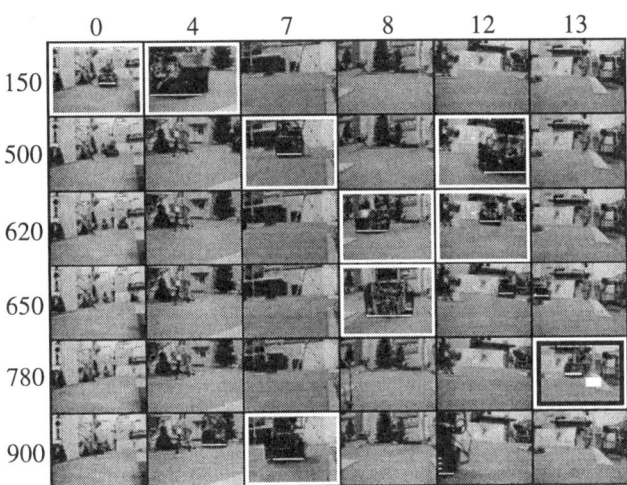

Figure 8: Images taken by VAs

a complex environment. That is, the DVS solves the attention control problems for the autonomous robots discussed in Section 1 with a different but more robust manner.

## 5 Conclusion and Research Plan Toward PI$^2$

The DVS is an alternative approach for realizing robust behaviors of intelligent robots. By organizing VAs, the DVS tightly couples robot actions and observations by the VAs. The previous autonomous robot has constraints of the hardware, the size and the number of sensors. On the other hand, the DVS does not have such constants and it is a promising approach for realizing real systems.

The purposes in this research approach are to solve the fundamental problems of the DVS, to develop real systems and to establish key technologies of PI$^2$. Especially, to develop systems for real applications is important. It will extend possibility of robotics and computer networks and be able to make users realize usefulness of computer vision techniques. The plan of this research is as follows:

1. Develop a DVS for navigating robots with a model town as a testbed (See Section 4).
2. Study the fundamental issues of mobile robot navigation (See Section 3).
3. Develop a DVS for observing and supporting human behaviors.
4. Study the fundamental issues of the human behavior support.
5. Extend the DVS to the PI$^2$ which supports various real world agents and develop a PI$^2$ in the university campus.

Results for the fundamental problems can be estimated with traditional evaluation criterion, such as originality and applicability. On the other hand, evaluations for the developed infrastructure is not so easy since the system should be collectively evaluated with various criterion. For the evaluations, the author considers it is important to disclose the details of the system development, which cannot be reported in technical papers, with the world-wide web (See http://www.lab7.kuis.kyoto-u.ac.jp/vision/).

Recent developments of multimedia computing environments have established huge number of cameras and computers in offices and towns. They are expected to be more intelligent systems as the PI$^2$ discussed in this paper. The PI$^2$ is a key issue in the next decade.

## Acknowledgment

The author would like to thank Prof. Toru Ishida for his stimulating discussions and constructive criticism, and Mr. Goichi Tanaka for his programming work.

## References

[Ballard 89] D. H. Ballard, Reference frames for animate vision, Proc. IJCAI, pp. 1635-1641, 1989.

[Durfee 91] E. H. Durfee and V. R. Lesser, Partial global planning: A coordination framework for distributed hypothesis formation, IEEE Trans. SMC, Vol. 21, No. 5, pp. 1167-1183, 1991.

[Hosoda 94] K. Hosoda and M. Asada, Versatile Visual Servoing without Knowledge of True Jacobian, Proc. IROS, pp. 186-193, 1994.

[Lesser 83] V. R. Lesser, and D. D. Corkill, The distributed vehicle monitoring testbed: A tool for investigating distributed problem solving networks, AI Magazine, pp. 15-33, 1983.

[Moezzi 96] S. Moezzi, An emerging Medium: Interactive three-dimensional digital video, Proc. Int. Conf. Multimedia, pp. 358-361, 1996.

[Pinhanez 96] C. S. Pinhanez and A. F. Bobick, Approximate world models: Incorporating qualitative and linguistic information into vision systems, Proc. AAAI, pp. 1116-1123, 1996.

[Mizoguchi 96] H. Mizoguchi, T. Sato and T. Ishikawa, Robotic office room to support office work by human behavior understanding function with networked machines, Proc. ICRA, pp. 2968-2975, 1996.

[Tanaka 97] G. Tanaka, H. Ishiguro and T. Ishida, Mobile robot navigation by distributed vision agents, Proc. ICCIMA, pp. 86-90, 1997.

# AI CHALLENGES

## Challenge 3

# Challenges in bridging plan synthesis paradigms

Subbarao Kambhampati[*]
Department of Computer Science and Engineering
Arizona State University, Tempe AZ 85287-5406
rao@asu.edu; http://rakaposhi.eas.asu.edu/rao.html

**Abstract**

In the last three years, several "radically new" and promising approaches have been developed for tackling the plan synthesis problem. Currently, these approaches exist in isolation as there is no coherent explanation of their sources of strength *vis a vis* the traditional refinement planners. In this paper, I provide a generalized view of refinement planning, that subsumes both traditional and newer approaches to plan synthesis. I will interpret the contributions of the new approaches in terms of a new subclass of refinement planners called "disjunctive planners". This unifying view raises several intriguing possibilities for complementing the strengths of the various approaches. I will identify and pose these as challenges to the planning community.

## 1. Introduction

Most traditional approaches to plan generation, developed over the last twenty years, work by searching in the space of partial plans, extending a plan incrementally until it becomes a solution, and backtracking when a plan can no longer be fruitfully refined. More recently, several approaches have been developed that relate plan synthesis to constraint satisfaction. These include Graphplan [Blum & Furst, 1995], SATPLAN [Kautz & Selman, 1996], COPS [Ginsberg, 1996], Descartes [Joslin & Pollack, 1996] and UCPOP-D [Kambhampati & Yang, 1996]. Graphplan and SATPLAN approaches, in particular, have demonstrated impressive scale-up potential in practice.

At present, there exists a huge gulf between these new breed of algorithms and the traditional refinement planning approaches. An important challenge for the planning community is to bridge these strands of research to see if their strengths can be complemented. To this end, I present a generalization of the refinement planning framework developed in our previous work [Kambhampati, Knoblock & Yang, 1995; Kambhampati & Srivastava, 1996] that covers most of the currently known approaches.

According to my general framework, partial plans are shorthand notations for sets of action sequences (called the candidate set of the plan). Plan synthesis involves a "*split and prune*" search [Pearl, 1980]. The pruning is carried out by applying refinement operations to a given set of partial plans (called a *planset*). Pruning attempt to incrementally get rid of non-solutions from the candidate set. The splitting part involves pushing the component partial plans of a planset into different branches of the search tree. Its main aim is to reduce the cost of applying refinements and termination check to the planset. Termination test involves checking if a solution can be extracted from the (possibly exponential number of) minimal candidates of the current planset. The extraction process can in general be cast as a constraint satisfaction search.

Within this framework, traditional planners can be seen as doing both pruning (refinement) and full splitting (pushing each component of the planset into a different search branch). They thus refine and terminate on individual partial plans (rather than plansets). The scale-up problems associated with these planners can be related to their full splitting. An obvious way of curing this malady involves avoiding splitting or controlling it intelligently. To implement this idea--which I call "**disjunctive planning**"-- we need to find ways of effectively refining and terminating on large sets of plans. All the newer planners can be seen as providing potential solutions to these two problems. Graphplan (and COPS) can be seen as clarifying the issues involved in compactly representing and reasoning with plansets. SATPLAN demonstrates how the problem of extracting solutions from a planset can be posed as a SAT problem, so that it can be solved by the new breed of efficient SAT solvers [Kautz et. al., 1992; Crawford & Auton, 1996; Bayardo & Schrag, 1997]. Finally, Descartes and UCPOP-D can be seen as exploring the effect of controlled (rather than complete or no) splitting of the plansets.

Viewing existing planners as points in a spectrum of possible disjunctive planners opens several exciting avenues of focused research: Could the missing planners corresponding to the other points in the spectrum be more efficient? What are the tradeoffs governing the efficiency of these spectrum of planners? I will identify specific short-term and intermediate-term challenges that have to be undertaken to answer these questions.

The rest of the paper is organized as follows. Section 2 presents my generalized framework for refinement planning. Section 3 motivates the idea of disjunctive planning, and interprets several newer approaches as providing guidelines for implementing disjunctive planners. Section 4 relates SATPLAN family of algorithms to disjunctive planning. Section 5 lists and motivates a set of challenge problems that arise from this unifying view. Section 6 briefly discusses the logistics of coordinating the research into the challenge problems.

## 2. Refinement Planning: Overview

Since a solution for a planning problem is ultimately a sequence of actions, plan synthesis in a general sense involves sorting our way through the set of *all* action sequences until we end up with a sequence that is a solution. This is the essential idea behind refinement planning. The

---

[*] This research is supported in part by the NSF NYI award IRI-9457634, the ARPI Initiative grant F30602-95-C-0247 and the ARPA AASERT grant DAAH04-96-1-0231. I would like to thank Bart Selman, Laurie Ihrig, Amol Mali and the anonymous reviewers for critical comments on a previous draft.

sets of action sequences are represented and manipulated in terms of **partial plans** which can be seen as a collection of constraints. The action sequences denoted by a partial plan, i.e., those that are consistent with its constraints, are called its **candidates**. For technical reasons that will become clear later, we find it convenient to think in terms of *sets* of (instead of single) partial plans. A set of partial plans is called a **planset** with its constituent partial plans referred to as the **components**. The candidate set of a planset is defined as the union of the candidate sets of its components.

A refinement (pruning) operation narrows the candidate set of a planset by adding constraints to its component plans. If no solutions are eliminated in this process, we will eventually progress towards the set of all solutions. Termination can occur as soon as we can pick up a solution using some bounded time operation -- called the *solution extraction function*. To make these ideas precise, we shall now look at the syntax and semantics of partial plans and refinement operations.

## 2.1 Partial Plan Representation: Syntax

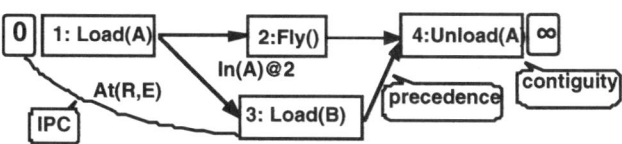

*Figure 1. An example (partial) plan in rocket domain.*

A partial plan can be seen as any set of constraints that together delineate which action sequences belong to the plan's candidate set and which do not. One commonly used representation models partial plans as a set of steps, ordering constraints between the steps, and auxiliary constraints (we shall discuss alternative representations, such as that proposed by Ginsberg [1996], later). The ordering constraints may require steps to precede each other or be contiguous. The auxiliary constraints demand the preservation of a condition over a time interval between two steps (called IPCs) or the truth of a condition at a time point (called point truth constraints). An example plan in this representation from the one-way rocket domain (involving transportation of two packages from earth to moon using a single one-way rocket) is shown in Figure 1.

## 2.2 Partial Plan Representation: Semantics

The semantics of the partial plans are given in terms of candidate sets. A candidate can be seen as a model of the partial plan constraints. An action sequence belongs to the candidate set of a partial plan if it contains the actions corresponding to all the steps of the partial plan, in an order consistent with the ordering constraints on the plan, and it also satisfies all preservation constraints. For the example plan shown in Figure 1, the action sequences shown on the left in Figure 2 are candidates, while those on the right are non-candidates.

Notice that the candidates may contain more actions than are present in the partial plan. Because of this, a plan's candidate set can be potentially infinite. We define the notion of "**minimal candidates**" to let us restrict our attention to a finite subset of the possibly infinite candidate set. Specifically, minimal candidates are candidates that only contain the actions listed in the partial plan (thus their length is equal to the number of steps in the plan other than 0 and $\infty$). The top candidate on the left of Figure 2 is a minimal candidate while the bottom one is not. There is a one-to-one correspondence between the minimal candidates and the syntactic notion of safe linearizations of a plan, where a safe linearization is a permutation of plan steps that satisfies the auxiliary preservation constraints. For example, the minimal candidate on the top left of Figure 2 corresponds to the safe linearization 0-1-3-2-4-$\infty$ (as can be verified by translating the step names in the latter to corresponding actions).

| Candidates ($\in$ «P») | Non-Candidates ($\notin$ «P») |
|---|---|
| [Load(A),Load(B),Fly(),Unload(A)] | [Load(A),Fly(),Load(B),Unload(B)] |
| Minimal candidate. Corresponds to the safe linearization [ 01324∞ ] | Corresponds to unsafe linearization [ 01234∞ ] |
| [Load(A),Load(B),Fly(), Unload(B),Unload(A)] | [Load(A),Fly(),Load(B), Fly(),Unload(A)] |

*Figure 2. Candidate set of a plan*

## 2.3 Refinement Strategies

A refinement strategy $R$ maps a planset $P$ to another planset $P^R$ such that the candidate set of $P^R$ is a subset of the candidate set of $P$. $R$ is said to be **complete** if $P^R$ contains all the solutions of $P$. It is said to be **progressive** if the candidate set of $P^R$ is a strict subset of the candidate set of $P$. $R$ is said to be **strongly progressive** if the length of the minimal candidates increases after the refinement. The degree of **progressiveness** is measured in terms of the reduction in the candidate set size. It is said to be **systematic** if no action sequence falls in the candidate set of more than one component of $P^R$.

Completeness ensures that we don't lose solutions by the application of refinements. Progressiveness ensures that refinement narrows the candidate set (i.e., has pruning power). Strongly progressive refinement strategies simultaneously shrink the candidate set of the plan, and increase the length of its minimal candidates. This provides an incremental way of exploring the (potentially infinite) candidate set of a planset for solutions: *Examine the minimal candidates (corresponding to safe linearizations) of the planset after each refinement to see if any of them correspond to solutions*. Systematicity ensures that we never consider the same candidate more than once, if we were to explore the components of the planset separately.

Refinements are best seen as canned inference procedures that compute the consequences of the meta-theory of planning, and the domain theory (in the form of actions), in the specific context of the current partial plan constraints. Traditional planners use four types of refinement strategies -- forward state space, backward state space, plan space and task-reduction [Kambhampati, 1997]. Figure 3 shows a forward state space refinement of a partial plan in the one-way rocket domain. It takes the null planset, corresponding to all action sequences and maps it to a planset containing 3 components. In this case, given the theory of planning, which says that solutions must have actions that are executable in their respective states, and the current planset constraint that the state of the world before the first step is the initial state, the forward state space refinement infers that the only actions that can come as the second step in the solution are Load(A), Load(B) and Fly().

This particular refinement is complete since no solution to the rocket problem can start with any other action for the given initial state. It is progressive since it eliminated the action sequences not beginning with Load(A), Load(B) or Fly() from consideration. Finally, it is systematic since no action sequence will belong to the candidate set of more than one component (the candidates of the three components will differ in the first action).

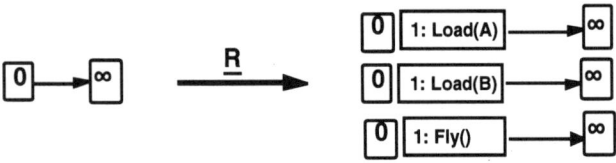

*Figure 3. An example refinement strategy (Forward State Space)*

## 2.4 Plan Synthesis

We are now in a position to present the general refinement planning template, which we do in Figure 4. If the current planset has an extractable solution—which is checked by inspecting its minimal candidates to see if any of them is a solution— we terminate. If not, we select a refinement strategy $R$ and apply it to the current planset to get a new planset. This is the pruning step. The third step is the splitting step, and involves pushing the components of the refined planset into different branches. The splitting is controlled by the parameter $k$. If $k$ equals 1, no splitting is done. If $k$ equals the number of components of the refined planset, full splitting is done. Intermediate values of $k$ correspond to intermediate levels of splitting. After the splitting step, one of the search branches is selected non-deterministically, and pruning and splitting is applied to it recursively.

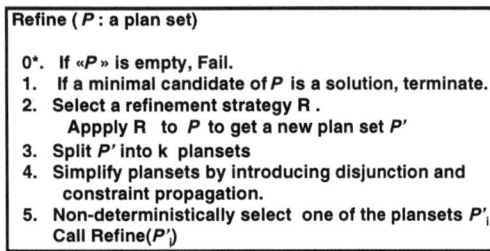

*Figure 4. Refinement planning with pruning and splitting*

As long as the selected refinement strategy $R$ is complete, the algorithm will never lose a solution. As long as the refinements are strongly progressive, for solvable problems, the algorithm will eventually reach a planset one of whose minimal candidates will be a solution.

The solution extraction process involves searching through the possibly exponential number of minimal candidates of a planset for a solution. As $k$ increases, individual plansets have fewer components, and consequently the cost of solution extraction reduces.

The algorithm template in Figure 4 covers both traditional and newer plan synthesis approaches (see [Kambhampati, 1997] for a more elaborate discussion). Traditional refinement planners, such as UCPOP, Prodigy, SNLP, etc. correspond to complete splitting (i.e., $k$ equals the number of components of the planset) differing mainly in the type of refinements they employ. The newer approaches such as Graphplan, COPS, Descartes and UCPOP-D can be seen as handling plansets without splitting. The first two do not do any splitting, while the last two do controlled splitting. Finally, SATPLAN approaches can be understood as posing the solution extraction as a SAT problem.

## 3. Disjunctive planning

We saw that traditional planners do refinement planning with full splitting, differing mainly in terms of the specific refinement strategies they use. Unfortunately, these planners tend to generate huge search spaces and have in practice shown disappointing scale-up potential. Viewing the newer approaches such as Graphplan and SATPLAN as instances of our framework suggests that a promising general solution involves handling plansets without splitting. This idea, referred to as disjunctive planning, raises three immediate issues: *How do we (a) represent, (b) refine and (c) extract solutions from large plansets?* I will discuss the first two issues in this section. The most promising approach for solution extraction at present seems to be to pose it as a CSP/SAT problem. I will discuss this further in Section 4, in the context of SATPLAN.

### 3.1 Disjunctive Representations

First off, keeping plansets together may lead to very unwieldy data structures. The way to get around this is to "internalize" the disjunction in the plansets so that we can represent them more compactly. The general idea of disjunctive representations is to allow disjunctive step, ordering, and auxiliary constraints into a plan. Figure 5 and Figure 6 show two examples the idea. The three plans on the left in Figure 5 can be combined into a single disjunctive step, with disjunctive contiguity constraints. Similarly, the two plans in Figure 6 can be combined by using a single disjunctive step constraint, a disjunctive precedence constraint, a disjunctive interval preservation constraint and a disjunctive point truth constraint.

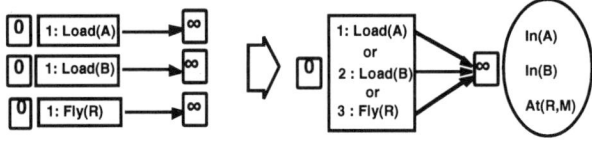

*Figure 5. Disjunction over state-space refinements.*

Candidate set semantics for disjunctive plans follow naturally from the fact that the presence of the disjunctive constraint $c_1 \vee c_2$ in a partial plan constrains its candidates to be consistent with either $c_1$ or $c_2$. So, solution extraction on disjunctive plans can still be posed as the problem of searching through minimal candidates for a solution.

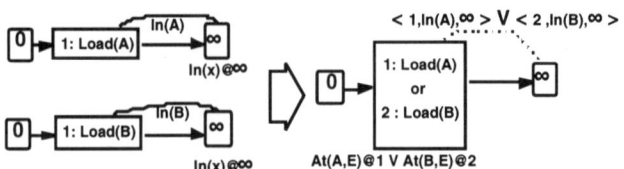

*Figure 6. Disjunction over plan-space refinements*

### 3.2 Refining Disjunctive Plans

In refining the disjunctive plans, we face a tradeoff between the pruning power of the refinement and its cost. The maximal pruning power is obtained by externalizing the disjunction, applying the traditional refinements to each of the resulting non-disjunctive partial plans, and combining them back into a disjunctive plan. For example, to refine the plans on the left in Figures 5 and 6, we simply go back to the plans on the right, refine them, and combine the results into a new disjunctive plan. This approach is however infeasible in practice since it makes the refinement stage exponentially hard. We need to generalize the traditional refinement strategies so that they can apply directly to the disjunctive plans. I will now attempt to outline the critical issues in doing this.

Since refinement strategies essentially derive the consequences of planning and domain theory in the context of the current plan constraints, it is clear that internal disjunction in the partial plan may lead to the derivation of weaker consequences (thus reducing their pruning power). For example, for the disjunctive plan on the right in Figure 5, we don't know which of the three steps will actually occur in the eventual solution. Consequently, we don't exactly know what the state of the world will be after the disjunctive step. This means that the forward state space refinement will not be able to infer exactly which actions can appear in the second step of the solution. Similarly, for the disjunctive plan in Figure 6, we don't know whether steps 1 or 2 or both will be present in the eventual solution. Thus a plan space refinement won't know whether it should introduce steps relevant to At(A,E) precondition, those relevant to At(B,E) precondition, or both.

All is not lost however, as the refinements can still infer a superset of the relevant actions. For example, for the plan in Figure 5, although we do not know the exact state after the first (disjunctive) step, we know that it can only be a subset of the union of conditions in the effects of the three steps. Knowing that only Load(A), Load(B) or Fly(R) can be the first steps in the plan tells us that the state after the first step can only contain the conditions In(A), In(B) and At(R,M). We can thus generalize forward state space refinement to add only those actions whose preconditions are subsumed by the union of propositions comprising the effects of the three steps. Similarly, plan-space refinement can be generalized to introduce actions to achieve all parts of a disjunctive precondition.

These "naive" generalizations of the standard refinements are still complete, but have far less pruning power than the standard refinements operating on non-disjunctive plansets. In the case of the generalized forward state space refinement, even if the preconditions of an action are in the union of effects of the preceding disjunctive step, there may be no real way for that action to actually be take place. For example, in Figure 5, although the preconditions of "unload at moon" action may seem satisfied, it is actually never going occur as the second step in any solution because Load() and Fly() cannot be done at the same time. In fact, this naive generalization may allow an exponential number of additional actions that will not be considered by traditional forward state space refinement operating on non-disjunctive plans [Kambhampati & Lambrecht, 1997].

Although the loss of progressiveness in refining a disjunctive plan cannot be completely avoided, it can be reduced to a significant extent by *extending the types of consequences inferred by the refinements*. Traditional refinement strategies concentrate exclusively on inferring the identities of new actions that must be part of any eventual solution. We can widen the focus in the context of disjunctive plans. For example, in Figure 5, using the domain and planning theory, we can recognize that actions Load(A) and Fly(R) cannot both occur in the first step (since their preconditions and effects are interacting). Propagating this information tells us that the second state may either have In(A) or At(R,M), but not both. Here the interaction between the steps 1 and 3 propagates to make the conditions In(A) and At(R,M) "mutually exclusive" in the next disjunctive state. Thus any action which needs both In(A) and At(R,M) can be ignored at the next level. The particular strategy described here is similar to the one employed by Blum and Furst's [1995] Graphplan algorithm, and has been shown to be a major source of its efficiency [Kambhampati & Lambrecht, 1997]. Similar techniques need to be developed for plan-space refinements.

To summarize, disjunctive plans can be refined directly and efficiently, at the expense of some of the progressiveness (pruning power) of the refinement. The loss of pruning power can be countered by widening the scope of the refinements to infer more than just the potential actions in the solution.

## 4. Relating SATPLAN to disjunctive planning

The SATPLAN approach involves generating a SAT encoding, all models of which will correspond to $k$-length solutions to the problem (for some fixed integer $k$). Model-finding is done by efficient SAT solvers [Selman et. al., 1992; Crawford & Auton, 1996; Bayardo & Schrag, 1997]. Kautz et. al. propose to start with some arbitrary value of $k$, and increase it if they do not find solutions of that length. They have considered a variety of ways of generating the encodings, corresponding loosely to different traditional planning algorithms.

In the context of the general refinement planning framework, we can offer a rational basis for the generation of the various encodings. Specifically, the natural place where SAT solvers can be used in refinement planning is in the "solution extraction phase". As illustrated in Figure 7, after doing $k$ "complete" and "strongly progressive" refinements on a null plan, we get a planset whose minimal candidates contain all $k$-length solutions to the problem. So, picking a solution boils down to searching through the minimal candidates-- which can be cast as a SAT problem. This account naturally relates the character of the encodings to the type of refinements used in coming with the $k^{th}$-level planset and how the plansets themselves are represented.

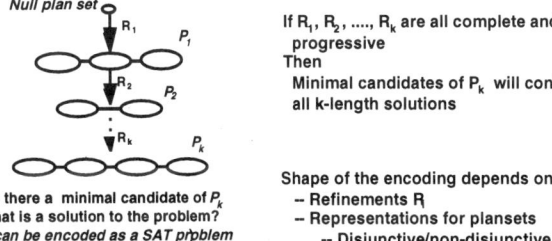

*Figure 7. Relating refined plan at k-th level to SATPLAN encodings.*

Kautz et. al. concentrate primarily on direct translation of planning problems to SAT encodings, sidelining refinement planning [Kautz et. al., 1996] (the only exception is their Graphplan based encoding; see below). I believe that such an approach confounds two orthogonal issues: (1) Reaching a disjunctive plan whose minimal candidates contain all the solutions and (2) Posing the solution extraction as a SAT problem.

It is best to separate these issues and see the main contributions of SATPLAN as pertaining to the second one. The issues guiding the first are by and large specific to planning, and are best addressed in terms of refining disjunctive plans. The issues guiding the second are general heuristics regarding the effective ways of compiling a combinatorial search problem into a SAT instance, and are only loosely tied to planning. The main concern here is to end up with a SAT instance that is "small" (e.g., in terms of number of variables). The techniques used to achieve this involve syntactic optimizations such as converting $n$-ary predicates into binary predicates, or compiling out dependent variables.

Separating planset construction and its compilation into a SAT instance allows SATPLAN techniques to exploit, rather than *re-invent*, (disjunctive) refinement planning. It also removes the arbitrariness involved in "guessing" the length of the solution as a pre-requisite for coming up with a SAT encoding, since solution extraction is interleaved with refinements. It may also be possible to develop heuristics that attempt to predict whether it is worth doing solution extraction on the current plan (as for example, Graphplan does).

In addition, I speculate that basing encodings on $k^{th}$ level plansets may also lead to SAT instances that are "smaller" on the whole. Specifically, the pruning done by disjunctive refinements can also lead to a planset with fewer number of minimal candidates than can be achieved by direct encodings. This can in turn lead to a tighter SAT encoding. My speculation is supported to some extent by the results reported in [Kautz and Selman, 1996] which show that SAT encodings based on $k$-length planninggraphs generated by Graphplan, can be solved more efficiently than the "linear" encodings generated by direct translation.[1]

## 5. Challenge Problems

The framework of relations that we have outlined in this paper gives rise to several open issues that are worth investigating. I pose them below as challenge problems to the researchers in the planning community:

### 5.1 Disjunctive refinements

In Section 3, I argued that the success of Graphplan can be interpreted in terms of the idea of disjunctive planning. The only implemented and widely tested disjunctive planner is Graphplan, and it is based on forward state space refinements. Given the known disadvantages of forward state-space refinements *vis a vis* the other goal directed refinements, it seems likely that disjunctive planners based on these other refinements may scale-up even better. This leads to the first challenge:

**Challenge 1.** *Develop disjunctive planners based on (1.a) backward state space, (1.b) plan-space and (1.c) task reduction refinements, either working independently or in combination.*

Section 3 explicates some of the critical issues involved in implementing disjunctive planners, but a lot remains to be done. An easy first step in this direction would be to have a version of Graphplan algorithm that uses backward state space refinements (which can in principle out-perform Graphplan by being goal-directed). A more interesting step is to develop a disjunctive planner based on plan-space refinement, and see if the advantages of plan-space refinement over state-space ones in the traditional planners carry over to disjunctive planners. Given the dominance of task reduction refinements in practical planners, it is also imperative to

develop disjunctive planners based on task reduction refinement. Finally, since interleaving refinements is known to have advantages in traditional planners [Kambhampati & Srivastava, 1996], it would be worth considering disjunctive planners that interleave refinements. The key in all these cases will be to handle the tradeoff between cost of the refinement and the loss of pruning power by generalizing the corresponding refinements to make them infer more than just action constraints. The discussion in Section 3 may suggest some ideas in this regard. Progress on the first challenge can put us at a good vantage point to answer a more global one:

**Challenge 2.** *Characterize tradeoffs offered by different refinements in supporting disjunctive planning.*

The most interesting issues to be resolved here include whether and how the tradeoffs between the various refinements change when we shift from traditional refinement planners to disjunctive planners (c.f. [Barrett & Weld, 1994]): Would plan-space and task-reduction refinements maintain their advantages over in disjunctive planning algorithms too, or would they be less effective (presumably because they do not support effective refinement and constraint propagation)? The answer may depend on how effectively the corresponding refinements can be generalized to disjunctive plans.

### 5.2 Issues in SAT compilation

Given that SAT compilation seems to be the most promising way of doing solution extraction on disjunctive plans, development of effective disjunctive planners will require attention to the SAT compilation issues. The problem-independent tradeoffs in SAT compilation are the subject of a challenge paper by Selman et. al. [1997], and will not be discussed here. Our interest here is to understand the specific compilation tradeoffs directly related to planning:

**Challenge 3.** *Characterize the tradeoffs offered by SAT-encodings based on disjunctive plans derived by different types of refinements (alone or working in an interleaved combination).*

The issues to be resolved here include understanding (1) how easy would it be to solve the SAT encodings resulting from different refinements and (2) how to effectively exploit the shared structure of the SAT encodings corresponding to successive levels of disjunctive plans. The first issue is especially interesting in light of the current experience that SNLP encodings, despite being more compact, are harder to solve [Selman et. al., 1997]. The second issue is motivated by the fact that solution extraction is interleaved with incremental refinement operations, and thus the disjunctive plans at successive levels share a significant amount of substructure. Indeed, one of the sources of efficiency for Graphplan algorithm is the way the backward search phase on successive planning-graphs can use the information computed in the previous levels (e.g. memoizing).

It would also be interesting to understand why SAT compilation is so effective in the first place. Kautz and Selman's [1996] original arguments seemed to imply that the effectiveness has a lot to do with stochastic search methods used in solving SAT problems. This implication been called into question by the recent results of Bayardo and Schrag [1997], which show that systematic approaches can do just as well. So, a pertinent question is: Is the compilation into SAT justified purely by modularity concerns (in that different planning algorithms do not have to worry about specialized solution extraction procedures), or is there any inherent ad-

---

[1] Kautz and Selman interpret the superior performance of graphplan encodings over linear encodings to the fact that the former allow action parallelism. However, as shown in [Kambhampati & Lambrecht, 1997], the real contribution of graphplan is not supporting parallel actions, but rather allowing forward state space refinements over disjunctive plans.

vantage to SAT representation? A partial attempt at resolving this issue would be to enrich the backward search (solution extraction) phase of the Graphplan algorithm with the CSP techniques such as dependency directed backtracking, to see if it could rival the SAT performance.

## 5.3 The utility of controlled splitting

In terms of our general refinement planning template, traditional planners split each of their component plans into a different search branch, while Graphplan and SATPLAN can be thought of in terms of not splitting the plansets at all. Research in constraint satisfaction and operations research shows that pruning and splitting can have synergistic interactions. In fact, the best CSP search strategies combine low-level constraint propagation (pruning) techniques, such as forward checking, with splitting [Bayardo & Schrag, 1997]. This raises the possibility that best disjunctive planners may also do controlled splitting of plansets (rather than no splitting at all) to facilitate refinements with more pruning power. Controlled splitting may also be helpful in scenarios where planning and execution need to be interleaved of necessity. Although some initial exploration has been done in this direction by Descartes and UCPOP-D planners, a more systematic analysis of tradeoffs is still needed. This leads to the next challenge:

**Challenge 4.** *Develop and explore the tradeoffs offered by planners that split some of the disjunction into the search space, while keeping the rest together in disjunctive format.*

One immediate question is exactly how many branches should a planset be split into? Since the extent of propagation depends on the amount of shared sub-structure between the disjoined planset components, one way of controlling splitting may be to keep plans with shared sub-structure together. Of particular theoretical interest would be exhibiting domains and problems where full disjunction is counter-productive. A first step in this direction would involve experimenting with versions of Graphplan algorithm that maintain multiple planning-graphs at each level [Kambhampati & Lambrecht, 1997].

## 5.4 Exploring alternative Representations

We have noted in Section 2.2 that there is nothing sacrosanct about the particular representation for partial plans that is used by most existing planners. Alternative representations may either be dictated by necessity (for example, the time-map based plan representations used in planners like HSTS [Muscettola, 1993] to support the use of expressive resource constraints in planning), or as more flexible ways of representing and handling sets of action sequences (for example, the representation used in COPS [Ginsberg, 1996]). The general framework and ideas described in this paper are independent of the specific representations used for partial plans. For example, Ginsberg's COPS planner can be seen as an instance of the disjunctive planning approach, that uses a very different partial plan representation (albeit with the same candidate set semantics). The approach shares the ideas of direct refinement of disjunctive plans and solution extraction by enumerating minimal candidates. Representations can however have a significant impact on the kinds of refinements that can be supported. Ginsberg describes a type of refinement on his plan representation that has the ability compute both upper and lower approximations on the set of solutions. This leads us to an interesting, but somewhat open-ended challenge.

**Challenge 5.** *Develop and investigate the utility of alternative plan representations on the efficiency of plan generation.*

While development of new representations is an open-ended task, a first feasible step in this direction would be to do a direct empirical comparison of currently available alternative representations, such as those used in COPS or HSTS. It would also be interesting to undertake each of the previous challenges in the context of the alternative representations.

## 6. Conclusion

In this paper, I presented a generalized view of refinement planning that bridges several hither-to distinct strands of planning research. I have outlined how the insights from this unification can help both traditional and SAT-based planners. Finally, I have proposed as a challenge to the community, a set of research problems, tackling which can lead to a significantly improved understanding of the issues involved in efficient plan synthesis.

To support research into these challenge problems, I intend to maintain a web site that will act as a clearing house for information on our evolving understanding. Its URL will be *http://rakaposhi.eas.asu.edu/challenge.html*. This site will contain references to other relevant literature, pointers to benchmark problems and test-domains, as well as more fleshed out versions of the challenges and their up-to-date status.

## References

Barrett, A. and Weld, D. 1994. Partial Order Planing: Evaluating possible efficiency gains. *Artificial Intelligence*, 67(1):71-112.

Bayardo, R and Schrag, R. 1997. Using CSP look-back techniques to solve real-world SAT instances. In Proc. AAAI-97.

Blum, A. and Furst, M. 1995. Fast planning through plan-graph analysis. In Proc. IJCAI-95.

Crawford, J. and Auton, L. 1996. Experimental results on the crossover point in random 3SAT. *Artificial Intelligence*, 81.

Ginsberg, M. 1996. A new algorithm for generative planning. In Proc. KR-96.

Joslin, D and Pollack, M. 1996. Is least commitment always a good idea? In Proc. AAAI-96.

Kambhampati, S. and Srivastava, B. 1996. Unifying classical planning approaches. ASU CSE TR 96-006. (Preliminary version appeared in Proc. 3rd European workshop on planning).

Kambhampati, S., Knoblock, C., and Yang, Q. 1995. Planning as refinement search: A unified framework for evaluating design tradeoffs in partial order planning. *Artificial Intelligence*, 76(1-2):167-238.

Kambhampati, S. and Yang, X. 1996. On the role of disjunctive representations and constraint propagation in refinement planning. In Proc. KR-96.

Kambhampati, S and Lambrecht, E. 1997. Why does Graphplan work? ASU CSE TR 97-005. (Poster at IJCAI-97)

Kambhampati, S. 1997. Refinement planning as a unifying framework for plan synthesis. AI Magazine. Summer issue.

Kautz, H. and Selman, B. 1996. Pushing the envelope: Planning Propositional Logic and Stochastic Search. In Proc. AAAI-96.

Kautz, H., McAllester, D. and Selman, B. 1996. Encoding plans in propositional logic. In Proc. KR-96.

Muscettola, N. 1993. HSTS: Integrating planning and Scheduling. In: *Intelligent Scheduling.* M. Fox and M. Zweben (eds). Morgan Kautmann.

Pearl, J. *Heuristics.* Addison-Wesley. 1984.

Selman, B., Levesque, H.J., and Mitchell, D. 1992. GSAT: a new method for solving hard satisfiability problems. In Proc. AAAI-92.

Selman, B., Kautz, H. and McAllester, D. 1997. Computational challenges in propositional reasoning and search. In Proc. IJCAI-97.

# Ten Challenges in Propositional Reasoning and Search

**Bart Selman, Henry Kautz,** and **David McAllester**
AT&T Laboratories
600 Mountain Avenue
Murray Hill, NJ 07974
{selman, kautz, dmac}@research.att.com
http://www.research.att.com/~selman/challenge

## Abstract

The past several years have seen much progress in the area of propositional reasoning and satisfiability testing. There is a growing consensus by researchers on the key technical challenges that need to be addressed in order to maintain this momentum. This paper outlines concrete technical challenges in the core areas of systematic search, stochastic search, problem encodings, and criteria for evaluating progress in this area.

## 1 Introduction

Propositional reasoning is a core problem in many areas of artificial intelligence. In recent years this area has seen growing interest and activity, due to advances in our ability to solve large problem instances, including ones that encode real-world problems such as planning and diagnosis. Contributions to the area of propositional deduction and satisfiability testing have come from research communities in artificial intelligence, operations research, and theoretical computer science. A set of key technical challenges have begun to emerge from interactions within and among these research groups. We will describe some of these challenges and discuss why they are important for progress in the field.

For many years the problem of propositional reasoning received relatively little attention. It appeared that nothing could be done to improve upon the performance of the Davis-Putnam procedure (1960), which held its position as the most efficient satisfiability testing algorithm for three decades. Furthermore, most researchers in artificial intelligence felt that the representational power of propositional logic was too limited, and turned their attention to first-order logic or even more powerful formalisms, such as various modal and non-monotonic logics. Several factors contributed to a renewed interest in propositional reasoning. First, new algorithms were discovered, including ones based on stochastic local search as well as systematic search, that have better scaling properties than the basic Davis-Putnam algorithm. Second, improvements in machine speed, memory size, and implementations extended the range of the algorithms. Third, researchers began to develop and solve propositional encodings of interesting, real-world problems such as planning and diagnosis, with others on the horizon, such as natural language processing and machine learning. Such encodings were not even being considered a few years ago because they were thought to be far too large to be handled by any method. Between 1991 and 1996 the size of hard satisfiability problems that could be feasibly solved grew from ones involving less than 100 variables to ones involving over 10,000 variables.

Progress has been spurred by the interaction of researchers in AI, operations research, and theory, in making available benchmark problems (DIMACS, see Trick and Johnson 1996), holding joint workshops (Ginsberg 1995), and sharing algorithms and code. There is a growing consensus, however, that certain key technical challenges must be addressed in order to continue to increase the range of problems that can be practically solved. Work on propositional reasoning can be grouped in three core areas: systematic search; stochastic search; and problem encodings. For each area, we will discuss concrete challenges that have arisen. We stress that these challenges are not particularly our own invention, but rather summarize the issues that frequently arise in informal discussions among researchers. Progress on any of these challenges would directly extend the usefulness of the propositional reasoning approach to problems in AI.

Because of the large number of references involved, we have not attempted to cite all related literature. However, a web page for these challenge problems has been constructed with a more complete bibliography, further details on the challenge problems, and pointers to existing satisfiability testing procedures. See http://www.research.att.com/~selman/challenge.

With each challenge, we give our best estimate of a time frame within which we think the challenge can be met. For the harder challenges, even partial solutions will constitute a significant contribution to the field.

## 2 Evaluation Criteria

The main evaluation criteria that has been adopted in the satisfiability testing (SAT) community is empirical performance on shared benchmark problems. We believe that this remains the best way to evaluate responses to any of the algorithmic challenges we will present. The alternative, evaluation by theoretical analysis, is problematic. Worst-case complexity results are all usually exponential. Furthermore, theoretical average case results are difficult to obtain, and can even be misleading because of the difficulty of formally characterizing problem distributions. Even when a sound theoretical analysis is devised, it may only characterize the asymptotic performance of the algorithm, and that asymptote may lie too far out (*e.g.*, to problems containing trillions of variables) to be of practical consequence. Of course, theoretical analysis can be a very useful *complement* to empirical evaluation, and can help us gain insights into *why* an algorithm works well or poorly on a given problem distribution. (In fact, one of our non-algorithmic challenges, as we will see, is to develop a theoretical analysis that explains *why* local search works well on certain problem distributions.)

Reasonably good benchmark collections of satisfiability problems, mainly in CNF form, are publically available from a number of sources. These include the DIMACS collection, which was used as a testbed for a large number of algorithms as reported in (Trick and Johnson 1996); a collection of circuit diagnosis problems encoded as SAT (Larrabee 1992); the planning problems Kautz and Selman (1996) devised for their SATPLAN system; hardware and software verification problems from the model checking community; and others. All these collections have the property that published papers exist specifying the best known algorithms and running times for solving each of their instances.

Progress on any of our challenges (when applicable) will be measured by showing that the proposed method outperforms all other known methods on at least *one* set of benchmark problems. Some of our challenges in the area of problem encodings involve finding better ways to represent a real-world problem as SAT. In these cases, the empirical evaluation should involve comparing performance of the best SAT algorithms on the proposed encoding against the best published results for *any* algorithm that can solve the original (*i.e.*, unencoded) problem instances. In such a comparison it is usually not possible to *outperform* the specialized algorithms (although it is sometimes possible, see for example (Kautz and Selman 1996)), but one should provide evidence that solving the encoded problem with general purpose Boolean satisfiability procedures is at least competitive.

Whenever possible, comparisons should also be made to the difficulty of solving the problems when alternative encodings are used, such as may be found in some of the benchmark sets. In cases where alternative encodings are available, one should show that the new encoding yields problems that are easier to solve by at least one state-of-the-art SAT procedure. (For example, it would be counted as progress if one developed a new encoding of planning problems that was good for stochastic search procedures, even if the encoding did not help systematic search procedures.)

In order to keep the criteria for comparison as objective as possible, it is important that results include total running time. This may be adjusted for machine speed, but it is not sufficient to *only* report certain characteristics of the execution, such as "number of nodes expanded" (Johnson 1996). There are many ways to shift the computational effort in search algorithms — for example, to visit fewer nodes by doing more work at each node — and an objective evaluation must consider the entire picture. Comparisons should cite the best results from the Operations Research and computer science theory literature, as well as the AI literature.

Tests should be done on a variety of problem sizes, up to the hardest available instances that the method can solve. It is important to show the limits of the proposed method, that is, where it becomes highly exponential or otherwise fails. Often methods that look good on small instances break down on larger ones (Johnson 1996).

## 3 Challenging SAT Problems

We shall begin the list of challenges by citing two specific open SAT problems. The first is to develop a way to *prove* that unsatisfiable 700 variable 3-CNF formulas, randomly generated in the "hard" region where the ratio of variables to clauses is 4.3, are in fact unsatisfiable (Mitchell *et al.* 1992; Crawford and Auton 1993). Randomly-generated *satisfiable* formulas of this size are regularly solved by stochastic search algorithms such as GSAT, but they are unable to prove unsatisfiability; and no systematic algorithm has been able to solve hard random formulas of this size. A generator for these hard random problem instances can be found at the DIMACS benchmark archive (Trick and Johnson 1996).

**CHALLENGE 1:** (1-2 yrs) *Prove that a hard 700 variable random 3-SAT formula is unsatisfiable.*

The second challenge problem is satisfiable. It is an encoding of a 32-bit parity problem, that appears in the DIMACS benchmark set. It appears to be too large for current systematic algorithms. It also defeats the hill-climbing techniques used by current local search algorithms.

**CHALLENGE 2:** (2-5 yrs) *Develop an algorithm that finds a model for the DIMACS 32-bit parity problem.*

For the second challenge, of course, the algorithm should not be told in advance the known solution! Given the amount of effort that has been spent on these two instances, any algorithm solving one or both will have to do something significantly different from current methods.

## 4  Challenges for Systematic Search

In the previous section we described two challenging SAT problems. In this section, and the next one, we describe promising ideas whose utility has not yet been demonstrated. In each case the challenge is to demonstrate the utility of a currently known, but as yet unproven, approach to solving SAT problems. In keeping with the discussion in section 2, to demonstrate the utility of an idea one must construct a procedure which uses that idea in an essential way in solving at least one class of problems more effectively than any other known approach.

Our next challenge is using proof systems stronger than resolution. All of the best systematic methods for propositional reasoning are based on creating a resolution proof tree. This includes depth-first search algorithms such as the Davis-Putnam procedure, where the proof tree can be recovered from the trace of the algorithm's execution, but is not explicitly represented in a data structure (the algorithm only maintains a single branch of the proof tree in memory at any one time). Most work on systematic search concentrates on heuristics for variable-ordering and value selection, all in order to the reduce size of the tree.

However, there are known fundamental limitations on the size of the shortest resolution proofs for certain problems (Haken 1985; Chvatal and Szemeredi 1988). For example, "pigeon hole" problems (showing that $n$ pigeons cannot fit in $n-1$ holes) are intuitively easy, but shortest resolution refutation proofs are of exponential length. Shorter proofs do exist in more powerful proof systems. Examples of proof systems more powerful than resolution include extended resolution, which allows one to introduce new defined variables, and resolution with symmetry-detection, which uses symmetries to eliminate parts of the tree without search. Assuming $NP \neq co-NP$, even the most powerful propositional proof systems would require exponential long proofs worst case — nonetheless, such systems provably dominate resolution in terms of minimum proof size.

However, at this point in time, attempts to mechanize these more powerful proof systems usually yield no computational savings, because it is *harder* to find the small proof tree in the new system, than to simply crank out a large resolution proof. In essence, the overhead in dealing with the more powerful rules of inference consumes all the potential savings. There is promising work in this area (Crawford *et al.* 1996; de la Tour and Demri 1995), but not yet convincing empirical results on a variety of benchmark problems; such evidence would meet our third challenge:

**CHALLENGE 3:** (2-5 yrs) *Demonstrate that a propositional proof system more powerful than resolution can be made practical for satisfiability testing.*

Our next challenge is the use of integer programming techniques in solving SAT problems. It is straightforward to translate SAT problems into 0-1 integer programming problems (Hooker 1988), and thus it has been argued that integer programming techniques should be useful for propositional reasoning. In fact, however, this has not been shown to be the case. For example, one of the main techniques in integer programming is to compute the *linear relaxation* of the problem, and then to use the (easily-found) solution of the relaxed problem to guide the selection of values in solving the integer problem. However, in most formulations the solution to the linear relaxation of *any* SAT problem simply sets all the variables to the value "1/2" (modulo unit propagation), thus yielding no guidance at all. Therefore we offer a challenge to show that the well-developed body of tools and techniques from Operations Research *does* in fact have something new to offer for propositional reasoning.

**CHALLENGE 4:** (2-5 yrs) *Demonstrate that integer programming can be made practical for satisfiability testing.*

## 5  Challenges for Stochastic Search

Stochastic local search has been shown to be a powerful alternative to systematic search for finding models of satisfiable CNF formulas. Stochastic algorithms are inherently incomplete, because if they fail to find a model for a formula one cannot be *certain* that the formula is unsatisfiable. This has led to an asymmetry in our ability to solve satisfiable and unsatisfiable instances drawn from the same problem distribution. Stochastic algorithms can solve hard random satisfiable formulas containing thousands of variables, but we cannot solve unsatisfiable instances of the same size.

Can local search be made to work for proving unsatisfiability? This apparently would require searching in space of refutation proofs, rather than in space of truth assignments. Each state would be an incomplete proof tree, *e.g.*, a proof tree that rules out some fraction of the truth assignments. The neighborhood of a state would be similar proof trees. The step in the local search would try to transform the proof into one that rules out a larger fraction of assignments.

**CHALLENGE 5:** (5-10 yrs) *Design a practical stochastic local search procedure for proving unsatisfiability.*

Our next challenge is that of distinguishing "dependent" from "independent" variables. SAT encodings of structured problems such as planning and diagnosis often contain large numbers of variables whose values are constrained to be a simple Boolean function of other variables. We call these *dependent* variables. Variables whose values can not be easily determined to be a simple function of other variables are called independent. For a given SAT problem there may be many different ways to classify the variables as dependent and independent. But for most SAT encodings of real-world problems there is a natural division between dependent and independent variables. Since an assignment to the independent variables determines a truth value for each dependent variable, the number of assignments that need be considered in a systematic search is at most $2^n$ where $n$ is the number of *independent* variables.

We belief that it fairly easy to empirically demonstrate that identification of dependent variables improves the performance of systematic search. It appears to be more difficult to establish that identification of dependent variables improves stochastic search. The challenge, therefore, is to improve local search for problems with dependent variables: the search should concentrate only (or mainly) on the independent variables, and some fast mechanism should then set the dependent variables.

**CHALLENGE 6:** (1-2 yrs) *Improve stochastic local search on structured problems by efficiently handling variable dependencies.*

As we have noted, systematic and local search procedures outperform each other on different problem classes. Even putting the issue of incompleteness aside, there exist classes of satisfiable problems for which one or the other approach is clearly the winner. This leads to the general question: Can we develop a procedure that leverages the strengths of each? The obvious way, of course, is to simply run good implementations of each approach in parallel. But is there a more powerful way of combining the two? Recently there has been some intriguing work on using local search to implement the variable ordering heuristic for systematic search (Boufkhad 1996; Mazure et al. 1996). These methods and other ways of combining systematic and stochastic search need to be further developed and compared with previous approaches on a range of benchmark problems.

**CHALLENGE 7:** (1-2yrs) *Demonstrate the succesful combination of stochastic search and systematic search techniques, by the creation of a new algorithm that outperforms the best previous examples of both approaches.*

## 6 Challenges for Problem Encodings

The value of research on propositional reasoning ultimately depends on our ability to find suitable SAT encodings of real-world problems. As discussed in the introduction, there has been significant recent progress along this front. Examples include classical constraint-based planning (Blum and Furst 1995; Kautz and Selman 1996), problems in finite algebra (Fujita et al. 1993), verification of hardware and software, scheduling (Crawford and Baker 1994), circuit synthesis and diagnosis (Larrabee 1992), and many other domains. Experience has shown that different encodings of the same problem can have vastly different computational properties. For example, in planning, "causal" encodings appear to be harder to solve than "state-based" encodings.

A challenge, therefore, is to develop a general characterization of encodings that can be *efficiently* solved. This characterization may involve, for example, understanding the relationship between the encoding and the shape of its search space. Note that the characterization cannot be as simple as stating that the search space has no local minima, because realistic problems will almost certainly have local minima. For example, the SAT encodings of blocks-world planning problems do have deep local minima, yet can be solved by local search. We need to understand why search can escape from local minima in some encodings but not in others. Perhaps one can find easily measurable statistical properties of the encoding that predict the behavior of various search algorithms on a given instance.

**CHALLENGE 8:** (5-10 yrs) *Characterize the computational properties of different encodings of a real-world problem domain, and/or give general principles that hold over a range of domains.*

Note that evaluation of progress on this last challenge cannot be purely empirical. There will be some subjective judgement necessary in determining whether the characterization is useful and enlightening. *Predictions of the theory can be put to an empirical test*: in the best case, the general principles would suggest new encodings which can be solved more easily than previous ones.

One problem with all of the propositional encodings for real-world problems that have been suggested in the literature is that they are extremely "brittle". If we look at formulations where models that satisfy the encoding correspond to solutions to the original problem instance, we see little relationship between models that "almost" satisfy the encoded problem and candidate solutions that "almost" satisfy the original problem instance. For example, Kautz and Selman (1992) noted that it was easy to find truth-assignments that satisfied all but a single clause of SAT encodings of blocks-world planning problems. These "near models" corresponded to making a series of random motions, and then, in the last state the blocks "magically" (violating physical constraints) arrange themselves in the correct position.

The robustness of encodings, and in particular, the robustness of local search algorithms applied to those encodings, could be improved by finding ways to more closely align the semantics of the SAT instance with the semantics of the source domain:

**CHALLENGE 9:** (1-2 yrs) *Find encodings of real-world domains which are robust in the sense that "near models" are actually "near solutions".*

As we discussed earlier, benchmarks have been an important driving force behind work on propositional reasoning algorithms. Ideally one would have access to a very large number of real-world problem encodings, so that one could develop solid statistical evidence of the performance of various techniques. In practice the number of real problems that can be obtained is limited. It is extremely time-consuming and knowledge-intensive work to manually create such instances, and the largest and potentially most useful examples are often part of some proprietary project, and thus cannot be shared. Hard randomly-generated problems have proven to be a good alternative for testing — so far, the algorithms that are the best on randomly generated instances are also best on structured problems. However, there is the concern that we may be reaching a point where this is no longer the case, and that of the simple random distributions now used for testing may be driving us in the wrong direction in our research (Johnson 1996). Therefore we present a final challenge, which if answered would provide vital tools for ensuring progress in this field.

**CHALLENGE 10:** (2-5 yrs) *Develop a generator for problem instances that have computational properties that are more similar to real-world instances.*

It is necessary to provide concrete evidence, empirical and/or theoretical, that the problem distribution so generated closely matches some set of real-world domains.

## 7 Conclusions

We have presented a series of technical challenge problems in the area of propositional reasoning and search. We believe that progress towards solving these problems will directly extend the usefulness of the propositional reasoning approach to problems in artificial intelligence, and computer science in general.

## References

Blum, A. and Furst, M.L. (1995). Fast planning through planning graph analysis. *Proc. IJCAI-95*, Canada.

Boufkhad, Y. (1996) Aspects probabilistes et algorithmiques du probleme de satisfiabilite. Ph.D. Thesis, Univ. of Paris, 1996.

Cook, S. and Mitchell, D., Finding Hard Instances of the Satisfiability Problem: A Survey. *DIMACS Series in Discr. Math. and Theoretical Comp. Sci..* (to appear)

Crawford, J.M. and Auton, L.D. (1993) Experimental results on the cross-over point in satisfiability problems. *AAAI-93* (1993) 21–27. (Ext. version in Artif. Intel.)

Crawford, J. and Baker, A.B. (1994). Experimental results on the application of satisfiability algorithms to scheduling problems. *Proc. AAAI-94*, Seattle, WA.

Crawford, J.M. , Ginsberg, M., Luks, E., and Roy, A. (1996). *Proc. KR-96*, Boston, MA, 148–158.

Chvatal, V. and Szemeredi, E. (1988). Many hard examples for resolution. *J. of the ACM*, 35(4) (1988) 759–208.

Davis, M. and Putnam, H. (1960) A computing procedure for quantification theory. *J. of the ACM*, 7 (1960).

de la Tour, T. and Demri, S. (1995). On the complexity fo extending ground resolution with symmetry rules. *Proc. IJCAI-95*, Montreal, Canada, 289–295.

Dubois, O., Andre, P., Boufkhad, Y., and Carlier, J. SAT versus UNSAT. *DIMACS Series in Discr. Math. and Theoretical Comp. Sci.*, Vol. 26, 1996, 415–433.

Fujita, M., Slaney, J., and Bennett, F. (1993). Automatic Generation of Results in Finite Algebra *Proc. IJCAI*, 1993.

Ginsberg, M. (1995). First Intl. Workshop on AI&OR. Timberline, Oregon, OR (1995).

Haken, A. (1985). The intractability of resolution. *Theoretical Computer Science* 39 (1985) 297–308.

Hooker, J.N. (1988). Hooker, J.N., Resolution vs. cutting plane solution of inference problems: Some computational experience. *Oper. Res. Letter*, 7(1) (1988) 1–7.

Johnson, D. (1996). Experimental Analysis of Algorithms: The Good, the Bad, and the Ugly. Invited Lecture, *AAAI-96*, Portland, OR. See also http://www.research.att.com/ dsj/papers/exper.ps.

Larrabee, T. (1992) Efficient generation of test patterns using Boolean satisfiability, *IEEE Trans. on CAD*, vol 11, 1992, pp 4-15.

Kautz, H. and Selman, B. (1992) Planning as Satisfiability. *Proceedings ECAI-92*, Vienna, Austria, 1992, 359–363.

Kautz, H. and Selman, B. (1996) Pushing the envelope: planning, propositional logic, and stochastic search. *Proc. AAAI-96*, Portland, OR, 1996.

Mazure, B. Sais, L. Gregoire E. (1996). Boosting complete techniques thanks to local search methods. *Proc. Math & AI*, 1996.

Mitchell, D., Selman, B., and Levesque, H.J. (1992) Hard and easy distributions of SAT problems. *Proc. AAAI-92*, San Jose, CA (1992) 459–465.

Papadimitriou, C.H. (1993). *Computational Complexity.* Addison Wesley, 1993.

Selman, B. , Kautz, H., and Cohen, B. (1994). Noise Strategies for Local Search. *Proc. AAAI-94*, Seattle, WA, 1994, 337–343.

Selman, B., Levesque, H.J., and Mitchell, D. (1992) A new method for solving hard satisfiability problems. *Proc. AAAI-92*, San Jose, CA, 440–446.

Trick, M. and Johnson, D. (Eds.) (1996) Cliques, Coloring and Satisfiability, *DIMACS Series in Discrete Mathematics and Theoretical Computer Science*, vol. 26. For benchmarks, see URL: http://dimacs.rutgers.edu/Challenges/index.html.

Williams, B.C. and Nayak, P. (1996) A model-based approach to reactive self-configuring systems. *Proceedings AAAI-96*, Portland, OR. 971–978.

# Challenge: How IJCAI 1999 can Prove the Value of AI by Using AI

James Geller
Department of Computer and Information Sciences
New Jersey Institute of Technology
Newark, NJ 07102
U. S. A.

## Abstract

The challenge formulated in this paper is a direct reaction to the reviewer selection process of IJCAI 1997, which relied on an unintelligent string matching program. We challenge the AI community to implement an intelligent knowledge-based reviewer selection program for IJCAI 1999.

## 1 Introduction

In this paper we will formulate a challenge that should result in the design and implementation of a reviewer selection program to be used by the organizing committee of IJCAI 1999. The motivation behind this problem is partially rooted in a dissatisfaction with the assignment process of papers to reviewers that was apparently used for IJCAI 1997. The challenge is that the process of assigning papers to reviewers should be automated and performed by a flexible and intelligent AI program. We will explain this motivation in more detail, based on a metaphor from the business of telephone system development. Then we will outline some of the major problems that are part of this challenge. Finally, we will recapitulate the importance of this challenge problem. Indeed, in our opinion, this challenge problem can well compete with any other challenge problem that is likely to be presented at IJCAI 1997.

## 2 Motivation

It is helpful to recount very briefly a personal experience of this author to create a clear picture of the motivation for this challenge problem. After graduating from college, this author looked for his first entry level job in software development. At the time and place where this occurred, the only game in town were the phone companies. There were four of them, let's call them AB, CD, EF, and YZ. The author was eventually hired by the AB phone company.

One of the main activities of all four companies was to build large PBX phone systems. These systems would connect to a number of outside lines and make them available to 300 or more local extensions. Such PBX systems would be bought by businesses in town. In fact, this was the accepted way businesses supplied their employees with a connection to the outside world. Naturally, there was fierce competition among the four phone companies to provide cheap and feature-rich PBX systems.

However, the AB company had a special secret. Something they would not have wanted their customers to know. The AB company used a PBX system made by the YZ company for their own in-house communication needs! This system was eventually replaced by an AB system, but still, there was this period when one could have imagined a TV advertisement of the YZ company, sounding like this: "YZ PBX systems – they are so good, even our competition is using them."

Recently, when writing a(nother) paper for IJCAI 1997, the author was reminded of this incident. After having finished the bulk of the paper, it was time to check the list of Content Areas contained in the Call for Papers. (Content Areas are more commonly called Keywords and denote the subareas of AI into which a given paper can be classified.) There were two obvious Content Areas, "semantic networks" and "parallel AI," describing our paper. Thus, the author consulted the IJCAI 1997 Call for Papers list of Content Areas in search of them.

However, both those areas were not contained in the official list of Content Areas. Obviously, "semantic networks" has been a main stream area in AI almost since the beginning of the field [Quillian, 1968]. Parallel AI is less common, but it has been a field of investigation for many years; see e.g., [Geller, 1991; Kanal et al., 1994; Geller et al., 1997].

A quick inquiry with the authorities of IJCAI confirmed that no new Content Areas could be added to the Call for Papers of IJCAI 1997 at such a late date.

The reader might think that this should cause no problems. Unfortunately, a preamble to the list of Content Areas stated:

> Please use the exact phrases below; do not make up new keywords, since they will be ignored by the paper classification software.

What conclusions do we have to draw from this statement? First of all, papers will be assigned (partially?) to reviewers by a program. Secondly, this program is so unintelligent that it cannot even deal with variant spellings, much less Content Areas that are not known to it. What else could the admonishment to use "exact phrases" mean?

This sounds like a phone company that does not have a phone system itself.

## 3 Current Reviewer Assignment

After acceptance of this challenge paper, the author contacted the IJCAI 1997 program chair to find out more about the reviewer assignment program that had been used. The program was developed by Ramesh Patil at USC/ISI[1] for use by the National Conference on Artificial Intelligence, commonly known as "AAAI." Due to the different committee setup at IJCAI and AAAI, a considerable amount of human post-processing was necessary for IJCAI 1997 reviewer assignment [Pollack, 1997].

An inquiry at ISI revealed that the AAAI reviewer assignment program has been used for five years. It relies on a system of bidding for papers, in combination with stored profiles of expertise. Details of this program have not been published, as the author considers them "not algorithmically novel." However, future versions of the reviewer assignment program might be published [Patil, 1997].

## 4 The Challenge Problem

The following is a suggestion of what should appear in the Call for Papers of IJCAI 1999:

> Authors must supply a list of Content Areas from any subarea of Artificial Intelligence that they consider appropriate. These Content Areas will be used by a flexible and intelligent program to assign papers to reviewers. Only in cases where this program reports insufficient confidence in its own results, will human editors make the assignment of papers to reviewers.

Actions speak louder than words. Having the major world-wide conference on Artificial Intelligence in 1997 issue a Call for Papers that admits that it relies on exact string matching for reviewer preselection speaks very loudly about the state-of-the-art of Artificial Intelligence. We think that this is a statement that we as members of the AI community do not want to make.

What would be involved in the kind of processing required for an intelligent reviewer assignment program (which we will call REVIEWER-SELECT)? We need at least a Natural Language Processing system that is able to parse noun phrases. We need a knowledge base that associates in a flexible way content areas with reviewers. We need a second Natural Language Processing system that is powerful enough to extract possible content areas from given source text, as will be explained below. Finally, we need an expert system that selects reviewers, keeping in mind certain overall constraints which will be elaborated in more detail shortly.

To summarize, the challenge for IJCAI 1999 one more time succinctly:

*Papers should be assigned to reviewers by an intelligent program that does not impose any limitations, except that the Content Areas supplied by paper authors need to be cogent to the eyes of a human expert.*

## 5 Sources of Information

In order to create the same base-line conditions for all the teams that want to take on this challenge, the IJCAI organization is asked to do the following. Due to the electronic submission process of abstracts through the world-wide web, the paper abstracts of most IJCAI 1997 submissions will be available on-line. The relatively few abstracts that are not submitted electronically should be typed in by IJCAI volunteers. This set of text sources should be made available through the world-wide web to the world AI community immediately after the IJCAI 1997 conference. The (second) Natural Language Processing system mentioned above should be used by research teams to extract candidate Content Areas from the stored abstracts. Some human filtering and/or augmentation might still be necessary to create a comprehensive list of Content Areas.

An important issue to address in preparing this source of information is how to treat the abstracts of rejected IJCAI 1997 papers. Copyright reasons as well as the large number of rejected papers make it difficult to deal with them as part of the source data. On the other hand, omitting the abstracts of rejected papers might bias the reviewer selection process towards papers in the areas that were accepted at IJCAI 1997. A solution to this problem needs to be found and publicized quickly, to create an even playing field for all future paper submitters.

Secondly, after IJCAI 1997, authors of papers should be asked to supply a list of five to ten individuals that they would have considered good reviewers for their own

---

[1] University of Southern California – Information Systems Institute

papers. Authors might also be asked to supply all the references cited in their own 1997 papers in electronic form to the IJCAI organization, adding an additional source of knowledge for the selection of reviewers. Another source of information is already there, and available to anybody with a browser, and that is the world-wide web itself. How the web could be used will be hinted at below.

One additional point should be made, although it is hopefully obvious at this stage. A response to this challenge will not consist of a Call for Papers containing a 10 page list of Content Areas. Rather, there might be no Content Areas listed at all. Anybody who is capable of submitting a paper to IJCAI should be aware of what constitutes an acceptable Content Area of AI. On the other hand, it appears useful to make the knowledge base of REVIEWER-SELECT itself publicly accessible at some point in time. Whether this should be before or after the IJCAI 1999 dead-line is open to discussion.

## 6 Problems

The problems posed by designing and implementing the REVIEWER-SELECT program are definitely non-trivial. This includes all aspects of it, but we want to stress the overall control that has to be exercised by this program. For instance, the REVIEWER-SELECT program should not come back with

1. Reviewers with the same affiliation as the author (conflict of interest);
2. The author of a paper as a suggested reviewer (extreme conflict of interest);
3. The doctoral advisor of the author as a suggested reviewer (another extreme conflict of interest);
4. Reviewers that the author has recently written a paper with (potential conflict of interest);
5. Assignment of fifty papers to one reviewer, and one paper each to fifty other reviewers (uneven load for reviewers);
6. Assignment of thirty reviewers to one paper, and no reviewers to thirty other papers (uneven load for papers).

The problem of identifying conflicts of interest could be attacked by using the world-wide web. As many institutions now have home pages on the world-wide web, information about affiliations could be retrieved by appropriate agents and softbots. Even potential conflicts of interest could be detected when authors have their vitae available on-line.

## 7 Limitations and Success Measurements

In order to make this challenge self-contained and its solution feasible in the allotted time, we must allow for some Content Areas that will not be contained in the knowledge base of REVIEWER-SELECT. In such cases the system will need to ask a human editor for help. It will be a measure of the success of any program taking on this challenge, how many (or how few) requests for human intervention will be required in processing all IJCAI 1999 submissions.

To ensure both fairness to IJCAI 1999 submissions and an alternative evaluation process of several programs that might take on this challenge (assuming that there will be several) the reviewer assignments should be rated blindly by a committee of human experts. By computing precision and recall measures compared to the ratings of human experts, a numeric evaluation of different programs will be performed.

A second, more expensive, way to evaluate the success of a REVIEWER-SELECT program would be to perform the reviewer selection process exactly as it was done in 1997, and to use the results of this process as the gold standard. This appears attractive for IJCAI 1999 because confidence in a completely automated review process might still be low in the AI community at large. If the evaluation is successful, then IJCAI 2001 might opt for completely automatic reviewer selection.

A third way to assess the success of a REVIEWER-SELECT program would be to compare its results with the names of reviewers suggested by the authors of IJCAI 1999 papers.[2] If this route is chosen, then the IJCAI 1999 Call for Papers must state that authors need to propose reviewers for their papers at submission time.

Whatever choice is made for evaluation, two conditions must be maintained. (1) If there are several teams taking on this challenge, then all of them must get the same starting conditions, including access to relevant materials collected by AAAI. (2) It must be assured that no REVIEWER-SELECT program has a built-in bias towards papers of its own designers. The author of this challenge paper will gladly serve as a coordinator of such evaluation activities throughout the whole life-time of this effort.

## 8 Acknowledgment of Debt to Previous Research

The idea of doing "reviewer selection driven research" is not new. Gio Wiederhold, a leading researcher in databases and other fields, and previously the editor of the most prestigious database journal, has mentioned the problem of automatically finding reviewers for his journal in conference presentations [Wiederhold, 1992]. However, we feel that in our formulation as a challenge problem, the difficulties that will be encountered are by

---

[2]This idea is due to one of the anonymous reviewers of this paper.

an order of magnitude larger. On the other hand, this challenge problem appears still small enough and well defined enough to lead to real practical results within the given time frame.

## 9 Why this Problem is Important

We have no doubt that there are many other challenge problems that make sense at the current state-of-the-art. Kitano's Grand Challenge problems have not been solved [Kitano et al., 1993]. Those problems are, however, in a different league, as opposed to the relatively small and well defined problems sought in this year's Challenge Paper track. The Turing Test, in whichever guise, is still the grand challenge problem to all of AI, but there is little hope to see it solved by IJCAI 1999.

Among problems in the same league as the reviewer assignment problem, we feel that our challenge is very important for a simple reason. It is the same reason why a phone company should first build a phone system for in-house use, before starting to sell it to other people. It looks very bad when an enterprise does not produce products that are good enough for its own needs. If IJCAI 1999 will have to rely on string matching for assigning reviewers to papers, it might be better to return to manual assignments, and to never even mention the use of a program in the Call for Papers. On the other hand, any solution to this challenge problem would considerably improve the credibility of AI as a whole.

## 10 The Real Challenge – A Final Remark

The challenge formulated in this paper is very limited, compared to the real challenge of making AI programs superior to average human practitioners in a field. From this point of view, why are authors of papers asked to supply Content Areas at all? Why can't an intelligent program "just read" the abstract of a paper and decide what the Content Areas are? A program that can assign papers to reviewers without being given Content Areas and that can do this for the whole range of AI problems would be a supreme achievement, as very few human individuals can perform this task. Such a program would probably build on many current subareas of AI, including data mining, machine learning, advanced natural language processing, knowledge representation, reasoning, etc. Research on summarizing text might be a valuable starting point. Maybe we will be able to pose this task as a challenge problem for IJCAI 2009.

## Acknowledgment

Mike Halper has improved the English of this paper.

## References

[Geller et al., 1997] J. Geller, H. Kitano, and C.B. Suttner, editors. *Parallel Processing for Artificial Intelligence 3*. North-Holland Elsevier, Amsterdam, 1997.

[Geller, 1991] J. Geller. Upward-inductive inheritance and constant time downward inheritance in massively parallel knowledge representation. In *Proceedings of the Workshop on Parallel Processing for AI at IJCAI 1991*, pages 63–68, Sydney, Australia, 1991.

[Kanal et al., 1994] L. Kanal, V. Kumar, H. Kitano, and C. Suttner, editors. *Parallel Processing for Artificial Intelligence*. Elsevier Science Publishers, Amsterdam, 1994.

[Kitano et al., 1993] H. Kitano, L. Hunter, B. Wah, W. von Hahn, R. Oka, and T. Yokoi. Grand challenge AI applications. In *Proceedings of the Thirteenth International Joint Conference on Artificial Intelligence*, pages 1677–1683. Morgan Kaufmann, San Mateo, CA, 1993.

[Patil, 1997] R. Patil. Personal communication, 1997.

[Pollack, 1997] M. Pollack. Personal communication, 1997.

[Quillian, 1968] M. R. Quillian. Semantic memory. In M. L. Minsky, editor, *Semantic Information Processing*, pages 227–270. The MIT Press, Cambridge, MA, 1968.

[Wiederhold, 1992] G. Wiederhold. Model-free optimization. In *DARPA Software Technology Conference*, pages 83–96. Meridien Corp., Arlington, VA, 1992.

# AUTOMATED REASONING

# AUTOMATED REASONING

Automated Reasoning 1: Belief Revision

# Qualitative Relevance and Independence: A Roadmap

**Didier Dubois, Luis Fariñas del Cerro, Andreas Herzig and Henri Prade**
Institut de Recherche en Informatique de Toulouse (IRIT) – Université Paul Sabatier – CNRS
118 route de Narbonne – 31062 Toulouse Cedex 4 – France

### Abstract

Several qualitative notions of epistemic dependence between propositions are studied. They are closely related to the ordinal notion of conditional possibility. What this paper proposes is a systematic investigation of how the fact of learning a new piece of evidence individually affects previous beliefs. Namely a new piece of information A can either leave a previous belief untouched, or cancel it from the set of accepted beliefs, or even refute it. On the contrary, A can justify a new belief, not previously held, or fail to justify it. We provide axiomatizations of epistemic independence and relevance and show the close links between qualitative independence and the theory of belief change. It turns out that qualitative independence and AGM belief change operations have the same expressive power. Lastly, it is briefly suggested how qualitative independence can be applied to plausible reasoning.

## 1. Introduction

It has been known for some time that the AGM revision theory (named from Alchourrón, Gärdenfors and Makinson; see Gärdenfors (1988)), and the preferential approach to nonmonotonic reasoning are two sides of the same coin. In recent years we have shown that this tight link could be explained in the setting of possibility theory, using set-functions satisfying the single axiom $\Pi(A \vee B) = \max(\Pi(A), \Pi(B))$, together with qualitative conditioning. The contribution of this paper is to show that there is a "third side" of the coin, viz. qualitative independence. The notion of epistemic independence naturally arises in the framework of reasoning under uncertainty and belief change. Most prominently, probabilistic conditional independence (between variables) plays a key role in Bayesian nets. More recently several authors (Delgrande and Pelletier, 1994; Benferhat et al., 1994; Dubois et al., 1994) have advocated the interest of qualitative independence notions for nonmonotonic reasoning. Gärdenfors (1990) and Fariñas del Cerro and Herzig (1996) have investigated the complementary notion of relevance in relation to belief change. The aim of the paper is to provide an exhaustive typology of the forms that independence and relevance can assume in the setting of an ordinal approach to uncertainty, like the one underlying major belief change and nonmonotonic inference theories.

In the paper, A, B, C,... stand for events belonging to a Boolean algebra of subsets of a set W. T and F are propositional constants denoting the true and false events respectively. Let us assume that our representation framework enables us to distinguish between three states of cognitive attitudes regarding C: i) C is an accepted belief, ii) ¬C is an accepted belief (i.e., C is refuted), iii) neither C nor ¬C is accepted (i.e., total ignorance about C). Hence C and ¬C cannot be held as accepted beliefs simultaneously. Intuitively, an event C is said to be independent of another event A when one's opinion about C is not affected by learning A.

Any definition of independence or relevance in such a setting can be expressed in terms of five basic notions corresponding to the possible effects of learning A on the belief status of C (Table 1). As already said an important distinction has to be made between propositions C that are a priori believed and those which are a priori ignored. Independence may then refer either to the lack of influence of A on a believed proposition C that remains accepted (line 1), or on the contrary, to the lack of influence of A on an ignored proposition that remains ignored (line 5). We shall speak of "qualitative independence" in the former case, and of "uninformativeness" in the latter.

|  |  | C | C given A |
|---|---|---|---|
| C is qualitatively independent of A | | accepted | accepted |
| A is qualitatively relevant for C | A cancels C | accepted | ignored |
| | A refutes C | accepted | refuted |
| A justifies C | | ignored | accepted |
| A does not inform about C | | ignored | ignored |

Table 1. Forms of epistemic independence and relevance

Relevance then may mean either that A negatively affects an agent's belief in C, or that A makes the agent start to believe C. This covers three situations: C was an accepted belief and upon learning A, C becomes ignored (line 2), or refuted (rejected) (line 3); if C was previously ignored, C may become accepted when learning A (line 4). In the following we reserve the name "qualitative relevance" for the first two cases, keeping the third one apart since the belief change then takes the opposite direction. It can be checked that all the other situations can be obtained from these 5 cases changing C into ¬C, or exchanging C and A.

What are the properties of such independence and relevance notions? Can they be characterized in a precise way? How are they related to theories of uncertainty? It is the purpose of this paper to answer these questions. In particular, it turns out that these notions are generally non-symmetric and sensitive to negation. However, they can be preserved via conjunction or disjunction, and this behavior is the one that was found natural by philosophers of probability such as Keynes or Gärdenfors. It is also worth noticing that due to the ternary structure of states of belief, it cannot be expected that independence and relevance be complementary notions.

For the sake of brevity, proofs of propositions are omitted. Moreover, definitions of independence and relevance are not given here with respect to a given context. However, the extension to ternary relations ("A is relevant/independent w.r.t. to C, given evidence E") is straightforward, and used in Section 5.

## 2. Relevance and Independence

### 2.1. The Probabilistic Framework

Suppose that a cognitive state is represented by means of a probability-like set-function. The standard definition of probabilistic independence is in terms of invariance with respect to conditioning: C is independent of A iff Prob(C | A) = Prob(C). It follows from the axioms of probability theory that independence then satisfies:

(symmetry) If C is independent of A then A is independent of C.
(negation) If C is independent of A then C is independent of ¬A.
(truth) A and T are independent.

Probabilistic independence has been criticized quite early by several authors such as Keynes: its symmetry, and the lack of properties with respect to conjunction and disjunction have been found debatable in an epistemic perspective. Following Keynes, Gärdenfors (1978) has discussed *conjunction criteria for dependence* and *independence*.

(CCD$\ell$) If C depends on A, and C depends on B, and $A \wedge B \neq F$, then C depends on $A \wedge B$
(CCI$\ell$) If C is independent of A, and C is independent of B then C is independent of $A \wedge B$.

Later, Gärdenfors (1990) has proposed that the concept of *relevance* should satisfy four minimal requirements (up to a fifth axiom, superfluous in our representational setting, stating that relevance is syntax-independent):

R1: A is relevant for C iff C is not independent of A
R2: If A is relevant for C then ¬A is relevant for C
R3: T is independent of C
R4: If C is contingent (= neither T nor F) then C is relevant for C.

These postulates equate relevance with dependence (i.e., the complement of independence), and insist on negation insensitivity (so that F is not relevant for C). Gärdenfors (1978) shows that under R1-R4, CCI$\ell$ + CCD$\ell$ leads to trivialization in the probabilistic framework.

An alternative attitude is, rather than rejecting CCD$\ell$, to accept regularities w.r.t. conjunction and disjunctions (such as CCD$\ell$ and CCI$\ell$) and drop R1 and R2. Namely, we shall object to negation insensitivity in some contexts, and we shall question the postulate that there is no middle way between relevance and independence. We shall choose a representation framework where the three cognitive attitudes (acceptance, rejection and ignorance) can be distinguished, viz. qualitative possibility theory, an ordinal setting for representing uncertainty in a way that directly extends classical logic with levels of acceptance. This framework has strong connections with belief change (Dubois and Prade, 1991) and ordering-based nonmonotonic reasoning (Gärdenfors and Makinson, 1994; Benferhat et al., 1992).

### 2.2. The Possibilistic Framework

In this section, we briefly recall the notions of possibility measure and distribution (Zadeh, 1978) and of conditional possibility (Dubois and Prade, 1988). Possibility theory provides a simple uncertainty representation setting where ordinal information about events derives from a complete transitive ordering of elementary events (the interpretations of a language). Dual rankings of events (or formulas) are induced in terms of possibility and certainty.

A function $\Pi$ from a set of events into any finite totally ordered set L (with top 1 and bottom 0) is a *possibility measure* if it satisfies the following decomposability axiom: $\Pi(A \vee B) = \max(\Pi(A), \Pi(B))$. This axiom enables an ordering on events to be recovered from an ordering of elementary events. The quantity $N(A) = 1 - \Pi(\neg A)$ is called the necessity of A, and represents a level of certainty (or acceptance) of A. $1 - (\cdot)$ is just a notation for the order-reversing function on L (if L = $\{1 = \ell_1 > \ell_2 > ... > \ell_n = 0\}$, $1 - (\ell_i) = \ell_{n+1-i} \, \forall \, i$). In can be checked that min(N(A), N(¬A)) = 0; A is said to be *accepted* iff N(A) > N(¬A), or equivalently N(A) > 0. And we have the reasonable axiom of acceptance saying that if A is accepted and so is B, then $A \wedge B$ is accepted too, since $N(A \wedge B) = \min(N(A), N(B))$ holds. If A is not accepted (N(A) = 0), it does not entail that it is rejected (N(¬A) > 0). This makes it clear that possibility theory can express the three possible attitudes that we want to distinguish.

Every possibility measure can be viewed as an encoding of a comparative possibility relation on events "$\geq$" defined by $A \geq B$ if and only if $\Pi(A) \geq \Pi(B)$ (Dubois and Prade, 1991). As shown in (Fariñas del Cerro and Herzig, 1991, Hajek et al., 1994; Boutilier, 1994), such a notion of comparative possibility is equivalent to that in (Lewis, 1973).

Conditional possibility can be defined similarly to conditional probability, changing the Bayes identity $P(A \wedge C) = P(C | A) \cdot P(A)$ into a more qualitative counterpart:

$$\Pi(A \wedge C) = \min(\Pi(C | A), \Pi(A)).$$

The use of minimum is justified by the ordinal nature of the possibility scale. The *conditional possibility* $\Pi(C | A)$ is then defined as the maximal solution of the above equation. This choice of the maximal solution is due to the *principle of minimal specificity* which urges to select the least informative or committed possibility measure, i.e., the one which allows each event to have the greatest possibility level:

$\Pi(C \mid A) = 1$ if $\Pi(A) = \Pi(A \wedge C)$ and $C \neq F$
$\Pi(C \mid A) = \Pi(A \wedge C)$ otherwise.

By duality the conditional necessity is $N(\neg C \mid A) = 1 - \Pi(C \mid A)$. Hence

$$N(C \mid A) = \begin{cases} 0 \text{ if } N(\neg A) = N(A \vee C) \text{ and } C \neq T \\ N(A \vee C) \text{ if } N(\neg A) < N(A \vee C) \end{cases}$$

The following notable property expresses that C is accepted in context A iff $A \wedge C$ is more plausible than $A \wedge \neg C$:

$$N(C \mid A) > 0 \text{ iff } \Pi(A \wedge C) > \Pi(A \wedge \neg C).$$

Note that if $\Pi(A \wedge C) > \Pi(A \wedge \neg C)$ then $N(C \mid A) = N(\neg A \vee C) = 1 - \Pi(A \wedge \neg C) \geq 1 - \Pi(\neg C) = N(C)$. Hence if $N(C) > 0$, the situation $N(C) > N(C \mid A) > 0$ (attenuation of acceptance) may never happen. So the input information A either confirms C or totally destroys our confidence in it. This is typical of the ordinal conditioning.

## 2.3. A Typology

It is tempting to define independence via conditioning in possibility theory, in a way similar to probability theory, namely to define C as independent of A when the (conditional) measure of C given A is equal to the unconditional measure of C. We can define independence either as $\Pi(C \mid A) = \Pi(C)$ or as $N(C \mid A) = N(C)$. If $\Pi(C \mid A) = \Pi(C) < 1$ then we are in the situation where C is plausibly rejected (since $\Pi(\neg C) = 1 > \Pi(C)$). It means that learning that A is true does not affect the plausible rejection of C. This expresses the negative statement that rejecting C is independent of A. It suggests to use $N(C \mid A) = N(C)$ in order to express the positive statement that A is independent of (the level of acceptance of) C. It turns out that this notion of independence is not uniform because it expresses the disjunction of two distinct forms of irrelevance:

$N(C \mid A) = N(C)$ is equivalent to
(i) $1 = \max(\Pi(\neg A \wedge \neg C), \Pi(A \wedge \neg C))$ and
$\Pi(A \wedge \neg C) \geq \Pi(A \wedge C)$,
or (ii) $\Pi(A \wedge C) > \Pi(A \wedge \neg C) \geq \Pi(\neg A \wedge \neg C)$.

The two situations (i) and (ii) correspond to (almost) reversed orderings of interpretations. Case (i) corresponds to the situation where $N(C \mid A) = N(C) = 0$, that is, C is either ignored or rejected both a priori and in the context A, which is again a composite situation. Now, in possibility theory, the full knowledge about C is expressed by the pair $(N(C), N(\neg C))$ and it covers the three situations where C is accepted, rejected or unknown. This leads to recognize three situations of independence in the absolute form:

– absolute independence of C vis-à-vis A
$N(C \mid A) = N(C) > 0$ (hence $N(\neg C \mid A) = N(\neg C) = 0$)
– absolute independence of $\neg C$ vis-à-vis A
$N(\neg C \mid A) = N(\neg C) > 0$ (hence $N(C \mid A) = N(C) = 0$)
– uninformativeness
$N(C \mid A) = N(C) = N(\neg C \mid A) = N(\neg C) = 0$.

The first (resp.: second) condition means that believing C (resp.: $\neg C$) is not affected by A, while the third condition means that A does not inform about C. In the first (resp.: second) situation we shall say that *believing C (resp.: $\neg C$) is absolutely independent* of A, where the term "absolute" refers to the stability of the level of acceptance, and the expression "believing C" indicates that C is an accepted a priori belief. The latter situation, which cannot be expressed in the probabilistic framework, means that in the presence of A, the piece of belief C, which was originally ignored, is still ignored. In this case, we shall speak of *uninformativeness* of A about C (or $\neg C$, equivalently), a notion that is negation-insensitive with respect to C. This is formalized by the following definitions:

**Definition 1.** *Believing C is absolutely independent of* A (noted $A \perp_{>} C$) iff $N(C \mid A) = N(C) > 0$.

**Definition 2.** A *does not inform about* C iff $N(C \mid A) = N(C) = N(\neg C \mid A) = N(\neg C) = 0$.

Now in order to investigate the opposite notions of relevance, simply taking the complement of the absolute independence or uninformativeness relation is not satisfactory. For instance the negation of "believing C is absolutely independent of A" is "either $N(C) = 0$, or $N(C \mid A) \neq N(C) > 0$". But it is hard to see why $N(C) = 0$ alone would mean that A is relevant to C. So in the possibilistic framework, we must give up the idea that "relevance" is just the negation of "independence". If we investigate relevance, we must keep the acceptance condition ($N(C) > 0$) and only negate the other equality condition. So A is said to be absolutely *relevant to believing C* iff $N(C) > 0$ but $N(C \mid A) \neq N(C)$. Again, this situation splits into three cases

– $N(C) > 0$ and $N(C \mid A) > N(C)$ (confirmation)
– $N(C) > 0$ and $N(C \mid A) = N(\neg C \mid A) = 0$ (cancellation)
– $N(C) > 0$ and $N(\neg C \mid A) > 0$ (refutation).

In the first situation, learning A confirms C by increasing its level of acceptance. In the second case, learning A leads us to forget about C and we say that A cancels C. In the third case, the agent's belief in C is reversed: we say that A refutes C. (Remember that the missing case $N(C) > N(C \mid A) > 0$ (attenuation) cannot occur here.)

The two cases when A confirms C, and believing C is absolutely independent of A, are those where learning A neither cancels nor refutes the agent's a priori acceptance of C. In the purely ordinal case where levels of belief are represented in a relative fashion only, it is not really meaningful to distinguish confirmation from absolute independence. This argument is reinforced by the fact that attenuation of acceptance can only occur in a drastic way: namely if A confirms C, but B subsequently does not and on the contrary weakens our belief in C, then $A \wedge B$ either cancels or refutes C. To sum it up, it means that there will not be any compensation effect between the confirmation of C by A and the subsequent negative effect of B. The latter will prevail in any case. So it is legitimate to consider the disjunction of the two cases when A neither cancels nor refutes C (in other words either A confirms C or believing C is absolutely independent to A) as expressing a single form of qualitative independence of C w.r.t. A. This leads to purely ordinal notions of relevance and independence:

**Definition 3.** *Believing C is qualitatively independent of A* (denoted A ≠> C) iff N(C) > 0 and N(C | A) > 0.

**Definition 4.** *A is qualitatively relevant to believing C*, denoted (A ≈> C), iff N(C) > 0 and N(C | A) = 0.

Hence A is qualitatively relevant for C iff A cancels or refutes our belief in C. Mind that relevance can but negatively affect beliefs. A last form of dependence is the one obtained when neither C nor ¬C is an accepted belief but C becomes accepted in the context where A is true. This is a form of direct relevance of A for C akin to causality, at least an epistemic form of it, since it means that A is a reason for starting to believe C.

**Definition 5.** *A justifies C* iff N(C) = N(¬C) = 0 and N(C | A) > 0.

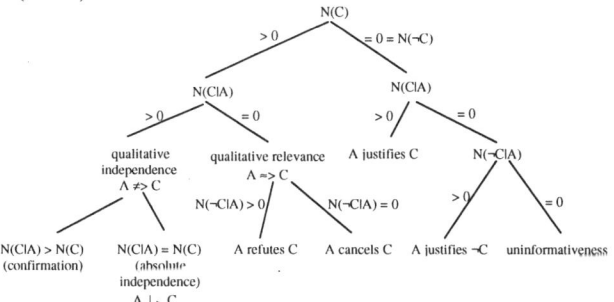

**Figure 1**

Figure 1 is the counterpart of Table 1. It exhaustively summarizes the various cases of relevance and independence that can be expressed in an ordinal setting. (Dual cases can be expressed in terms of those which appear in the table, changing A into ¬A or C into ¬C.) Figure 1 lays bare the fact that two distinct concepts of independence exist: uninformativeness, and a qualitative independence that expresses that an accepted belief C resists an input information A. The latter notion has absolute independence as a particular case.

Independence expresses a property of invariance (under conditioning) of the ordering of situations where C is true. This is similar to the case of probabilistic independence, that expresses numerical invariance under conditioning.

## 3. Properties and Representation Theorems

We now examine the properties of the above notions with respect to conjunction and disjunction. We shall first do it for the two notions of independence (Definitions 2 and 3) and three notions of "non-relevance" understood as the negations of each relevance notions (cancellation, refutation, justification). Then we also do it for these three relevance notions and the negation of the two independence notions. There are eight possible properties as follows, here formulated in terms of qualitative independence (≠>) and its negation (=>, qualitative dependence).

(CCIℓ)   If A ≠> C and B ≠> C then A ∧ B ≠> C
(DCIℓ)   If A ≠> C and B ≠> C then A ∨ B ≠> C
(CCIr)   If A ≠> B and A ≠> C then A ≠> B ∧ C
(DCIr)   If A ≠> B and A ≠> C then A ≠> B ∨ C
(CCDℓ)  If A => C and B => C then A ∧ B => C
(DCDℓ)  If A => C and B => C then A ∨ B => C
(CCDr)  If A => B and A => C then A => B ∧ C
(DCDr)  If A => B and A => C then A => B ∨ C

In the above properties named XCYz, X stands for disjunction or conjunction, C for criterion, Y for independence or dependence and z stands for left or right. Whether or not a criterion holds for a given notion is listed in Table 2 below. Grey slots refer to the negation of the notion on the corresponding line.

|  | DCIℓ | CCIr | DCIr | DCDℓ | CCDr | DCDr |
|---|---|---|---|---|---|---|
| C qual. independent of A (N(C) > 0, N(C\|A) > 0) | y | y | y | y | y | n |
| A does not inform about C (N(C)=N(¬C)=N(C\|A)= N(¬C\|A)=0) | y | n | n | n | n | n |
| A cancels C (N(C) > 0, N(C\|A) = N(¬C\|A) = 0) | y | n | n | n | n | n |
| A refutes C (N(C) > 0, N(¬C\|A) > 0) | y | n | n | y | y | y |
| A justifies C (N(C) = N(¬C) = 0, N(C\|A) > 0) | y | n | n | y | n | n |

**Table 2.** Properties of independence and relevance relations

All of the relations satisfy DCIℓ, but the properties CCIℓ and CCDℓ are never satisfied and do not appear in the table. This apparent paradox will be explained in Section 4. In the case of uninformativeness and cancellation DCIℓ is the only property that holds (formally, this is due to the presence of negations in the definition of the former). "Qualitative independence" and "uninformativeness" are the most regular notions, due to their simple definition. Lastly, it can be proved that independence and relevance are related via negation as follows: If A ≈> C then both ¬A ≠> C and C ≠> ¬A.

The above properties do not completely characterize the respective notions. It can be proved that qualitative independence and qualitative relevance can be axiomatized in such a way as to recover qualitative possibility theory. We have established the following results:

**Theorem 1** (axiomatic equivalence of ≠> with possibility theory). Let ≠> be any relation on events such that

QI1  T ≠> T (Tautologies do not undermine tautologies)
QI2  if A ≠> C then A ≠> B ∨ C (Right weakening)
QI3  if T ≠> C then C ≠> C (If C is believed, then it cannot undermine itself)
QI4  if A ≠> C then T ≠> C (If there is A that does not undermine C then C is a priori believed)
QI5  A ≠> ¬A never holds
QI6  if A ≠> C and B ≠> C then A ∨ B ≠> C (left OR rule, i.e., DCIℓ)

QI7 if $A \vee B \not\Rightarrow C$ then either $A \not\Rightarrow C$ or $A \vee B \not\Rightarrow \neg A$ or both (similar to rational monotony)

QI8 if $A \not\Rightarrow B$ and $A \not\Rightarrow C$, then $A \not\Rightarrow B \wedge C$ (stability under conjunction for acceptance, i.e., CCIr)

and let N be any mapping from the set of events to [0,1] such that $A \not\Rightarrow C$ iff $N(C \mid A) \geq N(C) > 0$. Then $\not\Rightarrow$ is a qualitative independence relation iff N is a non-trivial necessity measure.

Lastly, it is possible to axiomatize qualitative relevance $A \approx\!\!> C$. Note that $A \approx\!\!> C$ is not equivalent to $\neg(A \not\Rightarrow C)$, so that Gärdenfors' R1 does not hold. We have that $A \approx\!\!> C$ iff $\neg(A \not\Rightarrow C)$ and $T \not\Rightarrow C$. It is thus easy to see that $A \vee C \approx\!\!> \neg C$ iff $\Pi(\neg C) > \Pi(C) \geq \Pi(A)$. So the axiomatization of relevance does not follow immediately from that of independence.

**Theorem 2** (axiomatic equivalence of $\approx\!\!>$ with possibility theory). Let $\approx\!\!>$ be any relation on events satisfying

QR1 it does not hold that $A \approx\!\!> A \vee C$
QR2 $F \approx\!\!> A$ iff $A = T$ or there exists C such that $\neg A \approx\!\!> C$
QR3 if $A \approx\!\!> C$ and $\neg B \approx\!\!> B$ then $A \approx\!\!> B \wedge C$
QR4 if $\neg A \approx\!\!> C$ then $\neg A \approx\!\!> A$
QR5 if $A \vee B \approx\!\!> C$ then $A \approx\!\!> C$ or $B \approx\!\!> C$ (DCI$\ell$)
QR6 if $A \vee B \approx\!\!> A$ and $A \approx\!\!> C$ then $A \vee B \approx\!\!> C$ (restricted transitivity)
QR7 if $A \approx\!\!> B \wedge C$ then $A \approx\!\!> B$ or $A \approx\!\!> C$
(contraposed CCIr)

and let N be any mapping from the set of events to L such that $A \approx\!\!> C$ iff $N(C \mid A) = 0$ and $N(C) > 0$. Then $\approx\!\!>$ is a qualitative relevance relation iff N is a non-trivial necessity measure.

Although Gärdenfors' R1 does not hold, QR1 is related to QI1, QR3 to QI2, and QI5 is related to QR4. QR5-QR7 are contraposed forms of QI6-QI8. The contraposition of QI3-QI5 does not hold for $\approx\!\!>$.

## 4. Qualitative Independence and Belief Change

Several notions of independence and relevance studied above, among which qualitative independence, can be fully expressed in the framework of revision of propositional theories also called belief sets (Gärdenfors, 1988). Revising a belief set K by a sentence A means to add A to K and to restore consistency so as to keep A. Gärdenfors (1990) proposes the following criterion for the revision of a belief set: *If a belief state K is revised by a sentence A, then all sentences in K that are independent of the validity of A should be retained in the revised state of belief*. This seems to be a very natural requirement for belief revision operations, as well as a useful tool when it comes to implementing belief change operations. As noted by Gärdenfors, "a criterion of this kind cannot be given a technical formulation […] in a simple propositional language because the notion of relevance is not available in such a language." However the above criterion does make sense in the ordinal setting of possibility theory.

Given a belief set K, that is, a set of propositional formulas closed under deduction, and a revision operation *, K * A represents the result of revising K by a formula A. As stated in Gärdenfors (1988), if the revision operation satisfies the AGM postulates, then K and * can be represented equivalently by an epistemic entrenchment ordering, which in turn is nothing else than a qualitative necessity ordering (Dubois and Prade, 1991). Conversely, any qualitative necessity ordering leads to an AGM revision operation. Namely, given a necessity function N, the set K= {C, N(C) > 0} is a belief set that is, K is closed under conjunction and logical consequence. Moreover, it can be proved that the revision operation * can be defined in terms of possibility theory as follows: $C \in K * A$ is equivalent to $N(C \mid A) > 0$ (Dubois and Prade 1991). If we translate the various definitions of independence and relevance in terms of revision we get the following facts:

1. C is qualitatively independent of A ($A \not\Rightarrow C$) iff $C \in K$ and $C \in K * A$
2. A cancels C iff $C \in K$ and $C \notin K * A$ and $\neg C \notin K * A$
3. A refutes C iff $C \in K$ and $\neg C \in K * A$
4. A is qualitatively relevant for C ($A \approx\!\!> C$) iff $C \in K$ and $C \notin K * A$
5. A justifies C iff $C \notin K$ and $C \in K * A$
6. A does not inform about C iff $C \notin K$, $\neg C \notin K$, $\neg C \notin K * A$ and $C \notin K * A$.

Qualitative independence exactly expresses Gärdenfors' above requirement for independence-based revision.

The operation opposite to revision is contraction. Contracting a belief set K by a sentence A means to delete A from K, as well as those sentences that enable A to be derived so as to obtain a belief set K − A that does not contain A. The Harper Identity (Gärdenfors, 1988) defines contraction in terms of revision as follows: $K - A = K \cap (K * \neg A)$, i.e., first revise K to accept $\neg A$ and then keep only those formulas in K. Conversely $K * A = Cn((K - \neg A) \cup \{A\})$, where Cn is the consequence operation. This is the Levi Identity whereby revising by A means deleting $\neg A$ first and then adding A. Companion definitions of qualitative independence and relevance relations $\not\Rightarrow_c$ and $\approx\!\!>_c$ can be associated to a contraction operation "−" via the following definition:

$A \not\Rightarrow_c C$ iff $C \in K$ and $C \in K - A$
$A \approx\!\!>_c C$ iff $C \in K$ and $C \notin K - A$

where $A \not\Rightarrow_c C$ reads: forgetting A does not affect the belief in C and $A \approx\!\!>_c C$ reads forgetting A destroys the belief in C. It is easy to check that Levi and Harper Identities can be written in terms of independence relations between events as follows:

$A \not\Rightarrow_c C$ iff $\neg A \not\Rightarrow C$ ; $A \approx\!\!>_c C$ iff $\neg A \approx\!\!> C$.

Fariñas del Cerro and Herzig (1996) have proved the equivalence between $\approx\!\!>_c$ satisfying QR1-QR7 (where A is changed into $\neg A$) and AGM contraction operators.

Similarly, postulating the equivalence between $\neg A \not\Rightarrow C$ and $C \in K - A$, it can be proved that axioms QI1-QI8 are equivalent to the AGM postulates. Indeed, with the Harper Identity, C is qualitatively independent of A ($A \not\Rightarrow C$) when $C \in K$ and $C \in K - \neg A$ (because $K - \neg A$ is $K \cap K * A$). Due to the set inclusion of $K - \neg A$ in K, this is just equivalent to C ∈ K − ¬A, which makes this independence notion parti-cularly simple: in fact, we are able to express that C ∈ K by F ≠> C. This permits to obtain a complete axiomatization of qualitative independence ≠> by just rewriting the AGM postulates for contraction, turning A into ¬A.

If A either cancels or refutes C then C ∈ K and C ∉ K ∗ A. With the Harper Identity this is equivalent to C ∈ K and C ∉ K − ¬A. This corresponds to a dependence notion proposed in (Fariñas del Cerro and Herzig, 1996). If we had presented relevance and independence this way in terms of belief contractions instead of revisions, the properties CCD$\ell$ and CCI$\ell$ would have been fulfilled whenever the corresponding revision-based notion (via Levi or Harper Identity) satisfies DCI$\ell$ and DCD$\ell$, respectively. In this way, we can recover the original Keynes-Gärdenfors criteria (absent in Table 2).

## 5. Conclusion

We have established the equivalence between the statements "the agent's belief in C is independent of proposition A" (A ≠> C) and "the agent still believes C if his belief set is revised by A". This notion of independence can be expressed in terms of possibilistic conditioning, thus laying bare the analogy with probabilistic independence.

This paper indicates that the notions of ordinal independence introduced here can be modelled as extra constraints on the ordering of interpretations of a language, and have the same expressive power as the AGM theory of belief revision. However the latter is in turn equivalent to the rational monotony approach to plausible inference (Lehmann and Magidor, 1992; Gärdenfors and Makinson, 1994) and to possibility theory. Namely any AGM-like revision operation ∗ on a belief set there corresponds to a possibility measure $\Pi$ such that $\Pi(A \wedge B) > \Pi(A \wedge \neg B)$ iff B ∈ K ∗ A iff A ⊢ B (A plausibly infers B). So, a revision operation generates a conditional knowledge base $\Delta^* = \{A \vdash B : B \in K * A\}$. The nice interaction between the basic laws of plausible inference (embedded in Lehmann's axiomatic framework), belief revision, and possibilistic independence, augmented with results of this paper, suggest that independence assumptions can be expressed by means of supplementary conditional assertions, provided that the above introduced independence notions are extended to conditional independence:

*Qualitative conditional independence*:
(A ≠> C)$_D$ iff N(C | D) > 0 and N(C | A ∧ D) > 0.

So if a piece of knowledge comes under the form "C is independent of A in the context D", it can be expressed by the set of default rules {D ⊢ C, A ∧ D ⊢ C}. Note that the corresponding conditional relevance (A ≈> C)$_D$ (i.e., A refutes or cancels C in the context D) corresponds to the idea of Delgrande and Pelletier (1994) that A is relevant to a conditional assertion D ⊢ C. However their definition is more specific than ours: it can be expressed in our terminology by "A refutes C in context D or A refutes ¬C in context D". They do not seem to consider the possibility of a mere cancellation of C.

Noticeably, the rational monotony axiom RM: D ⊢ C implies D ⊢ ¬A or A ∧ D ⊢ C (underlying rational closure), does express a condition for conditional qualitative independence: In the context D, C is qualitatively independent of A as soon as ¬A is not accepted in this context. These results are promising in the scope of exception-tolerant inference, because they suggest that conditional knowledge bases not delivering expected answers can be repaired by means of suitable conditional independence assumptions. It opens the road to a sound, feasible and computationally reasonable treatment of exception-tolerant plausible inference (Benferhat et al., 1996a, b) that can cope with most, if not all, counterexamples to rational closure.

## References

S. Benferhat, D. Dubois and H. Prade. 1992. Representing default rules in possibilistic logic. In *Proc. KR'92*, 673-684.

S. Benferhat, D. Dubois and H. Prade, 1994. Expressing independence in a possibilistic framework and its application to default reasoning. In *Proc. ECAI'94*, 150-154.

S. Benferhat, D. Dubois and H. Prade. 1996a. Beyond counter-examples to nonmonotonic formalisms: A possibility-theoretic analysis. In *Proc. ECAI'96*, 652-656.

S. Benferhat, D. Dubois and H. Prade. 1996b. Coping with the limitations of rational inference in the framework of possibility theory. In *Proc. UAI'96*, 90-97.

C. Boutilier. 1994. Modal logics for qualitative possibility theory. *Int. J. Approximate Reasoning*, 10:173-201.

J.P. Delgrande and J. Pelletier. 1994. A formal approach to relevance. Tech. Report, Simon Fraser University, Burnaby, BC, Canada.

D. Dubois, L. Fariñas del Cerro, A. Herzig and H. Prade. 1994. An ordinal view of independence with application to plausible reasoning. In *Proc. UAI'94*, 195-203.

D. Dubois and H. Prade. 1991. Epistemic entrenchment and possibilistic logic. *Artificial Intelligence*, 50:223-239.

L. Fariñas del Cerro and A. Herzig. 1991. A modal analysis of possibility theory. *In:* LNCS 535, Springer Verlag, 11-18.

L. Fariñas del Cerro and A. Herzig. 1996. Belief change and dependence. In *Proc. TARK'96*, 147-161.

P. Gärdenfors. 1978. On the logic of relevance. *Synthese*, 37:351-367.

P. Gärdenfors. 1988. *Knowledge in Flux*. The MIT Press, Cambridge, MA.

P. Gärdenfors. 1990. Belief revision and irrelevance. *PSA*, 2:349-356.

P. Gärdenfors and D. Makinson. 1994. Nonmonotonic inference based on expectations. *Artificial Intelligence*, 65:197-245.

P. Hajek, D. Harmancova and R. Verbruggen. 1994. A qualitative fuzzy possibilistic logic. *Int. J. of Approximate Reasoning*:2, 1-19.

D. Lehmann and M. Magidor. 1992. What does a conditional knowledge base entail? *Artificial Intelligence*, 55:1-60.

D. Lewis. 1973. *Counterfactuals*. Basil Blackwell, Oxford. 2nd edition, Billing & Sons Ltd., Worcester, 1986.

L.A. Zadeh. 1978. Fuzzy sets as a basis for a theory of possibility. *Fuzzy Sets and Systems*, 1:3-28.

# The Complexity of Belief Update

## Paolo Liberatore
Dipartimento di Informatica e Sistemistica
Universitá di Roma "La Sapienza"
Via Salaria 113, 00198 Rome, Italy
Email: liberato@dis.uniroma1.it
WWW: http://www.dis.uniroma1.it/~liberato/

## Abstract

Belief revision and belief update are two different forms of belief change, and they serve different purposes. In this paper we focus on belief update, the formalization of change in beliefs due to changes in the world. The complexity of the basic update (introduced by Winslett [1990]) has been determined in [Eiter and Gottlob, 1992]. Since then, many other formalizations have been proposed to overcome the limitations and drawbacks of Winslett's update. In this paper we analyze the complexity of the proposals presented in the literature, and relate some of them to previous work on closed world reasoning.

## 1 Introduction

The study of belief change has received considerable attention from the AI, databases and philosophy communities. Belief change deals with the incorporation of new facts into an agent's beliefs. There are two basic forms of belief change: belief revision and belief update. The difference lies in what is the source of incorrectness (if any) in the previous agent's beliefs.

The old beliefs of the agent may be mistaken or incomplete: in this case the usual approach is that of belief revision, captured by the AGM postulates [Alchourrón et al., 1985]. On the other hand, an agent's beliefs, while correct at one time, may become obsolete due to changes in the world. The basic treatment of updates is given in [Winslett, 1990], while a general framework is proposed in [Katsuno and Mendelzon, 1991].

Many drawbacks and limitations of the initial proposal of Winslett have been discovered. One regards how it treats disjunctive information: as in many approaches of non-monotonic reasoning, the update treats incorrectly a new piece of information that is in disjunctive form (such as $x_1 \vee x_2$). The next example shows one of these situations.

**Example 1** *The RCH company deals with computers. It stores computers not yet sold in a warehouse composed by three separated rooms. The first room contains PCs, the second MacIntoshes and the third Suns, X-terminals etc. By accident, a fire is set up in the room that contains PCs, and all the items it contains are moved to the other two rooms. The old database is (for our purposes)* $\neg contains(PC, R2) \wedge \neg contains(PC, R3)$, *which means only the first room contains PCs. After the fire destroyed that room, all we know is that the PCs has been moved in the rooms R2 and R3. Some items may have been moved to a room and some to the other one. The new piece of information to be incorporated is* $\mu = contains(PC, R2) \vee contains(PC, R3)$, *which denotes the fact that at least one room (possibly both) must have PCs in it. The intended result of the update should be* $\mu$ *itself, but Winslett's gives instead* $contains(PC, R2) \neq contains(PC, R3)$. *This is not the intuitive result, since the possibility that some items have gone to R2 and some to R3 should not be ruled out.*

Another drawback of the original formulation of update is that it never allows changes to the previous states: sometimes updates lead agents to revise their knowledge about the previous state. In these cases, update is related with revision as intended in the AGM framework [Alchourrón et al., 1985; Dalal, 1988]. The next example, shamelessly stolen from [Boutilier, 1995], shows this fact.

**Example 2** *We have a beaker with an unknown liquid in it. We want to determine if it is an acid or a base. Initially, all we know is* acid $\neq$ base: *the content of the beaker may be an acid or a base, but not both.*

*In order to determine which kind of substance we are dealing with, we drop a litmus paper in the beaker. After that, we discover that the liquid is an acid. Updating the old knowledge base we obtain (correctly)* acid $\wedge \neg$base. *The problem is that this fact is true not only after the experiment, but was also true before it: the liquid was acid before the test, although we did not know this. As a result, the initial state should be modified accordingly. However, Winslett's approach modifies only the new state, and does not make any change on the belief about the previous state: using this update, we can say, about the liquid before the test, only* acid $\neq$ base.

In this paper we analyze the complexity of some of the proposals introduced so far. The complexity of the

PMA approach (Possible Models Approach, also known as Winslett's update) has been proved by Eiter and Gottlob [1992] to be at the second level of the polynomial hierarchy, namely $\Pi_2^p$. The problem considered there is the basic entailment, that is

> given a knowledge base $\phi$, an update $\mu$ and a propositional formula $\psi$, decide whether $\psi$ is implied by $\phi *_W \mu$, the updated knowledge base

The paper is organized as follows: in the next section we give the basic definitions of propositional calculus, belief update, and the various definitions of update proposed so far. In section 3 we study the updates proposed for solving the problem of disjunctive updates: we start showing how some of the semantics of update are related to previous work on closed world reasoning, and use the results on complexity of closed world reasoning to prove some complexity results of update. In section 4 we show the complexity of the remaining proposals. In section 5 we draw some conclusions.

## 2  Definitions

Throughout this paper, we assume a propositional language $\mathcal{L}$ over an alphabet of atoms $X$. Any piece of information (such as previous agents' beliefs, updates, etc.) is represented by a propositional formula, where not otherwise specified. An interpretation is a truth assignment of the atoms, that is, a function from atoms into the set {true, false}. We extend this assignment to propositional formulas in the usual way. An interpretation $I$ is a model of a formula $\phi$ if and only if the formula is true in that interpretation. We denote interpretations and models by sets of atoms (those mapped into true). The set of all the possible interpretations over the given alphabet is denoted by $\mathcal{M}$. A formula is said to be complete if it has exactly one model. Given a formula $\phi$, we denote by $Var(\phi)$ the set of atoms it contains.

We denote by $Mod(\phi)$ the set of all the models of the formula $\phi$. We use also $Form$ to denote the inverse function of $Mod$, that is, $Form(A)$ is a propositional formula whose set of models is $A$. This function is not unique, since equivalent formulas have the same set of models; however, this is unimportant for our purposes.

Given two interpretations $I$ and $J$, we denote by $Diff(I,J)$ their symmetric difference, that is, $Diff(I,J) = I \backslash J \cup J \backslash I$. Intuitively, $Diff(I,J)$ is the set of atoms to which the interpretations $I$ and $J$ assign a different truth value. Given a set $A$ and an ordering $<$, we denote by $\min(A, <)$ the set of the minimal elements of $A$ w.r.t. the ordering $<$.

We sometimes use substitutions. A substitution is a set of pairs $atom/formula$. Given a substitution $S = \{x_i/\psi_i\}$, and a propositional formula $\mu$, we denote by $\mu[S]$ the formula obtained from $\mu$ by replacing any occurrence of $x_i$ with the corresponding $\psi_i$.

### 2.1  Winslett's Update

Consider a propositional formula $\phi$ representing the state of the world. This information is assumed to be correct, but not (necessarily) complete. When a change in the world occurs, this description of the world must be modified. The assumption behind belief update is that what we know about the change is a propositional formula $\mu$ that is true in the new situation.

Winslett's update operates on a model by model base. Let $I$ be an interpretation, and define $\leq_I$ to be the ordering on interpretations defined as

$$J \leq_I Z \text{ iff } Diff(I,J) \subseteq Diff(I,Z)$$

The update of the old state $\phi$ when a new formula $\mu$ becomes true after a change is

$$\phi *_W \mu = Form(\bigcup_{I \in Mod(\phi)} \min(Mod(\mu), \leq_I))$$

This update was initially proposed in the context of databases, in which one can safely assume that all the updates are atomic, that is, the formula $\mu$ is always a literal. In AI this is often considered to be too restrictive. When $\mu$ is allowed to be any formula, and mainly when it is a disjunction of literals, the result of the Winslett's update may be intuitively incorrect. The example 1 of the last section shows a situation of this kind.

### 2.2  Minimal Change with Exceptions

This approach, introduced in [Zhang and Foo, 1996], attempts to overcome the non-intuitive treatment of disjunctive information of the original formulation. The definition is similar to that of Winslett's update, except for a set of atoms $EXC$ that are not considered in the computation of the symmetric difference between interpretations.

First, define $D(\mu)$ as the set of all the minimal clauses implied by $\mu$, that is

$$D(\mu) = \left\{ d \;\middle|\; \begin{array}{l} \mu \models d, \text{ and there is no other clause } d' \\ \text{s.t. } Var(d') \subset Var(d) \text{ and } \mu \models d' \end{array} \right\}$$

Then define $EXC(I, \mu)$, where $I$ is an interpretation and $\mu$ a formula, as

$$EXC(I, \mu) = \bigcup_{\substack{d \in D(\mu) \\ I \notin Mod(d)}} Var(d)$$

Then, define $\leq_I^R$, where $R$ is a set of variables, as the ordering among interpretations such that

$$J \leq_I^R Z \text{ iff } Diff(I,J) \backslash R \subseteq Diff(I,Z) \backslash R$$

finally, the update is defined as

$$\phi *_{MCE} \mu = Form(\bigcup_{I \in Mod(\phi)} \min(Mod(\mu), \leq_I^{EXC(I,\mu)}))$$

## 2.3 Minimal Change with Maximal Disjunctive Inclusion

Given a model $I$ and a set of models $S$, define [Zhang and Foo, 1996]

$$Dis(I,S) = \left\{ S' \middle| \begin{array}{l} S' \in Mod(\mu) \text{ and} \\ \text{i. } \forall S_i \in S,\ Diff(I,S_i) \subseteq Diff(I,S') \\ \text{ii. there are no other } S'' \text{ satisfying i.} \\ \quad \text{but } Diff(I,S'') \subset Diff(I,S') \end{array} \right\}$$

then define the result of the update as

$$\phi *_{MCD} \mu = \bigcup_{\substack{I \in mod(\phi) \\ S \subseteq \min(Mod(\mu), \leq_I)}} Dis(I,S)$$

This approach is different from the previous one. Let $\phi$ and $\mu$ be as follows.

$$\begin{aligned} \phi &= \neg x_1 \wedge \neg x_2 \wedge \neg x_3 \wedge \neg x_4 \\ \mu &= ((x_1 \neq x_2) \wedge \neg x_3 \wedge \neg x_4) \vee (x_1 \wedge x_2 \wedge x_3) \end{aligned}$$

The initial knowledge base $\phi$ has only one model $I = \emptyset$ (the interpretation that assigns false to all the variables). The only element of $D(\mu)$ that has not $I$ as a model is $x_1 \vee x_2$. As a result, $EXC(I,\mu) = \{x_1, x_2\}$ and thus $\phi *_{MCE} \mu = (x_1 \neq x_2) \wedge \neg x_3 \wedge \neg x_4$.

On the other hand, $\phi *_{MCD} \mu = \mu \wedge \neg x_4$. Indeed, $\min(Mod(\mu), \leq_I) = \{\{x_1\}, \{x_2\}\}$, that has three subsets. The first is $\{x_1\}$: we have $Dis(I, \{\{x_1\}\}) = \{\{x_1\}\}$. The second is $\{x_2\}$, for which $Dis(I, \{\{x_2\}\}) = \{\{x_2\}\}$. The last subset is $\{\{x_1\}, \{x_2\}\}$ itself. One can verify that $Dis(I, \{\{x_1\}, \{x_2\}\}) = \{\{x_1, x_2, x_3\}\}$. Thus, the models of the updated knowledge base are $\{x_1\}$, $\{x_2\}$, and $\{x_1, x_2, x_3\}$. The formula $\mu \wedge \neg x_4$ represents this set of models.

## 2.4 Update with Dependence Function

Assume that there is a function $DEP$ from atoms to set of atoms. We use $DEP(x_i)$ to denote the set of atoms whose value depends on that of $x_i$. This means that a change in the value of $x_i$ may affect the value of a $x_j \in DEP(x_i)$ (see [Herzig, 1996], where the operator is introduced, for more details.) It is assumed that for any atom $x_i$, it holds $x_i \in DEP(x_i)$. Define $DEP(\mu)$ as the union of $DEP(x_i)$ for each $x_i \in Var(\mu)$.

Now, define

$$CTX(\mu) = \{l_1 \wedge \cdots \wedge l_k \mid l_i = x_i \text{ or } \neg x_i,\ x_i \in DEP(\mu)\}$$

and the update is defined as

$$\phi *_{He} \mu = \bigvee_{\gamma \in CTX(\mu)} (\phi *_W \gamma) \wedge \mu$$

## 2.5 Generalized Update

This operator was introduced in [Boutilier, 1995] to formalize scenarios where an update may lead to revising the knowledge about the initial state. Define a ranking $\kappa$ as a partial function from interpretations to integers, such that there is at least one interpretation $I$ such that $\kappa(I) = 0$. The meaning of this ranking is that if $\kappa(I) < \kappa(J)$ then the interpretation $I$ is believed to be more plausible than $J$, the world described exactly by $I$ is more likely than the world described by $J$, from the agent's perspective. The models of $\phi$ must be exactly those ranked 0. When $\kappa(I)$ is not defined, the model $I$ is considered unplausible. In this case, we write $\kappa(I) = \infty$.

We formalize the possible changes in the world as a set of possible *events* that may occur. For each event, we must specify what is the plausibility that the possible world represented by a model $I$ will be transformed into the world represented by another model $J$. This plausibility is a function $\bar{e}$ from triples $\langle event, model, model \rangle$ to integers. It can be a partial function, and $\bar{e}(e, I, J) = \infty$ means that in the event $e$, the transition from $I$ to $J$ is considered implausible.

A generalized update model is a 5-tuple $\langle \mathcal{M}, \kappa, \bar{e}, E, m \rangle$, where $\mathcal{M}$ is as usual the set of all the interpretations, $\kappa$ is a ranking, $E$ is a set of events, $\bar{e}$ is as said above a function from triples $\langle event, model, model \rangle$ to integers, and $m$ is a function from pairs $\langle event, model \rangle$ to integers, representing the plausibility of the occurrence of an event in the world represented by a given model.

Define the new rank of the interpretations after the update as

$$\kappa^\diamond(J) = \min_{I \in \mathcal{M}, e \in E} \{\bar{e}(e, I, J) + m(e, I) + \kappa(I)\}$$

This function $\kappa^\diamond(J)$ induces an ordering in a natural way: $I \leq^\diamond J$ if and only if $\kappa^\diamond(I) \leq \kappa^\diamond(J)$. One way to define the update is

$$\phi *_{GU} \mu = Form(\min(Mod(\mu), \leq^\diamond))$$

that is, the models of the result are the models of $\mu$ whose rank $\kappa^\diamond$ is minimal.

## 2.6 Possible Causes Approach

This approach was proposed in [Li and Pereira, 1996] and is based upon the idea that changes in the world are caused by *actions*. Thus, update can be formalized with ad-hoc languages such as $\mathcal{A}$ (introduced in [Gelfond and Lifschitz, 1993]) with an appropriate semantics. The main difference between reasoning about actions and belief update formalisms is that in the formers, actions are often assumed to be known in advance, and the aim of the theory is to understand what is true after they are performed. In belief update, actions are always assumed unknown, and the only evidence they have happened is their consequences.

The language $\mathcal{A}$ is built over an alphabet composed of fluent names (that are the facts, or what in the previous formalizations are the atoms of the propositional language), and three special symbols **after**, **causes** and **if**. A fluent expression is a fluent name or a fluent name preceded by $\neg$. Propositions are of two types: value propositions $F$ **after** $A_1; \ldots; A_m$, where $F$ is a fluent expression and $A_1; \ldots; A_m$ are actions; and effect propositions $A$ **causes** $F$ **if** $P_1, \ldots, P_m$, where $A$

is an action and $F, P_1, \ldots, P_m$ are fluent expressions. If $m = 0$ the after proposition $F$ `after` $A_1, \ldots, A_m$ is written `initially` $F$, and is called initially proposition.

In order to formalize the belief update in this framework, we must introduce a finite set of (linearly ordered) temporal points $T$, and two other kinds of propositions, the *happens* propositions and the *holds* propositions. Given a time point $t$ and an action $A$, a happens proposition is written $A$ `happens at` $t$, and means that the action $A$ happens in the instant $t$. Given a fluent expression $F$ and a time point $t$, a holds proposition has the form $F$ `holds at` $t$, and means that $F$ is true in the time point $t$.

A domain description is a set of value, effect, and happens propositions. Given a domain description $D$, it may happen that new information, represented by a holds proposition $H$, has to be incorporated. Due to the lack of space, we cannot introduce formally the semantics of $\mathcal{A}$ neither that of the belief update based on it. We refer to [Li and Pereira, 1996], where the update is introduced, for a more detailed explanation. The new holds proposition $H$ could be not implied by $D$. In this case, we must find an explanation, i.e., a possible cause of the change. Formally, an explanation is a set of happens and initially propositions $P$ such that $D \cup P$ implies $H$. To decide whether a proposition $H'$ is implied by the updated description, we check whether $D \cup P$ implies $H'$ for each minimal explanation $P$.

### 2.7 Closed World Reasoning

In the sequel we use two formalizations of the closed world assumption, namely the generalized world closed assumption and the *CURB*.

Given a propositional formula $\mu$, the set of its free for negation atoms, written $FFN(\mu)$, is the set of the atoms that are false in all the minimal (w.r.t. set inclusion) models of $\mu$. The generalized closed world assumption of $\mu$ is the formula $GCWA(\mu) = \mu \wedge FFN(\mu)$.

The *CURB* is introduced in [Eiter *et al.*, 1993] as a variant of the circumscription. In this paper we use a simplified version of it, the $CURB^1$, also introduced in the paper above. Given a formula $\mu$, the $CURB^1(\mu)$ is defined as the formula whose models are the models $J$ of $\mu$ such that there exists a set of minimal models $S$ of $\mu$ such that a) $J$ contains all the models of $S$, and b) there is no other $J'$ of $\mu$, contained in $J$, with the same property.

### 2.8 Computational Complexity

We assume that the reader is familiar with the basic concepts of computational complexity. We use the standard notation of complexity classes that can be found in [Johnson, 1990]. Namely, the class P denotes the set of problems whose solution can be found in polynomial time by a *deterministic* Turing machine, while NP denotes the class of problems that can be resolved in polynomial time by a *non-deterministic* Turing machine. The class coNP denotes the set of decision problems whose complement is in NP. We call NP-hard a problem G if any instance of a generic problem NP can can reduced to an instance of G by means of a polynomial-time (many-one) transformation (the same for coNP hard).

Clearly, P $\subseteq$ NP and P $\subseteq$ coNP. We assume, following the mainstream of computational complexity, that these containments are strict, that is P $\neq$ NP and P $\neq$ coNP.

We also use higher complexity classes defined using oracles. In particular $P^A$ ($NP^A$) corresponds to the class of decision problems that are solved in polynomial time by deterministic (nondeterministic) Turing machines using an oracle for $A$ in polynomial time (for a much more detailed presentation we refer the reader to [Johnson, 1990]). The classes $\Sigma_k^p$, $\Pi_k^p$ and $\Delta_k^p$ of the polynomial hierarchy are defined by $\Sigma_0^p = \Pi_0^p = \Delta_0^p = P$, and for $k \geq 0$,

$$\Sigma_{k+1}^p = NP^{\Sigma_k^p}, \quad \Pi_{k+1}^p = co\Sigma_{k+1}^p, \quad \Delta_{k+1}^p = P^{\Sigma_k^p}.$$

## 3 Disjunctive Information

In this section we analyze the definitions proposed to manage the problem of disjunctive information in belief update. The reason of their introduction is that updating $\neg x \wedge \neg y$ with $x \vee y$ one does not want, in general, to obtain $x \neq y$, that is the result given by Winslett's update.

This problem is analogous of that of other forms of non monotonic reasoning, notably circumscription and closed-word reasoning.

We start with the two proposals by Zhang and Foo. The minimal change with exception update have some similarities with the generalized closed world assumption. Indeed, $GCWA$ can be reduced to MCE, and vice versa.

**Theorem 1** *For any propositional formulas $\phi$ and $\mu$, it holds*

$$GCWA(\mu) = \left(\bigwedge_{x_i \in X} \neg x_i\right) *_{MCE} \mu$$

$$\phi *_{MCE} \mu = \bigvee_{I \in Mod(\phi)} GCWA(\mu[S])[S]$$

*where $X$ is the set of atoms of $\mu$ and $S$ is the substitution $\{x_i/\neg x_i | x_i \in I\}$.*

This proves that $*_{MCE}$ is based upon the concept of generalized closed world assumption. As a result of the first statement, the update $*_{MCE}$ is $\Pi_2^p$ hard, since the generalized closed world assumption is so, [Eiter and Gottlob, 1993], and it can be reduced polynomially to $*_{MCE}$. About the membership in that class, we note that the second statement gives a non-polynomial reduction. The following theorem shows the complexity of $*_{MCE}$.

**Theorem 2** *The update $*_{MCE}$ is $\Pi_2^p$ complete.*

This can actually proved in a very simple manner using the result of $\Pi_2^p$ membership of $GCWA$ given by Nebel [1996].

While the minimal change with exception is based upon the generalized closed world assumption, the minimal change with maximal disjunctive inclusion update is similar to the variant of circumscription called $CURB^1$. About the relation between $*_{MCD}$ and $CURB^1$, we have the following theorem.

**Theorem 3** *For any propositional formulas $\phi$ and $\mu$, it holds*

$$CURB^1(\mu) = \left(\bigwedge_{x_i \in X} \neg x_i\right) *_{MCD} \mu$$

$$\phi *_{MCE} \mu = \bigvee_{I \in Mod(\phi)} CURB^1(\mu[S])[S]$$

*where $S$ is the substitution $\{x_i/\neg x_i | x_i \in I\}$.*

This theorem shows also that the $*_{MCD}$ is $\Pi_2^p$ hard, since this is the complexity of $CURB^1$, as proved in [Eiter et al., 1993].

The second reduction is not polynomial. However, with a little effort it is possible to obtain the exact complexity of $*_{MCD}$.

**Theorem 4** *The update $*_{MCD}$ is $\Pi_2^p$ complete.*

The update with dependence function is based on a function $DEP$ that represents the causal relations between literals of the given alphabet. The easiest case is having all literals independent from each other, that is $DEP(x_i) = \{x_i\}$. In this case, the following theorem shows that the Herzig's proposal coincides with the standard semantics update $*_{SSU}$ (for an explanation of the standard semantics update see [Winslett, 1990]).

**Theorem 5** *When $DEP$ is defined as $DEP(x_i) = \{x_i\}$ for each $x_i$, then for each pair of propositional formulas $\phi$ and $\mu$ it holds $\phi *_{He} \mu = \phi *_{SSU} \mu$.*

A more general statement can be given. Recall that a way to represent the logical consequences of $\phi *_{SSU} \mu$ is $\phi[\{x_i/z_i | x_i \in Var(\mu)\}] \wedge \mu$, where $z_i$ are new variables appearing nowhere else. For $*_{He}$ we have the following lemma.

**Lemma 1** *For each triple of propositional formulas $\phi$, $\mu$ and $\psi$, it holds $\phi *_{He} \mu \models \psi$ if and only if $\phi[\{x_i/z_i | x_i \in DEP(\mu)\}] \wedge \mu \models \psi$, where $z_i$'s are variables that do not appear in any of the three formulas.*

Assuming that verifying whether $x_i \in DEP(\mu)$ is a polynomial task, we have the following result.

**Theorem 6** *The update with dependence function is coNP complete.*

## 4 Complexity of Generalized Updates

In this section, we show the complexity of the two other updates, the generalized update by Boutilier [1995] and the possible causes approach by Li and Pereira [1996]. The former turns out to be simpler than the basic (Winslett's) update, while the latter is one level higher in the polynomial hierarchy.

About Boutilier's generalized update, we have to make some computational assumptions on the functions involved. Namely, we still write $\phi *_{GU} \mu$, and we implicitly assume that $E$ is part of the input, but not $\mathcal{M}$ (this is the set of all the interpretations over the given alphabet). Also, we assume that the functions $\kappa$, $\overline{e}$ and $m$ of the revision model can be calculated in polynomial time. This means that given an event $e$ and two models $I$ and $J$, it must be possible to determine the integer $\overline{e}(e, I, J)$ in polynomial time. The same for $\kappa$ and $m$.

Under these assumptions, the entailment problem for this update is a $\Delta_2^p$ problem.

**Theorem 7** *The update $*_{GU}$ is $\Delta_2^p$.*

We give a very short explanation of this result. As for Dalal's revision, the check $\phi *_{GU} \mu \models \psi$ can be done in two steps: first, determine the value of the minimum $\kappa(I) + \overline{e}(e, I, J) + m(e, I)$. Using this value $k_{\phi,\mu}$, verifying whether $\phi *_{GU} \mu \models \psi$ is a coNP complete problem. Indeed, guess a model $I$ of $\phi$, a model $J$ of $\mu$ and an event $e$ such that $\kappa(I) + \overline{e}(e, I, J) + m(e, I) = k_{\phi,\mu}$ and $J \not\in Mod(\psi)$. If this is possible, then $\phi *_{GU} \mu \not\models \psi$.

The first step, the determination of $k_{\phi,\mu}$, can be done with a polynomial number of calls to an NP oracle. Start with a guess for $k_{\phi,\mu}$, and verify (with the oracle) whether there exist $I$, $J$ and $e$ such that $\kappa(I) + \overline{e}(e, I, J) + m(e, I)$ is less than this number. If this is true, the value of this $k_{\phi,\mu}$ must be decreased (otherwise, it must be increased).

About the hardness of this update, we note that one of the revision introduced, namely Dalal's revision [Dalal, 1988], can be polynomially reduced to the generalized update.

**Theorem 8** *Let $\overline{e}$ be the function such that $\overline{e}(e, I, J)$ is the number of elements in $Diff(I, J)$. Then, using the revision models $\langle \mathcal{M}, \kappa, \overline{e}, \{e\}, m \rangle$ where $k(I) = 0$ if $I \in model(\phi)$ and $\kappa(I) = \infty$ otherwise, and $m(e, I) = 0$, then $\phi *_{GU} \mu = \phi *_D \mu$, where $*_D$ is the Dalal's revision.*

As a corollary, since the revision of Dalal is $\Delta_2^p[\log n]$ complete, the complexity of the generalized update is $\Delta_2^p[\log n]$ hard. This result can be strengthened.

**Theorem 9** *The generalized update is $\Delta_2^p$ complete.*

The Generalized Update has a lower complexity than Winslett's update. Thus (unless the polynomial hierarchy collapses) it is impossible to reduce the latter to the former in polynomial time.

We now consider the Possible Causes Approach. The problem we analyze is still to decide whether a formula is entailed by the old knowledge (in this case, a domain description) updated with a holds proposition.

**Theorem 10** *Entailment under the Possible Causes Approach is $\Pi_3^p$ complete.*

This is the only update considered in this work that turns out to be more complex than Winslett's. We try to explain this result.

Consider the definition of the possible causes approach restricted with the following hypothesis.

a. There are no holds proposition in the domain description (the only holds proposition is the update).

b. Given the set of fluents $\{F_1, \ldots, F_n\}$, there are only $n$ actions $\{A_1, \ldots, A_n\}$, and the only effect propositions in the domain description are

$$A_i \text{ causes } F_i \text{ if } \neg F_i$$
$$A_i \text{ causes } \neg F_i \text{ if } F_i$$

c. Explanations are composed only of after propositions (initially propositions are not allowed in explanations).

In the holds propositions $F$ holds at $t$ we allow $F$ to be a boolean formula on the alphabet of the fluent names (instead of a fluent expression). Under these restrictions, the update is still $\Pi_3^p$ complete.

The reason of the increase of the complexity is that, in the Winslett's approach, each model of $\phi$ is updated separately, and this makes easy to verify if the result of an action entails the new piece of information.

Given a model $I \in Mod(\phi)$, in order to decide if $J \in \min(Mod(\mu), \leq_I)$, we have to check whether $Diff(I, J)$ is minimal. Each element of $Diff(I, J)$ can be interpreted as an action of the kind of (b) above, that changes the value of a literal. As a result, an explanation of the change is a minimal set of actions that maps $I$ into a model of $\mu$. Given a model $I$ and a set of those actions, deciding whether the resulting interpretation $J$ is a model of $\mu$ is a polynomial problem.

In the PCA, the possible causes affect the initial state $D$ altogether: a possible explanation is a set of propositions $P$ such that $D \cup P$ implies the new fact to be incorporated. This is more complex than verifying a possible explanation in Winslett's approach.

## 5 Conclusions

In this paper we have investigated some of the semantics that have been introduced to formalize changes in the world of interest. The complexity of most of them turns out to be at the second level of the polynomial hierarchy. This is quite a remarkable result: indeed, these frameworks have been proposed to overcome drawbacks and limitations of the original semantics of update introduced by Winslett, but this does not seem to introduce any increase in the computational complexity.

Two approaches turn out to have a different complexity than Winslett's. The first is Boutilier's Generalized Update, that is easier ($\Delta_2^p$ complete), and the second is the Possible Causes Approach by Li and Pereira, that is more difficult ($\Pi_3^p$ complete). The first one is a generalization of Dalal's revision, and has a slightly higher complexity. The reason of the higher complexity of the Possible Causes Approach seems to be the globality of this operator: a change is considered not to affect each single possible initial state, but all the possible initial states altogether. This makes NP complete a subproblem that is polynomial in Winslett's approach.

## Acknowledgments

We would like to thank Marco Schaerf for revising a draft version of this paper. Most of the work reported here has been done while the author was visiting AT&T Bell Labs. We thank Henry Kautz for his support during this period.

## References

[Alchourrón et al., 1985] C. E. Alchourrón, P. Gärdenfors, and D. Makinson. On the logic of theory change: Partial meet contraction and revision functions. *J. of Symbolic Logic*, 50:510–530, 1985.

[Boutilier, 1995] C. Boutilier. Generalized update: belief change in dynamic settings. In *Proc. of IJCAI-95*, pages 1550–1556, 1995.

[Dalal, 1988] M. Dalal. Investigations into a theory of knowledge base revision: Preliminary report. In *Proc. of AAAI-88*, pages 475–479, 1988.

[Eiter and Gottlob, 1992] T. Eiter and G. Gottlob. On the complexity of propositional knowledge base revision, updates and counterfactuals. *AIJ*, 57:227–270, 1992.

[Eiter and Gottlob, 1993] T. Eiter and G. Gottlob. Propositional circumscription and extended closed world reasoning are $\Pi_2^p$-complete. *Theor. Comp. Sci.*, 114:231–245, 1993.

[Eiter et al., 1993] T. Eiter, G. Gottlob, and Y. Gurevich. CURB your theory! In *Proc. of IJCAI-93*, pages 634–639, 1993.

[Gelfond and Lifschitz, 1993] M. Gelfond and V. Lifschitz. Representing action and change by logic programs. *J. of Logic Programming*, 17:301–322, 1993.

[Herzig, 1996] A. Herzig. The PMA revised. In *Proc. of KR-96*, pages 40–50, 1996.

[Johnson, 1990] D. S. Johnson. A catalog of complexity classes. In J. van Leeuwen, editor, *Handbook of Theoretical Computer Science*, volume A, chapter 2. Elsevier, 1990.

[Katsuno and Mendelzon, 1991] H. Katsuno and A. O. Mendelzon. On the difference between updating a knowledge base and revising it. In *Proc. of KR-91*, pages 387–394, 1991.

[Li and Pereira, 1996] R. Li and L. Pereira. What is believed is what is explained. In *Proc. of AAAI-96*, pages 550–555, 1996.

[Nebel, 1996] B. Nebel. How hard is it to revise a knowledge base? Technical Report 83, Albert-Ludwigs-Universität Freiburg, Institut für Informatik, 1996.

[Winslett, 1990] M. Winslett. *Updating Logical Databases*. Cambridge University Press, 1990.

[Zhang and Foo, 1996] Y. Zhang and N. Foo. Updating knowledge bases with disjunctive information. In *Proc. of AAAI-96*, pages 562–568, 1996.

# Anytime Belief Revision

Mary-Anne Williams
Information Systems
Department of Management
University of Newcastle
Australia
maryanne@frey.newcastle.edu.au

## Abstract

Belief Revision is a ubiquitous process underlying many forms of intelligent behaviour. The AGM paradigm is a powerful framework for modeling and implementing belief revision systems based on the principle of Minimal Change; it provides a rich and rigorous foundation for computer-based belief revision architectures.

Maxi-adjustment is a belief revision strategy for theory bases that can be implemented using a standard theorem prover, and one that has been used successfully for several applications.

In this paper we provide an *anytime decision procedure* for maxi-adjustments, and study its complexity. Furthermore, we outline a set of guidelines that serve as a protomethodology for building belief revision systems employing a maxi-adjustment. The algorithm is under development in the belief revision module of the CIN Project.

## 1 Introduction

Belief Revision underlies many forms of intelligent behaviour. An intelligent agent must be adept at revising its beliefs in a rational way. The AGM paradigm, so named after its founders Alchourron, Gärdenfors and Makinson [1985], is a powerful theoretical framework for modeling and implementing belief revision systems; it provides a rich and rigorous foundation for principled computer-based architectures that endow agents with the ability to change their beliefs in a coherent and rational fashion.

Gärdenfors and Makinson [1988] provided a constructive means for defining revision functions based on an epistemic entrenchment ordering of a reasoning agent's beliefs. Furthermore, they showed that there is a one-to-one relationship between revision functions and epistemic entrenchment orderings.

Iterated revision can be achieved by transmuting epistemic entrenchment orderings where the emphasis is not exclusively on acceptance and removal of beliefs from a theory, but also on raising and lowering of the degree of acceptance of beliefs. Raising the degree of acceptance of a belief corresponds to a revision, whilst lowering it corresponds to a contraction.

Maxi-adjustment is a specific strategy for implementing belief revision systems. It strives for maximal inertia of information under change, and was devised by Williams [1996]. It has been shown to be successful in applications where the systems designer or knowledge engineer is able to specify dependencies among beliefs [MacNish and Williams, 1996]. In essence, when incoming information is inconsistent with the agent's knowledge, a maxi-adjustment retracts only those minimally entrenched beliefs that are inconsistent with the new information.

In this paper we provide an anytime decision procedure for maxi-adjustments. Furthermore, we outline a set of guidelines that serve as a protomethodology for building belief revision systems. The algorithm is under development in the belief revision module of the CIN Project; a project that is seeking to develop an intelligent information management toolkit [Antoniou and Williams, 1996].

Section 2 outlines belief revision in the AGM paradigm. Section 3 discusses several important modeling problems that arise when AGM change functions are used in practice, it thence outlines how maxi-adjustments overcome them. Section 4 describes subsumption removal; an optional feature of maxi-adjustment that can be used to enhance its performance. In section 5 we give an anytime algorithm for maxi-adjustments, and in section 6 we discuss its complexity. In section 7 we make several methodological remarks concerning the design and development of belief revisions systems that employ the maxi-adjustment strategy.

## 2 Belief Revision

An intelligent information system must possess the ability to revise its knowledge base when it receives new information. The AGM paradigm has become one of the standard frameworks for modeling change.

Let us begin with some technical preliminaries: $\mathcal{L}$ will denote a (possibly first-order) language which contains a complete set of Boolean connectives. We will denote sentences in $\mathcal{L}$ by lower case Greek letters. We assume $\mathcal{L}$ is governed by a logic that is identified with its consequence relation $\vdash$ which is assumed to satisfy the following conditions [Gärdenfors, 1988]: (a) if $\alpha$ is a truth-functional tautology, then $\vdash \alpha$, (b) if $\vdash (\alpha \rightarrow \beta)$ and $\vdash \alpha$, then $\vdash \beta$ (*modus ponens*), (c) $\vdash$ is consistent, i.e. $\not\vdash \bot$, where $\bot$ denotes the inconsistent theory, (d) $\vdash$ satisfies the deduction theorem, and (e) $\vdash$ is compact.

The set of all logical consequences of a set $T \subseteq \mathcal{L}$, i.e. $\{\alpha : T \vdash \alpha\}$, is denoted by $Cn(T)$. A *theory* of $\mathcal{L}$ is any subset of $\mathcal{L}$ closed under Cn. We let $\mathcal{L}^\bowtie$, pronounced 'elbow', denote the set of *contingent* sentences.

Within the AGM paradigm a body of information is represented as a theory, and informational changes are regarded as transformations on theories. The principal AGM functions are contraction and revision. They can be described using the well known AGM rationality postulates, and individual contraction and revision functions can be uniquely determined by any of the several standard constructions, e.g. the epistemic entrenchment ordering construction. Both the postulates and the constructions attempt to encapsulate the principle of Minimal Change. The *magnitude of change* may not be based on set inclusion measures; sometimes the most rational response to avoiding inconsistency is to forfeit more than the minimal number of beliefs, e.g. it may be better to retract several weakly held beliefs than to surrender a single strongly held belief.

A *revision*, $T_\alpha^*$, attempts to change a theory $T$ to incorporate $\alpha$ so that the resultant theory is consistent provided $\alpha$ itself is consistent. A *contraction*, $T_\alpha^-$, involves the removal of a set of sentences from $T$ so that a nontautological sentence $\alpha$ is no longer implied. A *withdrawal* function [Makinson, 1987] is a generalised contraction function in that it satisfies all but the most notorious postulate for contraction, namely *recovery*; the property $T = \text{Cn}(T_\alpha^- \cup \alpha)$. It has been argued in the literature that recovery is not always appropriate for a limited reasoner, however it is one of the most important postulates for capturing the notion of *minimal change* when information is given up.

## 3 Implementing Belief Revision

For the purpose of developing an implementation of AGM change functions, Gärdenfors and Makinson's [1988] work was a significant breakthrough. They showed that an epistemic entrenchment ordering (certain total preorder on the sentences in the language) can uniquely determine how the system will react to the pressures of impinging information. In order to develop computational models based on the entrenchment construction two obvious problems must be overcome: first an epistemic entrenchment ordering has to be propagated by the change function, and second a finite representation for epistemic entrenchment orderings is needed.

We use partial entrenchment rankings[1] as our representation of well-ranked epistemic entrenchment orderings, and we model iterated belief revision by propagating these rankings using a maxi-adjustment; a procedure described in Williams [1996].

### 3.1 Partial Entrenchment Rankings

Finite partial entrenchment rankings will be sufficient for present purposes. They represent *finite* epistemic entrenchment orderings of $\mathcal{L}$ where the elements of a finite (a not necessarily closed) set of sentences are mapped to the natural numbers.

**Definition:** *A finite partial entrenchment ranking is a function* $\mathbf{B}$ *from a finite subset of sentences in $\mathcal{L}$ into the*

[1] Partial entrenchments were defined in [Williams 1995], and essentially identical representations can be found in [Dubios et al 1994, Rott 1992, Williams 1992], and elsewhere.

*natural numbers $\mathcal{N}$ such that the following conditions are satisfied for all $\alpha \in \text{dom}(\mathbf{B})$:*

**(PER1)** *If $\not\vdash \alpha$ then $\{\beta \in \text{dom}(\mathbf{B}) : \mathbf{B}(\alpha) < \mathbf{B}(\beta)\} \not\vdash \alpha$.*
**(PER2)** *If $\vdash \neg\alpha$ then $\mathbf{B}(\alpha) = 0$.*
**(PER3)** $\vdash \alpha$ *if and only if $\mathbf{B}(\alpha) = \max(\mathbf{B}(\text{dom}(\mathbf{B})))$.*

The higher the integer assigned to a sentence by a partial entrenchment ranking the more firmly it is held. **(PER2)** says that inconsistent information is assigned zero, and **(PER3)** says that the tautologies alone are assigned the highest rank. The most piquant property of a ranking, is given in **(PER2)**, namely a nontautological sentence $\alpha$ cannot be entailed by sentences ranked strictly higher than $\alpha$ itself.

We refer to $\mathbf{B}(\alpha)$ as the degree of acceptance of $\alpha$. The intended interpretation of a partial entrenchment ranking is that sentences assigned a degree of acceptance greater than zero form the agent's *explicit* beliefs of the system, and their logical closure form his *implicit* beliefs.

**Definition:** *Define the* explicit information content *of a ranking* $\mathbf{B}$ *to be* $\{\alpha \in \text{dom}(\mathbf{B}) : \mathbf{B}(\alpha) > 0\}$, *and denote it by* $\exp(\mathbf{B})$. *Define the* implicit information content *of $\mathbf{B}$ to be* $\text{Cn}(\exp(\mathbf{B}))$, *and denote it by* **content**$(\mathbf{B})$.

Typically, we are not only interested in the degree of acceptance of explicit information, but also in the degree of acceptance of sentences they entail. A partial entrenchment ranking represents a system's incomplete preferences from which a complete entrenchment ranking can be generated. There could, well, be an infinite number of entrenchment rankings which are compatible with a given partial specification. The following function 'degree' derives the minimum possible degree of acceptance for implicit information as specified by a partial entrenchment ranking.

**Definition:** *Let $\alpha \in \mathcal{L}^\bowtie$. If $\mathbf{B}$ is a finite partial entrenchment ranking, then define* **degree**$(\mathbf{B}, \alpha)$ *to be*

$$\begin{cases} \text{largest } j \text{ such that } \{\beta \in \exp(\mathbf{B}) : \mathbf{B}(\beta) \geq j\} \vdash \alpha \\ \qquad \qquad \text{if } \alpha \in \textbf{content}(\mathbf{B}) \\ 0 \qquad \qquad \text{otherwise} \end{cases}$$

**Example:** If $\mathbf{B}(\alpha \to \beta) = 3$, $\mathbf{B}(\alpha) = 2$, $\mathbf{B}(\gamma) = 1$, then we can compute **degree**$(\mathbf{B}, \beta) = 2$, **degree**$(\mathbf{B}, \alpha \wedge \beta) = 2$, **degree**$(\mathbf{B}, \alpha \vee \gamma) = 2$, **degree**$(\mathbf{B}, \alpha \wedge \gamma) = 1$, **degree**$(\mathbf{B}, \delta) = 0$, **degree**$(\mathbf{B}, \neg\delta) = 0$.

### 3.2 Maxi-Adjustments

The $(\alpha, i)$–maxi-adjustment of the partial entrenchment ranking $\mathbf{B}$, is denoted $\mathbf{B}^\star(\alpha, i)$, it modifies $\mathbf{B}$ by assigning $\alpha$ the new ranking $i$ whilst maintaining the properties of a partial entrenchment ranking. It is formally defined in [Williams, 1996].

The input $i$ can be specified as an integer, as a number in a specified range, e.g. $[0,1]$, as $\mathbf{B}(\beta)$ for some $\beta$ which would mean give the new information $\alpha$ the same rank as $\beta$, or in linguistic terms depending on the application at hand.

Maxi-adjustments are motivated by Spohn's notion of a *reason*. According to Spohn [1983] $\beta$ is a *reason* for $\alpha$ if and only if raising the epistemic rank of $\beta$ would raise the epistemic rank of $\alpha$.

Maxi-adjustment is based on the idea that if you believe a reason your coffee cup is leaking is that it has a hole, then whereupon closer inspection you discover its not leaking at all, it would seem a rational response to retract *your cup has a leak* and *your cup has a hole*. Contrariwise, if you are unaware that holes are reasons for leaks, then you could quite happily continue to believe *your cup has a hole* even when you discover that it is not leaking. In other words, common sense suggests that information should be retracted only if there is good reason to do so.

To evaluate reasons, it seems eminently sensible to require that raising the rank of $\alpha$ should disturb the agent's background ranking as little as possible. It is easy to show that whenever $\alpha$ and $\beta$ are in **exp(B)**, if we simply enforce the properties of a partial entrenchment ranking then $\beta$ is a reason for $\alpha$ if and only if **degree**$(\mathbf{B}, \beta \to \alpha) > \mathbf{B}(\alpha)$. Maxi-adjustments aim to use a closed world assumption with respect to reasons: if it cannot be derived that $\beta$ is a reason for $\alpha$, then it is assumed that $\beta$ is not a reason for $\alpha$. The system designer specifies reasons by ensuring that **degree**$(\mathbf{B}, \beta \to \alpha)$ is strictly larger than **degree**$(\mathbf{B}, \alpha)$. All nontautological reasons are defeasible, and there may be reasons for reasons.

Like most revision procedures on theory bases maxi-adjustments are syntactically dependent on **exp(B)**. Despite this syntax dependence it can be shown that (i) $\mathbf{B}^\star(\alpha, i)$ is a partial entrenchment ranking in which $\alpha$ is assigned the degree of acceptance $i$, (ii) if $i$ is greater than zero then **content**$(\mathbf{B}^\star(\alpha, i))$ is an AGM revision (**content**$(\mathbf{B}))^\star_\alpha$, (iii) **content**$(\mathbf{B}^\star(\alpha, 0))$ is an AGM withdrawal function, and (iv) $Cn(\mathbf{B}^\star(\alpha, 0) \cup \neg \alpha) \cap$ **content**$(\mathbf{B})$ is an AGM contraction (**content**$(\mathbf{B}))^-_\alpha$.

The change functions defined above by maxi-adjustment may not be the same as those obtained via Gärdenfors and Makinson's construction using the epistemic entrenchment ordering derived in the obvious way from the relative ordering given by **B** using the function **degree**. Ordinary adjustment [Williams 1995] is a procedure that complies exactly with the standard entrenchment construction. There exist several variants of maxi-adjustment, all are minor variations of the main algorithm described in the next section. Our implementation [Williams and Williams, 1997] offers anytime algorithms for all of the variants. By performing a so-called hybrid-adjustment (an adjustment followed by a maxi-adjustment) we can guarantee that at least as much information is retained as would be using the standard entrenchment procedure. In practice hybrid-adjustments and maxi-adjustments retain a great deal more information than the standard recipe.

Other properties of maxi-adjustment include: (i) every ranking is *reachable* in a finite language via a finite number of maxi-adjustments, (ii) maxi-adjustments only use the relative ranking of sentences, (iii) maxi-adjustments *preserve finiteness*; $|\exp(\mathbf{B}^\star(\alpha, i))| \le |\exp(\mathbf{B})| + 1$, (iv) maxi-adjustment does not reassign explicit beliefs holding ranks greater than $\max(\{i, \mathbf{B}(\alpha), \mathbf{B}(\neg \alpha)\})$, i.e. it preserves information that is more inportant than the incoming information, (v) moving a sentence up $j$ ranks in one step results in the same ranking as 'jacking it up' $j$ ranks one rank at a time, and (vi) similarly, moving a sentence down $j$ ranks in one step results in the same ranking as moving it down $j$ ranks one rank at a time.

Finally, based on the well known connections between belief revision, possibilistic reasoning and nonmonotonic reasoning, maxi-adjustments can be applied to possibilistic knowledge bases [Dubois *et al*, 1994], and nonmonotonic knowledge bases [Gärdenfors and Makinson 1994, Lehmann 1992].

## 4 Subsumption Removal

The idea of using reasons to determine what information to retract during the contraction of a sentence $\alpha$ works well for all beliefs except for some ranked at **degree**$(\mathbf{B}, \alpha)$. For example, if $\beta \to \alpha$, $\beta$ and $\alpha$ are equally ranked, then neither $\beta \to \alpha$ nor $\beta$ is a Spohnian reason for $\alpha$, yet belief in both cannot be maintained if $\alpha$ is to be contracted. In this case the Gärdenfors and Makinson construction would remove both $\beta \to \alpha$ and $\beta$. A standard maxi-adjustment is designed to do the same, however it also provides an *optional* procedure that uses *subsumption removal* which if selected removes all the beliefs subsumed by the information to be contracted. For the case in question above, this would result in the retraction of $\beta \to \alpha$ in preference to $\beta$ because $\beta \to \alpha$ is subsumed by $\alpha$.

Subsumption removal is not always appropriate, has a computational cost, and should be used with caution. When contracting (or lowering) a belief $\alpha$ using a maxi-adjustment we justify removing beliefs subsumed by $\alpha$ on the following grounds:

(i) If **B** is interpreted to be a partial specification of an agent's epistemic entrenchment, then it will, in general, be compatible with several epistemic entrenchments (which could, in principle, be generated using a different mechanism for **degree**). If $\mathbf{B}(\beta \to \alpha) = \mathbf{degree}(\mathbf{B}, \alpha)$ then there may exist a compatible epistemic entrenchment ordering that would retain $\beta$ using the standard Gärdenfors and Makinson construction, but no such compatible ordering could exist that would retain $\beta \to \alpha^2$. If $\alpha \vee \beta$ is explicit and $\mathbf{B}(\alpha \vee \beta) = \mathbf{degree}(\mathbf{B}, \alpha)$ then we may prefer not to use subsumption removal because it could be interpreted to mean that the user definitely wants $\beta$ removed when $\alpha$ is - this situation can be trivially detected, and built into the algorithm.

(ii) Since $\alpha \vdash (\beta \to \alpha)$ the sentence $\beta \to \alpha$ does not add any *epistemic power* to the ranking if $\alpha$ itself is an explicit belief. In other words, the same epistemic entrenchment ordering would be generated using the function **degree** irregardless of whether $\beta \to \alpha$ is explicit. If subsumption removal is applied then whenever $\alpha$ is an implicit but not explicit belief the maxi-adjustment algorithm makes it explicit. In other words, reducing the ranking of $\alpha$ will not remove $\alpha$ even if it wasn't previously explicit, except of course when the new rank is zero.

## 5 Anytime Maxi-Adjustments

An anytime algorithm is one that if interrupted has constructed a partial solution that approximates the actual solution, and the longer it runs the better the approximation. The proposed algorithm for maxi-adjustment is

---

[2] In other words, if $\alpha = \beta \to \alpha$, then $\alpha \vee \beta > \alpha$ is possible, but $\alpha \vee (\beta \to \alpha) > \alpha$ would violate the postulates of entrenchment.

*anytime* because it constructs the desired resultant partial entrenchment ranking in a top down iterative fashion refining the partly constructed ranking until it converges on the actual ranking. Furthermore, it captures important beliefs before less important beliefs, by rebuilding the ranking from the highest rank to the lowest.

Put simply, the algorithm consists of two main phases: First find the largest cut of the ranking that is consistent with the sentence to be moved, and second salvage as much of the remainder of the ranking by giving preference to higher ranked beliefs.

An important and endearing feature of the algorithm is that after completing its first phase the content of the ranking constructed so far is guaranteed to satisfy the AGM postulates for withdrawal and revision. Furthermore, there is enough information to construct a contraction function as well.

The second phase refines the ranking, in the sense that it recaptures as many beliefs as possible by removing only the minimal subsets at each rank that together with the higher ranked beliefs to be kept entail the information to be contracted.

For simplicity of exposition, and without loss of generality, nonbeliefs and tautologies are not explicit in our rankings, in addition we focus on the principle case of movement of contingent beliefs within the ranking.

The algorithm requires the services of a standard theorem prover to implement references to entailment in the function **degree**, and in the generation of minimal subsets that entail sentences to be moved down the ranking via the **movedown** procedure. Before the procedure can decide whether to raise or to lower the degree of acceptance of the sentence $\alpha$ it computes its current degree of acceptance. If its degree is to decrease, then **movedown** is performed. If its degree is to increase then $\neg\alpha$ is decreased first (if necessary) using **movedown**, and then $\alpha$ is moved up the ranking using **moveup**. The procedure **maxi-adjustment**$(\alpha, i, \mathbf{B}, \mathbf{newB})$ is the $(\alpha, i)$-maxi-adjustment of $\mathbf{B}$ resulting in the formation of the new ranking **newB**. The highest integer given to any sentence in the domain of $\mathbf{B}$ is assigned to the global variable **max_degree**.

If the **maxi-adjustment** procedure is interrupted then **newB** constructed so far is returned and if $i > 0$ then we check that $\alpha$ has been added to rank $i$, if not we do so.

**PROCEDURE: maxi-adjustment**$(\alpha, i, \mathbf{B}, \mathbf{newB})$
**Input:** A partial entrenchment ranking $\mathbf{B}$, a sentence $\alpha$, and an integer $i$ which represents the new desired rank of $\alpha$.

if $\alpha \in \mathrm{dom}(\mathbf{B})$ then degree_$\alpha := \mathbf{B}(\alpha)$
                else degree_$\alpha := \mathbf{degree}(\mathbf{B}, \alpha)$
**end if**
**max_degree** $= \max(\mathbf{B}(\mathrm{dom}(\mathbf{B})))$
**case of** degree_$\alpha$:
  greater than $i$:
    **movedown**$(\alpha, $ degree_$\alpha, i, \mathbf{B}, \mathbf{newB})$.
  lower than $i$:
    **if** degree_$\alpha = 0$
      **then** degree_not_$\alpha := \mathbf{degree}(\mathbf{B}, \neg\alpha)$
      **else** degree_not_$\alpha = 0$
    **end if**
    **if** degree_not_$\alpha > 0$ **then**
      **movedown**$(\alpha, $ degree_not_$\alpha, 0, \mathbf{B}, \mathbf{newB})$
    **end if**
    $\mathbf{B} := \mathbf{newB}$
    **moveup**$(\alpha, i, \mathbf{B}, \mathbf{newB})$
  otherwise:
    $\mathbf{newB} := \mathbf{B}$
**end case**
**Output:** The new partial entrenchment ranking **newB**.

**FUNCTION: degree**$(\mathbf{B}, \alpha)$
**Input:** A partial entrenchment ranking $\mathbf{B}$, and a sentence $\alpha$.

  degree := **max_degree**$+1$
  **do** degree := degree $-1$
  **until** $\{\beta : \mathrm{degree} \geq \mathbf{B}(\beta)\} \vdash \alpha$ or degree $= 0$
  **assign** degree$(\alpha, \mathbf{B}) := $ degree

**Output:** The degree of $\alpha$ in the ranking $\mathbf{B}$.

**PROCEDURE: movedown**$(\alpha, j, i, \mathbf{B}, \mathbf{newB})$
**Input:** A partial entrenchment ranking $\mathbf{B}$, a sentence $\alpha$, an integer $j$ representing the current degree of $\alpha$ in $\mathbf{B}$, and an integer $i$ representing the desired new degree of $\alpha$.

  **for** $k = $ **max_degree** down to $j + 1$ **do**
    $\mathbf{newB}(\beta) := k$
    for all sentences $\beta$ such that $\mathbf{B}(\beta) = k$.
  **for** $k = j$ down to $i$ **do**
    - generate minimal subsets of sentences at rank $k$ that together with $\mathrm{dom}(\mathbf{newB})$ entail $\alpha$
    - $\mathbf{newB}(\beta) := k$ for all $\beta$ not in any such minimal subset.

**Output:** The new partial entrenchment ranking **newB**.

**PROCEDURE: moveup**$(\alpha, i, \mathbf{B}, \mathbf{newB})$
**Input:** A partial entrenchment ranking $\mathbf{B}$, a sentence $\alpha$, a integer $i$ representing the desired new degree of $\alpha$.

  **for** $k = $ **max_degree** down to $i + 1$ **do**
    $\mathbf{newB}(\beta) := k$
    for all sentences $\beta$ such that $\mathbf{B}(\beta) = k$.
  $\mathbf{newB}(\alpha) := i$
  **for** $k = i - 1$ down to $1$ **do**
    $\mathbf{newB}(\beta) := \mathbf{degree}(\mathbf{newB}, \alpha \rightarrow \beta)$.

**Output:** The new partial entrenchment ranking **newB**.

*Subsumption removal* is an optional feature of the maxi-adjustment procedure. There are several ways to use it: (i) eliminate subsumed beliefs before the minimal subsets are generated, or (ii) eliminate subsumed beliefs from the minimal subsets after they have been found. The first approach will, in general, result in more beliefs being removed than the second, and will also reduce the number of minimal subset that must be generated. To incorporate the first approach (the second is equally obvious) we insert the following instruction between the first and second **for** loop constructs in the **movedown** procedure, so that subsumed sentences are removed before the minimal subsets are generated: **remove** sentences in $\mathrm{dom}(\mathbf{B})$ if $\mathbf{B}(\beta) = \mathbf{degree}(\alpha, \mathbf{B})$ and $\alpha \vdash \beta$. Then after the second **for** construct we add: **if** $i > 0$ **then** $\mathbf{newB}(\alpha) := i$

# 6 Complexity

The procedure **maxi-adjustment**$(B, \alpha, i, \textbf{newB})$ returns a revised ranking **newB** for any allocation of computational time. The longer the algorithm runs the closer the ranking **newB** approximates $\textbf{B}^\star(\alpha, i)$.

First-order logics satisfy the conditions required of the underlying logic given in section 2, hence the AGM framework and the proposed algorithm supports changes to rich knowledge bases, in principle. However, it is well known that satisfiability in first-order languages is undecidable, consequently nontrivial belief revision algorithms that are guaranteed to terminate cannot be constructed. The maxi-adjustment decision procedure described herein is an anytime algorithm, so it can be used to generate an infinite sequence of better and better approximations to revision and contraction functions.

As noted in the previous section the maxi-adjustment procedure essentially consists of two main phases.

The first phase is computationally easier than the second and if it is completed then we are guaranteed to satisfy the AGM postulates for revision and withdrawal, and notably we have enough information to construct a contraction function, if desired. It is interesting to note that for query evaluations, such as is $\beta \in \textbf{B}^\star(\alpha, i)$, only the first phase need be carried out.

The function **degree** is the workhorse of the first phase. Computing **degree**$(\textbf{B}, \alpha)$ is NP-hard. If a polynomial fragment of propositional logic is used, then computing **degree** is polynomial. Our anytime algorithm uses a top-down strategy that, if interrupted will always err on the side of overestimating the degree of sentences which in turn will never lead to inconsistency. A purely bottom up procedure would not exhibit this behaviour. However a hybrid strategy that combined a top-down and a bottom-up binary search would be more efficient on average, than a purely top-down linear search for the degree of $\alpha$, and if **B** has $n$ natural partitions then it requires $\lceil log_2\ n \rceil$ satisfiability checks [Lang, 1997]. An interpolation strategy that used information about ranking's history, or information available from the application at hand would also improve the performance of the function **degree**. Hybrid and informed techniques have been investigated in Lang [1997]. If we adopt a hybrid algorithm for **degree** in our anytime algorithm then should the program be interrupted we simply use the most recent upper bound to calculate the resultant ranking.

The first phase determines the most important core of the ranking to survive the change, the second phase refines it by maintaining as many other beliefs as possible based on the original ranking. The second stage of the maxi-adjustment is in $\Delta_2^P$, and hence solvable with a polynomial number of calls to an NP oracle (c.f. [Nebel 1991, Eiter and Gottlob 1992]). The worse case arises when all explicit beliefs are equally ranked. The computational cost decreases as the number of ranks increase. So the more discerning the agent the easier it is for him to modify his beliefs using a maxi-adjustment. This property concords with our intuition, i.e. it seems psychologically plausible, but not all revision strategies exhibit it. For example, ordinary adjustment which is based on the standard entrenchment construction does not.

# 7 Methodological Remarks

## 7.1 Contraposition

The use of maxi-adjustment presupposes that the knowledge engineer is able to identify reasons. The inability to identify reasons simply means that more information than is perhaps intended is retained in practice. If $\beta$ is a reason for $\alpha$ then $\beta \rightarrow \alpha$ should be placed higher in the ranking than $\alpha$. Using material implication in this way has an important ramification, namely *contraposition of reasons*, i.e. whenever $\alpha$ is a (whole) reason for $\beta$, then $\neg\beta$ is a (whole) reason for $\neg\alpha$. For example, if one of the reasons my hang-guilder is ascending is that the up-lift is sufficient to overcome the effects of gravity, then one of the reasons it is not ascending is that the up-lift is insufficient. Consequently, reasons *do not capture causality* in a broader context.

Another effect of contraposition is that if $\beta$ is a reason for $\alpha$, then when an agent revises by adding $\neg\alpha$ he will, if only implicitly, accept $\neg\beta$. Maxi-adjustment can also be used to model applications in which contraposition of reasons is not an appropriate assumption. This is achieved by breaking down the changes to the ranking into more primitive operations, and composing a *transaction* on the ranking!

## 7.2 Some Guidelines

Knowledge Engineers and System Designers are accustomed to the syntax sensitivity present in prevalent information modelling methods, such as (restricted) logical languages, entity-relationship models, conceptual graphs, etc. Methodologies have been developed for these traditional techniques: they guide the development process to a faithful, hopefully optimal, representation of the application at hand.

Maxi-adjustment is also syntax dependent, and whilst a methodology for using maxi-adjustments is not yet available the following application independent guidelines have helped in the development of several belief revision applications.

(1) Important information should be explicit.
(2) Information should be in its simplest logical form.
(3) The number of ranks should be maximised if incoming information is expected to be inconsistent with highly entrenched information.
(4) Conjunction can be used to bind information items together, e.g. if the application calls for $\alpha$ to be removed whenever $\beta$ is and vice versa, then $\alpha \wedge \beta$ can be used. If the conjuncts themselves are not explicit, or not derivable from other explicit beliefs then they will stand and fall together.
(5) Represent sentences that do not need to be bound as independent sentences, e.g. if $\alpha$ and $\beta$ are not related then using $\alpha$ and $\beta$ is preferable to the compound sentence $\alpha \wedge \beta$.
(6) Irredundant rankings are preferable.
(7) If the set $\{\beta_1, \beta_2, \ldots, \beta_n\}$ constitutes a reason for $\alpha$ (i.e. their simultaneous satisfaction would mean that $\alpha$ must hold), then the sentence $\beta_1 \wedge \beta_2 \wedge \ldots \wedge \beta_n \rightarrow \alpha$ is placed higher in the ranking than $\alpha$.
(8) Subsumption at the same rank should be avoided, and should only be used to satisfy guideline (1).

Guidelines 4, 5, and 6 are related to data normalisation in database design; a process used to transform a database into a representation that minimises update anomalies. As is commonly the case with sets of guidelines, there are exceptions and some guidelines may be in conflict with one another for a particular application. For example, following (1) may lead to a redundant ranking which clearly offends (6).

Methodologies will have to be developed to support the effective use of belief revision in real-world applications. Their development will be facilitated through experience with implemented prototype systems out in the field. Our aim is, not only, to develop a robust belief revision system but also to assist the user in making design choices. This is partially achieved in our system [Williams and Williams, 1997] by making as many of the consequences of a users ranking representation visible. For example beliefs are highlighted before being moved, changes can be stepped through, reasons can be queried for in advance (since they are determined by the current ranking), and rankings can be unwound and saved using *commit* and *rollback* mechanisms. Rankings can also be modified hypothetically during development and testing.

The anytime algorithm for maxi-adjustment is under development in the CIN Project: a project that is seeking to develop an Intelligent Information Management Toolkit. It offers a suite of sophisticated methods for default reasoning and belief revision. The system is currently founded on an objected-oriented design. The core of the belief revision system is implemented in $C^{++}$ using a state-of-the-art tableau theorem prover, and a Java based Graphical User Interface that provides facilities for *dropping* and *dragging* sentences up and down a ranking. Several rankings can be manipulated simultaneously.

## 8 Discussion

Iterated belief revision can be achieved by transmuting a partial entrenchment ranking. In this paper we described an anytime decision procedure for iterated belief revision.

We discussed the complexity of our anytime algorithm. In essence, it possesses two main phases: (i) determine the degree of acceptance for the information to be moved down the ranking, and (ii) remove minimal sets of sentences at each rank that entail the information to be moved down the ranking. The first phase is computationally simpler than the second, and if the first phase is completed before the algorithm is interrupted then we are guaranteed to be able to identify a theory base whose closure satisfies the AGM postulates for revision and withdrawal. Furthermore, the ranking so far contains enough information to construct a theory satisfying the postulates for contraction.

The proposed anytime algorithm has been used for several applications, and as a result of the experienced gained in using our system we were able to provide a domain independent protomethodology for developing belief revision applications based on maxi-adjustment.

## References

Alchourrón, C., Gardenfors, P., and Makinson, D. [1985], *On the logic of theory change: Partial meet functions for contraction and revision*, Journal of Symbolic Logic 50: 510 - 530.

Antoniou, G. and Williams, M.A. [1996], *Reasoning with Incomplete and Changing Information*, in N. Terashima and E. Altman (eds), *Advanced Information Technology Tools*, Chapman and Hill, 395 -401.

Dubois, D., Lang, J., and Prade, H. [1994], *Possibilistic Logic*, Handbook of Logic in Artificial Intelligence and Logic Programming, Gabbay, D., Hogger, C, and Robinson, J. (eds), Claredon Press, Oxford.

Eiter, T and Gottlob, G. [1992], *The Complexity of Propositional Knowledge Base Revision*, Artificial Intelligence 57: 227 - 270.

Gärdenfors, P. [1988], *Knowledge in Flux*, A Bradford Book, The MIT Press.

Gärdenfors, P., and Makinson, D. [1988], *Revisions of Knowledge Systems using Epistemic Entrenchment*, Proceedings of TARK, 83 – 96.

Gärdenfors, P. and Makinson, D. [1994], *Nonmonotonic Inference Based on Expectations*, Artificial Intelligence 65: 197 – 245.

Lang, J. [1997], *Possibilistic Logic: Algorithms and Complexity* in J. Kohlas and S. Moral (eds), *Handbook of Algorithms for Uncertainty and Defeasible Reasoning*, Kluwer Academic Publishers.

Lehmann, D. [1992], *What should a Knowledge Base Entail*, in Principles of Knowledge Representation and Reasoning, Morgan Kaufmann, 1992.

MacNish, C.K. and Williams M.A. [1996], *From Belief Revision to Design Revision: Applying Theory Change to Changing Requirements*, in Reasoning about Incomplete and Changing Information, Lecture Note Series, Springer Verlag, 1997.

Makinson, D. [1987], *On the Status of the Postulate of Recovery in the Logic of Theory Change*, Journal of Philosophical Logic, 16: 383 – 394.

Nebel, B., [1991] *Belief Revision and Default Reasoning*, in Principles of Knowledge Representation and Reasoning, Morgan Kaufmann, 417 - 428.

Rott, H. [1992] *Preferential Belief Change using Generalised Epistemic Entrenchment*, Journal of Logic, Language and Information, 1: 45 - 78.

Spohn, W., [1983] *Deterministic and Probabilistic Reasons*, Erkenntis 19: 371 – 396.

Williams M.A. [1992], *Two functions for theory base change*, in the Proceedings of the Australian Joint Conference on Artificial Intelligence, p259 – 265.

Williams, M.A. [1995] *Iterated Theory Base Change: A Computational Model*, IJCAI-95, Morgan Kaufmann, 1541 – 1550.

M.A. Williams, [1996] *A Practical Approach to Belief Revision: Reason-Based Change*, in L. Aiello, and S. Shapiro (eds), Principles of Knowledge Representation and Reasoning: Proceedings of the Fifth International Conference, Morgan Kaufmann, San Mateo, CA, 412 - 421, 1996.

Williams, M.A., and Williams, D.M., *A Belief Revision System for the World Wide Web*, IJCAI Workshop on Artificial Intelligence and the Internet, 1997. http://u2.newcastle.edu.au/webworld/papers/ai-internet.html

# AUTOMATED REASONING

Automated Reasoning 2: Belief Revision

# Towards Generalized Rule-based Updates

**Yan Zhang**
Department of Computing
University of Western Sydney, Nepean
Kingswood, NSW 2747, Australia
E-mail: yan@st.nepean.uws.edu.au

**Norman Y. Foo**
School of Computer Science and Engineering
University of New South Wales
NSW 2052, Australia
E-mail: norman@cse.unsw.edu.au

## Abstract

Recent work on rule-based updates provided new frameworks for updates in more general knowledge domains [Marek and Truszczński, 1994; Baral, 1994; Przymusinski and Turner, 1995]. In this paper, we consider a simple generalization of rule-based updates where incomplete knowledge bases are allowed and update rules may contain two types of negations. It turns out that previous methods cannot deal with this generalized rule-based update properly. To overcome the difficulty, we argue that necessary preferences between update rules and inertia rules must be taken into account in update specifications. From this motivation, we propose prioritized logic programs (PLPs) by adding preferences into extended logic programs [Gelfond and Lifschitz, 1991]. Formal semantics of PLPs is provided in terms of the answer set semantics of extended logic programs. We then show that the procedure of generalized rule-based update can be formalized in the framework of PLPs. The minimal change property of the update is also investigated.

## 1 Introduction

Marek and Truszczński's recent work on rule-based updates [Marek and Truszczński, 1994] provided a new framework for updates in more general knowledge domains. Generally, they addressed the following problem: given an initial *knowledge base* $\mathcal{B}$, i.e. a set of ground atoms, and a set of *update rules* $\mathcal{P}$ with the forms[1]:

$$in(A) \leftarrow in(B_1), \cdots, in(B_m), out(C_1), \cdots, out(C_n), \quad (1)$$

$$out(A) \leftarrow in(B_1), \cdots, in(B_m), out(C_1), \cdots, out(C_n), \quad (2)$$

[1] $\mathcal{P}$ was called a *revision program* in [Marek and Truszczński, 1994].

where $A, B_1, \cdots, B_m, C_1, \cdots, C_n$ are ground atoms, what is the resulting knowledge base $\mathcal{B}'$ after updating $\mathcal{B}$ by $\mathcal{P}$? The intuitive meaning of (1) (or (2)) is that if $B_1, \cdots, B_m$ are *in* the knowledge base, and $C_1, \cdots, C_n$ are *not in* the knowledge base, then $A$ should be (or not be) in the knowledge base.

For example, given a knowledge base $\mathcal{B} = \{A, D\}$ and a set of update rules $\mathcal{P} = \{in(C) \leftarrow in(A), out(B), out(D) \leftarrow in(C), out(B)\}$, where $A, B, C$ and $D$ are ground atoms, then after updating $\mathcal{B}$ by $\mathcal{P}$, according to Marek and Truszczński's approach, we would expect to have a resulting knowledge base $\mathcal{B}' = \{A, C\}$.

Relationships between rule-based updates and logic programming have been studied by Baral [Baral, 1994] and Przymusinski and Turner [Przymusinski and Turner, 1995]. In particular, they showed that Marek and Truszczński's formal procedure of specifying $\mathcal{B}'$ can be reduced to a computation of the answer set of a corresponding extended logic program (we will review this procedure in next section). However, there are two limitations with Marek and Truszczński's rule-based update: the initial knowledge base should be complete, i.e. any ground atom not in the initial knowledge base is treated as its negation; and update rules only contain classical negations. For instance, following ideas of [Baral, 1994] and [Przymusinski and Turner, 1995], rules (1) and (2) are translated into the following inference rules respectively in the corresponding extended logic program:

$$A \leftarrow B_1, \cdots, B_m, \neg C_1, \cdots, \neg C_n,$$
$$\neg A \leftarrow B_1, \cdots, B_m, \neg C_1, \cdots, \neg C_n.$$

In this paper, we consider a simple generalization of rule-based updates where a knowledge base can be *incomplete*, eg. a set of ground literals, and update rules have the following form:

$$L_0 \leftarrow L_1, \cdots, L_m, \text{not } L_{m+1}, \cdots, \text{not } L_n,$$

where each $L_i$ ($0 \leq i \leq n$) is a literal, and *not* represents *negation as failure* (or called weak negation). As a literal can be a negative atom, the above rule actually may contain two types of negations, i.e. classical and weak negations. The intuitive semantics of this rule can be interpreted as follows: if facts $L_1, \cdots, L_n$ are true in the knowledge base, and there are no explicit representations saying that facts $L_{m+1}, \cdots, L_n$ are true in the knowledge base, then fact $L_0$ should be true in the knowledge base.

Such generalized rule-based update is important in many applications. For example, in a secure computer system, a formal specification of users' access rights is usually required, and the access policy of the system can be represented by a knowledge base. An update of this knowledge base must be performed whenever new access control rules are applied to the system. Generally, two types of negations are needed to specify access control rules. Let $\mathcal{B} = \{Member(A,G), Member(B,G), Access(A,F), \neg Access(B,F)\}$ represent the current access policy of the system, where $Member(A,G)$ and $Member(B,G)$ mean that users $A$ and $B$ are members of group $G$, and $Access(A,F)$ and $\neg Access(B,F)$ indicate that user $A$ can access file $F$ and user $B$ cannot access file $F$ respectively. Suppose that new users $C$ and $D$ are added into group $G$ and a global access control rule is now applied to each member of group $G$:

$$Access(x,F) \leftarrow Member(x,G), not \neg Access(x,F).$$

This rule actually says that any user belonging to group $G$ can access file $F$ unless it is explicitly stated that the user is not allowed to access $F$. Updating $\mathcal{B}$ by $\mathcal{P} = \{Member(C,G) \leftarrow, Member(D,G) \leftarrow, Access(x,F) \leftarrow Member(x,G), not \neg Access(x,F)\}$, from our intuition, we would expect that $Access(C,F)$ and $Access(D,F)$ are obtained, while facts $Access(A,F)$ and $\neg Access(B,F)$ remain persistent.

As we will see next, previous rule-based update approaches are not suitable to deal with this generalized rule-based update properly. To overcome the difficulty, we argue that necessary preferences between update rules and inertia rules with respect to the update must be taken into account. From this motivation, we propose prioritized logic programs (PLPs) by adding preferences into extended logic programs. Formal semantics of PLPs is provided in terms of the answer set semantics of extended logic programs. We then show that the procedure of generalized rule-based update can be formalized in the framework of PLPs. The minimal change property of the update is also investigated.

## 2 A Motivating Example

In this section we first review the concept of extended logic programs proposed by Gelfond and Lifschitz [Gelfond and Lifschitz, 1991] and then discuss an example of generalized rule-based update in the framework of extended logic programs.

### 2.1 Preliminaries

A language $\mathcal{L}$ of extended logic programs is determined by its object constants, function constants and predicates constants. *Terms* are built as in the corresponding first order language; *atoms* have the form $P(t_1, \cdots, t_n)$, where $t_i$ $(1 \leq i \leq n)$ is a term and $P$ is a predicate symbol of arity $n$; a *literal* is either an atom $P(t_1, \cdots, t_n)$ or a negative atom $\neg P(t_1, \cdots, t_n)$. A *rule* is an expression of the form:

$$L_0 \leftarrow L_1, \cdots, L_m, not L_{m+1}, \cdots, not L_n, \quad (3)$$

where each $Li$ $(0 \leq i \leq n)$ is a literal. $L_0$ is called the *head* of the rule, while $L_1, \cdots, L_m, not\ L_{m+1}, \cdots, not\ L_n$ is called the *body* of the rule. Obviously, the body of a rule could be empty. A term, atom, literal, or rule is *ground* if no variable occurs in it. An *extended logic program* $\Pi$ is a collection of rules. The following is an example of extended logic program $\Pi_0$:

$$\neg Employed(x) \leftarrow Student(x),$$
$$\qquad not\ Employed(x),$$
$$Employed(x) \leftarrow Age(x, > 25), \neg Student(x),$$
$$\qquad not\ \neg Employed(x).$$

To evaluate a extended logic program, Gelfond and Lifschitz proposed the answer set semantics for extended logic programs. For simplicity, we treat a rule $r$ in $\Pi$ with variables as the set of all ground instances of $r$ formed from the set of ground literals of the language of $\Pi$. In the rest of paper, we will not explicitly declare this assumption whenever there is no ambiguity in our discussion.

Let $\Pi$ be an extended logic program not containing *not* and *Lit* the set of all ground literals in the language of $\Pi$. The *answer set* of $\Pi$, denoted as $Ans(\Pi)$, is the smallest subset $S$ of *Lit* such that

(i) for any rule $L_0 \leftarrow L_1, \cdots, L_m$ from $\Pi$, if $L_1, \cdots, L_m \in S$, then $L_0 \in S$;

(ii) if $S$ contains a pair of complementary literals, then $S = Lit$.

Now let $\Pi$ be an extended logic program. For any subset $S$ of *Lit*, let $\Pi^S$ be the logic program obtained from $\Pi$ by deleting

(i) each rule that has a formula *not L* in its body with $L \in S$, and

(ii) all formulas of the form *not L* in the bodies of the remaining rules.

We define that $S$ is an *answer set* of $\Pi$, denoted $Ans(\Pi)$, iff $S$ is an answer set of $\Pi^S$, i.e. $S = Ans(\Pi^S)$.

Consider an extended logic program $\Pi_1$ obtained from $\Pi_0$ by adding other two rules in $\Pi_0$: $\Pi_1 = \Pi_0 \cup \{\neg Student(Peter) \leftarrow, Age(Peter, > 25) \leftarrow\}$.

It is not difficult to see that $\Pi_1$ has a unique answer set set: $\{\neg Student(Peter), Age(Peter, > 25), Employed(Peter)\}$.

## 2.2 An Example

As we mentioned earlier, in previous formulations a specification of rule-based update can be represented by an extended logic program [Baral, 1994; Przymusinski and Turner, 1995]. By illustrating a simple example here, we will show that these methods are not suitable for specifying generalized rule-based updates.

**Example 1** Suppose $\mathcal{B} = \{\neg A, B, C\}$ is a knowledge base, and $\mathcal{P} = \{\neg B \leftarrow not\ B, A \leftarrow C\}$ is a set of update rules. Consider an update of $\mathcal{B}$ by $\mathcal{P}$. Obviously, fact $\neg A$ should change to $A$ by applying the second rule of $\mathcal{P}$. Fact $B$, on the other hand, seems persistent because $B$ is true in the initial knowledge base, and $\neg B$ can only be derived from the first rule of $\mathcal{P}$ if fact $B$ is absent from the current knowledge base. Therefore, from our intuition, the resulting knowledge base should be $\{A, B, C\}$.

Now we follow the principle of Baral and Przymusinski and Turner's methods [Baral, 1994; Przymusinski and Turner, 1995] to specify the above update procedure within an extended logic program[2]. Firstly, we need to extend the language of our domain by adding new propositional letters with the form $New\text{-}L$ if $L$ is a propositional letter in the original language[3]. In this example, the extended language will include propositional letters $A, B, C, New\text{-}A, New\text{-}B$ and $New\text{-}C$. Then an extended logic program $\Pi(\mathcal{B}, \mathcal{P})$ is formed by the following rules:

*Initial knowledge rules:*
$\quad \neg A \leftarrow,$
$\quad B \leftarrow,$
$\quad C \leftarrow,$

*Inertia rules:*
$\quad New\text{-}A \leftarrow A, not\ \neg New\text{-}A,$
$\quad New\text{-}B \leftarrow B, not\ \neg New\text{-}B,$
$\quad New\text{-}C \leftarrow C, not\ \neg New\text{-}C,$
$\quad \neg New\text{-}A \leftarrow \neg A, not\ New\text{-}A,$
$\quad \neg New\text{-}B \leftarrow \neg B, not\ New\text{-}B,$
$\quad \neg New\text{-}C \leftarrow \neg C, not\ New\text{-}C,$

*Update rules:*
$\quad \neg New\text{-}B \leftarrow not\ New\text{-}B,$
$\quad New\text{-}A \leftarrow New\text{-}C.$

Generally speaking, an answer set of program $\Pi(\mathcal{B}, \mathcal{P})$ represents a possible resulting knowledge base after updating $\mathcal{B}$ by $\mathcal{P}$, where literal $New\text{-}L$ in the answer set

---
[2]The formalism used here, of course, is different from theirs.

[3]For simplicity, here we restrict the language to be propositional.

denotes the persistence of literal $L$ if $L \in \mathcal{B}$, or a change of $L$ if $\neg L \in \mathcal{B}$ or $L \notin \mathcal{B}$ with respect to this update.

It is easy to see that the above $\Pi(\mathcal{B}, \mathcal{P})$ has two answer sets: in one $New\text{-}B$ is true where in the other $New\text{-}B$ is false. Obviously this solution is not consistent with our previous observation. ∎

Observing program $\Pi(\mathcal{B}, \mathcal{P})$, it is not difficult to see that a conflict occurs between inertia rule $New\text{-}B \leftarrow B, not\ \neg New\text{-}B$ and update rule $\neg New\text{-}B \leftarrow not\ New\text{-}B$, that is, applying $New\text{-}B \leftarrow B, not\ \neg New\text{-}B$ will defeat $\neg New\text{-}B \leftarrow not\ New\text{-}B$, and *vice versa*. This conflict leads $\Pi(\mathcal{B}, \mathcal{P})$ to have two different answer sets with an indefiniteness of $New\text{-}B$.

On the other hand, it seems that the inertia rule $New\text{-}B \leftarrow B, not\ \neg New\text{-}B$ should *override* the update rule $\neg New\text{-}B \leftarrow not\ New\text{-}B$ during the evaluation of $\Pi(\mathcal{B}, \mathcal{P})$ in order to obtain the desired solution. But this preference information cannot be expressed in Gelfond and Lifschitz's extended logic programs.

From the above discussion, we argue that to represent such generalized rule-based update properly, necessary preferences between inertia rules and update rules have to be taken into account during the evaluation of the update. We approach this problem from a general ground: we will first propose prioritized logic programs where preferences between rules can be explicitly expressed, and then formalize the generalized rule-based update in the framework of prioritized logic programs.

## 3 Prioritized Logic Programs (PLPs)

In this section we propose prioritized logic programs (PLPs) which extend Gelfond and Lifschitz's extended logic programs [Gelfond and Lifschitz, 1991] by adding preference information into programs. We first describe the syntax of PLPs and then provide an answer set semantics for PLPs.

### 3.1 Syntax

The language $\mathcal{L}^P$ of PLPs is a language $\mathcal{L}$ of extended logic programs just with the following augments:

- *Names:* $N, N_1, N_2, \cdots$.
- A strict partial ordering (i.e. antireflexive, antisymmetric and transitive) $<$ on names.
- A naming function $\mathcal{N}$, which maps a rule to a name.

Terms, atoms, literals and rules in PLPs are defined as the same in extended logic programs. For the naming function $\mathcal{N}$, we require that for any rules $r$ and $r'$ in a PLP (see the following definition), $\mathcal{N}(r) = \mathcal{N}(r')$ iff $r$ and $r'$ indicate the same rule.

A *prioritized logic program* (PLP) $\mathcal{P}$ is a triplet $(\Pi, \mathcal{N}, <)$, where $\Pi$ is an extended logic program, $\mathcal{N}$ is a naming function mapping each rule in $\Pi$ to a name, and

$<$ is a relation representing all strict partial orderings on names.

The following is an example of prioritized extended logic program.
$\mathcal{P}_1 = (\{P \leftarrow not\ Q, not\ R, Q \leftarrow not\ P, R \leftarrow not\ P\}$, $\{\mathcal{N}(P \leftarrow not\ Q, not\ R) = N_1, \mathcal{N}(Q \leftarrow not\ P) = N_2$, $\mathcal{N}(R \leftarrow not\ P) = N_3\}, \{N_1 < N_2, N_2 < N_3\})$. To simplify our presentation, we usually represent $\mathcal{P}_1$ as the following form:

$\mathcal{P}_1$:
$N_1 : P \leftarrow not\ Q, not\ R,$
$N_2 : Q \leftarrow not\ P,$
$N_3 : R \leftarrow not\ P,$
$N_1 < N_2, N_2 < N_3.$

We also use notations $\mathcal{P}_1(\Pi)$, $\mathcal{P}_1(\mathcal{N})$, and $\mathcal{P}_1(<)$ to denote the sets of rules, naming function's values and $<$-relation of $\mathcal{P}_1$ respectively.

Consider the following program:

$\mathcal{P}_2$:
$N_1 : P \leftarrow not\ Q, not\ R,$
$N_2 : Q \leftarrow not\ P,$
$N_3 : R \leftarrow not\ P,$
$N_1 < N_2, N_2 < N_3, N_1 < N_3.$

Obviously, the only difference between $\mathcal{P}_1$ and $\mathcal{P}_2$ is that there is one more relation $N_1 < N_3$ in $\mathcal{P}_2$. As we mentioned earlier, $<$ is a strict partial ordering (i.e., antireflexive, antisymmetric and transitive), we would expect that $\mathcal{P}_1$ and $\mathcal{P}_2$ are identical in some sense. Furthermore, if we rename rules in $\mathcal{P}_2$ as follows,

$\mathcal{P}'_2$:
$N'_1 : P \leftarrow not\ Q, not\ R,$
$N'_2 : Q \leftarrow not\ P,$
$N'_3 : R \leftarrow not\ P,$
$N'_1 < N'_2, N'_2 < N'_3, N'_1 < N'_3,$

$\mathcal{P}'_2$ would be also identical to $\mathcal{P}_2$ and hence to $\mathcal{P}_1$ too from our intuition. To make this precise, we first introduce $<$-closure as follows.

**Definition 1** *Given a program $\mathcal{P} = (\Pi, \mathcal{N}, <)$. $\mathcal{P}(<^+)$ is the $<$-closure of $\mathcal{P}$ iff $\mathcal{P}(<^+)$ is the smallest set containing $\mathcal{P}(<)$ and closed under transitivity.*

We also need to define a renaming function as follows. A *renaming function* $Rn$ maps a PLP $\mathcal{P} = (\Pi, \mathcal{N}, <)$ to another PLP $\mathcal{P}'$, i.e. $Rn(\mathcal{P}) = \mathcal{P}' = (\Pi', \mathcal{N}', <')$, such that (i) $\mathcal{P}(\Pi) = \mathcal{P}'(\Pi')$; (ii) for each rule $r \in \mathcal{P}(\Pi)$[4], $\mathcal{N}(r) = N \in \mathcal{P}(\mathcal{N})$ iff $\mathcal{N}'(r) = N' \in \mathcal{P}'(\mathcal{N}')$ ($N$ and $N'$ are not necessarily different); (iii) for any rules $r_1$ and $r_2$ in $\mathcal{P}(\Pi)$, $\mathcal{N}(r_1) = N_1, \mathcal{N}(r_2) = N_2 \in \mathcal{P}(\mathcal{N})$, and $N_1 < N_2 \in \mathcal{P}(<)$ iff $\mathcal{N}'(r_1) = N'_1, \mathcal{N}'(r_2) = N'_2 \in \mathcal{P}'(\mathcal{N}')$, and $N'_1 < N'_2 \in \mathcal{P}'(<')$. It is easy to see that applying a renaming function to a PLP will only change the names of rules in the PLP.

Two prioritized extended logic programs $\mathcal{P}_1$ and $\mathcal{P}_2$ are *identical* iff there exists a renaming function $Rn$, mapping $\mathcal{P}_2$ to $\mathcal{P}'_2$ such that $\mathcal{P}_1(\Pi) = \mathcal{P}'_2(\Pi')$, $\mathcal{P}_1(\mathcal{N}) = \mathcal{P}'_2(\mathcal{N}')$, and $\mathcal{P}_1(<^+) = \mathcal{P}'_2(<'^+)$.

We have defined that a prioritized extended logic program is an extended logic program by associating with a partial ordering $<$ to it. Intuitively such ordering represents a preference of applying rules during the evaluation of a query of the program. In particular, if in a program $\mathcal{P}$, relation $\mathcal{N}(r) < \mathcal{N}(r')$ holds, rule $r$ would be preferred to apply over rule $r'$ during the evaluation of $\mathcal{P}$ (i.e. rule $r$ is more preferred than rule $r'$). Consider the following classical example represented in our formalism:

$\mathcal{P}_3$:
$N_1 : Fly(x) \leftarrow Bird(x), not\ \neg Fly(x),$
$N_2 : \neg Fly(x) \leftarrow Penguin(x), not\ Fly(x),$
$N_3 : Bird(Tweety) \leftarrow,$
$N_4 : Penguin(Tweety) \leftarrow,$
$N_2 < N_1.$

Obviously, rules $N_1$ and $N_2$ conflict with each other as their heads are complementary literals[5], and applying $N_1$ will defeat $N_2$ and *vice versa*. However, as $N_2 < N_1$, we would expect that rule $N_2$ is preferred to apply first and then defeat rule $N_1$ after applying $N_2$ so that the desired solution $\neg Fly(Tweety)$ could be derived.

### 3.2 Answer Sets for PLPs

Now we are ready to provide the semantics of PLPs. The semantics of PLPs is defined in terms of the answer set semantics of extended logic programs described earlier.

In program $\mathcal{P}_3$, we have seen that rules $N_1$ and $N_2$ conflict with each other. Since $N_2 < N_1$, we try to solve the conflict by applying $N_2$ first and defeating $N_1$. However, in some programs, even if one rule is more preferred than the other, these two rules may not affect each other at all during the evaluation of the program. In this case, the preference relation between these two rules does not play any role in the evaluation and should be simply ignored. This is illustrated by the following program:

$\mathcal{P}_4$:
$N_1 : P \leftarrow not\ Q_1,$
$N_2 : \neg P \leftarrow not\ Q_2,$
$N_1 < N_2.$

Although heads of $N_1$ and $N_2$ are complementary literals, applying $N_1$ will not affect the applicability of $N_2$ and *vice versa*. Hence $N_1 < N_2$ should not be taken into account during the evaluation of $\mathcal{P}_4$. The following definition provides a formal description for this intuition.

---

[4]Of course, $r$ is also in $\mathcal{P}'(\Pi')$.

[5]Precisely, $N_2$ is the name of rule $\neg Fly(x) \leftarrow Penguin(x), not\ Fly(x)$. Whenever there is no confusion in the context, we just simply refer a rule by its name.

**Definition 2** Let $\Pi$ be an extended logic program and $r$ a rule with the form $L_0 \leftarrow L_1, \cdots, L_m, \text{not } L_{m+1}, \cdots, \text{not } L_n$ ($r$ does not necessarily belong to $\Pi$). Rule $r$ is defeated by $\Pi$ iff for any answer set $Ans(\Pi)$ of $\Pi$, there exists some $L_i \in Ans(\Pi)$, where $m+1 \leq i \leq n$.

Now our idea of evaluating a PLP is described as follows. Let $\mathcal{P} = (\Pi, \mathcal{N}, <)$. If there are two rules $r$ and $r'$ in $\mathcal{P}(\Pi)$ and $\mathcal{N}(r) < \mathcal{N}(r')$, $r'$ will be ignored in the evaluation of $\mathcal{P}$, only if keeping $r$ in $\mathcal{P}(\Pi)$ and deleting $r'$ from $\mathcal{P}(\Pi)$ will result in a defeat of $r'$, i.e. $r'$ is defeated by $\mathcal{P}(\Pi) - \{r'\}$. By eliminating all such potential rules from $\mathcal{P}(\Pi)$, $\mathcal{P}$ is eventually reduced to an extended logic program in which the partial ordering $<$ has been removed. Our evaluation for $\mathcal{P}$ is then based on this extended logic program.

Let us consider program $\mathcal{P}_3$ once again. Since $N_2 < N_1$ and $N_1$ is defeated by $\mathcal{P}_3 - \{N_1\}$ (i.e. the unique answer set of $\mathcal{P}_3 - \{N_1\}$ is $\{Bird(Tweety), Penguin(Tweety), \neg Fly(Tweety)\}$), rule $N_1$ should be ignored during the evaluation of $\mathcal{P}_3$. For program $\mathcal{P}_4$, on the other hand, although $N_1 < N_2$, relation $N_1 < N_2$ will not affect the solution of evaluating $\mathcal{P}_4$ as $\mathcal{P}_4(\Pi) - \{N_2\}$ does not defeat $N_2$ (i.e. the unique answer set of $\mathcal{P}_4(\Pi) - \{N_2\}$ is $\{P\}$).

**Definition 3** Let $\mathcal{P} = (\Pi, \mathcal{N}, <)$ be a prioritized extended logic program. We define a reduct of $\mathcal{P}$ with respect to $<$, denoted as $\mathcal{P}^<$, as follows.

(i) $\Pi_0 = \Pi$;

(ii) $\Pi_i = \Pi_{i-1} - \{r_1, \cdots, r_k \mid \text{there exists } r \in \Pi_{i-1} \text{ such that } \mathcal{N}(r) < \mathcal{N}(r_i) \in \mathcal{P}(<^+) \ (i = 1, \cdots, k) \text{ and } r_1, \cdots, r_k \text{ are defeated by } \Pi_{i-1} - \{r_1, \cdots, r_k\}\}$;

(iii) $\mathcal{P}^< = \bigcap_{i=0}^{\infty} \Pi_i$.

In above definition, clearly $\mathcal{P}^<$ is an extended logic program obtained from $\Pi$ by eliminating some rules from $\Pi$. In particular, if $\mathcal{N}(r) < \mathcal{N}(r')$ and $\Pi - \{r'\}$ defeats $r'$, rule $r'$ is eliminated from $\Pi$. This procedure is continued until a fixed point is reached. Note that due to the transitivity of $<$, we need to consider each $\mathcal{N}(r) < \mathcal{N}(r')$ in the $<$-closure of $\mathcal{P}$. It is also not difficult to note that the reduct of a PLP may not be unique generally.

**Example 2** Using Definition 1 and 3, it is not difficult to conclude that $\mathcal{P}_1$, $\mathcal{P}_3$ and $\mathcal{P}_4$ have unique reducts as follows respectively:

$\mathcal{P}_0^< = \{P \leftarrow \text{not } Q\}$,
$\mathcal{P}_1^< = \{P \leftarrow \text{not } Q, \text{not } R\}$,
$\mathcal{P}_3^< = \{\neg Fly(x) \leftarrow Penguin(x), \text{not } Fly(x),$
$\quad\quad Bird(Tweety) \leftarrow, Penguin(Tweety) \leftarrow\}$,
$\mathcal{P}_4^< = \mathcal{P}_4(\Pi)$.

∎

Now it is quite straightforward to define the answer set for a prioritized extended logic program.

**Definition 4** Let $\mathcal{P} = (\Pi, \mathcal{N}, <)$ be a PLP and Lit the set of all ground literals in the language of $\mathcal{P}$. For any subset $S$ of Lit, $S$ is an answer set of $\mathcal{P}$, denoted as $Ans^P(\mathcal{P})$, iff $S = Ans(\mathcal{P}^<)$.

**Example 3** Immediately from Definition 4 and Example 2, we have the following solutions:

$Ans^P(\mathcal{P}_0) = \{P\}$,
$Ans^P(\mathcal{P}_1) = \{P\}$,
$Ans^P(\mathcal{P}_3) = \{Bird(Tweety),$
$\quad\quad Penguin(Tweety), \neg Fly(Tweety)\}$,
$Ans^P(\mathcal{P}_4) = Lit$,

which, respectively, are also consistent with our intuitions. ∎

### 3.3 Basic Properties of PLPs

We now discuss some properties of PLPs. To simplify our presentation, let us introduce some useful notations. Let $\Pi$ and $\mathcal{P}$ be an extended logic program and a PLP respectively. We use $ANS(\Pi)$ to denote the classes of answer sets of $\Pi$. Suppose $\mathcal{P} = (\Pi, \mathcal{N}, <)$ is a PLP. From Definition 3, we can see that a reduct $\mathcal{P}^<$ of $\mathcal{P}$ is generated from a sequence of extended logic programs: $\Pi = \Pi_0, \Pi_1, \Pi_2, \cdots$. We use notation $\{\Pi_i\}$ $(i = 0, 1, 2, \cdots)$ to denote this sequence and call it a reduct chain of $\mathcal{P}$. Then we can prove the following useful solutions[6].

**Theorem 1** Let $\mathcal{P} = (\Pi, \mathcal{N}, <)$ be a PLP, and $\{\Pi_i\}$ $(i = 0, 1, 2, \cdots)$ a reduct chain of $\mathcal{P}$. Suppose each $\Pi_i$ has answer set(s). Then for any $i$ and $j$ where $i < j$, $ANS(\Pi_j) \subseteq ANS(\Pi_i)$.

**Theorem 2** Let $\mathcal{P} = (\Pi, \mathcal{N}, <)$ be a PLP. Then a subset $S$ of Lit is an answer set of $\mathcal{P}$ iff $S$ is an answer set of each $\Pi_i$ for some reduct chain $\{\Pi_i\}$ $(i = 0, 1, 2, \cdots)$ of $\mathcal{P}$, where each $\Pi_i$ has answer set(s).

## 4 Generalized Rule-based Update

Consider a language $\mathcal{L}$ of extended logic programs as described in section 2.1. We specify that a knowledge base $\mathcal{B}$ is a set of ground literals of $\mathcal{L}$ and $\mathcal{P}$ is a set of rules of $\mathcal{L}$ with form (3) that are called update rules. Note that we allow a knowledge base to be incomplete. That is, a literal not in a knowledge base is treated as unknown.

We will use a prioritized logic program to specify an update of $\mathcal{B}$ by $\mathcal{P}$. For this purpose, we first need to extend language $\mathcal{L}$ by the following way. We specify $\mathcal{L}_{new}^P$ to be a language of PLPs based on $\mathcal{L}$ as described in section 3.1 with one more augment: For each predicate symbol $P$ in $\mathcal{L}$, there is a corresponding predicate symbol $New\text{-}P$ in $\mathcal{L}_{new}^P$ with the same arity of $P$.

---

[6] All proofs of theorems presented in this paper were given in our manuscript [Zhang and Foo, 1997].

To simplifying our presentation, in $\mathcal{L}_{new}^P$ we use notation $New\text{-}L$ to denote the corresponding literal $L$ in $\mathcal{L}$. For instance, if a literal $L$ in $\mathcal{L}$ is $\neg P(x)$, then notation $New\text{-}L$ simply means $\neg New\text{-}P(x)$. We use $Lit_{new}$ to denote the set of all ground literals of $\mathcal{L}_{new}^P$. Clearly, $Lit_{new} = Lit \cup \{New\text{-}L \mid L \in Lit\}$. Now we are ready to formalize our generalized rule-based update.

**Definition 5** *Let $\mathcal{B}$, $\mathcal{P}$, $\mathcal{L}$ and $\mathcal{L}_{new}^P$ be defined as above. The specification of updating $\mathcal{B}$ by $\mathcal{P}$ is defined as a PLP of $\mathcal{L}_{new}^P$, denoted as $Update(\mathcal{B}, \mathcal{P}) = (\Pi(\mathcal{B},\mathcal{P}), \mathcal{N}, <)$, as follows:*

1. *$\Pi(\mathcal{B}, \mathcal{P})$ consists of following rules:*
   *Initial knowledge rules: for each $L$ in $\mathcal{B}$, there is a rule $L \leftarrow$;*
   *Inertia rules: for each predicate symbol $P$ in $\mathcal{L}$, there are two rules:*
   $New\text{-}P(x) \leftarrow P(x), not\ \neg New\text{-}P(x)$[7]*, and*
   $\neg New\text{-}P(x) \leftarrow \neg P(x), not\ New\text{-}P(x),$
   *Update rules: for each rule*
   $L_0 \leftarrow L_1, \cdots, L_m, not\ L_{m+1}, \cdots, not\ L_n$ *in $\mathcal{P}$, there is a rule*
   $New\text{-}L_0 \leftarrow New\text{-}L_1, \cdots, New\text{-}L_m,$
   $\qquad not\ New\text{-}L_{m+1}, \cdots, not\ New\text{-}L_n$;

2. *Naming function $\mathcal{N}$ assigns a unique name $N$ for each rule in $\Pi(\mathcal{B}, \mathcal{P})$;*

3. *For any inertia rule with name $N$ and update rule with name $N'$, there is a partial ordering between $N$ and $N'$: $N < N'$.*

Comparing Definition 5 with the update specification described in Example 1, we can see that the difference between these two approaches is that in our formulation preference relations between inertia and update rules are explicitly expressed. We specify inertia rules to be more preferred than update rules in $Update(\mathcal{B}, \mathcal{P})$.

The intuitive idea behind this is that a preference ordering between an inertia rule and an update rule in $Update(\mathcal{B}, \mathcal{P})$ will affect the evaluation of $Update(\mathcal{B}, \mathcal{P})$ *only if* these two rules conflict with each other, eg. applying one rule causes the other inapplicable. On the other hand, a fact in the initial knowledge based $\mathcal{B}$ is always preferred to persist during an update whenever there is no violation of update rules[8]. Therefore, when conflicts occur between inertia and update rules, inertia rules should override the corresponding update rules. Otherwise, the preference ordering does not play any role in the evaluation of $Update(\mathcal{B}, \mathcal{P})$. Also note that there will be at most $2k \cdot l$ instances of relation $<$ in $Update(\mathcal{B}, \mathcal{P})$, where $k$ is the number of predicate symbols of $\mathcal{L}$ and $l$ is the number of update rules in $\mathcal{P}$.

[7] $x$ might be a tuple of variables.
[8] Note that an update rule in $Update(\mathcal{B}, \mathcal{P})$ is defeasible if it contains a weak negation *not* in the body.

Finally, on the basis of Definition 5, we can formally define a knowledge base $\mathcal{B}'$ resulting from updating $\mathcal{B}$ by $\mathcal{P}$ in a straightforward way.

**Definition 6** *Let $\mathcal{B}, \mathcal{P}, \mathcal{L}$ and $\mathcal{L}_{new}^P$ be specified as before, and $Update(\mathcal{B}, \mathcal{P})$ the specification of updating $\mathcal{B}$ by $\mathcal{P}$ as defined in Definition 5. A set of ground literals of $\mathcal{L}$, $\mathcal{B}'$, is called a possible resulting knowledge base with respect to the update specification $Update(\mathcal{B}, \mathcal{P})$, iff $\mathcal{B}'$ satisfies the following conditions:*

1. *if $Update(\mathcal{B}, \mathcal{P})$ does not have an answer set, then $\mathcal{B}' = \mathcal{B}$;*

2. *if $Update(\mathcal{B}, \mathcal{P})$ has a consistent answer set, say $Ans^P(Update(\mathcal{B}, \mathcal{P}))$, then*
   $\mathcal{B}' = \{L \mid New\text{-}L \in Ans^P(Update(\mathcal{B}, \mathcal{P}))\};$

3. *$\mathcal{B}' = Lit$ if $Ans^P(Update(\mathcal{B}, \mathcal{P})) = Lit_{new}$.*

**Example 4** *Example 1 continued. Let $\mathcal{B} = \{\neg A, B, C\}$ and $\mathcal{P} = \{\neg B \leftarrow not\ B,\ A \leftarrow C\}$. From Definition 5, the specification of updating $\mathcal{B}$ by $\mathcal{P}$, $Update(\mathcal{B}, \mathcal{P})$, is as follows:*

*Initial knowledge rules:*
$N_1: \neg A \leftarrow,$
$N_2: B \leftarrow,$
$N_3: C \leftarrow,$
*Inertia rules:*
$N_4: New\text{-}A \leftarrow A, not\ \neg New\text{-}A,$
$N_5: New\text{-}B \leftarrow B, not\ \neg New\text{-}B,$
$N_6: New\text{-}C \leftarrow C, not\ \neg New\text{-}C,$
$N_7: \neg New\text{-}A \leftarrow \neg A, not\ New\text{-}A,$
$N_8: \neg New\text{-}B \leftarrow \neg B, not\ New\text{-}B,$
$N_9: \neg New\text{-}C \leftarrow \neg C, not\ New\text{-}C,$
*Update rules:*
$N_{10}: \neg New\text{-}B \leftarrow not\ New\text{-}B,$
$N_{11}: New\text{-}A \leftarrow New\text{-}C,$
$<:$
$N_4 < N_{10}, N_5 < N_{10}, N_6 < N_{10},$
$N_7 < N_{10}, N_8 < N_{10}, N_9 < N_{10},$
$N_4 < N_{11}, N_5 < N_{11}, N_6 < N_{11},$
$N_7 < N_{11}, N_8 < N_{11}, N_9 < N_{11}.$

*Now from Definitions 3 and 4, it is not difficult to see that $Update(\mathcal{B}, \mathcal{P})$ has a unique answer set: $\{\neg A, B, C, New\text{-}A, New\text{-}B, New\text{-}C\}$. Note that in $Update(\mathcal{B}, \mathcal{P})$, only ordering $N_5 < N_{10}$ is used in $Update(\mathcal{B}, \mathcal{P})$'s evaluation, while other orderings are useless (see Definition 3). Hence, from Definition 6, the only resulting knowledge base $\mathcal{B}'$ after updating $\mathcal{B}$ by $\mathcal{P}$ is: $\{A, B, C\}$* ∎

**Example 5** *Let us consider the secure computer system domain described in section 1 again. Let*

$\mathcal{B} = \{Member(A, G), Member(B, G),$
$\qquad Access(A, F), \neg Access(B, F)\}$ *and*
$\mathcal{P} = \{Member(C, G) \leftarrow, Member(D, G) \leftarrow,$

$$Access(x, F) \leftarrow Member(x, G),$$
$$not \neg Access(x, F)\}.$$

Consider the update of $\mathcal{B}$ by $\mathcal{P}$. Ignoring the detail, using the approach presented above, we get a unique resulting knowledge base

$$\mathcal{B}' = \{Member(A, G), Member(B, G),$$
$$Member(C, G), Member(D, G),$$
$$Access(A, F), \neg Access(B, F),$$
$$Access(C, F), Access(D, F)\}.$$

■

## 5  Update and Minimal Change

In this section we investigate the minimal change property for the generalized rule-based update described previously. Let $\mathcal{B}$ be a consistent knowledge base and $r$ a rule with the form (3). $\mathcal{B}$ *satisfies* $r$ iff if facts $L_1, \cdots, L_m$ are in $\mathcal{B}$ and fact $L_{m+1}, \cdots, L_n$ are not in $\mathcal{B}$, then fact $L_0$ is in $\mathcal{B}$. Let $\mathcal{P}$ be a set of rules with the form (3). $\mathcal{B}$ satisfies $\mathcal{P}$ if $\mathcal{B}$ satisfies each rule in $\mathcal{P}$.

Let $\mathcal{B}$ and $\mathcal{B}'$ be two knowledge bases. We use $Diff(\mathcal{B}, \mathcal{B}')$ to denote the set of different ground atoms between $\mathcal{B}$ and $\mathcal{B}'$, i.e.

$$Diff(\mathcal{B}, \mathcal{B}') = \{|L| \mid L \in (\mathcal{B} - \mathcal{B}') \cup (\mathcal{B}' - \mathcal{B})\},$$

where notation $|L|$ indicates the corresponding ground atom of ground literal $L$, and $Min(\mathcal{B}, \mathcal{P})$ to denote the set of all consistent knowledge bases satisfying $\mathcal{P}$ but with minimal differences from $\mathcal{B}$, i.e.

$$Min(\mathcal{B}, \mathcal{P}) = \{\mathcal{B}' \mid \mathcal{B}' \text{ satisfies } \mathcal{P} \text{ and}$$
$$Diff(\mathcal{B}, \mathcal{B}') \text{ is minimal with}$$
$$\text{respect to set inclusion}\}.$$

Then we have the following result.

**Theorem 3** *Let $\mathcal{B}$ be a knowledge base, $\mathcal{P}$ a set of update rules, and $Update(\mathcal{B}, \mathcal{P})$ the specification of updating $\mathcal{B}$ by $\mathcal{P}$ as defined in Definition 5. If $\mathcal{B}'$ is a consistent resulting knowledge base with respect to the update specification $Update(\mathcal{B}, \mathcal{P})$, then $\mathcal{B}' \in Min(\mathcal{B}, \mathcal{P})$.*

The above theorem guarantees that our generalized rule-based update satisfies the principle of minimal change. However, it should be noted that not every element of $Min(\mathcal{B}, \mathcal{P})$ could be a resulting knowledge base of updating $\mathcal{B}$ by $\mathcal{P}$. The following example illustrates the case.

**Example 6** Let $\mathcal{B} = \{A, B\}$ and $\mathcal{P} = \{\neg A \leftarrow B, \neg B \leftarrow not\ B\}$. Using our approach, updating $\mathcal{B}$ by $\mathcal{P}$ will give us a unique resulting knowledge base $\mathcal{B}' = \{\neg A, B\}$. However, it is not difficult to see that $Min(\mathcal{B}, \mathcal{P})$ also contains another element: $\mathcal{B}'' = \{A, \neg B\}$. Although $\mathcal{B}''$ satisfies $\mathcal{P}$ and has minimal difference from $\mathcal{B}$, it does not represent an intuitive solution of updating $\mathcal{B}$ by $\mathcal{P}$ according to our previous discussion. ■

## 6  Concluding Remarks

In this paper we considered generalized rule-based updates in the framework of prioritized logic programs. We should mention that Marek and Truszczński's rule-based update is embeddable into our framework. For example, if our update rules only contain classical negations, our formulation is reduced to Przymusinski and Turner's described in [Przymusinski and Turner, 1995].

Finally, we have noticed that the issue of logic programs with preferences has also been explored recently by some other researchers (eg. [Brewka, 1996]). In fact, we can further extend our prioritized logic programs by associating with *dynamic preference* so that our prioritized logic programs can be used as a more general tool in broad areas of knowledge representation and reasoning. A detailed comparison between our PLPs and others' work and other applications of PLPs in reasoning about change is beyond the scope of this paper and was represented in our full version manuscript [Zhang and Foo, 1997].

## Acknowledgements

This research is supported in part by a grant from the Australian Research Council. We thank to Chitta Baral for many valuable comments on an earlier draft of this paper.

## References

[Baral, 1994] C. Baral. Rule based updates on simple knowledge bases. In *Proceedings of the Eleventh National Conference on Artificial Intelligence (AAAI'94)*, pages 136–141. AAAI Press, 1994.

[Brewka, 1996] G. Brewka. Well-founded semantics for extended logic programs with dynamic preferences. *Journal of Artificial Intelligence Research*, 14:19–36, 1996.

[Gelfond and Lifschitz, 1991] M. Gelfond and V. Lifschitz. Classical negation in logic programs and disjunctive databases. *New Generation Computing*, 9:365–386, 1991.

[Marek and Truszczński, 1994] M. Marek and M. Truszczński. Update by means of inference rules. In *Proceedings of JELIA'94, Lecture Notes in Artificial Intelligence*, 1994.

[Przymusinski and Turner, 1995] T.C. Przymusinski and H. Turner. Update by means of inference rules. In *Proceedings of LPNMR'95, Lecture Notes in Artificial Intelligence*, pages 156–174, 1995.

[Zhang and Foo, 1997] Y. Zhang and N.Y. Foo. Prioritized logic programming and reasoning about change. Technical report, University of Western Sydney, Nepean, 1997.

# Representation Theorems for Multiple Belief Changes

Dongmo Zhang[1,2] Shifu Chen[1] Wujia Zhu[1,2] Zhaoqian Chen[1]

[1]State Key Lab. for Novel Software Technology
Department of Computer Science and Technology
Nanjing University, Nanjing, 210093, China
[2]Department of Computer Science, Nanjing University
of Aeronautics and Astronautics, Nanjing, 210016, China
e-mail:aics@nuaa.edu.cn

## Abstract

This paper aims to develop further and systemize the theory of multiple belief change based on the previous work on *the package contraction*, developed by [Fuhrmann and Hansson 1994] and *the general belief changes*, developed by [Zhang 1996]. Two main representation theorems for general contractions are given, one is based on partial meet models and the other on nice-ordered partition models. An additional principle, called *Limit Postulate*, for the general belief changes is introduced which specifies properties of infinite belief changes. The results of this paper provides a foundation for investigating the connection between infinite non-monotonic reasoning and multiple belief revision.

## 1 Introduction

Belief change is the process through which a rational agent acquires new beliefs or retracts previously held ones. A very influential work on belief change goes back to Alchourrón, Gärdenfors and Makinson [Alchourrón et al. 1985], who developed a formal mechanism for the revision and the contraction of beliefs, which has been now widely referred to as the AGM theory. For a set of existing beliefs, represented by a deductively closed set $K$ of propositional sentences, and a new belief, represented by a propositional sentence $A$, three kinds of belief change operations are considered in the AGM theory: expansion, contraction and revision, denoted by $K + A$, $K - A$ and $K * A$, respectively. A set of rationality postulates for belief contractions and belief revisions, based on the idea of minimal change, are given and two different tools, partial meet model and epistemic entrenchment ordering, for constructing belief change operations have been developed in [Alchourrón et al. 1985] and [Gädenfors and Makinson 1988], respectively. Although AGM's belief change operators appear to capture of what is required of an ideal system of belief change, they are not suitable to characterize changes of beliefs with sets of new beliefs, especially with infinite sets. A number of studies on extending and generalizing these operations so as to enable a treatment of belief change by sets of sentences then come out [Fuhrmann 1988] [Niederée 1991] [Rott 1992] [Hansson 1992] [Fuhrmann and Hansson 94] [Zhang 1995] [Zhang 1996]. The extended operators for expansion, contraction and revision are usually called multiple ones while the original operators are referred to as singleton ones. A framework for multiple belief changes is not only interesting but also useful. We will benefit from it at least in the following aspects:

- The new information an agent accepts often involves simultaneously more than one belief, or even infinitely many, especially when the underlying language is extended to the first-order logic.

- It has been found that there are fundamental differences between iterated belief changes and simultaneous belief changes. The revisions of a belief set by a sentence $A$ and then by a sentence $B$ are by no means identical to the revision simultaneously by the set $\{A, B\}$. A framework for multiple changes will provide a possibility to describe the relationship between two sorts of belief changes (see [Zhang 1995]). A ready example is the supplementary postulates for multiple revisions(see subsection 2.2 of this paper).

- Connections between belief change and non-monotonic reasoning have been widely investigated in the literature [Makinson and Gärdenfors 1991] [Brewka 1991] [Nebel 1992] [Gärdenfors and Makinson 1994] [Zhang 1996]. The key idea is translating $B \in K * A$ into $A \hspace{-0.5ex}\mid\hspace{-1.5ex}\sim\hspace{0.5ex} B$ and vice versa. As claimed in [Gärdenfors and Makinson 1994] this translation makes sense only on the finite level. 'The idea of infinite revision functions seems to make good intuitive sense.' (see [Makinson and Gärdenfors 1991 ]P.190)

There have been several proposals for multiple belief changes ([Fuhrmann 1988] [Rott 1992] [Hansson 1992] [Fuhrmann and Hansson 94] [Zhang 1995] [Zhang 1996]). This paper is by no means to present an alternative one. Instead of that, we attempt to combine these approaches and develop some necessary tools to improve and systemize them. We will outline, in section 2, the two main paradigms of multiple belief changes: *package contraction*, developed by [Fuhrmann and Hansson 1994], and *general belief change operations*, developed by

[Zhang 1996]. It seems, however, that both paradigms fail to capture full characterization of multiple belief changes. The former succeeds in specifying the basic properties of multiple contractions but fails to give the generalization of the supplementary postulates, whereas the latter presents a full extension of AGM's postulates for belief changes but without providing a representation theorem for its framework. In section 3 we will devote to present a representation theorem for the general contraction, partially using the similar result of the package contraction. In section 4 we will argue with a counterexample that the postulates available for general contractions are not strong enough to characterize the multiple contractions. An additional principle, called *Limit Postulate*, is introduced, which reflects the relationship between the contraction by an infinite set and the ones by its finite subsets. The related representation result for Limit Postulate is then given. Section 5 will conclude the paper with a discussion on the application of this research to non-monotonic reasoning.

Unfortunately, space limitations do not allow a full presentation. All the proofs of lemmas and theorems and some of the lemmas which lead to the main results of the present paper are omitted.

Throughout this paper, we consider the first-order language $\mathcal{L}$ with the standard logical connectives $\neg, \vee, \wedge, \rightarrow$, and $\leftrightarrow$. The set of all subsets of $\mathcal{L}$ is denoted by $\mathcal{F}$. If $F = \{A_1, \cdots, A_n\}$ is a set of sentences, $\wedge F$ is an abbreviation of $A_1 \wedge \cdots \wedge A_n$. We shall assume that the underlying logic includes classical first-order logic and is compact. The notation $\vdash$ means classical first-order derivability and $Cn$ the corresponding closure operator. We call a set $K$ of sentences a *belief set*, which means that $K = Cn(K)$. The set of all belief set in $\mathcal{L}$ is denoted by $\mathcal{K}$. The notation $K + F$ will denote $Cn(K \cup F)$.

## 2 Postulates for Multiple Belief Change

In this section we try to give a survey of the current research on multiple belief change. Two types of multiple contraction and one type of multiple revision are discussed and their relationships is then established.

### 2.1 Multiple Contraction

[Fuhrmann and Hansson 1994] introduced two types of multiple contraction operations: *package contraction* and *choice contraction*[1]. They may all be viewed as generalizations of AGM contraction operation, but the former seems more acceptable. The so-called package contraction means contracting a belief set by removing all members of a set of sentences from it. For characterizing this operation, *six basic postulates* as generalizations of the corresponding basic postulates for AGM contraction are given as follows:

---
[1]Similar formalisms are also introduced by [Rott 1992] and [Hansson 1992].

**(K[-]1)** $K[-]F = Cn(K[-]F)$.

**(K[-]2)** $K[-]F \subseteq K$.

**(K[-]3)** If $\phi \not\vdash F$, then $F \cap (K[-]F) = \phi$.

**(K[-]4)** If $\phi \vdash F$, then $K \subseteq K[-]F$.

**(K[-]5)** If $A \in K \setminus K[-]F$, then there is some subset $S$ of $K$ such that $K[-]F \subseteq S$ and $S \not\vdash F$, but $S \cup \{A\} \vdash F$.

**(K[-]6)** If $F_1 \equiv_K F_2$, then $K[-]F_1 = K[-]F_2$.

Here $\Gamma \vdash F$ represents that there is some $A \in F$ such that $\Gamma \vdash A$; $F_1 \equiv_K F_2$ represents $\forall X \subseteq K(X \vdash F_1 \leftrightarrow X \vdash F_2)$.

A model for package contractions based on partial meet method has been constructed and the representation theorem for these basic postulates was also given in [Fuhrmann and Hansson 1994]. Although two tentative generalizations of AGM's supplementary postulates were given in the same paper, unfortunately, one of them was found to be inconsistent with the basic ones(personal communication).

Another kind of multiple contractions, called *the general contraction* introduced by [Zhang 1996], is motivated by a quite different idea. It seems to lay more emphasis on 'contracting' rather than 'removing'. The principal idea of general contractions is contracting a belief set so that the resulting set is consistent with a set of sentences.

Formally, for a given belief set $K$, a function $K \ominus : \mathcal{F} \rightarrow \mathcal{F}^2$, is a *general contraction operation* over $K$ if it satisfies the following postulates:

$(K \ominus 1)$ $K \ominus F = Cn(K \ominus F)$.

$(K \ominus 2)$ $K \ominus F \subseteq K$.

$(K \ominus 3)$ If $F \cup K$ is consistent, then $K \ominus F = K$.

$(K \ominus 4)$ If $F$ is consistent, then $F \cup (K \ominus F)$ is consistent.

$(K \ominus 5)$ $\forall A \in K(F \vdash \neg A \rightarrow K \subseteq K \ominus F + A)$.

$(K \ominus 6)$ If $Cn(F_1) = Cn(F_2)$, then $K \ominus F_1 = K \ominus F_2$.

$(K \ominus 7)$ $K \ominus F_1 \subseteq K \ominus (F_1 \vee F_2) + F_1$.

$(K \ominus 8)$ If $F_1 \cup (K \ominus (F_1 \vee F_2))$ is consistent, then $K \ominus (F_1 \vee F_2) \subseteq K \ominus F_1$.

Here $F_1 \vee F_2 = \{A \vee B : A \in F_1 \wedge B \in F_2\}$.

Among the above postulates, $(K \ominus 1)$-$(K \ominus 4)$ and $(K \ominus 6)$ are direct generalizations of the AGM postulates $(K-1)$-$(K-4)$ and $(K-6)$, respectively. The postulate $(K \ominus 5)$, being claimed as the generalization of *AGM's Recovery* (the most controversial among the AGM postulates for contractions), seems to be stronger than its original. Here we provide an equivalent property, called *Saturation*, which somewhat supports the postulate.

**Lemma 2.1** *If $\ominus$ satisfies $(K \ominus 1)$-$(K \ominus 4)$, then $(K \ominus 5)$ is equivalent to the following property:*

---
[2]In the following, $K$ will be omtted from '$K \ominus$' for convenience.

$(\ominus Sat)$ $(K \ominus F + F) \cap K \subseteq K \ominus F$ *(Saturation)*.

Saturation expresses the idea that if a piece of knowledge could be kept in the new knowledge base, it does not need to be abandoned when the contraction is conducted.

The postulate $(K \ominus 6)$ seems to be too weak when we consider the relationship between general contractions and package contractions. Instead, the following stronger principle is suggestible:

$(\ominus 6_S)$ If $\forall A \in K(F_1 \vdash \neg A \leftrightarrow F_2 \vdash \neg A)$, then $K \ominus F_1 = K \ominus F_2$.

It is obvious that $(K \ominus 6_S)$ implies $(K \ominus 6)$, but the inverse needs the presence of $(K \ominus 7)$ and $(K \ominus 8)$.

The postulates $(K \ominus 7)$ and $(K \ominus 8)$ are clearly non-intuitive. A slight improvement may be done by giving the following alternatives:

$(K \ominus 7')$ If $F_1 \subseteq F_2$, then $K \ominus F_2 \subseteq K \ominus F_1 + F_2$.

$(K \ominus 8')$ If $F_1 \subseteq F_2$ and $F_2 \cup K \ominus F_1$ is consistent, then $K \ominus F_1 \subseteq K \ominus F_2$.

In fact, $(K \ominus 7)$ and $(K \ominus 8)$ are equivalent to $(K \ominus 7')$ and $(K \ominus 8')$, respectively, by noting the fact that $F_1 \vee F_2 \dashv\vdash Cn(F_1) \cap Cn(F_2)$.

In this paper we will call $(\ominus 1)$-$(\ominus 5)$ and $(\ominus 6_S)$ *the basic postulates* for general contractions, whereas $(\ominus 7')$ and $(\ominus 8')$ *the supplementary postulates*.

An explicit construction of a general contraction has been given in [Zhang 1996] (also see section 4 of this paper), which shows that the set of postulates for general contractions is consistent, but it does not lead to a representation theorem for this sort of multiple contractions.

Despite the differences in motivation for two types of multiple contractions, they are closely related. In fact, the general contraction can be defined by the package contraction and, inversely, the latter can be partially defined in terms of the former. To show this, let us start with two notations. For any $F \in \mathcal{F}$, $\overline{F} = \{A : \exists B_1 \cdots B_n \in F(A = \neg B_1 \vee \cdots \vee \neg B_n)\}$ and $\overline{F} = \{\neg A : A \in F\}$. Then we have

**Proposition 2.2** *Let $K \in \mathcal{K}$ and '$[-]$' be a package contraction function over $K$. Define a general contraction function '$\ominus$' over $K$ as follows: for any $F \in \mathcal{F}$,*

$$K \ominus F \stackrel{Def}{=} K[-]\overline{F}$$

*If '$[-]$' satisfies all the basic postulates for package contractions, then '$\ominus$' satisfies all the basic postulates for general contractions.*

**Proposition 2.3** *Let $\mathcal{F}_\vee = \{F \in \mathcal{F} : \forall A, B \in F(A \vee B \in F)\}$ and $\ominus$ be a general contraction function over $K$. Define a package contraction function $[-] : \mathcal{F}_\vee \to \mathcal{F}_\vee$ over $K$ as follows: For any $F \in \mathcal{F}_\vee$,*

$$K[-]F = K \ominus \overline{F}$$

*If $\ominus$ satisfies all the basic postulates for general contractions, then $[-]$ satisfies all the basic postulates for package contractions.*

## 2.2 Multiple Revision

There are few investigations for multiple revision. A reason for this may be that it is widely agreed that revisions can be reduced to contractions. As a kind of generalizations of AGM revision operation, [Zhang 1995] [Zhang 1996] introduced a multiple revision function '$\otimes$', called *general revision*. Formally, for any belief set $K$, a function $\otimes : \mathcal{F} \to \mathcal{F}$ is a *general revision function over $K$* if it satisfies the following postulates:

$(K \otimes 1)$ $K \otimes F = Cn(K \otimes F)$.

$(K \otimes 2)$ $F \subseteq K \otimes F$.

$(K \otimes 3)$ $K \otimes F \subseteq K + F$.

$(K \otimes 4)$ If $K \cup F$ is consistent, then $K + F \subseteq K \otimes F$.

$(K \otimes 5)$ $K \otimes F$ is inconsistent if and only if $F$ is inconsistent.

$(K \otimes 6)$ If $Cn(F_1) = Cn(F_2)$, then $K \otimes F_1 = K \otimes F_2$.

$(K \otimes 7)$ $K \otimes (F_1 \cup F_2) \subseteq K \otimes F_1 + F_2$.

$(K \otimes 8)$ If $F_2 \cup (K \otimes F_1)$ is consistent, then $K \otimes F_1 + F_2 \subseteq K \otimes (F_1 \cup F_2)$.

In analogy with *Levi identity* and *Harper identity* in the AGM framework, the relationship between the general contraction and the general revision has been established by the following definitions:

(Def $\otimes$) $K \otimes F \stackrel{def}{=} (K \ominus \overline{F}) + F$.

(Def $\ominus$) $K \ominus F \stackrel{def}{=} (K \otimes \overline{F}) \cap K$

**Theorem 2.4** *[Zhang 1996] If $\ominus(\otimes)$ satisfies $(K \ominus 1)$-$(K \ominus 8)((K \otimes 1)$-$(K \otimes 8))$, then $\otimes(\ominus)$ obtained from $Def \otimes (Def \ominus))$ satisfies $(K \otimes 1) - (K \otimes 8)((K \ominus 1) - (K \ominus 8))$.*

This result enables us to lay more emphasis on the task of characterizing contraction operations.

## 3 Partial Meet Model for General Contractions

As mentioned above, the general contraction is successful in generalizing the AGM supplementary postulates, but fails to give a representation result whereas the package contraction is just opposite. In this section we try to give a representation theorem for the general contraction.

According to the relationship of package contractions and general contraction, a partial meet model for the general contraction can readily be constructed. The only problem is whether and how this kind of model can be suitably restricted so that the supplementary postulates for general contractions are also satisfied. This section will try to give an answer for it.

In the AGM theory the notation $K \perp A$ represents the set of maximal subsets of $K$ that does not imply $A$. This notation can be easily generalized to the following form:

**Definition 3.1** *For $K \in \mathcal{K}$ and $F \in \mathcal{F}$, $K' \in K\|F$ if and only if*
1. $K' \subseteq K$;
2. $F \cup K'$ *is consistent, and*
3. $\forall K'' \subseteq K(K' \subset K'' \to K' \cup F$ *is inconsistent).*

It is easy to see that $K\|\{A\} = K \perp \neg A$. The following notations are useful.
$\mathcal{U}_K = \bigcup\{K\|F : F \in \mathcal{F}\}$ and $\mathbf{U}_K = \{K\|F : F \in \mathcal{F}\}$

**Definition 3.2** *For any $K \in \mathcal{K}$, a selection function for $K$ is a function $S : \mathbf{U}_K \to 2^{\mathcal{U}_K}$ such that*

$$\forall H \in \mathbf{U}_K(S(H) \subseteq H \wedge (H \neq \phi \to S(H) \neq \phi))$$

**Definition 3.3** *An operation $K \ominus : \mathcal{F} \to \mathcal{F}$ is a partial meet contraction over $K$ if and only if there exists a select function $S$ such that for any $F \in \mathcal{F}$,*

$$K \ominus (F) = \begin{cases} K & \text{if } F \text{ is inconsistent;} \\ \bigcap S(K\|F) & \text{otherwise.} \end{cases}$$

We will omit $K$ form '$K\ominus$'.

A similar proof of representation theorem for the package contraction leads to the following representation result for the general contraction on the basic postulates.

**Theorem 3.4** *For any belief set $K$, $\ominus$ satisfies all the basic postulates for general contractions if and only if it is a partial meet contraction.*

**Definition 3.5** *Let $K \in \mathcal{K}$. A selection function $S$ for $K$ is complete if for all $H \in \mathbf{U}_K$*

$$S(H) = \{K' \in H : \bigcap S(H) \subseteq K'\}$$

*A partial meet contraction function is complete if it can be generated by such a selection function.*

**Definition 3.6** *Let $K \in \mathcal{K}$. A selection function $S$ for $K$ is (transitively) rational when there is a (transitive) relation $\leq$ on $\mathcal{U}_K$ such that for any $H \in \mathbf{U}_K$,*

$$S(H) = \{X \in H : \forall Y \in H(Y \leq X)\}$$

*The contraction function generated from such $S$ is called a (transitively) relational partial meet contraction function.*

One of the main results in this paper is the following representation theorem.

**Theorem 3.7** *(The first representation theorem) For any belief set $K$, $\ominus$ satisfies postulates $(K \ominus 1) - (K \ominus 8)$ if and only if $\ominus$ is a complete transitively rational partial meet contraction function.*

It should be remarked that there is an important difference between representation theorems of general contractions and of singleton contractions that the completeness of selection functions is needed just one direction in singleton contractions rather than both.

The following lemmas are found to be critical for the proof of the theorem.

**Lemma 3.8** *Let $K \in \mathcal{K}$ and $S$ be a complete select function for $K$. For any $\Delta \in \mathcal{U}_K$ and $F \in \mathcal{F}$, if $F$ is consistent and $\bigcap S(K\|F) \subseteq \Delta$, then $\Delta \in S(K\|F)$ or $\Delta = K$.*

**Lemma 3.9** *Let $K \in \mathcal{K}$, $F \in \mathcal{F}$ and $\Gamma$ be a closed subset of $K$. If $F \cup K$ is inconsistent but $\Gamma \cup F$ is consistent, then*

$$\bigcap\{\Delta \in K\|F : \Gamma \subseteq \Delta\} + F = \Gamma + F$$

**Lemma 3.10** *If $\ominus$ is a complete relational partial meet contraction, then $(K \ominus 7)$ holds.*

**Lemma 3.11** *Let $S$ be a complete selection function for $K$. If $F_1 \subseteq F_2$ and $F_2 \cup (\bigcap S(K\|F_1))$ is consistent, then $(K\|F_2) \cap S(K\|F_1) \neq \phi$*

**Lemma 3.12** *Let $S$ be a transitively rational selection function. If $F_1 \subseteq F_2$ and $(K\|F_2) \cap S(K\|F_1) \neq \phi$, then*

$$S(K\|F_2) \subseteq S(K\|F_1)$$

**Lemma 3.13** *If $\ominus$ is a complete transitively rational partial meet contraction, then $(K \ominus 8)$ holds.*

There are two limiting cases of singleton contractions: *maxichoice contractions* and *full meet contractions*, being investigated in the AGM theory, which are viewed as the lower and upper bounds of partial meet contractions, and therefore, are useful sometimes for understanding contraction operations sometimes. Here we present two similar representation results for general contractions. Assuming that the maxichoice contractions and the full meet contraction for general contractions are defined as usual, then we have

**Proposition 3.14** *A function $\ominus : \mathcal{F} \to \mathcal{F}$ is a maxichoice contraction over a belief set $K$ if and only if it satisfies $(K \ominus 1)$ -$(K \ominus 5)$, $(K \ominus 6_S)$ and the condition:*
*If $A \in K \setminus K \ominus F$, then there exists $B \in K$ such that $F \vdash \neg B$ and $A \to B \in K \ominus F$.*

**Proposition 3.15** *A function $\ominus : \mathcal{F} \to \mathcal{F}$ is a full meet contraction over a belief set $K$ if and only if it satisfies $(K \ominus 1)$ -$(K \ominus 5)$, $(K \ominus 6_S)$ and the condition:*
*If $F_1 \subseteq F_2$ and $F_1 \cup K$ is inconsistent, then $K \ominus F_1 \subseteq K \ominus F_2$.*

# 4 Limited Postulation for General Contractions

In this section, we use the approach of nice-ordered partition models, developed by [Zhang 1996], to show why we think that the postulates available for the general contraction are insufficient to characterize infinite belief changes. An additional postulate for the general contraction is introduced and its representation theorem is provided.

In order to construct a model for the general contraction operation, [Zhang 1996] introduced the notion of nice-ordered partition with the motivation of capturing the idea of degrees of reliability of information.

**Definition 4.1** *[Zhang 1996] For any belief set $K$, let $\mathcal{P}$ be a partition of $K$ and $<$ a total-ordering relation*

$cn$ $\mathcal{P}$. The triple $\Sigma = (K, \mathcal{P}, <)$ is called a *total-ordered partition (TOP)* of $K$. For any $p \in \mathcal{P}$, if $A \in p$, $p$ is called the rank of $A$, denoted by $b(A)$.

A total-ordered partition $\Sigma = (K, \mathcal{P}, <)$ is a *nice-ordered partition (NOP)* if it satisfies the following Logical Constraint:

If $A_1, \cdots, A_n \vdash B$, then $\sup\{b(A_1), \cdots, b(A_n)\} \geq b(B)$.

The idea underlying this definition is that different pieces of knowledge in a knowledge base are accepted with different degrees of reliability. It could be supposed that the knowledge base has been grouped according to degrees of reliability of knowledge. The rank of a piece of knowledge represents the group it belongs to. According to the above definition the less rank a sentence is, the higher degree of reliability it has. The relationship between NOP and epistemic entrenchment has been studied in [Zhang 1996]. Roughly speaking, the latter can be viewed as a special case of the former in some sense.

An explicit construction for multiple contraction functions is then given as follows:

**Definition 4.2** *[Zhang 1996] Let* $\Sigma = (K, \mathcal{P}, <)$ *be an NOP of a belief set $K$. Define a function* $\dot{\ominus} : \mathcal{F} \to \mathcal{F}$, *called NOP contraction over $K$, as follows: for any $F \in \mathcal{F}$*

i). If $F \cup K$ is consistent, then $K \dot{\ominus} F = K$; otherwise,
ii). $B \in K \dot{\ominus} F$ if and only if $B \in K$ and there exists $A \in K$ such that $F \vdash \neg A$ and

$\forall C \in K(A \vdash C \wedge F \vdash \neg C \to (b(A \vee B) < b(C) \vee \vdash A \vee B))$

*In particular, when $F$ is finite, then $B \in K \dot{\ominus} F$ if and only if*

$$B \in K \text{ and } b(\neg(\wedge F) \vee B) < b(\neg(\wedge F)) \vee \vdash \neg(\wedge F) \vee B \quad (1)$$

The following result expresses the fact that an NOP contraction must be a general contraction and, therefore, the set of postulates for general contractions is consistent.

**Theorem 4.3** *[Zhang 1996] IF $\dot{\ominus}$ is an NOP contraction over $K$, then it satisfies* $(K \ominus 1) - (K \ominus 8)$.

A natural question is now whether a general contraction function must be an NOP contraction. For the finite case the following lemma gives an affirmative answer.

**Lemma 4.4** *If '$\ominus$' is a general contraction function over a belief set $K$, then there exists a unique NOP, $\Sigma = (K, \mathcal{P}, <)$, of $K$ such that for any finite set $F$ of sentences, $B \in K \ominus F$ iff $B \in K$ and*

$b(\neg(\wedge F) \vee B) < b(\neg(\wedge F)) \vee \vdash \neg(\wedge F) \vee B$

For the infinite case, however, the answer is negative. Before we illustrate why, let us first introduce a variant of NOP contraction functions.

**Definition 4.5** *Let* $\Sigma = (K, \mathcal{P}, <)$ *be an NOP of a belief set $K$. Define a contraction function* $\hat{\ominus} : \mathcal{F} \to \mathcal{F}$ *over $K$ as follows: for any $F \in \mathcal{F}$*

i). If $F \cup K$ is consistent, then $K \hat{\ominus} F = K$, otherwise;
ii). $B \in K \hat{\ominus} F$ if and only if $B \in K$ and there exists $A \in K$ such that $F \vdash \neg A$ and

$\forall C \in K(A \vdash C \wedge F \vdash \neg C \to (b(C \vee B) < b(C) \vee \vdash C \vee B))$

It is easy to see that two types of NOP contractions agree on finite arguments. The following example shows that they branch in the infinite case.

**Example 4.1** We consider a propositional language $\mathcal{L}$ only including propositional variables $p_0, p_1, \cdots$. Let $K = Cn(\{p_0, p_1, \cdots\})$. Constructing an NOP $\mathcal{P}$ of $K$ as follows:
1. $\mathcal{P} = \{P_{-\infty}\} \cup \{P_{-n} : n \in \omega\}$, where

i) $P_{-\infty} = Cn(\phi)$.

ii) For any $A \in K$ such that $A \notin Cn(\phi)$,

    **a)** If $A = p_{i_1} \wedge \cdots \wedge p_{i_n}$, then $A \in P_{-\min\{i_1, \cdots, i_n\}}$, specially, $p_i \in P_{-i}$; otherwise,

    **b)** If $A'$ is the unique complete disjunctive normal form of $A$, then each disjunctive branch of $A'$ is of the form:

$$*p_{i_1} \wedge \cdots \wedge *p_{i_k}$$

where $*$ is empty or negative '$\neg$'. Since $A \in K$, there exists a disjunctive branch of $A'$ which belongs to $K$. Suppose that all such branches have been assigned into $P_{-j_1}, \cdots, P_{-j_m}$, respectively, then $A \in P_{-\max\{j_1, \cdots, j_m\}}$.

2. For any $n \in \omega$, $P_{-\infty} < P_{-n}$. For any $m, n \in \omega$, $P_{-m} < P_{-n}$ if and only if $m > n$.

It is not difficult to show that $\Sigma = (K, \mathcal{P}, <)$ is an NOP of $K$.

According to definitions of $\dot{\ominus}$ and $\hat{\ominus}$, we have
(1). If $F = \{\neg p_{i_0}, \cdots, \neg p_{i_n}\}(i_0 < \cdots < i_n)$, then $K \dot{\ominus} F = K \hat{\ominus} F = Cn(\{p_{i_{n+1}}, p_{i_{n+2}}, \cdots\} \cup F) \cap K$.
(2). If $F = \{\neg p_{i_0}, \neg p_{i_1}, \cdots\}(i_0 < i_1 < \cdots)$, then $K \dot{\ominus} F = K \hat{\ominus} F = Cn(F) \cap K$.

Now we change $\Sigma$ into $\Sigma'$ in this way of bringing all the formulas with the form $p_0 \vee p_i (i = 1, 2, \cdots)$ into $P_{-\infty}$ and doing the corresponding changes in order to make the resulting partition an NOP.

It is not difficult to check that if $F = \{\neg p_{i_0}, \neg p_{i_1}, \cdots\}$ $(0 < i_0 < i_1 < \cdots)$, then (with respect to the NOP $\Sigma'$):

$$K \dot{\ominus} F = Cn(F) \cap K \quad (2)$$

$$K \hat{\ominus} F = Cn(F \cup \{p_0\}) \cap K \quad (3)$$

The above example and Lemma 4.4 tell us the fact that the available postulates for general contractions are not sufficient to uniquely determine an NOP contraction as a general contraction function is given. We need some additional principles to describe the relationship between a belief set contracted by a infinite set of sentences and by its finite subsets. In fact, we can easily see from the above example that there is a situation in which an infinite set is inconsistent with all the contractions by its finite subsets (noting that it must consistent with the contraction by itself). In this case the available postulates for general contractions are insufficient to restrict the relation between the contraction by the infinite set and the ones by its finite subsets.

A reasonable thought is to assume that the contraction by a infinite set is some limiting case of the ones by its finite subset. This motivates us to introduce the following *Limit Postulate* for the general contraction:

Let $\mathcal{C}_F = \{\bar{F} : \bar{F} \subseteq Cn(F) \text{ and } \bar{F} \text{ is finite }\}$,

$$(K \ominus LP) \qquad K \ominus F = \bigcup_{\bar{F} \in \mathcal{C}_F} \bigcap_{\substack{\bar{F} \subseteq \bar{F}' \\ \bar{F}' \in \mathcal{C}_F}} K \ominus \bar{F}'$$

In other words, $B \in K \ominus F$ if and only if there exists a finite subset $\bar{F}$ of $Cn(F)$ such that for each finite subset $\bar{F}'$ of $Cn(F)$, if $\bar{F} \subseteq \bar{F}'$, then $B \in K \ominus \bar{F}'$.

An equivalent assumption for the general revision is readily given in terms of $(Def\ominus)$ and $(Def\otimes)$ as follows (an easy check is necessary):

$$(K \otimes LP) \qquad K \otimes F = \bigcup_{\bar{F} \in \mathcal{C}_F} \bigcap_{\substack{\bar{F} \subseteq \bar{F}' \\ \bar{F}' \in \mathcal{C}_F}} K \otimes \bar{F}'$$

Before we switch to present the representation theorem for the Limit Postulate we have to make a choice between the two types of NOP contractions (see Definition 4.2 and 4.5). By comparing Equations 2 and 3 we find that $\hat{\ominus}$ is more acceptable than $\dot{\ominus}$. Note that even though the proposition $p_0$ has the lowest degree of reliability in the old knowledge base, the agent believes extremely in $p_0 \vee p_i (i = 1, 2, \cdots)$. Therefore, when the fact tells that some $p_i'$s are false, he will conclude that $p_0$ is true. In the remaining part of this paper, an NOP contraction will only refer to $\hat{\ominus}$.

**Theorem 4.6** *If $\hat{\ominus}$ is an NOP contraction function, then $\hat{\ominus}$ satisfies $(K \ominus LP)$.*

**Theorem 4.7** *Let $\ominus$ be a general contraction function over $K$. If $\ominus$ satisfies $(K \ominus LP)$, then there exists an NOP, $\Sigma = (K, \mathcal{P}, <)$, of $K$ such that $\ominus$ is exactly the NOP contraction generated by $\Sigma$.*

The Lemma 4.4 and the above theorems lead to another main result of this paper:

**Theorem 4.8** *(The second representation theorem) For any belief set $K$, $\ominus$ satisfies $(K \ominus 1) - (K \ominus 8)$ and $(K \ominus LP)$ if and only if $\ominus$ is an NOP contraction over $K$.*

## 5 Conclusions and Future Work

Two representation theorems for the general contraction are presented in this paper. The first one illustrates that the AGM partial meet contraction can be smoothly extended to the case of multiple changes. The second one persuades us that confining our attention only to the generalization of postulates for singleton belief change would fail to get a full characterization of infinite belief changes. This result also provides a powerful tool to investigate the connection between belief revision and non-monotonic reasoning. In [Zhang et al. 1997], we introduced a non-monotonic logic the semantic of which bases on the theory of multiple belief revision and one of the inference rules in which, *finite supracompactness*, is just the counterpart of the Limit Postulate.

## References

[Alchourrón et al. 1985] C. E. Alchourrón, P. Gärdenfors and D. Makinson, On the logic of theory change: partial meet contraction and revision functions, *The Journal of Symbolic Logic* 50(2)(1985), 510-530.

[Brewka 1991] G. Brewka, Belief revision in a framework for default reasoning, in: A. Fuhrmann and M. Morreau eds. *The Logic of Theory Change*, (LNCS 465, Springer-Verlag, 1991) 185-205.

[Fuhrmann 1988] A. Fuhrmann, *Relevant Logics, Modal Logics, and Theory Change*, Phd thesis, Canberra: Australian National University.

[Fuhrmann and Hansson 1994] A. Fuhrmann and S. O. Hansson, A survey of multiple contractions, *Journal of Logic, Language, and Information* 3:39-76, 1994.

[Gädenfors and Makinson 1988] P. Gärdenfors and D. Makinson, Revisions of knowledge systems using epistemic entrenchment, in: M. Vardi, ed., *Proceedings of the Second Conference on Theoretical Aspects of Reasoning about Knowledge*, edited by M. Vardi (Morgan Kaufmann Publ., Los Altos, CA, 1988) 83-95.

[Gärdenfors and Makinson 1994] P. Gärdenfors and D. Makinson, Nonmonotonic inference based on expectations, *Artificial Intelligence* 65(1994), 197-245.

[Gärdenfors and Rott 1995] P. Gärdenfors and H. Rott, Belief revision, in: D. M. Gabbay C. J. Hogger and J. A. Robinson eds., *Handbook of Logic in Artificial Intelligence and Logic Programming*, Clarendon Press, Oxford, 1995, 35-132.

[Hansson 1992] S. O. Hansson, A dyadic representation of belief, in: P. Gärdenfors ed., *Belief Revision* (Cambridge University Press, 1992) 89-121.

[Makinson and Gärdenfors 1991] D. Makinson and P. Gärdenfors, Relations between the logic of theory change and nonmonotonic logic, in: A. Fuhrmann and M. Morreau eds. *The Logic of Theory Change*, (LNCS 465, Springer-Verlag, 1991) 185-205.

[Nebel 1992] B. Nebel, Syntax based approaches to belief revision, in: P. Gärdenfors ed., *Belief Revision* (Cambridge University Press, 1992) 52-88.

[Niederée 1991] R. Niederèe, Multiple contraction: a further case against Gärdenfors' principle of recovery, in: A. Fuhrmann and M. Morreau eds. *The Logic of Theory Change*, (LNCS 465, Springer-Verlag, 1991) 322-334.

[Rott 1992] H. Rott, Modellings for belief change: base contractions, multiple contractions, and epistemic entrenchment, *Logic in AI* (LNAI 633, Springer-Verlag 1992) 139-153.

[Zhang 1995] D. Zhang, A general framework for belief revision, in: *Proc. 4th Int. Conf. for Young Computer Scientists*(Peking University Press, 1995) 574-581.

[Zhang 1996] Zhang Dongmo, Belief revision by sets of sentences, *Journal of Computer Science and Technology*, 1996, 11(2), 1-19.

[Zhang et al. 1997] Dongmo Zhang, Shifu Chen, Wujia Zhu and Hongbing Li, Nonmonotonic reasoning and multiple belief revision, *this volume*.

# Nonmonotonic Reasoning and Multiple Belief Revision

Dongmo Zhang[1,2]  Shifu Chen[1]  Wujia Zhu[1,2]  Hongbing Li[1]

[1]State Key Lab. for Novel Software Technology
Department of Computer Science and Technology
Nanjing University, Nanjing, 210093, China
[2]Department of Computer Science, Nanjing University
of Aeronautics and Astronautics, Nanjing, 210016, China
e-mail:aics@nuaa.edu.cn

## Abstract

The aim of the present paper is to reveal the interrelation between general patterns of nonmonotonic reasoning and multiple belief revision. For this purpose we define a nonmonotonic inference frame in which individual inference rules have been proposed in the literature but their combination as a system has not been investigated. It is shown that such a system is so strong that almost all the rules (including the supracompactness) suggested for nonmonotonic inference relations in the literature hold in it. We prove that this nonmonotonic inference frame is strictly correspondent with multiple belief revision operation. On the basis of this result we analyse a specific paradigm of defult theory which satisfies all the rules under consideration and discuss limitations of methods based on consequence relations for the study of nonmonotonic reasoning.

## 1 Introduction

In recent years much work has been done on the relationship between nonmonotonic reasoning and belief revision [Makinson and Gärdenfors 1991] [Brewka 1991] [Nebel 1992] [Cravo and Martins 1993] [Li 1993][Gärdenfors and Makinson 1994] [Boutilier 1994] [Gärdenfors and Rott 1995] [Zhang 1996]. A very close correspondence between them has been found based on the following formal translation:

$$A \mathrel{\vert\!\sim}_K C \text{ iff } C \in K * A$$

The main idea is to identify revision of a belief set $K$ by a proposition $A$ with nonmonotonic inference from $A$ under the guidance of the background knowledge $K$. With this connection, it has been shown in [Makinson and Gärdenfors 1991] [Gärdenfors and Rott 1995] that each postulate for the belief revision function $*$ can be translated into a plausible conditions on the nonmonotonic inference relation $\mathrel{\vert\!\sim}$; conversely, almost all the plausible conditions on the nonmonotonic inference relation in the literature can also be translated into conditions on $*$ that are consequences of the postulates for the revision function. In fact, it is not difficult to verify that the revision function $*$ satisfies all eight postulates in [Gärdenfors 1988] if and only if $\mathrel{\vert\!\sim}$ satisfies the following five inference rules:

1. If $A \vdash B$, then $A\mathrel{\vert\!\sim} B$ (Supraclassicality).
2. If $A\mathrel{\vert\!\sim} \bot$, then $A \vdash \bot$ (Consistency Preservation).
3. If $A\mathrel{\vert\!\sim} B_i$ for all $B_i \in \Gamma$, $\Gamma \vdash C$, then $A\mathrel{\vert\!\sim} C$ (Closure).
4. If $A \wedge B\mathrel{\vert\!\sim} C$, then $A\mathrel{\vert\!\sim} B \to C$ (Conditionalization).
5. If $A \mathrel{\vert\!\not\sim} \neg B$ and $A\mathrel{\vert\!\sim} C$, then $A \wedge B\mathrel{\vert\!\sim} C$ (Rational Monotony).

This translation may be extended to the finite case. If $\Gamma$ is a finite set of propositions, written by $\{A_1, \cdots, A_n\}$, then:

$$\Gamma \mathrel{\vert\!\sim}_K A \text{ iff } A \in K * (A_1 \wedge \cdots \wedge A_n)$$

As mentioned in [Makinson 1993], however, this extension muddies the 'neat' distinction between $A_1, \cdots, A_n\mathrel{\vert\!\sim} A$ and $A_1 \wedge \cdots \wedge A_n\mathrel{\vert\!\sim} A$. A possible improvement is to replace the revision operation with some sort of multiple revision function. Suppose we have had a multiple revision function $\otimes$ such that $K \otimes F$ represents the result of revising a belief set $K$ with a set $F$ of propositions. The translation given below would be more natural:

$$\Gamma \mathrel{\vert\!\sim}_K A \text{ iff } A \in K \otimes \Gamma$$

This extension is also essential because it enables a treatment of inference relation in which premises are arbitrary sets of propositions, including infinite sets.

The questions arises naturally now that:

- how the nonmonotonic inference rules on $\mathrel{\vert\!\sim}$ are extended to the infinite level so that they are still plausible for nonmonotonic reasoners;
- how an infinite revision framework is constructed so that it is a natural generalization of the original one;
- whether the strict correspondence between belief revision and nonmonotonic reasoning can be preserved in the setting of the extended frameworks.

Fortunately, the first question has been widely investigated in the literature [Makinson 1989] [Freund 1990]

[Makinson 1993] [Herre 1994], only the presentation of the extended rules is mostly in the Tarski-style's inference operation $C$.

As far as the generalization of belief revision are concerned, [Zhang 1996] presented a kind of multiple revision framework, called *general revision*, which enables a treatment of revisions of belief set by arbitrary set of sentences. [Zhang et al. 1997] further developed the framework by providing two presentation theorems and suggesting an additional postulate to characterize the infinite properties of revision operations.

This paper is devoted to the last question. In the next section, we combine some of the nonmonotonic inference rules which have been suggested in the literature into a system of nonmonotonic reasoning, called **RN**, and discuss its properties. Section 3 outlines the general belief revision, and then, section 4 investigates the relationship between the system **RN** and the general belief revision. Section 5 presents a specific system of default reasoning which satisfies all the inference rules of **RN**. The last section discusses the inference power of **RN** and concludes the paper.

## 2 Rational Nonmonotonic Frame

This section will define a nonmonotonic frame of inference through combining generalized rules of the five nonmonotonic relations of inference mentioned above into a system, named **RN**. Although each of the generalized rules has been suggested in the literature, their properties as a whole have not been investigated. We start with the syntax of **RN** and then discuss its properties and derived rules.

We shall restrict the language of the indented system within any propositional language $\mathcal{L}$ with the standard logical connectives $\neg$, $\vee$, $\wedge$ and $\rightarrow$. Elements of $\mathcal{L}$ are called formulas which are denoted by $A,B,C$. Sets of formulas are denoted by $\Gamma$, $\Delta$, $F$ and etc. There are two relations of inference between premises on the left and conclusions on the right: $\vdash$, denoting the classical propositional derivability, and $\mid\sim$, used for a nonmonotonic relation of inference. An associated Tarski-style's consequence operation may be defined by each of the relations of inference in such manner:
$Cn(\Gamma) = \{A : \Gamma \vdash A\}$
$C(\Gamma) = \{A : \Gamma \mid\sim A\}$.

It is presupposed that the inference relation $\vdash$ satisfies all the inference rules of the classical propositional logic so it is compact:
$\Gamma \vdash A$ iff there exists a finite subset $\Gamma_0$ of $\Gamma$ such that $\Gamma_0 \vdash A$.

A set $\Gamma$ of formulas is said to be *closed* if $\Gamma = Cn(\Gamma)$. $\Gamma\mid\sim(\vdash)\Delta$ indicates that $\Gamma\mid\sim(\vdash)A$ for all $A \in \Delta$ ($\Delta$ may be empty); $\Gamma \not\mid\sim A$ indicates that $\Gamma\mid\sim A$ does not hold.

**Definition 2.1** *A system* **RN** $= (\mathcal{L}, \mid\sim)$ *is said to be a rational nonmonotonic frame if $\mathcal{L}$ is a language of classical propositional logic at least including propositional connectives ($\neg, \wedge, \vee$ and $\rightarrow$) and $\mid\sim$ is a relation from $2^{\mathcal{L}}$ to $\mathcal{L}$, called the rational nonmonotonic inference relation, if it satisfies:*

**(RN1)** *If $\Gamma \vdash A$, then $\Gamma\mid\sim A$ (Supraclassicality).*

**(RN2)** *If $\Gamma\mid\sim \bot$, then $\Gamma \vdash \bot$ (Consistency Preservation).*

**(RN3)** *If $\Gamma\mid\sim\Delta \vdash A$, then $\Gamma\mid\sim A$ (Closure or Weak Transitivity).*

**(RN4)** *If $\Gamma \cup \Delta\mid\sim A$ and $\Delta \neq \phi$, then there are $A_1, \cdots, A_n \in \Delta$ such that $\Gamma\mid\sim (A_1 \wedge \cdots \wedge A_n) \rightarrow A$ (Infinite Conditionalization).*

**(RN5)** *If $\Gamma \not\mid\sim \neg(A_1 \wedge \cdots \wedge A_n)$ for all $A_1, \cdots, A_n \in \Delta$, then $\Gamma\mid\sim A$ implies $\Gamma \cup \Delta\mid\sim A$ (Infinite Rational Monotonicity).*

*Furthermore, a rational nonmonotonic frame is said to be finite supracompact if it satisfies:*

**(RN6)** *$\Gamma\mid\sim A$ iff there exists a finite subset $\Gamma_0$ of $\Gamma$ such that $\Gamma_0 \cup \Gamma'\mid\sim A$ for every finite subset $\Gamma'$ of $Cn(\Gamma)$ (Finite Supracompactness).*

The name 'rational' follows from [Lehmann and Magidor 1992] [Herre 1994] but the rational inference relation here is stronger because the consistency preservation is added.

For those who are familiar with Tarski-style's nonmonotonic consequence operations, the following equivalent presentation of the conditions (**RN1**) − (**RN6**) would be preferential.

1. $Cn(\Gamma) \subseteq C(\Gamma)$(Supraclassicality).
2. If $Cn(\Gamma) \neq \mathcal{L}$, then $C(\Gamma) \neq \mathcal{L}$(Consistency Preservation).
3. $Cn(C(\Gamma)) \subseteq C(\Gamma)$(Closure).
4. $C(\Gamma \cup \Delta) \subseteq Cn(\Gamma \cup C(\Delta))$(Infinite Conditionalization).
5. IF $\Delta \cup C(\Gamma) \neq \mathcal{L}$, then $C(\Gamma) \subseteq C(\Gamma \cup \Delta)$(Rational Monotony).
6. $\Gamma\mid\sim A$ iff there exists a finite subset $\Gamma_0$ of $\Gamma$ such that $\Gamma_0 \cup \Delta\mid\sim A$ for every finite subset $\Delta$ of $Cn(\Gamma)$(Finite Supracompactness).

It should be noted that none of the above conditions is the authors' invention. They all have been suggested for nonmonotonic reasonings in the literature. In fact, the conditions 1-4 were presented in [Makinson 1993] and the last two conditions are found in [Herre 1994][1]

---

[1]In [Herre 1994] the finite supracompactness refers to that $\Gamma \mid\sim A$ *if* there exists a finite subset $\Gamma_0$ of $\Gamma$ such that $\Gamma_0 \cup \Delta \mid\sim A$ for every finite subset $\Delta$ of $Cn(\Gamma)$. However the complete $\Delta_2$-compactness is just the meaning of the finite supracompactness in this paper.

In order to reveal the power of **RN**, we shall show that most of the inference rules for nonmonotonic reasoning suggested in the literature are derived rules of **RN**.

**Lemma 2.2** *The following rules are derived rules of* **RN**:

(1). $\Gamma\mathrel{|\!\sim}\Gamma$ *(Reflexivity)*

(2). *If* $\Gamma, A\mathrel{|\!\sim}B, \neg B$, *then* $\Gamma\mathrel{|\!\sim}\neg A$ *(Reductio ad Absurdum)*.

(3). *If* $\Gamma, A\mathrel{|\!\sim}B$, *then* $\Gamma\mathrel{|\!\sim}A \to B$ *(Deduction Theorem)*.

(4). *If* $\Gamma\mathrel{|\!\sim}A \to B$ *and* $\Gamma\mathrel{|\!\sim}A$, *then* $\Gamma\mathrel{|\!\sim}B$ *(Modus Ponens)*.

**Proof:** (1) follows (**RN1**). (2) follows (**RN2**) and (**RN3**). (3) is the special case of (**RN4**) where $\Delta = \{A\}$. For (4), since $\Gamma\mathrel{|\!\sim}A \to B$ and $\Gamma\mathrel{|\!\sim}A$, so $\Gamma\mathrel{|\!\sim}\{A \to B, A\} \vdash B$. By (**RN3**) we get $\Gamma\mathrel{|\!\sim}B$. □

The above theorem shows that $\mathrel{|\!\sim}$ satisfies all the formal inference rules of classical propositional logic except for the following deductive transitivity:

If $\Gamma\mathrel{|\!\sim}\Delta\mathrel{|\!\sim}A (\Delta \neq \phi)$, then $\Gamma\mathrel{|\!\sim}A$.

**Lemma 2.3** *The following rules are derived rules of* **RN**:

(1). *If* $\Gamma\mathrel{|\!\sim}\Delta$ *and* $\Gamma \cup \Delta\mathrel{|\!\sim}A$, *then* $\Gamma\mathrel{|\!\sim}A$ *(Cumulative Transitivity)*

(2). *If* $\Gamma\mathrel{|\!\sim}\Delta$ *and* $\Gamma\mathrel{|\!\sim}A$, *then* $\Gamma \cup \Delta\mathrel{|\!\sim}A$ *(Cautious Monotony)*

(3). *If* $\Gamma\mathrel{|\!\sim}\Delta$ *and* $\Delta\mathrel{|\!\sim}\Gamma$, *then* $\Gamma\mathrel{|\!\sim}A$ *if and only if* $\Delta\mathrel{|\!\sim}A$ *(Reciprocity)*

(4). *If* $\Gamma \dashv\vdash \Delta$, *then* $\Gamma\mathrel{|\!\sim}A$ *if and only if* $\Delta\mathrel{|\!\sim}A$. *(Left Logical Equivalence)*

(5). *If* $A \dashv\vdash B$, *then* $\Gamma\mathrel{|\!\sim}A$ *if and only if* $\Gamma\mathrel{|\!\sim}B$. *(Right Logical Equivalence)*

**Proof:** For (1), suppose that $\Gamma\mathrel{|\!\sim}\Delta (\Delta \neq \phi)$ and $\Gamma \cup \Delta\mathrel{|\!\sim}A$. Then, by (**RN4**), there exists $A_1, \cdots, A_n \in \Delta$ such that $\Gamma\mathrel{|\!\sim}(A_1 \wedge \cdots \wedge A_n) \to A$. Since $A_1, \cdots, A_n \in \Delta$ implies $\Delta \vdash A_1 \wedge \cdots \wedge A_n$, hence we obtain $\Gamma\mathrel{|\!\sim}A$ by (**RN3**) and Theorem 2.2 (4).

For (2), suppose that $\Gamma\mathrel{|\!\sim}\Delta$ and $\Gamma\mathrel{|\!\sim}A$. If there are $A_1, \cdots, A_n \in \Delta$ such that $\Gamma\mathrel{|\!\sim}\neg(A_1 \wedge \cdots \wedge A_n)$, since $\Gamma\mathrel{|\!\sim}\Delta$ implies $\Gamma\mathrel{|\!\sim}A_1 \wedge \cdots \wedge A_n$ by (**RN3**), then we have $\Gamma\mathrel{|\!\sim}\bot$ again by (**RN3**). It follows from (**RN2**) that $\Gamma \vdash \bot$. By the compactness of the classical propositional logic, $\Gamma \cup \Delta \vdash \bot$, so $\Gamma \cup \Delta \vdash A$. By the Supraclassicality, we have $\Gamma \cup \Delta\mathrel{|\!\sim}A$. If $\Gamma\mathrel{|\!\not\sim}\neg(A_1 \wedge \cdots \wedge A_n)$ for any $A_1, \cdots, A_n \in \Delta$, then by (**RN5**) and $\Gamma\mathrel{|\!\sim}A$, we have $\Gamma \cup \Delta\mathrel{|\!\sim}A$ as desired.

(3) follows from (1) and (2). (4) follows from (**RN1**) and (3). (5) follows from (**RN3**). □

As shown by [Makinson 1993], the Infinite Conditionalization along with other rules implies the following Distributivity.

**Lemma 2.4** *If* $\Gamma \cup \Delta_1\mathrel{|\!\sim}A$, $\Gamma \cup \Delta_2\mathrel{|\!\sim}A$, *then* $\Gamma \cup (\Delta_1 \bigvee \Delta_2)\mathrel{|\!\sim}A$.
*Specially, if* $\Delta_1\mathrel{|\!\sim}A$ *and* $\Delta_2\mathrel{|\!\sim}A$, *then* $\Delta_1 \bigvee \Delta_2\mathrel{|\!\sim}A$ *(Distribution)*.
*where* $\Delta_1 \bigvee \Delta_2 = \{A \vee B : A \in \Delta_1 \text{ and } \Delta_2\}$.

It is well-known that compactness is a very important property of the classical logic which provides a bridge between inferences of finite and infinite premises: $\Gamma \vdash A$ iff $\Gamma_0 \vdash A$ for some finite subset $\Gamma_0$ of $\Gamma$. But such equivalence implies monotony, so this kind of compactness must fail in any nonmonotonic logic. This does not mean that there are no properties of compactness for the nonmonotonic logic. In fact there are a number of alternative versions of compactness for nonmonotonic reasoning proposed([Freund 1990] [Makinson 1993] [Herre 1994]).

[Freund 1990] suggested the following *Supracompactness* for nonmonotonic inference:
$\Gamma\mathrel{|\!\sim}A$ iff there exists a finite subset $\Gamma_0$ of $\Gamma$ such that for any set of formulas $\Delta$, $\Gamma\mathrel{|\!\sim}\Delta$ implies $\Gamma_0 \cup \Delta\mathrel{|\!\sim}A$.

The following theorem shows that such supracompactness follows from the finite supracompactness. This was also noted by [Freund 1990] and [Makinson 1993] with a little different setting.

**Theorem 2.5** *Any finite supracompact rational inference relation satisfies the Supracompactness.*

**Proof:** It is enough to show that if $\Gamma\mathrel{|\!\sim}A$ then there exists a finite subset $\Gamma_0$ of $\Gamma$ such that $\Gamma\mathrel{|\!\sim}\Delta$ implies $\Gamma_0 \cup \Delta\mathrel{|\!\sim}A$. For this, let $\Gamma\mathrel{|\!\sim}A$. By Finite Supracompactness, there exists a finite subset $\Gamma_0$ of $\Gamma$ such that $\Gamma_0 \cup \Gamma'\mathrel{|\!\sim}A$ for every finite subset $\Gamma'$ of $Cn(\Gamma)$. Suppose that $\Gamma\mathrel{|\!\sim}\Delta$. Since $Cn((\Gamma_0 \cup \Delta) \bigvee \Gamma) \subseteq Cn(\Gamma)$, the finite supracompactness implies that $\Gamma_0 \cup \Gamma'\mathrel{|\!\sim}A$ for every finite subset $\Gamma'$ of $Cn((\Gamma_0 \cup \Delta) \bigvee \Gamma)$. Again by the finite supracompactness, we have $Cn((\Gamma_0 \cup \Delta) \bigvee \Gamma)\mathrel{|\!\sim}A$ (noting that $\Gamma_0 \subseteq Cn((\Gamma_0 \cup \Delta) \bigvee \Gamma)$). It follows by the left logical equivalence that $(\Gamma_0 \cup \Delta) \bigvee \Gamma\mathrel{|\!\sim}A$. On the other hand, by $\Gamma\mathrel{|\!\sim}\Delta$ and the supraclassicality as well as Lemma 2.4, it is not difficult to verify that $(\Gamma_0 \cup \Delta) \bigvee \Gamma\mathrel{|\!\sim}\Delta$. Thus by the cautious monotony we have $\Delta \cup ((\Gamma_0 \cup \Delta) \bigvee \Gamma)\mathrel{|\!\sim}A$. Noting that $\Delta \cup ((\Gamma_0 \cup \Delta) \bigvee \Gamma) \dashv\vdash \Gamma_0 \cup \Delta$, we conclude from the left logical equivalence that $\Gamma_0 \cup \Delta\mathrel{|\!\sim}A$ as desired. □

On the basis of Makinson's work on general patterns in nonmonotonic reasoning, it is not difficult to see that any rational nonmonotonic relation of inference also satisfies conditions such as Absorption, Cut, Cumulativity, Loop, Negation Rationality(see [Makinson 1989] and [Makinson 1993]).

## 3 Multiple Belief Revision

This section recalls definitions and results on the multiple belief revision. [Zhang 1995][Zhang 1996] introduced and further developed by [Zhang et al. 1997] a framework for multiple belief changes through extending the AGM theory([Gärdenfors 1988]). The extended revision function was called the general revision. Formally, a function $K \otimes$ [2] $: 2^{\mathcal{L}} \to 2^{\mathcal{L}}$ with respect to a given belief set $K$ is said to be a *general revision function over $K$* if it satisfies the following nine postulates:

($K \otimes 1$) $K \otimes F = Cn(K \otimes F)$.

($K \otimes 2$) $F \subseteq K \otimes F$.

($K \otimes 3$) $K \otimes F \subseteq K + F$.

($K \otimes 4$) If $K \cup F$ is consistent, then $K + F \subseteq K \otimes F$.

($K \otimes 5$) $K \otimes F$ is inconsistent iff $F$ is inconsistent.

($K \otimes 6$) If $Cn(F_1) = Cn(F_2)$, then $K \otimes F_1 = K \otimes F_2$.

($K \otimes 7$) $K \otimes (F_1 \cup F_2) \subseteq K \otimes F_1 + F_2$.

($K \otimes 8$) If $F_2 \cup (K \otimes F_1)$ is consistent, then $(K \otimes F_1) + F_2 \subseteq K \otimes (F_1 \cup F_2)$.

($K \otimes LP$) $K \otimes F = \bigcup_{\bar{F} \in \mathcal{C}_F} \bigcap_{\substack{\bar{F} \subseteq \bar{F}' \\ \bar{F}' \in \mathcal{C}_F}} K \otimes \bar{F}'$

where $\mathcal{C}_F = \{\bar{F} : \bar{F} \subseteq Cn(F) \text{ and } \bar{F} \text{ is finite}\}$.

The postulates ($K \otimes 1$)-($K \otimes 8$) were presented in [Zhang 1996] and the last one, called *the Limit Postulate*, was introduced by [Zhang et al. 1997]. The representation theorem for all nine postulates was given in [Zhang et al. 1997] based on the following notions:

For any set $\Gamma$ of formulas, let $\mathcal{P}$ be a partition [3] of $\Gamma$ and $<$ a total-ordering(well-ordering) relation on $\mathcal{P}$. For any $p \in \mathcal{P}$, if $A \in p$, $p$ is called the *rank* of $A$, denoted by $b(A)$.

The triple $\Sigma = (\Gamma, \mathcal{P}, <)$ is called a *nice-ordered partition*(NOP) (*perfect-ordered partition*(POP)) of $\Gamma$ if it satisfies the following *Logical Constraint*:

If $A_1, \cdots, A_n \vdash B$, then $\sup\{b(A_1), \cdots, b(A_n)\} \geq b(B)$.

Now let $K$ be a closed set of formulas and $\Sigma = (K, \mathcal{P}, <)$ a nice-ordered partition. A function $\otimes : 2^{\mathcal{L}} \to 2^{\mathcal{L}}$ is said to be *the revision function generated by* $\Sigma$ if for any $F \subseteq \mathcal{L}$,

i). if $F \cup K$ is consistent, then $K \otimes F = K + F$; otherwise,

ii). $B \in K \otimes F$ if and only if $B \in K + F$ and there exists $A \in K$ such that $F \vdash \neg A$ and

$\forall C \in K(A \vdash C \land F \vdash \neg C \to (b(C \lor B) < b(C) \lor \vdash C \lor B))$

The original presentation of the representation theorem is based on the contraction function(see [Zhang et al. 1997]). The following theorem is obtained by using the interrelation of revision and contraction.

**Theorem 3.1** *For any closed set $K$ of formulas, a revision function $\otimes$ satisfies ($K \otimes 1$) - ($K \otimes 8$) as well as ($K \otimes LP$) if and only if there exists a nice-ordered partition $\Sigma = (K, \mathcal{P}, <)$ such that $\otimes$ is the revision function generated by $\Sigma$.*

---

[2] In the present paper, $K \otimes$ is also written as $\otimes_K$ or $\otimes$ if without confusion.

[3] A partition of a set $\Gamma$ is a disjoint family $\mathcal{P}$ of subsets of $\Gamma$ such that $\Gamma = \bigcup\{p : p \in \mathcal{P}\}$.

## 4 Representation Theorem

In order to reveal the interrelation between **RN** and the multiple belief revision, we shall take revision operations as the semantic of **RN** rather than follow the traditional approach of Shoham's preferential models.

**Theorem 4.1** *(Soundness) Let $\mathcal{L}$ be a language of propositional logic and $K$ a consistent closed set in $\mathcal{L}$. Let $\otimes$ be a general revision function over $K$. Define a relation $\mathrel{|\!\sim} \subseteq 2^{\mathcal{L}} \times \mathcal{L}$ as follows: for any set $\Gamma \subseteq \mathcal{L}$ and any formula $A \in \mathcal{L}$,*

$$\Gamma \mathrel{|\!\sim} A \text{ iff } A \in K \otimes \Gamma$$

*then $(\mathcal{L}, \mathrel{|\!\sim})$ is a finite supracompact rational nonmonotonic frame.*

**Proof:** We need to show $\mathrel{|\!\sim}$ satisfies the rules (**RN1**)-(**RN6**). For (**RN1**), assume that $\Gamma \mathrel{|\!\sim} A$. Since $\Gamma \subseteq K \otimes \Gamma$ and $K \otimes \Gamma$ is closed, thus $A \in K \otimes \Gamma$, that is, $\Gamma \mathrel{|\!\sim} A$.

For (**RN2**), assume that $\Gamma \mathrel{|\!\sim} \bot$, i.e., $\bot \in K \otimes \Gamma$, which means that $K \otimes \Gamma$ is inconsistent. It follows by ($K \otimes 5$) that $\Gamma$ is inconsistent. Thus $\Gamma \vdash \bot$.

For (**RN3**), assume that $\Gamma \mathrel{|\!\sim} \Delta \vdash A$ which means that $\Delta \subseteq K \otimes \Gamma$ and $A \in Cn(\Delta)$. By ($K \otimes 1$), we have $A \in K \otimes \Gamma$, i.e., $\Gamma \mathrel{|\!\sim} A$.

For (**RN4**), assume that $\Gamma \cup \Delta \mathrel{|\!\sim} A$, that is, $A \in K \otimes (\Gamma \cup \Delta)$. By ($K \otimes 7$), $K \otimes (\Gamma \cup \Delta) \subseteq K \otimes \Gamma + \Delta$, so we have $A \in K \otimes \Gamma + \Delta$. There exist then $A_1, \cdots, A_n \in \Delta$ such that $(A_1 \land \cdots \land A_n) \to A \in K \otimes \Gamma$, that is $\Gamma \mathrel{|\!\sim} (A_1 \land \cdots \land A_n) \to A$.

For (**RN5**), if for any $A_1 \land \cdots \land A_n \in \Delta$, $\Gamma \mathrel{|\!\not\sim} \neg(A_1 \land \cdots \land A_n)$, or $\neg(A_1 \land \cdots \land A_n) \notin K \otimes \Gamma$, then $\Delta \cup (K \otimes \Gamma)$ is consistent. Therefore, when $\Gamma \mathrel{|\!\sim} A$, or $A \in K \otimes \Gamma$, we conclude by ($K \otimes 8$) that $A \in K \otimes (\Gamma \cup \Delta)$, so $\Gamma \cup \Delta \mathrel{|\!\sim} A$ as desired.

For the finite supracompactness, suppose that $\otimes$ satisfies ($K \otimes LP$), that is, $A \in K \otimes \Gamma$ iff there exists a finite subset $\Gamma_1$ of $Cn(\Gamma)$ such that for any finite subset $\Gamma_2 \subseteq Cn(\Gamma)$, $\Gamma_1 \subseteq \Gamma_2$ implies $A \in K \otimes \Gamma_2$. It is easy to see that we only need to show that $\Gamma \mathrel{|\!\sim} A$ implies that there exists a finite subset $\Gamma_0$ of $\Gamma$ such that $\Gamma_0 \cup \Delta \mathrel{|\!\sim} A$ for every finite subset $\Delta$ of $Cn(\Gamma)$. To this end, assume that $\Gamma \mathrel{|\!\sim} A$, that is $A \in K \otimes \Gamma$. By ($K \otimes LP$), there exists a finite subset $\Gamma_1$ of $Cn(\Gamma)$ such that for any finite subset $\Gamma_2$ of $Cn(\Gamma)$, if $\Gamma_1 \subseteq \Gamma_2$, then $A \in K \otimes \Gamma_2$. Let $\Gamma_0$ be a finite subset of $\Gamma$ such that $\Gamma_1 \in Cn(\Gamma_0)$. For any finite subset $\Delta$ of $Cn(\Gamma)$, since $\Gamma_0 \cup \Gamma_1 \cup \Delta$ is finite and also a subset of $Cn(\Gamma)$, we have $A \in K \otimes (\Gamma_0 \cup \Gamma_1 \cup \Delta)$. It follows from ($K \otimes 6$) that $A \in K \otimes Cn(\Gamma_0 \cup \Gamma_1 \cup \Delta)$. On the other hand, $\Gamma_1 \subseteq Cn(\Gamma_0)$ implies $Cn(\Gamma_0 \cup \Delta) = Cn(\Gamma_0 \cup \Gamma_1 \cup \Delta)$. Thus we obtain that $A \in K \otimes Cn(\Gamma_0 \cup \Delta)$. It follows from ($K \otimes 6$) again that $A \in K \otimes (\Gamma_0 \cup \Delta)$, that is $\Gamma_0 \cup \Delta \mathrel{|\!\sim} A$. □

**Theorem 4.2** *(Completeness) Let $(\mathcal{L}, \mathrel{|\!\sim})$ be a finite supracompact rational nonmonotonic frame. Let $K = \{A \in \mathcal{L} : \phi \mathrel{|\!\sim} A\}$. Define a function $\otimes_K : 2^{\mathcal{L}} \to 2^{\mathcal{L}}$ as follows: for any $F \subseteq \mathcal{L}$,*

$$\otimes_K(F) = \{A \in \mathcal{L} : F \mathrel{|\!\sim} A\}$$

Then $\otimes_K$ is a general belief revision function over $K$.

**Proof:** We first prove that $K$ is closed and consistent. The consistency of $K$ follows easily from **(RN2)**. To show that $K$ is closed, let us assume that $K \vdash A$. There are then $A_1, \cdots, A_n \in K$ such that $A_1, \cdots, A_n \vdash A$. Hence $\phi \hspace{-0.5mm}\mid\hspace{-2mm}\sim A_1, \cdots, \phi \hspace{-0.5mm}\mid\hspace{-2mm}\sim A_n$, that is $\phi \hspace{-0.5mm}\mid\hspace{-2mm}\sim \{A_1, \cdots, A_n\}$. By **(RN3)**, we see $\phi \hspace{-0.5mm}\mid\hspace{-2mm}\sim A$ and then $A \in K$.

We now turn to show that $\otimes_K$ satisfies all nine postulates for the general belief revision.

Proof of $(K \otimes 1)$ is similar to that of closeness of $K$. $(K \otimes 2)$ follows immediately from the Reflexivity. $(K \otimes 3)$ and $(K \otimes 4)$ are special cases of $(K \otimes 7)$ and $(K \otimes 8)$, respectively. $(K \otimes 5)$ follows directly from **(RN2)**. $(K \otimes 6)$ follows from the Reciprocity.

For $(K \otimes 7)$, assume that $A \in \otimes_K(F_1 \cup F_2)$, or $F_1 \cup F_2 \hspace{-0.5mm}\mid\hspace{-2mm}\sim A$. Then by **(RN4)**, there are $A_1, \cdots, A_n \in F_2$ such that $F_1 \hspace{-0.5mm}\mid\hspace{-2mm}\sim (A_1 \wedge \cdots \wedge A_n) \to A$, that is, $(A_1 \wedge \cdots \wedge A_n) \to A \in \otimes_K(F_1)$. Consequently we have $A \in \otimes_K(F_1) + F_2$. Therefore, $\otimes_K(F_1 \cup F_2) \subseteq \otimes_K(F_1) + F_2$.

For $(K \otimes 8)$, assume that $F_2 \cup \otimes_K(F_1)$ is consistent, which means that for any $A_1, \cdots, A_n \in F$, $\neg(A_1 \wedge \cdots \wedge A_n) \notin \otimes_K(F_1)$, or $F_1 \hspace{-0.5mm}\mid\hspace{-2mm}\not\sim \neg(A_1 \wedge \cdots \wedge A_n)$. Now suppose $A \in \otimes_K(F_1) + F_2$, then there exist $B_1, \cdots, B_m \in F_2$ such that $(B_1 \wedge \cdots \wedge B_m) \to A \in \otimes_K(F_1)$, or $F_1 \hspace{-0.5mm}\mid\hspace{-2mm}\sim (B_1 \wedge \cdots \wedge B_m) \to A$. It follows from **(RN5)** that $F_1 \cup F_2 \hspace{-0.5mm}\mid\hspace{-2mm}\sim (B_1 \wedge \cdots \wedge B_m) \to A$. Since $B_1, \cdots, B_m \in F_2$ implies $F_1 \cup F_2 \hspace{-0.5mm}\mid\hspace{-2mm}\sim B_1 \wedge \cdots \wedge B_m$, we conclude, by Theorem 2.2(4), that $F_1 \cup F_2 \hspace{-0.5mm}\mid\hspace{-2mm}\sim A$, that is, $A \in \otimes_K(F_1 \cup F_2)$. Therefore we have proven that $\otimes_K(F_1) + F_2 \subseteq \otimes_K(F_1 \cup F_2)$ as desired.

The proof of the limit postulate is similar to that of the soundness. □

## 5  A Paradigm of Default Reasoning

Following the general considerations of the previous sections, we now look at a specific approach to nonmonotonic reasoning. We aim to seek a 'natural' system of nonmonotonic logic which satisfies all the inference rules for the rational nonmonotonic frame. On the basis of Makinson's 'satisfaction table' in [Makinson 1993], only Poole's system without constraints based on finite set of defaults in the systems of nonmonotonic logic considered in that paper satisfies all the inference rules of **RN** except the rational monotony. There is a disadvantage of Poole's approach, however, that it does not allow to represent priorities between defaults, which causes that the inference relations generated by Poole's system happen to collapse into the classical one when the default set is closed. [Nebel 1992] developed a system of default logic, called *ranked default theory* (RDT), which efficiently overcame this shortage. We here reformulate Nebel's system in a more general fashion.

Let $(F, D)$ be a default theory, where $F$ and $D$ are both sets of propositions, interpreted as 'facts' and 'defaults', respectively. $(F, D)$ is said to be a *perfect-ordered partitioned default theory* (POP DT) w.r.t. $\Sigma$ if $\Sigma = (D, \mathcal{P}, <)$ is a perfect-ordered partition(see section 3). The order-type $\eta$ of $\mathcal{P}$ is called the type of $(F, D)$, denoted by $\eta_D$. The partition $\mathcal{P}$ is denoted as $\{D_\alpha : \alpha < \eta_D\}$.

A set $E$ of propositions is a *syntax-based extension* of $(F, D)$ if $E = Cn((\bigcup_{\alpha < \eta_D} R_\alpha) \cup F)$ such that for all $\alpha < \eta_D$, $R_\alpha \subseteq D_\alpha$ and $R_\alpha$ is maximal (with respect to set-inclusion) among the subsets of $D_\alpha$ such that $(\bigcup_{\gamma \leq \alpha} R_\gamma) \cup F$ is consistent.

A proposition A is *strongly provable* in $(F, D)$, denoted by $F \hspace{-0.5mm}\mid\hspace{-2mm}\sim_D A$, iff for every extension $E$ of $(F, D)$, $A \in E$.

It is easy to see that Poole' system without constraints is a limiting case of POP DT when $\mathcal{P} = \{D\}$ and Nebel' RDT is the special case when $\eta_D$ is finite. Unfortunately, as pointed out by [Nebel 1992], the inference relation $\hspace{-0.5mm}\mid\hspace{-2mm}\sim_D$ generated by syntax-based extensions still fails to satisfy the rational monotony. [Zhang 1996] modified the definition of extensions into the following form:

a set $E$ is a *syntax-independent extension* of $(F, D)$ if $E = Cn((\bigcup_{\alpha < \eta_D} R_\alpha) \cup F)$ such that for all $\alpha < \eta_D$, $R_\alpha \subseteq Cn(\bigcup_{\gamma \leq \alpha} D_\gamma)$ and $R_\alpha$ is maximal among the subsets of $Cn(\bigcup_{\gamma \leq \alpha} D_\gamma)$ such that $(\bigcup_{\gamma \leq \alpha} R_\gamma) \cup F$ is consistent.

This approach, though slightly complicated, can yet be regarded as 'natural'. The only difference between two types of extensions is that the former does not satisfies the principle of irrelevance of syntax but the latter does.

On the basis of the notion of syntax-independent extensions, we have the following result:

**Theorem 5.1** *Let $D$ be a set of formulas in a language $\mathcal{L}$. For any perfect-ordered partition $\Sigma$ of $D$, $(\mathcal{L}, \hspace{-0.5mm}\mid\hspace{-2mm}\sim_D)$ is a finite supracompact rational nonmonotonic frame.*

The limited space does not afford a direct proof of the theorem. An indirect one may be done by using the result in [Zhang 1996] that $F \hspace{-0.5mm}\mid\hspace{-2mm}\sim_D A$ iff $A \in Cn(D) \otimes F$.

## 6  Discussions and Conclusions

We have established a very close connection between the general patterns of nonmonotonic reasoning and the multiple belief revision. This enables us to take the strategy to use methods from belief revision, set-theoretical, to contribute to a better understanding of nonmonotonic reasoning. We have seen that **RN** is such a strong system that almost all the rules suggested for nonmonotonic inference in the literature are the derived rules of **RN**. One may think that much more consequences would be derived in **RN** than in the classical logic from the same premises. This is clearly false when none of the pieces of background knowledge is available. Precisely specking, we have

**Proposition 6.1** Let $(\mathcal{L}, \hspace{-2pt}\mid\hspace{-4pt}\sim)$ be a rational nonmonotonic inference frame. If $K = \{B : \phi \hspace{-2pt}\mid\hspace{-4pt}\sim B\} = Cn(\phi)$, then

$$\Gamma \hspace{-2pt}\mid\hspace{-4pt}\sim A \quad \text{iff} \quad \Gamma \vdash A.$$

Furthermore, even though we equip with the whole background knowledge, the upshot is still less optimistic.

**Proposition 6.2** For any propositional language $\mathcal{L}$, there is a rational nonmonotonic frame $(\mathcal{L}, \hspace{-2pt}\mid\hspace{-4pt}\sim)$ such that for any $\Gamma \subseteq \mathcal{L}$ and any formula $A \in \mathcal{L}$,
i). if $K \cup \Gamma$ is consistent, then $\Gamma \hspace{-2pt}\mid\hspace{-4pt}\sim A$ iff $K \cup \Gamma \vdash A$;
ii). if $K \cup \Gamma$ is inconsistent, then $\Gamma \hspace{-2pt}\mid\hspace{-4pt}\sim A$ iff $\Gamma \vdash A$.
where $K = \{B : \phi \hspace{-2pt}\mid\hspace{-4pt}\sim B\}$.

This means that we can not always count on entailing more information from nonmonotonic inference rules alone than from classical ones. For example, even if we are told that $\phi \hspace{-2pt}\mid\hspace{-4pt}\sim p \to q$ and $\phi \hspace{-2pt}\mid\hspace{-4pt}\sim \neg p$, we still can not conduct the inference $p \hspace{-2pt}\mid\hspace{-4pt}\sim q$. There are two ways to surmount this obstacle. One is to construct some sort of ordering for the background knowledge such as nice-(perfect-)ordered partition, epistemic entrenchment or expectation ordering. The other is to transform the background knowledge into a conditional knowledge base as [Kraus *et al.* 1990] and [Lehmann and Magidor 1992] have already done. After all, the less we know, the less we can do.

# References

[Boutilier 1994] Craig Boutilier, Unifying default reasoning and belief revision in a modal framework, *Artificial Intelligence*, 68(1994), 33-85.

[Brewka 1989] Gerhard Brewka, Preferred subtheories: an extended logical framework for default reasoning, in: *Proceedings IJCAI-89*, (Detroit, Mich., 1989) 1034-1048.

[Brewka 1991] Gerhard Brewka, Belief revision in a framework for default reasoning, in: A. Fuhrmann and M. Morreau eds. *The Logic of Theory Change*, (LNCS 465, Springer-Verlag, Berlin, Germany, 1991) 185-205.

[Cravo and Martins 1993] Maria R. Cravo and João P. Martins, A unified approach to default reasoning and belief revision, in: M. Filgueiras and L. Damas eds., *Progress in Artificial Intelligence*, (LNAI 727, Springer-Verlag, 1993), 226-241.

[Freund 1990] Michael Freund, Supracompact inference operations, in: J. Dix, K. P. Jantke and P. H. Schmitt, eds. *Non-monotonic and Inductive Logic*, (LNAI 543, Springer-Verlag, 1990), 59-73.

[Gärdenfors 1988] P. Gärdenfors, *Knowledge in Flux: Modeling the Dynamics of Epistemic States* (The MIT Press, 1988).

[Gärdenfors and Makinson 1994] Peter Gärdenfors and David Makinson, Nonmonotonic inference based on expectations, *Artificial Intelligence* 65(1994), 197-245.

[Gärdenfors and Rott 1995] P. Gärdenfors and H. Rott, Belief revision, in: D. M. Gabbay C. J. Hogger and J. A. Robinson eds., *Handbook of Logic in Artificial Intelligence and Logic Programming*, Clarendon Press, Oxford, 1995, 35-132.

[Herre 1994] Heinrich Herre, Compactness properties of nonmonotonic inference operations, in: C. MacNish, D. Pearce and L. M. Pereira eds., *Logics in Artificial Intelligence*, (LNAI 838, Springer-Verlag, 1994), 19-33.

[Kraus *et al.* 1990] Sarit Kraus, Daniel Lehmann and Menachem Magidor, Nonmonotonic reasoning, preferential models and cumulative logics, *Artificial Intelligence*, 44(1990), 167-207.

[Lehmann and Magidor 1992] Daniel Lehmann and Menachem Magidor, What does a conditional knowledge base entail?, *Artificial Intelligence*, 55(1992), 1-60.

[Li 1993] Wei Li, An open logic system, *Scientia Sinica* (Series A), March ,1993.

[Makinson 1989] David Makinson, General theory of cumulative inference, in: M. Reinfrank, J. de Kleer, M. L. Ginsberg and E. Sandewall, eds., *Non-monotonic Reasoning*, (LNAI 346, Springer-Verlag, 1989), 1-18.

[Makinson 1993] David Makinson, General patterns in nonmonotonic reasoning, in: D. Gabbay, ed., *Handbook of Logic in Artificial Intelligence and Logic Programming*, (Oxford University Press, 1993), 35-110.

[Makinson and Gärdenfors 1991] David Makinson, Peter Gärdenfors, Relations between the logic of theory change and nonmonotonic logic, in: A. Fuhrmann and M. Morreau eds. *The Logic of Theory Change*, (LNCS 465, Springer-Verlag, Berlin, Germany, 1991) 185-205.

[Nebel 1992] Bernhard Nebel, Syntax based approaches to belief revision, in: P. Gärdenfors ed., *Belief Revision* (Cambridge University Press, Cambridge, 1992) 52-88.

[Poole 1988] David Poole, A logical framework for default reasoning, *Artificial Intelligence*, 36(1988), 27-47.

[Ryan 1991] Mark Ryan, Defaults and revision in structured theories, in: *1991 IEEE 6th Annual Symposium on Logic in Computer Science*, IEEE Computer Society Press, Los Alamitos, California, 1991.

[Zhang 1995] Dongmo Zhang, A general framework for belief revision, in:*Proc. 4th Int. Conf. for Young Computer Scientists* (Peking University Press, 1995) 574-581.

[Zhang 1996] Zhang Dongmo, Belief revision by sets of sentences, *Journal of Computer Science and Technology*, 1996, 11(2), 1-19.

[Zhang *et al.* 1997] Dongmo Zhang, Shifu Chen, Wujia Zhu, and Zhaoqian Chen, Representation theorems for multiple belief change, *this volume*.

# AUTOMATED REASONING

Automated Reasoning 3: Theorem Proving

# High Performance ATP Systems by Combining Several AI Methods

**Jörg Denzinger**
Fachbereich Informatik
Universität Kaiserslautern
67663 Kaiserslautern
Germany

**Marc Fuchs**
Fakultät für Informatik
TU München
80290 München
Germany

**Matthias Fuchs**
Fachbereich Informatik
Universität Kaiserslautern
67663 Kaiserslautern
Germany

## Abstract

We present a design for an automated theorem prover that controls its search based on ideas from several areas of artificial intelligence (AI). The combination of case-based reasoning, several similarity concepts, a cooperation concept of distributed AI and reactive planning enables a system to learn from previous successful proof attempts. In a kind of bootstrapping process easy problems are used to solve more and more complicated ones. We provide case studies from two domains in pure equational theorem proving. These case studies show that an instantiation of our architecture achieves a high grade of automation and outperforms state-of-the-art conventional theorem provers.

## 1 Introduction

Research concerned with achieving more efficient (fully automated) theorem provers focuses on three directions: higher inference rates, eliminating unnecessary inferences, and better control of the search. Although all these directions can indeed lead to more efficiency, better control of the search offers the highest gains, but also causes the most problems and has some risks. Nearly all approaches to improving search control involve the use of techniques and methods from other areas of artificial intelligence (AI), as for example knowledge representation, case-based reasoning (CBR), learning, planning, or multi-agent systems. In most of the known works only ideas from one of these areas are exploited.

One area that should—from the human point of view—be the most promising for high efficiency gains is learning. But the use of machine-learning techniques for improving automated theorem provers faces several severe problems. Learned knowledge has to be stored, retrieved, and very often must be combined. So, the focus of attention should not be restricted to the area of machine learning. Other areas of AI must also contribute in order to successfully apply the results the learning techniques produce.

In this paper we present an approach to controlling the search of an automated theorem prover that combines techniques from several areas of AI to overcome the problems that arise when trying to learn and to use control knowledge. The central idea is to utilize a known (i.e., learned) proof of a so-called *source problem* solved previously in order to guide the search for a proof of the *target problem* at hand. To this end we employ a method called *flexible re-enactment* (cp. [9]).

Source problems must of course satisfy certain similarity criteria with respect to the target problem. Our techniques for maintaining a database of source problems and our mechanisms for selecting source problems that are the most similar to the target are inspired by CBR. Unfortunately, one of the important premises of CBR, namely that "*small differences between problems result in small differences of their solutions*", is not fulfilled in automated theorem proving.

We cope with this uncertainty by applying the TEAMWORK method ([4]), a multi-agent approach to distributed search. TEAMWORK reduces the risk of deploying an inappropriate heuristic by having a *team* of heuristics (agents) guide the search *concurrently* and *cooperatively*. The reactive planning capabilities of a further agent, namely the supervisor, are made use of to compose a suitable team (cf. [7]). Moreover, the selection of the most suitable source problem required by flexible re-enactment can also be integrated with TEAMWORK in form of a specialized agent.

The combination of all these AI methods allowed us to build a theorem prover for pure equality reasoning that is fully automated, in both learning and proving, and is able to solve hard problems by using a kind of bootstrapping process that starts with easy problems and uses their proofs to gradually solve harder and harder problems. Besides providing the problems, no interaction with the system is required. Our experiments validate and substantiate the achievements of our system.

We cannot provide as many details as some readers (and we) would like. These readers may refer to [6].

## 2 Equational Reasoning

Equational reasoning deals with the following problem: Given a finite set $E$ of equations (of terms over a fixed signature $sig$) and a goal $u = v$. The question is whether the goal equation is a logical consequence of $E$, i.e., $E \models$

$u = v$. Unfailing completion (e.g., [2]) has proven to be quite successful for solving such a *proof problem* $\mathcal{A} = (E, u = v)$. The method is also a good example for so-called generating calculi that are based on generating new facts until a fact describing the goal is reached.

The inference rules of a generating theorem prover can be divided into two classes: *expansion* and *contraction* rules (see [3]). Completion uses the expansion *critical-pair-generation* and the contractions *reduction* and *subsumption*. Basis for the completion procedure is a so-called reduction ordering $\succ$ that is used to restrict the applicability of the inference rules and to avoid cycles.

An algorithmic realization of the inference rules of a generating theorem prover can be characterized as follows. There are two sets of facts (equations in our case): the set $F^A$ of *active facts* and the set $F^P$ of *passive facts*. The algorithm centers on a main loop with the following body: At first a fact $\lambda$ is selected and removed from $F^P$ ("activate $\lambda$"). After that $\lambda$ is normalized resulting in a fact $\lambda'$. (Normalization denotes the application of contraction rules to a fact until none of these rules is applicable anymore.) If $\lambda'$ is neither trivial nor subsumed by an active fact, all elements of $F^A$ are normalized, $\lambda'$ is added to $F^A$, and all facts that can be generated with $\lambda'$ and other elements of $F^A$ are added to $F^P$. A proof is found if the normalization of the two terms of the goal leads to the same term.

Assuming that there is a given order in which contracting inference rules are applied, normalization of a fact is a deterministic process. Hence, the remaining indeterminism is to determine which fact should be activated next. In order to eliminate this indeterminism, selection strategies and heuristics are used (see, e.g., [5]). In section 4 we present such a selection heuristic that is based on re-enacting a successful proof attempt for a problem that is somewhat *similar* to the problem at hand.

Since we want to learn from (successful) proof attempts, we have to obtain, represent, and store an actual proof run produced by the algorithm and a selection heuristic. $\mathcal{S}_\mathcal{A}$ denotes the sequence of facts activated during a proof run for problem $\mathcal{A}$ using a fixed heuristic $\mathcal{H}$. The actual proof to $\mathcal{A}$ is denoted by $\mathcal{P}_\mathcal{A}$ and it is obtained by eliminating from $\mathcal{S}_\mathcal{A}$ all facts that did not contribute to the proof. We refer to the facts occurring in $\mathcal{P}_\mathcal{A}$ also as the set $P_\mathcal{A}$ of *positive facts*. The other facts in $\mathcal{S}_\mathcal{A}$ form the set $N_\mathcal{A}$ of negative facts that is needed for some learning approaches (see [8; 9]). A successful proof attempt is stored as the quadruple $(\mathcal{A}, \mathcal{H}, \mathcal{S}_\mathcal{A}, \mathcal{P}_\mathcal{A})$ (or $(\mathcal{A}, \mathcal{C}, \mathcal{S}_\mathcal{A}, \mathcal{P}_\mathcal{A})$ with $\mathcal{C}$ denoting the teams used) that allows the use of various approaches for learning from previous proof experience.

## 3 Teamwork

The TEAMWORK method is a knowledge-based distribution method for certain search processes ([4]). Equational deduction by completion, as well as for example first-order deduction by (hyper-) resolution, is a member of this class of search processes. In a TEAMWORK-based system there are four different types of agents: experts, specialists, referees, and a supervisor. Experts and specialists are the agents that work on really solving a given problem. *Experts* form the core of a team. They are problem solvers (in our case theorem provers) that use the same inference mechanism (in our case unfailing completion), but different selection strategies for the next inference step to do. *Specialists* can also search for a solution (using other inference mechanisms) or they can help the supervisor, for example by analyzing and classifying the given problem like PES (see section 5.2). Each expert/specialist needs its own computing node. Therefore, the supervisor determines the subset of experts/specialists that are active during a working period.

After a working period a *team meeting* takes place. In the *judgment phase*, each active expert and specialist is evaluated by a referee. Each referee has two tasks: judging the whole work of the expert/specialist of the last working period and selecting outstanding results. The first task results in a *measure of success*, an objective measure that allows the supervisor to compare the experts. The second task is responsible for the cooperation of the experts and specialists, since each selected result will be part of the common start search state of the next working period. The referees send the results of their work to the supervisor.

In the *cooperation phase* the supervisor has to construct a new starting state for the next working period, select the members of the team for this next period and determine the length of the period. The new start state for the whole team consists of the whole search state of the best expert enriched by the selected results of the other experts and the specialists. The supervisor determines the next team with a reactive planning process involving general information about components and problem domains (*long-term memory*) and actual information about the performance of the components (*short-term memory*). The long-term memory suggests a *plan skeleton* that contains several small teams for different phases of a proof attempt. These suggested teams are reinforced with appropriate experts/specialists (if more computing nodes are available). During each team meeting the plan has to be updated. This means that adjustments are made according to the actual results (see [7]).

TEAMWORK allows for synergetic effects that result in enormous speed-ups and in finding solutions to problems that are beyond the possibilities of the single experts and specialists. While the competition of the experts directs the whole team into interesting (and promising) parts of the search space, the cooperation provides the experts with excellent facts they are not able to come up with alone. Thus gaps in their derivations towards the goal can be closed. This makes TEAMWORK the ideal basis for a learning theorem prover.

## 4 Flexible Re-enactment

Similarity between two proof problems $\mathcal{A}$ and $\mathcal{B}$ can occur in many variations. One possible kind of similarity is that a considerable number of the facts that contribute to a proof of $\mathcal{A}$ are also useful for proving $\mathcal{B}$ (or vice versa). This means in our terminology that the associated sets of positive facts $P_\mathcal{A}$ and $P_\mathcal{B}$ or the proofs $\mathcal{P}_\mathcal{A}$ and $\mathcal{P}_\mathcal{B}$ "have a lot in common" or, in other words, share many facts. ("$P_\mathcal{A} \cap P_\mathcal{B}$ is almost equal to $P_\mathcal{A}$ and $P_\mathcal{B}$.") Our goal is to think up a heuristic that is able to exploit such a similarity.

Given $\mathcal{I} = (\mathcal{A}_S, \mathcal{H}, \mathcal{S}_{\mathcal{A}_S}, \mathcal{P}_{\mathcal{A}_S})$ as past experience regarding a source problem $\mathcal{A}_S$, and assuming that a target problem $\mathcal{A}_T$ is similar to $\mathcal{A}_S$ in the way just described, it is reasonable to concentrate on mainly deducing facts when attempting to prove $\mathcal{A}_T$ that also played a role in finding the source proof $\mathcal{P}_{\mathcal{A}_S}$, namely the positive facts $P_{\mathcal{A}_S}$. We therefore design a heuristic FlexRE which—when trying to prove $\mathcal{A}_T$—makes use of $\mathcal{I}$ by giving preference to facts that were important for finding $\mathcal{P}_{\mathcal{A}_S}$. Such facts will henceforth be referred to as *focus facts*. Note that focus facts are facts inferred or inferable in connection with $\mathcal{A}_T$. They must be distinguished from the positive facts $P_{\mathcal{A}_S}$ belonging to the source problem $\mathcal{A}_S$, since it might be the case that some $\lambda \in P_{\mathcal{A}_S}$ is not deducible at all in connection with $\mathcal{A}_T$ (due to a different axiomatization). $P_{\mathcal{A}_S}$ is merely used to determine if some fact $\lambda$ inferable in connection with $\mathcal{A}_T$ is a focus fact. To put it another way, the use of $P_{\mathcal{A}_S}$ is effected by FlexRE on a strictly heuristic basis, meaning that $P_{\mathcal{A}_S}$ only influences the selection of facts from $F^P$, not, for instance, $F^P$ itself. That is, $P_{\mathcal{A}_S}$ is a guideline that FlexRE tries to follow if possible.

Depending on how strongly focus facts are preferred, FlexRE will re-enact (parts of) $\mathcal{P}_{\mathcal{A}_S}$ more or less quickly. Some of the focus facts, though useful for proving $\mathcal{A}_S$, may be irrelevant regarding the proof $\mathcal{P}_{\mathcal{A}_T}$ of $\mathcal{A}_T$ eventually found. But these irrelevant focus facts are not a big problem. The crucial difficulty is to find those (non-focus) facts that have to supplement the relevant focus facts in order to obtain a proof $\mathcal{P}_{\mathcal{A}_T}$. It is very likely that these (few) missing facts are descendants of relevant focus facts. Consequently, FlexRE should also favor descendants of focus facts. Favoring descendants should weaken with their "distance" from focus facts, since it cannot be assumed that the few missing facts are located very deeply relative to focus facts.[1]

Preferring descendants of focus facts in addition to giving preference to focus facts themselves justifies the attribute 'flexible' in the term 'flexible re-enactment' which summarizes the working method of FlexRE (see [9] or [6]).

## 5 Learning and CBR in the Teamwork Environment

In sections 3 and 4 we concentrated on how to use knowledge learned from previous successful proofs (in form of flexible re-enactment) and on how to overcome the problems such a use might cause (in form of the TEAMWORK method with cooperation with other experts, assessment of experts and results, and reactive planning to adapt to the problem at hand using long- and short-term memory). The problems that remain are how to find a proof that should be re-enacted in order to solve a given target problem and how to structure, build, and maintain the long-term memory from proof run to proof run.

The first problem will be tackled by a specialist PES that is providing the supervisor with information about known proof problems that are similar to the given target problem (see subsection 5.2). The second problem naturally depends on how the proof problems are presented to the system. Found proofs have simply to be extracted, analyzed and stored (the latter depending on how specialist PES will perform its retrieval). As we shall see in the next subsection, the necessary components are already provided in form of TEAMWORK agents.

### 5.1 The Basic Learning Cycle

Systems that use learning techniques for solving their tasks can be (very) roughly divided into two groups: systems that have a clearly defined learning phase after which (usually) no further learning takes place, and systems that always learn. In automated theorem proving, systems of the first type may be usable in clearly defined situations (see, for example, [8]), but in general learning should never stop.

Nevertheless, one can observe times in the use of a (learning) theorem prover in which new domains are explored, and other times in which one is interested in proving one particular problem. When exploring a new domain, typically there is a set of problems to be solved, and when starting the exploration no knowledge in the prover will be triggered. In the following, we will first concentrate on the exploration of a new domain and then we will point out how the one-problem case is handled.

When exploring a new domain the ordering of the problems given to a prover may influence its success. In order to deal with this problem we decided to let the prover handle the ordering of the given set of problems and also allow the prover to make several attempts to solve a problem. The latter is necessary since each solved problem may result in new knowledge that allows for solving some other problems that could not be solved so far ("*bootstrapping*"). Note that the set of problems given to the prover has to include easy and typical problems of a domain that the prover can use to get fundamental knowledge about this domain.

As already stated, in a TEAMWORK-based system the long-term memory that represents knowledge about domains is the responsibility of the supervisor. When confronted with a set of example problems of a new domain, the supervisor controls not only the single proof attempts, but a whole series of proof attempts that are to result in solving as many of the problems as possible.

Since the supervisor has no appropriate information when being confronted with a new domain, the first step

---

[1] "Distance" and (relative) depth basically refer to the number of inference steps separating two facts, one of these facts contributing to the deduction of the other.

is to try to solve the given problems with conventional means, i.e., without the use of experts and specialists that employ learned knowledge. This is accomplished by using a pre-defined team for a few cycles for each of the given problems (in a separate run). After generating $\mathcal{I} = (\mathcal{A}, \mathcal{C}, \mathcal{S}_{\mathcal{A}}, \mathcal{P}_{\mathcal{A}})$ for each solved problem $\mathcal{A}$, this data is integrated into a database of past proof experience that is part of the long-term memory. This database is essentially organized as case base. Hence, the structure and retrieval processes regarding this database are strongly related to CBR techniques (cp. [10] and subsection 5.2). Then the supervisor tries to solve the remaining problems (again imposing a time limit on each run), but now the teams are different. The team of the first working period is again pre-defined, and contains, besides good general purpose experts, the specialist PES.

After the first working period, the supervisor uses its reactive planning process to adapt the team to the problem. If PES was not able to report to the supervisor a problem from the case base that is similar to the target problem at hand, then the supervisor proceeds according to its standard procedure. Otherwise, the expert FlexRE will become a member of the next team, utilizing the reported problem from the case base as source problem.[2] Problems that can now be solved are analyzed so as to produce new data for the case base. For the remaining unsolved problems this whole process is repeated until no more new problems can be proven. Note that after the initial round that uses the pre-defined team without components using learned knowledge, in all other rounds each solved problem is immediately added to the case base so that it already can be used for the proof attempt of the next problem. Thus the number of rounds is often reduced.

In the one-problem case, i.e., in the presence of fundamental domain knowledge, the supervisor immediately employs a team that includes PES, and the supervisor plans and controls the whole proof attempt as described above. If the system is successful, the data of the run is also added to the case base.

## 5.2 Specialist PES

As described before, specialist PES retrieves one or possibly several source problems that are similar to a given target problem $\mathcal{A}_T$, and transmits information on these problems to the supervisor. More exactly, after receiving a target problem $\mathcal{A}_T = (Ax_T, \lambda_T)$ given over a fixed signature $sig_T$, PES returns information $R_{\text{PES}}(\mathcal{A}_T)$ on similar problems $\mathcal{A}_S$, where $R_{\text{PES}}(\mathcal{A}_T) = \{(\sigma, P_S) : cond_T(\sigma(\mathcal{A}_S))\}$. The data is determined by $cond_T$ and comprises the set of positive facts $P_S$ associated with source problem $\mathcal{A}_S$, and a signature match $\sigma$ from $\mathcal{A}_S$ to $\mathcal{A}_T$. ($\sigma$ provides an appropriate renaming of the function symbols occurring in $sig_S$.) $cond_T$ denotes that a problem $\mathcal{A}_S$ (translated from $sig_S$ to $sig_T$ by applying $\sigma$) is most similar to $\mathcal{A}_T$ (there are no other problems more similar to $\mathcal{A}_T$ than $\mathcal{A}_S$), but also that the similarity between $\mathcal{A}_S$ and $\mathcal{A}_T$ seems to be sufficient.

In order to construct such a predicate we shall use a quasi-ordering $\succeq_T$ that allows us to compare the similarity between proof problems and a target problem. It should hold true that $\mathcal{A}_1 \succ_T \mathcal{A}_2$ if (and only if) $\mathcal{A}_T$ is more similar to $\mathcal{A}_1$ than to $\mathcal{A}_2$. Furthermore, an absolute measure of similarity is needed in order to construct a *minimal similarity* predicate $ms$ that estimates whether the similarity between the target and a source problem seems to be sufficient or not.

Basically, there are two possibilities to construct $\succeq_T$ in order to estimate whether a problem $\mathcal{A}_1$ can provide a more suitable source proof than a problem $\mathcal{A}_2$: On the one hand it is possible to compare the problem descriptions, and on the other hand one can compare the search effort spent on solving the two problems. Considering the working method of FlexRE it is reasonable to consider $\mathcal{A}_1$ as providing a more suitable source proof than $\mathcal{A}_2$, if the target $\mathcal{A}_T$ is more similar to problem $\mathcal{A}_1$ than to $\mathcal{A}_2$ w.r.t. its problem description. Nevertheless some information on the search conducted to solve a problem, namely the length of the search protocol $\mathcal{S}$, can be consulted. Since some of the positive facts of more difficult problems (problems with a longer search protocol) were quite difficult to reach it seems to be sensible to force the activation of these probably useful, but hard to reach facts by using them as focus facts (cp. section 4). Thus more difficult problems should be favored.

Similarity between problem descriptions is surely also suitable for constructing an absolute similarity value in order to decide in favor of or against sufficient similarity. Information on the search protocol is difficult to use for this particular purpose and should only be used as a criterion to compare different problems.

In order to assess the differences between the problem descriptions of two proof problems we employ the similarity measure $sim_T$. As discussed before, this measure is useful to construct $\succeq_T$ and $ms$. The design of $sim_T$ is motivated by the fact that a target problem $\mathcal{A}_T = (Ax_T, \lambda_T)$ is proved by virtue of a proof of a source problem $\mathcal{A}_S = (Ax_S, \lambda_S)$, if all axioms of $Ax_S$ are subsumed by axioms of $Ax_T$ and the target goal $\lambda_T$ is subsumed by the source goal $\lambda_S$. Hence, subsumption criteria will play the major role in our approach. Furthermore, refinements by using other criteria (e.g., subsumption modulo $AC$) are imaginable (see below). These refinements have proven their usefulness in experiments, although naturally a simple proof replay often is impossible if $Ax_S$ and $Ax_T$ are similar in such a way.

In order to realize measure $sim_T$ we introduce a (asymmetric) similarity rating $sim_T^{eq}$ defined on equations over $sig_T$. Let $ax_1$ and $ax_2$ be two equations. We define $sim_T^{eq}(ax_1, ax_2) \in [0; 1]$ by $sim_T^{eq}(ax_1, ax_2) = \max\{rating(\rho_i) : \rho_i(ax_1, ax_2), 1 \leq i \leq m\}$. $\rho_1, \ldots, \rho_m$

---

[2]If PES provides more than one possible source problem, then the supervisor can react in various ways: It can select the most similar source problem and discard all others (which was done in section 6), or it may use several experts FlexRE, each using a different source problem, or it may supply the single FlexRE with the source problems provided by PES in succession.

are similarity criteria defined on equations. The function *rating* judges the reliability of a measure in order to judge similarity between two equations.

Possible similarity criteria are the relations $\triangleleft$, $\triangleleft_A$, and $\triangleleft^H \cup \triangleright^H$ where $\triangleleft$ denotes "plain" subsumption, and $\triangleleft_A$ subsumption modulo the theory given by a set of equations $A$. (Here, we shall always use $A = AC$). $ax_1 \triangleleft^H ax_2$ stands for a homeomorphic embedding of $ax_2$ in $ax_1$. We used the ratings 1, 0.8, and 0.2 for the criteria $\triangleleft$, $\triangleleft_A$, and $\triangleleft^H \cup \triangleright^H$, respectively. Hence, subsumption is considered to be very important, whereas homeomorphic embedding is considered to be a very weak similarity criterion. Note that we actually refined $sim_T^{eq}$ by adding further similarity criteria thus increasing the ability to produce distinctive measures. But a description of these technical details is beyond the scope of this paper.

With the help of this measure we are able to construct a similarity measure defined on proof problems. In the following let $\mathcal{A}_T = (Ax_T, \lambda_T)$ be a target problem, and $\mathcal{A}_S = (Ax_S, \lambda_S)$ be a source problem given over $sig_T$. Let $Ax_T = \{ax_1, \ldots, ax_m\}$, $Ax_S = \{ax'_1, \ldots, ax'_n\}$ $(n, m > 0)$. The similarity of target and source problem is $sim_T(\mathcal{A}_T, \mathcal{A}_S) = (s_1, s_2, s_3) \in [0; 1]^3$, where

$$s_1 = \frac{1}{n} \cdot \sum_{i=1}^{n} \max\{sim_T^{eq}(ax, ax'_i) : ax \in Ax_T\}$$

$$s_2 = \frac{1}{m} \cdot \sum_{i=1}^{m} \max\{sim_T^{eq}(ax_i, ax) : ax \in Ax_S\}$$

$$s_3 = sim_T^{eq}(\lambda_S, \lambda_T)$$

Thus, $s_1$ judges the degree of "coverage" of $Ax_S$ through similar axioms of $Ax_T$. For example, we have $s_1 = 1$ if all source axioms are subsumed by target axioms. $s_1$ decreases if only weaker similarity criteria are fulfilled. The value $s_2$ represents the percentage of target axioms that have a similar counterpart in $Ax_S$. Additional axioms in $Ax_T$ do not prevent the source proof from being applicable (in the case $s_1 = 1$), but may complicate the search for this proof. Finally, $s_3$ measures the similarity between target and source goal $\lambda_T$ and $\lambda_S$. We have $s_3 = 1$ if $\lambda_S$ subsumes $\lambda_T$.

With the help of $sim_T$ we are now able to estimate if the similarity between $\mathcal{A}_T$ and $\mathcal{A}_S$ seems to be sufficient. To this end, we check if the predicate $ms$ is fulfilled with $ms(sim_T(\mathcal{A}_T, \mathcal{A}_S))$ iff $c_1 \cdot s_1 + c_2 \cdot s_2 + c_3 \cdot s_3 \geq min$, where $c_1, c_2, c_3 \in \mathbb{R}$, and $min \in \mathbb{R}$ is the threshold. In our implementation we use $c_1 = 3$, $c_2 = 1$, $c_3 = 2$, and $min = 1$. Hence additional axioms in $Ax_T$ are considered quite harmless, while a good coverage of $Ax_S$ is considered to be important. Since we use $min = 1$, a subsumption of one third of the axioms of $Ax_S$, or no superfluous target axioms, or a subsumed target goal each suffice alone to reach the threshold.

$sim_T$ is also employed to define $\succeq_T$. We define $\succeq_T$ as a lexicographic combination of two quasi-orderings $\geq_S$ and $\geq_D$, where $\geq_S$ compares the similarity of two problem descriptions with respect to the target problem (using $sim_T$), and $\geq_D$ compares the length of the search

protocols (cp. [6]). Using $\succeq_T$ and $ms$, we can define $cond_T$. We apply

$$cond_T(\mathcal{A}_S) \quad \text{iff} \quad ms(sim_T(\mathcal{A}_T, \mathcal{A}_S)) \wedge \neg \exists \mathcal{A} : \mathcal{A} \succ_T \mathcal{A}_S.$$

In the subsequent section we shall discuss the experimental results we obtained with DISCOUNT by using specialist PES and expert FlexRE.

## 6 Experimental Results

Finding appropriate test sets of problems for our learning prover was not easy, because in most publications only hard problems are given. But we must solve at least one problem with conventional means in order to start the bootstrapping process. Fortunately, the TPTP library ([13]) contains two domains that include related problems of various degrees of difficulty. While we could use the domain groups (GRP) without any changes, the domain logic calculi (LCL) consists of several subdomains (some of which contain only hard problems, again) and is not given in a pure unit-equality axiomatization. Therefore we had to transform the problems of one subdomain (the CN calculus) for our experiments.

The achievements of a learning approach can only be observed when results from conventional provers are also provided. Besides two of our experts (AddWeight, or Add for short, and Occnest, see [5]) we will also use OTTER (version 3.0, using the *autonomous mode*, see [11]) to allow a comparison. Since OTTER has won the CADE-13 theorem prover competition ([12]) in the category "Unit Equality Problems", we think that a comparison with the current state-of-the-art is thus provided.

Our learning approach is integrated into our prover DISCOUNT (see [1]), that is implemented in C on Unix machines and has an old and slow inference engine. We limited each distributed run of it to 3 minutes, while the other provers had 10 minutes. The ordering in which DISCOUNT tried the problems is the lexicographical ordering of the names of the problem. All experiments were performed on SUN Sparc-10 machines, the team runs employing two of them. There are 125 problems in the GRP domain and 24 in the LCL (CN) domain (for the transformation into equality problems, for more results and a broader analysis of them, see [6]).

Table 1 shows that our main goal, developing a prover that is able to automatically learn and therefore to solve more problems than other provers, has definitely been achieved. In both domains, our learning team clearly outperformed OTTER and the single experts by at least 15 percent. Table 2 highlights some of the problem solution chains that are produced by our prover. In the GRP domain, problem 179-1 was solved conventionally and used in the next round to solve problem 179-2, which was basis for solving problem 183-1 (in the same round). Using problem 183-1 DISCOUNT was able to solve problem 167-3, which was then used to solve problem 167-1. In the LCL domain, problem 047-1 was used to solve 048-1 which allows for solving 050-1 that then is the source for problem 051-1. Due to the immediate usage of solved

Table 1: Comparison learning team vs. OTTER vs. best experts

| Domain | number of problems | learning team | | OTTER | | Add | | Occnest | |
|---|---|---|---|---|---|---|---|---|---|
| | | # solved | % | # solved | % | # solved | % | # solved | % |
| GRP | 125 | 113 | 90 | 93 | 74 | 86 | 67 | 91 | 73 |
| LCL (CN) | 24 | 17 | 71 | 11 | 46 | 12 | 50 | 6 | 25 |

Table 2: Selection of results

| target | source | runtime | OTTER | Add | Occnest |
|---|---|---|---|---|---|
| 179-1 | — | 12s | — | — | — |
| 179-2 | 179-1 | 37s | — | — | — |
| 183-1 | 179-2 | 40s | — | — | — |
| 167-3 | 183-1 | 129s | — | — | — |
| 167-1 | 167-3 | 32s | — | — | — |
| 047-1 | — | 21s | 137s | 27s | — |
| 048-1 | 047-1 | 14s | 138s | 27s | 35s |
| 050-1 | 048-1 | 32s | — | — | — |
| 051-1 | 050-1 | 15s | — | — | — |

problems when trying to solve the next one, the chain of the GRP domain is the result of 4 rounds, while the chain of the LCL domain was produced in only 3 rounds of the bootstrapping process. Note that any other ordering of the problems would produce the same Table 1 and the same chains in Table 2. Only the number of rounds needed may be different.

Among all the problems there is only one problem that OTTER can solve and our learning team cannot, but 27 problems that our learning team can solve and OTTER cannot.

In general, our experiments show that our concept of a learning theorem prover clearly outperforms current conventional theorem provers if the learning prover is provided with enough "exercise" in the domain it has to work in. In this case even the learning process is accomplished by the prover without help from the user.

## 7 Conclusion and Future Work

We presented a concept for a learning theorem prover that uses methods from several areas of AI. AI methods like planning and CBR are combined with the TEAMWORK multi-agent architecture, resulting in a theorem prover that clearly outperforms renowned provers. A prerequisite for the success of our system—as for a human student—is the presentation of problem domains in a "learnable" way, meaning that the presented problems cover the whole spectrum of difficulty ranging from easy to challenging. In a kind of bootstrapping process the system is able to solve harder and harder problems without any interaction on the parts of a user ("teacher").

Despite the success of our system, nevertheless most components are only first ideas that leave much room for improvements and extensions. Also, there are other concepts that can be used (e.g., [8]) or are at least worth investigating (e.g., the division of problems into easier sub-problems on the basis of learning), that will provide a wider range in the use of learned knowledge.

## Acknowledgments

This work was supported by the *Schwerpunktprogramm Deduktion* of the *Deutsche Forschungsgemeinschaft (DFG)*.

## References

[1] Avenhaus, J.; Denzinger, J.; Fuchs, M.: DISCOUNT: A System For Distributed Equational Deduction, *Proc. 6th RTA*, Kaiserslautern, LNCS 914, 1995, pp. 397–402.

[2] Bachmair, L.; Dershowitz, N.; Plaisted, D.: Completion without Failure, *Coll. on the Resolution of Equations in Algebraic Structures*, Austin (1987), Academic Press, 1989.

[3] Dershowitz, N.: A maximal-Literal Unit Strategy for Horn Clauses, *Proc. 2nd CTRS*, Montreal, LNCS 516, 1990, pp. 14–25.

[4] Denzinger, J.: Knowledge-Based Distributed Search Using Teamwork, *Proc. ICMAS-95*, San Francisco, AAAI-Press, 1995, pp. 81–88.

[5] Denzinger, J.; Fuchs, M.: Goal-oriented equational theorem proving using teamwork, *Proc. 18th KI-94*, Saarbrücken, LNAI 861, 1994, pp. 343–354.

[6] Denzinger, J.; Fuchs, Marc; Fuchs, M.: High Performance ATP Systems by Combining Several AI Methods, SEKI-Report SR-96-09, University of Kaiserslautern, 1996. (ftp://ftp.uni-kl.de/reports_uni-kl/computer_science/SEKI/1996/Denzinger.SR-96-09.ps.gz)

[7] Denzinger, J.; Kronenburg, M.: Planning for Distributed Theorem Proving: The Teamwork Approach, *Proc. KI-96*, Dresden, LNAI 1137, 1996, pp. 43–56.

[8] Fuchs, M.: Learning proof heuristics by adapting parameters, *Proc. 12th ICML*, Tahoe City, CA, USA, 1995, pp. 235–243.

[9] Fuchs, M.: Experiments in the Heuristic Use of Past Proof Experience, *Proc. CADE-13*, New Brunswick, LNAI 1104, 1996, pp. 523–537.

[10] Kolodner, J.L.: An Introduction to Case-Based Reasoning, *Artificial Intelligence Review* 6:3–34, 1992.

[11] McCune, W.W.: OTTER 3.0 Reference manual and Guide, Tech. rep. ANL-94/6, Argonne National Laboratory, 1994.

[12] Sutcliffe, G.; Suttner, C.B.: The Design of the CADE-13 ATP System Competition, *Proc. CADE-13*, New Brunswick, LNAI 1104, 1996, pp. 146–160.

[13] Sutcliffe, G.; Suttner, C.B.; Yemenis, T.: The TPTP Problem Library, *Proc. 12th CADE*, Nancy, LNAI 814, 1994, pp. 252–266.

# Equational Reasoning using AC Constraints

David A. Plaisted and Yunshan Zhu
Computer Science Department
University of North Carolina
Chapel Hill, NC 27599-3175
{plaisted,zhu}@cs.unc.edu
Fax: (919)962-1799

## Abstract

Unfailing completion is a commonly used technique for equational reasoning. For equational problems with associative and commutative functions, unfailing completion often generates a large number of rewrite rules. By comparing it with a ground completion procedure, we show that many of the rewrite rules generated are redundant. A set of consistency constraints is formulated to detect redundant rewrite rules. We propose a new completion algorithm, consistent unfailing completion, in which only consistent rewrite rules are used for critical pair generation and rewriting. Our approach does not need to use flattened terms. Thus it avoids the double exponential worst case complexity of AC unification. It also allows the use of more flexible termination orderings. We present some sufficient conditions for detecting inconsistent rewrite rules. The proposed algorithm is implemented in PROLOG.

## 1 Introduction

Knuth-Bendix completion [Knuth and Bendix, 1970] and its extensions [Bachmair et al., 1989] have been widely used for equational reasoning. One of the main bottlenecks of completion-based inference strategies is the large number of critical pairs generated. This is particularly evident when dealing with equational problems involving associative and commutative functions, i.e. AC equational problems. In this paper, we present a technique for reducing redundant equational inferences using constraints.

AC equational problems represent an important class of problems in theorem proving. Many mathematical functions of interest are associative and commutative. For example, the union and intersection operations in set theory and the addition operations in ring structures are all associative and commutative. Furthermore, addition and multiplication in arithmetic are both AC. Some form of AC equational reasoning is necessary if a term rewriting system is used to perform integer arithmetic.

AC equational reasoning is often difficult. One of the most recognized recent successes of automated reasoning, perhaps of AI in general, is the solution of the Robbins Problem. The Robbins Problem is in fact formulated as an equational problem with AC functions [McCune, 1996]. Most term rewriting systems do inferences by generating rewrite rules using Knuth-Bendix completion or unfailing completion. Associativity axioms and commutativity axioms often cause an explosion of new rewrite rules in the completion process. When overlapped with the AC axioms, an existing rewrite rule can generate an exponential number of AC equivalent rewrite rules. Since commutativity axioms cannot be oriented in any termination ordering, many of these new rewrite rules generated from the AC axioms cannot be simplified.

Special completion procedures are developed to handle AC equational problems [Lankford and Ballantyne, 1977; Peterson and Stickel, 1981]. Most of the approaches in the literature use flattened terms. For example, $+(a, +(b, c))$ is represented as $+(a, b, c)$, and terms $+(a, c, b)$ and $+(a, b, c)$ are considered identical. A flattened term represents all terms equivalent up to the AC axioms. Since flattening breaks the original term structure, special unification techniques are needed. AC unification algorithms [Stickel, 1981; Domenjoud, 1991; Boudet et al., 1996] are developed to compute all possible most general unifiers of two terms up to associativity and commutativity axioms. Special AC termination orderings are also needed to show termination of AC rewriting systems. Many commonly used orderings, such as recursive path ordering and lexicographic path ordering, are no longer well founded when flattened terms are used. Several AC termination ordering have been devised [Dershowitz et al., 1983; Bachmair and Plaisted, 1985; Kapur et al., 1990].

We study the problem of AC equational reasoning with a different approach. We propose a procedure called consistent unfailing completion in which only consistent rules and equations are used for critical pair generation and rewriting. A consistent unfailing completion procedure can be regarded as the lifted version of a ground completion procedure. In a standard unfailing completion procedure, some rewrite rules do not correspond to any rewrite rules in the ground procedure. We call such rules inconsistent. In consistent unfailing completion, inconsistent rules are replaced by its consistent permutations. For example, $x*y*x \to x$ is inconsistent with respect to the lexicographic ordering. This is because $a_1*a_1*a_2$, an arbitrary instance of $x*y*x$, is either simplified to $a_1*a_1*a_2$ or $a_2*a_1*a_1$ depending on the ordering of $a_1$ and $a_2$. $y*x*x \to x$ and $x*x*y \to x$ are the only two consistent permutations of $x*y*x \to x$. We derive two sufficient conditions for detecting inconsistent rules. Consistency checking based on these conditions greatly reduces the number of critical pairs generated. In the examples that we tested, it shows a factor of 2 to 4 reduction. Our implementation of the algorithm is able to solve all six of equality benchmark problems proposed by [Lusk and Overbeek, 1985].

Consistent unfailing completion does not require the use of flattened terms. AC unification is therefore not needed. Detecting inconsistency takes quadratic time and generating consistent permutations takes single exponential time. Thus the worse case complexity of an inference rule in consistent unfailing completion is single exponential. The worse case complexity for AC unification is double exponential. A special AC termination ordering is not needed either. Consistent unfailing completion works for most of the commonly used orderings, such as recursive path ordering, lexicographic path ordering, etc. In fact, we have a built-in ordering in our implementation, and all of our examples are solved using the same ordering. The built-in ordering enables us to fully automate the equational reasoning process and incorporate it in a general first-order theorem prover.

The rest of the paper is organized as follows. We provide some background in the next section. We then present the consistent unfailing completion algorithm and show its completeness. We then show how the algorithm can be refined and implemented. We give some test results in the end.

## 2 Background

In this section, we define some relevant concepts. For surveys on equational reasoning, see [Plaisted, 1993; Dershowitz and Jouannaud, 1990; Klop, 1992]. For introductions to the area of theorem proving, see [Chang and Lee, 1973; Loveland, 1978; Wos et al., 1984]. We follow the conventions of [Plaisted, 1993] in this paper.

We use the standard definitions of term, substitution, instance, the most general unifier etc. Readers may consult the aforementioned references for more details. We use $f, g, h, \ldots$ as function symbols, $a, b, c, \ldots$ as constant symbols, and $x, y, z, \ldots$ as variables.

An equation is an expression of the form $s = t$ where $s$ and $t$ are terms. An equational system is a set of equations. Assume $E$ is an equational system, we use $E \models s = t$ to represent that $s = t$ is a logical consequence of $E$. The problem of equational reasoning can be reduced to the problem of first-order theorem proving. $E \models s = t$ iff $\{E \cup A_E\} \vdash_L s = t$, where $A_E$ is the set of equality axioms and L is any sound and complete first-order strategy. In practice, term rewriting based approaches are often much more efficient.

A rewrite rule is written as $r \to s$, where $r$ and $s$ are terms. A rule $r \to s$ indicates that an instance of $r$ can be replaced by an instance of $s$, but not vice versa. $r \leftrightarrow s$ if $r \to s$ or $s \to r$. Sometimes $r \to s$ is also rewritten as $s \leftarrow r$. A term rewriting system $R$ is a set of rewrite rules. We use $\to_R$ to represent the rewrite relation. For example, a term rewriting system $R$ is $\{f(x) \to b, g(b) \to a\}$. Then $g(f(x)) \to_R g(b)$. We define $\to_R^*$ as the reflexive transitive closure of $\to_R$. Thus $g(f(x)) \to_R^* a$. We define $\leftrightarrow_R^*$ as the reflexive transitive closure of $\leftrightarrow_R$. Suppose $R$ is a term rewriting system $\{r_1 \to s_1, \ldots, r_n \to s_n\}$, $R^=$ is defined as the associated equational system $\{r_1 = s_1, \ldots, r_n = s_n\}$. By Birkhoff's theorem [Birkhoff, 1935], $R^= \models r = s$ iff $r \leftrightarrow_R^* s$. This essentially shows the soundness and completeness of equational reasoning based on term rewriting. However, the rewriting system is not yet satisfactory for theorem proving purposes: to show $r \leftrightarrow_R^* s$, rewriting has to be done in both directions, and it is highly non-deterministic and inefficient.

A term $r$ is reducible if there is a term $s$ such that $r \to s$, otherwise $r$ is irreducible. If $r \to^* s$ and $s$ is irreducible then we call $s$ a normal form of $r$. We write $r \downarrow s$ if there is a term $u$ such that $r \to^* u$ and $s \to^* u$. We write $r \uparrow s$ if there is a term $u$ such that $u \to^* r$ and $u \to^* s$. We say a rewriting system $R$ is confluent if for all terms $r$ and $s$, $r \uparrow s$ implies $r \downarrow s$. It can be shown that if $R$ is confluent and $r \leftrightarrow_R^* s$ then $r \downarrow s$. A rewriting system $R$ is terminating if it has no infinite rewriting sequence. If there exists a termination ordering $>$ such that for all rules $l \to r$ in R, $l > r$, then $R$ is terminating. If $R$ is confluent and terminating, $r \leftrightarrow_R^* s$ can be decided by rewriting $r$ and $s$ to normal forms and compare the normal forms.

**Definition 2.1** *Suppose that $l_1 \to r_1$ and $l_2 \to r_2$ are two rewrite rules with no common variables. Suppose $u$ is a non-variable subterm of $l_1$, (if $l_1$ is a variable, $u$ may be a variable). Suppose $u$ and $l_2$ are unifiable and $\theta$ is the most general unifier. We call the pair $(l_1[u \leftarrow r_2]\theta, r_1\theta)$ a critical pair for rules $l_1 \to r_1$ and $l_2 \to r_2$. $l_1[u \leftarrow r_2]$ denotes the term obtained by replacing a specified occurrence of the subterm $u$ in $l_1$ with $r_2$.*

**Theorem 2.2** *Suppose a term rewriting system $R$ is terminating. $R$ is confluent iff for all critical pairs $(s, t)$ in $R$, $s \downarrow t$.*

If a rewriting system is not confluent, an equivalent rewriting system might be obtained by adding new rewriting rules. Theorem 2.2 can be used to generate such new rules. The idea is to add rule $s \to t$ or $s \leftarrow t$ to $R$ while preserving termination for all critical pairs $(s, t)$ in $R$. The algorithm was initially proposed in [Knuth and Bendix, 1970], and is often called Knuth-Bendix completion, or simply, completion. Sometimes, it is not possible to orient an equation, i.e. $f(x, y) =$

$f(y,x)$, in either direction while maintaining termination. Thus Knuth- Bendix completion is extended to unfailing completion[Bachmair et al., 1989]. In unfailing completion, if a critical pair $(s,t)$ can not be oriented then it is added as an equation $s = t$. Equation $s = t$ may participate in further critical pair generation. It is regarded as containing rules $s \to t$ and $t \to s$. However, the use of equation in rewriting is restricted : $s\theta$ can be replaced by $t\theta$ only if $s\theta > t\theta$, where $>$ is the termination ordering.

We give names to a set of inference rules. We follow the convention used in [Plaisted, 1993]. The rule $UCP(\to, \to)$ is the operation of generating critical pairs between two rewrite rules. If a critical pair can be oriented, a new rewrite rule is added to the rewrite system; otherwise, an equation is added. $UCP$ stands for Unfailing Critical Pair generation. Similarly, $UCP(=, \to)$ generates critical pairs between a rewrite rule and an equation; $UCP(=, =)$ generates critical pairs between two equations. Inference rule $SIMPRULE$ is the operation of simplifying a rewrite rule or an equation with other existing rewrite rules or equations. For example, if a rewriting system $R$ contains $\{f(x) \to x, f(b) \to f(a)\}$, the second rewrite rule can be simplified to $b \to a$, and thus $R$ contains $\{f(x) \to x, b \to a\}$ after rule $SIMPRULE$ is applied. $E \models s = t$ iff $E \cup \{s' \neq t'\}$ is unsatisfiable, where $s'$ and $t'$ are skolemized terms of $s$ and $t$, respectively. Thus we may include $\{s' \neq t'\}$ as an inequation in a rewriting system $R$ and show a contradiction can be derived from $R$. $SIMPGOAL$ is the operation of simplifying an inequation using existing rewrite rules or equations. The rule $CONTRA$ derives $FALSE$ when the inequation $s \neq s$ is derived. In applying these inference rules, the concept of *fairness* is needed. Fairness means that every henceforth possible inference will eventually be performed.

**Theorem 2.3** *Fair unfailing completion starting from a set of $R$ of rules, equations and ground inequation and using the inference rule $\{UCP(\to,\to), UCP(=,\to), UCP(=,=), SIMPRULE, SIMPGOAL, CONTRA\}$ will eventually generate $FALSE$ if $R^=$ is unsatisfiable.*

Theorem 2.3 show the completeness of an equational reasoning strategy based on unfailing completion. We will also use the fact that $SIMPRULE$ is inessential to the completeness of the strategy.

In this paper, we are concerned with equational systems that involve associative and commutative functions, or AC functions. An AC function $f$ satisfies the following axioms $\{f(f(x,y),z) = f(x,f(y,z)), f(x,y) = f(y,x)\}$.

## 3 AC-Consistent Completion

In this section, we introduce the concept of consistency for AC equational problems. We present a new completion algorithm in which only consistent rules can be used for critical pair generation and simplification. We call the new algorithm consistent unfailing completion. We show the correspondence between a consistent unfailing completion and a ground completion procedure. The correspondence helps us in deriving the consistency constraints and establishing the completeness proof of the consistent unfailing completion algorithm.

**Definition 3.1** *Suppose that $>$ is a termination ordering and $>$ is total on ground terms. A ground term $s$ is an AC minimal term if $s$ is the minimal term of its AC-equivalence class with respect to the ordering $>$. A rewrite rule $s \to t$ is consistent if there exists an instance $s\theta \to t\theta$ such that both $s\theta$ and $t\theta$ are AC minimal terms, and that $\theta$ assigns distinct ground terms to distinct variables in $s$ and $t$. The consistency of an equation (a critical pair) is similarly defined.*

**Example 3.2** *In this example, we use the lexicographic ordering as the termination ordering. We assume $f$ is an AC function and $g$ is not an AC function. $f(a, f(b,c))$ is an AC minimal term. $f(f(a,b),c)$ is not an AC minimal term, because $f(f(a,b),c) > f(a,f(b,c))$ and they are AC equivalent. We sometimes say $f(a,f(b,c))$ is the AC minimal term of $f(f(a,b),c)$. The rewrite rule $g(f(x,f(y,z))) \to a$ is consistent, $g(f(a,f(b,c))) \to a$ is a consistent instance of the rule. Rewrite rule $g(f(x,y)) \to f(y,x)$ is not consistent, neither are $f(g(x),g(y)) = f(y,x)$ and critical pair $(f(x,a), f(a,x))$.*

**Definition 3.3** *Inference rule $CUCP(\to,\to)$ is the operation that generates critical pairs between two consistent rewrite rules. The critical pairs generated are oriented as rewrite rules or added as equations. $CUCP(\to,\to)$ is the same as $UCP(\to,\to)$ except that only consistent rewrite rules are used for critical pair generation in $CUCP$. $CUCP(=,\to)$ and $CUCP(=,=)$ are similarly defined. We say the extended set of AC axioms is the set $\{f(f(x,y),z) = f(x,f(y,z)), f(x,y) = f(y,x), f(x,f(y,z)) = f(y,f(x,z))\}$ for all AC functions $f$. Inference rule $UCP_{AC}$ is the operation that generates critical pairs between a rule or an equation and an equation from the extended set of AC axioms.*

**Theorem 3.4** *Fair consistent unfailing completion starting from a set of $R$ of rules, equations and ground inequation and using the inference rule $\{CUCP(\to,\to), CUCP(=,\to), CUCP(=,=), UCP_{AC}, SIMPGOAL, CONTRA\}$ will eventually generate $FALSE$ if $R^=$ is unsatisfiable.*

Theorem 3.4 outlines the consistent unfailing completion procedure. The soundness of the procedure is obvious, as each inference rule generates only logical consequences. We show that consistent unfailing completion can be regarded as a lifted version of the ground completion procedure outlined in Theorem 3.6, and thus establish its completeness proof.

**Definition 3.5** *Inference rule $INST$ is the operation of generating an instance of a rewrite rule or an equation and orienting it as a new rewrite rule. Inference rule $GCP(\to,\to)$ is the operation of generating critical pairs* between two ground rewrite rules and orienting them as new rewrite rules. $GSIMP_{AC}$ is the operation of replacing a ground rewrite rule $s \to t$ by a new rewrite rule $s' \to t'$, where $s'$ and $t'$ are AC minimal terms of $s$ and $t$, respectively.

**Theorem 3.6** *Fair instance-based completion starting from a set of $R$ of rules, equations and ground inequation and using the inference rule $\{INST, GCP(\to, \to), GSIMP_{AC}, SIMPGOAL, CONTRA\}$ will eventually generate $FALSE$ if $R^=$ is unsatisfiable.*

PROOF: If $R^=$ is unsatisfiable, by Herbrand's theorem, there is a finite unsatisfiable set of instances $R_I^=$ of $R^=$. By the fairness assumption, all instances in the set $R_I^=$ will be generated by $INST$. As a consequence of Theorem 2.3, $\{GCP(\to, \to), GSIMP_{AC}, SIMPGOAL, CONTRA\}$ will eventually generate $FALSE$ from the set $R_I^=$.

**Remark 3.7** *The consistent unfailing completion procedure in Theorem 3.4 is a lifted version of the instance-based completion procedure in Theorem 3.6.*

We now give an example to show the effect of the consistency constraints.

**Example 3.8** *Consider a rewriting system with the following rules, $\{(x*y)*z \to x*(y*z), x*y = y*x\}$. We use an ordering in which the associativity axiom can be oriented* [1]. *We refer to the associativity axiom as $R_1$, and the commutativity axiom as $R_2$.*

*Applying $UCP(\to, =)$ on $R_1$ and $R_2$, equations $(y*x)*z = (x*y)*z$ and $z*(x*y) = (x*y)*z$ will be generated. Applying rule $SIMPRULE$, they are simplified to $R_3$ $y*(x*z) = x*(y*z)$ and $R_4$ $z*(x*y) = x*(y*z)$. Applying $UCP(=, =)$ and $SIMPRULE$ on $R1$ and $R3$, equation $R_5$ $x*(y*(z*w)) = y*(x*(z*w))$ will be generated.*

*Now let's consider the permutations of the variables on each side of the equation $R_5$. In unfailing completion, all such permutations, e.g. $x*(y*(z*w)) = y*(x*(w*z))$, will eventually be generated from $R_1$ and $R_2$. There are 24 such permutations up to the renaming of variables. Since none of the permutations except $x*(y*(z*w)) = x*(y*(z*w))$ can be simplified, 23 equations will be added to the rewriting system. However, none of the permuted equations are consistent, and $x*(y*(z*w)) = x*(y*(z*w))$ is an instance of $x = x$ and can be deleted. Thus in consistent unfailing completion, no permutations of the equation $R_5$ will be added to the rewriting system.*

## 4 Refinements

In this section, we present some refinements of Theorem 3.4. Theorem 4.2 outlines the refined consistent unfailing completion algorithm. We then show how the algorithm can be efficiently implemented.

**Definition 4.1** *Inference rule $PERM$ is the operation that generates all consistent AC equivalent rewrite rules*

[1] one such ordering is defined in Definition 4.3.

*(equations) of an existing rewrite rule (equation). Inference rule $CUCP_A$ is the operation that generates critical pairs between a consistent rule or equation and the associativity axioms $\{f(f(x,y),z) = f(x,f(y,z))\}$.*

The inference rule $PERM$ and $CUCP_A$ combined are equivalent to the rule $UCP_{AC}$ in Theorem 3.4, and they are more efficient. $SIMPRULE$ often greatly reduces the number of rewrite rules and equations in a rewriting system. $SIMPRULE$ can be added to the consistent unfailing completion procedure in Theorem 3.4 without affecting its completeness. The proof involves extending the instance-based completion procedure as in Theorem 3.6 with a restricted version of $SIMPRULE$ and then lifting it to the nonground case.

**Theorem 4.2** *Fair consistent unfailing completion starting from a set of $R$ of rules, equations and ground inequation and using the inference rule $\{CUCP(\to, \to), CUCP(=, \to), CUCP(=, =), PERM, CUCP_A, SIMPRULE, SIMPGOAL, CONTRA\}$ will eventually generate $FALSE$ if $R^=$ is unsatisfiable.*

We now define an ordering that's used in our implementation.

**Definition 4.3** *We define size lexicographic path ordering $>_{slpo}$ on ground terms as follows. Suppose $s$ and $t$ are ground terms, $s = f_s(a_1, \ldots, a_m)$ and $t = f_t(b_1, \ldots, b_n)$, and $>_l$ is the lexicographic ordering. $s >_{slpo} t$ iff*
*1) $size(s) > size(t)$, or*
*2) $size(s) = size(t)$ and $f_s >_l f_t$, or*
*3) $size(s) = size(t)$, $f_s = f_t$, $a_k = b_k$ $\forall k$ from $1$ to $i-1$, and $a_i >_{slpo} b_i$.*

$size(s)$ is defined as the length of $s$ written as a character string(excluding commas and parentheses). The size lexicographic path ordering is extended to nonground terms as follows, $s >_{slpo} t$ iff $\forall \theta$ $s\theta >_{slpo} t\theta$. It can be difficult to order two nonground terms with respect to $>_{slpo}$. In practice, we use sufficient conditions based on special cases such as $g(x) >_{slpo} x$ and $f(x,x) >_{slpo} g(x)$. It can be showed that size lexicographic path ordering is a termination ordering.

**Definition 4.4** *Suppose that $f$ is an AC function, we call $f(s_1, s_2, \ldots, s_n)$ the flattened term of $f(s_1, f(s_2, \ldots f(s_{n-1}, s_n), \ldots))$, where $s_i$ does not contain $f$ as a top level function symbol. Suppose that $s$ and $t$ are subterms of an expression $e$, we say $s$ AC-precedes $t$ in $e$, or $s << t$, if the flattened term of $e$ contains a subterm of the form $f(\ldots, s, \ldots, t, \ldots)$, where $f$ is an AC function.*

We now describe two sufficient conditions for detecting the inconsistency of rewrite rules(or equations).

**Theorem 4.5** *Suppose a size lexicographic path ordering $<_{slpo}$ is used as the termination ordering. A rewrite rule $s \to t$ is inconsistent if*
*1) $u_1 <_{slpo} u_2$ and $u_2 << u_1$*
*or 2) $u_1 << u_2$ and $u_2 << u_1$, where $u_1$ and $u_2$ are subterms of $s$ and $t$, respectively.*

PROOF: For case 1): Equation $s = t$ contains a subterm $f(\ldots, u_2, \ldots, u_1, \ldots)$. Since $u_1 <_{slpo} u_2$, $f(\ldots, u_2, \ldots, u_1, \ldots)\theta >_{slpo} f(\ldots, u_1, \ldots, u_2, \ldots)\theta$, for all $\theta$. That is, no instances of $f(\ldots, u_2, \ldots, u_1, \ldots)$ can be an AC minimal term. $s \rightarrow t$ is thus inconsistent. The proof for case 2) is similar.

The inference rule $PERM$ generates all consistent AC-equivalent rewrite rules or equations of an existing rewrite rule or equation. It essentially involves solving a constraint satisfaction problems. The simplest solution is to enumerate and check. Namely, for each rewrite rule $s \rightarrow t$, all AC equivalent rules of $s \rightarrow t$ are enumerated, and then Theorem 4.5 is used to filter out the inconsistent rules. More specifically, to apply $PERM$ on a rewrite rule $s \rightarrow t$, we flatten $s \rightarrow t$, collect new rewrite rules by permuting the arguments of AC functions in $s \rightarrow t$, delete all inconsistent rules, and finally unflatten the remaining rewrite rules in the collection. For a term $f(a_1, \ldots, a_n)$ where $f$ is an AC function, unflattening the term generates $f(a_1, f(\ldots, f(a_{n-1}, a_n)))$. To unflatten a term or an expression, unflattening is done recursively for all its subterms. For example, $g(f(a, h(b), x)) \rightarrow c$ is unflattened to $g(f(a, f(h(b), x))) \rightarrow c$, assuming that $f$ is the only AC function. Note that flattened terms are only used as an intermediate representation for detecting inconsistent rules, and they are not used in the rewrite rules and equations of the rewriting system.

Theorem 4.5 provides only sufficient conditions for inconsistency, and thus some inconsistent rules can also be generated. More deliberate approaches based on constraint satisfaction might be able to compute the exact set of consistent rules, and thus further reduce the number of critical pairs and rewrite rules generated in the completion procedure.

## 5 Test Examples

It is fairly straightforward to implement the consistent unfailing completion algorithm. We implemented the algorithm with several hundred lines of Prolog code. We tested a number of pure equality problems. The results are listed in Table 1.

E1-E6 is a set of benchmark problems for equality proposed by [Lusk and Overbeek, 1985]. E3 and E6 involve AC functions. wos21(equality version), RNG015-6, RNG023-6 and RNG024-6 are problems from ring theory, all four of them involve AC functions. For non-AC problems, the procedure is same as unfailing completion. For AC problems, we study the effect of consistency checking by comparing the result from unfailing completion and consistent unfailing completion. Note that consistency checking significantly reduces the number of critical pairs generated, the number of rules kept and the time needed to obtain a proof. Our implementation serves the purpose of demonstrating the effect of consistency checking. A state of the art equality prover can solve most of the listed problems in tens of seconds[McCune, 1990]. Efficient data structures can be used to improve our current implementation.

| Problem | without AC constraints | | | with AC constraints | | |
|---|---|---|---|---|---|---|
| | CPs | Rules | Time | CPs | Rules | Time |
| E1(GRP001-2) | 261 | 26 | 2.8 | n/a | n/a | n/a |
| E2(GRP022-2) | 191 | 18 | 1.9 | n/a | n/a | n/a |
| E4(GRP002-4) | 90830 | 401 | 801.1 | n/a | n/a | n/a |
| E5(BOO002-1) | 16550 | 119 | 83.6 | n/a | n/a | n/a |
| E3(RNG008-7) | 23970 | 260 | 865.9 | 12330 | 193 | 494.2 |
| E6(RNG009-7) | - | - | >360000 | 628693 | 2590 | 182510 |
| wos21 | 12288 | 84 | 192.0 | 5928 | 64 | 84.4 |
| RNG015-6 | 42443 | 326 | 881.3 | 20257 | 205 | 410.4 |
| RNG023-6 | - | - | >15000 | 137138 | 591 | 3880.8 |
| RNG024-6 | - | - | >15000 | 138154 | 589 | 3995.7 |

Table 1: Timing on a set of equality problems. Size lexicographic path ordering is used for all problems. Column "CPs" shows the total number of critical pair generated. Column "Rules" shows the number of rewrite rules kept. Time is measured in seconds on a SPARC-20. n/a means that the problem has no AC functions. – means that time out occurred.

It is particularly encouraging that we are able to prove E6, which says a ring with $x^3 = x$ is commutative. The problem has a long history. Wos had the following remarks on the problem, "if one succeeds in having a reasoning program prove this theorem, and it can be shown that the success is the result of a new technique, then one has solid evidence of the potential value of the new idea"[Wos, 1988]. Veroff obtained a proof of the problem using AURA [Veroff, 1981]. However, the input contained many additional clauses that facilitated the proof. Stickel was the first to prove the problem with a natural set of input equations [Stickel, 1984]. The proof took over 14 hours. Zhang and Kapur could find a proof in a few minutes using RRL [Zhang and Kapur, 1990]. Both [Stickel, 1984] and [Zhang and Kapur, 1990] used approaches based on AC unifications. The use of the cancellation law for additional group was important for them to get the proof efficiently.

## 6 Discussion

In term rewriting systems with AC functions, a term can have an exponential number of AC equivalent terms, and a rewrite rule can generate an exponential number of new rules when combined with AC axioms for critical pairs generation. A common approach for avoiding the combinatorial explosion is to represent all AC equivalence terms as a single term. Flattened terms are used to represent terms equivalent up to associativity, and terms with arguments permuted are considered identical. AC unification algorithm is used to unify two flattened terms. Flattening breaks the well-foundedness of most termination orderings. Special AC termination orderings are needed to handle flattened terms.

On the other hand, the combinatorial explosion occurs only because all AC equivalent terms of a term are kept in a rewriting system. If they are simplified to a single normal form, the explosion will not occur. This is exactly the case for ground completion procedures: all ground AC equivalent terms can be rewritten to a normal form using AC axioms when a total termination ordering on ground terms is used. Our approach is based on the observation that a nonground completion procedure can be constructed by lifting a ground completion procedure, and thus avoid much of the redundancy in representing AC equivalent terms. Our approach does not need a special AC unification. It does not need a special AC termination ordering either. The latter facilitates the complete automation of the equational reasoning process. We used a single termination ordering for all of our test examples. We were also able to use the equational prover as a component in a general first order theorem prover.

There are some previous works on reducing unnecessary equational inferences, both for AC equational problems and equational problems in general[Zhang and Kapur, 1990; Bachmair et al., 1992]. Most of them are fundamentally different from our approach. They reduce the number of critical pairs generated by blocking out certain positions for overlapping two rewrite rules. Some of these works might be combined with our approach. It would also be interesting to study the applicability of our approach to other equational theories.

# References

[Bachmair and Plaisted, 1985] L. Bachmair and D. Plaisted. Termination orderings for associative-commutative rewriting systems. *J. Symbolic Computation*, 1:329–349, 1985.

[Bachmair et al., 1989] Leo Bachmair, N. Dershowitz, and D. Plaisted. Completion without failure. In Hassan Aït-Kaci and Maurice Nivat, editors, *Resolution of Equations in Algebraic Structures 2: Rewriting Techniques*, pages 1–30, New York, 1989. Academic Press.

[Bachmair et al., 1992] L. Bachmair, H. Ganzinger, C. Lynch, and W. Snyder. Basic paramodulation and basic strict superposition. In *Proceedings of the 11th International Conference on Automated Deduction*, pages 462–476, 1992.

[Birkhoff, 1935] G. Birkhoff. On the structure of abstract algebras. *Proc. Cambridge Philos. Soc.*, 31:433–454, 1935.

[Boudet et al., 1996] A. Boudet, E. Contejean, and C Marche. AC-complete unification and its application to theorem proving. In *Proceedings of the 7th International Conference on Rewriting Techniques and Applications*, pages 18–32, July 1996.

[Chang and Lee, 1973] C. Chang and R. Lee. *Symbolic Logic and Mechanical Theorem Proving*. Academic Press, New York, 1973.

[Dershowitz and Jouannaud, 1990] N. Dershowitz and J.-P. Jouannaud. Rewrite systems. In J. van Leeuwen, editor, *Handbook of Theoretical Computer Science*. North-Holland, Amsterdam, 1990.

[Dershowitz et al., 1983] Nachum Dershowitz, J. Hsiang, N. Josephson, and David A. Plaisted. Associative-commutative rewriting. In *Proceedings of the Eighth International Joint Conference on Artificial Intelligence*, pages 940–944, August 1983.

[Domenjoud, 1991] E. Domenjoud. Ac-unification through order-sorted ac1-unification. In *Proceedings of the 4th International Conference on rewriting techniques and applications*. Springer-Verlag, 1991.

[Kapur et al., 1990] D. Kapur, G. Sivakumar, and H. Zhang. A new method for proving termination of ac-rewrite systems. In *Proc. of Tenth Conference on Foundations of Software Technology and Theoretical Computer Science*, pages 133–148, December 1990. Springer Verlag LNCS 472.

[Klop, 1992] Jan Willem Klop. Term rewriting systems. In S. Abramsky, D. M. Gabbay, and T. S. E. Maibaum, editors, *Handbook of Logic in Computer Science*, volume 2, chapter 1, pages 1 – 117. Oxford University Press, Oxford, 1992.

[Knuth and Bendix, 1970] D.E. Knuth and P.B. Bendix. Simple word problems in universal algebras. In *Computational Problems in Abstract Algebra*, pages 263–297. Pergamon, Oxford, U.K., 1970.

[Lankford and Ballantyne, 1977] D. Lankford and A.M. Ballantyne. Decision problems for simple equational theories with commutative-associative axioms: Complete sets of commutative-associative reductions. Technical Report Memo ATP-39, Department of Mathematics and Computer Science, University of Texas, Austin, TX, 1977.

[Loveland, 1978] D. Loveland. *Automated Theorem Proving: A Logical Basis*. North-Holland, New York, 1978.

[Lusk and Overbeek, 1985] E. Lusk and R. Overbeek. Non-horn problems. *Journal of Automated Reasoning*, 1:103–114, 1985.

[McCune, 1990] William W. McCune. *OTTER 2.0 Users Guide*. Argonne National Laboratory, Argonne, Illinois, March 1990.

[McCune, 1996] W. McCune. Solution of the robbins problem. draft, 1996.

[Peterson and Stickel, 1981] G.E. Peterson and M.E. Stickel. Complete sets of reductions for some equational theories. *J. Assoc. Comput. Mach.*, 28(2):233–264, 1981.

[Plaisted, 1993] D. Plaisted. Equational reasoning and term rewriting systems. In D. Gabbay, C. Hogger, J. A. Robinson, and J. Siekmann, editors, *Handbook of Logic in Artificial Intelligence and Logic Programming*, volume 1, pages 273–364. Oxford University Press, 1993.

[Stickel, 1981] M.E. Stickel. A unification algorithm for associative-commutative functions. *Journal of the Association for Computing Machinery*, 28:423–434, 1981.

[Stickel, 1984] M Stickel. A case study of theorem proving by the knuth-bendix method:discovering that $x^3 = x$ implies ring commutativity. In *Proceedings of the 7th International Conference on Automated Deduction*, pages 248–258, 1984.

[Veroff, 1981] R.L. Veroff. Canonicalization and demodulation. Technical Report ANL-81-6, Argonne National Laboratory, Argonne, IL, 1981.

[Wos et al., 1984] L. Wos, R. Overbeek, E. Lusk, and J. Boyle. *Automated Reasoning: Introduction and Applications*. Prentice Hall, Englewood Cliffs, N.J., 1984.

[Wos, 1988] L. Wos. *Automated Reasoning: 33 Basic Research Problems*. Prentice Hall, Englewood Cliffs, N.J., 1988.

[Zhang and Kapur, 1990] H. Zhang and D. Kapur. Unnecessary inferences in associative-commutative completion procedures. *Mathematical Systems Theory*, 23:175–206, 1990.

# Strategies in Rigid-Variable Methods

Andrei Voronkov*

## Abstract

We study complexity of methods using rigid variables, like the method of matings or the tableau method, on a decidable class of predicate calculus with equality. We show some intrinsic complications introduced by rigid variables. We also consider strategies for increasing multiplicity in rigid-variable methods, and formally show that the use of intelligent strategies can result in an essential gain in efficiency.

## 1 Introduction

Automated reasoning methods for first-order classical logic can generally be divided in two classes.

Methods of the first class use *universal variables* (resolution [Robinson, 1965], the inverse method [Maslov, 1983]). Variables in these methods are local to a clause (formula, sequent) and can be considered as universally quantified in this clause (respectively formula or sequent). [Maslov et al., 1983; Maslov, 1987] characterized these methods as *local methods* (see also [Mints, 1990]).

Methods of the second class use *rigid variables* (the tableau method [Beth, 1959], the mating or the connection method [Andrews, 1981; Bibel, 1981], model elimination [Loveland, 1968], SLD-resolution [Kowalski, 1974; Apt and van Emden, 1982], SLO-resolution [Demolombe, 1989; Rajasekar, 1989]). Variables in these methods are local to a set of clauses (formulas, sequents) and can be considered as universally quantified in this set of clauses (respectively formulas or sequents). [Maslov et al., 1983; Maslov, 1987] characterized these methods as *global methods*. In this paper, we shall call such methods *rigid-variable methods*.

Both kinds of methods have their advantages and disadvantages which are well-known. There are papers comparing resolution and tableau-like calculi, for example [Eder, 1988; 1991] (see also [Bibel and Eder, 1993]).

---
*Computing Science Department, Uppsala University, Box 311, S-751 05 Uppsala, Sweden. Supported by a TFR grant. URL page: http://www.csd.uu.se/~voronkov. Email: voronkov@csd.uu.se.

Recently, there have been proposals combining both kinds of methods in one calculus, for example the equality elimination method [Degtyarev and Voronkov, 1995b; 1995a; 1996a] or a modification of model elimination [Moser et al., 1995].

Although there are many implementations of rigid-variable methods, there are almost no papers investigating the intrinsic complexity of problems arising in rigid-variable methods. In this paper we study several complexity problems related to rigid-variable methods on a decidable fragment of predicate calculus with equality. We show that the use of such methods can introduce essential complications even for this relatively simple fragment. We also demonstrate that methods with rigid variables can gain from the intelligent use of *strategies for multiplicity*. In addition, we show that a recent result of [Voda and Komara, 1995] on the Herbrand skeleton problem is related to an inadequate formulation of the problem, and pose a new open problem.

## 2 Preliminaries

A term or atomic formula is *ground* iff it has no variables. The symbol $\vdash$ denotes provability in first-order logic. When we write $\varphi_1, \ldots, \varphi_n \vdash \varphi$, where $\varphi_1, \ldots, \varphi_n, \varphi$ are formulas, it means provability of the formula $\varphi_1 \wedge \ldots \wedge \varphi_n \supset \varphi$. *Substitutions* of terms $t_1, \ldots, t_n$ for variables $x_1, \ldots, x_n$ are denoted $\{t_1/x_1, \ldots, t_n/x_n\}$. The *application of such a substitution $\theta$ to a term $t$*, is the operation of simultaneous replacement of all occurrences of $x_i$ by $t_i$. The result of the application is the term denoted $t\theta$. We shall also apply substitutions to formulas, equations and sets of equations and use the same notation for the result of the application.

We shall consider first-order predicate logic with equality. The equality predicate is denoted by $\simeq$, in order to distinguish it from the metasymbol $=$ used to denote the identity of two expressions. The symbol $\rightleftharpoons$ means "equal by definition". Atomic formulas $s \simeq t$ are called *equations*, and their negations $\neg s \simeq t$, denoted $s \not\simeq t$, *disequations*. We do not distinguish an equation $s \simeq t$ from the equation $t \simeq s$. For a technical convenience, we can also use equations and disequations between atomic formulas, for example $P(s_1, \ldots, s_m) \simeq Q(t_1, \ldots, t_n)$.

1. Choose a positive integer $\mu$.
2. Construct the formula $\varphi_\mu = (\varphi(\bar{x}_1) \vee \ldots \vee \varphi(\bar{x}_\mu))$ and check whether there is a substitution $\theta$ such that the formula $\varphi_\mu \theta$ is provable.
3. If such a substitution exists, then the goal formula $\exists \bar{x} \varphi(\bar{x})$ is provable. Otherwise, increase $\mu$ and come back to step (2).

Figure 1: The Procedure: a typical procedure for rigid-variable methods

Let $s_1, t_1, \ldots, s_n, t_n, s, t$ be terms. We write $s_1 \simeq t_1, \ldots, s_n \simeq t_n \vdash s \simeq t$ to denote that the formula $\forall (s_1 \simeq t_1 \wedge \ldots \wedge s_n \simeq t_n \supset s \simeq t)$ is true, i.e. it is provable in first-order logic. Equivalently, we can say that $s$ and $t$ lie in the same class of the congruence generated by $\{s_1 \simeq t_1, \ldots, s_n \simeq t_n\}$.

A *rigid equation* is an expression $\mathcal{E} \vdash_\forall s \simeq t$, where $\mathcal{E}$ is a finite set of equations. $\mathcal{E}$ is called the *left hand side* of this rigid equation, and the equation $s \simeq t$ — its *right hand side*. A *solution to a rigid equation* $\{s_1 \simeq t_1, \ldots, s_n \simeq t_n\} \vdash_\forall s \simeq t$ is any substitution $\sigma$ such that $s_1 \sigma \simeq t_1 \sigma, \ldots, s_n \sigma \simeq t_n \sigma \vdash s\sigma \simeq t\sigma$. A *system of rigid equations* is a finite set of rigid equations. A *solution to a system of rigid equations* $\mathcal{R}$ is any substitution that is a solution to every rigid equation in $\mathcal{R}$. The problem of solvability of rigid equations is known as *rigid E-unification*. The problem of solvability of systems of rigid equations is known as *simultaneous rigid E-unification*, or SREU for short.

We shall denote sets of equations by $\mathcal{E}$, systems of rigid equations by $\mathcal{R}$ and rigid equations by $R$. We shall sometimes write the left hand side of a rigid equation as a *sequence* of equations, for example $x \simeq a \vdash_\forall g(x) \simeq x$ instead of $\{x \simeq a\} \vdash_\forall g(x) \simeq x$.

For simplicity, in this paper we consider the provability problem for closed prenex existential formulas, i.e. formulas of the form $\exists \bar{x} \varphi(\bar{x})$, where $\varphi(\bar{x})$ is a quantifier-free formula. There is a provability-preserving polynomial time translation of arbitrary formulas to closed prenex existential formulas by means of Skolemization and prenexing.

According to the Herbrand theorem, such a formula is provable if and only if there exists a positive integer $\mu$ and a substitution $\theta$ such that the formula $(\varphi(\bar{x}_1) \vee \ldots \vee \varphi(\bar{x}_\mu))\theta$ is provable. This fact is used in several automated reasoning methods, for example in the method of matings, in the way shown in Figure 1. The procedure shown in that figure will simply be called *the Procedure*.

In the method of matings, before step (2) the formula $\varphi_\mu$ is represented in the form of a *matrix $M$* and the provability of $\varphi_\mu \theta$ means that any vertical path in $M\theta$ is inconsistent. The tableau method represents the formula in the form of a tree and uses branches of the tree instead of vertical paths.

The number $\mu$ used in the Procedure (the number of copies of $\varphi(\bar{x})$ which can be used) is usually called *multiplicity*. Of course, there are various modifications of the Procedure, for example, the goal formula may be non-prenex. In this case the notion of multiplicity is more complicated. Our results can also be generalized to more complex notions of multiplicity. However, we shall only consider prenex existential formulas, for which the notion of multiplicity is defined as a positive integer number.

We informally call a *strategy for multiplicity* any procedure which selects the initial multiplicity and increases multiplicity in the Procedure. The *standard strategy for multiplicity* is the strategy which sets $\mu$ initially to 1 and increments it by 1 on any further step. A strategy for multiplicity is called *formula-independent* iff it does not depend on the input formula.

There are various algorithms for checking, for a given formula $\varphi_\mu$, whether there is a substitution making this formula provable. Instead of studying concrete procedures, we shall study the intrinsic complexity of the problem which can be formulated as follows.

**Problem 1 (Herbrand Skeleton)** *Given a quantifier-free formula $\varphi(\bar{x})$ and a positive integer $\mu$, are there term sequences $\bar{t}_1, \ldots, \bar{t}_n$ such that the formula $\varphi(\bar{t}_1) \vee \ldots \vee \varphi(\bar{t}_\mu)$ is provable?*

[Degtyarev et al., 1996a] give an informal survey of several decision problems arising from the Herbrand theorem, including the Herbrand Skeleton problem. It is clear that Problem 1 is decidable if and only if the following problem is decidable.

**Problem 2 (Formula Instantiation)** *Given a quantifier-free formula $\varphi(\bar{x})$, is there a term sequence $\bar{t}$ such that the formula $\varphi(\bar{t})$ is provable?*

Note that the formula instantiation problem is repeatedly used in the procedures used by the method of matings or the tableau method.

The decidability of these problems is equivalent to the decidability of SREU. Unfortunately, it turned out that SREU has almost no decidable fragments which are general enough. Some known results on SREU are the following.

- SREU is undecidable [Degtyarev and Voronkov, 1996b].
- SREU with ground left hand sides is undecidable [Plaisted, 1995].
- SREU with ground left hand sides and two variables is undecidable [Veanes, 1996].
- SREU with one variable is DEXPTIME-complete [Degtyarev et al., 1997].

The case of one variable is hardly useful in automated reasoning. When all function symbols have arity $\leq 1$, Formula Instantiation is equivalent to *monadic SREU*, i.e. SREU in the signature where all function symbols have arity $\leq 1$. The decidability of monadic SREU is an open problem. The following facts are known about monadic SREU.

- Monadic simultaneous rigid *E*-unification with one function symbol is decidable (this fact has a rather non-trivial proof in [Degtyarev et al., 1996b]).
- Monadic simultaneous rigid *E*-unification with more than one function symbol is equivalent to a non-trivial extension of word equations [Gurevich and Voronkov, 1997a].

Some other decidable fragments of monadic SREU are considered in [Gurevich and Voronkov, 1997a], but they are hardly of much use for automated reasoning.

Since predicate calculus is undecidable, the undecidability of SREU does not add much to the complexity of predicate calculus. In this paper we consider the behavior of methods based on rigid variables on a decidable fragment of predicate calculus.

## 3 Ground-negative fragment of predicate calculus

A formula $\varphi$ is called *positive* if all atomic subformulas of $\varphi$ are positive. A closed formula $\varphi$ is *ground-negative* iff any occurrence of a variable in $\varphi$ is either an occurrence in a positive atomic subformula of $\varphi$ or is bound by an essentially universal quantifier (i.e. an universal quantifier occurring in $\varphi$ positively or an existential quantifier occurring in $\varphi$ negatively)[1]. [Kozen, 1977] proves the following result.

**Theorem 1** *The class of provable formulas of the form $A_1 \wedge \ldots \wedge A_n \supset \varphi$, where $A_1, \ldots, A_n$ are ground atomic formulas and $\varphi$ is a positive formula, is NP-complete.*

Using this result, one can prove the following:

**Theorem 2** *The provability problem for ground-negative formulas is in $\Pi_2^p$.*

Since we only consider prenex existential formulas, such a formula $\exists \bar{x} \varphi(\bar{x})$ is ground-negative if and only if all negative atomic subformulas of $\varphi(\bar{x})$ are ground. So we consider the Herbrand skeleton problem and the formula instantiation problem for ground-negative formulas. We assume that $\exists \bar{x} \varphi(\bar{x})$ is a fixed ground-negative existential prenex formula.

**Theorem 3** *The formula instantiation problem for ground-negative formulas is undecidable.*

*Proof.* We shall use the result proved in [Plaisted, 1995] that SREU with ground left hand sides is undecidable. Consider any system $\mathcal{R}$ of rigid equations with ground left hand sides:

$$\begin{array}{ll} s_{11} \simeq t_{11}, \ldots, s_{1n_1} \simeq t_{1n_1} & \vdash_\forall \quad s_1 \simeq t_1 \\ \quad \cdots & \\ s_{m1} \simeq t_{m1}, \ldots, s_{mn_m} \simeq t_{mn_m} & \vdash_\forall \quad s_m \simeq t_m \end{array}$$

---
[1] Variables occurring in essentially universal quantifiers can also be characterized as eigenvariables. Thus, a formula $\varphi$ is ground-negative if and only if every variable occurring in a negative atom in $\varphi$ is an eigenvariable.

Consider also the following formula $\varphi$:

$$\begin{array}{rl} (s_{11} \simeq t_{11} \wedge \ldots \wedge s_{1n_1} \simeq t_{1n_1} & \supset \quad s_1 \simeq t_1) \quad \wedge \\ & \cdots \qquad \wedge \\ (s_{m1} \simeq t_{m1} \wedge \ldots \wedge s_{mn_m} \simeq t_{mn_m} & \supset \quad s_m \simeq t_m) \end{array}$$

It is straightforward that a substitution $\theta$ is a solution to $\mathcal{R}$ if and only if the formula $\varphi\theta$ is provable. Note that all negative atoms in $\varphi\theta$ are ground. Hence, SREU with ground left hand sides is effectively reducible to the formula instantiation problem for ground-negative formulas. Thus, the formula instantiation problem for ground-negative formulas is undecidable. □

If we consider which systems of rigid equations arise from ground-negative formulas (e.g. according to the procedures of [Gallier et al., 1990; 1992]), we shall find out that these are precisely all systems of rigid equations with ground left hand sides.

Theorem 3 shows that a straightforward use of rigid-variable methods can create unnecessary complications, for example, the necessity to solve an intermediate undecidable subproblem in order to solve a problem in $\Pi_2^p$. This theorem can be reformulated as a statement about the standard strategy for multiplicity:

**Theorem 4** *For the standard strategy for multiplicity, a subproblem arising at step (2) of the Procedure is undecidable for the class of ground-negative formulas.*

*Proof.* Indeed, the subproblem with $\mu = 1$ arising at the first iteration of the algorithm is equivalent to the formula instantiation problem which is undecidable by Theorem 3. □

Hence, the use of the standard strategy may introduce unnecessary complications into rigid-variable methods. We can prove that the same holds for arbitrary formula-independent strategies. To this end, we shall use a result proven in [Veanes, 1997]. First, we cite a result by [Voda and Komara, 1995] which generalizes the undecidability of SREU.

We call a *specialization* of the Herbrand skeleton problem for any fixed $\mu$ the following problem[2]:

Given a quantifier-free formula $\varphi(\bar{x})$, are there term sequences $\bar{t}_1, \ldots, \bar{t}_\mu$ such that the formula $\varphi(\bar{t}_1) \vee \ldots \vee \varphi(\bar{t}_\mu)$ is provable?

The following result is proved in [Voda and Komara, 1995]:

**Theorem 5** *The specialization of the Herbrand skeleton problem for any fixed $\mu$ is undecidable.*

This result has recently been improved in [Veanes, 1997], where they have shown that it also holds for ground-negative formulas:

---
[2] The Herbrand skeleton problem described in [Voda and Komara, 1995] is precisely this specialization.

**Theorem 6** *The specialization of the Herbrand skeleton problem for ground-negative formulas and any fixed multiplicity $\mu$ is undecidable.*

Theorem 4 about the standard strategy for multiplicity can be generalized as follows.

**Theorem 7** *For any formula-independent strategy, a subproblem arising at step (2) of the Procedure for the class of ground-negative formulas is undecidable.*

*Proof.* Since the strategy is formula-independent, some $\mu$ independent of the input formula will be selected at the first iteration of step (2). Then continue similar to the proof of Theorem 4 but using Theorem 6 instead of Plaisted's result. □

We shall consider an intelligent strategy for multiplicity in Section 4.

## 4 The Herbrand Skeleton Problem and intelligent strategies for multiplicity

Intelligent strategies for multiplicity have always been considered of paramount importance for rigid-variable methods. However, essentially no formal results are known about such strategies. Existing systems based on rigid-variable methods use some heuristic methods for incrementing multiplicity and universal variables whenever possible (see e.g. [Hähnle et al., 1994]). In this section we show that there is an efficient formula-dependent strategy for multiplicity for the class of ground-negative formulas.

In this section $\psi = \exists \bar{x} \varphi(\bar{x})$ will denote a fixed ground-negative existential prenex formula. Let $A_1, \ldots, A_n$ be all its negative atomic subformulas. Denote by $N$ the set of all subsets of $\{1, \ldots, n\}$. Introduce $2^n$ sets of ground atoms $\mathcal{E}_I \rightleftharpoons \{A_i \mid i \in I\}$, for every $I \in N$. Using transformations similar to those used in the conjunctive normal form translation, we can assume that $\varphi(\bar{x})$ has the form

$$\bigwedge_{I \in N} ((\bigwedge_{i \in I} A_i) \supset \varphi_I),$$

where $\varphi_I$ are formulas constructed from atomic formulas using $\wedge, \vee$. Without loss of generality we can assume that for every $I, J$, if $J \subseteq I$, then $\vdash \varphi_I \supset \varphi_J$. Indeed, if this is not true, we can replace $\varphi_I$ by $\varphi_I \wedge \varphi_J$, then $\psi$ will be replaced by an equivalent formula.

The notion of the least Herbrand model of a set of formulas is standard and can be found in e.g. [Lloyd, 1987] or [Apt, 1990].

**Lemma 4.1** *Let $\mathfrak{A}$ be a set of ground atomic formulas and $\varphi$ be a closed formula constructed from atomic formulas using only $\wedge, \vee$ and $\exists$. Let $\mathfrak{M}$ be the least Herbrand model of $\mathfrak{A}$. If $\mathfrak{M} \models \varphi$, then $\mathfrak{A} \vdash \varphi$.*

*Proof.* Straightforward, by induction on $\varphi$. □

A more general variant of Lemma 4.1 also holds, where $\mathfrak{A}$ is a set of universal quasy-identities (see [Makowski, 1986] or [Sheperdson, 1988]).

**Lemma 4.2** *Let $\vdash \psi$. Then for every $I \in N$ there is a substitution $\theta$ such that $\mathcal{E}_I \vdash \varphi_I \theta$.*

*Proof.* Denote by $\mathfrak{M}_I$ the least Herbrand model of $\mathcal{E}_I$. We have $\mathfrak{M}_I \models \psi$. Since $\mathfrak{M}_I$ is a Herbrand model, we have $\mathfrak{M}_I \models \varphi(\bar{x})\theta$ for some substitution $\theta$ making all variables in $\bar{x}$ ground. We prove that $\theta$ satisfies the claim.

Indeed, we have $\mathfrak{M}_I \models (\bigwedge_{i \in I} A_i) \supset \varphi_I \theta$. Since $\mathfrak{M}_I \models (\bigwedge_{i \in I} A_i)$, we have $\mathfrak{M}_I \models \varphi_I \theta$. By Lemma 4.1 we have $\mathcal{E}_I \vdash \varphi_I \theta$. □

For every $I \in N$, we denote by $\theta_I$ some substitution satisfying Lemma 4.2.

**Theorem 8** *Let $\vdash \psi$. Then $\vdash \bigvee_{I \in N} \varphi(\bar{x})\theta_I$.*

*Proof.* Let $\mathfrak{M}$ be an arbitrary model. Consider $I \rightleftharpoons \{i \mid \mathfrak{M} \models A_i\}$. Since $\mathfrak{M} \models \bigwedge_{i \in I} A_i$, by Lemma 4.2 we have $\mathfrak{M} \models \varphi_I \theta_I$. Consider an arbitrary $J \in N$. Let us prove that $\mathfrak{M} \models (\bigwedge_{i \in J} A_i) \supset \varphi_J \theta_I$. Consider two cases:

1. $J \subseteq I$. Then we have $\vdash \varphi_I \supset \varphi_J$. It follows that $\vdash \varphi_I \theta_I \supset \varphi_J \theta_I$. Hence, $\mathfrak{M} \models \varphi_J \theta_I$.

2. $J \not\subseteq I$. By the choice of $\mathfrak{M}$ we have $\mathfrak{M} \not\models \bigwedge_{i \in J} A_i$.

In both cases we have $\mathfrak{M} \models (\bigwedge_{i \in J} A_i) \supset \varphi_J \theta_I$. Since $J$ was arbitrary, we have

$$\bigwedge_{J \in N} ((\bigwedge_{i \in J} A_i) \supset \varphi_J) \theta_I,$$

i.e. $\mathfrak{M} \models \varphi(\bar{x})\theta_I$. Then we have $\mathfrak{M} \models \bigvee_{I \in N} \varphi(\bar{x})\theta_I$. Since $\mathfrak{M}$ is arbitrary, we have $\vdash \bigvee_{I \in N} \varphi(\bar{x})\theta_I$. □

**Theorem 9** *The following problem is in $\Pi_2^p$ (compare it with the Herbrand Skeleton problem).*

*Given any ground-negative formula $\exists \bar{x} \varphi(\bar{x})$ and any positive integer $\mu \geq 2^n$, where $n$ is the number of negative atomic subformulas of $\varphi(\bar{x})$, are there term sequences $\bar{t}_1, \ldots, \bar{t}_\mu$ such that the formula $\varphi(\bar{t}_1) \vee \ldots \vee \varphi(\bar{t}_\mu)$ is provable?*

*Proof.* By Theorem 8, if the formula $\exists \bar{x} \varphi(\bar{x})$ is provable, then such term sequences $\bar{t}_1, \ldots, \bar{t}_\mu$ exists. Obviously, the converse is also true. Hence, the problem defined in the theorem is polynomial-time equivalent to the provability problem for ground-negative formulas. By Theorem 2, the provability problem for such formulas is in $\Pi_2^p$. □

Thus, for ground-negative formulas we have an interesting phenomenon. For small values of multiplicity $\mu$ the Procedure should solve an undecidable subproblem at step (2); for large enough values of $\mu$, this subproblem is in $\Pi_2^p$. Thus, we have formally shown that for the class of ground-negative formulas, formula-dependent strategies for multiplicity can result in a huge gain in efficiency.

Theorem 9 also shows that the result of Voda and Komara is not related to formula-dependent strategies for multiplicity. In order to formally define a formula-dependent strategy, we can represent the value of $\mu$ on the $n$th iteration of the Procedure as a function $f$ of two arguments: the input formula $\exists \bar{x}\varphi(x)$ and the number $n$. Since $\mu$ should be increased with each next iteration, we have $f(\exists \bar{x}\varphi(x), n+1) > f(\exists \bar{x}\varphi(x), n)$.

Then we can formally define a *strategy for multiplicity* as any function $f$ such that

1. The first argument of $f$ ranges over prenex existential formulas $\exists \bar{x}\varphi(x)$;
2. The second argument of $f$ ranges over positive integers;
3. for every positive integers $k > m$ and every prenex existential formula $\psi$ we have $f(\psi, k) > f(\psi, m)$.

The following problem arises:

**Problem 3** *Is there a strategy for multiplicity $f$ such that*

1. *$f$ is computable;*
2. *The following problem is decidable. Given a number $k$ and a prenex existential formula $\exists \bar{x}\varphi(x)$, are there term sequences $\bar{t}_1, \ldots, \bar{t}_{f(\exists \bar{x}\varphi(x),k)}$ such that the formula $\varphi(\bar{t}_1) \vee \ldots \vee \varphi(\bar{t}_{f(\exists \bar{x}\varphi(x),k)})$ is provable?*

This problem is still open. We conjecture that such function does not exist, but the proof of this fact would require some non-trivial diagonalization.

Even if this problem has a negative solution, there are still at least two known ways for the use of rigid-variable methods for logic with equality. One way is to augment rigid-variable methods by universal-variable parts, as for example in [Degtyarev and Voronkov, 1995b; 1996a] or [Moser *et al.*, 1995]. Another way is to use incomplete but terminating algorithms on step (2) of the Procedure, as demonstrated in [Plaisted, 1995] or [Degtyarev and Voronkov, 1996c].

**Acknowledgments**

I thank Anatoli Degtyarev, Yuri Gurevich and Margus Veanes. Some ideas of this paper appeared as a result of our discussions. I thank an anonymous referee who pointed out serious inaccuracies in the preliminary version of this paper.

## References

[Andrews, 1981] P.B. Andrews. Theorem proving via general matings. *Journal of the Association for Computing Machinery*, 28(2):193–214, 1981.

[Apt, 1990] K.R. Apt. Logic programming. In J. Van Leeuwen, editor, *Handbook of Theoretical Computer Science*, volume B: Formal Methods and Semantics, chapter 10, pages 493–574. Elsevier Science, Amsterdam, 1990.

[Apt and van Emden, 1982] K. Apt and M. van Emden. Contributions to the theory of logic programming. *Journal of the Association for Computing Machinery*, 29(3), 1982.

[Beth, 1959] E.W. Beth. *The Foundations of Mathematics*. North Holland, 1959.

[Bibel, 1981] W. Bibel. On matrices with connections. *Journal of the Association for Computing Machinery*, 28(4):633–645, 1981.

[Bibel and Eder, 1993] W. Bibel and E. Eder. Methods and calculi for deduction. In D.M. Gabbay, C.J. Hogger, and J.A. Robinson, editors, *Handbook of Logic in Artificial Intelligence and Logic Programming*, volume 1, chapter 3, pages 67–182. Oxford University Press, 1993.

[Degtyarev and Voronkov, 1995a] A. Degtyarev and A. Voronkov. Equality elimination for the inverse method and extension procedures. In C.S. Mellish, editor, *Proc. International Joint Conference on Artificial Intelligence (IJCAI)*, volume 1, pages 342–347, Montréal, August 1995.

[Degtyarev and Voronkov, 1995b] A. Degtyarev and A. Voronkov. General connections via equality elimination. In M. De Glas and Z. Pawlak, editors, *Second World Conference on the Fundamentals of Artificial Intelligence (WOCFAI-95)*, pages 109–120, Paris, July 1995. Angkor.

[Degtyarev and Voronkov, 1996a] A. Degtyarev and A. Voronkov. Equality elimination for the tableau method. In J. Calmet and C. Limongelli, editors, *Design and Implementation of Symbolic Computation Systems. International Symposium, DISCO'96*, volume 1128 of *Lecture Notes in Computer Science*, pages 46–60, Karlsruhe, Germany, September 1996.

[Degtyarev and Voronkov, 1996b] A. Degtyarev and A. Voronkov. The undecidability of simultaneous rigid $E$-unification. *Theoretical Computer Science*, 166(1-2):291–300, 1996.

[Degtyarev and Voronkov, 1996c] A. Degtyarev and A. Voronkov. What you always wanted to know about rigid $E$-unification. In J.J. Alferes, L.M. Pereira, and E. Orlowska, editors, *Logics in Artificial Intelligence. European Workshop, JELIA'96*, volume 1126 of *Lecture Notes in Artificial Intelligence*, pages 50–69, Évora, Portugal, September/October 1996.

[Degtyarev *et al.*, 1996a] A. Degtyarev, Yu. Gurevich, and A. Voronkov. Herbrand's theorem and equational reasoning: Problems and solutions. In *Bulletin of the European Association for Theoretical Computer Science*, volume 60, page ??? October 1996. The "Logic in Computer Science" column.

[Degtyarev *et al.*, 1996b] A. Degtyarev, Yu. Matiyasevich, and A. Voronkov. Simultaneous rigid $E$-unification and related algorithmic problems. In *Eleventh Annual IEEE Symposium on Logic in Computer Science (LICS'96)*, pages 494–502, New Brunswick, NJ, July 1996. IEEE Computer Society Press.

[Degtyarev *et al.*, 1997] A. Degtyarev, Yu. Gurevich, P. Narendran, M. Veanes, and A. Voronkov. The decidability of simultaneous rigid $E$-unification with one variable. UPMAIL Technical Report 139, Uppsala University, Computing Science Department, March 1997.

[Demolombe, 1989] R. Demolombe. An efficient strategy for non-Horn deductive databases. In G.X. Ritter, editor, *Information Processing 89*, pages 325–330. Elsevier Science, 1989.

[Eder, 1988] E. Eder. A comparison of the resolution calculus and the connection method, and a new calculus generalizing both methods. In E. Börger, G. Jäger, H. Kleine Büning, and M.M. Richter, editors, *CSL'88 (Proc. 2nd Workshop on Computer Science Logic)*, volume 385 of *Lecture Notes in Computer Science*, pages 80–98. Springer Verlag, 1988.

[Eder, 1991] E. Eder. Consolution and its relation with resolution. In *Proc. International Joint Conference on Artificial Intelligence (IJCAI)*, pages 132–136, 1991.

[Gallier et al., 1990] J. Gallier, P. Narendran, D. Plaisted, and W. Snyder. Rigid $E$-unification: NP-completeness and applications to equational matings. *Information and Computation*, 87(1/2):129–195, 1990.

[Gallier et al., 1992] J. Gallier, P. Narendran, S. Raatz, and W. Snyder. Theorem proving using equational matings and rigid $E$-unification. *Journal of the Association for Computing Machinery*, 39(2):377–429, 1992.

[Veanes, 1997] M. Veanes. *On Simultaneous Rigid E-Unification*. PhD thesis, Uppsala University, 1997.

[Gurevich and Voronkov, 1997a] Yu. Gurevich and A. Voronkov. Decision problems for logic with unary function symbols. In *ICALP'97*, page 12, 1997. To appear.

[Gurevich and Voronkov, 1997b] Yu. Gurevich and A. Voronkov. Monadic simultaneous rigid $E$-unification and related problems. UPMAIL Technical Report 134, Uppsala University, Computing Science Department, January 1997.

[Hähnle et al., 1994] R. Hähnle, B. Beckert, and S. Gerberding. The many-valued tableau-based theorem prover $_3T^AP$. Technical Report 30/94, Universität Karlsruhe, Fakultät für Informatik, November 1994.

[Kowalski, 1974] R.A. Kowalski. Predicate logic as a programming language. In *Proc. IFIP'74*, pages 569–574. North Holland, 1974.

[Kozen, 1977] D. Kozen. Complexity of finitely presented algebras. In *Proc. of the 9th Annual Symposium on Theory of Computing*, pages 164–177, New York, 1977. ACM.

[Lloyd, 1987] J.W. Lloyd. *Foundations of Logic Programming (2nd edition)*. Springer Verlag, 1987.

[Loveland, 1968] D.W. Loveland. Mechanical theorem proving by model elimination. *Journal of the Association for Computing Machinery*, 15:236–251, 1968.

[Makowski, 1986] J.A. Makowski. Why Horn formulas matter in computer science: Initial structures and generic examples. In H. Ehrig et al., editor, *TAPSOFT'86*, volume 185 of *Lecture Notes in Computer Science*, pages 374–387. Springer Verlag, 1986.

[Maslov, 1983] S.Yu. Maslov. An inverse method for establishing deducibility of nonprenex formulas of the predicate calculus. In J.Siekmann and G.Wrightson, editors, *Automation of Reasoning (Classical papers on Computational Logic)*, volume 2, pages 48–54. Springer Verlag, 1983.

[Maslov, 1987] S.Yu. Maslov. *Theory of Deductive Systems and its Applications*. MIT Press, 1987.

[Maslov et al., 1983] S.Y. Maslov, G.E. Mints, and V.P. Orevkov. Mechanical proof-search and the theory of logical deduction in the USSR. In J.Siekmann and G.Wrightson, editors, *Automation of Reasoning (Classical papers on Computational Logic)*, volume 1, pages 29–38. Springer Verlag, 1983.

[Mints, 1990] G. Mints. Gentzen-type systems and resolution rules. Part I. Propositional logic. In P. Martin-Löf and G. Mints, editors, *COLOG-88*, volume 417 of *Lecture Notes in Computer Science*, pages 198–231. Springer Verlag, 1990.

[Moser et al., 1995] M. Moser, C. Lynch, and J. Steinbach. Model elimination with basic ordered paramodulation. Technical Report AR-95-11, Fakultät für Informatik, Technische Universität München, München, 1995.

[Plaisted, 1995] D.A. Plaisted. Special cases and substitutes for rigid $E$-unification. Technical Report MPI-I-95-2-010, Max-Planck-Institut für Informatik, November 1995.

[Rajasekar, 1989] A. Rajasekar. *Semantics for Disjunctive Logic Programs*. PhD thesis, University of Maryland, 1989.

[Robinson, 1965] J.A. Robinson. A machine-oriented logic based on the resolution principle. *Journal of the Association for Computing Machinery*, 12(1):23–41, 1965.

[Sheperdson, 1988] J.C. Sheperdson. Negation in logic programming. In J. Minker, editor, *Foundations of Deductive Databases and Logic Programming*, pages 19–88. Morgan Kaufmann, 1988.

[Veanes, 1996] M. Veanes. Uniform representation of recursively enumerable sets with simultaneous rigid $E$-unification. UPMAIL Technical Report 126, Uppsala University, Computing Science Department, 1996.

[Voda and Komara, 1995] P.J. Voda and J. Komara. On Herbrand skeletons. Technical report, Institute of Informatics, Comenius University Bratislava, July 1995.

# AUTOMATED REASONING

Automated Reasoning 4: Propositional KBs

# Tractable Cover Compilations*

Yacine Boufkhad[1], Éric Grégoire[2], Pierre Marquis[2],
Bertrand Mazure[2], Lakhdar Saïs[2,3]

[1] LIP6
Université Paris 6
4, place Jussieu
F-75252 Paris Cedex 05, FRANCE
boukhad@laforia.ibp.fr

[2] CRIL   [3] IUT de Lens
Université d'Artois
Rue de l'Université, S.P. 16
F-62307 Lens Cedex, FRANCE
{gregoire,marquis,mazure,sais}@cril.univ-artois.fr

## Abstract

*Tractable covers* are introduced as a new approach to equivalence-preserving compilation of propositional knowledge bases. First, a general framework is presented. Then, two specific cases are considered. In the first one, partial interpretations are used to shape the knowledge base into tractable formulas from several possible classes. In the second case, they are used to derive renamable Horn formulas. This last case is proved less space-consuming than prime implicants cover compilations for every knowledge base. Finally, experimental results show that the new approaches can prove efficient w.r.t. direct query answering and offer significant time and space savings w.r.t. prime implicants covers.

## 1 Introduction

Different approaches have been proposed to circumvent the intractability of propositional deduction. Some of them restrict the expressive power of the representation language to tractable classes, like the Horn, reverse Horn, binary, monotone, renamable Horn, q-Horn, nested clauses formulas [Dowling and Gallier, 1984; Lewis, 1978; Boros *et al.*, 1994; Knuth, 1990]. Unfortunately, such classes are not expressive enough for many applications. Contrastingly, *compilation approaches* apply to full propositional-logic knowledge bases (KBs for short). Thanks to an off-line pre-processing step, a KB $\Sigma$ is compiled into a formula $\Sigma^*$ so that on-line query answering can be performed tractably from $\Sigma^*$. Many approaches to compilation have been proposed so far, mainly [Reiter and De Kleer, 1987; Selman and Kautz, 1991; 1994; del Val, 1994; Dechter and Rish, 1994; Marquis, 1995; del Val, 1995; 1996; Marquis and Sadaoui, 1996; Schrag, 1996].

*This work has been supported in part by the Ganymède II project of the Contrat État/Région Nord–Pas-de-Calais.

In this paper, a new approach to equivalence-preserving compilation, called *tractable covers*, is introduced. In short, a tractable cover of $\Sigma$ is a finite set $\mathcal{T}$ of tractable formulas $\Phi$ (disjunctively considered) s.t. $\Sigma \equiv \mathcal{T}$. Tractable covers of $\Sigma$ are equivalence-preserving compilations of $\Sigma$: a clause $c$ is a logical consequence of $\Sigma$ iff for every $\Phi$ in $\mathcal{T}$, $c$ is a logical consequence of $\Phi$. Since $\Phi$ is tractable, each elementary test $\Phi \models c$ can be computed in time polynomial in $|\Phi| + |c|$. The point is to find out tractable $\Phi$s that concisely represent (i.e. cover) the largest sets of models of $\Sigma$, so that $|\mathcal{T}|$ remains limited. To some extent, the present work could then be related to other model-based approaches to knowledge representation and reasoning, like [Khardon and Roth, 1996].

First, a general framework is presented, which can take advantage of most tractable classes, simultaneously. In many respects, it generalizes the prime implicants cover technique recently used for compilation purpose [Schrag, 1996]. Then, the focus is laid on tractable covers that can be computed and intensionally represented thanks to (partial) interpretations. Two specific cases are considered. In the first one, partial interpretations are used to shape the KB into formulas from several possible classes. In the second one, they are used to derive renamable Horn formulas. The last one is proved less space-consuming than prime implicants covers [Schrag, 1996] for every KB. Since tractable covers of $\Sigma$ are equivalence-preserving compilations, their size may remain exponential in $|\Sigma|$ unless $NP \subseteq P/poly$ [Selman and Kautz, 1994], which is very unlikely. However, experimental results show that the new approaches can prove efficient w.r.t. direct query answering and offer significant time and space savings w.r.t. prime implicants covers.

## 2 Formal Preliminaries

A literal is a propositional variable or a negated one. A clause (resp. a term) is a finite set of literals, representing their disjunction (resp. conjunction). A Horn (resp. reverse Horn) clause contains at most one literal that is positive (resp. negative). A binary clause contains at most two literals. A KB $\Sigma$ is a finite set of propositional formulas, conjunctively considered. $Var(\Sigma)$ is the set of variables occurring in $\Sigma$. Every KB can be rewritten into a conjunction of clauses (into CNF for short) while preserving equivalence. A KB $\Sigma$ is said renamable Horn iff there exists a substitution $\sigma$ over literals $l$ built up from $Var(\Sigma)$, s.t. $\sigma(l) = \bar{l}$ and $\sigma(\Sigma)$ is Horn.

Interpretations and models are defined in the usual way. Let us stress that, quite unconventionally, (partial) interpretations will be represented as terms, i.e. as satisfiable sets of literals (vs. sets of variables). An implicant of a KB $\Sigma$ is a partial interpretation $\alpha$ s.t. $\alpha \models \Sigma$. A prime implicant of $\Sigma$ is an implicant $\pi$ of $\Sigma$ s.t. for all implicants $\alpha$ of $\Sigma$ s.t. $\pi \models \alpha$, we have $\alpha \models \pi$. The set of all prime implicants of $\Sigma$ (up to logical equivalence) is noted $PI(\Sigma)$. A prime implicant cover of $\Sigma$ is any subset $S$ of $PI(\Sigma)$ (disjunctively considered) s.t. $S \equiv \Sigma$.

## 3 Tractable Cover Compilations

### 3.1 The General Framework

First, let us make precise what classes of tractable formulas will be considered:

**Definition 3.1** *A list $Cs$ of classes $C$ of tractable propositional formulas (w.r.t. cover compilation) is s.t.:*

- *There exists a polytime algorithm $TRACTABLE?$ for checking whether any formula $\Phi$ belongs to $C$.*
- *There exists a polytime algorithm $QUERY?$ for checking whether $\Phi \models c$ holds for any $\Phi$ in $C$ and any clause $c$.*
- *For every $C$ in $Cs$, for every formula $\Phi$ of $C$ and every term $p$, there exists $C'$ in $Cs$ s.t. $(\Phi \wedge p) \in C'$.*

Since classes of tractable formulas are not finite sets in the general case, they are intensionally represented by ordered pairs of decision procedures $\langle TRACTABLE?, QUERY? \rangle$.

Interestingly, the great majority of classes of formulas tractable for SAT (the well-known propositional satisfiability decision problem) are also tractable for cover compilations. Especially, this is the case for the Horn, reverse Horn, binary, renamable Horn, q-Horn and nested clauses classes.

We are now ready to define the notion of a tractable cover of a propositional KB.

**Definition 3.2** *Given a CNF-KB $\Sigma$ and a finite list $Cs$ of classes of tractable formulas w.r.t. cover compilation (represented intensionally), a tractable cover of $\Sigma$ w.r.t. $Cs$ is a finite set $\mathcal{T}$ of satisfiable formulas (disjunctively considered) s.t.:*

- *For every $\Phi_i \in \mathcal{T}$, there exists $C$ in $Cs$ s.t. $\Phi_i$ belongs to $C$, and*
- *$\mathcal{T} \equiv \Sigma$.*

In the following, we will assume that $Cs$ contains at least the class $\{t \text{ s.t. } t \text{ is a term}\}$. This ensures that there always exists at least one tractable cover of $\Sigma$, namely a prime implicants one.

In this paper, we focus on tractable covers that can be *intensionally represented*, using (partial) interpretations. The corresponding explicit covers can be generated on-line from the intensional ones in polynomial time.

### 3.2 Carver-Based Tractable Covers

A first way to derive covers consists in *simplifying* the KB using partial interpretations. For any partial interpretation $p = \{l_1, \ldots, l_n\}$, the KB $\Sigma$ simplified w.r.t. $p$ is the set $\Sigma_p = (((\Sigma_{l_1})_{l_2}) \ldots)_{l_n}$, where $\Sigma_l = \{c \setminus \{\neg l\}$ s.t. $c \in \Sigma$ and $c \cap \{l\} = \emptyset\}$.

**Definition 3.3** *Given a CNF-KB $\Sigma$ and a finite list $Cs$ of classes of tractable formulas (represented intensionally):*

- *A carver of $\Sigma$ w.r.t. $Cs$ is a partial interpretation $p$ s.t. there exists $C$ in $Cs$ s.t. $\Sigma_p$ belongs to $C$ and $\Sigma_p$ is satisfiable.*
- *A carver-based tractable cover of $\Sigma$ w.r.t. $Cs$ is a set $PC$ of carvers of $\Sigma$ w.r.t. $Cs$ (disjunctively considered) s.t. $(PC \wedge \Sigma) \equiv \Sigma$.*
- *A carver-based cover compilation of $\Sigma$ is a pair $\mathcal{C} = \langle \Sigma, PC \rangle$, where $PC$ is a carver-based tractable cover of $\Sigma$ w.r.t. $Cs$.*

Interestingly, the space required to store a carver $p$ is always lower or equal to the space needed by the corresponding tractable formula $\Phi = p \wedge \Sigma_p$ (with a $O(|\Sigma|)$ factor in the worst case). At the on-line query answering stage, the price to be paid is a $O(|\Sigma \wedge p|)$ time complexity overhead per carver $p$ but this does not question the tractability of query answering.

As an example, let $\Sigma = \{\bar{x}_1 \vee \bar{x}_2 \vee \bar{x}_3 \vee x_4, x_1 \vee x_2 \vee x_4\}$ and let $Cs$ contain the Horn and the binary classes. The set $PC = \{\{x_1\}, \{\bar{x}_1\}\}$ is a carver-based tractable cover of $\Sigma$ w.r.t. $Cs$ since $\Sigma_{\{x_1\}} = \{\bar{x}_2 \vee \bar{x}_3 \vee x_4\}$ is Horn and $\Sigma_{\{\bar{x}_1\}} = \{x_2 \vee x_4\}$ is binary. If $Cs$ reduces to the renamable Horn class, $\{\emptyset\}$ is a carver-based tractable cover of $\Sigma$ w.r.t. $Cs$ since $\Sigma$ belongs to the renamable Horn class.

Clearly enough, carver-based cover compilations are equivalence-preserving compilations:

**Proposition 3.1** *Let $\Sigma$ be a CNF-KB, $\mathcal{C} = \langle \Sigma, PC \rangle$ be a carver-based cover compilation of $\Sigma$ and let $c$ be a clause. $\Sigma \models c$ iff for every $p$ in $PC$, $((p \models c)$ or $(\Sigma_p \models c))$, which can be decided in time polynomial in $|\mathcal{C}| + |c|$.*

Although the class(es) to which $\Sigma_p$ belongs can be polynomially recognized on-line for each carver $p$ of $PC$, in practice, they are determined once only at the off-line compiling stage. Accordingly, each carver $p$ found during the compiling process is indexed with a reference $label(p)$ to the concerned class in $Cs$. A significant amount of time in on-line query answering can be saved via such indexing.

Interestingly, carver-based cover compilations can lead to exponential space savings w.r.t. prime implicants cover compilations for some KBs. An extreme case consists of tractable KBs $\Sigma$ that remain invariant under carver-based cover compilation but are such that every prime implicant cover is exponential in the size of $\Sigma$.

## 3.3 Hyper-Implicant Covers

Formulas in covers must be tractable and exhibit as many models of $\Sigma$ as possible. Interestingly, several classes $C$ of tractable formulas admit model-theoretic characterizations and features that can be exploited. For the binary, Horn, reverse Horn, renamable Horn classes, among others, class membership is equivalent to the existence of some specific interpretation(s). Formally, a CNF-KB $\Sigma$ is renamable Horn [Lewis, 1978] iff there exists an interpretation $p$ over $Var(\Sigma)$ s.t. for every clause $c$ of $\Sigma$, at most one literal of $c$ does not belong to $p$. Interestingly, the renamable Horn class includes both the Horn, reverse Horn, and satisfiable binary KBs as proper subsets.

As the semantical characterizations above indicate it, literals that belong both to $p$ and to $\Sigma$ play a central role that we can take advantage of. In order to derive satisfiable renamable Horn formulas, every clause $c$ of $\Sigma$ is shortened using a model $p = m$ of $\Sigma$: only the literals occurring in $m$ are kept, plus an additional literal of $c$ (when possible). In this way, every shortened clause is such that every literal of $m$ (except possibly one) belongs to it. Accordingly, the resulting set of clauses, called *hyper-implicant* of $\Sigma$, is satisfiable and renamable Horn.

Formally, let $m$ be a model of $\Sigma$. For every clause $c$ in $\Sigma$, let $c^m$ be the clause consisting of the literals common to $c$ and $m$. Let $P(\Sigma, m)$ denote the set of clauses of $\Sigma$ s.t. for every clause $c$ in $P(\Sigma, m)$, $c$ and $c^m$ are identical. Let $N(\Sigma, m)$ be $\Sigma \setminus P(\Sigma, m)$. For every clause $c$ in $N(\Sigma, m)$, let $l_c^m$ denote a literal of $c$ s.t. $\overline{l_c^m}$ belongs to $m$. Obviously, several candidates $l_c^m$ may exist in the general case.

**Definition 3.4** *Given a CNF-KB $\Sigma$:*

- *A hyper-implicant of $\Sigma$ (w.r.t. to a model) $m$ of $\Sigma$ is a formula $\Sigma^m = (\bigwedge_{c \in P(\Sigma,m)} c) \wedge (\bigwedge_{c \in N(\Sigma,m)} (c^m \vee l_c^m))$.*
- *A hyper-implicant cover $\mathcal{H}$ of $\Sigma$ is a set of hyper-implicants of $\Sigma$ (disjunctively considered) s.t. $\mathcal{H} \equiv \Sigma$.*

Importantly, the size of any hyper-implicant $\Sigma^m$ of $\Sigma$ is strictly lower than the size of $\Sigma$, except when $\Sigma$ is renamable Horn.

Hyper-implicant covers are equivalence-preserving compilations:

**Proposition 3.2** *Let $\Sigma$ be a CNF-KB, $\mathcal{H}$ be a hyper-implicant cover of $\Sigma$ and $c$ be a clause. $\Sigma \models c$ iff for every hyper-implicant $\Sigma^m$ of $\mathcal{H}$, we have $\Sigma^m \models c$. This can be decided in time linear in $|\mathcal{H}| + |c|$.*

Assuming that the clauses in $\Sigma = \{c_1, \ldots, c_n\}$ are totally ordered, hyper-implicant covers can be represented intensionally by $\Sigma$ and sets of pairs $\langle m, [l_{c_1}^m, \ldots, l_{c_n}^m] \rangle$ where $m$ is a model of $\Sigma$ and $l_{c_i}^m$ ($i \in 1..n$) is *false* if $c_i \in P(\Sigma, m)$ and is the literal of $c_i \setminus m$ which is kept, otherwise. Each $\Sigma^m$ can be derived on-line from $\Sigma$ and $\langle m, [l_{c_1}^m, \ldots, l_{c_n}^m] \rangle$ in linear time.

As an example, let $\Sigma = \{x_1 \vee x_3 \vee x_4, x_1 \vee \bar{x}_2 \vee x_3, x_1 \vee x_2 \vee \bar{x}_3 \vee \bar{x}_4, \bar{x}_1 \vee \bar{x}_2 \vee \bar{x}_3 \vee \bar{x}_4, \bar{x}_1 \vee x_2 \vee \bar{x}_3 \vee x_4\}$. The hyper implicant cover represented by $\Sigma$ and the two pairs $\langle \{x_1, \bar{x}_2, \bar{x}_3, x_4\}, [x_3, x_3, x_2, \bar{x}_1, \bar{x}_1] \rangle$ and $\langle \{\bar{x}_1, x_2, x_3, \bar{x}_4\}, [x_1, \bar{x}_2, \bar{x}_3, \bar{x}_2, x_4] \rangle$, contains two hyper-implicants: $\{x_1 \vee x_4 \vee x_3, x_1 \vee \bar{x}_2 \vee x_3, x_1 \vee \bar{x}_3 \vee x_2, \bar{x}_2 \vee \bar{x}_3 \vee \bar{x}_1, \bar{x}_3 \vee x_4 \vee \bar{x}_1\}$ and $\{x_3 \vee x_1, x_3 \vee \bar{x}_2, x_2 \vee \bar{x}_4 \vee \bar{x}_3, \bar{x}_1 \vee \bar{x}_4 \vee \bar{x}_2, \bar{x}_1 \vee x_2 \vee x_4\}$.

Intuitively, any hyper-implicant $\Sigma^m$ of $\Sigma$ is a concise representation of some prime implicants of $\Sigma$. To be more specific, the prime implicants of $\Sigma$ which are entailed by $m$ are prime implicants of $\Sigma^m$.

**Proposition 3.3** *For every model $m$ of $\Sigma$, let $PI_m(\Sigma)$ be the set of prime implicants of $\Sigma$ s.t. for every $\pi$ in $PI_m(\Sigma)$, we have $m \models \pi$. Then, $PI_m(\Sigma) \subseteq PI(\Sigma^m)$.*

Consequently, $\Sigma^m$ covers every model of $\Sigma$ covered by a prime implicant of $\Sigma$ entailed by $m$.

Hyper-implicants covers are economical representations w.r.t. prime implicants ones. Not withstanding the fact that hyper-implicants of $\Sigma$ are smaller than $\Sigma$, the number of hyper-implicants in a cover is lower than the number of implicants in a cover:

**Proposition 3.4** *For every prime implicant cover of $\Sigma$ containing $t$ prime implicants there exists a hyper-implicant cover of $\Sigma$ containing at most $\lfloor \frac{t+1}{2} \rfloor$ hyper-implicants.*

In the example above, the hyper-implicant cover contains 2 formulas, while the smallest prime implicant cover of $\Sigma$ consists of 6 prime implicants. Actually, experiments show significant savings w.r.t. the sizes of the prime implicants covers for most KBs (cf. Section 5).

## 4 Computing Compilations

(Partial) interpretations giving rise to intensionally-represented tractable covers are computed using systematic search, thanks to a Davis/Putnam-like procedure DPTC. This procedure is closely related to Schrag's DPPI algorithm [Schrag, 1996]. Cs is empty for the hyper-implicant case.

```
DPTC(Σ,Cs):
1    PC ← ∅;
2    DP*(Σ,Cs,∅);
3    Return(⟨Σ,PC⟩).

DP*(Σ,Cs,p):
1    If not PRUNING(Σ,p) then
2        UNIT_PROPAGATE(Σ,p);
3        If ∅ ∈ Σ then return;
4        If (p ⊨ Σ)
```

```
5        then PROCESS_IMPLICANT(p,Σ,Cs)
6        return;
7    l← CHOOSE_BEST_LITERAL(Σ);
8    DP*(Σ,Cs,p ∪ {l});
9    DP*(Σ,Cs,p ∪ {¬l}).
```

The CHOOSE_BEST_LITERAL branching rule and UNIT_PROPAGATE procedures are standard Davis/Putnam features. In our experiments, the branching rule by [Dubois et al., 1996] is used. The main role of DP* is to find implicants of Σ in the whole search tree. PRUNING and PROCESS_IMPLICANT depend on the considered approach. From the found implicants, (partial) interpretations (carvers and implicit representations of hyper-implicants) are derived thanks to PROCESS_IMPLICANT; they are collected into the global variable PC.

### 4.1 The Carver-Based Case

PRUNING$_C$(Σ,new_p):
```
1    If there exists p ∈ PC s.t. (new_p ⊨ p)
2        then return(true) else return(false).
```

PROCESS_IMPLICANT$_C$(new_p,Σ,Cs):
```
1    Carver ← DERIVE_C(new_p,Σ,Cs);
2    For every p ∈ PC do
3        If (p ⊨ Carver) then remove p from PC;
4    Put Carver in PC.
```

DERIVE$_C$(p,Σ,Cs):
```
1    Prime ← ONE_PRIME(p,Σ);
2    For every literal l ∈ Prime do
3        For every ⟨TRACTABLE?,QUERY?⟩ in Cs do
4            If TRACTABLE?(Σ_{Prime\{l}})
                 then Return(DERIVE_C(Prime\{l},Σ,Cs));
5    Return(Prime).
```

Whenever an implicant $p$ of Σ is found, a prime implicant Prime of Σ s.t. $p \models$ Prime is extracted from $p$, thanks to the ONE_PRIME procedure. ONE_PRIME considers every literal $l$ of $p$ successively, check whether $l$ is necessary (i.e. if there exists a clause $c$ of Σ s.t. $c \cap p = \{l\}$), and remove $l$ from $p$ if $l$ is not necessary (see [Castell and Cayrol, 1996][Schrag, 1996] for details). These prime implicants are then tentatively simplified; all the literals of each Prime are considered successively; for every literal $l$ of Prime, $\Sigma_{Prime\setminus\{l\}}$ is checked for tractability using the recognition procedures given in Cs. If $\Sigma_{Prime\setminus\{l\}}$ is found tractable, $l$ is ruled out from Prime and is kept otherwise. Once all the literals of Prime have been considered, the resulting simplified Prime (indexed by the label of the corresponding class in Cs) is checked against the set PC of carvers collected so far. Only maximal carvers w.r.t. ⊨ are kept in PC. Interestingly, the set PC of carvers stored during the traversal of the DP search tree is used to prune this tree, thanks to the PRUNING$_C$ procedure; indeed, whenever a candidate (partial) interpretation new_p is elected, it can be immediately removed if new_p entails one of the current carvers.

Clearly enough, both the literal ordering and the recognition procedure ordering used in DERIVE$_C$ can greatly influence the cover generated in this way.

### 4.2 The Hyper-Implicant Case

PRUNING$_H$(Σ,new_p):
```
1    If there exists p ∈ PC s.t. (Σ^{new_p} ⊨ Σ^p)
2        then return(true) else return(false).
```

PROCESS_IMPLICANT$_H$(new_p,Σ,_):
```
1    Model ← DERIVE_H(new_p,Σ);
2    For every p ∈ PC do
3        If (Σ^{Model} ⊨ Σ^p) then remove p from PC;
4    Put Model in PC.
```

DERIVE$_H$(p,Σ):
```
1    Model ← p;
2    For every variable v ∈ Var(Σ) do
3        If ({v} ∩ p = ∅) and ({¬v} ∩ p = ∅)
4            then put v with its most frequent
                 sign in Σ to Model;
5    Return(Model).
```

Each time an implicant $p$ of Σ is found, a model Model of Σ s.t. Model ⊨ $p$ is derived from $p$. The variables that do not occur in $p$ are added with the sign occurring the most frequently in Σ. When $\Sigma^{Model}$ is computed for the first time, a list $[l_{c_1}^m, \ldots, l_{c_n}^m]$ is attached to Model. The DERIVE$_H$ procedure is both simpler and more efficient than DERIVE$_C$ (roughly, it is not more time-consuming than the prime implicant extraction achieved by ONE_PRIME). The tractable formulas $\Sigma^p$ corresponding to the models $p$ collected in PC are used to prune the search tree (PRUNING$_H$). Only the $p$s giving rise to the logically weakests $\Sigma^p$ are kept in PC. Since the renamable Horn formulas $\Sigma^{Model}$ and $\Sigma^p$ stem from the same KB Σ, checking whether ($\Sigma^{Model} \models \Sigma^p$) can be performed in a very efficient way.

Both the literals chosen to extend the found implicants to models $p$ of Σ, and the literals $l_c^m$ selected in clauses of $\Sigma^p$ may have a significant impact on the hyper-implicants $\Sigma^p$ that are generated.

## 5 Experimental Results

In contrast to SAT, only few benchmarks for knowledge compilation can be found in the literature (with the well-developed experimental framework of [Schrag, 1996] as an exception). Actually, no comprehensive analysis of what should be the nature of meaningful benchmarks for evaluating compilation approaches has ever been conducted. Clearly, benchmarks must be hard for query answering since the goal of knowledge compilation is to overcome its intractability. However, in contrast to [Schrag, 1996], we do not focus on hard SAT instances,

| problems | #var | #cla | $\alpha_\mathcal{P}$ $\alpha_\mathcal{C}$ $\alpha_\mathcal{H}$ | $\beta_\mathcal{P}$ $\beta_\mathcal{C}$ $\beta_\mathcal{H}$ | $|\mathcal{P}|$ $|\mathcal{C}|$ $|\mathcal{H}|$ |
|---|---|---|---|---|---|
| adder | 21 | 50 | 5.08 2.09 0.20 | — — 78.75 | 5376 1044 305 |
| history-ex | 21 | 17 | 2.00 0.75 0.14 | — 1000.00 11.66 | 172 234 77 |
| two-pipes | 15 | 54 | 0.66 0.60 0.12 | 125.00 125.00 20.00 | 255 234 192 |
| three-pipes | 21 | 82 | 0.47 0.42 0.27 | 45.45 <1 17.50 | 441 373 284 |
| four-pipes | 27 | 110 | 0.36 0.25 0.20 | 23.80 18.51 29.16 | 675 528 380 |
| regulator | 21 | 106 | 0.51 0.29 0.17 | 26.31 20.83 29.99 | 861 530 422 |
| selenoid | 11 | 19 | 2.16 1.40 0.20 | — — 17.49 | 297 115 94 |
| valve | 13 | 50 | 0.85 0.66 0.16 | <1 <1 21 | 312 297 246 |

Table 1: Experimental results.

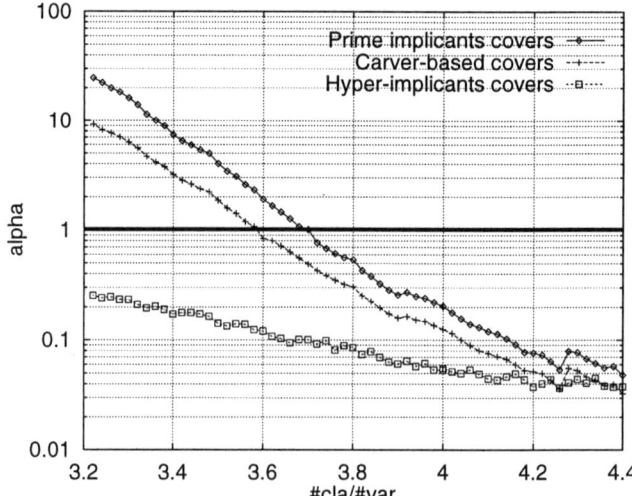

Figure 1: Ratios $\alpha$.

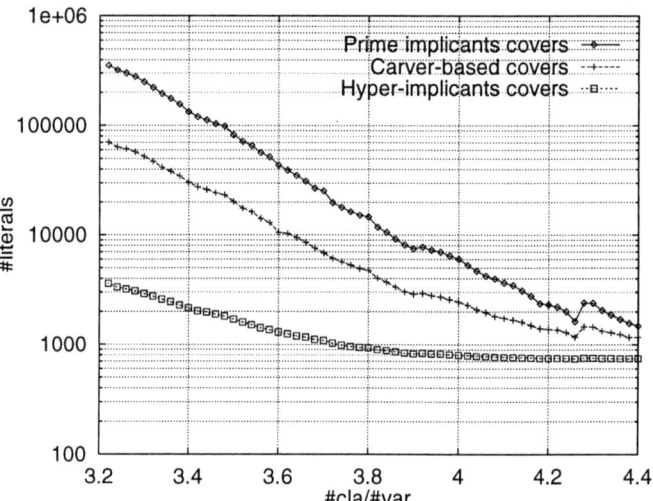

Figure 2: Sizes of the compilations.

only. Hard SAT instances (with respect to current algorithms) should be considered hard for query answering since at least one query (namely, the empty clause) is difficult. However, easy SAT instances can exhibit hard queries that differ from the empty clause.

Accordingly, we tested the tractable covers and the prime implicants approaches w.r.t. many KBs, including "standard" structured problems, taken from [Forbus and De Kleer, 1993], and random k-SAT problems [Dubois et al., 1996], varying the #cla(use)/#var(iable) ratio from the easy to the hard regions. Each KB $\Sigma$ has been compiled using the 3 techniques. Then, 500 queries have been considered. In order to check the usefulness of the compilation process, we also answered these queries from $\Sigma$, using a direct, uncompiled, Davis/Putnam-based approach [Dubois et al., 1996]. For each problem $\Sigma$ and each compilation technique, the ratios $\alpha = Q_C/Q_U$ and $\beta = C/(Q_U - Q_C)$ have been computed. $Q_C$ (resp. $Q_U$) is the time needed by the compiled (resp. uncompiled) approach to answer all the queries, and $C$ is the compilation time. $\alpha$ (resp. $\beta$) tells us how much query time improvement we get from compilation (resp. how many queries are required to amortize the cost of compilation).

Table 1 reports some results of our extensive experiments. For each problem [Forbus and De Kleer, 1993], it lists results for the prime implicants, carvers, and hyper-implicants covers, successively. Especially, it gives the ratios $\alpha$ and $\beta$ and the size (in literals) of the corresponding cover. The size of any tractable cover compilation is the size of $\Sigma$ plus the size of the set of (partial) interpretations used as an implicit representation. For the carver-based approach, only the Horn, reverse Horn and binary classes have been considered. For the hyper-implicant approach, simplification of the cover (i.e. lines 2 to 4 of the PROCESS_IMPLICANT$_\mathcal{H}$ procedure) has not been implemented.

Results obtained on 50 variables random 3-SAT problems, where the ratio #cla/#var varies from 3.2 to 4.4, are reported on the two next figures. 50 problems have been considered per point and the corresponding scores averaged. Figure 1 (resp. Figure 2) gives aggregate values of ratios $\alpha$ (resp. sizes in literals) obtained for each compilation technique. $\alpha = 1$ separates the region for which compilation is useful from the region for which it is not.

At the light of our experiments, tractable covers prove better than prime implicants covers, both for structured and random k-SAT problems. Significant time savings w.r.t. query answering and significant space savings are obtained. Especially, tractable covers prove useful for many KBs for which prime implicants covers are too large to offer improvements w.r.t. query answering. More, the tractable cover approach allows the compilation of KBs which have so huge prime implicants covers $\mathcal{P}$ that $\mathcal{P}$ cannot be computed and stored. This coheres with the theoretical results reported in [Boufkhad and Dubois, 1996], showing that the average number of prime implicants of k-SAT formulas is exponential in their number of variables.

## 6 Conclusion

Both theoretical and experimental results show the tractable cover approach promising and encourage us to extend it in several directions. A first issue for further research is how to determine efficiently the best suited classes of tractable formulas for a given KB $\Sigma$. On the experimental side, an extensive evaluation of the carver-based technique equipped with more expressive tractable classes must be done. Extending the hyper-implicant approach to other tractable classes, especially the q-Horn one [Boros et al., 1994], is another interesting perspective. Finally, fragments of tractable covers of $\Sigma$ can serve as approximate compilations (lower bounds) of $\Sigma$ in the sense of [Selman and Kautz, 1991; 1994; del Val, 1995; 1996]. Since the tractable cover approach allows disjunctions of tractable formulas from several classes, better approximations could be obtained.

## References

[Boros et al., 1994] E. Boros, P.L. Hammer, and X. Sun. Recognition of q-horn formulae in linear time. *Discrete Applied Mathematics*, 55(1):1–13, 1994.

[Boufkhad and Dubois, 1996] Y. Boufkhad and O. Dubois. Length of prime implicants and number of solutions of random r-cnf formulas. *(submitted)*, 1996.

[Castell and Cayrol, 1996] T. Castell and M. Cayrol. Computation of prime implicates and prime implicants by the davis and putnam procedure. In *Proc. ECAI'96 Workshop on Advances in Propositional Deduction*, pages 61–64, Budapest, 1996.

[Dechter and Rish, 1994] R. Dechter and I. Rish. Directional resolution: The davis-putnam procedure, revisited. In *Proc. KR'94*, pages 134–145, Bonn, 1994.

[del Val, 1994] A. del Val. Tractable databases: How to make propositional unit resolution complete through compilation. In *Proc. KR'94*, pages 551–561, 1994.

[del Val, 1995] A. del Val. An analysis of approximate knowledge compilation. In *Proc. IJCAI'95*, pages 830–836, Montreal, 1995.

[del Val, 1996] A. del Val. Approximate knowledge compilation: The first-order case. In *Proc. AAAI'96*, pages 498–503, Portland (OR), 1996.

[Dowling and Gallier, 1984] W. Dowling and J. Gallier. Linear time algorithms for testing the satisfiability of propositional horn formulae. *Journal of Logic Programming*, 1(3):267–284, 1984.

[Dubois et al., 1996] O. Dubois, P. Andre, Y. Boufkhad, and J. Carlier. Sat vs. unsat. In $2^{nd}$ *DIMACS Implementation Challenge*, volume 26 of *DIMACS Series*, pages 415–436. American Mathematical Society, 1996.

[Forbus and De Kleer, 1993] K.D. Forbus and J. De - Kleer. *Building Problem Solvers*. MIT Press, 1993.

[Khardon and Roth, 1996] R. Khardon and D. Roth. Reasoning with models. *Artificial Intelligence*, 87:187–213, 1996.

[Knuth, 1990] D.E. Knuth. Nested satisfiability. *Acta Informatica*, 28:1–6, 1990.

[Lewis, 1978] H.R. Lewis. Renaming a set of clauses as a horn set. *JACM*, 25:134–135, 1978.

[Marquis and Sadaoui, 1996] P. Marquis and S. Sadaoui. A new algorithm for computing theory prime implicates compilations. In *Proc. AAAI'96*, pages 504–509, Portland (OR), 1996.

[Marquis, 1995] P. Marquis. Knowledge compilation using theory prime implicates. In *Proc. IJCAI'95*, pages 837–843, Montreal, 1995.

[Reiter and De Kleer, 1987] R. Reiter and J. De Kleer. Foundations of assumption-based truth maintenance systems: Preliminary report. In *Proc. AAAI'87*, pages 183–188, Seattle (WA), 1987.

[Schrag, 1996] R. Schrag. Compilation for critically constrained knowledge bases. In *Proc. AAAI'96*, pages 510–515, Portland (OR), 1996.

[Selman and Kautz, 1991] B. Selman and H. Kautz. Knowledge compilation using horn approximations. In *Proc. AAAI'91*, pages 904–909, 1991.

[Selman and Kautz, 1994] B. Selman and H. Kautz. Knowledge compilation and theory approximation. *JACM*, 43(2):193–224, 1994.

# A Four-Valued Fuzzy Propositional Logic

**Umberto Straccia**
Istituto di Elaborazione della Informazione
Via S. Maria, 46 56126 Pisa (PI) ITALY

## Abstract

It is generally accepted that knowledge based systems would be smarter and more robust if they can manage inconsistent, incomplete or imprecise knowledge. This paper is about a four-valued fuzzy propositional logic, which is the result of the combination of a four-valued logic and a fuzzy propositional logic. Besides the nice computational properties, the logic enables us also to deal both with inconsistency and imprecise predicates in a simple way.

## 1 Introduction

The management of uncertainty in inference systems is an important issue due to the imperfect nature of real world information. There are several fields in which this information has to do with *vague concepts*, *i.e.* concepts without clear definition. The key fact about vague concepts is that while they are not well defined, propositions involving them may be quite well defined. For instance, the boundaries of the Mount Everest are ill defined, whereas the proposition stating that the Mount Everest is the highest mountain of the world is definite, and its definiteness is not compromised by the ill-definitess of it's exact boundaries. Propositions of this kind are called *fuzzy propositions*. Each fuzzy proposition may have a degree of truth between $[0, 1]$. On the other hand, there exists propositions which are true or false, but due to the lack of precision of the available information we can in general only estimate to what extend it is possible or necessary that they are true. This kind of propositions are called *uncertain propositions*. For example, the concept triangle is well defined, but we can only estimate to what extend it is possible that *e.g.* a shape in a picture is a triangle if the segments are not exactly bounded. Certainly, any combination of the two is possible, *e.g.* uncertain fuzzy propositions are fuzzy propositions for which the available reference information is not precise.

In this paper we will concentrate our attention to (certain) fuzzy propositions. In particular, fuzzy proposition we will handle are of the form $[A \geq n]$ (where $A$ is a proposition and $n \in [0,1]$) and have intended meaning "it is *certain* that the degree of truth of $A$ is *at least* $n$". But, rather mapping $[A \geq n]$ as usual into *true* or *false* (as *e.g.* in [Chen and Kundu, 1996]), we will give to it a four-valued semantics. This will be done by mapping $[A \geq n]$ into an element of $2^{\{t,f\}}$, where $\{t\}$, $\{f\}$, $\emptyset$ and $\{t,f\}$ stand for the four truth values *true*, *false*, *unknown* and *contradiction*, respectively, as in [Levesque, 1984]. A first consequence of this semantics is that in certain "useful" circumstances the deduction process is tractable from a computational point of view. A second consequence is that the semantics enables us to deal with inconsistencies as the four-valued logic we will adopt is known to be paraconsistent (see, *e.g.* [Wagner, 1991]).

Our four-valued fuzzy semantics has been shown to be useful in the area of content-based retrieval of multimedia data [Meghini *et al.*, 1997]. In this context the (semantic) content of *e.g.* an image region $r$ is described by means of fuzzy propositions like "$r$ represents the Mount Everest with degree $\geq 0.8$". Since images (or any other media) are the subjective work of their authors, contradictions could arise among their content representations (possibly together with domain knowledge), which typically may not be the subject of a belief revision process.

This paper is organised as follows. In the next section we will briefly resume some aspects of the four-valued logic we are based on and in Section 3 we will extend it to the fuzzy case. In Section 4 we will extend our logic by allowing a sort of conditional reasoning[1]. Calculi for deciding entailment will be given for all logics presented and Section 5 concludes.

## 2 Four-valued propositions

The four-valued logic we will base our work on is essentially [Belnap, 1977; Levesque, 1984]. Let $\mathcal{L}$ be the language of propositional logic, with connectives $\wedge, \vee$ and $\neg$. We will use metavariable $A, B, C, \ldots$ and $p, q, r, \ldots$ for propositions and propositional letters, respectively[2]. *Negation Normal Forms* (NNF) and *Conjunctive Normal Forms* (CNF) are defined as usual.

A *four-valued interpretation* $\mathcal{I}$ maps a proposition into an element of $2^{\{t,f\}}$ and has to satisfy the following equa-

---
[1]Notice that in our basic logic modus ponens is not a valid rule of inference.

[2]All metavariables could have an optional subscript.

tions: $t \in (A \wedge B)^{\mathcal{I}}$ iff $t \in A^{\mathcal{I}}$ and $t \in B^{\mathcal{I}}$; $f \in (A \wedge B)^{\mathcal{I}}$ iff $f \in A^{\mathcal{I}}$ or $f \in B^{\mathcal{I}}$; $t \in (A \vee B)^{\mathcal{I}}$ iff $t \in A^{\mathcal{I}}$ or $t \in B^{\mathcal{I}}$; $f \in (A \vee B)^{\mathcal{I}}$ iff $f \in A^{\mathcal{I}}$ and $f \in B^{\mathcal{I}}$; $t \in (\neg A)^{\mathcal{I}}$ iff $f \in A^{\mathcal{I}}$ and $f \in (\neg A)^{\mathcal{I}}$ iff $t \in A^{\mathcal{I}}$. It is worth noting that a two-valued interpretation is just a four-valued interpretation $\mathcal{I}$ such that $A^{\mathcal{I}} \in \{\{t\},\{f\}\}$, for each $A$. We might characterise the distinction between two-valued and four-valued semantics as the distinction between *implicit* and *explicit* falsehood: in a two-valued logic a formula is (implicitly) false in an interpretation iff it is not true, while in a four-valued logic this need not be the case. Our truth conditions are always given in terms of belongings $\in$ (and never in terms of non belongings $\notin$) of truth values to interpretations.

Let $\mathcal{I}$ be an interpretation, let $A, B$ be two propositions and let $\Sigma$ be a set of propositions, called *Knowledge Base* (KB): $\mathcal{I}$ satisfies (is a model of) $A$ iff $t \in A^{\mathcal{I}}$; $A$ and $B$ are *equivalent* (written $A \equiv_4 B$) iff they have the same models; $\mathcal{I}$ satisfies (is a model of) $\Sigma$ iff $\mathcal{I}$ is a model of $A$, for all $A \in \Sigma$; $\Sigma$ *entails* $A$ (written $\Sigma \models_4 A$) iff all models of $\Sigma$ are models of $A$. Without loss of generality, we can restrict our attention to propositions in NNF only, as $\neg\neg A \equiv_4 A$, $\neg(A \wedge B) \equiv_4 \neg A \vee \neg B$ and $\neg(A \vee B) \equiv_4 \neg A \wedge \neg B$ hold. For ease of notation, we will often omit braces, thus writing e.g. $A, B \models_4 C$ in place of $\{A, B\} \models_4 C$ and $\models_4 A$ in place of $\emptyset \models_4 A$.

**Relation 1** *The following relations can easily be verified:* 1. $A \equiv_4 B$ *does not imply* $\neg A \equiv_4 \neg B$ *(and viceversa)*, 2. $A \wedge B \models_4 A$, 3. $A \models_4 B$ *and* $B \models_4 C$ *implies* $A \models_4 C$, $A \models_4 A \vee B$, $A \wedge (\neg A \vee B) \not\models_4 B$ *and* $\Sigma \models_4 A$ *implies* $\Sigma \models_2 A$, *where* $\models_2$ *is the classical two-valued entailment relation. Note that there are no tautologies, i.e. there is no $A$ such that $\models_4 A$. Moreover, every KB is satisfiable. Hence, $A \wedge \neg A \not\models_4 B$, as there is a model $\mathcal{I}$ ($A^{\mathcal{I}} = \{t,f\}$, $B^{\mathcal{I}} = \emptyset$) of $A \wedge \neg A$ not satisfying $B$.* ∎

Certainly, not allowing modus ponens is penalizing. But, we will include this form of inference in the extended language $\mathcal{L}^+$ described in Section 4.

## 2.1 Deciding entailment in $\mathcal{L}$

Effectively deciding whether $\Sigma \models_4 A$ requires a calculus. A well known algorithm for deciding entailment in $\mathcal{L}$ is Levesque's algorithm [Levesque, 1984]: in order to check whether $A \models_4 B$, we put $A$ and $B$ into an equivalent CNF (say $C$ and $D$) and verify whether for each conjunct $D_j$ of $D$ there is a conjunct $C_i$ of $C$ such that $C_i \subseteq D_j$, where $C_i$ and $D_j$ are clauses. Hence, entailment between two propositions $C$ and $D$ in CNF can be verified in time $O(|C||D|)$, whereas checking whether $A \models_4 B$ is a coNP-complete problem in the general case. We propose an alternative calculus which (i) does not require any transformation into CNF, (ii) has the same polynomial complexity for the CNF case and (iii) is easy extensible to the treatment of conditional reasoning which will be the topic of Section 4. The calculus we have developed is one inspired on the calculus **KE** [D'Agostino and Mondadori, 1994]. The calculus, a semantic tableaux, is based on *signed propositions of type* $\alpha$ ("conjunctive propositions") and of *type* $\beta$ ("disjunctive propositions") and on their *components* which are defined as usual [Smullyan, 1968][3]:

| $\alpha$ | $\alpha_1$ | $\alpha_2$ | $\beta$ | $\beta_1$ | $\beta_2$ |
|---|---|---|---|---|---|
| T$A \wedge B$ | T$A$ | T$B$ | T$A \vee B$ | T$A$ | T$B$ |
| NT$A \vee B$ | NT$A$ | NT$B$ | NT$A \wedge B$ | NT$A$ | NT$B$ |

T$A$ and NT$A$ are called *conjugated signed propositions* and with $\beta_i^c$ we indicate the *conjugate* of $\beta_i$. An interpretation $\mathcal{I}$ *satisfies* T$A$ iff $\mathcal{I}$ satisfies $A$, whereas $\mathcal{I}$ *satisfies* NT$A$ iff $\mathcal{I}$ does not satisfy $A$. A set of signed propositions is *satisfiable* iff each element of it is satisfiable. Therefore, $\Sigma \models_4 A$ iff T$\Sigma \cup \{$NT$A\}$ is not satisfiable, where T$\Sigma = \{$T$A : A \in \Sigma\}$. The calculus is based on the rules:

$$(A) \; \frac{\alpha}{\alpha_1, \alpha_2} \qquad (PB) \; \frac{}{\text{T}A \mid \text{NT}A}$$

$$(B1) \; \frac{\beta, \beta_1^c}{\beta_2} \qquad (B2) \; \frac{\beta, \beta_2^c}{\beta_1}$$

An instance of e.g. rule $(B1)$ is $\frac{\text{T}A \vee B, \text{NT}A}{\text{T}B}$. Notice that the only branching rule is $(PB)$ (called *Principle of Bivalence*). As usual, a deduction is represented as a tree, called *deduction tree*. A branch $\phi$ in a deduction tree is closed iff for some proposition $A$, both T$A$ and NT$A$ are in $\phi$. With $S^\phi$ we indicate the set of signed propositions occurring in $\phi$. A set of signed propositions $S$ has a *refutation* iff in each deduction tree all branches $\phi$ are closed. Furthermore, we will restricted the proof procedure to the so-called *canonical* form [D'Agostino and Mondadori, 1994, p. 299]: a proposition is *AB-analysed* in a branch $\phi$ if either (i) it is of type $\alpha$ and both $\alpha_1$ and $\alpha_2$ occur in $\phi$; or (ii) it is of type $\beta$ and (iia) if $\beta_1^c$ occurs in $\phi$ then $\beta_2$ occurs in $\phi$, (iib) if $\beta_2^c$ occurs in $\phi$ then $\beta_1$ occurs in $\phi$. A branch is *AB-completed* if all the propositions in it are AB-analysed. A proposition of type $\beta$ is *fulfilled* in a branch $\phi$ if either $\beta_1$ or $\beta_2$ occurs in $\phi$. We say that a branch $\phi$ is *completed* if it is AB-completed and, every proposition of type $\beta$ occurring in $\phi$ is fulfilled. A deduction tree is *completed* if all its branches are completed. The procedure $Sat(S)$ below determines whether $S$ is satisfiable or not.

**Algorithm 1** ($Sat(S)$)
$Sat(S)$ starts from the root labelled $S$ and applies the rules until the resulting tree is either closed or completed. If the tree is closed, $Sat(S)$ returns false, otherwise true. At each step of the construction the following steps are performed:

1. select a branch $\phi$ which is not yet completed;
2. expand $\phi$ by means of the rules $(A)$, $(B1)$ and $(B2)$ until it becomes AB-completed, generating branch $\phi'$;
3. if $\phi'$ is neither closed nor completed then
   (a) select a proposition of type $\beta$ which is not yet fulfilled in the branch;
   (c) apply rule $(PB)$ with $\beta_1$ and $\beta_1^c$ as PB-formulae and go to step 1.
   otherwise, go to step 1. ∎

The following proposition can be shown.

---
[3] T and NT play the role of "True" and "Not True", respectively. In classical calculi NT may be replaced with F ("False").

**Proposition 1** *Let $S$ be a set of signed propositions in $\mathcal{L}$. Then $Sat(S)$ iff $S$ is satisfiable.* ⊣

**Example 1** It can easily verified that a canonical proof of $p \vee (q \wedge r) \models_4 (p \vee q) \wedge (p \vee r)$ starts with $S = \{\text{T}p \vee (q \wedge r), \text{NT}(p \vee q) \wedge (p \vee r)\}$ and generates two branches $\phi_1$ and $\phi_2$, by using $\beta_1 = \text{NT}(p \vee q)$ and $\beta_1^c = \text{T}(p \vee q)$ as PB-formulae, such that $S^{\phi_1} = S \cup \{\text{NT}p \vee q, \text{NT}p, \text{NT}q, \text{T}q \wedge r, \text{T}q, \text{T}r\}$ and $S^{\phi_2} = S \cup \{\text{T}p \vee q, \text{NT}p \vee r, \text{NT}p, \text{NT}r, \text{T}q \wedge r, \text{T}q, \text{T}r\}$. Both $\phi_1$ and $\phi_2$ are closed. ■

If $\Sigma$ and A are in CNF[4], then rule $(PB)$ is *not* needed, *i.e.* we can eliminate Step 3. from *Sat*. Hence, any deduction tree will have one branch. As a consequence, by observing that $\Sigma \models_4 A_1 \wedge \ldots A_n$ iff for each $1 \leq i \leq n$ $\Sigma \models_4 A_i$, it can be shown that

**Proposition 2** *If $\Sigma$ and A are in CNF then checking $\Sigma \models_4 A$ can be done in time $O(|\Sigma||A|)$ using Sat.* ⊣

Two-valued soundness and completeness is obtained by extending signed propositions as usual: (i) $\text{T}\neg A$ is of type $\alpha$ and $\text{NT}A$ is its $\alpha_1$ and $\alpha_2$ component; (ii) $\text{NT}\neg A$ is of type $\alpha$ and $\text{T}A$ is its $\alpha_1$ and $\alpha_2$ component. Just notice that in this case *Sat* is exactly the canonical procedure for **KE** [D'Agostino and Mondadori, 1994].

## 3 Fuzzy propositions

Now, we extend our propositional language $\mathcal{L}$ to the fuzzy case. A *fuzzy valuation* is a function mapping propositions into $[0, 1]$. Consistently with our approach of distinguishing explicit from implicit falsehood (*i.e.* distinguishing $f \in A^{\mathcal{I}}$ from $t \notin A^{\mathcal{I}}$) we will use two fuzzy valuations, $|\cdot|^t$ and $|\cdot|^f$: $|A|^t$ will naturally be interpreted as the *degree of truth* of $A$, whereas $|A|^f$ will analogously be interpreted as the *degree of falsity* of $A$. Classical "two-valued" fuzzy propositions $|\cdot|^t$ and $|\cdot|^f$ are such that $|A|^f = 1 - |A|^t$, for each $A$. In our case, instead, we might well have $|A|^t = 0.6$ and $|A|^f = 0.8$. This is a natural consequence of our four-valued approach.

A (certain) *fuzzy proposition* is an expression of type $[A \geq n]$, where $A$ is a proposition in $\mathcal{L}$ and $n \in [0, 1]$; $\mathcal{L}^f$ is just the set of fuzzy propositions. For instance, $[\text{ItsCold} \geq 0.7]$ is a fuzzy proposition meaning that it is certain that the degree of truth of ItsCold is at least 0.7, while $[\text{ItsCold} \geq 1]$ means that it is definitely cold. On the other hand $[\neg\text{ItsCold} \geq 0.7]$ means that it is certain that it is likely to be not cold, while $[\neg\text{ItsCold} \geq 1]$ may be interpreted as saying that it is definitely not cold. $[A \geq n]$ is in CNF whenever $A$ is.

A *fuzzy interpretation* $\mathcal{I}$ is a triple $\mathcal{I} = ((\cdot)^{\mathcal{I}}, |\cdot|^t, |\cdot|^f)$, where $|\cdot|^t$ and $|\cdot|^f$ are fuzzy valuations and $(\cdot)^{\mathcal{I}}$ maps each fuzzy proposition into an element of $2^{\{t,f\}}$. Additionally, $(\cdot)^{\mathcal{I}}, |\cdot|^t$ and $|\cdot|^f$ have to satisfy the following equations: $|A \wedge B|^t = \min\{|A|^t, |B|^t\}$; $|A \wedge B|^f = \max\{|A|^f, |B|^f\}$; $|A \vee B|^t = \max\{|A|^t, |B|^t\}$; $|A \vee B|^f = \min\{|A|^f, |B|^f\}$; $|\neg A|^t = |A|^f$; $|\neg A|^f = |A|^t$; $t \in [A \geq n]^{\mathcal{I}}$ iff $|A|^t \geq n$; and $f \in [A \geq n]^{\mathcal{I}}$

iff $|A|^f \geq n$. It is easy to see that, *e.g.* $|A \wedge B|^t = |\neg A \vee \neg B|^f$. Similarly for $|A \vee B|^t$.

It is worth noting that there is a simple connection between the four-valued semantics given in Section 2 and the fuzzy counterpart. In fact, the above conditions can be reformulated as *e.g.* $t \in [A \wedge B \geq n]^{\mathcal{I}}$ iff $t \in [A \geq n]^{\mathcal{I}}$ and $t \in [B \geq n]^{\mathcal{I}}$; and $f \in [A \wedge B \geq n]^{\mathcal{I}}$ iff $f \in [A \geq n]^{\mathcal{I}}$ or $f \in [B \geq n]^{\mathcal{I}}$. If $|A|^f = 1 - |A|^t$ and $[A \geq n]^{\mathcal{I}} \in \{\{t\}, \{f\}\}$, classical "two-valued" fuzzy logic is obtained.

Fuzzy satisfiability, fuzzy equivalence and fuzzy entailment are defined as the natural extensions of the non fuzzy case. We will use the relation $\approx_{(\cdot)}$ in place of $\models_{(\cdot)}$ whenever we refer to the fuzzy case (*e.g.* $\Sigma \approx_4 [A \geq n]$). Since $\approx_4 [A \geq 0]$, we will not consider those $[A \geq n]$ for $n = 0$. Given a KB $\Sigma$ and a proposition $A$, we define the *maximal degree of truth* of $A$ with respect to $\Sigma$ (written $Maxdeg(\Sigma, A)$) to be $\sup\{n > 0 : \Sigma \approx_4 [A \geq n]\}$ ($\sup \emptyset = 0$). Notice that $\Sigma \approx_4 [A \geq n]$ iff $Maxdeg(\Sigma, A) \geq n$.

There is a strict relation between fuzzy propositions and propositions. Given a KB $\Sigma$, let $\overline{\Sigma}$ be the (crisp) KB $\{A : [A \geq n] \in \Sigma\}$.

**Proposition 3** *Let $\Sigma$ be a KB and let $[A \geq n]$ be a fuzzy proposition. If $\Sigma \approx_4 [A \geq n]$ then $\overline{\Sigma} \models_4 A$.* ⊣

Proposition 3 states that there cannot be fuzzy entailment without entailment. Hence, $\{[A \geq 0.7], [\neg A \geq 0.5]\} \not\approx_4 [B \geq n]$, for all $n > 0$. In fact, consider an interpretation $\mathcal{I}$ such that $|A|^t = 0.7$, $|A|^f = 0.5$, $|B|^f = 0$ and $|B|^t = \frac{n}{2}$.

**Example 2** Let $\Sigma$ be the set $\Sigma = \{[p \geq 0.1], [p \wedge q \geq 0.5] [q \vee r \geq 0.6]\}$. Let $A$ be $p \vee r$. One may check that $\Sigma \approx_4 [A \geq 0.5]$ and $Maxdeg(\Sigma, A) = 0.5$. $\overline{\Sigma} \models_4 A$ is easily verified, thereby confirming Proposition 3. ■

### 3.1 Deciding fuzzy entailment in $\mathcal{L}^f$

The calculus is a straightforward extension of the procedure *Sat*. In fact, just consider the following *fuzzy signed propositions* and the obvious extension of the definition of satisfiability:

| $\alpha$ | $\alpha_1$ | $\alpha_2$ |
|---|---|---|
| $\text{T}[A \wedge B \geq n]$ | $\text{T}[A \geq n]$ | $\text{T}[B \geq n]$ |
| $\text{NT}[A \vee B \geq n]$ | $\text{NT}[A \geq n]$ | $\text{NT}[B \geq n]$ |
| $\beta$ | $\beta_1$ | $\beta_2$ |
| $\text{T}[A \vee B \geq n]$ | $\text{T}[A \geq n]$ | $\text{T}[B \geq n]$ |
| $\text{NT}[A \wedge B \geq n]$ | $\text{NT}[A \geq n]$ | $\text{NT}[B \geq n]$ |

By considering *Sat* extended to the fuzzy case, where $\text{T}[A \geq n]$ and $\text{NT}[A \geq m]$ are *conjugated* whenever $n \geq m$, we obtain

**Proposition 4** *Let $S$ be a set of signed fuzzy propositions in $\mathcal{L}^f$. Then $Sat(S)$ iff $S$ is satisfiable.* ⊣

**Example 3** The application of *Sat* to $\Sigma$ and $[A \geq 0.5]$, as in Example 2, starts with $S = \text{T}\Sigma \cup \{\text{NT}[A \geq 0.5]\}$ and generates a closed deduction tree with unique branch $\phi$ such that $S^{\phi} = S \cup \{\text{T}[p \geq 0.5], \text{T}[q \geq 0.6], \text{NT}[p \geq 0.5], \text{NT}[r \geq 0.5]\}$. Hence, $\Sigma \approx_4 [A \geq 0.5]$. ■

---
[4] A set is in CNF iff each component of it is.

Using Proposition 4, fuzzy entailment and entailment may be shown to be in the same complexity class.

**Proposition 5** *Checking $\Sigma \not\approx_4 [A \geq n]$ is a coNP-complete problem, as is $\Sigma \models_4 A$. Given $\Sigma$, $[A \geq n]$ in CNF, checking $\Sigma \not\approx_4 [A \geq n]$ can be done in time $O(|\Sigma||A|)$.* ⊣

One can notice that, any successful refutation of $\mathbf{T}\Sigma \cup \{\mathbf{NT}[A \geq n]\}$ does not rely on those $[B \geq m] \in \Sigma$ such that $m < n$. Hence, if we let $\Sigma$ be a KB and consider the set $\Sigma^n = \{[A \geq m] \in \Sigma : m \geq n\}$, then

**Proposition 6** *Let $\Sigma$ be a $\mathcal{L}^f$ KB. Then $\Sigma \approx_4 [A \geq n]$ iff $\overline{\Sigma^n} \models_4 A$.* ⊣

As a consequence, *fuzzy entailment inherits all the properties of entailment* seen in Section 2 (Relation 1):

**Proposition 7** *Let $\Sigma$ be a $\mathcal{L}^f$ KB. If $\overline{\Sigma} \models_4 A$ then there is a $n > 0$ such that $\Sigma \approx_4 [A \geq n]$.* ⊣

which completes Proposition 3. Just note that Proposition 6 does not hold for $\approx_2$. In fact, consider $\Sigma_1 = \{[p \geq 0.2], [\neg p \geq 0.3]\}$ and $\Sigma_2 = \{[p \geq 0.2], [\neg p \geq 0.9]\}$. It can easily be verified that $\overline{\Sigma_1^{0.1}} \models_2 q$ and $\Sigma_1 \not\approx_2 [q \geq 0.1]$, whereas $\Sigma_2 \approx_2 [q \geq 0.3]$ and $\overline{\Sigma_2^{0.3}} \not\models_2 q$.

Proposition 6 gives us a way for computing $Maxdeg(\Sigma, A)$ in the style of the method proposed in [Hollunder, 1994]. This is important, as computing $Maxdeg(\Sigma, A)$, is in fact the way to answer a query of type "to which degree is $A$ (at least) true, given the facts in $\Sigma$?" The method, which requires an algorithm for computing (crisp) entailment (*e.g. Sat*), is based on the observation that $Maxdeg(\Sigma, A) \in \{0\} \cup N_\Sigma$, where $N_\Sigma = \{n : [A \geq n] \in \Sigma\}$, and that $\Sigma^m \supseteq \Sigma^n$ if $n \geq m$.

**Algorithm 2**
Let $\Sigma$ be a KB and $A$ a proposition. Set $Min = 0$, $Max = 2$.

1. Pick $n \in N_\Sigma$ such that $Min < n < Max$. If there is no such $n$, then set $Maxdeg(\Sigma, A) := Min$ and exit.

2. Check if $\overline{\Sigma^n} \models_4 A$. If so, then set $Min = n$ and go to Step 1. If not so, then set $Max = n$ and go to Step 1.

By a binary search on $N_\Sigma$ the value of $Maxdeg(\Sigma, A)$ can be determined in $O(\log|N_\Sigma|)$ entailment tests. Hence, if $\Sigma$ and $A$ are in CNF, the complexity of determining $Maxdeg(\Sigma, A)$ is $O(|A||\Sigma|\log|\Sigma|)$.

**Example 4** Consider Example 2: $N_\Sigma = \{0.1, 0.5, 0.6\}$. By binary search, let $n := 0.5$. $\overline{\Sigma^{0.5}} = \{p \wedge q, q \vee r\} \models_4 A$ holds. Thus, $Min := 0.5$; pick $n := 0.6$. Now, $\overline{\Sigma^{0.6}} = \{q \vee r\} \not\models_4 A$ holds. Thus, $Max := 0.6$. Since there is no $Min < n < Max$ such that $n \in N_\Sigma$, the procedure stops. Hence, $Maxdeg(\Sigma, A) = 0.5$ as expected. ∎

A drawback, which Algorithm 2 inherits is that checking entailment several times is generally not feasible from a practical point of view, as it could be exponential in time and $N_\Sigma$ can be $O(\Sigma)$. In Section 4.1 we will present a method where computing $Maxdeg(\Sigma, A)$ "corresponds" to performing the entailment test only *once*.

## 3.2 Relations to Possibilistic Logic

There is a strict connection between our logic and (necessity-valued) possibilistic logic [Dubois and Prade, 1986], which allows the expression of uncertain propositions. In possibilistic logics, the expressions are of type $(A, \mathbf{P}n)$ and $(A, \mathbf{N}n)$. A weight $\mathbf{P}n$ (resp. $\mathbf{N}n$) attached to $A$ models to what extent $A$ is *possible* (resp. *necessarily*) true. The semantics is given in terms of fuzzy sets of interpretations, *i.e.* to each propositional interpretation $\mathcal{I}$ a weight $\pi(\mathcal{I}) \in [0,1]$ is assigned. The possibility and necessity of a proposition is then given by $\Pi(A) = \max\{\pi(\mathcal{I}) : \mathcal{I} \text{ satisfies } A\}$ and $N(A) = 1 - \Pi(\neg A)$. An interpretation satisfies an expression of type $(A, \mathbf{P}n)$ (resp. $(A, \mathbf{N}n)$) if $\Pi(A) \geq n$ (resp. $N(A) \geq n$).

A closer look to Proposition 6 reveals that it is similar to Hollunder's Theorem 3.4 in [Hollunder, 1994]:

**Theorem 1 (Hollunder)** *Let $\Sigma$ be a set of possibilistic propositions and $n > 0$. Then $\Sigma \models_2^{pos} (A, \mathbf{N}n)$ iff $\Phi_n \models_2 A$, where $\models_2^{pos}$ is the possibilistic entailment relation and $\Phi_n = \{A : (A, \mathbf{N}m) \in \Sigma \text{ and } m \geq n\}$.* ⊣

As a consequence, let $\Sigma$ be a $\mathcal{L}^f$ KB and let $\hat{\Sigma}$ be $\{(A, \mathbf{N}n) : [A \geq n] \in \Sigma\}$. Since $\models_4 \subset \models_2$, from Proposition 6 and Theorem 1 it follows that

**Proposition 8** *Let $\Sigma$ be a $\mathcal{L}^f$ KB. If $\Sigma \approx_4 [A \geq n]$ then $\hat{\Sigma} \models_2^{pos} (A, \mathbf{N}n)$.* ⊣

The converse of Proposition 8 is not true, *i.e.* $\hat{\Sigma} \models_2^{pos} (A, \mathbf{N}n)$ does not imply $\Sigma \approx_4 [A \geq n]$. For instance, $[p \geq 0.6], [\neg p \geq 0.7] \not\approx_4 [q \geq 0.6]$, whereas $(p, \mathbf{N}0.6), (\neg p, \mathbf{N}0.7) \models_2^{pos} (q, \mathbf{N}0.6)$ and, thus, $\approx_4 \subset \models_2^{pos}$ holds. Neither Proposition 8 nor the converse of it holds for $\approx_2$. We can confirm this by considering $\Sigma_1$ and $\Sigma_2$ of the previous section. Therefore, $\Sigma_2 \approx_2 [q \geq 0.3]$ and $\hat{\Sigma}_2 \not\models_2^{pos} (q, \mathbf{N}0.3)$, whereas $\Sigma_1 \not\approx_2 [q \geq 0.1]$ and $\hat{\Sigma}_1 \models_2^{pos} (q, \mathbf{N}0.1)$.

## 4 Conditionals

In Section 2 we have seen that modus ponens is not a valid inference rule in $\mathcal{L}$ and, thus, in $\mathcal{L}^f$. In order to deal with conditional reasoning we introduce a new connective $\Rightarrow$. Let $\mathcal{L}_+$ be $\mathcal{L}$ plus the set of propositions involving connective $\Rightarrow$. From a semantic point of view, an interpretation $\mathcal{I}$ has also to satisfy the following conditions: $t \in (A \Rightarrow B)^\mathcal{I}$ iff $t \in A^\mathcal{I}$ implies $t \in B^\mathcal{I}$, whereas $f \in (A \Rightarrow B)^\mathcal{I}$ iff $t \in A^\mathcal{I}$ and $f \in B^\mathcal{I}$. Notice, that now $A, A \Rightarrow B \models_4 B$ holds. Moreover, $\neg(A \Rightarrow B) \equiv_4 A \wedge \neg B$, $A \Rightarrow B \not\equiv_4 \neg B \Rightarrow \neg A$ (no contraposition) and $\models_4 A \Rightarrow (B \Rightarrow A)$ hold. Hence, there are tautologies in $\mathcal{L}_+$. As for $\mathcal{L}$, every $\mathcal{L}_+$ KB is satisfiable. A complete calculus with respect to $\mathcal{L}_+$ is obtained by extending the definition of signed propositions of type $\alpha$ and type $\beta$ to the cases $\mathbf{NT}A \Rightarrow B$ and $\mathbf{T}A \Rightarrow B$, respectively, in a similar way as in [D'Agostino and Mondadori, 1994]. Furthermore, we extend *Sat* with Step 3b: if $\beta_1$ is of type $\mathbf{T}A \Rightarrow B$ then let $S' = (S^{\phi'} \cup \{\mathbf{NT}A\}) \setminus \{\mathbf{T}A \Rightarrow B\}$ and if not $Sat(S')$ then expand $\phi'$ by means of one children node labelled $\mathbf{T}B$ and go to step 1.

**Proposition 9** *Let $S$ be a set of signed propositions in $\mathcal{L}_+$. Then $Sat(S)$ (with Step 3b) iff $S$ is satisfiable.* ⊣

Checking whether $\Sigma \models_4 A$ is a coNP-complete problem in $\mathcal{L}_+$. But, if we restrict $\mathcal{L}_+$ to $\overline{\mathcal{L}}_+$ we obtain a tractable logic. $\overline{\mathcal{L}}_+$ is defined inductively as follows: $\overline{\mathcal{L}}_+$ is the minimal set such that (i) every proposition in $\mathcal{L}$ in CNF is in $\overline{\mathcal{L}}_+$; (ii) if $A, A_1, \ldots, A_n$ and $B, B_1, \ldots, B_m$ are literals in $\mathcal{L}$, then both $A_1 \wedge \ldots \wedge A_n \Rightarrow B$ and $A \Rightarrow B_1 \vee \ldots \vee B_m$ are in $\overline{\mathcal{L}}_+$. By considering that Step 3. can be eliminated (rule $(PB)$ is *not* necessary), we have

**Proposition 10** *Let $\Sigma$ be a $\overline{\mathcal{L}}_+$ KB and let $A$ be in $\overline{\mathcal{L}}_+$. Checking if $\Sigma \models_4 A$ can be done in time $O(|\Sigma||A|)$.* ⊣

Now, let $\mathcal{L}_+^f$ be the extension of $\mathcal{L}_+$ to the fuzzy case. An interpretation $\mathcal{I}$ has now also to satisfy the semantic clauses $|A \Rightarrow B|^t = \min\{1, \frac{|B|^t}{|A|^t}\}$ (Gödel implication) and $|A \Rightarrow B|^f = \min\{|A|^t, |B|^f\}$. It is worth noting that $|A \Rightarrow B|^f = |A \wedge \neg B|^t$, whereas $|A \Rightarrow B|^t \neq |\neg B \Rightarrow \neg A|^t$ (no contraposition). The clause for $|A \Rightarrow B|^t$ models a sort of conditional $Cond(B|A) = \frac{\|B \cap A\|}{\|A\|}$. It is easily verified that the above conditions are equivalent to: $t \in [A \Rightarrow B \geq n]^{\mathcal{I}}$ iff $\forall m \in [0,1]$, if $t \in [A \geq m]^{\mathcal{I}}$ then $t \in [B \geq n \cdot m]^{\mathcal{I}}$; $f \in [A \Rightarrow B \geq n]^{\mathcal{I}}$ iff $t \in [A \geq n]^{\mathcal{I}}$ and $f \in [B \geq n]^{\mathcal{I}}$, which are similar to the non fuzzy case. The semantics for $\mathcal{L}_+^f$ enables thus a simple form of modus ponens: $[A \geq m], [A \Rightarrow B \geq n] \approx_4 [B \geq n \cdot m]$. Just notice that if $|\cdot|^t, |\cdot|^f \in \{0,1\}$ then classical two-valued $\Rightarrow$ is obtained.

**Example 5** Let $\Sigma = \{[J \Rightarrow S \wedge A \geq 0.9], [B \Rightarrow T \geq 0.4], [G \Rightarrow A \geq 0.9], [A \Rightarrow T \geq 0.8], [K \Rightarrow C \geq 0.7], [C \Rightarrow T \geq 0.2], [S \Rightarrow T \geq 0.5]\}$, where $J, S, A, B, T, G, K$ and $C$ stand for *Jon, Student, Adult, Boy, Tall, Gil, Karl* and *Child*, respectively. Then $\Sigma \cup \{[G \geq 0.8]\} \approx_4 [T \geq 0.576]$, $\Sigma \cup \{[J \geq 0.7]\} \approx_4 [T \geq 0.504]$ and $\Sigma \cup \{[G \vee K \geq 0.8]\} \approx_4 [T \geq 0.112]$ (the values are maximal). ∎

Note that $\models_4 [A \Rightarrow (B \Rightarrow A) \geq 1]$ holds, whereas if $C$ is $(p \Rightarrow q) \vee ((p \Rightarrow q) \Rightarrow q)$ then $\models_4 C$ and $\not\models_4 [C \geq n]$, for all $n > 0$. In fact, let $\mathcal{I}$ be an interpretation such that $|p|^t = \frac{n}{2}$ and $|q|^t = \frac{n^2}{3}$. $|C|^t = \frac{2n}{3} < n$ holds. As a consequence, Proposition 7 is not valid in $\mathcal{L}_+^f$, whereas Proposition 3 and Proposition 8 remain valid[5].

### 4.1 Deciding entailment in $\mathcal{L}_+^f$

Unfortunately, finding a calculus for entailment in $\mathcal{L}_+^f$ is not as easy as for $\mathcal{L}^f$, since Algorithm 2 does not work in the context of $\mathcal{L}_+^f$: Proposition 6 does not hold and $Maxdeg(\Sigma, A)$ may be not in $\{0\} \cup N_\Sigma$.

First, we generalise fuzzy propositions to the form $[A \geq \lambda]$, where $\lambda$ is a fuzzy value defined as follows. Let $\mathcal{X}$ be a new alphabet of *fuzzy variables* (with metavariable $x$). A *multiset*, (with metavariable $L$) is a finite

---

[5]Note that $(p, \mathbf{N}m), (p \Rightarrow q, \mathbf{N}n) \models_2^{pos} (q, \mathbf{N}\min\{m,n\})$ holds and $n \cdot m \leq \min\{m,n\}$.

set of fuzzy variables in which a variable $x$ can occur more than once. A *fuzzy value* (with metavariable $\lambda$) is a pair $(n, L)$ where $n \in [0,1]$ and $L$ is a multiset. An interpretation $\mathcal{I}$ is such that $x^{\mathcal{I}} \in [0,1]$, $\{x_1, \ldots, x_n\}^{\mathcal{I}} = x_1^{\mathcal{I}} \cdot \ldots \cdot x_n^{\mathcal{I}}$, $\emptyset^{\mathcal{I}} = 1$ and $(n, L)^{\mathcal{I}} = n \cdot L^{\mathcal{I}}$. We extend the multiplication function $\cdot$ to fuzzy values by defining $(n, L_1) \cdot (m, L_2)$ as $(n \cdot m, L_1 \cup L_2)$. For ease of notation we will write $0$, $n$ and $x$ in place of $(0, L)$, $(n, \emptyset)$ and $(1, \{x\})$, respectively. Moreover, we will allow fuzzy values $\frac{\lambda_1}{\lambda_2}$ and $\lambda^{\frac{m}{n}}$ ($n, m$ are positive integers) with obvious semantics. The greater equal relation $\geq$ is extended to fuzzy values as follows: $\lambda_1 \geq \lambda_2$ iff for all interpretations $\mathcal{I}$, $\lambda_1^{\mathcal{I}} \geq \lambda_2^{\mathcal{I}}$. Similarly for the relation $>$. Checking whether $\lambda_1 \geq \lambda_2$ can be done by observing that $(n, L) \geq (m, L')$ iff $n \geq m$ and if $m \neq 0$ then $L \subseteq L'$. The reader can verify that it is determinable whether $\frac{\lambda_1}{\lambda_2} \geq \frac{\lambda_3}{\lambda_4}$ and $\lambda_1^{\frac{m_1}{n_1}} \geq \lambda_2^{\frac{m_2}{n_2}}$. For instance, $0.7 \cdot x_1 \geq 0.6 \cdot x_1 \cdot x_2$, whereas $(0.9 \cdot x_1)^{\frac{1}{3}} \geq (0.5 \cdot x_1 \cdot x_2)^{\frac{1}{2}}$, since $(0.9 \cdot x_1)^2 \geq (0.5 \cdot x_1 \cdot x_2)^3$.

In what follows, we will use the obvious extension of the definition of satisfiability with the following clauses on signed propositions involving $\Rightarrow$: (i) $\mathbf{NT}[A \Rightarrow B \geq \lambda]$ is of type $\alpha$ and $\mathbf{T}[A \geq x]$ and $\mathbf{NT}[B \geq \lambda \cdot x]$ are its $\alpha_1$ and $\alpha_2$ components (for a "new" fuzzy variable $x$), and (ii) $\mathbf{T}[A \Rightarrow B \geq \lambda_1]$ is of type $\beta$ and $\mathbf{NT}[A \geq \lambda_2]$ and $\mathbf{T}[B \geq \lambda_1 \cdot \lambda_2]$ are its $\beta_1$ and $\beta_2$ components (for a all $\lambda_2$). Moreover, $\mathbf{T}[A \geq \lambda_1]$ and $\mathbf{NT}[A \geq \lambda_2]$ are called *conjugated signed propositions* if $\lambda_1 \geq \lambda_2$.

Algorithm $MaxVal(S, A)$ below computes the set of maximal fuzzy values $\lambda_1, \ldots, \lambda_n$ for which $S \cup \{\mathbf{NT}[A \geq \lambda_i]\}$ is not satisfiable. If $MaxVal(\mathbf{T}\Sigma, A) = \{n\}$, where $n \in [0,1]$, then $Maxdeg(\Sigma, A) = n$, otherwise $Maxdeg(\Sigma, A) = 0$.

Let $\phi_i$ be a not closed and completed branch of a deduction tree and $x$ a fuzzy variable. Let $N_i$ be the set of all fuzzy values $\lambda$ such that (i) both $\mathbf{T}[A \geq x^{n_1} \cdot \lambda_1]$ and $\mathbf{NT}[A \geq x^{n_2} \cdot \lambda_2]$ are in $S^{\phi_i}$, where $n_1 + n_2 \geq 1$ and $n_1 \neq n_2$; (ii) if $n_1 < n_2$ then $\lambda = (\frac{\lambda_1}{\lambda_2})^{\frac{1}{n_2 - n_1}}$; (iii) if $n_1 > n_2$ then $\lambda = (\frac{\lambda_2}{\lambda_1})^{\frac{1}{n_1 - n_2}}$. $N_i$ is just the set of fuzzy values $\lambda$ such that $\phi_i$ is closed whenever $x$ is substituted by $\lambda$, i.e. $\mathbf{T}[A \geq x^{n_1} \cdot \lambda_1]$ and $\mathbf{NT}[A \geq x^{n_2} \cdot \lambda_2]$ will be a conjugated pair. It can be verified that $0 \leq \lambda \leq 1$.

**Algorithm 3** ($MaxVal(S, A)$)
*Let the root node be labelled with $S \cup \{\mathbf{NT}[A \geq x]\}$, where $x$ is a new fuzzy variable. At each step of the construction of a deduction tree the following steps are performed*[6]:

1. *select a branch $\phi$ which is not yet completed;*
2. *expand $\phi$ by means of the rules (A), (B1) and (B2) until it becomes AB-completed. Let $\phi'$ be the resulting branch;*
3. *if $\phi'$ is neither closed nor completed then*
   (a) *select a proposition of type $\beta$ which is not yet fulfilled in the branch;*

---

[6]The branches $\phi$ will be maintained maximal, i.e. not both $\mathbf{T}[A \geq \lambda_1]$ and $\mathbf{T}[A \geq \lambda_2]$ are in $S^\phi$ with $\lambda_2 > \lambda_1$. Moreover, $\mathbf{T}[A \geq 0] \notin S^\phi$. Similarly for case $\mathbf{NT}$.

(b) if $\beta_1$ is of type $\text{T}[A \Rightarrow B \geq \lambda]$ then let $\{\lambda_1, \ldots, \lambda_l\}$ be $MaxVal(S^{\phi'}\setminus\{\text{T}[A \Rightarrow B \geq \lambda]\}, A)$, expand $\phi'$ by means of one children node labelled $\text{T}[B \geq \lambda \cdot \lambda_1], \ldots, \text{T}[B \geq \lambda \cdot \lambda_l]$ and go to step 1;
   
   (c) otherwise apply rule $(PB)$ with $\beta_1$ and $\beta_1^c$ as PB-formulae and go to step 1;
   
   otherwise, go to step 1.

4. for all not closed and completed branches $\phi_i$ ($1 \leq i \leq h$) let $n_i := \max N_i$ ($\max \emptyset = 0$); for all closed and completed branches $\phi_i$ ($h+1 \leq i \leq k$), let $n_i := 1$. $MaxVal(S, A) := \min\{n_1, \ldots n_k\}$. ∎

Just notice that Step 3b is not needed in $\mathcal{L}^f$. Moreover, the procedure can be improved by performing Step 4. during Step 2. - 3. It can be shown that

**Proposition 11** Let $\Sigma$ be a $\mathcal{L}^f_+$ KB and $A \in \mathcal{L}_+$. $Maxdeg(\Sigma, A) = n > 0$ iff $MaxVal(\text{T}\Sigma, A) = \{n\}$, and checking $Maxdeg(\Sigma, A) \geq n$ is a coNP-complete problem. ⊣

**Example 6** Consider Example 2. $MaxVal(\text{T}\Sigma, A)$ generates two branches $\phi_1$ and $\phi_2$, where $S^{\phi_1} := S \cup \{\text{T}[q \geq 0.6]\}$, $S^{\phi_2} := S \cup \{\text{NT}[q \geq 0.6], \text{T}[r \geq 0.6]\}$ and $S = \{\text{T}[p \geq 0.5], \text{T}[q \geq 0.5], \text{NT}[p \geq x], \text{NT}[r \geq x]\}$. Now, $N_1 = \{0.5\}$, $n_1 = 0.5$, $N_2 = \{0.5, 0.6\}$ and $n_2 = 0.6$. Hence, $Maxdeg(\Sigma, A) = \min\{n_1, n_2\} = 0.5$. ∎

**Example 7** Let $A$ be $(p \Rightarrow q) \vee ((p \Rightarrow q) \Rightarrow q)$. We have seen that $\models_4 A$, whereas $\not\models_4 [A \geq n]$, for all $n > 0$. Compute $MaxVal(\emptyset, A)$. The computation generates an unique branch $\phi$ such that $S^\phi$ contains $\text{T}[p \geq x_1]$, $\text{NT}[q \geq x \cdot x_1]$, $\text{T}[p \Rightarrow q \geq x_2]$ and $\text{NT}[q \geq x \cdot x_2]$. By Step 3b, a recursive call $MaxVal(S^\phi \setminus \{\text{T}[p \Rightarrow q \geq x_2]\}, p)$ will be performed answering with $\{x_1\}$, i.e. the set of maximal degrees of $p$ with respect to $S^\phi \setminus \{\text{T}[p \Rightarrow q \geq x_2]\}$. Hence, the computation proceeds with branch $\phi'$, where $S^{\phi'}$ is $S^\phi \cup \{\text{T}[q \geq x_2 \cdot x_1]\}$. Finally we will have MaxVal$(\emptyset, A) = \{x_1, x_2\}$. Therefore, $Maxdeg(\emptyset, A) = 0$. ∎

Finally, let $\overline{\mathcal{L}}^f_+$ be the extension of $\overline{\mathcal{L}}_+$ to the fuzzy case. In this case, it can be shown that $MaxVal(S, A)$, without Step 3. and such that $N_i$ is computed during Step 2., can be modified in such a way that it runs in polynomial time. Roughly, given e.g. $\{\text{T}[A \wedge B \Rightarrow C \geq 0.6], \text{T}[A \geq 0.7], \text{T}[B \geq 0.8], \text{NT}[C \geq x]\}$, rule $(B1)$ can be applied and we add $\text{T}[C \geq \min\{0.7, 0.8\} \cdot 0.6]$ to the branch and finally we get $x = 0.42$.

**Proposition 12** Let $\Sigma$ be a $\overline{\mathcal{L}}^f_+$ KB and $A \in \overline{\mathcal{L}}_+$. Computing $Maxdeg(\Sigma, A)$ can be done in time $O(|\Sigma||A|)$. ⊣

Note that the KB in Example 5 is a $\overline{\mathcal{L}}^f_+$ KB.

## 5 Conclusions

There are two main contributions in this paper. The first one is an alternative procedure to Levesque's algorithm for deciding entailment in $\mathcal{L}$ (with same complexity on propositions in CNF), but which works too for $\mathcal{L}_+$, i.e. $\mathcal{L}$ with modus ponens. The second one is the definition of the logic $\mathcal{L}^f_+$ for reasoning in presence of vague concepts and inconsistencies with an expressively powerful and computationally tractable case. These two parts can be furthermore combined, without affecting the computational complexity, by combining fuzzy propositions $[A \geq n]$ with the operators $\wedge, \vee, \neg$ and $\Rightarrow$ and, thus, allowing fuzzy propositions of type, e.g. $[A \geq n] \vee [B \geq m]$ and $[A \geq n] \Rightarrow [B \geq m]$. A decision procedure can simply be obtained by combining the algorithms for deciding entailment in $\mathcal{L}_+$ and the one for $\mathcal{L}^f_+$.

Uncertain fuzzy propositions can be obtained by allowing expressions of type $(\gamma, \text{P}n)$ and $(\gamma, \text{N}n)$, where $\gamma$ is a fuzzy proposition. The development of both a precise semantics within our four-valued framework and a calculus for automated reasoning in it, can be seen as interesting topics of further research.

## Acknowledgements

This work is funded by the European Community ESPRIT project FERMI 8134. Thanks go to C. Meghini and F. Sebastiani for their suggestions and comments.

## References

[Belnap, 1977] N. D. Belnap. A useful four-valued logic. In G. Epstein and J. M. Dunn, ed., *Modern uses of multiple-valued logic*, pages 5–37. Reidel, Dordrecht, NL, 1977.

[Chen and Kundu, 1996] J. Chen and S. Kundu. A sound and complete fuzzy logic system using Zadeh's implication operator. In *Proc. of the 9th Int. Symp. on Methodologies for Intelligent Systems*, LNAI 1079, pages 233–242. Springer, 1996.

[D'Agostino and Mondadori, 1994] M. D'Agostino and M. Mondadori. The taming of the cut. Classical refutations with analytical cut. *Journal of Logic and Computation*, 4(3):285–319, 1994.

[Dubois and Prade, 1986] D. Dubois and H. Prade. Possibilistic logic. In D. M. Gabbay and C. J. Hogger, ed., *Handbook of Logic in Artificial Intelligence*, 3:439–513. Clarendon Press, Oxford, Dordrecht, NL, 1986.

[Hollunder, 1994] B. Hollunder. An alternative proof method for possibilistic logic and its application to terminological logics. In *10th Annual Conference on Uncertainty in Artificial Intelligence*, Seattle, WA, 1994.

[Levesque, 1984] H. J. Levesque. A logic of implicit and explicit belief. In *Proc. of AAAI-84*, pages 198–202, Austin, TX, 1984.

[Meghini et al., 1997] C. Meghini, F. Sebastiani, and U. Straccia. Reasoning about the form and content for multimedia objects (extended abstract). In *Proc. of AAAI 1997 Spring Symp. on Intelligent Integration and Use of Text, Image, Video and Audio*, pages 89–94, Stanford University, CA, 1997.

[Smullyan, 1968] R. M. Smullyan. *First Order Logic*. Springer, 1968.

[Wagner, 1991] Gerd Wagner. Ex contradictione nihil sequitur. In *Proc. of IJCAI-91*, pages 538–543, Sydney, Australia, 1991.

# AUTOMATED REASONING

Automated Reasoning 5: Description Logic

# Autoepistemic Description Logics

**Francesco M. Donini** and **Daniele Nardi** and **Riccardo Rosati**

Dipartimento di Informatica e Sistemistica
Università di Roma "La Sapienza"
Via Salaria 113, I-00198 Roma, Italy
email: ⟨lastname⟩@dis.uniroma1.it

## Abstract

We present Autoepistemic Description Logics (ADLs), in which the language of Description Logics is augmented with modal operators interpreted according to the nonmonotonic logic $MKNF$. We provide decision procedures for query answering in two very expressive ADLs. We show their representational features by addressing defaults, integrity constraints, role and concept closure. Hence, ADLs provide a formal characterization of a wide variety of nonmonotonic features commonly available in frame-based systems and needed in the development of practical applications.

## 1 Introduction

Description Logics (DL) have been studied in the past years to provide a formal characterization of frame-based systems. However, while the fragment of first-order logic which characterizes the most popular constructs of these languages has been clearly identified (see for example [Woods & Schmolze, 1992]), there is not yet consensus on the features of frame-based systems that cannot be formalized in a classical first-order setting. In fact, frame systems, as well as DL-based systems [Brachman *et al.*, 1990, MacGregor, 1988], admit forms of nonmonotonic reasoning, such as defaults and closed world reasoning, and procedural features, e.g. rules. These issues have been addressed in the recent literature (see for example [Baader & Hollunder, 1995, Donini *et al.*, 1992, Donini, Nardi, & Rosati, 1995, Padgham & Zhang, 1993, Quantz & Royer, 1992]), but the proposals typically capture one of the above mentioned aspects.

In addition, most implementations of DL-based systems are object-centered, which enables them to perform efficient reasoning on the properties of individuals. Such behaviour can be naturally justified if one can restrict the reasoning to the individuals that are known to the knowledge base (i.e. individuals that have an explicit name). Based on this intuition, an epistemic extension of DLs with a modal operator **K**, interpreted in terms of minimal knowledge has been proposed in [Donini *et al.*, 1992]. In that formalism one can express a form of closed-world reasoning, as well as integrity constraints in the form of epistemic queries (as proposed in [Reiter, 1990]); in addition, by admitting a simple form of epistemic sentences in the knowledge base, one can formalize the so-called procedural rules.

In this paper we propose a new framework of Autoepistemic Description Logics (ADLs) which follows the lines of [Donini *et al.*, 1992], extending it in two respects: we rely on the nonmonotonic modal logic $MKNF$ [Lifschitz, 1994] and we consider several kinds of epistemic sentences to be used in the knowledge base. In $MKNF$ one can formalize Default Logic, Autoepistemic Logic, Circumscription and Logic Programming, i.e. many of the best known formalisms for nonmonotonic reasoning. With $MKNF$ we can naturally extend the previous approach to modal DLs, by introducing a second modal operator interpreted as autoepistemic assumption. Moreover, reasoning methods are available for deduction in propositional $MKNF$ [Rosati, 1997].

As for the representational features of the framework, we show that ADLs are able to capture a large variety of non-first-order features. In addition to procedural rules and epistemic queries, the formalism accounts for defaults, integrity constraints inside the KB, role and concept closure, which are addressed in the paper. Moreover, the whole representational power of $MKNF$ becomes available, thus making it feasible to consider new features, like autoepistemic reasoning, that are not implemented in current DL-based systems.

As for reasoning in ADLs we define methods for query answering which provide sound and complete reasoning procedures in DL-based systems admitting the above mentioned non-first-order features. It turns out that the proposed deduction methods constitute interesting decidable extensions of propositional nonmonotonic reasoning.

Based on the above considerations we argue that ADLs can indeed be considered as a unified framework for the logical reconstruction of frame-based systems.

The paper is organized as follows. We first present the extension of DLs obtained by adding the epistemic operators of $MKNF$. We then discuss the representational features of ADLs by considering several forms of nonmonotonic reasoning and integrity constraints. Finally, we address reasoning in these logics by providing reasoning methods for two rather general cases.

## 2 Autoepistemic Description Logics

Autoepistemic Description Logics (ADLs) are defined as an extension of DLs, in which the modal operators **K** and **A** are allowed in the formation of concept and role expressions. The meaning of modal sentences is given according to the logic *MKNF* [Lifschitz, 1994, Lin & Shoham, 1992].

Let $\mathcal{DL}$ be a generic description logic. Then, $\mathcal{DLK}_{\mathcal{NF}}$ stands for the description logic $\mathcal{DL}$ augmented with the modal operators **K** and **A**. We say that $C$ is a $\mathcal{DLK}_{\mathcal{NF}}$-*concept* if $C$ is a concept expression of $\mathcal{DLK}_{\mathcal{NF}}$. Analogously, $R$ is a $\mathcal{DLK}_{\mathcal{NF}}$-*role* if $R$ is a role expression in $\mathcal{DLK}_{\mathcal{NF}}$.

Below we present the epistemic DL $\mathcal{ALCK}_{\mathcal{NF}}$, namely we refer to the DL $\mathcal{ALC}$, although some of the results concern more expressive DLs. The syntax of $\mathcal{ALCK}_{\mathcal{NF}}$ is as follows:

$$C ::= \top \mid \bot \mid C_a \mid C_1 \sqcap C_2 \mid C_1 \sqcup C_2 \mid \neg C \mid \exists R.C \mid \forall R.C \mid \mathbf{K}C \mid \mathbf{A}C$$
$$R ::= P \mid \mathbf{K}P \mid \mathbf{A}P$$

where $C_a$ denotes an atomic concept, $C$ (possibly with a subscript) denotes a concept, $P$ denotes an atomic role, and $R$ (possibly with a subscript) denotes a role.

The semantics is obtained by interpreting concepts and roles on *MKNF* structures. With respect to the original semantics for *MKNF* in the first-order case, we introduce two changes: (i) following the approach of [Reiter, 1990, Donini et al., 1992] the semantics of ADLs is based on the *Rigid Term Assumption*: for every interpretation the mapping from the individuals into the domain elements is fixed; (ii) the semantics of ADLs is also based on the following *Common Domain Assumption*: in each model, every interpretation is defined over the same, fixed, countable-infinite domain of individuals $\Delta$. Hence, we define an *epistemic interpretation* as a triple $(\mathcal{I}, \mathcal{M}, \mathcal{N})$ where $\mathcal{I}$ is a $\mathcal{DL}$-interpretation (a possible world) and $\mathcal{M}, \mathcal{N}$ are sets of interpretations defined over the domain $\Delta$.

Atomic concepts and roles are interpreted as subsets of $\Delta$ and $\Delta \times \Delta$, respectively. $\top$ is interpreted as $\Delta$ and $\bot$ as $\emptyset$. Non-epistemic concepts and roles are given the standard semantics of DLs; conversely epistemic sentences are interpreted on epistemic interpretations, as follows.

$$\begin{aligned}
(\neg C)^{\mathcal{I},\mathcal{M},\mathcal{N}} &= \Delta \setminus (C)^{\mathcal{I},\mathcal{M},\mathcal{N}} \\
(C_1 \sqcap C_2)^{\mathcal{I},\mathcal{M},\mathcal{N}} &= (C_1)^{\mathcal{I},\mathcal{M},\mathcal{N}} \cap (C_2)^{\mathcal{I},\mathcal{M},\mathcal{N}} \\
(C_1 \sqcup C_2)^{\mathcal{I},\mathcal{M},\mathcal{N}} &= (C_1)^{\mathcal{I},\mathcal{M},\mathcal{N}} \cup (C_2)^{\mathcal{I},\mathcal{M},\mathcal{N}} \\
(\exists R.C)^{\mathcal{I},\mathcal{M},\mathcal{N}} &= \{d \in \Delta \mid \exists d'.(d,d') \in (R)^{\mathcal{I},\mathcal{M},\mathcal{N}} \\
&\quad \text{and } d' \in (C)^{\mathcal{I},\mathcal{M},\mathcal{N}}\} \\
(\forall R.C)^{\mathcal{I},\mathcal{M},\mathcal{N}} &= \{d \in \Delta \mid \forall d'.(d,d') \in (R)^{\mathcal{I},\mathcal{M},\mathcal{N}} \\
&\quad \text{implies } d' \in (C)^{\mathcal{I},\mathcal{M},\mathcal{N}}\} \\
(\mathbf{K}C)^{\mathcal{I},\mathcal{M},\mathcal{N}} &= \bigcap_{\mathcal{J} \in \mathcal{M}} (C)^{\mathcal{I},\mathcal{M},\mathcal{N}} \\
(\mathbf{A}C)^{\mathcal{I},\mathcal{M},\mathcal{N}} &= \bigcap_{\mathcal{J} \in \mathcal{N}} (C)^{\mathcal{I},\mathcal{M},\mathcal{N}} \\
(\mathbf{K}P)^{\mathcal{I},\mathcal{M},\mathcal{N}} &= \bigcap_{\mathcal{J} \in \mathcal{M}} (P)^{\mathcal{I},\mathcal{M},\mathcal{N}} \\
(\mathbf{A}P)^{\mathcal{I},\mathcal{M},\mathcal{N}} &= \bigcap_{\mathcal{J} \in \mathcal{N}} (P)^{\mathcal{I},\mathcal{M},\mathcal{N}}
\end{aligned}$$

For example, an individual $d \in \Delta$ is an instance of a concept $\mathbf{K}C$ (i.e. $d \in (\mathbf{K}C)^{\mathcal{I},\mathcal{M},\mathcal{N}}$) iff $d \in C^{\mathcal{J},\mathcal{M},\mathcal{N}}$ for all interpretations $\mathcal{J} \in \mathcal{M}$. In other words, an individual is "known" to be an instance of a concept if it belongs to the concept interpretation of every possible world in $\mathcal{M}$. An individual $d \in \Delta$ is an instance of a concept $\mathbf{A}C$ (i.e. $d \in (\mathbf{A}C)^{\mathcal{I},\mathcal{M},\mathcal{N}}$) iff $d \in C^{\mathcal{J},\mathcal{M},\mathcal{N}}$ for all interpretations $\mathcal{J} \in \mathcal{N}$. In other words, an individual is "assumed" to be an instance of a concept $C$ if it belongs to $C$ in all possible worlds of $\mathcal{N}$. Similarly, an individual $d \in \Delta$ is an instance of a concept $\exists \mathbf{K}R.\top$ iff there is an individual $d' \in \Delta$ such that $(d,d') \in R^{\mathcal{J},\mathcal{M},\mathcal{N}}$ for all $\mathcal{J} \in \mathcal{M}$.

The truth of inclusion statements in an epistemic interpretation $(\mathcal{I}, \mathcal{M}, \mathcal{N})$ is defined in terms of set inclusion: $C \sqsubseteq D$ is satisfied in $(\mathcal{I}, \mathcal{M}, \mathcal{N})$ iff $(C)^{\mathcal{I},\mathcal{M},\mathcal{N}} \subseteq (D)^{\mathcal{I},\mathcal{M},\mathcal{N}}$. Assertions are interpreted in terms of set membership: $C(a)$ is satisfied in $(\mathcal{I}, \mathcal{M}, \mathcal{N})$ iff $a \in (C)^{\mathcal{I},\mathcal{M},\mathcal{N}}$ and $R(a,b)$ is satisfied in $(\mathcal{I}, \mathcal{M}, \mathcal{N})$ iff $(a,b) \in (R)^{\mathcal{I},\mathcal{M},\mathcal{N}}$. A $\mathcal{DLK}_{\mathcal{NF}}$-*knowledge base* $\Psi$ is defined as a pair $\Psi = \langle \mathcal{T}, \mathcal{A} \rangle$, where $\mathcal{T}$ (called *TBox*) is a finite set of inclusion statements (intensional knowledge) of the form $C \sqsubseteq D$, where $C, D$ are $\mathcal{DLK}_{\mathcal{NF}}$-concepts, and $\mathcal{A}$ (called the *ABox*) is a finite set of membership assertions (extensional knowledge) of the form $C(a)$ or $R(a,b)$, where $C$ is a $\mathcal{DLK}_{\mathcal{NF}}$-concept, $R$ is a $\mathcal{DLK}_{\mathcal{NF}}$-role, and $a, b$ are individuals in $\Delta$. We call $\mathcal{O}_\Sigma$ the set of individuals occurring in $\Sigma$.

An inclusion $C \sqsubseteq D$ is *satisfied* by a structure $(\mathcal{M}, \mathcal{N})$ (denoted by $(\mathcal{M}, \mathcal{N}) \models \Psi$) iff each interpretation $\mathcal{I} \in \mathcal{M}$ is such that $(\mathcal{I}, \mathcal{M}, \mathcal{N})$ satisfies $C \sqsubseteq D$. An assertion $C(a)$ (resp. $R(a,b)$) is satisfied by $(\mathcal{M}, \mathcal{N})$ (denoted by $(\mathcal{M}, \mathcal{N}) \models \Psi$) iff each interpretation $\mathcal{I} \in \mathcal{M}$ is such that $(\mathcal{I}, \mathcal{M}, \mathcal{N})$ satisfies $C(a)$ (resp. $R(a,b)$). A $\mathcal{DLK}_{\mathcal{NF}}$-knowledge base $\Psi$ is *satisfied* by a structure $(\mathcal{M}, \mathcal{N})$ (denoted by $(\mathcal{M}, \mathcal{N}) \models \Psi$) iff each interpretation $\mathcal{I} \in \mathcal{M}$ is such that every sentence (inclusion or membership assertion) of $\Psi$ is true in the epistemic interpretation $(\mathcal{I}, \mathcal{M}, \mathcal{N})$.

A set of interpretations $\mathcal{M}$ is a *model* for $\Psi$ iff the structure $(\mathcal{M}, \mathcal{M})$ satisfies $\Psi$ and, for each set of interpretations $\mathcal{M}'$, if $\mathcal{M} \subset \mathcal{M}'$ then $(\mathcal{M}', \mathcal{M})$ does not satisfy $\Psi$. Roughly speaking, such a preference semantics gives a minimal knowledge interpretation to the modality **K**, while the operator **A** is interpreted in terms of autoepistemic assumption (see [Lifschitz, 1994, Lin & Shoham, 1992] for further details).

The $\mathcal{DLK}_{\mathcal{NF}}$-knowledge base $\Psi$ is *satisfiable* if there exists a model for $\Psi$, *unsatisfiable* otherwise. $\Psi$ logically implies an inclusion assertion $C \sqsubseteq D$ (where $C, D$ are $\mathcal{DLK}_{\mathcal{NF}}$-concepts), written $\Psi \models C \sqsubseteq D$, if $C \sqsubseteq D$ is true in every model for $\Psi$. Analogously, *instance checking* in $\Sigma$ of a membership assertion $C(a)$ (where $C$ is a $\mathcal{DLK}_{\mathcal{NF}}$-concept and $a \in \mathcal{O}_\Sigma$) is defined as follows: $\Psi \models C(a)$ iff $C(a)$ is satisfied by every model of $\Psi$.

## 3 Reconstruction of frame-based systems

In this section we show that the expressive capabilities of ADLs allow for the reconstruction of several nonmonotonic features of KR systems. In particular, we focus on: defaults, integrity constraints, role and concept closure.

**Defaults.** Some studies on formalizing defaults in frame-based systems in DLs [Baader & Hollunder, 1995, Quantz & Royer, 1992] propose the extension of DLs through the use of Default Logic. We argue that, in order to provide a unified framework to the formalization of several forms of KB closure, it is more convenient to treat defaults as epistemic sentences. A first attempt in this direction is presented in [Donini, Nardi, & Rosati, 1995], where it is shown that one can translate defaults in a logic of minimal knowledge, but based on a different (and less intuitive) semantics for the modal operator **K**. In the following, we only address the "closed" semantics for defaults, but it is worth noticing that also different forms of "open" semantics (see e.g. [Kaminski, 1995]) can be formalized in ADLs. We call $\mathcal{DL}$-default a default rule of the form $d = \frac{\alpha : \beta_1, \ldots, \beta_n}{\gamma}$, where $\alpha, \beta_i, \gamma$ are $\mathcal{DL}$-concepts and $n \geq 0$.

The semantics proposed in [Baader & Hollunder, 1995] for defaults in $\mathcal{DL}$-KBs restricts the application of defaults only to the individuals explicitly mentioned in the ABox. Notice that this semantics can be viewed as the natural extension of the semantics of procedural rules given in [Donini et al., 1992], where rules are applied only to the known individuals in the KB. Under this assumption, we are able to translate default rules in terms of $\mathcal{DLK}_{\mathcal{NF}}$ inclusions in the framework of ADLs. To this purpose, the translation of defaults into MKNF [Lifschitz, 1994] provides a modular and faithful translation of default rules into $\mathcal{DLK}_{\mathcal{NF}}$. More specifically, a $\mathcal{DL}$-default $d$ is translated as:

$$\tau_{DK}(d) = \mathbf{K}I \sqsubseteq \neg\mathbf{K}\alpha \sqcup \mathbf{A}\neg\beta_1 \sqcup \ldots \sqcup \mathbf{A}\neg\beta_n \sqcup \mathbf{K}\gamma$$

where $I$ is an atomic concept not appearing in the KB.

Let $(\Sigma, \mathcal{D})$ be a pair such that $\Sigma = \langle \mathcal{T}, \mathcal{A} \rangle$ is a $\mathcal{DL}$-KB and $\mathcal{D}$ is a set of $\mathcal{DL}$-defaults. Then, $\tau_{DK}(\Sigma, \mathcal{D}) = \langle \mathcal{T}', \mathcal{A}' \rangle$, where $\mathcal{T}' = \mathcal{T} \cup \{\tau_{DK}(d) : d \in \mathcal{D}\}$, $\mathcal{A}' = \mathcal{A} \cup \{I(a) : a \in \mathcal{O}_\Sigma\}$, i.e. for each individual $a \in \mathcal{O}_\Sigma$, the assertion $I(a)$ is added to $\mathcal{A}$. The condition $\mathbf{K}I$ corresponds to adding an extra prerequisite for the application of the default, which expresses the fact that the individual must be in $\mathcal{O}_\Sigma$. Such an extra condition is needed to realize the closed semantics for prerequisite-free defaults. Indeed, without imposing the prerequisite $I$, a prerequisite-free default would be applicable to *each* individual of $\Delta$. The resulting modular translation based on $\tau_{DK}$ is faithful.

**Theorem 3.1** *Given a $\mathcal{DL}$-KB with defaults $\langle \Sigma, \mathcal{D} \rangle$, where $\Sigma$ is a $\mathcal{DL}$-KB and $\mathcal{D}$ is a set of $\mathcal{DL}$-defaults, the $\mathcal{DLK}_{\mathcal{NF}}$ KB $\tau_{DK}(\Sigma, \mathcal{D})$ is such that $\langle \Sigma, \mathcal{D} \rangle \models C(a)$ iff $\tau_{DK}(\Sigma, \mathcal{D}) \models C(a)$ for each $\mathcal{DL}$-concept $C$ and each $a \in \mathcal{O}_\Sigma$.*

**Integrity Constraints.** In this section we study the problem of representing integrity constraints (IC) in ADLs. [Reiter, 1990] pointed out the *epistemic* nature of ICs: they are not statements about the world, they are statement about what the KB is said to know. Generally speaking, the satisfaction of ICs in Reiter's approach is checked in the following way: let $\mathcal{P}$ be a property that the KB must satisfy. Find a suitable epistemic query $Q_\mathcal{P}$ formalizing $\mathcal{P}$. Then, check whether the KB (interpreted under a closure assumption) entails $Q_\mathcal{P}$.

Previous work [Donini et al., 1992] has shown that Reiter's approach can be realized in $\mathcal{DL}$-KBs by endowing the query language with epistemic abilities. Nevertheless, it would be very desirable to express ICs as any other piece of information on the domain of interest, i.e. as sentences *inside* the KB. The difficulty that arises is precisely in the formalization of the notions of closure underlying these forms of integrity constraints. Notably, ICs do not add "objective information": rather, they impose conditions on the consistency of the KB. This is true in the case of first-order KBs, or KBs with a single model as in the case under consideration. In the case of KBs with multiple models, ICs can be viewed as an *a fortiori* check which establishes which of the models are actually allowed. In particular, many forms of ICs impose properties that must hold for the *known* individuals of the KB. The modal operator **K** appears as an appropriate way to formalize this intuition. Moreover, conditions imposed by ICs are consistency conditions which *cannot change the content of the KB*. In other words, augmenting the KB with ICs can only have one of the following two possible effects: either the model of the KB remains unchanged (it satisfies the ICs), or the KB becomes inconsistent (since its model does not satisfy the ICs). As we shall see, the modal operator **A** turns out to be well-suited to this purpose.

We now show that the combination of the modalities **K** and **A** provides for the formalization in ADLs of sophisticated constraints on the KB content.

**Example 1.** Let us consider the IC "each known person must be known to be either male or female", which is meant to avoid any situation where an individual has been added to the KB without specifying her/his sex. One might attempt to formalize it through the $\mathcal{DLK}_{\mathcal{NF}}$ inclusion $I_1 = \mathbf{K}\texttt{person} \sqsubseteq (\mathbf{K}\texttt{male} \sqcup \mathbf{K}\texttt{female})$. However, this formalization is incorrect, since such an assertion *forces* in the KB the knowledge about the sex of each known person. Instead, the meaning of the IC is to *check* whether the sex of each known person is known to the KB. This difference can be better explained as follows. Suppose $\Sigma$ contains only one assertion, $\texttt{person(Bob)}$. Of course, $\Sigma$ does not satisfy the IC. Now, if we add $I_1$ to $\Sigma$, we obtain two models for $(\Sigma, \{I_1\})$: one in which $\texttt{male(Bob)}$ holds, and another in which $\texttt{female(Bob)}$ holds. On the contrary, we would like $(\Sigma, \{I_1\})$ to be inconsistent, since $\Sigma$ does not satisfy the IC. The solution to the above problem lies in the use of the autoepistemic belief operator **A**. Indeed, if we add to a KB the $\mathcal{DLK}_{\mathcal{NF}}$ inclusion $I'_1 = \mathbf{K}\texttt{person} \sqsubseteq (\mathbf{A}\texttt{male} \sqcup \mathbf{A}\texttt{female})$ we formalize the intended meaning of the IC. The difference between $I_1$ and $I'_1$ lies in the fact that $I'_1$ does not force any new knowledge on known persons. In our example, since $\texttt{person(Bob)}$ holds, the assertion $\mathbf{A}\texttt{male} \sqcup \mathbf{A}\texttt{female(Bob)}$ must hold. Now, since there is no reason to conclude either $\texttt{male(Bob)}$ or $\texttt{female(Bob)}$ from $\Sigma$, the autoepistemic beliefs $\mathbf{A}\texttt{male(Bob)}$ and $\mathbf{A}\texttt{female(Bob)}$ are not consistent with the objective knowledge of $\Sigma$, therefore $\mathbf{A}\texttt{male} \sqcup \mathbf{A}\texttt{female(Bob)}$ is inconsistent with $\Sigma$. □

We remark that there is a precise correspondence between the **A** operator and Moore's $L$ operator of Autoepistemic Logic [Rosati, 1997]. From this correspondence, the above example can also be understood as a variation of the "classical" inconsistent autoepistemic theory $\{L\varphi\}$. Thus, the idea in the formalization of ICs is precisely to represent an IC as a believed sentence: if such a belief is not "supported" by the objective knowledge, then an inconsistency arises.

**Example 2.** The IC "Each known employee must have a known social security number, which must be known to be valid" can be correctly formalized by the set of $\mathcal{ALCK_{NF}}$ assertions $I_2 = \{\neg\mathbf{K}\texttt{emp} \sqcup \exists\texttt{KSSN}.\mathbf{A}\texttt{valid}(a) \mid a \in \mathcal{O}_\Sigma\}$. In fact, it can be shown that an $\mathcal{ALCK_{NF}}$ ABox does not satisfy the IC iff $\Sigma \cup I_2$ is inconsistent. □

**Role and concept closure.** Finally, we show how two particular forms of closed-world reasoning, namely role closure and concept closure, can be nicely formalized in the framework of $\mathcal{DLK_{NF}}$. These kinds of closure appear as very useful tools in knowledge representation.

Closure on roles is available both in CLASSIC [Brachman et al., 1990] and in LOOM [MacGregor, 1988]. The idea is to restrict universal role quantifications to the known individuals filling the role in the KB.

**Example (Role closure).** Let $\Psi_1$ be the following $\mathcal{ALC}$-KB: $\{\texttt{CHILD}(\texttt{Ann}, \texttt{Marc}), \texttt{CHILD}(\texttt{Ann}, \texttt{Paula})\}$, where doctor is an abbreviation for the concept $d \sqcap \neg l \sqcap \texttt{rich}$, and lawyer is an abbreviation for $l \sqcap \texttt{rich}$ (expressing that doctors and lawyers are disjoint concepts, and both are rich). Now, suppose we want to formalize the property: "one of the *known* children of Ann is *known* to be a doctor, and another one is *known* to be a lawyer". One would like to conclude that "all known children of Ann are known to be rich". It turns out that the correct formalization is provided by the use of both modalities **A** and **K**. Formally: let $B = \exists\texttt{ACHILD}.\mathbf{K}\texttt{doctor} \sqcap \exists\texttt{ACHILD}.\mathbf{K}\texttt{lawyer}(\texttt{Ann})$. Then, $\Psi_1 \cup \{B\} \models \forall\mathbf{K}\texttt{CHILD}.\mathbf{K}\texttt{rich}(\texttt{Ann})$. □

Epistemic operators make it natural to extend the notion of closure to concept expressions.

**Example (Concept closure).** Let $\Psi_2$ be the following $\mathcal{ALCK}$-KB: $\{\texttt{doctor}(\texttt{Paula}), \texttt{lawyer}(\texttt{Marc}), \forall\texttt{CHILD}.\texttt{hasBlueEyes}(\texttt{Ann})\}$. Suppose we want to add to $\Psi_2$ the following informal property $\mathcal{P}$: "One of Ann's children is one of the *known* doctors". Now, since Paula is the only known doctor, we want to be able to conclude that Paula is one of Ann's children, and hence $\Psi_2 \cup \{\mathcal{P}\} \models \texttt{hasBlueEyes}(\texttt{Paula})$. This can be obtained by formalizing $\mathcal{P}$ through the assertion $\exists\mathbf{K}\texttt{CHILD}.\mathbf{A}\texttt{doctor}(\texttt{Ann})$, since it can be shown that all models for the KB $\Psi_2 \cup \{\mathcal{P}\}$ either Paula or Marc is one of Ann's children. Therefore, $\Psi_2 \cup \{\mathcal{P}\} \models \texttt{hasBlueEyes}(\texttt{Paula})$. □

## 4 Reasoning in ADLs

In this section we study reasoning in ADLs. First, we study $\mathcal{DLK_{NF}}$-*simple KBs*, i.e. $\mathcal{DLK_{NF}}$-knowledge bases in which there is no occurrence of epistemic operators in the scope of quantifiers. We prove that decidability is preserved in simple theories. Furthermore, in many cases the worst-case complexity of deduction is not affected by such a nonmonotonic extension. We also provide an algorithm for instance checking in $\mathcal{DLK_{NF}}$-simple KBs which is parametric wrt the DL in which the KB is expressed. As we shall see, such an algorithm allows for reasoning in DLs with (closed) defaults.

Then, we address *subjectively quantified* $\mathcal{ALCK_{NF}}$-*ABoxes*, i.e. $\mathcal{ALCK_{NF}}$-ABoxes in which occurrences of epistemic operators in the scope of quantifiers are allowed (under some restrictions). We prove that instance checking in such ABoxes is decidable. This result allows us to prove decidability of reasoning in DLs with features like role and concept closure and integrity constraints, which can be expressed in $\mathcal{ALCK_{NF}}$-ABoxes with quantifying-in, as shown in the previous section.

Notably, in the case of $\mathcal{ALCK_{NF}}$ we can easily reduce a simple KB to a subjectively quantified ABox. On the other hand, the method for reasoning on simple KBs is applicable to *any* non-modal DL for which a procedure for instance checking is available.

### 4.1 Reasoning without quantifying-in

We start by defining the notion of $\mathcal{DLK_{NF}}$-simple KBs.

**Definition 4.1** *A $\mathcal{DLK_{NF}}$-concept $C$ is simple iff there are no occurrences in $C$ of an epistemic operator in the scope of quantifiers.*

**Definition 4.2** *A $\mathcal{DLK_{NF}}$-simple KB is a pair $(\Sigma, \Gamma)$ such that $\Sigma = \langle \mathcal{T}, \mathcal{A} \rangle$ is a $\mathcal{DL}$-KB and $\Gamma$ is a set of $\mathcal{DLK_{NF}}$-simple inclusions, i.e. inclusions of the form $\mathbf{K}C \sqsubseteq D$, where $D$ is a $\mathcal{DLK_{NF}}$-simple concept and $C$ is a $\mathcal{DL}$-concept such that $\Sigma \not\models \top \sqsubseteq C$.*

As in [Donini et al., 1992], the condition $\Sigma \not\models \top \sqsubseteq C$ corresponds to a "closed" semantics for the $\mathcal{DLK_{NF}}$-simple inclusion $\mathbf{K}C \sqsubseteq D$, in the sense that it corresponds to consider the application of such axioms only to the individuals in $\mathcal{O}_\Sigma$. This is precisely due to the minimal knowledge semantics of the modal operator **K**: in fact, due to the form of $\mathcal{DLK_{NF}}$-simple inclusions, there is no way to force the property $\mathbf{K}C$ on individuals $\notin \mathcal{O}_\Sigma$ (since $\Sigma \not\models \top \sqsubseteq C$). Now, from the definition of model in $\mathcal{DLK_{NF}}$ (that is, from the minimal knowledge semantics of the operator **K**) it is easy to see that only sets of $\mathcal{DL}$-interpretations in which $\neg\mathbf{K}C$ holds for each individual not in $\mathcal{O}_\Sigma$ are preferred. That is, the application on such individuals of the epistemic inclusion $\mathbf{K}C \sqsubseteq D$ has no effect. Therefore, the following property holds.

**Lemma 4.3** *Let $\Sigma = (\langle \mathcal{T}, \mathcal{A} \rangle, \Gamma)$ be a $\mathcal{DLK_{NF}}$-simple KB. Let $\Sigma' = \langle \mathcal{T} \cup \Gamma, \mathcal{A} \rangle$, and let $\Sigma'' = \langle \mathcal{T}, \mathcal{A} \cup \{\neg\mathbf{K}C \sqcup D(a) | \mathbf{K}C \sqsubseteq D \in \Gamma \text{ and } a \in \mathcal{O}_\Sigma\}\rangle$. Then, the sets of models of $\Sigma'$ and $\Sigma''$ coincide.*

We now present a method for computing instance checking in $\mathcal{DLK_{NF}}$-simple KBs, based on the procedure defined for propositional MKNF theories (see [Rosati, 1997]). However, the extension to the case of ADLs requires several preliminary notions. In particular, we have to define consistency of a partition of assertions wrt a $\mathcal{DLK_{NF}}$-simple KB.

**Definition 4.4** *Let $C$ be a $\mathcal{DLK_{NF}}$-concept expression. Let $P, N$ be sets of $\mathcal{DLK_{NF}}$-assertions of the form $\mathbf{K}D(a), \mathbf{A}D(a)$, such that $P \cap N = \emptyset$. Then, $C(a)(P, N)$ is the assertion obtained by substituting with $\top$ each occurrence in $C$ of a concept $D$ which is not within the scope of a modal operator, and such that $D(a) \in P$, and with $\bot$ each occurrence in $C$ of a concept $D$ which is not within the scope of a modal operator, and such that $D(a) \in N$.*

Informally, $C(a)(P, N)$ is the assertion representing the evaluation of $C(a)$ wrt the guess of the modal subformulas of $C(a)$ according to the partition $(P, N)$. Notice that, if each modal subexpression in $C(a)$ appears either in $P$ or in $N$, then $C(a)(P, N)$ is a non-modal assertion. We denote $P^K = \{C(a)(P, N) | \mathbf{K}C \in P \text{ and } a \in \mathcal{O}_\Sigma\}$, $P^A = \{C(a)(P, N) | \mathbf{A}C \in P \text{ and } a \in \mathcal{O}_\Sigma\}$, and $\mathbf{A}P^K = \{\mathbf{A}C(a) | C(a) \in P^K\}$.

In order to reason with $\mathcal{DLK}_{\mathcal{NF}}$-simple KB we need an effective method for identifying models. Let $C$ be a $\mathcal{DLK}_{\mathcal{NF}}$-concept. The set of *modal atoms* of $C$, denoted as $MA(C)$, is the set of role subexpressions or concept subexpressions of $C$ of the form $\mathbf{K}C'$ or $\mathbf{A}C'$. The set of modal atoms of an inclusion $C \sqsubseteq D$ is the union of the sets $MA(C)$ and $MA(D)$.

**Definition 4.5** *Let $\gamma$ be a $\mathcal{DLK}_{\mathcal{NF}}$-inclusion. Let $\mathcal{O}$ be a set of individuals. Let $MA(\gamma)$ be the set of modal atoms in $\gamma$. Then, the set of* instances of $MA(\gamma)$ in $\mathcal{O}$, denoted as $MI(\gamma, \mathcal{O})$, is the set $\{C(a) | a \in \mathcal{O} \text{ and } C \in MA(\gamma)\}$. *Moreover, if $\Gamma$ is a set of $\mathcal{DLK}_{\mathcal{NF}}$-inclusions, then $MA(\Gamma, \mathcal{O}) = \bigcup_{\gamma \in \Gamma} MA(\gamma, \mathcal{O})$.*

**Definition 4.6** *Let $(\Sigma, \Gamma)$ be a $\mathcal{DLK}_{\mathcal{NF}}$-simple KB. Let $(P, N)$ be a partition of $MI(\Gamma, \mathcal{O}_\Sigma)$. Then, $(P, N)$ is consistent with $(\Sigma, \Gamma)$ iff the following conditions hold:*
*i. for each $\gamma \in \Gamma$ and for each $a \in \mathcal{O}_\Sigma$, $\gamma(a)(P, N) \equiv \top$;*
*ii. the $\mathcal{DL}$-knowledge base $\langle \mathcal{T}, \mathcal{A} \cup P^K \rangle$ is satisfiable;*
*iii. the $\mathcal{DL}$-knowledge base $\langle \mathcal{T}, \mathcal{A} \cup P^A \rangle$ is satisfiable;*
*iv. for each $\mathbf{K}C(a) \in N$, $\Sigma \cup P^K \not\models C(a)(P, N)$;*
*v. for each $\mathbf{A}C(a) \in N$, $\Sigma \cup P^A \not\models C(a)(P, N)$.*

Notice that in the above definition both $P^K$ and $P^A$ are sets of $\mathcal{DL}$-assertions, since $(P, N)$ is a partition of $MI(\Gamma, \mathcal{O}_\Sigma)$. Therefore, $\langle \mathcal{T}, \mathcal{A} \cup P^K \rangle$ and $\langle \mathcal{T}, \mathcal{A} \cup P^A \rangle$ are $\mathcal{DL}$-KBs.

**Lemma 4.7** *Let $(P, N)$ be defined as above. $(P, N)$ identifies a model for $\langle \mathcal{T} \cup \Gamma, \mathcal{A} \rangle$ iff*
*i. $(P, N)$ is consistent with $(\Sigma, \Gamma)$;*
*ii. $\Sigma \cup P^K \models P^A$;*
*iii. for each partition $(P', N')$ of $MI(\Gamma', \mathcal{O}_\Sigma)$, where $\Gamma' = \Gamma \cup \{\mathbf{A}P^K\}$ either (a) $(P', N')$ is not consistent with $(\Sigma, \Gamma)$ or (b) $\Sigma \cup P^K \not\models P'^K$ or (c) $\Sigma \cup P'^K \models P^K$ or (d) $\Sigma \cup P^K \not\models P'^A$.*

The proof follows from Lemma 4.3 and properties of the logic $\mathcal{MKNF}$ [Rosati, 1997]. Based on the above lemma, we can provide a procedure for establishing whether a partition $(P, N)$ identifies a model for $\langle \mathcal{T} \cup \Gamma, \mathcal{A} \rangle$.

Figure 1 reports the algorithm Simple-Not-Entails for computing instance checking in $\mathcal{DLK}_{\mathcal{NF}}$-simple KBs. Notice that, in the case when the query $C(a)$ is non-modal, the method can be realized using a procedure for computing instance checking in $\mathcal{DL}$.

Based on the algorithm Simple-Not-Entails, one can prove decidability of instance checking in $\mathcal{DLK}_{\mathcal{NF}}$-simple KBs.

**Theorem 4.8** *Let $(\Sigma, \Gamma)$ be a $\mathcal{DLK}_{\mathcal{NF}}$-simple KB. Let $C(a)$ be a $\mathcal{DL}$-assertion. The problem $(\Sigma, \Gamma) \models C(a)$ is decidable iff instance checking in $\mathcal{DL}$ is decidable.*

We can provide a computational characterization of the instance checking problem in $\mathcal{DLK}_{\mathcal{NF}}$-simple KBs. In particular, the above theorem implies that adding $\mathcal{ALCK}_{\mathcal{NF}}$-simple inclusions to an $\mathcal{ALC}$-KB with empty TBox $\Sigma =$

---

**Algorithm** Simple-Not-Entails($\mathcal{DL}, (\Sigma, \Gamma), C(a)$)
**Input:** description logic $\mathcal{DL}$,
    $\mathcal{DLK}_{\mathcal{NF}}$-simple KB ($\Sigma = \langle \mathcal{T}, \mathcal{A} \rangle, \Gamma$),
    $\mathcal{DL}$-assertion (query) $C(a)$;
**Output:** true if $(\Sigma, \Gamma) \not\models C(a)$, false otherwise.
**begin**
$MA(\Gamma)$ = set of modal atoms in $\Gamma$;
$MI(\Gamma, \mathcal{O}_\Sigma) = \{D(a) | a \in \mathcal{O}_\Sigma \text{ and } D \in MA(\Gamma)\}$;
**if there exists** partition $(P, N)$ of $MI(\Gamma, \mathcal{O}_\Sigma)$
    such that
    $(P, N)$ identifies a model of $\langle \mathcal{T} \cup \Gamma, \mathcal{A} \rangle$ **and**
    $\Sigma \cup P^K \not\models C(a)$
**then return** true
**else return** false
**end**

Figure 1: Algorithm Simple-Not-Entails.

$\langle \emptyset, \mathcal{A} \rangle$ does not increase the worst-case complexity of instance checking of an $\mathcal{ALCK}$-assertion, which is PSPACE-complete in the case of a non-modal $\Sigma$ [Donini et al., 1992]. The same result holds in the case of an $\mathcal{ALC}$-KB $\Sigma = \langle \mathcal{T}, \mathcal{A} \rangle$ with $\mathcal{ALCK}_{\mathcal{NF}}$-simple inclusions (i.e. instance checking is EXPTIME-complete as in the non-modal case).

Notably, it is easy to see that the translation $\tau_{DK}(\Sigma, \mathcal{D})$ of a $\mathcal{DL}$-KB with defaults $(\Sigma, \mathcal{D})$ is equivalent to the $\mathcal{DLK}_{\mathcal{NF}}$-simple KB $(\Sigma, \Gamma)$, in which $\Gamma$ is the set $\{\tau_{DK}(d) : d \in \mathcal{D}\}$. Therefore, from Theorem 3.1 it follows that in many cases adding defaults to a $\mathcal{DL}$-KB does not increase the complexity of deduction: in particular, reasoning in $\mathcal{ALC}$-ABoxes with defaults is PSPACE-complete.

### 4.2 Reasoning with quantifying-in

We start by defining the notion of subjectively quantified $\mathcal{ALCK}_{\mathcal{NF}}$-ABoxes.

**Definition 4.9** *A subjectively quantified $\mathcal{ALCK}_{\mathcal{NF}}$-assertion is an $\mathcal{ALCK}_{\mathcal{NF}}$-assertion $C(a)$ where $C$ is a concept expression of $\mathcal{ALCK}_{\mathcal{NF}}$ in which each quantified subexpression is of the form $\exists P.D', \forall P.D', \exists \mathbf{M}P.\mathbf{M}'D, \forall \mathbf{M}P.\mathbf{M}'D$, where $\mathbf{M}, \mathbf{M}' \in \{\mathbf{K}, \mathbf{A}\}$, $D'$ is an $\mathcal{ALC}$-concept, $P$ is an atomic role.*

An $\mathcal{ALCK}_{\mathcal{NF}}$-ABox composed of subjectively quantified assertions is called *subjectively quantified $\mathcal{ALCK}_{\mathcal{NF}}$-ABox*. The method for reasoning on subjectively quantified ABoxes is based on a tableaux calculus following the lines of [Donini et al., 1996], where special closure conditions are defined to enforce the preference criterion on the models represented by the branches of the tableau. For ADLs, the lifting from propositional logic to DLs raises several difficulties that are addressed in the sequel. In particular, we sketch a calculus for characterizing the models of a subjectively quantified $\mathcal{ALCK}_{\mathcal{NF}}$-ABox in terms of $\mathcal{ALC}$-KBs, to which the procedure for query answering defined in [Donini et al., 1992] can be applied.

The tableaux rules include the standard rules for an S5 tableau [Fitting, 1983] for handling propositional connectives and epistemic expressions of the form $\mathbf{K}C, \mathbf{A}C$ and $\neg \mathbf{K}C, \neg \mathbf{A}C$. The tableaux calculus for generating the models for subjectively quantified $\mathcal{ALC}$-KBs behaves as usual, except for the fact that each rule is applied only to epistemic prefixed formulas.

First, we define the notion of modal atoms of a subjectively quantified $\mathcal{ALCK}_{\mathcal{NF}}$-assertion $C(x)$ as the set of assertions $C'(x)$, where $C'$ is a subexpression of $C$ of the form $\mathbf{K}C''$ or $\mathbf{A}C''$ or $\exists \mathbf{M}P.\mathbf{M}'C''$ or $\forall \mathbf{M}P.\mathbf{M}'C''$, where $\mathbf{M}, \mathbf{M}' \in \{\mathbf{K}, \mathbf{A}\}$. The set of modal atoms $MA(\Sigma)$ of a subjectively quantified $\mathcal{ALCK}_{\mathcal{NF}}$-ABox $\Sigma$ is the union of the sets of modal atoms of all the assertions it contains.

A branch $\mathcal{B}$ is a set of prefixed formulas of the form $\langle w : C(x)\rangle$. The tableau for $\Sigma$ starts with the set $\{\langle 1 : \mathbf{K}A\rangle \mid$ assertion $A \in \Sigma\}$. $\mathcal{O}_\mathcal{B}$ is the set of individuals in $\mathcal{B}$.

We concentrate on the rules that handle subjectively quantified expressions, and on a rule that saturates the branch wrt modal atoms of $\Sigma$. We phrase them using $\mathbf{K}$; the cases arising from the use of both modalities $\mathbf{K}$ and $\mathbf{A}$ are analogous. The rules are as follows:

$\forall$-rule: if $\langle w : \forall \mathbf{K}P.\mathbf{K}C(x)\rangle \in \mathcal{B}$, then for each $\langle 1 : \mathbf{K}P(x, y)\rangle \in \mathcal{B}$ add $\langle 1 : \mathbf{K}C(y)\rangle$ to $\mathcal{B}$.

$\exists$-rule: if $\langle w : \exists \mathbf{K}P.\mathbf{K}C(x)\rangle \in \mathcal{B}$ and there is no $y$ such that both $\langle 1 : \mathbf{K}P(x,y)\rangle$ and $\langle 1 : \mathbf{K}C(y)\rangle \in \mathcal{B}$, then add $\langle 1 : \mathbf{K}P(x,z)\rangle$ and $\langle 1 : \mathbf{K}C(z)\rangle$ to $\mathcal{B}$, where $z \in \mathcal{O}_\mathcal{B} \cup \{\iota\}$ and $\iota \notin \mathcal{O}_\mathcal{B}$.

mcut-rule: if $\mathbf{K}C(x) \in MA(\Sigma)$ then add $\langle 1 : \mathbf{K}C(x)\rangle$ or $\langle 1 : \neg \mathbf{K}C(x)\rangle$ to $\mathcal{B}$, if neither is present in $\mathcal{B}$.

A branch is *completed* if no rule is applicable to it; a branch is *open* if there is no pair of prefixed formulas in $\mathcal{B}$ of the form $\langle w : C(x)\rangle$ and $\langle w : \neg C(x)\rangle$.

An open completed branch $\mathcal{B}$ does not always represent a model. To select models according to the preference criterion, one needs to characterize the objective knowledge associated with $\mathcal{B}$. In particular, one has to distinguish between the objective knowledge implied by $\mathbf{K}$-prefixed and $\mathbf{A}$-prefixed modal atoms [Rosati, 1997]. To this aim, we remark that the mcut-rule forces a partition on the modal atoms of $\Sigma$ in $\mathcal{B}$. We call $(P_\mathcal{B}, N_\mathcal{B})$ such a partition. Based on Def. 4.4, we define the following $\mathcal{ALC}$-ABoxes:

$$OBJ_K(\mathcal{B}) = \{C(x)(P_\mathcal{B}, N_\mathcal{B}) | \mathbf{K}C(x) \in P_\mathcal{B}\}$$
$$OBJ_A(\mathcal{B}) = \{C(x)(P_\mathcal{B}, N_\mathcal{B}) | \mathbf{A}C(x) \in P_\mathcal{B}\}$$

We can now define the notion of preferred branch.

**Definition 4.10** *A branch $\mathcal{B}$ of the tableau for $\Sigma$ is preferred iff $\mathcal{B}$ is open and completed, $OBJ_K(\mathcal{B}) \models OBJ_A(\mathcal{B})$ and, for each open and completed branch $\mathcal{B}'$ of the tableau for $\Sigma \cup \{\mathbf{A}C(x) | C(x) \in OBJ_K(\mathcal{B})\}$, either $OBJ_K(\mathcal{B}) \not\models OBJ_K(\mathcal{B}')$ or $OBJ_K(\mathcal{B}') \models OBJ_K(\mathcal{B})$ or $OBJ_K(\mathcal{B}) \not\models OBJ_A(\mathcal{B}')$.*

The above notion of preferred branch allows us to identify *all* the models of $\Sigma$, up to renaming of individuals in $\mathcal{O}_\mathcal{B} - \mathcal{O}_\Sigma$.

**Theorem 4.11** *Let $\Sigma$ be a subjectively quantified $\mathcal{ALCK}_{\mathcal{NF}}$-ABox. Then, a branch $\mathcal{B}$ of the tableau for $\Sigma$ is preferred iff there exists a model $\mathcal{M}$ for $\Sigma$ and a mapping $\mu : \Delta \to \Delta$ such that $\mathcal{M} = \{\mathcal{I} | \mathcal{I} \models \mu(OBJ_K(\mathcal{B}))\}$. Moreover, let $C(a)$ be an $\mathcal{ALCK}$-assertion. Then, $\Sigma \models C(a)$ iff there exists a preferred branch $\mathcal{B}$ of the tableau for $\Sigma$ such that $OBJ_K(\mathcal{B}) \not\models C(a)$.*

It can be shown that the tableaux method above outlined always terminates. Moreover, since $OBJ_K(\mathcal{B}) \not\models C(a)$ can be checked by the algorithm presented in [Donini et al., 1992], we have proved decidability of the instance checking problem for subjectively quantified $\mathcal{ALCK}_{\mathcal{NF}}$-ABoxes.

## Acknowledgments

This research has been funded by Italian MURST, "Tecniche di ragionamento non monotono" and "Linguaggi per la modellizzazione concettuale dei requisiti", by EC Esprit LTR Project "Foundations of Data Warehouse Quality", and by ASI (Italian Space Agency).

## References

[Baader & Hollunder, 1995] F. Baader and B. Hollunder. Embedding defaults into terminological knowledge representation formalisms. *JAR*, 14:149–180, 1995.

[Brachman et al., 1990] R. J. Brachman, D. McGuinness, P. F. Patel-Schneider, L. A. Resnick. Living with CLASSIC. In *Principles of Semantic Networks*, J. Sowa ed., Morgan Kaufmann, 1990.

[Donini et al., 1992] F. M. Donini, M. Lenzerini, D. Nardi, W. Nutt, and A. Schaerf. Adding epistemic operators to concept languages. In *Proc. of KR-92*, 342–353, Morgan Kaufmann, 1992.

[Donini et al., 1996] F. M. Donini, F. Massacci, D. Nardi, and R. Rosati. A uniform tableau method for nonmonotonic modal logics. In *Proc. of JELIA'96*, LNAI 1126, 87–103, Springer-Verlag, 1996.

[Donini, Nardi, & Rosati, 1995] F. M. Donini, D. Nardi, and R. Rosati. Non-first-order features in concept languages. In *AI*IA-95*, LNAI 992, 91–102. Springer-Verlag, 1995.

[Fitting, 1983] M. Fitting. *Proof Methods for Modal and Intuitionistic Logics*. Reidel, 1983.

[Kaminski, 1995] M. Kaminski. A comparative study of open default theories. *AIJ*, 77:285–319, 1995.

[Lakemeyer, 1996] G. Lakemeyer. Limited reasoning in first-order knowledge bases with full introspection. *AIJ* 84:209–255, 1996.

[Lifschitz, 1994] V. Lifschitz. Minimal belief and negation as failure. *AIJ*, 70:53–72, 1994.

[Lin & Shoham, 1992] F. Lin and Y. Shoham. Epistemic semantics for fixed-point non-monotonic logics. *AIJ*, 57:271–289, 1992.

[MacGregor, 1988] R. MacGregor. A deductive pattern matcher. In *Proc. of AAAI-88*, 403–408, 1988.

[Padgham & Zhang, 1993] L. Padgham and T. Zhang. A terminological logic with defaults: a definition and an application. In *Proc. of IJCAI-93*, 662–668, 1993.

[Quantz & Royer, 1992] J. Quantz and V. Royer. A preference semantics for defaults in terminological logics. In *Proc. of KR-92*, 294–305, Morgan Kaufmann, 1992.

[Reiter, 1990] R. Reiter. What should a database know? *Journal of Logic Programming*, 14:127–153, 1990.

[Rosati, 1997] R. Rosati. Reasoning with minimal belief and negation as failure: algorithms and complexity. To appear in *Proc. of AAAI-97*.

[Woods & Schmolze, 1992] W.A. Woods and J.G. Schmolze. The KL-ONE Family. *Computers & Mathematics with Applications*, Volume 23, 2–9, 1992, 133–178.

# Reifying Concepts in Description Logics

Liviu Badea
AI Research Lab
Research Institute for Informatics
8-10 Averescu Blvd., Bucharest, Romania
e-mail: badea@roearn.ici.ro

## Abstract

Practical applications of description logics (DLs) in knowledge-based systems have forced us to introduce the following features which are absent from existing DLs:

- allowing a concept to be regarded at the same time as an individual (the instance of some other meta-level concept)
- allowing an individual to represent a collection (set) of other individuals.

The first extension, called *concept reification*, is more general and thus can cover the second one too. We argue that the absence of these features from existing DLs is an important reason for the lack of a unified approach to description logics and object-oriented databases.

We also show that concept reification cannot be dealt with by the standard DL semantics and propose a slightly modified semantics that takes care of the inherent higher-order features of reification in a first-order setting. A sound and complete inference algorithm for checking consistency in reified $\mathcal{ALCO}_\in$ knowledge bases is subsequently put forward.

## 1 Introduction

Description logics (DLs) are descendants of the famous KL-ONE system [Brachman and Schmolze, 1985] and can be viewed as formalizations of the frame-based knowledge representation languages.

Systems based on DLs are hybrid systems which separate the described knowledge in two distinct categories: *terminological* and *assertional* knowledge. The terminological knowledge is generic and refers to classes of objects and their relationships, while the assertional knowledge describes particular instances (individuals) of these classes. These two levels are completely disjoint since a given object cannot be at the same time a concept *and* an instance. (Description logics further distinguish between two kinds of terminological knowledge, namely concepts and roles. Concepts are essentially unary predicates interpreted as sets of individuals, while roles represent binary predicates interpreted as (binary) relations between individuals.)

An important limitation of current description logics is the clear-cut separation between the terminological (intensional) and the assertional (extensional) level. For example, concepts (representing intensional descriptions of sets of individuals) and their instances are stored at different levels and cannot be mixed under any circumstances.

In certain applications, however, it may be useful to be able to regard a given concept (class) as the instance of a higher level meta-concept (meta-class). This would allow us to *reuse* terminologies by constructing a unique generic terminology which could then be instantiated to produce several particular terminologies.

This paper presents an extension of description logics in which a given concept can be regarded as an individual (i.e. an instance of some other meta-level concept). This process, called *concept reification*, has not been extensively studied in the framework of description logics[1], mainly since it mixes the terminological and assertional levels and therefore spoils the simplicity of the currently used reasoning techniques. Also, reification introduces a form of *higher-order constructs* in description logics thereby complicating the issue of defining a proper semantics of the logic as well as the associated inference services.

In spite of these difficulties, reification is absolutely necessary whenever we want to achieve *reusability* in a knowledge-based system. The following example, taken from [Badea and Țilivea, 1996], deals with allocating the staff of some research institution. In such a setting we may want to introduce concepts like *manager*, *secretary*, *researcher* and instances like *Tom*, *Joan*, *Mary*, *Peter*, *Fred*, etc: $Tom \in manager$, $Joan \in secretary$, $Mary \in secretary$, $Peter \in researcher$, $Fred \in researcher$.

But now note that the concepts *manager*, *secretary* and *researcher* represent positions in the research institute. They are therefore not only concepts, but also *instances* of the (metal-level) concept

---

[1] The CLASSIC system [Brachman et al., 1991] already included meta-individuals (a pre-theoretic form of concept reification), but these are not taken into account in DL inferences.

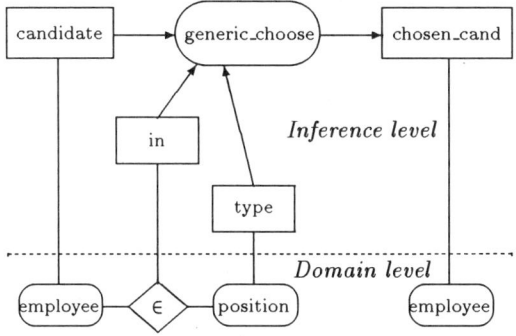

Figure 1: A generic inference using the relation $\in$.

*position*: *manager* $\in$ *position*, *secretary* $\in$ *position*, *researcher* $\in$ *position*.

The concept *position* should not be confused with the concept *employee*, which is a super-concept of *manager*, *secretary* and *researcher*: *manager* $\subset$ *employee*, *secretary* $\subset$ *employee*, *researcher* $\subset$ *employee*.

Now consider an employee allocation problem. In a typical knowledge-based system, we would have to write separate inferences for instance for retrieving managers, secretaries and researchers. This would amount to writing three separate pieces of code that are extremely similar (though not identical, as at least the positions of the employees to be retrieved would be different).

If we would like to avoid writing three separate pieces of code (inferences), we would have to write a *generic* inference that would be parametrized by the *position* of the employee we would like to choose. This can be accomplished by using an input parameter, called *type*, linked to the concept *position* and which is supposed to specify the position of the person to be chosen. Figure 1 is a graphical representation of such a generic inference in the KADS-based $E_xClaim$ system (see [Badea, 1996] for more details on $E_xClaim$). Note that we are making use of the built-in membership role $\in$, which links an instance $X$ with a concept $C$ whenever $X$ is an instance of $C$.

The predefined role $\in$ links a concept (*employee*) with a meta-concept (*position*) and allows therefore a kind of generic inferences which are essential for developing *domain-independent and reusable models*.[2]

Note that domain-independence and reusability of models could not have been achieved without concept reification and the role $\in$.

In order to be usable in real-life applications, description logics will also have to allow for an individual (instance) to represent a *collection* (set) of other individuals. However, existing (implemented) DL systems usually lack constructors for sets or lists of objects[3] and we therefore have to represent such collections outside the DL thereby affecting the completeness of the DL reasoning services.

Using concept reification, we can obtain individuals representing sets of other individuals by reifying concepts of the form $one\_of(i_1, \ldots, i_n)$. This observation allows us to concentrate in the following on "concept reification".

Note that the role $\in$ allows us to regard the assertional component of the DL (the ABox) as consisting of role tuples only, since instance assertions of the form $X : C$ can be viewed as tuples of the role $\in$.

Concept reification, as studied in this paper, is different from Kobsa's role reification implemented in SB-ONE [Kobsa, 1991]. More precisely, while the reification of a concept is an individual, Kobsa's reification of a role is a concept. (Kobsa's approach has been motivated by natural language applications in which a verb, for example, is regarded in some contexts as a role and in other contexts as a concept.) Therefore, while we are concerned with mixing the TBox and the ABox of a DL, Kobsa has dealt with mixing concepts and roles within the TBox (while keeping TBox and ABox disjoint).

As previously mentioned, concept reification involves a form of higher-order logic since the interpretations of value and existential restrictions on the $\in$-role employ a form of quantification over *concept-valued* variables. Therefore, whereas in ordinary DLs concepts exist only as named, terminological (*TBox*) level elements, in reified DLs concepts may be individuals ("data") as well.

Reification and the related higher-order features are also essential in object oriented databases (OODBs) [Beeri, 1990]. In classical database systems there are two distinct levels: *data* and *schema* (similar to ABox and respectively TBox in description logics).

In OODBs, meta-data (such as classes and functions), are frequently treated as data. Class objects acquire thereby a dual nature: on the one hand they are data and can be manipulated by the system; on the other hand, they are schema-level objects and thus part of the schema.

This situation is similar to concept reification as introduced in this paper. In fact, it is our opinion that the lack of a unified approach to description logics and OODBs is mainly due to the clear-cut separation of TBox and ABox (i.e. to the absence of reification) in DLs.

However, introducing concept reification in DLs is significantly harder than in OODBs since we have to modify the DL inference services (consistency and subsumption tests) to cope with the new construct. Since no analog inference services exist in OODBs and as long as we

---

[2] Not only is it cumbersome to have three identical pieces of code, but these pieces of code would depend on the domain level (the types of positions – *director*, *secretary* and *researcher* are domain-dependent; we cannot change the domain level, for example by introducing a new position, without having to modify the inference level too, since we would have to add a new inference for the new position type).

[3] Some description logics provide the $one\_of(i_1, \ldots, i_n)$ construct which denotes the concept whose extension is given by the set of instances $\{i_1, \ldots, i_n\}$. However, what we need is a concept construct whose *instances* denote sets or lists of other instances.

deal with reification in an explicit manner alone, there seem to be no complications in reasoning with the new construct in OODBs.

Another somewhat related formalism is *F-logic* [Kifer et al., 1995], which attempts to provide sound logical foundations of object-oriented as well as frame-based languages and can be considered as a declarative approach to deductive object-oriented databases.

F-logic provides a form of explicit reification, but it lacks the $\in$-role and the related DL inference services. Therefore, we can easily represent F-logic object models in reified DLs. For instance, an F-logic *non-inheritable property object[property $\rightarrow$ value]* would be represented in reified DLs as a tuple of the role *property* involving the individual *object*: $(object, value)$ : *property*, while an *inheritable* property of the form *object[property $\bullet\!\!\rightarrow$ value]* would be captured by a DL terminological axiom imposing a restriction on the fillers of the role *property* for the *instances* of *object*: $object \sqsubseteq \exists property.\{value\}$. F-logic signature (typing) expressions of the form *object[property $\Rightarrow$ type]* can also be represented in DLs as value restrictions $object \sqsubseteq \forall property.type$. Of course, set-valued attributes in F-logic correspond to DL roles, while normal F-logic attributes would be represented in DLs as functional roles (attributes). It is our opinion that the DL representation makes the distinction between the assertional and terminological properties of objects clearer.

It is worth noting that many frame-based knowledge representation systems (for example KEE) allow a class to be at the same time an instance, but this feature has usually no associated formal semantics.

## 2 Reifying concepts in description logics - semantical considerations

Traditional description logics separate the terminological (TBox) and assertional (ABox) levels completely by not allowing a concept to be regarded at the same time as an individual. This simplifies the semantics and corresponding reasoning algorithms.

In this paper we consider an extension to DLs that eliminates this restriction. Concept reification amounts to associating with each concept $C$ an individual $C'$.

Additionally, we allow the membership role $\in$ and its inverse $\ni$. The role $\in$ links an individual $X$ with some other reified concept $C'$ whenever $X$ is an instance of $C$ (regarded as a concept).

As already mentioned, concept reification involves a form of higher-order logic. For example, in defining the interpretation of $\forall \ni .C$:

$$(\forall \ni .C)^{\mathcal{I}} = \{x \in \Delta \mid \forall y . \, x \ni^{\mathcal{I}} y \rightarrow y \in C^{\mathcal{I}}\}$$

we quantify over *all concepts* $y$, not just the ones that are explicitly given. Since higher-order logics lack even a sound a complete axiomatization and in order to preserve the desired computational properties, we will restrict the semantics of the logic to a first-order semantics. This amounts to interpreting quantified concept variables as ranging over (explicitly given) reified individuals, or *intensions*, rather than over all concepts that potentially exist (or their *extensions*). Therefore, we will allow reification of *concept names* only.

Actually, we can drop the explicit reification construct $.'$ (and use $C$ instead of $C'$ for all concept names $C$) because we can determine the type of an object (whether it is a concept or an individual) from the context in which it is used. For example, $C$ in the concept term $\forall R.C$ is regarded as a concept, while if we use it in the assertional axiom $C : D$, then $C$ represents the reification of the corresponding concept (i.e. an individual). As previously mentioned, since we don't have an explicit reification operator, we will allow reification of concept names only. For instance, we will not allow ABox assertions like $(\forall R.C) : D$.

Also, since we interpret concept-valued variables over intensions rather than extensions, it may be that the reified counterparts of two equivalent concepts, $C$ and $D$, represent two different individuals.

We can use the membership role $\in$ to express all instance assertions $X : C$ (for concept *names* $C$) as tuple assertions of $X \in C$. Therefore, we can regard the ABox as consisting of tuple assertions only.

Reified description logics are quite expressive.

For example, it is easy to check that $\forall \ni .C$ represents, roughly speaking, the power-set $\mathcal{P}(C)$ of $C$ (the set of subset of $C$), while $\exists \in .C$ denotes the union $\bigcup C$ of the instances of $C$, regarded as sets.

Although reified DLs are capable of representing concepts like $\forall \ni .\top$ denoting "the set of all sets", we do not run into logical paradoxes, since $\forall \ni .\top$ is equivalent, as expected, to the top concept $\top$.

Since the union of all subsets of a set $C$ is equal to $C$, $\bigcup \mathcal{P}(C) = C$, we obtain the identity $\exists \in .\forall \ni .C = C$ which could be regarded as an axiom in reified DLs. However, only the "$\leftarrow$" direction is specific to $\in$, the "$\rightarrow$" direction being an instance of the axiom for role inverses.

Additionally, note that $C$ is an instance of $\exists \ni .D$ iff $C \sqcap D$ is consistent, and that $C$ is an instance of $\forall \ni .D$ iff $C \sqsubset D$. In reified DLs, subsumption is therefore reducible (within the language) to instance checking.

The following identities specific to reified DLs can also be easily checked[4]:

$$\exists \in .\top = \top \qquad (or\ its\ dual : \forall \in .\bot = \bot)$$
$$\forall \ni .\bot = \{\bot\} \qquad \exists \in .\{C\} = C.$$

Also, since $C$ is an instance of $\exists \ni .\{X\}$ iff $X \in C$, we can regard $\exists \ni .\{X\}$ as denoting the set of concepts $C$ for which $X$ is an instance. This observation shows that the "realization problem"[5] for $X$ has a solution expressible in the language, namely $\exists \ni .\{X\}$.

---

[4] $\{X\}$ is the singleton concept whose extension has only one element, $X$.

[5] retrieving the set of concepts $C$ that admit a given individual $X$ as an instance.

Observe that $X$ is an instance of $\forall \in .C$ iff $\exists \ni .\{X\} \subset C$.

The examples and observations above give a flavour of the intricate ways in which reification and the $\in$-role interact with one another and the other DL constructors.

After having informally presented concept reification in DLs, let us now try to formalize it. For reasons of simplicity, we are going to discuss reification in the $\mathcal{ALCO}_\in$ language ($\mathcal{ALC}$ of Schmidt-Schauß and Smolka [1991] extended with the *one-of* construct), but the results are easily extensible to more expressive languages.

The *syntax* of reified $\mathcal{ALCO}_\in$ deals with the following three sets of (syntactic) objects:

- *Names* denoted by $X, Y, \ldots$ (including $\bot$)
- *Concept-Terms* denoted by $C, D, \ldots$
- *Role-Terms* denoted by $R, Q, \ldots$

The set *Names* contains individuals and concept names occurring in the DL knowledge base. Since we want to allow for the reification of concept names, individuals and concept names will have to belong to a single syntactic category (*Names*) (as opposed to traditional DLs where they fall under syntactically disjoint categories).

*Concept-Terms* are terms built from concept names (belonging to *Names*) using the following $\mathcal{ALCO}_\in$ concept constructors:

$$C ::= X \mid C \sqcap D \mid C \sqcup D \mid \neg C \mid \forall R.C \mid \exists R.C \mid \{X\}.$$

($\{X\}$ is a singleton concept; general *one-of*$(X_1, \ldots, X_n)$ concepts can be represented as $\{X_1\} \sqcup \ldots \sqcup \{X_n\}$.)

*Role Terms* are built from role names (which are *not* in *Names*, since we do not reify roles) using the specific DL role constructors. Since $\mathcal{ALCO}$ admits only role names, $\mathcal{ALCO}_\in$ will admit only the following roles:

$$R ::= RN \mid \in \mid \ni .$$

The *semantics* of a DL allowing reification is also a bit different from the usual DL semantics.

Traditional DLs separate the terminological (TBox) and assertional (ABox) levels completely and define a polymorphic interpretation function $\cdot^\mathcal{I}$ which interprets concepts as sets of elements of some interpretation domain $\Delta$ and individuals as elements of $\Delta$:

$$\mathcal{I} : Concepts \to 2^\Delta \qquad \mathcal{I} : Individuals \to \Delta.$$

As long as individuals and concepts are distinct, this works fine. But as soon as we allow concept reification, a given object name $X$ can play both the role of an individual *and* of a concept name and we cannot use the above polymorphic interpretation function any more (because we wouldn't know how to define $X^\mathcal{I}$: as an element of $\Delta$, or as a subset of $\Delta$ ?).

Therefore, we are forced to introduce two different interpretation functions:

(1) one that interprets object names as elements of the interpretation domain $\Delta$ (i.e. regards them as individuals) $\nu : Names \to \Delta$ (the "name function") and

(2) one that associates an *extension* with concept and role terms (the "extension function"):

$$\varepsilon : Concept\_Terms \to 2^\Delta \qquad \varepsilon : Role\_Terms \to 2^{\Delta \times \Delta}.$$

The "name function" maps object names to elements of the interpretation domain $\Delta$. Such elements $x \in \Delta$ can be regarded either as individuals (if we use their names) or as concept names (if we use their extensions).

In order to retrieve the extension of an element $x \in \Delta$ regarded as a concept name, we need an additional function, the *"value function"* $\mathcal{V} : \Delta \to 2^\Delta$.

$\mathcal{V}$ associates with each $x \in \Delta$ the extension $\mathcal{V}(x) \subset \Delta$ of the concept name denoted by $x$, and is therefore uniquely determined by the extension of the $\in$-role:

$$\mathcal{V}(x) = \{y \in \Delta \mid (y, x) \in \varepsilon(\in)\} \qquad (1)$$

(in fact, $\mathcal{V}$ is a functional representation of the extension of the role $\ni$). $\varepsilon(\in)$ must verify the following (first-order) constraint:

$$\forall x \in \Delta. \exists y \in \Delta. [(x, y) \in \varepsilon(\in) \wedge \forall z. ((z, y) \in \varepsilon(\in) \to z = x)]$$

(i.e. $\forall x \in \Delta. \exists y \in \Delta. \mathcal{V}(y) = \{x\}$). This constraint says that each singleton $\{x\}$ must have an *intension* $y$ in $\Delta$ (this is needed, for example, to prove $C \to \exists \in .\forall \ni .C$). We also interpret the $\ni$-role as the inverse of the $\in$-role: $\varepsilon(\ni) = \varepsilon(\in)^{-1}$.

Using the "value function" and the "name function", we can now construct the extensions of concept names as

$$\varepsilon(X) = \mathcal{V}(X^\nu) \qquad \varepsilon(\bot) = \emptyset. \qquad (2)$$

The extension of concept terms is defined as usual depending on the particular DL concept constructors:

$$\begin{aligned}
\varepsilon(C \sqcap D) &= \varepsilon(C) \cap \varepsilon(D) \\
\varepsilon(C \sqcup D) &= \varepsilon(C) \cup \varepsilon(D) \\
\varepsilon(\neg C) &= \Delta \setminus \varepsilon(C) \\
\varepsilon(\forall R.C) &= \{x \in \Delta \mid \forall y \in \Delta.(x,y) \in \varepsilon(R) \to y \in \varepsilon(C)\} \\
\varepsilon(\exists R.C) &= \{x \in \Delta \mid \exists y \in \Delta.(x,y) \in \varepsilon(R) \wedge y \in \varepsilon(C)\} \\
\varepsilon(\{X\}) &= \{X^\nu\}.
\end{aligned} \qquad (3)$$

An interpretation $\mathcal{I}$ is uniquely determined by the "name function" $\nu : Names \to \Delta$ and the "extension function" restricted to role names (including $\in$) $\varepsilon : Role\_Names \cup \{\in\} \to 2^{\Delta \times \Delta}$. It can be extended to a full interpretation as follows:

- first, we extend $\varepsilon$ from role names to role terms
- then we use $\varepsilon(\in)$ to determine the "value function" $\mathcal{V} : \Delta \to 2^\Delta$ according to (1)
- this in turn helps us define the extension of concept names according to (2)
- finally, $\varepsilon$ extends naturally to concept terms as in (3).

Having two different interpretation functions $\nu$ and $\varepsilon$ applicable to a given object $X$ allows us to talk about the interpretation $X^\nu$ of $X$ as an individual *and* as a concept $\varepsilon(X)$ *at the same time*! This wasn't possible in the old setting, where we had a single polymorphic interpretation function $\cdot^\mathcal{I}$.

## 3 Reasoning in reified $\mathcal{ALCO}_\in$

Like in traditional description logics, the reasoning services in reified $\mathcal{ALCO}_\in$ are reducible to the knowledge base (KB) consistency test [Buchheit et al., 1993]. Therefore, we will concentrate in the following on checking consistency in $\mathcal{ALCO}_\in$ knowledge bases. The algorithm is a non-trivial extension of the algorithm for $\mathcal{ALC}$ and is based on a tableaux-like calculus operating on constraint systems.

Starting from an initial constraint system representing the KB, the calculus tries to construct a model of the knowledge base by applying a series of propagation rules. In doing so, it may discover obvious contradictions (clashes) and report the inconsistency of the original KB, or it may come up with a complete clash-free model, thus proving the satisfiability of the knowledge base.

The initial knowledge base to be tested for consistency is represented as a set of constraints of the form:

$$X:C, \quad (X,Y):R, \quad def(CN,C)$$

where $X$, $Y$ and $CN$ are names, $C$ is a concept term and $R$ a role term. We also assume that all the concepts and roles occurring in constraints have been previously brought to the *negation normal form*.

The KB consistency checking algorithm applies a series of propagation rules to a given constraint set $S$, until either an obvious contradiction (or clash) is generated (thereby proving the consistency of $S$), or no propagation rules are applicable any more (case in which the constraint system is called *complete* and can be used to construct an interpretation of $S$).

The propagation rules for $\mathcal{ALCO}_\in$, presented in Figure 2, can take the following two forms:

$$\alpha \rightarrow \beta \ if \ \gamma \qquad \alpha \Rightarrow \beta \ if \ \gamma.$$

Both forms fire only if the condition $\gamma$ holds and if the current constraint system contains constraints matching $\alpha$. After execution, the first deletes the constraints matching $\alpha$ from the constraint system, while the second keeps them. Both forms add the constraints from $\beta$ to the constraint system after firing.

The predicate $individual(X)$ succeeds on constant individuals, variables or singleton constructs $\{X\}$, while $role\_name(R)$ succeeds only on role names (excluding $\in$ and $\ni$).

The rule $(\neg\{\bot\})$ can be explained as follows: $C : \neg\{\bot\}$ holds iff $C \neq \bot$, i.e. $C$ is not the "empty" concept. $C$ is therefore consistent and admits an instance $X : C$. The special case $\bot : \neg\{\bot\}$ is avoided by this rule since it is dealt with by the $(clash_{\{\}})$ rule.

Note that the $(\forall_\in)$ rule asserts $Y$ to be an instance (of $C$) only if it is an individual (since we allow reification of concept names only).

Since $X$ is an instance of the singleton concept $\{X\}$, rules $(\forall_{\in\{\}})$ and $(\forall_{\ni\{\}})$ make sure that this is taken into consideration during constraint propagation.

| | |
|---|---|
| $(clash_\neg)$ | $X : \neg C, X : C \rightarrow fail$ |
| $(clash_{\{\}})$ | $X : \neg\{X\} \rightarrow fail$ |
| $(\bot)$ | $X : \bot \rightarrow fail$ |
| $(\sqcap)$ | $X : C \sqcap D \rightarrow X : C, X : D$ |
| $(\sqcup)$ | $X : C \sqcup D \rightarrow X : C \mid X : D$ |
| $(\exists)$ | $X : \exists R.C \Rightarrow (X,Y):R, Y:C \ (new \ variable \ Y)$ |
| $(\forall)$ | $X : \forall R.C, (X,Y):R \Rightarrow Y:C \ if \ role\_name(R)$ |
| $(\forall_\in)$ | $X : \forall \in .C, X:Y \Rightarrow Y:C \ if \ individual(Y)$ |
| $(\forall_{\in\{\}})$ | $X : \forall \in .C \Rightarrow \{X\}:C$ |
| $(\forall_\ni)$ | $X : \forall \ni .C, Y:X \Rightarrow Y:C$ |
| $(\forall_{\ni\{\}})$ | $\{X\} : \forall \ni .C \Rightarrow X:C$ |
| $(\{\})$ | $X : \{Y\} \rightarrow X = Y$ |
| $(\neg\{\bot\})$ | $C : \neg\{\bot\} \rightarrow X : C \ if \ C \neq \bot \ (new \ variable \ X)$ |
| $(\in)$ | $(X,C):\in \rightarrow X:C$ |
| $(\ni)$ | $(C,X):\ni \rightarrow X:C$ |
| $(def)$ | $def(CN,C), X:CN \Rightarrow X:C$ |

Figure 2: The propagation rules for reified $\mathcal{ALCO}_\in$

Finally, rule $(def)$ deals with acyclic[6] definitions $def(CN,C)$ of concept names $CN$ ($C$ being a concept term). Such definitions are interpreted semantically as $\varepsilon(CN) \subset \varepsilon(C)$. We shall not address the issue of more complex definitions (like general concept inclusions or equations) in this setting, since $\mathcal{ALCO}_\in$ is *"unstable"* due to the presence of the $\ni$ role (the inverse of $\in$). Instability amounts to the possibility that after having expanded all the constraints for some individual $X$, at some point in the future new constraints involving $X$ get discovered. For example, if we apply the propagation rules to $x : (c \sqcap \forall \in .\forall \ni .d)$ and expand all the constraints involving $x$, we obtain the constraints (1)-(3) below. But subsequent applications of propagation rules eventually discover a new constraint involving $x$ (namely (5)), thus proving the instability of $\mathcal{ALCO}_\in$.

(1) $x : (c \sqcap \forall \in .\forall \ni .d)$
(2) $x : c$      $(\sqcap) : (1)$
(3) $x : \forall \in .\forall \ni .d$      $(\sqcap) : (1)$
(4) $c : \forall \ni .d$      $(\forall_\in) : (2),(3)$
(5) $x : d$      $(\forall_\ni) : (2),(4)$

The same kind of problem occurs if we allow role inverses and general inclusions, for which the results in [Buchheit et al., 1993] are no longer applicable (because the stability lemma fails in the presence of role inverses). The problems posed by role inverses are deep and will not be tackled in the present paper since they are orthogonal to the issue of interest here (namely reification).

Note that the ABox assertion $\top : \forall \ni .C$ is equivalent

---

[6] Cycles "through" the $\in$ (or $\ni$) role are anyhow problematic from a semantical point of view. For example, since $\ni$ should be well-founded ($\neg repeat(\ni)$ in *repeat-PDL* notation), cycles involving $\ni$ should be interpreted probably w.r.t. least fixpoint semantics.

Also, such cycles don't seem to occur in practical applications anyway.

to the TBox axiom stating the validity of $C$ and can therefore be used to express general concept inclusions or equations. In order to avoid the above-mentioned problems with general inclusions, we will not allow $\top$ to be used as a concept name.[7]

The following sequence of constraints illustrates the consistency checking algorithm applied to the KB consisting of constraints (1) and (2) below.

| | | |
|---|---|---|
| (1) | $X : \forall R.C \sqcap \forall \in . \exists \ni .D$ | |
| (2) | $def(D, \exists R.\neg C)$ | |
| (3) | $X : \forall R.C$ | $(\sqcap):(1)$ |
| (4) | $X : \forall \in . \exists \ni .D$ | $(\sqcap):(1)$ |
| (5) | $\{X\} : \exists \ni .D$ | $(\forall_\in{\{\}}):(4)$ |
| (6) | $(\{X\}, Y) : \ni$ | $(\exists):(5)$ |
| (7) | $Y : D$ | $(\exists):(5)$ |
| (8) | $Y : \{X\}$ | $(\ni):(6)$ |
| (9) | $Y = X$ | $(\{\}):(8)$ |
| (10) | $X : \exists R.\neg C$ | $(def):(2),(7),(9)$ |
| (11) | $(X, Z) : R$ | $(\exists):(10)$ |
| (12) | $Z : \neg C$ | $(\exists):(10)$ |
| (13) | $Z : C$ | $(\forall):(11),(3)$ |
| (14) | fail | $(clash_\neg):(12),(13)$ |

### 3.1 Soundness and completeness

The *termination* and *soundness* of the algorithm are easy to prove. Its *completeness* is established by constructing a canonical interpretation $\mathcal{I}_S$ for each clash-free and complete constraint system $S$. Note that the propagation rules deal with so-called *"extended"* constraints, i.e. constraints of the form $X : Y$ and $(X, Y) : R$ where $X$ and $Y$ can be not just names, but also "extended" individuals represented by arbitrarily nested singletons like $\{\ldots\{Z\}\ldots\}$ (if we disallow extended individuals, we loose the completeness of the algorithm).

We can extend the "name function" $\nu$ to the set $Names^{\{\}}$ of "extended" individuals and thus talk about the name (intension) $\{X\}^\nu$ of an "extended" individual $\{X\}$. The extended name function must satisfy the constraint $\mathcal{V}(\{X\}^\nu) = \{X^\nu\}$.

The canonical interpretation $\mathcal{I}_S$ is defined by

$$(X^\nu, Y^\nu) \in \varepsilon(R) \quad \text{iff} \quad (X, Y) : R \text{ is in } S \quad (4)$$
$$(X^\nu, Y^\nu) \in \varepsilon(\in) \quad \text{iff} \quad (X : Y \text{ is in } S) \vee Y = \{X\} \quad (5)$$

for $X, Y \in Names^{\{\}}$ and $R \neq \in, \ni$. The extension of concept names can thus be obtained as:

$$\begin{aligned}\varepsilon(CN) &= \mathcal{V}(CN^\nu) = \{X^\nu \in \Delta \mid (X^\nu, CN^\nu) \in \varepsilon(\in)\} \\ &= \{X^\nu \in \Delta \mid (X : CN \text{ is in } S) \vee CN = \{X\}\}\end{aligned}$$

$\varepsilon$ extends naturally to concept terms according to (3).

Testing the consistency of $\mathcal{ALCO}_\in$ knowledge bases w.r.t. the proposed semantics is therefore decidable.

## 4 Conclusions

Extending description logics with concept reification is essential for developing domain-independent and reusable models. Nevertheless, it has not been extensively studied, mainly due to the semantical problems posed by its inherent higher-order features and because of the complications in the reasoning algorithms.

The semantical problems related to the higher-order features implicit in reification are solved by defining a first-order semantics which ensures the decidability of the main inference services. We have also described sound and complete inference algorithms for the reified terminological language $\mathcal{ALCO}_\in$ (but the algorithms can be extended to more expressive languages).

In our view, concept reification represents an essential element for bridging the gap between description logics and (deductive) object-oriented databases.

It also makes description logics expressive enough to be used for developing *generic* problem solving models [Badea and Țilivea, 1996] and even libraries of such models.

## Acknowledgments

Thanks are due to Doina Țilivea and Alon Levy for interesting discussions as well as to anonymous reviewers for their helpful comments and especially for pointing out the paper [Franconi, 1993] (which also introduces an explicit membership role, but does not provide a complete reasoning algorithm).

## References

[Badea, 1996] Badea L. *$E_x$Claim: a hybrid language for knowledge representation and reasoning using description logics.* Proc. ECAI'96 Workshop on Validation, Verification and Refinement of KBS, Budapest, 1996.

[Badea and Țilivea, 1996] Badea L., Țilivea D. *$E_x$Claim: a language for operationalizing CommonKADS expertise models using description logics.* PEKADS Report 4.5.1.

[Beeri, 1990] Beeri C. *A formal approach to object-oriented databases.* Data & Knowledge Eng. 5 (1990) 353-382.

[Brachman and Schmolze, 1985] Brachman R.J., Schmolze J.G. *An Overview of the KL-ONE Knowledge Representation System.* Cognitive Science 9(2), 171-216, 1985.

[Brachman et al., 1991] Brachman R.J., e.a. *Living with CLASSIC: When and How to Use a KL-ONE like Language,* in Sowa J.F. (ed) *Principles of Semantic Networks,* Morgan Kaufmann 1991.

[Buchheit et al., 1993] Buchheit M., Donini F.M., Schaerf A. *Decidable Reasoning in Terminological Knowledge Representation Systems.* J. of AI Research 1 (1993), 109-138.

[De Giacomo and Lenzerini, 1995] De Giacomo G., Lenzerini M. *What's in an aggregate: foundations for description logics with tuples and sets.* Proc. IJCAI'95, 801-807.

[Franconi, 1993] Franconi E. *A treatment of plurals and plural quantifications based on a theory of collections.* Minds and Machines, 1993 special issue on KR for NL, 453-474.

[Kobsa, 1991] Kobsa A. *Reification in SB-ONE.* In Proc. Int. Workshop on Terminological Logics, 72-74, DFKI D-91-13.

[Kifer et al., 1995] Kifer M., Lausen G., Wu J. *Logical foundations of object-oriented and frame-based languages.* Journal of the ACM, May 1995.

[Schmidt-Schauß and Smolka, 1991] Schmidt-Schauß M., Smolka G. *Attributive concept descriptions with complements.* Artificial Intelligence 48 (1), 1-26, 1991.

---

[7]We also disallow iff-definitions since these would enable us to define a concept name $D = E \sqcup \neg E$ equivalent to $\top$, and the above problems with expressing valid concepts $C$ using $D : \forall \ni .C$ reappear.

# AUTOMATED REASONING

Automated Reasoning 6: Nonmonotonism

# Circumscribing Inconsistency

**Philippe Besnard**
IRISA
Campus de Beaulieu
F-35042 Rennes Cedex

**Torsten H. Schaub**[*]
Institut für Informatik
Universität Potsdam, Postfach 60 15 53
D-14415 Potsdam

## Abstract

We present a new logical approach to reasoning from inconsistent information. The idea is to restore modelhood of inconsistent formulas by providing a third truth-value tolerating inconsistency. The novelty of our approach stems first from the restriction of entailment to three-valued models as similar as possible to two-valued models and second from an implication connective providing a notion of restricted monotonicity. After developing the semantics, we present a corresponding proof system that relies on a circumscription schema furnishing the syntactic counterpart of model minimization.

## 1 Introduction

The capability of reasoning in the presence of inconsistencies constitutes a major challenge for any intelligent system. This is because in practical settings it is common to have contradictory information. In fact, despite its many appealing features for knowledge representation and reasoning, classical logic falls in the same trap: A single contradiction may wreck an entire reasoning system, since it may allow for deriving any proposition. This comportment is due to the fact that a contradiction denies any classical two-valued model, since a proposition must be either true or false. We thus aim at providing a formal reasoning system satisfying the *principle of paraconsistency*: $\{\alpha, \neg\alpha\} \not\vdash \beta$ for some $\alpha, \beta$. In other words, given a contradictory set of premises, this should not necessarily lead to concluding all formulas. We address this problem from a semantic point of view. We want to counterbalance the effect of contradictions by providing a third truth-value that accounts for contradictory propositions. As already put forward by [Priest, 1979], this provides us with inconsistency-tolerating three-valued models. However, this approach turns out to be rather weak in that it invalidates certain classical inferences, even if there is no contradiction. Intuitively, this is because there are too many three-valued models, in particular those assigning the inconsistency-tolerating truth-value to propositions that are unaffected by contradictions.

Our idea is to focus on those three-valued models that are as similar as possible to *two-valued models* of the knowledge base. In this way, we somehow hand over the model selection process to the knowledge base by preferring those models that assign *true* to as many items of the knowledge base as possible. As a result, our approach reduces nicely to classical reasoning in the absence of inconsistency. (For the reader familiar with the work of [Priest, 1989] we note that ours is different from preferring three-valued models having the highest number of classical truth-values, which amounts to approximating two-valued interpretations while somehow discarding the underlying knowledge base.) The syntactic counterpart of our preferential reasoning process is furnished by an axiom schema, similar to the ones found in circumscription [McCarthy, 1980]. Another salient feature of our approach is driven by the desire to preserve existing proofs even though they may lead to contradictory conclusions. This is because proofs provide evidence for derived conclusions. We accomplish this by introducing an implication connective that reduces (inside the knowledge base) to classical implication in the absence of inconsistency, while its resulting inferences are conserved under inconsistency.

The paper is organized as follows. Section 2 lays the semantic foundations of our approach; it presents a novel three-valued logic comprising two special connectives: The aforementioned implication and a truth-value-indicating connective (used for later axiomatization of the model selection process). To a turn, we define our *paraconsistent inference relation* by means of a preference relation over the set of models obtained in this logic. Section 3 presents the syntactic counterpart by proposing a corresponding formal proof system. We present an axiomatization of the underlying three-valued logic and we furnish a circumscription axiom providing syntactic means for reasoning from preferred inconsistency-tolerating models.

## 2 Model theory

This section presents our semantic approach to reasoning from possibly inconsistent knowledge bases expressed in a propositional language. We use $\vdash$ for classical entailment wrt two-valued interpretations and $Cn_\vdash$ for classical deductive closure. For dealing with inconsistencies we rely on an extended

---
[*]Previously at LERIA, Université d'Angers, France.

propositional language:

**Definition 2.1** *Given a set $\mathcal{P}$ of propositional symbols let $\mathcal{L}$ be the set of all formulas generated from $\mathcal{P}$ using connectives $\top, \bot, \neg, \vee, \wedge, \rightarrow, \leftrightarrow, \flat, \leq$.*

The last two connectives serve as truth-value indicators. That is, $\flat \alpha$ means that $\alpha$ is true and $\alpha \leq \beta$ signifies that the truth value of $\alpha$ is less than that of $\beta$. This order is proper to this connective and is no intrinsic feature of the rest of the logic. We define $\top$ as $\flat\alpha \rightarrow \alpha$ and $\bot$ by $\neg\top$. Also, we define $\alpha \leftrightarrow \beta$ as $(\alpha \rightarrow \beta) \wedge (\beta \rightarrow \alpha)$. In fact, $\leq$ is also a defined connective (using $\flat$), whose discussion is deferred to Section 3.

**Definition 2.2** *An interpretation is a function*

$$v : \mathcal{P} \rightarrow \{t, f, o\} \quad \text{extending to} \quad \bar{v} : \mathcal{L} \rightarrow \{t, f, o\}$$

*according to the truth tables below.*

| $\neg$ | |
|---|---|
| $t$ | $f$ |
| $f$ | $t$ |
| $o$ | $o$ |

| $\wedge$ | $t$ | $f$ | $o$ |
|---|---|---|---|
| $t$ | $t$ | $f$ | $o$ |
| $f$ | $f$ | $f$ | $f$ |
| $o$ | $o$ | $f$ | $o$ |

| $\vee$ | $t$ | $f$ | $o$ |
|---|---|---|---|
| $t$ | $t$ | $t$ | $t$ |
| $f$ | $t$ | $f$ | $o$ |
| $o$ | $t$ | $o$ | $o$ |

| $\flat$ | |
|---|---|
| $t$ | $t$ |
| $f$ | $f$ |
| $o$ | $f$ |

| $\rightarrow$ | $t$ | $f$ | $o$ |
|---|---|---|---|
| $t$ | $t$ | $f$ | $o$ |
| $f$ | $t$ | $t$ | $t$ |
| $o$ | $t$ | $f$ | $o$ |

A model of a formula $\alpha$ is an interpretation that assigns either $t$ or $o$ to $\alpha$.

Modelhood extends to sets of formulas in the standard way. Observe that $\wedge$ and $\vee$ are de Morgan duals. Also, note that the truth-value of $\alpha \rightarrow \beta$ differs from that of $\neg \alpha \vee \beta$ only in the case of $v = \{\alpha : o, \beta : f\}$ resulting in $v(\alpha \rightarrow \beta) = f$ and $v(\neg \alpha \vee \beta) = o$. This difference is prompted by the fact that $t$ and $o$ indicate modelhood, which motivates the assignment of the same truth-values to $\alpha \rightarrow \beta$ no matter whether we have $\alpha : t$ or $\alpha : o$. This has actually to do with the difference between *modus ponens* (MP) and *disjunctive syllogism* (DS):

$$\frac{(\alpha \rightarrow \beta) \quad \alpha}{\beta} \text{ (MP)} \qquad \frac{(\alpha \vee \beta) \quad \neg\alpha}{\beta} \text{ (DS)}$$

The latter yields $B$ from $A \wedge \neg A \wedge \neg B$ because $A \vee B$ follows from $A$. The overall inference seems wrong because in the presence of $A \wedge \neg A$, $A \vee B$ is satisfied (by $A : o$) with no need for $B$ to be $t$. This is why we center our approach upon modus ponens.

We then obtain the following consequence relation:

**Definition 2.3** *Let $\Gamma$ be a set of formulas and $\gamma$ a formula. We define $\Gamma \Vdash \gamma$ iff each model of $\Gamma$ is a model of $\gamma$.*

The reader is warned that replacement of equivalents fails: Let $\gamma[\phi_1, \ldots, \phi_k]$ be the formula obtained from $\gamma[\psi_1, \ldots, \psi_k]$ by replacing all occurrences of $\psi_1, \ldots, \psi_k$ by $\phi_1, \ldots, \phi_k$. Then,

$$\Vdash \alpha \leftrightarrow \beta \not\Rightarrow \Vdash \gamma[\alpha] \leftrightarrow \gamma[\beta]$$

Letting $\alpha$ be $A \rightarrow \neg\neg A$, $\beta$ be $B \rightarrow \neg\neg B$, and $\gamma$ be $\neg\psi$ shows the failure of replacement of equivalents: $\Vdash (A \rightarrow \neg\neg A) \leftrightarrow (B \rightarrow \neg\neg B)$ but $\not\Vdash \neg(A \rightarrow \neg\neg A) \leftrightarrow \neg(B \rightarrow \neg\neg B)$.

We now turn to the key definition of our approach:

**Definition 2.4** *Let $v$ and $v'$ be two interpretations and $\Gamma$ a set of formulas. We define*

$$v \prec_\Gamma v' \text{ iff } \{\gamma \in \Gamma \mid v(\gamma) = o\} \subsetneq \{\gamma \in \Gamma \mid v'(\gamma) = o\}.$$

Observe that $\prec_\Gamma$ is a strict partial order on interpretations. Hence, we can speak of minimal models for a set of formulas $\Gamma$. This leads us to the following paraconsistent inference relation:

**Definition 2.5** *Let $\Gamma$ be a set of formulas and $\gamma$ a formula. We define $\Gamma \Vvdash \gamma$ iff each $\prec_\Gamma$-minimal model of $\Gamma$ is a model of $\gamma$.*

Definition 2.4 and 2.5 show that we focus on models of $\Gamma$ that assign $t$ (instead of $o$) to a maximal subset of $\Gamma$. Since $o$ accounts for inconsistency all this amounts to minimizing inconsistency. In fact, both aforementioned inference relations are paraconsistent: $\{A, \neg A\} \not\Vdash B$ and $\{A, \neg A\} \not\Vvdash B$.

Since we aim at modeling reasoning from knowledge bases expressed in a propositional language, we impose the following restriction: As modus ponens is a fairly uncontroversial reasoning mode, we take it as a basis for our approach. In particular, premises are required to be in conditional form prone to application of modus ponens:

**Definition 2.6** *Let $\mathcal{P}$ be a set of propositional symbols and $\mathcal{L}_\rightarrow$ the set of all expressions of the form*

$$L_1 \wedge \ldots \wedge L_m \rightarrow L_{m+1} \vee \ldots \vee L_n$$

*where $L_i \in \{\alpha, \neg\alpha \mid \alpha \in \mathcal{P}\}$ for $i = 1..n$ and $0 \leq m < n$.*

We refer to expressions in $\mathcal{L}_\rightarrow$ as *clauses*. For $m = 0$, such clauses reduce to $L_1 \vee \ldots \vee L_n$. As a whole, $\mathcal{L}_\rightarrow$ is generated from $\mathcal{P}$ using connectives $\neg, \vee, \wedge, \rightarrow$.

Consider the set of formulas

$$\Gamma = \{A \rightarrow B, A \rightarrow \neg B\}. \tag{1}$$

We obtain $\Gamma \Vvdash \neg A$. In fact, $\neg A$ is concluded for the reason that, if $A$ were true then there would be a contradiction about $B$. So, when it really is the case that there *is* a contradiction about $B$, the reason for $\neg A$ to be concluded no longer applies. That is, we have $\Gamma \cup \{A\} \not\Vvdash \neg A$ and $\Gamma \cup \{A\} \Vvdash A \wedge B \wedge \neg B$. Observe that this example violates unrestricted monotonicity (the relative theories must be both consistent or both inconsistent; cf. Theorem 2.2). This comportment can be verified in Table 1. An entry like $o/2$ in column $\Gamma$ means that interpreta-

| $A$ | $B$ | $\Gamma$ | $\Gamma\cup\{A\}$ | $\Gamma'$ | $\Gamma'\cup\{A\}$ | $\Gamma''$ | $\Gamma''\cup\{A\}$ |
|---|---|---|---|---|---|---|---|
| $t$ | $t$ | $f$ | $f$ | $f$ | $f$ | $f$ | $f$ |
| $t$ | $f$ | $f$ | $f$ | $f$ | $f$ | $f$ | $f$ |
| $t$ | $o$ | $o/2$ | $o/2$ | $o/2$ | $o/2$ | $f$ | $f$ |
| $f$ | $t$ | $t/0$ | $f$ | $t/0$ | $f$ | $t/0$ | $f$ |
| $f$ | $f$ | $t/0$ | $f$ | $t/0$ | $f$ | $t/0$ | $f$ |
| $f$ | $o$ | $t/0$ | $f$ | $t/0$ | $f$ | $t/0$ | $f$ |
| $o$ | $t$ | $f$ | $f$ | $o/1$ | $o/2$ | $f$ | $f$ |
| $o$ | $f$ | $f$ | $f$ | $o/1$ | $o/2$ | $f$ | $f$ |
| $o$ | $o$ | $o/2$ | $o/3$ | $o/2$ | $o/3$ | $o/4$ | $o/5$ |

Table 1: Truth tables for $\Gamma$, $\Gamma'$, and $\Gamma''$.

tion $v$, given in the first two rows, assigns $o$ to (the conjunction of) $\Gamma$, while $|\{\gamma \in \Gamma \mid v(\gamma) = o\}| = 2$. Such a number is however just an indication and should *not* be confused with the actual ordering relation on models which is based on set inclusion! A preferred model is indicated by boldface typesetting.

For a complement, take a look at clause set

$$\Gamma' = \{\neg A \vee B, \neg A \vee \neg B\}.$$

We have $\Gamma' \cup \{A\} \not\Vdash \neg B \wedge B$. We only have $\Gamma' \cup \{A\} \Vdash B \vee \neg B$. This illustrates the difference between an implication, like $A \to B$ and a disjunction, like $\neg A \vee B$. Unlike the latter, connective $\to$ allows us to construct a proof for $B \wedge \neg B$ from $\Gamma \cup \{A\}$. Compare this with the case of all contrapositives:

$$\Gamma'' = \{A \to B, \neg B \to \neg A, A \to \neg B, B \to \neg A\}.$$

This yields $\Gamma'' \cup \{A\} \Vdash A \wedge \neg A \wedge \neg B \wedge B$.

The previous examples have illustrated that whenever there are two-valued models, they are the only *relevant* minimal models. That is, in case of a 3-valued model $v$ assigning $t$, we find also all 2-valued models obtained by substituting $o$ in $v$ by $t$ and $f$, respectively. Hence, such 3-valued models are irrelevant. See columns $\Gamma, \Gamma', \Gamma''$ in Table 1. In fact, we have the following result showing that our mechanism amounts to classical (two-valued) logic, whenever we deal with a classically consistent theory.

**Theorem 2.1** *Let $\Gamma$ be a classically consistent set of clauses and $\gamma$ a formula whose connectives are among $\neg, \wedge, \vee$. Then,*

$$\Gamma \vdash \gamma \qquad \text{iff} \qquad \Gamma \Vdash \gamma.$$

This result does not extend to the underlying inference relation $\Vdash$. A counterexample is simply $\Gamma$ as in (1) and $\gamma$ being $\neg A$. Also, this theorem does not extend to conclusions containing $\to$, eg. $\not\Vdash (\neg A \vee B) \to (A \to B)$ although $(\neg A \vee B) \to (A \to B)$ is a classical tautology. Theorem 2.1 is neither expected to carry over to the case where $\Gamma$ is inconsistent.

A salient property of our approach is that it is *monotonic* on inconsistent premises:

**Theorem 2.2** *For sets of clauses $\Gamma$ and $\Delta$, we have*

$$\Gamma \Vdash \gamma \implies \Delta, \Gamma \Vdash \gamma$$

*whenever $\Gamma \Vdash \alpha \wedge \neg \alpha$ and $\forall \Gamma' \subsetneq \Gamma$. $\Gamma' \not\Vdash \alpha \wedge \neg \alpha$.*

We now need a few definitions on restricted alphabets:

$$\mathcal{P}_{\natural \Gamma} = \{P \in \mathcal{P} \mid \Gamma \Vdash \natural P \vee \natural \neg P\}$$

For an alphabet $\mathcal{P}_x$, let $\mathcal{L}_x$ denote the language generated from $\mathcal{P}_x$ using $\neg, \vee, \wedge, \to, \leftrightarrow$. Then, we have the following result showing that truthful parts of the knowledge base are closed under classical logic:

**Theorem 2.3** *For all sets of clauses $\Gamma$ and $\Delta$ such that $\Delta = \{\alpha \mid \Gamma \Vdash \natural \alpha\}$, we have*

$$\Delta \cap \mathcal{L}_{\natural \Gamma} = Cn_\vdash(\Delta) \cap \mathcal{L}_{\natural \Gamma}.$$

Moreover, we can show that truthful parts are never polluted by contradictions:

**Theorem 2.4** *Let $\Gamma = \Gamma_1 \cup \Gamma_2$ be a clause set such that $\mathcal{P}_1 \cap \mathcal{P}_2 = \emptyset$ where $\mathcal{P}_i$ is the set of propositional symbols occurring in $\Gamma_i$. We have for each $\alpha \in \mathcal{L}_1$,*

$$\Gamma \Vdash \alpha \qquad \text{iff} \qquad \Gamma_1 \vdash \alpha$$

*whenever $\Gamma \Vdash \natural \gamma$ for each $\gamma \in \Gamma_1$.*

As illustrated below, the last theorem extends in some cases to non-disjoint parts, as witnessed by $\Gamma_1, \Gamma_1^\to, \Gamma_1^\leftarrow$ below.

For further illustration, consider first the set of clauses

$$\Gamma_0 = \{\neg A, B, (\neg B \vee C)\}$$

Indeed $\Gamma_0$ has a single two-valued model $\{A : f, B : t, C : t\}$ (apart from 9 three-valued ones assigning $o$ to $\Gamma_0$). The former is clearly the only minimal model of $\Gamma_0$. We thus have

$$\Gamma_0 \vdash \neg A \wedge B \wedge C \quad \text{and} \quad \Gamma_0 \Vdash \neg A \wedge B \wedge C.$$

Adding $A$ to $\Gamma_0$ yields inconsistent theory $\Gamma_0' = \{A, \neg A, B, (\neg B \vee C)\}$ having only three-valued models left. In fact, all former models of $\Gamma_0$ with $A : f$ do now falsify $\Gamma_0'$. All remaining models of $\Gamma_0$ assign thus $o$ to $\Gamma_0'$ and $A$, the actual heart of the contradiction. Among the resulting models, we have a single minimal model, $\{A : o, B : t, C : t\}$, giving $\Gamma_0' \Vdash A \wedge \neg A \wedge B \wedge C$ by "applying" disjunctive syllogism to the consistent part of $\Gamma_0'$.

Next, consider the set of clauses

$$\Gamma_1 = \{A, \neg A, (\neg A \vee B)\}$$

This theory induces the truth-values given in Table 2. Among

| $A$ | $B$ | $\Gamma_1$ | $\Gamma_1^\to$ | $\Gamma_1^\leftarrow$ | $\Gamma_1'$ | $\Gamma_1'^\to$ | $\Gamma_1'^\leftarrow$ | $\Gamma_1'^\rightleftarrows$ | $\Gamma_1'^\leftrightarrows$ |
|---|---|---|---|---|---|---|---|---|---|
| $t$ | $t$ | $f$ | $f$ | $f$ | $f$ | $f$ | $f$ | $f$ | $f$ |
| $t$ | $f$ | $f$ | $f$ | $f$ | $f$ | $f$ | $f$ | $f$ | $f$ |
| $t$ | $o$ | $f$ | $f$ | $f$ | $f$ | $f$ | $f$ | $f$ | $f$ |
| $f$ | $t$ | $f$ | $f$ | $f$ | $f$ | $f$ | $f$ | $f$ | $f$ |
| $f$ | $f$ | $f$ | $f$ | $f$ | $f$ | $f$ | $f$ | $f$ | $f$ |
| $f$ | $o$ | $f$ | $f$ | $f$ | $f$ | $f$ | $f$ | $f$ | $f$ |
| $o$ | $t$ | $o/2$ | $o/2$ | $o/2$ | $o/3$ | $f$ | $o/3$ | $o/3$ | $f$ |
| $o$ | $f$ | $o/3$ | $f$ | $o/3$ | $o/3$ | $f$ | $o/3$ | $f$ | $o/3$ |
| $o$ | $o$ | $o/3$ | $o/3$ | $o/3$ | $o/4$ | $o/4$ | $o/4$ | $o/4$ | $o/4$ |

Table 2: Truth tables for $\Gamma_1$ and $\Gamma_1'$.

the three models of $\Gamma_1$, there is only one minimal one: $\{A : o, B : t\}$. As a consequence, we obtain

$$\Gamma_1 \Vdash A \wedge \neg A \wedge B.$$

For those familiar with [Priest, 1989], we note that this approach has $\{A : o, B : f\}$ as a second preferred model, which denies conclusion $B$. See Section 4 for details. The example illustrates further the aforementioned extendibility of Theorem 2.4: Despite the inconsistency of $A$, we derive $B$ from the consistent premises $A$ and $\neg A \vee B$.

Actually, things do not necessarily change by orienting the above disjunctions as implications:

$$\Gamma_1^\to = \{A, \neg A, (A \to B)\} \text{ and } \Gamma_1^\leftarrow = \{A, \neg A, (\neg B \to \neg A)\}$$

$\Gamma_1^{\rightarrow}$ and $\Gamma_1^{\leftarrow}$ have the same minimal model as $\Gamma_1$; thus offering the same conclusions. However, while $\Gamma_1^{\leftarrow}$ has the same models as $\Gamma_1$, interpretation $\{A:o, B:f\}$ falsifies $\Gamma_1^{\rightarrow}$.

Adding clause $A \vee \neg B$ to Theory $\Gamma_1$ yields

$$\Gamma_1' = \{A, \neg A, (\neg A \vee B), (A \vee \neg B)\}$$

$\Gamma_1'$ has two minimal models, both of which were models of $\Gamma_1$, yet only one of them was $\Gamma_1$-preferred. We thus get

$$\Gamma_1' \Vvdash A \wedge \neg A \quad \text{and} \quad \Gamma_1' \not\Vvdash B$$

illustrating that inferences by disjunctive syllogism are not always preserved.

For a complement, consider rule sets

$$\begin{aligned}
\Gamma_1'^{\rightarrow} &= \{A, \neg A, A \rightarrow B, \neg A \rightarrow \neg B\} \\
\Gamma_1'^{\leftarrow} &= \{A, \neg A, \neg B \rightarrow \neg A, B \rightarrow A\} \\
\Gamma_1'^{\rightleftarrows} &= \{A, \neg A, A \rightarrow B, B \rightarrow A\} \\
\Gamma_1'^{\leftrightarrows} &= \{A, \neg A, \neg B \rightarrow \neg A, \neg A \rightarrow \neg B\}
\end{aligned}$$

From these, we obtain after consulting Table 2:

$$\begin{aligned}
\Gamma_1'^{\rightarrow} &\Vvdash A \wedge \neg A \wedge B \wedge \neg B \\
\Gamma_1'^{\leftarrow} &\Vvdash A \wedge \neg A \\
\Gamma_1'^{\rightleftarrows} &\Vvdash A \wedge \neg A \wedge B \\
\Gamma_1'^{\leftrightarrows} &\Vvdash A \wedge \neg A \wedge \neg B
\end{aligned}$$

The derivability of $B$ and $\neg B$ illustrates the role of connective $\rightarrow$ as proof-provider: All proofs obtained from clauses by modus ponens are set in stone. This general property is reflected by the validity of $\Vdash (\alpha \wedge (\alpha \rightarrow \beta)) \rightarrow \beta$.

## 3 Proof theory

This section presents a formal proof system for our approach to circumscribing inconsistency. In analogy to the semantics, we first axiomatize $\Vdash$ and then we account for minimization by providing a syntactic axiom schema, so that the resulting system axiomatizes $\Vvdash$.

The axiomatization of $\Vdash$ consists of modus ponens as inference rule and the following axiom schemas:

$$\alpha \vee \neg \alpha \tag{1}$$
$$\alpha \wedge \beta \rightarrow \alpha, \; \alpha \wedge \beta \rightarrow \beta \tag{2}$$
$$\alpha \rightarrow \alpha \vee \beta, \; \alpha \rightarrow \beta \vee \alpha \tag{3}$$
$$\alpha \rightarrow (\beta \rightarrow (\alpha \wedge \beta)) \tag{4}$$
$$((\alpha \rightarrow \beta) \rightarrow \alpha) \rightarrow \alpha \tag{5}$$
$$(\alpha \rightarrow \gamma) \rightarrow ((\beta \rightarrow \gamma) \rightarrow (\alpha \vee \beta \rightarrow \gamma)) \tag{6}$$
$$\alpha \leftrightarrow \neg\neg\alpha \tag{7}$$
$$\neg(\alpha \vee \beta) \leftrightarrow \neg\alpha \wedge \neg\beta \tag{8}$$
$$\neg(\alpha \wedge \beta) \leftrightarrow \neg\alpha \vee \neg\beta \tag{9}$$
$$\alpha \rightarrow (\beta \rightarrow \alpha) \tag{10}$$
$$(\alpha \rightarrow (\beta \rightarrow \gamma)) \rightarrow ((\alpha \rightarrow \beta) \rightarrow (\alpha \rightarrow \gamma)) \tag{11}$$
$$(\alpha \rightarrow \beta) \rightarrow \neg\alpha \vee \beta \tag{12}$$
$$\alpha \wedge \neg\beta \rightarrow \neg(\alpha \rightarrow \beta) \tag{13}$$
$$\natural\alpha \rightarrow \alpha \tag{14}$$
$$\natural\alpha \rightarrow \natural\natural\alpha \tag{15}$$
$$\natural\neg\alpha \rightarrow \neg\natural\alpha \tag{16}$$
$$\neg\natural\alpha \leftrightarrow \natural\neg\alpha \tag{17}$$
$$\natural(\alpha \wedge \beta) \leftrightarrow \natural\alpha \wedge \natural\beta \tag{18}$$
$$\natural(\alpha \vee \beta) \leftrightarrow \natural\alpha \vee \natural\beta \tag{19}$$
$$\natural(\alpha \rightarrow \beta) \rightarrow (\natural\alpha \rightarrow \natural\beta) \tag{20}$$
$$\neg\alpha \rightarrow \natural(\alpha \rightarrow \beta) \tag{21}$$
$$\natural(\neg(\alpha \rightarrow \beta)) \rightarrow \alpha \wedge \natural\neg\beta \tag{22}$$
$$\natural((\alpha \rightarrow \beta) \rightarrow \alpha) \rightarrow \alpha \tag{23}$$
$$\natural\alpha \leftrightarrow \natural\beta \quad \text{for} \quad \alpha \leftrightarrow \beta \in \{(7), (8), (9)\} \tag{24}$$
$$\natural\alpha \rightarrow \natural\beta \quad \text{for} \quad \alpha \rightarrow \beta \in \{(10), \ldots, (13)\} \tag{25}$$
$$\natural\alpha \quad \text{for} \quad \alpha \in \{(14), \ldots, (20)\} \tag{26}$$

As can be shown, this proof system is sound and complete for $\Vdash$. We write $\gamma \in Cn_{\Vdash}(\Gamma)$ to indicate that $\gamma$ can be derived from $\Gamma$ by the above proof system.

Semantically, the move from $\Vdash$ to $\Vvdash$ amounts to minimizing the set of premises with truth-value $o$. That is, we prefer models that assign truth-value $o$ to a minimal set of premises. We can turn this idea into the syntax by using a connective indicating that a formula has a truth value which is less than the one of another formula. As anticipated in Section 2, such a connective can be defined as follows:

$$\alpha \leq \beta \;=_{\text{def}}\; (\natural\alpha \wedge \natural\beta) \vee (\natural\neg\alpha \wedge \natural\neg\beta) \vee (\neg\natural\beta \wedge \neg\natural\neg\beta)$$

This induces the following truth table corresponding to the poset of truth-values on the right hand side.

| $\leq$ | $t$ | $f$ | $o$ |
|---|---|---|---|
| $t$ | $t$ | $f$ | $t$ |
| $f$ | $f$ | $t$ | $t$ |
| $o$ | $f$ | $f$ | $t$ |

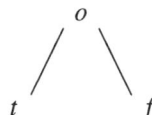

With this connective, we are now ready to express the following *circumscription schema* providing a syntactic account for preferring $\prec_\Gamma$-minimal models. For readability, we identify in the next definition clause set $\{\gamma_1, \ldots, \gamma_n\}$ with $\bigwedge_{i=1}^n \gamma_i$.

**Definition 3.1** *Let $\Gamma = \{\gamma_1, \ldots, \gamma_n\}$ be a finite set of formulas over alphabet $P_1, \ldots, P_k$ (abbreviated $\vec{P}$) so that $\gamma_i = \gamma_i[\vec{P}]$ and $\Gamma = \Gamma[\vec{P}]$. We define the three-valued paraconsistent circumscription schema $Circ_3 P(\Gamma)$ as*

$$\Gamma[\vec{\phi}] \wedge \left(\bigwedge_{i=1}^n \gamma_i[\vec{\phi}] \leq \gamma_i[\vec{P}]\right) \rightarrow \left(\bigwedge_{i=1}^n \gamma_i[\vec{P}] \leq \gamma_i[\vec{\phi}]\right)$$

Importantly, combining $Circ_3 P(\Gamma)$ with the proof system for $\Vdash$ captures the desired paraconsistent inference relation $\Vvdash$:

**Theorem 3.1** *Let $\Gamma$ be a set of clauses. Then, we have*

$$\Gamma \Vvdash \gamma \quad \text{iff} \quad \gamma \in Cn_{\Vdash}(\Gamma \cup \{Circ_3 P(\Gamma)\})$$

For illustration, let us return to our initial example
$$\Gamma = \{A \to B, A \to \neg B\}$$
We consider the following instance of $Circ_3 P(\Gamma)$ where $\phi_A = \bot$ and $\phi_B = B$:

$$\begin{aligned}&\big((\bot \to B) \wedge (\bot \to \neg B)\big)\\ \wedge\ &\big((\bot \to B \le A \to B) \wedge (\bot \to \neg B \le A \to \neg B)\big)\\ \to\ &\big((A \to B \le \bot \to B) \wedge (A \to \neg B \le \bot \to \neg B)\big)\end{aligned}$$

From $\Gamma$, we obtain the right hand side (RHS) of $Circ_3 P(\Gamma)$, that is, $(A \to B \le \bot \to B)$ and $(A \to \neg B \le \bot \to \neg B)$ after establishing the LHS by means of theorem $(\alpha \wedge (\beta \le \top)) \to (\beta \le \alpha)$. By applying transitivity of $\le$ to RHS and $(\bot \to B) \le \top$ and $(\bot \to \neg B) \le \top$, we then get $(A \to B) \le \top$ and $(A \to \neg B) \le \top$. So, we get $(\neg A \le \top) \vee ((\neg B \wedge B) \le \top)$ yielding $\neg A \le \top$, hence $\neg A$. Notably, it is the circumscription schema that reduces the three-valued consequence relation $\Vdash$ to its classical two-valued counterpart $\vdash$ (cf. Theorem 2.1).

For further illustration, consider $\Gamma \cup \{A\}$ along with the instance of $Circ_3 P(\Gamma \cup \{A\})$ obtained by taking $\phi_A = \top$ and $\phi_B = B \wedge \neg B$. We obtain $A \le \top$ and so $A$ using theorem $(\alpha \wedge (\gamma \le \beta)) \to (\gamma \le (\alpha \to \beta))$. Of course, not every $\Vdash$ conclusion necessitates the circumscription schema in order to be derived. For instance, $B$ and $\neg B$ are directly derived by modus ponens from $\Gamma \cup \{A\}$.

## 4 Related work

There are a number of proposals addressing inconsistent information. At first, there is the wide range of paraconsistent logics [Priest et al., 1989]. As opposed to our approach, such logics usually fail to identify with classical logic when the set of premises is consistent. There are also many approaches dealing with classical reasoning from consistent subsets. In a broader sense, this includes also belief revision and truth maintenance systems. A comparative study of the aforementioned approaches in general is given in [Besnard, 1991].

A system, at first sight closely related to ours, is $LP_m$ [Priest, 1989]; it was conceived to overcome the failure of disjunctive syllogism in $LP$ [Priest, 1979]. $LP$ amounts to the 3-valued logic obtained by restricting $\Vdash$ to connectives $\neg, \vee$ and $\wedge$ and defining $\alpha \to \beta$ as $\neg \alpha \vee \beta$. In $LP_m$ modelhood is then limited to models containing a minimal number of *propositional variables* being assigned $o$. As our approach, this allows for drawing "all classical inferences except where inconsistency makes them doubtful anyway" [Priest, 1989]. There are two major differences though: First, the aforementioned restriction of modelhood focuses on models as close as possible to 2-valued interpretations, while the one in our approach aims at models next to 2-valued *models* of the considered formula. The effects of making the formula select its preferred models can be seen by looking at $\Gamma_1$: While $LP_m$ yields two preferred models $\{A:o, B:t\}$ and $\{A:o, B:f\}$

from which one obtains $A \wedge \neg A$, $\Gamma_1$ makes our approach prefer the former over the latter, thus yielding $B$ as additional conclusion. Second, we have introduced implication as a primitive connective rather than a defined one. As a consequence, a modus ponens inference, like deriving $B$ from $A$ and $A \to B$, is preserved no matter what other premises are given; this fails in $LP_m$. Note that we get distinct truth-tables (and so different conclusions) for $\Gamma'_1$ and its variants $\Gamma'^{\rightrightarrows}_1, \ldots$, while $LP_m$ does not differentiate these variations. A resolution-based system close to $LP$ yet with a stronger disjunction is described in [Lin, 1987].

A whole variety of approaches uses lattices for dealing with inconsistency, eg. [Arieli and Avron, 1994; Belnap, 1977; Sandewall, 1985]. For instance, [Arieli and Avron, 1994; 1996] describes a system based on 4-valued logic that allows for constraining "the most consistent" models in the meta-level by a user-given set of propositions taking classical truth-values only. [Carnielli et al., 1991] proposes a translation-based approach to reasoning in the presence of contradictions that translates a logic into a family of other logics, eg. classical logic into 3-valued logics.

The difference between our approach and "reasoning from maximal consistent subsets of the premises" is that we still pay attention to one objection motivating relevant logics [Anderson and Belnap, 1975] and that is applying disjunctive syllogism to contradictory premises. However, we do not go as far as sanctioning any classical inference not using inconsistent subformulas. That is, we still follow the principle of relevant logics that an inference rule is a priori applicable to any premise. This is in contrast with the idea of restricted access logic [Gabbay and Hunter, 1993], where all classical inference rules are admitted with some special application conditions.

Among others, logic programming with inconsistencies was addressed in [Blair and Subrahmanian, 1988; 1989]. [Wagner, 1991] describes a procedural framework for handling contradictions that relies on the notions of "support" and "acceptance". The former avenue of research is further developed in [Grant and Subrahmanian, 1995], where it is shown how the approach of [Blair and Subrahmanian, 1988] can be extended by classical inferences, like reasoning by cases. Intuitively, the corresponding entailment relations amount to logic programming in a 3-valued (and 4-valued, respectively) logic. The major difference to our approach is that compared to classical entailment, these approaches are sound but not complete (even when the set of premises is consistent). As with other approaches, this is because they aim at paraconsistent reasoning in a logic programming setting that does not necessarily coincide with classical logic.

Our approach is clearly semantical in contrast to many other proposals to paraconsistency: (i) the idea of "forgetting" literals [Kifer and Lozinskii, 1989; Besnard and Schaub, 1996]; (ii) the idea of stratified theories [Benferhat et al., 1993]; (iii) the idea of reliability relation [Roos, 1992], (iv) and more generally the idea of reasoning from consistent subsets of the premises. In contrast to [Turner, 1990], where the baseline is to analyze propositions (so as to resolve paradoxes about truth, for instance), we simply apply a system of truth-values so that we can have non-trivial inconsistent premises. Moreover, our approach is purely deductive, as opposed to argumentation-based frameworks, like [Wagner, 1991; Elvang and Hunter, 1995]. An unusual approach to reasoning from inconsistency is due to [Lin, 1996], who introduces the notion of consistent belief by means of modal operators. This approach fails to satisfy reflexivity (not every premise is concluded).

## 5 Conclusion

We presented a semantical approach to dealing with inconsistent knowledge bases that is founded on the minimization of three-valued models. This was complemented by a formal proof system accomplishing model minimization by appeal to a circumscription axiom. The distinguishing features of our approach are (i) its desire to provide models making true (instead of true and false) as many as possible items of the knowledge base, (ii) its centering on inferences drawn by modus ponens by means of a primitive implication connective, and (iii) its property of restricted monotonicity. A major further development will be lifting the approach to the first-order case. In this context, we draw the reader's attention to the fact that our approach (unlike [Priest, 1989]) does not rely on the notion of an atomic proposition, which is always problematic when passing from the propositional case to the first-order case.

## References

[Anderson and Belnap, 1975] A. Anderson and N. Belnap. *Entailment: The Logic of Relevance and Necessity*. Princeton University Press, 1975.

[Arieli and Avron, 1994] O. Arieli and A. Avron. Logical bi-lattices and inconsistent data. In *Logic in Computer Science Conf.*, pp 468–476, 1994.

[Arieli and Avron, 1996] O. Arieli and A. Avron. Automatic diagnoses for properly stratified knowledge-bases. In *Int. Conf. on Tools with Artificial Intelligence*, pp 392–399. IEEE Press, 1996.

[Belnap, 1977] N. Belnap. A useful four-valued logic. In J. Dunn and G. Epstein, eds, *Modern Uses of Multiple-Valued Logic*. Reidel, 1977.

[Benferhat et al., 1993] S. Benferhat, D. Dubois, & H. Prade. Argumentative inference in uncertain and inconsistent knowledge bases. In *Int. Conf. on Uncertainty in Artificial Intelligence*, pp 411–419, 1993.

[Besnard and Schaub, 1996] P. Besnard and T. Schaub. A simple signed system for paraconsistent reasoning. In *European Workshop on Logics in Artificial Intelligence*, pp 404–416. Springer Verlag, 1996.

[Besnard, 1991] P. Besnard. Paraconsistent logic approach to knowledge representation. In *World Conf. on Fundamentals of Artificial Intelligence*, 1991.

[Blair and Subrahmanian, 1988] H. Blair and V.S. Subrahmanian. Paraconsistent foundations of logic programming. *Journal of Non-Classical Logics*, 5(2):45–73, 1988.

[Blair and Subrahmanian, 1989] H. Blair and V.S. Subrahmanian. Paraconsistent logic programming. *Theoretical Computer Science*, 68(2):135–154, 1989.

[Carnielli et al., 1991] W. Carnielli, L. Fariñas del Cerro, and M. Lima Marques. Contextual negations and reasoning with contradictions. In *Int. Joint Conf. on Artificial Intelligence*, pp 532–537. Morgan Kaufmann, 1991.

[Elvang and Hunter, 1995] M. Elvang and A. Hunter. Argumentative logics: reasoning with classically inconsistent information. *Journal of Knowledge and Data Engineering*, 16:125–145, 1995.

[Gabbay and Hunter, 1993] D. Gabbay and A. Hunter. Restricted access logics for inconsistent information. In *European Conf. on Symbolic and Quantitative Approaches to Reasoning and Uncertainty*. Springer Verlag, 1993.

[Grant and Subrahmanian, 1995] J. Grant and V.S. Subrahmanian. Reasoning in inconsistent knowledge bases. *IEEE Transactions on Knowledge and Data Engineering*, 7(1):177–189, 1995.

[Kifer and Lozinskii, 1989] M. Kifer and E. Lozinskii. RI: A logic for reasoning with inconsistency. In *Logic in Computer Science*, pp 253–262, 1989.

[Lin, 1987] F. Lin. Reasoning in the presence of inconsistency. In *AAAI Nat. Conf. on Artificial Intelligence*, pp 139–143. AAAI/MIT Press, 1987.

[Lin, 1996] J. Lin. A semantics for reasoning consistently in the presence of inconsistency. *Artificial Intelligence*, 86(1-2):75–95, 1996.

[McCarthy, 1980] J. McCarthy. Circumscription — a form of nonmonotonic reasoning. *Artificial Intelligence*, 13(1-2):27–39, 1980.

[Priest et al., 1989] G. Priest, R. Routley, and J. Norman, editors. *Paraconsistent Logics*. Philosophica Verlag, 1989.

[Priest, 1979] G. Priest. Logic of paradox. *Journal of Philosophical Logic*, 8:219–241, 1979.

[Priest, 1989] G. Priest. Reasoning about truth. *Artificial Intelligence*, 39:231–244, 1989.

[Roos, 1992] N. Roos. A logic for reasoning with inconsistent knowledge. *Artificial Intelligence*, 57:69–103, 1992.

[Sandewall, 1985] E. Sandewall. A functional approach to non-monotonic logic. *Computational Intelligence*, 1:80–87, 1985.

[Turner, 1990] R. Turner. *Truth and Modality for Knowledge Representation*. Pitman, 1990.

[Wagner, 1991] G. Wagner. Ex contradictione nihil sequitur. In *Int. Joint Conf. on Artificial Intelligence*, pp 538–543. Morgan Kaufmann, 1991.

# A default interpretation of defeasible network

*Xianchang Wang, Jia-Huai You, Li Yan Yuan*
Department of Computing Science
University of Alberta
Edmonton, Alberta, Canada T6G 2H1

## Abstract

This paper studies the semantics for the class of *all* defeasible (inheritance) networks, including *cyclic* and *inconsistent* networks using a transformation approach. First we show that defeasible networks can be translated, tractably, to default theories while preserving Horty's path-off credulous semantics for all consistent networks. Using the existing methods in dealing with the semantics of default logic, we are able to provide a tractable skeptical semantics, the *well-founded semantics*, and a new credulous semantics, the *regular semantics*, both of which are defined for *any* defeasible network. Furthermore, we show that these semantics are based on the same principle of specificity used by Horty in defining his credulous semantics of defeasible networks.

## 1 Introduction

Two fundamental problems are to be addressed in this paper. First, the semantics of defeasible networks has previously been studied mainly under the assumption that such a network is acyclic and consistent. There is little understanding of the semantics for the class of *all* defeasible networks. Researchers in the field have not been able to provide an acceptable semantics for any defeasible networks that may involve cycles and/or that may be inconsistent.

Second, although a significant body of knowledge has been accumulated providing us with a good understanding of general nonmonotonic formalisms, such as default logic, autoepistemic logic, circumscription, and logic programming with negation, and their relationships, little is known about how path-based reasoning is related to other forms of nonmonotonic reasoning. For this, Horty raised the question of whether it is possible, and if yes, how to specify the consequences of a network by interpreting it in some more standard nonmonotonic formalism [Horty, 1994].

A number of transformations from defeasible network to a more general form of nonmonotonic reasoning have been proposed [Etherington and Reiter, 1983; Gelfond and Przymusinska, 1990; Gregorie, 1989; Haugh, 1988; Lin, 1991; Reiter and Cirscuolo, 1987]. None of these proposed transformations preserves the path-off credulous semantics even for consistent, acyclic networks. Only recently there had been some breakthrough. E.g. by transforming consistent and acyclic defeasible network into an abstract argumentation framework [Dung and Son, 1995], Dung and Son argue that the credulous semantics of consistent and acyclic network can be expressed in Dung's argumentation framework. In the same paper they show that the answer set semantics of extended logic programming can be used to express the credulous semantics of consistent and acyclic networks. However, if a network is cyclic, their transformation could generate an infinite extended logic program. Later, they reformulated a new semantics for default logic based on the idea of path-based defeasible reasoning and provided a translation from consistent and acyclic networks to their default logic [Dung and Son, 1996].

Cycles in a network are indispensable in representing certain concepts.

The first three nets in Figure 1 summarize some of the situations where a cycle in a network may be formed. E.g. *Net 1* is the case where two properties may lead to each other. In the strict sense, it describes an if and only if relation. A simple case of this relation is that two different names describe the same property. In addition, since we are dealing with *defeasible* networks, a link from $p$ to $q$ could mean, *normally* the property $p$ leads to the property $q$. For example, a professor who teaches a course on logic programming usually also teaches a course on AI, and vice versa.

*Net 2* is about two properties being mutually exclusive; e.g. a male is not a female and vice versa.

*Net 3* shows a case where one concept leads to another which, usually, leads to the negation of the former. An example is that an Edmontonian is a North American

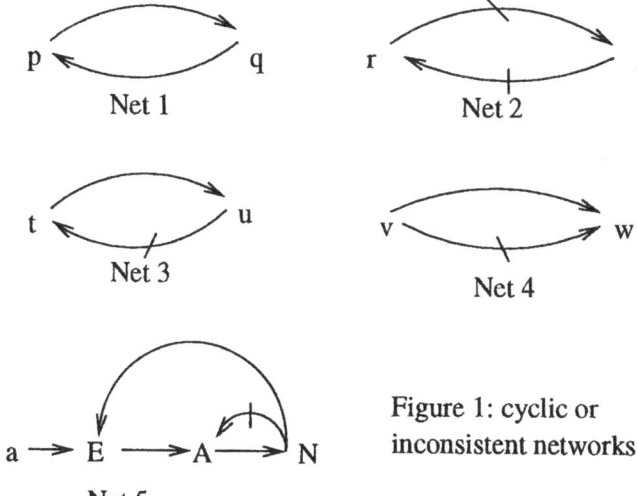

Figure 1: cyclic or inconsistent networks.

but a North American is usually not an Edmontonian. An extreme example is that every lottery winner is a person but a person is usually not a lottery winner.[1]

Note that, if a piece of information (such as, a person is usually not a lottery winner) is not presented in a network, no derivation is possible. This is because defeasible networks do not directly implement the Closed World Assumption.

*Net 4* shows the situation where inconsistency arises in Horty's definition of *defeasible inheritability*. Such a definition allows an extension to include contradictory conclusions (thus allowing derivations from contradicted information). Alternative definitions could avoid this problem; e.g. Touretzky's [Touretzky, 1986] allows two extensions, in this case, with consistent conclusions.

We stress that it is the cycles, not the inconsistency described above, that causes a network to lose all its extensions: Horty showed that any acyclic network possesses at least one credulous extension, but a network involving cycles may not have any credulous extension [Horty, 1994].

The problem that a network has no credulous extension is caused by the presence of *conflicting information*: there is a part of the network, no matter how it is interpreted there is another part that contradicts it.

*Net 5* illustrates a cyclic network that has no credulous extension. This could be understood when $a$ is an individual, $E$ is interpreted as Edmontonian, $A$ as Albertan, and $N$ as North American. Then, it is correct to say that a North American is usually not an Albertan ($N \not\to A$). However, this way of interpreting the network makes it erroneous to say that a North American is usually also an Edmontonian ($N \to E$).

The network could be interpreted differently; e.g. $E$, $A$, and $N$ are different names of the same property. Then, it is the link $N \not\to A$ that is erroneous.

In many application domains, the presence of only consistent information could be considered an exception rather than the norm. For example, in medical diagnosis, contradictions arise from many forms of knowledge incompleteness, e.g., lack of medical knowledge, lack of individual patient symptoms, and error or misinterpretation of both collective and individual data.

It has been argued by many authors that inconsistent information should not result in one of the two extremes: anything or nothing. In particular, inconsistent information should be localized. For example, the contradiction out of passing and failing a student should not affect the derivation that Houston is a city in Texas, and should not allow us to derive that Houston is a city in California. Though defeasible networks seem to provide a particularly suitable form of reasoning to accommodate a notion of local inconsistency, so far there has been no investigation into this possibility.

When we study the semantics of networks with cycles, what principle(s) should we follow? An important insight provided by this paper is that no new principle needs to be proposed. All we need is the principle of specificity that has been used all along in defining credulous semantics. *Net 6* illustrates this principle for an acyclic network: since the property $y$ has conflicting inheritance from the two nodes $u$ and $v$, the application of the path $\pi(x, \tau_1, v, \tau_2, u)$ blocks the path $u \to y$.

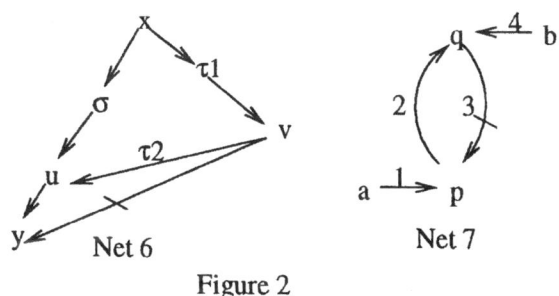

Figure 2

Under this principle, let us consider the network *Net 7* of Figure 2. From a conservative point of view, one may not be able to conclude anything from this net, since if we have $p(a)$, then we will get $q(a)$ and thus $\neg p(a)$. However, under the above principle, this net has a conservative extension $\{1, 12, 4, 43\}$ which concludes $p(a), q(a), q(b), \neg p(b)$. This is because the path 12 blocks the path 3 on object node $a$. That is, $p(a)$ is more specific than $\neg p(a)$. It would be easy to understand what the net could mean in a realistic situation if we represent $p$ as *Lottery_winner*, $q$ as *Person*, $a$ and $b$ as two individuals.

---

[1] Some of these examples can be better represented using both *defeasible* and *strict* links. The work has been extended to networks of mixed links in a forthcoming paper by the authors.

We choose a transformation approach to the semantics of all defeasible networks. First, we translate a network to a default theory and show that the translation preserves Horty's credulous semantics for all consistent networks (even if cyclic). This gives a default interpretation of defeasible networks. Then, we use the existing methods in dealing with the semantics of default logic to define a conservative, skeptical semantics, called the *well-founded semantics*, and a new credulous semantics, called the *regular semantics*, for arbitrary defeasible networks. The two semantics are named after their counterparts in logic programming [Van Gelder *et al.*, 1991; You and Yuan, 1995].

This paper is organized as follows: the next section describes the transformation from defeasible network to default theory, followed by a section on the new semantics for defeasible networks.

## 2 From Network to Default

In this section we give a translation from any defeasible network to a default theory and show a one-to-one correspondence between the credulous extensions of a consistent defeasible network [Horty, 1994] and Reiter's extensions of the translated default theory.

To simplify our discussion, we assume that a default theory is a set of defaults. A default $d$ is of the form $\frac{A:B_1,...,B_n}{C}$, where $n \geq 0$ and $A$, $B_i$, $C$ are formulas of the underlying language $\mathcal{L}$. We denote $A$ by $Pre(d)$, $\{B_1,...,B_n\}$ by $Just(d)$, and $C$ by $Cons(d)$. The definition of a default extension is standard [Reiter, 1980].

### 2.1 Defeasible network

A defeasible (inheritance) network $\Gamma$ is defined as a finite collection of positive and negative links between nodes. If $x, y$ are nodes then $y \leftarrow x$ (resp. $y \not\leftarrow x$) represents a positive (resp. negative) *link* from $x$ to $y$ where $x$ is called the *root* and $y$ is called the *head*. Nodes are divided into two disjoint classes: *object nodes* which are denoted as $a$, $a_1$, ..., and *property nodes* which are denoted as $p$, $q$, $p_1$, .... We assume that an object node can only be used as a root node. If a link's root is an object node, then this link is called an *object link*. Two links are said to be *in conflict* if they have the same head but one is positive and the other is negative. We say that $(x_1, x_2), ..., (x_{n-1}, x_n)$ (or simply $x_1...x_n$, $n \geq 2$) is a *general path* of $\Gamma$ if for every $i = 1, ..., n-1$, $(x_i, x_{i+1})$ is a link of $\Gamma$ whose root is $x_i$ and whose head is $x_{i+1}$. The path above is called a *general cycle* if $x_1 = x_n$. A network is said to be *acyclic* if it has no general cycle.

A network $\Gamma$ is *consistent* if there exist no two nodes $x, y$ such that both $y \leftarrow x$ and $y \not\leftarrow x$ belong to $\Gamma$.

A *path* of $\Gamma$ is either a link of $\Gamma$ or a sequence of $\Gamma$'s links $x_1 \to x_2, ..., x_n \to x_{n+1}$ (called a *positive path*), $n \geq 1$ (resp. $x_1 \to x_2, ..., x_{n-1} \to x_n, x_n \not\to x_{n+1}$, called a *negative path*). We simply denote the above positive path by $\pi(x_1, \sigma, x_{n+1})$ and negative path by $\overline{\pi}(x_1, \sigma, x_{n+1})$.

We now give the off-path credulous semantics of Horty [Horty, 1994].

**Definition 2.1** (Path constructibility and conflict in a path set) Suppose $\Gamma$ is a defeasible network, $\Phi$ is a path set of $\Gamma$. A path $\sigma$ of $\Gamma$ is *constructible* in a path set $\Phi$ if $\sigma$ is an object link or $\sigma = \pi(x_1, \sigma_1, x_n, x_{n+1})$ (resp. $\sigma = \overline{\pi}(x_1, \sigma_1, x_n, x_{n+1})$) and $\pi(x_1, \sigma_1, x_n) \in \Phi$.

A path $\sigma$ of $\Gamma$ is *conflicting* in $\Phi$ if $\sigma = \pi(x_1, \sigma_1, x_n)$ (resp. $\sigma = \overline{\pi}(x_1, \sigma_1, x_n)$) and $\sigma' = \overline{\pi}(x_1, \sigma'_1, x_n) \in \Phi$ (resp. $\sigma' = \pi(x_1, \sigma'_1, x_n) \in \Phi$). □

**Definition 2.2** (Preemption)
A positive path $\pi(x, \sigma, u, y)$ (resp. negative path $\overline{\pi}(x, \sigma, u, y)$) is *preempted* (see Net 6 of Figure 2) in $\Phi$ iff there is a node $v$ such that (i) $v \not\to y \in \Gamma$ (resp. $v \to y \in \Gamma$) and (ii) either $v = x$ or there is a path of the form $\pi(x, \tau 1, v, \tau 2, u) \in \Phi$. □

**Definition 2.3** (Defeasible inheritability)
Path $\sigma$ is *defeasible inheritable* in $\Phi$, written as $\Phi \vdash_d \sigma$, iff either $\sigma$ is an object link[2] or $\sigma$ is a compound path, $\sigma = \pi(x, \tau, y)$ (likewise for negative path) such that
 (i) $\sigma$ is constructible in $\Phi$;
 (ii) $\sigma$ is not conflicting in $\Phi$; and
 (iii) $\sigma$ is not preempted in $\Phi$. □

**Definition 2.4** (Credulous extension)
A set $\Phi$ of paths is a *credulous extension* of a net $\Gamma$ iff $\Phi = \{\sigma \mid \Phi \vdash_d \sigma\}$. □

**Example 2.5** Consider *Net 7* of Figure 2. This network allows one credulous extension: $\Phi_1 = \{1, 12, 4, 43\}$. □

### 2.2 Translation

For the presentation purpose, we first present a translation that is intuitive but not tractable, and then show how to modify it slightly to make it tractable.

Given a network $\Gamma$, a path $\pi(u, \sigma, x)$ is called a *simple path* if there is no path $\pi(u, \sigma', x)$ such that the set of links in $\pi(u, \sigma', x)$ is a proper subset of the links in $\pi(u, \sigma, x)$. We say that *a simple path $\pi(u, \sigma, x)$ causes conflict in node $r$* if two links from $u$ to $r$ and from $x$ to $r$ are in conflict in $\Gamma$. We say that *a simple path causes conflict* if it causes conflict in some node $r$. E.g. in *Net 7*, the simple path 12 causes conflict in node $p$, and it is the only simple path of *Net 7* that causes conflict.

We now relate the credulous semantics of a defeasible network with Reiter's extension semantics of its translated default theory. First, we transform a network to a default theory. In the following transformation, we use a predicate $in_l(a)$ to mean that *link $l$ is in a path from object node $a$*. The role played by $in_l(a)$ is similar to

---
[2]In [Horty, 1994], it is any *direct link*. The difference, however, is inessential.

the normality predicate in nonmonotonic reasoning; the link $l$ in a path to conclude a property of $a$ is normally accepted unless there is a reason to reject it.

**Definition 2.6** Let $\Gamma$ be an arbitrary defeasible network. We translate it into a default theory $\Pi(\Gamma)$ as:
$d_l = \frac{:B, in_l(a)}{B \wedge in_l(a)}$, for every object link $l \in \Gamma$, if $l = p \leftarrow a$, then $B = p(a)$; if $l = p \not\leftarrow a$, then $B = \neg p(a)$;
$d_l(x) = \frac{p(x):B, in_l(x)}{B \wedge in_l(x)}$ for every non-object link $l \in \Gamma$, if $l = q \leftarrow p$, then $B = q(x)$; if $l = q \not\leftarrow p$, then $B = \neg q(x)$;
$d_{(\pi(u,\sigma,v),l)}(x) = \frac{\wedge \{in_l(x) \mid l \in \pi(u,\sigma,v)\}:}{\neg in_l(x)}$, for every simple path $\pi(u, \sigma, v)$ and link $l$ from $v$ to $r$ such that it conflicts with a link from $u$ to $r$. □

Here, the underlying first order language of $\Pi(\Gamma)$ contains the predicate set of the property nodes of $\Gamma$ and the newly introduced predicate $in_l$ for every link $l$ of $\Gamma$; Its constant set is the set of the object nodes of $\Gamma$.

The first two kinds of defaults are translations directly from individual links. Let us denote the set of all these defaults by $\Pi_m(\Gamma)$. The last kind of defaults are translated from the simple paths that lead to a conflict. We denote it by $\Pi_p(\Gamma)$. Clearly, $\Pi(\Gamma) = \Pi_m(\Gamma) \cup \Pi_p(\Gamma)$.

We use the following examples to illustrate how a defeasible network is translated into a default theory and their relationships.

**Example 2.7** Continue with *Net 7* of Figure 2. The translated default theory $\Pi(Net\ 7)$ is:
$\Pi_m(Net\ 7) =$
$\{\frac{:p(a),in_1(a)}{p(a)\wedge in_1(a)}, \frac{p(x):q(x),in_2(x)}{q(x)\wedge in_2(x)}, \frac{q(x):\neg p(x),in_3(x)}{\neg p(x)\wedge in_3(x)}, \frac{:q(b),in_4(b)}{q(b)\wedge in_4(b)}\}$
$\Pi_p(Net\ 7) = \{\frac{in_1(x)\wedge in_2(x):}{\neg in_3(x)}\}$

The defaults in $\Pi_m(Net\ 7)$ are self-explainable. The default in $\Pi_p(Net\ 7)$ is due to the fact that links 1 and 3 are in conflict and the simple path 12 is from link 1's root to link 3's root. This default theory has one R-extension

$$E = Th(\{p(a), q(a), q(b), \neg p(b),$$
$$in_1(a), in_2(a), \neg in_3(a), in_4(b), in_3(b)\}),$$

which corresponds (after removing $in_l(x)$ literals) to the only credulous extension as given in Example 2.5. □

**Example 2.8** Consider the network *Net 8* adopted from [Dung and Son, 1996] in Figure 3 where $St(x)$ means $x$ is a student, $Yad(x)$ $x$ is a young adult, $Ad(x)$ $x$ is an adult and $Emp(x)$ $x$ is employed. This network has only one credulous extension $\Phi = \{1, 12, 15, 123\}$. Now we translate it into a default theory:
$\Pi_m(Net\ 8) = \{\frac{:St(a),in_1(a)}{St(a)\wedge in_1(a)}, \frac{St(x):Yad(x),in_2(x)}{Yad(x)\wedge in_2(x)},$
$\frac{Yad(x):Ad(x),in_3(x)}{Ad(x)\wedge in_3(x)}, \frac{Ad(x):Emp(x),in_4(x)}{Emp(x)\wedge in_4(x)}, \frac{St(x):\neg Emp(x),in_5(x)}{\neg Emp(x)\wedge in_5(x)}\}$
$\Pi_p(Net\ 8) = \{\frac{in_2(x)\wedge in_3(x):}{\neg in_4(x)}\}$

This default theory has only one R-extension, which implies $St(a), Yad(a), Ad(a), \neg Emp(a)$, along

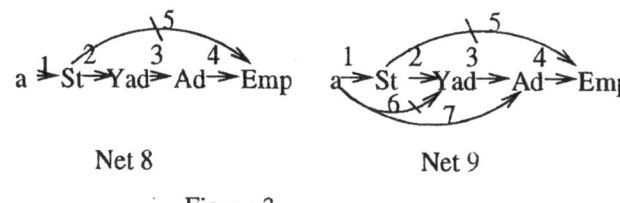

Net 8      Net 9

Figure 3

with $in_1(a), in_2(a), in_3(a), \neg in_4(a), in_5(a)$. It corresponds to the unique credulous extension of the network.

If we add two more rules $6: Yad \not\leftarrow a$ and $7: Ad \leftarrow a$ into the network *Net 8*, the new network, *Net 9*, has two credulous extensions: $\Phi_1 = \{1, 6, 7, 15\}$ and $\Phi_2 = \{1, 6, 7, 74\}$. When we translate it into a default theory $\Pi(Net\ 9)$, we get $\Pi_m(Net\ 9)$ and $\Pi_p(Net\ 9)$ where
$\Pi_p(Net\ 9) = \{\frac{in_1(x):}{\neg in_2(x)}, \frac{in_2(x)\wedge in_3(x):}{\neg in_4(x)}\}$.
$\Pi(Net\ 9)$ has two extensions, which imply, respectively,
$\{St(a), \neg Yad(a), Ad(a), \neg Emp(a)\}$,
$\{St(a), \neg Yad(a), Ad(a), Emp(a)\}$,
They correspond to the two credulous extensions of *Net 9*. □

We now show that the relationship demonstrated above holds for all consistent networks. First we explain some notation.

Let $\Gamma$ be a defeasible network, $E$ a first older theory under the language of default theory $\Pi(\Gamma)$, and $\Phi$ a path set of the network $\Gamma$. We use $E/\Gamma$ to mean the union of set $\{p(a) \mid p(a) \in E, p \text{ is a property node and } a \text{ is an object node}\}$ and set $\{\neg p(a) \mid \neg p(a) \in E, p \text{ is a property node and } a \text{ is an object node}\}$.

The path set $\mathcal{P}(E)$ of $\Gamma$ is defined as follows: $\mathcal{P}(E) = \{\pi(a, \sigma, p) \mid \forall l \in \pi(a, \sigma, p), in_l(a) \in E\} \cup \{\overline{\pi}(a, \sigma, p) \mid \forall l \in \overline{\pi}(a, \sigma, p), in_l(a) \in E\}$.

The consequence set of path set $\Phi$ of the net $\Gamma$ is defined by $Cons(\Phi) = \{p(a) \mid \exists \pi(a, \sigma, p) \in \Phi\} \cup \{\neg p(a) \mid \exists \overline{\pi}(a, \sigma, p) \in \Phi\}$.

**Theorem 2.9** *For any consistent network $\Gamma$, $\Phi$ is its credulous extension iff $\Pi(\Gamma)$ has an R-extension $E$ such that $\mathcal{P}(E) = \Phi$ and thus $E/\Gamma = Cons(\Phi)$.* □

### 2.3 A note on translation

Clearly, the number of defaults in a translated default theory depends on the number of the simple paths that lead to a conflict in the given network. In the worst case, this number is exponential to the node size. However, this still improves Dung and Son's translation [Dung and Son, 1995] from consistent and acyclic networks to argumentation frameworks, where the number of arguments could be infinite.

Selman and Levesque [Selman and Levesque, 1993] show that deciding whether a credulous extension exists for a defeasible network is NP-complete. Kautz and Selman [Kautz and Selman, 1991] show that the same decision problem for the simple default logic (restricted to literals) is also NP-complete. These results provide strong evidence that an extension preserving, polynomial transformation exists (note that these results only guarantee the existence of a polynomial transformation such that this same decision problem is preserved).

We now present such a transformation. The key idea is to use a predicate, $c_{p,q}$, to represent a positive path from node $p$ to $q$ where all the links on it have been *accepted*. Therefore, we modify $\Pi_p(\Gamma)$ to contain only the following defaults:

$\frac{in_{p \leftarrow a}(a):}{c_{a,p}(a)}$, where $p \leftarrow a$ is a positive link of the network.

$\frac{in_{q \leftarrow p}(x), c_{p',p}(x):}{c_{p',q}(x)}$, i.e., $c_{p,q}$ is transitive.

$\frac{c_{p,q}(x):}{\neg in_l(x)}$, where $l$ is a link from $q$ to $r$ and the network has another link from $p$ to $r$ that conflicts with $l$.

Clearly, the size of $\Pi_p(\Gamma)$ is polynomial to the node size of $\Gamma$. By encoding the connection relation between any two nodes, we can avoid enumerating all the possible paths that connect the two nodes.

An interesting implication of such a polynomial translation is that any decision problem for a defeasible network based on the credulous semantics is *no harder* than the corresponding decision problem for the simple default logic (cf. [Kautz and Selman, 1991]).

## 3 Improved Semantics

The default interpretation of defeasible networks presented above provides a way to understand, indirectly, the possible semantics of defeasible networks. Namely, any semantics of default logic yields a semantics of defeasible networks. If a semantics is defined for all default theories, then it is defined for all defeasible networks. In this section we propose two of such semantics to address the problems of no extension and inconsistency.

As we know, not every default theory has an extension. This is because an R-extension $E$ of a default theory $D$ is a fixpoint of the following *anti-monotonic* operator:

**Definition 3.1** (Anti-monotonic operator $\mathcal{R}$)

Suppose $D$ is a default theory. For any first order theory $E$, we define $\mathcal{R}(E)$ to be *the smallest formula set* that satisfies the following conditions:

1. $Th(\mathcal{R}(E)) = \mathcal{R}(E)$;
2. For every $d \in D$, if $Pre(d) \in \mathcal{R}(E)$ and for any $B \in Just(d)$, $\neg B \notin E$, then $Cons(d) \in \mathcal{R}(E)$. □

$\mathcal{R}$ is anti-monotonic because for any theories $E_1$ and $E_2$, $E_1 \subseteq E_2$ implies $\mathcal{R}(E_2) \subseteq \mathcal{R}(E_1)$. Thus, $\mathcal{R}^2$, the operator that applies $\mathcal{R}$ twice, is monotonic. It thus has a least fixpoint and maximal fixpoints. Such a fixpoint is called an *alternating fixpoint* of $\mathcal{R}$.

A number of researchers have proposed to use the technique of alternating fixpoints to define partial semantics for default logic and logic programs. [Baral and Subrahmanian, 1991; Przymusinska and Przymusinski, 1991].

**Definition 3.2** Let $D$ be a default theory. A fixpoint $E$ of $\mathcal{R}^2$ is said to be an *alternating extension* of $D$. It is said to be *normal* if $E \subseteq \mathcal{R}(E)$.

The *well-founded semantics* of $D$ is defined by the least alternating extension (which is necessarily normal). The *regular semantics* of $D$ is defined by the set of all maximal normal alternating extensions. □

We call the least normal alternating extension the *well-founded extension*, and a maximal normal alternating extension a *regular extension*.

However, this is not enough. It is well known that the well-founded extension defined this way is rather weak. Since in classic logic inconsistency is a global phenomenon, the possibility of deriving contradictory conclusion could nullify other expected conclusions. E.g. the default theory $\{\frac{:c}{c}, \frac{:\neg c}{\neg c}, \frac{:p}{p}\}$ has a well-founded extension $Th(\{\})$ and two regular extensions $Th(\{c,p\})$ and $Th(\{\neg c, p\})$. Obviously, we need to make $p$ as a conservative conclusion in the well-founded semantics. Thus, we modify the definition of closure $Th(E)$ as follows: $Th(E) = \{p \mid \exists \text{ consistent } E' \subseteq E \text{ such that } E' \models p\}$. Under this subtle modification, we can see that the preceding default theory has the well-founded extension $Th(\{p\})$ and the same regular extensions.

**Theorem 3.3** *For any network $\Gamma$, its translated default theory $\Pi(\Gamma)$ must have a unique, consistent well-founded extension $W$, and one or more consistent regular extension $E$. Conversely, the well-founded extension and any regular extension of the default theory $\Pi(\Gamma)$ must be consistent.* □

Based on this result, we define the well-founded extension of a defeasible network $\Gamma$ by $\mathcal{P}(W)$ and a regular extension of $\Gamma$ by $\mathcal{P}(E)$.

### 3.1 Examples

**Example 3.4** Consider *Net 11* in Fig. 5. Its default translation yields
$\{\frac{p(a), in_1(a)}{p(a) \wedge in_1(a)}, \frac{p(x):q(x), in_2(x)}{q(x) \wedge in_2(x)}, \frac{p(x):\neg q(x), in_3(x)}{\neg q(x) \wedge in_3(x)}, \frac{p(x):r(x), in_4(x)}{r(x) \wedge in_4(x)}\}$
Its well-founded extension is
$W = Th(\{p(a), r(a), in_1(a), in_4(a)\})$,
and the two regular extensions are
$E_1 = Th(\{p(a), r(a), q(a), in_1(a), in_4(a), in_2(a)\})$ and
$E_2 = Th(\{p(a), r(a), \neg q(a), in_1(a), in_4(a), in_3(a)\})$.

Thus, *Net* 11 has the well-founded extension, $\{1, 14\}$, and two regular extensions, $\{1, 12, 14\}$ and $\{1, 13, 14\}$. □

**Example 3.5** Consider the cyclic network *Net 12* of Fig. 5 (same as *Net 5* in Fig. 1). Its translated default theory is:

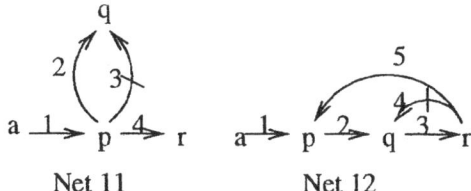

Figure 5

$\Pi_m(Net\ 12) = \{\frac{:p(a), in_1(a)}{p(a) \wedge in_1(a)}, \frac{p(x):q(x), in_2(x)}{q(x) \wedge in_2(x)},$
$\frac{q(x):r(x), in_3(x)}{r(x) \wedge in_3(x)}, \frac{r(x):\neg q(x), in_4(x)}{\neg q(x) \wedge in_4(x)}, \frac{r(x):p(x), in_5(x)}{p(x) \wedge in_5(x)}\}$

$\Pi_p(Net\ 12) = \{\frac{in_5(x):}{\neg in_2(x)}, \frac{in_2(x) \wedge in_3(x):}{\neg in_4(x)}\}$

The default theory $\Pi(Net\ 12)$ has the well-founded extension $Th(\{p(a), in_1(a)\})$ which is also its unique regular extension.

Thus, $Net\ 12$ has $\{1\}$ as its well-founded extension and regular extension. $\square$

### 3.2 Properties

The most important property is a soundness property for any well-founded, regular extension of a network $\Gamma$. Essentially, it says that any path constructed from such an extension cannot be preempted.

**Theorem 3.6** (Soundness property) *Let $\Gamma$ be any defeasible network, $\Phi$ is the well-founded or regular extension of $\Gamma$. Then,*

1. $\Phi \subseteq \{\sigma \mid \Phi \vdash_d \sigma\}$;
2. *For any path $\beta$ of $\Gamma$, if $\beta \not\subseteq \Phi$, and $\Phi \mid_d \beta$, then $\{\sigma \mid \Phi \vdash_d \sigma\} \not\vdash_d \beta$.* $\square$

If a network is consistent and acyclic, the correspondence between regular extensions and Horty's credulous extensions is one-to-one.

**Theorem 3.7** *Suppose $\Gamma$ is a consistent and acyclic defeasible network. $\Phi$ is a regular extension of $\Gamma$ iff $\Phi$ is a credulous extension of $\Gamma$.* $\square$

Since the least alternating fixpoint of a simple default theory can be tractably computed, we get

**Theorem 3.8** *The set of literals that are implied by a network under the well-founded semantics can be computed tractably.* $\square$

## References

[Baral and Subrahmanian, 1991] C. Baral and V. Subrahmanian. Dualities between alternative semantics for logic programming and nonmonotonic reasoning. In *Proc. of the First International Workshop on LPNMR*, pages 69–86. MIT Press, 1991.

[Dung and Son, 1995] P. Dung and T. Son. Nonmonotonic inheritance, argumentation and logic programming. In *Proc. of 3rd International Conference on LPNMR*, pages 317–329, 1995.

[Dung and Son, 1996] P. Dung and T. Son. An argument-theoretic approach to reasoning with specificity. In *KR-96*, pages 506–528, 1996.

[Etherington and Reiter, 1983] D. Etherington and R. Reiter. On inheritance hierarchies with exceptions. In *AAAI-83*, pages 104–108. Morgan Kaufmann, 1983.

[Gelfond and Przymusinska, 1990] M. Gelfond and H. Przymusinska. Formalization of inheritance reasoning in autoepistemic logic. *Fundamental Informaticae*, XIII:403–443, 1990.

[Gregorie, 1989] E. Gregorie. Skeptical theories of inheritance and nonmonotonic logics. *Methodologies for Intelligence System*, 4:430–438, 1989.

[Haugh, 1988] B. Haugh. Tractable theories of multiple defeasible inheritance in ordinary non-monotonic logics. In *AAAI-88*, pages 421–426. AAAI Press, 1988.

[Horty, 1994] J.F. Horty. Some direct theories of nonmonotonic inheritance. In M. Cabbay, C. Hogger, and J. Robinson, editors, *Handbook of Logic in Artificial Intelligence and Logic Programming, Vol 3: Nonmonotonic Reasoning and Uncertain Reasoning*, pages 111–187. Oxford University, 1994.

[Kautz and Selman, 1991] H. Kautz and B. Selman. Hard problems for simple default logics. *Artificial Intelligence*, pages 243–279, 1991.

[Lin, 1991] F. Lin. A study of nonmonotonic reasoning. PhD thesis, Stanford University, 1991.

[Przymusinska and Przymusinski, 1991] H. Przymusinska and T.C. Przymusinski. Stationary default extensions. Technical Report, California State Polytechnic and University of California at Riverside, 1991.

[Reiter and Cirscuolo, 1987] R. Reiter and G. Cirscuolo. On interacting defaults. In *Readings in Nonmonotonic Reasoning*, pages 94–100. Morgan Kaufmann, 1987.

[Reiter, 1980] R. Reiter. A logic for default reasoning. *Artificial Intelligence*, 13:81–132, 1980.

[Selman and Levesque, 1993] B. Selman and H. Levesque. The complexity of path-based defeasible inheritance. *Artificial Intelligence*, 62(2):303–340, 1993.

[Touretzky, 1986] D. Touretzky. *The Mathematics of Inheritance Systems*. Morgan Kaufmann, 1986.

[Van Gelder et al., 1991] Van Gelder, A. Ross, and J.S. Schlipf. The well-founded semantics for general logic programs. *Journal of the ACM*, 38(3):620–650, 1991.

[You and Yuan, 1995] J. You and L. Yuan. On the equivalence of semantics for normal logic programs. *J. Logic Programming*, 22:212–221, 1995.

# A Cumulative-Model Semantics for Dynamic Preferences on Assumptions

Ulrich Junker
ILOG
9, rue de Verdun, BP 85
F-94253 Gentilly Cedex
junker@ilog.fr

## Abstract

Explicit preferences on assumptions as used in prioritized circumscription [McCarthy, 1986; Lifschitz, 1985; Grosof, 1991] and preferred subtheories [Brewka, 1989] provide a clear and declarative method for defining preferred models. In this paper, we show how to embed preferences in the logical theory itself. This gives a high freedom for expressing statements about preferences. Preferences can now depend on other assumptions and are thus dynamic. We elaborate a preferential semantics based on Lehmann's cumulative models, as well as a corresponding constructive characterization, which specifies how to correctly treat dynamic preferences in the default reasoning system EXCEPT [Junker, 1992].

**Keywords:** nonmonotonic reasoning, common sense reasoning.

## 1 Introduction

In the absence of complete information, it is necessary to base decisions and conclusions on assumptions. If those assumptions were arbitrary, the resulting decisions and conclusions would be arbitrary as well. Depending on the given information, *best* assumptions are chosen.

Different ways for defining best assumptions (or default rules) have been studied in nonmonotonic reasoning. A sound and declarative method is provided by preferences on assumptions. They are e.g. used in prioritized circumscription [McCarthy, 1986; Lifschitz, 1985; Grosof, 1991] and for preferred subtheories [Brewka, 1989]. Preferences decide which assumptions will be selected first in presence of conflicts between assumptions. In absence of conflicts, they don't have any effect. Furthermore, preferences enable a preferential semantics leading to clear logical properties, as well as constructive characterizations in form of inductive definitions. Finally, they allow to express the important specificity principle in inheritance systems in a clear way. All these points are difficult to achieve in alternative approaches such as default logic. A problem, however, is how to specify preferences:

1. **Static preferences** are specified outside the logical theory to which they apply. They are given in form of priorities [McCarthy, 1986; Lifschitz, 1985] or in form of a partial order on assumptions [Brewka, 1989; Grosof, 1991]. Specifying such an ordering is a minutely work. It would be preferable to write down quantified and conditional statements on preferences in the logical theory itself.

2. **Implicit preferences** are used in conditional approaches [Geffner and Pearl, 1992; Kraus *et al.*, 1990]. Default rules of the form $\alpha_i \hspace{0.1em}\mid\hspace{-0.5em}\sim \gamma_i$ can be (partially) ordered by exploiting specificity relations between the contexts $\alpha_i$. However, other kinds of preference knowledge cannot be expressed. We refer the reader to [Brewka, 1994] who argues in favour of explicit preferences.

In order to allow a clear, explicit, and flexible specification of preferences, we embed them in the logical theory itself. As a consequence, preferences on assumptions can depend on (other) assumptions and thus become dynamic. We argue that those *dynamic preferences* are quite natural in human commonsense reasoning and illustrate this by the following example:

*Jim and Jane have the following habits:*

1. *Normally, Jim and Jane go to at most one attraction each evening.*
2. *Jim prefers the theatre to the night club.*
3. *Jane prefers the night club to the theatre.*
4. *If Jim invites Jane then he respects her preferences (and vice versa).*
5. *Normally Jim invites Jane.*
6. *An exception to 1 is Saturday.*
7. *An exception to 5 is Jim's birthday, where Jane invites Jim.*

*If no further information is given we conclude that Jim and Jane will go to the night club. When we learn that Jim has birthday we revise this and conclude that they go to the theatre. However, the day in question is a Saturday. Hence, they should go to both attractions. Finally the news tell that the theatre is closed for work. Thus we again conclude that they go to the night club.*

Dynamic preferences have been examined in the scope of the TASSO-project on graphic configuration under uncertainty. The default reasoning system EXCEPT II uses dynamic preferences for determining an order in which assumptions are inspected [Junker, 1992; Junker, 1995]. Problems are provided by cyclic preferences, as well as by new preferences that contradict the already chosen part of the order. Brewka succeeded to integrate dynamic preferences into default logic [Brewka, 1994] and logic programming [Brewka, 1996]. Defaults are applied in a certain order that is chosen initially. The dynamic preferences obtained as consequences of defaults must be consistent with this order.

Both approaches do not guarantee the existence of solutions. Furthermore, they miss a clear preferential semantics as well as a constructive characterization. In this paper, we present a solution to these problems:

1. In section 2, we show how to embed assumptions and preferences in a logical language.

2. We seek a preferential semantics for static preferences in section 3. We analyse limits of existing approaches and elaborate a preferential semantics based on cumulative models [Kraus et al., 1990].

3. We extend this semantics to dynamic preferences in section 4. The resulting nonmonotonic inference relation inherits all properties of Lehmann's system $\mathcal{C}$ [Kraus et al., 1990].

4. This semantics then points out how to modify the constructive approach in [Brewka, 1989; Junker and Brewka, 1991] to dynamic preferences.

Finally, we discuss a simple example in section 5, as well as related work in section 6.

## 2 Preferences in a Logical Language

In this section, we show how to express preferences on assumptions in a first-order language. For this purpose, assumptions must be named by ground terms. Similar to circumscription, we do not change the syntax of a logical language, but introduce special predicate symbols:

- a unary predicate symbol $c$ for *chosen* assumptions[1].
- a binary predicate symbol $\leftarrow$ for preferences.

If $t, t_1, t_2$ are ground terms then $c(t)$ means that an assumption of name $t$ is chosen and $t_1 \leftarrow t_2$ means that the assumption of name $t_1$ is preferred to the assumption of name $t_2$. Let $\mathcal{L}$ be a first-order language having the predicate symbols $c$, $\leftarrow$ and an unsatisfiable constant $\perp$.

Named default rules in the sense of [Poole, 1988] can easily be translated to this quite technical representation. A default rule $d : \alpha \supset \gamma$ of name $d$ means that $\alpha$ normally implies $\gamma$. It can be written as $\alpha \wedge c(d) \supset \gamma$.

We write our example directly in the translated form. We abbreviate night club by $nc$ and theatre by $th$. Furthermore, we introduce the terms $one_D$, $go_D(x)$, $inv_D$ for naming default rules:

0. $\forall x. c(go_D(x)) \supset go(x)$
1. $c(one_D) \wedge go(nc) \wedge go(th) \supset \perp$
2. $invites(jim, jane) \supset go_D(nc) \leftarrow go_D(th)$
3. $invites(jane, jim) \supset go_D(th) \leftarrow go_D(nc)$
4. $c(inv_D) \supset invites(jim, jane)$
5. $saturday \supset \neg c(one_D)$
6. $hasBirthday(jim) \supset \neg c(inv_D) \wedge invites(jane, jim)$

Since the assumptions $inv_D$, $one_D$ influence decisions about the assumptions $go_D(nc)$ and $go_D(th)$, but not vice versa, we put them into a level of higher priority by adding the following preferences:

7. $\forall x. one_D \leftarrow go_D(x)$
8. $\forall x. inv_D \leftarrow go_D(x)$

In order to simplify the discussion in this paper, we follow [Poole, 1988] and suppose that all possible assumptions are given explicitly. Furthermore, we suppose that the set of these assumptions is finite:

1. Let $\mathcal{N}$ be a *finite* set of ground terms of $\mathcal{L}$ that serve as names of assumptions.

2. Let $\mathcal{A} := \{c(t) \mid t \in \mathcal{N}\}$ be the corresponding set of assumptions.

Thus, only assumptions from $\mathcal{A}$ will be selected and only preferences between elements of $\mathcal{N}$ will be relevant.

Given a logical theory $\Gamma \subseteq \mathcal{L}$, we consider subsets of $\mathcal{A}$ that are consistent w.r.t. $\Gamma$. Let $\mathcal{C}_\Gamma := \{A \subseteq \mathcal{A} \mid A \cup \Gamma \not\models \perp\}$ be the set of these assumption sets. The following sections show how to define preferred elements of $\mathcal{C}_\Gamma$.

## 3 Static Preferences

In this section, we suppose that static preferences in form of a strict partial order $\prec \subseteq \mathcal{N} \times \mathcal{N}$ are given.

We first examine existing approaches for defining preferred assumption sets. The first one follows the idea of a preferential semantics [Shoham, 1987]. We lift the partial order $\prec$ on assumption names to a partial order $\prec_G$ on assumption sets and select the $\prec_G$-minimal elements of $\mathcal{C}_\Gamma$. We use an order proposed in [Geffner and Pearl, 1992; Grosof, 1991], where worse assumptions are exchanged by better assumptions:

**Definition 3.1** *Let* $\prec_G \subseteq 2^\mathcal{A} \times 2^\mathcal{A}$ *s.t.*
$A_1 \prec_G A_2$ *iff* $A_1 \neq A_2$ *and*
$$\forall c(t) \in A_2 - A_1 \exists c(t^*) \in A_1 - A_2 : t^* \prec t$$
$A \subseteq \mathcal{A}$ *is called* G-preferred assumption set *of* $\Gamma$ *iff 1.* $A \in \mathcal{C}_\Gamma$ *and 2.* $A^* \prec_G A$ *implies* $A^* \notin \mathcal{C}_\Gamma$.

The order $\prec_G$ is transitive and irreflexive [Geffner and Pearl, 1992]. Since the set $\mathcal{A}$ is finite the existence of G-preferred assumption sets is thus guaranteed.

The second approach considers a partial order an incomplete specification of a strict total order $< \subseteq \mathcal{N} \times \mathcal{N}$ [Brewka, 1989]. A strict total order on a finite set $\mathcal{N}$ uniquely defines an enumeration $t_1, \ldots, t_n$ of $\mathcal{N}$ that *respects* $<$ in the sense that $t_j < t_k$ iff $j < k$ for all $j, k = 1, \ldots, n$.

---
[1]In fact $c(t)$ corresponds to $\neg ab(t)$.

We inspect the assumptions in increasing order and select them if this selection is consistent w.r.t. the already chosen assumptions:

**Definition 3.2** *Let $< \subseteq \mathcal{N} \times \mathcal{N}$ be a strict total order and $t_1, \ldots, t_n$ be the enumeration of $\mathcal{N}$ that respects $<$. Let $A_0 := \emptyset$ and*

$$A_{i+1} := \begin{cases} A_i & \text{if } \Gamma \cup A_i \cup \{c(t_{i+1})\} \models \bot \\ A_i \cup \{c(t_{i+1})\} & \text{otherwise} \end{cases}$$

*Then $A_n$ is the selection of $<$. $A$ is a B-preferred assumption set iff $A$ is the selection of a strict total order $<$ which satisfies $\prec \subseteq <$.*

This constructive definition immediately gives rise to an algorithm for computing B-preferred assumption sets for decidable sub-languages $\mathcal{L}$. [Junker and Brewka, 1991] shows that every B-preferred assumption set is a G-preferred one, but the inverse is not true. A counterexample is the partial order $t_1 \prec t_3$, $t_2 \prec t_4$ and the theory $\Gamma_0 = \{\neg c(t_1) \vee \neg c(t_2), \neg c(t_1) \vee \neg c(t_3), \neg c(t_1) \vee \neg c(t_4), \neg c(t_2) \vee \neg c(t_3), \neg c(t_2) \vee \neg c(t_4)\}$. When determining B-preferred theories, we must start an enumeration by $t_1$ or $t_2$. Thus, we obtain two B-preferred sets, namely $\{c(t_1)\}$ and $\{c(t_2)\}$. The set $\{c(t_3), c(t_4)\}$ is G-preferred, but not B-preferred. Hence, G-preferred assumption sets cannot serve as a preferential semantics for B-preferred assumption sets.

Thus, we have two approaches for treating static preferences on assumptions. G-preferred assumption sets seem to be too weak since we would not accept the worst choices $\{c(t_3), c(t_4)\}$ as long as the better choices $\{c(t_1)\}$ and $\{c(t_2)\}$ are possible.

If we vary the example we observe that there is no strict partial order $\prec'$ on assumption sets that produces B-preferred assumption sets as minimal elements of $\mathcal{C}_\Gamma$. Consider the following theories and their B-preferred assumption sets:

$\Gamma_0$: $\quad\quad\quad\quad\quad\quad\quad\quad\quad \{c(t_1)\}, \{c(t_2)\}$
$\Gamma_1 := \Gamma_0 \cup \{\neg c(t_1)\}: \quad \{c(t_2)\}, \{c(t_3), c(t_4)\}$
$\Gamma_2 := \Gamma_0 \cup \{\neg c(t_2)\}: \quad \{c(t_1)\}, \{c(t_3), c(t_4)\}$
$\Gamma_3 := \Gamma_1 \cup \Gamma_2: \quad\quad\quad \{c(t_3), c(t_4)\}$

Assume that the $\prec'$-minimal elements of $\mathcal{C}_\Gamma$ coincide with the B-preferred assumption sets. From the last case, we infer that $\{c(t_3), c(t_4)\}$ is $\prec'$-smaller than the sets $\emptyset$, $\{c(t_3)\}$, and $\{c(t_4)\}$. The second and third cases show that neither $\{c(t_1)\}$, nor $\{c(t_2)\}$ are $\prec'$-smaller than $\{c(t_3), c(t_4)\}$. Hence, the set $\{c(t_3), c(t_4)\}$ must be a B-preferred set of $\Gamma_0$ which is a contradiction. We conclude that the simple semantical framework is not sufficient to give a preferential semantics to B-preferred assumption sets.

Rescue comes from Lehmann's more general framework [Kraus et al., 1990], which is based on structures of the form $C := (\mathcal{S}, l, \leftarrow)$ where $\mathcal{S}$ is a set of states and $\leftarrow \subseteq \mathcal{S} \times \mathcal{S}$ is an antisymmetric relation on states. Each state $s \in \mathcal{S}$ is labelled with a set $l(s)$ of worlds (i.e. logical interpretations). A state $s$ satisfies a theory $\Gamma$ if all interpretations in $l(s)$ satisfy $\Gamma$. Let $\mathcal{S}_\Gamma$ be the set of all states satisfying $\Gamma$. A state $s$ is a $\leftarrow$-minimal element of $\mathcal{S}_\Gamma$ iff $s \in \mathcal{S}_\Gamma$ and $s^* \leftarrow s$ implies $s^* \notin \mathcal{S}_\Gamma$. $\psi$ can nonmonotonically be inferred from $\Gamma$ (written as $\Gamma \hspace{0.2em}\sim\hspace{-0.9em}\mid\hspace{0.4em}_C \psi$) iff all $\leftarrow$-minimal elements of $\mathcal{S}_\Gamma$ satisfy $\psi$.

The existence of $\leftarrow$-minimal states is ensured by the following property: The order $\leftarrow$ is called *smooth* on a set $X$ iff for all $s \in X$ there exists a $\leftarrow$-minimal state $s^*$ in $X$ such that $s^* = s$ or $s^* \leftarrow s$. $C = (\mathcal{S}, l, \leftarrow)$ is called a *cumulative model* iff the order $\leftarrow$ is smooth on $\mathcal{S}_\Gamma$ for all theories $\Gamma$. Lehmann has shown that each cumulative model defines a nonmonotonic inference relation satisfying the following basic properties. Properties 4 and 5 together are called cumulativity:

1. $\alpha \hspace{0.2em}\sim\hspace{-0.9em}\mid\hspace{0.4em}_C \alpha$
2. if $\models \alpha \equiv \beta$, $\alpha \hspace{0.2em}\sim\hspace{-0.9em}\mid\hspace{0.4em}_C \gamma$ then $\beta \hspace{0.2em}\sim\hspace{-0.9em}\mid\hspace{0.4em}_C \gamma$
3. if $\models \alpha \supset \beta$, $\gamma \hspace{0.2em}\sim\hspace{-0.9em}\mid\hspace{0.4em}_C \alpha$ then $\gamma \hspace{0.2em}\sim\hspace{-0.9em}\mid\hspace{0.4em}_C \beta$
4. if $\alpha \wedge \beta \hspace{0.2em}\sim\hspace{-0.9em}\mid\hspace{0.4em}_C \gamma$, $\alpha \hspace{0.2em}\sim\hspace{-0.9em}\mid\hspace{0.4em}_C \beta$ then $\alpha \hspace{0.2em}\sim\hspace{-0.9em}\mid\hspace{0.4em}_C \gamma$
5. if $\alpha \hspace{0.2em}\sim\hspace{-0.9em}\mid\hspace{0.4em}_C \beta$, $\alpha \hspace{0.2em}\sim\hspace{-0.9em}\mid\hspace{0.4em}_C \gamma$ then $\alpha \wedge \beta \hspace{0.2em}\sim\hspace{-0.9em}\mid\hspace{0.4em}_C \gamma$

For a given partial order $\prec$ on assumption names, we will define a particular cumulative model producing the B-preferred assumption sets as minimal states. A B-preferred set is determined by choosing a total completion of the given partial order. We make this explicit by including this order in a state. Furthermore, a state contains an assumption set $A$ and a non-empty set of worlds that satisfy $A$. This set of worlds will serve as the label of a state.

**Definition 3.3** *Let $A \subseteq \mathcal{A}$ be an assumption set, $W$ be a non-empty set of worlds satisfying $A$, and $< \subseteq \mathcal{N} \times \mathcal{N}$ be a strict total order s.t. $\prec \subseteq <$. The triple $s := (W, A, <)$ is called a static state. Let $\mathcal{S}$ be the set of all static states.*

A static state $s := (W, A, <)$ is labelled with the set $l(s) := W$ of worlds. Let $\mathcal{S}_\Gamma$ be the set of all static states satisfying a theory $\Gamma$. The usage of a set of worlds instead of a single world means that static states represent incomplete information. If $\Gamma$ contains a disjunction $a \vee b$ then a state satisfying $\Gamma$ will contain worlds that satisfy $a$ and worlds that satisfy $b$.

If a state $s := (W, A, <)$ satisfies $\Gamma$ then all worlds $w$ in $W$ satisfy $\Gamma$, as well as $A$. Since $W$ is not empty, $\Gamma \cup A$ is consistent in this case:

$$\Gamma \cup A \not\models \bot \quad \text{if} \quad (W, A, <) \in \mathcal{S}_\Gamma$$

We compare two static states $(W_1, A_1, <_1)$ and $(W_2, A_2, <_2)$ if they have the same total order $< = <_1 = <_2$. This order $<$ gives rise to the following lexicographical order.

**Definition 3.4** *Let $<$ be a strict total order and $t_1, \ldots, t_n$ be the enumeration of $\mathcal{N}$ that respects $<$. Then $<^L \subseteq 2^{\mathcal{A}} \times 2^{\mathcal{A}}$ is defined as follows: $A_1 <^L A_2$ iff there exists an $k$ s.t. $c(t_k) \in A_1 - A_2$ and*

$$A_1 \cap \{c(t_1), \ldots, c(t_{k-1})\} = A_2 \cap \{c(t_1), \ldots, c(t_{k-1})\}$$

The order $<^L$ is a strict total order and it is smooth on $\mathcal{C}_\Gamma$ for all theories $\Gamma$. B-preferred sets are obtained in the following way:

**Lemma 3.1** *$A$ is the selection of $<$ iff $A$ is the $<^L$-minimal element of $\mathcal{C}_\Gamma$.*

We prefer a static state $(W_1, A_1, <)$ to $(W_2, A_2, <)$ if $A_1 <^L_1 A_2$:

**Definition 3.5** *Let $s_1 := (W_1, A_1, <_1)$ and $s_2 := (W_2, A_2, <_2)$ be two static states. The relation $\prec_S \subseteq \mathcal{S} \times \mathcal{S}$ is defined as follows: $s_1 \prec_S s_2$ iff 1. $<_1 = <_2$ and 2. $A_1 <^L_1 A_2$.*

The resulting relation $\prec_S$ is a strict partial order which is smooth on all $\mathcal{S}_\Gamma$. Therefore, $(\mathcal{S}, l, \prec_S)$ is a cumulative model and has all the desired properties. We now establish the link to B-preferred assumption sets:

**Theorem 3.2** *$A$ is a B-preferred assumption set of $\Gamma$ iff there exists a $\prec_S$-minimal state $s := (W, A, <)$ in $\mathcal{S}_\Gamma$.*

Thus, we established a preferential semantics for B-preferred assumption sets. Note that this result corrects the approach in [Roos, 1992] who gave a first trial for such a semantics.

## 4 Dynamic Preferences

In this section, we extend the preferential semantics and the constructive characterization to dynamic preferences. First, we modify the notion of a state. Given a triple $(W, A, <)$, we consider all preferences, i.e. ground formulas of the form $t_1 \leftarrow t_2$, that are satisfied by the state. A preference $t_1 \leftarrow t_2$ is satisfied by the state if it is satisfied by all worlds in $W$.

**Definition 4.1** *Let $A \subseteq \mathcal{A}$ be an assumption set, $W$ be a non-empty set of worlds satisfying $A$, and $< \subseteq \mathcal{N} \times \mathcal{N}$ be a strict total order such that $t_1 < t_2$ if all worlds $w \in W$ satisfy $t_1 \leftarrow t_2$ (for $t_1, t_2 \in \mathcal{N}$). The triple $s := (W, A, <)$ is called a dynamic state. Let $\mathcal{D}$ be the set of all dynamic states.*

A dynamic state $s := (W, A, <)$ is labelled with the set $l(s) := W$ of worlds. Let $\mathcal{D}_\Gamma$ be the set of dynamic states satisfying the theory $\Gamma$. If $s := (W, A, <)$ satisfies a theory $\Gamma$ then $\Gamma \cup A$ is consistent and $<$ respects all preferences that can be derived from $\Gamma \cup A$ in the following way:

$$\leftarrow_{\Gamma, A} := \{(t_1, t_2) \in \mathcal{N} \times \mathcal{N} \mid \Gamma \cup A \models t_1 \leftarrow t_2\}$$

Let $W_{\Gamma \cup A}$ be the set of all worlds satisfying $\Gamma \cup A$. Then:

**Lemma 4.1** *Let $A \subseteq \mathcal{A}$ and $< \subseteq \mathcal{N} \times \mathcal{N}$ be a strict total order. $s := (W_{\Gamma \cup A}, A, <)$ is a dynamic state iff $\Gamma \cup A \not\models \bot$ and $\leftarrow_{\Gamma, A} \subseteq <$.*

We compare two dynamic states $s_1 = (W_1, A_1, <_1)$ and $s_2 = (W_2, A_2, <_2)$ even if the orders $<_1$ and $<_2$ are different. $s_1$ is smaller than $s_2$ iff there is an $\alpha \in A_1 - A_2$ such that $<_1$ and $<_2$, as well as $A_1$ and $A_2$ agree on the elements that are $<$-smaller than $\alpha$:

**Definition 4.2** *Let $s_1 := (W_1, A_1, <_1)$ and $s_2 := (W_2, A_2, <_2)$ be two dynamic states. Let $u_1, \ldots, u_n$ and $v_1, \ldots, v_n$ be the two enumerations of $\mathcal{N}$ s.t. the first one respects $<_1$ and the second one respects $<_2$. Let $\prec_D \subseteq 2^\mathcal{D} \times 2^\mathcal{D}$ be defined as follows: $A_1 \prec_D A_2$ iff there exists a $k$ s.t. $c(u_k) \in A_1 - A_2$ and*

1. *$u_i = v_i$ for $i = 1, \ldots, k$*
2. *$A_1 \cap \{c(u_1), \ldots, c(u_{k-1})\} = A_2 \cap \{c(u_1), \ldots, c(u_{k-1})\}$*

**Lemma 4.2** *$\prec_D$ is a strict partial order.*

Thus, we obtained a simple preferential semantics for dynamic preferences. We require that the total order of a state respects the preferences that are satisfied by all worlds of the state and we use a kind of lexicographical order for comparing states with different base orders.

Before showing that $\prec_D$ is smooth and that $D := (\mathcal{D}, l, \prec_D)$ is a cumulative model, we give a constructive characterization of the $\prec_D$-minimal states of $\mathcal{D}_\Gamma$. We will, step by step, construct an assumption set $A$, as well as a corresponding order $<$. Since preferences are dynamic we have to avoid certain pitfalls.

1. We can obtain cyclic preferences such as $t_1 \leftarrow t_2$ and $t_2 \leftarrow t_1$. A relation $\leftarrow$ has a cycle iff its *transitive closure* $\leftarrow^+$ is not irreflexive.

2. We can obtain preferences $t_2 \leftarrow t_1$ although we have already chosen $t_1 < t_2$.

If the first case is obtained by the initial preferences which are derived from $\Gamma$ then there is no dynamic state satisfying $\Gamma$:

**Definition 4.3** *A theory $\Gamma$ is called D-consistent iff $\Gamma$ is consistent and $\leftarrow^+_{\Gamma, \emptyset}$ is irreflexive.*

**Lemma 4.3** *There exists a dynamic state satisfying $\Gamma$ iff $\Gamma$ is D-consistent.*

The question is what to do if any of the two problematic cases above is obtained after adding an assumption $\alpha$ to a set $A$ of already selected assumptions. The answer is quite simple: Don't select $\alpha$ because otherwise the current construction will not lead to a dynamic state. This modifies the constructive definition as follows:

**Definition 4.4** *Let $\Gamma$ be a consistent theory. Let $< \subseteq \mathcal{N} \times \mathcal{N}$ be a strict total order and $t_1, \ldots, t_n$ be the enumeration of $\mathcal{N}$ that respects $<$. Let $B_i := A_i \cup \{c(t_{i+1})\}$. We define $A_0 := \emptyset$ and*

$$A_{i+1} := \begin{cases} A_i & \text{if } \Gamma \cup A_i \cup \{c(t_{i+1})\} \models \bot \\ A_i & \text{if } \leftarrow^+_{\Gamma, B_i} \text{ is not irreflexive} \\ A_i & \text{if } t_j \leftarrow_{\Gamma, B_i} t_k \text{ for a } k, j \text{ s.t.} \\ & k \leq j \text{ and } k \leq i+1 \\ A_i \cup \{c(t_{i+1})\} & \text{otherwise} \end{cases}$$

*Then $A_i$ is the dynamic selection of $t_1, \ldots, t_i$ and $A_n$ is the dynamic selection of $<$. The sequence $t_1, \ldots, t_i$ is correct iff*

$$t_k \leftarrow_{\Gamma, A_j} t_{j+1} \quad \text{implies} \quad k < j+1$$

for $j = 0, \ldots, i-1$. The order $<$ is correct iff $t_1, \ldots, t_i$ is correct for all $i = 1, \ldots, n$.

$A$ is a D-preferred assumption set of $\Gamma$ iff $A$ is the dynamic selection of a correct strict total order $<$.

We now explore the properties of these definitions. First of all, dynamic selections are consistent and correct orders respect the dynamic preferences they produce:

**Lemma 4.4** Let $\Gamma$ be a consistent theory and $< \; \subseteq \mathcal{N} \times \mathcal{N}$ be a strict total order. Let $A$ be the dynamic selection of $<$. Then $\Gamma \cup A \not\models \bot$. If $<$ is correct then $\leftarrow_{\Gamma,A} \; \subseteq \; <$.

Correct sequences $t_1, \ldots, t_i$ can be constructed incrementally. In each step, we pick a best element $t_{i+1}$ among the non-enumerated assumptions and add it after $t_i$. $t_{i+1}$ must be a best element w.r.t. the preferences $\leftarrow_{\Gamma,A_i}$ in order to guarantee that $t_1, \ldots, t_i, t_{i+1}$ is correct. Let $R_i := \mathcal{N} - \{t_1, \ldots, t_i\}$:

$$t_{i+1} \in \{x \in R_i \mid \not\exists y \in R_i : y \leftarrow_{\Gamma,A_i} x\}$$

The existence of such best elements is insured since cyclic preferences are avoided:

**Lemma 4.5** Let $\Gamma$ be a D-consistent theory. Let $t_1, \ldots, t_i$ be a correct sequence of elements of $\mathcal{N}$ and $A_i$ its dynamic selection. Then there exists an enumeration $t_{i+1}, \ldots, t_n$ of $\mathcal{N} - \{t_1, \ldots, t_i\}$ s.t. $t_1, \ldots, t_j$ is correct for all $j = 1, \ldots, n$.

Consider a dynamic state $s := (W, A, <)$ in $\mathcal{D}_\Gamma$. If $A$ is the dynamic selection of $<$ then $<$ is correct and $s$ is a $\prec_D$-minimal state in $\mathcal{D}_\Gamma$. Otherwise, there exists a state that is $\prec_D$-smaller:

**Lemma 4.6** Let $\Gamma$ be a D-consistent theory and $s := (W, A, <)$ be a dynamic state in $\mathcal{D}_\Gamma$. Let $A^*$ be the dynamic selection of $<$.

1. If $A = A^*$ then $<$ is correct.
2. If $A = A^*$ and $s^* \prec_D s$ then $s^* \notin \mathcal{D}_\Gamma$.
3. If $A \neq A^*$ then $(W_{\Gamma \cup A^*}, A^*, <) \prec_D s$.

These lemmas allow to establish the two main theorems of the paper. First, $\prec_D$-minimal states have a constructive characterization:

**Theorem 4.7** Let $\Gamma \subseteq \mathcal{L}$ be a theory. $A$ is a D-preferred assumption set of $\Gamma$ iff there exists a dynamic state $(W, A, <)$ that is a $\prec_D$-minimal element of $\mathcal{D}_\Gamma$.

Second, we can now demonstrate the smoothness of $\prec_D$:

**Lemma 4.8** $\prec_D$ is smooth on $\mathcal{D}_\Gamma$ for $\Gamma \subseteq \mathcal{L}$.

**Theorem 4.9** $D := (\mathcal{D}, l, \prec_D)$ is a cumulative model.

Therefore, the nonmonotonic inference relation $\vdash_D$ has the five basic properties of system $\mathcal{C}$.

It is straightforward to adapt our approach to Lehmann's preferential models. A *preferential model* is a triple $C := (S, l, \prec)$ where $S$ is a set of states, $l$ a function mapping each state to a *single* world, and $\prec$ is a strict partial order on states that is smooth on all $S_\Gamma$, i.e. the set of states satisfying a theory $\Gamma$. Preferential models additionally support reasoning by cases:

6. $\alpha \vdash_C \gamma, \beta \vdash_C \gamma$ implies $\alpha \vee \beta \vdash_C \gamma$

We obtain a preferential model by restricting dynamic states $(W, A, <)$ to those where the set $W$ of worlds is a singleton, i.e. contains only one world. Further work is needed to adapt the constructive approach to this preferential-model semantics.

## 5 Example

We determine the D-preferred assumption sets of our initial example. Let $\Gamma_0$ be the set of formulas 0. - 8 and

$$\Gamma_1 := \Gamma_0 \cup \{birthday(jim)\}$$
$$\Gamma_2 := \Gamma_1 \cup \{saturday\}$$
$$\Gamma_3 := \Gamma_2 \cup \{\neg go(th)\}$$

We consider two correct strict total orders $<_1$ and $<_2$ where

$$go_D(nc) <_1 go_D(th)$$
$$go_D(th) <_2 go_D(nc)$$

Due to formulas 7. and 8., the assumptions $one_D$, $inv_D$ are smaller than the assumptions $go_D(nc)$ and $go_D(th)$. These formulas have been included to give the assumptions $one_D$, $inv_D$ a higher priority. Now we consider the dynamic selections $A_{i,j}$ of $\Gamma_i$ and $<_j$:

$$A_{0,1} = \{c(one_D), c(inv_D), c(go_D(nc))\} \quad *$$
$$A_{0,2} = \{c(one_D), c(inv_D), c(go_D(th))\}$$
$$A_{1,1} = \{c(one_D), c(go_D(nc))\}$$
$$A_{1,2} = \{c(one_D), c(go_D(th))\} \quad *$$
$$A_{2,1} = \{c(go_D(nc)), c(go_D(th))\}$$
$$A_{2,2} = \{c(go_D(th)), c(go_D(nc))\} \quad *$$
$$A_{3,1} = \{c(go_D(nc))\}$$
$$A_{3,2} = \{c(go_D(nc))\} \quad *$$

Since a normal invitation $c(inv_D)$ implies $go_D(nc) \leftarrow go_D(th)$ the order $<_2$ is not correct w.r.t. $\Gamma_0$. Since the theory $\Gamma_1$ implies $go_D(th) \leftarrow go_D(nc)$ the order $<_1$ is not correct w.r.t. $\Gamma_1$, $\Gamma_2$, and $\Gamma_3$. As a consequence, each $\Gamma_i$ has a unique D-preferred assumption set (marked with a *) and we obtain the following inferences:

$$\Gamma_0 \vdash_D go(nc) \qquad \Gamma_2 \vdash_D go(nc) \wedge go(th)$$
$$\Gamma_1 \vdash_D go(th) \qquad \Gamma_3 \vdash_D go(nc)$$

The conclusions change from $\Gamma_0$ to $\Gamma_1$ since the preferences change. The change from $\Gamma_1$ to $\Gamma_2$ is due to the removal of a conflict. The final change is due to a new inconsistency.

## 6 Related Work

Brewka has extended Reiter's default logic by dynamic preferences on defaults [Brewka, 1994]. As in our approach, defaults are named by constants and preferences between defaults are expressed by a binary predicate symbol. The additional expressiveness of default logic, however, makes it difficult to establish a preferential semantics. Even normal defaults as considered in [Brewka, 1994] do not have a cumulative-model semantics as shown by Makinson. In order to compare both approaches, we restrict our attention to normal defaults without prerequisites, which correspond to assumptions.

Brewka requires that a theory $\Gamma$ contains axioms stating that the predicate symbol $\leftarrow$ represents a strict partial order. These axioms ensure that no cyclic preferences are obtained. In our approach, we did not want to change the original theory $\Gamma$ and therefore required a corresponding property on the meta-level. Brewka determines preferred assumption sets as follows. An assumption set $A \subseteq \mathcal{A}$ is a BD-preferred iff 1. it is the (static !) selection of a strict total order $< \subseteq \mathcal{N} \times \mathcal{N}$ and 2. $\Gamma \cup A \cup \{t_1 \leftarrow t_2 \mid t_1 < t_2\}$ is consistent. Here, a total order on assumptions is chosen initially and verified in the end by comparing the chosen preferences with those that are implied by $\Gamma \cup A$. Unfortunately, there are examples that don't have BD-preferred assumptions:

$$c(t_1) \supset t_2 \leftarrow t_1, \; c(t_2) \supset t_1 \leftarrow t_2$$

The order $t_1 < t_2$ is not compatible with its (static) selection $\{c(t_1)\}$, but correct w.r.t. its dynamic selection $\{c(t_2)\}$. The selection of $c(t_1)$ fails in the second case since the preference $t_2 \leftarrow t_1$ contradicts the order $t_1 < t_2$. An analogue argument holds for $t_2 < t_1$.

The example shows that cyclic dependencies between preferences and assumptions make the search for preferred assumption sets quite difficult. Therefore, we interleave the construction of an order and an assumption set and we do not choose assumptions that have drawbacks on the already chosen part of the order.

## 7 Conclusion

We showed how preferences on assumptions can directly be expressed in a logical theory. The resulting system offers a high degree of freedom for "programming" preference rules: Preferences can be used in implications, in quantified statements, and can themselves depend on other assumptions.

Finding a clear mathematical treatment of dynamic preferences turned out to be a non-trivial task. We developed a preferential semantics based on Lehmann's cumulative models and an equivalent constructive characterization. The resulting nonmonotonic logic

1. allows to program preference rules,
2. satisfies all properties of Lehmann's system $\mathcal{C}$,
3. can be implemented for decidable sublanguages.

In order to keep the presentation simple and intuitive, we considered only finite assumption sets in this paper. In a long version of the paper, we will generalize the results to infinite assumption sets and well-founded orders on assumptions.

Thus, an important milestone in the design of an applicable and powerful nonmonotonic logic has been achieved. It can be applied to default reasoning in inheritance system, to diagnostic reasoning, and to decision making. Future work will concentrate on algorithms and applications. Furthermore, we will elaborate a variant of our approach in Lehmann's system $\mathcal{P}$ which additionally supports reasoning by cases.

## Acknowledgements

I would like to thank Gerd Brewka, Markus Junker, and the anonymous reviewers for helpful comments that improved the quality of this paper. The paper would not have been written without the moral support of my wife Isabelle, my son Kevin, and my daughter Céline.

## References

[Brewka, 1989] G. Brewka. Preferred subtheories: An extended logical framework for default reasoning. In *IJCAI-89*, pages 1043–1048, Detroit, MI, 1989. Morgan Kaufmann.

[Brewka, 1994] G. Brewka. Reasoning about priorities in default logic. In *AAAI-94*, 1994.

[Brewka, 1996] G. Brewka. Well-founded semantics for extended logic programs with dynamic preferences. *Journal of Artificial Intelligence Research*, 4, 1996.

[Geffner and Pearl, 1992] H. Geffner and J. Pearl. Conditional entailment: Bridging two approaches to default reasoning. *Artificial Intelligence*, 53:209–244, 1992.

[Grosof, 1991] B. Grosof. Generalizing prioritization. In *KR'91*, pages 289–300, Cambridge, MA, 1991. Morgan Kaufmann.

[Junker and Brewka, 1991] U. Junker and G. Brewka. Handling partially ordered defaults in TMS. In R. Kruse and P. Siegel, editors, *Symbolic and Quantitative Aspects for Uncertainty. Proceedings of the European Conference ECSQAU*, pages 211–218. Springer, LNCS 548, Berlin, 1991.

[Junker, 1992] U. Junker. *Relationships between Assumptions*. Doctoral thesis, University of Kaiserslautern, Kaiserslautern, 1992.

[Junker, 1995] U. Junker. Büroeinrichtung mit EXCEPT II. In F. di Primio, editor, *Methoden der Künstlichen Intelligenz für Graphikanwendungen*. Addison-Wesley, 1995.

[Kraus et al., 1990]
S. Kraus, D. Lehmann, and M. Magidor. Nonmonotonic reasoning, preferential models and cumulative logics. *Artificial Intelligence*, 44:167–208, 1990.

[Lifschitz, 1985] V. Lifschitz. Computing circumscription. In *IJCAI-85*, pages 121–127, Los Angelos, CA, 1985. Morgan Kaufmann.

[McCarthy, 1986] J. McCarthy. Applications of circumscription to formalizing common–sense knowledge. *Artificial Intelligence*, 28:89–116, 1986.

[Poole, 1988] D. Poole. A logical framework for default reasoning. *Artificial Intelligence*, 36:27–47, 1988.

[Roos, 1992] N. Roos. A logic for reasoning with inconsistent knowledge. *Artificial Intelligence*, 1992.

[Shoham, 1987] Y. Shoham. A semantical approach to nonmonotonic logics. In *Logics in Computer Science*, 1987.

# Compiling reasoning with and about preferences into default logic

**James P. Delgrande**
School of Computing Science
Simon Fraser University
Burnaby, B.C., Canada V5A 1S6
jim@cs.sfu.ca

**Torsten H. Schaub**[*]
Institut für Informatik
Universität Potsdam
Postfach 60 15 53, D–14415 Potsdam
torsten@cs.uni-potsdam.de

## Abstract

We address the problem of introducing preferences into default logic. Two approaches are given, one a generalisation of the other. In the first approach, an *ordered default theory* consists of a set of default rules, a set of world knowledge, and a set of fixed preferences on the default rules. This theory is transformed into a second, standard default theory, where, via the naming of defaults, the given preference ordering on defaults is respected. In the second approach, we begin with a default theory where preference information is specified as part of an overall default theory. Here one may specify preferences that hold by default, or give preferences among preferences. Again, such a theory is translated into a standard default theory. The approach differs from previous work in that we obtain standard default theories, and do not rely on prioritised versions, as do other approaches. In practical terms this means we can immediately use existing default logic theorem provers for an implementation. From a theoretical point of view, this shows that the explicit representation of priorities adds nothing to the overall expressibility of default logic.

## 1 Introduction

In many situations in nonmonotonic reasoning the application of one default is preferred to another. Perhaps the best known example is inheritance of properties, where an individual is assumed to have properties by default according to the most specific class(es) to which it belongs. Hence an individual that is a penguin (and so a bird) does not fly by default, since penguins typically don't fly, even though birds do typically fly. Preferences are also found in decision making and in scheduling. For example, in scheduling not all deadlines may be simultaneously satisfiable; preferences then may allow some compromise solution. In legal reasoning laws may apply by default, but the laws themselves may conflict; such conflicts may be adjudicated by higher-level principles.

Our goal in this paper is to explore preference orderings in nonmonotonic reasoning, specifically in default logic [Reiter, 1980]. In the next section we examine the general notion of preference orderings on defaults. We note that there is not a single way in which preferences should be applied. Rather what we will call *specificity* orderings, as are used in property inheritance, can be distinguished from *preference* or *priority* orderings. These notions have frequently been conflated previously in the literature; here our concerns lie solely with preference orderings.

In considering how preference orderings may be enforced in default logic, we consider first where a default theory consists of world knowledge and a set of default rules, together with (external) preference information between default rules. We show how such a default theory can be translated into a second theory wherein preference information is now incorporated into the theory. So with this translation we obtain a theory in "standard" default logic, rather than requiring machinery external to default logic, as is found in previous approaches. We subsequently generalise this approach so that preferences may appear arbitrarily as part of a default theory and, specifically, preferences among default rules may (via the naming of default rules) themselves be part of a default rule. We again show how such a generalised default theory can be translated into a "standard" default theory where preference information is incorporated into the theory.

Previous approaches have generally added machinery to an extant approach to nonmonotonic reasoning. We remain within the framework of standard default logic, rather than building a scheme on top of default logic, for several reasons. First, there exist theorem provers for default logic. Consequently our approach can be im-

[*]Previously at LERIA, Université d'Angers, France.

mediately incorporated in such a prover. Second, it is easier to compare differing approaches to handling such orderings. Third, by "compiling" preferences into default logic, and in using the standard machinery of default logic, we obtain insight into the notion of preference orderings. Thus for example we implicitly show that explicit priorities provide no real increase in the expressibility of default logic.

## 2 Preference Orderings

This section discusses preference orderings in general; while we employ default logic, the discussion is independent of any particular approach to nonmonotonic reasoning. We can use default rules[1] to express that $\beta$ follows by default from $\alpha$ by $\frac{\alpha:\beta}{\beta}$. Then we can write $\frac{\alpha:\beta}{\beta} < \frac{\gamma:\delta}{\delta}$ to express a preference between two defaults. Assume that we have an *ordered default theory*. The exact details are given in the next section; for the time being assume that we have a triple $(D, W, <)$, where $D$ is a set of default rules, $W$ a set of formulas, and $<$ is a strict partial order on the default rules in $D$.

Informally a higher-ranked default should be applied or considered before a lower-ranked default. But what exactly does this mean? Consider for example the defaults concerning primary means of locomotion: "animals normally walk", "birds normally fly", "penguins normally swim":

$$\frac{Animal:Walk}{Walk} < \frac{Bird:Fly}{Fly} < \frac{Penguin:Swim}{Swim}. \quad (1)$$

If we learn that some thing is penguin (and so a bird and animal), then we would want to apply the highest-ranked default, if possible, and only the highest-ranked default. Significantly, if the penguins-swim default is blocked (say the penguin in question has a fear of water) we *don't* try to apply the next default to see if it might fly. This then is standard inheritance of default properties.

The situation is very different in the next example. We have the defaults that "Canadians speak English by default", "Québecois speak French by default", "residents of the north of Québec speak Cree by default":

$$\frac{Can:English}{English} < \frac{Que:French}{French} < \frac{NQue:Cree}{Cree}. \quad (2)$$

Now if a resident of the north of Québec didn't speak Cree, it would be reasonable to assume that that person spoke French, and if they didn't speak French, then English.

Assume that we have a chain of defaults $\delta_1 < \delta_2 < \ldots < \delta_m$. Informally, we have the following possibilities with respect to how the defaults may be applied.

[1]Default logic is introduced shortly.

**P1** For the maximum $i$ for which $\delta_i$ is applicable, and for $j > 0$, no $\delta_{i+j}$ is denied,[2] apply $\delta_i$ if possible. No other default is considered. This is the situation in (1).

**P2** Apply $\delta_m$ if possible; apply $\delta_{m-1}$ if possible, continue in this fashion until no more than $k$ (for fixed $k$ where $1 \leq k \leq m$) defaults have been applied. This is the situation in (2) with $k = 1$.

An example of a general instance of **P2** is where a student wishes to take $k = 3$ computing courses, out of $m = 10$ possible courses, and so provides a list of preferences over the courses. There are two important subcases of **P2** corresponding to $k = 1$ and $k = m$. In the first case a maximum of one default is applied. In the second case one attempts to apply every default.

**P1** is essentially (default property) inheritance. The ordering on defaults reflects a relation of *specificity*; one attempts to apply the most specific default possible. In approaches such as [Touretzky et al., 1987; Pearl, 1990; Geffner & Pearl, 1992] specificity is determined implicitly, emerging as a property of the underlying system. [Reiter & Criscuolo, 1981; Etherington & Reiter, 1983; Delgrande & Schaub, 1994] have addressed adding specificity information in default logic. For incorporating preferences (as given in **P2**), [Boutilier, 1992; Brewka, 1994a; Baader & Hollunder, 1993] consider adding preferences in default logic while [McCarthy, 1986; Lifschitz, 1985; Grosof, 1991] do the same in circumscription. We note however that some of these latter papers include examples best interpreted as dealing with specificity (as given in (1)), and so would appear to conflate **P1** and **P2**.

Our concerns in this paper are with specifying preferences and priorities, as given in **P2**. We assume only that we are given a set of defaults and a priority policy on defaults, along with other world knowledge. We observe that the framework as given in **P2** is significantly more general that that of **P1**. For example it seems to be an intrinsic property of inheritance as given in **P1** that the ordering on defaults is determined by relative specificity of the prerequisites. So if $\delta_1 < \delta_2$ then the antecedent of $\delta_1$ is less specific than the antecedent of $\delta_2$. This is not the case for **P2** though. Consider a variation on (2) where in the north of Québec the first language is French, then English, then Cree: The resulting preference ordering is as follows.

$$\frac{NQue:Cree}{Cree} < \frac{Can:English}{English} < \frac{Que:French}{French}. \quad (3)$$

Indeed for preferences, one need not have any antecedent information. That one prefers something (say, a car) that is red, then green might be expressed as $\frac{:Green}{Green} <$

[2]i.e. we don't have that the prerequisite is true and the consequent false.

$\frac{:Red}{Red}$. In the most general case, we might have two defaults, with no relation between them, except for some given preference relation.

## 3 Default Logic and Ordered Default Logic

Default logic [Reiter, 1980] augments classical logic by *default rules* of the form $\frac{\alpha:\beta_1,\ldots,\beta_n}{\gamma}$. For the most part we deal with *singular* defaults for which $n = 1$. [Marek & Truszczyński, 1993] show that any default rule can be transformed into a set of singular defaults; hence our one use of a non-singular rule in Section 5 is for notational convenience only. A singular rule is *normal* if $\beta$ is equivalent to $\gamma$; it is *semi-normal* if $\beta$ implies $\gamma$. We sometimes denote the *prerequisite* $\alpha$ of a default $\delta$ by $Prereq(\delta)$, its *justification* $\beta$ by $Justif(\delta)$ and its *consequent* $\gamma$ by $Conseq(\delta)$. Empty components, such as no prerequisite or even no justifications, are assumed to be tautological. Defaults with unbound variables are taken to stand for all corresponding instances. A set of default rules $D$ and a set of formulas $W$ form a *default theory* $(D, W)$ that may induce a single or multiple *extensions* in the following way.

**Definition 3.1** *Let $(D, W)$ be a default theory and let $E$ be a set of formulas. Define $E_0 = W$ and for $i \geq 0$:*

$$\Gamma_i = \left\{ \frac{\alpha:\beta_1,\ldots,\beta_n}{\gamma} \in D \,\middle|\, \alpha \in E_i, \neg\beta_1 \notin E, \ldots, \neg\beta_n \notin E \right\}$$
$$E_{i+1} = Th(E_i) \cup \{Conseq(\delta) \mid \delta \in \Gamma_i\}$$

*Then $E$ is an extension for $(D, W)$ if $E = \bigcup_{i=0}^{\infty} E_i$.*

Any such extension represents a possible set of beliefs about the world at hand. The above procedure is not constructive since $E$ appears in the specification of $E_{i+1}$. We define[3] $\Gamma = \bigcup_{i=0}^{\infty} \Gamma_i$ as the set of default rules *generating* extension $E$.

For adding preferences among default rules, a default theory is usually extended with an ordering on the set of default rules. In analogy to [Baader & Hollunder, 1993; Brewka, 1994a], an *ordered default theory* $(D, W, <)$ is a finite set $D$ of default rules, a set $W$ of formulas, and a strict partial order $< \subseteq D \times D$ on the default rule. For simplicity we assume the existence of a default $\delta_\top = \frac{\top:\top}{\top} \in D$ where for every rule $\delta_i \neq \delta_\top$ we have $\delta_i < \delta_\top$. This gives us a (trivial) maximally preferred default that is always applicable to "start things off".

## 4 Static Preferences on Defaults

We show here how ordered default theories can be translated into standard default theories. Our strategy is to add sufficient "hooks" to a default rule in a theory to enable the control of rule application. We begin with an ordered default theory $(D, W, <)$ which is then translated into a standard default theory $(D', W')$ such that the explicit preferences in $<$ are "compiled" into $D'$ and $W'$.

To this end, we associate a unique name with each default rule. This is done by extending the original language by a set of constants $N$[4] such that there is a bijective mapping $n : D \to N$. We write $n_\delta$ instead of $n(\delta)$ (and we often abbreviate $n_{\delta_i}$ by $n_i$ to ease notation). For default rule $\delta$ along with its name $n$, we sometimes write $n : \delta$ to render naming explicit. To reflect the fact that we deal with a finite set of distinct default rules, we adopt a unique names assumption (UNA) and domain closure assumption (DCA) with respect to $N$. That is, for a name set $N = \{n_1, \ldots, n_m\}$, we add axioms $\forall x.\, name(x) \equiv (x = n_1 \vee \ldots \vee x = n_m)$ and $(n_i \neq n_j)$ for all $n_i, n_j \in N$ with $i \neq j$. For convenience, we then write $\forall x \in N.\, P(x)$ instead of $\forall x.\, name(x) \supset P(x)$. In the sequel, we use $\text{DCA}_N$ and $\text{UNA}_N$ to abbreviate domain closure and unique names axioms for $N$.

The use of names allows the expression of preference relations between default rules in the object language. So we can assert that default $n_j : \frac{\omega:\phi}{\varphi}$ is preferred to $n_i : \frac{\alpha:\beta}{\gamma}$ by $n_i \prec n_j$, where $\prec$ is a (new) predicate in the object language. Finally, in discussions of preference relations, we sometimes write $\delta_i < \delta_j$ or $\frac{\alpha:\beta}{\gamma} < \frac{\omega:\phi}{\varphi}$, to show a preference between two defaults; however it should be kept in mind that these latter expressions are not expressions *within* a default theory (as are given by $\prec$), but expressions *about* a default theory.

If we are given $\delta_i < \delta_j$, then we want to ensure that before $\delta_i$ is applied, that $\delta_j$ be applied or found to be inapplicable.[5] We do this by first translating default rules so that rule application can be explicitly controlled. For this purpose, we need to be able to, first, detect when a rule has been applied or when a rule is blocked, and, second, control the application of a rule based on other antecedent conditions. For a default rule $\frac{\alpha:\beta}{\gamma}$ there are two cases for it to not be applied: it may be that the antecedent is not known to be true (and so its negation is consistent), or it may be that the justification is not consistent (and so its negation is known to be true). For detecting this case, we introduce a new, special-purpose predicate $\mathsf{bl}(\cdot)$. Similarly we introduce a special-purpose predicate $\mathsf{ap}(\cdot)$ to detect the case where a rule has been applied. For controlling application of a rule we introduce predicate $\mathsf{ok}(\cdot)$. Then, a default rule $\delta = \frac{\alpha:\beta}{\gamma}$ is

---

[3] For simplicity, we refrain from parameterizing $\Gamma$ by $D$ and $E$.

[4] This is done also in [Brewka, 1994b]. Theorist [Poole, 1988] uses atomic propositions to name defaults.

[5] That is, we wish to exclude the case where $\delta_i \in \Gamma_n$ and $\delta_j \in \Gamma_m$ for $n \leq m$.

mapped to

$$\frac{\alpha \wedge \mathsf{ok}(n_\delta) : \beta}{\gamma \wedge \mathsf{ap}(n_\delta)}, \quad \frac{\mathsf{ok}(n_\delta) : \neg\alpha}{\mathsf{bl}(n_\delta)}, \quad \frac{\neg\beta \wedge \mathsf{ok}(n_\delta) :}{\mathsf{bl}(n_\delta)}.$$

These rules are sometimes abbreviated by $\delta_a, \delta_{b_1}, \delta_{b_2}$, resp.

None of the three rules in the translation can be applied unless $\mathsf{ok}(n)$ is true. Since $\mathsf{ok}(\cdot)$ is a new predicate symbol, it can be expressly made true in order to potentially enable the application of the three rules in the image of the translation. If $\mathsf{ok}(n)$ is true, the first rule of the translation may potentially be applied. If a rule has been applied, then this is "recorded" by assertion $\mathsf{ap}(n)$. The last two rules give conditions under which the original rule is inapplicable: either the negation of the original antecedent $\alpha$ is consistent (with the extension) or the justification $\beta$ is known to be false; in either such case $\mathsf{bl}(n)$ is concluded.

This translation clearly says nothing about which default precedes what in order of application. However, for $\delta_i < \delta_j$ we can now fully control the order of rule application: if $\delta_j$ has been applied (and so $\mathsf{ap}(n_j)$ is true), or known to be inapplicable (and so $\mathsf{bl}(n_j)$ is true), then it's ok to apply $\delta_i$. So we would have something like $(\mathsf{ap}(n_j) \vee \mathsf{bl}(n_j)) \supset \mathsf{ok}(n_i)$, but adjusted to allow for the fact that we might have other rules with higher priority than $\delta_i$. Further, given that $\delta_1 < \delta_2 < \delta_3$ we would want the order of application to be $\delta_3$ then $\delta_2$ then $\delta_1$. Given the predicates $\mathsf{bl}(\cdot)$ and $\mathsf{ap}(\cdot)$ it is a straightforward matter to also assert that a maximum of one default in a priority order can be applied, or in the general case that $k$ rules can be applied.

Taking all this into account, we obtain the following translation mapping ordered default theories in some language $\mathcal{L}$ onto standard default theories in the language $\mathcal{L}^+$ obtained by extending $\mathcal{L}$ by new predicates symbols $(\cdot \prec \cdot)$, $\mathsf{ok}(\cdot)$, $\mathsf{bl}(\cdot)$, and $\mathsf{ap}(\cdot)$, and a set of associated default names:

**Definition 4.1** *Given an ordered default theory $(D, W, <)$ over $\mathcal{L}$ and its set of default names $N = \{n_\delta \mid \delta \in D\}$, define $\mathcal{T}((D, W, <)) = (D', W')$ over $\mathcal{L}^+$ by*

$$D' = \left\{ \frac{\alpha \wedge \mathsf{ok}(n) : \beta}{\gamma \wedge \mathsf{ap}(n)}, \frac{\mathsf{ok}(n) : \neg\alpha}{\mathsf{bl}(n)}, \frac{\neg\beta \wedge \mathsf{ok}(n) :}{\mathsf{bl}(n)} \,\middle|\, n : \frac{\alpha : \beta}{\gamma} \in D \right\} \cup D_\prec$$

$$W' = W \cup W_\prec \cup \{DCA_N, UNA_N\}$$

*where*

$$D_\prec = \left\{ \frac{: \neg(x \prec y)}{\neg(x \prec y)} \right\}$$

$$W_\prec = \{n_\delta \prec n_{\delta'} \mid (\delta, \delta') \in <\} \cup \{\mathsf{ok}(n_\top)\}$$
$$\cup \{\forall x \in N. [\forall y \in N. (x \prec y) \supset (\mathsf{bl}(y) \vee \mathsf{ap}(y))] \supset \mathsf{ok}(x)\}$$

$W'$ contains prior world knowledge, together with assertions for managing the priority order $<$ on defaults.

The first part of $W_\prec$ specifies that $\prec$ is a predicate drawn from strict partial order $<$. $\mathsf{ok}(n_\top)$ asserts that it is ok to apply the maximally preferred (trivial) default. The third formula in $W_\prec$ controls the application of defaults: for every $n_i$, we derive $\mathsf{ok}(n_i)$ whenever for every $n_j$ with $n_i \prec n_j$, either $\mathsf{ap}(n_j)$ or $\mathsf{bl}(n_j)$ is true. This axiom allows us to derive $\mathsf{ok}(n_i)$, indicating that $\delta_i$ may potentially be applied, whenever we have for all $\delta_j$ with $\delta_i < \delta_j$ that $\delta_j$ has been applied or cannot be applied.

This alone however gives necessary but not sufficient conditions for rendering $\delta_i$ potentially applicable. If $(\delta_i, \delta_j) \notin <$ then $(n_i \prec n_j) \notin W_\prec$; however, for the last formula in $W_\prec$ to work properly we must be able to conclude (in the extension) that $\neg(n_i \prec n_j)$. This is addressed by adding the default rule in $D_\prec$ that renders the resulting theory complete with respect to priority statements. That is, for all resulting extensions $E$ we have that $(n_i \prec n_j) \in E$ or $\neg(n_i \prec n_j) \in E$. We also have $(n_\delta \prec n_\top) \in W'$ for every rule $\delta \neq \delta_\top$ by the definition of ordered default theories. Since $<$ is a strict partial order, $W'$ also includes the transitive closure of $\prec$ and no reflexivities, such as $n \prec n$.

As an example, consider the defaults:

$$n_1 : \frac{A_1 : B_1}{C_1}, \quad n_2 : \frac{A_2 : B_2}{C_2}, \quad n_3 : \frac{A_3 : B_3}{C_3}, \quad n_\top : \frac{\top : \top}{\top}$$

we obtain for $i = 1, 2, 3$:

$$\frac{A_i \wedge \mathsf{ok}(n_i) : B_i}{C_i \wedge \mathsf{ap}(n_i)} \quad \frac{\mathsf{ok}(n_i) : \neg A_i}{\mathsf{bl}(n_i)} \quad \frac{\neg B_i \wedge \mathsf{ok}(n_i) :}{\mathsf{bl}(n_i)}$$

and analogously for $\delta_\top$ where $A_i, B_i, C_i$ are $\top$. Given that $\delta_1 < \delta_2 < \delta_3$, we obtain $n_1 \prec n_2, n_2 \prec n_3, n_1 \prec n_3$ along with $n_k \prec n_\top$ for $k \in \{1, 2, 3\}$ as part of $W_\prec$. From $D_\prec$ we get $\neg(n_i \prec n_j)$ for all remaining combinations of $i, j \in \{1, 2, 3, \top\}$. It is instructive to verify that $\mathsf{ok}(n_3)$, along with $(\mathsf{ap}(n_3) \vee \mathsf{bl}(n_3)) \supset \mathsf{ok}(n_2)$, and $((\mathsf{ap}(n_2) \vee \mathsf{bl}(n_2)) \wedge (\mathsf{ap}(n_3) \vee \mathsf{bl}(n_3))) \supset \mathsf{ok}(n_1)$ is contained in $E_2$ in Definition 3.1; from this we get that $n_3$ must be applied first, followed by $n_2$ and then $n_1$.

Now, given $A_1, A_2, A_3$, we obtain subsequently $C_3 \wedge \mathsf{ap}(n_3) \in E_3$, $\mathsf{ok}(n_2) \in E_4$, $C_2 \wedge \mathsf{ap}(n_2) \in E_5$, $\mathsf{ok}(n_1) \in E_6$, $C_1 \wedge \mathsf{ap}(n_1) \in E_7$. Given additionally $C_3 \supset \neg B_2$ and $C_2 \supset \neg B_3$, we get $\mathsf{bl}(n_2) \in E_5$ instead of $C_2 \wedge \mathsf{ap}(n_2) \in E_5$ because $\neg B_2 \in E_4$. Suppose there is an extension containing $C_2 \wedge \neg B_3$ as opposed to $C_3 \wedge \neg B_2$. Then however we have neither $\mathsf{ap}(n_3) \in E_3$ nor $\mathsf{bl}(n_3) \in E_3$, which makes it impossible to derive $\mathsf{ok}(n_2)$ and thus $\neg B_3$ cannot belong to such an extension.

The following theorems summarize the major properties of our approach, and demonstrate that rules are applied in the desired order:

**Theorem 4.1** *Let $E$ be an extension of $\mathcal{T}((D, W, <))$ for ordered default theory $(D, W, <)$. We have for all $\delta, \delta' \in D$*

*1. $n_\delta \prec n_{\delta'} \in E$ or $\neg(n_\delta \prec n_{\delta'}) \in E$*

2. $\mathsf{ok}(n_\delta) \in E$

3. either $\mathsf{ap}(n_\delta) \in E$ or $\mathsf{bl}(n_\delta) \in E$

4. $\mathsf{ok}(n_\delta) \in E_i$ and $\mathit{Prereq}(\delta) \in E_j$ and $\neg \mathit{Justif}(\delta) \notin E$ implies $\mathsf{ap}(n_\delta) \in E_{\max(i,j)+1}$

5. $\mathsf{ok}(n_\delta) \in E_i$ and $\mathit{Prereq}(\delta) \notin E$ implies $\mathsf{bl}(n_\delta) \in E_{i+1}$

6. $\mathsf{ok}(n_\delta) \in E_i$ and $\neg \mathit{Justif}(\delta) \in E$ implies $\mathsf{bl}(n_\delta) \in E_j$ for some $j > i$

7. $\mathsf{ok}(n_\delta) \notin E_{i-1}$ and $\mathsf{ok}(n_\delta) \in E_i$ implies $\mathsf{ap}(n_\delta), \mathsf{bl}(n_\delta) \notin E_j$ for $j \leq i$

Notably, Theorem 4.1.5 allows us to detect blockage due to non-derivability of the prerequisite immediately after having the "ok" for the default at hand. For an extension $E$ and its generating default rules $\Gamma$, we trivially have $\delta_a \in \Gamma$ iff $\mathsf{ap}(n_\delta) \in E$, and $\delta_{b_1} \in \Gamma$ or $\delta_{b_2} \in \Gamma$ iff $\mathsf{bl}(n_\delta) \in E$.

**Theorem 4.2** *Let $E$ be an extension of $\mathcal{T}((D,W,<))$ for ordered default theory $(D,W,<)$ and $\Gamma, \Gamma_i$ be defined wrt $E$. Then, we have for all $\delta \in D$*

8. $\delta_a \in \Gamma$ or $\delta_{b_1} \in \Gamma$ or $\delta_{b_2} \in \Gamma$

9. $\delta_a \in \Gamma$ iff $(\delta_{b_1} \notin \Gamma$ and $\delta_{b_2} \notin \Gamma)$

*For all default rules $\delta, \delta' \in D$ such that $\delta < \delta'$, we have*

10. $\delta'_a, \delta'_{b_1}, \delta'_{b_2} \notin \Gamma_i$ implies $\delta_a, \delta_{b_1}, \delta_{b_2} \notin \Gamma_j$ for $j < i + 3$

11. $\delta'_a \in \Gamma_i$ or $\delta'_{b_1} \in \Gamma_i$ or $\delta'_{b_2} \in \Gamma_i$ implies $\delta_a \in \Gamma_j$ or $\delta_{b_1} \in \Gamma_j$ or $\delta_{b_2} \in \Gamma_j$ for some $j > i + 1$

12. $\delta_a \in \Gamma_i$ or $\delta_{b_1} \in \Gamma_i$ or $\delta_{b_2} \in \Gamma_i$ implies $\delta'_a \in \Gamma_j$ or $\delta'_{b_1} \in \Gamma_j$ or $\delta'_{b_2} \in \Gamma_j$ for some $j < i - 1$.

The minimum two-step delay between rules stemming from $\delta$ and those originated by $\delta'$ is due to to the fact that in Definition 3.1 the deductive closure of $E_i$ is determined at $E_{i+1}$. The important overall consequence of this series of propositions is that we have full control over default application.

Using the above properties, we can show that any extension of a translated default theory is a regular extension of the underlying *unordered* default theory:

**Theorem 4.3** *Let $E$ be an extension of $\mathcal{T}((D,W,<))$ for ordered default theory $(D,W,<)$ over $\mathcal{L}$. Then $E \cap \mathcal{L}$ is an extension of $(D,W)$.*

The approach is equivalent (modulo the original language) to standard default logic if there are no preferences:

**Theorem 4.4** *For a default theory $(D,W)$ over $\mathcal{L}$ and a set of formulas $E$, we have that $E$ is an extension of $\mathcal{T}((D,W,\emptyset))$ iff $E \cap \mathcal{L}$ is an extension of $(D,W)$.*

## 5 Dynamic Preferences on Defaults

We now consider situations where the presence of preferences is context-dependent. We deal with standard default theories $(D,W)$ over a language already including a predicate $\prec$ expressing a preference relation by means of default names. In order to keep a finite domain closure axiom, we restrict ourselves to a finite set of default rules $D$, being in one-to-one correspondence with a finite name set $N$.

Since preferences are now available dynamically by inferences from $W$ and $D$, we lack a priori complete information about the ordering predicate $\prec$, which was available in the rigid case by appeal to the explicit order $<$ between rules and the "closed world default" $\frac{:\neg(x \prec y)}{\neg(x \prec y)}$. This however leads to a problem. Consider where our only preference is given by $\frac{:n \prec m}{n \prec m}$. Either the default would apply or it would not; in either case we would expect one extension only. However we also have the "closed world" default for preferences, given in $D_\prec$, that asserts that if there is no known or derived preference between rules, then no preference exists. An instance of $D_\prec$ is $\frac{:\neg(n \prec m)}{\neg(n \prec m)}$. So if we simply have these two defaults then we run the risk of potentially having an unwanted extension where $\frac{:\neg(n \prec m)}{\neg(n \prec m)}$ applies over $\frac{:n \prec m}{n \prec m}$. Obviously we can't solve the problem by asserting that $\frac{:\neg(n \prec m)}{\neg(n \prec m)} < \frac{:n \prec m}{n \prec m}$ since our approach would now be circular. We address this issue by adding a new binary predicate $\not\prec$ indicating that for defaults $\delta$ and $\delta'$ neither $(n_\delta \prec n_{\delta'}) \in E$ nor $\neg(n_\delta \prec n_{\delta'}) \in E$. We add the following rule, where $x, y$ are variables ranging over default names:

$$\frac{:\neg(x \prec y), (x \prec y)}{(x \not\prec y)} \qquad (4)$$

This rule accounts for situations where neither $(x \prec y)$ nor $\neg(x \prec y)$ is derivable. That is for names $n$ and $n'$, the only time this rule will apply is when $n \prec n' \notin E$ and $\neg(n \prec n') \notin E$. So, since $\not\prec$ is an introduced predicate, the only time we have $n \not\prec n' \in E$ is when the default theory has no information on whether the two defaults are in a preference relation or not.

We now consider standard default theories in a language $\mathcal{L}$ including the set of default names and propositions formed by binary predicate $\prec$ applied to variables and default names only; these are mapped onto theories in the language $\mathcal{L}^\star$ obtained by extending $\mathcal{L}$ with new predicate symbols $(\cdot \not\prec \cdot)$, $\mathsf{ok}(\cdot)$, $\mathsf{bl}(\cdot)$, and $\mathsf{ap}(\cdot)$:

**Definition 5.1** *Given a default theory $(D,W)$ over $\mathcal{L}$ and its set of default names $N = \{n_\delta \mid \delta \in D\}$, we define $\mathcal{T}((D,W)) = (D', W')$ over $\mathcal{L}^\star$ by*

$$D' = \left\{ \frac{\alpha \wedge \mathsf{ok}(n) : \beta}{\gamma \wedge \mathsf{ap}(n)}, \frac{\mathsf{ok}(n) : \neg \alpha}{\mathsf{bl}(n)}, \frac{\neg \beta \wedge \mathsf{ok}(n) :}{\mathsf{bl}(n)} \;\middle|\; n : \frac{\alpha : \beta}{\gamma} \in D \right\} \cup D_\prec$$

$$W' = W \cup W_\prec \cup \{DCA_N, UNA_N\}$$

*where*

$$D_\prec = \left\{ \frac{:\neg(x \prec y), (x \prec y)}{(x \not\prec y)} \right\}$$

$$\begin{aligned}W_\prec =\ & \{\forall x \in N.\, \neg(x \prec x)\} \\ \cup\ & \{\forall xyz \in N.\, ((x \prec y) \wedge (y \prec z)) \supset (x \prec z)\} \\ \cup\ & \{\forall x \in N.\, (x \neq n_\top) \supset (x \prec n_\top)\} \cup \{\mathsf{ok}(n_\top)\} \\ \cup\ & \{\forall x \in N.\, (\forall y \in N.\, (x \not\prec y)\ \vee \\ & \quad [(x \prec y) \supset (\mathsf{bl}(y) \vee \mathsf{ap}(y))]) \supset \mathsf{ok}(x)\}\end{aligned}$$

In contrast to Definition 4.1, $D$ and $W$ now may contain preference information expressed by $\prec$ applied to default names. The first three axioms in $W_\prec$ account for information that was explicitly provided by ordered default theories in the rigid case. The last axiom is a straightforward extension of that found in the rigid case, now also accounting for the information provided by the default rule in $D_\prec$.

We note that Theorem 4.1 and Theorem 4.2 carry over to the general case except for Theorem 4.1.1. We get instead

$1'$. $n_\delta \prec n_{\delta'} \in E$ or $\neg(n_\delta \prec n_{\delta'}) \in E$ or $n_\delta \not\prec n_{\delta'} \in E$

In fact, ordered default theories are treated in the same way by our basic and general approach, except for different augmented languages:

**Theorem 5.1** *Let $(D, W, <)$ be an ordered default theory over $\mathcal{L}$. For each extension $E$ of $\mathcal{T}((D, W, <))$ there is an extension $E'$ of $\mathcal{T}((D, W \cup \{n_\delta \prec n_{\delta'} \mid (\delta, \delta') \in <\}))$ such that $E \cap \mathcal{L} = E' \cap \mathcal{L}$ and vice versa.*

As a consequence, our general approach yields all regular extensions (modulo the original language) if $(D, W)$ does not contain an occurrence of $\prec$.

Given Theorems 4.3 and 5.1, one would expect that ordered default theories would enjoy the same properties as standard default logic. This indeed is the case, but with one important exception: normal ordered default theories do not guarantee the existence of extensions. For example, the image of the ordered default theory (under our translation)

$$(\{n_1 : \tfrac{A:B}{B}, n_2 : \tfrac{B:C}{C}\}, \{A\}, n_1 < n_2) \qquad (5)$$

has no extension. The full paper describes formally why this is the case. Informally the problem is that we have a preference $n_1 < n_2$; however, given $W = \{A\}$ default $n_1$ applies first, and once it has applied, $n_2$ becomes applicable. Thus we have a preference ordering implicit in the form of the defaults and world knowledge, but where this implicit ordering is contradicted by the assertion $n_1 < n_2$. Not surprisingly then there is no extension. There are two immediate ways to resolve this problem. First, replace the rigid preference by $\frac{:n_1 \prec n_2}{n_1 \prec n_2}$; so we have this preference by default only. Second is to recognise that (5) is "buggy", in the same way that incorrect programs require modification. The lack of extension then indicates a problem in the specification of the original theory.

We conclude this section with the observation that our translation results in a manageable increase in the size of the default theory. For ordered theory $(D, W)$, the translation $\mathcal{T}((D, W))$ is only a constant factor larger than $(D, W)$.[6]

## 6 Discussion and Related Work

We have presented a very general framework for incorporating preferences into default logic. Via the naming of defaults we allow preferences to appear arbitrarily in $D$ and $W$ in a default theory. This allows preferences among preferences, preferences by default, preferences holding only in certain contexts, and so on. Strictly speaking, such generality isn't required: [Doyle & Wellman, 1991], building on work by Arrow, argue that in any preference-based default theory, for coherence, one requires a "dictator" to adjudicate preferences. That is, there must be, essentially, some way of determining a unique, complete, priority ordering. So in this sense, all one needs is what we have called the rigid approach of Section 4. We provide the more general framework of Section 5 for two reasons. First, it allows the more flexible specification of preferences, leaving it up to the user to ensure that there is no ambiguity in preferences. In the case where there is ambiguity in preferences (for example we might have $D \supseteq \{\frac{:n_1 \prec n_2}{n_1 \prec n_2}, \frac{:n_2 \prec n_1}{n_2 \prec n_1}\}$) one typically obtains multiple extensions. Second, we feel that the general approach is of technical interest: arbitrary defaults may be "compiled" into standard default theories, and so in a certain sense the explicit representation of priorities adds nothing to the fundamental power or expressibility of default logic.

Of other work in default logic treating preferences, we have argued that [Reiter & Criscuolo, 1981; Etherington & Reiter, 1983; Delgrande & Schaub, 1994] treat a separate problem, that of specificity orderings, as exemplified by (1). [Baader & Hollunder, 1993] and [Brewka, 1994a] present prioritised variants of default logic in which the iterative specification of an extension is modified. In brief, a default is only applicable at an iteration step (cf. Definition 3.1) if no $<$-greater default is applicable.[7] In contrast we translate priorities into standard default theories. There is insufficient space to fully compare approaches; see [Delgrande & Schaub, 1994] for a full discussion of these approaches with regard to how they address specificity in a theory.

---

[6]This assumes we count the default in $D_\prec$ as a single default.

[7]These authors use $<$ in the reverse order from us.

We conclude with an example from [Gordon, 1993], discussed in [Brewka, 1994b]. A person wants to find out if her security interest in a certain ship is "perfected", or legally valid. This person has possession of the ship, but has not filed a financing statement. According to the code UCC, a security interest can be perfected by taking possession of the ship. However, the federal Ship Mortgage Act (SMA) states that a security interest in a ship may only be perfected by filing a financing statement. Both UCC and SMA are applicable; the question is which takes precedence here. There are two legal principles for resolving such conflicts. *Lex Posterior* gives precedence to newer laws; here we have that UCC is more recent than SMA. But *Lex Superior* gives precedence to laws supported by the higher authority; here SMA has higher authority since it is federal law. Apart from $\delta_T$, we get:

$$ucc : \frac{possess : perf}{perf} \qquad sma : \frac{ship \wedge \neg finstmt : \neg perf}{\neg perf}$$

$$lp(x,y) : \frac{newer(y,x) : x \prec y}{x \prec y} \qquad ls(x,y) : \frac{stalaw(x) \wedge fedlaw(y) : x \prec y}{x \prec y}$$

To preserve finiteness, we restrict our attention to name set $N = N_0 \cup N_1$ where $N_0 = \{n_T, ucc, sma\}$ and $N_1 = \{lp(x,y), ls(x,y) \mid x,y \in N_0\}$, and the corresponding default instances. We have the facts: *possess*, *ship*, $\neg finstmt$, $newer(ucc, sma)$, $fedlaw(sma)$, $stalaw(ucc)$, $\forall x,y,u,v \in N_0.\ lp(x,y) \prec ls(u,v)$. [Brewka, 1994b] solves this problem by first generating 4 *complete* extensions. In a second step he rules out three of these extensions since they do not satisfy a certain priority criterion. In contrast, we obtain one extension, $E \supseteq \{\neg perf, ucc \prec sma\}$, and no other extension; this is the extension that Brewka ultimately obtains after ruling out non-preferred extensions.

## References

[Baader & Hollunder, 1993] F. Baader & B. Hollunder. How to prefer more specific defaults in terminological default logic. *Proc. Int'l Joint Conference on Artificial Intelligence*, p. 669-674, 1993.

[Boutilier, 1992] C. Boutilier. What is a default priority? *Canadian Conference on AI*, p. 140-147, 1992.

[Brewka, 1994a] G. Brewka. Adding priorities and specificity to default logic. *Proc. European Workshop on Logics in Artificial Intelligence*, p. 247-260. Springer, 1994.

[Brewka, 1994b] G. Brewka. Reasoning about priorities in default logic. *Proc. AAAI Conference on Artificial Intelligence*, p. 940-945, 1994.

[Delgrande & Schaub, 1994] J. Delgrande & T. Schaub. A general approach to specificity in default reasoning. *Proc. Fourth Int'l Conference on Knowledge Representation and Reasoning*, p. 146-157. Morgan Kaufmann, 1994.

[Doyle & Wellman, 1991] J. Doyle & M.P. Wellman. Impediments to universal preference-based default theories. *Artificial Intelligence*, 49(1-3):97-128, 1991.

[Etherington & Reiter, 1983] D.W. Etherington & R. Reiter. On inheritance hierarchies with exceptions. *Proc. AAAI Conference on Artificial Intelligence*, p. 104-108, 1983.

[Geffner & Pearl, 1992] H. Geffner & J. Pearl. Conditional entailment: Bridging two approaches to default reasoning. *Artificial Intelligence*, 53(2-3):209-244, 1992.

[Gordon, 1993] T. Gordon. *The pleading game: An Artificial Intelligence Model of Procedural Justice*. PhD thesis, University of Darmstadt, Germany, 1993.

[Grosof, 1991] B. Grosof. Generalizing prioritization. *Proc. Second Int'l Conference on Knowledge Representation and Reasoning*, p. 289-300. Morgan Kaufmann, 1991.

[Lifschitz, 1985] V. Lifschitz. Closed-world databases and circumscription. *Artificial Intelligence*, 27:229-235, 1985.

[Marek & Truszczyński, 1993] W. Marek & M. Truszczyński. *Nonmonotonic logic: context-dependent reasoning*. Springer, 1993.

[McCarthy, 1986] J. McCarthy. Applications of circumscription to formalizing common-sense knowledge. *Artificial Intelligence*, 28:89-116, 1986.

[Pearl, 1990] J. Pearl. System Z: A natural ordering of defaults with tractable applications to nonmonotonic reasoning. *Proc. of the Third Conference on Theoretical Aspects of Reasoning About Knowledge*, p. 121-135, 1990.

[Poole, 1988] D.L. Poole. A logical framework for default reasoning. *Artificial Intelligence*, 36(1):27-48, 1988.

[Reiter & Criscuolo, 1981] R. Reiter & G. Criscuolo. On interacting defaults. *Proc. Int'l Joint Conference on Artificial Intelligence*, p. 270-276, 1981.

[Reiter, 1980] R. Reiter. A logic for default reasoning. *Artificial Intelligence*, 13(1-2):81-132, 1980.

[Touretzky et al., 1987] D.S. Touretzky, J.F. Horty, & R.H. Thomason. A clash of intuitions: The current state of nonmonotonic multiple inheritance systems. *Proc. Int'l Joint Conference on Artificial Intelligence*, p. 476-482, 1987.

# AUTOMATED REASONING

Automated Reasoning 7:
Nonmonotonism for Logic Programming

# Learning Extended Logic Programs

**Katsumi Inoue**
Department of Electrical
and Electronics Engineering
Kobe University
Rokkodai, Nada-ku, Kobe 657, Japan
inoue@eedept.kobe-u.ac.jp

**Yoshimitsu Kudoh**
Division of Electronics
and Information Engineering
Hokkaido University
N-13 W-8, Sapporo 060, Japan
kudo@db.huee.hokudai.ac.jp

## Abstract

This paper presents a method to generate nonmonotonic rules with exceptions from positive/negative examples and background knowledge in Inductive Logic Programming. We adopt extended logic programs as the form of programs to be learned, where two kinds of negation—negation as failure and classical negation—are effectively used in the presence of incomplete information. While default rules are generated as specialization of general rules that cover positive examples, exceptions to general rules are identified from negative examples and are then generalized to rules for cancellation of defaults. We implemented the learning system LELP based on the proposed method. In LELP, when the numbers of positive and negative examples are very close, either parallel default rules with positive and negative consequents or nondeterministic rules are learned. Moreover, hierarchical defaults can also be learned by recursively calling the exception identification algorithm.

## 1 Introduction

*Inductive logic programming* (ILP) is a research area which provides theoretical frameworks and practical algorithms for inductive learning of relational descriptions in the form of logic programs [12, 10, 4]. Most previous work on ILP consider definite Horn programs or classical clausal programs in the form of logic programs to be learned. However, research work on knowledge representation in AI has shown that such monotonic programs are not adequate to represent our commonsense knowledge including notions of concepts and taxonomies. In this respect, there have been much work on *nonmonotonic reasoning* in AI. To learn default rules or concepts in taxonomic hierarchy, we thus need a learning mechanism that can deal with nonmonotonic reasoning.

On the other hand, recent advances on theories of logic programming and nonmonotonic reasoning have revealed that logic programs with *negation as failure* (NAF) is an appropriate tool for knowledge representation [3]. *Normal logic programs* (NLPs) are the class of programs in which NAF is allowed to appear freely in bodies of rules. NLPs are useful not only to represent default rules or rules with exceptions but also to write shorter and clearer programs than definite programs in many cases [5]. Learning NLPs has recently been considered in such as [2, 15, 5, 11].

While learning NLPs is an important step towards a better learning tool, there is still a limitation as a knowledge representation tool: NLPs do not allow us to deal directly with incomplete information [8]. NLPs automatically applies the *closed world assumption* (CWA) to all predicates, and any query is answered either *yes* or *no*, in which the latter negative answer is the result of CWA. In the context of inductive concept learning, the automatic application of CWA is not appropriate in the presence of both positive and negative examples. Positive examples represent instances of the target concept, while negative examples are non-instances. By CWA other objects are assumed non-instances, but then the role of negative examples is not clear because it is as if we supply a complete classification of all objects. This causes the paradox pointed out by De Raedt and Bruynooghe [6]: if everything is known, why should we still learn something? In the real world, we may not know whether some objects are positive or negative. But such incomplete information cannot be represented by NLPs.

To overcome the above problem of NLPs, we propose in this paper a new learning method which can deal with incomplete information in the form of *extended logic programs* (ELPs). ELPs are introduced by Gelfond and Lifschitz [8] to extend the class of NLPs by including *classical negation* (or explicit negation). The semantics of ELPs is given by the notion of *answer sets*, and is an extension of the *stable model semantics*. The answer to a ground query $A$ is either *yes*, *no*, or *unknown*, depending on whether the answer set contains $A$, $\neg A$, or neither. Using ELPs, the role of negative examples becomes clear, and any object not contained in either positive or negative examples is considered *unknown* unless the learned theory says that it must or must not be in that concept.

In this paper, we present a system, called **LELP** (Learning ELPs), to learn default rules with exceptions in the form of extended logic programs given incomplete positive and negative examples and background knowledge. LELP first generates candidate rules from positive examples (or negative examples if non-instances are much more than instances) and background knowledge in an ordinary ILP framework. Exceptions can be identified as negative examples (or positive examples if candidate rules have negative consequents) that are derived from the generated monotonic rules and background knowledge. Default rules with NAF are then computed by specializing candidate rules using the *open world specialization* (OWS) algorithm. This OWS algorithm is closely related to Bain and Muggleton's CWS algorithm [2], but works better in the three-valued semantics. Then, default cancellation rules are generated to cover exceptions using an ordinary ILP framework.

In the real world, it is not easy to know that a general default rule should have the positive or negative consequent. In LELP, it is determined according to the ratio of positive examples. Nevertheless, if it is still hard to know which is more general, LELP can generate nondeterministic rules in the context of the answer set semantics. Furthermore, by calling the OWS algorithm recursively, LELP can generate hierarchical default rules.

The rest of this paper is organized as follows. Section 2 outlines how our system LELP produces ELPs to learn simple default rules. Section 3 extends LELP to deal with complex concept structures with hierarchical exceptions. Section 4 presents related work, and Section 5 concludes the paper.

## 2 Learning Default Rules

This section shows how LELP learns default rules with exceptions. To clarify the underlying idea, we here consider a simple model that there are general rules and exceptions to them but there are no exceptions to exceptions. This model will be extended in Section 3.

The algorithm of LELP is summarized as follows.

**Algorithm 2.1** $LELP1(E^+, E^-, BG, T1, T2, T3)$
Input: positive examples $E^+$, negative examples $E^-$,
background knowledge $BG$
Output: default rules $T1$, exception rules $T2 \cup T3$

1. According to the ratio of $E^+$ or $E^-$ to all objects, determine whether the learned general rule is positive or negative;
2. Given $E^+$ (or $E^-$ if a negative rule is to be learned) and $BG$, generate general rules $T$ using an ordinary ILP technique;
3. Given $T$, $E^+$, $E^-$ and $BG$, compute default rules $T1$ and exceptions $AB$ using the OWS algorithm; Generate rules $T2$ deriving $E^-$ (or $E^+$) from $AB$ at the same time using an ordinary ILP technique;
4. Given $AB$ and $BG$, generate default cancellation rules $T3$ using an ordinary ILP techniques.

In LELP, the input positive examples are represented as *positive literals*, and negative examples are denoted as *negative literals*. We allow *rules* in the form of ELPs in background knowledge.

## 2.1 Extended Logic Programs

*Extended logic programs* (ELPs) were introduced in [8] as a tool for reasoning in the presence of incomplete information. They are defined as sets of rules of the form

$$L_0 \leftarrow L_1, \ldots, L_m, not\, L_{m+1}, \ldots, not\, L_n$$

where $L_i$'s ($0 \leq i \leq n$; $n \geq m$) are literals. Two kinds of negation appear in a program: *not* is the *negation as failure* operator, and $\neg$ is *classical negation*. Intuitively, the above rule can be read as: if $L_1, \ldots, L_m$ are believed and $L_{m+1}, \ldots, L_n$ are not believed then $L_0$ is believed. The semantics of ELPs are defined by the notion of *answer sets* [8], which are sets of ground literals representing possible beliefs. The class of ELPs are considered as a subset of *default logic* [14]: each rule of the above form in an ELP can be identified with the *default* of the form

$$\frac{L_1 \wedge \ldots \wedge L_m : \overline{L_{m+1}}, \ldots, \overline{L_n}}{L_0}$$

where $\overline{L}$ stands for the literal complementary to $L$. Then, each *answer set* is the set of atoms in an extension of the default theory. We say that a literal $L$ is *entailed* by an ELP $P$ if $L$ is contained in every answer set of $P$. While we adopt the answer set semantics in this paper, other semantics for ELPs may be applicable to our learning framework with minor modification.

We call a rule having a positive literal in its head *positive rule*, and a rule having a negative literal in its head *negative rule*. In the following, we denote classical negation $\neg$ as - and NAF *not* as \+ in programs.

The completeness and consistency of concept learning (see [10, 4] for instance) can be reformulated in the three-valued setting as follows. Let $BG$ be an ELP as background knowledge, $E$ a set of positive/negative literals as positive/negative examples, and $R$ a set of rules as hypotheses. $R$ is *complete* with respect to $BG$ and $E$ if for every $e \in E$, $e$ is entailed by $BG \cup R$ ($R$ covers $e$). $R$ is *consistent* with $BG$ and $E$ if for any $e \in E$, $\overline{e}$ is not entailed by $BG \cup R$ ($R$ does not cover $\overline{e}$). Note here that positive examples are not given any higher priority than negative ones. Namely, *both positive and negative examples are to be covered by the learned rules that are consistent with background knowledge and examples*. Thus, we will learn both positive and negative rules: no CWA is assumed to derive non-instances (see also [6]).

Although both positive and negative rules are generated by LELP, each default rule for the target concept should be either positive or negative. In LELP, it is determined according to the ratio of positive examples to all objects. In the following, we assume that positive rule is learned as a general rule unless otherwise specified.

## 2.2 Generating General Rules

In Algorithm 2.1, given positive (resp. negative) examples $E$ and background knowledge $BG$, LELP generates general rules $T$ to cover every example in $E$ using an ordinary ILP technique. We denote this part of algorithm as $GenRules(E, BG, T)$. In generating positive (resp.

negative) rules, no negative (resp. positive) example is used to specialize rules. The specialization of general rules is performed in the OWS algorithm (Section 2.3).

We do not assume any particular learning algorithm for the implementation of an "ordinary" ILP technique in $GenRules(E, BG, T)$. This part can be considered as a black box, and this paper is not concerned with the detail. In our real implementation, we used the notion of Plotkin's RLGG (relative least general generalization) with the bottom-up technique used in GOLEM [13]. When background knowledge contains rules with bodies, the *unfold* transformation in logic programming is also used to truncate literals in bodies of learned rules.

**Example 2.1** Suppose that positive examples $E^+$ are:
{flies(1),flies(2),flies(3),flies(4)},
and that background knowledge $BG$ is:
{bird(1),bird(2),bird(3),bird(4),bird(c),
   (bird(X) :- pen(X)),pen(a),pen(b)}.
Then, $GenRules(E^+, BG, T)$ generates:

    flies(X) :- bird(X).

## 2.3 Specializing Rules using NAF

The general rules computed to cover the positive (resp. negative) examples by $GenRules(E, BG, T)$ may also cover the complements of some of negative (resp. positive) examples. To specialize general rules, we propose the algorithm of *open world specialization* (OWS). The OWS algorithm is closely related to Bain and Muggleton's closed world specialization (CWS) [2]. Like CWS, OWS produce rules with NAF as default rules. Unlike CWS, however, OWS does not apply the closed world assumption (CWA) to identify non-instances of the target concept. In OWS, exceptions are identified as objects contained in negative examples (or positive examples if the general rule is negative) such that they are proved from the general rule with background knowledge and positive (or negative) examples.

In the following OWS algorithm, we assume here that each general rule in $T$ is positive.

**Algorithm 2.2** $OWS(T, E^+, E^-, BG, AB, T')$
Input: rules $T$, positive examples $E^+$, negative examples $E^-$, background knowledge $BG$
Output: default rules $T'$, set of exceptions $AB$
Let $T' := T$; $AB := \emptyset$;
for each $C_i = (H \text{ :- } B)$ in $T$ do:
  Find a literal $L$ such that $\overline{L} \in E^-$ and $L$ is entailed by $BG \cup \{C_i\} \cup E^+$ (we call $L$ an *exception* to $C_i$ and $C_i$ a *rule with exceptions*);
  $\theta :=$ the answer substitution for $C_i$ in proving $L$;
  if $B$ contains the literal $\backslash$+ $N$ then
    $AB := AB \cup \{N\theta\}$
  else $N := \text{ab}_i(V_1, \ldots, V_n)$,
    where $\{V_1, \ldots, V_n\}$ is the domain of $\theta$;
    $T' := (T' \setminus \{C_i\}) \cup \{(H \text{ :- } B, \backslash\text{+} N)\}$;
    $AB := AB \cup \{N\theta\}$.

In the real implementation of LELP, we used Prolog's top-down proof is used to get answer substitutions.

**Example 2.2** (cont. from Example 2.1)
$BG$, $E^+$: the same as Example 2.1,
$T = \{(\text{flies}(X) \text{ :- bird}(X))\}$,
$E^- = \{-\text{flies}(a), -\text{flies}(b), -\text{flies}(c)\}$.
Then, the exceptions to $C1 = (\text{flies}(X) \text{ :- bird}(X))$ are computed as $\{\text{flies}(a), \text{flies}(b), \text{flies}(c)\}$. In this case, substitutions $\theta$'s are X/a, X/b, X/c, and $N$ is ab1(X). Hence,
$T' = \{(\text{flies}(X) \text{ :- bird}(X), \backslash\text{+ ab1}(X))\}$,
$AB = \{\text{ab1}(a), \text{ab1}(b), \text{ab1}(c)\}$.

## 2.4 Negative Rules for Exceptions

Since we use OWS, we need rules to derive negative examples (or positive examples if the default rule is negative). Given negative examples $E^-$ (resp. positive examples $E^+$) and the set $AB$ of exceptions, LELP generates negative (resp. positive) rules $R$ to derive exceptions as $GenRules(E^-, AB, R)$ (resp. $GenRules(E^+, AB, R)$).

In the bird example, such a rule is generated as:

    -flies(Y) :- ab1(Y).

## 2.5 Cancellation Rules

In the OWS algorithm, the set $AB$ of exceptions is output as a set of ground atoms. However, if exceptions have some common properties, this expression is not informative and *rules about exceptions* are useful. These rules work as *default cancellation rules*.

After applying OWS, each exception is in the form of ground atom whose predicate is $\text{ab}_i$. Rules about exceptions have such abnormal predicates in their heads and are results of generalizations of some abnormal atoms. When such a common rule cannot be generated or there are some exceptions that cannot be covered by such a rule, those exceptions are left as they are.

Since exceptions are not anticipated in general, rules about exceptions should be used to derive only exceptions. In fact, exceptions are usually minimized in non-monotonic reasoning. To this end, we apply a limited form of CWA here. If a rule about exceptions is too general, that is, it derives negative facts more than expected, it should be rejected. This test can be done easily using a bottom-up model generation procedure. The algorithm to generate rules about exceptions is as follows.

**Algorithm 2.3** $Cancel(AB, BG, R)$
Input: set of exceptions $AB$, background knowledge $BG$
Output: default cancellation rules $R$
1. $GenRules(AB, BG, T)$;
2. For each $C \in T$, compute the set $L$ of $\text{ab}_i$ literals that are entailed by $BG \cup \{C\} \cup AB$;
   if $L \supset AB$ then $R := T \setminus \{C\}$
       else $R := T$.

**Example 2.3** (cont. from Example 2.2)
$BG$: the same as Example 2.1,
$AB = \{\text{ab1}(a), \text{ab1}(b), \text{ab1}(c)\}$,
$E = \{\text{flies}(1), \text{flies}(2), \text{flies}(3), \text{flies}(4)\}$.
Suppose that $GenRules$ outputs $T$ that contains rules
(ab1(X) :- pen(X)) and (ab1(Y) :- bird(Y)).

Here, the rule (ab1(Y) :- bird(Y)) enables us to derive ab1(1), ab1(2), ab1(3), ab1(4) besides the set $AB$, so it is removed by Algorithm 2.3. The atom ab1(c) represents an exception that is not a penguin, and hence cannot be generalized. The final rules are:
$$R = \{(\text{ab1}(Z) \text{ :- } \text{pen}(Z)), \text{ab1}(c)\}$$

While in this section we apply CWA to generate rules about exceptions in LELP, this assumption is not appropriate for exceptions to exceptions. We will extend the framework in this respect in Section 3.2.

## 2.6 Properties

Given consistent background knowledge $BG$ and positive and negative examples $E = E^+ \cup E^-$, let $R$ be the output hypotheses learned by LELP. Then, the next theorem is proved in [9].

**Theorem 2.1** *R is complete with respect to BG and E, and is consistent with BG and E.*

## 2.7 Example

The LELP program is implemented in Prolog and is called by

lelp(Examples,Background_Knowledge,Result).

In Examples, atoms preceded by + represents positive examples, and those with - are negative examples. In the following example, if the ratio of positive (resp. negative) examples exceeds 50%, positive (resp. negative) rules are generated.

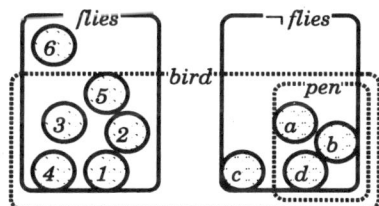

Figure 1: Bird Example

**Example 2.4** (Bird Example Summary: Figure 1)

```
| ?- lelp([+flies(1),+flies(2),+flies(3),
    +flies(4),+flies(5),+flies(6),
    -flies(a),-flies(b),-flies(c),-flies(d)],
    [bird(1),bird(2),bird(3),bird(4),bird(5),
    bird(c),(bird(X) :- pen(X)),
    pen(a),pen(b),pen(d)],Rules).

RLGG/LL of flies(1) and flies(2) is
    flies(_4594):-bird(_4594)
Covered examples: flies(1), flies(2), flies(3),
    flies(4), flies(5)
RLGG/LL of -flies(a) and -flies(b) is
    -flies(_42885):-pen(_42885),bird(_42885)
Covered examples: -flies(a), -flies(b), -flies(d)
RLGG/LL of ab1(a) and ab1(b) is
    ab1(_85217):-pen(_85217),bird(_85217)
Covered examples: ab1(a), ab1(b), ab1(d)
RLGG/LL of -flies(a) and -flies(b) is
    -flies(_127580):-ab1(_127580)
Covered examples: -flies(a), -flies(b), -flies(c),
    -flies(d)
Unfolding of ab1(_410):-pen(_410),bird(_410)
with bird(_410):-pen(_410) is
    ab1(_410):-pen(_410)

Running Time: 280 (msec)

Rules =
[(flies(A):-bird(A),\+ab1(A)),flies(6),
 (-flies(B):-ab1(B)),
 (ab1(C):-pen(C)),ab1(c)] ?
```

## 3 Extension

In this section, we extend LELP to learn more complex concept structures.

### 3.1 Nondeterministic Rules

When the number of positive examples is close to that of negative examples, it is difficult to judge whether the general rule should be positive or negative. Two solutions can be considered to this problem: (1) *parallel default rules*, and (2) *nondeterministic rules*. Parallel default rules are generated when exceptions exist for both positive and negative rules in parallel (e.g., mammals normally do not fly except bats, and birds normally fly except penguins). Nondeterministic rules are generated when some object is proved to be positive and negative by a program such that a contradiction occurs. An extension of Algorithm 2.1 is shown in Section 3.2, where hierarchical defaults can also be learned.

In the following example, if the ratio of positive examples is between 40% and 60%, parallel default rules or nondeterministic rules are generated.

**Example 3.1** (Learning Nondeterministic Rules)

```
| ?- lelp([+flies(1),+flies(2),+flies(3),+flies(4),
    -flies(5),-flies(6),-flies(7),-flies(8)],
    [bird(1),bird(2),bird(3),bird(4),
    bird(5),bird(6),bird(7),bird(8)],Rules).
```

Here, the ratio of positive examples is 50%, and hence both positive and negative rules are generated. If we use the previous method, the following rules are computed:

```
(flies(A):-bird(A),\+ab1(A)),
(-flies(B):-ab1(B)),
ab1(5),ab1(6),ab1(7),ab1(8),
(-flies(C):-bird(C),\+ab2(C)),
(flies(D):-ab2(D)),
ab2(1),ab2(2),ab2(3),ab2(4)
```

The above rules look like correct, but the bodies of both rules are the same as bird(X) except the $\text{ab}_i$ literals. Then, if bird(9) is added to background knowledge, neither ab1 nor ab2 is proved so that a contradiction occurs. In this case, the nondeterministic rules are generated:

```
(flies(A):-bird(A),\+ab1(A),\+ -flies(A)),
(-flies(B):-ab1(B)),ab1(5),ab1(6),ab1(7),ab1(8),
(-flies(C):-bird(C),\+ab2(C),\+flies(C)),
(flies(D):-ab2(D)),ab2(1),ab2(2),ab2(3),ab2(4)
```

This form of nondeterministic rules are shown in [3] to represent default rules in ELPs. For the added ground term 9, the ground instances of general rules are:

```
flies(9) :- bird(9), \+ab1(9), \+ -flies(9).
-flies(9) :- bird(9), \+ab2(9), \+flies(9).
```

Hence, for the bird 9, two answer sets exist, one concluding `flies(9)` and the other `-flies(9)`.

Here are two examples of Nixon Diamond.

**Example 3.2** (Pacifist Nixon)

```
| ?- lelp([+pacifist(c),+pacifist(d),
      -pacifist(a),-pacifist(b),+pacifist(nixon)],
    [republican(a),republican(b),
     republican(nixon),quaker(nixon),
     quaker(c),quaker(d)],Rules).
```

The result shows that parallel rules are generated in which only Nixon is an exception of republicans:

```
(pacifist(A):-quaker(A)),
(-pacifist(B):-republican(B),\+ab1(B)),
ab1(nixon)
```

**Example 3.3** (Nixon's Trouble)

```
| ?- lelp([+pacifist(c),+pacifist(d),
      -pacifist(a),-pacifist(b)],
    [republican(a),republican(b),
     republican(nixon),quaker(nixon),
     quaker(c),quaker(d)],Rules).
```

The result implies two answer sets for Nixon, and neither `pacifist(nixon)` nor `-pacifist(nixon)` is entailed:

```
(pacifist(A):-quaker(A),\+ -pacifist(A)),
(-pacifist(B):-republican(B),\+pacifist(B))
```

## 3.2 Hierarchical Defaults

Here, we further extend LELP to deal with hierarchical structures of concepts and exceptions. First, we modify Algorithm 2.3 to generate default cancellation rules.

**Algorithm 3.1** $Cancel2(AB, BG, R)$
Input: set of exceptions $AB$, background knowledge $BG$
Output: default cancellation rules $R$
1. $GenRules(AB, BG, T)$;
2. For each $C \in T$, compute the set $L$ of $ab_i$ literals that are entailed by $BG \cup \{C\} \cup AB$;
   if $|L - AB| \geq |AB|$ then $R := T \setminus \{C\}$
   else $R := T$.

The new condition $|L - AB| \geq |AB|$ in Step 3 above replaces the CWA condition $L \supset AB$ in Algorithm 2.3, and represents the *monotone assumption* that "in every level of the hierarchy, the number of exceptions is less than that of instances with default properties".

To learn hierarchical default rules, after removing over-general rules in Algorithm 3.1, we need to call algorithms of $OWS$ and $Cancel$ recursively. The procedure stops when there are no more exceptions or no more objects to be generalized. The extended algorithm LELP2 is as follows. In the following, we denote sets of opposite examples as $E_1, E_2$ instead of positive examples $E^+$ and negative examples $E^-$.

**Algorithm 3.2** $LELP2(E_1, E_2, BG, R)$
Input: positive examples $E_1$, negative examples $E_2$, background knowledge $BG$
Output: $R := R_3 \cup R_4 \cup R_5 \cup R_6$
1. According to the ratio of $E_1$ (or $E_2$) to all objects, determine to learn either (1) default rules for $E_1$, (2) default rules for $E_2$, (3) parallel default rules, or (4) nondeterministic rules;
   In case of (1), perform Steps 2, 4, 7, put $R_4 = R_6 = \emptyset$, and return;
   In case of (2), perform Steps 3, 5, 8, put $R_3 = R_5 = \emptyset$, and return;
   Otherwise, execute every step after 2 in order;
2. $GenRules(E_1, BG, R_1)$;
3. $GenRules(E_2, BG, R_2)$;
4. $OWS(R_1, E_1, E_2, BG, AB_1, R_3)$;
5. $OWS(R_2, E_2, E_1, BG, AB_2, R_4)$;
6. If there are contradictory rules in $R_1$ and $R_2$, transform $R_3$ and $R_4$ into nondeterministic rules, put $R_5 = R_6 = \emptyset$, and return;
7. if $AB_1 \neq \emptyset$ then $ABs(BG, AB_1, E_1, E_2, R_5)$;
   else $R_5 := \emptyset$;
8. if $AB_2 \neq \emptyset$ then $ABs(BG, AB_2, E_2, E_1, R_6)$;
   else $R_6 := \emptyset$.

**Algorithm 3.3** $ABs(BG, AB_1, E_1, E_2, R)$
1. $Cancel2(AB_1, BG, R_1)$;
2. $OWS(R_1, E_2, E_1, BG, AB_2, R_2)$;
3. if $AB_2 \neq \emptyset$ then $ABs(BG, AB_2, E_2, E_1, R_3)$;
   else put $R := R_1$, and return;
4. $GenRules(AB_2, BG, R_4)$;
   put $R := R_2 \cup R_3 \cup R_4$, and return.

**Example 3.4** (Learning Hierarchical Defaults)

```
| ?- lelp([-flies(1),-flies(2),+flies(3),+flies(4),
     +flies(5),-flies(6),-flies(7),-flies(8),
     -flies(9),-flies(10),-flies(11),-flies(12)],
    [pen(1),pen(2),bird(3),bird(4),bird(5),
     (bird(X) :- pen(X)),animal(6),animal(7),
     animal(8),animal(9),animal(10),animal(11),
     animal(12),(animal(X) :- bird(X))],Rules).
```

Here, the ratio of positive examples is 25%, and hence the negative rule `(-flies(A):-animal(A))` is firstly generated. Using OWS, the set of exceptions in the first level {`ab1(3), ab1(4), ab1(5)`} and the default rule `(-flies(A):-animal(A),\+ab1(A))` are then computed. This set of exceptions is generalized to the cancellation rule `(ab1(C):-bird(C))` and the positive rule for exceptions `(flies(B):-ab1(B))` are also generated. Nextly, after applying OWS to the cancellation rule, the exceptions in the second level {`ab2(1), ab2(2)`} and the default rule are computed, and then its cancellation rule and the negative rule for exceptions are generated. At this point, no more exception is left and the algorithm stops. The final **Rules** are as follows.

```
(-flies(A):-animal(A),\+ab1(A)),
(flies(B):-ab1(B)),
(ab1(C):-bird(C),\+ab2(C)),
(-flies(D):-ab2(D)),
(ab2(E):-pen(E))
```

## 4 Related Work

Bain and Muggleton's CWS algorithm [2] has been applied to non-monotonic versions of CIGOL and GOLEM in [1] and a learning algorithm that can acquire hierarchical programs in [15]. CWS produces default rules with NAF in stratified NLPs. Since CWS is based on CWA in the two-valued setting, it regards every ground atom that is not contained in an intended model as an exception. In LELP, on the other hand, OWS is employed instead of CWS, and incomplete information can be represented in ELPs with the three-valued semantics.

$TRACY^{not}$ by Bergadano et al. [5] learns stratified NLPs using trace information of SLDNF derivations. Since this system needs the hypothesis space in advance, it does not invent a new predicate like $ab_i$ expressing exceptions, and hence seems more suitable for learning rules with negative knowledge and CWA rather than learning defaults. Martin and Vrain [11] use the three-valued semantics for NLPs in their inductive framework. Since they do not adopt ELPs, CWA is still employed and two kinds of negation are not distinguished.

While no previous work adopts full ELPs in the form of learned programs, a limited form of classical negation has been used in [6, 7]. De Raedt and Bruynooghe [6] firstly discussed the importance of the three-valued semantics in ILP. However, since they did not allow NAF, an explicit list of exceptions is necessary for each rule, which causes the *qualification problem* in AI. Wrobel [16] also used exception lists to specialize over-general rules, but their underlying language is monotonic first-order. Dimopoulos and Kakas [7] propose a learning method that can acquire rules with hierarchical exceptions. They also do not use NAF to represent defaults, but adopt their own nonmonotonic logic. Moreover, using the approach of [7], one has to determine whether each negative information should be used in the usual specialization process or in the exception identification process. In our approach, such distinction can be clearly done by an appropriate usage of NAF and classical negation.

Finally, in any previous work, nondeterministic rules cannot be generated, and hence commonsense knowledge with multiple extensions cannot be learned.

## 5 Conclusion

This paper proposed new techniques to learn nonmonotonic rules with exceptions, and introduced the learning system LELP. Extended logic programs are adopted as program forms, in which two kinds of negation are effectively used in the presence of incomplete information. Default rules are generated using OWS, and their exceptions are then generalized to cancellation rules. LELP can also learn parallel/nondeterministic rules and hierarchical defaults within the three-valued semantics.

In this paper, we treated every explicit negative information as an exception to a positive hypothesis. In the real world, however, negative knowledge may often be irrelevant to the concepts to be learned. In this respect, a method of separation of noise from exceptions has been proposed in [15]. Another approach is that we may add information that each concept can have exceptions or not or that CWA can be applied or not. These extensions can easily be accommodated within LELP.

## References

[1] Michael Bain. Experiments in non-monotonic first-order induction. In: [12], pages 423–436.

[2] Michael Bain and Stephen Muggleton. Non-monotonic learning. In: [12], pages 145–161.

[3] Chitta Baral and Michael Gelfond. Logic programming and knowledge representation. *Journal of Logic Programming*, 19/20:73–148, 1994.

[4] Francesco Bergadano and Daniele Gunetti. *Inductive Logic Programming: From Machine Learning to Software Engineering*. MIT Press, 1996.

[5] F. Bergadano, D. Gunetti, M. Nicosia and G. Ruffo. Learning logic programs with negation as failure. In: Luc De Raedt, editor, *Proceedings of ILP-95*, pages 33–51, K. U. Leuven, 1995.

[6] Luc De Raedt and Maurice Bruynooghe. On negation and three-valued logic in interactive concept-learning. In: *Proceedings of ECAI '90*, pages 207–212, Pitman, 1990.

[7] Yannis Dimopoulos and Antonis Kakas. Learning non-monotonic logic programs: learning exceptions. In: Nada Lavrač and Stefan Wrobel, editors, *Proceedings of ECML-95*, pages 122–137, LNAI 912, Springer, 1995.

[8] Michael Gelfond and Vladimir Lifschitz. Classical negation in logic programs and disjunctive databases. *New Generation Computing*, 9(3,4):365–385, 1991.

[9] Katsumi Inoue and Yoshimitsu Kudoh. Learning default rules in extended logic programs. Submitted for publication, 1997 (in Japanese).

[10] Nada Lavrač and Sašo Džeroski. *Inductive Logic Programming: Techniques and Applications*. Ellis Horwood, 1994.

[11] Lionel Martin and Christel Vrain. A three-valued framework for the induction of general logic programs. In: Luc De Raedt, editor, *Advances in Inductive Logic Programming*, pages 219–235, IOS Press, 1996.

[12] Stephen Muggleton, editor. *Inductive Logic Programming*. Academic Press, London, 1992.

[13] Stephen Muggleton and Cao Feng. Efficient induction of logic programs. In: [12], pages 281–298.

[14] Raymond Reiter. A logic for default reasoning. *Artificial Intelligence*, 13:81–132, 1980.

[15] Ashwin Srinivasan, Stephen Muggleton and Michael Bain. Distinguishing exceptions from noise in nonmonotonic learning. In: *Proceedings of ILP-92*, ICOT, 1992.

[16] Stefan Wrobel. On the proper definition of minimality in specialization and theory revision. In: Pavel B. Brazdil, editor, *Proceedings of ECML-93*, pages 65–82, LNAI 667, Springer, 1993.

# Compiling Prioritized Circumscription into Extended Logic Programs

Toshiko Wakaki
Dept. of Computational Intelligence
and Systems Science
Tokyo Institute of Technology
Nagatsuda, Midori-ku, Yokohama, Japan

Ken Satoh
Division of Electronics and
Information Engineering
Hokkaido University
N13W8, Kita-ku, Sapporo 060, Japan

## Abstract

We propose a method of compiling circumscription into Extended Logic Programs which is widely applicable to a class of parallel circumscription as well as a class of prioritized circumscription. In this paper, we show theoretically that any circumscription whose theory contains both the domain closure axiom and the uniqueness of names axioms can always be compiled into an extended logic program Π, so that, whether a ground literal is provable from circumscription or not, can always be evaluated by deciding whether the literal is true in all answer sets of Π, which can be computed by running Π under the existing logic programming interpreter.

## 1 Introduction

Circumscription [McCarthy, 1980; Lifschitz, 1985] was proposed to formalize the commonsense reasoning under incomplete information.

So far, many studies have been proposed to explore the approach of the use of logic programming for the automation of circumscription based on the relationship between the semantics of circumscription and the semantics of logic programs.

Gelfond and Lifschitz [1988a] was the first to consider a computational method for some restricted class of prioritized circumscription which compiles circumscriptive theories into a stratified logic program. Though their method is computationally efficient, the applicable class is too limited.

So we proposed the extension of Gelfond and Lifschitz's method which also compile prioritized circumscription into a stratified logic program [Wakaki and Satoh, 1995]. With keeping the efficiency of Gelfond and Lifschitz's method, our method expands the applicable class of circumscription by making use of the result [Lifschitz, 1985] about parallel circumscription of a solitary formula. However, as far as a class of stratified logic programs is considered as the target language to which circumscription is compiled, the applicable class is limited within a class of circumscription which has a unique minimal model since every stratified logic program has a unique perfect model [Przymusinski, 1987]. But there are many examples of nonmonotonic reasoning whose intended meaning cannot be represented by a unique model such as *multiple extension problem*, a class of circumscription which has fixed predicates, and so on.

Recently Sakama and Inoue [1995; 1996] proposed two methods both of which compile circumscription into classes of more general logic programs whose semantics are given by stable models for the first one [Sakama and Inoue, 1995] and by preferred answer sets for the second one [Sakama and Inoue, 1996]. Though both of their methods can handle the multiple extension problem as well as circumscription with fixed predicates, the first one is only applicable to parallel circumscription but not to prioritized circumscription. On the other hand, the second one is applicable to prioritized circumscription, but it gives only the semantic aspects and is lack of the feasible logic programming interpreter for prioritized logic programs proposed by them as the target language into which prioritized circumscription is compiled.

In this paper, we propose a new method of compiling circumscription into *extended logic programs* proposed by Gelfond and Lifschitz [1991] as its target language. It is widely applicable to a class of parallel circumscription as well as a class of prioritized circumscription. Showing the semantic correspondence between circumscription with fixed predicates and Reiter's default theory which generalizes Theorem 2 in [Etherington, 1987], we can give not only the semantic relationship between a class of parallel circumscription and a class of extended logic programs but also the one between a class of prioritized circumscription and a class of extended logic programs. As a result, any circumscription whose theory contains both the domain closure axiom and the uniqueness of names axioms can always be compiled into an extended logic program Π, so that, whether a ground literal is provable from circumscription, can always be evaluated by deciding whether the literal is true in all answer sets of Π. This can be computed by running Π under the existing logic programming interpreter such as Satoh and Iwayama's top-down query evaluation procedure for abductive logic programming [Satoh and Iwayama, 1992].

Finally, we present that our approach exploiting classical negation $\neg$ can also give an extension of Sakama and Inoue's first method [1995] to make it possible to compute prioritized circumscription.

The structure of the paper is as follows. In section 2, we give preliminaries related to circumscription and extended logic programs. In section 3, we provide two syntactical definitions of extended logic programs into which parallel circumscription and prioritized circumscription are compiled respectively. Then, we present theorems and corollaries on which our method is theoretically based, along with examples. We finish section 4 with comparing our method with related researches.

## 2 Preliminaries

We review parallel circumscription and prioritized circumscription as well as the syntax and semantics of extended logic programs (ELP) [Gelfond and Lifschitz, 1991] in a slightly extended form.

Let $T(P, Z)$ be a sentence, $P$ be a tuple of minimized predicates and $Z$ be a tuple of variable predicates. $Q$ denotes the rest of the predicates occurring in $T$, called the fixed predicates. Then parallel circumscription of $P$ in $T(P, Z)$ with variable $Z$ is denoted by $Circum(T; P; Z)$. If $P$ is broken into parts $P^1, \ldots, P^k$, then the circumscription assigning a higher priority to the members of $P^i$ than to the members of $P^j$ for $i < j$ is denoted by the following prioritized circumscription:

$$Circum(T; P^1 > P^2 > \ldots > P^k; Z).$$

Prioritized circumscription can be represented by using parallel circumscription as follows:

$$Circum(T; P^1 > \ldots > P^k; Z)$$
$$\equiv \bigwedge_{i=1}^{k} Circum(T; P^i; P^{i+1}, \ldots, P^k, Z) \quad (2.1)$$

Extended logic programs have the expressive power to represent classical negation ($\neg$) along with negation as failure ($not$) which enables us to represent incomplete information.

A extended logic program is a set of rules of the form:

$$L \leftarrow L_1, \ldots, L_m, not L_{m+1}, \ldots, not L_n, \quad (2.2)$$

or of the form:

$$\leftarrow L_1, \ldots, L_m, not L_{m+1}, \ldots, not L_n, \quad (2.3)$$

where $n \geq m \geq 0$, and $L$ and $L_i$'s are literals. Each rule of the form (2.3) is called an *integrity constraint*.

The semantics of an extended logic program is an extension of the *stable model semantics*[Gelfond and Lifschitz, 1988b] and is given by the *answer sets* [Gelfond and Lifschitz, 1991]. The answer sets of an extended logic program are defined in the following two steps.

1. Let $\Pi$ be a set of ground rules of an extended logic program not containing $not$ and $Lit$ be the set of ground literals in the language. The answer set, $\alpha(\Pi)$, of $\Pi$ is the smallest subset $S$ of $Lit$ satisfying the conditions:

   (a) For any rule $L \leftarrow L_1, \ldots, L_m$ in $\Pi$, if $L_1, \ldots, L_m \in S$, then $L \in S$;
   
   (b) For any integrity constraint $\leftarrow L_1, \ldots, L_m$ in $\Pi$, if $L_1, \ldots, L_m \in S$, then $S = Lit$;
   
   (c) if $S$ contains a pair of complementary literals, then $S = Lit$.

2. Let $\Pi$ be any extended logic program without variables. For any set $S \subseteq Lit$, let $\Pi^S$ be the set of rules without $not$ obtained from $\Pi$ by deleting

   (a) every rule containing a formula $not L$ in its body with $L \in S$, and
   
   (b) every formula $not L$ in the bodies of the remaining rules.

   Then, $S$ is an answer set of $\Pi$ if $S$ is the answer set of $\Pi^s$, that is, $S = \alpha(\Pi^S)$.

Let $\Sigma$ be a set of clauses. Then $Th(\Sigma)$ stands for a set of clauses which are theorems of $\Sigma$. A clause in $Th(\Sigma)$ which is not properly subsumed by any theorem in $Th(\Sigma)$ is called a *characteristic clause*. $\mu Th(\Sigma)$ denotes the set of all characteristic clauses in $Th(\Sigma)$ [Inoue, 1992].

## 3 Translation from Circumscription into Extended Logic Programs

We give a translation from circumscription to extended logic programs. Parallel circumscription is translated into ELP $\Pi^\alpha$ and prioritized circumscription is translated into ELP $\Pi^\beta$ respectively.

In the following Definition 3.1 and Definition 3.2, we restrict a first order theory $T$ to the one which is function-free and contains both the *domain closure axiom* (DCA) and *uniqueness of names axioms* (UNA). We consider $T$ as a set of clauses:

$$T \stackrel{def}{=} \Sigma \cup \text{DCA} \cup \text{UNA}.$$

We suppose that every clause in $\Sigma$ containing variables is replaced by all its ground instances obtained by substituting every ground term in DCA for each variable. $t$ stands for a tuple of ground terms occurring in DCA.

**Definition 3.1**

Given parallel circumscription $Circum(\widetilde{\forall} T; P; Z)$ where $\widetilde{\forall}$ is a universal closure, ELP $\Pi^\alpha$ is constructed as follows:

1. For any minimized predicate $p$ from $P$ and any $t$,

$$\neg p(t) \leftarrow not\ p(t).$$

2. For any fixed predicate $q$ from $Q$ and any $t$,

$$\neg q(t) \leftarrow not\ q(t),$$
$$q(t) \leftarrow not\ \neg q(t).$$

3. For any clause $C \stackrel{def}{=} \ell_1 \vee \ell_2 \vee \ldots \vee \ell_n$ in $\mu Th(\Sigma)$ and any contrapositive of $C$ such as $\neg \ell_{i_2} \wedge \ldots \wedge \neg \ell_{i_n} \supset \ell_{i_1}$ and $\neg \ell_{i_1} \wedge \neg \ell_{i_2} \wedge \ldots \wedge \neg \ell_{i_n} \supset false$,

$$\ell_{i_1} \leftarrow \neg \ell_{i_2}, \ldots, \neg \ell_{i_n},$$

$$\leftarrow \neg \ell_{i_1}, \neg \ell_{i_2}, \ldots, \neg \ell_{i_n}.$$

where $\ell_{i_j} \in \{\ell_1, \ldots, \ell_n\}$ (for $j = 1, \ldots, n$).
As a special case, if $C$ is a unit literal $\ell$,

$$\ell \leftarrow, \qquad \leftarrow \neg \ell.$$

**Definition 3.2**
Given prioritized circumscription as follows:

$$Circum(\widetilde{\forall} T(P^1 \ldots P^k, Z, Q); P^1 > P^2 > \ldots > P^k; Z),$$

where $P^r(1 \leq r \leq k), Z, Q$ are tuples of predicate symbols such as $[(P^r)_1, \ldots, (P^r)_{\ell_r}], [Z_1, \ldots, Z_m], [Q_1, \ldots, Q_n]$, ELP $\Pi^\beta$ is constructed in the following two steps:

1. According to (2.1), a given prioritized circumscription is represented by the conjunction of $k$ parallel circumscriptions. So, let every $i$th ($1 \leq i \leq k$) parallel circumscription,

   $$Circum(\widetilde{\forall} T(P^1 \ldots P^k, Z, Q); P^i; P^{i+1}, \ldots, P^k, Z)$$

   be transformed in such a way that all predicate symbols occurring in it are renamed by using $Pi^1, \ldots, Pi^k, Zi, Qi$ instead of $P^1, \ldots, P^k, Z, Q$, which leads to as follows:

   $$Circum(\widetilde{\forall} Ti; Pi^i; Pi^{i+1}, \ldots, Pi^k, Zi), \quad (3.1)$$

   where $Ti$ denotes $T(Pi^1, \ldots, Pi^k, Zi, Qi)$, and $Pi^r, Zi, Qi$ are tuples of predicate symbols such as $[(Pi^r)_1, \ldots, (Pi^r)_{\ell_r}], [(Zi)_1, \ldots, (Zi)_m], [(Qi)_1, \ldots, (Qi)_n]$.

2. ELP $\Pi^\beta$ consists of all rules from $(\Pi^\alpha)_1, \ldots, (\Pi^\alpha)_k$ and $\Pi_\gamma$ where

   (a) each $(\Pi^\alpha)_i$ is an extended logic program (ELP) which is constructed from the $i$th renamed parallel circumscription (3.1) according to Definition 3.1.

   (b) $\Pi_\gamma$ is an extended logic program which consists of the following rules:
   For any predicate $u$ from $P, Z, Q$ and any $t$,

   $$u(t) \leftarrow u_i(t),$$
   $$\neg u(t) \leftarrow \neg u_i(t). \quad (1 \leq i \leq k)$$

   where $u$ and $u_i$ stand for any predicate symbol of $(P^r)_f, Z_g, Q_h$ and any one of $(Pi^r)_f, (Zi)_g, (Qi)_h$ respectively.
   $(1 \leq f \leq \ell_r, 1 \leq g \leq m, 1 \leq h \leq n)$

First of all, we show the following theorem which generalizes Theorem 2 in [Etherington, 1987].

**Theorem 3.3** *Assume that $T$ is a first order theory including DCA and UNA. Then, those formulas true in every default extension of the default theory:*

$$\left( \left\{ \frac{:\neg p(x)}{\neg p(x)}, \frac{:\neg q(x)}{\neg q(x)}, \frac{:q(x)}{q(x)} \middle| p \in P, q \in Q \right\}, T \right)$$

*are precisely those theorems of $Circum(T; P; Z)$ where $P, Q$ and $Z$ are tuples of minimized, fixed and variable predicate symbols respectively.*

**Proof:(sketch)** According to Theorem 1 in [de Kleer and Konolige, 1989], circumscription including fixed predicates $q_i$ can be logically equivalently transformed into circumscription without fixed predicates where $q_i$ and $\neg q_i$ are added to circumscribed predicates. As a result, this theorem is proved by applying Etherington's result [Etherington, 1987] about correspondence between circumscription without fixed predicates and Reiter's default theory [Reiter, 1980] to the transformed circumscription.

Based on Theorem 3.3 as well as a 1-1 correspondence between the extensions of a default theory and the answer sets of an extended logic program shown in [Gelfond and Lifschitz, 1991], the semantic relationships between circumscription and the translated ELPs $\Pi^\alpha, \Pi^\beta$ are given as follows.

**Theorem 3.4** *For any ground literal $G$ of the language of $T$ whose predicate symbol is not equality, it holds that,*

$$Circum(\widetilde{\forall} T; P; Z) \models G$$

*iff $G$ is true in all answer sets of ELP $\Pi^\alpha$,*

*where ELP $\Pi^\alpha$ is constructed from $Circum(\widetilde{\forall} T; P; Z)$ by using definition 3.1.*

**Theorem 3.5** *For any ground literal $G$ of the language of $T$ whose predicate symbol is not equality, it holds that,*

$$Circum(\widetilde{\forall} T; P^1 > P^2 > \ldots > P^k; Z) \models G$$

*iff $G$ is true in all answer sets of ELP $\Pi^\beta$.*

*where ELP $\Pi^\beta$ is constructed from*

$$Circum(\widetilde{\forall} T; P^1 > P^2 > \ldots > P^k; Z),$$

*by using Definition 3.2.*

**Remarks.** Theorem 3.4 and Theorem 3.5 can be easily generalized for a ground formula $F$ instead of for a ground literal $G$.

Satoh and Iwayama [1992] show that whether a ground atom $G$ is true in all stable models of a normal logic program $\Pi$, can be decided by running their top-down query evaluation procedure for abductive logic programs where an abductive framework is given by $\langle \Pi, A \rangle$ in which $A$ is a set of predicate symbols called abducible predicates. Their result is given as follows:

Suppose that a normal logic program $\Pi$ is consistent, which means that there exists a stable model of $\Pi$, and all ground rules obtained by replacing all variables in each rule in $\Pi$ by every element of its Herbrand universe are finite. Then it holds that,

$G$ *is true in all stable models of* $\Pi$
  iff *derive*$(not\ G, \{\})$ *fails*
     *under the abductive framework* $\langle \Pi, \{\} \rangle$.

where *derive* is a procedure given by them and *not* in its first argument denotes the negation-as-failure operator.

Since all ground rules in ELP $\Pi^\alpha$ and $\Pi^\beta$ constructed by using Definition 3.1 and Definition 3.2 respectively are finite, we can make use of their result in a slightly extended form, and Theorem 3.4 and Theorem 3.5 can be said in other words by the following corollaries.

**Corollary 3.6** *Suppose that $T$ is consistent and function-free. For any ground literal $G$ of the language of $T$ whose predicate symbol is not equality, it holds that,*

$$Circum(\widetilde{\forall}T; P; Z) \models G$$

    iff  $derive(not\ G, \{\})\ fails$
          $under\ the\ abductive\ framework\ \langle \Pi^\alpha, \{\}\rangle,$

*where* ELP $\Pi^\alpha$ *is constructed from* $Circum(\widetilde{\forall}T; P; Z)$ *by using Definition 3.1.*

**Corollary 3.7** *Suppose that $T$ is consistent and function-free. For any ground literal $G$ of the language of $T$ whose predicate symbol is not equality, it holds that,*

$$Circum(\widetilde{\forall}T; P^1 > P^2 > \ldots > P^k; Z) \models G$$

    iff  $derive(not\ G, \{\})\ fails$
          $under\ the\ abductive\ framework\ \langle \Pi, \{\}\rangle.$

*where* $\Pi$ *consists of all rules of* ELP $\Pi^\beta$ *in Definition 3.2 plus rules,* $\leftarrow u(t), \neg u(t)$, *for every predicate $u$ in $T$.*

**Remarks.** There is Ginsberg's early work [Ginsberg, 1989] on evaluating circumscription using abduction. But his method does not exploit the logic programming.

**Example 3.8**
Our compilation can handle circumscription representing *multiple inheritance*. Consider parallel circumscription:

$$Circum(\widetilde{\forall}T; ab_1, ab_2; pacifist) \quad (1)$$

where $T \stackrel{def}{=} \Sigma \cup$ DCA $\cup$ UNA, $\Sigma$ consists of the following clauses and DCA is $x = Nixon$:

$$pacifist(x) \vee ab_1(x) \vee \neg republican(x),$$
$$\neg pacifist(x) \vee ab_2(x) \vee \neg quaker(x),$$
$$republican(Nixon), \quad quaker(Nixon).$$

A set $Q$ of fixed predicates is $\{republican, quaker\}$ in this example. Hereafter we abbreviate *pacifist, republican, quaker* and *Nixon* to *pac, rep, quak* and $N$ respectively.

After replacing a variable $x$ in every clause in $\Sigma$ by a ground term $N$ in DCA, $\mu Th(\Sigma)$ is obtained as follows:

$$pac(N) \vee ab_1(N),$$
$$\neg pac(N) \vee ab_2(N),$$
$$ab_1(N) \vee ab_2(N),$$
$$rep(N), \quad quak(N).$$

According to Definition 3.1, let us construct ELP $\Pi^\alpha$ from circumscription (1).

1. For minimized predicates $ab_1, ab_2$ and a constant $N$,
$$\neg ab_1(N) \leftarrow not\ ab_1(N),$$
$$\neg ab_2(N) \leftarrow not\ ab_2(N),$$

2. For fixed predicates $rep, quak$ and a constant $N$,
$$\neg rep(N) \leftarrow not\ rep(N),$$
$$rep(N) \leftarrow not\ \neg rep(N),$$
$$\neg quak(N) \leftarrow not\ quak(N),$$
$$quak(N) \leftarrow not\ \neg quak(N),$$

3. For all contrapositives of all clauses of $\mu Th(\Sigma)$,
$$ab_1(N) \leftarrow \neg pac(N),$$
$$pac(N) \leftarrow \neg ab_1(N),$$
$$\leftarrow \neg pac(N), \neg ab_1(N),$$
$$ab_2(N) \leftarrow pac(N),$$
$$\neg pac(N) \leftarrow \neg ab_2(N),$$
$$\leftarrow pac(N), \neg ab_2(N),$$
$$ab_1(N) \leftarrow \neg ab_2(N),$$
$$ab_2(N) \leftarrow \neg ab_1(N),$$
$$\leftarrow \neg ab_1(N), \neg ab_2(N),$$
$$rep(N) \leftarrow, \quad \leftarrow \neg rep(N),$$
$$quak(N) \leftarrow, \quad \leftarrow \neg quak(N),$$

As a result, $\Pi^\alpha$ has two answer sets:
$$\{\neg ab_1(N), ab_2(N), pac(N), rep(N), quak(N)\},$$
$$\{ab_1(N), \neg ab_2(N), \neg pac(N), rep(N), quak(N)\}.$$

**Example 3.9** Let us compile the following prioritized circumscription:

$$Circum(\{p \vee q, q \vee r\}; p > q; r).$$

Since $\Sigma = \{p \vee q, q \vee r\}$, it holds that $\Sigma = \mu Th(\Sigma)$. Contrapositives of $p \vee q$ are as follows:

$$\neg q \supset p, \quad \neg p \supset q, \quad \neg p \wedge \neg q \supset false.$$

and those of $q \vee r$ are as follows:

$$\neg r \supset q, \quad \neg q \supset r, \quad \neg q \wedge \neg r \supset false.$$

According to (2.1), it holds that

$$Circum(\{p \vee q, q \vee r\}; p > q; r)$$
$$\equiv Circum(\{p \vee q, q \vee r\}; p; q, r)$$
$$\wedge Circum(\{p \vee q, q \vee r\}; q; r).$$

Then according to Definition 3.2, predicate symbols occurring each parallel circumscription are renamed as follows:

$$Circum(\{p1 \vee q1, q1 \vee r1\}; p1; q1, r1), \quad (2)$$
$$Circum(\{p2 \vee q2, q2 \vee r2\}; q2; r2). \quad (3)$$

Therefore ELPs $(\Pi^\alpha)_1, (\Pi^\alpha)_2$ constructed from (2) and (3) respectively as well as $\Pi_\gamma$ are obtained as follows:

$(\Pi^\alpha)_1:$
$$\neg p1 \leftarrow not\ p1,$$
$$p1 \leftarrow \neg q1, \quad q1 \leftarrow \neg p1,$$
$$\leftarrow \neg p1, \neg q1,$$
$$q1 \leftarrow \neg r1, \quad r1 \leftarrow \neg q1,$$
$$\leftarrow \neg q1, \neg r1,$$

$(\Pi^\alpha)_2:$
$$\neg q2 \leftarrow not\ q2,$$
$$\neg p2 \leftarrow not\ p2,$$
$$p2 \leftarrow not\ \neg p2,$$
$$p2 \leftarrow \neg q2, \quad q2 \leftarrow \neg p2,$$
$$\leftarrow \neg p2, \neg q2,$$
$$q2 \leftarrow \neg r2, \quad r2 \leftarrow \neg q2,$$
$$\leftarrow \neg q2, \neg r2,$$

$$\Pi_\gamma: \quad \begin{aligned} & p \leftarrow p1, & & p \leftarrow p2, \\ & q \leftarrow q1, & & q \leftarrow q2, \\ & r \leftarrow r1, & & r \leftarrow r2, \\ & \neg p \leftarrow \neg p1, & & \neg p \leftarrow \neg p2, \\ & \neg q \leftarrow \neg q1, & & \neg q \leftarrow \neg q2, \\ & \neg r \leftarrow \neg r1, & & \neg r \leftarrow \neg r2. \end{aligned}$$

ELP $\Pi^\beta$ has the only one answer set:

$$\{\neg p, q, \neg p1, q1, \neg p2, q2\},$$

where $\Pi^\beta \stackrel{def}{=} (\Pi^\alpha)_1 \cup (\Pi^\alpha)_2 \cup \Pi_\gamma$.
Thus according to Theorem 3.5, we can conclude that

$$Circum(\{p \lor q, q \lor r\}; p > q; r) \models \neg p,$$
$$Circum(\{p \lor q, q \lor r\}; p > q; r) \models q,$$

but neither $r$ nor $\neg r$ is provable from this prioritized circumscription.

In the following, we give an extension of Sakama and Inoue's first method [1995]. According to their method, parallel circumscription is translated into a *general disjunctive program* (GDP) whose semantics is given by stable models. Alternatively, they also show the translation of parallel circumscription into an *Extended disjunctive program* (EDP) whose semantics is given by the answer sets, which just corresponds to the stable models of the translated GDP. Each of their methods is applicable to a class of parallel circumscription, but not to prioritized circumscription. Our target language ELP has the expressive power of classical negation $\neg$, which enables us to compute prioritized circumscription as is shown in Theorem 3.5. Since the classical negation is also available to EDP, we can make use of our method to extend their alternative method whose target language is EDP, so that it may become applicable to a class of prioritized circumscription as follows.

**Definition 3.10** [Sakama and Inoue, 1995]
Given parallel circumscription $Circum(\tilde\forall T; P; Z)$, EDP $\Pi^\alpha$ is constructed as follows, where DCA and UNA are incorporated in a first order theory $T$ since only its Herbrand models are considered, and $p_1, \ldots, p_\ell, z_1, \ldots, z_m$, and $q_1, \ldots, q_n$ are used to denote atoms from $P, Z$ and $Q$ respectively.

1. For any clause in $T$ of the form:

$$p_1 \lor \ldots \lor p_\ell \lor z_1 \lor \ldots \lor z_m \lor q_1 \lor \ldots \lor q_n \lor \neg p_{\ell+1} \lor \ldots$$
$$\lor \neg p_s \lor \neg z_{m+1} \lor \ldots \lor \neg z_t \lor \neg q_{n+1} \lor \ldots \lor \neg q_u,$$

$\Pi^\alpha$ has the rule:

$$z_1|\ldots|z_m|q_1|\ldots|q_n \leftarrow p_{\ell+1},\ldots,p_s, z_{m+1},\ldots,z_t, q_{n+1},$$
$$\ldots, q_u, not\, p_1, \ldots, not\, p_\ell.$$

2. For every clause in $\mu Th(\Sigma)$ of the form:

$$p_1 \lor \ldots \lor p_\ell \lor q_1 \lor \ldots \lor q_n \lor \neg q_{n+1} \lor \ldots \lor \neg q_u,$$

$\Pi^\alpha$ has the rule:

$$p_1|\ldots|p_\ell|q_1|\ldots|q_n \leftarrow q_{n+1}, \ldots, q_u,$$

3. For any atom $p, z, q$, $\Pi^\alpha$ has the rule:

$$\neg p \leftarrow not\, p,$$
$$z \mid \neg z \leftarrow, \qquad q \mid \neg q \leftarrow.$$

**Remarks.** It is shown in [Sakama and Inoue, 1995] that there is a 1-1 correspondence between the models of parallel circumscription and the answer sets of the translated EDP $\Pi^\alpha$.

**Definition 3.11**
Given prioritized circumscription:

$$Circum(\tilde\forall T(P^1 \ldots P^k, Z, Q); P^1 > P^2 > \ldots > P^k; Z),$$

EDP $\Pi^\beta$ is constructed in the following two steps:

1. This is the same as the first step given by Definition 3.2.

2. EDP $\Pi^\beta$ consists of all rules from $(\Pi^\alpha)_1, \ldots, (\Pi^\alpha)_k$ and $\Pi_\gamma$ where

   (a) each $(\Pi^\alpha)_i$ is an extended disjunctive program (EDP) which is constructed from the $i$th renamed parallel circumscription (3.1) according to Definition 3.10.

   (b) $\Pi_\gamma$ is the same set given by Definition 3.2.

Then the relationship between prioritized circumscription and the translated EDP $\Pi^\beta$ are given as follows.

**Theorem 3.12** *Let $F$ be a ground formula of the language of $T$ and equality does not occur as a predicate symbol in $F$. Then, it holds that, for any $F$,*

$$Circum(\tilde\forall T; P^1 > P^2 > \ldots > P^k; Z) \models F$$

*iff $F$ is true in all answer sets of EDP $\Pi^\beta$.*

where EDP $\Pi^\beta$ is constructed from

$$Circum(\tilde\forall T; P^1 > P^2 > \ldots > P^k; Z),$$

*by using Definition 3.11.*

**Example 3.13**
Consider prioritized circumscription given in Example 3.9. We apply Theorem 3.12 instead of Theorem 3.5 to it. Then according to Definition 3.11, the translated EDPs $(\Pi^\alpha)_1$ and $(\Pi^\alpha)_2$ are as follows:
Since $P = \{p1\}$, $Z = \{q1, r1\}$, $Q = \phi$ in the renamed circumscription: $Circum(\{p1 \lor q1, q1 \lor r1\}; p1; q1, r1)$,

$$(\Pi^\alpha)_1: \quad \begin{aligned} & q1 \leftarrow not\, p1, \\ & q1 \mid r1, \\ & \neg p1 \leftarrow not\, p1, \\ & q1 \mid \neg q1 \leftarrow, \qquad r1 \mid \neg r1 \leftarrow. \end{aligned}$$

Since $P = \{q2\}$, $Z = \{r2\}$, $Q = \{p2\}$ in the renamed circumscription: $Circum(\{p2 \lor q2, q2 \lor r2\}; q2; r2)$,

$$(\Pi^\alpha)_2: \quad \begin{aligned} & p2 \leftarrow not\, q2, \\ & r2 \leftarrow not\, q2, \\ & p2 \mid q2 \leftarrow, \\ & \neg q2 \leftarrow not\, q2, \\ & r2 \mid \neg r2 \leftarrow, \qquad p2 \mid \neg p2 \leftarrow. \end{aligned}$$

$\Pi_\gamma$ is obtained as the same one shown in Example 3.9. As a result, EDP $\Pi^\beta$ has the following two answer sets:

$$\{\neg p, q, r, \neg p1, q1, r1, \neg p2, q2, r2\},$$
$$\{\neg p, q, \neg r, \neg p1, q1, \neg r1, \neg p2, q2, \neg r2\},$$

where $\Pi^\beta$ consists of all rules in the above $(\Pi^\alpha)_1$, $(\Pi^\alpha)_2$ and $\Pi_\gamma$. Both $\neg p$ and $q$ are true in all of these answer sets, but so is neither $r$ nor $\neg r$. Notice that this is the same evaluation result as the one in Example 3.9.

## 4 Related Works and Conclusion

In this paper, we present a method of compiling circumscription into extended logic programs which is widely applicable to parallel circumscription as well as prioritized circumscription. Our method always enables us to compute any circumscription whose theory includes both DCA and UNA by compiling it into ELP $\Pi$, which can be evaluated by using the existing logic programming interpreter such as Satoh and Iwayama's top-down query evaluation procedure for abductive logic programming.

In the following, we compare our method with related researches from the viewpoints of the applicable class as well as the computational efficiency and feasibility.

- Gelfond and Lifschitz's method [1988a] as well as our previous method [Wakaki and Satoh, 1995] are the most efficient since they are as efficient as the evaluation of a stratified logic program. But their applicable classes are limited within a class of circumscription which has a unique model as mentioned in the introduction of this paper.

- In Sakama and Inoue's first method [1995], parallel circumscription is translated into a general disjunctive program (GDP). This method is applicable to a wide class of parallel circumscription, but inapplicable to prioritized circumscription though a class of GDP has the expressive power of the *positive occurrences* of negation as failure [Inoue and Sakama, 1994]. The most important difference between their GDP and our ELP is whether classical negation $\neg$ is available or not. Theorem 3.12 shows that our approach exploiting classical negation can also give an extension of their alternative method whose target language is EDP, so that it may become applicable to prioritized circumscription.

- In Sakama and Inoue's second method [1996], parallel circumscription as well as prioritized circumscription is translated into prioritized logic programs proposed by them whose declarative meaning is given by preferred answer sets defined by them. Their method, however, is immature for the purpose of the automation of circumscription since their prioritized logic program is not feasible because their method gives only the semantic aspects, but procedural issues for the query evaluation are left as their future works.

Our future work is to implement our method proposed in this paper.

## References

[de Kleer and Konolige, 1989] de Kleer, J. and Konolige, K. Eliminating the Fixed Predicates from a Circumscription. *Artificial Intelligence 39*, pages 391-398, 1989.

[Etherington, 1987] David W. Etherington. Relating Default Logic and Circumscription. *Proc. IJCAI-87*, pages 489-494, 1987.

[Gelfond and Lifschitz, 1988a] Gelfond, M. and Lifschitz, V. Compiling Circumscriptive Theories into Logic Programs. *Proc. AAAI-88*, pages 455-459. Extended version in: *Proc. 2nd Int. Workshop on Nonmonotonic Reasoning*, LNAI 346, pages 74-99, 1988.

[Gelfond and Lifschitz, 1988b] Gelfond, M. and Lifschitz, V. The Stable Model Semantics for Logic Programming. *Proc. LP'88*, pages 1070-1080, 1988.

[Gelfond and Lifschitz, 1991] Gelfond, M. and Lifschitz, V. Classical Negation in Logic Programs and Disjunctive Databases. *New Generation Computing 9*, pages 365-385, 1991.

[Ginsberg, 1989] Matthew L. Ginsberg, A Circumscriptive Theorem Prover, *Artificial Intelligence 39*, pages 209-230, 1989.

[Inoue, 1992] Inoue, K. Linear Resolution for Consequence Finding, *Artificial Intelligence 56*, pages 301-353, 1992.

[Inoue and Sakama, 1994] Inoue, K. and Sakama, C. On Positive Occurrences of Negation as Failure. *Proc. KR'94*, pages 293-304, 1994.

[Lifschitz, 1985] Lifschitz, V. Computing Circumscription. *Proc. IJCAI 85*, pages 121-127, 1985.

[McCarthy, 1980] McCarthy, J. Circumscription - a Form of Non-monotonic Reasoning. *Artificial Intelligence 13*, pages 27-39, 1980.

[Przymusinski, 1987] Przymusinski, T. On the Declarative Semantics of Deductive Databases and Logic Programs. in *Foundations of Deductive Databases and Logic Programming* (J. Minker, Ed.), Morgan Kaufmann, pages 193-216, 1987.

[Reiter, 1980] Reiter, R. A Logic for default reasoning. *Artificial Intelligence 13*, pages 81-132, 1980.

[Sakama and Inoue, 1995] Sakama, C. and Inoue, K. Embedding Circumscriptive Theories in General Disjunctive Programs. *Proc. 3rd International Conference on Logic Programming and Nonmonotonic Reasoning*, 1995, LNAI 928, pages 344-357, Springer.

[Sakama and Inoue, 1996] Sakama, C. and Inoue, K. Representing Priorities in Logic Programs. *Proc. Joint International Conference and Symposium on Logic Programming*, pages 82-96, 1996.

[Satoh and Iwayama, 1992] Satoh, K. and Iwayama, N. A Query Evaluation Method for Abductive Logic Programming. *Proc. Joint International Conference and Symposium on Logic Programming*, pages 671-685, 1992.

[Wakaki and Satoh, 1995] Wakaki, T and Satoh, K. Computing Prioritized Circumscription by Logic Programming. *Proc. 12th International Conference on Logic Programming*, pages 283-297, 1995.

# AUTOMATED REASONING

Automated Reasoning 8: Modal Logic

# Prefixed Tableaux Systems for Modal Logics with Enriched Languages*

**Philippe Balbiani**
Laboratoire d'informatique de Paris-Nord,
Avenue Jean-Baptiste Clément,
93430 Villetaneuse, France.

**Stéphane Demri**
Laboratoire LEIBNIZ,
46 Avenue Félix Viallet,
38031 Grenoble, France.

## Abstract

We present sound and complete prefixed tableaux systems for various modal logics with enriched languages including the "difference" modal operator $[\neq]$ and the "only if" modal operator $[-R]$. These logics are of special interest in Artificial Intelligence since their expressive power is higher than the standard modal logics and for most of them the satisfiability problem remains decidable. We also include in the paper decision procedures based on these systems. In the conclusion, we relate our work with similar ones from the literature and we propose extensions to other logics.

## 1 Introduction

The definition of logical formalisms that model cognitive and reasoning processes has been always confronted to two issues: how to decrease the expressive power of existing untractable logics in order to obtain tractable fragments and how to increase the expressive power of decidable logics while preserving decidability - this includes for instance the extension of known decidable fragments of the classical logic. These fragments include various modal logics (see e.g. [Hughes and Cresswell, 1984]) if one translates them in the standard way to classical logic. The modal logics have been recognized in the Artificial Intelligence community as serious candidates to capture different aspects of reasoning about knowledge (see e.g. [Fagin et al., 1995]). However the standard modal logics have a restricted expressive power (for instance the class of irreflexive frames is not definable by a modal formula of the logic K).

That is why in the literature various modal logics with enriched languages have been defined. Most of the work done for these logics has been dedicated to study their expressive power (see e.g. [Goranko and Passy, 1992; Rijke, 1993]). In the paper our aim is to analyze various features related to the mechanization of numerous modal logics with enriched languages. To do so, we define prefixed tableaux which are known to be close to the semantics of the logics and they allow a user-friendly presentation of the proofs. Moreover, the use of prefixes

*Work supported by C.N.R.S., France.

(see e.g. [Fitting, 1983; Wallen, 1990; Massacci, 1994; Governatori, 1995]) is known to take advantage of the computational features of the logics. Namely, each prefix occurring at some stage of the proof contains some information about part of the current proof. However we ignore whether a matrix characterization of the logics treated herein exist in order to avoid some redundancies in the tableaux proof search - *notational redundancy*, *irrelevance* and *non-permutability* [Wallen, 1990].

The logics treated in the paper contain various operators that differ from the standard necessity operator □ (also noted $[R]$):

- the difference operator $[\neq]$ that allows to access to the worlds different from the current world (see e.g. applications of its use in [Segerberg, 1981; Sain, 1988; Koymans, 1992; Rijke, 1993])
- the complement operator $[-R]$ that allows to access to the worlds not accessible from the current world (see e.g. [Humberstone, 1983; Goranko, 1990a; Levesque, 1990; Lakemeyer, 1993])
- and by a side-effect the universal operator $[U]$ that allows to access to any world of the model (see e.g. [Goranko and Passy, 1992]). $[U]A$ can be defined in various ways: for instance $[U]A =_{def} A \wedge [\neq]A$ or $[U]A =_{def} [R]A \wedge [-R]A$.

Adding these operators to standard modal logics can significantly increase their expressive power. For instance every finite cardinality is definable in a modal logic whose language contains $[\neq]$ [Koymans, 1992]. Most of the logics dealt with in the paper have a decidable satisfiability problem and we shall provide decision procedures based on our systems. However because of the expressive power of the logics our calculi have two original features: a current information $\mathcal{C}$ is associated to each branch of a tableau and a restricted cut rule is included in various calculi that can be viewed as a modal variant of the cut rule in the d'Agostino's calculi [d'Agostino, 1993].

The rest of the paper is structured as follows. Section 2 presents the logics considered in the paper. The sections 3, 4, 5 and 6 present the calculi for the various logics as well as the decision procedures. Because of lack of space we have omitted part of the proofs as well as the possible extensions where the accessibility relations satisfy standard conditions (reflexivity, symmetry, transitivity,...). Section 7 compares our calculi with existing ones for other modal logics and concludes the paper by presenting possible extensions.

## 2 Enriched multi-modal logics

### 2.1 Syntax and semantics

A modal language L is determined by three sets that are supposed to be pairwise disjoint: a set $\text{For}_0 = \{p, q, \ldots\}$ of *propositional variables*, a set $\{\neg, \wedge\}$ of propositional operators (the connectives $\vee, \Rightarrow, \Leftrightarrow$ are defined as for the propositional calculus) and a (possibly finite) countable set $OP = \{[i] : i \in I\}$ of *modal operators*. The set of formulae For of the language L is defined by the following grammar: $A ::= p \mid \neg A \mid A \wedge B \mid \oplus A$ where $p \in \text{For}_0$, $A, B \in \text{For}$ and $\oplus \in OP$. In the sequel we assume that $OP$ is finite and as usual $\langle i \rangle A =_{def} \neg [i] \neg A$. A *frame* is a structure $(W, (R_i)_{i \in I})$ where $W$ is a non-empty set of *worlds* (sometimes also called *knowledge states*) and $(R_i)_{i \in I}$ is a family of binary relations on $W$. A *model* $\mathcal{M}$ is a structure $(W, (R_i)_{i \in I}, V)$ where $(W, (R_i)_{i \in I})$ is a frame and $V$ is mapping $\text{For}_0 \to \mathcal{P}(W)$, the power set of $W$. For each set $W$, we write $id_W$ (resp. $dif_W$) to denote the binary relation $\{\langle w, w \rangle : w \in W\}$ (resp. $W \times W \setminus id_W$). Let $\mathcal{M} = (W, (R_i)_{i \in I}, V)$ be a model. As usual, we say that a formula $A$ is *satisfied* by the world $w \in W$ (denoted by $\mathcal{M}, w \models A$) when the following conditions are satisfied:

- $\mathcal{M}, w \models p$ iff $w \in V(p)$ for all $p \in \text{For}_0$,
- $\mathcal{M}, w \models \neg A$ iff not $\mathcal{M}, w \models A$,
- $\mathcal{M}, w \models A \wedge B$ iff $\mathcal{M}, w \models A$ and $\mathcal{M}, w \models B$,
- $\mathcal{M}, w \models [i]A$ iff for all $w' \in W$ such that $(w, w') \in R_i$, we have $\mathcal{M}, w' \models A$.

In the sequel by a *logic* $\mathcal{L}$ we understand a pair $\langle \text{For}, \mathcal{S} \rangle$ such that For is a set of formulae from a given language and $\mathcal{S}$ is a set of models. A formula $A$ is said to be $\mathcal{L}$-*valid* iff for all models $\mathcal{M} \in \mathcal{S}$ and all $w \in W$, $\mathcal{M}, w \models A$. A formula $A$ is said to be $\mathcal{L}$-*satisfiable* iff $\neg A$ is not $\mathcal{L}$-valid.

### 2.2 Logics in the paper

In the paper we shall consider numerous logics that admit interactions between the modal operators:

1. $K_I = \langle \text{For}, \mathcal{S} \rangle$ is the logic such that $\mathcal{S}$ is the set of all the models. The $K_I$-satisfiability problem is **PSPACE**-complete (see e.g. [Fagin *et al.*, 1995]).

2. $\mathcal{L}([R], [-R]) = \langle \text{For}, \mathcal{S} \rangle$ (see e.g. [Goranko, 1990a]) is the logic such that $I = \{1, 2\}$ and $\mathcal{M} = (W, R_1, R_2, V) \in \mathcal{S}$ iff $R_1 = W \times W \setminus R_2$. The satisfiability problem is decidable and **EXPTIME**-hard [Spaan, 1993]. Similar modal logics are considered in the context of knowledge representation and reasoning (see e.g. [Lakemeyer, 1993]).

3. $\mathcal{L}([\neq]) = \langle \text{For}, \mathcal{S} \rangle$ (see e.g. [Segerberg, 1981]) is the logic such that $I = \{1\}$ and $\mathcal{M} = (W, R_1, V) \in \mathcal{S}$ iff $R_1 = dif_W$. The $\mathcal{L}([\neq])$-satisfiability problem is **NP**-complete when $\text{For}_0$ is infinite and in **P** otherwise (see e.g. [Spaan, 1993; Demri, 1996]).

4. $K_I([\neq]) = \langle \text{For}, \mathcal{S} \rangle$ is the logic such that $1 \in I$ (a distinguished element of $I$), $card(I) \geq 2$ and $\mathcal{M} = (W, (R_i)_{i \in I}, V) \in \mathcal{S}$ iff $R_1 = dif_W$. Axiomatization of $K_I([\neq])$ has been studied in [Rijke, 1993; Balbiani, 1997]. For $I = \{1, 2\}$, the $K_I([\neq])$-satisfiability problem is decidable and **EXPTIME**-complete [Rijke, 1993].

The models for $\mathcal{L}([R], [-R])$ satisfy $(\star)$ $R_1 = W \times W \setminus R_2$. If we require $(\star\star)$ $R_1 = dif_W$ then $[2]A \Leftrightarrow A$ is valid in this new logic. $\mathcal{L}([\neq])$ can be seen as $\mathcal{L}([R], [-R])$ except that the models satisfy $(\star)$ and $(\star\star)$ and only $[1]$ is in the language. Moreover, $K_I([\neq])$ is obtained from $\mathcal{L}([\neq])$ by adding the operators $\{[i] : i \in I \setminus \{1\}\}$ that behave as in $K_I$. The notion of complementary relations is therefore crucial in the semantics of the logics.

It is not the purpose of this section to recall all the features of the expressive power of the abovementioned logics (see e.g. [Goranko, 1990a; Koymans, 1992; Rijke, 1993]). By way of example we consider the logic $K_I([\neq])$ with $I = \{1, 2\}$. As usual, a class $\mathcal{F}$ of frames $(W, R_1, R_2)$ is said to be $K_I([\neq])$-*definable* iff there exists a $K_I([\neq])$-formula $A$ such that for all frames $(W, R_1, R_2)$, $(W, R_1, R_2) \in \mathcal{F}$ iff $(W, R_1, R_2) \models A$ (i.e. for all valuations $V$ and all $w \in W$, $(W, R_1, R_2, V), w \models A$). A similar notion of definability can be naturally defined for other logics.

**Fact 2.1.** [Goranko, 1990b; Koymans, 1992]

- All universal first-order conditions on $R, =$ are $K_I([\neq])$-definable.
- Every finite cardinality is $\mathcal{L}([\neq])$-definable.
- Each universal first-order formula on $R$ is $\mathcal{L}([R], [-R])$-definable.

The statements of Fact 2.1 do not hold for the logic $K_I$: for example the class of irreflexive frames is not $K_I$-definable.

## 3 Tableaux for $K_I$

The calculus defined for $K_I$ in this section can be easily obtained from existing ones in the literature (see e.g. [Fitting, 1983]) but it will be the opportunity to introduce various definitions smoothly.

We shall define prefixed tableaux following the methodology described in [Fitting, 1983]. We make substantial use of the uniform notation for modal formulae defined in [Fitting, 1983]. Four types of formulae are usually distinguished: $\nu$ (necessity), $\pi$ (possibility), $\alpha$ (conjunction) and $\beta$ (disjunction). For $i \in I$, we introduce the types $\nu^i$ and $\pi^i$. For instance, $\neg \langle i \rangle A$ and $[i]A$ are of type $\nu^i$ ($\nu_0^i$ denotes the formulae $\neg A$ and $A$ respectively) and $\neg[i]A$ and $\langle i \rangle A$ are of type $\pi^i$ ($\pi_0^i$ denotes the formulae $\neg A$ and $A$ respectively).

A prefixed formula is a triple of the form $\sigma : A\ [\mathcal{C}]$ where $\sigma$ is a *prefix*, i.e. $\sigma$ is a finite sequence of natural numbers possibly superscripted by some $i \in I$, $A$ is a formula and $\mathcal{C}$ is a couple $\langle \mathcal{C}_1, \mathcal{C}_2 \rangle$. Each $\mathcal{C}_i$ is a set of pairs of prefixes. When the context is clear we omit $\sigma$ or $[\mathcal{C}]$. The condition $\mathcal{C}$ is the current information on the branch

$$\frac{\sigma : \alpha \; [\mathcal{C}]}{\substack{\sigma : \alpha_1 \; [\mathcal{C}] \\ \sigma : \alpha_2 \; [\mathcal{C}]}} \; \alpha\text{-}rule$$

$$\frac{\sigma : \beta \; [\mathcal{C}]}{\sigma : \beta_1 \; [\mathcal{C}] \mid \sigma : \beta_2 \; [\mathcal{C}]} \; \beta\text{-}rule$$

$$\frac{\sigma : \pi^i \; [\mathcal{C}]}{\sigma k^i : \pi_0^i \; [\mathcal{C}]} \; \pi^i\text{-}rule, \; \text{new } k \in \omega \text{ on the branch}$$

$$\frac{\sigma : \nu^i \; [\mathcal{C}]}{\sigma' : \nu_0^i \; [\mathcal{C}]} \; \nu^i\text{-}rule$$

if $\sigma'$ is already on the branch and for some $k \in \omega$, $\sigma' = \sigma k^i$.

Figure 1: Tableaux system for $K_I$

that is stored during its development. At each step of the development of a branch, $\mathcal{C}$ is identical for all the prefixed formulae on that branch, i.e. $\mathcal{C}$ is an attribute for branches. We refer to a prefixed formula as *atomic* if it is of the form $\sigma : p \; [\mathcal{C}]$ or $\sigma : \neg p \; [\mathcal{C}]$ when $p$ is an atomic formula. Figure 1 presents the prefixed tableau system for the logic $K_I$. Observe that the condition $[\mathcal{C}]$ is of no use in this calculus.

In the sequel we omit the *presentation* of the $\alpha$-rule (decomposition of conjunctions) and the $\beta$-rule (decomposition of disjunctions) but these rules are included in any forthcoming calculus. A branch is *closed* if it contains contradictory prefixed formulae (for any formula $A$, $\sigma : A$ and $\sigma : \neg A$ are contradictory). A tableau is *closed* if every branch is closed. A formula $A$ is said to *have a closed tableau* iff there is a closed tableau which root is $0 : \neg A \; [\langle \emptyset, \emptyset \rangle]$. Termination occurs when no operation is possible. A branch is *open* if it is not closed and a tableau is *open* if at least one branch is such.

**Theorem 3.1.** A formula $A$ is $K_I$-valid iff $A$ has a closed tableau built with the rules presented in Figure 1.

The proof of Theorem 3.1 can be easily obtained from existing ones from the literature [Fitting, 1983].

## 4 Tableaux for $\mathcal{L}([R], [-R])$

Instead of defining a sound and complete calculus for the logic $\mathcal{L}([R], [-R])$ we define a sound and complete calculus for the logic $K_{1,2}^-$ ($I = \{1, 2\}$) characterized by the models $(W, R_1, R_2, V)$ where $R_1 \cup R_2 = W \times W$ (we do not require $R_1 \cap R_2 = \emptyset$). It is known that $\mathcal{L}([R], [-R])$ and $K_{1,2}^-$ have the same class of valid formulae [Goranko, 1990a] and we shall provide a decision procedure for the set of $K_{1,2}^-$-valid formulae based on our tableaux approach. Actually from the calculus for $K_{1,2}^-$ the careful reader will observe that a calculus for $\mathcal{L}([R], [-R])$ can be easily defined. However the calculus for $K_{1,2}^-$ is more adequate to define a decision procedure. The rules for the logic $K_{1,2}^-$ are those in Figure 2 where

- $\mathcal{C}' = \langle \mathcal{C}_1 \cup \{\langle \sigma, \sigma' \rangle\}, \mathcal{C}_2 \rangle$, $\mathcal{C}'' = \langle \mathcal{C}_1, \mathcal{C}_2 \cup \{\langle \sigma, \sigma' \rangle\} \rangle$,
- $\mathcal{C}''' = \langle \mathcal{C}_1 \cup \{\langle \sigma, \sigma' \rangle\}, \mathcal{C}_2 \cup \{\langle \sigma, \sigma' \rangle\} \rangle$.

$$\frac{\sigma : \nu^i \; [\mathcal{C}]}{\sigma' : \nu_0^i \; [\mathcal{C}]} \; \nu^i\text{-}rule, \; i \in \{1, 2\}$$

if $\mathtt{C}_i(\langle \sigma, \sigma' \rangle, \mathcal{C})$ holds and $\sigma'$ already occurs on the branch.

$$\frac{\sigma : \pi^i \; [\mathcal{C}]}{\sigma k^i : \pi_0^i \; [\mathcal{C}]} \; \pi^i\text{-}rule, \; \text{new } k \in \omega \text{ on the branch}$$

if there is no $\sigma'$ such that $\sigma' : \pi_0^i$ on the branch and either $\mathtt{C}_i(\langle \sigma, \sigma' \rangle, \mathcal{C})$ or (for all $\sigma : \nu^i$ on the branch, $\sigma' : \nu_0^i$ is on the branch).

$$\frac{\sigma'' : A \; [\mathcal{C}]}{\sigma'' : A \; [\mathcal{C}'] \mid \sigma'' : A \; [\mathcal{C}''] \mid \sigma'' : A \; [\mathcal{C}''']}$$

$\sigma, \sigma'$ not already applied with this rule

Figure 2: Tableaux system for $K_{1,2}^-$

For the logic $K_{1,2}^-$, $\mathtt{C}_i(\langle \sigma, \sigma' \rangle, \mathcal{C})$ holds ($i \in \{1, 2\}$) iff either $\langle \sigma, \sigma' \rangle \in \mathcal{C}_i$ or $\sigma' = \sigma k^i$ for some $k \in \omega$. Intuitively, $\mathtt{C}_i$ encodes the accessibility relation $R_i$. The condition $\mathcal{C}$ could be deleted in the definition of the calculus since it only stores some information about the way the rules have been applied on the branch. However, if one wishes to implement our calculi, the actual presentation is well-suited for this purpose. For instance the $\nu^i$-rule can be read as follows. If the formula $\sigma : \nu^i$ occurs on the branch and if the current information on the branch is $\mathcal{C}$ then add $\sigma' : \nu_0^i$ on the branch and $\mathcal{C}$ remains unchanged. It is worth observing that the cut rule cannot be deleted unless completeness is lost. This property is also shared by the cut rule in the calculi defined in [d'Agostino, 1993]. It is also worth noting that the condition of the restricted cut rule in Figure 2 is equivalent to: either not $\mathtt{C}_1(\langle \sigma, \sigma' \rangle, \mathcal{C})$ or not $\mathtt{C}_2(\langle \sigma, \sigma' \rangle, \mathcal{C})$. Moreover, by applying the restricted cut rule, the current information $\mathcal{C}$ on the branch is updated.

### 4.1 Soundness

Let $X$ be a set of prefixed formulae having the same condition $\mathcal{C}$ (what happens at a current stage of the development of a given branch). Let $\mathcal{M} = (W, R_1, R_2, V)$ be a $K_{1,2}^-$-model. By an *interpretation of $X$ in $\mathcal{M}$* we mean a mapping $\mathcal{I} : \{\sigma : \sigma : A \in X\} \to W$ such that if $\sigma, \sigma'$ occur in $X$, then $\mathtt{C}_i(\langle \sigma, \sigma' \rangle, \mathcal{C})$ implies $\langle \mathcal{I}(\sigma), \mathcal{I}(\sigma') \rangle \in R_i$ ($i = 1, 2$). We say that $X$ is $K_{1,2}^-$-*satisfiable under the interpretation* $\mathcal{I}$ if for each $\sigma : A \in X$, $\mathcal{M}, \mathcal{I}(\sigma) \models A$. We say that $X$ is $K_{1,2}^-$-satisfiable if $X$ is $K_{1,2}^-$-satisfiable under some interpretation. We say that a branch of a tableau is $K_{1,2}^-$-*satisfiable* if the set of prefixed formulae on it is $K_{1,2}^-$-satisfiable. A tableau is $K_{1,2}^-$-*satisfiable* if some branch is.

**Lemma 4.1.** Suppose $\mathtt{T}$ is a prefixed tableau that is $K_{1,2}^-$-satisfiable. Let $\mathtt{T}'$ be the tableau that results from a single tableau rule being applied to $\mathtt{T}$. Then $\mathtt{T}'$ is also $K_{1,2}^-$-satisfiable.

**Proof:** By an easy verification. **Q.E.D.**

**Proposition 4.2.** (soundness) If $A$ has a closed tableau built with the rules in Figure 2 then $A$ is $K_{1,2}^-$-valid.

**Proof:** Similar to the proof of Theorem 3.2 in [Fitting, 1983] (p.400). **Q.E.D.**

## 4.2 Completeness

Let $A$ be a formula. As done in [Fitting, 1983], we define a systematic attempt to produce a proof of $A$. The procedure is in stages and the stage 1 consists in placing $0 : \neg A \ [\langle \emptyset, \emptyset \rangle]$ at the root. Now suppose $n$ stages of the construction have been done. If the tableau is closed then we stop. Similarly if every occurrence of a prefixed formula is *finished* (see the definition of 'finished' below) then we stop. Otherwise we go on. If $n + 1$ is even, $\sigma, \sigma'$ satisfies the condition of the cut rule on some open branch **BR** (chosen in some *fair* way) and $\langle \sigma, \sigma' \rangle$ is the smallest pair (for some encoding in the set of natural numbers $\omega$) satisfying this property then split the end of branch **BR** in three sub-branches by applying the restricted cut rule with $\langle \sigma, \sigma' \rangle$. Otherwise ($n + 1$ odd) any stage $n + 1$ consists in choosing an occurrence of a prefixed formula $\sigma : B \ [\mathcal{C}]$ as high up in the tree as possible (as close to the origin as possible) that has not been finished. If $\sigma : B \ [\mathcal{C}]$ is atomic then the occurrence is declared finished. This ends the stage $n + 1$ otherwise we extend the tableau as follows. For each open branch **BR** through the occurrence of $\sigma : B \ [\mathcal{C}]$ (*under the proviso the conditions to apply the rules hold*):

P1 If $\sigma : B \ [\mathcal{C}]$ is of the form $\sigma : \alpha \ [\mathcal{C}]$ add $\sigma : \alpha_1 \ [\mathcal{C}]$ and $\sigma : \alpha_2 \ [\mathcal{C}]$ to the end of **BR**.

P2 If $\sigma : B \ [\mathcal{C}]$ is of the form $\sigma : \beta \ [\mathcal{C}]$ split the end of **BR** and add $\sigma : \beta_1 \ [\mathcal{C}]$ to the end of one *sub-branch* and $\sigma : \beta_2 \ [\mathcal{C}]$ to the end of the other one.

P3 If $\sigma : B \ [\mathcal{C}]$ is of the form $\sigma : \nu^i \ [\mathcal{C}]$ then for all $\sigma'$ satisfying the condition of the $\nu^i$-rule add $\sigma' : \nu_0^i \ [\mathcal{C}]$ to the end of **BR**, after which add a fresh occurrence of $\sigma : \nu^i \ [\mathcal{C}]$ to the end of **BR**.

P4 If $\sigma : B \ [\mathcal{C}]$ is of the form $\sigma : \pi^i \ [\mathcal{C}]$ then add $\sigma k^i : \pi_0^i \ [\mathcal{C}]$ to the end of **BR**. Moreover for $\sigma : \nu^i \ [\mathcal{C}]$ on the branch add $\sigma k^i : \nu_0^i \ [\mathcal{C}]$ to the end of **BR** (applications of the $\nu^i$-rule)

Having done this for each branch **BR** through the particular occurrence of $\sigma : B \ [\mathcal{C}]$ being considered, declare that occurrence of $\sigma : B \ [\mathcal{C}]$ finished. This ends stage $n + 1$.

**Definition 4.1.** Let $X$ be a set of prefixed formulae and $\mathcal{C}$ be a condition. We say $X$ is *downward-saturated with respect to $\mathcal{C}$* iff:

C1 For all $\sigma, \sigma' \in X$, (C1.1) either $\mathtt{C}_1(\langle \sigma, \sigma' \rangle, \mathcal{C})$ or $\mathtt{C}_2(\langle \sigma, \sigma' \rangle, \mathcal{C})$ and, (C1.2) for all $p \in \mathtt{For}_0$, $\{\sigma : p, \sigma' : \neg p\} \subseteq X$ implies $\sigma \neq \sigma'$.

C2 if $\sigma : \alpha \in X$ then $\{\sigma : \alpha_1, \sigma : \alpha_2\} \subseteq X$.

C3 if $\sigma : \beta \in X$ then either $\sigma : \beta_1 \in X$ or $\sigma : \beta_2 \in X$.

C4 if $\sigma : \nu^i \in X$ then for all $\sigma'$ in $X$ satisfying the condition of the $\nu^i$-rule, we have $\sigma' : \nu_0^i \in X$.

C5 if $\sigma : \pi^i \in X$ then there is $\sigma'$ such that $\sigma' : \pi_0^i \in X$ and, either $\mathtt{C}_1(\langle \sigma, \sigma' \rangle, \mathcal{C})$ or (for all $\sigma : \nu^i \in X$, $\sigma' : \nu_0^i \in X$).

$\triangledown$

**Lemma 4.3.** If $X$ is downward-saturated with respect to $\mathcal{C}$ then $X$ is $K_{1,2}^-$-satisfiable.

**Proof:** Assume $X$ is downward-saturated wrt $\mathcal{C}$. Let $\mathcal{M} = (W, R_1, R_2, V)$ be the structure such that $W = \{\sigma : \sigma : B \in X\}$, for all $p \in \mathtt{For}_0$ $V(p) = \{\sigma : \sigma : p \in X\}$ and for all $\sigma, \sigma'$ in $X$ and $i \in \{1, 2\}$ $\sigma R_i \sigma'$ iff either $\mathtt{C}_i(\langle \sigma, \sigma' \rangle, \mathcal{C})$ or $\{\nu_0^i : \sigma : \nu^i \in X\} \subseteq \{B : \sigma' : B \in X\}$. One can easily check that the definition of $\mathcal{M}$ is correct, i.e. $\mathcal{M}$ is a $K_{1,2}^-$-model. It can be shown by induction on the structure of the formulae that for every formula $B$ and every prefix $\sigma$, if $\sigma : B \in X$ then $\mathcal{M}, \sigma \models B$ (and therefore $X$ is $K_{1,2}^-$-satisfiable). **Q.E.D.**

**Proposition 4.4.** (completeness) If $A$ is $K_{1,2}^-$-valid then $A$ has a closed tableau built with the rules presented in Figure 2.

**Proof:** Suppose $A$ has no closed prefixed tableau. So the systematic procedure does not generate a closed tableau. We build a tableau with this procedure by considering $0 : \neg A \ [\langle \emptyset, \emptyset \rangle]$ at the root. If the procedure terminates then the tableau contains a non-closed branch. If the procedure does not terminate, by König's Lemma, there is an infinite non-closed branch. The systematic procedure guarantees that the non-closed branch **BR** is downward-saturated wrt some $\mathcal{C}$. By Lemma 4.3, **BR** is $K_{1,2}^-$-satisfiable. Since $0 : \neg A \in$ **BR**, there is a $K_{1,2}^-$-model $\mathcal{M}$ and a world $w$ such that $\mathcal{M}, w \models \neg A$, which leads to a contradiction. **Q.E.D.**

In the systematic procedure, we require that if $\sigma : B$ is a conclusion of some inference of the $\nu^i$-rule and if an occurrence of $\sigma : B$ has already been introduced on the branch then no new occurrence is added on the branch. The systematic procedure still guarantees completeness but it terminates since the $\pi^i$-rule can be applied only a finite number of times. Actually, each $\pi^i$-rule is applied at most $mw^i(A) \times 2^{card(\{B, \neg B : B \ subformula \ of \ A\})}$ times on a branch where $mw^i(A)$ is the number modal operators of the form $[i]$ or $\langle i \rangle$ occurring in $A$. The other rules do not introduce new prefixes which guarantees termination since their applications are restricted (while insuring completeness). The systematic procedure above is therefore a decision procedure for the $\mathcal{L}([R], [-R])$-validity problem.

## 5 Tableaux for $\mathcal{L}([\neq])$

For any finite set $X$ of pairs we write $X(a, b)$ to denote that $\langle a, b \rangle$ belongs to the smallest equivalence relation containing $X$. The rules for $\mathcal{L}([\neq])$ are those for $K_{1,2}^-$

except that the $\pi^1$-rule becomes

$$\frac{\sigma : \pi^1 \ [\mathcal{C}]}{\sigma k^1 : \pi_0^1 \ [\langle \mathcal{C}_1, \mathcal{C}_2 \cup \{\langle \sigma, \sigma k^1 \rangle\} \rangle]} \text{ new } k \in \omega \text{ on the branch}$$

and the restricted cut rule is replaced by:

$$\frac{\sigma'' : A \ [\mathcal{C}]}{\sigma'' : A \ [\langle \mathcal{C}_1 \cup \{\langle \sigma, \sigma' \rangle\}, \mathcal{C}_2 \rangle] \mid \sigma'' : A \ [\langle \mathcal{C}_1, \mathcal{C}_2 \cup \{\langle \sigma, \sigma' \rangle\} \rangle]}$$

$\sigma, \sigma'$ occur on the branch and neither $\mathtt{C}_1(\langle \sigma, \sigma' \rangle, \mathcal{C})$ nor $\mathtt{C}_2(\langle \sigma, \sigma' \rangle, \mathcal{C})$ holds.

The definitions of $\mathtt{C}_1$ and $\mathtt{C}_2$ are modified as follows: $\mathtt{C}_1(\langle \sigma, \sigma' \rangle, \mathcal{C})$ holds iff either $\mathcal{C}_1(\sigma, \sigma')$ holds or $\sigma = \sigma'$ and $\mathtt{C}_2(\langle \sigma, \sigma' \rangle, \mathcal{C})$ holds iff there exist $\sigma_1$ and $\sigma_1'$ such that $\{\langle \sigma_1, \sigma_1' \rangle, \langle \sigma_1', \sigma_1 \rangle\} \cap \mathcal{C}_2 \neq \emptyset$, $\mathtt{C}_1(\langle \sigma, \sigma_1 \rangle, \mathcal{C})$ and $\mathtt{C}_1(\langle \sigma', \sigma_1' \rangle, \mathcal{C})$. For instance $\mathtt{C}_1(\langle \sigma, \sigma_1 \rangle, \mathcal{C})$ can be interpreted by "$\sigma$ and $\sigma_1$ are equal modulo $\mathcal{C}$". A branch is *closed* if there exist prefixed formulae $\sigma : A$ and $\sigma' : \neg A$ on that branch such that $\mathtt{C}_1(\langle \sigma, \sigma' \rangle, \mathcal{C})$ holds. This calculus for $\mathcal{L}([\neq])$ strongly differs from the one in [Demri, 1996] due to the machinery associated to $\mathcal{C}$ and to the restricted cut rule.

**Theorem 5.1.** (soundness and completeness) A formula $A$ is $\mathcal{L}([\neq])$-valid iff $A$ has a closed tableau built with the rules for $\mathcal{L}([\neq])$.

In order to provide a decision procedure for $\mathcal{L}([\neq])$ it is sufficient to consider the decision procedure in Section 4 adequately modified for $\mathcal{L}([\neq])$ except that the following conditions are required to apply the $\pi^1$-rule:

$\rho 1$ it is not possible to apply the restricted cut rule (that is the restricted cut rule is *saturated* before applying the $\pi^1$-rule),

$\rho 2$ there is no $\sigma' : \pi_0^1$ on the branch such that $\mathtt{C}_2(\langle \sigma, \sigma' \rangle, \mathcal{C})$,

$\rho 3$ there are no $\sigma_1 : \pi_0^1$ and $\sigma_2 : \pi_0^1$ on the branch such that $\mathtt{C}_2(\langle \sigma_1, \sigma_2 \rangle, \mathcal{C})$.

It is possible to show that the calculus is sound and complete and the systematic procedure defined above always terminates (each formula $\pi^1$ occurring in $\neg A$ can be used at most twice as a premise of a $\pi^1$-rule inference on a given branch). Actually, at most $1 + 2 \times mw(A)$ different prefixes can occur on a given branch where $mw(A)$ is the so-called *modal weight* of $A$, i.e. the number of modal operators occurring in $A$. Hence the above systematic procedure constructs a polynomial-size $\mathcal{L}([\neq])$-model for $\neg A$ (with respect to the *size* of $A$) if $A$ is not $\mathcal{L}([\neq])$-valid.

## 6 Tableaux for $K_I([\neq])$

The conditions $\mathtt{C}_1$ and $\mathtt{C}_2$ are defined as in Section 5 as well as the closure conditions. The tableaux rules for $K_I([\neq])$ are given in Figure 3. Let $X$ be a set of prefixed formulae having the same condition $\mathcal{C}$ and $\mathcal{M} = (W, (R_i)_{i \in I}, V)$ be a $K_I([\neq])$-model. By an *interpretation of $X$ in $\mathcal{M}$* we mean a mapping $\mathcal{I} : \{\sigma : \sigma : A \in X\} \to W$ such that if $\sigma, \sigma'$ occur in $X$, then

- $\sigma' = \sigma k^i$ for some $k^i$ implies $\langle \mathcal{I}(\sigma), \mathcal{I}(\sigma') \rangle \in R_i$,

$$\frac{\sigma : \pi^i \ [\mathcal{C}]}{\sigma k^i : \pi_0^i \ [\mathcal{C}]} \pi^i\text{-rule, new } k \in \omega, i \in I \setminus \{1\}$$

$$\frac{\sigma : \pi^1 \ [\mathcal{C}]}{\sigma k^1 : \pi_0^1 \ [\langle \mathcal{C}_1, \mathcal{C}_2 \cup \{\langle \sigma, \sigma k^1 \rangle\} \rangle]} \pi^1\text{-rule, new } k \in \omega$$

$$\frac{\sigma : \nu^i \ [\mathcal{C}]}{\sigma' : \nu_0^i \ [\mathcal{C}]} \nu^i\text{-rule, } i \in I \setminus \{1\}$$

if there exist $\sigma_1, \sigma_1 k^i$ on the branch such that $\mathtt{C}_1(\langle \sigma, \sigma_1 \rangle, \mathcal{C})$ and $\mathtt{C}_1(\langle \sigma', \sigma_1 k^i \rangle, \mathcal{C})$.

$$\frac{\sigma : \nu^1 \ [\mathcal{C}]}{\sigma' : \nu_0^1 \ [\mathcal{C}]} \nu^1\text{-rule, if } \mathtt{C}_2(\langle \sigma, \sigma' \rangle, \mathcal{C})$$

$$\frac{\sigma'' : A \ [\mathcal{C}]}{\sigma'' : A \ [\langle \mathcal{C}_1 \cup \{\langle \sigma, \sigma' \rangle\}, \mathcal{C}_2 \rangle] \mid \sigma'' : A \ [\langle \mathcal{C}_1, \mathcal{C}_2 \cup \{\langle \sigma, \sigma' \rangle\} \rangle]}$$

$\sigma, \sigma'$ on the branch and neither $\mathtt{C}_1(\langle \sigma, \sigma' \rangle, \mathcal{C})$ nor $\mathtt{C}_2(\langle \sigma, \sigma' \rangle, \mathcal{C})$ holds.

Figure 3: Tableaux system for $K_I([\neq])$

- $\mathtt{C}_1(\langle \sigma, \sigma' \rangle, \mathcal{C})$ implies $\mathcal{I}(\sigma) = \mathcal{I}(\sigma')$ and $\mathtt{C}_2(\langle \sigma, \sigma' \rangle, \mathcal{C})$ implies $\mathcal{I}(\sigma) \neq \mathcal{I}(\sigma')$.

Lemma 4.1 can be shown to hold for $K_I([\neq])$ associated with the calculus presented in Figure 3: if $A$ has a closed tableau built with the rules in Figure 3 then $A$ is $K_I([\neq])$-valid. We also use the systematic procedure defined in Section 4.2 (with the binary restricted cut rule) except that (P4) is replaced by:

P4' If $\sigma : B \ [\mathcal{C}]$ is of the form $\sigma : \pi^i \ [\mathcal{C}]$ with $i \neq 1$ (resp. $\sigma : \pi^1 \ [\mathcal{C}]$) then add $\sigma k^i : \pi_0^i \ [\mathcal{C}]$ (resp. $\sigma k^1 : \pi_0^1 \ [\langle \mathcal{C}_1, \mathcal{C}_2 \cup \{\langle \sigma, \sigma k^1 \rangle\} \rangle]$) to the end of $\mathtt{BR}$.

Similarly, we say $X$ is *downward-saturated wrt $\mathcal{C}$* iff:

C1' For all $\sigma, \sigma' \in X$, (C1'.1) $\mathtt{C}_1(\langle \sigma, \sigma' \rangle, \mathcal{C})$ iff not $\mathtt{C}_2(\langle \sigma, \sigma' \rangle, \mathcal{C})$ (note the difference with C1.1 in Section 4) and, (C1'.2) for all $p \in \mathtt{For}_0$, $\{\sigma : p, \sigma' : \neg p\} \subseteq X$ implies $\mathtt{C}_2(\langle \sigma, \sigma' \rangle, \mathcal{C})$.

- Conditions C2,C3 from Section 4.2 and C4 for $i \neq 1$

C5' if $\sigma : \pi^i \in X$ with $i \neq 1$ then there exist $\sigma' : \pi_0^i \in X$ and $\sigma k^i$ in $X$ such that $\mathtt{C}_1(\langle \sigma', \sigma k^i \rangle, \mathcal{C})$.

C6 if $\sigma : \nu^1 \in X$ then for all $\sigma'$ in $X$ such that $\mathtt{C}_2(\langle \sigma, \sigma' \rangle, \mathcal{C})$, we have $\sigma' : \nu_0^1 \in X$

C7 if $\sigma : \pi^1 \in X$ then there is $\sigma'$ such that $\sigma' : \pi_0^1 \in X$ and $\mathtt{C}_2(\langle \sigma, \sigma' \rangle, \mathcal{C})$.

**Lemma 6.1.** If $X$ is downward-saturated wrt $\mathcal{C}$ then $X$ is $K_I([\neq])$-satisfiable.

**Proof:** Assume $X$ is downward-saturated wrt $\mathcal{C}$. Let $\mathcal{M} = (W, (R_i)_{i \in I}, V)$ be the structure such that,

- $W = \{|\sigma| : \sigma : B \in X\}$ where $|\sigma| = \{\sigma' : \sigma : B \in X, \mathtt{C}_1(\langle \sigma, \sigma' \rangle, \mathcal{C})\}$.
- for all $p \in \mathtt{For}_0$ $V(p) = \{|\sigma| : \sigma : p \in X\}$.
- $R_1 = dif_W$ and for all $\sigma, \sigma'$ in $X$, $|\sigma| R_i |\sigma'|$ ($i \neq 1$) iff $\exists \sigma_1, \sigma_1 k^i$ in $X$, $\mathtt{C}_1(\langle \sigma, \sigma_1 \rangle, \mathcal{C})$ and $\mathtt{C}_1(\langle \sigma', \sigma_1 k^i \rangle, \mathcal{C})$.

$\mathcal{M}$ is a $K_I([\neq])$-model. It can be shown (by induction on $B$) that if $\sigma : B \in X$ then $\mathcal{M}, |\sigma| \models B$. **Q.E.D.**

**Proposition 6.2.** (completeness) If $A$ is $K_I([\neq])$-valid then $A$ has a closed prefixed tableau built with the rules presented in Figure 3.

In order to obtain a decision procedure, take the systematic procedure, incorporate the restrictions $\rho 1$, $\rho 2$ and $\rho 3$ from Section 5 and for $i \neq 1$, add the following restriction to the $\pi^i$-rule: there is no $\sigma' : \pi_0^i$ on the branch such that $\mathtt{C}_1(\langle \sigma k^i, \sigma' \rangle, \mathcal{C})$ holds for some $k \in \omega$.

## 7 Concluding remarks

The use of *prefixes* for tableaux systems dedicated to modal logics has been thoroughly developed in [Fitting, 1983] whereas our treatment of the condition $\mathcal{C}$ (see e.g. Sections 4, 5, 6) can be viewed as a means to parametrize our calculi by the *theory* of the accessibility relations. Hence, the idea of *theory resolution* [Stickel, 1985] in which a theory is separately dealt with from the rest of the calculus is present in our calculi. This idea is not new in the realm of the mechanization of modal logics (see e.g. [Frisch and Scherl, 1990; Gent, 1993]) but the originality of our work is related to the conditions satisfied by the accessibility relations of the models.

The second important feature of our calculi is the use of a restricted cut rule. Recently, various works have *tamed* the cut rule for calculi dedicated to modal logics (see e.g. [d'Agostino, 1993; Governatori, 1995]). However our calculi do not have a cut rule with a branching for formulae. In that sense, the cut rule in our calculi is even more restricted than the one in [Governatori, 1995].

We have defined sound and complete prefixed tableaux calculi for the logics $\mathcal{L}([R], [-R])$, and $K_I([\neq])$ (also for $K_I$ and $\mathcal{L}([\neq])$) and decision procedures have been designed from these systems. It is worth noting that the expressive power of the modal logics with enriched languages is attractive in the Artificial Intelligence community since for instance the operator $[\neq]$ has already been shown to be useful to reason about time [Sain, 1988; Koymans, 1992] or space [Balbiani et al., 1997].

Future work could be oriented towards the incorporation of our calculi into existing tableaux-based theorem provers for modal logics and towards the definition of other prefixed tableaux for modal logics with enriched languages including for instance, the logics in the paper where standard conditions for the accessibility relations are required -reflexivity, symmetry, ....

**Acknowledgments**: the authors thank Luis Fariñas del Cerro for his encouragements.

# References

[Balbiani et al., 1997] Ph. Balbiani, L. Fariñas del Cerro, T. Tinchev, and D. Vakarelov. Modal logics for incidence geometries. *Journal of Logic and Computation*, 7(1):59-78, 1997.

[Balbiani, 1997] Ph. Balbiani. Inequality without irreflexivity, 1997. http://www-lipn.univ-paris13.fr/.

[d'Agostino, 1993] M. d'Agostino. The taming of the cut. Classical refutations with analytic cut. *Journal of Logic and Computation*, 4(3), 1993.

[Demri, 1996] S. Demri. A simple tableau system for the logic of elsewhere. In *TABLEAUX-6*, pages 177-192. LNAI 1071, Springer-Verlag, 1996.

[Fagin et al., 1995] R. Fagin, J. Halpern, Y. Moses, and M. Vardi. *Reasoning about Knowledge*. The MIT Press, 1995.

[Fitting, 1983] M. Fitting. *Proof methods for modal and intuitionistic logics*. Reidel, 1983.

[Frisch and Scherl, 1990] A. Frisch and R. Scherl. A constraint logic approach to modal deduction. In *JELIA*, pages 234-250. Springer-Verlag, LNAI 478, 1990.

[Gent, 1993] I. Gent. Theory matrices (for modal logics) using alphabetical monotonicity. *Studia Logica*, 52(2):233-257, 1993.

[Goranko and Passy, 1992] V. Goranko and S. Passy. Using the universal modality: gains and questions. *Journal of Logic and Computation*, 2(1):5-30, 1992.

[Goranko, 1990a] V. Goranko. Completeness and incompleteness in the bimodal base $\mathcal{L}(R, -R)$. In P. Petkov, editor, *Mathematical Logic*, pages 311-326. Plenum Press, 1990.

[Goranko, 1990b] V. Goranko. Modal definability in enriched languages. *Notre Dame Journal of Formal Logic*, 31:81-105, 1990.

[Governatori, 1995] G. Governatori. Labelled tableaux for multi-modal logics. In *TABLEAUX-5*, pages 79-94. LNAI 918, Springer-Verlag, 1995.

[Hughes and Cresswell, 1984] G. Hughes and M. Cresswell. *A companion to modal logic*. Methuen, 1984.

[Humberstone, 1983] L. Humberstone. Inaccessible worlds. *Notre Dame Journal of Formal Logic*, 24(3):346-352, 1983.

[Koymans, 1992] R. Koymans. *Specifying message passing and time-critical systems with temporal logic*. LNCS 651, Springer-Verlag, 1992.

[Lakemeyer, 1993] G. Lakemeyer. All they know: a study in multi-agent autoepistemic reasoning. In *IJCAI-13*, pages 376-381. Morgan Kaufmann, 1993.

[Levesque, 1990] H. Levesque. All I know : a study in autoepistemic logic. *Artificial Intelligence*, 42:263-309, 1990.

[Massacci, 1994] F. Massacci. Strongly analytic tableaux for normal modal logics. In *CADE-12*, pages 723-737. Springer Verlag, LNAI 814, 1994.

[Rijke, 1993] M. de Rijke. *Extending modal logic*. PhD thesis, ILLC, Amsterdam University, December 1993.

[Sain, 1988] I. Sain. Is 'some-other-time' sometimes better than 'sometime' for proving partial correctness of programs. *Studia Logica*, 47(3):278-301, 1988.

[Segerberg, 1981] K. Segerberg. A note on the logic of elsewhere. *Theoria*, 47:183-187, 1981.

[Spaan, 1993] E. Spaan. *Complexity of Modal Logics*. PhD thesis, ILLC, Amsterdam University, March 1993.

[Stickel, 1985] M. Stickel. Automated deduction by theory resolution. *Journal of Automated Reasoning*, 1:333-355, 1985.

[Wallen, 1990] L. Wallen. *Automated Deduction in Nonclassical Logics*. MIT Press, 1990.

# A Set-Theoretic Approach to Automated Deduction in Graded Modal Logics

**A. Montanari** and **A. Policriti**
Dipartimento di Matematica e Informatica, Università di Udine
Via delle Scienze 206, 33100 Udine, Italy

## Abstract

In the paper, we consider the problem of supporting automated reasoning in a large class of knowledge representation formalisms, including terminological and epistemic logics, whose distinctive feature is the ability of representing and reasoning about finite quantities. Each member of this class can be represented using graded modalities, and thus the considered problem can be reduced to the problem of executing graded modal logics. We solve this problem using a set-theoretic approach that first transforms graded modal logics into polymodal logics with infinitely many modalities, and then reduces derivability in such polymodal logics to derivability in a suitable first-order set theory.

## 1 Introduction

The general theme of this paper is the description of a novel approach to the problem of supporting the automation of reasoning in a family of knowledge representation formalisms. Such a family is characterized by the fact that its members need to represent and reason about finite quantities, and it includes terminological logics, epistemic logics, universal modalities. van der Hoek and de Rijke have shown that all these languages can be represented using graded modalities [Fattorosi-Barnaba and De Caro, 1985] (cf. [Hoek and de Rijke, 1995] for a complete description of this kind of reductions). In this paper, we propose an approach to automated deduction in graded modal logics which is based on a set-theoretic translation method introduced by D'Agostino et al. in [D'Agostino et al., 1995] to support derivability in propositional modal logic.

Most inference systems for modal logic are defined in the style of sequent or tableaux calculi, e.g. [Fitting, 1983; Wansing, 1994]. As an alternative, a number of *translation* methods for modal logic into classical first-order logic have been proposed in the literature (for a comprehensive survey, cf. [Ohlbach, 1993]). Such methods allow the use of Predicate Calculus mechanical theorem provers to implement modal theorem provers. Compared with the direct approach of finding a proof algorithm for a specific class of modal logics, the translation methods have the advantage of being *independent* of the particular modal logic under consideration: a single theorem prover may be used for any translatable modal logic.

In the standard approach, the first-order language $\mathcal{L}$ into which the translation is carried out contains a constant $\tau$ denoting the initial world in the frame, a binary relation $R(x,y)$ denoting the accessibility relation, and a denumerable number of unary predicates $\mathtt{P_i}(x)$. The translation function $\pi$ is defined by induction on the structural complexity of the modal formula as follows:

- $\pi(P_j, x) \equiv \mathtt{P_j}(x)$;
- $\pi(-, x)$ commutes with the boolean connectives;
- $\pi(\Box \psi, x) \equiv \forall y(xRy \rightarrow \pi(\psi, y))$.

Efficiency concerns have motivated further investigations on the above (relational) translation method. Such studies (e.g. [Ohlbach, 1991]) suggested a "functional" semantics for modal logic and resulted in a family of more efficient and general translation methods. From the computational point of view, the functional translation may still cause some problem when using a first-order theorem prover, due to the presence of equalities in the translation of the axioms. A method for limiting the complexity induced by the introduction of equality using a mixed relational/functional translation is proposed in [Nonnengart, 1993].

A common feature of all the methods mentioned above is that, in order to be applied directly, the underlying modal logic must have a first-order semantics. All attempts to apply them to logics not having a first-order semantics have required *ad-hoc* techniques. Moreover, if the logic has a first-order semantics, but it is only specified by Hilbert axioms, a preliminary step is necessary to find the corresponding first-order axioms. The question of automatically solving this last problem has been extensively studied and algorithms have been proposed, e.g. [Benthem, 1985; Gabbay and Ohlbach, 1992].

The above analysis can be easily tailored to the case of graded modalities. The semantics of graded modalities is very natural and intuitive, but it has a disadvantage: the inference systems based on it deal with $\Diamond_n$ and $\Box_n$ operators by generating a number of terms that, in general, can be very large. This problem can be overcome by using a Hilbert-style axiomatic system, which allows one to perform arithmetic symbolic reasoning; in such a case, however, the search space for proving even very simple theorems can grow very much and it is usually rather unstructured. In view of the previous points, a translational approach to automated reasoning with graded modalities has been considered by Ohlbach et al. (cf. [Ohlbach et al., 1995]). Such an approach provides the possibility of using a standard deductive system — thereby guaranteeing symbolic reasoning — for which optimizations and good implementations are available.

In this paper, we exploit an alternative translation method whose basic idea is to map modal formulae into set-theoretic terms. Such a method works for all normal complete finitely axiomatizable modal logics, regardless of the first-order axiomatizability of their semantics. It also works if the modal logic under consideration is only specified by Hilbert axioms. Furthermore, it can be easily generalized to polymodal logics with finitely many modalities [D'Agostino et al., 1995].

Even though graded modal logics can be seen as polymodal logics ([Ohlbach et al., 1995]), the set-theoretic translation method cannot be applied directly, because the number of modalities involved in their translation is infinite. In the following, we show how to adapt the set-theoretic translation for polymodal logics with finitely many modalities to encompass an infinite number of accessibility relations (each one corresponding to a different "grade"). As a matter of fact, graded modal logics are treated as a special case of a more general technique able to deal with polymodal logics with infinitely many modalities.

## 2 Graded modal logics

Graded modal logics have been introduced in the 60's by Goble [Goble, 1970], who proposed a logic with a fixed number of modalities, each one associated with a natural number and representing a different degree of necessity. As an example, the formula $N_n\varphi \wedge N_m\psi$, with $m > n$, espresses the fact that both $\varphi$ and $\psi$ are necessary, but $\psi$ is *more* necessary than $\varphi$. This approach has been later generalized by Fine (cf. [Fine, 1972]) who, inspired by Tarskian numerical quantifiers, introduced modal operators associated with natural numbers: the so-called *graded* modal operators $\Box_n$ and $\Diamond_n$, with $n \in \mathbb{N}$. Finally, in the 80's, Fattorosi-Barnaba, De Caro, and Cerrato provided sound and complete axiomatizations of graded modal logics, together with some interesting decidability results [Fattorosi-Barnaba and De Caro, 1985; De Caro, 1988; Cerrato, 1990; 1992].

Graded modal logics allow one to express conditions on the number of objects satisfying a given property such as: "at least $n$ elements (satisfying relation $R$) have property $y$". Formally, the basic system of graded modal logic $\overline{K}$ is an extension of $K$ obtained by adding graded modalities. The language of $\overline{K}$ is obtained from the standard language of pure modal logic by substituting $\Box_n$ and $\Diamond_n$ ($n \in \mathbb{N}$) for $\Box$ and $\Diamond$. Formulae of $\overline{K}$ are:

$$Form(\overline{K}, \Phi) = p, q, \ldots, \neg\varphi, \varphi \vee \psi, \Diamond_n\varphi,$$

where $p, q, \ldots$ stand for propositional letters, and modal formulae are defined inductively as usual.

It is convenient to define the following abbreviation:

$$\Diamond!_n\varphi = \begin{cases} \neg\Diamond_0\varphi & \text{if } n = 0; \\ \Diamond_{n-1}\varphi \wedge \neg\Diamond_n\varphi & \text{otherwise,} \end{cases}$$

whose intuitive meaning is that $\varphi$ holds at exactly $n$ $R$-accessible worlds.

$\overline{K}$ is a normal modal logic with respect to $\Box_0$, and it is characterized by the following axioms (cf. [Fattorosi-Barnaba and De Caro, 1985]):

$A_1$ $\vdash_{\overline{K}} \varphi$ for any propositional tautology $\varphi$;

$A_2$ $\Box_n\varphi \to \Box_{n+1}\varphi$;

$A_3$ $\Box_0(\varphi \to \psi) \to (\Diamond_n\varphi \to \Diamond_n\psi)$;

$A_4$ $\Box_0\neg(\varphi \wedge \psi) \to ((\Diamond!_n\varphi \wedge \Diamond!_m\psi) \to \Diamond!_{n+m}(\varphi \vee \psi))$,

and by the rules of modus ponens, substitution and necessitation:

MP $\vdash_{\overline{K}} \varphi \to \psi, \vdash_{\overline{K}} \varphi \Rightarrow \vdash_{\overline{K}} \psi$;

SUB $\vdash_{\overline{K}} \alpha \leftrightarrow \beta \Rightarrow \vdash_{\overline{K}} \varphi \leftrightarrow [\alpha|\beta]\varphi$;

N $\vdash_{\overline{K}} \varphi \Rightarrow \vdash_{\overline{K}} \Box_0\varphi$.

The semantics of $\overline{K}$ is given in terms of Kripke frames. In particular, the satisfiability relation is defined as usual over atomic formulae and boolean connectives, while the clauses for graded modalities are the following ones:

$$w \models_{\overline{K}} \Diamond_n\varphi \Leftrightarrow |\{v \in W : wRv \wedge w \models_{\overline{K}} \varphi\}| > n;$$

$$w \models_{\overline{K}} \Box_n\varphi \Leftrightarrow |\{v \in W : wRv \wedge w \models_{\overline{K}} \neg\varphi\}| \leq n.$$

It is worth noting that the standard modal operators $\Box$ and $\Diamond$ correspond to $\Box_0$ and $\Diamond_0$, respectively, thereby showing that $\overline{K}$ is an extension of $K$.

It can be showed that $\overline{K}$ is sound and complete with respect to the class of all Kripke frames.

**THEOREM 2.1** *For any formula $\varphi \in Form(\overline{K}, \Phi)$, it holds that*

$$\vdash_{\overline{K}} \varphi \Leftrightarrow \models \varphi.$$

Soundness is proved by induction on the structural complexity of $\varphi$, while completeness is proved by using an argument *à la* Henkin.

The soundness and completeness proof given for $\overline{K}$ can be immediately generalized to deal with all those graded modal logics whose accessibility relation either excludes pairwise *distant* worlds (e.g. $\overline{K}, \overline{D}, \overline{T}$) or is transitive (e.g. all systems of graded modal logic over $\overline{S5}$). In order to prove the completeness of the remaining graded modal logics, it is necessary to work with a weakened notion of canonical model (cf. [Cerrato, 1990]).

## 3 A set-theoretic translation method

In [D'Agostino *et al.*, 1995] D'Agostino et al. proposed a set-theoretic translation method ($\Box$-as-*Pow* translation, from now on) to execute modal logics. The main idea underlying the $\Box$-as-*Pow* translation is to formalize the notion of validity in Kripke frames by a set-theoretic formula that is provable in the underlying set theory if and only if the original formula is modally derivable. According to the $\Box$-as-*Pow* translation, any Kripke frame *is* the set of its worlds and any world in a frame *is* the set of those worlds accessible from it. The theory driving the translation is a very weak, finitely (first-order) axiomatizable set theory, called $\Omega$ [D'Agostino *et al.*, 1995], whose axioms, in the language with relational symbols $\in$ and $\subseteq$, and functional symbols $\cup, \setminus$, and *Pow*, are:

$x \in y \cup z \iff x \in y \lor x \in z;$

$x \in y \setminus z \iff x \in y \land x \notin z;$

$x \subseteq y \iff \forall z(z \in x \to z \in y);$

$x \in Pow(y) \iff x \subseteq y.$

A peculiarity of the technique is the weakness of the theory $\Omega$: it consists of only four axioms describing the most rudimentary and basic among the operators of naive set theory. In particular, notice that neither the extensionality axiom nor the axiom of foundation are in $\Omega$. Given a modal formula $\phi(P_1, ..., P_n)$, its translation is defined as the set-theoretic *term* $\phi^*(x, x_1, ..., x_n)$, with variables $x, x_1, ..., x_n$, built using $\cup, \setminus$, and *Pow*. Intuitively, the term $\phi^*(x, x_1, ..., x_n)$ represents the set of those worlds (in the frame $x$) in which the formula $\phi$ holds. The inductive definition of $\phi^*(x, x_1, ..., x_n)$ is the following:

- $P_i^* = x_i$;
- $(\phi \lor \psi)^* = \phi^* \cup \psi^*$;
- $(\neg \phi)^* = x \setminus \phi^*$;
- $(\Box \phi)^* = Pow(\phi^*)$.

For all modal formulae $\phi, \psi$, the following results hold, showing, respectively, the completeness and the soundness of the translation [D'Agostino *et al.*, 1995]:

$$\phi \vdash_{K_s} \psi \Rightarrow$$
$$\Omega \vdash \forall x(Trans(x) \land \forall \vec{z}(x \subseteq \phi^*(x, \vec{z})) \to \forall \vec{z}(x \subseteq \psi^*(x, \vec{z}))),$$
and
$$\Omega \vdash \forall x(Trans(x) \land \forall \vec{z}(x \subseteq \phi^*(x, \vec{z})) \to \forall \vec{z}(x \subseteq \psi^*(x, \vec{z})))$$
$$\Rightarrow \phi \models_f \psi,$$

where $Trans(x)$ stands for $\forall y (y \in x \to y \subseteq x)$. It is immediate to see that, for frame-complete theories, the above translation captures exactly the notion of $K_s$-derivability.

### 3.1 Polymodal logics with finitely many modalities

In [D'Agostino *et al.*, 1995], D'Agostino et al. also show how to generalize the $\Box$-as-*Pow* translation to polymodal logics. The basic idea is to mimic a polymodal frame, provided with finitely many accessibility relations, with a set, provided with the membership relation only. To this end, D'Agostino et al. defined an alternative semantics for polymodal logic, called p-semantics, that replaces the plurality of accessibility relations $\triangleleft_1, ..., \triangleleft_k$ by a single accessibility relation $R$ and $k$ copies $U_1, ..., U_k$ of the universe $U$. The p-semantics is formally defined as follows.

**DEFINITION 3.1** *A p-frame $\mathcal{F}$ is a $(k+2)$-tuple $(U, U_1, ..., U_k, R)$, where $U, U_1, ..., U_k$ are sets and $R$ is a binary relation on $U \cup U_1 \cup ... \cup U_k$, such that, for all $u, v, t$ in $U \cup U_1 \cup ... \cup U_k$, if $u \in U$, $uRv$ and $vRt$, then $t \in U$ (we will denote this property by $Trans^2(U)$).*

The intuition behind the notion of p-frame is the following one. Consider two worlds $w, w' \in U$ such that $w \triangleleft_i w'$ in the original polymodal frame. Since only one accessibility relation is available in p-frames, we cannot directly access $w'$ from $w$ (via $R$) anymore. However, we can follow a two-step path: first we move to (the unique) $w_i \in U_i$ $R$-accessible from $w$; then we move from $w_i$ to $w'$ via $R$.

A *p-valuation* $\models_p$ is a subset of $U \times \Phi$, where $\Phi$ is the set of propositional variables. In the case of boolean combinations, the p-valuation $\models_p$ may be lifted to the set of all polymodal formulae in the canonical fashion. In the case of $\Box_i$, with $i = 1, ..., k$, for all $u \in U$ we put

$$u \models_p \Box_i \phi \Leftrightarrow \forall v(uRv \land v \in U_i \to \forall t(vRt \to t \models_p \phi)).$$

A polymodal formula $\phi$ is p-valid in a p-frame $(U, U_1, ..., U_k, R)$ if and only if for all p-valuations $\models_p$ and all worlds $u \in U$, $u \models_p \phi$ holds.

The link between polymodal frames and p-frames is formally expressed by the following theorem [D'Agostino *et al.*, 1995]:

**THEOREM 3.2** *If $\psi, \phi$ are polymodal formulae, then $\psi \models \phi$ if and only if $\phi$ is p-valid in all p-frames in which $\psi$ is p-valid.*

A set-theoretic counterpart of p-frames can be easily given in the standard way. As far as the translation of the modal operators is concerned, on the ground of the definition of $\models_p$, the set-theoretic semantics of $\Box_i \phi$ becomes:

$$(\Box_i \phi)^* \equiv Pow((\bar{x} \setminus y_i) \cup Pow(\phi^*)),$$

where $\bar{x} = x \cup y_1 \cup ... \cup y_k$.

**THEOREM 3.3** *Let $H$ be a $k$-dimensional polymodal logic extending $K \otimes ... \otimes K$ with the axiom schema $\psi(\alpha_{j_1}, ..., \alpha_{j_m})$. For any polymodal formula $\phi$,*

**(soundness)**

$$\Omega \vdash \forall x \forall y_1 ... \forall y_k (Trans^2(x) \land Axiom_H(x, y_1, ..., y_k) \to \forall \vec{z}(x \subseteq \phi^*(x, y_1, ..., y_k, \vec{z}))) \Rightarrow \psi \models \phi$$

**(completeness)**

$$\vdash_H \phi \Rightarrow \Omega \vdash \forall x \forall y_1 ... \forall y_k (Trans^2(x) \land Axiom_H(x, y_1, ..., y_k) \to \forall \vec{z}(x \subseteq \phi^*(x, y_1, ..., y_k, \vec{z}))),$$

where
$Axiom_H(x, y_1, \ldots, y_k)$ is $\forall \vec{y}(x \subseteq \psi^*(x, y_1, \ldots, y_k, \vec{y})))$,
and $Trans^2(x)$ stands for $\forall y \forall z \ (y \in z \land z \in x \rightarrow y \subseteq x)$,
that is, $x \subseteq Pow(Pow(x))$.

## 4 Translating graded modalities

The general scheme followed for applying the set theoretic translation is the one suggested by [Ohlbach et al., 1995]: a two-step translation that first transforms graded modal logic into a polymodal logic with infinitely many modalities, and then reduces derivability in such a polymodal logic to derivability in a suitable first-order set theory.

A graded modal logic expresses properties of different (infinitely many) modalities which are all referring to the same accessibility relation. In other words, infinitely many Kripke semantics are provided over the same accessibility relation scheme. The task of the first step of the translation is that of rewriting the semantics of a graded modal logic in such a way to introduce a different accessibility relation for each different modality. Once this step has been performed, the next task is to generalize the existing translation for polymodal logics with finitely many modalities, to the case of infinitely many ones.

The advantage of using the $\square$-as-$Pow$ translation is its ability of dealing with non-first-order axiomatizable polymodal logics. This fact has two important consequences: on the one hand, it allows one to naturally translate the polymodal counterpart of $\overline{K}$ ($\overline{K}_E$), which is indeed non-first-order axiomatizable; on the other hand, it can be applied to non-first-order axiomatizable logics over $\overline{K}$. A further advantage is the fact that the technique introduced is very general and can in fact be employed to translate polymodal logics with infinitely many modalities.

### 4.1 Polymodal logics with infinitely many modalities

According to Kripke semantics, a $\overline{K}$-formula of the form $\diamond_n \varphi$ is true at a given world $x$ of a frame $\mathcal{F} = (W, R)$ if and only if there exists $Y \subseteq R(x)$ of cardinality greater than $n$ and such that $\varphi$ is true at any world $y \in Y$.

An alternative interpretation for $\overline{K}$ can be obtained introducing a new class of worlds, denoted by $W_Y$, representing *sets* of accessible worlds ($W_Y \subseteq Pow(W)$). The single accessibility relation $R$ can now be replaced by the denumerable set of relations $\{R_n : n \in \mathbb{N}\}$, where $R_n$ associates a given world $x$ with those elements of $W_Y$ having cardinality greater than $n$. A further accessibility relation $E$ will be used to associate a given element of $W_Y$ with its elements. The situation is described by the following picture:

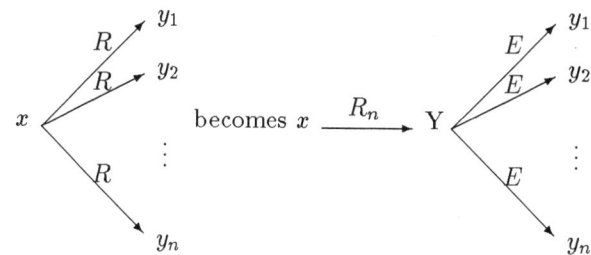

The alternative semantics described above suggests the introduction of the following modal logic $\overline{K}_E$. The language of $\overline{K}_E$ is obtained from that of $\overline{K}$ by substituting $\langle n \rangle$, $[n]$, $\diamond$, and $\square$ for $\diamond_n$ and $\square_n$. Formulae are defined as usual, and denoted by:

$$Form(\overline{K}_E, \Phi) = p, q, \ldots, \neg\varphi, \varphi \lor \psi, \langle n \rangle \varphi, \diamond \varphi.$$

The intuitive meaning of the newly introduced symbols is the following:

- $\langle n \rangle \varphi$ is true at a given world if and only if there exists an $R_n$-accessible world where $\varphi$ holds;
- $\diamond \varphi$ is true at a given world if and only if there exists an $E$-accessible world where $\varphi$ holds.

It is immediate to see that $\diamond_n p$ and $\square_n p$ correspond to $\langle n \rangle \square p$ and $[n] \diamond p$, respectively. Clearly, on the ground of this definition, $\overline{K}_E$ can turn out much more expressive than $\overline{K}$, since one can combine the modal operators of $\overline{K}_E$ arbitrarily.

**DEFINITION 4.1** *Axioms and rules of $\overline{K}_E$ are the following:*

*N1 propositional logic axioms together with modus ponens;*

*N2 the axioms of $K$ for $[n]$ and $\square$:*
$$[n](\varphi \rightarrow \psi) \rightarrow ([n]\varphi \rightarrow [n]\psi),$$
$$\square(\varphi \rightarrow \psi) \rightarrow (\square\varphi \rightarrow \square\psi);$$

*N3 the rule of necessitation for $[n]$ and $\square$:*
$$\text{if } \vdash_{\overline{K}_E} \varphi \text{ then } \vdash_{\overline{K}_E} [n]\varphi,$$
$$\text{if } \vdash_{\overline{K}_E} \varphi \text{ then } \vdash_{\overline{K}_E} \square\varphi;$$

*N4 $[0]\diamond\varphi \rightarrow [n]\diamond\varphi$;*

*N5 $\langle n \rangle \square \varphi \rightarrow \langle n \rangle \diamond \varphi$;*

*N6 $[n]\varphi \rightarrow [n+1]\varphi$;*

*N7 $\langle n+m \rangle \square(\varphi \lor \psi) \rightarrow (\langle n \rangle \square \varphi \lor \langle m \rangle \square \psi)$;*

*N8 $(\langle n \rangle \square(\varphi \land \psi) \land \langle m \rangle \square(\varphi \land \neg \psi)) \rightarrow \langle n+m+1 \rangle \square \varphi$.*

The presence of axioms N1-N3 ensures the possibility to give a semantics to $\overline{K}_E$ simply considering it as a particular polymodal logic.

It can be shown (cf. [Ohlbach et al., 1995]) that $\overline{K}_E$ is not first-order axiomatizable.

The next question to answer is relative to the soundness and completeness of $\overline{K}_E$ with respect to the chosen semantics. The soundness of $\overline{K}_E$ can be easily established. As for completeness, a partial result can be obtained making use of the following translation function:

**DEFINITION 4.2** Let $\Pi : \mathcal{L}(\overline{K}, \Phi) \to \mathcal{L}(\overline{K}_E, \Phi)$ be the function that maps formulae of $\overline{K}$ in formulae of $\overline{K}_E$ according to the following rules:

$\Pi(p) = p$ for all $p \in \Phi$;

$\Pi(\neg \varphi) = \neg \Pi(\varphi)$;

$\Pi(\varphi \circ \psi) = \Pi(\varphi) \circ \Pi(\psi)$ per $\circ \in \{\wedge, \vee, \to, \leftrightarrow\}$;

$\Pi(\Diamond_n \varphi) = \langle n \rangle \Box \Pi(\varphi)$;

$\Pi(\Box_n \varphi) = [n] \Diamond \Pi(\varphi)$.

The above translation function is sound and complete (cf. [Ohlbach et al., 1995]):

**THEOREM 4.3** *For any formula $\varphi \in \mathcal{L}(\overline{K}, \Phi)$*

$$\vdash_{\overline{K}} \varphi \text{ if and only if } \vdash_{\overline{K}_E} \Pi(\varphi).$$

From the above result, we derive the completeness of $\overline{K}_E$ for the fragment of the language consisting of the translation of $\overline{K}$-formulae (cf. [Ohlbach et al., 1995]):

**THEOREM 4.4** *For any formula $\varphi \in \mathcal{L}(\overline{K}, \Phi)$*

$$\text{if } \vdash_{\overline{K}_E} \Pi(\varphi) \text{ then } \vdash_{\overline{K}_E} \Pi(\varphi).$$

The above theorem guarantees the possibility of using the system $\overline{K}_E$ as an intermediate system for the translation described in the next section.

### 4.2 The set-theoretic translation of graded modal logics

The logic $\overline{K}_E$ can be seen as the extension of a normal modal logic $K_{\overline{\mathcal{R}}}$, where $\overline{\mathcal{R}}$ is the set of accessibility relations $\{R_i : i \in \mathbb{N}\} \cup \{E\}$, with the axiom $\Psi = N4 \wedge \ldots \wedge N8$. From a general point of view, the problem we want to solve is to design a (set-theoretic) translation method that can be applied to a logic $\tilde{H}$ extending $K_{\overline{\mathcal{R}}}$ with a (possibly non-first-order) axiom $\psi(p_{j_1}, \ldots, p_{j_m})$. $\overline{K}_E$ will thus be considered as a particular case in which $\psi(p_{j_1}, \ldots, p_{j_m})$ is $\Psi$.

A frame for $\tilde{H}$ is a structure $\mathcal{F} = (W, \{R_i\}_{i \in \mathbb{N}^E})$, where $\mathbb{N}^E$ stands for $\mathbb{N} \cup \{E\}$. Since $\tilde{H}$ is a polymodal logic with infinitely many modalities, in order to apply the set-theoretic translation method is necessary to define a semantics that allows us to consider only one accessibility relation, to be interpreted by the membership relation $\in$. The fact that we must deal with infinitely many modalities implies that we cannot use the technique introduced in [D'Agostino et al., 1995]. The key definition for the proposed semantics is the following notion of *p-frame*.

**DEFINITION 4.5**
*Given a frame $\mathcal{F} = (W, \{R_n\}_{n \in \mathbb{N}^E})$ for $\tilde{H}$, a p-frame is a pair $(S, R)$, where:*

- $S = W \cup \bigcup_{i \in \mathbb{N}^E} U_i \cup \mathbb{N}^E$, where the $U_i$'s are pairwise distinct copies of $W$;
- $R$ is a binary relation on $S$ such that

  i. $\forall w \in W, t \in S \setminus (W \cup \mathbb{N}^E), x \in S \ (wRt \wedge tRx \to x \in W \cup \mathbb{N}^E)$;

  ii. $\forall w \in W, i \in \mathbb{N}^E$ *it is not the case that* $wRi$ *and* $\exists t \in S \setminus (W \cup \mathbb{N}^E) \ (wRt \wedge tRi)$;

  iii. $\forall w \in W, t, u, \in S \setminus (W \cup \mathbb{N}^E), i \in \mathbb{N}^E \ (wRt \wedge wRu \wedge tRi \wedge uRi \to t = u)$;

  iv. $\forall t \in S \setminus (W \cup \mathbb{N}^E), i, j \in \mathbb{N}^E \ (tRi \wedge tRj \to i = j)$.

For example, the elements $x_1, \ldots, x_4$ below are those in relation $R_i$ with $w$. The element $t$ of $U_i$ is introduced to simulate such a relation, and the element $i$ is used to determine the index of the relation.

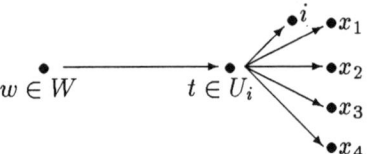

As for the semantics described in the case of finitely many modalities, a p-valuation $\models_p$ is a subset $W \times \Phi$ assigning truth values to propositional variables at $W$-worlds only. The extension of $\models_p$ to all $\tilde{H}$-formulae is inductively defined as follows:

$w \models_p \neg \varphi$ if and only if $w \not\models_p \varphi$;

$w \models_p \varphi \vee \psi$ if and only if $w \models_p \varphi$ or $w \models_p \psi$;

$w \models_p [i]\varphi$ if and only if $\forall t(wRt \wedge tRi \to \forall v(tRv \wedge v \neq i \to v \models_p \varphi))$.

The following results guarantee that the proposed semantics can be safely used in place of the usual one:

**LEMMA 4.6** *Given a p-frame $(S, R)$, there exists a frame $(W, \{R_i\}_{i \in \mathbb{N}^E})$ in which are valid all the formulae p-valid in $(S, R)$.*

**LEMMA 4.7** *For any frame $(W, \{R_i\}_{i \in \mathbb{N}^E})$, there exists a p-frame $(S, R)$ in which are p-valid all the formulae valid in $(W, \{R_i\}_{i \in \mathbb{N}^E})$.*

Any p-frame for $\tilde{H}$ can be embedded in a model of a suitable set theory in such a way that $W$ is a set, and any $w \in W$ is mapped in a set of elements of $w_i$'s for $i \in \mathbb{N}^E$. Any $w_i$ is of the form $w_i = \{v : wR_iv\} \cup \{i\}$.

The translation for propositional letters and boolean connectives is defined as usual:

$P_i^* = x_i$;

$(\neg \varphi)^* = x \setminus \varphi^*$;

$(\varphi \vee \psi)^* = \varphi^* \cup \psi^*$.

The translation of the different modalities must be given according to the p-semantics and replacing the accessibility relation by the membership relation:

$w \in ([i]\varphi)^* \Leftrightarrow \forall t(t \in w \wedge i \in t \to \forall v(v \in t \setminus \{i\} \to v \in \varphi^*))$,

from which it can be easily checked that the translation of $[i]\varphi$, for $i \in \mathbb{N}^E$, must be defined as follows:

$([i]\varphi)^* = Pow((S \setminus Rng(\{i\} \times_\in S)) \cup Pow(\varphi^* \cup \{i\}))$.

Let $\overline{\Omega}$ be the sub-theory of the theory $\Omega_c$ introduced in [Benthem et al., 1995] and defined as follows:
$$\overline{\Omega} = \Omega + \{\} + \times_\in + Rng.$$
The following two theorems state the completeness and the soundness of the proposed translation with respect to $\overline{\Omega}$:

**THEOREM 4.8 (Completeness)** *For any modal formula $\varphi \in \mathcal{L}(K_{\overline{\mathcal{R}}})$ we have that*
$$\vdash_{\tilde{H}} \varphi \Rightarrow \overline{\Omega} \vdash \forall y \forall x (Trans^2(x,y) \wedge Axiom_{\tilde{H}}(x,y) \to$$
$$\forall \vec{z}(x \subseteq \varphi^*(x,y,\vec{z}))),$$
*where $Trans^2(x,y)$ stands for the conjunction of the clauses on $R$ introduced in the definition of p-frame, and $Axiom_{\tilde{H}}(x,y)$ is the translation of the formula consisting of the conjunction of the axioms of $\tilde{H}$.*

**THEOREM 4.9 (Soundness)** *For any modal formula $\varphi \in \mathcal{L}(K_{\overline{\mathcal{R}}})$*
$$\overline{\Omega} \vdash \forall y \forall x (Trans^2(x,y) \wedge Axiom_{\tilde{H}}(x,y) \to$$
$$\forall \vec{z}(x \subseteq \varphi^*(x,y,\vec{z}))) \Rightarrow \models_{\tilde{H}} \varphi.$$

Once we have defined the set-theoretic translation in the context of a generic extension $\tilde{H}$ of $K_{\overline{\mathcal{R}}}$, the result for $\overline{K}_E$ follows as a special case:
$$\overline{\Omega} \vdash \forall y \forall x (Trans^2(x,y) \wedge Axiom_{\overline{K}_E}(x,y)$$
$$\to \forall \vec{z}(x \subseteq \varphi^*(x,y,\vec{z}))) \Rightarrow \models_{\overline{K}_E} \varphi,$$
$$\vdash_{\overline{K}_E} \varphi \Rightarrow \overline{\Omega} \vdash \forall y \forall x (Trans^2(x,y) \wedge Axiom_{\overline{K}_E}(x,y)$$
$$\to \forall \vec{z}(x \subseteq \varphi^*(x,y,\vec{z}))),$$
where $Axiom_{\overline{K}_E}(x,y)$ is the translation of $\Psi$, namely $\forall \vec{z}(x \subseteq \Psi^*(x,y,\vec{z}))$.

Notice that the soundness of the translation is stated with respect to the validity in $\overline{K}_E$, since we do not have the completeness of $\overline{K}_E$ with respect to derivability. However, using the translation function $\Pi$ we can prove the following theorem:

**THEOREM 4.10** *For any formula $\varphi \in \mathcal{L}(\overline{K})$ we have that:*
$$\vdash_{\overline{K}} \varphi \Leftrightarrow \overline{\Omega} \vdash \forall x \forall y (Trans^2(x,y) \wedge Axiom_{\overline{K}_E}(x,y) \to$$
$$\forall \vec{z}(x \subseteq (\Pi(\varphi))^*(x,y,\vec{z}))).$$

## 5 Conclusions and further directions

In this paper, we generalized the $\Box$-as-$Pow$ translation, proposed by D'Agostino et al. in [D'Agostino et al., 1995], to apply it to graded modal logic. The resulting method allows us to support automated reasoning in a large class of knowledge representation formalisms that can be reduced to graded modal logic. It can actually be applied to polymodal logics with infinitely many modalities. Indeed, there are no axioms in the underlying set theory constraining the behavior of the different modalities; such a behavior is governed by (the translation of) the axioms of the considered polymodal logic. As an example, it can be exploited to execute two-sorted metric temporal logics [Montanari and de Rijke, 1995], provided that they are reinterpreted as (a special kind of) propositional dynamic logics.

## References

[Benthem et al., 1995] J. van Benthem, G. D'Agostino, A. Montanari, and A. Policriti. Modal deduction in second-order logic and set theory. Technical Report ML-95-02, Institute for Logic Language and Information, 1995.

[Benthem, 1985] J. van Benthem. *Modal Logic and Classical Logic*. Bibliopolis, Napoli, Italy, 1985.

[Cerrato, 1990] C. Cerrato. General canonical models for graded normal logics (graded modalities IV). *Studia Logica*, 49:241–252, 1990.

[Cerrato, 1992] C. Cerrato. Decidability by filtration for graded modal logics (graded modalities V). *Studia Logica*, 53:61–74, 1992.

[D'Agostino et al., 1995] G. D'Agostino, A. Montanari, and A. Policriti. A set-theoretic translation method for polymodal logics. *Journal of Automated Reasoning*, 15:317–337, 1995.

[De Caro, 1988] F. De Caro. Graded modalities II (canonical models). *Studia Logica*, 47:1–10, 1988.

[Fattorosi-Barnaba and De Caro, 1985] M. Fattorosi-Barnaba and F. De Caro. Graded modalities I. *Studia Logica*, 44:197–221, 1985.

[Fine, 1972] K. Fine. In so many possible worlds. *Notre Dame Journal of Formal Logic*, 13:516–520, 1972.

[Fitting, 1983] M. Fitting. *Proofs Methods for Modal and Intuitionistic Logics*. D. Reidel Pub. Comp., Dordrecht, Boston, and Lancaster, 1983.

[Gabbay and Ohlbach, 1992] D. M. Gabbay and H. J. Ohlbach. Quantifier elimination in second-order predicate logic. In Morgan Kaufmann, editor, *Proc. of the $4^{th}$ International Conference on Principles of Knowldedge Representation and Reasoning, KR'92*, pages 425–436, 1992.

[Goble, 1970] L. F. Goble. Grades of modality. *Logique et Analyse*, 13:323–334, 1970.

[Hoek and de Rijke, 1995] W. van der Hoek and M. de Rijke. Counting objects. *Journal of Logic and Computation*, 5(3):325–345, 1995.

[Montanari and de Rijke, 1995] A. Montanari and M. de Rijke. Completeness results for two-sorted metric temporal logics. In Springer Verlag, editor, *Proc. of the $4^{th}$ International Conference on Algebraic Methodology and Software Technology, LNCS 936*, pages 385–399, 1995.

[Nonnengart, 1993] A. Nonnengart. First-order modal logic theorem proving and functional simulation. In *Proc. of $13^{th}$ International Joint Conference on Artificial Intelligence, IJCAI-93*, pages 80–85, Chambery, France, 1993.

[Ohlbach et al., 1995] H. J. Ohlbach, A. Schmidt, and U. Hustadt. Translating graded modalities into predicate logic. Technical Report MPI-I-95-2-008, Max-Planck-Institut für Informatik, Saarbrüken, Germany, 1995.

[Ohlbach, 1991] H. J. Ohlbach. Semantic-based translation methods for modal logics. *Journal of Logic and Computation*, 1(5), 1991.

[Ohlbach, 1993] H. J. Ohlbach. Translation methods for non-classical logics: An overview. *Bull. of the IGLP*, 1(1):69–89, 1993.

[Wansing, 1994] H. Wansing. Sequent calculi for normal modal propositional logics. *Journal of Logic and Computation*, 4(2):125–142, 1994.

# On evaluating decision procedures for modal logic

Ullrich Hustadt and Renate A. Schmidt
Max-Planck-Institut für Informatik, 66123 Saarbrücken, Germany
{hustadt,schmidt}@mpi-sb.mpg.de

## Abstract

This paper investigates the evaluation method of decision procedures for multi-modal logic proposed by Giunchiglia and Sebastiani as an adaptation from the evaluation method of Mitchell *et al.* of decision procedures for propositional logic. We compare three different theorem proving approaches, namely, the Davis-Putnam-based procedure KSAT, the tableaux-based system $\mathcal{KRIS}$ and a translation approach combined with first-order resolution. Our results do not support the claims of Giunchiglia and Sebastiani concerning the computational superiority of KSAT over $\mathcal{KRIS}$, and an easy-hard-easy pattern for randomly generated modal formulae.

## 1 Introduction

There are a variety of automated reasoning approaches for the basic propositional multi-modal logic $K(m)$ and its syntactical variant, the knowledge representation formalism $\mathcal{ALC}$. Some approaches utilize standard first-order theorem proving techniques in combination with translations from propositional modal logic to first-order logic [Ohlbach and Schmidt, 1995]. Others use Gentzen systems [Goble, 1974]. Still others use tableaux proof methods [Baader and Hollunder, 1991].

Usually, the literature on theorem provers for modal logic confines itself to a description of the underlying calculus and methodology accompanied with a consideration of the worst-case complexity of the resulting algorithm. Sometimes a small collection of benchmarks is given as in [Catach, 1991]. However, there have not been any exhaustive empirical evaluations or comparisons of the computational behavior of theorem provers based on different methodologies.

Giunchiglia and Sebastiani [1996a; 1996b] changed that. They report on an exhaustive empirical analysis of a new theorem prover, called KSAT, and the tableaux system $\mathcal{KRIS}$. KSAT is an adaptation for the multi-modal logic $K(m)$ of a SAT-procedure for checking satisfiability in propositional logic. The evaluation of Giunchiglia and Sebastiani has some shortcomings which we address. The random generator used to set up a benchmark suite produces formulae containing a substantial amount of tautologous and contradictory subformulae. It favours the SAT-procedure KSAT which utilizes a preprocessing routine that eliminates trivial tautologies and contradictions from the formulae. This property of the random formulae mislead Giunchiglia and Sebastiani in their analysis and comparison of KSAT and $\mathcal{KRIS}$. We show the random generator does not produce challenging unsatisfiable modal formulae.

The paper is structured as follows. In Sections 2, 3 and 4 we briefly describe the inference mechanisms of KSAT, $\mathcal{KRIS}$ and the translation approach. Section 5 describes the evaluation method of Giunchiglia and Sebastiani. The main part is Section 6 which evaluates the test method.

## 2 The SAT-based procedure KSAT

By definition, a *formula* of the multi-modal logic $K(m)$, where $m$ is a natural number, is a boolean combination of propositional and modal atoms. A *modal atom* is an expression of the form $\Box_i\psi$, where $i$ is such that $1 \leq i \leq m$ and $\psi$ is a formula of $K(m)$. $\Diamond_i\psi$ is an abbreviation for $\neg\Box_i\neg\psi$. The semantics of $K(m)$ is given by the usual Kripke semantics.

KSAT tests the satisfiability of a given formula $\phi$ of $K(m)$. Its basic algorithm, called KSAT0, is based on the following two procedures:

KDP: Given a modal formula $\phi$, this procedure generates a truth assignment $\mu$ for the propositional and modal atoms in $\phi$ which renders $\phi$ true propositionally. This is done using a decision procedure for propositional logic.

KM: For a given $\phi$ and $\mu$ computed by KDP, let $\Box_i\psi_{ij}$ denote any modal atom in $\phi$ with $\mu(\Box_i\psi_{ij}) = \bot$ and $\Box_i\phi_{ik}$ any modal atom with $\mu(\Box_i\phi_{ik}) = \top$. The procedure checks for each index $i$, $1 \leq i \leq m$, and each $j$ whether the formula $\varphi_{ij} = \bigwedge_k \phi_{ik} \wedge \neg\psi_{ij}$ is satisfiable. This is done with KDP. If at least one of the formulae $\varphi_{ij}$ is not satisfiable, then KM *fails on* $\mu$, otherwise it *succeeds*.

KSAT0 starts by generating a partial truth assignment $\mu$ for $\phi$ using KDP. If KM succeeds on $\mu$, then $\phi$ is $K(m)$-satisfiable. Otherwise, we have to generate a new truth assignment for $\phi$ using KDP. If no further truth assignment is found, then $\phi$ is $K(m)$-unsatisfiable.

The decision procedure KDP for propositional logic can be described by a set of transition rules on ordered pairs $P \triangleright S$ where $P$ is a sequence of pairs $\langle \phi, \mu \rangle$, and $S$ is a set of satisfying truth assignments.

$$\texttt{dp\_sol:} \quad \frac{\langle \top, \mu \rangle \mid P \triangleright S}{P \triangleright S \cup \{\mu\}}$$

$$\texttt{dp\_clash:} \quad \frac{\langle \bot, \mu \rangle \mid P \triangleright S}{P \triangleright S}$$

$$\texttt{dp\_unit:} \quad \frac{\langle \phi[c], \mu \rangle \mid P \triangleright S}{\langle \phi', \mu \cup \{c = \top\} \rangle \mid P \triangleright S}$$

if $c$ is a unit clause in $\phi$ and $\phi'$ is the result of replacing all occurrences of $c$ and $\bar{c}$ by $\top$ and $\bot$, respectively, followed by boolean simplification.

$$\texttt{dp\_split:} \quad \frac{\langle \phi[m], \mu \rangle \mid P \triangleright S}{\langle \phi[m] \wedge p, \mu \rangle \mid \langle \phi[m] \wedge \neg p, \mu \rangle \mid P \triangleright S}$$

if $\texttt{dp\_unit}$ cannot be applied to $\langle \phi[m], \mu \rangle$, and $m$ is a propositional or modal atom.

The symbol $|$ denotes concatenation. $\bar{\phi}$ and $\phi$ are complementary, e.g. $\overline{\neg p} = p$ and $\overline{\Box_i p} = \Diamond_i \neg p$.

Starting with $\langle \phi, \emptyset \rangle \triangleright \emptyset$, exhaustively applying the inference rules will result in $\emptyset \triangleright S$ where $S$ is a complete set of partial truth assignments making $\phi$ true.

Note that the transition rules form a variant of the Davis-Putnam procedure for propositional formulae not in conjunctive normal form. The crucial nondeterminism of the procedure is the selection of the splitting 'variable' $m$ in the transition rule $\texttt{dp\_split}$. KSAT employs the heuristic that selects an atom with a maximal number of occurrences in $\phi$.

## 3  The tableaux-based system $\mathcal{KRIS}$

While KSAT abstracts from the modal part of formulae to employ decision procedures for propositional logic, $\mathcal{KRIS}$ manipulates modal formulae (in variant notation) directly. More precisely, the inference rules of $\mathcal{KRIS}$ are relations on sequences of sets of labeled modal formulae of the form $w{:}\psi$, where $w$ is a label chosen from a countably infinite set of labels $\Gamma$ and $\psi$ is modal formula. For improved readability we write $w{:}\psi, C$ instead of $\{w{:}\psi\} \cup C$.

$$\bot\texttt{-elim:} \quad \frac{w{:}\bot, C \mid S}{S}$$

$$\top\texttt{-elim:} \quad \frac{w{:}\top, C \mid S}{C \mid S}$$

$$\wedge\texttt{-clash:} \quad \frac{w{:}\phi, w{:}\bar{\phi}, C \mid S}{S}$$

$$\wedge\texttt{-elim:} \quad \frac{w{:}\phi \wedge \psi, C \mid S}{w{:}\phi, w{:}\psi, C \mid S}$$

$$\vee\texttt{-elim:} \quad \frac{w{:}\phi \vee \psi, C \mid S}{w{:}\phi, C \mid w{:}\psi, C \mid S}$$

if $w{:}\phi \vee \psi, C$ has been simplified by

$\vee\texttt{-simp}_0{:} \quad w{:}\phi \vee \psi, w{:}\phi, D \to w{:}\phi, D$

$\vee\texttt{-simp}_1{:} \quad w{:}\phi \vee \psi, w{:}\bar{\phi}, D \to w{:}\psi, w{:}\bar{\phi}, D.$

$$\Diamond_i\texttt{-elim:} \quad \frac{w{:}\Diamond_i \phi, D, C \mid S}{v{:}\phi \wedge \psi_1 \wedge \ldots \wedge \psi_n, D, C \mid S}$$

if $D = w{:}\Box_i\psi_1, \ldots, w{:}\Box_i\psi_n$, $C$ does not contain any $w{:}\Box_i\psi$, none of the other rules can be applied to $C$, and $v$ is a new label from $\Gamma$.

The application of the $\vee$-elim rule to any labeled formula $w{:}\phi \vee \psi$ is preceeded by the application of the simplification rules to that formula. In no other situation the simplification rules are invoked. Given a formula $\phi$, the input sequence for $\mathcal{KRIS}$ is $w_0{:}\phi'$, where $w_0$ is a new label chosen from $\Gamma$ and $\phi'$ is the modal negation normal form of $\phi$. If $\mathcal{KRIS}$ arrives at a sequence $C \mid S$ such that no transformation rule can be applied to $C$, then the original formula $\phi$ is satisfiable. Otherwise the transformation rules will eventually reduce $w_0{:}\phi'$ to the empty sequence and $\phi$ is unsatisfiable.

## 4  The translation approach

The *translation approach* (TA) is based on the idea that modal inference can be done by translating modal formulae into first-order logic and conventional theorem proving. We use the *optimised functional translation* approach of Ohlbach and Schmidt [1995]. It has the property that ordinary resolution without any refinement strategies is a decision procedure for $K(m)$ [Schmidt, 1997]. The translation maps modal formulae into a logic, called *basic path logic*, which is a monadic fragment of sorted first-order logic with one binary function symbol $\circ$ that defines accessibility. A formula of path logic is further restricted in that its clausal form may only contain Skolem terms that are constants.

The optimised functional translation does a sequence of transformations. The first transformation $\Pi_f$ maps a modal formula $\phi$ to its so-called functional translation defined by $\Pi_f(\phi) = \forall x \, \pi_f(\phi, x)$. For $K(m)$, $\pi_f$ is defined by

$$\begin{aligned} \pi_f(p, s) &= P(s) \\ \pi_f(\Box_i \phi, s) &= \mathit{def}_i(s) \to \forall \alpha_i \, \pi_f(\phi, s \circ \alpha_i). \end{aligned}$$

$p$ is a propositional variable and $P$ is a unary predicate uniquely associated with $p$, $\mathit{def}_i$ is a special unary predicate of sort $i$, and $\alpha_i$ denotes a variable of sort $i$. For the propositional connectives $\pi_f$ is a homomorphism. The second transformation applies the so-called quantifier exchange operator $\Upsilon$ which moves existential quantifiers inwards over universal quantifiers using the rule '$\exists\alpha\forall\beta\,\psi$ becomes $\forall\beta\exists\alpha\,\psi$'. Ohlbach and Schmidt prove $\Upsilon\Pi_f$ preserves satisfiability.

Our aim is to test the satisfiability of a given modal formula $\phi$. This can be achieved by testing the satisfiability of the clausal form of $\neg\Upsilon\Pi_f(\neg\phi)$. The theorem prover we use is SPASS Version 0.55 developed by Weidenbach et al. [1996] which is an ordered resolution-based theorem prover for sorted first-order logic.

## 5 The test method

The evaluation method adopted by Giunchiglia and Sebastiani follows the approach of Mitchell et al. [1992]. To set up a benchmark suite for Davis-Putnam-based theorem provers Mitchell et al. generate propositional formulae using the fixed clause-length model. Giunchiglia and Sebastiani modify this approach for $K(m)$.

There are five parameters: the number of propositional variables $N$, the number of modalities $M$, the number of modal subformulae per disjunction $K$, the number of modal subformulae per conjunction $L$, the modal degree $D$, and the probability $P$. Based on a given choice of parameters random modal $K$CNF formulae are defined inductively as follows. A *random (modal) atom* of degree 0 is a variable randomly chosen from the set of $N$ propositional variables. A *random modal atom* of degree $D$, $D>0$, is with probability $P$ a random modal atom of degree 0 or an expression of the form $\Box_i\phi$, otherwise, where $\Box_i$ is a modality randomly chosen form the set of $M$ modalities and $\phi$ is a random modal $K$CNF clause of modal degree $D-1$ (defined below). A *random modal literal* (of degree $D$) is with probability 0.5 a random modal atom (of degree $D$) or its negation, otherwise. A *random modal $K$CNF clause* (of degree $D$) is a disjunction of $K$ random modal literals (of degree $D$). Now, a *random modal $K$CNF formula* (of degree $D$) is a conjunction of $L$ random modal $K$CNF clauses (of degree $D$).

For the comparison of the performance of KSAT and $\mathcal{KRIS}$, Giunchiglia and Sebastiani proceed as follows. They fix all parameters except $L$, the number of clauses. For example, they choose $N=3$, $M=1$, $K=3$, $D=5$, and $P=0.5$. The parameter $L$ ranges from $N$ to $40N$. For each value of the ratio $L/N$ a set of 100 random modal $K$CNF formulae of degree $D$ is generated. We will see that for small $L$ the generated formulae are most likely to be satisfiable and for larger $L$ the generated formulae are most likely to be unsatisfiable. For each generated formula $\phi$ they measure the time needed by one of the decision procedures to determine the satisfiability of $\phi$. Since checking a single formula can take arbitrarily long in the worst case, there is an upper limit for the CPU time consumed. As soon as the upper limit is reached,

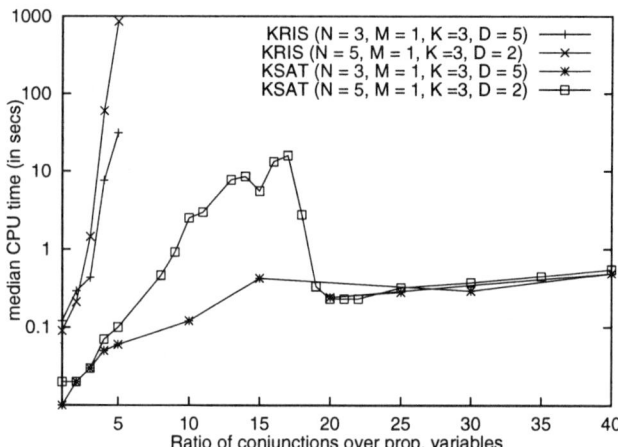

Figure 1: Performance comparison of $\mathcal{KRIS}$ and KSAT

the computation for $\phi$ is stopped. If the computation for more than 50% of the formulae of a set associated with a value of $L$ has been abandoned, then the computation for the set is discontinued. Now, the median CPU time over the ratio $L/N$ is presented. For example, the graphs of Figure 1 show the performance of $\mathcal{KRIS}$ and KSAT on the parameter settings **PS0** ($N=3$, $M=1$, $K=3$, $D=5$) and **PS1** ($N=5$, $M=1$, $K=3$, $D=2$). Our tests have been run on a Sun Ultra 1/170E with 196MB main memory using a time-limit of 1000 CPU seconds. Altogether Giunchiglia and Sebastiani [1996b] present graphs for ten different parameter settings. Based on their graphs including Figure 1 they come to the following conclusions:

(1) KSAT outperforms by orders of magnitude the previous state-of-the art decision procedures.

(2) All SAT-based modal decision procedures are intrinsically bound to be more efficient than tableaux-based decision procedures.

(3) There is partial evidence of an easy-hard-easy pattern on randomly generated modal logic formulae independent of all the parameters of evaluation considered.

We show that the situation is more complex and does not justify such strong claims. For our analysis it suffices to focus on the settings **PS0** and **PS1** of Figure 1.

## 6 Analysis of the test method

Selecting good test instances is crucial when evaluating and comparing the performance of algorithms empirically. We address the question whether the random generator and the parameter settings chosen by Giunchiglia and Sebastiani [1996b; 1996a] are appropriate for this purpose and actually support claims (1) to (3).

It is important to note that the claim of Giunchiglia and Sebastiani [1996b, p. 307] that for $D=0$ random modal 3CNF formulae coincide with random 3SAT

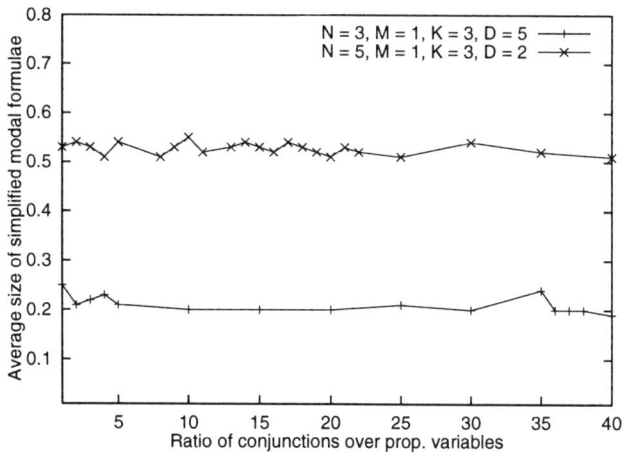

Figure 2: Effect of simplifying modal 3CNF formulae

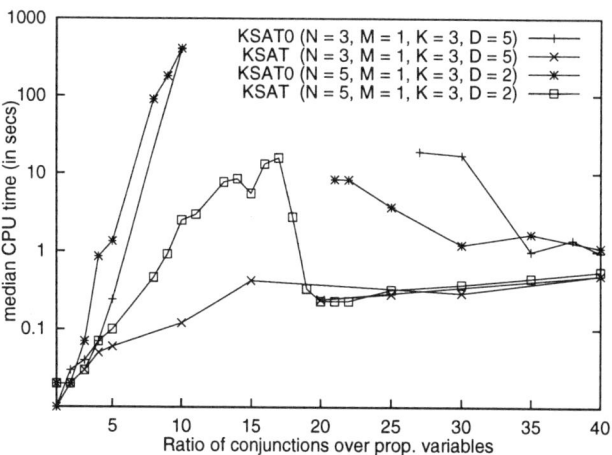

Figure 3: Performance comparison of KSAT and KSAT0

clauses as defined in Mitchell et al. is wrong: To generate a random 3SAT clause we have to randomly generate a *set* of three propositional variables and negate each member of the set with probability 0.5. In contrast, to generate a random modal 3CNF clause of degree 0, we have to randomly generate a *multiset* of three propositional variables and negate each member of the multiset with probability 0.5. For example, $p \vee q \vee \neg r$ is a 3SAT clause and also a modal 3CNF clause of degree 0. The clauses $p \vee \neg p \vee p$ and $p \vee p \vee q$ are not random 3SAT clauses, but both are random modal 3CNF clauses of degree 0. As we move to random modal 3CNF formulae of higher degree, such clauses may occur within the scope of a modal operator. For example, expressions like $\neg \Box_1 (p \vee \neg p \vee p)$ may occur which are contradictory. Consequently, random modal $K$CNF formulae contain tautological and contradictory subformulae. It is straightforward to remove these subformulae without affecting satisfiability. The extent to which the size of the random modal 3CNF formulae can be reduced by such simplifications is reflected by the graphs of Figure 2. They depict the average ratio of the size of the simplified random modal 3CNF formulae over the size of the original formulae. For the random modal 3CNF formulae generated using three propositional variables only, on average, the size of a simplified formula is only 1/4 of the size of the original formula. For the second parameter setting we see a reduction to 1/2 of the original size. In other words, one half to three quarters of the random modal 3CNF formulae is "logical garbage" that can be eliminated at little cost.

KSAT utilizes a form of preprocessing that removes duplicate and contradictory subformulae of an input formula. That is, KSAT performs exactly the simplification whose effect we have just described. $\mathcal{KRIS}$ on the other hand does not perform a similar simplification. We consider how KSAT performs if we remove the preprocessing step from its code. In Figure 3 KSAT0 denotes this modified form of KSAT, since it is actually identical to the algorithm described in Section 2. We see that the behavior of KSAT0 differs from the behavior of KSAT by orders of magnitude. Since the preprocessing is not an intrinsic part of the decision procedures, for the comparison of the procedures, either both KSAT and $\mathcal{KRIS}$ should utilize the preprocessing or none of them should. Simplification of the generated formulae is reasonable, so we have added the preprocessing function to $\mathcal{KRIS}$. This modified version of $\mathcal{KRIS}$ will be denoted by $\mathcal{KRIS}^*$. The graphs in Figure 4 show the performance of KSAT and $\mathcal{KRIS}^*$. Although the performance of KSAT is still better than that of $\mathcal{KRIS}^*$, KSAT is no longer qualitatively better than $\mathcal{KRIS}$ with preprocessing.

We now address claim (2) that, intrinsically, SAT-based modal decision procedures are bound to be more efficient than tableaux-based decision procedures. The explanation is based on the work by D'Agostino [1992], who shows that in the worst case algorithms using the

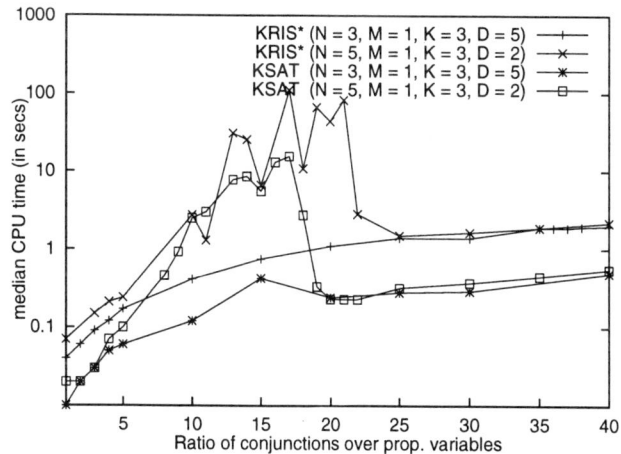

Figure 4: The performance of KSAT and $\mathcal{KRIS}^*$

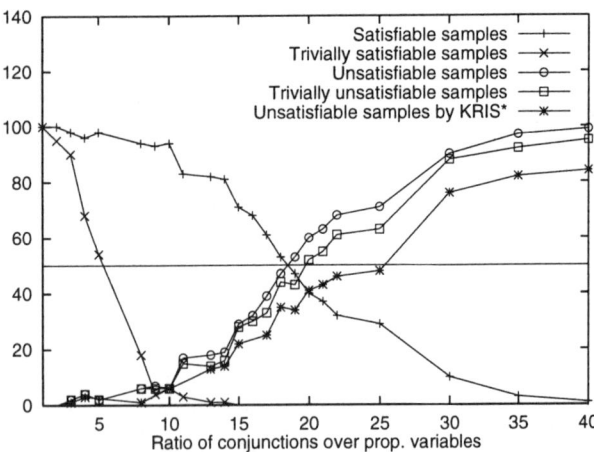

Figure 5: Quality of the test set **PS1**

∨-elim rule cannot simulate truth tables in polynomial time. Instead one has to replace ∨-elim by the rule

∨-elim': $\dfrac{w{:}\phi \vee \psi, C \mid S}{w{:}\phi, C \mid w{:}\psi, w{:}\neg\phi, C \mid S}$

This rule ensures that the two generated subproblems $w{:}\phi, C$ and $w{:}\psi, w{:}\neg\phi, C$ are mutually exclusive.

We have just seen that a major cause of the difference in computational behavior of the two algorithms is actually the absence of the preprocessing step in $\mathcal{KRIS}$. To explain the remaining difference we study the quality of the random modal 3CNF formulae. Suppose that we want to test a random modal 3CNF formula $\phi$ generated using $N$ propositional variables for satisfiability in a Kripke model with only one world. We have to test at most $2^N$ truth assignments to the propositional variables. Since $N{\leq}5$ for the modal formulae under consideration, this is a trivial task, even if we use the truth table method. We say a random modal 3CNF formula $\phi$ is *trivially satisfiable* if $\phi$ is satisfiable in a Kripke model with only one world. We also say a random modal 3CNF formula $\phi$ is *trivially unsatisfiable* if the conjunction of the purely propositional clauses of $\phi$ is unsatisfiable. Again, testing whether $\phi$ is trivially unsatisfiable requires the consideration of $2^N$ truth assignments only.

The graphs of Figure 5 show the percentage of satisfiable, trivially satisfiable, unsatisfiable, trivially unsatisfiable, and unsatisfiable samples detected by $\mathcal{KRIS}^*$ for the parameter setting **PS1**. We see that almost all unsatisfiable formulae are trivially unsatisfiable. We have verified that this also holds for all the other parameter settings used by Giunchiglia and Sebastiani. This indicates, none of these parameter settings is suited to generate challenging unsatisfiable modal formulae.

If we consider Figure 4 and 5 together, for ratios $L/N$ between 19 and 21 we observe the graph of $\mathcal{KRIS}^*$ deviates a lot (by a factor of more than 100) from the graph of KSAT. This is the area near the crossover point where the percentage of trivially unsatisfiable formulae rises above 50%, however, the percentage of unsatisfiable formulae detected by $\mathcal{KRIS}^*$ is still below 50% in this area. $\mathcal{KRIS}^*$ does not detect all trivially unsatisfiable formulae within the time-limit which explains the deviation in performance from KSAT. The reason for $\mathcal{KRIS}^*$ not detecting all trivially unsatisfiable formulae within the time limit, can be illustrated by the following example. Let $\phi$ be a simplified modal 3CNF formula

$$p \wedge q \wedge (m_{11} \vee m_{12} \vee m_{13})$$
$$\ldots$$
$$\wedge (m_{n1} \vee m_{n2} \vee m_{n3}) \wedge (\neg p \vee \neg q)$$

where the $m_{ij}$, with $1{\leq}i{\leq}n$, $1{\leq}j{\leq}3$, are modal literals different from $p$, $q$, $\neg p$, and $\neg q$. Evidently, $\phi$ is trivially unsatisfiable. KSAT does the following: Since $p$ and $q$ are unit clauses in $\phi$, it applies the rule `dp_unit` twice to $\phi$. The rule replaces the occurrences of $p$ and $q$ by ⊤, it replaces the occurrences of $\neg p$ and $\neg q$ by ⊥, and it simplifies the formula. The resulting formula is ⊥. At this point only the rule `dp_clash` is applicable and KSAT detects that $\phi$ is unsatisfiable. In contrast, $\mathcal{KRIS}^*$ proceeds as follows. First it applies the ∧-elim rule $n+2$ times, eliminating all occurrences of the ∧ operator. Then it applies the ∨-elim rule to all disjunctions, starting with $(m_{11} \vee m_{12} \vee m_{13})$ and ending with $(m_{n1} \vee m_{n2} \vee m_{n3})$. This generates $3^n$ subproblems. Each of these subproblems contains the literals $p$ and $q$ and the disjunction $\neg p \vee \neg q$. The rule ∨-simp$_1$ eliminates $\neg p \vee \neg q$ and a final application of the ∧-clash rule exhibits the unsatisfiability of each subproblem. Obviously, for $n$ large enough, $\mathcal{KRIS}^*$ will not be able to finish this computation within the time-limit.

Note, it makes no difference whether $\mathcal{KRIS}^*$ eliminates disjunctions by the ∨-elim rule or the ∨-elim' rule. The reason for $\mathcal{KRIS}^*$ not finishing within the time-limit is that it does not apply the simplification rules ∨-simp$_1$ and ∨-simp$_2$ and the ∧-clash rule eagerly before any application of the ∨-elim rule.

Finally, we consider claim (3) conjecturing an easy-hard-easy pattern, independent of all the parameters of evaluation, in randomly generated modal logic formulae. We have seen in Figure 1 that the mean CPU time consumption of KSAT decreases drastically at the ratio $L/N{=}17.5$ for the second sample. This is almost the point, where 50% of the sample formulae are satisfiable. This decline resembles the behavior of propositional SAT decision procedures on randomly generated 3SAT problems. Figure 6 compares the performance of KSAT with the performance of the translation approach on two parameter settings, where the easy-hard-easy pattern is most visible for KSAT. The translation approach does not show the peaking behavior of KSAT. The median CPU time grows monotonically with the size of

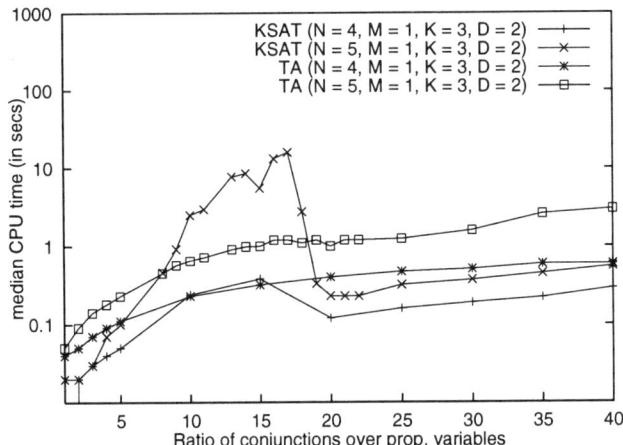

Figure 6: Performance comparison of KSAT and TA

modal formulae. Thus, the phase transition observed by Giunchiglia and Sebastiani is an artificial phenomenon of KSAT (and $\mathcal{KRIS}$), and not an intrinsic property of the generated modal formulae.

Observe that the peaking behavior occurs in the area where the number of trivially satisfiable sample formulae approaches zero. The following example tries to explain this. Let $\psi$ be a simplified modal 3CNF formula of the form

$$\neg \Box_1 s \land \Box_1 (p \lor r) \land (\Box_1 \neg r \lor \Box_1 q) \land (\neg \Box_1 p \lor \Box_1 r)$$
$$\land (m_{11} \lor m_{12} \lor m_{13}) \land \ldots \land (m_{n1} \lor m_{n2} \lor m_{n3})$$

where the $m_{ij}$, with $1 \leq i \leq n$, $1 \leq j \leq 3$, are modal literals different from the modal literals in the first three conjunctions of $\psi$. Assume, $\psi$ is satisfiable. $\Box_1 \neg r$ is false in any model of $\psi$. In the situation that $\Box_1 \neg r$ is one of the first split literals chosen by KSAT, it generates a huge search tree without finding a satisfying truth assignment before it eventually turns to the case where $\Box_1 \neg r$ is assigned $\bot$. This explains the bad behaviour of KSAT on those sample formulae where satisfiability tests in the non-propositional contexts are essential. $\mathcal{KRIS}^*$ behaves similarly.

In contrast, the translation approach proceeds as follows. It generates a clause set for $\psi$ containing

$$def_1(\underline{\iota})$$
$$\neg S(\underline{\iota} \circ \underline{a})$$
$$\neg def_1(\underline{\iota}) \lor P(\underline{\iota} \circ x) \lor R(\underline{\iota} \circ x),$$
$$\neg def_1(\underline{\iota}) \lor \neg R(\underline{\iota} \circ x) \lor \neg def_1(\underline{\iota}) \lor Q(\underline{\iota} \circ y),$$
$$\neg P(\underline{\iota} \circ \underline{b}) \lor \neg def_1(\underline{\iota}) \lor R(\underline{\iota} \circ x)$$

where $\underline{\iota}$, $\underline{a}$ and $\underline{b}$ denote Skolem constants and $x$ and $y$ are variables. SPASS applies unit propagation to the first clause followed by subsumption. Three resolvents can be derived: $P(\underline{\iota} \circ x) \lor Q(\underline{\iota} \circ y)$, $\neg P(\underline{\iota} \circ \underline{b}) \lor Q(\underline{\iota} \circ y)$, and $R(\underline{\iota} \circ \underline{b}) \lor R(\underline{\iota} \circ x)$. A factoring step on the last resolvent yields $R(\underline{\iota} \circ \underline{b})$. This means $\Box_1 \neg r$ is false in any model. An additional inference step computes the unit clause $Q(\underline{\iota} \circ y)$. No further inference is possible on this subset.

## 7 Conclusion

We have pointed out a number of problems with evaluating the performance of different algorithms for modal reasoning. Our investigations show benchmarking needs to be done with great care. A crucial factor is the quality of the randomly generated problems, which we think are too easy. Further investigations are required concerning the parameter settings and fundamental properties of modal $K$CNF formulae before we can come to safe conclusions about different theorem proving approaches for modal logic.

A longer version of this paper is available as Research Report MPI-I-97-2-003.

## References

[Baader and Hollunder, 1991] F. Baader and B. Hollunder. A terminological knowledge representation system with complete inference algorithms. In *Proc. PDK '91*, LNAI 567, pages 67–86. Springer, 1991.

[Catach, 1991] L. Catach. Tableaux: A general theorem prover for modal logics. *J. Automated Reasoning*, 7(4):489–510, 1991.

[D'Agostino, 1992] M. D'Agostino. Are tableaux an improvement on truth-tables? *J. Logic, Language, and Information*, 1:235–252, 1992.

[Giunchiglia and Sebastiani, 1996a] F. Giunchiglia and R. Sebastiani. Building decision procedures for modal logics from propositional decision procedures: Case study of modal K. In *Proc. CADE-13*, LNAI 1104, pages 583–597, 1996.

[Giunchiglia and Sebastiani, 1996b] F. Giunchiglia and R. Sebastiani. A SAT-based decision procedure for $\mathcal{ALC}$. In *Proc. KR'96*, pages 304–314, 1996.

[Goble, 1974] L. F. Goble. Gentzen systems for modal logic. *Notre Dame J. Formal Logic*, 15:455–461, 1974.

[Mitchell et al., 1992] D. Mitchell, B. Selman, and H. Levesque. Hard and easy distributions of SAT problems. In *Proc. AAAI-10*, pages 459–465, 1992.

[Ohlbach and Schmidt, 1995] H. J. Ohlbach and R. A. Schmidt. Functional translation and second-order frame properties of modal logics. Res. Report MPI-I-95-2-002, MPI für Informatik, Saarbrücken, 1995.

[Schmidt, 1997] R. A. Schmidt. Resolution is a decision procedure for many propositional modal logics: Extended abstract. To appear in *Proc. AiML'96*, 1997.

[Weidenbach et al., 1996] C. Weidenbach, B. Gaede, and G. Rock. SPASS & FLOTTER version 0.42. In *Proc. CADE-13*, LNAI 1104, pages 141–145, 1996.

# AUTOMATED REASONING

Automated Reasoning 9: Analogy

# Prediction: A Common Form of Induction and Analogy

Jun Arima
Fujitsu Laboratories Ltd.,
Fujitsu Kyushu R & D center,
2-2-Momochihama, Sawara-ku,Fukuoka-shi 814,
Japan.

## Abstract

Deduction, induction, and analogy pervade all our thinking. In contrast with deduction, understanding logical aspects of induction and analogy is still an important and challenging issue of *artificial intelligence*. This paper describes a logical formalization, called *preduction*, of common conjectural reasoning of both induction and analogy. By introduction of preduction, analogical reasoning is refined into "preduction + deduction" and (empirical) inductive reasoning is refined into "preduction + mathematical induction". We examine generality of preduction through applications to various examples on induction and analogy.

## 1 Introduction

Deduction, induction, and analogy are most common patterns of our thinking. While *deduction* infers a property about a specific individual from a general property which every individual satisfies, *inductive reasoning* infers an unknown property which every individual will satisfy commonly from specific properties about individuals. *Analogical reasoning* infers an unknown property about an individual from known properties about similar others. Because of their generality and importance in our intelligent activities, understanding their reasoning processes is indispensable for embodying *artificial intelligence*.

More formally, inferences by deduction, induction, and analogy are typically represented as Table 1[1]. Deduction(D.1) expresses "(D.1.1) $a$ is $P$. (D.1.2) all $P$-things are $Q$. Therefore (D.1.3) $a$ is $Q$." Induction(I.1) which we call *mathematical* expresses the usual axiom schema of induction in the arithmetic axioms. "(I.1.1) the case of 0 satisfies $P$. (I.1.2) if the case of $x$ satisfies $P$, the succeeding case of $x$ also satisfies $P$. Consequently, (I.1.3) any case will satisfy $P$." Induction (I.2) expresses more *empirical* reasoning than (I.1); the same consequence is inferred not from a general assertion like (I.1.2) but from an observation that (I.2.2) every case of 0 to $n$ satisfies $P$. Analogy (A.1) and (A.2) express "(A.*.1) the *base case* $b$ satisfies $P$. (A.*.2) the *target case* $t$ is similar to $b$. Thus, (A.*.3) $t$ also satisfies $P$." They are different in that the *similarity* between a target and a base is explicit as a property $S$ in (A.2).

| Deduction(D.1) | |
|---|---|
| (D.1.1) | $P(a)$ |
| (D.1.2) | $\forall x(P(x) \supset Q(x))$ |
| (D.1.3) | $Q(a)$ |

| Induction(I.1) | | Analogy(A.1) | |
|---|---|---|---|
| (I.1.1) | $P(0)$ | (A.1.1) | $P(b)$ |
| (I.1.2) | $\forall x(P(x) \supset P(s(x)))$ | (A.1.2) | $t \sim b$ |
| (I.1.3) | $\forall x.P(x)$ | (A.1.3) | $P(t)$ |

| Induction(I.2) | | Analogy(A.2) | |
|---|---|---|---|
| (I.2.1) | $P(0)$ | (A.2.1) | $P(b)$ |
| (I.2.2) | $P(1) \wedge \cdots \wedge P(n)$ | (A.2.2) | $S(b) \wedge S(t)$ |
| (I.2.3) | $\forall x.P(x)$ | (A.2.3) | $P(t)$ |

Table 1: Deduction, induction, and analogy

In this paper, we especially focus on two logical aspects of induction and analogy; their consistency and their relationship on inference. In contrast with (D.1) and (I.1), each inference rule of (I.2), (A.1), and (A.2) has at least one critical logical defect. The former, each of (D.1) and (I.1), *preserves consistency* (i.e., only consistent theorems are inferred from consistent axioms), while the latter, each of (I.2), (A.1), and (A.2) does not. In spite of this fact, the latter inference rules seem to be natural to our common sense. For example, when we infer a general rule from individual observations, where knowledge as (I.1.2) comes from? Can we directly recognize the knowledge from observations? Although we may recognize knowledge such as (I.2.2) from our environments, we do not recognize such an (I.1.2) at least from our daily life. From our view, it is the heart of our empirical induction to infer (I.1.2) from (I.2.1) and (I.2.2), and it is the process which we should investigate and formalize.

Conjectural reasoning which always brings a consistent conclusion at present but possibly inconsistent in

---

[1] Deduction (D.1) is a derivative rule from ∀-elimination and modus ponens.

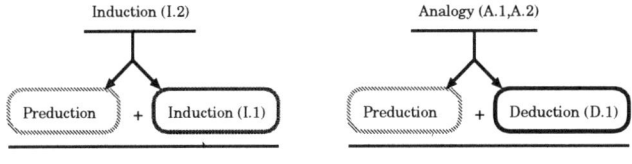

Figure 1: Preduction + (Deduction/Induction)

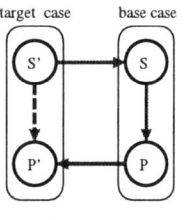
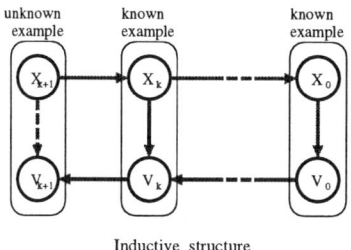

Figure 2: Common structure to analogy and induction

the future immediately implies that it is *non-monotonic*. Analogy and empirical induction often bring us conclusions turned to be wrong after we know more. In such cases, their conclusions are invalidated in our belief. This non-monotonicity is quite common to our reasoning. McCarthy introduced *circumscription*[McCarthy, 1980] for the purpose of formalizing non-monotonicity in common-sense reasoning. Circumscription of a predicate $P$ makes its extension minimized; anything is not $P$ unless it is stated $P$ by given axioms. Helft[Helft, 1988] used minimization of all predicates to formalize inductive reasoning. By means of minimizing all predicates, although preferably inferred when $P(a) \wedge S(a)$, a generalization $\forall x(S(x) \supset P(x))$ is no more inferred when $P(a) \wedge S(a) \wedge S(b)$ that normally happens whenever we want to deduce a useful conclusion from the generalization. Inductive reasoning (and analogical reasoning) should be naturally interpreted as a particular expansion of the extension of a predicate rather than minimization. Our approach follows circumscription just for formalization of non-monotonicity, but does not inherit the idea of minimality.

Relation between induction and analogy is another point to be investigated. Analogy (A.2) can be viewed as a two-step argument[Davies and Russell, 1987]; $\forall x(S(x) \supset P(x))$ by a *single-instance* induction from $S(b)$ in (A.2.2) and $P(b)$ in (A.2.1), and then by deduction with $S(t)$ in (A.2.2), we obtain (A.2.3). A similar idea is suggested in [Peirce, 1932; Mostow, 1983]. Unlike to their views, we consider there is a common inferential structure behind induction (I.2), analogy (A.1), and (A.2). Analogy includes a projection of information from a similar known object to an unknown object. Induction similarly includes a projection of information from known previous cases. We discuss more about this common denominator in the next section. We formalize this type of projection based on a transitive relation between objects. We call the formalization *preduction* because the introduction of preduction allows empirical induction and analogy to be broken into a stable part, (I.1) and (D.1), and the *preceding* more conjectural part corresponding to the preduction (Figure 1).

Formalizing consistent *preduction* has at least two significant points. One is (of course) to allow us to use empirical induction and analogy free from inconsistency. If consistency of preduction guaranteed, because of consistency of deduction and mathematical induction, the whole processes of empirical induction and analogy would become consistent.

The other is to bring us a better understanding of logical aspects of induction and analogy by stepping into their insides. Their inferences are reduced into two clear pieces; a common denominator, preduction, and each of their residues, mathematical induction and deduction that are well investigated. It allows us to focus on an unexplored central part of thinking related to induction and analogy by removing differential and well-explored parts from them.

This paper is organized as follows. Section 2 gives a formal view about the common denominator between induction and analogy. Section 3 proposes a form of preduction. Section 4 shows its generality by applying the form to various examples over induction and analogy. Section 5 proposes its model theory and shows that the form preserves consistency. Section 6 concludes this.

## 2 Common Denominator

Figure 2 illustrates two typical processes by analogy and by induction from another point of view than their logical aspects. In analogical reasoning, an unknown property P' about a *target* case is inferred by finding, based on a property S' of the target case, a *base* case which satisfies a corresponding property S to S', and by projecting a relevant base case property P to the target case. In inductive reasoning, we see a similar inference which is made possible by a result of induction rather than induction itself. By a result of induction, an unknown attribute value $V_{k+1}$ of an example indexed by $X_{k+1}$ becomes possible to be inferred based on the known value $V_k$ of its *preceding* example indexed by $X_k$. (e.g., the value, $V_{k+1}$, of factorial of $(k+1)$, $X_{k+1}$, is computed from the preceding case, the value of factorial of $k$.) To get the unknown value $V_{k+1}$, we may have to trace back (recursively) to some example indexed by $X_0$ whose value is known (e.g., the value of factorial of 0 is 1). Let us identify this known example in induction with a *base* case in analogy, and the unknown example indexed by $X_{k+1}$ with a *target* case. Then, although the number of repetition of inference is different, we can identify these structures where information is mapped from a base case into a target case.

Now, in order to formalize a more general common structure related to both analogy and induction, we abstract these in two ways: number of repetition and parameters of their cases/examples. Let us represent *inner-relation* among the $n$-parameters inside a case/example by an $n$-ary predicate $P$ ($n \geq 1$), and their parameters by a tuple of $n$-arguments of $P$. In this representation, we do not distinguish a parameter for an index from a parameter for a value. This abstraction is needed to capture an inference about a concept which is naturally represented by a predicate without parameter for value (e.g., a predicate, "Is a number"). Let us denote *outer-relation* between the parameters in a case/example and the parameters in a preceding case/example by a $2 \cdot n$-ary predicate $.R.$. Then, letting $x$ and $y$ be $n$-tuples of variables, the following sentence absorbs the number of repetition,

$$\forall x, y (P(x) \land xRy \supset P(y)),$$

which we call the *R-expansile sentence of P*. This implies that, if there is a $P$-thing, any entities which relate to it by $R$ are also $P$-things. That is, by this sentence the property $P$ will be (recursively) projected from a case/example known to be $P$ to the unknown successive case/example.

Returning to the schemas in Table 1, let us see how this sentence relates with them. If we substitute $y = s(x)$ for $xRy$, we get

$$\forall x, y (P(x) \land y = s(x) \supset P(y)) \quad \cdots (\text{i})$$

which is equivalent to (I.1.2) by the nature of $=$, and if $x \sim y$ and $S(x) \land S(y)$ for $xRy$, we get

$$\forall x, y (P(x) \land x \sim y \supset P(y)), \quad \cdots (\text{ii})$$

$$\forall x, y (P(x) \land S(x) \land S(y) \supset P(y)) \quad \cdots (\text{iii})$$

with which (A.1.3) and (A.2.3) can be deduced from their premises, respectively. Now let us assume an inferential schema, *preduction*, which can conclude each $R$-expansile sentence of (i), (ii), and (iii) from the premises of (I.2), (A.1), and (A.2), respectively. Then by this new schema, Induction (I.2) can be broken down into two steps; *1)* from the premises (I.2.1) and (I.2.2), conclude the expansile sentence (i) by preduction, and *2)* from (I.2.1) and (i), conclude (I.2.3) by Induction (I.1). Each of Analogy (A.1) and (A.2), on the other hand, becomes a derivative rule from preduction and deduction; *1)* from the premises, conclude each $R$-expansile sentence (ii)/(iii) by preduction, and *2)* from (ii)/(iii) together with their premises, conclude (A.1.3)/(A.2.3) by deduction, respectively. Thus, preduction is a common inference of (I.2), (A.1), and (A.2) (Figure 1).

## 3 Preduction

Preduction is a formal representation of the following concept: *"If every entity known to be $P$ can be traced back to some roots of $P$ along a relation $R$, then the unknown descendants of their roots will satisfy $P$ similarly."* Here, a set of *roots* corresponds to a set which includes a base case/an example $x_0$ in the previous section.

Before describing a form of the concept, we introduce some terminology. For a transitive relation $\prec$ and a pair of objects $e$ and $e'$, when $e \prec e'$, we say $e$ is an *ancestor* of $e'$ and $e'$ is a *descendant* of $e$ with respect to $\prec$. An $n$-ary *predicate* $U$ is generally expressed by $\lambda x Q$, where $x$ is a tuple of $n$ object variables, $Q$ is a sentence in which no object variables except variables in $x$ occur free.

For a $2 \cdot n$-ary predicate $.R.$ and $.\prec_R.$, let $Tr(R; \prec_R)$ express that $\prec_R$ is the transitive closure of $R$, that is, $\prec_R$ is the minimal predicate which satisfies

$$\forall x, y (xRy \supset x \prec_R y)$$
$$\land \forall x, y, z (x \prec_R z \land z \prec_R y \supset x \prec_R y).$$

Let $P$ be an $n$-ary predicate symbol and $A$ a first order sentence. Let $.\prec.$ be a $2 \cdot n$-ary transitive predicate in which $P$ does not occur. The *preductive sentence* of $P$ on $\prec$ in $A$, written $Pd(A; P; \prec)$, is

$$\exists \Phi ( \forall x (\Phi(x) \supset P(x)) \land \exists x, y (\Phi(x) \land x \prec y \land P(y))$$
$$\land A[\lambda x (\Phi(x) \lor \exists z (\Phi(z) \land z \prec x))] )$$
$$\supset \forall x, y (P(x) \land x \prec y \supset P(y)),$$

where $A[\lambda x (\Phi(x) \lor \exists z (\Phi(z) \land z \prec x))]$ expresses the result of substituting the predicate $\lambda x (\Phi(x) \lor \exists z (\Phi(z) \land z \prec x))$ for all occurrences of $P$ in $A$.

In the left side of the implication of the preductive sentence, a predicate variable $\Phi$ represents a concept of *root* of $P$-things with respect to a relation $\prec$. The first conjunct expresses that $\Phi$ is a sub-class of $P$. The second conjunct confirms that there is an entity of $\Phi$ that is an ancestor (with respect to $\prec$) of an entity of $P$. The third conjunct expresses that $P$ can be interpreted as a set of entities who have their roots in $\Phi$. If a predicate $\Phi$ satisfies all these three conditions, $Pd(A; P; \prec)$ tells us every $\prec$-descendants of an entity satisfying $P$ satisfies $P$ similarly.

Let $P$ do not occur in $R$. Then, the *preduction of $P$ on $R$ from $A$* is the sentence,

$$A \land Tr(R; \prec_R) \land Pd(A; P; \prec_R),$$

denoted by $Preduct(A; P; R)$, where $\prec_R$ is a new predicate symbol which does not occur in $A$.

Preduction is guaranteed to maintain consistency, which will be proved in Section 5. The following properties are straightforward and used in the next section. The conclusion of a preductive sentence, $\forall x, y (P(x) \land x \prec_R y \supset P(y))$, implies the $R$-expansile sentence of $P$, $\forall x, y (P(x) \land xRy \supset P(y))$, because of $Tr(R; \prec_R)$. $Tr(R; \prec_R)$ can be expressed formally by a second order sentence which *circumscribes* $\prec_R$ alone in the above two sentences of $Tr(R; \prec_R)$ [McCarthy, 1980]. If $R$ is transitive, the transitive closure of $R$ is $R$ itself. Thus, in this case $Preduct(A; P; R)$ is simplified to $A \land \forall x, y (xRy \equiv x \prec_R y) \land Pd(A; P; R)$.

## 4 Examples: Induction and Analogy

In the following examples, we implicitly use a classical logic with equality axioms and with the *unique name axioms* [Clark, 1978] which tells us that different ground terms denote different objects (or you may always add these axioms to $A_i$ in these examples).

## 4.1 Induction

**Example 1.** Let $A_1$ and $S$ be
$$A_1 \equiv N(0) \wedge N(s(s(0)))$$
and
$$xSy \equiv y = s(x).$$
Let $\prec_S$ denote the transitive closure of $S$ by $Tr(S; \prec_S)$. Then, $Pd(A_1; N; \prec_S)$ is

$$\exists \Phi(\ \forall x(\Phi(x) \supset N(x))$$
$$\wedge \exists x, y(\Phi(x) \wedge x \prec_S y \wedge N(y))$$
$$\wedge (\Phi(0) \vee \exists z(\Phi(z) \wedge z \prec_S 0))$$
$$\wedge (\Phi(s(s(0))) \vee \exists z(\Phi(z) \wedge z \prec_S s(s(0)))))$$
$$\supset \forall x, y(N(x) \wedge x \prec_S y \supset N(y)).$$

The sentence scoped by $\exists \Phi$, if substituted $\Phi(x) \equiv x = 0$ ("$x$ is a root iff $x$ is 0"), becomes
$$N(0) \wedge \exists y(0 \prec_S y \wedge N(y)) \wedge 0 \prec_S s(s(0))$$
and follows from $A_1 \wedge Tr(S; \prec_S)$. Thus, as the left side of the implication of $Pd(A_1; N; \prec_S)$ hold, the preduction of $P$ on $S$ from $A_1$ concludes
$$\forall x, y(N(x) \wedge x \prec_S y \supset N(y)),$$
and, by $\forall x, y(y = s(x) \supset x \prec_S y)$ (in $Tr(S; \prec_S)$) and by the equality axioms,
$$\forall x(N(x) \supset N(s(x))).$$

As this example shows, the preduction from "0 and 2 are natural numbers" gives "the successor of each natural number is a natural number" and thus, $N(1), N(3), \cdots$ are also theorems of the preduction. It shows the preduction expands the extension of $N$.

The preduction is non-monotonic with respect to $A$, that is, the theorems of some preduction from $A$ does not always include the theorems of the preduction from a sub-set theory of $A$. The following shows an example where a theorem of a preduction from a sub-set $A_1$ of $A_2$ is not the theorem from $A_2$.

**Example 2.** Let $A_2 \equiv A_1 \wedge \neg N(s(0))$ and $S$ is the same as in Example 1. $Pd(A_2; N; \prec_S)$ is the result obtained just by attaching the following sentence to the left side of the implication of $Pd(A_1; N; \prec_S)$ in the scope of $\exists \Phi$;
$$\neg(\Phi(s(0)) \vee \exists z(\Phi(z) \wedge z \prec_S s(0))),$$
which, by distribution of the negation, becomes
$$\neg \Phi(s(0)) \wedge \neg \exists z(\Phi(z) \wedge z \prec_S s(0))).$$

This second conjunct telling us "any less than 1 does not satisfy $\Phi$". This contradicts $(\Phi(0) \vee \exists z(\Phi(z) \wedge z \prec_S 0))$ in the left side of the implication of $Pd(A_1; N; S)$ by the transitivity of $\prec_S$. Thus, for $\Phi$, no predicate satisfies the left side of the implication of $Pd(A_2; N; S)$, and $Pd(A_2; N; S)$ is seen to be tautology.

**Example 3.** In Inductive Logic Programming (ILP), the inference of the following form is most commonly used in generalizing a clause (*absorption* [Muggleton, 1991], *saturation* [Rouveirol and Puget, 1990], *folding* [Lu and Arima, 1996], and etc.):

Induction(I.3)
(I.3.1)    $\forall x(\alpha(x) \supset P(x))$
(I.3.2)    $\forall x(\alpha(x) \supset \beta(x))$
(I.3.3)    $\forall x(\beta(x) \supset P(x))$,

where $P$ is a predicate symbol and $\alpha, \beta$ are conjunctions of literals. Unfortunately, this rule also does not maintain consistency. Thus, to check consistency is always necessary to discard an inconsistent conclusion produced by this rule. (It corresponds to *over-generalization* in ILP where a *negative example* $e$ of $P$ is covered by the newly obtained generalized clause $H$, that is, $A \vdash \neg P(e)$ but $A, H \vdash P(e)$.) The preduction can work as a consistent version of this rule.

**Proposition 1** *Let $P$ do not appear in $\beta$. Let $xBy$ be $\beta(x) \wedge \beta(y)$. Then the following is a theorem of $Preduct(A; P; B)$:*
$$\forall x(\alpha(x) \supset P(x)) \wedge \forall x(\alpha(x) \supset \beta(x))$$
$$\wedge \exists x \alpha(x) \wedge A[\beta]$$
$$\supset \forall x(\beta(x) \supset P(x)),$$

**Proof.** $B$ is transitive. Thus, $Pd(A; P; B)$ is a theorem of $Preduct(A; P; B)$. Assuming $\forall x(\alpha(x) \supset P(x))$, $\forall x(\alpha(x) \supset \beta(x))$, and $\exists x \alpha(x)$, it is sufficient to prove that $A[\beta] \supset \forall x(\beta(x) \supset P(x))$ is a theorem of $Pd(A; P; B)$.

Consider a preductive sentence $Pd(A; P; B)$ where $\alpha$ substituted for $\Phi$. The first condition of the sentence is just the same as the first assumption $\forall x(\alpha(x) \supset P(x))$. The second condition,
$$\exists x, y(\alpha(x) \wedge \beta(x) \wedge \beta(y) \wedge \alpha(y)),$$
is equivalent to $\exists x(\alpha(x) \wedge \beta(x))$, which follows from the second and the third assumptions. The third condition of the preductive sentence becomes simply $A[\beta]$ under the second and the third assumptions, because
$$(\alpha(x) \vee \exists z(\alpha(z) \wedge \beta(z) \wedge \beta(x))) \equiv \beta(x).$$
Consequently, the conditions of the preductive sentence is equivalent to $A[\beta]$. Also, these assumptions give
$$\exists x(P(x) \wedge \beta(x)),$$
which simplifies the conclusion of the preductive sentence $\forall x, y(P(x) \wedge \beta(x) \wedge \beta(y) \supset P(y))$ to $\forall x(\beta(x) \supset P(x))$.

As Proposition 1 shows, the preduction proposes two more conditions than the two premises of Induction(I.3). The former condition $\exists x \alpha(x)$ relates to the *justification* of induction (I.3). If $\exists x \alpha(x)$ is not required like Induction(I.3), it allows the case $A \vdash \neg \exists x \alpha(x)$, where (I.3.1) and (I.3.2) hold always, because these preconditions are always false. Thus Induction(I.3) yields, no matter what $\beta$ is, $\forall x(\beta(x) \supset P(x))$. There would be no reason to justify such an inference. The latter condition keeps consistency. This will be shown in Theorem 2.

## 4.2 Analogy

A consistent version of Analogy (A.2) is also deduced from a preduction. In the next proposition, as in (A.2), $S$ corresponds to similarity and $P$ a projected property by the similarity.

**Proposition 2** *Let $P$ do not appear in $S$. Let $xCy$ be $S(x) \wedge S(y)$. Then the following is a theorem of $Preduct(A; P; C)$:*

$$\exists z(P(z) \wedge S(z)) \wedge A[S] \supset \forall x(S(x) \supset P(x)),$$

**Proof.** $C$ is transitive. Thus, $Pd(A; P; C)$ is a theorem. Assuming $P(v) \wedge S(v)$, consider the sentence obtained by substituting $\lambda x(x = v)$ for $\Phi$ in $Pd(A; P; C)$. The first condition $\forall x(x = v \supset P(x))$ of the sentence is satisfied because $P(v)$, and the second condition $\exists y(S(v) \wedge S(y) \wedge P(y))$ is satisfied because of $v$ as $y$. The third condition becomes simply $A[S]$, because $(x = v \vee vCx) \equiv S(x)$ from $S(v)$. The conclusion of the preductive sentence is simplified to $\forall x(S(x) \supset P(x))$ because $\exists x(P(x) \wedge S(x))$ from the assumption. Consequently, $Preduct(A; P; C) \wedge P(v) \wedge S(v)$ gives

$$A[S] \supset \forall x(S(x) \supset P(x)),$$

which proves the proposition.

**Example 4.** Let $A_3$ be

$$human(t) \wedge mortal(b) \wedge human(b).$$

We define $C$ by their similarity, $human(= S)$, that is,

$$xCy \equiv human(x) \wedge human(y).$$

Then, as Proposition 2 shows, $Preduct(A_3; mortal; C)$ yields

$$\exists z(mortal(z) \wedge human(z)) \wedge A_3[human]$$
$$\supset \forall x(human(x) \supset mortal(x)).$$

Substituted the target case $b$ for $z$, the left side of the implication is

$$mortal(b) \wedge human(b)$$
$$\wedge human(t) \wedge human(b) \wedge human(b)$$

which follows from $A_3$. Thus, $Preduct(A_3; mortal; C)$ yields

$$\forall x(human(x) \supset mortal(x)),$$

which tells us that every human is mortal. Therefore, the preduction of *mortal* on $C$ can derive an analogical conclusion $mortal(t)$ by the fact $human(t)$.

Attempts to understand analogical reasoning are rigorously continued in Philosophy, Cognitive Science, and Artificial Intelligence[Helman, 1988]. The schemas (A.1) and (A.2) are too simple description of the process of analogy. Davies and Russell clearly argue that there should be more premises in their schemas by posing the following example: we will not infer that one ($t$) of two cars of the same model ($S$) is painted red ($P$) just because the other ($b$) is painted red, although we may guess a price ($P'$) of the one just because the other same model car is valued at the price ($P'$)[Davies and Russell, 1987]. That is, we prefer the latter inference, ($P'(t)$ just for $P'(b) \wedge S(b) \wedge S(t)$) than the former ($P(t)$ just for $P(b) \wedge S(b) \wedge S(t)$) by the difference of the properties ($P$ and $P'$) although there is no difference in applying the schemas to the cases. The missing premise should be weaker than $\forall x(S(x) \supset P(x))$, because, if otherwise, $S(t)$ is enough for the conclusion $P(t)$ and thus information about the base case ($b$) becomes unnecessary. Davies and Russell proposed the following premise:

$$\forall p, s(\exists x(\Sigma(x, s) \wedge \Pi(x, p)) \supset \forall y(\Sigma(y, s) \supset \Pi(y, p))),$$

which allows a particular pair of $s$ (as a similarity) and $p$ (as a projected property) to work in analogical reasoning based on a base case ($x$). (e.g., $\Sigma = Model$ and $\Pi = Price$). They call this sentence *determination rule*.

Although weaker than $\forall x(S(x) \supset P(x))$, this preduction is not an answer for the query of the missing premise in analogical schema. Instead, preduction provides a consistent way to infer determination itself. Given a certain triplet, a sentence $A$, a predicate $P$ and a relation $R$, the preduction can yield the determination rule as a theorem. This would be enough by showing the fact that the determination rule is an expansile sentence.

**Example 5.** In the $R$-expansile sentence, substituting $\lambda x, p\Pi(x, p)$ for $P$, $\lambda x, p, y, p'\exists s(\Sigma(x, s) \wedge \Sigma(y, s) \wedge p = p')$, for $R$, $\{x, p\}$ for $x$ and $\{y, p'\}$ for $y$ results in

$$\forall x, y, p(\Pi(x, p) \wedge \exists s(\Sigma(x, s) \wedge \Sigma(y, s)) \supset \Pi(y, p))$$

by the nature of $=$, which is arranged to the determination rule.

## 5 Model Theory

Let $U$ and $\varphi$ be sets such that $\varphi \subseteq U$. Let $\prec$ be a transitive relation on $U$. We say a set $\psi$ is *the expansile set of $\varphi$ with respect to $\prec$* if $\psi$ is the union of $\varphi$ with the (ascending) segment by $\varphi$ with respect to $\prec$, that is,

$$\psi = \varphi \cup \{e \mid \text{ there exists } z \in \varphi, z \prec e\}.$$

For a predicate $P$ and a structure $\mathcal{M}$, we write $P^{\mathcal{M}}$ for the extension of $P$ in $\mathcal{M}$. Let $\mathcal{M}$ and $\mathcal{N}$ be structures. For a predicate $P$ and a predicate $R$ in which $P$ does not occur, we say that $\mathcal{M}$ is an *$R$-expansile structure of $\mathcal{N}$ in $P$*, if i) $\mathcal{M}$ and $\mathcal{N}$ have the same universe, ii) all other predicate symbols and function symbols besides $P$ have the same extensions in $\mathcal{M}$ and $\mathcal{N}$ (thus, $R^{\mathcal{M}} = R^{\mathcal{N}}$), iii) but, for some subset $\phi$ of $P^{\mathcal{N}}$ such that an entity $e \in \phi$ and an entity $e' \in P^{\mathcal{N}}$ satisfy $e \prec_R e'$, $P^{\mathcal{M}}$ is the expansile set of $\phi$ with respect to $\prec_R$, where $\prec_R$ is the transitive closure of $R^{\mathcal{N}}$.

Let $\mathcal{M}$ be a model of $A$. We say that $\mathcal{M}$ is a *preductive model* of $A$ in $P$ on $R$, if i) any $R$-expansile structure of $\mathcal{M}$ is not a model of $A$, or otherwise ii) $\mathcal{M}$ is an $R$-expansile structure of a model of $A$.

**Theorem 1** We write $A \models_R^P f$ if a sentence $f$ is true in all preductive models of $A$ in $P$ on $R$. Then,

$$A \models_R^P f \text{ if } Preduct(A; P; R) \vdash f$$

**Proof.** Let $\mathcal{M}$ be a preductive model of $A$ in $P$ on $R$. Let $\varphi$ be a predicate which, for the transitive closure $\prec_R$ of $R$, satisfies the left side of the implication of $Pd(A; P; \prec_R)$ when substituted for $\Phi$. Then the extension of $\varphi$ is a subset of the extension of $P$ and includes an entity which is an ancestor of an entity of $P$ with respect to $\prec_R$, and its expansile set with respect to $\prec_R$ is an alternative extension of $P$. Thus, there is a model of $A$ that is an $R$-expansile structure of $\mathcal{M}$. It implies that $\mathcal{M}$ should also be an $R$-expansile structure of a model of $A$ by the definition of the preductive model. If the right side of $Pd(A; P; R)$ were not true on $\mathcal{M}$, there would exist a pair of entities, $(e_1, e_2)$, such that they satisfies $R$ and that $e_1$ is an entity of $P$ but $e_2$ is not of $P$. This would contradict that $\mathcal{M}$ is an $R$-expansile structure.

**Theorem 2** *(consistency): A predicate $P$ does not occur in a predicate $R$. If $A$ is consistent, the preductive sentence $Preduct(A; P; R)$ is consistent.*

**Proof.** Assume that $A$ is consistent but that $Preduct(A; P; R)$ is inconsistent. $A \wedge Tr(R; \prec_R)$ is consistent because $Tr(R; \prec_R)$ just defines a new predicate $\prec_R$ to be the transitive closure of $R$. By the assumption, for any model $\mathcal{N}$ of $A \wedge Tr(R; \prec_R)$, $\mathcal{N} \models \neg Pd(A; P; \prec_R)$. Thus, because the left side of $Pd(A; P; \prec_R)$ is true on $\mathcal{N}$, there is a predictive model of $A \wedge Tr(R; \prec_R)$ that is an $R$-expansile structure of $\mathcal{N}$. Let this be $\mathcal{M}$. Now, by the assumption, the negation of the right side of $Pd(A; P; \prec_R)$ is also true on $\mathcal{M}$, that is,

$$\mathcal{M} \models \exists x, y (P(x) \wedge x \prec_R y \wedge \neg P(y)).$$

$\mathcal{M}$ agrees with $\mathcal{N}$ on all symbols (including $R$ and $\prec_R$) except $P$ and, on $P$, agrees with the expansile set of the extension of $\Phi$ on $\mathcal{N}$ with respect to $\prec_R$. Therefore, letting $\Phi^{\mathcal{N}} = \phi^{\mathcal{M}}$,

$$\mathcal{M} \models \exists x, y((\phi(x) \vee \exists z(\phi(z) \wedge z \prec_R x))$$
$$\wedge x \prec_R y \wedge (\neg \phi(y) \wedge \neg \exists z(\phi(z) \wedge z \prec_R y))).$$
$$\mathcal{M} \models \exists x, y((\phi(x) \wedge x \prec_R y \vee \exists z(\phi(z) \wedge z \prec_R x \wedge x \prec_R y))$$
$$\wedge \neg \exists z(\phi(z) \wedge z \prec_R y)).$$

The both disjuncts in the first conjunct contradict the second conjunct by the transitivity of $\prec_R$. Thus, there is no such a model $\mathcal{M}$. This contradicts the assumption.

## 6 Conclusion

This paper proposes a common form of analogy and (empirical) induction. The form, called *preduction*, preserves consistency and brings a view that analogy and induction come from the same type of inference and diverge depending on types of sequent inference, deduction and mathematical induction, respectively. The generality of this form is verified by its broad application ranging over analogy and induction in the logical approach of artificial intelligence. Although the form will not contribute directly to a design of a general inference machine, we hope that this promotes us to devise it by better understanding logical aspects of analogy and induction.

## Acknowledgments

I would like to thank Norman Foo, his research members and Takeshi Ohtani for their helpful comments.

## References

[Clark, 1978] Clark,K.L.. Negation as Failure. in *Logic and Databases*, H.Gallaire and J.Minker (Eds). Plenum Press, New York, 1978, 293–322.

[Davies and Russell, 1987] Davies,T. and Russell,S.J.. A logical approach to reasoning by analogy. in *IJCAI-87*, 1987, 264–270.

[Helft, 1988] Helft,N.. Induction As Nonmonotonic Inference. in *Proceedings of the 1st International Conference on Principles of Knowledge Representation and Reasoning*. Morgan Kaufmann, 1989, 149–156.

[Helman, 1988] Helman,D.H., editor. *Analogical Reasoning – Perspectives of Artificial Intelligence, Cognitive Science, and Philisophy*. Kluwer Academic Publishers, Dordrecht, 1988.

[Lu and Arima, 1996] Lu,J. and Arima,J.. Inductive Logic Programming Beyond logical Implication. in *Proceedings of the 7th International Workshop, ALT96*, (Arikawa S. and Sharma A.K. Eds). LNAI 1160, Springer, 1996, 185–198.

[McCarthy, 1980] McCarthy,J.. Circumscription–A Form of non-monotonic reasoning. in *Artificial Intelligence 13*. North-Holland, 1980, 27–39.

[Mostow, 1983] Mostow,J.. International Machine Learning Workshop: An Informal Report, *SIGART Newsletter 86*, acm, 1983, 24–31.

[Muggleton, 1991] Muggleton,S.. Inverting the Resolution Principle. in *Machine Intelligence 12*. Oxford University Press, 1991.

[Peirce, 1932] Peirce,C.S.. *Elements of Logic*, in: C. Hartshorne and P. Weiss (eds.), *Collected Papers of Charles Sanders Peirce*, Volume 2 Harvard University Press, Cambridge, MA, 1932.

[Rouveirol and Puget, 1990] Rouveirol,C. and Puget,J.F.. Beyond inversion of resolution. in *Proceedings of the seventh international conference on machine learning*. Morgan Kaufmann, 1990, 122–130.

[Lifschitz, 1986] Lifschitz,V.. On the Satisfiability of Circumscription. in *Artificial Intelligence 28*. 1986, 17–27.

# Analogy and Abduction in Automated Deduction

## Gilles Défourneaux and Nicolas Peltier
46 Avenue Félix Viallet
38031 Grenoble FRANCE
Tel: (+33) (0)4-76-57-46-59
Gilles.Defourneaux@imag.fr, Nicolas.Peltier@imag.fr

## Abstract

A method is presented for analogical reasoning in Automated Deduction. We focus on the abductive aspects of analogy and give a unified treatment for theorems and non-theorems. Abduction allows to deal with *partial* analogies thus strongly increasing the application field of the method. It also allows to detect "bad analogies" in several cases. Explanatory examples as well as more realistic examples quantifying the effects of using analogy (for theorem-proving *and* for counter-example building) are given.

## 1 Introduction

Analogy is in the very heart of human reasoning, in particular in Mathematics. Roughly speaking, *reasoning by analogy* consists in using informations deduced from the solving of a given problem *or set of problems* (the *source problems*) for solving a new one (the *target problem*). In Artificial Intelligence and Automated Deduction, the mechanization of this approach is a crucial issue (see for example [Bledsoe, 1977; Plaisted, 1981; Hall, 1989]). Moreover analogy is also an intrinsically interesting way of reasoning: discovering similarities between existing proofs or theorems can be of highest importance (for instance in mathematical practice and teaching). As far as we know, in all the existing works in theorem-proving, analogy is used for *finding the proof* of a given target theorem from an existing one. This view of analogy is very limited since analogy can obviously be also useful if the target or source problems are *not* theorems: in this case one can try to find a counter-example of the target conjecture by using counter-examples of the source non-theorems. In this paper we give an unifying treatment of these two cases (theorems and non-theorems). *No assumption is made about the way these proofs and counter-examples are generated* (any existing method for search for proofs or counter-examples can be used: (hyper)resolution or tableaux-based methods, connexion method, but also (finite) model builder such as FINDER [Slaney, 1993], SEM [Zhang and Zhang, 1995] or the method RAMC [Bourely *et al.*, 1994] looking *si-multaneously* for a proof or a counter-example of a given formula $\mathcal{F}$.

Other approaches have been proposed to tackle the problem of analogy by second-order means. In [Boy de la Tour and Caferra, 1987], the paradigm of "propositions as types" is used and proofs are represented as terms. Higher-order functions are then applied to transform the base proof into the target one. In [Kolbe and Walther, 1995], higher-order evaluation techniques are used to refine the problem and ultimately have its premises match with axioms of the calculus, allowing lemma speculation as a side effect. However, both approaches only deal with proofs (not counter-examples). To the best of our knowledge there is no other approaches allowing to deal with model building by analogy in first-order logic.

To make the presentation of the method shorter, we assume that the problems are specified in clausal form and that we use a refutational approach. Therefore, formulae are sets of clauses, proofs are *refutations* and counter-examples are *models*.

## Analogy and abductive reasoning

According to Peirce [Hartshorne *et al.*, ], *analogy* can be seen as an *induction* and an *abduction* followed by a *deduction*. Our approach to analogy follows directly these steps. It can be summarized as follows. We assume a knowledge base $\mathcal{K}$ containing theorems with their proofs and non-theorems associated with counter-examples.

**Generalization step.** The first step consists in a *generalization* occurring at the presentation of a new source formula $\mathcal{F}_S$ to $\mathcal{K}$. $\mathcal{F}_S$ is transformed into a *more general formula* and stored into the knowledge base. This corresponds to the *inductive* part of analogy.

**Matching step.** The second step applies when a new target conjecture $\mathcal{F}_T$ is considered. It consists in trying to find one (or more) "analogical" formulae in $\mathcal{K}$. Then the proof or the counter-example of the conjecture $\mathcal{F}_T$ is built *from* the proofs/counter-examples of the formulae in $\mathcal{K}$. This corresponds to the *deductive* part of analogy.

However in most cases $\mathcal{F}_T$ cannot be directly (dis)proved by using only the information in $\mathcal{K}$ (this is obviously possible only if $\mathcal{F}_T$ is an instance of a problem in $\mathcal{K}$). However, even if a proof or a counter-example of $\mathcal{F}_T$ cannot be straightforwardly deduced from the known formulae, the use of the informations stored in $\mathcal{K}$ will very likely provide interesting hints for finding a proof or a counter-example of $\mathcal{F}_T$. For example it can guide lemmata generation that could be completed later by using any existing theorem prover or model builder. This corresponds to the *abductive* part of analogy: finding hypotheses that allows to prove *or to disprove* the target conjecture.

The *generalization step* has been precisely defined in [Bourely et al., 1996], where an algorithm is given to transform any formula $\mathcal{F}$ into a "more general formula" $\mathcal{F}_{gen}$. The *matching* step is defined in [Défourneaux and Peltier, 1997]. *In this paper we focus mainly on the abductive part of analogical reasoning, that is to say on the generation of lemmata*. We propose a *partial* matching algorithm especially devoted to finding such lemmata, i.e. abduction will be incorporated to step 2.

Abductive reasoning is the process of generating the *explanations* of a given fact (see for example [Polya, 1973; Pople, 1973; Console et al., 1991; Hobbs et al., 1993]). It has deep connections with other forms of reasoning such as model building [Inoue et al., 1993; Console et al., 1991]. Aristotle calls *abduction* a syllogism "in which the major is sure and the minor only probable" (see [Lalande, 1980], page 1). More recently, it is defined by Peirce [Peirce, 1955] as the process of finding the *minor premise* from a *major premise* and the *conclusion*: for example "infer" $A$ from $A \Rightarrow B$ and $B$. Peirce clearly points out the importance of abductive reasoning in science (he gave as a paradigm the discovery of Kepler's laws). From a deductive point of view, this inference is clearly *not* sound. However it provides interesting informations for proving $B$, since $A$ can be considered as a *lemma*, whose proof immediately yields a proof of $B$.

The aim of this paper is to show how to use analogy for generating such lemmata. Since our approach deals with simultaneous search for proofs and counter-examples, the notion of *lemma* is *much more general* than the standard one: it can be either a conjecture that (if it is true) is sufficient for proving $B$ or a *partial counter-example* that must be completed and extended for finding a counter-example of $B$.

## 2 Basic notions

We assume the reader is familiar with the usual terminology of First-Order Logic and Automated Deduction. We briefly review most of the basic notions used throughout this work.

Let $\Sigma$ be a set of functional symbols, $\Omega$ be a set of predicate symbols and $\mathcal{X}$ be an (countable) infinite set of variables. Let $a$ be a function mapping each symbol in $\Sigma \cup \Omega$ to a natural number (the *arity* of the symbol). Function symbols of arity 0 are called *constants* (denoted by $a$, $b$, ...). The set of *terms* $\tau(\Sigma, \mathcal{X})$ is defined as usual over the alphabet $\Sigma, \mathcal{V}$ (where $\mathcal{V} \subseteq \mathcal{X}$). If $\mathcal{V}$ is empty, $\tau(\Sigma, \mathcal{X})$ is denoted by $\tau(\Sigma)$. An *atom* is of the form $P(t_1, \ldots, t_n)$, where $P \in \Omega$, $arity(P) = n$ and $\forall i \in [1..n].t_i \in \tau(\Sigma, \mathcal{X})$. A *literal* is either an atom or the negation of an atom. If $p$ is a literal, $\neg p$ denotes the literal with the same predicate symbol and the same arguments than $p$ but with different sign. A *clause* is a finite set (or disjunction) of literals. First-order formulae are built as usual over atoms by using the logical symbols $\vee, \wedge, \neg, \exists, \forall \ldots$. By $\mathcal{V}ar(E)$ we denote the set of (free) variables occurring in the expression (term, clause, atom, literal...) $E$. A term or a clause containing no variables is called *ground*. The notion of *substitution* is defined as usual. The result of applying a substitution $\sigma$ to a term $t$ is noted $t\sigma$. The domain of a substitution $\sigma$ is the set of variables $x$ such that $x\sigma \not\equiv x$ (noted $\mathcal{D}om(\sigma)$). If for all variables $x \in \mathcal{D}om(\sigma)$, $x\sigma$ is ground then $\sigma$ is called *ground*.

For any set of clauses $S$ we denote by $\mathcal{S}(S)$ the set of ground instances of clauses in $S$.

As is well known, a ground clause $C'$ subsumes a ground clause $C$ (noted $C' \leq_s C$) iff $C' \subseteq C$. If $C, C'$ are two clauses, $C \leq_s C'$ iff for all $D' \in \mathcal{S}(C')$, $\exists D \in \mathcal{S}(C)/D \leq_s D'$.

### 2.1 Higher order formulae

Step 2 of the method (see Introduction) needs an algorithm transforming a given set of clauses $S$ into a more general one. The latter is represented by second order terms and clauses. Higher-order terms and formulae are built as usual from a signature $\mathcal{V}, \Sigma, \Omega$ by using *application* and *abstraction* (see [Huet, 1975]).

The set of types $\mathcal{S}_k$ of order $k$ is defined as follows. If $k = 1$, $\mathcal{S}_k = \{T\}$ (base type). If $\forall i \in [1..n].t_i \in \mathcal{S}_{k_i}$ and $s \in \mathcal{S}_k$, and $k = 1 + \max(k_1, \ldots, k_i)$, then $t_1 \times \ldots \times t_n \to s \in \mathcal{S}_k$.

Let $\mathcal{S} = \bigcup_{i=1}^{\infty} \mathcal{S}_i$. Let $(\mathcal{V})_{t \in \mathcal{S}}$ a set of disjoint infinite (countable) sets of variables (called variables of type $t$). Let $\mathcal{V} = \bigcup_{t \in \mathcal{T}} \mathcal{V}_t$, let $(\Sigma)_{t \in \mathcal{S}}$ a collection of disjoint set of constants of type $t$. We assume that $\mathcal{V} \cap \Sigma = \emptyset$.

The sets of terms $\tau_t(\Sigma, \mathcal{V})$ of type $t$ are built as usual from $\mathcal{V}, \Sigma$ using *application* or *abstraction*. We denote by $\tau(\Sigma, \mathcal{V}) = \bigcup_{t \in \mathcal{T}} \tau_t(\Sigma, \mathcal{V})$, the whole set of terms. The definition of formulae, equational formulae, substitutions, etc. can be extended straightforwardly to higher-order term. We do not give all the definitions (see for example [Lugiez, 1995]).

Roughly speaking sets of clauses will be represented by sets of *generalized clauses* i.e. clauses where the function symbols are replaced by *variables* of order 2.

A *generalized formula* is a formula such that each bound variable is of order 1, and each free variable of $\mathcal{F}$ is of order 1 or 2 and occurs in a term of the form $X(\bar{t})$. A *generalized clause* is of the form: $\forall \bar{x}.C$ where $C$ is a generalized formula of the form $\bigvee_{i=1}^{n} P_i$ (where $P_i$ are generalized literals and $\bar{x}$ a $n$-uple, possibly empty, of variables of order 1. A *generalized sequence* is a finite sequence of generalized formulae. The set of *instances* of $S$ is the set of sequences of clauses $S\sigma$ such that $\sigma$ is a ground substitution of $\mathcal{V}ar(S)$.

## 3 An order among formulae

Before presenting our method we must clarify the notion of generalization and give a precise definition of it. A definition of this notion was given in [Bourely et al., 1996]. We give here a new refined definition of the generalization order that takes into account the *semantic* aspect of the clauses (i.e. the set of ground clauses denoted by the set) rather than its syntax.

The underlying ideas of this ordering are the following. Informally, a given theorem $S$ will be said to be "more general" than a theorem $S'$ iff the hypotheses of $S$ are *weaker* than the one of $S'$ or if the conclusion of $S$ is *stronger* than the ones of $S'$. Indeed, it is obvious that $S$ provides more information than $S'$. If sets of clauses are considered, the set of hypotheses is the set of clauses $S$ itself, and the conclusion is $\Box$ (the empty clause). Hence $S$ will be *more general* than $S'$ iff each clause in $S$ belong to $S'$, i.e. if $S \subseteq S'$. However, what is important here is *not* the sets of clauses $S$ and $S'$ by themselves but rather the set of ground clauses they denote. Therefore, we must take into account the two following possibilities:

1. The set of clauses $S'$ does not explicitly contain $S$ but *any ground clause* denoted by $S$ belongs to the set of ground clauses denoted by $S'$, i.e. $\mathcal{S}(S) \subseteq \mathcal{S}(S')$. In this case $S$ can be said to be "more general" than $S'$.

2. $C \in \mathcal{S}(S)$ and $C \notin \mathcal{S}(S')$ but $C$ is *subsumed* by a clause $C'$ in $\mathcal{S}(S')$.

In order to capture these two cases we introduce the following relation $\sqsubseteq$ between sets of clauses.

**Definition 1** *let $S$ and $S'$ be sets of clauses. We note $S \sqsubseteq S'$ iff for all $C \in \mathcal{S}(S)$ there exists $C' \in \mathcal{S}(S')$ such that $C' \leq_s C$.*

**Lemma 1** *If $S \sqsubseteq S'$ then $S' \models S$.*

**Proof 1** *Assume that $S \sqsubseteq S'$. Let $\mathcal{I}$ be a model of $S'$. By definition if $C \in \mathcal{S}(S')$ then $\mathcal{I} \models C$. Let $D \in S$. We have $S \sqsubseteq S'$ hence by definition there exists $C \in S'$ such that $C \subseteq D$. $\mathcal{I} \models C$ (since $C \in S'$) hence $\mathcal{I} \models D$.*

An **unsatisfiable** set of clauses $S$ is said to be *more general* than $S'$ iff $S \sqsubseteq S'$. A **satisfiable** set of clauses $S$ will be said to be *more general* than $S'$ iff the hypotheses of $S$ are stronger than the one of $S'$ (this implies that $S$ is false in more interpretations than $S'$). Consequently a satisfiable set of clauses $S$ will be said to be *more general* than a set $S'$ iff $S' \sqsubseteq S$.

Please note that these two notions of generalization are *not* equivalent. The following definition formalizes this idea.

**Definition 2** *Let $T_1$ and $T_2$ be two sequences of sets of generalized clauses. $T_1$ is said to be more general than $T_2$ (noted $T_1 \succeq T_2$) iff for all substitution $\sigma_2$ of $Var(T_2)$, there exists a substitution $\sigma_1$ of $Var(T_1)$ such that*

- $T_1$ *is unsatisfiable and* $T_1\sigma_1 \sqsubseteq T_2\sigma_2$
- $T_1$ *is satisfiable and* $T_2\sigma_2 \sqsubseteq T_1\sigma_1$.

**Proposition 1** $\succeq$ *is a pre-order.*

**Example 1** *Let $t_1 : \forall x.P(x) \wedge \neg P(a)$, $t_2 : P(a) \wedge \neg P(a)$. $t_1$ and $t_2$ are two unsatisfiable sets. The set of ground clauses denoted by $t_1$ and $t_2$ are respectively $\{P(t)/t \in \tau(\Sigma, \mathcal{V})\} \cup \{\neg P(a)\}$ and $\{P(a), \neg P(a)\}$. We have $t_2 \sqsubseteq t_1$, hence $t_2$ is more general than $t_1$.*

**Example 2** *Let $t_1 : \forall x.P(f(x)) \wedge \neg P(a)$ and $t_2 : P(f(a)) \wedge \neg P(a)$. $t_1$ and $t_2$ are satisfiable. We have $t_2 \sqsubseteq t_1$, hence $t_2$ is less general than $t_1$. $t_1$ is indeed falsified by more interpretations than $t_2$ (any model of $t_1$ is a model of $t_2$).*

## 4 Matching

We call "matching" the process of finding in the knowledge base the formulae analogous to a given target formula. It is inductively defined as follows.

**Definition 3** *A matching problem is of the form $t = s$ (where $t$ and $s$ are terms or formulae), $S \sqsubseteq S'$ (where $S$ and $S'$ are sets of clauses), $C \leq_s C'$ (where $C, C'$ are clauses), $\mathcal{F} \vee \mathcal{G}$, $\mathcal{F} \wedge \mathcal{G}$, $\forall x.\mathcal{F}$, $\exists x.\mathcal{F}$, where $\mathcal{F}, \mathcal{G}$ are matching problems.*

In order to give a semantics to matching problems we only have to choose the semantics of atomic formulae.

**Definition 4** *Let $\mathcal{F}$ be a matching problem and let $\mathcal{I}$ be an interpretation. A substitution $\sigma$ is a semantic solution of $\mathcal{F}$ w.r.t. $\mathcal{I}$ iff:*

- $\mathcal{F}$ *is of the form $t = s$ (where $t, s$ are terms) and $\mathcal{I} \models t\sigma = s\sigma$.*
- $\mathcal{F}$ *is of the form $\phi = \psi$ (where $\phi, \psi$ are formulae) and $\mathcal{I} \models \phi\sigma \Leftrightarrow \psi\sigma$.*
- $\mathcal{F}$ *is of the form $S_1 \sqsubseteq S_2$ and there exists a set of clauses $S_3$ such that $S_3 \sqsubseteq S_2\sigma$ and $\mathcal{I} \models (S_1\sigma \Leftrightarrow S_3)$.*
- $\mathcal{F}$ *is of the form $C \leq_s D$ and there exists a clause $C'$ such that $C' \leq_s D\sigma$ and $\mathcal{I} \models (C\sigma \Leftrightarrow C')$.*

*The set of semantic solutions of $\mathcal{P}$ w.r.t. $\mathcal{I}$ is noted $\mathcal{S}_\mathcal{I}(\mathcal{P})$.*

Notice that the semantics of matching problem takes into account the *semantics* of the sets of clauses rather than their syntax. This is very important since analogy must focus on the semantic information contained in a theorem rather than in its statement.

Let $S_S, S_T$ be two sets of clauses. Finding a substitution $\sigma$ such that $S_S\sigma \sqsubseteq S_T$ (unsatisfiable case) or $S_T \sqsubseteq S_S\sigma$ (satisfiable case) is equivalent to finding a solution of the matching problem $S_S \sqsubseteq S_T$ (unsatisfiable case) or $S_T \sqsubseteq S_S$ (satisfiable case). Both cases are noted $\mathcal{P}rob(S_S, S_T)$ in the following.

Obviously finding the solutions of a matching problem is undecidable hence we cannot hope to get a general solution to this problem. In the present paper we only give a set of rules allowing to find the solutions of some matching problem. We *do not specify* here the strategy guiding the application of the rules (several different strategies can be proposed). These rules are *sound*: any solution of the obtained problem is a solution of the initial one. However they are obviously not complete.

## Matching problems transformation rules
### Clausal transformation rules
$$\begin{array}{rcl} P \vee Q & \to & Q \vee P \\ \forall x, y.P & \to & \forall y, x.P \\ P & \to & \neg \neg P \\ P & \to & P \vee \bot \\ P & \to & P \vee P \end{array}$$

### Higher-order unification rules
The following rules are simply the standard higher-order unification rules (see [Huet, 1975]).
$$\begin{array}{rcll} t = t & \to & \top & \\ f(\bar{t}) = f(\bar{s}) & \to & \bar{t} = \bar{s} & \\ f(\bar{t}) = g(\bar{s}) & \to & \bot & \text{if } f \neq g \\ x = t & \to & \bot & \text{if } x \in t \end{array}$$

### Imitation
$$\{X(t_1, \ldots, t_n) = f(s_1, \ldots, s_m)\}$$
$$\to \{X = \lambda x_1, \ldots, x_m f((X_1 x_1 \ldots x_n), \ldots, (X_m x_1 \ldots x_n))\}$$
$$\cup \bigcup_{i=1}^{m} \{X_i(t_1 \ldots t_n) = s_i\}$$
where $X_i$ ($1 \leq i \leq n$) are new free variables.

$$\{X(t_1, \ldots, t_n) = x\}$$
$$\to \{X = \lambda x_1, \ldots, x_n.x\}$$

### Projection
$$\{X(t_1, \ldots, t_n) = f(s_1, \ldots, s_m)\}$$
$$\to \{X = \lambda x_1, \ldots, x_n.x_i, t_i = f(s_1, \ldots, s_m)\}$$

### Replacement
$x = t \wedge \mathcal{P} \to x = t \wedge \mathcal{P}\{x \to t\}$ if $x \notin t$

### ∨-Elimination rule
$P \vee Q \to P$ If $P, Q$ are two matching problems.

### Matching rules
The following rules are new. They allow to eliminate the symbols $\sqsubseteq$ and $\leq_s$ from matching problems.

#### $\sqsubseteq$-Elimination rules
$$\begin{array}{lrcl} \sqsubseteq\text{-}E_1 & S_1 \cup S_2 \sqsubseteq S & \to & S_1 \sqsubseteq S \wedge S_2 \sqsubseteq S \\ \sqsubseteq\text{-}E_2 & \emptyset \sqsubseteq S & \to & \top \\ \sqsubseteq\text{-}E_3 & S \sqsubseteq \emptyset & \to & S = \emptyset \\ \sqsubseteq\text{-}E_4 & \{\forall \bar{x}.C\} \sqsubseteq S & \to & \forall \bar{x}.(\{C\} \sqsubseteq S) \\ \sqsubseteq\text{-}E_5 & \{C\} \sqsubseteq S_1 \cup S_2 & \to & \{C\} \sqsubseteq S_1 \vee \{C\} \sqsubseteq S_2 \\ \sqsubseteq\text{-}E_6 & \{C\} \sqsubseteq \{D\} & \to & D \leq_s C \\ \sqsubseteq\text{-}E_7 & \{A \vee B\} \sqsubseteq S & \to & \{A\} \sqsubseteq S \vee \{B\} \sqsubseteq S \end{array}$$

#### $\leq_s$-Elimination rules
$$\begin{array}{lrcl} \leq_s\text{-}E_1 & C \leq_s \forall y.D & \to & \forall y.(C \leq_s D) \\ \leq_s\text{-}E_2 & \forall x.C \leq_s D & \to & \exists x.(C \leq_s D) \\ \leq_s\text{-}E_4 & C \leq_s D_1 \vee D_2 & \to & (C \leq_s C_1) \vee (C \leq_s C_2) \\ \leq_s\text{-}E_5 & \bot \leq_s C & \to & \top \\ \leq_s\text{-}E_6 & P \leq_s Q & \to & P = Q \\ \leq_s\text{-}E_7 & C_1 \vee C_2 \leq_s D & \to & C_1 \leq_s D \wedge C_2 \leq_s D \end{array}$$

Next example illustrates the application of these transformation rules.

**Example 3** Let the source theorem be the set of clauses $S_S = \{c_1, c_2, c_3\}$ with $c_1 \equiv \forall x.P(x) \vee Q(x)$, $c_2 \equiv \neg P(A)$, $c_3 \equiv \neg Q(A)$ and the target theorem be the set of clauses $S_T = \{c_4, c_5\}$, where $c_4 \equiv \forall y.l(f(y)) \vee l(y)$ and $c_5 \equiv \forall y.\neg l(y)$. We solve the matching problem $\mathcal{P}: S_S \sqsubseteq S_T$ ? *Remark:* $P, Q, A$ denote variables (of order $2, 2, 1$ respectively) and $l, f$ denote constant symbols (of order $2, 1$).

$\mathcal{P}$
$$\begin{array}{ll} \to_{\sqsubseteq\text{-}E_1} & \{c_1\} \sqsubseteq S_T \wedge \{c_2\} \sqsubseteq S_T \wedge c_3 \sqsubseteq S_T \\ \to_{\sqsubseteq\text{-}E_5} & \{c_1\} \sqsubseteq \{c_4\} \wedge \{c_2\} \sqsubseteq \{c_5\} \wedge c_3 \sqsubseteq \{c_5\} \\ \to_{\sqsubseteq\text{-}E_7} & c_4 \leq_s c_1 \wedge c_5 \leq_s c_2 \wedge c_5 \leq_s c_3 \\ \to_{\leq_s\text{-}E_1} & \forall x.(c_4 \leq_s P(x) \vee Q(x)) \\ & \wedge c_5 \leq_s c_2 \wedge c_5 \leq_s c_3 \\ \to_{\leq_s\text{-}E_2} & \forall x.\exists y.l(f(y)) \vee l(y) \leq_s (P(x) \vee Q(x)) \\ & \wedge \exists y.\neg l(y) \leq_s \neg P(a) \wedge \exists y.\neg l(y) \leq_s \neg Q(a) \\ \to_{\leq_s\text{-}E_6} & \forall x.\exists y.(P(x) \vee Q(x) = l(f(y)) \vee l(y)) \\ & \wedge \exists y.P(A) = l(y) \wedge \exists y.Q(A) = l(y) \\ \to_{unification} & P = \lambda y.l(f(y)) \wedge Q = \lambda y.l(y) \end{array}$$

We obtain a solution of the initial problem:
$$P \to \lambda y.l(f(y)), Q \to \lambda y.l(y)$$

**Remark:** *of course, other solutions could be obtained.*

## 5 Lemma generation

In this section we identify a class of matching problems from which the solutions can be obtained automatically and we show how the proof (resp. the counter-example) of the target theorem can be automatically built from the one of the source (non-)theorem. In most cases, analogy will not directly give a proof (counter-example) of the target formula from the source formula. Our method can cope with this case by generating *lemmata* that have to be proved in order to complete the proof/counter-example of the target theorem. This is done by performing a *partial resolution* of matching problems. We do not solve the problem completely, but only *partially*. Formulae that cannot be matched correspond to additional hypotheses that must be proven in order to complete the proof or the counter-example. The derivation can then be completed in two different ways. Either by proving the lemma using a theorem-prover or by a *recursive call* of the matching method (unsatisfiable case). Or by showing that the lemmata are *compatible* with the target set, i. e. that the adding of those lemmata to the set preserves satisfiability of the set (satisfiable case).

A matching problem is in normal form iff it is of the form $\mathcal{P}: \bigwedge_{i=1}^{n} x_i = t_i \wedge \mathcal{F} = \bot$, where, for all $i \leq n$, $x_i$ occurs once in $\mathcal{P}$ and $\mathcal{F}$ is a first-order formula. The formula $\mathcal{F}$ corresponds to a formula that *cannot* be matched. We define new rules in order to generate problems in normal form.

### Lemma generation rules
$$\begin{array}{rcl} \forall y.(C = \bot) & \to & (\forall y.C) = \bot \\ \exists y.(C = \bot) & \to & (\exists y.C) = \bot \\ C = \bot \wedge D = \bot & \to & C \vee D = \bot \\ C = \bot \vee D = \bot & \to & C \wedge D = \bot \\ C = \top \vee D = \top & \to & C \vee D = \top \\ C = \top \wedge D = \top & \to & C \wedge D = \top \\ \mathcal{F} = \bot & \to & \neg \mathcal{F} = \top \\ S = \emptyset & \to & S = \top \end{array}$$

We note **matching** the system composed by the **unification and matching rules**, the usual **transformation rules for equational problems** (see for example [Comon and Lescanne, 1989; Lugiez, 1995]) and the **lemma generation** rules.

**Theorem 1** *(Soundness) Let* $\mathcal{P} \twoheadrightarrow_{matching} \mathcal{P}'$. *For all interpretation* $\mathcal{I}$, $\mathcal{S}_{\mathcal{I}}(\mathcal{P}') \subseteq \mathcal{S}_{\mathcal{I}}(\mathcal{P})$.

**Theorem 2** *Let $S_S$ be a generalized sequence and $S_T$ a set of clauses. Assume that $Prob(S_S, S_T) \twoheadrightarrow_{matching} \mathcal{P} \wedge \mathcal{F} = \perp$. Let $\sigma$ be a solution of $\mathcal{P}$.*

1. *If $S_S\sigma$ is unsatisfiable and if for every interpretation $\mathcal{I}$, $\mathcal{I} \models S_T \Rightarrow \mathcal{I} \models \neg \mathcal{F}\sigma$, then $S_T$ is unsatisfiable. On the other hand, there exists an algorithm that computes a refutation of $S_T$ from a refutation of $S_S\sigma$ and $\mathcal{F}\sigma \wedge S_T$.*

2. *If $S_S\sigma$ is satisfiable and if there exists a model $\mathcal{I}$ of $S_S\sigma$ such that $\mathcal{I} \models \neg \mathcal{F}\sigma$, then $S_T$ is satisfiable and $\mathcal{I} \models S_T$.*

In both cases, satisfiability or unsatisfiability of $S_T$ is deduced straightforwardly from the one of $S_S$. To show that $\mathcal{F}\sigma$ is false in all models of $S_T$ amount to prove that $\mathcal{F}\sigma \wedge S_T$ is unsatisfiable. We can therefore use whatever theorem prover to prove that. The proof can also be obtained with a *recursive call* of the procedure of analogy-proving. If the source formula is satisfiable, one must (according to theorem 2) find a model $\mathcal{I}$ of $S_S$ falsifying $\mathcal{F}\sigma$. In this aim, we try to find an extension of the (partial) model of $S_S\sigma$ falsifying $\mathcal{F}\sigma$. The difference with the unsatisfiable case is that the considered interpretation $\mathcal{I}$ is unknown a priori before the beginning of the search. In both cases, the derivation leading to the empty clause (resp. to a model) from the target set of clauses can be rebuilt straightforwardly from the source derivation and the proofs (or counter-example building) corresponding to $\mathcal{F}$.

**Example 4** *Consider the formula $S_S$ and $S_T$ of Example 3. Once we have found the solution $\sigma$ of the problem $S_S \sqsubseteq S_T$, it suffices according to theorem 2 to instantiate the proof of $S_S$ in order to obtain a proof of $S_T$.*

*Now assume that we replace clause $c_4$ in $S_T$ by: $c_6 \equiv \forall y. l(f(y)) \vee l(y) \vee l(g(y))$. In this case the literal $l(g(y))$ cannot be matched by the algorithm. If we apply the matching problem transformation rules, we get the problem in normal form:*

$$P = \lambda y. l(f(y)) \wedge Q = \lambda y. l(y) \wedge \exists y. l(g(y)) = \perp$$

*In order to rebuild the proof of $S_T$ we only have to prove the lemma: $\exists y. \neg l(g(y))$, i.e. to prove that $S_T \cup \{l(g(y))\}$ is not satisfiable. Any theorem prover can do that.*

## 6 Examples

In this section we give two more realistic examples.

**Example 5** *(Unsatisfiable set of clauses) Let us consider the problems* SYN310-1 *and* SYN312-1 *of* TPTP *[Suttner and Sutcliffe, 1996].*

Problem $S_S$ is:
$$\begin{aligned} &\neg p(x2, x1, x) \vee p(x, x1, x2) \\ &\neg p(x1, x, x2) \vee p(x, x1, x2) \\ &\neg p(x, x1, g(x2)) \vee p(x, x1, x2) \\ &\neg p(f(x), x1, x2) \vee p(x, x1, x2) \\ &\neg p(a, b, c) \\ &p(f(g(a)), f(g(b)), f(g(c))) \end{aligned}$$

$S_T$ is:
$$\begin{aligned} &p'(x, x3, x2) \vee \neg p'(x, x1, x2) \vee \neg p'(x1, x3, x2) \\ &p'(x2, x1, x) \vee \neg p'(x, x1, x2) \\ &p'(x1, x, x2) \vee \neg p'(x, x1, x2) \\ &p'(x, x1, f'(x2)) \vee \neg p'(x, x1, x2) \\ &p'(g'(x), x1, x2) \vee \neg p'(x, x1, x2) \\ &p'(a', f'(b'), c') \\ &p'(f'(b'), d, c') \\ &\neg p'(g'(f'(a')), g'(f'(d)), g'(f'(c'))) \end{aligned}$$

$S_S$ is very easy (it can be proven in 0.17 seconds by OTTER 3.0 [McCune, 1995] with "automatic" mode). $S_T$ is more difficult (OTTER takes 536.31 s to prove it in automatic mode). We assume that a proof of $S_S$ is known and we try to build automatically a proof of $S_T$ by analogy with the one of $S_S$.

We apply our algorithm in order to solve the problem $S_S \sqsubseteq S_T$? No general solution can be found. Indeed $S_T$ is not a direct instance of $S_S$. However we get the following partial solution: $\{p = \lambda x_1, x_2, x_3. \neg p'(x_3, x_2, x_1), g = \lambda x. f'(x), f = \lambda x. g'(x), a = a', b = b', c = c', d = d'\}$ with the lemmata:

$$\{p'(a, b, c), \forall x_1, x_2, x_3. p'(x_1, x_2, x_3) \vee \neg p'(x_1, x_3, x_1)\}$$

These two formulae are clauses belonging to $S_S$ that cannot be matched by $S_T$.

According to Theorem 2 it only remains to prove the lemmata. The first one can be proven (very easily) by adding the negation of $p'(a, b, c)$ to $S_T$ (OTTER takes 0.37 s to prove it).

In order to transform the second lemma into a set of clauses we need to use skolemization. We get $\{\neg p'(a_1, a_2, a_3), p'(a_1, a_3, a_2)\}$. This set is added to $S_T$. The corresponding set can again be proven very easily (OTTER takes 0.18 s).

This example suggests that reasoning by analogy can enhance significantly the power of theorem-provers. Indeed the total amount of time required to prove the lemmata is 0.55 s and the time required to prove the whole theorem without using analogy is 536.31 s. Please note that the proofs of the lemmata does not require any user-interaction. However the matching process is not yet implemented, hence must be done by hand.

**Example 6** *(Satisfiable set of clauses) Let $S_S$ be the following satisfiable set of clauses (adapted from [Church, 1940]).*

$$\begin{aligned} \neg f(x) \vee \neg p(x) \qquad & p(x) \vee f(x) \\ \neg f(y(x)) \vee \neg p(x) \qquad & p(x) \vee f(y(x)) \end{aligned}$$

The finite model builder FMC developed in our inference laboratory builds the following model (in 0.01 s).

```
Model:
f(0)= true    f(1)= false   p(0)= false
p(1)= true    y(0)=1        y(1)=0
```

Now, let $S_T$ be the following set of clauses $\{\neg f'(x) \vee \neg(f'(y'(x))), f'(x) \vee f'(y'(x)))\}$. Here, if we replace $f', y'$ by $f, y$, we have obviously $S_S \models S_T$. Hence we try to build a model of $S_T$ by analogy with the one of $S_S$. We first solve the matching problem: $\mathcal{P}rob(S_S, S_T) = S_T \sqsubseteq S_S$. We obtain the following solution: $y = \lambda x.y'(x), f = \lambda x.f'(x)$, with the lemma: $\forall x.p(x) \vee \neg p(x)$. According to theorem 2 it only remains to check if the lemma holds in the model of $S_S$. This is obviously the case since $\forall x.p(x) \vee \neg p(x)$ is a tautology. Therefore the following interpretation is a model of $S_T$: $\{f'(0) = true, f'(1) = false, y'(0) = 1, y'(1) = 0\}$.

# 7 Discussion and perspectives

We have presented a calculus for the discovery and handling of analogy between sets of clauses. It is able in particular to handle *partial* analogy between statements by generating *lemmata*. Our work is the first step toward a system for simultaneous search for refutations and models using analogy and abductive reasoning. Main lines of future research are:

- Define strategies or heuristics for our calculus *matching* and study their properties (termination, efficiency, completeness for some classes of problems etc.).
- Extend this approach to first-order formulae (not in clausal form). In particular we have to give a new definition of the generalization order.

A comparative analysis with the lemmata produced by other techniques such as [Kolbe and Walther, 1995] will be performed in the future. We are currently working on the implementation of the calculus and on the definition of the knowledge base.

# Acknowledgements

We thank Ricardo Caferra and the anonymous referees for their pertinent and precise comments.

# References

[Bledsoe, 1977] W. W. Bledsoe. Non-resolution theorem proving. *Artificial Intelligence*, 9:1–35, 1977.

[Bourely et al., 1994] Ch. Bourely, R. Caferra, and N. Peltier. A method for building models automatically. Experiments with an extension of Otter. In *Proceedings of CADE-12*, pages 72–86. Springer, 1994. LNAI 814.

[Bourely et al., 1996] C. Bourely, G. Défourneaux, and N. Peltier. Building proofs or counterexamples by analogy in a resolution framework. In *Proceedings of JELIA 96, LNIA 1126*, pages 34–49. Springer, LNAI, 1996.

[Boy de la Tour and Caferra, 1987] Th. Boy de la Tour and R. Caferra. Proof analogy in interactive theorem proving: A method to express and use it via second order pattern matching. In *Proceedings of AAAI 87*, pages 95–99. Morgan Kaufmann, 1987.

[Church, 1940] A. Church. A formulation of the simple theory of types. *Journal of Symbolic Logic*, 5(1):56–68, 1940.

[Comon and Lescanne, 1989] H. Comon and P. Lescanne. Equational problems and disunification. *Journal of Symbolic Computation*, 7:371–475, 1989.

[Console et al., 1991] L. Console, D. Theseider Dupre, and P. Torasso. On the Relationship between Abduction and Deduction. *Journal of Logic and Computation*, 1(5):661–690, 1991.

[Défourneaux and Peltier, 1997] G. Défourneaux and N. Peltier. Partial matching for analogy discovery in proofs and counter-examples. In Springer, editor, *Proceedings of CADE 14*, 1997.

[Hall, 1989] R.P. Hall. Computational approaches to analogical reasoning: A comparative analysis. *Artificial Intelligence*, pages 39–120, 1989.

[Hartshorne et al., ] Hartshorne, Weiss, and Burks. *Collected Papers of C.S. Peirce (1930-1958)*. Harward U. Press.

[Hobbs et al., 1993] J. Hobbs, M. Stickel, D. Appelt, and P. Martin. Interpretation as abduction. *Artificial Intelligence*, 63:69–142, 1993.

[Huet, 1975] G. Huet. A unification algorithm for typed $\lambda$-calculus. *Theorical Computer Science*, 1:27–57, 1975.

[Inoue et al., 1993] K. Inoue, Y. Ohta, R. Hasegawa, and M. Nakashima. Bottom-up abduction by model generation. In *Proc. IJCAI-93*, volume 1, pages 102–108, Morgan Kaufmann, 1993.

[Kolbe and Walther, 1995] Th. Kolbe and Ch. Walther. Second-order matching modulo evaluation - A technique for reusing proofs. In Chris S. Mellish, editor, *Proceedings of IJCAI 95*, pages 190–195. IJCAI, Morgan Kaufmann, 1995.

[Lalande, 1980] A. Lalande. *Vocabulaire technique et Critique de la Philosophie*. Presses Universitaires de France, 1980.

[Lugiez, 1995] D. Lugiez. Positive and negative results for higher-order disunification. *Journal of Symbolic Computation*, 1995.

[McCune, 1995] W. McCune. *Otter 3.0 Reference Manual and Guide*. Argonne National Laboratory, August 1995. Revision A.

[Peirce, 1955] C.S. Peirce. *Philosophical Writings of PEIRCE*, chapter Abduction and induction, pages 150–156. Dover Books, 1955.

[Plaisted, 1981] D.A. Plaisted. Theorem proving with abstraction. *Artificial Intelligence*, 16:47–108, 1981.

[Polya, 1973] G. Polya. *How to Solve It, a New Aspect of Mathematical Method*. Princeton University Press. Second Edition, 1973.

[Pople, 1973] H. Pople. On the Mechanization of Abductive Logic. In *Proc. of the IJCAI'73*, pages 147–152, 1973.

[Slaney, 1993] J. Slaney. SCOTT: a model-guided theorem prover. In *Proceedings IJCAI-93*, volume 1, pages 109–114. Morgan Kaufmann, 1993.

[Suttner and Sutcliffe, 1996] Ch. Suttner and G. Sutcliffe. The TPTP problem library. Technical report, TU München / James Cook University, 1996. V-1.2.1.

[Zhang and Zhang, 1995] J. Zhang and H. Zhang. SEM: a system for enumerating models. In *Proc. IJCAI-95*, volume 1, pages 298–303. Morgan Kaufmann, 1995.

# CASE-BASED REASONING

# CASE-BASED REASONING

Case-Based Reasoning 1

# How Similar is VERY YOUNG to 43 Years of Age?
## On the Representation and Comparison of Polymorphic Properties

Werner Dubitzky[‡§], Alfons Schuster[§], John G. Hughes[§], David A. Bell[§], Kenneth Adamson[§]

University of Ulster, Jordanstown
Northern Ireland Bio-Engineering Centre[‡]
Faculty of Informatics[§]
Co. Antrim BT37 OQB Northern Ireland

## Abstract

Intelligent computer systems rely on more or less complex computational entities that *represent* occurrences and events in the real world. Usually, such entities are formed from representational primitives called properties, attributes, features, etc. To reflect varying degrees of uncertainty, originating from human judgement and the intrinsic nature of the world, such property values occur as more or less vague linguistic symbols or exact numeric expressions. Determining similarity between two properties is usually done on either the symbolic or the numeric level. This seems to be too restrictive for case-based reasoning and similar approaches as these often face mixed specifications. In this paper we propose a flexible and systematic scheme for representing crisp properties and two types of fuzzy properties. It also provides a consistent mechanism to establish similarity scores for the various instance combinations.

## 1. Introduction

The main thread of this paper is that of *conceptual similarity* (or distance). Two quotes by I. Kant and W.V. Quine succinctly convey the essence and significance of this idea. 'A concept without a percept is empty; a percept without a concept is blind.', and 'There is nothing more basic to thought and language than our sense of similarity; our sorting of things into kinds.'

Many knowledge and information representation models employ the notion of a *property* by some means or other. For example, object-oriented database systems (*attribute* or *atomic object*), multiple criteria decision making (*criterion*), relational databases (*field* or *attribute*), statistics (*statistical variable*), frame-based systems (*slot*), case-based reasoning (*feature*), concept theory (*property*), and so on. Properties are usually viewed as representational primitives that provide a 'vehicle' to form more complex and more abstract structures like *classes, complex objects, alternatives, relations, records, tables, frames, cases, concepts, concept exemplars,* and so forth. It is these higher level entities that are used in computer systems to represent and capture the essence of real-world occurrences and events. Frequently, the underlying processes of knowledge-based and other systems require the (conceptual) comparison of such computational units. Comparison methods of this kind establish similarity scores based on the similarity of such atomic properties [Chen and Hwang, 1992].

Intelligent computer systems involve application-geared people at both the knowledge acquisition stage as well as at the problem-solving and decision-making stage. The reasoning quality of such systems hinges upon the adequacy and appropriateness with which experts and users define the characteristics of properties and specify property instances. Inevitably, because of human involvement and the intrinsic nature of real-world circumstances, uncertainty is manifest in the definitions and specifications of many properties [Zadeh, 1973; Klir et al., 1988].

This work proposes the concept of a *polymorphic property*. It enables the specifier of instances of the *same* property type to choose from three value representations according to prevailing uncertainty and his or her level of expertise. Further, to enable proximity-based reasoning models to take advantage of this representation device, a consistent method is presented that allows the computation of similarity between the various value combinations. This method has been developed with four objectives in mind: it should be intuitive, intellectually satisfying, easy to use, and computationally efficient.

The approach has been applied within a case-based reasoning (CBR) framework in the Coronary Heart Disease risk assessment domain. The major results of this study were that 1) Unassisted subjects (users) are able to specify their own case data in the presence of varying degrees of uncertainty. 2) the system can consistently process the typical CBR inference steps on cases whose property values were provided in crisp numeric and vague linguistic values. And 3) A benchmark test with 10,000 cases shows that the method is highly efficient.

It is claimed that the method's mechanisms are applicable to a wider range of systems and applications, especially in practical knowledge engineering environments.

The paper is organised as follows: Section 2 briefly discusses some related approaches and illustrates the need for more flexible uncertainty handling. Section 2.1 introduces and defines the notion of polymorphic property values. Sections 3.1 to 3.3 develop the various comparison methods. Section 4 briefly mentions the CBR application in which this approach has been tested, and a performance experiment that has been carried out. Section 5 ends with conclusions.

## 2. Multiple-Format Property Values

In CBR systems uncertainty is pervasive in, amongst other things, the features used to describe the cases [Dutta *et al.*, 1991]. In such and other systems uncertainty may originate from various sources [Zadeh, 1973; Chen and Hwang, 1992; Baldwin *et al.*, 1995]:

- **unquantifiable information** (judgement of individual): e.g., a patient's *age* can be easily determined while the *criticality* of his or her condition is not readily quantifiable,
- **incomplete information** (inexact measurement): e.g., a patient's *blood pressure* may be measured as 'about 145 mmHg', but not as 'exactly 145 mmHg',
- **non-obtainable information** (data too expensive to obtain): sometimes crisp data is in principle obtainable, but the cost is too high, e.g., a patient's *family history of high cholesterol*. It may, however, be possible to get a useful approximation of that data,
- **partial ignorance** (partially known facts about a phenomenon): some facts may only be partially known, e.g., the (average) *number of cigarettes per day* a patient smoked over the last 20 years.

Various data/knowledge models handle some of these forms of uncertainty in some way or another. For example, multiple attribute decision making systems [Chen and Hwang, 1992] (fuzzy sets), relational data models [Buckles *et al.*, 1982; Petry, 1996] (fuzzy sets, rough sets), object-oriented information models [George *et al.*, 1993] (fuzzy sets), and [Baldwin *et al.*, 1995] (fuzzy sets, Evidence Theory). And in CBR [Tirri *et al.*, 1996] (Bayesian), [Dubitzky *et al.*, 1996a/b] (fuzzy sets, Evidence Theory), and [Dutta *et al.*, 1991] (fuzzy sets).

Common to these approaches is that a particular property may only take values of a single type. Usually these values are expressed by *linguistic symbols* (associated with a fuzzy set or a similarity matrix), *or* [exclusive or!] *crisp numbers*. This restriction, however, does not reflect the practices in many real-world problem-solving and decision-making environments, where data are provided (mainly by people) in whatever format is appropriate or available at the time. This may result in values of a particular property type occurring in both fuzzy and crisp formats.

Consider, for example, a printer product support help desk scenario where customers are asked how many times a certain failure had occurred. Some may have actually counted the breakdowns, others may only be able to give an approximate estimate. Or imagine a healthcare system asking advice-seeking users to state their blood cholesterol concentrations. Some users may have recently had their cholesterol checked, and remember exact readings (or indeed a value like *low* reported to them by their medic). Others may only be able to recall, vaguely or accurately, measurements given to them in the past. Yet another group may simply come up with a more or less informed guess which is likely to be expressed linguistically rather than numerically.

To facilitate properties that consistently represent and process (similarity assessment) such polymorphic instance values, the notion of a property *concept frame* is introduced.

### 2.1 Property Concept Frame

In order to equip a property (type) with the capability to have instances of different form, the property (type), and its instances, are associated with a *concept frame*. The concept frame provides the representation platform to model the relationships between the various value formats, thus enabling the computation of cross-format similarity scores.

To be more precise, a concept frame serves as a unifying representation formalism for three property value formats: *real number*, *linguistic term*, and *fuzzy predicate*. These provide a means to express property instances at various levels of certainty and expertise. The concept frame of the property *cholesterol* (abbreviated by *chol*) depicted in Figure 1 should help to illustrate this idea.

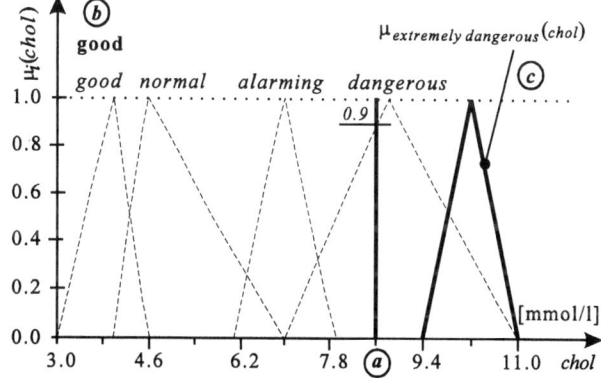

Figure 1: Property concept frame and three actual property values.

Basically, a concept frame consists of a finite *universe of discourse* (in Figure 1, the blood cholesterol concentration range 3.0...11.0 mmol/l), and a set of pre-defined concepts. These concepts are represented via *linguistic symbols* and the corresponding membership functions which essentially define *fuzzy sets*. Membership functions are referred to as *v-functions*; *v* stands for value.

So the concepts in the *cholesterol* concept frame example (see diagram) are defined by the linguistic terms *good*, *normal*, *alarming*, and *dangerous*, and the v-functions (dashed triangles in Figure 1) $\mu_{good}(chol)$, $\mu_{normal}(chol)$, $\mu_{alarming}(chol)$, and $\mu_{dangerous}(chol)$.

By means of the linguistically expressed concepts, whose semantics are reflected via the location and shape of the corresponding v-functions, a domain expert effectively conceptualises the universe of discourse according to his or her experience. Thus, the universe of discourse together with the linguistic terms and v-functions form the conceptual frame of discernment, or concept frame, within which *all* instances of the associated property are interpreted.

Figure 1 also demonstrates three examples of actual cholesterol *values* (objects depicted in bold style) indicated by *a*, *b*, and *c*. The value *a* is given in real number format, $a := \langle 8.6 \rangle$ [mmol/l], *b* in linguistic term format, $b := \langle good \rangle$, and *c* in fuzzy predicate format, $c := \langle extremely\ dangerous, \mu_{extremely\ dangerous}(chol) \rangle$.

The real number format is employed where quantifiable data is readily available. However, as it will be seen later, real numbers are reflected upon the concepts (v-functions) mapped out on a concept frame. For example, the crisp value $a := \langle 8.6 \rangle$ maps onto the *dangerous* concept as follows: $\mu_{dangerous}(a) = 0.90$. Note, on the conceptual plane the value *a* is perceived as $\langle dangerous, 0.90 \rangle$. Real number format values must be drawn from the concept frame's universe of discourse.

To specify a property value linguistically, the specifier of the value may chose any of the linguistic terms associated with the v-functions of a particular concept frame. For example, $\langle dangerous \rangle$. At first glance this may appear rather restrictive. However, the concepts defined on a concept frame are carefully established according to an expert's decision-making and problem-solving experience with the property and application in question. This ensures some degree of consistency, especially when inexperienced users specify values. An additional format (see below) provides more freedom and flexibility in cases where crisp data is not available and the set of pre-defined linguistic symbols does not seem to include an appropriate symbol.

Finally, the fuzzy predicate format. It provides a flexible mechanism to freely specify a value by introducing and defining a *new* concept and its semantics. He or she does so by stating a linguistic expression, naming the actual concept, and defining the corresponding membership function which characterises the meaning of the concept. An example for a fuzzy predicate value is given by the value *c* depicted in Figure 1. The intention is to equip the more experienced user (including the domain experts themselves) with a flexible and powerful representation 'vehicle' to express values more appropriately in accordance with the prevailing uncertainty.

**Definition 1** A concept frame associated with a polymorphic property defines a *universe of discourse U* and a *concept domain C*. *C* is composed of linguistic symbol/v-function pairs as follows:

$$C \underset{def}{=} \{(A_1, \mu_{A_1}(x)), (A_2, \mu_{A_2}(x)), \ldots, (A_n, \mu_{A_n}(x))\}$$

where $A_i$ denotes a linguistic label, and each $\mu_{A_i}(x)$ is a membership function or v-function defining a fuzzy set over $U$, such that $\mu_{A_i}(x) : U \to [0,1]$, for $x \in U \subseteq \mathbf{R}$, and $1 \leq n \leq 9$. (note, all v-functions must define *normalised* and *convex* fuzzy sets)

**Definition 2** The *value* formats by means of which instances of a polymorphic property may be specified comprise the real number (*RN*), the linguistic term (*LT*), and the fuzzy predicate format (*FP*). These are succinctly defined as follows:

$$RN \underset{def}{=} \langle x \rangle,\ LT \underset{def}{=} \langle A \rangle,\ FP \underset{def}{=} \langle B, \mu_B(x) \rangle$$

where *A* denotes a linguistic symbol that must be drawn from the terms defining the concept domain, i.e., $A \in \{A_1, A_2, \ldots, A_n\}$, *B* denotes a linguistic term naming a concept whose semantics is defined by the membership function $\mu_B(x)$, such that $\mu_B(x) : U \to [0,1]$, and $x \in U$, $B \notin \{A_1, A_2, \ldots, A_n\}$.

## 3. Comparing Polymorphic Property Values

Many decision-making and problem-solving models require the (conceptual) comparison of more or less complex entities involved in the process [Chen and Hwang, 1992; Dubitzky et al., 1996b]. Eventually, such algorithms come down to establishing similarity or distance between the constituent properties used to describe the entities in question. So providing a system or knowledge engineer with a powerful representation mechanism, such as the polymorphic property discussed above, is not enough. One must also put at his or her disposal a scheme that allows the systematic comparison of instances of such properties.

Because of symmetry, i.e., $sim(x,y) = sim(y,x)$, a total of six possible value format combinations need to be considered, namely (*LT,LT*), (*FP,FP*), (*LT,FP*), (*RN,RN*), (*RN,LT*), and (*RN,FP*). (where *RN* = real number, *LT* = linguistic term, and *FP* = fuzzy predicate format) These can be grouped into 1) fuzzy/fuzzy format comparison: (*LT,LT*), (*FP,FP*), and (*LT,FP*), 2) crisp/crisp: (*RN,RN*), and 3) crisp/fuzzy: (*RN,LT*) and (*RN,FP*). As the method for each group is in principle the same, only one combination per group has to be investigated.

### 3.1 Comparing Fuzzy Properties

Here, a fuzzily formatted property value is represented via a linguistic term and a associated membership function (essentially defining a fuzzy set). This is the case for both the linguistic term format as well as the fuzzy predicate format. Since fuzzy sets, or for that matter, fuzzy numbers represent many possible real numbers (with different membership degrees) they do not always yield a totally ordered set. This makes comparison a non-trivial affair. Various approaches have been proposed in the literature [Chen and Hwang, 1992]. Because of its simplicity, intuitive appeal, and effectiveness, Chen and Hwang's *crisp score method* [Chen and Hwang, 1992, 465-486] is adopted here to defuzzify fuzzy sets.

Given the universe of discourse $U$, a fuzzy set $A$ is characterised by a membership function $\mu_A(x)$, such that $\mu_A(x) : U \to [0,1]$, where $x \in U$. Then, depending on shape/location of $\mu_A(x)$ in $U$, the crisp score method derives a crisp score $s(A)$ from $A$, such that $s(A) \in [0,1]$. And in case of fuzzy numbers $U \subseteq \mathbf{R}$.

Now, based on crisp scores, the idea of a *basic concept distance* is defined. It will play a part in the computation of concept distances for most of the possible value combinations.

**Definition 3** Let $A$ and $B$ denote linguistic terms naming the corresponding concepts (linguistic term, fuzzy predicate values) on a concept frame, then the *basic concept distance* $\delta(A,B)$ between $A$ and $B$ is defined by the crisp scores $s(A)$ and $s(B)$ as follows:

$$\delta(A,B) = |s(A) - s(B)| \quad (1)$$

such that $A, B \in \{A_1, A_2, ..., A_n\}$ for linguistic term values, and $A, B \in$ {newly introduced labels} for fuzzy predicate values.

For example, the basic concept distance between the linguistic *cholesterol* values *good* and *alarming* yields $\delta(good, alarming) = |s(good) - s(alarming)| = 0.56$.

As the (general) concept distance $d(A,B)$ between two fuzzy value format values (linguistic term, fuzzy predicate) is identical to the corresponding basic concept distance $\delta(A,B)$, i.e., $d(A,B) = \delta(A,B)$, it is not defined separately.

## 3.2 Comparing Real Number Properties

The real world can be thought of as consisting of objects and events which are characterised by continuous numeric values. People, on the other hand, represent and process their knowledge by means of symbols. In order to make it possible to compare real number format values with any of the other three formats they are 'mapped' onto the concept arrangement (essentially *expressed* via symbols) of the corresponding concept frame. The vehicle to relate the symbolic and numeric levels is provided by the fuzzy sets (v-functions) used to represent the meaning of the linguistically described concepts. This means, for example, that the value $a := \langle 8.6 \rangle$ becomes $\langle dangerous/0.90 \rangle$.

Essentially, the computation of a *conceptual* distance measure $d(a,b)$ between two real number values $a$ and $b$ is determined by the basic concept distances of the concepts on the concept frame that are 'affected' by $a$ and $b$, *and* the degree to which these concepts are reflected.

First, the situation is considered where the two values $a$ and $b$ map onto a single concept $A$. The initial observation is that, independent of $a$ and $b$, the basic concept distance $\delta(A,A)$ is zero; see Definition 3. Clearly then, with respect to concept $A$, the conceptual difference between $a$ and $b$ is measured on how $a$ and $b$ differ with respect to their conformance with concept $A$. This, of course, is computed by $|\mu_A(a) - \mu_A(b)|$. A crucial consequence of this is that, conceptually, two crisp numeric values $a$ and $b$ may be identical, i.e., concept distance $d(a,b) = 0$, if $\mu_A(a) = \mu_A(b)$ holds, in spite of $a \neq b$.

Second, what happens if the value $a$ maps onto a concept $A$, and $b$ onto a *different* concept $B$ (i.e., $a \neq b$ and $A \neq B$)? Here, according to Definition 3, the basic concept distance $\delta(A,B)$ is always greater than zero. Somehow, one would think, the concept distance $d(a,b)$ between $a$ and $b$ should depend on $\delta(A,B)$. Now let $a$ and $b$ take on values such that $\mu_A(a) = 1$ and $\mu_B(b) = 1$, indicating full compliance of $a$ and $b$ with the concepts $A$ and $B$ respectively. In this special case, the concept distance $d(a,b)$ should be equal to the basic concept distance $\delta(A,B)$, that is, $d(a,b) = \delta(A,B)$. This state of affairs is brought about by treating $\langle A/1.00 \rangle$ as $\langle A \rangle$, and $\langle B/1.00 \rangle$ as $\langle B \rangle$. To put it another way: $(\mu_A(x_1) = 1) \to A$, and $(\mu_B(x_2) = 1) \to B$. This effectively increases the degree of uncertainty in the system, for the sake of being able to compare crisp numeric with vague linguistic values.

Preserving these two principles, the comparison method is based on two functions which are derived from the v-functions describing the concepts on a concept frame. These functions are $\delta^+(x)$ (solid graph in Figure 2b) and $\delta^-(x)$ (dashed graph in Figure 2b).

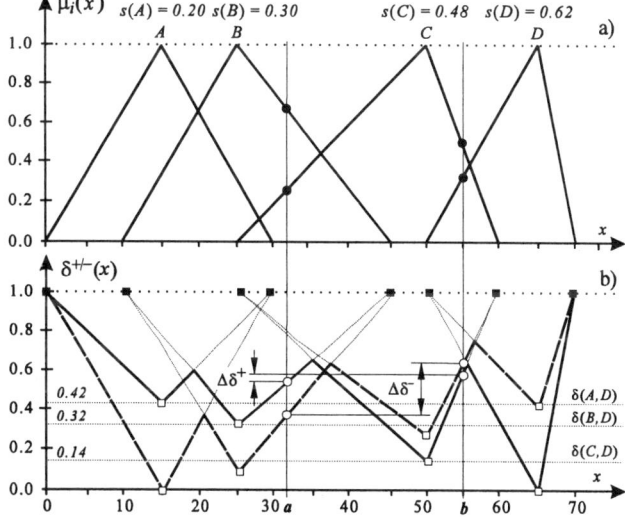

Figure 2: Concept frame and $\delta^+(x)$ and $\delta^-(x)$.

Function $\delta^+(x)$ describes the *directed conceptual distance* measured from the concept with the lowest crisp score ($A$) to the one with the highest ($D$), and $\delta^-(x)$ from that with the highest crisp score ($D$) to the one with the lowest ($A$). The functions are constructed from the (directed) basic concept distances $\delta(A,D)$, $\delta(B,D)$, $\delta(C,D)$, $\delta(D,D)$, for $\delta^+(x)$, and $\delta(D,A)$, $\delta(C,A)$, $\delta(B,A)$, $\delta(A,A)$, for $\delta^-(x)$ (white squared dots). For example,

$\delta(A,D) = 0.42$, $\delta(B,D) = 0.32$, $\delta(C,D) = 0.14$, $\delta(D,D) = 0$. The concept distance between two values $a$ and $b$ is then computed via the *differentials* $\Delta\delta^+(a,b)$ and $\Delta\delta^-(a,b)$ as: $d(a,b) = (\Delta\delta^+(a,b) + \Delta\delta^-(a,b))/2$, where $\Delta\delta^+(a,b) = |\delta^+(a) - \delta^+(b)|$, and $\Delta\delta^-(a,b) = |\delta^-(a) - \delta^-(b)|$.

**Definition 4** Let $a$ and $b$ denote real number format values in $U$, then the concept distance $d(a,b)$ is determined by the directed conceptual distances $\delta^+(x)$ and $\delta^-(x)$ as follows (see also Definition 5):

$$d(a,b) = \tfrac{1}{2}[\,|\delta^+(a) - \delta^+(b)| + |\delta^-(a) - \delta^-(b)|\,] \quad (2)$$

With the example values $a = 32$ and $b = 55$, depicted in Figure 2, the concept distance works out to $d(32,55) = 0.12$.

Figure 3 below illustrates the behaviour of the concept distance between two crisp numeric values. There, crisp *age* values from 0 to 100 are compared with the value 65 years. To demonstrate the relationship with the *age* concept frame, the v-functions (dashed graphs) and their corresponding linguistic terms are indicated in the diagram.

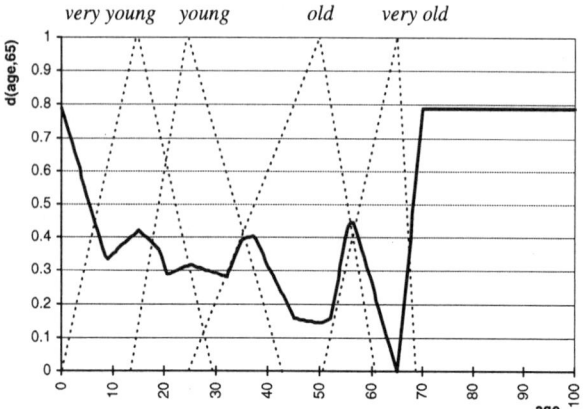

Figure 3: Concept distance between *age* and 65 years.

**Definition 5** The functions $\delta^+(x)$ and $\delta^-(x)$ are defined by the functions $\rho_{A_1}(x), \rho_{A_2}(x), \ldots, \rho_{A_i}(x), \ldots, \rho_{A_n}(x)$, and $\lambda_{A_1}(x), \lambda_{A_2}(x), \ldots, \lambda_{A_i}(x), \ldots, \lambda_{A_n}(x)$ as follows:

$$\delta^+(x) = min[\,\rho_{A_1}(x), \rho_{A_2}(x), \ldots, \rho_{A_i}(x), \ldots, \rho_{A_n}(x)\,] \quad (3a)$$

$$\delta^-(x) = min[\,\lambda_{A_1}(x), \lambda_{A_2}(x), \ldots, \lambda_{A_i}(x), \ldots, \lambda_{A_n}(x)\,] \quad (3b)$$

where each $\rho_{A_i}(x)$ and $\lambda_{A_i}(x)$ relates to and is derived from the corresponding v-function $\mu_{A_i}(x)$ on the concept frame ($A_1$ is the concept with the lowest crisp score, and $A_n$ the highest). In Figure 2b, $\rho_{A_i}(x)$ and $\lambda_{A_i}(x)$ are indicated by the lines connecting a white squared dot with two black squared dots.

The functions $\rho_{A_i}(x)$ and $\lambda_{A_i}(x)$ are obtained from the corresponding v-functions $\mu_{A_i}(x)$ by the linear mappings $r$ and $l$ as follows:

$$r : \mu_{A_i}(x) \to \rho_{A_i}(x) \quad (4a)$$

$$l : \mu_{A_i}(x) \to \lambda_{A_i}(x) \quad (4b)$$

such that $(x_L, \mu_{A_i}(x_L)=0) \to (x_L, \rho_{A_i}(x_L)=1)$, $(\hat{x}, \mu_{A_i}(\hat{x})=1) \to (\hat{x}, \rho_{A_i}(\hat{x})=\delta(A_i,A_n))$, $(x_R, \mu_{A_i}(x_R)=0) \to (x_R, \rho_{A_i}(x_R)=1)$, and $(x_L, \mu_{A_i}(x_L)=0) \to (x_L, \lambda_{A_i}(x_L)=1)$, $(\hat{x}, \mu_{A_i}(\hat{x})=1) \to (\hat{x}, \lambda_{A_i}(\hat{x})=\delta(A_i,A_1))$, $(x_R, \mu_{A_i}(x_R)=0) \to (x_R, \lambda_{A_i}(x_R)=1)$, for $x_L, x_R, \hat{x} \in U$. Where $x_L$ and $x_R$ refer to the left and right zero point and $\hat{x}$ to the height point of $\mu_{A_i}(x)$.

For example, from $\mu_C(x)$, third v-function of the concept frame shown in Figure 2a, the functions $\rho_C(x)$ and $\lambda_C(x)$ are obtained as:

$$\rho_C(x) = \begin{cases} \frac{1-\delta(C,D)}{(x_L=25)-(\hat{x}=50)}(x-x_L)+1, & \text{for } x_L \le x \le \hat{x} \\ \frac{1-\delta(C,D)}{(x_R=60)-(\hat{x}=50)}(x-x_R)+1, & \text{for } \hat{x} \le x \le x_R \end{cases}$$

$$\lambda_C(x) = \begin{cases} \frac{1-\delta(C,A)}{(x_L=25)-(\hat{x}=50)}(x-x_L)+1, & \text{for } x_L \le x \le \hat{x} \\ \frac{1-\delta(C,A)}{(x_R=60)-(\hat{x}=50)}(x-x_R)+1, & \text{for } \hat{x} \le x \le x_R \end{cases}$$

### 3.3 Comparing Crisp with Fuzzy Properties

Essentially, fuzzy predicate and linguistic term format values are represented, albeit not expressed, in the same way, namely via a fuzzy set (v-function, 'free' membership function). It is therefore sufficient to focus on the comparison between the crisp value format and linguistic term format values.

A crucial property of such a comparison must be that it never yields zero as a result. For $d(very\ young, 43) = 0$ would imply identity and therefore suggest that *very young* = 43! This, however, would assert certainty where there is in fact no certainty, and would therefore render the scheme rather flawed.

The graph in Figure 4 illustrates the behaviour of the method by comparing the crisp *age* values 0 to 100 years with the vaguely specified value *very young*. It is interesting to note that the concept distances $d(age, very\ young)$ are minimal around the *very young* region, but never reach zero! By way of digression, the white dot in the diagram indicates the answer to the question posed in the title of the paper, since $similarity(43, very\ young) = 1 - d(43, very\ young) \approx 0.77$.

Figure 4: Concept distance between crisp *ages* and *very young*.

Defining the concept distance for this fuzzy-against-crisp case, three requirements are considered: 1) there should be no identity such that d(fuzzy,crisp) = 0.0, 2) the general tendency should be preserved; this means that the distance should decrease, if non-monotonously, for values closer to the *support* of the fuzzy set representing the fuzzy value, and 3) the method, obviously, should be sensitive to the shape/location of the membership function associated with the vague value.

Criterion 1) is achieved by deriving the distance form the sum of the (directed) basic concept distances $\delta(A_i,A_n)$ and $\delta(A_i,A_1)$ of the involved fuzzy value $A_i$. (see also (4a) and (4b) ) Without formal proof it is evident that $\delta(A_i,A_n) + \delta(A_i,A_1) > 0$ always holds (see Figure 2b).

Criterion 2) is inherently met by the definition of $\delta(A_i,A_n)$ and $\delta(A_i,A_1)$ which already reflect tendency or directedness.

And to satisfy criterion 3), the crisp score $s(A_i)$—which captures the membership function's shape and location on the concept frame—of the fuzzy value $A_i$ is explicitly used.

**Definition 6** Let $A_i$ be a fuzzy property value (linguistic term or fuzzy predicate), and $x$ a real number value. Then the concept distance $d(x,A_i)$ between $A_i$ and $x$ is computed by (see also (2), (3a), and (3b) ):

$$d(x, A_i) = \frac{1}{2}\left[\left|\delta^+(x) - \frac{s(A_i)+\delta(A_i,A_n)}{2}\right| + \left|\delta^-(x) - \frac{s(A_i)+\delta(A_i,A_1)}{2}\right|\right] \quad (5)$$

where $s(A_i)$ denotes $A_i$'s crisp score, $\delta(A_i,A_n)$ and $\delta(A_i,A_1)$ denote the (directed) basic concept distances with respect to $A_i$, and $\delta^+(x)$ and $\delta^-(x)$ denote the directed conceptual distances, such that $x \in U$, for example:

$$d(43, very\ young) = \frac{1}{2}\left[\left|0.38 - \frac{0.20+0.42}{2}\right| + \left|0.48 - \frac{0.20+0}{2}\right|\right] = 0.23$$

## 4. A CBR Application and Some Benchmarks

The method has been tested in an experimental CBR system to give initial advice to subjects on their coronary heart disease (CHD) risk. The main point of this study was to have the subjects use the system and provide CHD-relevant data *without* first consulting a medic to establish various parameters such as *blood pressure*, *cholesterol*, or *anxiety*. To establish some idea about the method's performance 10,000 cases described by 21 polymorphic properties (7 concept frames with 5 v-functions, 7 with 7, and 7 with 9) have been seeded randomly with values of the three formats. On 100 different value distributions over all cases, the average time to compare all 10,000 cases was 11.74 seconds (minimum 7.22 sec, maximum 16.63 sec). The test configuration was a 90 MHz Pentium PC, 32 MB RAM on a Windows NT4 platform.

## 5. Conclusions

The polymorphic property approach presented in this paper led to encouraging results. It seems pertinent to applications (e.g., tele-medicine or help desk systems) involving a great deal of data that is provided in crisp as well as in fuzzy format. Further, performance tests indicate that the method does not incur exceptional overheads. Finally, the method is expected to have some appeal to practitioners as it is easy to use and understand.

A valid criticism of the method might be that distance scores never reach 1 (but a maximum $m < 1$) indicating total dissimilarity. Also, non-technical users, if they want to make use of the fuzzy predicate format, require to obtain some knowledge on the basics of fuzzy sets.

## References

[Baldwin et al., 1995] J.F. Baldwin, T.P. Martin, B.W. Pilsworth, *Fril—Fuzzy and Evidential Reasoning in Artificial Intelligence*, Research Studies Pr., 1995.

[Buckles et al., 1982] B.P. Buckles, F.E. Petry, "A Fuzzy Model for Relational Databases", in *Fuzzy Sets and Systems*, 7:213-226, 1982.

[Chen and Hwang, 1992] S-J. Chen, C-L. Hwang, *Fuzzy Multiple Attribute Decision Making, Methods and Applications*, Springer Verlag, 1992.

[Dubitzky et al., 1996a] W. Dubitzky, J.G. Hughes, D.A. Bell, "Multiple Opinion Case-Based Reasoning and the Theory of Evidence", in *6th Int. Conference on Information Processing and Management of Uncertainty in Knowledge-Based Systems*, 447-452, 1996.

[Dubitzky et al., 1996b] W. Dubitzky, J.G. Hughes, D.A. Bell, "Case Memory and the Behaviouristic Model of Concepts", in *Advances in Case-Based Reasoning, 3rd European Workshop*, EWCBR-96, 120-134, 1996.

[Dutta et al., 1991] S. Dutta, P.P. Bonissone, "Integrating Case Based and Rule Based Reasoning: the Possibilistic Connection", in *Uncertainty in Artificial Intelligence 6*, 281-290, N.-Holland, 1991.

[George et al., 1993] R. George, B.P. Buckles, F.E. Petry, "Modelling Class Hierarchies in the Fuzzy Object-Oriented Data Model", in *Fuzzy Sets and Systems*, 60(3):259-272, 1993.

[Klir et al., 1988] G.J. Klir, T.A. Folger, *Fuzzy Sets, Uncertainty and Information*, Prentice Hall, Englewood Cliffs, NJ, 1988

[Petry, 1996] F.E. Petry, *Fuzzy Databases: Principles and Applications*, Kluver Academic Pub., MA, 1996.

[Tirri et al., 1996] H. Tirri, P. Kontkanen, P. Myllymäki, "A Bayesian Framework for Case-Based Reasoning", in *Advances in Case-Based Reasoning, 3rd European Workshop*, EWCBR-96, 113-127, 1996.

[Zadeh, 1973] L.A. Zadeh, "Outline of a New Approach to the Analysis of Complex Systems and Decision Processes", in *IEEE Transactions on Systems, Man, and Cybernetics*, 3(1), 28-44, 1973.

# The Competence of Sub-Optimal Theories of Structure Mapping on Hard Analogies

**Tony Veale,**
School of Computer Applications,
Dublin City University,
Dublin, Ireland.

**Mark Keane,**
Dept. of Computer Science,
University of Dublin, Trinity College,
Dublin, Ireland.

## Abstract

Structure-mapping is a provably NP-Hard problem which is argued to lie at the core of the human metaphoric and analogical reasoning faculties. This NP-Hardness has meant that early attempts at optimal solutions to the problem have had to be augmented with sub-optimal heuristics to ensure tractable performance. This paper considers various grounds for qualifying the competence of such heuristic approaches, and offers an evaluation of the sub-optimal performance of three different models of analogy, SME, ACME and Sapper.

## 1. Introduction

Metaphor interpretation and Analogical reasoning are two, closely related, cognitive faculties which rely upon structure mapping to generate coherent and systematic correspondences between two domains of discourse. But since structure-mapping is clearly a *graph-isomorphism process* which must consider a combinatorial number of such correspondences to generate an optimal mapping, it is both intuitively and provably an NP-hard problem.

A variety of computational approaches to the problem have been described in the AI literature, such as the *Structure Mapping Engine* (SME) of [Falkenhainer *et al.* 1989], the *Analogical Constraint Mapping Engine* (ACME) of [Holyoak and Thagard 1989] and the *Sapper* model of [Veale *et al* 1996a,b]. The first of these, SME, provided an optimal (and thus potentially exponential) solution to the problem, followed by a heuristic, sub-optimal greedy-merge approach (see [Oblinger and Forbus 1990]) and later, an incremental approach (see [Forbus *et al.* 1994]). Falkenhainer *et al.* [1989] provide a complexity analysis of SME that identifies several factors leading to factorial explosion, but argued that analogies producing such difficulties were unlikely to occur. At its heart the original SME is a forest-matching mechanism, which extends known results regarding the $O(N^2)$ complexity of determining sub-tree isomorphism (e.g., see [Garey & Johnson 1979], [Akutsu, 1992]) to forests of inter-tangled tree representations. Layered on top of this forest matcher is a factorial merge process which combines the results of the polynomial sub-tree matching phase (called partial maps, or *pmaps*) into larger, global mappings (*gmaps*). This merge process is clearly $O(2^N)$ where N is the number of pmaps (isomorphic sub-tree matches) involved. SME's designers state that flat representations (i.e., non-nested) that cause N to be large will cause SME to be overly factorial, but that analogies leading to this situation would be rare or incoherent *'jumbles of unconnected expressions'* (p28, 1989).

However, this is shown not to be the case. Veale *et al.* [1996b] demonstrate that many concepts—most notably those that underlie nouns (such as Composer and War) but also story-based or narrative-structured concepts—are essentially *object-centred*, and are most naturally represented as a multitude of shallow trees. These trees *are* highly-connected in a coherent manner by means of shared arguments (common leaves). A mapping between two such domains is illustrated in Figure 1 below:

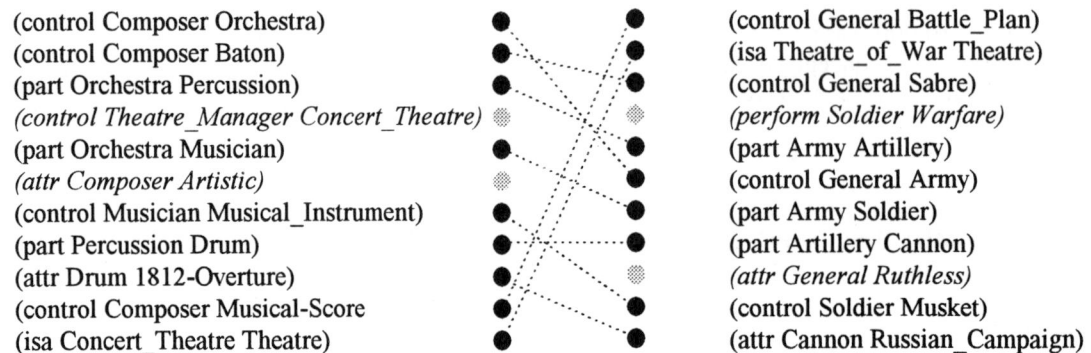

Figure 1: Partial domain descriptions relating to the concepts **Composer** and **General**. Noisy predications which do not contribute to the metaphor "Composers are Generals" are shown in italics.

This type of commonly occurring metaphor exacerbates SME's original $O(2^N)$ complexity. For example, the metaphor *Surgeons are Butchers* requires less that 15 seconds of processing time in Sapper, yet generates enough pmaps to keep an optimal SME busy for many billions of years. Veale *et al.* report the results of Figure 2 for an empirical test involving over 100 object-centred representations (drawn from the domain of professions):

| Aspect | optimal-SME | ACME | Sapper |
|---|---|---|---|
| Avg. # pmaps | 386 per metaphor | 12,657 | 18 |
| Avg. Time per Metaphor | N/A - worst case $O(2^{386})$ seconds | N/A | 12.5 seconds |

*Figure 2: Evaluation of Sapper, SME and ACME. The unavailability of times for SME and ACME reflects the inability of these models to run in a matter of days.*

Sapper out-performs SME in these domains because it is designed to seek out structure *laterally* from shallow trees that are connected via common elements, while SME seeks structure vertically from the hierarchical nesting of deep trees. Sapper can also encode the kind of verb-centred story analogues upon which SME demonstrates its strongest competence (such as the *Karla the Hawk* and *Fortress: Tumor* stories). This paper builds upon these results to show that while heuristic, sub-optimal *Greedy-SME* and *Incremental-SME* avoid factorial time performance, they are still very sensitive to tree organization, producing poor mappings when dealing with object-centred representations. We also apply our intuitions to the Sapper and ACME models, demonstrating that while the latter is from the outset a sub-optimal model, it also exhibits diminished competence on hard problems. This discussion will allow us to outline in greater detail exactly what we mean by a *hard* analogical problem.

## 2. Cognitive Theories of Structure Mapping

ACME approaches the structure-mapping problem from a different perspective than either SME or Sapper, pursuing what might be called a *natural computation approach* to analogy and metaphor. ACME models structure-mapping as a problem of parallel constraint satisfaction, in which the demands of 1-to-1 coherence and structural systematicity are coded as soft constraints, or pressures, on the system. Ultimately, it is a sub-optimal approach which offers no guarantees of mapping quality.

ACME employs a Hopfield-style connectionist network to encode mapping constraints (see [Hopfield and Tank, 1985]). Every structure-mapping hypothesis—either between a source and target predicate or between a source and target entity—is coded as a distinct neuron. Likewise, structural entailments among these hypotheses are coded as bi-directional excitatory links between the corresponding nodes, while inhibitory links are used to connect mutually exclusive hypotheses.

Such an arrangement is the connectionist equivalent of a 2-CNF SAT formula, raising the question of ACME's logical soundness. Indeed, it happens that the use of bi-directional linkages in ACME—which makes all implications *mutual* implications—means that an ACME representation is logically unsound. Because argument mappings can dictate predicate mappings, ACME is sound only when the source and target structures are trees (hence argument mappings *do* imply predicate mappings), but as noted, structure mapping is polynomially bounded anyway in such situations.

Overall, the complexity prognosis of ACME is not good: as a feedback-based neural network, there is no guaranteed polynomial bound on its time performance. Yet, because the network size is polynomially-bounded (i.e., $O(n^2)$ nodes and $O(n^4)$ linkages, where n is the number of distinct symbols in the source domain), the theoretical results of [Bruck and Goodman 1990] apply, who prove that a Hopfield-style network of polynomial size can only optimally solve NP-hard problems if NP = P. So, since an ACME network realistically embodies a polynomial algorithm, why should it be allowed to consume an exponential amount of time doing so?

### 2.1. Sapper: A Memory-Situated Model

The Sapper model of [Veale *et al.* 1996a,b] views semantic memory as a localist graph in which nodes represent distinct concepts, and arcs between those nodes represent semantic / conceptual relations between those concepts. Memory management under Sapper is pro-active toward structure mapping, that is, it employs rules of structural similarity—called Triangulation and Squaring—to determine if any two nodes may at some future time be placed in systematic correspondence in a metaphoric context. If so, Sapper notes this fact by laying down a *bridge relation* between these nodes, to be exploited in some future structure-mapping session.

**Triangulation Rule:** *If memory already contains two linkages $L_{ij}$ and $L_{kj}$ of semantic type L forming two sides of a triangle between the concept nodes $C_k$, $C_i$ and $C_j$, then complete the triangle and augment memory with a new bridge linkage $B_{ik}$.*

> Spread Activation from node T in long-term memory to a horizon H
> Spread Activation from node S in long-term memory to a horizon H
>
> When a wave of activation from T meets a wave from S at a bridge T':S'
>     linking the target domain concept T' to the source domain concept S' Then:
>     Find a chain of semantic relations R that links both T' to T and S' to S
>     If R is found, then the bridge T':S' is balanced relative to T:S, so Do:
>         Generate a partial interpretation (pmap) $\pi$ of the metaphor T:S as follows
>             For every target concept t between T' and T as linked by R Do
>                 Put t in alignment with the equivalent concept s between S' and S
>                 Thus, $\pi \leftarrow \pi \cup \{<t: s>\}$
>         Let $\Phi \leftarrow \Phi \cup \{\pi\}$
>
> Once the set $\Phi$ of all pmaps within the horizon H have been found, Do
>     Evaluate the richness of each pmap $\pi \in \Phi$
>     Sort the collection $\Phi$ of pmaps in descending order of richness.
>     Pick the first (richest) interpretation $\Gamma \in \Phi$ as a seed for overall interpretation.
>     Visit every other pmap $\pi \in (\Phi - \Gamma)$ in descending order of richness
>         If it is coherent to merge $\pi$ with $\Gamma$ (i.e., without violating 1-to-1ness) then
>             $\Gamma \leftarrow \Gamma \cup \pi$
>         Otherwise discard $\pi$
>     When $\Phi$ is exhausted, $\Gamma$ will contain the overall Sapper interpretation of T:S

*Figure 3: The Sapper Algorithm, as based on the exploitation of cross-domain bridge-points in semantic memory.*

**Squaring Rule:** *If $B_{jk}$ is a bridge, and if there already exist the linkages $L_{ij}$ and $L_{lk}$ of the semantic type L, forming three sides of a square between the concept nodes $C_i$, $C_j$, $C_k$ and $C_l$, then complete the square and augment memory with a new bridge $B_{il}$.*

At some future time, if Sapper wishes to determine a structural mapping between a target domain rooted in the concept node T (for Target) and one rooted in the node S (Source), it applies the algorithm of Figure 3.

The Sapper algorithm comprises two main phases: the first of these seeks out the set $\Phi$ of all well-formed and balanced semantic pathways (of length $\leq$ 2H) that originate at the root node of the target (T), and terminate at the root node of the source (S), crossing a single conceptual bridge (i.e., the domain cross-over point) at its mid-point. Each such pathway corresponds to a partial interpretation (a pmap in SME parlance) of the metaphor/analogy. The second phase coalesces this collection of pmaps $\Phi$ into a coherent global whole; it does this using a *seeding algorithm* (see [Keane and Brayshaw, 1988]) which starts with the structurally richest pmap $\Gamma$ as its seed, and then attempts to fold each other pmap into this seed, if it is coherent to do so, in descending order of the richness of those pmaps. This seeding phase is directly equivalent to the greedy merge phase of Greedy-SME (see [Oblinger & Forbus 1990]).

## 3. Proof: Structure-Mapping is NP-Hard

In this section we place our arguments on a solid footing by proving the NP-Hardness of the structure mapping problem. Though the known NP-complete problem LCS (Largest Common Sub-Graph) is perhaps a more immediate match, we instead employ here 3DM (3-Dimensional Matching) as a proof basis, a problem which seeks to obtain a non-overlapping matching of points in a 3-D space. A consideration of 3DM will shed light on the worst case scenario as encountered by the greedy heuristics employed by greedy-SME. Garey and Johnson ([1979]) define 3DM as follows:

*Unique 3-Dimensional Matching (3DM):* Given a set M of points in 3-D space, i.e., $M \subseteq X \times Y \times Z$, where X, Y and Z are disjoint sets of integers and $|X|=|Y|=|Z| = q$, find the largest set $M' \subseteq M$ such that no two elements of $M'$ agree in any coordinate.

**Proof**: To reformulate 3DM as a problem of structure-mapping, it is necessary to represent each 3-D point $<X_i, Y_i, Z_i> \in M$ as a pair of predicates, one in each of the source S and target T domains, such that these predicates are only allowed to map onto each other. Furthermore, any isomorphic mapping must not contain two different predicate matches that arise from two points sharing one or more coordinates. We can ensure this using the following polynomial transformation:

∀ <$X_i, Y_i, Z_i$> ⊆ M Do

  add $P_{XYZ}(X_i, Z_i)$ to S

*and* add $P_{XYZ}(Y_i, \Omega^* X_i + Y_i)$ to T

*where* $\Omega = \max(\max(X \cup Y \cup Z), |\min(X \cup Y \cup Z)|)$

Now, because the predicate P is uniquely tagged with the subscript XYZ which ties it to a particular 3-D point, these two predicate structures can map only to each other. When so mapped during the analogy process, such a mapping results in the creation of the following structure (a *pmap* in SME parlance):

$$\text{map}_i = \{<X_i, Y_i>, <Z_i, \Omega^* X_i + Y_i>\}$$

In this manner a root mapping will be created for each point in M. Note also that $\Omega^* X_i + Y_i$ is unique for each pairing of $X_i$ and $Y_i$, thus $X_i$, $Y_i$ and $Z_i$ are *tied together* and cannot be cross-mapped with any other point coordinate. Suppose we have two such pmaps, $\text{map}_i = \{<X_i, Y_i>, <Z_i, \Omega^* X_i + Y_i>\}$ and $\text{map}_k = \{<X_k, Y_k>, <Z_k, \Omega^* X_k + Y_k>\}$, arising out of the two points <$X_i, Y_i, Z_i$> and <$X_k, Y_k, Z_i$> which share a Z-coordinate $Z_i$. These maps cannot therefore be merged to create a larger mapping as such a merge results in $Z_i$ being mapped to both $\Omega^* X_i + Y_i$ and $\Omega^* X_k + Y_k$, a clear violation of mapping isomorphism.

Once a maximal gmap is found for the analogy, each pair <$X_i, Y_i$> and <$Z_i, \Omega^* X_i + Y_i$> of this gmap can then be decomposed and reassembled (in polynomial time) to recreate a point <$X_i, Y_i, Z_i$> that is added to M'. Since the gmap is maximal, so is M'. Because it solves 3DM, structure-mapping is thus NP-Hard. □

## 4. Problem Reorganization for Tractability

A large body of problem instances may nevertheless be tractably amenable to an optimal Sapper variant. If an optimal Sapper solution can be obtained for a large enough body of problem examples, these solutions can be used as ceilings to measure the competence of sub-optimal heuristics like greedy merging / seeding.

The domain descriptions in the Sapper profession corpus contain on average over 120 predications each. Test metaphors in the profession corpus thus generate too many partial mappings to make optimal evaluation tractable. Yet, some problem re-organization can be applied to reduce the number of pmaps to frequently make an Optimal-Sapper interpretation feasible, without losing the combinatorial scope of the interpretation. This reorganization process, whereby redundant areas of the combinatorial search space are pruned, is the equivalent of *arc-consistency testing* in satisfaction problems to a priori remove contradictory variable assignments (see [Mohr and Henderson, 1986]).

For each metaphor (whose pmap set is denoted $\Phi$) a conflict graph is constructed in $O(|\Phi|^2)$ time, by determining for each pmap the set of other pmaps with which it cannot be combined. This set is similar to the set of *NoGoods* calculated by the SME algorithm, though it used differently to achieve more extensive reductions in performance time. The conflict set $CF_i$ for a particular pmap $\pi_i \in \Phi$ is thus defined as:

$$CF_i = \{\pi_k \mid k \neq i \land \neg\text{systematic}(\pi_i, \pi_k)\}$$

Compatibility between pmaps can thus be defined as:

$$\text{compatible}(\pi_i, \pi_k) \text{ iff } CF_i \subseteq CF_k \land \pi_i \notin CF_k$$

In contrast to SME, Optimal-Sapper uses this information to recognize any compatibility-based redundancies, and redistribute them accordingly *before* entering the punishing factorial merge-stage, as follows:

$$\forall \pi_i \pi_j, i \neq j, \text{ if compatible}(\pi_i, \pi_j) \text{ then}$$

$$\forall \pi_k \in CF_j - CF_i \text{ do}$$

$$\pi_k \leftarrow \pi_k \cup \pi_i$$

$$\Phi \leftarrow \Phi - \pi_i$$

Given that the combinatorial merge stage of an Optimal-Sapper algorithm is $O(2^{|\Phi|})$, each such pmap factored out a priori lowers the eventual cost another exponential notch. On our corpus of profession metaphors, we have found that problem reduction of this form reduces the number of pmaps for each metaphor by an average of 60%, pruning the search space of the most intractable instance, Generals are Surgeons, from $O(2^{39})$ to one more manageable by Optimal-Sapper, $O(2^{17})$.

## 5. Experiment: Sapper V<u>s.</u> Greedy-SME

We can now quantify the competence of sub-optimal heuristics such as seeding and greedy-search as a percentage of optimal performance. But first, we consider the nature of the interpretations that structure-mapping algorithms will generate for these test metaphors. The mapping of Figure 1 is the Sapper interpretation of the metaphor Composers are Generals, while the mapping of Figure 4 is that returned by greedy-SME for the same metaphor.

Since an official implementation of greedy-SME is not yet publicly available ([Forbus, 1996]), we therefore simulate greedy-SME by feeding the pmaps generated by the available optimal-SME through the Sapper seeding stage. This a computationally equivalent process.

| **If Composer is like General** |
| :--- |
| *Then* Drum is like Cannon |
| *and* Powerful is like Loud |
| *and* Loud is like Powerful |
| *and* Conductor_Baton is like Sword |
| *and* Tchaikovsky is like Napoleon |
| *and* Libretto is like Plan |
| *and* *Narrow is like Dangerous* |
| *and* *19th_Century is like French* |
| *and* Music_Recital is like Cavalry_Charge |
| *and* *Long is like Sharp* |
| *and* Orchestra is like Army |
| *and* Listener is like Soldier |
| *and* W_A_Mozart is like George_Patten |
| *and* Percussion is like Artillery |
| *and* *Theatre is like Influential* |
| *and* *Russian is like 19th_century* |
| *and* Music_Composition is like Bomb_Raid |
| *and* *Musical is like Healthy* |
| *and* Music_Note is like Enemy_Soldier |
| *and* *Sudden is like Dead* |
| *and* *Piano is like Snub_Fighter* |
| *and* *Fictional is like On_Target* |
| *and* *Character is like Smart_Bomb* |
| *and* *18th_Century is like Arrogant* |
| *and* Symphony is like Military_Propaganda |
| *and* Violin is like Musket |
| *and* Musical_Score is like Enemy_Army |
| *and* Operatic_Act is like Medal |
| *and* Opera is like Military_Uniform |
| *and* *Inspiration is like Corpse* |

*Figure 4: Simulated Greedy-SME interpretation of "Composers are Generals".*

A selection of the mappings in Figure 4 above are displayed in an italics face to convey their 'ghost' status: 'ghosts' are essentially noisy mappings that might work in another metaphoric context but which are not systematic here. But why does greedy-SME generate so many ghosts while Sapper produces none, when both employ equivalent merge processes? To see why, consider that SME and Sapper agree on three tacit assumptions for seeding: first, that a goodness ordering can be placed upon the set of pmaps; secondly, that the pmap chosen as seed for the merge is rich enough to justify its own inclusion in the global mapping; and thirdly, that this seed is rich enough to nudge the overall merge process toward a good to optimal global mapping. However, Greedy-SME—unlike Sapper—does not generate sufficiently rich (and thus differentiable) pmaps in object-centred domains to make these assumptions work. As these domains are best represented as a broad forest of many shallow trees rather than a tight forest of few, deep trees (see [Veale *et al.* 1996]), the pmaps generated by SME for object-centred metaphors are equally shallow and numerous. In fact, these impoverished SME pmaps resemble the geometric pmaps generated in section 3 when reposing 3DM as structure-mapping. One clearly would not expect a greedy approach to work in this geometric context as no one pmap would have enough structure to successfully guide the merge process to a good solution.

The competence of Sapper and greedy-SME has been determined over the test corpus of 100+ profession metaphors, where the optimal-Sapper of section 4 is used as a savant: a mapping of a sub-optimal interpretation is considered valid if it is also contained in the optimal Sapper interpretation. The sub-optimal competence of Sapper and greedy-SME is thus calculated as **100 * (No. of valid mappings) / (Total No. of mappings)**. If this validity criterion seems overly strict and *all-or-nothing*, it needs to be for tractability reasons. If one were to evaluate a noisy interpretation on the basis of its largest systematic subset, the partition of the interpretation into signal and noise would in itself be an intractable problem of combinatorial dimensions. Comparative results are displayed in Figure 5 below:

| Rating | Sapper | Greedy SME | Optimal | Random |
| :--- | :---: | :---: | :---: | :---: |
| Competence | 95.2% | 18.7% | 100% | 80.5% |
| % of Times Optimal | 77% | 0% | 100% | 45% |

*Figure 5: Comparative trials of Sapper and Sub-Optimal greedy-SME with a random control.*

Greedy-SME performs disappointingly on these trials, significantly trailing even the random control trial, in which a random merging of coherent Sapper pmaps is generated as an interpretation for each metaphor. These results speak for the importance of structurally rich pmaps, for when these are rich enough even a random coalescence of pmaps will generate a good interpretation. But if the set of pmaps is structurally impoverished, as with SME in object-centred domains, not even a best-first sorting will compensate. These random trials indicate that a system's true competence is to be found in the processes which generate pmaps, more so than in those which combine them.

## 6. Where the Hard Analogies Are

What do the results of section 5 say about the identifiable qualities of *hard* analogies/metaphors? Clearly, when employing an optimal mapping algorithm, the number of distinct roots in the forest-of-trees representation of each domain is a direct indicator of the exponential requirements of the algorithm. What can be said of the hardness of analogies as perceived by sub-optimal approaches such as Sapper and SME?

In complexity terms no problem instance is—strictly speaking—*hard* to a sub-optimal structure matcher, as the number of pmaps is largely irrelevant in a $O(N^2)$ greedy merging / seeding process. However, if one measures hardness in terms of the likelihood of generating a quality (i.e., ghost-free and accurate) interpretation, the best indicator of hardness is the average structural richness of each pmap (i.e., the average number of mappings in each pmap). The lower this average richness, the more probable it is that any two pmaps can be coherently merged, and thus the more likely that the final interpretation will be noisy and ghost-ridden. In contrast, the higher this average, the more probable it is that a final interpretation will be near optimal, and less likely to contain ghosts (as each pmap merge operation will have a greater chance of failure).

If we have side-lined ACME's sub-optimal approach to structure-mapping in this paper, it is due to the belief that ACME represents an excessive approach to the problem. Recall that ACME can be characterized as a 2-SAT problem, where network nodes mirror SAT variables, and network linkages mirror SAT clauses. From the ratio of ACME nodes to linkages for any given metaphor/analogy, we can determine the equivalent SAT ratio of clauses to variables as $O(N^2)$, thus making an ACME problem hugely over-constrained (see [Mitchell et al. 1992]). Given the large networks which ACME can construct for a hard problem (> 12,000 nodes), existing relaxation techniques based on constraint prioritization do not seem practical (see [Bakker et al. 1993]). ACME thus reduces to a difficult subclass of *maximal 2-SAT*, with the size of that subset of clauses it must leave unsatisfied growing exponentially with the extent of network over-constraint, which itself grows quadratically with metaphor size. In this case, sub-optimality certainly thus not imply tractability.

In closing, we note that the profession corpus upon which our experiments are based is available from the following URL, in Sapper, ACME and SME formats:
*http://www.compapp.dcu.ie/~tonyv/metaphor.html*

## References

[Akutsu, 1992] T. Akutsu. An (RNC) Algorithm for Finding a Largest Common Subtree of Two Trees, *IEICE Transactions on Information and Systems*, 75-D, pages 95-101, 1992.

[Bakker et al. 1993] R. Bakker, F. Dikker, F. Templeman, and P. Wognum. Diagnosing and Solving Over-determined constraint satisfaction problems, in the Proceedings of IJCAI'93, the thirteenth International Joint Conference on Artificial Intelligence. Morgan Kauffman, 1993.

[Bruck and Goodman, 1990]. J. Bruck and J. W. Goodman. On the Power of Neural Networks for Solving Hard Problems, *Journal of Complexity* 6, pages 129-135, 1990.

[Falkenhainer et al. 1989] B. Falkenhainer, K. D. Forbus, and D. Gentner. Structure-Mapping Engine. *Artificial Intelligence*, 41, pages 1-63, 1989.

[Forbus et al. 1994]. K. D. Forbus, R. Ferguson and D. Gentner. Incremental Structure-Mapping, in *the Proceedings of the Sixteenth Annual Meeting of the Cognitive Science Society*, Atlanta, Georgia. Hillsdale, NJ: Lawrence Erlbaum, 1994.

[Forbus, 1996]. K. D. Forbus, personal communication, August 1996.

[Garey and Johnson, 1979] M. R. Garey and D. S. Johnson. *Computers and Intractability: A Guide to the Theory of NP-Completeness*. Freeman, NY, 1979.

[Holyoak and Thagard, 1989] K. J. Holyoak and P. Thagard. Analogical Mapping by Constraint Satisfaction, *Cognitive Science* 13, pp 295-355, 1989.

[Hopfield and Tank, 1985] J. J. Hopfield and D. W. Tank. "Neural" Computation of Decisions in Optimization Problems. *Biological Cybernetics* 52, pp 141-152, 1985.

[Keane and Brayshaw, 1988] Keane, M. T. and M. Brayshaw. The Incremental Analogical Machine: A computational model of analogy. In D. Sleeman (Ed.), *European Working Session on Learning*. Pitman, 1988.

[Mitchell et al. 1992] D. Mitchell, B. Selman and H. J. Levesque. Hard and Easy distributions of SAT problems, in the proceedings of AAAI'92, the 1992 conference of the American Association for AI, 1992.

[Mohr and Henderson, 1986] R. Mohr, and T. Henderson. Arc and Path Consistency Revisited. *Artificial Intelligence* 25, pages 65-74, 1986.

[Oblinger and Forbus, 1990] D. Oblinger, and K. D. Forbus. Making SME Pragmatic and Greedy, in *the Proc. of the Twelfth Annual Meeting of the Cognitive Science Society*. Lawrence Erlbaum, 1990.

[Veale et al. 1996a] T. Veale, B. Smyth, D. O'Donoghue and M. T. Keane. Representational Myopia in Cognitive Mapping, in the Proc. of the AAAI workshop on Source of the Power in Cognitive Theories, Portland, 1996.

[Veale et al. 1996b] T. Veale, D. O'Donoghue and M. T. Keane. Computability as a Limiting Cognitive Constraint: Complexity Concerns in Metaphor Comprehension, *Cognitive Linguistics: Cultural, Psychological and Typological Issues (forthcoming)*.

# An Average-Case Analysis of the $k$-Nearest Neighbor Classifier for Noisy Domains

**Seishi Okamoto**
Fujitsu Laboratories Ltd.
2-2-1 Momochihama, Sawara-ku
Fukuoka 814, Japan
seishi@flab.fujitsu.co.jp

**Yugami Nobuhiro**
Fujitsu Laboratories Ltd.
2-2-1 Momochihama, Sawara-ku
Fukuoka 814, Japan
yugami@flab.fujitsu.co.jp

## Abstract

This paper presents an average-case analysis of the $k$-nearest neighbor classifier($k$-NN). Our analysis deals with $m$-of-$n/l$ concepts, and handles three types of noise: relevant attribute noise, irrelevant attribute noise, and class noise. We formally compute the expected classification accuracy of $k$-NN after a certain fixed number of training instances. This accuracy is represented as a function of the domain characteristics. Then, the predicted behavior of $k$-NN for each type of noise is explored by using the accuracy function. We examine the classification accuracy of $k$-NN at various noise levels, and show how noise affects the accuracy of $k$-NN. We also show the relationship between the optimal value of $k$ and the number of training instances in noisy domains. Our analysis is supported with Monte Carlo simulations.

## 1 Introduction

The $k$-nearest neighbor classifier ($k$-NN) is one of the most widely applied learning algorithms. Although $k$-NN is a powerful algorithm and has been studied by many researchers, it is *not* clear how noise affects the classification accuracy of $k$-NN. Moreover, it is also unclear which value should be chosen for $k$ to maximize accuracy in noisy domains. These are crucial problems in $k$-NN applications, because there are few noise-free problems in practical domains.

Variants of $k$-NN have been proposed to tolerate noise (*e.g.*, [Aha and Kibler, 1989]), and to choose an appropriate value of $k$ (*e.g.*, [Creecy et al., 1992]). These proposals exhibit the high performance of $k$-NN by empirical evaluations. However, the noise effects on the accuracy of $k$-NN and on the optimal value of $k$ still remain unclear. It is therefore important to understand the noise effects on $k$-NN and the optimal $k$ by theoretical evaluations.

There have been several theoretical analyses of $k$-NN. The upper bound of $k$-NN error rate (risk) is twice the optimal Bayes risk under the assumption of an infinite number of training instances [Cover and Hart, 1967]. Moreover, $k$-NN risk converges to the optimal Bayes risk as $k$ approaches infinity [Cover, 1968]. For a finite set of training instances, the new bounds of 1-NN risk are given using Bayes risk [Drakopoulos, 1995]. Aha *et al.* [1991] analyze 1-NN with a similar model to PAC (Probably Approximately Correct) learning, and this analysis is generalized to $k$-NN [Albert and Aha, 1991]. Although these theoretical results are important and give some insights into the behavior of $k$-NN, all of these studies assume noise-free instances.

An average-case analysis is a useful theoretical framework to understand the behavior of learning algorithms [Pazzani and Sarrett, 1992]. This framework is based on the formal computation of the expected accuracy of a learning algorithm for a certain fixed class of concepts. Using the result of this computation, we can explore the predicted behavior of an algorithm. There have been some average-case analyses of $k$-NN. Langley and Iba [1993] analyzed 1-NN for conjunctive concepts, and we analyzed $k$-NN for $m$-of-$n$ concepts without irrelevant attribute [Okamoto and Satoh, 1995]. However, these studies assumed noise-free instances. Recently, we presented an average-case analysis of 1-NN for $m$-of-$n$ concepts with irrelevant attribute in noisy domains [Okamoto and Yugami, 1996]. This paper generalizes our recent study for 1-NN to $k$-NN.

In this paper, we present an average-case analysis of $k$-nearest neighbor classifier for noisy domains. Our analysis handles $m$-of-$n$ concepts with $l$ irrelevant attributes, and deals with three types of noise: relevant attribute noise, irrelevant attribute noise, and class noise. First, we formally compute the expected classification accuracy (*i.e.*, predictive accuracy) of $k$-NN after $N$ training instances are given. This accuracy is represented as a function of the domain characteristics: $k$, $N$, $m$, $n$, $l$, the probabilities of occurrence for relevant and irrelevant attributes, and noise rates. Using the accuracy function, we explore the predicted behavior of $k$-NN in noisy domains. We describe the predictive accuracy of $k$-NN at various noise levels, and show the effects of noise on the accuracy of $k$-NN. We also show the relationship between the optimal value of $k$ and the number of training instances in noisy domains. Our theoretical analysis is supported with Monte Carlo simulations.

## 2 Problem Description

Our analysis deals with $m$-of-$n/l$ concepts defined over the threshold $m$, $n$ relevant and $l$ irrelevant Boolean attributes [Murphy and Pazzani, 1991]. These concepts classify an instance as positive if and only if at least $m$ out of $n$ relevant attributes occur (*i.e.*, take the value 1) in its instance.

Our analysis handles three types of noise. Each type of noise is independently introduced by the following common definition. Relevant (irrelevant, *resp.*) attribute noise flips an arbitrary relevant (irrelevant, *resp.*) attribute value in each instance with a certain probability $\sigma_r$ ($\sigma_i$, *resp.*). Class noise replaces the class label for each instance with its negation with a certain probability $\sigma_c$.

We investigate a $k$-nearest neighbor classifier using hamming distance (*i.e.*, the number of attributes on which two instances differ) as a distance measure. For the distribution over the instance space, our analysis assumes every relevant and irrelevant attribute independently occurs with a certain probability $p$ and $q$. Each training instance is independently drawn from the instance space. After the effects of each type of noise, all training instances are stored into memory to allow for duplication. When a test instance is given, $k$-NN classifies the test instance into a majority class (positive or negative) among its $k$ nearest training instances. If the number of positive instances equals that of negative instances among its $k$ nearest neighbors, then $k$-NN randomly determines the class of the test instance (this situation can occur only when $k$ is an even number).

## 3 Predictive Accuracy

We formally compute the predictive accuracy of $k$-NN for $m$-of-$n/l$ target concepts after $N$ training instances are given. The predictive accuracy is represented as a function of the domain characteristics: $k$, $N$, $m$, $n$, $l$, $p$, $q$, $\sigma_r$, $\sigma_i$, and $\sigma_c$. However, to avoid complicated notation, we will not explicitly express these characteristics as parameters of the accuracy function with the exception of $k$.

We compute the predictive accuracy in the case where each type of noise affects only training instances. After this computation, we also give the accuracy function in the case where noise affects both test and training instances.

To compute the predictive accuracy, we use a set of instances in which $x$ relevant attributes and $y$ irrelevant attributes simultaneously occur (we denote this set with $I(x,y)$). Let $P_{\text{occ}}(x,y)$ be the probability that an arbitrary noise-free instance belongs to $I(x,y)$. This probability is given by

$$P_{\text{occ}}(x,y) = \binom{n}{x}\binom{l}{y} p^x(1-p)^{n-x} q^y(1-q)^{l-y}.$$

Under our assumptions given in Section 2, $k$-NN has the same expected probability of correct classification for an arbitrary test instance in $I(x,y)$. Hence, we can represent the predictive accuracy of $k$-NN after $N$ training instances as

$$A(k) = \sum_{y=0}^{l}\left\{\sum_{x=0}^{m-1} P_{\text{occ}}(x,y)\left(1-P_{\text{pos}}(k,x,y)\right) + \sum_{x=m}^{n} P_{\text{occ}}(x,y)P_{\text{pos}}(k,x,y)\right\},$$

where $P_{\text{pos}}(k,x,y)$ represents the probability that $k$-NN classifies an arbitrary test instance in $I(x,y)$ as positive.

Let $t(x,y)$ be an arbitrary test instance in $I(x,y)$. To represent $P_{\text{pos}}(x,y)$, we compute the appearance probability for an arbitrary training instance with distance $e(0 \leq e \leq n+l)$ from $t(x,y)$. Let $P_{\text{dp}}(x,y,e)$ ($P_{\text{dn}}(x,y,e)$, *resp.*) be this probability for an arbitrary training instance with the positive (negative, *resp.*) class label. $P_{\text{dp}}(x,y,e)$ and $P_{\text{dn}}(x,y,e)$ were computed using Eq.(14) and Eq.(15) in our previous paper [Okamoto and Yugami, 1996]. Hence, we simply state the computation of these probabilities here.

$P_{\text{dp}}(x,y,e)$ and $P_{\text{dn}}(x,y,e)$ are given by

$$P_{\text{dp}}(x,y,e) = \sum_{X=0}^{n}\sum_{Y=0}^{l} P_{\text{p}}(X,Y) P_{\text{dis}}(x,y,X,Y,e),$$

$$P_{\text{dn}}(x,y,e) = \sum_{X=0}^{n}\sum_{Y=0}^{l} P_{\text{n}}(X,Y) P_{\text{dis}}(x,y,X,Y,e).$$

In these equations, $P_{\text{p}}(X,Y)$ ($P_{\text{n}}(X,Y)$, *resp.*) represents the probability that an arbitrary training instance belongs to $I(X,Y)$ and has the positive (negative, *resp.*) class label. Moreover, $P_{\text{dis}}(x,y,X,Y,e)$ denotes the probability that an arbitrary training instance in $I(X,Y)$ has distance $e$ from $t(x,y)$.

First, we represent $P_{\text{p}}(X,Y)$ and $P_{\text{n}}(X,Y)$ by considering the effects of each type of noise on the training instances. These probabilities are represented as

$$P_{\text{p}}(X,Y) = (1-\sigma_c)P_{\text{p}_0}(X,Y) + \sigma_c P_{\text{n}_0}(X,Y),$$
$$P_{\text{n}}(X,Y) = \sigma_c P_{\text{p}_0}(X,Y) + (1-\sigma_c)P_{\text{n}_0}(X,Y),$$

where $P_{\text{p}_0}(X,Y)$ ($P_{\text{n}_0}(X,Y)$, *resp.*) denotes the appearance probability for an arbitrary positive (negative, *resp.*) training instance in $I(X,Y)$, before the effect of class noise. $P_{\text{p}_0}(X,Y)$ and $P_{\text{n}_0}(X,Y)$ are given by

$$P_{\text{p}_0}(X,Y) = \sum_{X_0=m}^{n}\sum_{Y_0=0}^{l} P_{\text{occ}}(X_0,Y_0) P_{\text{nr}}(X_0,X) P_{\text{ni}}(Y_0,Y),$$

$$P_{\text{n}_0}(X,Y) = \sum_{X_0=0}^{m-1}\sum_{Y_0=0}^{l} P_{\text{occ}}(X_0,Y_0) P_{\text{nr}}(X_0,X) P_{\text{ni}}(Y_0,Y),$$

where $P_{\text{nr}}(X_0,X)$ ($P_{\text{ni}}(Y_0,Y)$, *resp.*) represents the probability that the number of relevant (irrelevant, *resp.*) attributes occurring in an arbitrary training instance is changed from $X_0$ ($Y_0$, *resp.*) to $X$ ($Y$, *resp.*) by the effect of relevant (irrelevant, *resp.*) attribute noise. These probabilities are represented as

$$P_{\mathrm{nr}}(X_0, X) = \sum_{s=\max(0, X_0-X)}^{\min(X_0, n-X)} \left\{ \binom{X_0}{s} \binom{n-X_0}{X-X_0+s} \right.$$
$$\left. \times \sigma_{\mathrm{r}}^{X-X_0+2s}(1-\sigma_{\mathrm{r}})^{n-(X-X_0+2s)} \right\},$$

$$P_{\mathrm{ni}}(Y_0, Y) = \sum_{t=\max(0, Y_0-Y)}^{\min(Y_0, l-Y)} \left\{ \binom{Y_0}{t} \binom{l-Y_0}{Y-Y_0+t} \right.$$
$$\left. \times \sigma_{\mathrm{i}}^{Y-Y_0+2t}(1-\sigma_{\mathrm{i}})^{l-(Y-Y_0+2t)} \right\}.$$

Next, we represent $P_{\mathrm{dis}}(x, y, X, Y, e)$. Let $z_{\mathrm{r}}$ ($z_{\mathrm{i}}$, resp.) be the number of relevant (irrelevant, resp.) attributes which occur in both $t(x,y)$ and an arbitrary training instance in $I(X, Y)$. Then, $P_{\mathrm{dis}}(x, y, X, Y, e)$ is given by

$$P_{\mathrm{dis}}(x, y, X, Y, e) = \sum_{(z_{\mathrm{r}}, z_{\mathrm{i}}) \in \mathcal{S}} \frac{\binom{x}{z_{\mathrm{r}}}\binom{n-x}{X-z_{\mathrm{r}}}}{\binom{n}{X}} \frac{\binom{y}{z_{\mathrm{i}}}\binom{l-y}{Y-z_{\mathrm{i}}}}{\binom{l}{Y}},$$

where $\mathcal{S}$ is a set of a pair of $z_{\mathrm{r}}$ and $z_{\mathrm{i}}$, that satisfies all conditions of

$$\max(0, x+X-n) \leq z_{\mathrm{r}} \leq \min(x, X),$$
$$\max(0, y+Y-l) \leq z_{\mathrm{i}} \leq \min(y, Y),$$
$$z_{\mathrm{r}} + z_{\mathrm{i}} = \frac{x+y+X+Y-e}{2}.$$

We have represented $P_{\mathrm{dp}}(x, y, e)$ and $P_{\mathrm{dn}}(x, y, e)$. Using these probabilities, we compute $P_{\mathrm{pos}}(k, x, y)$ in the accuracy function. For this computation, we consider the distance from $t(x,y)$ to the $k$-th nearest neighbor. We denote this distance with $d(0 \leq d \leq n+l)$. When the $k$-th nearest neighbor has distance $d$ from $t(x,y)$, then exactly $a(0 \leq a \leq k-1)$ out of $N$ training instances have the distance less than $d$ from $t(x,y)$, and exactly $b$ training instances have distance $d$. Here, we have $(k-a) \leq b \leq (N-a)$ from $k \leq (a+b) \leq N$. We use $P_{\mathrm{num}}(x, y, d, a, b)$ to denote the probability that this situation occurs. We also use $P_{\mathrm{sp}}(k, x, y, d, a, b)$ to designate the probability that $k$-NN classifies $t(x,y)$ as positive in this situation. Combining $P_{\mathrm{num}}(x, y, d, a, b)$ and $P_{\mathrm{sp}}(k, x, y, d, a, b)$, $P_{\mathrm{pos}}(k, x, y)$ can be computed as

$$P_{\mathrm{pos}}(k, x, y) = \sum_{d=0}^{n+l} \sum_{a=0}^{k-1} \sum_{b=k-a}^{N-a} P_{\mathrm{num}}(x, y, d, a, b) P_{\mathrm{sp}}(x, y, d, a, b).$$

We can represent $P_{\mathrm{num}}(x, y, d, a, b)$ as

$$P_{\mathrm{num}}(x, y, d, a, b) = \binom{N}{a}\binom{N-a}{b} P_{\mathrm{l}}(x, y, d)^a$$
$$\times P_{\mathrm{d}}(x, y, d)^b (1 - P_{\mathrm{l}}(x, y, d) - P_{\mathrm{d}}(x, y, d))^{N-a-b},$$

where $P_{\mathrm{l}}(x, y, d)$ and $P_{\mathrm{d}}(x, y, d)$ denotes the probability that an arbitrary training instance has the distance less than and equal to $d$ from $t(x,y)$, respectively. $P_{\mathrm{l}}(x, y, d)$ and $P_{\mathrm{d}}(x, y, d)$ are given by

$$P_{\mathrm{d}}(x, y, d) = P_{\mathrm{dp}}(x, y, d) + P_{\mathrm{dn}}(x, y, d),$$
$$P_{\mathrm{l}}(x, y, d) = \sum_{e=0}^{d-1} \{P_{\mathrm{dp}}(x, y, e) + P_{\mathrm{dn}}(x, y, e)\}.$$

Finally, we compute $P_{\mathrm{sp}}(k, x, y, d, a, b)$ by considering the following situations. When exactly $a$ training instances have the distance less than $d$ from $t(x,y)$, we let exactly $u$ out of these $a$ instances have the positive class label. We use $P_{\mathrm{lp}}^{\mathrm{u}}(a, u)$ to denote the probability that this situation occurs. To get $k$ nearest neighbors for $t(x,y)$, $k$-NN selects exactly $(k-a)$ out of $b$ training instances with distance $d$ from $t(x,y)$. We let $w$ out of these $(k-a)$ instances have the positive class label. Under these situations, if we have $u+w > k/2$, then $k$-NN always classifies $t(x,y)$ as positive, and if $u+w < k/2$, then always classifies $t(x,y)$ as negative. Moreover, if $u+w = k/2$, then $t(x,y)$ is classified as positive with the probability of $1/2$. Hence, we can represent $P_{\mathrm{sp}}(k, x, y, d, a, b)$ as

$$P_{\mathrm{sp}}(k, x, y, d, a, b) = \sum_{u=0}^{a} \sum_{v=0}^{b} P_{\mathrm{lp}}^{\mathrm{u}}(a, u) P_{\mathrm{dp}}^{\mathrm{v}}(b, v)$$
$$\times \left\{ \sum_{w=\lceil \frac{k+1}{2} \rceil - u}^{v} P_{\mathrm{dp}}^{\mathrm{w}}(k, a, b, v, w) + \frac{1}{2} P_{\mathrm{dp}}^{\mathrm{w}}\left(k, a, b, v, \frac{k}{2} - u\right) \right\}.$$

In this equation, $P_{\mathrm{dp}}^{\mathrm{v}}(b, v)$ denotes the probability that exactly $v$ out of $b$ training instances with distance $d$ from $t(x,y)$ have the positive class label. Moreover, $P_{\mathrm{dp}}^{\mathrm{w}}(k, a, b, v, w)$ denotes the probability that exactly $w$ out of $(k-a)$ training instances, selected by $k$-NN from $b$ instances with distance $d$ from $t(x,y)$, have the positive class label. Note that $P_{\mathrm{dp}}^{\mathrm{w}}(k, a, b, v, \frac{k}{2} - x)$ becomes zero, when $k$ is an odd number.

We can represent $P_{\mathrm{lp}}^{\mathrm{u}}(a, u)$ as

$$P_{\mathrm{lp}}^{\mathrm{u}}(a, u) = \binom{a}{u} \left(\frac{P_{\mathrm{lp}}(x, y, d)}{P_{\mathrm{l}}(x, y, d)}\right)^u \left(\frac{P_{\mathrm{ln}}(x, y, d)}{P_{\mathrm{l}}(x, y, d)}\right)^{a-u},$$

where $P_{\mathrm{lp}}(x, y, d)$ ($P_{\mathrm{ln}}(x, y, d)$, resp.) is the probability that an arbitrary instance has the distance less than $d$ from $t(x,y)$ and has the positive (negative, resp.) class label. $P_{\mathrm{lp}}(x, y, d)$ and $P_{\mathrm{ln}}(x, y, d)$ are give by

$$P_{\mathrm{lp}}(x, y, d) = \sum_{e=0}^{d-1} P_{\mathrm{dp}}(x, y, e),$$
$$P_{\mathrm{ln}}(x, y, d) = \sum_{e=0}^{d-1} P_{\mathrm{dn}}(x, y, e).$$

In a similar manner, we can represent $P_{\mathrm{dp}}^{\mathrm{v}}(b, v)$ as

$$P_{\mathrm{dp}}^{\mathrm{v}}(b, v) = \binom{b}{v} \left(\frac{P_{\mathrm{dp}}(x, y, d)}{P_{\mathrm{d}}(x, y, d)}\right)^v \left(\frac{P_{\mathrm{dn}}(x, y, d)}{P_{\mathrm{d}}(x, y, d)}\right)^{b-v}$$

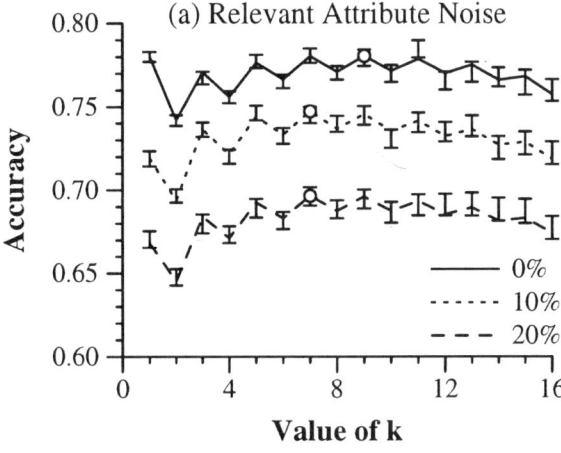
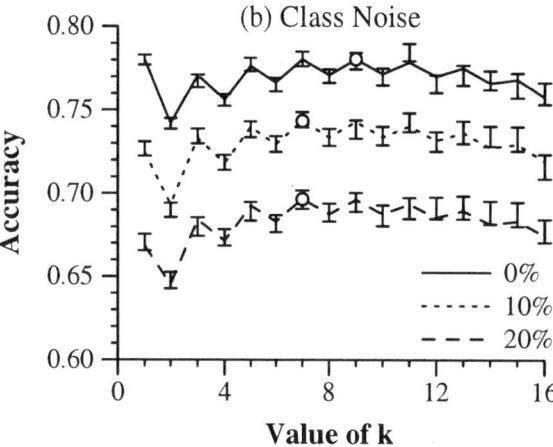

Figure 1: The predictive accuracy of $k$-NN against the value of $k$ for a 3-of-5/2 concept. The lines and the error bars represent the theoretical results and the empirical results of Monte Carlo simulations. Each circle denotes the accuracy for the optimal $k$ at the corresponding noise level. The number of training instances is fixed at 32.

When $k$-NN selects exactly $(k-a)$ out of $b$ training instances with distance $d$ from $t(x,y)$, these $(k-a)$ training instances comprise exactly $w$ out of $v$ instances with the positive class label and exactly $(k-a-w)$ out of $(b-v)$ instances with the negative class label. Hence, $P_{\mathrm{dp}}^{\mathrm{w}}(k,a,b,v,w)$ is given by

$$P_{\mathrm{dp}}^{\mathrm{w}}(k,a,b,v,w) = \frac{\binom{v}{w}\binom{b-v}{k-a-w}}{\binom{b}{k-a}}.$$

We have computed the predictive accuracy of $k$-NN in the case where each type of noise affects only the training instances. When noise affects test instances, the appearance probability for an arbitrary test instance with the positive (negative, *resp.*) class label in $I(x,y)$ is $P_{\mathrm{p}}(x,y)$ ($P_{\mathrm{n}}(x,y)$, *resp.*). Hence, when noise affects both test and training instances, the predictive accuracy of $k$-NN after $N$ training instances can be represented as

$$A(k) = \sum_{x=0}^{n}\sum_{y=0}^{l} \{P_{\mathrm{n}}(x,y)\left(1 - P_{\mathrm{pos}}(k,x,y)\right) + P_{\mathrm{p}}(x,y)P_{\mathrm{pos}}(k,x,y)\}.$$

## 4 Predicted Behavior

Using the accuracy function described in Section 3, we explore the predicted behavior of $k$-NN. Although the accuracy function was obtained for both noise-free and noisy test instances, our exploration deals with only noise-free test instances for lack of space. Moreover, we investigate the effects of each individual noise type on $k$-NN.

For irrelevant attribute noise, we can formally prove the following claim from the accuracy function (the proof is omitted here due to space limitations).

**Claim 1**
*If the probability of occurrence for irrelevant attribute is 1/2, then the predictive accuracy of $k$-NN for m-of-n/l concepts is entirely independent of the noise rate for irrelevant attributes.*

From this claim, we can expect that irrelevant attribute noise does *not* greatly affect the classification accuracy of $k$-NN, *nor* the optimal value of $k$. Therefore, the following discussions focus on the effects of relevant attribute noise and class noise. Throughout our exploration, we set the probabilities of occurrence for both relevant and irrelevant attributes to 1/2.

In addition to the theoretical results from the accuracy function, we give the results of Monte Carlo simulations to confirm our analysis. For each case, 500 training sets are randomly generated in accordance with each noise rate, then the data is collected as the classification accuracy measured over the entire space of noise-free instances. For each case, we report a 95% confidence interval for the mean accuracy of 500 data items. In the following figures, the error bar indicates this confidence interval.

### 4.1 Accuracy against Value of $k$

First, we report the predicted behavior of $k$-NN against the value of $k$ at several levels of noise, as shown in Figure 1. In this figure, the number of training instances is fixed at 32, and the target is a 3-of-5/2 concept. The lines indicate the theoretical results from the accuracy function, and the error bars represent the empirical results of Monte Carlo simulations. The theoretical results agree well with the empirical ones for both relevant attribute noise and class noise.

Figure 1 shows that the predictive accuracy of $k$-NN markedly drops off for each noise level when $k$ is an even number. This negative influence of an even number for $k$ on the accuracy is caused by a random determination of class when a tie occurs. This negative influence suggests that a choice of even number for $k$ is undesirable when applying $k$-NN.

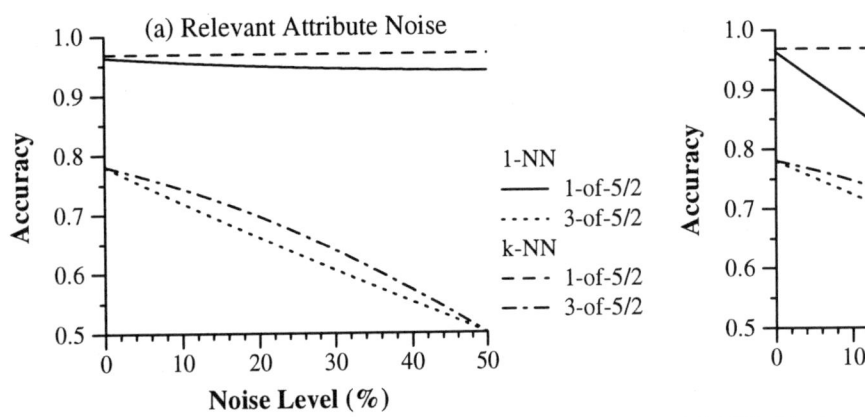
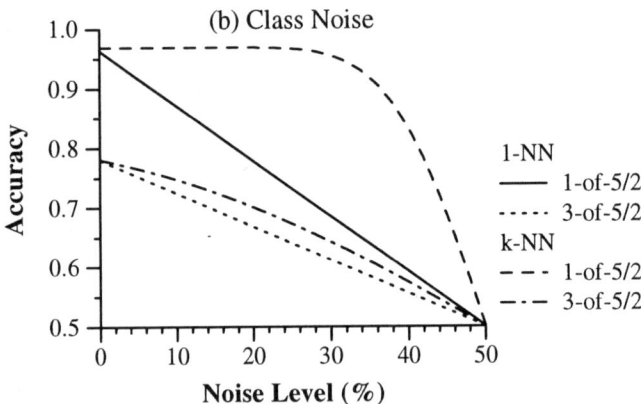

Figure 2: The effects of noise on the predictive accuracy of $k$-NN. Each curve for $k$-NN indicates the accuracy of $k$-NN with the optimal value of $k$. The number of training instances is fixed at 32.

In Figure 1, each circle represents the predictive accuracy for the optimal value of $k$ at the corresponding noise level. For each odd number for $k$, the accuracy of $k$-NN for a 0% noise level has two peaks. One appears at $k = 1$, while the other appears at the optimal value of $k$. In contrast, the accuracy for a 10% noise level and a 20% level have one peak at the corresponding optimal $k$. This is because the peak at $k = 1$ disappears due to the effect of noise.

### 4.2 Effects of Noise on the Accuracy

We further investigate the effects of noise on the predictive accuracy of $k$-NN, as shown in Figure 2. In this figure, the number of training instances is fixed at 32, and all curves come from the theoretical results. Each curve for $k$-NN represents the predictive accuracy of $k$-NN with the optimal value of $k$ at each noise level.

Figure 2(a) shows the effects of relevant attribute noise on the predictive accuracies of 1-NN and the optimal $k$-NN. When the noise level is 0%, the accuracy of 1-NN is comparable to that for the optimal $k$-NN, for both 1-of-5/2 and 3-of-5/2 concepts. However, the predictive accuracy of 1-NN almost linearly decreases with an increase in the noise level. For a 50% noise level, the accuracy of 1-NN equals that of a random prediction algorithm which predicts the same class as that for a randomly selected training instance. These observations suggest that 1-NN is strongly sensitive to relevant attribute noise. In contrast, the predictive accuracy of the optimal $k$-NN exhibits slower degradation. For the disjunctive concept (1-of-5/2 concept), the accuracy of the optimal $k$-NN is *not* greatly changed as the noise level increases.

Figure 2(b) shows the effects of class noise on the predictive accuracies of 1-NN and the optimal $k$-NN. For the 3-of-5/2 concept, both 1-NN and the optimal $k$-NN exhibit similar behavior to the corresponding tests with relevant attribute noise. However, the effects of class noise on the accuracy differ entirely from ones of relevant attribute noise for the disjunctive concept. The predictive accuracy of 1-NN linearly decreases to 0.5. In contrast, the optimal $k$-NN's accuracy does *not* substantially change until about a 30% noise level, whereafter it rapidly decreases to 50%.

These observations show that the predictive accuracy of 1-NN is strongly affected by both relevant attribute noise and class noise. Also, they suggest that we can restrain the degradation in the predictive accuracy of $k$-NN caused by an increase in noise level by optimizing the value of $k$.

### 4.3 Optimal Value of $k$

Finally, we give the relationship between the optimal value of $k$ and the number of training instances in noisy domains, as shown in Figure 3. In this figure, the optimal value of $k$ comes from the theoretical results, and the target is a 3-of-5/2 concept. In the following discussions, we use $N$ to refer to the number of training instances.

For a 0% noise level, the optimal value of $k$ remains $k = 1$ until $N = 28$. There is a rapid increase in the optimal $k$ at $N = 32$, and then the optimal $k$ almost linearly increases with an increase of $N$. This rapid increase is caused by the change of the peak given the highest accuracy from $k = 1$ to another (as mentioned in Section 4.1, $k$-NN's predictive accuracy has two peaks).

For each level (5%, 10%, and 30%) for both relevant attribute noise and class noise, the optimal value of $k$ is changed from $k = 1$ to another at small $N$. This observation can be explained by the strong sensitivity of the accuracy of 1-NN to both relevant attribute noise and class noise (as mentioned in Section 4.2). That is, the peak at $k = 1$ disappears due to the effect of noise, even though $N$ is a small number. After changing from $k = 1$ to another, the optimal value of $k$ almost linearly increases with an increase of $N$.

These observations from Figure 3 show that the optimal value of $k$ almost linearly increases with an increase of $N$ after the optimal $k$ is changed from $k = 1$ to another, regardless of the noise level for both relevant attribute noise and class noise. That is, the optimal value of $k$ strongly depends upon the number of training instances in noisy domains.

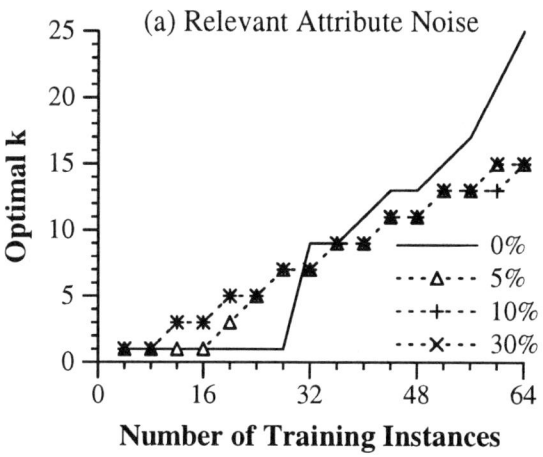
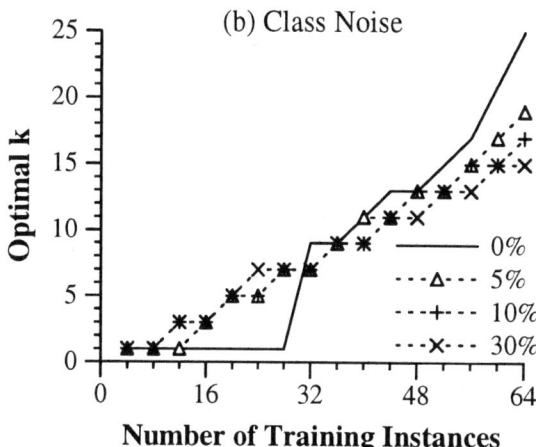

Figure 3: The optimal value of $k$ against the number of training instances for a 3-of-5/2 concept.

## 5 Conclusion

In this paper, we presented an average-case analysis of the $k$-nearest neighbor classifier ($k$-NN) for $m$-of-$n/l$ target concepts in noisy domains. Our analysis dealt with three types of noise: relevant attribute noise, irrelevant attribute noise, and class noise.

We formally defined the predictive accuracy of $k$-NN as a function of the domain characteristics. Using the accuracy function, we explored the predicted behavior of $k$-NN for each type of noise. The predictive accuracy of $k$-NN was given at various levels of noise, then the noise effects on $k$-NN's accuracy were shown. We also show that the optimal value of $k$ almost linearly increases with an increase in the number of training instances in noisy domains. Our analysis was supported with Monte Carlo simulations.

In the future, we will extend the framework of average-case analysis to relax many restrictions such as Boolean attributes, a fixed class of target concepts, and a fixed distribution over the instance space. Using the extended framework, we would like to analyze learning algorithms to give more useful insights into their practical applications.

## References

[Aha and Kibler, 1989] Aha, D. and Kibler, D. Noise-Tolerant Instance-Based Learning Algorithms. In *Proceedings of IJCAI-89*, pages 794–799. Morgan Kaufmann.

[Aha et al., 1991] Aha, D., Kibler, D., and Albert, M. Instance-Based Learning Algorithms. *Machine Learning*, 6:37–66.

[Albert and Aha, 1991] Albert, M. and Aha, D. Analyses of Instance-Based Learning Algorithms. In *Proceedings of AAAI-91*, pages 553–558. AAAI Press/MIT Press.

[Cover and Hart, 1967] Cover, T. and Hart, P. Nearest Neighbor Pattern Classification. *IEEE Transactions on Information Theory*, 13(1):21–27.

[Cover, 1968] Cover, T. Estimation by the Nearest Neighbor Rule. *IEEE Transactions on Information Theory*, 14(1):50–55.

[Creecy et al., 1992] Creecy, H., Masand, M., Smith, J., and Waltz, D. Trading Mips and Memory for Knowledge Engineering. *Communications of the ACM*, 35(8):48–63.

[Drakopoulos, 1995] Drakopoulos, J. Bounds on the Classification Error of the Nearest Neighbor Rule. In *Proceedings of ICML-95*, pages 203–208. Morgan Kaufmann.

[Langley and Iba, 1993] Langley, P. and Iba, W. Average-Case Analysis of a Nearest Neighbor Algorithm. In *Proceedings of IJCAI-93*, pages 889–894. Morgan Kaufmann.

[Murphy and Pazzani, 1991] Murphy, P. and Pazzani, M. ID2-of-3: Constructive Induction of $M$-of-$N$ Concepts for Discriminators in Decision Trees. In *Proceedings of IWML-91*, pages 183–187. Morgan Kaufmann.

[Okamoto and Satoh, 1995] Okamoto, S. and Satoh, K. An Average-Case Analysis of $k$-Nearest Neighbor Classifier. In *Proceedings of ICCBR-95* (Veloso, M. and Aamodt, A. Eds., *LNAI*, 1010), pages 243–264. Springer-Verlag.

[Okamoto and Yugami, 1996] Okamoto, S. and Yugami, N. Theoretical Analysis of the Nearest Neighbor Classifier in Noisy Domains. In *Proceedings of ICML-96*, pages 355–363. Morgan Kaufmann.

[Pazzani and Sarrett, 1992] Pazzani, M. and Sarrett, W. A Framework for Average Case Analysis of Conjunctive Learning Algorithms. *Machine Learning*, 9:349–372.

[Rachlin et al., 1994] Rachlin, J., Kasif, S., Salzberg, S., and Aha, D. Toward a Better Understanding of Memory-Based Reasoning Systems. In *Proceedings of ICML-94*, pages 242–250. Morgan Kaufmann.

# CASE-BASED REASONING

Case-Based Reasoning 2

# Learning to Integrate Multiple Knowledge Sources for Case-Based Reasoning[*]

**David B. Leake, Andrew Kinley, and David Wilson**
Computer Science Department
Lindley Hall 215, Indiana University
Bloomington, IN 47405, U.S.A.
{leake, akinley, davwils}@cs.indiana.edu

## Abstract

The case-based reasoning process depends on multiple overlapping knowledge sources, each of which provides an opportunity for learning. Exploiting these opportunities requires not only determining the learning mechanisms to use for each individual knowledge source, but also how the different learning mechanisms interact and their combined utility. This paper presents a case study examining the relative contributions and costs involved in learning processes for three different knowledge sources—cases, case adaptation knowledge, and similarity information—in a case-based planner. It demonstrates the importance of interactions between different learning processes and identifies a promising method for integrating multiple learning methods to improve case-based reasoning.

## 1 Introduction

The case-based reasoning (CBR) process solves new problems by retrieving records of problem solving for similar prior problems and adapting their solutions to fit new needs. Learning by acquiring new cases is an integral part of this process: each problem-solving episode itself provides a new case to save for future reuse. However, learning new cases is only one of many ways to learn within the CBR framework. CBR systems rely on at least four types of knowledge: the case base, indexing scheme, similarity criteria, and case adaptation knowledge. Each of these types of knowledge provides an opportunity for learning. Consequently, a multistrategy learning approach [Michalski and Tecuci, 1994] that improves multiple types of knowledge is promising for improving case-based reasoning. Because the information content of the different types of knowledge in a CBR system may overlap [Richter, 1995], learning that augments one type of knowledge can even help overcome deficiencies in the others. For example, learning new cases might reduce the need for case adaptation knowledge, by enabling the system to start from more relevant cases; conversely, learning new case adaptation knowledge might enable a system to solve a wider range of problems with its existing cases.

Developing the requisite learning methods for each knowledge type requires addressing questions about the learning mechanisms to use, how to integrate them, and the overall utility of adding them to the CBR process. A simple approach is to develop learning strategies for each knowledge type individually and then add them all to the CBR system. Learning methods exist, for example, for refining indexing criteria (see [Kolodner, 1993] for an overview); learning methods have also been applied to case adaptation knowledge [Hanney, 1997; Sycara, 1988]; and some CBR systems already combine multiple forms of learning [Hammond, 1989].

However, simply combining methods may not achieve the desired overall benefits, even if each method is effective individually. For example, Leake, Kinley, and Wilson [1996] describe tests in which case learning, and learning about case adaptation, each independently made solution generation much more effective, but when case learning was added to adaptation learning, the addition yielded minimal improvement over adaptation learning alone. One possible explanation would be that in these tests, adaptation learning alone was almost sufficient for optimal performance, leaving little room for improvement. However, tests described in this paper show that adding learning for another knowledge source can actually degrade performance: when the system learned both new cases and new adaptations, it was unable to retrieve the cases it needed in order to take full advantage of the learned adaptations. This interaction raises questions about how a CBR system can best exploit learning for each of its multiple knowledge sources.

---

[*]This work was supported in part by the National Science Foundation under Grant No. IRI-9409348.

This paper presents a case study examining the relationship of case learning, learning to refine case adaptation, and learning to refine similarity judgments in a case-based planning system. It considers two sets of issues: the requirements for each of the individual learning methods to be effective, and the requirements for realizing their full potential for improving overall system performance. It demonstrates the tight coupling of knowledge sources for CBR and shows that linking similarity assessment to learned adaptation knowledge can yield important benefits for exploiting both case and adaptation learning.

## 2 Motivations and Issues

This study grew out of research on learning to refine case adaptation. Case adaptation remains the least understood part of case-based reasoning, and experts agree that the state of the art in case adaptation is inadequate for automatic case adaptation to be included in fielded applications of CBR [Barletta, 1994; Mark et al., 1996]. One possible way to alleviate this problem is to develop new methods for automatic learning of case adaptation knowledge. The DIAL system, our testbed case-based planner, uses a hybrid approach to learning adaptations [Leake et al., 1996], building initial adaptations by reasoning from scratch and then reusing adaptations by case-based reasoning. Learning of adaptation cases takes place in tandem with learning of plan cases to be reused by the normal case-based planning process [Hammond, 1989].

Unlike previous case-based approaches to case adaptation (e.g., [Sycara, 1988]), DIAL's method reuses adaptations by derivational analogy [Carbonell, 1986; Veloso, 1994], replaying the *derivations* of previous adaptations to generate analogous adaptations, rather than transforming the solutions to prior adaptation problems. When the rationale for a problem-solving process is available, derivational approaches can increase problem-solving efficiency for the broad class of problems with similar derivations [Veloso, 1994]. In addition to recording and replaying the traces of adaptations done from scratch, DIAL also stores traces of user-performed adaptations for problems it cannot adapt, increasing the range of adaptation problems it can solve.

Both case learning and adaptation learning would be expected to reduce the effort expended on case adaptation. Case learning should increase the range of plans available as the starting point for reasoning, reducing the need to reason from distant plans requiring more adaptation. Adaptation learning should increase the availability of relevant adaptation knowledge, reducing the amount of effort required for each adaptation. Prior tests showed that as expected, the learning methods, used individually, each produced a marked improvement in the speed of case adaptation. Surprisingly, however, adding case learning to adaptation learning (the method that performed best individually) produced only small additional speedup when compared with the best of the individual learning methods (adaptation learning) [Leake et al., 1996].

We hypothesized that the problem might be caused by a mismatch between the system's similarity assessment criteria and the system's changing case adaptation abilities. To facilitate adaptation, similarity criteria should reflect adaptability [Birnbaum et al., 1991; Leake, 1992a; Smyth and Keane, 1996]. Thus when new adaptations are learned, similarity criteria should be modified to reflect changed adaptation abilities, in order to select the cases that will be easiest to adapt. However, early versions of DIAL—like most other CBR systems—relied on static similarity criteria. As a result, when it learned both new plan cases and new adaptations, the new plan cases it selected as most similar might be more difficult to adapt than plans that appeared less similar, but that involved problems it had learned how to adapt.

To link similarity assessment directly to adaptation knowledge, we developed a simple similarity assessment method called RCR (for Re-application costs and relevance) [Leake et al., 1996]. RCR estimates the cost of performing adaptations by using simple case-based reasoning about the costs of previous adaptations. Such a method makes learning to refine similarity a natural side-effect of adaptation learning, but also has two potential drawbacks: either generating inaccurate similarity judgments (if the costs of the previous adaptations retrieved turn out to be poor predictors), or imposing excessive computational overhead, because of embedding another case-based reasoning process within the main CBR cycle. Consequently, we asked four questions:

1. Whether the linkage between similarity and adaptation knowledge provided by RCR similarity assessment can markedly decrease case adaptation effort when case learning and adaptation learning are used together.

2. How the *overall* planning efficiency of DIAL is affected by RCR and adaptation learning.

3. How the total planning cost breaks down into costs of RCR similarity assessment versus case adaptation.

4. How adaptation learning and case learning affect the range of problems that the system can solve.

After a synopsis of the learning methods investigated and how they are applied, this paper examines these four issues. It briefly addresses the first issue, which is considered in depth in Leake, Kinley, & Wilson [1997], and focuses primarily on the remaining three.

# 3 Task domain and basic processing sequence

DIAL's task domain is disaster response planning: the strategic planning used to guide damage assessment, evacuations, etc., in response to natural and man-made disasters such as earthquakes and chemical spills. Human disaster response planners appear to depend heavily on prior experiences when they address new problem situations [Rosenthal et al., 1989], making it a natural task domain for case-based reasoning. For example, when generating a response plan to bring help to an isolated area, a previously-generated plan for another isolated area may provide helpful information for planning emergency transportation.

DIAL generates disaster response plans for disasters reported in simple (1-2 line) news stories. The system includes a simple schema-based story understanding component that processes conceptual representations of news stories describing the initial events in a disaster, and a retrieval component that selects a prior response plan expected to be easily adaptable to the new disaster. Problems in the retrieved plan are detected by a simple evaluator for candidate response plans (based on the problem-detection process described in [Leake, 1992b], and supplemented by inputs from a human user).

When problems are found, a description of the problem in a pre-defined problem vocabulary is provided to the adaptation component. That component can either build up adaptations from scratch or by case-based reasoning starting from previous adaptations. During adaptation, DIAL learns by storing traces of its case adaptation process and of the memory search process used to find needed information. For example, if it performs a substitution to replace an unavailable object (e.g., supplies were previously delivered by the Red Cross but there is no Red Cross in the country where the new disaster occured), the stored memory search trace records the path it followed to find a substitution (e.g., moving from a memory node for Red Cross to the memory node for its abstraction of relief organizations, and then moving to specifications of that node). More complete descriptions of the system are available in [Leake et al., 1996].

# 4 Types of learning

**Response plan case learning:** DIAL begins its processing supplied with a small library of hand-coded disaster response plans, using a representation analogous to that used by CHEF [Hammond, 1989]. When new disasters are encountered, these response plan cases are reapplied by transformational analogy, changing components as needed to fit new constraints. The results are then stored for future reuse, adding to the case library. Because this process is a standard part of case-based planning systems, we will not discuss it further. When DIAL is unable to generate a suitable plan autonomously, its plan library can be augmented by user-generated plans, increasing the range of problems the system can solve autonomously, as described in the following paragraphs.

**Adaptation case learning:** As described in [Leake et al., 1996], DIAL's initial case adaptation knowledge is a small set of abstract transformation rules and a library of domain-independent "weak methods" for memory search (e.g., the "local search" strategy to find related concepts by considering nearby nodes in memory). When presented with a new adaptation problem, DIAL first selects a transformation rule to apply and then performs memory search to find the information needed to operationalize the transformation rule and apply it to the problem at hand (for example, if a *substitution* transformation is selected, to find what to substitute). Once a successful adaptation has been generated, the system saves a trace of the steps used in solving the adaptation problem for future reuse. In this way, the system learns specific adaptation procedures starting from domain-independent adaptation methods when no specific knowledge is available. Adaptation cases may themselves be "adapted" in a simple way: If the derivation does not identify a solution, "local search" considers alternatives near the one suggested by the derivation, terminating its search after reaching a user-defined limit on the number of nodes visited. When the process terminates without finding an adaptation, the user can guide the system through the adaptation process to generate the new plan. Both the new adaptation and the resulting plan are stored for future use.

DIAL's adaptation cases have two basic parts: indexing information and adaptation information. The indexing information includes a representation of the type of problem to adapt and information about the response plan for which the adaptation case was generated. The problem description information is similar in spirit to the problem vocabularies used to guide adaptation in other CBR systems (e.g., [Leake, 1992b]), and serves as an index to guide retrieval of adaptation cases to use for new adaptation problems. The problem vocabulary divides problems according to categories such as UNAVAILABLE-FILLER and LACK-OF-ACCESS. Each problem type is associated with a structure to be filled by a fixed range of descriptive information (e.g., the particular role, filler, and attempted action involved). To streamline access to relevant adaptation cases, stored adaptation cases are organized in memory by the types of problems they address.

The adaptation information packages a transformation type (e.g., substitute, add, delete) and a pointer to a memory search case containing the memory search steps used to find the information needed to apply the transformation. The memory search steps are described in terms of a vocabulary of standard memory operations, such as extracting a role-filler or moving up the abstraction hierarchy in memory.

The adaptation information is used both to guide future adaptations and to estimate their cost. Once an adaptation case has been retrieved, the cost of memory search dominates all other costs involved in adaptation. Consequently, the cost can be approximated by the memory search cost involved in replaying the stored memory search trace.

**Similarity learning:** The RCR similarity assessment method predicts the cost of adapting a problem in a case-based way, using learned adaptation knowledge. Given a new disaster situation and a candidate response plan with applicability problems, RCR first retrieves the adaptation cases most relevant to the current problem types, one for each problem to adapt, using the problem description as an index into the library of adaptation cases. It next estimates the cost to re-apply each of the adaptation cases retrieved, based on the length of its adaptation derivation.

Ideally, in similar future contexts, replaying the derivation will lead to an analogous result that applies to the new context, so that the length of the stored derivation suggests the re-application cost. However, differences between the old and new problems may prevent the prior derivation from being directly applicable, increasing the cost of adaptation. Consequently, the estimated cost is multiplied by a "dissimilarity" factor based in a simple way on the semantic similarity of old and new situations. To calculate the dissimilarity factor DIAL simply sums semantic distances between role-fillers in the problem descriptions, according to its memory hierarchy. The benefits of RCR compared to alternative methods are discussed in Leake, Kinley, and Wilson [1997].

Note that because RCR focuses on the difficulty of adapting problems, a response plan that requires several simple adaptations could be chosen over a response plan that requires a single difficult adaptation. Because this similarity learning method focuses on finding the cases that are easiest to adapt (those with the least important differences), it differs from learning methods such as Prodigy/Analogy's [Veloso, 1994] "foot-print" similarity metric that are aimed at learning situations with the most relevant similarities. RCR is in the spirit of Smyth and Keane's [1996] adaptation-guided retrieval, but learns about the difficulty of adaptations from experience rather than using static criteria to estimate adaptation cost.

Thus DIAL's learning mechanisms include response plan learning, by CBR/transformational analogy; adaptation learning, by CBR/derivational analogy applied to traces of internal processing or user adaptations; and similarity learning, by CBR/transformational analogy applied to previous adaptations. The combination of methods allows different lessons to be drawn from a single episode and reapplied independently in new contexts.

## 5 Effects of Individual and Combined Learning Strategies

To answer the questions listed in section 2, we performed a series of tests. These tests compared DIAL's performance under five conditions: No learning of either cases or adaptations (NL); case learning, of plan cases only—the standard learning of case-based planners (CL); adaptation learning, in which only adaptation cases are stored (AL); learning of both response plan cases and adaptation cases (AL+CL); and learning of both response plan cases and adaptation cases, using the RCR method to base similarity assessment during plan retrieval on learned adaptation cases (AL+CL+RCR). Each condition except the last used traditional semantic similarity for case retrieval, with ties broken by a simple count of the number of problems in a plan case requiring adaptation.

The initial memory for the trials included nodes for 1264 concepts and an initial case library containing 5 response plans for earthquake, air quality, flood, and fire disasters. During testing, DIAL processed conceptual representations of 18 news stories (7 floods, 5 earthquakes, 4 forest fires, and 2 industrial air quality problems). Generating response plans for these disasters required generating 119 adaptations, each of which was stored as a new adaptation case. These experiments extended the trials reported in [Leake et al., 1996], which processed 5 stories, resulting in 30 adaptation cases.

Test runs were divided into two sets. Processing of the first third of the adaptation problems was treated as a learning phase to build up initial knowledge sources, and statistics were gathered on the remaining two thirds of the adaptations.

**Effects of linking similarity and adaptation knowledge on adaptation efficiency:** The measure used for adaptation efficiency was memory search effort, calculated by two machine-independent measures: the number of memory nodes visited, and the number of primitive memory search operations performed. Figure 1 shows that both case learning and adaptation learning individually provide large efficiency increases over no learning (as expected), while adaptation efficiency with AL+CL provides smaller gains over adaptation learning alone. The results for the first four cases are consistent with those of [Leake et al., 1996].

The fifth result, for AL+CL+RCR, suggests the potential benefits of directly linking similarity judgments to learned adaptation knowledge. The tests do not,

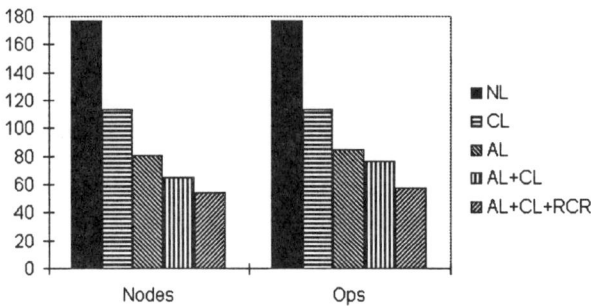

Figure 1: Average adaptation costs.

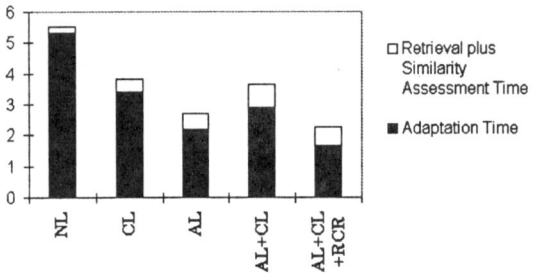

Figure 2: Overall processing costs.

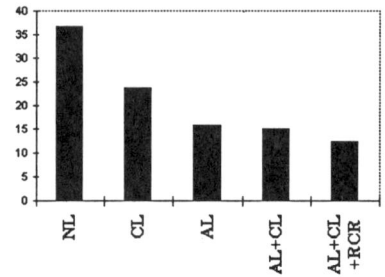

Figure 3: Failure rates for the adaptation process.

however, address another crucial question: whether the combination of AL+CL+RCR improves overall planning performance. RCR similarity assessment involves retrieving adaptation cases applicable to all the problems in a set of candidate cases, possibly imposing considerable overhead on the retrieval/similarity assessment process. This led us to examine the overall efficiency of generating response plans, measured by execution time.

**Overall Planning Efficiency and Cost Breakdown:** Figure 2, illustrating execution time in CPU seconds, shows that AL+CL+RCR in fact provided some improvement over the other conditions in terms of overall planning time. The light bands at the top of the bars show the portion of the execution time for the retrieval/similarity assessment process. Interestingly, although AL+CL decreased the machine-independent measures of adaptation effort compared to AL (as shown by figure 1), it resulted in a noticeably worse total execution time than AL. This is partially due to increased retrieval time of retrieving from growing sets of plan and adaptation cases. However, it also appears that the machine-independent measures do not completely capture the factors affecting adaptation time.

**Effects on the Range of Problems the System Can Solve:** DIAL's initial domain theory is incomplete, but its ability to store and reuse user-provided solutions (both disaster response plans and adaptations) enables it to augment its knowledge. Consequently, its learning affects not only efficiency, but also the range of problems that it can solve. Figure 3 shows the percentage of the trial problems the system is unable to solve autonomously after the learning phase on the test set of problems. The combination of case learning and adaptation learning performed better than either method alone, and the improved case selection of AL+CL+RCR increased the proportion of problems that the system could solve compared to AL+CL based on static semantic similarity criteria. However, these differences are small and possibly insignificant.

## 6 Lessons and Directions

These results suggest the importance of adjusting similarity criteria—and hence the selection of learned cases—to keep pace with adaptation learning. By enabling more effective use of two types of learned knowledge, integrating different types of learning improved both the speed of processing and the range of problems that DIAL could solve. These pilot experiments raise many issues for future study: the relative importance of adaptation and case learning in different domains, the learning curve for each type of knowledge, the utility of case-based similarity assessment methods (like RCR) as the number of adaptation cases increases, whether it is possible to partially alleviate the utility problem by retaining only a subset of the many adaptation cases that are generated, and the tradeoffs and utility of alternative methods for retrieving and applying adaptation knowledge during case selection. However, the results support the potential value of coordinating different types of learning in CBR and the need for further investigation.

# 7 Conclusion

Case-based reasoning exploits multiple knowledge sources. Consequently, it provides an opportunity for multistrategy learning to refine each of those knowledge sources. Our studies of multistrategy learning in the case-based planner DIAL provide—to our knowledge—the first empirical demonstrations of the complementary roles that can be played by these multiple learning processes. However, they also show that the learning strategies must be coordinated to realize their potential benefit. Similarity criteria for selecting cases must change as adaptation knowledge is learned; neither coverage of the case library, nor case adaptation abilities, can be judged in isolation from the other knowledge sources. Likewise, in developing a combined method, how one type of learning affects the efficiency of one component of the CBR process is secondary to the efficiency effects of the learning on the CBR process as a whole. The close coupling of multiple processes and knowledge sources in CBR complicates the application of learning to each one, but also provides a new motivation for combined learning: Combined learning can enable a CBR system to better exploit the relationships between multiple types of knowledge.

# References

[Barletta, 1994] R. Barletta. A hybrid indexing and retrieval strategy for advisory CBR systems built with ReMind. In *Proceedings of the Second European Workshop on Case-Based Reasoning*, pages 49–58, Chantilly, France, 1994.

[Birnbaum et al., 1991] L. Birnbaum, G. Collins, M. Brand, M. Freed, B. Krulwich, and L. Pryor. A model-based approach to the construction of adaptive case-based planning systems. In R. Bareiss, editor, *Proceedings of the DARPA Case-Based Reasoning Workshop*, pages 215–224, San Mateo, 1991. Morgan Kaufmann.

[Carbonell, 1986] J. Carbonell. Derivational analogy: A theory of reconstructive problem solving and expertise acquisition. In R. Michalski, J. Carbonell, and T. Mitchell, editors, *Machine Learning: An Artificial Intelligence Approach*, volume 2, pages 371–392. Morgan Kaufmann, Los Altos, CA, 1986.

[Hammond, 1989] K. Hammond. *Case-Based Planning: Viewing Planning as a Memory Task*. Academic Press, San Diego, 1989.

[Hanney, 1997] K. Hanney. Learning adaptation rules from cases. Master's thesis, Trinity College, Dublin, 1997.

[Kolodner, 1993] J. Kolodner. *Case-Based Reasoning*. Morgan Kaufmann, San Mateo, CA, 1993.

[Leake et al., 1996] D. Leake, A. Kinley, and D. Wilson. Acquiring case adaptation knowledge: A hybrid approach. In *Proceedings of the Thirteenth National Conference on Artificial Intelligence*, pages 684–689, Menlo Park, CA, 1996. AAAI Press.

[Leake et al., 1997] D. Leake, A. Kinley, and D. Wilson. Case-based similarity assessment: Estimating adaptability from experience. In *Proceedings of the Fourteenth National Conference on Artificial Intelligence*. AAAI Press, 1997.

[Leake, 1992a] D. Leake. Constructive similarity assessment: Using stored cases to define new situations. In *Proceedings of the Fourteenth Annual Conference of the Cognitive Science Society*, pages 313–318, Hillsdale, NJ, 1992. Lawrence Erlbaum.

[Leake, 1992b] D. Leake. *Evaluating Explanations: A Content Theory*. Lawrence Erlbaum, Hillsdale, NJ, 1992.

[Mark et al., 1996] William Mark, Evangelos Simoudis, and David Hinkle. Case-based reasoning: Expectations and results. In D. Leake, editor, *Case-Based Reasoning: Experiences, Lessons, and Future Directions*, pages 269–294. AAAI Press, Menlo Park, CA, 1996.

[Michalski and Tecuci, 1994] R. Michalski and G. Tecuci, editors. *Machine Learning: A Multistrategy Approach*. Morgan Kaufmann, San Mateo, CA, 1994.

[Richter, 1995] Michael Richter. The knowledge contained in similarity measures. Invited talk, the First International Conference on Case-Based Reasoning, Sesimbra, Portugal., October 1995.

[Rosenthal et al., 1989] U. Rosenthal, M. Charles, and P. Hart, editors. *Coping with crises: The management of disasters, riots, and terrorism*. C.C. Thomas, Springfield, IL, 1989.

[Smyth and Keane, 1996] B. Smyth and M. Keane. Design à la Déjà Vu: Reducing the adaptation overhead. In D. Leake, editor, *Case-Based Reasoning: Experiences, Lessons, and Future Directions*. AAAI Press, Menlo Park, CA, 1996.

[Sycara, 1988] K. Sycara. Using case-based reasoning for plan adaptation and repair. In J. Kolodner, editor, *Proceedings of the DARPA Case-Based Reasoning Workshop*, pages 425–434, San Mateo, CA, 1988. Morgan Kaufmann.

[Veloso, 1994] M. Veloso. *Planning and Learning by Analogical Reasoning*. Springer Verlag, Berlin, 1994.

# Aggregating Features and Matching Cases on Vague Linguistic Expressions

Alfons Schuster[*], Werner Dubitzky[*§], Philippe Lopes[**], Kenneth Adamson[*], David A. Bell[*],
John G. Hughes[*], John A. White[***]

University of Ulster, [*]Faculty of Informatics, [§]Northern Ireland Bio-Engineering Centre
Co. Antrim, BT37 0QB
Northern Ireland

[**]University of Wales
Dep of Physical Education, Sport and Exercise Science Unit
Aberystwyth, Dyfed, SY23 3DE
Wales

[***]Department of Public Health Medicine and Epidemiology
Queen's Medical Centre
Nottingham NG7 2UH
England

## Abstract

Decision making based on the comparison of multiple criteria of two or more alternatives, is the subject of intensive research. In many decision making situations, a single criterion consists of more than one piece of information, and therefore might be regarded as a lump of aggregated information. This paper proposes a general method for aggregating information. To accomplish information aggregation we have developed a fuzzy expert system. Results from an application of our approach in the domain of Coronary Heart Disease Risk Assessment (CHDRA) indicate the value of the information aggregation process of the system. We also show in this paper, how a case-based reasoning (CBR) system can greatly benefit—in its time performance and ability to manage uncertainty—from the information aggregation method.

## 1 Introduction

Decision making situations very frequently require an ability to compare multiple criteria of two or more alternatives [Chen and Hwang, 1992]. In many cases a single criterion has a complex structure, but even if the meaning of such a complex criterion can be represented in terms of simpler ones, the need for a higher level entity persists [Wilensky, 1986].

As an example, consider a job application scenario, where a manager is confronted with the decision to choose between two candidates, A and B. Let us suppose the manager decides to employ candidate A. Then, when asked to explain his decision, he might say that comparing the two candidates, A has *better prepared* and also showed a *better personality profile*. A closer look at the applicants' documents will probably show, that candidate A indeed has *better references, more experience* and *better working skills* than B. Furthermore, asked in more detail about the *better personality profile*, the manager might answer that candidate A had *better communication skills* and was *more confident* during the interview—therefore his decision was right.

So:

- All information about the candidates is expressed by rather vague or imprecise linguistic terms.

- To explain his decision, the manager uses the linguistic terms *better prepared* and *better personality profile*, rather than details (*better references, more experience, better working skills, better communication skills, more confidence*).

- The manager would possibly have arrived at the same choice (candidate A) if his only information was that candidate A is *better prepared* and has a *better personality profile* than candidate B.

Our observations are:

(1) Many decision making situations require the capability to manage and process vague or imprecise information, perhaps via linguistic terms.

(2) In many decision making situations, information can be crudely but usefully classified as *higher level information*, and *lower level information* respectively.

In CBR for example, complex case features represent higher level information, and primitive case features represent lower level information respectively. Higher level information can be composed of: (a) lower level information, (b) other higher level information, or (c) a mixture of both information types. In our job application scenario, the higher level information *better preparation* aggregates the lower level information *better references, more experience* and *better working skills*.

And so: It is possible to arrive at useful decisions using information at various levels.

A study of this analysis might lead to a hierarchical structure of information as it is shown in Figure 1.

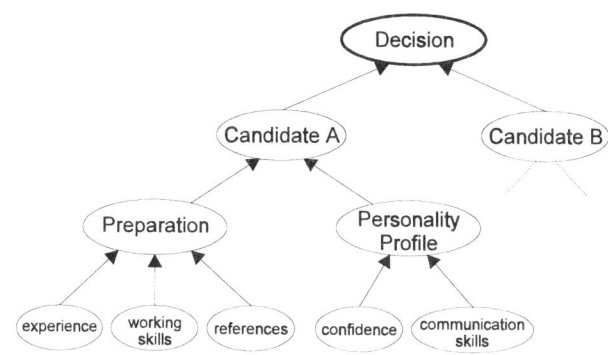

Figure 1: Hierarchical structure of information.

The purpose of this paper is to propose a method that aggregates available lower level information to units of higher level information. In many cases, the lower level information will be imprecise or vague, and so the proposed method should be able to manage uncertainty [Bonissone, 1985]. We therefore have used an expert system shell to develop a fuzzy expert system that accomplishes such aggregation. The applicability and usefulness of this approach was tested in the domain of CHDRA. We show that our fuzzy expert system is able to manage the uncertainty contained in the aggregation process, and that its reasoning process (information aggregation) leads to meaningful, consistent and valuable outcomes. Furthermore, we show how a CBR system can benefit from an implementation of this information aggregation method. For example, we model primitive case features and complex case features via fuzzy sets. This, significantly increases the CBR system's ability to manage uncertainty.

The remainder of this paper is organized as follows: Section 2 briefly describes the advantages that fuzzy primitive and fuzzy complex case features provide for CBR. In Section 3, we describe our fuzzy expert system and the information aggregation process in more detail. The results of applying the approach are outlined in Section 4. Finally, in Section 5, we finish with a discussion, conclusions and future work.

## 2 CBR, Fuzzy Primitive Case Features and Fuzzy Complex Case Features

CBR is a problem-solving model that allows reasoning to be performed by using past experience [Brown, 1992; Kolodner, 1993, Riesbeck and Schank, 1989]. Past experience—i.e., knowledge about situations that have been solved in the past—is represented by entities, called *cases*. These cases are stored and organized in a memory-like construct called a *case knowledge base* or simply *case base*. Reasoning in CBR systems is accomplished by retrieving the base case(s) most relevant to a new situation or problem at hand, called the *query case*, and then adapting the solution(s) to the actual problem. Because the knowledge contained in a case base is basically determined by its constituents (the stored base cases) the representation of base cases is an important issue in CBR. We describe cases in a compact, characteristic fashion by *abstract* or *salient features*, here simply referred to as *features*. There exist two types of features, primitive features and complex features. Complex features are composites of several (primitive or complex) features.

For example, in the domain of CHDRA the features Smoking and Cholesterol have been identified (among other factors) to be main risk factors for myocardial infarction and subsequent sudden death. In the assessment process, Smoking is regarded as a primitive feature, used to indicate the number of cigarettes a person smokes per day, whereas the complex feature Cholesterol is a composite of the three primitive feature cholesterol types: TOTAL cholesterol, LDL cholesterol and HDL cholesterol.

Cholesterol travels in the blood in distinct particles called lipoprotein. The two major types of lipoproteins are low-density lipoproteins (LDL) and high-density lipoproteins (HDL). LDL, often called 'bad' cholesterol, delivers the cholesterol to the arterial walls with the ultimate consequence of narrowing the arteries [Slyper, 1994]. HDL, often called 'good' cholesterol, protects against heart disease by removing excess cholesterol from the blood [Gordon et al., 1989]. In a fasting blood test, a clinician first finds out what a person's TOTAL cholesterol level is. If the TOTAL cholesterol level is too high then further measurements of LDL and HDL are required (note: a *high* HDL value 'compensates' a *high* TOTAL cholesterol value, and therefore, a person's cholesterol can be still described as *normal*). In this paper we use 'cholesterol' when we are discussing generally, and Cholesterol when we talk about a complex case feature; but they mean the same thing—an aggregate or composite of three cholesterol type values.

Possible instances of Smoking and Cholesterol might be given by [Smoking/<40cigarettes per day>], and [Cholesterol/<TOTAL 4.6 mmol/l >, <LDL 3.0 mmol/l >, <HDL 1.0 mmol/l >].

Frequently, it is not possible to obtain or assess a value of a feature precisely [Dubitzky et al., 1995]. In situations like this, use is often made of linguistic terms. For example, it is not possible to 'measure' a person's cholesterol value, because it is a composite of three cholesterol type values. But, asked about it, the doctor might describe the person's cholesterol to be simply as *normal*, rather than state: TOTAL 4.6, LDL 3.0 mmol/l, and HDL 1.0 mmol/l. Even in situations where precise values are obtainable, humans often fall back upon to use vague or imprecise linguistic terms. For example, the doctor might describe a LDL value of 3.0 mmol/l simply as *normal*, and one of

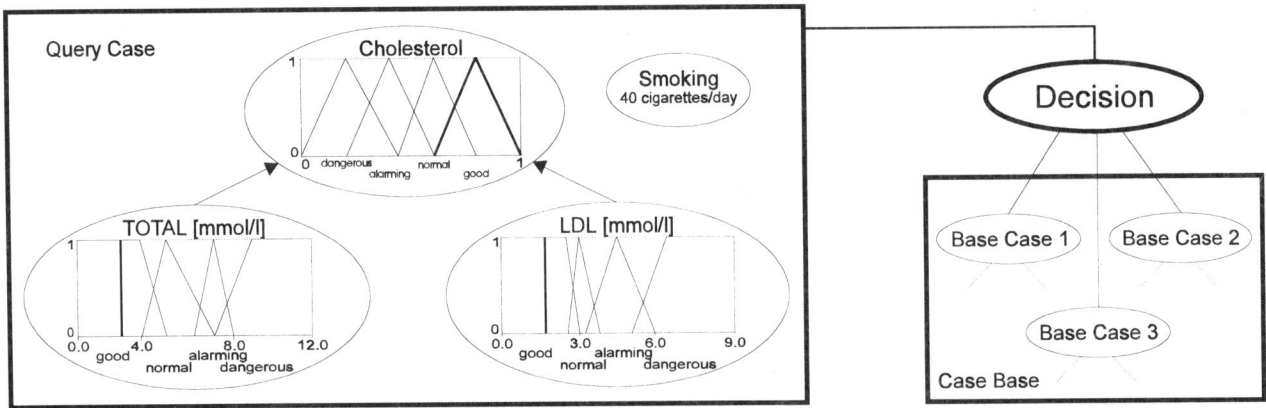

Figure 2: Fuzzy primitive and fuzzy complex case features.

4.5 mmol/l simply as *alarming*. Furthermore, there is no exact boundary between a *normal* and an *alarming* LDL value; that is, the transition between *normal* and *alarming* is gradual or fuzzy, rather than abrupt or crisp [Zadeh, 1973; Klir and Folger, 1988]. In this work we use fuzzy set theory to model primitive features and complex features, and therefore those features will be called fuzzy primitive features, and fuzzy complex features respectively [Na and Park, 1996; Dubitzky et al., 1995]. Figure 2 illustrates the aforementioned CHDRA example in a CBR context. It shows, a query case consisting of the primitive feature Smoking and the fuzzy complex feature Cholesterol, where the fuzzy complex feature Cholesterol is a composite of the fuzzy primitive features Total and LDL (for the sake of simplicity the fuzzy primitive feature HDL is omitted in Figure 2). The cases: Base Case 1, Base Case 2 and Base Case 3 in Figure 2 constitute a simplified case base.

The reasoning process performed should retrieve the base case(s) most relevant to the query case. Therefore, corresponding feature-value pairs for Smoking, Total and LDL of the query case and the base case(s) have to be compared and to be aggregated to an overall similarity score.

## 3 Information Aggregation via a Fuzzy Expert System

Instead of performing the reasoning task by comparing each single feature value (Smoking, TOTAL, LDL and HDL) of the query case and the base case(s), our approach allows us to compare features on a fuzzy complex feature level. This means that only the feature values Smoking and Cholesterol of the query case and the base case(s) are used. This reasoning process has two main advantages: firstly, for a very large case base the promise of better performance; secondly, sometimes data at primitive feature level is not available, but a description is available on complex feature level. For example, patients may not know the value of each cholesterol type, but possibly remember that during their last health test the cholesterol was *normal*.

To make information on fuzzy complex feature level available, we have developed an inference process based on fuzzy set theory that maps (aggregates) fuzzy primitive feature values to fuzzy complex feature level. For example, the two fuzzy primitive feature values <Total/3.0> and <LDL/2.0> in Figure 2, might map on fuzzy complex feature level to <Cholesterol/(good)>. Such a mapping (aggregation) should satisfy the following requirements:

(1) The aggregated values on complex feature level should be intuitively appealing to an expert's understanding of the problem in question.

(2) Using aggregated values on complex feature level in a decision making process should lead to meaningful, justifiable and consistent results.

To manage the proposed information aggregation process, the normal steps of knowledge acquisition, knowledge representation, and design of an inference engine were realized.

Within the knowledge acquisition process for our application the knowledge engineer and the domain expert were involved to extract the domain knowledge for its use in the fuzzy expert system (e.g., establishing the various fuzzy sets for the different cholesterol types). The basis for the knowledge acquisition was a data set, consisting of 133 records. One record for each person initially held values for TOTAL cholesterol, LDL cholesterol and HDL cholesterol, as well as the two ratios TOTAL/HDL and LDL/HDL. These two ratios are also important because they provide more meaningful indicators of coronary heart disease risk than TOTAL cholesterol per se [Kinosian et al., 1994]. The expert was asked to provide expertise for determining each person's cholesterol value, and so was asked to indicate one of the fields (*dangerous, alarming, normal* and *good*) for each data record as illustrated in Table 1.

| Nr. | Cholesterol Data | | | | | Expert's Decision | | | |
|---|---|---|---|---|---|---|---|---|---|
| | Total | LDL | HDL | Total/HDL | LDL/HDL | dangerous | alarming | normal | good |
| 1 | 6.5 | 4.9 | 1.2 | 5.4 | 4.1 | | X | | |
| 2 | 6.3 | 4.7 | 1.0 | 6.3 | 4.7 | | X | | |
| ... | ... | ... | ... | ... | ... | | | | |
| 133 | 6.4 | 4.4 | 0.6 | 10.7 | 7.3 | X | | | |

Table 1: Cholesterol values, taken from 133 persons. Associated with each data record is an expert's decision, representing the expert's interpretation of the person's cholesterol value.

Typically the category that a cholesterol type value or ratio value belongs to is expressed in intervals [Pyorala et al., 1994]. For example, a TOTAL/HDL ratio between 4 and 4.5 is considered as *good*, and one below 4 is regarded to be even *better*. There is no doubt that such a representation is not intuitive to a human's understanding of the problem. In our understanding, the transition from *good* to *better* should be gradual, rather than abrupt. To represent such categories, the three different cholesterol types, and the two ratios are modeled via fuzzy sets. As an example, Figure 3 shows the fuzzy sets for the cholesterol type TOTAL, and for the ratio TOTAL/HDL.

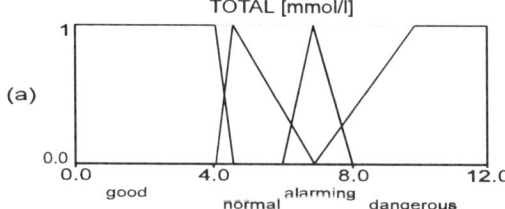

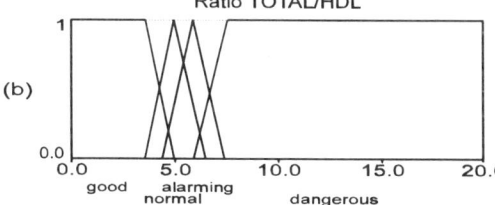

Figure 3: Fuzzy sets for (a) the cholesterol type TOTAL and (b) the ratio TOTAL/HDL.

The knowledge representation scheme used for the proposed information aggregation process was that of production rules, formulated as if-then statements, where the if-part of a rule (the antecedent) is the input, and the then-part of the rule (the consequent) is the output of the fuzzy expert system. Here, the rule base consisted of four rules only. As an example, Figure 4 shows a typical rule used in the fuzzy expert system.

A crucial concept of the proposed fuzzy expert system is, that all rules apply at all times, but some may have more influence than others. This means that if more than one rule is active, the separate responses have to be combined to a composite output. This idea is central to fuzzy logic systems.

**If** (Total is dangerous) or (LDL is dangerous) or (HDL is dangerous) or (Ratio_Total_HDL is dangerous) or (Ratio_LDL_HDL is dangerous)

**Then** Cholesterol is dangerous

Figure 4: Example rule.

Therefore, the inference process performed by the fuzzy expert system consists of three sub-processes: (a) scaling of the fuzzy input, (b) combination of the output, and (c) defuzzification of the output. There exist different methods for all three sub-processes, and it is part of the knowledge engineer's work to find the methods appropriate to the actual problem. Scaling was done via the *correlation-product* encoding, the combination step via *sum combination* and finally, for the defuzzification of the output, the *center of gravity* method is applied. As an example, Figure 5a shows the fuzzy sets for the system's output (cholesterol), and Figure 5b shows a possible output activation.

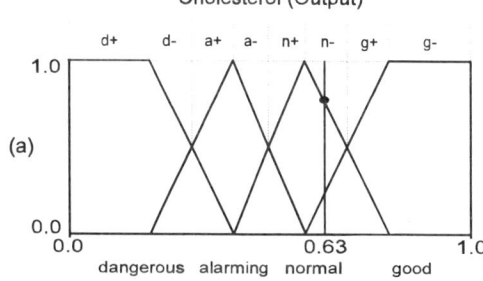

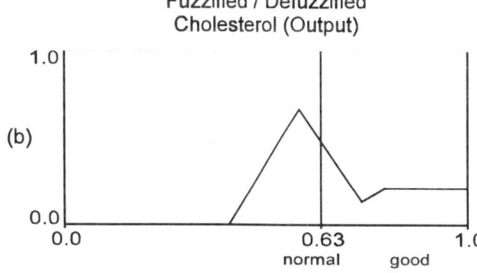

Figure 5: (a) Fuzzy sets for the cholesterol output, and (b) activated cholesterol output and defuzzification via the center of gravity method.

| Nr. | Cholesterol Data | | | | | Expert's Decision | | | | System Output | |
|---|---|---|---|---|---|---|---|---|---|---|---|
| | Total | LDL | HDL | Total/HDL | LDL/HDL | dangerous | alarming | normal | good | COG | Cholesterol |
| 1 | 6.5 | 4.9 | 1.2 | 5.4 | 4.1 | | X | | | 0.63 | a+ |
| 2 | 6.3 | 4.7 | 1.0 | 6.3 | 4.7 | | X | | | 0.49 | d− |
| ... | ... | ... | ... | ... | ... | | | | | | ... |
| 133 | 6.4 | 4.4 | 0.6 | 10.7 | 7.3 | X | | | | 0.28 | d+ |

Table2: Expert's decision and system output for each cholesterol data record.

Figure 5b also illustrates that the two fuzzy sets *normal* and *good* have been activated by the rules, and that the defuzzification of the output via the center of gravity method results in an output value of 0.63. The location of this value in Figure 5b also shows, that the aggregated cholesterol should be interpreted as *normal* 'with a tendency' to *good*. The tendency of an output is indicated here by a plus (+) or a minus (−) sign, attached to the corresponding fuzzy set and derived as illustrated in Figure 5a. The output value (0.63) intersects with the fuzzy sets *normal* and *good*. When the value intersects with two or more fuzzy sets like this, we take and qualify by 'tendency' the fuzzy set where the output value intersects with the highest score. In this example, the system's output would look like <Cholesterol/n−>, and should be interpreted as: 'The aggregated cholesterol value of the person is *normal* with a tendency to *good*'. Such a result is intuitively appealing and close to a human expert's explanation in such a situation. In the next section, we investigate the usefulness, the validity and the consistency of the system's output.

## 4 Results

After the inference process was accomplished for all 133 data records, the fuzzy expert system output of each record was compared with the expert's judgment of the record in question. Table 2 is similar to Table 1, but additionally contains two columns for the system's output. The first column displays the center of the gravity (COG) of the system output, and the second column shows, the system's decision on the cholesterol for the corresponding record.

The results have been evaluated in two steps. In the first step the number of direct matches was computed, and in the second step the number of 'tendency' matches. A direct match was considered to be the case when the expert and the system evaluated the data record belonging to the same category. For example, this is the case for the first and the last record in Table 2. Both the expert and the system evaluated the first record to be *alarming* and the last record to be *dangerous*. Record two in Table 2 represents a tendency match. The expert considers the cholesterol of record 2 to be *alarming*, whereas the system's response is d−, which means *dangerous* with a tendency to *alarming* (see Figure 5a). This is a meaningful result because, as pointed out before, the transition from *alarming* to *dangerous* is gradual.

Our approach led to the following results. A direct match happened 83 times, and a tendency match 34 times. Therefore, the system derived 118 meaningful results, i.e. in 88.7% of the sample. The inference process was not to be expected to establish a direct match of 100% for a number of reasons. Firstly, asked about the same situation or problem twice (e.g. repeated after some weeks), even a single expert's decision-making diverges very often. Secondly, when several experts are available, it is very likely, that they will disagree in some cases. Thirdly, during knowledge acquisition, the expert was enforced to chose one of the four categories (*dangerous*, *alarming*, *normal*, *good*) for a record invoking one of the weaknesses of a discrete choice; very often it is not possible to express intermediate values.

## 5 Discussion, Conclusions and Future Work

A general method to aggregate information has been presented. Based on fuzzy set theory and fuzzy logic the aggregation process was implemented in a fuzzy expert system. The aggregated information, derived by the fuzzy expert system is meaningful, valuable and consistent. According to [Hall and Kandel, 1992], the proposed fuzzy expert system displays most of the characteristics of 'class one' expert systems; e.g.: (a) the domain of the problem is limited and very well defined, (b) an expert was available during the development, (c) the complexity of the problem is not extreme in the eyes of the knowledge engineer, and (d) the uncertainty prevailing in the domain was manageable.

We applied this process to a specified problem, but its applicability to similar problems is manifest (e.g. at the moment the approach is tested on a data set of cancer patients), and therefore, its potential is obvious. The integration of the proposed information aggregation method into a CBR environment is very promising because the reasoning process in CBR can benefit in two ways: (1) in cases with a lack of data (e.g. unavailable data at the primitive feature level) the higher level information, available at the fuzzy complex feature level, can be used in the reasoning process, and so the CBR system's capability to handle uncertainty increases significantly; and (2) CBR systems with a large case base will improve in their performance in the time domain.

record carries data about a patient suffering a cancerous disease. The support the proposed information aggregation method can provide to the CBR system will be investigated from the point of view of management of uncertainty and performance. There is also work underway to use the aggregation method in a multiple expert scenario and to relate this work with the theory of evidence.

## References

[Bonissone, 1985] Piero P. Bonissone. Editorial: Reasoning with uncertainty in expert systems. *International Journal Man-Machine Studies*, 22:241-250, 1985.

[Brown, 1992] Mike Brown. Case-Based Reasoning: principles and potential. *AI Intelligence*, January 1992.

[Chen and Hwang, 1992] Shu J. Chen and Ching L Hwang. *Fuzzy multiple attribute decision making, methods and applications.* Springer Verlag, Berlin, Heidelberg, 1992.

[Dubitzky et al., 1995] Werner Dubitzky, Alfons Schuster, John G. Hughes, and David A. Bell. Conceptual distance of numerically specified case features. In *Proceedings of the Second New Zealand International Two-Stream Conference on Artificial Neural Networks and Expert Systems*, pages 210-213, Dunedin, New Zealand, 1995.

[Gordon et al., 1989] D.J. Gordon, J.L. Probstfield, R.J. Garrison. High density lipoprotein cholesterol and cardiovascular disease. *Circulation*, 79(8):8-15, 1989.

[Hall and Kandel, 1992] Lawrence O. Hall and Abraham Kandel. The evolution from expert systems to fuzzy expert systems. Abraham Kandel, editor. *Fuzzy Expert Systems.* pages 3-21, CRC Press, Boca Raton, Florida, 1992.

[Kinosian et al., 1994] B. Kinosian, H. Glick, G. Garland. Cholesterol and coronary heart disease - predicting risks by levels and ratios. *Annals of Internal Medicine*, 121(9):641-647, 1994.

[Klir and Folger, 1988] George J. Klir and Tina A. Folger. *Fuzzy Sets, Uncertainty and Information.* Prentice Hall, Englewood Cliffs, New Jersey, 1988.

[Kolodner, 1993] Janet Kolodner. *Case-Based Reasoning.* Morgan Kaufmann, San Mateo, California, 1993

[Na and Park, 1996] Selee Na and Seog Park. Management of fuzzy objects with fuzzy attribute values in a fuzzy object-oriented data model. In *Proceedings of Flexible Query-Answering Systems*, pages 19-40, Rosklide, Denmark, 1996.

[Pyorala et al., 1994] Kaveli Pyorala, Guy De Backer, Ian Graham, Philip Poole-Wilson, and Wood David. Prevention of coronary heart disease in clinical practice: Recommendations of the Task Force of the European Society of Cardiology, European Atherosclerosis Society and European Society of Hypertension. *Atherosclerosis*, 110(2):21-61, 1994

[Riesbeck and Schank, 1989] Christopher K. Riesbeck and Roger C. Schank. Inside Case-Based Reasoning. Lawrence Erlbaum Associates, Hillsdale, New Jersey, 1989.

[Schneider and Kandel, 1992] Mordechay Schneider and and Abraham Kandel. General purpose fuzzy expert systems. Abraham Kandel, editor. *Fuzzy Expert Systems.* pages 23-41, CRC Press, Boca Raton, Florida, 1992.

[Slyper, 1994] A.H. Slyper. Low-density-lipoprotein density and atherosclerosis - unravelling the connection. *JAMA*, 272(4):305-308, 1994.

[Wilensky, 1986] Robert Wilensky. Knowledge representation—A critique and a proposal. In J. Kolodner and K. Riesbeck, editors, *Experience, Memory, And Reasoning*, pages 15-28, Lawrence Erlbaum Associates, Hillsdale, New Jersey, 1986.

[Zadeh, 1973]. Lotfi A. Zadeh. Outline of a new approach to the analysis of complex systems and decision processes. *IEEE Transactions on Systems, Man, and Cybernetics*, SMC-3(1):28-45, January 1973.

# Using Case-Based Reasoning in Interpreting Unsupervised Inductive Learning Results *

**Tu Bao Ho**
School of Information Science
Japan Advanced Institute
of Science and Technology
Tatsunokuchi, Ishikawa 923-12, Japan

**Chi Mai Luong**
Institute of Information Technology
National Center for
Natural Science and Technology
Nghiado, Tuliem, Hanoi, Vietnam

## Abstract

The objective of this work is to interpret inductive results obtained by the unsupervised learning method OSHAM. We briefly introduce the learning process of OSHAM, that extracts concept hierarchies from unlabelled data, based on a representation combining the classical, prototype and exemplar views on concepts. The interpretive process is considered as an intrinsic part in OSHAM and is carried out by a combination of case-based reasoning with matching approaches in inductive learning. An experimental comparative study of some learning methods in terms of knowledge description and prediction is given.

## 1 Introduction

Though the interpretation of induction results has a significant role in machine learning applications, there have been little work in inductive learning, particularly in unsupervised learning, associated with interpretation procedures, e.g., [Bergadano et al., 1992], [Wu, 1996].

There are three broad classes of logical, threshold, and competitive interpreters for intensional concept descriptions [Langley, 1996]. Naturally, three types of outcomes occur when matching logically an unknown instance with learned concepts: no-match, single-match, and multiple-match. Most work dealing with the cases of no-match and multiple-match employ a probabilistic estimation, e.g., [Michalski et al., 1986]. However, it is not always possible to obtain such an estimation in unsupervised learning when it requires using the class information. Moreover, the logical match of the concept intent does not always provide a prediction with enough satisfaction, particularly in boundary regions of concepts.

---

*This work is supported by Kokusai Electric Co., Ltd (Japan) to JAIST (Japan Advanced Institute of Science and Technology), and by the National Research Programme on Information Technology KC01-RD08 (Vietnam).

One of two styles of case-based reasoning (CBR) is interpretive by which new situations are evaluated in the context of old situations [Kolodner, 1992]. Rather than classifying new cases using the intensional concept descriptions, CBR typically does classification by using the nearest neighbor methods which have been demonstrated to be able to work often as well as other inductive learning techniques [Aha et al., 1991]. However, one limitation of the CBR is that it does not provide the concept description which is the main advantage of inductive learning about the knowledge understandability.

This paper highlights the intrinsic role of the interpretive process in unsupervised inductive learning and proposes a procedure that combines CBR with matching approaches in inductive learning to interpret the concepts learned by method OSHAM [Ho, 1996], [Ho, 1997]. The reason for this combination lies in the fact that the use of results in unsupervised learning, obtained by a non-exhaustive searching for regularities, can be suported by the nearest neighbor rule. The paper is organized as follows. Section 2 briefly resumes the learning phase in OSHAM consisting of an extended representation of concepts in the Galois lattice, and the essential ideas of the learning algorithm for extracting concept hierarchies from unsupervised data. Section 3 presents the interpretation phase of OSHAM to classify unknown cases using learned knowledge. Section 4 presents an experimental comparative study of four learning methods and the discussion. Section 5 is a short conclusion.

## 2 Learning Concept Hierarchies

### 2.1 Concept representation and extraction

Among *views on concepts* in cognitive science and machine learning, the classical, prototype and exemplar ones are widely known and used. Main strengths and limitations of these views on concepts have been widely recognized, e.g., [Van Mechelen et al., 1993], [Wrobel, 1994]. Moreover, without the class information, unsupervised learning systems often compose solutions by employing one or more of three main *categorization con-*

*straints* based on similarity, feature correlation, and syntactical structure of conceptual knowledge.

Recently, several concept learning systems have been developed using the classical view on concepts in the Galois lattice structure [Wille, 1982]. In [Godin and Missaoui, 1994], [Carpineto and Romano, 1996], the authors generate incrementally all possible concepts in the Galois lattice. In [Ho, 1995], an alternative approach to hierarchical conceptual clustering was proposed that extracts a part of the Galois lattice in the form of a concept hierarchy. Although the Galois lattice provides a powerful structure for learning concepts, the classical view on concepts in this framework has some considerable limitations, such as it does not capture typicality effects and vagueness. Otherwise, to find all possible concepts is not always tractable as in the worse case the number of concepts can be exponential in the size of datasets, e.g., even for the small well-known dataset of Congressional voting (435 instances × 17 attributes), the Galois lattice has about 150,000 nodes [Carpineto and Romano, 1996].

As analysed in [Van Mechelen et al., 1993], each system relies on a single view on concepts can be limited in capturing the rich variety of conceptual knowledge. Therefore, hybrid systems attempt to improve the concept learning process by combining fairly different theoretical views on concept and categorization constraints. In [Ho, 1996], the learning method OSHAM was improved by an extension of the classical view of concepts in the Galois lattice. Instead of characterizing a concept only by its intent and extent, OSHAM represents each concept $C_k$ in a concept hierarchy $\mathcal{H}$ by a 10-tuple

$$< l(C_k), f(C_k), s(C_k), i(C_k), e(C_k),$$
$$d(C_k), p(C_k), d(C_k^r), p(C_k^r \mid C_k), q(C_k) > \quad (1)$$

where

- $l(C_k)$ is the level of $C_k$ in $\mathcal{H}$;
- $f(C_k)$ is the list of direct superconcepts of $C_k$;
- $s(C_k)$ is the list of direct subconcepts of $C_k$;
- $i(C_k)$ is the intent of $C_k$ which is the set of all common properties of instances of $C_k$;
- $e(C_k)$ is the extent of $C_k$ which is the set of all instances satisfying properties of $i(C_k)$;
- $d(C_k)$ is the dispersion between instances of $C_k$;
- $p(C_k)$ is the occurrence probability of $C_k$;
- $d(C_k^r)$ is the dispersion of local instances of $C_k$ which are not classified into subconcepts of $C_k$;
- $p(C_k^r \mid C_k)$ is the conditional probability of these unclassified instances of $C_k$;
- $q(C_k)$ is the quality estimation of splitting $C_k$ into subconcepts $C_{k_i}$.

The induction and interpretation of components in (1) employ the following distance metrics. Denote by $\lambda(X)$ for any instance set $X$ the largest set of properties common to all elements of $X$, and by $\rho(S)$ for any property set $S$ the set of all instances satisfying $S$. The *distance* $\delta(o_p, o_q)$ between two instances $o_p$ and $o_q$ is defined as an extension of Jaccard distance

$$\delta(o_p, o_q) = 1 - \frac{\sum_{a \in \lambda(\{o_p, o_q\})} \gamma(a)}{\sum_{a \in \lambda(\{o_p\}) \cup \lambda(\{o_q\})} \gamma(a)} \quad (2)$$

where $\gamma(a) \in \mathbb{Z}^+$ are positive integer weights of attributes $a$ (with value 1 by default). The *dispersion* $d(C_k)$ between instances in $e(C_k)$, considered as the inverse of the homogeneity of $e(C_k)$, is defined as the average distance between all pairs of instances in $e(C_k)$

$$d(C_k) = \frac{2 \times \sum_{o_p, o_q \in e(C_k)} \delta(o_p, o_q)}{\mid e(C_k) \mid \times (\mid e(C_k) \mid -1)} \quad (3)$$

If $C_k$ is a non-leaf concept, its local instances in $C_k^r$ can be considered to be more typical and representative than its instances classified into subconcepts $C_{k_i}$. As an instance $o$ is member of different concepts along a branch in the concept hierarchy, the concepts $C_k$ that $o \in C_k^r$ is of particular interest. If $C_k$ is a leaf concept, we have $e(C_k) = C_k^r$ and all of its instances are considered with the same representative role.

The extent of all direct subconcepts $C_{k_1}, C_{k_2}, ..., C_{k_n}$ of $C_k$ and the set of local instances $C_k^r$ form a partition $P$ of $e(C_k)$. Denote by $W(C_k)$ the average of all $d(C_{k_i})$ and $d(C_k^r)$. The dissimilarity between subconcepts of $C_k$, denoted by $B(C_k)$, is defined as the average of distances $\Delta(c(C_{k_i}), c(C_{k_j}))$ between all pairs $C_{k_i}, C_{k_j}$ in P, where the distance $\Delta(c(C_{k_i}), c(C_{k_j}))$ is defined as the smallest distance among distances of all pairs $o_p \in C_{k_i}, o_q \in C_{k_j}$

Table 1: Brief description of learning process

---

1. While $C_k$ is still splittable, find a new subconcept of it that corresponds to the hypothesis minimizing the quality function $q(C_k)$ among $\eta$ hypotheses generated by the following steps

    (a) Find a "good" attribute-value pair concerning the best cover of $C_k$.
    (b) Find a closed attribute-value subset $S$ containing this attribute-value pair.
    (c) Form a subconcept $C_{k_i}$ with the intent is $S$.
    (d) Evaluate the quality function with the new hypothesized subconcept.

    Form intersecting concepts corresponding to intersections of the extent of the new concept with the extent of existing concepts excluding its superconcepts.

2. If one of the following conditions holds then $C_k$ is considered as unsplittable

    (a) There exist not any closed proper feature subset.
    (b) The local instances set $C_k^r$ is too small.
    (c) The local instances set $C_k^r$ is homogeneous enough.

3. Apply recursively the procedure to concepts generated in step 1.

$$\Delta(c(C_{k_i}), c(C_{k_j})) = Min_{o_p \in C_{k_i}, o_q \in C_{k_j}} \delta(o_p, o_q) \quad (4)$$

The *quality* of splitting a concept $C_k$ into subconcepts in the next level, denoted by $q(C_k)$, is measured by

$$q(C_k) = W(C_k)/B(C_k) \quad (5)$$

The basis of learning in OSHAM is a generate-and-test procedure to split a concept $C$ into subconcepts at a higher level of $\mathcal{H}$. Starting from the root concept of the Galois lattice with the whole set of training instances, it extracts the concept hierarchy $\mathcal{H}$ recursively in a top-down direction. This algorithm is originally designed for discrete attributes with unordered nominal values. In the current version, continuous attributes are discretized before learning process by k-means clustering [Hartigan, 1975]. In fact, for each continuous attribute the k-means algorithm is applied to cluster its values into $k$ groups ($k = 1, 2, ..., K$). A criterion similar to (5) with the Euclidean distance is used to choose a value of $k$ that corresponds to the best partition according to this criterion. The basic idea of learning algorithm in OSHAM, described fully in [Ho, 1996], is resumed in Table 1.

## 2.2 About learned concept hierarchies

By using different constraints for 1.(a) in the learning algorithm, OSHAM is able to extract both overlapping or disjoint concepts depending the user's interest [Ho, 1997]. OSHAM has been implemented in the X Window on a Sparcstation with the direct manipulation style of interaction which allows the user to participate actively in the learning process. The user can initialize parameters to cluster data, visualize the concept hierarchy gradually, observe the results and the quality estimation, manually modify the parameters when necessary before the system continues to go further to cluster subsequent data or backtrack to regrow branches of the concept hierarchy with respect to the categorization scheme. Figure 1 shows a main screen of the interactive OSHAM with a hierarchy of overlapping concepts learned from the Wisconsin breast cancer dataset. A full description of concept 43 in this figure is given below

```
CONCEPT 43
Level = 5
Super_Concepts = {29}, Sub_Concepts = {52, 53}
Features = (Uniformity of Cell Size, 1) ∧ (Bare Nuclei, 1)
∧ (Bland Chromatin, 1) ∧ (Uniformity of Cell Shape, 2)
Local_instances/Covered_instances = 6/25
Local_instances = {8, 127, 221, 236, 415, 661}
Concept_probability = 0.041666
Local_instance_conditional_probability = 0.240000
Concept_dispersion = 0.258848
Local_instance_dispersion = 0.055556
Subconcept_partition_quality = 0.519719
```

There is a considerable distinction in the concept description of OSHAM in contrast to those of other methods such as the supervised learning system C4.5 [Quinlan, 1993], the unsupervised learning systems COBWEB [Fisher, 1987] and AUTOCLASS [Cheeseman and Stutz, 1996]. C4.5 induces decision trees in which concepts are represented by their intent associated with a predicted error rate, and it has not to maintain intermediate concepts. COBWEB represents each concept $C_k$ as a set of attributes $a_i$ associated with a set of their possible values $v_{ij}$, the occurrence probability, and the conditional probability $P(a_i = v_{ij} \mid C_k)$ associated with each value $v_{ij}$. A *classification* in AUTOCLASS is defined as a set of classes, the probability of each class, and two additional probabilities for each hypothesized model: the model probability $P(H)$ and the conditional parameter probability distribution $P(p \mid H)$.

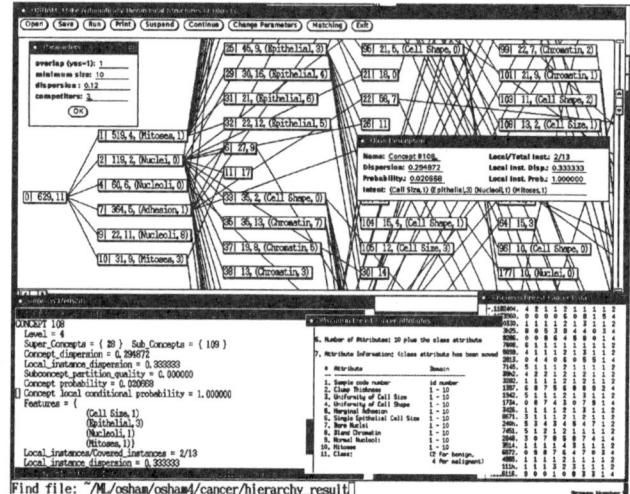

Figure 1: A screen of the interactive OSHAM

We share the opinion in [Langley, 1996] that the interpretive process is a central issue in learning. An intensional representation has no meaning (e.g., no extension) without some associated interpreters and different interpreters can yield different meaning for the same representation. Next section describes the second phase in OSHAM - its interpretive process.

## 3 Interpreting concept hierarchies

### 3.1 Interpretive CBR

Interpretive case-based learning is a process of evaluating situations in the context of previous experience. One way a case-based classifier works is to ask whether the unknown case is enough like another one known. It does classification by trying to find the closest matching case in its case base to the new case rather than using intensional concept descriptions. Many studies have pointed out the strong points of CBR (e.g., simplicity, relatively robust, often excellent performance, etc.) and its weak points, e.g., [Aha et al., 1991], [Kolodner, 1992].

In inductive learning, the *logical* interpretation approach carries out an "all or none" matching process depending on whether the unknown instance satisfies the concept intent. The *threshold* approach carries out a partial matching process and employs some threshold to determine an acceptable degree of match. The *competitive* approach also carries out a partial matching process and selects the best competitor based on estimated degrees of match [Langley, 1996]. The interpretation of inductive learning results is commonly understood as the process of comparing an unknown case to the learned concepts. In OSHAM, as the generality decreases along branches of the concept hierarchy, we say that a concept $C_k$ *matches* the unknown instance $e$ if $C_k$ is the most specific concept in a branch that matches $e$ intensionally (though all superconcepts of $C_k$ match $e$). Naturally, there are three types of outcomes when matching logically an unknown instance $e$ with the learned concepts: only one concept that matches $e$ (*single-match*), many concepts that match $e$ (*multiple-match*), and no concept that matches $e$ (*no-match*). We believe in the alternative roles of CBR and generalization in inductive learning. As noted in [Kolodner, 1992], rules could be used when they matched cases exactly, while cases would be used when rules were not immediately applicable. In fact, the nearest neighbor of $e$ in the training set and the learned concept to which it belongs, denoted by $NN(e)$ and $c[NN(e)]$, provide useful information which could be used in all cases of single-match, multiple-match and no-match.

## 3.2 Interpretation of induction results

We develop an interpretation procedure for concept hierarchies that uses the concept intent, the hierarchical structure information, the probabilistic estimations and the nearest neighbors of unknown instances. This interpretation procedure consists of two stages: (1) find all concepts on the concept hierarchy that match $e$ intensionally, and (2) decide among these concepts which matches $e$ best. This procedure shares the same scheme of the system POSEIDON [Bergadano et al., 1992], but functions differently. In the second stage, it determines the best matched concept of $e$ with some satisfying degree of prediction.

Consider the multiple-match case when we have to decide among the competitors the best matched concept. To do it we need to determine and compare the degree of match of competitors. From various experiment casestudies we note that a logically matched concept $C_k$ will match $e$ well (with low error rate) if it satisfies a majority of the following conditions: $l(C_k)$ is high, $C_k$ is a leaf concept, $p(C_k) \times p(C_k^r)$ is high, $d(C_k)$ is low, $d(C_k^r)$ is low, and generally none of these factors has a clearly higher priority than the others. Formally, the following functions $\tau_N, \tau_L, \tau_P, \tau_D, \tau_R$ can be used to compare these factors, respectively, between two concepts $C_k$ and $C_h$ which matched $e$ intensionally

$$\tau_N(C_k, C_h) = \begin{cases} 1, & \text{if } l(C_k) > l(C_h) \\ 0, & \text{if } l(C_k) = l(C_h) \\ -1, & \text{if } l(C_k) < l(C_h) \end{cases} \quad (6)$$

$$\tau_L(C_k, C_h) = \begin{cases} 1, & \text{if } C_k = \text{leaf} \wedge C_h \neq \text{leaf} \\ 0, & \text{if } C_k = \text{leaf} \wedge C_h = \text{leaf} \\ & \vee C_k \neq \text{leaf} \wedge C_h \neq \text{leaf} \\ -1, & \text{if } C_k \neq \text{leaf} \wedge C_h = \text{leaf} \end{cases} \quad (7)$$

$$\tau_P(C_k, C_h) = \begin{cases} 1, & \text{if } p(C_k) \times p(C_k^r \mid C_k) \\ & > p(C_h) \times p(C_h^r \mid C_h) \\ 0, & \text{if } p(C_k) \times p(C_k^r \mid C_k) \\ & = p(C_h) \times p(C_h^r \mid C_h) \\ -1, & \text{if } p(C_k) \times p(C_k^r \mid C_k) \\ & < p(C_h) \times p(C_h^r \mid C_h) \end{cases} \quad (8)$$

$$\tau_D(C_k, C_h) = \begin{cases} 1, & \text{if } d(C_k) < d(C_h) \\ 0, & \text{if } d(C_k) = d(C_h) \\ -1, & \text{if } d(C_k) > d(C_h) \end{cases} \quad (9)$$

$$\tau_R(C_k, C_h) = \begin{cases} 1, & \text{if } d(C_k^r) < d(C_h^r) \\ 0, & \text{if } d(C_k^r) = d(C_h^r) \\ -1, & \text{if } d(C_k^r) > d(C_h^r) \end{cases} \quad (10)$$

The following heuristic function $\tau$ compares the degree of match of $C_k$ and $C_h$. We say that $C_k$ has a *higher satisfying degree than* $C_h$ in matching $e$ if

$$\tau(C_k, C_h) = \theta_N \times \tau_N(C_k, C_h) + \theta_L \times \tau_L(C_k, C_h) + \theta_P \times \tau_P(C_k, C_h) + \theta_D \times \tau_D(C_k, C_h) + \theta_R \times \tau_R(C_k, C_h) > 0 \quad (11)$$

Table 2: Interpretation procedure

*Input*  concept hierarchy $\mathcal{H}$, unknown instance $e$.
*Result*  best matched concept $c[e]$, satisficing degree $\phi$.
*Variables*  $\sigma$ is a given threshold.

*Procedure Interpretation*($\mathcal{H}, e, c[e], \phi$)

If there is only one concept $C_k \in \mathcal{H}$ that matches $e$ intensionally then

  if $c[NN(e)] = C_k$ then $c[e] \leftarrow C_k, \phi \leftarrow S_1$
  else if $\tau(C_k, c[NN(e)]) \geq 0)$ then $c[e] \leftarrow C_k, \phi \leftarrow S_2$
  else $c[e] \leftarrow c[NN(e)], \phi \leftarrow S_3$.

If there are $m$ concepts $C_{i_1}, ..., C_{i_m} \in \mathcal{H}$ that match $e$ intensionally then

  Choose $i_K \in \{i_1, ..., i_m\}$ satisfying $\tau(C_{i_K}, C_{i_k}) \geq 0$
  when comparing $C_{i_K}$ with $C_{i_k}$ for all $i_k \in \{i_1, ..., i_m\}$.
  If $C_{i_K} = c[NN(e)]$ then $c[e] \leftarrow C_{i_K}, \phi \leftarrow M_1$
  else if $\tau(C_{i_K}, c[NN(e)]) \geq 0)$ then $c[e] \leftarrow C_{i_K}, \phi \leftarrow M_2$
  else $c[e] \leftarrow c[NN(e)], \phi \leftarrow M_3$.

If there is not any concept that match $e$ intensionally then
  if $\delta(NN(e), e) \leq \sigma$ then $c[e] \leftarrow c[NN(e)], \phi \leftarrow N_1$
  else $c[e] = \varnothing, \phi \leftarrow N_2$.

where $\theta_N, \theta_L, \theta_P, \theta_D, \theta_R$ are positive weights for the importance of the level, leaf concept, local instance conditional probability, concept dispersion, and local instance dispersion (all with value 1 by default).

Denote by $c[e]$ the best matched concept that $e$ is finally decided to belong to, and by $\phi$ the *satisficing degree* of prediction. Table 2 presents the interpretation procedure in OSHAM based on the function $\tau$ and the nearest-neighbor principle. Essentially, this procedure makes final decision by comparing the best concept matching $e$ with the concept $c[NN(e)]$ containing the nearest neighbor $NN(e)$ of $e$ regarding the function $\tau$. In this interpretation procedure, different symbolic values are assigned to the satisficing degree of prediction $\phi$.

Values of $\phi$ indicate the decreasing rank of prediction satisfaction. For example, we may consider $S_1$ as "best prediction", $M_1$ as "strong prediction" while $N_1$ as "weakly accepted prediction" and $N_2$ as "no prediction". The interpretation for different values of $\phi$ depends on the judgment of the user or domain experts.

## 4 Evaluation

### 4.1 Experimental Results

A way to evaluate unsupervised learning system is to employ supervised data but hide the class information in the whole learning and interpreting phases and use the class information only to estimate the predictive accuracy. We employ this way to evaluate unsupervised learning systems where the predicted name of each learned concept $C_k$ is determined by the most frequently occurring name of instances in $e(C_k)$. With this predicted name of learned concepts, the error rate of an unsupervised learning system can be estimated as the ratio of the number of testing instances correctly predicted regarding the predicted name over the total number of testing instances. It is worth mentioning that multiple train-and-test experiments are much more computationally expensive but give more reliable evaluation than a single train-and-test experiment.

Experiments are carried out on ten datasets from the UCI repository of machine learning databases, including the Wisconsin breast cancer (breast-w), Congressional voting (vote), Mushroom (mushroom), Tic-tac-toe (tictactoe), Glass identification (glass), Ionosphere (ionosphere), Waveform (waveform), Pima diabetes (diabetes), Thyroid (new) disease (thyroid), and Heart disease cleveland (heart-c). The numbers of attributes (discrete and continuous), instances and "natural" classes of these datasets are given in columns 2-5 of Table 3. All experiments on these datasets are carried out with 10-fold cross validation by four programs C4.5 [Quinlan, 1993], CART-like [Breiman et al., 1984], AUTOCLASS and OSHAM in the same condition, i.e., the same randomly divided datasets into subsets. For AUTOCLASS, we use the public version AUTOCLASS-C implemented in C and run three steps of *search, report* and *predict* with the default parameters. The predicted name and predictive accuracy of AUTOCLASS and OSHAM are obtained as mentioned above. Columns 6-8 report the predictive accuracies of C4.5, CART-like and AUTOCLASS, respectively.

For OSHAM we carried out experiments for two interpretation procedures: OSHAM-NN (using only the nearest neighbors) and OSHAM-NN+$\tau$ (using the nearest neighbors and matching regarding the function $\tau$ by the procedure described in Table 2). Experimental results are reported in columns 9-10, respectively. In order to avoid a biased evaluation of OSHAM, although with each dataset parameters can be adjusted to obtain the most suitable concept hierarchy, we fixed values $\alpha = 1\%$ of the size of the training set, $\beta = 15\%$ and $\sigma = 10\%$ of the number of attributes, and the beam size $\eta = 3$, commonly to all datasets. Two last columns in Table 3 give the average size of concept hierarchies (number of concepts) and CPU time (in second) of OSHAM learned from these datasets.

### 4.2 Discussion

The predicted name obtained in OSHAM and AUTOCLASS by the majority of occurring name of instances in concepts is different from the concept name obtained in supervised learning (e.g., C4.5) using the pruning threshold based on the class information. An unsupervised concept in the worse case may contain nearly equal numbers of instances belonging to different natural classes, and an unsupervised classification may be failed in distinguishing very similar instances. It explains that while the predictive accuracies between these supervised and unsupervised methods look not so different, they are slightly different in nature. Note that the recent release 8 of C4.5 [Quinlan, 1996] treats the continuous attribute better than the release we used in this work.

The predictive accuracies of OSHAM and AUTOCLASS in these experiments are only slightly different. In these first trials, each system is better in several datasets and these two systems can be considered having comparable performance. One advantage of OSHAM is its concept hierarchies can be easily understood by its extended classical view on concepts and the graphical support.

Empirical results with OSHAM-NN and OSHAM-NN+$\tau$ illustrate that these strategies are both good for interpreting unsupervised induction results. We believe that in general CBR can also be used to interpret results of unsupervised learning and this topic is worth for a further investigation.

Table 3: Datasets, Predictive accuracies of methods, Average size and CPU time of OSHAM

| Datasets | attributes | | inst. | class | C4.5 | CART-like | AUTO-CLASS | OSHAM | | | |
|---|---|---|---|---|---|---|---|---|---|---|---|
| | disc | cont | | | | | | NN | NN+$\tau$ | concept | class |
| breast-w | 9 | – | 699 | 2 | 93.3 | 94.1 | 96.6 | 92.1 | 92.6 | 98 | 134 |
| vote | 17 | – | 435 | 2 | 94.5 | 93.8 | 91.2 | 93.0 | 93.7 | 69 | 43 |
| mushroom | 23 | – | 8125 | 2 | 100.0 | 100.0 | 86.5 | 92.3 | 88.2 | 63 | 336 |
| tictactoe | 9 | – | 862 | 9 | 88.0 | 86.2 | 82.3 | 93.2 | 92.6 | 204 | 133 |
| glass | – | 9 | 214 | 6 | 66.6 | 66.0 | 55.7 | 63.3 | 65.3 | 37 | 6.5 |
| ionosphere | – | 35 | 351 | 2 | 91.5 | 88.3 | 91.5 | 86.8 | 84.6 | 110 | 64 |
| waveform | – | 21 | 300 | 3 | 72.4 | 72.5 | 59.2 | 73.0 | 73.0 | 215 | 278 |
| diabetes | – | 8 | 768 | 2 | 71.2 | 72.2 | 68.2 | 72.1 | 72.7 | 39 | 72 |
| thyroid | – | 6 | 215 | 3 | 91.1 | 90.3 | 89.3 | 78.6 | 84.6 | 19 | 5 |
| heart-c | 8 | 5 | 303 | 2 | 59.5 | 52.4 | 49.2 | 61.0 | 60.8 | 65 | 13 |

## 5  Conclusion

In this paper we first briefly resumed the main ideas of the unsupervised learning method OSHAM in terms of description and extraction of concepts. We then described how the nearest neighbor rule is combined with domain knowledge to interpret the induction results. Careful experiments with different datasets have demonstrated that this combination provides a good solution to the use of unsupervised learned knowledge in prediction. The main conclusions can be drawn from this research are (1) the interpretive process needs to be a part of a unsupervised learning method, and (2) the CBR can be used to interpret results obtained from the non-exhaustive search in unsupervised inductive learning. Our near future research concerns further investigations on the effects of k-nearest neighbors and the discretization of continuous attributes to the interpretation of unsupervised induction results, based on experiments with a larger number of datasets.

## References

[Aha et al., 1991] D.W. Aha, D. Kibler, M.K. Albert. Instance-based learning algorithms. *Machine Learning*, Vol. 6 (1991), 37–66.

[Bergadano et al., 1992] F. Bergadano, S. Matwin, R.S. Michalski, J. Zhang. Learning two-tiered descriptions of flexible concepts: the POSEIDON system. *Machine Learning*, Vol. 8 (1992), 5–43.

[Breiman et al., 1984] L. Breiman, J. Friedman, R. Olshen, C. Stone. *Classification and Regression Trees*. Belmont, CA: Wadsworth, 1984.

[Carpineto and Romano, 1996] C. Carpineto and G. Romano. A lattice conceptual clustering system and its application to browsing retrieval. *Machine Learning*, Vol. 10 (1996), 95–122.

[Cheeseman and Stutz, 1996] P. Cheeseman and J. Stutz. Bayesian classification (AutoClass): Theory and results. In *Advances in Knowledge Discovery and Data Mining*, U.M. Fayyad et al. (Eds.). AAAI Press/MIT Press, 1996, 153–180.

[Fisher, 1987] D. Fisher. Knowledge acquisition via incremental conceptual clustering. *Machine Learning*, Vol.2 (1987), 139–172.

[Godin and Missaoui, 1994] R. Godin and R. Missaoui. An incremental concept formation approach for learning from databases. *Theoretical Computer Science*, 133 (1994), 387–419.

[Hartigan, 1975] J.A. Hartigan. *Clustering Algorithms*. Wiley, New York, 1975.

[Ho, 1995] T.B. Ho. An approach to concept formation based on formal concept analysis. *IEICE Trans. Information and Systems*. Vol. E78-D (1995), No.5, 553–559.

[Ho, 1996] T.B. Ho. A hybrid model for concept formation. In *Information Modelling and Knowledge Bases* VII, Y. Tanaka et al. (Eds.). IOS Publisher, 1996, 22–35.

[Ho, 1997] T.B. Ho. Discovering and Using Knowledge From Unsupervised Data. In *Decision Support Systems*, June 1997 (in press).

[Kolodner, 1992] J. Kolodner. An introduction to Case-Based Reasoning. *Artificial Intelligence Review*, Vol. 6 (1992), 3–34.

[Langley, 1996] P. Langley. *Elements of Machine Learning*. Morgan Kaufmann, 1996.

[Michalski et al., 1986] R.S. Michalski, I. Mozetic, J. Hong, N. Lavrac. The multi-purpose incremental learning systems AQ15 and its testing application to three medical domains. In *Proceedings of AAAI 1986*, 1041–1045.

[Quinlan, 1993] J.R. Quinlan. *C4.5: Programs for Machine Learning*. Morgan Kaufmann, 1993.

[Quinlan, 1996] J.R. Quinlan. Improved use of continuous attributes in C4.5. *Journal of Artificial Intelligence Research*, Vol. 4 (1996), 77–90.

[Van Mechelen et al., 1993] I. Van Mechelen, J. Hampton, R.S. Michalski, P. Theuns (Eds.)., *Categories and Concepts. Theoretical Views and Inductive Data Analysis*. Academic Press, 1993.

[Wille, 1982] R. Wille. Restructuring lattice theory: an approach based on hierarchies of concepts. In *Ordered Sets*, I. Rival (Ed.). Reidel, 1982, 445–470.

[Wrobel, 1994] S. Wrobel. *Concept Formation and Knowledge Revision*. Kluwer Academic Publishers, 1994.

[Wu, 1996] X. Wu. Hybrid interpretation of induction results. In *Advanced IT Tools*, N. Terashima and E. Altman (Eds.). Chapman & Hall, 1996, 497–506.

# COGNITIVE MODELING

# COGNITIVE MODELING

Cognitive Modeling 1

# Acquisition of Human Feelings in Music Arrangement

Masayuki Numao  Masashi Kobayashi  Katsuyuki Sakaniwa*
Department of Computer Science, Tokyo Institute of Technology
2-12-1, O-okayama, Meguro-ku 152, JAPAN
Email: numao@cs.titech.ac.jp

## Abstract

We often make decisions based on our feelings, which are implicit and very difficult to express as knowledge. This paper details an attempt to acquire feelings automatically. We assume that some relations or constraints exist between impressions felt and *situations*, which consist of an object and its environment. For example, in music arrangement, the object is a music score and its environment contains listeners, etc. Our project validates this assumption through three levels of experiments. At the first level, a program simply mimics human arrangements in order to transfer their impressions to another arrangement. This implies that the program is capable of distinguishing patterns that result in some impressions. At the second level, in order to produce a music recognition model, the program locates relations and constraints between a music score and its impressions, by which we show that machine learning techniques may provide a powerful tool for composing music and analyzing human feelings. Finally, we examine the generality of the model by modifying some arrangements to provide the subjects with a specified impression.

## 1 Introduction

KANSEI (Japanese; lit. *human feelings*) [Tsuji, 1995] has been analyzed using quantitative psychological analysis methods, such as the semantical differential (SD) method and multivariate analysis, which analyze human feelings. Such analyses can isolate feelings associated with known objects, but cannot predict feelings for a new object, nor create a new object for the purpose of generating in the subject a specific feeling. We would like to introduce and discuss a system capable of predicting feelings and creating new objects based on *seed* structures that have been extracted and are perceived as favorable by the test subject, such as patterns and

---
*Presently at Matsushita Electric Industrial Co., Ltd., Kadoma, Osaka, 571, JAPAN

Figure 1: Melody, Chords and their Functions

colors for pictures, or spectrums and their transition for sounds. Until now, such emergent structures have been obtained only through a random and intractable combination of elements. In this paper, we explain how this difficulty is overcome using machine learning techniques.

As a representative medium, we focus on a MIDI-based music arrangement system which is applied to automatic selection or arrangement in an online KARAOKE system. This system is used for automatic downloading of MIDI data as requested by a user, or can be used more generally for information retrieval or filtering based on emotional data.

## 2 Melody and Chords

We attempt to extract a musical structure based on melody and chords as shown in Figure 1. In a musical piece, a function — *tonic* (T), *dominant* (D), *subdominant* (S) or *subdominant minor* (SDm) — is assigned to each chord. This paper discusses the extraction of two aspects of the structure (i.e., each chord and a sequence of functions) from which the system derives constraints for assigning chords to a melody (supplemented by functions).

## 3 Mimicking Arrangements

To investigate the feasibility of generating arrangements automatically, the authors constructed a system that mimics human arrangements as shown in Figure 2.

The chord analyzer assigns a function to each chord by parsing chord progression, and translates scores arranged by corresponding human composers into training examples. Each example consists of a chord and its

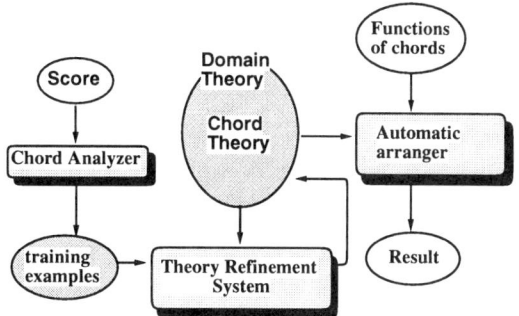

Figure 2: Automatic Arrangement System

corresponding function, from which the system learns (derives) constraints for automatic arrangement. The system is initially provided with a general theory of harmony, which is refined by a learner called a *theory refinement system* [Tangkitvanich and Shimura, 1992; Numao and Shimura, 1989; Mooney and Ourston, 1994]. Each note in the score has a function, to which the automatic arranger assigns a chord.

Figure 3 shows an example of such an arrangement. The system can assign chords based on the general theory of harmony, even though they are mediocre. Theory refinement introduces some *decorations* to the chords and refines the techniques used in the training examples, improving the arrangement.

The present authors prepared a set of training and test examples by the same composer. Figure 4 compares the arrangements produced by the initial and refined theories with those of the human composer, which shows that refinement based on the training examples increases the number of matched chords in the test examples.

Verbal reports from the subjects characterized the arrangements produced by the initial theory as *simple, not interesting* and *flavorless*, while those produced using the refined theory are described as *refreshing* and *novel*. Learning using multiple scores enhances *variety*, but the results lack uniformity.

There is more demand today for professional MIDI arrangers than there was several years ago, since digital synthesizers, electronic pianos, piano players, online KARAOKE systems, etc. have become more popular. This type of arrangement system will satisfy these needs by learning and rivaling the skills of professional arrangers.

## 4 Music Recognition Model

The system outlined above only mimics arranged scores and does not consider human feeling, so we have attempted to develop a system for acquiring music recognition model. Such a model is necessary to achieve automatic arrangement based on situation-dependent human feelings. The authors prepared some musical pieces of 8 bars, and then played them for some subjects in order to get their impressions. This was accomplished using the semantic differential (SD) method. The results were

Training Example:

An Arrangement composed using the Initial Theory:

An Arrangement composed using the Refined Theory:

Figure 3: An Example of Arrangement

utilized by an inductive logic programming (ILP) system [DeRaedt, 1996] capable of deriving a model for music recognition.

### 4.1 Training Examples

In order to categorize responses by subjects based on the musical pieces they listened to, the authors selected the following 5 pairs of complementary adjectives: *bright - dark, clear - unclear, fast - slow, favorable - unfavorable, stable - unstable*. A musical piece is evaluated as one of 7 grades for each pair, such that 1 indicates most *bright*, and 7 indicates most *dark* in the spectrum presented by a bright-dark pair.

Let evaluation of a piece $P$ for a pair of adjectives $A$ be $E_A^P$ for a subject ($1 \leq E_A^P \leq 7$). The system generalizes melody by analyzing notes in each bar based on background knowledge described in Section 4.2. For each piece $P$ consisting of 8 bars: $Bar_1, Bar_2, \cdots, Bar_8$, the system learns to evaluate $A$ from the following 8 training examples:

$$(Bar_1, E_A^P), (Bar_2, E_A^P), \ldots, (Bar_8, E_A^P)$$

where $Bar_i$ is a sequence of (*pitch_name, length*). An example is considered positive if $5 \leq E_A^P \leq 7$, and negative if $1 \leq E_A^P \leq 4$. The system also weighs each learned clause according to the values $E_A^P$ of the examples used.

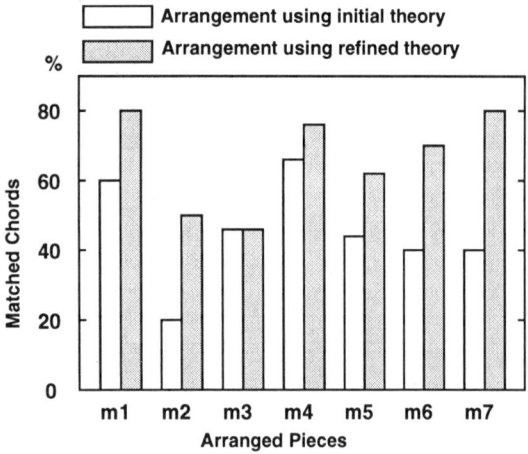

Figure 4: Matched Chords

The system generalizes chord progression by analyzing successive chords. For each piece $P$ with a sequence of $n$ chords: $Chord_1, Chord_2, \cdots, Chord_n$, the system learns to evaluate based on the following $n-1$ training examples:

$$(Chord_1, Chord_2, E_A^P)$$
$$(Chord_2, Chord_3, E_A^P)$$
$$\cdots$$
$$(Chord_{n-1}, Chord_n, E_A^P)$$

where $Chord_i$ is a combination of the following notes and their function:

**Root:** i, ♭ii, ii, ♭iii, iii, iv, ♭v, v, ♭vi, vi ♭vii, vii
**Third:** major, minor, suspended fourth
**Fifth:** fifth, augmented, diminished
**Seventh:** none, major sixth, minor seventh, major seventh
**Function:** T, D, S, SDm

which constructs $12 \times 3 \times 3 \times 4 \times 4 = 1728$ combinations. The model describes the relationship among two successive chords and their functions. The high number combinations and relations results in neither neural networks nor decision trees being appropriate as learning tools. Instead, what is required is inductive logic programming that learns not only attributes and propositional descriptions, but also predicates for finding useful relations in obscure combinations. In the experiment, the authors use a learner similar to FOIL [Quinlan, 1990] except that its background knowledge may also be described using Horn clauses.

### 4.2 Background Knowledge for ILP

The model is constructed based on background knowledge — definitions of predicates that describe melody and chords. Melody is considered to be a sequence of notes that have pitch and duration described by the following predicates:

- Average duration
- Minimum (lowest) pitch
- Maximum (highest) pitch
- Difference between minimum and maximum pitch
- Pitch transition (rising or falling)

Chords are analyzed based on background knowledge as follows:

root_i(Chord) : The root of Chord is i.
root_v(Chord) : The root of Chord is v.
major(Chord) : Chord has the major third.
fifth(Chord) : Chord has the perfect fifth.
seventh(Chord) : Chord has the minor seventh.
tonic(Chord) : The function of Chord is T.
dom(Chord) : The function of Chord is D.
subdom(Chord) : The function of Chord is S.
succ(Chord1,Chord2) : Chord1 and Chord2 are successive chords.

Using the background knowledge for each adjective pair, some predicates *adjective_pair(Example,Weight)* are derived. E.g. bright-dark(*Example of Chords, Weight*) to detect bright or dark bar is learned as follows:

```
bright-dark((C1,C2,_),7)
 :- succ(C1,C2),major(C1),subdom(C1),dom(C2).
bright-dark((C1,_,_),6)
 :- major(C1),dom(C1),root_v(C1).
```

### 4.3 Predicting an evaluation

After the system creates a recognition model for a subject by processing the results of his/her evaluation of various pieces, it predicts the subject's evaluation of subsequent pieces, which are transformed into a set of examples for analyzing melody and chords. If an *adjective_pair(Example,$W_j$)* is satisfied by $k_j$ examples, the evaluation for the adjective pair is calculated by

$$\frac{\sum_j W_j k_j}{\sum_j k_j}.$$

### 4.4 Experiments in Recognition

The present authors prepared 100 well-known music pieces, from which they extracted 8 successive bars without modulation. The subject evaluates 85 of the 100 pieces, and the results for adjective pairs are studied by the system to predict evaluations for the other 15 pieces. They then take an average of the results by 8 different subjects.

Let $F_A^p$ be a prediction of the evaluation of a piece $p$ for adjective pair $A$, and $n$ be the number of pieces. The difference between the prediction and the evaluation by a subject is shown to be:

$$\mathit{diff}_A = \frac{1}{n}\sum_{p=1}^{n}|E_A^p - F_A^p|.$$

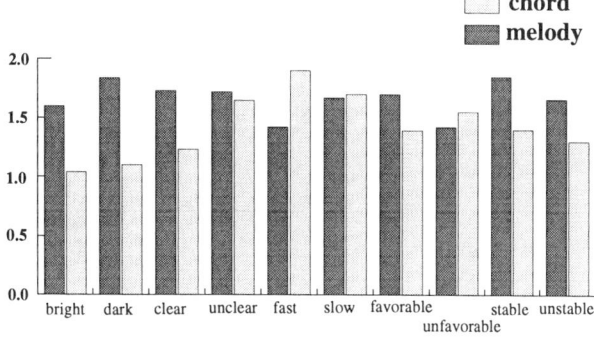

Figure 5: Average Difference

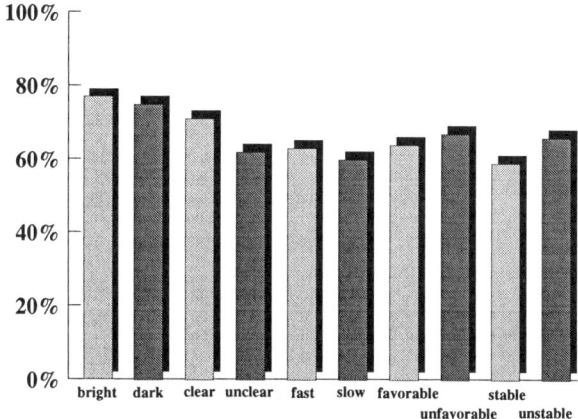

Figure 6: Prediction Accuracy

Figure 5 shows the average difference for the subjects: $\frac{1}{8}\sum_{s=1}^{8} \mathit{diff}_A$, when the prediction is based on chords or melody. Figure 6 shows the percentage of correct predictions. The results show that the differences are between 1.0 and 2.0 in the 7-grade evaluation, and that on average 70% of predictions are correct, which indicates that the system is capable of predicting evaluation of new pieces by the subjects very well. For criteria *fast*, *slow* and *unfavorable*, the difference increases when the prediction is based on chords. This means that these impressions are based mainly on melody and particularly from the length of each note. The system predicts responses within the criteria *bright* and *dark* from chords very well. These impressions are dependent mainly on whether the third is major or minor. Although the system cannot discern these impressions in melody, the authors are now attempting to detect them by analyzing sequences of notes.

Figure 7 shows the variance of impressions among the subjects, which is relatively large in the adjective pairs fast-slow and stable-unstable. These subject-dependent pairs are learned by the system very well according to Figures 5 and 6.

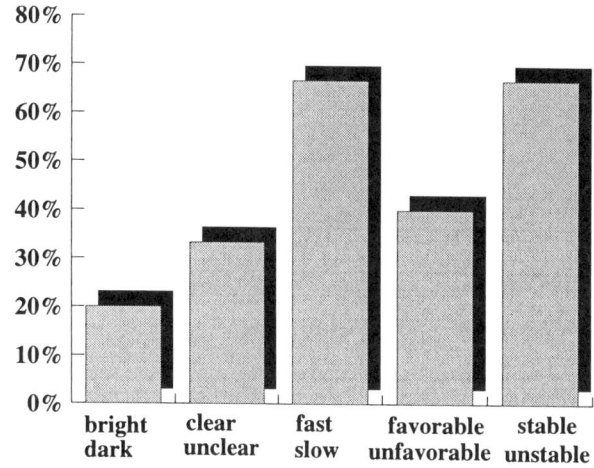

Figure 7: Pieces with Variance > 1

This level of arrangement will be employed in KARA-OKE systems for generating a user model of emotional data to extract an appropriate piece from a MIDI database, or to select suitable background pictures or video for each title.

## 5 Arrangement based on the Model

Utilizing the music recognition model, the system controls the arrangement process based on its overall mood. To improve the arrangement, we assume that a human arranger ( or the system shown in section 3 ) composes the original arrangement, which is then modified slightly to result in a specified impression.

### 5.1 Modifying a Score to Change its Impression

Before arrangement, the system develops a music recognition model for the user. Information about the mood of the user is provided in the form of a 7-grade scale for each of the adjective pairs. The chord progression of a score is modified according to the following priorities:

1. Remove any chord or chord progression whose impression is described by the opposite adjective in the pairs.

2. To minimize differences from the original score, avoid:

    (a) modifying any chord and chord progression that satisfy the given adjective,

    (b) changing any functions, and

    (c) modifying any chord and chord progression that is not in opposition to the given adjective.

3. Modify chords to minimize differences in the evaluation.

### 5.2 Experiments in Arrangement

The authors prepared 94 well-known music pieces without modulation, from which they extracted 4 or 8 suc-

Table 1: Percentage of pieces with matching impressions

| subject | bright | stable | favorable | average |
|---|---|---|---|---|
| A | 45 | 55 | 50 | 50 |
| B | 50 | 75 | 60 | 63 |
| C | 71 | 71 | 50 | 65 |
| D | 78 | 54 | 33 | 55 |
| E | 91 | 64 | 78 | 77 |
| F | 67 | 55 | 58 | 58 |
| G | 45 | 54 | 70 | 56 |
| H | 56 | 83 | 80 | 74 |
| average | 67 | 62 | 61 | 63 |
| standard deviation | 16.2 | 10.7 | 14.9 | 8.7 |

Table 2: Percentage of pieces with improved impressions

| subject | brightness | stable | favorable | average |
|---|---|---|---|---|
| A | 60 | 33 | 37 | 41 |
| B | 45 | 54 | 63 | 54 |
| C | 50 | 78 | 44 | 57 |
| D | 70 | 64 | 43 | 61 |
| E | 82 | 56 | 56 | 68 |
| F | 57 | 50 | 75 | 62 |
| G | 40 | 75 | 75 | 64 |
| H | 50 | 100 | 71 | 74 |
| average | 58 | 62 | 59 | 60 |
| standard deviation | 13.0 | 18.9 | 14.3 | 9.3 |

cessive bars. The subject evaluates 80 of the 94 pieces in 3 pairs of adjectives: *bright - dark, favorable - unfavorable, stable-unstable*, and the results of the evaluations are processed by the system to modify the chord progressions of 6 of the 14 pieces in 6 criteria — bright, dark, favorable, unfavorable, stable and unstable. The subject evaluates the modified $6 \times 6 = 36$ pieces and the original 6 pieces without being notified of how each piece has been modified.

The present authors repeated the above experiment using 8 subjects. Table 1 shows the percentage of arrangements whose modifications corresponded to the intended changes, as well as their standard deviation. According to the table, intended arrangements are produced on average 60% of the time. Table 2 shows the percentage of arrangements for which the subjects' impression is improved and the standard deviation.

Figure 8 shows the average percentage of pieces for which the impression is improved by the modification. According to the figure, 60% of arrangements are improved for the criteria brightness or darkness. In general, if an arranger is less than proficient, the results tend to be evaluated as unstable or unfavorable. Therefore, it is more likely that a piece will be arranged and evaluated as unstable or unfavorable than as stable or favorable.

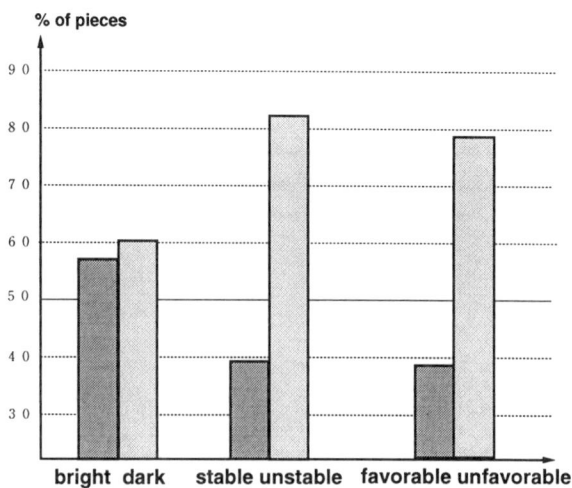

Figure 8: Improvement in impression

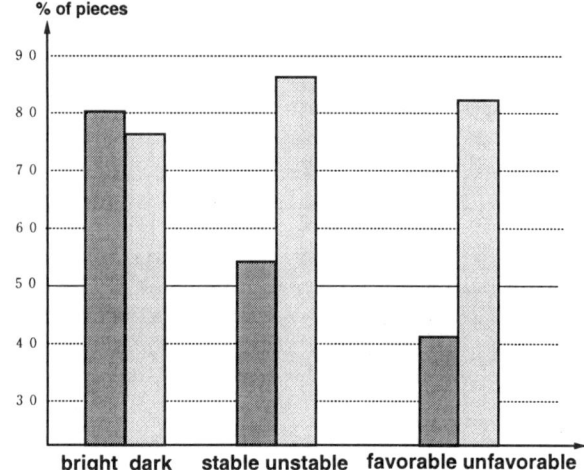

Figure 9: Impression modified (opposing criteria)

According to the figure, some arrangements are evaluated as stable or favorable, although these make up less than half of the total. This suggests that an arranger based on learning personal feelings is effective.

If an original arrangement was evaluated as very bright, it would be difficult to rearrange the piece for a higher level of brightness. It would be easier to rearrange the piece to less bright. Figure 9 shows the average percentage of pieces successfully rearranged to cause the opposite impression. According to the figure, 80% of arrangements are rearranged in this fashion.

## 6 Related Works

Widmer [1994] proposed a method of accomplishing explanation-based learning by attaching harmonies — chord symbols to the notes of a melody. This task is closely related to that described in Section 3. The present paper further discusses a means of controlling the process based on learned feelings.

Katayose, Imai and Inokuchi [1988] approached the understanding of music based on the following rules:

Melody:
 Both the fourth and the seventh do not exist
 → Oriental mood

Our system not only uses this kind of rule but also creates it and applies it through a process that considers its weight.

Most applications of machine learning related to music investigate approaches to interpretation of a music score for playing, for example, measuring the loudness of each note in a score [Widmer, 1993; 1994], or acquiring playing viola [Furukawa, 1997; 1996].

Neural Networks have been used for dealings with problems in music [Todd and Loy, 1991]. The present authors believe that inductive logic programming offers a better means of describing music scores, in which structures are very important. To quantify feelings in music, we introduce a weight $W_j$ to each clause. A better solution is to combine logic programming and neural networks by weighting links in dynamic logical networks, as described in [Numao et al., 1997].

## 7 Conclusion

We present an approach for utilizing human feeling in listening to and composing music. This approach to feelings is an interesting test of the application of machine learning techniques, and should be developed further due to the lack of practical tests to date. If background knowledge of the learner incorporates some important theories in musicology and psychology [Meyer, 1956; Hiraga, 1987; Longuet-Higgins, 1987], we may obtain a more powerful tool for composing music and analyzing human feelings.

## Acknowledgments

An earlier version of the system was implemented by Takashi Shirai (Nintendo), Koji Yamaguchi (Matsushita Electric) and Masatake Saito (SONY).

## References

[DeRaedt, 1996] L. DeRaedt, editor. *Advances in Inductive Logic Programming*. IOS Press / Ohmsha, Amsterdam, Oxford, Tokyo, Washington, DC, 1996.

[Furukawa, 1996] K. Furukawa. Towards verbalization of tacit knowledge by inductive logic programming. In *Keio International Workshop on Verbalization of Tacit Knowledge based on Inductive Inference*. 1996.

[Furukawa, 1997] K. Furukawa. A framework for verbalizing unconscious knowledge based on inductive logic programming. In *Machine Intelligence 15*. Oxford University Press, 1997.

[Hatano, 1987] G. Hatano, editor. *Music and Cognition (in Japanese)*. University of Tokyo Press, Tokyo, 1987.

[Hiraga, 1987] Y. Hiraga. A knowledge representation for recognition of music (in Japanese). In *In [Hatano, 1987]*, chapter 4, pages 97–130. 1987.

[Katayose et al., 1988] H. Katayose, M. Imai, and S. Inokuchi. An approach to extract sentiments in music (in Japanese). *Journal of Japanese Society for Artificial Intelligence*, 3(6):748–754, 1988.

[Longuet-Higgins, 1987] H. C. Longuet-Higgins. Music. In *Mental Processes*, chapter II, pages 57–188. The MIT Press, Cambridge, MA, 1987.

[Meyer, 1956] L. B. Meyer. *Emotion and meaning in music*. University of Chicago Press, 1956.

[Michalski and Tecuci, 1994] R. S. Michalski and G. Tecuci, editors. *Machine Learning: A Multistrategy Approach (Vol. IV)*. Morgan Kaufmann, San Francisco, CA, 1994.

[Mooney and Ourston, 1994] R. J. Mooney and D. Ourston. A multistrategy approach to theory refinement. In *In [Michalski and Tecuci, 1994]*, chapter 5, pages 141–164. 1994.

[Numao and Shimura, 1989] M. Numao and M. Shimura. Explanation-based acceleration of similarity-based learning. In *Proceedings of the Sixth International Workshop on Machine Learning*, pages 58–60, Palo Alto, CA, 1989. Morgan Kaufmann.

[Numao et al., 1997] M. Numao, S. Morita, and K. Karaki. A learning mechanism for logic programs using dynamically shared substructures. In *Machine Intelligence 15*. Oxford University Press, 1997.

[Quinlan, 1990] J. R. Quinlan. Learning logical definitions from relations. *Machine Learning*, 5:239–266, 1990.

[Tangkitvanich and Shimura, 1992] S. Tangkitvanich and M. Shimura. Refining a relational theory with multiple faults in the concept and subconcept. In *Machine Learning: Proc. 9th International Workshop*, pages 436–444, 1992.

[Todd and Loy, 1991] P. M. Todd and D. G. Loy, editors. *Music and Connectionism*. The MIT Press, Cambridge, MA, 1991.

[Tsuji, 1995] S. Tsuji, editor. *KANSEI Information Processing (in Japanese)*. Grant in Aid for Scientific Research of Ministry of Education, Science, Culture, and Sports of Japan under Grant-in-Aid for Special Project Research 04236107, 1995.

[Widmer, 1993] Gerhard Widmer. Understanding and learning musical expression. In *Proc. International Computer Music Conference*, pages 268–275, 1993.

[Widmer, 1994] G. Widmer. Learning with a qualitative domain theory by means of plausible explanations. In *In [Michalski and Tecuci, 1994]*, chapter 25, pages 635–655. 1994.

# Using Data and Theory in Multistrategy (Mis)Concept(ion) Discovery

Raymund Sison, Masayuki Numao and Masamichi Shimura
Department of Computer Science
Tokyo Institute of Technology
2-12-1 Ohokayama, Meguro
Tokyo 152, Japan

## Abstract

Most conceptual clustering systems rely solely on data to form concepts without supervision; the few that exploit causalities in the background knowledge do so only after the completion of a similarity-based learning phase. In this paper, we describe a multistrategy misconception discovery system, MMD, that utilizes data and theory in a more tightly coupled way. The integration of similarity- and causality-based learning in MMD is shown to be essential for the automatic construction of accurate and meaningful misconceptions that account for errors in novice behavior.

## 1 Introduction

The primary method for unsupervised *concept formation* in AI is *conceptual clustering,* which is the grouping of unlabeled *objects* into *categories* for which conceptual descriptions (i.e., *concepts*) are formed. Although conceptual clusterers differ in the way they address six key dimensions,[1] they largely share the characteristic of relying solely on data to form concepts, a situation that likewise characterizes concept learning research in cognitive psychology [Komatsu, 1992]. Recent works (e.g., [Barsalou, 1991; Rips and Collins, 1993; Wisniewski and Medin, 1994]), however, reveal an increasing dissatisfaction over similarity-based models with their almost exclusive reliance on data and show an increasing interest in the role of theories and goals in concept formation.

While there may be a couple of AI systems that use data (similarity-based learning (SBL)) and theory (explanation-based learning (EBL)) to form concepts without supervision, these systems treat SBL and EBL as phases that are performed one after the other. The unsupervised learning systems of [Lebowitz, 1986] and [Pazzani, 1993], for example, first apply SBL on objects to form clusters which are then fed into an EBL or EBL-like component for further processing. The supervised learners of [Flann and Dietterich, 1989], [Mooney and Ourston, 1989] and [Yoo and Fisher, 1991], on the other hand, operate in the reverse fashion: the output of the EBL phase is sent to an SBL component. However, [Wisniewski and Medin, 1994] argue cogently that such loosely coupled approaches to using data and theory, while undoubtedly useful, remain inadequate as models of concept formation.

In this paper, we describe MMD, a <u>m</u>ultistrategy concept (or, more specifically, <u>m</u>isconception) <u>d</u>iscovery system that utilizes data and theory in a more tightly coupled way than previous systems have. As a system that incrementally constructs and revises a hierarchy of possibly overlapping categories of relational descriptions, MMD is also unique in the manner in which it addresses the key dimensions of conceptual clustering earlier mentioned.

**Misconception Discovery as a Special Form of Concept Formation** Errors in novice behavior, such as bugs in a program written by a novice programmer, can be represented as logic formulas that describe specific relations, i.e., *discrepancies*, between the incorrect behavior and an ideal (Table 1 illustrates). Such sets of discrepancies are analyzed in order to uncover the underlying misconceptions that cause them. Knowledge of the causes of bugs will enable a tutor to both present a lesson and remediate a student more effectively.

In *misconception discovery*, therefore, the usual problem of concept formation is further complicated by the fact that conceptual descriptions of clusters of discrepancies can hardly be considered misconceptions — unless causal explanations for them are found.

In what follows, we first present a basic similarity-based algorithm for clustering relational descriptions and then describe how causal relationships in the background knowledge can be exploited to construct or correct descriptions of concepts/misconceptions while they are being formed. We then report experimental results showing that the approach to concept discovery embodied in

---

[1]These dimensions are: data processing mode (incremental or batch), data description (attributional or relational), concept description (definitional, prototypical or probabilistic), category organization (hierarchical or flat), category overlap (overlapping or disjoint), and clustering criterion (ranging from ad hoc to Bayesian measures).

Table 1: Discrepancies in Behavior

---

Ideal behavior:
```
reverse([H|T],R):-
   reverse(T,T1),
   append(T1,[H],R).
```

Buggy behavior:
```
reverse([H|T],[T1|H]):-
   reverse(T,T1).
```

Discrepancies:
```
replace(head,R,[T1|H]),
remove(append_subgoal).
```

---

Table 2: Basic Clustering Algorithm

---

1. From the children $N_1, \ldots, N_m$ of a given node $N$ of the concept hierarchy, determine those that *match* the set of input discrepancies, $O$. The *match* function computes, for every child node $N_i$:

   - the set of commonalities, $Com(N_i, O)$, between a node, $N_i$, and $O$, and
   - the degree of similarity, $Sim(N_i, O)$, between $N_i$ and $O$

   and determines whether $Sim$ exceeds a system threshold, $\gamma$.

2. If no match is found, place $O$ under $N$. Otherwise, for every $N_i$ that matches $O$, perform one of the following depending on the value of $Com$:

   - increase the weight counter of $N_i$;
   - replace $N_i$ with $O$ and insert $(N_i - O)$ under $O$;
   - cluster $(O - N_i)$ under $N_i$ (i.e., repeat the procedure, this time matching $(O - N_i)$ with the children of $N_i$);
   - create a new node, $Com$, under $N$, representing the commonalities of $O$ and $N_i$, and place their differences under this node.

3. Nodes whose (*weight·height*) values fall below a system parameter may be discarded on a regular or demand basis.

---

MMD enables the automatic construction of meaningful misconceptions from theory and data.

## 2 Incremental Clustering of Relational Descriptions

### 2.1 Basic Similarity Measure

Basically an object $O$ is classified into a category with concept description $C$ with which it has more similarities than differences. To measure this degree of similarity/dissimilarity we use as our basis Tversky's contrast model [Tversky, 1977]:

$$Sim(C,O)) = \theta f(C \cap O)) - \alpha f(C - O) - \beta f(O - C)$$

which expresses the similarity between two sets of features (in our case, discrepancies), $C$ and $O$, as a function of the weighted measures of their common $(C \cap O)$ and distinctive $(C - O, O - C)$ features.

We compute the commonalities between two sets of relational descriptions $C$ and $O$ using:

$$(C \cap O) = Com(C,O) = \bigcup_{i=1}^{m} \bigcup_{j=1}^{n} lgg(C_i, O_j)$$

where $lgg(x,y)$ is the least general generalization [Plotkin, 1970; Muggleton and Feng, 1990] of atomic formulas $x$ and $y$ in the function-free first-order logic, and $m$ and $n$ are the number of atoms in $C$ and $O$, respectively.

### 2.2 Basic Relational Clustering Algorithm

The basic similarity-based clustering algorithm classifies a sequence of objects into a hierarchy which it inductively revises in the process. The algorithm is *incremental*, so it takes one object at a time and classifies this object recursively into the nodes that match it to a certain degree. Each node in the hierarchy denotes a *concept*, which is either (a) a generalization (intersection or variableization) of the subconcepts below it, or (b) a record of an instance, or both. A counter is used to store the number of instances associated with a node. Table 2 describes the basic algorithm.

The above algorithm is similar to UNIMEM [Lebowitz, 1987] and COBWEB [Fisher, 1987], which are both incremental conceptual clusterers.[2] UNIMEM's similarity measure, however, considers only the differences between two sets of features. Furthermore, UNIMEM retrieves only a set of "potentially relevant" nodes to compare against the new object (rather than examining every child of a given node), and maintains a total of 13 different parameters.

COBWEB, on the other hand, uses a probabilistic concept representation (rather than set theoretic) and a corresponding probabilistic similarity measure (category utility [Gluck and Corter, 1985; Corter and Gluck, 1992]), and can only produce disjoint clusters (but see the probabilistic clusterer in [Martin and Billman, 1994]). In terms of explaining errors in novice behavior, disjoint clusters mean that a set of discrepancies can only be classified under one "misconception", though it may well be symptomatic of several.

Both UNIMEM and COBWEB deal only with attributional descriptions, though one of COBWEB's variants, Labyrinth [Thompson and Langley, 1991], extends COBWEB to handle structured objects. Like its predecessor,

---

[2] The above algorithm and the $Com(C, O)$ function are presented in greater detail in [Sison and Shimura, 1996a], where the algorithm is called RC. Said report also provides a more in-depth discussion of the similarities and differences among UNIMEM, COBWEB, and RC.

however, Labyrinth can only produce disjoint clusters. We also mention here CLUSTER/S [Stepp and Michalski, 1986], which is an early algorithm that handles structured descriptions by first transforming these into attribute-value form, then feeding these into an attributional clusterer (CLUSTER/2 [Michalski and Stepp, 1983]) that is nonincremental.

## 3 Using Causal Relations to Strengthen Coherence of Concept Descriptions

### 3.1 Causality in Background Knowledge

Similarity-based clusterers like the ones described or mentioned above form categories on the basis of regularities (e.g., co-occurrence, frequency) among features in the data, but ignore qualitative relationships among these same features. We argue that the presence of a qualitative, particularly causal relationship between features of a concept serves at least two purposes:

- First, causal relationships strengthen the coherence of a conceptual description, and can warrant the splitting of a concept or an object when some regularities are coincidental.

- Second, and more important for the problem of misconception discovery, causal relationships explain the regularities in the data. By examining causal relationships it is possible to gain a better understanding of the causes of raw discrepancies.

### 3.2 Causality heuristics

Causal relationships between features can be induced or deduced in a variety of ways. Lebowitz [1986], for example, suggests first using the frequency of occurrence of a feature in other concepts as a heuristic indicator of whether the feature is a cause or an effect, and then forward-chaining from the causative features to the other features using heuristic, low-level, causal domain rules. In [Pazzani, 1993], there are only two "kinds" of features, namely, actions and state changes, and actions are always the causative features. Determining which state changes are caused by which actions is achieved by instantiating general causal patterns.

In our case, we use causal relationships among components of the ideal behavior, together with the following heuristics:

- *Component-level causality:* Causal (or enabling or determination) relationships among the components of the ideal behavior that are present in a set of discrepancies suggest causal relationships among these discrepancies.

- *Concept-level causality:* A causal relationship between two discrepancies in a generalization node, where one is an intersection generalization and the other a variableization, suggests that the former causes the latter.

- *Subconcept-level causality:* Causal relationships between a parent node and its child suggests that the latter causes the former.

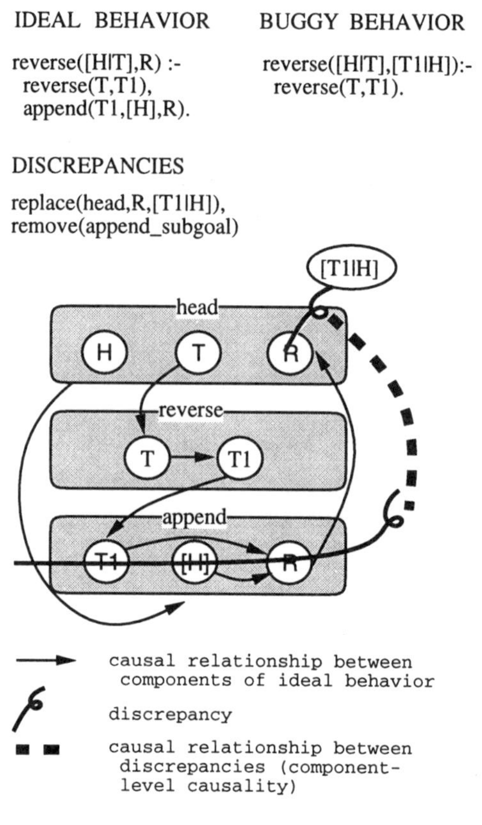

**IDEAL BEHAVIOR**

reverse([H|T],R) :-
 reverse(T,T1),
 append(T1,[H],R).

**BUGGY BEHAVIOR**

reverse([H|T],[T1|H]):-
 reverse(T,T1).

**DISCREPANCIES**

replace(head,R,[T1|H]),
remove(append_subgoal)

Figure 1: Causal relationships

The second and third causality heuristics assume the existence of component-level causal relationships, and are used to determine the direction of casuality.

To illustrate the first heuristic, recall the ideal program for `reverse/2` in Table 1 (reproduced in Figure 1). The ideal program states that the reverse of a list is the concatenation of the reverse of its tail and its head. Note that this can be viewed as describing relationships among four objects, namely, the head $H$ and the tail $T$ of the list to be reversed, the reversed list $R$, and a temporary entity $T1$ representing the reverse of $T$; and the relations `reverse/2` and `append/3`. These relationships are illustrated graphically in Figure 1.

The *component-level* causality heuristic suggests that if two discrepancies $d1$ and $d2$ involve two features $f1$ and $f2$, respectively, both of which involve a common object $c$ (i.e., $c$ in $f1$ causes, enables or determines $c$ in $f2$, or vice versa), then $d1$ and $d2$ are causally related. Thus, since both discrepancies in Figure 1 involve the object $R$ ($R$ in the relation in the second discrepancy causes (enables) the $R$ in the first), then the two discrepancies are, according to this heuristic, causally related; that is, the student's use of the construct [|] is related to the conspicuous absence of the `append/3` subgoal in his program. The drawing in Figure 1 illustrates.

The *concept-level* causality heuristic suggests that if

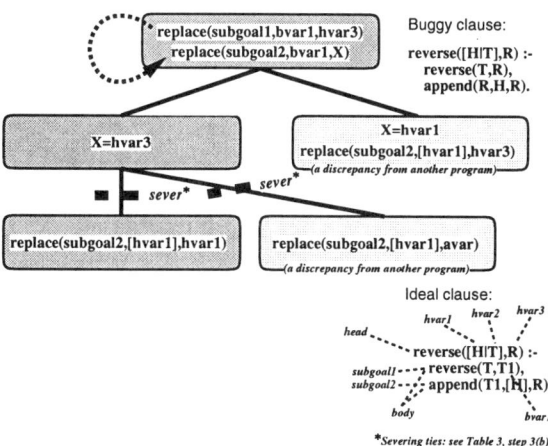

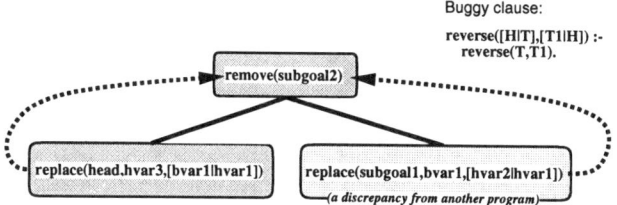

Figure 2: Direction of causality between discrepancies in a node

Figure 3: Direction of causality between discrepancies along a link

Table 3: Incorporating Causality into Concept Formation

---

1. Same as in Table 2, with the addition that causality relationships among discrepancies are to be determined using the component-level heuristic.
2. Same as in Table 2.
3. For every new node created in (2),
   (a) If concept-level causality exists among discrepancies in this node, record the direction of causality. If no such causality exists, retain the node nevertheless. (If the node needs to be split, it will be split by step (2) on another occasion.)
   (b) If subconcept-level causality exists between this node and its parent, record the direction of causality. If no such causality exists, sever the link between this child and its parent, linking it instead with its grandparent, then recheck for subconcept-level causality (i.e., step 3b).
4. Same as step (3) in Table 2.

---

two discrepancies $d1$ and $d2$ are causally related and belong to the same concept/node, and if $d1$ is an intersection generalization and $d2$, a variableization generalization, then $d1$ causes $d2$. This is because the intersection would be the discrepancy that is present together with all the various possible instantiations of the variableization. Thus, for example, students who are not confident in introducing variables in the body of a clause would omit the variable $T1$ in the recursive subgoal of reverse/2, and, as a result, use possibly different variables from the head (e.g., $R$ or $H$) in place of $T1$ in the append/3 subgoal. Figure 2 illustrates.

Finally, the *subconcept-level* causality heuristic suggests that if two discrepancies $d1$ and $d2$ are causally related, and $d1$ is a parent of $d2$, then $d2$ is a probable cause of $d1$, i.e., based on empirical data, the child implies the parent. This is because a child is more likely to be encountered or seen with its parent, than the parent with a particular child. For instance, it is possible that the student in the example in Table 1 (and in Figure 1) put [T1|H] in the head because he/she purposely omitted the append/3 subgoal in the body of his/her clause. This suggests that the student knew about the append/3 relation but decided that using [|] was better, at least in this case. However, it is more likely that the student omitted the append/3 subgoal as a result of putting [T1|H] in the head (Figure 3). This means that the student thought, incorrectly, that the [|] construct could be used to prepend a list to an object, and having dealt with the necessary concatenation, had no further need for a concatenation subgoal in the body of the clause.

## 3.3 A Similarity- and Causality-Based Clustering Algorithm

Existing approaches (e.g., [Lebowitz, 1986; Pazzani, 1993]) to using data and theory (causality) in concept formation use separate SBL and EBL components one after the other. In MMD, SBL and EBL are tightly coupled in the concept formation process. This entails two revisions to the basic algorithm presented in the previous section (rather than a separate algorithm altogether).

- Causal relationships are to be determined using the component-level causality heuristic.
- The directions of causalities are to be determined whenever possible using the concept and subconcept-level heuristics. This may lead to the severing of ties between a parent node and its child when the two are in fact unrelated.

These revisions are found in Table 3. Note that step (3b) effectively functions as a reorganization operator that is causality-based. Step (4) likewise reorganizes (prunes) the hierarchy, though it does this based on frequencies. Reorganization operators are especially important for incremental learners to mitigate ordering effects.

## 4 Evaluation

For a preliminary empirical evaluation of the ability of MMD to discover actual misconceptions in real-world behavior, a sufficiently large corpus of incorrect novice behavior, here in the form of buggy Prolog programs, had to be compiled. A total of 64 buggy reverse/2 and 56 buggy sumlist/2 programs (for the naive reversal of lists and for summing up the elements of a list of numbers, respectively) were obtained from third-year undergraduate students who have learned basic Prolog concepts, and then submitted for expert (teacher) analysis of the underlying misconceptions. The discrepancies between the buggy programs and their associated ideal programs were also computed and then fed into MMD. A point-by-point comparison of the misconception hierarchies generated by the expert and by MMD is presented elsewhere (see [Sison, Numao and Shimura, 1997]); here we compare instead the accuracies of the misconception/classification hierarchies generated by:

a) the basic UNIMEM-like similarity-based algorithm in Table 2[3] (called SMD in Figure 4); and

b) MMD

given *worst-case* orderings of the objects in the reverse/2 and sumlist/2 datasets. A misconception or classification generated by MMD or SMD is considered accurate if it matches that of the expert.

The results obtained are very encouraging (Figure 4). MMD was able to correctly classify most of the bugs in the student programs. The lower accuracy of the hierarchies generated by SMD were mainly due to incoherent groupings and multiple bugs, which SMD is insensitive to. The bugs which MMD (and of course SMD) was not able to classify correctly were primarily due to discrepancies which could be transformed to other, "more meaningful" discrepancies. For MMD to classify these bugs correctly, two options are possible. One option would be to give MMD the ability to recognize discrepancies between discrepancies (i.e., to transform one discrepancy to another). Alternatively, this task could be given to the preprocessor which computes discrepancies between buggy programs and an ideal. The second option is preferable since MMD's primary task is clustering discrepancies rather than transforming them.

## 5 Concluding Remarks

A similarity-based approach to misconception discovery is important because it reveals regularities in the data, which in turn may indicate the existence of underlying causalities. Moreover, in the absence of feature relationships deducible from the background knowledge, an SBL-generated node with high confidence can be learned as a new, though yet unexplainable, misconception. On the other hand, an explanation(causality)-based approach is necessary because concepts based solely on regularities might not be coherent and because some features in concept descriptions can be irrelevant. Furthermore, a similarity-based learner can only roughly classify an erroneous program but not specify the cause(s) of its errors.

---
[3]Using $\theta = \alpha = \beta = 1, \gamma \geq 0$.

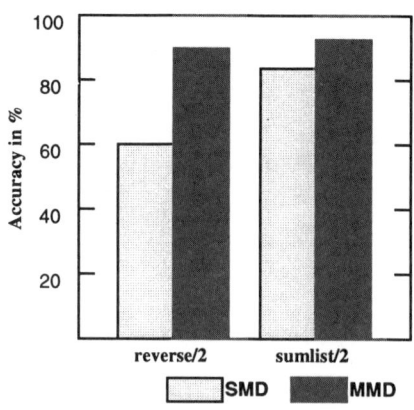

Figure 4: Accuracy of hierachies generated by SMD and MMD given *worst-case* orderings of objects in the reverse/2 and sumlist/2 datasets

The tight integration of similarity- and causality-based learning in the multistrategy unsupervised concept discovery system MMD has been shown to be useful, if not essential, for the the automatic construction of meaningful misconceptions that can be used to account for discrepant behavior in student programs. The applicability of MMD's approach should extend naturally to similar domains, i.e., domains in which causal relationships exist among components of behavior in the background knowledge. Future work will involve investigating mechanisms for generalizing misconceptions across problems, for using domain semantics to articulate causal relationships, and for exploiting and dynamically choosing other qualitative relationships (e.g., goals) that might exist in the background knowledge. MMD is a step toward the automatic discovery of (Prolog programming) misconceptions [Sison, 1997] and their use in multistrategic student modeling [Sison and Shimura, 1996b].

## Acknowledgment

The first author thanks Ethel Chua Joy and Philip Chan for their assistance in compiling and analyzing the programs in the reverse/2 and sumlist/2 datasets.

## References

[Barsalou, 1991] L. Barsalou. Deriving categories to achieve goals. *The Psychology of Learning and Motivation*, 27:1-64, 1991.

[Corter and Gluck, 1992] J. Corter and M. Gluck. Explaining basic categories: Feature predictability and information. *Psychological Bulletin*, 111: 291-303, 1992.

[Fisher, 1987] D. Fisher. Knowledge acquisition via incremental conceptual clustering. *Machine learning*. 2:139-172, 1987.

[Flann and Dietterich, 1989] N. Flann and T. Dietterich. A study of explanation-based methods for inductive learning, *Machine Learning,* 4:187-226, 1989.

[Gluck and Corter, 1985] M. Gluck and J. Corter. Information, uncertainty, and the utility of categories, In *Proceedings of the Annual Conference of the Cognitive Science Society,* pages 283-287. Lawrence Erlbaum, 1985.

[Komatsu, 1992] L. Komatsu. Recent views of conceptual structure, *Psychological Bulletin,* 112(3):500-526. 1992.

[Lebowitz, 1986] M. Lebowitz. Integrated learning: Controlling explanation, *Cognitive Science,* 10:219-240, 1986.

[Lebowitz, 1987] M. Lebowitz. Experiments with incremental concept formation. *Machine Learning,* 2:103-138, 1987.

[Martin and Billman, 1994] J. Martin and D. Billman. Acquiring and combining overlapping concepts. *Machine Learning,* 16:121-155, 1994.

[Michalski and Stepp, 1983] R. Michalski and R. Stepp. Learning from observation: Conceptual clustering. In R. Michalski, J. Carbonell, and T. Mitchell, editors, *Machine Learning: An Artificial Intelligence Approach.* Tioga, Palo Alto, CA, 1983.

[Mooney and Ourston, 1989] R. Mooney and D. Ourston. Induction over the unexplained: Integrated learning of concepts with both explainable and conventional aspects, In *Proceedings of the Sixth International Workshop on Machine Learning,* pages 5-7. Morgan Kaufmann, 1989.

[Muggleton and Feng, 1990] S. Muggleton and C. Feng. Efficient induction of logic programs, In *Proceedings of the First Conference on Algorithmic Learning Theory,* Tokyo Ohmsha, 1990.

[Murphy and Medin, 1985] G. Murphy and D. Medin. The role of theories in conceptual coherence, *Psychological Review,* 92(3):289-316, 1985.

[Pazzani, 1993] M. Pazzani. Learning causal patterns: Making a transition from data-driven to theory-driven learning, *Machine Learning,* 11(2/3):173-194, 1993.

[Plotkin, 1970] G. Plotkin. A node on inductive generalization, *Machine Intelligence,* 5:153-163, 1970.

[Rips and Collins, 1993] L. Rips and A. Collins. Categories and resemblance, *Journal of Experimental Psychology: General,* 122(4):468-486, 1993.

[Sison, 1997]. R. Sison. Toward the automatic discovery of misconceptions. To appear in *Proceedings of the International Joint Conference on Artificial Intelligence,* 1997.

[Sison and Shimura, 1996a] R. Sison and M. Shimura. Incremental clustering of relational descriptions. Technical Report TR96-0011. Department of Computer Science, Tokyo Institute of Technology, 1996.

[Sison and Shimura, 1996b] R. Sison and M. Shimura. The application of machine learning to student modeling: Toward a multistrategic learning student modeling system. In *Proceedings of the European Conference on Artificial Intelligence in Education,* pages 87-93, 1996.

[Sison, Numao and Shimura, 1997] R. Sison, M. Numao and M. Shimura. Discovering misconceptions using multistrategic conceptual clustering. To appear in *Proceedings of the World Conference on Artificial Intelligence in Education,* 1997.

[Stepp and Michalski, 1986] R. Stepp and R. Michalski. Conceptual clustering of structured concepts: A goal-oriented approach, *Artificial Intelligence,* 28:43-69, 1986.

[Thompson and Langley, 1991] K. Thompson and P. Langley. Concept formation in structured domains. In D. Fisher, M. Pazzani and P. Langley, editors, *Concept Formation: Knowledge and Experience in Unsupervised Learning.* Morgan Kaufmann, 1991.

[Tversky, 1977] A. Tversky. Features of similarity, *Psychological Review,* 84(4):327-352, 1977.

[Wisniewski and Medin, 1994] E. Wisniewski and D. Medin. On the interaction of theory and data in concept learning, *Cognitive Science,* 18:221-281, 1994.

[Yoo and Fisher, 1991] J. Yoo and D. Fisher. Concept formation over problem-solving experience. In D. Fisher, M. Pazzani and P. Langley, editors, *Concept Formation: Knowledge and Experience in Unsupervised Learning.* Morgan Kaufmann, 1991.

# An Aggregation Procedure for Building Episodic Memory

**Olivier Ferret**
LIMSI-CNRS
BP 133
91 403 Orsay CEDEX
FRANCE

**Brigitte Grau**
LIMSI-CNRS and IIE-CNAM
BP 133
91 403 Orsay CEDEX
FRANCE

## Abstract

When dealing with narrative texts, a system must possess a strong domain theory, and especially knowledge about situations occurring in the world. Otherwise the system must envisage comprehension as a complex process including learning from the texts themselves to improve its capabilities. This requires managing past solutions and completing them when analoguous situations happen in other texts in order to create general situations. We propose a system, MLK (Memorization for Learning Knowledge), that organizes specific situations in an episodic memory by aggregating the similar ones in a single unit. This aggregation process leads to a progressive enrichment and generalization of the overall situations and of their specific features. MLK is a system conceived to allow the emergence of structures, their accessing being realized by a propagation process. Therefore, with MLK, we are able to address the problem of understanding and learning even when a domain theory is lacking.

## 1 Introduction

Modelling and using the experience of a system is central for text understanding as the machine does not possess all the necessary background knowledge. So if a system can take advantage of the information present in texts and reuse it, it will be able to improve itself. Psychological studies [Vygotsky, 1962; Bartlett, 1932] show that, when experiences are memorized, they are grouped according to resemblance criteria to form a unit, and then abstract classes emerge progressively. The possibility of using known situations related to a new one for understanding it, allows a human to complete his own knowledge about this situation and to improve his capacities of understanding. A system applying such a method is permanently evolving and as a result, a same story may be analyzed in different ways at two different times. We are mostly interested in pragmatic knowledge about concrete situations occurring in the world, traditionally represented by schemata. This kind of knowledge is necessary to integrate sentences in a context and to infer chains of causal relations between events.

Many studies have been carried out on memorizing, understanding and learning as components of a whole process. These systems learn new specialized situations, SWALE [Schank and Leake, 1989] and AQUA [Ram, 1993] by explanation based learning (EBL) and OCCAM [Pazzani, 1988] by integrating empirical learning and EBL. A common point in these systems is the presence of general knowledge to guide the explanation process.

In contrast, our goal is to build a system able to work even when lacking general pragmatic knowledge about concrete situations. In this context, text analysis relies on linguistic clues, such as causal and temporal information, and on other known similar situations if they exist. These weak methods do not lead to a causal explanation of all the events of a text, and thus their pertinence is not fully justified. The presence of an event in two particular cases is not sufficient to conclude if it belongs to a possible general situation. Only the conjunction of its recurrence and the presence of causal clues in different texts will make its abstraction possible. Furthermore, all the events describing a situation are not specified in a text, but those which are relevant to the story. That is why a learning process must recognize similar situations to complete them text after text and not just memorize these situations independently of each other.

To allow the emergence of pertinent features from all similar situations, our system, MLK, uses incremental learning. When memorizing, MLK groups similar situations together to form an aggregated structure, in which similar features are reinforced while rare events become weaker. This aggregation process leads to a progressive enrichment and generalization of the overall situations and of their specific features. This kind of memorization is conceived to build structures on which abstraction of general units will be based and to provide an adaptative retrieval process by propagating activation. The retrieval of situations analoguous to a new one is always dependent on the state of the memory. This approach has several advantages related to the management and the retrieval of similar cases and provides a more psychologically-plausible model of memorization and learning when general background knowledge is lacking.

## 2 The MLK system

MLK is initially given a semantic representation of sentences organized in situations. The segmentation is currently done by hand. We use linguistic criteria (causal clues, order of sentences and temporal markers) plus the type of the predicates in each clause. Situations are named thematic units (TUs) because they group the specific knowledge about a topic. Representations of texts form episodes: a set of TUs related by thematic links, such as topic shift (another subject) and topic deviation (a particular point). The episodes define more global contexts containing specific situations. A TU is a causal structuration of events. Features contained in each slot are represented by conceptual graphs [Sowa, 1984]. They are derived from the semantic knowledge of the system which is represented by a lattice of types of concepts associated to canonical and definition graphs. The canonical graphs contain semantic constraints on the casual roles fillers. In definition graphs, more precision about the roles is available, as well as about the predicate consequences.

The aggregation process is based on similarities between situations. Two situations are similar if their causal structure and their features are analoguous. A similarity measure has been defined on the basis of the structure itself and of its features. The process entails grouping similar features and reinforcing them by augmenting their weight. Other features are simply added. The reinforcement of some elements implies the weakening of others. By iterating this process, a situation will be progressively completed and the most recurrent features describing it will emerge. They will be the candidates for the description of the abstracted situation.

We consider short narrative texts, such as those found in newspapers, that may reference more than one situation. For example, a murder attempt, an aggression, a quarrel, political murders: they all describe analoguous situations that lead to progressively build a single aggregated TU about an attempted murder, described in figure 3 (section 3).

The aggregation process also accounts for the case retrieval problem. In most CBR systems, cases are indexed by predetermined features. In [Kolodner and Simpson, 1989], cases are classified according to general situations and the access mode is based on the traversal of a network of discriminating indices. In AQUA, cases are indexed by several features, especially by the general situations (MOPs) involved. But when lacking general situations, the system does not know what it is learning about and therefore no indices can be used. Another approach is to perform the retrieval of cases by a propagation process, as presented in REMIND [Lange and Wharton, 1993]. Propagation requires weighing the features of the structure. In REMIND, this task is made by hand when the domain knowledge is given. In MLK, the aggregation leads to computing all the weights and to dynamically modifying them according to the experience. Retrieval criteria evolve while the memory content changes without maintaining an explicit structure of indexation features. The problem of deciding what features are relevant for retrieving a structure does not occur. The spreading activation network encodes knowledge about concepts and situations. Concepts can be viewed as indices for situations, because they are part of the conceptual graphs that represent the features of the TUs. Furthermore they create links between TUs other than the thematic links.

When analyzing a text, sentences are decomposed in clauses and the MLK retrieval process is given each concept of a clause. Units in the network representing these concepts are clamped to high level of activation. Activation then spreads into the network. As the activation function encodes the prior context, situations that fit with the current clause and are coherent with the context are highly activated.

Thus the aggregation mechanism provides an answer to the organization and structuration of cases in an episodic memory for the purpose of learning from texts, even when general situations do not guide the process. Our model satisfies two constraints: it allows incremental learning independently of the domain and is a case basis for CBR.

## 3 The aggregating episodic memory

### 3.1 Representation of TUs and encoding of episodes

A text is represented by an episode which is a structured set of TUs. One of them encodes the main topic of the episode and the others are linked to this main situation by thematic links. TUs can be considered as instances of schemata. They encode causal and temporal knowledge between events of a specific situation issued from a text analysis. Their representation formalism is derived from the schema representation chosen, which is close to MOPs [Schank, 1982]. Figure 1 shows the segmentation of a text, the distribution of the clauses in the TUs and details one TU.

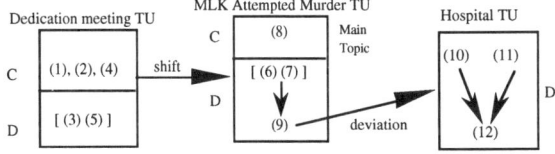

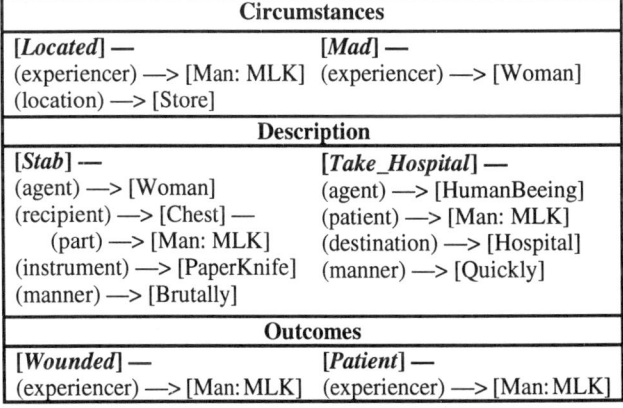

Figure 1: A text and its representation

A TU is described by three slots: Circumstances, Description and Outcomes. The Circumstances slot is valued by a set of conceptual graphs that represent the states known when the situation occurs. Conceptual graphs in the Description slot represent the events which take place and are partially ordered by chronological links. The Outcomes slot contains the resulting states and is also valued by a set of graphs. We can see that some conceptual graphs are issued from semantic inferences, as Patient and Wounded. The conceptual graphs are build around a predicate. The predicate is the central concept when describing an event; its type is a subtype of action, process or state. The other concepts fill in the casual roles linked to the predicate. Their types must be specializations of the types coming from the canonical graphs (we call these latter types 'reference types').

## 3.2 Similarity

The aggregation process requires that MLK find known situations in memory similar to the new TUs. Situations in memory result from preceding aggregations and are named aggregated TUs. The aggregation takes place at two levels. At the episode level, its purpose is to make the global contexts emerge by the means of recurrent sets of situations. At the TU level, its purpose is to learn about a situation by incrementally completing it while recurrent events emerge. At both levels, the aggregation process has to evaluate the similarity of TUs with aggregated ones.

For this purpose, the similarity measure we use [Ferret and Grau, 1995] is decomposed in two steps. Global similarity is based on the number of similar events, according to the slots they belong to. The exact degrees of similarity of slots and events are examined when this global similarity gives a medium result. Two events are similar if their predicates contain the same type and if at least one of their concepts filling casual roles is similar. This occurs if their types share a supertype inferior to the reference type of the casual role. The degree of similarity of events and TUs is computed respectively on the basis of the weights of concepts and graphs. An example is given in section 5.

This similarity measure results in grouping features that belong to a same topic without consideration of a description level. Situations in texts which refer to the same topic are often different specializations in which some events are the same but roles are different. An aggregated TU is conceived to be the basic structure on which abstraction will rest. This latter process will be in charge to eventually split a memorized situation into several units, general and specialized schemata, according to the weights of the events and concepts. If the similarity criteria were more strict, each aggregated TU would be too specific and the common points of a general unit would not emerge.

## 3.3 Episode aggregation

At the episode level, MLK reinforces configurations of similar situations. Aggregation in these cases consists in the reinforcement of the common thematic links and the aggregation of the related TUs. At the TU level, the graphs of events that have the same predicate are aggregated; the others are just added. Same causal or temporal relations between graphs are also reinforced. Aggregation of graphs leads to progressively generalizing their concepts.

As previously mentioned, aggregated TUs result from successive aggregations in the episodic memory. So, they are structures like the TUs, whose features are weighed aggregated graphs. This kind of graphs is defined as an extension of conceptual graphs in such a way as to maintain the capability of applying the basic derivational operations. We have added to them the similarity computation and the aggregation process. The aggregation process is defined as a maximal join where predicates must join: same casual roles are joined and their types of concepts are abstracted as in figure 2. New relations and concepts are then added.

```
[Stab ](1.0) —
    (agent) (1.0) —>      [Man] (1.0)
      (agent) [1,2]         Soldier [1], Young-Man [2]
    (recipient) (1.0) —>   [Body] (1.0) —
      (recipient) [1,2]     Arm [1], Stomach [2]
    (part) (1.0) —>        [Man] (1.0)
      (part) [1,2]          Head-of-State [1], Young-Man [2]
    (instrument) (1.0) —>  [Knife] (1.0)
      (instrument) [1,2]    Bayonet [1], Flick knife [2]
```

Figure 2: An aggregated graph

Types of aggregated concepts are computed at each aggregation. They are the most specific concept abstraction (msca), different from the reference type, of all the aggregated instances. If such a type does not exist in the lattice, we choose the msca that groups the maximum of instances if it is possible. To compute it, each aggregated concept keeps the instances it comes from. For each of these instances, we also keep the specific episode it is part of (as [1] in the figure). Thus, we can rebuild all the specific situations which have contributed to the formation of an aggregated TU. This capability is preserved to enable the retrieving of specific links between graphs. It will be useful to justify a reasoning based on the aggregated situation, TU aggregation being done regardless of shared instances between graphs. For example, if an agent of an event is also the patient of another event in the same situation, this piece of knowledge is not maintained at the aggregation level since each aggregated concept type evolves inside its own graph regardless of the evolution in the other graphs. This kind of knowledge is also basic to schemata abstraction; it prepares the formation of roles. The last characteristic added on the aggregated concepts is a counter to compute their weight. Weights on the predicates represent the importance of the event inside the TU and weights on the other concepts, the importance of the concept in the graph. These counters are equal to the number of aggregations. Thus, the weight of an event is computed by dividing the counter of the predicate by the number of times the TU has been aggregated. Similarly, the weight of a concept is the counter of the concept divided by the number of aggregations of the predicate. Weights on relations between two graphs are the number of aggregations divided by the number of common episodes between the two graphs.

The following example, in which five TUs have been aggregated, only shows the predicates of the graphs with their weights and the episodes they belong to. For presentation purpose, chronological relations between events in the

| Circumstances (C) | |
|---|---|
| *(a) Located* (**0.2**) [1] | *(b) Disagree* (**0.4**) [2,3] |
| *(c) Live* (**0.4**) [3,5] | *(d) Have_Bath* (**0.2**) [3] |
| *(e) Threatened* (**0.2**) [3] | *(f) Commanding* (**0.2**) [4] |
| *(g) Sleeping* (**0.2**) [5] | |
| Description (D) | |
| *(a) Attack* (**0.4**) [1,5] | *(b) Stumble* (**0.2**) [1] |
| *(c) Stab* (**1.0**) [1,2,3,4,5] | *(d) Arrest* (**0.4**) [1,2] |
| *(e) Hit* (**0.2**) [2] | *(f) Enter* (**0.4**) [3,5] |
| *(g) Lose* (**0.2**) [4] | *(h) Attach* (**0.2**) [5] |
| *(i) Tear* (**0.2**) [5] | |
| Outcomes (O) | |
| *(a) Imprisoned* (**0.5**) [1,2,5] | *(b) Wounded* (**0.4**) [1,5] |
| *(c) Dead* (**0.5**) [2,3,4] | *(d) Guillotined* (**0.2**) [3] |

Inter-graphs relations:   D.d -> O.a (**1.0**) [1, 2]
  D.c -> O.b (**1.0**) [1,5], O.c (**1.0**) [2,3,4]

Figure 3: The aggregated TU Attempted_Murder

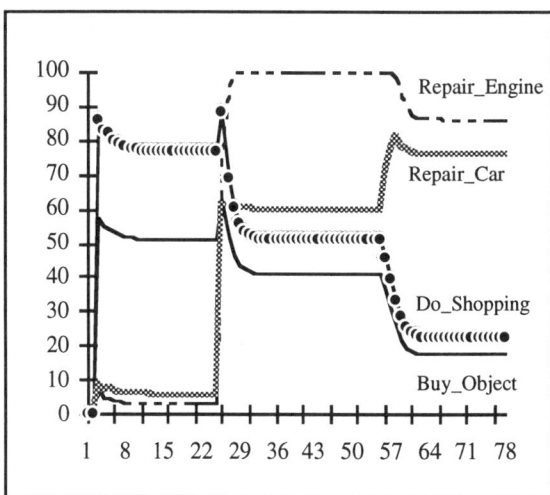

Figure 4: Activation level of aggregated graphs

description slot and the aggregated concepts have not been mentioned. Only causal relations between actions and their results have been precised. States setting up a situation are extremely various in texts, and often without explicit causal relations; so they are all considered as possible circumstances. At this state of the memorization, none emerge completely, but we can note that some are semantically close: *Located* vs *Live* and *Disagree* vs *Threatened*. Events in the description slot are more interesting with a characteristic event always reinforced, *Stab*. Others have to be confirmed before deciding their relevance, like *Arrest*. Notice that the reinforcement of a causal relation between two events is also an indication of their interest. Weak weights on events can be interpreted as anecdotes (*Stumble*) or events that do not really belong to the situation. Outcomes are often linked to an action and are significant. A discussion about the formation of this TU is presented in section 5.

## 4 The retrieval process

A spreading activation mechanism makes possible the use of episodic memory as an associative memory in order to select the relevant knowledge for processing the clauses coming from the texts. The figure 4 shows activation of four aggregated graphs resulting from the network after it successively receives as input the clauses of the following little story:

*Yesterday, I went out to do shopping. I bought a spare part to repair my car. But finally, I let the mechanic repair it.*

The two aggregated graphs, BuyObject and Shopping, are part of an aggregated TU called *GoingSupermarket*. RepairCar and RepairEngine are part of the *RepairingCar* aggregated TU. In accordance with the thematic shift in the story (going from shopping to repairing a car), there is a change among the most activated elements of the memory. After clamping the concepts of the first sentence, the most activated elements are part of the *GoingSupermarket* TU. On the contrary, the graphs of the *RepairingCar* aggregated TU are not activated because nothing is mentioned about the situation of repairing. When the second sentence is introduced (after cycle 25), the graphs of the *GoingSupermarket* TU are still highly activated. This is the result of two influence streams. First, the previous activation state of the memory created a context which constrains its evolution. Second, concepts coming from the sentence (in this case, the concept Buy) confirm the previous topic. But we can see also that others concepts (SparePart, Car, Repair) make another topic appear. This explains why the graphs of the *RepairingCar* TU, and the TU itself, begin to be activated. After the processing of the third sentence (after cycle 56), the context effect of the memory for the first topic is not significant any more (the previous confirmation was not very strong) and no element of this third sentence is specifically linked to it. On the contrary, it strengthened the second topic. This explains why we can observe that the activation of the elements of the *GoingSupermarket* TU decreases rather strongly and that the opposite effect happens for the elements of the *RepairingCar* TU.

The spreading activation mechanism which supports this process is divided in two steps. Firstly, it defines a subset of the memory, in which the selection will be done. The episodic memory may be very large if the system has a great amount of experiences. Hence it is not realistic from a computational point of view to involve the whole memory in the core selection mechanism. Moreover, this could be a source of noise and disturb the selection process. It is of course necessary in this task to bypass some similarity failures: the above little story does not tell us about a supermarket, yet the *GoingSupermarket* TU, which is the best thematically related TU, has been nevertheless retrieved. This is due to the spreading of the activation flow towards more general or more specific concepts than the ones initially clamped. But this flow has to stay around the initial concepts. Thus it is not interesting to select a surtype of Supermarket as PublicPlace because we do not want to activate all the situations which take place in a public place.

The definition of this subset of the memory is based on the propagation of a constant flow of activation in the network, a flow that starts from the concepts that make up the clauses of the texts. As the global amount of activation that can be used for the propagation is always the same, the activation level of the units is less and less high as these units are more and more distant. When the activation level of an unit is under a given threshold, the propagation towards its next units is stopped. In the lattice of concept types, a specific mechanism makes more activation going towards subtypes of a concept type than towards surtypes.

The second step of the spreading activation mechanism aims at selecting the aggregated TUs in the episodic memory that are the most relevant with regard to the clauses which are considered at a given stage of the analysis of a text. This process is akin to the *evidential activation* in REMIND. When this working mode is active, the network mentioned above, which is a recurrent one due to the symmetry of the connections, has a two-phases dynamics: at the beginning, it has great liberty to explore the space of possible states and this liberty is restrained progressively in order to make the network converge on one state. This is achieved locally by the activation function of the units through its divisor term:

$$A_i(t+1) = \frac{\sum_j w_{ij} \cdot A_j(t)}{t^2} + A_i(t)$$

with  $A_i(t)$: activation level of unit i after t cycles
$w_{ij}$: weight of the connection between units i and j

The second term of this function is particularly significant since it is the support of the context effect mentioned above which makes the network take into account the result of the previous spreading activation sessions.

## 5 Discussion

### 5.1 Processing of texts

We give here a complete example processed by MLK. The memory contains in particular the TU *Attempted_Murder* described in figure 3 resulting from the processing of five texts. The creation of the TU comes from the first episode, Airport_Murder_Attempt because none similarity has been found with existing TUs. The other texts, a Quarrel [2], two Politic_Murders [3] and [4] and an Aggression [5], contain a situation matching with the evolutionary aggregated *Attempted_Murder* TU. With the episodes [2] and [3], the similarity results from a deep comparison of the graphs while within the two last episodes, it results more from the global measure. Thus, the more the TU is growing, the more the similarity becomes obvious. The new processed text is given in figure 1. We will focus on the processing of the TU MLK_Attempted_Murder. Each graph causes a propagation and entails highly activating the *Attempted_Murder* unit. The Circumstances and Outcomes slots are strongly similar while the Description slot has been found similar after a deep evaluation. Therefore aggregation is done; results are given in figure 5.

The aggregated TU has been enriched by new events. *Located* has been reinforced while *Mad* is added in the Circumstances slot. In the Description slot, *Stab* is reinforced and *Take_Hospital* is added. In the third slot, *Wounded* is reinforced and *Patient* added. A new causal relation appears

| Circumstances | |
|---|---|
| (a) *Located* (0.33) [1,6] | (b) *Disagree* (0.33) [2,3] |
| (c) *Live* (0.33) [3,5] | (d) *Have_Bath* (0.16) [3] |
| (e) *Threatened* (0.16) [3] | (f) *Commanding* (0.16) [4] |
| (g) *Sleeping* (0.16) [5] | (h) *Mad* (0.16) [6] |
| Description | |
| (a) *Attack* (0.33) [1,5] | (b) *Stumble* (0.16) [1] |
| (c) *Stab* (1.0) [1,2,3,4,5,6] | (d) *Arrest* (0.33) [1,2] |
| (e) *Hit* (0.16) [2] | (f) *Enter* (0.33) [3,5] |
| (g) *Lose* (0.16) [4] | (h) *Attach* (0.16) [5] |
| (i) *Tear* (0.16) [5] | (j) *Take_Hospital* (0.16) [6] |
| Outcomes | |
| (a) *Imprisoned* (0.5) [1,2,5] | (b) *Wounded* (0.5) [1,5,6] |
| (c) *Dead* (0.5) [2,3,4] | (d) *Guillotined* (0.16) [3] |
| (e) *Patient* (0.16) [6] | |

Figure 5: A resulting TU after the aggregation process

between *Take_Hospital* and *Patient*. Comparison of the weights before and after aggregation, shows that they increase with reinforcement while decreasing in the opposite case. Even with only six examples, a structure begins to emerge and the overall frame becomes meaningful. Our method groups events that are thematically linked and, even if misunderstandings occur when segmenting texts, the aggregation will fix them in the long term. Aggregation provides a strong process where some noises are allowed.

### 5.2 Towards abstraction

At this state of structuration, we can notice in the evolution of the aggregated TU some indications about further abstraction. Strongly weighed events will belong to a general situation, as *Stab*; events as *Arrest* have to be confirmed. Events such as *Dead* and *Wounded* are not simultaneously present in the different episodes but are causally linked to the same action. This will suggest creating a generalized event for them in the general structure, with two specializations including these specific events. Thus, an aggregated structure will lead to create schemata hierarchically structured.

Generalization of concepts also prepares the abstraction phase. Information that will be used are weights, but also the distribution of the episodes and the aggregated relations.

### 5.3 A reasoning medium

The episodic memory is build for the purpose of learning and improving the comprehension. In this latter perspective, we have developed a thematic analysis that segments a text into TUs by using the memory and its selection mechanism [Ferret and Grau, 1997]. The accumulated experience of MLK is then reused to process new texts.

Enrichment of TUs will allow to improve understanding and make inferences. For example, in the fifth episode, the fact that the aggressor is imprisoned is explicitly mentioned. Because this kind of fact has been memorized as the outcome of the *Arrest* event with an important weight, the arrest of the aggressor could be inferred. Such an inference can be justified because the roles are identical in the specific situations. MLK makes this kind of reasoning possible due to the case basis dimension of its memory.

# 6 Comparison with other models

AQUA is the closest system to MLK in terms of their goals. It learns explanation patterns (XPs), even if they are incomplete. These XPs are completed when the lacking explanations are found in further texts. But the two processes are quite different. AQUA needs a strong domain theory to build its explanations. This knowledge is represented by MOPs and abstract XPs. AQUA learns new specializations of these XPs. Learning and generalization occur at each new case. Generalization relies on the explicative structure even if some actions need further explanations. In MLK, to avoid the need of general descriptions in all domains, causal links come from linguistic clues, from the assumption that texts are coherent by themselves and from semantic knowledge. If these links are recurrent, they will be confirmed. On the other hand, if they are unusual or result from a misinterpretation, they will disappear by non-reinforcement. So generalization is done in two steps. The aggregation principle allows relevant events to appear with their inter-relations. It replaces the explanation mechanism needed to justify events and solves the incremental aspect of learning. The second step of abstraction, where events and roles are abstracted to produce general situations, must be realized by another process, when situations in the episodic memory are stabilized.

Case retrieval is quite different in MLK and problems of misindexation solved by AQUA do not occur. It is closer to REMIND which integrates episodic memory retrieval and comprehension in the same propagation process. Comparison is only done on the episode retrieval results, since comprehension is quite different when dealing with incomplete knowledge. Differences come from the integration and the structuration of episodes. First, updating of the weights in MLK is done automatically. Second, frames in REMIND are intermediate between our semantic graphs and the TUs. Episodes in REMIND are analogous to our TUs but TUs are much more structured units where all the frames explaining an episode are grouped. In MLK, weights on events code the importance of the frame in the episode. Thus, this upper level affects the spreading of the activation and selects the TUs not only superficially (as they can be in REMIND) but also thematically close.

# 7 Conclusion

In this article, we have presented MLK, an episodic memory model which has been designed to support a comprehension process tightly tied to learning. In this model, text representations built by the comprehension process are stored in memory in order to be used later by this same process to analyze other texts. But unlike traditional CBR systems, the memorization in MLK does not consist in classifying a new element according to an existing frame. By aggregating the text representations, the similar situations to which they refer form aggregated units containing events having a weight which characterizes their recurrence degree. This allows a new sort of knowledge to emerge. Weights are used in order to support the retrieval process in the memory: a spreading activation mechanism selects the aggregated units that are the most relevant according to both the current input and the context set by the previous inputs. So, even when a strong domain theory is absent, MLK is able to memorize and organize the result of a comprehension process in such a way that it is able to recall it later in a contextually relevant way. Moreover, this knowledge evolves progressively as new text representations are memorized.

At its present state, our memory model is implemented in Smalltalk and has been tested with a set of 1939 conceptual frames (concept types and the graphs associated to them) and 19 aggregated TUs resulting from 26 texts, i.e. 65 TUs. A linguistic based module whose goal is to build the internal representation of the TUs and a generalization process are being studied. So, in the future, we intend to integrate all these aspects in order to build a complete system integrating learning and comprehension.

# References

[Bartlett, 1932] Frederic C. Bartlett. *Remembering*. Cambridge University Press, Cambridge, 1932.

[Ferret and Grau, 1995] Olivier Ferret and Brigitte Grau. An episodic memory for understanding and learning. In *Proceedings of RANLP'95*, pp. 221-229, Bulgarie, 1995.

[Ferret and Grau, 1997] Olivier Ferret and Brigitte Grau. Une analyse thématique s'appuyant sur une mémoire épisodique. In *Proceedings of JST Francil 97*, pp. 161-168, France, 1997.

[Kolodner and Simpson, 1989] Janet Kolodner and R. L. Simpson. The MEDIATOR: Analysis of an early case-based problem solver. *Cognitive Science*, 13(4):507-549, 1989.

[Lange and Wharton, 1993] Trent E. Lange and Charles M. Wharton. Dynamic Memories: Analysis of an Integrated Comprehension and Episodic Memory Retrieval Model. In *Proceedings of the 13th International Joint Conference on Artificial Intelligence,* pp. 208-213, France, 1993.

[Pazzani, 1988] Michael J. Pazzani. Integrating explanation-based and empirical learning methods in OCCAM. In *Proceedings of the 3rd European Working Session on Learning*, pp. 147-165, Glasgow, 1988.

[Ram, 1993] Ashwin Ram. Indexing, elaboration and refinement: incremental learning of explanatory cases. *Machine Learning*, 10(3):7-54, 1993.

[Schank, 1982] Roger C. Schank. *Dynamic Memory: a Theory of Reminding and Learning in Computers and People*. N.Y. Cambridge University Press, 1982.

[Schank and Leake, 1989] Roger C. Schank and David B. Leake. Creativity and Learning in a Case-Based Explainer. *Artificial Intelligence*, 40(1-3):353-385, 1989.

[Sowa, 1984] John F. Sowa. *Conceptual Structures: Information Processing in Mind and Machine*. Addison Wesley, 1984.

[Vygotsky, 1962] L.S. Vygotsky. *Thought and Language*. MIT Press, Cambridge-Mass, 1962.

# COGNITIVE MODELING

Cognitive Modeling 2

# An Achievement Test for Knowledge-Based Systems: QUEM

Caroline Clarke Hayes
University of Illinois
Department of Computer Science
405 North Mathews Avenue
Urbana, Illinois 61801 U.S.A.

Michael I. Parzen
University of Chicago
Graduate School of Business
1101 East 58th Street,
Chicago, Illinois, 60637 U.S.A.

## Abstract

This paper describes QUEM, a method for assessing the skill level of a knowledge-based system based on the quality of the solutions it produces. QUEM is demonstrated by using it to assess the performance of a particular knowledge-based system, $P^3$. QUEM can be viewed as an achievement or job placement test given to knowledge-based systems to help system designers determine how the system should be used, and in what capacity by what level of users. In general, it is difficult to find useful metrics for assessing a system's over-all performance. Most literature on evaluation deals with validation, verification and testing in which the primary concern is the correctness and consistency in the databases and rulebases. However, these properties alone may not be sufficient to determine how well a system performs its task. QUEM allows software developers to assess their system's performance by constructing a skill function based on human performance data that relates experience and solution quality. QUEM can be used to gauge the experience level of an individual system, compare two systems, or compare a system to its intended users. This represents an important advance in quantitative measures of over-all system performance that can be applied to a broad range of systems.

## 1 Introduction

When evaluating knowledge-based systems (KBSs) it is often difficult to find useful metrics for assessing a system's overall performance. Most literature on KBS evaluation deals with validation, verification and testing [Nazareth and Kennedy, 1993] in which the primary concern is with the correctness and consistency in the databases and rule-bases. Other systems address modifiability, ease of use and cost of the system. However, these properties alone may not be sufficient to determine how *well* a system performs its task. For example,
a complete and consistent KBS may not necessarily create high quality solutions. It would be useful to have a method to estimate how well a KBS's performs its task on some absolute scale that would allow comparisons between systems to be made. However, it is not immediately obvious what type of scale should be used. In this paper we present QUEM (Quality and Experience Metric, pronounced "kwem"), a method for evaluating the *experience level* of a knowledge-based system and the *quality* of its solutions. In other words, QUEM estimates the performance of a system in terms of the years of experience a human would require to generate solutions of the same quality; experience level is the scale on which quality is assessed. QUEM can be considered to be an *achievement test* for KBSs since it estimates the level that a KBS has achieved. We use expert judges to assess the quality of solutions generated by human experts and KBS's. We then construct a "skill function" for the human data relating experience and solution quality. This skill function is used to estimate KBS experience level.

QUEM provides a *quantitative* way to estimate the experience level of a KBS, compare two KBS's, or compare the experience level of a KBS to that of its users. This type comparison is of particular importance if a KBS is to be used as an aid to human users. Understanding the skill level of the KBS relative to its users is important in determining how the system should be used and in predicting whether users will accept it. For example, it is often necessary that the skill level of the KBS equal or exceed that of its users. If the KBS produces solutions of lower sophistication and quality than the user can produce on his or her own, the user may consider the system to be a hindrance. Additionally, estimation of a KBS's experience level is one measure of how well developers have succeeded in capturing the domain expertise.

The development of QUEM arose from our desire to measure the quality of various knowledge-based systems which were under development. We were particularly interested in having a way to assess the quality of the problem solving abilities of prototype systems because it would be of great assistance in making development decisions. Before continuing a large programming effort to improve a given system, we wanted to have some assurance that the approach used in the prototype was a reasonable one. If evaluation were to show that the best solutions produced by a given system were of low quality, then we would conclude that the approach was not a reasonable one and our efforts should be focused on re-structuring the approach and problem solving architecture. On the other hand, if the solutions produced proved to be of high quality then we could feel more confident that the approach was reasonable and that it would be worthwhile to put efforts into further development of the current system architecture.

In this paper we will outline the QUEM method developed for assessing average solution quality in terms of experience level, and demonstrate the use of QUEM to estimate the solution quality and experience level of a KBS for creating manufacturing plans. We used these results to assess the competence of our KBS problem solving approach, and to determine if we should continue on the same approach in future developments.

## 1.1 Challenges

Our first challenge was to identify a method for assessing solution quality in complex domains. Quality is the perceived utility that an artifact has to some set of people in a given context. A utility function is often used to provide a precise number to estimate quality, just as a watch provides a precise number to estimate the true time. Solution quality is in general hard to measure because it can be hard to quantify. This is particularly true in domains that are very rich such as architectural design, military battle planning, and manufacturing. There are usually many, sometimes conflicting factors that determine solution quality such as cost and accuracy, and esthetics. Even if one can generate a utility function to describe quality in a given domain, it may be hard to quantify the component factors that determine quality.

In our initial attempts to estimate solution quality for the manufacturing domain, we tried to construct a quality utility function composed of factors which our experts believed to be important. In the manufacturing domain, important factors included plan cost, feasibility and reliability. We attempted to generate a quality function for this domain based on these factors. However, we soon found this approach to be inadequate; it was not feasible to construct an accurate mathematical quality function because some important component factors, such as reliability, were very difficult to quantify.

To describe the example further, *plan reliability* is the likelihood that the operations within the plan will fail or will produce marginal results. Plans can fail in catastrophic ways resulting in physical damage to agents or equipment executing the plans, or in subtler ways, such as when the resulting product does not meet requirements. Predicting reliability requires knowledge of a wide variety of situations which are hard to capture without a large body of empirical data. Because of these difficult to quantify component factors, the task of constructing a reasonably accurate mathematical quality metric for KBS solutions is very difficult in practice, for many (if not most) rich and complex domains.

However, after some initial disappointment on finding that a good mathematical quality function was not feasible in our domain, it occurred to us that we did not actually need a quality function because we had a number of fairly robust quality measuring devices readily available to us: human experts. Human experts can succeed in assessing quality where a quality function may fail because experts are able to estimate hard-to-quantify quality factors, such as reliability, based on their broad empirical experience. Additionally, we found that although there was some variation in how judges assessed quality, their quality assessments were usually fairly close. Experts with similar experience tend to make similar quality assessments of a given solution. Furthermore, those assessments correlate very strongly with the experience of the problem solver. Thus, although there is a perception that humans are unreliable, we found that human experts were fairly consistent in assessments, and that their variability can be measured (for example, by having several experts independently rate the same solution) and taken into account. We decided look for a way to use the solution quality assessments produced by human experts to assess KBS solution quality.

Our second challenge was to devise a scoring system in which human judges could report their quality assessments. The scoring system must allow the quality assessments of different judges to be compared. Initially, we considered having the judges assign quality scores between 1 and 10, like Olympic sports judges, indicating the absolute quality of each plan. However, we decided against such an approach because experts do not have a standard or agreed upon method for assigning quality measures to plans. We were concerned that it might be difficult to compare scores assigned by two judges; if 10 is the best quality score, an enthusiastic judge might give many 10's while a conservative judge may rarely give a score better than 6. However, the first judges's 10 may mean the same thing as the second judges's 6. We decided that it would be more appropriate to have the judges rank order solutions from best to worst, rather than to assign them scores.

## 2 Related Work

As mentioned earlier, most literature on knowledge-based system (KBS) evaluation deals with validation, verification and testing (VVT) [Nazareth and Kennedy, 1993] in which the primary concern is with correctness, circularity, inconsistency, redundancy, modifiability, ease of use and cost [Lane, 1986], [Liebowitz, 1986]. However, these properties alone may not be enough to describe a system's competence in solving problems effectively. [Clancey, 1993] describes four perspectives useful for evaluating a system's competence: performance, articulation, accuracy and completeness. Other parameters important to system competence are: solution feasibility, solution quality, problem solving range, computer effort, and user effort.

Most competence evaluations provide *relative* measures of system performance. These evaluations provide the information that system x works better than system y, or human z. For example, when Aikins [Aikins, 1981] evaluated her system, Puff, a medical diagnostic system for cardio-pulmonary diseases, she compared the performance of her system against the diagnostic performance of three human doctors. She found that Puff's diagnosis agreed with the average diagnosis more often than did any of the individual doctors. From this she concluded that not only could Puff perform competently, but it was also more accurate on average than any of the individual experts in the study. Dixon, et al. evaluated their system, Dominic [Dixon et al., 1987], by comparing its results against those of two other KBS's and a human expert. From this comparison they concluded that "Dominic is a reasonably capable designer ... although the two domain specific programs produced slightly superior performance."

However, simply knowing that one KBS produces better quality solutions than another does not necessarily tell the KBS developers if either produces particularly good solutions. For this reason we also felt it was necessary to develop a *quantitative* measure of KBS experience level which would allow one to make statements such as, "My KBS is estimated to have captured *n* years of experience." Such measures can better aid system developers in assessing whether their KBS is sophisticated enough for their purposes.

## 3 General Method

The QUEM procedure requires one or more *knowledge-based systems* for comparison, a set of *problems*, several *subjects* of various experience levels, and two or more *expert judges*. The expert subjects should have differing levels of experience. The expert judges should have experience equal to or greater than that of all subjects. The judges should not double as subjects in order for this test to produce meaningful results. Additionally, the domain of experience for the KBS, judges and subjects, must all be very similar, otherwise they may judge quality by very different criteria. The QUEM procedure for rating KBS experience level is:

1. **Solve.** Have all subjects and all KBSs each solve all problems in the problem set.

2. **Sort.** For each problem, put all solutions together in a group. If there are 3 problems, there will be 3 solution groups.

3. **Rank.** Have the expert judges independently rank order all of the solutions in each group from worst quality to best quality. Label the worst solution in each group as number 1. Successively number each solution, assigning the highest number to the best solution.

4. **Adjust Ranks.** If a judge ranks several solutions as having equal quality, the ranks must be normalized so that they can be compared to other rankings. For example, suppose Judge 1 is given 6 solutions which he ranks 1 through 6, while Judge 2 is given the same six solutions but she ranks two solutions as worst, three as intermediate, and one as best, producing the ranks of 1, 1, 2, 2, 2, and 3. The rankings of Judge 2 must be adjusted if they are to be compared to Judge 1's rankings. To adjust the rankings, they must be divided in to tied groups. Judge 2's rankings would be divided into three groups: (1, 1) (2, 2, 2) (3). All data points must be renumbered starting from the lowest number, such that each has a separate consecutive rank: (1, 2) (3, 4, 5) (6). Next, the average rank of each group is computed, and each member of a group is assigned the value of its group average. Thus, Judge 2's adjusted rankings would be: 1.5, 1.5, 4, 4, 4, and 6.

5. **Compute subject averages.** Compute the average quality ranking for each subject and KBS across all problems using the adjusted rankings.

6. **Plot subject averages.** Put the KBS data aside for a moment. Plot each human subject's experience on the y axis and his or her average quality ranking on the x axis.

7. **Fit a skill function to the data.** Fit a line or curve to these data using linear regression or other method appropriate to the data. Call this the *skill function*.

8. **Construct confidence bands** to indicate the amount of variation one can expect in individual performances at any given experience level. A point estimate of experience is not useful without some idea how accurate the estimate is. To compute these bands let $x_m$ denote the average quality rank of a KBS. Using the linear regression model described above, our experience estimate of the KBS is $y_m = b_0 + b_1 x_m$. A 95% confidence interval for this estimate is given by

$$y_m \pm t_{(n-2,.025)} \sqrt{(\frac{1}{n} + \frac{(x_m - \overline{x})^2}{\sum_{i=1}^{n} x_i^2 - n\overline{x}^2}) s_e^2}$$

where $t_{(n-2,.025)}$ is the 95% confidence coefficient based on the $t$−distribution and $s_e^2$ is an estimate of the amount of variability in the relationship between experience level and average quality rank. All these quantities are standard output results from statistics packages.

9. **Construct an experience estimate and interval.** For each KBS in the study,

    (a) Plug the KBS's average quality ranking (x) into the skill function to obtain the *experience estimate* for the KBS.

    (b) Plug the KBS's average quality ranking (x) into the equation for the upper confidence band. Repeat for the lower confidence band. The two numbers produced represent the *experience interval* for the KBS.

The results of this process are:

- A *skill function* for humans, relating length of experience to solution quality.
- *Confidence bands* around the skill function showing the expected range of skill for practitioners having a given length of experience.
- An *experience estimate* for the KBS. This value indicates the most likely value for the experience-level of the KBS.
- An *experience interval* showing the range of human experience levels expected to achieve the same average quality as the KBS, with 95% confidence.

## 3.1 Applications of QUEM

QUEM can be used in a variety of ways:

1. **Estimate the experience level of a single KBS.** When applied in this way, solutions created by a single KBS are ranked along with solutions created by a range of humans.

2. **Identify a change in experience level** between an old and a new version of a KBS. Solutions of two or more versions of the same KBS are ranked along with solutions created by humans.

3. **Compare two or more KBSs in the same domain** using the same method described in (2).

4. **Compare two unrelated KBSs** that operate in different domains. In order to compare two unrelated KBS's, two separate QUEM tests must be performed, and the resulting experience levels compared. A separate group of judges and subjects with appropriate domain knowledge must be selected for each test.

5. **Estimate the amount by which a computer assistant raises the skill level of a user.** Run two problem solving sets of the same difficulty on the same user: one set without the aid of the KBS and one with the KBS. A separate problem set must be used for each trial to avoid learning effects. Use the same analysis method as (2) – treat the user's two trials as one would treat two different versions of the same KBS.

## 3.2 Selecting judges, subjects, problems

In order to perform a test, the experimenter will need to take some care in selecting judges and a range of subjects. Selection of problems turned out to be a less difficult issue. We found that in the domains which we studied (manufacturing and software development) even very simple problems were of sufficient complexity to show strong differences between practitioners ranging between 0 and 10 years of experience. This is probably true for most complex domains, although it may not be true for very simple or toy domains.

**Subject and judge selection.** The judges should preferably have 10 or more years of experience. (MacMillan [MacMillan *et al.*, 1993] refers to such experts as "super experts.") However, given the rarity of highly experienced experts, one may have to settle for what one can get. The subjects' and the judges' experience area should closely match the domain of the KBS being evaluated.

**Range of subjects.** Ideally, one would like to select subjects so that the experience level of the KBS falls within the subjects' experience range. The method will still work even if the KBS falls slightly outside the range of the subjects' experience, however if it falls too far outside their range then the experience interval may become too broad to supply a useful experience estimate. For example, if the experience level of the KBS is 5 years, one may want to select subjects ranging from 2 to 10 years of experience. Unfortunately, before applying QUEM, one does not know the experience level of the KBS, so one must make an initial educated guess as to what the range of experience levels should be for the subjects. It may be necessary to conduct 1 or more pilot studies to identify the appropriate experience range for the subjects. The first time we tested the manufacturing KBS (described in the later example), we did not guess the subject range correctly. We selected subjects having between 2 and 5 years of experience, but found that the KBS's experience level was above the range of these subjects. This provided useful information, but it did not allow us to put an upper bound on the system's experience level. After two additional years of development on the system, we conducted a second test in which we selected subjects between 2 and 24 years of experience. This time we found that our KBS's experience level did fall inside the range of the subjects (approximately at the 8 year mark). These two previous studies enabled us to select the correct range of subjects (2 - 10 years) for the study in this paper.

**Graceful degradation.** There are many sources of variation in the data. Variations may arise from differences in the way judges make assessments, motivation levels of the subjects, and other factors. The total variation in the data from all such sources is reflected in the width of the confidence bands and experience interval. This representation of variability makes QUEM robust to noise to an extent, and provides QUEM with the property of graceful degradation. Thus, if the experimenter accidentally introduces additional variation by poor selection of one judge or subject, it may broaden the experience interval (reducing the precision of the answer) but it will not greatly change the result.

## 3.3 Limitations of the method

QUEM can provide useful information for a domain only when practitioners in the field show a distinct improvement in solution quality over time. An example of a domain where this relationship is known to exist is management planning of software development projects[Fiebig, 1997]. However, experience may not bring quality improvements in all domains. The existence of such a relationship can easily be determined by testing if a simple function can be found which fits the data well. The converse, that no relation exists, is harder to determine. If

| Problem Solver | Years of Experience | Judge 1 | | | Judge 2 | | | Average Solution Rank |
|---|---|---|---|---|---|---|---|---|
| | | P1 | P2 | P3 | P1 | P2 | P3 | |
| Subject 1 | 2 | 2 | 2 | 8 | 1 | 1 | 1 | 2.50 |
| Subject 2 | 2 | 1 | 1 | 5 | 2 | 5 | 5 | 3.17 |
| Subject 3 | 5 | 3 | - | 4 | 7 | - | 2 | 4.00 |
| Subject 4 | 5 | 5 | 3 | 7 | 4 | 4 | 4 | 4.50 |
| Subject 5 | 7 | 4 | 5 | 6 | 3 | 3 | 3 | 4.50 |
| Subject 6 | 8 | 8 | 8 | 1 | 8 | 8 | 7 | 6.67 |
| Subject 7 | 10 | - | 7 | 9 | - | 6 | - | 7.33 |
| KBS | * | 7 | 4 | 2 | 6 | 7 | 8 | 5.67 |

Figure 1: Quality rankings assigned to solutions

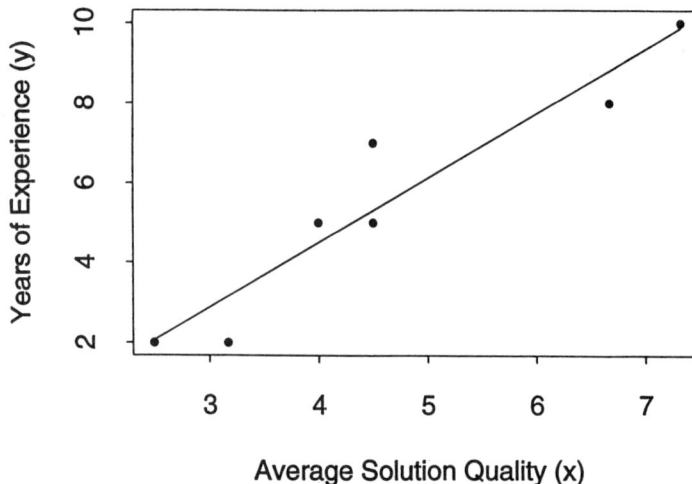

Figure 2: Average quality rankings and skill function.

no clear relationship is found in the data, it does not necessarily mean that one does not exist. It could also mean that the subjects or judges were not chosen well, the range of experience levels was too narrow, or increased skill manifests itself in ways other than through increased solution quality (such as increased speed in producing a solution).

## 4 Example: Evaluation of a KBS

QUEM was used to evaluate a manufacturing KBS, MACHINIST (later called $P^3$ [Hayes, 1996]). Experts in this domain are highly skilled and require as much as 8 to 10 years of intensive practice to achieve master level status. Seven subjects and two judges were selected. The subjects ranged between 2 and 10 years of experience. They had 2, 2, 5, 5, 7, 8, and 10 years of experience respectively. The two expert judges had 15 and 18 years of experience respectively. We prepared 3 problems for the subjects to solve, P1, P2 and P3, all of approximately the same difficulty level.

### 4.1 Evaluating a KBS with QUEM

1. **Solve:** We had each of the subjects and the KBS solve all three problems. We wrote up all solutions in a uniform format and handwriting (to disguise their source).

2. **Sort:** We sorted the solutions into 3 groups: each group contained all solutions to a specific problem.

3. **Rank:** We had two expert judges independently rank the plans in each group, from worst to best. The worst plan was given a score of 1. The ranks assigned to each plan are shown in Figure 1. P1, P2 and P3 are problems 1, 2 and 3. The missing data points resulted when subjects were unable to complete all three problems when they called away due to immediate job demands.

4. **Adjust Ranks.** This step was not necessary for these data because each plan had a unique rank.

5. **Compute Subject Averages.** The average quality ranking received by each subject across all three problems was computed. These values are shown in the last column of Figure 1. The lowest average score, 2.50, was received by the subject 1 who had only two years of experience. The highest average score of 7.33 was received by subject 7 who had 10 years of experience. The KBS received a quality ranking of 5.67. A factorial analysis performed on the data showed experience to be statistically significant, but not judge nor part (which confirmed our expectations).

6. **Plot.** The average quality rankings or the human subjects were plotted on the graph shown in Figure 2.

7. **Fit a Skill Function to the Data.** Several types of curves were fit to these data, including a logarithmic function and several types of polynomials. However, a simple linear regression fit the data best. The regression yielded the following equation for the model: $y = -1.97 + 1.67x$. We use this as the *skill function,* shown in Figure 2 as a heavy diagonal line.

8. **Construct confidence bands.** 95% confidence bands are shown in Figure 3 as curved bands flanking the skill function.

9. **Plot the KBS average quality rank.** The average quality rank for the KBS, $x(m)$, was plotted on the quality ($x$) axis.

10. **Construct an experience estimate and interval.** The average plan quality rating of the KBS, 5.67, is plugged into the skill function. This produces a value which estimates the KBS's experience level at 7.20 years of experience. This is the *experience estimate.* Using the equations for the confidence bands, it was determined that the *experience interval* is 6.03 to 8.36 years of experience. This means that the true experience level of the KBS lies

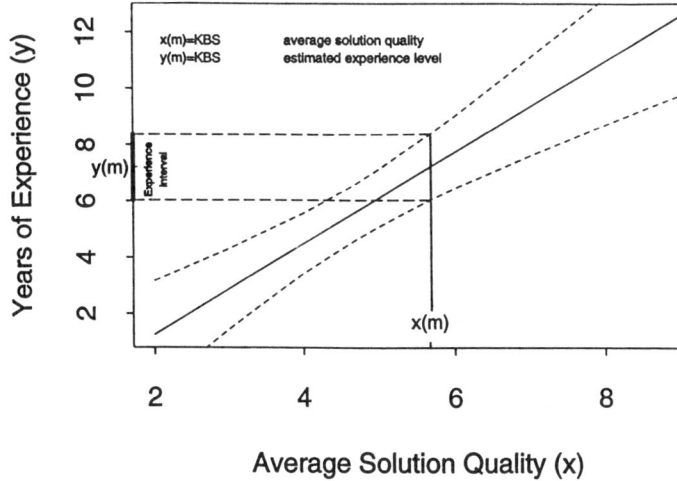

Figure 3: The KBS experience estimate and interval.

somewhere between 6.03 and 8.36 years, with 95% confidence.

Figure 3[1] shows the experience estimate and experience interval for the KBS on the y axis. From this evaluation we concluded that our system exhibits a very high experience level. On the basis of these results we confirmed that our problem solving approach was a reasonable and effective one. We decided that our basic approach was sound and that we should proceed with development along the same lines. Information on how to change the system to improve it further was derived from further knowledge engineering and protocol analysis.

## 5 Summary and Conclusions

In this paper we present QUEM, a general method for measuring the experience level of a KBS, and assessing the quality of its solutions relative to human practitioner. This method allows researchers to answer the question, "How expert is my expert system?" Assessing the solution quality and experience level of a KBS system is important in helping developers decide if their approach is sufficient, how the system should be used, and with what level of user it should interact.

The method utilizes expert judges to rank order solutions produced by both KBSs and human subjects. The rankings for the human subjects are used to construct a skill function describing the relationship between experience and solution quality. Lastly this skill function, and its confidence bands can be used to estimate the KBSs' experience level, and to bound the true value.

Previous methods for evaluating a KBS performance involve *qualitative* comparisons, such as "system x performs *better than* system y," which does not say if either system performs well at all. The QUEM procedure allows a system developer to make a *quantitative* assessment of the solution quality and experience level of a KBS. This measure allows system developers to answer the questions such as, "*How much* better is system x than system y?" or "*How many* years of experience does my KBS capture?"

QUEM can be used in any domain in which increased experience leads to measurably increased solution quality (which is presumably most complex domains). Additionally, it can be used to assess a partially complete KBS which can construct solutions but which may not be complete or correct in all aspects. It can be used to measure the experience level of an individual KBS, to compare several KBSs which operate in the same domain, or to compare the experience levels of several KBS's that operate in unrelated domains. QUEM represents an important advance in providing quantitative measures of system performance that can be applied to a broad range of complex domains in which solution quality may otherwise be hard to quantify.

## References

[Aikins, 1981] J. S. Aikins. Representation of control knowledge in expert systems. In *Proceedings of the First AAAI*, pages 121–123, Stanford, CA., 1981.

[Clancey, 1993] W. J. Clancey. Acquiring, representing and evaluating a competence model of diagnostic strategy. In et al Buchanan, Bruce G., editor, *Readings in Knowledge Acquisition*, pages 178–215. Morgan Kaufman Publishers, San Mateo, California, 1993.

[Dixon et al., 1987] J. R Dixon, A. Howe, P. R. Cohen, and M. K. Simmons. Dominic 1: Progress toward domain independence in design by iterative redesign. *Engineering with Computers*, 2:137–145, 1987.

[Fiebig, 1997] C. Fiebig. The development of expertise in complex domains. Master's thesis, University of Illinois, Urbana, Illinois, May 1997.

[Hayes, 1996] C. C. Hayes. $p^3$: A process planner for manufacturability analysis. *IEEE Transactions on Robotics and Automation*, 12(2):220–234, April 1996.

[Lane, 1986] N. E. Lane. Global issues in evaluation of expert systems. *Proceedings of the 1986 IEEE International Conference on Systems, Man and Cybernetics*, pages 121–125, 1986.

[Liebowitz, 1986] J. Liebowitz. Useful approach for evaluating expert systems. *Expert Systems*, 3(2):86–96, 1986.

[MacMillan et al., 1993] J. MacMillan, E. B. Entin, and D. Serfaty. Evaluating expertise in a complex domain – measures based on theory. In *Proceedings of the Human Factors and Ergonomics Society*, pages 1152–1155, 1993.

[Nazareth and Kennedy, 1993] D. L. Nazareth and M. H. Kennedy. Knowledge-based system verification, validation and testing. *International Journal of Expert Systems*, 6(2):143–162, 1993.

---

[1]Figure 3 courtesy of M. C. P. Dorneich.

# A Functional Theory of Design Patterns

**Sambasiva R. Bhatta**
NYNEX Science & Technology
500 Westchester Ave.
White Plains, NY 10604, USA.

**Ashok K. Goel**
College of Computing
Georgia Institute of Technology
Atlanta, GA 30332-0280, USA.

## Abstract

Design patterns specify generic relations among abstract design elements. In the domain of physical devices, design patterns, called *generic teleological mechanisms* (or GTMs), specify generic functional relations and abstract causal structure of a class of devices. We describe a functional theory of acquisition, access, and use of GTMs, but focus on their use in analogical design. In this theory, GTMs are acquired by abstraction over known designs, accessed by goals of adapting a familiar design to meet new design requirements, and used by instantiation in the context of a familiar design. This account of design patterns is one part of a general theory of analogical design called *model-based analogy* (or MBA). The IDEAL system implements the MBA theory for conceptual design of physical devices and evaluates its account of design patterns.

## 1 Introduction

Design patterns specify generic relations among abstract design elements. The relations are generic in that they are independent of any specific design situation and the elements are abstract in that they do not refer to any specific physical structure. We focus on a specific kind of design patterns that specify generic functional relations and abstract causal structure of a class of physical devices. We call these functional and causal design patterns *generic teleological mechanisms* (or GTMs). The abstract concept of feedback in control systems is one example of a GTM; a generic mechanism for transforming translational motion into rotational motion is another example. The feedback GTM, for example, specifies both the generic function it achieves (e.g., regulation of a device output, given possible fluctuations in the device input) independent of any specific design situation, and the abstract causal structure that achieves it (e.g., transmission of information about fluctuations in the device output to a device control input) without reference to the physical structure of any particular device.

We describe a functional theory of acquisition, access and use of GTMs. In this theory, GTMs are acquired by abstraction over known designs, accessed by goals of adapting a familiar design to meet new design requirements, and used by instantiation in the context of a familiar design. In particular, we hypothesize that knowledge of structure-behavior-function (SBF) models of specific designs enables the acquisition, access, and use of GTMs. SBF model of a device specifies the internal causal behaviors of the device that explain how the device works, i.e., how the device structure delivers its functions. This account of design patterns is one part of a general normative theory of analogical design called *model-based analogy* (or MBA). The IDEAL system instantiates the MBA theory for conceptual design of physical devices and evaluates its account of design patterns. In this paper, we briefly describe the MBA theory, focusing on our hypothesis about SBF models enabling the access and use of GTMs in analogical design. [Bhatta and Goel, 1997] describes model-based learning of GTMs.

## 2 Model-Based Analogy

The process of MBA takes as input a specification of a target design problem in the form of the functional requirements and structural constraints on a desired design, and gives as output a solution in the form of a structure that realizes the specified function(s) and also satisfies the structural constraints. In addition, MBA gives an SBF model that explains how the structure realizes the desired function. Figure 1 illustrates a part of the MBA process that pertains to GTMs. A stored design analogue in this process specifies (i) the functions delivered by the known design, (ii) the structure of the design, and (iii) a pointer to the causal behaviors of the design (the SBF model). The design analogues are indexed both by the functions that the stored designs deliver and by the structural constraints they satisfy.

If a source analogue that exactly matches the target problem cannot be found in memory, MBA spawns reasoning goals for adapting the source design. Different types of functional differences between the target and the source lead to different types of adaptation goals, some requiring only simple modifications (such as parameter tweaks) and some others requiring more complex modifications (such as topological changes). In order to control the reasoning involved in making complex modifications, MBA requires knowledge that can encapsulate the relationships between candidate modifications and their causal effects. In device design, design patterns, and, in particular, GTMs, provide such knowledge. Therefore, MBA uses the knowledge of GTMs in modifying device topology

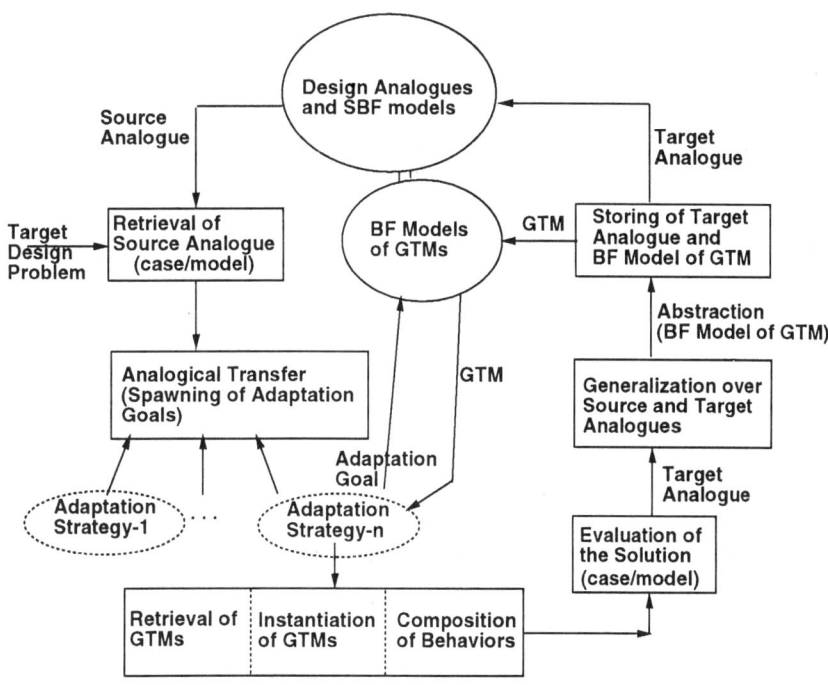

Figure 1: A Portion of the MBA Process Relating to GTMs

in the source design. MBA evaluates a modified design by qualitative simulation of its SBF model.

## SBF Device Models

IDeAL represents its comprehension of specific design cases (i.e., device models) in a structure-behavior-function (SBF) language [Goel et al., 1997]. This language provides conceptual primitives for representing and organizing knowledge of the structures, behaviors, and functions of a device. In this representation, the **structure** of a device is viewed as constituted of *components* and *substances.* Substances have *locations* in reference to the components in the device. They also have *behavioral properties,* such as *voltage* of *electricity,* and corresponding *parameters,* such as *1.5 volts, 3 volts,* etc. Figure 2(a, b, c) illustrates the SBF model of a simple design of gyroscope follow-up: the structure, its function, and the behavior that achieves the function are shown.

A **function** in the SBF models is a behavioral abstraction, and is represented as a schema that specifies the behavioral state the function takes as input, the behavioral state it gives as output, and a pointer to the internal causal behavior of the design that achieves the function. The pair of states indicated by GIVEN and MAKES in Figure 2 (b) shows the function of the simple gyroscope follow-up. Both the input state and the output state are represented as *substance schemas.* Informally, the function specifies that the device takes as input angular momentum of magnitude $L_i$ and clockwise direction at the input (gyroscope) location, and produces a proportional angular momentum of magnitude $L_o$ and of clockwise direction at the output shaft location. $L_o$ fluctuates over a large range, i.e., $L_o = L_{avg} \pm \Delta$, where $\Delta$ is large. Note that while the representation of a specific design may specify fluctuations in terms of quantitative tolerance limits (e.g., $L_o$ and $\Delta$ may be numbers), the representation of a design pattern would specify fluctuations in terms of qualitative abstractions such as *small, medium* or *large,* independent of specific quantitative values.

The internal causal **behaviors** in the SBF model of a device explicitly specify and explain how the functions of structural elements in the device get composed into device functions. The annotations on the state transitions express the *causal, structural,* and *functional contexts* in which the transformation of state variables, such as substance, location, properties, and parameters, can occur. Figure 2(c) shows the causal behavior that explains how angular momentum from the input gyroscope location is transferred to the output shaft location. The functional context specified by the annotation USING-FUNCTION in $transition_{3-4}$ indicates that the transition occurs due to the primitive function "CREATE Angular Momentum" of Hydraulic-Motor.

## Design Patterns

IDeAL represents GTMs as BF (Behavior-Function) models using a subset of the SBF language as above. The SBF representation of a GTM encapsulates two types of knowledge: knowledge about the patterns of differences between the functions of known designs and desired designs that the GTM can help reduce; and knowledge about patterns of modifications to the internal causal behaviors of the known designs that are necessary to reduce the differences. That is, it specifies relationships between patterns of functional differences and patterns of behavioral modifications to reduce those functional differences. For example, Figures 3 (a) & (b) respectively show these two types of knowledge for a partial model of

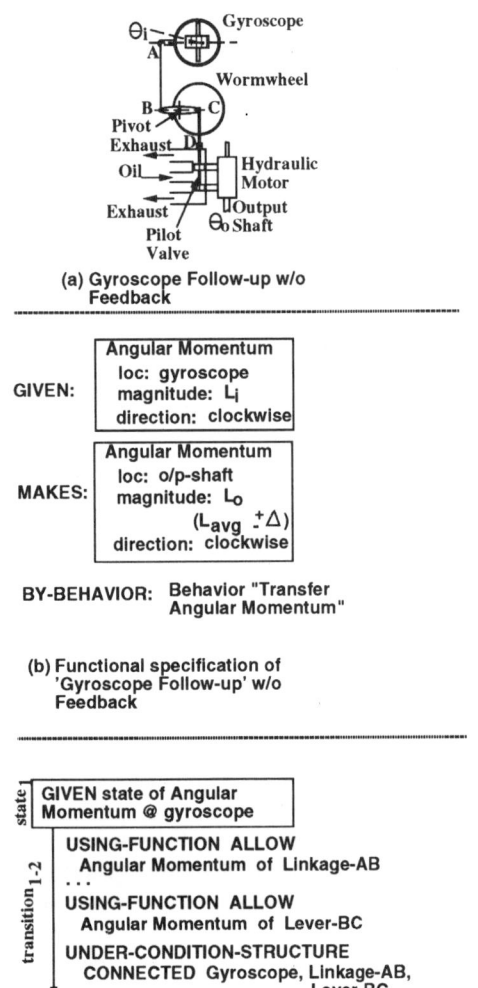

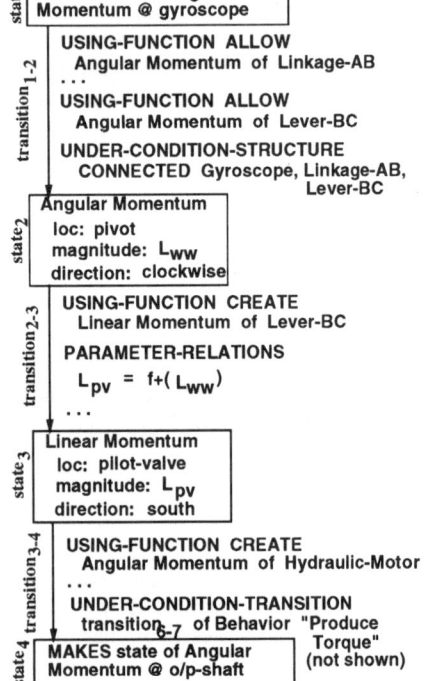

Figure 2: Simple Design of Gyroscope Follow-up (without feedback control)

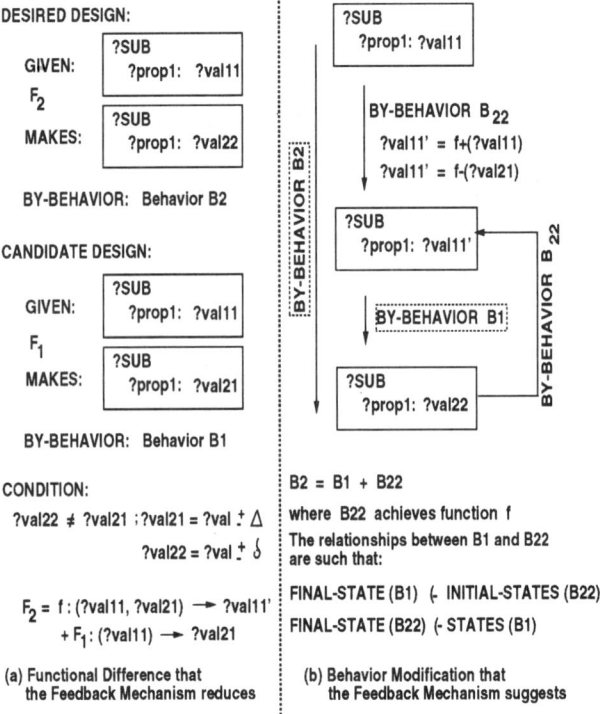

Figure 3: Feedback Mechanism IDEAL Learns

the feedback mechanism.[1] Figure 3 (a) shows the patterns of functions $F_1$ and $F_2$ respectively of a candidate design available and the desired design, and the conditions underwhich the mechanism is applicable. Because of the tasks for which they are used in MBA, the GTMs are indexed by the patterned functional differences such as shown in Figure 3 (a) (i.e., the fluctuations in the output substance property values in the candidate design function and the desired design function respectively are large and small). The model of the feedback indicates that the desired behavior ($B_2$) can be achieved by modifying the candidate behavior ($B_1$) through setting up the indicated causal relationships between the latter and the additional behaviors (that achieve the subfunctions of $F_2$ other than $F_1$ characterized in the conditions of the mechanism). In particular, the feedback mechanism suggests the addition of a causal link from a change in the output substance state to a change in an earlier state (input state or intermediate state) in the candidate behavior so that the effective input to the device is modified. Figure 3 (b) shows the relationships in the model of the feedback that IDEAL learns from two designs of amplifiers (one without feedback and the other with feedback). [Bhatta and Goel, 1996; 1997] provide a detailed account of the learning task and method.

---

[1]Feedback can be open loop or closed loop. The feedback mechanism described here is one type of closed-loop feedback in which the output substance, fedback substance, and the input substance are all same.

## 3 Access and Use of GTMs

Let us now consider a design problem in the domain of mechanical controllers presented to IDEAL. The new problem has a functional specification that given the substance angular momentum with a magnitude of $L_i$ and clockwise direction at an input location (gyroscope), the device needs to produce the angular momentum with a magnitude $L_o'$ proportional to the input and the same direction at a specified output location. It also specifies the constraint that the output cannot fluctuate much around an average value (i.e., $L_o' = L_{avg} \pm \delta$, where $\delta$ is small). This is the problem of designing a gyroscope follow-up [Hammond, 1958].

Let us consider the knowledge condition in which the design of a device (Figure 2) which transfers angular momentum from a gyroscope to an output shaft location is available in IDEAL's analogue memory (or is given explicitly as part of the *adaptive* design problem). Given an input angular momentum of magnitude $L_i$ and clockwise direction at the input (gyroscope) location, this device produces a proportional angular momentum of magnitude $L_o$ and of clockwise direction at the output shaft location; however, $L_o$ fluctuates over a large range, i.e., $L_o = L_{avg} \pm \Delta$, where $\Delta$ is large. IDEAL retrieves (if not given explicitly) the design of gyroscope control system available in its memory because the desired function matches with the function of this design. That is, the function of this available device is similar to the function of the desired design in that the input states are identical and the output states differ in a parameter value and a constraint on that value.

Now, the task for IDEAL is to modify the available design of gyroscope control system (Figure 2) to deliver the desired function. Simple modifications such as replacing a component in the given design analogue will not result in a device that can solve the new design problem because there is no single component in the device that seems responsible for the large fluctuations and that which may be selected for modification. The issue becomes if and how IDEAL can modify the device topology using the knowledge of GTMs.

IDEAL first retrieves the relevant GTM: it uses the difference in the functions of the candidate and desired designs as a probe into its memory because it indexes the mechanisms by the functional differences and the decomposability conditions on the desired functions. It retrieves the feedback mechanism because the current functional difference, namely, the fluctuation in the output property being large vs. small (i.e., $\Delta$ vs. $\delta$), matches with the difference that the feedback mechanism reduces which is specified in a device-independent manner. Then, it tries to match the decomposability condition on the desired function in the feedback mechanism (see Figure 3(a) for the condition $F_2 = f : (?val11, ?val21) \rightarrow ?val11' + F_1 : (?val11) \rightarrow ?val21)$ with the desired function in order to find the subfunctions $f$ (or $g$) that need to be designed for and composed with the candidate function. By performing this match, as guided by the language of SBF models, IDEAL finds the subfunction $f:(L_i, L_o') \rightarrow L_{ww}'$, i.e., it needs to design for a structure that takes two inputs, angular momentum with a magnitude of $L_i$ and angular momentum with a magnitude of $L_o'$, and gives as output an angular momentum of $L_{ww}$ in the opposite direction at the location of pivot in the candidate design.

Next IDEAL instantiates the retrieved GTM in the context of the target problem. The algorithm for IDEAL's process of instantiating the GTMs is shown in Figure 4. When the abstractions are GTMs, this process involves designing for the subfunction(s) determined by matching the applicability conditions of the mechanism (in steps 2 & 3 of the algorithm) and composing the new sub-behavior(s) with the behavior of the candidate design as per the relationships specified in the retrieved mechanism (in step 5). Let us walk through the algorithm as it applies to the current example. Step 1 is to select the behavior relevant for the function of the available design. Since the function in an SBF model of a device directly points to the behavior relevant for that function, this step is trivial. In the current example, $B_1$ is the behavior shown in Figure 2(c). Step 2 is to identify bindings for variables in the retrieved GTM, in particular, in the subfunctions to be designed for. Some of the bindings for the state variables are obtained while doing the matching for the retrieval of the GTM itself. As described above, in the current example, IDEAL finds the subfunction $f$ to be $(L_i, L_o') \rightarrow ?val11'$ because ?val11 is the value of the property (whose output values in the desired and retrieved functions are different) in the initial state of $B_1$ and ?val21 is the value in the final state of $B_1$. Like in this example, even after step 2, some other variables such as $?val11'$ still need to be bound with specific values from the behavior of the available design. Step 3 is exactly for doing that: the idea is to trace the relevant behavior of the available design, $B_1$, backwards from the final state to the initial state, and identify the intermediate states that are possible candidates for the states of the subfunctions. In the current example, IDEAL needs to find a candidate state from $B_1$ that could be the output state of the subfunction $f$. As it traces back the behavior shown in Figure 2, the first state to be considered is $state_3$. But since it describes a substance (linear momentum) different from what the substance (angular momentum) is from the bindings in step 2, this state cannot be a candidate. Next, it considers $state_2$ which is the only state left and which is a candidate for the output state of $f$. If $state_3$ were to describe angular momentum, it would also have been a candidate. In such a case, IDEAL would have chosen $state_3$, the state nearest to the final state of $B_1$. The rationale in this is that the modification selected should cause as minimal disturbance in the candidate behavior as possible, which means modifying the state as near to the final state as possible in order to solve the problem. Since $state_2$ is selected in the current example, IDEAL gets the binding for $?val11'$ from this state, and it has all parts of the subfunction specified.

Since the subfunction has multiple states, step 4 is relevant. In the current design scenario, the subfunction IDEAL needs to design really has two parts (as it takes two inputs and produces one output): one that specifies the need for transferring angular momentum from the input location to the pivot location, and the other for transferring angular momentum from the output shaft location to the pivot location. Applying step 4, we can find that $transition_{1-2}$ really covers the transfor-

**Input:**
- $M_1$, the SBF Model of the Design Analogue, and its Function $F_1$.
- $F_2$, the desired function.
- $G$, a GTM (retrieved by matching $F_2 \sim F_1$).

**Output:**
- $M_2$, the SBF Model of the new device that achieves $F_2$.

**Procedure:**
begin
(1) Select the behavior $B_1$ in $M_1$ relevant to $F_1$.
(2) Bind the initial & final states of $B_1$ to the appropriate GIVEN and MAKES states of the subfunctions $f$ and $g$ in $G$.
(3) **if** $\exists$ an unbound state variable in $f$ or $g$
   **then** backtrace $B_1$ to find states in $B_1$ that may be modified, considering the bindings from step **2**.
   (3.1) **if** $\exists$ multiple candidate states for modification
      **then** Select the state that is nearest to the final state in $B_1$.
   (3.2) Compute values of unbound state variables in $f$ and $g$ based on the selected state, ($F_2 \sim F_1$), and PARAMETER-RELATIONS in $B_1$.
(4) **if** $\exists$ multiple GIVEN or MAKES states in $f$ or $g$
   **then** Check if $\exists\ b \in B_1$ that achieves the transformation from any of the GIVEN states to any of the MAKES states in $f$ or $g$.
   (4.1) **if** *yes*
      **then** $f'$ = rest of the transformation in $f$.
      (i.e., < (GIVEN-states($f$) - initial-state($b$)), (MAKES-states($f$) - final-state($b$)) >.)
      $g'$ = rest of the transformation in $g$.
      (i.e., < (GIVEN-states($g$) - initial-state($b$)), (MAKES-states($g$) - final-state($b$)) >.)
(5) Retrieve subdesigns for $f'$ and $g'$.
   (5.1) **if** $\exists$ no subdesigns for $f'$ or $g'$ **then** FAIL.
   (5.2) **else**
      (5.2.1) Adapt the retrieved subdesigns for $f'$ and $g'$ (if necessary).
      (5.2.2) Compose $B_{f'}$, the behavior for $f'$, and $B_{g'}$, the behavior for $g'$, with $B_1$ as per the relationships in $G$.
      (5.2.3) Propagate the resulting changes in state variables forward in $B_1$ and in the dependent behaviors in $M_1$.
**end.**

Figure 4: IDEAL's Method for Instantiating A GTM

mation $?val11 \rightarrow ?val11'$ and the remaining transformation ($f'$) in $f$ is $?val21 \rightarrow ?val11'$. That is, the first part is already designed for in the candidate design as the behavior segment $state_1 \rightarrow state_2$ (Figure 2(c)) achieves it. Therefore, in successfully instantiating the mechanism in the candidate design of gyroscope follow-up, IDEAL only needs to find a behavior (and a structure) that accomplishes the second part of the subfunction ($f'$) given the context of the first transformation.

Let us consider the knowledge condition in which IDEAL has the knowledge of a component (called *worm*) whose function is to transfer an input angular momentum to an output location with the magnitude proportional to the output component and the direction dependent on the direction of threading on the worm. This component reverses the direction of the input angular momentum. In step 5, given the subfunction $f'$, IDEAL retrieves that component because the desired part of the subfunction matches with the component's function. It substitutes the appropriate parameters in the behavior of the retrieved design (i.e., worm) to generate a behavior for the desired subfunction. Then it composes that behavior (i.e., $B_{22}$) with the behavior of the candidate design (i.e., $B_1$) as per the specification of the causal relationships in the feedback mechanism (as in Figure 3(b)) to propose a behavior (shown in Figure 5(b)) for achieving the desired function. Note that the resulting modification is non-local in that it modifies the device topology (see the structure of the desired device in Figure 5(a)). It finally propagates the changes in states resulting from composing the subdesign's behavior with $B_1$ forward to the final state or until a state is revisited.

## 4 Evaluation

IDEAL provides a testbed for experimenting with the MBA theory. We conducted several kinds of experiments with IDEAL that evaluate the MBA theory for its acquisition, access, and use of GTMs. One kind of experiment contained two steps. The first step involved giving IDEAL a pair of designs, one without any instance of GTM and the other with an instance of a GTM, and testing IDEAL's ability to learn a BF representation of the GTM instantiated in one of the two input designs. In the second step, IDEAL is given a design problem, from a different domain in some cases, such that it would need to access and use a previously learned GTM in order to solve the given problem. We verified if it can autonomously recognize the applicability of a GTM and successfully access and use it to solve the given problem. We conducted 12 such experiments with different combinations of design sources and target problems from 4 different design domains involving 28 distinct designs. The largest design had about 10 structural elements and 10 structural relations, and 3 inter-dependent behaviors in its SBF model. We tested IDEAL's learning of 6 different GTMs and its use of 3 of them. In all these cases, IDEAL was successful in learning GTMs, and in accessing and using them in solving design problems. The behavior of IDEAL in these experiments led us to conclude the following four results:

**(1) Computational feasibility and efficacy:** IDEAL successfully addresses the multiple tasks in the MBA theory, for

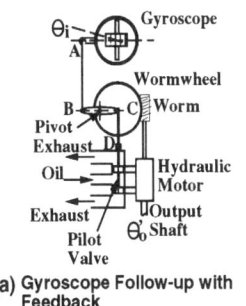

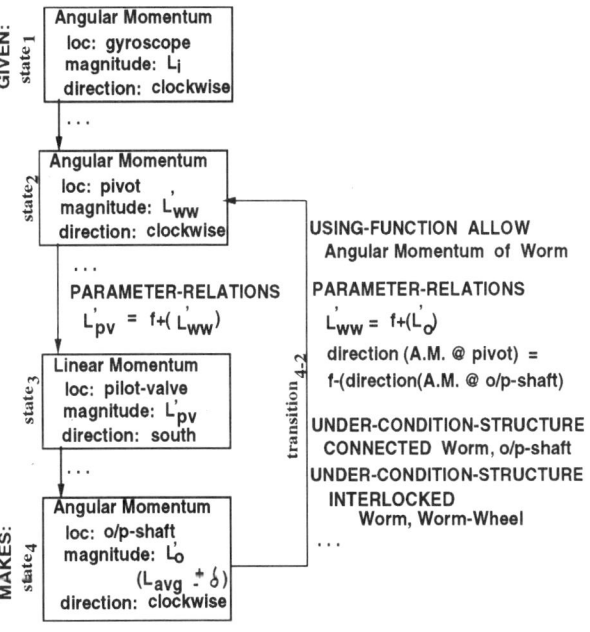

Figure 5: The Design of Gyroscope Follow-up Instantiating the Feedback Mechanism

example, the tasks of learning, accessing, transferring and using GTMs.

**(2) Uniformity of representations:** The different tasks in the MBA theory impose constraints on one another. They also impose different constraints on the design knowledge representations. IDEAL uses the same SBF language for addressing the different tasks. The GTMs, for example, are represented in a BF language that is a subset of the SBF language. The design analogues too are indexed in the SBF vocabulary.

**(3) Generality of domains:** As mentioned above, IDEAL presently contains about thirty design analogues from four different device domains, namely, the domains of simple electric circuits, heat exchangers, electronic circuits, and complex mechanical devices (such as momentum controllers and velocity controllers). This includes the design problem used as an illustrative example in this paper, which was taken from a classical textbook on mechanical design [Hammond, 1958].

**(4) Generality in terms of different GTMs:** IDEAL presently covers six different GTMs: cascading, four different types of feedback, and one type of feedforward.

## 5 Related Research

The notion of design patterns can be traced at least as far back as Christopher Alexander's [1964] *Notes on the Synthesis of Form* in which he provides a conceptual analysis of evolutionary design of village centers in rural India in terms of topological design patterns. Recently Gamma *et al.* [1995] have analyzed designs of object-oriented programs in terms of patterns of data and control flow, and described a library of twenty three reusable patterns for supporting interactive object-oriented programming. But insofar as we have been able to determine, at present there is no computational theory (other than MBA) of automated acquisition, access and use of design patterns. The MBA theory not only identifies GTMs as a useful class of design patterns in the domain of physical devices, but also shows that knowledge of SBF device models is sufficient for enabling automated acquisition, access and use of GTMs.

Our theory of design patterns is also related to Gero's theory of design prototypes [1990]. Like our design patterns, his design prototypes too specify functional relations and causal structures in a class of devices, but, unlike a design pattern, a design prototype also specifies the generic physical structure of the device class. While a design prototype is a generalization over design cases such that a case is an instance of a prototype, a design pattern is an abstraction over design prototypes such that a design prototype is a subclass of a design pattern. IDEAL too contains design prototypes as an intermediate abstraction between design cases and design patterns, but uses only the latter for analogical transfer. If the SBF model of the retrieved design case indicates that a local (i.e., parametric or componential) modification would be sufficient for meeting the new design requirements, then the system simply adapts the retrieved design. But if the SBF model does not identify a local modification, then the system accesses and instantiates a relevant design pattern to make a "non-local" (i.e., topological) modification. Furthermore, Gero's theory of analogical design uses a process similar to that of the structure-mapping engine (SME) [Falkenhainer *et al.*, 1989] to abstract causal behaviors at transfer time. In contrast, in MBA, design patterns are abstracted at storage time and acquired for potential reuse, and the process is different.

PHINEAS [Falkenhainer, 1989], which evolves from SME, also uses high-level abstractions for establishing correspondence between the source and the target situations. But it provides neither any content account of the high-level abstractions nor a process account of their acquisition. MBA provides both a content account of generic abstractions in relation to device design, and also a process account of the acquisition, access, and use of the abstractions. As in [Kedar-Cabelli, 1988], the generic abstractions in MBA are purpose-directed. In particular, in MBA the design patterns are indexed by the problem-solving goals stated in terms of functional differences between two design situations.

Case-based theories of design mostly involve direct transfer of the structure of familiar designs to new design situations. That is, the transfer is not mediated by high-level abstractions. In some case-based theories (e.g., Shinn, 1988), however, high-level abstractions do enable case reminding, but still play little role in analogical transfer. Also, the high-level abstractions in these case-based theories are generalizations over features of a problem, and do not specify relations that characterize a problem and its solution. Finally, design adaptations in case-based design are in general limited to local (typically parametric) design modifications.

The MBA theory evolves from an earlier theory of case-based design called *adaptive modeling* [Goel 1991a; 1991b]. The adaptive-modeling theory described case-specific structure-behavior-function (SBF) models. It showed how case-specific SBF models enable local (i.e., parametric or componential) modifications to source designs for solving target design problems. It also showed how case-specific SBF models of new designs can be acquired by adapting the models of known designs. Stroulia and Goel [1992] described how case-independent generic models enable topological modifications to source designs in the same domain as that of target problems. Bhatta and Goel [1996; 1997] showed how generic models can be acquired by abstraction over case-specific SBF models. The MBA theory completes the circle by showing how generic models mediate analogical transfer of design knowledge from the source domain to a target domain (e.g., from amplifiers to gyroscopes).

## 6 Conclusions

Design patterns in general specify generic relations among abstract design elements. GTMs, that specify generic functional relations and abstract causal structure of a class of devices, are a kind of design patterns useful for conceptual device design. In particular, GTMs mediate analogical transfer of design knowledge from one device domain to another, and enable topological modifications to familiar designs for meeting new functional requirements. SBF model of a specific device specifies the internal causal behaviors of the device which explain the functioning of the device. The SBF ontology provides a language for representing GTMs and a vocabulary for indexing them. In addition, knowledge of SBF models of familiar designs appears sufficient for acquisition, access, and use of GTMs.

## Acknowledgments

This paper has benefited from numerous discussions with members of the Intelligence and Design research group at Georgia Tech. This work has been supported in part by research grants from NSF (IRI-92-10925 and DMI-94-20405) and ONR (research contract N00014-92-J-1234).

## References

[Alexander, 1964] C. Alexander. *Notes on the Synthesis of Form*. Harvard University Press, 1964.

[Bhatta and Goel, 1996] S. Bhatta and A. Goel. From design experiences to generic mechanisms: Model-based learning in analogical design. *Artificial Intelligence for Engineering Design, Analysis and Manufacturing*, 10:131–136, 1996.

[Bhatta and Goel, 1997] S. Bhatta and A. Goel. Learning generic mechanisms for innovative strategies in adaptive design. *The Journal of the Learning Sciences*, 1997. Forthcoming.

[Falkenhainer et al., 1989] B. Falkenhainer, K. Forbus, and D. Gentner. The structure-mapping engine: Algorithm and examples. *Artificial Intelligence*, 41:1–63, 1989.

[Falkenhainer, 1989] B. Falkenhainer. *Learning from Physical Analogies: A Study in Analogy and the Explanation Process*. PhD thesis, University of Illinois, Department of Computer Science, Urbana, IL, 1989.

[Gamma et al., 1995] E. Gamma, R. Helm, R. Johnson, and J. Vlissides. *Design Patterns: Elements of Reusable Object-Oriented Software*. Addison-Wesley, 1995.

[Gero, 1990] J.S. Gero. Design prototypes: A knowledge representation schema for design. *AI Magazine*, 11(4):26–36, 1990.

[Goel et al., 1997] A. Goel, S. Bhatta, and E. Stroulia. Kritik: An early case-based design system. In Mary Lou Maher and Pearl Pu, editors, *Issues in Case-Based Design*. Erlbaum, Hillsdale, NJ, 1997.

[Goel, 1991a] A. Goel. A model-based approach to case adaptation. In *Proc. of the Thirteenth Annual Conf. of the Cog. Sci. Soc.*, pages 143–148, Chicago, August 1991.

[Goel, 1991b] A. Goel. Model revision: A theory of incremental model learning. In *Proc. of the Eighth Intl. Conf. on Machine Learning*, pages 605–609, Chicago, June 1991.

[Hammond, 1958] P. H. Hammond. *Feedback Theory and Its Applications*. The English Univ. Press Ltd., London, UK, 1958.

[Kedar-Cabelli, 1988] S.T. Kedar-Cabelli. Toward a computational model of purpose-directed analogy. In R.S. Michalski, J.G. Carbonell, and T.M. Mitchell, editors, *Machine Learning II: An Artificial Intelligence Approach*, pages 284–290. Morgan Kaufmann, Los Altos, CA, 1988.

[Shinn, 1988] H. S. Shinn. Abstractional analogy: A model of analogical reasoning. In Janet Kolodner, editor, *Proc. of the DARPA Workshop on Case-Based Reasoning*, pages 370–387, Clearwater Beach, FL, May 1988.

[Stroulia and Goel, 1992] E. Stroulia and A. Goel. Generic teleological mechanisms and their use in case adaptation. In *Proc. of the Fourteenth Annual Conf. of the Cog. Sci. Soc.*, pages 319–324, Bloomington, IN, August 1992.

# Mental Tracking: A Computational Model of Spatial Development

**Kazuo Hiraki**
Presto, JST / ETL, MITI
1-1-4 Umezono Tsukuba-shi
Ibaraki, 305 Japan
khiraki@etl.go.jp

**Akio Sashima** and **Steven Phillips**
ETL, MITI
1-1-4 Umezono Tsukuba-shi
Ibaraki, 305 Japan
{sashima,stevep}@etl.go.jp

## Abstract

Psychological experiments on children's development of spatial knowledge suggest experience at self-locomotion with visual tracking as important factors. Yet, the mechanism underlying development is unknown. We propose a robot that learns to *mentally track* a target object (i.e., maintaining a representation of an object's position when outside the field-of-view) as a model for spatial development. Mental tracking is considered as prediction of an object's position given the previous environmental state and motor commands, and the current environment state resulting from movement. Following Jordan and Rumelhart's (1992) forward modeling architecture the system consists of two components: an inverse model of sensory input to desired motor commands; and a forward model of motor commands to desired sensory input (goals). The robot was tested on the "three cups" paradigm (where children are required to select the cup containing the hidden object under various movement conditions). Consistent with child development, without the capacity for self-locomotion the robot's errors are self-center based. When given the ability of self-locomotion the robot responds allocentrically.

## 1 Introduction

This research challenges the traditional approach of theory construction in cognitive development by using the framework of robot learning. Traditionally, researchers in cognitive development (e.g., developmental psychologist) have focused on general and abstract descriptions of experimental data as explanations for their observations. However, developmental psychology is intrinsically limited with respect to the question *"how does development occur?"*, because of difficulties in the methodology (e.g., scientists should not open an infant's head to check for internal representations, and should not control their everyday experiences). Instead of real infants, we need a substitute that can be used for testing the theory and controlling conditions without ethical limitation. Consequently, the requirement of a computer simulation can no longer be ignored.

Over the past few decades several studies have been conducted on computational models of cognitive development. For example, Klahr and Wallance developed a computer model of acquisition of number conservation[1] using *self-modifying production system* [Klahr and Wallance, 1976]. Drescher proposed a *schema mechanism* to elaborate and test Piaget's theory from a constructivist's perspective [Dresher, 1991]. However, what is lacking in these approaches is an account of the interaction between children and environment. Consequently, models based on these approaches sometimes lack realism. We should pay more attention to the dynamics of cognitive development in the real world.

In contrast to these approaches, we propose using autonomous robots as the subject of cognitive development, and constructing computer programs by which robots can develop or learn analogously to infants. The advantage of using robots is twofold. First, we can utilize a robot's vision sensors and actuators as the inputs and outputs of the model. This forces us to use the same input stimuli and action goals as those of the infant, whereas the input and output representations of a computer simulation must be assumed. Second, we can construct a theory absorbing *activeness* in cognitive development. Recently, researchers have emphasized the importance of activeness (i.e., mobility) of infants during development [Thelen and Smith, 1994]. However, the theory derived from this stream needs

---

[1] Conservation is a term introduced by Piaget for the child's understanding that quantitative aspects of a set of materials are not changed or affected by transformations of the display itself.

to be tested and refined in more detail. We believe that using a robot leads us to a more concrete theory. More recently, Elman et. al. published an exciting book on development from a connectionist perspective [Elman *et al.*, 1996]. We follow their approach, but concentrate much more on interaction between individuals and environment.

As a first step to constructing a complete computational theory of cognitive development, we address the question of how infants relate to their spatial environments, and how this changes as the infant matures. To explore these issues, we focus on the change of *mental tracking*: the ability to update spatial relations between self and object without real (visual) tracking during the locomotion. We modeled the development of mental tracking as a learning task for a simulated robot, and conducted experiments simulating an infant's experience of locomotion.

The following sections describe our first results of modeling infant's spatial development. In Section 2, psychological evidence for spatial development is introduced. In Section 3, we elaborate our model for the development of mental tracking. Section 4 describes an empirical experiment with a simulated robot. In Section 5, we discuss the implications of our approach and future work.

## 2 Psychological Evidence for Spatial Development

### 2.1 Egocentrism in early infants

Piaget suggested that before infants are 1 year old, they exhibit a kind of sensorimotor *egocentrism* [Piaget, 1971]. Although the term egocentrism refers to young children's general tendency to view the world solely from their own perspective[2], we focus on infant's egocentric behavior in the *spatial environments* and how egocentric behavior changes into the *allocentric behavior* that normal adults exhibit. In other words, we address the question of how infants relate to their spatial environment, and how this changes as the infant matures.

Figure 1 shows an experiment designed to investigate infant's spatial searching [Bower, 1979]. A doll (prize) was put inside one of three cups, in this case the middle one, and then the infant moved around the table. The doll's relative position from the infant's view was changed from 'middle' to 'right'. Thus, the infant should look for the doll under the right cup. However, early infants frequently fail to

[2] Piaget used the word egocentrism referring not only to spatial behavior but also to more general aspects of young children such as *egocentic communication*.

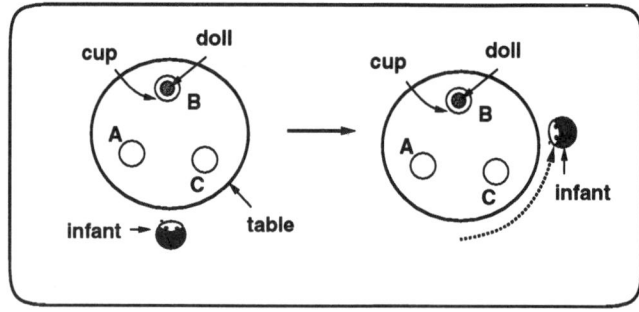

Figure 1: Self centered representation of space.

compensate for changes in their own spatial position. They continue to turn in the direction that previously led them toward the target (middle cup). The egocentric behavior in the searching task can be observed up to the age of about 18-months, but older children consistently show non-egocentric behavior on spatial tasks that involve searching for objects from different view points.

In the searching task the containers (cups) used to hide the prize were the same size and color. However, the actual environment provides much more information than position. In fact, infants can search the prize more correctly when distinctive cues (landmarks) are provided [Acredelo, 1979].

Phenomena concerning egocentrism are quite controversial and there are still many on-going studies. However, it's important to note that egocentic encoding of the target's position is not so irrational for infants who cannot move around. Pre-walking infants don't have to take into account changes in their own position in order to search for the target.

### 2.2 Effect of Locomotor Experience

So far, we have seen an interesting developmental change in infants spatial behavior. 12-year-olds behave egocentrically and 18-year-olds do not. What is the difference between 12-year-olds and 18-year-olds? How does egocentric behavior change into allocentric behavior? Experience in moving around environments seems to be one of the factors that effects the difference in the searching task. Kermoian and Campos suggested the importance of locomotor experience [Kermoian and Campos, 1988]. Infants who have experience in a walker or who can crawl are superior in spatial tasks to infants who have no such experiences.

Acredelo, Asams and Goodwyn conducted experiments to test the role of self-locomotion as opposed to passive transport concerning infant's spatial cognition [Acredelo *et al.*, 1984]. Their results suggested the importance of active movement with visual tracking. When 12-month-olds walked to the other side of a layout and have the opportunity to continually looked in the direction of a hidden prize, they looked in that direction more often and subsequently did better at turning toward the object from the new location than children who were carried. In contrast, when they could not see the prize as they walk from one position to the other, they were subsequently no better in turning toward it than children who were carried. Based on these results they hypothesized self-produced motion leads to more effective deployment of visual attention.

## 3 Learning to Mentally Track

The psychological experiments mentioned previously suggest two important factors for the spatial development:

- self-locomotor experience; and
- visual tracking.

Yet, it is still unclear what information is central in promoting the change from egocentric behavior to allocentric behavior. In Acredelo's experiments, 18-month-olds can behave correctly without visual tracking of the target object. This leads us to the necessity of modeling the mechanism of the change, taking into account of the effect of locomotion experience.

In this section, we focus on the change of the ability of updating the spatial relation between self and target object without visual tracking during locomotion. We call this ability *mental tracking*, and propose a learning architecture for mental tracking by which robots can learn it analogously to infants. Firstly, we present our assumptions and identify the information that is available during the experience of self-locomotion.

### 3.1 Formalization as a Robot Learning Task

**The Robot**

Figure 2 shows a robot that was used for modeling infant spatial development. The robot is based on Nomad 200 (Nomadic Technologies,Inc.). It can control two wheels and trunk orientation. The robot is equipped with a movable stereo-camera (Sony EVID30 x 2) that is connected with a vision processing unit that uses a Fujitsu tracking module and DSP board(TMS320C40 x 2) for accelerating image processing. Using these facilities, the robot can detect relative distance and orientation to the target.

Figure 2: A robot for modeling cognitive development.

**Locomotion Experience**

Suppose that the above robot is the infant who has just started toddling. What types of information can the robot receive from walking? We assume that the self-locomotion experience of robots can be characterized by applying a next-state function $f$ and an output function $g$ successively. At time step $n-1$ the robot produces motor command $u[n-1]$. In conjunction with the state of the environment $x[n-1]$, the motor command determines the next state:

$$x[n] = f(x[n-1], u[n-1]). \quad (1)$$

Corresponding to each state $x[n]$ there is also a *sensation* $y[n]$:

$$y([n]) = g(x[n]). \quad (2)$$

We assume that the robot has access to the state of the environment: $y([n])$ can be seen as visual information directly obtained from the camera. The formalism is analogous to a standard state-action loop of mobile robot.

**The Learning Task for Visual Tracking**

Now we model the experience of locomotion with visual tracking. Locomotion experience with visual tracking can be modeled with a robot that moves while tracking the target with its movable camera and trunk. In other words, the task of visual tracking can be seen as generation of motor commands to the camera and the trunk to keep the target object on the center of the visual image. Note that we should consider two types of motor commands; one for moving and the other for visual tracking.

Let $y^*([n])$ be a desired sensation, and in this case the target object in the center of the visual image.

Let $u_m[n]$ be a motor command for moving[3], and $u_v[n]$ be a command for visual tracking. Given the state $x[n-1]$ (representing the current posture of the robot), $y^*([n])$ and $u_m[n]$, the robot produces an action $u_v[n-1]$:

$$u_v[n-1] = h(x[n-1], u_m[n-1], y^*([n])). \quad (3)$$

The learning task for visual tracking with locomotion is to make appropriate adjustments to the input-to-action mapping $h$ based on data obtained from interaction with the environment. Note that we assume $u_m[n-1]$ is also given. This is because the robot should know how to move to the next position. The robot produces $u_m[n-1]$ independently of visual tracking.

### 3.2 The Learning Architecture for Mental Tracking

So far, we have defined the learning task for visual tracking with locomotion: Nonetheless, what we need is a model of mental tracking. It must be noted here that *mental tracking can be accomplished by mentally simulating visual tracking*. In other words, if the robot can learn to track the target while in motion, the robot can mentally track the target by applying input-to-action mapping $h$ successively as an internal process. First, we present a learning architecture for visual tracking with locomotion, and then describe how to use the acquired knowledge for mental tracking.

**Learning Inverse Model**

As mentioned above, the learning task for visual tracking is to determine a proper command $u_v[n-1]$ given $x[n-1], u_m[n-1], y^*([n])$. This is analogous to the so called *inverse model* in control system design. A controller receives the desired sensation $y^*([n])$ as input and must find actions that cause actual sensations to be as close as possible to the desired sensation. The controller must invert the transformation from actions to sensations.

We developed this mechanism based on the neural network architecture proposed by Jordan and Rumelhart [Jordan and Rumelhart, 1992]. There are several reasons for using their approach. One of the advantages of this architecture is that we don't need an explicit teacher. The robot can use the difference between predicted position of the target and the next input of the target position as training data. Another reason is that the architecture is capable of addressing the *many-to-one mapping* problem from actions to sensations. The robot shown in

Figure 2 must control at least two parameters, one for the camera and the other for the trunk to track the target. So there are infinite number of possible inverse models[4].

Using these features, the mechanism learns to produce appropriate motor command $u_v[n-1]$ to keep the target in the center of the visual image, given the current state $x[n-1]$ and a motor command to move to the next position $u_m[n-1]$.

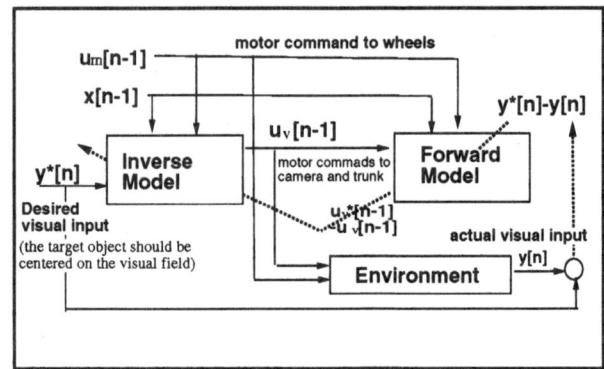

Figure 3: The learning mechanism for visual tracking with locomotion.

Figure 3 shows the learning architecture for visual tracking. $y^*([n])$ denotes a desired position of the target in the visual image, $u_v^*[n-1]$ denotes a proper command to the movable camera. $y([n])$ denotes the actual position of the target in the visual image, and $u_v[n-1]$ denotes the actual motor command for the movable camera.

In order to learn the inverse model to keep the target in the center of the field-of-view, we need the difference between the proper command $u_v^*[n-1]$ and actual command $u_v[n-1]$ to adjust the motor command:

$$u_v^*[n-1] - u_v[n-1] \quad (4)$$

We used the method described in [Jordan and Rumelhart, 1992]. Firstly the robot learns a forward model based on the difference between $y^*[n]$ (the output forward model) and $y[n]$. Here the difference (4) can be acquired by backpropagating the difference between $y^*[n]$ and $y[n]$ through the forward model. Then, the robot learns the inverse model based on the difference (4).

**The Network Architecture**

We implemented the above learning architecture using a feedforward network based on the block diagram shown in Figure 3. The network is composed

---
[3]The motor command can be seen as the command to move the robot's two wheels.

[4]See [Jordan and Rumelhart, 1992] for more details.

of two subnetworks: one for the inverse model; and the other for the forward model. The inverse model consists of 5 input, 15 hidden and 1 output units. The forward model consists of 6 input, 15 hidden and 3 output units. The output of the inverse model is taken as input to the forward model.

**Mental Tracking via Acquired Knowledge**

Figure 4 illustrates the way to mentally track the target using the learned visual tracking. The shaded portion denotes the acquired knowledge for visual tracking. The command for moving around in the environment is denoted as $u_m[n-1]$.

Note that the robot uses the output of the forward model as the current state $x^+[n-1]$, instead of actual input from the camera (environment). In order to mentally track the target with locomotion, the robot produces virtual command $u_v^+[n-1]$ to the forward model as an internal process.

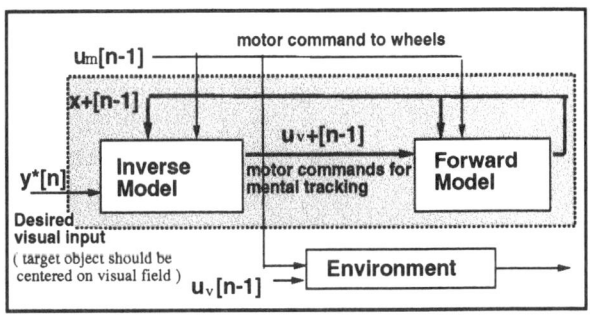

Figure 4: The architecture for the mental tracing.

## 4 Empirical Experiments with a Simulated Robot

### 4.1 The Setup

In the following experiments, we simulated three stages of a child's development of motor skills with the robot by varying its permitted actions. In stage 1, the robot is only permitted head rotation. In stage 2, the robot can rotate both head and body. Finally, in stage 3, the robot is also permitted self-locomotion, whereas in stages 1 and 2, locomotion was performed by an external agent.

For each stage, the forward and inverse models of the network were trained until:

**Forward model** Prediction error was less than 0.0005 for 50 training steps, or 50000 training steps were completed.

**Inverse model** The difference between the proper command $u_v^*[n-1]$ and actual command $u_v[n-1]$ was less than 0.0005 for 50 training steps, or 50000 training steps were completed.

### 4.2 The Three Cups Task

Following the "three cups" paradigm [Bower, 1979] discussed previously, the robot is placed in front of three cups and shown which cup hides the target object[5]. The robot moves (or is moved) to a new position from which it must predict which cup hides the target object.

### 4.3 Results

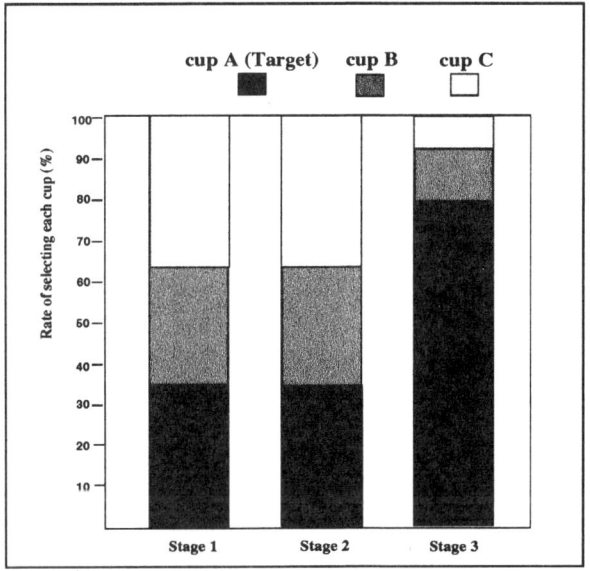

Figure 5: Prediction rate after each stage (average over 10 runs).

Figure 5 shows the rate of selection one of three cup after training for each of the three stages. The target cup (cup A) is black. The robot's performance at stages 1 and 2 was at chance level (35%). However, analysis showed that responses were consistent with an egocentric based prediction, not random choice. For example, from an allocentric perspective, a cup on the left-hand side of one's field of view would appear on the right-hand side if one views the cups from behind. From an egocentric perspective, however, one would predict the target as being on the left-hand side, which was the robot's prediction for stages 1 and 2. Since the cups were arranged in the shape of an equilateral triangle, only 35% of positions will yield a correct prediction based on egocentric knowledge. Under random choice there is no correlation between the relative

---

[5]This was done by labeling the target cup at the initial location.

positions of the selected cups before and after movement. In stage 3, when the robot also had control of translational movement, its predictive accuracy was above chance and egocentric levels, and more consistent with an allocentric based choice. Thus, locomotion experience was important for learning to predict the target's position.

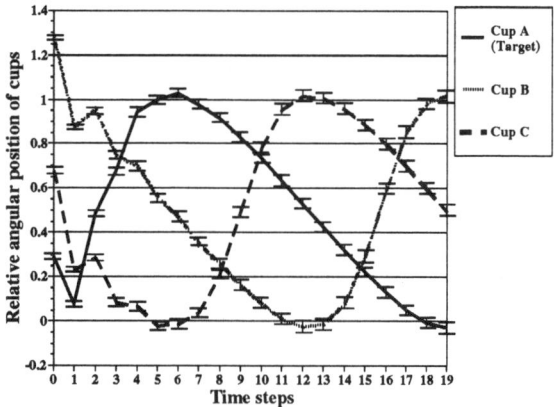

Figure 6: Relative angular position of cups as a function of robot location at the end of stage 1.

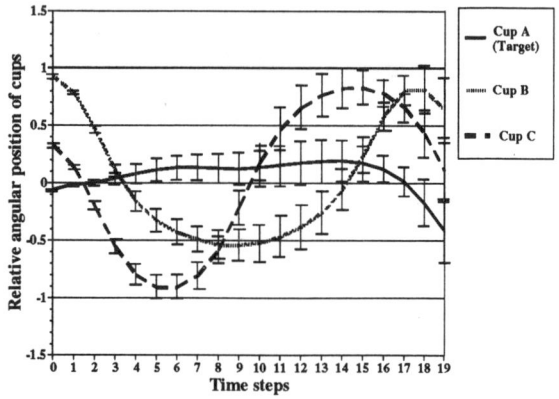

Figure 7: Relative angular position of cups as a function of robot location at the end of stage 3.

The contrast between egocentric behaviour (at stage 1) and allocentric behaviour (at stage 3) is made clearer by plotting cup position in the robot's field-of-view as a function of the robot's location.

Figure 6 shows the angular position of each cup relative to the center of the robot's field of view at various locations around the cups. The robot was moved along the circumference of a circle enclosing the cups (see Figure 1), and cup positions were recorded at 1/20th intervals. For example, 0 on the x-axis corresponds to the robot's initial position. For the y-axis, negative and positive values correspond to the left and right halves of the field-of-view (respectively). For each location, the cup with the smallest angular position (in magnitude) is the selected cup. For example, at position 5, cup C was selected. As evident from Figure 6, the robot always selected the leftmost cup as the target. In other words, the robot adjusted its head so that the leftmost cup (relative to the robot) was positioned at the center of the field-of-view (0 angular position). Consequently, the other two cups appeared in the right half of the field-of-view. This is to say that the robot behaved egocentrically. Crossovers on the graph (e.g., position 8) occurred because there are 6 locations on the circumference of the enclosing circle for which one cup is occluded by another.

Figure 7 shows mental tracking at the end of stage 3. As can be seen from the graph the target cup (cup A) remains closest to the center of the robot's field-of-view for most locations (i.e., it behaved allocentrically). Again, the exceptional cases (i.e., crossovers) are due to occlusions.

## 5 Discussion and Future Work

So far we have described mental tracking as a computational model of infant's spatial development. The simulation results support the evidence found in developmental psychology: the importance of self-locomotion. Furthermore, the results of our simulation suggest that experience of self-locomotion *with visual tracking* can accelerate spatial development. This offers the key to an understanding of *how* egocentric behavior changes into allocentric behavior.

As for mental tracking, Tani proposed a similar idea in the context of robot navigation [Tani, 1995]. He developed a robot that is capable of *mentally simulating* action plans based on a forward modeling scheme using recurrent network learning. Although he was not concerned with cognitive development, he did suggest its relevance to cognition.

Perhaps the closest approach to ours is the Cog Project of Brooks and colleagues [Brooks and Lynn, 1993]. They have been developing human-body like robots, *humanoid*. The idea of creating humanoids to investigate human cognition is very attractive in a sense that *intelligence cannot come without body*. We believe that the key concept of using robot for modeling cognitive development can be achieved even with simple mobile robots.

Current simulations were limited to one step back in time (i.e., target visible from the previous time step). For more complex environments, the target will be outside the field-of-view for indefinite periods. An obvious (and elegant) solution is to incorporate recurrent connections and have the network learn to remember positional information (e.g., [Elman, 1990]). However, learning to maintain information over long periods in the absence of additional input is difficult without special learning techniques (e.g., incremental learning, [Elman, 1993]). The extent to which mental tracking is maintenance of internal representations, or the search for alternative visual cues is an interesting research issue. And one that can best be addressed in real-world active environments such as we have proposed with the use of robots.

In our simulations, we divided child's development of motor skills into three stages. In general, however, motor skills develop more gradually, and interact with spatial development more tightly. More likely is that the development of motor skills and spatial knowledge interact both ways [Thelen and Smith, 1994]. We need to explore this kind of interaction in future work.

## 6 Conclusion

In this paper, we addressed the question of how infants relate to their spatial environment, and how this changes as the infant matures. To explore these issues, we introduced mental tracking as a key concept, and propose a learning architecture for mental tracking analogous to infants. Although there is much work to be done, we believe that the idea of using autonomous robots as the subject of development will open a new approach to modeling cognitive development. We take inspiration from recent work in robot vision, where the problem of making a robot see generated predictions leading to discoveries in insect vision [Franceschini et al., 1992]. We expect similar results for cognitive development.

## Acknowledgement

We thank Hideki Asoh for his comments on initial stage of this work. We also thank Motoi Suwa, Kazuhisa Niki and Hideyuki Nakashima for their support.

## References

[Acredelo et al., 1984] L.P. Acredelo, A. Adams, and S.W. Goodwyn. "The role of self-produced movement and visual tracking in infant spatial orientation". *Journal of Experimental Child Psychology*, 38:312–327, 1984.

[Acredelo, 1979] L.P. Acredelo. "Laboratory versus home: The effect of environment on the 9-month-old infant's choice of spatial refence system". *Developmental Psychology*, 15:666–667, 1979.

[Bower, 1979] T.G.R. Bower. *"Human Development"*. W.H. Freeman and Company, 1979.

[Brooks and Lynn, 1993] R. A. Brooks and A. S. Lynn. "Building Brains for Bodies". Technical report, MIT. AI Lab, 1993.

[Dresher, 1991] G.L. Dresher. *"Made-up Minds: A Constructivist Approach to Artificial Ingelligence"*. MIT Press, 1991.

[Elman et al., 1996] J.L. Elman, E.A. Bates, M.H. Johnson, A. Karmiloff-Smith, D. Parisi, and K. Plunkett. *"Rethinking Innateness: A Connectionist Perspective on Development"*. MIT Press, 1996.

[Elman, 1990] J. L. Elman. Finding structure in time. *Cognitive Science*, 14:179–211, 1990.

[Elman, 1993] J. L. Elman. Learning and development in neural networks: The importance of starting small. *Cognition*, 48:71–99, 1993.

[Franceschini et al., 1992] N. Franceschini, M. Pichon, and C. Blanes. From insect vision to robot vision. *Phil. Trans. R. Sac. Lond.*, B337:283–294, 1992.

[Jordan and Rumelhart, 1992] M.I. Jordan and D.E. Rumelhart. "Forward Models: Supervised learning with a distal teacher". *Cognitive Science*, 16:307–354, 1992.

[Kermoian and Campos, 1988] R. Kermoian and J.J. Campos. "Locomotor experience: A facilitator of spatial cognitive development". *Child Development*, 59:908–917, 1988.

[Klahr and Wallance, 1976] D. Klahr and J.G. Wallance. *"Cognitive development: An information processing view"*. NJ: Erlbaum, 1976.

[Piaget, 1971] J. Piaget. *"The construction of reality in the child"*. New York: Ballantine., 1971.

[Tani, 1995] J. Tani. "Self-Organization of Symbolic Processes through Interaction with the Physical World". In *Proc. of IJCAI'95*, pages 112–118, 1995.

[Thelen and Smith, 1994] E. Thelen and L.B. Smith. *"A Dynamic Systems Approach to the Development of Cognition and Action"*. MIT Press, 1994.

# COGNITIVE MODELING

Cognitive Modeling 3

# In the Quest of the Missing Link

Guilherme Bittencourt*
Laboratório de Controle e Microinformática
Departamento de Engenharia Elétrica – Universidade Federal de Santa Catarina
88040-900 – Florianópolis – SC – Brazil – E-mail: gb@lcmi.ufsc.br

## Abstract

This paper presents a generic model for a cognitive agent based on the hypothesis that the cognitive activity has three main characteristics: self-organization, evolutionary nature and history dependence. According to this model, a cognitive agent presents three levels: reactive, instinctive and cognitive. Each level, together with its lower levels, is intended to model a complete agent, each new level just increasing the behavior complexity. The generic model is instantiated into a computational architecture that integrates connectionist, evolutionary computation and symbolic approaches.

## 1 Introduction

During its forty years of existence, *Artificial Intelligence (AI)* research produced an heterogeneous set of methods adapted to solve problems in some, usually rather specific, domains. Efforts geared towards an unified theory have not succeeded, even the fundamental research is divided among several uncompatible approaches, e.g., *Physical Symbol Systems* [Newell, 1980], *Connectionism* [Rumelhart and McClelland, 1986] and *Evolutionary Computation* [Goldberg, 1989]. Although methods from different approaches have been successfully combined in some systems (e.g., neural networks and expert systems [Fu, 1994]), no general theory of *Hybrid Systems* is presently available.

This paper defines a generic model for a *cognitive agent* and proposes a computational architecture, coherent with this model, that integrates all the approaches mentioned above. The model is based on the following basic hypothesis: (i) Cognition is an emergent property of a cyclic dynamic self-organizing process [Morin, 1991] based on the interaction of a large number of functionally independent units of a few types [Changeux, 1983]. (ii) Any model of the cognitive activity should be epistemologically compatible with the Theory of Evolution. That applies not only to the "hardware" components of this activity but also to its "psychological" aspects [Wright, 1994]. (iii) Learning and cognitive activity are closely related and, therefore, the cognitive modeling process should strongly depend on the cognitive agent's particular history [Piaget, 1963].

The paper is organized as follows. In Section 2, we introduce the generic model for a cognitive agent. In Section 3, we present a computational architecture coherent with the proposed model. In Sections 4, 5 and 6, we discuss some details of the proposed architecture that are relevant to the three adopted hypothesis. Finally, in Section 7, we summarize our proposal.

## 2 Generic Model

The proposed generic model for a cognitive agent presents three levels: *reactive*, *instinctive* and *cognitive*. Functionally, these three levels are similar to the *reactive*, *deliberative* and *meta-management* components of Sloman's architectures for human-like agents [Sloman, 1996]. The model also presents some similarities with the Rasmussens's models [Rasmussen, 1991]. The reactive level consists of an evolutionary environment, whose elements are *patterns*, extracted from perceptive information about some external world, *effector controls* that can produce some action in the same external world and a population of *reactive agents* that tie together perception and action. This evolutionary environment is submitted to a *Natural Selection* process where the *fitness function* is associated with the *emotions* of the agent, defined as a global response reflecting the agent's present state [Kitano, 1995]. The environment complexity and the agent variety are not bounded, phenomena such as co-evolution, arms race, parasitism, symbiosis, etc are expected to occur. In particular, the agents can organize themselves in co-evolutionary groups analogous to the *agencies* proposed by Minsky in his *Society of Minds*

---

*The author is partially supported by *Conselho Nacional de Desenvolvimento Científico e Tecnológico (CNPq)*, grant No. 352017/92-9.

[Minsky, 1986]. At this level, processing is totally parallel and is characterized by a rapid perception/action cycle. At the end of each cycle, the best agents in the community, according to the fitness function, are allowed to *act*, i.e., to emit control commands to the body. This first level is intended to model simple animals, such as insects. The reactive level definition has many points in common with the *Enactive* theory proposed by Varela et al. [Varela *et al.*, 1991].

The instinctive level introduces a *long term memory* into the model. As the evolutionary process at the reactive level proceeds and the situations repeat themselves in the world, it is possible to identify the populations of agents in the environment responsible for a useful action in a given situation. If we take, from these populations, the best and the worst agents, according to the fitness function, it is possible to abstract their properties and to obtain a *general description* of a given population, a kind of "genetic reserve" indexed by situation. We claim that long term memory is composed essentially by these descriptions and that the act of "remembering" corresponds to the introduction at the reactive level of *new populations*, whose agents are *genetically encoded* according to these general descriptions. Once there, these new populations are, up to a certain limit, able to recreate their original enviroment, from which they were generalized. The long time effect of memory in such a model is analogous to the "breeding" and "taming" of reactive agent populations. At this level, processing is less massively parallel and is characterized by a much longer cycle that needs many repetitions of the same situation to be completed. The two lower levels together are intended to model more complex animals, such as mammals.

Finally, the cognitive level is concerned with the manipulation of the general descriptions generated at the instinctive level. Its functions are the usual cognitive functions: *deduction*, *abduction* and *induction*. The cognitive level is based on two complementary activities: the learning of descriptions of relevant situations – a synthetic, holystic activity – and the generation of new strategies of action – an analytical, local activity. The *contents* of the cognitive activity, i.e., the contents of the *short term memory*, are defined to be exactly the relations between these two complementary activities.

Although the cognitive level could be based on any adequate symbolic logic formalism, it is "embodied" in a strict sense: on the one hand, its symbolic expressions refers to "lived" situation descriptions, giving it a "real" semantic. Moreover, these descriptions can be used to generate specific behavior patterns in a real world environment or in some "hypothetical" environment, simulated through the *sentiments* of the agent, defined as emotions induced through memories generated by cognitive activities. On the other hand, the results of its infer-

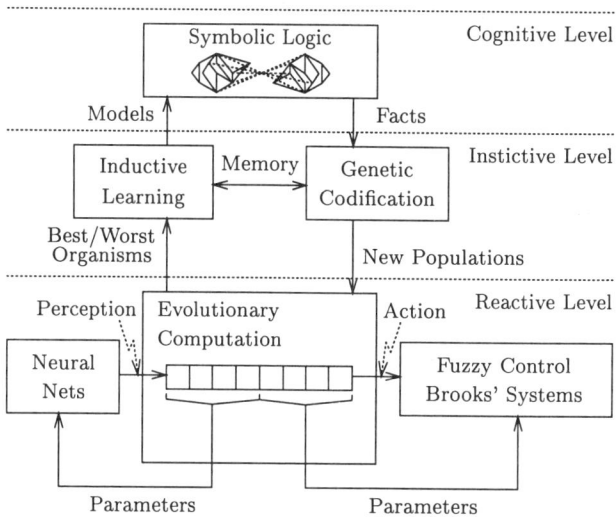

Figure 1: The Architecture

ence mechanisms are intended to be useful as the seeds for new useful behaviors and not simply true theorems. These two interpretations of the symbolic expressions in the cognitive level – (i) representations of the properties of the instintive level descriptions and (ii) seeds for action skills – are similar to the *representational* and *functional* roles of Smith's *impressions* [Smith, 1987].

It is interesting to note that the instintive and cognitive levels together can be thought of as a more complex version of a reactive level agent, also functionally conecting perception and action. But in this case, the agent refers to an internal world: the reactive level environment. We claim that a necessary condition for the development of cognition is that this high level agent (which in some sense is the only one that we can really call "cognitive agent") be itself "embodied" in a community of similar cognitive agents. Moreover, we propose that the structure of long term memory reflects a *reconstruction*, at the reactive level, of the relevant structures of the environment in which the cognitive agent is itself embodied, particularly its *social* structures [Neumann, 1968]. The agents in this cognitive environment would have two communication channels: (i) *experience*, they (partially) agree about the effects of actions in the external world and about the contents of emotions and sentiments, i.e., they believe they have similar reactive levels; and (ii) *language*, they can share, through the representations at the cognitive level, knowledge about experience, specially that knowledge that would give rise to useful skills at the reactive level.

## 3 Architecture

The instantiation of the above model for the human cognition would rise several questions about its coherence with respect to the known facts about human evolution, physiology and psychology. We avoid these questions and, heeding Sloman's advice [Sloman, 1995], try to explore the *design space* through a computational architecture for cognitive agents based on that model. The proposed architecture, presented in figure 1, has the same three levels and is intended to be a minimal implementation of the generic model.

The central activity of the machine is performed by the *Evolutionary Computation* module at the reactive level. It computes some kind of genetic algorithm where the genotype of the population members encode control parameters of the perception and action mechanisms, represented in figure 1 as the *Neural Nets* and *Fuzzy Control/Brooks' Systems* modules, respectively. The function of the *Neural Nets* module is to account for perception and the *Fuzzy Control/Brooks' Systems* module represents the effectors of the agent. Both these technologies have interesting properties for our purposes: neural nets are able to learn how to extract patterns from raw information, even without an external supervisor [Kohonen, 1987], reducing the complexity of the situation identification process. Moreover, they can be easily coded into a genotype structure through the weights associated with the network links. Fuzzy control and Brooks' systems are able to control quite complex behaviors in a distributed way and can also be easily coded, because their mechanisms depend on a limited number of parameters.

The agents in the reactive level environment can be seen as "atomic potential actions" or "skill fragments" joining together the perception and action mechanisms and the actual functional capability for an effective guided action in the external world. The genotype of each of these agents encodes three types of information: (i) the weights to be used in the input neural nets, (ii) the parameters that define the action mechanism, e.g., the fuzzy set definitions in the case of a fuzzy control mechanism, and (iii) a functional definition of the connection between the neural net output and the effector control input. Such an agent, when it is "plugged" into the appropriate perception/action circuit, uses its encoded weights and parameters to tune the perception/action mechanisms according to its necessity, and applies its encoded function to drive the relevant parts of the external effectors through the desired action according to the present perceptual input.

The model supposes a fitness function that attributes a grade to each reactive level agent and can be associated with some measure of the internal state of the machine. The real *actions* occur at the reactive level: after the genetic algorithm is running for some generations since the last action, the best agents according to the fitness function are activated. There are no central control and there are lots of things happening at the same time. Each action changes the internal state of the agent and the external environment, as perceived through the neural net module, and the cycle restarts. The model does not represent an "information processing machine", the perceptual input represents a restriction on the generated actions but does not determine them. Furthermore, the machine may continue its activities independent of input and output, it is enough to *simulate* a previously learned fitness function to create a hypothetical situation. In fact, we claim that much of the machine activity would be performed during the periods of inactivity, in which it "sleeps" and "dreams".

To learn a fitness function, a memory mechanism is necessary. If we observe the "arena" at the reactive level and, as a situation repeats itself, collect the best and the worst agents in handling such situation, then the genetic material in these agents can be used as examples and counterexamples in an *Inductive Learning* mechanism (e.g., [Salzberg, 1991]). The results of this inductive mechanism are general descriptions of populations, that can be associated with appropriate fitness function values. The complexity of these descriptions is proportional to the generality of the inductive learning mechanism and its underlying representation language. If this language is general enough to be interpreted as a logical description of the world situations, then we can use it to give a semantic account to a symbolic logic module, where the learned descriptions are interpreted as *models* of the external world.

Finally, the cognitive level is defined as a *theorem proving mechanism*, although it presents some unusual features in such systems: its expressions refer to well defined structures, the general descriptions of the instinctive level, and its inference results can be used to genetically encode new populations at the reactive level, and therefore, their value as representations of the world can be tested in "real action" conditions. In our minimal model, the genetic codification module corresponds to a fuzzy control system generator, but in a more general specification it could be any control system generator, reintroducing the *Cybernetics* view that *Control Theory* is an important part of cognitive activity. Coherently with the generic model, we suppose that the machine would be "embodied" in a world where human beings and cognitive machines are joined in communities.

## 4 Self-Organizing Logic

The connection between a symbolic theorem proving mechanism and the elements in the instinctive level of the proposed architecture deserves a more detailed treatment. The cognitive level activity could be described as follows. Initially, the learning mechanism, at the instinctive level, selects the best reactive level agents. The genetic material of these agents encode situation descriptions and adequate actions to be performed in these situations and it is used to derive a *logical description* of the relevant behaviors in these situations. The process that generates these descriptions is called *conceptualization*. At the cognitive level, the logical descriptions are interpreted as *possible models* of the world. The main activity of the cognitive level is to assume that the learned models are *all* the possible models for these situations and to derive an associated *factual description* that has exactly those models. This mechanism is very similar to McCarthy's *Circumscription* [McCarthy, 1980].

The usual theorem proving methods – *Resolution* [Robinson, 1965] and *Semantic Tableaux* [Smullyan, 1968] – present some properties that make them not adequate for our purposes: (i) they present external, i.e., meta-logical, inference rules, (ii) they are not designed to generate a factual description from a set of possible model descriptions, only to the reverse, and (iii) their proof methods are local mechanisms. To avoid these problems, we propose a new inference method, based on the transformation between conjunctive and disjunctive canonical normal forms, the *dual transformation*, that involves only internal properties of the underlying logical system (i.e., there is no inference rule external to the logical language) and that has a global and concurrent nature [Bittencourt, 1997].

Given a logical formula $W$, we call a *theory* the two sets $\Phi$ and $\Psi$ that contain, respectively, the clauses and dual clauses associated with the canonical normal forms of $W$. The two sets $\Phi$ and $\Psi$ are a kind of "holographic" representation of each other. Each clause in $\Phi$ consists of a combination of all dual clauses in $\Psi$ and, conversely, each dual clause in $\Psi$ consists of a combination of all clauses in $\Phi$.

Intuitively, the proposed inference algorithm consists in the elimination of the contradictory dual clauses, the combination of the associated substitution fragments into a set of independent substitutions, the application of these substitutions to the dual clause set and, finally, the generation, through the dual transformation, of the clause sets associated with these instances of the dual clause set. The clauses in these instances are new theorems which can be added to the original clause set and the cycle may be repeated. Clearly this mechanism does not present external inference rules, all that is done is to exploit the duality of the canonical representations and their semantics.

The inference method is totally symmetric. It does not matter if we begin with clauses, i.e., a factual account of a situation, or dual clauses, i.e., a description of the possible models of the situation. Analogously to the elimination of contradictory dual clauses, in the clause form it is the tautologic clauses that can be eliminated.

Therefore, the repeated transformation between canonical representations is not only able to infer new theorems but it is also able to *refine* the theory.

These theoretic properties of the inference method are enough to eliminate the two first restrictions to theorem proving methods. The third one must necessarily take into account implementation concerns. Clearly, the most expensive operation of the proposed inference method is the dual transformation. To improve its efficiency, we developed a concurrent algorithm for the dual transformation, based on a geometrical representation, that, on the one hand, generates only dual clauses that are not subsumed by any other and, on the other hand, is naturally concurrent and therefore easy to parallelize.

According to this algorithm, a theory is represented through a $n$ dimensional *hypercube*, where $n$ is the number of predicate symbols that occur in the theory. Each vertex of this hypercube can be associated with a certain combination of predicates given by those coordinates of the vertex that are not zero. A (dual) clause is associated with the vertex labeled exactly with those predicates that appear in its literals. The literals in each dual clause can be thought to *represent* some clauses where they appear or where there are literals subsumed by them. This information is stored in the geometric representation as integer sets associated with each literal that indicate in which (dual) clauses the literal is present. The result of this construction is that the dual transformation and the manipulations necessary to the inference/refinement mechanism can be done through local communication between neighbor (dual) clauses in the hypercube. The holographic properties of the canonical representation allow for a global effect in one form to be calculated locally in the other form [Bittencourt, 1996].

This representation allows the integration of the instinctive and cognitive levels: the learning mechanism could be thought of as concurrently "feeding" each separate node of the hypercube with the appropriate dual clause representation of some situation. After some refinement/inference steps the new inferred model descriptions can be directly verified through the learning mechanism. Furthermore, the clause representation can be used to generate new reactive level populations, whose action effects could again be verified through the learning mechanism. It is interesting to note that this kind of integration does not eliminate the *robustness* of the reactive level environment. In the model, the cognitive activity can *influence* the global behavior, but its scope is limited.

Another interesting property of the representation is that it easily supports a special kind of higher-order logic, where the theories, with their predicate and function symbols as parameters, are considered as new atomic formulas. This property has much in common with Peirce's *semiosis* process [Peirce, 1974]. According to Peirce, a *sign (S)* represents an *object (O)* for another sign, the *interpreter (I)*. This last sign – called an *habit* or *mental law* – may determine another interpreter $I_1$, which may determine an $I_2$, etc. The actualization of the potentially infinite sequence $I, I_1, I_2, \ldots$ is called semiosis. This triadic relation can be found in all three levels of the model. In particular, at the cognitive level, the memorized descriptions are the signs that represent possible models to theories and theories are signs that represent new populations to the instinctive level. Moreover, theories as signs may represent possible (hypothetical) models for other theories, mediated by a meta-conceptualization process.

## 5 Evolution

In the proposed model, we are double committed with the evolutionary approach: on the one hand, the most important component of the model is in itself an evolutionary mechanism – the reactive level – and, on the other hand, the model is designed in such a way that the components at each level could have evolved one after the other, because both, the reactive level and the pair reactive/instinctive levels, are complete models of simpler agents.

The choice of an evolutionary mechanism as the base of the model is inspired by the fact that the function of the central nervous system is to integrate the most different perception and action mechanisms. This integration could be done through a genetically "wired" functional mechanism, but none would be flexible enough to adapt itself to a variable environment. A more robust solution is to use a *simulation* of the relevant characteristics of the external world, under the same natural selection principles, to make more secure decisions about future actions. If the real-time simulation is reliable enough, even without a memory, a reactive level agent would be able to "learn" how to act in a given situation, i.e., the adequate reactive level agents will be (slowly) selected through interactions with the real world. The problem is that, without a memory, the range of experiences is limited to the actually lived situations and that implies that certain situations, e.g., fatal ones, cannot be simulated.

In this sense, long term memory is an *improvement* to a previously existing mechanism. An improvement which allows *off-line* simulations, because memory allows the simulation of the internal fitness function associated with *hypothetical* situations. These hypothetical simulations greatly accelerate the evolutionary process at the reactive level. The cognitive level is another improvement of the mechanism, because it allows the *creation*, through inference, of models of situations that never occurred in the agents' experience, further accelerating the evolutionary process at the reactive level. But its full strength comes from the fact that it also allows the establishment of a second communication channel between agent and world: language. Language may be used in the description of *abstract* situations, in particular, it allows one to conceptualize social situations and to reconstruct the relevant social relations into a coherent internal theory that can be suitably simulated at the reactive level.

## 6 Learning

One of the central issues of the model is the fitness function to be provided and its relation with the meaning of the words "emotions" and "sentiments", loosely introduced in the model description. More formally, we call emotions an internally generated feedback mechanism, based on perception and memory, able to *guide* the evolution of the reactive level environment in such a way that the global agent *survives*. Emotions are the keys for long term memory, only emotionally important learned descriptions are memorized and the same emotions are able to fetch them from memory.

According to the model, long term memory existed prior to symbolic cognitive activity and it was in some sense *adapted* to cognitive purposes. As cognitive products, models and factual descriptions, do not have in principle any emotional contents, to memorize them it is necessary to artificially create these emotional contents. These cognitively generated emotions we call sentiments. Because sentiments are associated with cognitive states and therefore to language constructions, they are fundamental for the appropriate learning of social behaviors, mainly transmitted through language. The coherence between sentiments and emotions is the base for the cognitive development. If the descriptions we learn lead in fact, at the reactive level, to the emotional effects we were told they should, we can have the "sentiment" that we learned an useful thing.

Although the theories at the cognitive level are strictly grounded in the learned descriptions, the real "meaning" of these descriptions is different for each cognitive agent, because it is defined as the effect of these descriptions when their associated populations are introduced in the reactive level, and this effect is determined by the particular history of the reactive level environment, i.e., by the history of emotions and sentiments that guided its development.

## 7 Conclusion

In their book *The Embodied Mind*, Varela et al. [Varela et al., 1991] propose three questions about cognition and answer them from three points of view: the cognitivist research program, the emergence (connectionist) program and their own *Enaction* theory. To summarize our proposal, we answer the same three questions from the point of view of the proposed model:

- *What is cognition ?* The articulation of two parallel evolutionary processes: one that occurs in an internal environment where a history of structural coupling brings forth a (physical) world, i.e., the reactive level, and, another that occurs in an external sentimental/social/cultural environment where a history of symbolic exchanges, mainly through language, between cognitive agents brings forth, in each one, a (sentimental) world, i.e., the cognitive level. The integration between the environments is mediated by the long term memory.
- *How does it work ?* The reactive level is based on an evolutionary process guided by the emotions and the cognitive level is based on a cyclic process that transforms factual descriptions into models and vice versa. The models are interpreted as memorized general descriptions of situations and the factual descriptions are used to generate new reactive level populations. Cognitive theories can be communicated through language. The cognitive level can also generate sentiments, i.e., artificial emotions, that allow it to influence memory contents.
- *How do I know when a cognitive system is functioning adequately ?* When it becomes part of an ongoing existing physical and sentimental/social/cultural world.

## Acknowledgments

The author is grateful to all those that helped in the development of this paper, in particular the LCMI team and the anonymous referees.

## References

[Bittencourt, 1996] G. Bittencourt. Boxing theories (abstract). *Journal of the Interest Group in Pure and Applied Logics (IGPL)*, 4(1):479–481, 1996.

[Bittencourt, 1997] G. Bittencourt. Concurrent inference through dual transformation. *Journal of the Interest Group in Pure and Applied Logics (IGPL), in press*, 1997.

[Changeux, 1983] J.-P. Changeux. *L'Homme Neuronal.* Collection Pluriel, Librairie Arthème Fayard, 1983.

[Fu, 1994] L. Fu. Rule generation from neural networks. *IEEE Transactions on Systems, Man and Cybernetics*, 24(8):1114–1124, August 1994.

[Goldberg, 1989] D.E. Goldberg. *Genetic Algorithms in Search, Optimization, and Machine Learning.* Addison-Wesley Publishing Company, Reading, MA, 1989.

[Kitano, 1995] H. Kitano. A model for hormonal modulation of learning. In *Proceedings of IJCAI 14*, 1995.

[Kohonen, 1987] T. Kohonen. *Content-Addressable Memories.* Springer-Verlag, Berlin, 1987.

[McCarthy, 1980] J. McCarthy. Circumscription - a form of non-monotonic reasoning. *Artificial Intelligence*, 13(1,2):27–39, 1980.

[Minsky, 1986] M.L. Minsky. *The Society of Mind.* Simon and Schuster, New York, 1986.

[Morin, 1991] E. Morin. *La Méthode 4, Les Idées.* Editions du Seuil, Paris, 1991.

[Neumann, 1968] E. Neumann. *Ursprungsgeschichte des Bewusstseins.* Kindler Verlag GmbH, München, 1968.

[Newell, 1980] A. Newell. Physical symbol systems. *Cognitive Science*, 4:135–183, 1980.

[Peirce, 1974] C.S. Peirce. *The Collected Papers of C.S. Peirce.* Harvard University Press, Cambridge, Mass., 1974.

[Piaget, 1963] J. Piaget. *The Origins of Intelligence in Children.* Norton, New York, 1963.

[Rasmussen, 1991] Steen Rasmussen. Aspects of information, life, reality, and physics. In C. Langton, C. Taylor, J.D. Farmer, and S. Rasmussen, editors, *Artificial Life II.* Addison-Wesley, 1991.

[Robinson, 1965] J.A. Robinson. A machine-oriented logic based on the resolution principle. *Journal of the ACM*, 12(1):23–41, January 1965.

[Rumelhart and McClelland, 1986] D.E. Rumelhart and J. McClelland, editors. *Parallel Distributed Processing: Explorations in the Microstructure of Cognition 1: Foundations*, volume 1. M.I.T. Press, Cambridge, MA, 1986.

[Salzberg, 1991] S.L. Salzberg. A nearest hyperrectangle learning method. In *Machine Intelligence 6*, pages 251–276. 1991.

[Sloman, 1995] A. Sloman. A philosophical encounter. In *Proceedings of IJCAI 14*, 1995.

[Sloman, 1996] A. Sloman. What sort of architecture is required for a human-like agent ? In *Proceedings of AAAI-96*, August 1996.

[Smith, 1987] B.C. Smith. The correspondence continuum. *Center for the Study of Language and Information*, January 1987. Report No. CSLI 87-71.

[Smullyan, 1968] R.M. Smullyan. *First Order Logic.* Springer-Verlag, 1968.

[Varela *et al.*, 1991] F.J. Varela, E. Thompson, and E. Rosch. *The Embodied Mind: Cognitive Science and Human Experience.* MIT Press, Cambridge, MA, 1991.

[Wright, 1994] R. Wright. *The Moral Animal.* Vintage Books, New York, 1994.

# Implementing BDI-like Systems by Direct Execution

**Michael Fisher**
Department of Computing
Manchester Metropolitan University
Manchester M1 5GD, United Kingdom
EMAIL: M.Fisher@doc.mmu.ac.uk

## Abstract

While the Belief, Desire, Intention (BDI) framework is one of the most influential and appealing approaches to rational agent architectures, a gulf often exists between the high-level BDI model and its practical realisation. In contrast, the Concurrent METATEM language, being based upon executable formal specifications, presents a close link between the theory and implementation, yet lacks some of the features considered central to the representation of rational agents. In this paper, we introduce a hybrid approach combining the direct execution of Concurrent METATEM with elements of rationality from the BDI framework. We show how this system can capture a range of agent behaviours, while retaining many of the advantages of executable specifications.

## 1 Introduction

The inability of traditional software to handle complex modern applications, together with the rapid expansion of infrastructure such as the INTERNET, has led to the introduction of a new technology, termed *agent-based systems*. An *agent* is a semi-autonomous process, typically communicating via message-passing and cooperating with other agents in order to achieve common goals. This technology has been particularly successful in producing distributed information systems where centralised control is either impractical or undesirable. Here, not only is the ability of agents to act autonomously vital, but such agents are often required to dynamically adapt to unforeseen circumstances and to work cooperatively with other agents in order to overcome problems. These facets make agent-based systems ideal for complex tasks in real-world applications and, consequently, this technology has been applied in a wide variety of areas, from industrial process control to cooperative information retrieval.

In spite of the rapid spread of agent technology, there are, as yet, relatively few high-level programming languages for agent-based systems. Although considerable research has been carried out concerning the development of theories of agency and cooperation, the lack of appropriate high-level logic-based programming languages often means that implemented systems have very little connection with these high-level theories (an exception to this is Shoham's *Agent-Oriented Programming* work [Shoham, 1993]). Traditional programming languages typically lack the flexibility to handle high-level concepts such as an agent's dynamic control of its own behaviour. Consequently, it is widely regarded as important that high-level languages be provided, which can support the principled development of multi-agent systems, from logical theory to implemented system.

In this paper, we consider two approaches to the high-level representation of agent-based systems, namely Rao and Georgeff's BDI model [Rao and Georgeff, 1991] and Fisher's Concurrent METATEM language [Fisher, 1993]. In spite of their differing backgrounds we show that they have a great deal in common and introduce, in §4, a hybrid approach. Whilst not as general as the BDI model, nor as simple as the Concurrent METATEM language, this hybrid approach can capture a range of agent behaviours, while retaining many of the advantages of executable specifications.

## 2 The BDI Framework

Rao and Georgeff [1991] considered a particular agent framework whereby individual rational agents incorporate certain "mental attitudes" of *Belief*, *Desire* and *Intention* (BDI). These are used to represent, respectively, the information, motivational and deliberative states of the agent, and together effectively determine the system's behaviour. This framework is both appealing and influential, being used in a number of practical systems [Jennings and Wooldridge, 1995].

### 2.1 Agents

The core components of the BDI framework are rational agents [Wooldridge and Jennings, 1995]. These are autonomous entities, which execute independently, and have complete control over their own internal behaviour. The core elements of BDI agents are as follows. *Beliefs* correspond to the information that the agent has assimilated about the world; the set of beliefs is typically incomplete and may be incorrect with respect to the true situation. *Desires* are the agent's high-level goals, which need not be achievable simultaneously, but are usually consistent. *Intentions* essentially represent a subset of the agent's desires that it has committed to achieve.

## 2.2 Executing Agent Descriptions

The key process in executing BDI systems is *deliberation*. This consists of two aspects:

- deciding which desires will become intentions;
- deciding how to achieve those intentions.

Thus, an abstract execution cycle usually involves following steps [Rao and Georgeff, 1991].

1. Update beliefs based upon observations and actions.
2. Based on these beliefs, generate new desires.
3. Select a subset of desires to act as intentions.
4. Select a step to perform based upon intentions and the current state of the agent.

Thus, intentions are only completed one step at a time. The step selected corresponds either to performing an action, modifying a belief, or generating a subgoal and, once this step has been completed, the execution cycle begins again.

## 2.3 Real BDI Systems

Since its inception, many real-world agent-based systems have been based upon the BDI philosophy, most notably the Procedural Reasoning System (PRS) [Georgeff and Ingrand, 1989], but also systems such as INTERRAP [Fischer et al., 1996]. In the PRS, the selection of a step to be performed involves searching a plan library for plans which can achieve the selected intention. This library consists of pre-constructed plans; planning from first principles is not undertaken. In the PRS, *intention structures*, which are essentially partial orders of dependencies, provide linkage between all related intentions. For example, such structures record the fact that some intentions may be suspended, some may be deferred and some depend on further intentions.

Although both PRS and its successor, dMARS, have been successfully used in a number of areas, such as Air Traffic Control [Rao and Georgeff, 1995], the link between such systems and the BDI framework is often tenuous. In particular, while the practical systems incorporate elements termed beliefs and intentions, these are distinct from the formally defined beliefs and intentions of BDI model.

Consequently, there is a requirement for mechanisms linking the high-level BDI model with its low-level realisation. Several different approaches have been considered. For example, a number of BDI-based programming languages have been proposed [Weerasooriya et al., 1995; Rao, 1996a]. However, these have either been too low-level or not expressive enough to capture the key elements of BDI systems.

More recently, object-oriented development methodologies have been adapted in order to provide, through a form of *agent-oriented* development methodology, a closer link between the model and its realisation [Kinny and Georgeff, 1997]. Since this practical development approach still does not consider the maintenance of formal properties, the need for high-level logic-based languages capturing the key components of the BDI model remains.

## 3 The Concurrent METATEM Framework

Fisher [1993] introduced Concurrent METATEM, an agent-based programming language comprising two elements:

1. The representation of each individual agent's behaviour using a temporal specification.
2. An operational framework providing both asynchronous concurrency and broadcast message-passing.

Temporal logic provides a means of declaratively specifying agent behaviour. It not only represents the dynamic aspects of an execution, but also contains a mechanism for representing and manipulating the goals of the agent. The operational model of Concurrent METATEM is both general purpose and intuitively appealing. The use of *broadcast* message-passing provides both a general and flexible communication model for concurrent objects [Birman, 1991] and a natural interpretation of distributed deduction [Fisher, 1997b]. These features together provide a coherent and consistent programming model within which a variety of agent applications can be represented [Fisher, 1994].

### 3.1 Agents

As in the BDI framework, the basic elements of Concurrent METATEM are agents, although these need not necessarily be rational. There are two elements to each agent: its *interface definition* and its *internal definition*.

The definition of which messages an agent recognises, together with a definition of the messages that an agent may itself produce, is provided by the interface definition, which may be given as follows.

```
car()
  in:  go,stop,turn
  out: empty,overheat
```

Here, {go, stop, turn} is the set of messages the 'car' agent recognises, while the agent itself is able to produce the messages {empty, overheat}.

The internal definition of each agent is given by a temporal logic specification [Manna and Pnueli, 1992]. Temporal logic is seen as classical logic extended with various modalities for representing temporal aspects of logical formulae [Emerson, 1990]. The temporal logic used here is based on a linear, discrete model of time. Thus, time is modelled as an infinite sequence of discrete 'moments', with an identified starting point, called 'the beginning of time'. Classical formulae are used to represent constraints within individual moments, while temporal formulae represent constraints *between* moments. Examples of temporal operators are given below.

- $\Diamond \varphi$ is true now if $\varphi$ is true at *some* moment in the future.
- $\Box \varphi$ is true now if $\varphi$ is true *always* in the future.
- $\varphi \,\mathcal{U}\, \psi$ is true now if $\varphi$ is true from now *until* a future moment when $\psi$ is true.
- $\bigcirc \varphi$ is true now if $\varphi$ is true at the *next* moment in time.
- **start** is only true at the *beginning* of time.

As an agent's behaviour is represented by a temporal formula, this can be transformed into the temporal normal form, SNF [Fisher, 1997a]. This process not only removes the majority of the temporal operators, but also translates the formula into a set of *rules* suitable for either execution or verification. Each of these rules is of one of the following varieties.

$$\textbf{start} \Rightarrow \bigvee_{j=1}^{r} m_j$$

$$\bigwedge_{i=1}^{q} k_i \Rightarrow \bigcirc \bigvee_{j=1}^{r} m_j$$

$$\bigwedge_{i=1}^{q} k_i \Rightarrow \Diamond l$$

where each $k_i$, $m_j$ or $l$ is a literal. Note that SNF is just as expressive as the full temporal logic. This logical form provides the core elements for describing basic dynamic execution:

- a description of the current moment;
- a description of transitions that *might* occur between the current and the *next* moment;
- a description of situations that will occur at some, unspecified, moment in the future.

Thus, using this approach, the behaviour of an agent can be represented now, in transition to the next moment in time and at some time in the future [Fisher, 1995].

### 3.2 Executing Agent Descriptions

An agent's temporal specification can be implemented in a number of ways, for example through refinement to traditional programming languages. However, since temporal logic represents a powerful, high-level notation, a viable alternative, at least for prototyping purposes if not for full implementation, is to animate the agent by directly executing its temporal specification [Fisher, 1996].

In the case of Concurrent METATEM, the set of SNF rules is executed using the *imperative future* paradigm [Barringer *et al.*, 1996]. Here, a *forward-chaining* process is employed, using information about both the history of the agent and its current set of rules in order to constrain its future execution.

As an example of a simple set of rules which might be part of the `car` agent's description, consider the following.

$$\begin{aligned}
\textbf{start} &\Rightarrow \neg\texttt{moving} \\
\texttt{go} &\Rightarrow \Diamond\texttt{moving} \\
(\texttt{moving} \wedge \texttt{go}) &\Rightarrow \bigcirc(\texttt{overheat} \vee \texttt{empty})
\end{aligned}$$

Here, `moving` is false at the beginning of time and whenever `go` is true (for example, if a `go` message has just been received), a commitment to eventually make `moving` true is given. Similarly, whenever both `go` and `moving` are true, then either `overheat` or `empty` will be made true in the next moment in time.

The operator used to represent basic temporal indeterminacy is the *sometime* operator, '$\Diamond$'. When a formula such as '$\Diamond\varphi$' is executed, the system must attempt to ensure that $\varphi$ *eventually* becomes true. As such eventualities might not be able to be satisfied immediately, a record of the outstanding eventualities must be kept, so that they can be re-tried as execution proceeds. The standard heuristic used is to attempt to satisfy as many eventualities as possible, starting with the oldest outstanding eventuality [Fisher and Owens, 1992].

An important consideration with respect to the practical implementation of Concurrent METATEM is that, although it *can* be made into a complete theorem-prover for temporal logic (at least in the propositional case), this is rarely done. Thus, the complexity of full theorem-proving is usually avoided [Fisher, 1996].

### 3.3 Concurrency, Communication and Grouping

We note that such asynchronously executing agents, communicating via broadcast message-passing, are very useful for developing open systems [Hewitt, 1991], while the notion of agent *groups* [Maruichi *et al.*, 1991] is essential both for restricting the extent of broadcast messages and for structuring the agent space.

### 3.4 Applications

The combination of executable temporal logic, asynchronous message-passing and broadcast communication provides a powerful and flexible basis for the development of agent-based systems. Consequently, Concurrent METATEM has been applied in a number of areas, including distributed artificial intelligence, concurrent theorem-proving, artificial agent societies and transport systems [Fisher, 1994].

## 4 A Hybrid Approach

In this section, we show how an extended version of Concurrent METATEM can be used to implement the core facets of the BDI model. As we are keen to represent only the *fundamental* elements, there are a variety of details that we ignore. For example, we will not consider real-time constraints, complex plan structures/libraries, or complex intention structures from the PRS, nor will we consider first-order temporal specifications, concurrent actions, multiple threads, cloning

| BDI | METATEM |
|---|---|
| initialise-state<br>**repeat**<br>  options = option-generator(event-queue)<br>  selected-options = deliberate(options)<br>  update-intentions(selected-options)<br>  execute-step()<br>  get-new-external-events()<br>**end repeat** | initialise-state<br>**repeat**<br>  futures = generate-choices(state)<br>  choice = choose(futures,outstanding)<br>  outstanding = update(outstanding,choice)<br>  execute-step(choice)<br>  state = message-update(choice)<br>**end repeat** |

Figure 1: A Comparison of Interpreter Cycles.

or grouping from Concurrent METATEM. In spite of these restrictions, we will show that the key elements of simple BDI architectures can be concisely and consistently represented and implemented using extended Concurrent METATEM.

### 4.1 Informational Aspects

As, in practical BDI architectures such as the PRS, beliefs are usually coded by ground instances of first-order predicate calculus, then these can be represented directly as (internal agent) facts in Concurrent METATEM. However, as Concurrent METATEM does not incorporate persistence, then this necessitates the addition of meta-level frame axioms, such as

$$[bel(\varphi) \wedge \neg change(\varphi)] \Rightarrow \bigcirc bel(\varphi).$$

Thus, an alternative approach is to directly add a new modal operator for belief to the logic executed within Concurrent METATEM. As in BDI, this is a modal dimension satisfying the KD45 axioms and the execution mechanism itself is modified to handle the persistence of beliefs, i.e. both the above frame axiom and the KD45 axioms are 'built in' to the execution mechanism.

Thus, this is the key logical extension provided to Concurrent METATEM. The logic executed is now a multi-modal logic consisting of both temporal and doxastic dimensions. Note that SNF has been extended to such logics elsewhere [Fisher et al., 1996], and rules now incorporate the new belief modality, '**B**'.

### 4.2 Motivational Aspects

The key observation we use regarding the BDI framework is that, in practice, there is often little fundamental difference between desires and intentions. The former represent all the goals that the agent wishes to achieve; the latter represent those that it is actively pursuing at present. Indeed, in the logical foundations of the BDI model [Rao, 1996b], both desires and intentions are represented by the same type of modal logic (KD), while beliefs are represented by a different one (KD45). Thus, a very natural (although Bratman [1987] takes a different view) mechanism for representing both intentions and desires is to use temporal eventualities. In particular, these are required to be satisfied eventually (if consistent), can be conflicting (e.g. $\Diamond \varphi$ and $\Diamond \neg \varphi$), and the system must choose between them in order to generate further execution.

### 4.3 Deliberative/Execution Aspects

Before considering the detailed execution within our approach and its relationship to BDI architectures, it is useful to examine the basic interpretation cycles for both the PRS [Rao and Georgeff, 1995] and core METATEM execution [Fisher and Owens, 1992]. These are presented in Fig. 1. Note the close similarities, the main difference being the more complex (and stratified) deliberation phase in the PRS.

Thus, the key modification of the execution mechanism concerns the implementation of deliberation. The subset of eventualities selected, the order in which they are attempted, and the mechanism for achieving them, are all the concern of deliberation. If, for the moment, we assume that desires and intentions are trivially achievable (this restriction is considered further below), then we can describe the key decision mechanisms of the two approaches as follows. As deliberation is concerned with weighing up each of the possible choices and choosing a subset to actually undertake, then in Concurrent METATEM, the eventualities (desires/intentions) are simply ordered by age and the oldest one is attempted first, while in BDI systems, each is examined in turn and a specific cost function is used to generate an ordering.

In order to provide added flexibility within the language, we allow the user to redefine the priority functions used within deliberation. Thus, the system requires a *desire priority* function, $\rho_d$, which takes as arguments a list of eventualities (desires), and the history of the execution, and returns an ordered list of eventualities (intentions). This is then used to control the part of deliberation that extracts intentions from desires. In standard Concurrent METATEM, this would simply involve comparing the age of each eventuality; in the PRS this function would typically be much more complex, incorporating a variety of aspects of the current and past state of the agent.

If we were to remove the above simplification whereby intentions are trivially achieved, we could also allow a user defined *intention priority* function, $\rho_i$, which takes as arguments a list of eventualities (intentions), the history of the execution, and a list of plan definitions, and returns an ordered list of eventualities (intentions) to be executed immediately. In the PRS, the list returned would typically only contain one element (as only one step is taken at each cycle), though this framework can obviously accommodate more complex systems.

Thus, the simple deliberation mechanism in Concurrent METATEM (see Fig. 1) is replaced by a two stage process whereby the function $\rho_d$ is used to generate a list of intentions and the function $\rho_i$ is used to choose appropriate actions, based upon this intention list. The implementation of these priority functions may be carried out in a number of ways, for example via meta-level METATEM rules [Barringer et al., 1991] directly provided by the user.

In this way a language for implementing simple BDI-like systems can be provided by extending Concurrent METATEM with belief elements and user definable priority functions for use in deliberation. Note that the more complex elements of practical BDI systems may also be represented in this language. For example, in [Mulder et al., 1996], an implementation of a simplified version of the PRS is provided, incorporating a range of low-level details, such as the manipulation and interpretation of plan components. Simply, dependencies between plans can be provided by asserting that if any sub-plan has not been constructed, the plan itself can not be completed, e.g.

$$\left(\bigvee_i \neg subplan_i\right) \Rightarrow \neg plan.$$

However, motivation for the work described in this paper is provided by the observation that such a representation of the PRS can be improved, simply by extending Concurrent METATEM with elements from the BDI model.

### 4.4 Analysis

There are five notable points to make regarding this hybrid approach of extending executable temporal logic with aspects of the BDI model.

Firstly, since the key elements added are the KD45 belief modality and the user defined priority functions, the Concurrent METATEM execution mechanism remains relatively unchanged. The efficiency of such an approach depends primarily on the complexity of the priority functions provided. For example, if a simple age-ordering function is used, then the speed is comparable with that of Concurrent METATEM.

A variety of deliberation mechanisms can be represented by the use of user definable priority functions within the execution mechanism. The constraints upon this are the complexity of evaluating such functions and the expressive power of the logical notation. Regarding the latter, the logical rule form is as expressive as full temporal logic and so a wide variety of functions can potentially be provided.

For this hybrid approach a formal semantics can be developed, though it would be parameterised by the semantics of the priority functions. If only sequences of actions undertaken are considered, then executions represent appropriate models for the BDI logics [Rao, 1996b].

This approach provides increased expressive power. Not only can a variety of priority functions be defined, but additional elements of Concurrent METATEM, such as multi-threading and disjunctive rules, can potentially be incorporated.

Finally, while there are no explicit axioms linking, say, beliefs and intentions (as are often found in theoretical works concerning BDI), rules in the program can be used to partially provide these, for example $B\varphi \Rightarrow \Diamond \xi$ links beliefs and desires (eventualities).

## 5 Conclusions and Future Work

In this paper, we have described an extension to Fisher's Concurrent METATEM language that presents the possibility of implementing BDI-like systems using a form of executable multi-modal logic. While we do not claim that all BDI systems can be implemented using this approach, we believe a significant range of applications can be, providing the priority functions defined are not excessively complex.

The key observations from this work are that, in BDI systems, beliefs are primitive and distinct, while desires and intentions can essentially be represented as the same kind of entities. This, together with the observation that desires/intentions correspond, in many respects, to eventualities in temporal logic, allows the definition of basic BDI entities within Concurrent METATEM. Thus, this work separates the basic elements of the model (beliefs and desires) from the mechanisms for manipulating these and deciding how to act (deliberation).

While this approach will not be as efficient as directly implemented BDI systems, such as the PRS, it provides the opportunity at least for prototyping of BDI systems. The execution of temporal logic through Concurrent METATEM is well understood and, although the extended language requires the execution of a multi-modal logic, it remains a relatively inexpensive extension. As execution at an attempt to construct a model, rather than a complete theorem-proving process, it represents a relatively cheap mechanism for animating a specification. At the other extreme, it is difficult to see how the multi-modal BDI logic from [Rao, 1996b] could be directly executed.

Our future work concerns three areas: extended languages; refined priority functions; and expanded formal semantics. As mentioned above, there are elements of Concurrent METATEM that are not required in the BDI model. It is thus interesting to consider how these elements could be used in the framework described here. Also, once such a system is produced, it is relatively easy to extend it with the other elements of Concurrent METATEM, such as cloning and grouping. The definition of a range of detailed deliberation strategies, via appropriate priority functions, is another obvious direction. Finally, the temporal semantics of Concurrent METATEM must be adapted to fully incorporate the above extensions.

# References

[Barringer et al., 1991] H. Barringer, M. Fisher, D. Gabbay, and A. Hunter. Meta-Reasoning in Executable Temporal Logic. In *Proceedings of the International Conference on Principles of Knowledge Representation and Reasoning (KR&R)*. Morgan Kaufmann, 1991.

[Barringer et al., 1996] H. Barringer, M. Fisher, D. Gabbay, R. Owens, and M. Reynolds, editors. *The Imperative Future: Principles of Executable Temporal Logics*. Research Studies Press, Chichester, United Kingdom, 1996.

[Birman, 1991] K. P. Birman. The Process Group Approach to Reliable Distributed Computing. Techanical Report TR91-1216, Department of Computer Science, Cornell University, July 1991.

[Bratman, 1987] M. E. Bratman. *Intentions, Plans, and Practical Reason*. Harvard University Press, 1987.

[Emerson, 1990] E. A. Emerson. Temporal and Modal Logic. In *Handbook of Theoretical Computer Science*. Elsevier, 1990.

[Fischer et al., 1996] K. Fischer, J. Müller and M. Pischel. A Pragmatic BDI Architecture. In *Intelligent Agents II (LNAI 1037)*. Springer-Verlag, 1996.

[Fisher and Owens, 1992] M. Fisher and R. Owens. From the Past to the Future: Executing Temporal Logic Programs. In *LPAR (LNCS 624)*. Springer-Verlag, 1992.

[Fisher, 1993] M. Fisher. Concurrent METATEM — A Language tor Modeling Reactive Systems. In *Parallel Architectures and Languages, Europe (LNCS 694)*. Springer-Verlag, June 1993.

[Fisher, 1994] M. Fisher. A Survey of Concurrent METATEM — The Language and its Applications. In *First International Conference on Temporal Logic (LNCS 827)*. Springer-Verlag, July 1994.

[Fisher, 1995] M. Fisher. Representing and Executing Agent-Based Systems. In *Intelligent Agents*. Springer-Verlag, 1995.

[Fisher, 1996] M. Fisher. An Introduction to Executable Temporal Logics. *Knowledge Engineering Review*, 11(1), March 1996.

[Fisher, 1997a] M. Fisher. A Normal Form for Temporal Logic and its Application in Theorem-Proving and Execution. *Journal of Logic and Computation*, 7(4), July 1997.

[Fisher, 1997b] M. Fisher. An Open Approach to Concurrent Theorem-Proving. In *Parallel Processing for Artificial Intelligence III*. North-Holland, 1997.

[Fisher et al., 1996] M. Fisher, M. Wooldridge and C. Dixon. A Resolution-Based Proof Method for Temporal Logics of Knowledge and Belief. In *Proceedings of the International Conference on Formal and Applied Practical Reasoning (FAPR)*. Springer-Verlag, June 1996.

[Georgeff and Ingrand, 1989] M. P. Georgeff and F. F. Ingrand. Decision-Making in an Embedded Reasoning System. In *Proceedings of the Eleventh International Joint Conference on Artificial Intelligence (IJCAI)*. Morgan Kaufmann, 1989.

[Hewitt, 1991] C. Hewitt. Open information systems for distributed artificial intelligence. *Artificial Intelligence*, 47, 1991.

[Jennings and Wooldridge, 1995] N. R. Jennings and M. Wooldridge. Applying Agent Technology. *Applied Artificial Intelligence*, 9(4), 1995.

[Kinny and Georgeff, 1997] D. N. Kinny and M. P. Georgeff. Modelling and design of multi-agent systems. In *Intelligent Agents III (ATAL-96)*. Springer-Verlag, 1997.

[Manna and Pnueli, 1992] Z. Manna and A. Pnueli. *The Temporal Logic of Reactive and Concurrent Systems: Specification*. Springer-Verlag, New York, 1992.

[Maruichi et al., 1991] T. Maruichi, M. Ichikawa, and M. Tokoro. Modelling Autonomous Agents and their Groups. In *Proceedings of the $2^{nd}$ European Workshop on Modelling Autonomous Agents and Multi-Agent Worlds (MAAMAW '90)*. Elsevier/North Holland, 1991.

[Mulder et al., 1996] M. Mulder, M. Fisher and J. Teur. A Comparison of Concurrent METATEM and DESIRE. Internal Report. Department of Mathematics and Computer Science, Vrije Universiteit Amsterdam, 1996.

[Rao and Georgeff, 1991] A. S. Rao and M. P. Georgeff. Modeling Agents within a BDI-Architecture. In *International Conference on Principles of Knowledge Representation and Reasoning (KR)*, Cambridge, Massachusetts, April 1991. Morgan Kaufmann.

[Rao and Georgeff, 1995] A. S. Rao and M. Georgeff. BDI Agents: from theory to practice. In *Proceedings of the First International Conference on Multi-Agent Systems (ICMAS-95)*, San Francisco, CA, June 1995.

[Rao, 1996a] A. Rao. Agentspeak(l): BDI Agents Speak out in a Logical Computable Language. In *MAAMAW-96 (LNAI 1038)*. Springer-Verlag, 1996.

[Rao, 1996b] A. S. Rao. Decision procedures for propositional linear-time Belief-Desire-Intention logics. In *Intelligent Agents II (LNAI 1037)*. Springer-Verlag, 1996.

[Shoham, 1993] Y. Shoham. Agent-oriented programming. *Artificial Intelligence*, 60(1), 1993.

[Weerasooriya et al., 1995] D. Weerasooriya, A. Rao, and K. Ramamohanarao. Design of a concurrent agent-oriented language. In *Intelligent Agents*. Springer-Verlag, 1995.

[Wooldridge and Jennings, 1995] M. Wooldridge and N. R. Jennings. Agent theories, architectures, and languages: A survey. In *Intelligent Agents*. Springer-Verlag, 1995.

# Managing decision resources in plan execution

Michael Freed and Roger Remington
NASA Ames Research Center, MS 262-2
Moffett Field, CA 94035 USA

## Abstract

We describe an approach to the problem of managing resources in routine decision-making tasks. The central feature of this approach is the use of reusable RAP-like plans to generate decisions. This allows our system, APEX, to take advantage of the flexibility in scheduling and method selection provided by execution mechanisms and thereby minimize or circumvent resource conflicts. We then discuss an application of APEX for simulating a human air traffic controller in order to aid in the evaluation of radar display designs.

## 1 Introduction

In this paper, we describe an approach to managing resources in *routine decision tasks* and apply this approach to a practical problem. Routine decisions are choices that occur regularly in an agent's everyday tasks. For example, drivers are often faced with decisions such as whether to slow down for a yellow light and whether to turn at an often-encountered intersection. Making such decisions involves several resource-demanding activities including acquiring decision-relevant information (internally or from the task environment) and making inferences. In time-pressured conditions, or when multiple tasks compete for the same computational and perceptual resources, the ability to manage scarce resources becomes an important determiner of agent performance.

Researchers have taken a variety of approaches to managing resources, especially computational resources, when deciding action in realistically complex, dynamic task environments. One approach is to eliminate certain expensive computations. Reactive planners, for example, use only current perceptions to conditionalize action choice, thus avoiding the expensive computations required to construct plans and retrieve items from memory. Some systems avoid specific classes of inference such as the prediction of future states and deductive retrieval [Firby89]. Others allow expensive operations when they are most likely to prove essential but otherwise avoid them [Chien91;Hayes-Roth95].

A second approach is to delay decisions until relevant information can be acquired cheaply (or at all), thus avoiding the computationally expensive process of conditionalizing decisions on a large number of possible future states. Systems that employ this approach [Firby89; Gat96;Simmons94;Pell97] are referred to as execution systems since they interleave planning (deciding action) with plan execution.

Our system's plan execution component, like the RAP system [Firby89] on which it is based, combines these approaches but differs from previous systems in how it handles routine decisions. In particular, routine decision tasks are treated in a uniform manner with tasks of other kinds — i.e. they are carried out by general-purpose task-execution mechanisms in accordance with task-specific, variablized plan structures called *methods*. Methods consist of steps, each corresponding either to a primitive action or to some non-primitive that must be decomposed into substeps by selecting a more specific method.

## 2 Decision-methods

The use of specialized decision-plans has a number of advantages. In particular, plan execution mechanisms can begin, abort, retry, interrupt, resume, specify and terminate decision tasks as needed to handle situational constraints and coordinate resource use with other tasks. Consider, for example, the method below for deciding between alternative routes home from work.

```
(method-25 (decide-route-from-work-to-home)
   (step1 (acquire-info:time-of-day => ?time)
       (priority +1))
   (step2 (acquire-info:day-of-week => ?day))
   (step3 (infer rule-57 ?time ?day => ?best-route)
       (wait-for step1 step2)))
```

In this case, the decision is based solely on whether or not the agent is likely to face rush hour traffic, a function of the time of day and day of week. The two information acquisition steps are carried out in parallel, each by one of several methods. For instance, day of the week information is acquired either by checking a calendar or querying memory for the current day. When these steps complete, an inference rule is applied to decide the best route. Information acquisition tasks can fail and thus fail to provide a value to a decision-relevant variable— e.g. when no calendar can be found and no information on the current day can be acquired from memory. Similarly, execution may omit an information-acquisition task in response to time-pressure or other factors. In these cases, the inference-rule will rely on a default value — e.g. that today is a weekday.

This framework provides two means for managing resource use in routine decision tasks. First, execution mechanisms can control the timing of resource-demanding tasks by interruption, delaying task initiation, or delaying resumption. For example, if finding a calendar would take perceptual resources (gaze) away from a higher priority task, execution mechanisms can delay this action until the higher priority task completes. Second, since different information-acquisition methods will generally differ in type and amount of required resources, methods can be selected in order to allocate resources them most effectively. For instance, if finding a calendar is prevented by a higher priority use of the gaze resource, execution could try retrieving the information from memory or using the *null-method* (i.e. omitting information acquisition and forcing reliance on a default) which requires minimal resources.

Method-selection in our system is handled by *method-selection-rules* (MSRs) whose syntax mirrors that of the COND macro in LISP. MSRs employ both transient and long-term knowledge. Relevant long-term knowledge can include: the expected interval during which resources must be allocated to a method for it to complete; the usual level of competition for resources from other tasks present during execution of the decision-task (expected workload); the usual amount of time available to complete the decision-task (expected urgency); and the likelihood that the default value associated with an information-acquisition task will prove accurate. Such factors determine the expected utility of alternative information acquisition methods and thus determine a stable preference between alternatives. Taken together, the method preferences for all steps of a decision-method constitute a baseline *decision strategy*.

Transient information can be used to adapt a decision strategy to an agent's current situation. The current model supports adaptation from several kinds of information including especially: *subjective workload* and *default counterevidence*. Subjective workload corresponds to an agent's evaluation of its overall "busyness" compared to expected workload. During periods of unusually high workload for a given decision task, decision strategy is biased in favor of the least resource-demanding methods. In some cases, this will cause decisions to rely on defaults when more reliable information acquisition methods would normally be selected.

Default counterevidence is knowledge that the default value for some decision-relevant factor is likely to prove incorrect for some time into the future. For example, an agent may tend to assume that today is a weekday when deciding a route home from work. Deciding to go into work on a weekend day invalidates this assumption and should (temporarily) reduce the system's tendency to rely on it — i.e. it should bias execution mechanisms to avoid using the default. This function is carried out by task-specific *bias rules*. Since avoiding reliance on valid defaults wastes resources, bias rules must specify a duration, after which their effect expires. The length of this interval depends on the expected duration of the non-default condition and the expected interval between successive observations of the condition if it persists.[1] The system will thus tend to rely on invalid defaults when the default condition lasts for an unusually long time or when an unusually long period has passed since the condition was last observed.

## 3 Application: user interface evaluation

We have incorporated this approach to routine decision making into action selection mechanisms of our human operator model, APEX.[2] The model consists primarily of two components: action selection mechanisms based on the RAP plan execution system, and a resource architecture which describes limitations on perceptual, cognitive, and motor resources and constrains action selection mechanisms to operate within those limits.

### 3.1 To err is human, to prevent error is good design

APEX is intended to address a fundamental problem in the design of user interfaces. In particular, newly designed equipment and procedures often inadvertently facilitate human error. Techniques for identifying error facilitations in design tend to be either ineffective or very expensive. For example, one of the most effective ways to test new designs is to hire human operators to carry out tasks using prototyped equipment, and then observe their performance in a wide range of operating conditions. In our domain, air

---

[1] See [Freed97] for information on how the bias duration parameter is set.

[2] Architecture for Procedure EXecution

traffic control, such tests typically require hiring highly paid expert controllers as subjects, often for extended periods. The limited amount of testing that results from high cost can stifle innovation and compromise safety.

One way to get some of the benefits of a "human in the loop" study at much lower cost is to use a computer to simulate all elements of such a study including the equipment, human operators, and experimental observers. Human simulation has been used successfully by others to guide design (e.g. [John90,Corker95]). However, ours appears to be the first system to employ the powerful and versatile action selection mechanisms provided by AI plan execution systems, and thus the first able to function effectively in inherently complex, dynamic, and uncertain domains such as air traffic control. By employing action selection mechanisms designed for robot control, our model overstates human capabilities in some ways, but can operate in domains where predicting human error would be most useful.

Though not specifically designed to make errors of any kind, our approach to managing the resource cost of routine decision-making enables APEX's plan execution component to help predict a type of error sometimes referred to as a "habit capture" [Reason90]. Habit captures are defined by their apparent cognitive cause. In particular, people make such errors when, instead expending resources to acquire reliable information, they act in accordance with a false but usually reliable default assumption. Habit captures are reported quite frequently in naturalistic studies of error [Reason82]. For example:

> "I went to the bedroom to change in to something more comfortable for the evening, and the next thing I knew I was getting into my pyjama trousers, as if to go to bed.
>
> "I had decided to cut down my sugar consumption and wanted to have my cornflakes without it. But the next morning, however, I sprinkled sugar on my cereal just as I always do."

In our view, much of people's tendency to rely on default assumptions can be explained as adaptations to regularities in the task environment. For instance, people will be more likely to rely on a default if, apriori, it is especially likely to be true or if the environment reliably provides default counterevidence when it is false. This view provides a basis for predicting when people will rely on false defaults and make habit capture errors as a result.

## 3.2 An Example

At a TRACON air traffic control facility, one controller will often be assigned to the task of guiding planes through a region of airspace called an arrivals sector. This task involves taking planes from various sector entry points and getting them lined up at a safe distance from one another on landing approach to a particular airport. Some airports have two parallel runways. In such cases, the controller will form planes up into two lines.

Occasionally, a controller will be told that one of the two runways is closed and that all planes on approach to land must be directed to the remaining open runway. A controller's ability to direct planes exclusively to the open runway depends on remembering that the other runway is closed. How does the controller remember this important fact? Normally, the diversion of all inbound planes to the open runway produces an easily perceived reminder. In particular, the controller will detect only a single line of planes on approach to the airport, even though two lines (one to each runway) would normally be expected.

However, problems may arise in conditions of low workload. With few planes around, there is no visually distinct line of planes to either runway. Thus, the usual situation in which both runways are available is perceptually indistinguishable from the case of a single closed runway. The lack of perceptual support would then force the controller to rely on memory alone, thus increasing the chance that the controller will accidentally direct a plane to the closed runway.

## 3.3 Simulation

When the simulated controller hears that the left runway is closed, interpretation mechanisms cause a propositional representation of this fact to be encoded in memory. The encoding event generates bias (default counterevidence) according to the following rule:

```
(bias-rule-17
  (if (closed ?rwy)
      (create-bias method-27 step5 (10 minutes))))
```

Newly generated bias is represented explicitly in memory along with a notation indicating when the bias will expire if not renewed. In this case, bias lasting 10 minutes causes decision mechanisms to consider the possibility of runway closure (step5 below) in cases where the usual state — all runways open — might otherwise be assumed.

When a plane approaches its airspace, the simulated controller initiates a routine plane-handling method involving accepting responsibility for the plane, determining where the plane is headed, and then guiding it to its destination. If the plane's destination is Los Angeles airport (LAX), guiding it to its destination will involve selecting between the airport's two parallel runways. For highly routine decisions such as runway selection, human controllers can reasonably be expected to know which factors to consider in making the decision and how to appropriately weight each factor. This knowledge is incorporated into the following decision method:

```
(Method-27 (select-runway ?plane)
  (step1 (id-rwy-with-fewer-planes => ?fewer))
  (step2 (id-rwy-fastest-approach ?plane =>
      ?fastest))
  (step3 (id-rwy-easiest-for-me => ?easiest))
  (step4 (id-rwy-better-microclimate =>?climate))
  (step5 (id-available-runways => ?available))
  (step6 (id-safest-rwy ?plane => ?safest))
  (step7 (infer rule-19 ?fewer ?fastest …)
     (wait-for step1 step2 … step6)))
```

In most cases, more than one method will be available for acquiring information about a factor. In this example, the controller could determine runway availability by retrieving information from memory, asking another controller, or by assuming the most likely condition — that the runway is open. Since runways closures are rare and memory retrieval is expensive [Carrier95;Stein93], the decision strategy underlying this method (along with associated MSRs) prescribes reliance on the default assumption unless transient bias (default counterevidence) promotes a more effortful alternative.

In the described scenario, bias produced after learning of the runway closure causes the agent to temporarily avoid reliance on the default. Instead, for some time thereafter, the runway's availability is verified by retrieving information from working memory whenever a runway selection task occurs. Eventually, the initial bias expires. To select a runway for a newly arrived plane, the agent's decisions will once again conform to the default assumption. Other factors will then determine which runway is selected. For example, the controller may choose to direct a heavy plane to the longer left runway which, in normal circumstances, would allow the plane an easier and safer landing. With the left runway closed, actions following from this decision result in error.

Avoiding error requires maintaining appropriate bias. In a variation of the described scenario in which no error occurs, visually perceived reminders of the runway closure cause bias to be periodically renewed. In particular, whenever visual attention mechanisms attend to plane icons on an approach path to the airport, interpretation mechanisms note the absence of a line of planes to the left runway and signal an expectation failure.

```
(expectation-generation-rule-5
  (if (and
         (not (visual-group plane-icons
                              lft-rwy-path))
         (visual-group plane-icons rt-rwy-path))
     (assert-anomaly (rwy-imbalance right))))
```

In general, whenever an expectation failure occurs, a task to explain the observed anomaly is initiated. The first step in such a task is to try to match the anomaly to a known explanation-pattern (XP) [Schank86]. A match results in a task to verify the XP hypothesis.

```
(explanation-pattern-5
  (:anomaly    (rwy-imbalance ?rwy))
  (:candidate-explanation   (closed ?rwy)))
```

In principle, verifying a hypothesis could involve mental and physical actions of any kind. In the current model however, the only way to verify a hypothesis is to check for a match in working memory. In this case, the contents of working memory are adequate; the anomalous absence of planes on approach to the left runway is explained as a result of the left runway's closure.

Bias renewal occurs whenever the working memory item that originally produced the bias is re-encoded or retrieved. Thus, retrieving the proposition (closed runway left) triggers the bias generation rule just as if the proposition had been encoded for the first time. Thus, the unusual arrangement of planes on the radar scope acts as a constant reminder, preventing the agent from reverting to the use of its default assumption and thereby preventing error.

## 3.4 Aiding user interface design

Interface designers often overlook aspects of an interface that facilitate user errors, though in many instances, design problems are obvious once pointed out [Norman88]. The problem of noticing these design problems becomes especially difficult in domains such as air traffic control where interfaces must mediate complex tasks carried out in diverse operating conditions. By employing a model of error-prone human behavior, we hope to partially automate the process of predicting design-facilitated errors.

The basis for these predictions arises from an analysis of how agents generally, and humans in particular, can manage limited resources in decision-making. Employing this analysis, it is possible not only to simulate the influence of interface attributes on human tendencies to err, but also to provide causal explanations for predicted errors that indicate ways to repair the design. For example, explaining the described error scenario to a designer as an indirect consequence of low workload indicates a clever fix: runway closures can be visually signaled, but only in low workload conditions when the added screen clutter would cause little distraction.

## 4 Discussion

The process of deciding action can make substantial demands on limited computational and physical resources. To cope with time-pressure and competing demands, our system decides action on the basis of flexible strategies incorporated into RAP-like reusable plans and other structures. When carried out by the same task execution mechanisms used to carry out non-decision tasks, these decision plans provide two means for managing resources.

First, execution mechanisms can delay or interrupt decision subtasks to give higher priority tasks preferential access to resources. Second, execution can select between alternative methods for subtasks on the basis of duration or compatibility between their different resource requirements and demands from other tasks. Selecting a method for its resource-demand characteristics will sometimes entail trading off against some other desirable attribute such as reliability. In particular, execution may rely on a default assumption in making a decision rather than engage in time- and resource-demanding efforts to acquire more reliable information.

Resource-management strategies that prescribe reliance on defaults make an agent vulnerable to habit-capture errors when assumptions underlying the strategy do not hold in its current environment. In the described example, the strategy of assuming runway availability in the absence of default counterevidence makes the simulated air traffic controller vulnerable in conditions of low workload when counterevidence is unavailable.

The systematicity of such errors makes our approach useful for predicting circumstances in which aspects of an interface design might inadvertantly facilitate error. By alerting designers to the potential for such errors early in the design process, we hope to reduce the cost of evaluation and thereby speed the safe introduction of new technology.

## Acknowledgements

Thanks to Jim Johnston, Eric Ruthruff, Mark van Selst, and Mike Shafto for many useful discussions.

## References

[Carrier95] Carrier, L.M. and Pashler, H. Attentional limitations in memory retrieval. *Journal for experimental psychology: learning, memory, & cognition*, 21, 1339-1348, 1995.

[Corker93] Corker, K.M. and Smith, B.R. An architecture and model for cognitive engineering simulation analysis. *Proceedings of the AIAA Computing in Aerospace 9 Conference*, San Diego, CA, 1993.

[Firby89] R.J. Firby. Adaptive execution in complex dynamic worlds. Ph.D. thesis, Yale University, 1989.

[Freed97] Freed, M. and Shafto, M. Human-system modelling: some principles and a pragmatic approach. *Proceedings of the Fourth International Workshop on the Design, Specification, and Verification of Interactive System*. Granada, Spain.

[Gat96] Gat, Erann. The ESL User's Guide. Unpublished. Available at: www-aig.jpl.nasa.gov/ home.gat/esl.html

[John94] John, B.E. and Kieras, D.E. The GOMS Family of Analysis Techniques: Tools for Design and Evaluation. Carnegie Mellon University. School of Computer Science, TR CMU-CS-94-181, 1994.

[Norman88] Norman, Donald A. *The Psychology of Everyday Things*. Basic Books, New York, N.Y., 1988.

[Pell97] Pell, B., Bernard, D.E., Chien, S.A.., Gat, E., Muscettola, N., Nayak, P.P., Wagner, M., and Williuams, B.C. An autonomous agent spacecraft prototype. *Proceedings of the First International Conference on Autonomous Agents*, ACM Press, 1997.

[Reason90] Reason, J.T. *Human Error*. Cambridge University Press, New York, N.Y., 1990.

[Reason82] Reason, J.T. and Mycielska, K. *Absent-minded? The psychology of mental lapses and everyday errors*. Englewood Cliffs, N.J., Prentice Hall, 1982.

[Schank86] Schank, Roger C. *Explanation Patterns*. Lawrence Earlbaum Associates, Hillsdale, N.J., 1986.

[Simmons94] Simmons, R. Structured control for autonomous robots. *IEEE Transactions on Robotics and Automation*. 10(1), 1994.

[Stein93] Stein, Earl S. and Garland, Daniel. Air traffic controller working memory: considerations in air traffic control tactical operations. FAA technical report DOT/FAA/CT-TN93/37, 1993.

# COMPUTER-AIDED EDUCATION

# Use of Abstraction and Complexity Levels in Intelligent Educational Systems Design

**Ruddy Lelouche**
Département d'informatique, Université Laval
Québec   G1K 7P4   Canada
Tel.: (418) 656-2131, ext. 2597   Fax: (418) 656-2324
e-mail: `lelouche@ift.ulaval.ca`

**Jean-François Morin**
Département d'informatique, Université Laval
Québec   G1K 7P4   Canada
Tel.: (418) 843-5193   Fax: (418) 843-9245
e-mail: `morin@ift.ulaval.ca`

## Abstract

We are interested in problem-solving domains, like engineering and most "exact science" disciplines. In these domains, the knowledge to be acquired by the student is twofold: the domain knowledge itself, but also and mainly the knowledge necessary to solve problems in that domain. As a result, an education-oriented system in such a domain must encompass three knowledge types: the domain knowledge and the problem-solving knowledge, constituting the knowledge to be acquired and mastered by the student, and the tutoring knowledge, used by the system to facilitate the student's learning process. In this paper, we show how these three knowledge types can be modelled, how they should interact with one another in order to fulfil the system educational purpose, and above all how the abstraction and complexity levels can shed a uniformizing light on the system operation and make it more user-friendly. We thus hope to bring some contribution to the general and important problem of finding a generic architecture to intelligent tutoring systems.

## Introduction

Besides the complexity of teaching itself, the domains taught vary considerably in complexity, type of interactions, type of evaluation, relationships between theory and practice, etc. This way, teachable domains can be classified according to the type of knowledge to be acquired by the student: "know", "know-how", and "know-how-to-be". Examples of such knowledge types are respectively: anatomy or a language grammar, the skill to solve a mathematical or medical problem, and the capability to adapt to one's environment or to deal with personal relationships. Cost engineering knowledge is of the second type.

Moreover, almost all teachable domains vary in complexity, from simple basics to relatively complex problems to solve. Thus, a student should learn and master the basics of such a domain before he is taught wider notions. And when a human tutor detects errors or misunderstandings, he usually draws the student's attention on a small subset of the involved knowledge, so that he may correct his errors and/or misunderstandings. Such an interaction is focused on either a given set of the domain knowledge or the scope of knowledge involved by a given problem.

Problem-solving or know-how domains are the ones in which we are interested here. In such a domain, hereafter called a KH-domain, the knowledge to be acquired by the student is twofold: the domain knowledge itself, but also and mainly the knowledge necessary to solve problems in that domain. As a result, an education-oriented system in such a domain, which we here call a KH-ITS, must encompass three knowledge types: the domain knowledge and the problem-solving knowledge, constituting the knowledge to be acquired and mastered by the student, and the tutoring knowledge, used by the system to facilitate the student's learning process.

This paper has a double purpose:

- we present each of the three types of knowledge involved in a KH-ITS;
- for each type of knowledge, we show how abstraction and complexity levels appear and how it is possible to deal with them.

In order to do so, we present in section 1 our domain knowledge modelling and how we exemplify it in the case of cost engineering. Next, in section 2, we focus on the advantage of separating the problem-solving knowledge from the domain knowledge in a KH-ITS, and we present some problem-solving activities in cost engineering as an example. In section 3, we briefly describe some principles of tutoring knowledge modelling in a KH-ITS. In each of these three sections, we show how to use abstraction and complexity levels, exemplifying the principles in the cost engineering domain. Finally, section 4 presents the educational interests of using abstraction and complexity levels when modelling the three types of knowledge involved in a KH-ITS.

## 1 Domain knowledge

In order to describe the domain knowledge, we first present its characteristics in a general KH-ITS (§ 1.1). We then show how we model it in the cost engineering domain (§ 1.2), and how such an approach lets us introduce the notions of abstraction and complexity levels (§ 1.3).

## 1.1 General

The first type of knowledge involved in every ITS, the *domain knowledge* (DK), contains all theoretical and factual aspects of the knowledge to be taught to the student. Although its specific structure can be varied, it typically may include concepts, entities, and relations about the domain [Brodie & al., 1984], object classes and instances [Kim & Lochovsky, 1989], possible use restrictions, facts, rules, [Kowalski, 1979; Clocksin & Mellish, 1981], semantic or associative networks [Findler, 1979; Sowa, 1984], etc.

The main system activities centred on this knowledge type are:

- theoretical explanations about the various knowledge elements and their relationships in the teaching domain;
- providing the other modules of the ITS, i.e. problem-solving and tutoring, with the necessary background of domain knowledge that they need.

## 1.2 Application to cost engineering

In the particular domain of cost engineering, we chose to represent this type of knowledge with concepts, relations, and a special case of relations modelled as concepts, the factors. The model is presented with more details in [Lelouche & Morin, 1996].

*Concepts* can be basic entities like:

- *Investment:* It is an amount on which interest is computed; depending on the context or the type of problem, it may also be called capital, loan, or present value.
- *Interest:* It is the amount exceeding the initial investment, received by the lender from the borrower at the moment of its repayment.
- *Investment duration:* It is the total time during which the interest is computed over the investment; this duration is uninterrupted.
- *Future value:* It is the total amount to be repaid at the term of an investment; it is equal to the total accumulated interest added to the initial investment amount or present value.

With the notion of compounding, some other concepts can be defined:

- *Compounding:* If the investment duration is long enough, a partial interest may be periodically added to the principal; that amount itself then produces interest for the remaining of the investment duration; this principle is called interest compounding; when the interest is compounded before the term of the investment, the principal then augments accordingly, and becomes greater than the initial investment.
- *Compounding period:* It is the time elapsed between two consecutive interest compoundings; unless otherwise specified, the interest is compounded yearly.
- *Number of periods:* It is the number of interest compoundings during the investment duration.
- *Interest rate:* It is the amount of interest payable for one period depending on the principal:

$$\text{Interest rate} = \frac{\text{Interest amount for one period}}{\text{Capital amount}}$$

- *Annuity:* amount paid or received periodically (annually unless otherwise specified); although the periodic amounts may increase or decrease over time, we refer only to uniform (constant) annuities in this paper.

Concepts are linked to one another by various types of *relations*. The relations between concepts can be either usual knowledge-representation relations, like *subclass of*, *element of*, *sort of*, etc., or numerical relations represented by formulæ. Such a formula is given above. Another one is:

$$F = P \times (1 + i)^n \qquad (1)$$

which, given the present value $P$ of an investment over $n$ periods at rate $i$, computes the corresponding future value $F$ of that investment.

A formula such as (1) can be rewritten as:

$$F = P \times \Phi_{PF,i,n} \text{ where } \Phi_{PF,i,n} = (1 + i)^n \qquad (2)$$

$$P = F \times \Phi_{FP,i,n} \text{ where } \Phi_{FP,i,n} = (1 + i)^{-n} \qquad (3)$$

thus introducing the *factors* $\Phi_{PF,i,n}$ and $\Phi_{FP,i,n}$. Factors allow us to separate their definition (rightmost equalities above, a quantitative aspect) from their possible uses in the application domain (leftmost equalities, a qualitative aspect).

Similarly, the factor $\Phi_{AP,i,n}$ converts a series of identical annual amounts $A$ into an unique present value $P$:

$$P = A \times \Phi_{AP,i,n} \quad \text{where} \quad \Phi_{AP,i,n} = \frac{(1+i)^n - 1}{i(1+i)^n} \qquad (4)$$

Actually, $\Phi_{AP,i,n}$ is a sum of $\Phi_{FP}$ factors. The factor $\Phi_{PA,i,n}$ does the reverse process:

$$A = P \times \Phi_{PA,i,n} \quad \text{where} \quad \Phi_{PA,i,n} = \frac{i(1+i)^n}{(1+i)^n - 1} \qquad (5)$$

There exist other factors converting gradient and geometrical series of amounts into a present or future value; such factors are also computed as a sum of $\Phi_{FP,i,n}$ factors.

Although the formulæ related to these factors essentially involve quantitative aspects, the similarities and differences between them, and the circumstances regulating the use of either one, are of a deeply qualitative ground. If the *value* of a factor is indeed calculated from two or three numerical parameters, the *context* in which they are defined depends on whether we have to timewise move an unique amount or a series of amounts, identical or not, or conversely to compute an equivalent annual amount, etc. In fact, this context corresponds to the type of conditions that govern the investment, or *investment conditions type*, without respect to the amounts and durations involved, and is thus essentially qualitative.

This characteristic allows us to qualitatively describe relationships between factors which are basically quantitative: the *transitivity* ($\Phi_{AF,i,n} = \Phi_{AP,i,n} \times \Phi_{PF,i,n}$) and *inversion* ($\Phi_{FP,i,n} \times \Phi_{PF,i,n} = 1$) relations. Moreover, this qualitative description (including the notation) leads to a significant improvement of the pedagogical approach used to teach this subject to university students.

## 1.3 Abstraction and complexity levels

As it turns out, every factor introduces an *intermediate abstraction level* between the concepts implied the equation defining it. For example, in the case of formula (1), or equivalently formulæ (2) and (3), we have (see figure 1):

- at the bottom of the hierarchy, the interest rate and the number of periods, basic concepts "making technicalities explicit";
- above them, concepts more fundamentally related to the cost engineering problem being solved, namely the present and future values of the investment at hand;
- between these two levels, the intermediate one brought by the introduction of the factor $\Phi_{FP}$ or $\Phi_{PF}$.

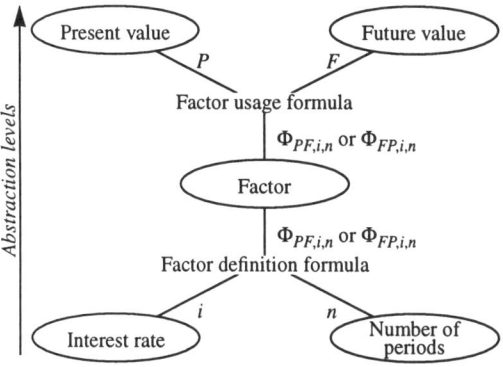

**Figure 1** — Representation of a factor as a concept.

That intermediate status of the factor, originally just an intermediate variable in calculations [see formulæ (2) and (3)], makes it appear as a pedagogically oriented concept, which clearly separates

- the computational, quantitative aspect of the factor definition, i.e. the interest rate and the number of periods
- from the practical, qualitative aspect of the factor usage in a domain problem, i.e. the present and future values.

This follows the theory [Lenat & al., 1979; Malec, 1989] according to which the use of multiple *abstraction levels* eases the modelling process and simplifies inferences which may be made on the domain concepts.

Most interestingly, our scaffolding approach can be made more general: we may present and use higher-level factors built upon these first ones. Indeed, "above" $\Phi_{FP}$ and $\Phi_{PF}$, we can define factors to express the present and future values of a series of identical amounts (and vice versa), which are a first way to generalize this concept hierarchy. For example, the $\Phi_{AP}$ factor is indeed a sum of $\Phi_{FP}$ factors:

$$\Phi_{AP, i, n} = \sum_{k=1}^{n} \Phi_{FP, i, k} = \sum_{k=1}^{n} (1+i)^{-k} = \frac{(1+i)^n - 1}{i(1+i)^n}$$

where the last expression results from computing the geometrical series shown. This example constitutes a proof of (4), but also and mainly shows that the $\Phi_{AP}$ factor is at a higher level than $\Phi_{FP}$. Note that this refers to a *complexity level* rather than an abstraction level, since it is due to the way the $\Phi_{AP}$ factor is defined and computed.

In the sequel of the cost engineering course, other abstraction or complexity levels appear when introducing the factors related to investments made of periodic amounts in arithmetic or geometric progression, and the factors related to various depreciation types.

## 2 Problem-solving knowledge

In order to describe the problem-solving knowledge, we now present the general characteristics regarding problem-solving knowledge modelling in a KH-ITS (§ 2.1). As we did in section 1, we then present our model in the cost-engineering domain (§ 2.2).

### 2.1 General

The second type of knowledge is specific to KH-domains, henceforth to KH-ITSs. We call it *problem-solving knowledge* (PSK). It contains all inferential processes used to solve a problem resulting from the instantiation of a practical situation based on the domain knowledge [Kowalski, 1979; Patel & Kinshuk, 1996]. In other words, in order to be able to solve a problem, the problem-solving knowledge needs a theoretical background, which is found in the domain knowledge. The processes stored in PSK may be represented in various ways, using either or all of logic [Kowalski, 1979], procedural networks [Brown & Burton, 1978], semantic networks with procedural attachments, (augmented) transition networks, production rules [Goldstein, 1979; Anderson & Reiser, 1985], etc.

The main system activities centred on this knowledge type are:

- providing the inferential tools for problem solving;
- providing the inferential tools for coaching a student in a problem-solving session.

The main advantage of separating the problem-solving knowledge from the domain knowledge is that it emphasizes the distinction between the domain itself and the skills used to solve a practical problem in that domain, thus simplifying the learning process. That knowledge separation into DK and PSK is common to all KH-domains; this is why we believe that KH-ITSs, which are aimed at helping the student to learn how to solve problems, should display the same knowledge separation.

Besides, we can use — we believe in a novel way — that separation between DK and PSK to define four *generic operating modes* in a KH-ITS, based on the type of knowledge involved (DK or PSK), and on who "generates" it (the system or the student).

- In *domain-presentation mode*, the student asks the system some information about a domain theoretical element, and the system reacts by transferring to the student the required information or knowledge. The knowledge involved in this category is always DK, system-generated.
- In *demonstration mode*, the student asks the system to solve a practical problem or to coach him while he solves a problem. In the first case, the problem typically comes from the student himself, whereas in the latter one the problem typically comes from the tutoring sys-

tem. In either case, the main level of knowledge involved is PSK, system-generated.

- In *domain-assessment mode*, the system prompts the student to develop a domain element, and the student thus expresses his understanding of that element. If judged necessary, the system may then intervene to correct that understanding. The knowledge involved in this mode is essentially DK, student-generated.
- Finally, in *exercising mode*, the system prompts the student to solve a practical problem. The student then solves it step by step, showing what he understands of the involved problem-solving knowledge and of the associated domain knowledge. If necessary, the system may decide to intervene in order to help him reach his goal or to correct it. The knowledge involved in this mode is naturally PSK, student-generated.

## 2.2 Application to cost engineering

In the particular case of cost engineering, various problem-solving activities can be identified:

1. identify the given problem data;
2. identify the expected result;
3. draw a temporal diagram to represent the relevant events;
4. apply a formula;
5. compare amounts located at the same date;
6. compare amounts located at different dates;
7. add amounts situated at the same date;
8. add amounts situated at different dates;
9. choose a reference date;
10. move an amount from one date to another;
11. collapse a series of periodic amounts into one single amount;
12. explode an amount into a series of periodic amounts;
13. etc.

As it turns out, many problem-solving tasks may be divided into smaller ones, letting us introduce the notion of *complexity levels* in these tasks. For example, the comparison of two amounts situated at different dates implies:

- first, choosing a reference date at which to make the comparison;
- then, moving either (or both) amount(s) from its present date to the reference date;
- finally, comparing the amounts now both located at the same reference date.

These subactivities (of types 9, 10, and 5 respectively in the sample list above) thus appear to be of a lower complexity level than the initial one (of type 6). However, it is interesting to note that, although activity 6 turns out to be more complex than activity 5 (the latter is part of the former), both are stated using the same *abstraction level*.

It may also happen that some lower-level activities can *only* appear as components of a higher-level one. For example, the activity "drawing a temporal diagram" (type 3 above) implies the following tasks, which can only be accomplished as part of that activity (hence their identification from 3a to 3d):

3a. draw a timeline to encompass all periods implied by the problem data;
3b. draw arrows representing the amounts involved in the problem data;
3c. if necessary, split an amount (or each amount in a series) to simplify the computations;
3d. qualitatively draw as a special arrow the expected result of the computations to be made.

## 3 Tutoring knowledge

We now briefly present the *tutoring knowledge* (TK) in order to help the reader to better apprehend the relationships of that knowledge with DK and PSK. This third type of knowledge contains all tutoring processes enclosed in the ITS. It is not directly related to the teaching domain or to problem solving, but helps the student to understand, assimilate, and master the knowledge included in DK and PSK [Gagné & Trudel, 1996].

The main system activities using TK are:

- ordering and formatting the topics to be presented to the student;
- monitoring a tutoring session, i.e., triggering the various tutoring processes according to the system tutoring goal and the student's actions; such monitoring may imply giving explanations, asking questions, changing to another type of interaction, etc.;
- in a KH-domain, while the student is solving an exercise, monitoring the student's problem-solving activities: understanding and assessing these activities, giving advice to correct or optimize them, giving hints or partly solving the exercise at hand (as required by the student or by the tutoring module), etc.;
- continuously analysing the student's progress in order to optimize the tutoring process.

The advantage of separating the tutoring knowledge from the knowledge of the domain to be taught has been emphasized long ago [Goldstein, 1977; Sleeman & Brown, 1982; Clancey, 1986; Wenger, 1987], and lies in the reusability of TK in various domains. In the case of KH-domains, the domain to be taught clearly encompasses both DK and PSK; indeed, the term "domain knowledge" applies to DK if referring to the knowledge type, and to DK + PSK if referring to the knowledge to be acquired. Therefore, as shown earlier, in a KH-ITS, knowledge ends up being separated into three categories rather than two.

We believe that the tutoring processes are triggered by tutoring goals which depend on the current educational setting and learning context. In the current state of our research, our assumption is therefore the following: the underlying hierarchy or hierarchies governing the way they interact with one another is not related to these processes *per se*, but rather to the goal to be attained when they are invoked. If that assumption turns out to hold, then the *dynamic structure of educational goals and subgoals* — which itself depends on the student's desires or abilities, the main underlying objective of the tutoring system, the student's state (e.g. of tiredness, etc.) and performance, etc. — will determine the succession of tutoring processes activated and tutoring interactions taking place. To our knowledge, the use of abstraction levels to induce a dynamic hierarchy of tutoring goals is new, as is the assumption that such a hierarchy will play a major role in activating the various tutoring processes and student–system interactions.

## 4 Educational interests of the abstraction and complexity levels

In the above sections, we have described a complexity- and abstraction-level approach to help model the three types of knowledge involved in a KH-ITS. In this section, we present the educational interests of our model. Sections 4.1 to 4.3 focus on the type of knowledge respectively presented in sections 1 to 3. Section 4.4 summarizes that discussion with some overall pedagogical interests of our approach.

### 4.1 Domain modelling

The factor hierarchy described in section 1 lets us derive an *order for the presentation of factors* to the student, from the lowest (simplest) level up to the highest, i.e. with increasing understanding complexity. That does not imply that such an order is unique, or even the best (e.g. a student's personal interests might make another order more motivating for him), but it is justified by our model. This presentation order may itself induce a possible *order for prerequisites*; e.g., if a student experiments difficulties to deal with $\Phi_{AP}$, has he well mastered $\Phi_{FP}$, a conceptually simpler factor?

Moreover, the intermediate abstraction levels will permit our ITS to exhibit a *sharper error modelling*. For example, the source of an understanding error concerning either relation in equation (2) or (3) is much easier to identify using the corresponding factor, either as a definition error or as a usage error, than an error concerning the global equation (1), where the definition and application relationships are not made explicit, and are therefore impossible to distinguish. Similarly, an error using a $\Phi_{AP}$ factor may be diagnosed as resulting from an insufficient mastery of the simpler factor $\Phi_{FP}$ as concept (which in turn will be diagnosed as related either to the definition, or to the usage of that concept).

Finally, in a first development stage, this modularity will ease our *defining the exercise types* to be implemented into the ITS, and will ease the tutor module task of *choosing the exercise type* to challenge the student with. Later, once the basic system is operational, that modularity will help us develop an automatic *exercise generator* dealing with the domain elements to be mastered by the student. That approach will then help the student to acquire a better critical mind about the relative importance of *know-how* knowledge vs. *domain* knowledge.

### 4.2 Problem-solving modelling

The problem-solving activities briefly presented in section 2 naturally display abstraction and complexity levels. Indeed, a standard cost engineering problem can be divided, possibly in more than one way, into major steps, which can then be split into simpler substeps. Such "relatively complex" cost engineering activities shown in section 2.2 were activity 3 (draw a temporal diagram) and activity 6 (compare amounts situated at different times). Using activities 5 and 6 as examples, we also made a distinction between the complexity level, based on the execution process, and the abstraction level, based on expressiveness or scope.

Such abstraction and complexity levels will let us introduce multiple levels of explanations, which will ease their tailoring, vary the answers to the student's questions, and adapt to the theoretical and practical reminders needed by the student.

Moreover, such an approach will lead the student to focus specifically on the activities for which he needs more tutoring, with the abstraction and complexity levels appropriate to his individual case.

That tutoring may take the form of explanations, guidance, hinting, even partially solving the exercise on which the student is currently working.

Table 2 — Characteristics of the four typical functioning modes of an ITS based on a problem-solving domain.

| Functioning mode | | Domain-presentation mode | Demonstration mode | Domain-assessment mode | Exercising mode |
|---|---|---|---|---|---|
| Main type of knowledge involved | | Domain knowledge | Problem-solving knowledge | Domain knowledge | Problem-solving knowledge |
| Student's main goal | | To learn | | To assess his learning | |
| Direction of the knowledge transfer | | System → Student | | Student → System | |
| Typical interaction | Trigger (start) | The student asks the system... | | The system prompts the student... | |
| | | some information about a domain theoretical element | to solve a practical problem or to coach him while he solves a problem | to develop a domain element | to solve a practical problem |
| | Knowledge exchange | The system presents... | | The student presents his view of... | |
| | | the requested element | a possible solution to the requested problem | the requested element | a possible solution to the given problem |
| | Result (closure) | The student expresses his understanding... | | The system assesses the student's answers, and possibly corrects them | |
| | | of the element | of the problem solution | | |

## 4.3 Tutoring modelling

As presented in section 2, the distinction between DK and PSK leads to four natural functioning modes: the domain-presentation mode, the demonstration mode, the domain-assessment mode, and the exercising mode. Their main characteristics are recalled in Table 2.

The abstraction levels of the tutoring goals aimed at by the system when invoking the tutoring processes is likely to result in a chain of recursive calls of these processes. This recursivity will or will not be direct, depending on the tutoring interaction types being chained: the system might freely decide to temporarily change between interaction types. However, the potential complexity of this chain is only apparent: because of the tutoring goal abstraction hierarchy, each newly invoked process will be called with a *narrower scope*, which naturally eliminates the risk of "forgetting" the initial tutoring goal or running into an infinite loop.

## 4.4 Overall interests of these abstraction and complexity levels

Abstraction levels are certainly not new. What we think is new is to use them in a systematic way to shed a uniformizing light on the system design and operation, and to make it more user-friendly once implemented.

First, they may help to better tailor the system tutorial interventions to fulfil the student's needs and the system tutoring goals, thus improving its conviviality and efficiency.

Then, all the capabilities presented above should result in smoother, more "natural", human-like interactions with the student. This improved ability to reproduce a human teacher's behaviour contributes again to make the system more user-friendly.

Finally, although that aspect is not in the scope of this paper, our refining of the three types of knowledge as described in sections 1 to 3 paves the way for the implementation of a structured error model, and eventually a student model.

# Conclusion

This presentation of a possible knowledge structure for KH-domains, which emphasizes the separation between domain knowledge and problem-solving knowledge, shows how a general functioning theory of such an ITS — namely the four functioning modes described in section 4 — can naturally be derived.

Moreover, the abstraction and complexity levels highlighted throughout this paper can be used as a common paradigm to help finding an appropriate representation for each knowledge type, and thus help creating more efficient ITSs. More generally, this paradigm can shed a uniformizing light on the system design, although it has never been used in a systematic way in the design or implementation of an ITS.

We thus hope to bring some contribution to the general and important problem of finding a generic architecture for intelligent tutoring systems.

# Bibliographic references

Anderson John R. (1986) *Cognitive Modelling and Intelligent Tutoring*. National Science Foundation (Washington, DC).

Anderson, John R. & B. J. Reiser (1985) "The LISP Tutor". *Byte* **10**, no. 4, p. 159–175.

Brodie, Michael L., John Mylopoulos & Joachim W. Schmidt, eds. (1984) *On Conceptual Modelling, Perspectives from Artificial Intelligence, Databases, and Programming Languages*. Springer Verlag (New York).

Brown, John S. & Richard R. Burton (1978) "Diagnostic models for procedural bugs in mathematical skills". *Cognitive Science* **2**, p. 155–192.

Clancey, William J. (1986) "Qualitative student models". *Annual Review of Computer Science* **1**, p. 381–450.

Clocksin, William F. & Christopher S. Mellish (1981) *Programming in Prolog*. Springer-Verlag (Berlin).

Díaz de Ilarraza Sánchez, Arantza & Isabel Fernández de Castro, eds. (1996) *Computer-Aided Learning and Instruction in Science and Engineering*, Proceedings of the *Third International Conference, CALISCE'96*, San Sebastian (España), 29–31 July 1996. LNCS 1108, Springer (Berlin).

Findler, Nicholas V., ed. (1979) *Associative Networks*. Academic Press (Orlando, FL).

Frasson, Claude, Gilles Gauthier & Alan Lesgold, eds. (1996) *Intelligent Tutoring Systems*, Proceedings of the *Third International Conference, ITS'96*, Montréal (Canada), 12-14 June 1996. LNCS 1086, Springer (Berlin).

Gagné, Denis & André Trudel (1996) "A highly flexible student-driven architecture for computer-based instruction". [Frasson & al., 1996], p. 66–74.

Goldstein, Ira P. (1977) *The Computer as Coach: an Athletic Paradigm for Intellectual Education*. AI Memo 389, AI Laboratory, Massachusetts Institute of Technology (Boston, MA).

Goldstein, Ira P. (1979) "The genetic epistemology of rule systems". *International Journal of Man–Machine Studies* **11**, no. 1, p. 51–77.

Kim, Won & Frederick H. Lochovsky, eds. (1989) *Object-Oriented Concepts, Databases, and Applications*. ACM Press, Addison-Wesley Publ. (Reading, MA).

Kowalski, R. (1979) *Logic for Problem Solving*. North-Holland (Berlin).

Lelouche, Ruddy & Jean-François Morin (1996) "The formula: a relation? Yes, but a concept too!". [Díaz de Ilarraza Sánchez & Fernández de Castro, 1996], p. 176–185.

Lenat, D., F. Hayes-Roth & P. Klahr (1979): *Cognitive Economy*. Working Paper HPP-79-15, Stanford Heuristic Programming Project. Stanford University (CA), June, 46 pages.

Malec, J. (1989) "Knowledge elicitation during dynamic scene description". *ACM–SIGART Newsletter: special issue on knowledge acquisition*, no. 108, April, p. 162–163.

Patel, Ashok & Kinshuk (1996) "Applied artificial intelligence for teaching numeric topics in engineering disciplines". [Díaz de Ilarraza Sánchez & Fernández de Castro, 1996], p. 132–140.

Sleeman, D. H. & John Seely Brown, eds. (1982) *Intelligent Tutoring Systems*. Academic Press (London).

Sowa, John F. (1984) *Conceptual Structures*. Addison-Wesley (Reading, MA).

Wenger, Étienne (1987) *Artificial Intelligence and Tutoring Systems*. Morgan Kaufmann (Los Altos, CA).

# Reasoning Symbolically About Partially Matched Cases

## Kevin D. Ashley and Vincent Aleven

Learning Research and Development Center, Graduate Program in Intelligent Systems Program, and School of Law
University of Pittsburgh
Pittsburgh, PA 15260 USA
+1 (412) 624-7496
ashley+@pitt.edu, aleven+@pitt.edu

## Abstract

In teaching case-based argumentation skills, the CATO program, an intelligent learning environment, guides students' assessments of partial matches between problems and cases by generating alternative interpretations of the similarities and differences. CATO's Factor Hierarchy captures information about the significance of similarities and differences given the normative purposes of the domain classification. Its algorithms for emphasizing or downplaying significance tailor interpretations to the comparison context, block interpretations strongly contradicted by other factors and strategically determine how and how abstractly to characterize a difference. An empirical evaluation confirmed CATO's effectiveness in teaching basic argumentation skills.

## 1 Introduction

Traditionally, researchers in case-based and analogical reasoning have regarded as an obstacle the fact that cases can be described at various levels of abstraction. Carbonell abandoned a transformational approach to analogy in favor of a derivational approach because of the problems of recognizing analogous problems despite apparent differences in the language or levels of abstraction of their descriptions [Carbonell, 1983]. Describing cases at different levels of abstraction also frustrates isomorphism in structure-mapping approaches to analogical reasoning. Forbus acknowledges the obstacle as a limitation in MAC/FAC. Either the cases all have to be described in the same way and at exactly the same levels of abstraction or else the mappings will fail [Forbus, et al. 1994, p. 198].

In argumentational or interpretive CBR applications, however, multilevel case descriptions are not just an inconvenient obstacle for computerizing analogical reasoning. They are an integral part of why and how experts reason with cases. Experts assess cases and similarity in terms of a general domain model, to the extent there is one. They fit new cases into the domain model and into the body of cases, grouping a new case with similar examples and reconciling it with negative examples. The level of abstraction with which to describe a case is an important degree of freedom in fitting it into a general model and reconciling it with other cases. The importance of similarities and differences varies as cases are characterized abstractly in different ways and levels; the general model helps experts assess the importance of shared and unshared case features viewed more abstractly or less. By choosing the groupings and characterizations carefully, an expert may even extend the general domain model.

This is certainly true of legal case-based reasoning in which the general principles and policies of a legal domain inform a determination of what case features and which analogical similarities and differences are important [Burton, 1985]. Since quantitative weighting schemes are either inappropriate for or not sufficiently context-sensitive to be applied in legal argument [Ashley & Rissland, 1988], one needs to reason explicitly in terms of some model of why similarities and differences are significant. In law, however, the general model of a domain is often weak enough, and there is enough uncertainty about how cases fit the model and each other, that characterizing cases at different levels of abstraction is often done strategically; the level of abstraction an arguer chooses in asserting that cases are the same or different reflects the arguer's purpose.

In designing a program, CATO, to teach first year law students basic skills of making case-based legal arguments to classify new cases, we have partially addressed three of four problems associated with comparing cases at multiple levels of abstraction, the problems of: (1) representing cases at multiple levels, (2) reasoning symbolically about partially matched cases in light of the possible abstract descriptions, and (3) modeling the strategic uses of case description. Students learn how to argue that the complaining party should win (or lose) its claim in a new case by drawing analogies to past cases where the corresponding side won (or lost) and by distinguishing or otherwise counteracting similar cases with the opposite result. CATO teaches students, among other things, how to characterize a problem abstractly in arguing that a particular side should win and how to characterize the significance of particular differences between the problem and past cases so as to emphasize or downplay that significance. The lessons should help us address the remaining problem (4): identifying and retrieving cases described at different levels of abstraction.

In making competing arguments that a partially matched case is close enough to a problem or not, CATO alternatively interprets the differences in terms of more abstract characterizations contained in its domain model, the Factor Hierarchy. This graph captures information about the plausible significance of similarities and differences given the domain's normative purposes. It enables the program to determine which abstractions help an argument, hurt the argument, or are indifferent. Although CATO's case representations do not include multiple levels of abstraction, CATO does compare cases at multiple levels of abstraction. Its algorithms (for downplaying and emphasizing differences and for making issue-based arguments) select, from among all the abstractions at multiple levels and along multiple paths which may apply to a case, just the right ones (i.e., the focal abstractions) for characterizing the case in various argument moves. CATO's algorithm for selecting focal abstractions implements strategic criteria so that the resulting arguments avoid contradictions and exposure to counterattacks. When there is conflicting evidence whether an abstraction applies, CATO applies its general knowledge in the Factor Hierarchy to resolve the conflict by blocking certain interpretations, if possible. If not, its case-based arguments play out the conflict.

As compared with either CASEY or GREBE, CATO employs an alternative approach and a different kind of model to guide interpretation of partial matches. CASEY reasons symbolically about differences between a problem and an explained, diagnosed case in terms of the strong causal diagnostic model used to generate the past case's explanation and evidence principles for characterizing whether differences are important. [Koton, 1988]. GREBE represents cases at multiple levels of abstraction in a structural sense. Each case's explanation is like a small structural model indicating which features are relevant to which conclusions. GREBE attempts to bridge gaps in the mapping of explanations by importing inferences from other case explanations and legal rules [Branting, 1991].

Neither CASEY nor GREBE have anything like a Factor Hierarchy that represents *how* features strengthen or weaken a conclusion and *why*, information which informs CATO's arguments emphasizing or downplaying differences. Although CATO's model cannot solve problems by itself as CASEY's can, that is to be expected given the differences between CASEY's medical domain and CATO's legal one. While GREBE may compare cases at multiple levels of structural abstraction, it does not make sense to ask whether it strategically selects a level of abstraction at which to characterize similarity or differences as CATO does in selecting focal abstractions. GREBE uses a different kind of model to make a different kind of argument.

## 2 CATO's Factor Hierarchy

Students use CATO to analyze argumentation problems in a traditional casebook chapter on trade secret misappropriation, a legal regime protecting confidential commercial information from unfair access and use by competitors. Students may retrieve cases from CATO's database of 147 trade secret cases. A special query language helps students formulate constraints that a case must satisfy to be useful in an argument and translate them into queries to retrieve relevant cases.

### 2.1 Factor Representation

In CATO, legal cases are represented by a short textual description of the facts and decision (i.e., the case squib) and a set of indexing factors. Factors represent stereotypical collections of facts which tend to strengthen or weaken a plaintiff's legal claim [Ashley, 1990]. CATO's representation is illustrated in Figure 1, showing a squib's description of the facts of the *MBL* case, and Figure 2, box 1 indicating the set of six factors indexing the case.

**Facts**: Chemi-Flex (a division of plaintiff) had developed, over a number of years, a molding process for manufacturing urethane belts. Defendant Diekman was a former employee. When he first worked for plaintiff, Diekman had signed an employer-employee agreement which contained a confidentiality and non-competition clause. However, when he was rehired in 1979, he refused to sign such an agreement. Following his second tenure with Chemi-Flex, defendant formed two companies and began to design centrifugal molding machines and designs for other belt-producing equipment. Plaintiff's employees were not aware of what information plaintiff considered to be confidential. Some employees, but not all, signed nondisclosure agreements. Plaintiff's plans, designs and customer data were not locked up; customer names and orders were not marked confidential; process formulas and machines were shown, without restriction, to employees, outside consultants and others; no licensing or confidentiality agreements were signed by outside parties with access to the process and formulas; and the entire process, formula and machinery were open to a team of engineers from a stockholder corporation of plaintiff. Customer names and specifications were on orders and requisitions located in various places in the plant. The use of plaintiff's process was known in the industry, although the process is not specifically discussed in any industry literature.

Figure 1: Squib's Facts for *MBL (USA) Corp. v. Diekman*

Typically, legal disputes involve some factual strengths for the plaintiff's claim, the pro-plaintiff factors, and some weaknesses, the pro-defendant factors. In *MBL*, some factors favor the plaintiff (p): the defendant had entered into nondisclosure and noncompetition agreements (at least, at one time), F4 and F13, evidence that plaintiff took at least one kind of measure to protect the security of its confidential information, F6. On the other hand, the agreement did not specifically indicate what information was regarded as secret, F5, some of the information had been disclosed to outsiders, F10, and in any event, some of the information was already known in the industry, F20.

In law, there is no algorithmic or statistical technique for combining strengths and weaknesses to come to a decision. Instead, lawyers make arguments by analogy to past cases presenting the same combinations of strengths and weaknesses, arguing for a similar result in the problem.

### 2.2 Constructing Arguments

CATO's recipe for constructing such an argument is illustrated in Figure 3, which shows on the right side a set of steps for justifying a favorable decision for a side on a legal issue associated with his claim. On the left is an outline of an argument in favor of plaintiff's trade secret misappropriation claim in the *MBL* case focusing on one of two issues: that the information is a trade secret. Pursuant to the recipe, each issue argument attempts to capitalize on the related factual strengths and overcome the weaknesses. Accomplishing each goal requires finding and citing appropriate cases satisfying certain constraints. As the capitalized text indicates, in order to emphasize the strengths and downplay the weaknesses, the arguer needs to find cases which satisfy the specified constraints.

| | |
|---|---|
| 1. *MBL* = F4 Agreed-Not-To-Disclose (p)<br>        * F5 Agreement-Not-Specific (d)<br>        = F6 Security-Measures (p)<br>        * F10 Secrets-Disclosed-Outsiders (d)<br>        = F13 Noncompetition-Agreement (p)<br>        * F20 Info-Known-To-Competitors (d) | 2. *Elcor* (p) = F4 Agreed-Not-To-Disclose (p)<br>              = F6 Security-Measures (p)<br>             = F13 Noncompetition-Agreement (p)<br>             * F15 Unique-Product (p)<br>             * F18 Identical-Products (p) |
| 3. *MBL* = F4 Agreed-Not-To-Disclose (p)<br>        * F5 Agreement-Not-Specific (d)<br>        = F6 Security-Measures (p)<br>        * F10 Secrets-Disclosed-Outsiders (d)<br>           F13 Noncompetition-Agreement (p)<br>        * F20 Info-Known-To-Competitors (d) | 4. *Sperry* (p) * F2 Bribe-Employee (p)<br>                F3 Employee-Sole-Developer (d)<br>             = F4 Agreed-Not-To-Disclose (p)<br>             = F6 Security-Measures (p)<br>             * F7 Brought-Tools (p)<br>             * F15 Unique-Product (p)<br>             * F18 Identical-Products (p)<br>             * F26 Deception (p) |
| = shared factor   * distinction | |

Figure 2: Factor Comparison of *MBL* Case with the *Elcor* Case (Boxes 1, 2) and with the *Sperry* Case (Boxes 3, 4)

CATO's Factor Hierarchy provides information relating factual strengths and weaknesses (i.e., factors) to those legal issues for which they are relevant. CATO uses this information in constructing its descriptions of the constraints on cases which would be useful in an argument (i.e., the capitalized text in Figure 3.) Excerpts of the Factor Hierarchy are shown in Figure 4. Base-level factors at the bottom of the Hierarchy are linked through intermediate legal concerns to high-level legal issues at the top. The nodes representing intermediate concerns and top-level issues are called "abstract factors". Each represents two opposite conclusions, one favoring plaintiffs and one favoring defendants. The links represent a (defeasible) support relation between the nodes. Links may be strong (thick) or weak (thin), indicating the level of support they provide, and are marked as to whether they support the same side as the primary conclusion of the parent node (+) or the opposing conclusion (-). CATO uses the link strength to determine whether to block certain inferences (discussed below.) Currently, the Factor Hierarchy contains 26 base-level factors for trade secret law, 16 abstract factors (5 of which are legal issues) and 50 links.

CATO's query language helps students translate argument constraints directly into queries for relevant cases from its database. For example, students could find cases relevant for plaintiff's argument on the trade secret issue using queries such as "(and F4 F6)" and "(or F4 F6) (or F10 F20)". Among the cases returned by the queries are the *Elcor* and *Sperry* cases. Figure 2 shows a factor comparison between the *MBL* case and each of the *Elcor* and *Sperry* cases which confirms that they satisfy the first query's constraints and thus are eligible to be used in the argument of Figure 3.

Deciding whether to include these cases in the argument necessitates reasoning about partial matches. As Figure 2 indicates, each case only partially matches *MBL*. Each case has numerous distinctions with respect to *MBL*. Distinctions are those unshared factors that tend to make the cited cases (*Elcor* or *Sperry*) stronger for plaintiff than *MBL*. In particular, if the case is employed in the argument, the opponent could respond to it by distinguishing it, that is, pointing out these relevant differences. Are the problem and cited case really the same or different? Are the distinctions between them really important? Do they warrant not including the cases in the argument?

## 3. Reasoning Techniques for Partial Matches

CATO's Factor Hierarchy and techniques for emphasizing or downplaying distinctions help students frame answers to these questions, and thus reason about partial matches, by generating examples of arguments to consider. Figure 5 shows several arguments CATO makes to emphasize or downplay a particular distinction (F15 Unique-Product (p)) between *MBL* (or versions thereof) and the *Elcor* and *Sperry* cases. In both *Elcor* and *Sperry*, the plaintiff's products were unique, no other competitors marketed products like that. Since pro-plaintiff F15 was not in *MBL*, the defendant could distinguish these cases, pointing out that *MBL* was correspondingly weaker for plaintiff. This distinction is, indeed, important. As CATO's arguments in boxes 1 and 3 of Figure 5 indicate, CATO can make strong arguments on behalf of defendant emphasizing the distinction but cannot make any argument for the plaintiff downplaying it.

In emphasizing a distinction, one has to find abstract interpretations to use as "focal points" for characterizing the two cases as very different. One interprets the distinction's significance in a case, shows factors in that case which support the interpretation (i.e., "corroborating factors"), and shows factors in the other case supporting a contrary characterization (i.e., "contrasting factors"). In downplaying a distinction, by contrast, the goal is to dismiss the distinction, finding an abstract interpretation to use as a focal point for characterizing the two cases as the same.

In CATO's algorithm for emphasizing a distinction, the key is selecting the focal point for emphasizing a distinction D in case C1 in favor of side S (either plaintiff or defendant.) A focal point is a 3-tuple comprising an abstract factor P to use as a focal abstraction for characterizing the distinction's significance, a set X of contrasting factors, and a set Y of corroborating factors. Set X comprises con-S factors in the other case C2. Set Y comprises pro-S factors in C1. P is the most specific common ancestor in the Factor Hierarchy of D and the factors in X and Y.

**Argument for Plaintiff in the MBL (USA) Corp. v. Diekman problem**

Plaintiff should win a claim of trade secrets misappropriation. Plaintiff's information is a trade secret [F101] and a confidential relationship existed between plaintiff and defendant [F114].

**Plaintiff's information is a trade secret [F101]**

Plaintiff's information is a trade secret [F101]. Restatement 1st of Torts s 757, and Comment b, factors 1-6 (1939). In the current fact situation, plaintiff and defendant entered into a nondisclosure agreement [F4] and plaintiff took security measures to protect the information [F6]. This shows that plaintiff took efforts to maintain the secrecy of its information [F102]. [TO EMPHASIZE STRENGTHS: NEED CASES WON BY PLAINTIFF WITH F4 AND/OR F6.]

The fact that plaintiff disclosed its product information to outsiders [F10] does not preclude a conclusion that plaintiff's information is a trade secret [F101], given that plaintiff took measures to keep its information secret [F6] and plaintiff and defendant entered into a nondisclosure agreement [F4]. The factual strengths favoring plaintiff warrant the conclusion that plaintiff's information is a trade secret [F101], even though plaintiff's information was known to competitors [F20]. [TO DOWNPLAY WEAKNESSES: NEED CASES WHERE PLAINTIFF WON IN SPITE OF F10 AND/OR F20, PREFERABLY CASES WITH F4 OR F6]

**A confidential relationship existed between plaintiff and defendant [F114]** ...

To justify a favorable decision on an issue:
- Point to strengths related to issue; say why they matter.
- Show cases in which these strengths led to favorable outcome.
- Discuss weaknesses related to issue. Point to strengths that may compensate.
- Show cases that had favorable outcome in spite of weaknesses.

Figure 3: Sample CATO Issue-based Argument (left) and Recipe for Making It (right).
(Factors and Abstract Factors are indicated in square brackets.)

**To emphasize distinction D of case C1 as compared to C2 [where D favors side S]:**

1. **Select candidate focal points to emphasize D:**
   A. Find all combinations of a focal abstraction P and a set of contrasting factors X in C2 such that P is a most specific common ancestor of D and the factors in X.
   B. Find all combinations of a focal abstraction P and a set of corroborating factors Y in C1 such that P is a most specific common ancestor of D and the factors in Y.
2. **Organize candidate focal points**
   A. Join candidate focal points with same focal abstraction P.
   B. Filter unsuitable factors and focal points. For each candidate:
   • Check if contrast exists between C1 and C2 for P's ancestors.
   • If candidate has no contrasting factors, check if closed-world assumption is appropriate that pro-S conclusion associated with P is absent in C2.
   • Remove contrasting or corroborating factors shared by C1 and C2. (Shared factors not useful to draw contrast.)
   C. Consolidate candidate focal points whose focal abstractions are ancestor and descendant.
   D. Order the focal points by estimated strength: (1) X is not empty. (2) Y contains other factors beside D. (3) The rest.
3. **Generate text for surviving focal points with templates.**

The algorithm (and the one for downplaying a distinction in [Aleven and Ashley, 1996]) supports three "smart" techniques for reasoning symbolically about the significance of similarities and differences: (1) Tailor interpretations to fit the context of comparison. (2) Block interpretations strongly contradicted by other factors. (3) Strategically interpret similarities and differences.

**Tailoring Interpretations**: CATO achieves a measure of context sensitivity in comparisons by virtue of the Factor Hierarchy's structure in which individual factors may relate to more than one abstract factor and by the Emphasize Distinction Algorithm's identifying abstract factors (P) as focal points which are the most specific common ancestors (msca) of the distinction to be emphasized (D), the contrasting factors, and the corroborating factors. (See [Kolodner, 1993] for other uses of msca.) By "context sensitivity", we mean knowing which similarities and differences are most salient in different circumstances and why: which should a reasoner focus upon, how should it characterize them, and which should it ignore. Although a normative concern may imply generally that a particular factual circumstance is significant, in the context of a particular problem and case, that assessment may be affected by the co-occurrence of other factual circumstances and other concerns, the arguer's rhetorical viewpoint and the dialectical role in which the arguer is engaged.

The effect of tailoring an interpretation of a difference may be seen in Figure 5. Each box shows an argument emphasizing the same distinction involving F15 between the *MBL* case (and two variations of *MBL*) and the *Elcor* and *Sperry* cases, respectively. The arguments focus on different interpretations of the distinction in terms of abstract factors F106, F120, or F104. Depending on the context, one or other of these abstract factors is the focal abstraction. For instance, in Box 1, CATO interprets F15's significance in *Elcor* (C1) in terms of abstract factor F106 as showing that in *Elcor*, plaintiff's information was not known outside its business, whereas in *MBL* (C2) it was. In Box 3, F15 is significant in *Sperry* (C1) because it suggests the defendant in *Sperry* got the information through improper means, unlike *MBL* (C2). In comparing *MBL* and *Elcor*, the algorithm finds that F106 is the most specific common ancestor of F15 and the contrasting factors in *MBL*, F10 and F20. In comparing *MBL* and *Sperry*, however, the algorithm finds that by selecting the more abstract F120, additional corroborating factors can be brought to bear in a broader contrast distinguishing *Sperry* as a case turning upon improper means, namely F2, F7, and F26. The decision, to adopt F120 as the interpretation, is made in the algorithm's consolidation step, 2C.

Other evidence of CATO's context sensitive interpretations is in Figure 5. Comparing Boxes 1 and 2 shows the effect on the *MBL/Elcor* comparison of making *MBL* stronger for plaintiff by adding F8, so that the defendant saved product development time or expense by accessing plaintiff's information. CATO now makes an argument downplaying the F15 distinction interpreting it in terms of F104, a new focal abstraction indicating that the information was valuable. Boxes 3 and 4 show the effect on the *MBL / Sperry* comparison of making *MBL* stronger for plaintiff by adding F7, so that the defendant brought plaintiff's product-development tools with him to the defendant. CATO again says the F15 distinction shows possible improper means in *Sperry*, but this time CATO downplays the distinction, arguing on the basis of F7 that the *MBL* hypo and *Sperry* are the same.

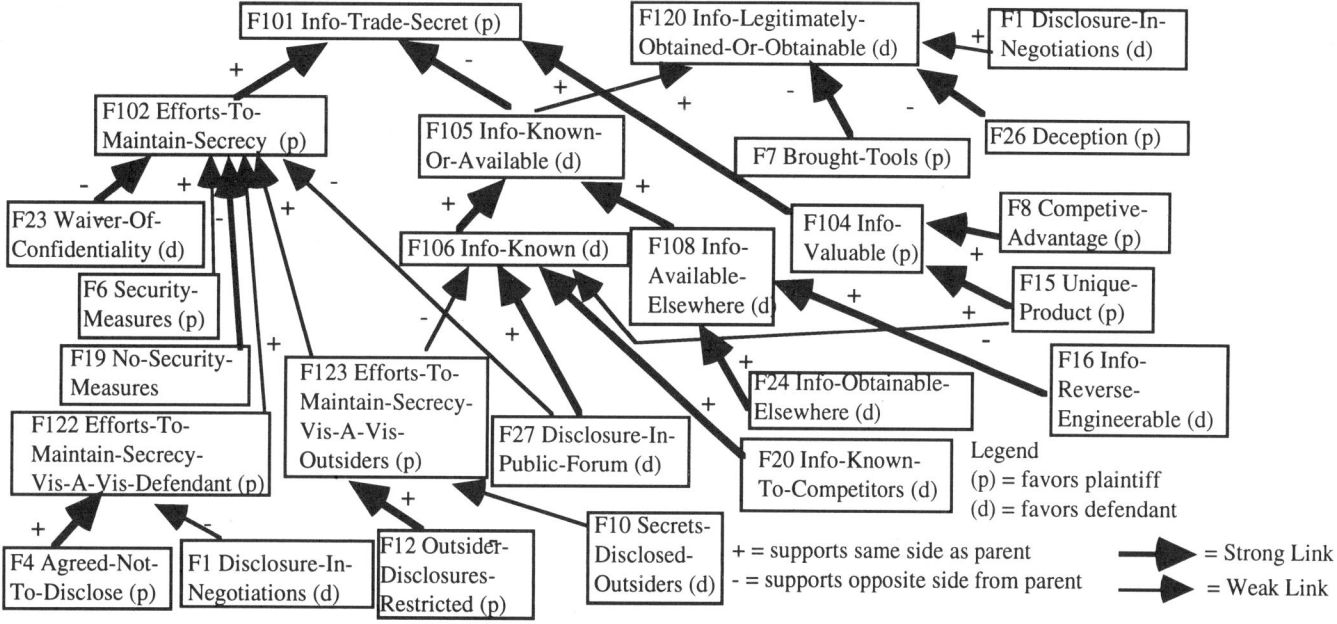

Figure 4: Excerpts from CATO's Factor Hierarchy

**Blocking Interpretations:** Link strengths in the Factor Hierarchy are designed to block certain interpretations in appropriate circumstances. The Hierarchy will characterize a similarity or difference in terms of only those abstractions not strongly contradicted by other factors. Evidence of blocking can be seen in Figure 5, Box 4, involving the *MBL* problem as modified by adding F7. As a result of blocking, CATO, arguing for defendant, does not characterize the *MBL* factors as evidence that defendant's means were not improper, as it does in Box 3. Where defendant brought plaintiff's product development tools, one does not want to argue that no improper means were used (F120) solely on the basis of plaintiff's disclosures to outsiders (F10) and competitors' having knowledge of the information (F20). F7 is itself evidence of improper means. F7 should block the interpretation that F10 and F20 are evidence of no improper means. In the Factor Hierarchy, Figure 4, a strong negative link from F7 to F120 blocks the weak positive link from F105 through which pro-defendant factors F10 and F20 in *MBL* are related to F120.

**Strategic Interpretation:** CATO's algorithm for emphasizing distinctions (and for downplaying them) takes strategic considerations into account in deciding which abstraction path to follow in the Factor Hierarchy and how abstractly to characterize a distinguishing factor. One wants to characterize a distinction broadly enough to take advantage of all relevant contrasting and corroborating factors. One also wants to focus on distinctions that hold up at more abstract levels and avoid characteriziations so broad as to destroy the contrast or allow obvious objections.

An example of a strategic choice using the Factor Hierarchy is choosing a narrower interpretation of a distinction where a broader interpretation does not lead to a contrast, as shown by comparing Boxes 3 and 4 in Figure 5. After hypothetically modifying the *MBL* problem so that the defendant took product development information (F7), CATO can still emphasize the F15 distinction but this time does not focus on the improper means aspect (F120), as it did before in the *MBL/Sperry* comparison (Box 3). This would not be an effective way to draw a contrast, since in the modified problem there is also evidence of improper means (as CATO now points out in the new argument downplaying F15). Instead, CATO again interprets F15 in terms of F106, an abstract factor "lower down" in the Factor Hierarchy (see Figure 4). CATO eschews a broader interpretation in favor of a narrower one that more clearly contrasts with the other case, even though that means certain corroborating factors in the broader interpretation cannot be used.

How does CATO use strategic considerations to decide which path to use and how far up or down the Factor Hierarchy to go in interpreting a distinction? The answer lies in heuristic policies embodied in CATO's algorithm for emphasizing a distinction. The policies ensure that a sufficient contrast between the cases exists, with respect to both the focal abstraction itself (steps 1.A and 1.B) and its ancestors in the Hierarchy (step 2.B).

As a result, CATO decides not to use F120 in Box 4 for emphasizing the distinction, because there is no suitable contrast with respect to this abstract factor. F120 must either not be supported for plaintiff in *MBL*, or supported for defendant as well. However, it is supported for plaintiff in *MBL* (by F7), but not supported for defendant. As noted above, pro-plaintiff factor F7 in *MBL* blocks the pro-defendant inference from F10 and F20 to F120. Abstract factor F106, however, does satisfy the requirements of both heuristic policies. The two cases have opposing factors related to F106 itself (pro-plaintiff F15 in *Sperry*, pro-defendant F10 and F20 in *MBL*). Also, they contrast suitably with respect to ancestors of F106: The path from F106 to F105 to F101 does not contain abstract factors supported for plaintiff in *MBL*, satisfying the second heuristic policy.

| 1. *MBL* / *Elcor* / F15 Unique-Product (p) <br> **Plaintiff's argument downplaying distinction F15 in *Elcor*:** <br> None. <br> **Defendant's argument emphasizing distinction F15 in *Elcor*:** <br> In *Elcor*, plaintiff was the only manufacturer making the product [F15]. This was not so in *MBL*. This is a marked distinction. It shows that in *Elcor*, the information apparently was not known outside plaintiff's business [F106], whereas in *MBL*, plaintiff's information was known outside plaintiff's business [F106]: Plaintiff disclosed its product information to outsiders [F10] and plaintiff's information was generally known in the industry [F20]. | 2. *MBL* + F8 Competitive-Advantage (p) / *Elcor* / F15 Unique-Product (p) <br> **Plaintiff's argument downplaying distinction F15 in *Elcor*.** <br> In *Elcor*, plaintiff was the only manufacturer making the product [F15]. This was not so in *MBL*. However, this does not amount to an important distinction. In *MBL*, defendant's access to plaintiff's information enabled it to develop its product in less time or at lower cost [F8]. It follows that in both cases, plaintiff's information was valuable for plaintiff's business [F104]. <br> **Defendant's argument emphasizing distinction F15 in *Elcor*:** <br> (Same as in Box 1) |
| --- | --- |
| 3. *MBL* / *SPERRY* / F15 Unique-Product (p) <br> **Plaintiff's argument downplaying distinction F15 in *Sperry*:** <br> None. <br> **Defendant's argument emphasizing distinction F15 in *Sperry*.** <br> In *Sperry*, plaintiff was the only manufacturer making the product [F15]. This was not so in *MBL*. This is a clear and striking distinction. It shows that in *Sperry*, defendant may have acquired plaintiff's information through improper means [F120]. Other facts in *Sperry* also support this: Defendant offered a salary increase or bonus to induce plaintiff's employee to work for them and bring plaintiff's trade secrets [F2], plaintiff's former employee brought product development information to defendant [F7], and defendant obtained plaintiff's information through deception [F26]. By contrast, in *MBL*, defendant obtained or could have obtained its information by legitimate means [F120]: Plaintiff made disclosures to others outside its business [F10] and plaintiff's information was generally known in the industry [F20]. | 4. *MBL* + F7 Brought-Tools (p) / *Sperry* / F15 Unique-Product (p) <br> **Plaintiff's argument downplaying distinction F15 in *Sperry*.** <br> In *Sperry*, plaintiff was the only manufacturer making the product [F15]. This was not so in *MBL*. However, this difference is not significant. In *MBL*, plaintiff's former employee took documents, blueprints, or tools to defendant [F7]. In both cases, therefore, defendant may have acquired plaintiff's information through improper means [F120]. <br> **Defendant's argument emphasizing distinction F15 in *Sperry*.** <br> In *Sperry*, plaintiff was the only manufacturer making the product [F15]. This was not so in *MBL*. This distinction is highly significant. It shows that in *Sperry*, the information apparently was not known outside plaintiff's business [F106]. In *MBL*, by contrast, plaintiff's information was known outside plaintiff's business [F106]: Plaintiff made disclosures to others outside its business [F10] and plaintiff's information was known to competitors [F20]. |

Figure 5: CATO's Arguments Emphasizing / Downplaying Distinctions

## 4 Evaluation

An experiment with 30 first-year law students compared CATO's instructional effectiveness to that of an accomplished legal writing teacher teaching the same material in a traditional way [Aleven & Ashley, 1997]. The subjects were randomly assigned to experimental and control groups. All subjects read a traditional trade secret law casebook chapter.

The control group attended six classroom session of 50 minutes in small groups of about eight students. The human instructor discussed the casebook cases and presented a framework for analyzing trade secret problems. During two sessions, students made oral arguments about two problems in a moot court setting in which the instructor played the role of "judge". Students prepared for the moot court sessions in two 75 minute practice sessions outside of class.

In the experimental group's nine 50-minute sessions, students learned to use CATO's tools to address the argumentation problems at the end of each section of the casebook following instructions in four workbooks. Students worked with CATO in pairs. After sessions introducing students to CATO's tools and to factors and their interpretations, students used CATO to analyze problems in terms of factors, retrieved cases to test certain hypotheses about the importance of certain factors, and then worked with a legal dispute much like the *MBL* case. They determined which factors apply, retrieved and compared cases as in Figure 2, practiced argument moves with the problem like those in the argument recipe of Figure 3 and in emphasizing and downplaying distinctions as in Figure 5, organized and wrote a multi-case argument and compared it with CATO's argument (as in Figure 3 but incorporating actual cases.)

All subjects took a pre-test and post-test of Basic Argument Skills. We also administered a more advanced take-home post-test involving a memo-writing assignment, considerably beyond the sophistication of the CATO instruction, using a previous semester writing assignment grade as a control. The legal writing instructor graded the exams in a blind test.

|  | Basic Argument Skills | | | | Memo Writing | | |
| --- | --- | --- | --- | --- | --- | --- | --- |
|  | Pre-Test | | Post-Test | | Prev. | Post-Test | |
| Exper. Gr. Avg. | 60 | C- | 70 | C+ | 63 | 70 | B- |
| Control Gr. Avg. | 55 | D | 68 | C | 63 | 79 | B+ |

Table 1: Pre-test and post-test scores (maximum is 100)

As shown in Table 1, on the Basic Argument Skills tests, in both groups the improvement from pre- to post-test was statistically significant (t-test, p < .05). There was no significant difference between the two groups, and their improvement scores (i.e., post-test score minus pre-test score) did not differ significantly between the two groups (t-test, p > .05). On the memo-writing assignment however, the control group did better, and the difference was statistically significant. The previous semester's assignment showed no statistically significant difference.

## 5 Discussion and Conclusion

CATO instruction lead to a significant improvement in students' basic argumentation skills, comparable to that achieved by a legal writing instructor teaching small groups. The subjects were all from a special program for students judged most in need of individualized attention, and the human instructor, the director of that program, was experienced and successful in teaching such students.

The memo-writing post-test shows a limitation on CATO's efficacy. While both groups learned basic argumentation skills, the control group integrated these skills better in a complex assignment. The human taught skills indirectly and holistically by engaging students in oral arguments without focusing on elements of argumentation. By contrast, CATO taught students basic elements with examples gradually combining more elements. CATO's exercises to help students integrate the components came at the end; few students had enough time to practice them. As a result, we need additional time and techniques for integration. For example, with an LCD projector, instructors could demonstrate complex arguments and techniques with CATO in class which students could then practice with CATO at home.

CATO tackles some general problems of abstraction in case description. It illustrates, and the evaluation lends some support to the utility of, three techniques in comparing partially matched cases at multiple levels of abstraction, the use of: (1) factors to represent relevant similarities and differences, (2) a Factor Hierarchy to represent the significance of factors abstractly in terms of the purposes of the classification, and (3) a set of criteria for selecting focal abstractions. Implemented in CATO's algorithms for emphasizing and downplaying differences and for making issue-based arguments, these criteria enable CATO to select the right path and level of abstraction for characterizing a case.

CATO's approach helps frame the remaining problems of representing, recognizing and retrieving cases described at different levels of abstraction. Given the range of choice in characterizing a case abstractly, and the underlying strategic considerations, any multilevel case representation must preserve flexibility of interpretation. Multilevel descriptions also complicate the problem of evaluating which candidate cases are best. The value of a case lies in its uses in arguments. Multilevel characterizations expand the ways a case may be used in arguments and responses. Those possible uses need to be folded into the retrieval process. BankXX's heuristically-guided argument search employing complex argument evaluation functions [Rissland, et al. 1996], or factoring adaptation cost into retrieval [Smyth and Keane, 1994] may provide models. In an educational context like CATO's, however, the uses of retrieved cases in arguments need to be played out explicitly for students to understand and learn from them.

## References

[Aleven & Ashley, 1997] V. Aleven and K. Ashley "Teaching Case-Based Argumentation through a Model and Examples: Empirical Evaluation of an Intelligent Learning Environment" To appear in *Proc. AI in Education*. Kobe, Japan. August.

[Aleven and Ashley, 1996] V. Aleven and K. Ashley. How Different is Different? Arguing About the Significance of Similarities and Differences. In Smith and Faltings, ed. *Advances in CBR*. Lecture Notes in AI 1168. pp. 1-15. Springer: Berlin.

[Ashley, 1990] K. Ashley, *Modeling Legal Argument, Reasoning with Cases and Hypotheticals*. MIT Press: Cambridge.

[Ashley and Rissland, 1988]. K. Ashley and E. Rissland. Waiting on Weighting: A Symbolic Least Commitment Approach. In *Proc., AAAI-88*. pp. 239-244. Morgan Kaufmann: San Mateo, CA.

[Branting, 1991]. Branting, L. K. *Integrating Rules and Precedents for Classification and Explanation: Automating Legal Analysis*. Ph.D. Dissertation. AI Lab., Tech. Rep. AI90--146. The University of Texas at Austin.

[Burton, 1985] Steven J. Burton. *An Introduction to Law and Legal Reasoning*. Little, Brown: Boston.

[Carbonell, 1983] Carbonell, J. G. "Derivational Analogy and Its Role in Problem Solving". In *Proc. AAAI83*. pp. 64-69. William Kaufmann: Los Altos, CA.

[Forbus, et al. 1994] K. Forbus, D. Gentner, and K. Law "MAC/FAC: A Model of Similarity-based Retrieval", *Cognitive Science* 19, 141-205.

[Kolodner, 1993] J. Kolodner. *Case-Based Reasoning*. Morgan Kaufmann: San Mateo, CA.

[Koton, 1988] P. Koton. *Using Experience in Learning and Problem Solving*. Ph.D. Thesis, MIT.

[Rissland, et al. 1996] E. Rissland, D. Skalak, and M. Friedman. "BankXX: Supporting Legal Arguments through Heuristic Retrieval". *AI and Law*. V. 4 No. 1. pp. 1-71. Kluwer: Dordrecht.

[Smyth & Keane, 1994] Smyth, B. and Keane, M. "Retrieving Adaptable Cases: The Role of Adaptation Knowledge in Case Retrieval". In *Topics in Case-Based Reasoning*. S. Wess, K-D. Althoff, M. Richter, eds. pp. 209-220 Lecture Notes in AI series. Springer: Berlin.

# Task Ontology Makes It Easier To Use Authoring Tools

## Mitsuru Ikeda  Kazuhisa Seta  Riichiro Mizoguchi
The Institute of Scientific and Industrial Research
Osaka University
8-1, Mihogaoka, Ibaraki, Osaka, 567,
Japan

## Abstract

The main purpose of this paper is to illustrate the characteristics of ontology-based authoring tools for Computer Based Training (CBT) systems. It has two major advantages as follows. (A) It provides human-friendly primitives in terms of which users can easily describe their own model of a task(descriptiveness, readability). (B) It can simulate the abstract behavior of the model in terms of conceptual level primitives (conceptual level operationality). In this paper, we will discuss the basic issues on the concept of task ontology and then describe the design principle of an ontology-based authoring tool for Computer Based Training (CBT) systems.

## 1. Introduction

Recently much attention has been paid to the notion of "ontology" in the expectation that it can serve as the new, strong foundation of knowledge engineering. In the conventional approach to theory of knowledge, to give the operational semantics of knowledge representation has been regarded as of major importance and the analysis of contents of knowledge has been considered to be subordinate to it. To solidify the foundation of knowledge engineering, however, many researchers, especially in the field of knowledge sharing and reuse, has strongly felt necessity of the change of such a way of thinking. The key to the problem is to understand the essential interaction between "form" and "contents" on equal importance. This implies that deep understanding of "content" will give us new insight into design of knowledge representation. The notion of "ontology" can be key to this issue.

The ultimate goal of research on ontology is to give the full picture of theory of knowledge. To make improvements in the study of this difficult issue, of course, it is important to accumulate huge amount of "contents", and develop sophisticated ontology representation language as fundamental "form" of knowledge.

The same thing applies to the field of intelligent educational systems (IES). Building an IES requires a lot of work. At the present situation, however, it is always built from scratch. Few functional components are reusable and we cannot compare or assess the existing systems. Only existing contribution to the solution of the problem can be found in study of the authoring tools for educational systems. However, it is considered questionable whether substantial benefit for the authors engaged in the complex task may be expected or not, since most of existing authoring tools do not satisfy the requirements for the authoring tools as shown below.

- To provide human-friendly primitives in terms of which authors can easily describe define their own skeleton of IES.
- To give appropriate guidance to authors based on the established principle of the educational task by checking the rationality of the skeleton of IES.
- To show the dynamic behavior of the IES in conceptual level by which the authors can examine its validity.

We think the key to the solution of the problem is intelligent support based on "task ontology" which serves as a theory of vocabulary/concepts used as building blocks for knowledge-based systems. The issues here also include how to represent what we know about the fundamental characteristics of an IES as "task ontology" and how to integrate it into intelligent authoring tools. Our solution is integration of an ontology construction environment CLEPE as a part of the authoring tool we have developed. CLEPE provides us with all the functions needed to satisfy the requirements shown above.

The most important role of CLEPE is to lay the theoretical foundation for IES development process. It maintains continuity from author's conceptual understanding of educational task to the computational semantics of IESs. It provides human friendly vocabulary for authors to describe the educational task. For the authoring tools, on the other hand, it specifies the computational semantics of vocabulary and also provides a set of components represented in terms of both conceptual primitives and object-oriented code fragments.

The goals of our research on task ontology are to exemplify the benefits of task ontology through the development of an ontology based authoring tool for Computer Based Training (CBT) systems. In this paper, we will discuss the basic issues on the concept of task ontology and then describe the design principle of an ontology-based authoring tool for Computer Based Training (CBT) systems.

## 2. What is task ontology

Now let us go into the detail of task ontology. Generally ontology is composed of two parts, that is, taxonomy and axioms. Taxonomy is a hierarchical system of concepts and axioms are rules, principles, or constraints among the concepts. From the viewpoint of the ontology use, axioms specify the competence of an ontology. In other words, a class of the questions to which the answers can be derived from the axiom specify the competence of the ontology.

Following the analogy to natural language processing, we can easily understand the role of task ontology as a system of semantic features to represent the meaning of the problem solving description. The advantages of the integration of task ontology is as follows:
  A. Task ontology provides human-friendly primitives in terms of which users can easily describe their own task (descriptiveness, readability).
  B. The system can simulate the problem solving processes at the conceptual level and show users the execution process in terms of conceptual level primitives (conceptual level operationality).
  C. The system makes the task description runnable by translating it into symbol level code (symbol level operationality).

For the moment, it may be useful to look more closely at the functional feature of task ontology. Here, let us introduce three models M(A), M(B), and M(C), which embody the functions A, B, and C listed above, respectively. According to the analogy of natural language again, M(A) corresponds to sentences of natural language, M(B) is an internal model of intended meaning represented by the sentences, and M(C) has a capability to simulate the dynamic, concrete story implied by the sentences.

From now on, M(A), M(B) and M(C) are called "lexical level model", "conceptual level model", and "symbol level model", respectively. Lexical level model mainly deals with the syntactic aspect of the problem solving description, and conceptual level model captures conceptual level meaning of the description. Symbol level model corresponds to runnable program and specifies the computational semantics of the problem solving. Ontology at the top layer is called lexical level ontology in terms of which M(A) is represented. Ontology at the middle layer is called conceptual level ontology which specifies computational semantics of M(B). Lexical level ontology specifies the language in terms of which users externalize their own knowledge of the target task, while conceptual level ontology is an ontology which represents the contents of knowledge in their minds.

Figure 1 shows a taxonomy of terms included in lexical level ontology. Because of the space limitation, here we just show a part of the CBT ontology which we are building now (see [Mizoguchi et al., 1996] for a more detailed account of IES ontology). All the terms of lexical level ontology are organized into word classes, such as, generic verb, generic noun, etc. :
  - Generic nouns[N] representing objects reflecting their roles appearing in the problem solving process,
  - Generic verbs[V] representing unit activities appearing in the problem solving process,
  - Generic adjectives modifying the objects.

Task ontology for computer-based training, for example, looks as follows:
  Nouns: "Problem", "Scenario", "Answer", "Example", "Accident", "Operation", "Hint", etc.
  Verbs: "Provide", "Show", "Ask", "Simulate", "Give", etc.
  Adjectives: "Unsolved", "Easy", "Correct", "Counter" etc.
Verbs are defined as a set of procedures representing its operational meaning. So, they collectively serve as a set of reusable components for building IESs.

In the conceptual level ontology, the concepts to represent our perception of problem solving are organized into generic concepts, such as, activity, object, status, and so on. On top of that, a collection of symbol level CLOS code fragments are organized in symbol level ontology. There are some prerequisite relations among these three levels. Intuitively generic verb, generic noun, and generic adjective in lexical level ontology correspond to activity, object, and status in the conceptual level one, respectively. For each activity, at least one code fragment is prepared in the symbol level.

Thus, task ontology provides primitives in terms of which authors can describe their own educational task model. Using their own model, authors can easily put a piece of teaching material into the appropriate educational context, since it provides authors with abstract educational roles of various objects which could be easily instantiated to domain-specific objects. In the above examples, "problems", "examples", and "hints" represent such objects with educational roles.

One of the most important characteristics of task ontology is that meanings of verbs are also defined at the symbol level, that is, at least one executable code is associated with each verb to enable semiautomatic generation of runnable educational application.

Thus, roles of IES task ontology can be summarized as follows:
(1) To provide vocabulary/concepts in terms of which one can compare and assess existing IESs.
(2) To formalize training tasks.
(3) To specify the tutoring/training context which contributes to making it easy to put domain knowledge into a right context, since it provides us with abstract roles of various objects which could be instantiated to domain-specific objects.
(4) To provide reusable components for IES design and development.
(5) To enable translation of the knowledge-level description of the skeleton of IES into symbol-level executable code.

## 3. An Authoring tool based on CBT task Ontology.

Authors' work of IES systems can be roughly sketched out in the following three stages.
1. Design the skeleton of the CBT process of interest.
2. Compile training material or training aids, for example, notes, pictures, sounds, simulator, etc. into the skeleton.
3. Adjust the control of CBT process in detail.

Of course, the distinction among the stages is not necessarily important for authors. Rather strict distinction sometimes becomes disadvantage of authoring tools. The reason for making the distinction is to focus our discussion to the author's work which enjoy the ontology support.

The existing authoring tools in the literature [Inui et al, 1990], [Major, 1995], [Murray and Woolf, 1992] seems to cover all the three stages. Especially, some authoring tools for ITSs( for example, [Van Marcke and Vedelaar, 1995] ) provide sophisticated set of descriptive primitives and corresponding computational components of IES for authors. When looking carefully at the support functions provided by them, however,

```
Tutoring Actions                     Explanation                                      a known schema[N]
    Help[V]                              Deep/Shallow knowledge level[N]              Creating a new problem solving
        Answer (+ to the questions/requests)[V]  Knowledge necessary for solving          schema[N]
        Help (+learner to overcome               the problem[N]                   Knowledge state
impasses)[V]                             Prerequisite knowledge[N]                    Coverage of concepts/skills mastered[N]
    Getting learners motivated[V]        Simulation result[N]                         Degree of mastery[N]
        Encourage[V]                     Hints                                        Belief[N]
        Compliment[V]                        Deductive hints                          Connection strength of relations[N]
    Exercise                                     Verification process[N]              Location of bugs
        Select (+problems) [V]                   Intermediate solution[N]                 Conditions/Premises/Factors/etc.
        Give (+problems) [V]                     Abstract examples[N]                     Actions/Conclusions/Hypotheses/etc.
    Guide                                    Inductive hints                              Order of rules[N]
        Show/Prvesent[V]                         Kinds of bugs[N]                     Types of bugs
        Explain [V]                              Counter examples[N]                      Lack of concepts/relations/rules[N]
        Simulate [V]                             Extreme examples    [N]                  Insertion of concepts/relations[N]
        Suggest(+Hints[N]) [V]           Learner's state                                  Incorrect concepts/relations[N]
        Asking [V]                           Phase in learning process                    Incorrect order of operations[N]
    Evaluate/assess[V]                       Phase of concepts learning           Teaching material knowledge
Objects of Training task                         Learning of new concepts[N]          Domain knowledge
    Scenario[N]                                  In-depth understanding of                Nodes(category)
        process-oriented scenario[N]                 the new concepts[N]                      Concepts[N]
        topic-oriented scenario[N]               Assimilation of the concept[N]               Rules/Problem solving schema[N]
    Problems                                 Phase of problem solving learning            Principles[N]
        Similar problems[N]                      basic schema[N]                          Strategy[N]
        Easier problems[N]                   Learning advanced schema[N]              ....
        More difficult problems[N]           Learning solving a problem by using
```

Figure 1. Lexical level CBT task ontology (Partial)

we found that little effort has been done to give active guidance based on the explicit first principle on education. The reason of this could be lack of explicit "ontology" as design rationale of IESs. To embody active support of authoring process, the authoring tool should know the rationale in the form of meta description of the design.

Based on the above consideration, we have tried to build IES ontology. Although the step we have made is not large, we believe that its implications to the future research in IESs area is not small. CBT ontology we discuss here uses the IES ontology as its core. We organize a set of concepts specific to training system under the IES ontology. In the following sections, we try to show the significance of the CBT ontology. First, we present brief overview of *SmartTrainer* which have been developed as a prototype of Intelligent CBT system, then we will show the image of the authoring tool based on the CBT ontology built based on the analysis of *SmartTrainer*'s task.

### 3.1 *SmartTrainer*

*SmartTrainer* is a CBT System in the area of electric power systems. The target task of *SmartTrainer* is mainly to recover the accidents of substations in the electric power systems. When an accident happens, the electric power transmission will be interrupted, and the operators should recover it as quickly as possible. The operators should find the spot of the accident, continue to supply the electric power to some special places such as hospital, police station at once by borrowing some power from the other substations, find the causes of the accident and recover it within the limited time.

The goal of the training conducted by *SmartTrainer* is to improve capability of not only skill-based or rule-based reasoning but also knowledge-based reasoning. The set of the scenarios incorporated into *SmartTrainer* has been designed by the experienced trainers. In order to let the trainee master the principled knowledge, *SmartTrainer* let them do practice first and then teach them the first principle behind it adaptively to their mistakes, and finally, check their learning result by practice again. This is a form of "learning by doing."

*SmartTrainer* is basically composed of three models, that is, training process model, teaching materials model, learner model, which are part of our CBT task ontology.

### 3.2 An Ontology-based Authoring Tool

Among three stages of authors' work listed above, we have concentrated on the first one, that is, design process of the skeleton of CBT process. Hereinafter, we sometimes call the skeleton of the CBT as the *model* of a CBT system to emphasize the importance of existence the ontology which governs it. Therefore, the term *model* implies two ideas, one is that it has a set of general axioms to follow and the other is, in an ordinary sense, it abstracts the skeleton from the real CBT systems.

Basic teaching material prepared by the authors of CBT are the following

A) knowledge needed for target operation,

B) training scenario, and

C) simulator of a target system.

The authors compile these three items into the skeleton of a CBT system at the first stage. The typical work of the compilation is to put a fragment of knowledge(A) into the target operation context(B) appropriate for training. The most desired thing for authors is the established guideline on the compilation task, for example, the fundamental structure of the scenario, the typical style of questions, appropriate timing and method of learner modeling, and so on. As we discussed in the previous section, we have defined a set of terms to express the training task of the electric power system operation in the lexical level task ontology and the meaning of the terms in conceptual level task ontology. The author's task is to build a model of the intended training task in terms of the lexical level task ontology as a result of the compilation. Such terms and concepts are provided by an ontology maintenance system, which maintains the consistency between the task ontology

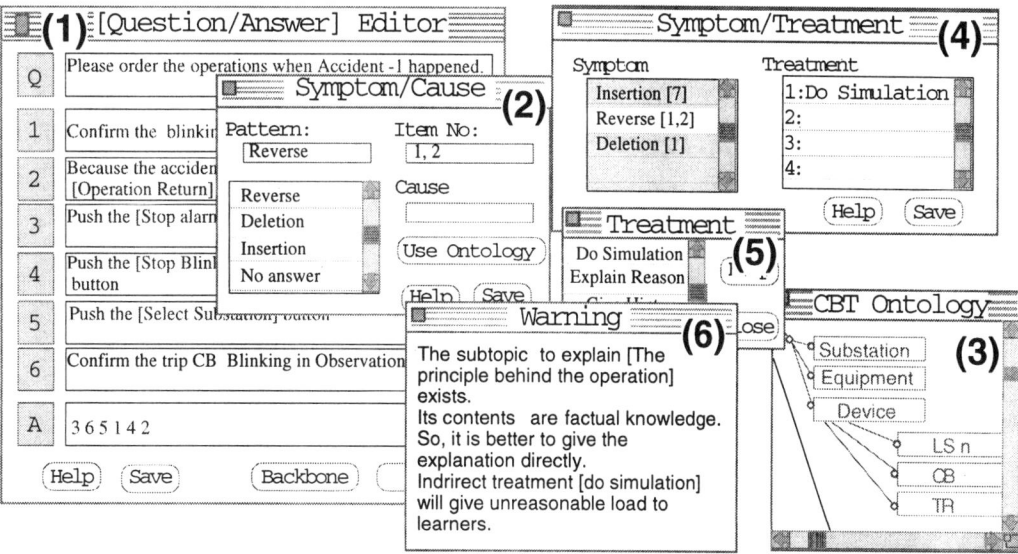

Figure 2 Interface of an Ontology based Authoring Tool

and the model built based on the ontology. As a part of our ontology engineering project, an ontology maintenance environment, called CLEPE, has been developed. Further details of CLEPE will be described later.

On the top of CLEPE, we have been developing an authoring tool for CBT systems. Figure 2 shows the image of the authoring tool interface. The interface is designed in full consideration of the intuitively clear way to express the skeleton of training tasks. In other words, the structure of the interface windows reflects the training task ontology and acts as the guideline on the author's task.

Figure 2 shows the snapshot of interface of the authoring tools we have developed. When the authoring tool is activated, the author is asked to create a training scenario which is a series of question and answer sessions. The window (1) is for editing a question/answer session. Before the window appears, the author was asked to select one from several question/answer types defined by CBT task ontology. In this case, the author selected "ordering question." And then he/she inputs the components of the question/answer, that is, "question", "items to be ordered" and "correct answer." After inputting the question, the author edits a set of learner's errors, called "symptom" in the window (2). The author inputs the possible symptoms as one of "reversing", "missing", "superfluity", which are defined as subcategories of the symptom for ordering question in CBT task ontology. In addition, the author is asked to clarify the cause of the symptom by selecting one node from the domain knowledge shown in the window (3). Since *SmartTrainer* adopts overlay learner model, the cause is formalized as missing of the selected node. Although we assume that domain knowledge is prepared in advance, the author can modify it at anytime when needed. Once the author complete inputting a symptom, the symptom/treatment window (4) will appear. In this window he/she specifies the treatment for the symptom. The possible treatments are listed in the window (5) based on tutoring strategy concept defined in CBT task ontology. The treatment selected by the author is "let him/her do operation". However, the selection violates the constraint on symptom/treatment rationality. It says that direct explanation is better than indirect treatment when the question asked to the student is not very difficult. The window (6) shows the rationality and recommends the "explain the principle" treatment instead.

From figure 2 one might get impression that the description of the question is too detail for the first stage of the author's work. Of course, the boundary between the first two stages of author's work is gray. Our intention of this is, however, that the textual information acts as just labels of the entities of questions. Without such labels, authors cannot structure the skeleton in their mind and also may lack the consistency of their design. Note that the textual information in the question frame is meaningless for ontology management environment at this stage.

### 3.3 Layers of Ontology

As we have discussed in section 2, task ontology is divided into three complementary partitions. From now on, we will focus on the two of them, that is, lexical level and conceptual level, which are the most important for authoring support function at the first stage of author's work. In figure 3, the two partitions are arranged in depth. The two partitions in depth are allotted different functions to each. Lexical level ontology specifies syntactical aspects and conceptual level specifies semantics of vocabulary.

In addition to that, task ontology has three layers arranged in vertical direction in order of the degree of task-domain dependency. That is, core task ontology with the least dependency and arranged at the lowest layer, task-type specific ontology at middle, and task-domain ontology at the highest.

Core task ontology lays foundation for upper layers by defining inherent concept needed for modeling all the types of problem solving. Task-type specific ontology is a system/theory of vocabulary/concepts for describing task models of a certain type of task, and task-domain ontology is one for describing domain models from the task-type viewpoint. We have defined the CBT(Computer Based Training) ontology which organizes a set of concepts specific to training system as a task-specific ontology. Based on the topmost layer ontology, authors build the model of task of interest. In our case, the model means a skeleton of an CBT system(*SmartTrainer*).

The relationship between ontology and model is not absolute one but relative one. For example, when building a task-specific ontology, it can be regarded as a model, while the core task ontology as an ontology for the world. That is to say, an object can be considered as an ontology, meanwhile it can also be considered as a model based on the relatively lower layer. In principle, a descriptive primitive of upper layer has richer meaning in the sense of human understanding than that of lower layer. Needless to say, the rigid computational semantics defined for the primitives at the lowest layer.

Thus, we could say an intended model of CBT systems is described based on the upper, front part of task ontology and the computational semantics of it is finally defined at the lowest, rear part.

To characterize each layer from the above viewpoint, we can ideologically divide the authors concerned with the ontology into four types.
 (1) the authors who develop training systems (end authors).
 (2) The authors who build task-domain ontology (task-domain ontology authors).
 (3) The authors who build task-specific ontology (task-specific ontology authors)
 (4) The authors who build core task ontology-based on core-task ontology (Core task ontology authors).
In our case,
 (1) End author: the instructor of electric power system operation (instructor)
 (2) Task-domain ontology author: staff of the software development division of electric power company.
 (3) Task-specific ontology author: IES/CBT ontology researcher.
 (4) Core task ontology author: general task ontology researcher.

Note that each type of author needs appropriate guidelines corresponding to their jobs. Intuitively, the authors working in the upper layer are more constrained than those in the lower layer. Thus, the authoring tools is desired to have the capability to switch the ontology layer to provide the appropriate guidelines for each type of the authors. The capability is realized by integration of CLEPE into the authoring tool. In the next section, we will look into the role of CLEPE in detail.

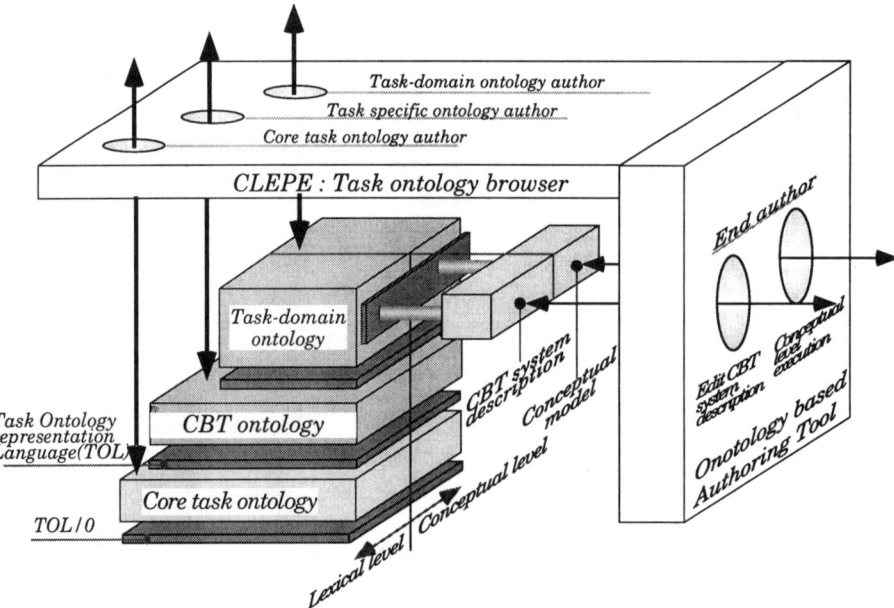

Figure 3 Overview of a Task Ontology based Authoring Tool

## 4. Task Ontology Management : CLEPE

CLEPE supports both ontology authors who construct task ontology and end authors who develop application, CBT systems in our case. Figure 3 shows the overview of CLEPE and the ontology-based authoring tool we have developed. Ontology authors are arranged above side and end authors right side.

Thin planes stand for languages. The language for task specific ontology author is called Task Ontology representation Language(TOL), whose semantics is specified by core task ontology and the base language TOL/0. TOL/0 provides description primitives for ontology authors and defines semantics of upper-layer languages. Therefore all the semantics of task ontology described with TOL is specified ultimately at the level of TOL/0. An ontology author of core task ontology specifies the lexical entities and conceptual ones using TOL/0 primitives, e.g. Define-Core-Lexical, Define-Core-Concept, etc. By reading the specification of core task ontology into CLEPE, a set of conceptual primitives at the TOL level is introduced. Task type specific ontology author specifies the concepts appearing in the target task type with TOL.

The main work of an end author are as follows:
 (1) To describe an intended model of a training task in terms of CBT task ontology (edit CBT system description),
 (2) to make sure that his/her description is correctly interpreted by the system (trace the conceptual level execution process),
 (3) to modify the description if necessary (debug the description).

In preparation for interpretation of the CBT task description written by an end author, CLEPE reads task ontology description represented with TOL. The task ontology description is translated into internal form by TOL-parser and stored into ontology base. Ontology Manager manages the ontology base and deals with the requests related to the ontology made by other modules, for example, inquiries for class information, creation of a class instance and so on. Once an end author completes editing his/her own description, CLEPE initiates the interpretation process. The model compiler compiles an internal form of the description and generates the conceptual model. Finally, he/she can *run* the conceptual model with executor.

## 4.1 Ontology Management

As we have discussed in 3.3, the relatively lower ontology provides the guideline on the construction of the relatively upper ontology. For example, the core task ontology act as a specification for the task-specific ontology. Definition of each concepts appearing in task-specific ontology layer should follow the regulation enforced by core task ontology. "Verb word class" definition in the core task ontology (meta-layer), for example, specifies that all the verbs defined in task-specific ontology layer (base-layer) have "input-slot", "output-slot" and "effect slot" in its definition. Furthermore, a verb definition of "give" in task specific ontology layer specifies that the input slot of each instance of "give" appearing in the model layer for end authors should be filled with limited noun instances such as, "Hint", "Problem", and so on.

Ontology management function of CLEPE is to support the authors work in a certain layer based on the next lower ontology. When it detects the inconsistency between model and ontology, it gives the authors some warning messages along with suggestions to encourage them to revise their models based on the ontology. The major merit of this function is that the authors could be guided kindly in their model building and the models is guaranteed to be consistent with the established ontology.

## 4.2 Conceptual level execution

Once CLEPE generates a conceptual model, an end author can inquire some questions about the dynamic behavior of the model, for example, "how student model updated", or "then what kind of tutoring action will be taken place." CLEPE can produce answers based on the conceptual model, for example, "the student model will be updated based on the buggy knowledge attached to the learner's answer to the question," or "system will show visual simulation to make learner notice the crucial breakdown." Since the answer to the inquiry is described at conceptual level, he/she could easily understand it. By keeping the continuity from the symbol level program code to conceptual level model, CLEPE can explain the symbol level execution result with conceptual level annotation.

Current implementation of conceptual level execution is limited to static analysis of the task model. The plan for the future development includes dynamic analysis of the task model. The basic idea is to introduce the concept of pseudo-learner. Training task ontology includes a set of terms and concepts to specify the status of the learner using vocabulary shown in figure 1. By specifying certain property of learner, for example, "a learner who prefers deductive way of thinking to inductive one" or "a learner who is good at skill-based operation", the author can examines whether the training model behave well for the learner or not by monitoring the results of conceptual level execution. The executor of the training model will assume the learner's responses to the questions along the training scenario, and activate the tutoring strategies based on the leaner model. Showing the dynamic behavior of the model to the author, the executor can check the validity of the training model based on the task ontology. When some defects or shortcoming of the model are identified, the executor informs the authors and suggest a desirable way to make alternation to the model.

## 5. Conclusion

We have investigated the inherent characteristic of CBT through the development of *SmartTrainer* and tried to build CBT task ontology. The main purpose of this paper is to illustrate the characteristics of ontology-based authoring tools. It provides two major advantages as follows. (A) It provides human-friendly primitives in terms of which users can easily describe their own model of a task (descriptiveness, readability). (B) It can simulate the abstract behavior of the model in terms of conceptual level primitives (conceptual level operationality).

We have implemented a prototype of ontology-based authoring tool and systematically accumulating CBT ontology and training scenarios. The most important, interesting part of CBT ontology is explicit description of the first principle of training such as "learning by doing." Since the accumulation of the ontology has been in steady progress, it is planned to pursue a close investigation of this issue in the near future.

## References

[Inui et al. ,1990] Inui, M., Miyasaka, N., Matsubara, A., Fujita, M. Development of A Model-based Intelligent Training System for Plant Operations. *In Proceedings of International Conference on Advanced Research on Computers in Education, ARCE90,* pages 89-94 , Tokyo, July 1990.

[Major, 1995] Major,M.P. REDEEM: Creating Reusable Intelligent Courseware, *In Proceedings of International Conference on Artificial Intelligence in Education, AI-ED95*, pages 75-82, Washington, DC, August1995.

[Mizoguchi et al., 1996] Mizoguchi, R. , Sinitsa,K., Ikeda,M. Knowledge Engineering of Educational Systems for Authoring System Design — A preliminary results of task ontology design —, *In Proceedings. of EuroAI-ED '96*, pages.329-335, Lisbon, 1996.

[Murray and Woolf, 1992] Murray, T, Woolf,B : Tools for Teacher Participation in ITS Design, *In Proceedings of Intelligent Tutoring Systems ITS'92 , Lecture Notes in Computer Science, 608.. Springer Verlag,* pages.593-600, Montreal, June 1992.

[Van Marcke and Vedelaar, 1995] Van Marcke, K. and Vedelaar, H. Learner adaptivity in generic instructional strategies. *In Proceedings of International Conference on Artificial Intelligence in Education, AI-ED95*, pages 323-333, Washington, DC, August1995.

# CONSTRAINT SATISFACTION

# CONSTRAINT SATISFACTION

## Constraint Satisfaction 1: Constraint Programming

# Semiring-based Constraint Logic Programming

Stefano Bistarelli, Ugo Montanari*, Francesca Rossi

University of Pisa
Computer Science Department
Corso Italia 40, 56125 Pisa, Italy
E-mail: {bista,ugo,rossi}@di.unipi.it

## Abstract

We extend the Constraint Logic Programming (CLP) formalism in order to handle semiring-based constraint systems. This allows us to perform in the same language both constraint solving and optimization. In fact, constraint systems based on semirings are able to model both classical constraint solving and more sophisticated features like uncertainty, probability, fuzzyness, and optimization. We then provide this class of languages with three equivalent semantics: model-theoretic, fixpoint, and proof-theoretic, in the style of CLP programs.

## 1 Introduction

Classical constraint satisfaction problems (CSPs) [Mac92] are a very expressive and natural formalism to specify many kinds of real-life problems. However, they also have evident limitations, mainly when they are used to represent real-life scenarios where the knowledge is not completely available nor crisp. In fact, in such situations, the ability of stating whether an instantiation of values to variables is allowed or not is not enough or sometimes not even possible. Recently, a proposal which extends classical CSPs in this direction has been developed [BMR95; BMR97], which is able to model many desired features, like fuzzyness [DFP93], probability [FL93], uncertainty, partiality [FW92], hierarchy [BMMW89], and optimization. This framework is based on the observation that a semiring (that is, a domain plus two operations satisfying certain properties) is all what is needed to describe many constraint satisfaction schemes. In fact, the domain of the semiring provides the levels of consistency (which can be interpreted as cost, or degrees of preference, or probabilities, or others), and the two operations define how to combine constraints together. In particular, from one of the operations we can derive a partial order $\leq$ among the elements

---
*On leave at SRI International, Menlo Park, CA.

of the semiring which allows us to compare different elements: if $a \leq b$ then it means that $b$ is *better* than $a$. This is crucial in situations which involve some kind of optimization. Constraint problems described according to this framework are called SCSP (for Semiring-based Constraint Satisfaction Problems).

Constraint logic programming (CLP) [JL87] languages extended logic programming (LP) by replacing term equalities with constraints and unification with constraint solving. Programming in CLP means choosing a constraint system for a specific class of constraints (for example, linear arithmetic constraints, or finite domain constraints) and embedding it into a logic programming engine. This approach is very flexible since one can choose among many constraint systems without changing the overall programming language, and has shown to be very successful in specifying and solving complex problems in terms of constraints of various kind. However, it can handle only classical constraint solving. Thus it is natural to try to extend the CLP formalism in order to be able to handle also SCSP problems. We will call such an extension SCLP (for Semiring-based CLP).

In passing from CLP to SCLP languages, we will replace classical constraints with the more general SCSP constraints. By doing this, we also have to modify the notions of interpretation, model, model intersection, and others, since we have to take into account the semiring operations and not the usual CLP operations. For example, while CLP interpretations associate a truth value (either *true* or *false*) to each ground atom, here ground atoms must be given one of the elements of the semiring. Also, while in CLP the value associated to an existentially quantified atom is the *logical or* among the truth values associated to each of its instantiations, here we have to replace the *or* with another operation which refers to one of the semiring operations.

After describing the syntax of SCLP programs, we will define three equivalent semantics for such languages: model–theoretic, fixpoint, and operational. These semantics are conservative extensions of the corresponding ones for LP, since by choosing a particular semiring (the one with just two elements, *true* and *false*, and the logical *and* and *or* as the two semiring operations) we get exactly the LP semantics. The extension is in some cases predictable but it possesses some crucial new features. For example, the presence of a partial order among the semiring elements (and not a *total* order like it is in the LP/CLP case, where we just have two comparable elements) brings some conceptual complexity in some aspects of the semantics. In fact, in the operational semantics there could be two refutations for a goal which lead to different semiring elements which are not comparable in the partial order. In this case, these elements have to be combined in order to get the solution corresponding to the given goal, and their combination could be not reachable by any derivation path in the search tree. This means that any constructive way to get such a solution by visiting the search tree would have to follow all the incomparable paths before being able to find the correct answer.

A related approach is HCLP (Hierarchical CLP) [BMMW89], where each constraint has a level of importance (like *strong, weak, required*), and these levels are used to decide which constraints to satisfy. However, a constraint can only be satisfied or not, and thus HCLP is a *crisp* formalism. Moreover, their treatment is only algorithmic, and they do not provide their language with a fixpoint or a model-theoretic semantics.

A recent approach to multi valued logic programming [MPS97] uses bilattices with two orderings to model both truth and knowledge levels. The resulting logic programming semantics is just operational and fixpoint, while no model-theoretic semantics is presented. Moreover, the presence in our approach of just one ordering (modelling truth levels) is not a restriction, since the vectorization of several semirings is still a semiring (see [BMR95; BMR97]) and thus optimization based on multiple criteria can be cast in our framework as well.

## 2 Semiring-based CSPs

Here we give the basic notions about constraint solving over semirings, introduced in [BMR95; BMR97].

**Definition 1 (semiring)** *A semiring is a tuple* $\langle A, +, \times, \mathbf{0}, \mathbf{1} \rangle$ *such that*

- *$A$ is a set and $\mathbf{0}, \mathbf{1} \in A$;*
- *$+$, called the additive operation, is a closed (i.e., $a, b \in A$ implies $a + b \in A$), commutative (i.e., $a + b = b + a$) and associative (i.e., $a + (b + c) = (a + b) + c$) operation such that $a + \mathbf{0} = a = \mathbf{0} + a$ (i.e., $\mathbf{0}$ is its unit element);*
- *$\times$, called the multiplicative operation, is a closed and associative operation such that $\mathbf{1}$ is its unit element and $a \times \mathbf{0} = \mathbf{0} = \mathbf{0} \times a$ (i.e., $\mathbf{0}$ is its absorbing element);*
- *$\times$ distributes over $+$ (i.e., $a \times (b + c) = (a \times b) + (a \times c)$).* □

**Definition 2 (c-semiring)** *A c-semiring is a semiring* $\langle A, +, \times, \mathbf{0}, \mathbf{1} \rangle$ *such that*

- *$+$ is idempotent (i.e., $a \in A$ implies $a + a = a$);*
- *$\times$ is commutative;*
- *$\mathbf{1}$ is the absorbing element of $+$.* □

The following is a list of properties about c-semirings which will be used in this paper:

- Given any c-semiring $S = \langle A, +, \times, \mathbf{0}, \mathbf{1} \rangle$, the relation $\leq_S$ over $A$ such that $a \leq_S b$ iff $a + b = b$ is a partial order.

- Since $\mathbf{0}$ is the unit element of the additive operation, it is the minimum element of the ordering. Thus, for any $a \in A$, we have $\mathbf{0} \leq_S a$.

- Both the additive and the multiplicative operation are monotone on the ordering $\leq_S$.

- Since $\mathbf{1}$ is also the absorbing element of the additive operation, then $a \leq_S \mathbf{1}$ for all $a$. Thus $\mathbf{1}$ is the maximum element of the partial ordering. This implies that the $\times$ operation is *intensive*, that is, that $a \times b \leq_S a$. This is important since it means that combining more constraints leads to a worse (w.r.t. the $\leq_S$ ordering) result.

- Given a c-semiring $S = \langle A, +, \times, \mathbf{0}, \mathbf{1} \rangle$, and its partial order $\leq_S$, $\langle A, \leq_S \rangle$ is a complete lattice[1]. Moreover, for any $a, b \in A$, we have $a \vee b = a + b$, where $\vee$ is the lub operation of the lattice.

- Given a c-semiring $S = \langle A, +, \times, \mathbf{0}, \mathbf{1} \rangle$, consider the corresponding lattice $\langle A, \leq_S \rangle$. If $\times$ is idempotent, then we have that:
  1. $+$ distributes over $\times$;
  2. $\times = \wedge$, where $\wedge$ is the glb operation of the lattice;
  3. $\langle A, \leq_S \rangle$ is a distributive lattice.

**Definition 3 (SCSPs)** *A constraint problem based on semirings (SCSP) consists of a set of variables with a finite domain $D$ and a set of constraints. Each constraint, which connects a subset of the variables $V$, is defined by associating an element of the semiring with each tuple of values of $D$ for the variables in $V$.* □

Note that the elements of the chosen semiring can be interpreted in many ways: cost, level of preference, certainty, probability, etc. Note also that the intuitive

---

[1] Actually, for this result to hold we must assume that there exists the sum of an infinite number of elements.

meaning of the partial order $\leq_S$ is to state when an element is better than another one: if $a \leq_S b$ then we mean that $b$ is better than $a$. Finally, it is interesting to notice that classical CSPs are just SCSPs where the semiring has just two values: *true* and *false*, and the two operations are *logical and* and *logical or*. That is, the semiring is $S_{CSP} = \langle \{true, false\}, \vee, \wedge, false, true \rangle$.

## 3 Syntax of SCLP programs

SCLP(S,D) programs are just Constraint Logic Programming (CLP) programs [JL87] where constraints are handled according to the chosen semiring $S = \langle A, +, \times, \mathbf{0}, \mathbf{1} \rangle$ and variables can be assigned values over a finite set $D$. As usual, a program is a set of *clauses*. Each clause is then composed by a *head* and a *body*. The head is just an atom and the body is either a collection of atoms, or a value of the semiring. Finally, a *goal* is a collection of atoms. The BNF for this syntax follows.

$P :: CL \mid CL, P$
$CL :: H : -B$
$H :: AT$ where $AT$ is the category of atoms
$LAT :: \square \mid LAT, AT$
$B :: LAT \mid a$ where $a \in A$
$G :: \quad : -LAT$

As an example, consider the following SCLP(S,D) program where $S = \langle -N \cup \{-\infty\}, max, +, -\infty, 0 \rangle$, $-N$ is the set of non-positive integers, and $D = \{a, b, c\}$. Note that the ordering $\leq_S$ in this semiring coincides with the $\leq$ ordering over integers.

```
s(x)    :- p(x,y).
p(a,b)  :- q(a).
p(a,c)  :- r(a).
q(a)    :- t(a).
t(a)    :- -2.
r(a)    :- -3.
```

The intuitive meaning of a semiring value like $-3$ associated to the atom $r(a)$ is that $r(a)$ costs 3 units. Thus the set $-N \cup \{-\infty\}$ contains all possible costs, and the choice of the two operations $max$ and $+$ implies that we intend to maximize the sum of the costs. This gives us the possibility to select the atom instantiation which gives the minimal cost overall.

## 4 Model-theoretic semantics

An interpretation is a function which takes a predicate and an instantiation of its arguments (that is, a ground atom), and returns an element of the semiring: $I : \bigcup_n (P_n \to (D^n \to A))$, where $P_n$ is the set of n-ary predicates. The notion of interpretation can be used to associate elements of the semiring also to formulas which are more complex than ground atoms. More precisely, we can define the function $value_I$ which takes any formula appearing in a program (not a clause) and returns an element of the semiring:

- The value associated to a non-ground atom of the form $F = \exists x.A(x)$ is computed by considering the lub of the values associated to all the ground atoms $A(x/d)$, where $d$ is any domain element. That is, $value_I(F) = lub\{I(A(d))$, for all $d \in D\}$. Formulas of this kind occur in SCLP languages since variables appearing in the body of a clause but not in its head are considered to be existentially quantified. For example, in the special case of logic programming the clause `p(a) :- q(x,a)` is just a shorthand for the formula $p(a) \leftarrow \exists x.q(a, x)$.

- The value associated to a conjunction of atomic formulas of the form $A \wedge B$ is the product of the values associated to $A$ and $B$: $value_I(A \wedge B) = value_I(A) \times value_I(B)$.

- For any semiring element $a$, $value_I(a) = a$.

Note that the meaning associated to formulas by function $value_I$ coincides with the usual logic programming interpretation [Llo93] when considering the semiring $S_{CSP}$. In fact, in this case the ordering $\leq_S$ is defined by $false \leq_S true$, the lub operation of the lattice $\langle \{true, false\}, \leq_S \rangle$ is $\vee$, and the glb is $\wedge$. Thus, for example, $value_I(\exists x.A(x)) = lub\{I(A(d))$, for all $d \in D\} = \vee\{I(A(d))$, for all $d \in D\}$. Thus it is enough that one of the $A(d)$ is assigned the value *true* that the value associated to the whole formula $\exists x.A(x)$ is *true*. Note also that in this special instance the lub and glb of the lattice coincide with the two semiring operations, but this is not true in general (see Section 2).

Given a clause of the form $H : -B$ and an interpretation $I$, we say that the clause is *satisfied* in $I$ if and only if $value_I(H) \geq value_I(B)$. This is consistent with the usual treatment of clauses in logic programming, where a clause is considered to be satisfied if the body logically implies the head, and by noting that logical implication in the semiring $S_{CSP}$ coincides with the ordering $\leq_{S_{CSP}}$.

For example, the clause `p(a) :- q(b)` is satisfied in $I$ if $I(p(a)) \geq I(q(b))$; the clause `p(x) :- q(x,a)` is satisfied if $\forall x.(I(p(x)) \geq I(q(x, a)))$; the clause `p(a) :- q(x,a)` is satisfied if $I(p(a)) \geq value_I(\exists x.q(x, a))$; the clause `p(x) :- q(x,y)` is satisfied if $\forall x.(I(p(x)) \geq value_I(\exists y.q(x, y)))$.

An interpretation is a *model* for a program $P$ if all clauses of $P$ are satisfied in $I$. Given a program and all its models, one would like to identify a unique single model as the representative one. In logic programming this is done by considering the minimal model [Llo93], which is obtained by intersecting all the models of the program. This works because models in logic programming are assimilable to sets of ground atoms, those with associated value *true*. Here we follow the same approach, but we have to generalize the notion of intersection of two models, written as "∘", as their glb in the lattice $\langle A, \leq \rangle$.

**Definition 4 (model intersection)** *For every ground atomic formula $A$ and two models $I_1$ and $I_2$, $I_1 \circ I_2(A) = glb(I_1(A), I_2(A))$.* □

**Theorem 5 (model intersection)** *Consider two models $M_1$ and $M_2$ for a CLP(S,D) program $P$. Then $M_1 \circ M_2$ is a model for $P$ as well.*

**Proof**: Since $M_i$ is a model for $P$ for $i = 1, 2$, it must be that, for every clause $H : -B$, $M_i(B) \leq M_i(H)$. Consider now the model $M = M_1 \circ M_2$. We need to prove that, for all $H : -B$, also $M(B) \leq M(H)$ holds. Without loss of generality, assume that $B = A_1, A_2$. Thus $M_i(B) = M_i(A_1) \times M_i(A_2)$ for $i = 1, 2$ and $M(B) = M(A_1) \times M(A_2) = glb(M_1(A_1), M_2(A_1)) \times glb(M_1(A_2), M_2(A_2))$. Also, $M(H) = glb(M_1(H), M_2(H))$. Since $glb(M_1(A_1), M_2(A_1)) \leq M_1(A_1)$ and $glb(M_1(A_2), M_2(A_2)) \leq M_1(A_2)$ by definition of glb, and recalling that $\times$ is monotone, we have that $M(B) \leq M_1(B)$. The same reasoning applies also for proving that $M(B) \leq M_2(B)$. By transitivity of $\leq$, we thus get $M(B) \leq M_1(H)$ and $M(B) \leq M_2(H)$. Since $M(H)$ is the glb of $M_1(H)$ and $M_2(H)$, and since the glb of two elements if the greatest among the elements which are smaller than both, we have that $M(B) \leq M(H)$. □

It is easy to see that the operation of model intersection is associative, idempotent, and commutative. Thus its application can be extended to more then two models. Given a set of models $MS$, we will write $\circ(MS)$ as the model obtained by intersecting all models in $MS$.

Given a program $P$ and the set of all its models, its *minimal model* is obtained by intersecting all models: $MM_P = \circ(\{M \mid M \text{ is a model for } P\})$. The *model-theoretic semantics* of a program $P$ is its minimal model, $MM_P$.

Consider the program $P$ described at the end of last section. The minimal model $MM_P$ for such a program must assign a non-positive integer to each ground atom, and it is the following function:
$MM_P(t(a)) = -2$
$MM_P(q(a)) = -2$
$MM_P(r(a)) = -3$
$MM_P(p(a,c)) = -3$
$MM_P(p(a,b)) = -2$
$MM_P(s(a)) = max(-2, -3) = -2$
For each other atom different from the ones considered above, $MM_P$ returns $-\infty$.

## 5 Fixpoint semantics

We define now the operator $T_P$ which extends the one used in logic programming [Llo93] by following the same approach as in the previous section. The resulting operator maps interpretations into interpretations, that is, $T_P : IS_P \to IS_P$, where $IS_P$ is the set of all interpretations for $P$. Given an interpretation $I$ and a ground atom $A$, assume that program $P$ contains $k$ clauses defining the predicate in $A$. Clause $i$ is of the form $A : -B_1^i, \ldots, B_{n_i}^i$. Then

$$T_P(I)(A) = \sum_{i=1}^{k} (\prod_{j=1}^{n_i} I(B_j^i)).$$

This function coincides with the usual immediate consequence operator of logic programming when considering the semiring $S_{CSP}$.

Consider now an ordering $\preceq$ among interpretations which respects the semiring ordering.

**Definition 6 (partial order of interpretations)** *Given a program $P$ and the set of all its interpretations $IS_P$, we define the structure $\langle IS, \preceq \rangle$, where for any $I_1, I_2 \in IS$, $I_1 \preceq I_2$ if $I_1(A) \leq_S I_2(A)$ for any ground atom $A$.* □

It is easy to see that $\langle IS, \preceq \rangle$ is a complete partial order, whose greatest lower bound coincides with the glb operation (suitable extended to interpretations). It is also possible to prove that function $T_P$ is monotone and continuous over the complete partial order $\langle IS, \preceq \rangle$.

By using these properties, classical results on partial orders [Tar55] allow us to conclude that

- $T_P$ has a least fixpoint, $lfp(T_P)$, which coincides with $glb(\{I \mid T_P(I) \preceq I\})$;
- the least fixpoint of $T_P$ can be obtained by computing $T_P \uparrow \omega$. This means starting the application of $T_P$ from the bottom of the partial order of interpretations, called $I_0$, and then repeatedly applying $T_P$ until a fixpoint.

Consider again the program at the end of Section 3. We recall that in this specific case the semiring is $S = \langle -N \cup \{-\infty\}, max, +, -\infty, 0 \rangle$ and $D = \{a, b, c\}$. Thus function $T_P$ is:

$$T_P(I)(A) = max\{\sum_{j=1}^{n_1} I(B_j^1), \ldots, \sum_{j=1}^{n_k} I(B_j^k)\}.$$

In this semiring the bottom interpretation $I_0$ is the interpretation which maps each semiring element into itself and each ground atom into the bottom of the lattice associated to the semiring, that is, $-\infty$. Note that we slightly abused the notation since interpretations are functions whose domain contains only ground atoms (see Section 4), while here we also included semiring elements. This simplifies the definition of $I_0$; however, it is possible to obtain the same result with a more complex definition of $I_0$ which satisfies the definition of interpretation. Given $I_0$, we obtain $I_1$ by applying function $T_P$ above. For example, $I_1(r(a)) = -3$. Instead, $I_1(p(a,c)) = -\infty$, and $I_2(p(a,c)) = I_1(r(a)) = -3$. The following table gives the value associated by the interpretations $I_i$ with each ground atom. Some of the atoms are not listed because each interpretation $I_i$ gives them value $-\infty$. All interpretation $I_i$ with $i > 4$ coincide with $I_4$, thus $I_4$ is the fixpoint of $T_P$.

|        | $I_1$     | $I_2$     | $I_3$     | $I_4$     |
|--------|-----------|-----------|-----------|-----------|
| t(a)   | -2        | -2        | -2        | -2        |
| r(a)   | -3        | -3        | -3        | -3        |
| q(a)   | $-\infty$ | -2        | -2        | -2        |
| p(a,c) | $-\infty$ | -3        | -3        | -3        |
| p(a,b) | $-\infty$ | $-\infty$ | -2        | -2        |
| s(a)   | $-\infty$ | $-\infty$ | -3        | -2        |
| s(b)   | $-\infty$ | $-\infty$ | $-\infty$ | $-\infty$ |
| s(c)   | $-\infty$ | $-\infty$ | $-\infty$ | $-\infty$ |

The most interesting case is the computation of the value associated to $s(a)$. In fact, $I_3(s(a)) = max\{I_2(p(a,a)), I_2(p(a,b)), I_2(p(a,c))\} = max\{-\infty, -\infty, -3\} = -3$. Instead, $I_4(s(a)) = max\{I_3(p(a,a)), I_3(p(a,b)), I_3(p(a,c))\} = max\{-\infty, -2, -3\} = -2$. Note that the clause s(x) :- p(x,y) is considered equivalent to all its instantiations. In particular, when $x = a$, we have the three clauses s(a) :- p(a,a), s(a) :- p(a,b), and s(a) :- p(a,c). These are the clauses to consider when computing $I(s(a))$.

We will now prove that the least fixpoint of function $T_P$ coincides with the minimal model of program $P$. To do that, we need an intermediate result which shows that the models of a certain program $P$ are the solutions of the equation $T_P(I) \preceq I$.

**Theorem 7 (models and $T_P$)** *Given any interpretation $I$ for a program $P$, $I$ is a model for $P$ if and only if $T_P(I) \preceq I$.*

**Proof**: Consider any ground atom $H$ and assume there are two clauses with $H$ as their head: $H : -B_1$ and $H : -B_2$. By definition of model, each clause $H : -B_i$ is satisfied in $I$. Thus $I(H) \geq I(B_i)$. Now, function $T_P$ assigns to $H$ the sum of the values assigned by $I$ to $B_1$ and $B_2$, thus $T_P(I)(H) = I(B_1) + I(B_2)$. But the + operation coincides with the lub of the semiring, thus any value of the semiring which is greated than both $I(B_1)$ and $I(B_2)$ is also greater than their sum. Therefore $T_P(I)(H) \leq I(H)$. A similar reasoning works also for proving that if $T_P(I)(H) \leq I(H)$ for any ground atom $H$ then $I$ is a model. $\square$

**Theorem 8 (model and fixpoint semantics)**
*Given a SCLP(S,D) program $P$, we have that $MM_P = lfp(T_P)$.*

**Proof**: By definition of minimal model, $MM_P = glb(\{I \mid I \text{ is a model for } P\})$. By Theorem 7, we get $MM_P = glb(\{I \mid T_P(I) \preceq I\})$. By the classical results cited above, this coincides with the least fixpoint of $T_P$. $\square$

## 6 Proof-theoretic semantics

We will define here a proof-theoretic semantics based on a resolution rule, just like in CLP [JL87]. However, we need first to rewrite the program into a form which is more suitable to our semantics.

First, we rewrite each clause so that the head is an atom whose arguments are only variables. This means that we must explicitely specify the substitution that was written in the head, by inserting it in the body. That is, given a clause $p(t_1, \ldots, t_n) : -B$ we transform it into $p(x_1, \ldots, x_n) : -\langle B, \theta \rangle$ where $\theta = \{x_1/t_1, \ldots, x_n/t_n\}$. Thus bodies now have the following form: $B1 :: \langle B, \theta \rangle$. We recall that $B$ can be either a collection of atoms or a value of the semiring. To give a uniform representation to bodies, we can define them as triples containing a collection of atoms (possibly empty), a substitution, and a value of the semiring (possibly, $\mathbf{1}$). Thus bodies are now of the form $B2 :: \langle LAT, \theta, a \rangle$. If we have a body belonging to the syntactic category $B1$ of the form $\langle a, \theta \rangle$, we get $\langle \square, \theta, a \rangle$. If instead we have $\langle C, \theta \rangle$, where $C$ is a collection of atoms, we get $\langle C, \theta, \mathbf{1} \rangle$. Thus clauses have now the form $CL1 :: H : -B2$. Initial goals need to be transformed as well: given a goal $G = (: -C)$, where $C$ is a collection of ground atoms, we get the goal $G' = (: -\langle C, \varepsilon, \mathbf{1} \rangle)$. The reason why we write the empty substitution and the value $\mathbf{1}$ of the semiring is that both these elements are the unit elements w.r.t. the operations we want to perform on them, that is, composition of substitution and constraint combination.

In summary, given a SCLP(S,D) program, we get a program in an intermediate language, whose syntax is as follows:
$B2 :: \langle LAT, \theta, a \rangle$
$CL1 :: H : -B2$
$P1 :: CL1 \mid CL1, P1$
$G1 :: B2$

Consider again the example at the end of Section 3. The transformed program is then

```
s(x)   :-   〈 p(x,y), ε, 0 〉.
p(x,y) :-   〈 q(a), {x=a,y=b}, 0 〉.
p(x,y) :-   〈 r(a), {x=a,y=c}, 0 〉.
q(x)   :-   〈 t(a), {x=a}, 0 〉.
t(x)   :-   〈 □, {x=a}, -2 〉.
r(x)   :-   〈 □, {x=a}, -3 〉.
```

Once we have transformed the given SCLP program into a program in the syntax just given, we can apply the following semantic rule. This rule defines the transitions of a nondeterministic transition system whose states are goals (according to the syntactic category G1).

If the current goal contains an atom which unifies with the head of a clause, then we can replace that atom with the body of the considered clause, performing a step similar to the resolution step in CLP. The main difference here is that we must update the third element of the goal, that is, the semiring value associated to the goal: if before the transition this value is $a$ and the transition uses a clause whose body has value $a_1$, then the value associated to the new goal is $a \times a_1$. The reason for using the $\times$ operation of the semiring is that this is exactly the operation used when accumulating constraints in the SCSP framework.

$$\frac{\begin{array}{c} C = A, Cr \\ A' := \langle C_1, \theta_1, a_1 \rangle \text{ is a clause} \\ \exists \theta' = mgu(A\theta, A'\theta_1) \end{array}}{\langle C, \theta, a \rangle \longrightarrow \langle (C_1, Cr), \theta' \circ \theta_1, a \times a_1 \rangle}$$

A *derivation* is a finite or infinite sequence of applications of the above rule. A *refutation* is a finite derivation whose final goal is of the form $\langle \Box, \theta, a \rangle$.

Let us now consider the set
$S = \{ \langle C, a \rangle \mid \langle C, \varepsilon, \mathbf{1} \rangle \Rightarrow^* \langle \Box, \theta, a \rangle \text{ for some } \theta \}$
which contains all pairs representing all refutations for the given program. Notice that we can forget about the derivation $\theta$ accumulated during the refutations, since we assumed to always start with a ground goal. Thus $\theta$ only refers to variables introduced during the derivation.

Now we are ready to define function $OS_P$ which, given a ground atom, returns a value of the semiring. More formally, function $OS_P : AT \to A$, where $AT$ is the set of ground atoms and $A$ is the semiring set, is defined as follows: $OS_P(C) = \sum_{a_i \mid \langle C, a_i \rangle \in S} a_i$. Notice that, if the set of all $a_i$ such that $\langle C, a_i \rangle$ is in $S$ is empty, $OS_P(C)$ returns the unit element for $+$, that is, $\mathbf{0}$.

For example, by considering the goal $\langle s(a), \varepsilon, \mathbf{1} \rangle$, we get two refutations, one represented by the pair $\langle s(a), -2 \rangle$, and the other one by $\langle s(a), -3 \rangle$. Thus $OS_P(s(a)) = max(-2, -3) = -2$.

**Theorem 9 (model and operational semantics)**
*Given a SCLP(S,D) program $P$, we have that $MM_P = OS_P$.*

**Proof (sketch):** The statement can be proved by induction on the length $n$ of the refutations, and considering at step $n$ the set $S_n = \{ \langle C, a \rangle \mid \langle C, \varepsilon, \mathbf{1} \rangle \to^n \langle \Box, \theta, a \rangle$ and such that there is no refutation of length greater than $n$ for $\langle C, a \rangle \}$. The proof is similar to that used in [Llo93] for logic programming, although we have to generalize because of the presence of semiring values. $\Box$

For lack of space and sake of readability, in this paper we treated only the case of goals consisting of a ground atom. However, our results can be extended also to the general case of non-atomic and/or non-ground goals.

## 7 Future work

We are now studying how to efficiently implement our framework. Techniques related to heuristically guided search can be useful to cut some paths in the search tree. However, when the partial order is not total, we may have to consider more than one path at a time, as noted in the previous section. Other techniques that we are investigating are based on dynamic programming [BMR97] or on other methods to efficiently compute the solutions in a bottom-up way.

## References

[BMMW89] A. Borning, M. Maher, A. Martindale, and M. Wilson. Constraint hierarchies and logic programming. In Martelli M. Levi G., editor, *Proc. 6th International Conference on Logic Programming*. MIT Press, 1989.

[BMR95] S. Bistarelli, U. Montanari, and F. Rossi. Constraint Solving over Semirings. In *Proc. IJCAI95*. Morgan Kaufman, 1995.

[BMR97] S. Bistarelli, U. Montanari, and F. Rossi. Semiring-based Constraint Solving and Optimization. *Journal of the ACM*, 1997. To appear.

[DFP93] D. Dubois, H. Fargier, and H. Prade. The calculus of fuzzy restrictions as a basis for flexible constraint satisfaction. In *Proc. IEEE International Conference on Fuzzy Systems*. IEEE, 1993.

[FL93] H. Fargier and J. Lang. Uncertainty in constraint satisfaction problems: a probabilistic approach. In *Proc. European Conference on Symbolic and Qualitative Approaches to Reasoning and Uncertainty (ECSQARU)*. Springer-Verlag, LNCS 747, 1993.

[FW92] E. C. Freuder and R. J. Wallace. Partial constraint satisfaction. *AI Journal*, 58, 1992.

[JL87] J. Jaffar and J.L. Lassez. Constraint logic programming. In *Proc. POPL*. ACM, 1987.

[Llo93] J. W. Lloyd. *Foundations of Logic Programming*. Springer Verlag, 1993.

[Mac92] A.K. Mackworth. Constraint satisfaction. In Stuart C. Shapiro, editor, *Encyclopedia of AI (second edition)*, volume 1, pages 285–293. John Wiley & Sons, 1992.

[MPS97] Bamshad Mobasher, Don Pigozzi, and Giora Slutzki. Multi-valued logic programming semantics: An algebraic approach. *Theoretical Computer Science*, 1997. to appear.

[Tar55] A. Tarski. A lattice-theoretical fixpoint theorem and its applications. *Pacific Journal of Mathematics*, 5:285–309, 1955.

# Computational Complexity of Multi-way, Dataflow Constraint Problems

**Gilles Trombettoni** and **Bertrand Neveu**
Projet Contraintes, CERMICS/INRIA,
2004 route des lucioles, 06902 Sophia-Antipolis Cedex, B.P. 93, France
Gilles.Trombettoni, Bertrand.Neveu@sophia.inria.fr

## Abstract

Although it is acknowledged that multi-way dataflow constraints are useful in interactive applications, concerns about their tractability have hindered their acceptance. Certain local propagation algorithms that solve these constraints are polynomial, others (such as Sky-Blue) are exponential. Every system handles a specific problem and the influence of any particular restriction on the computational complexity is not yet precisely determined. In this paper, we present three theoretical results that allow us to classify existing multi-way constraint problems. Especially, we prove that the problem handled by SkyBlue is NP-hard.

## 1 Introduction

*Dataflow* constraints are rapidly gaining popularity in interactive applications because they simplify the programming task. They are conceptually simple and easy to understand, and are capable of expressing relationships over multiple data types, including numbers, strings, booleans, bitmaps, fonts, and colors [Vander Zanden, 1996]. Dataflow constraint solvers are used in numerous interactive systems, such as graphical user interfaces, spreadsheets, graphical layout systems and animation.

Dataflow constraints are divided into two main categories. A *one way* dataflow constraint has *one* associated function for recovering its consistency. This function calculates output variables using the current value of input variables. The *spreadsheet model*, more formally known as the *dependency graph* model [Hoover, 1987], only takes into account one-way constraints. This model is widely used in interactive systems, mainly because the solving process is based on an efficient *incremental evaluation* phase that topologically sorts the functions to execute.

A *multi-way* dataflow constraint has *several* functions (called *methods*) that may be used to satisfy it. The solving process of problems that contain multi-way constraints needs an additional *planning phase* that assigns a method to each constraint, before the evaluation phase.

Although multi-way constraints are more expressive than one-way constraints, they are less recognized because of concerns about their tractability. Every solving algorithm handles a specific problem and the conditions that allow us to decide whether it is computationally difficult are not clear by now.

This paper aims at giving a classification for the computational complexity of the main existing multi-way constraint problems.

## 2 Background

A *multi-way dataflow constraint* system can be denoted as $(V, C, M)$. $V$ is a set of variables with a current value. $C$ is a set of dataflow constraints and $M$ is a set of *methods* that can satisfy the constraints.

**Definition 1** *A* **multi-way dataflow constraint** *is an equation that has one or more methods associated that may be used to satisfy the equation.*

*A* **method** *consists of zero or more inputs, one or more outputs, and an arbitrary piece of code that computes the output variables based on the current value of the input variables [Vander Zanden, 1996]. A* **single-output** *method determines only one variable.*

A (dataflow) constraint system is often represented by a *constraint graph* $G_c$ as shown in Figure 1 (a).

Local propagation is the technique used to solve multi-way constraint systems, typically when new constraints are incrementally added. It works in two phases:

- The *planning phase* directs the edges in $G_c$ by assigning one method to each constraint. The result of this phase (i.e., the solution of the corresponding problem) is a *valid* graph $G_m$ called *method graph* (see Figure 1 (b) and (c)).

**Definition 2** *A method graph $G_m$ is* **valid** *if (1) every constraint has one method associated with it in $G_m$, and (2) $G_m$ has no* **variable conflicts**, *that is,* each variable is the output of, at most, one method (i.e., has at most one incoming edge).

- When the method graph $G_m$ contains no directed cycles, the *evaluation phase* executes the methods in some topological order. When a method is executed, it sets the output variables to values such that the constraint is satisfied. When $G_m$ is cyclic, strongly connected components are collected and generally passed to external solvers to be satisfied as a whole.

Ignoring the operations involved in method execution and cycle solving, the evaluation problem is in the class $P$ of polynomial problems. Indeed, topological sort is $O(d \times |C|)$ where $d$ is the maximum number of methods associated to one constraint. We concern ourselves with the computational complexity of the problem solved by the planning phase that will be called *constraint planning problem* in the following.

Planning algorithms can be divided into three main categories. (1) DeltaBlue [Freeman-Benson *et al.*, 1990] and SkyBlue [Sannella, 1994] work by propagating the conflicts from the perturbations to the leaves of the constraint graph. (2) The propagation of the degrees of freedom scheme (in short PDOF) selects the methods in the reverse order (*i.e.*, executing first the methods that were chosen last). This algorithm has been used in SketchPad [Sutherland, 1963] and QuickPlan [Vander Zanden, 1996]. (3) A third approach is related to the classical problem of graph matching. It gives the *Maximum-matching* algorithm [Gangnet and Rosenberg, 1992].

## 3 Different types of constraint planning problems

Existing local propagation algorithms solve different planning problems that imply various tradeoffs between expressiveness and performance:

- *required* constraints only, or both required and *preferential* constraints that are satisfied if possible
- single-output constraints only, or multi-output ones
- acyclic constraint graphs only, or cycles allowed
- *method restriction* imposed or relaxed

**Definition 3** The **method restriction** imposes that every constraint method of a given problem must use all of the variables in the constraint either as an input or an output variable [Vander Zanden, 1996].

Moreover, some systems allow directed cycles in the method graph, whereas others do not, which leads in fact to two different (and incomparable) problems. Indeed, a general computational result states that a restriction

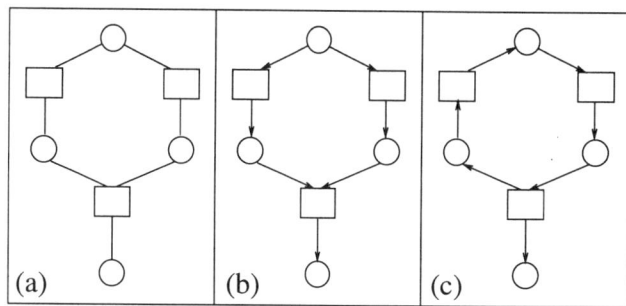

Figure 1: A *constraint graph* is a bipartite graph whose nodes are constraints and variables, respectively represented by rectangles and circles. Each constraint is connected to its variables. This is illustrated in Figure (a). Figures (b) and (c) show two possible *method graphs*. The method selected for each constraint is symbolized by directed edges from the constraint to the output variables, and from the input variables to the constraint. The method graph in (b) contains no directed cycles, as opposed to the method graph in (c).

imposed on a solution does not necessarily make the corresponding problem easier [Papadimitriou, 1994].

Table 1 shows the computational complexity of selected existing constraint planning problems.

## 4 An NP-complete planning problem with method restriction

Disjunctive constraints do not respect the method restriction (*e.g.*, constraint $a \vee b$ has two methods that output to either variable $a$ or $b$ with no input). However, these constraints are not usually needed in interactive systems. Therefore, all of the propagation algorithms impose the method restriction. The problems handled by these algorithms are in $P$, except $cp_2$ (see Table 1) which has not been yet analyzed.

The only known NP-completeness results aim at problems $acp_3$ and $acp_4$ [Maloney, 1991] which are not very interesting in practice because they relax the method restriction.

The following theorem states an NP-completeness result about the problem $cp_2$ for which the method restriction holds and which is handled by SkyBlue.

**Theorem 1** *Let $G$ be a dataflow constraint system for which the method restriction holds.*

*Then proving the existence of a valid method graph (cyclic or not) corresponding to $G$ is NP-complete.*

The proof and the polynomial reduction involved in it are described in following paragraphs.

### 4.1 Polynomial reduction

In Section 4.2, we prove that the known NP-complete problem "*Exact Cover by 3-Sets*" [Papadimitriou, 1994]

| problem | method restriction | single-output | complexity | proof | algorithms |
|---|---|---|---|---|---|
| $acp_1$ | yes | yes | $P$ | [Sutherland, 1963] | PDOF, DeltaBlue |
| $acp_2$ | yes | no | $P$ | [Vander Zanden, 1996] | QuickPlan |
| $acp_3$ | no | yes | $NPC$ | [Maloney, 1991] 3 | – |
| $acp_4$ | no | no | $NPC$ | [Maloney, 1991] 1 and 2 | – |
| $cp_1$ | yes | yes | $P$ | [Gangnet & Rosenberg, 1992] | Maximum-matching |
| **$cp_2$** | **yes** | **no** | **??** | – | **SkyBlue** |
| **$cp_3$** | **no** | **yes** | **??** | – | – |
| **$cp_4$** | **no** | **no** | **??** | – | – |

Table 1: Computational complexity of constraint planning problems. Cycles in the constraint graph are allowed. Constraints are *required* (not *preferential*). Every problem depends on three characteristics: (1) a problem that only accepts *a*cyclic method graphs is designed by $acp_i$ ($i \in \{1..4\}$), whereas a problem that accepts both acyclic and cyclic solutions is designed by $cp_i$; (2) the method restriction; (3) the presence of single-output constraints only. The complexity of problems $cp_2$, $cp_3$ and $cp_4$ is not yet known, especially $cp_2$ which is handled by SkyBlue.

can be reduced to "*constraint-planning*" (*i.e.*, $cp_2$). This reduction will be called *planning reduction*.

**Definition 4 (Exact Cover by 3-Sets)** *Let $X$ be a finite set, such that $|X| = 3q$ for some integer $q$. Let $E$ be a family of sets that contain 3 elements of $X$ each. Every element of $X$ belongs to at least one 3-set of $E$.*[1]

*Does $E$ contain an exact cover for $X$, that is, a subset $S$ of $E$ such that every element of $X$ belongs to exactly one 3-set of $S$?*

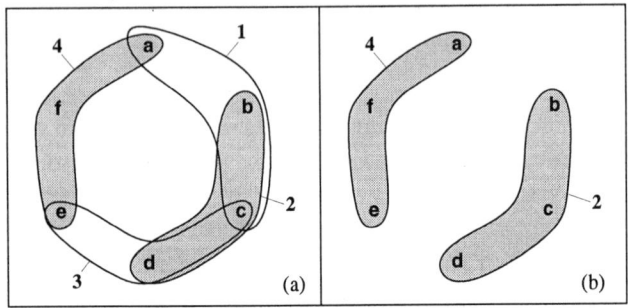

Figure 2: An instance of the "*Exact Cover by 3-Sets*" problem. $X = \{a...f\}$. The 3-sets in $E = \{1...4\}$ are represented by hyper-arcs, as shown in (a). The unique solution $S = \{2, 4\}$ is shown in (b).

Let $G = (X, E)$ be an instance of "*Exact Cover by 3-Sets*". Let $G' = (V, C, M)$ be an instance of "*constraint-planning*" obtained by a planning reduction applied to $G$ as follows:

- Variables in $V$ are divided into two sets *VarX* and *VarE*. Constraints in $C$ are divided into two sets *ConstX* and *ConstE*.

- Each element $i$ of $X$ corresponds to one variable *i-at-most* of *VarX*.

---
[1]This additional hypothesis discards trivial instances while keeping the problem NP-complete.

- A 3-set $p = \{a, b, c\}$ in $E$ corresponds to a constraint $set_p$ of *ConstE* connecting six variables. $set_p$ has two triple output methods which indicate whether the 3-set $p$ is either present or absent in the solution. The *present method* outputs to the three variables *a-at-most*, *b-at-most*, *c-at-most* of *VarX* (It ensures that no other 3-set will cover the corresponding elements.) The *absent method* outputs to the three variables *a-p*, *b-p*, and *c-p* of *VarE*.

- When element $i$ of $X$ can be covered by $n$ different 3-sets $\{p_1...p_n\}$ of $E$, $n$ variables $\{i\text{-}p_1...i\text{-}p_n\}$ of *VarE* are constructed. One constraint *i-at-least* of *ConstX* is also built, that connects these variables. *i-at-least* has $n$ single-output methods, one for each variable in the constraint. The method that outputs to variable $i\text{-}p_k$ ensures that element $i$ of $X$ is covered by (at least) the 3-set $p_k$.

Figure 2 shows an instance of "*Exact Cover by 3-Sets*" that is reduced to the "*constraint-planning*" instance of Figure 3.

The planning reduction is based on the following intuition. One element $i$ of $X$ appears in exactly one 3-set of solution $S$. Thus, $i$ appears *at most once* and *at least once* in a 3-set of $S$. This is translated into the planning problem as follows:

- (at most once) If the 3-set $p$ belongs to solution $S$, then the *present* method of $set_p$ is selected in the corresponding constraint planning problem. Thus, the variables determined by $set_p$ ensure that no other 3-set than $set_p$ covers them, as this would lead to variable conflicts.

- (at least once) As said above, a constraint *i-at-least* directed onto variable *i-p* ensures that element $i$ of $X$ is covered by the 3-set $p$ in the solution. In fact, no method can be selected for constraint *i-at-least* (thus involving no solution) if *every* connected variable is the output of an *absent* method.

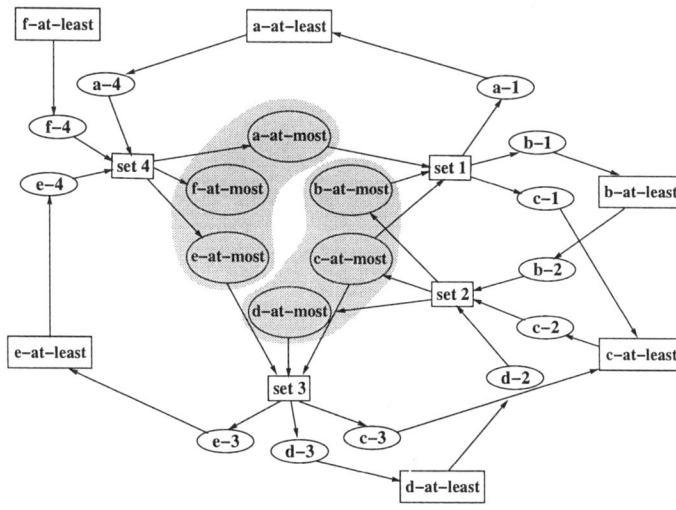

Figure 3: Valid method graph after transforming the instance of "*Exact Cover by 3-Sets*" (Figure 2).

## 4.2 Proof of Theorem 1

First, the planning reduction is $O(|X|+|E|)$. Indeed, one 3-set of $E$ corresponds to one constraint, five methods, nine edges, and three variables in the planning problem. One element of $X$ corresponds to one constraint and one variable.

Second, "*constraint-planning*" is in *NP* since verifying that a method graph issued from a planning reduction is valid is $O(|V|+|C|)$. Indeed, it just has to be verified that every variable is determined by at most one constraint, and that every constraint has one method selected for it.

Finally, the two following paragraphs prove the equivalence between $(i)$ a solution $S$ for any instance $G$ of "*Exact Cover by 3-Sets*" and $(ii)$ a solution $S'$ for the instance $G'$ of "*constraint-planning*" obtained by applying to $G$ the planning reduction.

$(i) \rightarrow (ii)$ Based on $S$, $S'$ is obtained by selecting present methods for the constraints $set_p$ in *ConstE* when the 3-set $p$ is in $S$. The absent method is selected for the 3-sets that are not in $S$. Every constraint $i$-at-least in *ConstX* is directed onto variable $i$-$p$ of *VarE* when the 3-set $p$ is in $S$. Variable $i$-$p$ is an input of the present method selected for the constraint $set_p$ in *ConstE*. By construction, every constraint has one method selected for it.

By hypothesis, every element of $X$ belongs to exactly one 3-set of $S$. Since there is no intersection between any two 3-sets in $S$, this construction does not generate conflicts on variables $i$-at-most of *VarX*.

Every element $i$ of $X$ is covered by (at least) one 3-set $p$ of $S$. By construction, $set_p$ is activated with the present method that outputs to $i$-at-most. By construction, variable $i$-$p$ is determined by constraint $i$-at-least, thus there is no variable conflict generated on $i$-$p$. The other variables of constraint $i$-at-least do not provide conflicts because they are linked only to two constraints and are not determined by $i$-at-least. Thus, for every element $i$ of $X$, no corresponding variable in the constraint planning problem can cause a variable conflict.

$(ii) \rightarrow (i)$ Based on $S'$, $S$ is built by collecting a 3-set $\{a,b,c\}$ when the present method that outputs to variables $a$-at-most, $b$-at-most, and $c$-at-most is selected. Since the method graph is valid, the intersection of any two 3-sets in $S$ is empty.

Let us consider every constraint $i$-at-least of *ConstX* in $S'$. Let $i$-$p$ be the variable determined by $i$-at-least. $i$-$p$ is necessarily an input variable of constraint $set_p$ that determines variable $i$-at-most, otherwise a variable conflict would occur on $i$-$p$. By construction, $i$ necessarily belongs to a 3-set in $S$. □

## 4.3 Complexity of 2-output constraint planning problems

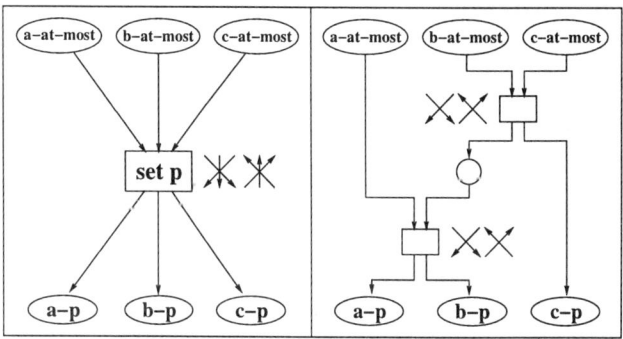

Figure 4: A 3-output constraint $set_p$ transformed into two 2-output constraints. The motifs next to a constraint indicate the possible methods.

We know that when "*constraint-planning*" is restricted to single-output constraints, the problem complexity comes down to P ($cp_1$). The planning reduction shows that "*constraint-planning*" is NP-complete with 3-output constraints. A natural question is therefore, whether the 2-output constraint restriction would yield a polynomial problem or not.

**Theorem 2** *Let $G$ be a dataflow constraint system for which the method restriction holds. $G$ contains methods that have at most two outputs.*

*Then proving the existence of a valid method graph (cyclic or not) corresponding to $G$ is NP-complete.*

**Proof.** Every 3-output constraint $set_p$ can easily be transformed into two 2-output constraints and a

| problem | method restriction | single-output | complexity | proof | algorithms |
|---------|-------------------|---------------|------------|-------|------------|
| $acp_1$ | yes | yes | $P$ | [Sutherland, 1963] | PDOF, DeltaBlue |
| $acp_2$ | yes | no | $P$ | [Vander Zanden, 1996] | QuickPlan |
| $acp_3$ | no | yes | $NPC$ | [Maloney, 1991] 3 | – |
| $acp_4$ | no | no | $NPC$ | [Maloney, 1991] 1 and 2 | – |
| $cp_1$ | yes | yes | $P$ | [Gangnet and Rosenberg, 1992] | Maximum-matching |
| **$cp_2$** | **yes** | **no** | **$NPC$** | **Theorems 1 and 2** | **SkyBlue** |
| **$cp_3$** | **no** | **yes** | **$P$** | **Theorem 3 and $cp_1$** | **Maximum-matching** |
| **$cp_4$** | **no** | **no** | **$NPC$** | **$cp_2$** | **SkyBlue** |

Table 2: Computational complexity of constraint planning problems. The contributions of this paper are bold-faced.

"dummy" variable, as shown in Figure 4. The global behavior remains exactly the same[2]. □

## 5 Influence of the method restriction on problems with cyclic solutions

The following theorem states that the method restriction has no influence on the computational complexity of the constraint planning problem when cyclic solutions are allowed.

**Theorem 3** *Let $C$ be a class of dataflow constraint systems and let $P_C$ be the problem of existence of a valid method graph (cyclic or not) for any instance in the class $C$. Let $P'_C$ be the restriction of $P_C$ to constraint systems that satisfy the method restriction.*

*Then $P_C$ and $P'_C$ are polynomially (actually LOG-space) equivalent.*

The proof is based on the *method transformation*.

**Definition 5** *Let $G_1 = (V_1, C_1, M_1)$ be a constraint system. Based on $G_1$, the method transformation provides a constraint system $G_2 = (V_2, C_2, M_2)$ such that: (1) $V_1 = V_2$, (2) $C_1 = C_2$, (3) methods in $M_1$ for which the method restriction holds, occur unchanged in $M_2$, and (4) every method $m_1$ in $M_1$ for which the method restriction does not hold is replaced by a method $m_2$ in $M_2$ for which the method restriction holds: $m_2$ has the same output variables as $m_1$ and has all of the other variables of the associated constraint as input.*

Note that this trivial transformation is LOG-space. Thus, Theorem 3 can be applied to problems that are in $P$ or are NP-complete.

**Proof of Theorem 3.** First, $P'_C$ can be reduced to $P_C$ since $P'_C$ is a restriction of $P_C$. Second, $P_C$ can be reduced to $P'_C$ thanks to the method transformation that reduces a constraint system $G_1$ into a constraint system $G_2$ for which the method restriction holds. We prove the equivalence between $(i)$ a solution of $G_1$ and $(ii)$ a solution of $G_2$.

[2] Figure 3 illustrates that every variable is determined by a constraint. Thus, no solution can be found for 2-output constraint problems when the two 2-output constraints are selected in opposite directions.

$(ii) \rightarrow (i)$ A valid method graph of $G_2$ can be transformed into a valid method graph of $G_1$ since $G_1$ is the same as $G_2$ without certain edges. This does not induce variable conflicts.

$(i) \rightarrow (ii)$ A valid method graph of $G_1$ can be transformed into a valid method graph of $G_2$ since every edge added by the method transformation connects a constraint and one of its *input* variable. This does not generate variable conflicts. □

Note that if directed cycles are not allowed in the solution, the last implication is false because adding an input edge could introduce a directed cycle.

## 6 Synthesis

These three theorems allow us to deduce the three missing computational complexity results, as shown in Table 2. Since $cp_2$ (which is NP-complete) is a restriction of $cp_4$, and $cp_4$ is in $NP$, $cp_4$ is also NP-complete[3]. Theorem 3 proves that $cp_1$ and $cp_3$ have the same computational complexity.

Since the table is now complete, we can highlight interesting points about the constraint planning problems handled by existing algorithms.

**SkyBlue** Problem $cp_2$ assumes that the constraints must be *required*, whereas SkyBlue [Sannella, 1994] can handle one type of *preferential* constraints. Therefore, the SkyBlue problem is NP-hard. The NP-completeness result given by Theorem 1 makes the exponential worst case time complexity of SkyBlue less surprising. However, [Sannella, 1994] has proven that SkyBlue could reach this worst case complexity even on problems $acp_2$ and $cp_1$ that are in $P$.

**QuickPlan** QuickPlan [Vander Zanden, 1996] cannot be extended without making the corresponding problem NP-complete. Indeed, the gap between $acp_2$ (in $P$) and $acp_4$ (NP-complete) is due to the method restriction, as described in [Vander Zanden, 1996]. Moreover, the gap between $acp_2$ (in $P$) and $cp_2$ (NP-complete) is due also to cyclic solutions being accepted or not.

[3] Theorem 3 applied to $cp_2$ and $cp_4$ also proves this result.

The companion paper [Trombettoni and Neveu, 1997] proves that there exist instances of "*constraint-planning*" issued from the planning reduction which only have cyclic solutions (otherwise $P$ would be equal to $NP$).

**Maximum-matching**  The Maximum-matching problem is in $P$ [Gangnet and Rosenberg, 1992]. Since the gap between $cp_1$ and $cp_2$ lies in the *single-output constraint* restriction, Maximum-matching cannot be extended to multi-output constraints.

Note that Maximum-matching can also solve problem $cp_3$. Indeed, Theorem 3 can easily be extended to the problem of *finding* a solution, thanks to a reverse transformation. So one needs to (1) transform an instance of $cp_3$ into one of $cp_1$ with the *method transformation*, (2) call Maximum-matching on the $cp_1$ instance and (3) retrieve the solution (if any) with the reverse method transformation.

## 7  Complexity of problems with acyclic constraint graphs

We know that an acyclic constraint graph cannot yield a method graph with directed cycles. The restriction of the two problems $acp_i$ and $cp_i$ ($i \in \{1..4\}$) to acyclic constraints graphs is then a unique problem $p'_i$.

Problems $p'_1$ and $p'_2$ are in $P$ because they are restrictions of $acp_1$ and $acp_2$. In the same way, $p'_3$ is in $P$ since it is a restriction of $cp_3$. $p'_2$ and $p'_4$ can be seen as problems where cyclic solutions are allowed (in fact, all solutions are acyclic and one does not need to disallow cyclic solutions). They satisfy the conditions of Theorem 3 so that $p'_4$ is in $P$. We can then conclude that all of the restrictions to acyclic constraint graphs are in $P$.

## 8  Conclusion

We have proven new computational complexity theorems. First, the constraint planning problem handled by SkyBlue is NP-hard. We do not know yet whether it is in $NP$, when handling *constraint hierarchies* that are a widely used type of preferential constraints. Second, the computational complexity is insensitive to the method restriction when cyclic solutions are allowed. Based on the theoretical results presented in this paper, the following simple rule gives sufficient conditions to determine if a given constraint planning problem is in $P$.

---
   **if** the constraint graph contains no cycle **then**
      the problem is in $P$
   **else if** an acyclic solution is expected **then**
      the problem is in $P$ if the method restriction is imposed
   **else**
      the problem is in $P$ if it only contains single-output constraints

---

This rule highlights the importance of the "cyclic/acyclic solution" condition. When directed cycles are not allowed in the solution, the gap between problems in $P$ and NP-complete problems comes from the method restriction, and not from the single-output constraint restriction. The problem complexity has exactly the opposite behavior when cyclic solutions are allowed. Finally, the companion paper shows that the polynomial complexity of problem $acp_1$, $acp_2$, $cp_1$, or $cp_3$ is not lost when handling constraint hierarchies.

We believe that these results will help designers to conceive multi-way constraint systems that provide a good balance between expressiveness and performance.

## Acknowledgements

Special thanks to N. Chleq and T. Schiex whose comments were very helpful. Also thanks to C. Bliek, C. Lottaz, N. Prcovic, and R. Stalker.

## References

[Freeman-Benson et al., 1990] Bjorn Freeman-Benson, John Maloney, and Alan Borning. An incremental constraint solver. *Communications of the ACM*, 33(1):54–63, January 1990.

[Gangnet and Rosenberg, 1992] Michel Gangnet and Burton Rosenberg. Constraint programming and graph algorithms. In $2^{nd}$ *International Symposium on Artificial Intelligence and Mathematics*, January 1992.

[Hoover, 1987] Roger Hoover. *Incremental Graph Evaluation*. PhD thesis, Cornell University, Ithaca, 1987.

[Maloney, 1991] John Maloney. *Using Constraints for User Interface Construction*. PhD thesis, Department of Computer Science and Engineering, University of Washington, Seattle, 1991. Published as Technical Report 91-08-12.

[Papadimitriou, 1994] Christos H. Papadimitriou. *Computational Complexity*. Addison-Wesley, 1994.

[Sannella, 1994] Michael Sannella. *Constraint Satisfaction and Debugging for Interactive User Interfaces*. PhD thesis, Department of Computer Science and Engineering, University of Washington, Seattle, 1994.

[Sutherland, 1963] Ivan Sutherland. *Sketchpad: A Man-Machine Graphical Communication System*. PhD thesis, Department of Electrical Engineering, MIT, 1963.

[Trombettoni and Neveu, 1997] Gilles Trombettoni and Bertrand Neveu. Computational complexity of multiway, dataflow constraint problems. Technical Report 97-86, CERMICS, January 1997.

[Vander Zanden, 1996] Bradley Vander Zanden. An incremental algorithm for satisfying hierarchies of multiway, dataflow constraints. *ACM Transactions on Programming Languages and Systems*, 18(1):30–72, January 1996.

# CONSTRAINT SATISFACTION

## Constraint Satisfaction 2: SAT

# Heuristics Based on Unit Propagation for Satisfiability Problems

## Chu Min Li & Anbulagan
LaRIA, Univ. de Picardie Jules Verne, 33, Rue St. Leu, 80039 Amiens Cédex, France
fax: (33) 3 22 82 75 02, e-mail: {cli@laria.u-picardie.fr, Anbulagan@utc.fr}

## Abstract

The paper studies new unit propagation based heuristics for Davis-Putnam-Loveland (DPL) procedure. These are the novel combinations of unit propagation and the usual "Maximum Occurrences in clauses of Minimum Size" heuristics. Based on the experimental evaluations of different alternatives a new simple unit propagation based heuristic is put forward. This compares favorably with the heuristics employed in the current state-of-the-art DPL implementations (C-SAT, Tableau, POSIT).

## 1 Introduction

Consider a propositional formula $F$ in Conjunctive Normal Form (CNF) on a set of Boolean variables $\{x_1, x_2, ..., x_n\}$, the satisfiability (SAT) problem consists in testing whether clauses in $F$ can all be satisfied by some consistent assignment of truth values (1 or 0) to the variables. If it is the case, $F$ is said satisfiable; otherwise, $F$ is said unsatisfiable. If each clause exactly contains $r$ literals, the subproblem is called $r$-SAT problem.

SAT problem is fundamental in many fields of computer science, electrical engineering and mathematics. It is the first NP-Complete problem [Cook, 1971] with 3-SAT as the smallest NP-Complete subproblem.

The Davis-Putnam-Loveland procedure (DPL) [Davis et al., 1962] is a well known complete method to solve SAT problems, roughly sketched in Figure 1.

DPL procedure essentially constructs a binary search tree, each recursive call constituting a node of the tree. Recall that all leaves (except eventually one for a satisfiable problem) of a search tree represent a dead end where an empty clause is found. The branching variables are generally selected to allow to reach as early as possible a dead end, i.e. to minimize the length of the current path in the search tree.

The most popular SAT heuristic actually is Mom's heuristic, which involves branching next on the variable having Maximum Occurrences in clauses of Minimum Size [Dubois et al., 1993; Freeman, 1995; Pretolani, 1993; Crawford and Auton, 1996; Jeroslow and Wang, 1990]. Intuitively these variables allow to well exploit the power of unit propagation and to augment the chance to reach an empty clause. Recently another heuristic based on Unit Propagation (UP heuristic) has proven useful and allows to exploit yet more the power of unit propagation [Freeman, 1995; Crawford and Auton, 1996; Li, 1996]. Given a variable $x$, a UP heuristic examines $x$ by respectively adding the unit clause $x$ and $\bar{x}$ to $F$ and independently makes two unit propagations. The real effect of the unit propagations is then used to weigh $x$.

**procedure DPL(F)**
**Begin**
```
if F is empty, return "satisfiable";

F:=UnitPropagation(F); If F contains an empty
clause, return "unsatisfiable".

/* branching rule */
select a variable x in F according to a heuristic
H, if the calling of DPL(F ∪ {x}) returns
"satisfiable" then return "satisfiable", otherwise
return the result of calling DPL(F ∪ {x̄}).
```
**End.**

**procedure UnitPropagation(F)**
**Begin**
```
While there is no empty clause and a unit clause l
exists in F, assign a truth value to the variable
contained in l to satisfy l and simplify F.
Return F.
```
**End.**

Figure 1: DPL Procedure

However, since examining a variable by two unit propagations is time consuming, two major problems remain open: should one examine all free variables by unit propagation at every node of a search tree? if not, what are the variables to be examined at a search tree node?

In this paper we try to experimentally solve these two problems to obtain an optimal exploitation of UP heuristic. We define a $PROP$ predicate at a search tree node whose denotational semantics is the set of variables to be examined at a search tree node, i.e. $x$ is to be examined if and only if $PROP(x)$ is true. By appropriately changing $PROP$, we experimentally analyse the behaviour of different UP heuristics. We write 12 DPL procedures which are different only in $PROP$ and run these procedures on a very large sample of hard random 3-SAT problems.

We begin in section 2 by describing the 12 DPL procedures and summarizing the experimental results on these programs. In section 3 we compare a pure UP heuristic and a pure Mom's heuristic and show the superiority of UP heuristics. In section 4 we study different restrictions of UP heuristics. In section 5 we discuss the related work and compare the best DPL procedure in our experimentation with three state-of-the-art DPL procedures. Section 6 concludes the paper.

## 2 UP Heuristics Driven by $PROP$

Let $diff(F_1, F_2)$ be a function which gives the number of clauses of minimum size in $F_1$ but not in $F_2$, we show a generic branching rule in Figure 2, where the equation defining $H(x)$ is suggested in [Freeman, 1995] and the weight 5 for uniformizing clauses of different length is empirically optimal.

```
For each free variable x such that PROP(x) is
true do
let F' and F'' be two copies of F
Begin
    F' := UnitPropagation(F' ∪ {x});
    F'' := UnitPropagation(F'' ∪ {x̄});
    If both F' and F'' contain an empty clause then
        return "F is unsatisfiable".
    If F' contains an empty clause then x := 0,
    F := F'' else if F'' contains an empty
    clause then x := 1, F := F';
    If neither F' nor F'' contains an empty clause
        then let w(x) denote the weight of x
        w(x) := diff(F', F) and w(x̄) := diff(F'', F)
End;

If all variables examined above are valued or
PROP(x) is false for every x then
For each free variable x in F do
    let r_i be the length of the clause C_i
    w(x) := Σ_{x̄∈C_i} 5^{-r_i}  and  w(x̄) := Σ_{x∈C_i} 5^{-r_i}

For each variable x do
    H(x) := w(x̄) * w(x) * 1024 + w(x̄) + w(x)

Branching on the free variable x such that H(x) is
the greatest.
```

Figure 2: A Generic Branching Rule Driven by $PROP$

The essential reason to use UP heuristics instead of Mom's one is that Mom's heuristic may not maximize the effectiveness of unit propagation, because it only takes binary clauses (if any) into account to weigh a variable, although some extensions try to also take longer clauses into account with exponentially smaller weights (e.g. 5 ternary clauses are counted as 1 binary clauses). A UP heuristic allows to take all clauses containing a variable and their relations into account in a very effective way to weigh the variable. As a secondary effect, it allows to detect the so-called *failed literals* in $F$ which when satisfied falsify $F$ in a single unit propagation. However since examining a variable by two unit propagations is time consuming, it is natural to try to restrict the variables to be examined. For this purpose we use $PROP$ predicate defined at a search tree node.

The success of Mom's heuristic suggests that the larger the number of binary occurrences of a variable is, the higher its probability of being a good branching variable is, implying that if one should restrict UP heuristics by means of $PROP$, he should restrict UP heuristics to those variables having a sufficient number of binary occurrences. For this reason, all restrictions on $PROP$ studied in this paper are defined according to the number of binary occurrences of a variable, so that the resulted UP heuristics rely on combinations of unit propagation and Mom's heuristics.

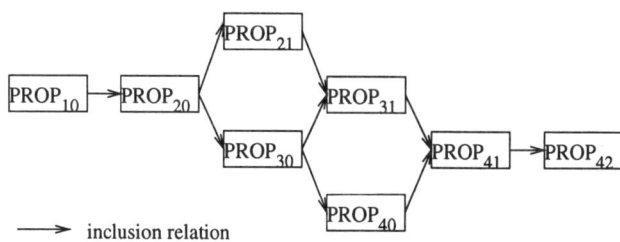

Figure 3: $PROP$ predicate hierarchy by inclusion relation. $PROP_{ij}(x)$ is true iff $x$ has $i$ binary occurrences of which at least $j$ negative and $j$ positive.

The first $PROP$ predicate in our experimentation is called $PROP_a$ and has empty denotational semantics, the resulted branching rule using a pure Mom's heuristic, and the second is called $PROP_0$ whose denotational semantics is the set of all free variables, the resulted UP heuristic is in its pure form and plays its full role:

$PROP_a$: $PROP_a(x)$ is false for every free variable $x$;

$PROP_0$: $PROP_0(x)$ is true for every free variable $x$.

Between $PROP_a$ and $PROP_0$, there are many other possible $PROP$ predicates. Figure 3 defines 8 predicates constituting a hierarchy by inclusion relation of their denotational semantics.

$PROP_{3141}$ is defined to be $PROP_{31}$, but for nodes under a fixed depth of a search tree, it is defined to be

$PROP_{41}$. Let $T$ be a constant, the last $PROP$ predicate is named $PROP_z$ and is defined to be the first of the three predicates $PROP_{41}$, $PROP_{31}$ and $PROP_0$ (in this order) which has at least $T$ variables in its denotational semantics.

Each $PROP$ predicate results in a DPL procedure, $PROP_a$, $PROP_0$, $PROP_{3141}$ and $PROP_z$ respectively giving $Sata$, $Sat0$, $Sat_{3141}$ and $Satz$, and $PROP_{ij}$ giving $Sat_{ij}$. These programs are different only in $PROP$ predicate, except $Sat0$ which need not count the occurrences of variables.

We run the 12 programs (compiled using gcc with optimization) on a PC with a 133 Mhz Pentium CPU under Linux operating system on a very large sample of random 3-SAT problems generated by using the method of Mitchell et al.[Mitchell et al., 1992]. Given a set $V$ of $n$ Boolean variables $\{x_1, x_2, ..., x_n\}$, we randomly generate $m$ clauses of length 3. Each clause is produced by randomly choosing 3 variables from $V$ and negating each with probability 0.5. Empirically, when the ratio $m/n$ is near 4.25 for a 3-SAT formula $F$, $F$ is unsatisfiable with a probability 0.5 and is the most difficult to solve. We vary $n$ from 140 variables to 340 variables incrementing by 20, for each $n$ the ratio clauses-to-variables ($m/n$) is set to 4.0, 4.1, 4.2, 4.25, 4.3, 4.4, 4.5. At each ratio and by each program, if $n < 280$ then 1000 problems are solved, if $280 \leq n \leq 300$ then 500 problems are solved, if $n = 320$ then 300 problems are solved, and if $n = 340$ then 100 problems are solved. A problem is solved successively by all the 12 DPL procedures before another to ensure the same environment to all programs. Due to the lack of space, we only present the experimental results for the ratio $m/n = 4.25$ in Figures 4, 5, and 6, where the DPL procedures corresponding to the curves are listed in the same order from top to bottom. The experimental results on the other ratios give exactly the same conclusions.

## 3 A Pure UP Heuristic Versus a Pure Mom's Heuristics: $Sat0$ vs $Sata$

$Sat0$ systematically examines all the variables by unit propagation at all nodes, using a pure UP heuristic, while $Sata$ does not examine any variable so and employs a pure Mom's heuristic. One might believe that $Sat0$ would be simply too slow, but it is not the case. *$Sat0$ is much faster than $Sata$*. In fact from Figures 4 and 5, all DPL procedures using a UP heuristic in our experimentation are substantially better than $Sata$ in terms of search tree size and real run time.

Note that Mom's heuristic used in $Sata$ is similar to the so-called two-sided Jeroslow-Wang rule [Hooker and Vinay, 1995], with the only difference that a clause of length i is counted as 5 clauses of length i+1 instead of 2. Our experiments suggest that 5 is better than 2. 5 is also similar to the exponential factors in C-SAT [Dubois et al., 1993] where 5.71 ternary clauses are counted as 1 binary clause.

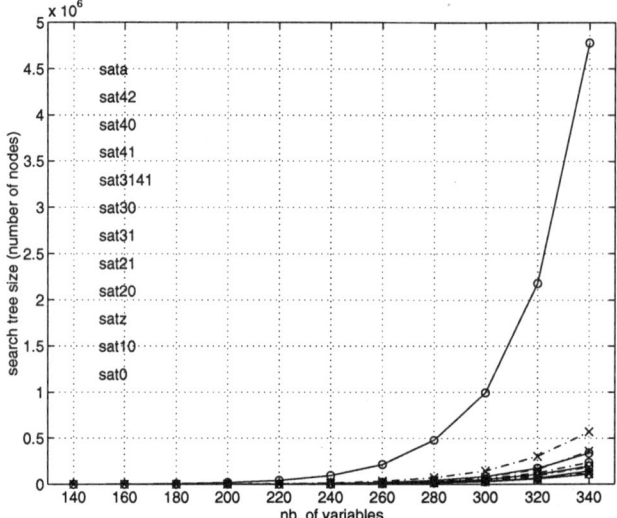

Figure 4: Mean search tree size of each program as a function of $n$ for hard random 3-SAT problems at the ratio $m/n = 4.25$

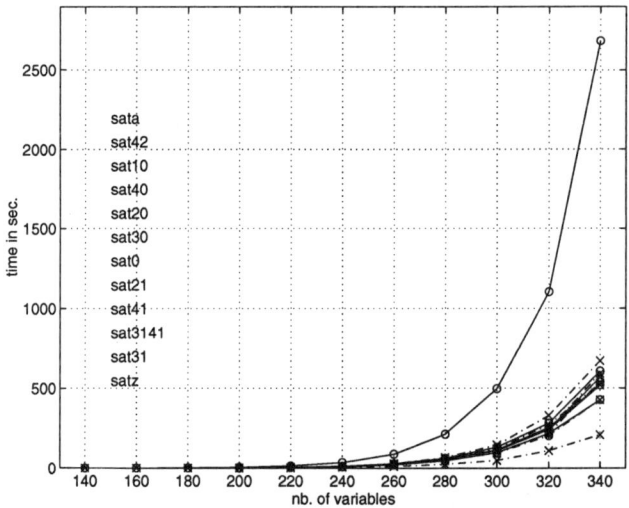

Figure 5: Mean run time of each program as a function of $n$ for hard random 3-SAT problems at the ratio $m/n = 4.25$

$Sat0$ actually is slower than five other programs based on balanced restrictions of variables to be examined by unit propagation, but not substantially so (except $Satz$). The surprisingly good performance of $Sat0$ confirms the power of UP heuristics for selecting the next branching variable and suggests that its effect for detecting failed literals is only secondary.

## 4 Restricted UP Heuristics

Figure 6 illustrates the number of variables examined by different restricted UP heuristics at a node.

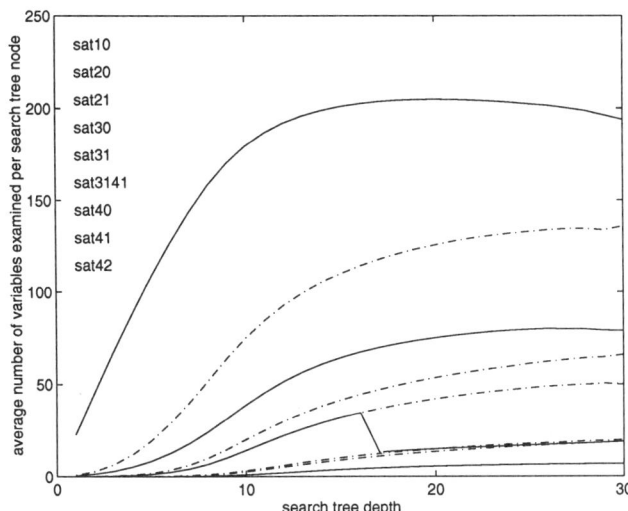

Figure 6: Average number of variables examined at a search tree node in a given depth when solving hard random 3-SAT problems of 300 variables and 1275 clauses (500 problems are solved) for 9 programs

## 4.1 Restriction by total number of binary occurrences of a variable

Four programs $Sat_{10}$, $Sat_{20}$, $Sat_{30}$ and $Sat_{40}$ realize this type of restrictions. While a classical Mom's heuristic selects *the* next branching variable having maximum binary occurrences, the restricted UP heuristics examine a set of variables having more binary occurrences than others, including the variable having maximum binary occurrences. From Figure 4, it is clear that the more variables are examined, the smaller the search tree size is.

## 4.2 Balanced restriction by total number of binary occurrences of a variable

Four programs $Sat_{21}$, $Sat_{31}$, $Sat_{41}$ and $Sat_{42}$ realize this type of restrictions. The $PROP$ predicates require that a variable occurs both positively and negatively in binary clauses to balance the search tree. We compare the duet $Sat_{i0}$ and $Sat_{i1}$ (i=2, 3, 4) and observe that $Sat_{i1}$ examines strictly fewer variables than $Sat_{i0}$ and is faster than it in spite of a slightly larger search tree. In particular, $Sat_{41}$ and $Sat_{40}$ examine almost the same number of variables (see Figure 6), but the balanced restriction gives a faster DPL procedure.

We pay special attention to $PROP_{31}$ and $PROP_{41}$ since they seem to be the best balanced restrictions.

## 4.3 Dynamic restriction as a function of search tree depth

$Sat_{3141}$ realizes this restriction. A general observation when solving 3-SAT problems using a DPL procedure is that there are more and more binary clauses when descending from the search tree root and the denotational semantics of a $PROP$ predicate such as $PROP_{31}$ becomes larger and larger. Furthermore, the nodes are more numerous near the leaves and the branching variables play a less important role there. It appeared that one could restrict more the variables to be examined by unit propagation near the leaves without important loss on the search tree size so as to obtain some gain in terms of real run time.

POSIT's UP heuristic (called BCP-based heuristic) [Freeman, 1995] realizes this idea: under the level 9 of a search tree, at most 10 variables are examined by unit propagation.

$Sat_{3141}$ uses $PROP_{31}$ from the top of a search tree, but under the depth empirically fixed to $n*4/70$, it uses $PROP_{41}$, where $n$ is the number of variables in the initial input 3-SAT problem. Note that if $n \geq 160$, $n*4/70 \geq 9$, so $Sat_{3141}$ generally strengthens the restriction later than POSIT.

From Figures 4 and 5 $Sat_{3141}$ is not better than $Sat_{31}$, although it makes many fewer unit propagations to examine variables (see Figure 6), suggesting that the search tree depth is rather irrelevant to the restriction of UP heuristics.

## 4.4 Dynamic restriction by number of variables to be examined

The relatively poor performance of $Sat_{42}$ seems due to the small number of variables examined at each node (see Figure 6), though these variables have many binary occurrences. A careful analysis shows that even $Sat_{31}$, the best one up to now, examines few or no variables at some nodes, especially near the root where there are few binary clauses, although these nodes are more determinant for the final search tree size. $PROP_z$ is then introduced to ensure that at least $T$ variables are examined at each node, $T$ being empirically fixed to 10. Near the root, all free variables are examined to exploit the full power of UP heuristic. As soon as the number of variables occurring both negatively and positively in binary clauses and having at least 4 (3) binary occurrences is larger than $T$, only these variables are examined to select the next branching variable.

## 5 Related Work

C-SAT [Dubois et al., 1993] examines some variables by unit propagations (called local processing) near the bottom of a search tree to rapidly detect failed literals there. Pretolani also uses a similar approach (called pruning method) based on hypergraphs in H2R [Pretolani, 1993]. But the local processing and the pruning method as are respectively presented in [Dubois et al., 1993] and [Pretolani, 1993] do not contribute to the heuristic to select the next branching variable. We find the first effective exploitation of UP heuristic in POSIT [Freeman, 1995] and Tableau [Crawford and Auton, 1996] which use a similar idea as in C-SAT to determine the variables to be examined at a node by unit propagation: *x is to be examined iff x is among the k most weighted variables by a Mom's heuristic.*

The main difference of *Satz* with Tableau and POSIT is that *Satz* does not specify a upper bound $k$ of the number of variables to be examined at a node by unit propagation. Instead, *Satz* specifies a lower bound. In fact, *Satz examines many more variables by an optimal combination of unit propagation and Mom's heuristics.*

Given the depth of a node, Table 1 illustrates the average number of variables examined ($\#examined\_vars$) at the node by *Satz*, with the depth of the root being 0. In order to compare with C-SAT, Tableau and POSIT we also give the theoretical value of $k_C$ (for C-SAT), $k_T$ (for Tableau) and $k_P$ (for POSIT) at the node, respectively according to the definitions of $k$ in [Dubois et al., 1993; Crawford and Auton, 1996; Freeman, 1995].

| depth | #free_vars | #examined_vars | $k_C$ | $k_T$ | $k_P$ |
|---|---|---|---|---|---|
| 1 | 298.24 | 298.24 | 0 | 263 | 265 |
| 2 | 296.52 | 296.52 | 0 | 227 | 230 |
| 3 | 294.92 | 293.89 | 0 | 193 | 198 |
| 4 | 292.44 | 292.21 | 0 | 141 | 149 |
| 5 | 288.60 | 282.04 | 0 | 61 | 72 |
| 6 | 285.36 | 252.14 | 0 | 0 | 10 or 3 |
| 7 | 281.68 | 192.82 | 0 | 0 | 10 or 3 |
| 8 | 277.54 | 125.13 | 0 | 0 | 10 or 3 |
| 9 | 273.17 | 71.51 | 0 | 0 | 10 or 3 |
| 10 | 268.76 | 40.65 | 0 | 0 | 10 or 3 |
| 11 | 264.55 | 26.81 | 0 | 0 | 10 or 3 |
| 12 | 260.53 | 21.55 | 0 | 0 | 10 or 3 |
| 13 | 256.79 | 19.80 | 0 | 0 | 10 or 3 |
| 14 | 253.28 | 19.24 | 0 | 0 | 10 or 3 |
| 15 | 249.96 | 19.16 | 0 | 0 | 10 or 3 |
| 16 | 246.77 | 19.28 | 0 | 0 | 10 or 3 |
| 17 | 243.68 | 19.57 | 0 | 0 | 10 or 3 |
| 18 | 240.68 | 19.97 | 0 | 0 | 10 or 3 |
| 19 | 237.73 | 20.46 | 0 | 0 | 10 or 3 |
| 20 | 234.82 | 20.97 | 0 | 0 | 10 or 3 |

Table 1: Average number of variables examined in *Satz* at a node in a given depth when solving a hard random 3-SAT problem of 300 variables and 1275 clauses (500 problems are solved) compared with theoretical value of $k$ in C-SAT, Tableau and POSIT

It is clear that *Satz* examines many more variables at each node than any of C-SAT, Tableau or POSIT. Near the root, *Satz* examines all free variables. Elsewhere *Satz* examines a sufficient number ($T$) of variables.

We compare C-SAT, Tableau, POSIT and *Satz* on a large sample of hard random 3-SAT problems on a SUN Sparc 20 workstation with a 125 MHz CPU. The 3-SAT problems are generated from 3 sets of $n$ variables and $m$ clauses at the ratio $m/n = 4.25$, $n$ steping from 300 variables to 400 variables by 50.

We use an executable of C-SAT dated July 1996. The version of Tableau used here is called *3tab* and is the same used for the experimentation presented in [Crawford and Auton, 1996]. POSIT is compiled using the provided *make* command on the SUN Sparc 20 workstation from the sources named *posit*-1.0.*tar.gz*[1]. Table 2 shows the performances of the 4 DPL procedures on problems of 300, 350, and 400 variables, where *time* standing for the real mean run time is reported by the unix command /usr/bin/time and *t_size* standing for search tree size (number of nodes) is reported (or computed from number of branches reported) by the DPL procedures.

| | 300 vars 300 problems | | 350 vars 250 problems | | 400 vars 100 problems | |
|---|---|---|---|---|---|---|
| System | time | t_size | time | t_size | time | t_size |
| C-SAT | 77 | 49567 | 512 | 275303 | 3818 | 1624869 |
| Tableau | 79 | 43041 | 558 | 253366 | 4544 | 1524551 |
| POSIT | 57 | 61797 | 474 | 400588 | 3592 | 2751611 |
| *Satz* | 34 | 32780 | 203 | 174337 | 1207 | 916569 |

Table 2: Mean run time (in second) and mean search tree size of C-SAT, Tableau, POSIT and *Satz* on ratio $m/n$=4.25

Table 2 shows that *Satz* is faster than the above cited versions of C-SAT, Tableau and POSIT, *Satz*'s search tree size is the smallest, and *Satz*'s run time and search tree size grow more slowly. Table 3 shows the gain of *Satz* compared with the cited version of C-SAT, Tableau and POSIT at the ratio $m/n$=4.25. Each item is computed from Table 2 using the following equation:

$$gain = (value(system)/value(Satz) - 1) * 100\%$$

where *value* is real mean run time or real mean search tree size and *system* is C-SAT, Tableau or POSIT. From Table 3, it is clear that the gain of *Satz* grows with the size of the input formula.

| | 300 vars 300 problems | | 350 vars 250 problems | | 400 vars 100 problems | |
|---|---|---|---|---|---|---|
| System | time | t_size | time | t_size | time | t_size |
| C-SAT | 126% | 51% | 152% | 58% | 216% | 77% |
| Tableau | 132% | 31% | 175% | 45% | 276% | 66% |
| POSIT | 68% | 89% | 133% | 130% | 198% | 200% |

Table 3: The gain of *Satz* vs. C-SAT, Tableau and POSIT in terms of run time and search tree size on the ratio $m/n$=4.25 computed from Table 2

The central strategy of *Satz* is to try to reach an empty clause as early as possible. Further along the line, we make two relatively small resolvents-driven improvements in *Satz*. The first improvement is the preprocessing of the input formula by adding some resolvents of length $\leq 3$. The second improvement consists in refining yet more the heuristic $H$ in the nodes where all free variables are examined by unit propagation. Refer to Figure 2, when $PROP_z$ is equal to $PROP_0$ we define $w(x)$ as the number of resolvents the newly produced binary clauses would result in in $F'$ by a single step of resolution. $w(\bar{x})$ is similarly defined.

---

[1] publicly available via anonymous ftp to ftp.cis.upenn.edu in pub/freeman/ directory

*Satz* improved in this way solves many real-world or structured SAT problems where previous heuristics were not successful. For example, Table 4 shows the performance of the 4 DPL procedures on the well-known Beijing challenging problems[2], where a problem that can not be solved in less than 2 hours is marked by "> 7200" and the version of Tableau is called *ntab*[3]. It is clear that *Satz* is much more efficient and solves many more problems in less than two hours.

| Problem | Satz | C-SAT | Posit | ntab |
|---|---|---|---|---|
| 2bitadd_10 | > 7200 | > 7200 | > 7200 | > 7200 |
| 2bitadd_11 | 201 | > 7200 | 0.3 | > 7200 |
| 2bitadd_12 | 0.4 | 6379 | 0.05 | > 7200 |
| 2bitcomp_5 | 0.03 | 0.1 | 0.01 | 0.4 |
| 2bitmax_6 | 0.07 | 3.7 | 0.01 | 1.6 |
| 3bitadd_31 | > 7200 | > 7200 | > 7200 | > 7200 |
| 3bitadd_32 | 4512 | > 7200 | > 7200 | > 7200 |
| 3blocks | 2.0 | 4.3 | 1.8 | 1468 |
| 4blocksb | 8.2 | 118 | 49 | > 7200 |
| 4blocks | 1542 | > 7200 | > 7200 | > 7200 |
| e0ddr2-10-by-5-1 | 215 | > 7200 | > 7200 | > 7200 |
| e0ddr2-10-by-5-4 | 232 | > 7200 | 3508 | 236 |
| enddr2-10-by-5-1 | > 7200 | > 7200 | > 7200 | > 7200 |
| enddr2-10-by-5-8 | 229 | > 7200 | > 7200 | 92 |
| ewddr2-10-by-5-1 | 339 | > 7200 | 283 | > 7200 |
| ewddr2-10-by-5-8 | 279 | > 7200 | > 7200 | 119 |

Table 4: Run time (in sec.) of Beijing challenging problems

## 6 Conclusion

We found that UP heuristic is substantially better than Mom's one even in its pure form realized by $PROP_0$ where all free variables are examined at all nodes. In its restricted forms based on combinations of unit propagation and Mom's heuristics, the more variables are examined, the smaller the search tree is, confirming the advantages of UP heuristic, but too many unit propagations slow the execution. The combinations realized by $PROP_{41}$ and $PROP_{31}$ represent good compromises.

A dynamic restriction such as $PROP_{3141}$ which strengthens the restriction under a fixed depth of a search tree fails to work better than the static restriction $PROP_{31}$. We design the dynamic restriction along another line: $PROP_z$ ensures that at least $T$ candidates are examined by unit propagation at *every* node of a search tree by successively using $PROP_{41}$, $PROP_{31}$ and $PROP_0$, giving the very efficient and very simple DPL procedure called *Satz*.

*Satz* is favorably compared with several current state-of-the-art DPL implementations (C-SAT, Tableau and POSIT) on a large sample of hard random 3-SAT problems and the recent Beijing SAT benchmarks. The good performance of *Satz* on the structured or real-world SAT problems shows that UP heuristic can tackle new problems or problem domains where Mom's heuristics were not successful and enhances the belief that if a DPL procedure is efficient for random SAT problems, it should be also efficient for a lot of structured ones.

## Acknowledgments

We thank Olivier Dubois, James M. Crawford and Jon W. Freeman for kindly providing us their DPL procedures and anonymous referees for their comments which helped improve this paper.

## References

[Chvatal and Szemeredi, 1988] V. Chvatal and E. Szemeredi. Many Hard Examples for Resolution. *Journal of ACM*, 35(4):759–768, October 1988.

[Cook, 1971] S. A. Cook. The Complexity of Theorem Proving Procedures. In *3rd ACM Symp. on Theory of Computing*, pages 151-158, Ohio, 1971.

[Crawford and Auton, 1996] J. M. Crawford and L. D. Auton. Experimental Results on the Crossover Point in Random 3-SAT. *Artificial Intelligence*, 81, 1996.

[Davis et al., 1962] M. Davis, G. Logemann, and D. Loveland. A machine program for theorem proving. *Communication of ACM*, 5(7):394-397, July 1962.

[Dubois et al., 1993] Olivier Dubois, P. Andre, Y. Boufkhad and Jacques Carlier. SAT versus UNSAT. *Second DIMACS Challenge: Cliques, Coloring and Satisfiability*, Rutgers University, NJ, 1993.

[Freeman, 1995] Jon W. Freeman. Improvements to Propositional Satisfiability Search Algorithms. Ph.D. thesis, Department of computer and Information science, Univ. of Pennsylvania, Philadelphia, PA, 1995.

[Hooker and Vinay, 1995] J. N. Hooker and V. Vinay. Branching Rules for Satisfiability. *Journal of Automated Reasoning*, 15:359-383, 1995.

[Jeroslow and Wang, 1990] R. Jeroslow and J. Wang. Solving Propositional Satisfiability Problems. *Annals of Mathematics and AI*, 1:167-187, 1990.

[Li, 1996] ChuMin LI. Exploiting Yet More the Power of Unit Clause Propagation to Solve 3-SAT Problem. In *ECAI'96 Workshop on Advances in Propositional Deduction*, pages 11-16, Budapest, Hungary, 1996.

[Mitchell et al., 1992] D. Mitchell, B. Selman, H. Levesque. Hard and Easy Distributions of SAT Problems. In *AAAI'92*, pages 459–465, San Jose, CA, 1992.

[Pretolani, 1993] Daniele Pretolani. Satisfiability and hypergraphs. Ph.D. thesis, Dipartimento di Informatica, Università di Pisa, 1993.

---

[2] available from http://www.cirl.uoregon.edu/crawford/beijing
[3] available from http://www.cirl.uoregon.edu/crawford/

# Hidden *Gold* in Random Generation of SAT Satisfiable Instances

**Thierry Castell**
IRIT - Université Paul Sabatier
118, route de Narbonne
31062 Toulouse Cedex
France

**Michel Cayrol**
IRIT - Université Paul Sabatier
118, route de Narbonne
31062 Toulouse Cedex
France

## Abstract

Evaluation of incomplete algorithms that solve SAT requires to generate hard satisfiable instances. For that purpose, the kSAT uniform random generation is not usable. The other generators of satisfiable instances generate instances that are not intrisically hard, or exhaustive tests have not been done for determining hard and easy areas. A simple method for generating random hard satisfiable instances is presented. Instances are empirically shown to be hard for three classical methods: the "Davis-Putnam" procedure (which is complete), and the two incomplete methods: GSAT and the Break Out Method. Moreover, a new method for escaping from local minima is presented.

## 1 Introduction

A crucial problem in practical AI development is the problem of satisfiability (called SAT) of a finite set of propositional clauses. SAT is a NP-complete problem. Which AI system does not use a satisfiability test or equivalent (Constraint Satisfaction Problem, Graph coloring, ...)? Formal reasoning is limited because there is no efficient propositional theorem prover. On one hand, the complete methods that solve SAT have not really progressed since the resolution method and Davis-Putnam procedure, except that it has been shown that the choice of the heuristics is central [Dubois *et al.*, 1993]. On the other hand, the incomplete methods have been really developed thanks to the results of GSAT for the random instances [Selman *et al.*, 1993].

A threshold phenomenom has been brought to light for the generation of random instances [Cheeseman *et al.*, 1991], making possible the generation of random hard instances. Satisfiable instances are needed for evaluating the incomplete methods. But only 50% of the hardest instances are satisfiable. It has been shown [Chvàtal and Szemerédi, 1988] that the unsatisfiable instances are hard (for the resolution) but it has not been shown [Cha and Iwama, 1995] that the satisfiable instances are difficult (for the local search for example). In works on evaluation of incomplete methods, [Konolige, 1994] showed the inefficiency of GSAT on random structured instances, compared with Davis-Putnam procedure. But with few changes, GSAT becomes efficient [Kask and Dechter, 1995]. May be it is due to the fact that these structured instances are not *intrisically hard*. But, what is it the definition of *intrisically hard*? So, a main problem about SAT is to determine if a set of instances is hard. There are no general definition and determination method of the complexity of a set of instances (except when a polynomial class is recognized). For this reason, empirical methods are used for showing the difficulty of the instances produced by a generator. A *hard* set of instances is a set of instances which are hard for the best known algorithms, namely the "Davis-Putnam" procedure [Davis *et al.*, 1962], GSAT with random walk [Selman and Kautz, 1993] and the Break Out Method [Morris, 1993].

## 2 The kSAT Uniform Random Generation

An instance of kSAT is produced by the uniform random generation if each clause of the instance is randomly and independently picked out. These instances are generated from two parameters: the number of variables and the ratio. The ratio is defined as the number of clauses divided by the number of variables. The uniform generation of kSAT instances is an interesting theoretical problem. A threshold phenomenom has been brought to light for the probability to pick out a satisfiable instance as a function of the ratio [Cheeseman *et al.*, 1991] [Crawford and Auton, 1993]. The threshold seems to appear when the probability to generate a satisfiable instance is equal to 1/2. The value of this critical ratio has been determined (empirically) as 4.25 for 3SAT, 9.8 for 4SAT etc [Mitchell *et al.*, 1992] [Dubois *et al.*, 1993]. A problem is to determine theoretically this threshold. Near the threshold, if the ratio decreases, the probability to pick out a satisfiable instance quickly tends to 1; if the ratio increases, the probability to pick out a satisfiable instance quickly tends to 0. The more important the number of variables, the more quickly the probability tends to 0 (or 1). Moreover, it has been shown (empirically) that the hardest instances are generated for a ratio value equal to the threshold. For other values, the instances are easier.

This generation method is used for evaluating complete or incomplete theorem provers. In the hardest region, it has been proved that inconsistent instances are difficult (for the resolution) [Chvàtal and Szemerédi, 1988]. But there is no theoretical result for the difficulty of the satisfiable instances. Futhermore, for generating hard instances, the higher the number of variables, the more the value for the ratio is precise. A little variation on the ratio can produce only satisfiable or only insatisfiable instances.

For example, if the theoretical value of the critical ratio for 1000 variables is equal to 4.255 then the uniform generation would produced, for 1000 variables and with a ratio equal to 4.25, more than 50% of satisfiable instances. In this case, the evaluation of a local search algorithm under the assumption "50% satisfiable" is a big mistake.

With a lot of variables, the complete methods are unusable. Thus it is impossible to know the rate of success of the incomplete methods. The uniform generation is not a good means for evaluating the incomplete methods if the *theoretical precise value* of the threshold is not known.

## 3 Satisfiable Instances

The algorithm kSAT_GEN presented in [Cha and Iwama, 1995] produces only random satisfiable instances. The clauses are randomly constructed in order to be satisfied by a given model. There are several generation parameters: number of variables, ratio and literal distribution (i.e. the number of occurrences of each literal in the generated set of clauses). In [Cha and Iwama, 1995], there is no result on the relation between the parameters and the difficulty of the instances. For not producing only easy instances, the studies on the uniform random generation have shown the importance of the value of the parameters. There are a lot of parameters for kSAT_GEN, then it will be difficult to make exhaustive tests.

Two other methods are proposed [Cha and Iwama, 1995] for generating always inconsistent sets of clauses or sets of clauses having one and only one model. The basic steps of these generators are the reversal resolution principle and the reversal subsumption simplification (a clause C1 subsumes a clause C2 iff C1 ⊆ C2). If the generation is reversed, a resolution proof can be easily obtained. The time of the generation of the sets of clauses must be limited. So there is a short proof, by resolution, of the empty clause for the inconsistent sets of clauses. And there is a short proof, by resolution, for each prime implicate for the sets of clauses having only one model. The smaller the generation time, the smaller the resolution proofs! This kind of instances is not intrinsically hard, because the resolution proof of the prime implicates is short. A very good resolution-based theorem prover would be able to solve easily this kind of instances.

Let x be a propositional symbol; x is called a *positive* literal and ¬x is called a *negative* literal. One possibility for generating only consistent sets of clauses is to leave out the clauses that have only positive litterals, from a set of clauses generated by uniform random generation (this idea appears in [Morris, 1993] [Rauzy, 1995]). For these instances, a model is obtained by assigning false to all the variables. Unfortunately, this method often produces easy instances. It is due to the fact that the clauses with only positive literals are left out, so there more negative literals than positive literals. Consequently, the Davis-Putnam procedure easily finds a model (with a good heuristic).

Moreover, [Rauzy, 1995] has noticed that only "nearly Horn-renamable" instances are generated. To generate hard instances, new parameters must be introduced.

Each clause has a negative literal. An "intelligent" algorithm would be able to show this property. To solved that problem, some literals of the instances have to be renamed (swap x and ¬x for some x), and then some clauses without negative literal appear.

## 4 How to be Hard

Let us consider the case where the clauses without negative literals are left out. In order to control the proportion of positive and negative literals, [Rauzy, 1995] introduces parameters for determining the probability to pick out a given sign of a clause. For 3SAT, there are four kinds of sign for a clause: (i) all the literals in the clause are positive, (ii) two literals are positive and one is negative, (iii) one literal is positive and two are negative, (iv) all the literals are negative. So there are four parameters for the generation: $\delta$ the probability to pick out a clause (i), $\gamma$ for a clause (ii), $\beta$ for a clause (iii) and $\alpha$ for a clause (iv).

The probability to pick out each sign of a clause is $\alpha=\delta=1/8$ and $\beta=\gamma=3/8$ for the uniform random generation. With a probability equal to $\delta=0$ to pick out a clause which sign is (i), and with given probabilities for the others, [Rauzy, 1995] proposed two equations for generating instances (before renaming) with approximatively the same number of positive and negative literals:

(1) $\alpha+\beta+\gamma=1$ (since clauses with only positive literals are rejected)
(2) $3\alpha+\beta=\gamma$ (for the equilibrium between positive and negative literals)

For the sake of simplicity, a restrictive strategy is chosen in our work. Only one parameter is introduced: the probability to pick out a positive literal, called *posp*. The following algorithm is proposed:

```
function kSAT_satisfiable(k, nbv, ratio, posp)
input: k = number of literals per clause, nbv = number of
variables, ratio = value for number of variables/number of
clauses, posp = probability to pick out a positive literal
(0 ≤ posp < 1).
output: a satisfiable instance of kSAT
B := ∅; nbc := 0;
while (nbc < nbv*ratio) do
    {x_1, ..., x_k} := a random part of the set of the nbv
    variables;
    Clause := ∅;
    for i := 1 to k do
        let p be a random number among [0,1[;
        if (p ≤ posp) then Clause := Clause∪{x_i};
        else Clause := Clause∪{¬x_i};
    if a negative literal belongs to Clause then
        B := B∪{Clause}; nbc := nbc+1;
rename B; %swap x and ¬x for some propositional symbol x %
return B;
```

## 5 Hidden Gold

In order to obtain instances of 3SAT consisting of sets of clauses with the same expected total number of positive and negative literals, in the same method as [Rauzy, 1995], *posp* must be equal to $\left(\sqrt{5}-1\right)/2 \approx 0.61803$.

Indeed, with the generator, the values of α,β,γ in function of *posp* are: α=$(1-posp)^3/(1-posp^3)$, β=$3posp.(1-posp)^2/(1-posp^3)$ and γ=$3posp^2(1-posp)/(1-posp^3)$.

In order to satisfy equations (1) and (2) there is only one solution for *posp*: $\left(\sqrt{5}-1\right)/2$. We've just found the expected gold, this value is equal to

$$1/the\_golden\_ratio$$

## 6 Search Algorithms

The generator kSAT_satisfiable has been tested on three classical different methods. This generator has also been used to evaluate two new personal versions of the Break Out method.

### 6.1 Complete Method

We use the classical "Davis-Putnam" procedure (DP). It is an enumeration of the interpretations. With sophisticated heuristics, it is the most efficient complete algorithm for the uniform random instances of kSAT. For our tests, an implementation with heuristics of C-SAT is used [Dubois et al., 1993]. C-SAT generally solves in few seconds hard instances (from uniform random generation) having 200 variables. Instances having 300 variables are solved in less than one hour. Instances having more than 500 variables require some days, years or much more ...

### 6.2 Incomplete Methods

These methods are incomplete because they cannot always find a model. But they can solve very large problems [Selman et al., 1992]. At the present time, the best algorithms for SAT are based on the hill-climbing algorithm. For a given interpretation, these methods try to decrease, by local changes on the interpretation, the number of falsified clauses. A local change in the interpretation is an inversion of the truth value of a variable, called *flip* of the variable. For these methods, the crucial problem is to escape from a local minimum.

We used the Break Out Method (BOut) [Morris, 1993] and GSAT with random walk [Selman and Kautz, 1993]. BOut escapes from a local minimum by weigthing the falsified clauses on that local minimum. The results presented in [Cha and Iwama, 1995] show that this method seems to be better than GSAT which is supposed to be the best incomplete method for solving SAT. GSAT [Selman et al., 1992] is the more popular incomplete algorithm for SAT. There exists some refinements, in particular a notion of weighting [Selman and Kautz, 1993] as for BOut. For the tests, we used the version with random walk. A random walk is an elementary improvement for escaping from local minima: with a given probability, a variable that appears in a falsified clause is flipped.

### 6.3 Escaping from a Local Maxima

**Break Out Method with Jump**

Now an improvement of BOut is presented. It is called the Break Out Method with Jump (BOJ). On a local minimum very often the same clauses are falsified. To avoid that, on a local minimum, *all the variables that appear in the falsified clauses are flipped*. Thus several flips have to be done for falsifying again these clauses.

Another change is introduced. In some cases, the current interpretation tends to the opposite of a model; in order to solve that, the following property is used: when the sum of the weights of the clauses that are totally satisfied (each literal of the clause is satisfied) is lower than the sum of the weights of the falsified clauses, if all the variables are flipped, the new interpretation is better.

The Break Out Method with Jump is represented by the following algorithm:

```
function BOJ(B) : boolean
input : a set B of clauses.
output : true if B is satisfiable.
let I an initial random interpretation;
until (I becomes a model of B) do
    if (the sum of the weights of the totally satisfied
        clauses is lower than the sum of the weights of the
        falsified clauses)
    then all the variables of I are flipped;
    if (I is not a local minimum) then
        Poss_flips:=the set of the variables that minimize
        the sum of the weights of the falsified clauses;
        flip on I a variable from Poss_flips;
    else
        add a weight to the falsified clauses;
        flip on I all the variables that appear in the
        falsified clauses;% jump %
return True;
```

**A Specialized Algorithm: *Mirror***

This variation of BOut called *Mirror* is presented because its efficiency is a mystery for us. Its principle is simple. It is BOut with only one new operation. On a local minimum, falsified clauses are normally weigthed and immediately after *all the variables are flipped* (it is the reason for the name *Mirror*). In the hardest areas of the other incomplete algorithms, this method is amazingly efficient. But it has a lot of problems on obvious areas. Moreover this method seems to be efficient only with our generator.

## 7 Empirical Results

The figures appear at the end of the paper.

kSAT_satisfiable has been tested with a lot of values for nbv, ratio and *posp*, for the 3SAT case. Each point is calculated from a sample of 100 instances. Only general results for 100 variables are presented (the results are similar for 50, 150 and 200 variables). Here we are interested in the behaviour of the different methods as well as the easy and hard instances of the generator. Comparing the algorithms is outside the scope of the paper.

The complexity of Davis-Putnam procedure is measured in terms of the sum of the average number of calls, the number of unit propagations and the number of pure literal simplifications (a literal is a pure literal for a set of clauses if its complementary literal does not appear in this set). The complexity of the incomplete methods is measured in terms of the average number of tested interpretations. This measurement is more precise than the number of flips. Indeed, in order to choose a best neighbour, a local search method inspects the interpretations in the neighbourhood of the current interpretation (explicitly or implicitly). Therefore, this measurement enables us to differenciate algorithms which do not have the same kind of neighbourhood, or which do not explore the search space in the same way.

The most classical parameters are chosen for the incomplete algorithms. The initial weight of the clauses is fixed to one and a unit weight is added on local minima. The probability of the random walk for GSAT is equal to 1/2. In order to avoid a too quick restart, no restart is realized. The programs are stopped after 100 000 flips, but GSAT with random walk is stopped after 200 000 flips. It is a high value compared to values used in experimentations with uniform random generation. Experimentations indicate that GSAT with random walk needs in average 27 654 flips and almost no restart in order to prove the consistency of the instances composed of 200 variables generated at the threshold by the standard model [Selman et al., 1994].

## 7.1 The Hardest Areas

As for the ratio of instances solved (figure 1), for DP it is equal to 100%, which is not surprising since DP is a complete method. The performances of GSAT with random walk are poor: close to 0% of instances solved for a ratio equal to 8 and $posp = 0.8$. BOut behaves better but still have big weaknesses. BOJ behaves quite well, almost all instances are solved. Moreover, BOJ has the same behaviour of the other algorithms, but the failure starts only at around 200 variables. As for complexity (figure 2), GSAT, BOut and BOJ have similary curves. The complexity of these algorithms grows with the ratio. It is more simple for DP, the form of the complexity is easy-hard-easy. It is like the uniform random generation. The hardest areas are located at a ratio$\approx$4.4 and $posp\approx$0.64. However, the same "ridge" of hardness is present in DP: when $ppos$ increases with the ratio.

We could conclude that this model for generating consistent instances generates hard instances and that the region of the hardest instances follows a ridge which is common to all methods: when $ppos$ increases with the ratio.

But we are immediately going to check that this conclusion is erroneous.

## 7.2 Calling the conclusions into question

We must not conclude too quickly, even if some evidences are showed. We are going to show that the presence of a common ridge in the complexity curves is not directly linked with the hardness of instances but with the methods used. To show this, we are going to make again the tests with *Mirror*, figure 3. Moreover, when describing the DP procedure used in the experimentations, we omitted an important factor: the presence of a heuristics in the choice of the first subtree explored (heuristics of S-SAT [Dubois et al., 1993]). We are also going to make again the tests with DP, but with a choice of the first subtree explored which is opposed to that proposed by the heuristics and which is random, figure 4.

The curves always show the presence of a ridge, but here *posp* decreases when the ratio increases! For a high *posp* value and a high ratio, we observe that the generated instances are easy for *Mirror*: 100% of the instances are solved quickly. And for the hardest area of *Mirror* (low *posp* value and a high ratio), the generated instances are easy for BOut, BOJ and GSAT: 100% are solved quickly.

When the first subtree explored by DP is selected randomly then the two ridges appear, figure 4.

## 8 Conclusion

With the kSAT-consistent function, it is easy to generate hard satisfiable instances. These instances turn out to be practically much harder than instances generated by the standard random model. 50 variables are enough to make GSAT+random walk and the Break Out Method little efficient. The generated instances have no structure and yet they are very hard to solved (by almost the most popular algorithms). As for the efficiency of local search, the results of *Mirror* bring into light gaps of classical approaches. There exists instances generally simple to solve (by*Mirror*) which may be extremely difficult for methods such as GSAT and the Break Out Method. An important result of our tests is that incomplete algorithms must be evaluated with all the different areas, and not only on the hardest areas of a given algorithm. If *Mirror* was tested only on the hardest area of BOut or GSAT, a mistake would be made on its efficiency.

New operations must be integrated to local search in order to make it more robust. The flip of the whole set of variables (mirror) is one of our proposals, but up to now this operation gives interesting results on this model of generation only. In order to capture the good characteristics of BOut, GSAT and *Mirror*, the BOJ algorithm has been developed. On our generator, BOJ improves BOut and seems also better on other kinds of instances.

We venture to conjecture that for the kSAT-consistent function, the generated instances which are the hardest to solve are, independently from the method used, located on a ratio $\approx$ 4.4 and $posp \approx$ 0.64. For a given algorithm, the presence of a zone harder than this point brings into light the gaps of the evaluated method.

Generally speaking, the kSAT-consistent function shows that, as complete methods, local search has weaknesses and allows for instances that are not solvable in a reasonable amount of time. As for the uniform generation, the simplified model generation of satisfiable instances must be intensively studied.

## Acknowledgments

To Elsa, Claudette for helping us with English, Jérôme for helping us finishing the paper and the members of RESSAC.

## References

[Cha and Iwama, 1995] B. Cha and K. Iwama. Performance Test of Local Search Algorithms Using New Types of Random CNF Formulas. *Proc. of IJCAI-95*, pages 304-310, 1995.

[Cheeseman *et al.*, 1991] P. Cheeseman, B. Kanefsky and W.M. Taylor.Where the Really Hard Problems Are. *Proc. of IJCAI-91*, pages 331-337, 1991.

[Chvàtal and Szemerédi, 1988] V. Chvàtal and E. Szemerédi. Many Hard Examples for Resolution. *JACM*, 35 (4):759-768, 1988.

[Crawford and Auton, 1993] J.M. Crawford and L.D. Auton. Experimental Results on the Cross over Point in Satisfiability Problems. *Proc. of AAAI-93*, pages 21-27, 1993.

[Davis *et al.*, 1962] M. Davis, G. Logemann and D. Loveland. A Machine Program for Theorem Proving. *JACM*, (5):394-397, 1962.

[Dubois *et al.*, 1993] O. Dubois, P. Andre, Y. Boufkhad and J. Carlier. SAT versus UNSAT. *2nd DIMACS Challenge Workshop*, 1993.

[Kask and Dechter, 1995] K. Kask and R. Dechter. GSAT and Local Consistency. *Proc. of IJCAI-95*, pages 616-621, 1995.

[Konolige, 1994] K. Konolige. Easy to be Hard: difficult problems for greedy algorithms. *Proc. of KR-94*, pages 374-378, 1994.

[Mitchell *et al.*, 1992] D. Mitchell, B. Selman.and H. Levesque. Hard and Easy Distribution of SAT problems. *Proc. of AAAI-92*, 1992.

[Morris, 1993] P. Morris. The Breakout Method For Escaping From Local Minima. *Proc. of AAAI-93*, 1993.

[Rauzy, 1995] A. Rauzy. On the Random Generation of 3-SAT Instances. *Technical Report 1060-95, LaBRI-URA CNRS 1304*, Université Bordeaux I, 1995.

[Selman and Kautz, 1993] B. Selman and H.A. Kautz. Domain Independent Extensions to GSAT: Solving Large Structured Satisfiability Problems. *Proc. of IJCAI-93*, pages 290-295, 1993.

[Selman *et al.*, 1992] B. Selman, H. Levesque and D. Mitchell. A new Method for Solving Hard Satisfiability Problems. *Proc. AAAI-92*, pages 337-343, 1992.

[Selman *et al.*, 1994] B. Selman, H.A. Kautz, B. Cohen. Noise strategies for improving local search. *Proc. of AAAI-94*, 1994.

## Figures

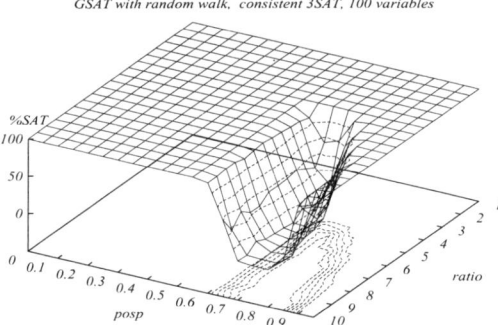
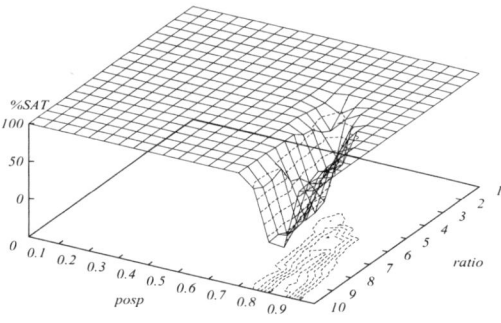
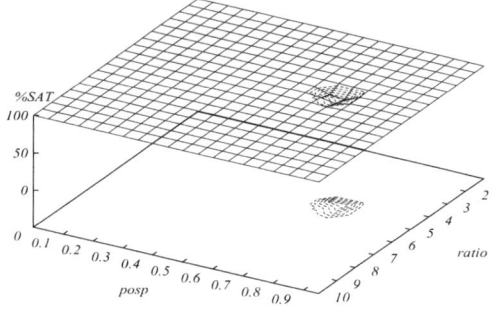
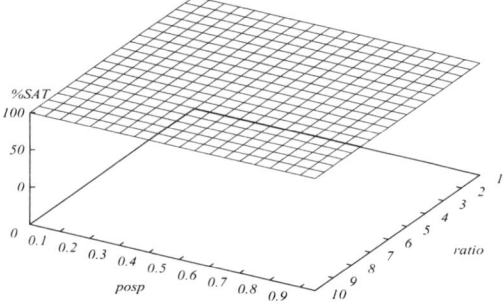

**Figure 1:** number of solved instances

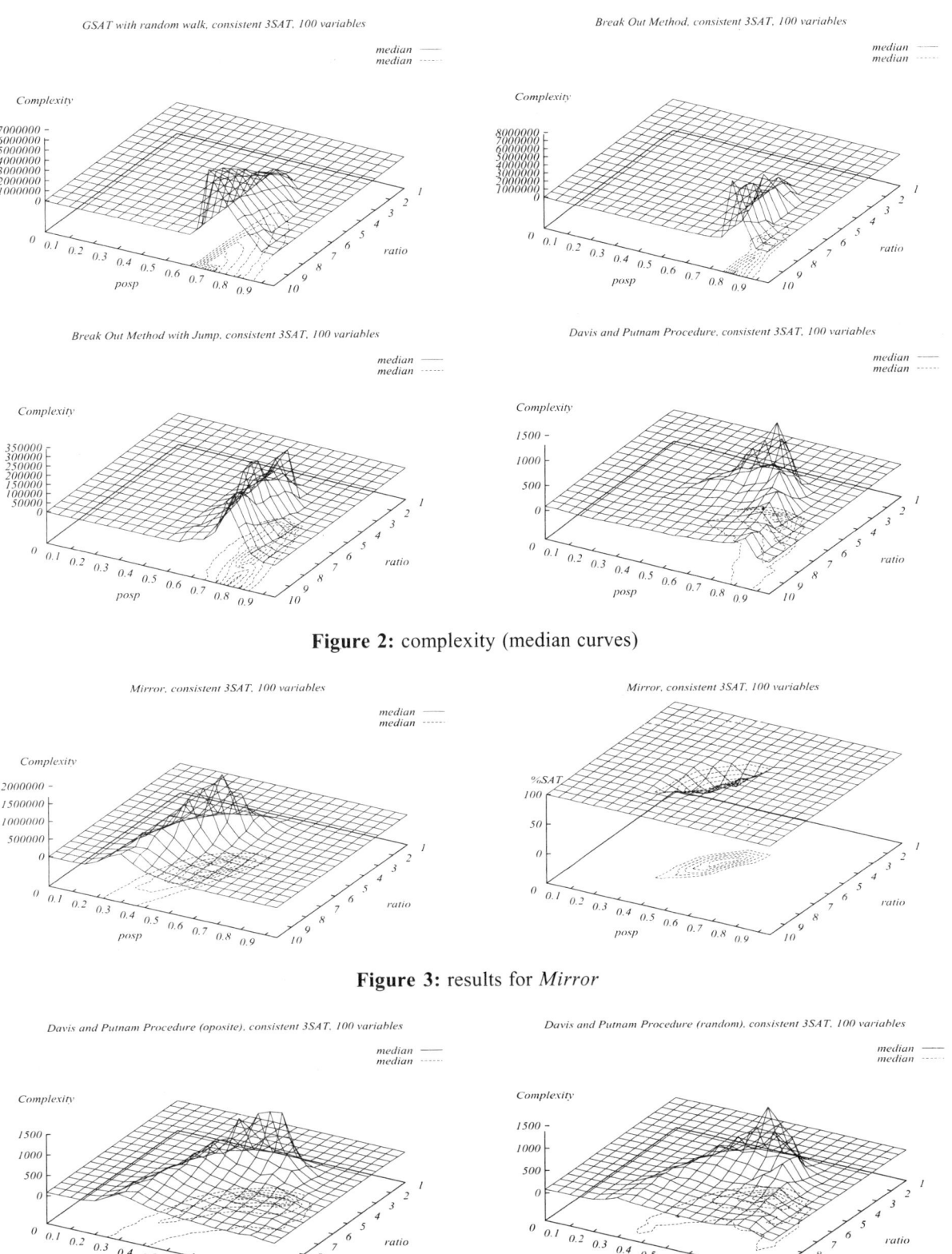

**Figure 2:** complexity (median curves)

**Figure 3:** results for *Mirror*

**Figure 4:** Davis-Putnam with opposite and random choice for the first subtree explored

# Discrete Lagrangian-Based Search for Solving MAX-SAT Problems *

Benjamin W. Wah and Yi Shang
Coordinated Science Laboratory
University of Illinois at Urbana-Champaign
1308 West Main Street
Urbana, IL 61801, USA
{wah, shang}@manip.crhc.uiuc.edu
URL: http://manip.crhc.uiuc.edu

## Abstract

Weighted maximum satisfiability problems (MAX-SAT) are difficult to solve due to the large number of local minima in their search space. In this paper we propose a new discrete Lagrangian based search method (DLM) for solving these problems. Instead of restarting from a new point when the search reaches a local minimum, the Lagrange multipliers in DLM provide a force to lead the search out of the local minimum and move it in a direction provided by the multipliers. Since DLM has very few parameters to be tuned, it can be made deterministic and the results, reproducible. We compare DLM with GRASP in solving a large set of test problems, and show that it finds better solutions and is substantially faster. DLM has a solid theoretical foundation that can be used as a systematic approach for solving general discrete optimization problems.

## 1 Introduction

The satisfiability (SAT) problem is defined as follows. Given a set of $n$ clauses $\{C_1, C_2, \cdots, C_n\}$ on $m$ variables $x = (x_1, x_2, \cdots, x_m)$, $x_i \in \{0, 1\}$, and a Boolean formula in conjunctive normal form (CNF),

$$C_1 \wedge C_2 \wedge \cdots \wedge C_n, \qquad (1)$$

find an assignment to the variables so that (1) evaluates to be *true*, or derive its infeasibility if (1) is infeasible.

The *weighted maximum satisfiability problem* (MAX-SAT) is a general case of SAT. In MAX-SAT, each clause $C_i$ is associated with weight $w_i$. The objective is to find an assignment to the variables that maximizes the sum of the weights of satisfied clauses,

$$\max_{x \in \{0,1\}^m} \sum_{i=1}^{n} w_i S_i(x), \qquad (2)$$

where $S_i$ equals 1 if logical assignment $x$ satisfies $C_i$, and 0 otherwise. This objective is equivalent to minimizing the sum of the weights of unsatisfied clauses.

MAX-SAT problems are difficult to solve for the following reasons. First, they have a large number of local minima in their search space, where a local minimum is a state whose local neighborhood does not include any state that is strictly better. Second, the weights in the objective function of a MAX-SAT problem can lead to a much more rugged search space than the corresponding SAT problem. When a MAX-SAT problem is satisfiable, existing SAT algorithms [Gu, 1989; Selman and Kautz, 1993] developed to find a satisfiable assignment to the corresponding SAT problem can be applied, and the resulting assignment is also optimal for (2). However, this approach does not work when the MAX-SAT problem is not satisfiable. In this case, existing local search SAT methods have difficulties in overcoming the rugged search space.

Methods for solving MAX-SAT can be classified as incomplete and complete, depending on whether they can find the optimal assignment. Complete algorithms can determine optimal assignments. They include mixed integer linear programming methods [Resende *et al.*, 1997] and various heuristics [Hansen and Jaumard, 1990; Goemans and Williamson, 1995; Joy *et al.*, 1997]. They are generally computationally expensive and have difficulties in solving large problems. On the other hand, incomplete methods are usually faster and can solve some large problems that complete methods cannot handle. Many incomplete local search methods have been designed to solve large SAT problems of thousands of variables [Gu, 1989; Selman and Kautz, 1993; Wah and Shang, 1996; Shang and Wah, 1997] and have obtained promising results in solving MAX-SAT problems [Jiang *et al.*, 1995; Resende *et al.*, 1997].

---

*Research supported by National Science Foundation Grant MIP 96-32316 and National Aeronautics and Space Administration Grant NAG 1-613.
Source code of DLM is at http://manip.crhc.uiuc.edu.

Another approach to solve MAX-SAT is to transform it into continuous formulations. Discrete variables in the original problem are relaxed into continuous variables in such a way that solutions to the continuous problem are binary solutions to the original problem. This transformation is potentially beneficial because an objective in continuous space may smooth out some local minima. Unfortunately, continuous formulations require computationally expensive algorithms, rendering them applicable only to small problems.

A discrete Lagrangian-based search method was developed previously for solving SAT problems [Wah and Shang, 1996; Shang and Wah, 1997]. Traditional Lagrangian theory for solving continuous problems was extended to discrete constrained optimization problems and incorporated in *DLM* (*Discrete Lagrangian method*) that works on discrete variables. Instead of restarting from a new starting point when a search reaches a local trap, the Lagrange multipliers in DLM provide a force to lead the search out of the local minimum and move it in the direction provided by the Lagrange multipliers. Hence, DLM can escape from local minima in a continuous trajectory without restarts, avoiding a break in the trajectory as in methods based on restarts. This is advantageous when the trajectory is already in the vicinity of a local minimum, and a random restart may bring the search to a completely different search space. Moreover, DLM has very few algorithmic parameters to be tuned by users, and the search procedure can be made deterministic and the results, reproducible. When applied to the DIMACS SAT benchmarks, DLM generally performs better than the best competing methods and can achieve an order-of-magnitude speedup for many problems.

In this paper we extend the discrete Lagrange-multiplier-based search method to solve MAX-SAT problems. We formulate a MAX-SAT problem in (2) as a discrete constrained optimization problem.

$$\min_{x \in \{0,1\}^m} \quad N(x) = \sum_{i=1}^{n} w_i\, U_i(x) \qquad (3)$$
$$\text{subject to} \quad U_i(x) = 0 \quad \forall i \in \{1, 2, \ldots, n\}.$$

where $w_i > 0$, and $U_i(x)$ equals 0 if the logical assignment $x$ satisfies $C_i$, and 1 otherwise. We propose a mechanism to control the growth rate and magnitude of Lagrange multipliers, which is essential in solving difficult MAX-SAT problems. Our method belongs to the class of incomplete methods that attempt to find approximate solutions in a fixed amount of time.

This paper is organized as follows. Section 2 presents our discrete Lagrangian algorithm. Section 3 discusses issues and alternatives in implementing DLM. Finally, Section 4 presents experimental results in applying DLM to solve a large set of test problems.

## 2 Discrete Lagrangian Methods (DLM)

In this section we first summarize previous work on Lagrangian methods for solving continuous constrained optimization problems. We then extend continuous Lagrangian methods to discrete constrained optimization problems, and apply it to solve MAX-SAT problems.

### 2.1 Continuous Lagrangian Methods

Lagrangian methods are classical methods for solving continuous constrained optimization problems [Luenberger, 1984]. Define an equality constrained optimization problem as follows:

$$\min_{x \in E^m} \quad f(x) \qquad (4)$$
$$\text{subject to} \quad g(x) = 0$$

where $g(x) = (g_1(x), g_2(x), \cdots, g_n(x))$ are $n$ constraints. Lagrangian function $F$ is defined by

$$F(x, \lambda) = f(x) + \sum_{i=1}^{n} \lambda_i g_i(x) \qquad (5)$$

where $\lambda = (\lambda_1, \cdots, \lambda_n) \in E^n$ are Lagrange multipliers.

A *saddle-point* $(x^*, \lambda^*)$ of Lagrangian function $F(x, \lambda)$ is defined as one that satisfies the following condition.

$$F(x^*, \lambda) \leq F(x^*, \lambda^*) \leq F(x, \lambda^*) \qquad (6)$$

for all $(x^*, \lambda)$ and all $(x, \lambda^*)$ sufficiently close to $(x^*, \lambda^*)$.

**(Continuous) Saddle-Point Theorem** [Luenberger, 1984]. $x^*$ is a local minimum to the original problem defined in (4) if and only if there exists $\lambda^*$ such that $(x^*, \lambda^*)$ constitutes a saddle point of the associated Lagrangian function $F(x, \lambda)$.

Based on the Saddle Point Theorem, numerical algorithms have been developed to look for saddle points corresponding to local minima in a search space. One typical method is to do descents in the original variable space of $x$ and ascents in the Lagrange-multiplier space of $\lambda$. The method can be written as a set of ordinary differential equations as follows:

$$\frac{dx}{dt} = -\nabla_x F(x, \lambda) \quad \text{and} \quad \frac{d\lambda}{dt} = \nabla_\lambda F(x, \lambda) \qquad (7)$$

where $t$ is an autonomous variable and $\nabla$ is a gradient.

### 2.2 General DLM

Little work has been done in applying Lagrangian methods to solve discrete constrained combinatorial search problems. The difficulty lies in the requirement of a differentiable continuous space. To apply Lagrangian methods to discrete optimization problems, we need to develop the counterpart of gradient in discrete space.

Similar to (4), a discrete optimization problem is defined as follows.

$$\min_{x \in D^m} \quad f(x) \qquad (8)$$
$$\text{subject to} \quad g(x) = 0$$

where $D$ is the set of integers. We can also define a discrete Lagrangian function similar to (5).

A saddle point $(x^*, \lambda^*)$ of $F(x, \lambda)$ is defined as one that satisfies the following condition:

$$F(x^*, \lambda) \le F(x^*, \lambda^*) \le F(x, \lambda^*) \qquad (9)$$

for all $\lambda$ sufficiently close to $\lambda^*$ and for all $x$ whose Hamming distance between $x^*$ and $x$ is 1.

**Discrete Saddle-Point Theorem.** $x^*$ is a local minimum of (8) if and only if there exists some $\lambda^*$ such that $(x^*, \lambda^*)$ constitutes a saddle point of the associated discrete Lagrangian function $F(x, \lambda)$.

The proof is quite straightforward and is based on showing that $U(x^*) = 0$ in both directions. Based on the Discrete Saddle-Point Theorem, we have developed the following DLM to find discrete saddle points.

**Discrete Lagrangian Method (DLM) $\mathcal{A}$.**

$$x^{k+1} = x^k \ominus \Delta_x F(x^k, \lambda^k) \qquad (10)$$
$$\lambda^{k+1} = \lambda^k + U(x^k) \qquad (11)$$

where $\ominus$ represents exclusive OR (XOR). We define the *discrete gradient* $\Delta_x F(x, \lambda)$ with respect to $x$ such that $\Delta_x F(x, \lambda) = (\delta_1, \cdots, \delta_m) \in \{0, 1\}^m$ with at most one non-zero $\delta_i$, and it gives $F(x, \lambda)$ the greatest reduction in the neighborhood of $x$ with Hamming distance 1.

DLM provides a theoretical foundation and generalization of local search schemes that optimize the objective alone and clause-weight schemes that optimize the constraints alone. In contrast to local search methods that restart from a new starting point when a search reaches a local trap, the Lagrange multipliers in DLM provide a force to lead the search out of a local minimum and move it in the direction provided by the Lagrange multipliers. In contrast to constraint-weight schemes that rely only on the weights of violated constraints to escape from local minima, DLM also uses the value of the objective function to provide further guidance.

In most existing clause-weight schemes, weights are updated dynamically based on which constraints are violated after each flip or after a series of flips. However, they do not use any objective function that may be important in some stages of the search. For instance, when the number of violated constraints is large, then it is important to reduce the number of violated constraints by choosing proper variables to flip. In this case, the weights on the violated constraints may be relatively small as compared to the number of violated constraints and, hence, play a minor role in determining the variables to flip. The advantage of the objective is more apparent in applications in which its value may be large, such as the MAX-SAT problems studied in this paper. The dynamic shift in emphasis between the objective and the constraints, depending on their relative values, is the key of DLM.

### 2.3 DLM Formulation of MAX-SAT

Using the MAX-SAT problem in (3), the Lagrangian function has the same form as in (5):

$$L(x, \lambda) = N(x) + \lambda^T U(x) = \sum_{i=1}^{n}(w_i + \lambda_i)U_i(x) \qquad (12)$$

where $x \in \{0,1\}^m$, $U(x) \in \{0,1\}^n$, and $\lambda^T$ (the transpose of $\lambda = (\lambda_1, \lambda_2, \ldots, \lambda_n)$) denotes the Lagrange multipliers.

There are three important properties about DLM.

- Since $U(x) = 0$ when $\mathcal{A}$ converges to a discrete saddle point, $x$ is globally (not just locally) optimal when $\mathcal{A}$ converges. If any of the constraints in $U(x)$ is not satisfied, then $\lambda$ will continue to evolve to handle the unsatisfied constraints, and the search continues.

- Since $U_i(x)$ is weighted by $w_i + \lambda_i$ in (12), the minimization of $L(x, \lambda)$ by DLM while $\lambda$ is changing is equivalent to perturbing the weights in the original objective function defined in (3). In contrast, existing MAX-SAT methods explore the search space without modifying the weights. This is the major difference between existing methods and DLM.

- The minimization of (12) is equivalent to a clause-weight scheme. This happens because the objective and the constraints in (3) are dependent. In general, when the objective and the constraints are independent, descents in the original-variable space and ascents in the Lagrange-multiplier space form two counteracting forces to converge to saddle points.

## 3 Implementation Issues of DLM

In this section, we discuss various considerations in implementing DLM. Figure 1 shows the pseudo code of $\mathcal{A}_1$, a generic DLM implementing (10) and (11). Define one *iteration* as one pass through the while loop. In the following, we describe some of our design considerations.

(a) *Initial Points* (Lines 1-2). DLM is started from either the origin or from a random initial point generated using a fixed random seed. Further, $\lambda$ is always set to zero. The fixed initial points allow the results to be reproducible easily.

1. Set initial $x$ randomly by a fixed random seed
2. Set initial $\lambda$ to be zero
3. **while** $x$ is not a feasible solution, *i.e.*, $N(x) > 0$
4.     update $x$: $x \longleftarrow x \ominus \Delta_x L(x, \lambda)$
5.     update incumbent if $N(x)$ is better than the current incumbent
6.     **if** condition for updating $\lambda$ is satisfied **then**
7.        update $\lambda$: $\lambda \longleftarrow \lambda + c \times U(x)$
8.     **end if**
9. **end while**

Figure 1: Generic DLM $\mathcal{A}_1$ for solving SAT problems.

(b) *Descent and Ascent Strategies* (Line 4). There are two ways to calculate $\Delta_x L(x, \lambda)$: greedy and hill-climbing, each involving a search in the range of Hamming distance one from the current $x$. In a *greedy strategy*, the assignment leading to the maximum decrease in $L(x, \lambda)$ is selected to update the current assignment. In *hill-climbing*, the first assignment leading to a decrease in $L(x, \lambda)$ is selected to update the current assignment. Depending on the order of search and the number of assignments that can be improved, hill-climbing strategies are generally much faster than greedy strategies.

Among various ways of hill-climbing, we used two alternatives in our implementation: flip the variables one by one in a predefined order, or maintain a list of variables that can improve the current $L(x, \lambda)$ and just flip the first variable in the list. The first alternative is fast when the search starts. By starting from a randomly generated initial assignment, it usually takes very few flips to find a variable that improves the current $L(x, \lambda)$. As the search progresses, there are fewer variables that can improve $L(x, \lambda)$. At this point, the second alternative becomes more efficient and should be applied.

(c) *Conditions for updating* $\lambda$ (Line 6). The frequency in which $\lambda$ is updated affects the performance of a search. The considerations here are different from those of continuous problems. In a discrete problem, descents based on discrete gradients usually make small changes in $L(x, \lambda)$ in each update of $x$ because only one variable changes. Hence, $\lambda$ should not be updated in each iteration of the search to avoid biasing the search in the Lagrange-multiplier space of $\lambda$ over the original variable space of $x$.

Experimental results show that a good strategy is to update $\lambda$ only when $\Delta_x L(x, \lambda) = 0$. At this point, a local minimum in the original variable space is reached, and the search can only escape from it by updating $\lambda$.

(d) *Amount of update of* $\lambda$ (Line 7). A parameter $c$ controls the magnitude of changes in $\lambda$. In general, $c$ can be a vector of real numbers, allowing non-uniform updates of $\lambda$ across different dimensions and possibly across time. In our experiments, $c = 1$ has been found to work well for most of the benchmarks tested. How-

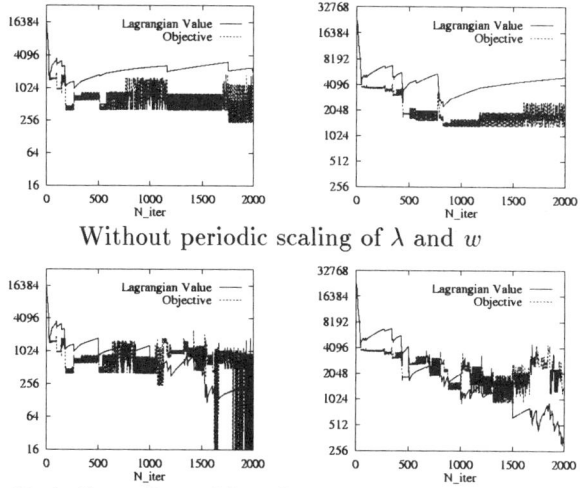

Without periodic scaling of $\lambda$ and $w$

Periodic scaling of $\lambda$ and $w$ every 500 iterations

Figure 2: Execution profiles of *jnh1* (left) and *jnh10* (right). *jnh1* is satisfiable and *jnh10* is not.

ever, for some larger problems, a larger $c$ could result in shorter search time and better solutions.

The update rule in Line 7 results in nondecreasing $\lambda$ because $U(x)$ is either 0 or 1. In contrast, in continuous problems $\lambda_i$ of constraint $g_i(x) = 0$ increases when $g_i(x) > 0$ and decreases when $g_i(x) < 0$.

From previous experience in solving SAT problems using DLM, some $\lambda$ values for some difficult problems can become very large after tens of thousands of iterations. At this point, the Lagrangian search space (12) becomes very rugged, and the search has difficulty in identifying an appropriate direction to move. To cope with this problem, $\lambda$ should be reduced periodically to keep them small and to allow DLM to better identify good solutions.

The situation is worse in MAX-SAT because the weights of clauses, $w$, can be in a large range, making $L(x, \lambda)$ in (12) even more rugged and difficult to search. Here, $\lambda + w$ in MAX-SAT has a similar role as $\lambda$ in SAT. Without any mechanism to reduce $\lambda + w$, $L(x, \lambda)$ can become very large and rugged as the search progresses.

This situation is illustrated in the first two graphs of Figure 2 that show the behavior of DLM when it was applied to solve two MAX-SAT problems. The graphs show that the Lagrangian values vary in a very large range, and the Lagrangian space is difficult to search.

One way to overcome this problem is to reduce $\lambda$ and $w$ periodically. For instance, in the last two graphs of Figure 2, $\lambda + w$ was scaled down by a factor of 2 every 500 iterations. This strategy reduces $L(x, \lambda)$ and restricts the growth of the Lagrange multipliers, leading to faster solutions for many test problems.

Figure 3 shows $\mathcal{A}_2$, our implementation of DLM to

```
Set initial x
Set λ = 0
Set c = 1
Set κ = n/3, where n is the number of variables
Set λ reduction interval I_λ (e.g. 500)
Set reduction ratio r (e.g. 2.0)
Set base λ value λ_b (e.g. 1)
Set base weight value w_b (e.g. 1)
while x is not a feasible solution, i.e., N(x) > 0
    if number of iterations ≥ κ then
        Maintain a list, l, of variables such that
            if one of them is flipped, L(x, λ) will improve.
        if l is not empty then
            Update x by flipping the first element of l
        else
            Update λ: λ ⟵ λ + c · U(x)
        end if
    else
        if ∃ variable v s.t. L(x', λ) < L(x, λ) when flipping v
            in a predefined order in x to get x' then
            x ⟵ x'
        else
            Update λ: λ ⟵ λ + c · U(x)
        end if
    end if
    if iteration index mod I_λ = 0 then
        Reduce λ and weights for all clauses if possible,
        e.g. λ ⟵ max(λ_b, λ/r), w_i ⟵ max(w_b, w_i/r)
    end if
    if x is better than incumbent, keep x as incumbent.
end while
```

Figure 3: $\mathcal{A}_2$: An efficient implementation of DLM.

solve MAX-SAT. $\mathcal{A}_2$ uses the first strategy of hill-climbing in descents in the beginning and switches to the second strategy later. We use $\kappa$ to control the switch from the first strategy to the second. We found that $\kappa = n/3$ works well and used it in our experiments.

$\mathcal{A}_2$ updates $\lambda$ only when a local minimum is reached. In our experiments, we have used a simple scheme that increases $\lambda$ by a constant across all clauses.

$\mathcal{A}_2$ controls the magnitude of $L(x, \lambda)$ by periodically reducing $\lambda$ and $w$. More sophisticated adaptive mechanisms can be developed to lead to better performance.

## 4 Experimental Results

In this section we evaluate DLM using some benchmarks and compare it with previous results obtained by GRASP [Resende and Feo, 1996; Resende et al., 1997]. GRASP is a greedy randomized adaptive search procedure that has been shown to quickly produce good-quality solutions for a wide variety of combinatorial optimization problems, including SAT and MAX-SAT.

The 44 MAX-SAT problems tested by GRASP were derived from DIMACS SAT benchmark jnh that have 100 variables and between 800 and 900 clauses. Some of them are satisfiable, while others are not. Their weights are integers generated randomly between 1 and 1000.[1]

Our DLM code was written in C. The experiments were run on a 50-MHz Sun SparcStation 10/51. The code was compiled using gcc with the "-O" option.

In our experiments, we first studied the effect of reducing $\lambda$ and $L(x, \lambda)$. We compared the results with and without periodic reduction of $\lambda$. We ran DLM between 5 to 20 times from random initial points, each for 10,000 iterations, for each test problem. We then took the best solution of the multiple runs as our final solution.

When $\lambda$ was not reduced periodically, DLM found 25 optimal solutions out of the 44 problems (based on the best results of 10 runs). When $\lambda$ was reduced periodically, we tried several values of $I_\lambda$, the reduction interval, and $r$, the reduction ratio. Using $I_\lambda$ of 100, 1000, 2000 and 5000 and $r$ of 1.5, 2 and 4, we found significant improvement in solution quality when $\lambda$ was reduced periodically (except for the case when $I_\lambda = 5000$). The best combination is when $I_\lambda = 500$ and $r = 2$, in which 12 more optimal solutions were found than without $\lambda$ reduction, 18 test problems were improved, and no solution is worse. Overall, we found optimal solutions in 39 out of the 44 test problems (based on best of 20 runs).

Table 1 compares the solutions obtained by DLM and by GRASP. The results of GRASP are from [Resende et al., 1997] that were run for 10,000 iterations on a 250-MHz SGI Challenge with MIPS R4400. In contrast, DLM was run between 5 and 20 times from random initial points using 10,000 iterations each. Our results show significant improvement over GRASP in 43 of the 44 test problems. It is important to note that, although the differences between GRASP's results and the optimal solutions are relatively small, improvements in this range are the most difficult for any search algorithm.

Table 2 shows the average time in seconds for 10,000 iterations of GRASP and DLM. Since GRASP was run on a faster machine, we estimate that one iteration of DLM takes around three orders of magnitude less time than that of GRASP on the same machine.

GRASP found better results when the number of iterations is increased. Using the ten test problems reported in [Resende et al., 1997] that were run for 100,000 iterations of GRASP, Table 3 compares the results of DLM and GRASP with longer runs. It shows that DLM still found better solutions in all the problems.

Increasing the number of iterations of DLM could improve the best solutions found. DLM was able to find 40 out of 44 optimal solutions when the number of iterations is increased to 100,000 (best of 10 runs).

Our experimental results show that DLM solved these benchmark problems much faster and found better solutions for a majority of the test problems.

---

[1] The 44 benchmark problems can be obtained from ftp://netlib.att.com/math/people/mgcr/data/maxsat.tar.gz.

Table 1: Comparison of results of DLM and those of GRASP with respect to the optimal solutions. (jnh1-19: 850 clauses, $\sum_i w_i = 420925$; jnh201-220: 800 clauses, $\sum_i w_i = 394238$; jnh301-310: 900 clauses, $\sum_i w_i = 444854$. GRASP was run 10,000 iterations on a 250-MHz SGI Challenge with MIPS R4400 whereas DLM was run 5-20 times, each for 10,000 iterations, from random starting points on a Sun SparcStation 10/51. For DLM, $I_\lambda = 500$ and $r = 2$.)

| Problem ID | \# Runs of DLM 5 | 10 | 20 | GRASP | Optimal Solutions |
|---|---|---|---|---|---|
| jnh1 | 0 | 0 | 0 | -188 | 420925 |
| jnh4 | -41 | -41 | -41 | -215 | 420830 |
| jnh5 | -164 | -131 | 0 | -254 | 420742 |
| jnh6 | 0 | 0 | 0 | -11 | 420826 |
| jnh7 | 0 | 0 | 0 | 0 | 420925 |
| jnh8 | 0 | 0 | 0 | -578 | 420463 |
| jnh9 | -7 | -7 | -7 | -514 | 420592 |
| jnh10 | 0 | 0 | 0 | -275 | 420840 |
| jnh11 | -8 | 0 | 0 | -111 | 420753 |
| jnh12 | 0 | 0 | 0 | -188 | 420925 |
| jnh13 | 0 | 0 | 0 | -283 | 420816 |
| jnh14 | -77 | 0 | 0 | -314 | 420824 |
| jnh15 | -79 | -79 | 0 | -359 | 420719 |
| jnh16 | 0 | 0 | 0 | -68 | 420919 |
| jnh17 | 0 | 0 | 0 | -118 | 420925 |
| jnh18 | -98 | 0 | 0 | -423 | 420795 |
| jnh19 | 0 | 0 | 0 | -436 | 420759 |
| jnh201 | 0 | 0 | 0 | 0 | 394238 |
| jnh202 | 0 | 0 | 0 | -187 | 394170 |
| jnh203 | 0 | 0 | 0 | -310 | 394199 |
| jnh205 | 0 | 0 | 0 | -14 | 394238 |
| jnh207 | 0 | 0 | 0 | -137 | 394238 |
| jnh208 | 0 | 0 | 0 | -172 | 394159 |
| jnh209 | 0 | 0 | 0 | -207 | 394238 |
| jnh210 | 0 | 0 | 0 | 0 | 394238 |
| jnh211 | 0 | 0 | 0 | -240 | 393979 |
| jnh212 | 0 | 0 | 0 | -195 | 394238 |
| jnh214 | 0 | 0 | 0 | -462 | 394163 |
| jnh215 | 0 | 0 | 0 | -292 | 394150 |
| jnh216 | -149 | 0 | 0 | -197 | 394226 |
| jnh217 | 0 | 0 | 0 | -6 | 394238 |
| jnh218 | 0 | 0 | 0 | -139 | 394238 |
| jnh219 | 0 | 0 | 0 | -436 | 394156 |
| jnh220 | 0 | 0 | 0 | -185 | 394238 |
| jnh301 | 0 | 0 | 0 | -184 | 444854 |
| jnh302 | **-510** | **-338** | **-338** | -211 | 444459 |
| jnh303 | -152 | -143 | -143 | -259 | 444503 |
| jnh304 | -106 | 0 | 0 | -319 | 444533 |
| jnh305 | -196 | -194 | -194 | -609 | 444112 |
| jnh306 | 0 | 0 | 0 | -180 | 444838 |
| jnh307 | 0 | 0 | 0 | -155 | 444314 |
| jnh308 | -103 | -60 | 0 | -502 | 444724 |
| jnh309 | 0 | 0 | 0 | -229 | 444578 |
| jnh310 | 0 | 0 | 0 | -109 | 444391 |

Table 2: Comparison of average CPU time in seconds for 10,000 iterations of DLM and GRASP.

| Problem | \# Clauses | GRASP | DLM |
|---|---|---|---|
| jnh1-19 | 850 | 799.8 | 1.7 |
| jnh201-220 | 800 | 692.9 | 1.7 |
| jnh301-310 | 900 | 937.8 | 2.0 |

Table 3: Comparison of solutions found by DLM and GRASP with longer iterations.

| Problem Id | GRASP 10K | 100K | DLM 10K iter | Optimal Solutions |
|---|---|---|---|---|
| jnh1 | -188 | -77 | 0 | 420925 |
| jnh10 | -275 | -259 | 0 | 420840 |
| jnh11 | -111 | -111 | 0 | 420753 |
| jnh12 | -188 | -54 | 0 | 420925 |
| jnh201 | 0 | 0 | 0 | 394238 |
| jnh202 | -187 | -141 | 0 | 394170 |
| jnh212 | -195 | -50 | 0 | 394238 |
| jnh304 | -319 | 0 | 0 | 444533 |
| jnh305 | -609 | -368 | -194 | 444112 |
| jnh306 | -180 | -63 | 0 | 444838 |

## References

[Goemans and Williamson, 1995] M. X. Goemans and D. P. Williamson. Improved approximation algorithms for maximum cut and satisfiability problems using semidefinite programming. *JACM*, 42(6):1115–1145, 1995.

[Gu, 1989] J. Gu. *Parallel Algorithms and Architectures for Very Fast AI Search*. PhD thesis, Dept. of Computer Science, University of Utah, August 1989.

[Hansen and Jaumard, 1990] P. Hansen and R. Jaumard. Algorithms for the maximum satisfiability problem. *Computing*, 44:279–303, 1990.

[Jiang et al., 1995] Y. Jiang, H. Kautz, and B. Selman. Solving problems with hard and soft constraints using a stochastic algorithm for MAX-SAT. In *Proc. of 1st Int'l Joint Workshop on Artificial Intelligence and Operations Research*, Timberline, Oregon, 1995.

[Joy et al., 1997] S. Joy, J. Mitchell, and B. Borchers. A branch and cut algorithm for MAX-SAT and weighted MAX-SAT. In *Satisfiability Problem: Theory and Applications. DIMACS Series on Discrete Mathematics and Theoretical Computer Science*. AMS (to appear), 1997.

[Luenberger, 1984] D. G. Luenberger. *Linear and Nonlinear Programming*. Addison-Wesley Publishing Company, 1984.

[Resende and Feo, 1996] M.G.C. Resende and T. Feo. A GRASP for satisfiability. In D. S. Johnson and M. A. Trick, editors, *Cliques, Coloring, and Satisfiability: Second DIMACS Implementation Challenge*, volume 26, pages 499–520. DIMACS Series on Discrete Mathematics and Theoretical Computer Science, AMS, 1996.

[Resende et al., 1997] M.G.C. Resende, L.S. Pitsoulis, and P.M. Pardalos. Approximate solution of weighted MAX-SAT problems using GRASP. In *Satisfiability Problem: Theory and Appl. DIMACS Series on Discrete Mathematics and Theoretical Computer Science*. AMS, 1997.

[Selman and Kautz, 1993] B. Selman and H. Kautz. Domain-independent extensions to GSAT: Solving large structured satisfiability problems. In *Proc. of 13th Int'l Joint Conf. on Artificial Intelligence*, pages 290–295, 1993.

[Shang and Wah, 1997] Y. Shang and B. Wah. Discrete lagrangian methods for solving satisfiability problems. *Journal of Global Optimization (to appear)*, 1997.

[Wah and Shang, 1996] B. W. Wah and Y. Shang. A discrete lagrangian-based global-search method for solving satisfiability problems. In *Proc. DIMACS Workshop on Satisfiability Problem: Theory and Applications*. AMS, March 1996.

# Learning Short-Term Weights for GSAT

Jeremy Frank *
frank@ptolemy.arc.nasa.gov
Caelum Research Corp.
NASA Ames Research Center Mail Stop 269-2
Moffett Field, CA. 94035-1000

## Abstract

We investigate an improvement to GSAT which associates a weight with each clause. We change the objective function so that GSAT moves to assignments maximizing the weight of satisfied clauses, and each clause's weight is changed when GSAT moves to an assignment in which this clause is unsatisfied. We present results showing that this version of GSAT has good performance when clause weights are reduced geometrically throughout the course of a single try. We conclude that clause weights are best interpreted as short-term, context sensitive indicators of how hard different clauses are to satisfy.

## 1 Introduction

Local search procedures are an alternative to complete search algorithms for solving combinatorially expensive search problems. GSAT is a local search algorithm which can often find solutions to satisfiable SAT problems in Conjunctive Normal Form (CNF) quickly [SLM92]. GSAT operates by changing a complete assignment of variables into one in which the maximum possible number of clauses are satisfied by flipping a single variable. GSAT can quickly solve satisfiable 3-CNF problem instances of up to 2000 variables which are known to have solutions, while the best known complete search procedures typically are able to solve problems with 500 variables or less.

A feature of GSAT is that often the best move it can make leaves the number of unsatisfied clauses the same. These moves are known as "sideways" moves, and GSAT typically searches regions of sideways moves for some time before finding a move which reduces the number of unsatisfied clauses. This means GSAT must randomly search these "plateaus" until it finds a way off. Much research into GSAT has focused on reducing the impact of plateau search, typically by encouraging exploration of parts of the space that GSAT has not explored yet.

Several modifications of GSAT have been developed to reduce the impact of searching the plateaus. Tabu search [Glo89], [HdW87] is one such scheme; when a variable is flipped, it is placed on a *tabu list* and can't be flipped again for a constant amount of time. This is a primitive learning scheme which remembers actions taken in the recent past and does not reverse them. Tabu search has been augmented with a variety of schemes to allow reversing flips under certain circumstances.

Selman and Kautz [SK93] added random walk to GSAT, enabling GSAT to make backwards moves in an attempt to escape plateaus. Gent and Walsh [GW95] developed a wide variety of modifications to GSAT and tested many of these with random walk. Spears [Spe95] discusses a modification of GSAT that employs simulated annealing to allow escape from local minima. In this modification the change in the number of satisfied clauses and a "temperature" parameter are used to compute the probability that some variable flip is accepted; as more flips are done the probability that backwards moves are selected decreases.

A modification of local search that has received recent attention is the "clause weights" scheme proposed in [SK93] and in [Mor93]. In this version, each clause has an associated weight, and these weights are modified during search in order to better inform the variable selection heuristic. The criterion for variable selection becomes "maximize the weight of satisfied clauses". This variant of GSAT has been investigated by Cha and Iwama as a stand-alone modification[CI95] and in conjunction with a method to add additional clauses to problem instances[CI96]. Davenport et. al. use a similar scheme in GENET applied to a connectionist architecture [DTWZ94]. Frank [Fra96] analyzed several variants

---

*This research was supported by NSF CCR-94-0365. I would like to thank Peter Cheeseman, John Stutz and John Allen of NASA Ames Research Center for their input during this research, and also Ian Gent of the University of Strathclyde, Toby Walsh of IRST and the IJCAI reviewers for their comments.

based on Selman and Kautz' approach.

We present a modification of weighted GSAT and describe different algorithms for modifying the clause weights. We interpret the process of modifying weights as *learning* how difficult a clause is to satisfy, and by extension which variables to focus on at each flip. We show that the best learning algorithm retains a context-sensitive weight for each clause rather than allowing clause weights to grow unbounded. Our interpretation is that as the assignments GSAT examine change, the weight should represent the difficulty of satisfying the clauses in the context of recent assignments. We also demonstrate a tradeoff between the rate of learning and the amount of contextual information preserved.

In §2 we discuss variants of weighted GSAT. In §3 we discuss the experiments we ran to gather empirical data on the performance of our variants, and our analysis is presented in §4. In §5 we discuss conclusions and future work.

## 2 GSAT and Variants

We give the familiar algorithm outline for GSAT in figure 1. GSAT is a local search algorithm which typically employs greedy hill-climbing [SLM92] designed to solve CNF formulae. GSAT begins with a randomly generated initial truth assignment, then hill-climbs by reversing or "flipping" the assignment of the variable which increases the number of satisfied clauses the most. If our CNF $\Sigma$ has $C$ clauses, then $eval = \sum_{i=1}^{C} Sat_i(A)$ where $Sat_i(A) = 0$ if clause $i$ is unsatisfied in assignment $A$ and 1 if it is satisfied. If PossFlips contains more than one variable, the algorithm must now pick one variable to flip. Different GSAT variants control how many and which variables are examined to determine the best flip and tie-breaking schemes to pick among equally good flips.

Selman and Kautz [SK93] describe a modification in which GSAT associates a weight with each clause. The weights of all clauses that remains unsatisfied at the end of a try are incremented. A clause that has remained unsatisfied through many tries will have a higher weight than a clause which has been satisfied in most of those tries. Each clause's weight is initialized to 1, so the first try is controlled as it would be in GSAT. GSAT then picks the flip that satisfies clauses with the highest total weight. So if $W_i$ denotes the weights of the clauses then our function $eval$ can be written $eval = \sum_{i=1}^{C} W_i * Sat_i(A)$.

We saw several modifications which we felt might improve the algorithm's performance and illuminate the effect that weights have on GSAT's performance, leading to even better improvements in the future. In particular, we feel that clause weights can be used to *learn* which parts of the problem instance are hard to solve,

```
procedure GSAT(Σ,MaxFlips, MaxTries)
    for i=1 to MaxTries
        A = gen_assignment
        for j = 1 to MaxFlips
            if solved_problem(A)
                return A
            else PossFlips = select(Σ,A)
                V=pick(PossFlips)
                A=A with V's value flipped
            end else
        end for
    end for
end

procedure select(Σ,A)
PossFlips=ε
Best = -∞
    for i=1 to Number of Vars
        A=A with Vᵢ's value flipped
        if eval(A, Σ) =Best
            PossFlips=PossFlips ∪ V
        else if eval(A, Σ)>Best
            PossFlips=V
            Best = eval(A, Σ)
        A=A with Vᵢ's value flipped
    end for
end
```

Figure 1: GSAT Algorithm Sketch.

which will improve the performance of weighted GSAT. We briefly describe these modifications below.

### 2.1 When to Change the Weights?

Selman and Kautz originally suggested altering the weights after each try. This has the effect of allowing GSAT to drive the number of satisfied clauses to a very low number relative to the initial number; as a result, few weights change after each try and feedback must wait until the end of a try. We decided to update the clause weights after each *flip* instead of after each try. Unsatisfied clauses provide feedback to the variable selection heuristic every flip. This has the effect of emphasizing variables in unsatisfied clauses almost immediately, without waiting for the try to complete. If we view the weight modification process as learning, we see that this version starts learning immediately rather than waiting until each try is complete, thereby acquiring information about clauses more rapidly. After each flip we update the clauses in the following way: if $W_{i,t}$ denotes the weight of clause $i$ at time $t$ then $W_{i,t} = W_{i,t-1} + \delta$ if clause $i$ not satisfied and $W_{i,t} = W_{i,t-1}$ otherwise. This scheme resembles that used by Morris [Mor93], who changes weights when the best move leaves the value of the objective function unchanged. If a try is unsuccessful then we reset the weights to 1 before starting the new try. We shall refer to this algorithm as WGSAT, not to be confused with WSAT, the random walk version of GSAT.

Different learning rates may make a difference in

WGSAT's performance [SK93]. Suppose a clauses' weight increases to $x$ after the first flip. This clause now "outweighs" $x$ clauses of weight 1. The heuristic prefers to satisfy clauses of higher weights, and increasing the increment size can result in different moves, and therefore varying performance of the algorithm. In a learning context, a larger increment implies "fast" learning or "focusing", in that larger weights mean that the algorithm rewards satisfaction of those highly weighted clauses more quickly than if the clauses are incremented at a more moderate rate. If the rate of learning is too fast then WGSAT may "thrash" by making moves which almost always unsatisfy more clauses than necessary. However, if the rate of learning is too slow then WGSAT will require more flips to solve problem instances.

## 2.2 Short Term Weights

One observation we made about clause weights is that their *relative magnitudes* control which state is best to move to. If a single clause has a high weight, it may be able to sway the heuristic to cause several clauses of lower weight to become unsatisfied, but can't sway a clause of higher weight than its own. The original WGSAT algorithm never decreases the weights once they are incremented, so after a time it becomes very difficult to change the relative magnitudes of the weights. In WGSAT clauses accumulate weight in the context of assignments which are constantly changing. It may be that weights acquired during one part of search will hinder later parts of the search.

To combat this problem we implemented two schemes to reduce weights throughout local search. The first reduces clause weights if the clause was satisfied in an assignment. For example, if a clause was satisfied in the current assignment we could reduce its weight by 1, while if it was unsatisfied then we could increase its weight by 5. We envision a large increment when a clause is unsatisfied so that the clauses weight will endure for a time, even if it is satisfied in the next several assignments. This gives the clause weight scheme a feel similar to that of tabu search [Glo89], [HdW87], where variables are not allowed to be flipped for some fixed duration. This scheme is more general than tabu search, because it provides a continuously weakening incentive not to flip the variable instead of disallowing the flip. We can analogize decrementing weights to "refocusing" on other clauses which have not been satisfied by changing the relative weight of satisfied clauses faster. The weight update scheme has the form $W_{i,t} = W_{i,t-1} - \delta_s$ if clause $i$ is satisfied and $W_{i,t} = W_{i,t-1} + \delta_u$ if clause $i$ not satisfied. We refer to $\delta_u$ as the refocusing rate to distinguish it from $\delta_s$ the learning rate for this version. In our example above, $\delta_u = 5$ and $\delta_s = 1$. Unfortunately, this scheme was not successful and due to space considerations we do not discuss it further here.

The second scheme only stores the increments which took place *recently* instead of keeping all clause weight changes. In this scheme clause weights reflect how difficult this clause was to satisfy only in the context of recent assignments. The general form of this scheme is $W_{i,t} = p * W_{i,t-1} + \delta$, where we refer to $p$ as the decay rate. We restrict $p < 1$ and the learning rate $\delta \geq 0$. With this scheme the maximum weight is $\frac{\delta}{1-p}$, which has the side effect of preventing unbounded growth of weights due to monotone increasing as in the previous schemes. The choice of decay rate will dictate how extensive the context is, and at what rate older weight information is "forgotten".

## 3 Experiments

We tested our algorithms on problems found near the phase transition for 3-CNF [CKT91]; unless otherwise stated, the algorithms were tested on problems for which the number of clauses $C$ and the number of variables $N$ obeyed the constraint $\frac{C}{N} = 4.3$. We created the problem instances randomly; that is, we generated each clause of each problem by selecting 3 of $N$ literals without replacement and negating each literal with probability $\frac{1}{2}$. We generated test suites of problems which were proven to have solutions using a variant of the Davis-Putnam procedure.

We employed two different metrics when analyzing the performance of the algorithm variants. The first metric is the proportion of problem instances from the test suite which the variant solved in a single try. This metric is appropriate for the situation where a practitioner is solving problems from an unknown distribution. The second metric is the minimum mean time to solve all problem instances in the test suite. Gent and Walsh analyze versions of GSAT by finding the value of MaxFlips minimizing the mean time to solve a test suite of known solvable problems [GW95]. This metric is appropriate when a practitioner knows something about the distribution of problems and has tuned the value of MaxFlips to minimize the time required to solve problem instances from the distribution. HSAT [GW93], which selects the variable flipped longest ago if there are several variables resulting in the best change in the number of satisfied clauses, has among the best performance with MaxFlips tuned. We therefore use HSAT as a benchmark algorithm and test our variants against the performance of HSAT.

### 3.1 Effectiveness of Decaying Weights

We implemented WGSAT with decay in order to test its performance. We found that the learning rate makes little difference in the single try performance of WGSAT without decay, so we only present results on combining different learning rates with decay. As we stated before, if weights are updated according to the formula $W_{i,t} = p * W_{i,t-1} + \delta$ then the maximum weight is $\frac{\delta}{1-p}$ and the maximum weight after $F$ flips is $\frac{\delta(1-p^F)}{1-p}$. We tested WGSAT with decay rates $p = 0.95, 0.99, 0.999$ and learning rates of $\delta = 1, 8, 16, 24$. The plots in figures 2 to 4 present the data for fixed decay rates; for comparison we also include the plot of WGSAT without decay in each figure. We see in figure 2 that when $p$ is close to 1, in this case 0.999, that even these small differences disappear and the number of solved problems rises for all values of MaxFlips. When the decay rate grows further

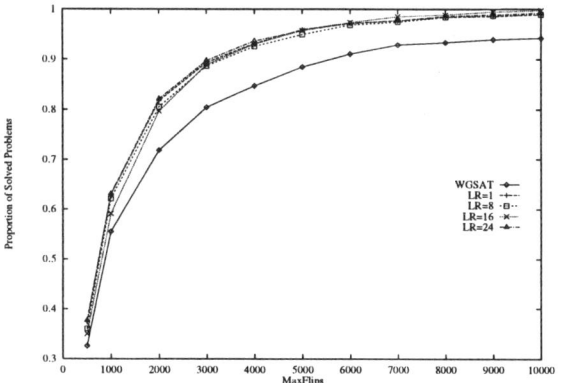

Figure 2: Satisfiability Performance of WGSAT With Varying Learning Rates and Decay Rate of 0.999.

from 1, as shown in figures 3 and 4, we notice that for some pairs of learning rates and decay rates the number of solved problems decreases. For instance in figure 4 we see that the performance of the variant with $\delta = 1$ solves more problems for small numbers of MaxFlips than the other variants. We believe that the other variants emphasize clauses too quickly, resulting in "thrashing" until the weights settle. However, as MaxFlips increases, the

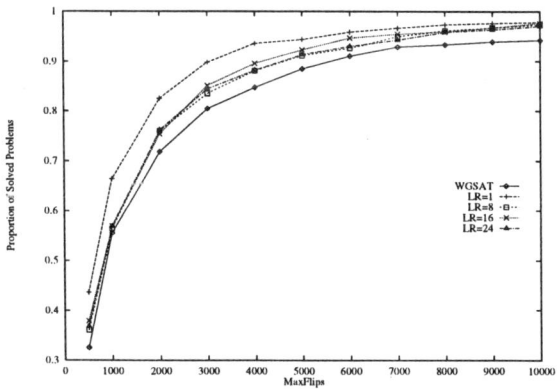

Figure 3: Satisfiability Performance of WGSAT With Varying Learning Rates and Decay Rate of 0.99.

variant with $\delta = 1$ becomes much worse than the rest of the variants. This indicates that when the learning rate is higher, WGSAT can tolerate a faster loss of weights before losing performance. By contrast, in figure 3 we see that the variant with $\delta = 1$ performs *better* than the rest of the variants. Our explanation is that with higher $\delta$ and $p$ very close to 1, weights accumulate too fast resulting in over-emphasizing clauses too early in search. We see that the performance of WGSAT with the decay rate very close to 1 is the best version. Figure 5 shows the performance of WGSAT with $\delta = 1$ for different decay rates and for no decay. Notice that when $p = 0.95$ and MaxFlips is low WGSAT outperforms the other variants, but as MaxFlips increases the number of solved problems for $p = 0.95$ is smaller than for $p$ closer to 1. This implies that it is important to discount weights acquired early in the process, but more important to preserve weights acquired later.

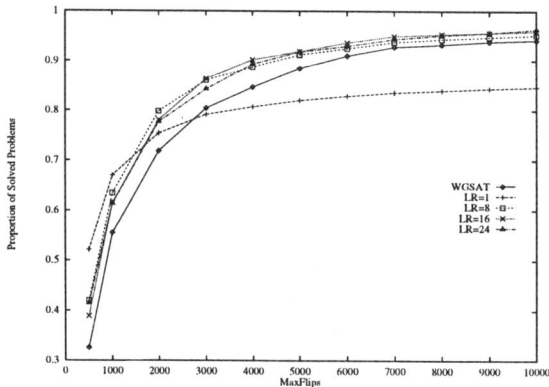

Figure 4: Satisfiability Performance of WGSAT With Varying Learning Rates and Decay Rate of 0.95.

Our analysis indicates that using a decay scheme with WGSAT might improve the optimal performance of WGSAT. Gent and Walsh [GW95] show that GSAT variants require variable amounts of time to solve problems for different values of MaxFlips. Accordingly we tested 3 variants of GSAT to find the value of MaxFlips which minimizes the mean number of flips to solve a test suite of 1000 known solvable problems. We allowed each variant of GSAT enough tries to guarantee that all problems were solved, then computed the mean number of flips required to solve the suite. We tested WGSAT without decay, WGSAT using a decay rate of 0.999, and HSAT. Both variants of WGSAT used a learning rate of 1. We ran our experiments on problems of 100, 125, 150 and 175 variables. The results are presented in figure 6. We see that the performance of WGSAT and WGSAT with decay are roughly comparable until we solved problems of 150 variables. This is somewhat surprising given that for a fixed number of flips WGSAT

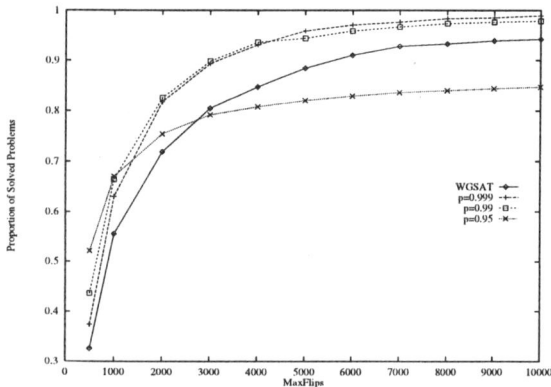

Figure 5: Satisfiability Performance of WGSAT With Varying Decay Rates and Learning Rate of 1.

| Vars | Variant | Opt MaxFlips | Mean | SDev |
|---|---|---|---|---|
| 100 | WGSAT | 1250 | 1591.95 | 2346.2 |
|  | WGSAT+Decay | 1000 | 1616.84 | 2322.05 |
|  | HSAT | 200 | 2437.27 | 4269.09 |
| 125 | WGSAT | 1750 | 3454.23 | 5159.49 |
|  | WGSAT+Decay | 1700 | 3511.1 | 5721.7 |
|  | HSAT | 300 | 5480.54 | 10220 |
| 150 | WGSAT | 2000 | 6982.04 | 12366.2 |
|  | WGSAT+Decay | 1900 | 6733.4 | 11510.6 |
|  | HSAT | 550 | 9836.46 | 17697 |
| 175 | WGSAT | 2800 | 13169.2 | 37702.7 |
|  | WGSAT+Decay | 2600 | 12081.2 | 21157.6 |
|  | HSAT | 700 | 22283.2 | 40605.1 |

Figure 6: Optimal Performance of GSAT Variants.

with decay solved more problems than WGSAT for problems of 100 variables and 125 variables. However, we notice that the best performance of WGSAT with decay occurs at a lower value of MaxFlips than the optimal value of MaxFlips for WGSAT with no decay. For problems of 150 variables WGSAT with decay begins to show a slight improvement in the mean number of flips as well as the variance. The best improvement appears for problems of 175 variables; we see that WGSAT with decay shows a definite improvement over WGSAT. Both versions are considerably better than HSAT for all problem sizes. We note that HSAT is somewhat better than versions of GSAT with Random Walk (results supporting this conclusion come from [GW95]), so improvements upon HSAT are strong results of the effectiveness of our variants.

### 3.2 Structured Problems

Numerous researchers have also analyzed the performance of GSAT and GSAT variants on structured SAT problem instances from planning and scheduling applications. We analyzed WGSAT and WGSAT with decay on several problem instances taken from standard problem instance sets used by other researchers. Due to the performance differences observed for problems of variable size we tested problems of a variety of different sizes. The instances are: satisfiability encodings of a simple logistics problem with 331 variables and 2246 clauses (Rocket) and a large blocks world problem with 459 variables and 4675 clauses (Large) from the SAT-PLAN library [KS96], a satisfiability encoding of the 3-disk Tower of Hanoi problem with 158 variables and 992 clauses (Tower), and a circuit diagnosis problem with 331 variables and 712 clauses from the DIMACS archive (SSA). For each problem instance we ran WGSAT with $\delta = 1$, prior=1 and $\alpha = 1$ and WGSAT with decay with $\delta = 1$ and $p = 0.999$. We set MaxFlips at 10000 and ran 100 tries; we present the number of successful tries and the mean and variance in the number of flips required to solve the problem instances for those successful tries.

| WGSAT | | | |
|---|---|---|---|
| Problem | Solved | Average flips | Sdev flips |
| Tower | 28 | 2900 | 3881 |
| SSA | 41 | 374 | 73 |
| Large | 76 | 3587 | 4324 |
| Rocket | 70 | 2626 | 3269 |
| WGSAT with Decay | | | |
| Problem | Solved | Average flips | Sdev flips |
| Tower | 32 | 3167 | 3622 |
| SSA | 44 | 370 | 68 |
| Large | 70 | 3799 | 4491 |
| Rocket | 63 | 2764 | 3551 |

Figure 7: Performance of WGSAT Variants on Structure SAT Problems.

GSAT and HSAT were unable to solve these problems in 100 tries for many settings of MaxFlips. Figure 7 shows the proportion of tries out of 100 that WGSAT variants solved the problems, the mean number of flips required to solve the problems on successful attempts, and the standard deviation, also on successful attempts. WGSAT variants can solve these problem instances with some regularity, so the modifications clearly have merit beyond applicability to randomly generated problems. WGSAT and WGSAT with decay perform comparably on these problem instances. WGSAT appears to be slightly better on the SAT encoding of the 3-disk Tower of Hanoi problem, while WGSAT with decay appears to be better on the Circuit Diagnosis problem, but the differences are slight.

## 4 Analysis

The results show that gradually decaying weights can improve the performance of WGSAT compared to not decaying weights for randomly generated problems as the problem size increases. We noticed that when the learning rate was high we can decay rates more rapidly and maintain good performance, but the best version seems to be a decay rate close to 1, and at this value the learning rate appears irrelevant. Our conclusion is that clause weights are best interpreted as a context dependent measurement of the difficulty of satisfying a clause, and as search moves away from the current context the weight should represent the difficulty of satisfying the clauses in the new context. However, since the context changes slowly the weights must also be decayed slowly. The results showing that rapid decay is effective early in the search indicates that WGSAT can suffer from prematurely weighting clauses, but the reduced effectiveness of rapid decay for higher MaxFlips indicates that we must walk a fine line between no decay and too much decay.

## 5 Conclusions and Future Work

We have presented a modification of an update to GSAT which uses clause weights to modify the GSAT variable selection heuristic continuously throughout local search. We have demonstrated that decaying these weights gradually throughout search results in a more effective version of GSAT than several previous versions. We have also performed experiments which assess the impact of different weight updating schemes and concluded that gradual decay of weights gives the best results. We interpreted these methods as learning the best way to conduct local search by discovering the hardest clauses to satisfy relative to recent assignments.

Currently none of our variants of WGSAT carry weights over to subsequent tries. We would like to combine information from previous tries in new tries for WGSAT. Our limited experiments found that using the weights found on a previous try can lead to occasional spectacular failures resulting in millions of flips required to solve instances. One idea is to "discount" the weights from previous tries, in effect using the previous try as a non-uniform prior weight. An important consideration in new versions of GSAT is the tradeoff between reduced numbers of flips required to solve problems and the CPU time required to run these schemes.

Many of the schemes presented in this paper show improvements over GSAT and HSAT, but require floating-point manipulations where GSAT and HSAT use only integer (or even bit) computations. Cache and memory performance issues can also play a role, since schemes like HSAT and WGSAT require variable amounts of memory. We are conducting investigations into the impact these issues have on the real time performance of GSAT variants.

## References

[CI95] B. Cha and K. Iwama. Performance tests of local search algorithms using new types of random cnf formulas. *14th International Joint Conference on Artificial Intelligence*, pages 304–310, 1995.

[CI96] B. Cha and K. Iwama. Adding new clauses for faster local search. *Proceedings of the 13th National Conference on Artificial Intelligence*, 1996.

[CKT91] P. Cheeseman, B. Kanefsky, and W. Taylor. Where the *really* hard problems are. *IJCAI*, pages 163–169, 1991.

[DTWZ94] A. Davenport, E. Tsang, C. J. Wang, and K. Zhu. Genet: A connectionist architecture for solving constraint satisfaction problems by iterative improvement. *Proceedings of the 12th National Conference on Artificial Intelligence*, pages 325–330, 1994.

[Fra96] J. Frank. Weighting for godot: Learning heuristics for gsat. *Proceedings of the 13th National Conference on Artificial Intelligence*, 1996.

[Glo89] F. Glover. Tabu search part i. *ORSA Journal on Computing*, 1989.

[GW93] I. Gent and T. Walsh. Towards an understanding of hill-climbing procedures for sat. *AAAI*, pages 28–33, 1993.

[GW95] I. Gent and T. Walsh. Unsatisfied variables in local search. In J. Hallam, editor, *Hybrid Problems, Hybrid Solutions*. IOS Press, 1995.

[HdW87] A. Hertz and D. de Werra. Using tabu search techniques for graph coloring. *Computing*, 39:345–351, 1987.

[KS96] H. Kautz and B. Selman. Pushing the envelope: Planning, propositional logic and stochastic search. *Proceedings of the 13th National Conference on Artificial Intelligence*, 1996.

[Mor93] P. Morris. The breakout method for escaping from local minima. *AAAI*, 1993.

[SK93] B. Selman and H. Kautz. Domain independent versions of GSAT solving large structured satisfiability problems. *IJCAI*, 1993.

[SLM92] B. Selman, H. Levesque, and D. Mitchell. A new method for solving hard satisfiability problems. *AAAI*, pages 440–446, 1992.

[Spe95] W. Spears. Simulated annealing for hard satisfiability problems. In D. Johnson and M. Trick, editors, *2d. DIMACS Implementation Challenge*, DIMACS Series in Theoretical Computer Science, 1995.

# CONSTRAINT SATISFACTION

## Constraint Satisfaction 3: Local Consistency

# Local consistency for ternary numeric constraints

Boi Faltings and Esther Gelle
Artificial Intelligence Laboratory (LIA)
Swiss Federal Institute of Technology (EPFL)
IN-Ecublens, 1015 Lausanne, Switzerland
faltings/gelle@lia.di.epfl.ch

## Abstract

We consider algorithms of the Waltz type for computing local consistency (also called arc-consistency) for constraints over numeric domains. Many commonly used propagation rules do not in fact enforce local consistency. We extend the propagation rule given by Faltings [Faltings, 1994] to the case of ternary constraints. Since any general n-ary continuous constraint can be represented as a collection of ternary ones, this also covers n-ary constraints in general. We show how the propagation can be implemented efficiently. The new algorithm gives significantly tighter labellings than previous propagation algorithms on most problems.

## 1 Introduction

The constraint satisfaction problem (CSP) is ubiquitous in many practical applications. We consider in particular CSPs with continuous-valued variables, given by:

- a set of *variables* $\{X_i\}$.
- a set of ternary *constraints* $C_{X_i, X_j, X_k}$, defined as inequalities $f(x_i, x_j, x_k) \geq 0$. Note that an equality $f = 0$ can be modelled as a conjunction of two inequalities $f \geq 0$ and $f \leq 0$. Binary and unary constraints are defined similarly with the corresponding number of variables.

A *solution* to a CSP consists of value assignments $\{X_i = x_i\}$ to all variables such that all constraints are satisfied. Throughout this paper, we use upper case letters for variables and lower case letters for particular values.

For linear constraints, there are efficient optimization algorithms for finding *single* solutions to CSPs. For non-linear constraints, optimization algorithms only apply to special cases. More importantly, there are many applications in configuration, diagnosis or planning where it is important to know not a *single* solution, but the space of *all* possible solutions. Algorithms for determining the space of all solutions also find applications in numerical analysis ([Benhamou et al., 1994]). A solution space can be represented by a set of *labels* defining the consistent values or value combinations. In numeric domains, it is very difficult to represent sets of value combinations. However, sets of values for individual variables can be represented efficiently by collections of *intervals*. We consider algorithms for computing labellings which are *locally consistent* according to the following definition:

**Definition 1** *A labelling is* locally consistent *if for every $X_i, X_j$ and every value $x_i \in L_i$, there exists a value $x_j \in L_j$ such that for every constraint $C_{X_i, X_j, X_k}$ involving $X_i$ and $X_j$ and possibly another variable $X_k$, there exists a value $x_k \in L_k$ such that $C_{X_i, X_j, X_k}$ is satisfied for $X_i = x_i, X_j = x_j, X_k = x_k$:*

$$(\forall i, j) \quad (\forall x_i \in L_i)(\exists x_j \in L_j)$$
$$(\forall C_{X_i X_j X_k})(\exists x_k \in L_k) C_{X_i X_j X_k}(x_i, x_j, x_k)$$

Note that this definition is different from others (e.g., [Davis, 1987]) in that it requires that one single value $x_j$ can satisfy *all* constraints. This ensures that the locally consistent labelling of the ternary system is at least as tight as the locally consistent labelling of its binary projection. Note that our work distinguishes itself from others, for example [Hyvönen, 1992] or [Lhomme, 1993], in that we are only considering efficient and general algorithms for *local* consistency, without any attempt to impose global consistency.

## 2 Local consistency with numeric constraints

A locally consistent labelling can be computed using the Waltz algorithm. It consists of applying a *propagation rule* to the labels of every variable pair until it results in no more change. While local consistency is very often applied for CSP with discrete domains, in continuous domains Davis ([Davis, 1987]) has reported a number of very negative results, including the fact that the Waltz algorithm does not guarantee a locally consistent labelling and may often fail to terminate. [Faltings, 1994] has shown that many of his negative results are due only to an overly straightforward formulation of the propagation rule used in the Waltz algorithm:

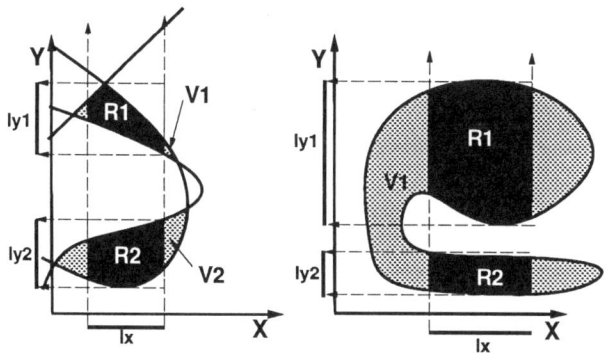

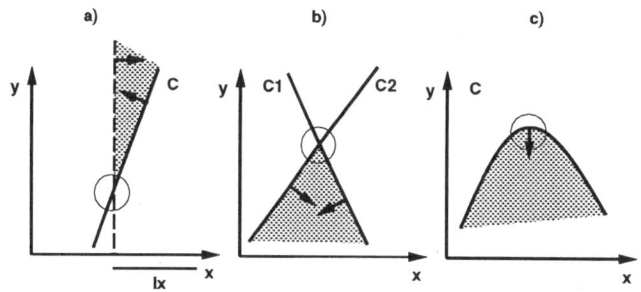

Figure 1: *Two examples of total constraints. The constraint on the left consists of the two feasible regions $V1$ and $V2$. When propagating from $X$ to $Y$, the interval $Ix$ generates the restricted regions $R1$ and $R2$ which project into interval $Iy1$ and $Iy2$. The example on the right shows that multiple restricted regions $R1$ and $R2$ can result from a single consistent region $V1$.*

Figure 2: *The three different types of local extrema: a minimum a) and two maxima b) and c).*

- Davis assumes that only interval boundaries are propagated, but with nonlinear constraints local extrema and intersections between constraints may also become interval bounds.

- Davis assumes that each constraint between the same set of variables is propagated individually, thus creating implicit cycles in the constraint network when several constraints involve the same variables.

The shortcomings introduced by these assumptions are already apparent in the following simple example of a parabola and a sphere:

$$X_1^2 + 1/2 * X_2 + 2 * (X_3 - 6) \geq 0$$
$$X_1^2 + X_3^2 + X_4^2 - 25 \leq 0$$

with initial intervals of $X_i$ $i = 1, 2, 3, 4$ of $[-10, 10]$. Here, Davis' propagation rule results in no refinement whatsoever of the initial labels. However, upon closer consideration, we find that:

- $X_3$ can never become greater than 5. This bound is not found since it is reached at a local extremum of a constraint.

- $X_3$ can never become smaller than -3.36. This bound is not found since it is given by the intersection of the two constraints.

A more powerful propagation rule can be obtained by generalizing the propagation rules used in discrete domains to the infinite, continous case. To do this, it is necessary to consider all constraints involving the same pair of variables $X$ and $Y$ simultaneously rather than individually; we call their conjunction the *total constraint* between $X$ and $Y$. It may consist of several regions $V_1, .., V_k$ of feasible value combinations (Figure 1).

In a propagation step from $X$ to $Y$, the label of $Y$ is modified to exclude those values which are not compatible with any value in the label of $X$. This can be accomplished by first restricting the feasible regions of the total constraint to those within the label of $X$, called the *restricted* regions (Figure 1) $R_1, ..., R_l$. Projecting these restricted regions onto the $Y$-axis gives the intervals of $Y$-values admitted by the propagation. [Faltings, 1994] shows that this propagation rule guarantees a locally consistent labelling upon termination.

Implementing such a rule poses some difficulties, since the regions making up the total constraint can become quite complex (see Figure 1). [Faltings, 1994] gives a method for propagation without explicitly considering the regions formed by the constraints. It is based on propagating *extrema* of boundaries of restricted regions. There are three types of extrema (see Figure 2):

a) intersections between constraints and interval bounds

b) constraint intersections

c) local extrema of constraint curves

We consider the propagation of the label of $X$ to that of $Y$ through a restricted region of a constraint between $X$ and $Y$. We let $B(R)$ be the continous boundary of a restricted region $R$. Extrema are then the local maxima and minima in $Y$ of $B(R)$ with respect to $X$. Note in particular that extrema here refer to the boundary $B(R)$, not the region $R$ itself. As a notation, we let $max_Y(B(R), y_0)$ be true if and only if $B(R)$ has a local maximum in $Y$ with $Y$-coordinate $y_0$, and $min_Y(B(R), y_0)$ if it has a local minimum. Note that if $R$ is bounded by a set of constraint curves, the extrema of its boundary $B(R)$ are a subset of the extrema and intersections of these curves. Only those which satisfy all other constraints bounding $R$ are valid, the others are ignored. Thus, the set of extrema can be found by purely local considerations.

We now define the function $\alpha_Y(R, y)$ to be the difference in the number of maxima and the number of minima at $y$-coordinates greater than or equal to $y$:

$$\alpha_Y(R, y) = |\{y_0 | max_Y(B(R), y_0) \land y_0 > y\}| - |\{y_0 | min_Y(B(R), y_0) \land y_0 > y\}|$$

It is then possible to prove ([Faltings, 1994]) that:

- if there is a point $(x^*, y^*) \in R$, $\alpha_Y(R, y^*) > 0$, (Lemma 2.9) and

- if there is no point $(x^*, y^*) \in R$, $\alpha_Y(R, y^*) = 0$ (Lemma 2.7)

The propagation rule should eliminate a value $y^*$ from the label $L_Y$ if and only if it is in no region, i.e. $\alpha_Y(R_i, y^*) = 0$ for all $R_i$. Since $\alpha_Y(R_i, ...) \geq 0$ for any $R_i$, this can be conveniently expressed as $\sum_i \alpha_Y(R_i, y^*) = 0$. By rewriting this as:

$$\sum_i \alpha_Y(R_i, y^*) = \sum_i |\{y_0 | max_Y(B(R_i), y_0) \wedge y_0 > y^*\}|$$
$$- \sum_i |\{y_0 | min_Y(B(R_i), y_0) \wedge y_0 > y^*\}|$$

we only require the total number of extrema in *all* regions taken together, without any consideration of the regions they belong to. The set of illegal $y^*$, and conversely the set of legal $y$, can thus be characterized without knowing which extrema belong to which region!

The following propagation rule computes the maximal intervals of admissible $y$ and thus achieves local consistency (for the proof, see [Faltings, 1994]):

1. $I_y \leftarrow \{\}$
2. compute the set $MAX$ of local maxima and the set $MIN$ of local minima of all restricted regions $r \in R(Q, I_x), Q \in C^t(x, y)$.
3. Filter both $MAX$ and $MIN$ to keep only those extrema which satisfy all other constraints, i.e. actually bound a legal region. Order the two sets of points according to their y-coordinate (these sets are obtained directly by consideration of individual constraints $c(x, y)$).
4. set $index \leftarrow 0$, consider the extrema $e$ in $MIN$ and $MAX$ in decreasing order of their y-coordinate, where elements of $MAX$ are always considered before elements of $MIN$, and do:
   (a) if $e \in MAX$, $index \leftarrow index + 1$. If $e \in MIN$, $index \leftarrow index - 1$.
   (b) if $index$ has just changed from 0 to 1, set $y_{lower} \leftarrow y - coordinate(e)$.
   (c) if $index$ has just changed from 1 to 0, set $y_{upper} \leftarrow y-coordinate(e)$ and add the interval $[y_{lower}..y_{upper}]$ to $I_y$.
5. return $I_y$.

We assume that all local extrema of constraints as well as constraint intersections have been precomputed before running the propagation, so that the computation of $MAX$ and $MIN$ in step 2 only requires evaluating the constraints at interval bounds; the amount of constraint mainpulation required in each propagation step is exactly the same as that of other more straightforward propagation rules based only on interval bounds.

## 3 Generalization to ternary constraints

In practice, constraint networks are almost never binary. The main novel contribution of this paper is to generalize the propagation rule of [Faltings, 1994] to ternary

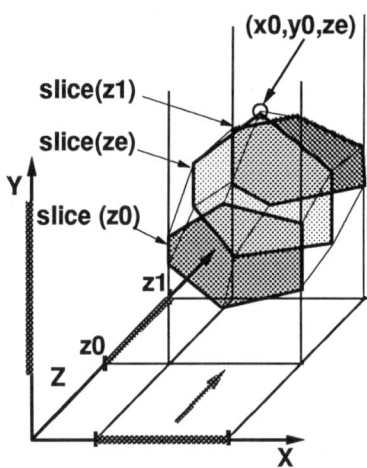

Figure 3: *Propagating intervals through a set of ternary constraints. The point $(x_0, y_0, z_e)$ is an extremum in Y of the ternary constraint.*

constraints over continous variables. The generalization again makes use of the property that the propagation rule only needs to know the complete set of extrema of all feasible regions, but not what regions there are.

**Propagation with a single third variable $Z$** We first consider the case of propagating the label of $X$ to $Y$ through a set of constraints $C_{XYZ}$ involving a single third variable $Z$ (see Figure 3). The consistency condition of Definition 1 for the label of $Y$ is now:

$$(\forall y \in L_Y)(\exists x \in L_X)(\exists z \in L_Z) C_{XYZ}(x, y, z)$$

We define the *projection* of the ternary constraint:

$$\mathcal{P}_{XY}(C_{XYZ}) = \{(x, y) | (\exists z \in L_Z) C_{XYZ}(x, y, z)\}$$

so that the local consistency condition for $L_Y$ for the ternary constraint $C_{XYZ}$ becomes the consistency condition for the binary constraint $\mathcal{P}_{XY}(C_{XYZ})$. Thus, propagation through the projection will give a consistent labelling.

In order to apply the propagation rule, we need to determine the extrema in the boundaries of this projection. The following lemma indicates where to find them:

**Lemma 1** *Assume that $(x_e, y_e)$ is an extremum of $\mathcal{P}_{XY}(C_{XYZ})$. Then, there must exist a $z_e \in L_Z$ such that $(x_e, y_e, z_e)$ is an extremum of $C_{XYZ}$ restricted to values of Z within its label $L_Z$.*

**Proof:**
Since $(x_e, y_e)$ is in $\mathcal{P}_{XY}(C_{XYZ})$, there must be a $z_e$ such that $(x_e, y_e, z_e)$ is in $C_{XYZ}$. On the other hand, since $(x_e, y_e)$ is an extremum in y of $\mathcal{P}_{XY}(C_{XYZ})$, there is a small interval of $x'$ around $x_e$ such that there is no point $(x_e, y') \in \mathcal{P}_{XY}(C_{XYZ})$ with $y' > y_e$ (maximum) or $y' < y_e$ (minimum). But this means that there is no $z'$ such that $(x', y', z') \in C_{XYZ}$, which is just the condition for $(x_e, y_e, z_e)$ to be an extremum of the ternary constraint. QED.

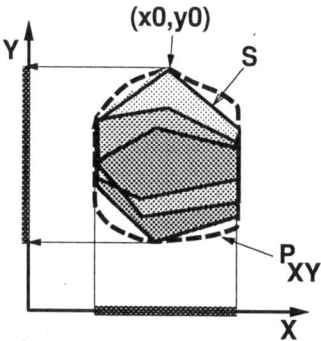

Figure 4: *The three slices generate the set S which mirrors all extrema of $P_{XY}$ even though it does not necessarily cover $P_{XY}$ (shown by dashed lines) exactly.*

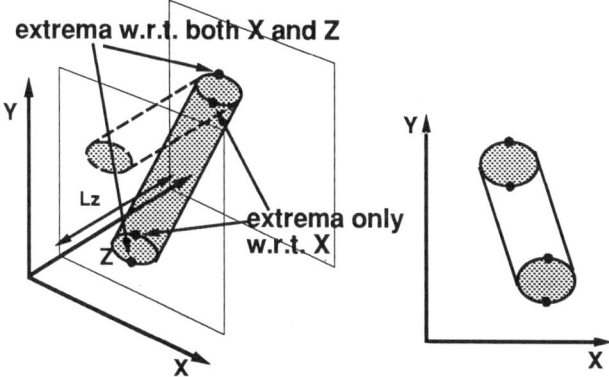

Figure 5: *The figure on the left defines a cylinder rising in Z. The two slices leave a gap which is not present in $P_{XY}$. The projection of the cylinder on axes X and Y with the slices hatched is shown on the right.*

Extrema of ternary constraints can fall on interval bounds, local extrema, constraint intersections, but also combinations thereof. For example, the intersection of two constraints involving the same variables forms a curve which can have an extremum, or a constraint curve can have a local extremum in one coordinate and reach interval bound in the other. Let $Z = \{z_{e1}, z_{e2}, ..., z_{ek}\}$ be the set of Z-coordinates of all extrema. Each $z_{ei}$ defines a "cut" through the space spanned by all constraints involving X and Y, called a *slice* (Figure 3). Now consider the set S:

$$S = \bigcup_{z_{ej} \in Z} \{(x,y) | C_{XYZ}(x, y, z_{ej})\}$$

which is the union of all slices at extrema of Z (Figure 4). The set S does not necessarily cover $P_{XY}(C_{XYZ})$, but because of Lemma 1 it contains all extrema. However, it is not true that the union of all extrema of slices in S will give the correct propagation, since it also contains *interior* extrema which would lead to false gaps during propagation. For example, the propagation shown in Figure 5 generates disjoint slices even though the projection should be one continuous region.

These interior extrema are characterized by the fact that they are not also extrema of $C_{XYZ}$ with respect to Z. Thus, the set of extrema E of $P_{XY}(C_{XYZ})$ is the set of all extrema of some slice which are also extrema with respect to Z. We can show the following:

**Lemma 2** *Updating the label of Y by propagating $L_X$ through E will give the same result as propagating it through $P_{XY}(C_{XYZ})$, and thus through $C_{XYZ}$ itself.*

**Proof:**
Interval bounds of $P_Y(C_{XYZ})$ can only fall on extrema of $P_{XY}(C_{XYZ})$ and thus of $C_{XYZ}$ itself. By construction, E contains all extrema in Y of $P_{XY}(C_{XYZ})$, with the interior ones filtered out. Thus, the projection of S onto the Y-axis will be identical to that of $P_{XY}(C_{XYZ})$.
QED

and we have a correct propagation rule for a set of constraints with a single third variable.

**Propagating constraints with several third variables** When there are several third variables $Z_1, Z_2, ..., Z_k$ with constraints $C_{XYZ_i}$, labels must be propagated through their *intersection*. Since the third variables are all distinct, the projection of their intersection is the intersection of the projections, i.e.:

$$\bigcap_{i=1..k} P_{XY}(C_{XYZ_i})$$

This intersection can be carried out implicitly by (i) filtering all extrema $(x_e, y_e)$ such that for some third variable $Z_i$, there does not exist a $z_{ei}$ such that $(x_e, y_e, z_{ei})$ satisfies $C_{XYZ_i}$, and (ii) adding all extrema resulting from intersections of $C_{XYZ_i}$ and $C_{XYZ_j}$.

**Identifying slices** Slices need to identify all points which could possibly give rise to extrema of the projection $P_{X_iX_j}$ of a constraint. Consider thus a total constraint $C_{X_iX_j}^t = \{C_{X_iX_jZ_1}^1, ..., C_{X_iX_jZ_k}^k\}$ involving k other variables $Z_1..Z_k$.

The first set of slices to be considered are those which fall at *interval bounds* of some $Z_i$. The second set of slices to be considered are *local extrema* of individual constraint curves. Consider a single ternary constraint $C_{XYZ_i}$. It has a local extremum in Y at $(x_e, y_e, z_e)$ if both:

- either $\delta C/\delta X(x_e, y_e, z_e) = 0$ or $x_e$ is an interval bound
- either $\delta C/\delta Z(x_e, y_e, z_e) = 0$ or $z_e$ is an interval bound

The set of relevant slices is obtained by considering all constraints which involve X and Y and collecting all $z_e$ satisfying the above conditions and falling into the label of $Z_i$, $L_{Z_i}$.

The third set of slices is generated by *intersections* between constraints. Here, the situation can be quite

The is the case exactly if the normal vectors of both constraints lie in a plane that also includes the $Y$-axis, i.e. $n(\vec{C_1}) + \alpha n(\vec{C_2})$ is a vector parallel to the $Y$ axis, for some $\alpha$. This leads to a system of equations for such extrema which can be solved using numerical methods.

The binary propagation rule is thus adapted to ternary constraints by replacing step 2) as follows:

2. $MIN \leftarrow \{\}, MAX \leftarrow \{\}$
   $\mathcal{Z} \leftarrow \{Z_i | \exists C_{XYZ_i}\}$
   for all $Z_i \in \mathcal{Z}$ do
   $\mathcal{S} \leftarrow$ all slices of $Z_i$
   for every slice $(z_{e1}, z_{e2}...) \in \mathcal{S}$ do

   1. add to $MAX$ all local maxima and to $MIN$ all local minima of all restricted regions $R_i$ in the slice $C^t(x,y,z_{ei})$, obtained directly by consideration of individual inequalities and their intersections.

   2. filter from $MAX$ and $MIN$ all those extrema which are not also extrema with respect to $Z_i$.

   3. for all other constraints $C_{XYZ_k}, Z_k \in \mathcal{Z}$ filter from $MAX$ and $MIN$ all those extrema which are not satisfiable by the constraint $C_{XYZ_k}$, and add all extrema which correspond to intersections between a constraint $C_{XYZ_i}$ and $C_{XYZ_k}$.

The following theorem shows that using this propagation rule in the Waltz algorithm results in a locally consistent labelling:

**Theorem 1** *The ternary propagation rule reaches quiescence only when the labels are locally consistent.*

**Proof:**
Follows from the correctness of the binary propagation rule (Theorem 2.1 in [Faltings, 1994]) and the equivalence of propagating through $S$ and $P_{XY}$.
QED

**Example of a propagation step.** We consider the propagation of $X_1$ to $X_3$ in the system of constraints already mentioned in section 2:

$$X_1^2 + 1/2 * X_2 + 2 * (X_3 - 6) \geq 0$$
$$X_1^2 + X_3^2 + X_4^2 - 25 \leq 0$$

with four variables $X_1, .., X_4$ and initial labels of $[-10, 10]$ for $X_1$ to $X_4$. The total constraint is thus formed by $\{C_1, C_2\}$. $C_2$ has a maximum in $X_3$ at $\{X_1 = 0, X_3 = 5, X_4 = 0\}$ and a minimum in $X_3$ at $\{X_1 = 0, X_3 = -5, X_4 = 0\}$. The projections on $X_2$ and $X_4$ of combinations of extrema (individual extrema and interval bounds) in $X_3$ give rise to the following potential slices:

$$X_2 \in \{-10, 10\} \quad \text{and} \quad X_4 \in \{-10, 0, 10\}$$

For each of the slices defined by substituting combinations of these values in the constraints, the label of $X_1$, $[-10, 10]$, is propagated to $X_3$. Only the slices $(X_2, X_4) = (-10, 0)$ and $(10, 0)$ give rise to valid extrema

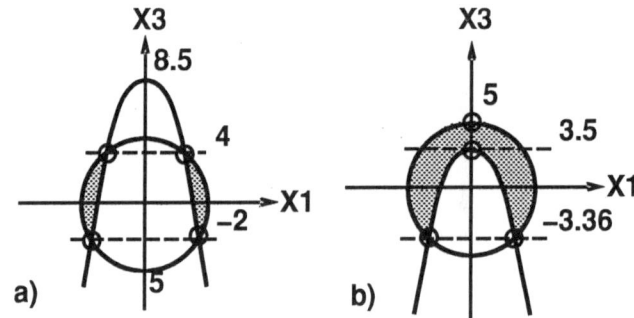

Figure 6: *The propagation of $X_1 = [-10, 10]$ through a) the slice at $\{X_2 = -10, X_4 = 0\}$ and b) through the slice at $\{X_2 = 10, X_4 = 0\}$ produces two maxima and two minima for each slice.*

(i.e. satisfying all other constraints); they are shown in Figure 6. The result propagating $X_1$ to $X_3$ through the total constraint formed by $C_1$ and $C_2$ is then obtained by collecting the valid extrema produced by these slices. The first slice with $(X_2, X_4) = (-10, 0)$ has two MAXIMA in $X_3$ at $\{X_1 = -3, X_3 = 4\}$ and $\{X_1 = 3, X_3 = 4\}$ and two MINIMA in $X_3$ at $\{X_1 = -4.583, X_3 = -2\}$ and $\{X_1 = 4.583, X_3 = -2\}$ all of them resulting from intersections. The second slice $(X_2, X_4) = (10, 0)$ has two local MAXIMA at $\{X_1 = 0, X_3 = 3.5\}$ and $\{X_1 = 0, X_3 = 5\}$ as well as two MINIMA produced by intersections at $\{X_1 = -3.7, X_3 = -3.359\}$ and $\{X_1 = 3.7, X_3 = -3.359\}$. Ordering all extrema in $X_3$ and applying step 4 of the propagation rule gives the single consistent interval $[-3.359, 5]$. In this example, the Davis method does not alter the labels ($[-10..10]$ for all variables) at all, whereas our propagation rule gives $X_1 = [-5, 5]$, $X_2 = [-10, 10]$, $X_3 = [-3.359, 5]$ and $X_4 = [-5, 5]$.

## 4 Performance on random examples

In this section, we would like to compare our algorithm with the propagation of individual constraints ([Davis, 1987]) and show on randomly generated examples why our algorithm is better than Davis'. Numeric values in these examples are all rounded on 2 decimals.

Supposing that the restricted region always falls within the propagated intervals, we applied Davis propagation to the following examples:

**Constraint set 1**
$C_1$    $-4 * X_5 - X_6 + 6 * X_9 + 6 > 0$
$C_2$    $4 * X_4 - 9 * X_5 - 5 * X_8 - 5 > 0$
$C_3$    $-9 * X_2 + 7 * X_3 + 4 * X_5 + 4 > 0$
$C_4$    $-X_6 + 10 * X_9 - 4 * X_{10} - 4 > 0$
$C_5$    $X_4 + X_7 - X_8 - 1 > 0$
$C_6$    $-4 * X_4 + X_7 - X_8 + 1 > 0$
$C_7$    $-6 * X_3 - 2 * X_5 + 6 * X_6 + 6 > 0$
$C_8$    $6 * X_2 - X_4 + 7 * X_5 + 7 > 0$
$C_9$    $8 * X_1 - 5 * X_2 - 8 * X_3 - 8 > 0$

| $X_i$ | Local C. | Davis | Simplex |
|---|---|---|---|
| $X_1$ | $[-10, 10]$ | $[-10, 10]$ | $[-10, 10]$ |
| $X_2$ | $[-10, 7.2]$ | $[-10, 10]$ | $[-8, 85, 7.2]$ |
| $X_3$ | $[-10, 10]$ | $[-10, 10]$ | $[-10, 10]$ |
| $X_4$ | $[-10, 5.25]$ | $[-10, 5.25]$ | $[-10, 5.25]$ |
| $X_5$ | $[-6.86, 7.33]$ | $[-10, 7.33]$ | $[-5.35, 7.33]$ |
| $X_6$ | $[-10, 10]$ | $[-10, 10]$ | $[-9.84, 10]$ |
| $X_7$ | $[-9.4, 10]$ | $[-10, 10]$ | $[-9.4, 10]$ |
| $X_8$ | $[-10, 9.4]$ | $[-10, 10]$ | $[-10, 7.16]$ |
| $X_9$ | $[-4.6, 10]$ | $[-4.6, 10]$ | $[-3.41, 10]$ |
| $X_{10}$ | $[-10, 10]$ | $[-10, 10]$ | $[-10, 10]$ |

Table 1: *Solutions for constraint set 1.*

| $X_i$ | Local Consistency | Davis |
|---|---|---|
| $X_1$ | $[-10, 10]$ | $[-10, 10]$ |
| $X_2$ | $[-10, 7.15]$ | $[-10, 10]$ |
| $X_3$ | $[-10, 5.32]$ | $[-10, 10]$ |
| $X_4$ | $[-10, 10]$ | $[-10, 10]$ |
| $X_5$ | $[-4.6, 7.16]$ | $[-10, 9.44]$ |
| $X_6$ | $[-4.37, 2.79]$ | $[-10, 10]$ |
| $X_8$ | $[-10, 10]$ | $[-10, 10]$ |
| $X_9$ | $[-6.35, 4.24]$ | $[-10, 10]$ |
| $X_{10}$ | $[-6.91, 10]$ | $[-10, 10]$ |

Table 2: *Solutions for constraint set 2.*

**Constraint set 2**

$C_1$ $-16 + \frac{7}{10} * X_9 * X_5 + \frac{7}{5} * X_6 - \frac{6}{5} * X_5 - \frac{6}{5} * X_9 - \frac{13}{50} * X_9 * X_6 - \frac{13}{50} * X_5 * X_6 + \frac{3}{10} * X_6 * X_6 + \frac{3}{5} * X_5 * X_5 + \frac{3}{5} * X_9 * X_9 < 0$

$C_2$ $4 * X_4 - 9 * X_5 - 5 * X_8 - 5 > 0$

$C_3$ $-9 * X_2 + 7 * X_3 + 4 * X_5 + 4 > 0$

$C_4$ $-\frac{4}{5} * X_{10} - \frac{4}{5} * X_6 + \frac{1}{10} * X_{10} * X_9 + \frac{1}{10} * X_6 * X_9 + \frac{1}{10} * X_6 * X_{10} + \frac{17}{10} * X_9 + \frac{1}{50} * X_9 * X_9 + \frac{1}{4} * X_6 * X_6 + \frac{1}{4} * X_{10} * X_{10} - \frac{22}{3} < 0$

$C_5$ $-6 * X_3 - 2 * X_5 + 6 * X_6 + 6 > 0$

$C_6$ $6 * X_2 - X_4 + 7 * X_5 + 7 > 0$

$C_7$ $8 * X_1 - 5 * X_2 - 8 * X_3 - 8 > 0$

A comparison of the results of Davis' algorithm and the local consistency rule (Table 1) reveals that local consistency prunes labels more drastically as it considers intersections between constraints defined on the same pairs of variables. $X_5$ is an example of such a variable involved in several constraints. When compared with the simplex algorithm, which computes the globally optimal solution for constraint set 1, the labels of $X_2$, $X_5$, $X_7$ and $X_8$ show that local consistency is better than the Davis rule although it does not achieve global consistency. In the second example, two nonlinear constraints $C_1$ and $C_4$ have been added to a set of linear inequalities. Here, the local consistency algorithm is far better because, additionally to label boundaries, it takes into account the individual extrema of nonlinear constraints (Table 2).

We randomly generated about 300 mixed constraints sets similar to constraint set 2. For each solution we evaluated the average number of labels different from the initial label and how much the labels had been reduced on average. Davis' algorithm reduced 3 out of 10 labels whereas local consistency reduced nearly 7 labels. The amount of pruning was 7.48% for Davis and 25.97 % for local consistency.

## 5 Conclusions

Among all techniques developed for constraint-based reasoning, local consistency and its weaker variants such as forward checking are by far the most widely used techniques in practice. This is due to the fact that it can be computed efficiently but still provides dramatic pruning of the search space, especially when some variables are already assigned values.

In continuous domains, straightforward implementations of Waltz propagation does not guarantee useful results ([Davis, 1987]). A modified propagation rule which guarantees local consistency has been given in [Faltings, 1994] for binary constraints. In this paper, we have shown how the rule can be generalized to ternary and thus n-ary constraints. The Waltz algorithm using the new propagation rule gives significantly better results than the commonly used propagation rule analyzed by Davis. This result should allow significant improvements in many applications of local consistency.

## References

[Benhamou et al., 1994] **F. Benhamou, D. McAllester, P. Van Hentenryck:** "CLP(Intervals) Revisited," Technical Report No. CS-94-18, Université Aix-Marseille, April 1994

[Davis, 1987] **Ernest Davis:** "Constraint Propagation with Interval Labels," *Artificial Intelligence* **32**, 1987, pp. 281-331

[Faltings, 1994] **Boi Faltings:** "Arc-consistency for continous variables," *Artificial Intelligence* **65**, 1994, pp. 363-376

[Freuder, 1982] **Eugene Freuder:** "A sufficient condition for backtrack-free search," *Journal of the ACM* **29**(1), 1982, pp. 24-32

[Hyvönen, 1992] **E. Hyvönen:** "Constraint reasoning based on interval arithmetic: the tolerance propagation approach," *Artificial Intelligence* **58**(1992), pp. 71-112

[Lhomme, 1993] **O. Lhomme:** "Consistency Techniques for Numeric CSPs," *Proceedings of the 13th IJCAI*, 1993, pp. 232-238

[Sam-Haroud and Faltings, 1996] **Djamila Sam-Haroud, Boi Faltings:** "Consistency Techniques for Continous Constraints," *Constraints* **1**(1), 1996, pp. 85-118

[Waltz, 1975] **David Waltz:** "Understanding line drawings of scenes with shadows," in P.H. Winston(ed.): *The Psychology of Computer Vision*, McGraw-Hill, NY 1975

# Arc consistency for general constraint networks: preliminary results

**Christian Bessière**
LIRMM–CNRS (UMR 5506)
161 rue Ada
34392 Montpellier cedex 5, France
Email: bessiere@lirmm.fr

**Jean-Charles Régin**
ILOG S.A.
9 rue de Verdun
94253 Gentilly Cedex, France
Email: regin@ilog.fr

## Abstract

Constraint networks are used more and more to solve combinatorial problems in real-life applications. Much activity is concentrated on improving the efficiency of finding a solution in a constraint network (the constraint satisfaction problem, CSP). Particularly, arc consistency caught many researchers' attention, involving the discovery of a large number of algorithms. And, for the last two years, it has been shown that maintaining arc consistency during search is a worthwhile approach. However, results on CSPs and on arc consistency are almost always limited to binary constraint networks. The CSP is no longer an academic problem, and it is time to deal with non-binary CSPs, as widely required in real world constraint solvers. This paper proposes a general schema to implement arc consistency on constraints of any arity when no specific algorithm is known. A first instantiation of the schema is presented here, which deals with constraints given by a predicate, by the set of forbidden combinations of values, or by the set of allowed ones.

## 1 Introduction
### 1.1 Overview

Constraint satisfaction problems (CSPs) occur widely in artificial intelligence. They are used more and more in real-life applications, such as resource allocation, crew scheduling, time tabling, frequency allocation, car sequencing, etc. The CSP involves finding a solution in a constraint network, i.e. finding values for problem variables subject to constraints.

The general task of solving a CSP being NP-hard, many researchers have concentrated on improving the efficiency of finding a solution in a constraint network. Particularly, arc consistency caught many researchers' attention, involving the discovery of a large number of algorithms [Mackworth, 1977a], [Mohr and Henderson, 1986], [Bessière, 1994], [Bessière et al., 1995]. And, recently, the value of these studies on arc consistency increased since it has been shown that maintaining arc consistency during search is definitely a worthwhile approach when solving large and hard problems [Sabin and Freuder, 1994], [Bessière et al., 1995], [Bessière and Régin, 1996], [Grant and Smith, 1996].

However, results about CSP solving and about arc consistency are almost always limited to *binary* constraint networks, justified by the fact that any non-binary constraint network can be translated into an equivalent binary one with additional variables [Rossi et al., 1990], [Dechter, 1990]. But, in practical cases, it is often inconceivable to translate a non-binary constraint into an equivalent set of binary ones with the technique cited above because of the underlying computational and memory cost. When a constraint is *representable* [Montanari, 1974]), it is sometimes possible to replace it by a set of binary constraints without introducing new variables and without generating explicitly the set of allowed tuples of each binary constraint. For example, a constraint enforcing that the three variables $x_1$, $x_2$, and $x_3$ must all take different values can equivalently be replaced by the three binary constraints $x_1 \neq x_2$, $x_1 \neq x_3$, and $x_2 \neq x_3$. But, as it has been shown in [Régin, 1994], non-binary constraints lose a part of their semantics when encoded into a set of binary ones in this way. This leads, for example, to less pruning for arc consistency algorithms handling them.

Hence, if we no longer want to consider the CSP as an academic problem we must be able to deal with any kind of constraints, and so, with non-binary constraints, as widely required in real-world constraint solvers.

### 1.2 Previous work

In the constraint satisfaction community, the number of works involving non-binary constraint networks is rather small. For the particular case of arc consistency, we know two algorithms capable of achieving it. Mackworth [Mackworth, 1977b] proposed the algorithm CN, which is a kind of generalization of AC-3 to non-binary constraints. As AC-3, that algorithm has a bad worst-case time complexity ($O(er^2d^{r+1})$, with $e$ the number of constraints in the network, $r$ the maximal arity of the constraints, and $d$ the size of the largest domain). Mohr and Massini [Mohr and Masini, 1988] proposed GAC4. It is based on the same idea as AC-4 (computing the number of supports for each value in each domain and removing those with this number equal to zero). Thus, it gets rid of the huge worst-case time complexity of CN (GAC4 is in $O(ed^r)$), but has the same drawbacks as AC-4: space complexity (because of the lists of supported values), and average time complexity. The respective drawbacks of CN and GAC4 are even more important than on their binary versions, so that CN can only be applied on ternary constraints and very small domains, and GAC4 on very tight constraints, where the number of allowed tuples of values is very small.

Finally, we can point out the work of van Beek and Dechter [van Beek and Dechter, 1995], who proposed another definition for arc consistency on non-binary constraint networks, namely *relational arc consistency*. This definition is much stronger than the classical one since for each constraint it requires global consistency on the underlying subnetwork (i.e. the network involving the variables of the given constraint, and all the "smaller" constraints defined on some of these variables).

### 1.3 Our purpose

Our general aim is to propose a new schema in order to perform arc consistency on any real world constraint network with a reasonably low time and space complexity. In industrial applications, a constraint can be given in many different forms: by the set of allowed tuples in extension (generally when the constraint is very tight), by the set of forbidden tuples in extension (when the constraint is very loose), by a conjunctive constraint, by an arithmetic relation, or by any predicate for which no particular semantics is known (data base query, user's context-dependent constraint, etc.). Thus, a suitable arc consistency algorithm should be able to efficiently handle any of these constraints.

The particular aim of this paper is to present the first steps of our work. First, we will present our general schema for arc consistency on non-binary constraint networks. This schema is based on the AC-7 schema given in [Bessière et al., 1995]. It makes use of the "current support" idea, and of "multidirectionality" (the generalization of bidirectionality to non-binary constraints) in order to save as many constraint checks as possible. Second, we will instantiate this schema with some of the most frequently occurring forms of constraint representations.

Constraints defined by a predicate for which no particular semantics is known are one of the most important forms that have to be handled. Indeed, if we except the predicates for which specific algorithms are known (arithmetic relations [Van Hentenryck et al., 1992], cardinality constraints [Régin, 1996], etc.), the only available algorithms for non-binary constraints are CN and GAC4. GAC4 deals only with constraints given in extension, and then is not capable of handling a predicate, even if, theoretically, it is always possible to build the whole set of allowed tuples by checking all the combinations of values for the variables involved in the constraint. It would be impracticable, at least because of space requirements. Moreover, asking the value of a predicate for a particular combination of values (a constraint check) can be very costly in practice, when the answer is contained in a database, or requires heavy computation. This definitively eliminates CN, with its huge time complexity, doing and doing again many times the same constraint checks. Therefore, we will propose an instantiation of the schema that will efficiently handle predicates, minimizing the number of constraint checks. However, this schema is certainly not competitive on predicates for which a specific algorithm is available.

Afterwards, we will instantiate the schema for constraints defined by a set of forbidden tuples. This kind of representation is of practical interest when constraints are very loose.

Finally, we will show that the schema can also deal with constraints defined by the list of allowed tuples given in extension. It leads to an improvement on GAC4, which was already written for that kind of constraint.

## 2 Preliminaries

**Constraint network.** A finite *constraint network* $\mathcal{P}$ is defined as a set of $n$ *variables* $X = \{x_1, \ldots, x_n\}$, a set of current *domains* $\mathcal{D} = \{D(x_1), \ldots, D(x_n)\}$ where $D(x_i)$ is the finite set of possible *values* for variable $x_i$, and a set $\mathcal{C}$ of *constraints* between variables. We introduce the particular notation $\mathcal{D}_0 = \{D_0(x_1), \ldots, D_0(x_n)\}$ to represent the set of initial domains of $\mathcal{P}$. Indeed, we consider that any constraint network $\mathcal{P}$ can be associated with an initial domain $\mathcal{D}_0$ (containing $\mathcal{D}$), on which constraint definitions were stated. A total ordering $<_d$ can be defined on $D(x_i), \forall x_i \in X$, without loss of generality.

**Constraints.** A constraint $C$ on the ordered set of variables $X(C) = (x_{i_1}, \ldots, x_{i_r})$ is a subset of the Cartesian product $D_0(x_{i_1}) \times \cdots \times D_0(x_{i_r})$ that specifies the *allowed* combinations of values for the variables $x_{i_1} \times \ldots \times x_{i_r}$. An element of $D_0(x_{i_1}) \times \cdots \times D_0(x_{i_r})$ is called a *tuple on* $X(C)$. Two tuples $\tau$ and $\tau'$ on $X(C)$ can be ordered by the natural lexicographic order $\prec_{lo}$ in which $\tau \prec_{lo} \tau'$ iff $\exists k/\tau[1..k-1] = \tau'[1..k-1]$ and $\tau[k] <_d \tau'[k]$ ($\tau[1..k]$ being the prefix of size $k$ of $\tau$, and $\tau[k]$ the $k^{th}$ value of $\tau$). The tuples of

$D_0(x_{i_1}) \times \cdots \times D_0(x_{i_r})$ not allowed by $C$ are called the *forbidden* tuples of $C$. Verifying whether a given tuple $\tau$ is allowed by $C$ or not is called a *constraint check*. $|X(C)|$ is the *arity* of $C$.

A constraint $C$ involving the subset of variables $X(C) = (x_{i_1}, \ldots, x_{i_r})$ can be defined by the set of allowed tuples (resp. the set of forbidden tuples) given in extension when the constraint is tight (resp. is loose), or by an arithmetic relation. More generally, it can be represented by any Boolean function $f_C$ defined on $D_0(x_{i_1}) \times \cdots \times D_0(x_{i_r})$.

**Solutions.** A *solution* of a constraint network is an instantiation of the variables such that all the constraints are satisfied.

**Notation.** A value $a$ for a variable $x$ is often denoted by $(x, a)$. $\text{var}(C, i)$ represents the $i^{th}$ variable of $X(C)$, while $\text{index}(C, x)$ is the position of variable $x$ in $X(C)$.

**Arc consistency.** Let $\mathcal{P} = (X, \mathcal{D}, \mathcal{C})$ be a constraint network, $C$ a constraint in $\mathcal{C}$.

A tuple $\tau$ of $X(C)$ is *valid* if $\forall (x, a) \in \tau, a \in D(x)$; otherwise, it is *rejected*.

A value $a \in D(x)$ is *consistent with* $C$ iff $x \notin X(C)$, or $\exists \tau$ allowed by $C$, such that $a = \tau[\text{index}(C, x)]$ and $\tau$ is valid. ($\tau$ is then called a *support* for $(x, a)$ on $C$.)

$C$ is *arc consistent* iff $\forall x_i \in X(C), D(x_i) \neq \emptyset$ and $\forall a \in D(x_i), a$ is consistent with $C$.

$\mathcal{P}$ is *arc consistent* iff all the constraints of $\mathcal{C}$ are arc consistent.

We achieve arc consistency in $\mathcal{P}$ by removing every value not consistent with at least one constraint in $\mathcal{C}$.

## 3 A general schema for arc consistency

As we pointed out in Section 1, our general aim is, on the one hand, to provide a schema sufficiently general to be instantiated with any kind of constraint. On the other hand, this schema must be powerful enough to avoid as many constraint checks as possible when achieving arc consistency.

We did not choose an AC-3 like or an AC-4 like schema for efficiency reasons. To be reasonably efficient, the schema has to be based on AC-6 or AC-7. They are both based on the search of a single support for each value, the worst-case time complexity being optimal in both algorithms. The difference between them is whether or not they deal with bidirectionality [Bessière et al., 1995].

In a non-binary constraint network, any constraint is multidirectional, as any constraint is bidirectional in a binary constraint network. *Multidirectionality*, indeed, is the fact that for any constraint $C$, a tuple $\tau$ on $X(C)$ is a support for the value $\tau[\text{index}(C, x)]$ ($x$ being a variable involved in $C$) iff $\forall y \in X(C), \tau$ is a support for $\tau[\text{index}(C, y)]$. We say that an algorithm "deals with" multidirectionality iff it never checks whether a tuple is a support for a value when it has already been checked for another value, and never looks for a support for a value on a constraint $C$ when a tuple supporting this value has already been checked.

**Algorithm 1:** function propagation

propagation (**in** $C$: constraint; **in** $x$: variable;
                 **in** $a$: value;
                 **in out** deletionStream: list): **boolean**

for each $\tau \in S_C(x, a)$ do
1    for each $(z, c) \in \tau$ do remove $\tau$ from $S_C(z, c)$;
2    for each $(y, b) \in S(\tau)$ do
       remove $(y, b)$ from $S(\tau)$;
       if $b \in D(y)$ then
3          $\sigma \leftarrow \texttt{seekInferableSupport}(C, y, b)$;
4          if $\sigma \neq \textbf{nil}$ then
            add $(y, b)$ in $S(\sigma)$;
         else
5            $\sigma \leftarrow \texttt{seekNextSupport}(C, y, b, last_C(y, b))$;
           if $\sigma \neq \textbf{nil}$ then
6              add $(y, b)$ in $S(\sigma)$;
              $last_C(y, b) \leftarrow \sigma$;
7              for $k$ from $1$ to $|X(C)|$ do
                 add $\sigma$ in $S_C(\text{var}(C, k), \sigma[k])$;
           else
8              remove $b$ from $D(y)$;
             if $D(y) = \emptyset$ then return **false**;
9              add $(y, b)$ in deletionStream;

return **true**;

And, dealing with multidirectionality appears to be promising on non-binary constraints. A tuple $\tau$, indeed, allowed by a constraint $C$, can support $|X(C)|$ values. Thus the possible savings are much more important than on binary constraints. AC-7 could save $d^2$ constraint checks thanks to bidirectionality, where $d$ is the domain size of the variables involved (see [Bessière et al., 1995]), while $d^r$ constraint checks can be saved thanks to multidirectionality on an $r$-ary constraint.

The framework we propose is then an AC-7 like schema in which search for support (function seekNextSupport) is instantiated differently depending on the type of the constraint involved. This schema (named GAC-schema) is able to handle any type of constraint, as soon as the corresponding function seekNextSupport is available. Before giving the seekNextSupport function associated with the types of constraints we will study in this paper (in the next section), let us describe the function propagation (see Alg. 1). The program including it must create and initialize the data structures ($S_C, S$, and $last_C$), and call propagation($C, x, a$, deletionStream) for each constraint $C$ involving $x$, each time a value $(x, a)$ is removed from $D(x)$, in order to propagate the consequences of this deletion.

$S_C$, $S$, and $last_C$ must be initialized in a way such that:

- $S_C(x,a)$ contains all the allowed tuples $\tau$ that are the current support for some value, and such that $\tau[index(C,x)] = a$.
- $S(\tau)$ contains all values for which $\tau$ is the current support.
- $last_C(y,b)$ is the last tuple returned by seekNextSupport as a support for $(y,b)$ if seekNextSupport$(C,y,b,-)$ has already been called; **nil** otherwise. $last_C(y,b)$ is perhaps no longer a support in $C$ for $(y,b)$ (if it is no longer valid and seekInferableSupport has found a new support for $(y,b)$). It is just here to give the point where seekNextSupport will have to restart the search for a support for $(y,b)$ on $C$ at the next call. There is an underlying ordering on the tuples, which is proper to seekNextSupport.

**Property 1** $\forall C \in \mathcal{C}, \forall$ allowed tuple $\tau : S(\tau) \neq \emptyset \Leftrightarrow \forall x \in X(C), \tau \in S_C(x, \tau[index(C,x)])$.

**Property 2** *Given any constraint $C$ of arity $r$, the space complexity of the data structure of GAC-schema for $C$ is $O(r^2 d)$.*

**Proof.** Each value has at most one support, then $\sum |S(\tau)| \leq rd$ and there are at most $rd$ tuples in memory. Each value is associated with one $last_C$, so the $last_C$ data structure requires $rd$ pointers. A tuple contains $r$ elements, thus the set of all tuples that support at least one value can be represented in $O(r^2 d)$. The number of elements that belong to $S_C$ is bounded above by $r^2 d$ by property 1 and because $\sum |S(\tau)| \leq rd$. Since an element of an $S_C$ list corresponds to a pointer to a tuple, the space complexity of the data structures of GAC-schema for $C$ is $O(r^2 d)$. ∎

Each time a value $(x,a)$ is removed from $D(x)$, we must propagate this deletion to each constraint $C$ involving $x$. So, for all the values $(y,b)$ that were supported by a tuple containing $a$ in position $index(C,x)$, we must find another support (line 2). To take multi-directionality into account, we first check (lines 3 and 4) if there exists a valid tuple containing $(y,b)$ that is already the current support for another value (function seekInferableSupport, Alg. 2). If not, the function seekNextSupport will look for another support for $(y,b)$, starting the search from $last_C(y,b)$ (line 5). If a new support $\sigma$ is found for $(y,b)$ on $C$, $(y,b)$ is recorded as being currently supported by $\sigma$ (line 6), and for each value contained in $\sigma$, we store the fact that $\sigma$ is the current support for some value (line 7); otherwise, $(y,b)$ has to be removed (lines 8 to 9).

## 4 The main types of constraint representations

In this section, we will present the instantiation of the schema in order to deal with three types of constraints.

---

**Algorithm 2:** function seekInferableSupport

seekInferableSupport (**in** $C$: constraint; **in** $y$: variable; **in** $b$: value): tuple

**while** $S_C(y,b) \neq \emptyset$ **do**
  $\sigma \leftarrow first(S_C(y,b))$ ;
  **if** $\exists k / \sigma[k] \notin D(var(C,k))$ **then** remove $\sigma$ from $S_C(y,b)$ ;
  **else return** $\sigma$   /* $\sigma$ is a support */ ;
**return nil** ;

---

Constraints given by a predicate for which we do not know specific algorithms is the most important type. Indeed, there does not exist any tool to process a predicate in reasonable time and space. Afterwards, we will present a type of constraint which can be viewed as a special case of predicate constraints: constraints given by the set of forbidden tuples. Finally, we will give the seekNextSupport function capable of processing constraints given by the set of allowed tuples.

### 4.1 Predicates

When a constraint $C$ is defined by a predicate for which no particular semantics is known, it is necessary to define an ordering on the tuples that will be followed when looking for a support for a value $(y,b)$ on $C$. By means of the function nextTuple (Alg. 3), the lexicographic order $\prec_{lo}$ defined in Section 2 is used to examine the valid tuples on $X(C)$. Given any tuple $\tau$ on $X(C)$, and any index $k \leq |X(C)|$, nextTuple$(C,y,b,\tau,k)$ will return the smallest valid tuple $\sigma$ greater than $\tau$ w.r.t. $\prec_{lo}$, such that the value of $y$ in $\sigma$ is $b$, and such that the prefixes of size $k$ of $\sigma$ and $\tau$ are not equal. The second parameter returned by nextTuple is the smallest index $k'$ on which $\sigma$ and $\tau$ have different values.

When the function seekNextSupport (see Alg. 4) is called to search for a new support for $(y,b)$ on $C$, we know from seekInferableSupport that there does not exist any tuple already checked that supports $(y,b)$ on $C$. However, we have to take care to avoid checking a tuple $\tau$ which has already been unsuccessfully checked for another value of another variable of $X(C)$. We name *candidate* a valid tuple which has never been checked and thus, which could be a support for $(y,b)$, as op-

---

**Algorithm 3:** function nextTuple

nextTuple (**in** $C$: constraint; **in** $y$: variable; **in** $b$: value; **in** $\tau$: tuple; **in** $k$: index): (tuple, index)

/* returns $(\sigma, k')$ where:
- $\sigma$ is the smallest valid tuple such that $\sigma[index(C,y)] = b$, $\tau \prec_{lo} \sigma$ and $\sigma[1..k] \neq \tau[1..k]$
- $k'$ is such that $\sigma[1..k'-1] = \tau[1..k'-1]$ and $\sigma[k'] \neq \tau[k']$

otherwise **returns** (**nil**,–)                         */

Note that $k'$ will always be less than or equal to $k$.

**Algorithm 4:** function seekNextSupport

seekNextSupport (**in** $C$: constraint; **in** $y$: variable; **in** $b$: value; **in** $\tau$: tuple): tuple

1  **if** $\tau \neq nil$ **then**
    | $(\sigma, dummy) \leftarrow \texttt{nextTuple}(C, y, b, \tau, |X(C)|)$;
  **else**
    | **for** $i$ from 1 to $|X(C)|$ **do**
      | $\sigma[i] \leftarrow first(\textsf{var}(C, i))$ ;
2     | $\sigma[\textsf{index}(C, y)] \leftarrow b$ ;
3  $\sigma \leftarrow \texttt{seekCandidateTuple}(C, y, b, \sigma, 1)$ ;
  $found \leftarrow false$;
4  **while** $(\sigma \neq nil)$ **and** $(not\ found)$ **do**
5     **if** $f_C(\sigma)$ **then** $found \leftarrow true$ ;
    **else**
     | $(\sigma, k) \leftarrow \texttt{nextTuple}(C, y, b, \sigma, |X(C)|)$;
6      | $\sigma \leftarrow \texttt{seekCandidateTuple}(C, y, b, \sigma, k)$ ;
  **return** $\sigma$;

**Algorithm 5:** function seekCandidateTuple

seekCandidateTuple (**in** $C$: constraint; **in** $y$: variable; **in** $b$: value; **in** $\sigma$: tuple; **in** $k$: index): tuple

  **while** $(\sigma \neq nil)$ **and** $(k \leq |X(C)|)$ **do**
1     /* $\sigma$ is candidate till index $k-1$ */;
    **if** $last_C(\textsf{var}(C, k), \sigma[k]) \neq nil$ **then**
2       | $\lambda \leftarrow last_C(\textsf{var}(C, k), \sigma[k])$ ;
      | $split \leftarrow 1$ ;
3       | **while** $\sigma[split] = \lambda[split]$ **do**
        | $split \leftarrow split + 1$ ;
4       | **if** $\sigma[split] < \lambda[split]$ **then**
5         | **if** $split < k$ **then**
6           | $(\sigma, k') \leftarrow \texttt{nextTuple}(C, y, b, \sigma, k)$ ;
7           | $k \leftarrow k' - 1$ ;
        **else**
8           | $(\sigma, k') \leftarrow \texttt{nextTuple}(C, y, b, \lambda, |X(C)|)$ ;
9           | $k \leftarrow min(k, k' - 1)$ ;
    | $k \leftarrow k + 1$ ;
  **return** $\sigma$ ;

posed to the valid tuples that have already been checked not to be allowed by $C$ (when we look for support for another value). Dealing with multidirectionality implies only checking the candidates. This can be done by means of the function seekCandidateTuple (see Alg. 5).

seekCandidateTuple$(C, y, b, \sigma, k)$ returns the smallest (w.r.t. $\prec_{lo}$) candidate greater than or equal to $\sigma$, where $\sigma$ is valid and $\sigma[1..k-1]$ has been verified to be a possible prefix for a candidate. For each index from $k$ to $|X(C)|$, seekCandidateTuple verifies whether $\sigma$ is greater than $last_C(\textsf{var}(C, k), \sigma[k])$, (denoted by $\lambda$) (lines 2 to 4). If $\sigma$ is smaller than $\lambda$, the search for a candidate has to jump forward: either to the smallest valid tuple following $\sigma$ with a prefix different from $\sigma[1..k]$ (if $\sigma$ and $\lambda$ were diverging before $k$) (line 6), or directly to the valid tuple following $\lambda$ (if $\sigma$ and $\lambda$ were diverging after $k$) (line 8). When we jump forward to the next valid tuple greater than $\sigma$ or $\lambda$, some values before index $k$ may have changed. Then, the value $k$ goes back to the smallest index where the value of $\sigma$ has changed (lines 7 and 9) to keep the property of line 1. When $k$ reaches $|X(C)| + 1$, $\sigma$ is a candidate and is returned.

The function seekNextSupport$(C, y, b, \tau)$ returns the smallest tuple greater than $\tau$ which is checked to be allowed by $C$. From line 1 to line 2, seekNextSupport assigns to $\sigma$ the smallest valid tuple following $\tau$ (depending on whether $\tau$ is **nil** or not). Afterwards, $\sigma$ is assigned to the smallest candidate (line 3). The search for a support for $(y, b)$ on $C$ is done in lines 4 to 6: we check $\sigma$ and jump to the next candidate until $f_C(\sigma)$ returns **true** (line 5).

**A sketch of proof.**
We will simply show here that seekCandidateTuple cannot miss any candidates when jumping forward in lines 6 and 8.

**line 6:** suppose there is a candidate $\sigma'$ between $\sigma$ and the tuple returned by nextTuple$(C, y, b, \sigma, k)$. This means that $\sigma'[1..k] = \sigma[1..k]$; otherwise $\sigma'$ would be the tuple returned by nextTuple. So, $\sigma'$ is smaller than $\lambda$ because of lines 4 and 5. $\sigma'[k]$ being equal to $\sigma[k]$, $\sigma'$ cannot be a candidate because $\lambda = last_C(\textsf{var}(C, k), \sigma[k])$ (see the definition of $last_C$).

**line 8:** suppose there is a candidate $\sigma'$ between $\sigma$ and $\lambda$. Then, $\sigma'[1..k]$ is equal to $\sigma[1..k]$ since $\sigma[1..k] = \lambda[1..k]$ (lines 3 to 5). Once again that is impossible because $\lambda = last_C(\textsf{var}(C, k), \sigma[k])$. Finally, $\sigma'$ cannot be equal to $\lambda$ since $\lambda$ is no longer a valid tuple. (Otherwise seekInferableSupport would have found it to support $(y, b)$.) ∎

### 4.2 Constraints given in extension

**Negative constraints**

In this case, the constraint $C$ is given in extension by the set of forbidden tuples, denoted by $\overline{T}(C)$. Even without loss of generality, we can assume that the constraint is very loose. (Otherwise, the space in memory needed prohibits us from using that representation.) Then $\overline{T}(C)$ contains few elements with regard to the set of tuples on $|X(C)|$. So, if we use the previous method with the predicate $f_C(\sigma)$ such that $f_C(\sigma) \Leftrightarrow \sigma \notin \overline{T}(C)$, then only a few constraint checks will be needed to find a new support or to prove there is none, because almost all valid tuples are candidates.

Such a predicate $f_C(\sigma)$ can be efficiently implemented by using a method like hashing [Sedgewick, 1990]. Hashing, indeed, permits us to find whether an element belongs to a set $S$, with an $O(|S|)$ space complexity, and an average time complexity close to a constant.

**Positive constraints**

In this case, the set of allowed tuples, denoted by $T(C)$, is explicitly given. Thus, for one constraint, the space

**Algorithm 6:** function seekNextSupport

```
seekNextSupport (in C: constraint; in y: variable;
                 in b: value; in dummy): tuple
  while elt(y,b) ≠ nil do
      σ ← Tuple(elt(y,b));
      if isValid(σ) then return σ ;
      elt(y,b) ← next(elt(y,b)) ;
  return nil ;
```

complexity will depend on the size of $T(C)$ similarly to GAC4. So, at first glance, it seems that the space complexity of GAC4 cannot be improved. But, in practice, we may have a problem in which some of the constraints appear many times involving different variables. For instance, in configuration problems, a configurator has to choose the most appropriate components among a catalog of predefined components and arrange them according to constraints that can be assembly rules, performance, etc. Most of the constraints involved with the same kind of components are repeated and are given by their common set of allowed tuples.

For a constraint $C$, the representation of the data structures of GAC4 corresponds to a particular representation of $T(C)$. So, if two constraints are given by $T(C)$, then GAC4 needs two distinct data structures, each with a $O(|T(C)|)$ space complexity. Hence, for $p$ repeated constraints the space complexity of GAC4 will be $O(p \cdot |T(C)|)$. The algorithm we presented only requires an explicit representation of $S$ and $S_C$ lists, and $last_C$ pointers, for each constraint. This leads to a $O(r^2 d)$ space complexity per constraint, $r$ being the arity of the constraint involved (see Property 2). If the representation of $T(C)$ can be shared by $p$ repeated constraints, then the global space complexity will be $O(|T(C)| + p.r^2 d)$, and a factor $p$ will be gained with regard to GAC4.

Such an algorithm can be defined by representing the common set $T(C)$ as GAC4 does for one constraint and by using for each repeated constraint and each value $(y,b)$ a pointer that indicates the last element reached in $T(C)$. We will denote it by $elt(y,b)$. The algorithm does no longer compute the tuples and test them; it only looks for a support directly in $T(C)$. The new algorithm is given by the function seekNextSupport (Alg. 6) which replaces the previous one. The function $next(elt(y,b))$ gives the next tuple in $T(C)$ containing $b$ for variable $y$.

## 5 Space and time analysis

Let $C$ be a constraint of arity $r$ with a tightness $t$ (proportion of forbidden tuples), and assume that all domains have a size $d$. Then, if a pointer is represented by 4 bytes, a good estimation of the memory required is:

- $16r \cdot |\{tuples\ on\ X(C)\ allowed\ by\ C\}| = 16rd^r(1-t)$ Bytes for GAC4
- $24r^2 d$ Bytes for the GAC-schema, according to property 2 ($C$ being a predicate[1])

The table below gives some results about the space requirements of GAC4 and the GAC-schema. The sizes are given in Megabytes. Only the configurations associated with bold numbers can be used in practice. ("−" means that more than 256 Mb are needed.) Bear in mind that these space requirements are given for only **one** constraint.

| $t$ | 0.001 | 0.02 | 0.2 | 0.5 | 0.8 | 0.98 | 0.999 |
|---|---|---|---|---|---|---|---|
| | \multicolumn{7}{c}{$r=8, d=10$} |
| GAC-s | \multicolumn{7}{c}{**0.015** for any tightness} |
| GAC4 | - | - | - | - | - | 256 | 12.8 |
| | \multicolumn{7}{c}{$r=5, d=20$} |
| GAC-s | \multicolumn{7}{c}{**0.012** for any tightness} |
| GAC4 | 256 | 251 | 205 | 128 | 51 | 5.1 | **0.26** |
| | \multicolumn{7}{c}{$r=4, d=50$} |
| GAC-s | \multicolumn{7}{c}{**0.019** for any tightness} |
| GAC4 | - | - | - | 200 | 80 | 8 | **0.4** |

Finally, to be exhaustive on complexities, we report the worst-case time complexities of the different versions of the GAC-schema presented in this paper. Let $C$ be a constraint of arity $r$. If $C$ is a predicate, the GAC-schema has a time complexity bounded above by $O(d^r)$. This is a gain of $r^2 d$ with respect to the $O(r^2 d^{r+1})$ of CN. If $C$ is a negative constraint, the complexity remains very close to $O(d^r)$ (see subsection 4.2). If $C$ is a positive constraint, the worst-case time complexity of the GAC-schema is $O(|T(C)|)$. This is the same as the one of GAC4.

## 6 A cryptogram as an example

In this section we briefly present a very small example (a cryptogram) to show some of the advantages of dealing with non-binary predicates. (This example is implemented in ILOG Solver 4.0, in which the GAC-schema has been inserted.) For a complete presentation of the example, see [ILOG, 1997].

The cryptogram is given in Fig. 1. The problem is to find a one to one mapping from letters $\{A, B, D, E, G, L, N, O, R, T\}$ to $\{0..9\}$ in such a way that the addition obtained by replacing each letter by its associated value is consistent. It is possible to solve this problem without the GAC-schema, by encoding it (for instance) as the following constraint network: a variable for each letter; a domain containing the numbers 0..9 for each variable; and two non-binary constraints for which efficient specific algorithms are known, "$100,000D + 10,000O + 1,000N + 100A + 10L + D +$

---
[1]For constraints given in extension we should add the size of the constraint representation (see subsection 4.2).

```
DONALD
+ GERALD
= ROBERT
```

Figure 1: The cryptogram

$100,000G + 10,000E + 1,000R + 100A + 10L + D = 100,000R + 10,000O + 1,000B + 100E + 10R + T$", and "all-different$(A, B, D, E, G, L, N, O, R, T)$". However, with this representation, some of the knowledge we have on the problem is not stated, while it could have been helpful in solving the problem efficiently. We know, for example, that in the rightmost column $D + D$ is necessarily equal to $T$ or $10 + T$ since there is no carry. We know that on the leftmost column $R$ is equal to $D + G$ or $D + G + 1$ since there is no letter on the left of $R$. More generally, for the third column for example, we know that $(N + R = B) \vee (N + R = 10 + B) \vee (N + R + 1 = B) \vee (N + R + 1 = 10 + B)$. All these constraints (six in the cryptogram of Fig. 1) are predicates, but no already known specific algorithm can handle them (they are not arithmetic relations, but disjunctions of several arithmetic relations).

The table below presents the number of backtracks (#bt) and cpu-time performances of solving the cryptogram with ILOG Solver 4.0 (maintaining arc consistency during search) on a Pentium Pro 200, with or without the GAC-schema and the 6 additional constraints, and with or without the use of the minimal domain dynamic variable ordering (DVO).

|  | without DVO | | with DVO | |
|---|---|---|---|---|
|  | #bt | time | #bt | time |
| without GAC-s | 4,612 | 0.72 sec. | 138 | 0.05 sec. |
| with GAC-s | 61 | 0.1 sec. | 1 | 0.00 sec. |

Of course it is a very small illustration that is given here with that sample problem. In this problem, indeed, the additional constraints are only added to improve the search. They are not necessary to encode the problem: the first two constraints were already ensuring that the solutions of the constraint network were the solutions of the cryptogram.

## 7 Conclusion

While arc consistency had been widely studied by the CSP community, there did not exist any algorithm that efficiently achieves arc consistency on non-binary constraints. Thus, we presented GAC-schema, which is built to take into account the last improvements available on binary constraints (AC-7 like schema). We saw that it is even more important than on binary constraints to use such improvements (e.g. multidirectionality). With specialized instantiations, this schema is capable of efficiently dealing with predicates, positive constraints, or negative constraints. The perspectives of this work are to propose other instantiations of the schema in order to improve arc consistency processing on some other types of constraint representations frequently occurring in industrial problems.

## References

[Bessière and Régin, 1996] C. Bessière and J.C. Régin. MAC and combined heuristics: two reasons to forsake FC (and CBJ?) on hard problems. In *CP'96*, pages 61–75, Cambridge MA.

[Bessière et al., 1995] C. Bessière, E.C. Freuder, and J.C. Régin. Using inference to reduce arc consistency computation. In *IJCAI'95*, pages 592–598, Montréal, Canada.

[Bessière, 1994] C. Bessière. Arc-consistency and arc-consistency again. *Artificial Intelligence*, 65:179–190, 1994.

[Dechter, 1990] R. Dechter. On the expressiveness of networks with hidden variables. In *AAAI'90*, pages 556–562, Boston MA.

[Grant and Smith, 1996] S.A. Grant and B.M. Smith. The phase transition behavior of maintaining arc consistency. In *ECAI'96*, pages 175–179, Budapest, Hungary.

[ILOG, 1997] ILOG. *User's manual*. ILOG Solver, 4.0 edition, 1997.

[Mackworth, 1977a] A.K. Mackworth. Consistency in networks of relations. *Artificial Intelligence*, 8:99–118, 1977.

[Mackworth, 1977b] A.K. Mackworth. On reading sketch maps. In *IJCAI'77*, pages 598–606, Cambridge MA.

[Mohr and Henderson, 1986] R. Mohr and T.C. Henderson. Arc and path consistency revisited. *Artificial Intelligence*, 28:225–233, 1986.

[Mohr and Masini, 1988] R. Mohr and G. Masini. Good old discrete relaxation. In *ECAI'88*, pages 651–656, Munchen, FRG.

[Montanari, 1974] U. Montanari. Networks of constraints: Fundamental properties and applications to picture processing. *Information Science*, 7:95–132, 1974.

[Régin, 1994] J.C. Régin. A filtering algorithm for constraints of difference in CSPs. In *AAAI'94*, pages 362–367, Seattle WA.

[Régin, 1996] J.C. Régin. Generalized arc consistency for global cardinality constraint. In *AAAI'96*, pages 209–215, Portland OR.

[Rossi et al., 1990] F. Rossi, C. Petrie, and V. Dhar. On the equivalence of constraint satisfaction problems. In *ECAI'90*, pages 550–556, Stockholm, Sweden.

[Sabin and Freuder, 1994] D. Sabin and E.C. Freuder. Contradicting conventional wisdom in constraint satisfaction. In *PPCP'94*, Seattle WA.

[Sedgewick, 1990] R. Sedgewick. *Algorithms in C*. Addison-Wesley Publishing Company, 1990.

[van Beek and Dechter, 1995] P. van Beek and R. Dechter. On the minimality and global consistency of row-convex constraint networks. *Journal of the ACM*, 42(3):543–561, 1995.

[Van Hentenryck et al., 1992] P. Van Hentenryck, Y. Deville, and C.M. Teng. A generic arc-consistency algorithm and its specializations. *Artificial Intelligence*, 57:291–321, 1992.

# Constraint Satisfaction over Connected Row Convex Constraints

Yves Deville     Olivier Barette
Université catholique de Louvain,
Pl. Ste Barbe 2,
B-1348 Louvain-la-Neuve, Belgium
{yde,barette}@info.ucl.ac.be

Pascal Van Hentenryck
Brown University
Box 1910
Providence, RI 02912, USA
pvh@cs.brown.edu

## Abstract

In this paper, we study constraint satisfaction over connected row convex (CRC) constraints, a large class of constraints subsuming, in particular, monotone constraints. We first show that CRC constraints are closed under composition, intersection, and transposition, the basic operations of path-consistency algorithms. This establishes that path consistency over CRC constraints produces a minimal and decomposable network, strenghtening the results of van Beek and Dechter [1995]. We then present a path-consistency algorithm for CRC constraints running in time $O(n^3d^2)$ and space $O(n^2d)$, where $n$ is the number of variables and $d$ is the size of the largest domain. This improves the traditional time complexity $O(n^3d^3)$ and space complexity $O(n^3d^2)$. Finally, we show that a solution can be found in time $O(n^2)$, once the graph is path-consistent.

## 1 Introduction

Constraint satisfaction techniques have been found useful in many areas such as Operations Research, hardware design, robotics, knowledge bases, and temporal resaoning to name a few. Some applications require to find one or all solutions, in which case consistency techniques such arc and path consistency are instrumental in reducing the size of the search space. Other applications require to put the graph of constraints in minimal form, e.g., to remove redundant information.

Increasing attention has been devoted recently to the study of special classes of constraints or contraint graphs. These studies are motivated both by practical considerations (e.g., constraint languages are based on a set of primitive constraints) and by theoretical considerations, since stronger results and more efficient algorithms can be obtained by exploiting special properties.

This paper considers the class of connected row convex (CRC) constraints. CRC constraints were motivated by van Beek's row convex constraints, their properties, and their applications to various tasks in artificial intelligence [van Beek and Dechter, 1995]. The class of CRC constraints include many constraints such as $ax + by + c \leq 0$, $ax + by + c \geq 0$, $axy + b \leq 0$, $axy + b \geq 0$, $af(x) + by + c \leq 0$, and $af(x) + by + c \geq 0$, where $a, b, c$ are rationals and $f(x)$ is a function whose derivative does not change sign in the considered domain as well as conjunctions of these constraints, some of which being non-monotone. We show that, contrary to row convex constraints, CRC constraints are closed under intersection, composition, and transposition, the basic operations of path-consistency algorithms. We then propose a generic path-consistency algorithm PC-GEN which mimics the generic arc-consistency algorithm AC-5 and we show that it can be instantiated to produce a path-consistency algorithm running in time $O(n^3d^2)$ and in space $O(n^2d)$ for CRC constraints. Finally, we show that finding a solution to a path-consistent graph of CRC constraints can be done in $O(n^2)$ time.

The rest of the paper is organized as follows. Section 2 introduces the necessary background on Constraint Satisfaction Problems (CSPs) and Section 3 discusses related work. Section 4 introduces CRC constraints and studies their properties Section 5 presents PC-GEN and Section 6 instantiates it for CRC constraints. Section 7 concludes the paper.

## 2 Background

In this paper, variables are represented by the natural numbers $1, \ldots, n$. Each variable $i$ has an associated finite domain $D_i$. All constraints are binary and relate two distinct variables. If $i$ and $j$ are variables ($i < j$), we assume, for simplicity, that there is at most one constraint relating them, denoted by $C_{ij}$. A constraint $C_{ij}$ denotes a set of couples ($C_{ij} \subseteq D_i \times D_j$). The fact that $(v, w) \in C_{ij}$ is also be denoted by $C_{ij}(v, w)$. We denote by $D$ the union of all domains and by $d$ the size of the largest domain. We assume the existence of a total ordering over $D$. Following Montanari [1974], a constraint $C_{ij}$ will also be seen as a (0,1)-matrix with $|D_i|$ rows and $|D_j|$ columns. Rows and columns are ordered according to the underlying order over $D$. A 1 (resp. 0) at position $(v, w)$ in the matrix means $(v, w) \in C_{ij}$ (resp. $(v, w) \notin C_{ij}$). To simplify the presentation, each domain $D_i$ is also represented by a (pseudo-binary) constraint $C_{ii}$ such that $C_{ii}(v, v)$ holds iff $v \in D_i$. Domain $D_i$ and constraint $C_{ii}$ can be used in an interchangeable way.

Consistency algorithms generally work on the graph representation of the CSP. We associate a graph $G$ to a CSP in the following way. $G$ has a node $i$ and an arc $(i, i)$ for each variable $i$. The constraint associated to arc $(i, i)$ is $C_{ii}$. For each constraint $C_{ij}$ relating variables $i$ and $j$ $(i < j)$, $G$ has two directed arcs, $(i, j)$ and $(j, i)$. The constraint associated to arc $(i, j)$ is $C_{ij}$ and the constraint associated to $(j, i)$ is $C_{ji}$, which is the transposition of $C_{ij}$. The graph is thus defined by the nodes, the arcs, and their associated constraints. We use $arc(G)$ and $node(G)$ to denote the set of arcs and the set of nodes of graph $G$. A CSP uniquely defines a graph and vice versa. The tuple $(v_1, \ldots, v_n) \in D^n$ is a solution of $G$ if $C_{ij}(v_i, v_j)$ holds for all $(i, j) \in arc(G)$. Two graphs $G$ and $G'$ are *equivalent* if $G$ and $G'$ have the same solutions.

Let us review some of the basics in the area of path consistency. A tuple $\langle v_{i_0}, v_{i_m} \rangle$ is *path-consistent* for path $(i_0, \ldots, i_m)$ wrt $G$ if $\exists v_{i_1}, \ldots, v_{i_{m-1}}$ such that $v_{i_k} \in D_{i_k}$ $(1 \leq k < m)$ and $C_{i_k i_{k+1}}(v_{i_k}, v_{i_{k+1}})$ $(0 \leq k < m)$. The path $p = (i_0, \ldots, i_m)$ in $G$ is *path-consistent wrt $G$* if for all $v_{i_0} \in D_{i_0}$ and $v_{i_m} \in D_{i_m}$ with $C_{i_0 i_m}(v_{i_0}, v_{i_m})$, $\langle v_{i_0}, v_{i_m} \rangle$ is path-consistent for $p$ wrt $G$. A graph $G$ is *path consistent* if for all paths $p$ in $G$, $p$ is path consistent wrt $G$. Montanari [1974] showed that a complete graph is path consistent iff all paths of length two are path-consistent. Typically, path-consistency algorithms work on complete graphs, and an incomplete graph can be easily transformed into a complete graph by adding $TRUE$ constraints. The objective of a path-consistency algorithm is thus, given a complete graph $G$, to compute new constraints which are path consistent and have the same solutions as $G$. Path-consistency algorithms are generally defined in terms of interesection and composition of the matrix representation defined as

$$(C_{ik} \cdot C_{kj})(v, w) = \vee_{u \in D} C_{ik}(v, u) \wedge C_{kj}(u, w)$$
$$(C_1 \cap C_2)(v, w) = C_1(v, w) \wedge C_2(v, w)$$

A graph $G$ is *minimal* if, for all $i, j \in node(G)$ and for all $v, w \in D$, $C_{ij}(v, w)$ implies that $(v, w)$ is part of some solution of $G$. A graph $G$ is *decomposable* if, for all $v_{i_1} \ldots v_{i_k}$ satisfying all the constraints relating nodes $i_1 \ldots i_k (1 \leq k < n)$ and for any new node $i_{k+1}$, there exists $v_{i_{k+1}}$ such that $v_{i_1} \ldots v_{i_k}, v_{i_{k+1}}$ satisfy all the constraints relating nodes $i_1 \ldots i_k, i_{k+1}$. A decomposable graph is also called strongly $n$-consistent [Freuder, 1982]. Decomposable graphs have thus the property that any consistent instantiation of some variables can be extended to a solution without backtracking. A decomposable graph is of course minimal. In a minimal graph, it is not possible to prune further the constraints without removing solutions.

## 3 Related Work

This research was motivated by van Beek's result on row convex constraint. A constraint $C_{ij}$ is *row convex* if, in each row of its matrix representation, all the ones are consecutive. Van Beek and Dechter [1995] show that, when the constraints of a path-consistent graph are row convex (or can be made row convex by permutation of values in the domain), then the graph is minimal and decomposable. One can thus compute a solution without backtracking in $O(n^2 d)$. Solving the CSP can then be done in $O(n^3 d^3)$, the time complexity of the PC algorithm. Unfortunately, row convex constraints are not closed under composition and intersection. As a consequence, no conclusion can be drawn a priori for a graph of row convex constraints, since its path consistent subgraph may or may not be row convex. CRC constraints remove this problem, since a graph of CRC constraints, after application of a path consistency algorithm, is still CRC and is thus minimal and decomposable.

In [1995], Jeavons and Cooper identify the class of max-closed constraints that can be solved in polynomial time ($O(n^4 d^4)$ for binary constraints). Our class of CRC constraints, which can be solved in $O(n^3 d^2)$, intersects with max-closed constraints, but is not a subset.

The idea of row convexity has also been exploited in the context of continuous constraints [Haroud and Faltings, 1996]. They start from the result that, when constraints are convex and binary, path-consistency is sufficient to ensure decomposability. They show that for continous domain, this result can be generalized to ternary and $n$-ary constraints using some other notion of consistency ((3,2)-relational consistency).

The class of CRC constraint is also related to discrete temporal reasoning [van Beek, 1992b]. Valdès-Peres [Valdès-Pérez, 1987] shows that path-consistency algorithms find the minimal network for a subclass of Allen's interval algebra [Allen, 1983]. Such a result has also been proposed in the context of point algebra [Vilain and Kautz, 1986; van Beek, 1989].

Montanari [Montanari, 1974] already shows that a path-consistent tree or distributive networks are minimal. He also shows that path consistency of (total) monotone constraints produces a decomposable network. Note that CRC constraints are not distributive and generalize the total monotone functions of Montanari.

Finally, recall that best path-consistency algorithms [Chmeiss, 1996; Han and Lee, 1988; Mohr and Henderson, 1986; Singh, 1995] run in time $O(n^3 d^3)$ and in space $O(n^3 d^2)$ (PC-6). These algorithms do not take advantage of properties of the constraints. A generic arc-consistency algorithm AC-5 has been proposed in [Van Hentenryck et al., 1992] and it can be instantiated to produce linear algorithms for different classes of constraints such as functional and monotone constraints.

## 4 Connected Row Convex Constraints

Row convex constraints exhibits two problems during path-consistency algorithms. First, when a row convex constraint is composed of disjoint blocks of 1s, its composition with another row convex constraint may not be row convex. Second, even if disjoint blocks are forbidden, intersection may create empty rows and columns and

thus disjoint blocks. Here is an illustration of these two problems:
$$\begin{pmatrix} 1 & 1 & 1 \\ 0 & 1 & 1 \\ 0 & 0 & 1 \end{pmatrix} \begin{pmatrix} 1 & 1 & 0 \\ 1 & 0 & 0 \\ 0 & 0 & 1 \end{pmatrix} = \begin{pmatrix} 1 & 1 & 1 \\ 1 & 0 & 1 \\ 0 & 0 & 1 \end{pmatrix}$$
$$\begin{pmatrix} 1 & 1 & 1 \\ 0 & 0 & 1 \\ 0 & 0 & 1 \end{pmatrix} \cap \begin{pmatrix} 1 & 0 & 0 \\ 1 & 0 & 0 \\ 1 & 1 & 1 \end{pmatrix} = \begin{pmatrix} 1 & 0 & 0 \\ 0 & 0 & 0 \\ 0 & 0 & 1 \end{pmatrix}$$

CRC constraints avoid both problems. Informally, a constraint is CRC if, after removing the empty rows, it is row convex and connected (two successive rows either intersect or are consecutive).

**Definition 1** The *reduced form* of a constraint $C_{ij}$, denoted by $C_{ij}^*$, is obtained by removing all the empty rows and columns in its matrix representation. The *domain* of $i$ through the constraint $C_{ij}$, denoted by $D_i(C_{ij})$, is the set $\{v \in D \mid \exists w : \langle v, w \rangle \in C_{ij}\}$.

**Definition 2** Let $C_{ij}$ be a row convex constraint and $v \in D_i(C_{ij})$. The *image* of $v$ in $C_{ij}$ is the set $\{w \mid \langle v, w \rangle \in C_{ij}^*\}$. Because of the row convexity of $C_{ij}$, this set is represented as an interval $[w_1, w_m]$ (over the domain $D_j(C_{ji})$) and we denote $w_1$ and $w_m$ by $min(C_{ij}, v)$ and $max(C_{ij}, v)$ respectively. We also denote by $succ(w)$ and $pred(w)$ the successor and the predecessor of $w$ in $D_j(C_{ji})$.

**Definition 3** A row convex constraint $C_{ij}$ is *connected* if the images $[a, b]$ and $[a', b']$ of two consecutive rows in $C_{ij}^*$ is such that $b' \geq pred(a) \wedge a' \leq succ(b)$.

**Definition 4** A constraint $C_{ij}$ is *connected row convex* (CRC) if $C_{ij}^*$ and $C_{ji}^*$ are both row convex and connected.

We assume that $C_{ij}$ is always the transposition of $C_{ji}$. Notice that CRC constraints are not necessarily row convex (because of empty rows) and that row convex constraints are not necessarily CRC (not connected rows).

The main result of this section is the fact that CRC constraints are closed under composition, intersection and transposition.

**Lemma 1** The deletion of rows and columns in a CRC constraint produces a CRC constraint.

Connectivity in Lemma 1 can be proven by showing that all the columns between two non-connnected rows are necessarily empty and must thus be deleted when considering the connectivity property.

**Lemma 2** Let $C_{ij}$ be a CRC constraint. Let $v_1, v, v_2$ be in $D_i(C_{ij})$ such that $v_1 < v < v_2$ and their respective images are $[a_1, b_1]$, $[a, b]$ and $[a_2, b_2]$ in $C_{ij}$.

$b_2 < a_1 \Rightarrow [a, b] \cap [b_2, a_1] \neq \emptyset$
$a_2 > b_1 \Rightarrow [a, b] \cap [b_1, a_2] \neq \emptyset$
$b_2 \geq a_1 \wedge a_2 \leq b_1 \Rightarrow [a_1, b_1] \cap [a_2, b_1] \subseteq [a, b]$

**Theorem 5** The intersection and composition of two CRC constraints and the transposition of a CRC constraint are CRC constraints.

**Proof** (Sketch) Transposition is obvious. For intersection and composition ($C = A \cap B$ or $C = A.B$), delete the empty rows and columns in the operands to give $A^+$ and $B^+$. By Lemma 1, it is sufficient to show that $C^+ = A^+ \cap B^+$ or $C^+ = A^+.B^+$ are CRC. Row convexity is obvious for intersection and results from Lemma 2 for composition. Connectivity can be proven by showing that all the columns between two non-connnected rows are necessarily empty and must thus be deleted when considering the connectivity property. □

**Theorem 6** Let $G$ be composed of CRC constraints. The application of a path-consistency algorithm to $G$ produces a minimal and decomposable graph.

## 5  PC-GEN: Generic Path Consistency

We now present a path-consistency algorithm for CRC constraints. We proceed in two steps. We first present a generic path-consistency algorithm PC-GEN which is parametrized in a way similar to AC-5. We then instantiate PC-GEN to CRC constraints.

---

**procedure** PRUNE(**in** $\Delta, i, j$)
*Pre:* $i, j \in node(G)$.
*Post:* $C_{ij} = C_{ij_0} \setminus \{\langle v, w \rangle \mid \langle v, w \rangle \in \Delta\}$,
$C_{ji} = C_{ji_0} \setminus \{\langle w, v \rangle \mid \langle v, w \rangle \in \Delta\}$.

**procedure** INITQUEUE(**out** $Q$)
*Post:* $Q = \{\}$.

**function** EMPTYQUEUE(**in** $Q$): Boolean
*Post:* EMPTYQUEUE $\Leftrightarrow (Q = \{\})$.

**procedure** DEQUEUE(**inout** $Q$, **out** $i, k, j, \langle v, u \rangle$)
*Post:* $\langle i, k, j, \langle v, u \rangle \rangle \in Q_0$ and $Q = Q_0 \setminus \{\langle i, k, j, \langle v, u \rangle \rangle\}$.

**procedure** ENQUEUE($i, j, \Delta$, **inout** $Q$)
*Pre:* $\Delta \subseteq C_{ij}$.
*Post:* $Q = Q_0 \cup \{\langle i, j, k, \langle v, w \rangle \rangle \mid k \in node(G)$
  and $j \neq k$ and $\langle v, w \rangle \in \Delta\}$
  $\cup \{\langle j, i, k, \langle w, v \rangle \rangle \mid k \in node(G)$
  and $j \neq i \neq k$ and $\langle v, w \rangle \in \Delta\}$.

Let $PC_{ikj}(v, w) = \exists u : \langle v, u \rangle \in C_{ik}$ and $\langle u, w \rangle \in C_{kj}$.
$PC'_{ikj}(v, w) = \exists u : (\langle v, u \rangle \in C_{ik} \vee \langle i, k, j, \langle v, u \rangle \rangle \in Q)$
  and $(\langle u, w \rangle \in C_{kj} \vee \langle j, k, i, \langle w, u \rangle \rangle \in Q)$

**procedure** PATHCONS(**in** $i, k, j$, **out** $\Delta$)
*Pre:* $i, k, j \in node(G)$.
*Post:* $\Delta = \{\langle v, w \rangle \in C_{ij} \mid \neg PC_{ikj}(v, w)\}$.

**procedure** LOCALPATHCONS(**in** $i, k, j, \langle v, u \rangle$, **out** $\Delta$)
*Pre:* $i, k, j \in node(G)$, and $\langle v, u \rangle \notin C_{ik}$.
*Post:* $\Delta_1 \subseteq \Delta \subseteq \Delta_2$, with
$\Delta_1 = \{\langle v, w' \rangle \in C_{ij} \mid \langle u, w' \rangle \in C_{kj}^{init}$ and $\neg PC_{ikj}(v, w')\}$
$\Delta_2 = \{\langle v', w' \rangle \in C_{ij} \mid \neg PC_{ikj}(v', w')\}$.

---

Figure 1: Subproblems for for PC-GEN

The specification of the main operations in PC-GEN are given in Figure 1. In all specifications, a parameter $p$ subscripted with 0 ($p_0$) represents the value of $p$ at call time and $C_{ij}^{init}$ is the original set of constraint tuples between $i$ and $j$. This justifies the restriction $i \leq j$ in line 2 of PC-GEN. As is traditional, PC-GEN uses a queue to drive the algorithm. Procedure ENQUEUE is required to take $O(s)$ time, where $s$ is the number of new elements to insert in the queue, and procedure DEQUEUE must take constant time. The deletion of tuples is performed by procedure PRUNE, which removes tuple $\langle v, w \rangle$ from $C_{ij}$, and $\langle w, v \rangle$ from $C_{ji}$. Hence, $\langle v, w \rangle \in C_{ij} \Leftrightarrow \langle w, v \rangle \in C_{ji}$ is an invariant of the algorithm.

PC-GEN is parametrized by two procedures, PATH-CONS and LOCALPATHCONS whose implementations are left open. Procedure PATHCONS computes the set $\Delta$ of tuples in $C_{ij}$ which are not path consistent for the path $(i, k, j)$. Because of the relationship between $C_{ij}$ and $C_{ji}$, $\Delta$ is also the set of tuples (in reverse order) of $C_{ji}$ that are not path consistent for path $(j, k, i)$.

Procedure LOCALPATHCONS returns in $\Delta$ a set of tuples of $C_{ij}$ that are not path consistent for $(i, k, j)$ after the tuple $\langle v, u \rangle$ has been removed from the constraint $C_{ik}$. The set $\Delta$ is also the set of tuples (in reverse order) of $C_{ji}$ that are not path consistent in path $(j, k, i)$ after tuple $\langle u, v \rangle$ has been removed from $C_{ki}$.

The size of $\Delta$ computed by LOCALPATHCONS can vary. The set $\Delta_1$ contains the tuples in $C_{ij}$ that become path inconsistent for $(i, k, j)$ due to the removal of the tuple $\langle v, u \rangle$ from $C_{ik}$. Specifically, a tuple $\langle v, w' \rangle$ is included in $\Delta_1$ if $u$ was a support (i.e., $(u, w') \in C_{kj}^{init}$) and $(v, w')$ is not supported anymore. In some cases, it is possible, but not always desirable, to prune a larger set of tuples. As an extreme case, $\Delta_2$ prunes all tuples in $C_{ij}$ which are not path inconsistent wrt $(i, k, j)$ at call time, regardless of whether they can be supported by $\langle v, u \rangle$. In fact, the specifications can be made even less restrictive by replacing $PC_{ikj}$ by $PC'_{ikj}$ in the definition of $\Delta$ in PATHCONS, and in the definition of $\Delta_1$ in LOCALPATHCONS. Our instantiation of PC-GEN to CRC constraints uses this flexibility to improve the time complexity of traditional path-consistency algorithms.

---

**Algorithm** PC-GEN
*Post*: $G$ is the largest path-consistent graph for $G_0$.
    **begin**
1      INITQUEUE($Q$);
2      **for each** $i, k, j \in node(G)$ with $i \leq j$ **do**
3      **begin**
4         PATHCONS($i,k,j,\Delta$);
5         ENQUEUE($i,j,\Delta,Q$);
6         PRUNE($\Delta,i,j$)
7      **end**;
8      **while not** EMPTYQUEUE($Q$) **do**
9      **begin**
10     DEQUEUE($Q,i,k,j,\langle v,u\rangle$);
11     LOCALPATHCONS($i,k,j,\langle v,u\rangle,\Delta$);
12     ENQUEUE($i,j,\Delta,Q$);
13     PRUNE($\Delta,i,j$)
14    **end**
    **end**

Figure 2: The Path Consistency Algorithm PC-GEN

PC-GEN is depicted in Figure 2 and mimics AC-5. In the loop on lines 2-7, procedure PATHCONS identifies the path-inconsistent tuples with respect to each path of length two. The inconsistent tuples are enqueued and processed in the second loop, on lines 8-14, where procedure LOCALPATHCONS is used to prune tuples of $C_{ij}$ which become inconsistent after the removal of a tuple from $C_{ik}$. The removal of the tuple $\langle v, w \rangle$ in $C_{ij}$ and $\langle w, v \rangle$ in $C_{ji}$ induces to reconsider all length-two paths involving either $(i, j)$ or $(j, i)$ as the first or as the second arc. It is however unnecessary to explicitly consider the involvement as a second arc (in the ENQUEUE procedure) since LOCALPATHCONS($i, j, k, \ldots$) will cover both paths $(i, j, k)$ and $(k, j, i)$, and LOCALPATHCONS($j, i, k, \ldots$) will cover paths $(j, i, k)$ and $(k, i, j)$.

It can be shown that PC-GEN is correct and that it enqueues and dequeues at most $O(n^3 d^2)$ elements.

**Theorem 7** *Given a time complexity of $O(d^2)$ for procedure PATHCONS and a time complexity of $O(\Delta)$ for procedure LOCALPATHCONS, the time complexity of algorithm PC-GEN is bounded by $O(n^3 d^2)$.*

PC-GEN can be instantiated for general constraints to produce a path-consistency algoritm with a time complexity of $O(n^3 d^3)$ and a space complexity of $O(n^3 d^2)$. We now turn to its instantiation for CRC constraints.

## 6 PC-GEN for CRC Constraints

---

Let $D = \{b, \ldots, B\}$.
Let $C_{ij} = \{\langle v_1, v_1 \rangle, \ldots, \langle v_m, v_m \rangle\}$ if $i = j$
       $= \{\langle v_1, w_1 \rangle, \ldots, \langle v_m, w_m \rangle\}$ if $i \neq j$ (where $v_k, w_k \in D$)

**Data Structure**
  **Syntax**
    $C_{ij}.supmin$: array $[b..B]$ of element $\in D$.
    $C_{ij}.supmax$: array $[b..B]$ of element $\in D$.
    $C_{ij}.first$: element $\in D$.
    $C_{ij}.succ$: array $[b..B]$ of element $\in D$.
    $C_{ij}.pred$: array $[b..B]$ of element $\in D$.
  **Semantics**
    $C_{ij}.supmin[v] = min(C_{ij}, v)$
    $C_{ij}.supmax[v] = max(C_{ij}, v)$
    $C_{ij}.first = min\{v \in D_i(C_{ij})\}$
    $C_{ij}.succ[v] = succ(v)$ in $D_i(C_{ij})$
    $C_{ij}.pred[v] = pred(v)$ in $D_i(C_{ij})$
  **Invariant**
    $C_{ij} = C_{ji}^T$
    $C_{ij}.supmin[v] \in D_j(C_{ji})$
    $C_{ij}.supmax[v] \in D_j(C_{ji})$

**Interface**
  **function** EMPTYSUPPORT(**in** v,w, i,k,j): Boolean
  *Post*: EMPTYSUPPORT(v,w, i,k,j) = $\neg PC'_{ikj}(v, w)$

  **function** FIRST(**in** i,j): Integer
  *Post*: FIRST(i,j) = $min\{v \in D_i(C_{ij})\}$

  **function** MIN(**in** v, i,j): Integer
  *Post*: MIN(v, i,j) = $min(C_{ij}, v)$

  **function** MAX(**in** v, i,j): Integer
  *Post*: MAX(v, i,j) = $max(C_{ij}, v)$

  **function** SUCC(**in** v, i,j): Integer
  *Post*: SUCC(v, i,j) = $succ(v)$ in $D_i(C_{ij})$

  **function** PRED(**in** v, i,j): Integer
  *Post*: PRED(v, i,j) = $pred(v)$ in $D_i(C_{ij})$

Figure 3: The CRC CONSTRAINT Module

CRC constraints can be represented in space $O(d)$ as shown in Figure 3. It is necessary to keep a description of $D_i(C_{ij})$, since row convexity is only enforced on the reduced form. Figure 3 also specifies the operations on CRC constraints which are all implemented in constant time. For instance, EMPTYSUPPORT($v, w, i, k, j$) can be implemented by $b' \geq a \land a' \leq b$ with $a =$ MIN($v, i, k$), $b =$ MAX($v, i, k$), $a' =$ MIN($w, j, k$), and $b' =$ MAX($w, j, k$).

```
procedure PATHCONS(in i, k, j, out Δ)
   begin
1     Δ := ∅;
2     for each v ∈ D_i(C_ij) do
3     begin
4        LOCALPATHCONS(i, k, j, v, Δ_v);
5        Δ := Δ ∪ Δ_v;
6     end
   end

procedure LOCALPATHCONS(in i, k, j, v, out Δ)
Pre: i, k, j ∈ node(G).
Post: Δ_1 ⊆ Δ ⊆ Δ_2, with
      Δ_1 = {⟨v, w'⟩ ∈ C_ij | ¬PC'_ikj(v, w')}.
      Δ_2 = {⟨v', w'⟩ ∈ C_ij | ¬PC_ikj(v', w')}.
   begin
1     BOUNDEDMIN(i, k, j, ⟨v, MAX(v, i,j)⟩, Δ', w_min);
2     if w_min = MAX(v, i,j) then Δ := Δ'
3     else
4     begin
5        BOUNDEDMAX(i, k, j, ⟨v, MIN(v, i,j)⟩, Δ'', w_max);
6        PROPAGATE(i,j,k,⟨v,w_min⟩, BOUNDEDMIN, PRED, Δ_1);
7        PROPAGATE(i,j,k,⟨v,w_min⟩, BOUNDEDMIN, SUCC, Δ_2);
8        PROPAGATE(i,j,k,⟨v,w_max⟩, BOUNDEDMAX, PRED, Δ_3);
9        PROPAGATE(i,j,k,⟨v,w_max⟩, BOUNDEDMAX, SUCC, Δ_4);
10       Δ := Δ' ∪ Δ'' ∪ Δ_1 ∪ Δ_2 ∪ Δ_3 ∪ Δ_4;
11    end
   end
```

Figure 4: PATHCONS and LOCALPATHCONS for CRC constraints.

```
procedure PROPAGATE(in i, k, j, ⟨v, w⟩, BOUNDED, NEXT, out Δ)
Let v_k = NEXT^k(v),
    w_k and Δ_k such that BOUNDED(i, k, j, ⟨v_k, w⟩, Δ_k, w_k),
    m = max{k | Δ_k ≠ ∅ ∧ w_k = w}.
Post: Δ = ⋃_{1≤k≤m+1} Δ_k
   begin
1     Δ := ∅;
2     v_calc := v;
3     repeat
4        v_calc := NEXT(v_calc);
5        BOUNDED(i, k, j, ⟨v_calc, w⟩, Δ_calc, w_calc);
6        Δ := Δ ∪ Δ_calc;
7     until (w_calc ≠ w);
   end

procedure BOUNDEDMIN(in i, k, j, ⟨v, w⟩, out Δ, w_min)
Post: w_min = max{w ∈ D_j(C_ji) |∀w' ∈ [MIN(v,i,j), w] :
                    EMPTYSUPPORT(v, w', i, k, j)}
      Δ = { ⟨v, w'⟩ | w' ∈ [MIN(v,i,j), w_min] }
   begin
1     Δ := ∅;
2     w_2 := MIN(v, i, j);
3     while (w_2 ≤ w) ∧ ¬EMPTYSUPPORT(v,w_2,i,k,j) do
4     begin
5        Δ := Δ ∪ {⟨v, w_2⟩};
6        w_2 := SUCC(w_2);
7     end ;
8     w_min := PRED(w_2);
   end

procedure BOUNDEDMAX(in i, k, j, ⟨v, w⟩, out Δ, w_max)
Post: w_max = min{w ∈ D_j(C_ji) |∀w' ∈ [w, MAX(v,i,j)] :
                    EMPTYSUPPORT(v, w', i, k, j)}
      Δ = { ⟨v, w'⟩ | w' ∈ [w_max, MAX(v,i,j)] }
   begin
1     Δ := ∅;
2     w_2 := MAX(v, i, j);
3     while (w_2 ≥ w) ∧ ¬EMPTYSUPPORT(v,w_2,i,k,j) do
4     begin
5        Δ := Δ ∪ {⟨v, w_2⟩};
6        w_2 := PRED(w_2);
7     end ;
8     w_max := SUCC(w_2);
   end

procedure PRUNE(in Δ, i, j)
Pre: i, j ∈ node(G),
     C_ij is a CRC constraint,
     C_ij \ Δ is a CRC constraint.
Post: C_ij = C_ij_0 \ {⟨v, w⟩ | ⟨v, w⟩ ∈ Δ},
      C_ji = C_ji_0 \ {⟨w, v⟩ | ⟨v, w⟩ ∈ Δ}.
```

Figure 5: Subproblems for PC-CRC.

An implementation of Procedures PATHCONS and LOCALPATHCONS is given in Figure 4. Note that the specification of LOCALPATHCONS has been relaxed further by removing parameter $u$ and that PATHCONS can now be expressed in terms of LOCALPATHCONS. In LOCALPATHCONS, BOUNDEDMIN computes the interval $\Delta'$ to be removed on the left of the interval in row $v$ while BOUNDEDMAX computes the interval $\Delta''$ to be removed on the right of the interval in row $v$. Although this pruning is sufficient, it may destroy the CRC property. To preserve the property, it is necessary to perform additional pruning on the rows above or below $v$. This is the role of the PROPAGATE instructions. The specifications and implementation of the procedures are given in Figure 5 and the intuition behind LOCALPATHCONS is captured in Figure 6. Because $C_{ij} := C_{ij} \cap C_{ik}.C_{kj}$ produces a CRC constraint, the implementation is guaranteed to keep $C_{ij}$ connected row convex. Note that PROPAGATE works from $v$ to the exterior, while BOUNDEDMIN and BOUNDEDMAX work from the exterior to the interior.

**Complexity** PRUNE can be performed in $O(\Delta)$ assuming the elements of $\Delta$ are ordered to preserve the CRC property. The ordering can be performed during the construction of $\Delta$ during LOCALPATHCONS without incurring any cost. An implementation of $\Delta$ as a doubly-linked list is sufficient for this purpose given the way $\Delta$ is constructed as mentioned in the previous section. The complexity of Procedures PROPAGATE, BOUNDEDMIN and BOUNDEDMAX is obviously $O(\Delta)$. Hence LOCALPATHCONS is $O(\Delta)$. By Theorem 7, the time complexity of PC-GEN is $O(n^3 d^2)$. The space complexity per con-

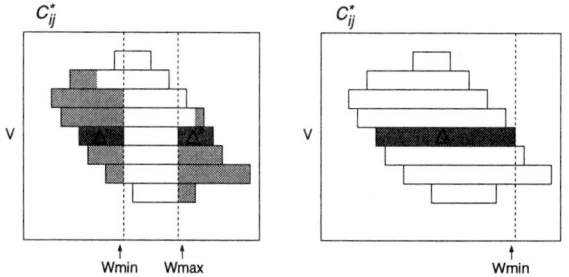

Figure 6: Illustrating LOCALPATHCONS for CRC constraints: two possible cases.

straint is $O(d)$ and $O(nd)$ for all the constraints. The space complexity of the queue is bounded by $O(n^2d)$ because procedure LOCALPATH does not use parameter $u$ and elements in the queue can be grouped as tuples of the form $\langle i, j, E, v \rangle$, where the set $E$ is initially $node(G) \backslash \{j\}$. The set $E$ can be shared by all elements of the queue except the first one.

**Theorem 8** For CRC constraints, PC-GEN has a time complexity of $O(n^3d^2)$ and a space complexity of $O(n^2d)$.

The above theorem is valid for incomplete graphs of CRC constraints as well, since the completion of the graph introduces $TRUE$ constraints which are CRC.

**Finding a Solution** A path-consistent graph with CRC constraints is decomposable due to Helly's theorem (e.g., [Haroud and Faltings, 1996]). The proof in [van Beek and Dechter, 1995] is constructive and the author proposes a $O(n^2d)$ algorithm to find a solution. We propose in Figure 7 an INSTANTIATE procedure with a time complexity of $O(n^2)$ for CRC constraints. It is based on van Beek's algorithm, but takes advantage of the data structure.

```
procedure INSTANTIATE(in G, out ⟨x₁,...,xₙ⟩)
Pre: G has only CRC constraints, and is path-consistent,
     Dᵢ ≠ ∅ (1 ≤ i ≤ n)
Post: ⟨x₁,...,xₙ⟩ is a solution of G.
    begin
1     for i := 1 to n do
2       begin
3         L := FIRST(i,i);
4         for j := 1 to i-1 do L := max( L, MIN(xⱼ,j,i) ) ;
5         xᵢ := L
6       end
    end
```

Figure 7: INSTANTIATE for CRC Constraints.

The total complexity to detect inconsistency or to find a solution of a graph composed with CRC constraints is thus $O(n^3d^2)$, the time complexity of the path-consistency algorithm.

## 7 Conclusion

In this paper, we studied constraint satisfaction over connected row convex (CRC) constraints, a large class of constraints subsuming, in particular, monotone constraints. We showed that, contrary to row convex constraints, CRC constraints are closed under composition, intersection, and transposition, establishing that path consistency over CRC constraints produces a minimal and decomposable network and strenghtening the results of van Beek and Dechter [1995]. We then presented a path-consistency algorithm for CRC constraints running in time $O(n^3d^2)$ and space $O(n^2d)$, where $n$ is the number of variables and $d$ is the size of the largest domain, as an instantiation of a generic path-consistency algorithm. This improves the traditional time complexity $O(n^3d^3)$ and space complexity $O(n^3d^2)$. Finally, we show that a solution can be found in time $O(n^2)$, once the graph is path-consistent.

Further research will be devoted to studying how to improve path consistency and its approximation to continuous domains, since path consistency has been shown instrumental in speeding search considerably for some transistor modelling problems, and to other classes of discrete domains.

**Acknowledgment** We thank anonymous reviewers for helpful comments. This research is partially supported by the *Actions de recherche concertées (ARC/95/00-187)* of the Direction générale de la Recherche Scientifique – Communauté Française de Belgique, by the Office of Naval Research (ONR Grant N00014-94-1-1153 and a NSF National Young Investigator Award with matching funds of Hewlett-Packard.

## References

[Allen, 1983] J.F. Allen. Maintaining Knowledge About Temporal Reasoning. *J.ACM*, 26:832–843, 1983.

[Chmeiss, 1996] A. Chmeiss. Sur la consistance de chemin et ses formes partielles. In *Actes du Congrès AFCET-RFIA'96, Rennes*, 1996.

[Freuder, 1982] E.C. Freuder. A Sufficient Condition for Backtrack-Free Search. *J.ACM*, 29:24–32, 1982.

[Han and Lee, 1988] C.C. Han and C.H. Lee. Comments on Mohr and Henderson's Path Consistency Algorithm. *Artif. Intel.*, 36:125–130, 1988.

[Haroud and Faltings, 1996] D. Haroud and B. Faltings. Consistency techniques for continuous constraints. *Constraints, An International Journal*, 1:85–118, 1996.

[Mohr and Henderson, 1986] R. Mohr and T.C. Henderson. Arc and Path Consistency Revisited. *Artif. Intel.*, 28:225–233, 1986.

[Jeavons and Cooper, 1995] P.G. Jeavons and M.C. Cooper. Tractable constraints on ordered domains. *Artif. Intel.*, 79:327-339, 1995.

[Montanari, 1974] U. Montanari. Networks of constraints: Fundamental properties and applications to picture processing. *Information Science*, 7(2):95–132, 1974.

[Singh, 1995] M. Singh. Path consistency revisited. In *IEEE-ICTAI'95, Washington DC*, 1995.

[Valdès-Pérez, 1987] R.E. Valdès-Pérez. The satisfiability of temporal constraint network. In AAAI-87, Seattle, pages 745-750, 1987.

[van Beek, 1989] P. van Beek. Approximation algorithms for temporal reasoning. In IJCAI-89, pages 745–750, 1989.

[van Beek, 1992b] P. van Beek. Reasoning About Qualitative Temporal Reasoning. *Artif. Intel.*, 58:297–326, 1992.

[van Beek and Dechter, 1995] P. van Beek and R. Dechter. On the Minimality and Global Consistency of Row Convex Networks. *J.ACM*, 42:543-561, 1995.

[Van Hentenryck et al., 1992] P. Van Hentenryck, Y. Deville, and C. Teng. A generic arc-consistency algorithm and its specializations. *Artif. Intel.*, 57(2–3):291–321, 1992.

[Vilain and Kautz, 1986] M. Vilain and Kautz. Constraint propagation algorithms for temporal reasoning. In AAAI-86, Philadelphia, pages 132–144, 1986.

# CONSTRAINT SATISFACTION

Constraint Satisfaction 4

# Some Practicable Filtering Techniques for the Constraint Satisfaction Problem

## Romuald Debruyne and Christian Bessière
LIRMM (UMR 5506 CNRS)
161 rue Ada
34392 Montpellier Cedex 5 - France
Email: {debruyne, bessiere}@lirmm.fr

## Abstract

Filtering techniques are essential in order to efficiently solve constraint satisfaction problems (CSPs). A blind search often leads to a combinatorial explosion, the algorithm repeatedly finding the same local inconsistencies. Maintaining a local consistency can strongly reduce the search effort especially on hard and large problems. A good illustration are the good time performances on such problems of maintaining arc consistency during search compared to forward checking which maintains a lower level of local consistency. On the one hand, arc consistency (2-consistency) is the most used filtering technique because it cheaply removes some values that cannot belong to any solution. On the other hand, other k-consistencies (k≥3) have important space and time requirements because they can change the set of constraints. They can only be used on very small CSPs. Thus, in this paper, we study and compare the filtering techniques that are more pruningful than arc consistency while leaving unchanged the set of constraints. As arc consistency, they only remove inconsistent values in the domains, and so, can deal with large CSPs.

## 1 Introduction

Constraint satisfaction problems (CSPs) are used more and more to solve combinatorial problems arising in artificial intelligence. They involve finding an assignment of values to variables subject to constraints. This is a NP-hard task. So, the time needed by a primitive backtrack search to find such an assignment will probably be strongly affected by thrashing because it will fall many times into the same local inconsistencies. Thus filtering techniques are used to remove some local inconsistencies before or during the search. To find the suitable level of consistency to achieve we have to both consider the pruning efficiency of the filterings involved, and their time and space complexities. Obviously, the overhead caused by removing local inconsistencies has to be outweighted by its gain, which can change depending on the problem to solve. On very easy and very small CSPs, filtering techniques are useless because we can quickly find a solution or prove the inconsistency. But it can be advantageous to use an algorithm that wastes a few seconds on easy CSPs if it can save many minutes (hours ?) on hard problems. A good illustration is the efficiency of maintaining arc consistency during search (MAC) on hard CSPs [Bessière and Régin, 1996]. On these problems, MAC can outperform forward checking (FC) which maintains a lower level of local consistency. The harder a CSP is, the more useful filtering techniques are. So, maintaining a local consistency during search is essential in order to deal with large and hard problems. Obviously, whatever the level of local consistency maintained may be, some pathological CSPs which are hard to solve can be found. But the more pruningful the filtering technique is, the more seldom these CSPs are.

Arc consistency (AC) filtering is widely used on binary CSPs because it removes some values that cannot belong to any solution with a low space and time complexity. On the other hand, strong path consistency can remove more inconsistencies but is never used because of its huge complexity and because it changes the structure of the problem by adding constraints. Indeed, path consistency algorithms remove compatible pairs of values that cannot be extended to a consistent instantiation including any third variable. If a path inconsistent pair of values between two independent variables is found, it is removed from the corresponding universal relation and so the structure of the constraint graph is changed. This is a real drawback, especially when we deal with sparse problems because the resulting constraint graph is often complete. The second important drawback is the lost of semantics. In real applications, a constraint is rarely represented by a boolean matrix or a set of compatible pairs of values. It is represented by a predicate which has a particular semantics ($\neq$, $\leq$, ...), and this semantics is lost when some allowed pairs of values are removed. This leads to the generation of the boolean matrix, and so to space intractability.

The aim of this paper is to study the filtering techniques that can be used in practice. So, we only care about those that do not change the constraints and that do not require to have the constraints in extension, i.e. those that only remove values in the variable domains.

## 2 Definitions and notations

A *binary CSP* $P = (\mathcal{X}, \mathcal{D}, \mathcal{C})$ is a set $\mathcal{X} = \{i, j, \ldots\}$ of $n$ variables, each taking value in its respective finite *domain* $D_i, D_j, \ldots$ elements of $\mathcal{D}$, and a set $\mathcal{C}$ of $e$ binary constraints. $d$ is the size of the largest domain.

A *binary constraint* $C_{ij}$ is a subset of the cartesian product $D_i \times D_j$ that denotes the compatible pairs of values for $i$ and $j$. We note $C_{ij}(a, b) = true$ to specify that $((i, a), (j, b)) \in C_{ij}$. We then say that $(j, b)$ is a *support* of $(i, a)$ on $C_{ij}$. With each CSP we associate a *constraint graph* in which nodes represent variables and arcs connect pairs of variables which are constrained explicitly. The *neighborhood* of $i$ is the set of variables linked to $i$ in the constraint graph. A domain $\mathcal{D}' = \{D'_i, D'_j, \ldots\}$ is a *sub-domain* of $\mathcal{D} = \{D_i, D_j, \ldots\}$ if $\forall i, D'_i \subseteq D_i$. An *instantiation* of a set of variables $S$ is an indexed set of values $\{I_j\}_{j \in S}$ s.t. $\forall j \in S \ \ I_j \in D_j$. An instantiation $I$ of $S$ satisfies a constraint $C_{ij}$ if $\{i, j\} \not\subseteq S$ or $C_{ij}(I_i, I_j)$ is true. An instantiation is *consistent* if it satisfies all the constraints. A *solution* of $P = (\mathcal{X}, \mathcal{D}, \mathcal{C})$ is a consistent instantiation of $\mathcal{X}$. A value $(i, a)$ is *consistent* if there is a solution $I$ such that $I_i = a$, and a CSP is consistent if it has at least one solution. In the following we denote by $P \mid_{D_i = \{a\}}$ the CSP obtained by restricting $D_i$ to $\{a\}$ in $P$.

## 3 Overview on filtering techniques

Filtering techniques may detect inconsistency or remove some local inconsistencies in order to make the search easier. They look for partial instantiations that cannot be extended to a solution and remove them.

Fig. 1 presents the properties corresponding to the filtering techniques studied. The best AC algorithms are AC-6 [Bessière, 1994] with the $O(ed^2)$ optimal worst-case time complexity and $O(ed)$ worst-case space complexity, and AC-7 [Bessière *et al.*, 1995], which deals with the bidirectionality of constraints to improve AC-6.

$k$-consistency, i.e. $(k-1, 1)$-consistency [Freuder, 1985], removes the consistent instantiations of length $k-1$ that cannot be extended to a consistent instantiation including any additional $k^{th}$ variable. The time and space complexity of $k$-consistency algorithms is polynomial in $k$. Furthermore, if $k \geq 3$, $k$-consistency changes the structure of the problem. So, only 2-consistency (i.e.

- A binary CSP is $(i, j)$-**consistent** iff $\forall k \in \mathcal{X}$, $D_k \neq \emptyset$, and any consistent instantiation of $i$ variables can be extended to a consistent instantiation including any $j$ additional variables.
- A domain $D_i$ is **arc consistent** iff, $\forall a \in D_i$, $\forall j \in \mathcal{X}$ s.t. $C_{ij} \in \mathcal{C}$, there exists $b \in D_j$ s.t. $C_{ij}(a, b)$. A CSP is **arc consistent** ((1, 1)-consistent) iff $\forall D_i \in D$, $D_i \neq \emptyset$ and $D_i$ is arc consistent.
- A pair of variables $(i, j)$ is path consistent iff $\forall (a, b) \in C_{ij}$, $\forall k \in \mathcal{X}$, there exists $c \in D_k$ s.t. $C_{ik}(a, c)$ and $C_{jk}(b, c)$. A CSP is **path consistent** ((2, 1)-consistent) iff $\forall i, j \in \mathcal{X}$, $(i, j)$ is path consistent.
- A binary CSP is **strongly path consistent** iff it is node-consistent, arc consistent and path consistent.
- A binary CSP is **path inverse consistent** iff it is (1, 2)-consistent i.e. $\forall (i, a) \in D \ \ \forall j, k \in \mathcal{X}$ s.t. $j \neq i, k \neq j$, $\exists (j, b) \in D$ and $(k, c) \in D$ s.t. $C_{ij}(a, b) \wedge C_{ik}(a, c) \wedge C_{jk}(b, c)$
- A binary CSP is **neighborhood inverse consistent** iff $\forall (i, a) \in D$, $(i, a)$ can be extended to a consistent instantiation including the neighborhood of $i$.
- A binary CSP is **restricted path consistent** iff $\forall i \in \mathcal{X}$, $D_i$ is a non empty arc consistent domain and, $\forall (i, a) \in D$, for all $j \in \mathcal{X}$ s.t. $(i, a)$ has an unique support $b$ in $D_j$, for all $k \in \mathcal{X}$ linked to both $i$ and $j$, $\exists c \in D_k$ s.t. $C_{ik}(a, c) \wedge C_{jk}(b, c)$.
- A binary CSP $P$ is **singleton arc consistent** iff $\forall i \in \mathcal{X}$, $D_i \neq \emptyset$ and $\forall (i, a) \in D$, $P \mid_{D_i = \{a\}}$ has an arc consistent sub-domain.
- A binary CSP $P$ is **singleton restricted path consistent** iff $\forall i \in \mathcal{X}$, $D_i \neq \emptyset$ and $\forall (i, a) \in D$, $P \mid_{D_i = \{a\}}$ has a restricted path consistent sub-domain.

Figure 1: The local consistencies studied.

arc consistency) can be used in practice. However, path consistency (3-consistency) being widely studied, we will compare strong path consistency (enforcing both arc and path consistency) with the other filtering techniques removing values. The best PC algorithms are PC-5 [Singh, 1995] with $O(n^3d^3)$ worst-case time complexity and $O(n^3d^2)$ worst-case space complexity, and PC-8 [Chmeiss and Jégou, 1996] which requires smaller space ($O(n^2d)$ [1]) but with $O(n^3d^4)$ worst-case time complexity.

Arc consistency being a low level of consistency and path consistency leading to important drawbacks, Berlandier proposed the restricted path consistency (RPC) in [Berlandier, 1995]. This filtering technique removes the arc inconsistent values but in addition it checks the path consistency of the pairs of values involving a weakly supported value. The idea is that if a value $(i, a)$ has an unique support $(j, b)$ on $C_{ij}$, it may be advantageous to check the path consistency of $((i, a), (j, b))$ because the deletion of this pair leads to the arc inconsistency of $(i, a)$. So, RPC uses path inconsistency to remove values but does not delete any pair of values. This allows to remove more values than AC while avoiding the drawbacks of PC. The RPC algorithm in [Berlandier, 1995] has $O(end^3)$ worst-case time complexity and $O(end + ed^2)$ worst-case space complexity. But this algorithm is not optimal in time.

---
[1] Remark that we still need a $O(n^2d^2)$ data structure for the constraints representation.

Another way to avoid the drawbacks of $k$-consistency is to consider the "opposite" property. $k$-inverse consistency (i.e. $(1, k-1)$-consistency [Freuder, 1985]) removes the values that cannot be extended to a consistent instantiation including any $k - 1$ additional variables. The constraint graph is unchanged by this filtering technique and the space required to enforce it is linear. However, as for $k$-consistency, the worst-case time complexity is polynomial in $k$, and so, inverse consistency quickly becomes prohibitive when $k$ grows. The first level removing more values than arc consistency is path inverse consistency ($k = 3$). Obviously, the more $k$ is important, the more pruningful $k$-inverse consistency is, but we cannot reasonably remove the $n$-inverse inconsistent values. A good compromise is to make sure that each value can be extended to a consistent instantiation including its neighborhood. This filtering technique called neighborhood inverse consistency (NIC)[Freuder and Elfe, 1996] adapts the level of local inverse consistency to the number of constraints involving the variable. This is an efficient method on sparse problems but its exponential worst-case time complexity cannot guarantee a reasonable cpu time.

Singleton consistency is a new class of filtering techniques. It is based on the fact that if a value $(i, a)$ is consistent, then the CSP obtained by restricting the domain of $i$ to the singleton $\{a\}$ is consistent. So, if $P\mid_{D_i=\{a\}}$ is inconsistent, we can delete $(i, a)$. Obviously finding if $P\mid_{D_i=\{a\}}$ is consistent may be costly. Therefore, we only check if a local consistency holds in $P\mid_{D_i=\{a\}}$. For example, a CSP is singleton arc consistent if it does not have any empty domain, and $\forall(i, a) \in \mathcal{D}$, $P\mid_{D_i=\{a\}}$ has a non empty arc consistent sub-domain. Any AC algorithm can be used to determine whether achieving arc consistency in $P\mid_{D_i=\{a\}}$ can yield a domain wipe out, but a lazy approach (such as LAC7 [Schiex et al., 1996]) is sufficient. The singleton consistency can be used with any algorithm removing inconsistent values as those studied in this paper. If the local consistency can be enforced in a polynomial time in $P\mid_{D_i=\{a\}}$, the corresponding singleton consistency has a polynomial worst-case time complexity too.

Since AC algorithms can incrementally propagate the deletion of a value, we propose the algorithm of Fig. 2. It can easily be adapted to enforce a singleton consistency with a stronger pruning efficiency, such as SRPC or SPIC.

## 4 Pruning efficiency
### 4.1 Qualitative study

In order to compare the pruning efficiency of the local consistencies we study, we introduce the following transitive relation. We say that a local consistency $LC$ is *stronger* than another local consistency $LC'$ if in any

**Procedure SingletonAC( $P$ )**
$P \leftarrow AC(P)$;
**Repeat**
  Changed $\leftarrow false$;
  **for** all $(i, a) \in D$ **do**
    **if** the propagation of the deletion of $\{b \in D_i \; s.t. \; b \neq a\}$
    in $P$ leads to a wipe-out$^{(*)}$ **then**
      $D_i \leftarrow D_i\backslash\{a\}$;
      Propagate the deletion of $(i, a)$ in $P$ to achieve AC;
      Changed $\leftarrow true$;
**until** Changed=$false$;

$^{(*)}$ This test only determines if achieving AC in $P\mid_{D_i=\{a\}}$ leads to a domain wipe out.

Figure 2: A SAC algorithm.

CSP in which $LC$ holds, $LC'$ holds too. For example and by definition, RPC is stronger than AC. Therefore, a RPC algorithm removes at least all the arc inconsistent values. Consequently, the domain of $RPC(P)$ is a sub-domain of the domain of $AC(P)$ whatever $P$ may be. A local consistency $LC$ is *strictly stronger* than another local consistency $LC'$ if $LC$ is stronger than $LC'$ and there is at least one CSP in which $LC'$ holds and $LC$ does not hold.

**Theorem 4.1** *If $\mid \mathcal{X} \mid \geq 3$, path inverse consistency is stronger than restricted path consistency.*

**Proof.** First, let us show that PIC is stronger than arc consistency if $\mid \mathcal{X} \mid \geq 3$. For all $(i, a) \in D$ and $C_{ij} \in \mathcal{C}$, let $k$ be any third variable. PIC implies the existence of $(j, b) \in D$ and $(k, c) \in D$ such that $C_{ij}(a, b)$, $C_{ik}(a, c)$ and $C_{jk}(b, c)$. So, for all $(i, a) \in D$ and $C_{ij} \in \mathcal{C}$, there exists a support $(j, b) \in D$. The second property of RPC is a trivial consequence of PIC. If PIC holds, any value $(i, a) \in D$ can be extended to a consistent instantiation including any two other variables. So, if $(i, a)$ has an unique support $(j, b)$ on $C_{ij}$ we are sure that whatever $k$ is, $(i, a)$ can be extended to a consistent instantiation $I$ including $j$ and $k$ s.t. $I_j = b$. ∎

**Theorem 4.2** *If $\mid \mathcal{X} \mid \geq 3$ and the constraint graph is connected, neighborhood inverse consistency is stronger than path inverse consistency.*

**Proof.** see [Debruyne and Bessière, 1997] ∎

**Theorem 4.3** *If $\mid \mathcal{X} \mid \geq 3$, singleton arc consistency is stronger than path inverse consistency.*

**Proof.** Suppose that a singleton arc consistent value $(i, a)$ is not path inverse consistent. Let $j$ and $k$ be two variables s.t. $(i, a)$ cannot be extended to a consistent instantiation including $j$ and $k$. $(i, a)$ being singleton arc consistent, let $D'$ be an arc consistent sub domain of $P\mid_{D_i=\{a\}}$ and $b$ a value in $D'_j$. There exists a value $c \in D'_k$ supporting $(j, b)$ on $C_{jk}$. Since $b$ and $c$ are in an arc consistent sub-domain of $P\mid_{D_i=\{a\}}$, $C_{ij}(a, b)$ and $C_{ik}(a, c)$. We have therefore a contradiction since $(a, b, c)$ is a consistent instantiation of $\{i, j, k\}$. ∎

**Theorem 4.4** *Singleton restricted path consistency is stronger than singleton arc consistency.*

**Proof.** Trivial since restricted path consistency is stronger than arc consistency. ∎

**Theorem 4.5** *Strong path consistency is stronger than singleton arc consistency.*

**Proof.** We have to prove that if a value $(i, a)$ is singleton arc inconsistent, a strong path consistency algorithm removes it. First, let us show that if a value $(j, b)$ is arc inconsistent in $P \mid_{D_i = \{a\}}$ and $(a, b) \in C_{ij}$ then $(a, b)$ is path inconsistent w.r.t. $P$ (by induction on the number of arc inconsistent values deleted before $(j, b)$). If $(j, b)$ is the first value deleted by an AC algorithm because it has no support in $D_k$, $(a, b)$ cannot be extended to a consistent instantiation of $\{i, j, k\}$. If after the deletion of $m$ arc inconsistent values, $(j, b)$ has to be removed because it no longer has a support in $D_k$, all the supports of $(j, b)$ on $C_{jk}$ have already been deleted in $P \mid_{D_i=\{a\}}$. Thus, by induction hypothesis $(a, c)$ is path inconsistent in $P$ for all $(k, c)$ compatible with $(j, b)$. So, $(a, b)$ is path inconsistent in $P$.

According to this lemma, if a value $(i, a)$ is singleton arc inconsistent because AC clears the domain of $j$ in $P \mid_{D_i=\{a\}}$, path consistency removes $(a, b)$ from $C_{ij}$ for all $b \in D_j$ and AC deletes $(i, a)$. ∎

SAC and strong PC are strongly related. In [Mc Gregor, 1979], Mc Gregor proposed a strong path consistency algorithm that can be modified to achieve SAC. This algorithm repeatedly enforces arc consistency on $P \mid_{D_i=\{a\}}$ for all $(i, a) \in D$. If a wipe out occurs, $(i, a)$ is removed, otherwise $(a, c)$ is deleted from $C_{ik}$ for all $(k, c)$ removed by arc consistency in $P \mid_{D_i=\{a\}}$. The algorithm stops when it can no longer detect inconsistent values or path inconsistent pairs of values. A SAC algorithm is obtained by omitting deletions of pairs of values in this algorithm. So, strong PC removes more values only when the propagation of the deletions of pairs of values leads to some additional value deletions.

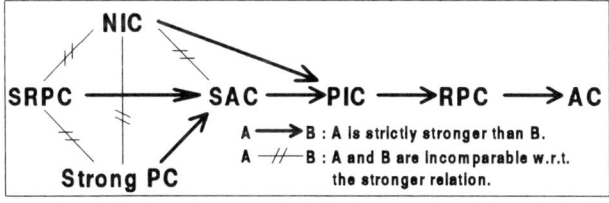

Figure 3: Relations between the local consistencies.

Fig. 3 shows the relations between the filtering techniques studied. A continuous arrow from $A$ to $B$ means that the local consistency $A$ is strictly stronger than $B$. There is a crossed line between $A$ and $B$ if $A$ and $B$ are incomparable w.r.t. the "stronger" relation. If $A$ is not stronger than $B$ ($B$ is strictly stronger than $A$ or $A$ and $B$ are incomparable), a CSP in which $A$ holds and $B$ does not hold can be found in [Debruyne and Bessière, 1997]. The "stronger" relation does not induce a total ordering. For instance, NIC and SRPC are incomparable.

## 4.2 Experimental evaluation

The study in 4.1 gives no quantitative information. A local consistency can be stronger than another while performing few additional value deletions. Furthermore, a local consistency $LC$ can remove more values than another local consistency $LC'$ on most of the CSPs while being incomparable with $LC'$ because of some particular CSPs.

Fig. 4 shows on which random CSPs the local consistencies are useful to detect inconsistency. The CSP generator involves four parameters : $n$ the number of variables, $d$ the common size of all the initial domains, $p_1$ the proportion of constraints in the network ($p_1 = 1$ corresponds to the complete graph) and $p_2$ the proportion of forbidden pairs of values in a constraint (the tightness)[2]. The problems of Fig. 4 have 20 variables, 10 values in each domain, and for each possible pair $(p_1, p_2)$, 500 problems were generated. The continuous line shows for each density the minimal tightness for which the 500 random CSPs are all inconsistent. For each filtering technique, Fig. 4 presents the pairs $(p_1, p_2)$ for which the given local inconsistency is detected on all the generated problems. For example, for AC the limit is 0.69 at density 0.5. This means that for a smaller tightness, there is at least one of the 500 problems on which AC does not remove all the values.

Restricted path consistency is very close to PIC although it requires much less computational effort. To achieve RPC, we have to try to extend a value $(i, a)$ to a consistent 3-tuple involving $\{i, j, k\}$ only if $(i, a)$ has an unique support in $D_j$. It is sufficient to remove most of the values deleted by PIC. As said in section 4.1, SAC is close to strong path consistency which removes more inconsistent values only when there is an important propagation of pairs of values removed. Although SRPC and strong PC are incomparable, SRPC can detect more inconsistent values on most of the CSPs. All these polynomial filterings have a similar behavior. They show difficulties in inconsistency detection on dense problems with loose constraints. It was predicted in [van Beek, 1994]. NIC has a very different behavior. On very dense CSPs, checking if a value can be extended to a consistent instantiation including its neighborhood is close to checking the consistency of the value. However, on the generated problems with a density between 0.2 and

---
[2]See [Frost et al., 1996] for a complete description of the uniform random CSP generator we used.

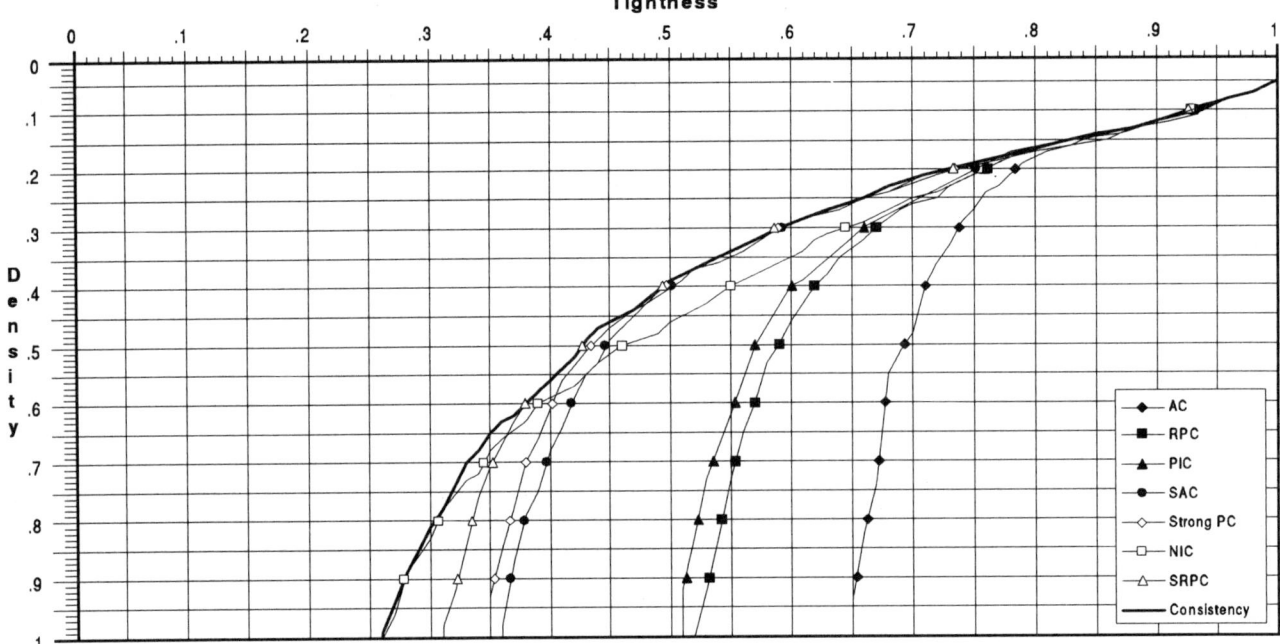

Figure 4: Evaluation of inconsistency detection on random problems with $n=20$ and $d=10$.

0.55, NIC prunes fewer values than SAC, strong PC, and SRPC.

## 5 Time efficiency

The same generator is used to evaluate the cpu time required to enforce the local consistencies. All the generated problems have 100 variables and 10 values in each initial domain. Fig. 5 shows the results for both CSPs having relatively few constraints, and dense CSPs. The algorithms used are AC-6, the RPC algorithm in [Berlandier, 1995], the PIC and NIC algorithms in [Freuder and Elfe, 1996] (NIC using forward checking with minimal domain heuristic), a strong path consistency algorithm based on PC-8 and AC-6, the singleton arc consistency algorithm of Fig. 2 based on AC-6, and a SRPC algorithm based on the RPC algorithm in [Berlandier, 1995]. All these algorithms have been modified to stop as soon as a wipe out occurs. For each tightness, 15 instances were generated and Fig. 5 presents mean values. For some problems at density 0.25, NIC is very time consuming. Therefore, we set a 2 hours limit. No result on NIC is given if one of the 15 instances required more than 2 hours.

Although NIC and SRPC remove more values, strong path consistency is the most expensive filtering technique on most of the generated problems. The consequences of the exponential worst case time complexity of NIC can be observed at density 0.25. For a tightness between 0.18 and 0.30, some problems were not finished after 2 hours. NIC cannot avoid the combinatorial explosion, and so, is unusable on these problems. In addition, NIC is the less efficient (if we exclude strong PC) on CSPs involving many constraints or having tight constraints. The most promising filtering techniques are SAC and RPC. SAC removes most of the strong path inconsistent values while requiring less cpu time than PIC. The RPC algorithm used is non optimal and does not propagate the value deletions as soon as possible. This explains the relatively bad time performances for tightness between 0.5 and 0.8 at density 0.25. However, RPC may remove much more values than AC.

## 6 Conclusion

In this paper we studied the filtering techniques that are practicable on large CSPs, i.e. those that do not change the constraints. When the aim is detecting inconsistency, SRPC and NIC are the most powerful local consistencies. However, they are expensive in time and the exponential worst case time complexity of NIC makes it unusable on some dense CSPs. Two promising local consistencies are SAC and RPC, SAC having a good cpu time to number of deleted values ratio, and RPC requiring little additional cpu time compared to AC while removing most of the path inverse inconsistent values.

## References

[Berlandier, 1995] P. Berlandier. Improving Domain Filtering using Restricted Path Consistency. In *Proceedings of the IEEE CAIA-95*, Los Angeles CA, 1995.

[Bessière and Régin, 1996] C. Bessière and J.C. Régin. MAC and Combined Heuristics: Two Reasons to Forsake FC (and CBJ ?) on Hard Problems. In *Proceedings of CP-96*, pages 61–75, Cambridge, MA, USA, 1996.

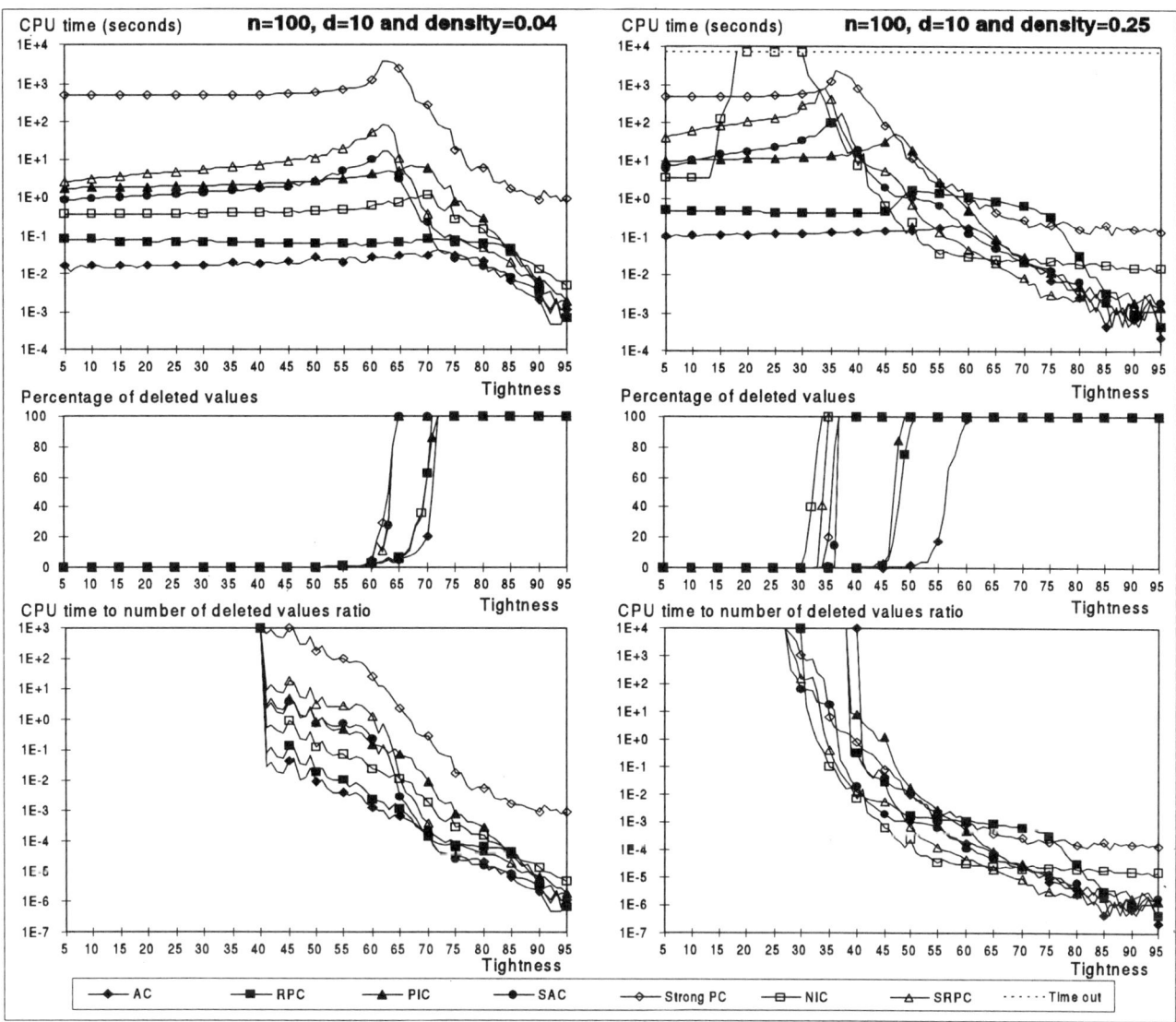

Figure 5: Comparison of the filtering techniques on random CSPs with $n=100$ and $d=10$.

[Bessière et al., 1995] C. Bessière, E.C. Freuder, and J.C. Régin. Using inference to reduce arc-consistency computation. In *Proceedings of the 14th IJCAI*, Montréal, Canada, 1995.

[Bessière, 1994] C. Bessière. Arc-consistency and arc-consistency again. *Artificial Intelligence 65*, pages 179–190, 1994.

[Chmeiss and Jégou, 1996] A. Chmeiss and P. Jégou. Two New Constraint Propagation Algorithms Requiring Small Space Complexity. In *Proceedings of the 8th IEEE ICTAI*, pages 286–289, Toulouse, France, 1996.

[Debruyne and Bessière, 1997] R. Debruyne and C. Bessière. Some Practicable Filtering Techniques for the Constraint Satisfaction Problem. Technical Report 97035, LIRMM, Montpellier, France, 1997.

[Freuder and Elfe, 1996] E. Freuder and C.D. Elfe. Neighborood Inverse Consistency Preprocessing. In *Proceedings of AAAI-96*, pages 202–208, Portland, Oregon, 1996.

[Freuder, 1985] E. Freuder. A sufficient condition for backtrack-bounded search. *Journal of the ACM*, 32(4):755–761, 1985.

[Frost et al., 1996] D. Frost, C. Bessière, R. Dechter, and J.C. Régin. Random Uniform CSP Generators. http://www.ics.uci.edu/~dfrost/csp/generator.html, 1996.

[Mc Gregor, 1979] J.J. Mc Gregor. Relational consistency algorithms and their application in finding subgraph and graph isomorphisms. *Information Sciences 19*, pages 229–250, 1979.

[Schiex et al., 1996] T. Schiex, J.C. Régin, C. Gaspin, and G. Verfaillie. Lazy arc consistency. In *Proceedings of AAAI-96*, pages 216–221, Portland, Oregon, 1996.

[Singh, 1995] M. Singh. Path Consistency Revisited. In *Proceedings of the 7th IEEE ICTAI*, Washington D.C., 1995.

[van Beek, 1994] P. van Beek. On the inherent level of local consistency in constraint networks. In *Proceedings of AAAI-94*, pages 368–373, Seattle WA, 1994.

# Structuring Techniques for Constraint Satisfaction Problems

Rainer Weigel and Boi V. Faltings
Artificial Intelligence Laboratory
Swiss Federal Institute of Technology in Lausanne (EPFL)
IN-Ecublens, CH-1015 Lausanne
Switzerland

## Abstract

We present a structuring method for discrete constraint satisfaction problems (CSPs). It takes advantage of interchangeabilities to represent sets of equivalent values by meta-values and thus obtain more compact representations. Strongly related variables are clustered into meta-variables to create occurrences of interchangeabilities. By iterative application, a CSP can be transformed into an hierarchy of equivalent CSPs, where each problem is significantly simpler than the original one. This structure is particularly advantageous when a large set of possible solutions must be inspected.

## 1 Introduction

We consider binary CSPs defined by $P = (X, D, C, R)$, where $X = \{X_1, \ldots, X_n\}$ is a set of variables, $D = \{D_1, \ldots, D_n\}$ a set of finite domains associated with the variables, $C = \{C_1, \ldots, C_m\}$ a set of constraints, and $R = \{R_{ij} \subseteq D_i \times D_j$ for $C_{ij}$ applicable to $X_i$ and $X_j\}$ a set of relations that define the value combinations allowed by the constraints. Solving such a CSP amounts to finding value assignments to variables subject to constraints. In this paper we address CSPs with many solutions which are to be inspected to select the most suitable one. *Structuring* this space of solutions means constructing a more efficient representation which simplifies such a selection. An example of such a structure is the crossproduct representation [Hubbe and Freuder, 1992]. Interchangeability [Freuder, 1991] is an important concept for structuring CSPs. It involves replacing equivalent values by single *meta values*, thus collapsing solutions with either value to a single one. Crossproduct representations can be constructed from interchangeabilities. The definition from Freuder for neighborhood interchangeability is recalled below. Neighborhood interchangeability captures the idea that values might satisfy exactly the same constraints and thus it is enough

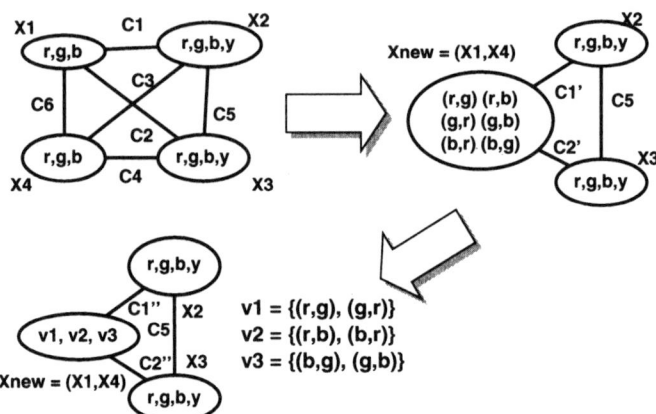

Figure 1: *Coloring example*

to consider only a single value which represents all the equivalent ones.

**Definition 1** *Neighborhood interchangeability:* A value $b$ for a CSP variable $V$ is neighborhood interchangeable (NI) with a value $c$ for $V$ iff for every constraint $C$ on $V$:

$$\{i \mid (b, i) \text{ satisfies } C\} = \{i \mid (c, i) \text{ satisfies } C\}$$

However, interchangeability is rarely valid for complete CSPs. It can often be found only for certain contexts, or for combinations of several variables. We call such combinations *meta variables*. The coloring example in Fig. 1 illustrates this fact. Assume that variables $X1$ and $X4$ are clustered into a new meta variable $X_{new}$. Then there are 6 values for $X_{new}$, namely the solutions of the subproblem induced by $X1$ and $X4$. $C1$, $C3$ and $C2$, $C4$ are merged into $C1'$ and $C2'$ respectively. The constraint $C6$ will no longer appear since it is already accounted in the value combinations allowed for $X_{new}$. The new structured value $(X1 = r, X4 = g)$ for $X_{new}$ is then compatible with $b$ and $y$ for variable $X3$. It turns out that by calculating interchangeability for values of $X_{new}$, we can find three equivalence classes each of size two, namely $v1 = \{(r, g), (g, r)\}$, $v2 = \{(r, b), (b, r)\}$ and $v3 = \{(b, g), (g, b)\}$. These are the meta values for $X_{new}$. One solution of the transformed CSP is $(X_{new} =$

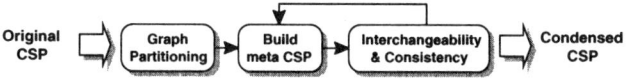

Figure 2: *The Structuring Process*

$v1, X2 = b, X3 = y)$ and this solution represents the two detailed solutions $(X1 = r, X2 = b, X3 = y, X4 = g)$ and $(X1 = g, X2 = b, X3 = y, X4 = r)$. This small example shows that one can find interchangeability in problems after solving a subproblem, although no interchangeability could be found in the original problem formulation. In [Freuder, 1991], Freuder gives an algorithm for computing neighborhood interchangeability which runs in $O(n^2 d^2)$ time, where $n$ is the number of variables and $d$ is the maximal domain size. This algorithm and its extension to *k-interchangeability* are the only polynomial time algorithms to find all interchangeable values in CSPs known so far. Finding all fully interchangeable values [Freuder, 1991] of a CSP, for example, requires finding all its solutions and is thus intractable in general. Thus we only consider neighborhood interchangeability. In order to compact the search space, we calculate neighborhood interchangeability in meta CSPs, and identify thereby *meta-interchangeability* as described in [Freuder, 1991]. We will first present the structuring algorithm. Then we describe an algorithm to calculate interchangeability for values of meta variables and illustrate in 4 the structuring algorithm with an example. In sec. 5 we describe a model for CSPs with clustered constraint graphs and conclude after an empirical evaluation of the algorithm.

## 2 The Structuring Algorithm

The various steps in the structuring algorithm are depicted in Fig. 2. A clustered structure of a constraint graph can be made explicit in the first step using graph partitioning algorithms. After having identified the clustered structure, it can be exploited for finding interchangeability and consistency in the meta CSPs that are generated iteratively. The iteration of calculating interchangeability and running consistency in meta CSPs continues until a single meta variable is created for each cluster identified by the graph partitioning step. Thus the amount of preprocessing that needs to be invested in the iteration depends on how many clusters the graph is decomposed.

**Graph Partitioning:** Given a CSP with clustered constraint graph $G(V, E)$, we apply a graph decomposition method called the *Recursive Bisection Partitioning* [Ciarlet et al., 1994] that allows us to identify a set of edges $E_p$, which, if removed, decompose the constraint graph into several connected components. For clustered graphs, one can expect that the set $E_p$ corresponds to the intercomponent edges of the graph, and each connected component that remains after removing $E_p$, corresponds to a cluster with higher connectivity. The problem of finding a minimum cut bisection of a graph is a well known NP-complete problem [MacGregor, 1978], and among the heuristics for finding an approximation to the problem the Recursive Bisection Method [Ciarlet et al., 1994] is a very successful one. It recursively applies the median cut partitioning method that could be described as follows:

1. Build the Laplacian matrix of the graph G, which is basically the adjacency matrix of the graph G and in the diagonal we put $-|adj(X_i)|$. Then calculate the second smallest eigenvalue of the matrix and the corresponding eigenvector $f$, which is called the fiedler vector [Fiedler, 1973].

2. Let $x_l$ be the median value of the components of the fiedler vector $f$. Node $X_i$ will be put in cluster $C1$ if its corresponding position $f_i$ is smaller than $x_l$, otherwise node $X_i$ belongs to cluster $C2$.

Consider the example from Fig. 3. The median cut procedure partitions the graph into clusters $C1 = V1, ..., V8$ and $C2 = V9, ..., V16$ indicated by hachted edges $E_P$. A stopping criteria for the recursive part of the algorithm needs to be defined. The recursion stops when the number of edges that should be removed in each step is bigger or equal to the number of edges that rest in each component. In an interactive mode, the user might be prompted and asked for the decision.

For every cluster we can identify a subset of nodes $V_E$ that are linked with nodes in other clusters, and a set of nodes $V_I$ having only internal links. The nodes $V_E$ can act like a set of articulation nodes that limit the interaction of assignments to the nodes $V_I$ with the other clusters. At the same time, these nodes are intuitively the part of the cluster through which value interaction is transmitted among the different clusters. Thus by finding consistent assignments to the variable in $V_E$, we can localize the effects of assignments to the other variables of the CSP. The clustered structure is therefore a major source for interchangeability. In the example described in sec. 4, we cluster the variables $V13, V14, V15, V16$ into a new meta variable $AV4$. Any set of values for $AV4$ that have the **same** assignments for the variables in $V_E = V13, V14, V15$ are interchangeable wrt. the rest of the CSP.

**Meta CSPs:** Variables, inducing subproblems of the CSP, are clustered together and treated as a single meta-variable. Each solution to a subproblem is a composed label of the size of the subproblem and becomes a value of the new meta-variable. Adjusting the constraints must in turn be done for all the values of each new variable. Values of the new variables are compatible wrt. the new constraint if all pairs of the compound labels are compatible wrt. the original constraints. The result is called a meta CSP in [Freuder, 1992]. There are many reasons why it can be interesting to cluster variables: the resulting meta CSP might have a nice structure so that it can be solved efficiently. This is the case in the tree clustering procedure [Dechter and Pearl, 1989] where variables are clustered together such that the resulting meta CSP is a tree. The ultimate goal, pursued here, is to find clusters of variables such that their values are meta interchangeable.

There are many possible subproblems that could be clustered together. The clustering procedure presented here will satisfy the following two conditions: 1) the nodes should be tightly connected because the domain size for the new variable should be kept reasonably small [1], and 2) the nodes should belong to the same cluster identified by the graph partitioning method. Therefore we use a greedy maximal clique decomposition algorithm, described next, applied on the graph $G(V, E \setminus E_p)$. A *greedy clique decomposition* algorithm of a graph $G$ partitions the nodes of $G$ into cliques [2]. To find such a decomposition, one starts with a single node and builds successively a bigger clique by putting nodes into the clique that are adjacent to all the nodes already in the clique until the clique cannot be extended further i.e. no other node is adjacent to all the nodes in the clique. Having found such a maximal clique one removes all the nodes in the clique from the graph. The algorithm finally stops when no nodes are left in $G$. There are many possible clique decompositions and a heuristic that always selects a node with the highest degree to be put into the current clique is used. An example of such a decomposition can be found in the example in Fig. 3, where the cliques are indicated with bold edges.

Using the clique decomposition as a heuristic to cluster variables can be justified by the fact that these problems are highly connected and thus might not allow many solutions that appear as new values for the meta variable in the next step. This heuristic corresponds partly to the maximum degree variable ordering heuristic [Dechter and Meiri, 1994] in backtrack search.

**Interchangeability and Consistency:** Calculating interchangeability for the values of the meta variables allows us to reduce the size of the CSP when sets of values can be treated as a single value. From the definition of neighborhood interchangeability one can see that the chance for finding these interchangeable values increases with a decreasing number of constraints that have to be checked. Thus the heuristic to choose cliques for clustering described above might result in more interchangeable sets of values compared to a random clustering.

When clustered variables have been replaced by meta variables, arc-consistency might find additional inconsistencies which were no apparent in the original problem. These pruned meta values are actually partial solutions of the original CSP that cannot be extended to a global solution.

## 3 Calculating Interchangeability

In this section we present another algorithm to identify interchangeable values for meta variables. The basic idea is to take advantage of the set of articulation nodes $V_E$ already presented in section 2. In order to identify interchangeability of two values $v1$ and $v2$ of a meta variable $X$ with Freuder's algorithm, one needs to iterate over all constraints that link $X$ with its neighbors and check if exactly the same tuples are valid for $v1$ and $v2$. For two values having the same assignments for the variables in $V_E$, we do not even have to check the constraint, because the same assignments to $V_E$ imply the same allowed tuples for the constraints. Thus these values are interchangeable. The following algorithm runs faster than Freuder's algorithm, but might not find all sets of interchangeable values.

**Algorithm 1** *Interchangeability of Meta Values*
Let $X$ be a meta variable created by clustering the set of variables $V$ ($V = V_I \cup V_E$):

1. Identify $V_E \subseteq V$.

2. Unless $V_E = V$, iterate over the values of a meta variable $X$ and use the projection of the compound labels of each value onto the set $V_E$ to discriminate between the different equivalence classes of interchangeable values.

One can show that the interchangeable sets for values in a meta CSP found by algorithm 1 are always subsets of the interchangeable sets found by Freuder's algorithm. Thus, to find all interchangeable values, it is proposed to run algorithm 1 followed by Freuder's algorithm. Although this results in the same interchangeable sets as when running Freuder's algorithm only, it can be much faster. Algorithm 1 runs in $O(dn)$ time[3]. Thus the complexity of Freuder's algorithm can be calculated by $O(D^2 n^2)$ where $D$ is the number of equivalence classes returned by algorithm 1. Most experiments done so far indicate that the sets not found by our algorithm are rare. This is not very surprising knowing that interchangeability applied directly on CSPs is also rare.

---

[1] The new domain size is exponential in the size of the subproblem.

[2] In the literature, however, clique decompositions often partition the edges and not the nodes into cliques.

[3] $d$, $n$ are the domainsize and the number of variables of the $metaCSP$.

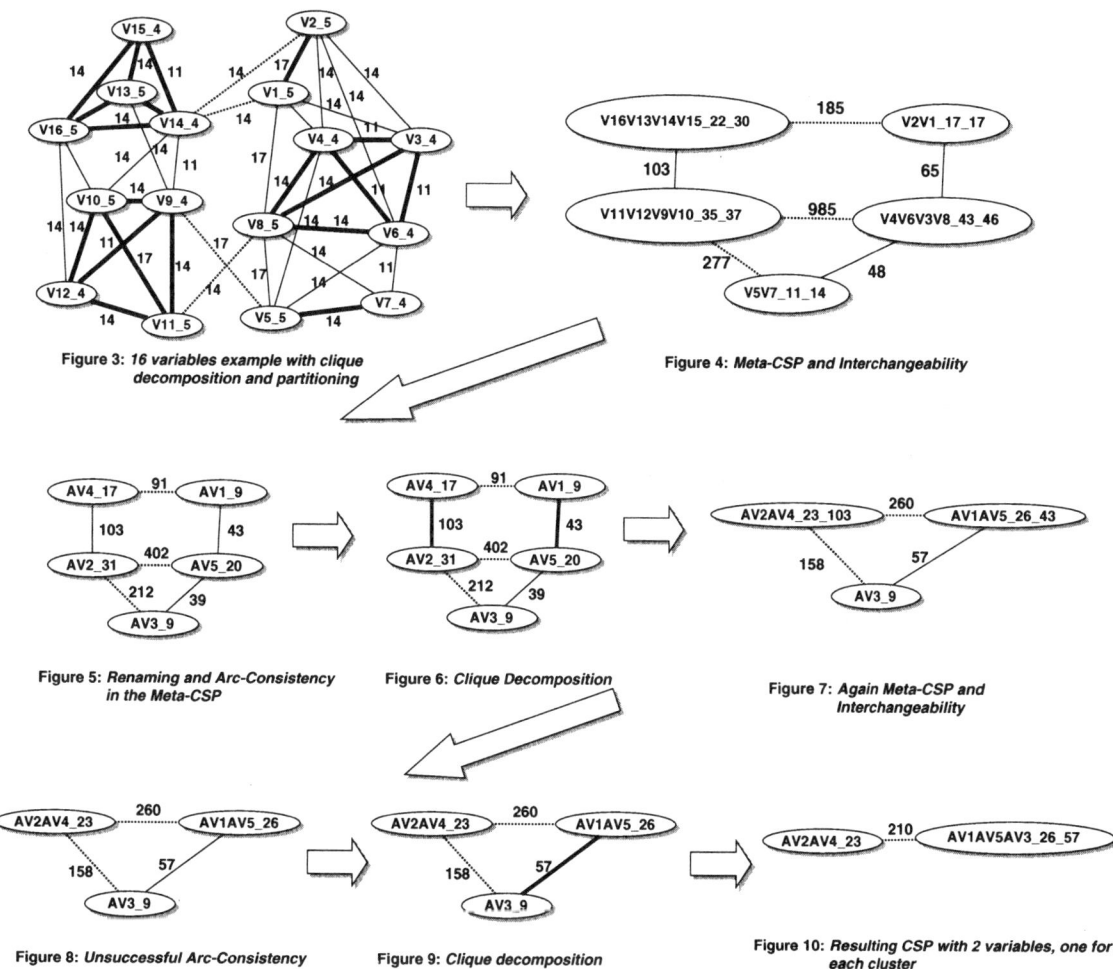

Figure 3: 16 variables example with clique decomposition and partitioning

Figure 4: Meta-CSP and Interchangeability

Figure 5: Renaming and Arc-Consistency in the Meta-CSP

Figure 6: Clique Decomposition

Figure 7: Again Meta-CSP and Interchangeability

Figure 8: Unsuccessful Arc-Consistency

Figure 9: Clique decomposition

Figure 10: Resulting CSP with 2 variables, one for each cluster

## 4 Example

Fig. 3 shows the constraint graph of a CSP with 16 variables. Every variable name is followed by its domain size and every constraint is labeled with the number of tuples allowed by the constraint. The median cut partitioning finds 4 edges between the nodes $(V14, V2), (V14, V1), (V9, V5)$ and $(V11V8)$ that partition the graph into cluster $V1, V2, ..., V8$ and a second cluster $V9, V10, ..., V16$. The partitioning algorithm is then stopped and the iteration on building the meta CSP calculating interchangeability and running consistency is repeated until a single meta variable is created for each cluster. The greedy clique decomposition algorithm determines in the first run 5 subproblems with complete graphs indicated by the bold edges in Fig. 3. Every subproblem is solved individually and appears in Fig. 4 as a new meta variable. The variables $V13, V14, V15$ and $V16$ for example are clustered together into the meta variable, named "$V16V13V14V15$". The number of solutions of this subproblem is 30, i.e. the last number of the node label. Interchangeability identifies 8 equivalence classes of values each of size 2. Thus the domain size can be reduced to 22. As an example of a set of interchangeable values consider $S1 = (V16 = 0, V13 = 4, V14 = 2, V15 = 3)$ and $S2 = (V16 = 0, V13 = 4, V14 = 2, V15 = 2)$. One can see how the nodes $V16, V13$ and $V14$ act as the articulation nodes $V_E$. Algorithm 1 finds the set $V_E$, the projection of $S1$ and $S2$ onto $V_E$ is $(V16 = 0, V13 = 4, V14 = 2)$ for both values, and thus $S1$ and $S2$ are interchangeable. Renaming $V16V13V14V15$ to $AV4$ ($AV$ stands for Abstract Variable) and applying arc-consistency to the CSP is shown in Fig. 5. With arc-consistency one can remove 5 values from the domain of $AV4$. The four edges found by the graph partitioning in the first step corresponds now to the three edges $(AV1, AV4), (AV2, AV5), (AV2, AV3)$ that are not allowed to appear in a clique in the next step. Consequently two new clusters $AV1AV5$ and $AV2AV4$ are created and $AV3$ is not changed (Fig. 8). In a third step "$AV1AV5$" and $AV3$ are clustered together resulting finally in a CSP with two nodes $AV2AV4$ with domain size 23 and $AV1AV5AV3$ with domain size 26. These two nodes corresponds to the two clusters $V1, V2, ..., V8$

and $V9, V10, ..., V16$ determined in the first step. 210 tuples are allowed by the constraint between the two nodes and these tuples correspond to the solutions for the CSP on this level (Fig. 10). Every meta-value in a solution of the resulting CSP can be replaced by the set of values it represents. In this way one obtains the cross product representation [Hubbe and Freuder, 1992] of solutions for the next lower level. Replacing meta values for each solution in the crossproduct continues until the lowest level is reached. The 210 solutions can be refined to 438 on the next lower level then to 2910 and finally to the 5366 solutions of the original CSP. It is worth to mention that the size of the equivalence classes of values plays a multiplicative role when determining the number of solutions on the next lower level. If a solution $S$ contains $t$ meta values that represent classes of size $s_i$ with $i = 0, ..., t-1$, then the number of solutions on the next level that are represented by $S$ equals $\prod_{i=0}^{t-1} s_i$.

## 5 Clustered Constraint Graphs

Constraint graphs of problems that arise for example from physical interactions often resemble clustered graphs. We shortly describe a simplified model for generating clustered graphs, however details can be found in [Hogg, 1996]. The nodes of the graph to be generated will appear as the leaves of a balanced binary tree. Thus to generate graphs with $2^D$ number of nodes, we build a balanced tree with depth $d = D - 1$. The tree induces a ultrametric distance between pairs of nodes, defined as the number of levels up in the tree from the nodes to reach a common ancestor. For a given $p$ with $0 < p < 1$, edges in the graph between nodes at distance $t$ ($0 < t \leq d$) can then be selected with a relative probability $p^t$. As an example consider the clustered constraint graph in Fig. 3. The nodes $V1, ..., V16$ are the leaves of a balanced tree with depth 3 arranged from left to right in increasing order. Given $p = 0.5$ the probability of having an edge in between two nodes from the set $V1, ..., V8$ and from the set $V9, ..., V16$ is $0.5^3 = 0.125$. An edge in between two nodes in $V1, ., V4$ and $V5.., V8$ appears with probability 0.25 etc.. Having build the clustered graph one can easily construct a CSP with clustered constraint graph by randomly selecting domain sizes and allowed tuples for the constraints.

## 6 Evaluation

Any evaluation of a technique depends on the objectives it is used for. Our objective is to structure a CSP so that its solution space can be easily inspected. Thus, a reasonable measure would be the average number of choices a user has to make to select one particular solution among all possibilities. The results for a small number of test cases generated with $n = 32$, $p = 0.4$, $domainsize = 3$,

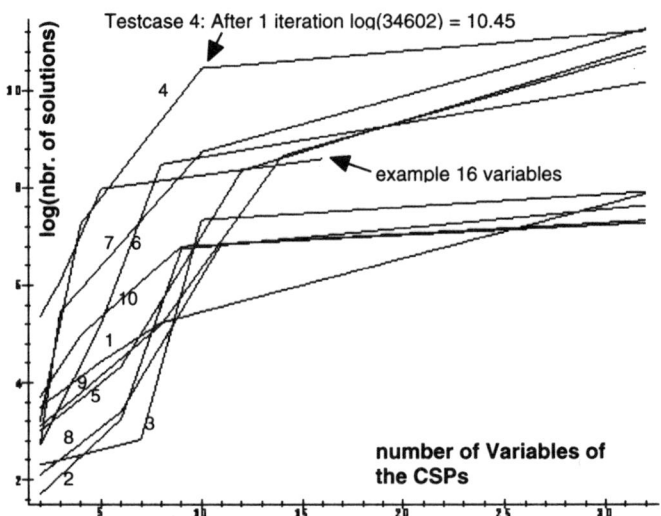

Figure 11: *Number of solutions on the different levels for the Testcases on a logarithmic scale*

$number\ of\ constraints = 110$, 70% from all possible tuples for each constraint are allowed, can be found in Table: 1 and in Fig. 11. The problems are tranformed until CSP with two variables are returned. Over-constrained problems and problems with less that 100 solutions are omitted. We define a condensing factor of the solution space $\rho_i = log|solutions\ level_i|/log|all\ solutions|$ with $1/log|all\ solutions| \leq \rho_i \leq 1$ and for different abstraction levels $i = 0, ..., k$. Using this measure, we can discriminate between CSPs where the algorithm returns good and bad results[4]. A factor close to 0 implies a very condensed representation, while a factor of 1 could not condense the solution space much. A series of runs confirm the intuition that for clustered CSPs with "many" solutions, the number of solutions of the resulting meta-CSPs is condensed considerably, often by a large factor (see. Fig. 11). Testcase 2 for example could be condensed from 2066 solutions to 6 solutions after 3 iterations. Resulting in a $\rho_3 = log6/log2066 = 1.7917595/7.6333694 = 0.234$. Thus, the structured problem will considerably reduce the effort required to search through the solution space of such a CSP.

The evaluation method needs to enumerate all solutions for the condensed CSP. This, however, is only possible for small CSPs. For large CSPs (i.e. when the resulting CSP has still > 100 variables), we will use a probabilistic approach that determines the expected number of solutions of the CSPs (based on [Knuth, 1975]) on the different levels to determine $\rho_i$.

---

[4] $\rho$ is not defined for CSPs with only a single solution

| Solutions on level: | Example and Testcases 1 - 10 | | | | | | | | | | |
|---|---|---|---|---|---|---|---|---|---|---|---|
| | Ex.(16 vars) | 1 | 2 | 3 | 4 | 5 | 6 | 7 | 8 | 9 | 10 |
| 3 | 210 | 36 | 6 | 11 | 26 | 22 | 15 | 15 | 9 | 24 | 43 |
| 2 | 438 | 84 | 26 | 17 | 1428 | 76 | 184 | 235 | 30 | 180 | 146 |
| 1 | 2910 | 188 | 862 | 1550 | 34602 | 4260 | 4832 | 6220 | 951 | 5712 | 904 |
| original CSP | 5366 | 2632 | 2066 | 2750 | 74718 | 54280 | 25872 | 77828 | 1512 | 48972 | 1460 |
| $\rho_3$ | 0.62 | 0.44 | 0.23 | 0.29 | 0.28 | 0.27 | 0.26 | 0.23 | 0.28 | 0.28 | 0.50 |

Table 1: *Condensing the solution space of some examples. The size of the CSPs on the various levels is omitted.*

## 7 Future Work and Conclusion

In this paper we have presented a structuring method for CSPs. Applying the method to a CSP results in a more condensed representation that allows the enumeration of sets of similar solutions and thus facilitates the selection of single solutions. One should contrast this with search algorithms for solutions in the original CSP: They find one solution after the other without showing the similarity in between them. Furthermore, in many applications it is not enough to present a human decision maker with a single solution, because this single solution might violates other criteria which the user has not made explicit to the problem. Thus one should give the user a chance to integrate hidden knowledge and unquantifiable preferences in the decision process by proposing sets of similar solutions from which he can make the selection.

The evaluation of our algorithm imposes many difficulties and a lot of work needs still to be done. It is for example not enough to know how many interchangeable sets we get and how big they are; it is very important to know in how many solutions a meta value will finally appear. This always requires us to look at all solutions of a CSP, a set which is impossible to generate for large CSP. In the future work we will also look at hard random problems. Basically one would like to compact the search space such that certain instantiations might have to be made only once for each equivalence class of values, thus reducing the search effort. Similarly these hard problems often exhibit a high degree of consistency and thus low consistency algorithms will not find inconsistencies. Running arc-consistency on a meta CSP, which is related to restricted form of $k$-*consistency* with $k$ equal to the sum of size of two adjacent clusters, might therefore be very interesting for hard CSPs.

## References

[Ciarlet et al., 1994] P. Ciarlet, F. Lamur, and B.F. Smith. On the influence of the partitioning schemes on the efficiency of overlapping domain decomposition methods. Technical Report 94-23806, UCLA Computitional and Applied Mathematics, Los Angeles, CA 90024-1555, 1994.

[Dechter and Meiri, 1994] R. Dechter and I. Meiri. Experimental evaluation of preprocessing algorithms for constraint satisfaction problems. *Artificial Intelligence*, 68:211–241, 1994.

[Dechter and Pearl, 1989] Rina Dechter and Judea Pearl. Tree Clustering for Constraint Networks. *Artificial Intelligence*, 38:353–366, 1989.

[Fiedler, 1973] M. Fiedler. Algebraic connectivity of graphs. *Czechoslovak Mathematical Journal*, 23 (98):298–305, 1973.

[Freuder, 1991] Eugene C. Freuder. Eliminating Interchangeable Values in Constraint Satisfaction Problems. In *Proc. of AAAI-91*, pages 227–233, Anaheim, CA, 1991.

[Freuder, 1992] Eugene C. Freuder. Constraint solving techniques. In E. Tyngu B. Mayoh and J. Penjaen, editors, *Constraint Programming of series F: Computer and Sytems Sciences*, pages 51–74. NATO ASI Series, 1992.

[Hogg, 1996] T. Hogg. Refining the phase transition in combinatorical search. *Artificial Intelligence*, 81:127–154, 1996.

[Hubbe and Freuder, 1992] P.D. Hubbe and E.C. Freuder. An efficient Cross Product Representation of the Constraint Satisfaction Search Space. In *Proc. of AAAI-92*, pages 421–427, San Jose, CA, 1992.

[Knuth, 1975] D. E. Knuth. Estimating the efficiency of backtrack programs. *Mathematics of Computations*, 29:121–136, 1975.

[MacGregor, 1978] R. M. MacGregor. *On Partitioning a Graph* . PhD thesis, University of California, Berkeley, 1978.

# Merging constraint satisfaction subproblems to avoid redundant search*

## Javier Larrosa

Dep. Llenguatges i Sistemes Informàtics
Universitat Politècnica de Catalunya
Jordi Girona Salgado, 1-3
E-08034 Barcelona, Spain

### Abstract

When solving a *constraint satisfaction problem* we may find similar subproblems. Algorithms not exploiting this similarity are condemned to duplicate some work. In this paper we introduce a technique for merging sibling subproblems which avoids redundancy in search associated to their similarity. We show that, when *forward checking* is enhanced with this capability, its performance may be increased. Experimental results on hard crossword puzzles support the practical validity of our approach.

## 1 Introduction

*Constraint Satisfaction Problems* (CSPs) involve finding values for problem variables subject to constraints on which combinations of values are permitted. They have many applications in *artificial intelligence* and other areas of *computer science*.

It is believed that some real CSPs have different values for the same variable that behave in a similar manner [Freuder, 1991; Haselbock, 1993]. *Depth-first* algorithms (such as *forward checking* [Haralick and Elliott, 1980]) not exploiting this similarity are condemned to do some redundant work. Figure 1a (left upper corner) shows a simple colouring graph problem used to illustrate this idea. The problem is to assign a colour to each node in such a way that adjacent nodes have different colours. Colour choices for each node are represented by letters and included in the nodes. Figure 1b shows the search tree traversed by forward checking before finding that the problem is unsolvable (variables and values are selected lexicographically).

It can be observed that the first two subtrees (after assigning A and B to $x_1$) have a large similar area (shadowed). The reason for this is that values A and B behave in a very similar manner. Thus, their assignment produces similar subproblems. This fact can be observed in Figure 1a, which also shows the three subproblems obtained after assigning the three alternative colours to $x_1$ and updating the domains. As can be seen, the first and the second have more in common than the third.

We claim that solving similar subproblems without exploiting their similarities is inefficient since an important part of the work is, somehow, duplicated. Obviously, it is desirable to avoid the unnecessary repetition of as much work as possible. This paper aims in that direction. We show that, if sibling subproblems are close enough, it is worth to *merge* their subtrees, and only distinguish to which value they are committed at deeper levels of the tree.

We introduce a technique for merging sibling subproblems based in the concept of *weak assignment*. Briefly, a weak assignment is the reduction of a variable

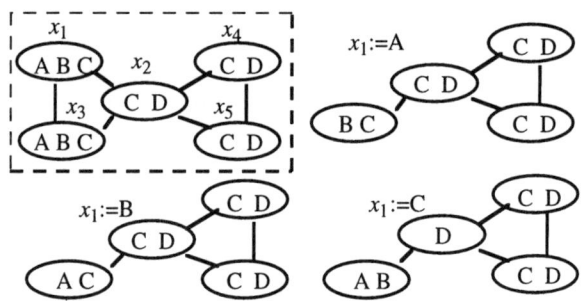

(a). A sample colouring problem (left upper corner) and the three first-level subproblems.

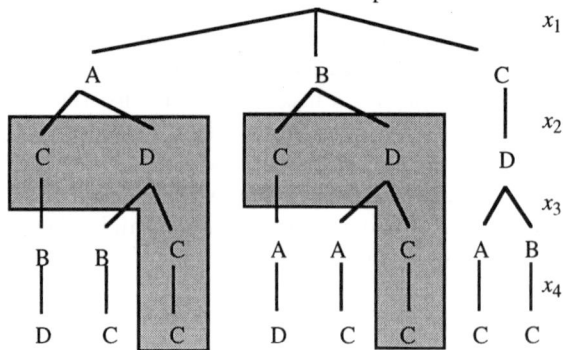

(b). search tree traversed by *forward checking* (common paths in the first and second subproblem are shadowed).

Figure 1. A colouring graph and the search tree traversed by *forward checking*.

---

*This research is supported by the Spanish CICYT project TIC96-0721-C02-02.

domain to a subset of its feasible values. The resulting subproblem merges into one the set of subproblems that would be obtained if the subset of values had been assigned in the classical way. If one delays the decision of what value is definitively assigned to the variable, the resulting search tree is the fusion of the set of trees. We show that this technique is especially convenient to merge similar subproblems.

We have introduced weak assignments into forward checking and we show that, for specific domains, its performance is significantly increased. Our experimental results on hard random crossword puzzles support the practical validity of our approach.

This paper is structured as follows. In Section 2 we introduce the notation and definitions that will be used throughout the paper. In Section 3 we present the concept of weak assignment and analyze its effectivess for merging subtrees. In Section 4 we introduce a modification to forward checking enhancing it with merging capabilities. In Section 5 we report our experiments on crossword puzzles. In Section 6 we review and compare work related to ours. Finally, in Section 7 we give the conclusions of our work.

## 2 Definitions and Notation

A discrete binary CSP is defined by a set of $n$ variables $\{x_i\}$ taking values on discrete and finite domains $\{d_i\}$ under a set of binary constraints $\{R_{ij}\}$ ($i,j=1,..,n$). A constraint $R_{ij}$ is a subset of $d_i \times d_j$, containing the permitted values for $x_i$ and $x_j$. A solution to the CSP is an assignment of values to variables satisfying every constraint. This paper is devoted to the task of finding one or all solutions to a CSP.

Given a CSP, a subproblem is what remains after the assignment of some of its variables. Assigned variables are called *past* variables and denoted by P. Unassigned variables are called *future* variables and denoted by F. A *subproblem* S is defined by a tuple S=<P,F,D> where P=$\{x_j:=k_j\}$ $\forall j \notin F$ are the assignments to past variables, F is the set of future variables and D=$\{d_j\}$ $\forall j \in F$ are the non empty sets of future values which remain feasible.

*Forward checking* (FC) is a simple, yet powerful algorithm for constraint satisfaction. Given a subproblem S, FC selects an unassigned variable $x_i$ and transforms S into a set of simpler subproblems $\{S^l \mid l \in d_i\}$. Each subproblem $S^l$=<$P^l,F^l,D^l$> is defined by $P^l=P\cup\{x_i:=l\}$, $F^l=F-\{x_i\}$, $D^l=\{d_j^l\}$ $\forall j \in F^l$. Domain $d_j^l$ is the updating of $d_j$ after removing those values unfeasible with the new assignment $x_i := l$. Variable $x_i$ is the *current variable* and it changes from being a future variable in S to a past variable in $S^l$.

FC can be seen as the depth-first traversal of a search tree where each subproblem is a tree node. Two nodes are *siblings* if they have the same parent. If all solutions are required, the whole tree is traversed. If one solution is wanted, the algorithm stops when it finds one. We denote by T the subtree rooted by node S. An important parameter related to the computational effort needed to solve a problem is the number of nodes that the algorithm requires to visit when traverses the whole search tree (|T|). It is especially appropriate when: (i) all solutions are required, or (ii) only one solution is wanted, but the problem is not trivial and a significant part of the search tree has to be traversed before finding a solution.

To evaluate the validity of our approach we suppose that forward checking uses a *static variable ordering* (the current variable selection depends on the tree level where it takes place). However, our approach is still valid when *dynamic variable orderings* are used, as will be shown in section 5.

Throughout this paper we will refer to a generic subproblem, S, and to its successors $\{S^l \mid l \in d_i\}$ (being $x_i$ the current variable). Specific successors will be denoted by $S^a$, $S^b$,..., where their index corresponds to the value assigned to $x_i$. To differentiate children elements from their parent we will use their index (i.e: $S^a$=<$P^a,F^a,D^a$>, $T^a$, etc.).

**Definition 1:** *Given S, $S^a$, $S^b$, and two of its descendants $S^\alpha \in T^a$ and $S^\beta \in T^b$, we say that $S^\alpha$ and $S^\beta$ follow* common paths *after S if their assignments to past variables ($P^\alpha$ and $P^\beta$) only differ in the assignment made to $x_i$.*

Therefore, common paths are those paths only differing in the assignment that caused their separation (for instance, in Fig. 1b (A,D,C,C) and (B,D,C,C) are common paths). Since we are assuming a static variable ordering, proceeding upwards from $S^\alpha$ and $S^\beta$ level by level, one encounters pairs of nodes which follow common paths, too.

Finally, *similarity* among subproblems is a graded relation which in this work is restricted to sibling nodes. For our purposes it is enough to give an intuitive definition. Similarity must measure how close are future variable domains. Thus, two subproblems have the lowest similarity if they have no common values in their domains. Two subproblems have the highest similarity if they have exactly the same values in their domains.

## 3 Merging Subproblems

### 3.1 Weak Assignments

Common paths between two subproblems define two subtree areas with identical arboreous structure (shadowed subtree areas in Fig. 1b correspond to common paths between the first and second subproblem). Each subtree node has its corresponding node at the other subtree which only differs in: (i) one assignment and (ii) the effect of that assignment on future variable domains. Common paths are, somehow, redundant since, except for one assignment, their nodes represent the same situation. It would be desirable to generate them only once and employ them for both subproblems. In what follows, we introduce an approach to this idea based on the concept of *weak assignment*.

**Definition 2:** *Given a subproblem S=<P,F,D>, a variable $x_i$, and a subset $K \subset d_i$ such that* card(K)>1, *the weak assignment $x_i := K$ produces a new simpler subproblem $S^K$=<$P,F,D^K$> where*

(i) $d_i^K = K$
(ii) $d_j^K = \{a \in d_j \mid a$ *consistent to at least one value in* $K\}$

In words, a weak assignment reduces the domain of an unassigned variable and prunes future values that are inconsistent with all values considered for $x_i$. Observe that $x_i$ does not become a past variable after the weak assignment because more than one value is still considered for it. Figure 2a shows the subproblem $S^{\{A,B\}}$ obtained after the weak assignment $x_1:=\{A,B\}$ in the colouring problem presented in Figure 1a.

The set of solutions of $S^K$ is equal to the set of solutions of $\{S^l \mid l \in K\}$. Therefore, one can either solve $S^K$ or the set $\{S^l \mid l \in K\}$. Solving $S^K$, one can delay the decision of what value (among K) is assigned to $x_i$ to deeper levels of the tree (unless otherwise indicated, we will assume that, after a weak assignment, the variable remains unassigned until the last tree level). The advantage of solving $S^K$ is that all common paths in $\{T^l \mid l \in K\}$ occurring before the last tree level are *merged* into a unique path, which considers, simultaneously, the different assignments to $x_i$. However, look ahead after a weak assignment has a lower pruning power. Therefore, the detection of domain wipe-outs may be delayed.

Figure 2b shows the search tree traversed by forward checking with the problem presented in Figure 1a, if the first and second subproblems are merged with the weak assignment $x_1:=\{A,B\}$. Observe that two subtrees are merged into one where their common paths occur only once. In this particular case, the delayed assignment of $x_1$ never takes place because the algorithm always backtracks before the last tree level.

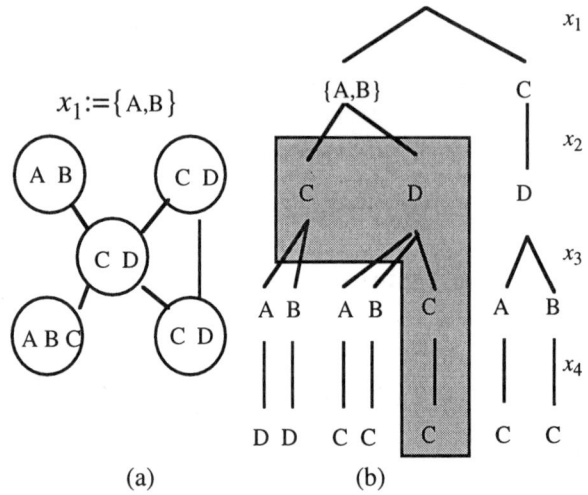

Figure 2. (a) Resulting problem after the weak assignment $x_1:=\{A,B\}$ which merges the first two subproblems. (b) The search tree traversed by *forward checking* with the weak assignment.

### 3.2 Effectiveness of Merging Subproblems

In this subsection we analyze the benefit that can be obtained when subproblems are merged. For the sake of clarity, we restrict our discussion to the case of *two-valued* weak assignments.

To evaluate the effectiveness of merging two subproblems $S^a$ and $S^b$ with a weak assignment $x_i:=\{a,b\}$ we compare the size of the subtree underneath $S^{\{a,b\}}$ (i.e.: $|T^{ab}|$) with the size of the subtrees under $S^a$ and $S^b$ ($|T^a|+|T^b|$). As a first step, we partition the set of descendant nodes of $S^a$ (and $S^b$) into three groups: $V^a$ are those nodes which are problem solutions, $W^a$ are those nodes having a node in $T^b$ with which follow common paths, and $Z^a$ are the remaining nodes (the same for $T^b$). For instance, if we denote by $S^A$ the subproblem after assigning A to $x_1$ in the graph colouring problem (first subtree in Fig. 1b): $V^A=\emptyset$, $W^A=\{AC, AD, ADC, ADCC\}$ and $Z^A=\{ACB, ACBD, ADB, ADBC\}$. The following expression trivially holds:

$$|T^a|+|T^b| = |V^a|+|W^a|+|Z^a|+|V^b|+|W^b|+|Z^b|$$

**Observation 1:** *By definition $|W^a|=|W^b|$, therefore the former expression can be rewritten as*

$$|T^a|+|T^b| = |V^a|+|Z^a|+|V^b|+|Z^b|+2|W^a|$$

We also use this partition to establish the following mappings between nodes (proofs are straightforward under the assumption that a static variable ordering is used):

**Observation 2:**

(i) *for each $S^\alpha \in V^a$ there is a unique $S^\gamma \in T^{ab}$ such that $S^\alpha$ and $S^\gamma$ are identical.*

(ii) *for each $S^\alpha \in Z^a$ there is a unique $S^\gamma \in T^{ab}$ such that $P^\alpha$ and $P^\gamma$ only differ in that $P^\gamma$ does not include $x_i:=a$.*

(iii) *for each $S^\alpha \in W^a$ and $S^\beta \in W^b$ following common paths, there is a unique $S^\gamma \in T^{ab}$ such that $P^\gamma$ only differs from $P^\alpha$ and $P^\beta$ in that it does not include any assignment to $x_i$.*

These observation establishes a one-to-one mapping between some nodes in $T^{ab}$ and nodes in $T^a$ and $T^b$. An important consequence is that any node in $T^a$ or $T^b$ has its associated node in $T^{ab}$. However, observation (iii) shows that non solution nodes in the common paths of $T^a$ and $T^b$ share their associated node in $T^{ab}$ (this is why the fusion occurs).

**Observation 3:** *$T^{ab}$ may have nodes neither mapped to $T^a$, nor to $T^b$.*

**Proof:** Consider a node $S^\gamma \in T^{ab}$, whose assignments to past variables include two assignments, $x_p:=l_p$ and $x_q:=l_q$ such that $l_p \in d_p^a - d_p^b$ and $l_q \in d_q^b - d_q^a$. This node is not mapped to $T^a$, nor to $T^b$.

**Observation 4:** *Combining observation 2 and observation 3 the following relation can be obtained:*

$$|T^{ab}| \geq |V^a|+|V^b|+|Z^a|+|Z^b|+|W^a|$$

Therefore, comparing *observation* 1 with *observation* 4, one may conclude that merging two subproblems with a weak assignment, $x_i:=\{a,b\}$, is effective (i.e: $|T^{ab}| < |T^a|+|T^b|$) if the number of nodes in the common paths of $T^a$ and $T^b$ (whose duplication is avoided in $T^{ab}$) surpasses the number of extra nodes in $T^{ab}$ that are not mapped to any node in $T^a$ or $T^b$.

If $S^a$ and $S^b$ are very similar, it is natural to expect $T^a$ and $T^b$ to have many common paths. It is so, because the only chance that $S^\alpha$, descendent of $S^a$, does not have its common node $S^\beta$, descendent of $S^b$, is due to the dissimilarities between $S^a$ and $S^b$. It may happen for two reasons: (i) $S^\alpha$ has a value assigned to a variable which is unfeasible in $S^b$, or (ii) assignments made to $S^\alpha$ would produce a domain wipe-out if made at $S^b$.

## 3.3 Nested Weak Assignments

Tree size reductions obtainable by merging subproblems can be increased if more than one weak assignment is performed through a tree path (*nested* weak assignments). Thus, if a weak assignment $x_2:=\{c,d\}$ is performed after a weak assignment $x_1:=\{a,b\}$, descendant nodes merge those subproblems considering the four possible combination for $x_1$ and $x_2$. In general, $u$ nested weak assignments of $v$ values each, can merge $v^u$ nodes differing in $u$ assignments.

However, the pruning power of the look ahead decreases with nested weak assignments because weaker information for pruning is available. Therefore, the detection of domain wipe-outs may be delayed to deeper nodes, producing the addition of new nodes. The benefit of nested weak assignments depends on the balance of these two factors.

## 4 FC with Weak Assignments

In this section we modify forward checking adding to it the capability of merging subproblems with nested weak assignments. The resulting algorithm, $FC_w$, is outlined in Figure 3. At each node, $FC_w$ selects a variable $x_i$ and partitions its domain into groups. Groups with more that one element are weakly assigned producing a new subproblem where $x_i$ remains being a future variable, groups with a single value are assigned in the classical way producing a new subproblem where $x_i$ becomes a past variable (therefore, if the domain is partitioned into groups of one value each, the algorithm behaves as FC). The look-ahead function removes from future variable domains those values inconsistent with *all* values in the group under consideration (therefore, if the group has only one value, this procedure behaves as the standard look-ahead).

The *select_var* function implements the variable ordering heuristic of choice, which can be either static or dynamic. However, some criterion must be added to delay the definitive assignment of variables that are weakly assigned. In our implementation, when a variable is selected and weakly assigned, it is not selected again until the rest of variables have been selected.

The *domain_partition* function is a key element for the algorithm efficiency. This function must group values producing sufficiently similar subproblems. The implementation of this function raises two important questions: (i) how to measure subproblem similarity, and (ii) what is the similarity threshold to cluster values after which $FC_w$ is cost effective. We believe that the answer to these questions is domain dependent (as well as the existence of similar subproblems). For instance, in *random problems* (a typical benchmark for CSP algorithms) weak assignments seem not to be useful because their random nature makes unprobable the occurrence of very similar subproblems. Moreover, they do not have any structure, therefore no domain knowledge can be used to evaluate similarity. On the other hand, more structured problems may be more appropriate for our approach because some domain specific knowledge can be used to detect subproblem similarities. We will come back to this point in the next section.

```
Procedure FC_w(subproblem <P,F,D>)
if(empty_domain(D)) then backtrack
if(F=∅) then P is solution
else
    x_i := select_var(F)
    Π := domain_partition(d_i)
    for all π_k ∈ Π do
        D' := look_ahead_w(x_i, π_k, F, D)
        if(|π_k|=1) then /* strong assign. */
            FC_w(P∪{x_i:=π_k}, F-{x_i}, D')
        else /* weak assignment */
            d'_i = π_k
            FC_w(P, F, D')
EndProcedure

Function look_ahead_w(x_i, π_k, F, D)
for all x_j ∈ F do if x_j ≠ x_i then
    for all a ∈ d_j do
        d'_j = ∅
        for all b ∈ π_k do
            if(consistent(x_i, x_j, b, a)) then d'_j = d'_j ∪ {a}
Return(D')
```

Figure 3. *Forward checking* with weak assignments.

## 5 Experimental Results

To evaluate the practical validity of our approach a set of experiments was devised using *crossword puzzles*. A crossword puzzle is characterized by a set of words (called dictionary) and a grid that may have some of its cells labelled as void. Consecutive non void cells (horizontally or vertically) define a slot. The task is to place a word to each horizontal slots in such a way that vertical slots form dictionary words, too ([Ginsberg *et al.*, 1990]). They are of practical interest because they can be seen as an abstraction of configuration tasks where the components of an assembly have to be decided in such a way that interdependency restrictions are fulfilled.

Crossword puzzles can be represented as binary CSPs with variables associated to grid slots. For each slot, its domain is the set of dictionary words having the appropriate length. Intersecting slots are constrained, restricting pairs of words to those matching in their intersection.

In our experiments we used random puzzles with square grids and without void cells. In addition, we tightened constraints preventing rows being equal from columns in order to avoid symmetric solutions (they were found to make problem instances significantly easier).

Thus, our random puzzles are characterized by a tuple $<l,m,v>$ where $l$ is the words length and the grid dimension, $m$ is the dictionary cardinality and $v$ is the alphabet cardinality. Random instances were generated by selecting $m$ words out of the $v^l$ choices using a uniform probability distribution. We experimented on the following problem classes: <5,20-99,5>, <7,20-62,3> and <9,15-31,2> ($m$ was used as a varying parameter). For each parameter setting, samples of 50 random instances were generated forming a total of 7,000 instances. We observed that, for small $m$, problems are overconstrained and do not have any solution. If $m$ is increased, there is a point after which problems become abruptly underconstrained and the number of solutions grows very fast. If only one solution is sought, this point corresponds to a peak on average problem difficulty similar to that observed in other domains [Cheeseman et al., 1991]. In our three problem classes the rank for $m$ was chosen to coincide with the peak. Therefore, our experiments were on *hard* instances having few, or no solution.

An interesting feature of crossword puzzles is that constraints are implicit in the letters forming the words. Words having many matching letters behave similarly in the problem. More precisely, the pruning effect of assigning two words to a slot is exactly the same with respect to slots intersecting in their matching letters and totally different with respect to slots intersecting in their different letters. In our Implementation of $FC_w$ for crossword puzzles, the *domain_partition* function (Fig. 3) partitions domains into groups where every word differs from the rest in at most $\delta$ letters. Delta was set to one for the first class of problems ($l=5$) and to two for the other two classes ($l=7$ and $l=9$). These values for $\delta$ were obtained by sampling the dictionaries and observing that more similarity rarely occurred and less similarity gave a too coarse partition.

Each problem instance was solved using FC and $FC_w$ finding one and all solutions. In all cases the *minimum remaining values* variable ordering heuristic was used [Haralick and Elliott, 1980]. Figure 4 summarizes the results of our experiment. We report average computational effort versus dictionary size for finding one and all problem solutions. Computational effort is given as the number of consistency checks (a consistency check occurs each time that the algorithm evaluates the consistency of assigning two words to two constrained variables). The number of consistency checks is a more appropriate measure for computational effort than the number of visited nodes, since visiting some nodes may require more computation than others. However, we observed a strong correlation between these two parameters.

Each graph in Figure 4 corresponds to a problem class and provides the information for the two tasks under consideration. It can be observed that, for the three problem classes, through the whole rank of $m$, and for the two tasks, $FC_w$ clearly outperforms FC. The gain of $FC_w$ goes up to a factor of 2.3 (up to 3.1 and 2.0 in terms of visited nodes and CPU time, respectively). The greatest gains are obtained on the hardest instances for each task. That is, at the peak if one solution is wanted, and for the largest dictionaries if all solutions are wanted.

It should be noted that $FC_w$ increases its advantage as the words length increases and the alphabet cardinality decreases. An explanation is that, for longer words, search trees are deeper and similar problems have longer common paths. Thus, it is more profitable merging similar problems. In addition, with smaller alphabets similar words are more likely to occur. Therefore, weak assignments of larger groups of words can be performed.

## 6 Related Work

The importance of values having similar behavior in a problem was stressed in [Freuder, 1991] and related to the concept of value *interchangeability*. Some forms of local interchangeability have been found especially useful. Two values are *neighborhood interchangeable* if they are inconsistent with *exactly* the same values of the rest of variables. Assigning neighborhood interchangeable values to a variable results on identical subproblems. In [Haselbock, 1993] an algorithm which merges search branches of neighborhood interchangeable values is presented. In addition, neighborhood interchangeability with respect single constraints is used to save consistency checks. Neighborhood interchangeability is a too strong condition for some real problems. For this reason weaker forms of interchangeability were defined. One value $a$ is *neighborhood substitutable* for another value $b$ if the set of values inconsistent with $a$ is a subset of the values inconsistent with $b$. Assigning neighborhood substitutable values produces subproblems one included in the other. In [Bellicha et al, 1994], algorithms exploiting neighborhood substitubility are developed. Neighborhood substitubility is still a strong condition. For instance, in the crossword domain there is not neighborhood substitubility (two different words have a completely different behaviour in, at least, one constraint). In this domain, neither Haselbock's algorithm, nor Bellicha's algorithm produce any tree reduction.

Our work extends this line of research in the sense that it does not require values to be *exactly* neighborhood interchangeable or substitutable to avoid unnecessary search. $FC_w$ would certainly be cost effective in those cases, but it can also obtain tree reductions when values have a *not so structured* similar behaviour, or when values are close to accomplish interchangeability or substitutability.

Avoiding unnecessary duplication of search has been the subject of intense research. In [Frost and Detchter, 1994], unsolvable subproblems found during search are recorded as nogoods and they are skipped if they are found again in the future. Alternatively, [Freuder and Hubbe, 1995] present a method for decomposing subproblems and propose an algorithm exploiting it. Roughly, when a subproblem is to be solved, its intersection with one of its previous siblings is computed and factored out. As in our approach, [Freuder and Hubbe, 1995] only consider similarities among sibling nodes, since they only differ in the effect of one assignment and are more likely to be similar. However, our approach is different in spirit. While their algorithm tries to detect similarities with previously unsuccessfully solved subproblems, $FC_w$ groups similar subproblems and merges them into a single problem. A performance comparison between these two methods in different domains remains as future work.

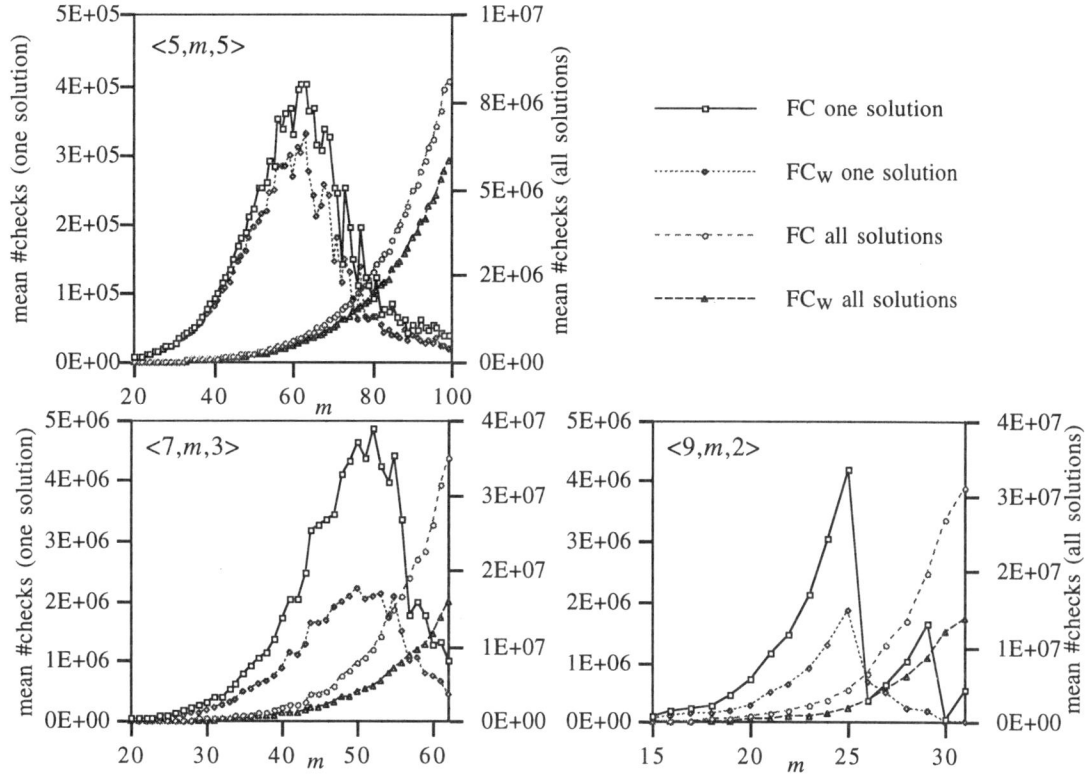

Figure 4. Experimental results on three classes of *random crossword puzzles*. Each graph reports mean computational effort (in number of consistency checks).

## 7 Conclusion

We have introduced a method, based in the concept of weak assignment, for merging constraint satisfaction subproblems. It avoids some search redundancy associated to subproblem similarities. We have incorporated this method to forward checking for the tasks of finding one and all problem solutions. Our experiments on hard random crossword problems show that forward checking can increase significantly its performance when enhanced with merging capabilities.

Our method extends previous work on value interchangeability in the sense that it requires weaker conditions for subproblem similarity. It can be combined with other CSP algorithms different from FC, keeping a higher consistency level among future variables. The study of these combinations and their applicability to other domains are open questions requiring further research.

## Acknowledgments

I am in debt to Pedro Meseguer for his help on carrying out this research. I thank Ricard Gavaldà for valuable comments on an earlier version of this paper.

## References

[Bellicha *et al.*, 1994] A. Bellicha, C. Capelle, M. Habib, T. Kökény, M. Vilarem. CSP techniques using partial orders on domain values. In *ECAI-94 Workshop on CSP issues raised by practical applications,* 47-56, 1994.

[Cheeseman *et al.*, 1991] P. Cheeseman, B. Kanefsky, W.M. Taylor. Where the really hard problems are. In *Proceedings of IJCAI-91* pages 331-337, 1991.

[Freuder, 1991] E. C. Freuder. Eliminating interchangeable values in constraint satisfaction problems. In *Proceedings of AAAI-91* pages 227-233, 1991.

[Freuder and Hubbe, 1995] Eugene C. Freuder and Paul D. Hubbe. Extracting constraint satisfaction subproblems. In *Proceedings of IJCAI-93*, pages 548-555, 1993.

[Frost and Detchter, 1994] Daniel Frost and Rina Dechter. Dead-end driven learning. In *Proceedings of AAAI-94*, pages 294-300, 1994.

[Ginsberg *et al.*, 1990] Matthew L. Ginsberg, Michael Frank, Michael P. Halpin, and Mark C. Torrance. Search lessons learned from crossword puzzles. In *Proceedings of AAAI-90*, pages 210-215, 1990.

[Haralick and Elliott, 1980] R. M. Haralick and G. J. Elliot. Increasing tree search efficiency for constraint satisfaction problems. *Artificial Intelligence*, 14:263-313, 1980.

[Haselbock, 1993] Alois Haselbock. Exploiting Interchangeabilities in constraint satisfaction problems. In *Proceedings of IJCAI-93*, pages 282-287, 1993.

# DIAGNOSIS AND QUALITATIVE REASONING

# DIAGNOSIS AND QUALITATIVE REASONING

Diagnosis 1

# Locating Faults in Tree-Structured Networks

## Christopher Leckie and Michael Dale
Telstra Research Laboratories
770 Blackburn Rd, Clayton
Victoria 3168, AUSTRALIA
Email: {c.leckie, m.dale}@trl.telstra.com.au

## Abstract

An important problem in fault diagnosis is how to locate faulty components by analysing performance measurements from those components. In this paper, we present an algorithm that uses the information-theoretic Minimum Message Length (MML) principle to locate faults in tree-structured networks. Tree-structured networks are an important application for fault diagnosis due to their use as distribution networks in telecommunication and power industries. The main advantage of using a MML approach is that we can model both the complexity of an explanation for a set of measurements, as well as the accuracy of the explanation. We demonstrate the effectiveness of this approach through an empirical evaluation on a range of simulated data.

## 1 Introduction

An important problem in fault diagnosis is how to locate a faulty component in a system by analysing a set of performance measurements from different parts of the system [Doyle 1995, El Fattah and Dechter 1995, de Kleer 1991]. An example of this type of problem is locating faults in tree-structured networks. Networks with a tree topology are widely used in applications such as telecommunications, power distribution and water reticulation, where material is to be distributed from a central site to a large number of consumers. The central site forms the root of the tree, while the consumers are located at the leaves of the tree. In such a network, a fault in one component of the network may affect all consumers connected downstream from that component. By analysing measurements of the quality of service received by each consumer we can infer the location of faults within the network. In this paper, we present an algorithm that uses the information-theoretic Minimum Message Length (MML) principle to locate faults in tree-structured networks.

The complexity of this fault diagnosis problem arises from the need to deal with continuous-valued measurements, and the uncertainty of what measurements are generated by a fault. Fault localisation is trivial when a network component fails completely, since all consumers will report a total loss of service. However, less severe faults can be more common in practice, and there are several complicating factors that arise from these faults: (1) many types of fault cause a degradation of the quality of service received by the consumers, rather than a total failure; (2) the extent of the service degradation may vary between consumers; (3) even when there is no fault, there can be random fluctuations in the quality of service due to external effects; and (4) at any time, there can be more than one fault affecting the network. Thus, we need an algorithm that can cope with these uncertainties and infer a set of faults that best explains the measurement data.

The main contribution of our work is an algorithm for locating faults in tree-structured networks using the MML principle. The main advantage of using a MML approach is that we can model both the *complexity* of an explanation for a set of measurements, as well as the *accuracy* of the explanation [Georgeff and Wallace 1984]. This means we can determine how many faults are occurring (the complexity of the explanation), in addition to where the faults have occurred (the accuracy of the explanation). In this paper we describe how the MML principle can be applied to this diagnosis problem, and provide an empirical evaluation of the benefits of this approach. We begin in the next section by describing the fault diagnosis problem in more detail. In Section 3, we provide an overview of the MML approach. In Section 4, we describe our fault localisation algorithm and how it uses the MML approach. We then provide the results of an empirical evaluation of our algorithm in Sections 5 and 6. Finally, in Section 7 we compare our approach to related work in the area of fault diagnosis.

## 2 Monitoring Tree-Structured Networks

In tree-structured networks there is one path from each node to the root of the tree. Figure 1 shows a simple example of such a network containing eight nodes. Consumers are attached to each of the leaf nodes, and the distribution of material is from the root node to the leaves, e.g., from node $n_1$ to nodes $n_3$, $n_5$, $n_6$, $n_7$ and $n_8$. In general, a fault can occur at any link or node in the network. However, since each node is fed by a single link, a fault in a link is indistinguishable from a fault in a node. Consequently, we can describe the fault diagnosis problem as a search for faults at nodes.

A fault at a node affects the distribution of material through the node, and can thus affect the quality of service

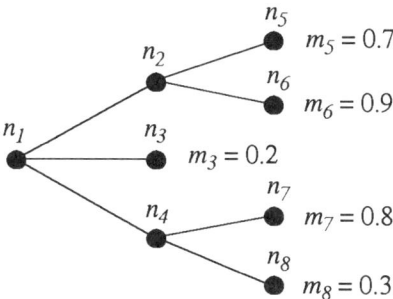

Figure 1: An Example of a Tree-Structured Network

provided to consumers downstream from that node. We say that a node $n_d$ is downstream from a node $n_u$ if there is a path from $n_u$ to $n_d$ and $n_u$ is closer to the root of the tree than $n_d$. For example, a fault at node $n_2$ in Figure 1 can affect consumers at nodes $n_5$ and $n_6$ but does not affect consumers at any other leaves. We assume that we can measure the quality of service provided to each consumer using monitoring devices at the leaves of the network. These measurements can then be reported back via a separate monitoring network such as the telephone network. We also assume that measurements are taken only at the leaves, and not at any of the internal nodes in the network, e.g., nodes $n_1$, $n_2$ or $n_4$. If measurements can be made at the internal nodes then the diagnostic task is considerably simplified. However, this is often not possible in practice, and thus we focus on the problem of locating faults using measurements only at the leaves.

In general, we can assume that an arbitrary number of different types of measurements can be simultaneously made at each leaf. Let $m_{ij}$ represent the $j$-th measurement at node $n_i$. We assume that we know the range of values that can be taken by $m_j$, i.e., the continuous interval $[m_j^L, m_j^U]$, as well as the probability density of these measurements under normal operation $f_j^N(m_{ij})$, as well as when the node is affected by a fault $f_j^F(m_{ij})$. In practice, these density functions can be compiled from historical measurements of normal and fault-affected networks.

In Figure 1 we have shown a set of measurements at the leaves of our example tree. For simplicity, we assume that each leaf node in this example has one measurement that is a real-valued number in the range [0, 1], where 0 implies perfect service and 1 implies total disruption of service. These measurements indicate that nodes $n_5$, $n_6$ and $n_7$ are experiencing a problem.

We need to determine whether these abnormal measurements have been caused by a single fault or multiple faults. For the moment, let us assume that a fault can affect all leaves downstream from the faulty node. One possible explanation for the measurements is that there are three faults in the network $\{F_5, F_6, F_7\}$, where $F_i$ represents a fault at node $n_i$. An alternative explanation is that there are two faults $\{F_2, F_7\}$, where $F_2$ causes the abnormal measurements at $n_5$ and $n_6$. Assuming faults are independent, we would assume that the second explanation is more plausible than the first since two faults are more likely than three.

However, the extent to which the measurements at the downstream leaves are affected depends on the corresponding probability distribution. This means that some leaves affected by a fault may occasionally return a satisfactory performance measurement. Consequently, another possible explanation is that there is a single fault $F_1$. This explanation is less likely since nodes $n_3$ and $n_8$ are reporting satisfactory measurements, but these measurements can still occur when there is a fault, albeit with a low probability. In terms of the number of faults this explanation is more likely since it involves only one fault, but in terms of the measurements it is less likely because of the measurements at $n_3$ and $n_8$.

The aim of this fault diagnosis problem is to find a set of faulty nodes that best explains the pattern of performance measurements at the leaves. The key issue in this problem is how to decide whether one set of possible faults provides a better explanation of the measurements than another set of faults. This decision function needs to be able to cope with the underlying randomness of the measurements, as well as the probability of having a given number of faults occur. In the following section, we describe the MML approach and its advantages for implementing such a decision function.

## 3 The Minimum Message Length Approach

The MML approach is a method that uses information theory to evaluate the effectiveness of different theories $T_i$ in explaining a given set of measurements or observations $M$ of the world. It was developed by Wallace as a method of inductive inference [Wallace and Boulton 1968, Georgeff and Wallace 1984], and has similarities to both Rissanen's Minimum Description Length technique [Rissanen 1984] and Bayesian techniques [Cheeseman et al. 1988]. In this section, we provide an introduction to the key principles of the MML approach. A detailed tutorial can be found in [Oliver and Hand 1994].

In order to assess how effectively a given theory $T_i$ explains a set of measurements $M$, we construct a message that describes both the theory, and the measurements in terms of the theory. A theory $T_1$ is said to be better than another theory $T_2$ if the total length of the message using $T_1$ is shorter than the message using $T_2$. The justification for this approach is as follows. A good theory should help explain the measurements, and hence the description of the measurements will be shorter than if no theory was used. By adding a description of the theory to the message, we can take into account the complexity of the theory. If the theory is too simplistic, then the theory will have a short description, but the length of the measurements' description will be long. As the complexity of the theory is increased to reflect any structure in the measurements, the length of the measurements' description will shrink. If the theory becomes too complex, the increased length of the theory's description will outweigh any savings in the description of the measurements.

We can encode the complete message as a binary string. By choosing an efficient coding scheme such as Huffman codes, the length of the measurements will reflect the probability of the measurements given the theory, i.e.,

$$length(M) = -\log_2 P(M \mid T_i).$$

Similarly, if we can estimate the *a priori* probability of each theory, then the length of the optimum code word for the

theory will be
$$length(T_i) = -\log_2 P(T_i).$$
Hence, the length of the combined message is
$$length(T_i + M) = -\log_2 [P(T_i) P(M \mid T_i)].$$
Note that by Bayes rule, $P(M) P(T_i \mid M) = P(T_i) P(M \mid T_i)$. Since $P(M)$ is constant for a given set of measurements, the theory that minimises the combined message length is effectively the theory that has the highest probability given the measurements. In practice, we do not need to construct the actual message since we are interested in the length of the message rather than the message itself.

The MML approach thus provides us with a method for comparing two different explanations of the same set of measurements, assuming that we have an efficient procedure for enumerating the search space of possible theories. Its key advantage is that it allows us to take into account both the effectiveness of the theory in explaining the measurements, as well as the complexity of the theory itself. This addresses the key issues of our diagnosis problem, namely, deciding how many faults are appropriate, and determining the effectiveness of these faults in explaining the measurements.

## 4 FLoc: An Algorithm for Locating Faults

In this section, we describe our fault localisation algorithm called FLoc. We define in Section 4.1 our structure for the message corresponding to a given set of measurements and possible faults. The length of this message is then used to guide our search for the best combination of faults to explain the measurements (Section 4.2). Let us begin by describing the assumptions FLoc makes about (1) the structure of the trees that it is given, (2) the types of faults that can be located, and (3) the types of measurements generated by these faults.

Trees can have an arbitrary number of nodes $N$ and an arbitrary number of children per node. We have modelled two different types of faults that can affect the nodes of a tree. The first type of fault ($F_i^A$) corresponds to a major failure at node $n_i$ that affects all nodes downstream. The second type of fault ($F_i^L$) corresponds to a failure at node $n_i$ that affects only the leaves directly connected to $n_i$. Such a fault can occur at joints where connections to customers are tapped from the main trunk. For example, if node $n_1$ in Figure 1 experienced a fault $F_1^L$, then node $n_3$ would be the only leaf affected. We denote the default situation where node $n_i$ is functioning normally as $F_i^{OK}$. The aim of the algorithm is to assign one of these three labels $\{F^A, F^L, F^{OK}\}$ to each node in the network. We assume that any node can experience either fault, although the two faults are equivalent at a leaf node. We also assume that $F^A$ and $F^L$ are equiprobable.

Each leaf has a set of $M$ measurements that are affected by the state of the nodes upstream from the leaf, and we assume that the range $[m_j^L, m_j^U]$ of each type of measurement $m_j$ is known. We also assume that two probability density functions are given for each type of measurement $m_j$. The first density function $f_j^N(.)$ describes the distribution of possible values for measurement $m_j$ when the leaf node is not affected by a fault. The second density function $f_j^F(.)$ describes the distribution of values for $m_j$ when the leaf node is affected by a fault either directly or indirectly. We assume that the distribution $f_j^F(.)$ can be used for both types of faults ($F^A$ and $F^L$).

### 4.1 Message Structure

To describe completely a given explanation for a set of measurements, we would need to describe the structure of the network, the locations of the faults, and the encoding of the observed measurements given these fault locations. In practice, the structure of the tree is fixed. This means that the tree structure can be ignored since it is a constant overhead in the total message and has no effect in minimising the message length. Consequently, we can divide our message into two parts, namely, the hypothesised faults and the observed measurements.

*Faults* - When describing a set of hypothesised faults, we can assume that the default status of each node is $F^{OK}$, i.e., it does not have a fault. This means we need to describe only those nodes where we suspect a fault has occurred. For each node where a fault is suspected, we must encode the identifier for that node and the type of fault that has occurred at the node. If there are $N$ nodes in the tree, then the length of the unique identifier for each node is $\log_2 N$ bits long. Note that since we are interested in the length of the message rather than the message itself, we can allow fractional numbers of bits. If the faulty node is a leaf, then the type of fault is redundant since $F^A$ is equivalent to $F^L$. Otherwise, we require one additional bit for each faulty node to indicate whether the fault is $F^A$ or $F^L$.

*Measurements* - When describing each of the observed measurements, we need to take into consideration whether the measurement is affected by a fault. With this knowledge we can then determine the probability of the measurement by using the appropriate density function, i.e., $f^F(.)$ or $f^N(.)$. Given that we cannot represent a real-valued measurement with infinite accuracy, we encode each type of measurement $m_j$ to a finite accuracy $\varepsilon_j$ [Wallace and Boulton 1968]. This is done by dividing the range of each type of measurement $[m_j^L, m_j^U]$ into a set of subintervals of width $\varepsilon_j$. We can then approximate a measurement $m_{ij}$ at a node $n_i$ by the value $m_{ij}^*$, which is the midpoint of the subinterval that contains $m_{ij}$. We can approximate the probability of $m_{ij}$ as

$$P(m_{ij} \mid no\ fault) \approx \varepsilon_j f_j^N(m_{ij}^*) \text{ or } P(m_{ij} \mid fault) \approx \varepsilon_j f_j^F(m_{ij}^*).$$

Thus, for each node $n_i$, the length of message describing the $j$-th measurement is either

- $-\log_2[\varepsilon_j f_j^F(m_{ij}^*)]$ if there is a fault affecting $m_{ij}$, or
- $-\log_2[\varepsilon_j f_j^N(m_{ij}^*)]$ otherwise.

### 4.2 Algorithm

For a given set of measurements, our algorithm recursively selects the most appropriate fault status for each subtree, starting at the leaves (see Figure 2). At each leaf there are two possible messages: one with a fault and one without. We select the fault status for the leaf based on the shorter of these two messages. We then work our way up the tree by selecting a fault status for each interior node $n_i$, based on the message lengths of each of the subtrees rooted at the children of $n_i$. Let us now examine how the message length at an

```
Algorithm FLoc(n_i)
{ if n_i is a leaf node
    { calculate the following messages:
        message(F_i^{OK}):
            add measurements m_{ij} using f_j^N(.);
        message(F_i^A):
            add fault F_i^A;
            add measurements m_{ij} using f_j^F(.);
        select fault label ∈ {F_i^{OK}, F_i^A}
            with the shortest corresponding message;
    }
    else
    { for each child n_c of n_i
        call FLoc(n_c);
      calculate the following messages:
        message(F_i^{OK}):
            for each child n_c of n_i
                add faults from the subtree at n_c;
                add measurements from the subtree at n_c;
        message(F_i^A):
            add fault F_i^A;
            for each child n_c of n_i
                add measurements from subtree at n_c
                    using f_j^F(.);
        message(F_i^L):
            add fault F_i^L;
            for each child n_c of n_i
                if n_c is a leaf node
                    add measurements from n_c using f_j^F(.);
                else
                    add faults from subtree at n_c;
                    add measurements from the subtree at n_c;
        select fault label ∈ {F_i^{OK}, F_i^A, F_i^L}
            with the shortest corresponding message;
        if fault label is F_i^A
            set all fault labels in the descendants of n_i to F^{OK};
    }
}
```

Figure 2: Fault Localisation Algorithm

interior node changes based on the fault status that is assigned, i.e., no fault ($F_i^{OK}$), a fault affecting all children ($F_i^A$), or a fault affecting only leaf children ($F_i^L$).

*No fault* - This is the null hypothesis. If we assume there is no fault at $n_i$ ($F_i^{OK}$), then the message length at $n_i$ is simply the sum of the message lengths from each child of $n_i$.

*Fault affecting all children* - If we assign node $n_i$ the fault status $F_i^A$, then this fault overrides all faults that were inferred at any of the descendants of $n_i$. The corresponding message at $n_i$ comprises the new fault label $F_i^A$ and the measurements from all leaves downstream from $n_i$ using the distribution $f^F(.)$. Note that if faults had been inferred at most of the children of $n_i$, then a single fault at $n_i$ is a more concise explanation of the measurements.

*Fault affecting only leaf children* - If we assign node $n_i$ the fault status $F_i^L$, then this fault overrides all faults that were inferred at any leaf children of $n_i$. The corresponding

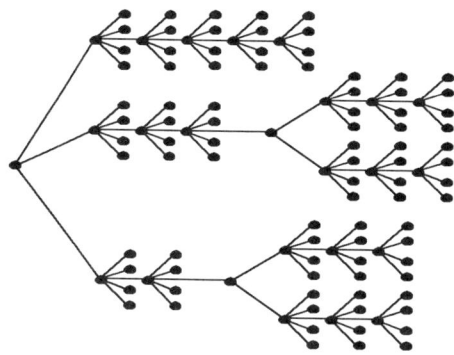

Figure 3: Test Network

message at $n_i$ comprises:
- the new fault label $F_i^L$,
- the measurements from all of the leaf children of $n_i$ using the distribution $f^F(.)$, and
- the messages from each of the non-leaf children of $n_i$.

In this way, we provide a more concise explanation for the leaf children of $n_i$ without affecting our explanations for the subtrees belonging to the non-leaf children.

Once we have calculated the lengths of these three messages, we can select the fault status for $n_i$ corresponding to the shortest message. While this fault status is locally optimal for the subtree rooted at $n_i$, it may be viewed as suboptimal at a higher level node when evidence from the siblings of $n_i$ is taken into account. For example, if we have assigned faults to nodes $n_2$ and $n_3$ in Figure 1, we may decide that a more appropriate location for the fault is node $n_1$. Assigning a fault $F_1^A$ will override the faults assigned to $n_2$ and $n_3$, and those lower-level faults are then discarded. To simplify the calculation of messages at higher level nodes, we record what the change in message length would be at each node if it is affected by a higher level fault. In this way, the algorithm does not have to backtrack, and its complexity is linear in the number of nodes in the tree.

## 5 Evaluation

### 5.1 Aims of Testing

In order to evaluate the performance of our algorithm, we have investigated how the accuracy of FLoc changes under different levels of noise. We used the network shown in Figure 3 as the basis for our tests. It contained 113 nodes, of which 88 were leaf nodes. The structure of this network reflects the types of networks we have encountered in practice. Our two control variables in these tests were the number of measurements per leaf node and the probability density of each measurement.

All leaf nodes have the same number of measurements, and all measurements have the same range and underlying probability density functions. The range of each measurement is the interval [0, 1], with 0 corresponding to normal operation, and 1 corresponding to a total disruption of service. If a node is unaffected by a fault, its measurements have a Gaussian distribution with a peak at 0 and a standard deviation $\sigma$,

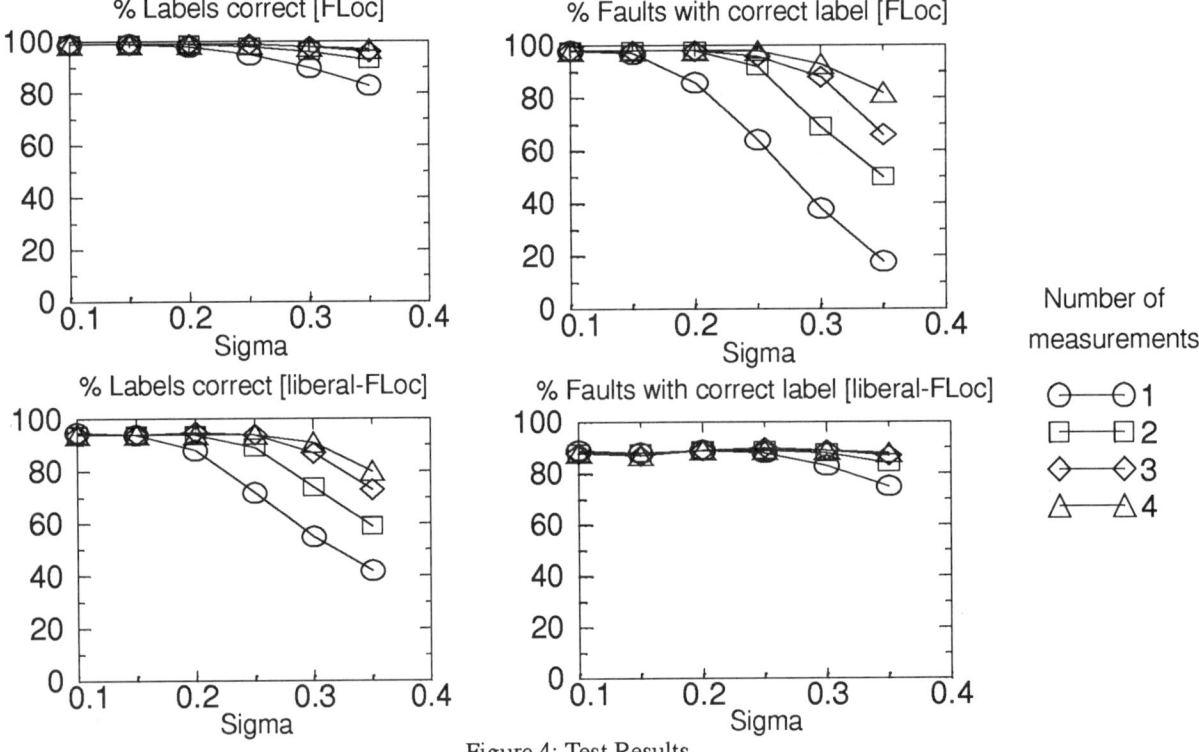

Figure 4: Test Results

$$f^N(m) = \frac{1}{\sigma\sqrt{2\pi}} e^{-m^2/(2\sigma^2)} \qquad 0 \le m \le 1.$$

The density function for fault-affected measurements is a Gaussian distribution with the peak at 1,

$$f^F(m) = \frac{1}{\sigma\sqrt{2\pi}} e^{-(1-m)^2/(2\sigma^2)} \qquad 0 \le m \le 1.$$

In our messages, we assume that measurements are encoded with an accuracy $\varepsilon = 0.1$, although the results were generally insensitive to the exact choice of $\varepsilon$.

These measurements were generated by randomly simulating faults in the test network. The probability of any node experiencing a fault of type $F^A$ was set at 0.05, and similarly for faults of type $F^L$. This allowed us to test the system with multiple simultaneous faults in the network.

By varying $\sigma$, we were able to test the sensitivity of our algorithm to noise. A small value of $\sigma$ means that normal measurements have little similarity to fault-affected measurements. In contrast, a large value of $\sigma$ means that there is a much greater overlap between the two distributions, thus making it easier to mistake a normal measurement as faulty, and vice versa. We tested FLoc under 24 different parameter settings, allowing $\sigma$ to vary from 0.1 to 0.35 and the number of measurements per node to vary from 1 to 4. We performed 10,000 trials for each combination of parameter settings, which was found to keep the uncertainty in the results to less than 1%.

To provide a basis for comparison, we also tested a less conservative version of FLoc called liberal-FLoc. This version of the algorithm did not include fault labels for leaves in the message it constructs. Instead, it selects the fault status of a leaf solely on the length of the encoding of its measurements. This makes liberal-FLoc more likely to assign faults to leaves than the null hypothesis of no fault.

**5.2 Results**

Figure 4 shows the results of these tests for the FLoc and liberal-FLoc respectively. In these graphs, we use the term "label" to describe an inferred fault at a node. In each case, we show (1) the percentage of inferred labels where the location and the type of label correctly matched the actual fault ($F^A$ or $F^L$), and (2) the percentage of faults with a correct label in the correct position. The first graph indicates how accurate are the labels reported by each algorithm, while the second graph indicates what proportion of faults are correctly detected. On average, each batch of 10,000 trials required 30 minutes of CPU time on a Sun Sparcserver 20 using an implementation of the algorithm written in Java.

## 6 Discussion

Let us begin by analysing the performance of Floc. The vast majority of labels generated by FLoc were consistently correct. When $\sigma$ was small, nearly all faults were correctly detected. As $\sigma$ increased, the accuracy of the labels was maintained, but fewer faults were detected. This was due to the increase in the number of faults with borderline measurements, especially at leaf faults. The proportion of faults detected decreased with the number of available measurements, since with fewer measurements, it is more likely that all of a fault's measurements have uncharacteristic values. Thus with noisy inputs, FLoc had a low number of errors of commission but a high number of errors of omission.

In comparison, when σ was small liberal-FLoc performed 4% worse in terms of the accuracy of its labels, and 9% worse in terms of the number of faults detected. This was mostly due to the tendency of liberal-FLoc to blame leaf faults on the parent of the leaf. This happens because liberal-FLoc is more likely to consider leaf nodes as faulty, and then combine imagined leaf faults with real leaf faults. As σ increased, liberal-FLoc manages to detect most faults, but at the price of generating a large number of labels where there is no fault. This was to be expected, since liberal-FLoc does not assign labels to faults in the messages at leaf nodes. Consequently, it is more willing to accept borderline measurements as faulty. Thus with noisy inputs, liberal-FLoc has the opposite behaviour to FLoc, i.e., a high number of errors of commission and a low number of errors of omission.

Each type of error has a different penalty. Errors of omission mean that consumers suffering from degraded service go undetected. However, these omissions are usually leaf faults that affect only a single consumer. In contrast, errors of commission mean that we are dispatching maintenance staff unnecessarily, which can be a significant expense. Hence, FLoc is clearly preferable when the cost of dispatching staff is high. One possibility is to use both algorithms and give priority to the output from FLoc, then as maintenance staff are available we can investigate the additional faults from liberal-FLoc.

## 7 Related Work

Senjen et al. [1993] have developed a hybrid approach to locating faults in tree-structured networks using a mixture of rule-based and connectionist techniques. They use preset thresholds to control the propagation of fault probabilities through the network, as well as rules that encode the relevance of each type of measurement to different faults. The main difficulty in their approach is how to select appropriate threshold levels to control the propagation of fault probabilities. By using an MML approach, we have been able to replace these heuristic thresholds with a more systematic algorithm that uses measurable distributions for the probability of each type of measurement.

El Fattah and Dechter [1995] have also addressed the problem of diagnosis for tree-structured networks. Their approach uses constraint satisfaction and optimisation techniques. However, their approach is designed for problems with a symbolic representation, e.g., boolean logic circuits. In contrast, our approach is designed for sub-symbolic domains that involve continuous and probabilistic representations. de Kleer [1991] has developed a probabilistic approach to focus the attention of a diagnostic system for boolean logic circuits. However, it not designed for domains involving continuous measurements.

Doyle [1995] uses a different approach to diagnosing continuous and probabilistic domains based on historical measurements and causal dependencies. He has developed a causal distance measure to detect and relate abnormal frequency distributions of measurements. This technique is specifically designed for time-series measurements, which we have not addressed. However, Doyle does not address the problem of multiple simultaneous faults. Thus, the two approaches are complementary, and the ability to analyse time-series measurements is an important future direction that we intend to incorporate into our work.

## 8 Conclusion and Further Work

We have developed an algorithm for locating faults in tree-structured networks, and demonstrated its effectiveness under a variety of test conditions. By using a MML approach for diagnosis, we are able to find the combination of faults that provide the best explanation for a set of continuous and probabilistic measurements. At present, our diagnosis is based on a "snapshot" of measurements at a single point in time. Our next step is to generalise this approach so that we can incorporate historical information in the form of time-series measurements into our model.

### Acknowledgements

We hereby acknowledge the permission of the Director, Telstra Research Laboratories, to publish this work.

### References

[Cheeseman et al., 1988] P. Cheeseman, M. Self, J. Kelly, J, Stutz, W. Taylor and D. Freeman. Bayesian Classification. In *Proceedings of the Seventh National Conference on Artificial Intelligence (AAAI-88)*, pp. 607-611, 1988.

[Doyle, 1995] R.J. Doyle. Determining the Loci of Anomalies Using Minimal Causal Models. In *Proceedings of the Fourteenth International Joint Conference on Artificial Intelligence (IJCAI-95)*, pp. 1821-1827, 1995.

[El Fattah and Dechter, 1995] Y. El Fattah and R. Dechter. Diagnosing Tree-Decomposable Circuits. In *Proceedings of the Fourteenth International Joint Conference on Artificial Intelligence (IJCAI-95)*, pp. 1742-1748, 1995.

[Georgeff and Wallace, 1984] M.P. Georgeff and C.S. Wallace. A General Criterion for Inductive Inference. In *Proceedings of the 6th European Conference on Artificial Intelligence (ECAI-84)*, 1984, pp. 473-482, 1984.

[de Kleer, 1991] J. de Kleer. Focusing on Probable Diagnoses. In *Proceedings of the Ninth National Conference on Artificial Intelligence (AAAI-91)*, pp. 842-848, 1991.

[Oliver and Hand, 1994] J. Oliver and D. Hand. *Introduction to Minimum Encoding Inference*. Technical Report 205, Department of Computer Science, Monash University, Clayton 3168, Australia, 1994.

[Rissanen, 1984] J. Rissanen. Universal coding, information, prediction and estimation. In *IEEE Transactions on Information Theory*, IT-30(4), pp. 629-636, 1984.

[Senjen et al., 1993] R. Senjen, M. de Beler, C. Leckie and C. Rowles. Hybrid Expert Systems for Monitoring and Fault Diagnosis. In *Proceedings of the Ninth IEEE Conference on AI for Applications*, pp. 235-241, 1993.

[Wallace and Boulton, 1968] C.S. Wallace and D.M. Boulton (1968). An Information Measure for Classification. In *Computer Journal* 11(2), pp. 185-194, 1968.

# Diagnosing Tree Structured Systems

## Markus Stumptner and Franz Wotawa
Institut für Informationssysteme
Technische Universität Wien
Paniglgasse 16, A-1040 Vienna, Austria, Europe
email: {mst,wotawa}@dbai.tuwien.ac.at

## Abstract

This paper introduces the algorithm **TREE_DIAG** for computing minimal diagnoses for tree structured systems. Diagnoses are computed by descending into the tree, enumerating the input combinations that might be reponsible for a given incorrect observation, and combining the diagnoses for the subtrees generating these inputs into diagnoses for the whole system. We prove soundness and correctness of the algorithm and show experimental results that indicate that it compares favorably to Reiter's hitting-set-based algorithm and El Fattah and Dechter's SAB. Extensions of the algorithm related to general acyclic systems, use of fault modes and the practical application to the software diagnosis domain are discussed.

*Keywords:* Model-Based Diagnosis, Algorithms

## 1 Introduction

Since the beginning of model-based diagnosis research, several attempts have been made to make model-based diagnosis of large systems feasible. This has been done by introducing probability measurements ([dK91]), by computing a component focus as in [FF92; PG93], or by using special control strategies (e.g., [SD89; GRST89]). However, fault probabilities are not available in all domains, and complete discrimination between diagnoses is also not possible in all examples.

Therefore in the past years, much effort has been spent on the development of faster diagnosis algorithms. In [FD95], an algorithm based on constraint satisfaction for tree structured diagnosis systems was introduced whose performance, as the accompanying experimental evaluation showed, compared favorably with the algorithms of Reiter [Rei87; GSW89] and de Kleer [dKW87]. A new algorithm for the diagnosis of general diagnosis systems was also presented in [Dar95] (who also presents some results on an abductive variant). It was shown that the algorithm behaves best if the system is tree structured, but without experimental results. Other diagnosis algorithms such as [AFK88] are limited to single fault diagnosis and/or to a specific data structure such as boolean algebra.

This paper presents a simple diagnosis algorithm designed for use on tree-structured systems. The algorithm description is given using standard model-based terminology and we show results based on testing them with automatically generated examples. The experiment showed that the improvement compared to Reiters diagnosis algorithm depends on the ratio between number of components and number of inputs. Since there exists an inverse relation between the number of inputs and the number of diagnoses, the performance depends also on the number of computed diagnoses. Systems with many inputs can be diagnosed faster than systems with only one input.

The paper is organized as follows. In section 2, we give the basic definitions. We present the **TREE_DIAG** algorithm in Section 3, discuss its complexity and show some experimental results in Section 4. Section 5 discusses the extension of **TREE_DIAG** to handle more general system topologies and fault modes. Finally, we summarize the results and discuss some open problems. Proofs have been omitted for brevity, see [SW96b] for details.

## 2 Basic Definitions

We define a diagnosis system as in [Rei87], except that we differentiate between components with inputs and those without inputs.

**Definition 2.1 (System)** *A system is a tuple $(SD, COMP \cup LEAF)$ where $SD$ denotes the system description, $COMP$ the set of components, and $LEAF$ the set of components not connected to the outputs of another component.*

$LEAF$ components represent the input ports of the system and we assume they always behave correctly.

A function describing the behavior in terms of input and output ports is associated with every component. We start out by assuming that every component has only one output port. The function associated with the component can be accessed using $func : COMP \mapsto FUNC$. The arity of a function is given by $arity :$

$FUNC \mapsto \mathbb{N}$. (For brevity we will write $arity(C)$ instead of $arity(func(C))$.) Input ports of $C$ are written as $in_i(C)$ and the output port as $out(C)$. Ports are assumed to have values, expressed by the function $val : PORTS \mapsto VALUES$. Because there exists only one output for every component, we use $val(C)$ instead of $val(out(C))$ for brevity. The system description uses the defined functions to express a diagnosis system. The behavior of components can be expressed by the rule

$$ok(C) \Rightarrow \bigl(val(C) = \\ func(C)(val(in_1(C)), \ldots, val(in_{arity(C)}(C)))\bigr). \quad (1)$$

Connections between ports of components are also element of the system description:

$$out(C) = in_i(C'). \quad (2)$$

Every connection allows a value propagation:

$$out(C) = in_i(C') \Rightarrow val(C) = val(in_i(C')). \quad (3)$$

The set of all components connected to input ports is denoted by $in(C) = \{C'|out(C') = in_i(C) \in SD\}$. Note that for all leaf components $L$, $in(L) = \emptyset$.

A diagnosis problem related to a given system can be stated by including a set of observations.

**Definition 2.2 (Diagnosis Problem)** *Let $(SD, COMP \cup LEAF)$ be a system and $OBS$ be observations. The tuple $(SD, COMP \cup LEAF, OBS)$ is said to be a diagnosis problem. We assume that all elements of $LEAF$ have an associated observation.*

The observations are given in terms of a function returning the observed value of the component:
$observed : COMP \cup LEAF \mapsto VALUES$.

A system is correct if the observed value is equal to the derived value for every component. We therefore introduce the rule

$$observed(C) = val(C). \quad (4)$$

Diagnoses are defined in the usual manner.

**Definition 2.3 (Diagnosis)** $\Delta$ *a subset of $COMP$ is a diagnosis for $(SD, COMP \cup LEAF, OBS)$ iff $SD \cup OBS \cup \{ok(C)| \in COMP \setminus \Delta\} \cup \{\neg ok(C)|C \in \Delta\}$ is consistent.*

Because leaf components only store input values for the system, they are not considered in $\Delta$. The set of top components $tops(SD, COMP \cup LEAF)$ includes all components that are not used as input of other components.

**Definition 2.4 (Path)** *Let $C_1, C_n$ be two components. A path connecting $C_1$ with $C_n$ is a sequence of components $[C_1, \ldots, C_n]$ where each $C_{i+1}$ is connected to an input of $C_i$.*
$in_i(C_1) = out(C_2) \in SD \Rightarrow path(C_1, C_2) = [C_1, C_2]$
$in_i(C_1) = out(C_2) \in SD \wedge path(C_2, C_n) = [C_2, \ldots, C_n]$
$\Rightarrow path(C_1, C_n) = [C_1, C_2, \ldots, C_n]$

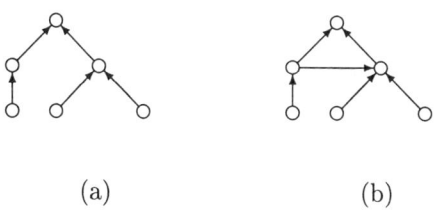

(a)          (b)

Figure 1: (a) A tree structured system, (b) an acyclic system

We can consider the diagnosis system as a directed graph whose vertices are components and whose edges are connections from an output of a component to the input of another component. We call the diagnosis system acyclic if this graph contains no directed cycles. The set of all paths from one component to another is defined by $paths(C_1, C_2) = \{p|p = path(C_1, C_2)\}$. We say a component $C'$ is connected to $C$ if there is a path from $C'$ to $C$ in the graph.

Tree structured systems have only one top component and from every other component to the top component there exists only one path. Figure 1 (a) shows a tree structured system, and (b) an acyclic one.

**Definition 2.5 (Tree Structured Systems)** *A system $(SD, COMP \cup LEAF)$ is tree structured iff $tops(SD, COMP \cup LEAF) = \{C\} \wedge$ $\forall C' \in COMP \cup LEAF : |paths(C', C)| = 1$.*

**Corollary 2.1** *Every tree structured system is acyclic.*

## 3 TREE_DIAG: An Algorithm for tree structured Systems

In this section we introduce the **TREE_DIAG** diagnosis algorithm for tree-structured systems. The basic idea is as follows. If the observed value of a component is not equal to the output of the component (as computed using the assumption that all components work correctly), there must be an incorrect component in the system. First, the component itself can behave incorrectly, i.e., it is a single fault. Secondly the component can be correct, but another component connected to the input can behave incorrectly. If this is the case, the inputs must have values leading to the observed output value. Therefore we introduce a function $INPUTS$ which computes all input values producing the observed value and call the diagnosis function recursively on the input components using those values: $INPUTS_f(val) = \{\vec{x} \in Def(f)|f(\vec{x}) = val\}$[1]. All diagnoses must be combined to give the overall diagnosis.

It is obvious that we do not need to compute all input tuples leading to the wrong output since we are only

---

[1] Note that in many cases it is possible (and necessary) to impose further restrictions on the definition, depending on the properties and domain of $f$.

interested in minimal diagnoses. We must only use the tuples which are as close as possible to the original input tuple. For example, if we have an **and** gate with the input tuple $(1,1)$ and the observed output $0$, we use the tuples $(1,0)$ and $(0,1)$ for computing diagnoses.

Before introducing the algorithm in detail, we define some auxiliary operators. For two diagnosis sets $S, S_1, S_2$, the operator $\times$ is defined as:

1. $S \times \{\} = \{\} \times S = \{\}$
2. $S_1 \times S_2 = \bigcup_{\substack{x_1 \in S_1 \\ x_2 \in S_2}} \{x_1 \cup x_2\}$

This definition can be extended to the general case, where $n$ sets should be combined.

1. $\times_{i=1}^{n} S_i = S_1 \times (\times_{j=2}^{n} S_j)$
2. $\times_{i=n}^{n} S_i = S_n$

We also want to restrict diagnosis sets to contain only diagnoses up to a given size. For a diagnosis set $S$ and $ds \in \mathbb{N}$, we write $S \rceil ds = \{D | D \in S \wedge |D| \leq ds\}$

In addition, we assume a function *checked* answering whether a component has been used for diagnosis or not. Before executing **TREE_DIAG**, *checked* answers *false* for all components. The **TREE_DIAG** algorithm uses a global variable, *diags*, to collect diagnoses as they are constructed. The variable *diags* is initialized to $\{\{\}\}$. For a given diagnosis problem $(SD, COMP, LEAF, OBS)$, the algorithm is called by TREE_DIAG($top(SD, COMP)$).

**Algorithm TREE_DIAG(C)**
*Evaluates every component, i.e., computes their output values and calls the diagnosis algorithm if the computed value is not equal to an observed value. It is assumed that observations exist for all leaf components.*
$C \ldots$ *current Component, initially the top component*
*Global constants: $SD, COMP, LEAF, OBS$*
*Global variables: diagSize (prespecified max. diagnosis size) and diags (Diagnosis set, initialized to $\{\{\}\}$)*

IF $C \in LEAF$ THEN
    $val(C) \leftarrow observed(C)$);
    $checked(C) \leftarrow true$
ELSE
    FOR $i \in \{1, \ldots, arity(C)\}$ DO
        $V_i \leftarrow$ TREE_DIAG($in_i(C)$)
    END FOR;
    $V \leftarrow func(C)(V_1, \ldots, V_{arity(C)})$;
    IF $exists(observed(C))$ AND
        $V \neq observed(C)$ THEN
        $diags \leftarrow (diags \times$
            DIAG($C, observed(C)$)) $\rceil diagSize$;
        $val(C) \leftarrow observed(C)$
    ELSE
        $val(C) \leftarrow V$
    END IF
END IF

**Algorithm DIAG(C,E)**
*Compute diagnosis by explaining differences between the computed and the observed behavior.*
$C \ldots$ *current Component*
$E \ldots$ *Observed value*
*Global constants: $SD, COMP, LEAF, OBS$*
*Global variables: diagSize (Diagnosis size) and diags (Diagnosis set)*

IF $checked(C)$ THEN
    IF $val(C) = E$ THEN
        RETURN $\{\{\}\}$
    ELSE
        RETURN $\{\}$
    END IF
ELSE
    $checked(C) \leftarrow true$
    IF $E = val(C)$ THEN
        RETURN $\{\{\}\}$
    ELSE
        $new\_diags \leftarrow \{\{C\}\}$;
        FOR $(v_1, \ldots, v_{arity(C)}) \in$ INPUTS(SD,C,E) DO
            $collected\_diags \leftarrow \{\{\}\}$;
            FOR $i \leftarrow 1$ TO $arity(C)$ DO
                $collected\_diags \leftarrow (collected\_diags \times$
                    DIAG($in_i(C), v_i$)) $\rceil diagSize$
            END FOR;
            $new\_diags \leftarrow new\_diags \cup collected\_diags$
        END FOR;
        RETURN $new\_diags$
    END IF
END IF

We explain the operation of **TREE_DIAG** using a small example. Inverters are denoted by $Ix$ and **or** gates by $Ox$. Figure 2(1) shows the initialization of the system with observed values. The input observations are propagated to the output causing a discrepancy (see Figure 2(2)). In Figure 2(3) the misbehavior is explained by assuming that $O1$ behaves incorrectly. In Figure 2(4), $O1$ is assumed to behave correctly, leading to the input vector $(0,1)$ which can be explained by assuming that $I2$ is faulty. In Figure 2(5) backward propagation is done by assuming that $I2$ behaves correctly, causing $I1$ to be incorrect.

The algorithm is sound and complete with regard to the definition of diagnosis.

**Theorem 3.1 (Termination of TREE_DIAG)** *The* **TREE_DIAG** *algorithm halts on every tree structured diagnosis system.*

**Theorem 3.2 (Soundness and Completeness)**
*The* **TREE_DIAG** *algorithm is sound wrt. the diagnosis definition, i.e., every element $\Delta \in diags$ after the algorithm has terminated is a diagnosis. Also, the algorithm computes all minimal diagnoses.*

We conclude the section with a brief discussion of the time complexity of the **TREE_DIAG** algorithm.

**Theorem 3.3 (Time Complexity of TREE_DIAG)**
*The maximum time complexity of the* **TREE_DIAG** *algorithm is of order $o(COMP \cdot 2^{|COMP|})$.*

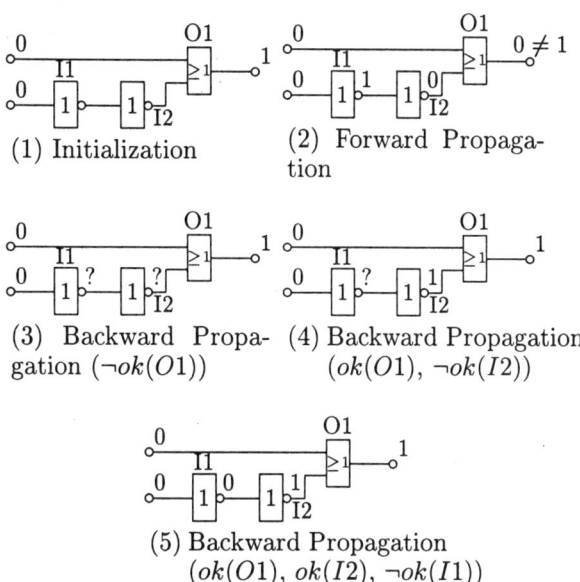

(1) Initialization

(2) Forward Propagation

(3) Backward Propagation ($\neg ok(O1)$)

(4) Backward Propagation ($ok(O1), \neg ok(I2)$)

(5) Backward Propagation ($ok(O1), ok(I2), \neg ok(I1)$)

Figure 2: How **TREE_DIAG** works

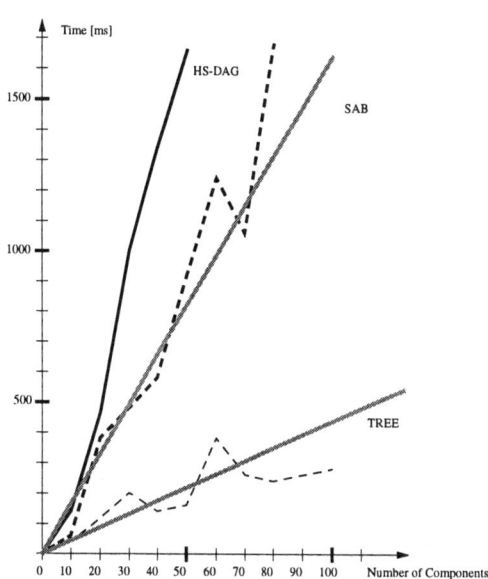

Figure 3: Performance of **TREE_DIAG**, HS-DAG, and SAB for $\frac{|COMPS|}{2}$ inputs

It is worth noting that this worst time complexity is very unlikely to occur, because we only consider minimal diagnoses, and a limit on the desired diagnosis size can be specified. Practical application of **TREE_DIAG** to the diagnosis of digital circuits has shown that the algorithm is highly efficient and can be used even for large examples (see Section 4). In the individual case, the time complexity of **TREE_DIAG** depends on the component functions, the structure of the system, and the diagnosis size.

**Theorem 3.4 (Time Complexity, Single Fault)**
*The maximum time complexity of the **TREE_DIAG** algorithm is of order $o(|COMP|)$ if only single faults are searched for.*

## 4 TREE_DIAG Implementation Results

**TREE_DIAG** was implemented in ParcPlace Visualworks/Smalltalk 2.5. The experiments were run on a SPARCStation 10-30 with 96 MB main memory. Examples were generated automatically using only **and**, **or**, and **not** gates. The number of components and number of inputs, i.e., leaf components, serve as input parameters. All listed figures are based on searching for single faults. There are no further restrictions on the internal structure of the circuit (i.e., no symmetry or constant depth of leaves). To allow a direct comparison of **TREE_DIAG** with other existing MBD algorithms, Reiter's **HS-DAG** and the SAB algorithm from [FD95] were also implemented.

Figure 3 shows the amount of diagnosis time as a function of the number of components for circuits with the number of inputs set at half the number of internal components. Results were derived by averaging over 100 different example runs. Comparing the results leads to the conclusion that **TREE_DIAG** outperforms the two other MBD algorithms.

Figure 4 shows the performance of **TREE_DIAG** and SAB algorithm for systems with up to 10,000 diagnosis components. We can see that **TREE_DIAG** has a good response time even for very large systems. The maximum diagnosis time was always less than 13 seconds.

From the evaluation of **TREE_DIAG**, it is clear that the diagnosis time not only depends on the number of components, but also varies with the number of system inputs. Figure 5 shows the runtime behavior for the number of components fixed at 100 and examples generated for varying number of inputs. It can be seen that we obtain the best performance for **TREE_DIAG** if the number of inputs is equal to the number of components. In the worst case (1 input) the diagnosis system degenerates to a row of inverters where diagnosis discrimination is not possible without using additional measurements. Note that the time scale for this figure (and Figure 4) is significantly larger than for Figure 3. The time required for finding single faults is not zero, but merely stayed clearly below 1 second in all cases.

In the case where the number of inputs is small compared with the number of components, the number of diagnoses returned is large, therefore many diagnoses are added to the diagnosis set of **TREE_DIAG** causing computational overhead. However, in summary our experiments indicate that **TREE_DIAG** always performs better than the HS-DAG and SAB algorithm.

## 5 Extending TREE_DIAG

The TREE_DIAG algorithm can be extended in a number of ways. First, we can release the restriction that

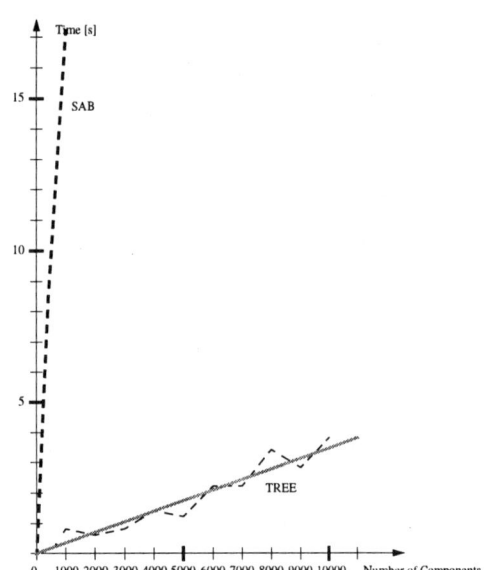

Figure 4: Performance of **TREE_DIAG** for $\frac{|COMPS|}{2}$ inputs

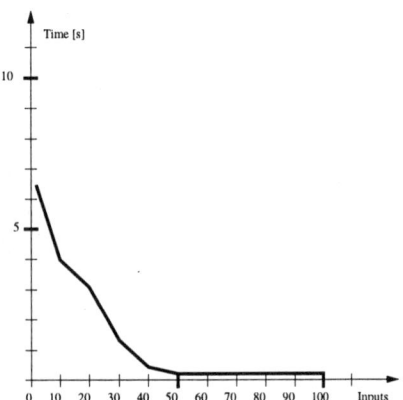

Figure 5: Influence of input number on **TREE_DIAG**, for $|COMPS| = 100$

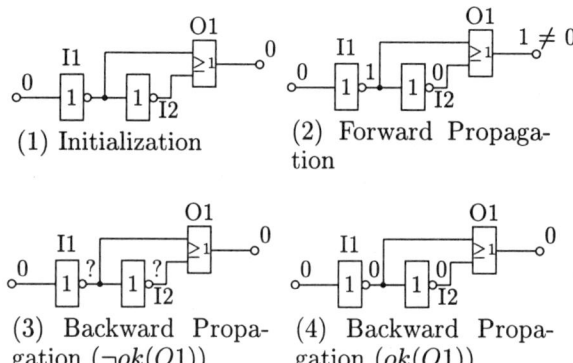

(1) Initialization

(2) Forward Propagation

(3) Backward Propagation ($\neg ok(O1)$)

(4) Backward Propagation ($ok(O1)$)

Figure 6: A simple Example explaining **DAG_DIAG**

a system may only possess one output. Second, by extending the algorithm from trees to directed acyclic systems, including systems that contain components with multiple outputs. Finally, by combining directed acyclic systems with system descriptions using fault modes.

The assumption that a system possesses only one output can be easily avoided, because every system with $n$ outputs can be transformed to a system with one output by adding a component $\hat{C}$ with one output and $n$ inputs with the following behavior:

$out(\hat{C}) = [val(in_1(\hat{C})), \ldots, val(in_n(\hat{C}))]$

$\hat{C}$ will under all circumstances be assumed to work correctly.

From now we will therefore assume that the diagnosis system has only one output. We will not present the **DAG_DIAG** algorithm in detail, but briefly discuss its basic characteristics. The structure of the algorithm is similar as for **TREE_DIAG**, except that in the general case, back propagation alone will not be sufficient. Instead, in some cases additional forward propagation must be performed and included into the final results. Consider the example of Figure 6.

Phase (1) shows the initialization of the system with observed values. The input values are propagated to the output causing a discrepancy (see (2)). In phase (3), the misbehavior is explained by assuming that $O1$ behaves incorrectly, in (4) it is explained by the assumption that $O1$ behaves correctly, but that both inputs of $O1$ must be 0. In this case, we see that the value of 0 on the output $I1$ causes a discrepancy and that this happens also if 0 is propagated through $I2$. Therefore we can derive one single fault $\{O1\}$ and one double fault $\{I1, I2\}$. Crucial to this example is that the value of 0 propagated back through $O1$ must afterwards be propagated forward through the cross connection via $I2$.

**DAG_DIAG** behaves similar to **TREE_DIAG**, except in the case where the output of a component $C$ is connected to more than one input. In this case, a new value is computed because the assumption that the component behaves correctly must be also propagated to the inputs of all components depending on $C$. If this propagation causes a discrepancy, the resulting diagnoses must be added to the diagnoses derived from back propagation.

The next extension is the handling of components with multiple outputs. Basically, a component $C$ with two outputs $out_1(C)$ and $out_2(C)$ whose values are described by two functions such that $val(out_1(C)) = func_1(val(\vec{inp}(C)))$ and $val(out_2(C)) = func_2(val(\vec{inp}(C)))$ respectively can be decomposed into two components $C'$ and $C''$, where $val(C') = func_1(val(\vec{inp}(C')))$ $val(C'') = func_2(val(\vec{inp}(C'')))$, and $inputs(C') = inputs(C'') = inputs(C)$. A diagnosis $\Delta'$ of the decomposed system is mapped into a diagnosis $\Delta$ of the original system by stating that $C \in \Delta$ if $C' \in \Delta'$ or $C'' \in \Delta'$.

To introduce fault modes, the main alteration is to the $INPUTS$ function in DIAG, which receives a fault mode assumption for the current component as an additional argument and computes the possible input vectors for that particular fault mode.

Finally, note that the SAB algorithm was also applied to cyclic systems by using a preprocessing stage based on tree clustering [FD95], and the same process can be applied for the TREE_DIAG algorithm. However, in our particular application domain, software diagnosis [SW96a], the utility of this approach will often be limited, since significant structural alterations can be expected between diagnosis runs for different versions of a program, so that almost every diagnosis run will require a separate preprocessing stage. For cases where preprocessing is not desired, the DAG_DIAG algorithm is also superior to the Reiter algorithm (see [SW96b] for experimental results). TREE_DIAG is sufficient, however, for the diagnosis of expressions in functional languages.

## 6 Conclusion

In this paper, we have presented a new diagnosis algorithm. The **TREE_DIAG** algorithm uses the tree structure of systems to subdivide the system and combine the subdiagnoses afterwards. Experimental results using automatically generated examples have shown that the new algorithm improves on the HS-DAG diagnosis algorithm and the SAB algorithm [FD95], with performance of the new algorithm (not too surprisingly) depending on the structure of the diagnosed system. In our case, a conventional system description is used, in conjunction with the requirement that a function $INPUTS$ must exist to compute inverses. There are no other requirements on $INPUTS$ (e.g., it could also be based on tables).

In addition, we have identified the number of system inputs as performance relevant factor: systems with many inputs can be diagnosed very quickly. Future work will include the implementation of the extended version of the algorithm that handles fault modes and the application of the algorithm to software debugging [SW96a].

## References

[AFK88] Magdy S. Abadir, Jack Ferguson, and Thomas E. Kirkland. Logic design verification via test generation. *IEEE Transactions on Computer-Aided Design*, 7(1), 1988.

[Dar95] Adnan Darwiche. Model-Based Diagnosis using Causal Networks. In *Proc. IJCAI*, pages 211–217, 1995.

[dK91] Johan de Kleer. Focusing on probable diagnoses. In *Proceedings AAAI*, pages 842–848, Anaheim, July 1991. Morgan Kaufmann Publishers, Inc.

[dKW87] Johan de Kleer and Brian C. Williams. Diagnosing multiple faults. *Artificial Intelligence*, 32(1):97–130, 1987.

[FD95] Yousri El Fattah and Rina Dechter. Diagnosing tree-decomposable circuits. In *Proc. IJCAI*, pages 1742–1748, 1995.

[FF92] Hartmut Freitag and Gerhard Friedrich. Focusing on independent diagnosis problems. In *Proc. KR '92*, Cambridge, MA, October 1992.

[GRST89] Massimo Gallanti, Marco Roncato, Alberto Stefanini, and Girogio Tornielli. A diagnostic algorithm based on models at different levels of abstraction. In *Proc. IJCAI*, pages 1350–1355, Detroit, August 1989. Morgan Kaufmann Publishers, Inc.

[GSW89] Russell Greiner, Barbara A. Smith, and Ralph W. Wilkerson. A correction to the algorithm in Reiter's theory of diagnosis. *Artificial Intelligence*, 41(1):79–88, 1989.

[PG93] Nooteboom P. and Leemeijer G.B. Focusing based on the structure of a model in model-based diagnosis. *International Journal of Man-Machine Studies*, 28:455–474, 1993.

[Rei87] Raymond Reiter. A theory of diagnosis from first principles. *Artificial Intelligence*, 32(1):57–95, 1987.

[SD89] Peter Struss and Oskar Dressler. Model-based Diagnosis with the Default-based Diagnosis Engine: Effective Control Strategies that Work in Practice. In *Proc. IJCAI*, pages 1318–1323, Detroit, August 1989. Morgan Kaufmann Publishers, Inc.

[SW96a] Markus Stumptner and Franz Wotawa. A model-based approach to software debugging. In *Proceedings on the Seventh International Workshop on Principles of Diagnosis*, Val Morin, Canada, 1996.

[SW96b] Markus Stumptner and Franz Wotawa. Two algorithms for model-based diagnosis. Technical Report DBAI-MBD-TR-96-08, Technische Universität Wien, September 1996.

# Event-Based Reasoning for Short Circuit Diagnosis in Power Transmission Networks

**Gianfranco Lamperti**
Dipartimento di Elettronica per l'Automazione
Università degli Studi di Brescia
via Branze 38, I-25123 Brescia, Italy
lamperti@bsing.ing.unibs.it

**Paolo Pogliano**
CISE s.p.a.
via Reggio Emilia, 39
I-20090 Segrate (Milano), Italy
paolo@sia.cise.it

## Abstract

This document introduces an original, model-based technique for short circuit diagnosis in power transmission networks. The diagnostic task is viewed as a bottom-up interpretation process of available observations. Three different interpretation layers have been defined: *local interpretation*, *global interpretation*, and *heuristic interpretation*. This paper presents the first two layers. The local interpretation focuses on the behavior of single protection components that are distributed in the power network and operate when a short circuit occur. The global interpretation provides a global behavior of the protection apparatus by combining consistent local behaviors so that given *interface constraints* among components are met. Finally, the heuristic interpretation is meant both to shrink the cardinality of the global interpretation on the basis of application-dependent *heuristic constraints*, and to eventually localize the short circuit and possibly faulty protection components within the transmission network. The implementation of the proposed technique is under way. The resulting system will be tested by ENEL, the Italian electricity board.

## 1 Introduction

To avoid service interruptions caused by short circuits in a power transmission network, it is necessary to isolate faults as soon as possible. For this reason, there are a number of protection components distributed over the network. The protection system is in charge of detecting dangerous conditions, of disconnecting a component (such as a line, a bus, a transformer or a generation group) as soon as its operating starts to be dangerous, and of keeping in operation non-faulty components as much as possible, in order to avoid a black-out. This is achieved by tripping the circuit breaker associated with each protection. Each protection has to protect mainly one component, but must also operate as a backup to other protections nearby.

When intervening, all the protections and the breakers send logical signals reporting their operation: these are recorded by an event recorder and transmitted to a *Regional Control Center* (RCC) where they are used for fault localization. Records, called *messages*, consist of a unique address of the event source, an event code, and possibly a timestamp. Operators of the RCC have to decide within one minute where the fault is located and what recovery actions have to be applied. Normally, due to time constraint reasons, only the current status of the breakers is considered, but this is not generally sufficient for fault localization. It is recognized that as a result of this an automated tool for fault diagnosis is an important requirement.

To this end, a technique based on the modeling of the protection system and on the interpretation of the reaction of the protection system to the occurrence of a short circuit has been defined. Three different interpretation layers are involved: the *local interpretation*, the *global interpretation*, and the *heuristic interpretation*. The focus of this paper is on the first two layers.

## 2 Modeling the protection system

Each protection component is modeled by means of a finite state machine (FSM), called *reactive model*, which is driven by a series of *input* events and generates some other *output* events when transitions between (*reactive*) *states* occur. For example, a distance protection is started when the impedance on the relevant line goes under a certain threshold (input event). After that, the protection is expected to change state at fixed time intervals (clock input events) and, possibly, to trip the associated breaker (tripping output event). Note that the tripping output event for the protection generates an input (tripping) event for the breaker: we say that the output event is *exported* by the protection and is *imported* by the breaker. Reactive states can be either *steady* or *unsteady*. This classification improves the semantics of reactive models and indirectly poses a number of constraints on the behavior of protection components which can be conveniently exploited by the interpretation algorithm. A *null event* $\epsilon$ is a formal notation to specify the absence of events.

Formally, a reactive model $M$ is a record of four elements: $M = (\Sigma, I, O, \delta)$, where $\Sigma$ is the set of the reactive states, $I$ is the set of input events, $O$ is the set of output events, and $\delta$ is the transition function: $\delta : \Sigma \times (I \cup \epsilon) \times (2^O \cup \epsilon) \to \Sigma$, ($2^O$ is the power-set of $O$). This means that a transition $T$ from state $S_1$ to state $S_2$ is in general triggered by an input event $i$ and generates, before changing state, the list of output events $< o_1, o_2, \ldots, o_n >$. This is denoted by $T = S_1 \xrightarrow{i \mid o_1, o_2, \ldots, o_n} S_2$. Furthermore, $\Sigma = \Sigma_s \cup \Sigma_u$, $\Sigma_s \cap \Sigma_u = \emptyset$, where $\Sigma_s$ and $\Sigma_u$ denote the set of steady and unsteady states respectively.

The specification of the reactive model of each type of protection component and the instantiation of such classes of components into a given network topology yields the whole model of the diagnosed system.

Each reactive model describes both the correct and faulty behavior of a class of protection components. Furthermore, the model allows for uncertainty due to possible loss of messages. Specifically, if during a transition $T_1 = S_1 \xrightarrow{i \mid m} S_2$, message $m$ may be lost, then the $\delta$ function will include an additional transition $T_2 = S_1 \xrightarrow{i \mid \epsilon} S_2$.

In the sequel, we assume the completeness of each reactive model, that is to say, the reactive model is assumed to describe each possible reaction.

## 3 Misbehaviors, observations, histories

When a short circuit occurs, we say there is a *misbehavior* $\mu$ of the transmission network. The set of protective components reacting to a misbehavior $\mu$ is called the *misbehavior extent* of $\mu$, and is denoted by *extent*$(\mu)$. After the specification of each reactive model, it is possible to interpret a given sequence of observations, the messages, in order to eventually find out the sequence of transitions covered by the involved protection components during $\mu$. When a short circuit occurs, each actual protection component is expected to react in a way that corresponds to an instantiation of the model, called the *local history* of the component. As such, a local history is a sequence of transitions within the reactive model. The initial and final states of the history are required to be steady. The local history is derived on the basis of the *local observation* of the component $P$, namely a list of messages relevant to $P$.

**Definition 1** (*local history*) Let $M = (\Sigma, I, O, \delta)$ be a reactive model of a protective component. A *local history* of $M$ is a (possibly empty) sequence $h = < T_1, T_2, \ldots, T_n >$ of transitions in $M$ so that $h$ conforms to the following *morphology constraints*:

1. *Determinism.* Each transition $T_i \in h$ is adorned with at most one allowed input event.

2. *Contiguity.* For each pair of contiguous transitions $T_i, T_{i+1}$ in $h$, the final state of $T_i$ coincides with the initial state of $T_{i+1}$.

3. *Stability.* Both the initial state of $T_1$ and the final state of $T_n$ are in $\Sigma_s$.

If the sequence of transitions is empty, $h$ is called a *null history*, and is denoted by $\epsilon$.

**Definition 2** (*local observation*) Let $M$ be a reactive model of a protection component $P$. A *local observation* *obs* of $P$, $obs(P) = < o_1, o_2, \ldots, o_n >$, is a sequence of temporally ordered observable output events, generated by a local history of $M$. A local observation which is empty is called a *null observation* and is denoted by $\epsilon$.

## 4 Local Interpretation

The first step towards the detection of the short circuit is represented by the local interpretation of every local observation. A local interpretation algorithm (*lia*) has been defined. The algorithm takes as input a local observation $obs(P)$ and generates a set $h^*$ of consistent local histories on the basis of the relevant reactive model. In general, a local interpretation gives rise to several consistent local histories, namely: $h^* = lia(obs(P)) = [h_1, h_2, \ldots, h_n]$.

Interestingly, a reactive model $M$ of $P$ can be formally interpreted as the grammar of a regular language $L$ where an observation *obs(P)* represents a phrase of $L$ [Hopcroft and Ullman, 1979]. Consequently, the problem of determining the local interpretation of *obs(P)* corresponds to the recognition of a string of tokens (the messages) as a phrase $f$ of $L$. What is essential in the local interpretation is to keep track of the sequence $h$ of transitions which are relevant to the recognition of $f$.

However, a model $M$ may possibly be non-deterministic, involving transitions which are not associated to any observable event. These are called *silent transitions*. Moreover, silent transitions may possibly form *silent cycles* so that $h^* = lia(obs(P))$ may potentially include an unlimited number of local histories, all of them being consistent with $obs(P)$. For example, if $h_1 = S_1 \xrightarrow{i_1 \mid m_1} S_2 \xrightarrow{i_2 \mid m_2} S_1$ is consistent with $obs(P)$, and the model of $P$ includes the silent transition $S_2 \xrightarrow{i \mid \epsilon} S_2$, then $h_2 = S_1 \xrightarrow{i_1 \mid m_1} S_2 \xrightarrow{i \mid \epsilon} S_2 \xrightarrow{i_2 \mid m_2} S_1$ is consistent with $obs(P)$ as well, since the silent transition is immaterial for the observation.

In the *lia* algorithm, a non-deterministic reactive model $M_n$ is first translated into a deterministic model

$M_d$, in which a connected subgraph of $M$ involving only silent transitions (called a *silent subgraph*) is collapsed into a new reactive state. The interpretation of $obs(P)$ is therefore performed on the basis of the new deterministic model $M_d$, instead of the original non-deterministic model $M_n$. However, once a local history $h$ is determined, for example $h = S_1 \xrightarrow{i_1 \mid m_1} S \xrightarrow{i_2 \mid m_2} S_2$, if $S$ corresponds to a collapsed silent subgraph, then the actual local history relevant to the original model $M_n$ will include a so-called *hypertransition*, namely the totality of the paths in $S$ which start at an initial state $S_i$ and end at a final state $S_f$ such that $S_1 \xrightarrow{i_1 \mid m_1} S_i$ and $S_f \xrightarrow{i_2 \mid m_2} S_2$ are legal transitions in $M_n$. A hypertransition from $S_i$ to $S_f$ is denoted by $S_i \Rightarrow S_f$. A local history involving a hypertransition is called a *hyperhistory*. In the above example, the hyperhistory will be: $S_1 \xrightarrow{i_1 \mid m_1} S_i \Rightarrow S_f \xrightarrow{i_2 \mid m_2} S_2$. A *hyperhistory graph* is the graph representation of the hyperhistory and is obtained by connecting the graph representation of the involved (hyper) transitions to each other.

## 5 Global Interpretation

Once observations of protective components reacting to a misbehavior $\mu$ are interpreted by $lia$, the set of consistent global histories must be generated.

**Definition 3** (*global history*) Let $[P_1, P_2, \ldots, P_m] = extent(\mu)$ be the set of protective components involved in a misbehavior $\mu$. A *global history* $H$ of $\mu$, $H(\mu) = (h_1, h_2, \ldots, h_m)$, is the aggregation of the local histories $h_1, h_2, \ldots, h_m$ relevant to $P_1, P_2, \ldots, P_m$ respectively.

**Definition 4** (*global history domain*) Let $[obs(P_1), obs(P_2), \ldots, obs(P_n)]$ be the set of local observations relevant to $extent(\mu)$. The *global history domain* of $\mu$, $H_d(\mu)$, is the cartesian product of the local interpretations of $P_1, P_2, \ldots, P_n$, namely: $H_d(\mu) = lia(obs(P_1)) \times lia(obs(P_2)) \times \cdots \times lia(obs(P_n))$.

**Definition 5** (*global interpretation*) The *global interpretation* of $\mu$, $H^*(\mu)$, is a relation among the local interpretations $[lia(obs(P_1)), lia(obs(P_2)), \ldots, lia(obs(P_n))]$, namely a subset of the global history domain $H_d(\mu)$: $H^*(\mu) \subseteq H_d(\mu) = lia(obs(P_1)) \times lia(obs(P_2)) \times \cdots \times lia(obs(P_n))$ so that the following conditions hold:

1. $\forall H = (h_1, h_2, \ldots, h_n) \in H^*(\mu)$, $H$ is globally consistent with respect to the interface constraints, and

2. $\neg \exists$ a globally consistent $H' \in H_d(\mu)$ such that $H' \notin H^*(\mu)$.

A systematic approach for selecting non spurious global histories is to apply a *balance* function to every pair of local histories $(h_1, h_2)$, where $h_1 \in lia(obs(P_1))$, $h_2 \in lia(obs(P_2))$, and $P_1$ is connected to $P_2$. In the simplest case in which both $h_1$ and $h_2$ are plain local histories (not hyperhistories), *balance* is a boolean function checking the consistency of $h_1$ and $h_2$ with respect to the interface constraints relevant to $h_1$ and $h_2$ only. Informally, each event of $h_1$ exported by $P_1$ to $P_2$ is required to be imported by $h_2$ and vice versa.

To this end, a *consistency matrix* is used, where the result of each application of *balance* is recorded. A consistency matrix $\Xi_\mu$ is relevant to a misbehavior $\mu$ and is composed of rows and columns associated with local interpretations relevant to $extent(\mu)$. Thus, $\Xi_\mu(h_1, h_2)$ denotes the element of the matrix which corresponds to histories $h_1$ and $h_2$. The *inconsistency set* of $\Xi_\mu$, denoted by $\Im(\Xi_\mu)$ is the set composed of those local history pairs $(h_1, h_2)$ for which $balance(h_1, h_2) = false$. On the basis of the inconsistency set we are allowed to remove the inconsistent global histories from the global history domain $H_d(\mu)$. This operation corresponds to the deletion of the global histories which include a pair of the inconsistency set. Therefore, $H \in H_d(\mu)$ is a consistent global history if and only if $\neg \exists (h_1, h_2) \in \Im(\Xi_\mu)$ so that $(h_1, h_2) \subseteq H$.

However, the trouble arises when the approach based on the consistency matrix is extended to deal with hyperhistories: in this case, *balance* is expected to return more information than a single boolean value. For example, if $h$ and $h^*$ denote a local history and a hyperhistory respectively, three different cases are possible for $balance(h, h^*)$, namely:

1. $balance(h, h^*) = \emptyset$. This is the easiest case since it establishes that every candidate global history involving both $h$ and $h^*$ is to be discarded. $\Xi_\mu(h, h^*)$ is marked with the empty set symbol.

2. $balance(h, h^*) = [h_1, h_2, \ldots, h_n]$. In that case the hyperhistory is *resolved* and a finite number of local histories is determined (possibly a singleton). $\Xi_\mu(h, h^*)$ is marked with a reference to this set.

3. $balance(h, h^*) = h'^* = [h_1, h_2, \ldots, h_n, h_{n+1}, \ldots]$, $h'^* \subseteq h^*$. The hyperhistory is *reduced* but not resolved. A reference to the hyperhistory graph of $h'^*$ is put in $\Xi_\mu(h, h^*)$.

But the real problem is how to extend the global history generation process once the $\Xi_\mu$ has been filled with this information. When all the local interpretations are composed of a limited number of local histories, we are able to enumerate all the candidate global histories $H \in H_d(\mu)$. In that case we say that we have an *extensional* representation. By contrast, the possible inclusion of hyperhistories in the local interpretation forces us to maintain a so called *intensional* representation, by means of the hyperhistory graph.

**Definition 6** (*hyper global history*) Let $H \in H_d(\mu)$ be a candidate global history. If there exists a hyperhistory $h^* \in H$, then $H$ is called a *hyper global history*.

Extending the scope of *balance* to cope with hyperhistories, requires $\Xi_\mu$ to record a more complex set of information. Specifically, the codomain of *balance* must be represented by a set $\nabla$ including elements which can be classified as follows ($h^\star$ denotes either a hyperhistory or a finite set of local histories): $\nabla = [true, false, \emptyset, h^\star, (\emptyset, \emptyset), (h^\star, \emptyset), (\emptyset, h^\star), (h_1^\star, h_2^\star)]$ which in turn can be seen as the union of three parts: $\nabla = \nabla_{h,h'} \cup \nabla_{h,h^\star} \cup \nabla_{h^\star,h'^\star}$, whereby: $\nabla_{h,h'} = [true, false]$, $\nabla_{h,h^\star} = [\emptyset, h^\star]$, and $\nabla_{h^\star,h'^\star} = [(\emptyset, \emptyset), (h^\star, \emptyset), (\emptyset, h^\star), (h_1^\star, h_2^\star)]$, corresponding respectively to the domain of two local histories, a local history and a hyperhistory, and two hyperhistories.

**Definition 7** (*resolvable hyper global history*) Let $H^*$ be a candidate hyper global history involving a number of infinite hyperhistories: $H^* = (h_1, h_2, \ldots, h_k, h_{k+1}^*, h_{k+2}^*, \ldots, h_n^*) = (h_1, h_2, \ldots, h_k) \times h_{k+1}^* \times h_{k+2}^* \times \cdots \times h_n^* = [H_1, H_2, \ldots, H_i, H_{i+1}, \ldots]$. If it is possible to enforce the interface constraints so that $H^*$ is reduced to a finite set $H^+ = [H_1', H_2', \ldots, H_p'] \subset H^*$, then $H^*$ is called a *resolvable hyper global history*.

**Definition 8** (*symbolic intersection*) Let $h_1^*$ and $h_2^*$ be two hyperhistories relevant to the same protective component $P$. The hyperhistory graph corresponding to $h^* = h_1^* \cap h_2^*$ is called the *intersection graph* of $h_1^*$ and $h_2^*$. The process of deriving the intersection graph is called *symbolic intersection*.

**Definition 9** (*residue*) Let $H^*$ be a candidate hyper global history. Let $h^* \in H^*$ be a hyperhistory embraced by $H^*$. Let $h_1^*, h_2^*, \ldots, h_n^*$ be the list of reduced hyperhistories relevant to $balance(h_i^*, h^*)$, $h_i^* \in H^*$, $h_i^* \neq h^*$. The *residue* of $h^*$ in the context of $H^*$, denoted by $\Re(h^*, H^*)$, is the symbolic intersection of $h_1^*, h_2^*, \ldots, h_n^*$, namely $\Re(h^*, H^*) = h_1^* \cap h_2^* \cap \ldots \cap h_n^*$.

### 5.1 Global Interpretation Algorithm

The *gia* global interpretation algorithm can be concisely described as a function having as input a global history domain $H_d(\mu)$ and returning a global interpretation $H^*(\mu) \subseteq H_d(\mu)$ by means of the following steps:

1. $H^*(\mu) \leftarrow H_d(\mu)$;

2. create the consistency matrix $\Xi_\mu$;

3. by applying *balance*, associate with each element of $\Xi_\mu$ a relevant instantiation of a symbol in $\nabla$;

4. build the inconsistency set $\Im(\Xi_\mu)$ as composed of the pairs of (possibly hyper) histories $(h_i, h_j)$ for which: $balance(h_i, h_j) = false \lor balance(h_i, h_j) = \emptyset \lor balance(h_i, h_j) = (\emptyset, h_{j'}) \lor balance(h_i, h_j) = (h_{i'}, \emptyset) \lor balance(h_i, h_j) = (\emptyset, \emptyset)$;

5. remove from the global interpretation those global histories $H$ including a pair $(h_i, h_j) \in \Im(\Xi_\mu)$, namely: $H^*(\mu) \leftarrow H^*(\mu) - [H \mid \exists (h_i, h_j) \subseteq H, (h_i, h_j) \in \Im(\Xi_\mu)]$;

6. for each hyperhistory $h^*$ of every hyper global history $H^* \in H^*(\mu)$, replace $h^*$ with the corresponding residue, namely: $\forall H^* \in H^*(\mu)$, $\forall h^* \in H^*$, $h^* \leftarrow \Re(h^*, H^*)$.

## 6 Comparison with related work

Several attempts to develop automated support tools for fault diagnosis in power transmission networks are reported in the literature. Somewhat surprisingly, they are based on a variety of different technological approaches including neural networks [Kezunovic *et al.*, 1994], fuzzy expert systems, [Cho *et al.*, 1994], Petri nets [Wang and Trecat, 1994], model-based diagnosis [Tornielli *et al.*, 1994; Beschta *et al.*, 1993], and temporal reasoning techniques [Baroni *et al.*, 1995].

In [Kezunovic *et al.*, 1994] a neural network is trained to recognize typical fault patterns, by processing in input the values of voltage and current measured in a given substation. Therefore this system aims mainly to provide a detailed characterization of a local fault, within the context of a specific substation and of its electrical and structural configuration, rather than to perform a diagnosis at the level of the overall network, considering the behavior of the protection system. Moreover, even within this limited context, it seems that a major difficulty of this approach is related to the huge amount of preliminary work required, since the neural network must be trained on a wide number of different cases each time it has to be applied to a different substation configuration.

In [Cho *et al.*, 1994], explicit relations between the location of a fault in the networks and the operation of protections and breakers are represented through sagittal diagrams. Temporal behavior of the components is not considered, however in order to encompass the fact that some protections should operate before others when a fault is located in a given position, a numerical label is associated with each relation, representing the possibility that each protection operates when the fault is in the considered position. Different diagnostic hypotheses can be generated starting from observations and are ranked according to their possibility value. This approach, not including an explicit model of the diagnosed system, suffers from all the well-known limitations of first-generation, rule-based, diagnostic systems (see for instance a discussion in [Beschta *et al.*, 1993]).

An alternative approach [Wang and Trecat, 1994] resorts to Petri nets in order to build a model of the protection system behavior. However this model is oversimplified: temporal aspects are not considered and the overall protection system behavior is modeled as a simple two steps activity. More detailed models are used in [Tornielli *et al.*, 1994; Beschta *et al.*, 1993], where a classical model-based diagnosis approach is adopted, using the GDE+ diagnostic engine [Struss and Dressler, 1989]. The approaches differ in the nature of the model adopted. In [Tornielli *et al.*, 1994], the behavior of network components is defined in terms of equations between the admittance values seen from different points, whereas the behavior of the components of the protection system is related to the impedance value seen by the component itself. The approach of [Beschta *et al.*, 1993] is quite similar but a qualitative evaluation of the fault distance is used instead of the admittance value in the component models. A major problem with these approaches is that the component models adopted, being quite abstract and not very detailed, do not take into account the rather complex behavior of the protection system and are not able to produce detailed diagnoses about its malfunctions.

Detailed models of component behaviors, including temporal features, are used in [Baroni *et al.*, 1995], where an original diagnostic algorithm for time-varying systems is proposed and applied to power networks. This algorithm is able to produce a detailed temporal reconstruction of the events occurred in the network, but it relies on the assumption that the timestamp of each message received is available, that is not always the case in practice. By contrast, the approach presented in this paper, while retaining the advantages of using detailed component models, does not rely on the availability of temporal information.

An approach whose rationale has some significant similarities with the work presented here has been developed in the frame of diagnosis of communication protocols [Riese, 1993]. In this context the problem consists of verifying, starting from a set of observations, if an implementation of a communication protocol is compliant with its formal definition. The protocol is modeled through a FSM and observations are matched with the possible evolution of the FSM, simulated on a discrete time scale. If none of the possible evolutions matches with the observations, the algorithm searches for modifications of the FSM that can explain the observations: each modification corresponds to a lack of compliance with respect to the standard. A detailed comparison with this work is beyond the scope of the present paper. However it can be noted that, even though the nature of the diagnostic problem is rather different, we share the use of FSM-based techniques, involving observation matching, propagation, and interpretation, in order to deal with complex diagnostic problems of time-varying systems.

This choice differs with respect to other approaches proposed in literature for diagnosis of dynamic systems, a topic that has been addressed by many researchers in recent years [Dvorak and Kuipers, 1989; Gluckenbiehl and Schäfer-Richter, 1990; Hamscher, 1991; Ng, 1991; Lackinger and Nejdl, 1991; Dressler and Freitag, 1994].

As a matter of fact, even though the importance of dynamic system diagnosis and the need of specific techniques for this problem was recognized early [Pan, 1984; Hamscher and Davis, 1984], the *classical* theory of diagnosis [Reiter, 1987; de Kleer and Williams, 1987] has been conceived for static systems only. Subsequent attempts to overcome this limitation have been, however, strongly influenced by the approach initially adopted for static systems. In fact, most of these approaches [Dvorak and Kuipers, 1989; Hamscher, 1991; Ng, 1991; Lackinger and Nejdl, 1991, Friedrich and Lackinger, 1991, Dressler and Freitag, 1994] share a common rationale: Reiter's algorithm is simply applied iteratively to subsequent instantaneous snapshots of the behavior of the system, generated through a suitable dynamic model. This method presents however some significant drawbacks, since it is unable to deal with faults whose manifestation involves several time instants and does not exploit efficiently knowledge about system dynamic behavior. An approach closer to our ideas is presented in [Gluckenbiehl and Schäfer-Richter, 1990] where the diagnostic activity for a dynamic system is seen as the task of reconstructing the history of the system starting from some temporally located information about system attributes and using a model of system behavior. However the behavioral representation adopted is made up of *if-then* rules including temporal information and suffers therefore from limited expressiveness.

## 7 Discussion and conclusions

Although featuring an original approach to model-based diagnosis of event-based dynamic systems, the proposed method relies on the availability of correct observations. More precisely, besides the intrinsic unavailability of messages, the method can cope with lost messages, but it assumes that those messages which are available are correct. The knowledge about possible missing messages must be specified in the model, as illustrated in Section 2. Intuitively, the richer is the set of observations, the smaller is the number of diagnostic alternatives produced. Furthermore, the algorithm is clearly designed for domains where discrete (and complete) models of correct and faulty behaviors are available.

Another important point concerns the resolvability of hyper global histories, which can be guaranteed by enforcing some *safety* constraints on a reactive model $M$. Intuitively, safety may be achieved when either no silent cycles are incorporated in $M$ or when, under additional assumptions, interface events are associated with silent transitions within silent cycles.

Finally, a formal analysis of the complexity of the

method and real benchmarks of the implementation are expected for the future. However it is already evident that the performance of the algorithm will strongly depend on the complexity of the reactive models exploited, and the extent to which they allow an accurate interpretation of the available observations. It is also worth pointing out that local interpretations might in principle be generated in parallel, either by a single parallel machine or by several machines installed in different substations.

## Acknowledgments

The present work is partially supported by the Commission of the European Union under the ongoing Esprit III Project 8491 *Timely* (Time-Constrained Integrated Management of Large-Scale Systems).

## References

[Baroni et al., 1995] P. Baroni, U. Canzi, and G. Guida. SHORT: a knowledge-based system for fault diagnosis in power transmission networks. In *Proc. of the Eighth Int. Symp. on Artificial Intelligence ISAI 95*, pages 199-208, Monterrey, MX, 1995.

[Beschta et al., 1993] A. Beschta, O. Dressler, H. Freitag, M. Montag, and P. Struss. A model-based approach to fault localisation in power transmission networks. *Intelligent Systems Engineering*, pages 3-14, Spring 1993.

[Cho et al., 1994] H.-J. Cho, J.-K. Park, and H.-J Lee. A fuzzy expert system for fault diagnosis of power systems. In *Proc. of the Int. Conf. on Intelligent System Application to Power Systems ISAP 94*, pages 217-226, Montpellier, F, 1994.

[Dressler and Freitag, 1994] O. Dressler and H. Freitag. Prediction sharing across time and contexts. In *Proc. of the Twelfth Nat. Conf. on Artificial Intelligence AAAI-94*, pages 1136-1141, Seattle, WA, 1994.

[Dvorak and Kuipers, 1989] D. Dvorak and B. Kuipers. Model-based monitoring of dynamic systems. In *Proc. of IJCAI-89*, pages 1238-1243, Detroit, MI, 1989.

[Friedrich and Lackinger, 1991] G. Friedrich and F. Lackinger. Diagnosing temporal misbehavior. In *Proc. of IJCAI-91*, pages 1116-1122, Sydney, Australia, 1991.

[Gluckenbiehl and Schäfer-Richter, 1990] T. Gluckenbiehl and G. Schäfer-Richter. SIDIA: Extending prediction based diagnosis to dynamic models. *Working Notes of the First Int. Workshop on Principles of Diagnosis*, pages 74-82, Stanford, CA, 1990.

[Hamscher, 1991] W. Hamscher. Modeling digital circuits for troubleshooting. *Artificial Intelligence*, 51 (1-3):223-271, 1991.

[Hamscher and Davis, 1984] W. Hamscher and R. Davis. Diagnosing circuits with state: an inherently underconstrained problem. In *Proc. of the Fourth Nat. Conf. on Artificial Intelligence AAAI-84*, pages 142-147, Austin, TX, 1984.

[Hopcroft and Ullman, 1979] J.E. Hopcroft and J.D. Ullman. *Introduction to Automata Theory, Languages and Computation*, Addison-Wesley, Reading, MA, 1979.

[de Kleer and Williams, 1987] J. de Kleer and B.C. Williams. Diagnosing multiple faults, *Artificial Intelligence*, 32(1):97-130, 1987.

[Kezunovic et al., 1994] M. Kezunovic, I. Rikalo, and D.J. Sobajic. Neural network applications to real-time and off-line fault analysis. In *Proc. of the Int. Conf. on Intelligent System Application to Power Systems ISAP 94*, pages 29-36, Montpellier, F, 1994.

[Lackinger and Nejdl, 1991] F. Lackinger and W. Nejdl. Integrating model-based monitoring and diagnosis of complex dynamic systems. In *Proc. of IJCAI-91*, pages 1123-1128, Sydney, Australia, 1991.

[Ng, 1991] H.T. Ng. Model-based, multiple-fault diagnosis of dynamic, continuous physical devices. *IEEE Expert*, 6(6):38-43, 1991.

[Pan, 1984] J. Y.-C. Pan. Qualitative reasoning with deep-level mechanism models for diagnosis of mechanism failures. In *Proc. of the First IEEE Conf. on AI Applications*, pages 295-301, Denver, CO, 1984.

[Reiter, 1987] R. Reiter. A theory of diagnosis from first principles. *Artificial Intelligence*, 32(1):57-95, 1987.

[Riese, 1993] M. Riese. Diagnosis of communicating systems: dealing with incompleteness and uncertainty. In *Proc. of IJCAI-93*, pages 1480-1485, Chambery, F, 1993.

[Struss and Dressler, 1989] P. Struss and O. Dressler. Physical negation-integrating fault models into the general diagnostic engine. In *Proc. of IJCAI-89*, pages 1318-1323, Detroit, Michigan, 1989.

[Tornielli et al., 1994] G. Tornielli, L. Capetta, S. Cermignani, A.S. Fabiano, and R. Schinco. An interval-based approach to fault diagnosis of power systems. In *Proc. of the Int. Conf. on Intelligent System Application to Power Systems ISAP 94*, pages 809-816, Montpellier, F, 1994.

[Wang and Trecat, 1994] J.P. Wang and J. Trecat. A parallel fault diagnosis expert system for one dispatch center. In *Proc. of the Int. Conf. on Intelligent System Application to Power Systems ISAP 94*, pages 201-208, Montpellier, F, 1994.

# DIAGNOSIS AND QUALITATIVE REASONING

Diagnosis 2

# Exploiting domain knowledge for approximate diagnosis

**Annette ten Teije**
SWI
University of Amsterdam
annette@swi.psy.uva.nl

**Frank van Harmelen**
Dept. of Math. and CS
Vrije Universiteit Amsterdam
frankh@cs.vu.nl

## Abstract

The AI literature contains many definitions of diagnostic reasoning most of which are defined in terms of the logical entailment relation. We use existing work on approximate entailment to define notions of approximation in diagnosis. We show how such a notion of approximate diagnosis can be exploited in various diagnostic strategies. We illustrate these strategies by performing diagnosis in a small car domain example.

## 1 Motivation

The AI literature contains many definitions of diagnostic reasoning. However, there are many reasons why we should not search for *the* appropriate definition of diagnosis, but instead search for alternative definitions, and investigate how they relate to each other. There exists a whole space of reasonable notions of diagnosis. These notions can be seen as mutual approximations.

Strategies for approximate diagnosis can be used (1) to choose another, related notion of diagnosis when one definition of diagnosis fails (e.g. too many diagnoses, no diagnosis), (2) to reduce the cost of diagnosis using an anytime algorithm, (3) to deal with incompleteness of data and knowledge (4) to find an appropriate definition suited for the purpose and circumstance of performing diagnosis. See [van Harmelen and ten Teije, 1995] for more motivation for approximations in diagnosis.

In the literature, the definition of diagnosis is usually characterised using the logical entailment relation (Sec. 2 of this paper). In this paper we use existing notions of approximate entailment [Schaerf and Cadoli, 1995] (Sec. 3) to define notions of approximation in diagnosis (Sec. 4). In Sec. 5 we give four strategies for exploiting approximations in diagnosis using approximate entailment, and we illustrate these strategies in a small car domain example. The final section discusses related work and concludes.

## 2 Definition of Diagnosis

We use a common definition of diagnosis that is widespread in the literature. We follow [Console and Torasso, 1991] and combine in our definition both abductive and consistency based diagnosis, which accounts for a large variety of diagnostic systems from the literature.

**Definition 1 (Diagnosis problem and solution)**
Given a behaviour model $BM$ (a logical theory in clausal form), and two sets of observations $O^+$ and $O^-$ (both sets of literals read as conjunctions), a solution to a diagnostic problem is a set of literals $E$ ("$E$" for explanation, again read as a conjunction), which satisfies the following:

$$\text{ABD}: \quad BM \cup E \vdash O^+ \quad (1)$$
$$\text{ABD}: \quad BM \cup E \not\vdash \bot \quad (2)$$
$$\text{CBD}: \quad BM \cup E \cup O^- \not\vdash \bot \quad (3)$$

We will write OBS for the set of all possible observables from which the letters of $O^+$ and $O^-$ must be taken, and require that $E$ is disjoint from OBS. $O^+$ is the set of observations that must be explained abductively (i.e. they must be implied by the explanation $E$), while $O^-$ only needs to be consistent with the explanation $E$.

Formulae (1) and (2) constitute the abductive part of our notion of diagnosis (ABD), and (3) the consistency based part (CBD). Although (2) directly follows from (3) for classical entailment we include both conditions explicitly, because the central idea of our method of approximations in diagnosis is to parameterise the notion of diagnosis over different approximations of the entailment relation (in particular of Schaerf & Cadoli's approximate entailment relations).

We emphasise that our particular definition of a diagnostic problem and its solution is not of crucial importance to our *central* message that approximate entailment can be usefully exploited for diagnostic reasoning to obtain interesting and efficient results.

## 3 Summarising approximate entailment

In this section we will summarise the work in [Schaerf and Cadoli, 1995], which defines the approximate entailment relations that we will exploit for our work on diagnoses. Schaerf and Cadoli define two approximations of classical entailment, named $\vdash_1$ and $\vdash_3$ which are either unsound but complete ($\vdash_1$) or sound but incomplete ($\vdash_3$). By analogy, they sometimes write $\vdash_2$ for classical entailment. Both of these approximations are parameterised over a set of predicate letters $S$ (written $\vdash_1^S$ and $\vdash_3^S$) which determines their accuracy. We repeat some of the basic definitions from [Schaerf and Cadoli, 1995]:

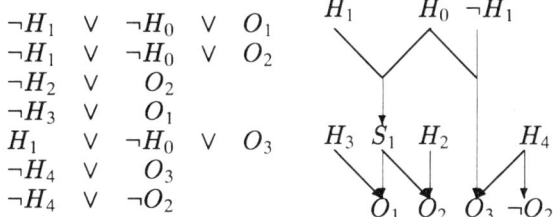

$$
\begin{aligned}
\neg H_1 &\vee \neg H_0 \vee O_1 \\
\neg H_1 &\vee \neg H_0 \vee O_2 \\
\neg H_2 &\vee O_2 \\
\neg H_3 &\vee O_1 \\
H_1 &\vee \neg H_0 \vee O_3 \\
\neg H_4 &\vee O_3 \\
\neg H_4 &\vee \neg O_2
\end{aligned}
$$

Figure 1: An example behaviour model (BM) formalised in clausal notation. In the formalisation of the network intermediate nodes (in this case only $S_1$) have been removed.

**Definition 2 (1-$S$-assignment, 3-$S$-assignment)**
A 1-$S$- and 3-$S$-assignment are defined as follows:
- If $x \in S$ then $x$ and $\neg x$ get opposite truth values
- If $x \notin S$ then
  - for a 1-$S$-assignment, $x$ and $\neg x$ both become 0.
  - for a 3-$S$-assignment, $x$ and $\neg x$ do not both become 0.

In other words: for letters in $S$, these assignments behave as classical truth assignments, while for letters $x \notin S$ they make either all literals false (1-$S$-assignments) or make one or both of $x$ and $\neg x$ true (3-$S$-assignments).

Satisfaction of a clause by a 1-$S$- or 3-$S$-assignment, and the notions of 1-$S$-entailment and 3-$S$-entailment are defined in the same way as classical satisfaction and entailment.

Intuitively, for 3-$S$-entailment the predicates outside $S$ are deemed irrelevant for deduction, while for 1-$S$-entailment these predicates are taken as false. The following syntactic notions can be used to clarify these definitions. For a theory in clausal form, 1-$S$-entailment corresponds to classical entailment, but after removing from every clause any literals with a letter outside $S$. When this results in an empty clause, the theory becomes the inconsistent theory $\bot$. Similarly, 3-$S$-entailment corresponds to classical entailment, but after removing every clause from the theory that contains a literal with a letter outside $S$. This may result in the empty theory $\top$.

The main result of [Schaerf and Cadoli, 1995] is:

**Theorem 1 (Approximate entailment)**

$$\vdash_3^\emptyset \Rightarrow \vdash_3^S \Rightarrow \vdash_3^{S'} \Rightarrow \vdash_2 \Rightarrow \vdash_1^{S'} \Rightarrow \vdash_1^S \Rightarrow \vdash_1^\emptyset$$

where $S \subseteq S'$. (Everywhere primed letters are a superset of the unprimed letters).

This states that $\vdash_3^S$ is a sound but incomplete approximation of the classical $\vdash_2$. The counterpositive of the second half of the theorem (reading $\not\vdash_1^S \Rightarrow \not\vdash_1^{S'} \Rightarrow \not\vdash_2$) states that $\not\vdash_1^S$ is a sound but incomplete approximation of $\not\vdash_2$.

**Example 1 (Illustrating $\vdash_3^S$ and $\not\vdash_1^S$)**
We illustrate these notions with figure 1. We can see that $\vdash_3^S$ is incomplete with respect to $\vdash_2$, since in the theory $BM$ of figure 1 we have that classically $BM \cup \{H_3\} \vdash_2 O_1$, but if we restrict $S$ to $\text{LET}(BM) \setminus \{H_3\}$, where $\text{LET}(BM)$ stands for all the predicate letters in $BM$, we do not have that $BM \cup \{H_3\} \vdash_3^S O_1$. This is so because taking $S = \text{LET}(BM) \setminus \{H_3\}$ amounts to removing $H_3 \rightarrow O_1$ from $BM$.

Similarly, $\not\vdash_1^S$ is incomplete with respect to $\not\vdash_2$ (or equivalently, $\vdash_1^S$ is unsound w.r.t. $\vdash_2$) since, for example, if $S = \text{LET}(BM) \setminus \{H_0\}$, then $BM \cup \{H_1\} \not\vdash_2 O_1$, but $BM \cup \{H_1\} \vdash_1^S O_1$. This is so because taking $S = \text{LET}(BM) \setminus \{H_0\}$ amounts to removing $H_0$ as a conjunct from $H_1 \wedge H_0 \rightarrow O_1$.

Furthermore, with increasing $S$, the accuracy of these approximations improves, until the approximate versions coincide with classical entailment when all letters are included in $S$.

Schaerf and Cadoli also give incremental algorithms for computing $\vdash_1^S$ and $\vdash_3^S$ when $S$ increases. They have obtained attractive complexity results which state that even when computing $\vdash_2$ through iterative computation of $\vdash_3^S$, the total cost of the iterated computation is not larger than the direct computation of $\vdash_2$ (and similarly for $\not\vdash_1^S$ to compute $\not\vdash_2$). However, the iterative computation of the approximate entailment has as important advantage that the iteration may be stopped when a confirming answer has already obtained for a smaller value of $S$. This yields a potentially drastic reduction of the computational costs. The size of these savings depend on the appropriate choice for $S$.

Although the summary above is based on a propositional calculus, the theory that we will apply the approximations to in this paper is first-order (Fig. 3). In [Schaerf and Cadoli, 1995] they show how the propositional results can be extended to the first-order case in a straightforward way.

## 4 Summarising approximate entailment in diagnosis

In this section we summarize the results [ten Teije and van Harmelen, 1996] on applying $\vdash_1^S$ and $\vdash_3^S$ in diagnosis. We use $\vdash_1^S$ and $\vdash_3^S$ in both the ABD-part of our definition of diagnosis (written as $\text{ABD}_1^S$, $\text{ABD}_3^S$) and the CBD-part (written as $\text{CBD}_1^S$, $\text{CBD}_3^S$). Since we write $\vdash_2$ for the classical entailment relation, we will also write $\text{ABD}_2$ and $\text{CBD}_2$.

The main intuitions behind using $\vdash_1^S$ and $\vdash_3^S$ in diagnosis are as follows. By using $\vdash_1^S$, candidate solutions more easily satisfy part (1) of our definition of diagnosis, because $\vdash_2 \Rightarrow \vdash_1^S$. Similarly, by using $\vdash_3^S$, candidate solutions more easily satisfy parts (2) and (3) of our definition of diagnosis, since $\not\vdash_2 \Rightarrow \not\vdash_3^S$.

We will write $\text{ABD}_i^S$ when we intend both $\text{ABD}_1^S$ and $\text{ABD}_3^S$, and similarly for $\text{CBD}_i^S$. Furthermore, we write $\mathcal{ABD}_i^S$ for the set of all diagnoses $E$ which satisfy $\text{ABD}_i^S(E)$, and similarly for $\mathcal{ABD}_2$, $\mathcal{CBD}_i^S$ and $\mathcal{CBD}_2$.

There are two important relations $\subseteqq\rightarrow$ and $\subseteqq\leftarrow$ for relating the $\text{ABD}_i^S$ and $\text{CBD}_i^S$ diagnoses.

**Definition 3** For any set of sets $P$ and $P'$, $P \subseteqq\rightarrow P'$ and $P \subseteqq\leftarrow P'$ are defined by:

$$P \subseteqq\rightarrow P' \equiv \forall p \in P \, \exists p' \in P' : p \subseteq p'$$
$$P \subseteqq\leftarrow P' \equiv \forall p' \in P' \, \exists p \in P : p \subseteq p'$$

Notice that these relations are relations between sets of sets. The required superset and subset relation is among the elements (sets) of these sets of sets.

| diagnosis definition | change of $S$ | new superset diagnosis | new subset diagnosis | nr. |
|---|---|---|---|---|
| $ABD_1^S$ | $S \to S'$ | yes | no | more |
| $ABD_1^{S'}$ | $S' \to S$ | no | only | less |
| $ABD_3^S$ | $S \to S'$ | no | yes | more |
| $ABD_3^{S'}$ | $S' \to S$ | only | no | less |
| $CBD_1^S$ | $S \to S'$ | only | no | more |
| $CBD_1^{S'}$ | $S' \to S$ | no | no | less |
| $CBD_3^S$ | $S \to S'$ | no | no | less |
| $CBD_3^{S'}$ | $S' \to S$ | no | only | more |

Figure 2: Summarising some results of using approximate entailment in the diagnosis definition [ten Teije and van Harmelen, 1996]. *"yes"* means that using the new $S$ results in superset/subset diagnoses, and similarly for *"no"*. *"only"* means that all the new computed diagnoses are superset/subset diagnoses. *more* and *less* means that the number of diagnoses increases and decreases respectively.

**Example 2 (Examples of $\Leftrightarrow$ and $\Leftarrow$ )**

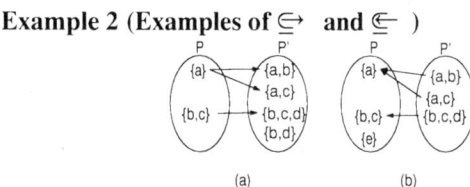

*Figure (a)*: $P \Leftrightarrow P'$, *figure (b)*: $P \not\Leftarrow P'$.

For the set of abductive diagnoses we have the following relation:

**Theorem 2 (Relations between $\mathcal{ABD}_i^S$)**

$$\emptyset = \mathcal{ABD}_1^\emptyset \subseteq \mathcal{ABD}_1^S \Leftrightarrow \mathcal{ABD}_1^{S'} \Leftrightarrow \mathcal{ABD}_2$$
$$\mathcal{ABD}_2 \Leftarrow \mathcal{ABD}_3^{S'} \Leftarrow \mathcal{ABD}_3^S \Leftarrow \mathcal{ABD}_3^\emptyset = \emptyset$$

This states that $ABD_1^S$ diagnoses consist of parts of $ABD_2$ diagnoses, and that $ABD_3^S$ diagnoses contain $ABD_2$ diagnoses. Another result is on the number of diagnoses:

**Theorem 3 (Sizes of $\mathcal{ABD}_i^S$)**

$$0 = |\mathcal{ABD}_1^\emptyset| \leq |\mathcal{ABD}_1^S| \leq |\mathcal{ABD}_1^{S'}| \leq |\mathcal{ABD}_2|$$
$$|\mathcal{ABD}_2| \geq |\mathcal{ABD}_3^{S'}| \geq |\mathcal{ABD}_3^S| \geq |\mathcal{ABD}_3^\emptyset| = 0$$

We have analogous results for the CBD-part:

**Theorem 4 (Relations between $CBD_i^S$)**

$$\emptyset = \mathcal{CBD}_1^\emptyset \subset \mathcal{CBD}_1^S \Leftrightarrow \mathcal{CBD}_2 \Leftrightarrow \mathcal{CBD}_3^S \Leftrightarrow \mathcal{CBD}_3^\emptyset = \mathcal{E}$$
$$\emptyset = \mathcal{CBD}_1^\emptyset \subset \mathcal{CBD}_1^S \subset \mathcal{CBD}_2 \subset \mathcal{CBD}_3^S \subset \mathcal{CBD}_3^\emptyset = \mathcal{E}$$
$$0 = |\mathcal{CBD}_1^\emptyset| < |\mathcal{CBD}_1^S| < |\mathcal{CBD}_2| < |\mathcal{CBD}_3^S| < |\mathcal{CBD}_3^\emptyset|$$

where $\mathcal{E}$ stands for any consistent set of literals whose letters are taken from $LET(BM) \setminus OBS$. The first sequence of inclusions states that $CBD_1^S$ diagnoses consist of parts of classical CBD-diagnoses, and that every classical diagnosis is contained in at least one $CBD_3^S$ diagnosis. In [ten Teije and van Harmelen, 1996] we also have theorems about the type (superset or subset) of new diagnoses that can be found by changing $S$. *"New"* means that using the new value of $S$, we compute at least one superset/subset diagnosis which was not present for the old value of $S$. These theorems are summarized in Fig. 2.

# 5 Strategies for approximate diagnosis

## 5.1 General strategies

We can use approximate diagnosis (results of Sec. 3 and 4) for solving problems of too many, too few, too large and too small diagnoses. When such a problem occurs with a particular notion of diagnosis, we could choose another related notion. In our case of using Schaerf & Cadoli's approximate entailment relation, this means changing the parameter $S$.

In this section we only consider increasing $S$, because this allows us to use the incremental algorithm of Schaerf & Cadoli. They show that the total cost of the iterated computation is not larger than directly computing classical entailment. However, the iterative computation of the approximate entailment may be stopped when a satisfactory answer is already obtained for a smaller value of $S$.

Focusing on increasing $S$ results in the following general strategies:

**Solutions for the problem of too few diagnoses:**
- Shifting from $ABD_1^S$ to $ABD_1^{S'}$
- Shifting from $ABD_3^S$ to $ABD_3^{S'}$ (see strategy III)
- Shifting from $CBD_1^S$ to $CBD_1^{S'}$ (see strategy II).

**Solutions for the problem of too many diagnoses**
- Shifting from $CBD_3^S$ to $CBD_3^{S'}$ (see strategy I)

**Solutions for the problem of too small diagnoses**
- Shifting from $ABD_1^S$ to $ABD_1^{S'}$ (see strategy IV)
- Shifting from $CBD_1^S$ to $CBD_1^{S'}$

**Solution for the problem of too large diagnoses**
- Shifting from $ABD_3^S$ to $ABD_3^{S'}$

These strategies are general in the sense that they do not exploit specific properties of the behaviour model. Using such properties would be more attractive because this enables us to be more precise about how we could extend $S$ and determine the characteristics of the diagnoses that will be computed. In Sec. 5.3 we give such specific strategies.

## 5.2 Example Behaviour Model (BM)

The example that we use for our strategies is taken from [Dupré, 1994] and is shown in Fig. 3. This figure shows a partial causal model of a car. The causal network contains 42 nodes and 40 causal links. We transform the causal network of Fig. 3 to an equivalent two layered network, because we use the results of [ten Teije and van Harmelen, 1996], and some of them are restricted to a two layered network.

In causal networks [Console and Torasso, 1990], states are represented as predicates, and the fact that state $S_i$ necessarily causes state $S_j$ is represented as $S_i \to S_j$. The fact that $S_i$ *possibly* causes state $S_j$ can be written as $S_i \wedge \alpha_{ij} \to S_j$, where $\alpha_{ij}$ (called the incompleteness assumption) is interpreted as the unknown condition required for $S_i$ to cause $S_j$.

The letters of the two layered version of the network from Fig. 3 are the initial causes (written as the set $\mathcal{H}$), the incompleteness assumptions (the set $\mathcal{A}$), and the observables (the set $\mathcal{O}$).

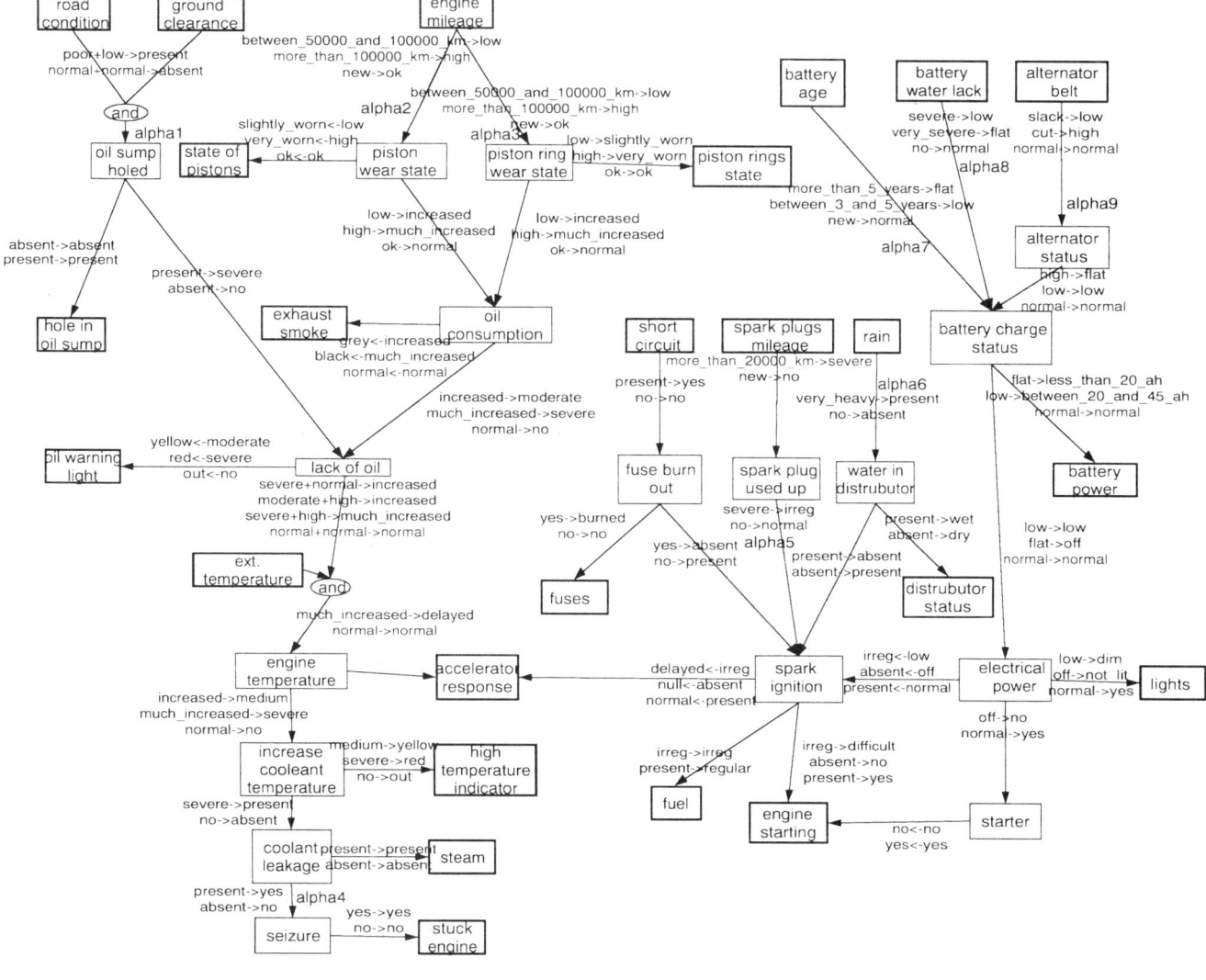

Figure 3: Behaviour model of a car from [Dupré, 1994]. The bold lined boxes are initial causes and observables. For example, the top-right-most causal link corresponds to the formulae: $alternatorbelt(slack) \wedge \alpha_9 \rightarrow alternatorstatus(low)$, $alternatorbelt(cut) \wedge \alpha_9 \rightarrow alternatorstatus(high)$ $alternatorbelt(normal) \wedge \alpha_9 \rightarrow alternatorstatus(normal)$

## 5.3 Strategies Dependent on properties of BM

In this section we give four strategies of approximations in diagnosis which depend on properties of the behaviour model.

Strategy (I) and (II) use the specificity of observables, strategy (III) and (IV) use the necessity of the causal relations. Besides using different properties of the behaviour model, the strategies are distinct examples of approximating diagnoses. Strategy (I) and (II) are examples of changing the CBD-part. Strategy (III) and (IV) are examples of changing the ABD-part. Also, they are strategies to deal with different types of problems with the diagnoses: strategy (I) reduces the number of diagnoses, strategy (II) and (IV) increase the number of diagnoses, strategy (III) increases the size of the diagnoses.

**Using Specificity of Observations**

Strategy (I) and (II) are based on the specificity of observables. We use a whole spectrum from specific to a-specific observables. We call an observable more specific if it has fewer possible causes. For the car example this spectrum is shown in Fig. 4.

**Strategy (I).** This strategy can solve the problem of "too many diagnoses". The strategy is based on the idea that an explanation cannot be a diagnosis if one of its specific observables is not observed. We approximate this by using the spectrum from the specific to the a-specific end. We apply definition 1 using $CBD_3^S$, $ABD_2$, and the initial value of $S$ ($S_{init}$) is only all the possible causes and the incompleteness assumptions, and none of the observables. Because the consistency part of the ABD-formula (2) implies $CBD_3^{S_{init}}$, any ABD-explanation will be a diagnosis. Extending $S$ with the most specific observables means that a cause of such an observable can only remain part of the diagnosis when its specific observables are observed. We extend $S$ with increasingly less specific observables until one diagnosis is left or no diagnosis is left. In the latter case we consider the diagnoses of the previous step as

| most specific | | | | least specific |
|---|---|---|---|---|
| 1 poss. cause ($Spec_1$) | 2 poss. causes ($Spec_2$) | 3 poss. causes ($Spec_3$) | 6 poss. causes ($Spec_4$) | 9 poss. causes ($Spec_5$) |
| hole in oil sump state of pistons piston ring state fuses distributor status | exhaust smoke | oil warning light battery power lights high temp.indic. steam stuck engine | fuel engine starting | accel. response |

Figure 4: Specificity of the observables in the car example. This spectrum is from specific observables (1 possible cause) to a-specific observables (9 possible causes).

the best diagnoses.

**Example 3 (Strategy (I))** *(In the examples we give only the subset minimal diagnoses.)* We illustrate this strategy using the following values:

$$O^+ = \{oil\ warning\ light(red)\}$$
$$O^- = \{\neg exhaust\ smoke(black),$$
$$\neg accelerator\ response(delayed)\}$$
$$S_{init} = \mathcal{H} \cup \mathcal{A}$$

| $S$ | $Diagnoses$ |
|---|---|
| $S_{init}$ | $\{engine\ mileage(> 100000km), \alpha_2\},$ |
| | $\{engine\ mileage(> 100000km), \alpha_3\},$ |
| | $\{road\ cond(poor),\ gr.\ clearance(low), \alpha_1\}$ |
| $+Spec_1$ | $\{engine\ mileage(> 100000km), \alpha_2\},$ |
| | $\{engine\ mileage(> 100000km), \alpha_3\},$ |
| | $\{road\ cond(poor),\ gr.\ clearance(low), \alpha_1\}$ |
| $+Spec_2$ | $\{road\ cond(poor),\ gr.\ clearance(low), \alpha_1\}$ |

For $S_{init}$ each ABD-explanation is $\text{CBD}_3^S$-consistent with $O^-$, because $S$ contains no observables. We do not introduce inconsistency by extending $S$ with $Spec_1$, in contrast with extending $S$ further with $Spec_2$. This extension introduces inconsistency between $\neg exhaust\ smoke(black)$ and each of the explanations $\{engine\ mileage(> 100000km), \alpha_2\}$ and $\{engine\ mileage(> 100000km), \alpha_3\}$. We stop at $S_{init}+Spec_1+Spec_2$ because we have just one diagnosis left.

Our final explanation is $\text{CBD}_3^S$-consistent with the negative observation $\neg$ accelerator response(delayed) *(for $S = S_{init} + Spec_1 + Spec_2$)*. Under $\text{CBD}_2$ it would have been inconsistent with this observation, but we consider this is of less importance because of the low specificity of this observable.

**Strategy (II).** This strategy can solve the problem of "too few diagnoses". The strategy is based on the idea that a cause has to be excluded if a specific observable is not observed. We apply definition 1 using $\text{CBD}_1^S$ and $\text{ABD}_2$. If an observable is not in $S$, then no possible cause of this observable can be part of the diagnosis. We allow non-observed observables to be implied by the diagnosis only if these additional observables are non-specific. Therefore the strategy uses the spectrum of specificity of observables from a-specific to specific.

We initialise $S$ with all possible causes, plus $O^+$ and $O^-$. Excluding $O^+$ from $S$ would result in inconsistency of all the explanations, because each possible cause of an $O^+$ observable would be excluded by $\text{CBD}_1^S$. Excluding $O^-$ would mean inconsistency with $O^-$ and $BM$. We increasingly extend $S$ with more specific observables. This introduces possibly more diagnoses, because more causes will be enabled by the observables in $S$.

**Example 4 (Strategy (II))** We illustrate this strategy using the following values:

$$O^+ = \{fuel(irreg)\}$$
$$O^- = \{\neg fuses(burned)\}$$
$$S_{init} = \mathcal{H} \cup \mathcal{A} \cup O^+ \cup O^-$$

| $S$ | $Diagnoses$ |
|---|---|
| $S_{init}$ | none |
| $+Spec_5$ | none |
| $+Spec_4$ | $\{spark\ plugs\ mileage(> 200000km), \alpha_5\}$ |
| $+Spec_3$ | $\{spark\ plugs\ mileage(> 200000km), \alpha_5\},$ |
| | $\{battery\ age(between\ 3\ and\ 5\ years), \alpha_7\},$ |
| | $\{battery\ water\ lack(severe), \alpha_8\},$ |
| | $\{alternator\ belt(slack), \alpha_9\}$ |

$S_{init}$ gives no diagnoses because every cause still has an observable which is not in $S_{init}$. $S_{init}$ extended with $Spec_5$ and $Spec_4$ results in the first diagnosis. All observables of the causes spark plugs mileage and $\alpha_5$ are in $S$. If we continue extending $S$ more diagnoses could be found. This example illustrates that one would prefer the diagnosis whose observables are observed (consistent) or (if they are not observed) not very specific.

**Using Necessity of Causal Relations**

Strategies (III) and (IV) are based on the necessity of the causal relation.

Our behaviour model contains necessary causal relations ($S_i \rightarrow S_j$) and possible causal relations ($S_i \wedge \alpha_{ij} \rightarrow S_j$). However, for our strategies we will use a whole spectrum of the necessity of the causal relation by dividing the incompleteness assumptions $\alpha_i$ in several groups and ordering them. This ordering is meant to indicate the degree of necessity of the causal relation. This ordering in the car domain is as follows:

| Necessary ($Poss_0$) | ($Poss_1$) | ($Poss_2$) | Possible ($Poss_3$) |
|---|---|---|---|
| No $\alpha_i$ in | $\alpha_5$ | $\alpha_2, \alpha_3, \alpha_4$ | $\alpha_1$ |
| in causal link | $\alpha_9$ | $\alpha_7, \alpha_8$ | $\alpha_6$ |

Note that this spectrum is domain specific knowledge, whereas the specificity of observables can be determined syntactically.

**Strategy (III).** This strategy can solve the problem of "too few diagnoses". The strategy is based on the idea of taking as little notice as possible of non-necessary relations. We use the spectrum from the necessary-side to the possible-side. This amounts to first completely ignoring the possible relations and introducing them increasingly starting from the most necessary ones. We apply definition 1 using $ABD_3^S$, $CBD_2$, and the initial value of $S$ is all symbols of $BM$ without the $\alpha_i$. This results in only diagnoses which use necessary causal relations. If we extend $S$ with $\alpha_i$, we also use less necessary causal relations.

**Example 5 (Strategy (III))** *We illustrate this strategy using the following values:*

$$O^+ = \{battery\ power(between\ 20\ and\ 45\ ah)\}$$
$$O^- = \emptyset$$
$$S_{init} = \mathcal{H} \cup \mathcal{O}$$

| $S$ | $Diagnoses$ |
|---|---|
| $S_{init}$ | $none$ |
| $+Poss_1$ | $\{alternator\ belt(slack), \alpha_9\}$ |
| $+Poss_2$ | $\{battery\ age(between\ 3\ and\ 5\ years), \alpha_7\}$, |
|  | $\{battery\ water\ lack(severe), \alpha_8\}$, |
|  | $\{alternator\ belt(slack), \alpha_9\}$ |

*In the first step only necessary relations are used, and no diagnosis is found. Extending $S$ with the incompleteness assumptions with the highest degree of necessity ($Poss_1$) enables the use of the causal relations with $\alpha_5$ and $\alpha_9$ giving the diagnosis* {alternator belt(slack),$\alpha_9$}. *Adding more incompleteness assumptions allows us to use more possible relations and results in two extra diagnoses.*

**Strategy (IV).** This strategy can solve the problem of "too small diagnoses". The strategy is based on the idea to start with paying no attention to the necessity of relations, i.e. to use all relations as necessary. All diagnoses will be without the incompleteness assumptions even though they might be using non-necessary links. These diagnoses can later be extended using the necessity of the causal relations. We apply definition 1 using $ABD_1^S$, $CBD_2$, and the initial value of $S$ is all letters of $BM$ excluding the $\alpha_i$. Extending $S$ with those $\alpha_i$ which are in the spectrum at the *possible* end, results in detailed diagnoses if such relations are used in the explanation. This means that those diagnoses with the most unreliable causal links will be the first to get extended with the appropriate $\alpha_i$.

We do not illustrate this strategy because of lack of space.

## 6 Discussion

The main message of this paper is that we apply approximation strategies to diagnosis, and that the approximation strategies are informed by particular properties of the domain knowledge. The strategies (I)–(IV) all deal with problems concerning the size and number of the diagnoses. However, our approximation techniques can also be used to model many of the general focusing strategies described in the literature. For instance, by dividing the behaviour model in sub-models, and by choosing for $S$ only the letters of a particular sub-model, we effectively obtain a focusing strategy based on the use of these sub-models. Other focusing strategies can be dealt with in the same way.

A first obvious task for future work would be to apply our proposed algorithms to a more realistic application. A candidate for this could be a domain where explanations can be ordered based on their urgency. According to theorem 2, using $ABD_1^S$ and initialising $S$ with the most urgent explanation candidates, and adding gradually less urgent candidates computes only urgent subsets of classical diagnosis, and computes only non-urgent abductive diagnoses when $S$ is increased if resources allows. This yields an anytime algorithm that performs well for urgent diagnoses under time pressure. A second obvious task is to study the efficiency behaviour of our approximation algorithms in larger behaviour models then we have presented here.

Finally, our approach to diagnostic strategies should be compared with other approaches, in particular [Böttcher and Dressler, 1994], based on [Struss, 1992]. In this approach, so called "working hypotheses" indicate various restrictions on, or preferences for potential diagnoses. The set of active working hypotheses is then modified to switch from one set of diagnoses to another. We claim that defining $S$ as the set of all letters from the behaviour model plus the active working hypotheses would yield an alternative formalisation.

## References

[Böttcher and Dressler, 1994] C. Böttcher and O. Dressler. A framework for controlling model-based diagnosis systems with multiple actions. *Annals of Mathematics and Artificial Intelligence*, 11(1-4), 1994.

[Console and Torasso, 1990] L. Console and P. Torasso. Hypothetical reasoning in causal models. *Int. J. of Intelligent Systems*, 5(1):83–124, 1990.

[Console and Torasso, 1991] L. Console and P. Torasso. A spectrum of logical definitions of model-based diagnosis. *Computational Intelligence*, 7(3):133–141, 1991.

[Dupré, 1994] D.Theseider Dupré. *Characterizing and Mechanizing Abductive Reasoning*. PhD thesis, Universita di Torino, 1994.

[Schaerf and Cadoli, 1995] M. Schaerf and M. Cadoli. Tractable reasoning via approximation. *Artificial Intelligence*, 74(2):249–310, April 1995.

[Struss, 1992] P. Struss. What's in SD? Towards a theory of modeling for diagnosis. In L. Console, J.H. de Kleer, and W.C. Hamscher, editors, *Readings in Model-based Diagnosis*, pages 419–449. Morgan Kaufmann, 1992.

[ten Teije and van Harmelen, 1996] A. ten Teije and F. van Harmelen. Computing approximate diagnoses by using approximate entailment. In *Proc. of the Int. Conf. on Principles of Knowledge Representation and Reasoning (KR'96)*, Boston, Massachusetts, November 1996.

[van Harmelen and ten Teije, 1995] F. van Harmelen and A. ten Teije. Approximations in diagnosis: motivations and techniques. In A. Levy and P. Nayak, editors, *Proc. of SARA-95, Symposium on Abstraction, Reformulation, and Approximation*, pages 149–155, Quebec, Aug 1995.

# Semantically Guided Theorem Proving for Diagnosis Applications

**Peter Baumgartner**
Univ. Koblenz
Inst. f. Informatik
peter@infko.uni-koblenz.de

**Peter Fröhlich**
Universität Hannover
froehlich@kbs.uni-hannover.de

**Ulrich Furbach**
Univ. Koblenz,
Inst. f. Informatik
uli@infko.uni-koblenz.de

**Wolfgang Nejdl**
Universität Hannover
nejdl@kbs.uni-hannover.de

## Abstract

In this paper we demonstrate how general purpose automated theorem proving techniques can be used to solve realistic model-based diagnosis problems. For this we modify a model generating tableau calculus such that a model of a correctly behaving device can be used to guide the search for minimal diagnoses. Our experiments show that our general approach is competitive with specialized diagnosis systems.

## 1 Introduction

In this paper we will demonstrate that model generation theorem proving is very well suited for solving consistency-based diagnosis tasks. More precisely, we want to emphasize two aspects:

i) Theorem proving techniques are very well applicable to realistic diagnosis problems, as they are contained in diagnosis benchmark suites.

ii) Semantic information from a specific domain, can be used to significantly improve performance of a theorem prover.

According to Reiter ([10]) a simulation model of the technical device under consideration is constructed and is used to predict its normal behavior. By comparing this prediction with the actual behavior it is possible to derive a diagnosis.

This work was motivated by the study of the diagnosis system DRUM-2 [5; 9]. The basic idea is to start with an initial model of a correctly functioning device. This model is revised, whenever the actual observations differ from the predicted behavior.

We will use a proof procedure, which is an implementation of the hyper tableaux calculus presented in [1]. We adapt the idea from DRUM-2 to this tableaux calculus, which yields *semantic hyper tableaux*. The resulting system approximates the efficiency of the DRUM-2 system.

The use of semantics within theorem proving procedures have been proposed before. There is the well-known concept of semantic resolution ([3]) and, more recently, there are approaches by Plaisted ([4] and Ganzinger ([6]). Plaisted is arguing strongly for the need of giving semantic information for controlling the generation of clauses in his instance-based proof procedures, like hyper-linking. Ganzinger and his coworkers are presenting an approach where orderings are used to construct models of clause sets. Indeed, they even relate their approach to SATCHMO-like theorem proving, which is an instance of the hyper tableau calculus. However, the semantics have to be given by orderings or alternatively, by Horn subsets of the set of clauses. In cases where the initially given semantics is not compatible with orderings or is not expressible by Horn subsets it is unclear how to proceed. We will show, that in the case of diagnosis an initial semantics, which is naturally given by a model of the correct behavior of the device under consideration, can improve performance significantly. Our proof procedure does not impose any restrictions on these initial models.

We assume that the reader is familiar with the basic concepts of propositional logic. *Clauses*, i.e. multisets of literals, are usually written as the disjunction $A_1 \vee \cdots \vee A_m \vee \neg B_1 \vee \cdots \vee \neg B_n$ or as an implication $A_1, \ldots, A_m \leftarrow B_1, \ldots, B_n$ ($m \geq 0$, $n \geq 0$). With $\overline{L}$ we denote the complement of a literal $L$. Two literals $L$ and $K$ are *complementary* if $\overline{L} = K$.

## 2 Model–Based Diagnosis

In model-based diagnosis [10] a simulation model of the device under consideration is used to predict its normal behavior, given the observed input parameters. This approach uses an logical description of the device, called the system description ($SD$), formalized by a set of first–order formulas. The system description consists of a set of axioms characterizing the behavior of system components of certain types. The topology is modeled separately by a set of facts. The diagnostic problem is described by system description $SD$, a set $COMP$ of components and a set $OBS$ of observations (logical facts). With each component we associate a behavioral mode: $Mode(c, Ok)$ means that component $c$ is behaving correctly, while $Mode(c, Ab)$ (abbreviated by $Ab(c)$) denotes that $c$ is faulty.

**Definition 2.1 (Reiter 87)** A *Diagnosis* of $(SD, COMP, OBS)$ is a set $\Delta \subseteq COMP$, such that $SD \cup OBS \cup \{Mode(c, Ab) | c \in \Delta\} \cup \{\neg Mode(c, Ab) | c \in COMP - \Delta\}$ is consistent. $\Delta$ is called a *Minimal Diagnosis*, iff it is the minimal set (wrt. $\subseteq$) with this property. □

The set of all minimal diagnoses can be large for complex technical devices. Therefore, stronger criteria than minimality are often used to further discriminate among the minimal diagnoses. These criteria are usually based on the probability or cardinality of diagnoses. In the remainder of this paper we will use restrictions on the cardinality of diagnoses. We say that a diagnosis satisfies the *n-fault assumption* iff $|\Delta| \leq n$. Other minimality criteria which prefer e.g. more probable diagnoses are conceivable as well, but are not treated in the present paper.

A widely used example of the $n$-fault assumption is the 1-fault assumption or *Single Fault Assumption*. Many specialized systems for technical diagnosis have the Single Fault Assumption implicit and are unable to handle multiple faults.

In model–based diagnosis systems the Single Fault Assumption can be activated explicitly in order to provide more discrimination among diagnoses and to speed up the diagnosis process.

As a running example consider the simple digital circuit on the right consisting of an or–gate (*or1*) and two inverters (*inv1* and *inv2*). The system description *SD* is given by the following propositional clauses:[1].

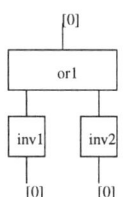

OR1:  $ab(or1), high(or1, i1), high(or1, i2) \leftarrow high(or1, o)$
      $ab(or1), high(or1, o) \leftarrow high(or1, i1)$
      $ab(or1), high(or1, o) \leftarrow high(or1, i2)$

INV1: $ab(inv1) \leftarrow high(inv1, o), high(inv1, i)$
      $ab(inv1), high(inv1, i), high(inv1, o) \leftarrow$

INV2: $ab(inv2) \leftarrow high(inv2, o), high(inv2, i)$
      $ab(inv2), high(inv2, i), high(inv2, o) \leftarrow$

$high(or1, i1) \leftarrow high(inv1, o)$    $high(or1, i2) \leftarrow high(inv2, o)$
$high(inv1, o) \leftarrow high(or1, i1)$    $high(inv2, o) \leftarrow high(or1, i2)$

We observe that both inputs of the circuit have low voltage and the output also has low voltage, i.e. the clause set of *OBS* is given by $\{\leftarrow high(inv1, i), \leftarrow high(inv2, i), \leftarrow high(or1, o)\}$.

The expected behavior of the circuit given that both inputs are low would be high voltage at the outputs of both inverters and consequently high voltage at the output of the or–gate. This model of the correctly functioning device, namely $I_0 = \{high(inv1, o), high(inv2, o), high(or1, i1), high(or1, i2), high(or1, o)\}$, can be computed very efficiently even for large devices by domain–specific tools, e.g. circuit simulators.

---

[1]These formulas can be obtained by instantiating a first order description of the gate functions with the structural information. For instance, the clauses for the OR1 gate stem from the formula $\forall OR\_gate : (\neg ab(OR\_gate) \to (high(OR\_gate, o) \leftrightarrow (high(OR\_gate, i1) \lor high(OR\_gate, i2))))$

## 3 Hyper Tableaux Calculus

In [1] a variant of clausal normal form tableaux called "hyper tableaux" is introduced. We briefly recall the ground version here.

From now on $S$ always denotes a finite ground clause set, and $\Sigma$ denotes its signature, i.e. the set of all predicate symbols occurring in it. We consider finite ordered trees $T$ where the nodes, except the root node, are labeled with literals. In the following we will represent a branch $b$ in $T$ by the sequence $b = L_1, \ldots, L_n$ ($n \geq 0$) of its literal labels, where $L_1$ labels an immediate successor of the root node, and $L_n$ labels the leaf of $b$. The branch $b$ is called *regular* iff $L_i \neq L_j$ for $1 \leq i, j \leq n$ and $i \neq j$, otherwise it is called *irregular*. The tree $T$ is *regular* iff every of its branches is regular, otherwise it is *irregular*. The set of *branch literals* of $b$ is $lit(b) = \{L_1, \ldots L_n\}$. For brevity, we will write expressions like $A \in b$ instead of $A \in lit(b)$. In order to memorize the fact that a branch contains a contradiction, we allow to label a branch as either *open* or *closed*. A tableau is *closed* if each of its branches is closed, otherwise it is *open*.

**Definition 3.1 (Hyper tableau)** A literal set is called *inconsistent* iff it contains a pair of complementary literals, otherwise it is called *consistent*. Hyper tableaux for $S$ are inductively defined as follows:

**Initialization step:** The empty tree, consisting of the root node only, is a hyper tableau for $S$. Its single branch is marked as "open".

**Hyper extension step:** If (1) $T$ is an open hyper tableau for $S$ with open branch $b$, and (2) $C = A_1, \ldots, A_m \leftarrow B_1, \ldots, B_n$ is a clause from $S$ ($m \geq 0$, $n \geq 0$), called *extending clause* in this context, and (3) $\{B_1, \ldots, B_n\} \subseteq b$ (equivalently, we say that $C$ is *applicable to* $b$) then the tree $T'$ is a hyper tableau for $S$, where $T'$ is obtained from $T$ by *extension of $b$ by $C$*: replace $b$ in $T$ by the *new* branches

$$(b, A_1) \ldots, (b, A_m), (b, \neg B_1) \ldots, (b, \neg B_n)$$

and then mark every inconsistent new branch as "closed", and the other new branches as "open". We say that a branch $b$ is *finished* iff it is either closed, or else whenever $C$ is applicable to $b$, then extension of $b$ by $C$ yields some irregular new branch. □

The applicability condition of an extension expresses that *all* body literals have to be satisfied by the branch to be extended (like in hyper *resolution*). From now on we consider only regular hyper tableaux. This restriction guarantees that for finite clause sets no branch can be extended infinitely often.

**Definition 3.2 (Branch Semantics)** As usual, we represent an interpretation $I$ for given domain $\Sigma$ as the set $\{A \in \Sigma \mid I(A) = true, A \text{ atom}\}$. *Minimality* of interpretations is defined via set-inclusion.

Given a tableau with consistent branch $b$. The branch $b$ is mapped to the interpretation $[\![b]\!]_\Sigma := \text{lit}(b)^+$, where $\text{lit}(b)^+ = \{A \in \text{lit}(b) \mid A \text{ is a positive literal}\}$. Usually, we write $[\![b]\!]$ instead of $[\![b]\!]_\Sigma$ and let $\Sigma$ be given by the context. □

Figure 1 contains a hyper tableau for the clause set from our running example. Each open branch for the clause set $SD \cup OBS$ corresponds to a partial model. The highlighted model can be understood as an attempt to construct a model for the whole clause set, without assuming unnecessary $ab$-predicates. Only for making the clauses from $OR1$ true is it necessary to include $ab(or1)$ into the model. $high(or1,o)$ cannot be assumed, as this contradicts the observation $\leftarrow high(or1,o)$.

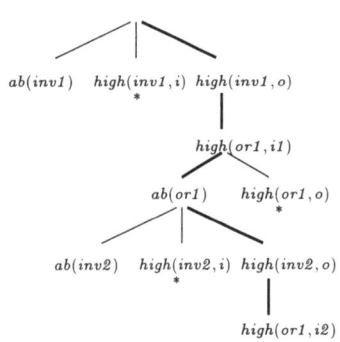

Figure 1: Hyper tableau.

A refutational completeness result for hyper tableaux was given in [1]. For our purposes of computing diagnosis (i.e. models), however, we need a (stronger) model completeness result:

**Theorem 3.3 (Model Completeness of Hyper Tableaux.)** Let $T$ be a hyper tableau for $S$ such that every open branch is finished. Then, for every minimal model $I$ of $S$ there is an open branch $b$ in $T$ such that $I = [\![b]\!]$.

This theorem enables us to compute in particular minimal diagnosis by simply collecting all $Ab$-literals along $[\![b]\!]$, because every minimal diagnosis must be contained in some minimal model of $S$.

### 3.1 Lessons from the Specialized Diagnosis System DRUM-2

In order to make the generation of models efficient enough for the large benchmark circuits used in this paper, additional knowledge has to be used to guide this model generation. This is done by starting from a model of the correct behavior of the device $SD$ and revising the model only where necessary. This idea of a "semantically guided" model generation has been introduced first in the DRUM-2 system [5; 9]. The basic idea of DRUM-2 is to start with a model of the correct behavior of the device under consideration, i.e. with an interpretation $I_0$, such that $I_0 \models SD \cup \{\neg Ab(c) | c \in COMP\}$. Then the system description $SD$ is augmented by an observation of abnormal behavior $OBS$, such that the assumption that all components are working correctly is no longer valid. Thus, $I_0$ is no model of $SD \cup OBS$, however it is used to guide the search for models of $SD \cup OBS$.

Note, that the initial model $I_0$ usually already takes into account a part of the observations which correspond to the inputs of the device under consideration. These input values are needed to simulate the correct behavior of the device. In our example, the initial model reflects the fact that the inputs of the inverters both have low voltage ($I_o$ was given at the end of Section 2).

## 4 Formalizing the Diagnosis Task with Semantic Hyper Tableaux

In this section we discuss how to incorporate initial interpretations into the calculus. Our first technique by *cuts* should be understood as the semantics of the approach; an efficient implementation by a *compilation technique* is presented afterwards.

### 4.1 Initial Interpretations via Cuts

The use of an initial interpretation can be approximated in the hyper tableau calculus by the introduction of an additional inference rule, the *atomic cut rule*.

**Definition 4.1** The inference rule **Atomic cut (with atom $A$)** is given by: if

$T$ is an open hyper tableau for $S$ with open branch $b$,

then the literal tree $T'$ is a hyper tableau for $S$, where $T'$ is obtained from $T$ by extension of $b$ by $A \vee \neg A$ (cf. Def. 3.1). □

Note that in regular tableaux it cannot occur that a cut with atom $A$ is applied, if either $A$ or $\neg A$ is contained on the branch. As a consequence it is impossible to use the "same cut" twice on a branch.

We approximate initial interpretations by applying atomic cuts at the beginning of each hyper tableau:

**Definition 4.2** An *initial tableau* for an interpretation $I_0$ is given by a regular tableau which is constructed by a applying atomic cuts with atoms from $I_0$ as long as possible. □

The branches of an initial tableau for an interpretation $I_0$ consist obviously of all interpretations with atoms from $I_0$. A part of the initial tableau for the initial interpretation $I_0$ given at the end of Section 2 is depicted in Figure 2. Note that the highlighted branch corresponds to the highlighted part in Figure 1 (a negative literal

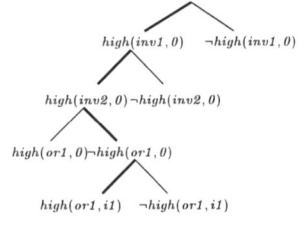

Figure 2: Initial tableau.

in an initial tableaux, such as $\neg high(or1,o)$, is represented implicitly in a hyper tableau as the *absence* of the complementary positive literal). If this branch is extended in successive hyper extension steps, the diagnosis $ab(or1)$, which was contained in the model from Figure 1 can be derived as well.

**Definition 4.3 (Semantic Hyper Tableau – SHT)** A *semantic hyper tableau* for $I_0$ and $S$ is a hyper tableau which is generated according to Definition 3.1, except that the empty tableau in the initialization step is replaced by an initial tableau for $I_0$. □

It is easy to derive an open tableau starting from the initial tableau for $I_0$ in Figure 2, such that it contains the model from Figure 1.

**Proposition 4.4 (Model Completeness of SHT)** Let $T$ be a semantic hyper tableau for $I_0$ and $S$, such that every open branch is finished. Then, for every minimal model $I$ of $S$ there is an open branch $b$ in $T$ such that $I = [\![b]\!]$.

### 4.2 Initial Interpretations via Renaming

The just defined "semantical" account for initial interpretations via cut is unsuited for realistic examples. This is, because all the $2^n - 1$ possible deviations from the initial interpretation will have to be investigated as well. Hence, in this section we introduce a compilation technique which implements the deviation from the initial interpretation only by need.

Assume we have an initial interpretation $I_0 = \{a\}$ and a clause set which contains $b \leftarrow$ and $c \leftarrow a \wedge b$. By the only applicable atomic cut we get the initial tableau with two branches, namely $\{a\}$ and $\{\neg a\}$. The first branch can be extended twice by an hyper extension step, yielding $\{a, b, c\}$. The second branch can be extended towards $\{\neg a, b\}$. No more extension step is applicable to this tableau. Let $T_{cut}$ be this tableau.

Let us now transform the clause set with respect to $I_0$, such that every atom from $I_0$ occurring in a clause is shifted to the other side of the $\leftarrow$ symbol and complemented. In our example we get the clause $c \vee \neg a \leftarrow b$; the fact $b \leftarrow$ remains, because $b$ is not in $I_0$. Using $b \leftarrow$ we construct a tableau consisting of the single branch $\{b\}$, which can be extended in an successive hyper step by using the renamed clause. We get a tableau consisting of two branches $\{b, c\}$ and $\{b, \neg a\}$. Let $T$ be that tableau. Now, let us interpret a branch in $T$ as usual, except that we set an atom from $I_0$ to *true* if its negation is not contained in the branch. Under this interpretation the branch $\{b, c\}$ in $T$ corresponds to the usual interpretation of $\{a, b, c\}$ in $T_{cut}$. Likewise, the second branch $\{b, \neg a\}$ in $T_{cut}$ corresponds to the second model in $T$.

Note that by this renaming we get tableaux where atoms from $I_0$ occur only *negatively* on open branches; such cases just mean deviations from $I_0$. In contrast to the cut approach, these deviations are now brought into the tableau by need.

The following definition introduces the just described idea formally. Since we want to avoid unnecessary changes to the hyper calculus, a new predicate name $neg\_A$ instead of $\neg A$ will be used.

**Definition 4.5 (I-transformation)** Let $C = L_1 \vee \cdots \vee L_n$ be a clause and $I$ be a set of atoms. The *I-transformation* of $C$ is the clause obtained from $C$ by replacing every positive literal $A$ with $A \in I$ by $\neg neg\_A$, and by replacing every negative literal $\neg A$ with $A \in I$ by $neg\_A$. The *I-transformation* of $S$, written as $S^I$, is defined as the $I$-transformation of every clause in $S$. □

It is easy to see that every $I$-transformation preserves the models of a clause set, in the sense that every model for the non-transformed clause set constitutes a model for the transformed clause set by setting $neg\_A$ to *true* iff $A$ is *false*, for every $A \in I$, and keeping the truth values for atoms outside of $I$. More formally we have:

**Proposition 4.6 (Model Preservation of I-transformation)** For every interpretation $J$: $J \models S$ iff $rename_I(J) \models S^I$, where $rename_I(J)(neg\_A) = \overline{J(A)}$ iff $A \in I$, and else $rename_I(J)(A) = J(A)$.

As explained informally above, the branch semantics of tableaux derived from a renamed, i.e. $I$-transformed clause set, is changed to assign *true* to every atom from $I$, unless its negation is on the branch. This is a formal definition:

$$[\![b]\!]^I = (I \setminus \{A \mid neg\_A \in \mathrm{lit}(b)\}) \cup \\ (\mathrm{lit}(b) \setminus \{neg\_A \mid neg\_A \in \mathrm{lit}(b)\})$$

The connection of semantic hyper tableaux to hyper tableaux and renaming is given by the next theorem.

**Theorem 4.7** Let $T$ be a semantic hyper tableau for $S$ and $I$ where every open branch is finished; let $T^I$ be a hyper tableau for the $I$-transformation of $S$ where every open branch is finished. Then, for every open branch $b^I$ in $T^I$ there is an open branch $b$ in $T$ such $[\![b^I]\!]^I = [\![b]\!]$. The converse does not hold.

The theorem tells us that with the renamed clause set we compute some deviation of the initial interpretation. The value of the theorem comes from the fact that the converse does not hold in general. That is, not every possible deviation is examined by naive enumeration of all combinations.

In order to see that the converse does not hold, take e.g. $S = \{a \leftarrow\}$ and $I_0 = \{b\}$. There is only one semantic hyper tableau of the stated form, namely the one with the two branches $\{b, a\}$ and $\{\neg b, a\}$. On the other side, the $I_0$-transformation leaves $S$ untouched, and thus the sole hyper tableau for $S$ consists of the single branch $\{a\}$ with semantics $[\![\{a\}]\!]^{I_0} = \{a, b\}$. However, the semantics of the branch $\{\neg b, a\}$ in the former tableau is different. The set of models which are computed in $T^I$ can be characterized by an ordering, $\geq_{I_0}$, which takes into account the deviation from the initial interpretation. If $b_1$ and $b_2$ are branches in a semantic hyper tableau for $S$ and $I$, $b_1 \geq_{I_0} b_2$ iff $[\![b_1]\!] \cap I \supseteq [\![b_2]\!] \cap I$ and $[\![b_1]\!] \setminus I = [\![b_2]\!] \setminus I$.

**Theorem 4.8** Let $T$ and $T^I$ be given as in Theorem 4.7. Then, for every open $\geq_{I_0}$-maximal branch $b$ of $T$ there is an open branch $b^I$ in $T^I$, such that $[\![b^I]\!]^I = [\![b]\!]$.

To conclude, the semantic hyper tableau approach serves us as a tool for *understanding* the effects of initial interpretations. For *efficient computing* we rely on the next theorem:

**Theorem 4.9 (Minimal Diagnosis Completeness)** Let $I_0$ be an interpretation such that $I_0 \cap \{Ab(c) \mid c \in COMP\} = \emptyset$, and $T$ be a hyper tableau for $S^{I_0}$ such that every open branch is finished. Then, for every minimal diagnosis $\Delta \subseteq COMP$ there is an open branch $b$ in $T$ such that

$$\Delta = \{c \in COMP \mid \mathit{rename}_{I_0}([\![b]\!]) \models Ab(c)\}$$
$$= \{c \in COMP \mid Ab(c) \in b\} \ .$$

**Proof.** (Sketch) We need Theorem 3.3 and Proposition 4.6: suppose $\Delta$ is a minimal diagnosis. Consider all atom sets $M$ such that $M \cup \{Ab(c) \mid c \in \Delta\} \models S$. As a consequence of the model correspondence expressed in Proposition 4.6, thereby using the facts that $I_0$ does not contain $Ab$-literals and that $\Delta$ is a minimal diagnosis, we can find an $M$ such that $J := \mathit{rename}_{I_0}(M \cup \{Ab(c) \mid c \in \Delta\})$ is a minimal model for $S^{I_0}$. Hence, by Theorem 3.3 this model, and in particular this diagnosis $\Delta$, will be computed along some finished open branch $b$. Q.E.D.

## 5 Implementation and Experiments

We have implemented a proof procedure for the hyper tableaux calculus of [1], modified it slightly for our diagnosis task, and applied it to some benchmark examples from the diagnosis literature.

**The Basic Proof Procedure.** A basic proof procedure for the plain hyper tableaux calculus for the propositional case is very simple, and coincides with e.g. SATCHMO [8]. Initially let $T$ be a tableau consisting of the root node only. Let $T$ be the tableau constructed so far. *Main loop:* if $T$ is closed, stop with "unsatisfiable". Otherwise select an open branch $b$ from $T$ (branch selection) which is not labeled as "finished" and select a clause $H \leftarrow B$ (extension clause selection) from the input clause set such that $B \subseteq b$ (applicability) and $H \cap b = \{\}$ (regularity check). If no such clause exists, $b$ is labeled as "finished" and $[\![b]\!]$ is a model for the input clause set. In particular, the set of literals on $b$ with predicate symbol $Ab$ (simply called $Ab$-literals) constitutes a (not necessarily minimal) diagnosis. If every open branch is labeled as "finished" then stop, otherwise enter the main loop again.

In the diagnosis task it is often demanded to compute every (minimal) diagnosis. Hence the proof procedure does not stop after the first open branch is found, but only marks it as "finished" and enters the main loop again.

**Adaption for the Diagnosis Task.** While incorporation of the initial interpretation is treated by renaming predicates in the input clause set (Section 4), and thus requires no modification of the prover, implementing the minimality restriction is dealt with by the following new inference rule: "any branch containing $n+1$ (due to regularity necessarily pairwise different) $Ab$-literals is closed immediately".

Notice that this inference rule has the same effect as if the $\binom{|COMP|}{n+1}$ clauses $\leftarrow ab(C_1), \ldots, ab(C_{n+1})$ (for $C_i \in COMP$, $C_i \neq C_j$, where $1 \leq i, j \leq n+1$ and $i \neq j$) specifying the $n$-faults assumption would be added to the input clause set. Since even for the smallest example (c499) and the 1-fault assumption the clause set would blow up from 1600 to 60000 clauses, the inference rule solution is mandatory.

**Computing Minimal Diagnoses** From Theorem 4.9 we know that for each minimal diagnosis $\Delta$ the hyper tableau contains an open finished branch. However, there are also branches corresponding to non–minimal extensions of $Ab$. Since the number of non–minimal diagnoses is exponential in the number of components of the device, our goal is to avoid extension of branches which correspond to non–minimal diagnoses. This can be achieved by a combined iterative deepening/lemma technique. It can be basically realized by a simple outer loop around the just described proof procedure. The outer loop includes a counter $N = 0, 1, 2, \ldots$, which stands for the cardinality of the diagnosis computed in the inner loop. The *invariant* for the outer loop is the following: *all minimal diagnosis with cardinality $\leq N - 1$ have been computed*, say it is the set $\Delta_{N-1}$, and every such diagnosis $\{c_1, \ldots, c_n\} \in \Delta_{N-1}$ has been added to the input clause set as a lemma clause $\neg Ab(c_1) \vee \cdots \vee \neg Ab(c_n)$. Before entering the proof procedure in the inner loop we set $\Delta_N := \Delta_{N-1}$, and the proof procedure is slightly modified according to the following rule: whenever a finished open branch $b$ is derived, a new diagnosis $\Delta_b = \{c \mid Ab(c) \in b\}$ is found. Hence we set $\Delta_N := \Delta_N \cup \{\Delta_b\}$ and add $\Delta_b$ as a lemma clause to the input clause set, as just described. No more modifications are necessary.

Notice that the lemma clauses are purely negative clauses and hence can be used to close a branch. Since we give preference to negative clauses, no diagnosis will be computed more than once, because as soon as a diagnosis is computed the first time, it is turned into a (negative) lemma clause which will be used to immediately close all branches containing the same diagnosis. Furthermore, we compute only *minimal* diagnosis as an immediate consequence of the iterative deepening over $N$: any branch containing a non-minimal diagnosis would have been closed by a lemma clause stemming from a diagnosis with strictly smaller cardinality, which must be contained in the input clause set due to the invariant for some value $< N$. Thus, the invariant holds for all $N$.

Although this procedure computing minimal diagnosis under the $n$-fault assumption is so simple, it has some nice prop-

| Name | # Gates | # Clauses | # Diagnosis | Time (sec.) | # Steps | All? |
|---|---|---|---|---|---|---|
| C499 | 202 | 1685 | 2 | 5 | 3015 | no |
| 2-fault: | | | 67 | 50 | 27323 | yes |
| C880 | 383 | 2776 | 19 | 2 | 161 | yes |
| C1355 | 546 | 3839 | 5 | 47 | 24699 | no |
| 2-fault: | | | 5 | 2948 | 1284454 | no |
| C2670 | 1193 | 8260 | 31 | 6 | 533 | yes |
| C3540 | 1669 | 10658 | 3 | 10853 | 1473572 | yes |
| C5315 | 2307 | 16122 | 5 | 13 | 3071 | yes |

Figure 3: ISCAS' 85 Circuits and runtime results.

erties. First, it can be implemented easily by slight modifications to the basic hyper tableau proof procedure. Second, in the hyper tableau proof procedure we look at one branch at a time, which gives a polynomial upper bound for the memory requirement for the tableau currently constructed. Admittedly, there are possibly exponentially many minimal diagnosis wrt. $|COMP|$, but we assume that those will be kept anyway. Further, it is important to notice that we compute minimal models only wrt. the extension of the $Ab$-predicate, but not of minimal models wrt. all predicates. Since $|COMP|$ will be usually much smaller than the number of atoms in the translation $(SD, COMP, OBS)$ to propositional logic, computing minimal models wrt. all predicates would usually require considerably more memory (such an approach to minimal model reasoning was proposed in [2]).

**Experiments.** We implemented a prover according to the proof procedure as outlined above. It is a prover (written in SCHEME) for first-order logic, and thus carries some significant overhead for the propositional case. For our experiments we ran parts of the ISCAS-85 benchmarks [7] from the diagnosis literature. This benchmark suite includes combinatorial circuits from 160 to 3512 components. Table 3 describes the characteristics of the circuits we tested. The observations which were used can be obtained from the authors.

The results are summarized in Table 3. *# Clauses* is the number of input clauses stemming from the problem description. We ran our prover in 1-fault and 2-fault assumption settings. *Time* denotes proof time proper in seconds, and thus excludes time for reading in and setup (which is less than about 10 seconds in any case). The times are taken on a Sparc-Station 20. *# Steps* denotes the number of hyper extension steps to obtain the final tableau, and *# Diag* denotes the number of diagnosis. When two rows for a circuit are given, the upper one is for the 1-fault assumption, and the lower one is for the 2-fault assumption (recall that the 2-fault diagnosis *include* the 1-fault diagnosis). We emphasize that the results refer to the clause sets with renamed predicates according to Section 4. Without renaming, and thus taking advantage of the initial interpretation, in the 1-fault assumption only c499 was solvable (in 174 seconds); all other examples could not be solved within 2 hours, whatever flag settings/heuristic we tried!

In the "All?" column a "yes" entry means that there are no diagnosis with cardinality $> 1$, resp. $> 2$ for the 2-fault rows. That is, only for c1355 there are possible diagnosis with cardinality $> 2$ which we did not compute, for all other examples all minimal diagnosis were computed. We can detect if all minimal diagnoses were found by checking if some tableau branch was closed due to the $n$-fault assumption. To our impression, increasing $n$ usually drastically increases the time to compute diagnosis.

## 6 Conclusions

In this paper we analyzed the relationship between logic-based diagnostic reasoning and tableaux based theorem proving. We showed how to implement diagnostic reasoning efficiently using a hyper tableaux based theorem prover. We identified the use of an initial model as the main optimization technique in the diagnostic reasoning engine DRUM-2 and showed how to apply this technique within a hyper tableaux calculus. We slightly modified an implementation of the basic hyper tableau calculus by augmenting it with a combined iterative deepening/lemma technique. This guarantees completeness for minimal diagnosis computation.

## References

1. P. Baumgartner, U. Furbach, and I. Niemelä. Hyper Tableaux. In *JELIA 96*. European Workshop on Logic in AI, Springer, LNCS, 1996.
2. F. Bry and A. Yahya. Minimal Model Generation with Positive Unit Hyper-Resolution Tableaux. In P. Moscato, U. Moscato, D. Mundici, and M. Ornaghi, editors, *Theorem Proving with Analytic Tableaux and Related Methods*, volume 1071 of *Lecture Notes in Artificial Intelligence*, pages 143–159. Springer, 1996.
3. C. Chang and R. Lee. *Symbolic Logic and Mechanical Theorem Proving*. Academic Press, 1973.
4. H. Chu and D. Plaisted. Semantically Guided First-Order Theorem Proving using Hyper-Linking. In A. Bundy, editor, *Automated Deduction — CADE 12*, LNAI 814, pages 192–206, Nancy, France, June 1994. Springer-Verlag.
5. P. Fröhlich and W. Nejdl. A model–based reasoning approach to circumscription. In *Proceedings of the 12th European Conference on Artificial Intelligence*, 1996.
6. H. Ganzinger, C. Meyer, and C. Weidenbach. Soft Typing for Ordered Resolution. Unpublished, 1996.
7. The ISCAS-85 Benchmarks. http://www.cbl.ncsu.edu/www/CBL_Docs/iscas85.html, 1985.
8. R. Manthey and F. Bry. SATCHMO: a theorem prover implemented in Prolog. In *Proc. 9th CADE*. Argonnee, Illinois, Springer LNCS, 1988.
9. W. Nejdl and P. Fröhlich. Minimal model semantics for diagnosis – techniques and first benchmarks. In *Seventh International Workshop on Principles of Diagnosis*, Val Morin, Canada, Oct. 1996.
10. R. Reiter. A theory of diagnosis from first principles. *Artificial Intelligence*, 32:57–95, 1987.

# A Static Model–Based Engine for Model–Based Reasoning

**Peter Fröhlich and Wolfgang Nejdl**
University of Hannover
Lange Laube 3, 30159 Hannover, Germany
{froehlich, nejdl}@kbs.uni-hannover.de

## Abstract

Most systems for model–based reasoning record justifications during the search for solutions. In the popular ATMS systems this can lead to an explosion of recorded information when solving complicated problems. We propose an engine for model–based reasoning which works directly on logical models, uses static precompiled information on the structure of the underlying theory and does no further recording during search. This engine (DRUM–II) solves large complicated problems with attractive time and space complexity. To demonstrate the efficiency of the engine we give a new characterization of a popular benchmark suite, solve it with our engine and compare the performance to previous results.

## 1 Introduction

Since Reiter's seminal paper [Reiter, 1987], several generic systems for model–based diagnosis have been developed using logical inference, assumption based truth maintenance and conflicts as their underlying principles (see [de Kleer and Williams, 1987] and many others). Efficiency problems due to administration overhead inherent to this approach have only recently been solved [Raiman et al., 1993].

DRUM–II follows a different line of research which investigates reasoning through direct manipulation of logical models. Chou and Winslett [Chou and Winslett, 1994] have developed model–based algorithms for the minimal change semantics. Fröhlich and Nejdl took up the ideas of Chou and Winslett and developed algorithms for minimal model computation and query answering under circumscription [Fröhlich and Nejdl, 1996].

In this paper we present a simpler and more efficient variant of this minimal model computation algorithm and apply it to diagnosis. First we introduce the basic algorithm which performs local search to find minimal models of the system description (and thus minimal diagnoses). Although this algorithm is already efficient enough for a wide range of applications (like alarm management in telecom networks) large complex problems require additional focusing techniques. In most systems focusing is supported by recording actual dependencies among the literals in the system description during runtime. We introduce a new approach in which possible dependencies among the literals are detected statically during the compilation of the system description and used for focusing the algorithm. To demonstrate the efficiency of the engine we give a new characterization of a popular benchmark suite and discuss why the examples are hard to solve. We run our engine on these examples and compare its performance to previous results.

## 2 The Model–Based Framework

### 2.1 Relating Diagnoses and System Models

The device under consideration is described by a logical system description $SD$, denoted as a set of clauses. To avoid confusion we use the term model only for logical Herbrand models represented by sets of atoms in this paper (and not for the system description). The observed behaviour is characterized by a set of observations $OBS$. As usual, the set of system components is called $COMP$. With each component $c \in COMP$ we associate at least two behavioral modes: a correct mode denoted by $Ok(c)$ and an abnormal mode denoted by $Ab(c) \equiv \neg Ok(c)$. Our framework is suitable for expressing a wide range of diagnostic concepts. First we reconstruct Reiter's concept of minimal diagnosis.

**Definition 1** *A Diagnosis of* $(SD, COMP, OBS)$ *is a set* $\Delta \subseteq COMP$, *such that* $SD \cup OBS \cup \{Ab(c) | c \in \Delta\} \cup \{\neg Ab(c) | c \in COMP - \Delta\}$ *is consistent.* $\Delta$ *is called a* Minimal Diagnosis, *iff it is the minimal set (wrt.* $\subseteq$*) with this property.*

Next we characterize this diagnosis concept by the logical models of $SD \cup OBS$. Given a model $M$ of $SD \cup OBS$, the extension of the $Ab$–predicate (denoted by $M|Ab|$) tells us which components are considered faulty by $M$. Thus, every model $M$ of $SD \cup OBS$ corresponds to a (possibly non–minimal) diagnosis $M|Ab|$. To characterize the minimal diagnoses using models, we define an order $<_{Ab}$ on the models, based on the abnormal components they contain.

**Definition 2** *Given models $M_1$ and $M_2$, we define $M_1 \leq_{Ab} M_2$, iff $M_1|Ab| \subseteq M_2|Ab|$ and $M_1 \sim_{Ab} M_2$, iff $M_1|Ab| = M_2|Ab|$. Furthermore we define $M_1 <_{Ab} M_2$, iff $M_1 \leq_{Ab} M_2$ and not $M_1 \sim_{Ab} M_2$*

The minimal diagnoses of $(SD, COMPS, OBS)$ correspond to the $\leq_{Ab}$–minimal models of $SD \cup OBS$.

**Proposition 3** *A set $\Delta$ of components is a minimal diagnosis of $(SD, COMP, OBS)$, iff there exists a $\leq_{Ab}$–minimal model $M$ of $SD \cup OBS$, such that $\Delta = M|Ab|$.*

The following example shows that there can be several system models corresponding to a single diagnosis. Consider a system component $C$ (an integrated digital circuit) with two output ports $X$ and $Y$.

Suppose the system description predicts the values $X = 1$ and $Y = 0$, given the current values of $I_1, I_2$ and $I_3$. If we observe $Y = 1$, we have to assume that $C$ is faulty. Since $C$ is behaving abnormally, we cannot predict the value of $X$. Thus we have the following two $\leq_{Ab}$–minimal models corresponding to the minimal diagnosis $\{C\}$: $M_1 = \{Value(X, 0), Value(Y, 1), Ab(C)\}$ and $M_2 = \{Value(X, 1), Value(Y, 1), Ab(C)\}$.

$M_1$ and $M_2$ are equivalent ($M_1 \sim_{Ab} M_2$), because they correspond to the same diagnosis. DRUM–II therefore computes only one model out of every $\sim_{Ab}$–equivalence class. Such a representative set of minimal models is called a *Transversal*.

**Definition 4** *Let $\mathcal{M}$ be a set of models and $\mathcal{M}/\sim_{Ab} = \{[M]|M \in \mathcal{M}\}$ be a set of equivalence classes. A set $\mathcal{M}'$ is called a Transversal of $\mathcal{M}/\sim_{Ab}$, iff $\mathcal{M}'$ contains exactly one member out of every equivalence class in $\mathcal{M}/\sim_{Ab}$.*

The following theorem shows that computing a transversal of the minimal models is a correct implementation of model–based diagnosis.

**Theorem 5** *Let $\mathcal{M} := \{M | M \models SD \cup OBS$ and $\not\exists N : (N \models SD \cup OBS$ and $N <_{Ab} M)\}$ be the set of all $\leq_{Ab}$–minimal models. Let $\mathcal{M}'$ be a transversal of $\mathcal{M}/\sim_{Ab}$. Then $D := \{M|Ab| \mid M \in \mathcal{M}'\}$ is the set of all minimal diagnoses of $(SD, COMP, OBS)$.*

While previous results on the relation of diagnosis and circumscription [Besnard and Cordier, 1994; Raiman, 1990] focus on formalizing stronger forms of explanation (than provided by consistency–based diagnosis), this theorem establishes an interesting connection between consistency–based diagnosis and circumscription, because $\mathcal{M}$ is the set of all models obtained by circumscribing $SD \cup OBS$ in $Ab$, while varying all other predicates.

## 2.2 Repairing Inconsistent Models

The following algorithm was motivated by Chou and Winslett, who have implemented a system (IMMORTAL) for model–based belief revision [Chou and Winslett, 1994]. Consider a system description $SD$ and a model $M$ which contains no abnormals, such that $M \models SD$. Circuits, for example, can be modeled by axioms describing the function of their gates. We split the observations into observations of input and output values respectively $OBS = OBS_{in} \cup OBS_{out}$. A model $M^0$ of the correctly functioning circuit can be obtained by simulating the circuit's correct behavior given the input observations. More generally, monotonic first order logics can be used to construct a model of $SD \cup \{Ok(C)|C \in Comp\} \cup OBS_{in}$. We denote the model which is obtained by incorporating the observations $OBS$ into the model $M^0$ by $Ins(OBS, M^0)$. It is obtained by inserting the positive literals from $OBS$ into $M^0$ and deleting the negative literals in $OBS$ from $M^0$.[1]

**Definition 6** *For a model $M$ we define $Ins(OBS, M) := OBS \cup M \setminus \{x | \neg x \in OBS\} \setminus \{\neg x | \neg x \in OBS\}$*

The changed model $M^1 = Ins(OBS, M^0)$ possibly contradicts some of the axioms in the system description. Our algorithm however uses the inconsistent model $M^1$ to construct a model of $SD \cup OBS$. It does so by inverting the truth values of literals in $M^1$ which contradict the axioms in $SD$.

**Example 7** *Consider $SD = \{a \rightarrow c \vee d, d \rightarrow \neg e\}$, in clause form $\{\{\neg a, c, d\}, \{\neg d, \neg e\}\}$ together with the initial model $M^0 = \{e\}$ and the observations $OBS = \{a\}$.*

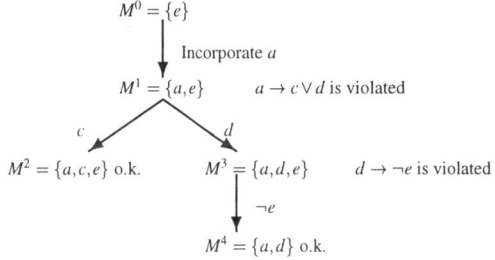

The model $M^1$ contradicts the axiom $a \rightarrow c \vee d$. We can change it in two ways in order to make the axiom hold again. Either we assume $c$ or $d$, because these are the other literals in the corresponding clause, which can make it true again. While assuming $c$ creates a consistent model ($M^2$) of $SD \cup OBS$, assuming $d$ violates the axiom $d \rightarrow \neg e$. Thus, we have to delete $e$ in order to obtain the second model $M^4$.

To formalize the iterative model repair algorithm motivated in the above example, we need the concept of *Committed Literals*. The positive and negative literals, which have been inverted in model $M^i$ compared to the initial model $M^0$ are called committed literals $Comm(M^i)$.

**Definition 8**
$Comm(M^i) := \{l | \text{literal } l \text{ and } M^i \models l \text{ and } M^0 \not\models l\}$

---

[1]Note, that inserting $OBS$ into $M^0$ has the same effect as inserting only $OBS_{out}$, since in our case $M^0$ already reflects the input observations. In general, any model without abnormals can serve as initial model, so the notation $Ins(M, OBS)$ is more general.

```
Repair_Model(SD, M⁰, OBS)
    Solutions := ∅;
    M¹ := Ins(OBS,M⁰);
    Models := {M¹};
    WHILE Models ≠ ∅ DO
        M := Select(Models);
        Models := Models\{M};
        IF M ⊨ SD∪OBS THEN
            Solutions := Solutions∪{M};
        ELSE
            FOR ALL Flippings f DO
                Models := Models∪{Ins(¬f,M)};
            END FOR;
        END IF;
    END WHILE;
```

Figure 1: Model Repair Algorithm

In our example $Comm(M^3) = \{a,d\}$. To avoid cycles in the model repair algorithm, a committed literal is not considered for inversion again. We consider a literal for inversion, if it occurs in a violated clause and neither the literal itself nor its negative counterpart are committed. Such a literal is called a *Flipping*.

**Definition 9** *Let $M^i$ be a model and $l$ a literal. $l$ is called a Flipping, iff*

1. *There exists a clause C in SD such that $l \in C$ and $M^i \not\models C$ and*

2. $l, \neg l \notin Comm(M^i)$.

Our iterative repair algorithm, which is summarized in figure 1 iteratively applies all possible flippings to the inconsistent models.[2]

### 2.3 Obtaining Minimal Models

During the repair process we maintain a set of partially repaired models (The set *Models* in figure 1). We have to select a model from this set and apply the next repair step to it. Since we want to compute minimal models, it is important to choose a model which is $<_{Ab}$–minimal.

An earlier solution to the problem of model selection, which we presented in [Fröhlich and Nejdl, 1996] and applied successfully to reasoning about change–problems, is a best–first algorithm. The list of models is organized as a priority queue. This priority queue is sorted by the number of *Ab*–literals in the models, and so the first element of the priority queue is always $<_{Ab}$–minimal. Although this model selection policy is sufficiently efficient to solve benchmark problems from the reasoning about action and change domain, it

---

[2]Our idea of applying local changes to models is similar to GSAT [Selman *et al.*, 1992]. However, in contrast to GSAT, our method performs a deterministic exhaustive search for minimal models. For this task no probabilistic algorithms have been proposed so far.

proved not suitable for larger diagnostic applications. The main problem here is the management of multiple models and copying of models which is necessary after each repair step.

In DRUM–II we therefore use an iterative deepening algorithm for obtaining the minimal models. For this algorithm the model set is organized as a stack with the usual operations Push and Pop. A stack has a better locality than a priority queue, because all operations only affect its first element. When using a stack, we can implement the push and pop operations without really copying the models. However, since its top element is just some arbitrary model, we must implement other mechanisms to avoid the generation of non–minimal models. The iterative deepening algorithm first defines an initial upper bound *Cutoff* = 1 on the number of *Ab*–literals in a model. If a partially repaired model contains more *Ab*–literals than *Cutoff* it is deleted, but a flag is set to indicate the deletion. The algorithm computes all consistent models with one *Ab*–literal and then sets *Cutoff* to the smallest number of *Ab*–literals of a model which was deleted. The algorithm proceeds until either no model is deleted or the bound *Cutoff* has reached an externally defined maximum (e.g. "only search for single and double faults").

It is easy to see that all models generated by this algorithm are indeed minimal models. In the first iteration we only search for single faults. All single faults found are minimal diagnoses. Then we set the search limit to the next smallest cardinality of a diagnosis which was pruned. During the search we always compare the models to the solutions already found. Thus, no non–minimal diagnoses are generated.

### 2.4 Formal Properties

The algorithm developed in the previous section is a simpler and more efficient variant of the algorithms presented in [Fröhlich and Nejdl, 1996] and fits in the framework for deciding entailment under circumscription presented in that paper. In the current paper we will only give the results relevant to diagnosis.

**Definition 10** *Let M and N be sets. $Diff(M,N) := (M\backslash N) \cup (N\backslash M)$*

Under the uniqueness of names assumption (UNA) and the domain closure assumption (DCA) (see [Lukaszewicz, 1990]) we can show that the set of models computed by `ID_Repair_Model` is a superset of those models of $SD \cup OBS$ which differ minimally from the initial model $M^0$.

**Lemma 11** *Let SD be a system description without function symbols, for which we assume UNA and DCA. Let $M^0$ be a model such that $M^0 \models SD$ and $M^0|Ab| = \emptyset$. Let OBS be a set of literals. Let $\mathcal{M}$ be the set of models returned by* `ID_Repair_Model`*(SD, $M^0$, OBS). Then $\mathcal{M} \supseteq \{M'|M' \models SD \cup OBS$ and $Diff(M^0,M')$ is minimal$\}$.*

Since $M^0|Ab| = \emptyset$ it is easy to show that $\mathcal{M}$ contains a model for each minimal extension of *Ab*. Thus $\mathcal{M}$ contains a transversal of the minimal models and using theorem 5 we can conclude, that `ID_Repair_Model` is a correct implementation of diagnosis:

**Proposition 12** *Let $SD, OBS, M^0$ and $\mathcal{M}$ be given as in lemma 11. For every diagnosis $\Delta$ there is a model $M \in \mathcal{M}$ such that $M|Ab| = \Delta$.*

## 3 Static Dependencies

We will now present a theory of dependencies between literals in sets of horn clauses. These dependencies (and corresponding indepencies) can be used to focus the search for diagnoses. The results presented below are applicable to all system descriptions, where the correct model can be formalized by propositional horn clauses. We do not assume that the whole system description horn since some clauses are typically not horn, e.g. relations among fault models like $Ok \equiv \neg Ab$.

### 3.1 Exploiting Indepence of Literals

Let us consider a set of Horn Clauses $\mathcal{S}$. Let $\{a_1, \ldots, a_n\}$ be the set of atoms occurring in $\mathcal{S}$. We say that $a_j$ depends directly on $a_i$, if there is a clause containing $a_j$ and $\neg a_i$.

**Definition 13** $a_j$ *depends directly on $a_i$, denoted by $a_i \to^1 a_j$, iff there is a clause $C \in S$ with $\{a_j, \neg a_i\} \subseteq C$.*

We define the transitive closure of the direct dependency relation $\to^1$ and call it $\to^+$ (depends on).

**Definition 14** *The relation $\to^+$ is defined inductively by*

1. *if $a_i \to^1 a_j$, then $a_i \to^+ a_j$.*
2. *If $a_i \to^1 a_k$ and $a_k \to^+ a_l$, then $a_i \to^+ a_l$.*

*We say $a_j$ depends on $a_i$, iff $a_i \to^+ a_j$.*

We want to study the effect of certain atoms on the other atoms of the theory. In particular, we want to see which changes are caused by assuming different sets $A, B$ of atoms. If we know that an atom $a$ was entailed by $\mathcal{S} \cup A$ (i.e. $\mathcal{S} \cup A \models a$), we want to know if $a$ is still entailed if we assume the set of atoms $B$ instead.

**Proposition 15** *Let $A$ and $B$ be sets of atoms such that $\mathcal{S} \cup A \not\models \bot$ and $\mathcal{S} \cup A \not\models \bot$. Let $a$ be an atom. If $\mathcal{S} \cup A \models a$ and there is no $b \in \text{Diff}(A, B)$ such that $b \to^+ a$, then $\mathcal{S} \cup B \models a$.*

**Proof Sketch:** The *positive* literal $a$ follows from $\mathcal{S} \cup A$, iff it holds in the minimal model of $\mathcal{S} \cup A$. This minimal model coincides with the minimal model of $\mathcal{S}^+ \cup A$ ($\mathcal{S}^+$ consists of those clauses of $\mathcal{S}$, which contain a positive literal). Thus, $\mathcal{S}^+ \cup A \models a$. Thus, there is a proof for $a$, only based on the clauses in $\mathcal{S}^+$. If we delete all unneccessary steps from this proof, only those steps remain, which derive $a$ itself or literals on which $a$ depends. This proof remains possible, if we replace $A$ by $B$ since we assume that $A$ and $B$ do not disagree on atoms, on which $a$ depends. Thus, $\mathcal{S}^+ \cup B \models a$ and, since $\mathcal{S} \cup B$ is consistent, we conclude $\mathcal{S} \cup B \models a$. #

The knowledge from the above proposition can be used to focus the search for diagnoses.

### 3.2 Application to Diagnosis

In diagnosis, the horn clauses $\mathcal{S}$ are given by the system description $SD$. The following proposition is a special case of proposition 15.

**Proposition 16** *Let $SD$ be a system description, $OBS = OBS_{in} \cup OBS_{out}$ an observation, $a$ an atom and $\Delta$ a candidate (a set of components considered faulty).*

*If $SD \cup OBS_{in} \models a$ and for all $c \in \Delta : Ok(c) \not\to^+ a$, then $SD \cup \{Ok(c) | c \in COMP \setminus \Delta\} \cup OBS \models a$.*

To proof this proposition, we must assume, that the observations $OBS_{out}$ are really the outputs of the device, i.e. no further atom depends on the output observations. The above proposition allows us to delete any model, which contains changes not influenced by the current candidate.

**Corollary 17** *Let $M$ be a model and $a \in M$ an atom occurring negatively in $Comm(M^i)$. Let $\Delta$ be a candidate. If for all $c \in \Delta : Ok(c) \not\to^+ a$ then $M \not\models SD \cup OBS \cup \{Ok(c) | c \in COMP \setminus \Delta\}$ and thus model $M$ can be deleted.*

The method described above enables us to eliminate a large number of candidates, which do not influence all abnormal outputs of the device as well as a large number of models which contain changes not influenced by the current candidate. The relation $\to^+$ needs only to be computed once for each system description. We compute $\to^+$ when the system description changes and store it in a (bit-) matrix, which is loaded into memory by the DRUM–II engine. The space complexity as well as the time complexity for computing the matrix is quadratic in the number of clauses of the underlying system description. The offline computation of the matrix is rather efficient. For a large circuit, the c6288 (see section 4) our current implementation uses approximately one minute to compute $\to^+$. While we have not yet compared the various optimizations of ATMS-based systems to our approach, a few comparisons can be made already. First, while an ATMS computes actual dependencies of propositions at runtime, our approach computes possible dependencies at compile time. Second, as these possible dependencies are correct for each truth value of the input propositions, they obviously cannot exploit specific values of input propositions. An ATMS-environment is therefore more specific than our dependency set, and can exploit specific input values of the circuit. On the other hand, it seems impossible to precompile ATMS-dependencies since they depend on the actual test vector used. Third, as ATMS's usually store environments as bit vectors, the storage requirement for one environment is the same as for one of our dependency sets. Fourth, while a proposition usually can be deduced in different ways, and the ATMS therefore has to store more than one environment for these propositions, we have exactly one dependency set for each proposition, which avoids any explosion of these dependency sets. An interesting issue for further exploration would be to use our possible dependencies in an ATMS instead of the specific justifications computed by the ATMS at run-time in order to speed up ATMS-based diagnosis systems.

## 4 Combinatorial Benchmark Circuits

The ISCAS–85 benchmark suite contains combinatorial circuits with 160 to 3512 components. In recent years it has been regarded as a challenge for diagnostic engines in several papers [de Kleer, 1991; Raiman *et al.*, 1993; Nejdl and Fröhlich, 1997; Williams and Nayak, 1997] due to the size of the problems and their inherent complexity. Before we discuss the performance of DRUM–II on these problems and compare it to previous results, we will give a characterization of the kind of complexity present in these problems and discuss why some of these circuits are hard to solve.

### 4.1 Why are these Problems so hard?

For satisfiability and constraint satisfaction benchmarks consisting of random problems are widely used. Random problems are often characterized by a set of parameters $\langle n, m, p_1, p_2 \rangle$ [Smith and Dyer, 1996], where $n$ is the number of variables, $m$ is the number of possibles values per variable, $p_1$ is the constraint density (the probabilty that a constraint exists between two variables) and $p_2$ is the constraint tightness (the probability that two variables related by a constraint have incompatible values). The constraint tightness is not applicable to the circuits because the constraints (clauses) are not all binary. Other parameters for the ISCAS circuits are given in table 1. The estimated constraint density of the problems decrease when problem size increases[3]. This suggests that the problems are hard in a dimension not yet captured by random problems or at least by their usual numerical characterization. Figure 2 shows the graph structure of a small part of the c6288 circuit (all successors of a particular component). We can observe the phenomenon of reconvergent fanout, which was already noted by de Kleer [Raiman *et al.*, 1993]: Although a value occurs only in a small number of constraints, there are long chains of constraints, through which nearly all variables in the graph are finally related. For an algorithm this means that it is easy to find a locally consistent solution but after a long computation it will eventually turn out that the solution is not globally consistent. Consequently, the circuits are not easily solvable by brute force search alone and focusing is necessary. Moreover, there is a large number of long propagation paths making straightforward dependency recording algorithms run out of memory.

### 4.2 Experimental Results

In this section we include experimental results for DRUM–II on the benchmark circuits. We have run 5 test vectors for each circuit. The running time is given in table 2. The times given are elapsed times in seconds on a lightly loaded SUN Ultra-1/170. They do not include model setup time, which is,

---

[3]Due to the special structure of the system description the actual constraint density is even lower than these estimates

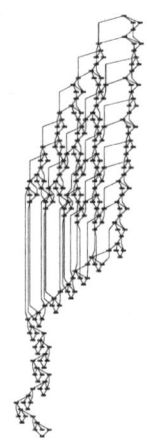

Figure 2: A Small Subgraph of the c6288 Circuit

| Circuit | Gates | $n$ | $m$ | $p_1$ |
|---|---|---|---|---|
| c499  | 202  | 894   | 2 | 0.02  |
| c880  | 383  | 1615  | 2 | 0.008 |
| c1355 | 546  | 2238  | 2 | 0.007 |
| c2670 | 1193 | 4928  | 2 | 0.002 |
| c3540 | 1669 | 6377  | 2 | 0.002 |
| c5315 | 2307 | 9356  | 2 | 0.002 |
| c6288 | 2406 | 9696  | 2 | 0.002 |
| c7552 | 3512 | 13582 | 2 | 0.001 |

Table 1: Statistical Parameters of the Circuits

however, very short (using a domain-specific forward chainer for generating the initial model).

| Circuit | Running Time (sec) | | | | |
|---|---|---|---|---|---|
|       | t1     | t2    | t3     | t4      | t5    |
| c499  | 0.013  | 0.036 | 0.040  | 0.036   | 0.013 |
| c880  | 0.023  | 0.003 | 0.028  | 0.027   | 0.004 |
| c1355 | 0.047  | 0.047 | 0.046  | 1.255   | 0.040 |
| c2670 | 0.061  | 0.083 | 0.222  | 0.077   | 0.226 |
| c3540 | 0.236  | 0.050 | 0.750  | 0.143   | 1.586 |
| c5315 | 0.025  | 0.131 | 0.051  | 0.080   | 0.019 |
| c6288 | 89.517 | 0.562 | 23.788 | 160.623 | 0.221 |
| c7552 | 0.056  | 2.093 | 1.598  | 0.126   | 2.261 |

Table 2: Running Times of DRUM–II for 5 test vectors

The test vectors used are available from the authors. Note, that the timing results show a large variation depending on the test vector used. This effect is striking with respect to the c6288. A reason for this is the different number of components which can explain the abnormal observation. Table 3 shows the number of propositions (possibly) influencing the abnormal observation for the different test vectors of the c6288. In the difficult cases t1 and t4 nearly the whole circuit influences the abnormal observation. We expect a large variation in running time for ATMS-based systems, too, though no such results have been published so far.

| Circuit | t1 | t2 | t3 | t4 | t5 |
|---|---|---|---|---|---|
| c6288 | 6316 | 795 | 2545 | 5804 | 375 |

Table 3: Propositions influencing the Abnormal Observation

In table 4 we compare the median running time of DRUM–II to previous results from the literature. DRUM–II is always the fastest system except for the c6288 where IMPLODE is faster. Since however the test vector used is not given in [Raiman et al., 1993] it is not clear, whether IMPLODE's critical reasoning techniques provide still better focusing on the special structure of the c6288 for all test vectors.

| Circuit | DRUM-II Median | de Kleer AAAI91 | IMPLODE IJCAI–93 | Williams DX–96 |
|---|---|---|---|---|
| c499 | 0.036 | 7.9 | 0.4 | 4.5 |
| c880 | 0.023 | 6.2 | 0.8 | 4.0 |
| c1355 | 0.047 | 242 | 1.4 | 12.3 |
| c2670 | 0.083 | 33 | 3 | 28.8 |
| c3540 | 0.236 | 1545 | 6 | 113.3 |
| c5315 | 0.051 | 1215 | 7 | 61.2 |
| c6288 | 23.788 | – | 8 | – |
| c7552 | 1.598 | 1028 | 14 | 61.5 |

Table 4: Comparison of Running Times

## 5 Conclusion and Further Work

In this paper we describe the reasoning algorithm DRUM-II for diagnosis and other non-monotonic reasoning problems, which is based on the direct manipulation of logical models. It uses static precompiled information on the structure of the underlying theory and does no further recording at runtime. Running DRUM-II on a set of circuit diagnosis benchmarks with more than 3000 components shows that its efficiency is in most cases much better than the performance of previous systems. The paper also contains a short characterization of these benchmarks and their inherent complexity

Currently, we are extending DRUM to handle stronger definitions of diagnosis (abductive diagnosis and kernel diagnosis). For these optimizations the filtering operator defined in [Fröhlich and Nejdl, 1996] is used, which decides, if the truth value of a given literal is entailed.

## 6 Acknowledgement

We thank Peter Baumgartner as well as an anonymous referee for their contributions to section 3

## References

[Besnard and Cordier, 1994] Philippe Besnard and Marie-Odile Cordier. Explanatory diagnoses and their characterization by circumscription. *Annals of Mathematics and Artificial Intelligence*, 11:75–96, 1994.

[Chou and Winslett, 1994] T. S-C. Chou and M. Winslett. A model–based belief revision system. *Journal of Automated Reasoning*, 12:157–208, 1994.

[de Kleer and Williams, 1987] Johan de Kleer and Brian C. Williams. Diagnosing multiple faults. *Artificial Intelligence*, 32:97–130, 1987.

[de Kleer, 1991] Johan de Kleer. Focusing on probable diagnoses. In *Proceedings of the National Conference on Artificial Intelligence (AAAI)*, pages 842–848, Anaheim, July 1991. Morgan Kaufmann Publishers, Inc.

[Fröhlich and Nejdl, 1996] Peter Fröhlich and Wolfgang Nejdl. A model–based reasoning approach to circumscription. In *Proceedings of the 12th European Conference on Artificial Intelligence*, 1996.

[Lukaszewicz, 1990] W. Lukaszewicz. *Non–monotonic reasoning: formalization of commonsense reasoning*. Ellis Horwood, 1990.

[Nejdl and Fröhlich, 1997] Wolfgang Nejdl and Peter Fröhlich. Minimal model semantics for diagnosis – techniques and first benchmarks. In *7th International Workshop on Principles of Diagnosis*, Val Morin, Canada, 1997.

[Raiman et al., 1993] Olivier Raiman, Johan de Kleer, and Vijay Saraswat. Critical reasoning. In *Proceedings of the International Joint Conference on Artificial Intelligence*, pages 18–23, Chambery, August 1993.

[Raiman, 1990] Olivier Raiman. Circumscribing diagnosis engines: Exploiting the alibi principle to avoid blind search. In *Proceedings of the International Workshop on Expert Systems in Engineering*, Vienna, September 1990. Springer-Verlag. Lecture Notes in AI.

[Reiter, 1987] Raymond Reiter. A theory of diagnosis from first principles. *Artificial Intelligence*, 32:57–95, 1987.

[Selman et al., 1992] Bart Selman, Hector Levesque, and David Mitchell. A new method for solving hard satisfiability problems. In *Proceedings of the National Conference on Artificial Intelligence (AAAI)*, pages 440–446, San Jose, July 1992. Morgan Kaufmann Publishers, Inc.

[Smith and Dyer, 1996] Barbara M. Smith and Martin E. Dyer. Locating the phase transition in binary constraint satisfaction problems. *Artificial Intelligence*, 81:155–181, 1996.

[Williams and Nayak, 1997] Brian C. Williams and Pandurang Nayak. A model–based approach to reactive self–configuring systems. In *7th International Workshop on Principles of Diagnosis*, Val Morin, Canada, 1997.

# DIAGNOSIS AND QUALITATIVE REASONING

Diagnosis 3

# Polynomial Temporal Band Sequences for Analog Diagnosis

**Etienne LOIEZ**[*][#]

[*]ISEN
41 Bd Vauban
59046 Lille (France)

[#]LAIL
USTL Bat. P2
59562 Villeneuve d'Ascq
elo@isen.fr

**Patrick TAILLIBERT**[+]

[+]Dassault Electronique
55 Quai Marcel Dassault
92214 Saint Cloud (France)
Patrick.Taillibert@dassault-elec.fr

## Abstract

In [Loiez and Taillibert 96a] we proposed a new way to model components and physical quantities in the framework of Model-based Diagnosis applied to analog systems. Components were represented by equations of the form " $f+g'=0$ " where f and g are functions of observable quantities of the system to be diagnosed; observations were represented by temporal band sequences (sets of consecutive trapezoids encompassing the observations); consistency checking was done by integration.

The present paper generalizes the observation representation to temporal bands defined by any polynomial function. It shows that this extension not only allows a better approximation of observations but also that models of components can now be represented by equations relating derivatives of any order ($f+g'+h''+...=0$) and involving any polynomial functions, which improves dramatically the ease of modeling analog systems.

## 1 Introduction

The diagnosis of analog electronic circuits, and more generally of analog systems, raises many specific difficulties which justify the development of specific methods. Thus, the most significant problems are:
- a very large number of possible failures (each abnormal value of a parameter can lead to a particular fault mode),
- the inaccuracy of parameter and measurement values,
- the importance of time to describe system behaviour,
- the small amount of useful information brought by the knowledge of the nominal behaviour of the system. Each component failure can provoke the other components (though perfectly correct) to shift well away from their nominal behaviour. This is especially the case when the system contains a feedback loop,
- multiple faults are not a rare occurrence, since the failure of one component can have a knock-on effect on others.

With the model based diagnosis framework [de Kleer and Williams 87; Reiter 87], many difficulties have naturally been overcome, but some of them, for instance the inaccuracy of parameters and measurements and also the management of time, have required the development of specific methods. Some purely qualitative approaches have been tested [Pan 84, Ng 90, Dague et al 87], but the ambiguity and the inaccuracy introduced by such a modeling method do not allow the detection of some failures (the change of a component parameter often leads to a simple distortion of observed signals, with no effective qualitative changing). In CATS [Dague et al 91], a first solution to the problems of inaccuracy and time management was proposed for systems whose observations are "repeatable"[1] and whose fault status does not change during the diagnosis process.

In CATS, physical quantities are represented, thanks to a sampling process, by arrays of intervals. The models of correct behavior of components are described by functional relations between physical quantities. Numeric intervals represent the value of component parameters and their inaccuracy. Some temporal properties of the components are expressed by using the first derivative of quantities. The diagnosis procedure consists of an interval propagator coupled with an ATMS.

CATS has been tested on many circuits of various sizes and complexity in many domains like electronics [Dague et al 91], electricity [Devès et al 92] or chemical engineering (HINT Esprit project) [Alarcon et al 94].
Experiments point out two difficulties:
- The first one concerns the use of arrays of intervals to represent values, which induces a huge amount of information (hundred samples per quantity) which is not used in an optimal manner.
- The second difficulty is induced by the fact that a numeric approximation of derivatives is required but, in general, no information about the derived functions is available (because of the fault, any kind of signal might appear in the system). Moreover, the limitation to first-order derivatives

---

1. For such systems, either the measurements of the different quantities are made simultaneously or, - as is often the case in electronics - the same system state occurs periodically, allowing sequential measurements.

makes it difficult to model complex components.

A possible solution to the first problem consists in using temporal intervals rather than samples, that is, expressing quantities in terms of their concise histories made up of episodes, maximal time intervals during which the (interval) value of the quantity remains constant (see TCP [Williams *et al* 86] and TINT [Hamscher 89]). This approach is difficult to transpose to continuous systems because avoiding to consider too many episodes requires a too abstract view of physical quantities which prevents finding many faults, especially these due to a slight variation of component parameters.

In [Loiez and Taillibert, 96a] we proposed a new way to model components and physical quantities. Components or sets of components are represented by equations of the form "f+g'=0" where f and g are functions of observable quantities of the system to be diagnosed; observations are represented by temporal band sequences (sets of consecutive trapezoids encompassing the observations). This approach allows a sparser representation of observations (at least than CATS) and does not require a numeric approximation of derivatives.

The present paper generalizes the observation representation to temporal bands defined by any polynomial function. It shows that this extension not only allows a better approximation of observations but also that models of components can now be represented by equations relating derivatives of any order (f+g'+h''+...=0) and involving any polynomial functions, which improves dramatically the ease of modeling analog systems.

Section 2 presents the temporal band sequence formalism, section 3 describes a consistency checking procedure and section 4 gives some hints to use it in a diagnosis system.

## 2 Observations modeling

We now describe how we propose to model observations, defined on a unique temporal window and representing a numeric quantity which varies continuously over time.

We first define the concept of "temporal band" which will allow us to approximate the value of physical quantities on the temporal interval [t1, t2[.

**Definition 1:** *"Temporal band"*
A temporal band is a tuple $(t_i, t_s, p_i, p_s)$ where $t_i, t_s$ are real numbers, $p_i, p_s$ are the polynomials

$$p_i(t) = \sum_{k=0}^{m} a_{ik} t^k \quad \text{and} \quad p_s(t) = \sum_{k=0}^{n} a_{sk} t^k \quad \text{such that:}$$

$$t_i < t_s \wedge \forall t \in [t_i, t_s[, \quad p_i(t) \leq p_s(t)$$

The interval $[t_i, t_s[$ is the *temporal extent* of the band and the highest degree of the polynomials $p_i, p_s$ (i.e. max(m, n)) is the *degree* of the band.

We will denote by $B(t) = [p_i(t), p_s(t)]$ the value (an interval) of a band $B = (t_i, t_s, p_i, p_s)$ at the instant t.
Example: the next figures show two bands of degree 1 and 2

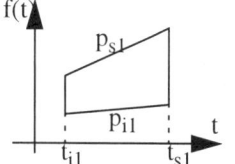 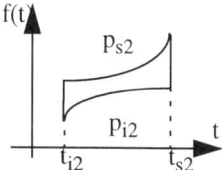

A high degree temporal band is, in general, mandatory to approximate an observation well. As we will see in section 3, our motivation to introduce polynomials of any degree in the temporal band definition is more to be able to process derivatives of higher order than to get a better approximation of an observation. That is the reason why we now define "temporal band sequences":

**Definition 2:** *"Temporal band sequence"*
A temporal band sequence SB is a finite ordered set of temporal bands $\{B_1, ..., B_n\}$ such that:

$$\forall k \in [1, n-1], t_s(B_k) = t_i(B_{k+1})$$

where $t_i(B)$ is the initial instant of the band B and $t_s(B)$ the final instant.

The interval $[t_i(B_1), t_s(B_n)[$ is the *temporal extent* of the band sequence and the highest degree of the bands is the *degree* of the temporal band sequence.

We will denote by $SB(t) = B(t)$ the value (an interval) of the temporal band sequence SB at the instant t (B is the temporal band of SB whose temporal extent contains t).

We can now define how to represent an observation:

**Definition 3:** *"Representation of an observation"*
A temporal band sequence SB whose temporal extent is $[t_1, t_2[$ is a representation of an observation v(t) over a time interval $[t_1, t_2[$ iff the observation is always encompassed within the bands, which means:

$$\forall t \in [t_1, t_2[, \exists (t_i, t_s, p_i, p_s) \in SB, \text{ such that:}$$
$$t \in [t_i, t_s[ \wedge p_i(t) \leq v(t) \leq p_s(t)$$

Example: the band sequences $SB_1$ and $SB_2$ are two possible representation of v(t)

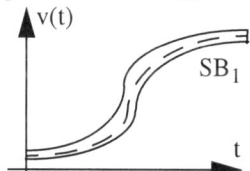 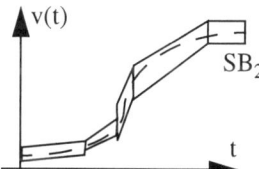

We have develop a procedure capable of computing a temporal band sequence given the information provided by the sensors (for example, a sampled signal). This procedure takes into account sensor inaccuracy in order to always properly represent the observation. It also makes a trade-off between the number of bands in the sequence, the inaccuracy introduced (characterized by the width of the band) and the degree of the bands. In case of observations made sequentially, it considers the temporal extent of the bands already computed in order to reduce the number of new bands possibly introduced by the "synchronization" operation defined below (Definition 5). So far, this procedure, based on the least squares method, generates band sequences whose degree is at most one; that does not make higher-degree bands irrelevant since such bands occur when combining first-degree bands together.

**Definition 4:** *"Equivalence of temporal band sequences"*
Two band sequences $SB_1$ and $SB_2$ of same temporal extent $[t_1, t_2[$ are equivalent iff:
$\forall t \in [t_1, t_2[, \exists (t_{i1}, t_{s1}, p_{i1}, p_{s1}) \in SB_1, \exists (t_{i2}, t_{s2}, p_{i2}, p_{s2}) \in SB_2,$
$p_{i1}(t) = p_{i2}(t) \wedge p_{s1}(t) = p_{s2}(t)$

Hence, two equivalent temporal band sequences differ only in their temporal splitting. For example:

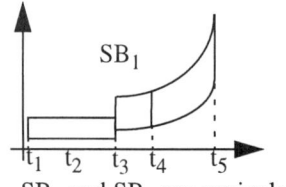 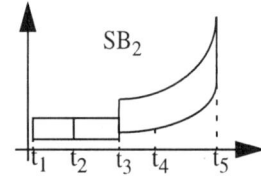

$SB_1$ and $SB_2$ are equivalents.

**Definition 5:** *"Synchronization of temporal band sequences"*
The synchronization of two temporal band sequences $SB_1$ and $SB_2$ whose temporal extents are the same consists in finding two temporal band sequences $SB_1'$ and $SB_2'$ such that:
- $SB_1$ and $SB_1'$ (resp. $SB_2$ and $SB_2'$) are equivalent,
- $|SB_1'| = |SB_2'| = n$
- $\forall k \in [1, n], \begin{cases} t_i(SB_1'[k]) = t_i(SB_2'[k]) \\ t_s(SB_1'[k]) = t_s(SB_2'[k]) \end{cases}$

where $SB[i]$ denote the $i^{th}$ band of $SB$.
For example, $SB'1$ and $SB'2$ have been made synchronous:

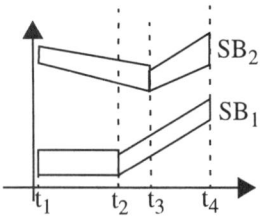 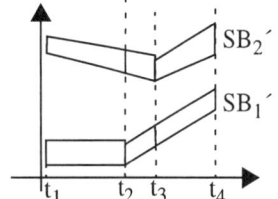

**Property 1:**
The minimal synchronization of two temporal band sequences whose temporal extents are the same always exists and is unique.

We now define the necessary operations to use the temporal band sequences formalism for components modeling.

**Definition 6:** *"Sum of temporal band sequences"*
The sum of two synchronous band sequences $SB_1$ and $SB_2$ of size n is the band sequence SB of size n such that:
$\forall k \in [1, n], SB[k] = SB_1[k] + SB_2[k]$,
where the sum of two bands $(t_i, t_s, p_{i1}, p_{s1})$ and $(t_i, t_s, p_{i2}, p_{s2})$ is the band $(t_i, t_s, p_{i1} + p_{i2}, p_{s1} + p_{s2})$

**Property 2:**
The sum of two band sequences maintains the property of representation; this means that, if any functions $g_1(t)$ and $g_2(t)$ are represented by temporal band sequences $SB_1$ and $SB_2$ then the function $g_1(t) + g_2(t)$ is represented by $SB_1 + SB_2$.

Computing the sum of two temporal band sequences which are not synchronous consists in synchronizing these sequences and then computing their sum.

This definition is easily extended to the sum of a temporal band sequence and an interval.

We now define the integral of a temporal band sequence. Our aim is to define a temporal band sequence representing the function $\int_\theta^t g(x)dx$ from the representation of the function $g(t)$.

**Definition 7:** *"Integral of a temporal band"*
The integral of a temporal band $B = (t_i, t_s, p_i, p_s)$ from any instant $\theta$ belonging to the open interval $(t_i, t_s)$ is a temporal band sequence $\{(t_i, \theta, P_s, P_i), (\theta, t_s, P_i, P_s)\}$, noted $\int_\theta B$, such that:

$$P_i(t) = \int_\theta^t p_i(x)\,dx \quad \text{and} \quad P_s(t) = \int_\theta^t p_s(x)\,dx$$

We extend this definition for $t_i$ and $t_s$ such that
$$\int_{t_i} B = \{(t_i, t_s, P_i, P_s)\} \quad \text{and} \quad \int_{t_s} B = \{(t_i, t_s, P_s, P_i)\}$$

It can easily be proved that the constraint of the temporal band definition holds, that is:
$\forall t \in [t_i, \theta[, P_s(t) \leq P_i(t)$ (for the first band) and
$\forall t \in [\theta, t_s[, P_i(t) \leq P_s(t)$ (for the second).

We now recursively define the integral of a temporal band sequence.

**Definition 8:** *"Integral of a temporal band sequence"*
The integral from any instant $\theta$ belonging to $]t_i(B_j), t_s(B_j)[$ of a temporal band sequence $SB = \{B_1, ..., B_n\}$ of size n and degree p is a temporal band sequence $\{B'_1, ..., B'_{j-1}\} \cup \{B'_{j1}, B'_{j2}\} \cup \{B'_{j+1}, ..., B'_n\}$ of size n+1 and degree p+1, noted $\int_\theta SB$, such that

- $\{B'_{j1}, B'_{j2}\} = \int_\theta B_j$,

- $B'_{j-1} = B'_{j1}(t_s(B_{j-1})) + \int_{t_s(B_{j-1})} B_{j-1}$
$\wedge \;\; \forall k \in [1, j-2], B'_k = B'_{k+1}(t_s(B_k)) + \int_{t_s(B_k)} B_k$,

- $B'_{j+1} = B'_{j2}(t_i(B_{j+1})) + \int_{t_i(B_{j+1})} B_{j+1}$
$\wedge \;\; \forall k \in [j+2, n], B'_k = B'_{k-1}(t_i(B_k)) + \int_{t_i(B_k)} B_k$.

The constant $B'_{k+1}(t_s(B_k))$ corresponds to the integration of the intermediate bands ($B_j$ to $B_k$).

Thus, computing the integral of a band sequence SB consists in computing the integral of the $j^{th}$ band of SB, then computing the integrals of the $j+1^{th}$ and $j-1^{th}$ bands from the integral of the $j^{th}$, ...

Example:

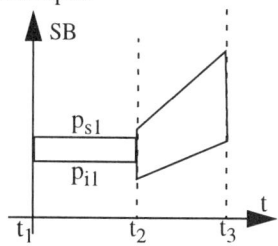 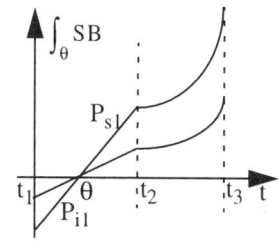

**Property 3:**
The integration of a temporal band sequence maintains the property of representation; this means that, if any function g(t) is represented by a temporal band sequence SB then the function $\int_\theta^t g(x)\,dx$ is represented by $\int_\theta SB$.

In the same way, we define the product of two temporal band sequences and the product of an interval and a temporal band sequence. In both cases, the property of representation is kept but the number of bands in the sequence and/or the degree of bands can increase. Complete definitions and proofs can be found in [Loiez 97].

Any polynomial operation can be built from the previous definitions (sum and product) at the cost of an increase of inaccuracy when the same quantity occurs several times. Primitive operations can be defined to avoid this loss of accuracy as it is done with real functions in [Moore 66].

## 3 Diagnosing with Temporal Band Sequences

### 3.1 Principle

The diagnosis approach we propose takes place in the model based diagnosis theory [Reiter, 87]: each model is associated (by a conjunction of $\neg ab(.)$ predicates) with the set of components which allows one to assume the model of a component or a block of components holds. Diagnosing, then, consists in detecting the inconsistent models to determine the conflict sets (nogoods).

**Definition 9:** *"Behaviour model of components"*
We call correct behavior model of the components $c_1, c_2, ..., c_p$, an expression of the form:

$$\boxed{\begin{array}{l}\neg ab(c_1) \wedge ... \wedge \neg ab(c_p) \\ \Rightarrow \forall t, f(t) + g'(t) + h''(t) + k'''(t) + ... = 0 \quad (1)\end{array}}$$

where f, g, h, k, ... are (any) temporal functions.

Allowing derivatives of any order in the model equation eases components modeling. In particular, the use of simulation models for diagnosing can be considered and also complex block modeling.
For example possible models are:

a resistor r whose value is R:

$\neg ab(r) \Rightarrow \forall t, \; vr(t) - R\,ir(t) = 0 \quad (f = vr - R\,ir)$

a block composed of a capacitor c and an inductance l:

$(f = vl/LC, h = vc)$

$\neg ab(c) \wedge \neg ab(l) \Rightarrow \forall t, \; \dfrac{vl(t)}{LC} - vc''(t) = 0$

Inverting definition 9 gives:
$\exists t, f(t) + g'(t) + h''(t) + k'''(t) + ... \neq 0 \Rightarrow ab(c_1) \vee ... \vee ab(c_n)$
Diagnosing then consists in determining for each model if:

$$\boxed{\exists t, f(t) + g'(t) + h''(t) + k'''(t) + ... \neq 0 \quad (2)}$$

Our main result is that there exists a procedure to make this computation in an efficient way when the temporal functions are represented by synchronous temporal band sequences. We shall now present this procedure.

### 3.2 An inconsistency detection procedure

Our idea is to represent each model expression $f(t) + g'(t) + h''(t) + k'''(t) + ...$ by a temporal band sequence thanks to temporal band sequence operations defined in section 2. Directly representing the model expression would require an approximation at every instant, not only of observations but also of their derivatives, which is not easy to obtain for the first derivative (and is a limitation of the CATS system) and is unrealistic for higher order derivatives.

To overcome this problem, the model expression is transformed into an equivalent expression in order to be able to represent it by a temporal band sequence without having to approximate the derivatives at every instant. For that, we will take advantage of the fact that integrals of temporal band sequence are easy to compute. As the procedure depends on the degree of the equations, we present it for different cases:

<u>Equations without any derivative:</u>
Let a model equation $f(t) = 0$ and SB a temporal band sequence representing f(t). As the value of the function f(t) is known (0), it is possible to verify if f(t) is really represented by SB. If it is not the case, the model is inconsistent. Thus, directly using equation (2) gives:

- $\exists (t_i, t_s, p_i, p_s) \in SB, \exists t \in [t_i, t_s[, p_i(t) > 0 \vee p_s(t) < 0 \quad (3)$

The procedure consists in verifying if this relation holds.

Model equations without any derivative are not a rare occurrence. For example, checking the inconsistency of the equation $vr = R\,ir$ of a resistor model whose quantities vr and ir are respectively represented by the temporal band sequences SBV and SBI consists in computing the temporal band sequence SBV - R SBI and checking if the relation (3) holds.

<u>First Order Equation:</u>
Let a model equation $f(t) + g'(t) = 0$ and SBf, SBg the temporal band sequences representing respectively f(t) and g(t). It is not possible to compute a temporal band sequence representing g'(t) from SBg. For that reason, we integrate the model equation. This gives:
$\forall t, f(t) + g'(t) = 0 \Leftrightarrow \forall t, \int_\theta^t f(x)dx + g(t) - g(\theta) = 0$
Thus, the procedure consists in:
- choosing an integration instant $\theta$ (this choice will be discussed in section 4.1),

- computing the temporal band sequence
$\int_\theta SBf + SBg - SBg(\theta)$ which represents the function:
$$k(t) = \int_\theta^t f(x)dx + g(t) - g(\theta),$$

- checking the consistency of the equation without any derivative $k(t) = 0$.

Hence, checking the consistency of a first order equation consists in integrating once the equation and then checking the consistency of an equation without any derivatives.

Moreover, the fact that all the operations maintain the property of representation allows us to ensure that no false conflict will be detected.

<u>Second Order Equation</u>:
Let a model equation $f(t) + g'(t) + h''(t) = 0$ and SBf, SBg, SBh the temporal band sequences representing respectively $f(t)$, $g(t)$ and $h(t)$. For the same reason as before, we integrate the equation from an instant $\theta$:
$$\forall t, f(t) + g'(t) + h'' = 0$$
$$\Leftrightarrow \forall t, \int_\theta^t f(x)dx + g(t) - g(\theta) + h'(t) - h'(\theta) = 0$$
Thus, the procedure consists in:
- choosing an integration instant $\theta$,
- approximating the constant $h'(\theta)$ by an interval H,
- computing the temporal band sequence
$\int_\theta SBf + SBg - SBg(\theta) - H$ which represents the function
$k(t) = \int_\theta^t f(x)dx + g(t) - g(\theta) - h'(\theta),$
- checking the consistency of the first order equation $k(t) + h'(t) = 0$.

Hence, checking the consistency of a second order equation consists in integrating once the equation and then checking the consistency of a first order equation.

For the same reason as before, no false conflict set will be detected. However, the consistency of a second order equation $f(t) + g'(t) + h''(t) = 0$ can be checked only if an approximation of $h'$ at time $\theta$ can be obtained; this point is discussed in section 4.

<u>Generalization</u>:
In the same manner, equations of any order can be checked.

The procedure to check the consistency of a $n^{th}$-order equation consists in:
- choosing an integration instant $\theta$,
- integrating once the equation,
- approximating the values of the derivatives which appear,
- computing the temporal band sequence which represents the function involving no derivatives,
- checking the consistency of the resulting $n-1^{th}$-order equation.

## 4 Discussion

The use of the consistency checking procedure just described for diagnosing analog systems raises two main questions related to the integration instant $\theta$ and the compositionality of the approach. We discuss these two points here.

### 4.1 Integration instant

For first-order equations any integration instant can be chosen; nonetheless, inconsistency of the equation $k(t) = 0$ can be missed since the integration process adds inaccuracies and the (possibly) only instant when this inconsistency can be found out might be far from the chosen $\theta$. Hopefully, when only temporal bands of degree 1 are considered, an analytic procedure can be devised for checking equation $k(t) = 0$, as shown in [Loiez and Taillibert, 96a]. This procedure computes whether there exists at least one $\theta$ for which the equation is inconsistent; it can be extended for bands of higher degree at the cost of some efficiency since it then requires solving numerically polynomial equations.

Another difficulty occurs for equations with second order derivatives: an approximation of the first derivative of function $h(t)$ at time $\theta$ is mandatory. This approximation can not be computed from the temporal band sequence representing $h(t)$, but must be obtained from observations (as is the case in CATS). We have said that it is difficult to approximate a function derivative at every instant, but the fact that we just have to approximate it at only one instant ($\theta$) makes that problem less critical. For example, it is often possible to choose an instant for which the system is stable, i.e. an instant when all derivatives are zero. The same reasoning applies for equations with higher order derivatives. Hence, for these equations, the choice of the integration instant $\theta$ is imposed by the need to have an approximation of derivatives at that time. That may be a cause of incompleteness since this constraint can conflict with the one mentionned just above. Experiments carried out so far do not suggest that this problem could be a serious impediment for a practical use of our method.

### 4.2 Compositionality

Thanks to the temporal band sequences formalism just described, we can represent observation of quantities which continuously evolve in time. The modeling we have suggested allows the description of components by equations like $f(t) + g'(t) + h''(t) + ... = 0$ where f, g, h, ... are functions involving quantities of the system to be diagnosed and sum and product operations.

However, the procedure we have presented can only detect inconsistencies - and hence find out conflicts - if a temporal band representation of all the functions of an equation is available. But, in general, not all the physical quantities of the system to diagnose are measurable. (For example, currents in electronics). If these quantities are involved in the functions of the equation, the consistency of the corresponding model will not be checked.

Moreover, the consistency checker tests the models one by one and independently of one another; to detect all the possible conflict sets with as few measurements as possible we need some composition of models.

To solve these problems, a classical propagation procedure is applicable: when temporal band sequences for a quantity and its derivative are predicted -say v and v'-, consistency is checked with the procedure described in section 3-2 for f+g'=0 (by making f=v' and g=-v).

In DOGS, the program we developed to experiment Temporal Band Sequences, we tried another approach by generating off-line all possible equations of the form f(t) + g'(t) + h''(t) + ... = 0 and such that the quantities present in the functions f, g, h, ... are measurable; all possible combination of these equations eliminating some quantity are also computed[1].

We notice that the theoretical complexity of such an algorithm is huge (more than $2^n$ generated equations where n is the number of components). But, in practice, the number of equations remains manageable when circuits are weakly connected (few equations contain the same quantity) or when the number of measurable quantities is low compared to the total number of quantities (about 50% in electronics). Moreover, if we impose a limit to the size of the conflicts to be detected, the number of the resulting equations should remain "reasonable".

## 5 Conclusion

The approach that we have presented has been essentially motivated by our need to represent observations such as signals which vary continuously over time and to use them for diagnosis purpose.

In [Loiez and Taillibert, 96a], Temporal Band sequences were limited to first order temporal bands (trapezoid) which was already a real improvement w.r.t. the result obtained with the CATS system, especially when fast varying signals were considered. The extension to polynomial temporal bands seems quite natural but we have had the pleasure to find out that it allowed not only a better approximation of signals, but mainly, that it also allowed one to take into account derivatives of higher order and more complex operations, and hence, ease the component modeling. The only constraint is to get an approximation of derivatives at one instant for which approximation is safe.

The temporal band sequence operations process can increase the number of "episodes" but, this drawback is limited since we do not propagate these new episodes.

Various extensions to the procedure generating the models can be imagined; for instance, we associate preconditions to models in order to allow a piecewise linear description and hence to ease modeling of complex components such as diodes or transistors[2].

DOGS has been tested with first degree bands, on the monostable circuit chosen as a reference problem by the diagnostic community [Dague and Taillibert, 96]. The results obtained by CATS on the same circuit [Dague et al, 90] are the same with DOGS without the risk of faulty diagnosis due to the computation of derivatives made by CATS in the fast-varying part of signals. Moreover, DOGS has given good results with second order circuits like filters.

## Acknowledgments

E.Loiez work is supported by the "Conseil General du Nord-Pas de Calais". A preliminary version of that work was presented in [Loiez and Taillibert, 96b].

## References

[Alarcon et al., 1994] I.Alarcon et al, "Heterogeneous Integration Architecture for Intelligent Control Systems", *Int. Journal of AI in Engineering,1994* .

[Dague et al., 1987] P. Dague, P. Devès, and O. Raiman, "Troubleshooting: when Modeling is the Trouble," *Proceedings of AAAI Conference, Seattle, July 1987.*

[Dague et al., 1990] P. Dague, P. Devès, P. Luciani, and P. Taillibert, "Analog System Diagnosis," *Proceedings of the 9th ECAI, Stockholm, August 1990.*

[Dague et al., 1991] P. Dague, P. Devès, P. Luciani, O. Jehl and P. Taillibert, "When oscillators stop oscillating," *Proceedings of the twelfth IJCAI, Sydney, August 1991.*

[Dague and Taillibert, 1996] P. Dague, P. Taillibert, "The Monostable: A reference Problem for Analog Diagnosis," *DX'96, 7th Int. Workshop on Principles of Diagnosis, Val Morin, Quebec, Canada, October 1996.*

[Devès et al., 1991] P. Devès, C. Fischer, P. Taillibert, "An Alternative to Expert Systems for Electrical Diagnosis", in Future Generation Computer Systems, Vol 7 91/92.

[De Kleer and Williams, 1987] J. de Kleer and B.C. Williams, "Diagnosis Multiple Faults,' *Artificial Intelligence 32, 1987.*

[Hamscher, 1989] W. Hamscher, "Temporally Coarse Representation of behaviour for Model-based Troubleshooting of Digital Circuits," *IJCAI, Detroit, August 1989.*

[Loiez and Taillibert, 1996a] E. Loiez, P. Taillibert, "Analog Systems Diagnosis: Modeling with Temporal Band,", *CESA96 IMACS-IEEE/SM, July 1996, Lille, FRANCE.*

[Loiez and Taillibert, 1996b] E. Loiez, P. Taillibert, "Polynomial Temporal Band Sequences for Analog Diagnosis,", *DX'96 7th Int. Workshop on Principles of Diagnosis, October 13-16, 1996, Val Morin, CANADA.*

[Loiez 1997] E. Loiez, "Contribution au Diagnostic de Systèmes Analogiques", *Thèse de l'Université des Sciences et Technologies de Lille -in french-, March 3, 1997, Lille, FRANCE.*

[Moore, 1966] R.E.Moore, "Interval Analysis", Prentice Hall, New Jersey, 1966

[Ng, 1990] H.T. Ng, "Model-based Multiple Fault Diagnoses of Time Varying, Continuous Physical Devices," *Proceedings of the sixth IEEE Conference on AI Applications, Santa Barbara, March 1990.*

[Pan, 1984] Y. Pan, "Qualitative reasoning with deep-level mechanism models for diagnoses of mechanism failures," *Proceedings of the first IEEE Conference on AI Applications, Denver, 1984.*

[Reiter, 1987] R. Reiter, "A Theory of Diagnosis from first Principles," *Artificial Intelligence 32, 1987.*

[Williams, 1986] B.C. Williams, "Doing Time: Putting Qualitative Reasoning on Firmer Ground," *Proceedings of AAAI Conference, Philadelphia, August 1986.*

---

1. DOGS processes only linear functions; it eliminates unmeasurable quantities by a gaussian elimination and generates all the equations by considering all the possible pivotal elements
2. As in CATS, equations modeling such components are given by experts and rely upon high-level parameters (gain, ...).

# Fundamentals of Model-Based Diagnosis of Dynamic Systems

## Peter Struss

TU Munich, Orleansstr.34, D-81667 Munich, Germany
OCC'M Software GmbH, Gleissentalstr. 22, D-82041 Deisenhofen, Germany
struss@informatik.tu-muenchen.de

### Abstract

The paper discusses theoretical foundations and practical aspects of applying model-based diagnosis (particularly consistency-based diagnosis) to dynamic systems. Many approaches to this task take it for granted that it requires simulation of the system being diagnosed. We present conditions for avoiding the often prohibitively expensive step of simulation, which are stated as properties of the model and the predictive algorithm and the observability of the system. The results provide design criteria for models and diagnostic systems and a foundation for tackling new significant types of applications. This fact is illustrated by a case study on diagnosis of the hydraulic circuit of an anti-lock braking system.

## 1 Introduction

Model-based diagnostic systems are becoming fairly successful and starting to address industrial applications. Many, if not most, systems can be regarded as some variant of the *General Diagnosis Engine* (GDE), following the principle of *consistency-based diagnosis* ([Dressler-Struss 96]):
- In order to perform fault detection and fault localization, check whether the observations about the actual device behavior are consistent with the behavior predicted by a model of the correct device (or some part of it).
- For fault identification, check consistency of observations with models of faulty behavior.

For static systems (or, rather, systems represented by a static model), such as combinatorial circuits, this amounts to checking satisfiability of a set of constraints, representing the observed state(s) and the state constraints of the device components.

In principle, of course, consistency-based diagnosis also applies to dynamic systems, but it requires checking the consistency of device **behaviors over time** with behaviors allowed by a dynamic model of the device. At a first glance, this inevitably demands
- tracking of the actual behavior over time and
- simulation of the modelled behavior

in order to check consistency of the results. This idea underlies several approaches (e.g. MIMIC, [Dvorak - Kuipers 92]) and has, on the other hand, probably prevented practical solutions to challenging problems. The reason for the latter is that the simulation task and the comparison of behavior sets can often be prohibitively expensive. This holds, in particular, when qualitative simulation yields ambiguous results and when fault identification requires simulation of many fault scenarios.

Other approaches, such as [Dressler 96], avoid simulation and generate diagnostic candidates based on checking consistency of the model with observed **states** only, (as opposed to observed **behaviors**, i.e. sequences of states); but so far a formal analysis of their preconditions and consequences is lacking.

Consistency of observed states with the model is a necessary condition for the observed behavior to be consistent with the model. Are there conditions under which this is **sufficient**? [Malik-Struss 96] states a condition for the equivalence of state-based and simulation-based diagnosis without giving a proof.

This paper discusses theoretical foundations of model-based predictors and consistency-based diagnosis (section 2) in order to derive criteria for the utility and equivalence of the approaches. This is done taking **practical** aspects of diagnosis into account, in particular measurability (section 3), and with the goal of designing diagnostic **algorithms** appropriately (section 4). This is illustrated by a case study on diagnosis of the hydraulic subsystem of an anti-lock braking system (section 5).

Many of the results are fairly fundamental. In fact, so fundamental that it is surprising they have never been spelled out in the literature, even more since they have a considerable **practical** impact on diagnostic systems.

## 2 Theoretical Foundations

### 2.1 Consistency-based diagnosis

As pointed out in the introduction, the key lies in checking whether or not a set of observations, OBS, contradicts the models of certain *behavior modes* of the device, the correct behavior (to perform fault detection) or faulty behaviors (for fault identification) which result from certain component faults and/or structural faults in the device. If such a behavior model of a mode, *model(mode)*, and the set OBS are considered as logical theories, the task is to check their joint consistency:

$$model(mode) \cup OBS \overset{?}{\vdash} \bot .$$

If the behavior model does not capture temporal aspects, it is a set of constraints, *state-constraints*, that restricts the set of states possible under the respective mode. Alternatively, we can think of it as being represented by this set, *STATES(mode)*. For our purposes, we have to extend the concepts.

## 2.2 Model-based prediction of behaviors

To characterize the evolution of a behavior over time, the behavior model does not only have to constrain the states, but also **relations between states across time**:

$$model(mode) = state\text{-}constraints(mode) \cup temp\text{-}constraints(mode).$$

The domain, $DOM(\underline{v})$, of the vector $\underline{v}$ of all variables describes the set of all (theoretically) possible states. We call the triple

$$(\underline{v}, DOM(\underline{v}), T),$$

where T is a universe of time instances, a *representation*(al space) for modeling. A *behavior* is seen as specifying the state at each time instance and, hence, can be defined in this representation as a mapping

$$behvr: T \rightarrow STATES \subseteq DOM(\underline{v}),$$

or, alternatively, as the graph of this mapping

$$\{(t, \underline{v}(t)) \mid t \in T\}.$$

Then *state-constraints* restricts STATES, whilst *temp-constraints* restricts the possible mappings, *behvr*, or their graphs.

We need a clear understanding of what kinds of constraints go into the different parts of the model. For instance, if we choose a set of ordinary differential equations (ODEs) to model a behavior mode, then this constitutes the *state-constraints*. This may seem surprising, since, after all, ODEs are supposed to express temporal behavior. However, the equations themselves only restrict the values of the involved variables and their derivatives at each time point. Restrictions on the temporal evolution of the described system are based on rules that capture properties of **continuity, integration, and derivatives** (CID). Usually, these restrictions are only implicitly represented in procedures, e.g. for numerical integration. For our analysis, we have to make them explicit in a set which we will call *CID-constraints*. In qualitative simulation, e.g. the system QSIM used in MIMIC, the qualitative versions of these restrictions are often captured by so-called transition rules which list the admissible pairs of neighboring states.

It is important to characterize the different form and contents of the different parts of the model: *state-constraints* captures the **specifics of the device** under a certain behavior **mode**. In contrast, *CID-constraints* comprises **general** constraints that are **independent** not only of the mode, but even **of the device** (and of time). Regarding the form, the former constraints relate **different variables** (including derivatives) at **one time** instance, whereas the latter, at least in their pure form, constrain values and derivatives of **only one variable across time**.

This splitting of dynamic models into two orthogonal sets of constraints that limit the possible states and the possible evolution of states, respectively, suggests that consistency of a behavior with a model can be checked by checking both aspects independently (as indeed done in QSIM) and motivates the following definition.

**Definition (Separable Model)**

A dynamic model *model(mode)* is called separable, if the following holds:

A behavior, *behvr*, is an admissible behavior under the model

if and only if

its states are admissible under *state-constraints* and
it satisfies its *temp-constraints*.

Formally:

$\{behvr\} \cup model(mode) \not\vdash \bot$ iff
$\{behvr\} \cup state\text{-}constraints(mode) \not\vdash \bot$ and
$\{behvr\} \cup temp\text{-}constraints(mode) \not\vdash \bot$.

A behavior model's *temp-constraints* can contain more than the *CID-constraints*. This is the case when it specifies changes of a variable over time which are independent of the *CID-constraints* or even contradict them. The former case is illustrated by a valve's state switching between *OPEN* and *CLOSED*, the latter happens when the same situation is modeled by the valve's opening area discontinuously changing from 0 to a positive value and vice versa. The *CID-constraints* for the directly or indirectly affected variables have to be suspended while such a change occurs. Such constraints on transitions which are specific for the device and usually also vary with the behavior mode will be denoted

*trans-constraints(mode)*,

and we obtain a partitioning of the dynamic part of a model:

$$temp\text{-}constraints(mode) = trans\text{-}constraints(mode) \cup CID\text{-}constraints,$$

where the latter possibly hold only with exceptions introduced by the *trans-constraints*.

Often, *trans-constraints* will violate the conditions of the above definition of a separable model, because they may mix state restrictions and temporal restrictions. For instance, difference equations, which belong to *trans-constraints*, usually do so. We will define a continuous behavior model through the absence of such constraints.

**Definition (Continuous Behavior Model/System Description)**

A behavior model *model(mode)* is called strictly continuous, if

it is separable and
*trans-constraints(mode)* = $\emptyset$, i.e.
*model(mode)*=
*state-constraints(mode)* $\cup$ *CID-constraints*.

A system description {*model(mode)*} is called strictly continuous, if all its models are.

From the above discussion, it follows that we consider the condition of a separable model in this definition as naturally satisfied if the dynamic part is confined to the *CID-constraints*. Strictly continuous system descriptions have an interesting property: since the *CID-constraints* do not depend on the mode, all models share the same *temp-constraints*. We call this property homogeneity.

**Definition (Homogeneous Dynamic Models/System Description)**
Two models
$model(mode_i) = state\text{-}constraints(mode_i) \cup temp\text{-}constraints(mode_i)$, i=1,2,
are called homogenous, if
$temp\text{-}constraints(mode_1) = temp\text{-}constraints(mode_2)$.
A system description $\{model(mode)\}$ is called homogeneous, if any two models are homogeneous.

This property turns out to have an important impact on discrimination among modes and, hence, diagnostic algorithms. This is indicated by the following proposition.

**Proposition (State and Behavior Equivalence)**
Two homogeneous models, and, hence also two strictly continuous models in the same representation, $model(mode_1)$, $model(mode_2)$,
share all states if and only if they share all behaviors:
$STATES(model(mode_1)) = STATES(model(mode_2))$
iff
$BEHAVIORS(model(mode_1)) = BEHAVIORS(model(mode_2))$.

The proof is trivial: if the STATES, i.e. the logical models of $state\text{-}constraints(mode_i)$, are equal, then
$state\text{-}constraints(mode_1) \Leftrightarrow state\text{-}constraints(mode_2)$,
and, due to homogeneity,
$model(mode_1) \Leftrightarrow model(mode_2)$,
which implies equal sets of behaviors. The other direction is obvious.

## 2.3 Consistency-based Diagnosis of Dynamic Systems

The proposition states a fundamental property of many model-based predictors and, in particular, qualitative simulation systems which, to the best of my knowledge, has not yet been pointed out in the literature. This is worth while, because the proposition indicates the possibility we are interested in: in order to discriminate between different modes (that is what diagnosis is about), we need not check for different behaviors; it suffices to check for the existence of different states.

However, we do not want to compare entire sets of states and behaviors. Using a few behaviors, or even only one (the observed one), should do. The foundation for this is provided by the following proposition and theorem.

**Proposition (Model Discrimination by State Checking)**
Let $model(mode_1)$, $model(mode_2)$ be two strictly continuous models in the same representation (or; more generally, separable and homogeneous models).
If a behavior, $behvr$, is admissible under $model(mode_1)$, then
it is admissible under $model(mode_2)$ iff it contains only states admissible under $state\text{-}constraints(mode_2)$:
if $\{behvr\} \cup model(mode_1) \not\vdash \bot$
then
$\{behvr\} \cup model(mode_2) \not\vdash \bot$
iff $\{behvr\} \cup state\text{-}constraints(mode_2) \not\vdash \bot$.

This follows directly from the definitions:
$\{behvr\} \cup model(mode_1) \not\vdash \bot$
implies
$\{behvr\} \cup temp\text{-}constraints(mode_1) \not\vdash \bot$.
Because the models are homogeneous,
$temp\text{-}constraints(mode_1) = temp\text{-}constraints(mode_2)$,
and $model(mode_1)$ is separable, the conclusion is obtained.
We are still comparing behaviors under different **models**. However, in a diagnostic setting, we have to check behaviors of the **real device** for consistency with one or more models. Especially for fault detection, the only model we want to use is the model of correct behavior. But the models of faulty behavior do not really have to be present and used in the diagnostic system. If $behvr$ is the description of a real behavior under a particular behavior mode in a representation ($\underline{v}$, DOM($\underline{v}$), T) and satisfies the *CID-constraints* for this representation, this suffices to exploit a separable model we compare $behvr$ with. With this step, the above proposition yields the following theorem.

**Theorem (Mode Discrimination by State Checking)**
Let $model(mode)$ be a strictly continuous model and $behvr$ the description of a real behavior in the same representation, satisfying the *CID-constraints*.
Then $behvr$ is admissible under $model(mode)$
iff
it contains only states admissible under $state\text{-}constraints(mode)$.

This means, all the diagnostic system has to do is to perform a consistency check of all states of the real behavior with the *state-constraints* of the model. The temporal constraints can be ignored, and, in particular, **no simulation of the modeled behavior is required. It simply could not reveal additional contradictions**. Also, it suffices to detect **a single state** to be inconsistent with the model, which means we do not have to rely on a temporally dense description of the behavior. All this sounds too wonderful to be true and to contradict many practical experiences. Indeed, we have to consider a pragmatic precondition for diagnosis that has been ignored in our theoretical considerations, so far.

## 3 Temporal Constraints and Measurability

Our definition of states and behaviors or, more precisely, of their description in a particular representation requires completeness: a state assigns a value to each variable (and derivative) that occurs in the representation. In many applications, limited measurability of the device or process to be diagnosed provides us only with a partial description of states. A dynamic system, by definition, has internal states that depend on previous input and states we may have no information about, and it is likely that the state descriptions are fairly incomplete.

There are three relevant limitations to measurability:
- The measurements of variables have limited precision.
- Only a subset of the variables in the respective representation can be measured, i.e. only a sub vector $\underline{v}_{obs} = p_{obs}(\underline{v})$, where $p_{obs}$ is a projection.
- We can obtain measurements only for a subset $T_{obs} \subset T$ of the temporal universe.

While the first limitation indeed affects both state checking

and behavior checking, there is a chance that using *CID-constraints* may help to detect more inconsistencies. Let $behvr_{obs}$ be the partial description of an actual behavior:
$behvr_{obs} = p_{obs} \circ behvr \mid T_{obs} : T_{obs} \to DOM(\underline{v}_{obs})$,
where „∘" denotes composition of mappings and „|" restriction of the mapping. But how should simulation be able to overcome limitations imposed by measurability? If
$\{behvr_{obs}\} \cup state\text{-}constraints(mode) \not\vdash \bot$,
how can joining *CID-constraints* change the situation to
$\{behvr_{obs}\} \cup state\text{-}constraints(mode) \cup CID\text{-}constraints$
$\vdash \bot$ ?

Basically, the *CID-constraints* can do so by improving the behavior description in two respects:

- Since they constrain variable values and derivative over time, they could complement the existing **partial state descriptions**. This is particularly true for derivatives which are difficult to measure, but might be determined (or estimated) based on variable values of adjacent states.
- They may complement the **partial behavior description** by inferring (partial) descriptions of states that were not directly observed, i.e. by extending $T_{obs}$. The mean value theorem can derive information about intermediate, unobserved states from values measured at time instances of $T_{obs}$.

In this sense, **exploiting the temporal constraints can compensate for limited measurability** of a device and be superior to simple consistency checking of states. However, it is necessary to ensure for any class of devices and measurement conditions whether or not this is actually true. In [Malik-Struss 96], we state a sufficient condition for the case where *CID-constraints* will not improve the diagnosis result.

The first part of this condition states that the sampling rate suffices to guarantee that no state in the evolution of a behavior is missed by the observation ("observations without gaps"). This may be fulfilled for qualitative behavior descriptions. Obviously, this precondition denies that the *CID-constraints* can reveal inconsistencies by providing information about unobserved states.

The second part formulates that measurements need not provide a complete state description, but only have to be "complete enough" to make the important distinctions between different modes visible. We restate and reformulate the definition.

**Definition (Measurability for Fault Detection)**

Let $\{model(mode)\}$ be a strictly continuous system description. Let the measurability of a device be characterized by
$p_{obs}: DOM(\underline{v}) \to DOM(\underline{v}_{obs})$.
The condition of measurability for fault detection is satisfied if at least one distinctive state can be measured: For any behavior *behvr* under a fault mode the following statement holds
If $\{behvr\} \cup (state\text{-}constraints(mode_{correct})) \vdash \bot$
then $\{behvr_{obs}\} \cup (state\text{-}constraints(mode_{correct})) \vdash \bot$.

(A similar condition can be stated for fault identification). If this precondition is satisfied, application of *CID-constraints* would be prevented from detecting additional inconsistencies based on completion of partial state descriptions. Thus, we obtain the following proposition stated in [Malik-Struss 96].

**Proposition (Fault Detection by Checking Measured States)**

Let $model(mode_{correct})$ be a strictly continuous model and *behvr* the description of a real behavior in the same representation, satisfying the *CID-constraints*. If the observations are without gaps and the condition of measurability for fault detection is satisfied then

*behvr* is admissible under $model(mode_{correct})$

iff

$behvr_{obs}$ contains only states admissible under $state\text{-}constraints(mode_{correct})$.

(Again, a respective proposition can be stated for the problem of fault identification.) In other words, under the conditions of this theorem, **diagnosis based on checking state consistency yields results equivalent to diagnosis based on checking state and temporal constraints**, which means simulation is needless.

An important question is whether we have a chance to determine if the condition of measurability for fault detection holds. We are able to formulate valid models of the relevant fault behaviors in the representation used by $model(mode_{correct})$, we can establish a necessary condition, namely whether or not $p_{obs}$ preserves the **distinctive states between the models:**

Measurability for fault detection is not satisfied, if there exists a *mode* such that the measurements of all states that are admissible under *model(mode)* but not under $model(mode_{correct})$ are consistent with $model(mode_{correct})$, i.e.

$\forall s \in STATES(model(mode)) \setminus STATES(model(mode_{correct}))$
$p_{obs}(s) \in p_{obs}(STATES(model(mode_{correct})))$.

This provides a criterion that can be checked by analyzing sets of states and applying a projection. Note that the models of fault modes are only used for this analysis and need not be represented in the diagnostic system. Because it cannot be excluded that a fault model contains whose projection to observables is inconsistent with the model of correct behavior but which does not occur in any real behavior, the proposition yields only a necessary condition.

Yet, there are fast processes that may not allow observations without gaps, and measurability may be bad enough to violate the second criterion. In such cases it is still worth while to check whether exploiting the *CID-constraints* actually improves results of consistency checking. But even if there is evidence of this possibility, there are different ways to achieve it. Performing simulation is not the only one.

## 4 State-based vs. Simulation-based Diagnosis

So far, we analyzed when and why behavior models can omit temporal constraints without impairing the results of the diagnosis. Now, let us assume we are certain that we have to deal with a situation that forces us to exploit the temporal constraints. Here, we consider strictly continuous systems again. The task of the diagnostic algorithm to be devised is to check consistency of observations with both *state-constraints* and *CID-constraints* :

$$\textit{state-constraints(mode)} \cup \textit{CID-constraints} \cup \text{OBS} \overset{?}{\vdash} \bot,$$

possibly for different modes if we are interested in fault identification. There are two extreme ways for performing this task, and, certainly, a number of mixed forms. Both ways could start by pruning the modes through

$$\textit{state-constraints(mode)} \cup \text{OBS} \overset{?}{\vdash} \bot.$$

Then, we can first compute the results of

$$\textit{state-constraints(mode)} \cup \textit{CID-constraints},$$

which basically means simulation (through integration) or constructing an envisionment for one or more modes. This corresponds to **extending information about the possible behaviors** which can then be checked for consistency with the observed behaviors in OBS. This can be expensive for several reasons, for instance, if we have to consider many modes, or if there is no metric information about time and it is not obvious how many simulation steps have to be carried out. Also, comparing behaviors involves both checking states and state transitions.

Alternatively, we can use *CID-constraints* to **extend information about the actual behavior** by computing deductions from

*CID-constraints* ∪ OBS

mainly by estimating derivatives and applying rules like the mean value theorem and then check with the newly derived information for states inconsistent with *state-constraints(mode)* for the respective modes. The advantage lies in applying *CID-constraints* only once and in the restricted context of OBS, as opposed to many modes. Also, no transition checking is required. This has an intuitive appeal, because, after all, it is the limitations in OBS that lead us to seeking the help of the *CID-constraints*. Hence, using them to enhance OBS seems appropriate. A potential source of problems is a situation where values of a variable vary strongly across neighboring samples and combination of results of the mean value theorem for different variables creates many intermediate states. Also, temporal information about derived intermediate states will tend to be weak. As stated above, there are mixed versions, for instance, in guiding the simulation task tightly by the incoming observations etc.

Although there is some evidence that the second scheme is advantageous for many situations, we are far from suggesting one single best approach to this task. We present the discussion to show that the often advocated simulation-based approach is by no means compelling, but, on the contrary, highly questionable. Much more work is needed to develop good designs and criteria for their utility. This will require a more detailed analysis of the form and contents of the *CID-constraints* and their possible applications and relating their results to various limitations in measurability.

## 5 Case Studies

The motivation for this work and the expectations are generated by the attempt to build systems that tackle industrial applications beyond the scope of previous model-based systems. One example is (off-board) diagnosis of the hydraulic circuit of an anti-lock braking system (ABS) used in cars. In this section, we first try to summarize key features of this problem, convey the basic ideas underlying our solution, relate them to the issues raised in this paper, and report some experimental results obtained. Details are presented in [Struss et. al. 96].

The purpose of an ABS is to prevent the wheels of the vehicle from locking up in order to enable the driver to be able to steer the car while using the brakes. This is achieved by controlling the pressure which is exerted on the wheel brake cylinders by pushing the pedal via the hydraulic circuit. The speed of each wheel is measured, and when the measurements indicate a tendency of a wheel to lock up, because the (negative) acceleration is too strong, the Electronic Control Unit reduces the pressure for some time, before increasing it again for the next de-acceleration phase. Fig. 1 shows one subsystem of the hydraulic circuit which typically affects two diagonally opposite wheels. For each wheel, an increase in pressure is achieved by an open inlet valve and a closed outlet valve. For maintaining pressure level, the inlet valve is closed, and for reduction of the pressure the outlet valve opened. The latter step is supported by a reservoir chamber that fills quickly in this phase. Also the pump starts immediately to transport the liquid back towards the main cylinder, and the next cycle may start, if necessary.

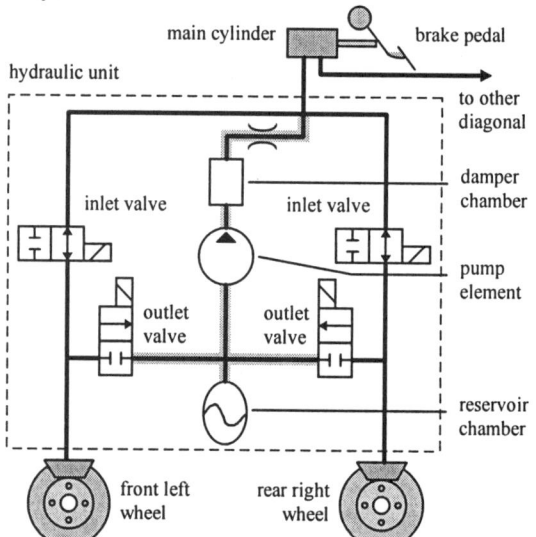

**Figure 1: Hydraulic Circuit of ABS**

Diagnosing this subsystem requires solving a number of challenging problems:
- Obviously, this is a system with **complex dynamics**.
- Observations are sparse: there is **not a single sensor** in the hydraulic circuit (the level sensor for the braking liquid is of no help). Information about pressures can only be obtained indirectly from the (de-)acceleration of the wheels. However:
- For **off-board diagnosis**, available observations are inherently **qualitative** in nature **and not temporally specified**. Typical observations would be "one wheel

tends to lock up" (indicating too high pressure in the respective brake cylinder) or "the brake pedal is too soft" (as a result of an unusually low pressure in the main cylinder). Even more precise values of speed and acceleration present on-board do not help too much, since:
- **Measurement** of strongly **influential exogenous factors** (e.g. the state of the road surface) is **impossible**.

In response to these conditions, we adopted an approach described in [Malik-Struss 96] using models that are stated in terms of qualitative deviations of variables and parameters from some unspecified and potentially changing nominal value.

It turns out that the "measurements" characterized above, enable the models to infer deviations in the pressure in different parts of the circuit. This trivially suffices to establish measurability for fault detection, but not measurability for fault identification and localization. The reason lies in the lack of information about the derivatives of pressures which cannot be provided by the observations and the state-constraints. Basically, this information would help to detect significant inconsistencies because resistive elements like valves relate (deviations of) flow to (deviations of) pressure, whereas pipes and other containers link (deviations of) flow and (deviations of) derivatives of pressure.

The solution was indeed provided by exploiting *CID-constraints*. However, they were not used for simulation of correct or faulty behavior modes, but to complement the observations in the following manner:

For each phase in the cycle (determined by valve positions), strict continuity can be assumed. Furthermore, we assume that there were no deviations when the phase was entered (this limits the applicability of the solution). If a deviation of a variable v was zero initially and different from zero after a while, then the *CID-constraints* tell us that there must exist a time interval during which both the deviation of v occurred and the derivative of v has a deviation with the same sign. In other words, *CID-constraints* were combined with OBS to deliver information about derivatives.

This turned out to be fairly successful: for a sample of component faults, such as clogged or punctured valves, defect pump, and air included, fault localization was successful: the respective faults were included in the set of single fault diagnoses, sometimes as only possible ones.

Other case studies that use related approaches and shed a light on its utility, are reported in the literature: [Dressler 96] empirically discovered the possibility of fault detection via state checking in a prototype for diagnosing ballast water tank systems in off-shore platforms (a fairly sensor-rich system). [Chantler et al. 96] reports results on equivalence of integration-based and differentiation-based algorithms for the special case of numerical models. [Williams-Nayak 96] pursues a transition-oriented, as opposed to simulation-oriented, approach to diagnosis and reconfiguration applied to the propulsion system of a space craft. [Malik-Struss 96] covers fault detection and identification in a simplified system (a controlled electric motor) with limited observability, but measurable derivatives based only on state checking.

# 6 Summary

As [Malik-Struss 96] states, "Diagnosis of dynamic systems does not necessarily require simulation". This paper presented theoretical foundations and practical considerations supporting model-based diagnostic systems that are confined to using state constraints for consistency checking and do without temporal behavior constraints, or use them for purposes other than simulation. There are empirical results that provide evidence of the utility of such approaches, but also attempt steps towards identifying their limitations, particularly limitations that result from limited observability of the system. We hope that our analysis
- discourages overly "straightforward" approaches to model-based diagnosis,
- encourages more steps towards a systematic and thorough investigation leading to the design of diagnosis systems that are both well-founded and efficient,
- shows that such steps require a specific and detailed analysis of properties of models and predictive engines on the one hand and their interrelationship with observability on the other hand.

The benefit of these efforts will be beyond theoretical insight and result in widening the scope of feasible industrial applications which often involve dynamic systems.

## References

[Chantler et al. 96] M. J. Chantler, S. Daus, T. Vikatos and G. M. Coghill, *The Use of Quantitative Dynamic Models and Dependency Recording for Diagnosis*, in: Workshop Notes of the Seventh International Workshop on Principles of Diagnosis (DX-96), Montreal, 1996.

[Dressler 96] Oskar Dressler, *On-line Diagnosis and Monitoring of Dynamic Systems based on Qualitative Models and Dependency-based Diagnostic Engines*, in: Proceedings of the European Conference on Artificial Intelligence (ECAI-96), John Wiley & Sons, 1996.

[Dressler-Struss 96] Oskar Dressler and Peter Struss, *The Consistency-based Approach to Automated Diagnosis of Devices*, in: Brewka (ed.), Principles of Knowledge Representation, CSLI, 1996.

[Dvorak-Kuipers 92] Daniel Dvorak and Benjamin Kuipers, *Model-based Monitoring of Dynamic Systems*, in: W. Hamscher et al. (eds.), Readings in Model-based Diagnosis, Morgan Kaufmann Publishers, 1992.

[Malik-Struss 96] Andreas Malik and Peter Struss, *Diagnosis of Dynamic Systems Does Not Necessarily Require Simulation*, in: Workshop Notes of the Seventh International Workshop on Principles of Diagnosis (DX-96), Montreal, 1996.

[Struss et al. 97] Peter Struss, Andreas Malik and Florian Dummert, *Diagnosing a Dynamic System with (almost) no Observations*, in: Workshop Notes of the Eleventh International Workshop on Qualitative reasoning (QR-97), Cortona, 1997.

[Williams-Nayak 96] Brian Williams and Pandurang Nayak, *A Model-based Approach to Reactive Self-Configuring Systems*, in: Workshop Notes of the Seventh International Workshop on Principles of Diagnosis (DX-96), Montreal, 1996.

# Comparative Analysis of Structurally Different Dynamical Systems

H. de Jong and F. van Raalte
Knowledge-Based Systems Group, Department of Computer Science
University of Twente, P.O. Box 217, 7500 AE Enschede, the Netherlands

## Abstract

CEC* is an implemented method for comparative analysis (CA) which is able to deal with intra-model and inter-model CA problems within a single framework. The algorithm constructs a comparative envisionment showing (1) possible effects of differences in the antecedent conditions or (2) possible causes of differences in the response of the two systems. CEC* has been proven sound and incomplete.

## 1 Introduction

Qualitative reasoning (QR) is concerned with predicting and explaining the qualitative behavior of dynamical systems. An important QR problem is the comparison of behaviors of two systems, that is, an assessment of the relative value of system variables at chosen time-points. More specifically, this *comparative analysis (CA)* task is directed at (1) predicting possible consequences of differential antecedent conditions and (2) finding possible causes of differential responses of the two systems.

CA questions arise in a variety of contexts. Consider an engineer diagnosing an electrical circuit by comparing the observed faulty behavior with a reference behavior. Or scientists estimating the impact of a disturbing process on a quantity measured in an experiment.

With few exceptions, QR methods for comparative analysis have focussed on the comparison of systems with the same structure: *intra-model* CA [Weld, 1990; Chiu and Kuipers, 1992; Neitzke and Neumann, 1994; de Jong et al., 1996]. The comparison of dynamical systems with a different structure, *inter-model* CA, is considered to be a "terribly difficult" problem in general [Weld, 1990]. The limitation to intra-model CA is a serious shortcoming, since many interesting questions can be mapped onto inter-model CA problems, including the examples given above.

In this paper we introduce CEC*, an implemented method for comparative analysis which treats intra-model CA *and* inter-model CA within a single framework [de Jong, 1996]. The exposition starts with the definition of some basic concepts and a formulation of the method as a propagation process (sec. 2). The constraints employed in the propagation process form the topic of sec. 3. Sec. 4 shows how CEC* deals with an example of a predictive and a diagnostic CA question. The discussion in sec. 5 gives correctness guarantees and pays attention to the generality of the method. Brief notes on further research conclude the paper.

## 2 Basic concepts and outline of method

### 2.1 Pairs of comparison

In a comparative analysis a qualitative behavior of a first system is compared with a behavior of a second system. We assume that the qualitative behaviors have been generated by the QSIM algorithm [Kuipers, 1994], although the principles underlying our method do not depend on this choice.

Fig. 1 shows a system of two cascaded tanks. The (constant) input variables are the inflow $i$ and the sizes $r_u$, $r_l$ of the orifices of the upper and lower tank. The amounts of water $a_u$, $a_l$ in the tanks form the state variables. One of the qualitative behaviors produced by QSIM for the cascaded tanks is summarized in fig. 1(c). As can be seen, the upper tank reaches equilibrium *before* the lower tank.

Now suppose we compare this system with a second cascaded-tanks system that has a *leaky* upper tank. The leaky system is structurally different from the watertight one: the netflow into the upper tank is given by $\hat{n}_u = \hat{i} - \hat{o}_u - \hat{o}_h$, where $\hat{o}_h = h(\hat{a}_u)$, $\partial h / \partial \hat{a}_u > 0$.[1] The qualitative behavior in which the upper tank reaches equilibrium first is also a behavior of the leaky system.

We will compare the watertight and leaky cascaded tanks when both exhibit the behavior of fig. 1(c). The behaviors of the two systems are topologically equal [Weld, 1990] with respect to the shared variables (a situation not obtaining in general, see sec. 4). The two behaviors are compared at so-called *pairs of comparison*. A pair of comparison $pc$ is a pair of distinguished time-points $\langle t, \hat{t} \rangle$ from the first and the second behavior, respectively.

---

[1] Following [Weld, 1990], the hat accent $\hat{\ }$ will be used to denote the variables of the second system.

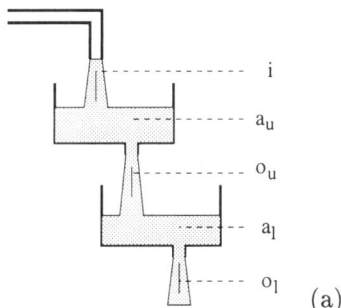 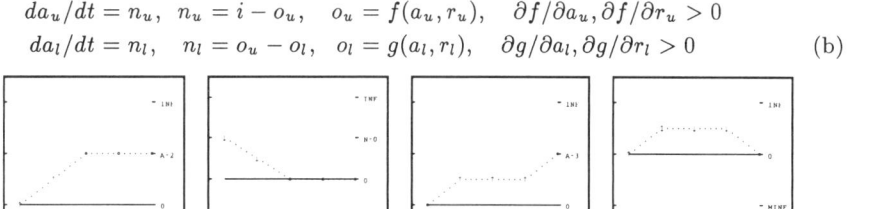

Figure 1: Qualitative description of the structure and behavior of a watertight cascaded-tanks system (adapted from [Kuipers, 1994]). The variables have the following interpretation: $a$ water amount, $r$ size of orifice, $o$ outflow, $i$ inflow, and $n$ netflow. The subscripts $_u$, $_l$, and $_h$ refer to the upper tank, lower tank, and leak, respectively. The system (a) has the qualitative differential equation (QDE) (b). A qualitative behavior of the system is shown in (c).

**Def. 1 (Meaningful pairs of comparison)** Two behaviors are given with sequences of distinguished time-points $t_0, \ldots, t_n$ and $\hat{t}_0, \ldots, \hat{t}_m$. The pair of comparison $pc = \langle t, \hat{t} \rangle$ is a *meaningful* pair of comparison, if

1. $t$ and $\hat{t}$ are the initial time-points $t_0$ and $\hat{t}_0$ or the final time-points $t_n$ and $\hat{t}_m$;

2. $t$ and $\hat{t}$ are time-points at which variables $p$ and $\hat{p}$ reach the same basic landmark value $0$, $\infty$, or $-\infty$.

A particular pair of comparison could be singled out by both of the criteria in the definition. For the behaviors of the watertight vs. the leaky cascaded-tanks system we find three meaningful pairs of comparison: $pc_0 = \langle t_0, \hat{t}_0 \rangle$, $pc_1 = \langle t_1, \hat{t}_1 \rangle$, $pc_2 = \langle t_2, \hat{t}_2 \rangle$ (see fig. 1(c)).

A partial ordering relation can be defined on the pairs of comparison which formalizes the intuitive notion that one pair of comparison occurs before another. For two pairs of comparison $pc_1 = \langle t_1, \hat{t}_1 \rangle$ and $pc_2 = \langle t_2, \hat{t}_2 \rangle$ it holds that $pc_1 \preceq pc_2$, iff $t_1 \leq t_2$ and $\hat{t}_1 \leq \hat{t}_2$. If $pc_1$ and $pc_2$ are different pairs of comparison and $pc_1 \preceq pc_2$, then $pc_1$ is said to be a *predecessor* of $pc_2$ (and $pc_2$ a *successor* of $pc_1$). A set of pairs of comparison ordered according to the $\preceq$-relation will be called an *ordered pairs of comparison (OPC) structure*. This structure always has a top element, the first pair of comparison, and a bottom element, the last pair of comparison. In the example we have the simple OPC structure $pc_0 \preceq pc_1 \preceq pc_2$.

Two successive pairs of comparison in an OPC structure mark out what we will call a pair of *behavior fragments*, one for each system. The behavior fragments defined by the successive pairs of comparison $pc_1 = \langle t_1, \hat{t}_1 \rangle$ and $pc_2 = \langle t_2, \hat{t}_2 \rangle$ are the closed intervals $[t_1, t_2]$ and $[\hat{t}_1, \hat{t}_2]$. If the interval of a behavior fragment does not contain any distinguished time-points except for its boundaries, we will say it is a *primitive* behavior fragment, otherwise a *composite* behavior fragment. Combining the three pairs of comparison identified for the cascaded-tanks example with the qualitative behavior in fig. 1(c), we directly see that all behavior fragments ($[t_0, t_1]$, $[t_1, t_2]$, $[\hat{t}_0, \hat{t}_1]$, $[\hat{t}_1, \hat{t}_2]$) are primitive.

## 2.2 Relative values

Pairs of comparison have been introduced to specify when the values of variables in the two systems can be meaningfully compared. In order to express how the values of a shared variable $p$ in the first and second system relate at a pair of comparison, the notion of *relative value (RV)* is introduced.

**Def. 2 (Relative value)** Given a variable $p$ and a pair of comparison $pc = \langle t, \hat{t} \rangle$. The relative value (RV) at $pc$ is defined as:

$$p \Uparrow_{pc}, \text{ iff } \hat{p}(\hat{t}) > p(t),$$
$$p \|_{pc}, \text{ iff } \hat{p}(\hat{t}) = p(t),$$
$$p \Downarrow_{pc}, \text{ iff } \hat{p}(\hat{t}) < p(t).$$

Often the RV of a variable at a pair of comparison $pc$ is denoted by $RV(p)_{pc}$, and $RV(p)_{pc} = \Uparrow$ is equivalent to $p \Uparrow_{pc}$. Remark that a relative value is only defined for variables that are shared by the two systems. A special kind of RV is the relative duration of the behavior fragments defined by two successive pairs of comparison. Written as $RV(T)_{pc_1 \to pc_2}$, this RV expresses whether the intervening behavior fragment of the second system has a longer, shorter, or equal duration compared to that of the first system.

A few basic (in)equality relations and algebraic operations for RVs can be introduced. This simple algebra is used in the definition and evaluation of propagation constraints for comparative analysis that are discussed below.

**Def. 3 (Equality and inequality)** For two relative values the equality $RV(p) = RV(q)$ and inequality $RV(p) < RV(q)$ relations are defined as follows:

| | $RV(q)$ | | | | | $RV(q)$ | | |
|---|---|---|---|---|---|---|---|---|
| = | $\Uparrow$ | $\|$ | $\Downarrow$ | | < | $\Uparrow$ | $\|$ | $\Downarrow$ |
| $\Uparrow$ | T | F | F | | $\Uparrow$ | T | F | F |
| $RV(p)$ $\|$ | F | T | F | $RV(p)$ | $\|$ | T | F | F |
| $\Downarrow$ | F | F | T | | $\Downarrow$ | T | T | T |

Similar definitions can be given for $>$, $\geq$, $\leq$, and $\gtreqless$.

**Def. 4 (Basic algebraic operations)** For two relative values the addition $RV(p) + RV(q)$ and subtraction $RV(p) - RV(q)$ operations are defined as follows:

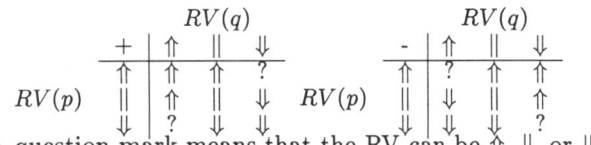

A question mark means that the RV can be $\Uparrow$, $\|$, or $\Downarrow$.

By means of the concepts introduced above comparative analysis can be formalized as a propagation problem. The input of CEC* consists of a pair of qualitative behaviors, the QDEs giving rise to them, and a set of relative values at either the first or the last pair of comparison, that is, the top or the bottom of the OPC structure. The analysis proceeds by propagating the initial RVs through the OPC structure, from each pair of comparison to its successor(s) or predecessor(s). When the direction of propagation is from top to bottom, the method is *predicting* the consequences of differential antecedent conditions. The reverse direction corresponds with the *explanation* or *diagnosis* of differential responses of the two systems.

For this propagation process to succeed we need *propagation* or *RV constraints*. These constraints relate RVs at the same pair of comparison or at successive pairs of comparison.

## 3 Constraints on relative values

### 3.1 Variables at a pair of comparison

A direct way to infer the RV of a variable $p$ at a pair of comparison $pc = \langle t, \hat{t} \rangle$ is to examine the qualitative value of the continuously differentiable functions $p$ and $\hat{p}$ at $t$ and $\hat{t}$. The qualitative value of a variable $p(t)$ is either a landmark value $l_j$ or an open interval between two landmark values $]l_j, l_{j+1}[$ in which $p(t)$ falls. If we extend the applicability of $>$, $=$, and $<$ to intervals, the following result is obtained:[2]

**Prop. 1 (Qualitative values)** Let $pv$ and $\hat{pv}$ be qualitative values of $p$ and $\hat{p}$, respectively, at pair of comparison $pc = \langle t, \hat{t} \rangle$. For the difference $\hat{p}(\hat{t}) - p(t)$ we have:

|  | $\hat{p}(\hat{t}) - p(t)$ | $> 0$ | $= 0$ | $< 0$ |
|---|---|---|---|---|
| $\hat{pv}$ | $> 0$ | ? | $> 0$ | $> 0$ |
|  | $= 0$ | $< 0$ | $= 0$ | $> 0$ |
|  | $< 0$ | $< 0$ | $< 0$ | ? |

Knowing the sign of the difference $\hat{p}(\hat{t}) - p(t)$ immediately gives the relative value of $p$ at $pc$, $RV(p)_{pc}$. For example, in fig. 1(c) we see that at $pc_0$ $a_u = 0$ and $\hat{a}_u = 0$, so that $RV(a_u)_{pc_0} = \|$.

### 3.2 Functions at a pair of comparison

Suppose that a variable $p$ shared by the two systems is a continuously differentiable function of a vector of variables $\boldsymbol{r}$ in the first system and $\boldsymbol{s}$ in the second system:

$$p = f(\boldsymbol{r}) \quad \text{and} \quad \hat{p} = g(\hat{\boldsymbol{s}}) \tag{1}$$

The problem is that $p(t)$ and $\hat{p}(\hat{t})$ cannot be straightforwardly compared, because in general the vectors $\boldsymbol{r}$ and

---
[2]For proofs of the propositions, see [de Jong, 1996].

$\boldsymbol{s}$ and the functions $f$ and $g$ will be different. To make the expressions for $p$ and $\hat{p}$ comparable, define $\boldsymbol{q}$ as the vector of variables occurring both in $\boldsymbol{r}$ and in $\boldsymbol{s}$, and $\boldsymbol{a}$ as the vector of variables occurring either in $\boldsymbol{r}$ or in $\boldsymbol{s}$, but not in both. We can now give a criterion for the comparability of $p$ and $\hat{p}$:

**Def. 5 (Comparable functions)** Suppose $p$ and $\hat{p}$ are defined by continuously differentiable functions $f$ and $g$, as in (1). The functions $f$ and $g$ are called *comparable* with respect to a continuously differentiable function $h$ and vectors $\boldsymbol{a^0}$ and $\hat{\boldsymbol{a}}^0$, if

$$p = h(\boldsymbol{q}, \boldsymbol{a^0}) \quad \text{and} \quad \hat{p} = h(\hat{\boldsymbol{q}}, \hat{\boldsymbol{a}}^0)$$

The variable $a_i^0$ ($\hat{a}_i^0$) equals either a variable in $\boldsymbol{r}$ ($\hat{\boldsymbol{s}}$) or, if not occurring in $\boldsymbol{r}$ ($\hat{\boldsymbol{s}}$), has a *comparison value* $c_i$ ($\hat{c}_i$).

An example will clarify this definition. From the cascaded-tanks QDEs we take $n_u = i - o_u$ and $\hat{n}_u = \hat{i} - \hat{o}_u - \hat{o}_h$. The vector $\boldsymbol{q}$ of shared variables is $[i \ o_u]'$ and $\boldsymbol{a} = o_h$. Now let $n_u = h([i \ o_u]', 0) = i - o_u$ and $\hat{n}_u = h([\hat{i} \ \hat{o}_u]', \hat{o}_h) = \hat{i} - \hat{o}_u - \hat{o}_h$, then $h$ makes $f$ and $g$ comparable. The comparison value $0$ is chosen for the flow out of the leak in the first system, i.e. $o_h = 0$.

With the generalized mean value theorem for differentiation we find:

**Prop. 2 (Functional relations)** The variables $p$ and $\hat{p}$ are defined by functions $f$ and $g$. If $f$ and $g$ are comparable with respect to a function $h$, then at pair of comparison $pc = \langle t, \hat{t} \rangle$:

$$\hat{p}(\hat{t}) - p(t) = \boldsymbol{d_q}(\hat{\boldsymbol{q}}(\hat{t}) - \boldsymbol{q}(t)) - \boldsymbol{d_a}(\hat{\boldsymbol{a}}^0(\hat{t}) - \boldsymbol{a^0}(t))$$

$\boldsymbol{d_q}$ and $\boldsymbol{d_a}$ are vectors with $d_{q,i} = \frac{\partial}{\partial q_i} h(\boldsymbol{\psi}, \boldsymbol{\alpha}^0)$ and $d_{a,i} = \frac{\partial}{\partial a_i} h(\boldsymbol{\psi}, \boldsymbol{\alpha}^0)$. Each $\psi_j$ lies between $q_j(t)$ and $\hat{q}_j(\hat{t})$, and each $\alpha_j^0$ between $a_j^0(t)$ and $\hat{a}_j^0(\hat{t})$.

In the cascaded-tanks example we have at $pc_0$

$$\hat{n}_u(\hat{t}_0) - n_u(t_0) = (\hat{i}(\hat{t}_0) - i(t_0))$$
$$- (\hat{o}_u(\hat{t}_0) - o_u(t_0)) - (\hat{o}_h(\hat{t}_0) - 0)$$

Since the tanks are filled from empty, $\hat{o}_h(\hat{t}_0) = 0$ and the last term disappears. The expression can be easily transformed into an RV constraint:

$$RV(n_u)_{pc_0} = RV(i)_{pc_0} - RV(o_u)_{pc_0}$$

Suppose that the inflow and outflow are known to be equal in the two systems ($i\|_{pc_0}$, $o_u\|_{pc_0}$), then we can readily infer with defs. 3 and 4 that the net flow is also equal ($n_u\|_{pc_0}$).

Prop. 2 is equally applicable to intra-model and inter-model CA problems. In the former case, the qualitative models of the two systems, and thus all corresponding functional relations, are identical. Consequently, $\boldsymbol{a} = \boldsymbol{0}$ and $\hat{p}(\hat{t}) - p(t) = \boldsymbol{d_q}(\hat{\boldsymbol{q}}(\hat{t}) - \boldsymbol{q}(t))$.

## 3.3 State variables between pairs of comparison

The question to be addressed now is how the relative values of variables at a pair of comparison $pc_1 = \langle t_1, \hat{t}_1 \rangle$ relate to those at a predecessor pair of comparison $pc_0 = \langle t_0, \hat{t}_0 \rangle$. For this, we have to take into account how the variables of the two systems change during the respective behavior fragments between the pairs of comparison.

We will assume that the two behavior fragments start simultaneously, i.e. $t_0 = \hat{t}_0$. If this is not the case, the behavior fragments have to be synchronized first by a suitable time translation. $T = t_1 - t_0$ and $\hat{T} = \hat{t}_1 - \hat{t}_0$ represent the *durations* of the two behavior fragments.

We can distinguish several cases in the comparison of two behavior fragments by considering how their durations relate. When comparing primitive behavior fragments, we have $\hat{T} < T$, $\hat{T} = T$, or $\hat{T} > T$, and we can rewrite the difference in $x$ at $pc_1$ as follows:

$$\hat{x}(\hat{t}_1) - x(t_1) = \begin{cases} x(\hat{t}_1) - x(t_1) + \hat{x}(\hat{t}_1) - x(\hat{t}_1) & , \hat{T} < T \\ \hat{x}(\hat{t}_1) - x(\hat{t}_1) & , \hat{T} = T \\ \hat{x}(\hat{t}_1) - \hat{x}(t_1) + \hat{x}(t_1) - x(t_1) & , \hat{T} > T \end{cases}$$

Consider the case $\hat{T} < T$. The difference $\hat{x}(\hat{t}_1) - x(t_1)$ is split up into $x(\hat{t}_1) - x(t_1)$ and $\hat{x}(\hat{t}_1) - x(\hat{t}_1)$. The sign of the former part directly follows from the qualitative behavior of $x$, whereas the latter part must be determined.

For the general problem of establishing the difference $\hat{x}(t_b) - x(t_b)$ in terms of the difference of variables at $t_a$, we have to study the QDEs which define the dynamical behavior of the two systems. First, we introduce three classes of variables: shared state variables $x$, shared input variables $u$, and auxiliary variables $a$. A variable $v$ is (1) a *shared state* variable if it is shared and a state variable in at least one of the systems, (2) a *shared input* variable if it is shared and an input variable in at least one of the systems, or (3) an *auxiliary* variable if it is not shared. The QDEs of the two systems give rise to the state equations:

$$\frac{dx(t)}{dt} = f_1(x(t), u(t), a(t)), \quad \frac{d\hat{x}(t)}{dt} = f_2(\hat{x}(t), \hat{u}(t), \hat{a}(t)) \quad (2)$$

We can now define a *comparison system* and transform it into a familiar linear form:

**Prop. 3 (Linear comparison system)** Consider the *comparison system* obtained by subtracting the state equations (2). If the functions $f_1$ and $f_2$ are comparable through a function $h$, such that $f_1(x, u, a) = h(x, u, a^0)$ and $f_2(\hat{x}, \hat{u}, \hat{a}) = h(\hat{x}, \hat{u}, \hat{a}^0)$, then we can write the state equation of the comparison system as:

$$\frac{d}{dt}(\hat{x}(t) - x(t)) = A(t)(\hat{x}(t) - x(t)) + B(t)(\hat{u}(t) - u(t)) + E(t)(\hat{a}^0(t) - a^0(t)) \quad (3)$$

$A(t)$, $B(t)$, and $E(t)$ are time-varying matrices of continuous functions $a_{ij}(t) = \frac{\partial}{\partial x_j} h_i(\xi(t), \nu(t), \alpha^0(t))$, $b_{ik}(t) = \frac{\partial}{\partial u_k} h_i(\xi(t), \nu(t), \alpha^0(t))$, and $e_{il}(t) = \frac{\partial}{\partial a_l} h_i(\xi(t), \nu(t), \alpha^0(t))$. $\xi_j(t)$ lies between $\hat{x}_j(t)$ and $x_j(t)$, $\nu_k(t)$ between $\hat{u}_k(t)$, and $u_k(t)$, and $\alpha_l^0(t)$ between $\hat{a}_l^0(t)$ and $a_l^0(t)$.

The proposition follows directly with the definition of comparable functions and the application of prop. 2 to each $h_i$.

For the comparison of watertight and leaky cascaded tanks, the relevant vectors of variables are $x = [a_u \; a_l]'$, $u = [i \; r_u \; r_l]'$, and $a = o_h$. With the QDE in fig. 1 we determine the following matrices in state equation (3):

$$A(t) = \begin{bmatrix} -a_{11}(t) & 0 \\ a_{21}(t) & a_{22}(t) \end{bmatrix}, \; E(t) = \begin{bmatrix} -1 \\ 0 \end{bmatrix},$$
$$B(t) = \begin{bmatrix} 1 & -b_{12}(t) & 0 \\ 0 & b_{22}(t) & -b_{23}(t) \end{bmatrix}, \quad (4)$$

where the matrix elements are given by the partial derivatives $\partial h_i / \partial x_j$, $\partial h_i / \partial u_k$, and $\partial h_i / \partial a_l$. When considering the comparison system on an interval in which $x, \hat{x}, u, \hat{u}, a^0, \hat{a}^0$ have signs as between $pc_0$ and $pc_1$, we find that $a_{ij}, b_{ij} > 0$.

Again, it is useful to point out how intra-model CA is treated as a special case of inter-model CA. When comparing two structurally identical systems, the term $E(t)(\hat{a}^0(t) - a^0(t))$ drops out of (3).

The formulation of a linear comparison system is a pivotal move in deriving RV constraints between successive pairs of comparison. An explicit solution to (3) is a basic result in linear system theory [Rugh, 1996].

**Prop. 4 (State variables)** Suppose two primitive behavior fragments are compared. The difference between the shared state variables $x$ at $pc_1 = \langle t_1, \hat{t}_1 \rangle$ can be expressed in terms of differences at $pc_0 = \langle t_0, \hat{t}_0 \rangle$ as:

$$\hat{x}(\hat{t}_1) - x(t_1) = \begin{cases} x(\hat{t}_1) - x(t_1) + F(\hat{t}_1, t_0) & , \hat{T} < T \\ F(t_1, t_0) & , \hat{T} = T \\ \hat{x}(\hat{t}_1) - \hat{x}(t_1) + F(t_1, t_0) & , \hat{T} > T \end{cases} \quad (5)$$

where the function $F$ is defined as follows:

$$F(t_b, t_a) = \Phi(t_b, t_a)(\hat{x}(t_a) - x(t_a)) + \int_{t_a}^{t_b} \Phi(t_b, \tau) [B(\tau)(\hat{u}(t_a) - u(t_a)) + E(\tau)(\hat{a}^0(\tau) - a^0(\tau))] \, d\tau \quad (6)$$

The usefulness of prop. 4 hinges on the possibility to determine the transition matrix $\Phi(t_b, \tau)$ for all $\tau$ on an interval $[t_a, t_b]$. Explicitly solving for $\Phi(t_b, \tau)$ is generally a very difficult task, but for our purposes it suffices to determine the *signs* of the elements of $\Phi(t_b, \tau)$. A direct way to find such a qualitative transition matrix is to trace the zero-input solutions of (3) in the phase space for initial states consisting of the column vectors of the unit matrix $I$ (see, e.g., [Rugh, 1996]). Here we restrict ourselves to stating the resulting transition matrix for the cascaded tanks between $pc_0$ and $pc_1$:

$$\Phi(t_b, \tau) = \begin{bmatrix} \phi_{11}(t_b, \tau) & 0 \\ \phi_{21}(t_b, \tau) & \phi_{22}(t_b, \tau) \end{bmatrix},$$

with $\phi_{ij}(t_b,\tau) > 0$ for all $\tau$ in $]t_a,t_b]$.

Armed with the vectors $\boldsymbol{x}$, $\boldsymbol{u}$, $\boldsymbol{a^0}$ and matrices $\boldsymbol{B}$, $\boldsymbol{E}$, $\boldsymbol{\Phi}$ determined above, we can use prop. 4 to express differences at $pc_1$ in terms of differences at $pc_0$. Evaluating (5) for $\hat{T} < T$, we obtain for state variable $a_u$:

$$\hat{a}_u(\hat{t}_1) - a_u(t_1) = a_u(\hat{t}_1) - a_u(t_1) + c_1(\hat{a}_u(\hat{t}_0) - a_u(t_0))$$
$$+ c_2(\hat{i}(\hat{t}_0) - i(t_0)) - c_3(\hat{r}_u(\hat{t}_0) - r_u(t_0)) - c_4 \quad (7)$$

and $c_1, \ldots, c_4 > 0$. The constant factors are transition matrix elements (e.g., $c_1 = \phi_{11}(\hat{t}_1, t_0)$) or integrals involving transition matrix elements (e.g., $c_3 = \int_{t_0}^{\hat{t}_1} \phi_{11}(\hat{t}_1, \tau) b_{12}(\tau) d\tau$). From the qualitative behavior in fig. 1(c) we see that the amount of water in the upper tank is increasing between $pc_0$ and $pc_1$, so $a_u(\hat{t}_1) - a_u(t_1) < 0$ (since $\hat{t}_1 < t_1$).

Eq. (7) can be directly translated into a constraint:

if $RV(T)_{pc_0 \to pc_1} = \Downarrow$, then
$$RV(a_u)_{pc_1} < RV(a_u)_{pc_0} + RV(i)_{pc_0} - RV(r_u)_{pc_0}$$

The RV constraint implies, by contraposition, that from $a_u\|_{pc_0}$, $a_u\|_{pc_1}$, $i\|_{pc_0}$, $r_u\|_{pc_0}$ we must conclude that the relative duration from $pc_0$ to $pc_1$ cannot be shorter.

### 3.4 Constants between pairs of comparison

Besides state variables, constants also have their difference propagated between pairs of comparison.

**Prop. 5 (Constants)** Suppose two behavior fragments defined by $pc_0 = \langle t_0, \hat{t}_0 \rangle$ and $pc_1 = \langle t_1, \hat{t}_1 \rangle$ are compared. If the variable $c$ has a constant value during the behavior fragments, then $\hat{c}(\hat{t}_1) - c(t_1) = \hat{c}(\hat{t}_0) - c(t_0)$.

The constancy of the inflow in the example leads to the constraint $RV(i)_{pc_1} = RV(i)_{pc_0}$.

## 4 Results

The CEC* algorithm outlined in sec. 2 has been implemented in Common Lisp, as a module interacting with the QSIM program. At each pair of comparison a constraint satisfaction problem is generated consisting of the variables shared by the two systems, domains $\{\Downarrow, \Uparrow, \|\}$ of their possible relative values, and constraints relating the RVs of variables at the current and at predecessor pairs of comparison. The RV constraints are instantiated from a database with templates derived from props. 1, 2, 4, 5. A consistent combination of RVs for the variables at a pair of comparison is called a *comparative state*. There may be several comparative states at a particular pair of comparison. The comparative states at a pair of comparison are linked to comparative states of predecessor and successor pairs of comparison to form a *comparative envisionment* [de Jong et al., 1996]. Branching in the comparative envisionment is possible, because at the pairs of comparison an RV is either $\Downarrow$, $\|$, or $\Uparrow$.

An example of a comparative envisionment produced by CEC* is shown in fig. 2(a). The envisionment answers the predictive CA question: How does the equilibrium state of a cascaded-tanks system with a leaky upper tank differ from that with a watertight upper tank when the systems are considered under the same conditions? Each path from the first to the last comparative state in the envisionment forms a *comparative behavior*. Although ambiguities arise with respect to $a_l, n_l$ at $pc_1$, the five possible comparative behaviors predict a single outcome: the leak tends to lower the amount of water in the tanks at equilibrium ($CS_{pc_2,0}$). No unambiguous conclusion can be drawn about the relative durations to reach the equilibrium state.

Fig. 2(b) illustrates a diagnostic CA question: When the relative amounts of water at $pc_1$ and $pc_2$ have been measured for watertight and leaky cascaded tanks filled from empty, and the orifices have the same size, which differences could account for the lower equilibrium water amounts in the leaky system? CEC* finds three possible causes: (1) a leaky upper tank and lower inflow ($CS_{pc_0,0}$), (2) a leaky upper tank and higher inflow ($CS_{pc_0,2}$), and (3) a leaky upper tank only ($CS_{pc_0,1}$). The leak, making the second system structurally different from the first, tends to lower the equilibrium amounts at $pc_2$. An additional difference in initial conditions may strengthen ($i\Downarrow_{pc_0}$) or attenuate ($i\Uparrow_{pc_0}$) this effect.

The CEC* program has been tested on a range of intra-model and inter-model CA questions concerning simple and more advanced QR systems, such as bathtubs, watertanks filled by a pump, masses on a spring, heat exchangers, sliding blocks, and tensile fracture tests. These examples generalize upon the cascaded-tanks example by requiring CEC* to deal with topologically different qualitative behaviors, region transitions, and composite behavior fragments between pairs of comparison. We are currently embedding CEC* in a system for the integration of measurements from a scientific knowledge base.

## 5 Discussion and related work

The CEC* algorithm has been proven sound and incomplete [de Jong, 1996]. Analogously to [Kuipers, 1994], soundness is taken to mean that for any two real behaviors of the systems that are in conformity with CEC*'s input, a comparative behavior describing their differences is included in the comparative envisionment. On the other hand, a comparative envisionment may contain comparative behaviors that do not correspond to any two real behaviors in conformity with CEC*'s input. CEC* then fails to derive a stronger conclusion and is consequently incomplete.

The proof of soundness is enabled by the mathematical foundation of the RV constraints. The propositions in sec. 3 also help in judging the generality of the method. In principle, the theorems apply to systems of arbitrary order and structure. A practical problem in evaluating (5) of prop. 4, however, is that it may be difficult to establish whether a transition matrix function $\phi_{ij}(t_b,\tau)$ changes sign during $[t_a, t_b]$, and hence to establish the sign of the elements of $\boldsymbol{\Phi}(t_b, t_a)$ and $\int_{t_a}^{t_b} \boldsymbol{\Phi}(t_b, \tau) \ldots d\tau$ in

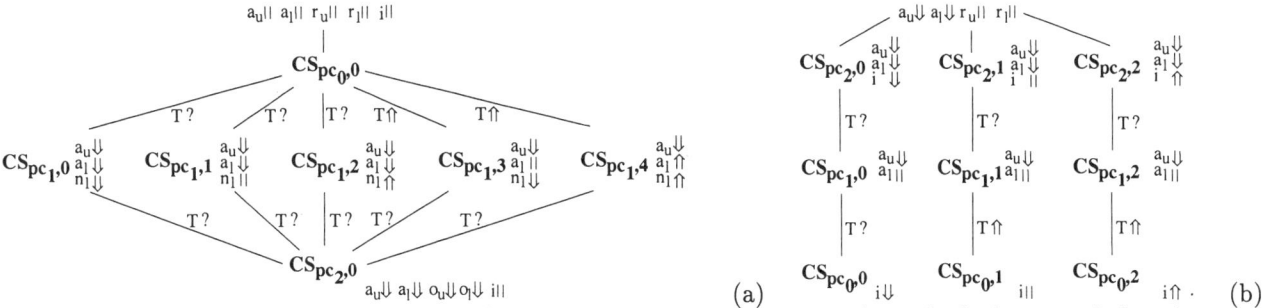

Figure 2: Comparative envisionment arising from the comparison of a watertight and a leaky cascaded-tanks system when both show the qualitative behavior of fig. 1(c). Comparative envisionment (a) is produced in response to a predictive CA question, whereas (b) arises from a diagnostic CA question. A few distinctive RVs are indicated at the comparative states (*CS*). The links between comparative states are labelled with the relative durations.

$F(t_b, t_a)$. The formulation of the difficulty in terms of concepts from linear system theory has the advantage of making this problem well-defined and of suggesting possible solutions (see [de Jong, 1996]).

The computational complexity of CEC* is of the order $\mathcal{O}(n3^q)$, where $q$ is the (maximum) number of shared variables and $n$ the number of pairs of comparison in the OPC structure. In the examples we have studied, the average-case behavior turned out to be more agreeable and could be improved by integrating additional RV constraints (e.g., measurements of variables at a pair of comparison) during the analysis.

CEC* borrows ideas on the mathematical formulation and the propagation approach from existing CA methods, in particular DQ analysis [Weld, 1990] and CEC [de Jong et al., 1996]. A notable exception to the rule that existing CA methods are restricted to intra-model CA is the model sensitivity approach described in [Weld, 1992].[3] In this approach, a behavior arising from a simple model is compared with a behavior from a more complex model. The simple model must be a *fitting approximation* of the more complex model, that is, the complex model must have an independent variable, such that, when this parameter is taken to a certain approximation limit, the RVs of the shared variables become ||. By constructing fitting approximations, the *inter*-model CA problem can be reduced to an *intra*-model CA problem.

The method's applicability thus depends on the opportunity to formulate a fitting approximation, which is not always possible (e.g., in the cascaded-tanks example). In addition, the implementation of the model sensitivity approach solves the resulting intra-model CA problem by DQ analysis. Unlike CEC*, DQ analysis terminates when confronted with ambiguities and can only reason in one direction through the OPC structure.

## 6 Conclusions

CEC* is an implemented method for predictive and diagnostic comparative analysis that is able to deal with intra-model and inter-model problems within a single framework. The constraints it uses in RV propagation have a solid mathematical foundation. This makes it possible to proof CEC*'s soundness and incompleteness.

Further work on CEC* concerns improvements upon the derivation of RV constraints from the models of the systems, especially those relating state variables across pairs of comparison. If the qualitative models and behaviors are extended to include quantitative information, a form of semi-quantitative comparative analysis could be considered as well.

**Acknowledgments** The authors would like to thank B. de Jong, N. Mars, P. van der Vet, and H. Zwart for their contributions.

## References

[Chiu and Kuipers, 1992] C. Chiu and B. Kuipers. Comparative analysis and qualitative integral representations. In B. Faltings and P. Struss, eds., *Recent Advances in Qualitative Physics*, p. 313–328. MIT Press, 1992.

[de Jong et al., 1996] H. de Jong, N.J.I. Mars, and P.E. van der Vet. Cec: Comparative analysis by envisionment construction. In W. Wahlster, ed., *Proceedings ECAI-96*, p. 476–480. John Wiley & Sons, 1996.

[de Jong, 1996] H. de Jong. Comparative envisionment construction. Memorandum UT-KBS-96-14, University of Twente, Enschede, 1996.

[Kuipers, 1994] B. Kuipers. *Qualitative Reasoning*. MIT Press, 1994.

[Neitzke and Neumann, 1994] M. Neitzke and B. Neumann. Comparative simulation. In *Proceedings AAAI-94*, p. 1205–1210. AAAI Press/MIT Press, 1994.

[Rugh, 1996] W.J. Rugh. *Linear System Theory*. Prentice Hall, 2nd ed., 1996.

[Weld, 1990] D.S. Weld. *Theories of Comparative Analysis*. MIT Press, 1990.

[Weld, 1992] D.S. Weld. Reasoning about model accuracy. *Artificial Intelligence*, 56:255–300, 1992.

---

[3] RSIM+ [Neitzke and Neumann, 1994] allows for a modest inter-model CA feature, namely differences in the monotonicity of functional relations.

# DIAGNOSIS AND QUALITATIVE REASONING

## Qualitative Reasoning 1: Modeling Support

# A Web-Based Compositional Modeling System for Sharing of Physical Knowledge

### Yumi Iwasaki, Adam Farquhar, Richard Fikes, James Rice
Knowledge Systems Laboratory, Department of Computer Science
Stanford University
Stanford, CA 94305 USA

## Abstract[1]

This paper describes a compositional modeling system called CDME (Collaborative Device Modeling Environment) for constructing domain theories of physical systems, composing models of devices, and simulating their behavior. We have implemented the system with the goal of encouraging sharing as well as the collaborative construction of knowledge bases describing physical domains. To maximize the chance of sharing and reuse of knowledge, CDME is implemented as a collection of network services on the World Wide Web. Knowledge is represented at three distinct levels: the physical, ontological, and logical. We describe the levels of representation, and how the system enables knowledge sharing at each level.

## 1 Introduction

Compositional modeling [1] is an effective method for automatically formulating a behavior model of a complex physical system. In compositional modeling, a system is provided with a knowledge base about the physical world, including knowledge of the behavior of a variety of devices and physical processes. The system formulates a model of a given situation by selecting pieces of knowledge that are applicable and composing them together.

The compositional modeling approach can lower the cost of analyzing physical systems if the cost of constructing the knowledge base can be amortized over many physical devices whose models can be quickly and easily formulated using the knowledge base. Thus, the advantages of compositional modeling hinge on one's ability to reuse knowledge.

Reuse and sharing is important in the task of representing a sufficient quantity of knowledge to make a viable compositional modeling system. Since compiling a large quantity of knowledge covering a wide variety of types of physical phenomena and devices requires a large amount of resources, one would like to be able to do so collaboratively, allowing a community of users to reuse and share existing knowledge bases or new ones constructed by others. The desire for sharing knowledge has lead some in the qualitative reasoning community to design a modeling language called CML (Compositional Modeling Language) in an attempt to unify similar languages used by different groups [2, 3]. This effort also reflects the awareness in the larger AI community of the need for knowledge reuse and sharing [4].

We have implemented a compositional modeling system called CDME (Collaborative Device Modeling Environment) for compositional modeling with the explicit goal of enabling collaborative construction as well as sharing of knowledge bases. We took two avenues towards facilitating knowledge sharing:

- The knowledge used by the system is represented at three levels: the physical, ontological and logical levels. Knowledge can be shared at each level.

- We made an explicit decision to implement the system as a network service. This allows the system to be used by a large community of users. More importantly, this allows a library of knowledge bases to be constructed collaboratively and in a modular fashion. The knowledge, as well as the result of inferences made by the system, can be shared.

---

[1] This project is supported in part by DARPA and the National Institute for Standards and Technology, Rapid Development, Exploration and Optimization (RaDEO) program, under cooperative agreement 70NANB6H0075.

This paper describes the multiple levels of representation in CDME and how sharing and collaborative construction of a knowledge base is supported at each level.

## 2 CDME, a Collaborative Device Modeling Environment

CDME automatically formulates the behavior model of a given device and simulates its behavior. To predict the behavior of a physical system in some domain, CDME requires knowledge, encoded in CML, about the physics of the domain. Such a knowledge base, called a *domain theory* in CML consists mainly of *model fragment* definitions, which are descriptions of classes of relevant objects, phenomena and systems. The domain theory of chemical processing plants, for example, might include descriptions of physical phenomena such as mass and heat flows, boiling, evaporation, and condensation; it would also include chemical reactions, the effects of catalysts, and models of components such as reaction vessels, pumps, controllers, and filters. A specific system or situation being modeled is described in a *scenario*. A scenario specifies the objects that exist in a given situation to be analyzed, their relations, and the initial conditions.

Given a scenario, CDME formulates a model by identifying model fragment instances whose operating conditions are satisfied in the scenario, and extracts a mathematical equation model from those instances. The equations are solved to produce a behavioral description. Because the conditions under which the model fragments hold are explicit in the domain theory, the system is able automatically to construct additional behavior models that describe the device as it moves into new operating regions. The functionality of CDME is similar to previous compositional modeling systems [1, 5-8], and the behavior of its direct predecessor is described in detail in [9].

### 2.1 Multi-layered representation

A modeling system such as CDME requires a rich repository of knowledge about a subject domain in order to be useful. In developing the system as well as its user interface, we often experienced a conflict between the desire to ensure reusability of the knowledge and the need to provide a rich environment in which users can express knowledge efficiently. Using a formal language such as the first-order predicate logic and making sure that all concepts used are logically defined is an important step towards ensuring knowledge can be reused. However, a formal logical language is not the most convenient language in which to state knowledge. The fine granularity and context-insensitivity of logical representations, the very characteristics that make formal representations suitable as vehicles for knowledge sharing, make such a representation inappropriate to use for stating a large amount of domain-specific knowledge.

To allow the efficient compilation of a large amount of knowledge, a language needs both semantic constraints and syntactic tools, such as those offered by a number of object-oriented representation languages, to help in the efficient capture of the structure of the knowledge. If knowledge is represented in a specialized language, however, the community of people that can use and understand the language is likely to be smaller, and consequently, the ability to share it will be reduced.

Though using a common knowledge representation formalism goes a long way towards enabling sharing, it is not the only issue. Equally important is making one's *ontology* explicit. By ontology, we mean the set of words used in describing a domain along with their definitions. In order for knowledge to be communicated accurately, one must state what the words are and what they mean. The ontology is also a useful body of knowledge to be shared in its own right since the set of concepts and their relations comprising an ontology reveal one's conceptual model of the domain of discourse.

We, therefore, have conflicting requirements: on the one hand, the desire to maximize the chance for reuse and sharing, in different forms and portions, and, on the other, the desire to enable the efficient encoding of a large amount of knowledge. To satisfy these requirements, we implemented CDME with three levels of representation: the physical, the ontological and the logical levels. Each level has its own representation language, which is independently designed to facilitate sharing. We built a web-based user interface for the physical and ontological levels to support encoding of knowledge at the respective level. Table 1 shows the three levels with the representation language, the authoring tool, and the types of objects manipulated. The following sections describe each level in detail.

### 2.2 Physical Level

The physical level is the level at which knowledge about physical phenomena and objects can be stated by users constructing domain theories. Using CML as the representation language, one defines the classes

| Level | Objects manipulated | Language | Authoring tool |
|---|---|---|---|
| Physical | model fragments, entities, quantities | CML | CML Editor |
| Ontological | concepts, concept definitions | Ontolingua | Ontolingua Editor |
| Logical | logical symbols, axioms | KIF | |

Table 1: Levels of Representation in CDME

of objects and physical phenomena along with their attributes that must be considered in modeling the domain.

CML is a general declarative modeling language for representing knowledge of physical domains for compositional modeling. CML was designed to facilitate model sharing between research groups, many of which have long been using similar languages. These languages are based primarily on the language originally defined by Qualitative Process Theory [5] and include the languages used for the Qualitative Physics Compiler [6, 7], compositional model formulation [1] and the Device Modeling Environment [8]. CML has been formulated by researchers involved in these projects to provide a clean redesign of these languages. CML is described in more detail in [2, 3].

The basic representational unit in CML is the *model fragment*. Each model fragment describes one aspect of a component's behavior or a physical process. Each model fragment is a self-contained assertion whose applicability to a given situation is decided separately to generate the model of an entire situation. Following is an example of a model fragment definition of contained-fluid.

```
(defModelFragment Contained-Stuff
 :subclass-of (physical-object)
 :participants
  ((sub  :type substance)
   (ctnr :type fluid-container))
 :conditions
  ((> (amount-of-in sub ctnr) 0))
 :quantities
  ((pressure
    :dimension pressure-dimension)
   (mass
    :dimension mass-dimension))
 :consequences
  ((== mass (amount-of-in sub ctnr))))
```

To facilitate sharing of CML knowledge bases, CML is fully translatable to the Knowledge Interchange Format (KIF). KIF is the proposed specification for a standard format for first-order predicate logic developed for the purpose of knowledge interchange [10], and indeed CDME automatically translates CML into KIF. Though one could simply use KIF, CML and the CML Editor in CDME provide more structure than pure KIF to simplify stating knowledge about physical phenomena.

To support collaborative construction of knowledge bases in CML, we have implemented a web-based CML editor as part of CDME. Figure 1 shows some user interactions with the editor. Once a domain theory is constructed and a scenario to be analyzed is given, a user can push the "Compose Model" button to instruct CDME to formulate a model for the scenario, and to simulate the model's behavior. CDME also includes facilities to help comprehension of simulation results, including causal explanations and plots of time-varying variables.

### 2.3 The ontological level

Having a common knowledge representation format is an important step in enabling communication. Equally important is to establish an explicit vocabulary that will be used in stating the content of knowledge. In our system, users as well as the implementers of CDME can work at the Ontological Level to construct fundamental ontologies for stating knowledge about the world.

An ontology is a formal specification that defines the representational vocabulary in some domain of discourse. In our case, the basic vocabulary of CML, including words such as *model fragment* and words naming parts of its definition (shown in **bold** in the example in Section 2.2), are all defined at the ontological level. General mathematical and physical concepts such as mathematical operators and the dimensions and units of quantities are also defined at this level. In general, concepts such as *systems*, *subsystems*, *components*, and *connections*, which are useful in describing physical domains in general are defined at the ontological level, while more domain-specific concepts such as *isothermal expansion* are defined at the physical level.

To support sharing and construction of ontologies, we have implemented the Ontolingua Editor, which provides a full-distributed

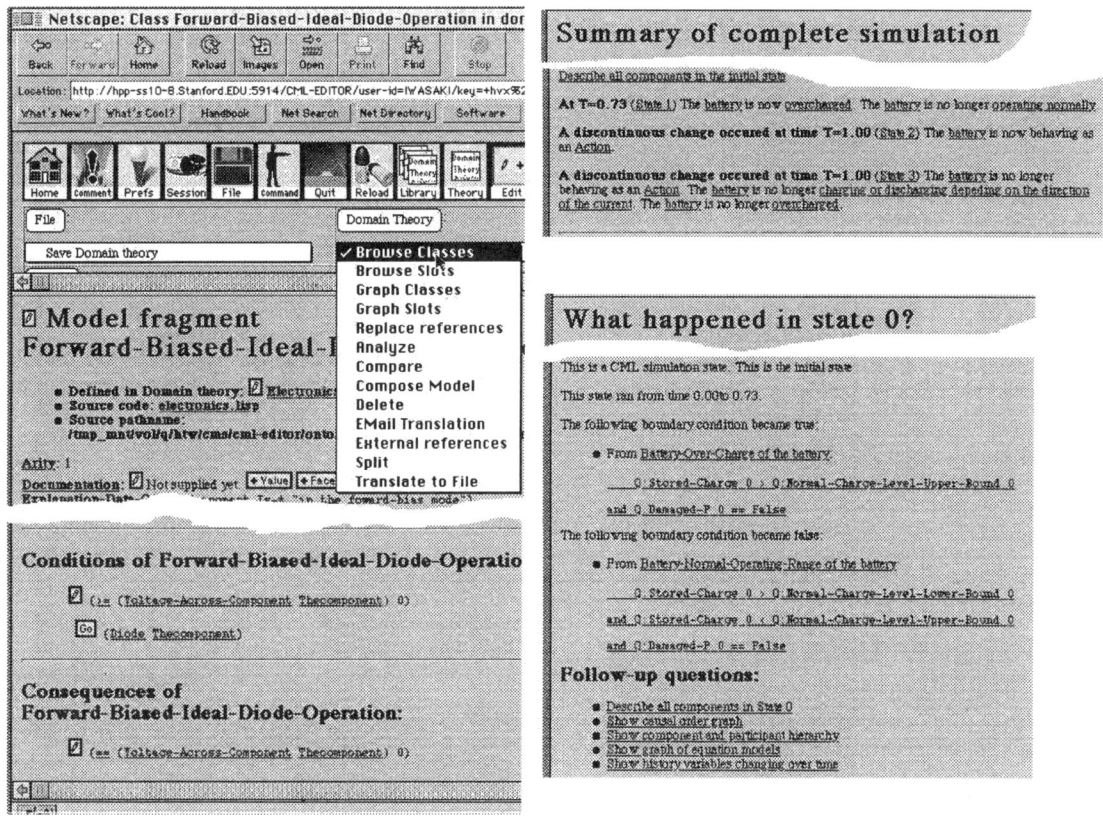

Figure 1: CML Editor: The left page shows a model fragment, *Forward-biased-ideal-diode-operation*. The user interface uses a row of icons for frequently used operations, and pop-up menus with submit buttons as labels. The two pages on the right summarize the simulated behavior and show the state of the device in State-0.

collaborative editing environment for constructing a library of ontologies [11]. To enable distributed, collaborative development of ontologies, the user-interface of the Ontolingua Editor is implemented using off-the-shelf Web-browsers as the interaction medium. Our Ontolingua Editor assists the user in the basic development tasks of browsing, creating, maintaining, sharing, and using ontologies. The Ontolingua Editor also supports modular construction of ontologies by allowing an ontology to include other ontologies. The CML Editor shares these features.

We use an ontology specification language called *Ontolingua* [12] for constructing ontologies. Ontolingua extends KIF by providing a rich set of constructs for modularizing and grouping axioms together. Ontolingua itself was designed to support sharing of knowledge, and a number of translators have been constructed between Ontolingua and various representation languages. While our ontologies are used for knowledge sharing and agent inter-operability [13], they share many properties in common with class libraries, object-oriented specifications, object-oriented simulation components, database schema definitions, and knowledge-bases.

Ontology construction usually is a distributed and collaborative activity [14], because in many domains no single person has the expertise necessary to construct an ontology for an entire domain. Therefore, it is essential for experts to collaborate on ontology construction. Furthermore, it is important for prospective users of an ontology, who may be geographically distributed, to examine, evaluate, and critique the ontology.

The CML and Ontolingua Editors are unusual among Web services because they allow users to create and edit objects, rather than simply retrieve them. The implemented interface provides a full, distributed collaborative editing environment with over a hundred user commands, context sensitive help, context sensitive user feedback and bug-report collection, multi-level undo/redo, multi-user sessions, and so on.

## 2.4 The logical level

This is the level of the formal, logical representation into which all the knowledge expressed at higher levels are automatically translated. We use KIF as the representation formalism at this level. Since the users of CDME are not expected to interact with the system at this level, no special purpose authoring tool is provided at this level. However, when necessary, the ontology authors can step outside of the frame language supported by the Ontolingua Editor and write axioms using the full expressiveness of KIF. Knowledge expressed at the physical and ontological levels are automatically translated into KIF by the system. As KIF is the proposed specification for a standard format for knowledge interchange, those who do not use Ontolingua or CML can still share the knowledge in CML knowledge bases through KIF.

## 2.5 Implementation

We now briefly describe the transformations between levels that take place in CDME. CDME requires several very different sorts of inference. A mixture of symbolic logic and mathematics is required during model formulation. Numeric integration may be used to simulate the systems behavior and monitor for boundary conditions. Logical inference is required again to determine the state of the system after a boundary condition has been reached, so that analysis can continue in the subsequent operating region.

Figure 2 shows a schematic view of the transformations. The left side of the figure illustrates domain theory construction, while the right half illustrates model formulation and simulation. As described above, the ontological knowledge is specified in the Ontolingua language. The physical knowledge is specified in CML, employing terms defined in the ontologies. To unify these representations, CDME translates the CML definitions into Ontolingua. This provides a uniform logical level representation for both the ontological and physical knowledge. This representation is the source for a variety of computationally efficient forms.

To perform model instantiation, the unified Ontolingua representation is translated into a restricted language (REL in the figure) supported by a forward chaining rule engine whose rules are compiled into efficient executable code. Most of the model instantiation work is done by the rule engine. Some inferences, such as reasoning about inequalities and mathematical functions, however, are not efficiently handled by forward chaining. These inferences are performed by a mixture of backward chaining (controlled by iterative deepening) and a separate mathematical solution engine. This engine also solves the equations extracted from the rule engine's knowledge-base and predicts behavior in terms of the variable values over time. When a model reaches its boundary of applicability, the rule engine produces a new model and simulation continues. In Figure 2, each state contains a representation of a behavior model and the variable values over the time interval of the state.

External agents can access CDME's analysis results, knowledge-base contents, domain theories, and ontologies by using the Generic Frame Protocol (GFP), which is an application program interface for accessing knowledge bases stored in frame knowledge representation systems [15].

## 2.6 Status

CDME has been implemented and available as a service on WWW. The Ontolingua Editor has been operational for over two years and been used by users around the world to construct ontologies for various purposes. The core part of CDME, the CML Editors, the model formulation and simulation mechanisms as well as some of the explanation facilities, is operational and has been used by us and a limited number of friendly users. The system is not yet robust enough for use by the general public. The devices for which we have used CDME include subsystems of micro-satellites being developed in the Aeronautics and Astronautics Department at Stanford.

Our limited experience with the system shows the beginning of knowledge reuse. Because of the modularity of domain theories, some of the fundamental ontologies (e.g. the ontologies of algebra, component connection) and some domain theories (e.g. kinematics, motors) are reused often. However, as a CML domain theory gets larger, it becomes more difficult to understand its context and implicit assumptions, and thus harder to reuse. This is a general problem with any knowledge base or collection of information. More research is needed to address the issue of making a large body of information more understandable through various types of aggregation, abstraction, and interface tools.

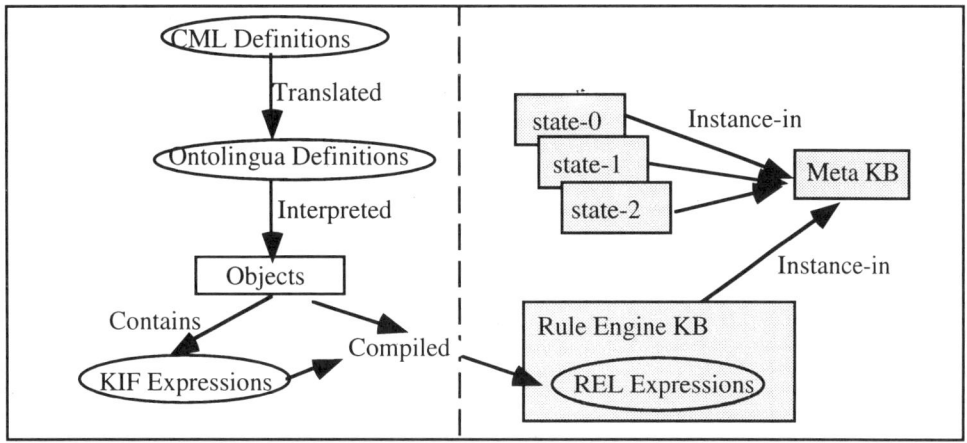

Figure 2: Transformation of knowledge in the system

## 3 Conclusions

CDME is a web-based compositional modeling system implemented with the explicit goal of enabling sharing and collaborative construction of knowledge bases. The system, which is accessible on the WWW, includes a full, distributed collaborative editing environment to construct CML domain theories. To encourage sharing, knowledge is represented at multiple levels. We have described the different representation levels, and how each level allows different groups of users to share and reuse knowledge in a different form.

Accessibility through the web allows a potentially large community of users not only to reuse but also to contribute to the construction of domain theories. We hope that the availability of CDME and similar services will encourage the development of a significant public library of domain theories by the research community and that it will spur further research and development in the field.

## Acknowledgment

The authors would like to thank the past and the present members of the How Things Work project at the Knowledge Systems Laboratory for their assistance in designing and implementing various features of CDME.

## References

1. Falkenhainer, B. and Forbus, K., *Compositional modeling: finding the right model for the job.* Artificial Intelligence, 1991. **51**(1-3).

2. Falkenhainer, B., et al. *CML: A Compositional Modeling Language.* 1994. Technical report KSL-94-16, Knowledge Systems Laboratory, Stanford University.

3. Bobrow, D., et al. A Compositional Modeling Language. In *Proceedings of The Tenth International Workshop on Qualitative Reasoning.* 1996. Fallenleaf Lake, CA. AAAI Press.

4. Neches, R., et al., *Enabling Technology for Knowledge Sharing,* in *AI Magazine.* 1991, p. 36-56.

5. Forbus, K.D., *Qualitative Process Theory.* Artificial Intelligence, 1984. **24**(1-3).

6. Crawford, J., Farquhar, A., and Kuipers, B. QPC: A Compiler from Physical Models into Qualitative Differential Equations. In *Proceedings of The Eighth National Conference on Artificial Intelligence.* 1990.

7. Farquhar, A. A Qualitative Physics Compiler. In *Proceedings of Proceedings, Twelfth National Conference on Artificial Intelligence.* 1994. Seattle, Washington. The AAAI Press/The MIT Press.

8. Iwasaki, Y. and Low, C.M. Device Modeling Environment: An Integrated Model-Formulation and Simulation Environment for Continuous and Discrete Phenomena. In *Proceedings of Conference on Intelligent Systems Engineering.* 1992. Edinburgh, Scotland.

9. Low, C.M. and Iwasaki, Y., *Device modelling environment: an interactive environment for modelling device behaviour.* Intelligent Systems Engineering, 1993. **1**(2): p. 115-145.

10. Genesereth, M.R. and Fikes, R.E. *Knowledge Interchange Format, Version 3.0 Reference Manual.* 1992. Technical report Logic-92-1, Stanford University Logic Group.

11. Rice, J., *et al.* Using the Web Instead of a Window System. In *Proceedings of CHI 96.* 1996. ACM Press.

12. Gruber, T.R. *Ontolingua: A Mechanism to Support Portable Ontologies.* 1991. Technical report KSL 91-66, Knowledge Systems Laboratory, Stanford University.

13. Fikes, R., *et al. Knowledge Sharing Project Overview.* 1991. Technical report KSL-91-71, Knowledge Systems Laboratory, Department of Computer Science, Stanford University.

14. Grudin, J., *Groupware and Cooperative Work: Problems and Prospects,* in *The art of human-computer interface design,* B. Laurel, Editor. 1990, Addison-Wesley: Reading, Massachusetts. p. 171-185.

15. Karp, P.D., Myers, K.L., and Gruber, T. The Generic Frame Protocol. In *Proceedings of the Fourteenth International Joint Conference on Artificial Intelligence.* 1995.

# A Causal Time Ontology for Qualitative Reasoning

### Yoshinobu Kitamura, Mitsuru Ikeda and Riichiro Mizoguchi
The Institute of Scientific and Industrial Research
Osaka University
8-1, Mihogaoka, Ibaraki, Osaka 567
Japan

## Abstract

Aiming at explicit description of temporal meaning of causal relations generated by qualitative reasoning systems, this article proposes a causal time ontology which defines a set of general time concepts in qualitative models, called *causal time scales*. Each of them associated with a modeling technique represents a temporal granularity and/or an ontological viewpoint. They allow us to specify temporal performance of the reasoning engines and to identify a general causal reasoning scheme together with sophisticated feedback analysis. Lastly, we present a causal time resolution required to derive causal relations in fluid-related systems and a reasoning system satisfying it.

## 1 Introduction

Causality plays a crucial role in human understanding of behavior of physical systems. A lot of research has been carried out on qualitative reasoning systems in order to derive causal relations from the models of the target systems, e.g.,[de Kleer and Brown, 1984]. Human recognition of causal relations is based on recognition of time delay (i.e., time interval) between the cause and the effect. Little, however, is known concerning temporal meaning of causal relations generated by the reasoning systems, that is, how long (or short) the time intervals in the causal relations in the real physical behavior as discussed in [Iwasaki et al., 1995]. There are the following two explanations for it. First, there are many modeling techniques and representations, each of which implies several temporal relations among variables. Secondly, such models are interpreted by the reasoning engines on the basis of their own time concepts behind their reasoning procedures. For example, using the same qualitative differential equations[1], QSIM [Kuipers, 1994] and the causal ordering procedure proposed in [Iwasaki and Simon, 1994] generate different causal relations together with different temporal meanings. As a consequence of the implicit existence of several time concepts, the temporal meaning of generated causal relations is not clear for the users of the reasoning engines.

[1] Strictly speaking, the causal ordering procedure [Iwasaki and Simon, 1994] needs additional information.

The goal of this article is to reveal the structure of causal time underlying the qualitative models and the causal reasoning engines. We propose a set of general time concepts in qualitative models, called *causal time scales*. Each causal time scale associated with a modeling technique represents a temporal granularity and/or an ontological viewpoint. In other words, the set of the causal time scales aims to enumerate all possible temporal meanings of the models, that is, an ontology of time for causal reasoning. Ontologies are explicit specifications of concepts [Mars, 1995], which can specify assumptions underlying knowledge-based systems [Mizoguchi and Ikeda, 1996].

We have identified 13 causal time scales shown in Table 1. They are classified into four categories each of which represents a modeling technique. They generalize time concepts in the some previous frameworks [de Kleer and Brown, 1984; Iwasaki and Simon, 1994; Kuipers, 1994; Rose and Kramer, 1991].

The utility of the causal time scales includes the following: First, causal relations generated by the reasoning engines can be categorized into one of the causal time scales. It clarifies not only the temporal meaning of the causal relations but also the performance of the reasoning engines with respect to causal ordering, called *causal time resolutions*. For example, causal relations generated by QSIM are categorized into the causal time scale named $Ta3$ associated with the mathematical integral operation. On the other hand, some of those generated by the causal ordering procedure are categorized into the time scale $Ta2$ which is a finer-grained time concept than $Ta3$. Thus, the causal time resolution of the causal ordering procedure is finer than that of QSIM. The time resolutions of other reasoning methods will be shown in section 3.1.

Secondly, we also identified a general causal reasoning scheme which can cope with multiple time scales. It can explain essential parts of the conventional reasoning methods. It will be shown section 3.2.

Lastly, fine-grained time scales enable sophisticated analysis of causality in feedback loop. According to the time scale associated with a feedback loop, the reasoning engine can suppress causal relations without physical meaning and ambiguities of reasoning results as discussed in section 3.3.

In section 4, we discuss a causal time resolution required to derive the causal relations in fluid-related systems. The constituents of the model for the time resolution is also discussed.

In this article, we do not discuss formal ontology based on axiomatization, aiming at getting on agreement on the content and the terminology. Next, we concentrate on ontological issues. For the details of model representation, reasoning engine and its evaluation, see other articles[Kitamura *et al.*, 1996a; Kitamura *et al.*, 1996b].

## 2 A Causal Time Ontology

### 2.1 Theoretical Foundation

In our causal time ontology, behavior over time generated by the reasoning engine is represented in terms of events and links among the events, in a similar way in the history model[Forbus, 1984]. An event $e \in E$ represents instantaneous changes of qualitative values of parameters and their resultant values at a time point. Changes of quantitative values are assumed to be continuous and differentiable. A new event $e_2$ is generated by applying an operators $o \in O$ to an old event $e_1$ according to the model $M$. A link $l_1 \in L$ between $e_1$ and $e_2$ represents a causal relation according to the model $M$. There is an open time-interval $t_1$ between $e_1$ and $e_2$, corresponding to the causal relation $l_1$. The roles of operators $o \in O$ are to propagate changes and to generate new events, time intervals and hence partial temporal relations. Note that the symbol 't' always represents not a time point but a time interval in this article. Although events correspond to time points, we concentrate on time intervals in which changes propagate.

The causal time ontology provides categories of such time intervals, called *causal time scales*. A causal time scale represents a concept of time interval for propagation of effect. The notation $\tau(l) = T$ denotes a time interval $t$ of a causal relation $l$ is categorized into a time scale $T$. We can say that "the causal relation $l$ is represented on the time scale $T$".

The ordinal relation $T_1 \prec T_2$ representing a time scale $T_1$ is shorter (finer-grained) than $T_2$ is defined as follows;

$$T_1 \prec T_2 \leftrightarrow \forall t_1 \in T_1, \forall t_2 \in T_2, t_1 < t_2$$

In other words, $T_1$ represents faster events than that $T_2$ does. This relation is transitive. The relation between $Ta2$ and $Ta3$ where $Ta2 \prec Ta3$ is shown in Figure 1. Although the figure shows the relation among concrete time scales due to limitation of space, we explain the general relation here. When a certain condition becomes true in the reasoning process on a shorter time scale $Ta2$, the reasoning shifts to a neighboring longer time scale $Ta3$. Such a condition is called as a *boundary condition* of $Ta2$ or a *precondition* of $Ta3$. The set of events grouped by the condition $e_{(1,1)}, e_{(1,2)}, ..., e_{(1,4)}$ on $Ta2$ is treated as the instantaneous events $e_{(2,1)}$ on $Ta3$. Then, the *reasoning operator* of $Ta3$ is applied to the event $e_{(2,1)}$. Each time scale has an operator. The resultant values on $Ta3$ can be treated as the initial values on $Ta2$. The same applies to $Ta3 \prec T_4$ cases recursively. In summary, a time scale $T$ can be defined by a tuple of three elements, $<$Pc, Op, Bc$>$, where these denote a precondition, an operator, a boundary condition, respectively. The elements of $T_1$ are denoted by $T_1$:Pc, $T_1$:Op and $T_1$:Bc, respectively.

### 2.2 Physical Meaning of Causal Relations

The relations $l$ generated by the reasoning engine do not always make sense from the physical viewpoints. There are

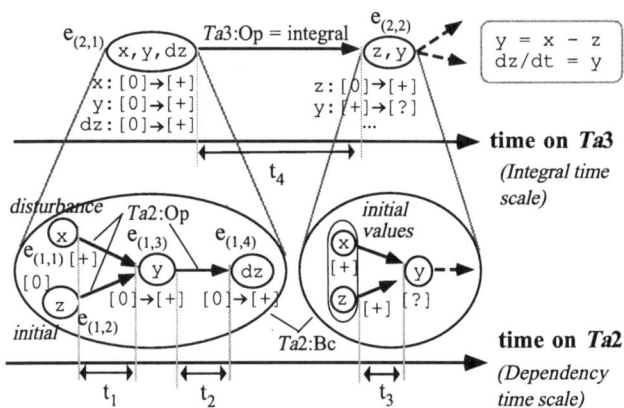

Figure 1: Relation between the time scales $Ta2 \prec Ta3$

such cases where a link $l$ represents an operational order which is not justified by the physical sense. In order to clarify *the physical meaning* of the causal relations, we will discuss two aspects of the physical meaning of each time scale, that is the *interval-meaning* and the *ordinal-meaning*. The former represents a physical justification of existence of the time intervals on the time scale. The latter represents that of order of events on the time scale.

### 2.3 Causal Time Scales

This section defines 13 causal time scales shown in Table 1. The time scales are classified into four categories each of which represents a modeling technique together with particular modeling rationales. The *direct modeling* is to describe models using the mathematical differential equations which directly represent dynamic behavior over time. The *time constant modeling* is to qualitatively categorize the time constants for modeling of phenomena. The *component structure modeling* is to introduce the concept of "component", aiming at causal relations reflecting the physical structures of the target systems. The *modeling of periods of interest* such as initial responses allows the reasoning engine to neglect changes of no interest. The notation "$Tx\#$" denotes a time scale, where '$x$' denotes a category (a,b,c or d) and the number '#' represents ascending order in each category. In Table 1, each condition denoted by a notation "c$x$#" represents the boundary condition of the time scale listed above and the precondition of that listed below.

**(a) Direct Modeling**

In the direct modeling, temporal characteristics of the phenomena are represented directly by the mathematical aspect of the models. The precondition of the time scale $Ta3$ is that a set of parameters are *completely determined* where every parameter in the set has values which satisfy all constraints. When the condition holds, the reasoning engine applies the integral operator and hence generates a new event. The integral operator embodies the qualitative mean value theorem $x_{new} = x_{old} + dx/dt$ [de Kleer and Brown, 1984]. The time intervals between the old events and the new events are categorized to $Ta3$:*integral time scale*. The time in QSIM[Kuipers, 1994] corresponds to $Ta3$. Furthermore, $Ta3$ is categorized

Table 1: The causal time scales
**(a) Direct Modeling**
    ca1: changes of parameter values
  $Ta1$: *Mutual Dependency time scale*
    ca2: a set of inherently simultaneous equations are satisfied.
  $Ta2$: *Dependency time scale*
    ca3: a set of constraints are completely determined.
  $Ta3$: *Integral time scale*
      $Ta3p$: *Integral-from-equality time scale*
      $Ta3i$: *Integral-to-equality time scale*
    ca4: a set of parameters reaches equilibrium.
  $Ta4$: *Equilibrium time scale*
**(b) Time Constant Modeling**
  $Tb1$: *A Faster Mechanism time scale*
    cb2: a faster mechanism reaches equilibrium.
  $Tb2$: *A Slower Mechanism time scale*
    cb3: a slower mechanism reaches equilibrium.
**(c) Component Structure Modeling**
  $Tc1$: *Intra-component time scale*
    cc2: all parameters in a component are determined.
  $Tc2$: *Inter-component time scale*
    cc3: all parameters in a global structure are determined.
  $Tc3$: *Global time scale*
    cc4: all parameters in the whole system are determined.
  $Tc4$: *The Whole System time scale*
**(d) Modeling of Periods of Interest**
  $Td1$: *Initial Periods time scale*
    cd1: the first event happens on a time scale.
  $Td2$: *Intermediate Transitional time scale*
    cd2: the last event happens on a time scale.
  $Td3$: *Final Periods time scale*

into two types; $Ta3p$ and $Ta3i$. The former represents the time intervals for integration from the same value as the landmark values to the interval of the landmark values. The latter represents those from the interval to the landmark values. $Ta3p \prec Ta3i$ holds.

On the other hand, until a set of parameters are completely determined, the time intervals are categorized to $Ta2$:*dependency time scale*. The precondition of $Ta2$ is that a set of inherently simultaneous equations[2] are satisfied. The time scale of the causal ordering theory [Iwasaki and Simon, 1994] corresponds to $Ta2$. Until the inherently simultaneous equations are satisfied, the time intervals are categorized to $Ta1$:*mutual dependency time scale*. Although this time scale has the interval-meaning mentioned in section 2.2, it has no ordinal-meaning. On the other hand, $Ta2$ and $Ta3$ can have both kinds of the physical meaning. When a set of parameters achieve its equilibrium, the reasoning shifts to $Ta4$:*equilibrium time scale*.

For example, consider a simple system modeled by the direct modeling, $y = x - z, dz/dt = y$. A variable takes one of the three qualitative values, $[+], [0]$ and $[-]$, where the landmark value is 0. In the initial state, all variables take $[0]$ except for a disturbance $x = [+]$. Figure 1 shows causal relations generated on $Ta2$ and $Ta3$. In this case, the method of constraint

---

[2]This term represents such simultaneous equations which cannot be solved by substitution alone, borrowed from [de Kleer and Brown, 1984].

satisfaction is simple propagation of values. First, the value of $y$ becomes greater than 0 (denoted by $[0] \rightarrow [+]$ in the figure) according to $y = x - z$. Next, the value is propagated to the derivative of $z$ (denoted by $dz$ in the figure). At this point, every parameter has a value which satisfies all constraints, that is, the precondition of $Ta3$ becomes true. Then the reasoning shifts to the longer time scale $Ta3$. On the scale $Ta3$, the integral operator is applied to $z$, then $z$ becomes greater than 0. Next, on $Ta2$, the new value of $z$ is propagated to $y$ and so on. Note that the change of $y$ and that of $dz$ happen at the same time point on the time scale $Ta3$ and then there is no causal relation between $y$ and $dz$ on $Ta3$, while the change of $y$ causes that of $dz$ after a small time interval $t_2$ on the time scale $Ta2$.

When model builders describe a phenomenon in terms of differential equations, the modeling rationale is to capture dynamic changes in the transitional behavior in $Ta3$ to its equilibrium. In general, it implies that the time interval to achieve its equilibrium is longer than the other phenomena.

**(b) Time Constant Modeling**
In order to represent differences in time constants, this modeling technique divides the target system into such parameter sets where the time intervals to achieve equilibrium $Ta4$ are extremely different from each other. In such a model, the (relatively) faster mechanism firstly reaches the equilibrium on the time scale $Tb1$. Then the reasoning shifts to the time scale $Tb2$. This kind of modeling is found in [Iwasaki and Simon, 1994; Kuipers, 1994]. This modeling has an advantage in reasoning efficiency because of separation of the reasoning space.

**(c) Component Structure Modeling**
This modeling is to divide the whole system into subparts according to component structures based on the device ontology [de Kleer and Brown, 1984]. In this article, devices in the minimum grain size are called "components". $Tc1$ represents internal behavior in components, while $Tc2$ represents behavior between neighboring components. Interactions between the global structures containing components are represented by $Tc3$. Those between more coarse-grained global structures are also represented by $Tc3$. $Tc4$ represents that the whole system eventually reaches equilibrium. The ordinal relations among these time scales reflect structural distances.

Although $Tc2$ and $Tc3$ have the interval-meaning, the connection information alone cannot give the ordinal-meaning to them. We will discuss additional knowledge for the ordinal-meaning later. On the other hand, $Tc1$ has no physical meaning in any sense. This modeling technique implies such modeling rationales that the causal relations should reflect functioning components and the medium flow along the structures.

**(d) Modeling of Periods of Interest**
This modeling allows the reasoning engine to treat only particular temporal periods of interest such as initial behavior. The time scales constrain not length but the number of time intervals. For example, QUAF [Rose and Kramer, 1991] reasons only the initial changes $Td1$ and the final responses $Td3$ without the intermediate transient behavior. This technique

Table 2: Causal time scales in reasoning systems

| QSIM [Kuipers, 1994] | $T1_{qs} : Ta3, T2_{qs} : Ta4$ |
|---|---|
| QSEA [Kuipers, 1994, ch.7] | $T_{qa} : Ta4$ |
| Time-Scale [Kuipers, 1994, ch.12] | $T1_{ts} : Ta3 \,\&\, Tb1,$ $T2_{ts} : Ta4 \,\&\, Tb1,$ $T3_{ts} : Ta3 \,\&\, Tb2,$ $T4_{ts} : Ta4 \,\&\, Tb2$ |
| QUAF [Rose and Kramer, 1991] | $T1_{qf} : Ta3 \,\&\, Td1,$ $T2_{qf} : Ta3 \,\&\, Td3$ |
| Mythical Time [de Kleer and Brown, 1984] | $T1_{mt} : Ta1 \,\&\, Tc1,$ $T2_{mt} : Ta1 \,\&\, Tc2,$ $T3_{mt} : Ta3 \,\&\, Tc4$ |
| Causal Ordering [Iwasaki and Simon, 1994] | $T1_{co} : Ta2,$ $T2_{co} : Ta3$ |
| Abstraction [Iwasaki and Simon, 1994] | $T1_{ab} : Ta2 \,\&\, Tb1,$ $T2_{ab} : Ta3 \,\&\, Tb1,$ $T3_{ab} : Ta2 \,\&\, Tb2,$ $T4_{ab} : Ta3 \,\&\, Tb2$ |

contributes to disambiguation of reasoning results and avoiding reasoning costs.

# 3 Causal Time Scales in Reasoning Systems

## 3.1 Causal Time Resolutions

Let us characterize some of the existing reasoning systems in terms of the causal time scales. In general, a time resolution of a reasoning system is specified by a set of combinations of the primitive time scales discussed thus far. The notation $T_1 : Tx_1 \& Tx_2$ represents that the time scale $T_1$ consists of $Tx_1$ and $Tx_2$. Table 2 shows the time scales which can be treated by some conventional qualitative reasoning systems. For example, QSIM[Kuipers, 1994] can cope with behavior on $Ta3$ and $Ta4$. QSIM uses only mathematical differential equations and adopts a kind of generate-and-test method for constraint satisfaction. Thus, no causal relation among transitional behavior to $Ta3$ is identified. The time of QSIM is corresponds to $Ta3$. QSEA[Kuipers, 1994, ch.7] treats only equilibrium states represented by $Ta4$. The time-scale abstraction[Kuipers, 1994, ch.12] is a kind of the time constant modeling represented by $Tb$. QUAF[Rose and Kramer, 1991] reasons only the initial changes $Td1$ and the final responses $Td3$ on the integral time scale $Ta3$.

The method proposed in [de Kleer and Brown, 1984] can generate causal relations among more fine-grained time scale $Ta1$, called "mythical time", on the basis of the concept of device. Causal relations on $T1_{mt}$, however, do not always have the physical meaning because $T1_{mt}$ consists of $Ta1$ and $Tc1$. On the other hand, in order to give the ordinal-meaning to $T2_{mt}$, de Kleer and Brown employ general heuristics representing physical intuitions. Causal relations generated by them, however, are ambiguous due to the arbitrariness of heuristics application.

The causal ordering theory [Iwasaki and Simon, 1994] derives causal relations on $Ta2$, which have the ordinal-meaning representing mathematical dependency. The theory, however, does not try to derive those on $Ta1$. Two abstraction techniques corresponding to $Tb$ are also discussed.

## 3.2 Primitive Reasoning Scheme

The primitive reasoning scheme of a reasoning system can be specified by the set of time scales which the system can cope with. Let $TS$ be such a set and $E_c$ be a current set of events to be carried out. The generic reasoning scheme for a current time scale $T_c$ and neighboring time scales $T_1$ and $T_2$ where $T_1 \prec T_c \prec T_2$ is defined as below.

1. On the time scale $T_c$, if an event $e_1 \in E_c$ satisies the precondition $T_c$:Pc, the operator $T_c$:Op is applied to $e_1$ and then a new event $e_2$ and a new link $l$ between $e_1$ and $e_2$ are generated. $\tau(l) = T_c$ holds.

2. The reasoning process shifts to the shorter time scale $T_1$. $T_c \leftarrow T_1$ and $E_c \leftarrow e_2$ and go to step 1 recursively[3].

3. If $e_2$ does not satisfies the boundary condition $T_c$:Bc, go back to step 1 and $E'_c \leftarrow E_c - \{e_1\} + \{e_2\}$.

4. If $e_2$ satisfies the boundary condition $T_c$:Bc, the reasoning process shifts to the longer time scale $T_2$. All events in $T_c$ are transferred to the event $e_3$ on $T_2$. Go to step 1 recursively.

The reasoning process starts with the minimum time scale $T_{min}$ in $TS$, given the initial value $E_c$. This reasoning process repeats recursively until the boundary condition of the maximum time scale holds. There are such cases that $T_{min}$ needs a special operator to satisfy the precondition of $T_{min}$.

The reasoning processes of the conventional systems can be explained by their time scales shown in Table 2. For example, the reasoning method called time-scale abstraction [Kuipers, 1994, ch.12] starts with the minimum time scale $T1_{ts}$. Since $T1_{ts}$ contains $Ta3$, the operator for $T1_{ts}$ is the integration[4]. When the boundary condition of $T1_{ts}$ becomes true, i.e., the faster system reaches equilibrium, the reasoning process shifts to $T2_{ts}$. Because the system is in equilibrium, no reasoning is carried out in $T2_{ts}$. Then, the reasoning process in $T3_{ts}$ starts and then the slower behavior is generated. In principle, the reasoning process at $T3_{ts}$ backs to the shorter time scales $T1_{ts}$ and $T2_{ts}$. In this case, however, because $T2_{ts}$ is in equilibrium and hence has no more events, only checks of values are needed. The primitive scheme of the algorithm shown in [Kuipers, 1994] is identical with this one.

The reasoning result consists of a set of events $E$ and a set of links $L$ each of which has a time scale $T \in TS$ associated with it where $\tau(l) = T$. If there is a (transitive) causal relation between $e_1$ and $e_2$, $\tau(e_1, e_2)$ denoting the time scale representing the time interval between $e_1$ and $e_2$ is defined as follows;

$$\tau(e_1, e_2) = \max_{l \in L_e} \tau(l)$$

where $L_e \subset L$ consists of the links between $e_1$ and $e_2$. This implies that a chain of time intervals represented on a time scale can be represented on the same time scale. In other words, time intervals on a time scale $T_1$ can never become longer enough to be categorized into the longer time scales

---

[3]The symbol '$\leftarrow$' denotes substitution

[4]Strictly speaking, the operator of QSIM is not identical with integration. It represents possible transitions over time for reasoning efficiency.

than $T_1$ unless the boundary condition is satisfied. In the cases of no causal relation, if $\tau(e_0, e_1) \prec \tau(e_0, e_2)$ where $e_0$ represents the last common event (i.e., the junction event), we only can say that $e_1$ happens before $e_2$. If not, there is no temporal order between such events.

## 3.3 Feedback and Causal Time Scales

Such phenomena that the effect of an event of a parameter is eventually propagated to the parameter itself are called as feedback. The time delay along the feedback loop plays a crucial role in human understanding of feedback. For example, in the cases where the time delay along a feedback is very short and then the modeler has no interest in the transitional behavior of the feedback, it is no need to generate causal relations among events in the feedback loop and to trace the changes of parameter values. Therefore, the reasoning engine can treat feedback according to the following heuristics.

**Feedback heuristics :** Whether or not a phenomenon is recognized as feedback depends on the time delay for the propagation loop according to the pre-defined threshold values $T_{s1}$ and $T_{s2} \in TS$. Let $L$ be a set of the links contained in the propagation loop and $T_l$ be the time scale for the time delay along the loop.

1. If $T_l \preceq T_{s1}$ then the phenomenon is not treated as feedback. The orders of events in $L$ have no physical meaning. If the new value after the feedback is different from the original value, that is viewed as contradiction at the same time point.

2. If $T_l \succ T_{s1}$ and $T_l \preceq T_{s2}$ then the phenomenon is treated as *semi* feedback. The orders of events in $L$ have the physical meaning. If there is a conflict between the old and new values then the new value is neglected.

3. If $T_l \succ T_{s2}$ then the phenomenon is treated as feedback. The orders of events in $L$ have the physical meaning. The values will be changed after the feedback.

The last one corresponds to the usual feedback. The first two are paraphrased as "the feedback is virtual, produced by the sequential operations of the reasoning method" and "there is no feedback which suppresses the original change instantaneously", respectively.

## 4 Time Scales for Fluid Systems

This section discusses a causal time resolution required to derive the causal relations in the fluid-related systems. A finer-grained time resolution than those of the conventional systems is required. Our reasoning system satisfying the required time resolution is also mentioned.

## 4.1 Required Time Resolution

Table 3 shows a time resolution, i.e., a set of time scales, required to derive causal relations in fluid systems based on the device ontology. The necessity to distinguish among these time units is justified by human recognition of causality or some assumptions. Firstly, the device ontology requires the discrimination between $Tc1$ of the $T1$:*inter-component time scale* and $Tc2$ of the $T2$ *inter-component time scale*. Secondly, in order to cope with global phenomena such as

Table 3: Time Scales required for Fluid Systems

| Name of time scale | Definition |
|---|---|
| $T1$:*Intra-Component time scale* | $Ta1/2$ & $Tb1/2$ & $Tc1$ |
| $T2$:*Inter-Component time scale* | $Ta1/2$ & $Tb2$ & $Tc2$ |
| $T3$:*Global time scale* | $Ta1/2$ & $Tb2$ & $Tc3$ |
| $T4$:*Globally Simultaneous time scale* | $Ta1/2$ & $Tb1$ & $Tc3$ |
| $T5$:*Integral time scale* | $Ta3$ & $Tb2$ & $Tc3$ |
| $T6$:*Partial Equilibrium time scale* | $Ta4$ & $Tb2$ & $Tc3$ |
| $T7$:*Complete Equilibrium time scale* | $Ta4$ & $Tb2$ & $Tc4$ |
| $T1 \prec T4 \prec T2 \prec T3 \prec T5 \prec T6 \prec T7$ ||

changes in temperatures caused by global heat balances, hierarchical structure ($Tc3$ of $T3$ and $T4$) is needed. The length of time interval of $T3$ is longer than that of $T2$ because of the structural distance represented by $Tc2$ and $Tc3$. There are, however, such cases where changes in non-neighboring components are simultaneous, called *globally simultaneous phenomena*. For example, on the assumption that fluid is incompressible, flow rate of such fluid at each component changes at the same time. Thus, $T4$:*globally simultaneous time scale* which is combination of $Tc3$ and $Tb1$ is needed. Since it is assumed that there is only one level of faster mechanisms which is represented by $T4$, the other scales are on $Tb2$. Because $Tc1$ in $T1$ represents the most primitive concept in the device ontology, $T1 \prec T4$ holds. Because $T4$ represents almost simultaneous phenomena, $T4 \prec T2$ holds.

## 4.2 A Reasoning System for Fluid Systems

We have developed a reasoning system which can cope with the above seven time scales finer than those of the existing systems [Kitamura *et al.*, 1996a; Kitamura *et al.*, 1996b]. In general, a main issue to discuss is what contents we have to describe in order to build such a model that generates causal relations having the physical meaning. For the required time resolution, we employ the modeling schemes such as hierarchical components modeling, description of time constants and causal characteristics of components. The last knowledge enables the reasoning engine to give the physical meaning of causal relations among components on the $Tc2$ of $T2$. As discussed thus far, additional knowledge is needed for the physical meaning on $Tc2$. Considering components have their own causal characteristics, our approach is to explicitly describe inherent causal characteristics of each parameter in components, called *causal specifications*, context-independently. Although such a description is prone to dependent on context as discussed in [de Kleer and Brown, 1984], categories of causal relations [Kitamura *et al.*, 1996b] helps capture causal characteristics context-independently.

The reasoning method of our reasoning system is based on the general reasoning scheme discussed in section 3.3. The reasoning of feedback is based on the feedback heuristics. In our system, since a part of causal relations in $T1$ have no physical meaning and $T4$ represents a very fast mechanism, the threshold values $T_{s1}$ and $T_{s2}$ are set to $T1$ and $T4$, respectively.

The reasoning system has been successfully applied to a power plant [Kitamura *et al.*, 1996b]. The model of the whole system consists of 27 components, 143 parameters and 102 constraints. All the reasoning results matched those obtained by a domain expert including their ambiguities.

## 5 Related Work

The time concept in QSIM is discussed in [Kuipers, 1994] from the mathematical viewpoint, which is categorized into $Ta$ or $Tb$. Iwasaki and Simon show a causal ordering theory for hierarchical sets of variables and discuss how to generate such hierarchical sets according to time scale and strength of interaction among variables [Iwasaki and Simon, 1994]. The causal time ontology allows us to clarify the modeling rationales underlying such sets from the physical viewpoint.

Ontologies of time itself have been discussed elsewhere such as [Allen, 1984] where Allen has identified primitives for representing time itself, and categorized of logical relationship between them. The causal time ontology provides cognitive categories of time intervals from the viewpoint of causal ordering of physical systems.

In [de Kleer and Brown, 1984; Top and Akkermans, 1991], although general causal properties of devices have been identified, causal relations generated by their methods are ambiguous in the case of inherently simultaneous equations. The TQ analysis [Williams, 1984] provides heuristics to analyze limited kinds of feedback according to time delay. A part of our causal specification corresponds to the descriptions of "exogenous parameters" [Iwasaki and Simon, 1994] of each component.

In [Forbus, 1984; Washio, 1989], causal characteristics of physical processes are described. One of our global constraints corresponds to an energy constraint (a global filter) for QSIM [Fouché and Kuipers, 1992].

## 6 Summary

We have proposed a causal time ontology containing a set of causal time scales shown in Table 1 to reveal the structure of causal time underlying the qualitative models and the causal reasoning engines. Some conventional reasoning systems have been characterized with respect to causal ordering using the time scales shown in Table 2. Furthermore, we present a reasoning system which can generate finer-grained causal relations than the existing systems.

We confined the topic to continuous changes. A discrete model of a phenomenon is, however, often the result of modeling according to such a rationale that the phenomenon is extremely faster than other phenomena, as discussed in [Iwasaki et al., 1995; Nishida and Doshita, 1987]. Thus, such discrete models can be viewed as another kind of temporal modeling techniques discussed in this article. Investigation on such discrete changes remains as future work.

As discussed in section 4, the causal time scales enable us to specify temporal performance required to derive desired causal relations. They can be viewed as specification of goal of design. They will govern the constituents of the models and the reasoning procedures. Investigation on design methodology of causal reasoning systems based on the causal time scales is in progress.

## Acknowledgments

The authors would like to thank Shinji Yoshikawa and Kenji Ozawa, Power Reactor and Nuclear Fuel Development Corp., and Munehiko Sasajima, I.S.I.R., Osaka University, for their help and comments on building the model of the power plant. The authors are grateful to the anonymous reviewers for valuable comments.

## References

[Allen, 1984] J. F. Allen. Towards a general theory of action and time. *Artificial Intelligence*, 23:123-154, 1984.

[de Kleer and Brown, 1984] J. de Kleer and J. S. Brown. A qualitative physics based on confluences. *Artificial Intelligence*, 24:7-83, 1984.

[Forbus, 1984] K. D. Forbus. Qualitative process theory. *Artificial Intelligence*, 24:85-168, 1984.

[Fouché and Kuipers, 1992] P. Fouché and B. J. Kuipers. Reasoning about energy in qualitative simulation. *IEEE Trans. on Systems, Man, and Cybernetics*, 22(1):47-63, 1992.

[Iwasaki and Simon, 1994] Y. Iwasaki and H. A. Simon. Causality and model abstraction. *Artificial Intelligence*, 67:143-194, 1994.

[Iwasaki et al., 1995] Y. Iwasaki, A. Farquhar, V. Saraswat, D. Bobrow, and V. Gupta. Modeling time in hybrid systems: how fast is "instantaneous"? *Proc. of IJCAI-95*, pages 1773-1780, 1995.

[Kitamura et al., 1996a] Y. Kitamura, et al. A method of qualitative reasoning for model-based problem solving and its application to a nuclear plant. *Expert Systems with Applications*, 10(3/4):441-448, 1996.

[Kitamura et al., 1996b] Y. Kitamura, M. Ikeda, and R. Mizoguchi. A qualitative reasoning based on an ontology of fluid systems and its evaluation on a power plant. *Proc. of PRICAI-96*, pages 288-299, 1996.

[Kuipers, 1994] B. J. Kuipers. *Qualitative Reasoning*. MIT Press, Boston, MA, 1994.

[Mars, 1995] N. J. I. Mars, editor, *Towards Very Large Knowledge Bases*. IOS Press, 1995.

[Mizoguchi and Ikeda, 1996] R. Mizoguchi and M. Ikeda. Towards Ontology Engineering. *Technical Report AI-TR-96-1*, I.S.I.R., Osaka University, Osaka, Japan, 1996.

[Nishida and Doshita, 1987] T. Nishida and S. Doshita. Reasoning about discontinuous change. *Proc. of AAAI-87*, pages 643-648, 1987.

[Rose and Kramer, 1991] P. Rose and M. A. Kramer. Qualitative analysis of causal feedback. *Proc. of AAAI-91*, pages 817-823, 1991.

[Top and Akkermans, 1991] J. Top and H. Akkermans. Computational and physical causality. *Proc. of IJCAI'91*, pages 1171-1176, 1991.

[Washio, 1989] T. Washio. Causal ordering methods based on physical laws of plant systems. MITNRL-033, MIT, 1989.

[Williams, 1984] B. C. Williams. Qualitative analysis of MOS circuits. *Artificial Intelligence*, 24:281-346, 1984.

# DIAGNOSIS AND QUALITATIVE REASONING

Qualitative Reasoning 2: Perception and Belief

# Qualitative Reasoning about Perception and Belief

**Alvaro del Val**[*]
Departamento de Ingeniería Informática
Universidad Autónoma de Madrid
28049 Madrid, Spain
delval@ai.ii.uam.es

**Pedrito Maynard-Reid II**[†]
Computer Science Department
Stanford University
Stanford, CA 94305
pedmayn@cs.stanford.edu

**Yoav Shoham**
Computer Science Department
Stanford University
Stanford, CA 94305
shoham@cs.stanford.edu

## Abstract

We present a qualitative model for reasoning about perceptions, sensors, and belief, and a logic to reason about this model. Basic to our model is a distinction between precision and accuracy, for both of which we provide qualitative definitions. In our logic this distinction gives rise to two modal operators—$P$ for actual perception, and $C_p$ for perceptual capability, which is captured as a set of possible percepts. Adding to these operators the standard $B$ operator to model belief, we end up with a logic combining standard Kripke-style semantics with the almost-standard 'neighborhood semantics.' We define various agent types in the logic, from agents who believe all and only what the sensors tell them, to much more skeptical agents. We define each agent both axiomatically and model-theoretically, and provide soundness and completeness results relating the two types of definitions.

## 1 Introduction

A great deal of attention in AI has been devoted to modeling states of information, whether qualitatively through, e.g., logics of knowledge and belief (cf. [Moore, 1985; Konolige, 1986; Fagin *et al.*, 1995]), or quantitatively. In contrast, relatively little attention has been paid to modeling information *sources*, on the basis of which the various information states are reached. In particular, there has been very little work on the relationship between belief and *perception*, the most common source of information in everyday life. This is the topic of this paper. (We hasten to add that there certainly has been some work on this topic in AI, notably

by Davis [1989] and Bacchus, *et al.* [1995]; we discuss this work in section 5.)

The connection between belief and perception is multifaceted. Here we concentrate on a particular set of issues, namely, the precision and accuracy of percepts and of sensors that generate them. This distinction is important but easily glossed over; we discuss it in detail in the next section.

The contributions of this paper are as follows:

- A qualitative model of precision and accuracy.

- A logic with which to reason about certain aspects of the model. The logic contains three different but related modalities: $P$ (for actual perception), $C_p$ (for the set of possible percepts), and $B$ (for belief). They are each defined in what by now is a standard fashion, but the combination is novel.

- A category of agent types, each defined both axiomatically and model-theoretically; the types differ either on the relationship between the agent's percepts and the world, or (and this is the most interesting case) on the relationship between the agent's percepts and his beliefs.

- Completeness results, relating the axiomatic systems and the semantic models.

The paper is organized as follows. In section 2 we present our model of precision and accuracy, and explain it with examples. In section 3 we present the logic **CP0** for modeling and reasoning about perception, and present our first completeness result. We also define the first two types of agent, *accurate* and *observant*. In section 4 we present the logic **BCP0** for modeling and reasoning about perception and belief, and present many more agent types. We provide completeness results for **BCP0** as well as for the various agent types. We discuss related work in section 5, and make some concluding remarks in section 6.

---

[*]This work performed at Stanford University's Computer Science Department.
[†]This work partly supported by a National Physical Science Consortium Fellowship.

## 2 Qualitative model of accuracy and precision

Our starting point is a distinction between *precision* and *accuracy*, both between the precision and accuracy of a particular percept, and the precision and accuracy of a sensor.

We illustrate these notions through an example. Consider a sensor that is supposed to detect the location of an object on the real line. A percept (or reading) of this sensor would be some interval such as $[1,7]$, meaning that the object is somewhere in that interval. Let's denote this percept by $P_1$. Another percept might be $P_2 = [3,5]$. In this case $P_2$ is said to be more precise than $P_1$. Suppose that the *actual* location is 2; then $P_1$ is said to be more accurate than $P_2$ since $P_1$ is right and $P_2$ is wrong. Now consider $P_3 = [3,4]$. Clearly $P_3$ is more precise than $P_2$, but which of them is more accurate? The intuitive answer is not obvious, since they are both wrong, but in our model $P_3$ will come out more accurate. Intuitively, $P_2$ allows locations that are "more wrong" than does $P_3$. Note that in order to capture "more wrong" we need to appeal to the notion of a "similarity ordering"—which of two points are closer to a third—but not necessarily an exact metric specifying the distance between two points.

The three example percepts discussed so far are related by set inclusion, but now consider $P_4 = [4,6]$; is it more or less precise than $P_2$? The answer depends on whether we can appeal to metric properties or not. If we can then we can judge the two percepts equally precise. However, since we wish to capture qualitative notions, in our model the precisions of $P_2$ and $P_4$ will come out as non-comparable. But what about their accuracies? In our model $P_2$ will come out as more accurate; the intuition is again that $P_4$ allows locations that are even more wrong than does $P_2$.

Let us now begin to develop a formal model; in the following definitions assume a set S (of worlds).

**Definition 1** *A percept is a subset of $S$.*

**Definition 2** *Percept $X$ is said to be as precise as percept $Y$ iff $X \subseteq Y$.*

**Definition 3** *Percept $X$ is accurate at $s \in S$, abbreviated s-accurate, iff $s \in X$.*

**Definition 4** *Given a set of allowable percepts $A$, percept $X$ is as accurate as percept $Y$ at $s \in S$ wrt $A$ iff $\forall Z \in A$, if $s \in Z$ and $Y \subseteq Z$ then $X \subseteq Z$.*

Thus $X$ is as accurate as $Y$ iff every accurate percept which includes $Y$ must also include $X$. This definition is intuitive but often quite weak. Therefore we consider ways to impose more structure upon possible perceptions, in particular a notion of "convexity" of the space derived from a similarity preorder.

**Definition 5** *A similarity ordering on $S$ at $s \in S$ is a binary relation $\leq_s \subseteq S \times S$; the intuitive reading of $\leq_s (s_1, s_2)$, which we abbreviate $s_1 \leq_s s_2$, is "$s_1$ is at least as close as $s_2$ to $s$". $<_s$ denotes the strict version of $\leq_s$. We require that $\leq_s$ be a preorder (that is, reflexive and transitive), and that if $s \neq s'$ then $s <_s s'$.*

**Definition 6** *Given $\leq_s$ as above, percept $X$ is convex at $s \in S$, abbreviated s-convex, iff $\forall x, y \in X$, if $x \leq_s z \leq_s y$ then $z \in X$. A set of percepts $A$ is s-convex iff every $X \in A$ is s-convex.*

**Definition 7** *The accurate extension of a percept $X$ at $s \in S$ is the percept $X^+(s) = \{y \in S \mid \exists x \in X : y \leq_s x\}$.*

**Definition 8** *A set of percepts $A$ is strongly s-convex iff $A$ is convex and $\forall X \in A : X^+(s) \in A$.*

**Proposition 1** *$X^+(s)$ is s-convex and s-accurate, and for any s-convex and s-accurate percept $Z \in A$, if $X \subseteq Z$ then $X^+ \subseteq Z$.*

In the following propositions, assume $A$ to be strongly s-convex and $X, Y \in A$.

**Proposition 2** *$X$ is as accurate as $Y$ at $s \in S$ wrt $A$ iff $X^+(s) \subseteq Y^+(s)$.*

**Proposition 3** *$X$ is as accurate as $Y$ at $s$ wrt $A$ iff $\forall x \in X \ \exists y \in Y : x \leq_s y$.*

**Proposition 4** *Let $\leq_s$ be a total preorder. $X$ is as accurate as $Y$ at $s \in S$ wrt $A$ iff $sup(X) \leq_s sup(Y)$, where $sup(Z)$ is an element of set $Z$ that is a least upper bound wrt $\leq_s$.*

Under the assumptions of this last proposition, therefore, the relation "being as accurate as" is a total preorder. Also, if $A$ is convex but not strongly convex, then the right hand sides of these three propositions are sufficient but not necessary for $X$ to be as accurate as $Y$.

So far we have discussed the relative precision and accuracy of percepts; let us now turn to comparing sensors. We model a sensor by its *perceptual capabilities*, which, loosely speaking, means the set of percepts it is capable of delivering in any particular circumstance. Because sensors represent sets of percepts, their relative precision and accuracy can be defined in a variety of ways depending on the application. Two possibilities are as follows: (1) Sensor A is considered more precise/accurate than sensor B in some situation if every percept of which sensor A is capable is more precise/accurate than some percept sensor B could deliver. (2) Sensor A is more precise/accurate than sensor B in a situation if every percept sensor A could return is more precise/accurate than any percept sensor B could return. The first case corresponds to a comparison of the worst case scenarios of the two sensors, whereas the second case corresponds to a guarantee on the relative precision/accuracy of the percepts actually returned. The second is obviously more restrictive, and implies the first.

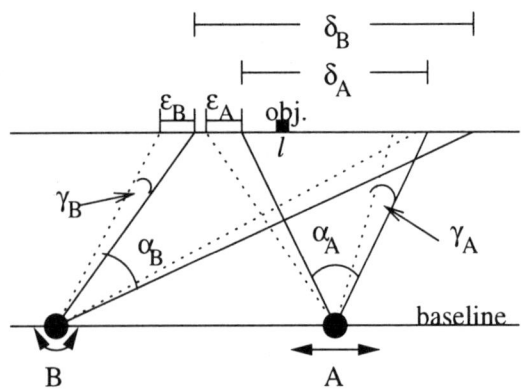

Figure 1: Two-sensor example.

Again, we illustrate these notions through an example. We will also demonstrate that not only can a given sensor's capabilities change with the world, but so can the relative relationship between the capabilities of different sensors. Consider two sensors, A and B, which report on the location of some object on a line. Imagine the two sensors as located on another line (the *baseline*) parallel to the line the object is on (see Figure 1).

The capabilities of sensor $x$ (where $x$ is A or B) are characterized by two parameters, $\alpha_x$ and $\gamma_x$. $\alpha_x$ is the angle of resolution of sensor $x$, whereas $\gamma_x$ is the maximum angle of divergence between the orientation implied by the particular reading and the sensor's actual orientation. Given these parameters, the distance of the baseline from the real line, and the location and orientation of sensor $x$, we can compute $\delta_x$, the size of the interval it returns, and $\epsilon_x$, the maximum displacement of the reading from the true interval. (We assume that $\gamma_x$ is small enough compared to $\alpha_x$ that the variation in interval length due to $\gamma_x$ is negligible.) Let us suppose that $\alpha_B < \alpha_A$ and $\gamma_B < \gamma_A$.

The object's location determines the world, and at a world each sensor returns one percept. To determine the percept it returns at a world, sensor A maintains a fixed orientation perpendicular to the real line, and travels back and forth along the baseline, periodically taking a reading until the object appears in its vision. Because its orientation is constant (within $\gamma_A$), both $\delta_A$ and $\epsilon_A$ are constant. Imagine, however, that it is more expensive for sensor B to move around, so instead it remains in one location and varies its orientation by rotating, randomly sampling the real line until the object appears in its vision. In this case both $\delta_B$ and $\epsilon_B$ vary with the sensor's orientation, so both its accuracy and precision decrease the farther away the object is.

Thus the specific set of possible percepts which sensor $x$ might deliver when the object is located at $l$ consists of the intervals $[r, r + \delta_x]$ for all $l - \delta_x - \epsilon_x \leq r \leq l + \epsilon_x$ (corresponding to the extreme cases where the the interval is $\epsilon_x$ to the left and right of the object, respectively) for $r \in \Re$ (the real numbers). Consider the world where the object is located directly in front of sensor B. Sensor B is more precise than sensor A because all the percepts sensor B might return are more precise than all those sensor A might return. It is also more accurate (under the worst-case interpretation) than sensor A because the least accurate percept it is capable of is more accurate than some of sensor A's possible percepts.

Now suppose the object is far away from sensor B. The size of the interval returned by sensor B and the magnitude of its maximum displacement will be much larger. In this case sensor B will be both less precise and less accurate (in the worst case) than sensor A because it will have possible percepts that are less precise and less accurate than all the percepts sensor A might return.

The definitions below extend our framework to include sensors. Assume a set of allowable percepts $A$ as before.

**Definition 9** *A sensor is a function $f : S \to 2^{2^S}$ that describes the set of possible percepts in a world.*

**Definition 10** *Sensor $f$ is* worst-case as precise as *sensor $f'$ at $s \in S$ iff $\forall X \in f(s) \; \exists Y \in f'(s)$ s.t. $X$ is as precise as $Y$ at $s$. It is* guaranteed as precise as *sensor $f'$ at $s$ wrt $A$ iff $\forall X \in f(s) \; \forall Y \in f'(s)$, $X$ is as precise as $Y$ at $s$ wrt $A$.*

**Definition 11** *Sensor $f$ is* worst-case as accurate as *sensor $f'$ at $s \in S$ wrt $A$ iff $\forall X \in f(s) \; \exists Y \in f'(s)$ s.t. $X$ is as accurate as $Y$ at $s$ wrt $A$. It is* guaranteed as accurate as *sensor $f'$ at $s$ wrt $A$ iff $\forall X \in f(s) \; \forall Y \in f'(s)$, $X$ is as accurate as $Y$ at $s$ wrt $A$.*

## 3 A logic for perception

We now aim to construct a logic with which to reason about these aspects of perception. As the above discussion illustrates, at a basic level reasoning about perception involves the counterfactual notion of what *could be perceived*, and it is this notion we will concentrate on. We capture it formally by introducing a logic with two modal operators, $P$ and $C_p$, which represent actual perception and possible perception, respectively.

This gives us a way to capture the perceptual limitations of an agent. Next we want to represent the degree to which the agent is aware of these limitations. What an agent believes given some percept may depend on the agent's level of trust in the reliability of her sensors, which in turn may depend to some degree on the actual percepts themselves. To be able to describe complex interactions of this sort between the agent's mental state and her perceptions and possible perceptions, we also add to the formalism a modal operator, $B$, for belief. We then use this model to define several agent types with various interesting perceptual-belief characteristics.

## 3.1 The logic CP0

The formulas of the language $\mathcal{L}$ are obtained by closing a set of propositional symbols under the boolean connectives, and under the operators $P$ and $C_p$. Intuitively $P\phi$ means "the agent perceives $\phi$", where $\phi$ is some arbitrary formula. $C_p\phi$ indicates that "the agent can perceive $\phi$", meaning that her perceptual apparatus could have produced a percept that validates $\phi$. We also ensure, of course, that what is perceived can be perceived.

For semantics, we define a P-structure as a tuple $\mathcal{M} = (S, N, I, \pi)$ where $S$ is a set of states or worlds, $\pi$ is a truth assignment $\pi : S \to 2^\mathcal{L}$, $N$ is a relation $N : S \to 2^{2^S}$ mapping each world to a set of sets of worlds, and $I$ is an accessibility relation $I : S \to 2^S$ mapping each world to a set of worlds.

The satisfaction relation is defined as usual. Only the truth conditions for the operators $P$ and $C_p$ are of interest. They are:

$\mathcal{M}, s \models C_p \phi$ iff $\exists Z \in N(s)$ s.t. $\forall t \in Z, \mathcal{M}, t \models \phi$,

$\mathcal{M}, s \models P\phi$ iff $\forall t \in W$, if $t \in I(s)$ then $\mathcal{M}, t \models \phi$.

Thus $P$ uses the standard semantics for necessity operators in modal logic, whereas $C_p$ uses a variant of neighborhood (or minimal model) semantics [Chellas, 1980]. For $s \in S$, we use the notation $N(s) = \{X \subseteq S \mid (s, X) \in N\}$, and $I(s) = \{t \in S \mid (s, t) \in I\}$. The idea is that $N(s)$ represents the set of possible percepts at a world, and the set $A$ of allowable percepts referred to in the previous section includes all these possible percepts from all worlds. In a domain that is strongly $s$-convex, $A$ would also include all the accurate extensions of these percepts, although these extensions may not be actually perceivable from any world.

We are interested in the class $\mathcal{M}_{CP}$ of P-structures for perception satisfying the following conditions:

CP1. $N(s) \neq \emptyset$.

CP2. $\emptyset \notin N(s)$.

CP3. $I(s) \in N(s)$

Condition (CP3) is the most interesting one. It requires every actual percept to be a possible percept, and corresponds to axiom A4 below. (CP1) and (CP2) correspond, in turn, to axioms A2 and A3, respectively.

Let **CP0** be the following set of axioms and rules of inference:

A1. All tautologies of propositional calculus.

A2. $C_p true$

A3. $\neg C_p false$

A4. $P\phi \supset C_p \phi$

A5. $(P\psi \wedge P(\psi \supset \phi)) \supset P\phi$.

R1. From $\psi$ and $\psi \supset \phi$ infer $\phi$.

R2. From $\psi \supset \phi$ infer $C_p \psi \supset C_p \phi$.

R3. From $\psi$ infer $P\psi$.

**Theorem 1** **CP0** *is a sound and complete axiomatization of the class $\mathcal{M}_{CP}$ of P-structures for perception.*

**Proposition 5** *These formulas are valid in $\mathcal{M}_{CP}$:*

1. $C_p(\phi \wedge \psi) \supset (C_p \phi \wedge C_p \psi)$.

2. $\neg P false$.

Thus **CP0** includes only some weak consistency conditions on $C_p$, namely A2 and A3, and on $P$, in addition to the aforementioned restriction that actual percepts must be possible percepts. On the other hand, the system does *not* sanction the axiom $(C_p\psi \wedge C_p\phi) \supset C_p(\psi \wedge \phi)$: for example, it may be possible for me to perceive some object as either blue or red depending on lightning conditions, but this does not mean that I can perceive the object as both red and blue.

## 3.2 Accurate and observant agents

Various agent types can be defined to capture varying assumptions about the application domain. Most definitions refer to the relationship between perception and belief, a topic we have not yet addressed. However, some types are defined in terms of the relationship between perception and objective facts in the world, independent of the belief modality. Thus we give them here.

We first consider *observant* agents, those from whose perception nothing escapes. Formally, the agent is said to be observant if the following schema is valid:

$$\phi \supset P\phi.$$

Similarly, the agent is said to be *accurate* if the schema

$$P\phi \supset \phi$$

is valid.

The schema corresponding to the observant agent is clearly very strong, unlikely to be applicable to agents in more complex environments, since nothing perceptually escapes the agent. Yet it is possible to think of simple agents for which it makes sense, such as, for example, a device continuously and accurately monitoring some variable, and whose model of the world consists exclusively of the possible values the variable can take.

We can also consider similar properties for the $C_p$ modality. The agent is *potentially observant* iff she can perceive any (true) fact, i.e. iff the following schema is valid:

$$\phi \supset C_p \phi.$$

An agent is said to be *necessarily accurate*, on the other hand, if anything she can perceive is true, i.e. iff the following schema is valid:

$$C_p \phi \supset \phi.$$

# 4 A logic for perception and belief

We are now interested in what the agent should believe, given what she perceives. It might be tempting to make the assumption that anything perceived is also believed. However, in (common) cases where the sensors might not be completely accurate, and in which the agent might know this, the assumption would be inappropriate. Sometimes the converse assumption is more adequate: one doesn't believe in anything without having perceived it. Keeping this in mind, we begin with the most general case, in which the relation between perception and belief is unconstrained, and which can be captured in a straightforward manner. We will then identify a number of special cases with further properties.

## 4.1 The logic BCP0

We expand **CP0** to include an operator $B$. A PB-structure is a tuple $\mathcal{M} = (S, N, I, R, \pi)$, where $S$, $N$, and $I$ and $\pi$ are as before (in particular (CP1)–(CP3) are satisfied), and $R$ is a serial, transitive and Euclidean relation over $S$. That is, we are characterizing belief with the standard KD45 system. In addition, we require the agent to be aware of her perceptions, by imposing the following semantic condition (which corresponds to axioms A10 and A11 below):

BCP1. $\forall s, t$, if $t \in R(s)$ then $I(s) = I(t)$.

Let $\mathcal{M}_{BCP}$ be the class of such structures. The semantics of the $B$ operator are given as usual by:

$\mathcal{M}, s \models B\psi$ iff $\forall t \in S$, if $t \in R(s)$ then $\mathcal{M}, t \models \psi$.

The system **BCP0** consists of all the axioms and rules of **CP0** together with:

A6. $(B\psi \wedge B(\psi \supset \phi)) \supset B\phi$.

A7. $\neg B(false)$.

A8. $B\psi \supset BB\psi$.

A9. $\neg B\psi \supset B\neg B\psi$.

A10. $P\psi \supset BP\psi$.

A11. $\neg P\psi \supset B\neg P\psi$.

R4. From $\psi$ infer $B\psi$.

The following result should be obvious:

**Proposition 6** **BCP0** *is a sound and complete axiomatization of the class $\mathcal{M}_{BCP}$ of PB-structures for perception and belief.*

Note that the formula $BP\psi \equiv P\psi$ follows from A10 and A11, and is therefore a theorem of **BCP0**. Although one can easily think of counterexamples, e.g. ambient noise, or sensory overload, where percepts are filtered out so that the agent is not aware of what she is perceiving, we have elected to adopt this assumption to simplify the remaining discussion. It is easy to weaken the formalism to deal with this situation if desired.

Given this logic, we can make additional inferences about perception and belief. For example, the formula

$$P\phi \wedge B(\psi \supset \neg C_p\phi) \supset B\neg\psi$$

is valid in **BCP0**. We can read this as saying that if the agent perceives $\phi$ and she believes she could not have perceived that in a world in which $\psi$ holds, then she believes that $\psi$ is false. We can obtain stronger results by imposing additional conditions on the agent. This motivates the definition of a number of agent types according to the connections between the various modalities. We do this in the next section.

## 4.2 More agent types

By placing additional joint restrictions on $I$, $N$, and $R$, it is possible to capture different ways in which perceptions and perceptual capabilities can be related to the agent's beliefs and to the environment. There are a number of basic dimensions to consider. In a previous section we appealed to the relationship between perception and the environment in defining the accurate and observant agents, and also between perceptual capabilities and the environment; we also decided to build into the logic the introspective abilities of the agent with respect to perception through axioms A10 and A11. In addition we may now consider the introspective abilities of the agent with respect to her perceptual capabilities, and the confidence that the agent places in her perceptions. As was discussed in the previous section, a third dimension—the correctness of the agent's beliefs about her perceptions—is built into the logic. Table 1 defines the various agent types along these dimensions in terms of the validity of certain (not necessarily independent) axiom schemas, and Table 2 identifies certain classes of PB-structures for perception and belief in terms of restrictions on $I$, $N$, and $R$. The correspondence between the axiom schemas and the corresponding classes of structures is given in the soundness and completeness results in the next theorem, where **BCP0**+$i$ refers to the system composed of **BCP0** and the $i$th axiom schema in Table 1.

**Theorem 2** **BCP0** $+ 1$ *(respectively $+ 2$, $+ 3$[1], $+ 4$, $+ 5$, $+ 6$, $+ 7$, $+ 8$, $+ 9$) is a sound and complete axiomatization of the class of PB-structures for perception and belief $\mathcal{M}_{observant}$ (respectively $\mathcal{M}_{accurate}$, $\mathcal{M}_{Cp-observant}$, $\mathcal{M}_{Cp-accurate}$, $\mathcal{M}_{Cp-aware(+)}$, $\mathcal{M}_{Cp-aware(-)}$, $\mathcal{M}_{confident}$, $\mathcal{M}_{cautious}$, $\mathcal{M}_{sib}$).*

---

[1] The completeness proof for the potentially observant agent assumes a finite propositional vocabulary.

| Agent Type | Axiom schema |
|---|---|
| Observant agent | 1. $\psi \supset P\psi$ |
| Accurate agent | 2. $P\psi \supset \psi$ |
| Potentially observant | 3. $\psi \supset C_p\psi$ |
| Necessarily accurate | 4. $C_p\psi \supset \psi$ |
| Aware (+) of perceptual capab. | 5. $C_p\psi \supset BC_p\psi$ |
| Aware (−) of perceptual capab. | 6. $\neg C_p\psi \supset B\neg C_p\psi$ |
| Perceptually confident | 7. $P\psi \supset B\psi$ |
| Cautiously confident | 8. $P\psi \supset \neg B\neg\psi$ |
| SIB (seeing-is-believing) | 9. $\neg P\psi \supset \neg B\psi$ |

Table 1: Agent types: axiomatic conditions.

| Class of structures | Restriction |
|---|---|
| $\mathcal{M}_{observant}$ | 1. $\forall s,t : t \in I(s) \Leftrightarrow s = t$ |
| $\mathcal{M}_{accurate}$ | 2. $\forall s : s \in I(s)$ |
| $\mathcal{M}_{Cp-observant}$ | 3. $\forall s : \{s\} \in N(s)$ |
| $\mathcal{M}_{Cp-accurate}$ | 4. $\forall s : s \in \bigcap N(s)$ |
| $\mathcal{M}_{Cp-aware(+)}$ | 5. $\forall s,t : t \in R(s) \Rightarrow N(s) \subseteq N(t)$ |
| $\mathcal{M}_{Cp-aware(-)}$ | 6. $\forall s,t : t \in R(s) \Rightarrow N(t) \subseteq N(s)$ |
| $\mathcal{M}_{confident}$ | 7. $\forall s : R(s) \subseteq I(s)$ |
| $\mathcal{M}_{cautious}$ | 8. $\forall s : R(s) \cap I(s) \neq \emptyset$ |
| $\mathcal{M}_{sib}$ | 9. $\forall s : I(s) \subseteq R(s)$ |

Table 2: Agent types: semantic conditions.

## 5 Related work

There has been relatively little work relating perception and belief; here are the closest of which we are aware.

First, related to our paper is Davis' earlier work [Davis, 1989] on belief and perceptual acuity. Davis' work, however, like some other work in related fields (cf., [Parikh, 1983]), is firmly based on modeling metric spaces, and is thus quantitative in nature. Davis' logic contains only notions analogous to our $P$ and $B$, and does not have a way of specifying the perceptual capabilities of the sensor. He was thus not able to differentiate in the logic between precision and accuracy, and ran into some paradoxes as a result.

Recently, Bacchus, et al. [1995] have worked on representing the effects of readings from noisy sensors on the belief state and subsequent actions of agents. They employ a formalism that combines a probabilistic definition of belief with a situation calculus framework. However, our approach, while remaining amenable to probabilistic extensions, is able to capture qualitative effects of perception as well. Their work also does not address precision.

## 6 Discussion

We have presented a formalism to reason about perception and belief, an under-explored area of investigation. We started by distinguishing between precision and accuracy, notions that are often conflated. We proposed what to our knowledge is the first qualitative definition of these notions.

We then provided a model in which to represent an agent's perceptions and beliefs, and a logic with which to reason about the model. We explored possible relations between an agent's perceptions, perceptual capabilities, and beliefs, and between these and the environment. Because no type of agent among those defined is likely to be useful in all scenarios, we have mentioned some circumstances in which different types might be used.

The formalism we have presented seems natural and well-behaved, but more research is needed on applying it, whether in general formulations of common-sense reasoning or in more specific applications. It would also be interesting to study extensions of the language that incorporate more specific domain structure (such as a metric space), and relate them to perceptual indistinguishability. Temporal and multi-agent aspects are also of interest, as well as the problem of updating the beliefs of the agent in the light of perceptually acquired information. Indeed, we view sensors as a source of information, and precision/accuracy as measures of that source's competence. We intend to apply this theory more generally to reason about multiple, partially-trusted information sources.

**Acknowledgements** We would like to thank the members of the Nobots research group at Stanford University for their helpful comments, as well as the referees for constructive critique on the paper.

## References

[Bacchus et al., 1995] F. Bacchus, J. Halpern, and H. Levesque. Reasoning about noisy sensors in the situation calculus. In *Proc. 14th Conf. IJCAI*, pages 1933–1940, 1995.

[Chellas, 1980] B. Chellas. *Modal Logic: An Introduction*. Cambridge University Press, 1980.

[Davis, 1989] E. Davis. Solutions to a paradox of perception with limited acuity. In *Proc. of the 1st Intl. Conf. on Knowl. Rep. and Reas.*, 1989.

[Fagin et al., 1995] R. Fagin, J. Halpern, Y. Moses, and M. Vardi. *Reasoning about Knowledge*. MIT Press, 1995.

[Konolige, 1986] K. Konolige. *A Deduction Model of Belief*. Pittman Research Notes in AI, 1986.

[Moore, 1985] R. Moore. Semantical considerations on non-monotonic logic. *Art. Int.*, 25:75–94, 1985.

[Parikh, 1983] R. Parikh. The problem of vague predicates. In R. Cohen and M. Wartofsky, editors, *Language, Logic, and Method*, pages 241–261. D. Reidel, 1983.

# Rule-based Contact Monitoring Using Examples Obtained by Task Demonstration

**Pavan Sikka** and **Brenan J. McCarragher**
Department of Engineering, F. E. I. T.
The Australian National University
Canberra, ACT 0200, Australia

## Abstract

This paper presents a new rule-based hybrid-system approach to contact monitoring. The rules are formulated in terms of temporal sequences of the contact force and they recognize temporal patterns of force associated with contact states and transitions. The rule-base is built using inductive learning techniques on force data obtained by human demonstration. This approach is suitable for monitoring robotic as well as human tasks. An advantage of this approach is that it allows process monitors for different tasks to be built quickly and easily by learning new sets of rules from a demonstration of the tasks. Experimental results are presented to demonstrate the effectiveness of this approach.

## 1 Introduction

Robotic systems are prone to errors due to uncertainties in the model of the process and of the environment. This makes process monitoring an important part of a robotic system. Process monitoring is based on continuously observing and interpreting the data from sensors that report on the system state, and is used for error detection, recovery, control, adaptation and learning. The focus of this paper is the design of process monitors using inductive learning techniques within a hybrid dynamic systems framework.

Robotic tasks can often be modeled as hybrid dynamic systems [Stiver and Antsaklis, 1992]. A task level description is based on discrete event dynamic systems [Ramadge and Wonham, 1989], while the physical aspects of the system are modeled using continuous variable dynamic systems. In this hybrid dynamic systems framework, a process monitor is defined as a mapping that transforms the physical state of the continuous system into discrete states and events at the task level. This approach can be contrasted with statistical change detection techniques for contact tasks. For example, Eberman and Salisbury [Eberman and Salisbury, 1994] consider a temporal sequence of the force signal to determine contact type; however, there is no consideration of a discrete-event task model and their approach is focused on the continuous variable representation. This paper presents a rule-based method for contact monitoring where temporal sequences of quantized continuous variables are matched with rules to determine the states and transitions of the discrete-event system.

Task demonstration has been used for task-level robot programming. Delson and West [Delson and West, 1996] considered the problem of robot programming from demonstrated motion during one task-level state and did not address the problem of contact monitoring. Wang et al. [Wang et al., 1996] present a method to derive compliant motion programs from human demonstration. However, an explicit discrete-event model of the task is not considered, and hence the subtasks cannot always be identified with task-level states. In our approach, the rules are obtained from data generated by several demonstrations of the task being considered. The data obtained from demonstration is clustered into a small number of groups corresponding to quantized values of the continuous variables. Inductive learning techniques are then applied to the discretized data to obtain rules for monitoring and classifying the discrete states and events of the hybrid system. Since the rules pertain to both discrete states and events, this approach exploits the dynamics of the task, as opposed to previous work in assembly based on the quasi-static assumption which ignores dynamic effects (see, for example, [Whitney, 1982; Bruyninckx et al., 1995]).

Our approach to obtaining the rules from examples can be contrasted with the model-based approach in [McCarragher and Asada, 1993] where qualitative reasoning techniques are used to obtain simple templates for reconizing contact state transitions. There is only one template for each contact state transition. The templates are based on the current values of quantized variables and do not consider temporal sequences. Furthermore, qualitative reasoning techniques are difficult to apply consistently to continuous dynamic models since it is difficult to obtain consistent qualitative algebras [Williams and de Kleer, 1991].

There are several advantages of this approach. Since the analysis is based on real sensory data, the approach has the advantage of taking into account various task parameters such as friction, which are difficult to model analytically. Most important, the approach can be adapted easily and quickly to different tasks by simply learning new sets of rules from the data obtained by task demonstrations. Since the learning phase is based on a demonstration of the task, this approach is applicable both to robotic process monitoring and to the study of human demonstration of tasks. Finally, our approach is general in that it can be applied to one or more sensory modalities taken together, for example position and force.

## 2 Contact Monitoring using a Hybrid System Model

We consider contact tasks, such as assembly, to illustrate our approach to process monitoring. An example of modeling an assembly task as a hybrid dynamic system is presented in [McCarragher and Asada, September 1995]. These tasks can be characterized by the different configurations of contact between the parts being assembled. The configurations are described geometrically in terms of contact between the edges and surfaces of the objects involved and are called *contact states*. The contact states are organized into a graphical structure $G = (\Gamma, E)$, called the *contact state network*, where $\Gamma$ is the set of nodes representing the contact states $\gamma$ and $E$ is the set of arcs connecting contact states and representing the allowed changes in contact states [Asada and Hirai, 1989]. This is illustrated in Figure 1, which shows a part of the contact state network corresponding to the planar peg-in-hole task. At the plant level, the contact states determine the geometrical constraints which, together with the equations of motion, determine the plant behavior in terms of state variables such as position, velocity, acceleration, and force.

The contact monitor tracks the state of the continuous plant by monitoring variables of interest such as the position of the robot and the force acting on the robot. The organization of the contact monitor within the hybrid-dynamic framework is shown in Figure 2. The system is modeled as a combination of a continuous-domain system interacting with a discrete-event system through an interface. The interface acts as a mapping between the two sub-systems, transforming and inter-connecting the discrete and continuous inputs and outputs between the two sub-systems. Within this framework, the transformation between the discrete event system and the con-

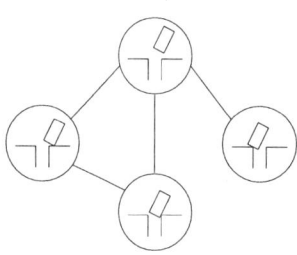

Figure 1: *A part of the contact-state network for the peg-in-hole task. Each circle denotes a contact-state, with the figure inside the circle illustrating the corresponding peg-in-hole configuration. Each arc represents an allowed change in contact-state.*

tinuous system becomes a key issue, and the process monitor is represented as a part of the interface. The vector $\mathbf{z}(t)$ denotes the state of the plant as seen by the process monitor, and consists of variables that are of interest to the process monitor. For example, in the planar case with force and changes in force being the variables of interest, $\mathbf{z}(t) = [f_x(t), f_y(t), \dot{f}_x(t), \dot{f}_y(t)]^t$, where $f_x$ and $f_y$ denote the planar components of force.

The contact monitor works in two steps. In the first step, the process monitor samples the variable $\mathbf{z}$ and then *quantizes* or *discretizes* it into the variable $\mathbf{z}_q$. For example, in an assembly task, $\mathbf{z}_q = [f_{xq}, f_{yq}, \dot{f}_{xq}, \dot{f}_{yq}]^t$. The discretization is performed using discriminant functions $g_{\gamma_c}$ based on the current contact state $\gamma_c$. The discriminant functions are also obtained from the data generated by human demonstration. The details of this step can be found in [Sikka and McCarragher, 1996].

The second step of the process monitor consists of using a rule-based system [Barr and Feigenbaum, 1981] to identify the process state $\gamma$ based on a sequence of discretized variables. This process is described in detail in the next section.

## 3 Rule-based Contact Monitoring

Each contact state transition can be characterized by temporal patterns of force. These patterns can easily be encoded in the form of rules for identifying the corresponding contact state transitions. This observation is the basis for our approach to rule-based contact state and transition identification.

A rule is defined as a *condition-action* pair where the condition part specifies the conditions under which the rule can be applied and the action part specifies the actions that are taken when the rule is applied. For example, in the context of contact monitoring, the following general rule can be used for recognizing a contact state transition:

**IF**

    state = $\gamma_1$   **AND**

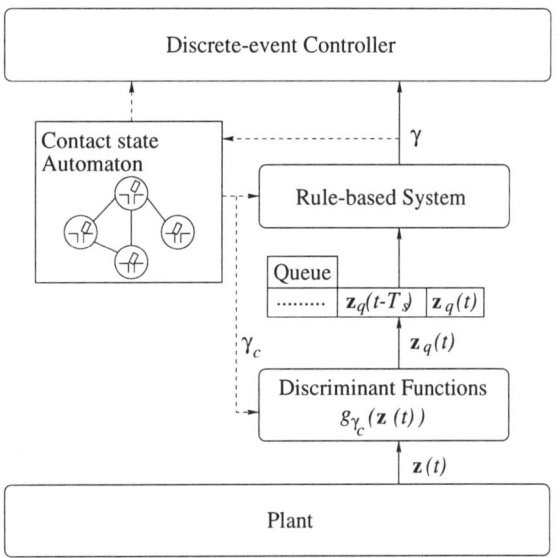

Figure 2: *The process monitor within a hybrid system framework. The dotted line encloses the main functional blocks of the process monitor.*

$$\begin{aligned}
\dot{f}_{xq}(k) &= d_1 \textbf{ AND} \\
\dot{f}_{xq}(k-1) &= d_1 \textbf{ AND} \\
\dot{f}_{xq}(k-2) &= d_2 \textbf{ AND} \\
\dot{f}_{xq}(k-3) &= d_3 \textbf{ AND} \\
\dot{f}_{xq}(k-4) &= d_1
\end{aligned}$$
**THEN**
$$\text{state} = \gamma_2$$

Here, $k$ refers to the current sampling period, $d_1 \cdots d_3$ refer to the discrete values of the quantized variables and $\gamma_1, \gamma_2 \in \Gamma$ denote contact states. To simplify notation, the remainder of this paper uses the corresponding integers $i$ to refer to the discrete values $d_i$.

The rule shown above checks for a sequence of discretized values of the rate of change of force to recognize a transition from contact state $\gamma_1$ to $\gamma_2$. In general, several variables are monitored and the rule-based process monitor consists of sets of rules for each variable and for each contact transition. This leads to a very general system for monitoring contact.

There are several important issues for rule-based systems. The matching of rules to data is computationally expensive [Barr and Feigenbaum, 1981]. A large rule-base coupled with a large database can make the system too inefficient to be of use. Another important issue for rule-based systems is the integrity of the rule-base. A rule-base with conflicting rules can lead to ambiguous results, and hence the rule-base must be checked for conflicts. These problems are addressed in the paper by the choice of finite-sate machines as the representation for rules and by the organization of the rule-base using the contact state network.

### 3.1 Finite-state machines: a representation for rules

A finite-state machine (FSM) is defined as the structure $M = \{S, \Sigma, f, s_0, F\}$, where $S$ is the set of states, $\Sigma$ is the set of input symbols, $f : S \times \Sigma \to S$ is the state transition function, $s_0 \in S$ is the initial state, and $F \subseteq S$ is the set of final, accepting states. A FSM operates as follows: The state of the machine is initialized to $s_0$. The input to the machine consists of a sequence $\sigma$ of symbols taken from the input alphabet $\Sigma$ of the FSM. The input string is processed sequentially. Thus, if $\sigma_i$ is the next unprocessed symbol in the input, and $s$ is the current state of the FSM, then the FSM changes its state to $f(s, \sigma_i)$ and $\sigma_{i+1}$ becomes the next symbol to be processed. The FSM stops after processing the last symbol in the input string. Let $s_t$ be the state of the FSM when it stops. If $s_t \in F$, then the FSM is said to *accept* the string $\sigma$.

A set of rules can be represented by a FSM under certain conditions. In particular, pattern classification rules can be easily implemented as a set of FSMs with one FSM for each set of patterns representing a class. The input alphabet of the FSM consists of the symbols used to express the patterns. A path from the initial state to a final state of the FSM corresponds to a pattern belonging to the corresponding class, and hence to a rule for that class.

The representation of a set of rules by a FSM significantly reduces the computational effort required to determine the rules that apply to the current information in the database. For a given pattern in the database, the task of finding the matching rule is reduced to determining if the FSM accepts the sequence representing the pattern.

### 3.2 The contact monitor

Let $\{\mathbf{z}_q(k)\}$ denote a temporal sequence of the discretized plant state vector. When this sequence corresponds to a contact state transition, it can be used to characterize the particular transition. In general, a contact state transition can be characterized by a set of temporal sequences or patterns of the discretized plant state vector $\mathbf{z}_q$. Thus, each component of $\mathbf{z}_q$ contributes an associated set of patterns for a given contact state transition. Each pattern becomes a rule for recognizing the corresponding contact state transition and these rules together are encoded as a FSM for recognizing the given transition.

The contact monitoring system is built around the contact state network $G$. Let $e_1, \cdots, e_p$ represent the transitions allowed from a contact state $\gamma$. As before, let the discretized plant state vector be defined by: $\mathbf{z}_q(k) = \mathbf{f}_q(k) = [f_{xq}(k), f_{yq}(k), \dot{f}_{xq}(k), \dot{f}_{yq}(k)]^t$. Then, for a transition $e$ from $\gamma$, a set of 4 FSMs are defined, one for each discretized variable: $M^e_{f_{xq}}, M^e_{f_{yq}}, M^e_{\dot{f}_{xq}}$ and $M^e_{\dot{f}_{yq}}$.

The rule-based contact monitor works as follows: Let $\{\cdots, \mathbf{f}_q(j), \cdots, \mathbf{f}_q(k)\}$ denote the sequence of values of the discretized plant state vector. Then, the input strings for the FSMs consist of the sequence $\{\mathbf{f}_q(k-m\Delta), \cdots, \mathbf{f}_q(k-\Delta), \mathbf{f}_q(k)\}$, where $\Delta$ is an integer representing a system-dependent delay and $m$ is an integer representing the the system-dependent sequence length. The sequence of each discretized variable is used as the input for the corresponding FSM. The transition for which all the FSMs accept the input sequence is taken as the transition occurring at instant $k$. The parameters $m$ and $\Delta$ in the input sequence are chosen based on the considerations such as the computing power available and the sampling rate. For example, in the implementation described later in the paper, the sampling rate is 120 Hz and typically about 60 samples make up a transition. The computing resources limit $m$ to 5. Hence, to represent the transitions, $\Delta$ is chosen to be 12.

Organizing the FSMs according to the contact state network has the advantage of simplifying the problem of determining the consistency of rules. The rules corresponding to transitions from a contact state can be examined independently of transitions from other contact states. Furthermore, for a given state, conflicts may exist among rules for a particular variable as long as there are no conflicts when the rules for all the variables are considered together.

## 4 Contact monitoring and human demonstration

This section addresses the problem of constructing a contact monitor for a given task. This involves several steps. First, the contact state network must be constructed and the variables making up the plant state vector must be specified. Then, for each state, the discriminant functions for quantizing the plant state variables need to be specified. Finally, the FSMs corresponding to the rules associated with each contact state transition must be constructed. In this section, we describe the use of task demonstration to construct the discriminant functions and the FSMs for the given task. The problem of constructing the contact state network for a given task is not addressed in this paper.

A person demonstrates the task several times. Variables constituting $\mathbf{z}(t)$, like position and force, are recorded during each demonstration of the task. Thus, human demonstration of the task yields a set of traces of the plant state $\mathbf{z}$, denoted by $\{y_1, \cdots, y_p\}$, where $y_i = \{\mathbf{z}(k)\}$. Each trace $y_i$ is used to generate sub-traces $y_{ij}$ corresponding to the individual contact state transitions present in the trace. This must be done manually, although automatic change detection algorithms, for example [Eberman and Salisbury, 1994], can be used to aid in this process. These sub-traces are then organized according to contact states. Thus, the set $X_{\gamma_1}$ contains all the subtraces corresponding to contact state transitions from the contact state $\gamma_1$.

The next step in setting up the contact monitor is to discretize the traces obtained from human demonstration. The discretized sequences are then used to obtain the discriminant functions as well as the rules for the contact monitor. For each state $\gamma_i$, the sub-traces in $X_{\gamma_i}$ are combined into one large set of data points $X$. Using clustering techniques, the points in $X$ are then grouped into a small number of classes for each component of the plant state vector $\mathbf{z}$. These classes correspond to the discretized values for the variable. In this manner, the clusters are then used to learn the discriminant functions for each of the variables and for each contact state. This process is described in detail in [Sikka and McCarragher, 1996].

Finally, the rules for contact state and transition recognition must be obtained from the clustered data. First, for each state $\gamma_i$, the clustered data is regrouped into sub-traces corresponding to the sub-traces in the set $X_{\gamma_i}$ to obtain the set of sub-traces $X_{q\gamma_i}$. Then, for each variable in $\mathbf{z}$, the sub-traces in $X_{q\gamma_i}$ are used to obtain examples of the corresponding contact state transition. This step is due to computational considerations and is required to reduce the length of long temporal sequences generated as a result of high sampling rate. Thus, we finally obtain the sets $X'_{q\gamma_i}$ corresponding to each contact state $\gamma_i$ and containing examples of contact state transitions.

In the last step for setting up the contact monitor, the sets $X'_{q\gamma_i}$ are used to learn the FSMs for each variable in $\mathbf{z}$. The FSM generated for each variable recognizes exactly the set of examples used. The FSM is constructed using the algorithm described in [Biermann and Feldman, 1972]. This algorithm constructs an FSM given a set of examples in an iterative manner, with each constructed FSM closer to the given set. The algorithm is based on progressively identifying longer and longer prefixes that are present in the given set of examples.

The approach is illustrated in Figures 3 and 4. Figure 3(a) shows the raw signal corresponding to a gain of contact. The first-difference for this signal, corresponding to the time-derivative, is shown in Figure 3(b). This signal is then clustered into 4 discrete values to obtain the discretized first-difference signal shown in Figure 3(c). Finally, Figure 3(d) shows the input string of length 5 obtained by further sampling the discretized signal, 13111. Such input strings are obtained for each demonstration of the particular contact state transition to generate a set of input strings, for example, {13111,

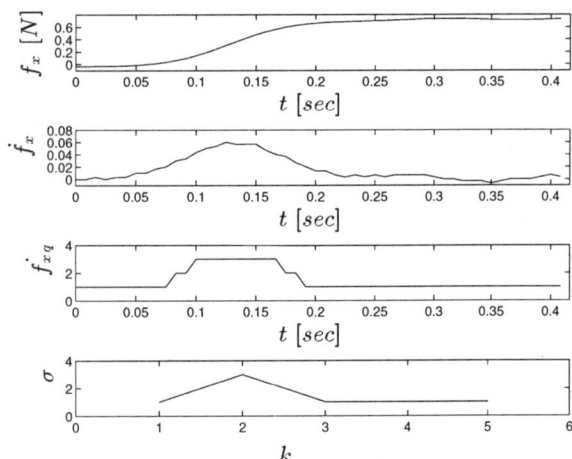

Figure 3: *The transformation of a continuous signal $f_x$ corresponding to a transition into a discrete sequence $\sigma$ of five symbols, 13111.*

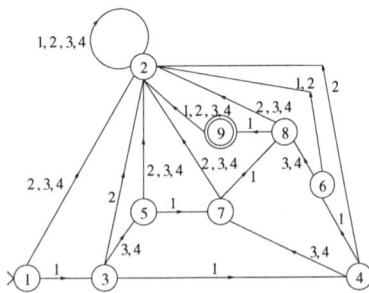

Figure 4: *The automaton obtained from the training set $\{13111,\ 14111,\ 11311,\ 11411,\ 11131,\ 11141\}$.*

14111, 11311, 11411, 11131, 11141}. Figure 4 shows the FSM obtained for this set using the algorithm described in [Biermann and Feldman, 1972]. The FSM consists of 9 states, with the state labeled 1 being the initial state $s_0$, and the state labeled 9 being the only accepting state, i.e., $F = \{s_9\}$.

## 5 Experimental Results

The rule-based process monitor was tested on a task consisting of several contact states resulting from contact between a rectangular peg and a solid block. These contact states are illustrated in Figure 5 and are denoted by $\gamma_1$, $\gamma_2$ and $\gamma_3$. Figure 5 also shows the contact state network for the experimental setup, and the allowed transitions $\tau_1 \cdots \tau_6$ between the contact states. The task used to test the process monitor consists of the sequence of contact states $\gamma_1 \rightarrow \gamma_2 \rightarrow \gamma_3 \rightarrow \gamma_1 \rightarrow \gamma_3 \rightarrow \gamma_2 \rightarrow \gamma_1$.

The task was demonstrated several times by a person and the data from three demonstrations was used to set up the rule-base of the process monitor as described in Section 4. Figure 6 shows the output of the two stages of the process monitor used during a task demonstration.

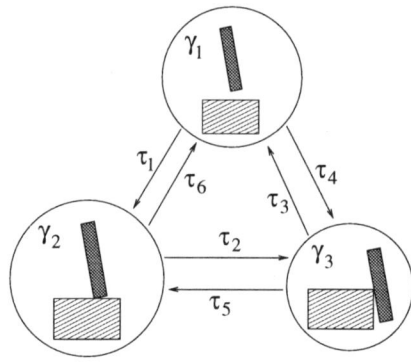

Figure 5: *The setup used to demonstrate the process monitoring techniques.*

The topmost graph in Figure 6 shows $\dot{f}_x$, the derivative of the force acting tangential to the top surface of the solid block. The second graph shows the discretized variable $\dot{f}_{xq}$ produced by the process monitor. Similarly, the third and fourth graphs relate to the force in the normal direction. The graph at the bottom of Figure 6 shows the discrete events recognized by the rule-based process monitor.

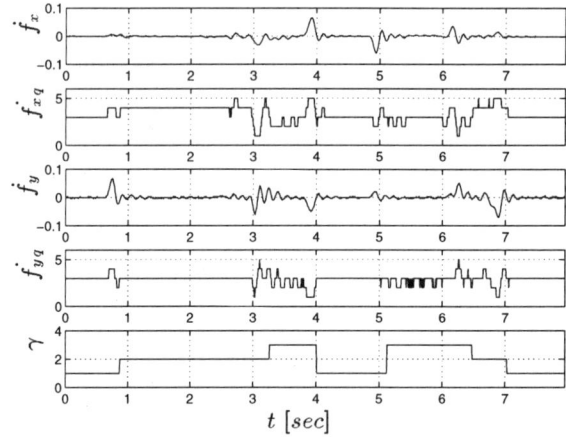

Figure 6: *The results obtained from the process monitor for a demonstration of the task.*

The performance of the contact monitor was measured by using it to monitor 25 demonstrations of the task described above. The percentage of transitions recognized correctly by the rule-based process monitor is summarized in Table 1 which shows the overall recognition accuracy for the task and the recognition accuracy for each individual transition. The first set of numbers in Table 1 show the performance after initially training the rulebase using 3 demonstrations. However, note that the recognition accuracy of the process monitor for transition $\tau_2$ from contact state $\gamma_2$ to $\gamma_3$ is very poor at 48 %.

Closer examination of the data revealed qualitative

Performance (%)

| | $\tau_1$ | $\tau_2$ | $\tau_3$ | $\tau_4$ | $\tau_5$ | $\tau_6$ | All |
|---|---|---|---|---|---|---|---|
| After initial training | | | | | | | |
| | 100 | 48 | 75 | 68 | 93 | 100 | 79 |
| After retraining | | | | | | | |
| | 100 | 72 | 61 | 83 | 93 | 95 | 84.5 |

Table 1: *The performance of the process monitor.*

differences corresponding to this transition. The contact monitor was therefore changed by replacing one of the demonstrations in the training set and then obtaining a new set of rules for the rule-base. The second set of numbers in Table 1 show the recognition accuracy of the process monitor based on the new rule-base. The Table shows that the recognition accuracy of the process monitor for transition $\tau_2$ improved to 72 % and the overall recognition accuracy also improved to 84.5 %. This demonstrates that the process monitoring system is very flexible and changes can easily be made to improve performance.

## 6 Conclusion

We have presented a new rule-based method for process monitoring within the framework of hybrid dynamic systems, based on human demonstration. The experimental results indicate that this method is effective and captures the dynamic nature of the sensory data corresponding to gain and loss of contact. Thus, rule-based transition detection, together with pattern recognition based techniques of discriminant functions and clustering methods, provide a flexible and robust method for robotic process monitoring.

# References

[Asada and Hirai, 1989] H. Asada and S. Hirai. Towards a symbolic-level feedback: Recognition of assembly process states. In *5-th International Symposium on Robotics Research*, Tokyo, 1989.

[Barr and Feigenbaum, 1981] Avron Barr and Edward A. Feigenbaum, editors. *The Handbook of Artificial Intelligence: Vol I*. William Kaufman, Inc., 1981.

[Biermann and Feldman, 1972] A. W. Biermann and J. A. Feldman. On the synthesis of finite-state machines from samples of their behavior. *IEEE Transactions on Computers*, C-21:592–597, June 1972.

[Bruyninckx et al., 1995] H. Bruyninckx, S. Dutré, and J. De Schutter. Peg-on-hole: A model based solution to peg and hole alignment. In *IEEE International Conference on Robotics and Automation*, pages 1919–1924, 1995.

[Delson and West, 1996] Nathan Delson and Harry West. Robot programming by human demonstration: Adaptation and inconsistency in constrained motion. In *IEEE International Conference on Robotics and Automation*, 1996.

[Eberman and Salisbury, 1994] Brian Eberman and J. Kenneth Salisbury, Jr. Application of change detection to dynamic contact sensing. *The International Journal of Robotics Research*, 13(5):369–394, October 1994.

[McCarragher and Asada, 1993] Brenan J. McCarragher and Haruhiko Asada. Qualitative template matching using dynamic process models for state transition recognition of robotic assembly. *The ASME Journal of Dynamic systems, Measurement and Control*, 115(2A):261–275, June 1993.

[McCarragher and Asada, September 1995] Brenan J. McCarragher and Haruhiko Asada. The discrete event modeling and planning of robotic assembly tasks. *The ASME Journal of Dynamic systems, Measurement and Control*, 117(3):394–400, September September 1995.

[Ramadge and Wonham, 1989] Peter J. G. Ramadge and W. Murray Wonham. The control of discrete event systems. *Proceedings of the IEEE*, 77:81–98, 1989.

[Sikka and McCarragher, 1996] Pavan Sikka and Brenan J. McCarragher. Monitoring contact using clustering and discriminant functions. In *IEEE International Conference on Robotics and Automation*, 1996.

[Stiver and Antsaklis, 1992] James A. Stiver and Panos J. Antsaklis. Modeling and analysis of hybrid control systems. In *Proceedings of the 31st Conference on Decision and Control*, pages 3748–3751, 1992.

[Wang et al., 1996] Qi Wang, Joris De Schutter, Wim Witvrouw, and Sean Graves. Derivation of compliant motion programs based on human demonstration. In *IEEE International Conference on Robotics and Automation*, pages 2616–2621, April 1996.

[Whitney, 1982] Daniel E. Whitney. Quasi-static assembly of compliantly supported rigid parts. *The ASME Journal of Dynamic systems, Measurement and Control*, 1982.

[Williams and de Kleer, 1991] B C Williams and J de Kleer. A theory of interactions: Unifying qualitative and quantitative algebraic reasoning. *Artificial Intelligence*, 51:1–79, 1991.

# DIAGNOSIS AND QUALITATIVE REASONING

Qualitative Reasoning 3:
Geometric and Spatial Reasoning

# On the Complexity of Qualitative Spatial Reasoning: A Maximal Tractable Fragment of the Region Connection Calculus *

## Jochen Renz  Bernhard Nebel

Institut für Informatik, Albert-Ludwigs-Universität Freiburg
Am Flughafen 17, D-79110 Freiburg, Germany

## Abstract

The computational properties of qualitative spatial reasoning have been investigated to some degree. However, the question for the boundary between polynomial and NP-hard reasoning problems has not been addressed yet. In this paper we explore this boundary in the "Region Connection Calculus" RCC-8. We extend Bennett's encoding of RCC-8 in modal logic. Based on this encoding, we prove that reasoning is NP-complete in general and identify a maximal tractable subset of the relations in RCC-8 that contains all base relations. Further, we show that for this subset path-consistency is sufficient for deciding consistency.

## 1 Introduction

When describing a spatial configuration or when reasoning about such a configuration, often it is not possible or desirable to obtain precise, quantitative data. In these cases, qualitative reasoning about spatial configurations may be used.

One particular approach in this context has been developed by Randell, Cui, and Cohn [1992], the so-called *Region Connection Calculus* (RCC), which is based on binary topological relations. One variant of this calculus, RCC-8, uses eight mutually exhaustive and pairwise disjoint relations, called base relations, to describe the topological relationship between two regions (see also Egenhofer [1991]).

Some of the computational properties of this calculus have been analyzed by Grigni et al. [1995] and Nebel [1995]. However, no attempt has yet been made to determine the boundary between polynomial and NP-hard fragments of RCC-8, as it has been done for Allen's [1983]

---
*This research was partially supported by DFG as part of the project FAST-QUAL-SPACE, which is part of the DFG special research effort on "Spatial Cognition".

interval calculus [Nebel and Bürckert, 1995]. We address this problem and identify a maximal fragment of RCC-8 that is still tractable and contains all base relations.

As in the case of qualitative temporal reasoning, this proof relies on a computer generated case-analysis that cannot be reproduced in a research paper.[1] Further, we show that for this fragment path-consistency is sufficient for deciding consistency.[2]

## 2 Qualitative Spatial Reasoning with RCC

RCC is a topological approach to qualitative spatial representation and reasoning where *spatial regions* are subsets of topological space [Randell et al., 1992]. Relationships between spatial regions are defined in terms of the relation $C(a, b)$ which is true iff the closure of region $a$ is connected to the closure of region $b$, i.e. if they share a common point. Regions themselves do not have to be internally connected, i.e. a region may consist of different disconnected parts. The domain of *spatial variables* (denoted as $X, Y, Z$) is the whole topological space.

In this work we will focus on RCC-8, but most of our results can easily be applied to RCC-5, a subset of RCC-8 [Bennett, 1994]. RCC-8 uses a set of eight pairwise disjoint and mutually exhaustive relations, called *base relations*, denoted as DC, EC, PO, EQ, TPP, NTPP, TPP$^{-1}$, and NTPP$^{-1}$, with the meaning of *DisConnected, Externally Connected, Partial Overlap, EQual, Tangential Proper Part, Non-Tangential Proper Part*, and their converses. Examples for these relations are shown in Figure 1. In RCC-5 the boundary of a region is not taken into account, i.e. one does not distinguish between DC and EC and between TPP and NTPP. These relations are combined to the RCC-5 base relations DR for *DiscRete* and PP for *Proper Part*, respectively.

Sometimes it is not known which of the eight base relations holds between two regions, but it is possible

---
[1] The programs can be obtained from the authors.
[2] Full proofs can be found in our technical report [Renz and Nebel, 1997].

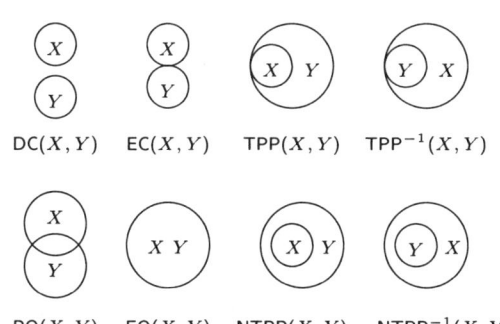

Figure 1: Two-dimensional examples for the eight base relations of RCC-8

to restrict to some of them. In order to represent this, unions of base relations can be used. Since base relations are pairwise disjoint, this results in $2^8$ different relations, including the union of all base relations, which is called *universal relation*. In the following we will write sets of base relations to denote these unions. Using this notation, DR, e.g., is identical to $\{DC, EC\}$. *Spatial formulas* are written as $XRY$, where $R$ is a spatial relation. Apart from union ($\cup$), other operations are defined, namely, converse ($\smile$), intersection ($\cap$), and composition ($\circ$) of relations. The formal definitions of these operations are:

$$\forall X, Y: \quad X(R \cup S)Y \leftrightarrow XRY \vee XSY,$$
$$\forall X, Y: \quad X(R \cap S)Y \leftrightarrow XRY \wedge XSY,$$
$$\forall X, Y: \quad XR^{\smile}Y \leftrightarrow YRX,$$
$$\forall X, Y: \quad X(R \circ S)Y \leftrightarrow \exists Z: (XRZ \wedge ZSY).$$

The compositions of the eight base relations are shown in Table 1. Every entry in the composition table specifies the relation obtained by composing the base relation of the corresponding row with the base relation of the corresponding column. Composition of two arbitrary RCC-8 relations can be obtained by computing the union of the composition of the base relations.

A *spatial configuration* can be described by a set $\Theta$ of spatial formulas. One important computational problem is deciding *consistency* of $\Theta$, i.e. deciding whether it is possible to assign regions to the spatial variables in a way that all relations hold. We call this problem RSAT. When only relations of a specific set $\mathcal{S}$ are used in $\Theta$, the corresponding reasoning problem is denoted RSAT($\mathcal{S}$). In the following $\widehat{\mathcal{S}}$ denotes the closure of $\mathcal{S}$ under composition, intersection, and converse.

## 3 Encoding of RCC-8 in Modal Logic

In this work we use Bennett's [1995] encoding of RCC-8 in propositional modal logic.[3] Bennett obtained this encoding by analyzing the relationship of regions to the universe $\mathcal{U}$. He restricted his analysis to closed regions

---
[3]We assume in the remainder that the reader is familiar with modal logic as presented, e.g., by Fitting [1993].

| $\circ$ | DC | EC | PO | TPP | NTPP | TPP$^{-1}$ | NTPP$^{-1}$ | EQ |
|---|---|---|---|---|---|---|---|---|
| DC | * | DR PO PP | DR PO PP | DR PO PP | DR PO PP | DC | DC | DC |
| EC | DR PO PP$^{-1}$ | DR,EQ PO,TPP TPP$^{-1}$ | DR PO PP | EC PO PP | PO PP | DR | DC | EC |
| PO | DR PO PP$^{-1}$ | DR PO PP$^{-1}$ | * | PO PP | PO PP | DR PO PP$^{-1}$ | DR PO PP$^{-1}$ | PO |
| TPP | DC | DR | DR PO PP | PP | NTPP | DR,EQ PO,TPP TPP$^{-1}$ | DR PO PP$^{-1}$ | TPP |
| NTPP | DC | DC | DR PO PP | NTPP | NTPP | DR PO PP | * | NTPP |
| TPP$^{-1}$ | DR PO PP$^{-1}$ | EC PO PP$^{-1}$ | PO PP$^{-1}$ | PO,EQ TPP TPP$^{-1}$ | PO PP | PP$^{-1}$ | NTPP$^{-1}$ | TPP$^{-1}$ |
| NTPP$^{-1}$ | DR PO PP$^{-1}$ | PO PP$^{-1}$ | PO PP$^{-1}$ | PO PP$^{-1}$ | PO,EQ PP PP$^{-1}$ | NTPP$^{-1}$ | NTPP$^{-1}$ | NTPP$^{-1}$ |
| EQ | DC | EC | PO | TPP | NTPP | TPP$^{-1}$ | NTPP$^{-1}$ | EQ |

Table 1: Composition table for the eight base relations of RCC-8, where $*$ specifies the universal relation

| Relation | Model Constraints | Entailment Constraints |
|---|---|---|
| DC | $\neg(X \wedge Y)$ | $\neg X, \neg Y$ |
| EC | $\neg(\mathbf{I}X \wedge \mathbf{I}Y)$ | $\neg(X \wedge Y), \neg X, \neg Y$ |
| PO | — | $\neg(\mathbf{I}X \wedge \mathbf{I}Y), X \to Y,$ $Y \to X, \neg X, \neg Y$ |
| TPP | $X \to Y$ | $X \to \mathbf{I}Y, Y \to X, \neg X, \neg Y$ |
| TPP$^{-1}$ | $Y \to X$ | $Y \to \mathbf{I}X, X \to Y, \neg X, \neg Y$ |
| NTPP | $X \to \mathbf{I}Y$ | $Y \to X, \neg X, \neg Y$ |
| NTPP$^{-1}$ | $Y \to \mathbf{I}X$ | $X \to Y, \neg X, \neg Y$ |
| EQ | $X \to Y, Y \to X$ | $\neg X, \neg Y$ |

Table 2: Encoding of the base relations in modal logic

that are connected if they share a point and overlap if they share an interior point. If, e.g, $X$ and $Y$ are disconnected, the complement of the intersection of $X$ and $Y$ is equal to the universe. Further, both regions must not be empty, i.e. the complements of both $X$ and $Y$ are not equal to the universe. In this way the eight base relations can be represented by constraints of the form $(m = \mathcal{U})$, called *model constraints*, and $(m \neq \mathcal{U})$, called *entailment constraints*, where $m$ is a set-theoretic expression containing perhaps the topological interior operator $i$. Any model constraint must hold, whereas no entailment constraint must hold [Bennett, 1994].

The model and entailment constraints can be encoded in modal logic, where spatial variables correspond to propositional atoms and the interior operator $i$ to a modal operator $\mathbf{I}$ (see Table 2). The axioms for $i$ must also hold for the modal operator $\mathbf{I}$, which results in the following axioms [Bennett, 1995]:

1. $\mathbf{I}X \to X$,   3. $\mathbf{I}\top \leftrightarrow \top$ (for any tautology $\top$),
2. $\mathbf{II}X \leftrightarrow \mathbf{I}X$,   4. $\mathbf{I}(X \wedge Y) \leftrightarrow \mathbf{I}X \wedge \mathbf{I}Y$.

Axioms 1 and 2 correspond to the modal logics T and 4, axioms 3 and 4 already hold for any modal logic K, so **I** is a modal S4-operator.

The four axioms specified by Bennett are not sufficient to exclude non-closed regions. In order to account for that, we add two formulas for each atom, which correspond to topological properties of closed regions. A closed region is the closure of an open region and the complement of a closed region is an open region:

$$X \leftrightarrow \neg \mathbf{I} \neg \mathbf{I} X, \qquad \neg X \leftrightarrow \mathbf{I} \neg X.$$

In order to combine the different model and entailment constraints, Bennett [1995] uses another modal operator $\Box$. $\Box m$ is interpreted as $m = \mathcal{U}$ and $\neg \Box m$ as $m \neq \mathcal{U}$. Any model constraint $m$ can be written as $\Box m$ and any entailment constraint as $\neg \Box m$. If $\Box X$ is true in a world $w$ of a model $\mathcal{M}$, written as $(\mathcal{M}, w \Vdash \Box X)$, then $X$ must be true in any world of $\mathcal{M}$. So $\Box$ is an S5-operator with the constraint that all worlds are mutually accessible. Therefore Bennett [1995] calls it a strong S5-operator. So the encoding of RCC-8 is done in multi-modal logic with an S4-operator and a strong S5-operator.

Let $\Theta$ be a set of RCC-8 formulas and $Reg(\Theta)$ be the set of spatial variables used in $\Theta$, then $m(\Theta)$ specifies the modal encoding of $\Theta$, where

$$m(\Theta) = \left( \bigwedge_{XRY \in \Theta} m_1(XRY) \right) \wedge \left( \bigwedge_{X \in Reg(\Theta)} m_2(X) \right).$$

$m_1(XRY)$ is a disjunction of the conjunctively connected model and entailment constraints for the base relations in $R$. $m_2$ results from the axioms of the **I**-operator and the additional properties of closed regions:

$$m_2(X) = \Box(\mathbf{I}X \to X) \wedge \Box(\mathbf{I}X \to \mathbf{II}X)$$
$$\wedge \Box(\neg X \to \mathbf{I} \neg X) \wedge \Box(X \to \neg \mathbf{I} \neg \mathbf{I} X).$$

$\Box(\mathbf{II}X \to \mathbf{I}X)$, $\Box(\mathbf{I} \neg X \to \neg X)$ and $\Box(\neg \mathbf{I} \neg \mathbf{I} X \to X)$ are entailed by the other formulas and can be ignored. As follows from the work by Bennett [1995], $\Theta$ is consistent iff $m(\Theta)$ is satisfiable.

In order to refer to the single model and entailment constraints, we will introduce some abbreviations.

**Definition 3.1** *Abbreviations for the model constraints:*

| | | | |
|---|---|---|---|
| $\alpha_{xy}$ | $\equiv \Box(\neg(X \wedge Y))$ | $A_{xy}$ | $\equiv \Box(\neg(\mathbf{I}X \wedge \mathbf{I}Y))$ |
| $\beta_{xy}$ | $\equiv \Box(X \to Y)$ | $B_{xy}$ | $\equiv \Box(X \to \mathbf{I}Y)$ |
| $\gamma_{xy}$ | $\equiv \Box(Y \to X)$ | $C_{xy}$ | $\equiv \Box(Y \to \mathbf{I}X)$ |

As the entailment constraints are negations of the model constraints, they will be abbreviated as negations of the above abbreviations. When it is obvious which atoms are used, the abbreviations will be written without indices. The abbreviations can be regarded as "propositional atoms". Then it is possible to write the modal encoding $m_1(XRY)$ of every relation $R$ of RCC-8 as a "propositional formula" of abbreviations. We will call this formula the *abbreviated form* of $R$. In the remainder we will use the encoding of $m_1(XRY)$ such that the abbreviated form is in conjunctive normal form (CNF).

## 4 Computational Properties of RCC-8

In this section we prove that reasoning with RCC-8 as well as RCC-5 is NP-hard. A similar but weaker result has been proven by Grigni et al. [1995] (see Section 8).

In this paper NP-hardness proofs for different sets $\mathcal{S}$ of RCC-8 relations will be carried out. All of them use a reduction of a propositional satisfiability problem to RSAT($\mathcal{S}$) by constructing a set of spatial formulas $\Theta$ for every instance $\mathcal{I}$ of the propositional problem, such that $\Theta$ is consistent iff $\mathcal{I}$ is a positive instance. These satisfiability problems include 3SAT, NOT-ALL-EQUAL-3SAT where every clause has at least one true and one false literal, and ONE-IN-THREE-3SAT where exactly one literal in every clause must be true [Garey and Johnson, 1979].

The reductions have in common that every literal as well as every literal occurrence $L$ is reduced to two spatial variables $X_L$ and $Y_L$ and a relation $R = R_t \cup R_f$, where $R_t \cap R_f = \emptyset$ and $X_L R Y_L$. $L$ is true iff $X_L R_t Y_L$ holds and false iff $X_L R_f Y_L$ holds. Additional "polarity" constraints have to be introduced to assure that for the spatial variables $X_{\neg L}$ and $Y_{\neg L}$, corresponding to the negation of $L$, $X_{\neg L} R_t Y_{\neg L}$ holds iff $X_L R_f Y_L$ holds, and *vice versa*. Using these polarity constraints, spatial variables of negative literal occurrences are connected to the spatial variables of the corresponding positive literal, and likewise for positive literal occurrences and negative literals. Further, "clause" constraints have to be added to assure that the clause requirements of the specific propositional problem are satisfied in the reduction.

**Theorem 4.1** RSAT(RCC-5) *is NP-hard.*

**Proof Sketch.** Transformation of NOT-ALL-EQUAL-3SAT to RSAT(RCC-5) (see also Grigni et al. [1995]). $R_t = \{\mathsf{PP}\}$ and $R_f = \{\mathsf{PP}^{-1}\}$. Polarity constraints:

$$X_L\{\mathsf{PP},\mathsf{PP}^{-1}\}X_{\neg L}, Y_L\{\mathsf{PP},\mathsf{PP}^{-1}\}Y_{\neg L},$$
$$X_L\{\mathsf{PO}\}Y_{\neg L}, Y_L\{\mathsf{PO}\}X_{\neg L}.$$

Clause constraints for every clause $c = \{i, j, k\}$:

$$X_i\{\mathsf{PP},\mathsf{PP}^{-1}\}X_j, X_j\{\mathsf{PP},\mathsf{PP}^{-1}\}X_k, X_k\{\mathsf{PP},\mathsf{PP}^{-1}\}X_i,$$
$$X_i\{\mathsf{PO}\}Y_k, X_j\{\mathsf{PO}\}Y_i, X_k\{\mathsf{PO}\}Y_j. \qquad \blacksquare$$

Since RCC-5 is a subset of RCC-8, this result can be easily applied to RCC-8.

**Corollary 4.2** RSAT(RCC-8) *is NP-hard.*

In order to identify the borderline between tractability and intractability, one has to examine all subsets of RCC-8. We limit ourselves to subsets containing all base relations, because these subsets still allow to express definite knowledge, if it is available. Additionally, we require the universal relation to be in the subset, so that it is possible to express complete ignorance. This reduces the search space from $2^{256}$ subsets to $2^{247}$ subsets. We proved a property that has likewise been used in identifying the maximal tractable subset of Allen's calculus [Nebel and Bürckert, 1995] that can be used to further reduce the search space.

**Theorem 4.3** $\mathsf{RSAT}(\widehat{S})$ *can be polynomially reduced to* $\mathsf{RSAT}(S)$

**Corollary 4.4** *Let $S$ be a subset of* RCC-8.

1. $\mathsf{RSAT}(\widehat{S}) \in \mathsf{P}$ *iff* $\mathsf{RSAT}(S) \in \mathsf{P}$.
2. $\mathsf{RSAT}(S)$ *is NP-hard iff* $\mathsf{RSAT}(\widehat{S})$ *is NP-hard.*

The first statement of Corollary 4.4 can be used to increase the number of elements of tractable subsets of RCC-8 considerably. With the second statement of Corollary 4.4, NP-hardness proofs of RSAT can be used to exclude certain relations from being in any tractable subset of RCC-8. The NP-hardness proof of Theorem 4.1, e.g., only uses the relations {PO} and {PP, PP$^{-1}$}. So for any subset $S$ with the two relations contained in $\widehat{S}$, $\mathsf{RSAT}(S)$ is NP-hard. The following NP-hardness results can be used to exclude more relations.

**Lemma 4.5** *Let $S$ be a subset of* RCC-8 *containing all base relations. If any of the relations* {TPP, NTPP, TPP$^{-1}$, NTPP$^{-1}$}, {TPP, TPP$^{-1}$}, {NTPP, NTPP$^{-1}$}, {NTPP, TPP$^{-1}$} *or* {TPP, NTPP$^{-1}$} *is contained in* $\widehat{S}$, *then* $\mathsf{RSAT}(S)$ *is NP-hard.*

**Proof Sketch.** When $R_f \cup R_t$ is replaced by {TPP, NTPP, TPP$^{-1}$, NTPP$^{-1}$}, {TPP, TPP$^{-1}$} or {NTPP, NTPP$^{-1}$}, the transformation of Theorem 4.1 can be applied. For {NTPP, TPP$^{-1}$} and {TPP, NTPP$^{-1}$} ONE-IN-THREE-3SAT has to be used. ∎

By computing the closure of all sets containing the eight base relations together with one additional relation, the following lemma can be obtained.

**Lemma 4.6** $\mathsf{RSAT}(S)$ *is NP-hard for any subset $S$ of* RCC-8 *containing all base relations together with one of the 72 relations of the following sets:*

$$\mathcal{N}_1 = \{R \mid \{\mathsf{PO}\} \not\subseteq R \text{ and } (\{\mathsf{TPP}, \mathsf{TPP}^{-1}\} \subseteq R \text{ or } \{\mathsf{NTPP}, \mathsf{NTPP}^{-1}\} \subseteq R)\},$$

$$\mathcal{N}_2 = \{R \mid \{\mathsf{PO}\} \not\subseteq R \text{ and } (\{\mathsf{TPP}, \mathsf{NTPP}^{-1}\} \subseteq R \text{ or } \{\mathsf{TPP}^{-1}, \mathsf{NTPP}\} \subseteq R)\}.$$

## 5 Transformation of RSAT to SAT

For transforming RSAT to propositional satisfiability (SAT) we will transform every instance $\Theta$ of RSAT to a propositional formula in CNF that is satisfiable iff $\Theta$ is consistent. We will start from $m(\Theta)$, the modal encoding of $\Theta$, and show that whenever $m(\Theta)$ is satisfiable it has a Kripke model of a specific type. This model will then be used to transform $m(\Theta)$ to a propositional formula.

$m(\Theta)$ is satisfiable if it is true in a world $w$ of a Kripke model $\mathcal{M} = \langle W, \{R_1 = W \times W, R_2 \subseteq W \times W\}, \pi \rangle$, where $W$ is a set of worlds, $R_1$ the accessibility relation of the $\Box$-operator, $R_2$ the accessibility relation of the **I**-operator, and $\pi$ a truth function that assigns a truth value to every atom in every world. The truth conditions for $\mathcal{M}, w \Vdash m(\Theta)$ can be specified as a combination of truth conditions of the single atoms according to the form of $m(\Theta)$. In this way $\mathcal{M}, w \Vdash \mathbf{I}\varphi$, e.g., can be written as $(\forall u : wR_2u.\mathcal{M}, u \Vdash \varphi)$ and $\mathcal{M}, w \Vdash \neg \mathbf{I}\varphi$ as $(\exists u : wR_2u.\mathcal{M}, u \Vdash \neg\varphi)$. We will call this form of writing $\mathcal{M} \Vdash m(\Theta)$ the *explicit form* of $m(\Theta)$.

Before transforming $m(\Theta)$ to a propositional formula, we have to show that there is a Kripke model of $m(\Theta)$ that is polynomial in the number of spatial variables $n$.

**Definition 5.1** *Let $u \in W$ be a world of the model $\mathcal{M}$.*

- $u$ *is a world of level 0 if $vR_2u$ only holds for $v = u$.*
- $u$ *is a world of level $l+1$ if $vR_2u$ holds for a world $v$ of level $l$ and there is no world $v \neq u$ of level $> l$.*

We assume that every occurrence of a sub-formula of $m(\Theta)$ of the form $\neg\Box\varphi$, where $\varphi$ contains no $\Box$ operators, introduces a new world of level 0. As these sub-formulas correspond to entailment constraints, the number of worlds of level 0 is polynomial in $n$.

For every spatial variable $X$ and every world $w$ there might be sub-formulas that force the existence of a world $u$ with $wR_2u$ where $X$ is true or where $\neg X$ is true. Because there are $n$ different spatial variables, $2n$ different worlds $u$ with $wR_2u$ are sufficient for each world $w$.

**Definition 5.2** *An* RCC-8-*frame* $\mathcal{F} = \langle W, \{R_1, R_2\}\rangle$ *has the following properties:*

1. *$W$ contains only worlds of level $0, 1$ and $2$.*
2. *For every world $w$ of level $k$ ($k = 0, 1$) there are exactly $2n$ worlds $u$ of level $k + 1$ with $wR_2u$.*
3. *For every world $w$ of level $k$ there is exactly one world $u$ for every level $0 \leq l \leq k$ with $uR_2w$.*

*An* RCC-8-*model is based on an* RCC-8-*frame.*

**Lemma 5.3** $m(\Theta)$ *is satisfiable iff* $\mathcal{M}, w \Vdash m(\Theta)$ *for an* RCC-8-*model $\mathcal{M}$ with polynomially many worlds.*

Now it is possible to transform the explicit form of $m(\Theta)$ to a propositional formula $p(m(\Theta))$ in CNF such that $p(m(\Theta))$ is satisfiable iff $m(\Theta)$ is satisfiable in a polynomial RCC-8-model $\mathcal{M}$. For this purpose, propositional atoms $X_w$ are introduced which stand for the truth of atom $X$ in world $w$ of the RCC-8-model $\mathcal{M}$. Further, universally quantified truth conditions are transformed into conjunctions and existentially quantified truth conditions are transformed into truth conditions on particular worlds, which can be determined using the structure of the RCC-8-frame and the modal formula.

**Theorem 5.4** RSAT(RCC-8) *can be polynomially reduced to* SAT.

With Corollary 4.2 this leads to the following theorem.

**Theorem 5.5** RSAT(RCC-8) *is* NP-*complete*.

## 6 Tractable Subsets of RCC-8

In order to identify a tractable subset of RCC-8, we analyze which relations can be expressed as propositional Horn formulas, as satisfiability of Horn formulas (HORN-SAT) is tractable.

**Proposition 6.1** *Applying the transformation p to the model and entailment constraints, to the axioms for* **I**, *and to the properties of closed regions leads to Horn formulas*.

Since the model constraints $\alpha$ and $A$ are transformed to indefinite Horn formulas, the transformation of any disjunction of these constraints with any other constraint is also Horn. All relations with an abbreviated form using only abbreviations or disjunctions of abbreviations transformable to Horn formulas can be transformed to Horn formulas. In this way 64 different relations can be transformed to Horn formulas. We call the subset of RCC-8 containing these relations $\mathcal{H}_8$.

**Theorem 6.2** RSAT($\mathcal{H}_8$) *can be polynomially reduced to* HORNSAT *and therefore* RSAT($\widehat{\mathcal{H}}_8$) $\in$ P.

**Theorem 6.3** $\widehat{\mathcal{H}}_8$ *contains the following* 148 *relations*:

$$\widehat{\mathcal{H}}_8 = \text{RCC-8} \setminus (\mathcal{N}_1 \cup \mathcal{N}_2 \cup \mathcal{N}_3)$$

*with* $\mathcal{N}_1$ *and* $\mathcal{N}_2$ *as defined in Lemma 4.6 and*

$\mathcal{N}_3 = \{R | \{\text{EQ}\} \subseteq R \text{ and } ((\{\text{NTPP}\} \subseteq R, \{\text{TPP}\} \not\subseteq R)$
$\text{or } (\{\text{NTPP}^{-1}\} \subseteq R, \{\text{TPP}^{-1}\} \not\subseteq R))\}$.

For proving that $\widehat{\mathcal{H}}_8$ is a maximal tractable subset of RCC-8, we have to show that no relation of $\mathcal{N}_3$ can be added to $\widehat{\mathcal{H}}_8$ without making RSAT intractable.

**Lemma 6.4** *The closure of every set containing* $\widehat{\mathcal{H}}_8$ *and one relation of* $\mathcal{N}_3$ *contains the relation* {EQ, NTPP}.

Therefore it is sufficient to prove NP-hardness of RSAT($\widehat{\mathcal{H}}_8 \cup \{\text{EQ, NTPP}\}$) for showing that $\widehat{\mathcal{H}}_8$ is a maximal tractable subset of RCC-8.

**Lemma 6.5** RSAT($\widehat{\mathcal{H}}_8 \cup \{\text{EQ, NTPP}\}$) *is* NP-*hard*.

**Proof Sketch.** Transformation of 3SAT to RSAT($\widehat{\mathcal{H}}_8 \cup \{\text{EQ, NTPP}\}$). $R_t = \{\text{NTPP}\}$ and $R_f = \{\text{EQ}\}$. Polarity constraints:
$X_L\{\text{EC, NTPP}\}X_{\neg L}, Y_L\{\text{TPP}\}Y_{\neg L}$,
$X_L\{\text{TPP, NTPP}\}Y_{\neg L}, Y_L\{\text{EC, TPP}\}X_{\neg L}$,
Clause constraints for each clause $c = \{i, j, k\}$:
$Y_i\{\text{NTPP}^{-1}\}X_j, Y_j\{\text{NTPP}^{-1}\}X_k, Y_k\{\text{NTPP}^{-1}\}X_i$. ∎

**Theorem 6.6** $\widehat{\mathcal{H}}_8$ *is a maximal tractable subset of* RCC-8.

It has to be noted that there might be other maximal tractable subsets of RCC-8 that contain all base relations.

As $\widehat{\mathcal{H}}_8$ is tractable, the intersection of RCC-5 and $\widehat{\mathcal{H}}_8$ is also tractable. We will call this subset $\widehat{\mathcal{H}}_5$.

**Theorem 6.7** $\widehat{\mathcal{H}}_5$ *is the only maximal tractable subset of* RCC-5 *containing all base relations*.

## 7 Applicability of Path-Consistency

As shown in the previous section, RSAT($\widehat{\mathcal{H}}_8$) can be solved in polynomial time by first transforming a set of $\widehat{\mathcal{H}}_8$ formulas to a propositional Horn formula and then deciding it in time linear in the number of literals. This way of solving RSAT does not appear to be very efficient.

As RSAT is a Constraint Satisfaction Problem (CSP) [Mackworth, 1987], where variables are nodes and relations are arcs of the constraint graph, algorithms for deciding consistency of a CSP can also be used. A correct but in general not complete $O(n^3)$ algorithm for deciding inconsistency of a CSP is the path-consistency method [Mackworth, 1977] that makes a CSP path-consistent by successively removing relations from all edges using $\forall k : R_{ij} \leftarrow R_{ij} \cap (R_{ik} \circ R_{kj})$, where $i, j, k$ are nodes and $R_{ij}$ is the relation between $i$ and $j$. If the empty relation occurs while performing this operation, the CSP is not path-consistent, otherwise it is.

In this section we will prove that path-consistency decides RSAT($\widehat{\mathcal{H}}_8$). This is done by showing that the path-consistency method finds an inconsistency whenever positive unit resolution (PUR) resolves the empty clause from the corresponding propositional formula. As PUR is refutation-complete for Horn formulas, it follows that the path-consistency method decides RSAT($\widehat{\mathcal{H}}_8$). The only way to get the empty clause is resolving a positive and a negative unit clause of the same variable. Since the Horn formulas that are used contain only a few types of different clauses, there are only a few ways to resolve unit clauses using PUR.

**Definition 7.1**

- $R_K$ denotes the set of relations of $\mathcal{H}_8$ with the conjunct $K$ appearing in their abbreviated form.
- $R_{K_1,K_2}$ denotes $R_{K_1} \cup R_{K_2}$.
- $R_\Gamma$ denotes $R_\gamma \cup R_{\alpha \vee \gamma} \cup R_{A \vee \gamma} \cup R_C \cup R_{\alpha \vee C} \cup R_{A \vee C}$.
- An $R_K$-chain $R_K^*(X, Y)$ is a path from region $X$ to region $Y$, where all relations between successive regions are from $R_K$.

**Lemma 7.2** *Let* $\Theta$ *be a set of* $\mathcal{H}_8$-*formulas*.

- *A positive unit clause* $\{X_w\}$ *can only be resolved from* $\{Y_w\}$ *and a clause resulting from* $XR_\Gamma Y \in \Theta$. *When such a resolution is possible,* $XR_{\alpha,A}Y$ *cannot hold, so* $XR_{\gamma,C}Y$ *must hold*.

- A negative unit clause $\{\neg X_w\}$ can only be resolved from $\{Y_w\}$ and a clause resulting from $XR_{\alpha,A}Y \in \Theta$.

**Lemma 7.3** *If the positive unit clause $\{X_w\}$ can be resolved with PUR using an $R_\Gamma$-chain from $X$ to $Y$, the path-consistency method results in $XR_{\gamma,C}Y$.*

Using Lemma 7.3, it can be proven that the path-consistency method decides $\mathsf{RSAT}(\widehat{\mathcal{H}}_8)$. Using the proof of Theorem 4.3, it is possible to express every relation of $\widehat{\mathcal{H}}_8$ as a Horn formula. Then the following theorem can be proven.

**Theorem 7.4** *The path-consistency method decides $\mathsf{RSAT}(\widehat{\mathcal{H}}_8)$.*

Another interesting question is whether the path-consistency method computes minimal possible relations on $\widehat{\mathcal{H}}_8$. As the following example demonstrates, this is not the case even for the set $\widehat{\mathcal{H}}_5$. $A\{\mathsf{PP}\}D$ is impossible although the constraint graph is path-consistent:

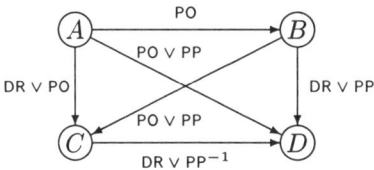

## 8 Related Work

Nebel [1995] showed that $\mathsf{RSAT}(\widehat{\mathcal{B}})$ can be decided in polynomial time, where $\mathcal{B}$ is the set of the RCC-8 base relations. Since $\mathcal{B} \subseteq \mathcal{H}_8$, our result is more general. Further, $\widehat{\mathcal{B}}$ contains only 38 relations, whereas $\widehat{\mathcal{H}}_8$ contains 148 relations, i.e. about 58% of RCC-8.

Grigni et al. [1995] proved NP-hardness of problems similar to RSAT. For instance, they considered the problem of *relational consistency*, which means that there exists a path-consistent refinement of all relations to base relations, and showed that this problem is NP-hard. While our NP-hardness result on RSAT implies their result, the converse implication follows only using the above cited result by Nebel [1995].

In addition to this syntactic notion of consistency, Grigni et al. [1995] considered a semantic notion of consistency, namely, the *realizability* of spatial variables as internally connected planar regions. This notion is much more constraining than our notion of consistency. It is also computationally much harder.

## 9 Summary

We analyzed the computational properties of the qualitative spatial calculus RCC-8 and identified the boundary between polynomial and NP-hard fragments. Using a modification of Bennett's encoding of RCC-8 in a multimodal propositional logic, we transformed the RCC-8 consistency problem to a problem in propositional logic and isolated the relations that are representable as Horn clauses. As it turns out, the fragment identified in this way is also a maximal fragment that contains all base relations and is still computationally tractable. Further, we showed that for this fragment path-consistency is sufficient for deciding consistency.

As in the case of qualitative temporal reasoning, our result allows to check whether the relations that are used in an application allow for a polynomial reasoning algorithm. Further, if the application requires an expressive power beyond the polynomial fragment, it can be used to speed up backtracking algorithms. Assuming that the relations are uniformly distributed, the average branching factor is reduced from 4.0 to 1.4375 using $\widehat{\mathcal{H}}_8$ instead of $\mathcal{B}$ to split the relations (see also [Nebel, 1997]).

## References

[Allen, 1983] J.F. Allen. Maintaining knowledge about temporal intervals. *Comm. of the ACM*, 26(11), 1983.

[Bennett, 1994] B. Bennett. Spatial reasoning with propositional logic. In *Proc. KR'94*, Morgan Kaufmann, 1994.

[Bennett, 1995] B. Bennett. Modal logics for qualitative spatial reasoning. *Bull. IGPL*, 4(1), 1995.

[Egenhofer, 1991] M.J. Egenhofer. Reasoning about binary topological relations. In *Proc. SSD'91*, LNCS 525, pp.143–160. Springer, 1991.

[Fitting, 1993] M.C. Fitting. Basic modal logic. In *Handbook of Logic in Artificial Intelligence and Logic Programming - Vol. 1*, pp.365–448, Oxford, 1993.

[Garey and Johnson, 1979] M.R. Garey and D.S. Johnson. *Computers and Intractability—A Guide to the Theory of NP-Completeness*. Freeman, 1979.

[Grigni et al., 1995] M. Grigni, D. Papadias, and C. Papadimitriou. Topological inference. In *Proc. IJCAI'95*, pp.901–906, 1995.

[Mackworth, 1977] A.K. Mackworth. Consistency in networks of relations. *Artificial Intelligence* 8, 1977.

[Mackworth, 1987] A.K. Mackworth. Constraint satisfaction. In S. C. Shapiro (ed), *Encyclopedia of Artificial Intelligence*, pp.205–211. Wiley, 1987.

[Nebel and Bürckert, 1995] B. Nebel and H.-J. Bürckert. Reasoning about temporal relations: A maximal tractable subclass of Allen's interval algebra. *Journal ACM*, 42(1), pp.43–66, 1995.

[Nebel, 1995] B. Nebel. Computational properties of qualitative spatial reasoning: First results. In *KI'95*, LNCS 981, pp.233–244, Springer, 1995.

[Nebel, 1997] B. Nebel. Solving hard qualitative temporal reasoning problems: Evaluating the efficiency of using the ORD-Horn class. *Constraints*, 3(1), pp.175–190, 1997.

[Randell et al., 1992] D.A. Randell, Z. Cui, and A.G. Cohn. A spatial logic based on regions and connection. In *Proc. KR'92*, pp.165–176, 1992.

[Renz and Nebel, 1997] J. Renz and B. Nebel. On the complexity of qualitative spatial reasoning: A maximal tractable fragment of the Region Connection Calculus. Tech. report 87, Inst. f. Informatik, Univ. Freiburg, 1997.

# Automation of Diagrammatic Reasoning

## Mateja Jamnik, Alan Bundy, Ian Green
Department of Artificial Intelligence, 80 South Bridge
Edinburgh, EH1 1HN, UK
matejaj@dai.ed.ac.uk, A.Bundy@ed.ac.uk, I.Green@ed.ac.uk

## Abstract

Theorems in automated theorem proving are usually proved by logical formal proofs. However, there is a subset of problems which humans can prove in a different way by the use of geometric operations on diagrams, so called diagrammatic proofs. Insight is more clearly perceived in these than in the corresponding algebraic proofs: they capture an intuitive notion of truthfulness that humans find easy to see and understand. We are identifying and automating this diagrammatic reasoning on mathematical theorems. The user gives the system, called DIAMOND, a theorem and then interactively proves it by the use of geometric manipulations on the diagram. These operations are the "inference steps" of the proof. DIAMOND then automatically derives from these example proofs a generalised proof. The constructive $\omega$-rule is used as a mathematical basis to capture the generality of inductive diagrammatic proofs. In this way, we explore the relation between diagrammatic and algebraic proofs.

## 1 Introduction

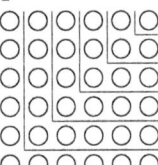

$$1 + 3 + 5 + \cdots + (2n - 1) = n^2$$

It requires only basic secondary school knowledge of mathematics to realise that the diagram above is a proof of a theorem about the sum of odd naturals.

It is an interesting property of diagrams that allows us to "see" and understand so much just by looking at a simple diagram. Not only do we know what theorem the diagram represents, but we also understand the proof of the theorem represented by the diagram and believe it is correct.

Is it possible to simulate and formalise this sort of diagrammatic reasoning on machines? Or is it a kind of intuitive reasoning particular to humans that mere machines are incapable of? Roger Penrose claims that it is not possible to automate such diagrammatic proofs.[1] We are taking his position as an inspiration and are trying to capture the kind of diagrammatic reasoning that Penrose is talking about so that we will be able to simulate it on a computer.

The importance of diagrams in many domains of reasoning has been extensively discussed by Larkin and Simon [Larkin and Simon, 1987], who claim that "a diagram is (sometimes) worth *ten* thousand words". The advantage of a diagram is that it concisely stores information, explicitly represents the relations among the elements of the diagram, and it supports a lot of perceptual inferences that are very easy for humans.

It is exactly these characteristics of diagrams that we wish to exploit in our project. In this paper we present a system (which is currently being developed) called DIAMOND (DIAgraMmatic reasONing and Deduction), which reasons with diagrams. With this system, the user inputs a theorem of mathematics to be proved, instructs the system what diagram to start the search for the proof from, and decides what geometric operations to perform during the proof search. To implement these components is a hard task. We do not aim to discover proofs, but to investigate the relation between formal algebraic proofs and more "informal" diagrammatic proofs. Usually, theorems are *formally* proved with the use of inference steps which often do not convey an intuitive notion of truthfulness to humans. The inference steps are just statements that follow the rules of some logic. The reason we trust that they are correct is that the logic has been previously proved to be sound. Following and applying the rules of such a logic guarantees us that there is no mistake in the proof. We might not have such a guarantee in DIAMOND, but will gain a more informal insight into the proof. Ultimately, the entire process of diagrammatically proving theorems will illuminate the issues of formality,

---

[1] Roger Penrose presented his position in the lecture at International Centre for Mathematical Sciences in Edinburgh, in celebration of the 50th anniversary of UNESCO on 8 November, 1995.

rigour, truthfulness and power of diagrammatic proofs.

In Section 2 we list some of the theorems that we aim to prove. Section 3 presents DIAMOND's architecture, some operations required, and the generalisation mechanism employed. Section 4 reports on some of our results and discusses future work. Section 5 discusses some of the related diagrammatic reasoning systems. Finally, we conclude by summarising the main points of this paper.

## 2 'Diagrammatic' Theorems

We are interested in mathematical theorems that admit diagrammatic proofs. In order to clarify what we mean by diagrammatic proofs we first list some example theorems. Then, we introduce a taxonomy for categorising these examples in order to be able to characterise the domain of problems under consideration.

### 2.1 Examples

**Pythagoras' Theorem**

Pythagoras' Theorem states that the square of the hypotenuse of a right angle triangle equals the sum of the squares of its other two sides. Here is one of the many different diagrammatic proofs of this theorem, taken from [Nelsen, 1993, page 3]:

$$a^2 + b^2 = c^2$$

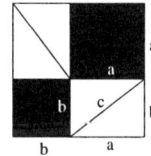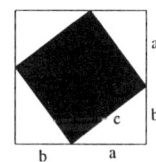

The proof consists of first taking any right angle triangle, completing a bigger square by joining to it identical triangles and squares along its sides, and then rearranging the triangles in a bigger square. For a more elaborate explanation, see [Jamnik *et al.*, 1996].

**Sum of Odd Naturals**

This example is also taken from [Nelsen, 1993, page 71]. The theorem about the sum of odd naturals states the following:

$$1 + 3 + \cdots + (2n - 1) = n^2$$

Note the use of parameter $n$. If we take a square we can cut it into as many L's (which are made up of two adjacent sides of the square) as the size of the side of the square. Note that one L is made out of two sides, i.e., $2n$, but the shared vertex has been counted twice. Therefore, one L has a size of $(2n - 1)$, where $n$ is the size of the square.

**Geometric Sum**

This example is also taken from [Nelsen, 1993, page 118]. A theorem about a geometric sum of $\frac{1}{2^n}$ states the following:

$$\frac{1}{2} + \frac{1}{4} + \frac{1}{8} + \cdots = 1$$

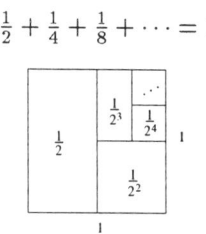

Note the use of ellipsis in the diagram. Take a square of unit size. Cut it down the middle. Now, cut one half of the previous cut square into halves again. This will create two identical squares making up a half of the original square. Take one of these two squares and continue doing this procedure indefinitely.

### 2.2 Classification

From the analysis of the examples that we presented above, and many others, three categories of proofs can be distinguished:

**Category 1:** Proofs that are not schematic: there is no need for induction to prove the general case. Simple geometric manipulations of a diagram prove the individual case. At the end, generalisation is required to show that this proof will hold for all $a, b$. Example theorem: *Pythagoras' Theorem*.

**Category 2:** Proofs that are schematic: they require no inductive step to prove the theorem for each concrete diagram (i.e., problem), but require induction for the general diagram of size $n$ (a concrete diagram cannot be drawn for this instance). The constructive $\omega$-rule (explained in more detail in Section 2.4) is used to generate a generalised proof from the individual proof instances. Example theorem: *Sum of Odd Naturals*.

**Category 3:** Proofs that are inherently inductive: for each individual concrete case of the diagram they need an inductive step to prove the theorem. Every particular instance of a theorem, when represented as a diagram requires the use of abstractions to represent infinity. Thus, the constructive $\omega$-rule (defined later) is not applicable here. Example theorem: *Geometric Sum*.

### 2.3 Problem Domain

We choose mathematics as our domain for theorems since it allows us to make formal statements about the reasoning, proof search, induction, generalisations, and such issues. Having introduced the examples and their categorisation, which is by no means exhaustive, we are now able to further restrict our domain of mathematical theorems.

First, we narrow down the domain to a subset of theorems that can be represented as diagrams without the need for abstraction (e.g., the use of ellipsis, as in the above example theorem for *geometric sum*). Conducting proofs and using abstractions in diagrams is problematic, since it is very difficult to keep track of these abstractions while manipulating the diagram during the proof procedure.

Second, we consider diagrammatic proofs that require induction to prove the general case (i.e., Category 2 above). Namely, diagrams can be drawn only for concrete situations and objects. We cannot draw, for example, an $n \times n$ square. Our challenge is to find a generalisation mechanism that does not require using abstractions in diagrams. The generality of the proof will be captured in an alternative way (by using the constructive $\omega$-rule; see Section 2.4).

It is clear that we will need a stronger problem domain definition which remains a subject of our research. One possibility is to consider theorems of arithmetic or number theory only. To date, DIAMOND is targeted to prove examples of Category 2, but we may also implement diagrammatic theorem proving of examples for Category 1.

## 2.4 Constructive $\omega$-Rule

As mentioned above we could use the constructive $\omega$-rule to prove theorems of Category 2. Siani Baker in [Baker et al., 1992] did some work on the constructive $\omega$-rule and schematic proofs for arithmetic theorems. Here, we explain the idea behind schematic proofs and how it can be applied to diagrammatic proofs.

**Schematic Proof**

Schematic proofs use the constructive $\omega$-rule which is an alternative to induction. The constructive $\omega$-rule allows inference of the sentence $\forall x P(x)$ from an infinite sequence $P(n)$ $n \in \omega$ of sentences.

$$\frac{P(0), P(1), P(2), \ldots}{\forall n. P(n)}$$

where "if each $P(n)$ can be proved *in a uniform way* (from parameter $n$), then conclude $\forall n P(n)$." The criterion for uniformity of the procedure of proof using the constructive $\omega$-rule is taken to be the provision of a general schematic proof, namely the proof of $P(n)$ in terms of $n$, where some rules $R$ are applied some function of $n$ (i.e., $f_R(n)$) times (a rule can also be applied a constant number of times). Now, $proof(n)$ is schematic in $n$, since we applied some rule $R$ $n$ times. The following procedure summarises the essence of using the constructive $\omega$-rule in schematic proofs:

1. Prove a few special cases (e.g., $P(2)$, $P(16)$, ...).
2. Generalise (guess) $proof(n)$ (e.g., from $proof(2)$, $proof(16)$, ...).
3. Prove that $proof(n)$ proves $P(n)$ by meta-induction on $n$.

The general pattern is extracted (guessed, to be exact) from the individual proof instances by (learning type) inductive inference. By meta (mathematical) induction we mean that we introduce system $PA_\omega$ (i.e., Peano Arithmetic with $\omega$-rule) such that:

$$proof(n) : P(n) \vdash_{PA_\omega} proof(s(n)) : P(s(n))$$

where ":" stands for "is a proof of", and $s(n)$ is a notation for successor of $n$. This essentially says that by using the rules on $P(s(n))$ we can reduce it to $P(n)$. For more information, see [Baker et al., 1992].

**Diagrams and Schematic Proofs**

We claim that we can extend Baker's work on schematic proofs in our diagrammatic proofs in that the generality of the diagrammatic proof is embedded in the schematic proof. Thus, we eliminate the need for abstractions in diagrams, and can generalise from manipulations on concrete diagrams.

The diagrammatic schematic proof starts with a few particular concrete cases of the theorem represented by the diagram. The diagrammatic procedures (i.e., operations) on the diagram are performed next, capturing the inference steps of the diagrammatic proof. In DIAMOND, this step (also referred to as the proof checking step) is done *interactively* with the user, and corresponds to the first step of the schematic proof procedure given in the previous section.

The second step is to generalise the operations involved in the schematic proof for $n$. Note that the generality is represented as a sequence of diagrammatic procedures (operations) and not as a general representation of a diagram. In DIAMOND, this step is done *automatically*. More precisely, the basic idea is to consider proofs for $n+1$ which can be reduced to proofs for $n$ (or conversely, such proofs for $n$ which can be extended to proofs for $n+1$ by adding to them some additional sequence of operations). The difference between the proof for $(n+1)$ and the proof for $n$, i.e., the additional sequence of operations in the proof for $(n+1)$ with respect to the proof for $n$ is referred to as the step case of the generalised proof.

The last step in the schematic proof procedure is to prove by meta-induction that the generalised diagrammatic schematic proof is indeed correct. It remains a subject of this research project to determine whether this will be considered at all or not. An alternative at this point could be to translate the diagrammatic proof to an algebraic proof.

## 2.5 Schematic Diagrammatic Proof for the Sum of Odd Naturals

Now we can attempt to structure the diagrammatic proofs in a more formal way. Here we list the proof for the theorem about the sum of odd naturals as a sequence of steps that need to be performed on the diagram:

1. Cut a square into $n$ L's, where an L consists of 2 adjacent sides of the square.
2. Cut each L into two segments.
3. For each L, join these two segments one on top of the other length-wise (note that one of the two segments is always one unit longer than the other, thus an L always consists of an odd number of units, i.e., $2n-1$).

Identifying the operations (i.e., geometric manipulations) that were required to prove the theorem will help us define a large repertoire of such operations which will be used in the diagrammatic proofs. The generality of the proof is captured by the use of the constructive $\omega$-rule, by which we take a few special cases of the diagram (say a square of size 15 and 16), and find the general pattern of the proof that will hold for each case (e.g., the schematic proof given above).

## 3 DIAMOND System

Clearly, an important issue in the development of DIAMOND is the *internal* representation of diagrams and operations on them. The hope is that these capture the intuitiveness, rigour and simplicity of human reasoning with diagrams. We aim to emulate human visual perception to enable a theorem prover to prove theorems using diagrammatic inference steps. There are several representations available to achieve this. In DIAMOND we use a mixture of Cartesian and topological representations.

The architecture of DIAMOND consists of two parts. The *diagrammatic component* forms and processes the diagram. The *inference engine* deals with the diagrammatic inference steps. It processes the operations on the diagram. An important submodule is the generalisation tool (i.e., implementation of the constructive $\omega$-rule).

The rest of this section presents the operations used to construct proofs, the structure of proofs and the generalisation mechanism used in DIAMOND. For more information see [Jamnik *et al.*, 1996].

### 3.1 Geometric Operations

Geometric operations (also referred to as manipulations or procedures) capture the inference steps of the proof. Thus, a sufficiently large number of such operations which are then available to the user in the search for the proof, needs to be identified and formalised. Since we are not generating, i.e., discovering diagrammatic proofs, but rather we are trying to understand them, we can expect from the user to input these operations. To date, a small number of such operations has been implemented and is available to the user.

We distinguish between two types of operations:

**Atomic operations:** they are basic one-step operations that will be combined into more complex operations. Examples of such operations are: rotate, translate, cut, join, project from 3D to 2D, remove, insert a segment.

**Composite operations:** they are more complex, typically recursive operations, composed from simple atomic ones. Perhaps, we can think of them as tactics or tacticals in automated reasoning. In the future we need to investigate several different recursive structures of diagrams. Depending on the theorem that we are proving, we use a different recursive composite operation. Ideally, the internal representation of the diagram should be pertinent to the composite operation that we are performing on it.

In the example of the theorem for the *sum of odd naturals* the proof consists of the following operations: *lcut, split_top_row, rotate90, join, rem_left_column, remove_dot*.

### 3.2 Constructing a Proof

DIAMOND's example proof consists of a sequence of applications of geometric operations on a diagram. The generalisation is then carried out automatically, if any such generalisation exists for the two given example proofs.[2] DIAMOND expects the example proofs to be formulated in a particular way where the order of operations in the user's formulation of the proofs is crucial. Both example proofs are expected to be given with the same order of operations, but with some extra operations in the case of the *proof(n + 1)* for some particular $n$.

Consider the example for the *sum of odd naturals*. The step cases for proofs for $n = 4$ and $n = 3$ look as follows:

The aim is to recognise automatically the structure of the proof from a linear sequence of applications of operations, so that the example proofs for $n$ and $n + 1$ can be reformulated in the general case into the following:

$$\begin{aligned} proof(n) &= \mathcal{A}(n)\mathcal{A}(n-1)\ldots\mathcal{A}(2)\mathcal{B}(1) \\ proof(n+1) &= \mathcal{A}(n+1)\mathcal{A}(n)\mathcal{A}(n-1)\ldots\mathcal{A}(2)\mathcal{B}(1) \end{aligned}$$

where for each $n$, $\mathcal{A}(n)$ is a step case consisting of a sequence of applications of some operations and $\mathcal{B}(1)$ is a base case for $n = 1$. Alternatively, we seek this recursive reformulation:

$$\begin{aligned} proof(n+1) &= \mathcal{A}(n+1) \; proof(n) \\ proof(1) &= \mathcal{B}(1) \end{aligned}$$

A further issue that we are investigating currently is to relax the requirement for a particular ordering of operations in formulating example proofs. Sets with partial ordering could be used as an alternative.

### 3.3 Generalisation

Given some example proofs DIAMOND needs to generalise from them, so that the final diagrammatic proof is not only for the cases of specific $n$'s, but holds for all $n$. Such a schematic proof consists of a general sequence of applications of some operations, where the number of application of each operation is dependent on $n$ or is a constant.

DIAMOND distinguishes between two types of example proofs: *destructive*, i.e., the example proofs which are formulated so that the base case operations are performed last (in a sense the initial diagram is "destructed" by the application of operations down to a trivial diagram, forming the proof along in this way); and *constructive*, i.e., the base case operations are performed first followed by the step case operations.

The proofs that have the same structure for all $n$ are called 1-homogeneous proofs. Proofs can be c-homogeneous; then there are $c$ cases of the proof. We say that if all concrete instances of the proof (for instances of numbers that "equal modulo $c$") have the same structure

---

[2]If the proof contains a case split for say, even and odd integers, and the two example proofs given are for two different cases, then DIAMOND cannot generalise from them. However, DIAMOND recognises that the example proofs were given for different cases, and requests the user to supply another example proof for each case, in order for it to be able to generalise. This will be further explained in Section 3.3.

and can be generalised, then the proof is $c$-homogeneous. If there are $c$ cases, then there are $c$ different generalised proofs, one for each case. The following theorem can be proved:

**Theorem 1:** If a proof is $c$-homogeneous, then it is also $(kc)$-homogeneous for every natural number $k > 0$.

The immediate consequence of **Theorem 1** is:

**Corollary 1:** If a proof is *not* $c$-homogeneous, then it is also *not* $f$-homogeneous for every factor $f$ of $c$.

In a $c$-homogeneous proof we will denote by $\mathcal{B}_r$ a base case for a branch of numbers which give remainder $r$ when divided by $c$. $\mathcal{B}_r$ is actually a proof for the smallest natural number that gives remainder $r$ when divided by $c$. A special case is a class of numbers divisible by $c$, where a base case is going to be $\mathcal{B}_c$, which is a proof for $n = c$.

The general representation of the destructive proof is formalised as follows – let:

- $n = kc + r$
- where $c = number\ of\ cases$ and $r < c$
- and $i \geq 1$.

Then the recursive definition of a general proof is:

for $r \neq 0$:
$$\begin{aligned} proof(ic+r) &= \mathcal{A}_r(ic+r)\ proof((i-1)c+r) \\ proof(r) &= \mathcal{B}_r(r) \end{aligned}$$

for $r = 0$:
$$\begin{aligned} proof(ic) &= \mathcal{A}_c(ic)\ proof((i-1)c) \\ proof(c) &= \mathcal{B}_c(c) \end{aligned}$$

where $\mathcal{A}_r$ is a step case and $\mathcal{B}_r$ is a base case for a class of proofs where $n \equiv r(mod\ c)$. The formalisation of generalised proof for *constructive* proofs is symmetric to the one given above.

### Generalising For All Linear Functions

As mentioned above, we aim to recognise the particular recursive structure of the given example proofs. More precisely, we want to extract the step case $\mathcal{A}$ and the base case $\mathcal{B}$ of the proof and then generalise them for all $n$. The general methodology employed for doing this can be demonstrated as:

```
given      n = 4           n = 3
           XY              Y  ——→  Y = A(3)Z ——→  ...
                                ?
              \           /
               [X] = A(4)
```

The first step of the generalisation algorithm is to extract the difference between the two given example proofs for $n_1$ and $n_2$ ($n_1 > n_2$), where $c = n_1 - n_2$, in the hope that this, when generalised, will be the step case $\mathcal{A}$ of the proof. This is done by associative matching which detects and returns the difference between the two example proofs. Now we have a concrete step case of the proof. This difference consists of a few operations $op_k$ each applied $x_{k,n_1}$ times for some natural $k$.

To make a step case general, we need to find the dependency function between every $x_{k,n_1}$ and $n_1$. This demands identifying a function of $n_1$, which would give a specific $x_{k,n_1}$, i.e., $f_k(n_1) = x_{k,n_1}$ for some $k$ and $n_1$. DIAMOND assumes that the dependency is linear: $an + b$. Thus, let us write for each $op_k$ a linear equation $an_1 + b = x_{k,n_1}$, where $n_1$ and $x_{k,n_1}$ are known.

The subsequent stage of the generalisation is to extract the next step case from the rest of the example proof for the corresponding new $n$ (i.e., $n_2$). If successful, continue extracting step cases for the corresponding $n$'s from the rest of the proof until only the base case is left.

Since we are dealing with inductive proofs, it is expected that every step case of a proof will have the same structure, i.e., will consist of the same sequence of application of operations, but a different number of times. Thus, we could in the same way as above for every operation $op_k$ write a linear equation $an_2 + b = x_{k,n_2}$. However, the number $x_{k,n_2}$ of applications of a particular operation $op_k$ in the next step case is not known. A possible value of $x_{k,n_2}$ is acquired by counting the number $x'$ of times every operation $op_k$ of the initial step case occurs in the rest of the proof. The actual value of the number of occurrences of each operation could be any number from 0 to $x'$. Thus, we do branching for all such values and thus we have:
$$\begin{aligned} an_1 + b &= x_{k,n_1} \\ an_2 + b &= x_{k,n_2} \end{aligned}$$
where $n_1, n_2, x_{k,n_1}$ and $x_{k,n_2}$ are known, so the equations can be solved for $a$ and $b$, and $x_{k,n_2}$ takes values from 0 to $x'$. This results in several possible potential generalisations of the step case. The aim is to eliminate those that are impossible. After checking if step cases for all $n$ down to base case are structurally consistent one hopes to be left with at least one possible generalisation of the example proofs. The step case is rejected when the sequence of operations in the subsequent step cases is impossible, i.e., the functions were wrong. This normally occurs when the dependency function gives a negative number of applications of a particular operation, when the calculated sequence is not identical to the rest of the example proof, or when there is no integer solution to our equations. Usually, there will be only one possible generalisation of the two given example proofs.

The example proof for the *sum of odd naturals* is generalised into the following step case and base case:

$\mathcal{A}(n)$ = {$lcut(1)$, $split\_top\_row(1)$, $rotate90(1)$, $join(1)$, $rem\_left\_column(n-1)$, $remove\_dot(1)$}

$\mathcal{B}(1)$ = {$remove\_dot(1)$}

where the function in parentheses indicates the number of times that the operations are applied for each particular $n$.

### f-Homogeneous Proof

Assume two example proofs for the *sum of odd naturals* (the example proof would consist of making $n$ lcuts, and then showing that each L consists of an odd number of dots). If the user supplies two example proofs for values of $n$ and $n+1$, for some concrete $n$, then there is no problem, so DIAMOND will generalise normally and determine that the proof is 1-homogeneous. However, should the user supply proofs for $n$ and $n + 2$ for some concrete $n$, the first stage of generalisation would determine that the step case consists of two lcuts. However, a complete recursive function for generalisation requires a step case to consist of one lcut only.

DIAMOND checks this by trying to split the step case into further $f$ structurally the same sequences of operations, for all factors $f$ of $c$ in order to obtain an $f$-homogeneous proof. If the method fails, then there is no such $f$-homogeneous further generalisation of the step case $\mathcal{A}(n)$. If the method succeeds, and DIAMOND finds a new generalisation of the step case, call this $\mathcal{A}'(n)$, then it also needs to find a new base case $\mathcal{B}'(r')$ if $r' \neq 0$, or $\mathcal{B}'(f)$ if $r' = 0$, where the previous $r$ for $c$ was such that $n = kc + r$ and $r < c$, and the new $r'$ is now such that $n = kf + r'$ and $r' < f$.

## 4 Results and Further Work

DIAMOND is implemented in Standard ML of New Jersey, Version 109. The code is available upon request to the first author.

So far, the interactive construction of proofs and automatic generalisation from example proofs have been implemented in DIAMOND. We can prove a few theorems: sum of odd naturals, sum of all naturals, sum of Fibonacci squares. We are working on more examples.

We want to relax the requirement for a particular formulation of example proofs. Partially ordered sets could be used. However, this would require recognition and generalisation of diagrams (as opposed to a sequence of operations).

There is also a possibility of allowing non-linear dependency functions.

Another issue that will be addressed is to prove that the derived generalisation of a diagrammatic proof is indeed correct.

Finally, some recognition and generalisation of diagrams using abstractions could be an interesting issue to consider. The difficulty is to keep track of these abstractions while manipulating the diagram.

## 5 Related Work

Several diagrammatic systems such as the Geometry Machine [Gelernter, 1963], Diagram Configuration model [Koedinger and Anderson, 1990], GROVER [Barker-Plummer and Bailin, 1992], and Hyperproof [Barwise and Etchemendy, 1991] have been implemented in the past and are of relevance to our system.

However, they all use diagrams to model algebraic statements, and use these models for heuristic guidance while searching for an *algebraic* proof. In contrast, proofs in our system are explicitly constructed by operations on diagrams.

Closer to our work, but not in the domain of diagrammatic reasoning, is work done by Siani Baker [Baker *et al.*, 1992] described in Section 2.4, whereby we exploit the uniform structure of inductive proofs to generalise from example proofs.

## 6 Conclusion

We presented a diagrammatic reasoning system, DIAMOND, which supports interactive construction of diagrammatic proofs. The system automatically generalises from examples to give a general proof for all $n$. The hope is that automating the 'informal' diagrammatic reasoning of humans will shed light on the issues of formality, informality, rigour and 'intuitive' understanding of the correctness of diagrammatic proofs. In the long run, tools such as DIAMOND could lead to e.g. improved maths courses in schools.

## Acknowledgements

We should like to thank Predrag Janičić for inspiring discussions about some of the work presented here. The research reported in this paper was supported by Artificial Intelligence Dept. grant, University of Edinburgh, and the Slovenian Scientific Foundation supplementary grant for the first author, and by EPSRC grant GR/L/11724 for the other two authors.

## References

[Baker *et al.*, 1992] S. Baker, A. Ireland, and A. Smaill. On the use of the constructive omega rule within automated deduction. In A. Voronkov, editor, *LPAR 92, St. Petersburg*, Lecture Notes in AI No. 624, pages 214–225. Springer-Verlag, 1992.

[Barker-Plummer and Bailin, 1992] D. Barker-Plummer and S. C. Bailin. Proofs and pictures: Proving the diamond lemma with the GROVER theorem proving system. In *Working Notes of the AAAI Symposium on Reasoning with Diagrammatic Representations*, Stanford, USA, March 1992.

[Barwise and Etchemendy, 1991] J. Barwise and J. Etchemendy. Visual information and valid reasoning. In W. Zimmerman and S. Cunningham, editors, *Visualization in Teaching and Learning Mathematics*, pages 9–24. Mathematical Association of America, 1991.

[Gelernter, 1963] H. Gelernter. Realization of a geometry theorem-proving machine. In E. Feigenbaum and J. Feldman, editors, *Computers and Thought*, pages 134–52. McGraw Hill, 1963.

[Jamnik *et al.*, 1996] M. Jamnik, A. Bundy, and I. Green. Automation of diagrammatic proofs in mathematics. In *Proceedings of the Third International Summer School in Cognitive Science, Sofia, Bulgaria*, 1996.

[Koedinger and Anderson, 1990] K.R. Koedinger and J.R. Anderson. Abstract planning and perceptual chunks. *Cognitive Science*, 14:511–550, 1990.

[Larkin and Simon, 1987] J.H. Larkin and H.A. Simon. Why a diagram is (sometimes) worth ten thousand words. *Cognitive Science*, 11:65–99, 1987.

[Nelsen, 1993] R. B. Nelsen. *Proofs Without Words: Exercises in Visual Thinking*. The Mathematical Association of America, 1993.

# Structural Inferences from Massive Datasets

Kenneth Yip
MIT AI Lab
Cambridge, MA 02139
U.S.A.

## Abstract

High-level understanding of data must involve the interplay between substantial prior knowledge with geometric and statistical techniques. Our approach emphasizes the recovery of basic structural elements and their interaction patterns in order to summarize and draw inferences about the significant features contained in the data. As a testbed for modeling how scientists analyze and extract knowledge of structure morphogenesis from data, we examine the datasets obtained from numerical simulation of turbulence. We describe a program that automatically extracts 3D structures, classifies them geometrically, and analyzes their spatial and temporal coherence. Our program is constructed by mixing and matching the aggregate, classify, and re-describe operators of the spatial aggregation language. The research is a continuation of the effort to investigate the role of imagistic reasoning in human thinking.

## 1 Introduction

An essential survival skill for humans is the ability to perceive objects, classify them, and predict their behavior. Much of the neural machinery is devoted to these tasks. We believe that the machinery that forms the foundation of intelligence is to be found in the mechanisms that support, for example, the vision, the language, and the sensorimotor faculties. Each faculty has to solve the figure-ground problem: To identify, interpret, manipulate, and track salient objects.

It is important to understand how each such modality contributes to overall intelligence. Each modality draws on proprietary representations and problem-solving competences. Each takes inputs from internal sources as well as from the outside world. For example, a linguistically represented problem can require us to harness the power of our visual system so as to engage the strategies of "imagistic reasoning" or visual imagination. What appears to be a high-level cognitive ability emerges from cooperative problem-solving as peripherals work out parts of a problem that have been reexpressed in terms of peripheral-specific representations.

If humans do have and make use of these peripheral representations, what is the form of these representations? What are the processes that manipulate them?

In this paper we focus on the development and new applications of imagistic reasoning. In particular, we propose a model to reason about spatially distributed, temporally evolving 3D structures in the domain of fluid dynamics. Our model aims to capture the process by which human scientists interpret numerical simulation data. Our approach follows the spatial aggregation framework [Yip and Zhao, 1996]: Programs are implemented by mixing and matching a few high-level operators such as aggregate, classify, and redescribe. We report on our progress in implementing the performance model.

The research is inspired by the practical needs for analyzing large amount of data and the intellectual curiosity to understand how humans discover regularities, summarize significant events, and debug conceptual models. We use scientific datasets as a testbed in order to exploit the vast amount of accumulated specialized knowledge. Scientists' reasoning about fluid dynamical objects rely on the same intuitive concepts of objects: cohesiveness and continuity [Spelke et al., 1995]. An object is cohesive if it is internally connected, externally bounded, and moves while maintaining its connectedness and boundedness. The motion of an object is not arbitrary; it traces one connected path in space-time and leaves no gaps.

Describing complex fluid phenomena in terms of such intuitive notions of cohesive objects[1] and their interaction patterns is useful for qualitative understanding and almost necessary for developing new conceptual models of dynamical mechanisms.

## 2 Human understanding

The entire theory of incompressible fluid flow is contained in the Navier-Stokes equations. However, progress in analyzing the equations analytically has been slow. Much of the understanding of flow mechanisms comes from an interplay between numerical simulation

---
[1] The term "coherent objects" is more commonly used in the turbulence research community.

and experimental data. To see an example of this kind of understanding that fluid dynamicists arrive at after detailed investigation of simulation data, let us quote a passage in the concluding section of [Robinson, 1991]:

> The main conclusion is that the self-maintaining cycle of turbulence production in the boundary layer is driven by the formation or regeneration of embedded vortical structures... A mature vortical arch gives rise to a trailing quasi-streamwise vortex... The quasi-streamwise vortex elements collect and lift low-speed near-wall fluid, leaving behind a persistent low-speed streak. Relatively high-speed fluid scrubbing the vortex-lifted low-speed fluid creates a shear layer which rolls up into a new vortical arch. The new arch grows outward by agglomeration and/or self-induction and circulation lift, and the cycle repeat itself.

See the diagram in Figure 1 accompanying the explanation.

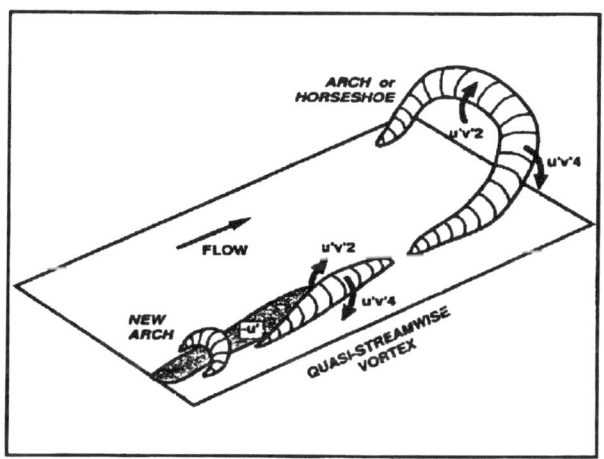

Figure 1: A conceptual model summarizing the morphogenesis of vortex structures near a turbulent boundary layer.

Much can be learned from this explanation. First, the explanation is conceptual: it does not provide any new equations. Second, the explanation refers to various kinds of structures: vortices [2], streaks, and shear layer. Third, the spatial character (e.g., arch-shaped, streamwise, persistent) of the structures are described and related. And finally the dynamical process is described in terms of structural morphogenesis: the growth and agglomeration of structures and their effect on others (e.g., lifting, rolling up).

---

[2] Vortex is an intuitive notion denoting some compact region of swirling fluid motion; it is related to but not the same as the vector quantity vorticity, the curl of the velocity, at a flow field point.

## 3 The Spatial Aggregation Framework

In recent years, a computational framework, *spatial aggregation*, has been developed to unify the description of a class of imagistic problem solvers. A program written in this framework has the following properties. It takes a continuous field and optional objective functions as input, and produces high-level descriptions of structure, behavior, or control actions. It computes a multi-layer of intermediate representations, called spatial aggregates, by forming equivalence classes and adjacency relations. It employs a small set of generic operators such as aggregation, classification, and localization to perform bidirectional mapping between the information-rich field and successively more abstract spatial aggregates. It uses a data structure, the neighborhood graph, as a common interface to modularize computations.

Our performance model is implemented in this style. In the following, we abstract away the details of the programming issues and focus on the incorporation of new aggregators and classifiers for manipulating 3D structures.

## 4 Performance Model

### 4.1 Objective and Strategies

The objective of the performance model is:

> Given a sampled flow-field, construct a conceptual model to summarize the significant events implicit in the numerical data.

Numerical simulation of turbulent flow generates massive amount of data. Even a modest sampled flow-field consists of tens of gigabytes of data describing quantities such velocity, pressure, and vorticity at each sampled point. From observing how some scientists work, we propose a 5-step strategy to summarize data:

1. Isolate and characterize interesting structures.
2. Determine the spatial relationships among structures.
3. Track the morphogenesis of structures backward and forward in time.
4. Describe the statistical and dynamical significance of each structural event.
5. Note correlations among structural events and assign cause-and-effect relationships.

Not all of these steps have been automated. In the following we describe the pieces that have been implemented.

### 4.2 Aggregating field points

Many categories of structures can be defined as the isosurface of a scalar field. An isosurface of a scalar field F is defined as the set of points (x,y,z) such that F(x,y,z) = c where c is a constant and is often called the threshold of the isosurface. Some examples are the low pressure region, the high-speed streak, and the eigenvalues of the velocity gradient.

A popular method to render isosurface is the marching cube algorithm [Lorensen and Cline, 1987]. The algorithm divides the field into little cubes. The intersection of the surface with the cube is determined by a table-lookup based on the signs of the values F(x,y,z) - c.[3] The algorithm is reasonably fast, but does not guarantee a topologically consistent surface when adjacent cubes share a face where the vertices on the two diagonals differ in sign.

Our aggregator extends the basic marching cube algorithm in four ways:

- *Recursive subdivision.* The aggregator recursively subdivides an ambiguous cube into 8 little cubes until the newly generated cubes are unambiguous. The subdivision method gives accurate results subject to the limit in sampling resolution. The values at the sub-sampled points are determined by polynomial interpolation using values from adjacent cubes.

- *Multiple isosurfaces.* It requires little overhead to compute surfaces corresponding to different thresholds during a single march over the entire flow field.

- *Connected components.* The connectivity of a surface patch intersecting a cube are precomputed and stored in each cube pattern. The global connected components of the surface are determined by aggregating individual patches.

- *Volume estimate.* The volume enclosed by an isosurface is computed from the contributions from all the cubes on and inside the surface. Each cube pattern is pre-partitioned into tetrahedra representing the volume enclosed by a surface patch.

Other structures are defined by integral curves of a vector field. For example, a vortex line, the integral curve of the instantaneous vorticity field, is the basic building block of a vortex structure. However, it is important to distinguish vorticity lines (lines everywhere parallel to the vorticity vector) and vortices (regions of nearly circular motions in the plane normal to the core of the vortex observed at the core speed). In particular, vortices are much more cohesive and less noisy than vorticity lines. We adopt the following working definition for vortices:

**Definition** A *coherent vortex* is a compact bundle of adjacent, high-intensity vortex lines that are geometrically similar.

Aggregating vortex lines to form coherent structures is not straightforward. Previous researchers [Moin and Kim, 1985; Robinson, 1991] have found that vortex lines are sensitive to initial conditions. Nearby vortex lines can diverge rapidly. If the initial conditions are not chosen carefully, the resulting vortex lines are likely to resemble badly tangled spaghetti wandering over the whole flow field, making the identification of organized structure extremely difficult. This might explain why vortex lines have not been widely used for structure identification.

Our search algorithm for vortices is adaptive[Yip, 1995]: it exploits local geometry (e.g., converging or diverging) of the vorticity lines to decide the direction and step size of integration.

### 4.3 Object Cohesiveness

There is a degree of arbitrariness in defining structures in terms of isosurfaces. The shape of the object depends on the threshold chosen, but it is often unclear how to set the appropriate thresholds. For example, what is the threshold for a low pressure region? Intuitively we prefer structures that are stable against small changes in the threshold. We compute a range of thresholds (the lower and upper bound of the thresholds are typically known) for each scalar field. Geometric properties of the re-described objects are plotted as functions of the thresholds. We choose the mean value in the largest threshold region in which the structures are stable as the desired threshold. Figure 2 shows an example of how cohesiveness of a vortex determines the choice of the threshold.

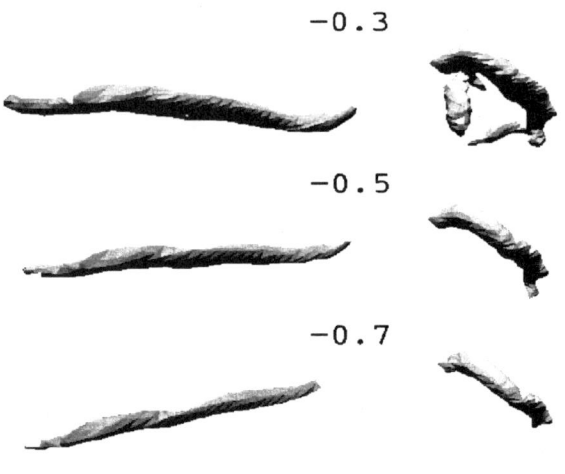

Figure 2: Cohesiveness of two vortices as a function of the isosurface threshold. Left: The shape of the vortex is stable in the range of threshold from -0.3 to -0.7. Right: The vortex is incohesive between -0.3 and -0.5.

To determine the stability of the shape of a collection of surface points for threshold selection, we find it sufficient to compute a gross geometric description of the points. The detailed shape representation is postponed until the surface classification stage (see next section). The algorithm is as follows. We compute a 3D bounding box that minimizes the thickness of the distribution of the points along the long axis of the box. The cloud of

---

[3]There are $2^8$ ways a cube intersects a surface. The number can be reduced to 14 distinct configurations using rotation and reflection symmetries. In our implementation, we find it more convenient to use the entire table of $2^8$ intersection patterns.

points is reoriented along the principal component directions. This reorientation is determined by rotating the points with the matrix of eigenvectors of the covariance matrix of the points. The reoriented points also serve as input to the detailed surface classification.

### 4.4 Classifying surfaces

The marching cube algorithm and the vortex line integration provide a surface representation of the fluid structures. To classify the structures, we need to recover their 3D geometry. We use generalized cylinders (GC) [Binford, 1990] as our basic shape model.

A generalized cylinder consists of a spline, a cross-section, and a sweeping rule. It is particularly useful for our problem because to a first approximation vortices are elongated slender structures which bend and turn.

Recovery of GC descriptions from images is a hard problem in computer vision because of noise and occlusion ([Zerroug and Nevatia, 1996]). Our problem is different and is much easier. We aim to recover the GC descriptions from 3D surface points. The recovery procedure has two stages: (1) recovery of the meridians of the GC, and (2) recovery of the cross sections and the spline of the GC.

The meridians of the GC are recovered as follows:

1. Reorient the surface points $S$ along the principal component directions.

2. Project the points onto the plane normal to the first principal component direction, i.e., the eigenvector associated with the smallest magnitude. Call the projected points $S_p$.

3. Compute a delaunay triangulation of $S_p$.

4. Remove exterior triangles of the triangulation by a sculpturing method [Boissonat, 1984]. See Figure 3.

5. Extract an oriented boundary (anticlockwise) of the triangulation.

6. Segment the boundary at the two points corresponding to the two largest local maximum curvatures of the boundary. Call these two segments $bd_1$ and $bd_2$.

7. Backproject $bd_1$ and $bd_2$ to obtain two meridians of the GC.

The surface points $S$ are rotated along the last principal component direction. The above procedure is then repeated to obtain additional meridians of the GC.

The cross-sections and the spline of the GC are recovered as follows:

1. The meridians are smoothed and parametrized by an normalized arc length $t$ from 0 to 1.

2. Align corresponding points of the meridians at fixed increments of $t$.

3. Fix an ellipse to these points to obtain the cross-section.

4. The center of the ellipses are smoothed to obtain the spline of the GC.

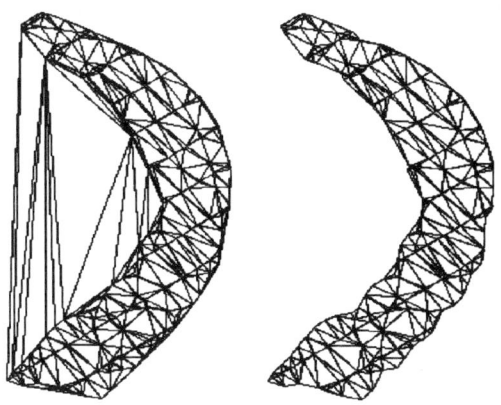

Figure 3: Sculpturing: Recover the shape of the 2D projection of a vortex by removing the exterior triangles from the delaunay triangulation. An exterior triangle is one whose circumcircle is much larger than the triangle (in terms of area). The meridians of the generalized cylinder representation of the vortex are found by back-projecting the boundary of the sculptured triangulation.

### 4.5 Re-describing Surfaces

A classified surface is a compact parametric description of a large number of field points. Like a lambda abstraction, the re-describe operator encapsulates the classified surface as a primitive object for further processing. For example, the center of mass, volume, orientation, and inertia tensor of the object are computed. From these quantities we estimate the distribution of characteristic objects (e.g, streamwise slender vortices) in interesting regions of the field (e.g., near the surface of a free surface flow).

### 4.6 Object Persistence

Structures typically persist over time. They evolve for a while and may grow or shrink or collide with other structures. Unlike structure recovery from 2D images, the analogous correspondence problem in recovery from 3D fields is much easier to solve. For example, we do not have to deal with occlusions.

Objects smoothly deform and move in connected paths. There is a lot of redundant information – size, shape, and relative positions – that carries over from one time slice to another. We track the largest structures (currently up to 20) and discard time slices which do not contain significant changes in the spatial character or distributions of the tracked structures. Conversely consecutive time slices that yield qualitatively different distributions are noted. These time slices are candidates for restart of numerical simulation to acquire better temporally resolved data.

### 4.7 Extensions

We have only automated the extraction of kinematics and statistics of a few turbulence structures (vortices and pressure regions). Much more remains to be done:

- More structures. Ejection zones, dissipation regions, high stress areas, pockets, large-scale motions – these are some of the more important structures that need to be tracked.

- Spatial correlations. We need to have efficient ways to compute pairwise or even triple correlations of spatial distributions.

- Dynamics. We need to incorporate qualitative rules of interaction to determine causal relations among structures. For example, a vortex accelerates velocity (lifts) on one side and slows down on another. A bent vortex exerts a self-induced motion along its binormal and the effect is proportional to the curvature of the bent [Arms and Hama, 1965].

## 5 Experiment

As a testbed, we choose a direct numerical simulation[4] of free surface turbulence generated by a shear flow in a $128^3$ rectangular box. The problem is important for several reasons. First, the mechanism for scalar transport into turbulent fluids across free-shear interfaces is important for the design and control of industrial equipments with free surfaces. Second, interest in environment and global warming raises the issue of how to accurately estimate mass transfer rate of $CO_2$ between atmosphere and ocean. Third, recent experiments have found evidence that low-speed streaks occur near free-slip surface under sufficient high shear. Visually these streaks resemble those found in wall-layer streaks. Without the complication of a solid boundary, studies in the formation of these streaks near free surface may shed additional light on the phenomenon.

It has been recently proposed that a definition of vortex in an incompressible flow in terms of the second eigenvalues of the symmetric tensor $S^2$ and $\Omega^2$ where $S$ and $\Omega$ are respectively the symmetric and antisymmetric parts of the velocity gradient tensor $\nabla u$ [Jeong and Hussain, 1995]. The symmetric part determines the contribution to local pressure minimum due to rotational motion alone.

This proposed definition is attractive computationally because it refers to the isosurfaces of a scalar field, the second eigenvalues of the symmetric tensor. We compare this definition with our working definition of vortex in terms of compact bundles of vorticity lines.

We have examined dozens of structures according to both definitions. We found a surprisingly good agreement between two definitions. Figure 4 is a typical result.[5] The figure shows a superposition of the vortex lines recovered by laborious integration and the surface patches quickly recovered by the extended marching cube algorithm. Because of results like these, we now routinely use the scalar isosurfaces as surrogates for vortices to develop our matching and correlation algorithms.

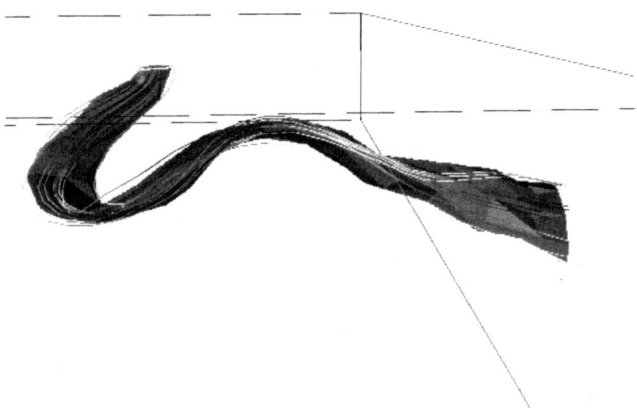

Figure 4: A superposition of vortex lines obtained by adaptive integration and the surface patches recovered by the marching cube algorithm. The agreement is surprisingly good.

## 6 Related Work

Our research shares some of the objectives with the visualization work in Rutgers [Silver and Zabusky, 1993]. The Rutgers group is interested in better interactive tools for helping scientists visualizing the datasets. In particular, they use ellipsoids to fit isosurfaces of vorticity, the modulus of the vorticity vector. Our effort emphasizes more on building a computational model of the process by which scientists understand those data. We ask questions about how scientists think visually and how this visual thinking is connected to the rather primitive knowledge of object cohesiveness, persistence, and continuity. We also use a more general shape model and explore the coincidence of different definitions of a coherent vortex.

Commonsense reasoning about fluids has been proposed as central problem in naive physics [Hayes, 1985]. The problem is hard because fluids do not conveniently divide into discrete pieces that can be easily combined. Collins and Forbus have partially implemented Pat Hayes' ideas for representing fluids using both the contained-liquid and piece-of-stuff ontologies [Collins and Forbus, 1987]. The work described here is closer to the scientific end of the formalization spectrum.

Our work builds on previous work by several groups on imagistic reasoning and spatial aggregation [Yip and Zhao, 1996]. Zhao[Zhao, 1994] extracts structures of trajectories from 3D phase space to guide control actions. Junker[Junker and Braunschweig, 1995] reconstructs the history of wavefront propagation from 2D seismic data.

---

[4] A direct simulation proceeds from the Navier-Stokes equations without extraneous modeling assumptions.

[5] We are exploring why there might be such a coincidence.

Nishida[Nishida et al., 1991] automates the qualitative analysis of 3D phase space trajectories. Our work extends the frontier to tackle full 3+1D spatial-temporal field. Unlike previous work in domains where one has a rather complete knowledge of what the interesting structures are, we need to rely on informal concepts of object cohesiveness and coincidence for structure discovery.

## 7 Discussion/Conclusion

We have described a framework for understanding flow fields and progress in implementation. The specific technical contributions are threefold:

- We show how aggregation algorithms (based on marching cube and adaptive line tracing) and classification algorithms (based on recovery of generalized cylinders) can be used to extract and characterize interesting 3D fluid structures.
- In collaboration with the programs, we found a substantial agreement between a recently proposed definition for vortex and the more intuitive definition of vortex as compact bundles.
- We extend the library of spatial aggregation routines by incorporating new aggregators and new classifiers.

Many of these ideas are generally useful for the recovery of spatial-temporal patterns in other physical domains.

We believe the current work also sheds light on the nature of human intelligence. Even in sophisticated scientific applications, some scientists appear to think visually and rely on commonsense intuitions of object cohesiveness, persistence, and continuity – not unlike those shown in recent cognitive psychology literature. This shared core of knowledge attests to the importance of recycling the peripheral machinery (such as vision) for solving otherwise symbolically-posed problems. Traditional AI has searched for the secrets of central intelligence for years. Perhaps the central is the peripherals.

## References

[Arms and Hama, 1965] R Arms and F.R. Hama. Localized-induction concept on a curved vortex and motion of an elliptic vortex ring. *Physics of Fluids*, 8, 1965.

[Binford, 1990] T.O. Binford. Generalized cylinder representation. In *Encyclopedia of Artificial Intelligence*. Wiley, 1990.

[Boissonat, 1984] Jean-Daniel Boissonat. Geometric structures for 3D shape representation. *ACM Transactions on Graphics*, 3(4), 1984.

[Collins and Forbus, 1987] John Collins and Kenneth Forbus. Reasoning about fluids via moecular collections. In *Proceedings AAAI-87*, 1987.

[Hayes, 1985] Patrick Hayes. The second naive physics manifesto. In J.R. Hobbs and R.C. Moore, editors, *Formal Theories of the Commonsense World*. Ablex Publishing Corp., 1985.

[Jeong and Hussain, 1995] Jinhee Jeong and Fazle Hussain. On the identification of vortex. *Journal of Fluid Mechanics*, 285, 1995.

[Junker and Braunschweig, 1995] U Junker and B Braunschweig. History-based interpretation of finite element simulations of seismic wave fields. In *Proceedings IJCAI-95*, 1995.

[Lorensen and Cline, 1987] W.E Lorensen and H.E. Cline. Marching cubes: A high resolution 3d surface construction algorithm. *Computer Graphics*, 21(4), 1987.

[Moin and Kim, 1985] P. Moin and J. Kim. The structure of the vorticity field in turbulent channel flow. *Journal of Fluid Mechanics*, 155, 1985.

[Nishida et al., 1991] Toyoaki Nishida, Kenji Mizutani, Atsushi Kubota, and Shuji Doshita. Automated phase portrait analysis by integrating qualitative and quantitative analysis. In *Proceedings AAAI-91*, 1991.

[Robinson, 1991] S.K. Robinson. The kinematics of turbulent boundary layer structure. Tech. Mem. 103859, NASA, 1991.

[Silver and Zabusky, 1993] D Silver and N Zabusky. Quantitfying visualizations for reduced modeling in nonlinear sciences: extracting structures from the datasets. *Journal of Visual Communication and Image Representation*, 4, 1993.

[Spelke et al., 1995] E Spelke, Grant Gutheil, and Gretchen Van de Walle. The development of objet perception. In *Visual Cognition*. MIT Press, 1995.

[Yip and Zhao, 1996] Kenneth Yip and Feng Zhao. Spatial aggregation: Theory and application. *Journal of Artificial Intelligence Research*, 1996.

[Yip, 1995] Kenneth Yip. Reasoning about fluid motion i: Finding structure. In *Proceedings IJCAI-95*, 1995.

[Zerroug and Nevatia, 1996] M. Zerroug and R. Nevatia. Three-dimensional descriptions based on the analysis of the invariant and quasi-invariant properties of some curved-axis generalized cylinders. *IEEE Transactions on PAMI*, 18, 1996.

[Zhao, 1994] Feng Zhao. Extracting and representing qualitative behvaiors of complex systems in phase space. *Artificial Intelligence*, 69, 1994.

# DIAGNOSIS AND QUALITATIVE REASONING

## Qualitative Reasoning 4: Causality

# Qualitative Analysis of Causal Graphs with Equilibrium Type-Transition

**Koichi Kurumatani**
Electrotechnical Laboratory
Umezono 1-1-4, Tsukuba,
Ibaraki 305, Japan
kurumatani@etl.go.jp

**Mari Nakamura**
LERC, Electrotechnical Laboratory
Nakoji 3-11-46, Amagasaki,
Hyogo 661, Japan
mari@osaka.etl.go.jp

## Abstract

In this paper, we present a method to qualitatively compute the global characteristics of causal graphs by the analysis of the underlying dynamical systems, rather than traditional qualitative simulations which suffer from intractability and difficulty in understanding their simulation results. The key idea is to translate a given causal graph into an autonomous dynamical system and to analyze equilibrium points in the system. The method requires no numerical information and it has the advantage of computing the conditions under which a certain equilibrium type holds and when equilibrium type-transitions occur. The method is firmly based on mathematical grounds, and the result is guaranteed to be valid for linear systems under the specified conditions.

## 1 Introduction

Traditional qualitative simulation methods are based on the qualitative value representation for each state variable. The behaviors of a target system are simulated on this discrete quantity, and the methods cannot avoid the ambiguity and intractability in their results.

In this paper, we present a qualitative method to analyze the global characteristics of a target system described as a causal graph without qualitative simulations, based on dynamical systems theory.

The key idea of the method is 1) to prepare a target system in the notation of autonomous linear dynamical system, 2) to analyze the type of equilibrium points, and 3) to determine the conditions when equilibrium type-transitions occur.

The method requires no numerical information. In addition to it, the method has the advantage of computing the conditions under which a certain equilibrium type holds and when equilibrium type-transitions occur. This kind of reasoning ability has not been achieved by previous works in the literature.

After the basic introduction of dynamical systems theory (section 2), we will analyze the basic simply-connected loops (section 3), followed by the key components of our method (section 4, 5, 6) and examples to show the ability of the method (section 7).

## 2 Basic Characteristics of Dynamical Systems

Linear dynamical systems are expressed by a differential equation:

$$d\mathbf{x}/dt = A\mathbf{x}, \quad \mathbf{x} \in R^n, A \in R^{n \times n}. \tag{1}$$

Each element of $\mathbf{x} = (x_1, x_2, \ldots, x_n)^t$ is a state variable, and the vector $\mathbf{x}$ is called state vector [1].

Equation (1) represents a linear dynamical system, which has one trivial equilibrium point $\mathbf{x} = 0$ and possibly has non-trivial ones.

Non-linear dynamical systems are represented by a differential equation:

$$d\mathbf{x}/dt = \mathbf{f}(\mathbf{x}), \quad \mathbf{x}, \mathbf{f} \in R^n. \tag{2}$$

Possibly there exist several equilibrium points given by $\mathbf{f}(\mathbf{x}) = 0$. The behavior in the neighborhood of an equilibrium point $\mathbf{x}_0$ is precisely governed by a linear equation:

$$d\mathbf{x}/dt = J\mathbf{x}, \quad J = (\partial f_i / \partial x_j).$$

Computing the Jacobian matrix $J$ can be done automatically by symbolic differentiation.

Equilibrium points are classified into *attractor, repellor* or *saddle*. Trajectories approaching to an attractor will finally reach the point. From a repellor, trajectories start and never return to the point. Trajectories approaching to a saddle are first attracted but finally repelled by the point.

The characteristics of an equilibrium point can be specified by the eigenvalues of the coefficient matrix (Table 1). In general, the eigenvalues consist of 1) real numbers and 2) pairs of conjugate complex numbers. If all the real parts of eigenvalues are positive or negative, the point is a repellor or an attractor respectively. Otherwise, the point is a saddle which is a combination of repellor and attractor. Exceptions are the cases where

---

[1] Throughout the paper, $(x_1, x_2, \ldots, x_n)^t$ represents $n$-tuple column vector.

Table 1: Eigenvalue and the type of equilibrium point.

| Eigenvalue: $\lambda = a \pm bi$ | Real Number: $b = 0$ | A Pair of Conjugate Complex: $b \neq 0$ |
|---|---|---|
| $a > 0$ | nodal repellor | spiral repellor |
| $a = 0$ | zero-eigenvalue | center |
| $a < 0$ | nodal attractor | spiral attractor |

Figure 1: A loop connected by influence relations.

the real parts of some eigenvalues are zero. Such a point is called *non-hyperbolic*.

About non-hyperbolic point [Guckenheimer and Holmes, 1990], if it is a pair of purely imaginary numbers, it is called *center*, which has eternal periodic trajectories around the point.

The eigenvalue whose value is exactly zero is called *zero-eigenvalue*. If there exist zero-eigenvalues, the behaviors around the point become strange ones. We ignore this special case for a while, before addressing structural stability in section 5.

## 3 Analysis of Causal Loops

In this section, we analyze basic simple loops connected by influence relations, i.e., causal loops with time-delay.

### 3.1 Illustrative Example

An example is a simple loop of three variables connected by direct influence relations [Forbus, 1984] (Figure 1). It is described by an equation:

$$d\mathbf{x}/dt = A\mathbf{x}, \quad \mathbf{x} \in R^3, A = \begin{pmatrix} 0 & 0 & a_{13} \\ a_{21} & 0 & 0 \\ 0 & a_{32} & 0 \end{pmatrix},$$

where $a_{21}, a_{32} > 0$ and $a_{13} < 0$.

Propagation-based qualitative simulation methods cannot be applied to this kind of example with loops, because a propagation will go back to its starting point.

The constraint-based filtering method, i.e., QSIM [Kuipers, 1986], can predict the behavior of the system. Starting with an initial state, e.g.,

$$(x_1, x_2, x_3) = ((+, dec), (+, inc), (+, inc)),$$

the method produces the unique successors and finds out that the twelfth state is qualitatively equivalent to the initial one, i.e., the behavior is cyclic, then stops.

There is basically no way to know which type of behavior it is (spiral attractor, repellor, or center) by QSIM. Although the energy filter [Fouché and Kuipers, 1992] and the envisionment-guided simulation [Clancy and Kuipers, 1992] have been proposed for that purpose, they have restrictions in applicability, e.g., the dimension of the state space.

Actually, this equilibrium point is a saddle which consists of a one-dimensional attractor and a two-dimensional spiral repellor, whenever $a_{21}, a_{32} > 0$ and $a_{13} < 0$. The trajectory will be repelled with rotating around the equilibrium point.

We will show a method of extracting these global characteristics of the system by qualitatively analyzing equilibrium points, instead of carrying out qualitative simulations.

### 3.2 Qualitative Analysis of Simple Loops

The general form of simple loops connected by direct influences are described by an equation:

$$d\mathbf{x}/dt = A\mathbf{x}, \quad \mathbf{x} \in R^n, A = (a_{ij}),$$
$$a_{ij} = \begin{cases} \neq 0 & \text{if } i = j+1 \vee (i = 1 \wedge j = n), \\ = 0 & \text{otherwise}. \end{cases} \quad (3)$$

The loop shown in Figure 1 is an instance ($n = 3$) of this class.

**Definition 1**
The *loop constant* of a loop by Equation (3) is defined as $L = a_{1n} a_{21} \ldots a_{n\,n-1}$. The loop constant should not be zero by definition. ∎

**Definition 2**
A loop is called *positive* or *negative*, iff its loop constant is greater or less than zero, respectively. ∎

The following lemma tells the relation between the eigenvalues and the trace of a matrix.

**Lemma 1 (Eigenvalues and Trace)**
The summation of the eigenvalues $\lambda_i$ of a matrix $A$ is equal to the trace $Tr(A) = \sum_{i=1}^{n} a_{ii}$, i.e.,

$$\sum_{i=1}^{n} \lambda_i = Tr(A).$$

**Proof :**
Eigenvalues are given by an equation:
$$(\lambda - \lambda_1)(\lambda - \lambda_2)\ldots(\lambda - \lambda_n) = 0.$$

By comparing the equation with the eigenpolynomial,

$$\lambda^n - Tr(A)\lambda^{n-1} + \ldots + (-1)^n Det(A) = 0, \quad (4)$$

where $Tr(A)$ is the trace and $Det(A)$ is the determinant, the proposition holds. ∎

Consequently, the summation of the real parts of all eigenvalues is also equal to the trace.

We can conclude the characteristics of simple loops with time-delay as follows.

**Theorem 1 (Characteristics of Simple Loops)**
The equilibrium point of a simple loop described by Equation (3) is a saddle when $n \geq 3$, except:

- The neighborhood of the equilibrium point of a positive loop includes a center as its subspace, iff both $n$ and $n/2$ are even numbers, e.g., $n = 4, 8, 12, \ldots$.

- The neighborhood of the equilibrium point of a negative loop includes a center as its subspace, iff $n$ is even and $n/2$ is odd, e.g., $n = 6, 10, 14, \ldots$.

**Proof:**
The eigenvalues are given by an equation $\lambda^n - L = 0$, where $L$ is the loop constant. Obviously, zero-eigenvalue is not a solution. Since the trace is zero, the sum of the real parts of all eigenvalues is zero. Assume that the loop is positive, i.e., $L > 0$. Since one eigenvalue is a positive real number $L^{1/n}$, there should exist at least one eigenvalue whose real part is negative. Consequently, the point is a saddle except the case where there exists a pair of purely imaginary eigenvalues $\lambda = \pm L^{1/n} i$ (a center). All conditions for this case are limited to exceptions shown above. Similarly the theorem holds when the loop is negative. ∎

This theorem tells that simply-connected causal loops should become a saddle when its length is greater than or equal to three, i.e., trajectories are first attracted to the equilibrium point but finally they are sure to move to the infinity point. The direction of the repelling move must include $(+, +, \ldots, +)^t$ when the loop is positive.

The exceptional case about a center does not have the importance from the actual application point of view, because only a very little perturbation will make it become a spiral attractor or repellor (in section 5).

Special cases in lower dimensions are as follows.

**Theorem 2**
A loop of length one is called *self-feedback*. A positive self-feedback has a repellor as its equilibrium point, and a negative self-feedback has an attractor.
A loop of length two is called *mutual-feedback*. A positive mutual-feedback has a saddle, and a negative mutual-feedback has a center. ∎

## 4 Convergence of General Systems

From the engineering point of view, the system is called *convergent* (or *stable*) when the equilibrium point is an attractor, because all the state variables, and consequently the system itself, are sure to converge on a stable state (the equilibrium point).

In this section, we discuss a method of verifying the convergence of general systems given by Equation (1).

### 4.1 System of Short Loops

The first way is to check whether a system consists of short loops whose lengths are less than three. It is straightforward to obtain the following theorem from the result in the previous section.

**Theorem 3 (Convergence of Short Loops)**
A system satisfying the following conditions is convergent, i.e., it has an attractor as its equilibrium point, while possibly the neighborhood includes a center.

- There exists no positive self-feedback, and there exists at least one negative self-feedback.
- There exists no positive mutual-feedback.
- There exists no loop whose length is greater than or equal to three. ∎

The conditions are same in essence as the results obtained by Ishida [Ishida, 1989].

The conditions required in this theorem are too strict to apply them to real systems, because many systems have loops whose lengths are greater than two.

### 4.2 System with Self-Feedbacks

The second way is to check whether each variable in the system has 'enough' self-feedbacks. This technique can be applied not only to short loops, but also to general complex systems.

The method is based on the relations between eigenvalues and diagonal elements in a matrix, proven in linear algebra [Chatelin, 1988].

**Definition 3 (Gershgorin Circle)**
For a matrix $A \in R^{n \times n}$, $A = (a_{ij})$, the *Gershgorin circle* in a complex plane for the $i$-th diagonal element $a_{ii}$ is defined as:

$$|z - a_{ii}| \le \sum_{j \ne i} |a_{ij}|. \tag{5}$$

∎

The center of the $i$-th Gershgorin circle in a complex plane is the place of the $i$-th diagonal element, and the radius is the summation of the elements in the $i$-th row. About the relations between the Gershgorin circles and the places where eigenvalues exist, the following theorem holds.

**Theorem 4 (Gershgorin Circle and Eigenvalue)**
Any eigenvalue of $A \in R^{n \times n}$ exists in (at least) one of the Gershgorin circles of $A$. ∎

This theorem tells that any eigenvalue exists in the area covered by Gershgorin circles, which can be used to reason about the type of equilibrium.

The condition for the convergence of a target system is given by the following theorem.

**Theorem 5 (Convergence by Self-Feedbacks)**
If all the diagonal elements are small enough, i.e., each state variable has enough negative self-feedback, the system is convergent. The condition is:

$$\forall i \, ( a_{ii} < -\sum_{j \ne i} |a_{ij}| ).$$

Similarly, the point becomes a repellor, when:

$$\forall i \, ( a_{ii} > \sum_{j \ne i} |a_{ij}| ). \quad \blacksquare$$

By this theorem, if all diagonal elements are smaller than the sum of other elements' absolute values in the same row, the system is convergent.

Intuitively speaking, it is rare for a complex system to become convergent, unless we add some special mechanisms to the system. The theorem shows such a mechanism which surely makes the system convergent.

## 5 Structural Stability and Equilibrium Type-Transition

Imagine a situation where we add a little perturbation to the vector field of a target system, i.e., to the elements of the coefficient matrix $A$ in Equation (1). The type of the equilibrium point (attractor, repellor, or saddle) will not change by the perturbation if its value is small enough. This attribute is called *structural stability* of the system [Guckenheimer and Holmes, 1990].

However, a system is structurally 'unstable' when it has a non-hyperbolic equilibrium point, i.e., the real parts of some or all eigenvalues are zero. Such a point is sure to transit to another type of equilibrium regardless of the value of a perturbation, because the real parts of eigenvalues are sure to become positive or negative.

The manner of 'type-transition' of a non-hyperbolic equilibrium point by a perturbation is as follows.

### Center to Spiral Repellor or Attractor

A pair of purely imaginary numbers will change into a pair of conjugate complex numbers whose real part is positive or negative. Thus a center will transit to a spiral repellor or attractor by a perturbation.

### Zero-Eigenvalue

If there exists only one eigenvalue which is exactly zero, i.e., zero-eigenvalue, the determinant of $A$ is zero by the eigenpolynomial of Equation (4). In this case, the rank of $A$ is $n-1$ and there exists a non-trivial solution of $A = 0$, which means that there exists a *equilibrium hyperline* in the phase space. The equilibrium line will change into a nodal repellor or attractor by a perturbation regardless of its value.

In order to determine the existence of a zero-eigenvalue, we have to compute the rank of $A$, e.g., to count the number of linearly independent rows or columns of $A$. Because we assume that $A$ is qualitative, i.e., the exact values of the elements of $A$ are not given, zero-eigenvalues exist only for specific sets of the numeric values. Consequently only the possibility of the existence can be pointed out. In special cases, however, we can show that a target system has zero-eigenvalues only by qualitative computation.

If there exist more than one zero-eigenvalues, it is usually very difficult to handle the equilibrium point and the behavior of the total system. We will not address this kind of points for simplicity.

## 6 Transforming General Graphs Including Monotonic Relations

Before applying our method to examples, we address the relations called indirect influences or monotonic relations, which are constraints on the manner of variable-value change, and show a way of transforming general graphs which include such relations into Equation (1).

We express indirect influence $Q^+$ [Forbus, 1984] by monotonic relation $M^+$ defined in [Crawford et al., 1990]:

$$y = Q^+(x_1) \equiv y = M^+(x_1) \equiv \\ (\, y = f(x_1, x_2, \ldots, x_n),\ \partial f / \partial x_1 > 0\,). \tag{6}$$

A monotonic relation $y = M_0^+(x)$ puts another constraint "$x = 0 \Rightarrow y = 0$" in addition to (6).

In order to decide $n$ in Equation (6), i.e., to decide how many variables $x_i$ indirectly influence $y$, we have to count the arcs flowing into $y$.

An arbitrary combination of monotonic relations in a causal graph can be mathematically meaningless. We assume that the condition for the change of variables concerning partial differentiation holds in a given causal graph, i.e., there is no loop which consists of monotonic relations. The exception is a simple monotonic loop whose all nodes have only one source, i.e., $n = 1$ in Equation (6) for all the nodes in the loop.

On this assumption, we can transform a given causal graph into a simpler form which can be handled by our method, using the following transformation rules.

### Transitive and Distributive Law

$$z = M^+(y) \wedge y = M^+(x) \Rightarrow z = M^+(x) \tag{7}$$

$$z = M^+(y + x) \Rightarrow z = M^+(y) \oplus M^+(x) \tag{8}$$

The notation '$\oplus$' means qualitative addition while '$+$' means exact addition. Similar relations hold for the combinations of $M^-, -$ and $\ominus$.

### Variable Identification by Monotonic Relations

When a monotonic relation $y = M^+(x)$ exists, we can identify a variable $x$ as $y$, with regard to the direction of variable change, i.e., the sign of time-derivate, iff there exists no other monotonic relation flowing into $y$, i.e.,

$$\neg \exists z\,(\,y = M(z)\,), \tag{9}$$

where $M$ is any of $M^+, M^-, M_0^+$, or $M_0^-$. Similarly, two variables connected by $M^-$ can be identified iff (9) holds, although the direction of change is opposite.

Two variables connected by $M_0^+$ or $M_0^-$ can be identified as a same qualitative variable in $\{+, 0, -\}$, although both the direction of change and the value are opposite in the case of $M_0^-$.

### Loop Including Monotonic Relations

The important issue is a loop which consists of both (direct) influences and monotonic relations. If there is no monotonic relation in the loop after variable identification by monotonic relations, the loop is handled in the way discussed in the previous sections. In this case, the sign of loop constant should be changed according to the number of monotonic relations $M^-, M_0^-$ appearing in the original loop.

When we cannot eliminate all monotonic relations in a loop, the behavior of the loop is separated into cases depending on the status of variables flowing into the loop.

## 7 Applications

We have applied our method to examples in the literature and found that many of the examples have small numbers of state variables in essence, e.g., from one to three variables. The method captures the characteristics of this class of the systems in lower dimensions which qualitative simulations can handle with less difficulty. In this section, we address three examples to show the ability of our method.

## 7.1 U-Tube

The first example is U-tube [Crawford et al., 1990]. Two tanks A and B are connected by a liquid flow path at the bottom of them [2]. In the initial state, tank A is filled with water and tank B is empty.

The system has six state variables (pressure at the bottom of tank $p_a, p_b$; mass in tank $m_a, m_b$; flow rate $r$; pressure difference $p_d$). The causal graph of the system is as follows.

$$p_a = M_0^+(m_a),\ p_b = M_0^+(m_b),$$
$$m_a = I^-(r),\ m_b = I^+(r),$$
$$r = M_0^+(p_d),\ p_d = p_a - p_b. \qquad (10)$$

After applying variable-identification rules, we obtain the following qualitative equation which consists of only two variables $p_a$ and $p_b$ ($\mathbf{x} = (p_a, p_b)^t$).

$$d\mathbf{x}/dt = A\mathbf{x},\ \mathbf{x} \in R^2,\ A = \begin{pmatrix} -a_1 & a_1 \\ a_2 & -a_2 \end{pmatrix},$$
$$a_1 > 0,\ a_2 > 0. \qquad (11)$$

The outline of reasoning process about the behaviors of this system is as follows. Because the system has a positive loop of length two, the convergence of short loops does not hold (theorem 3). Although all eigenvalues exist in two circles which are placed in the negative half-plane, the circles also include the origin $(0,0)^t$ (theorem 4). The convergence by self-feedbacks does not hold in the case where some eigenvalues exist on the origin, i.e., they are zero-eigenvalues (theorem 5).

Actually, this system is a special one which has one zero-eigenvalue, since 1) the second row is equal to $-a_2/a_1$ times the first row (linearly dependent), 2) the number of linearly independent rows is 1, and consequently, 3) the rank of $A$ is 1.

Non-trivial solution of $A = 0$ is $(k,k)^t$ where $k$ is an arbitrary real number, and the direction vector of equilibrium line is $(1,1)^t$. Because the dimension is lower, direct computation of another eigenvalue is possible. However, we can reason about it in a qualitative way, i.e., 1) another eigenvalue is a real number, but it is not zero because the rank of $A$ is 1, 2) all eigenvalues exist in the negative half-plane, and consequently, 3) it should be a negative real number.

By these reasoning results, the phase space is obtained as shown in Figure 2 (a). From any initial point, the trajectory will reach the equilibrium line and stop on it.

When a small perturbation is added to the system, this equilibrium line will transit to a nodal repellor or attractor. For instance, it will transit to a nodal attractor in Figure 2 (b), when a perturbation $\delta_1 > 0, \delta_2 > 0$ is added to $A$ as follows.

$$A = \begin{pmatrix} -a_1 & a_1 - \delta_1 \\ a_2 - \delta_2 & -a_2 \end{pmatrix}.$$

---

[2] We ignore the portal which is included in the original problem to clarify the essential points.

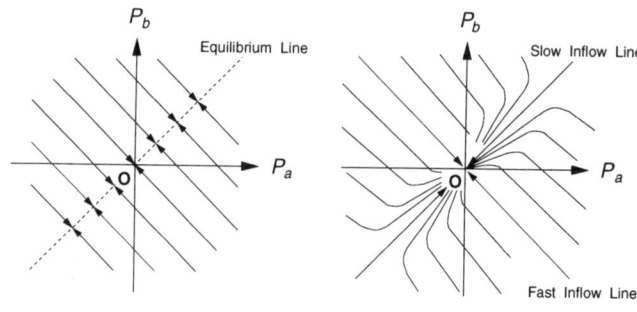

(a) Equilibrium Line.   (b) Nodal Attractor.

Figure 2: Phase space for U-tube behavior.

This perturbation physically means that a small leak exists in the liquid flow path. It is easy to know that both eigenvalues are negative real numbers. When the perturbation is relatively small, the trajectory will first approach the slow inflow line which the original equilibrium line above has changed into, then move along the slow inflow line, and finally reach the equilibrium point.

Relatively small perturbation in this case means that two negative eigenvalues have relatively different absolute values. Iwasaki proposed a method to analyze the system which has relatively different eigenvalues [Iwasaki and Bhandari, 1988], although the method is valid only for an attractor and its main purpose is to decompose a large-scale system into sub-components.

This kind of 'equilibrium type-transition' discussed here cannot be reasoned by comparative analysis [Weld, 1987], nor perturbation analysis [De Mori and Prager, 1989] [Rose and Kramer, 1991], since these methods are valid for the position move of an equilibrium point within a certain equilibrium type.

In the original problem [Crawford et al., 1990], the flow rate is assumed to be equal to the pressure difference as shown in Equation (10). This is the reason why the trajectory approaches the equilibrium state without oscillation. In real physical systems, however, the flow rate is not always equal to the pressure difference because of inertia of fluid and friction, which causes a damped oscillation before reaching the equilibrium state.

## 7.2 Damped Oscillation

The second example is an oscillator governed by an equation:

$$\ddot{y} + \mu \dot{y} + \omega^2 y = 0.$$

Letting $x_1 = y$, $x_2 = \dot{y}$, $\mathbf{x} = (x_1, x_2)^t$, we obtain the following equation.

$$d\mathbf{x}/dt = A\mathbf{x},\ \mathbf{x} \in R^2,\ A = \begin{pmatrix} 0 & 1 \\ -\omega^2 & -\mu \end{pmatrix}.$$

When $\mu = 0$, the system is an oscillator without damp. In this case, the system is a negative loop of length two,

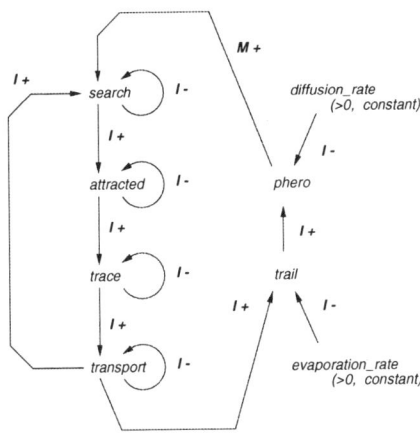

Figure 3: A Causal Graph of Ant Colony's Behavior.

whose equilibrium type is center (theorem 2). The trajectory is a closed cycle around the origin (eternal periodic oscillation).

Notice that two eigenvalues consist of a pair of conjugate purely imaginary numbers $\pm \alpha i$, and that $Tr(A)$ is zero. This non-hyperbolic equilibrium point changes its type into spiral attractor or repellor, when a little perturbation is added to the system.

When $\mu > 0$, the system is called damped oscillator. In such case, the eigenpolynomial $(\lambda^2 + \alpha^2)$, which corresponds to conjugate purely imaginary numbers, becomes $(\lambda^2 + \mu\lambda + \alpha^2)$, because the trace of $A$ becomes $-\mu$. Consequently, the real part of the eigenvalues becomes negative, i.e., the equilibrium type is spiral attractor. The trajectory approaches the origin while revolving around it.

### 7.3 Ant Colony's Macro-Behavior

The third example is a foraging behavior of an ant colony at macro-level [Kurumatani and Nakamura, 1996] [Kurumatani, 1995]. It is a coordinated group behavior achieved by a simple behavior of individual ants with a communication method using chemical material.

The key mechanism of the system is that 1) an ant changes its behavior-mode according to the internal/external situation, 2) an ant which transports bait puts chemical material called *trail* on the ground, 3) the trail evaporates and becomes *pheromone*, and 4) diffused pheromone induces other ants to gather at the bait-place.

The system has the state variables for the number of ants in each mode and the amounts of trail and pheromone. The causal graph is shown in Figure 3.

There are two loops and six variables in the graph. Although four variables have self-feedbacks, the rest do not. The trajectory is sure to be repelled in the direction of $(+, +, \ldots, +)^t$, if influences on the variables *phero* and *trail* by other variables are greater than ones by two constants.

This system is purely qualitative one, because there is no way to know the exact numeric coefficients. Our method, however, captures the characteristics of the behavior which is empirically analyzed [Nakamura and Kurumatani, 1996], where the positive influence loop really works.

## 8 Non-Linearity in Causal Graphs

Our method is based on the characteristics of linear dynamical systems. It cannot be applied directly to general non-linear systems.

However, the results are valid for the systems described by 'semi-linear' functions, e.g., influence, monotonic, addition, and subtraction, since these semi-linear functions do not change the type of equilibrium and the characteristics of equilibrium neighborhood discussed in this paper can be extended globally to the whole phase space, although trajectories might be curved by non-linearity of the functions. The examples in the previous section belong to this class.

It seems difficult to have purely qualitative methods of analyzing global topological characteristics of general non-linear systems. When a non-linear system is given, it is possible to qualitatively analyze each semi-linear segment of a phase space divided by piece-wise approximation and to connect them into a whole qualitative phase portrait. Methods in this direction of analyzing phase space [Sacks, 1987] [Yip, 1988] [Lee and Kuipers, 1993] and constructing the whole map [Nishida and Doshita, 1991] [Zhao, 1994] have been proposed. Applying our method in this direction is left for future work.

## 9 Conclusion

We have proposed a method to qualitatively analyze the characteristics of causal graphs without simulations.

The method consists of 1) pre-processing monotonic relations in a graph and reformulating it into a qualitative dynamical system, 2) analyzing the coefficient matrix and computing the type of equilibrium points, and 3) computing the conditions when equilibrium type-transitions occur.

The method can be regarded as direct qualitative analysis of equilibrium points and their neighborhoods without numerical information. It has the advantage of computing the conditions under which a certain equilibrium type holds and when equilibrium type-transitions occur.

Future work includes applications of our theory to global topology analysis of non-linear systems and to automatic categorization of equilibrium type-transition in such systems.

## Acknowledgement

A part of this work was carried out when the first author was staying at AI Lab. (LIA), Swiss Federal Institute of Technology, Lausanne (EPFL). We would like to thank Prof. Faltings and the members of LIA for fruitful discussions.

# References

[Chatelin, 1988] F. Chatelin. *Valeurs propres de matrices* (in French). Masson (Paris), 1988.

[Clancy and Kuipers, 1992] D. J. Clancy and B. J. Kuipers. Aggregating behaviors and tractable simulation. In *AAAI Design from Physical Principles Fall Symposium Working Notes*, pages 38–43, 1992.

[Crawford et al., 1990] J. Crawford, A. Farquhar, and B. J. Kuipers. QPC: A compiler from physical models into qualitative differential equations. In *Proc. of AAAI-90*, pages 365–372, 1990.

[De Mori and Prager, 1989] R. De Mori and R. Prager. Perturbation analysis with qualitative models. In *Proc. of IJCAI-89*, pages 1180–1186, 1989.

[Forbus, 1984] K. D. Forbus. Qualitative process theory. *Artificial Intelligence*, 24:85–168, 1984.

[Fouché and Kuipers, 1992] P. Fouché and B. J. Kuipers. Reasoning about energy in qualitative simulation. In *IEEE Transactions on Systems, Man and Cybernetics 22*, pages 47–63, 1992.

[Guckenheimer and Holmes, 1990] John Guckenheimer and Philip Holmes. *Nonlinear Oscillations, Dynamical Systems, and Bifurcations of Vector Fields (Applied Mathematical Sciences Vol. 42)*. Springer-Verlag, 1990.

[Ishida, 1989] Y. Ishida. Using global properties for qualitative reasoning: A qualitative system theory. In *Proc. of IJCAI-89*, pages 1174–1179, 1989.

[Iwasaki and Bhandari, 1988] Y. Iwasaki and I. Bhandari. Formal basis for commonsense abstraction of dynamic systems. In *Proc. of AAAI-88*, pages 307–312, 1988.

[Kuipers, 1986] B. J. Kuipers. Qualitative simulation. *Artificial Intelligence*, 29(3):289–338, 1986.

[Kurumatani and Nakamura, 1996] K. Kurumatani and M. Nakamura. Generating qualitative equations about macro-behaviors of foraging in ant colony. In *Proc. of German Conference on Bioinformatics, GCB'96* (Leipzig), pages 142–147, 1996.

[Kurumatani, 1995] K. Kurumatani. Generating causal networks for mobile multi-agent systems with qualitative regions. In *Proc. of IJCAI-95*, pages 1750–1756, 1995.

[Lee and Kuipers, 1993] W. Lee and B. J. Kuipers. A qualitative method to construct phase portraits. In *Proc. of AAAI-93*, pages 614–619, 1993.

[Nakamura and Kurumatani, 1996] M. Nakamura and K. Kurumatani. Formation mechanism of pheromone pattern and control of foraging behavior in an ant colony model. In *Proc. of Fifth Int'l Conf. on Artificial Life, ALife V* (Nara), pages 48–56, 1996.

[Nishida and Doshita, 1991] T. Nishida and S. Doshita. A geometric approach to total envisioning. In *Proc. of IJCAI'91*, pages 1150–1155, 1991.

[Rose and Kramer, 1991] P. Rose and M. Kramer. Qualitative analysis of causal feedback. In *Proc. of AAAI-91*, pages 817–823, 1991.

[Sacks, 1987] E. Sacks. Piecewise linear reasoning. In *Proc. of AAAI-87*, pages 655–659, 1987.

[Weld, 1987] D. S. Weld. Comparative analysis. In *Proc. of IJCAI-87*, pages 959–965, 1987.

[Yip, 1988] K. Yip. Generating global behaviors using deep knowledge of local dynamics. In *Proc. of AAAI-88*, pages 280–285, 1988.

[Zhao, 1994] F. Zhao. Extracting and representing qualitative behaviors of complex systems in phase space. *Artificial Intelligence*, 69:51–92, 1994.

# Action Localness, Genericity and Invariants in STRIPS

**Norman Y. Foo** and **Abhaya Nayak** and **Maurice Pagnucco**
Knowledge Systems Group
Department of Artificial Intelligence
School of Computer Science and Engineering
The University of New South Wales
NSW 2052 Australia

**Pavlos Peppas**
Knowledge Systems Group
School of MPCE
Macquarie University
NSW 2109, Australia

**Yan Zhang**
Knowledge Systems Group
Department of Computing
University of Western Sydney, Nepean
Kingswood, NSW 2747, Australia

## Abstract

The STRIPS specification for actions, despite criticisms of its shortcomings, remains popular in implementations because of its simplicity and intuitive appeal. Lifschitz showed in his logical account of STRIPS that states can safely be represented as sets of ground atom, and this is our starting point. Our aim is to provide logical foundations for design heuristics and procedural implementations of STRIPS. In particular, we formalize the notions of 'localness' in pre- and post-conditions of actions. This justifies and explicates the intuitions that underlie the diagrammatic reasoning typically used to design such conditions. A formal account is also given for 'generic' actions that have parameters. We conclude with a method for extracting candidate action invariants that distinguishes between two kinds of inertia implied by the action specifications.

## 1 Introduction

Among the earliest proposals for specifying actions is STRIPS, due to Fikes and Nilsson [Fikes and Nilsson 71]. STRIPS has the great merit of simplicity and ease of implementation. As an example and a quick review of it, here is how the action of picking up block A in a blocks world may be specified:

*action: pick-up(A)*
*pre-condition: on(A,X), clear(A)*
*post-condition:*
  *add-list: inhand(A)*
  *delete-list: on(A,X)*

In a Prolog setting, this is easily implemented by the rule: *pick-up(A) :- on(A,X), clear(A), retract(on(A,X)), assert(inhand(A))* assuming that the state of the blocks-world is represented as a collection of Prolog facts. An immediate observation is that both the specification and the implementation describe a *generic* action that has A as a parameter, and that X is implicitly existentially quantified. Even this example however is not as general as STRIPS can be, for it is possible to have first-order formulas in the pre-condition and post-condition. However, unless restrictions are observed, the inclusion of such formulas, as pointed out by Lifschitz [Lifschitz 86] can cause difficulties. Moreover, until that paper appeared, there was no proper logical semantics for STRIPS. Further, constraints or invariants (like "no block can be both inhand and on something") are observed either (a) by specification or programming of the add-list and delete-list to ensure non-violation, or (b) by Prolog rules of the form: *impossible :- inhand(A), on(A,X)* where *not(impossible)* is used as a guard in the body of action rules.

A logical semantics of STRIPS should build upon the insights of Lifschitz and also give an account of several other characteristics. Lifschitz showed that if states of the world are represented as sets of ground atoms, then his semantics of STRIPS is safe, and we will use this as a starting point because we are principally interested in understanding the efficacy of Prolog-like implementations that are based on such state representation. Designers of rules for such programs often draw pictures like that shown in figures 1 and 2 to reason about pre- and post-conditions, and to maintain constraints implicitly. The most striking characteristic of such pictures is the identification of 'localness' and 'genericity'. Localness has to do with the fact that an informal 'window' (see the dashed enclosures in the figures) is typically drawn around the objects relevant to the action instance. Genericity is the notion that each window to which an action can apply looks like any other similar window except for a change in parameters. Another characteristic, noted by Lifschitz [Lifschitz 86] is that the constraints like the *impossible* rule above are often maintained implicitly by clever design of the add- and delete-lists. In this paper we address these and related characteristics to provide a formal explication for why they work.

The organization of the paper is as follows. In section 2 we recapitulate a number of standard definitions, introduce a formal counterpart of the 'window' idea and show how it accounts for pre-condition 'localness'. We also provide the formal equivalent of 'genericity' for actions. In section 3 we examine state change due to actions and show that post-condition 'windows' have the same boundary as the pre-condition ones. We also identify two kinds of inertial assumptions that appear to be implicitly used by designers of action specifications. Section 4 considers intended design semantics in the add- and delete-lists, and shows how candidate action invariants can be derived from them depending on two kinds of inertial assumptions. Concluding the paper is section 5 which discusses what we are currently pursuing and what has yet to be done.

## 2 States, Actions and Windows

States of the world are describable formally in many ways, but a popular one is to represent them as collections of ground atoms. This is a particularly felicitous choice for databases, and for implementations in languages like Prolog. In this paper we accept this choice as it will permit us to focus on the issues that really matter without being distracted by technicalities that may attend alternative choices. It has the virtue of simplicity and is sanctioned by earlier investigations by Lifschitz. To this end therefore, we will adopt a first-order language with equality as our descriptive medium. Section 2.1 recalls the logic and model theory needed.

Actions are state transformers. In the STRIPS tradition, for an action to apply in a state its pre-condition must be satisfied by that state. Actions that are specified with parameters or variables have instances that are the result of instantiating or substituting constant names of objects in the state into the action pre-condition. If successfully applied, the post-conditions of the action are actuated, i.e., the add-list and delete-list operations are performed on the representation. If states are theories, then actions so modelled are theory change operators. Such operators have been studied extensively in the logics of action from many perspectives (an example is [Lin 95]), but it is not the purpose of this paper to introduce yet another. Rather, we wish to examine a well-known system that has wide practical acceptance to see if its design heuristics, particularly those with a diagrammatic flavor such as expositied by Barwise [Barwise 95], can be given formal justifications.

### 2.1 Basic Definitions and Constructions

Since ground atoms are the basis of state descriptions, in order to evaluate formulas with respect to states we need to know when an atom does not hold. The simplest assumption to achieve this is to use the version of the *closed world assumption* [Reiter 78] in the absence of constraints.

**Definition 1** *If $\Delta$ is a set of ground atoms of $\mathcal{L}$, then the closed world assumption applied to $\Delta$ is $CWA(\Delta) = \Delta \cup \{\neg L | L \notin \Delta$ and $L$ is a ground atom$\}$.*

This is actually a simplification of the original definition which involves state constraints, but here none are assumed to be (explicitly) available. A potential problem with this assumption is inconsistency when the description is *indefinite* [Zhang and Foo 96], e.g. there is a constraint of the form $p \vee q$ where $p$ and $q$ are atoms, and neither of them appear in the state description. The indefiniteness of actions in terms of their pre- or post-conditions is known to be problematic [Zhang and Foo 96]. For pre-conditions it is possible to avoid it by refining the action ontology to "split" the indefinite action into definite ones, but for post-conditions no such easy way is known. To circumvent this for the purpose of focusing on the main ideas, we shall therefore assume that all descriptions for states as well as for actions are *definite*. Then, using the closed world assumption, all states are *ground complete*, i.e., for each ground atom, either it or its negation holds in a state.

Formally, complete states are *structures* in model theory and logic, so we shall use the terms *state* and *structure* interchangeably. The next three definitions are paraphrased from [Enderton 72; Chang and Keisler 73]. (As it gets tedious to make distinctions between predicate (constant) symbols and the relations (domain elements) they denote, we will not observe these distinctions later.)

**Definition 2** *Suppose $\mathcal{L}$ is a first-order language with equality. An $\mathcal{L}$-structure $\mathcal{U}$ is a triple $\langle \mathcal{A}, \mathcal{R}, \mathcal{C} \rangle$ where $\mathcal{A}$ is a non-empty set called the domain of $\mathcal{U}$, $\mathcal{R}$ is a map that assigns relations of the appropriate arity on $\mathcal{A}$ to predicate symbols in $\mathcal{L}$, and $\mathcal{C}$ is a map that assigns elements of $\mathcal{A}$ to constant symbols in $\mathcal{L}$.*

**Definition 3** *If $\mathcal{U}_1$ and $\mathcal{U}_2$ are $\mathcal{L}$ structures with domains $\mathcal{A}_1$ and $\mathcal{A}_2$ respectively such that $\mathcal{A}_1 \subseteq \mathcal{A}_2$, and the relation and constant assignment maps of $\mathcal{U}_1$ are that of $\mathcal{U}_2$ restricted to $\mathcal{A}_1$, then $\mathcal{U}_1$ is a substructure of $\mathcal{U}_2$, and conversely $\mathcal{U}_2$ is an extension of $\mathcal{U}_1$.*

**Definition 4** *Two $\mathcal{L}$ structures $\mathcal{U}_1$ and $\mathcal{U}_2$ are elementarily equivalent, written $\mathcal{U}_1 \equiv \mathcal{U}_2$, if for every sentence $\phi \in \mathcal{L}$, $\mathcal{U}_1 \models \phi$ iff $\mathcal{U}_2 \models \phi$.*

Two elementarily equivalent structures satisfy exactly the same set of sentences, and are therefore indistinguishable by sentences. To look ahead a little, our interest in substructures arises from the intuition that they are the 'windows' within pictures of states to which actions are locally applied. The construction of these windows is determined by the arguments or parameters of the action pre-condition, and the objects of relevance are accumulated via a closure operation. This closure has the effect of accruing, to the constants that instantiate a pre-condition, other constants that may be needed to evaluate predicates in the pre-condition. For instance, if $\forall X P(X, b)$ is a component of a pre-condition and

$b$ is a constant instantiated into it as an action parameter, we would need to accrue all constants $c$ in the (domain of the) structure for which $P(c, b)$ holds in order to evaluate the component. This motivates the closure definition below.

**Definition 5** *Given a set $S$ of constants in $\mathcal{U}$, the expansion of $S$ by n-ary predicate $P$ in $\mathcal{U}$, denoted $closure(\mathcal{U}, S, P)$, is $\bigcup\{\{a_1, \ldots, a_n\} \mid$ some $a_i$ is in $S$, and $\mathcal{U} \models P(a_1, \ldots, a_n)\}$. The notation $closure(\mathcal{U}, S, P_1, \ldots, P_k)$ has the obvious generalization to several predicates.*

In fact, this closure is a special case of a more general closure which is suggested by the notation $closure(\mathcal{U}, S, P, k)$ where $k$ is the level of transitivity of $P$ used in the expansion of $S$, e.g., if $a \in S$ and $P$ is binary with $P(a, b)$ and $P(b, c)$ holding in $\mathcal{U}$ (while neither $b$ nor $c$ is in $S$), both $b$ and $c$ will be in $closure(\mathcal{U}, S, P, 2)$. Our closure is for $k = 1$, and we confine attention to this case simply because the key ideas in this paper are thereby not obscured by technical detail, and moreover for most practical applications $k = 1$ suffices. The generalization to arbitrary $k$ is conceptually easy. The construction of the closure implies the following lemma.

**Lemma 1** *Assume the notation as in definition 5. If $\mathcal{U} \models Q(b, X_1, \ldots, X_k, a_1, \ldots, a_m)$ for all $b$ in $closure(\mathcal{U}, \{a_1, \ldots, a_m\}, Q)$ then $\mathcal{U} \models \forall Y Q(Y, X_1, \ldots, X_k, a_1, \ldots, a_m)$.*

This lemma is the basis for an induction that establishes the corresponding lemma in which $Q$ above is replaced by general formulas called *D-formulas* in definition 6 below, where $D$ is $\{a_1, \ldots, a_m\}$ in the present context.

## 2.2 Blocks World Example

The blocks-world structure (state) illustrated by figures 1 and 2 will motivate much of the later development. All of figure 1 shows a world $W$ in which $\mathcal{L}$ has predicate symbol $on(\cdot, \cdot)$, constant symbols $\{a, b, c, d, e, table\}$. Letting the constant symbols name their domain elements and the predicate symbol name its relation, the theory $T_2$ (subtheory of $T_1$, to be introduced shortly) of the structure is axiomatizable by the set $\{on(a, b), on(c, d), on(e, table), on(b, table), on(d, table)\}$. The dashed part of figure 1 shows a substructure of $T_2$ with the theory $T_1$ axiomatizable by $\{on(a, b), on(b, table), on(e, table)\}$. Already, if one accepts that the dashed part is a sub-structure of the entire structure, from well-known *preservation theorems* [Chang and Keisler 73] we can conclude the following: the universal sentence $\forall x \neg on(x, a)$ holds in $T_2$, so by the fact above, it also holds in $T_1$; the (trivially) existential sentence $on(a, b)$ holds in $T_1$, so it also holds in $T_2$. However, it will turn out that these conclusions from standard model theory are special cases of the main theorem of this section. This example is much simplified, for in practice we may need more predicates than shown, e.g., *clear, block* etc. By the CWA the negative ground literals are implied and not shown. As a result, the formula $\forall X \neg on(X, a)$ holds in this structure. Observe that this is the intended meaning of

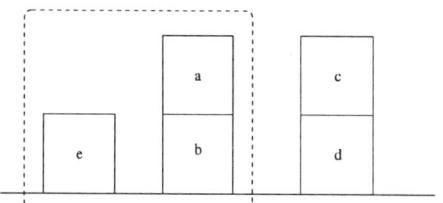

Figure 1: A pre-condition window

the predicate $clear(a)$ in practice, so $clear(\cdot)$ is *eliminable* in a language with the $on(\cdot, \cdot)$ predicate.

## 2.3 What Is A Window?

We are now ready to formalize the informal notion of a 'window' for an action instance. Let the pre-condition of action $A$ be $\phi(X_1, \ldots, X_n)$. If $A$ applies in state $\mathcal{U}$ with substitutions $a_1, \ldots, a_n$ for $X_1, \ldots, X_n$ respectively, construct a sub-structure $\mathcal{W}$ of $\mathcal{U}$ by letting the domain of $\mathcal{W}$ be $closure(\mathcal{U}, \{a_1, \ldots, a_n\}, P_1, \ldots, P_k)$ where $P_1, \ldots, P_k$ are the predicates that occur in $\phi$. In order to emphasize the origins of $\mathcal{W}$, it is convenient to also write it as $window(\phi[a_1, \ldots, a_n], \mathcal{U})$ or alternatively by $window(A[a_1, \ldots, a_n], \mathcal{U})$, being careful to remember that the domain of this structure includes constants beyond $a_1, \ldots, a_n$ by virtue of closure.

**Example 1** *Suppose the pre-condition of action $move(A, B)$ in the blocks world is $\forall X \neg on(X, A) \land \forall X \neg on(X, B) \land on(A, Y) \land on(B, Z)$ and it is applied to the structure $\mathcal{U}$ in figure 1 with the substitutions $A/a, B/e$, then the substructure $\mathcal{W}$ will have closure domain $\{a, e, b, table\}$, and relations restricted to these.*

This example explains the motivation for the name *window*. In figure 1 the objects in the dashed line box constitute the domain of $\mathcal{W}$. This window of the action $move(a, e)$ encloses all the objects that could possibly be relevant in a test of the applicability of this action. The extent of the window is determined here by the fact that its pre-condition is a $\{a, e\}$-formula, as specified in the definition below.

**Definition 6** *Given a set $D$ of domain constants in $\mathcal{L}$, the D-formulas are (partly) grounded formulas in which each predicate must have at least one argument that is from $D$.*

It appears that most practical pre-conditions of actions have as instances *D-formulas* where $D$ is the set of substituted constants in the particular instance of the action. For example, in the instance of the action $move(a, e)$, the pre-condition $\forall X \neg on(X, a) \land \forall X \neg on(X, e) \land on(a, Y) \land on(b, Z)$ is a $\{a, e\}$-formula. Note that this definition effectively enforces 1-transitivity in $P$. A 2-transitive formula would have (for a binary $P$) a subformula instance of the form $P(a, X) \land P(X, Y)$—or some such permutation thereof—where $a$ is in $D$, and unless $Y$ is instantiated to some $b$ in $D$, this subformula violates the definition of a $D$ formula since $P(X, Y)$ does not have a $D$ constant. However, if indeed $Y$ is some such $b$, the subformula reads $P(a, X) \wedge P(X, b)$ and is 1-transitive.

So, because of the above reasons, we shall *assume* the following:

**Assumption 1** (D-formula Assumption) *If $\phi$ is a pre-condition of action $A[a_1, \ldots, a_n]$, then $\phi[a_1, \ldots, a_n]$ is a $\{a_1, \ldots, a_n\}$-formula.*

The formal reason for why windows work in practice is provided by the next theorem.

**Theorem 1** (Pre-condition Conservative Extension) *Let $\phi[a_1, \ldots, a_n]$ be a pre-condition of an action instance to be applied to state $\mathcal{U}$. If $\alpha$ is any $\{a_1, \ldots, a_n\}$-formula, then $\mathcal{U} \models \alpha$ iff $window(\phi[a_1, \ldots, a_n], \mathcal{U}) \models \alpha$.*

**Proof:** We proceed by induction on the complexity of $\alpha$. For atomic $\alpha$ it suffices to consider by example the case of $P(a_1, X)$ as more general cases follow the same pattern. $\mathcal{U} \models P(a_1, X)$ iff $\mathcal{U} \models \forall X P(a_1, X)$ iff for all $b$ in the domain of $\mathcal{U}$, $\mathcal{U} \models P(a_1, b)$. By the construction of $window(\phi[a_1, \ldots, a_n], \mathcal{U})$ and lemma 1 the last assertion is equivalent to $window(\phi[a_1, \ldots, a_n], \mathcal{U}) \models P(a_1, X)$. The inductive steps in the proof of the equivalence is straightforward for the $\wedge$ connective. For the $\forall$ connective the inductive proof has the same structure as the atomic case. Finally $\mathcal{U} \models \neg \alpha$ iff $\mathcal{U} \not\models \alpha$ iff (by the inductive hypothesis) $window(\phi[a_1, \ldots, a_n], \mathcal{U}) \not\models \alpha$ iff $window(\phi[a_1, \ldots, a_n], \mathcal{U}) \models \neg \alpha$. □

In other words, the theory of $\mathcal{U}$ is a *conservative extension* [Chang and Keisler 73; Enderton 72] of that of $window(\phi[a_1, \ldots, a_n], \mathcal{U})$ for $\{a_1, \ldots, a_n\}$-formulas. This theorem is the essence of why the pictorial idea of a 'window' cut out of a picture of a state is intuitively appealing. We claim that it captures the notion of 'localness' in such pictures. The theorem states that an instantiated pre-condition is satisfied by the objects in the window if and only if it is satisfied by the entire picture.

## 3 State Change

Consider now figure 2 which intuitively represents the resultant state of the $move(a, e)$ action applied to the state in figure 1. Note that the window boundary is unchanged from figure 1. This is another characteristic of 'localness' that is captured. We now turn to a formal account of these observations based on the post-conditions of the action instance. Our first assumption is that all the objects and predicates of direct interest in the post-condition are already named in the pre-condition, i.e.,

**Assumption 2**
(Post-condition Assumption) *If $A[a_1, \ldots, a_n]$ is an action instance with pre-condition $\phi[a_1, \ldots, a_n]$, then all constants of this instance in the post-condition are among the $a_1, \ldots, a_n$, and all the predicates in the post-condition are among those of the pre-condition.*

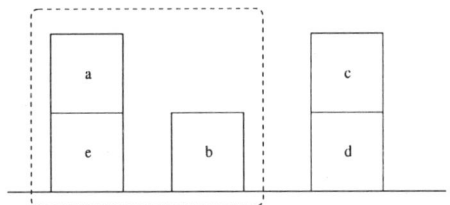

Figure 2: A post-condition window

The next assumption is the interpretation of what it means to add and delete atoms in the corresponding lists. Lifschitz [Lifschitz 86] proposes the very reasonable one that, for atomic add and delete lists, amounts to the:

**Assumption 3** (Inertia Assumption) *Let action instance $A$ transform state (structure) $\mathcal{U}$ to state (structure) $\mathcal{U}_1$. Then (i) $\mathcal{U} \models P$ implies $\mathcal{U}_1 \models P$ for any atom $P$ not in the delete-list of $A$ — this is the idea of inertia or persistence by default; (ii) if $Q$ is an atom in the add-list, then $\mathcal{U}_1 \models Q$.*

In practice, the second condition seems to apply by default to atoms $Q$ already known to be such that $\mathcal{U} \models \neg Q$. If the Inertia Assumption is understood to dominate pre-conditions to the extent that no atom that persists can be in the add-list (this is not required by the semantics of Lifschitz [Lifschitz 86]), and no atom that does not hold in $\mathcal{U}$ and $\mathcal{U}_1$ is in the delete-list, then we have:

**Assumption 4** (Strong Inertia Assumption) *This is as in the Inertia Assumption, except that in (i) we add the qualifier, "and only for atoms $P$ not in the delete-list", and in (ii) we strengthen the consequent to "$\mathcal{U}_1 \models Q$ and $\mathcal{U} \not\models Q$".*

For strong inertia then, nothing is in the add- or delete-list unless it has changed signs in the course of the action. Strong inertia was implicitly used by Reiter in [Reiter 91] in his formulas $\gamma_Q^+$ and $\gamma_P^-$ which combined explanation closure (due to Schubert) and result closure (due to Pednault) in his solution to the frame problem. It is tempting to think that this should be the natural default for designers of specifications. While inertia is a reasonable guide to the design of action specifications in practice, we shall see an example in section 4.2 that strong inertia does not appear to be the intended designer semantics in some STRIPS specifications, i.e. that sometimes (weak) inertia has significance.

For non-atomic add- and delete-lists, we might as well go the whole hog and consider post-conditions which are formulas. Bearing in mind the cautionary statements from [Lifschitz 86], it is prudent to consider only such formulas that are *definite* so that they in effect define a set of ground literals, whose predicate names and constants are subject to the Post-condition Assumption above. Then we apply the Inertia Assumption as before.

The result is that the Post-condition Assumption coupled with these remarks implies:

**Theorem 2** (Post-condition Conservative Extension) *If $\phi$ and $\psi$ are the pre- and post-conditions of action instance $A$ respectively, then with the notation above (i) subject to*

the Post-condition Assumption the domain of $window(\psi, \mathcal{U}_1)$ is a subset of the domain of $window(\phi[a_1, \ldots, a_n], \mathcal{U})$; (ii) in addition, if subject to the Inertia Assumption, the Precondition Conservative Extension theorem also holds for $window(\psi, \mathcal{U}_1)$ with respect to $\mathcal{U}_1$.

Assertion (i) is trivial. Assertion (ii) follows from the Inertia Assumption and reasoning similar to the Pre-condition Conservative Extension theorem. Together, the two theorems confirm that the window construction captures the intuitive notion of 'localness'. The 'genericity' of actions is captured by the following observation (see [Enderton 72; Chang and Keisler 73] for the definition of isomorphism between structures) and the fact that isomorphic structures are elementarily equivalent (see definition 4).

**Observation 1** (Isomorphism of Windows) *Any two instances of an applicable action has isomorphic pre-condition, and correspondingly isomorphic post-condition, windows.*

## 4 Action Invariants

Zhang and Foo [Zhang and Foo 96] have used persistence to derive action invariants from definite pre- and post-conditions. Since the Inertia Assumption is accepted in our formalization, one should expect that a similar derivation is possible here. If so, it would give a formal account of how STRIPS rule designers, for reasons of implementation efficiency, avoid having to specify the *impossible* predicates as action guards by clever selection of add- and delete-lists.

**Definition 7** *An invariant of an action A is a formula $\beta$ s.t., for all instances of A, if $\beta$ holds in the state (structure) before A is applied, $\beta$ will hold in the resulting state (structure).*

In the context of several actions (not merely different instances) constraints are simply the set of invariants common to all of them. So, it suffices to search for candidate invariants for each action in order to discover system constraints. In practice, designers of action specifications know many constraints, but they may not be aware of others that are implicit in the action specifications. Conversely, known constraints can be used to specify conditions for actions. In this section we indicate how invariants depend on action specifications.

The notions of pre- and post-condition windows introduced in the earlier sections provide the justification for localized reasoning about action invariants. While this section reports preliminary work toward the elimination of the *impossible* rules, the ideas proposed are prompted by the preceding formal development. We are currently investigating systematic algorithms to fulfill this work.

### 4.1 Ontological Primacy of Atoms

In many knowledge representation systems there is an implicit priority given to those facts of the world that can be expressed *positively* as ground atoms. This is a *linguistic* convention seldom breached, and no doubt there are strong psychological and cognitive reasons for it. Not being experts in these disciplines, we adduce evidence for what might be called a principle of the *ontological primacy of atoms* by recounting some facts that are widely known in our AI and KR community. First, databases (both deductive and otherwise) record atoms as the primary data. So do pure logic programs. Herbrand models are standard in KR semantics, and many of the characterization theorems of non-monotonic reasoning systems are expressed in terms of properties of such models. The bias toward positivity appears to be widespread even in formal approaches.

We therefore advance the following hypothesis, and then formalize it in the next definition.

**Assumption 5** (Positive Literal Bias) *In an action specification the add-list of atoms are determinants of the delete-list of atoms in the presence of constraints.*

**Definition 8** *Let $C$ be a set of formulas that express the constraints of the system. A set $\Gamma$ of atoms determines another set $\Lambda$ if $C \cup \Gamma \models \Lambda$.*

The Positive Literal Bias hypothesis then translates to this formal proposal. Let $A$ be an action that is applied to state $\mathcal{U}$ with resultant state $\mathcal{U}_1$, and with add-list $\Gamma$ of atoms, delete-list $\Lambda$ of atoms. Let $C$ be (possibly unknown) system constraints. Then $C \cup \Gamma \models \neg \bigvee \Lambda$, which is equivalent to $C \models \bigwedge \Gamma \supset \neg \bigvee \Lambda$. The notation $\bigvee \Lambda$ means the disjunction of the atoms in the set $\Lambda$. This means that $\bigwedge \Gamma \supset \neg \bigvee \Lambda$ is a candidate invariant for the action A if its constants are generalized. This generalization is possible because of observation 1 on isomorphisms of windows.

Our proposal is in the spirit of Explanation Closure attributed to Schubert [Reiter 91]. Indeed the assertion $C \cup \Gamma \models \neg \bigvee \Lambda$ has the structure of an abduction of the negative literals in terms of the positive ones with respect to $C$.

### 4.2 Generating Candidate Invariants

Let us now apply the above proposal to a few examples. Several caveats will be revealed in the process, which is to be thought of as a procedure for generating plausible guesses at invariants which will be correct if the Positive Literal Bias hypothesis holds along with some weak requirements.

For the $move(a, e)$ action, the add-list is $on(a, e)$ and the delete-list is $on(a, b)$, so by the proposal we have the candidate invariant $on(a, e) \supset \neg on(a, b)$, which generalizes (using unique names) to $(X \neq Y) \supset (on(A, X) \supset \neg on(A, Y))$ which is just a fancy way of saying that no block is in different places, i.e., the Prolog rule: *impossible :- on(A,X), on(A,Y), X $\neq$ Y* used as a guard for checking the move action. However, the generalization implies that every block has been moved before, since the generalization relies on an isomorphism of result windows and these are the effects of the action. This motivates the following definition.

**Definition 9** *A ground atom P is actualised by an action A if it was false in some initial window satisfying the pre-condition of A but true in the resulting window.*

By earlier remarks, we can replace 'window' by 'state', and equivalently say $P$ satisfies the post-condition of $A$. Then the proposed candidate invariant is plausible for every $A$-actualised atom. The *soundness* of the proposed invariant is not in question for the following reasons. Let $P_1, \ldots, P_n$ be the add-list of an instance of action $A$, and $Q_1, \ldots, Q_m$ be the delete-list. Then in the resultant state it is the case that $\bigwedge_{i=1}^{i=n} P_i \wedge \bigwedge_{i=1}^{i=m} \neg Q_i$ and this implies the candidate invariant in our proposal, which is $\bigwedge_{i=1}^{i=n} P_i \supset \bigwedge_{i=1}^{i=m} \neg Q_i$. In fact, if we assume strong inertia as in assumption 4, then all of $P_1, \ldots, P_n$ would have been false in the initial state, so the conjunctive formula $\bigwedge_{i=1}^{i=n} P_i \wedge \bigwedge_{i=1}^{i=m} \neg Q_i$ is false in it, and cannot be an invariant, while the weaker formula in question holds in it. To confirm that this is actually an invariant in our formalization an induction is necessary, but in [Zhang and Foo 96] a persistence logic has been used to verify it.

One final remark: the *clear* predicate used in some implementations has not been used in the blocks-world example here as it is eliminable. For technical reasons our proposal requires non-eliminable predicates.

### 4.3 Weak Inertia

In this rather preliminary subsection we show an example where strong inertia is not the intended semantics of a STRIPS specification.

**Example 2** *Consider a duopoly law for the ownership of a television station. Current owners may sell to others, but at all times there must be two owners. Here is the STRIPS specification for the action buy(A,E) meaning informally that owner A is selling his share of the station to E.*

> action: buy(A,E)
> pre-condition: own(B,T), own(E,T), $A \neq E$
> post-condition:
>   add-list: own(A,T)
>   delete-list: own(E,T)

Please note the formal similarity of this situation with that of *move(a,e)* that we discussed earlier. In any instance of $buy(a,e)$, the window of the action will contain in addition to $own(a,t)$ an instance $own(b,t)$ which will persist through the action but is not in the add-list. However, the candidate invariant using strong inertia is formally similar to that of the *move* action in the blocks world, viz., $(A \neq B) \supset (own(A,T) \supset \neg own(B,T))$ which is too weak to prevent a monopoly.

Consider now how *weak inertia* may be used here. If we change the add-list to $own(A,T), own(B,T)$, a candidate invariant is $own(A,T) \wedge own(B,T) \wedge (E \neq A \wedge E \neq B \wedge A \neq B) \supset \neg own(E,T)$ which is intuitively what the specifier intended under the unique names assumption. Evidently, weak inertia has different designer semantics.

## 5 Conclusion

We have provided a formalizations of the ideas of 'localness', 'genericity', and 'implicit constraints' in STRIPS-like specifications of actions. The first two are very faithful to the intuitions of designers, but the third is not quite settled as it appears that two different kinds of inertia are assumed. The first two formalizations have potential uses in diagrammatic reasoning, while the third suggests opportunities for identifying the psychology of action specifications with appropriate model classes for which inductive proofs of candidate invariants can be avoided. There is an increasing interest in diagrammatic reasoning [Barwise 95] in AI and we shall be extending the window semantics in this direction. We are currently pursuing the identification of dynamic system classes in the spirit of Sandewall [Sandewall 94] to see how invariants can be extracted completely from specifications without induction. There appear to be at least three notions of completeness for invariants that can be formalised.

## References

[Barwise 95] Barwise, J. and Etchemendy, J., "Heterogeneous logic", in *Diagrammatic Reasoning: Cognitive and Computational Perspective*, eds. J. I. Glasgow, et al. MIT Press, 1995, 209-232.

[Chang and Keisler 73] Chang, C.C., and Keisler, H.J., *Model Theory*, North Holland, 1973.

[Enderton 72] Enderton, H., *A Mathematical Introduction to Logic*, Academic Press, 1972.

[Fikes and Nilsson 71] Fikes, R.E. and Nilsson, N.J., " STRIPS: A New Approach to the Application of Theorem Proving to Problem Solving", *Artificial Intelligence*, 2, 1971, 189-208.

[Lin 95] Lin, F., "Embracing Causality in Specifying the Indirect Effects of Actions", Proceedings of the 14th International Joint Conference in Artificial Intelligence, IJCAI-95, 1995.

[Lifschitz 86] Lifschitz, V., "On the Semantics of STRIPS", in *Reasoning about Actions and Plans*, ed. M. Georgeff and A. Lansky, Morgan Kaufmann, 1986, 1-9.

[Reiter 78] Reiter, R., "On Closed World Data Bases", in *Logic and Database*, ed., Gallaire, H. and Minker, J., Plenum Press, 1978, 55-76.

[Reiter 91] Reiter, R., "The Frame Problem in the Situation Calculus: A Simple Solution (Sometimes)and a Completeness Result for Goal Regression", in *Artificial Intelligence and Mathematical Theory of Computation*, ed. V. Lifschitz, Academic Press, 1991, 359-379.

[Sandewall 94] Sandewall, E., *Features and Fluents; the Representation of Knowledge about Dynamical Systems, Vol. I*, Oxford University Press, 1994.

[Zhang and Foo 96] Zhang, Y. and Foo, N., "Deriving Invariants and Constraints from Action Theories", *Fundamenta Informaticae*, 30:23-41, 1996.

# Causality, Constraints and the Indirect Effects of Actions

Hector Geffner*
Dpto. de Computación
Universidad Simón Bolívar
Aptdo. 89000, Caracas 1080-A
Venezuela

## Abstract

The goal of this paper is the formulation of a well-founded solution to the ramification problem: the problem of determining the indirect effects of an action. The solution relies on a model of causal reasoning based on two simple ideas: that the values that a variable may take are a *function* of the values of its causes, and that the *form* of that function is determined by the rules acting upon the variable. The model is related to models of causal reasoning based on Bayesian Networks and Structural Equation Models but does not appeal to probabilities or rankings of any sort. It also combines (propositional) action rules and causal rules in a clear manner, and provides a simple way for determining when 'derived information' can be assumed to persist.

## 1 Introduction

The goal of this paper is the formulation of a well-founded solution to the ramification problem: the problem of determining the indirect effects of an action. The solution follows recent approaches that give causality a prominent role [McCain & Turner, 1997; Lin, 1995] but relies on a novel model of causal reasoning that is based on two simple ideas; namely that the values that a variable may take are a *function* of the values of its causes, and that the *form* of that function is determined by the causal rules acting upon the variable. The model is closely related to models of causal reasoning based on Bayesian Networks (e.g., [Goldszmidt & Pearl, 1992; Darwiche & Pearl, 1994; Geffner, 1996]) and Structural Equation Models [Simon, 1953; Druzdzel & Simon, 1993; Pearl, 1996]) but does not appeal to probabilities or rankings of any sort. It also combines (propositional) action rules and causal rules in a clear manner, provides a simple way for determining when 'derived information' can be assumed to persist [Myers & Smith, 1988;

---

*Mailing address from US and Europe: Hector Geffner, Bamco CCS 144-00, P.O.BOX 02-5322, Miami Florida 33102-5322, USA. E-mail: hector@usb.ve.

Giunchiglia & Lifschitz, 1995], and provides another perspective on the distinction between 'qualification' and 'ramification' constraints [Lin & Reiter, 1994].

The paper starts with the description of the causal model in Section 2, and proceeds in Section 3 with the model of actions and time. The two models are integrated in Section 4 and two extensions are considered in Section 5. A brief summary closes the paper in Section 6.

## 2 Causality

The causal model provides a semantics for collections of *causal rules* defined over a simple language made up of two sets of symbols: *variables* and *values*. All atoms are of the from x = $x$, where x is a variable and $x$ is a value. The language is the closure of all such atoms under the standard propositional connectives.

Each variable x has also a *domain* $D_x$ of values. For example, the variable temperature may take values from the domain $D_{temp} = \{high, medium, low\}$, distance may take values from $D_{dist} = \{1, 2, 3, \ldots, 10\}$, rain may take values from the boolean domain $D_{rain} = \{\mathbf{t}, \mathbf{f}\}$, etc.

A *state* $s$ is an interpretation that assigns to each variable x a value $x$ from its domain. We call $x$ the denotation of x in $s$ and refer to it as $x^s$. Atoms x = $x$ are true in a state $s$ when $x^s = x$, conjunctions $A \wedge B$ are true in $s$ when both $A$ and $B$ are true in $s$, disjunctions $A \vee B$ are true in $s$ when $A$ or $B$ are true, etc. A state thus provides an interpretation for *all* formulas in the language, and is a *model* of a set of formulas when it satisfies (i.e., makes true) each formula in the set.

*Causal rules* are of the form $F \rightarrow A$ where $F$ is a formula and $A$ is an atom. Causal rules express *mechanisms* by which the truth of $F$ *normally* causes the truth of $A$ [Simon, 1953; Druzdzel & Simon, 1993; Pearl, 1996]. Causal rules in which the truth of $F$ *always* causes the truth of $A$ are written as $F \supset A$. We call the former, *default* causal rules and the latter *strict* causal rules.

For example, using the expressions $x$ and $\neg x$ to abbreviate atoms of the form x = $\mathbf{t}$ and x = $\mathbf{f}$ for *boolean* variables x, a situation in which a car may fail to start due a dead battery can be modeled as:

$$turn\_key \rightarrow car\_starts$$

$$dead\_battery \supset \neg car\_starts$$
$$lights \to dead\_battery$$

where the last rule says that a battery dies when the lights are left on for a long time.

The *semantic problem* in causal reasoning is the problem of characterizing in a meaningful way the consequences that follow from a set $C$ of causal rules possibly extended with a set $O$ of observations (atoms). While the *models* of the resulting *causal theories* $T = \langle C, O \rangle$ are the states $s$ that satisfy the strict rules in $C$ and the formulas in $O$, not all such models can be considered. Defeasible causal rules in $C$ must play a role as well yet it hasn't been clear what such role actually is.

For instance, the simplest idea of focusing on the models that violate *minimal* sets of defaults does not work.[1] Such account correctly predicts *dead_battery* from the observation *lights* but incorrectly retracts the prediction in the presence of the additional but irrelevant piece of information *turn_key*. This problem is avoided by approaches in which causal rules are regarded as one-directional rules of inference.[2] However, those approaches do not explain *why* reasoning 'forward' is often the right thing to do, nor when reasoning 'forward' may fail (as when the three atoms above are all true). An interpretation of causal theories that sheds light on this question is [Goldszmidt & Pearl, 1992], where causal theories are seen as describing a family of Bayesian Networks [Pearl, 1988] that make explicit the independences in the theory. Our model is related to [Goldszmidt & Pearl, 1992] but is simpler and more transparent as it also appeals to notions of independence but does not rely on either Bayesian Networks or rankings of any sort.

## 2.1 The Model

The model is based on two ideas that are very simple and appear in one way or the other in several accounts of causal reasoning (e.g. [Goldszmidt & Pearl, 1992; Darwiche & Pearl, 1994; Geffner, 1996]):

1. the values that a variable x may take in a given state $s$ is a *function* of the value of the *causes* of x in $s$
2. the *form* of that function is determined by a standard interpretation of the rules that express causal influences over x

In order to formalize these two ideas, we express the set of *causal influences* acting on a variable x in a theory $T = \langle C, O \rangle$ by $T_x$; i.e. $T_x$ contains all the rules in $C$ of the form $F \supset x = x$ or $F \to x = x$. We call $T_x$ the (causal) *microtheory* for x.[3] The microtheories $T_x$ will be used to determine the possible values of x in a given state $s$. We assume that no variable is its own cause; i.e., no rule involves the same variable in both its body and its head.

We write $s/[x = x]$ for $x \in D_x$ to denote the state that makes $x = x$ true and is otherwise identical to $s$. We also say that a model $s/[x = x]$ of the microtheory $T_x$ is a *minimal model* of $T_x$ when no other model $s/[x = x']$ of $T_x$ violates a strictly smaller set of rules from $T_x$. The ideas above can then be expressed as follows:

1. for each variable x there is a function $\phi_x(s) \subseteq D_x$, such that $x \in \phi_x(s)$ means that $x$ is a possible value of x given the *causes* of x that are true in $s$
2. the function $\phi_x(s)$ is such that $x \in \phi_x(s)$ iff $s/[x = x]$ is a *minimal model* of the microtheory $T_x$

If $min$ is a function that takes a collection of states and returns the ones that violate a minimal set of defaults from $T_x$ but no strict rules, then the definition of the functions $\phi_x$ above can be expressed equivalently as:

$$\phi_x(s) \stackrel{def}{=} min^{-1}_{x \in D_x} s/[x = x] \quad (1)$$

Default rules of different priorities can be accommodated in this definition by doing the mimization lexicographically (stronger rules first).

The functions $\phi_x$, that we call *causal functions*, determine the values that a variable x can take given the value of its causes. We say that a state $s$ is a *causal interpretation* relative to a set of rules when the *actual* value of every variable x in $s$ is compatible with the value of its causes; i.e., when $x^s \in \phi_x(s)$. We also say that $s$ is a *causal model* of a causal theory $T = \langle C, O \rangle$ when $s$ is a causal interpretation relative to $C$ that satisfies the observations $O$. We assume throughout that causal theories are consistent in the sense that $\phi_x(s)$ is never empty.

The *consequences* of a causal theory $T$ are the formulas that hold in all its (causal) models.

**Example 1** The domain theory considered above involves four boolean variables: turn_key, lights, dead_battery and car_starts. The microtheories for the first two variables are empty and hence from (1):

$$\phi_{\mathsf{turn\_key}}(s) = \phi_{\mathsf{lights}}(s) = \{\mathbf{t}, \mathbf{f}\}$$

The microtheories for car_starts and dead_battery, on the other hand, are given by the first two rules and third rule respectively, and hence:

$$\phi_{\mathsf{dead\_battery}}(s) = \begin{cases} \{\mathbf{t}\} & \text{if } s \models lights \\ \{\mathbf{t}, \mathbf{f}\} & \text{otherwise} \end{cases}$$

$$\phi_{\mathsf{car\_starts}}(s) = \begin{cases} \{\mathbf{t}\} & \text{if } s \models turn\_key \land \neg dead\_battery \\ \{\mathbf{f}\} & \text{if } s \models dead\_battery \\ \{\mathbf{t}, \mathbf{f}\} & \text{otherwise} \end{cases}$$

Clearly the state $s$ that makes:

turn_key$^s$ = **t**, lights$^s$ = **t**, dead_battery$^s$ = **t**, car_start$^s$ = **f**

---

[1] Strict and defeasible rules $F \to A$ are violated in a state $s$ when $s$ satisfies $F$ but does not satisfy $A$.

[2] This directionality can be defined procedurally as in inheritance hierarchies and logic programs or by means of a formalism like Reiter's default logic.

[3] These microtheories are similar to the ones considered in [Darwiche & Pearl, 1994] even if their use here is different. Indeed, many of our microtheories are 'inconsistent' according to the criterion in [Darwiche & Pearl, 1994].

is a *causal interpretation*, while the state $s'$ that corresponds to the 'unintended' model:

turn_key$^{s'}=$ t, lights$^{s'}=$ t, dead_battery$^{s'}=$ f, car_start$^{s'}=$ t

is not as $\phi_{\text{dead\_battery}}(s') = \{t\}$ and thus dead_battery$^{s'} \notin \phi_{\text{dead\_battery}}(s')$. Indeed, the state $s$ is the *only* causal interpretation that satisfies the observations $O = \{turn\_key, lights\}$, and thus $s$ is the only causal model of the theory $T = \langle C, O \rangle$.

## 2.2 Related Models

**Theories of Action.** The model above is a refinement of the minimal model semantics in which a single global minimization over the whole set of rules is replaced by multiple minimizations over the local microtheories $T_x$. Moreover, the local minimizations are done over a a *fixed* state $s$ in which only the value of the variable x is allowed to change. Something similar occurs in the recent approaches to the frame problem. It is well known that if default rules are added for fluents and persistence violations are minimized, the resulting semantics is too weak [Hanks & McDermott, 1987]. The most recent approaches to the frame problem thus replace the global minimization by local minimizations in which the current state and actions are fixed, and the resulting state is allowed to change [Gelfond & Lifschitz, 1993; Reiter, 1991]. The result is that the action and persistence rules get compiled into state transition functions that determine the states that can follow a given state $s$ and action $a$, very much like causal rules above get compiled into the causal functions $\phi_x(s)$ that determine the states of the variable x that can 'follow' a given state of its causes.

**Bayesian Networks.** The definition of the causal interpretations $s$ above involves a fixed point condition: $s$ has to be such that the value x$^s$ of each variable x belongs to $\phi_x(s)$. This fixed point condition can be simplified when the causal rules in the theory do not involve cycles. Indeed, let us define the *causes* of x as any *minimal* set of variables $causes(\mathsf{x})$ that shield x from all other variables; i.e., $[causes(\mathsf{x})]^s = [causes(\mathsf{x})]^{s'}$ must imply $\phi_x(s) = \phi_x(s')$. Then it is easy to show that the function $\phi_x(s)$ can be replaced by an equivalent function $\psi_x(causes^s(\mathsf{x}))$ where the global state $s$ is replaced by the local state of x's causes. Then if there are no causal cycles and thus the variables can be ordered into a sequence $x_1, x_2, \ldots, x_n$ such no variable is preceded by one of its causes, the causal interpretations $s$ can be computed *iteratively* by setting the value of each variable $x_i$, from $i = 1$ to $n$, to any value $x_i \in \psi_{x_i}(causes^s(x_i))$. Indeed, the model that results in this case is exactly a Bayesian Network in which probabilities are replaced by $\kappa$ measures [Goldszmidt & Pearl, 1992] that are either 0 (plausible) or $\infty$ (impossible).

**Structural Equation Models.** In the presence of cycles, the resulting models cannot be interpreted as Bayesian Networks but as Structural Equation Models [Simon, 1953; Druzdzel & Simon, 1993]. Structural Equation Models are used widely in Economics and the Social Sciences and have the form of a set of simultaneous equations of the form:

$$x_i = f_i(x_1, x_2, \ldots, x_{i-1}, x_{i+1}, \ldots, x_n, \epsilon_i) \qquad (2)$$

where $x_i$ stands for the value of variable $x_i$, $\epsilon_i$ is an error variable, and $f_i$ is a deterministic function. Such equations are called *structural* as their *form* is important: equations that are *mathematically* equivalent are no necessarily *structurally* equivalent.[4] Structural equations represent mechanisms that propagate external perturbations on the right hand side variables to changes in the left hand side variable but not the other way around. For example, an structural equation like $spending = income/2 + \epsilon$ predicts that doubling the income may double the level of spending but does *not* predict that doubling the level of spending will double the income (see also [Pearl, 1996]).

The equation that defines our causal interpretations can be expressed in the form:

$$x_i^s \in \phi_{x_i}(x_1^s, x_2^s, \ldots, x_{i-1}^s, x_{i+1}^s, \ldots, x_n^s) \qquad (3)$$

that is very close to (2) except that the function $\phi_{x_i}$, unlike the function $f_i$ above, does not have to pick a *single* value for the variables $x_i$.

**Coherent Models.** Some approaches to causal reasoning replace the notion of *minimal models* by the notion of *coherent models* [Geffner, 1992; McCain & Turner, 1997]). Roughly, a model is *coherent* when all default violations are *causally explained*. In our language, we can consider the violation of a default rule $F \to A$ *explained* in a model $s$, when there is *conflicting* defeasible or strict rule $F' \to A'$ that is 'active' in $s$, i.e., whose body and head both hold in $s$. It is very easy to show then that causal and coherent models coincide. In this sense, coherence considerations which otherwise appear somewhat ad-hoc, can be justified by the ideas that define the causal models.

## 3 Time and Actions

Many causal relations such as the action of toggling a switch cannot be modeled without a notion of time or context. In this section we focus on such causal relations and consider the interpretation of *temporal* causal rules of the form $F \longrightarrow A$ where $F$ and $A$ are formulas and atoms as before. Strict temporal rules are written as $F \Longrightarrow A$. The difference between a (non-temporal) causal rule $F \to A$ and a temporal (causal) rule $F \longrightarrow A$ is that the former says that $A$ is normally true in the state $s$ when $F$ is true in $s$, while the latter says that $A$ is normally true in the state *following* $s$ when $F$ is true in $s$. The action of toggling a switch can be modeled by

---

[4] Note that the same holds for 'qualification' and 'ramification' constraints [Lin & Reiter, 1994]: they may be logically equivalent yet, for the same structural reasons, they behave in different ways. See Section 5.2.

the *temporal* rules of the form (the notation is similar to [Gelfond & Lifschitz, 1993]):

$$toggle \wedge \mathsf{switch} = on \longrightarrow \mathsf{switch} = \mathit{off}$$
$$toggle \wedge \mathsf{switch} = \mathit{off} \longrightarrow \mathsf{switch} = on$$

The semantics of temporal rules involves sequences of states $s_0, s_1, s_2, \ldots$, called *trajectories*, where $s_i$ refers to the state of the world at time $i$. As before, the *temporal interpretations* will be the trajectories that interpret all variables x in a way that is compatible with the causal function $\phi_\mathsf{x}^t$ determined by the causal microtheory $T_\mathsf{x}^t$ (we use the index $t$ to express that we are dealing now with *temporal* rules only).

The microtheories $T_\mathsf{x}^t$ are defined exactly as before as the collection of rules that express causal influences on x; namely the rules of the form $F \longrightarrow \mathsf{x} = x$ and $F \Longrightarrow \mathsf{x} = x$.

These microtheories are interpreted by means of *extended states* of the form $s//[\mathsf{x}] = x$, where $s$ represents a state at some time point and $\mathsf{x} = x$ represents the value of x at the *next* time point. More precisely, an *extended state* $s//[\mathsf{x} = x]$ satisfies a *temporal rule* $F \longrightarrow \mathsf{x} = x'$ if $F^s$ is false or $x = x'$ is true.

The function $\phi_\mathsf{x}^t(s)$ is defined as the set of values $x \in D_\mathsf{x}$ for which the extended state $s//[\mathsf{x} = x]$ satisfies all strict rules in $T_\mathsf{x}^t$ and violates a minimal set of default rules from $T_\mathsf{x}^t$. In analogy to (1):

$$\phi_\mathsf{x}^t(s) \stackrel{def}{=} min^{-1}_{x \in D_\mathsf{x}} s//[\mathsf{x} = x] \qquad (4)$$

Now $x \in \phi_\mathsf{x}^t(s)$ no longer means that $x$ is a possible value of x in the state $s$, but a possible value of x in the states *following* $s$. This suggests that the *temporal interpretations* could be defined as the trajectories $s_0, s_1, s_2, \ldots$ in which the condition $\mathsf{x}^{s_{i+1}} \in \phi_\mathsf{x}(s_i)$ holds for all variables x and all time points $i \geq 0$.

This actually captures the semantics of the rules but misses something important: the *law of inertia* that expresses the tendency of certain variables, usually called *fluents*, to retain their values.

The persistence of fluents can be captured by *extending* the causal functions $\phi_\mathsf{x}^t(s)$ into stronger functions $\phi_\mathsf{x}^*(s)$ in which the value $\mathsf{x}^s$ of fluents is allowed to persist when the rules in $T_\mathsf{x}$ provide no reason for a change.

More precisely we assume that the value of x persists when the rules in $T_\mathsf{x}$ do no discriminate among the values of x; i.e. when $\phi_\mathsf{x}^t(s) = D_\mathsf{x}$. The extended or *persistent causal function* $\phi_\mathsf{x}^*$ for *fluent* variables x is thus defined as follows:

$$\phi_\mathsf{x}^*(s) \stackrel{def}{=} \begin{cases} \phi_\mathsf{x}(s) & \text{if } \phi_\mathsf{x}^t(s) \subset D_\mathsf{x} \\ \{\mathsf{x}^s\} & \text{otherwise} \end{cases} \qquad (5)$$

For non-fluent variables, $\phi_\mathsf{x}^*(s) \stackrel{def}{=} \phi_\mathsf{x}(s)$.

With the functions $\phi_\mathsf{x}^*(s)$, the *temporal interpretations* can be defined in complete analogy to the *causal interpretations* as:

**Definition 1** *A state trajectory $s_0, s_1, \ldots$ is a* temporal interpretation *relative to a given set of temporal rules, if $\mathsf{x}^{s_{i+1}} \in \phi_\mathsf{x}^*(s_i)$ for each variable x and each time point $i \geq 0$.*

### 3.1 Temporal Theories

The *models* of a *temporal theory* $T = \langle D, A, O \rangle$ comprised of a set $D$ of *temporal rules*, a set $A$ of *actions*, and a set $O$ of *observations*, are the temporal interpretations (relative to $D$) that are *compatible with both the actions $A$ and the observations $O$*.

Actions and observations in temporal theories are encoded by means of *temporal atoms* of the form $\mathsf{x}[i] = x$, where x is an *action* or *fluent variable* respectively, $x$ belongs to $D_\mathsf{x}$, and $i$ is a time index (i.e., a non-negative integer). For boolean variables $\mathsf{x}[i] = \mathbf{t}$ and $\mathsf{x}[i] = \mathbf{f}$ are abbreviated as $x[i]$ or $\neg x[i]$. For simplicity, action variables are assumed boolean.

A trajectory $s_0, s_1, \ldots$ is *compatible with the observations* $O$ when $\mathsf{x}^{s_i} = x$ if $\mathsf{x}[i] = x$ is in $O$, and is *compatible with the actions* $A$ when for all action variables x, $\mathsf{x}^{s_i} = \mathbf{t}$ iff $x[i] \in A$.[5]

The *consequences* of a temporal theory are the temporal atoms $\mathsf{x}[i] = x$ that are true in all models, where $\mathsf{x}[i] = x$ is true in a model $s_0, s_1, \ldots$, when $\mathsf{x}^{s_i} = x$.

The model that results from these definitions is similar to several other models of action. Here we draw a simple comparison with the model $\mathcal{A}$ of Gelfond and Lifschitz [Gelfond & Lifschitz, 1993]. For simplicity we consider domain descriptions $D_\mathcal{A}$ comprised only of actions rules '$A$ **causes** $B$ **if** $C$' and initial conditions '**initially** $L$'.[6] Such description can be expressed as a set $D$ of temporal rules of the form $A \wedge C \longrightarrow B$ and a set $O$ of observations $L[0]$. Then, it is easy to show that:

**Theorem 1** *A value proposition $L$* **after** *$A_0, \ldots A_n$ is entailed by $D_\mathcal{A}$ iff $L[n+1]$ follows from the temporal theory $T = \langle D, A, O \rangle$, where $A = \{A_0[0], A_1[1], \ldots, A_n[n]\}$*

**Example 2:** The YSP [Hanks & McDermott, 1987] can be encoded by means of two temporal rules of the form:

$$load \longrightarrow loaded$$
$$shoot \wedge loaded \longrightarrow \neg alive$$

All variables are boolean, and *load* and *shoot* stand for actions. Given the observation *alive*[0] and the actions *load*[1] and *shoot*[3], $\neg alive$[4] follows. This is because $\phi_\mathsf{loaded}^*(s) = \{\mathbf{t}\}$ iff $load^s = \mathbf{t}$ or $loaded^s = \mathbf{t}$, while $\phi_\mathsf{alive}^*(s) = \{\mathbf{f}\}$ iff $shoot^s = loaded^s = \mathbf{t}$ or $alive^s = \mathbf{f}$. Then, since in all models $s_0, s_1, s_2, \ldots$, load and shoot are true in the states $s_1$ and $s_3$ and false in all other states, we get that $loaded^{s_1}$ must be true, $loaded^{s_2}$ and $loaded^{s_3}$ must also be true, and thus, $alive^{s_4}$ must be false.

---

[5] The double implication for actions captures that actions not included in $A$ must be false. A similar 'closed world assumption' for actions is implicit in approaches such as [Gelfond & Lifschitz, 1993; Reiter, 1991] in which actions are not represented as part of the state of the system.

[6] We assume that the domain description is consistent in the sense that no pair of rules associated with the same action have antecedents that are jointly satisfiable and consequents that are jointly unsatisfiable.

# 4 Causal and Temporal Rules Combined

We consider now the combination of temporal (causal) rules and (non-temporal) causal rules. Namely the *extended* temporal theories that we consider now are of the form $T = \langle D, C, A, O \rangle$ where $\langle D, A, O \rangle$ stands for a temporal theory as before, and $C$ stands for a set of (non-temporal) causal rules. Temporal and causal rules play the role of *effect axioms* and *ramification constraints* [Lin & Reiter, 1994] respectively.

For example, the scenario in [Kartha & Lifschitz, 1994] that involves two switches and a light that is 'on' when the switches are in the same position can be modeled by a theory with *temporal rules*:

$$toggle1 \wedge sw1 \longrightarrow \neg sw1$$
$$toggle1 \wedge \neg sw1 \longrightarrow sw1$$
$$toggle2 \wedge sw2 \longrightarrow \neg sw2$$
$$toggle2 \wedge \neg sw2 \longrightarrow sw2$$

and *causal rules*:

$$(sw1 \equiv sw2) \rightarrow on$$
$$\neg(sw1 \equiv sw2) \rightarrow \neg on$$

The semantics of temporal theories extended in this way must combine the causal and temporal rules into a single causal function $\phi_x^*$ for each variable x. For simplicity, we follow other formalisms and assume that *all temporal rules are strict*, and *all variables are boolean*. These two assumptions decouple the causal rules from the temporal rules, allowing us to define the persistent causal function $\phi_x^*$ for fluent variables x (that now takes two sucessive states $s_p$ and $s_c$ as arguments) as:[7]

$$\phi_x^*(s_p, s_c) \stackrel{\text{def}}{=} \begin{cases} \phi_x^t(s_p) & \text{if } \phi_x^t(s_p) \subset D_x \\ \phi_x^c(s_c) & \text{if } \phi_x^c(s_c) \subset D_x \\ \{x^{s_p}\} & \text{otherwise} \end{cases} \quad (6)$$

This definition assumes that fluent variables persist when neither the temporal or causal rules tell us otherwise. The restrictions above can be removed by simply defining a causal function $\phi_x(s_p, s_c)$ in terms of a single microtheory $T_x$ that includes both causal and temporal rules.

Provided with the functions $\phi_x^*$, and assuming that actions are exogenous (and hence that action variables do not occur in the head of rules), the resulting *temporal interpretations* can be defined as follows:

**Definition 2** *A trajectory $s_0, s_1, \ldots$ is a temporal interpretation relative to a set of temporal rules and causal rules if for each variable x, $x^{s_0} \in \phi_x^c(s_0)$, and $x^{s_{i+1}} \in \phi_x^*(s_i, s_{i+1})$ for each $i \geq 0$.*

The *models* of an *extended* temporal theory $T = \langle D, C, A, O \rangle$ are the temporal interpretations relative to the rules in $D$ and $C$ that are compatible with the actions $A$ and the observations $O$.

---
[7]The conditions are to be evaluated top down.

**Example 3.** Using the rules above, we prove that if $sw1$ and $sw2$ are initially true and false respectively, and the action $toggle2$ is done, then $sw1$ stays true, $sw2$ becomes true, and hence $on$ becomes true as well (see [Lin, 1995] for a similar example). Clearly, $\phi_{sw1}^*(s_i, s_{i+1}) = \{\mathbf{t}\}$ iff either $toggle1 \wedge \neg sw1$ or $\neg toggle1 \wedge sw1$ is true in $s_i$. Samething similar holds for $\phi_{sw2}^*(s_i, s_{i+1})$. Then since $toggle2[1]$ must be the only true action in the trajectory, sw1 must remain true at all points, sw2 must become true at time 2, and thus on must too become true at time 2.

# 5 Extensions

## 5.1 Non-persistent Causation

Rieger and Grinberg [Rieger & Grinberg, 1977] develop a language for representing causal mechanisms that, among other features, includes two different types of cause-effect relations: *continuous relations* in which the cause has to remain present in order to sustain the effect (e.g., $wet \supset hard\_to\_grasp$) and *one-shot relations* in which the effect remains true even after the cause that triggered the effect has vanished (e.g., $in\_water \supset wet$). In our model we have not made this distinction, yet this distinction is often needed.

Consider for instance a scenario from [Myers & Smith, 1988] cited in [Giunchiglia & Lifschitz, 1995]: if an object is put on the table it becomes safe for the baby, yet if removed from the table it won't necessarily remain safe. Such a scenario can be modeled by the temporal rules:

$$put \longrightarrow on\_table$$
$$remove \longrightarrow \neg on\_table$$

and the causal rule:

$$on\_table \rightarrow safe$$

When the action *put* is performed, *on_table* and *safe* become true, but if later *remove* is done, *on_table* becomes false but *safe* remains true. That's clearly anomalous. The problem, using the terminology of Rieger and Grinberg, is that we are modeling all causal rules as *one-shot* rules, yet the causal rule $on\_table \rightarrow safe$ is a *continuous* rule; i.e., *safe* is true *as long as on_table* is true. We call such rules, *non-persistent* rules.

Assuming now that all (non-temporal) causal rules are non-persistent, and for simplicity, that all rules are *strict*, the definition (6) of the function $\phi_x^*$ can be changed to:

$$\phi_x^*(s_p, s_c) \stackrel{\text{def}}{=} \begin{cases} \phi_x^t(s_p) \cap \phi_x^c(s_c) & \text{if } \phi_x^t(s_p) \cap \phi_x^c(s_c) \subset D_x \\ D_x & \text{if } \phi_x^c(s_p) \subset D_x \\ \{x^{s_p}\} & \text{otherwise} \end{cases}$$
$$(7)$$

where a variable x is *not* assumed to persist in states where it could have been caused by a non-persistent causal rule.[8]

---
[8]More accurate definitions are possible but the one above is simple and fits naturally into the format of the previous definition (6).

In the example above, the new definition means that *safe* won't be assumed to persist in states $s$ in which *on_table* holds. Yet if *on_table* persists so will be *safe* due of the rule *on_table* $\rightarrow$ *safe*.

## 5.2 Action Constraints

Strict *causal* rules of the form $F \supset A$ stand for a *mechanism* by which interventions to make $F$ true make $A$ true as well. Other type of expressions have the *same logical form* but do not have this property For example, the constraint *eat* $\supset$ *have_food* is always true yet we don't expect *have_food* to become true by performing the action *eat*. In other words, *eat* $\supset$ *have_food* does not stand for any mechanism. Something similar occurs with the so-called integrity constraints in databases like *high_salary* $\supset$ *is_a_boss*. Indeed, while causal rules express conditions that are self-enforcing in the model (i.e., they are assumed to be enforced by a mechanism), constraints like *eat* $\supset$ *have_food* and *high_salary* $\supset$ *is_a_boss* express conditions that must be enforced by the agent. For that reason we call the latter constraints *action constraints* (they correspond to the qualification constraints in [Lin & Reiter, 1994]).

In order to accommodate action constraints in our model, let us say that a theory $T = \langle D, C, A, O \rangle$ complies with a set $AC$ of *action constraints* if *every* state $s_i$ of *every* model $s_0, s_1, \ldots$ of $T$ satisfies those constraints. If so, we say that the theory $T' = \langle D, C, AC, A, O \rangle$ that results from incorporating those constraints is *consistent*. Otherwise we say that the theory is *inconsistent*.

Note that we do not prune the models that violate the constraints. When there are such models we simply declare the theory *inconsistent*. This is to comply with the so-called 'causality principle' in system theory [Padulo & Arbib, 1974] by which (exogenous) actions should not have an effect on the past. This principle holds for consistent theories but does not hold for inconsistent ones.

## 6 Summary

We have developed an approach to the ramification problem in terms of a simple and well-founded model of causal reasoning that is related to recent theories actions, Bayesian Networks and Structural Equation Models (see Section 2.2). A number of refinements and extensions that would be desirable are a more accurate model of non-persistent causation, constraints introducing *interactions* among mechanisms (e.g., a rigid bar among two moving blocks), and a more expressive language.

## References

[Darwiche & Pearl, 1994] Darwiche, A., and Pearl, J. 1994. Symbolic causal networks. In *Proceedings AAAI-94*, 238–244. Seattle, WA: MIT Press.

[Druzdzel & Simon, 1993] Druzdzel, M., and Simon, H. 1993. Causality in bayesian belief networks. In *Proceedings UAI-93*, 3–11.

[Geffner, 1992] Geffner, H. 1992. *Default Reasoning: Causal and Conditional Theories*. MIT Press.

[Geffner, 1996] Geffner, H. 1996. A formal framework for causal modeling and argumentation. In *Proceedings FAPR'96*, 208–222. Springer Verlag.

[Gelfond & Lifschitz, 1993] Gelfond, M., and Lifschitz, V. 1993. Representing action and change by logic programs. *J. of Logic Programming* 17:301–322.

[Giunchiglia & Lifschitz, 1995] Giunchiglia, E., and Lifschitz, V. 1995. Dependent fluents. In *Proceedings IJCAI-95*, 1964–1969.

[Goldszmidt & Pearl, 1992] Goldszmidt, M., and Pearl, J. 1992. Rank-based systems In *Proceedings KR'92*, 661–672.

[Hanks & McDermott, 1987] Hanks, S., and McDermott, D. 1987. Non-monotonic logics and temporal projection. *AIJ* 33.

[Kartha & Lifschitz, 1994] Kartha, G. N., and Lifschitz, V. 1994. Actions with indirect effects (preliminary report). In *Proceedings KR-94*, 341–350.

[Lin & Reiter, 1994] Lin, F., and Reiter, R. 1994. State constraints revisited. *J. of Logic and Computation* 4(5):655–678.

[Lin, 1995] Lin, F. 1995. Embracing causality in specifying the indirect effects of actions. In *Proceedings IJCAI-95*, 1985–1991.

[McCain & Turner, 1997] McCain, N., and Turner, H. 1997. Causal theories of action and change. In *Proceedings AAAI-97*.

[Myers & Smith, 1988] Myers, K., and Smith, D. 1988. The persistence of derived information. In *Proceedings AAAI-88*, 496–500.

[Padulo & Arbib, 1974] Padulo, L., and Arbib, M. 1974. *System Theory*. Hemisphere Publishing Co.

[Pearl, 1988] Pearl, J. 1988. *Probabilistic Reasoning in Intelligent Systems*. Morgan Kaufmann.

[Pearl, 1996] Pearl, J. 1996. Structural and probabilistic causality. *Motivation and Learning*.

[Reiter, 1991] Reiter, R. 1991. The frame problem in the situation calculus In Lifschitz, V., ed., *Artificial Intelligence and Mathematical Theory of Computation*. Academic Press.

[Rieger & Grinberg, 1977] Rieger, C., and Grinberg, M. 1977. The declarative representation and procedural simulation of causality in physical systems. In *Proceedings IJCAI-77*, 250-256.

[Simon, 1953] Simon, H. 1953. Causal ordering and identifiablity. In Hood, W., and Koopmans, T., eds., *Studies of the Econometric Method*. J. Wiley.

# DIAGNOSIS AND QUALITATIVE REASONING

Qualitative Reasoning 5

# Redesigning a Problem-Solver's Operators to Improve Solution Quality

**Eleni Stroulia**
Center for Applied Knowledge Processing
Helmholtzstr. 16
89081 Ulm, Germany
stroulia@faw.uni-ulm.de

**Ashok K. Goel**
College of Computing
Georgia Institute of Technology
Atlanta, GA 30332-0280
goel@cc.gatech.edu

## Abstract

The inability of a problem solver to produce solutions of a desired quality often may lie in the incorrect design of its operators. In this paper, we describe a method which, given a problem solver that produces a poor solution for a given problem, and given the desired solution for it, uses a model of the problem-solver's processing and knowledge to identify faults in the specification of its operators and to appropriately redesign them. This method is based on the structure-behavior-function (SBF) model of problem solving that explicitly captures the functional semantics of the problem-solver's tasks, the compositional semantics of its problem-solving methods that combine the operators' inferences into the outputs of the overall task, its domain knowledge, and the "causal" interdependencies between its tasks, methods and domain knowledge. We illustrate and evaluate this learning method through AUTOGNOSTIC.

## 1 Introduction

Research on unification of problem solving and learning has led to the development of several learning methods for improving problem-solving efficiency (e.g., [Mitchell *et al.*, 1981]). These methods typically assume that the set of available problem-solving operators is both complete and correct. The methods instead improve the control of processing, either by compiling sequences of operators into macro-operators or by acquiring better heuristics for their selection. In this paper, we are interested in the complementary learning goal of improving the quality of the solutions a problem solver generates. Modifying the control of processing is, in general, insufficient for attaining this goal because the reason why a problem solver produces poor solutions often is that its operators are incorrect. The specific research issue is as follows: given a problem solver that fails to produce solutions of a desired quality, and given desired solutions for problems for which it produced poor-quality solutions, what combination of knowledge and processing enables the identification of the incorrectly specified operators and their redesign, so that problem solver can deliver solutions of the desired quality?

To address this problem, we have developed a learning method based on a structure–behavior–function (SBF) model of a problem-solver's processing and knowledge. This model captures (i) the functional semantics of the problem-solver's tasks, (ii) the compositional semantics of its methods which recursively synthesize the inferences drawn by its operators into the outputs of its overall task, (iii) the domain knowledge available to it, and (iv) the "causal" interdependencies between its tasks, methods and domain knowledge. This model-based learning method is implemented in AUTOGNOSTIC, a "shell" which provides a language for representing SBF models of problem solvers, and mechanisms for monitoring the problem solving, receiving feedback on the result, and, in case of failure, assigning blame and repairing the problem solver. AUTOGNOSTIC's method for assigning blame for specific kinds of failure and identifying what operators to modify is described in [Stroulia and Goel, 1996]. In this paper, we focus on its method for redesigning the incorrect operators. AUTOGNOSTIC's integration with ROUTER, a path-planning system, is used to illustrate the method.

## 2 SBF Models of Problem Solving

SBF models analyze the problem-solver's task structure its domain knowledge and their interdependencies. In this framework, the problem-solver's tasks constitute the building blocks of its problem-solving mechanism. The methods that it employs decompose its complex overall tasks into simpler subtasks. These, in turn, get recursively decomposed into even simpler subtasks until they become elementary reasoning steps, i.e., *"leaf" tasks*, directly accomplished by the problem-solver's domain operators.

A task is specified as a transformation from an input to an output information state. It is characterized by the type(s) of information it consumes as input and produces as output, and the nature of the transformation it performs between the two. A task's *functional semantics* partially defines the task's intended correct behavior, by specifying the nature of the task's information transformation; it is expressed in terms of domain relations among the task's inputs and outputs. [1] No-

---
[1] The SBF language presently allows up to ternary relations and a single qualification level for the representation of functional semantics.

tice, that functional semantics is not simply a set of pre- and post-conditions for the application of an operator; it is rather a functional concept specifying how the task's output relates to its input. For a non-leaf task, the functional semantics of the subtasks into which the task is recursively decomposed, and the ordering relations that the decomposing methods impose over them, constitute a partial description of a correct reasoning *strategy* for this task.

Methods can be thought of as general plans for how the solutions of low-level subtasks get combined to deliver those of higher-level tasks. Each method captures the *semantics of the composition* of a set of lower-level subtasks into a higher-level task in terms of control, and information interdependencies. The tasks' control interdependencies are a set of relations partially ordering their executions; their information interdependencies are a set of information producer-consumer relations among them.

Finally, the SBF model of a problem solver captures the problem-solver's domain ontology, in terms of the types of objects that the problem solver knows about, and the relations applicable to them.

**The Case Study:** ROUTER, [Goel et al., 1994], the case-study problem solver which will be used in this paper to illustrate AUTOGNOSTIC's model-based learning method, is a path planner. Its spatial world model is organized in a neighborhood-subneighborhood hierarchy. High-level neighborhoods describe large spaces in terms of major streets and their intersections, and get refined into lower-level neighborhoods which describe both major and minor streets and their intersections but over smaller spaces. In addition to its world model, ROUTER contains a memory of past path-planning cases, also organized around the neighborhood-subneighborhood hierarchy. Figure 1 diagrammatically depicts part of ROUTER's SBF model. It illustrates the SBF specification of part of its task structure and domain knowledge.

ROUTER's task, *path-planning*, is to find a path from an initial to a goal location. To that end, ROUTER first identifies the neighborhoods in which these locations belong (*elaboration* subtask). Then, it searches in its path memory, for a path that is close to the current problem (*retrieval*). If the two given locations belong in the same neighborhood, ROUTER may use the *intrazonal-search* method to *search* for a path between them. This method is essentially a breadth-first search within the common neighborhood of the two locations. Initially, ROUTER sets up as its current location the initial location, and initializes its *tmp-path* to contain only this location. Then, by repeating the *path-increase* subtask, it incrementally expands the *tmp-path*. If, at some point, the *tmp-path* reaches the goal location, ROUTER assigns its value to the *path* and returns it as the solution. If the given locations do not belong in a common neighborhood, ROUTER can either perform a hierarchical search in its neighborhood organization (*interzonal-search* method) or it can use the path it retrieved from its memory, as the basis for solving the current problem (*case-based* method).

## 3 Redesigning Incorrect Operators

The inability of a problem solver to produce solutions of the desired quality is revealed when, for some particular problem, it receives as feedback a desired solution different from the one it actually produced. The first subtask of our model-based learning method is to assign blame for the failure to a specific operator so that it can then proceed to redesign it.

### 3.1 Assigning Blame

Blame assignment is performed through a successively focused examination of the tasks and methods involved in the production of the solution. Beginning with the problem-solver's overall task, the blame-assignment method evaluates whether the feedback and the task's actual input validate the task's functional semantics. If this is the case, it infers that the task in question should have produced the feedback, and the reason why it did not must lie within the subtasks in which it was decomposed. Based on the problem-solving trace, the last such subtask is identified and the blame assignment focuses on it. If the functional semantics of a particular task are violated by the feedback and its actual input, the blame-assignment method attempts to infer alternative inputs which would make the feedback value valid with respect to the task's semantics. If such alternative inputs can be found, the actual inputs are considered undesirable, and the blame-assignment focuses on the earlier subtask responsible for their production.

**Identifying Over-Constrained Operators:** If the blame assignment reaches a leaf task whose semantics is violated by the feedback solution and its actual input, and no alternative input can be found to make the feedback valid, then it postulates as one potential cause for the problem-solver's failure the *over-constrained semantics* of the task. The feedback solution exemplifies a set of quality requirements which, although in accordance with the overall behavior expected by the problem solver, conflict with the specification of the behavior delivered by a low-level operator. Thus, the feedback reveals an error in the information transformation that this particular operator was designed to accomplish in the context of the overall task, since it prevents the production of an acceptable and desired solution.

**Identifying Under-Specified Operators:** Alternatively, the blame-assignment may reach a leaf task whose semantics is validated by the output desired of it as well as its actual output. This situation implies that the problem-solver's task structure is not sufficiently specified to produce the right quality of solutions. Thus, the blame assignment postulates that the cause of the failure might be the *under-specified semantics* of the task, which allows both the actual and the desired values to be produced, when only the latter conforms with the requirements on the quality of the problem-solver's solutions, as exemplified by the feedback.

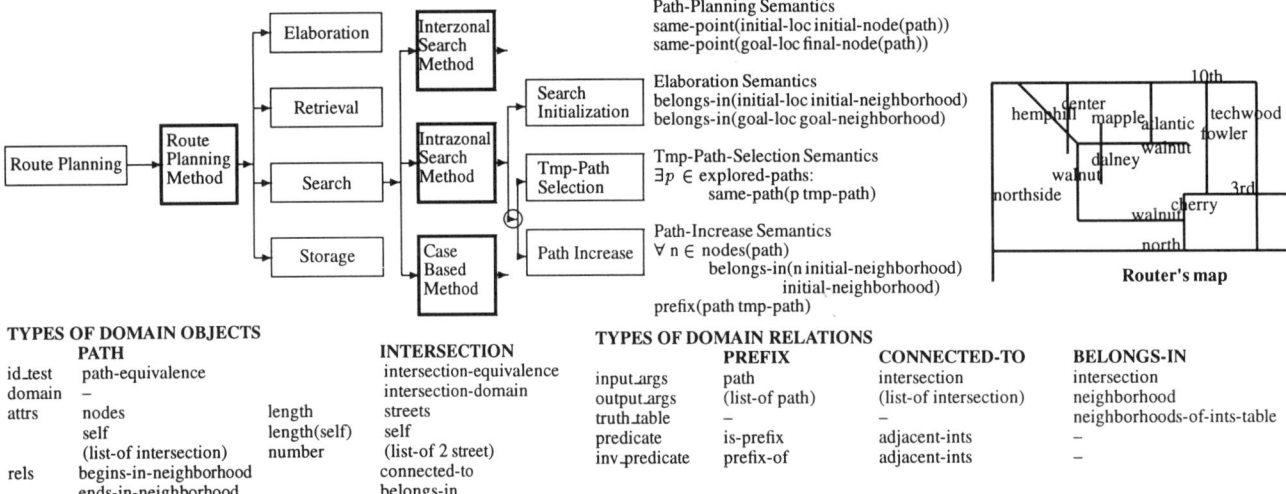

Figure 1: Fragment of ROUTER's SBF model: part of its task structure and domain knowledge.

## 3.2 Redesigning an Incorrect Operator

To address the learning tasks that arise from identifying an operator, *under-specified* or *over-constrained* with respect to the overall problem-solving task in service of which it is employed, a learning method must be able to formulate new functional semantics for this operator. That is, the learning method must be able to discover new functional concepts to characterize the relation of the operator's output to its input. In the former case, the new semantics will refine(specialize) the functionality of the faulty operator, in the latter one, it will extend(generalize) it. To that end, the model-based learning method we present in this paper relies first, on the examples of the behavior desired of the operator in the context of the overall task as exemplified by the feedback, and second, on its comprehension of the domain knowledge generally available to the problem solver.

Let us now see how the examples of the operator's desired behavior are generated. As it successively refines the focus of the investigation from the overall task to its increasingly specific subtasks, the blame assignment also "translates" the behavior desired of this overall task into behaviors desired of its subtasks. This translation occurs in two steps. In the first, when the investigation is focused from one task to the last subtask set up by the method used to accomplish it, the feedback becomes the output desired of the more specific subtask. In the second, when the investigation moves from a subtask to an earlier one, the feedback enables the inference of alternative inputs for the current task which, then, become the outputs desired of the earlier one. Thus, by the time the blame assignment reaches an incorrect (i.e., under-specified or over-constrained) operator, it has also produced an example of the behavior desired of this operator in the actual problem-solving session, namely an input-output tuple, which, if produced by the operator, could lead to the production of the feedback desired of the overall task. In the case of an under-specified operator, the *actual-input desired-output* tuple constitutes an example of the behavior desired of it, where the *actual-input actual-output* tuple constitutes an example of undesired behavior. In the case of an over-constrained operator, both the *actual-input desired-output* and the *actual-input actual-output* tuples constitute examples of desired behavior. Having such a set of positive and negative examples of the information transformation that the operator should perform, the next goal of the learning method becomes to discover a semantics specification to characterize them in an abstract way.

Here, let us discuss how the comprehension of the domain knowledge is used in the redesign of faulty operators. The SBF specification of the problem-solver's tasks' semantics is based on the domain relations known to the problem solver. In a sense, these domain relations provide the "terms" in which the information-transformation functions of the problem-solver tasks are expressed. By specifying the domain relations available to the problem solver, independently of whether or not they are used by its current task structure, the SBF model specifies the range of inferences that are possible in the problem-solver's domain. This specification of potential inferences enables our model-based learning method to recognize new uses for the problem-solver's knowledge in characterizing the alternative information-transformation functions desired of its currently incorrectly specified operators.

Figure 2 depicts the algorithm for discovering a new semantics with which to characterize a set of positive and negative examples of an operator's desired behavior. The learning method consists of two basic steps. First, it establishes a hypothesis space in which to search for possible alternative semantics for the incorrectly specified operator, and second, it successively evaluates the hypotheses in this space and collects these semantics that correctly characterize the behavior desired of the operator.

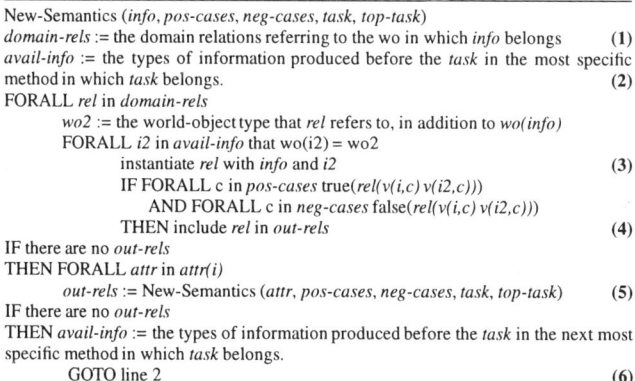

```
New-Semantics (info, pos-cases, neg-cases, task, top-task)
domain-rels := the domain relations referring to the wo in which info belongs   (1)
avail-info := the types of information produced before the task in the most specific
method in which task belongs.                                                    (2)
FORALL rel in domain-rels
    wo2 := the world-object type that rel refers to, in addition to wo(info)
    FORALL i2 in avail-info that wo(i2) = wo2
        instantiate rel with info and i2                                         (3)
        IF FORALL c in pos-cases true(rel(v(i,c) v(i2,c)))
           AND FORALL c in neg-cases false(rel(v(i,c) v(i2,c)))
        THEN include rel in out-rels                                             (4)
IF there are no out-rels
THEN FORALL attr in attr(i)
    out-rels := New-Semantics (attr, pos-cases, neg-cases, task, top-task)       (5)
IF there are no out-rels
THEN avail-info := the types of information produced before the task in the next most
specific method in which task belongs.
    GOTO line 2                                                                  (6)
```

Figure 2: The Algorithm for Discovering new functional semantics for incorrectly defined problem-solving operators

**Establishing a Hypothesis Space:** Given the representation of the functional semantics of a task in the SBF language, the formulation of the semantics hypothesis space requires first, the identification of the possible domain relations on which this semantics may be based, and second, the identification of the information types which, in addition to the operator's output, may be mentioned in the semantics.

To address the first requirement, the method uses the SBF specification of the operator's output information, to infer in which type of object it belongs. Next, based on the SBF specification of this object type, it identifies the set of domain relations relating this object type with other object types whose instances can be found among the set of available information types (line 1 of the algorithm shown in Figure 2). To address the second requirement, the method identifies the set of information types, available to the problem solver when the incorrect operator is applied. To that end, it identifies the most specific method which invokes the incorrect operator, and then these subtasks of the method which should have already been accomplished before the operator is invoked. Then, the types of information that can be used in the description of the new functional semantics are the types of information consumed and produced by these subtasks (line 2). The final step involves the formulation of hypotheses for the new semantics, using the set of available information types and the set of domain relations that may apply to them. For each domain relation, a semantics is formulated using as arguments the operator's output information and every other available information whose type conforms with the type of the relation's other argument (line 3).

**Evaluating the Semantics Hypotheses:** Having formulated a set of potential semantics for the incorrect operator, the learning method proceeds to evaluate which of them actually characterize the operator's desired behavior, by evaluating which ones are validated by the positive examples of the operator's desired behavior and fail for the negative examples (line 4). Any of the semantics relations that fulfill these requirements can be used to respecify the incorrect operator. If no such semantics relation is found, the learning method uses the SBF specification of the object type of the operator's output to infer its attributes and their corresponding values for the output actually produced and the output desired of it, and proceeds to repeat the above step sequence, in order to identify semantics in terms of the output-information's attributes (line 5). If again no semantics is found, the learning method extends the set of available information types by collecting the information types available in the context of a more general method subsuming the incorrect operator (line 6).

## 4  An Example in ROUTER

Let us now illustrate the above method with an example from ROUTER. ROUTER is given the problem of going from *(10th center)* to *(walnut dalney)*. It produces the path *(center 10th) (10th atlantic) (atlantic walnut) (walnut dalney)*, and is given the shorter path *(center 10th) (center mapple) (mapple dalney) (dalney walnut)* as feedback.

**Assigning Blame:** AUTOGNOSTIC first evaluates whether or not the feedback path is valid with respect to the functional semantics of ROUTER's overall *path-planning* task (as shown in Figure 1). The desired path begins at the initial intersection and ends at the final one, therefore, AUTOGNOSTIC infers that the desired value is within the class of solutions that ROUTER was intended to produce in this particular problem. Thus, it successively refines the focus of its investigation to increasingly specific subtasks of the *path-planning* task which produce the *path*, that is, *search* and subsequently *path-increase*.

As it evaluates the semantics of *path-increase*, AUTOGNOSTIC notices that the feedback path does not have as its subpath the task's input *tmp-path*. The *tmp-path* selected in the last repetition of the loop was *(center 10th) (10th atlantic) (atlantic walnut)*, and it is not a prefix of the desired path. At this point, AUTOGNOSTIC infers that the desired path could not possibly have been produced by the *path-increase* task given the information it received as input. To infer alternative values for the input *tmp-path* which could enable the task at hand to produce the desired path, AUTOGNOSTIC uses the predicate for the inverse *prefix-path* relation, and infers that *(center 10th) (center mapple) (mapple dalney)* is one prefix of the desired path, which, if given as *tmp-path* to the *path-increase* subtask, could result in the feedback's production. Thus, the cause of the failure must lie within the subtask that produced the "wrong" *tmp-path*, i.e., the task *tmp-path selection*.

AUTOGNOSTIC focuses its investigation towards identifying why this earlier subtask did not produce the desired *tmp-path*. It evaluates its functional semantics for the *tmp-path* desired of it, and notices that it is satisfied. Indeed, the desired *tmp-path* belongs in the set of *possible-paths*, and therefore, it could have been produced by *tmp-path selection*. Since this is a leaf task, accomplished by an operator, AUTOGNOSTIC infers that the specification of the *tmp-path selection* operator is too general with respect to the quality of solutions desired of the overall *path-planning* task.

**Redefining the *tmp-path selection* Operator:** Indeed, if "short length" is a quality desired of the paths that ROUTER produces, then its intrazonal-search method, which is essentially a breadth-first search within a neighborhood, is inappropriate. Instead, a depth-first search method should be used to actively prefer short paths, and in order for that to happen, the *tmp-path selection* should be redesigned to prefer the short paths in the *possible-paths* list. Let us now see how AUTOGNOSTIC reaches this conclusion.

The blame-assignment step has localized the underspecified *tmp-path selection* operator. It actually produced the *tmp-path (center 10th) (10th atlantic) (atlantic walnut)*, when it should have produced *(center 10th) (center mapple) (mapple dalney)* in order for the overall *path-planning* to produce the feedback.

The next step of the learning method is to collect the types of information available to ROUTER when the *tmp-path selection* is accomplished. This set includes the inputs and outputs of the *search-initialization* subtask, the only subtask of the *intrazonal-search* method (i.e., the method invoking the incorrect operator) which precedes the operator. Thus, the set consists of *initial-zone, initial-int, possible-paths*. These are the information types that can be used as terms in the new semantics of the *tmp-path selection* operator, and they are a *neighborhood*, an *intersection*, and a *path* respectively. Next, AUTOGNOSTIC collects the domain relations relating the above types of objects with *path*s, and formulates a set of potential semantics relations with these domain relations relating the *initial-zone*, or *initial-int* or *possible-paths* with *tmp-path*. Finally, it proceeds to evaluate whether any of these semantics characterizes the behavior desired of *tmp-path selection*. None of them does, so AUTOGNOSTIC proceeds to search for semantics based on the attributes of the available information types. In this effort, it identifies a single candidate for the role of *tmp-path selection* semantics, namely the relation *forall p in possible-paths, length(tmp-path)* $\leq$ *length(p)*. AUTOGNOSTIC proceeds to redefine the *tmp-path selection* operator to perform the function suggested by its new semantics, i.e., to select from the set of *possible-paths* the shortest one to expand [2]. Then, after presenting ROUTER with the same problem for a second time and producing the desired path as output, AUTOGNOSTIC verifies the appropriateness of the new semantics.

## 5 Experimentation and Evaluation

Two important issues for unified problem-solvers and learners, such as AUTOGNOSTIC, are how well do their learning methods accomplish their learning tasks in individual problem-solving episodes, and whether the learning converges to a stable design for the problem-solver. To test the efficacy of AUTOGNOSTIC's learning method in individual episodes, we gave AUTOGNOSTIC-on-ROUTER three (3) sequences of forty (40) randomly-generated path-planning problems. For each problem, we also gave as feedback the shortest path, where path length was measured by the number of discrete path segments in a path plan. We found that for all failed problem-solving episodes, in which ROUTER generated a path longer than the desired one, AUTOGNOSTIC succeeded in assigning blame and appropriately repairing ROUTER. However, since AUTOGNOSTIC is an incremental learner, it suffers from the familiar problem of overgeneralization (and over-specialization) of the functional semantics on the basis of a single example.

To test the long-term effectiveness of the learning process, after each of the above training sets, we gave AUTOGNOSTIC-on-ROUTER a sequence of one hundred and fifty (150) randomly generated path-planning problems. For each training set, and for each problem in the test set, we compared the quality of the solutions produced by the original and the modified ROUTER. The table below summarizes the results:

Table 1: Results on Quality-of-Solution Improvement.

| training set | problems with better path quality | problems with worse path quality | sign test | paired t-test |
|---|---|---|---|---|
| 1 | 55 | 13 | $1.3 \, 10^{-7}$ | 0.0002 (t=3.88) |
| 2 | 53 | 14 | $8.8 \, 10^{-7}$ | 0.0000 (t=5.17) |
| 3 | 44 | 30 | 0.06 | 0.14 (t=1.49) |

In these experiments, AUTOGNOSTIC modified ROUTER's *tmp-path selection* operator, as we described above, and two additional operators invoked by the *case-based* and the *interzonal-search* methods. This enabled ROUTER to perform a greedy search for short paths, which, for an almost canonical street network, returns the shortest path in most cases. In addition, AUTOGNOSTIC extended the size of ROUTER's neighborhoods to enable it to search through larger spaces in order to find shorter paths. That AUTOGNOSTIC modified the same three operators for all the different problems in the three experiments suggests that AUTOGNOSTIC is able to determine that only some operators are "faulty" in ROUTER's design. It also indicates that AUTOGNOSTIC does not keep on modifying the problem solver indefinitely, but stops when, after some exploration and redesign, a good enough design is found.

## 6 Related Research

Although much of past research on unification of problem solving and learning has focused learning methods for improving problem-solving efficiency, recent work in the Prodigy [Carbonell et al., 1989] framework has also explored learning of problem-solving operators. Perez [1994] describes a learning method for improving the quality of a problem-solver's solutions. This method uses both a trace of the problem solving that led to the solution of less than the desired quality and a trace of hypothetical problem solving that would lead to a solution of the desired quality. It assumes that the problem-solver's operators are correct and the cause of its failure to produce good solutions lies in its incorrect operator-selection heuristics. Gil [1994] describes a method for identifying missing pre- or post-conditions in a problem-solver's operator specification. This is similar to the

---

[2] The "reprogramming" of the operator is done manually.

learning task of AUTOGNOSTIC, but different, since it does not modify the information transformation the incorrect operators perform. Wang [1995] describes a method for acquiring operators from problem-solving traces, where the GPS-like operators again are specified in terms of pre- and post-conditions, without any functional specification of the input-output relations.

While the learning methods in the Prodigy framework are trace-based, the learning methods in our theory are model-based. In particular, AUTOGNOSTIC's learning methods use SBF models of problem solvers. We initially developed SBF models to capture the functional, compositional and causal semantics of physical devices [Goel 1991]. In AUTOGNOSTICwe have used the SBF modeling methodology to explicitly specify the tasks, methods, control of processing, and domain knowledge of problems solvers in a manner consistent with Chandrasekaran's framework of Generic Tasks and Task Structures [Chandrasekaran 1989]. CommonKADS [Wielinga *et al.* 1992] is another task-oriented framework for analyzing and describing problem solving. A major advantage of analyzing problem-solvers in terms of tasks and methods is that enable specification of the organization of problem solving, which in turns localizes the learning. Alternative model-based methods, such as that of Castle [Freed *et al.*, 1992], can only describe problem solvers performing a linear sequence of tasks, and, in addition, are limited in the learning tasks they address.

## 7 Conclusions

We have described a learning method based on SBF models of problem solvers which is able to identify incorrectly specified problem-solving operators, and to appropriately redesign them. This method relies on the specification of the functional semantics of the problem-solver's tasks, the compositional semantics of the methods that combine the inferences of its low-level operators into the outputs of its high-level tasks, and its domain knowledge.

The learning method uses the specification of the functional semantics of the problem-solver's tasks to establish the range of behaviors that the problem-solver's design is intended to deliver. Also, by evaluating the functional semantics of the problem-solver's subtasks relative to the solution desired of the problem solver, the learning method identifies which of the operators are incorrect with respect to the desired solution. This also enables the learning method to infer the behaviors desired of these subtasks in specific problems, which are then used as the positive examples in formulating their new functional semantics. Finally, the representation scheme for functional semantics in the SBF language enables the learning method to define the dimensions of the space of the new functions.

The SBF specification of the compositional semantics of the problem-solver's methods enables the learning method to limit the number of information interdependencies it examines, specifically by investigating only the tasks involved in the production of the undesired output. In addition, it enables the incremental specification of the information types in terms of which the new functional semantics of the incorrect operator can be specified. Finally, the specification of the problem-solver's domain ontology enables the learning method to establish the set of information relations that can be used in the specification of these new functions.

## Acknowledgments

This paper has benefited from numerous discussions with members of the Intelligence and Design research group at Georgia Tech. This work has been supported in part by research grants from NSF (IRI-92-10925 and DMI-94-20405) and ONR (research contract N00014-92-J-1234).

## References

[Carbonell *et al.*, 1989] Carbonell, J.G., Knoblock, C.A. and Minton, S. Prodigy: An Integrated Architecture for Planning and Learning. In *Architectures for Intelligence*, K. VanLehn (ed.), Hillsdale, NJ: Erlbaum.

[Chandrasekaran 1989] Chandrasekaran, B. Task Structures, Knowledge Acquisition and Machine Learning. *Machine Learning* 4:341-347, 1989.

[Freed *et al.*, 1992] Freed, M., Krulwich, B., Birnbaum, L. and Collins, G. Reasoning about performance intentions. In *Proceedings of the Fourteenth Conference of Cognitive Science Society*, pages 7-12, Hillsdale, NJ: Erlbaum.

[Gil, 1994] Gil, Y. Learning by Experimentation: Incremental Refinement of Incomplete Planning Domains. In *Proceedings of the Eleventh International Conference on Machine Learning*, pages 87-95. Los Altos, CA: Morgan Kaufmann.

[Goel 1991] Goel, A. Model Revision: A Theory of Incremental Model Learning. In *Proceedings of the Eighth International Conference on Machine Learning*, pages 605-609, Chicago, June 1991. Los Altos, CA: Morgan Kaufmann.

[Goel *et al.*, 1994] Goel, A., Ali, K., Donnellan, M., Gomez, A. and Callantine, T. Multistrategy Adaptive Navigational Path Planning. *IEEE Expert*, 9(6):57-65, 1994.

[Mitchell *et al.*, 1981] Mitchell, T.M., Utgoff, P.E., Nudel, B. and Banerji, R.B. Learning problem-solving heuristics through practice. In *Proceedings of the Seventh International Joint Conference on Artificial Intelligence*, pages 127-134.

[Perez, 1994] Perez, M.A. The goal is to produce better plans. In *Working Notes of AAAI Spring Symposium Goal-Driven Learning*, March 21-23, 1994, Stanford University; AAAI Press.

[Stroulia and Goel, 1996] Stroulia, E. and Goel, A. A Model-Based Approach to Blame Assignment: Revising the Reasoning Steps of Problem Solvers. In *Proceedings of the Thirteenth National Conference on Artificial Intelligence (AAAI-96)*, pages 959-965, Portland, Oregon, August 4-8. AAAI Press.

[Wang, 1995] Wang, X. Learning by Observation and Practice: An Incremental Approach for Planning Operator Acquisition. In *Proceedings of the Twelfth International Conference on Machine Learning*, pages 549-557. Los Altos, CA: Morgan Kaufmann.

[Wielinga *et al.* 1992] Wielinga, B.J., Schreiber, A.Th., and Breuker, J.A. KADS: A modelling approach to Knowledge Engineering. *Knowledge Acquisition*, 4(1):5-53.

# Formal Specifications for Hybrid Dynamical Systems

**Pieter J. Mosterman** and **Gautam Biswas**
Center for Intelligent Systems
Box 1679, Sta B
Vanderbilt University
Nashville, TN 37235.
`pjm,biswas@vuse.vanderbilt.edu`

## Abstract

Modeling abstractions in physical systems result in *hybrid models* which encompass continuous behaviors with discrete changes, causing discontinuities in system behavior generation which violate the physical laws of *conservation of energy* and *continuity of power*. This paper develops a formal specification for handling discrete model configuration changes at well-defined points in time, and a consistent transfer of the continuous system state from a previous model configuration to a new one based on the principle of *invariance of state*. Simulation algorithms designed to operate on hybrid models define behavior generation schemes that operate on the interval (continuous) to point (discrete) to interval (continuous) switches on the time line.

## 1 Introduction

Physical systems are inherently continuous and their behaviors are governed by the principles of conservation of energy and continuity of power [3]. Perceived discontinuities are in reality fast nonlinear continuous behaviors. For efficient analysis, the differences in time scale may be exploited so that the nonlinear behaviors can be abstracted to manifest as ideal discontinuities at *points in time*. An example is an ideal *elastic* collision between a body and a floor where the velocity of the body reverses instantaneously on impact. In reality, the collision occurs on a small time interval during which kinetic energy is converted into potential elastic energy, which then reverts back completely to kinetic energy for the body. Discontinuous effects can also be created by *parameter* abstraction [5]. Small parasitic physical effects that cause nonlinear continuous effects are abstracted away to simplify system description. For example, an *ideal non-elastic* collision between two bodies involves instantaneous discontinuous changes in velocity for the two bodies at the point of impact. A more precise model would have included small elasticity coefficients for the two bodies, and the period of impact would be a small but finite time interval, in which the change in velocities for the bodies would occur in a continuous manner. Models that combine continuous and discrete effects are called *hybrid* systems.

During discontinuous changes, physical laws of conservation of energy and continuity of power may be violated [5]. In such situations, the initial state vector following the discontinuous changes is computed using the principle of *conservation of state* along with explicitly modeled interactions with the environment. In previous work [5; 7], this theory of discontinuous configuration changes in physical system models has been developed into a *hybrid bond graph* modeling paradigm that combines traditional bond graph elements with *ideal switching* elements controlled by finite state automata. Formal schemes for verifying the correctness of models based on the principle of *divergence of time* have also been developed [7]. The hybrid bond graph formalism can be effectively applied to systematically design and analyze hybrid models of dynamic physical systems. This paper focuses on developing a formal semantics for analyzing systems with mixed continuous/discrete components.

## 2 A Hybrid Modeling Paradigm

Hybrid models operate in continuous modes (typical physical system behavior), but at points in time when signal values cross pre-defined thresholds or when explicit external (control) events are imposed on the system, changes in model configurations cause discrete changes in system behavior. An important observation is that the temporal trajectory of system behavior becomes piecewise continuous, where *simple discontinuities* can occur only at well-defined points in time. The key to developing a correct modeling paradigm is to ensure that interaction between the continuous and discrete modeling formalisms is unambiguous, rigorous, and consistent.

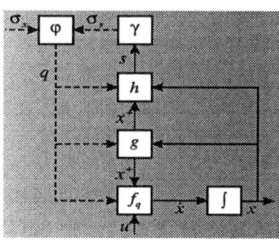

Figure 1: **A general hybrid system.**

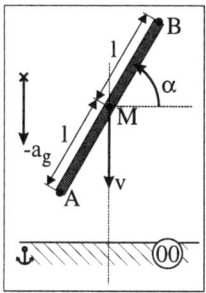 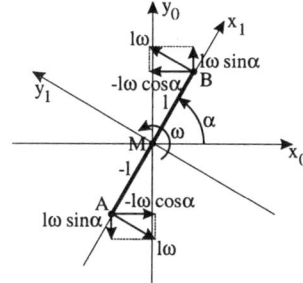

Figure 2: **A collision between a body and a floor.**

## 2.1 General Hybrid Dynamic System

The general architecture for a hybrid dynamic system model (Fig. 1) can be specified by the 9-tuple: $H = <I, \Sigma, \phi, X, U, f_\alpha, g, h, \gamma>$. Each mode of continuous behavior is given a unique state label $\alpha_k \in I$. Continuous behavior is governed by field $f_{\alpha_k}$ which determines the continuous state vector $x_{\alpha_k}$. The function $h$ computes signal ($s \in S$) values from the state vector $x_{\alpha_k}$ in mode $\alpha_k$, which may generate discrete events $\Sigma$ specified by a mapping $\gamma$. $\gamma$ is usually defined in terms of signal values *reaching* or *crossing* pre-specified threshold values. Occurrence of a discrete event suspends the continuous behavior mode $\alpha_k$, and a new mode, $\alpha_{k+1}$, is generated by the discrete transformation $\phi$. The function $g$ computes a new state vector, $x^+$, for the new operational mode $\alpha_{k+1}$ using values of the continuous state vector $x_{\alpha_k}$ in the previous operational mode $\alpha_k$.

## 2.2 The Continuous Model

Dynamic physical system models are best represented as a set of differential equations on the system state vector. For example, the falling rod in Fig. 2 can be described by three state variables, the rod's linear velocities, $v_x$ and $v_y$, and its angular velocity, $\omega$. When it is falling freely, only gravity acts on the center of mass and accelerates vertical movement. This can be described by the differential equation $\dot{\omega} = 0, \dot{v}_x = 0, \dot{v}_y = a_g$, where $a_g$ is the gravitational acceleration.

Differential equation state space models, supplemented by algebraic constraints (DAEs) directly reflect underlying physical principles such as Kirchhoff's laws and phenomenological relations like Ohm's law. Many model parameters have an immediate physical meaning and equations can be systematically derived from bond graphs, network representations, and block diagrams [2]. A general representation of an ODE model derived from DAEs is: $\dot{x}(t) = f_\alpha(x(t), u(t), t)$. The field, $f_\alpha$, describes continuous temporal evolution of system behavior in a mode of operation, $\alpha$, with the input vector, $u$, and the continuous state vector, $x$. Note that $f_\alpha$ is unique in mode $\alpha$.

## 2.3 The Discrete Model

Discrete events are modeled by a discrete indexing set, $I$ and a switching function, $\phi : I \times \Sigma \rightarrow I$. The set of discrete states corresponds to (a) *real* modes, where system behavior is governed by energy principles, and (b) *mythical* modes [5; 7], where the system behavior transitions are instantaneous. $\Sigma = \Sigma_s \times \Sigma_x$ captures the *event* set. $\Sigma_s$, is associated with closed loop control, and $\Sigma_x$ is governed by external, open loop control signals. The closed loop control is a function of the system's physical process variables. $\phi$, usually implemented with Petri-Nets or Finite State Automata, determines the next state after an event occurs.

## 2.4 Interactions

Interactions between the continuous and discrete modeling formalisms have to be specified correctly. For states that correspond to modes of continuous operation, ODEs determine behavior. A discrete event causes the system to change operational mode, and the correct state vector in the new mode is determined by the function $g : X \times I \rightarrow X^+$. $X$ defines state vector values just before switching occurs, and $X^+$ represents state vector values at the initial point in time when a switch or mode change has occurred. A function $h : X \times U \times I \rightarrow S$ determines signal values $S$ and $S^+$, computed by $h$ from $X$ and $X^+$, respectively. The function $\gamma : S \times S^+ \rightarrow \Sigma_s$ generates discrete events from the signal values. The interaction between the continuous and discrete part consists of (a) discrete events generated by the continuous signals, and (b) a change of operational mode by the discrete model, requiring a consistent mapping of the continuous state vector.

An example, adapted from [4], illustrates a rigid body collision of a rod falling to the floor (Fig. 2). On hitting the floor, the rod may disconnect after a point in time where contact occurred, slide along the floor and rotate about its point of contact, or just stick at the point of contact and rotate. Whether the rod sticks at the point of contact or slides is determined by a Coulomb friction coefficient, $\mu$, and whether the horizontal force exceeds a threshold value given by:

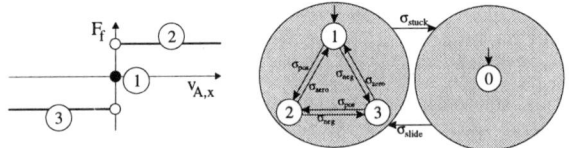

Figure 3: **Coulomb friction.**

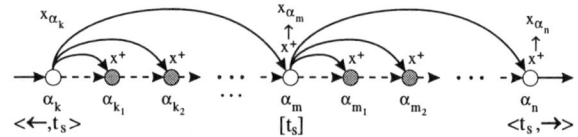

Figure 4: **System state is derived from the original state vector.**

$$\gamma : \begin{cases} |F_{A,x}| > \mu F_n \Rightarrow \sigma_{slide} \\ |v_{A,x}| \leq v_{th} \Rightarrow \sigma_{stuck}. \end{cases} \quad (1)$$

The friction force cannot be predetermined because it depends on the normal force at the surface. When the $\sigma_{slide}$ event becomes active, the friction force comes into effect and its direction is always opposed to the direction of velocity. The events, $\sigma_{zero}$, $\sigma_{pos}$, and $\sigma_{neg}$, correspond to states 1, 2, and 3 of the automata in Fig. 3, respectively. The events that cause these internal state changes are defined as:

$$\gamma : \begin{cases} v_{A,x} = 0 \Rightarrow \sigma_{zero} \\ v_{A,x} < 0 \Rightarrow \sigma_{pos} \\ v_{A,x} > 0 \Rightarrow \sigma_{neg}. \end{cases} \quad (2)$$

Since this behavior is piecewise continuous only *simple* behavior discontinuities occur at time points, which implies that operational modes have limit values at discontinuities. The complete event set for Coulomb friction is $\Sigma = \{\sigma_{slide}, \sigma_{stuck}, \sigma_{zero}, \sigma_{pos}, \sigma_{neg}\}$. A distinction is made between sliding with 0 velocity and being stuck, though the velocity of the rod at the surface is 0. In case the rod is stuck, the model does not have a degree of freedom in the $x$-direction.

As an example of a transfer of the continuous state vector between model configurations, consider the falling rod when it first makes contact with the floor, the model moves from mode $\alpha_{00}$ to mode $\alpha_{01}$ in Fig. 5. At this point it reaches a model configuration where $v_x$ and $v_y$ at the rod-tip are forced to 0. This requires the center of mass to move in the $x$ and $y$ direction with a velocity that is completely determined by the angular velocity. Conservation of momentum determines that the initial momentum in the $y$ direction is redistributed over the angular and linear momenta. Fig. 2 shows that the linear velocities can be represented in coordinate frame $(x_0, y_0)$ by $v_x = l\omega^+ sin\theta, v_y = -l\omega^+ cos\theta$. A detailed derivation (see [6]) yields the new state vector:

$$g_{\alpha_{01}} : \begin{cases} \omega^+ = \frac{1}{J+ml^2}(\omega J + ml(sin\theta v_x - cos\theta v_y)) \\ v_x^+ = l\omega^+ sin\theta \\ v_y^+ = -l\omega^+ cos\theta \end{cases} \quad (3)$$

## 3 Model Execution Semantics

A discontinuous change that occurs in given mode $\alpha_k$ has to happen at a point in time, say $t_s$. The state vector at this point, $x_{\alpha_k}$ labeled $x_{\alpha_k}^- = \lim_{t \uparrow \uparrow t_s} x_{\alpha_k}(t) = x_{\alpha_k}(t_s)$. This becomes the *a priori* vector for the state computation function $g$ that determines the initial state $x^+$ in the new mode $\alpha_{k+1}$. The state vector $x^+$ is referred to as the *a posteriori* vector computed by $g$. The new state vector, $x^+$, may immediately trigger further discrete events determined by $h$ and $\phi$, causing a sequence of discrete mode changes till a new operational mode, $\alpha_m$, is reached at which no further switching occurs. All the intermediate states traversed between two continuous modes are mythical [7]. The sequence of state and state vector changes is illustrated in Fig. 4. At mode $\alpha_m$ system behavior evolution in time resumes, with the state vector $x_{\alpha_m}(t_s) = x^+$. Sometimes, mode $\alpha_m$ may represent just a point of *continuous* operation (e.g, the point of contact in an elastic collision [7]). State vector changes from $x_{\alpha_k}^-(t_s)$ to $x^+$ in the new real mode may cause the $\gamma$ function to generate additional events resulting in another sequence of discrete state changes before the next continuous operational mode, $\alpha_n$, is arrived at (Fig. 4).

Consider the falling rod in Fig. 5. Initially, it is falling freely under gravity (mode $\alpha_{00}$). On hitting the floor it exerts a force with two components, $F_{A,y}$ and $F_{A,x}$ (mode $\alpha_{01}$). Since the floor surface has Coulomb friction, the rod immediately starts to slide if $|F_{A,x}| > \mu F_n$ (mode $\alpha_{11}$). Otherwise, it sticks and rotates around the point of contact (mode $\alpha_{01}$). When the rod starts to slide, the floor exerts an opposing friction force, $F_f$. In this case, the initial kinetic energy before contact is redistributed over the angular and vertical momentum to ensure the vertical velocity of the rod-tip, $v_{A,y}$, is 0. The horizontal velocity of the rod-tip, $v_{A,x}$, is determined by the angular velocity, $\omega$, and the horizontal velocity of the center of mass, $v_x$. Since $v_x$ is independent of $\omega$ and determined by $F_f$, it is initially 0 and the discontinuous change of $\omega$ results in a discontinuous change of $v_{A,x}$. Therefore, the system changes from the operational mode where $F_f = 0$ to mode $\alpha_{21}$ where $F_f = \mu F_n$.

The grayed modes of operation in Fig. 5 are mythical. They do not have physical meaning, therefore, no representation on the real time-line. However, they play

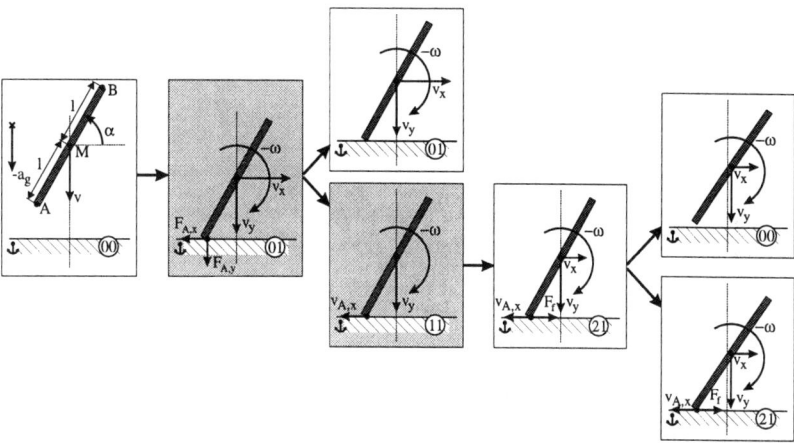

Figure 5: **Operational modes of falling rod.**

the role of transition points for locally defined switching functions.

## 3.1 Temporal Evolution of State

Discontinuities are abrupt point changes, caused by modeling abstractions. Discontinuities that persist in time intervals would violate continuity of power and conservation of energy principles. Further, asymmetry in temporal evolution ensures that the state vector in modes of continuous operation has to be *left closed* over the time intervals these modes are active. Mode changes and discontinuous changes in the continuous state vector can only occur at points in time $t_s$. We have shown in other work [7] that further continuous evolution may cause a mode change, $\alpha_m$ to $\alpha_n$, at the point of transfer, but no discontinuous change can occur in the continuous state vector between $x_{\alpha_m}(t_s)$ and $x_{\alpha_n}^+(t_s)$ since its initial value would be derived from $\lim_{t \downarrow t_s} x_{\alpha_n}(t)$ which requires knowledge of future behavior and conflicts with the assumption of causality in physical system models.

Consider the stiction force when the rod disconnects from the floor as it slides. If this force causes a discontinuous change in the vertical velocity of the rod, $\lim_{t \downarrow t_s} v_y(t)$ differs from the actual value $v_y(t_s)$. However, the value of $\lim_{t \downarrow t_s} v_y(t)$ may be such that its value indicates that the rod would have gotten stuck. This implies that in addition to the current state $v_y(t_s)$ and model configuration, the operational mode needs to know future modes and the limit values of state variables looking back in time. Such systems are acausal which is physically impossible and results in ill-defined models.

Since no discontinuous change of the state vector can occur, it is continuous over a left closed interval in time. This only requires the system state to operate continuously in left closed intervals but field $f$ is not required to be differentiable. Therefore, other derived system variables may still change discontinuously as a result of configuration changes. These jumps are well-defined by the continuous state vector and model configuration.

## 3.2 Invariance of State

A discontinuous change in the state vector may invoke further mode transitions. The state vector in a new mode is computed from the last continuous state vector, and the state vector in all new modes is computed from the last continuous state vector before switching started. This is the principle of *invariance of state* [7].

To illustrate, consider the falling rod in Fig. 2. When it hits the floor, its vertical momentum is distributed over its angular, horizontal and vertical momentum to ensure its rotation and translation of center of mass are such that the point of contact does not move (mode $\alpha_{01}$). In this situation, if the force at the rod-tip, $F_{A,x}$, exceeds a threshold value, it immediately starts to slide (mode $\alpha_{11}$). The rod-tip moves freely in the x-direction, and its initial vertical momentum is distributed only over its *a posteriori* angular momentum and vertical momentum to ensure the y-value does not change at the point of contact. If the continuous state vector in the sliding mode, $\alpha_{11}$, was computed from the previously inferred mode, $\alpha_{01}$, it would have a horizontal velocity associated with its center of mass which would keep the rod-tip from moving in the x-direction as well, which is incorrect. This demonstrates the importance of the proper computation of the state vector across a series of discontinuous changes.

## 3.3 Divergence of Time

Discontinuous configuration changes in system behavior are instantaneous so a model verification technique based on the principle of *divergence of time* ensures that the model does not end up in a loop of instantaneous changes where system behavior does not progress in time

[7]. In previous work, we have developed a multiple energy phase space analysis that establishes divergence of time before simulation is performed [5].

As an example, consider the falling rod when it starts to slide because its force in the vertical direction exceeds a threshold value. If the rod is specified to stick when the velocity of its rod-tip is below a certain threshold value, it may not have sufficient initial vertical momentum to maintain a high enough vertical velocity. Based on the specifications, this moves the model into the configuration where it sticks and rotates around the point of contact. However, in this configuration, based on the initial vertical momentum, its horizontal force causes it to start sliding and a loop of consecutive changes occurs.

## 3.4 Implementation

In previous work[5; 7], we have developed a hybrid bond graph modeling methodology that uses an *ideal switching element*, whose on-off conditions are governed by finite state automata, to dynamically construct model configurations as its behavior evolves in time. The bond graph model of the idealized thin rod and idealized floor, and the fragments dynamically generated by simulation[6] are shown in Fig. 6. The rod is assumed to have three degrees of freedom with associated buffers: angular velocity $\omega$ ( buffer $J$), and linear velocities $v_x$ (buffer $m_x$) and $v_y$ (buffer $m_y$). The relation between those velocities is modeled by a modulated transformer. Gravity is modeled by a constant effort source, $ma_g$, in the $y$ direction at the center of mass.

The $x$ and $y$ components of the forces and velocities at point A connect to the model at the $0_C$ junction. If the body is moving freely, this junction is *off*. If the body is in contact with the floor, $0_C$ is *on* and if no other elements are connected, it enforces a 0 velocity. The friction force, $F_f = \mu F_n$, in the $x$ direction is modeled as a piecewise continuous modulated source, $MS_e$, producing force values $0, F_f, -F_f$ at A opposite to the direction of the surface velocity.

The control specifications (CSPEC) of the switching junctions are specified by finite state automata, one for each controlled junction. The controlled junction is specified by a hierarchical finite state machine, which can be in one of several *on* states, depending on the bond graph signals. Depending on the specific state, a part of the piecewise continuous friction function is active. In its *off* state, the junction enforces 0 flow.

Initially, the rod is moving freely and controlled junctions $0_C$ and $1_S$ are *off*. This bond graph is mode $q_{00}$. The position of the rod-tip closest to the floor, $y_A$, is determined by the sum of the position of the center-point, $y_M = \int v_y$, and the distance of the rod-tip from the center point, $-l\sin\alpha$. When $x_A = 0$ the rod collides with the floor, $0_C$ comes *on* and the model transitions

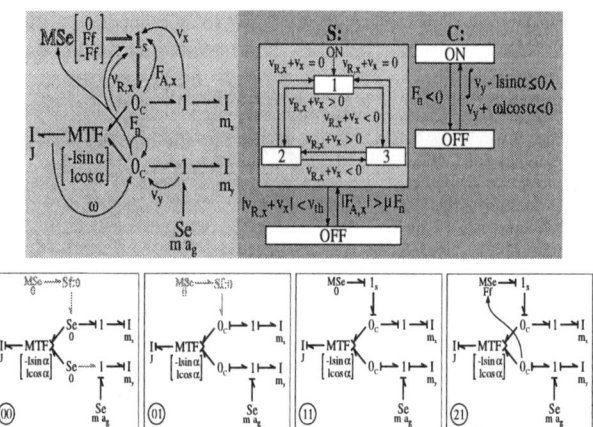

Figure 6: **Dynamically generated models.**

to mode $q_{01}$. If the rod-length and angle of collision are such that $1_S$ comes *on* (the model transitions into $q_{11}$), the rod begins to slide. The function, $g$, can be recalculated, and the piecewise continuous friction function may move into its $F_f$ area, mode $q_{21}$. The hybrid bond graph approach provides a seamless integration of configuration changes based on local switches. Details of the derived continuous system and discrete control models appear in [6]. Other examples of hybrid bond graph models are discussed in [7].

## 4 Hybrid System Simulation

The simulator operates in two modes: (i) continuous simulation during intervals of operation, and (ii) discrete changes during configuration changes. Numerical simulation schemes like Euler and Runge-Kutta can be used for continuous operation but discrete events generated by $\gamma$ triggers an event detection module to determine the switching time, $t_s$, within a margin of tolerance, $\epsilon$ (Fig 7). The continuous field, $f_k$, computes $x_k(t_s^-)$, then real time is suspended, and the meta-level control model, $\phi$, generates the discrete state transition. The original continuous state vector is then transferred to the newly found model configuration using $g$, and this may trigger further events. The resulting model configuration is established, and $x_k(t_s^-)$ is transferred to this model configuration. Again, discrete events may be generated and this process continues until no further transitions occur and the continuous system state is established as $x_m(t_s)$

If the new operational mode is valid for a point in time further events are generated till continuous simulation can be resumed for a model configuration valid over an interval of time. To detect possible configuration changes when the system evolves over an infinitesimal amount of time, the system, $f_m$, is considered continuous over an infinitesimal time, $\epsilon$, i.e., no configuration changes are allowed. Simulation up till $t_s + \epsilon$ may cause a new

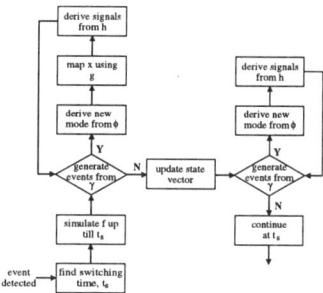

Figure 7: **Flow diagram of hybrid system simulation.**

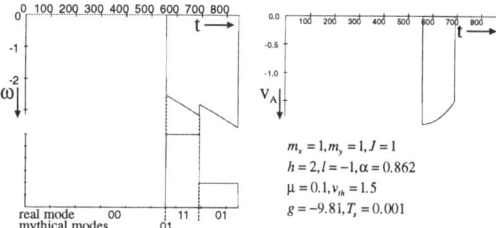

Figure 8: **Physically consistent simulation.**

series of configuration changes. These are executed by the discrete model using the $h$ function. There can be no more discontinuous changes so the state vector does not have to be recomputed. In the new continuous mode $q_n$, $f_n$, defines the simulation from time $t_s$ with initial vector $x_m(t_s)$. This implements simulation of $f_m$ at $t_s$ as a point in time and allows an energy redistribution that results from a function with discontinuities that are not simple. Note that the method only applies under the principle of temporal evolution of state.

A simulation run for one scenario is shown in Fig. 8. The rod falls down at a specific angle, hits the floor and moves into configuration 01. Based on the state vector an immediate configuration change to mode 11 occurs. In this configuration the rod slides with a velocity that decreases in magnitude due to the friction force acting on the rod-tip. At one time, this velocity falls below a preset threshold value and if the state vector is such that the rod gets stuck without immediately satisfying the condition to slide, the system moves into mode $q_{01}$. In this mode, it is stuck and rotates around the point of contact until it falls flat on the floor.

## 5 Conclusions

This work demonstrates a powerful hybrid system modeling scheme that incorporates modeling abstractions and embedded discrete control of physical systems. Configuration changes governed by local automata may produce discontinuities in system variables. The new state variables are then systematically derived using the principle of conservation of state combined with explicitly defined interactions with the environment. Global specifications are derived dynamically based on systematic principles of invariance of state, divergence of time, and temporal evolution of state. This simplifies the modeling task and truly demonstrates the use of *compositionality* in defining system models. This is in contrast with the approach by Alur et al. [1] which requires predefined global specifications of continuous system behavior in terms of differential equations. Furthermore, global knowledge in specifying discrete behavior is required to ensure no mythical modes exist. Also, unlike the hybrid bond graph modeling paradigm, there is no support for systematic modeling based on physical principles (e.g., conservation of state). The formal specifications are incorporated into a hybrid system simulation scheme that ensures the generation of correct system behavior. Overall, this is a systematic approach to abstracting physical system models: (i) time scale abstraction, and (ii) ignoring parasitic parameter effects that often cause sharp nonlinearities. The result is a truly hybrid behavior generation scheme, where the abstractions result in discrete qualitative behaviors (mode and configuration changes), otherwise system behavior evolves continuously. Future work will be directed toward applying this methodology in embedded (computer-based) control of physical systems.

## References

[1] R. Alur, et al., The algorithmic analysis of hybrid systems. J.W. Bakkers, C. Huizing, W.P. de Roeres, and G. Rozenberg (eds.), proc. *11th Intl. Conf. on Analysis and Optimization of Discrete Event Systems*, Springer, pp. 331–351, 1994.

[2] W. Borutzky. Exploiting differential algebraic system solvers in a novel simulation environment. *SAMS*, 17:165–178, 1995.

[3] P.C. Breedveld. Multibond graph elements in physical systems theory. *Jour. of the Franklin Institute*, 319(1/2):1–36, Jan./Feb. 1985.

[4] P. Lötstedt. Coulomb friction in two-dimensional rigid body systems. *Z. angew. Math. u. Mech.*, 61:605–615, 1981.

[5] P.J. Mosterman and G. Biswas. A formal hybrid modeling scheme for handling discontinuities in physical system models. *AAAI-96*, pp. 985–990, Portland, Oregon, 1996.

[6] P.J. Mosterman and G. Biswas. Hybrid modeling specifications for dynamic physical systems. *Intl. Conf. on Bond Graph Modeling and Simulation*, pp. 162-167, Phoenix, AZ, Jan. 1997.

[7] P.J. Mosterman and G. Biswas. A theory of discontinuities in dynamic physical systems. *Jour. of the Franklin Institute*, 334B(6), 1997.

# DISTRIBUTED ARTIFICIAL INTELLIGENCE

# DISTRIBUTED ARTIFICIAL INTELLIGENCE

Distributed AI 1: Interagent Communication

# Middle-Agents for the Internet*

**Keith Decker**
Computer and Information Sciences
University of Delaware
decker@cis.udel.edu

**Katia Sycara** and **Mike Williamson**
The Robotics Institute
Carnegie-Mellon University
(sycara,mikew)@cs.cmu.edu

## Abstract

Like middle-men in physical commerce, middle-agents support the flow of information in electronic commerce, assisting in locating and connecting the ultimate information provider with the ultimate information requester. Many different types of middle-agents will be useful in realistic, large, distributed, open multi-agent problem solving systems. These include *matchmakers* or *yellow page* agents that process advertisements, *blackboard* agents that collect requests, and *brokers* that process both. The behaviors of each type of middle-agent have certain performance characteristics—privacy, robustness, and adaptiveness qualities—that are related to characteristics of the external environment and of the agents themselves. For example, while brokered systems are more vulnerable to certain failures, they are also able to cope more quickly with a rapidly fluctuating agent workforce and meet certain privacy considerations. This paper identifies a spectrum of middle-agents, characterizes the behavior of three different types, and reports on initial experiments that focus on evaluating performance tradeoffs between matchmaking and brokering middle-agents, according to criteria such as load balancing, robustness, dynamic preferences or capabilities, and privacy.

## 1 Introduction

One of the basic problems facing designers of open, multi-agent systems for the Internet is the connection problem [Davis and Smith, 1983]—finding the other agents who might have the information or other capabilities that you need. There are two special types of information used in this process—*preferences* and *capabilities*. In multi-agent information systems, a preference is (meta) knowledge about what types of information have utility for a requester, both in form (John follows the price of SUNW) and other characteristics (John wants only free information; John wants stock quotes at least every 35 minutes). A capability is (meta) knowledge about what types of requests can be serviced by a provider (Mary can provide the current price of any NASDAQ stock, 15 minute-delayed, for free at a rate of 10 quotes per minute). While this paper will focus on information providing agents, providers might also accomplish other types of tasks.

From a privacy standpoint, preference information can flow from a requester to a provider, and capability information can flow the other way. Agents that deal with preference or capability information that are *neither* requesters *nor* providers (from the standpoint of the transaction under consideration) we call *middle-agents*. Different organizational solutions to the connection problem stop this flow of information at different points. We will discuss the full scope of design possibilities presented by our model and a full experimental implementation of two possible designs. In one, known variously as *matchmaker, yellow pages,* or *directory agent* systems [Genesereth and Ketchpel, 1994; Finin et al., 1994; Kuokka and Harada, 1995], capabilities come to be known by all including the requesters, but preferences are kept initially private. In the other, which we call in general *brokered* systems, only the broker comes to know both preferences and capabilities of a class of requesters and providers. Market-based programming systems [Wellman, 1993] are a popular subclass of brokered system.

Privacy, however, is only one concern when choosing a solution to the connection problem. A designer also needs to consider other characteristics, such as the efficiency with which requests are handled and resources are used, the vulnerability of the system to the failure of some component, and the ability to quickly adapt as an open system to changing preferences and capabilities. Our ongoing research aims to develop empirically-validated models of the relationships between the various performance characteristics and system parameters.

This paper will first present our privacy based model of the connection problem, and briefly lay out the space of organizational solutions and their characteristics. We then briefly describe how matchmaking and brokering behaviors can be defined in terms of well-understood communicative acts. By defining these behaviors and how they interact, we are able to better understand how our space of solutions to the connection

---

*This work was supported by ONR Grant #N-00014-96-1-1222.

| preferences initially known by | Capabilities initially known by | | |
| --- | --- | --- | --- |
| | provider only | provider + middle agent | provider + middle + requester |
| requester only | (broadcaster) | "front-agent" | matchmaker/yellow-pages |
| requester + middle agent | anonymizer | broker | recommender |
| requester + middle + provider | blackboard | introducer/bodyguard | arbitrator |

Table 1: Middle-agent roles in the solution space to the connection problem, categorized by explicit initial privacy concerns.

problem constrain agent architecture and behavior design.

Finally, we will examine some empirical results on the comparative performance of brokered and matchmade systems. The questions we will be examining include first the quantitative end-to-end response time advantages and disadvantages of matchmaking and brokering behavior. Second, we will examine characteristics of these behaviors with respect to robust and adaptive open systems, where agents might enter and exit the system at any time. Our experimental results were achieved using an implementation of the WARREN multi-agent financial portfolio management system [Sycara *et al.*, 1996; Decker *et al.*, 1996].

## 2 Definitions and Model

We will examine the connection problem from the standpoint of privacy considerations. In particular, we examine knowledge about requester agent preferences, and provider agent capabilities. A specific *request* is an instance of an agent's preferences, and a specific *reply* or action in service of a request is an instance of an agent's capabilities. Furthermore, an agent can have a mental state with respect to a particular specification of a preference or capability. An *advertisement* is a capability specification such that the agent creating the advertisement is committed to servicing any request that satisfies the advertisement's constraints. Symmetrically, a *want-ad* is a preference specification by a requester who is committed to accepting any reply that meets the constraints in the preference spec.

Preference information can initially be kept private at the requester, be revealed to some middle agent (neither the requester nor the ultimate provider), or be known by the provider itself. The same three possibilities exist for capability info (see Table 1). This leads to nine general middle-agent roles in information-gathering organizations, of which this paper has room to examine three. A **blackboard** is a middle-agent that keeps track of requests. Requesters post their problems; providers can then query the blackboard agent for events they are capable of handling. This class includes newsgroups and bulletin boards. A **broker** is a middle-agent that protects the privacy of *both* the requester and provider. The broker understands both the preferences and capabilities, and routes both requests and replies appropriately. Neither the requester nor provider ever knows directly about the other in a transaction. A **matchmaker/yellow-pages** is a middle agent that stores capability advertisements that can then be queried by requesters. The requesters then choose and contact any provider they wish directly.

Clearly whenever a requester and provider come into direct contact multiple times it is possible for one to learn the preferences or capabilities of the other. For example, some work describe agents that attempt to learn the preferences of people [Maes and Kozierok, 1993]. Other work reports on agents that learn capabilities [Kautz *et al.*, 1996]. Hybrid organizations can be used to protect both requesters and providers from learning agents; e.g., requesters can use an anonymizer middle-agent even if the nominal organization is blackboard-like. An anomynizer middle-agent knows the preferences of the requester, posting on the requester's behalf (such systems are really used on the Internet). Such hybrids add to communication costs and potential for failures, however. We will later discuss another hybrid, the combination of matchmade and brokered organizations, that avoids some of these problems and brings together robustness and efficiency characteristics.

We will focus our attention on the 3 diagonal boxes representing minimal information travel: the blackboard organization where preference commitments or individual requests are posted for everyone to see, but true capabilities remain hidden; a matchmaker organization where providers freely advertise their capabilities but requesters remain private; and brokered organizations where preferences/requests are joined to capabilities using either market allocation mechanisms, OS load balancing algorithms, or managerial task assignment schemes. Because our interest is in open systems, where all agents may come and go, and preferences and capabilities may change dynamically, we are limiting our discussion to middle-agent systems where the capabilities of all agents are not known and hard-coded at compile time.

### 2.1 Organizational Role Behaviors

Each organizational model relies on three basic roles: that of the requester, the middle-agent, and the provider. Any one agent in a domain system might take on multiple roles, for example an agent that requests basic info from several providers, does some complex integration, and then serves the integrated info to other requesters. In this model, communicative acts are limited to requests, replies, and commitments. This has two benefits: first, the semantics of requests and commitments are well-understood [Finin *et al.*, 1994; Cohen and Levesque, 1995], and second, such a model allows us to build simpler agents that can work in an open environment with hybrid behaviors (e.g., both matchmaking and brokering).

In a matchmade organization, providers *advertise*[1] their capabilities with a matchmaker[2]. If those capabilities change, or the agent exits the open system, the provider *unadvertises*. A matchmaker stores these advertisements in a local database. A requester wishing to ask a query first formulates a meta-query asking for advertisements from agents that could respond to the query. This meta-query is asked of a matchmaker, which responds with a set of matching advertisements. The requester can then use its full preferences to choose a provider, and make its request directly. Furthermore, if this type of query is asked often, then the requester can *subscribe* to updated advertisements from a matchmaker, and keep a local cache of the current advertisements (see Section 2.3). Blackboard organizations are extremely similar, but with requester and provider behavior reversed (e.g. requesters send *want-ads* to the blackboard). In brokered organizations, requester behaviors remain the same. In a *pure* brokered organization the brokers are generally known by all the agents, just like a matchmaker is. However, for practicality in an open system *hybrid* brokered organizations use a matchmaker so that providers and requesters can find the appropriate broker (see Section 2.5). Providers query a matchmaker to find an appropriate broker, and then advertise with one broker. Brokers advertise summary capabilities built from all the providers that have advertised with them; these capabilities are advertised in turn to the matchmaker. When a request comes in, the broker matches it with a provider and sends it on; the reply is then sent back to the original requester. The methods by which a broker assigns requests to providers can drawn from several areas of research, including OS load balancing, managerial task assignment, and market-based economics. Which methods to choose will depend on several environmental factors (such as whether the providers are self-interested or cooperative agents) and will be a subject for future work.

## 2.2 Efficiency

The main performance attribute which we have measured (Section 3) is $r$, the total elapsed time taken by a requester to satisfy a service objective. It includes: (1) time spent planning and scheduling by the requester, middle-agent, and provider, $S$ (2) time spent communicating between agents (given that we always use a middle-agent rather than compiling in fixed agent name, this feature of our agents, denoted $C$, includes four communication actions); (3) time spent by the provider providing the service $T$; (4) time spent waiting at a provider which is busy fulfilling prior requests, denoted $Q$ (this is a function of the request generation period, $P$, and of the number of providers, $N$).

Our system can be roughly described by a queuing network model [Lazowska et al., 1984]. According to queuing network theory, the total elapsed time to fulfill a request is $r = D + Q$, where $D$ is total computational demand of the request (in our case, $D = T + S + C$). Note that $r$, like $Q$, is a function of the request generation period and of the number of providers. If requests are generated at a rate greater than the maximum system throughput, i.e. if $P < \frac{D}{N}$ then the system will be *saturated* and $r$ will grow without bound. Otherwise, a fundamental result of queuing theory is that the expected elapsed time per request is:

$$r = \frac{D}{1 - \frac{D}{PN}} \quad (1)$$

This result depends on the service request load being equally distributed across all providers, or else the elapsed time per request will be greater. Since a brokered system can precisely balance the load on providers, while a matchmaker or blackboard organizations only stochastically does so, we would expect the broker to provide better elapsed times. Of course some decentralized load balancing can be done even in a matchmade organization, at the expense of extra communication.

## 2.3 Robustness

Decentralized organizations such as matchmade or blackboard orgs with caching are significantly more robust than centralized organizations such as brokered orgs. Malone [Malone, 1987] examines the basic vulnerability of decentralized and centralized markets, defined as the sum of the expected costs of each possible failure times the probability of that failure. Let $p_r, p_m$, and $p_p$ be the probability of failure of a requester, middle-agent, and provider respectively; and $C_p$ be the cost of reassigning a task upon provider failure, $C_r$ be the cost of losing all the requests at a single requester (this is a function of $r$ and $P$ above), and $C_m$ be the cost of losing all access to a single class of capabilities. Assuming $N$ providers, $A$ requesters, and $B$ brokers each brokering a single class of capabilities, the vulnerability of a brokered organization is

$$Ap_rC_r + Bp_mC_m + Np_pC_p$$

Interestingly, a matchmade system where every requester must check the matchmaker/yellow-pages every time, does no better! However, it is easy to have a requester cache matchmaker results so that if a matchmaker fails, the cached information can be used temporarily. This behavior changes the vulnerability of a matchmade system to

$$Ap_rC_r + Np_pC_p$$

which agrees with Malone's predictions for decentralized markets. Blackboard systems are similar to the matchmade systems, but it is more unlikely for cached *requests* to be useful, as opposed to cached *capabilities*.

## 2.4 Adaptivity

The final characteristic we will discuss is the effect of dynamically changing preferences and capabilities. An example of a preference change in the WARREN domain is the need to track a new stock, or an old stock at a new frequency. An example of

---

[1] All communications here are done via the appropriate KQML performatives [Finin et al., 1994].

[2] At some fixed location; there could be more than one; this can be handled like the Internet DNS.

a capability change is the entry or exit of a new provider. The ability of an organization to quickly adapt to new preferences or capabilities is a function of the distance that the information has to travel, and the costs of keeping that information up-to-date. For example, in a matchmade organization, capability information is available locally to the requesters, and so a change in preferences can be acted on instantaneously. The reverse holds for blackboard organizations and changing capabilities. In either case, the primary costs are in keeping the local cache of public information (capability info, in the matchmade case). Brokered organizations again represent a useful design alternative, requiring only that the broker be notified of changes, but requiring such messages for both capability and preference changes.

Since we are interested in an open system, let us look at what happens to the maximum service time $R$ when providers come and go. $R$ will obviously change as the number of providers $N$ changes, but Eqn 1 only holds when the system is not saturated. When the system *is* saturated, the queues of the remaining agents will begin to grow (along with $R$) until the lost providers are restored. If we define the excess capacity of the system at time $t$ as:

$$\text{Ex}(t) = \frac{N(t)}{D(t)} - \frac{1}{P(t)} \quad (2)$$

then the maximal queue length MQL at time $t$ is:

$$\text{MQL}(t) = \max(0, \text{MQL}(t-1) - \text{Ex}(t)) \quad (3)$$

and we can predict the maximal response time as the max of the steady state response time and the response time implied by the maximal (FIFO) queue length. As usual, this only holds for a brokered system where the load is perfectly balanced as providers come back on line. In a matchmade system, if the agents make no attempt at load balancing, the MQL may continue to grow even after providers come back on line, and when it falls it will do so more slowly than in the brokered system (Section 3).

### 2.5 Hybrid Organizations

Real solutions to real world problems will often require hybrid approaches. If capabilities and matching preferences can be clearly partitioned (creating different connection problem classes) then each class can use a different organization simultaneously. For example, how do agents find an appropriate broker? If we ignore precompiled solutions, one way is to use a matchmaker. Providers advertise their capabilities privately to a broker, and the broker in turn advertises a (usually) more abstract capability(ies) to a matchmaker.

Another way to use hybrid MM/Broker organizations is to take advantage of the best characteristics of each organization alone, at the cost of reduced privacy in emergency situations. Brokered organizations have higher efficiency because of their centralized load-balancing, but are critically non-robust to broker failure. By switching automatically to a matchmade organization (and thus losing provider privacy) the system as a whole becomes robust to broker failure.

Due to our careful definitions of roles earlier, such a hybrid system requires only a small change to the behavior of provider and broker agents: brokers must advertise their brokering as a capability to a matchmaker, and providers must upon initialization register with the matchmaker for broker-capability updates. When a broker is available, the provider uses it; if not, the provider advertises with a matchmaker until one becomes available. A broker failure (once detected via a KQML "Sorry" or timeout) can then transform into a matchmade organization.

## 3 Performance Tradeoffs

The decision to use matchmaking or brokering to solve the connection problem offers many performance tradeoffs. In our remaining space we present two short representative experiments.

System performance is dependent upon a large number of parameters, including the rate at which service requests are generated; the number of providers in the system; the time needed by each provider to fulfill a request; agent failure rates, and so on. We will consider two alternative systems. Each consists of some number of homogeneous providers and requesters. All agents run on serial processors, and the basic service action is non-interruptible. In the "Matchmade" system, each provider advertises itself to one *matchmaker* agent. Requesters query the matchmaker to obtain a current list of providers, choose one randomly, and send it a service request. In the "Brokered" system, providers advertise themselves to a distinguished *broker* agent. Requesters send all service requests directly to the broker, who farms them out to the providers— seeking to equalize the load among them.

Our implementation of these systems consists of real, implemented agents, who experience real communication and processor latencies, etc. These agents are part of the WARREN multi-agent financial portfolio management system. [Sycara *et al.*, 1996; Decker *et al.*, 1996] The broker and matchmaker are the same agents used in the actual portfolio management system. Providers and requesters are instances of WARREN providers and requesters, but we have standardized on a single abstract service to be provided (modeled after stock ticker services).

The actual service time for each request and the period between requests are generated randomly; the service time is distributed normally around the mean $T$, and the request generation period is distributed exponentially around the mean $P$. But because we are using real rather than simulated agents, some parameters are beyond our control, such as the inter-agent communication latency, the computational needs of the broker or matchmaker, and the amount of time spent by the providers and requesters on planning, scheduling, and other internal operations.

We make some further assumptions about the ranges of values that our system parameters will take on. First, we assume that the service time is relatively long compared to the computational overhead of the matchmaker, broker, and providers themselves. This is consistent with actual WARREN agents, which typically require 30 seconds or more to access Internet

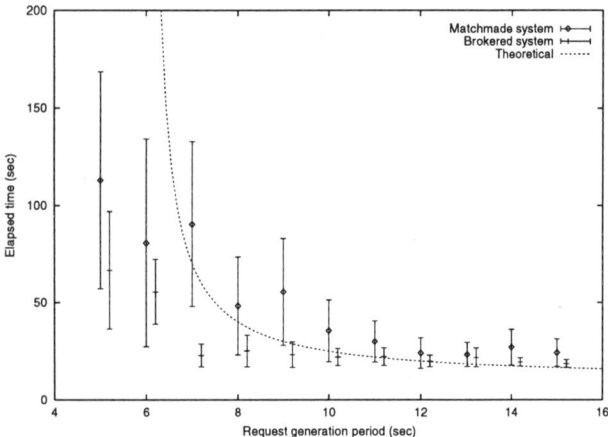

Figure 1: Mean time for 100 service requests as a function of request generation period

resources. Second, we assume that the number of providers is relatively small (we have experimented with systems of up to 20 providers), which is again consistent with the operational WARREN system.

### 3.1 Experiment One: Response Time

In our first experiment, we validate the theoretical model of Eq. 1, and empirically compare the brokered and matchmade systems. For a fixed service time $T$ and number of providers $N$, we vary the request generation period $P$. For each period, we generate 100 requests, and measure the mean and standard deviation of their elapsed times. Figure 1 shows the results as the request generation period varies between 5 and 15 seconds, for systems with 3 providers and a service time of 15 seconds.

Note that despite the crudeness of Eq. 1, it gives a good indication of the expected response time, especially for larger request generation periods[3]. It is clear that the load balancing of the brokered system confers a response time advantage over the matchmade system.

### 3.2 Experiment Two: Provider Failure and Recovery

In our second experiment, we investigate the effect of provider failure and recovery on our two systems. We begin with three providers, and fix the service time and request generation period at 15 and 10 seconds, respectively. After five minutes, we kill one of the providers, and after five more minutes, we kill a second one. Five minutes after that, we bring one of the servers back on line, and then ten minutes later the third one returns. When a provider dies, it sends a SORRY message for each outstanding request. Each of these requests must be reallocated (by either the broker or the original requester) to another provider.

---

[3]For shorter periods, when the system is more highly loaded—or even saturated—our measured values fall below the predictions because we are performing only 100 queries. The earlier requests experience less queuing time, and so skew the results downwards.

Figure 2 shows the results of this experiment. Each point represents the completion of a service request. The response-time superiority of the brokered system stems from the difference in behavior of the two systems when the failed providers come back online. When there is only one provider left operating, the system is saturated, so that provider begins to build up a large backlog of requests. When the second and third providers become available again, the requesters in the matchmade system continue to allocate one-half or one-third of their requests to the overloaded provider, so the backlog persists for a long time.[4] In the brokered system, on the other hand, all new requests are allocated to the new provider, allowing the backlog at the congested provider to quickly dissipate.

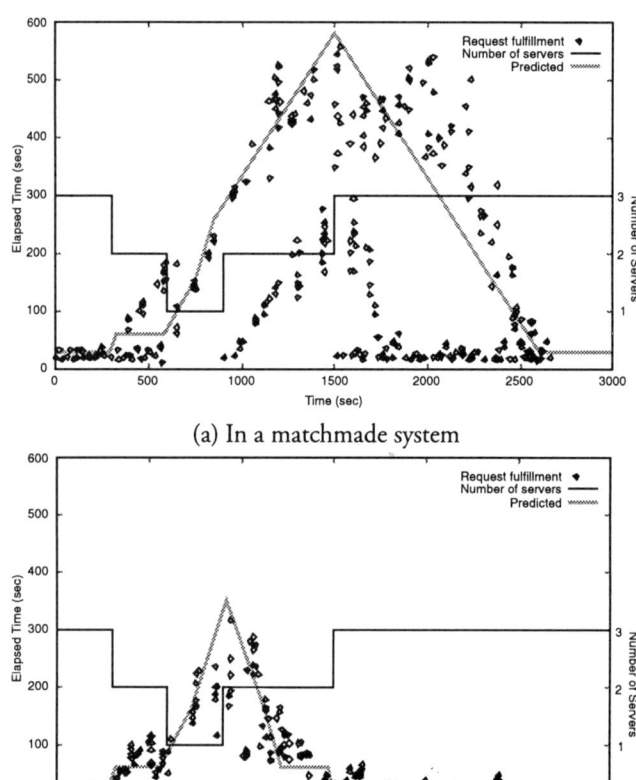

(a) In a matchmade system

(b) In a brokered system

Figure 2: Predicted and actual effect of provider failure and recovery

Figure 2 also shows the predicted response time (for the maximally loaded provider) in the brokered organization as derived from Eqs. 1, 2, and 3 in Section 2.4. Regression analysis shows that the predicted response matches the actual response of the maximally loaded provider with $R^2 = 0.77$ (77% of the variance explained).

Of course this prediction does not hold for the matchmade

---

[4]This effect could be reduced if requesters make an effort at active load balancing.

system. Rewriting Eq. 2 for matchmade organizations as a recurrence equation $\text{Ex}_M(t) =:$

$$\begin{cases} \text{Ex}_M(t-1) & N(t) = N(t-1) \\ \frac{N(t)}{D(t)} - \frac{1}{P(t)} & N(t) < N(t-1) \\ \text{Ex}_M(t-1) + \frac{1}{N}(\frac{N(t)}{D(t)} - \frac{1}{P(t)}) & N(t) > N(t-1) \end{cases}$$

This is the same as before except when adding new agents after being saturated, in which case the maximal queue length is reduced by an amount proportional to the optimum experienced by the brokered system. Using this excess capacity equation results in an regression $R^2$ value of 83% with respect to the measured experimental data.

## 4  Conclusions

This paper examined solutions to the agent connection problem using third-party "middle-agents." Using privacy as an organizing paradigm, we outlined 9 alternative types of middle-agents that result in corresponding organizational classes. We examined analytically the efficiency, robustness, and adaptiveness characteristics of three common alternatives. Finally, we validated two of our models using experimental implementations for matchmade and brokered organizations in a real multi-agent system, concentrating on the speed at which the alternatives adapt to provider failures and reappearances.

Although we did not have space to discuss all 9 alternatives, we hope to stimulate discussion about, and work on, these alternate approaches, each applicable to different situations. Although the concepts of matchmaking, recommending, and brokering have been described in general elsewhere, only Kuokka has provided any detailed experiences with matchmaking. This paper outlines a much larger playing-field for middle-agents, and includes implementation experiences with both matchmade and brokered systems. This is the first work to analyze some of the detailed characteristics of these organizations, drawing on queuing theory and Malone's modeling work with human organizations.

Matchmaker organizations can become elegant decentralized markets with the proper caching mechanisms. They offer preference privacy to requesters, and each requester keeps total control over its own control decisions (so adapting to changing preferences is immediate). Each agent needs to be smart enough to construct a meta-query and evaluate the resulting alternative provider choices. With local caching, matchmade organizations are highly resistant to total failure, but rather degrade gracefully. No simple optimal load balancing is possible; we described an analytical model that predicts system response times without load balancing. Using a matchmaker has slightly higher overhead due to the extra queries.

Brokered organizations include centralized economic markets or traditional bureaucratic managerial units. They can also handle the dynamic entry and exit of agents, and provide an easy way to do load balancing. They can protect the privacy of both requester preferences and provider capabilities. However, they suffer from the need of agents to have static knowledge of the brokers (in a pure, non-hybrid system). In addition, a broker is a communication bottleneck (since all requests and replies need to go through the brokers). Worst of all, they form a single point of failure that cannot be mitigated via local caching.

The solution that we are currently working on is a hybrid system with both matchmakers and brokers. By design, our specified agent behaviors work interchangeably with both organizational roles. A hybrid system allows us to capitalize on the lower overhead and efficient load balancing of a brokered system while retaining the dynamic naming capabilities and greater robustness of a matchmaker system. By trading off privacy in an emergency, a hybrid system can continue to function even in the face of broker failure.

## References

[Cohen and Levesque, 1995] P.R. Cohen and H.J. Levesque. Communicative actions for artificial agents. In *Proc. ICMAS-95*, pages 65–72. AAAI Press, June 1995.

[Davis and Smith, 1983] R. Davis and R. G. Smith. Negotiation as a metaphor for distributed problem solving. *Artificial Intelligence*, 20(1):63–109, January 1983.

[Decker et al., 1996] K. Decker, A. Pannu, K. Sycara, and M. Williamson. Designing behaviors for information agents. In *Proc. Autonomous Agents 97*, pages 404–413. ACM Press, February 1997.

[Finin et al., 1994] T. Finin, R. Fritzson, D. McKay, and R. McEntire. KQML as an agent communication language. In *Proc. CIKM-94*. ACM Press, November 1994.

[Genesereth and Ketchpel, 1994] M.R. Genesereth and S.P. Ketchpel. Software agents. *CACM*, 37(7):48–53,147, 1994.

[Kautz et al., 1996] Henry Kautz, Bart Selman, and Al Milewski. Agent amplified communication. In *Proc. AAAI-96*, pages 3–9, August 1996.

[Kuokka and Harada, 1995] D. Kuokka and L. Harada. On using KQML for matchmaking. In *Proc. ICMAS-95*, pages 239–245. AAAI Press, June 1995.

[Lazowska et al., 1984] E.D. Lazowska, J. Zahorjan, G.S. Graham, and K.C. Sevcik. *Quantitative System Performance*. Prentice Hall, 1984.

[Maes and Kozierok, 1993] P. Maes and R. Kozierok. Learning interface agents. *Proc. AAAI-93*, pages 459–465, July 1993.

[Malone, 1987] Thomas W. Malone. Modeling coordination in organizations and markets. *Management Science*, 33:1317–1332, 1987.

[Sycara et al., 1996] K. Sycara, K. Decker, A. Pannu, M. Williamson, and D. Zeng. Distributed intelligent agents. *IEEE Expert*, 11(6):36–46, December 1996.

[Wellman, 1993] Michael Wellman. A market-oriented programming environment and its application to distributed multicommodity flow problems. *JAIR*, 1:1–23, 1993.

# Semantics and Conversations for an Agent Communication Language *

Yannis Labrou and Tim Finin
Computer Science and Electrical Engineering Department,
University of Maryland, Baltimore County, Baltimore, MD 21250, USA

## Abstract

We address the issues of semantics and conversations for *agent communication languages* and the Knowledge Query Manipulation Language (KQML) in particular. Based on ideas from speech act theory, we present a semantic description for KQML that associates "cognitive" states of the agent with the use of the language's primitives (performatives). We have used this approach to describe the semantics for the whole set of *reserved* KQML performatives. Building on the semantics, we devise the conversation policies, *i.e.*, a formal description of how KQML performatives may be combined into KQML exchanges (conversations), using a *Definite Clause Grammar*. Our research offers methods for a speech act theory-based semantic description of a language of communication acts and for the specification of the protocols associated with these acts. Languages of communication acts address the issue of communication among software applications at a level of abstraction that is useful to the emerging *software agents* paradigm.

## 1 Introduction

Communication among software agents [Petrie, 1996; Nwana, 1996] is an essential property of agency [Wooldridge and Jennings, 1995]. Agent communication languages allow agents to effectively communicate and exchange *knowledge* with other agents despite differences in hardware platforms, operating systems, architectures, programming languages and representation and reasoning systems. We view an agent communication language

---

*This work was supported in part by the Air Force Office of Scientific Research under contract F49620–92–J–0174, and the Advanced Research Projects Agency monitored under USAF contracts F30602–93–C–0177 and F30602–93–C–0028 by Rome Laboratory.

as the medium through which the attitudes regarding the content of the exchange between agents are communicated; it suggests whether the content of the communication is an assertion, a request, a query, *etc.*

Knowledge Query and Manipulation Language (KQML) is such a language; it consists of primitives (called *performatives*) that express attitudes regarding the content of the exchange and allow agents to communicate such attitudes to other agents and find other agents suitable to process their requests. Our research provides semantics for KQML along with a framework for the semantic description of KQML-like languages for agent communication. We also address the issue of conversations, *i.e.*, of sequences of causally-related messages in exchanges between agents and present a method for the specification of conversations (*conversation policies*).

After an introduction to KQML, we describe our semantic framework and give the semantics for a small set of KQML performatives. We follow with our method for describing the protocols (conversations) associated with the primitives and present the resulting conversations for our set of performatives. [1] We end by summarizing our contributions regarding the semantics and the specification of the conversations.

## 2 KQML for Agent Communication

KQML is an abstraction, a collection of *communication* primitives (message types) that express an *attitude* regarding the actual expression being exchanged, along with the assumptions of a simple model for inter-agent communication and an abstract design for KQML-speaking agents. There is no such thing as an *implementation* of KQML, *per se*, meaning that KQML is not an *interpreted* or *compiled* language that is offered in some hardware platform or an abstract machine. Agents *speak* KQML in the sense that they use those primitives, this *library of communication acts*, with their reserved *meaning*. The application programmer is expected to provide

---

[1] Specifications for the full set of KQML performatives and associated policies are available in [reference omitted].

code that processes each one of the performatives for the agent's language or knowledge representation framework. This is a KQML message:

```
(ask-if :sender      A
        :receiver    B
        :language    prolog
        :ontology    bible-genealogy
        :reply-with  id1
        :content     ''spouse(adam,eve)'' )
```

In KQML terminology, *ask-if* is a *performative*.[2] The value of the `:content` is an expression in some language or another KQML message and represents the content of the communication act. The other parameters (*keywords*) introduce values that provide a context for the interpretation of the `:content` and hold information to facilitate the processing of the message. In this example, $A$ is querying $B$ (these are symbolic names for agents[3]), in *Prolog* (the `:language`), about the truth status of *spouse(adam,eve)*. Any response to this KQML message will be identified by *id1* (the `:reply-with`). The ontology[4] *bible-genealogy* may provide additional information for the interpretation of the `:content`. In an environment of KQML-speaking agents there are agents called *facilitators* (*mediators* or *brokers* [Decker et al., 1996]) denote similarly intended agents to whom agents *advertise* their services and ask for assistance in finding other agents that can provide services for them.

Our goal is to provide a semantic description for the language in a way that captures all the intuitions expressed in its existing documentation [ARPA Knowledge Sharing Initiative, 1993]. The lack of semantics for KQML has been a long-standing problem of KQML. Moreover, although agents engage into extended interactions with other agents (conversations), conversations is an issue that has received little attention with respect to KQML, or other agent communication languages (the few notable exceptions are [Barbuceanu and Fox, 1995; Kuwabara, 1995; Bradshaw et al., 1996; Parunak, 1996]). Building on the semantic description we explore the issue of specifying KQML conversations in a formal manner.

## 3 A Framework for the Semantics

We treat KQML performatives as speech acts. We adopt the descriptive framework for speech acts and particularly illocutionary acts suggested by Searle [Searle, 1969; Searle and Vanderveken, 1985]. The semantic approach we propose uses expressions, that suggest the minimum set of preconditions and postconditions that govern the use of a performative, along with conditions that suggest the final state for the *successful* performance of the performative; these expressions describe the states of the agents involved in an exchange and use propositional attitudes like *belief*, *knowledge*, *desire*, and *intention* (this *intentional description* of an agent is only intended as a way of viewing the agent) which have the following reserved meaning:

1. BEL, as in BEL(A,P), which has the meaning that $P$ is (or can be proven) true for A. $P$ is an expression in the native language of agent $A$.[5]
2. KNOW, as in KNOW(A,S), expresses knowledge for $S$, where $S$ is a state description (the same holds for the following two operators).
3. WANT, as in WANT(A,S), to mean that agent A desires the cognitive state (or action) described by $S$, to occur in the future.
4. INT, as in INT(A,S), to mean that A has every intention of doing $S$ and thus is committed to a course of action towards achieving $S$ in the future.

We also introduce two instances of actions:

1. PROC(A,M) refers to the action of $A$ processing the KQML message $M$. Every message after being *received* is *processed*, in the sense that it is a valid KQML message and the piece of code designated with processing the performative for the application indeed processes it. PROC(A,M) does not guarantee proper processing of the message (or conformance of the code with the semantic description).
2. SENDMSG(A,B,M) refers to the action of $A$ sending the KQML message M to $B$.

For an agent $A$ it is BEL(A,P) if and only if $P$ is true (in the *model-theoretic* sense) for $A$; we do not assume any axioms for BEL. Roughly, KNOW, WANT and INT stand for the psychological states of knowledge, desire and intention, respectively. All three take an agent's state description (either a cognitive state or an action) as their arguments. An agent can KNOW an expression that refers to the agent's own state or some other agent's state description if it has been communicated to it. So, KNOW(A,BEL(B,"foo(a,b)")) is valid, if BEL(B,"foo(a,b)") has been communicated to $A$ with some message, but KNOW(A,"foo(a,b)") is not valid because "foo(A,B)" is not a state description. Researchers have grappled for years with the problem of formally capturing the notions of *desire* and *intention*. Various formalizations exist but none is considered a definitive one. We do not adopt a particular one neither we offer a formalization of our own. It is our belief that any of the existing formalizations would accommodate the modest use of WANT and INT in our framework.

Our semantic description, which includes expressions with the mental attitudes and actions we described, provides the following: **(1)** a natural language description

---

[2]The term was first coined by Austin [Austin, 1962], to suggest that some verbs can be uttered so that they perform some action.

[3]We will use the term *agents* to indiscriminately refer to all kinds of KQML-speaking programs and applications.

[4]An ontology is a repository of semantic and primarily pragmatic knowledge over a certain domain.

[5]The *native* language of the application may or may not have modal operators but we do not assume any, here.

of the performative's intuitive meaning; **(2)** an expression which describes the content of the illocutionary act and serves as a formalization of the natural language description; **(3)** preconditions that indicate the necessary state for an agent in order to send a performative (**Pre(A)**) and for the receiver to accept it and successfully process it (**Pre(B)**); if the preconditions do not hold a *error* or *sorry* will be the most likely response; **(4)** postconditions that describe the states of both interlocutors after the *successful* utterance of a performative (by the sender) and after the receipt and processing (but before a counter utterance) of a message (by the receiver); the postconditions (**Post(A)** and **Post(B)**, respectively) hold unless a *sorry* or an *error* is sent as a *response* in order to suggest the unsuccessful processing of the message; **(5)** a completion condition for the performative (**Completion**) that indicates the final state, after possibly a conversation has taken place and the intention suggested by the performative that started the conversation, has been fulfilled; and **(6)** any explanatory comments that might be helpful. the performative.

## 4 Semantics for KQML Performatives

We present the semantics for five KQML performatives (*advertise*, *ask-if*, *tell*, *sorry* and *broker-one*) which can support some interesting agent conversations and illustrate our approach. [6] We first introduce our notation. For a KQML message **performative(A,B,X)**, **A** is the :sender, **B** is the :receiver and **X** is the :content of the performative (KQML message). Occasionally we use **M** to refer to an instance of a KQML message. Capital-case letters from the beginning of the alphabet (*e.g.*, $A$, $B$, etc.) are agents' names and letters towards the end of the alphabet (*e.g.*, $X, Y, Z$) are propositional contents of performatives. All underscores (_) are unnamed, universally quantified variables (they stand for performative parameters that do not have values in the KQML message). Capital case letters preceded by a question mark (?), *e.g.*, ?B, are existentially quantified variables.

All expressions in our language denote agents' states. Agents' states are either actions that have occurred (PROC and SENDMSG) or agents' mental states (BEL, KNOW, WANT or INT). Conjunctions ($\land$) and disjunctions ($\lor$) of expressions that stand for agents' states are agent's states, also, but we do not allow $\land$ and $\lor$ in the scope of KNOW, WANT and INT. Propositions in the agent's native language can only appear in the scope of BEL and BEL can only take such a proposition as its argument. BEL, KNOW, WANT, INT and actions can be used as arguments for KNOW (actions should then be interpreted as actions that have already happened). WANT and INT can only use KNOW or an action as ar-

---

[6] Semantics for the complete set appear in [Labrou, 1996].

guments. When actions are arguments of WANT or INT, they are actions to take place in the future.

A negation of a mental state is taken to mean that the mental state does not hold in the sense that it should not be inferred (we will use the symbol **not**). When ¬ qualifies BEL, *e.g.*, ¬ (BEL(A,X)), it is taken to mean that the :content expression $X$ is not true for agent $A$, *i.e.*, it is not provable in $A$'s knowledge base. Obviously, what "not provable" means is going to depend on the details of the particular agent system, for which we want to make no assumptions.

### advertise(A,B,M)

1. A states to B that A can and will process the message $M$ from B, if it receives one (A commits itself to such a course of action).
2. INT(A,PROC(A,M))
   where $M$ is the KQML message **performative_name(B,A,X)**.
3. **Pre(A)**: INT(A,PROC(A,M))
   **Pre(B)**: NONE
4. **Post(A)**:
   KNOW(A,KNOW(B,INT(A,PROC(A,M))))
   **Post(B)**: KNOW(B,INT(A,PROC(A,M)))
5. **Completion**: KNOW(B,INT(A,PROC(A,M)))
6. An *advertise* is a commisive act, in the sense that it commits its sender to process $M$, as suggested by the announcement of the intention to process. If B is a *facilitator* then B is interchangeable (in the semantic description) with the name of any agent the facilitator knows about.

### ask-if(A,B,X)

1. A wants to know what B believes regarding the truth status of the content $X$.
2. WANT(A,KNOW(A,S))
   where $S$ may be any of BEL(B,X), or ¬(BEL(B,X)).
3. **Pre(A)**: WANT(A,KNOW(A,S)) $\land$ KNOW(A,INT(B,PROC(B,M)))
   where $M$ is **ask-if(A,B,X)**
   **Pre(B)**: INT(B,PROC(B,M))
4. **Post(A)**: INT(A,KNOW(A,S))
   **Post(B)**: KNOW(B,WANT(A,KNOW(A,S)))
5. **Completion**: KNOW(A,$S'$) )
   where $S'$ is either BEL(B,X) or ¬(BEL(B,X)), but not necessarily the same instantiation of $S$ that appears in $Post(A)$, for example.
6. **Pre(A)** and **Pre(B)** suggest that a proper advertisement is needed to establish them (see *advertise* and our comments in Section 7).

### tell(A,B,X)

1. A states to B that A believes the content to be true.
2. BEL(A,X)

3. **Pre(A)**: BEL(A,X) ∧ KNOW(A,WANT(B,KNOW-(B,S)))
   **Pre(B)**: INT(B,KNOW(B,S))
   where $S$ may be any of BEL(B,X), or ¬(BEL(B,X)).
4. **Post(A)**: KNOW(A,KNOW(B,BEL(A,X)))
   **Post(B)**: KNOW(B,BEL(A,X))
5. **Completion**: KNOW(B,BEL(A,X))
6. The completion condition holds, unless a *sorry* or *error* suggests B's inability to acknowledge the *tell* properly, as is the case with any other performative.

**sorry(A,B,Id)**
1. A states to B that although it processed the message, it has no (possibly further) response to provide to the KQML message $M$ identified by the :reply-with value **Id** (some message identifier).
2. PROC(A,M)
3. **Pre(A)**: PROC(A,M)
   **Pre(B)**: SENDMSG(B,A,M)
4. **Post(A)**: KNOW(A,KNOW(B,PROC(A,M))) ∧ not($Post_M(A)$),
   where $Post_M(A)$ is the **Post(A)** for message $M$.
   **Post(B)**: KNOW(B,PROC(A,M)) ∧ not($Post_M(B)$)
5. **Completion**: KNOW(B,PROC(A,M))
6. The postconditions and completion conditions do not hold, even though A dispatched the performative to the appropriate function, because A could not (or did not want) to come up with a response that would result to their satisfiability. The *not* should be taken to mean that the mental state it qualifies should not be inferred to be true as a *result* of this particular message. This does not mean that for example $Post_M(B)$ does not hold if it has already been established by a previous message; it is up to B to decide (perhaps after using additional information) if and how it wants to alter its internal state with respect to the *sorry*.

**broker-one(A,B,performative(A,_,X))**
Let $D$ be an agent such that CANPROC$(D,performative(B,D,X))$ [7] and *performative* be a performative that entails a request (a *directive*); for the set of performatives presented here, only *ask-if* falls into this category. B sends **performative(B,D,X)** to $D$, receives some *response* (depending on the *performative*) from $D$, let us call it **response(D,B,X')**, and then B sends to A the message **forward(B,A,_,A,response(_,A,X'))**. [8]

Semantically this is a three-party situation. We break down the semantic description to the three (agent) pairs involved in the transaction.

**A and B** For A and B, the semantics are **not** those of a **performative(A,B,X)**, meaning that A is aware that whatever *response*, if any, comes from B is merely an "echo" of the utterance of the broker-ed agent $D$. So, the semantics is:

1. A wants B (a broker) to send the :content of the *broker-one* to some agent that can process it and eventually forward the response of the broker-ed agent back to A.
2. WANT(A,SENDMSG(B,D,M))
   where $M$ is **performative(B,D,X)** and D is an agent such that CANPROC(D,M).
3. **Pre(A)**: WANT(A,SENDMSG(B,D,M))
   **Pre(B)**: B has to be a *facilitator*; an agent can be a facilitator if and only if it can process performatives like *broker-one*, although it is usually more helpful to ascribe facilitator status to an agent in advance, so that agents can know which agent to contact for such requests.
4. **Post(A)**: KNOW(A,SENDMSG(B,D,M))
   **Post(B)**: SENDMSG(B,D,M))
5. **Completion**: SENDMSG(B,A,forward(B,A,_,A,M'))
   where M' is the message **response(_,A,X')** generated by the broker-ed agent's response to B, *i.e.*, **response(D,B,X')**.
6. To offer an example, if the :content of the *broker-one* was **ask-if(A,_,X)**, A understands that the (possible) *response* **forward(B,A,_,A,tell(_,A,X))** does not imply that BEL$(B,X)$, since D's response to B is wrapped in a *forward* and then sent to A. Also, D's name is omitted in the *forward*, so A does not know D's name.

**B and D** For B and D the semantics are those of **performative(B,D,X)**, meaning that as far as D knows of, the exchange has the meaning and repercussions of **performative(B,D,X)** (and whatever additional responses) being exchanged between B and D.

**A and D** For A and D the semantics are those of **performative(A,D,X)** (let us call it $M$) but with the major difference that this is an one-sided exchange. So, $Pre_M(D)$ and $Post_M(D)$ are empty because D does not know that it has this exchange with A. Additionally, A can have no prior knowledge (in $Pre_M(A)$) of its interlocutor's state. Finally, the applicable $Post_M(A)$ and $Completion_M$ lack the name of D. To show how this translates semantically, we present the semantics of **broker-one(A,B,ask-if(A,_,X))** for agent A and the broker-ed agent D.

1. A wants to know what some other agent believes regarding the truth status of the content $X$.

---

[7]CANPROC, as in CANPROC(A,M), stands for "A being able to process message $M$." It is always the case that if **advertise(A,B,M)** then CANPROC(A,M), but it could very well be the case that CANPROC(A,M) may be inferred in other ways (this is to be provided or inferred by B). CANPROC is entirely different from PROC; CANPROC suggests ability to process and PROC suggest that the agent will process (or has already processed) a performative, in the sense that it will (or did) dispatch the message to the appropriate piece of code for handling.

[8]The performative *forward* is not presented here. Its meaning is basically the intuitive one and the four parameters :from, :to, :sender and :receiver refer respectively to the originator of the performative in the :content, the final destination, the :sender of the *forward* and the :receiver of the *forward*.

2. WANT(A,KNOW(A,S))
   where $S$ may be any of BEL(?D,X), or $\neg$(BEL(?D,X)).
3. **Pre(A)**: WANT(A,KNOW(A,S))
   **Pre(D)**: NONE
4. **Post(A)**: INT(A,KNOW(A,S))
   **Post(D)**: NONE
5. **Completion**: KNOW(A,$S'$) )
   where $S'$ is either BEL(?D,X) or $\neg$(BEL(?D,X)), but not necessarily the same instantiation of $S$ that appears in $Post(A)$, for example.
6. In effect, $D$'s identity remains unknown to $A$ and $D$ is unaware that $A$ knows its belief regarding the truth status of X.

## 5 Describing Conversations

A *conversation* is a sequence of KQML messages that belong to the same thread of interaction between two or possibly more agents. We assume some sort of (intuitive) causal relation between messages that are taken to belong in the same *conversation* and we use the `:in-reply-to` value as the indicator of such linkage. *Conversation policies* are rules that describe permissible *conversations* among KQML-speaking agents. The *conversation policies* that we provide do not describe *all* possible *conversations* because more complex interactions (and thus conversations) are possible between KQML-speaking agents. The conversations we present can be used as building blocks for more complex interactions.

We use the Definite Clause Grammars (DCGs) formalism for the specification of the *conversation policies* for the KQML performatives. DCGs extend Context Free Grammars (CFGs) in the following way [Perreira and Warren, 1986]: 1) **Non-terminals** may be compound terms (instead of just atoms as in the CFG case), and 2) the body of a rule may contain **procedural attachments**, written within "{" and "}" (in addition to terminals and non-terminals), that express extra conditions that must be satisfied for the rule to be valid. For example, a DCG rule might look like
**noun(N)** $\longrightarrow$ **[W], {RootForm(W,N), is_noun(N)}**
with the possible meaning that "a phrase identified as the noun **N** may consist of the single word **W** ([**W**] is a terminal), where **N** is the root form of **W** and **N** is a noun" [Perreira and Warren, 1986].

### 5.1 DCGs & KQML conversation policies

*Conversation policies* describe both the sequences of KQML performatives and the constraints and dependencies on the values of the *reserved parameters* of the performatives involved in the conversations. In other words, we are not only interested in asserting that an *ask-if* might be followed by a *tell* (among other performatives) but we want to also capture constraints such as, the content$_{ask-if}$ being the same with the content$_{tell}$ or the reply$-$with$_{ask-if}$ being also the in $-$ reply $-$ to$_{tell}$. The DCG we provide in the next section fully describes the above in a declarative fashion.

Each KQML message is a *terminal* in the DCG. A terminal is a list of the following values: `performative_name`, `:sender`, `:receiver`, `:in-reply-to`, `:reply-with`, `:language`, `:ontology`, `IO` (if IO is set to 1 the message is an incoming message and if it is set to 0 the current message is an outgoing message), `:content`, and whenever the `:content` is a performative itself, then the `:content` is going to be a list itself. *Terminals* are enclosed in "[" and "]", so a terminal in our DCG will look like: **[[ask-if,A,B,id1,id2,prolog,bar,foo(X,Y)]]** In the DCG we present here, we omit the `:language` and `:ontology` values (we take them to remain unchanged throughout the same conversation).

The conversation policies we present are tied to the semantics in the sense that changes in the semantic description would result to different conversation policies. Our conversation policies technically are not inferred from the semantic description, but they define the *minimal* set of conversations that are consistent with the semantics when following these heuristics:

- If a performative has *preconditions* for the sender, then it cannot start a conversation if these preconditions have to be established by a communication act (see *tell*).
- If the *completion condition(s)* for a performative are not not a subset of the postconditions, then a performative cannot end a conversation since further (communicative) action has to take place to establish the *completion condition(s)* (see *ask-if*).
- A performative may be preceded by a performative that can (partially) establish its preconditions (*e.g.*, a *tell* may be preceded by an *ask-if*; compare **Post(A)** for *ask-if* and **Pre(A)** for *tell*).

## 6 Converation Policies, in detail

We present a complete DCG for the set of performatives presented in Section 4. This is a subset of the full DCG that describes the whole set of conversation policies (see [Labrou, 1996]) and is intended as a demonstration of how our method may be used.

**ask-if, tell**
S $\rightarrow$
    s(CC,P,S,R,IR,Rw,IO,C),
    {member(P,[advertise,broker-one])}
s(CC,ask-if,S,R,IR,Rw,IO,C) $\rightarrow$
    [[ask-if ,S,R,IR,Rw,IO,C]] |
    [[ask-if ,S,R,IR,Rw,IO,C]], {OI is abs(1-IO)},
        r(CC,ask-if,S,R,_,Rw,OI,C)
r(CC,ask-if,R,S,_,IR,IO,C) $\rightarrow$
    [[tell ,S,R,IR,Rw,IO,C]] |
    problem(CC,R,S,IR,_,IO)
The rules are organized into groups that describe the

sub-dialogues that may start with a performative, or a group of them and are written so that any sequence of messages that is reachable from the start is also a conversation that will be accepted by the DCG. Note that there is no notion of a *complete* KQML conversation, although it might be possible to define such conversations in some cases. Rules might be called by other rules.

As a result, an *advertise* of an *ask-if* is a conversation; if a proper *ask-if* follows the *advertise*, the sequence of *advertise* and *ask-if* is a conversation; and finally, if an appropriate *tell* follows the *ask-if*, the resulting sequence of the three messages will be a conversation that the DCG will accept. The values of the various terminals and non-terminals define what an *appropriate* follow-up is, at any point of a KQML exchange. We use the following variables for the various tokens that appear in the DCG (symbols that start with a capital-case letter are variables and those that start with small-case letters are constants): CC stands for the *current conversation* that the DCG handles; P is the *performative_name*; S is the :sender; R is the :receiver; IR is the :in-reply-to value; Rw is the :reply-with; IO and OI are the variables that indicates if a message is an incoming or outgoing one (they only take the values 0 and 1 and always have complimentary values) ; C is the :content; and [] is the *empty string*.

We take the position that all starting points for conversations are *advertise* performatives and the *broker-one* performative (when sent to, or processed by facilitators). *Ask-if* may follow an *advertise* and may be responded to (in this KQML subset) with a *tell*.[9] The :in-reply-to value of the response must equal the :reply-with of the *ask-if* for all performatives that act as a *response* or a *follow-up* to some other performative. Also, notice that the :content of a response is the same as the :content of the querying performative in the case of the *ask-if*.

### sorry

problem(CC,R,S,IR,Rw,IO) →
    [[sorry ,S,R,IR,Rw,IO,[]]]

A *problematic* or a *non-positive* response, *i.e.*, a *sorry* (or an *error*, not included here) is always a possibility and those two performatives may follow almost any performative (except for another *sorry* or *error*).

### advertise

s(CC,advertise,S,R,_,Rw,IO,_) →
    { OI is abs(1-IO) },
    [[advertise,S,R,_,Rw,IO,[P1,R,S,Rw,_,OI,C1]]] ,
        {member(P1,[ask-if])},
        c_adv(CC,P1,S,R,Rw,_,OI,C1)
c_adv(CC,P,R,S,Rw_adv,_,IO,C) →
    s(CC,P,S,R,Rw_adv,_,IO,C) |

---
[9]A response with a *sorry* or *error* (not included in this set) is always a possibility of course.

problem(CC,S,R,Rw_adv,_,IO) | []

The procedural attachment restricts the performatives that might appear in the :content of an *advertise*. The :content has the form of the expected follow-up to the *advertise*. This follow-up is given by the part of the DCG that starts the sub-dialogue for the embedded performative. Note that it is possible to have a *sorry* response to the *advertise* itself, as well to the follow-ups to the *advertise*.

### broker-one

s(CC,broker-one,S,R,IR,Rw,IO,C) →
    {OI is abs(1-IO)},
    [[broker-one,S,R,IR,Rw,IO,[P1,R,_,R,Rw,Rw1,_,C1]]] ,
        {member(P1,[ask-if])},
        c_brk_one(CC,P1,S,R,Rw,Rw1,OI,C1)
c_brk_one(CC,P,R,S,Rw_brk,Rw,1,C) → [] |
    problem(CC,S,R,Rw_brk,_,1) |
    r(CC,P,Brk,R,_,Rw,1,C)
c_brk_one(CC,P,R,S,Rw_brk,Rw,0,C) → [] |
    problem(CC,S,R,Rw_brk,Rw,0) |
    s(CC,P,S,Brk,Rw_brk,Rw,0,C),
        c_brk_one1(CC,P,S,R,Brk,Rw_brk,Rw,0,C)
c_brk_one1(CC,P,S,R,Brk,Rw_brk,Rw,IO,C) → [] |
    {OI is abs(1-IO), last(CC,[P1,Brk,S,Rw,Rw1,OI,C1]),
    assert(send_MSG([forward,S,R,Rw_brk,Rw2,OI,
            [P1,_Brk,R,Rw_brk,Rw1,OI,C1]]))} |
    [[forward,S,R,Rw_brk,Rw2,OI,
            [P1,_Brk,R,Rw_brk,Rw1,OI,C1]]]

The *broker-one* performative presents an interesting case because it involves a three-party interaction. The receiver$_{broker-one}$ sends the content$_{broker-one}$ (with the appropriate values) to some other agent and then passes the response(s) to it to the sender$_{broker-one}$. The last part of this exchange can be done automatically with a procedural attachment in the DCG instead of being taken care of by the handler function for *broker-one*. As the c_brk_one1 rule suggests, a sub-dialogue (a new conversation) with the third agent starts and the response (or follow-up), *i.e.*, the last message in the conversation being handled by the DCG with the expected values for :sender and :in-reply-to, is sent to sender$_{broker-one}$ (this is the meaning of the procedural attachment in the c_brk_one1 rule, that makes reference to predicates that are not a part of the DCG).

If the **local** agent sent a *broker-one*, the message expected is the prescribed response or follow-up to the *performative* in the :content. Technically this message (or messages) will arrive wrapped in a *forward* but from the DCG point of view will be stripped from their "forwarding" packaging. This performative is a prime example of how complicated interactions might be composed from the simpler building blocks.

# 7 Discussion

The issue of semantics for communication acts has received a fair share of attention. Cohen and Lesveque suggest a model for rational agents [Cohen and Levesque, 1990], which uses a *possible-worlds* formalism, that can in turn be used as a framework for the semantic description of illocutionary acts [Cohen and Levesque, 1995; Smith and Cohen, 1996]. Sadek [Sadek, 1992] has also taken on a similar task of defining rational agency and defining communicative acts on top of it. Finally, Singh proposes a model of agency [Singh, 1993a], which differs from that of Cohen and uses it as a framework for the semantic treatment of speech acts [Singh, 1993b].

In contrast, we draw directly from a high-level speech act account, although the resulting preconditions/postconditions framework is reminiscent of planning (but it could also be thought as operational semantics, *i.e.*, transitions on agents' states). Also, we provide no formal semantics (in a *possible-worlds* formalism or some similar framework) for the modal operators but we restrict the scope and use of these operators, so that they can be subsumed by similar modalities whose semantics could be provided by an intentional theory of agency. Apart from the complexity of possible-worlds–like formalisms which can be prohibiting for the intended audience of our semantic description that includes application developers that want to support KQML in their software agents, we want to avoid a tight coupling with a particular theory of agency. Another common element of the mentioned approaches is the strictly declarative definitions of the primitives. Instead, our preconditions, postconditions and completion conditions framework suggests a more operational approach which we hope will be useful to implementors that have to provide the code that processes the communication primitives.

By attempting a semantics for communication acts without a theory of agency, *i.e.*, formal semantics for the propositional attitudes (operators), we certainly give up interesting inferencing. For example, if an agent sends **tell(A,B,X)** and later **tell(A,B,X $\rightarrow$ Y)**, $B$ will not be able to infer that $\text{BEL}(A,Y)$ (since we do not even assume a universal *weak S4* model for BEL) based on the KQML semantics alone. Nothing is lost though, because the additional information of the agent theory that holds for the agent can be supplied as part of the KQML exchange (*e.g.*, in the :ontology value of a KQML message) and subsequently taken into consideration for further inferencing. In the end, we trade a formal semantics for the propositional attitudes, which inevitably define a *model of agency* that is unlikely to be universal for all agents, for a simpler formalism and agent theory independence.

Objections may be raised regarding some of our choices regarding the meaning we chose to attribute to some of the performatives. Our semantics for *tell*, for example, suggest that an agent can not offer unsolicited information to some other agent. This can be easily amended by introducing another performative, let us call it *proactive-tell* which has the same semantic description as *tell* with the following difference: **Pre(A)** is $\text{BEL}(A,X)$, and **Pre(B)** is empty. Similarly, an agent $A$ can send an *ask-if* to agent $B$ if and only if $A$ knows that $B$ is going to process such a request. Implicit in this choice, is our preference for a model where agents advertise their services so that other agents (with the help of *mediators* or *facilitators*) can find agents that can process requests for them. A "relaxed" version of *ask-if* can be introduced to allow for direct querying. The semantic description of this *proactive-ask-if* differs from that of *ask-if* as follows: **Pre(A)** is $\text{WANT}(A,\text{KNOW}(A,S))$, and **Pre(B)** is empty. Following KQML's tradition of an open standard, the KQML users' community should decide the performative names to be associated with whatever semantic description. Additionally, these two "new" performative could be starting points for conversations in our conversation policies.

Our description and implementation of the conversation policies using a DCG allows as to provide a description that would not be possible had we chosen a CFG or a Finite State Machine for the task. Another formalism that would probably provide us with the same flexibility is that of Augmented Transition Networks[10] (ATNs), but DCGs have the advantage that they can be expressed directly in a general purpose programming language like Prolog (in fact our DCG is a Prolog program). The conversation policies do not prescribe the only possible behavior for an agent but they rather define one which is consistent with the semantics. Such a specification is in no way a prescriptive one and thus does not constrain elaborate agents but it could be useful for simpler ones.

# 8 Conclusions

We have presented excerpts of a complete semantic description for the primitives in the agent communication language KQML. This specification uses a framework for the semantic description of KQML-like languages [11] for the linguistic communication among software agents along with a method for specifying the conversations that builds on our semantic description. We have used our approach to provide the semantics and conversation policies for the full set of KQML primitives and we have presented the framework and the semantic description along with the method and the conversation policies'

---

[10] Perreira and Warren claim that DCGs are at least as powerful of a formalism as ATNs ([Bates, 1979]), with DCGs having some considerable advantages over ATNs ([Perreira and Warren, 1986]).

[11] That is, languages of attitude-expressing communication primitives, modeled after speech acts.

specification for a handful of performatives.

The conversation policies present us with some attractive possibilities. They can be used to devise a software component that monitors an agent's incoming and outgoing messages and ensures that it only engages in valid KQML conversations of well-formed KQML messages. Such a component can keep track of an agent's multiple interactions (conversations) with other agents and offer ways to recover from unforeseen situations. Alternatively, one may view an agent as a collection of conversations that "unfold" concurrently as the agent interacts with other agents. Finally, the conversation policies can be used as building blocks for more complex interactions. In the end, we should keep in mind that agents do not use the primitives of a communication language statically, but in order to carry, often complex, interactions which the conversation policies can help describe.

# References

[ARPA Knowledge Sharing Initiative, 1993] ARPA Knowledge Sharing Initiative. Specification of the KQML agent-communication language. ARPA Knowledge Sharing Initiative, External Interfaces Working Group working paper., July 1993.

[Austin, 1962] J.L. Austin. *How to do things with words*. Harvard University Press, Cambridge, MA, 1962.

[Barbuceanu and Fox, 1995] M. Barbuceanu and M. S. Fox. COOL: a language for describing coordination in multi-agent systems. In *Proceedings of the 1st International Conference on Multi-agent systems (ICMAS'95)*, pages 17–24. AAAI/MIT Press, 1995.

[Bates, 1979] Madeleine Bates. The theory and practice of augmented transition network grammars. In Leonard Bolc, editor, *Natural Language Communication with Computers*, Lecture Notes in Computer Science, pages 191–260. Morgan Kaufmann, 1979.

[Bradshaw et al., 1996] Jeffrey M. Bradshaw, Stuart Dutfield, Pete Benoit, and John D. Woolley. Kaos: Toward an industrial-strength open agent architecture. In Jeffrey M. Bradshaw, editor, *Software Agents (in preparation)*. AAAI/MIT Press, 1996.

[Cohen and Levesque, 1990] Philip R. Cohen and Hector J. Levesque. Intention is choice with commitment. *Artificial Intelligence*, 42:213–261, 1990.

[Cohen and Levesque, 1995] Philip R. Cohen and H.J. Levesque. Communicative actions for artificial agents. In *Proceedings of the 1st International Conference on Multi-Agent Systems (ICMAS'95)*. AAAI Press, June 1995.

[Decker et al., 1996] Keith Decker, Mike Williamson, and Katia Sycara. Matchmaking and brokering. In *Proceedings of the 2nd International Conference on Multi-Agent Systems (ICMAS'96)*, December 1996.

[Kuwabara, 1995] K. Kuwabara. AgenTalk: coordination protocol description for multi-agent systems. In *Proceedings of the 1st International Conference on Multi-agent systems (ICMAS'95)*. AAAI/MIT Press, 1995.

[Labrou, 1996] Yannis Labrou. *Semantics for an Agent Communication Language*. PhD thesis, University of Maryland, Baltimore County, August 1996.

[Nwana, 1996] Hyacinth S. Nwana. Software agents: an overview. *Knowledge Engineering Review*, 11(3):1–40, September 1996.

[Parunak, 1996] H. Van Dyke Parunak. Visualizing agent conversations: Using enhanced dooley graphs for agent design and analysis. In *Proc. of the 2nd International Conference on Multi-Agent Systems (ICMAS'96)*, 1996.

[Perreira and Warren, 1986] F. Perreira and D. Warren. Definite clause grammars for language analysis. In Barbara J. Grosz, Karen Sparck Jones, and Bonnie Lynn Webber, editors, *Readings in Natural Language Processing*, pages 101–124. Morgan Kaufmann Publishers, 1986.

[Petrie, 1996] Charles Petrie. Agent-based engineering, the web, and intelligence. *IEEE Expert*, December 1996.

[Sadek, 1992] M.D. Sadek. A study in the logic of intention. In *Proceedings of the 3rd Conference on Principles of Knowledge Representation and Reasoning (KR'92)*, pages 462–473, Cambridge, MA, 1992.

[Searle and Vanderveken, 1985] J. Searle and D. Vanderveken. *Foundations of illocutionary logic*. Cambridge University Press, Cambridge, UK, 1985.

[Searle, 1969] John R. Searle. *Speech Acts*. Cambridge University Press, Cambridge, UK, 1969.

[Singh, 1993a] M.P. Singh. A logic of intentions and beliefs. *Journal of Philosophical Logic*, 22:513–544, 1993.

[Singh, 1993b] M.P. Singh. A semantics for speech acts. *Annals of Mathematics and Artificial Intelligence*, 8(I-II):47–71, 1993.

[Smith and Cohen, 1996] Ira A. Smith and Philip R. Cohen. Toward a semantics for an agent communications language based on speech-acts. In *Proceedings of the 13th National Conference on Artificial Intelligence*. AAAI/MIT Press, August 1996.

[Wooldridge and Jennings, 1995] M. Wooldridge and N.R. Jennings. Intelligent agents: Theory and practice. *Knowledge Engineering Review*, 10(2), 1995.

# Persuasion among Agents : An Approach to Implementing a Group Decision Support System Based on Multi-Agent Negotiation

Takayuki Ito and Toramatsu Shintani

Department of Intelligence and Computer Science
Nagoya Institute of Technology
Gokiso, Showa-ku, Nagoya, 466 JAPAN
E-Mail:{itota,tora}@ics.nitech.ac.jp

## Abstract

In this paper, we propose a persuasion mechanism for negotiation among agents for a group decision support system and implement a group choice design support system (GCDSS). GCDSS helps a group decision to make a reasonable choice from alternatives. In the system, each user manages a system for an Analytic Hierarchy Process (AHP) and an agent. Each user subjectively constructs a decision hierarchy and determines the various weights of alternatives by using AHP. Based on the hierarchy and weights, agents negotiate with each other on behalf of their users. During the negotiation, agents persuade one another. Adopting some of the features of AHP, we implement a new persuasion mechanism. We have implemented the GCDSS to see how effectively the persuasion mechanism can be used. The results of our current experiments demonstrated that the persuasion mechanism is an effective method for a group decision support system based on multi-agent negotiation.

## 1 Introduction

Group decision support systems (GDSSs)[Desanctis and Gallupe, 1987] are being investigated very actively in the field of operations research. The GDSS aims to improve the process of group decision-making by removing common communication barriers, providing techniques for structuring decision analysis, and systematically directing the pattern, timing or content of discussions.

We can classify GDSSs into three types according to their approach to supporting a group. Type 1 GDSSs improve the decision process by facilitating the exchange of information among members. Type 2 GDSSs provide decision-modeling and group-decision techniques aimed at reducing the uncertainty and noise that occur in the group's decision process. Type 3 GDSSs are characterized by machine-induced group communication support and often provide expert advice in the selection and arrangement of rules to be applied during a meeting. If we introduce Artificial Intelligence (AI) methods into the Type 3 GDSSs, we can expect to further enhance the intelligence of their support. For this reason, and in keeping with the current high expectations for effective GDSSs, we here implement a Type 3 GDSS using intelligent agents.

The term *agent*, used in a variety of ways, has recently commanded much attention in the field of AI. For purpose of this paper, an agent can act autonomously and cooperatively in a network environment on behalf of its users. Agents have many crucial functors, one of the important being the attainment of consensus. Reaching a consensus also is a group decision process that should be supported by the Type 3 GDSS.

Agent negotiation for group decision support has been studied widely. In particular, multi-agent meeting scheduling [Ephrati et al., 1994; Garrido and Sycara, 1996; Sen and Durfee, 1994] has been a current topic of research. [Sen and Durfee, 1994] has been focused on solving meeting scheduling problem using a central host agent. However, user preferences are not taken into account. [Ephrati et al., 1994] presented an alternative approach which is economic in flavor. They introduced the Clark Tax Mechanism as a method for removing manipulability from agents. [Garrido and Sycara, 1996] has been focused on decenterized meeting scheduling with user preferences taking into account. They did not, however, establish how to measure the subjective judgements of users. In this paper, in order to measure the judgements of users, we employ the Analytic Hierarchy Process (AHP) [Saaty, 1980].

In general, a group reaches consensus by use of a vote. But the result of voting are often inconsistent, largely due to the inconsistency of voting rules: Majority Rule, Single Voting Rule, etc. Arrow's impossibility theorem has shown that no voting method exists which satisfies all of the following four conditions: positive association of social and individual values, independence of irrelevant alternatives, citizen's sovereignty and no-dictatorship[Luce and Raiffa, 1985]. In this paper we propose a persuasion mechanism[Ito and Shintani, 1996a; 1996b] rather than voting methods for negotiation among agents.

In agent negotiation, the persuasion mechanism can be defined as follows. When agent A persuades agent B, agent A sends a persuasion message to agent B. Then, according to the message, agent B tries to change its belief. If agent B is able to change its belief, the persuasion is a success. However, if agent B cannot change its belief, the persuasion is a failure. A concrete method for implementing the persuasion mechanism is given in section 3.3.

Our agents are known as *software agents* [Kautz et al., 1994]. Their activities include carrying out tasks on behalf of their users, making suggestions to their users, and so on. Reliability and the ability to depute are important qualities of the software agents. It can be said that our agents have a high ability to depute users, since our agents negotiate with each other by use of the persuasion mechanism on behalf of users. In addition, our agents have an explanation mechanism. They can explain to their users why they have been persuaded, when they are persuaded, who persuaded them and how they are persuaded by use of graphical user interface. Because of their explanation mechanism, our agents are more reliable.

The aim of this paper is to present the persuasion mechanism among agents and to implement a Group Choice Design Support System (GCDSS) based on the persuasion mechanism. The paper consists of five sections. In section 2, we show the architecture of our system and the process of supporting a group decision. In section 3, we present an agent mechanism for our system. Our agent has a management mechanism for the user's hierarchy of AHP, a persuasion mechanism for negotiating with other agents, and an explanation mechanism for their reliability. In section 4, we show an implementation of the GCDSS using the persuasion mechanism and discuss the results of our current experiments. Some concluding remarks are presented in section 5.

## 2 Group Choice Design Support System

### 2.1 System Architecture

Figure 1 shows the architecture of our system. In our sys-

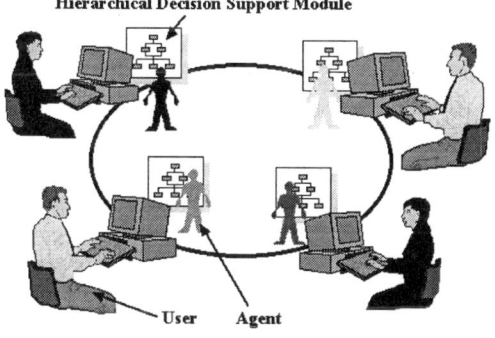

Figure 1: System architecture

tem, users' computers are connected by a network. Each

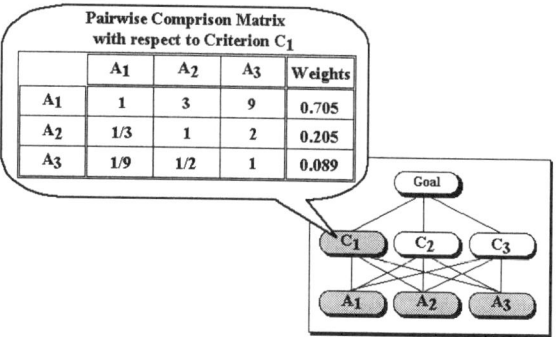

Figure 2: Analytic Hierarchy Process

user's decision-making is supported by their own hierarchical decision support module in our system. Agents manage their particular user's hierarchical decision support module, and negotiate based on the information it supplies. The agent's tasks are described in section 3. The hierarchical decision support module has functions to help generate alternatives, to make judgements for pairwise comparisons, and to construct a hierarchy. In order to lighten users' work load, we realize these functions using graphical user interfaces.

The process for supporting group decision-making in our system is described as follows: First, a host user proposes a topic to be decided. Second, users make and choose alternatives from alternative database they share. In order to get alternatives, users can employ a variety of methods, e.g., brainstorming. These methods are provided by the hierarchical decision support module. Third, each user constructs a decision hierarchy for AHP using the module. The hierarchy clarifies elements which should be considered in the decision making process. The module is used to quantify subjective judgements of decision makers by using AHP based pairwise comparisons. Fourth, agents negotiate with each other based on their users' subjective weights and decision hierarchy. Negotiation among agents is based on the persuasion mechanism described in section 3.2. Finally, the result of the negotiation is reported to all users.

### 2.2 Quantifying Subjective Judgements

In order to measure subjective judgements of users, we employ AHP in our system. The AHP is a method for making decisions that are hard to analyze quantitatively. It combines both systems approach and subjective judgements, and its primary purpose is to maximize the user's intuition and experience.

In the AHP, users decompose the problem into a hierarchy that consists of a goal, criteria (and possibly sub-criteria), and alternatives. The judgement of the pairwise comparison between factors (in Figure 2, alternatives $A_1, A_2$ and $A_3$) on a certain level is made with respect to the criterion that is a factor (in Figure 2, criterion $C_1$) on the upper level. By interpreting a set of values of judgements as a matrix (top left of Figure 2),

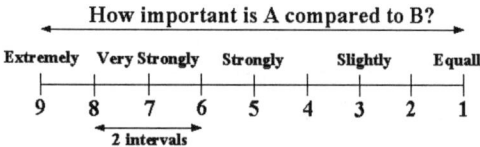

Figure 3: Scale of measurement for AHP

the weights (i.e., measurement of criteria) of factors are calculated analytically. To put it more concretely, we can achieve the weights of each factor as the eigen-vector for the max eigen-value of the pairwise comparison matrix. As a whole hierarchy, the weights of the alternatives can be calculated by composing the weights of the criteria.

A pairwise comparison matrix is characterized by the following: (1) diagonal values are 1; and (2) the values of elements in a matrix are $a_{ij} = 1/a_{ji}$. For judgements of pairwise comparisons in the AHP, we can use a 9 point scale consisting of five words (equally, slightly, strongly, very-strongly and extremely (Figure 3)) and four intermediate levels (e.g., between slightly and strongly)

AHP provides a measure the inconsistency in each set of judgements. This measure is called the inconsistency ratio (I.R.) and is defined as follows: $I.R. = (\lambda_{max} - n)/(n-1)$. Here, $\lambda_{max}$ is the max eigen-value described above and $n$ is the size of the pairwise comparison matrix. Ideally, a set of judgements in a pairwise comparison matrix will be consistent, and the inconsistency ratio will be 0. If the inconsistency ratio is no more than 0.1, this means that the pairwise comparison matrix is consistent.

In general, relative judgements are easier for users to make than absolute judgements. In the AHP, in order to lighten their work load, users can make comparisons between criteria using verbal and fuzzy expressions. Thus, the value of a pairwise comparison in the AHP is not a strict expression of a user's subjective judgement, but rather a rough approximation of that judgement. In our system, it may safely be assumed that a user's subjective judgement expresses 2 intervals with a certain value as the center point of the intervals on the 9 point scale (Figure 3). An interval is an unit of the 9 point scale. For example, while in the AHP the verbal expression "Very Strongly Important" means the value of 7 internally, in our system it means the values 6, 7, or 8.

In addition, we propose the expressions *fixed* and *assumed* to describe a user's belief in a judgement of pairwise comparison. For example, if the judgement is labeled fixed by the user, this means that the judgement value itself is fixed (i.e., the value is reliable). On the other hand, if the judgement is labeled assumed by the user, this means that the judgement value is also assumed (i.e., the value is unreliable). If an agent is persuaded in negotiation, the agent tries to adjust assumed judgements within 2 intervals in order to change the weights of the alternatives. These adjustments are made under the constraint of I.R. less than 0.1.

## 2.3 Public and Private Decision Hierarchy

In general, the AHP is employed in the common objective context (all users have the same objectives)[Dyer and Forman, 1992]. In this context, users decide one decision hierarchy among them first, then each user judges pairwise comparisons individually based on the decision hierarchy. In this case, because all users have the same objectives, i.e., criteria, the existing systems support users to construct only one decision hierarchy and to judge pairwise comparisons. On the other hand, our system supports group decision-making in the non-common objective context (each user has non-shared and sometimes hidden objectives). In the non-common objective context, it is difficult to construct a decision hierarchy among users, because the users have non-shared and hidden objectives (i.e., criteria). In our system, each user may construct a decision hierarchy individually. However, when there are same criteria among users, if the users can notice this fact, each can construct a decision hierarchy more effectively. Therefore, in our system, a decision hierarchy has public parts and private parts in order to make an effective group decision. The public parts can be referenced by all users. The private parts are hidden from other users. In the concrete system, the whole decision hierarchy is basically public. Individual users can designate as public or private each new criteria they create. The goal and alternatives must be public.

## 3 Agent's Behavior

### 3.1 Reducing the Pairwise Comparisons

Each agent manages the decision hierarchy and the pairwise comparison matrix constructed by its user. In general, the AHP requires too many judgements of its user. If there are n factors for a criterion, users have to make $n(n-1)/2$ judgements. If the number of levels of the decision hierarchy or the number of factors of the level is increased, more and more judgements are required. Acting on the user's behalf, the user's agent effectively reduces the number of judgements, which leads the user to make consistent judgements dynamically using the following methods.

In general, users have to judge all pairwise comparisons in the AHP. In fact, it is very hard for a user to judge all pairwise comparisons. In our system, the initial value of the elements of all pairwise comparison matrixes is 1 (i.e., "Equally Important") as an assumed value. Because there are initial values, the user changes only the value that the user wants to judge.

In the AHP, the value of an element $a_{ij}$ in the matrix equals $W_i/W_j$. Here, the $W_i$ expresses the weight of the factor $i$. If a pairwise comparison matrix is consistent (i.e., $I.R. = 0$), a certain element can be inferred from the other two elements. For example, in a matrix, because the value of an element $a_{12}$ is $W_1/W_2$ and the value of an element $a_{23}$ is $W_2/W_3$, we can infer that the value of an element $a_{13}$ is $a_{12} \times a_{23} = W_1/W_2 \times W_2/W_3 = W_1/W_3$. In the system, in order to reduce the number of judgements, agents infer the value of an element in a matrix using this feature of the AHP. When the agent infers the value of an element, if the element was judged by the user, the agent asks the user to change the judgement. If the element has never been judged, the agent changes the value of the element to the new value inferred. In this case, the new value is labeled as assumed. In general, in order to reduce the number of judgements, the Harker Method [Harker, 1987] is now widely employed. However, to use the Harker method, the covering condition must be satisfied on a matrix. In order to satisfy the covering condition, many judgements, from which all elements in the matrix must be inferred, are required. In a real system, it is also hard for the user to make judgements with the covering condition. If the user has a chance to satisfy the covering condition, the system exploits the Harker method instead of the simple method mentioned above to omit some redundant pairwise comparisons and check the consistency of each comparison dynamically. In a pairwise comparison matrix, when the value of $I.R.$ is more than 0.1, this means that the matrix is inconsistent and the user must remake all the judgements. Naturally, this remaking creates additional work load for the user. In our system, while the user is making judgements in a matrix, the agent is watching the $I.R.$ dynamically. When the $I.R.$ is more than 0.1, the agent requests that the user remake the present judgement.

### 3.2 The Persuasion Process

In the system, a negotiation among agents consists of persuasions between two agents. Figure 4 shows an example of negotiation among agents $a, b, c, d$, and $e$ in the system. First, agents pair-off into groups: agents $a$ and $b$ make a group and agents $c$ and $d$ make a group. Next, within each group one agent who selected randomly persuades the other. In Figure 4, agent $a$ persuades agent $b$ and agent $c$ persuades agent $d$. If these individual persuasions succeed, the persuading agents assume the representation of their respective groups. In Figure 4, each persuasion succeeds, and agent $a$ and agent $c$ are representative of their groups. If the persuasions had failed, the agents would change places. For example, if the agent $a$ failed to persuade the agent $b$, the agent $b$ persuades the agent $a$ next time. The groups are now $a, b$ and $c, d$, and the representatives advance to negotiate with each other singly. During negotiation between agents $a, c$, and $e$, agent $a$ persuades agent $e$. So that the groups become $a, b, e$ and $c, d$. Finally, agent $c$ persuades agent $a$, and the agents reach a consensus.

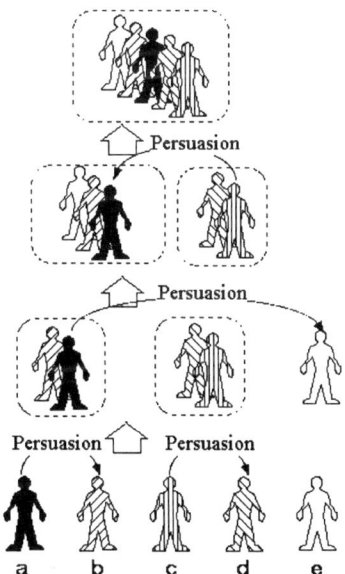

Figure 4: Negotiation among agents

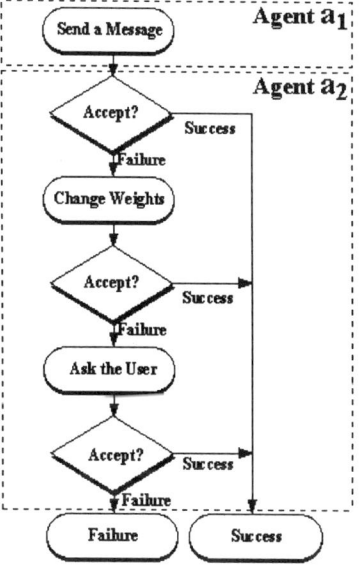

Figure 5: The process of persuasion (Agent $a_1$ persuades Agent $a_2$)

Figure 5 shows the process of persuasion between two agents. Suppose that agent $a_1$ and agent $a_2$ are in a group and agent $a_1$ persuades agent $a_2$. First, agent $a_1$ sends a persuasion message to agent $a_2$. The persuasion message is the most preferable alternative that has the highest weight and is decided by agent $a_1$'s user with the AHP. Secondly, agent $a_2$ accepts or rejects this message according to the following process.

First, agent $a_2$ checks whether its own most preferable alternative is the same as that in the persuasion message. If this alternative is mutual, the agent $a_2$ accepts the persuasion message and this persuasion is a success. If not, agent $a_2$ does not accept the message and the persuasion process advances to the next step.

Second, agent $a_2$ tries to change the preference order of alternatives by adjusting the judgements of matrixes in the decision hierarchy. Figure 6 shows an example. The top and bottom halves of Figure 6 show, respectively,

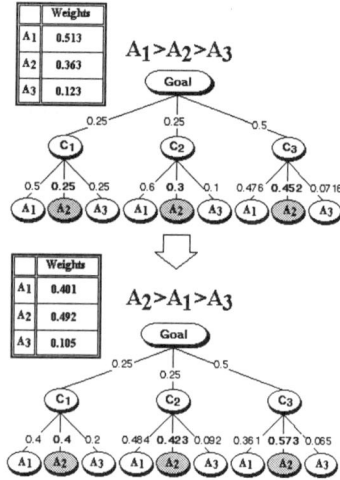

Figure 6: An example of adjusting judgements

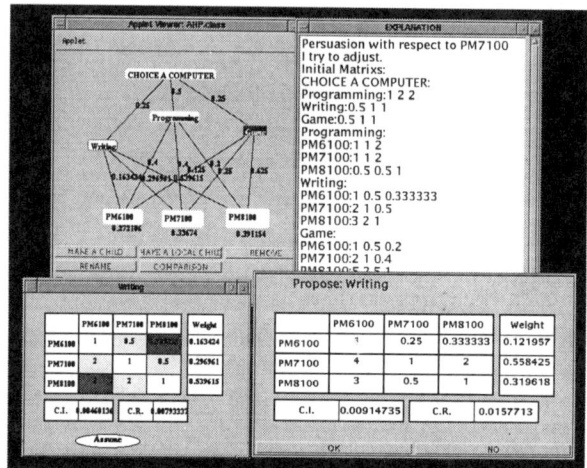

Figure 7: An example of adjusting judgements in a matrix

the hierarchy before and after adjustment. In the top half of the figure, alternative $A_1$ is more preferable than alternative $A_2$ or $A_3$. Now, suppose that agent $a_1$ proposes alternative $A_3$ as the persuasion message. Agent $a_2$ tries to adjust the judgements in order to change the preference order so that alternative $A_2$ is more preferable than alternative $A_1$.

The agents adjust the judgements in a matrix by employing the following method. From the feature of the AHP, in order to increase the weight of the alternative $I_i$, we increase the value of elements at $i$th row except a diagonal element. In the system, the agents increase the value of assumed elements of $i$th row of the matrix except a diagonal element within 2 intervals of the 9 point scale, in order to increase the weight of alternative $I_i$ and change the preference order of the alternatives. Figure 7 shows an example of such an adjustment. Suppose that all elements of the matrix are labeled *assumed*, and agent $a_2$ wants to increase the weight of alternative $A_2$. In this case, agent $a_2$ increases 1 interval of the value of elements (i.e., judgements) on alternative $A_2$ against the alternative $A_1$ and $A_3$. By this adjustment, alternative $A_2$ becomes most preferable. Agent $a_2$ asks the user for permission to change the weights in practice. If the user permits, agent $a_2$ changes the judgements. If not, agent $a_2$ tries to adjust again.

Third, agent $a_2$ checks whether the persuasion message can be accepted using his new preference order. If the persuasion message can be accepted, this persuasion is a success. If not, this persuasion is a failure.

Fourth, the agent asks the user to change his or her judgements. If the user agrees, agent $a_2$ indicates which judgement the user should change.

Fifth, agent $a_2$ again checks whether or not the persuasion message can be accepted.

Note that the reliability of the system is best enhanced by the persuasion mechanism's second and fourth steps.

Figure 8: An example

### 3.3 Reliability

The reliability is an important factor to implement a software agent. The agent briefs its users on the details of negotiation via graphical user interface during negotiation. The agent explains which agents persuaded, which element in a matrix should be changed, which group the agents participated in at the time, and so on. This explanation mechanism renders the agents more reliable.

## 4 An Example and Discussion

As an experiment, we used the system to choose a new computer for our laboratory. The number of members in our laboratory is 12, and each research group has different objectives. Therefore, the group decision-making context was a non-common objective context. The common objectives (i.e., criteria) were writing papers, programming, and so on. The non-common objectives were playing games, music, and so on.

Figure 8 shows an example of the persuasion mechanism of the system. Here, the agent received an alternative "PM7100" as a persuasion message. Then, the agent adjusted judgements and proposed the new judgements by showing the window at the bottom right of Figure 8. This shows the agent's explanation in the process of a persuasion. The agent asks the user's permission to change particular judgements. The user agrees by pushing the *OK* button, or disagrees by pushing the *NO* button. The top right window in Figure 8 gives a detailed explanation.

These experiments have yielded some interesting results that merit discussion. A consensus is sometimes disturbed by the user who makes arbitrary judgements. To deal with such cases, we must consider the following. (1) In the AHP, we generally make decisions constructively. For this reason, arbitrary judgements should be prohibited. This idea can be applied to many cases of group decision making, but it also restricts the user's judgements. (2) An arbitrary judgement can be regarded as an opinion of the user. In this case, it will be worth reaching a consensus among a sub group that excludes this user. In addition, this idea can be applied to cases in which a group does not need consensus among all members. For example, in deciding the destination for a trip, we do not require consensus among all members. We can assume that the member who makes arbitrary judgements does not want to go on the trip, and can thus be disincluded.

Finally, the necessity of the explanation mechanism, particularly with regard to the persuasion mechanism, should be discussed. The successful persuasion of an agent requires the compliance of that agent's user. At first, we implemented a system in which the users were removed from the negotiation of their agents. In this case, we found that even if the users knew the system architecture and the process of agent negotiation in advance, the users who were persuaded were not satisfied. In other words, the reliability of agents is an important factor in realizing a multi-agent system. For this reason, in our system, as we have proposed in section 3.3, each agent explains to the user the details of the negotiation, in order to gain the user's consent and make agents more reliable.

## 5 Conclusion

In this paper, we proposed a persuasion mechanism among agents and implemented a group decision support system GCDSS to see how effectively the persuasion mechanism can be used. The GCDSS helps a group decision to make a reasonable choice from alternatives. We found useful characteristics of AHP for multi-agent based group decision support systems. The characteristics are as follows. (1) We can quantify the subjective judgements of users by using AHP. Agents negotiate effectively based on the subjective judgements quantified. (2) The verbal and fuzzy measurement of AHP enables us to realize the persuasion protocol. Agents can adjust the decision hierarchy of their users. The results of our current experiments demonstrated that the persuasion protocol is a suitable method for reaching a consensus among agents in group decision support systems.

## References

[Desanctis and Gallupe, 1987] Gerardine Desanctis and R. Brent Gallupe. A foundation for the study of group decision support systems. *Management Science*, 33(5):589–609, May 1987.

[Dyer and Forman, 1992] Robert F. Dyer and Ernest Forman. Group decision support with the analytic hierarchy process. *Decision Support Systems*, 8:99–124, 1992.

[Ephrati et al., 1994] Eithan Ephrati, Gilad Zlotkin, and Jeffrey S. Rosenschein. A non-manipulable meeting scheduling system. In *The Thirteenth International Distributed Artificial Intelligence Workshop*, pages 105–125, July 1994.

[Garrido and Sycara, 1996] Leonardo Garrido and Katia Sycara. Multi-agent meeting scheduling: Preliminary experiment results. In *Proceedings of Second International Conference on Multi-Agent Systems(ICMAS-96)*, pages 95–102. AAAI Press, December 1996.

[Harker, 1987] Patrick T. Harker. Incomplete pairwise comparisons in the analytic hierarchy process. *Mathmatical Modelling*, 9:353–360, 1987.

[Ito and Shintani, 1996a] Takayuki Ito and Toramatsu Shintani. An approach to a multi-agent based scheduling system using a coalition formation. In *Proceedings of the Ninth International Conference on Industrial & Engineering Applications of Artificial Intelligence and Expart Systems (IEA/AIE-96)*, page 780. Gordon and Breach Science Publishers, June 1996.

[Ito and Shintani, 1996b] Takayuki Ito and Toramatsu Shintani. Implementing an agent negotiation protocol based on persuasion. In *Proceedings of Second International Conference on Multiagent Systems(ICMAS-96)*, page 443. AAAI Press, December 1996.

[Kautz et al., 1994] Henry Kautz, Bart Selman, Michael Coen, Steven Ketchpel, and Chris Ramming. An experiment in the design of software agents. In *AAAI-94*, pages 438–443, 1994.

[Luce and Raiffa, 1985] Robert D. Luce and Howard Raiffa. *Games and decisions*. Dover Publications, 1985.

[Saaty, 1980] Thomas L. Saaty. *The Analytic Hierarchy Process*. McGraw Hill, 1980.

[Sen and Durfee, 1994] Sandip Sen and Edmund H. Durfee. On the design of an adaptive meeting scheduler. In *The Tenth IEEE Conference on Artificial Intelligence for Applications*, pages 40–46, March 1994.

# DISTRIBUTED ARTIFICIAL INTELLIGENCE

Distributed AI 2: Coordination and Cooperation

# Cooperation Structures

**Mark d'Inverno**
School of Computer Science
University of Westminster
London W1M 8JS, UK
dinverm@wmin.ac.uk

**Michael Luck**
Dept. of Computer Science
University of Warwick
Coventry CV4 7AL, UK
mikeluck@dcs.warwick.ac.uk

**Michael Wooldridge**
Agent Systems Group
Zuno Ltd, International House
London W5 5DB, UK
mjw@dlib.com

## Abstract

In order to cooperate effectively with its peers, an agent must manipulate representations of the social structures in which it plays a part. The purpose of this paper is to investigate the mathematical and computational aspects of this social reasoning process. We begin by defining an abstract representation of *cooperation structures*, wherein agents cooperate to achieve goals on each other's behalf. We then investigate the question of whether or not cooperation is feasible with respect to an agent's goal, and we show that answering this question is an NP-complete problem. Finally, we investigate the conditions under which such structures can be composed to form larger structures.

## 1 Introduction

Cooperation is perhaps the paradigm example of social activity in both real and artificial social systems; it is certainly the best studied process in multi-agent systems research. Cooperation in human societies is an intricate and subtle activity, which has defied many attempts to formalise it. However, some progress has been made on understanding the types of situation in which cooperation can arise, and how it can proceed.

Central to the study of cooperation is the notion of a *social structure*. A social structure is a set of relations that hold between agents in a society. These relations define the dependencies that exist between agents (e.g., [Castelfranchi, 1990; d'Inverno and Luck, 1996a; 1996b]), and determine the rights and responsibilities of each agent in the society with respect to its peers. In order to cooperate effectively with its peers, an agent must *represent* any social structures in which it plays a part, and *reason* with these representations. This reasoning process is carried out in order to answer such questions as whether cooperation is possible, and to investigate how an agent stands in relation to other agents in the society. In multi-agent systems, the representation of social structures is a central research issue. For example, Durfee has developed representations of multi-agent activity known as *partial global plans* [Durfee and Lesser, 1987]. These structures can be manipulated by an agent in order to find more efficient routes to solving complex multi-agent problems.

Much work on representing social structures in multi-agent systems has been purely formal, with no obvious, direct route to implementation. Game-theoretic and economic-theoretic studies of social behaviour fall into this category [Rosenschein and Genesereth, 1985]. Although such work is central to our understanding of cooperation, it has little to tell us about the *computational* aspects of social reasoning, such as what types of social reasoning are tractable. Our aims in this paper are therefore threefold:

- to introduce an abstract symbolic representation of social structures;
- to identify and formally define some basic reasoning problems associated with these social structures; and finally
- to investigate the computational complexity of these problems.

We begin in the following section by introducing a simple, general formal framework, which can be used to represent a wide range of multi-agent scenarios. This framework defines agents as systems that have the ability to achieve certain goals, and that are capable of making independent decisions about how they will interact with other agents. We informally discuss the properties of cooperation and cooperation structures, and formally define cooperation structures within our framework. We then introduce COOPSAT, a key decision problem in social reasoning. COOPSAT is the problem of determining, given some society of agents and a particular agent's goal, whether or not cooperation is *in principle* possible to achieve the agent's goal. We show that the problem is NP-complete, and that it cannot therefore be answered in practice. We then address the issue of *manipulating* social structures, and the conditions under which they can be combined. Finally, we discuss related work, and present some conclusions and future research directions.

## 2 Cooperation Structures

Before we can define cooperation and cooperation structures, we need a formal framework within which we can express the definitions. A number of such formal frameworks have previously been developed, the most obvious of which being game-theory (e.g., [Rosenschein and Genesereth, 1985]) and multi-modal logic (e.g., [Wooldridge and Jennings, 1994]).

With respect to the former, the models derived are by nature quantitative rather than symbolic, and hence not well-suited to representation within a computer system. With respect to the latter, the models derived tend to be rather arcane, and do not easily lend themselves to, for example, complexity-theoretic analysis. For these reasons, we develop a simple formal framework for expressing multi-agent scenarios, using a notation based on the Z specification language [Spivey, 1992] which, in turn, is based on set theory and first-order logic. In Z, a relation $R$ is defined as a set of ordered pairs. The expression dom $R$ represents the set of the first elements of each of the ordered pairs of $R$, and ran $R$ represents the set of second elements. Also, $R^+$ is the transitive closure, and $R^*$ is the reflexive transitive closure of $R$.

We start by assuming a fixed, finite set $Ag$ of agents. We use $i, j, k$ as variables ranging over $Ag$. The main assumptions that we make with respect to agents are that they are *autonomous* (in that they are not benevolent), and that they have *capabilities* (in that they have the ability to achieve goals). The properties of agents are discussed in more detail elsewhere [Wooldridge and Jennings, 1995; Luck and d'Inverno, 1995].

Next, we assume a fixed, finite set $G$ of *goals*. We use $g$ with annotations ($g'$, $g_1$, ...) as variables ranging over $G$. Whereas the assumption that $Ag$ is finite seems intuitively reasonable, it may seem odd to assume that the set of goals is finite: it is common in AI to represent goals as logical formulae, and hence to allow the set of goals to be infinite in size. We make this assumption in the interests of simplicity. In this paper, we are not concerned with the question of what a goal *is* — the contents of $G$ are left undefined (but have been considered elsewhere [Luck and d'Inverno, 1995]). However, it seems essential to introduce some notion of *consistency* between goals. We thus assume a relation $con \subseteq G \times G$, such that if $(g, g') \in con$, then $g$ and $g'$ are said to be consistent with one another. We write $con(g, g')$ to indicate that $(g, g') \in con$. The intuitive meaning of $con(g, g')$ is that $g$ being satisfied does not preclude $g'$ being satisfied: the two goals are not mutually exclusive. We shall not give a formal semantics to $con$, but we do require that this relation satisfies the following properties:

- reflexive: $\forall g \in G \bullet con(g, g)$; and
- symmetric: $\forall g, g' \in G \bullet con(g, g') \Leftrightarrow con(g', g)$.

In addition to consistency, we assume that goals are related to each other through a *sub-goal* relation: $\leq \subseteq G \times G$. A sub-goal is a component goal of another, higher-order, goal. Thus if $(g, g') \in \leq$ (written $g \leq g'$), then $g$ is said to be a *sub-goal* of $g'$.

The $\leq$ relation must satisfy the following properties:

- reflexive:
  $$\forall g \in G \bullet g \leq g;$$
- transitive:
  $$\forall g, g', g'' \in G \bullet g \leq g' \wedge g' \leq g'' \Rightarrow g \leq g'';$$
- well-founded:
  $$\forall g \in G \bullet \#\{g' \mid g' \leq g\} \in \mathbb{N}.$$

The first two conditions are intuitively obvious; well-foundedness simply states that no goal has an infinite, non-terminating chain of sub-goals.

We define the *strict sub-goal* relation, $<$, in the obvious way: $g < g' \Leftrightarrow g \leq g' \wedge g \neq g'$. With the exception of reflexivity, $<$ enjoys all the properties of $\leq$ and, in addition, $<$ is asymmetric. Since $<$ is well-founded, some goals are primitive and have no strict sub-goals. We write $\bot < g$ to indicate that $g$ has no strict sub-goals. We assume that there is just one way to *achieve* a goal, by achieving all its strict sub-goals. We make this assumption in the interests of simplicity though, of course, the real-world is somewhat more complicated than this. If I have a goal of drinking tea, I have many different possible courses of action available to me: for example, I can make the tea myself, ask someone to make me a cup, or go to a cafe.

Another useful extension to $\leq$ is the *immediate sub-goal* relation: $\ll$. This relation is defined as follows: $g \ll g' \Leftrightarrow g < g' \wedge (\neg \exists g'' \in G \bullet ((g < g'') \wedge (g'' < g')))$. The function $isg : G \to \mathbb{P} G$ takes a goal and returns the set of all its *immediate* sub-goals: $isg(g) = \{g' \mid g' \ll g\}$.

We noted above that benevolence is not assumed in our framework. Crudely, the benevolence assumption states that agents will always attempt to do what is requested of them: they are not autonomous [Rosenschein and Genesereth, 1985]. While benevolence is reasonable for many distributed problem-solving systems, it is not an appropriate assumption in most multi-agent scenarios. In order to capture autonomy in our framework, we make use of a *will-adopt* function, $will : Ag \times Ag \to \mathbb{P} G$. The idea is that agent $i$ will adopt a goal $g$ on behalf of another agent $j$ iff $g \in will(i, j)$. Where there can be no confusion, we write $will(i, g, j)$ to indicate $g \in will(i, j)$. We do not give a formal semantics to this relation, as its properties will be domain specific[1].

The *capabilities* of agents are represented in a function $cap : Ag \to \mathbb{P} G$. The idea is that $cap(i)$ represents the set of goals that agent $i$ can achieve in isolation. We require the following invariant to hold between the $cap$ and $will$ functions: $will(i, g, j) \Rightarrow g \in cap(i)$.

The various sets and relations introduced above together comprise a *framework*.

**Definition 1** *A framework is a 6-tuple*

$$\langle Ag, G, con, \leq, will, cap \rangle$$

*with components as above. Let Fr be the set of all frameworks. We use F with annotations ($F'$, $F_1$, ...) as variables ranging over Fr.*

For most of this paper, the framework is assumed to be fixed and understood.

## 2.1 Defining Cooperation Structures

Previously, Luck and d'Inverno [1996] have taxonomised the types of interactions that occur between agents in a multi-

---

[1]There are certain properties that it seems reasonable to demand of *will*. For example, we might specify that if $i$ will adopt $g$ for $j$, then $i$ will also adopt any sub-goal of $g$ for $j$: $\forall i, j \in Ag \bullet \forall g, g' \in G \bullet (g \in will(i, j) \wedge g' \leq g) \Rightarrow g' \in will(i, j)$. However, none of these properties are essential for our framework.

agent system, distinguishing in particular between *engagements* of non-autonomous agents and *cooperation* between autonomous (motivated) agents. In this view, autonomous agents will only adopt goals if it is to their motivational advantage to do so, while non-autonomous agents may benevolently adopt goals. Though we focus on autonomous agents, the discussion of cooperation structures below does not refer to the *reasons* for goal adoption, nor to the autonomy or non-autonomy of the agents involved and, consequently, these structures are generic and may be applied to cooperations and engagements, as described here, equally.

We define cooperation by one agent with another to mean that the agent will adopt a goal on behalf of that other agent. These cooperations give rise to a graph structure, with nodes in the graph corresponding to agents, and arcs in the graph corresponding to cooperations, labelled with goals. This leads us to the definition of a *structure*.

**Definition 2** *A structure is a pair $(C, l)$ where:*

- $C \subseteq Ag \times Ag$ *is a binary* cooperates *relation;*
- $l : C \to G$ *labels each arc in $C$ with a goal.*

If $(C, l)$ is a structure, then we write $C(i, j)$ to indicate $(i, j) \in C$. The intuitive interpretation of $C(i, j)$ is that agent $i$ has delegated goal $l(i, j)$ to agent $j$, and thus that $j$ is cooperating with $i$ over this goal.

Not all structures are cooperation structures, however. For a structure to be a cooperation structure, there must first be at least two agents cooperating, and each agent in the structure must be connected to another through a cooperation over goals between them. Furthermore, agents cannot delegate goals to others who, in turn, delegate them back to the original agent. Hence there are no cycles, and no agent cooperates with itself. Also, if one agent $j$ cooperates with another agent $i$ over some goal $g$, and another agent $k$ cooperates with $j$ for goal $g'$ then $g'$ must be a sub-goal of $g$. Thus when agents delegate, they delegate sub-goals. Finally, all goals in a cooperation structure must be consistent with each other. These considerations lead to the following definition.

**Definition 3** *A structure $(C, l)$ is a* cooperation structure *iff:*

1. *$C$ is non-empty: $C \neq \emptyset$;*
2. *$C$ is weakly connected;*
3. *$C$ is acyclic: $\forall i \in Ag \bullet \neg C^+(i, i)$;*
4. *$C$ is irreflexive: $\forall i \in Ag \bullet \neg C(i, i)$;*
5. *agents delegate sub-goals: $\forall i, j, k \in Ag \bullet C(i, j) \wedge C(j, k) \Rightarrow l(j, k) \leq l(i, j)$;*
6. *goals within a structure are mutually consistent: $\forall g, g' \in \operatorname{ran} l \bullet con(g, g')$; and*
7. *$j$ cooperates with $i$ only if $j$ is willing to do so: $C(i, j) \Rightarrow will(j, l(i, j), i)$.*

*Let Coop be the set of all cooperation structures.*

Condition (3) may seem too strong. To see why, consider the following scenario: John asks Paul to make some tea. Paul agrees, but asks that John boil the kettle. This kind of scenario (where $i$ delegates a goal $g$ to $j$, and $j$ in return delegates a strict sub-goal of $g$ back to $i$) occurs frequently in

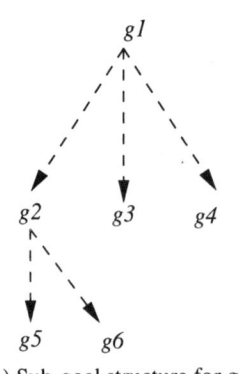
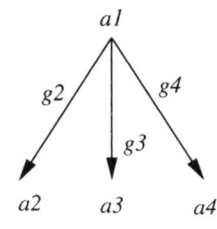

(a) Sub-goal structure for g1 | (b) Incomplete cooperation structure

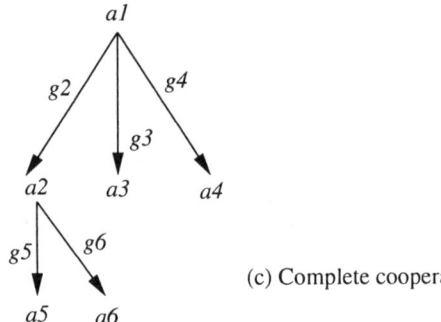

(c) Complete cooperation structure

Figure 1: Completeness

real life. However, the problem of determining whether an arbitrary, possibly cyclic structure was in fact a legal cooperation structure would then become much harder. In addition, determining whether it was possible to fuse two cooperation structures (a problem we consider below) would also be much more complicated. For these reasons, cooperation structures are required to be acyclic.

Finally, the function, $deleg : Ag \times Coop \to \mathbb{P}\, G$, returns the set of goals that an agent delegates in some cooperation structure: $deleg(i, (C, l)) = \{l(i, j) \mid C(i, j)\}$.

## 2.2 Complete Cooperation Structures

Consider the following scenario. Agent $a_1$ wants to achieve goal $g_1$, but $g_1 \notin cap(a_1)$. The complete sub-goal structure for $g_1$ is illustrated in Figure 1(a). Agent $a_1$ delegates goals $g_2$, $g_3$, and $g_4$ to agents $a_2$, $a_3$, and $a_4$ respectively. Agents $a_3$ and $a_4$ have the capabilities to achieve their respective goals (i.e., $g_3 \in cap(a_3)$ and $g_4 \in cap(a_4)$), but $a_2$ is not capable of $g_2$. It should therefore delegate the sub-goals $g_5$ and $g_6$ to other agents, but it does not. So $g_2$ will not be achieved, and hence neither will $g_1$. The cooperation structure in Figure 1(b) is thus in some sense *incomplete*.

This leads to the idea of a cooperation structure being *complete* for some agent-goal pair. Informally, a cooperation structure is said to be complete for agent $i$ and goal $g$ *iff* either:

1. agent $i$ has been delegated the goal $g$, and $i$ is capable of $g$; or else

2. agent $i$ has delegated each immediate sub-goal $g'$ of $g$ to some agent $j$, and $(C, l)$ is complete for agent $j$ and goal $g'$.

Completeness is hence a recursive notion, with the first clause as the base. It is not difficult to see that the cooperation structure in Figure 1(b) is incomplete according to this definition, but that Figure 1(c) *is* complete (assuming that $g_5 \in cap(a_5)$ and $g_6 \in cap(a_6)$). Formally, completeness is defined as follows.

**Definition 4** *A cooperation structure $c = (C, l)$ is said to be* complete *with respect to agent $i$ and goal $g$ iff either:*

1. $g \in cap(i)$, and for some $j \in Ag$, we have $C(j, i)$ and $l(j, i) = g$; or else
2. $isg(g) \subseteq deleg(i, (C, l))$, and $\forall j \in Ag$, if $C(i, j)$ and $l(i, j) \ll g$, then $(C, l)$ is complete for agent $j$ and goal $l(i, j)$.

In a complete cooperation structure, there are no sub-goals left dangling: all sub-goals are successfully delegated and hence, by the intuitive semantics for the sub-goal relation, every goal in the structure is achieved.

## 3 Is Cooperation Possible?

Suppose an agent has a goal that it wants to achieve, and further suppose that the agent either cannot achieve the goal in isolation (because it does not have the resources), or does not want to achieve it in isolation (because in so doing, it would clobber one of its other goals) [Wooldridge and Jennings, 1994]. The obvious question this agent should ask is: can I get other agents to help me with this goal? This is a *satisfiability* problem, similar in nature to the question of whether a formula of some particular logic is true under some interpretation. Formally, the problem can be stated as follows.

**Definition 5** *(The COOPSAT problem.) Given a framework $F = \langle Ag, G, con, \leq, will, cap \rangle$, an agent $i \in Ag$, and a goal $g \in G$, does there exist a cooperation structure over $F$ that is complete for $(i, g)$?*

**Theorem 1** COOPSAT *is* NP-*complete.*

**Proof:** Membership of NP is easy: given an instance $\langle\langle Ag, G, con, \leq, will, cap\rangle, i, g\rangle$ of the COOPSAT problem, simply guess a cooperation structure $(C, l)$ that is complete for $(i, g)$. The size of the structure is bounded above by $\#(Ag \times Ag)$ and, since $Ag$ is finite, guessing can be done in polynomial time. Verifying that $(C, l)$ is complete for $(i, g)$ can also be done in polynomial time.

For completeness, we must show that COOPSAT is in some sense no easier than all other NP-complete problems. To do this, it suffices to show that any instance $I$ of some known NP-complete problem can be transformed into an instance of $\tau(I)$ of COOPSAT such that the transformation can be done in polynomial time, and the transformed problem $\tau(I)$ has a solution only if the original problem $I$ has a solution. For COOPSAT, we define a reduction from a version of the well-known HAMILTONIAN CYCLE (HC) problem.

An instance of HC is determined by a graph $(N, A \subseteq N \times N)$. The aim is to answer 'yes' if $A$ has a cycle containing all nodes without repetition, 'no' otherwise. The idea behind the reduction is to encode the relation $A$ in the *will* relation, and the requirement for the cycle in the sub-goal relation $<$.

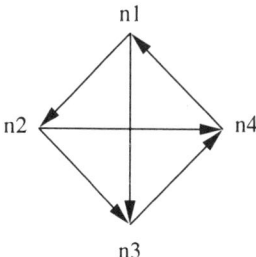

Figure 2: Illustrating the Reduction

To see how the reduction works, consider the following graph $G_1$ (illustrated in Figure 2):

$$G_1 = (\{n_1, n_2, n_3, n_4\}, \{(n_1, n_2), (n_2, n_3), \\ (n_3, n_4), (n_4, n_1), (n_1, n_3), (n_2, n_4)\})$$

This graph has a Hamiltonian cycle $(n_1, n_2, n_3, n_4)$. We transform $G_1$ into an instance $\tau(G_1)$ of COOPSAT. To do this, we first create five goals $g_0, \ldots, g_4$, and five agents, $n_1, \ldots, n_4, end$. We then create a linear sub-goal structure for $g_0$:

$$g_4 \ll g_3 \ll g_2 \ll g_1 \ll g_0$$

The *will* relation is then generated as follows.

|  |  | To |  |  |  |  |
|---|---|---|---|---|---|---|
|  |  | $n_1$ | $n_2$ | $n_3$ | $n_4$ | end |
| From | $n_1$ |  |  |  |  |  |
|  | $n_2$ | $g_1,\ldots,g_4$ |  |  |  |  |
|  | $n_3$ | $g_1,\ldots,g_4$ | $g_1,\ldots,g_4$ |  |  |  |
|  | $n_4$ |  | $g_1,\ldots,g_4$ | $g_1,\ldots,g_4$ |  |  |
|  | end |  |  |  | $g_1,\ldots,g_4$ |  |

Now consider the COOPSAT problem determined by this framework and the agent-goal pair $(n_1, g_0)$. The problem clearly has a solution, which will look like that below, where each question mark represents an agent.

$$n_1 \xrightarrow{g_1} ? \xrightarrow{g_2} ? \xrightarrow{g_3} ? \xrightarrow{g_4} ?$$

In fact, this problem has a simple solution:

$$n_1 \xrightarrow{g_1} n_2 \xrightarrow{g_2} n_3 \xrightarrow{g_3} n_4 \xrightarrow{g_4} end$$

The transformation from HC to COOPSAT is entirely automatic, (see Figure 3) and is polynomial. We leave it for the reader to see that the generated COOPSAT problem has a solution only if the original HC problem does, and so we are done. □

Suppose we had an algorithm that was guaranteed to give us the correct answer to an instance of the COOPSAT problem. This algorithm would take as input a framework, an agent and a goal and, some time later, would be guaranteed to generate as output of either 'yes' (indicating that a solution did indeed exist), or 'no' (indicating that the problem had no solution). Now, suppose the algorithm answered 'yes'. Then the agent would know that cooperation was possible. It would then have to start delegating goals to other agents. But the simple

```
let n = #N be the size of N
for each node i ∈ N, create a corresponding agent i
create a 'dummy' agent, called end
create n + 1 goals, g_0, ..., g_n
set con(g, g') for all goals g, g'
define < by g_n ≪ ··· ≪ g_1 ≪ g_0
for each agent j ≠ end
    for each agent k ≠ end
        if (j, k) ∈ A then
            if k = n_1 then
                will := will ∪ {(j, end) ↦ {g_n}}
            else
                will := will ∪ {(j, k) ↦ {g_1, ..., g_n}}
        else
            will := will ∪ {(j, k) ↦ ∅}
    end-for
end-for
cap := {end ↦ {g_n}}
```

Figure 3: Reducing HC to COOPSAT

'yes' answer does not indicate exactly who the agent should delegate *to*. A more useful algorithm would not only say 'yes', but would also *produce* a solution to the problem. This leads us to the following, closely related problem.

**Definition 6** (*The* COOPFIND *problem.*) *Given a framework* $F = \langle Ag, G, con, \leq, will, cap \rangle$, *an agent* $i \in Ag$, *and a goal* $g \in G$, *find a cooperation structure over F that is complete for* $(i, g)$ *if such a structure exists, or else answer that there is no solution.*

This problem is clearly related to many similar planning problems [Allen *et al.*, 1990]; it is also naturally viewed as a type of constraint satisfaction problem.

## 4 Composing Cooperation Structures

In a real society there will be many cooperation structures in existence at any given time. This may result in redundancy through different agents achieving the same goals in different contexts. In order to remove this redundancy it may be possible to *compose* two cooperation structures. Given two cooperation structures, $c = (C, l)$ and $c' = (C', l')$, we write $c \cup c'$ to denote the set-theoretic union $(C \cup C', l \cup l')$ of these structures. However, this composition can only occur in certain situations: we cannot expect the union of two arbitrary cooperation structures to be a legal structure. For example, if we have some arc, $(i, j)$, that appears in both $C$ and $C'$, but $l(i, j) \neq l'(i, j)$, then $l \cup l'$ is not a function (since $(l \cup l')(i, j)$ is not well-defined). This raises the question of what conditions are required for the union of two cooperation structures itself to be a cooperation structure. It turns out that we can define these conditions precisely. First, we extend the notion of *intra*-structure goal consistency to *inter*-structure goal consistency.

**Definition 7** *Let* $c = (C, l)$ *and* $c' = (C', l')$ *be cooperation structures. Then c and c' are said to be* consistent *(written* $cons(c, c')$*) iff* $\forall g \in ran\ l \bullet \forall g' \in ran\ l' \bullet con(g, g')$.

We can now define what it means for two structures to be *compatible*.

**Definition 8** *If* $c = (C, l)$ *and* $c' = (C', l')$ *are cooperation structures, then c and c' are said to be* compatible *(written* $compat(c, c')$*) iff:*

1. $C \cup C'$ *is weakly connected;*
2. $C \cup C'$ *is acyclic;*
3. *the two structures agree on labels:* $\forall i, j \in Ag \bullet C(i, j) \land C'(i, j) \Rightarrow l(i, j) = l'(i, j)$
4. $\forall i, j, k \in Ag,$
    (a) *if* $C(i, j) \land \neg C(j, k) \land \neg C'(i, j) \land C'(j, k)$ *then* $l'(j, k) \leq l(i, j);$
    (b) *if* $C'(i, j) \land \neg C'(j, k) \land \neg C(i, j) \land C(j, k)$ *then* $l(j, k) \leq l'(i, j);$ *and*
5. $cons(c, c').$

One might expect that compatibility is an equivalence relation over the set of all cooperation structures, but this is not in fact the case, as the following theorem establishes.

**Theorem 2** *The compatibility relation is reflexive and symmetric, but not transitive.*

**Proof:** Reflexivity and symmetry are obvious. For transitivity, it is easy to construct a counter-example. Suppose we had three cooperation structures $c, c', c''$. Further suppose that $compat(c, c')$ and $compat(c', c'')$, and that $c$ and $c'$ share a single agent $i$ in common, and $c'$ and $c''$ share a different single agent $j$ in common ($i \neq j$). Hence $c$ and $c''$ are disjoint, so $c \cup c''$ is not weakly connected. Hence $c$ and $c''$ are not compatible. □

Determining whether two cooperation structures are compatible is a tractable problem.

**Theorem 3** *It is possible to determine whether two cooperation structures are compatible in polynomial time — no worse than* $O(\#(Ag)^3)$.

**Proof:** The only non-trivial step involves showing that the resulting cooperation relation is acyclic, which requires checking the transitive closure of the relation — using Warshall's algorithm, the transitive closure can be computed in time $O(\#(Ag)^3)$ [van Leeuwen, 1990, pp540–544]. The overall time complexity is therefore no worse than $O(\#(Ag)^3)$. □

**Theorem 4** *If* $c = (C, l)$ *and* $c' = (C', l')$ *are cooperation structures, then* $c \cup c'$ *is a cooperation structure iff* $compat(c, c')$.

**Proof:** (Omitted due to lack of space.) □

Once we know that two cooperation structures are compatible, the problem of generating their union is computationally trivial.

## 5 Related Work

As we noted in Section 1, cooperation is a widely studied issue in multi-agent systems research. Despite this, we are aware of little other work that considers cooperation in the complexity-theoretic way proposed in this paper. Probably the most closely related work to ours is that of Shehory and Kraus on *coalition formation* [Shehory and Kraus, 1996]. Coalition formation is the process of devising a team of agents to work on a goal, and is rather similar to our COOPSAT problem. The most obvious differences between our work and that of Shehory and Kraus are that they assume benevolent agents, and they present algorithms (adapted from the *set covering problem*) to design coalitions. In addition, the work of Tennenholtz and Moses on the *multi-entity* model of multi-agent systems [Tennenholtz and Moses, 1989] is also closely related. This model is used to define the *cooperative goal achievement* (CGA) problem, which can crudely be stated as: given a set of benevolent agents, each with their own goals, is there some plan for the set that will achieve all their goals? Tennenholtz and Moses show that this problem is PSPACE-complete. Our framework most significantly differs from theirs in that they allow a richer representation of goals (as arbitrary propositional logic formulae) and, in addition, they also assume benevolence.

## 6 Conclusion

Cooperation is a key process for multi-agent systems research and, as such, it has received a considerable amount of attention in the multi-agent systems literature. However, mathematical treatments of cooperation have focussed primarily on either game-theoretic or modal logic formulations.

When many agents cooperate together, a cooperation structure emerges, which can be represented formally as a directed graph with certain properties. In this paper, we have formally defined the properties that must hold of such a graph to be considered as a cooperation structure. We have shown that, even when making simplifying and limiting assumptions about the world, the problem of determining whether cooperation structures are available to achieve an agent's goal is NP-complete. The problem of computational complexity and tractability has often been overlooked in the design of multi-agent systems. In future work we wish to use our framework in order to formally define other key social reasoning problems, and analyse the computational complexity of such problems. In addition, we aim to investigate algorithms for solving these problems.

## References

[Allen et al., 1990] J. F. Allen, J. Hendler, and A. Tate, editors. *Readings in Planning*. Morgan Kaufmann, San Mateo, CA, 1990.

[Castelfranchi, 1990] C. Castelfranchi. Social power. In Y. Demazeau and J.-P. Müller, editors, *Decentralized AI — Proceedings of the First European Workshop on Modelling Autonomous Agents in a Multi-Agent World (MAAMAW-89)*, pages 49–62. Elsevier Science Publishers B.V.: Amsterdam, The Netherlands, 1990.

[d'Inverno and Luck, 1996a] M. d'Inverno and M. Luck. A formal view of social dependence networks. In *Distributed Artificial Intelligence: Architecture and Modelling, Lecture Notes in Artificial Intelligence, 1087*, pages 115–129. Springer Verlag, 1996.

[d'Inverno and Luck, 1996b] M. d'Inverno and M. Luck. Formalising the contract net as a goal directed system. In W. Van de Velde and J. W. Perram, editors, *Agents Breaking Away: Proceedings of the Seventh European Workshop on Modelling Autonomous Agents in a Multi-Agent World (MAAMAW-96), Lecture Notes in Artificial Intelligence Volume 1038*, pages 72–85. Springer-Verlag, 1996.

[Durfee and Lesser, 1987] E. H. Durfee and V. R. Lesser. Using partial global plans to coordinate distributed problem solvers. In *Proceedings of the Tenth International Joint Conference on Artificial Intelligence (IJCAI-87)*, pages 875–883, 1987.

[Luck and d'Inverno, 1995] M. Luck and M. d'Inverno. A formal framework for agency and autonomy. In *Proceedings of the First International Conference on Multi-Agent Systems (ICMAS-95)*, pages 254–260. AAAI Press / MIT Press, 1995.

[Luck and d'Inverno, 1996] M. Luck and M. d'Inverno. Engagement and cooperation in motivated agent modelling. In *Distributed Artificial Intelligence: Architecture and Modelling, Lecture Notes in Artificial Intelligence, 1087*, pages 70–84. Springer-Verlag, 1996.

[Rosenschein and Genesereth, 1985] J. S. Rosenschein and M. R. Genesereth. Deals among rational agents. In *Proceedings of the Ninth International Joint Conference on Artificial Intelligence (IJCAI-85)*, pages 91–99, 1985.

[Shehory and Kraus, 1996] O. Shehory and S. Kraus. Formation of overlapping coalitions for precedence-ordered task-execution among autonomous agents. In *Proceedings of the Second International Conference on Multi-Agent Systems (ICMAS-96)*, pages 330–337. AAAI Press / MIT Press, 1996.

[Spivey, 1992] J. M. Spivey. *The Z Notation*. Prentice Hall, Hemel Hempstead, 2nd edition, 1992.

[Tennenholtz and Moses, 1989] M. Tennenholtz and Y. Moses. On cooperation in a multi-entity model: Preliminary report. In *Proceedings of the Eleventh International Joint Conference on Artificial Intelligence (IJCAI-89)*, pages 918–923, 1989.

[van Leeuwen, 1990] J. van Leeuwen. Graph algorithms. In J. van Leeuwen, editor, *Handbook of Theoretical Computer Science*, pages 525–631. Elsevier Science Publishers B.V., 1990.

[Wooldridge and Jennings, 1994] M. Wooldridge and N. R. Jennings. Formalizing the cooperative problem solving process. In *Proceedings of the Thirteenth International Workshop on Distributed Artificial Intelligence (IWDAI-94)*, pages 403–417, July 1994.

[Wooldridge and Jennings, 1995] M. Wooldridge and N. R. Jennings. Intelligent agents: Theory and practice. *The Knowledge Engineering Review*, 10(2):115–152, 1995.

# Exploration and Adaptation in Multiagent Systems: A Model-based Approach

## David Carmel and Shaul Markovitch
Computer Science Department,
Technion, Haifa 32000, Israel
{carmel,shaulm}@cs.technion.ac.il

## Abstract

Agents that operate in a *multi-agent system* can benefit significantly from adapting to other agents while interacting with them. This work presents a general architecture for a model-based learning strategy combined with an exploration strategy. This combination enables adaptive agents to learn models of their rivals and to explore their behavior for exploitation in future encounters. We report experimental results in the *Iterated Prisoner's Dilemma* domain, demonstrating the superiority of the model-based learning agent over non-adaptive agents and over reinforcement-learning agents. The Experimental results also show that exploration can improve the performance of a model-based agent significantly.

## 1 Introduction

The recent tremendous growth of the Internet has motivated a significant increase of interest in the field of *multi agent systems* (MAS). In such a system, a group of egocentric autonomous agents interact with each other to achieve their private goals. Generally, agents are designed to achieve their masters' goals by trying to maximize some given utility measurement. Designing an "effective" strategy for interaction is a hard problem because its effectiveness depends mostly on the strategies of the other agents involved. However, the agents are autonomous, hence their strategies are private. One way to deal with this problem is to endow agents with the ability to adapt their strategies based on their interaction experience [Weiß and Sen, 1996].

A common technique used by adaptive agents in MAS is *reinforcement learning* [Shoham and Tennenholtz, 1994; Sandholm and Crites, 1995]. The adaptive agent adapts its strategy according to the rewards received while interacting with other agents. A major problem with this approach is its slow convergence. An effective interaction strategy is acquired only after processing a large number of interaction examples. During the long period of adaptation the agent pays the cost of interacting sub-optimally.

In previous work [Carmel and Markovitch, 1996b] we describe a model-based approach that reduces the number of interaction examples needed for adaptation by investing more computational resources in deeper analysis of past interaction experience. This approach splits the learning process into two separate stages. In the first stage, the learning agent infers a model of the other agent based on past interaction. In the second stage the agent utilizes the learned model for designing effective interaction strategy for the future.

A *model-based* adaptive agent might converge to suboptimal behavior. Acting according to the current best-response strategy may leave unknown aspects of the opponent's strategy unexplored. Finding the balance between *exploitation* of the current model and *exploration* of other alternatives is a well-known dilemma for on-line adaptive agents. In this work we describe methods for combing exploration strategies with model-based learning. The exploring model-based agent sacrifices immediate rewards to explore the opponent behavior; the better model resulted will then yield a better interaction strategy.

The paper is organized as follows: First, we formalize a framework for describing interaction among agents in multi-agent systems. An encounter between agents is described as a *two-player game* and a sequence of encounters as a *repeated game*. We also present a general architecture for a model-based adaptive strategy for repeated games. Second, we describe some exploration methods for on-line learning and show how to combine them with model-based learning. Third, we show experimentally the superiority of a model-based adaptive agent over non-adaptive agent and over reinforcement-learning agent in the *Iterated Prisoner's Dilemma game*. We also report some experimental results comparing different exploration methods for model-based learning in repeated-games.

## 2 A general framework for a model-based interacting agent

To formalize the notion of interacting agents we consider a framework where an encounter between two agents is represented as a *two-player game* and a sequence of encounters as a *repeated game*. A *two-player game* is a tuple $G = \langle R_1, R_2, u_1, u_2 \rangle$, where $R_1, R_2$ are finite sets of alternative moves for the players (called *pure strategies*), and $u_1, u_2 : R_1 \times R_2 \to \Re$ are utility functions that define the utility of a joint move $(r_1, r_2)$ for the players. For example, the *Prisoner's dilemma* (PD) is a two-player game, where each player has two actions, cooperate (c) and defect(d), $R_1 = R_2 = \{c, d\}$. The utility functions, $u_1, u_2$, are described by the payoff matrix shown in Figure 1.

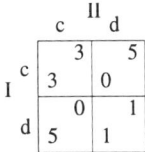

Figure 1: The payoff matrix for the Prisoner's dilemma game

A sequence of encounters among agents is described as a repeated game, $G^\#$, based on the repetition of $G$ an indefinite number of times. At any stage $t$ of the game, the players choose their moves, $(r_1^t, r_2^t) \in R_1 \times R_2$, simultaneously. A history, $h(t)$ of $G^\#$, is a finite sequence of joint moves chosen by the agents up to the current stage of the game, $h(t) = \langle (r_1^0, r_2^0), (r_1^1, r_2^1), \ldots, (r_1^{t-1}, r_2^{t-1}) \rangle$. $H(G^\#)$ is the set of all finite histories for $G^\#$.

A *strategy* $s_i : H(G^\#) \to R_i$ for player $i, i \in \{1, 2\}$, is a function that takes a history and returns an action. $\mathcal{S}_i$ is the set of all possible strategies for player $i$ in $G^\#$. A pair of strategies $(s_1, s_2)$ defines an infinite sequence of joint moves (a path) while playing the game $G^\#$, $g_{(s_1, s_2)}(t)$.

The *discounted-sum* function is common utility function for repeated games:

$$U_i^\gamma(s_1, s_2) = (1-\gamma) \sum_{t=0}^\infty \gamma^t u_i(s_1(g_{(s_1,s_2)}(t)), s_2(g_{(s_1,s_2)}(t))) \quad (1)$$

$0 \leq \gamma < 1$ is a discount factor for future payoffs of player $i$. It is easy to show that $U^\gamma(s_1, s_2)$ converges for any $\gamma < 1$.

**Definition 1** $s_i^{opt}(s_j, U_i)$ *will be called the* best response *for player $i$ with respect to strategy $s_j$ and utility $U_i$, iff* $\forall s \in \mathcal{S}_i, [U_i(s_i^{opt}(s_j, U_i), s_j) \geq U_i(s, s_j)]$.

The Iterated Prisoner's Dilemma (IPD) is an example of a repeated game based on the PD game that attracts significant attention in the game-theory literature. Tit-for-tat (TFT) is a simple, well known strategy for IPD that has been proven to be extremely successful in IPD tournaments [Axelrod, 1984]. It begins with cooperation and imitates the opponent's last action afterwards. The *best-response* against TFT with respect to $U^\gamma$ depends on the discount parameter $\gamma$ [Axelrod, 1984]:

$$s^{opt}(\text{TFT}, U^\gamma) = \begin{cases} \text{all-c} & \frac{2}{3} \leq \gamma \\ \text{all-d} & \gamma \leq \frac{1}{4} \\ \text{``Alternate between c and d''} & \text{otherwise} \end{cases}$$

One of the basic factors affecting the behavior of agents in MAS is the knowledge that they possess about each other. In this work we assume that each player is aware of the other player's actions, i.e. $R_1, R_2$ are *common knowledge*, while the players' preferences, $u_1, u_2$, are *private*. In such a framework, while the history of the game is *common knowledge*, each player predicts the future course of the game differently. The prediction of player $i$ is based on the player's strategy $s_i$ and on the player's belief about the opponent's strategy, $\hat{s}_j$. $\hat{s}_j$ will be called an *opponent model*.

How can a player acquire a model for its opponent's strategy? One source of information available for the player is the history of the game. Another possible source of information is observed games between the opponent and other agents. Note that any history of length $t$ provides a sample, $E_j(h(t)) = \{(h(k), r_j^k) | 0 \leq k \leq t\}$, of $t+1$ examples of the opponent's behavior.

Given a learning algorithm $L_i$ that infers an opponent model based on a sample of its behavior, and a utility function $U_i$, we can define the strategy $s_i^{U_i, L_i}$ of a model-based learning agent as

$$s_i^{U_i, L_i}(h(t)) = s_i^{opt}(L_i(E_j(h(t))), U_i)(h(t))$$

The above definition yields a model-based player (MB-agent) that adapts its strategy during the game. A MB-agent begins the game with an arbitrary opponent model $\hat{s}_j^0$, finds the *best response* $s_i^0 = s_i^{opt}(\hat{s}_j^0, U_i)$, and plays according to the *best response*, $r_i^0 = s_i^0(\lambda)$, where $\lambda$ is the null history. At any stage $t$ of the game, the MB-agent acquires an updated opponent model by applying its learning algorithm $L_i$ to the current sample of the opponent's behavior, $\hat{s}_j^t = L_i(E_j(h(t)))$. It then finds the best response against the current model, $s_i^t = s_i^{opt}(\hat{s}_j^t, U_i)$, and plays according to the best response $r_i^t = s_i^t(h(t))$. Figure 2 illustrates the general architecture of an on-line model-based learning agent for repeated games.

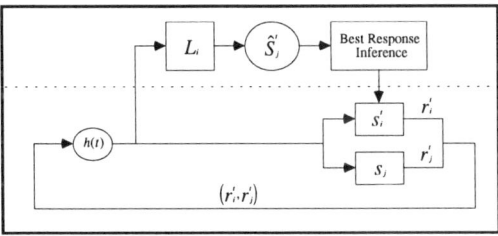

Figure 2: An architecture for a model-based learning agent in repeated games.

The efficiency of this adaptive strategy depends mainly on the two main processes involved: 1) The finding of the best response against a given model; 2) the learning process of the opponent model. In previous work [1996b] we show that when the players are restricted to *regular strategies*, i.e. strategies that can be represented by deterministic finite automata (DFA) [Rubinstein, 1986], the best-response problem can be solved efficiently. We also describe a learning algorithm for inferring a regular model based on the history of the game.

## 3 Exploring the opponent's strategy

One of the weaknesses of the *model-based* approach is that it might converge to sub-optimal behavior. The best-response strategy ignores the possibility that the current model is not identical to the opponent's strategy and that other actions might have a better utility (according to the actual opponent's strategy). The left part of Figure 3 shows an example of an opponent's strategy for IPD. If the player uses the model shown in the right part of the figure then the best-response corresponding to this model is "all-d" which yields sub-optimal utility. Furthermore, playing "all-d" prevents the player from observing counterexamples, therefore, the wrong model will never be corrected and the player will never discover the best-response for the actual opponent strategy – "play c and than all-d".

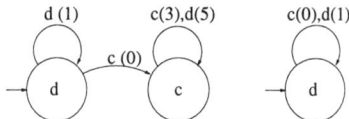

Figure 3: An example of an opponent model in local minimum. (Left): An Opponent's strategy. (Right): An opponent model. The model dictates the play "all-d" while the actual best-response is "play c and then all-d".

The example above demonstrates that it is better sometimes to play sub-optimally in order to explore the opponent's behavior. This phenomenon is known as the dilemma between exploration and exploitation. The learning player has to choose between the wish to exploit the current model maximally, and the desire to explore other alternatives to improve its future play.

In recent years a variety of exploration strategies have been proposed for reinforcement-learning agents. *indirect strategies* are based on incorporating randomness into the decision procedure of the learning agent. The key idea is to associate a positive probability for any state-action pair. The agent randomly selects an action according to the given distribution and therefore no action is neglected forever. A common exploration strategy in the reinforcement learning paradigm is to play according to a distribution that takes into account the expected utility of the actions and the learning stage. We will use similar method for the MB-agent with the discounted-sum utility function. The expected utility of an action $r$ is computed by summing the instant expected reward of executing $r$ and the discounted sum of rewards expected from following the optimal policy from then on.

Let $\hat{s}_j^t = L_i(h(t))$ be an opponent model inferred by the learning algorithm $L_i$ based on history $h(t)$. Let $s_i^t = s_i^{opt}(\hat{s}_j^t, U_i^\gamma)$ be the best-response for player $i$. The *expected utility* for action $r$ is defined by:

$$U_i'(\hat{s}_j^t, r) = u_i\left(r, \hat{s}_j^t(h(t))\right) + \gamma U_i^\gamma\left(s_i^t, \hat{s}_j^t\right)$$

$U_i'$ can be computed efficiently since any game between two automata converges to a finite cycle, and the discounted sum of rewards can be computed by analyzing the expected game-cycle of the two automata.

The Boltzmann distribution [Sutton, 1990] assigns a probability for any possible action according to its expected utility and according to a decreasing parameter $T$ called a temperature.

$$Pr(r) = \frac{e^{U_i'(\hat{s}_j^t, r)/T}}{\sum_{r' \in R_i} e^{U_i'(\hat{s}_j^t, r')/T}} \quad (2)$$

$T$ depends on the stage of the game and determines the rate of convergence. We use a common temperature function $T = \alpha^t$ where $\alpha < 1$ is the exploration parameter.

*Direct strategies* try to explore the environment more efficiently by using statistics based on the learning experience of the agent. These heuristic methods compute an exploration bonus for each possible action and incorporate this bonus into the decision procedure of the agent. Let $E_i'(\hat{s}_j^t, r)$ be the exploration bonus computed for an action $r$ according to the current model $\hat{s}_j^t$ and the history $h(t)$. The value for each action is computed according to the following formula:

$$V_i(\hat{s}_j^t, r) = (1-\alpha) \frac{U_i'(\hat{s}_j^t, r)}{\sum_{r' \in R_i} U_i'(\hat{s}_j^t, r')} + \alpha \frac{E_i'(\hat{s}_j^t, r)}{\sum_{r' \in R_i} E_i'(\hat{s}_j^t, r')} \quad (3)$$

As for indirect methods, $\alpha$ is the exploration parameter that determines the ratio between exploration and exploitation.

Statistics might include action-counts [Sato et al., 1991] where higher exploration bonuses are given to actions that have been chosen less frequently. Sutton [1990] suggests an exploration bonus based on the time that have passed since the action was taken. Kaelbling [1993] suggests an exploration bonus based on the upper bound of the confidence interval computed for the expected utility of the possible actions.

To summarize, at any stage $t$ of the game, an exploring MB-agent updates the opponent model $\hat{s}_j^t = L_i(E_j(h(t)))$. It then finds the best response against the model $s_i^t = s_i^{opt}(\hat{s}_j^t)$. When using an *indirect exploration method* it computes $Pr(r)$ for every action $r \in R_i$ and randomly selects an action according to this distribution. When using *directed methods* it computes the exploration bonus for every action $r$ based on the history statistics, and then selects the action with the maximal utility.

## 4 Experimentation: On-line learning in repeated-games

We conducted a set of experiments to test the capabilities of a model-based learner in repeated games. The stage-game used in these experiments is the PD-game. The first experiment tests the basic capabilities of an on-line learner against random opponents, the second compares indirect and direct exploration methods, and the third tests the MB-agent against non-random automata.

For each experiment, a random opponent automaton was generated by choosing a random transition function and a random output function. The MB-agent begins with a random DFA as a model of its opponent's strategy and plays according to the *best response* strategy combined with an exploration strategy. The opponent model is modified whenever it fails to predict the opponent's actual play by using the learning algorithm IT-US-$L^*$ with $(h_i(t), r_j^t)$ as a counterexample [Carmel and Markovitch, 1996b]. We use the average cumulative reward achieved by the player to measure the quality of its learning strategy:

$$\frac{\sum_{k=0}^{t-1} u_i(r_i^k, r_j^k)}{t}$$

Figure 4 shows the average cumulative reward attained during 400 stages of the IPD game, averaged over 100 trials against 100 different random automata of size 20. The lowest curve shows the results of the non-adaptive player TFT. TFT achieves only 2.25 points, which is the average payoff of the PD game, and this result is not improved during the game.

The three upper curves stand for MB-strategies combined with Boltzmann strategies with different exploration parameters, $T = 5.0 * \alpha^t$. All the MB-strategies achieve results that are far superior to those achieved by TFT. The exploring adaptive agents are far more successful than the non-exploring agent (with $\alpha = 0$). The graphs highlight an interesting phenomenon. At early stages of the game, the exploring agents pay for the suboptimal decisions induced by their "curiosity". The cost for exploration increases with the exploration parameter $\alpha$. However, the better models generated by the exploring agents pay off in later stages of the game, and the exploring strategies outperform the non-exploring strategy dominantly.

The second curve shows the results of a reinforcement-learning (RL) agent. RL is based on the idea that the tendency to produce an action should be reinforced if it produces a favorable results, and weakened if it produces unfavorable results. Q-learning [Watkins and Dayan, 1992] is a well-known RL algorithm that works by estimating the values of all state-action pairs. For repeated games, an entire history is needed for representing a game state. In such framework, any state is visited only once and no generalization can be made. A possible alternative is to use a fixed window of previous moves for representing a state. A too wide window can cause the agent to use a too sparse table and to disable convergence of the learning process in practical time. A too narrow window can cause perceptual aliasing, i.e., different states can appear identical and therefore can be represented by the same state.

Q-learning was tried in repeated games against Tit-for-tat (TFT) by Sandholm and Crites [1995]. The Q-agent succeeded in learning an optimal strategy against TFT using a window of width one, but it needed about 100,000 iterations for convergence. Similar results were obtained by us. In our experiments the Q-agent used a window of width 2 and a Boltzmann exploration strategy. It is quite clear that the performance of the Q-agent is much inferior to that of the MB-agent in the given domain. Increasing the width of the window did not help. The Q-agent managed to achieve results comparable to those achieved by the MB-agents only when the length of the game was increased up to 40,000.

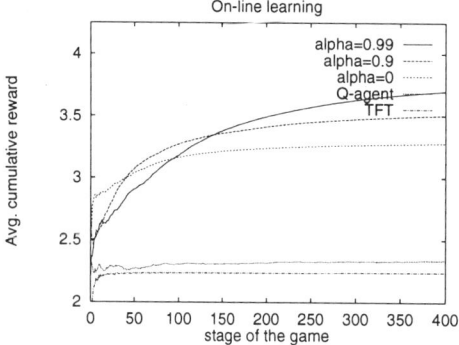

Figure 4: On-line learning: The average reward attained by different playing strategies against 100 random automata of size 20.

The results for the MB-agents address a question of how much exploration is needed during interaction. The answer depends on the "greed" of the learning agent. The more weight the agent gives to future payoffs, the more resources it should spend on exploration. This weight is expressed by the discount parameter $\gamma$. We repeated the last experiment for various values of $\gamma$. For each $\gamma$ we tried several values of exploration parameter $\alpha$ and recorded the one for which the agent achieved the best performance. The left part of Figure 5 shows the best $\alpha$ for each $\gamma$. As expected, as $\gamma$ increases, it is better to invest more on exploration by using larger $\alpha$.

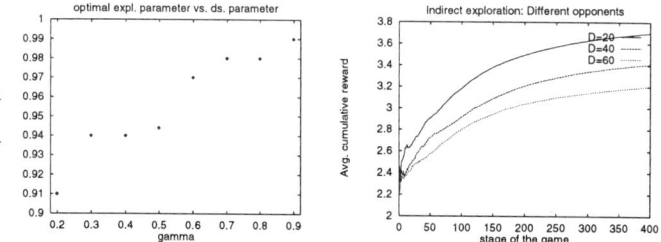

Figure 5: Left: The best exploration parameter for various discount parameters. Right: On-line learning of random automata with various sizes.

The right part of Figure 5 shows the effect of increasing the complexity of the opponent strategy on the learning process. The three curves show the average results attained by an MB-agent with $\alpha = 0.99$ against 100 random automata of sizes 20, 40, and 60. The results show that increasing the size of the automata deteriorates the rate of adaptation. Learning effective models for complex automata demands more examples and hence the learning process converges slower.

In the second experiment we compared indirect and direct exploration strategies. The indirect-MB agent uses Boltzmann distribution with temperature function $T = 5 * 0.99^t$. The direct-MB agent uses *recency-based* exploration [Sutton, 1990] – for every action $r$ it counts the number of stages that have passed since the last time that $r$ was taken, $\rho(\hat{s}_j^t, r)$, at the current state of the opponent model $\hat{s}_j^t$. It then computes an exploration bonus $E'(\hat{s}_j^t, r) = \sqrt{\rho(\hat{s}_j^t, r)}$ and measures the utility of action $r$ according to Equation 3, using $\alpha = 0.1$. For automata of size 20, the direct and indirect methods showed comparable performance. However, when the size of the automata was increased to 40, the direct strategies outperformed the indirect one. The results are shown in Figure 6.

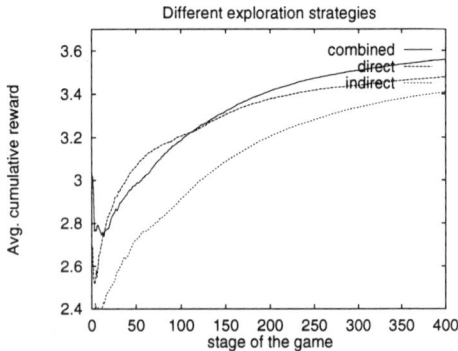

Figure 6: Different exploration strategies while playing IPD against 100 random automata of size 40.

The exploration behavior of the indirect strategy is sensitive only to the length of the history and not to its content. Therefore it does not modify its exploration behavior when playing against different opponents. The direct strategy, on the other hand, utilizes the content of the history to modify its exploration behavior. This is the reason why it outperforms the indirect method when increasing the complexity of the opponent. However, the direct method does not reduce its exploration tendency with time. Therefore, even after learning a perfect model of the opponent, it will still pick non-optimal actions.

We have designed a *combined* exploration strategy that uses a combination of the direct and indirect strategies: it uses *recency-based* exploration strategy but decreases the exploration parameter with time, $\alpha_t = 0.1 * 0.99^t$. Figure 6 shows that the combined strategy enjoys the advantages of both methods and outperforms them both.

In the third experiment we tested the MB-strategy against non-random opponents that were hand-crafted specifically for the IPD game. We repeated the famous tournament organized by Axelrod [1984] for IPD. In the original tournament, fifteen attendees (strategies) competed in a round-robin tournament, where every interaction was based on 200 repetitions of the PD game. The participated strategies were sent as computer programs by different researchers around the world. Most programs were designed to basically cooperate. They differ mainly in the way that they deal with defection of their rivals. For this experiment we allowed only deterministic regular strategies to participate. Figure 7 shows the models and the best-responses learned by the MB-agent for five different opponents. The MB-agent learned its opponents on-line during the 200 iterations using Boltzmann exploration with the temperature function $T = 5 * 0.99^t$. The best-response strategies were computed according to the discounted sum utility function with $\gamma = 0.9$.

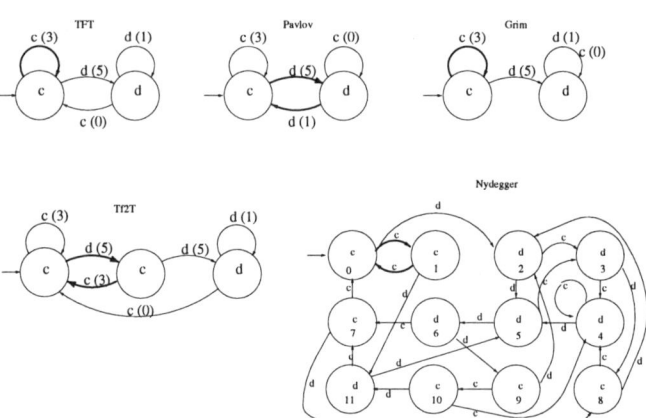

Figure 7: The models learned by a MB-agent after 200 iterations of the PD-game. The best-response cycles are highlighted.

The MB-agent succeeded very well against the strategies described above. The adaptive strategy begins without any knowledge about the game and only few iterations are needed to converge to the *best response*: Cooperate with TFT and exploit players like TF2T and Pavlov. The best-response for Nydegger which is "all-c" was found after about 100 iterations. The model given in the figure was learned after 200 iterations. It encapsulates many features of Nydegger. For example it cooperates after 3 mutual defections (see the paths 2-5-6-9, 4-5-6-9, 11-5-6-9). The MB-agent also succeeds in finding the right model for Grim. However, the exploring agent falls into the "defection sink" and is not being able to get back to the cooperation cycle. This problem is common to all exploring strategies that eventually will explore the "wrong" action and will fall into the sink. Grim is a typical example for strategies that are problematic for on-line adaptive players.

## 5 Conclusions

This work presents a model-based approach for learning interaction strategies. We present an architecture for a model-based learner that uses the *best-response* strategy combined with an exploration strategy to explore its opponent's behavior for exploitation in future encounters.

We conducted a set of experiments where an adaptive model-based agent played against randomly generated opponents. In these experiments, the MB-agent performed significantly better than non-adaptive agents and Q-agents. This result supports our claim that it is more effective to learn a model and use it for designing a best-response strategy than trying to directly learn such a strategy. We compared direct and indirect exploration methods and showed that a combination of these two methods can have an advantage. We also showed that the MB adaptive learner, without any background knowledge, outperforms hand-crafted opponents in the IPD domain.

The above experimental results demonstrate average-case performance of our learning strategy. In the worst-case, Fortnow and Whang [1994] show that for any learning algorithm there is an adversary DFA for which the learning process will converge to the *best response* in at least exponential number of stages. One way of dealing with this complexity problem is by limiting the space of automata available for the opponent [Mor et al., 1996]. Such methods are based on long exploration sequences during the course of the game. The high cost of such exploration sequences diminishes in infinite games for utility functions that measure only asymptotic performance. However, these methods may fail for the discounted-sum utility which takes into account also immediate rewards. The exploration methods described in this work take into account the cost of exploration and are therefore suitable for such utility functions as well.

One of the basic assumptions of our framework is that the opponent is stationary. When this assumption is removed, we may want to look at windows of the input rather than the complete history, and to increase the exploratory rate of the algorithms. An interesting class of non-stationary opponents, is the class of adaptive opponents. For example, we might assume that the opponent is a MB-adaptive agent. This can be further extended to a general framework of recursive modeling, where a player holds a model of the opponent which is also a modeling player. We have developed a similar framework for two-player zero-sum games [Carmel and Markovitch, 1996a].

## References

[Axelrod, 1984] R. Axelrod. *The Evolution of Cooperation*. Basic Books, New York, 1984.

[Carmel and Markovitch, 1996a] David Carmel and Shaul Markovitch. Incorporating opponent models into adversary search. In *Proceedings of thirteenth National Conference on Artificial Intelligence (AAAI 96)*, pages 120 – 125, Portland, Oregon, August 1996.

[Carmel and Markovitch, 1996b] David Carmel and Shaul Markovitch. Learning models of intelligent agents. In *Proceedings of thirteenth National Conference on Artificial Intelligence (AAAI 96)*, pages 62 –67, Portland, Oregon, August 1996.

[Fortnow and Whang, 1994] L. Fortnow and D. Whang. Optimality and domination in repeated games with bounded players. In *Proceedings of the 25th Annual ACM Symposium on Theory and Computing*, pages 741–749, 1994.

[Kaelbling, 1993] Leslie P. Kaelbling. *Learning in embedded Systems*. MIT Press, Cambridge, Mass, 1993.

[Mor et al., 1996] Yishay Mor, Claudia V. Goldman, and Jeffery S. Rosenschein. Learn your opponent's strategy (in polynomial time). In G. Weiß and S. Sen, editors, *Adaptation and Learning in Multi-agent Systems, Lecture Notes in AI*. Springer-Verlag, 1996.

[Rubinstein, 1986] A. Rubinstein. Finite automata play the repeated Prisoner's Dilemma. *Journal of Economic Theory*, 39:83–96, 1986.

[Sandholm and Crites, 1995] T. W. Sandholm and R. H. Crites. Multiagent reinforcement learning and the iterated Prisoner's Dilemma. *Biosystems Journal*, 37:147–166, 1995.

[Sato et al., 1991] Mitsuo Sato, Kenichi Abe, and Hiroshi Takeda. Learning control of finite markov chains with an explicit trade-off between estimation and control. In *IEEE Transactions on Systems, Man and Cybernetics*, volume 18 (5), September 1991.

[Shoham and Tennenholtz, 1994] Y. Shoham and M. Tennenholtz. Co-Learning and the evolution of social activity. Technical Report STAN-CS-TR-94-1511, Stanford University, Department of Computer Science, 1994.

[Sutton, 1990] Richard S Sutton. Integrated architectures for learning, planning, and reacting based on approximating dynamic programming. In *Proceedings of the 7th international conference on Machine Learning*, pages 216–224, San Mateo CA,, 1990. Morgan Kaufman.

[Watkins and Dayan, 1992] C. J. C. H. Watkins and P. Dayan. Technical notes: Q-learning. *Machine Learning*, 8:279–292, 1992.

[Weiß and Sen, 1996] G. Weiß and S. Sen. *Adaptation and Learning in Multi-agent Systems, Lecture Notes in AI 1042*. Springer-Verlag, 1996.

# The Effects of Runtime Coordination Strategies Within Static Organizations

**Edmund H. Durfee and Young-pa So**
AI Laboratory, EECS Department
University of Michigan,
Ann Arbor, MI 48109, USA
durfee@umich.edu

## Abstract

Much work in multiagent systems has been devoted to developing runtime mechanisms to coordinate computational agents. However, the efficacy of such mechanisms depends to a large extent on the opportunities for flexible choices of action imposed by the implicit or explicit organizational relationships among agents. This paper examines the dependencies between the design of static organizational structures and the efficacy of runtime coordination. We show that runtime coordination strategies can be viewed as transforming the configuration of an organization, with principled effects on the expected performance of the agent system. By understanding these effects in the context of organizations, we can begin to articulate the conditions under which the introduction of coordination mechanisms (and their associated costs) can be justified.[*]

## 1 Introduction

The debate over the efficacy of runtime coordination mechanisms has in many ways revisited the deliberative versus reactive debate of the planning community. At one extreme are researchers who consider runtime coordination time-consuming and unnecessary. By instituting enough structure in the system, the local decisions of agents are assured of being compatible. Social Laws [Shoham and Tennenholtz, 1992] are one example of such static structuring mechanisms. At the other extreme are researchers who strive toward endowing each agent in a complex system with the ability to dynamically maintain models of other agents and to use these models to make arbitrarily complex coordination decisions interleaved with execution (e.g. [Gmytrasiewicz et al., 1991]).

Between these extremes are the many techniques for coordinating agents within the bounds of some fixed commitments upon which they can all depend (e.g. [Decker, 1995; Durfee et al., 1987; Gasser et al., 1989]). But these techniques have generally been developed to meet the needs of particular applications. While many have been successful, it has seldom been the case that they are found to be easily transferred and effectively employed across many different application domains.

Why do runtime coordination mechanisms sometimes make a big difference, and sometimes have little impact? We argue that runtime coordination is situated contextually in ways that affect its impact. Simply, if an agent is placed in a world where it faces no indecision, then coordinating with others at runtime is unlikely to help it perform better. On the other hand, if it faces several choices, and making a good choice depends on knowing what others around it are doing, then runtime coordination is a major help.

The context within which an agent—especially a computational agent—works is controllable to a large extent. In particular, in multiagent systems, the designers of a system dictate many of the features of the context in which an agent will be operating. Of particular interest in this paper, for example, are the relationships between agents' roles that are defined by the authority structure and the inherent structure of the distributed tasks that they pursue. We call this role structure the **organization** within which the agents work.

We are investigating the relationship between static organizations (how roles are defined prior to execution) and runtime coordination (how agents choose to occupy and fulfill roles during execution). Our ultimate goals include being able to predict what kinds of coordination mechanisms agents should possess to be effective in a particular organization, and to provide design advice for the creation of organizations that are suited for agents with known coordination capabilities. In this paper, we describe steps toward this end. We provide a parametric characterization of a class of organizations that allows us to systematically generate and evaluate alternatives, and describe performance measures for comparison. We describe the effects that different basic coordination strategies have in different organizations under different operating conditions. With these observations, we can distill some general lessons in designing agents and organizations that they inhabit.

## 2 Organization Structure

We will confine our discussion to the *multi-level hierarchy* class of organization structures. Hierarchies capture many of the most common notions of organization, where author-

---

[*] This work was supported in part by NSF PYI award 91-58473.

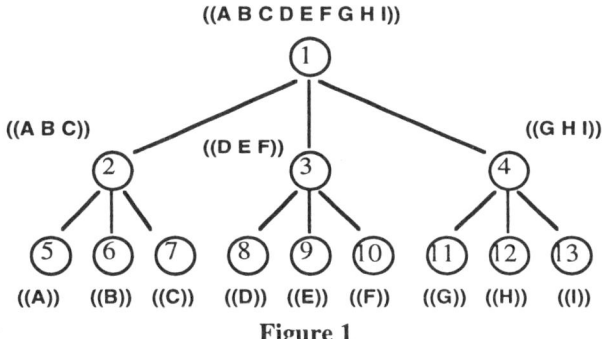

Figure 1

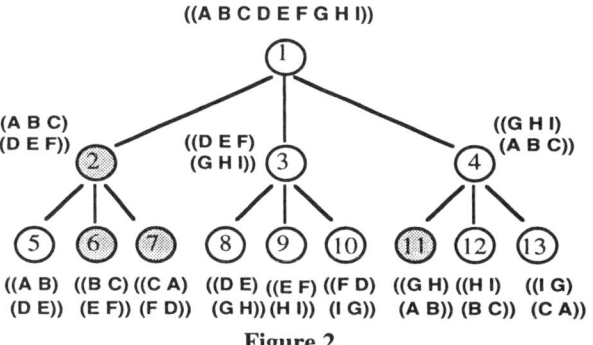

Figure 2

ity and breadth of perspective are concentrated in managers at higher levels of the organization, and tasks are decomposed and distributed down to the lowest level members. For example, this is the form of organization that emerges from the use of contracting protocols [Smith, 1980].

We characterize these structures in terms of the number of (primitive or supervisory) tasks assigned to members, and the branching factor of the organizational tree. For simplicity, we assume in this paper that these are constant in the structure (the organizational structure is uniformly branchy and the tasks are evenly distributed). For example, in Figure 1 is an organizational structure with a branching factor of 3 and that assumes that the "leaf" agents should perform one task (labeled alphabetically). Therefore, a middle manager's task is to supervise the performance of three tasks below it, and the top-level manager supervises three such supervisory tasks, meaning that it is indirectly overseeing 9 primitive tasks (A-I). The leaf-level agents complete their tasks, pass the results up to the middle managers, who synthesize them into a single result to pass to the top manager, who then generates from these the complete result. The solution time depends on the time required for each unit task and synthesis task (we assume the same values for these) and for the communication delays in distributing the subtasks and returning the results.

When uncertainty exists about the rate at which tasks will be achieved, imbalances of effort in the organization can degrade performance. A countermeasure is to give members of an organization overlapping responsibilities (tasks) such that if one is delayed in fulfilling its responsibilities (such as if it catastrophically fails), others can pick up the slack [Durfee et al., 1987]. For example, to make the previous organization more robust to agent failures, we could redundantly assign supervisory and primitive tasks among the agents. Note that the top-level agent is still a single point of failure, but we can do assignments beneath it such that the organization (or any subtree of it) can be assured of tolerating a certain number of failures among those being supervised. In the example of Figure 2, the redundant task assignments ensure that a (sub)tree succeeds so long as no more than one of the branches of that (sub)tree fail. A particular failure configuration is shown. The notation ((AB)(DE)) means the agent has 2 task sets, each with 2 tasks. In some coordination schemes, the first task set might be the (possibly prioritized) primary tasks, and the second might be secondary tasks (since they are another agent's primary tasks).

As has been discovered in other organizational approaches to cooperative problem solving, however, the reliability comes at the potential cost of duplicated effort [Durfee et al., 1987]. So, in cases where agents do not fail, they should coordinate their choices of which of their assigned tasks to each do to avoid redundancy. And despite the built-in reliability, if too many (or the wrong combination of) agents fail, the organization will still fail.

Runtime coordination, therefore, should be employed within these organizations to overcome these problems. When agents do not fail, runtime coordination should ensure that they are working on complementary tasks. When agents do fail, runtime coordination should ensure that the remaining agents are applied most effectively. Of course, if the organization involves no redundancy, then the first reason for runtime coordination will never arise; if the organization is completely redundant (all agents have all tasks), then the second reason will never arise. Thus the effects of runtime coordination mechanisms are highly dependent on the underlying organization structure among the agents.

## 3 Performance Measures

To study the effects of runtime coordination mechanisms in the context of organizational structures, we need to be more precise about what features of the performance we will measure to assess the effects. To keep our discussion simple, we will consider two measures: the response time of an organization (how quickly it completes the task set) and the reliability of the organization (how probable it is that it will successfully complete the entire task set).

### 3.1 Response Time

The response time of an organization is the time it takes for the agents to decompose and distribute the tasks to subordinates (who then decompose and distribute, and so on), for the agents doing the unit tasks to do them all, and for the results to be passed up and synthesized. We will omit the details of the calculation of the response time in this paper; we use the model described by [So and Durfee, 1997] where we are additionally assuming that the communication delay is equivalent to the unit task execution time. Thus, we characterize the response time of the organization when it is successful as $Ts(\gamma, N, k, m, o, f, s)$, where:

- $\gamma$ is the ratio of unit task execution time to communication delay (in this paper, $\gamma=1$)

- $N$ is the size of the overall task set
- $k$ is the organization's branching factor
- $m$ is the number of unit tasks per leaf role
- $o$ is the overlap, representing how many subtrees can fail while still being assured of success (note that $o < k$)
- $f$ is the probability of failure for each agent (we assume agents fail independently)
- $s$ is the coordination strategy outlined below

Of course, given the possibility of agent failures, the performance will depend on which agents do and do not fail. We refer to a particular arrangement of working (and non-working) agents as a **configuration**. The expected response time for successful configurations is thus the weighted (by probability that a configuration will occur given $f$) average of the response times of successful configurations. The response time of a successful configuration is itself generally an expected value, since a particular configuration will behave differently depending on the order in which redundantly assigned tasks are pursued.

As an example, for the organization in Figure 2, one configuration is where agents 2, 6, 7, and 11 have failed. Despite these failures, eventually all of the unit tasks will be done by agents that can supply results back to the top (by agents 8, 9, 10, 12, and 13). But note that the task (set) ordering represented in the figure means that task A will be done last of all (by agent 13). A reordering by 13 could decrease the delay for A (which only 13 can accomplish); in fact, agents 3 and 4 could reorder their subtask sets such that agent 4 would prioritize ABC over GHI, and this would yield better performance. Given random orderings, therefore, we will have a distribution of response times for this configuration, as shown in Figure 3.

To get the distribution of response times for the organization as a whole, we combine the distributions of the different configurations, weighted by their probabilities. This leads to another distribution, from which we can compute best-case, worst-case, and expected-case response times for the organization, *when it is successful*, given $f$ (Figure 4).

### 3.2 Reliability

Of course, if very unlikely configurations are the only ones that ever succeed, the organization is not a very good one no matter what its expected response time is conditioned on success. We can compute the probability of success by simply summing up the probabilities for the successful configurations. The probability of not succeeding (failing) is one minus the success probability, and for larger values of $f$ the failure probability can be substantial! We can thus evaluate organizations on likelihood of success independently of response time.

### 3.3 Overall Performance

Ultimately, to compare organizations with each other, we need a performance measure that combines the relevant features. In our case, we need to combine reliability and response time. Our way of doing this makes several assumptions that allow us to cast unreliability as delay in the achievement of goals. Simply put, our performance measure assumes that when an organization fails, it tries again and again until it succeeds. Thus, the organization times out when the worst-case response time across all successful configurations has elapsed. It then restarts the process, assuming that a new random configuration of the organization will be manifested, which might allow success. If the worst-case response time goes by again, it tries again, and so on until it finally succeeds.

Graphically, this means that the organization's response time distribution (Figure 4) is repeated over and over again, each time with a lower amplitude, such that the areas under all of the distributions sum to one (the area of the first distribution is the probability of success, of the second is the probability of failure times probability of success, of the third is the probability of failing twice before success, and so on). This is shown in Figure 5. We truncate this pattern when the remaining area is negligible, and we can compute the expected response time from this overall distribution.

## 4 Effects of Runtime Coordination

Given this distribution over performance, we can now be more precise about what runtime coordination is intended to do. Put simply, runtime coordination should move this distribution toward faster response times. There are two obvious ways that the distribution can be "moved" like this. One would be to increase the areas of the "humps" at faster response times, which means increasing the probability of success for any given execution of the organization. The other is to shift the expected value of the initial distribution to a faster time. That is, making the response time when successful faster will shift each of the repeated "humps" to

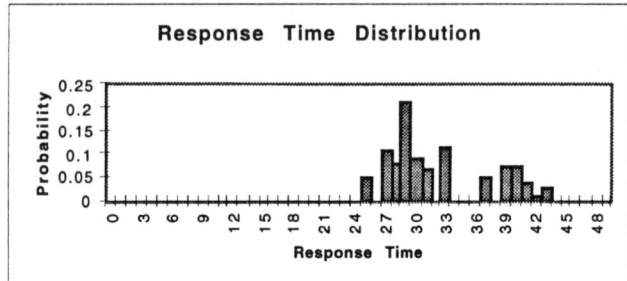

Figure 3

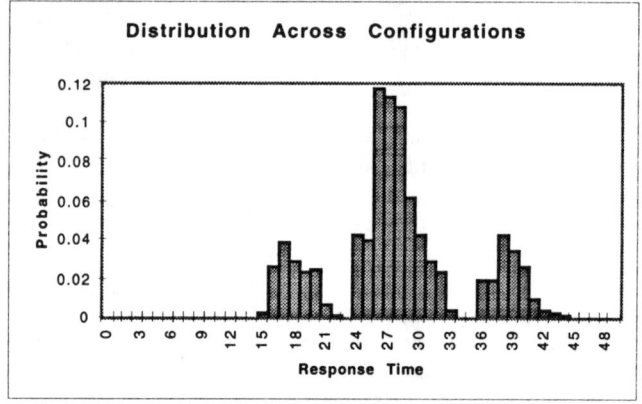

Figure 4

**Figure 5**

have faster expected response times, which will lead to an overall faster expected response time.

Thus, we are able to say something about the *potential* effects of runtime coordination without having to be specific about the mechanisms for realizing the coordination. Instead, we can characterize runtime coordination mechanisms in terms of their effects on the performance distributions, rather than by implementation details of the mechanisms. More substantially, as we shall see, we can talk about whether runtime coordination mechanisms that affect the distributions in particular ways would be expected to benefit particular organizational structures.

## 5 Coordination Strategies

In this paper, we will consider two main coordination strategies, corresponding to the transformations on response time distributions previously mentioned. Again, while the detailed design and implementation of these strategies are certainly of concern (as we discuss later), our goal here is to ascertain whether the runtime coordination through either (or both) of these strategies is potentially worthwhile however the strategies are implemented.

### 5.1 Role Reallocation

A **role** corresponds to a position in the tree structure of the organization. Agents occupy roles. If an agent fails, then the role that the agent plays in the organization is not filled. As we saw with our example in Figure 2, loss of roles affects performance. In fact, for particular configurations, the lost roles can render the organization ineffective.

Role reallocation is a runtime strategy whereby agents can reassign roles among themselves when they discover that certain critical roles are unfilled. Note that we assume that it is still the case that an agent can only occupy one role. But an agent can abandon a less important role to take on the responsibilities of a fallen comrade. For example, in Figure 2, agent 5 serves no useful purpose since the synthesis pathway for its results is blocked by agent 2's failure. 5 could therefore take on the role of failed agent 11, which (given a fortunate ordering of when tasks are done) can lead to faster organization response times than had it not done so.

More importantly, role reallocation (RR) can overcome agent failures that cause an organization to be unsuccessful. This strategy in principle allows agents at runtime to transform the current configuration (with some agents working and some failed) into any other configuration with the same number of working agents. Thus, they can transform the configuration into the best configuration with that many failed agents. Of course, this implies that the agents not only are able to recognize the current configuration, but can identify and agree on a better configuration. Examples of reconfiguration mechanisms exist in the multiagent systems literature (e.g. [Smith, 1980; So and Durfee, 1992]). Our investigation asks what kinds of organizations and environments could exploit agents that can do this.

### 5.2 Local Task Reordering

The second coordination strategy we look at here focuses on the local decisions agents make about which of their tasks to do next. This problem appears in many distributed applications, ranging from job-shop scheduling to distributed operating systems, with many approaches to supporting this kind of runtime coordination having been identified (e.g. [Decker, 1995; Liu and Sycara, 1996]. In general, task interdependencies and uncertainties about execution time complicate the ordering/scheduling of tasks in distributed systems.

In the context of our analysis, local task reordering (LTR) is a coordination strategy that lets agents take actions that complement those of other agents. In the simple scenarios we have outline here, this amounts to doing tasks that others will not have already done first, although our results should apply to more complex task relationships as well. Viewed as a transformation on the organization, LTR transforms a particular configuration into an instance of the same configuration but with the best possible orderings over the tasks of the agents. That is, if agents know which of them have failed, and thus which of the redundantly assigned tasks are now mandatory, they could reorder their tasks more optimally (as illustrated in section 3.1). This shifts the distribution of expected response times toward the faster end.

### 5.3 Combined Strategies

The two strategies outlined above are not mutually exclusive. At runtime, agents might be able to identify the configuration of empty roles (due to agent failures), and could both reassign roles among themselves and, within those roles, find the best local task reorderings. Both of these factors would be brought to bear on the coordination decisions: the agents should transform their configuration into the configuration (with the same number of failed agents) that has the best expected performance for the best ordering (compared to the RR strategy alone which transforms the configuration to the one with the best expected performance given random orderings).

### 5.4 Other Strategies

While the strategies outlined above cover the predominant runtime coordination mechanisms in the literature, they are not the only possible ones. Even better runtime coordination might occur, for example, if agents can redefine roles (swap tasks) or adopt multiple roles. However, as strategies like this begin to emerge, the notion of a "static" organizational structure begins to blur, since redefining roles, or dynamically creating new roles from merged pieces, amounts to runtime organizational redesign [Ishida et al., 1992]. Organizational redesign by agents is an important runtime strategy, but is beyond the scope of this paper.

## 6 Empirical Observations

By defining in a precise way a class of organizations, a set of performance measures, and the transformations on an organization caused by runtime coordination mechanisms, we can begin to understand the multiagent system design problem in integrated terms. To motivate this discussion, though, let us first look at some experimental results to illustrate some trends we see in studying the effects of runtime coordination in the context of organizational structures.

To keep things simple, the experiments we outline here are based on two organizations. One is a simple variation of Figures 1 and 2, where instead the organization was given $N=36$ tasks, and each "leaf" agent did 4 tasks ($m=4$). Thus, there were still 13 roles ($r=13$), with a branching factor $k=3$. The other organization has a branching factor $k=2$ and each "leaf" does $m=9$ of the 36 tasks so there are $r=7$ roles. In all cases, we assume $\gamma=1$.

In these experiments, we vary four parameters: the organization (as captured by the number of roles $r$), the organization's role overlap ($o$), the agent failure probability ($f$), and the coordination strategy ($s$) of the agents. The role overlap ranges from 0 to 2, the failure probability ranges from 0 up to 0.7, and the coordination strategy is one of four values: RR, LTR, BOTH, or NONE. (When there is no overlap, ALL are the same.) To gather the data, we ran each experiment that involved randomization at least 100 times. The results are summarized in Table 1. We have also experimented with many other organization structures, and the trends we emphasize in this paper appear to generalize.

### 6.1 Organization Structure

To begin with, the designer of an organization within the limited class we are considering can choose an organization size based on the branching factor and task sizes at the leaves. Not unexpectedly, an organization of 13 agents can solve the problem faster than an organization of 7 agents, so long as all agents are available. Our experiments show, however, that a smaller organization is inherently more robust, because the chances of any agent failing are smaller (Figure 6, $o=0$ cases). A simple change to the organization structure of introducing redundancy ($o$ greater than 0) can compensate at lower failure rates, and larger values of $o$ postpones organizational failures to higher values of $f$, but at a cost of being less efficient at lower $f$ values (Figure 6). This figure shows that at high $f$, a 13-role organization with $o=2$ is best (this can still succeed even losing 10 of 13 agents), but is inferior to the same organization with $o=1$ at low values of $f$. The 7-role organization with $o=1$ is better than the 13-role organization with $o=1$ at high $f$, needing only 3 of 7 working agents compared to 7 of 13.

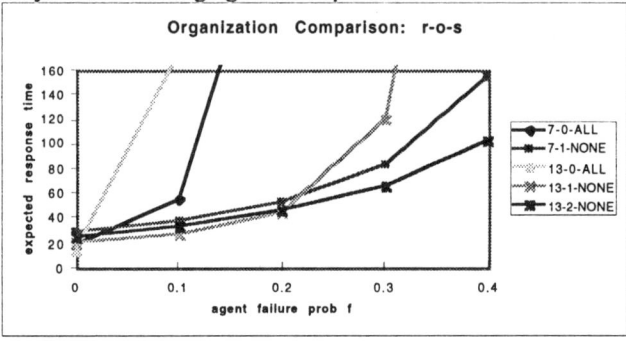

**Figure 6**

### 6.2 Runtime Coordination

Introducing redundancy illustrates a trend that we have recognized in all our other experiments, and that has intuitive appeal. Under low failure rates, what drags down response time is poor task selection, leading to duplication of effort on some tasks and delays in attending to others. Low failure rates mean that the redundant assignment seldom is needed, so the inefficiencies are incurred without the benefits unless an LTR strategy is employed For $r=13$ and $o=2$ (Figure 7), LTR can cut response time by about 40%. As failure rates increase, two things happen. First, there is less redundant activity because fewer agents with redundant tasks are active. Second, the dominant influence on response time is the repeated retrying of the problem. Thus, LTR has decreasing impact, while RR's effect becomes more pronounced (Figure 7) since, even for configurations for which the number of failed agents is high, there is often a very

**Table 1**

|      | r  | o | f=0  | f=.1 | f=.2 | f=.3 | f=.4 | f=.5 |
|------|----|---|------|------|------|------|------|------|
| ALL  | 13 | 0 | 14   | 176  | 4391 | $10^6$ |      |      |
| NONE | 13 | 1 | 20.2 | 26.2 | 44.5 | 120  | 534  | 3950 |
| LTR  | 13 | 1 | 14   | 21.6 | 40.9 | 117  | 532  | 3948 |
| RR   | 13 | 1 | 20.2 | 20.5 | 21.3 | 23.7 | 32.3 | 74.7 |
| BOTH | 13 | 1 | 14   | 15.6 | 17.1 | 20.3 | 29.6 | 72.5 |
| NONE | 13 | 2 | 25.3 | 33.7 | 45.5 | 65.2 | 104  | 198  |
| LTR  | 13 | 2 | 15   | 24.4 | 37.6 | 58.6 | 98.6 | 194  |
| RR   | 13 | 2 | 25.3 | 26.0 | 26.7 | 27.7 | 29.4 | 32.2 |
| BOTH | 13 | 2 | 15   | 16.2 | 17.9 | 19.9 | 22.6 | 26.6 |
| ALL  | 9  | 0 | 17   | 55.8 | 323  | 2317 | $10^5$ | $10^6$ |
| NONE | 9  | 1 | 28.6 | 37.7 | 52.7 | 82.9 | 156  | 369  |
| LTR  | 9  | 1 | 17   | 29.1 | 46.2 | 77.9 | 152  | 366  |
| RR   | 9  | 1 | 28.6 | 30.1 | 32.2 | 35.4 | 41.3 | 54.4 |
| BOTH | 9  | 1 | 17   | 21.2 | 25.0 | 29.5 | 36.6 | 50.7 |

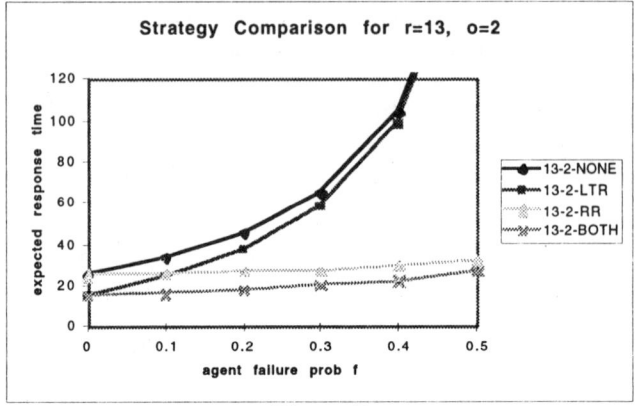

**Figure 7**

small number of configurations with the same number of failed agents where the organization can succeed. While the chances of randomly being in such configurations are small, the RR strategy allows these to be found and exploited.

### 6.3 Runtime Coordination and Role Overlap

Another important observation is that the coordination strategies are less effective as the role overlap decreases. That is, when the role overlap approaches 0 (no redundant task assignments), there is less opportunity for duplication of effort. The distribution for a single configuration (Figure 3) approaches a single spike, and the number of successful configurations (Figure 4) approaches the single "no failure" configuration, so LTR is less useful. In our data, LTR reduces runtime by about 40% with $o=2$ (for $r=13$), but only reduces runtime by about 30% with $o=1$, and of course has no effect at all for $o=0$. And, with smaller overlap, tolerance of *any* failure is reduced, so RR has no alternative configuration to move to. Thus, the efficacy of the coordination strategies not only varies with environmental factors like failure rate, but also the efficacy varies because strategies can only coordinate within the boundaries specified by the organizational structure. Structures that give less opportunity for agents to adapt to runtime conditions are less affected by runtime coordination mechanisms.

### 6.4 Runtime Coordination Costs

Of course, all of the preceding considered the potential advantages of runtime coordination without considering its costs. When there are no potential advantages, we can avoid introducing runtime coordination mechanisms that would not help even if they were overhead free.

More generally, when we accept that runtime coordination requires time (and other resources) that is taken away from performing tasks, then we need to trade off the costs and benefits of the runtime mechanisms. A thorough exploration of the costs is one of our future tasks. Here, we give a brief flavor of the kinds of results we are working toward.

A challenge in comparing coordination mechanisms is that the quality (or lack thereof) of implementation can obscure design comparisons. We sidestep this by introducing parametric models of coordination mechanisms, based on the parameters we have already described, such that we can assess performance under different assumptions about mechanism efficiency. For example, we might consider the cost of LTR at any given time to be dependent on the size of the space of possible collective orderings, which can be approximated by: $(m(o+1))^a$, where $a$ is the number of active agents in the organization. Note as more agents fail, the cost of LTR decreases dramatically (along with effective role overlap). The cost of RR might simply be based on periodic communication between all of the active agents. One simplistic model of coordination cost, therefore, is that it adds to the runtime a factor proportional to the actual runtime (assuming longer runs need more coordination):

$$\Delta RT = [\alpha*(m(o+1))^a + \beta*a^2] * RT$$

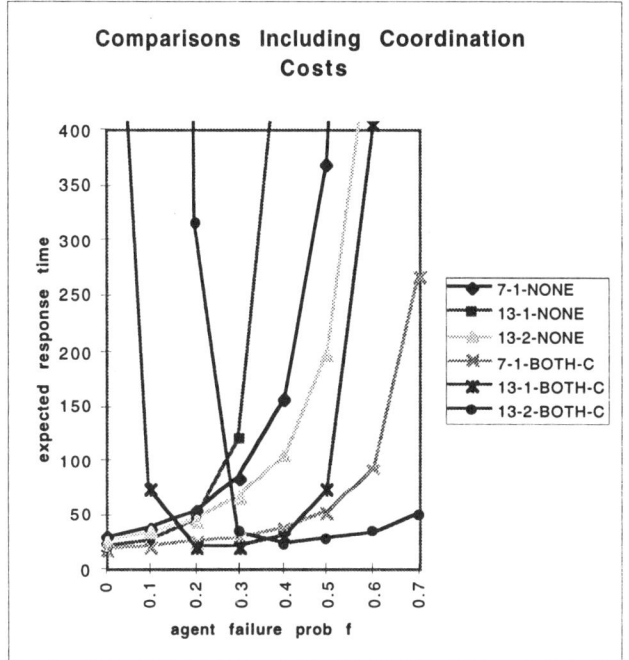

**Figure 8**

where $\alpha$ and $\beta$ are constants capturing in a gross form the quality of the coordination algorithm design and implementation. $\alpha$ and $\beta$ are typically small relative to task execution time, and for purposes of illustration we will assume assessing a single collective ordering is very fast ($\alpha=10^{-10}$) and selecting the right configuration is somewhat fast ($\beta=10^{-4}$) compared to the application tasks. For the 7 and 13 role organizations, approximating $a$ as $(1-f)r$, the comparison between using both or neither of the coordination strategies is shown in Figure 8. This illustrates that, for the given parameters, coordination overhead in the smaller (7-role) organization is cost-effective throughout. In the larger 13-role organization, the combinatoric costs swamp the benefits of runtime coordination until higher values of $f$, where the benefits are more substantial and the costs have decreased in this simple model because fewer "live" agents remain to coordinate. Note that for the 13-role case there is an interval for which less overlap is better than more, because more overlap increases coordination costs as well.

We should again emphasize that this example is simplistic and is intended to be only suggestive, rather than prescriptive. The combinatorics we have outlined for larger organizations have been addressed numerous ways in more practical coordination mechanisms, including reducing the frequency of coordination, coordinating agents in smaller local groups, and searching less of the collective task-ordering space (such as by using hill-climbing) [Durfee and Lesser, 1991]. These relaxations can lead to suboptimal coordination; using the model we have presented here, our research can work toward prescriptions about what relaxations make sense given the costs and the potential benefits for runtime coordination within an organization.

## 7 Designing Agents and Organizations

In this paper, we have taken initial steps toward characterizing how statically-defined organizational roles and relationships affect the potential utility of making runtime coordination decisions about what roles agents should fill and which of the tasks associated with a role each agent should perform next. From the agent-design perspective, our work emphasizes that building agents that are better at coordinating only makes sense in organizational structures that provide agents with sufficient latitude to exploit these capabilities. Agile organizations that are built to be robust under changing environmental conditions (such as when organization members fail to deliver) provide more opportunities for coordination mechanisms. Organizations that have evolved for stable conditions, however, might benefit little from better coordination mechanisms. In human terms, putting coordination technology on an assembly line would not be expected to help much, given our studies. Improving agent coordination capabilities should be considered in the context of the organization, whether agents are computational or human.

From the organization-design perspective, this tells us that we might well design organizations differently depending on the coordination abilities of the participants in the organization. Better responsiveness and reliability can both be attained with appropriate coordination mechanisms. But agents with such mechanisms should be organized differently for different environments, as we saw in Figure 8, where different organizational structures were better over different failure probability ranges. And, of course, if all we have are simpler agents, then we should tailor the organization to *not* depend on runtime coordination!

Most of our observations have been based on identifying *potential* for improvement. By looking at how runtime coordination might affect performance in the "best" case, we can validate beforehand whether there is sufficient opportunity to make investment in implementing coordination mechanisms for an application, and what class of coordination mechanisms (such as reconfiguration, or reordering, or both) hold the most promise. But we have also acknowledged that coordination comes at a cost, so that even when there are potential benefits from coordination, it might well be the case that the costs of the runtime mechanisms outweigh their benefits. To advance the design of multi-agent systems from an "art" into an engineering discipline, therefore, will require a careful articulation of the various design parameters for organizations and agents, and their impact on the performance measures [Carley and Prietula, 1994]. In this paper, we have taken steps in this direction so that we can begin to make recommendations for the restricted set of organizations outlined here, organizations for multiagent systems that have already been shown to be useful in multiagent applications including distributed sensor interpretation [Durfee and Lesser, 1991] and distributed network monitoring [So and Durfee, 1992].

## References

[Carley and Prietula, 1994] Carley, K. M., and Prietula, M. J., (Eds.), *Computational Organization Theory*, Hillsdale, NJ, Lawrence Earlbaum Associates.

[Decker, 1995] Decker, K., *Environment Centered Analysis and Design of Coordination Mechanisms*, PhD dissertation, University of Massachusetts.

[Durfee et al., 1987] Durfee, E., Lesser, V., and Corkill., D., "Coherent cooperation among communicating problem solvers." *IEEE Trans on Computers*, C36(11):1275-1291.

[Durfee and Lesser, 1991] Durfee, E., and Lesser, V., "Partial Global Planning: A coordination framework for distributed hypothesis formation." *IEEE Trans on Systems, Man, and Cybernetics*, SMC21(5):1167-1183.

[Gasser et al., 1989] Gasser, L., Rouquette, N. F., Hill, R. W., and Lieb, J., "Representing and Using Organizational Knowledge in Distributed AI Systems." In Gasser and Huhns (eds.) *Distributed Artificial Intelligence II*, San Mateo, CA, Morgan Kaufmann Publishers.

[Gmytrasiewicz et al., 1991] Gmytrasiewicz, P., Durfee, E., and Wehe, D., A decision-theoretic approach to coordinating multiagent interactions." In *IJCAI-91*, pp 62-68.

[Ishida et al., 1992] Ishida, T., Gasser, L., and Yokoo, M., Organization self-design of distributed production systems, *IEEE Trans on Knowl and Data Eng* DKE4(2):123-134.

[Liu and Sycara, 1996] Liu, J. and Sycara, K., Multiagent coordination in tightly coupled task scheduling, *International Conf on Multi-Agent Systems*, pp. 181-188.

[Shoham and Tennenholtz, 1992] Shoham, Y. and Tennenholtz, M., On the synthesis of useful social laws for artificial agent societies, *AAAI-92*, pages 276-281.

[Smith, 1980] Smith, R. The contract-net protocol, *IEEE Trans on Computers* C29(12):1104-1113.

[So and Durfee, 1997] So, Y. and Durfee, E., Designing span-of-control for computational agent organizations, *Journal of Computational and Mathematical Organization Theory* 2(3):219-246.

[So and Durfee, 1992] So, Y. and Durfee, E., A distributed problem-solving infrastructure for computer network management." *Int Journal of Intelligent and Cooperative Info Sys*, 1(2):363-392.

# DISTRIBUTED ARTIFICIAL INTELLIGENCE

Distributed AI 3: Multiagent Algorithms

# Dynamic Prioritization of Complex Agents in Distributed Constraint Satisfaction Problems

**Aaron Armstrong**
**Edmund Durfee**

Department of Electrical Engineering and Computer Science
University of Michigan
Ann Arbor, MI 48109 USA
{armst, durfee}@umich.edu

## Abstract

Cooperative distributed problem solving (CDPS) by loosely-coupled agents can be effectively modeled as a distributed constraint satisfaction problem (DCSP) where each agent has multiple local variables. DCSP protocols typically impose (partial) orders on agents to ensure systematic exploration of the search space, but the ordering decisions can have a dramatic effect on the overall problem-solving effort. In this paper, we examine several heuristics for ordering agents, and conclude that the best heuristics attempt to order agents based on the cumulative difficulty of finding assignments to their local variables. Less costly heuristics are sometimes also effective depending on the structure of the variables' constraints, and we describe the tradeoffs between heuristic cost and quality. Finally, we also show that a combined heuristic, with weightings determined through a genetic algorithm, can lead to the best performance.

## 1 Introduction

Cooperative distributed problem solving (CDPS) is often modeled as being done by a group of loosely-coupled computational agents involved in extensive local computations [Durfee et al. 1989; Luo et al. 1993]. Because these agents need to develop local solutions that together comprise one or more solutions to collective problems, they need to communicate intermittently about aspects of their local solutions to ensure compatibility. This may be usefully viewed as a distributed constraint satisfaction problem, where there are constraints between the local solutions of the different agents [Yokoo et al. 1992]. The agents want to exchange enough information to identify and to rectify violations of constraints. Rapid delivery of pertinent information is essential for the agents to avoid computationally expensive dead-ends. The challenge is in controlling this exchange so that it does not swamp the agents with messages, and so that it efficiently results in convergence to consistent solutions.

One way of ensuring systematic exchange of partial solutions and of ensuring the identification of constraint violations is to order the agents, such that some agents make commitments to particular solutions around which others must work. If a work-around cannot be found, the system backtracks by asking agents up the pecking order to try different commitments. This strategy is the multi-agent version of a centralized, backtracking search. In fact, it is possible for backtracking to exploit parallelism, in cases where constraints are not highly constraining, by asynchronous backtracking (ABT) [Yokoo et al. 1992]. With ABT, all agents in parallel pass their own variable assignments to relevant, lower priority agents and pass information on inconsistent combinations of value assignments (no-goods) to higher priority agents.

While instituting an ordering over the agents leads to systematic exploration, in the worst case there could still be an exhaustive search over the space of combinations of local solutions. To make this approach more effective, therefore, it can help if the agent ordering tends to focus search in more promising areas first. For example, highly constrained agents should have first choice.

Mapping this once again to the constraint satisfaction problem (CSP) framework, it would appear that ordering the agents is analogous to ordering the variables. In fact, this is the strategy that has generally been employed [Minton et al. 1990; Yokoo 1993], along with the typical assumption that each agent has one variable. The trouble in CDPS is that, to use communication bandwidth efficiently, the problem is distributed into a relatively small number of complex local problems—corresponding to a number of local CSPs. Realistically, the agents cannot be modeled as each having a single variable, but rather as each having multiple variables. In addition, considerations such as geographic distribution may suggest that we model the agents as each having a fixed set of variables, corresponding to a fixed problem decomposition (e.g. specified by resource availability). Now, even if variable ordering information is available, agents cannot be ordered strictly based on the variable ordering, because it is unclear how best to combine variable priorities to obtain a ranking of the agents. Further, even if good average-case methods for generating agent priorities were

---

This work has been supported, in part, by the National Science Foundation under PYI award 91-58473.

found, they could still be inferior to algorithms allowing dynamic priority assignment, since dynamic prioritization allows the CSP search process to discover and use additional information particular to the current problem. This research uses problems from a path planning domain to investigate the problems of how to prioritize dynamically and how to apply heuristics to ensembles of variables, so that DCSP algorithms can be efficiently used by the agents.

## 2  Distributed Constraint Satisfaction

In standard formulations of constraint-satisfaction, the problem is defined as one of instantiating an ordered set of variables V from a respective set of domains D such that a set of constraints C over the variables is satisfied. Figure 1 illustrates a CSP where $V = (x_1, x_2, x_3, x_4)$, $D = (\{1, 2, 3\}, \{1, 2\}, \{1, 3\}, \{1\})$, $C = \{(x_1 \neq x_2), (x_3 = 3), (x_3 = x_2 + 2), (x_3 \neq x_4)\}$. A solution is $(x_1, x_2, x_3, x_4) = (2, 1, 3, 1)$.

Backtracking search is a basic approach to solving CSPs. The variables are ordered and then the algorithm does a preorder walk of the implicit search tree. To reduce the size of the search space, we may use static or dynamic consistency methods to prune the tree, such as node, arc, and path consistency checks.

### 2.1  Asynchronous Backtracking

To allow us to study dynamic prioritization, we developed a DCSP protocol inspired by the asynchronous backtracking (ABT) [Yokoo et al. 1992] and weak-commitment (WC) protocols [Yokoo 1994]. A key idea of asynchronous backtracking is to distribute the search problem and then allow the agents to work concurrently on their local problems. This creates potential parallelism by allowing each agent to actively guess solutions and by allowing agents to discover no-goods simultaneously. In ABT, each agent is responsible for a single variable, and the agents are usually related by a fixed, total order (though the communication is only between mutually constrained agents). They use periodic communication to synchronize constraint checking information. Each constraint is checked by the lowest priority agent among the set of agents involved in the constraint. Continuing the example above, the constraint $(x_1 \neq x_2)$ would be checked by $x_2$, the constraint $(x_3 \neq x_4)$ would be checked by $x_4$, and the other constraints would be checked by $x_3$. (Assuming a total order of $x_1 > x_2 > x_3 > x_4$.)

The process begins by having each agent assign a value to its variable and then passing that value to agents who are constrained by the agent's variable. Each agent, on receiving values from higher priority agents, checks to see if its own choice is compatible. If not, it tries to pick a new value. If it finds a legal new value, it passes on this change to its dependents and otherwise tells its parent that its parent's value is no-good. (A no-good is defined by a subset of the agents and their variables' instantiations such that some lower priority agent cannot instantiate its variable(s) without violating some constraint.) If an agent receives a no-good message, it records the no-good as a new constraint and tells the other (more important) agents involved in the no-good to keep the agent who is processing the no-good informed of any changes to their variables. The agent then tries to find a new value for its variable, as above. The algorithm terminates when the lowest priority agent has found a value consistent with all higher-priority agents—success. It also terminates when an agent discovers that it has an empty domain (e.g. from new constraints)—failure.

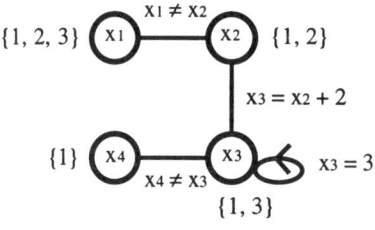

Figure 1

In the example above (figure 1), the agents choose $(x_1, x_2, x_3, x_4) = (1, 1, 3, 1)$. Note that node constraints can be locally processed and so $x_3$ does not choose value 1. Each agent passes its information to agents checking mutual constraints. Agent (variable) $x_2$ receives $x_1$'s message and changes its value to 2, sending along a message to $x_3$ stating the changes. Agent $x_3$ receives $x_2$'s message and responds saying that $x_2 = 2$ is a no-good for $x_2$. Agent $x_2$ accordingly adds a self-loop constraint that $x_2 \neq 2$. Agent $x_2$ informs $x_1$ that $x_1 = 1$ is a no-good for $x_1$ and informs $x_3$ that it has now chosen $x_2 = 1$. Agent $x_1$ adds a self-loop constraint that $x_1 \neq 1$, chooses $x_1 = 2$, and sends this on to $x_2$. Agent $x_2$ receives this message and has already chosen $x_2 = 1$. The other two agents likewise do not need to change their assignments (a slight time savings from parallelism). The agents find the solution $(x_1, x_2, x_3, x_4) = (2, 1, 3, 1)$.

### 2.2  Asynchronous Weak-Commitment Search

In (asynchronous) weak commitment search [Yokoo 1994, 1995], agents solve their local problems and check constraints in a manner similar to asynchronous backtracking, but whenever a no-good is discovered, the agent ordering is changed so that the agent who discovered the no-good now has highest priority. In weak-commitment search, the ordering is total and dynamic. Yokoo [1995] is not explicit in this regard, but many agents will need to be apprised of the situation whenever an agent is reprioritized so that constraints and no-goods will not be lost. Usually the lowest priority agent involved in a no-good stores the no-good information. Changing the order may then cause the no-good to be effectively forgotten, since another of the involved agents may become lower in the ranking and thus may eventually have to rediscover the no-good.

In our continuing example, the agents initially choose $(x_1, x_2, x_3, x_4) = (1, 1, 3, 1)$. Agent $x_2$ then receives $x_1$'s message ($x_1 = 1$) and changes its value to 2, sending along a message to $x_3$ stating the changes. Agent $x_3$ receives $x_2$'s message and responds saying that $x_2 = 2$ is a no-good for $x_2$. The ordering then changes to $(x_3, x_1, x_2, x_4)$. Agent $x_3$ sets its value to 3 and informs $x_1$. Agent $x_1$ retains its value of 1, causing $x_2$ to report a no-good of $x_1 = 1$. The priority then changes to $(x_2, x_3, x_1, x_4)$. The agents choose $(x_2, x_3, x_1, x_4) = (1, 3, 2, 1)$ and the problem is solved.

## 2.3 Summary

ABT and WC have similar benefits and a couple of problems. Their common benefit is the distribution of the problem and the result of possible parallelism. In addition, WC adds the ability to change bad orderings so that bad choices made by the high-priority agents will be discovered faster. One resulting problem of these distributed methods is the exponential space usage from the no-good storage. Another problem, the focus of our research, is the rigidness of the agent ordering. In ABT, the agent ordering is static and in WC, the ordering method is limited to a single heuristic.

## 3 Algorithm

Our algorithm was developed from the desire to use partial information from the search process to guide the remainder of the search, given our basic goal of investigating the problem of ordering agents with multiple variables. We generalized from ABT and WC to allow asynchronous local search with a more flexible reordering of the agent hierarchy.

### 3.1 Description

In our distributed algorithm, the variables and their domains are distributed among agents, each of which can check all constraints in which its variables are involved. There is also a central agent, which is responsible for starting and stopping the process and a no-good processor (possibly distributed), which keeps no-good information from being lost when the agents reorder.

The problem solving is divided into epochs, in which the central agent gives an initiation signal, the agents calculate and broadcast their priorities (thus establishing a total order of agents), and then the agents attempt to solve the problem. The epoch is terminated when a solution is discovered or when a no-good is discovered with fewer than $m$ agents involved. (The parameter $m$ is constant, typically small, during the search.) When $m = 0$, there will only be one epoch in the process, since a no-good involving zero agents implies that the problem has no solution. With this setting, the protocol reduces to a case of asynchronous backtracking in which communication is limited to an agent's immediate (typically two) neighbors in the total ordering. When $m > n$, where $n$ is the number of agents, every no-good will cause a re-prioritization. With this setting and with the prioritization determined by a rapidly decaying number-of-no-goods heuristic, the protocol reduces to weak-commitment search.

When a no-good is discovered, the variable assignments causing it and the IDs of the agents involved are sent to the no-good processor, which saves the no-good. Later, when an agent has found a tentative assignment for its variables, it consults the no-good processor to make sure that its assignment along with any known assignments of higher priority agents do not constitute a no-good. More on the no-good processor appears in the next subsection.

Each agent, after broadcasting and receiving priorities, constructs a tentative assignment of its variables. It consults with the no-good processor, then passes its assignment on to the next agent in the ranking. Gradually, information from higher ranking agents accumulates at each agent. This information is used to constrain the possible assignments of the agent's variables. All current information is passed on to the next agent in the ranking whenever a tentative assignment is made. In the case that the bottom agent is able to make an assignment to its variables and it has information from all higher ranking agents, the agent contacts the central agent to signal a solution to the problem. In the case that an agent cannot make an assignment to its variables, it has discovered a no-good. It will again contact the central agent, this time to signal a no-good (which may potentially trigger a new epoch). It will also signal the next-higher agent to find a new assignment for the higher agent's variables.

### 3.2 The No-Good Processor(s)

The issue of no-goods is crucially connected to completeness and to the space requirements of the DCSP algorithm. With asynchronicity, the algorithm could get stuck and continue to check and recheck the same variable assignments. To avoid this, as sections of the search tree are found not to contain a solution, data specifying the fruitless branches are recorded as no-goods to be avoided in the future. In asynchronous backtracking, when a no-good is discovered, the lowest priority agent involved in the no-good stores it in the form of an additional constraint.

In asynchronous weak-commitment search and in our protocol, using this method could result in the agents rediscovering and storing many copies of the same no-good, one for each time a different participating agent in the no-good had lowest priority. A no-good processor gives us the benefit of reduced storage costs (only one copy) and reduced search time (only discovered once). The tradeoff is in additional message passing. As mentioned before, we could distribute the no-good processor to reduce the computation and storage load on any particular process. In this case when an agent wants to check a collection of instantiations for no-goods, it sends the set of agents involved (and the variable assignments) to the no-good processors. The no-good processors are each responsible for mutually exclusive partitions of the power set of the agents. Each processor checks all subsets of the current list of agents corresponding to subsets in its piece of the partition. For example, with two agents and 2 no-good processors, processor A might store no-goods involving agents {1} and {2}, while processor B might store the no-goods for {1, 2}. If an agent wanted to check whether an assignment to the variables of agents {1, 2} was valid, processor A could check no-goods involving just agent 1 or just agent 2 and B could check no-goods involving both agents. Note further that if information from the higher-priority agents is consistent, the processors only have to check subsets involving the most recently added agent.

## 3.3 Performance Optimization

Now that the basic algorithm has been described and since it is a complete search technique, the next question is one of performance. Since CSP is NP-complete, we will merely attempt to improve the average case performance.

**Heuristics**

Obviously, to get good average performance from any search algorithm, we need to focus the search in more promising areas. These DCSP algorithms are focused by determining which agents have precedence and which values they prefer (choose first) for their variables. Value ordering heuristics, though important, were not the aim of this research and so were not used. We concentrated instead on the problem of deriving agent ordering heuristics from variable ordering heuristics which have been described in the literature, e.g. [Yokoo 1993; Minton et al. 1990]. In much of the previous work, there had been a trivial derivation: since each agent had one variable, an ordering of the variables constituted an ordering of the agents. With multiple variables for an agent, there must be a method of combining the ordering information of single variables to produce agent-ordering information.

To improve the performance of our algorithm on test problems in the path-planning domain (discussed later in the paper), we investigated various heuristics. For uniformity and control purposes, we used two null heuristics—one static and the other dynamic. The remaining heuristics attempted to quantify the degree of constraint on an agent, equating "more constrained" with "more important."

- Random (but static). This heuristic would initially assign a random total order to the agents and hold the order fixed throughout the search. This was our standard for comparison.
- Random (dynamic). This heuristic randomly generated a different order with each reprioritization.
- A pseudo-heuristic which ordered the agents the same way that the dynamic random heuristic had ordered them when it had finished solving the same problem. This allowed us to examine the importance of ordering vs. building up knowledge of no-goods.
- The number of no-goods discovered or a decaying average of the number of no-goods discovered. These heuristics gave priority to agents which had discovered larger numbers of no-goods. The hope was that this would dynamically determine the most constrained agents. An exponentially decaying average is the basic heuristic of Yokoo's weak-commitment search [1994].
- Total or average number of different values in a single agent's domains. These heuristics gave priority to agents with fewer choices for their variables.
- A weighted average of the domain sizes. This heuristic is similar to the last one, but it gave more weight to variables representing important choices (e.g. more likely to conflict). In the path planning domain, the size of the domains of variables representing agent location at the middle of the path were considered to be most informative and so were given the greatest weight.
- Number of local solutions. This heuristic requires exhaustive constraint-satisfaction internal to the agent to be performed first, effectively reducing the agent's set of variables to a single variable. If this is computationally infeasible, we could also estimate this number by checking some subset of the local problem for solutions. Another variation is to dynamically account for changes in the number of local solutions as no-goods accumulate and rule out solutions.

**Combining Heuristics**

In addition to investigating single heuristics, we also implemented an algorithm to automate performance tuning of combinations of heuristics. Priorities were assigned to agents by combining the heuristic values in a weighted sum. We used a genetic algorithm [Holland 1992] to search the space of heuristic weightings, to automatically discover which heuristics were effective, and if possible to exploit epistatic relations between them.

**Example**

In our toy example (figure 1), there are four problem-solving agents, each with one variable. (In our experiments, there are more variables for each agent—around 5-10.) The other two agents are the central agent and the no-good processor.

At the beginning, the central agent broadcasts an initiation message to all the agents. The agents calculate their priorities and broadcast them to each other. Suppose the ordering is $(A = x_1, B = x_2, C = x_3, D = x_4)$, where we equate agents with single variables. The agents then choose $(x_1, x_2, x_3, x_4) = (1, 1, 3, 1)$. Each checks with the no-good processor, which okays these choices. Agent A sends B a message that $x_1 = 1$. B discovers a conflict. It changes $x_2 = 2$ and passes $(x_1, x_2) = (1, 2)$ to C. Agent C discovers a conflict. C determines further that there is no selection for $x_3$ that is consistent with the assignments to $x_1$ and $x_2$. It tells the no-good processor that the ordered agent set (A, B) cannot use the assignments $(x_1, x_2) = (1, 2)$. Agent C sends a message to B telling it to try a different local solution and also sends a message to the central agent announcing a no-good involving (A, B).

If we assume the constant $m > n$, the central agent will start a new epoch for each no-good discovered. It instructs the agents to restart. They recalculate their priorities. Let us assume that agents who have discovered more no-goods and who have fewer solutions get higher priority. The ordering may then change to (C, D, B, A), where C's no-good increased its priority. The agents pick $(x_1, x_2, x_3, x_4) = (1, 1, 3, 1)$. The no-good processor then forces A to choose 2 instead. The agents pass on information until A has received information from the others. Agent A now has a consistent assignment, minimal priority, and solutions from all higher priority agents, so the CSP has been solved. Agent A notifies the central agent, which broadcasts a halt message.

As this example illustrates, we can potentially benefit from the parallelism of computation as in ABT. We can also benefit from a good initial ordering and any reordering of the variables, using dynamic information on the number of no-goods discovered and increasingly accurate estimates of the number of local solutions.

## 4 Experimental Evaluation

To evaluate our protocol and test our intuitions about various heuristics, we selected a simple family of problems and collected statistics on the performance of the algorithm using the different heuristics.

### 4.1 Problem Domain

We chose a problem in the domain of multiple-agent path planning as a source of the agents and their constraints. The $n$ agents inhabit nodes in an arbitrary graph. Each agent starts at some node and must arrive at some other node. There are $t$ time steps available for solving the problem. At each time step, each agent traverses an edge (possibly a self-loop). The solution to the problem consists of $n$ paths between the $n$ starts and goals. The paths must not conflict with one another (no simultaneous occupation of the same node or edge).

In figure 2 we have an example of a small problem. There are 3 agents, A, B, and C. This could potentially be a starting configuration. If the final configuration was (A, B, C) = (6, 4, 5), we could have a solution in which A uses path (1, 2, 3, 6); B uses path (7, 8, 5, 4); and C uses path (6, 9, 8, 5). In the CSP formulation, each agent might have 4 variables, corresponding to its position at times 1-4.

### 4.2 Topologies

We considered various topologies, since the topologies affect the kinds and numbers of constraints between variables more generally. We looked at randomly connected graphs, tree graphs, grid graphs, and hub graphs (each graph had multiple hubs). We did much of our work with grid graphs for ease of visualization and with hub graphs as a source of more difficult problems.

### 4.3 Experimental Analyses

We here summarize a few of our experimental results, focusing on the hub graph topology, but unless explicitly stated otherwise the reader can assume that the trends reported apply to the other topologies as well. We measured the impact of each agent-ordering heuristic on the time required by the agents to solve their path problems and on the amount of communication overhead incurred—the number of messages passed. The charts in figure 3 represent averages over 1200 randomly generated problems for the hub topology (with 12-16 nodes in each graph).

**Effects of Agent Ordering**

During the constraint satisfaction problem-solving, performance depends on the accumulation of additional constraints (discovery of no-goods) as well as on the ordering of agents. To get an appreciation of the relative influence of these factors, we can use the situation where the agents were given a fixed order for comparison. The order can be assigned randomly (in the charts, this is the "uniquifier" column) or can be based on the final ordering discovered by the dynamic random heuristic (in the chart,

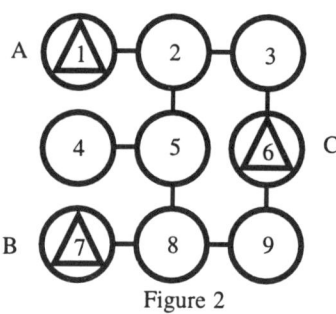

Figure 2

this is the "last run" heuristic). As shown in the charts (rightmost columns), a good ordering results in a time savings of 37% and a message savings of 32% compared to a random ordering. Similar savings occur in other topologies. Clearly, proper ordering can make a big difference.

**Effects of Dynamic Ordering**

As just seen, agent ordering can make a big difference. If a random order starts out badly, performance will suffer. If agents are given a chance of reordering dynamically, their performance might improve even if the reordering is random (the "random" column in the charts). Our experiments bear this hypothesis out.

**Performance of Heuristics Based on No-goods**

Yokoo's weak commitment search strategy would assume that an agent which discovers a no-good is highly constrained and should be moved to the front of the priority list. We evaluated this heuristic ("Decaying NG"), along with the variation of this heuristic that prioritized agents based on total no-goods discovered so far ("Num NGs"). Putting the most recently over-constrained agent first does shuffle the ordering to improve performance better than a random reshuffling, but we discovered that, at least for the DCSPs created in our domain across various topologies, it is better to consider more history: the total number of no-goods discovered heuristic generally performed better, and never performed worse, than the decaying no-goods heuristic that mimics Yokoo's.

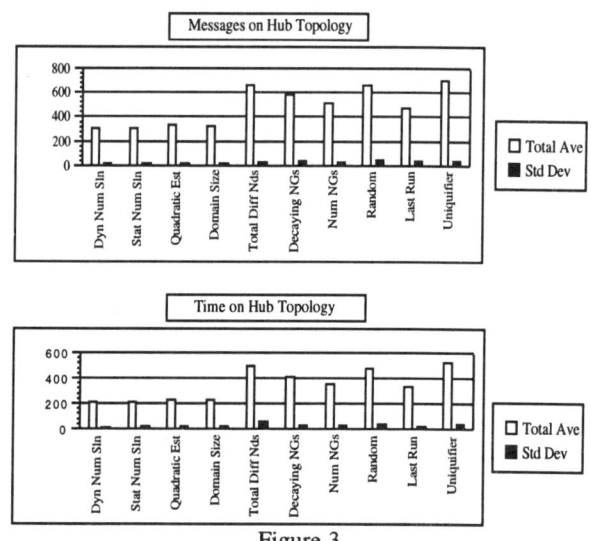

Figure 3

**Performance of Heuristics Based on Domain Size**

A widely used heuristic for ordering variables in CSPs is to assign the most-constrained variable first. Our heuristics approximated this by assessing the total sizes of the domains of all of an agent's variables ("Total Diff Nds"), the average domain size of an agent's variables ("Domain Size"), and a quadratically weighted average of the domain sizes ("Quadratic Est"). The first did poorly overall, probably because the different agents tend to have similar domain sizes (similar path characteristics), but the latter two generally did well, though slightly less well than the number of solutions, discussed next.

**Number of Local Solutions**

Of course, a better approximation of the most-constrained variable heuristic is to treat each agent as having a single variable (as in previous DCSP work). This amounts to having an agent generate the set of legal solutions to its local CSP, and then ordering the agents by their respective numbers of local solutions. We experimented with both static ("Stat Num Sln") and dynamic ("Dyn Num Sln") versions of this heuristic. Not surprisingly, it did very well across the board. Somewhat surprisingly, a static ordering did statistically as well as a dynamic ordering, but considering that usually constraints across groups of agents are discovered dynamically, we can see that dynamic changes to local solution sets may be minor (not enough new information).

**Combinations of Heuristics**

We used a generation-based genetic algorithm (GA) to study combinations of heuristics. Each of the 100 population members consisted of a concatenation of Grey-coded weights. The weights determined the amount of influence exerted by each heuristic. There was some difficulty with slow convergence because of the wide variation of the randomly generated problems.

After running the GA, we selected some of the top combination heuristics and ran them against several of our "pure" heuristics, measuring the number of time steps and the number of messages sent averaged over 1500 problems. Successful combinations placed high emphasis on the number of local solutions and on the number of no-goods. A high value was also accorded to the dynamic random heuristic. Combination heuristics proved to be modestly better than any pure heuristic.

In practice, such an algorithm could be used to gradually optimize the performance of a large CSP system. Heuristics and parameter settings could be guessed and as the system was used, it would adapt to the particular distribution of problems that it faced and identify the most useful heuristics with relatively little overhead.

## 5 Conclusions

While several researchers have recognized the similarities between cooperative distributed problem solving and distributed constraint satisfaction, the emphasis of the former on loosely-coupled agents solving substantial local problems has not been adequately addressed in the latter. In this paper, we describe a foray into this area of investigation. We have described a protocol that is assured of terminating and that generalizes asynchronous backtracking and weak commitment search to permit variations in the timing and criteria of agent reordering. Our empirical investigation using these capabilities has revealed that a good ordering is critical to performance, and that ordering based on the local solution spaces is most effective. However, because the computations for this heuristic amount to solving a substantial local CSP, more cost effective approximations are available that perform nearly as well. Yokoo's approach of placing the most recently over-constrained agent first is one possibility, although our results indicate that the total number of no-goods and the average domain size heuristics can be even more effective.

Many open problems remain, including characterizing the tradeoffs between local computation and the benefits of the heuristics. We also need to investigate further decentralization of the protocol and better storage of no-goods (cf. the polynomial space usage in [Ginsberg and McAllester 1994]). Finally, it might be the case that a good ordering of agents is not possible without redistribution of variables, leading to issues of negotiation and load-balancing.

## References

[Durfee et al. 1989] Durfee, E., Lesser, V., and Corkill, D. Cooperative Distributed Problem Solving. In A. Barr, P. Cohen, and E. Feigenbaum (eds.). *The Handbook of Artificial Intelligence*, Volume IV, Addison-Wesley.

[Ginsberg and McAllester 1994] Ginsberg, M. and McAllester, D. GSAT and Dynamic Backtracking. *Proc. of Princ. and Pract. of Constraint Prog.-1994*, 243-265.

[Holland 1992] Holland, J. *Adaptation in Natural and Artificial Systems: an Introductory Analysis with Applications to Biology, Control, and AC*, 1st MIT Press ed. Cambridge, MA: MIT Press.

[Luo et al. 1993] Luo, Q., Hendry, P., and Buchanan, J. Heuristic Search for Distributed Constraint Satisfaction Problems. Research report KEG-6-93, University of Strathclyde, UK.

[Minton et al. 1990] Minton, S., Johnston, M., Philips, A., and Laird, P. Solving Large-Scale Constraint Satisfaction and Scheduling Problems using a Heuristic Repair Method. *Proc. of AAAI-1990*, 17-24.

[Yokoo et al. 1992] Yokoo, M., Durfee, E., Ishida, T., and Kuwabara, K. Distributed Constraint Satisfaction for Formalizing Distributed Problem Solving. *12th IEEE Int. Conf. on Dist. Computing Sys.*, 614-621.

[Yokoo 1993] Yokoo, M. Dynamic Variable/Value Ordering Heuristics for Solving Large-Scale Distributed Constraint Satisfaction Problems. *Proc. of 12th Int. Workshop on Distributed Art. Int.*, 407-422.

[Yokoo 1994] Yokoo, M. Weak-Commitment Search for Solving Constraint Satisfaction Problems. *Proc. of the 12th National Conf. on Art. Int.*, 313-318.

[Yokoo 1995] Yokoo, M. Asynchronous Weak-Commitment Search for Solving Large-Scale Distributed Constraint Satisfaction Problems. *Proc. of the 1st Int. Conf. on Multi-Agent Systems*, 467.

# A dynamic theory of incentives in multi-agent systems (Preliminary report)

Yoav Shoham[*]
Stanford University
Computer Science Department
Stanford, CA 94305
U. S. A.

Katsumi Tanaka[†]
Toshiba Corporation
Kansai Research Laboratories
8-6-26 Motoyama-minami-cho
Higashinada-ku, Kobe 658
Japan

## Abstract

Motivated by problems that plague the acceptance of groupware products, as well as the challenge of designing effective multi-agent systems, we investigate the role of incentives in "public goods" settings, that is, settings in which the value of the system is supplied by its members. In such settings one encounters a prisoners-dilemma-like problem, in which each individual has a disincentive to contribute, but much benefit from the contributions of others. The natural result is that, unless special care is taken to engineer correct incentives, no contribution is made by anyone and the system has no value. Drawing on the literature on mechanism design and public goods in economics, we present a model in which to reason about such settings. Our model has several novel features, the most important of which is its dynamic component; the model describes how the value of the system is determined over time as a result of its use. We provide several results about the model, including the identification of a *critical mass* point, which is a value of the system that, if reached, is guaranteed to henceforth enable the system to engineer incentives that lead to contribution to (and thus added value of) the system.

## 1 Introduction

A fundamental tenet of rational behavior, and of much work in AI, is that agents are incentive driven. One way in which this assumption manifests itself in AI is in many attempts to endow programs with various notions of motivation – preferences, utilities, goals, and so on – as well as with a general mechanism (such as a planner) that acts in accordance with these motivations, the idea being that once this general mechanism is in place, one need only specify the motivational component in order to elicit the desired behavior. One can also turn the picture on its head, and ask how the view of *people* as rational beings should impact the design of effective programs, especially those that interact with their users. Of course, this too has been an important strand of AI research, perhaps best exemplified by application of plan- and goal-recognition techniques to question-answering systems (for example, Cohen *et al.*'s [Cohen *et al.*, 1982]).

Reasoning about incentives in a multi-agent setting is complicated, since the incentives offered to an agent may depend not only on its behavior but also on the behavior of others. The area of *mechanism design* [Fudenberg and Tirole, 1991] (also known as *implementation theory*) in economics studies the crafting of incentive structures that lead to desired behaviors. There has already been work in AI that embraces some ideas from mechanism design [Zlotkin and Rosenschein, 1993; Shoham and Tennenholtz, 1997]. Here we continue in that tradition, but concentrate on an important particular case, which in the economics jargon is referred to as *public goods*.

The prototypical example of public goods are tax-financed highway construction or national defense. In such situations there is some "public good" that is paid for by some segment of the population, but that can be enjoyed equally by everyone regardless of actual payment made (see caveat below). It is easy to see that in the absence of (e.g., judicial) external incentives, there is no incentive for any individual to pay taxes, but a great benefit from having other people pay *their* taxes. Turning to computational applications that motivate our work, consider a shared database, one that is populated by items contributed by the set of users themselves (these have been called *discretionary databases* [Connolly and Porter, 1990]). The various users (and contributors) might be people or software agents. In either case, each user derives benefit from having a rich database to draw from, but has a slight disincentive to actually contribute anything. The result is that the database remains empty, to the detriment of all users; this is of course a manifestation of a multi-person version of the prisoners dilemma.

Lest this sound too academic, let us mention that

---

[*]This work was supported by NSF grants IRI-9220645 and IRI-9503109.

[†]This work was carried out while the author was a visiting scholar at Stanford University.

the problem is very real. In the case of groupware it is well recognized, as documented for example in [Markus and Connolly, 1990]. We experienced it ourselves in our experiments with Fab, a collaborative recommendation system [Balabanovic and Shoham, 1997]. In addition, while it is still early days for software agents, we conjecture that the problem will be real there as well. Indeed, several researchers share the belief that future software agents will exhibit strategic interactions similar to those of people (cf. [Varian, 1995; Zlotkin and Rosenschein, 1993]).

The basic question we will ask is whether the incentives can be meddled with so that the users are incented to contribute. Stated this generally, this is the problem tackled by public-good theory in economics. However, as we shall see, our computational orientation will cause us to diverge from that literature. One divergence is the degree of freedom we allow in meddling with the incentives. Public-goods theory includes the notion of *exclusion*, which refers to the ability to withhold the benefits of the public good from some people. Given our intended applications, we will allow greater freedom. We will inherit from the public-good model the assumption that the incentives are derived entirely from the value of the system (e.g., the database), but will allow the system to offer each user any amount of benefit between zero and the entire value of the system, based on their contribution.

We inherit from the economic literature the assumption that, given a particular incentive structure, the users will contribute in a way that maximizes their rewards. But here is where we further diverge from classical work in mechanism design and public-good theory; we add a function that determines how the value of the database evolves as a result of the various contributions to it. This generates a dynamic model, which is absent from the model one finds in economics: At each stage the current value of the database determines the incentive structure, the incentive structure determines the various contributions, the various contributions together with the current value of the database determine its next value, and the process repeats.

Once we have this model in place, we can meaningfully ask various questions, such as: Does the current value of the database allow an incentive structure that incents at least one person to contribute? Is there some minimum value for the database (a "critical mass") above which we are guaranteed that the database value will grow monotonically in time? As it turns out, these are two of the questions we will answer once we have presented the formal model.

The paper is structured as follows. In section 2 we present the formal model, accompanied by intuitive explanations and some algebraic examples. In sections 3 and 4 we present our results that were derived within the model, concerning the existence of appropriate reward schemes that lead to continuing participation in the system; from the technical point of view, these sections contains the main "results" of the paper. We conclude in section 5 with various comments, discussion of related work, and our own planned future work in this area.

## 2 A dynamic model of public goods

In the following definitions we assume some fixed number $n$ of *agents*, a set $E = [0..\Omega] \subset \mathcal{R}$ of *possible effort levels*, and a set $\Theta = [0, \infty)$ of user types. In the following, $\mathcal{R}$ denotes the real numbers and $\mathcal{R}^+$ the non-negative reals.

We first define how the value of the system (e.g., database) evolves.

**Definition 1** *A (group-based)* system growth function *is a function* $f : E^n \times \mathcal{R}^+ \to \mathcal{R}^+$ *defined by*

$$f((e_1, ..., e_n), W) = w_1(\Sigma_{i=1}^n e_i) + w_2(W)$$

*where $w_1$ and $w_2$ are continuous and monotonic increasing functions.*

*Intuition:* The system's value is a function of its old value and the incremental efforts put into it by each of the $n$ agents. Note that the growth depends only on the sum on the efforts; in particular, it does not depend on the identities of the contributors. Also note that 'growth' can be negative.

This defines one direction of the dynamics, namely how the value of the system is determined by the agents. We now define the more involved way in which the system provides incentives to the agents and thus, indirectly, leads them to behave in certain ways.

**Definition 2** *A (system-based)* reward function *is a function* $r : E \times E^n \times \mathcal{R}^+ \to \mathcal{R}^+$ *and must satisfy* $r(e, (e_1, \ldots, e_i, \ldots, e_j, \ldots, e_n), W) = r(e, (e_1, \ldots, e_j, \ldots, e_i, \ldots, e_n), W)$ *for all* $i, j \in \{1, \ldots, n\}$

*Intuition:* Based on its own most recent effort level $e$, efforts of all agents $(e_1, \ldots, e_n)$, and on the current value $W$ of the system (e.g., the database), the agent is given a certain reward $r(e, (e_1, \ldots, e_n), W)$. Note that rewards are symmetric; they can not favor one agent over another based on any consideration other than the effort made by the agents.

**Definition 3** *A* disutility function *is a function* $d : E \times \Theta \to \mathcal{R}^+$, *such that*

*for all $\theta$, $d(\Omega, \theta) = \infty$,*

*for all $\theta$, $d(0, \theta) = 0$,*

*$d(e, \theta)$ is a continuous and twice differential function of $e$ and $\theta$,*

$\frac{\partial d}{\partial e} > 0$, $\frac{\partial^2 d}{\partial e^2} > 0$, $\frac{\partial d}{\partial \theta} < 0$, $\frac{\partial^2 d}{\partial \theta \partial e} < 0$

*Intuition:* Each agent experience a certain disutility from engaging in effort; the larger the effort, the greater the disutility. The disutility function is concave; disutility increases more rapidly the higher the effort, and is infinite at the level of maximum effort (one concrete interpretation is to view effort level as hours of work in a day, and $\Omega$ as 24). However, different agents experience different disutilities at different levels (again, for concreteness, think of different sleep requirements of people). This is reflected in the agent type; the higher the type, the lower is the disutility at any given effort level.

A reward function and a disutility function form the basis for a composite utility function:

**Definition 4** *A (system-based) utility function is a function* $u : E \times E^n \times \mathcal{R}^+ \times \Theta \to \mathcal{R}$ *defined by*

$$u(e, (e_1 \ldots, e_n), W, \theta) = r(e, (e_1 \ldots, e_n), W) - d(e, \theta)$$

*where $r$ is a (system-based) reward function, and $d$ is a disutility function.*

Note that for any $W$, $\theta$ and $e_1, \ldots, e_n$ other than $e$, we have $u(0, (e_1 \ldots, e_n), W, \theta) \geq 0$. $u(0, (e_1 \ldots, e_n), W, \theta)$ is called the *reservation value* of the agent; in order to get the agent to take any action at all the system must generate at least this great an incentive.

Let us now tie this all into a definition of a dynamic public-good system:

**Definition 5** *A dynamic public-good system (DPGS) is a tuple* $((\theta_1, \ldots \theta_n), r, d, f)$, *where $\theta_i \in \Theta$, $r$ is a reward function, $d$ is a disutility function, and $f$ is a growth function.*

*Intuition:* We have $n$ agents, where $\theta_i$ is the type of the $i$'th agent. They are all subject to the same reward and disutility functions, though, of course, they will experience different values due to their different types and the different efforts they will make.

**Definition 6** *A DPGS* $((\theta_1, \ldots \theta_n), r, d, f)$ *with* $f = w_1(\Sigma_{i=1}^n e_i) + w_2(W)$ *is said to be* endogenous *if* $r(e, (e_1 \ldots, e_n), W)$ *is constrained by* $f = w_1(\Sigma_{i=1}^n e_i) + w_2(W)$.

*Intuition:* In endogenous systems all the rewards come from within the system. However, agents can be "promised" a reward based on the future value of the system (e.g., people are persuaded to pay for a highway upgrade based on the value of the upgraded highway, not the existing one). This is crucial if we wish for the system value to increase with time.

In the sequel we will be concerned mostly with endogenous systems.

We are close to having the dynamics of the system completely specified. Clearly, given an initial value of the system, if we knew what effort levels would be selected by the each agent for each value of the system we would be able to predict a deterministic evolution of the system's value.

But rather than take these effort levels as given, we try to deduce them under the assumption that all the agents are economically rational. Specifically, we assume that agents select effort levels that maximize their utilities. The problem, of course, is that this is not well defined in general, since each agent's reward (and thus utility) depends in part on the efforts made by all the agents. Indeed, this situation defines a game in the economic sense, and we can apply the various solution concepts from game theory [Osborne and Rubinstein, 1994].

However, we are really not looking to solve a particular game, but are rather asking the inverse question, namely how can we craft a reward function that induces a game with desired properties. This is precisely the realm of mechanism design [Fudenberg and Tirole, 1991].

Traditional work on mechanism design assumes direct manipulation of the utilities. We have only roundabout (and, because of the restriction to endogenous systems, constrained) ways of doing so, via manipulation of the reward function. There are potentially many sorts of mechanisms we can attempt to set up here, but we will be looking at a very particular case, one based on *dominant strategies*.

**Definition 7** *A vector of efforts* $(e_1^*, \ldots, e_n^*)$ *is said to be* dominant *for a DPGS* $((\theta_1, \ldots \theta_n), r, d, f)$ *at $W$ iff for all $i \in \{1, \ldots, n\}$, and all $e_1, \ldots, e_n \in E$, it is the case that* $u(e_i^*, (e_1, \ldots, e_i^*, \ldots, e_n), W, \theta_i) \geq u(e_i, (e_1, \ldots, e_i, \ldots, e_n), W, \theta_i)$

*Intuition:* This is a strong requirement. It says that each agent can choose an effort level that is guaranteed to maximize his utility, no matter what the other agents do. If such an effort level exists, surely we can assume that the agent will adopt it. Therefore, if we are guaranteed that a given DPGS has a dominant effort vector for *any* value of the system, we can predict a deterministic evolution of the system value.

In section 3 we will look at conditions in which we can devise reward functions that lead to dominant strategies. Of course, the fact that a strategy is dominant doesn't imply that it is a good strategy, and in particular that it leads to growth in the value of the system; we will be looking for dominant strategies that do have this property.

We conclude the section with an example that illustrates the definitions.

**Example 1** Consider a community-based news clipping service, implemented as a discretionary database. During the day each user can annotate news articles s/he came across with some keywords to facilitate indexing, and place it in the database. At the end of the day each user can query the database; for simplicity, assume that each query returns a single item. We can model the situation with the following DPGS:

$e_i \in [0..\Omega]$: the number of items placed in the database by user $i$ during the day. Assume for simplicity that no two users submit the same article.

$W$: the number of items in the database at the end of the day.

$f$: the growth function which determined the value W' of the database at the end of the day, based on its value W yesterday and the items placed during the day: $f((e_1, \ldots, e_n), W) = \Sigma_{i=1}^n e_i + 0.5 \cdot W$. $f$ reflects an assumption that each day half the database items left over from the previous day become obsolete and are removed from the database automatically.

$r$: the number of articles the user $i$ is allowed to draw from database at the end of the day (assume that beyond that number queries result in an "account overdrawn" message): $r(e_i, (e_1, \ldots, e_n), W) = f((e_1 \ldots, e_n), W)$. $r$ reflects the fact that each user is given free access to the database regardless of contribution.

$d$: the disutility function $d(e, \theta) = exp(\frac{e}{(\Omega-e)\theta}) - 1$.

For simplicity, let us assume that there are two users with types $\theta_1 = 1$ and $\theta_2 = 0.5$.

Since $r(e, (e_1, e_2), W) = e_1 + e_2 + 0.5 \cdot W$, the utility function for the first user becomes $u(e_1, (e_1, e_2), W, \theta_1) = e_1 + e_2 + 0.5 \cdot W - exp(\frac{e_1}{(\Omega-e_1)}) + 1$, and similarly for second user. The derivative of our particular utility function for either agent is therefore $u'(e) = 1 - \frac{exp(\frac{e}{(\Omega-e)})\Omega}{(\Omega-e)^2}$, for any value of the effort level of the other agent. As for any legal utility function, since our utility function is convex it is maximized where the derivative is zero if such a point exists, or at 0 otherwise. Figure 1 plots the utility function for two values of $\Omega$; for $\Omega = 3$ the utility has a maximum at $e^* > 0$, and for $\Omega = 1$ the derivative is always negative and the utility is maximum at 0. The second is an extreme case of lack of incentives; under no circumstance could a user be incented to place items in the database. The first case does incent the users to exert some effort, but whether this effort is sufficient to sustain growth depends on the value of the database; we return to this issue in the next sections.

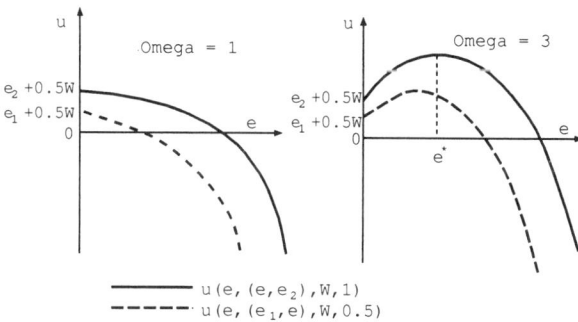

Figure 1: Utility functions

## 3 Reward design

As shown in example 1, careless reward schemes can lead to a prisoners-dilemma situation, in which no agent contributes to the system, and the system dies. We begin to investigate now the design of reward functions that avoid this situation. In this section we look at what it takes to design a value-increasing reward function at a particular value of the system. In the next section we'll look at conditions that guarantee that once such a reward can be offered, the dynamics of the system will ensure that henceforth this will be the case.

Our first theorem gives necessary and sufficient conditions on effort vectors such that the vectors are dominant strategies for some reward function. (All the proofs are omitted from this article.)

**Theorem 1** *Given:*
- *a DPGS* $((\theta_1, \ldots, \theta_n), r, d, f)$,
- *a system value $W$, and*
- *a vector $(e_1^*, \ldots e_n^*)$ such that $0 \leq e_i^* < \Omega$,*

*the following two conditions are equivalent:*

1. *There is a reward function $r$ such that $(e_1^*, \ldots, e_n^*)$ is a dominant strategy for DPGS $((\theta_1, \ldots \theta_n), r, d, f)$ at $W$.*

2. $e_i^* \geq e_j^* \equiv \theta_i \geq \theta_j$ *for all $i, j \in \{1, \ldots, n\}$.*

Intuitively, Theorem 1 states that it is possible to design a reward function that induces any desired effort levels on the part of the various agents, so long as the effort assigned to low-disutility agents is higher than that of high-disutility agents. Note that Theorem 1 does not restrict the reward function. It is perhaps not surprising that agents can be incented to do almost anything given arbitrary incentives. The following theorem looks at the effects of endogeny:

**Theorem 2** *Under the same conditions as in Theorem 1, and given an endogenous DPGS $D = ((\theta_1, \ldots \theta_n), r, d, f)$, the following two conditions are equivalent:*

1. *There is a reward function $r$ such that $(e_1^*, \ldots, e_n^*)$ is a dominant strategy for $D$ at $W$.*

2. $-f((0, \ldots, e_i, \ldots, 0), W) \leq d(e_i^*, \theta_i) - d(e_i, \theta_i) \leq f((0, \ldots, e_i^*, \ldots, 0), W)$ *forall $i \in \{1, \ldots, n\}, e_1, \ldots, e_n$.*

Theorem 2 states the strict relationship between system growth function and disutilities of agents for endogenous DPGS, whose intuition is that the disutility incurred by an agent at any given effort level should be compensated by system's value. Overall these two theorems tells us that as system designers we have quite a lot of power, even when working within endogenous *DPGS*. This is illustrated in the following example.

**Example 2** Let us modify Example 1 (the case of $\Omega = 1$) by adjusting the reward function $r$ such that
$r(e, (e, e'), W) = r(e, (e', e), W) =$
$$\begin{cases} \beta + 0.5 \cdot W & (e = \alpha, e' = \beta) \\ \alpha & (e = \beta, e' = \alpha) \\ \alpha + exp(\frac{\alpha}{1-\alpha}) - exp(\frac{\beta}{1-\beta}) & (e = \alpha, e' = \alpha) \\ \beta + 0.5 \cdot W - \\ \quad exp(\frac{2\alpha}{1-\alpha}) + exp(\frac{2\beta}{1-\beta}) & (e = \beta, e' = \beta) \\ \alpha + 0.5 \cdot W & (e = \alpha, e' \neq \alpha, \beta) \\ \alpha + 0.5 \cdot W - \\ \quad exp(\frac{2\alpha}{1-\alpha}) + exp(\frac{2\beta}{1-\beta}) & (e = \beta, e' \neq \alpha, \beta) \\ Min(exp(\frac{e}{1-e}) - exp(\frac{\alpha}{1-\alpha}), \\ \quad exp(\frac{2e}{1-e}) - exp(\frac{2\beta}{1-\beta})) & (otherwise) \end{cases}$$

If $\alpha \geq \beta$ there is a dominant strategy $(\alpha, \beta)$ since the following conditions are satisfied:
$u(\alpha, (\alpha, e_2), W) \geq u(e_1, (e_1, e_2), W)$ *forall* $e_1, e_2, \in [0..\Omega)$
$u(\beta, (e_1, \beta), W) \geq u(e_2, (e_1, e_2), W)$ *forall* $e_1, e_2 \in [0..\Omega)$.

In order to ensure that this *DPGS* is endogenous, $r$ must satisfy $0 \leq r(e, (e, e'), W) \leq f((e, e'), W)$ *forall $e$,*

$e'$, which corresponds to the following conditions:
$-e - 0.5 \cdot W \leq exp(\frac{\alpha}{1-\alpha}) - exp(\frac{e}{1-e}) \leq \alpha + 0.5 \cdot W$,
$-e - 0.5 \cdot W \leq exp(\frac{2\beta}{1-\beta}) - exp(\frac{2e}{1-e}) \leq \beta + 0.5 \cdot W$
forall $e$.

## 4 Critical mass

The previous section discussed our ability to engineer good incentives at a particular system value. We now take this a step further and ask whether we can have some assurance that from a certain point on system will continue to be used and to grow monotonically in value.

The key notion here is that of *critical mass*:

**Definition 8** *The critical mass of a DPGS* $((\theta_1, \ldots \theta_n), , r, d, f)$ *is the smallest system value* $W_c$ *such that*
$$W_c \in \mathcal{W} =_{def} \{W \in \mathcal{R}^+ : f((\Omega, \ldots, \Omega), W) > W\},$$
*and*

*for all* $W \in \mathcal{W}, W \geq W_c$ *there exists some function* $r'$ *such that*

1. $((\theta_1, \ldots \theta_n), r', d, f)$ *is endogenous.*
2. *there exists a strategy* $(e_1^*, \ldots, e_n^*)$ *that is dominant for* $((\theta_1, \ldots \theta_n), r', d, f)$ *at* $W$, *and such that* $f((e_1^*, \ldots, e_n^*), W) > W$.

The intuition is as follows. Clearly, system growth increases monotonically with the total effort made, and so positive growth is achieved by some combination of efforts if and only if it is achieved by maximal efforts by all agents. By definition, above the critical mass it must be the case that at all system values in which positive system growth is possible at all, agents can be endogenously induced to in fact exert sufficient effort to enable positive (though not necessarily maximal) growth.

Note that this definition could be stronger; it requires an appropriate reward function at every $W \in \mathcal{W}, W \geq W_c$, but these functions need not be the same. Furthermore, we give no guidance about how to compute these reward functions.

The critical mass does not always exist; here is a necessary and sufficient condition for its existence.

**Theorem 3** *Let the growth function of a DPGS D be* $w_1(\Sigma e_i) + w_2(W)$ *as in Definition 1. Then D has a critical mass* $W_c$ *if and only if there exists a* $W$ *such that*

*there exists a effort level* $\bar{e}$ *such that*
$$0 \leq \bar{e} < \Omega,$$
$$w_1(\bar{e} \cdot n) + w_2(W) > W, \text{ and}$$
$$d(\bar{e}, \theta_i) \leq w_1(\bar{e}) + w_2(W) \text{ forall } i \in \{1, \ldots, n\}$$

Intuitively, this says that there exists a critical mass iff there would be enough compensation for every agent's disutility in the case that maximum effort of each agent is exerted. Unfortunately, we provide no guidance here about how to compute this critical mass, or even how to determine whether the condition stated holds in any given circumstance.

However, we do end with one bit of good news. One might worry that if despite the meager instructions offered here one found a critical mass, the increments of the system value might not promised. The following theorem guarantees that this is not the case; once the critical mass is reached, the system's value will increase.

**Theorem 4** *Given an endogenous DPGS D, a system value W, and a dominant strategy* $(e_1^*, \ldots, e_n^*)$ *for D at* $W$, *then for every* $W'$ *there exists an endogenous DPGS* $D'$ *such that*

1. $D = D'$ *except possibly in the reward function,*
2. *there exists a dominant strategy* $(e_1^{*'}, \ldots, e_n^{*'})$ *for* $D'$ *and* $W'$, *and*
3. $\sum_{i=1}^n e_i^{*'} > \sum_{i=1}^n e_i^*$ *if* $W' > W$.

Recall that the growth function is a function of the current system value and the sum of the efforts of the agents. Therefore beyond the critical mass the system value will increase monotonically within $\mathcal{W}$.

We illustrate these definition and properties in the following example.

**Example 3** Let us modify Example 2 to realize system growth. Suppose there exists $W_c$ which is the smallest value such that all $W < 4$ (corresponding to $\mathcal{W}$) and $W \geq W_c$ satisfy the following conditions for all $\alpha < \bar{e}, \beta < \bar{e}$ such that $\alpha \geq \beta, \bar{e} < 1$.
$-e - 0.5 \cdot W \leq exp(\frac{\alpha}{1-\alpha}) - exp(\frac{e}{1-e}) \leq \alpha + 0.5 \cdot W$, and
$-e - 0.5 \cdot W \leq exp(\frac{2\beta}{1-\beta}) - exp(\frac{2e}{1-e}) \leq \beta + 0.5 \cdot W$ forall $e \in [0, \bar{e}]$.

Then clearly $W_c$ is a critical mass point. To see why, note that we can design a reward function which gives a dominant strategy $(\alpha, \beta)$ for an endogenous $DPGS$ at $W$ such that the next system value $W'$ will be $\alpha + \beta + W \cdot 0.5 > W$, because $W < 4 \cdot \bar{e}$ and $\alpha, \beta$ can be arbitrarily set in the range $0 \leq \alpha \leq \bar{e}, 0 \leq \beta \leq \bar{e}$. Furthermore, Theorem 4 says that there exists a dominant strategy $(\alpha', \beta')$ at $W'$ such that $\alpha' + \beta' > \alpha + \beta$. Obviously $\alpha' + \beta' + W' \cdot 0.5 > \alpha + \beta + W \cdot 0.5$ holds, which promises monotonic increasing system growth in the future.

## 5 Final remarks

We have provided a formal model of a multi-agent public-goods system. Given that our model draws heavily on work in economics, it is important that we identify the particular contributions of this paper. These contributions include the following:

- Allowing arbitrary reward functions that don't exceed the system's value.
- Defining a dynamic model in which agents' efforts and system value are mutually reinforcing.
- Providing various results about this model, which (to our knowledge) have no analogue in existing multi-agent models. These results include necessary and sufficient conditions for the existence of a dominant strategy.

- Defining in the model the notion of a critical mass, the point beyond which the continued well being of the system is made possible, and providing necessary and sufficient (albeit non-constructive) conditions for the existence of the critical mass.

Our model is based on notions from micro-economics, but also has several novel features. One is the total freedom to choose reward functions within the current system value. A second, more fundamental one is the construction of a dynamic model.

While the literature on public goods, agency theory, and mechanism design is too rich for us to cover here, the literature on applications of mechanism design and/or public goods in AI or software design is very limited. One notable exception is [Avery *et al.*, 1996]'s discussion of a pricing method for inducing optimal provision of evaluations (as in restaurant reviews, or Web-page ratings) from users. They regard the database of evaluations as a public good, and discuss implementability of payment scheme between evaluators and readers. They too consider a dynamic model, but unlike us they do not assume a variable effort but a fixed cost. [Varian, 1995]'s discussion of the application of agency theory to software design is related insofar as it too examines the application of agency theory – the asymmetric, two-person case of mechanism design – in the software domain. However, the similarity ends there, as he does not consider the public good aspect at all, let alone its dynamic version. Other work in generally in the area of application of economic mechanisms in AI includes Market Oriented Programming [Wellman, 1996], [Zlotkin and Rosenschein, 1993], and [Shoham and Tennenholtz, 1997]; however, there is no direct connection between the technical development nor the motivation here and any of these three.

Several challenges remain, in the areas of models, algorithms, and application. Our model makes several assumptions, which we mention throughout the construction. First, we have looked only at dominant pure strategies. It is natural to relax this strong requirement and look at general Nash equilibria among possibly mixed strategies. Second, we have assumed a particular additive form of the system's growth function, the result being that the system's value can grow unboundedly. This is unrealistic in any interesting domain; for example, in a discretionary database it is it is unrealistic to assume that no matter how many people post items to the database, these items are always distinct. It would be natural to allow more complex growth functions which model the phenomonon of diminishing returns. Moving beyond the limitations of the model itself, we note that we have given no algorithms. In particular, we have given no algorithm to compute the critical mass, nor to determine whether it exists. Finally, as was mentioned in the introduction, this work was motivated by real difficulties encountered in incenting users to participate in groupware. We hope to incorporate the lessons learned from this work in future groupware design.

**Acknowledgments.** The authors are grateful for discussion with members of the Nobots group and for comments by anonymous reviwers.

# References

[Avery *et al.*, 1996] Chris Avery, Paul Resnick, and Richard Zeckhauser. The market for evaluations. In Harvard University Mimeo, http://www.research.att.com/~presnick/papers/mfe/, 1996.

[Balabanovic and Shoham, 1997] Marko Balabanovic and Yoav Shoham. Combining content-based and collaborative recommendation. In *Communications of the ACM, Special Issue on Recommender Systems (to appear)*, 1997.

[Cohen *et al.*, 1982] P. R. Cohen, C. R. Perrault, and J. Allen. Beyond question-answering. In W. Lehnert and M. Ringle, editors, *Strategies for natural language processing*. L. Erlbaum Associates, 1982.

[Connolly and Porter, 1990] Terry Connolly and Alan. L. Porter. Discretionary databases in forecasting. *Journal of Forecasting*, 9:1–12, 1990.

[Fudenberg and Tirole, 1991] Drew Fudenberg and Jean Tirole. Bayesian games and mechanism design. In *Game Theory*. MIT Press, 1991.

[Markus and Connolly, 1990] M. Lynne Markus and Terry Connolly. Why CSCW applications fail: Problems in the adoption of interdependent work tools. In *CSCW '90*, pages 371–380, 1990.

[Osborne and Rubinstein, 1994] M. J. Osborne and A. Rubinstein. *Game Theory*. MIT Press, 1994.

[Shoham and Tennenholtz, 1997] Y. Shoham and M. Tennenholtz. Emergent conventions in a game theoretic setting. *AI Jouranal special issue on Economic Foundations of Multi-Agent Systems (to appear)*, 1997.

[Varian, 1995] Hal R. Varian. Mechanism design for computerized agents. In Usenix Workshop on Electronic Commerce, http://www.sims.berkeley.edu/~hal/people/hal/papers.html, July 1995.

[Wellman, 1996] Michael P. Wellman. Market-oriented programming: Some early lessons. In S.Clearwater, editor, *Market-Based Control: A Paradigm for Distributed Resource Allocation*. World Scientific, 1996.

[Zlotkin and Rosenschein, 1993] Gilad Zlotkin and Jeffrey S. Rosenschein. A domain theory for task oriented negotiation. In *Proc.of IJCAI-93*, pages 416–422, 1993.

# On the Gains and Losses of Speculation in Equilibrium Markets

Tuomas Sandholm
Department of Computer Science
Washington University
One Brookings Drive
St. Louis, MO 63130-4899
sandholm@cs.wustl.edu

Fredrik Ygge
EnerSearch and
Department of Computer Science (IDE)
University of Karlskrona/Ronneby
372 25 Ronneby
Fredrik.Ygge@enersearch.se

## Abstract

In computational markets utilizing algorithms that establish a market equilibrium (general equilibrium), *competitive behavior* is usually assumed: each agent makes its demand (supply) decisions so as to maximize its utility (profit) assuming that it has no impact on market prices. However, there is a potential gain from *strategic behavior* (via speculating about others) because an agent does affect the market prices, which affect the supply/demand decisions of others, which again affect the market prices that the agent faces.

This paper presents a method for computing the maximal advantage of speculative behavior in equilibrium markets. Our analysis is valid for a wide variety of known market protocols. We also construct demand revelation strategies that guarantee that an agent can drive the market to an equilibrium where the agent's maximal advantage from speculation materializes.

Our study of a particular market shows that as the number of agents increases, gains from speculation decrease—often turning negligible already at moderate numbers of agents. The study also shows that under uncertainty regarding others, competitive acting is often close to optimal, while speculation can make the agent significantly worse off—even if the agent's beliefs are just slightly biased. Finally, protocol dependent game theoretic issues related to multiple agents counterspeculating are discussed.

## 1 Introduction

*General equilibrium theory*, a microeconomic market framework, has recently been successfully adapted for and used in computational multiagent systems in many application domains, see e.g. [6; 16; 17; 9; 18; 2]. It provides a distributed method for efficiently allocating goods and resources among agents. Such a market can have two types of agents, *producers* and *consumers*. It has a finite number of *commodities*. The amount of each commodity is unrestricted, and each commodity is usually assumed arbitrarily divisible. Different elements within a commodity are not distinguishable, while elements from different commodities are. Each consumer has a *utility function* which encodes its preferences over different *consumption bundles*, i.e. over vectors. Each element of the vector describes how much of a given commodity the agent consumes. Each consumer also has an initial *endowment* of the different commodities. The producers can use some commodities to produce others. The *production vector* of a producer describes how much of each commodity the agent produces. Net usage of a commodity is denoted by a negative number. A producer's capability of turning inputs into outputs is characterized by its *production possibilities set*, which is the set of feasible production vectors. The producer's profits are divided among the consumers according to predetermined proportions which need not be equal (one can think of the consumers owning stocks of the producers).

The market is said to be in *general equilibrium* in terms of the prices on commodities, consumers' consumption decisions, and producers' production decisions if

I markets clear: for each commodity, production plus endowments equals consumption, and

II each consumer consumes a bundle of commodities such that the agent could not afford another bundle of higher utility given its initial endowments, the current prices, and the profits it receives from producers, and

III each producer uses the feasible production vector that maximizes its profits given the prices.

If the production possibilities sets are convex, and the consumers' preferences are continuous, non-decreasing and locally insatiable, then at least one such equilibrium *exists* [8; 5; 15]. [1] A sufficient condition for *uniqueness* of such an equilibrium is that the demand for each good is nondecreasing in the prices of the other goods.

---
[1] If agents act competitively (as opposed to strategically via speculation), each general equilibrium is Pareto efficient: no agent can be made better off—with any methodology for making decisions—without making some other agent worse off. Such competitive general equilibria are also stable in the sense of the *core* solution concept of coalition formation games: no subgroup of agents is motivated to pull out of the general equilibrium and form their own market.

The analysis in this paper is based on the assumption that a protocol is used that establishes a market price for each good such that supply meets demand, and that reallocation is performed *after* these prices have been established. There are many alternative algorithms that can be used to find such a general equilibrium. Clearly, if no such equilibrium exists, no algorithm can find it. In this paper we analyze the gains and losses of strategic behavior via speculation. We do this by analyzing the equilibrium. If an equilibrium does not exist, the agents will not achieve a resource reallocation, and in that case, the gains and losses of speculation are not well defined. The equilibrium-based analysis makes our results protocol independent—as long as the agents exchange goods only after an equilibrium has been reached. This allows our results to hold for most market algorithms that have been used to find an equilibrium. Some of these algorithms are now discussed.

To reach a general equilibrium, the *price tâtonnement process* is usually used. This is an iterative mechanism, and the trades, production, and consumption are assumed to occur only after the process has terminated. At each iteration, the *auctioneer* sets a vector of prices. Then all agents have to declare a vector of how much they are willing to buy and sell of each commodity at the current prices. Based on this information, the auctioneer updates the price vector for the next iteration. Under certain technical conditions, this process is guaranteed to converge to a general equilibrium [8]. Within computational multiagent systems, Wellman has developed a general equilibrium based software system called WALRAS. As example domains, he has used WALRAS in flow routing in a network [16], and in configuration design [17]. Wellman discusses how the latter application fails to meet the assumptions for existence of a general equilibrium. For example, because some design parameters are discrete, production possibilities sets cannot be convex. Mullen and Wellman have applied WALRAS to a distributed information network example [9]. The iterative market process of WALRAS differs from tâtonnement. Specifically, WALRAS uses asynchronous declarations by the agents, and the agents bid demand functions (of price) as opposed to just quantities. This process converges to a general equilibrium [2]. As in tâtonnement, trades in WALRAS only occur after the market process has terminated.

In addition to the price-based market mechanisms, resource-based mechanisms exist for reaching the general equilibrium. In such a resource-based protocol (*cf. quantity tâtonnement* [8]), the auctioneer sets the allocation of commodities to agents at each iteration, and agents report how much more or less they are willing to pay for each commodity. The auctioneer then takes these declarations into account in changing the allocation in the next iteration: agents that increase their willingness to pay get more, and others get less than on the previous iteration. The algorithm terminates when an equilibrium is reached. Kurose and Simha have developed a market mechanism—applied to file allocation—where at each iteration, the agents compute their marginal worths for resources [6]. Based on these worths, resources are reallocated at every iteration. This differs from most other equilibrium approaches where trades only occur after all iterations have been completed. In Kurose and Simha's approach, the solution gets better at every iteration, and is guaranteed to finally converge to the optimum. Recently, resource-based market mechanisms have been used and thoroughly studied in electricity distribution [18]. A sufficient condition for the applicability of our analysis to resource-based approaches is that reallocation is not performed until an equilibrium has been reached.

Classically in equilibrium markets, the agents are assumed to act *competitively*: they treat prices as exogenous. This means that each agent makes and reveals its demand (supply) decisions truthfully so as to maximize its utility (profit) given the market prices—assuming that it has no impact on those prices. The idea behind this *price-taking assumption* is that the market is so large that no single agent's actions affect the prices. However, this is paradoxical since the agents' declarations completely determine the prices. The price-taking assumption becomes valid as the number of agents approaches infinity: with infinitely many agents (of comparable size), each agent is best off acting competitively since it will not affect the prices.

However, in markets with a finite number of agents, an agent can act strategically, and potentially achieve higher utility by over/under representing [7, pp. 220-223], [4]. In doing so, the agent has to speculate how its misrepresentation affects the market prices, which are simultaneously affected by how other agents respond to the prices which changed due to the first agent's strategic actions. This paper addresses the question of how much an agent can gain or lose by such strategic behavior in such a complex setting of interrelationships.

The theory in this paper stems from well established principles in microeconomics—such as general equilibrium theory and the theory of optimal price setting in oligopoly and monopoly markets—and is extended for situations where agents have estimation errors in their beliefs regarding the behavior of other agents. First, a general, protocol independent study of the potential gains and losses from speculation is presented in Section 2. This is concretized for the case where only one agent is speculating under perfect information in Section 3, and under biased beliefs about others in Section 4. Simultaneous speculation by multiple agents is game theoretically discussed in Section 5. Finally, Section 6 concludes. Note that all numerical utility values in this paper have been multiplied by 100 for readability. In all, we believe that the methodology presented in this paper is important for builders of computational markets where the agents represent self-interested real world parties that can tailor their agents so as to take advantage of the other agents in the system.

## 2 A Method for Analyzing the Potential Gains from Speculation

The goal of a self-interested consumer is to find the consumption bundle that maximizes its utility. To find the optimal bundle when acting in an equilibrium market, the consumer must speculate how other agents respond to prices. This is because its demand decisions affect the prices, which affect the demand and supply decisions of others, which again affect the prices that the consumer faces. Using the model of other agents, the consumer computes its optimal demand decisions. Note that other agents might also be speculating (in the same way or some other, suboptimal way). That is included in the agent's model of the other agents.

The goal of a self-interested producer is to find the production vector that maximizes its profits.[2] Again, this requires a model of how others react to prices because the producer's production decisions affect the prices, which affect the demand and supply decisions of others, which again affect the prices that the producer faces.

Using standard notation from microeconomics, let there be $n$ commodities, and let the price vector be $\mathbf{p} = [p_1, p_2, \ldots, p_n]$, where $p_i$ is the price for good $i$. Let the resulting allocation for consumers be $\mathbf{x}_i(\mathbf{p}) = [x_{i1}(\mathbf{p}), x_{i2}(\mathbf{p}), \ldots, x_{in}(\mathbf{p})]^T$, where $x_{ij}$ is consumer $i$'s allocation of good $j$. Let the initial endowments be $\mathbf{e}_i = [e_{i1}, e_{i2}, \ldots, e_{in}]^T$, where $e_{ij}$ is consumer $i$'s endowment of good $j$. Now, consumer $i$'s excess (net) demand of good $j$ is $z_{ij}(\mathbf{p}) = x_{ij}(\mathbf{p}) - e_{ij}$. Let $u_i(\mathbf{x}_i)$ be consumer $i$'s utility as a function of its allocation, and let $\theta_{ih}$ be the fraction of producer $h$ that consumer $i$ owns. The producers' profits are divided among consumers according to these shares. However, the consumers are assumed to have no say-so in the producers' production decisions.

Furthermore, let $\mathbf{y}_i(\mathbf{p}) = [y_{i1}(\mathbf{p}), y_{i2}(\mathbf{p}), \ldots, y_{in}(\mathbf{p})]^T$ be the production vector of producer $i$, where $y_{ij}$ represents the amount produced, and a negative number means that the good is an input. Let $Y_i$ be the production possibilities set of producer $i$. The profit of producer $i$ is $\mathbf{p} \cdot \mathbf{y}_i(\mathbf{p})$, where $\mathbf{y}_i \in Y_i$. For presentation uniformity with the case of the consumer, we define the excess demand of producer $h$ to be $z_{hj}(\mathbf{p}) = -y_{hj}(\mathbf{p})$.

Let there be $k$ agents in addition to the speculating agent that we investigate. The excess demand of these $k$ agents for good $j$ is

$$z_j^k(\mathbf{p}) = \sum_{i=1}^{k} z_{ij}(\mathbf{p}). \qquad (1)$$

For the purposes of this section, we do not make any restricting assumptions about how these $k$ agents make their supply/demand decisions (which determine the excess demands). In particular, we do not assume that agents act competitively. The speculating agent that we investigate uses its information about $z_j^k(\mathbf{p})$ as the basis of its strategic behavior as is now described.

The total excess demand with the speculating agent included is

$$z_j(\mathbf{p}) = z_j^k(\mathbf{p}) + z_{sj}(\mathbf{p}). \qquad (2)$$

Once the market has reached a general equilibrium, supply meets demand, i.e. $z_j(\mathbf{p}) = 0$ for every good $j$.[3] Substituting this into (2) gives

$$z_{sj}(\mathbf{p}) + z_j^k(\mathbf{p}) = 0. \qquad (3)$$

### 2.1 Case A: Speculating Consumer

A solution to the following maximization problem gives the highest utility that a speculating consumer theoretically can obtain.

$$\max_{\mathbf{p}} u_s(\mathbf{x}_s(\mathbf{p})) \quad \text{s.t.} \qquad (4)$$

$x_{sj}(\mathbf{p}) \geq 0$ (consumer does not produce)

$x_{sj}(\mathbf{p}) = e_{sj} - z_j^k(\mathbf{p})$ (supply meets demand)

$\mathbf{p} \cdot \mathbf{z}_s(\mathbf{p}) = \sum_{h \in producers} \theta_{sh} \, \mathbf{p} \cdot \mathbf{y}_h(\mathbf{p})$ (budget constraint)

provided that the equilibrium is unique and the market protocol finds it (this is discussed further in Section 3.2). The last equality assumes that either the speculating consumer's utility does not decrease as the amount of any good is increased in the consumer's consumption bundle, or the consumer can freely dispose of any good. Otherwise, "=" should be changed to "$\leq$".

### 2.2 Case B: Speculating Producer

Similarly to the case of a speculating consumer, a solution to the following maximization problem gives the highest profit that a speculating producer can obtain.

$$\max_{\mathbf{p}} \mathbf{p} \cdot \mathbf{y}_s(\mathbf{p}) \quad \text{s.t.} \qquad (5)$$

$\mathbf{y}_s(\mathbf{p}) \in Y_s$ (feasible production plan)

$y_{si} = z_i^k(\mathbf{p})$ (supply meets demand)

provided that the equilibrium is unique and the market protocol finds it. The last equality turns into $y_{si} \geq z_i^k(\mathbf{p})$ if free disposal [8] for both inputs and outputs is possible for each commodity.

We call the solution to the applicable optimization problem above (depending on whether the speculator is a producer or a consumer) $\mathbf{p}^*$.

---

[2] This makes the standard assumption that the producer is able to alter its production plan costlessly during the search for equilibrium.

[3] This holds even if agents are strategic—assuming that the particular market algorithm finds an equilibrium. The equilibrium reflects how the speculating agent is acting strategically, and how the other agents have reacted to the new price vector that came about due to the strategic agent's actions.

# 3 Strategic Behavior with Perfect Information

In the previous section we obtained a method for determining the highest utility that an agent theoretically can obtain. This section shows how this method can be used in practice.

## 3.1 A Simple Example

To demonstrate the method, we show how it applies to a specific case. We choose a setting with no producers. In such *pure exchange markets*, the consumers just reallocate their initial endowments among themselves [15; 5]. We choose a set of agents that is similar to the one described by Hu and Wellman [3]. Specifically, we let every agent—except for the speculating one that we investigate—be a competitive agent with constant elasticity of substitution, i.e. a utility function of the form

$$u_i(\mathbf{x}) = \left(\sum_{j=1}^{n} \alpha_{ij} x_{ij}^{\rho}\right)^{\frac{1}{\rho}} \quad (6)$$

where we have chosen $\alpha_{ij} = 1$ and $\rho = \frac{1}{2}$. Since these agents act competitively, and the speculating agent is assumed to have perfect information, the analysis of this example is protocol independent as long as the resources are reallocated after equilibrium (3) has been reached.

For simplicity and readability, we use only two goods ($n = 2$). The endowments are the same for all the competitive agents and they are 2 for good 1 and 1 for good 2. We let the speculating agent have the utility function

$$u_s(\mathbf{x}) = \sum_{j=1}^{n} ln(x_{sj}), \quad (7)$$

and an endowment of 1 for both goods.

Most of this section is devoted to computing the gain from strategic behavior. For the casual reader, the main results of this section are condensed in Table 1 and Figure 1 below.

We get, e.g. from [17] and the definition of excess demand, that the excess demand of the competitive agents is

$$z_1^k(\mathbf{p}) = k\left(\frac{2p_1 + 1}{p_1(p_1 + 1)} - 2\right). \quad (8)$$

Because prices are only relative, we can set one of the prices arbitrarily, e.g. $p_n$ can be set to 1, i.e. $p_2 = 1$. From the budget constraint $(\mathbf{p} \cdot \mathbf{z}(\mathbf{p}) = 0)$ we then get $z_n(\mathbf{p}) = -\sum_{j=1}^{n-1} p_j z_j(\mathbf{p})$, i.e. $z_2^k(p_1) = -p_1 z_1^k(p_1)$. Using these and (7) and (8), we get

$$u_s(\mathbf{x}(p_1)) = \begin{array}{l} ln\left(1 - k\left(\frac{2p_1+1}{p_1(p_1+1)} - 2\right)\right) + \\ ln\left(1 + k\left(\frac{2p_1+1}{p_1+1} - 2p_1\right)\right). \end{array} \quad (9)$$

From (7) we see that $x_{sj}$ must be greater than zero. If the speculating agent chooses to minimize $p_1$, it should sell as much of $x_1$ as possible and thus, as seen from the expression for $x_{s1}$ in (9), and the requirement that $x_{s1} > 0$, we have $p_1 > p_1^{min} = -\frac{1}{2(2k+1)} + \sqrt{\frac{k}{2k+1} + \left(\frac{1}{2(2k+1)}\right)^2}$. Analogous reasoning for $x_{s2}$ shows that $p_1 < p_1^{max} = \frac{1}{4k} + \sqrt{\frac{1}{2}\left(1 - \frac{1}{k}\right) + \left(\frac{1}{4k}\right)^2}$. If $k$ approaches infinity, both $p_1^{min}$ and $p_1^{max}$ approach $\sqrt{\frac{1}{2}} \approx 0.707$. Therefore, with an infinite number of agents, the speculator cannot afford to affect the price in any way.

The first derivative of $u_s$ with respect to $p_1$ is

$$\frac{\partial u_s}{\partial p_1} = \\ \frac{-1}{1 - k\left(\frac{2p_1+1}{p_1(p_1+1)} - 2\right)} k\left(2\frac{1}{p_1(p_1+1)} - \frac{2p_1+1}{p_1^2(p_1+1)} - \frac{2p_1+1}{p_1(p_1+1)^2}\right) + \\ \frac{1}{1 - k\left(\frac{2p_1+1}{p_1+1} - 2p_1\right)} k\left(\frac{2}{p+1} - \frac{2p_1+1}{(p_1+1)^2} - 2\right). \quad (10)$$

It turns out that $\lim_{p_1 \to p_1^{min+}} \frac{\partial u_s}{\partial p_1} > 0$ and $\lim_{p_1 \to p_1^{max-}} \frac{\partial u_s}{\partial p_1} < 0$, and that the solution to $\frac{\partial u_s}{\partial p_1} = 0$ is unique in the interval $p_1^{min} < p_1 < p_1^{max}$. Therefore, the optimum, $p_1^*$, is obtained by solving $\frac{\partial u_s}{\partial p_1} = 0$.

The results of optimal strategic behavior are compared to the results of competitive behavior by the same agent. When the agent acts competitively, the excess demand (with $u_s$) is $z_{s1} = \frac{p_1+1}{2p_1} - 1$. Setting the excess demand to zero gives $k\left(\frac{2p_1+1}{p_1(p_1+1)} - 2\right) + \frac{p_1+1}{2p_1} - 1 = 0$. Solving for the competitive price gives $p_1^c = \sqrt{\frac{2k+1}{4k+1}}$. The results are shown in Table 1. [4]

| $k$ | $p_1^{min}$ | $p_1^{max}$ | $p_1^*$ | $p_1^c$ | $u_s(p_1^*)$ | $u_s(p_1^c)$ |
|---|---|---|---|---|---|---|
| 1 | 0.4343 | 1.281 | 0.7601 | 0.7746 | 1.746 | 1.626 |
| 2 | 0.5403 | 1.000 | 0.7402 | 0.7454 | 2.202 | 2.152 |
| 5 | 0.6303 | 0.8262 | 0.7227 | 0.7237 | 2.614 | 2.602 |
| 10 | 0.6667 | 0.7670 | 0.7154 | 0.7157 | 2.788 | 2.784 |
| 20 | 0.6863 | 0.7372 | 0.7114 | 0.7115 | 2.884 | 2.884 |
| 30 | 0.6931 | 0.7272 | 0.7100 | 0.7160 | 2.918 | 2.918 |
| 100 | 0.7029 | 0.7131 | 0.7080 | 0.7080 | 2.967 | 2.967 |

Table 1: *Acting strategically vs. acting competitively. $k$ is the number of agents acting competitively. $p_1^{min}$ is the price for $x_1$ in the market when the speculating agent sells as much $x_1$ as possible. $p_1^{max}$ is the price for $x_1$ in the market when the speculating agent sells as much $x_2$ as possible. $p_1^*$ is the market price as a consequence of strategic acting. $p_1^c$ is the market price as a consequence of competitive acting by the same agent. The values $u_s$ are the corresponding utilities for the agent under observation.*

In Figure 1, the utility is plotted for the situations where the agent acts strategically and where it acts competitively. As expected (see e.g. [11]), the larger the number of agents, the smaller the gain from strategic behavior, and the less reason not to act competitively. In this example, already when the number of competitive

---

[4]Since utility is ordinal, one should be careful when discussing degrees of improvement.

agents is around five, the gain from strategic behavior is negligible.

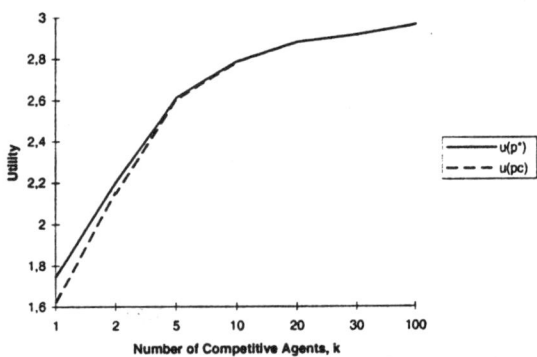

Figure 1: *Comparison of strategic and competitive utility, $u_s(p_1^*)$ and $u_s(p_1^c)$. The horizontal axis shows a number of interesting values rather than a specific scale. We see that the larger the number of agents, the smaller the gain from strategic behavior.*

### 3.2 Reaching Equilibrium

Obviously an agent's best strategy is to declare an excess demand function such that the market will converge to the prices that are optimal for the speculating agent. More formally, when perfect information is available, an agent's best strategy—even if the other agents are not acting competitively, and some of them may be producers—is to declare an excess demand function with the property

$$z_{sj}^*(\mathbf{p}^*) = -z_j^k(\mathbf{p}^*), \qquad (11)$$

for each good $j$, and which has a form such that the particular algorithm for searching for the market equilibrium converges to $\mathbf{p}^*$. Depending on the setting, the difficulty of finding such a function varies. As shown below, finding it for the simple example from above where all other agents' supply/demand functions can be taken as fixed (to do this, they do not have to be competitive in general) is normally not hard. However, finding it may be more difficult if the other agents try to act strategically via speculation as well. These game theoretic issues will be addressed in Section 5.

Having computed the optimal speculative solution, $u_s^*(p_1^*)$, we would like to describe the strategic behavior leading to this solution under any particular market protocol used. For example, if $p_1$ is established via an algorithm whose only requirement for finding the equilibrium is $\partial z_1(p_1)/\partial p_1 < 0$, [5] and we have that $\partial z_1^k(p_1)/\partial p_1 < 0$ (i.e. demand decreases as price increases—as is the case in our example above), we see that if

$$z_{s1}^*(p_1^*) = -z_1^k(p_1^*) \text{ and } \frac{\partial z_{s1}^*(p_1)}{\partial p_1} \leq 0, \qquad (12)$$

---
[5] An example of such an algorithm is binary search. The WALRAS market framework uses this implementation.

there is a single solution for $p_1 = p_1^*$, and that solution will be found by the algorithm. [6] It turns out that simple demand revelation strategies exist for the speculator which guarantee that an equilibrium will be reached where the speculator's maximal gain from speculation (derived earlier in this paper) materializes. The following linear function defines one such strategy—i.e. it fulfills (12):

$$z_{s1}(p_1) = \overline{p_1^*} - \overline{z_1^k(p_1^*)} - p_1, \qquad (13)$$

where $\overline{z_1^k(p_1^*)}$ and $\overline{p_1^*}$ are the speculator's (perfect) estimates of $z_1^k(p_1^*)$ and $p_1^*$ respectively. Another viable strategy is defined by the constant function

$$z_{s1}(p_1) = -\overline{z_1^k(p_1^*)}. \qquad (14)$$

## 4 Dealing with Uncertainty: Strategic Behavior under Biased Beliefs about Others

This section extends the discussion to include the impact of uncertainty on the speculating agent's strategy.

### 4.1 The General Setting

Above, when an excess demand function was chosen based on perfect information, the exact form of the function was unimportant as long as it fulfilled (12). However, if the speculating agent cannot estimate $z_1^k$ perfectly, its outcome will depend on the function chosen, i.e. the choice of an excess demand function will depend on the probability distribution that characterizes the uncertainty in the speculator's beliefs. Rather than analyzing how the excess demand is chosen for different uncertainty situations and how the speculator should try to reduce the uncertainty by learning about its environment in an efficient way, we describe how, for a specific choice of $z_{sj}$, a specific error in the estimation of the competitive agents' behavior affects the outcome of the market.

Independently of how the $z_{sj}$ function is chosen, the possible market outcomes can be determined by solving (3). If no solution exists, of course, no algorithm can find it, and if multiple equilibria exist, a protocol dependent analysis is required to find out which one will be chosen.

### 4.2 The Example Revisited

Returning to the above example, we first assume that $z_{s1}$ has been chosen as describe by (13), but that now there is an error in the estimation. We define this error

---
[6] The above reasoning extends easily to a market with more than two goods. If $\mathbf{p}$ is established via an algorithm whose only requirements for finding the equilibrium are $\partial z_i(\mathbf{p})/\partial p_i < 0$ and $\partial z_i(\mathbf{p})/\partial p_j \geq 0$, $i \neq j$ (e.g. WALRAS), and we have that $\partial z_i^k(\mathbf{p})/\partial p_i < 0$ and $\partial z_i^k(\mathbf{p})/\partial p_j \geq 0$, $i \neq j$, then (12) should be generalized to $z_{si}^*(\mathbf{p}^*) = -z_i^k(\mathbf{p}^*)$ and $\partial z_{si}^*(\mathbf{p})/\partial p_i \leq 0$ and $\partial z_{si}^*(\mathbf{p})/\partial p_j \geq 0$, $i \neq j$.

learn more about other agents' excess demand/supply functions. If many agents are involved in such probing, it seems that time becomes an important factor. Some agents might reveal progressively more of their competitive demands in order to speed up the convergence (as it might be urgent for them to get the resources traded), while others might extend the probing in order to maximize their benefit from the trade. [9]

Once the other agents' strategies are known (e.g. from a game theoretic equilibrium analysis), the methods of this paper can be used to analyze the speculating agent's strategy alternatives. The methods are protocol independent, and they can be used to estimate the potential gains from speculation in any particular setting, as well as to determine how far from the optimal strategy a particular strategy is—as long as the other agents' strategies can be fixed conceptually. This does not mean that they need to be known with certainty.

The methods of this paper can also be used when game theoretic analysis fails. Especially when speculation is based on expected actions of other agents—instead of a game theoretic equilibrium analysis of strategies—the theory of speculation under biased beliefs is highly applicable. In addition, the example market used in this paper suggests that if a market is of at least moderate size, the gain from speculation is so small, and the risk of being significantly worse off due to estimation error is so great, that one can expect that each agent will act close to its competitive behavior.

## 6 Conclusions

We presented a method for computing the maximal advantage of speculative strategic behavior in general equilibrium based markets. It is computed from the other agents' supply/demand functions (classic competitive behavior by the other agents is a special case of this). The method enables one to analyze how much an agent could gain or lose by speculating in a particular system, and it is also useful when evaluating different strategies since it allows one to determine how close to the optimal strategy they are. Our analysis is valid for a wide variety of known market protocols. We also constructed demand revelation strategies that guarantee that an agent can drive the market to an equilibrium where the agent's maximal advantage from speculation materializes.

Our study of a particular market shows that as the number of agents increases, the gains from speculation decrease—often turning negligible already at moderate numbers of agents. The study also shows that under

---

[9]Some work has addressed non-competitive behavior in WALRAS [3], although there was only one speculating agent in the experiments, and this agent was limited to simple linear price prediction about how its actions affect the prices. Further analysis is required to determine whether its optimal strategy can be captured in this model. This need not be the case because the optimal strategy may involve some more "aggressive" behavior, e.g. the probing described above.

uncertainty regarding other agents, competitive acting is often close to optimal, while speculation can make the agent significantly worse off—even if the agent's beliefs are just slightly biased.

We believe that computational agents representing self-interested real world parties will deviate from competitive behavior in practice only if the potential gain from speculation is sufficiently large compared to the cost of the computation or information gathering actions required for speculation. We also believe that an agent will not speculate when that incurs potentially large losses when estimation errors are present. Merging the above results with these beliefs suggests that speculation problems diminish as the market grows or the agents' information about others becomes less certain.

Finally, we discussed the protocol dependent game theoretic issues related to multiple agents counterspeculating. There are many open issues in this area and the difficulties should not be underestimated.

## Acknowledgments

The authors thank Arne Andersson, Junling Hu, John Nachbar, Eric Schenk, Fernando Tohmé, Michael Wellman, and Curt Wells for interesting discussions and comments. Fredrik also thanks Hans Akkermans, Rune Gustavsson and Hans Ottosson for all their support.

## References

[1] S. Barbera and M. O. Jackson. Strategy-proof exchange. *Econometrica*, 63(1):51–87, 1995.

[2] J. Q. Cheng and M. P. Wellman. The WALRAS algorithm: A convergent distributed implementation of general equilibrium outcomes. *Computational Economics*, 1997. To appear.

[3] J. Hu and M. P. Wellman. Self-fulfilling bias in multiagent learning. In *Proc. of the Second International Conference on Multi-Agent Systems (ICMAS)*, pages 118–125, Keihanna Plaza, Kyoto, Japan, Dec. 1996.

[4] L. Hurwicz. On informationally decentralized systems. In C. McGuire and R. Radner, editors, *Decision and Organization*, chapter 14, pages 297–336. University of Minnesota Press, 1986. Second edition.

[5] D. Kreps. *A Course in Microeconomic Theory*. Princeton U. Press, 1990.

[6] J. F. Kurose and R. Simha. A microeconomic approach to optimal resource allocation in distributed computer systems. *IEEE Transactions on Computers*, 38(5):705–717, 1989.

[7] E. Malinvaud. *Lectures on Microeconomic Theory*. North-Holland, 1985.

[8] A. Mas-Colell, M. Whinston, and J. R. Green. *Microeconomic Theory*. Oxford University Press, 1995.

[9] T. Mullen and M. P. Wellman. A simple computational market for network information services. In *Proceedings of the First International Conference on Multi-Agent Systems (ICMAS)*, pages 283–289, San Francisco, CA, June 1995.

[10] J. Nash. Equilibrium points in n-person games. *Proc. of the National Academy of Sciences*, 36:48–49, 1950.

[11] D. J. Roberts and A. Postlewaite. The incentives for price-taking behavior in large exchange economies. *Econometrica*, 44(1):115–127, 1976.

[12] J. S. Rosenschein and G. Zlotkin. *Rules of Encounter*. MIT Press, 1994.

[13] T. W. Sandholm. Limitations of the Vickrey auction in computational multiagent systems. In *Proceedings of the Second International Conference on Multi-Agent Systems (ICMAS)*, pages 299–306, Keihanna Plaza, Kyoto, Japan, Dec. 1996.

[14] T. W. Sandholm and V. R. Lesser. Advantages of a leveled commitment contracting protocol. In *AAAI-96*, pages 126–133, Portland, OR, Aug. 1996. Extended version: UMass Comp. Sci. technical report 95-72.

[15] H. R. Varian. *Microeconomic analysis*. New York: W. W. Norton, 1992.

[16] M. Wellman. A market-oriented programming environment and its application to distributed multicommodity flow problems. *Journal of Artificial Intelligence Research*, 1:1–23, 1993.

[17] M. Wellman. A computational market model for distributed configuration design. In *AAAI-94*, pages 401–407, Seattle, WA, July 1994.

[18] F. Ygge and J. Akkermans. Power load management as a computational market. In *Proceedings of the Second International Conference on Multi-Agent Systems (ICMAS)*, pages 393–400, Keihanna Plaza, Kyoto, Japan, Dec. 1996.

through[7]

$$\overline{p_1^*} - \overline{z_1^k(p_1^*)} = (1+e)(p_1^* - z_1^k(p_1^*)). \qquad (15)$$

Provided that the speculator stops learning about $z_1^k(p_1)$ at some point in time, and hereby fixes $z_{s1}(p_1)$, any market protocol whose only requirement for finding the equilibrium is that $\partial z_1(p_1)/\partial p_1 < 0$ (e.g. the binary search used in WALRAS) is guaranteed to converge independently of how large the estimation error is, since (12) is satisfied. Inserting the expressions from (8), (13), and (15) in (3) gives

$$(1+e)\left(p_1^* - k\left(\frac{2p_1^*+1}{p_1^*(p_1^*+1)} - 2\right)\right) - p_1 + \\ k\left(\frac{2p_1+1}{p_1(p_1+1)} - 2\right) = 0. \qquad (16)$$

This equation was solved for different errors $e$, and different numbers of competitive agents $k$. The corresponding utilities—normalized to the utility that would result from competitive behavior—are plotted in Figure 2.

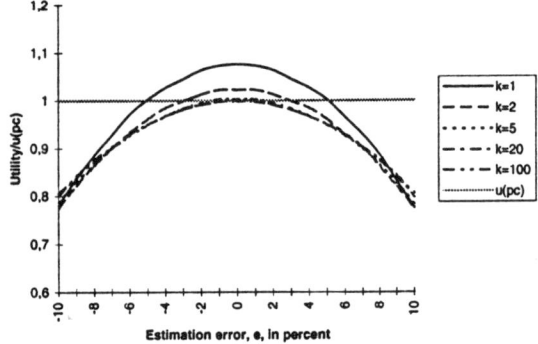

Figure 2: *The utility as a function of the estimation error. $k$ is the number of competitive agents.*

As seen from Figure 2, when the number of agents is small, the gain from speculating is large; even when there is an error involved, it often pays off to speculate. But as the number of agents increases, acting competitively is close to acting strategically with perfect information, and it is worse to speculate, even with surprisingly small errors. The break even points for 1, 2, 5, 20, and 100 agents are at an estimation error of approximately ±5%, ±3%, ±1%, ±0.4%, and ±0.05% respectively. Looking the other way around, with 100 competitive agents and a ±5% error (the break even error for 1 competitive agent), the loss compared to acting competitively is substantial.

## 5 Strategic Behavior by Multiple Agents

In our simple example, only one agent was speculating and the others were acting competitively. Even though

---

[7] If the speculating agent can learn about the other agents and change its excess demand during the market process, the error might decrease during this process. The error described here is the error remaining when the process terminates.

(1)–(5) and (11)–(15) are valid even in the case where every agent is speculating, determining market outcomes in a protocol independent way is not possible in such settings in general. The reason is that an agent cannot treat others' strategies (e.g. policies for revealing excess supply/demand *functions*) as fixed because the others would like to tailor their strategies to the specific strategy that the agent chooses. The strategies are in *Nash equilibrium* [10; 8; 5] if each agent's strategy is its best response to the others' strategies. This can be viewed as a necessary condition for system stability in settings where all agents act strategically.[8] In sequential protocols, one can also strengthen the Nash equilibrium solution concept in multiple ways by requiring that the strategies stay in equilibrium at every step of the game [8; 5]. Unlike our analysis, the Nash equilibrium outcome is specific to the market protocol. Important factors impacting the outcome are the order in which bids are submitted (see e.g. Stackleberg vs. Cournot models [8]), whether the bids are sealed or open [13], whether the protocol is iterative (the agents can change their excess demand between iterations) or not, whether the agents can decommit from their agreements by paying a penalty [14], *etc.*

In some games, no Nash equilibrium exists in pure (non-randomized) strategies. The following simple example illustrates this. Let there be two consumer agents, A and B, that engage in a market where they reveal their excess demand functions simultaneously and in a single round. Agent A can choose between two strategies (A1 and A2), and B can choose between B1 and B2. Provided that A knows that B will choose B1, A will choose A2, and A1 if B chooses B2. Provided that B knows that A will choose A2, B will choose B2, and B1 if A chooses A1. Now, from every possible pair of strategies, one agent would be motivated to deviate to another strategy, i.e. no Nash equilibrium exists. In general, existence and uniqueness of a general equilibrium (where agents act competitively) for a market does not imply existence and uniqueness of a Nash equilibrium (or an individually rational non-cooperative equilibrium [7]).

We argue that for each protocol proposed for implementation of equilibrium markets including self-interested computational agents, a thorough game theoretic analysis should be attempted [14; 13; 12]. However, as discussed above, some games lack a Nash equilibrium, or it may not be unique. In addition, some protocols are hard to analyze game theoretically. For example, in WALRAS, the agents might change their demand functions during the computation of the equilibrium. Then some agents may deliberately send false bids to generate more iterations of the market process in order to

---

[8] A stronger condition of stability is to require incentive compatibility, i.e. that an agent's strategy is optimal (for the agent) no matter what strategies others choose. Market protocols have been studied using this solution concept in [1].

# DISTRIBUTED ARTIFICIAL INTELLIGENCE

Distributed AI 4: Multiagent Algorithms

# The Use of Meta-level Information in Learning Situation-Specific Coordination*

M V Nagendra Prasad and Victor R Lesser
Department of Computer Science
University of Massachusetts, Amherst, MA 01003.
{nagendra,lesser}@cs.umass.edu

## Abstract

Achieving effective cooperation in a multi-agent system is a difficult problem for a number of reasons such as limited and possibly out-dated views of activities of other agents and uncertainty about the outcomes of interacting non-local tasks. In this paper, we present a learning algorithm that endows agents with the capability to choose the appropriate coordination algorithm from a set of available coordination algorithms based on meta-level information about their problem solving situations. We present empirical results that strongly indicate the effectiveness of the learning algorithm.

## 1 Introduction

Coordination is the act of managing interdependencies in a multi-agent system[Decker & Lesser, 1995]. Achieving effective coordination in a multi-agent system (MAS) is a difficult problem for a number of reasons. An agent's local control decisions about what activity to do next or what information to communicate and to whom or what information to ask others may be inappropriate or suboptimal due its limited view of the interactions between its own activities and those of the other agents. In order to make more informed control decisions, the agents have to acquire a view of the task structures of other agents. To the extent that this resolves agents' uncertainty about the non-local problem solving activities, they can act coherently. However, an agent has to expend computational resources in acquiring and exploiting such non-local views of other agents' activities. This involves communication delays and the computational cost of providing this information and assimilating the information from other agents. Given the inherent uncertainty in agents' activities and the cost of meta-level processing, relying on sophisticated coordination strategies to acquire non-local views of task structures may not be worthwhile for all problem-solving situations[Durfee & Lesser, 1988; Decker & Lesser, 1995; Nagendra Prasad et al., 1996]. For example, when the agents are under severe time pressure and the load of the activities at the agents is high, it is often difficult for them to rearrange their local activities so that they can exploit the results generated by other agents or to generate results that the other agents can exploit. In these situations, sophisticated agent coordination strategies may not pay-off. In this paper, we will be dealing with how agents can learn to dynamically choose the appropriate coordination strategy in different coordination problem instances. We empirically demonstrate that even for a narrow class of agent activities, learning to choose an appropriate coordination strategy based on meta-level characterization of the global problem solving state outperforms using any single coordination strategy across all problem instances.

In order to accomplish learning, we break the coordination problem into two phases. In the first phase, the agents exchange meta-level information not directly used for coordination. This information is used by the agents to derive a prediction of the effectiveness of various coordination mechanisms in the present problem solving episode. These mechanisms differ in the amount of non-local information they acquire and use, and in the complexity of analysis of interactions between activities at the agents. Agents choose an appropriate subset of the coordination mechanisms (or a coordination strategy) based on the meta-level information and enter Phase II. In this phase, the coordination strategy decides the types of information to be exchanged and the kind of reasoning about local and non-local activities the agents perform to achieve coherent activity. We call the meta-level information a *situation* and the two phase process *situation-specific coordination*. Learning situation-specific coordination involves associating appropriate views of the global situation with the knowledge learned about the effectiveness of the coordination mechanisms.

The rest of the paper is organized as follows. After placing our work in context, we briefly review the TÆMS task structure representation for coordination problems. We then describe our learning algorithm that learns to choose among three coordination modes of different levels of sophistication. We then present some of our experimental results and conclude.

## 2 Related Work

Much of the literature in multi-agent learning relies on reinforcement learning and classifier systems as learning algorithms. In Sen, Sekaran and Hale[Sen, Sekaran, & Hale, 1994] and Crites and Barto[Crites & Barto, 1996], the agents

---

*This material is based upon work supported by the National Science Foundation under Grant Nos. IRI-9523419. The content of this paper does not necessarily reflect the position or the policy of the Government, and no official endorsement should be inferred.

do not communicate with one another and an agent treats the other agents as a part of the environment. Weiss[Weiss, 1994] uses classifier systems for learning appropriate multi-agent hierarchical organization structuring relationships. In Tan[Tan, 1993], the agents share perception information to overcome perceptual limitations or communicate policy functions. In Sandholm and Crites[Sandholm & Crites, 1995] the agents are self-interested and an agent is not free to ask for any kind of information from the other agents. The MAS that this paper deals with contain complex cooperative agents (each agent is a sophisticated problem solver) and an agent's local problem solving control interacts with that of the other agents' in intricate ways. In our work, rather than treating other agents as a part of the environment and learning in the presence of increased uncertainty, an agent tries to resolve the uncertainty about other agents' activities by communicating meta-level information to resolve the uncertainty to the extent possible (there is still the environmental uncertainty that the agents cannot do much about). In cooperative systems, an agent can "ask" other agents for any information that it deems relevant to appropriately situate its learned local coordination knowledge. Agents sharing perceptual information as in Tan[Tan, 1993] or bidding information as in Weiss[Weiss, 1994] do not make explicit the notion of situating the local control knowledge in a more global abstract situation. The information shared is weak and they are studied in domains such as predator-prey[Tan, 1993] or blocks world [Weiss, 1994] where the need for sharing meta-level information and situating learning in it is not apparent.

Sugawara and Lesser[Sugawara & Lesser, 1993] also recognize the need for situation specificity in learning coordination, though they do have the notion of two phase coordination. They are concerned with learning to make the situations more discriminating to avoid using an inappropriate coordination strategy in the domain of distributed network diagnosis. Their learning relies on deep domain knowledge and agent homogeneity assumptions to learn to progressively refine situations based on failure-driven explanation and comparative analysis of problem solving traces. They test the theory on very limited number of coordination situations and the evidence was anecdotal. It is not clear how such knowledge-intensive learning can be generalized to other instances without significant knowledge engineering and the development of more sophisticated explanation-based learning techniques. Despite these limitations, combining their work on learning situation representations with the learning presented here on situation-based choice of coordination could have interesting implications for situation-specific learning.

## 3 Task Analysis, Environment Modeling, and Simulation

### 3.1 TÆMS

The TÆMS framework (Task Analysis, Environment Modeling, and Simulation) [Decker & Lesser, 1993] represents coordination problems in a formal, domain-independent way. A TÆMS model of a task environment specifies what actions are available to agents and how those actions relate to one another and to the performance of the system as a whole. A coordination problem instance is represented as a *task group* $\mathcal{T}$. Each task group has an arrival time $\text{Ar}(\mathcal{T})$, and a deadline $\text{D}(\mathcal{T})$. A task group, consisting of a set of computationally related actions, is represented by a directed acyclic graph. The quality of a task $T$ at a particular time t ($Q(T,t)$) is a function of the quality of its subtasks (in this paper this function is either minimum(AND-like) or maximum(OR-like)). *Quality* is used as a catch-all term representing acceptability characteristics like certainty, precision completeness, other than temporal characteristics. Leaf tasks are called *methods M* and they represent domain actions, such as executing a blackboard knowledge source, running an instantiated plan, or executing a piece of code with its data. Executable methods have base level quality and duration. A task may have multiple ways to accomplish it, represented by multiple methods, that trade off the time to produce a result for the quality of the result.

Besides tasks/subtask relationships, there can be other interrelationships between tasks in a task group[Decker & Lesser, 1993]. In this paper, we will be dealing with two such interrelationships:

- `facilitates` relationship or soft interrelationship: If the results of execution of Task A are available for Task B before it starts executing, then Task B will have increased quality and/or decreased duration.

- `enables` relationships or hard interrelationship: Task A must be executed before Task B can be executed.

A design-to-time scheduling[Garvey & Lesser, 1993] algorithm heuristically enumerates a promising subset of quality and time trade-offs to produce schedules that maximize quality given the deadlines. In a cooperative multi-agent system, the goal of the agents is to work together to produce the highest possible quality for as many task groups as possible.

### 3.2 Environment-Specific Coordination Mechanisms

In order to bring to bear different collections of coordination mechanisms for different multi-agent problem-solving environments, we use the Generalized Partial Global Planning (GPGP) approach[Decker & Lesser, 1995]. GPGP consists of several coordination mechanisms, each of which notices certain features in its local partial view of the task structures of other agents and their relationship with its own task structure and responds by taking certain communication or information gathering actions, or by proposing new commitments to other agents. In GPGP, a coordination strategy can be derived by activating a subset of the coordination mechanisms. Specifically we will be investigating the effect of three coordination strategies:

*Balanced (or dynamic-scheduling)*: Agents coordinate their actions by dynamically forming commitments. Relevant results are generated by specific times and communicated to the agents to whom corresponding commitments are made. Agents schedule their local tasks trying to maximize the accrual of quality based on the commitments made to it by the other agents, while ensuring that commitments to other agents are satisfied. The agents have the relevant non-local view of the coordination problem, detect coordination relationships, form commitments and communicate the committed results.

*Data Flow Strategy*: An agent communicates the result of performing a task to all the agents and the other agents can exploit these results if they still can. This represents the other extreme where there are no commitments from any agent to any other agent.

*Rough coordination*: This is similar to `balanced` but commitments do not arise out of communication between agents but are known *a priori*. Each agent has an approximate idea of when the other agents complete their tasks and communicate results based on its past experience. "Rough commitments" are a form of tacit social contract between agents about the completion times of their tasks.

The latter two coordination strategies are the alternatives normally used in the distributed data processing domain[Nagendra Prasad *et al.*, 1996]. [Decker & Lesser, 1995] proposed `balanced` as a sophisticated strategy that exploits a number of mechanisms to achieve coordination.

## 4 COLLAGE: Learning Coordination

### 4.1 Learning Coordination

Our learning algorithm, called COLLAGE, uses abstract meta-level information about coordination problem instances to learn to choose, for the given problem instance, the appropriate coordination strategy from the three strategies described previously. Learning in COLLAGE (**CO**ordination **L**earner for mu**L**tiple **AGE**nt systems) falls into the category of Instance-Based Learning algorithms[Aha, Kibler, & Albert, 1991] originally proposed for supervised classification learning. We, however, use the IBL-paradigm for unsupervised learning of decision-theoretic choice.

Learning involves running the multi-agent system on a large number of training coordination problem instances and observing the performance of different coordination strategies on these instances. When a new task group arises in the environment, each of the agents has its own partial local view of the task group. Based on its local view, each agent forms a *local situation* vector. A local situation represents an agent's assessment of the utility of reacting to various characteristics of the environment. Such an assessment can potentially indicate how to activate the various GPGP mechanisms and consequently has a direct bearing on the type of coordination strategy that is best for the given coordination episode. The agents then exchange their local situation vectors and each of the agents composes all the local situation vectors into a global situation vector. All agents agree on a choice of the coordination strategy and the choice depends on the kind of learning mode of the agents:

*Mode 1*: In this mode, the agents run all the available coordination strategies and note their relative performances for the each of the coordination problem instances. Thus, for example, agents run each of `data-flow`, `rough`, and `balanced` for a coordination episode and store their performances for each strategy.

*Mode 2*: In this mode, the agents choose one of the coordination strategies for a given coordination episode and observe and store the performance only for that coordination strategy. They choose the coordination that is represented the least number of times in the neighborhood of a small radius around the present global situation. This is done to obtain a balanced representation for all the coordination strategies across the space of possible global situations.

Mode 2 is quasi-online algorithm. In the initial stages it just explores and in the later stages it just exploits the learned information. A more typical online learning algorithm interleaves exploration and exploitation. Studying COLLAGE in this kind of a setup is high on our agenda of future work.

At end of each run of the coordination episode with a selected coordination strategy, the performance of the system is registered. This is represented as a vector of four performance measures: total quality, number of methods executed, number of communications, and termination time. Learning involves simply adding the new instance formed by the performance of the coordination strategy along with the associated problem solving situation to the "instance-base". Thus, the training phase builds a set of $\{situation, coordination\_strategy, performance\}$ triplets for each of the agents. Here the global situation vector is the abstraction of the global problem solving state associated with the choice of a coordination-strategy. Note that at the beginning of a problem solving episode, all agents communicate their local problem solving situations to other agents. Thus, each agent aggregates the local problem solving situations to form a common global situation. All agents form identical instance-bases because they build the same global situation vectors through communication.

**Forming a Local Situation Vector**

The situation vector is an abstraction of the coordination problem and the effects of the coordination mechanisms in GPGP. It is composed of six components:

(a) The first component represents an approximation of the effect of detecting soft coordination relationships on the quality component of the overall performance. An agent creates virtual task structures from the locally available task structures by letting each of the `facilitates` coordination relationships potentially affecting a local task to actually take effect and calls the scheduler on these task structures. In order to achieve this, the agent detects all the `facilitates` interrelationships that affect its tasks. An agent can be expected to know the interrelationships affecting its tasks though it may not know the exact tasks in other agents that affect it without communicating with them. The agent then produces another set of virtual task structures, but this time with the assumption that the `facilitates` relationships are not detected and hence the tasks that can potentially be affected by them are not affected in these task structures. The scheduler is again called with this task structure. The first component, representing the effect of detecting `facilitates` is obtained as the ratio of the quality produced by the schedule without `facilitates` relationships and the quality produced by the schedule with `facilitates` relationships.

(b) The second component represents an approximation of the effect of detecting soft coordination relationships on the duration component of the overall performance. It is formed using the same techniques discussed above for quality but using the duration of the schedules formed with the virtual task structures.

(c) The third and fourth components represent an approximation of the effect of detecting hard coordination interrelationships on the quality and duration of the local task structures at an agent. They are obtained in a manner similar to that described for `facilitates`.

(e) The fifth component represents the time pressure on the agent. In a design-to-time scheduler, increased time pressure on an agent will lead to schedules that will still adhere to the deadline requirements as far as possible but with a sacrifice in quality. Under time pressure, lower quality, lower duration methods are preferred over higher quality, higher duration methods for achieving a particular task. In order to get an estimate of the time pressure, an agent generates virtual task structures from its local task structures by setting the deadlines of the task groups, tasks and methods to ∞ (a large number) and scheduling these virtual task structures. The agents schedule again with local task structures set to the actual deadline. Time pressure is obtained as the ratio of the schedule quality with the actual deadlines and the schedule quality with large deadlines.

(f) The sixth component represents the load. It is obtained as the ratio of execution time under actual deadline and the execution time under no time pressure. It is formed using methods similar to that discussed above for time pressure but using the duration of the schedules formed with the virtual task structures.

In the work presented here, the cost of scheduling is ignored and the time for scheduling is considered negligible compared to the execution time of the methods. However, more sophisticated models would need to take into consideration these factors too. We view this a one of our future directions of research.

**Forming a Global Situation Vector**

Each agent communicates its local situation vector to all other agents. An agent composes all the local situation vectors: its own and those it received from others to form a global situation vector. We can have a number of composition functions but the one we used in the experiments reported here is simple: component-wise average of the local situation vectors. Thus the global situation vector has six components where each component is the average of all the corresponding local situation vector components.

For example, one global situation vector looks as follows: (0.82 0.77 0.66 0.89 1.0 0.87). Here the low value of the third component represents large quality gains by detecting and coordinating on hard interrelationships. Thus two of the more sophisticated coordination strategies called `balanced` and `tough`[Decker & Lesser, 1995] are found to be better performers in this situation. On the other hand, in a global situation vector such as (0.80 0.90 0.88 0.80 0.61 0.69) the low values of fifth and sixth components indicate high time pressure and load in the present problem solving episode. Even if the agents use sophisticated strategies to coordinate, they may not have the time to benefit from it. Hence, relatively simple coordination strategies like `simple` or `mute`[Decker & Lesser, 1995] do better in this scenario.

Note, however, that in most situation vectors, these tradeoffs are subtle and not as obvious as the above examples. It is difficult for a human to look at the situations and easily predict which strategy is the best performer. Hence, hand-coding the strategies by a designer is not a practical alternative.

### 4.2 Choosing a Coordination Strategy

COLLAGE chooses a coordination strategy based on how the set of available strategies performed in similar past cases. We adopt the notation from Gilboa and Schmeidler[Gilboa & Schmeidler, 1995]. Each case $c$ is triplet

$$\langle p, a, r \rangle \in C_i$$
$$C_i \subseteq P \times A \times R$$

where $p \in P$ and $P$ is the set of situations representing abstract characterization of coordination problems, $a \in A$ and $A$ is the set of coordination choices available, $r \in R$ and $R$ is the set of results from running the coordination strategies.

Decisions about coordination strategy choice are made based on similar past cases. Outcomes decide the desirability of the strategies. We define a similarity function and a utility function as follows:

$$s : P^2 \rightarrow [0, 1]$$
$$u : R \rightarrow \Re$$

In the experiments presented later, we use the Euclidean metric for similarity.

The desirability of a coordination strategy is determined by a similarity-weighted sum of the utility it yielded in the similar past cases in a small neighborhood around the present situation vector (in our experiments, the neighborhood radius was heuristically set to 0.05). We observed that such an averaging process in a neighborhood around the present situation vector was more robust than taking the nearest neighbor. Let $M$ be the set of past similar cases to problem $p_{new} \in P$ (greater than a threshold similarity).

$$m \in M \Leftrightarrow s(p_{new}, m) \geq s_{threshold}$$

For $a \in A$, let $M_a \equiv \{m = \langle p, \alpha, r \rangle \in M | \alpha = a\}$. The utility of $a$ is defined as

$$U_{(p_{new}, a)} = \frac{1}{|M_a|} \sum_{\langle q, a, r \rangle \in M_a} s(p_{new}, q) u(r)$$

## 5 Experiments

### 5.1 Experiments in the DDP domain

Our experiments on learning coordination were conducted in the domain of distributed data processing[Nagendra Prasad et al., 1996]. This domain consists of a number of geographically dispersed data processing centers (agents). Each center is responsible for conducting certain types of analysis tasks on streams of satellite data arriving at its site: "routine analysis" that needs to be performed on data coming in at regular intervals during the day, "crisis analysis" that needs to be performed on the incoming data but with a certain probability and "low priority analysis", the need for which arises at the beginning of the day with a certain probability. Low priority analysis involves performing specialized analysis on specific archival data. Different types of analysis tasks have different priorities. A center should first attend to the "crisis analysis tasks" and then perform "routine tasks" on the data. Time permitting, it can handle the low-priority tasks. The processing centers have limited resources to conduct their analysis on the incoming data and they have to do this within certain deadlines. Results of processing data at a center may need to be communicated to other centers due the interrelationships between the tasks at these centers. [Nagendra Prasad *et al.*, 1996] developed a graph-grammar-based stochastic task structure description language and generation tool for modeling task structures arising in a domain such as this. They present the results of empirical explorations of the effects of varying deadlines and crisis task group arrival probability. Based on the experiments, they noted the need for different coordination strategies in different situations to achieve good performance. In this section, we intend to demonstrate the power of COLLAGE in choosing the most appropriate coordination strategy in a given situation. We performed two sets of experiments varying the probability of the centers seeing crisis tasks. In the first set of experiments, the crisis task group arrival probability was 0.25 and in the second set it was 1.0. For both sets of experiments, low priority tasks arrived with a probability of 0.5 and the routine tasks were always seen at the time of new arrivals. A day consisted of a time slice of 140 time units and hence the deadline for the task structures was fixed at 140 time units. In the experiments described here, utility is the primary performance measure. Each message an agent communicates to another agent penalizes the overall utility by a factor called *comm_cost*. However, achieving a better non-local view can potentially lead to higher quality that adds to the system-wide utility. Thus, $utility = quality - total\_communication \times comm\_cost$. The system consisted of three agents (or data processing centers).

**Experiments**
For the experiments where crisis task group arrival probability was 0.25, COLLAGE was trained on 4500 instances in Mode 1 and on 10000 instances in Mode 2. For the case where crisis task group arrival probability was 1.0, it was trained on 2500 instances in Mode 1 and on 12500 instances in Mode 2. Figure 1 shows the average quality over 100 runs for different coordination strategies at various communication costs. The curves for both Mode 1 and Mode 2 learning algorithms lie above those for all the other coordination strategies for the most part in both the experiments. We performed a Wilcoxon matched-pair signed ranks analysis to test for significant differences (at significance level 0.05) between average performances of the strategies across communications costs upto 1.0 (as versus, pairwise tests at each communication cost). This test revealed significant differences between each of the learning algorithms (both Mode 1 and Mode II) and each of the other three coordination strategies, indicating that we can assert with a high degree of confidence that the performance of the learning algorithms across various communication costs is better than statically using any one of the family of coordination strategies[1]. As the communication costs go up, the mean performance of the coordination strategies go down. For crisis task group arrival probability of 0.25, the `balanced` coordination strategy performs better than the learning algorithms at very high communication costs because, learning algorithms use additional units of communication to form the global situation vectors. At very high communication costs, even the three additional meta-level messages for local situation communication (one for each agent) led to large penalties on utility of system. At communication cost of 1.0, Mode 1 learner and Mode 2 learner average at 77.72 and 79.98 respectively, whereas, choosing `balanced` always produces an average performance of 80.48. Similar behavior was exhibited at very high communication costs when the crisis task group arrival probability was 1.0. Figure 2 gives an example of situation-specific choice of coordination strategies for Mode 1 learner in 100 test runs when the crisis task group probability was 1.0. The Z-axis shows the number of times a particular coordination strategy was chosen in the 100 runs at a particular communication cost. X-axis shows the communication cost and the Y-axis shows the coordination strategy.

**When not to learn!**
In order to test COLLAGE on interesting scenarios with a range of characteristics, we created a number of "synthetic domain theories" using graph grammar formalisms[Nagendra Prasad *et al.*, 1996]. Space limitations do not permit us to discuss all the results but we would like to briefly talk about a very interesting result seen in the synthetic grammar "G3". We trained COLLAGE on the G3 domain in both Mode 1 and Mode 2 and tested them on 100 runs for different coordination strategies at various communication costs. We found that a coordination strategy called `tough`[2] coordination performs slightly better than COLLAGE. Upon closer examination of the problem instances, it was noted that `tough` was the best performer in 81% of the instances and other coordination strategies did better in the rest of the 19%. COLLAGE learns to choose the right coordination strategy in all the 100 instances. However, the agents require additional units of communication of meta-level information to form the global situation vector and decide that `tough` is the strategy of choice (in most cases). The lesson we learn from this grammar is that, if there is an overwhelming favorite for best performance in the family of strategies, then it may not pay to use COLLAGE to determine the best performer through additional situation communication. Sticking to the favorite without awareness of the nonlocal situation may yield as good a performance. However, if the few cases that warrant the choice of another strategy give far superior performance, then the gains from choosing a strategy can more than compensate for the additional communication. This, however, was not the case in environments produced by grammar G3.

### 5.2 Discussion
COLLAGE chooses an appropriate coordination strategy by projecting decisions from past similar experience into the newly perceived situation. COLLAGE agents performed better than using any single coordination strategy across all the

---

[1] Testing across communication costs is justified because in reality, the cost may vary during the course of the day.

[2] Definition of the `tough` coordination strategy is not important here.

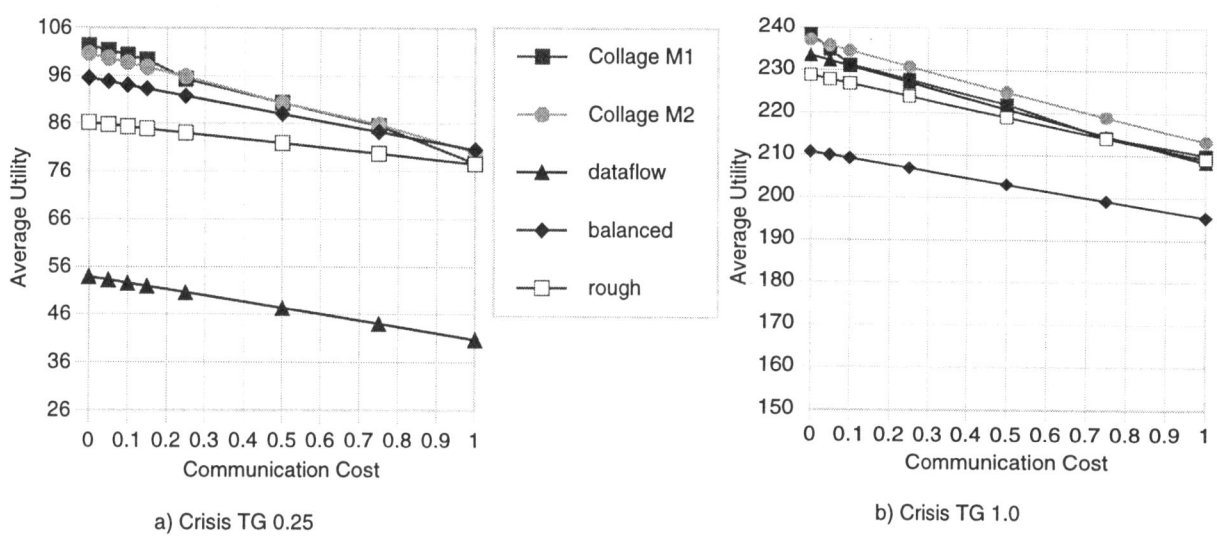

Figure 1: Average Quality versus Communication Cost

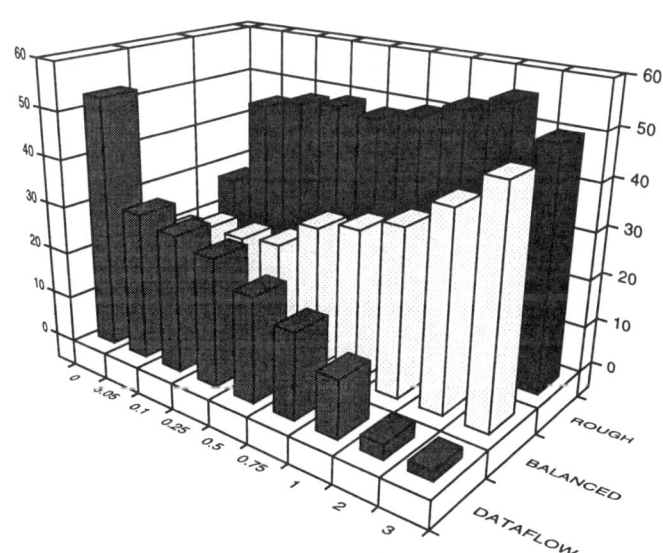

Figure 2: Strategies chosen by COLLAGE M1 for Crisis TG Probability 1.0 at various communication costs

100 instances in all domains we experimented with, except G3. In these domains, the cost incurred by additional communication for detecting global situation is offset by the benefits of choosing a coordination strategy based on globally grounded learned knowledge. Domain G3, however, is distinguished by the fact that there is little variance in the choice best coordination strategy and the best coordination strategy was almost always tough. This highlights the fact that learning is especially beneficial in more dynamic environments.

## 6 Conclusion

Many researchers have shown that no single coordination mechanism is good for all situations. However, there is little in the literature that deals with how to dynamically choose a coordination strategy based on the situation. In this paper, we presented a learning system, called COLLAGE, that uses meta-level information in the form of abstract characterization of the coordination problem instance to learn to choose the appropriate coordination strategy from among a class of strategies. Our experiments provide strong empirical evidence of the benefits of learning situation-specific coordination.

However, one important limitation of COLLAGE is its scalability. As the number of coordination alternatives become large in number, the learning phase could become computationally very intensive and the instance-base size could increase enormously with respect to Mode 2. We are looking at how to integrate methods for progressively refining situation vectors such as those in [Sugawara & Lesser, 1993], ways to organize the instance-base to access and detect regions where there is insufficient learning and also ways to do more directed experimentation during learning rather than randomly sampling the problem space.

In COLLAGE, all the agents form identical instance-bases. We could as well have done with one designated agent forming the instance-base and choosing the coordination strategy. However, our configuration was set up with a more general scheme in mind. Instead of all agents choosing the same coordination algorithm, they can choose pairwise or group-wise. A subset of the agents coordinate to choose the same strategy. This will lead to different case-bases at different agents and an agent may have more than one case-base if it is a part of more than one group. This leads us to another scalability issue: the number of agents. If there are a large number of agents, then common situation vectors may lose "too many" details about the situations. Pairwise or group-wise coordination may be a better option. However, we have to deal with issues such as inconsistent and conflicting knowledge among the case-bases, formation of appropriate groups, and different amounts of learning for different groups.

## Acknowledgments

We would like to thank Keith Decker and Alan Garvey for their help and input during the course of this work. Thanks also go to Daniel Neiman, Mike Chia and the anonymous referees for their comments and feedback on draft versions of this paper.

## References

[Aha, Kibler, & Albert, 1991] Aha, D. W.; Kibler, D.; and Albert, M. K. 1991. Instance-based Learning Algorithms. *Machine Learning* 6:37–66.

[Crites & Barto, 1996] Crites, R. H., and Barto, A. G. 1996. Improving elevator performance using reinforcement learning. In *Advances in Neural Information Processing Systems 8*.

[Decker & Lesser, 1993] Decker, K. S., and Lesser, V. R. 1993. Quantitative modeling of complex computational task environments. In *Proceedings of the Eleventh National Conference on Artificial Intelligence*, 217–224.

[Decker & Lesser, 1995] Decker, K. S., and Lesser, V. R. 1995. Designing a family of coordination algorithms. In *Proceedings of the First International Conference on Multi-Agent Systems*, 73–80. San Francisco, CA: AAAI Press.

[Durfee & Lesser, 1988] Durfee, E., and Lesser, V. 1988. Predictability vs. responsiveness: Coordinating problem solvers in dynamic domains. In *Proceedings of the Seventh National Conference on Artificial Intelligence*, 66–71.

[Garvey & Lesser, 1993] Garvey, A., and Lesser, V. 1993. Design-to-time real-time scheduling. *IEEE Transactions on Systems, Man and Cybernetics* 23(6):1491–1502.

[Gilboa & Schmeidler, 1995] Gilboa, I., and Schmeidler, D. 1995. Case-based Decision Theory. *The Quaterly Journal of Economics* 605–639.

[Nagendra Prasad et al., 1996] Nagendra Prasad, M. V.; Decker, K. S.; Garvey, A.; and Lesser, V. R. 1996. Exploring Organizational Designs with TAEMS: A Case Study of Distributed Data Processing. In *Proceedings of the Second International Conference on Multi-Agent Systems*. Kyoto, Japan: AAAI Press.

[Sandholm & Crites, 1995] Sandholm, T., and Crites, R. 1995. Multi-agent reinforcement learning in the repeated prisoner's dilemma. to appear in Biosystems.

[Sen, Sekaran, & Hale, 1994] Sen, S.; Sekaran, M.; and Hale, J. 1994. Learning to coordinate without sharing information. In *Proceedings of the Twelfth National Conference on Artificial Intelligence*, 426–431. Seattle, WA: AAAI.

[Sugawara & Lesser, 1993] Sugawara, T., and Lesser, V. R. 1993. On-line learning of coordination plans. In *Proceedings of the Twelfth International Workshop on Distributed AI*.

[Tan, 1993] Tan, M. 1993. Multi-agent reinforcement learning: Independent vs. cooperative agents. In *Proceedings of the Tenth International Conference on Machine Learning*, 330–337.

[Weiss, 1994] Weiss, G. 1994. Some studies in distributed machine learning and organizational design. Technical Report FKI-189-94, Institut für Informatik, TU München.

# Analysis of Inheritance Mechanisms in Agent-Oriented Programming

**Lobel Crnogorac**
Dept. of Computer Science
The University of Melbourne
Vic. 3052, Australia
E-mail: lobel@cs.mu.oz.au

**Anand S. Rao**
Australian AI Institute
Level 6, 171 La Trobe Street
Vic. 3000, Australia
E-mail: anand@aaii.oz.au

**Kotagiri Ramamohanarao**
Dept. of Computer Science
The University of Melbourne
Vic. 3052, Australia
E-mail: rao@cs.mu.oz.au

## Abstract

Research into the design methodologies for agent-oriented programming (AOP) is still in its infancy, with most of the work focused towards applying the well-known object-oriented concepts. Integration of inheritance into AOP seems a natural step in that direction. This paper addresses the notion of inheritance in AOP. We identify and formally define two problems which significantly reduce the usefulness of inheritance. We present a formal framework to investigate the notion of inheritance and the above problems. We define two alternative inheritance mechanisms for AOP and formally show the trade-offs involved. We hope that the results obtained from the analysis will help AOP language designers in their search for better inheritance mechanisms and a successful integration of inheritance within AOP.

## 1 Introduction

Multi-agent systems are concurrent systems based on the notion of autonomous, reactive, internally-motivated agents interacting in a constantly changing environment. Research in this field has been primarily concerned with conceptual and architectural issues of these systems and how they can be applied to real-world problems. More recently, there has been an increasing trend towards capturing the essential features of these systems and embodying them into language constructs, popularly called Agent-Oriented Programming (AOP) [Shoham, 1993; Wooldridge and Jennings, 1994], as well as developing a methodology for analysing and designing such systems, the emergence of Agent-Oriented Methodology (AOM) [Brazier et al., 1995; Kinny et al., 1996].

Naturally, work in AOP and AOM has been influenced by existing paradigms - especially the Object-Oriented Programming (OOP) paradigm. Although, AOP is often viewed as a special case of OOP [Shoham, 1993], very little attention has been given to one of the major aspects of OOP, namely, *inheritance*. The basic concept of inheritance allows the user to define a new class by extending an existing class with additional variables and methods. By sharing code among classes in this way, the total amount of code in a system can be reduced drastically.

Agents, unlike classical objects, need to synchronise their autonomous and reactive behaviour (multiple threads), they need to interact with the environment and other agents (the agent space is inherently concurrent) as well as commonly being required to satisfy real-time constraints. As a result, agents are usually significantly more complex than typical objects, both in their internal structure and in the behaviours they exhibit. Code re-use becomes highly important in such a context since it is much easier to design systems by extending already proven programs. Thus, the very complexity of agents makes the notion of inheritance a very useful and powerful tool for AOP. This paper addresses the notion of inheritance - its meaning, its potential benefits, and the problems that arise in introducing it within AOP.

An *inheritance mechanism* specifies exactly how existing agent descriptions may be inherited to produce new descriptions. One of the important decisions to be made by the language designer with respect to inheritance is deciding on its granularity, *i.e.*, what is the smallest unit of re-use within agents. This choice influences the usefulness of the concept, its expressiveness and its flexibility in aiding the development of re-usable agents.

The potential benefits of introducing inheritance in AOP are two-fold. First, it can reduce the complexity of the programming effort by enabling code sharing. In particular, an agent system may be designed by modelling the abstract behaviours of its constituent agents (*e.g.*, their roles). Agents exhibiting related abstract behaviours are expected to share a large part of their implementations. Second, it can facilitate the modular verification of multi-agent systems. Assume that we have proven certain behavioural properties of an agent family. It would be very useful, if rather than starting from first principles, one could prove properties of a derived agent family from provable properties of the original family.

However, the realisation of these potential benefits depends crucially on how one defines the inheritance mechanism. In particular, two types of problems could occur: the *lack of behaviour preservation* and the *presence of inheritance anomalies*. The lack of behaviour preservation leads to the need to prove properties about each agent implementation in an inheritance hierarchy from first principles (*i.e.*, the second benefit above is not realised). Inheritance anomalies are related to the expressive power of an inheritance mechanism. An inheritance anomaly occurs when the inheritance mechanism is not powerful enough to provide code sharing between agents implementing related abstract behaviours (*i.e.*, the first benefit above is not realised). These two problems can greatly reduce the usefulness of inheritance. Solving them requires the design of more complex inheritance mechanisms. We show that this usually leads to a trade-off, *i.e.*, designing more complex mechanisms may reduce the occurrence of inheritance anomalies while increasing the problems with behaviour preservation.

The key contributions of the paper are at two levels. At the technical level it presents a formal framework of inheritance, behaviour preservation, and inheritance anomaly. Using this framework, we prove that certain classes of inheritance mechanisms are behaviour preserving and anomaly-free. Examples of inheritance mechanisms are provided using the agent model of an implemented agent system – dMARS (distributed Multi-Agent Reasoning System). At the conceptual level, we believe the paper brings the technical rigour of analysis of classical programming languages to the field of agent-oriented programming. We hope that our analysis can help other researchers to make similar investigations of formal aspects of agent-oriented concepts before introducing them as language constructs in AOP languages.

## 2 Informal analysis

This section presents the dMARS programming environment and gives examples of typical problems that can arise when inheritance is introduced.

### 2.1 The agent setting

dMARS is an agent-oriented distributed real-time system based on the popular PRS model of agents [Georgeff and Lansky, 1986]. It provides a representational framework and reasoning mechanism for implementing agents.

An *agent family* $X$ is a tuple $\langle B_X, G_X, P_X \rangle$. $B_X$ is a set of belief atoms where each belief atom $b(\mathbf{v})$ ($b$ is a predicate symbol, $\mathbf{v} = (v_1, \ldots, v_n)$, $v_i$ is a term) provides information about the state of the environment as perceived by the agent. $G_X$ is a set of goals, where each goal describes the desired state of the system. If $g$ is a predicate symbol and $\mathbf{v}$ is as defined before, the

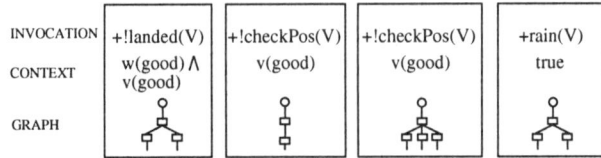

Figure 1: Plans of *SimplePlane*

goal $!g(\mathbf{v})$ states that the agent wants to achieve a state where $g(\mathbf{v})$ is a true belief. $P_X$ is a set of plans, where each plan specifies how to accomplish a goal or react to certain situations. An agent family is a set of *agent instances* (or simply *agents*) that have the same set of beliefs, achieve the same set of goals, and use the same set of plans to achieve all their goals. Agent instances are dynamic, run-time entities, while agent families are static descriptions corresponding to *classes* in OOP. At any point in time an agent contains a set of ground belief atoms. Agent reacts to events occurring in its environment by invoking its plans. Plans can also be invoked by events generated internally by the agent. Each plan consists of an invocation condition, context condition and a plan graph, *e.g.*, the plan $p$ is denoted by $\langle p_{inv}, p_{con}, p_{pg} \rangle$. The invocation condition identifies the event that must occur for this plan to be invoked. The events considered are the addition (+)/deletion (−) of beliefs/goals. Hence, $+b(\mathbf{v}), -b(\mathbf{v}), +!g(\mathbf{v}), -!g(\mathbf{v})$ are events. The context condition is a logical formula composed from belief atoms and the connectives $\wedge$ and $\vee$. The context condition must be satisfied by the set of ground belief atoms of the agent for the plan to be applicable. Plan graphs are acyclic graphs with their edges labelled with the sub-goals or primitive actions that have to be satisfied or performed in order for the main goal to be satisfied. We make a distinction between *public* goals and *private(internal)* goals. In particular, $G_X \subseteq \{!g(\mathbf{v}) : \exists p \in P_X \text{ such that } p_{inv} = +!g(\mathbf{v})\}$, is the set of all public goals that can invoke plans of an agent family $X$. A specific instance of an agent family acquires an initial set of ground beliefs, goals to be achieved, and the plans of the agent family. When the agent interacts with the environment, it adopts and commits to plans forming intentions. Details are provided elsewhere [Rao, 1996].

**Example 1** Figure 1 shows several plans of an agent family *SimplePlane* which describes agents simulating simple types of planes. The event +!*landed(base)* triggers the first plan which is designed to land the plane in good weather (*w(good)*) and good visibility (*v(good)*). The execution of this plan involves executing the sub-goal !*checkPos(V)* which then triggers the respective plans. The choice between the two plans for !*checkPos(V)* is based on a notion of priorities. The higher priority plan is tried first, and if it fails then the second plan is tried.

The plan for $+rain(V)$ reacts to a change in the environment. It is triggered when the new belief $rain(V)$ is added to the agent's set of beliefs. ∎

We now illustrate the problems that may arise when a simple inheritance mechanism is added to an agent-oriented language such as dMARS. Inheritance is introduced by following well-known OOP concepts. Hence, an agent (object) inherits all the plans (methods) of its parent, re-defines some plans and possibly adds new plans.

## 2.2 Behaviour preservation - an example

Agents, interacting with the environment generate behaviours. Characterising the interesting aspects of a behaviour is highly application specific and non-trivial. In some applications, behaviour can be characterised by considering *what* goals were achieved, while in some others one may want to capture *how* the goals were achieved, or *when* they were achieved (*e.g.*, real-time systems). In this section, we look at behaviour preservation with respect to the first characterisation, *i.e.*, what goals are achieved. Suppose we have implemented an agent family $X$, and proven that it achieves some desired behaviour. A new agent family $Y$ is defined by inheriting $X$ and adding some new plans. Does an arbitrary instance of $Y$ at least still achieve all the goals that an arbitrary instance of $X$ could achieve (if placed in the same environment)? In other words, can we guarantee some properties of $Y$ without needing to prove them explicitly?

**Example 2** Suppose instances of *SimplePlane* always land successfully in good weather and visibility. Consider a new agent family *AdvancedPlane* which inherits all the plans of *SimplePlane* and adds a new plan for the goal $!landed(V)$ enabling its instances to land in all weather and visibility conditions. We have inherited and extended *SimplePlane* with a new plan. Now we need to verify that an instance of *AdvancedPlane* can actually land in all conditions. However, we cannot even guarantee that it can land in good weather and visibility. If the new plan has higher priority than the inherited plan for $!landed(V)$ then the new plan is triggered first. The new plan could contain an error, causing *SimplePlane* to crash. Hence, a verification of the desirable behaviour for *AdvancedPlane* must be done from first principles. ∎

Note that there are always particular examples of inheritance which do in fact preserve behaviour. In Example 2, if the inherited plan has higher priority than the new plan then the behaviour of *SimplePlane* is preserved by *AdvancedPlane* (it is immediately known that both can land in good weather and visibility). More complex examples can be constructed that do not involve plan priority and yet fail to guarantee behaviour preservation. Such examples are based on the operational semantics of plan execution. Ideally, the inheritance mechanism should guarantee behaviour preservation.

## 2.3 Inheritance anomaly - an example

Suppose we have defined an agent family $X$ which implements our desired behaviour. Now, we want to build a specialised agent family $Y$ that can achieve all the goals that $X$ can achieve, and also some additional goals. We should be able to implement $Y$ by inheriting all the plans from $X$, while any behaviours not exhibited by $X$ are implemented by writing new plans. If it is not possible for $X$ and $Y$ to share the code in this way, then we have an occurrence of inheritance anomaly.

**Example 3** Suppose we define a new agent family *NewPlane* by inheriting *SimplePlane* and adding a new plan for $!checkPos(V)$. The new plan has the context condition $v(poor)$, enabling the agent to check its position in poor visibility. Thus, an instance of *NewPlane* should be able to land in good weather, under any visibility. The problem is that the context condition of the inherited plan for $!landed(V)$ still restricts the agent to land only in good visibility. This plan must be re-written with a new context condition, $w(good)$. Hence, even though there is some code sharing between *NewPlane* and *SimplePlane* (the plan for $+rain(V)$ and the existing two plans for $!checkPos(V)$), the amount of code sharing has been unnecessarily reduced. It is possible to construct examples where none of the plans can be shared. ∎

Re-defining plans of the parent family is error-prone, time consuming, and it violates the notions of modularity and encapsulation. Inheritance was designed to solve exactly this problem by code sharing. Thus, inheritance anomalies greatly decrease the usefulness of inheritance.

## 3 Formal analysis

In this section, we firstly define an inheritance mechanism to be the set of transitions from one syntactic specification to another. We provide two examples of such mechanisms with varying expressive power. Next, abstract behaviours of agent families are formalised and two alternative ways of defining behavioural equivalence (*i.e.*, when do two agent families exhibit the same behaviour) are considered. The language semantics determines how syntactic specifications are mapped into behaviours. The notions of behaviour preservation and inheritance anomaly are then formalised by analysing the behaviours of derived and original syntactic specifications. Behaviour preservation is similar to the notion of *soundness*: only the desirable behaviours should be derivable by the given inheritance mechanism. Inheritance anomaly is similar to the notion of *completeness*: all desirable behaviours should be derivable by the inheritance mechanism. Finally, we prove which of our inheritance mechanisms are behaviour preserving and anomaly-free.

## 3.1 Inheritance mechanism

First, we provide a general definition of an inheritance mechanism and then specialise it to AOP. An inheritance mechanism can be viewed as a transition relation between syntactic specifications. More formally we have:

**Definition 1** An *inheritance mechanism* is a tuple $(Spec, \Delta, \rightarrow, \dashrightarrow)$ where $\rightarrow \subseteq Spec \times \Delta \times Spec$. An element of $\rightarrow$, $(a, \delta, b)$ is called a *transition* where $a, b \in Spec$ and $\delta \in \Delta$. $Spec$ is the set of all syntactic specifications. The set $\Delta$ is the set of syntactic entities which specify the differences between $a$ and $b$. We write $a \xrightarrow{\delta} b$ for $(a, \delta, b) \in \rightarrow$. Furthermore, $a \xrightarrow{\delta \in \Delta^*} b$ is used to denote the transitive closure of $\rightarrow$, i.e., a sequence of transitions. Remark that the overloading of the notion $a \xrightarrow{\delta} b$ is quite harmless. The relation $\dashrightarrow \subseteq \rightarrow$ is the set of *incremental* transitions which maximise code sharing. ∎

In the context of AOP, a syntactic specification is simply the definition of an agent family. Intuitively, the transitions specify how inheritance can be used to move from a syntactic specification $a$ (e.g., *SimplePlane* in Example 2) to a new specification $b$ (a full definition of *AdvancedPlane*) by specifying the differences (the new plan for *!landed(V)*) in $\delta$. The general form of Definition 1 is specialised to AOP by instantiating $Spec$ to the set of all possible agent family definitions, i.e., the set including $X, Y, \ldots$ etc. Introducing inheritance to an agent language also means defining a particular instance of relations $\rightarrow$ and $\dashrightarrow$, i.e., giving a set of transitions. Given our definition of an agent family the following possibilities arise: We can derive an agent family specification $Y$ from an agent family $X$, by (a) adding or deleting belief atoms; (b) adding or deleting goals; and (c) adding, deleting, or modifying plans. While additions and deletions of belief atoms, goals, and plans in an agent family specification are straightforward, there are many ways in which a plan can be modified. A plan can be modified by modifying the invocation, context, or plan graph. We do not consider Case (a) in this paper. Now we consider a useful type of inheritance obtained by following the basic idea of inheritance as found in OOP (as in Section 2):

**Definition 2** $X \xrightarrow{\delta}_{mod} Y$ for some $\delta$ iff $|P_X| \leq |P_Y|$ and $B_X = B_Y$. It follows that $G_X \subseteq G_Y$. ∎

Agent family $Y$ is obtained from $X$ by having the same set of beliefs, but adding more goals, adding new plans for these goals, and modifying some or all of the existing plans for the existing goals. Definition 2 states that if an agent family $Y$ has more plans than an agent family $X$ then $Y$ can be obtained by inheritance from $X$. The point is that re-defining all the inherited plans may leave the specifications of $X$ and $Y$ totally unrelated (no code is shared) apart from their relative number of plans. The specifications of $X$ and $Y$ only share code if at least some inherited plans are not re-written in $Y$. Of course, the desirable situation arises when there is a maximal amount of code sharing in the system, i.e., when some new plans are added and none of the inherited plans is re-defined. This leads to the notion of *incremental* transitions. Incremental transitions only extend the existing family specifications without re-defining them in any way. We consider two useful inheritance mechanisms. The first mechanism (Definitions 2 and 3) allows incremental addition of plans for new and existing goals. The second mechanism (Definitions 2 and 4) allows incremental addition of plans only for new goals.

**Definition 3** $X \dashrightarrow^{\delta}_{plan} Y$ iff $X \xrightarrow{\delta}_{mod} Y$ and $P_X \subseteq P_Y$. ∎

The relation $\dashrightarrow_{plan}$ allows family specifications to be extended by the addition of new plans. The granularity of $\dashrightarrow_{plan}$ is a single plan since any new plan can be added. Another possible choice is the granularity of sets of plans, i.e., the set of plans for an existing goal cannot be incrementally extended. The relation $\dashrightarrow_{set}$ is the subset of all incremental transitions of $\rightarrow_{mod}$ at the granularity of sets of plans.

**Definition 4** $X \dashrightarrow^{\delta}_{set} Y$ iff $X \dashrightarrow^{\delta}_{plan} Y$ and $\forall p \in P_Y, p \notin P_X \implies p_{inv} \cap \{q_{inv} : q \in P_X\} = \emptyset$ where $p_{inv} = \pm!g(\mathbf{v})$. ∎

Under this definition only the plans written to achieve new goals can be added. For instance, a new plan with the invocation condition $+!landed(V)$, $+!checkPos(V)$ or $+rain(V)$ cannot be added in Example 1.

## 3.2 Behavioural equivalence

Next, we consider the notion of behaviour of an agent family. As discussed earlier, one can capture the notion of behaviour in many ways depending on the purpose of a particular application domain. We need some way of comparing the behaviours of derived agent families with the behaviours of the original agent families. In this paper, we consider two alternatives. We say that two agent families are behaviourally equivalent if: (a) they have the same public goals; or (b) their instances can achieve the same goals, when placed in the same environment. We consider each of these alternatives more formally.

**Definition 5** Let $X$ and $Y$ be agent families. $X \equiv_1 Y$ iff $G_X = G_Y$ ($X \leq_1 Y$ iff $G_X \subseteq G_Y$). ∎

Definition 5 is a simple definition of behavioural equivalence which states that agent families $X$ and $Y$ have equivalent behaviour ($X \equiv_1 Y$) if they have the same public goals. Hence, their instances have the potential to achieve the same goals. However, instances of $X$ and $Y$ may achieve the same goals under very different conditions (since their plans may be different). This definition corresponds to the OOP notion of type-equivalence based on interfaces. It can be shown that $\equiv_1$ is an equivalence relation, it is reflexive, symmetric and transitive.

The second definition of behavioural equivalence is concerned with more details about exactly under which conditions are particular goals achieved by an agent. It uses the notion of *traces*. Assume there is an external observer who records the sequence of all the events occurring in the environment of an agent. At every point in time the observer has noted down a finite trace of a possible evolution of this concurrent agent/environment system. The (infinite) set of all such finite sequences of events is called the set of *traces* of the system and is denoted by $traces(x||E)$ where $x$ is an agent instance, $E$ is the environment (non-deterministic) and $||$ is the concurrency operator. Placing an agent into an environment determines the agent's initial beliefs, its initial goals and its subsequent behaviour. Different instances of an agent family yield the same set of traces if placed in the same environment, but may exhibit different sets of traces if placed in different environments. We define $traces(X)$ as the union of all $traces(x||E)$ for all possible $E$ and $x$.

**Example 4** Two possible traces of the system in Example 1 are: $tr_1 = \langle\ +!landed(base), +pos(5), +sun(strong), +pos(3), +wind(weak), +pos(1), +landed(base)\ \rangle$ and $tr_2 = \langle\ +!landed(base), +rain(strong), +pos(5), +wind(strong), +crashed()\ \rangle$. ∎

**Definition 6** The trace $s \preceq t$ if $s$ is a prefix of $t$ ($\langle e_1 \rangle \preceq \langle e_1, e_2 \rangle$). Concatenation of two traces is denoted by $\hat{}$ ($\langle e_1 \rangle \hat{} \langle e_2, e_3 \rangle = \langle e_1, e_2, e_3 \rangle$). We use the standard *interleaving* semantics of concurrent events. ∎

**Definition 7** Let $x, y$ be arbitrary instances of $X, Y$. Let $goal\_traces_{!g(\mathbf{v})}(x||E) = \{s : s\hat{}\langle +g(\mathbf{v})\rangle \in traces(x||E)\}$ where $!g(\mathbf{v}) \in G_X$ is the goal to be achieved. Then, $X \leq_2 Y$ if for any environment $E$, whenever $s \in goal\_traces_{!g(\mathbf{v})}(x||E)$ either:

a) $\exists t \in goal\_traces_{!g(\mathbf{v})}(y||E)$ such that $t \preceq s$; or

b) $\exists t \in traces(y||E) : s \preceq t$, and for each such $t$ $\exists u \in goal\_traces_{!g(\mathbf{v})}(y||E) : s \preceq u \preceq t \lor t \preceq u$.

Also, $X \equiv_2 Y$ iff $X \leq_2 Y$ and $Y \leq_2 X$. Again, it can be checked that $\equiv_2$ is reflexive, symmetric and transitive. ∎

This definition of behavioural equivalence requires that all instances of two agent families can achieve the same goals if placed in the same environment (the goals may be achieved by adopting different plans, and thus generating different internal events). An observer records events occurring in the environment until the achievement of the goal (or failure). The set $goal\_traces_{!g(\mathbf{v})}$ contains all traces which lead to the achievement of goal $!g(\mathbf{v})$[1]. The trace $tr_1$ from Example 4 (without the last event $+landed(base)$) is an element of $goal\_traces_{!landed(base)}$ of an instance of *SimplePlane*, while $tr_2$ is not. An agent family $Y$ preserves and extends the behaviour of an agent family $X$ ($X \leq_2 Y$) if whenever an instance $x$ of $X$ achieves the goal then an instance $y$ of $Y$ (assuming $y$ is in the same environment as $x$) either: (a) has already achieved the goal; or (b) independently of what happens in the environment after this point, $y$ eventually achieves the goal.

**Example 5** Recalling the examples of Section 2 we have: $SimplePlane \leq_2 NewPlane$ ($NewPlane$ can land in poor visibility), and $NewPlane \leq_2 AdvancedPlane$ ($AdvancedPlane$ can also land in bad weather). ∎

### 3.3 Behaviour preservation

Having defined the inheritance mechanism and behavioural equivalence we examine their relationship by revisiting the notions of behaviour preservation and inheritance anomaly. Behaviour preservation expressed in a general form addresses the soundness issue. It ensures that an incremental transition from one syntactic specification to another leads to a related behaviour (the original behaviour is preserved and extended, as given by $\leq$).

**Definition 8** Let $a, b \in Spec, \delta \in \Delta$ and let $\leq$ be an ordering on behaviours. Inheritance mechanism $(Spec, \Delta, \rightarrow, \dashrightarrow)$ is *behaviour preserving* under $\leq$ if $\forall a, b$. $a \xrightarrow{\delta} b \implies a \leq b$. ∎

Definition 8 is specialised to AOP by considering specific inheritance mechanisms and specific behavioural equivalences. Behaviour preservation also depends on the exact semantics of the chosen language. Theorem 1 shows that the inheritance mechanism which allows new plans to be added for new and existing goals is behaviour preserving only under our simpler definition $\leq_1$ (for counter-example under $\leq_2$ see Example 2).

**Theorem 1** $(Spec_{AOP}, \Delta_{AOP}, \rightarrow_{mod}, \dashrightarrow_{plan})$ is behaviour preserving under $\leq_1$.

**Proof**: $X \xrightarrow{\delta}_{plan} Y \implies G_X \subseteq G_Y \implies X \leq_1 Y$. ∎

The inheritance mechanism where plans can be added only for new goals is behaviour preserving under both characterisations of behavioural equivalence.

**Theorem 2** *The mechanism* $(Spec_{AOP}, \Delta_{AOP}, \rightarrow_{mod}, \dashrightarrow_{set})$ *is behaviour preserving under* $\leq_1$ *and under* $\leq_2$.

**Proof**: Let $X \xrightarrow{\delta}_{set} Y$ and $s \in goal\_traces_{!g(\mathbf{v})}(x||E)$. The trace of $s$ contains $+!g(\mathbf{v})$ which triggered some plan $q \in P_X$ to execute. Since $P_Y = P_X \cup \delta$, and since $\{p_{inv} : p \in P_X\} \cap \{p_{inv} : p \in \delta\} = \emptyset$, the same plan $q \in P_Y$ was executed in $y$. It follows by similar arguments that $s \in goal\_traces_{!g(\mathbf{v})}(y||E)$. Hence, $goal\_traces_{!g(\mathbf{v})}(x||E) \subseteq goal\_traces_{!g(\mathbf{v})}(y||E)$. Also, $X \leq_2 Y$ by letting $t = s$ in Case (a) of Definition 7. The proof for $\leq_1$ is similar to Theorem 1. ∎

---
[1] We assume that an agent is never asked to achieve a goal which is not in the set of its public goals.

The results of Theorems 1 and 2 seem to suggest that the inheritance mechanism employing the incremental additions at the granularity of sets of plans is the better solution. The next section will make a similar comparison with respect to the problem of inheritance anomalies.

### 3.4 Inheritance anomaly

Inheritance anomaly, in its general form, addresses the completeness issue. An inheritance mechanism is anomaly-free if all related behaviours can be derived by incremental transitions, thus maximally sharing code.

**Definition 9** Let $a, b \in Spec$ and $\leq$ be an ordering relation on behaviours. Inheritance mechanism ($Spec, \Delta, \rightarrow, \dashrightarrow$) is *anomaly-free* under $\leq$ if $\forall a, b.\ a \leq b \implies \exists \delta \in \Delta^*$ and $c \in Spec$ such that $a \overset{\delta}{\dashrightarrow} c$ and $b \equiv c$, where $\equiv$ is the equivalence relation induced by $\leq$. ∎

Definition 9 requires that any related behaviour of $a$ must be reachable from $a$ by incremental transitions. The desired behaviour of $b$ is reached by inheriting and extending $a$ to obtain $c$. If this requirement cannot be satisfied for some $a$ and $b$ then we have an occurrence of the anomaly. Note that in general $c$ and $b$ are different implementations of the same behaviour. Similarly to behaviour preservation, the anomaly is highly language dependent, and also depends on the definition of $\equiv$.

**Theorem 3** *The inheritance mechanism $\rightarrow_{mod}$ is anomaly-free under $\leq_1$. This holds for both $\dashrightarrow_{plan}$ and $\dashrightarrow_{set}$. The proof follows from Definitions 3 and 4.* ∎

Both inheritance mechanisms suffer from inheritance anomalies under $\leq_2$, as shown in Example 3. However, the choice of the granularity of sets of plans ($\dashrightarrow_{set}$) leads to more anomalies than the finer granularity of $\dashrightarrow_{plan}$.

**Theorem 4** *Every inheritance anomaly that occurs in $\rightarrow_{mod}$ with $\dashrightarrow_{plan}$ also occurs in $\rightarrow_{mod}$ with $\dashrightarrow_{set}$. The converse does not hold. The proof follows from the fact that $\dashrightarrow_{set} \subseteq \dashrightarrow_{plan}$.* ∎

This theorem illustrates the trade-off in the choice of inheritance mechanisms. The granularity of sets of plans is behaviour preserving, but it leads to more anomalies.

## 4 Comparison and conclusion

The notion of inheritance has been studied extensively in OOP and is non-existent in AOP. The well-known inheritance mechanisms in OOP languages do not suffer from the problems described in the paper. Inheritance anomaly occurs, and has been researched, in concurrent object-oriented programming [Matsuoka and Yonezawa, 1993]. However, it has not been formally defined.

In this paper, we have introduced the notion of inheritance into AOP and examined the potential benefits it offers in terms of code sharing and modular verification of agent systems. However, there are potentially serious problems with inheritance in AOP, namely behaviour preservation and inheritance anomaly. In this paper we have provided a formal framework to examine these problems and the trade-offs involved in solving them. We analysed two inheritance mechanisms and two definitions of behavioural equivalence, and showed which of them were behaviour preserving and anomaly-free. The formal framework we provide is generic and can be applied to other agent-oriented languages, such as AGENT0 [Shoham, 1993], PLACA [Thomas, 1994], or CONGOLOG [Lesperance *et al.*, 1995], by altering the definition of specifications ($Spec$), transitions ($\Delta$), and behavioural equivalence. More importantly, we believe that our approach to the introduction of inheritance into AOP lays the preliminary groundwork for rigorous analysis of language issues, such as inheritance, within the framework of multi-agent system research.

## References

[Brazier *et al.*, 1995] F. Brazier, B. D. Keplicz, N. R. Jennings, and J. Treur. Formal specification of multi-agent systems: a real-world case. In *ICMAS-95*, pages 25–32, California, 1995. AAAI Press/MIT Press.

[Georgeff and Lansky, 1986] Michael P. Georgeff and Amy L. Lansky. Procedural knowledge. *Proceedings of the IEEE Special Issue on Knowledge Representation*, 74:1383–1398, 1986.

[Kinny *et al.*, 1996] D. Kinny, M. P. Georgeff, and A. S. Rao. A methodology and modelling technique for systems of BDI agents. In *MAAMAW'96*, LNAI 1038, pages 56–71, The Netherlands, 1996. Springer-Verlag.

[Lesperance *et al.*, 1995] Y. Lesperance, H. J. Levesque, F. Lin, D. Marcu, R. Reiter, and R. B. Scherl. Foundations of a logical approach to agent programming. In *IJCAI'95 Workshop (ATAL), LNAI 1037*, pages 331–346, Montréal, Canada, 1995. Springer-Verlag.

[Matsuoka and Yonezawa, 1993] Satoshi Matsuoka and Akinori Yonezawa. Analysis of inheritance anomaly in object-oriented concurrent programming languages. In *Research Directions in COOP*. MIT Press, 1993.

[Rao, 1996] Anand S. Rao. AgentSpeak(L): BDI Agents speak out in a logical computable language. In *MAAMAW'96*, LNAI 1038, pages 42–55, 1996.

[Shoham, 1993] Yoav Shoham. Agent-oriented programming. *Artificial Intelligence*, 60(1):51–92, 1993.

[Thomas, 1994] S. Rebecca Thomas. The PLACA agent programming language. In *ECAI/ATAL-94*, LNAI 890, pages 355–370, The Netherlands, 1994.

[Wooldridge and Jennings, 1994] M. J. Wooldridge and N. R. Jennings. Agent theories, architectures, and languages: A survey. In *ECAI/ATAL-94*, LNAI 890, pages 2–39, The Netherlands, 1994. Springer-Verlag.

# EXPERT SYSTEMS

# An Expert System Using Nonmonotonic Techniques for Benefits Inquiry in the Insurance Industry

**Leora Morgenstern**
IBM T.J. Watson Research Center
30 Saw Mill River Road
Hawthorne, NY 10532
leora@watson.ibm.com

**Moninder Singh**
Department of Computer & Information Science
University of Pennsylvania
Philadelphia, PA 19104-6389
msingh@gradient.cis.upenn.edu

## Abstract

This paper describes BenInq, an expert system for benefits inquiry in the insurance industry. BenInq is designed to be used by both customer service representatives, who answer customers' questions in real time, and policy modifiers, who update insurance products. The main challenges were the design of the KR structure — in particular, representing a huge knowledge base at varying levels of granularity, and the development of the reasoning methods — in particular, developing a method that determines which insurance regulations apply to an insurance service. This required nonmonotonic reasoning, which we modeled using an extension of inheritance methods. BenInq represents one of the few large scale industrial applications that explicitly uses formal nonmonotonic reasoning techniques.

## 1 Introduction

Benefits inquiry, the process of querying an insurance company to determine one's benefits, is becoming increasingly complex. Whereas years ago, a medical insurance company would offer only a small number of products, today companies have thousands of different insurance products, each of which contains a myriad of services and regulations. Furthermore, these products are always changing. The amount of information is difficult to keep up with; thus, the need for an expert system.

The expert system described in this paper, BenInq, was created for the application of benefits inquiry in the medical insurance domain. The paper is structured as follows. We first describe the existing technology for benefits inquiry. We next list the desiderata for our expert system. Subsequently, we discuss the knowledge representation and reasoning issues, focusing on the nonmonotonic techniques used. We then present BenInq, evaluate its usefulness, and discuss how the techniques can be generalized.

## 2 Existing Technology

Benefits inquiry has traditionally been performed by customer service representatives (CSRs) who rely on basic charts, detailed manuals, and implicit knowledge gained through common sense and job experience. As insurance products have proliferated, there have been attempts to harness computing power to help with the benefits inquiry task. Often, such efforts are limited to placing charts and manuals on-line. Other efforts include text-based and code-based systems, discussed below.

**Text-based Systems:** Text-based systems allow for some search and indexing of subject areas. The medical insurance company for which we consulted had a rudimentary text-based system. Information is divided into chunks or subject areas. A piece of text is associated with each subject area. Subject areas might include preventive care, immunizations, and maternity. The text associated with preventive care lists the different types of preventive care available, such as routine physicals and standard immunizations, as well as coverage rates and allowed frequency of services. Another screen may deal with one of the topics described in one screen; e.g., there may be a screen devoted to immunizations, a topic described in the preventive care screen. The CSR uses this system by pulling down a menu of topics, clicking on a topic, and reading the information on the associated screen which comes up.

The advantages to this system over on-line manuals are first, the system allows rudimentary search, and second, the information is partly organized in a modular fashion. But there is no reasoning; the system has merely pruned the amount of information to be read. Second, the system does not make explicit the many interconnections between subject areas. For example, nothing in the system indicates a connection between the screens on immunization and preventive care. The CSR must reason, e.g., that the schedule rate for preventive care most likely applies to routine immunizations. The lack of connection between related screens makes updating the system difficult. If both the preventive care and immunization screens do have schedule rate information and this schedule rate changes, the individual modifying the system must make changes on both screens. Third, there are a small number of screens relative to the number of types of questions a CSR may have to answer. Thus, the CSR may not get the level of information that she needs. This is not an artifact particular to this application but is due to the logistical difficulties of creating and continually modifying a large number of screens.

**Making inquiries via codes:** Many insurance companies have a code-based scheme for answering customers' questions. Whether automated or not, such schemes operate using a table-lookup methodology. When a customer calls to inquire if a particular service is covered, he is asked to provide the CSR with the *procedural* (CPT) code, which represents the service which is to be performed, and the *diagnostic* (ICD-9) code, which represents the diagnosis that is the reason for the service that is to be performed. The CSR then feeds this pair of codes into the system, which responds with the information that the service is or is not covered, along with the appropriate cost of the service.

The advantage of code-based inquiry is that often customers' question can be answered quickly and unambiguously. The disadvantages are: First, customers must know complete code information before they can call their CSRs; this slows down the inquiry process. Second, the code-based scheme only allows questions at a detailed level of granularity. But customers often wish to ask general questions, such as Are routine immunizations covered?. Third, code-based systems do not allow questions that rely on information in addition to codes such as If a 44-year-old woman had a mammogram two years ago, will she be covered for a routine mammogram now?, in which the treatment history of the patient is relevant. Fourth, updating code-based systems is an exceedingly difficult process. There are tens of thousands of CPT codes and tens of thousands of ICD-9 codes, and a large subset of the possible pairings of these codes must be considered.[1]

**Desiderata — Toward Replacing Existing Technology:** We aimed to develop an expert system that supports benefits inquiry but avoids the problems of both the text-based and code-based systems. In particular, we wished to develop a system that
- allows questions at varying levels of granularity
- gives clear, unambiguous answers to commonly asked questions
- allows representation of very large amounts of material and navigation around a large information space
- supports connections among related topics
- supports easy updates and modifications

The ability to modify is important because products change so frequently; an outdated benefits inquiry system is useless. Thus, the system had to be usable not only by CSRs but also by *policy modifiers* (PMs), the insurance company employees responsible for making changes within a particular insurance product.

---

[1]The nightmarish prospect of updating millions of code pairs is somewhat mitigated by the fact that CPT and ICD-9 codes have their own structure and support some rudimentary abstraction.

## 3 Knowledge Rep and Reasoning

Insurance employees such as CSRs and PMs must reason about services, benefits, coverage information, and business rules. A *service* represents some class of medical services, such as Surgery or X-rays. A *benefit* is some component of an insurance product which covers or excludes a particular service. For example, the Drug Benefit may generally cover Prescription Drugs, though it may exclude certain classes of prescription drugs, such as Experimental Drugs. *Coverage information* indicates which services are covered or excluded by which benefit, as in the above example. *Business rules* limit or modify the ways in which services are covered by benefits. Rules include *cost-share* rules, which specify the deductible and co-pay of a covered service, *access* rules, specifying whether it is required to go to a network provider, and *administrative* and *medical* rules specifying other restrictions. For example, there is a cost-share rule specifying a 20% co-pay for surgical services, and an administrative rule stating that patients in drug rehabilitation programs lose all rehab benefits for a year if they are non-compliant.

CSRs and PMs must deal with a large amount of information. There are tens of thousands of services of interest. There may be several thousand business rules for an insurance product; many of these rules (such as cost-share) will apply to a great number of services.

In developing BenInq, we needed to focus on the *external* representation—the representation that the CSRs and the PMs see, as well as the *internal* representation, which the inner workings of the system use. Since BenInq would be used primarily to answer questions about service coverage and applicable business rules, the internal representation had to facilitate efficient reasoning about these issues. In addition, to ease use by CSRs and PMs, the external representation had to be perspicuous to employees of the insurance company. Moreover, PMs needed to directly manipulate knowledge in the system in order to modify products. The easiest way to achieve this was to have a direct mapping between the internal and external knowledge representations. A complete direct mapping is not always possible.

The choice of representation resulted from the observation that many services can be organized in a taxonomic manner (see Figure 1). For example, the services of PET Scans, Lab Tests, EKG/EEG, and Genetic Testing are all subtypes of Diagnostic Services. The taxonomy generally is at least several layers deep. Benefits likewise can be arranged taxonomically (see Figure 2).

The structure, however, is not purely taxonomic. Certain services have multiple supertypes. For example, Genetic Testing is both a subtype of Diagnostic Services and of Family Planning Services. Thus, the structure is a dag rather than a tree.

The quasi-taxonomic structure led us to consider a semantic network—in particular, an inheritance network with exceptions—as our KR structure. There are efficient algorithms for reasoning with a semantic network.

Moreover, such networks are natural to laypersons such as CSRs and PMs; this naturalness guarantees a direct mapping between the external and internal knowledge representations. We needed to determine, however, whether a semantic network was capable of encoding all the knowledge that was needed for the benefits inquiry system, specifically, (1) coverage or exclusion of medical services and (2) rules limiting or modifying the ways in which a service was covered.

## 3.1 Representing Coverage Information

Coverage and exclusion information can easily be encoded in the semantic network. In addition to the subtype link, one introduces covers and excludes link which connect benefit and service nodes. Thus, to represent the information that Diagnostic Services are *covered* by the Diagnostic Benefit, one places a covers link between Diagnostic Benefit and Diagnostic Services; to represent the information that PET Scans are *excluded* by the Diagnostic Benefit, one places an excludes link between Diagnostic Benefit and PET Scans. The covers and excludes links propagate along the taxonomy, so that, e.g., we can reason that Diagnostic Benefit covers Blood Work, since Blood Work is a subtype of Diagnostic Services. The propagation is relative to specificity considerations, so we do not use propagation to conclude that PET Scans are covered by the Diagnostic Benefit: the path from PET Scans to Diagnostic Benefit along the excludes link is more direct than the path from Pet Scans to Diagnostic Benefit along the covers link.

It is obvious that there is a close connection between this structure and a standard inheritance network with exceptions [Horty et al., 1990], and in fact we can make this connection (an example of a direct mapping) precise in the following way: Replace each benefit node by a service node which represents all services covered by that particular benefit (so that, e.g., the Diagnostic Benefit node would be replaced by the node Services Covered by Diagnostic Benefit). Replace each covers link by a subtype (isa) link, and each excludes link by a cancels link. One then uses a standard traversal algorithm for inheritance networks with exceptions [Horty et al., 1990; Stein, 1992] to determine whether or not services are covered. Both representations are useful: the standard inheritance network with exceptions representation allows us to use well-documented algorithms without modification, while the explicit use of covers and excludes links is more accessible to CSRs and PMs.

The importance of the cancels links in the standard inheritance representation points out that much of the relevant information for benefits inquiry is nonmonotonic. It is rarely the case that a class of services is covered or excluded by some benefit; there are almost always exceptions. This nonmonotonicity is hardly surprising; proponents of nonmonotonic logic [Reiter, 1980] have argued since its inception that nonmonotonic reasoning is common in everyday reasoning (including the business world). What is surprising is how rarely nonmonotonicity has entered commercial applications, even

in well-understood forms such as inheritance with exceptions. Nonmonotonicity is also present in this application in the regulations that apply to a node; we discuss this further in section 3.4.

## 3.2 Representing Rules

Certain rules lend themselves to representation within a semantic network. Consider the rule:

There is a co-pay of 20% for diagnostic services

To represent this rule using the standard inheritance network model, one could have a node representing the services which have a 20% co-pay, and a subtype link between the Diagnostic Services node and this node.

On the other hand, a more complex rule such as

Patients in Drug Rehabilitation programs lose all rehab benefits for a year if they are non-compliant

cannot be so easily represented. One could posit a node that represents the services which have the property that if patients are non-compliant wrt that service, then they lose all rehab benefits for a year, and then have a subtype link between the Drug Rehab Services node and this node. But such a node appears quite artificial, and well outside of the spirit of a semantic network, where nodes are supposed to represent easily understood and expressible concepts.

Instead, we choose to represent rules as well-formed formulae in some classical logic. For example, the rule above could be translated as

Drug-rehab(p) $\wedge$ enrolled(x,p,i) $\wedge$ $\exists$ j sub(j,i) $\wedge$ non-compliant(x,p,j) $\Rightarrow$ ((Drug-rehab(p2) $\wedge$ enrolled(x,p2,i2) $\wedge$ time(start(i2)) - time(end(j)) $\leq$ 1year) $\Rightarrow$ $\neg$ covered(x,p2)). The wff translation of business rules exists in the internal representation.

These rules seem to be attached to specific nodes in the net. For example, the rule specifying that the diagnostic co-pay is 20% ought to be attached to the Diagnostic Services node in the network. This observation prompted the development of FANs (formula-augmented semantic networks). A FAN is a structure which allows attaching wffs to nodes in a semantic network. Formally, a FAN is a tuple consisting of a set of nodes, a set of wffs, a set of links (subtypes) on nodes, a (possibly empty) set of partial orderings on these links, allowing the prioritization of links in cases of multiple supertyping, and links connecting nodes and sets of wffs. Details are given in [Morgenstern, 96].

The semantics of a rule attached to a node is roughly: the rule attached to a node usually applies to the services represented by the node. Nonmonotonicity is again present; the rule may not apply to all subtypes of the node. Thus, a natural question is: are rules inherited along taxonomic lines? We discuss this issue in sec. 3.4.

## 3.3 Categorizing Nodes

Benefit nodes correspond to the types of benefits provided by a particular insurance product. There are about twenty to thirty benefit nodes in a semantic network representing a typical insurance product.

In contrast to benefit nodes, which are few in number, there are many service nodes. Medical care providers speak of tens of thousands of medical services. Even if we do not represent all of these, we need to have some broad classification of services. A first cut at classification is provided by the insurance industry, which categorizes services according to *(professional) procedure*, *condition* or *setting*. Examples of professional procedure services, which are categorized according to the professional who provides the services, are Surgical services, Therapy, and Diagnostic services. Examples of condition services, which are categorized according to a particular class of conditions, include Maternity and Mental Health. Examples of setting services are Hospital Ancillary services or Inpatient Hospice Room and Board.

Each of these classes of services itself contains a few major classifications, roughly corresponding to the way laymen speak of medical services. Formally, a classification is considered major if it is a root node in the service taxonomy. Examples are Surgical, Diagnostic, Preventive, and Hospital Services. There are between 30 and 40 major classifications for a typical insurance product.

If we decide to explicitly represent tens of thousands of services in the semantic network, each major classification will have a very large number of nodes, much too large to display or for a CSR to comprehend. We are faced with a trade-off of ease of display, navigation, and comprehensibility, versus granularity and comprehensiveness. Achieving arbitrarily fine levels of granularity for all types of services will result in a system that may be too unwieldy for a CSR to use.

Our approach is to generate nodes in the (external representation of the) semantic network according to the following principle. A subtype is explicitly given only if there is a reason: for example, a particular subtype may not be covered, or different rules may attach to the subtype. Thus, Endoscopic Surgery is given as a subtype of Surgery because there are different rules for Endoscopic Surgery than for standard Surgery; Routine Endoscopic Surgery is given as a subtype of Endoscopic Surgery because Routine Endoscopic Surgery is not covered. On the other hand, there are no explicit nodes for appendectomy or gall bladder surgery, because these are standard surgeries; one can always just look at the Surgery node to determine coverage and the rules.

Indeed, a fully comprehensive network would have considerably more than tens of thousands of nodes. As indicated above, insurance companies often speak of procedure/condition or procedure/setting pairs or even procedure/condition/setting triples. For example, Lab Tests performed in a hospital are generally paid in a different manner (different schedule of payment and different regulations) than Lab Tests performed in a doctor's office or home. Expanding the network to include these pairs and triples would mean that the network would have millions of nodes. This would not only yield an unwieldy network, but would also be extremely wasteful: most pairs and triples are not of interest. We deal with this problem in several ways. First, sometimes a pair or

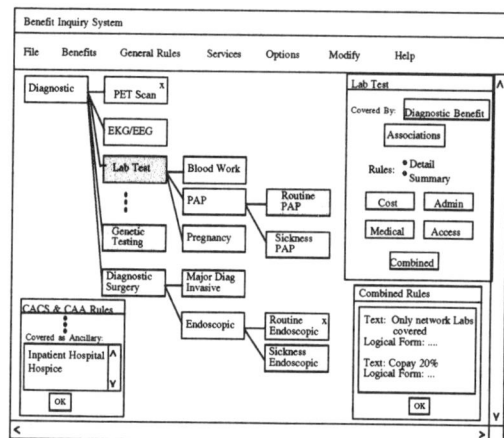

figure 1: CSR's view of BenInq — Diagnostic Services

triple, if common enough, is given its own node. Thus we have Routine Endoscopic and Sickness Endoscopic. [2] Second, each node is associated with a (possibly empty) list of settings which change the coverage or regulations of that node. For example, the node Lab Tests would have associated with it a list of settings including Hospital and Hospice, since payment for Lab Tests is different in those settings. We discuss this further in section 4.1.

### 3.4 Reasoning Methods

Two types of reasoning are needed for the semantic network:

1. Determining if a node is covered or excluded by some benefit.
2. Determining which rules apply to a node.

The first problem has been studied extensively in the literature [Horty *et al.*, 90; Stein 92]. BenInq implements Stein's (92) algorithm to solve the problem.

The second problem has been studied extensively by [Morgenstern, 96]. A short summary follows. The rules that *apply* to a node are not merely the rules that are *attached* to a node. For example, the rules that are attached to Diagnostic Services should also apply to the node Lab Tests. It might seem that given a node of interest (which we call the *focus node*), we can collect all rules attached to supertypes of that node[3]. But such an approach could lead to rampant inconsistency; it is very possible that one of the rules at Lab Tests is inconsistent with the rules at Diagnostic Services.

The issue of which rules apply to a node in a FAN is known as the *wff-inheritance problem*. The solution is to compute a *maximally consistent subset* (mcs) of all the rules attached to nodes to which there is a positive undefeated path from the focus node. Since there may be

---
[2] Of course, these are really sets of pairs; Sickness Endoscopic is the set of pairs (Sickness Endoscopic, x) where x stands for one of a large number of conditions for which endoscopic surgery is considered appropriate.

[3] Or more precisely to nodes to which there is a positive undefeated path [Horty *et al.*, 90] from the focus node.

many such maximally consistent subsets, one prefers the *preferred maximally consistent subsets* (pmcs's) based on criteria of specificity and path preference. For example, if there are two mcs's at a focus node, and they differ in that the first has a wff from a more specific node than the second, than the first is preferred to the second.

The algorithm to determine a pmcs at a focus node is given in [Morgenstern, 96]. It calls for first preprocessing the semantic network to remove conflicted and pre-empted edges. Then the network is traversed upward, starting at the focus node. As each node is visited, the node is marked, and a pmcs of the collected wffs along with the new wffs is computed. At each upward branching point, the branches are ordered according to the given prioritization, and the branches are recursively traversed according to that order.

The most computationally intensive part of this process is computing the pmcs, and in particular, determining if a set of wffs is consistent.[4] Our system uses the method of Selman and Kautz (1993) to randomly produce a valid truth assignment to a set of wffs. If it is consistent, this procedure generally works very rapidly. Thus, if after a short time no valid assignment has been produced, the system tries to show inconsistency in a more traditional manner.

## 4 System Description

BenInq has been implemented in VisualAge Smalltalk on an OS/2 platform. The system incorporates two tools: the *inquiry tool* which is used by CSRs to answer customers' questions, and the *authoring tool* which is used by PMs to modify products. BenInq has two main components, a graphical user interface and a reasoning engine, which are used by both tools. The graphical user interface is used to navigate through the large number of nodes and to display the dags that are rooted at specific benefit or service nodes. It consists mainly of functions to search through the network, draw nodes in the network structure, and display dialog boxes. We exploited the power of VisualAge Smalltalk, which comes with a library of drawing tools (particularly for menus and dialog boxes) and facilitates the creation of graphical interfaces.

The reasoning engine consists of two components: one for performing standard attribute inheritance and one for performing wff inheritance. The attribute-inheritance component uses the algorithm described in [Stein, 1992]. The wff inheritance component uses the algorithm described in [Morgenstern, 1996] and summarized in section 3.4.

The authoring tool uses both components of the reasoning engine as well as the graphical interface. The graphical interface is used to search for the node or portion of the network to be modified. If a node is inserted or deleted, or inheritance information is changed, the

---

[4]Determining consistency is only semi-decidable for the predicate case; it is decidable but intractable for the propositional case.

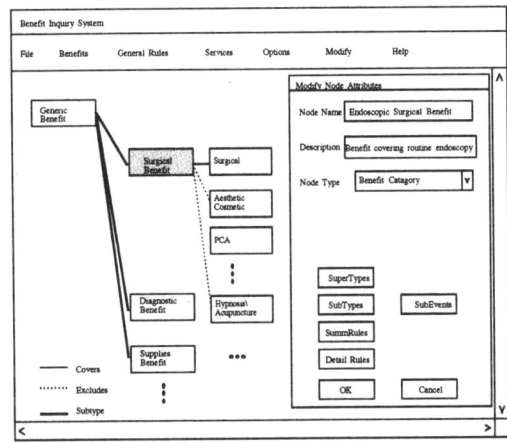

Figure 2: Surgical Benefit before updating

attribute-inheritance component recomputes the coverage or exclusion of the relevant nodes. If a wff is added, deleted, or modified at a node, the wff-inheritance component recomputes the wffs that apply to that node and the nodes beneath it in the dag.

The inquiry tool uses the attribute-inheritance component of the reasoning engine in addition to the graphical interface. Since the wff-inheritance component is by far the most computation-intensive portion of BenInq, we try to restrict its use within the system as much as possible. It is therefore not used by the inquiry tool. Thus, the set of wffs that apply to a node are stored during the authoring process.

Below, we briefly discuss how BenInq is used by its two classes of users, CSRs and PMs.

### 4.1 A CSR's view of BenInq

BenInq gives a CSR the ability to view various benefits and services by either selecting the appropriate item in one of several pull-down menus (organized by node type) or by using a node search mechanism. This allows the CSR to view all the subtypes of a chosen service or benefit, and to view all top-level services that are either covered or excluded by a chosen benefit. The system also gives the CSR the ability to determine which services are covered or excluded, and by which benefit. Moreover, the CSR can easily determine all the rules that apply to a particular service, including those inherited from its ancestors. The CSR can choose to view rules of a particular sort (e.g., cost-share or access) or all rules. BenInq also allows the CSR to view the *associations* at a particular node. Such associations allow the CSR to see how a procedure's coverage may change depending on the setting in which the procedure is performed.

Figure 1 illustrates how a CSR would use the system to answer questions concerning different Lab Tests. The CSR can view the relevant portion of the network either by first selecting Diagnostic Services using the pulldown menu and then clicking on Lab Tests, or by using the node search mechanism. The screen displayed provides

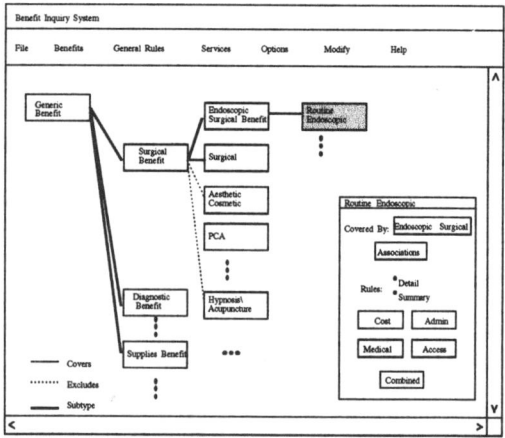

Figure 3: Surgical Benefit after updating

## 4.2 A Policy Modifier's view of BenInq

The authoring tool provides the Policy Modifier with several features to effectively modify benefits and services. The PM can add new benefits or services, modify existing services and benefits, and delete existing products. The PM can add a node either before or after an existing node in the network. She can also specify other supertypes, subtypes, and the rules that apply to the new node. For service nodes, the PM can specify whether the service is covered (or excluded), and by which benefit, or whether coverage/exclusion is inherited from another service node. Similarly, the PM can modify various attributes of any existing benefit or service including name, list of supertypes, prioritization of the supertypes, list of subtypes, set of rules that apply to that node, and for service nodes, the coverage/exclusion information and service classification. Finally, the PM can delete services or benefits from the system.

Figures 2 and 3 demonstrate how the authoring tool would be used by the PM to add a new benefit, Endoscopic Surgical Benefit, to explicitly cover the previously excluded Routine Endoscopic Services. Since the new benefit is a form of Surgical Benefit, she adds it as a subtype of Surgical Benefit. Figure 2 shows a portion of the dag displaying the Surgical Benefit. The PM adds a new node as a subtype of Surgical Benefit (by choosing the appropriate option from the Modify menu), and the dialog window pops up. She can then specify the name, type, subtypes, and supertypes as well as the rules that apply to the new node. Figure 3 shows the situation in which the new benefit, Routine Endoscopic Benefit has been added as a subtype of Surgical Benefit. The service node Routine Endoscopic has also been modified to change its coverage/exclusion information so that it is now covered by the new benefit (as opposed to being excluded by Diagnostic Benefit).

the CSR with much useful information at a single glance. First, it is immediately apparent which nodes are not covered; they have an x in the upper right corner. All other nodes are covered. There are visual cues indicating that a node is not covered by the expected benefit (in this example, Diagnostic Benefit). Thus, for example, the CSR finds out that Routine Pap (which is a different color than other nodes), although a lab test, is not covered by the Diagnostic benefit. To find out which benefit covers a particular node, the CSR clicks on that node. In Figure 1, since Lab Tests is currently selected, a dialog box shows, among other things, that Lab Tests is covered by Diagnostic Benefit. If the CSR clicks on the Routine Pap node, a similar dialog box comes up indicating that it is covered by the Preventive Benefit. Moreover, the CSR is given the option to switch views to a tree rooted at the Preventive benefit so that all information relevant to Routine Pap becomes visible.

In addition, the CSR may view various rules applicable to the node by clicking on the appropriate button. In Fig. 1, the CSR has selected the Combined button, indicating that all rules applying to the service should be displayed. The lower right portion of the screen shows the dialog box that pops us when the Combined button is selected. Note that both the text and logical forms of the rules are given. Similarly, the CSR can view the *associations* at a particular node by clicking on the appropriate button. The lower left portion of Figure 1 displays the dialog box which pops up when the CSR clicks on the associations button.[5] This box lets the CSR know that if lab tests are performed in an inpatient hospital (resp. hospice) setting, the procedure is covered by the Inpatient Hospital (resp. Hospice) Benefit and is subject to those rules. If the CSR wishes to see these rules, she clicks on the desired setting in the dialog box and the appropriate regulations appear.

## 5 Discussion
### 5.1 Evaluating the System

BenInq represents a significant advance in existing technology for benefits inquiry. The departure from a code-based system to a knowledge-based system is a big change for the insurance industry. This is, as far as we know, one of the few large-scale industry applications that explicitly models nonmonotonic reasoning and uses formal nonmonotonic reasoning techniques. This is clearly an advance for the insurance industry, which previously could deal with exceptions only in an ad-hoc manner. It also validates the work of the nonmonotonic reasoning community, which has often been accused of being insular and producing no useful results for the real world.

BenInq allows CSRs to immediately determine whether or not a service is covered and which rules apply to that service. It further provides PMs with an easy method of updating and modifying insurance products. The current version of BenInq was completed after our consulting engagement had ended. The medical in-

---

[5]In reality, only one of the rules/associations dialog boxes can be shown at one time. Both dialog boxes are shown for explanatory purposes.

surance company received an earlier version of BenInq, which incorporated some but not all of the reasoning methods described in this paper. This version supported standard inheritance with exceptions and attaching wffs to a node, but did not perform wff-inheritance. Thus PMs who modified the rules at a node had to modify the rules at all subtypes of the node as well in order to ensure integrity. Despite this drawback, the earlier version of BenInq received excellent reviews from the CSRs and PMs who used it. They needed virtually no training, surpassing our most optimistic expectations. They were delighted with the ease of finding out information and modifying products.

The desiderata outlined in section 2 were largely satisfied, although the last desideratum, supporting easy updates and modification, was only partly satisfied. In particular, while it is easy for a PM to add, delete, or modify links and nodes in the network, it is not always easy to modify or add wffs. The PM may easily modify the text version of business rules, but cannot modify the logical form of business rules.

This is a direct consequence of the fact that there is no direct mapping between the internal (logical form) and external (text) representations of business rules, as discussed in section 3. This is in direct contrast to all other parts of the semantic network, in which there is a natural, direct mapping between the internal and external representations. Fixing the problem requires the construction of a mapping between the text and logical form representations; this problem is in general difficult and beyond the scope of this work. However, one can identify certain types of rules for which there is a reasonable mapping. We have sketched out mappings for cost-share and access rules, as well as for administrative rules involving temporal and spatial (setting) constraints. The aim is the construction of templates which will allow the PM to directly enter rules in logical form, and will generate a reasonable text equivalent.

**What were the alternatives?** The success of BenInq was due to two facts: first, the realization that the bulk of the information was taxonomic, but that a significant chunk was not, and second, the construction of a knowledge structure, the FAN, which allowed the representation of both taxonomic and non-taxonomic information, and linkage between the two. Existing benefits inquiry systems had ignored the taxonomic nature of the medical insurance domain. This resulted in systems that were extremely spotty in their coverage (text-based systems) or were unwieldy (codes). Even a more sophisticated rule-based (or production) system that ignored the taxonomic nature of the domain would run into the problem of redundant information and the need for multiple updates. On the other hand, any attempt to force all of the information into a taxonomy would have resulted in highly artificial nodes with no clear semantics.

## 5.2 Generalizations and Extensions

The techniques described in this paper, while developed specifically for benefits inquiry in medical insurance, can be generalized to other domains as well. The construct of a FAN and its associated algorithms may prove useful in other domains which satisfy the following criteria:
- there exists a large amount of taxonomic information
- there exists a significant amount of non-taxonomic information, conceptually linked to the taxonomic information
- the non-taxonomic information can be mapped into wffs

The construction of an internal representation which closely parallels the external representation is a useful technique for domains where there is a non-technical class of users who must directly manipulate the knowledge structure. This technique is best suited to domains in which there is a natural translation between the user's conceptual model and a knowledge structure which lends itself to efficient reasoning methods.

Examples of such domains include other parts of the insurance industry, such as life, property, and casualty insurance; legal reasoning, especially case law (cases are often organized taxonomically, and different legal rulings are associated with cases); and medical reasoning.

We are currently using these techniques in the development of a system to dynamically configure property and casualty insurance products. In this system, the constructs of composition and subtyping are both important. The methods developed to reason with subtypes are also applicable for reasoning with composition, with appropriate modifications. This indicates that the techniques described in this paper are useful not only in domains in which taxonomic information is central but in domains in which composition is a useful construct. This has the potential to expand the applicability of this work. We hope to report on this research in the future.

**Acknowledgements:** We are grateful to Lynn Bricking, Kathleen Jamison, Bill Wynne, Tracy Edmonds, Terry Brown, and Kay Krimmer for their assistance in the knowledge engineering phase. An earlier version of BenInq was implemented with the assistance of Mike Ashwell and Jason Springer. Thanks also to Ernie Davis and Benjamin Grosof for helpful suggestions.

## References

[Horty et al., 1990] J. Horty, R. Thomason, and D. Touretzky. A Skeptical Theory of Inheritance in Nonmonotonic Semantic Networks. *Artif. Intell. 42*, 311-349.

[Morgenstern, 1996] L. Morgenstern. Inheriting Well-formed Formulae in a Formula-Augmented Semantic Network. *Proc. KR-96*, 268-279.

[Morgenstern, 1997] L. Morgenstern. Inheritance Comes of Age. Invited Talk, IJCAI-97.

[Reiter, 1980] R. Reiter. A Logic for Default Reasoning. *Artif. Intell. 13*, 81-132.

[Selman and Kautz, 1993] B. Selman and H. Kautz. Domain-Independent Extensions to GSAT. *Proc IJCAI*.

[Stein, 1992] L. Stein. Resolving Ambiguity in Nonmonotonic Inheritance Hierarchies, *Artif. Intell.*, 259-310.

# Can We Benefit from Metrics in KBS Development?

**Stefan Kramer** and **Hermann Kaindl**
Siemens AG Österreich
Geusaugasse 17
A-1030 Vienna, Austria
{stefan.kramer, hermann.kaindl}@siemens.at

**Stefan Schlee**
Softlab Austria
Moeringgasse 20
A-1150 Vienna, Austria
sts@softlab.co.at

## Abstract

A major issue with knowledge-based systems (KBS) is their maintenance. For dealing with this and other issues in software engineering for conventional software, there is a long tradition of using metrics. More recently, metrics have become popular also for object-oriented development. For the development of a KBS, however, no metrics are available yet.

This paper presents the core of the first metrics suite for KBS development. We focus on the fundamental concepts of coupling and cohesion, as well as on the impact of modifications. All these concepts are very important for ease of maintenance. The application of our metrics to a real-world KBS based on rules and frames helped pinpointing several maintenance problems. Therefore, we propose to use our and other yet to be developed metrics in order to improve KBS development.

## 1 Introduction

The maintenance of knowledge-based systems (KBS) is one of the major issues in their practical use. According to the experience with conventional software development, many maintenance problems stem from bad designs and implementations. So, improving these is an effective approach to address the maintenance problem.

Both in "traditional" and in object-oriented (O-O) software engineering, *metrics* are used (among other things) to determine current and to predict future qualities of a product relative to expected standards. In particular, it is possible to pinpoint those parts of a product with potential defects. In this way, it is easier to improve the quality of the product and therefore to improve its maintainability.

In order to address the maintenance problem of KBS, we devised the first metrics suite for KBS development.[1]

In this paper, we present three metrics of this suite dealing with frames and rules and their relationship in a knowledge base. These metrics represent quantitative models of the fundamental concepts *coupling* and *cohesion*, as well as the impact of modifications.

We have applied these metrics to a real-world KBS based on frames and rules. The numerical values computed there helped locating those frames and rules which would most likely pose problems during the maintenance of the KBS. So, the application of our metrics helped pinpointing maintenance problems.

This paper is organized as follows. First, we present some of the metrics that we devised for frames and rules. Then we show some empirical evidence with these metrics from real-world experience and from using a published knowledge base. After that, we shortly discuss previous work on the use of metrics. Finally, we discuss our approach and the use of metrics in KBS development more generally.

## 2 Metrics for Frames and Rules

We propose to use metrics for knowledge bases consisting of frames and rules in order to pinpoint potential problems with maintenance. Another possible use of metrics reported in the literature is the evaluation and comparison of alternatives. In both cases, metrics are useful as a means to make certain properties related to quality objective and quantifiable.

In the metrics literature there is a general distinction between *process metrics* (measuring certain aspects of the development process), and *product metrics* (measuring aspects of items produced during software development). In this paper we focus on product metrics related to knowledge bases consisting of frames and rules. In case of procedural attachment in the form of method slots, it would have been possible to simply transfer well-known metrics from the object-oriented paradigm. (For a comprehensive discussion of O-O met-

---

[1] The research reported has been carried out in the ViVa (Verification, Improvement and Validation of Knowledge-Based Systems) project, partially funded by the ESPRIT Programme of the Commission of the European Communities as project number 6125. The overall goal of the ViVa project is to provide a methodology and a supporting tool-set for V & V of knowledge-based systems in industry. More details can be found at http://www.cri.dk/ViVa — the *ViVaWeb* — where also our metrics suite can be tried out.

rics, see [Henderson-Sellers, 1996].) However, for knowledge bases consisting of frames and rules that are not attached to frames, a new set of metrics had to be derived.[2]

## 2.1 Procedure for the Derivation and Validation of Metrics

For the derivation and validation of the metrics, we adopted a procedure that is agreed upon in the software engineering literature (e.g. see [Hitz and Montazeri, 1996]). This procedure consists of three steps:

1. Identify the attributes of interest. In this context, attributes are properties of the item in question. For the attributes of interest, quantitative models are to be defined.

2. Derive the metrics for these attributes. The derivation starts with an intuitive account of the so-called *empirical relation system*, i.e., a kind of qualitative model of the attribute of interest. Then formalize the relations established in the empirical relation system.

3. Validate the metrics by determining the *scale type*, by careful application of some of *Weyuker's axioms* [Weyuker, 1988] and by checking the effects of so-called *atomic modifications* [Zuse, 1990].

We followed this procedure, and our description parallels its steps.

## 2.2 Identification of the Attributes of Interest

Regarding the selection of attributes of interest, we decided to focus on the concepts of coupling and cohesion, which are generally accepted to be relevant both in object-oriented and "traditional" software engineering. Coupling and cohesion can be explained in terms of elements (like modules, classes or frames) that are linked in some way (for instance by function calls, common usage of global variables, or message connections). The degree of connectivity *within* such an element is called cohesion, and the degree of connectivity *among* these elements is called coupling. In general, low coupling and high cohesion are indicators of good modularization and minimal interfaces, and thus contribute to maintainable software. In our context, the elements are frames, and they are linked by those rules where they are referenced together. Regarding cohesion, we are interested in the degree to which the *slots* belong together. This is different from the object-oriented version of cohesion, where the cohesion of *methods* is usually determined by their use of instance variables.

Besides coupling and cohesion, we are interested in measuring the impact that a change of a frame may have on the rule base. This additional attribute supplements coupling and cohesion, and is also related to maintenance.

[2] For a discussion of the concepts underlying object-oriented systems, frame systems and description logics see Appendix A.

| Frame $A$ |
|---|
| Slot $v$ |
| Slot $w$ |
| Slot $x$ |
| Slot $y$ |
| Slot $z$ |

| Frame $B$ |
|---|
| Slot $x$ |
| Slot $y$ |

| Frame $C$ |
|---|
| Slot $x$ |
| Slot $y$ |

| Frame $D$ |
|---|
| Slot $z$ |

Rule $r_1$: if $(A.w = 2)$ and $(A.x = 2)$ and $(B.x = 0)$ then $(C.y = 0)$

Rule $r_2$: if $(A.v = 2)$ and $(A.z = 1)$ then $(C.y = 1)$

Rule $r_3$: if $(A.w = 2)$ and $(A.x = 0)$ and $(A.y = 1)$ then $(B.y = 0)$

Rule $r_4$: if $(B.x = 0)$ and $(D.z = 0)$ then $(C.x = 1)$

Figure 1: Frames and rules of a hypothetical mini-KBS.

## 2.3 Derivation of the Metrics

The derivation will be illustrated by a running example, a hypothetical mini-KBS as shown in Fig. 1. It consists of four rules. The rules refer to frames ($A, B, C, D$) and their slots ($A.v, A.w, \ldots$).

First we want to derive a quantitative model for coupling. We focus on a single frame, and the degree of coupling between this frame and the other frames. The degree of coupling of a frame can be thought of as the "connectivity" with respect to other frames. The larger the number of other frames that are linked to a frame, the higher the coupling of the frame. As stated above, frames are linked by rules, but more precisely they are linked if and only if there is at least one rule where both frames are referred to. So, the degree of coupling of a frame may simply be defined as the number of other frames that are referred to in the same rules as the frame under consideration.

To formalize this notion of coupling of a frame, we define two auxiliary functions. Let $R$ be the set of rules in the knowledge base, and $F$ be the set of frames in the taxonomy. Then we define *Rules* as

$$Rules : F \to \mathcal{P}(R)$$

This function determines all rules that refer to a particular frame. For our running example, $Rules(A) = \{r_1, r_2, r_3\}$. Dually, we define *Frames* as

$$Frames : R \to \mathcal{P}(F)$$

This function simply returns the set of frames referred to by a given rule. For instance, $Frames(r_1) = \{A, B, C\}$.

Now we can define the metric for coupling, $DCpF$ (*Degree of Coupling of Frame*) as:

$$DCpF(f) = \left| \bigcup_{r_i \in Rules(f)} Frames(r_i) \right| - 1$$

In our running example, $DCpF(A) = 2$, since $\bigcup_{r_i \in \{r_1, r_2, r_3\}} Frames(r_i) = \{A, B, C\}$.

Another attribute of interest related to maintenance is

the impact that a modification of some frame has on the rule base.[3] In case of such a modification, the affected rules have to be checked if they are still valid. Given that we have to check those rules where the particular frame is referred to, the impact on the rule base may simply be measured by the number of rules using the frame. So we define the metric $IMF$ (*Impact of Modification of Frame*) as follows:

$$IMF(f) = |Rules(f)|$$

So, for our running example, $IMF(A) = 3$.

Next, we want to derive a quantitative model for cohesion. Our goal is to define a function measuring the degree to which the slots of a frame belong together (with respect to the rules). Intuitively, two slots belong together if they are connected by the rules, i.e. if they are used in the same rules. In other words, there is a connection between two slots if and only if they are both used by at least one rule. It is convenient to think of this in terms of graphs: for each frame we define a graph, where the nodes represent the slots, and an arc exists if the corresponding slots are both used in at least one of the rules. An obvious way to define the lack of cohesion is to count the number of "disconnected subgraphs", in graph-theoretical terminology the number of components. We define the metric for cohesion, $LChF$ (*Lack of Cohesion of Frame*) as follows:

$LChF(f)$ = Number of components of graph $G$, where the nodes of $G$ represent the slots of $f$, and an arc exists in $G$ if the corresponding slots are both used in at least one of the rules.

In our example, there are two components of the corresponding graph of frame $A$, one containing $\{w, x, y\}$, and one containing $\{v, z\}$ (see Fig. 2). Thus, $LChF(A) = 2$.

## 2.4 Validation of the Metrics

The soundness of these metrics can be checked by determining the scale type, by applying those of Weyuker's axioms that make sense in our context, and by checking the effects of atomic modifications. These "acid tests" are based on measurement theory and are established in the theory of metrics in the software engineering literature.

The *scale type* of all three metrics is absolute, since they are all counting some entities. This means that a full range of descriptive statistics is applicable, for instance, means and standard deviations may be calculated.

*Weyuker's axioms* state desirable properties of software complexity measures. For various reasons (explained in [Henderson-Sellers, 1996; Chidamber and Kemerer, 1994]) only a subset of Weyuker's axioms makes sense in our context. These properties can be paraphrased as follows:

---
[3]This attribute is different from coupling, since it relates elements of *different* sorts (frames and rules), while coupling just relates elements of the *same* sort (namely frames).

Figure 2: An example graph for computing $LChF(A)$.

- Property 1: not all possible frames have the same values for the metric.
- Property 3: for every frame there might exist another frame of the same complexity.
- Property 4: design decisions should be reflected in the values calculated by the metrics.
- Property 5: complexity should increase when adding something.

Properties 1, 3 and 4 are trivially satisfied. For property 5, we define the concatenation of frames as the merging of the slots into a new frame. The proof for property 5 is simple, but we cannot include it in this paper due to lack of space. In brief, all of these properties are satisfied by the metrics.

Finally, we check the effects of *atomic modifications* [Zuse, 1990]. Atomic modifications are the smallest possible changes in the model underlying the calculation of the metric. If the changes in the values of a metric caused by these modifications do not translate back to acceptable statements within the empirical relation system, the metric has to be rejected. In our context, the atomic modifications are additions or deletions of rules or frames. Since our metrics are all basically counting some entities, there is a direct correspondence between the modifications and the values of the metrics. Due to the straightforward derivation of the metrics, the increase in the metrics directly translates back to correct qualitative statements about the KBS. For instance, adding a rule that refers to a frame $A$ increases $IMF$ by 1, and the statement in the empirical relation system is that the impact of changing $A$ increases as well.

We briefly sketched three ways of validating the metrics for frames and rules. Summing up, our metrics turned out to be valid according to all three "acid tests".

## 3 Empirical Results

In this section we present the results obtained for two knowledge bases. First, we present some findings from a real-world project, where the knowledge base is implemented in Kappa [IntelliCorp, 1993], a hybrid frame system with traditional frames and rules. In the second application, we put the generality of the approach to the test, and calculated values for a knowledge base represented in CLASSIC [Brachman et al., 1991], a knowledge representation system based on description logic.

### 3.1 Results for a "Traditional" Frame System

In the following, we describe the application of the metrics to a real-world KBS implemented using a "traditional" frame system. Our role was to control the quality of

| Frame | $DCpF$ | $IMF$ | $LChF$ |
|---|---|---|---|
| $DataBoxes$ | 3 (60%) | 5 (8%) | 1 |
| $TMEntry$ | 1 (20%) | 2 (3%) | 1 |
| $StorageDisks$ | 3 (60%) | 14 (21%) | 1 |
| $Spindles$ | 2 (40%) | 28 (42%) | 1 |
| $Diagnosers$ | 5 (100%) | 66 (100%) | 1 |
| $AmfModels$ | 4 (80%) | 40 (60%) | 1 |

Table 1: Values obtained by the metrics ($DCpF$...Degree of Coupling of Frame, $IMF$...Impact of Modification of Frame and $LChF$...Lack of Cohesion of Frame) for 6 frames of a real-world knowledge base.

| Concept | $DCpF$ | $IMF$ | $LChF$ |
|---|---|---|---|
| $wine$ w/o s. | 23 (17%) | 148 (100%) | 7 |
| $wine$ w.s. | 1 (20%) | 148 (100%) | 1 |
| $meal\text{-}c.$ w/o s. | 0 (0%) | 0 (0%) | 2 |
| $meal\text{-}c.$ w.s. | 1 (20%) | 57 (39%) | 1 |
| $fruit\text{-}c.$ w/o s. | 1 (0.7%) | 2 (1.4%) | 2 |

Table 2: Values obtained by the metrics with and without substitution of "antecedent concepts" for the CLASSIC demo knowledge base "winery".

the KBS in question. So we were not involved in the development of the KBS, but in quality assurance. Table 1 summarizes the data obtained by the metrics. To improve the comprehensibility of the results, we also include percentages.

Applying the metric for coupling ($DCpF$) shows that the frame *Diagnosers* couples *all* frames (5 frames, 100%). In addition, the value for $IMF$ indicates that a change in this frame affects the *whole* rule base (66 rules, 100%). The second finding is that the frame *AmfModels* couples 4 out of 5 frames (80%), and that 60% of the rules (40 rules) would be affected by some change of this frame. These findings hinted at the possibility that both frames *AmfModels* and *Diagnosers* might have been pragmatically created for implementation considerations rather than designed as semantically meaningful units. So we asked the development team, and they confirmed our assumptions. In effect, the metrics helped us uncover a flaw that might have led to serious maintenance problems.

Checking the cohesion of the frames, we did not find indications for problems using the corresponding metric, since $LChF = 1$ for all frames. However, the applications of our other metrics helped pinpointing serious problems with maintenance. Our experience also suggests that metrics are useful as a means to determine certain properties of KBS in an objective and effective way, and that they help to focus the attention on potential problems.

### 3.2 Results for a CLASSIC Knowledge Base

We applied our metrics to the "winery" knowledge base that is distributed as a demo of the CLASSIC knowledge representation system. In order to apply our metrics to CLASSIC knowledge bases, it is first necessary to conclude which concept is the right-hand side concept of a rule. In Table 2, we summarize a few results for the "winery" knowledge base. We included values for its most fundamental concepts, *wine* and *meal-course*, and for a subconcept of *meal-course*: *fruit-course*. In addition, we performed a thought experiment, and calculated values for a slightly modified knowledge base: in CLASSIC, the antecedent of a rule has to refer to a defined concept. So if the antecedent of the rule is not yet defined as a concept, we have to define it to formulate the rule. We were interested in the values for the metrics if we could simply substitute the definitions of these "antecedent" concepts, and refer to them directly in the rules.

The results for the original winery knowledge base show that for the concept *wine* the degree of coupling ($DCpF$) is high (23 frames, 17%), and that the impact of modifications ($IMF$) is very high (148 rules, 100%). This is due to the fact that many rules (91 rules, 61%) are used to define contingent properties of wines (properties that do not belong to the concept itself), and that the rest of the rules (57 rules, 39%) make statements about the relationship between specific meal courses and sorts of wine. Note that the concept *meal-course* itself is never referred to in the rules (0 rules, 0%), but only specific meal courses like *fruit-course*.

Looking at the modified knowledge base, we find that the degree of coupling of the concept *wine* ($DCpF$) is low (1 concept, namely *meal-course*), but the value for $IMF$ is still very high (148 rules, 100%). Moreover, *meal-course* itself is referred to in the rules, but it couples only the concept *wine* (so $DCpF = 1$). Although coupling is low, the impact of modifications of this concept ($IMF$) is high (57 rules, 39%).

For the original knowledge base (without substitution of the antecedent definition), the cohesion of *wine* and *meal-course* is low (the lack of cohesion $LChF$ is high). The reason for this is that all the rules use only one attribute/role of these concepts. Interestingly, this is not the case if we substitute the antecedent definitions. The metric reflects that the substitution yields rules where attributes/roles of the same concept are used together.

Although the metrics have not been devised for description logics, they helped us make a few interesting observations.

## 4 Related Work

Coupling and cohesion are well-known and established concepts from traditional software engineering. Although the definition of the *concepts* of coupling and cohesion is clear and established in software engineering, there are various versions of metrics defined for these concepts (e.g., in [Heyliger, 1994]).

Chidamber and Kemerer [1994] were the first to define a metrics suite for object-oriented design and programming. Their metrics suite includes measures for coupling and cohesion. Coupling among objects is defined in terms of incoming and outgoing messages (method invocations) of an object. We defined coupling in analogy to this formulation in the object-oriented framework, where the rules take the role of the methods. Additionally, we defined a metric measuring the impact of modifications of a frame ($IMF$), that is specific for knowledge bases consisting of frames and rules.

Since Chidamber and Kemerer's metric for cohesion suffers several measurement-theoretic anomalies, we took a different approach. We adopted both the view of cohesion as proposed in [Li and Henry, 1993] and the graph-theoretic approach by [Hitz and Montazeri, 1996]. In a graph-theoretic view of Li and Henry's definition, the nodes represent the methods of a class, and they are connected if there is at least one instance variable that is used by both methods corresponding to the nodes. While in this view the measure for cohesion by Li and Henry also counts the number of components of a graph, in our approach the graph is defined in a dual way, i.e., the nodes represent slots, and the arcs are defined by "common" rules.

Besides coupling and cohesion, there are various metrics in the object-oriented literature measuring properties of taxonomies. (Some of them, like $DIT$, the depth of inheritance tree, are problematic from a measurement-theoretic view.) We also developed a few metrics for taxonomies, but in this paper we focus on coupling and cohesion.

In conventional and object-oriented software engineering there is a rapidly increasing number of publications about metrics. To the best of our knowledge, however, metrics have not yet been described in the published literature about KBS development.

## 5 Discussion

Now let us discuss our approach more generally. The question may arise, for instance, what such numbers signify. In some sense, this question relates to the utility of metrics in general. While this issue is surely beyond the scope of this paper, we claim to have developed our metrics suite according to the state of the art. In addition, we have validated them both according to acknowledged theoretical standards and through practical experience.

It must be understood that these like any other comparable metrics are just special quantitative models of "quality". So, it is very important for successful applications to beware of misunderstandings and misuse. Still, wherever some of these numbers indicate problems, it is worth a closer look.

More specifically, the concepts of coupling and cohesion and therefore also the corresponding metrics are related to the principle of minimal interfaces. While this principle is generally agreed to in the context of software engineering, it is not normally dealt with in frame systems and rule bases. A major effect of adhering to this principle is that we can achieve greater ease of maintenance. Since this should be a goal for the development based on frames and rules as well, and since also our third metric pursues this goal, we think that the metrics proposed in this paper should be useful for improving the development of KBS based on frames and rules.

Let us shortly point to an interesting difference of our approach compared to the corresponding approach for O-O software development (for the general correspondence between objects and frames see the Appendix A). While we measure how rules use the slots of frames, the usual O-O approach measures how the methods *within* an object use the instance variables of the same object. So, our approach measures more global dependencies between frames and rules that are *outside* of these frames. Normally, global dependencies are more critical for maintenance than local ones. So, we conjecture that our approach may be even more helpful in the context of KBS based on frames and rules than the approach for O-O software development.

Finally, our metrics approach relates to verification and validation of KBS consisting of rules and frames. In contrast to mainstream work in this area dealing with static KBS properties, we focus on the *relation between* rules and frames.

## 6 Conclusion

In this paper, we presented three metrics from our metrics suite for KBS development. They represent the core of this suite and measure the fundamental concepts of coupling and cohesion, as well as the impact of modifications. In essence, these metrics help pinpointing problems for the maintenance of knowledge bases consisting of frames and rules.

To the best of our knowledge, this is the first published account of a metrics suite for KBS development. Some of our metrics have been developed in analogy to existing metrics in software engineering, but we also developed completely new metrics: the metric measuring the impact of modifications of a frame has no analogue, and our measure for cohesion is dual to the one proposed by [Li and Henry, 1993].

The results from applying these metrics to a real-world KBS ("traditional" frames and rules) as well as to a demo for CLASSIC (description logic and rules) suggest the usefulness of these metrics. So, we think that this approach has the potential to improve the development of a KBS in the sense of enabling quantitative and objective judgements of quality properties relevant for the developer. Therefore, we propose to use our and other yet to be developed metrics in order to improve KBS development.

## Acknowledgments

We would like to thank Martin Hitz for useful discussions and comments on an earlier draft of this paper. Leslie Benjamin helped to improve the English. The implementation of our metrics in the *ViVaWeb* is due to Gert Schmeltz Pedersen.

# References

[Brachman et al., 1991] R. J. Brachman, D. L. McGuiness, P. F. Patel-Schneider, L. A. Resnick, and A. Borgida. Living with CLASSIC: when and how to use a KL-ONE-like language. In J. F. Sowa, editor, *Principles of Semantic Networks: Explorations in the Representation of Knowledge*, pages 401–456. Morgan Kaufmann, San Mateo, CA, 1991.

[Chidamber and Kemerer, 1994] S.R. Chidamber and C.F. Kemerer. A metrics suite for object oriented design. *IEEE Transactions on Software Engineering*, 20(6):476–493, 1994.

[Fikes and Kehler, 1985] R. Fikes and T. Kehler. The role of frame-based representation in reasoning. *Communications of the ACM*, 28(9):904–920, 1985.

[Henderson-Sellers, 1996] B. Henderson-Sellers. *Object-Oriented Metrics: Measures of Complexity*. Prentice Hall, Upper Saddle River, NJ, 1996.

[Heyliger, 1994] G.E. Heyliger. Coupling. In J.J. Marciniak, editor, *Encyclopedia of Software Engineering*, pages 220–228. John Wiley & Sons, Inc., 1994.

[Hitz and Montazeri, 1996] M. Hitz and B. Montazeri. Chidamber & Kemerer's metrics suite: A measurement theory perspective. *IEEE Transactions on Software Engineering*, 22(4), 1996.

[IntelliCorp, 1993] IntelliCorp. Kappa user's guide. Manual, IntelliCorp, Inc., Mountain View, CA, September 1993. Publication Number: K3.0-UG-2.

[Li and Henry, 1993] W. Li and S. Henry. Maintenance metrics for the object-oriented paradigm. In *Proc. First International Software Metrics Symp.*, pages 52–60, Los Alamitos, CA, May 1993. IEEE Comp. Soc. Press.

[Minsky, 1975] M. Minsky. A framework for representing knowledge. In P. Winston, editor, *The Psychology of Computer Vision*, pages 211–277. McGraw-Hill, New York, 1975.

[Weyuker, 1988] E.J. Weyuker. Evaluating software complexity measures. *IEEE Transactions on Software Engineering*, 14(9):1357–1365, 1988.

[Winston, 1977] P. H. Winston. *Artificial Intelligence*. Addison-Wesley, Reading, MA, 1977.

[Winston, 1984] P. H. Winston. *LISP*. Addison-Wesley, Reading, MA, 1984.

[Zuse, 1990] H. Zuse. *Software Complexity: Measures and Methods*. Walter de Gruyter, Berlin, 1990.

# A Relation Between O-O Languages and Frame Systems

In order to make this paper self-contained, we summarize the relation between object-oriented languages and frame systems, including systems based on description logics like CLASSIC [Brachman et al., 1991].

| Object-Oriented Languages | Traditional Frame Systems | Description Logics |
|---|---|---|
| (object) classes | (class) frames | concepts |
| (object) instances | (instance) frames | individuals |
| class / instance var. | slots | roles |

Table 3: Correspondence of key notions.

Objects and traditional frames have much in common. However, there are also some fundamental differences.

Although Winston heavily uses the notion of a frame due to [Minsky, 1975] in his AI textbook [Winston, 1977], his definition from the viewpoint of implementing frames in LISP simply says that a frame is a generalized property list [Winston, 1984, p. 311]. Traditional frame languages like, e.g., FRL, KEE [Fikes and Kehler, 1985] or Kappa [IntelliCorp, 1993], differ in that they provide frames in a way similar to records in Pascal. The general theme, however, is that a frame is a data structure that combines data stored in slots. While these definitions do not explicitly mention operations (in contrast to most of those for objects), these are typically representable via *procedural attachment*, just being stored in slots like data. But these definitions do not mention at all that data are hidden from the outside.

The principal element of systems based on description logics like CLASSIC [Brachman et al., 1991] is the *structural conceptual object*, or *concept*. The components of a concept are its subsuming[4] concepts (its superconcepts), and its local internal structure. This structure is expressed in terms of *roles*, which describe potential relationships between instances of the concept and those of other concepts (i.e., its properties, parts, etc.), and *structural descriptions*. In CLASSIC it is possible to model *concepts*, *roles* and *individuals*.

Table A summarizes the correspondences of these key notions in object-oriented languages and frame systems. (Object) classes and (class) frames roughly correspond to concepts (described in expressions according to a description logic). (Object) instances and (instance) frames correspond to individuals. Class / instance variables and slots have roles as their counterpart.

While it is possible to put these notions into correspondence, there are essential differences particularly between concepts in description logics and traditional frames as well as objects: firstly, they differ in the way subsumption is (formally) defined. Secondly, because the formal and well-defined descriptions of the former systems are based on a logic. Thirdly, the general intent of the former systems is to model the semantics and conceptual structure of a space of concepts, rather than to act as a data structure in a computer implementation. Still, the correspondence may be exploited for the purpose of this paper — to apply similar metrics.

---

[4] *Subsumption*: given two CLASSIC descriptions, an important question to consider is whether one *subsumes* the other—that is, whether an instance of one is always an instance of the other.

# Multi-Perspective Modelling of the Air Campaign Planning process*

**John Kingston**
AIAI, University of Edinburgh
Edinburgh EH1 1HN
Scotland

**Anna Griffith**
ISX Corporation
1165 Northchase Parkway
Marietta, GA USA 30067

**Terri Lydiard**
AIAI, University of Edinburgh
Edinburgh EH1 1HN
Scotland

## Abstract

This paper describes work performed to acquire knowledge about, and produce models of, the USAF Air Campaign Planning (ACP) process. The aim of this work was to produce a set of "knowledge models" which researchers in the area could refer to, rather than having each of them interview the expert planners.

It was decided that the models which were produced should be *multi-perspective* models; that is, a variety of models would be produced, each containing a particular type of knowledge about the air campaign planning process. The basis for this approach was the CommonKADS methodology for modelling organisational and expert knowledge. This paper describes the development of organisational, task and communication models to represent air campaign planning from various perspectives.

For some models, it was decided that CommonKADS' representations were not sufficiently rich, and so alternative modelling techniques (IDEF3 and Role Activity Diagrams) were used to represent the Task and Communication models. It was discovered that these techniques could be used without modification to represent CommonKADS models. An architecture is proposed, based on the Sowa/Zachman framework for Information Systems Architecture, to help determine the types of knowledge addressed by various modelling techniques

---

*This work is sponsored by the Defense Advanced Research Projects Agency (DARPA) and ISX Corporation under grant number F30602-95-C-0275. The U.S. Government is authorised to reproduce and distribute reprints for Governmental purposes notwithstanding any copyright annotation hereon. The views and conclusions contained herein are those of the authors and should not be interpreted as necessarily representing official policies or endorsements, either express or implied, of DARPA, Rome Laboratory or the U.S. Government.

## 1 Introduction

This paper describes work performed to acquire knowledge about, and produce models of, the USAF Air Campaign Planning (ACP) process. The aim of this work was to produce a set of "knowledge models" which researchers in the area could refer to, rather than having each of them interview the expert planners.

It was decided that the models which were produced should be *multi-perspective* models; that is, a variety of models would be produced, each containing a particular type of knowledge about the air campaign planning process (declarative, procedural, communications, etc), and thus representing the ACP process from a number of different viewpoints. The chief benefits of multi-perspective modelling are that it is easier to ensure complete acquisition of knowledge of each type, easier to maintain the knowledge, and easier to re-use individual models if another application should be developed.

The basis for this approach was the CommonKADS methodology for modelling organisational and expert knowledge. CommonKADS [Schreiber et al., 1994; Breuker & van de Velde, 1994] is a collection of structured methods for building knowledge based systems, analogous to structured methods for software engineering; as such, it provides an *enabling technology* for the analysis of acquired knowledge and the design of knowledge based systems. It was developed between 1983 and 1994 on two projects funded by the European Community's ESPRIT program, and has recently been identified as an technology with much potential by the Strategic Directions For Computing Research working group in North America [Computing AI Working Group, 1996].

For analysis of any knowledge-based application, CommonKADS recommends the development of one or more of the six models described below. The models, which are represented using node-and-arc diagrams, relate to each other as shown in Figure 1.

- An *organisational* model to represent the processes, structure and resources within an organisation;
- A *task* model to show the activities carried out in the course of a particular process;
- An *agent* model to represent the capabilities required of the agents who perform a process, and

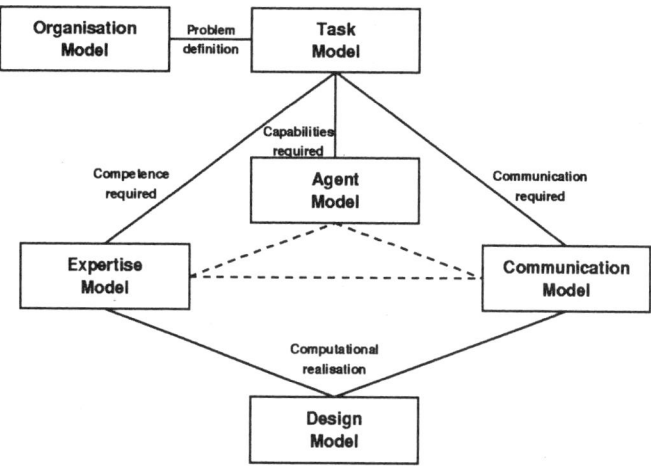

Figure 1: The CommonKADS Model Set

constraints on their performance;

- A *communication* model to show the communication required between agents during a process;
- An *expertise* model, which is a model of the expertise required to perform a particular task. This model is divided into three components:
  - declarative knowledge about the domain;
  - the inference processes required during problem solving;
  - a hierarchical classification and ordering of the inference processes.
- A *design* model, which culminates in the design of a knowledge based system to perform all or part of the process under consideration.

In order to represent the ACP process to a sufficient level of detail for other researchers, it was decided that the Organisational, Task and Communication models would be required; the Expertise model was also partially developed. The purpose of this paper is to describe the development of these models, with an emphasis on the use of different modelling techniques within a CommonKADS framework to represent the different perspectives. The paper concludes with discussion of a suggested framework for classifying and applying modelling techniques to multi-perspective modelling.

## 2 Knowledge acquisition

Knowledge acquisition was initially carried out using interviews. These interviews provided much useful knowledge, and also highlighted the existence of a number of relevant documents, from which a lot of knowledge was acquired. Two other knowledge acquisition techniques were also used: analysis of protocols acquired during a sample planning scenario provided useful information about the priorities, ordering, and necessary information for the planning process; and the repertory grid knowledge elicitation technique [Gaines & Shaw, 1993] was used at a later stage to determine which activities within the planning process might benefit from knowledge based system support.

The knowledge that was acquired showed that air campaign planning is hierarchically organised. When a crisis occurs, the Commander in Chief (CinC) provides planning guidance to the Joint Forces Commander (JFC). This guidance is communicated to the component commanders (e.g. the Air Component commander), who will in turn communicate the guidance to the air campaign planners. Based on the guidance, the planning staff will take between 3 days and 1 week to build a plan that may be executed. The acquired knowledge also showed that certain documents (such as the Master Attack Plan and the Air Tasking Order) form the outputs of key activities, and constitute a major method of communication within the planning process.

## 3 CommonKADS models

The acquired knowledge was then classified into appropriate CommonKADS models.

### 3.1 Organisational model

The CommonKADS organisational model [de Hoog et al., 1993b] is a collection of diagrams, each taking a different perspective on an organisation. CommonKADS recommends and suggests formats for diagrams of the organisational structure, activities, power/authority relationships and various resources. These models analyse each type of knowledge individually; "cross products" can also be developed which combine the knowledge from two different perspectives.

There may be various reasons for developing an organisational model:

- The model may be intended to describe the organisation in a way which facilitates identification of organisational needs and opportunities. By giving detailed but separate description of the activities, the structure which contains agents who perform activities, and the resources required for activities, it becomes easier to identify insufficient or excessive constraints on agents, under- or over-resourcing, and other potential areas for improvement.
- By emphasising knowledge-based activities in the modelling, this model can be used as a basis for *knowledge management* within an organisation. Knowledge management, which is a growing area of interest, involves surveying, categorising, analysing and synthesising knowledge-based activities, codifying and organising knowledge, appraising and evaluating knowledge and related activities, and deciding how to leverage, apply and control the knowledge [Wiig, 1993].
- The model can be used as part of a feasibility assessment of a proposed knowledge based system. A good example of this can be found in [de Hoog et al., 1993a], which describes a feasibility study into developing a KBS to support decision making within

a social security department, in order to reduce the time taken to process social security applications. A "cross product" of the organisational activities and organisational structure showed that decision making was distributed among several departments, and only occupied a small proportion of the total processing time for an application.

- The model may simply be required to provide a comprehensible, high level overview of an organisation for specialists who work in a narrowly focused area, and who need to know how their work relates to the rest of the organisation.

The modelling of air campaign planning was carried out for the final reason: to provide an overview of the air campaign planning process to ARPI researchers who are developing systems to perform knowledge-based planning and other tasks. The organisational model was also adapted according to the demands of the domain; specifically, the representation of power/authority relationships was considered superfluous by the experts, which is unsurprising in a model of a military organisation, where all authority is inherent in the organisational structure.

The resulting organisational model consisted of diagrams of activities (such as Figure 2), agents within an organisational structure, and resources. These were then combined to produce "cross products"; Figure 3 shows an early version of the cross product between activities and resources.

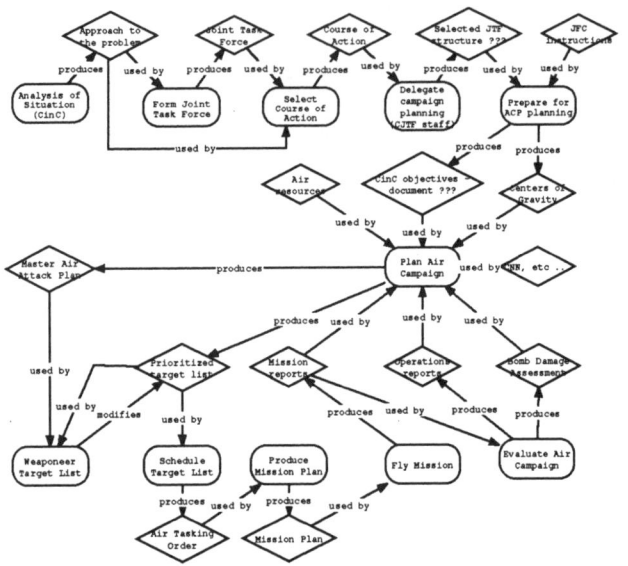

Figure 3: Air Campaign Planning: Cross Product of Activities and Resources

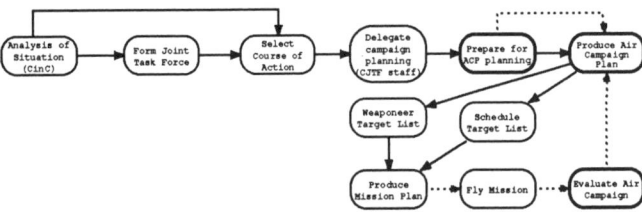

Figure 2: Air Campaign Planning: Top Level Activities

**Key:**

| Solid link | the first activity *precedes* the second |
|---|---|
| Dashed link | there is an *information flow* from the first activity to the second. |

With its combination of perspectives, the CommonKADS Organisational model provides a good overview of an organisation and its activities. However, there is a need for a more detailed analysis of the Air Campaign Planning process in order to supply researchers and developers of potential automated planning and decision support tools with sufficient information on where automated support is required, and therefore on where their tools might be used. As a result, it was decided that the CommonKADS Task and Communication Models needed to be developed as well.

**Key:**

| Labelled link | the activity *produces/consumes/uses/modifies* the resource |
|---|---|
| Diamond | Passive resource |
| Rounded rectangle | Activity |

### 3.2 Task Model

The CommonKADS Task Model [Duursma, 1993] expands one of the organisational activities identified in the organisational model into a more detailed view of the processes required to perform that activity. The underlying assumption is that this organisational activity has been identified as a source of an organisational weakness or opportunity, and so the activity should be investigated in more detail, with a view to automating certain sub-processes, or improving the knowledge management of sub-processes in other ways.

CommonKADS recommends a format for Task model diagrams. CommonKADS' recommended representation has the benefits of simplicity and clarity, but certain key information is relegated to textual annotations rather than being represented graphically. The representation of processes has been addressed by a number of different methods and techniques, each of which has its own strengths and weaknesses; for this project, where the primary aim was to communicate detailed information to other researchers, it was decided that CommonKADS' representation for the Task model was not adequate. The Process Flow Network recommended by the IDEF3 method was used instead.

IDEF3 is a process capture technique which was designed to be tolerant of incomplete and inconsistent descriptions, and to be flexible enough to deal with the incremental nature of the information acquisition process. It provides both a process-centred view of a system, via the *Process Flow Network*, and an object-centred view of a system via the *Object State Transition Network*.

A process flow network displays a sequence of *Units of Behaviour* (UOB) which represent activities, actions, processes or operations. These are linked together by precedence arcs. Where the process flow diverges (fan-out) or converges (fan-in) junction boxes are used. Junctions are of the AND, OR or Exclusive OR type and can be synchronous or asynchronous. This notation may impose timing constraints on the process flow. For example, a synchronous fan-in junction indicates that the incoming processes must complete simultaneously before the next UOB can begin.

In addition to UOBs and junctions, process flow networks can include referents, elaboration forms and UOB decompositions. Referents are used to indicate context-sensitive information and may refer to any other type of UOB such as an elaboration form, another process flow network, an object state transition network, an entirely different scenario, a note, or act as a GO-TO within the network. In some cases referents may impose timing constraints on the process so there is the option to be synchronous or asynchronous as needed. An elaboration form holds specific textual information for each UOB such as the object used by it, constraints acting on it, facts about it and a description of it. Decompositions enable each step of the process to be broken down into more detailed process descriptions, allowing descriptions to be held at varying levels of abstraction. This is indicated on the diagrams by a shadow on the parent UOB box.

An example of an IDEF3 process flow network can be seen in Figure 4.

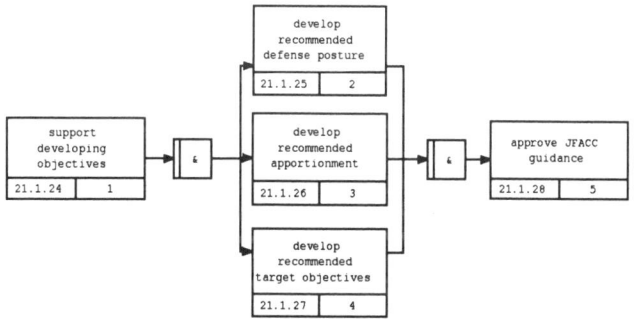

Figure 4: An IDEF3 Process Flow Network diagram

### 3.3 Agent and Communication Models

The CommonKADS Agent and Communication models support the Task model by identifying other information about the sub-processes being modelled. The Agent model represents the capabilities and skills of the agents (staff, clients, or computer programs) who perform each activity within a sub-process; the Communication model represents all communication which must take place between agents in order for a process to be completed.

One of the primary uses of the Agent model is to determine which roles can be performed by a human, which by a computer, and which by a human and computer working together. As this was not a major purpose of the modelling exercise, it was decided that an Agent model was not required. A Communication model was required, however, because effective transfer of information from one person or working group to the next is an important factor in the completion and efficiency of the planning process.

As with the Task model, CommonKADS' recommended format for communication modelling (described in [Waern *et al.*, 1994]) was not the richest technique available, and it was decided that a model rich in detail was required for this project. Role Activity Diagrams (RADs), a representation developed by a UK consultancy [Ould, 1992-3], were used instead. RADs are designed to represent all communication transactions succinctly; they represent the initiator of each communication; they link communication to activities, thus providing an explicit link with the Task Model; and they allow textual annotations to represent details. Part of the RAD which was developed for Air Campaign Planning can be seen in Figure 5.

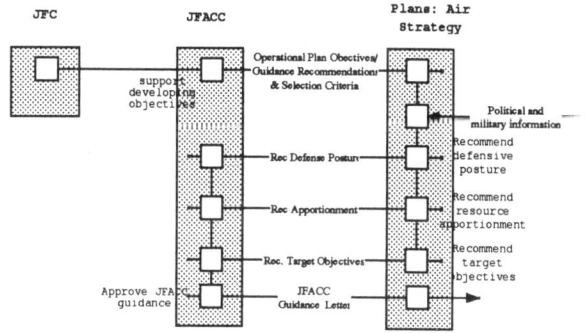

Figure 5: Role Activity Diagram showing initial communication in the air campaign planning process. Each shaded box represents a person or group of people.

### 3.4 Matching modelling techniques to perspectives

An important observation which was made was that IDEF3 Process Flow Networks and RADs could be used *without modification* to represent the CommonKADS Task model; this implies that these techniques model a single perspective on an organisational process. If this observation can be extended to other modelling techniques, then it should be possible for organisations to use CommonKADS as a modelling framework within which existing techniques can be used where appropriate, and CommonKADS' recommended techniques can be used for perspectives on knowledge which are not covered, or inadequately covered, by existing techniques. This would allow introduction of a CommonKADS approach without discarding work previously performed using other modelling techniques, and would reduce the learning curve for CommonKADS.

However, there is currently no systematic way of selecting a suitable technique to represent a particular perspective without having detailed knowledge of the capabilities of every possible technique.

As part of the ACP modelling effort, the authors of this paper met with a "modelling cluster" of researchers within the overall project who were interested in wider modelling issues. At the suggestion of one of the participants, the Sowa/Zachman framework for Information Systems Architecture [Zachman, 1987; 1992] was proposed as a basis for classification of modelling techniques.

The Information Systems Architecture (ISA) framework, which can be seen in Figure 6, classifies information and/or systems on two dimensions: *perspective* and *level of abstraction*. The six categories in the perspective classification are represented succinctly by the words *What*, *How*, *Who*, *When*, *Where* and *Why*. In other words, the ISA framework is suggesting that, to model an information system completely, it is necessary to determine what information will be used; how it will be processed; who will use it; when in the process each item of information is needed; the location at which each piece of information is needed; and the underlying strategy. These perspectives map closely to the perspectives recommended by CommonKADS:

- the Task model shows *how* information is processed;
- the Agent model shows *who* handles the information
- the Communication model shows *where* the information is needed
- part of the Expertise model defines the flow of control that specifies *when* activities occur
- both the Organisational Model and the Expertise model have components describing *what* information needs to be processed.

CommonKADS does not have an explicit "strategy" perspective; however, the library of "problem-solving methods" within the Expertise Model, and some aspects of the Design Model, are considered sufficient to support and represent *why* a particular way of representing knowledge has been chosen.

The second classification which appears within the ISA framework is *level of abstraction*. The framework suggests that each perspective should be modelled at six levels of abstraction, ranging from the *scope* of the system through *enterprise*, *system* and *technology* models to *components* and *functioning systems*. These terms are intended to be generic terms which can be customised for each particular application; for example, in the design and construction of a business system or building, the six levels of abstraction might correspond to the considerations of the *planner*, the *owner*, the *designer*, the *builder*, the *sub-contractor* and the *user*. By combining these different levels of abstraction with the different

|  | FOCUS | | | | | |
|---|---|---|---|---|---|---|
| Generic Framework<br>Element<br>Bond<br>Element | WHAT<br>(Data)<br>Entity<br>Relationship<br>Entity | HOW<br>(Function)<br>Process<br>Input-Output<br>Process | WHERE<br>(Network)<br>Node<br>Line<br>Node | WHO<br>(People)<br>Agent<br>Work<br>Agent | WHEN<br>(Time)<br>Event<br>Cycle<br>Event | WHY<br>(Rationale)<br>End<br>Means<br>End |
| SCOPE<br>(Planner) | Entity List | Process List | Location List | Organization List | Major Event List | Objective List |
| ENTERPRISE MODEL<br>(Owner) | Enterprise Entity<br>Enterprise Rule<br>Enterprise Entity | Enterprise Process<br>Resource<br>Enterprise Process | Enterprise Location<br>Enterprise Channel<br>Enterprise Location | Organization<br>Work<br>Organization | Enterprise Event<br>Enterprise Cycle<br>Enterprise Event | Objective<br>Strategy<br>Objective |
| SYSTEM MODEL<br>(Designer) | Entity Type<br>Relationship Type<br>Entity Type | System Process<br>User View<br>System Process | Site<br>Link<br>Site | Role<br>Presentation<br>Role | System Event<br>System Cycle<br>System Event | Criterion<br>Choice<br>Criterion |
| TECHNOLOGY MODEL<br>(Builder) | Data Structure<br>Referential Integrity<br>Data Structure | Application<br>Device Format<br>Application | Connection Point<br>Communication Line<br>Connection Point | User<br>Technical Interface<br>User | Technical Event<br>Technical Cycle<br>Technical Event | Condition<br>Action<br>Condition |
| COMPONENTS<br>(Sub-contractor) | Data Container<br>Acquisition<br>Data Container | Module/Object<br>Couple/Message<br>Module/Object | Address<br>Protocol<br>Address | Individual<br>Transaction<br>Individual | Component Event<br>Component Cycle<br>Component Event | Sub-condition<br>Step/Task<br>Sub-condition |
| FUNCTIONING SYSTEM<br>(User) | Information<br>Integrity<br>Information | Procedure<br>Request<br>Procedure | Client/Server<br>Access<br>Client/Server | Worker<br>Work Session<br>Worker | Operating Event<br>Operating Cycle<br>Operating Event | Target<br>Option<br>Target |

Figure 6: The Information Systems Architecture framework

perspectives on information, 36 possible classifications are created.

The four intermediate levels of abstraction identified by the ISA framework have direct correspondence with levels of abstraction identified by CommonKADS. The enterprise level corresponds with the Organisational Model, the system level with the Task, Agent & Communication models, the technology level with the Expertise Model, and the component level with the Design Model. The scope level, which defines what is inside and outside the boundaries of the information system, is implicitly represented by the presence or absence of knowledge items in CommonKADS models, while the functioning system level clearly corresponds to implemented KBS systems.

From this analysis, it can be seen that the CommonKADS approach to knowledge modelling can be mapped to the perspectives and levels of abstraction recommended by Sowa and Zachman in their framework for Information Systems Architecture. Given that CommonKADS is a proven enabling technology for KBS development, this analysis suggests that:

- CommonKADS could usefully be extended in certain areas to increase its coverage of the framework. Some of this work is already taking place; for example, the Communication model was originally seen as a way of modelling communication at the system level in order to identify communication bottlenecks and difficulties, but has been applied at the technology level (e.g. [Kingston, 1992] because it provides a useful tool for modelling inter-system communication, and for defining requirements for user interfaces).

- Other techniques could be used in conjunction with CommonKADS, to augment or extend its capabilities. The benefits of using known techniques in conjunction with CommonKADS include reduced learning curves, richness of representation, and reuse of existing models. This project has shown how

IDEF3 Process Flow Networks and Role Activity Diagrams can be used to augment certain aspects of CommonKADS modelling; it is not difficult to imagine circumstances where knowledge modelling would benefit from representation of the *what* perspective at the system level, which could be accomplished using IDEF3's Object State Transition Networks, or other techniques.

- The ISA framework could be used to inhibit techniques from being used where they are not appropriate. There is a danger that the known technique will be over-used, because people who are familiar with a technique believe that they can "hack" a solution; however, classifying techniques using the ISA framework promises to identify techniques that should be appropriate for modelling particular types of knowledge, as well as making explicit the limits of these techniques.

### 3.5 Summary

The CommonKADS methodology is an enabling technology for the analysis and design of knowledge based systems. It recommends analysis of knowledge by producing mutiple models showing individual perspectives on knowledge, and this multi-perspective approach is one of its main strengths. CommonKADS has proved to be very helpful for producing a well-defined model of air campaign planning; however, because of the richness of models required, CommonKADS was augmented with alternative modelling techniques that are able to replace the CommonKADS modelling techniques for certain perspectives.

The Sowa/Zachman framework for Information Systems Architecture takes a similar multi-perspective view on the representation of information and knowledge to the approach taken by CommonKADS. The ISA framework is capable of identifying specific areas where CommonKADS can be improved; if a suitable classification can be agreed, then there is potential for the ISA framework to identify modelling techniques which are suitable for representing different types of knowledge.

## References

[Breuker & van de Velde, 1994] Breuker, J., and van de Velde, W. 1994. *The CommonKADS Library: reusable components for artificial problem solving.* Amsterdam, Tokyo: IOS Press.

[de Hoog et al., 1993a] de Hoog, R.; Benus, B.; Metselaar, C.; et al. 1993a. Applying the CommonKADS organisational model. KADS-II/T1.1/UvA/RR/004/4.1, ESPRIT project P5248 KADS-II.

[de Hoog et al., 1993b] de Hoog, R.; Benus, B.; Metselaar, C.; and Vogler, M. 1993b. The Common KADS Organisational Model. ESPRIT Project P5248 KADS-II KADS-II/TM6/DM6.2/UvA, University of Amsterdam.

[Duursma, 1993] Duursma, C. 1993. Task Model definition and Task Analysis process. ESPRIT Project P5248 KADS-II KADS-II/M5/VUB/RR/004/1.1c, Vrije Universiteit Brussel.

[Gaines & Shaw, 1993] Gaines, B., and Shaw, M. 1993. Knowledge Acquisition Tools based on Personal Construct Psychology. *Knowledge Engineering Review* 8(1):49–85.

[Computing AI Working Group, 1996] Computing AI Working Group, S. D. 1996. in Strategic Directions in Artificial Intelligence. *ACM Computing Surveys* 28(4). Also available at URL http://www.medg.lcs.mit.edu/sdcr/ai/report/latest-draft.html#knowledge.

[Kingston, 1992] Kingston, J. 1992. Pragmatic KADS: A methodological approach to a small KBS project. *Expert Systems: The International Journal of Knowledge Engineering* 9(4). This paper is also available as AIAI Technical report AIAI-TR-110.

[Ould, 1992-3] Ould, M. 1992-3. Process Modelling with RADs. *IOPener: the newsletter of Praxis plc* 1-5 to 2-2.

[Schreiber et al., 1994] Schreiber, G.; Wielinga, B.; de Hoog, R.; Akkermans, H.; and van de Velde, W. 1994. CommonKADS: A Comprehensive Methodology for KBS Development. *IEEE Expert* 28–37.

[Waern et al., 1994] Waern, A.; Höök, K.; Gustavsson, R.; and Holm, P. 1994. The Common KADS Communication Model. ESPRIT Project P5248 KADS-II KADS-II/M3/TR/SICS, Swedish Institute of Computer Science.

[Wiig, 1993] Wiig, K. 1993. *Knowledge Management Foundations*. Arlington, Texas: Schema Press.

[Zachman, 1987] Zachman, J. 1987. A Framework for Information Systems Architecture. *IBM Systems Journal* 26(3).

[Zachman, 1992] Zachman, J. 1992. Extending and Formalizing the Framework for Information Systems Architecture. *IBM Systems Journal* 31(3).

# GAME PLAYING

# GAME PLAYING

Game Playing 1: Go

# A Model of Strategy for the Game of Go Using Abstraction Mechanisms

## Patrick RICAUD

LAFORIA - Université Pierre et Marie Curie - Paris VI
4, Place Jussieu - 75 252 PARIS cedex 05 - FRANCE

ricaud @ laforia.ibp.fr

## Abstract

While Chess programs are almost at the level of top world players, Go programs have yet to reach the beginner's one. Particularities of Go make it unsuitable for the arborescent approaches applied to Chess. The very large branching factor (200 to 350 possible moves), the length of the games (250 moves), and the difficulty to conceive global functions of evaluation adapted to it, are sufficient to explain the weak level of Go programs. Moreover, they totally lack the strategic dimension that is essential and crucial to the reasoning of the human player. Our work presents a new approach by using the possibility of abstraction to model the elementary strategy of Go. We have modeled, and implemented in the GOBELIN system, strategic mechanisms and basic concepts of Go. To this end, we have used mechanisms of abstraction that permit the search of a solution (a plan) in a simplified (abstracted) domain. This method allows us not only to avoid combinatorial crash, but also to provide the system with a reasoning level higher than the tactical one and a strategic coherent behavior close to, yet humbly, what the Go players name the global sense.

## 1 Introduction

### 1.1 A brief presentation of the game of Go

Go is an ancient game of strategy which originated in China and which has been played by Occidentals since the sixties. The game is played on a 19 x 19 board on which two players in turn place a stone of their colour. Starting with an empty board, each stone is placed on one of the vacant points of intersection of any two lines. The players can remove the opposing stones from the board by completely enclosing them. The object of the game is to have a greater number of stones on the board than the opponent at the end of the game. These basic rules differ from one country to the next, but all of them can be interpreted according to the above presentation. Playing Go is not essential to understand the issues which are dealt with in this paper.

### 1.2 The problems posed by Go with regard to Artificial Intelligence

As the characteristics of Go are not well known yet, we shall try and compare them to those of the game of Chess so as to bring to light the various problems which are encountered. Four words epitomize the four main problems which make the computerization of Go difficult : **Combinatory, Strategy, Perception,** and **Understanding**.

- In Go, the **branching factor** ranges from 200 to 350 moves whereas in the game of Chess it totals 40 moves. For 100 half moves in Chess, the **number of moves** in Go ranges from 200 to 250 for a match. The result of those two characteristics is an immense search space which makes the use of tree-search techniques impossible. It should be noted that these techniques were developed for the resolution of Chess problems.

- As a basic mechanism, the programs used in Chess problem resolution utilize an **evaluation function** not only on the number and nature of the pieces on the board, but also on their positions. The notions at stake in Go are not utterly compatible with such a computation and therefore various perspectives are often necessary to evaluate a position with respect to a plan of action. The **impossibility to significantly evaluate a position** at the beginning or in the middle of a party (in terms of points) makes it quite clear that there is no point in having a global evaluation function for Go [Victorri, 93].

- **Perception** also plays an important part in the concepts of aesthetics, equilibrium, and «natural form» which are commonly used by commentators of Go matches. If the problem at a local level seems quite simple to formalize, things are different at a global level where some forms more or less «work» with others towards achieving a goal (sometimes even several conflicting goals). Go players unconsciously use mechanisms of **recognition and interpretation at a very low level** to recognize, to qualify and use the afore said forms in their reasonings [Bouzy 94]. The formalization of such mechanisms, which might be compared to the human acquisition of reflexes, is one of the main obstacles in the creation of a Go program [Reitman & Wilcox, 79; Wilcox, 85].

- We can distinguish **several levels in the understanding** of a move in Go : There is a **tactical level** which enables one to recognize the basic forms. In addition this level is concerned with the structure and the tactical interactions of sets of adjacent pieces. There is also a **strategic level** which enables one to interpret the meaning of these forms in the particular context of a particular Go match. The polysemous characteristic of the entities at the tactical level in the interpretation of the strategic level makes us think that there may be some similarities between a) the cognitive mechanisms which are operated to understand a move or a position in Go, and b) the understanding of a sentence or a text in a natural language (with a very short semantic context for Go).

In short, the specific characteristics of Go cset this game at the frontier of the realm of reality and the formal domain. Indeed, the domain is not only **formal** with respect to its representation which allows no ambiguity -concerning its description at the lowest level- but also **real** with regard to the nature and complexity of the knowledge, as well as the concepts and the mechanisms which are put to work.

## 1.3 The significance of an abstract approach to the game of Go

Resolving the problems of Go through a standard combinatorial approach would be useless. Therefore, we have tried to change the problem by using a player's strategic approach. *«It is natural to simplify an intricate problem if its resolution can then be guided by the resolution of the simple problem. Abstraction itself consists in creating an easier problem that can be solved first, so that its solution can be used as the starting point for resolving the real problem»* [Pitrat, 90]. We believe that the possibility of using part of the strategic human reflection onto the framework of an abstraction is a natural approach. Consequently, thanks to our strategic knowledge of Go, we have created a domain which is parallel with this game but which has a smaller space search. We shall name this parallel domain the **Abstract-Go**.

Unlike the applications usually tested within the context of abstraction, [Giunchiglia 92], there is neither solution nor theorem to be found in Go but a possibly profitable position as it is the case in most of the games where the space search is large. A system using abstraction is generally made up of at least two levels : a **real level** and an **abstract level**. They respectively correspond to the initial domain and the simplified one. Each level has its own description of the states of the problem, a mechanism of exploration and generation of new states, and a mechanism of translation into the other representation.

## 2 General presentation of the system

The GOBELIN system is made up of four modules which can be thematically divided into the two following manners : **Real/Abstract & Description/Analysis**. There are two modules for the handling of real and abstract descriptions 1 and 2 and two modules of analysis -one for the research of plans 3 and the other for the tactical verification of the latter 4. The arrows ▷ signals the transitions between the different modules during the working cycle. The modules of reflection also takes into consideration the modules of description to create new states (in their respective level).

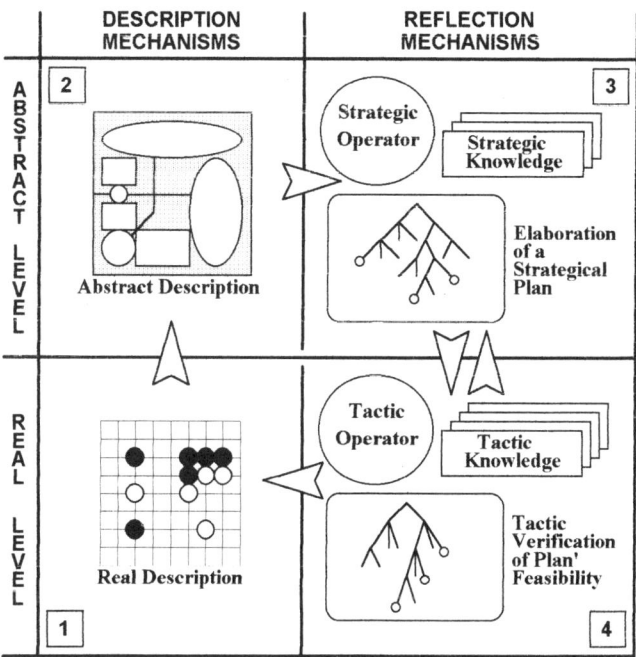

**Module 1    Handling of the real representation**

This module deals with the evolution of the real representation of the different game states. It can generate a subsequent position from a current position. It governs the rules of the game at the lowest level, that is to say the placing of a new stone, the capture, and the detection of illegal moves. It is the generator of states that is used in most computer game programs.

**Module 2    Calculation of the abstract representation**

From the real representation of the board position, this module builds an abstract representation of that position where the real objects are piled up and where the strategic relations are clarified in order to create new objects which are strategically more relevant. The system also determines the implicit objects which corresponds to strategically profitable moves. Thus, we are provided with a new representation of the board position as a network of object relations of a higher level than the real representation : Blocks, Chains, Groups and same colour stones are connected by a relation of proximity and of natural extension.

## Module 3    Abstract reflection

Following the abstract description which is created by the module 2, the mechanism of strategic reflection evaluates the position and seeks a profitable plan of action for the colour to be played. It also takes into consideration the various emergencies, postponable actions and possible threats. A move is comparable to a series of abstract operators (possibly a tree) representing operations which can be implemented by several elementary actions -for instance, splitting two groups, removing the heart of a colour group, defending a group which undergoes attacks, extending a group by natural extension, etc.)

## Module 4    Real verification

When a strategic move is chosen by module 3, the «tactical operator» 4 verifies the feasibility of the series of operators in the real representation. It does so by developing a reduced arborescence for each operator in order to validate or invalidate the initial move. In case of failure in the tactical feasibility of the strategic plan, a return to the abstract level 3 is operated (transition 4 ▷ 3) so as to allow the strategic plan to search for a new move with the new indications (notably on the impossibility of realizing the operators which are tactically unfeasible) which were provided by the tactical evaluation. Finally, when the tactical module delivers a positive verdict on the feasibility of a strategic move, the systems actualizes the first real move of the first operator of the plan.

## 4    An example

In this section we shall illustrate the methods that are used in GOBELIN with a relatively simple example of a beginning of a match with a handicap of five stones. In order not to cumbersome the exposé, we have not mentioned a vast number of details which, though useful, could not be accounted for in this paper. For further information see [Ricaud, 95; Ricaud, 96].

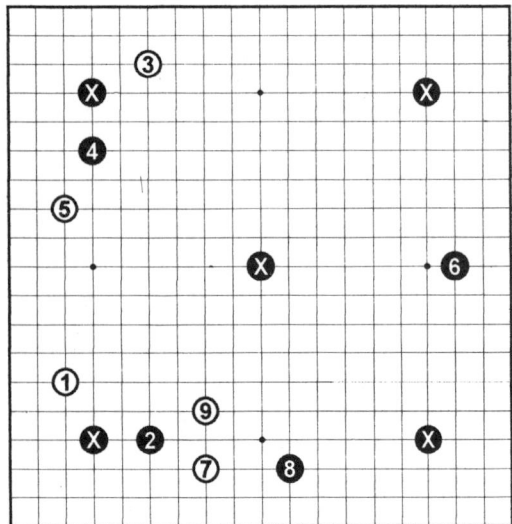

Real description of a position

In this example, Black has begun the game with five stones already on the board (the five stones denoted ⓧ), which gives it some lead over White. As White tries to build territories and areas, it tries to neutralize the potential territories and the areas which are under the control of Black.

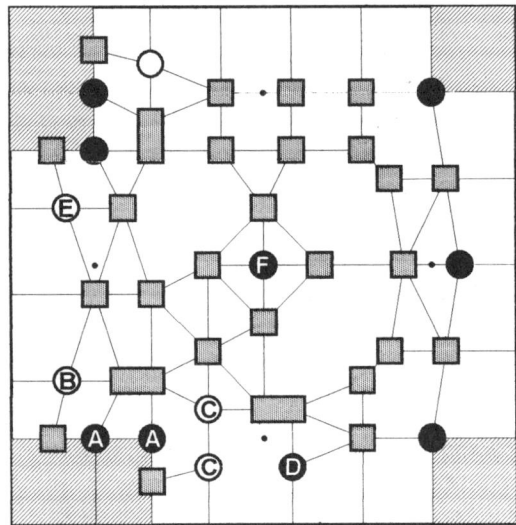

Abstract description of the same position

We shall illustrate the various methods which are implemented in GOBELIN by studying the reactions of the system with respect to the position following the move ⑨ that has just been played by White. Therefore, in our example, GOBELIN makes the decisions for Black. In this instance, the strategic search is focused on the group of stones Ⓐ, Ⓑ, Ⓒ and Ⓓ in the south-west corner because they have particular characteristics which make them vulnerable. The two areas that are controlled by the Black family Ⓐ are invadable through gliding, moreover this group has only one escape extension towards the center of the board.

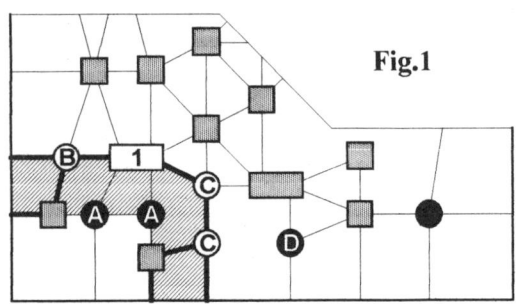

**Fig.1** If it were White's turn, it would be in its interest to play ☐1 for two reasons : the most important one is that it would enable White to connect its two groups Ⓑ and Ⓒ which are quite unsettled. It would also surround the Black group Ⓐ, threatening then to remove Ⓐ's heart through gliding in its controlled zones.

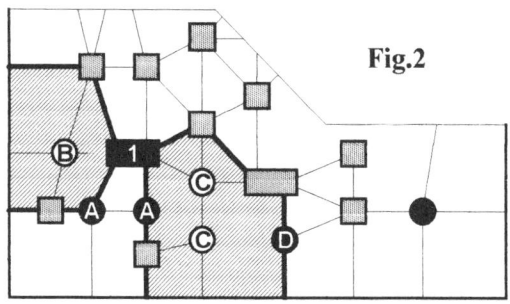

**Fig.2** If Black plays ■1, it prevents the previous surrounding from taking place and moreover it can accomplish two interesting attacks (the strategic equivalent of a fork). Thus, it is possible to surround the white stone Ⓑ in two moves. Another attack would be to surround group Ⓒ in three moves. This attack is less dangerous for White but quite inconvenient because White cannot stabilize this group by creating two boundary zones as it is the case for group Ⓑ. Group Ⓒ is thus named floating families.

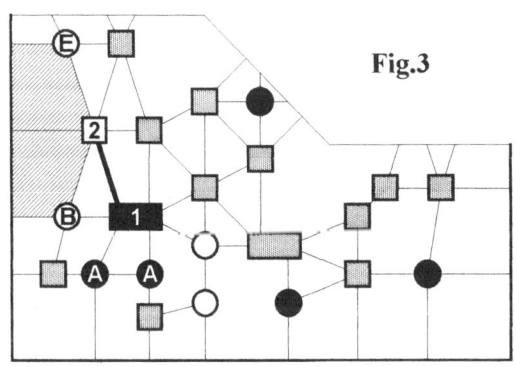

**Fig.3** The standard response to the Black move ■1 is White ☐2 because not only does it respond to the threat of surrounding but it also creates two white boundary zones which stabilizes the white stones of the western corner Ⓑ, ☐2 and Ⓔ.

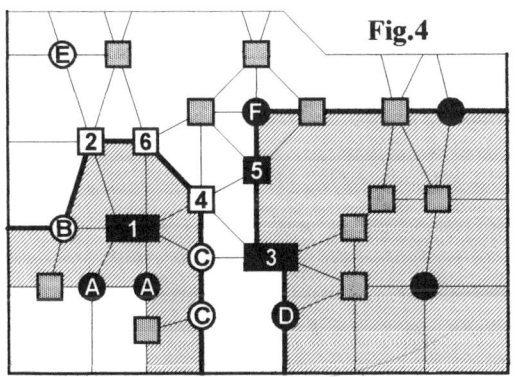

**Fig.4** A possible continuation to the previous exchange is the sequence from ■3 to ☐6 which would permit Black to develop an immense potential territory in the East-southern quarter part of the checkerboard. As a compensation, White would connect its very unstable group Ⓒ in the Southern corner to the stable group Ⓑ ☐2 Ⓔ in the Western corner; this would result in effectively stabilizing the whole body. In addition, White could surround the black group in the south-western corner and put it in an awkward position, though not in a desperate one. This sequence would be particularly profitable to Black and unacceptable to White.

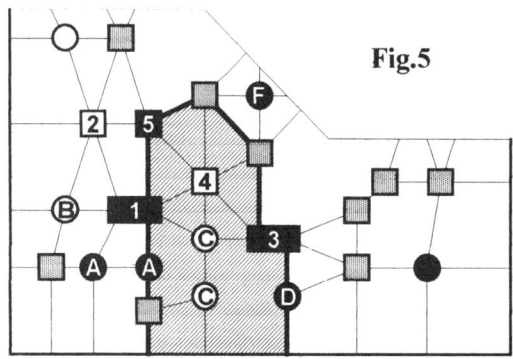

**Fig.5** A variant of the previous sequence consists in preventing White from connecting its two groups by playing Black ■5, which results in leaving the white group Ⓒ ☐4 to drift over the board with a very unstable status from a strategical point of view. Again, this variant would be quite profitable to Black. White should therefore try and find a sequence which would be less unprofitable.

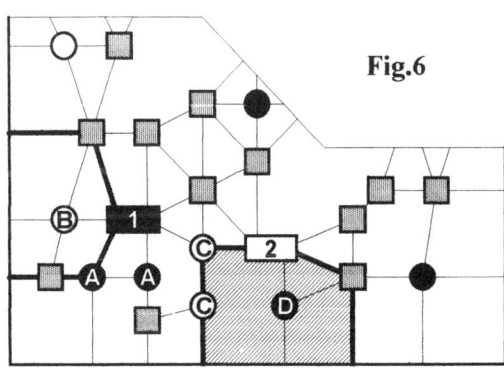

**Fig.6** Thanks to a standard back-track, the tree search resumes its activity after Black ■1 with the new move made by White ☐2. Because such a sequence could surround black group Ⓓ in one move, this move is a strong threat.

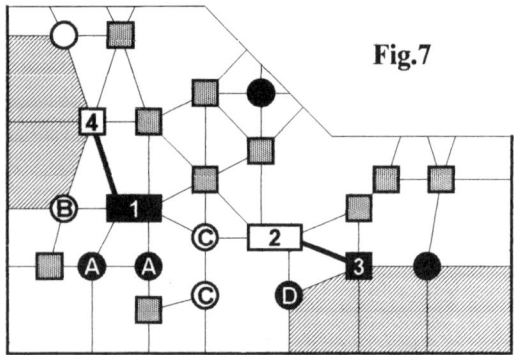

**Fig.7** The sequence which is less penalizing for White is the following: after **1**, White postpones the response to the threat we have just described by threatening of surround Black **2**. Black responds to the latter threat with **3**; then White can responds to the former threat of Black by playing **4**. Comparative to the previous variants, White managed a less disastrous exchange in so far as it is now difficult to surround group Ⓒ which starts to create the immense territory that could have been built by Black. After a similar analytic search on the black group in the North-west corner (which is also problematic), GOBELIN chooses to play move **1** of the previous sequence.

## 5  Validity of the abstract solution

The solution we found with respect to the Abstract-Go is not always valid in the Real-Go, and a stage of verification of the validity and feasibility of the abstract sequence is necessary (see above). The abstract description entails a loss of information which results in a loss of accuracy in the evaluation of game positions which are generated at the abstract level. Consequently, the longer the abstract solution is (in terms number of operators), the less accurate it is and the less probable it is that it shall be converted into a real solution. The tactical validation can be non-existent for some moves which are itemized as always valid. This is the case for sliding in certain situation. In the opposite case, a geographically limited tree, which is guided by a goal, is run on the real representation so as to test the possibility or impossibility of fulfilling the goal with respect to the particular context of the position.

## 5.1  The limits of the Approach with regard to the domain

The abstraction which is implemented to model the strategy operated in the game of Go is not a faithful representation of human players' reasonings. Indeed, players often use tactical sequences in the midst of their strategic reflections to immediately verify the feasibility of their reasoning and modify it in case of impossibility or major obstacles. GOBELIN should be able to **tactically zoom** in in its strategic research when a tactical opportunity occurs or when it is not sure of its strategic sequence.

Similarly, because the opponent has to respond to tactical threats by an easily predictable sequence of moves, giving a particular advantage to the player, every simple tactical threat has to be taken into consideration at the strategic level. In the game of Go, the Shibori (wringing out) is a classical example of such a situation. As we know, players can shift to a tactical exploration during the phase of strategic reflection. This fact allows them to develop an adaptable game which fails Go programs nowadays.

## 6  Conclusion

Strategic modules of present Go programs yields bad results, which strongly penalizes the general level of the game. The strategic dimension of Go is not at all mastered in these modules, and yet it is an essential part in the human player's reflection. We have used an applied abstraction to model the fundamental strategic mechanisms of the game of Go. Contrary to most of the domains which are dealt with on publications about abstraction, our abstraction deals with a non-logical domain. The mechanisms that make up abstraction ( a) the translation of a real description into an abstract description, b) the research of an abstract solution, c) the translation of the abstraction solution into a real solution) are implemented in the GOBELIN system.

GOBELIN is the end-result of a thesis which was carried out within the Metaknowledge team and under the direction of Jacques Pitrat at the Pierre et Marie Curie University (Paris 6). GOBELIN has been evaluated upon a basis of tests carried out at the beginning of a match. It yields responses the quality of which is assessed at about 9th kyu. Considering the modest, encouraging results of our approach, we hope to be able to improve our modelling by creating a new one which would take into account the interaction between strategy and tactics that is used by human players.

It is clear that human strategy is made up of more than the operators and knowledge that were introduced in our modelling, but we think that the strategic reactions we have obtained are encouraging. In addition, the concepts that were outlined are comparable to the concepts of elementary strategy that are used by the average player in game commentaries.

# References

## Abstraction

[Allis et al, 96] L.V. ALLIS, H.J. VAN DER HERIK, M.P.H. HUNTJENS. Go-Moku Solved by New Search Techniques. In Computational Intelligence, vol.12, num.1, 1996, pages 7-23.

[Culberson & Schaeffer, 96] Joseph C. CULBERSON, Jonathan SCHAEFFER. Searching with Pattern Databases. In Lecture Notes In Artificial Intelligence, num.1081, 1996, pages 402-416.

[Prieditis, 93] Armand E. PRIEDITIS. Machine Discovery of Effective Admissible Heuristics. In Machine Learning vol.12, num.1/2/3, August 1993, pages 117-141.

[Giunchiglia & Walsh, 92] F. GIUNCHIGLIA, T. WALSH. A Theory of Abstraction. In AI 57 1992, pages 323-389.

[Holte et al, 96] R.C. HOLTE, T. MKADMI, R.M. ZIMMER, A.J. MACDONALD. Speeding Up Problem Solving by Abstaction : a Graph Oriented Approach. In AI 85 1996, pages 321-361.

[Imielinski, 87] Tomasz IMIELINSKI. Domain Abstraction and Limited Reasoning. In IJCAI 87, Vol.2, p. 997-1003.

[Knoblock et al, 91] KNOBLOCK, TENENBERG, YANG Characterizing Abstraction Hierarchies for Planning. In AAAI 91, vol.2.

[Lowry, 87] Michael R. LOWRY. The Abstraction / Implementation Model of Problem Reformulation. In IJCAI 87, vol.2, pages 1004-1010.

[Pitrat, 90] Jacques PITRAT. Métaconnaissance. Futur de l'intelligence artificielle. Hermès ed., Paris, 1990.

[Sacerdoti, 74] Earl D. SACERDOTI. Planning in a Hierarchy of Abstraction Spaces. In AI, vol. 5, 1974.

[Unruh & Rosenbloom, 89] Amy UNRUH, Paul S. ROSENBLOOM. Abstraction in Problem Solving and Learning. Iin Proceedings of the 11' IJCAI, vol. 1, 1989, pages 681-687.

## Game of Go

[Berlekamp & Wolfe, 94] Elwin BERLEKAMP, David WOLFE. Mathematical Go Endgame. Ishi Press ed., 1994.

[Bouzy, 95] Bruno BOUZY. Modélisation Cognitive du Joueur de Go. Thèse de Doctorat Paris 6, Informatique, 95.

[Reitman & Wilcox, 79] Walter REITMAN, Bruce WILCOX. The structure and Performance of the INTERIM-2 Go Program. In proceedings of the IJCAI 6, 1979, pages 711-719

[Ricaud, 95] Patrick RICAUD. GOBELIN : une Approche Pragmatique de l'Abstraction Appliquée à la Modélisation de la Stratégie Elémentaire du Jeu de Go. Thèse de Doctorat Paris 6, Informatique, 1995.

[Ricaud, 96] Patrick RICAUD. Une Approche Pragmatique de l'Abstraction - Application à la Modélisation de la Stratégie Elémentaire du Jeu de Go. In proceedinds of RFIA'96, Rennes (France), 1996.

[Victorri, 93] Bernard VICTORRI. Eléments d'une théorie géométrique du jeu de Go. Second Cannes/Sophia-Antipolis Go Reseach Day, 1993.

[Wilcox, 85] Bruce WILCOX. Reflections on Building Two Go Programs. In CIGART Newsletter, ACM, n°94, Oct.85, pages 29-43.

# An Evolutionary Algorithm Extended by Ecological Analogy and its Application to the Game of Go

**Takuya Kojima**  **Kazuhiro Ueda**  **Saburo Nagano**

College of Arts and Sciences
The University of Tokyo
3-8-1 Komaba, Meguro-ku, Tokyo, 153, JAPAN
kojima@taikan.c.u-tokyo.ac.jp

## Abstract

The following two important features of human experts' knowledge are not realized by most evolutionary algorithms: one is that it is various and the other is that the amount of knowledge, including infrequently used knowledge, is large. To imitate these features, we introduce an activation value for individuals and a new variation-making operator, splitting, both of which are inspired by ecological systems. This algorithm is applied to the game of Go and a large amount of knowledge evaluated as appropriate by a human expert is acquired. Various kinds of Go knowledge may be acquired such as patterns, sequences of moves, and Go maxims, part of which has already been realized.

## 1 Introduction

The knowledge of human experts has two important features. One is that it is full of *variety*; humans can have many types of knowledge, such as verbal or visual, and there is little limitation in size and shape. The other is that it can be a *large amount*; it is estimated that a chess master has between 10,000 and 100,000 chunks of knowledge [Simon and Gilmartin, 1973]. As a consequence, experts also have knowledge which is *infrequently used*.

In contrast, knowledge acquired by most evolutionary algorithms, such as Genetic Algorithm (GA), has two features. One is that it is *inflexible*; knowledge representation is uniform such that the length of all strings is unique. The other is that in most cases acquired knowledge is limited to one kind; all rules tend to be similar [Goldberg, 1989] in most algorithms. Moreover, algorithms can only acquire knowledge which is frequently used. These two characteristics of algorithms are far from those of human experts.

Among evolutionary algorithms, Genetic Programming (GP) overcomes the first weakness; it can use programs as representation, which are far flexible representation than that used by other major evolutionary algorithms. However, GP does not overcome the second weakness; all the individuals tend to be similar.

We choose ecological systems as a new source of ideas. In ecological systems, many species coexist in environment. We introduce an activation value for individuals and a new variation-making operator, splitting, in order that different individuals coexist.

This research aims to acquire a large number of flexible rules, including useful but infrequently used ones. This algorithm may help us to build a learning model in an evolutionary way which learns as people do.

This algorithm is applied to the game of Go, which has recently attracted considerable attention from AI researchers as the target next to chess [Kitano and others, 1993; McCarthy, 1990]. Games have been used as a testbed for AI algorithms because they are well-defined and easy to use to test the algorithms. Almost all the games studied so far have been so simple that the game playing systems are on an expert level only by using the search-intensive approach. Go is the only exception. Since Go has a much larger search space than chess, taking the search-intensive approach used for chess is not enough to make Go systems play as well as human experts. Thus, another approach in AI, the knowledge-intensive approach, should be considered. Go playing systems of this approach have a large amount of knowledge and search a small space resembling human experts [Saito, 1996]. Thus, Go is considered to be an appropriate testbed for the knowledge-intensive approach.

The knowledge-intensive approach is difficult to take because of the difficulty of acquiring a huge amount of knowledge. This algorithm aims to acquire large numbers of rules, making it possible to take the knowledge-intensive approach.

## 2 Explanation of this Algorithm

Since it is inspired by ecological systems, this algorithm is explained using ecological terms.

### 2.1 Overview of the Algorithm

This algorithm acquires a large number of rules in the form of production rules. Each rule is considered as an *individual* and has an *activation value*. There are no rules in the initial state. A training datum is considered as *food* which is eaten by any rule that matches it and the activation value of the rule increases. If there are no rules which match a given datum, a new rule is created which matches the datum. Thus, the number of rules increases at an early stage. Rules whose activation value is over a certain threshold *split* into the original rule and a more specific rule. Every rule eats food during each step and decreases one activation value. Rules whose activation value is 0 *die*. The procedure of this algorithm is shown below.

**Algorithm 1** Outline of this Algorithm
```
1   step ← 1
    while step ≤ laststep
2       choose a datum from a training set
3       if no rule matches the datum
            then make a new rule
            else feed matched rules
        for all rules
4           if activation of a rule > {threshold}
                then split the rule
5           activation of a rule ← activation of a rule − 1
6           if activation of a rule = 0 then the rule dies
        end
7       step ← step + 1
    end
```

## 2.2 Details of the Algorithm

Rules take the form of production rules. They are Horn clauses; the condition part of each rule consists of a set of clauses and contains only $\wedge$ as a logical connective.

**Matching and Feeding Rules**

Matched rules are those whose condition part and action part match a given training datum. Those rules that do not have more specific rules than themselves are fed. An activation value equal to Parameter $FOOD$ is shared among the rules fed by one training datum. The following is an example.

Suppose that the following five rules are matched.

1. **IF** $C_1$ **THEN** $A_1$
2. **IF** $C_1 \wedge C_2$ **THEN** $A_1$
3. **IF** $C_2 \wedge C_3$ **THEN** $A_1$
4. **IF** $C_4 \wedge C_5$ **THEN** $A_1$
5. **IF** $C_2 \wedge C_3 \wedge C_4$ **THEN** $A_1$

Since Rules 1 and 3 have more specific rules than themselves (Rules 2 and 5), they are not fed. The others, Rules 2, 4 and 5, do not have more specific rules than themselves, so they share food and each gets one third of the food.

In summary, the more general the rules are, the more frequently they are matched, but the lower the probability is that they are fed when they are matched.

**Splitting and Making Rules**

A rule whose activation value is over a certain threshold splits into two rules; one is the original rule and the other is created by adding a new predicate to the condition part of the original (parent) rule. For example, a pattern, **IF** $C_1 \wedge C_2$ **THEN** $A_1$, splits into the original rule and a new pattern, **IF** $C_1 \wedge C_2 \wedge C_3$ **THEN** $A_1$. The new predicate, $C_3$, is randomly chosen as long as the newly created rule matches the given datum.

If no rule is matched, a new rule is created which has only one predicate chosen randomly in the condition part as long as the newly created rule matches the given datum.

When a new rule is created, its activation value is decided by a parameter, $INI\_ACT$. When a rule is split, the activation value of the newly created rule is determined by $INI\_ACT$ and is subtracted from the activation of the parent rule.

## 2.3 General Description of the System Behavior

The following is the general description of the behavior of this system. In the initial state, this system has no rules. A training datum is given and if no rules match the given datum, a new rule having only one predicate in the condition part is created, thus many rules with only one condition are created in the early stages. A rule whose activation value is over a certain threshold splits into the original rule and a more specific rule created by adding a new predicate to the condition, so rules that often match data split frequently and more specific rules are created. The frequency of matching of a newly created rule is always less or equal to that of its parent rule, because the condition of the new rule is always more specific than that of its parent. Rules split and evolve into more specific rules, as long as the frequency of being fed is higher than a certain threshold. Therefore, in the stable state, all rules are expected to match data with almost the same frequency.

The number of rules is expected to be almost equal to the amount of $FOOD$. Let $N$ be the number of rules. For each step, every rule loses one unit from its activation value, so the total loss in activation value is $N$; on the other hand, the total gain in activation value is $FOOD$.

# 3 Application to the Game of Go

We chose the game of Go as a testbed for the algorithm, because Go has a much larger search space than chess and knowledge is therefore indispensable for computer systems to play Go. Moreover, recent cognitive studies on Go players [Saito and Yoshikawa, 1996] have revealed that human experts use Go knowledge such as "patterns" and also search a game tree a little; this shows the importance of Go knowledge to human players. Therefore, studying acquisition of Go knowledge may help Go playing systems to play better.

Game knowledge can be classified into two categories: *strict knowledge* and *heuristic knowledge*. The former is the knowledge that can always be applied and the validity can easily be shown, such as the way to capture stones. The latter is the knowledge whose validity depends on situations, such as *tesuji*, a heuristic way of attack or defense. In Go, a search space is so huge and moves are so sensitive to situations that most of the knowledge used by human players is the heuristic knowledge.

In our previous studies [Kojima et al., 1994; Kojima, 1995], we built a system that deductively acquires Go patterns. Although the system acquires reliable patterns, it can only acquire strict rules, not heuristic ones.

The two big problems in the present systems that acquire heuristic knowledge are that the acquired knowledge is localized and small, and that the knowledge representation is fixed. For example, one system [Sei and Kawashima, 1994] collects from master's game records patterns, which are fixed diamond shape and whose area is within five Manhattan distances of the center. Our algorithm makes it possible to acquire more flexible knowledge; patterns with any shapes and of unlimited size can be acquired. Thus, patterns acquired by our algorithm are more similar to those which human players have than those acquired by the previous systems.

### 3.1 Details of Application to Go

Go is a game between two players, who take turns putting black and white stones on a board. The standard board size is 19x19; sometimes a smaller board, such as 13x13 or 9x9, is used usually by beginners.

Training examples are game records between professional players. Two kinds of training sets are used: one is 19x19 board games (total 600 games) and the other is 9x9 board games (total 200 games). Training sets are chosen as follows. One game is randomly chosen and all the moves of the game are taken as training data[1] from the first move to the last move. After all the moves are used as training data, another game is randomly chosen. This process is repeated. Each move is one time step in this algorithm.

There are many kinds of Go knowledge, such as patterns, sequences of moves, and maxims. A pattern is a rule whose condition part is a part of board configuration and whose action part is a move. A sequence of moves is a rule that suggests several moves. A maxim is a rule consisting of Go terms. This paper describes the acquisition of patterns, so the rules mentioned in the most of the paper are Go patterns.

Rules are described in relative terms so they can be executed by either players. A predicate of the condition part consists of objects and their coordinates (*coordinate=object*); the action point is the point to which a stone is placed and has the coordinates [0,0]; other coordinates are relative to the action point, and objects are either stones or edges of the board. Stones are either "SAME" (i.e. belonging to the active player) or "DIFF" (i.e. belonging to the opponent). The action part of a rule is always "place a stone to [0,0]", so it is omitted in the following examples.

For example, "**If** [2,1]=SAME ∧ [0,1]=DIFF ∧ [5,0]=EDGE" means that **if** a *same* color stone as the active player's is in the point [2,1] (relative to the action point), and if a *different* color stone exists in point [0,1] and *edge* exists in point [5,0] **then** put a stone on [0,0].

When rules split or are newly created, a new condition about stones or edges of the board is chosen.

### 3.2 Parameters

This algorithm has two parameters: one is a constant, $INI\_ACT$, which is the initial activation value for newly created rules, and the other is $FOOD$, a function of the number of games used as training data ($= i$), which is the total amount of activation given by one training datum.

In the experiment with 9x9 board games, $FOOD$ is defined as follows: if $i < 4000$, $FOOD = 400$, and if $i \geq 4000$, $FOOD = i \div 10$, and $INI\_ACT$ is 200. In the experiment with 19x19 games, $FOOD$ is defined as follows: if $i < 4000$, $FOOD = 2000$, and if $i \geq 4000$, $FOOD = i \div 2$, and $INI\_ACT$ is 1000.

The threshold of splitting is twice the value of $FOOD$.

Parameter $INI\_ACT$ is smaller than $FOOD$. A newly created rule is an assumption, thus it is often inappropriate and consequently dies, because a new condition is randomly chosen. Therefore, the activation value given to a newly created rule is smaller than that given to an existing rule.

The number of iterations is counted by the number of games used. It is 50,000 for 9x9 board games and 26,500 for 19x19 board games.

### 3.3 Experimental Results

**The Number of Acquired Rules and the Probability of Being Fed**

In the 9x9 board games, 4,684 rules were acquired, while in the 19x19 board games, 12,850 rules were acquired.[2] This shows that our algorithm, unlike the general evolutionary algorithms, enables a large number of rules to be acquired.

Rules that are fed over 100 times were selected for evaluation, because rules that are just created may be inappropriate and die after only a few steps. As a result, 3,211 rules were selected for 9x9 board games and 7,587 rules for 19x19 board games.

The average probability that the selected rules are fed[3] is calculated. The results were 1.58% for 9x9 board games and 0.188% for 19x19 board games, demonstrating that infrequently used rules, which are difficult for ordinary evolutionary algorithms to acquire, can be acquired by this algorithm.

**Quality of Acquired Rules**

Acquired rules from 9x9 board games were evaluated by an expert[4]. It is difficult to evaluate the patterns stored by previous studies, because they have a fixed shape and have too many indifferent stones to be called rules. The

---

[1] All the moves are used, because almost all the moves of professional players are taken as good moves for amateur players and only less than one percent of moves are bad.

[2] In this simulation, newly created rules that are the same as a rule already in the rule set is deleted. Thus, there is no duplication of rules.

[3] The average of the probability that each rule is fed during the learning process.

[4] 5 *dan* amateur player

rules acquired by our algorithm are sufficient for evaluation by human experts.

Of the 3,211 selected rules, the top 5% of the 161 rules (the order being determined by the number of being fed) were evaluated. Moves indicated by the rules were evaluated and categorized as good, average, or bad. Of the 161 rules, 67 (41.6%) were evaluated as good, 34 (21.1%) as average, and 60 (37.3%) as bad.

This result shows that about two thirds of the rules acquired by this algorithm are acceptable, showing that the algorithm works effectively, even without any domain specific heuristics.

**Automatic Context Separation**

This algorithm automatically acquires specialized rules, which are used only in a certain stage or context. Most evolutionary algorithms can acquire only general rules and require humans to classify training data into contexts in order to acquire specialized rules. Our algorithm, however, does not require the classification by humans and automatically separates situations or contexts. For example, Figure 1 shows a rule used only in the opening and Figure 2 a rule used only in the end game.

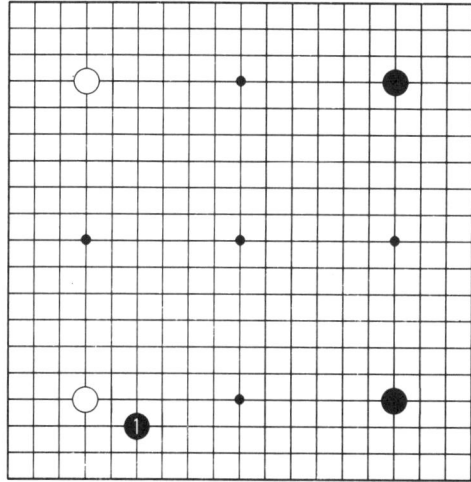

Figure 1: An example of a rule used in the opening.

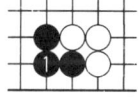

Figure 2: An example of a rule used in the end game.

**The Variety of Acquired Rules**

In previous studies, only small and fixed-shaped rules are acquired, whereas this algorithm permits the rules to be any size and shape.

Our algorithm acquires rules with a variety of the sizes, never realized in previous studies. Figure 1 is an example of a large (whole-board) rule. Figure 3 is an example of a small rule.

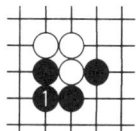 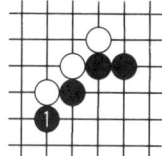

Figure 3: An example of a small rule.

Figure 4: An example of an irregular-shaped rule.

The variety in shape of the rules acquired by our algorithm is demonstrated by the following three examples. Figure 1 is an example of the rule with stones spread over a wide area. Figure 3 is an example of a compact, square rule. Figure 4 is an example of an irregularly shaped rule.

Rules that are used in different contexts can be acquired, as shown in "Automatic Context Separation".

The examples show the great variety in size, shape, and context of knowledge acquired by this algorithm.

**Learning Process**

In the early trials, general rules tend to be acquired and gradually specialized rules are acquired. This is because the more trials there are, the more food that is given. While frequently used rules can be acquired when the amount of food is small, less frequently used rules are not acquired until the amount of food is large.

Take Figures 1[5] and 3[6] for example.[7] While Figure 3 can occur in all stages of the game, Figure 1 appears only in the opening. The ages[8] of the rules are 1,139,957 and 335,825 respectively, showing that the more popular the rules are, the older they are.

## 4 Discussion and Related Work

### 4.1 Advantages of Introducing an Activation Value

One of the major differences between our algorithm and other evolutionary algorithms is that our algorithm gives each individual an activation value. Two advantages of introducing the activation value are discussed below.

---

[5] If [10,1]=SAME [10,13]=SAME [-2,1]=DIFF [-2,13]=DIFF [0,-3]=EDGE [-6,0]=EDGE

[6] If [1,0]=SAME [0,1]=SAME [2,1]=SAME [1,1]=DIFF [0,2]=DIFF [1,2]=DIFF

[7] Since both rules have six conditions, the probability that they are acquired is the same.

[8] the time steps from the birth of the rule

**Avoidance of Convergence**

In almost all evolutionary algorithms, individuals tend to become similar and converge at one point or in one set of a small number of rules. This convergence is avoided by introducing activation value to individuals. The reason is as follows.

Suppose many similar rules are created and they tend to match the same data. As explained before, among matched rules, those that do not have more specific rules than themselves are fed, so those that do have more specific rules matching the same data are not fed and die. Some individuals that are fed also die if there are too many similar rules being fed at the same time, because the food is shared among them. As a consequence, the appropriate number of individuals for the given food survives and rules acquired will not be similar as long as there is variety in the training data.

**Acquisition of Rules Infrequently Used and Specialized**

Ordinary evolutionary algorithms acquire the rules that are used at almost every time step, not those that are infrequently used, because all individuals are evaluated at every time step. By introducing the activation value, such rules are acquired if they are fed before activation value becomes zero. This is demonstrated by the average probabilities that rules are fed being 1.58% for 9x9 board games and 0.188% for 19x19 board games.

Automatic situation separation is also the consequence of introducing activation value. Opening rules, such as Figure 1, will appear only in the opening and not in other stages. Without the activation value such specialized rules would not be acquired.

### 4.2 Variation-making Operator

Genetic Algorithm (GA) and Genetic Programming (GP) adopt crossover as a variation-making operator. Crossover is a good operator for spreading good building blocks (*schema*); once good schema are found, they are spread rapidly into the population by the crossover operator, and individuals in the population tend to be similar. Therefore, it is quite useful for optimizing functions.

Instead, our algorithm adopts splitting, which is a new operator. Unlike crossover, this operator does not refer to other individuals in making a variation, and each individual makes independently new rules by adding one condition to itself. As a result, the probability of making a novel rule is expected to be higher than with crossover.

While it is said that crossover tends to make individuals similar to each other, splitting is expected to make individuals different. Since our purpose is to acquire various rules, we adopted splitting.

### 4.3 Comparison with Other Evolutionary Algorithms

Since our algorithm is a kind of an evolutionary algorithm, we compare it with other evolutionary algorithms in this section.

**Genetic Programming**

Among major evolutionary algorithms, only Genetic Programming (GP) [Koza, 1992] admits flexible knowledge representation. It admits more flexible knowledge representation than our algorithm; while our algorithm admits only Horn clauses, GP admits any type of programs in describing rules.

On the other hand, while GP acquires small sets of rules, our algorithm can acquire large numbers of rules and some infrequently used rules.

Therefore, our algorithm is appropriate for acquiring a large number of rules that can be described by Horn clauses, while GP is appropriate for acquiring a small number of optimal rules that are too complex to be described by Horn clauses.

**Classifier Systems**

Classifier Systems (CS) [Holland *et al.*, 1986] use many classifiers to solve problems, as does our algorithm. However, CS do not acquire the thousands of rules, which our algorithm does. Moreover, the probability that rules are matched should be high and rules which are infrequently used are not acquired by CS, whereas they are with our algorithm.

CS need a uniform format (i.e. the same length of strings) because they use crossover. Our algorithm adopts splitting and does not require the same length of strings, and the acquired rules are varied in the number of predicates in the condition part.

### 4.4 Comparison with Humans

The learning process of this algorithm can be compared with that of humans.

In the beginning of this algorithm, there are no rules and the simplest forms of rules are created. Gradually they evolve from simple rules to more complex ones. This is similar to the human learning process. Human beginners also learn simple rules first and then gradually learn exceptions and more complicated rules.

Food might be compared with the memory capacity of humans. While novices have much less memory capacity, their capacity gradually increases; experts have much larger capacity than novices[Chase and Simon, 1973]. In this algorithm, the amount of food given for one training datum is determined by the *FOOD* parameter. As mentioned in Section 3.2, *FOOD* increases as the iteration continues. This is similar to the process of human learning in expanding memory capacity.

### 4.5 Improving the Quality of Acquired Rules

The human expert's evaluation of the acquired rules was shown above. Most rules evaluated as average or bad lacked some conditions, or stones. The complexity of the rules depends on the number of iterations; the more the algorithm iterates, the more specific the rules are. The number of the rules acquired depends on the parameter *FOOD*. Therefore, changing the parameter and the number of iterations may improve the quality of the acquired rules. We are currently studying this assumption.

## 4.6 Future Work: Acquisition of other kinds of knowledge in Go

This paper focuses only on the acquisition of patterns. Since this algorithm is flexible, other kinds of rules may be acquired.

Human players are said to have not only patterns but also sequences of moves, which they consider important [Saito, 1996]. Since the condition part of this algorithm can take any predicates, indexes of previous moves (for example, two moves before) can be a part of the condition. Thus, sequences of moves have already been acquired, although they are not mentioned in this paper.

Furthermore, once Go terms are defined, Go maxims using these Go terms might be acquired. Since rules of this algorithm are Horn clauses, predicates of rules can be Go terms. In splitting, a new predicate which is a Go term can be added to the rule condition to make the rule more complex. In this way, many rules using Go terms, which we call *maxims*, may be acquired by this algorithm. Acquiring Go maxims, which human experts implicitly have, may help both computers and human novices to understand Go.

The the weakest point in present Go playing systems is said to be that they do not behave differently according to the situations they face, but it is very difficult to write rules for so many situations. Once situation terms in Go are defined, Go patterns or rules may be acquired for the situations using the terms, as with the Go maxims. This may enable computers to play a much stronger game of Go.

## 5 Conclusions

A new evolutionary algorithm inspired by ecological systems is introduced. Although previous evolutionary algorithms acquire only a few kinds of rules, this algorithm acquires a large number of them. It can acquire rules used as infrequently as 1% of the time, which have never been acquired by other evolutionary algorithms. It acquires Horn clauses, which are flexible compared to the fixed knowledge representation of most evolutionary algorithms.

The algorithm was applied to the game of Go. A large number of appropriate rules were acquired; this may enable Go playing systems to take knowledge-intensive approach. Automatic context separation was also observed.

This algorithm should enable systems to acquire sequences of moves and Go maxims, which are said to be important for human players, because it permits flexible knowledge representation. So this study may enable Go playing systems to be much stronger.

It is shown that the learning process of this algorithm is similar to that of humans.

This algorithm enables a large number of rules to be acquired in an evolutionary way and introduces the possibility of integrating evolutionary algorithms and knowledge engineering.

## Acknowledgements

The authors would like to thank Masahiro Okasaki for useful information, and Prof. Takao Terano, Masahiro Hachimori, and Kiyoshi Izumi for their helpful comments.

## References

[Chase and Simon, 1973] William G. Chase and Herbert A. Simon. Perception in chess. *Cognitive Psychology*, 4:55–81, 1973.

[Goldberg, 1989] David E. Goldberg. *Genetic Algorithms in Search, Optimization, and Machine Learning*. Addison-Wesley, 1989.

[Holland et al., 1986] J. H. Holland, K. J. Holyoak, R. E. Nisbett, and P. R. Thagard. *Induction: Processes of Inference, Learning, and Discovery*. MIT Press, 1986.

[Kitano and others, 1993] Hiroaki Kitano et al. Grand challenge AI applications. In *Proceedings of the Thirteenth International Joint Conference on Artificial Intelligence*, pages 1677–1683. Morgan Kaufmann, 1993.

[Kojima et al., 1994] Takuya Kojima, Kazuhiro Ueda, and Saburo Nagano. A case study on acquisition and refinement of deductive rules based on EBG in an adversary game: how to capture stones in Go. In *Game Programming Workshop in Japan '94*, pages 34–43. 1994.

[Kojima, 1995] Takuya Kojima. A model of acquisition and refinement of deductive rules in the game of Go. Master's thesis, The University of Tokyo, 1995.

[Koza, 1992] John R. Koza. *Genetic Programming*. MIT Press, 1992.

[McCarthy, 1990] J. McCarthy. Chess as the Drosophila of AI. In T. Anthony Marsland and Jonathan Schaeffer, editors, *Computers, Chess, and Cognition*, chapter 14, pages 227–237. Springer-Verlag, 1990.

[Saito and Yoshikawa, 1996] Yasuki Saito and Atsushi Yoshikawa. An analysis of strong go-players' protocols. In *Proceedings of Game Programming Workshop in Japan '96*, pages 66–75. 1996.

[Saito, 1996] Yasuki Saito. *Cognitive Scientific Study of Go*. PhD thesis, The University of Tokyo, 1996. in Japanese.

[Sei and Kawashima, 1994] Shinichi Sei and Toshiaki Kawashima. The experiment of creating move from "local pattern" knowledge in Go program. In *Game Programming Workshop in Japan '94*, pages 97–104, 1994. in Japanese.

[Simon and Gilmartin, 1973] Herbert A. Simon and Kevin Gilmartin. A simulation of memory for chess positions. *Cognitive Psychology*, 5:29–46, 1973.

# GAME PLAYING

## Game Playing 2

# Search Versus Knowledge in Game-Playing Programs Revisited*

**Andreas Junghanns, Jonathan Schaeffer**
University of Alberta
Dept. of Computing Science
Edmonton, Alberta
CANADA T6G 2H1
Email: {andreas, jonathan}@cs.ualberta.ca

## Abstract

Perfect knowledge about a domain renders search unnecessary and, likewise, exhaustive search obviates heuristic knowledge. In practise, a tradeoff is found somewhere in the middle, since neither extreme is feasible for interesting domains.

During the last two decades, the focus for increasing the performance of two-player game-playing programs has been on enhanced search, usually by faster hardware and/or more efficient algorithms. This paper revisits the issue of the relative advantages of improved search and knowledge. It introduces a revised search-knowledge tradeoff graph that is supported by experimental evidence for three different games (chess, Othello and checkers) using a new metric: the "noisy oracle".

Previously published results in chess seem to contradict our model, postulating a linear increase in program strength with increasing search depth. We show that these results are misleading, and are due to properties of chess and chess-playing programs, not to the search-knowledge tradeoff.

## 1 Introduction

Many experiments have been performed in game-playing programs that measure the benefits of improved knowledge and/or deeper search. In particular, chess has been a popular application for these experiments. The explicit or implicit message of these works is that the results for chess are generalizable to other games. There have been few studies that examined the impact of improved knowledge on program performance [Schaeffer and Marsland, 1985; Mysliwietz, 1994]. In contrast, the benefits of additional search are well documented: deeper search provides immediate performance gains (for example, [Thompson, 1982]).

*This research was supported by the German Academic Exchange Service (DAAD), the Natural Sciences and Engineering Research Council of Canada (NSERC) and the Killam Foundation.

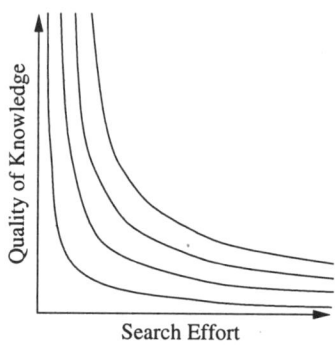

Figure 1: Proposed Search-Knowledge Relationship

Figure 1 has been hypothesized to represent the relationship between the quality of knowledge and search effort expended (first expressed in [Michie, 1977] and later refined in [Berliner *et al.*, 1990]). The curves represent various combinations of search and knowledge with equivalent performance. The figure illustrates that by increasing the search effort, less knowledge is required by the application to achieve the same level of performance, and vice versa.

Of the two dimensions, improvements in search are the easiest to address. Gains can often be achieved with little effort. One can redesign algorithms or rewrite code to execute faster or, even better, do nothing and just wait for a faster computer to become available.

The knowledge dimension, however, is nebulous. Whereas there are well-defined metrics for measuring search effort (such as search depth, execution time, and nodes examined), there is nothing comparable for knowledge.

This paper makes a number of contributions to our understanding of the relationship between search and knowledge in game-playing programs:

- Figure 1 is a hypothesis and has not been verified. In fact, it turns out to be misleading. Analytical and experimental data (from three game-playing programs: chess, Othello, and checkers) allows us to construct a new view of the search-knowledge tradeoff. This is shown in Section 2.

- To do the experiments, we needed a way of assessing

the quality of a program's knowledge. To do this, we introduce a new metric, the *noisy oracle*. Section 3 presents experimental data that yields new insights into the shapes of the curves.

- Figure 1 predicts decreasing benefits for increasing search effort—diminishing returns. This has been confirmed for both Othello and checkers by others. However, numerous papers suggest that for chess the relationship between search depth and performance is relatively constant and non-decreasing. In Section 4, we demonstrate that diminishing returns do indeed occur in chess, and that the reason for this discrepancy with the literature is rather surprising—the game length in combination with relatively high error rates.

## 2 Search Versus Knowledge: Theory

In this section, we use an idealized definition of knowledge. Knowledge is uniformly applicable throughout the search tree. This allows us to avoid thorny issues such as search pathology [Nau, 1983] and search-depth-dependent anomalies.

Figure 1 shows various performance levels for different combinations of search effort and quality of knowledge (isocurves). This graph has been hypothesized, but never verified. The shape of the isocurves comes largely from two known data points in the graph: no search and perfect knowledge, as well as exhaustive search with no knowledge—both yield a perfect program[1]. However, there is nothing in this data that implies that the isocurves should be concave down, or even that they should be curves at all. In fact, [Michie, 1977] and [Berliner *et al.*, 1990] provide no justification for their shape. However, experience suggests this shape to be likely (unproven). For now, we assume that they are concave down, and we examine this issue in the next section.

What does it mean to do no search and have perfect knowledge? In fact, this implies a minimal amount of search (1 ply) to evaluate all the moves and then choose the one leading to the highest outcome. What does it mean to perform exhaustive search (to depth $GL$, the maximum game length) with no knowledge? The "no knowledge" is misleading, because to play perfectly one must have some knowledge—in this case being able to identify and correctly back up the scores for wins, losses, and draws. This suggests a scale for these two data points. We let $WLD$ represent the knowledge about correctly backing up terminal nodes in the search. This is knowledge supplied by the application domain rules. The "perfect knowledge" program requires 100% of the domain-specific knowledge required to play flawlessly. The above distinction allows us to plot two data points on the $y$-axis. We can now

---

[1]Perfect is meant to imply that the program never makes a game-theoretic-value error. We do not consider the case where the program is also required to play the move that maximizes the chances for improving its expected outcome.

plot the isocurve for perfect programs, a concave down curve between the points $(1, +100)$ and $(GL, WLD)$.

Although we are interested in programs that have high performance, we can also consider the case where the quality of a program's knowledge is worse than $WLD$. The worst-case scenario is a program whose knowledge is $-100$: the program assesses positions inversely proportional to their worth. Thus, this *anti-perfect* program with a one-ply search will always choose the worst move on the board.

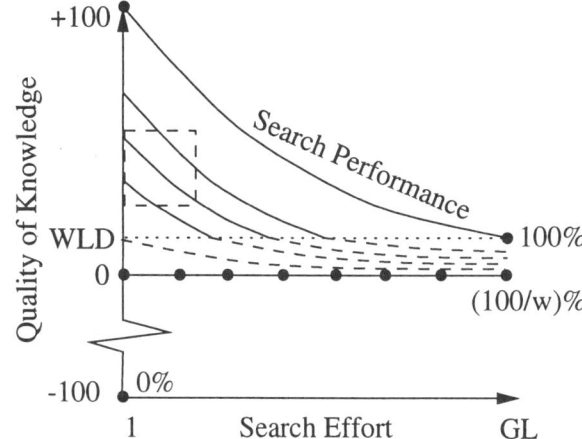

Figure 2: Search Versus Knowledge Revisited

One other data point of interest is the 0 knowledge program. Since the the program has no knowledge, its move choices are random. Given an average branching factor of $w$ move choices in a position, the program will make the right move $1/w$ of the time ($w = 40$ for chess, while only 8 in non-capture checkers positions). Here there are no benefits of search; the program will play the correct move $1/w$ of the time regardless of the search depth[2]. This program is clearly better than the anti-perfect program, but must be worse than a $WLD$ program. The latter follows since the $WLD$ program has positive knowledge about wins, losses, and draws, which can only improve the likelihood of selecting the right move.

These arguments allow us to construct a revised version of Figure 1, as shown in Figure 2 (ignore the dashed box for now). The $x$-axis is search depth, starting at 1 and increasing. The $y$-axis is the quality of knowledge, ranging from perfect positive knowledge to perfect negative knowledge. Each isocurve represents a fixed level of performance. If we measure performance by the percentage of correct move decisions made by the program, then the perfect program is 100% correct, and the anti-perfect program scores 0%. The random program alway scores $100/w\%$ (we ignore the minor differences that occur if $w$ varies during the game).

---

[2]We use the simplifying assumption of a uniform branching factor. As [Beal and Smith, 1994] showed, random evaluations can implicitly capture concepts like mobility in non-uniform trees.

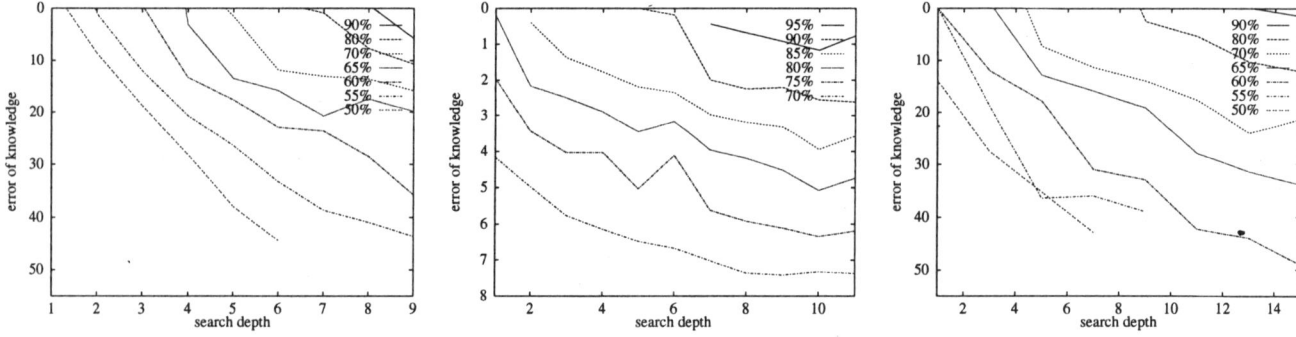

Figure 3: Search Knowledge Behavior in Chess (left), Othello (middle) and Checkers (right)

Note that we have not included any isocurves in the 0 to -100 range. Here the knowledge is worse than random, and one can expect to see search pathology [Nau, 1983]. Since this region is not of interest in practice, for reasons of brevity we ignore it. It is interesting to note that the anti-perfect program, which always makes the worst move with a 1-ply search, may play the right move given a 2-ply or larger search (albeit for the wrong reasons).

Consider the region in Figure 2 that is bounded by the perfect curve (100%) and the random line ($100/w\%$). All curves start at $depth = 1$, but are spaced out over a range from $+100$ to $0$ on the knowledge axis. They end up at $depth = GL$, in the smaller range from $WLD$ to $0$. Therefore the curves move closer together as the search depth increases. In other words, the isocurves do not have the same slope. The lower the performance, the flatter the curve—the extreme being the flat random line. The higher the performance, the steeper the curve—the extreme being the perfect performance isocurve. Hence, as one moves to higher performance levels, the slope of the isocurves increase. This implies that for shallow search depths, more knowledge is required to move to a higher isocurve than for deeper search depths.

## 3 Search Versus Knowledge: Practise

The difficulty in experimentally verifying Figure 2 lies in quantifying the knowledge axis. Perfect knowledge assumes an oracle, which for most games we do not have. However, we can approximate an oracle by using a high-quality, game-playing program that performs deep searches. Although not perfect, it is the best approximation available. Using this, how can we measure the quality of knowledge in the program?

A heuristic evaluation function, as judged by an oracle, can be viewed as a combination of two things: oracle knowledge and noise. The oracle knowledge is beneficial and improves the program's play. The noise, on the other hand, represents the inaccuracies in the program's knowledge. It can be introduced by several things, including knowledge that is missing, over- or under-valued, and/or irrelevant. As the noise level increase, the beneficial contribution of the knowledge is overshadowed.

By definition, an oracle has no noise. We can measure the quality of the heuristic evaluation in a program by the amount of noise that is added into it. To measure this, we add a random number to each leaf node evaluation ($N_L$).

In most games of skill, the value of a parent node is strongly correlated with the values of its children. Hence, our noise model should reflect this. Following the previous work of [Iida et al., 1995], we define the noise of a leaf node in a search to be $N_L = \sum_{i=1}^{d} r_i$, where $-R < r_i < R$, $R$ is an adjustable parameter, and $d$ is the depth in the tree of the leaf node. This simple representation comes closer to approximating the parent/child behavior. The resulting random numbers at the depth $d$ leaf nodes have a normal distribution with mean 0 and a standard deviation of $\sqrt{d * (R^2/3)}$. One should be careful: simulating tree behavior is fraught with pitfalls [Plaat et al., 1996].

The above discussion assumed we have a perfect oracle. For real games such as chess, Othello and checkers, the best we can do is use a high-quality, deep-searching program as our best approximation. In effect, this program is a *noisy oracle* with noise level $N_O$. We can now increase the noise level by increasing the distribution of random scores added to the evaluation ($N_O + N_L > N_O$).

To show the tradeoff between search and knowledge, we conducted experiments with chess, Othello, and checkers. The programs used were *TheTurk* (chess), *Keyano* (Othello) and Chinook (checkers)[3]. All three are well-known internationally. For each game, 256 positions from grandmaster play were selected. The noisy oracle would determine the best move in the position. Since the oracle is noisy, and evaluation functions differentiate positions by insignificant margins, all moves that were within 5 points (1/20th of a pawn/checker) in chess/checkers or 8 disc in Othello were considered as best moves. For each game, each position was searched to a variety of search depths with a variety of noise. The programs were searched with $R = 0, 5, 10, 15, 20, 25, 50, 100, 150$ for chess and checkers, and $R = 0, 1/8, 1/4, 1/2, 1, 2, 4, 8, 16$

---
[3]*TheTurk* is a chess program authored by Andreas Junghanns and Yngvi Bjornsson. *Keyano* is an Othello program developed by Mark Brockington.

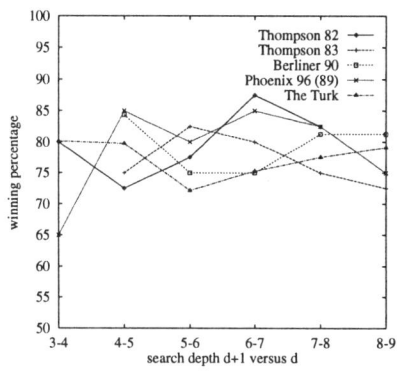 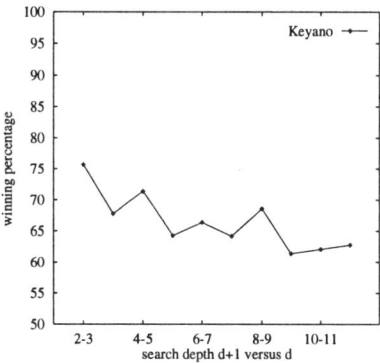 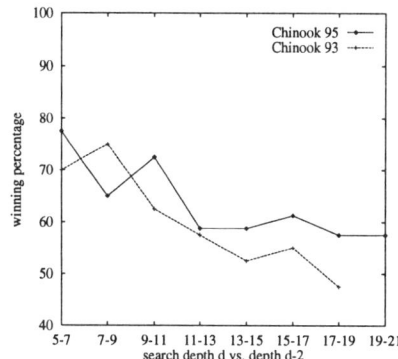

Figure 4: Self-Play Experiments in Chess (left), Othello (middle) and Checkers (right)

for Othello.

Figure 3 shows the results for the three games (only some of the $R$ values are shown). The $x$ axis is the search depth, ranging from 1 to 9-15 depending on the game. The $y$ axis measures the quality of the noisy oracle's knowledge, beginning at $R = 0$. The isocurves represent different levels of performance, where performance is measured as the percentage of times that the program makes the correct move selection in the test set.

All three programs exhibit similar behavior. The isocurves appear to be curved and concave down, although in many cases they are almost linear. The curves are not perfectly formed because of the statistical nature of the experiments. All three games show the curves leveling off, suggesting that for deeper searches, the benefits of additional knowledge (less noise) are more significant than for additional search.

In our experimental setting, we are restricted to a small range of possible values on the $x$ and $y$ axis. From the shape of the curves in Figure 3, we can approximate where this graph fits into the Figure 2 framework (shown by the dashed box).

When comparing the graphs for the different games, the reader should keep in mind that neither the search nor the knowledge axis are comparable, since it is not clear how close we are to perfect knowledge and exhaustive search depth. Although it is well-defined what it means to search an additional ply of search, it is not clear what it means to reduce the noise from, say, 20 to 10. In other words, although the $y$ axis is shown as a linear scale, the effort required to improve the program along this axis may not be linear.

## 4 The Chess Anomaly

The results from Sections 2 and 3 suggest that the benefits of additional search decline as the search depth increases—so-called *diminishing returns*. A number of papers have experimentally addressed this question. Figure 4 graphs some of those results. These graphs are the result of self-play experiments, where a program searching to depth $d$ plays matches against the same program searching to depth $d + \delta$, where $\delta = 1$ for chess and Othello, and $\delta = 2$ for checkers. The idea is that, for example, the winning percentage of a 3-ply program playing against a 2-ply program should be higher than for a 13-ply program playing a 12-ply program. At least in Othello (experiments with *Keyano*, and supported by data in [Lee and Mahajan, 1990]) and checkers [Schaeffer et al., 1993], this seems to be borne out.

However, the results for the game of chess are perplexing because, even though there is a logical argument for stating that the benefits obtained by deeper searching will gradually reduce, the experimental evidence does not substantiate this. Many publications consistently show a linear relationship between search depth and performance (for example, [Newborn, 1979; Thompson, 1982; Condon and Thompson, 1983; Newborn, 1985; Berliner et al., 1990; Mysliwietz, 1994]). Only [Condon and Thompson, 1983] shows a slight decline in performance with increased search depth, however this trend is still within the range of statistical noise. Intuitively, diminishing returns must exist, since eventually exhaustive search solves the problem and additional search effort would be entirely wasted.

Our new experiments with chess show that there are diminishing returns, further confirming the general shape of Figure 2. The reason that these results were not evident in previous work is twofold; one reason having to do with the quality of the program's knowledge, and the other having to do with a characteristic of the game.

**Decision Quality**

Searching to depth $(d + 1)$ pays off only if the deeper search results in a better move choice than is possible with a $d$-ply search. The smaller the probability that this happens, the better the $d$-ply search is a predictor of the $(d + 1)$-ply search. Note that the value of the search is irrelevant; only the move selection influences the game result (even if the right move is played for the wrong reasons).

We conducted an experiment to measure how the move choice changes as a function of search depth (similar to [Newborn, 1985]). One thousand opening positions were searched by a deep (9-ply) version of *TheTurk* (a noisy oracle) to determine the best move and value. Figure 5 shows in the top-most curve the percentage of times that a deeper search changes the move choice. A move change might not be a significant event if the value difference between the moves is small, as judged by the noisy oracle. The additional curves in the figure represent the percentage of *significant* move changes, according to the difference in move values (at least 10, 15, 20, 25, 50, 100 points), where 100 points is the equivalent of a pawn.

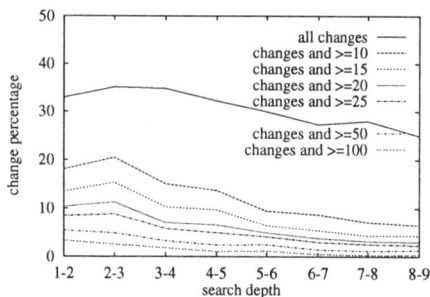

Figure 5: Move Changes from $d$ to $(d+1)$ Ply (Chess)

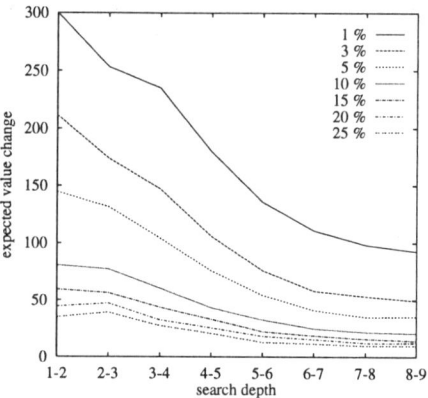

Figure 6: Value Changes from $d$ to $(d+1)$ Ply (Chess)

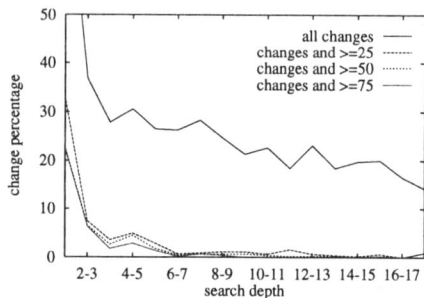

Figure 7: Move Changes from $d$ to $(d+1)$ Ply (Checkers)

The graph shows a reduction in error (or alternatively, an increase in prediction accuracy) with increasing search depth, but the error reduction slows down with deeper searches. Figure 6 shows a different view of the data. Here the change in value in going from $d$ to $d+1$ ply is plotted versus depth. The curves represent the percentage of moves that achieve a certain level of performance. For example, the top curve shows that 1% of the moves result in value changes of roughly 100 points (a pawn) when you search from 8 to 9 ply. The curves show a dramatic decrease in expected error and, again, exhibits a tapering off with deeper searches—an indication of diminishing returns.

The surprising feature of Figure 6 is the magnitude of the errors. In going from 8 to 9 ply, 10% of the moves result in at least a 25-point differential; usually a significant score swing. In other words, the error rates of even an 8-ply search in *TheTurk* are extremely high.

This data can be dramatically put into perspective by comparing it with the results of a similar experiment with Chinook. Chinook is the world's strongest checkers playing entity (man or machine). With its massive endgame databases (444 billion positions), the program is close to being an oracle. Figure 7 shows the percentage of move changes for checkers. The difference is clear: the error rates are much lower, an indication of how much better the evaluation quality of *Chinook* is as compared to *TheTurk*. With such low error rates, searching deeper in *Chinook* yields little benefits. In *TheTurk*, the error rates are still high enough to allow for significant improvements as search depth increases, which in return obscures the effect of diminishing returns in self-play games.

**Game Length**

The above suggests that the decision quality in chess is not as good as one would like (i.e. the noisy oracle is too noisy). Each move played by the $d$-ply program against the $d+1$-ply program is fraught with danger, since the deeper searching program has less probability of making a mistake. This suggests that the longer the games lasts, the greater the winning chances of the $d+1$-ply program.

To test this hypothesis, we conducted two experiments. First, we measured the average length of self-play games played by *TheTurk*. As the search depth of the programs increased, so did the length of the game. In other words, for shallow searches, the games tended to be shorter because the probability of an error was higher. As the search depth increased, the error probability dropped and, hence, the games lasted longer because the opponents were more evenly matched. Games played between 8- and 9-ply programs averaged out to be 29% longer than games between 3- and 4-ply programs.

This suggests that game length has something to do with chess self-play results. To test this hypothesis, we played a series of 80 self-play games where the game length was restricted. After a specified number of moves, the game was adjudicated.

Figure 8 shows a constant winning percentage for games of unrestricted length (top line) that would lead to the conclusion that diminishing returns do not exist in chess. How-

696  GAME PLAYING

ever, if we restrict the length of the games (to 10 through 45 moves), a decline in the winning percentage is visible, leading to the conclusion that diminishing returns exist in chess for truncated games.

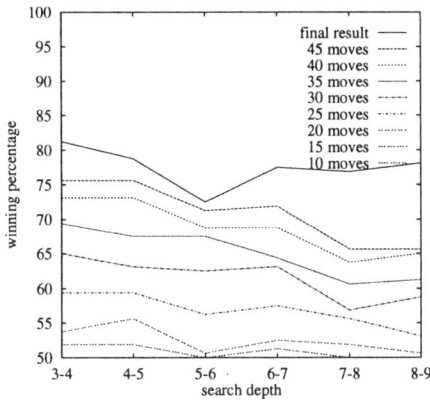

Figure 8: Winning % for Truncated Games (Chess)

Both Othello and checkers games are limited in the number of moves. Othello games are constrained to a maximum of 30 moves aside. Checkers, with its forced capture rule, tends to have similarly short games. In contrast, chess has no such limitations. Given that the $d+1$-ply searching program has an advantage over the $d$-ply searcher, the longer the game, the greater the likelihood that the advantage will manifest itself. Essentially, self-play experiments in chess suffer from the *gambler's ruin*, the reason why diminishing returns have remained hidden for almost 20 years.

## 5 Conclusion and Future Work

A new graph for the search-knowledge tradeoff was proposed and experimentally verified. This graph suggests diminishing returns for both the knowledge and search axis. We show that, contrary to the previous literature, there are diminishing returns in chess. This is due to two reasons. The first, decision quality, is not a surprise. The second, game length, is a new result that illustrates how sensitive experimental data can be to hidden properties of the search domain.

Diminishing returns for both increasing knowledge and search raises the question as to what the best way is for improving program performance. The answer depends on several factors including, for example, the application domain, the quality of the evaluation function, and the computational resources available. Future research is needed to understand the role played by each of these factors in program performance.

The designers of the current best chess program, *Deep Blue*, have concentrated their efforts on the search axis. In a typical search, 50 billion positions are considered. The *Deep Blue* chess knowledge is limited because it is implemented in silicon. Our results suggest that small improvements in their knowledge, even at the expense of some search effort, could greatly improve their performance.

## 6 Acknowledgements

This paper benefited from interactions with Yngvi Björnsson, Tony Marsland, Aske Plaat and Manuela Schöne. Special thanks are due to Mark Brockington for providing the data for Othello.

## References

[Beal and Smith, 1994] D. Beal and M. Smith. Random evaluations in chess. *ICCA Journal*, 17(1):3–9, 1994.

[Berliner et al., 1990] H. Berliner, G. Goetsch, M. Campbell, and C. Ebeling. Measuring the performance potential of chess programs. *Artificial Intelligence*, 43(1):7–21, April 1990.

[Condon and Thompson, 1983] J. Condon and K. Thompson. Belle. In P. Frey, editor, *Chess Skill in Man and Machine*, pages 201–210. Springer-Verlag, 1983.

[Iida et al., 1995] H. Iida, K.-I. Handa, and J.W.H.M. Uiterwijk. Tutoring strategies in game-tree search. *ICCA Journal*, 18(4):191–204, 1995.

[Lee and Mahajan, 1990] K.-F. Lee and S. Mahajan. The development of a world class Othello program. *Artificial Intelligence*, 43(1):21–36, 1990.

[Michie, 1977] D. Michie. A theory of advice. *Machine Intelligence 8*, pages 151–170, 1977.

[Mysliwietz, 1994] P. Mysliwietz. *Konstruktion und Optimierung von Bewertungsfunktionen beim Schach*. PhD thesis, University of Paderborn, 1994.

[Nau, 1983] D.S. Nau. Pathology on game trees revisited, and an alternative to minimaxing. *Artificial Intelligence*, 21(1–2):221–244, March 1983.

[Newborn, 1979] M. Newborn. Recent progress in computer chess. In M. Yovits, editor, *Advances in Computers*, volume 18, pages 59–117. Academic Press, 1979.

[Newborn, 1985] M. Newborn. A hypothesis concerning the strength of chess programs. *ICCA Journal*, 8(4):209–215, 1985.

[Plaat et al., 1996] A. Plaat, J. Schaeffer, W. Pijls, and A. de Bruin. Best-first fixed-depth minimax algorithms. *Artificial Intelligence*, 87(1–2):255–293, November 1996.

[Schaeffer and Marsland, 1985] J. Schaeffer and T. Marsland. The utility of expert knowledge. In *IJCAI'85*, pages 585–587, 1985.

[Schaeffer et al., 1993] J. Schaeffer, P. Lu, D. Szafron, and R. Lake. A re-examination of brute-force search. In *Games: Planning and Learning*, pages 51–58. AAAI, 1993. 1993 Fall Symposium, Report FS9302.

[Thompson, 1982] K. Thompson. Computer chess strength. In M.R.B. Clarke, editor, *Advances in Computer Chess 3*, pages 55–56. Pergamon Press, 1982.

# Learning Strategies in Games by Anticipation

## Christophe Meyer, Jean-Gabriel Ganascia and Jean-Daniel Zucker

LIP6 - CNRS, Pôle IA
Université Pierre et Marie Curie
Tour 46-0, 4 Place Jussieu
75252 Paris CEDEX, FRANCE
e-mail: meyer@laforia.ibp.fr

## Abstract

Game Theory is mainly seen as a mathematical theory which tries to replace pure chance and intuitive behavior in a competitive situation by calculations. This theory has been widely used to define computer programs. The aim of the research described here is to design an artificial system which is able to play efficiently certain games to which Game Theory cannot be applied satisfactorily (such as games with incomplete or imperfect information). When it cannot find a winning solution, the system is able to play through a process of anticipation. This is done by building and refining a model of the adversary's behavior in real time during the game. The architecture proposed here relies on two genetic classifiers, one of which models the adversaries' behaviors while the other uses the models thus built in order to play. The system's strategy learning ability has been tested on a simple strategic game. The results show the advantages of this approach over human and traditional artificial adversaries (simple probabilistic and adaptive probabilistic) and illustrate how the system learns the strategies used by its adversaries.

## 1. Introduction

Despite its name, Game Theory, which is the branch of operational research dealing with decision taking in a competitive situation, is not only about play activities such as chess or bridge. All strategic activities, so-called because they require the coordination and organization of forces against adversity, are involved, be they questions of diplomacy, politics, industrial choices, economic or military policy. Like all branches of operational research, Game Theory is mainly seen as a mathematical theory which tries to replace pure chance and intuitive behavior in a competitive situation by calculations. If we suppose that those in competition with each other are all trying to optimize their respective advantages, how should the most efficient lines of play be determined ? This is the generic question to which Game Theory tries to find an answer.

Having said this, Game Theory addresses different points of view depending on the type of game and the optimization criteria taken into account. Traditionally, games fall into two categories, depending on whether the information is complete or incomplete, i.e. whether or not the players have full knowledge about the game. Thus chess is a game with complete information while bridge is one with incomplete information. Another distinction is whether the information is perfect or imperfect. It is perfect if the players control all the parameters of the game, as in chess, and imperfect if, for example, two players have to make a move simultaneously without knowing what move the other is making. The optimization criteria also vary depending on whether the adversaries wish to minimize the risk of losing or maximize the chance of winning while trying to avoid a draw, if possible.

In the past, mathematical theories on games went back to the origin of probability theory for games based on pure chance, i.e. with no competitor other than fate. Mathematical theory on games was developed in the 20th century with the work of Emile Borel, followed by that of Von Neumann and Morgenstein in the forties [Neumann and Morgenstern, 1944], then Nash in the fifties. The last three invented the concepts of mini/max and the search for a point of equilibrium that are the basis of what today is called Theory of Game.

With the appearance of the first computers, Game Theory was seen to be an appropriate formalism for the design of game machines in the case of games with complete and perfect information. In this framework the task of the computer was to simulate the predictable futures, while supposing that the adversaries themselves play in a way that is totally rational and try to optimize their advantages. Such a simulation supposes the exploration of all (or most) possibilities in a search tree. When this simulation becomes impossible because of the size of the tree, heuristics are introduced to get round the combinatorial explosion and the machine must presume that the adversaries adopt similar heuristics. Thus most chess playing machines rely on the mini/max algorithm and a common assessment function which takes into account the state of the chess board as it would be after a near exhaustive exploration of a number of moves [Fürnkranz, 1996].

However, despite the undoubted success of this method, a number of problems exist. The first one is that games with complete and perfect information are not the only competitive situations which might benefit from the contribution of computer simulations. And yet, with the notable exceptions of the poker program designed by Waterman using production rule systems [Waterman and Hayes-Roth, 1978] and the one designed by Findler [Findler, 1977], most computer programs have tackled games with complete and perfect information. Note that existing simulations of games such as Mastermind or Scrabble are not strictly speaking games since there is no strategic dimension involving competition. The second is that nothing proves that when games with complete and perfect information are strongly combinatorial they benefit from Game Theory inspired mini/max processing methods. In this case computer simulation is far from covering all possibilities and therefore the presumed rationality of the adversaries is just an illusion. In a word, it would be a good idea to contemplate other types of games with methods other than those inspired by Game Theory. This would make it possible not only to build more efficient programs where machines currently operate but also to extend the empire of computer programs to competition situations such as the design of interactive interfaces which have not really been simulated before.

It is the above considerations that led to considering a new line of research in which adversarial behaviors are no longer computed from a more or less long list of possible behaviors but are inferred from the adversaries' past behavior. The progress over the last few years in the design of adaptive learning algorithms, with the help of genetic algorithms, has made it possible to make such inferences thanks to which the machines build for themselves an internal representation of their adversaries. We set apart from the mainstream approach in 'Learning in games' which consist in learning an evaluation function [Lorenz and Markovitch,1996].

Once this representation has been built it has to be used in order to act and this is done through the anticipation mechanism which is presented in this article. In fact, the article is divided into three sections and a conclusion. The first presents the general concepts of inference and anticipation on which the approach is based. The second describes the SAGACE system (Genetic Algorithmic Solution for the Anticipation of Evolving Behaviors) which implements this model with the help of genetic algorithms and the game ALESIA on which SAGACE will be assessed. The third reports on tests which show that when humans play against such a machine they are rapidly beaten by the ability of the machine to predict their reactions. Other tests assess different specially implemented strategies. In particular, it is shown that the behaviors predicted by Game Theory appear as the fixed point extremes of the behavior of SAGACE which, in this sense, has implemented a generalization of Game Theory to non rational adversaries.

## 2. The new approach

### 2.1 General overview

The aim of the research described here is to design an artificial system which is able to play efficiently certain games to which Game Theory cannot be applied satisfactorily. The system has been provided with two essential features. The first is that it does a comprehensive scan, to a certain depth, of all possible future states of play according to its actions and regardless of those of the adversary (thus making it possible to identify a possible short term winning strategy). The second is that when it cannot find a winning solution it plays according to a model of the adversary's behavior which it builds and refines in real time during the game. This modeling allows it to anticipate the adversary's actions. As opposed to the work of Carmel and Markovitch [Carmel and Markovitch, 1993], our approach doesn't consist in learning the evaluation function of an adversary (to be used in a mini/max algorithm) but to construct an real-time behavioral model of this adversary.

The system is built around SAGACE, an original learning method based on inference, modeling and anticipation which enables it to learn both an adversary's strategy and the way this strategy evolves.

### 2.2 SAGACE

This algorithmic method is implemented in the form of a classifier system [Holland et al., 1986], the rules of which evolve thanks to a genetic algorithm [Holland, 1975], [Goldberg, 1989], and is based on a purely behavioral approach. It predicts and anticipates the behavior of an individual (human or artificial) but does not claim to be able to discover the underlying cognitive or algorithmic mechanisms [Meyer and Ganascia, 1996a]

#### 2.2.1 General architecture

The system is made up of several components which interact in order to determine the most judicious action for each move.

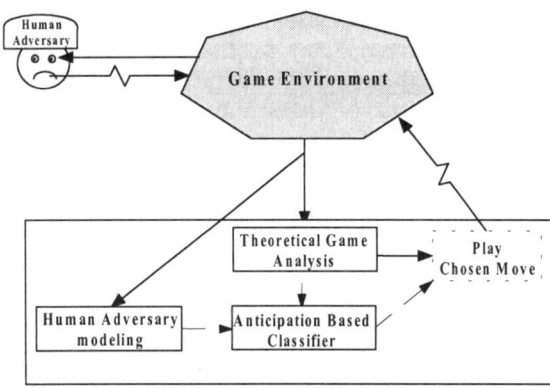

Figure 1. General Architecture

### 2.2.2 The theoretical analysis component

The role of this component is to manage the theoretical knowledge obtained mathematically using Game Theory. It is used to analyze a given game situation and to determine whether or not it corresponds to a winning position, i.e. whether a pure winning strategy exists. In this way, whenever a winning strategy is identified it is adopted by the system.

If no winning solution is identified, the system tries to use its model of the adversary.

### 2.2.3 The anticipation component

Each time the system takes a decision the anticipation component determines whether or not the model it has of its adversary is satisfactory, based on certain criteria concerning the accuracy and error rate of past predictions. If the degree of prediction is considered to be satisfactory (above a predefined anticipation confidence rate), the component provides a list of the adversary's most probable actions. Using the list, the component goes on to determine the most suitable move, with the help of a classifier system. Each classifier of the anticipation component is of the form:

'Game_situation & prediction_action_adversary
→ action_to_be_performed'.

Note that as classifiers these rules are genetically evolutive and therefore adaptive.

If the degree of prediction is considered to be unsatisfactory (below the confidence rate) the system determines its actions by considering that the adversary is likely to make any valid move. In this case the 'prediction_adversary_action' part of the classifiers is undetermined.

### 2.2.4 The modeling component

The modeling component records in real time all the decisions taken by the adversary according to the configuration of the game. Modeling is performed by creating the adversary's behavior rules and then adjusting them regularly so as to reflect the real attitude of the adversary as faithfully as possible. These rules are also managed like classifiers and have the following form:

'Game_situation → action_performed'.

When, in similar circumstances, the adversary repeats an action he has already performed, the modeling component increases a confidence coefficient linked to *this* action in *this* configuration. If the action is different, a new behavior rule is created.

Two coefficients are added to the behavior rules, their role being to give information about the frequency of use and success rates of the rules w.r.t to the modeled player. If, for example, a modeled player often performs a certain action in a given context but with a negative effect, then the coefficient of the frequency of use of the corresponding behavior rule will be high whereas that of the success rate will be low. In addition, the coefficients are given a time weighting. The fact that an action has been chosen recently is more important than if it had been chosen much earlier. Similarly, if the behavior of a modeled player was successful in the past but has recently turned out to be inappropriate, the success rate coefficient will depend more on recent results than on past ones and will certainly be low. Management of this time weighting can be parametrized according to the degree to which the built model of the adversary corresponds to his real observed behavior.

In any situation the system can thus send a request to this component in order to get some idea of how the adversary will play.

## 3. Tests

In order to assess this new approach, an artificial player called Sagace was designed to play against different types of adversary at a repeated game with complete and imperfect information. This game is called ALESIA and is described below.

### 3.1 Alesia

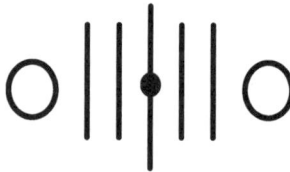

*Alesia is a game between two players who have to try and push the marker across the board to the other one's citadel. At best this can be done in three consecutive winning moves. At the beginning of the game each player receives a capital of 50 points. The players decide in secret and announce simultaneously how many points (at least one and at most their capital) they have decided to play during a given move. The player who announces the highest number pushes the marker*

*(initially at the center) one slot towards the adversary's citadel. If they both announce the same number, the marker remains where it is. In both cases the number of played points is deducted from the players' capital. A player wins if he manages to move the marker to the adversary's citadel. Otherwise it is a draw.*

This game contains a number of interesting features. The rules are very easy and can very quickly be understood by a human, the information is complete and imperfect and there is no general non-mixed winning strategy (this has been proved mathematically).

### 3.2 Implementation

The system has been implemented in Delphi 2.0 for Windows 95 and it is possible for two computer players to play 1000 games in about 30 seconds on a Pentium Pro PC. Thus a large number of series of tests have been run to assess system performance. For each adversary (human or simulated) of the system there is a modeling base and an anticipation base (to begin with it is the same for all adversaries but is likely to evolve differently for each).

The classifiers contained in the modeling bases are encoded in the form 'Condition -> action'. The 'condition' part is made up of intervals of possible values for the number of points of the two opponents and the position of the marker. The 'action' part indicates the number of points to be played.

The anticipation base classifiers are encoded in the same formalism with just an additional indication in the 'condition' which encodes the prediction of the adversary's action. These classifiers have, for the most part, been elaborated by a human player using introspection and then encoded in the formalism imposed by the form of the classifiers. Others have been built after looking at the behavior of good players and have in fact been taken from the modeling bases of certain particularly successful players. This idea has already been studied concerning the bootstrapping of a classifier system [Meyer, Ganascia 1996b].

## 4. Results

A test protocol has been defined involving three specific groups of adversaries for Sagace. This set of players seems to be sufficiently varied so as to represent the different strategies that a human player could try and adopt when playing Alesia. Sagace has therefore been set against real humans, simulated non-adaptive probabilistic players and simulated adaptive probabilistic players (the exact nature of which will be described below). Each of the tests measures the evolution in the percentage of games won since the beginning of the session.

### 4.1. Human adversaries

The system played against different human players and the most representative curve corresponding to the results of this series of tests was derived. Each human played a hundred games (Figure 2).

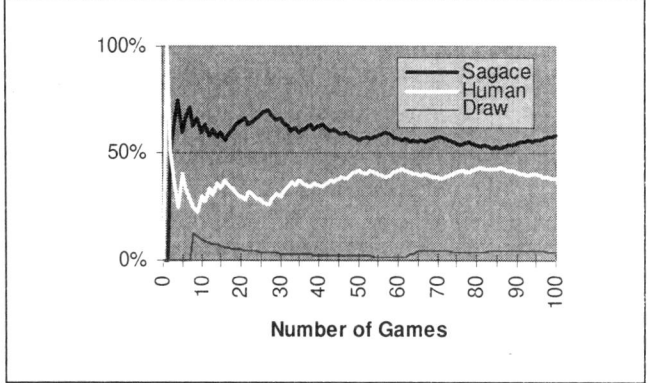

Figure 2. Human adversaries

As can be seen in figure 2, the progression in the rate of games won by the players is not monotonic. The two players seem to have continually adapted their way of playing to the other. After 85 games the human player no longer seems capable of imagining new strategies adapted to Sagace and the rate of games won by the latter rises slightly. Longer series of games against different players show that the curves end up stabilizing, in general around the same values as those shown in figure 2 (55% for Sagace and 45% for the human player). It is difficult to interpret this type of curve but what is clear is that both human and machine adapt very quickly to each other. It is also clear that the intensity of the variations decreases with the number of games.

If the system uses exactly the same components with the same classifiers when the adversary modeling component is disabled, the human wins more than 80% of the games. This result proves the importance of modeling and anticipating.

### 4.2 Probabilistic adversaries

In the second type of test the system played against simulated players with probabilistic strategies. Since these strategies rely on chance, they naturally come to mind when a human is trying to thwart the modeling he knows he may be the subject of. The first of these adversaries is a probabilistic player who plays at random a number between one and the number of points he has left.

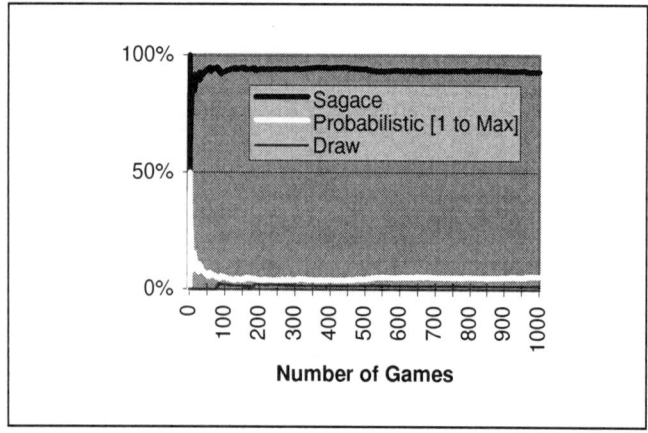

Figure 3. First probabilistic adversary

Very quickly Sagace analyzes the situation, generates a very reliable model of this adversary and adapts perfectly (Figure 3).

The second adversary is also probabilistic but always chooses a number between 1 and 50. If the number that is computed is greater than the number of points he has left, he plays all his remaining points.

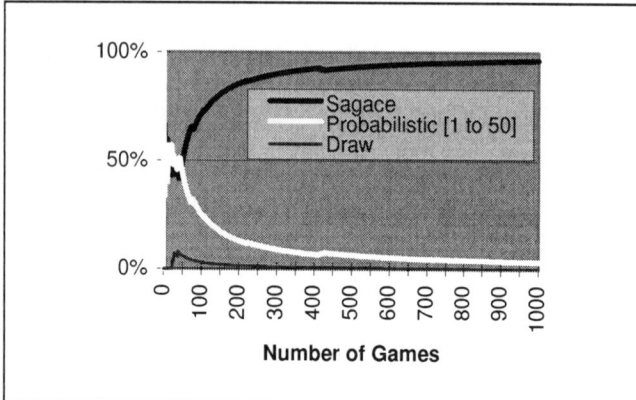

Figure 4. Second probabilistic adversary

It takes longer to model this player than the previous one and the results are not as good at the beginning but better at the end (Figure 4). Modeling is more tricky (convergence is slower) because there is no equiprobability in the choices of action as the higher moves are naturally favored. After fifty games, however, the model elaborated by Sagace is very reliable.

The third adversary simulates an expert theoretician and is able to recognize mathematically when a situation allows a winning strategy, which will always be played. Otherwise his behavior is determined by a set of probabilistic rules that have been introduced according to a games expert (e.g. 'never play more than one point above your adversary's remaining points').

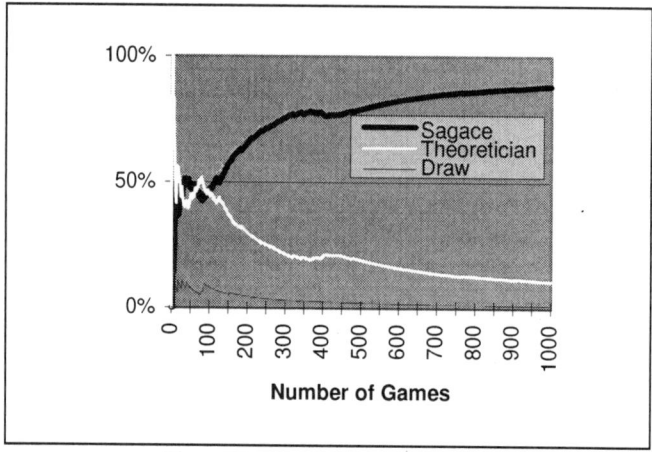

Figure 5. Theoretician adversary

Sagace needs at least 150 games in order to produce a reliable model of this player (Figure 5). Before that there is a period during which the players are equal, during the following 150 games Sagace plays very well, and then the percentage of games won stabilizes. If play continues, the percentage reaches its extreme limit of 85%.

### 4.3 adaptive adversaries

In the third type of test the system played against simulated players with adaptive strategies. These strategies naturally come to mind when a human is trying to thwart the modeling he knows he may be the subject of by dynamically modifying his strategies. The simulated adversary is a player who will adapt his strategies according to results. This player has been implemented in the form of a classifier system; it learns through the traditional method of positive and negative retribution of the rules governing choice of action ac cording to results [Holland et al., 1986].

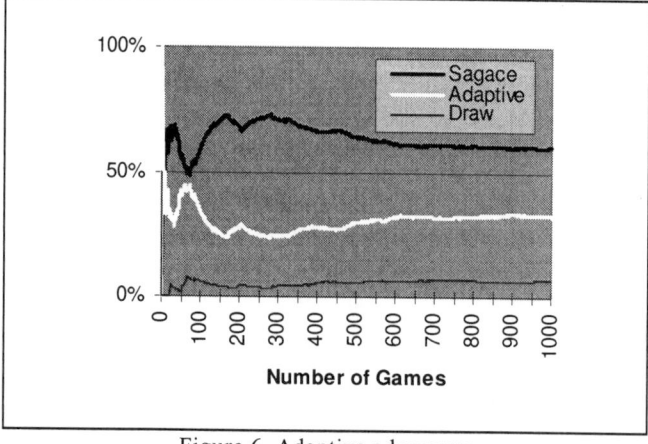

Figure 6. Adaptive adversary

During the first fifty games, Sagace manages to model the adaptive player relatively well and the percentage of games won increases (Figure 6). During the following fifty games the adaptive player adapts, then Sagace manages to determine a new consistent model and the percentage of games won increases again. After a certain number of these 'adaptation' swings the percentage of games won stabilizes (60% for Sagace, 40% for the adaptive player). Note that this type of player is less efficient than a good human player.

## 4. Discussion

Thanks to the architecture of this system it is possible to be free of the limitations of caution imposed by Game Theory while at the same time taking advantage of its mathematic results when they are optimal. Yet this method offers no theoretical guarantee of results, unlike Game Theory. In fact, it is a different way of apprehending the inequation between security, risk and results. Against any adversary with a fixed strategy this approach is formidable since it can determine the adversary's behavior rules and choose the appropriate strategies. Against an adaptive adversary, the efficiency of the system naturally depends on the degree of adaptability of the adversary. A priori, one would assume that a human is a highly adaptive adversary and yet human players are often beaten by the system. When a human player is informed of the type of learning process used by the system he often tries to adopt a strategy that is very different from his original one. Either he tries to make himself unpredictable by playing more or less at random (but this will often make him make moves that are not optimal and he will lose) or he will try and deceive the system by playing the same move several times over and then changing as soon as he thinks the system is going to play according to this information that he has deliberately given. In fact, these human players model the system that is trying to model them. This human strategy requires great perspicacity and heavy concentration but may be extremely successful.

## 5. Conclusions and future lines of research

Not only does this approach involving repeated games with complete and imperfect information not challenge mathematically valid results but it also offers a solution to address/apprehend complex situations in which it is important if not essential to have a reliable model of individuals or agents with which one interacts.
The system will be improved by giving it means to model itself so as to be able to tell when it has become predictable and thus change behavior accordingly. This improvement will also make it possible to analyze more rigorously the way the system learns.

In addition, work will be done to see how to adjust the probabilistic coefficients of an optimal mixed strategy that has been determined using Game Theory, according to the way SAGACE models the behavior of its adversaries.

Also, some of the ideas concerning the generalization of genetic rules in classifiers will be implemented so as to extend the predictions of an adversary's actions to situations that have not actually been observed but are similar.

Finally, extension to more complex games (chess, poker) and to cooperative games (iterated prisoner's dilemma) will be studied.

## References

[Carmel and Markovitch, 1996] David Carmel and Shaul Markovitch. Incorporating opponent models into adversary search. In Proceedings of the Thirteenth National Conference on Artificial Intelligence, Portland, Oregon, 1996.

[Findler, 1977] Nicholas V. Findler. *Studies in machine cognition using the game of poker.* Communications of the ACM, 20(4):230-245, 1977.

[Fürnkranz, 1996] Johannes Fürnkranz. Machine learning in computer chess: The next generation. ICCA Journal, 19(3):147-160, September 1996.

[Goldberg, 1989] David E. Goldberg. *Genetic Algorithms in Search, Optimization, and Machine learning.* Addison-Wesley, 1989.

[Holland, 1975] John H. Holland. *Adaptation in Natural and Artificial Systems.* University of Michigan Press, 1975.

[Holland et al., 1986] John H. Holland, K.J. Holyoak, P.R. Thagard, and R.E. Nisbet. *Induction : Processes of Inference, Learning, and Discovery.* MIT Press, Cambridge, 1986.

[Lorenz and Markovitch,1993] David H. Lorenz and Shaul Markovitch. *Derivative evaluation function learning using genetic operators.* In Epstein and Levinson, editors. Proceedings of the AAAI Fall Symposium on Intelligent Games: Planning and Learning, Menlo Park, CA. The AAAI Press, 1993.

[Meyer and Ganascia, 1996a] Christophe Meyer and Jean-Gabriel Ganascia. *S.A.G.A.C.E. Genetic Algorithmic Solution for the Anticipation of Evolving Behaviors.* Paris VI, LAFORIA N°96/32, 1996.

[Meyer and Ganascia, 1996b] Christophe Meyer and Jean-Gabriel Ganascia. *Utilization of imitation and anticipation mechanisms to bootstrap an evolutive distributed artificial intelligence system.* Animal Societies as an Alternative Metaphorical Basis for DAI - Workshop ICMAS, 1996.

[Neumann and Morgenstern, 1944] John von Neumann and O. Morgenstern. *Theory of Games and Economic Behavior*, Princeton University Press, 1944.

[Waterman and Hayes-Roth, 1978] D. Waterman and F. Hayes-Roth. *Pattern-Directed Inference Systems.* Academic Press, 1978.

# INFORMATION RETRIEVAL

# INFORMATION RETRIEVAL

Information Retrieval Distinguished Paper

# Translingual Information Retrieval: A Comparative Evaluation

Carbonell, Jaime G.; Yang, Yiming; Frederking, Robert E.;
Brown, Ralf D.; Geng, Yibing; Lee, Danny

Language Technologies Institute, Carnegie Mellon University
5000 Forbes Avenue, Pittsburgh, PA 15213 USA

## Abstract

Translingual information retrieval (TIR) consists of providing a query in one language and searching document collections in one or more different languages. This paper introduces new TIR methods and reports on comparative TIR experiments with these new methods and with previously reported ones in a realistic setting. Methods fall into two categories: query translation based, and statistical-IR approaches establishing translingual associations. The results show that using bilingual corpora for automated extraction of term equivalences in context outperforms other methods. Translingual versions of the Generalized Vector Space Model (GVSM) and Latent Semantic Indexing (LSI) perform relatively well, as does translingual pseudo relevance feedback (PRF). All showed relatively small performance loss between monolingual and translingual versions. Query translation based on a general machine-readable bilingual dictionary – heretofore the most popular method – did not match the performance of other, more sophisticated methods. Also, the previous very high LSI results in the literature were disconfirmed by more realistic relevance-based evaluations.

## 1 Introduction

Translingual information retrieval (TIR) is starting to receive considerable attention in recent years with the increased accessibility of ever-more-diverse on-line international text collections, including centrally the World Wide Web. In spite of some pioneering work [Salton, 1970; Dumais et al., 1996; Davis and Dunning, 1996; Hull and Grefenstette, 1996], evaluations of different TIR techniques in realistic retrieval tasks are rare. This paper reports our evaluation results of both newly developed TIR techniques and reimplementations of previously reported techniques.

Translingual information retrieval (aka "multilingual" or "crosslingual" IR) consists of providing a query in one language and searching document collections in one or more different languages. One can envision many ways to bridge the language barrier between query and collection. In this paper, we focus on query translation and methods based on automatically establishing translingual associations between queries and documents without need of translating either.

## 2 MT-Based Methods for TIR

The *machine translation methods* for TIR require that either the query be translated into the target language, and the translation be used to search the target-language collection, or the collection be translated into the source language, and the original query be used to search. Let us consider the pros and cons of each approach:

- *Translation Accuracy* – Both human and machine translation [Carbonell, 1985; Nirenburg et al., 1991] require context to achieve accuracy. Translating isolated words in a query is unreliable, due to unresolved lexical ambiguity. Translating documents should yield greater accuracy.

- *Retrieval Accuracy* – Since documents contain far more information than queries, random translation errors should cause less degradation for the IR task in documents than in queries. Hence for both this reason and the above, document translation is in principle preferable. In fact, preliminary findings by Dumais et al [Dumais et al., 1996] support this line of reasoning.

- *Practicality* – Many document collections are very large. Most are searched remotely. Some are proprietary; individual documents may be read or downloaded, but the entire collection may not be copied or translated. Even if these problems were surmountable, translating the collection may require inordinately long computation and massive storage, not to mention re-indexing the translated collection.

Because of the above practicality constraint, we report only on translating the query for TIR. If the query were formulated as phrases, as a full sentence, or as a paragraph, we could apply MT systems far more reliably. However, experience shows that users typically prefer to give isolated words, or at best short phrases to an IR system.[1] The question is how to best translate a set of isolated words. Full fledged MT is not applicable. Instead, we investigated three approaches:

1. *Dictionary Term Translation* – Look up each query term in a general-purpose bilingual dictionary, and use all its possible translations. This is a form of query expansion upon translation.

2. *Corpus-based Term Translation* – Use a sentence-aligned bilingual training corpus to find the terms that co-occur in context across languages, thus creating a corpus-based term-equivalence matrix.

3. *Corpus-based Term-to-Sentence* – Use the same type of aligned bilingual training corpus to extract full sentences that in the target language co-occur with query terms in the source language. Term-to-sentence expansion may enhance recall, but at a cost in precision.

All three MT-based methods used variations of the Pangloss Example-Based Machine Translation engine (PanEBMT) [Brown, 1996], however only corpus-based term translation (called EBT below, for "Example-Based Term" translation) is further described since it produced better results.

## 2.1 PanEBMT Translations

In general, EBMT systems [Brown, 1996; Nagao, 1984] use a large corpus of example pairs of previously translated sentences, in order to find close matches and translations of words and phrases in context. The PanEBMT parallel corpus was derived primarily from the Spanish and English portions of the UN Multilingual Corpus [Graff and Finch, 1994], with an admixture of texts from the Pan-American Health Organization and ARPA MT evaluations. The total corpus contains some 685,000 sentence pairs after removing duplicated Spanish sentences. PanEBMT translates by finding the set of matches to a new text string (word, phrase or sentence) in the indexed bilingual corpus. Then the translations corresponding to these matches are combined into candidate translations of the new text. Because queries contain more isolated terms than phrases or sentences, our query-translation experiment is unable to exploit the power of EBMT.

## 2.2 Corpus-based Term Translation

In order to create domain-specific or corpus-specific bilingual dictionaries automatically, we start from a large sentence-aligned bilingual corpus and generate a large term co-occurrence table combined with a thresholding scheme[Brown, 1997]. The result was used as the dictionary for corpus-based (example-based) term substitution MT (EBT).

Co-occurrence dictionary generation is performed in two phases: First the co-occurrence matrix (indexed by source-language words on one axis and target-language

[1]LYCOS reports that their typical user queries for general web search are only one to three words long, although they are occasionally reformulated into longer queries.

words on the other) is generated. Each cell in the matrix represents the number of times the source-language word occurred in the same sentence pair as the target-language word. Given this matrix, we compute the conditional probability that if the term occurs in one language its counterpart (i.e. its candidate translation) also occurs in the other language. If this probability is above a threshold, then the term translation is added into the dictionary. Should a term in one language co-occur with several terms in the other language with sufficient frequency to pass the conditional probability threshold, *all* are stored as candidate translations.

## 3 IR-based Methods for TIR

We extended three monolingual retrieval methods to translingual retrieval: pseudo-relevance feedback (PRF)[Buckley *et al.*, 1995], the general vector space model (GVSM)[Wong *et al.*, 1985], and the latent semantic indexing (LSI) approach[Deerwester *et al.*, 1990]. In each case, a translingual semantic correspondence between queries and documents is established based on a document-aligned bilingual training corpus, without requiring sentence-level alignment, bilingual dictionaries or machine translation.

### 3.1 Pseudo-Relevance Feedback

Relevance feedback (RF) is an approach to query expansion in monolingual text retrieval[Salton, 1989]. It requires a user to judge interactively which retrieved documents are relevant, and uses the relevance judgments to expand the original query for additional search. By "pseudo-relevance feedback" (PRF) we mean using the top-ranking documents obtained in an initial retrieval without human judgements, assuming that a significant fraction of top-ranked documents will be relevant. Both RF and PRF are query expansion techniques similar to case-based IR [Rissland and Daniels, 1995], and both typically improve performance in monolingual retrieval compared to not using them. The adaptation of PRF to translingual retrieval is relatively simple if a bilingual corpus is available. That is, we find the top-ranking documents for a query in the source language, substitute the corresponding documents in the target language, and use these documents to form the corresponding query in the target language. Figure 1 illustrates the data flow for translingual RF and PRF.

All the IR techniques discussed here, including the PRF approach are variants of the vector space model (VSM)[Salton, 1989] where both queries and documents are represented using vectors of term weights. To allow clear theoretical comparison of these IR-based methods, let us define the notation for VSM (including PRF):

$$\vec{q} = (q_1, q_2, \ldots, q_m)^t$$
$$\vec{d} = (d_1, d_2, \ldots, d_m)^t$$
$$sim(\vec{q}, \vec{d}) = \cos(\vec{q}, \vec{d}) = \frac{\sum_{i=1}^m q_i d_i}{\sqrt{\sum_{i=1}^m q_i^2}\sqrt{\sum_{i=1}^m d_i^2}}$$

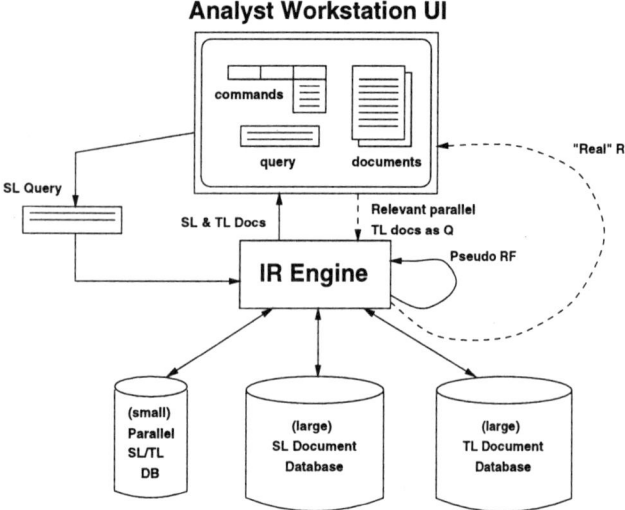

Figure 1: Data flow for translingual relevance feedback and pseudo-relevance feedback

where $\vec{q}$ is the query, $\vec{d}$ is a document in a corpus, $m$ is the number of unique terms (words or phrases) in the corpus, and $q_i$ and $d_i$ are the term weights in the query and the document, respectively.

### 3.2 Generalized Vector Space Model

A criticism of conventional VSM is that it uses terms as an orthogonal basis of the vector space, but terms are often not semantically independent. Wong *et al* proposed an alternative, namely the "generalized vector space model" (GVSM) [Wong *et al.*, 1985], also referred to as "the dual space" [Sheridan and Ballerini, 1996]. The idea is to use documents as the basis for representing terms instead of using terms.

Consider a term-document matrix, $A_{m \times n}$, as a training corpus where $m$ is the size of the vocabulary, and $n$ is the number of unique documents in this corpus. One can view this matrix as a way to represent documents (the columns) using terms, and to represent terms (the rows) using documents. The former view corresponds to the conventional vector space model, and the latter view corresponds to the dual space. Each row vector of $A$ reflects the term usage in the corpus, i.e., the *pattern* of this term distributed over documents.

Matrix $A$ can be used for query transformation by computing $\vec{q'} = A^t \vec{q}$ where $\vec{q}$ is the original query vector whose dimensions are unique terms, and $\vec{q'}$ is the transformed vector whose dimensions are unique documents. The transformation is equivalent to weighting the *distribution pattern* of each term using its weight in the original query, and summing up the weighted patterns to obtain a new representation of the query. Similar to query transformation, a document can also be transformed into a vector in the dual space by computing $\vec{d'} = A^t \vec{d}$ where $\vec{d}$ is the document vector in the conventional VSM. The retrieval criterion in GVSM for monolingual retrieval is defined to be:

$$sim(\vec{q}, \vec{d}) = \cos(A^t \vec{q}, A^t \vec{d}).$$

Here we propose a novel extension of the monolingual GVSM for translingual retrieval. Assuming a bilingual corpus for training, we form two matrices, $A$ and $B$, where $A$ is a term-document matrix for the training documents in the source language (also the language of the queries), $B$ is a term-document matrix for the training documents in the target language, and the corresponding columns of $A$ and $B$ are the matching pairs of documents in the bilingual corpus. We use $A$ for query transformation and $B$ target-language document transformation. The retrieval criterion is defined to be:

$$sim(\vec{q}, \vec{d}) = \cos(A^t \vec{q}, B^t \vec{d})$$

Since matrix $A$ and $B$ share the same dual space, the transformations $A^t \vec{q}$ and $B^t \vec{d}$ give the query and the document a common basis (presenting distribution patterns of terms over documents) on which they can be compared. This is how the translingual correspondence is established.

The computation in GVSM consists of the transformation ($A^t \vec{q}$ and $B^t \vec{d}$) and the cosine computation. The time complexity of the first part is similar to the computation in VSM. It is proportional to the number of non-zero elements in a query or document vector, $O(kn)$ where $k$ is the average number of unique terms per query or document, and $n$ is the number of document pairs in the bilingual training corpus. The time complexity in the second part, is $O(n)$ per document, or $O(nl)$ for a test corpus of $l$ documents. It is possible to significantly reduce this complexity in large problems by aggressively removing non-influential elements from the transformed document vectors [Yang, 1995].

### 3.3 Latent Semantic Indexing

Latent Semantic Indexing[Deerwester *et al.*, 1990] (LSI) is a one-step extension of GVSM. The claim is that neither terms nor documents are the optimal choice for the orthogonal basis of a semantic space, and that a reduced vector space consisting of the most meaningful linear combinations of documents would be a better representative basis for the content of documents.

In monolingual retrieval, LSI uses the term-document matrix ($A$) for training, the same as in GVSM. It computes the orthogonal dimensions ("the latent semantic structures") in matrix $A$, and selects the principal dimensions as the new basis for a reduce vector space. The monolingual LSI retrieval criterion is defined to be:

$$A = U \Sigma V^t$$
$$L = U \Sigma^{-1}$$
$$sim(\vec{q}, \vec{d}) = \cos(L^t \vec{q}, L^t \vec{d})$$

where matrices $U$ and $V$ contain a set of $p$ orthogonal singular vectors each (one for the representation of terms, and another for the representation of documents). Matrix $\Sigma$ is $p$-diagonal, containing the singular values indicating the importance of the corresponding singular vectors in matrices $U$ and $V$. Matrix $L$ contains the weighted singular vectors in $U$, where the weights are the corresponding singular values in $\Sigma$. This matrix can be viewed as a reduced version of matrix $A$. That is, both $A$ and $L$ use their row vectors to represent terms, but the term vectors in $L$ are much shorter than the term vectors in $A$. The dimensions in $L$ are linear combinations of documents, while the dimensions in $A$ are individual documents.

The translingual LSI model [Dumais et al., 1996] is similar to the model for monolingual LSI, except that a bilingual document corpus is needed for training instead of a monolingual corpus. Let $\vec{q}$ be a query in the source language, $\vec{d}$ be a document in the target language, and $\begin{bmatrix} A \\ B \end{bmatrix}$ be the matrix of bilingual document pairs where $A$ and $B$ are the same as defined in GVSM. Then the translingual LSI retrieval criterion is defined to be:

$$\begin{bmatrix} A \\ B \end{bmatrix} = U_2 \Sigma_2 V_2^t$$
$$L_2 = U_2 \Sigma_2^{-1}$$
$$sim(\vec{q}, \vec{d}) = \cos(L_2^t \vec{q}, L_2^t \vec{d})$$

where $U_2$, $V_2$, $\Sigma_2$ and $L_2$ are the matrices computed using the singular value decomposition of the bilingual input matrix.

LSI has a quadratic time complexity of $O(n'p)$ where $n' = \max\{m, n\}$ is the larger number between the size ($m$) of the joint vocabulary of both languages and the number ($n$) of document pairs in the bilingual training corpus; $p$ is the number of orthogonal dimensions (singular vectors) computed in the singular value decomposition. Thus the scalability of this method to a large corpus would be much more limited than the VSM or GVSM approach if a large number of singular vectors is necessary for good retrieval performance.

### 3.4 The Scientific Challenge

The similarities and differences between the three models mentioned above can be seen in their retrieval criteria:

$$VSM: \quad sim(\vec{q}, \vec{d}) = \cos(\vec{q}, \vec{d})$$
$$GVSM: \quad sim(\vec{q}, \vec{d}) = \cos(A^t \vec{q}, B^t \vec{d})$$
$$LSI: \quad sim(\vec{q}, \vec{d}) = \cos(L_2^t \vec{q}, L_2^t \vec{d})$$

The fundamental difference, in theory, is the choice of the basis for the similarity comparison between queries and documents. VSM assumes semantic independence of terms in its basis. GVSM uses documents instead, assuming documents are semantically independent. LSI computes the orthogonal dimensions in a training corpus, and chooses the principal dimensions as the basis of a reduced vector space. GVSM and LSI are close variants in the sense that both exploit the dual space. The only difference is whether to use the original dimensions (document vectors) or the reduced dimensions (the orthogonal singular vectors) as the basis for the vector space. Which model best represents the semantic space of documents and queries is a scientifically challenging question.

Given the methods, empirical validation is important. For monolingual retrieval, performance improvement of GVSM over VSM was observed on small collections [Wong et al., 1985]; improvement of LSI over VSM was observed sometimes but not always [Deerwester et al., 1990]. Until our work reported below, there has not been a comparison between GVSM and LSI, in either monolingual or translingual retrieval.

## 4 Empirical Evaluation

We carried out a comparative evaluation of the six translingual IR methods described above (the three term-based MT methods, PRF, GVSM, and LSI) on a realistic retrieval task. The large UN Multilingual Corpus[Graff and Finch, 1994] from the Linguistic Data Consortium was available to us, but, among other problems, there were no queries or human relevance judgements available for training and evaluation. We conducted our experiments on a subset of this corpus, consisting of 2255 document pairs pertaining to UNICEF reports and deliberations. Each document pair consists of an English document and its corresponding Spanish translation. 1134 document pairs were randomly selected and used for translingual training. The remaining 1121 pairs were set aside for testing. The average (monolingual) document is 9 paragraphs long. Altogether, the training and test sets in both languages consist of almost 2 million words of text.

We conducted our experiments as follows: First, we created 30 queries in English, germane to the UNICEF subcollection. The average query length was 11 words. Second, we obtained human relevance judgements on the cross product of the 30 queries and 1121 test documents (33,630 samples in all), and used these as our gold standard for testing[2]. Third, we trained each method to find translingual equivalences using paired documents, without queries; hence no relevance judgements were required for training. Fourth, we tested each method *monolingually* on the test set to obtain ranked lists of retrieved documents. Fifth, we applied the translingual version of our methods. Finally, we evaluated the results by comparing the retrieval degradation when moving from monolingual to translingual IR for all the methods.

We optimized each method for monolingual retrieval, with respect to its performance on 11-point average precision using the full human relevance judgements on the 30 queries for the test corpus. Optimizations include the settings on TF and IDF weights for cosine-similarity scoring, setting thresholds on pseudo-relevance feedback, setting cutoff levels for number of non-influential elements for GVSM (200 was optimal), and determining the optimal number of singular vectors in LSI (also 200, coincidentally). When each retrieval method was performing at optimum on monolingual retrieval, we tested

---

[2] An initial set of experiments was conducted before these judgements were available, as described below.

that method with exactly the same parameter settings on translingual retrieval.

We carried out two sets of experiments. The first comparison focuses on overlap between monolingually-retrieved documents and translingual retrievals, based on the parallel corpus to establish correspondences. The central tenet of the first evaluation was that perfect translingual retrieval would retrieve exactly the corresponding set of documents as monolingual retrieval, but in the target language. Hence, monolingual versus translingual comparison was our primary effectiveness measure. The second comparison is a direct evaluation of monolingual and of translingual retrieval using human relevance judgements and computing precision and recall scores in the traditional IR manner [Salton and McGill, 1983].

### 4.1 Initial Evaluation

Before we had relevance judgements available, we carried out an initial set of experiments, using two novel evaluation methods. Each of the retrieval methods was trained on the same bilingual training data. We then ran each of the translingual methods using English queries to retrieve Spanish documents, and compared the Spanish documents retrieved translingually to the English documents retrieved monolingually by each method.

The results were compared using both an 11-point average precision and an overlap measure. "Precision" in this case was measured by comparing the top $L2$ of the translingually-retrieved ranked list against the top $L1$ of the corresponding monolingually-retrieved ranked list, and taking the monolingually-retrieved documents as relevant to the query. The overlap measure was simply the percentage of identical documents present in the top $L1$ ranked documents retrieved monolingually and the top $L2$ translingually, for a given $L1$ and $L2$. We calculated both measures in order to minimize any artifact of our evaluation method.

The translingual results are presented in Figures 2 and 3. Figure 2 presents the 11-point average precision for each method plotted against the number of reference documents used ($L1$). $L2$ is four times $L1$ in each case, in an attempt to make the second ranked list sufficiently long. Figure 3 presents the overlap measure for each method, again versus $L1$. Here we use $L2 = L1$, as per our definition of overlap. As mentioned above, we only report the results of the corpus-based EBT method in these graphs, for brevity, given that it performed best among the MT methods.

From these figures, we see that GVSM outperformed the other methods in terms of these document overlap measures. All the methods have a large overlap between their monolingual and translingual retrievals, which is a good sign. Moreover, the graphs of the two evaluation methods are remarkably consistent.

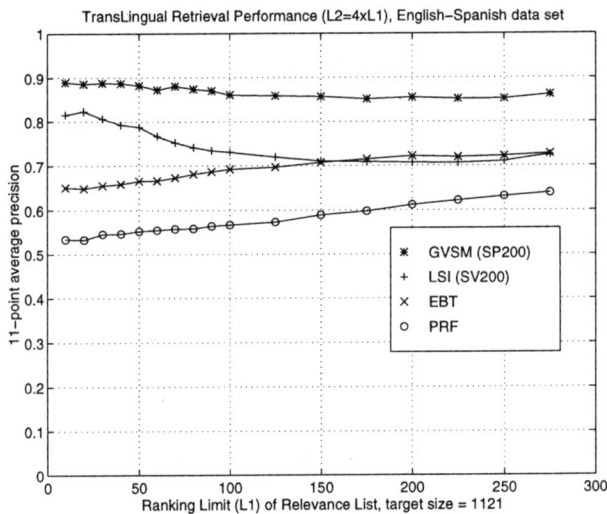

Figure 2: 11-pt. ave. precision vs. rank limit threshold

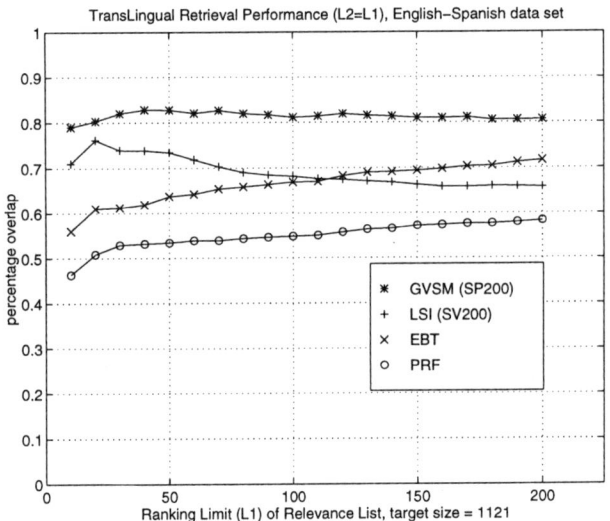

Figure 3: Overlap % vs. rank limit threshold

### 4.2 Evaluation with Human Judgements

For our second comparison, we evaluated each method, monolingually and translingually, using human relevance judgements. The corresponding 11-point average precision values in the table in figure 4 below. For comparison, we also include corresponding translingual results reported by other researchers. Because the methods have been run on different corpora with different queries, direct comparisons on absolute 11-point-precision recall figures are not meaningful. However, the ratio of translingual IR (TIR) over monolingual IR results may be more indicative of the relative power of the TIR methods. We encourage direct comparisons on the same corpus. We also present our results in the standard recall-precision graphs for monolingual and translingual IR in figures 5 and 6, respectively.

| Site | Method | (Monolingual) IR | TIR | TIR/MIR |
|------|--------|------------------|-----|---------|
| CMU | Dict Q-tran | .4721 (SMART.basic) | .2898 | 61% |
| CMU | EBT Q-tran | .4721 (SMART.basic) | .4318 | 91% |
| CMU | PRF Q-exp | .4487 | .4043 | 90% |
| CMU | GVSM QD-tran (SP=200) | .4008 | .3804 | 95% |
| CMU | LSI QD-tran (200 SVs) | .3689 | .3093 | 84% |
| UMASS | CorpusPhrase Q-exp | .20 | .1358 | 68% |
| ETH | LSI thesaurus Q-exp | .527 | .212-.278 | 40-53% |
|  | (LSI thesaurus + RF) | ? | ? | (68%) |
| XEROX | Dict Q-tran | .393 | .235 | 60% |
| NMSU | Dict Q-tran | ? | ? | 40-50%? |

Figure 4: Results of TIR methods (with ntc.ntc weighting)

As the table in figure 4 shows, example-based term (EBT) translation (a.k.a. corpus-based term translations), never before tried for TIR, exhibits top absolute performance, whereas general-purpose machine-readable dictionary (MRD) query-translation exhibits the worst performance. In spite of the similarity between these methods (both translate the query), the former is trained to the corpus and exploits context, and is therefore much superior. This result indicates that the most popular TIR method reported in the literature (MRD-based query translation) may be the simplest, but its performance leaves much to be desired.

Pseudo-relevance feedback also performed well in absolute terms, indicating that if the user were willing to provide true relevance judgements, full relevance feedback could become the top-performing method for TIR.

GVSM, never before tried for TIR, performed relatively well and showed the least degradation from MIR to TIR. LSI did not perform according to expectations from the literature. In earlier less realistic experiments (with queries formulated directly from documents), LSI had performed better [Dumais et al., 1996].

## 5 Conclusions

This paper reports a thorough evaluation of multiple methods for translingual retrieval in a query-based retrieval task. We believe that the evaluation methodology used here may be generally useful when costly human relevance judgements are unavailable. Our experimental results indicate that:

- Translingual retrieval is viable by a number of different techniques, ranging from term-based query translation and pseudo-relevance feedback to generalized vector spaces and latent semantic indexing.

- In the translingual retrieval test, example-based MT establishing corpus-based term equivalences performed best, followed by PRF, GVSM, LSI and MRD-based query translation. However, in terms of performance relative to monolingual retrieval, GVSM performed best.

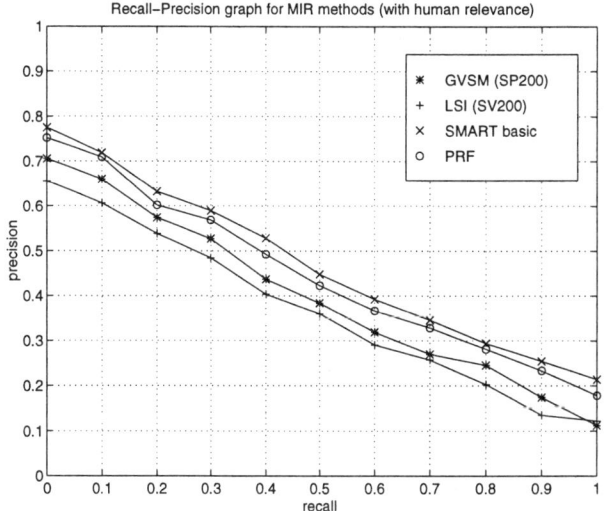

Figure 5: Recall-Precision performance of MIR methods

- MRD-based query translation, though popular in the literature, should be re-examined as the TIR method of choice given the results in this paper.

- It appears that Translingual LSI is not as good in a realistic setting with actual queries and 11-point-average precision evaluations as in the preliminary Dumais et al study, although LSI does perform better that simple MRD-based query translation.

It is worth noting that GVSM is simple to compute and easy to scale up, somewhat better than LSI, and its performance is not crucially dependent on the exact value of a tuned parameter (such as the number of singular vectors of LSI). More work is clearly called for in further evaluating the GVSM method and corpus-based term-translation in other realistic contexts, and investigating whether other forms of tunable MT-based translingual IR could be made to perform reasonably well, especially in situations where translating the collection does not pose serious problems.

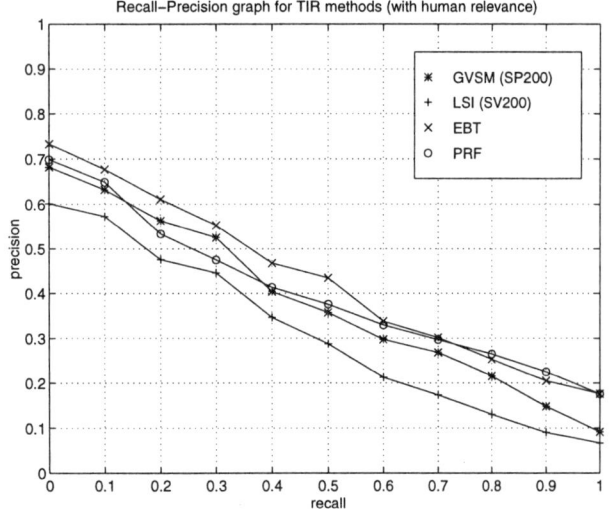

Figure 6: Recall-Precision performance of TIR methods

# References

[Brown, 1996] R.D. Brown. Example-Based Machine Translation in the Pangloss System. In *Proceedings of the Sixteenth International Conference on Computation Linguistics*, pages 169–174, 1996.

[Brown, 1997] R.D. Brown. Automated Dictionary Extraction for "Knowledge-Free" Example-Based Translation. In *Proceedings of the Seventh International Conference on Theoretical and Methodological Issues in Machine Translation*, 1997.

[Buckley et al., 1995] C. Buckley, G. Salton, J. Allan, and A. Singhal. Automatic Query Expansion Using SAMRT: TREC 3. In *Overview of the Third Text REtrieval Conference (TREC-3)*, pages 69–80, 1995.

[Carbonell, 1985] J. G. Carbonell. New Approaches to Machine Translation. In *Proceedings of the conference on Theoretical and Methodological Issues in Machine Translation of Natural Languages*, Hamilton, NY, 1985.

[Davis and Dunning, 1996] M. Davis and T. Dunning. A TREC evaluation of query translation methods for multi-lingual text retrieval. In *The 4th Text Retrieval Conference (TREC-4)*, 1996.

[Deerwester et al., 1990] S. Deerwester, S.T. Dumais, G.W. Furnas, T.K. Landauer, and R. Harshman. Indexing by Latent Semantic Analysis. In *J Amer Soc Inf Sci 1, 6*, pages 391–407, 1990.

[Dumais et al., 1996] S. Dumais, T. Landauer, and M. Littman. Automatic Cross-Linguistic Information Retrieval using Latent Semantic Indexing. In *Proceedings of SIGIR-96*, Zurich, August 1996.

[Graff and Finch, 1994] David Graff and Rebecca Finch. Multilingual Text Resources at the Linguistic Data Consortium. In *Proceedings of the 1994 ARPA Human Language Technology Workshop*. Morgan Kaufmann, 1994.

[Hull and Grefenstette, 1996] D.A. Hull and G. Grefenstette. Querying Across Languages: a Dictionary-based Approach to Multilingual Information Retrieval. In *19th Ann Int ACM SIGIR Conference on Research and Development in Information Retrieval (SIGIR'96)*, pages 49–57, 1996.

[Hutchins, 1986] W. J. Hutchins. *Machine Translation: Past, Present, Future*. Ellis Horwood, Chichdster, UK, 1986.

[Nagao, 1984] M. Nagao. A Framework of a Mechanical Translation between Japanese and English by Analogy Principle. In A. Elithorn and R. Banerji (eds), editors, *Artificial and Human Intelligence*. NATO Publications, 1984.

[Nirenburg et al., 1991] S. Nirenburg, J. G. Carbonell, M. Tomita, and K. Goodman. *Knowledge-Based Machine Translation*. Morgan Kaufmann Inc, San Mateo, CA, 1991.

[Rissland and Daniels, 1995] Edwina L. Rissland and Jody J. Daniels. Using CBR to Drive IR. In *Proceedings of the Fourteenth International Joint Conference on Artificial Intelligence (IJCAI-95)*, pages 400–407, 1995.

[Salton and McGill, 1983] G. Salton and M. J. McGill. *Introduction to Modern Information Retrieval*. McGraw-Hill Computer Science Series. McGraw-Hill, New York, 1983.

[Salton, 1970] G. Salton. Automatic Processing of Foreign Language Documents. *Journal of American Society for Information Sciences*, 21:187–194, 1970.

[Salton, 1989] G. Salton. *Automatic Text Processing: The Transformation, Analysis, and Retrieval of Information by Computer*. Addison-Wesley, Reading, Pennsylvania, 1989.

[Sheridan and Ballerini, 1996] P. Sheridan and J.P. Ballerini. Experiments in Multilingual Information Retrieval using the SPIDER System. In *19th Ann Int ACM SIGIR Conference on Research and Development in Information Retrieval (SIGIR'96)*, pages 58–65, 1996.

[Wong et al., 1985] S.K.M. Wong, W. Ziarko, and P.C.N. Wong. Generalized Vector Space Model In Information Retrieval. In *ACM SIGIR Conference on Research and Development in Information Retrieval (SIGIR'85).*, pages 18–25, 1985.

[Yang, 1995] Y. Yang. Noise Reduction in a Statistical Approach to Text Categorization. In *Proceedings of the 18th Ann Int ACM SIGIR Conference on Research and Development in Information Retrieval (SIGIR'95)*, pages 256–263, 1995.

# INFORMATION RETRIEVAL

Information Retrieval 1

# Adaptive Personal Information Filtering System that Organizes Personal Profiles Automatically

Toshiki Kindo   Hideyuki Yoshida   Tetsuro Morimoto   Taisuke Watanabe

Matsushita Research Institute, Tokyo Inc.
3-10-1 Higashimita Tama-ku Kawasaki, 214
JAPAN

## Abstract

An adaptive personal information filtering system is presented. The adaptive learning algorithm of the system is derived from the statistical consideration - what keywords are useful for retrieving items in which the user want to see. According to the algorithm, the system organizes a personal profile to reflect the user's interests from an empty profile through user-system interaction - the system displays items and the user returns the answer, "interesting" or "uninteresting". The system gives the user items according to his or her interest estimated using the user's personal profile. The density of the interesting items at the head of the sequence of items sorted by the system is as high as that in the items retrieved using the keyword matching method whose Boolean expression of keywords was made by a researcher who has much experience in patent retrieval. The number of the interesting items included the head of the sorted sequence is larger than that in the retrieved items. A field test is executed to ensure that the system responds to the varied interests of many users. The items used in the test are Japanese newspaper headlines which are supplied at a rate of 600 per day from information providers. Not only researchers and engineers in various technical fields, but also staff members participated in the test as users. The results of the test suggest that the system can be put to practical use.

## 1 Introduction

With the increasing availability of information, it becomes more important to provide methods of filtering information to reflect the user's needs. There are many difficulties in determining what information the user wants to see. The difficulties are based on the following facts: the words people use do not always clearly reflect the topic or content. A single word has more than one meaning in many documents[Foltz and Dumais, 1992]. Furthermore, not only the user's interest but also the value of the words used to retrieve items change, i.e. "internet" was a useful key word a few years ago. However, it is no longer useful today.

Many studies have been performed in an attempt to overcome the ambiguity of words. Latent Semantic Indexing is a methods that takes account of the ambiguity for document profiles[Foltz, 1990]. Information Tapestry is a method used to overcome it by means of collaborative filtering[Goldberg et al. , 1992].

Furthermore, adaptive learning is discussed as an approach to overcome changes in the user's interest[Asakura et al., 1995]. Unfortunately, the discussed adaptive learning algorithm has many heuristic parameters and its effects are not shown.

In this paper, we present an adaptive personal information filtering system with which users do not need to directly manipulate their profile. The system gives them items sorted in order of interest (priority). We call the present system "information filtering system sorting items in order of user's priority" or simply "INSOP". INSOP has only 2 parameters: one is a parameter to limit the size of user profile and the other is a parameter to control the method of manipulation of rare keywords. In Section 2, we derive an adaptive learning algorithm based on the statistical consideration - what key words are useful for retrieving the items in which only one user is interested. In Section 3, we demonstrate the effects of INSOP. We show that INSOP is a good information filtering system. In Section 4, we show the architecture and browser of the current INSOP system used in our field test.

## 2 Adaptive Learning Algorithm

In this section, we show the adaptive learning algorithm of INSOP. First, we consider which key words are useful for retrieving items in which the user is interested. Next, we give an adaptive learning algorithm for an ideal case and refine the algorithm for application to actual cases.

### 2.1 Classification of Key Words

If our subject is to supply a set of key words which are useful for retrieving items in which only one user is interested, then the intrinsic value of each key word depends on whether the key word reflects his or her interest or not. The semantics of each key word is not important in this context. Using this viewpoint, we classify key words (including Boolean combinations of key words) into 3 groups. The first group includes those whereby if one of them appears in an item, then the user has an interest in the item. We call such key words "positive key words". The second group is that whereby if one of the key words appears in an item, then the user dose not have an interest in the item. We call such key words "negative key words". The third group includes those which are not useful in determining whether that the user has an interest in the items with the key word or not. We call such key words "neutral key words".

## 2.2 Ideal Case

In order to explain the principal idea of INSOP clearly, we consider an ideal case in which we can manipulate items, keywords, and user's inputs by statistical methods.

Here we define notations as follows: $w_j$ is the $j$-th key word, $p^*$ is the true probability that an item is one in which the user has an interest, and $q^*(w_j)$ is the true probability that an item with the $j$-th key word is one in which the user has an interest.

The positive (negative) key words satisfy $q^*(w_j) > p^*$ ($q^*(w_j) < p^*$), and the neutral key words satisfy $q^*(w_j) \sim p^*$.

In an ideal case, we can define a statistical quantity to classify key words into the 3 groups as:

$$SKC(q^*(w_j), p^*) = q^*(w_j) \log \frac{q^*(w_j)}{p^*} \quad (1)$$
$$- (1 - q^*(w_j)) \log \frac{(1 - q^*(w_j))}{(1 - p^*)}.$$

$SKC(\cdot)$ is the "Statistical Key word Classifier". When the Statistical Key word Classifier receives one of positive, or negative, or neutral key words as an input, it outputs a positive, or negative value, or the value close to 0, respectively. The Statistical Key word Classifier is similar to Kullback-divergence, which is used to measure the "distance" between 2 distributions in statistics,

$$KD(q^*(w_j), p^*) = q^*(w_j) \log \frac{q^*(w_j)}{p^*} \quad (2)$$
$$+ (1 - q^*(w_j)) \log \frac{(1 - q^*(w_j))}{(1 - p^*)}$$

but the sign of its second term is different from that of the Kullback-divergence.

In addition, we define the "Statistical Item Classifier" as

$$SIC(item_a) = \sum_{j=1}^{n_a} SKC(q^*(w_j^a), p^*) \quad (3)$$

where $item_a$ is the $a$-th item, $n_a$ is the number of the key words of the $a$-th item, and $w_j^a$ is the $j$-th key word of them. When the $a$-th item has many positive (negative) key words, the output of the Statistical Item Classifier is a large positive (negative) value.

The Statistical Key word Classifier and the Statistical Item Classifier work well as tools for classifying items in which a user is interested or not, in the ideal case.

As a result, a system which sorts items in order of the output of the Statistical Item Classifier works as an information filtering system. This is the principal idea of INSOP.

## 2.3 Actual Case

If we apply the Statistical Key word Classifier and the Statistical Item Classifier to an actual information filtering system, we must estimate the probability that an item with the $j$-th key word is of interest based on sample items. The estimated probability, $q(w_j)$, is not equal to the true probability, $q^*(w_j)$, in actual cases. The accuracy of the estimated probability depends on the number of sample data. Accordingly, the following problems must be solved: (1) how to treat key words which are too rare to be manipulated by statistical methods, (2) how to initialize the system, (3) how to adapt user's interests and their modifications, (4) how to treat items which have both positive and negative key words. In this section, we provide the solutions to these problems by extending the Statistical Key word Classifier to the "Key word Classifier" for the treatment of both frequent and rare key words and by introducing an adaptive learning algorithm.

First, we extend the Statistical Key word Classifier in order to solve the first problem. The rare key word problem is that the probability estimated on the basis of a small mount of sample data is different from the true value. We consider a typical case where the $r$-th key word, which is a rare key word whose true probability, $q^*(w_r)$, satisfies $0 \ll q^*(w_r) \ll 1$, appears just once with an interest item. Note that a case in which a new key word, either is rare or not, appears is equivalent to this case. In the real world, there are many such cases. Then, the probability $q(w_r)$ is estimated as

$$q(w_r) = 1 \gg q^*(w_r). \quad (4)$$

As a result, the Statistical Key word Classifier input with the rare (new) $r$-th key word outputs a large but unreliable value. Here, we remember that the Statistical Key word Classifier input with one of the neutral key words, which are useless to estimate user's interest, outputs the value close to 0. Therefore, if we interpret "unreliable" as "useless", we are able to introduce the following "Key word Classifier" to treat frequent and rare key words,

$$KC(w_j) = \tanh\left(\frac{c(w_j)}{\alpha}\right) \cdot SKC(q(w_j), p), \quad (5)$$
$$p = \frac{P^i}{P^i + P^u}, \quad (6)$$
$$c(w_j) = Q_j^i + Q_j^u, \quad (7)$$
$$q(w_j) = \frac{Q_j^i}{Q_j^i + Q_j^u}, \quad (8)$$

where $P^i$ is the appearance count of interesting items, $P^u$ is the appearance count of uninteresting items, $Q_j^i$ is the appearance count of interesting items with the $j$-th key word, $Q_j^u$ is the appearance count of uninteresting items with the $j$-th key word, and $\alpha$ is the parameter, called the "rarity parameter", which controls the manipulation of rare key words. Accordingly, we define "Item Classifier" as:

$$IC(item_a) = \sum_{j=1}^{n_a} KC(w_j^a). \qquad (9)$$

The Key word Classifier outputs a value close to 0 when it receives a key word whose appearance count is less than the rarity parameter $\alpha$. Namely, the Key word Classifier classifies the key word as a useless key word. Consequently, the unfavorable influence of the rare key words to the output of the Item Classifier is small compared with that to the Statistical Key word Classifier. As a result, the Key word Classifier and the Item Classifier work well as tools for classifying the interesting, neutral, or uninteresting items in actual cases.

This is one solution to the first problem of rare key words.

The definitions of the Key word Classifier and the Item Classifier require that the personal profile of INSOP must be in the form given in Table 1.

**Table 1:** Format of INSOP personal profile

| user name | $P^i$ | $P^u$ |
|---|---|---|
| $w_1$ | $Q_1^i$ | $Q_1^u$ |
| $w_2$ | $Q_2^i$ | $Q_2^u$ |
| $\vdots$ | $\vdots$ | $\vdots$ |
| $w_K$ | $Q_K^i$ | $Q_K^u$ |

Accordingly, the adaptive learning algorithm of the personal profile is given as follows.

1) A user reads an item, $item_a$, with key words, $w_1^a, \cdots, w_{n_a}^a$.

2) If the user inputs "interesting" to INSOP, then

   2-1) $P^i \leftarrow P^i + 1$ and

   2-2) search user's personal profile for the same key word as $w_k^a, k = 1, \cdots, n_a$.

   2-3) If there is a key word which satisfies $w_j = w_k^a$, then $Q_j^i \leftarrow Q_j^i + 1$.

   2-4) Otherwise, add $w_k^a$ into user's personal profile as the $(K+1)$-th key word and set $Q_{K+1}^i = 1, \quad Q_{K+1}^u = 0$.

3) If the user inputs "uninteresting" to INSOP, then

   3-1) $P^u \leftarrow P^u + 1$ and

   3-2) search user's personal profile for the same key word as $w_k^a, k = 1, \cdots, n_a$.

   3-3) If there is the key word which satisfies $w_j = w_k^a$, then $Q_j^u \leftarrow Q_j^u + 1$.

   3-4) Otherwise, add $w_k^a$ into user's personal profile as the $(K+1)$-th key word and set $Q_{K+1}^i = 0, \quad Q_{K+1}^u = 1$.

Note that we do not need to embed the user's interest into the personal profile initially. INSOP can be started an empty profile. This is an advantage of INSOP and is also the solution to the second problem (how to initialize the system). There is, however, a disadvantage of the algorithm as follows: the algorithm continually increases the number of key words in the user's personal profile.

We overcome this disadvantage by adding an algorithm which selects the key words which give the "best" finite personal profile of INSOP. In INSOP, positive and negative key words are useful but neutral key words are useless. We introduce a function which outputs a large positive value for positive or negative key words but which outputs a value close to 0 for neutral key words in order to construct INSOP which selects a finite number of useful key words. Such a function is given by

$$KCH(w_j) = \tanh\left(\frac{c(w_j)}{\alpha}\right) \cdot KD(q(w_j), p). \qquad (10)$$

We call $KCH(\cdot)$ the "Key word Checker". When we design the INSOP system, we must give the limit of the number of key words in the personal profile, $N$. In order to make the "best" personal profile of INSOP under the limit, we select $N$ key words which give large outputs of the Key word Checker compared to other key words.

Next, we show that the rarity parameter $\alpha$ is related to the adaptive speed of INSOP. The properties of the Key word Checker are summarized in Table 2.

**Table 2:** Relations between input key word and the Key word Checker output

| input | output |
|---|---|
| positive | large |
| negative | large |
| rare/new | small $\sim$ large |
| neutral | zero |

The outputs of the Key word Checker following input of rare or new key words depend on the rarity parameter. If we give the rarity parameter a value close to 0, then the outputs become very large and INSOP manipulates rare and/or new key words as useful key words compared to positive and negative key words. As a result, INSOP retains new or rare key words whose appearance counts are small and releases old key words - positive or negative key words - whose appearance counts are large. As the rarity parameter increases, rare and new key words become useless and are easily released but positive and negative key words are held fast in INSOP. This shows that we can control the adaptive speed of INSOP according to our requirements using the rarity parameter. This is the solution to the third problem (how to adapt the user's interests and their modifications) and also that of the first problem (how to treat rare key words).

The remaining problem is how to treat items which have both positive and negative key words. We can determine the characteristic of such items but we have no direct solution for this problem yet. The characteristic is that the output of the Item Classifier is close to 0 but

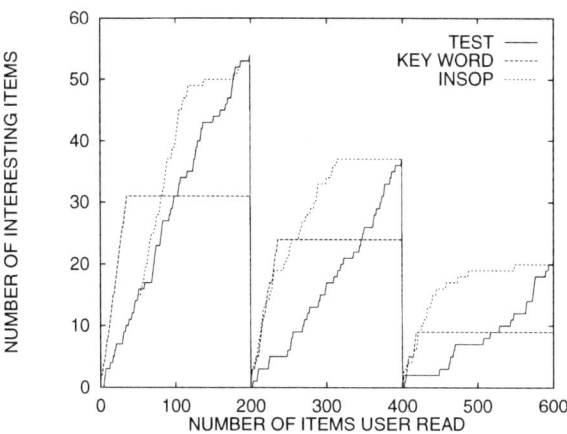

Figure 1: The distribution of items in which the testee has interests

the sum of the outputs of the Key word Checker is large when we input such an item. Accordingly, we designed the current INSOP system which displays the output of the Item Classifier as "needs" and the sum of the Key word Checker outputs as "reliance" in order to help users to identify such items.

## 3 Performance of INSOP

In this section, we demonstrate the performance of INSOP based on experiments conducted in collaboration with several researchers who cooperated as testees. All of them are highly experienced in patent retrieval. We set the limit of the number of key words in the personal profile to $N = 750$ in this experiment.

We obtained test sequences of items which were headlines of Japanese newspapers provided by information providers and the Boolean combination of key words for retrieving to the items in which each testee had interest as follows.

1) The tester requested each testee to mark the items in which they had interests and executed the interface program which displays one item without key words and receives the testee's input - "interesting" or "uninteresting".

2) The testee read 600 items and input his answers.

3) After the testee had input all the answers, the tester stopped the program and requested the testee to produce the Boolean combination of key words to retrieve the items which he marked "interesting".

In order to simulate a case where each user received three sets of items, we divided each test sequence into 3 short test sequences including 200 items.

In Figure 1, we show typical results of the experiments to evaluate the performance of INSOP. The horizontal axis indicates the number of items which the testees read. We refer to the number as $m$. The vertical axis indicates the number of interesting items including the read items. Three sequences of items are shown in Figure 1. The first sequence corresponding to a solid line is the test sequence. The second sequence corresponding to a dashed line is that of items retrieved by the key word matching method using the Boolean combination of key words set by the testee. The third sequence corresponding to a dotted line is that of items sorted by INSOP.

The solid line shows that the distribution of the marked items in the short test sequence is approximately uniform.

The dashed line stops to increase at $m = 36, 238$ and $421$. This shows that the key word matching method retrieves $10\% \sim 20\%$ items. About 60% marked items are included in the retrieved items. Consequently, the density of the marked items in the items retrieved using the key word matching method is about 3 times as high as that in the short test sequences. This rate is consistent with the value of $1.5 \sim 2.5$ reported by ref.[Foltz and Dumais, 1992]. Note that the key word matching method misses several of the marked items (about 40%). Namely, it is difficult to create a Boolean combination of key words that includes all our personal interests, which are indefinite, even if we have much experience in information retrieval. This result suggests that the key word matching method is not useful for retrieving the marked items in this test task.

However, based on this result, we are not suggesting that the key word matching method is useless for all types of information retrieval or filtering. In general, retrieval methods, into which the key word matching method is classified, in which personal profiles are directly manipulated by users, are still useful in cases where users understand the exact subjects to retrieve, i.e. patent retrieval.

The dotted line shows the performance of INSOP. In the experiment, adaptive learning of the personal profile is executed at $m = 50, 100, 250, 300, 450$ and $500$. The order of the first 50 items presented by an INSOP system whose personal profile is empty is the same as the order of the test sequence. At $m = 50$, INSOP executes the first adaptive learning of the personal profile and sorts the remaining 150 items into the first short sequence in the order of the testee's degree of interest estimated using the learned personal profile. As a result, the dotted line begins to go upward and deviates from the solid line at $m = 50$. This indicates that INSOP moves interesting items forward and uninteresting items backward. INSOP executes the second learning and sorting of the remaining items in the first short sequence at $m = 100$. At $m = 200$, INSOP sorts all the items in the second short sequence and executes the adaptive learning and sorting of the remaining items in the short sequence at $m = 250$.

The density of interesting items at the head of the sorted sequence is as high as that in the retrieved items using the key word matching method. The number of interesting items including in the head is larger than the number including in the retrieved items.

These graphs show that INSOP is a good information filtering system which sorts items in order of interest-

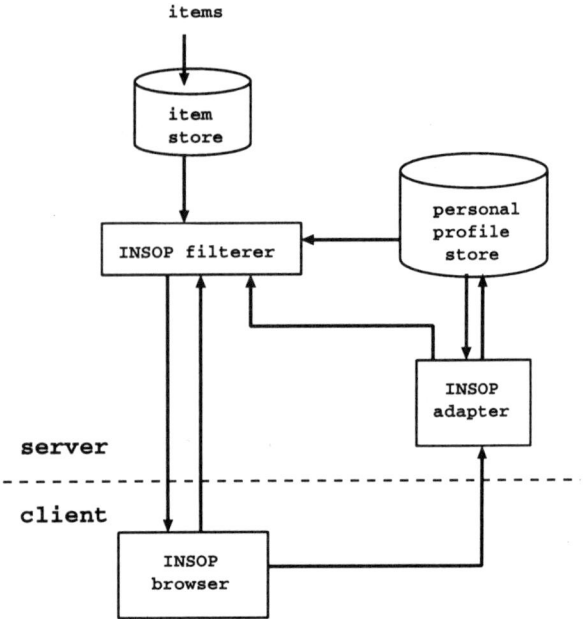

Figure 2: The flow of items through the current INSOP system

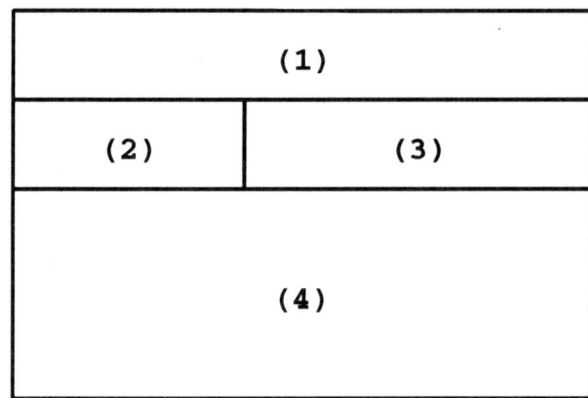

Figure 3: Four areas of the INSOP browser

ing, neutral and uninteresting items compared to the key word matching method.

## 4 INSOP in Field Test

In this section, we demonstrate the current INSOP system in our field test. First, we explain the architecture of the current INSOP system. Next, we show how to use the INSOP browser to receive the INSOP service, and we present the results of the field test.

### 4.1 Architecture

Figure 2 shows the flow of items through the following major architectural components of the current INSOP system.

- *Item store.* Provides storage for items sent by information providers.
- *Personal profile store.* Provides storage of a personal profile.
- *INSOP filterer.* Sorts items in order of user's interest.
- *INSOP browser.* Provides the user with an interface for accessing INSOP services. This includes facilities for displaying items and receiving user inputs.
- *INSOP adapter.* Organizes user's personal profile according to the adaptive learning algorithm described in the previous section.

The current INSOP system uses a client/server model. The client(INSOP browser) is installed into over 50 personal computers (PCs) and operates as an application software on each PC whose OS is Windows95. The server (INSOP filterer/adapter) operates as a daemon process on several UNIX machines. The clients and the servers send data (sorted items, user's input etc.) to each other according to our original protocol based on the TCP/IP.

### 4.2 Browser

If we execute the INSOP browser, the browser requests for us to input a server name, our user IDs, and our passwords. After receiving them, the browser opens the window in which there are 4 areas as shown in Figure 3. In area (1), the name of the opened file or the folder is shown. In area (2), the trees of the files and folders are shown according to the same format as Microsoft Explore. In area (3), accessible files or folders are shown. Area (4) displays items with the two values, "needs" and "reliance", defined at the end of Section 2.

We open an item folder and select a file using a PC mouse in the same manner as with Explore. The items included in the file are displayed in area (4) (we show a copy of the browser at Figure 4 where Japanese items are displayed). The bold frame indicates the item to which we should respond using the mouse buttons - its left button for "interesting" and its right button for "uninteresting" and is moved using the mouse. If we mark an item with "interesting" or "uninteresting", then the color of the left block in area (4) where "needs" and "reliance" are displayed changes from white to dark or light gray, respectively. If we attempt to open a new file after marking items using the buttons, the INSOP browser sends the set of our inputs and the key words with the items to the INSOP adapter. The INSOP adapter organizes our personal profiles according to the adaptive learning algorithm. Then, the INSOP filterer, based on our personal profiles, sorts the items of the new file and sends the sorted items to the INSOP browser. The INSOP browser displays them in area (4).

The calculation time from the start of the adaptive learning to the sorting items is less than 1s in the cases where the number of items is 100. The actual waiting time, depending on the load levels of the network and the server, is 1~5s on the local area network (LAN) of Matsushita Research Institute, Tokyo Inc. (MRIT).

## 4.3 Field Test

We executed the field test of the current INSOP system on the LAN. The purpose of the test was to ensure that INSOP responds to the various interests of many users. The items used in the test were headlines of Japanese newspapers which were supplied by information providers at a rate of about 600 per day. Over 50 users participated in the test including not only researchers and engineers in material, mechatronics, image processing, speech technology and mobile communication but also staff members.

We asked a set of questions to the users regarding their impression of INSOP and the sorting ability 3 times: when the INSOP filterer was operating (case 1), when the we stopped the INSOP filterer (case 2), and when the INSOP filterer was operating again (case 3).

Table 3 shows the result of a questionnaire on the current INSOP system.

**Table 3a** Impression of INSOP

|        | excellent |     | normal |     | poor |
|--------|-----------|-----|--------|-----|------|
| case 1 | 7%        | 50% | 39%    | 4%  | 0%   |
| case 2 | 0%        | 4%  | 35%    | 42% | 18%  |
| case 3 | 11%       | 68% | 18%    | 3%  | 0%   |

**Table 3b** Sorting ability of INSOP

|        | excellent |     | normal |     | poor |
|--------|-----------|-----|--------|-----|------|
| case 1 | 0%        | 79% | 21%    | 0%  | 0%   |
| case 2 | 0%        | 0%  | 14%    | 61% | 25%  |
| case 3 | 14%       | 75% | 11%    | 0%  | 0%   |

Furthermore, over 90% of users answered, "INSOP is useful reading the headlines of Japanese newspapers.".

## 5 Conclusion

In this paper, we presented an adaptive personal information filtering system, INSOP, whose personal profile is automatically organized from an empty profile by means of user-system interaction. The INSOP system is a good information filtering system which sorts items in order of interesting, neutral, and uninteresting items. The results of the experiment and field test suggest that INSOP can be put to practical use.

## Acknowledgments

We thank Electric Library Co. which allows us to demonstrate the INSOP system using the headlines of Japanese newspapers, Mr. Natsuki Oka, Mr. Seiji Kishimoto and Mr. Tatsuro Itoh for their useful comments, and the members of MRIT who cooperated with us as testees.

## References

[Foltz and Dumais, 1992] Peter W. Foltz and Susan T. Dumais. Personalized information delivery: an analysis of information filtering methods. *Communications of the ACM*,35(12):51–60, December 1992.

[Foltz, 1990] Peter W. Foltz. Using Latent Semantic Indexing for information filtering. *Proceedings of the ACM Conference on Office Information Systems*, pages 40-47, Boston, April, 1990.

[Goldberg et al. , 1992] David Goldberg, David Nichols, Brain M.Oki, and Douglas Terry. Using collaborative filtering to weave an information TAPESTRY. *Communications of the ACM*,35(12):61–70, December 1992.

[Asakura et al., 1995] Takayoshi Asakura, Koji Kida, Hiroyuki Tarumi, and Toshiaki Miyashita. Agent based infromation filtering *Information Processing Society of Japan SIG Notes* 95-20-7:49–55, March 1995.

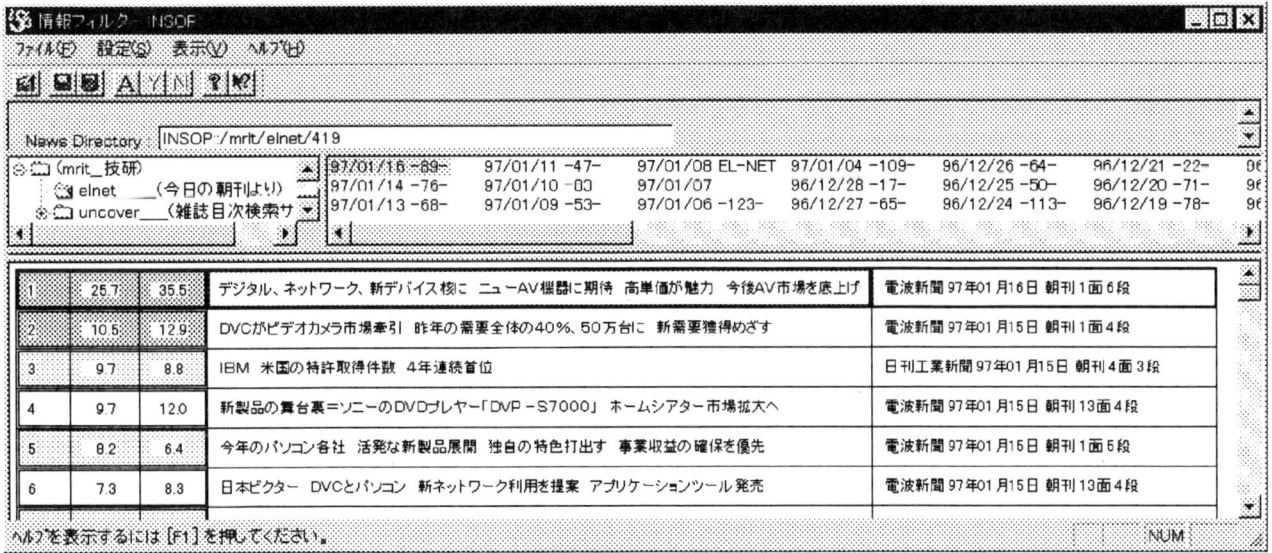

Figure 4:INSOP browser in the field test.

# An Index Navigator for Understanding and Expressing User's Coherent Interest

Yukio OHSAWA* and Masahiko YACHIDA
Dept. Systems and Human Science, Osaka University
1-3 Machikaneyama, Toyonaka, Osaka 560
Japan

## Abstract

It is important to guide a user to interesting documents in a large-scale document-database. However, when the user is not an expert of the area of interest, it is difficult for him/her to name precise keywords in which he/she is interested, nor to select areas of his/her own interest. This paper presents an Index Navigator which clarifies what areas the user is interested in, what keywords he/she should investigate, and what documents concern his/her interest. A tough problem for such a system is to understand interesting areas for the user, among other areas-sets which can explain his/her behavior. Our Index Navigator employs an inference method called Cost-based Cooperation of Multi-abducers (CCMA), for understanding user's interest from the history of the user's expression of interest in incomplete keywords, even if the changing speed of the user's interest is totally unknown. With this device, the Index Navigator guides the user to really important areas, keywords and documents.

## 1 Introduction

Recently, very large amounts of text have been stored in several types of media such as CD-ROMs, electronic libraries, and the Internet. In such situations, a user desiring important documents often loses his or her way because there are too many information resources.

In order to cope with this problem, methods have been presented to make a match between the user's interest and the documents. A major strategy is to represent the user's interest by a keyword-vector, in which each keyword is assigned a weight based on the frequency it was read or written by the user [Krulwich, 1995], or by the sheer keywords requested by the user. This vector is used to match keyword-vectors of documents in hand. Unfortunately, especially when the user is not familiar with the domain of the desired document, this matching becomes extremely difficult because of the discrepancy between the terms in the domain and the vocabulary of the user [Furnas et al, 1987]. Also, the user may fail to name precise keywords in which he/she is interested, because of the poor expressiveness due to haste.

In order to cope with this problem, we aim at improving the user's ability to express his/her interest. We propose a method to *infer* the areas of the user's interest from *incomplete* (poor) keywords he/she requested. Additional keywords are obtained from these areas (classifying the interest to areas is beneficial itself too, e.g. for searching documents in a library classified by areas, or investigating what academic society the user should subscribe). By seeing additional keywords, the user learns to express his/her own interest, which is useful when the user writes, reads, or searches documents from a larger set of documents in electronic libraries or the Internet.

A tough problem in this inference, however, is again the lack in the keywords requested by the user - areas cannot be uniquely determined from too few keywords. An effective approach to obtain unique and precise areas of the user's interest is to consider the history of requested keywords, i.e. inferring the user's current interest to be coherent with his/her recent interests. However, changes in the user's interest are difficult to understand. For example, a user who has been interested in using syntactic rules for machine translation may, one day, face a problem that sentences cannot be translated without considering semantics. He may change his interest from syntax to semantics. Such a change comes from newly encountered situations, unable to be expressed in the user's model acquired from the user's past behaviors [Armstrong et al, 1995; Krulwich, 1995]. In [Balabanovic and Shoham, 1995], the user's model is revised dynamically by changing weights of words based on their frequency in documents the user accessed recently. However, frequencies of incomplete keywords compose incomplete information about the change in the user's interest, e.g. shifting from the use of "Indonesian" to "English" does not mean that interest in "translation" from Indonesian to English disappeared.

In this paper, we present an *Index Navigator* which clarifies to a user what *areas*, *keywords*, and *documents* match his/her interest, based on the history of the keywords the user selected. For understanding the user in-

*Email: osawa@sys.es.osaka-u.ac.jp

terest transitions, the Index Navigator uses an inference method called *Cost-based Cooperation of Multi-abducers* (CCMA) [Ohsawa and Yachida 1997]. CCMA obtains a coherent explanation of user's history of selected keywords, to classify the user's current interest to areas. Since CCMA understands an event-sequence even if the changing speed is totally unknown, our Index Navigator allows for unpredictable changes in the user's interest. The Index Navigator presents areas, keywords, and documents matching the user's interest, so that the user learns to express his/her own interest.

## 2 The Outline of Index Navigator

In this paper, we present an Index Navigator, which supports a user who searches documents matching his/her interest, but cannot express his/her interest fully by keywords nor content areas.

A user of the Index Navigator selects keywords, i.e., index terms he/she can request, by clicking on a given *Input-Index List* (the largest sub-window in Figure 1) or by entering keywords in the upper left window in Figure 1. Each time the user selects a set of keywords, a set of areas are inferred corresponding to the user's current interest, from the history of selected keywords, by the *Interest Tracer*. These inferred areas are listed in the *Areas List*, in the middle right of Figure 1. Keywords expressing the user's interest are listed in the *Output-Index List*, the upper right of Fig.1. At the same time, documents corresponding to the user's interest are displayed in the *Document List*, at the bottom of Fig.1.

In the remainder of this paper, we first sketch the mechanism of the Index Navigator in Section 3 where the subsystems of the Index Navigator, each executing inference to present the Areas List, the Output-Index List, and the Document List, are stated. Then, we evaluate the performance of the Index Navigator experimentally in Section 4 and discuss about the results. We make conclusional remarks in Section 5.

## 3 The Mechanism of Index Navigator

The Index Navigator is composed of three subsystems. First, the *Interest Tracer* fills in the Areas List. Then, the Output-Index List and the Document List are filled in, in the second and the third subsystems respectively. The following subsections describe the details of these subsystems.

### 3.1 Interest Tracer

In order to consider changes in user's interest, we employ an inference system called *Cost-based Cooperation of Multi-abducers* (CCMA), which explains an event-sequence by hypotheses (corresponding to the causes), even if the changes in causal events occur at a totally unknown speed. In this subsection, let us show why and how we use CCMA for the Interest Tracer.

#### Coherent explanation of selected keywords

The history of the user's behaviors can be regarded as a coherent sequence of events (caused by the user's coher-

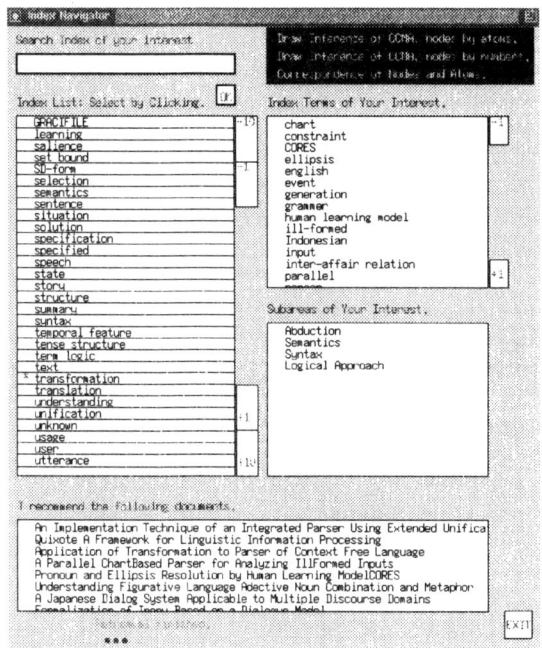

Figure 1: A window of the Index Navigator.

ent interest), which has been considered in many works. In probabilistic models, *transition probabilities* represent the coherence among events at nearby moments or places [Kashino et al, 1995; Chen et al, 1994]. However, a transition probability, i.e., the occurrence probability of an event after a fixed or unfixed interval from a preceding event cannot be determined if the situation changes at a totally unknown speed, though it may work to obtain probable areas of a user or a document given without history [Fung and Favero, 1995].

Coherence among sequential events has been formalized in [Ng and Mooney, 1990], where *Explanatory Coherence* (EC, hereafter) is defined by the number of common hypotheses shared by the solution hypotheses-sets of *abduction* (*hypothetical reasoning*). In abduction, causes of observed events (goals) are inferred as a solution hypotheses-set [Poole et al, 1987; Poole, 1993]. Abduction on EC obtains the solution hypotheses-sets of maximal explanatory coherence. EC is useful even if the speed of changes in events are unknown, because it does not rely on any predefined parameter of changing speed, like transition probabilities. However, EC cannot evaluate an explanation of observed events, if they do not have common causes, i.e. if the situation changes radically. On the other hand, in cost-based abduction (CBA, hereafter), a weight is assigned to each element hypothesis and the best solution hypotheses-set is defined as the hypotheses-set with the lowest cost, where cost is the sum of weights of element hypotheses in the solution hypotheses-set. The solution hypotheses-set of CBA is meaningful in that it corresponds to the MAP (maximum a-posteriori) explanation of one goal [Charniak and Shimony 1994]. However, CBA does not consider the coherence in the sequence.

For understanding an event-sequence changing in an unknown speed, both of these two criteria, i.e. the MAP explanation in CBA and the coherence considered in EC, are merged in CCMA.

**Process and Semantics of CCMA**

The fundamental mechanism and algorithm of CCMA was presented very recently [Ohsawa and Yachida 1997](*in Japanese*), as a new distributed reasoning framework. However, CCMA has been applied only to *non-real world* toy problems, e.g., artificially given varying fault-states (bad connections which is sometimes naturally fixed by unpredictable external forces) in electronic circuits such as small full-adders. This restriction in the previous applications of CCMA came from the fact that the semantics of CCMA was yet under construction and all the efforts have been devoted to analyzing "why and when CCMA works" by solving artificial (i.e. intentionally produced) problems where too complex elements in the *real-world* to take into account are not included (more detailed experimental data than [Ohsawa and Yachida 1997] about the analysis of CCMA on artificial problems will appear in future in English, which shows that CCMA catches state transitions better than previous distributed reasoning methods e.g. FA/C [Carver and Lesser, 1995]).

With our later progress in formulating the semantics of CCMA (as in Eq.(3) and Eq.(4) below), we came to deal with unpredictable changes in the *real world* such as the will of users of document search engines. The Interest Tracer is the first application of CCMA to a real world problem: understanding the changing interest of users – changing in unpredictable speed moved by complex elements in the real world (emotional occasions or accidentally encountered situations) which is hardly included in artificial problems such as ones we considered so far.

Here, let us introduce reader to CCMA's fundamental algorithm and the updated formulation of CCMA semantics.

In CCMA, multiple abduction systems (*abducers*, hereafter) of CBA are assigned to goals (events observed in an event-sequence), and each abducer obtains an optimal (lowest cost) solution hypotheses-set to explain its goal. Each abducer iterates cycles of revising weights of hypotheses according to the explanations of recent events by recent abducers and then explaining its own goal again, until the solution hypotheses-sets converge to a coherent explanation of the sequential events.

Formally, the problem of CCMA is described as finding solution-sets $h_j (j = 1, 2, ...N)$, which satisfies the following constraints in Eq.(1) where $N$ is the length of the events-sequence, i.e., the number of observed events. Here, knowledge is composed of *background knowledge* $\Sigma$ (knowledge given as truth in any environment) and *hypothetical knowledge* $H$ (a set of element hypotheses which may be false and may contradict others). $G_j$ is the $j$ th event in the sequence. The solution to be searched is a set of hypotheses-sets $h_j \in 2^H$ which, if combined with $\Sigma$, supports $G_j$ for each $j(j = 1, 2, ...N)$ without contradiction ($\phi$ denotes the empty clause).

$$\begin{aligned} h_j &\in 2^H, \\ h_j \cup \Sigma &\vdash G_j, \\ h_j \cup \Sigma &\not\vdash \phi. \end{aligned} \quad (1)$$

We deal with $\Sigma$ composed of causal rules expressed in Horn clauses as Eq.(2), which means that $y$ is true if all of $x_1, x_2, ...x_n$ are true, where variables $x_1, x_2, ...x_n$ and $y$ are atoms each of which corresponds to an event occurrence. When applied to our Index Navigator, $x$ is true if $x$, for example, means "semantics" and the user is really interested in semantics.

$$y : -x1, x2, ..., xn. \quad (2)$$

Also, a probabilistic causality can be described by assigning a conditional atom $x_{n+1}$ representing an event of probability $p(x_{n+1})$. The *weight* of $x_i$ is defined by $weight(x_i)$ equal to $-log\, p(x_i)$, by which the solution of CBA corresponds to the MAP hypotheses as we stated above.

For example, in Figure 2, leaves denote the given hypotheses (content areas of the user's interest) $H$ and other nodes denote goals (selected keywords). Edges in Figure 2 denote the causal links between areas and keywords, e.g., the left-most graph means a Horn clause

$$GPSG : -syntax, p1,$$

where $p_1$ is an atom representing an event of probability $p(p_1)$, i.e., (the number of documents including "GPSG") / (the number of documents in the area of syntax analysis)(note: GPSG is a grammatical analysis method). Also, the probabilities of "GPSG" and "syntax" are (the number of documents including "GPSG") / ($D$:=the number of documents in the database) and (the number of documents in the area of "syntax") / $D$, respectively. This implies that we approximate the interest of a user by the interest of the society, i.e. the number of accepted documents. Weights of hypotheses are defined on these probabilities. In Figure 2, "GPSG,""transformation,""understanding," and "translation" depict keywords that the user selected sequentially at time $t$ $(G_1), t + dt$ $(G_2), t + 2dt$ $(G_3)$, and $t+3dt$ $(G_4)$ respectively. Bold-face numbers depicted below the nodes in Figure 2 are the weights of corresponding hypotheses (virtual weights for a simple illustration). As a measure of coherence, the solution hypotheses-set for $G_i$ is selected by minimizing $C_{1j}$ in Eq.(3).

$$C_{1j} = \Sigma_\eta weight(\eta, s.t., \eta \in h_j, \eta \notin h_{j+1} \cup h_{j-1}), \quad (3)$$

i.e. hypotheses *not believed* in adjacent abducers are *least believed*. Here we have the notion of coherence in EC, where hypotheses *believed* in adjacent abducers are *strongly believed*, and also the notion of lowest-cost solution employed in CBA. If minimizing $C_{1j}$ is not enough

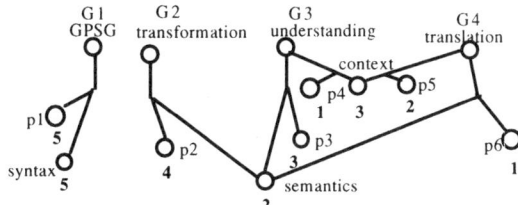

Figure 2: A knowledge-base for CCMA. Nodes at the bottom (i.e. leaves) depict areas, and nodes at the top depict requested keywords.

to define a unique solution, we should focus attention to more strongly believed hypotheses by minimizing

$$C_{2j} = \Sigma_\eta weight(\eta, s.t., \eta \in h_j, \eta \notin h_{j+1} \cap h_{j-1}). \quad (4)$$

In CCMA, the minimization of $C_{1j}$ and $C_{2j}$ are realized by distributed reasoning. Here, the $j$ th abducer is the abducer which explains the $j$ th goal $G_j$ from $\Sigma$ and $H$ by the lowest-cost solution hypotheses-set $h_j$. The solution hypotheses-set $h_j$ is a set of atoms in $H$ whose cost, defined in Eq.(5), is the lowest of all the feasible explanations of $G_j$. Here, $\eta_{ij}$ is the $i$ th element hypothesis in $H$ in the $j$ th abducer, whose weight is $w_{ij}$.

$$cost_j = \sum_{j|\eta_{ij} \in h_j} w_{ij}. \quad (5)$$

Variables appearing in the following procedure of CCMA, are defined as, $k$ : the count of distributed abduction cycles of the multiple abducers, $t_{ij}$ : the truth value of $\eta_{ij}$ defined as 1 for true, 0 for false. The item in Table 1 for $\{t_{i(j-1)} + t_{i(j+1)}$, (current) state of $\eta_{ij}\}$ denotes the next value of $w_{ij}$, i.e., "$w_{ij}$" in Table 1 means that $w_{ij}$ stays with the current value and "$w_{oij}$" denotes that $w_{ij}$ changes into $w_{oij}$, the value of $w_{ij}$ given initially when $k$ was 0. Refractory periods are explained later.

*The procedure of CCMA*

1. $k = 0$.
2. $k = k+1$. Obtain the lowest cost solution $h_j$ for $G_j$ by the $j$ th abducer. Put $\eta_{ij}$ to an *absolute refractory period* in the $k$ th cycle if $t_{ij}$ came to be 1 in the $k$ th cycle and to a *relative refractory period* if $t_{ij}$ came to be 0 in the $k$ th cycle but was 1 in the $(k-1)$ th. Otherwise put $\eta_{ij}$ to a *reactive period*.
3. Change $w_{ij}$ as shown in Table 1. **Stop** if none of $w_{ij}$ for $j = 1$ to $N$ is changed here. Otherwise Go to 2.

Briefly, changing a weight means a belief revision in the corresponding hypothesis. $w_{ij}$ is set to 0 in the $k-1$ th cycle which is a reactive period, if the corresponding hypotheses in adjacent abducers (i.e. $\eta_{i(j-1)}$ and $\eta_{i(j-1)}$) are inferred as true, by which $C_{1j}$ in Eq.(3) is minimized locally in the $j$ th abducer in the $k$ th cycle.

Refractory periods, in which weights are not reduced so easily as in a reactive period, prevents CCMA from

| State of $\eta_{ij}$ \ $t_{i(j+1)} + t_{i(j-1)}$ | 0 | 1 | 2 |
|---|---|---|---|
| reactive period | $w_{ij}$ | 0 | 0 |
| relative refractory period | $w_{ij}$ | $w_{ij}$ | 0 |
| absolute refractory period | $w_{oij}$ | $w_{ij}$ | $w_{ij}$ |

Table 1: Change of $w_{ij}$, for the state of $\eta_{ij}$ and value of $t_{i(j-1)} + t_{i(j+1)}$.

*limitless cycles*. For example, Figure 3 shows the process of CCMA for Figure 2. Each graph in Figure 3 is a subtree of Figure 2, extracted in a top-down manner beginning with each goal. In Cycle 1, abducers infer that the user is interested in $\{h_1:$ *syntax*, $h_2:$ *semantics*, $h_3:$ *context*, $h_4:$ *semantics*$\}$ because these are the lowest-cost hypotheses-sets in abducers 1, 2, 3 and 4 (for $G_1, G_2, G_3$ and $G_4$) respectively. Messages are sent as the dashed arrows in Figure 3, which change the weights of the hypotheses at the arrowheads, e.g., the arrow from "semantics" in abducer 4 to "semantics" in abducer 3 in Cycle 1 reduces the weight of the latter and enhances the belief in the interest in "semantics" of abducer 3 in Cycle 2. In Cycle 3, the procedure stops because the weights do not change any more. The result is that the user is coherently interested in "semantics", after the change from "syntax" to "semantics" at time $t + dt$. Thus, the coherent interest of the user and the radical change in the interest are obtained. Generally, the propagation of the weight-changing message from a hypothesis successfully stops at the abducer whose goal cannot be supported by the same hypothesis. However, suppose that "context" of $G_3$ receives a message from "context" of $G_4$ in Cycle 2, it becomes true again in Cycle 3 as well as in Cycle 1, so that abducers return to the same states as in Cycle 1 - hence limitless cycles.

Refractory periods extinguish the causes of limitless cycles (not completely but mostly) by two main effects. First, in an *absolute refractory period*, $\eta_{ij}$ totally ignores the belief of the adjacent abducers in $\eta_{i(j+1)}$ and $\eta_{i(j-1)}$, which may be due to a previous message from $\eta_{ij}$. Second, in a *relative refractory period* of $\eta_{ij}$, $\eta_{ij}$ is believed more strongly if it is believed by both of its neighboring abducers, than if it is believed by only one of them. This helps focusing the solution to the one of the minimal $C_{2j}$ (see Eq.(4)) and avoiding too many hypotheses exchanged for one observation, as "semantics" and "context" in Figure 3.

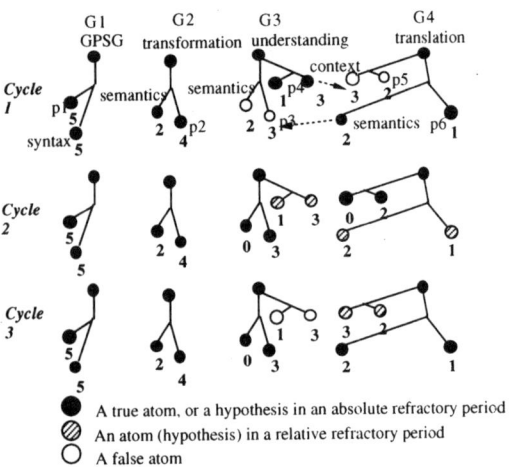

Figure 3: CCMA process for the sequence of $G1, G2, G3,$ and $G4$.

## 3.2 Generating the Output-Index List and the Document List

Keywords and documents relative to the intersection of areas in the Areas List are presented in the Output-Index List and the Document List respectively. For example, for the Horn-clause in Eq.(6), $keyword0$ is obtained if $area1$ and $area2$ are in the Areas List, with an assigned priority of $p0$. The priority of $keyword1$ becomes $p1 \cdot p(area3)$ in this case, which is normally less than $p0$ (if $p0$ is not far less than $p1$ and $p(area3)$ is far less than 1). Keywords in some area in the Areas List are sorted by priority values and listed in the Output-Index List,

$$keyword0 : -area1, area2, p0.$$
$$keyword1 : -area1, area3, p1. \quad (6)$$

A document in the Document List is selected similarly. The only difference is that another knowledge-base including rules as in Eq.(7) is used, where the probability of $p_0$ is given by 1/(number of documents in $area1 \cap area2$).

$$document0 : -area1, area2, p_0. \quad (7)$$

## 4 The Performance of the Index Navigator

The system is coded in C++ and the interface is in X-window, on a Pentium 166MHz computer with 64M RAM running Linux. In the experiment here, the Index Navigator is used with a knowledge base acquired from the data-base in the CD-ROM of the Journal of Japanese Society of Artificial Intelligence, in HTML format, in which 741 papers are stored, given keywords, and classified by areas. Ten students in the engineering field, who know nothing about the mechanism of CCMA, used this system for one week concentrating on retrieving interesting documents. These students were assigned the task of presenting reports at the end of the week, about the documents they read.

Four windows of Index Navigators, similar to Figure 1, were displayed, each employing an Interest Tracer realized by CCMA, CBA, EC, and Bayesian Networks. Users could select favorite windows without being bound by prejudice on inference methods, because these windows were marked simply "A", "B", "C" and "D" respectively (though the keywords selected are shared by all four methods, the user selected keywords by clicking on the Input Index List in his/her favorite window).

During the week, the user's selection of keywords and the inference results (presented lists) were recorded automatically. Also, in order to construct a Bayesian network to be compared with CCMA, probabilities were calculated for any transition from the interest in one area to another at the next time a user selects keywords (areas of the user's interest at a time were determined by the areas in the Areas List in the window he/she selected).

### 4.1 Evaluation of the Performance of the Index Navigator

**The usefulness of the Interest Tracer by CCMA**

Figure 4 shows the changes in the number of users who clicked on the Input Index List in each of the four windows for each three hours (the periodical increase and decrease are due to days and nights).

Multiple different ways to interpret the results in Figure 4 may be possible. First of all, the curves depict that the number of users of every method decreased with date to be none finally, which may seem to imply that all the tested methods were judged to be useless at last. However, this is not the case because the tested students were assigned the task of presenting a report about their retrieved documents. That is, they had to spend time reading the retrieved documents *carefully* and for finishing their reports in the last few days, rather than retrieving new documents and learning to express their interest in keywords. Secondly, those methods whose users came to be none early may seem to be so sophisticated that they satisfied users quickly, because users of one method did not increase with the decrease of other methods, according to Figure 4. However, this is neither the case – according to the detailed data obtained by monitoring users' behaviors, users tried all the methods (windows A, B, C, and D) in the early days of the week and quitted to use useless ones – it was very rare that a user suddenly began to use methods new to him/her (in other words, users seldom stopped their tasks on the early days by being satisfied with the information they obtained). Thus, we interpret a method to be the more useful, the longer used it was.

The performance of Bayesian network grew first due to its learning, but CCMA overwhelmed this afterward. According to the users' comments, the inferred areas by Bayesian network rapidly went out of their interests. For example, when a Bayesian network was used and a user's interest seldom stayed in "semantics" for a long time on the first day, it was inferred that his interest changed to

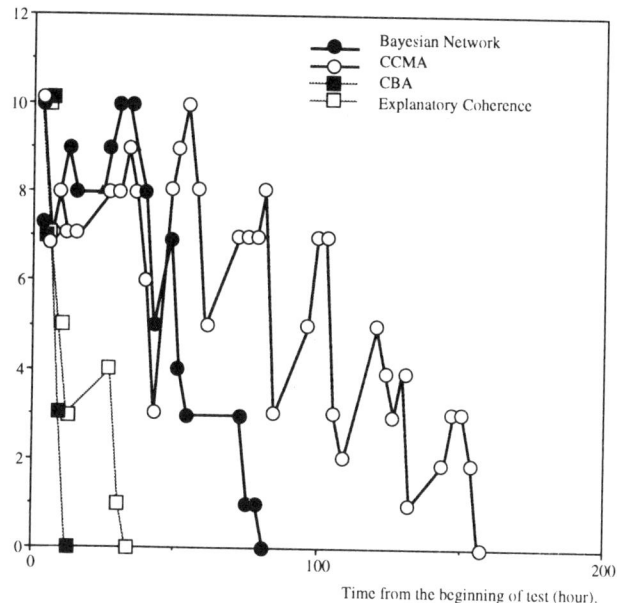

Figure 4: Number of users of Index Navigator employing variable methods

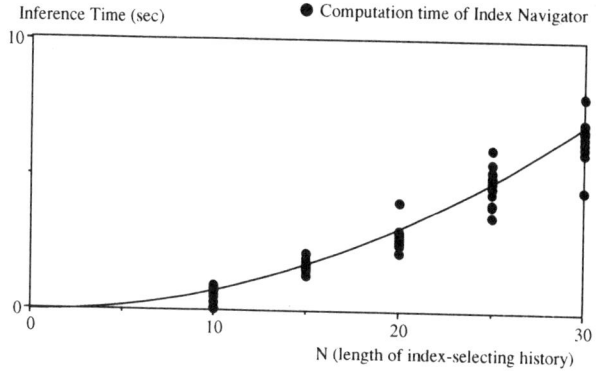

Figure 5: The time performance of CCMA.

area "syntax" even when he selected "translation" after "understanding", i.e. even though he was still interested in "semantics" on the second day. This is because the transition probabilities learned on the first day were used even though they no longer worked - the changing speed of user's interest varies on many factors, e.g., whether he is trying to catch the overall perspective of a field or to understand one concept deeply. On the other hand, by CCMA this user was inferred to be interested in the area "semantics" throughout the second day. Also, his interest was inferred as in "syntax" area when he selected "translation" after "GPSG", and this was a correct understanding on the first day. The performance of CBA (employing a CBA reasoner for one keywords-set selected at each moment independently of other moments, in the Interest Tracer) was the worst, because CBA does not consider the coherence at all. Finally, *too coherent* history of areas came out by EC (Explanatory Coherence). Whenever EC is used, areas which *are* now not, but *were* in the user's interest previously, came into the Areas List. This is especially troublesome when the user's interest changes from one area to another, because areas of his/her previous interest also appear in the areas-list and thus decreases the extracted keywords and documents.

### The keywords in the Output-Index List worked

Although the Index Navigator itself cannot be applied to the entire Internet yet, because not all the pages are classified by areas, the keywords learned by using the Index Navigator can be used as terms to search for interesting pages using existing search engines.

The pages extracted by the search engine TITAN (http://isserv.tas.ntt.jp/chisho/titan.html) after the testing week drastically came to fit the user's interest. Only 7.2% accessible pages (pages except those already closed or forbidden to access) in TITAN's output matched the tested users' interest before the testing week, which grew to 92.2% after the week.

### The speed of the Interest Tracer by CCMA

The most time-consuming component in the Index Navigator is the Interest Tracer. Here, we sped up the abducers in the CCMA of the Interest Tracer by the *Networked Bubble Propagation* (NBP) method, a high-speed CBA method. NBP obtains a near lowest cost solution of CBA within a polynomial-time of $O(|H|^2)$, where $|H|$ is the number of hypothesis in $H$ [Ohsawa and Ishizuka 1997]. We employed this approximate solution method, because multiple abducers compensate for inaccuracy (difference from the exact optimal solution) in each other. The CCMA calculation time is approximately $O(N^2 \cdot |H|^2)$, since each of $N$ abducers takes $O(|H|^2)$ in one cycle, and $N$ cycles run before all the messages propagate over $N$ abducers where $N$ is the length of the sequence.

The time taken for $N = 1$ to 30 of the Interest Tracer by CCMA is shown in Figure 5. The data fits $O(N^2)$, corresponding to the above anticipation of $O(N^2|H|^2)$ considering that $|H|$ is independent of $N$. The current Index Navigator considers keywords selected at 20 recent moments, to make the system respond within 5 second.

## 5 Conclusions

We presented an Index Navigator, which provides a user with keywords that express his/her interest. A user who learns to express his/her own interest in keywords by the Index Navigator becomes able to search for important documents in larger information sources. The successful results show that user interests are understood correctly, even if the user interest changes radically. CCMA, an inference method for understanding changing event-sequences, achieves good inference for this purpose, and also good time-performance, i.e. polynomial against the size of knowledge and the length of the sequence.

The Index Navigator is applied to a CD-ROM currently and users are indirectly aided to use the Internet, if they will, by learning to express their interests. In order to make the Index Navigator useful as a tool to directly search documents from a larger document collection e.g. an electronic library or the Internet, the inference speed of CCMA will be an important issue. In order to apply the Index Navigator to such large-scale document collections, our next plan is to re-install the distributed inference of CCMA on a parallel computer, which will further improve the inference-speed.

## Acknowledgments

We appreciate Prof. Mitsuru Ishizuka (the University of Tokyo) and Prof. Toru Ishida (Kyoto University) for helpful comments. Also, we express special thanks to Assoc. Prof. Seiji Yamada (Tokyo Institute of Technology), Assoc. Prof. Takashi Washio (Osaka University) and Dr. Tomohiro Yamaguchi for constructive discussions.

## References

[Armstrong et al, 1995] Armstrong,R., Freitag,D., Joachims,T. and Mitchell,T., WebWatcher: A Learning Apprentice for the World Wide Web, *AAAI Spring Sympo. Series on Information Gathering from Distributed, Heterogeneous Environments*, 1995[1].

[Balabanovic and Shoham, 1995] Balabanovic,M., Shoham,Y., Learning Information Retrieval Agents: Experiments with Automated Web Browsing, *AAAI Spring Sympo. Series on Information Gathering from Distributed, Heterogeneous Environments*, 1995.

[Carver and Lesser, 1995] Carver, N., Lesser, V., The DRESUN Tested for Research in FA/C Distributed Situation Assessment, Proc. Int'l Conf. Multiagent Systems, 33-40, 1995.

[Charniak and Shimony 1994] Charniak,E., Shimony, S.E., Cost-based abduction and MAP explanation, *Artif.Intell.* 66(2): 345-374, 1994.

[Chen et al, 1994] Chen, M.Y., Kundu, A. and Zou, J., Off-Line Handwritten Word Recognition Using a Hidden Markov Model Type Stochastic Network, *IEEE Trans. PAMI*, 16, 481-496, 1994.

[Fung and Favero, 1995] Fung,R., Favero,B.D., Applying Bayesian Networks to Information Retrieval, *Com. ACM*, 38(3): 42-48, 1995.

[Furnas et al, 1987] Furnas,G.W., Landauer,T.K., Gomez, L.M., and Dumais, S.T., The Vocabulary Problem in Human-System Communication, *Com. ACM*, 30(11): 964-971, 1987.

[Kashino et al, 1995] Kashino,K., Nakadai,K., Kinoshita,T. and Tanaka,H., Organization of Hierarchical Perceptual Sounds: Music Scene Analysis with Autonomous Processing Modules and a Quantitative Information Integration Mechanism, *Proc.IJCAI'95*, 158-164, 1995.

[Krulwich, 1995] Krulwich,B., Learning User Interests Across Heterogeneous Document Databases, *AAAI Spring Sympo. Series on Information Gathering from Distributed, Heterogeneous Environments*, 1995.

[Ng and Mooney, 1990] Ng,H., Mooney,R., On the Role of Coherence in Abductive Explanation, *Proc. AAAI'90*, 337-342, 1990.

[Ohsawa and Ishizuka 1997] Ohsawa,Y., Ishizuka,M., Networked Bubble Propagation: a Polynomial-time Hypothetical Reasoning Method for Computing Near-optimal Solutions, *Artif. Intell.* July 1997.

[Ohsawa and Yachida 1997] Ohsawa,Y., Yachida,M., Understanding Sequential Events by Cost-based Cooperation of Multiple Abducers, *J. Japanese Soc.AI*, 12(4), July 1997 (in Japanese).

[Poole et al, 1987] Poole,D., Aleliunas,R. and Goebel,R., Theorist : A Logical Reasoning System for Defaults and Diagnosis, *in the Knowledge Frontier : Essays in the Knowledge Representation*, Springer-Verlag, 1987.

[Poole, 1993] Poole,D., Probabilistic Horn Abduction and Bayesian Networks, *Artif. Intell.* 64(1): 81-130, 1993.

---

[1] *Papers referred here from AAAI Spring Sympo. Series on Information Gathering from Distributed, Heterogeneous Environments are available from http://www.isi.edu/sims/knoblock/sss95/proceedings.html*

# Wrapper Induction for Information Extraction

**Nicholas Kushmerick   Daniel S. Weld**
Department of Computer Science & Engineering
University of Washington, Box 352350
Seattle WA 98195-2350 USA
{nick, weld}@cs.washington.edu

**Robert Doorenbos**
NETbot, Inc.
4530 Union Bay Pl. NE, Suite 208
Seattle WA 98105 USA
bobd@netbot.com

## Abstract

Many Internet information resources present relational data—telephone directories, product catalogs, *etc.* Because these sites are formatted for people, mechanically extracting their content is difficult. Systems using such resources typically use hand-coded *wrappers*, procedures to extract data from information resources. We introduce *wrapper induction*, a method for automatically constructing wrappers, and identify HLRT, a wrapper class that is efficiently learnable, yet expressive enough to handle 48% of a recently surveyed sample of Internet resources. We use PAC analysis to bound the problem's sample complexity, and show that the system degrades gracefully with imperfect labeling knowledge.

## 1 Introduction

The Internet contains many sources of relational data. For example, when queried with a name, email address services return ⟨name, email⟩ pairs. But because these sites are designed for people, the content is formatted for human browsing (*e.g.* an HTML page), rather than for use by a program. Therefore, software systems using such resources (*e.g.*, heterogeneous database systems [Chawathe *et al.*, 1994; Arens *et al.*, 1996] or software agents [Etzioni & Weld, 1994; Kirk *et al.*, 1995]) must translate query responses to relational form.

*Wrappers* are commonly used as such translators. A wrapper is a procedure, specific to a single information resource, that translates a query response to relational form. Wrappers are typically hand-coded; unfortunately, hand-coding is tedious and error-prone.

We seek an automated solution to this problem of constructing wrappers. Natural language processing has been used for similar information-extraction tasks; see [Cowie & Lehnert, 1996] for a recent summary. But many information resources do not exhibit the rich grammatical structure such techniques are designed to exploit. Moreover, linguistic approaches tend to be slow, while ideally wrappers should execute quickly, because they are used on-line to satisfy users' queries.

*Wrapper induction* is a new technique for automatically constructing wrappers. Our system learns a wrapper by generalizing from example query responses. A PAC model bounds the number of examples needed to generate a satisfactory wrapper. The inductive algorithm requires an oracle to label examples; we solve this *labeling problem* [Etzioni, 1996] by composing oracles from heuristic knowledge, and we demonstrate that our system degrades gracefully with imperfect heuristics.

We identify HLRT, a class of wrappers which is efficiently learnable, yet expressive enough to handle numerous actual Internet information resources. HLRT is designed for resources that display their content in a tabular layout. HLRT wrappers scan their input for substrings that delimit the information to be extracted. Though our focus is on Internet resources, these learned delimiters need not be HTML tags, but can be arbitrary text.

HLRT corresponds essentially to a class of finite-state automata, so wrapper induction is similar to FSA induction (*e.g.*, [Angluin, 1982]). Since FSAs run in linear time, HLRT satisfies the desire that wrappers be fast. However, since wrappers are used for parsing (rather than just classification), the learned FSA must have a specific state topology. Existing FSA induction algorithms do not make such guarantees, so we have developed a new algorithm targeted specifically at HLRT.

We make the following contributions. *First*, we formalize the wrapper construction problem as that of inductive generalization. *Second*, we identify the HLRT wrapper class, which is efficiently learnable yet reasonably expressive. *Third*, we show how to compose the required oracle from (possibly imperfect) heuristics.

We proceed as follows. In Sec. 2, we describe wrappers. In Sec. 3, we cast wrapper construction as inductive generalization; we then spell out this framework by describing how to learn HLRT (Sec. 4), applying the PAC framework (Sec. 5), and presenting a modular approach to building oracles (Sec. 6). Sec. 7 provides an empirical evaluation of our approach. Finally, Sec. 8 describes related work.

## 2 Wrappers

A wrapper is a procedure for extracting tuples from a particular information source. Formally, a wrapper is a function from a page[1] to the set of tuples it contains.

---

[1] We use the term *page* generically, referring to whatever query response is returned by an information resource.

(a)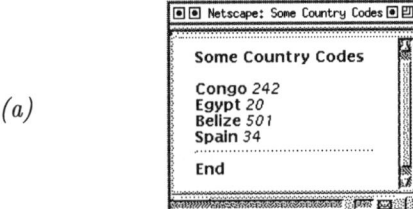

(b)
```
<HTML><TITLE>Some Country Codes</TITLE>
<BODY><B>Some Country Codes</B><P>
<B>Congo</B> <I>242</I><BR>
<B>Egypt</B> <I>20</I><BR>
<B>Belize</B> <I>501</I><BR>
<B>Spain</B> <I>34</I><BR>
<HR><B>End</B></BODY></HTML>
```

(c)
ExtractCCs(page $P$)
  skip past first occurence of <P> in $P$
  while next <B> is before next <HR> in $P$
    for each $\langle \ell_k, r_k \rangle \in \{\langle\text{<B>},\text{</B>}\rangle, \langle\text{<I>},\text{</I>}\rangle\}$
      skip past next occurence of $\ell_k$ in $P$
      extract attribute from $P$ to next occurence of $r_k$
  return extracted tuples

(d)
ExecuteHLRT($\langle h,t,\ell_1,r_1,\ldots,\ell_K,r_K \rangle$, page $P$)
  skip past first occurence of $h$ in $P$
  while next $\ell_1$ is before next $t$ in $P$
    for each $\langle \ell_k, r_k \rangle \in \{\langle \ell_1, r_1 \rangle, \ldots, \langle \ell_K, r_K \rangle\}$
      skip past next occurence of $\ell_k$ in $P$
      extract attribute from $P$ to next occurence of $r_k$
  return extracted tuples

Figure 1: (a) A fictitious example query response page; (b) the HTML from which it was rendered; (c) a wrapper for this resource; and (d) the HLRT wrapper template.

For example, consider a fictitious information resource that provides a tabular list of countries and their telephone country codes. Suppose that in response to a query, the resource responds as displayed in Fig. 1(a), which was rendered from the HTML shown in (b). Many kinds of wrappers could be written; we consider wrappers that use the positions of particular strings to delimit the extracted text. From (b), it appears that this resource renders tuples by surrounding countries with <B> and </B>, and country codes with <I> and </I>.[2] So one candidate wrapper relies on these four delimiters.

But note that this simple left-right (LR) strategy fails, because not all occurrences of <B>···</B> indicate a country. However, the string <P> can be used to distinguish the head of the page from the tuples proper. Similarly, <HR> separates the last tuple from the tail. Fig. 1(c) shows ExtractCCs, a wrapper based on this more sophisticated head-left-right-tail (HLRT) approach.

We focus on wrappers that are structurally similar to ExtractCCs. Fig. 1(d) shows the template for HLRT wrappers. Note that instantiating the template with the six

[2] Note that though this example involves HTML tags such as <B>, our system does not require the use of HTML; any text fragment (such as just B>) that reliably delimits the attribute is acceptable.

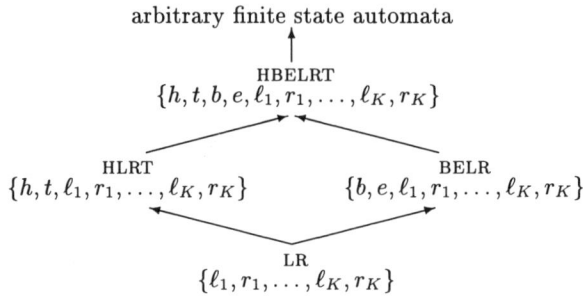

Figure 2: A partial hierarchy of wrapper biases. Arrows indicate that one bias is more expressive than another.

strings $\langle\text{<P>},\text{<HR>},\text{<B>},\text{</B>},\text{<I>},\text{</I>}\rangle$ yields ExtractCCs.

Formally, an HLRT wrapper for a domain with $K$ attributes per tuple is encoded as a vector of $2K+2$ strings $\langle h,t,\ell_1,r_1,\ldots,\ell_K,r_K \rangle$. One string ($h$) marks the end of the header, another ($t$) marks the start of the tail, and two strings ($\ell_k$ and $r_k$) delimit each of the $K$ attributes.

We focus on HLRT, but alternatives abound. Fig. 2 illustrates a partial hierarchy of wrapper classes. LR is less expressive than HLRT; for example, the country/code resource can be wrapped by HLRT but not LR. BELR's $b$ and $e$ mark the beginning and end of each tuple, rather than page's body. In the extreme, arbitrary finite-state automata could be used as wrappers. In [Kushmerick, 1997], we analyze this hierarchy in detail.

## 3 Constructing wrappers by induction

The *wrapper construction* problem is the following: given a supply of example query responses, learn a wrapper for the information resource that generated them. For the country/code resource, the problem is to induce the ExtractCCs procedure, given a supply of HTML pages similar to that shown in Fig. 1(b).

Induction thus provides a natural framework for formalizing wrapper construction. Induction is the task of generalizing from labeled examples to a hypothesis, a function for labeling instances. For our problem:

**Instances** correspond to pages—e.g. Fig. 1(b).

**Labels** correspond to pages' tuples—e.g. the example page is labeled as containing {⟨Congo,242⟩, ⟨Egypt,20⟩, ⟨Belize,501⟩, ⟨Spain,34⟩}.

**Hypotheses** correspond to HLRT wrapper template parameters—e.g. ⟨<P>,<HR>,<B>,</B>,<I>,</I>⟩ is the encoding of ExtractCCs.

**Oracles** correspond to sources of example query responses and their labels. We split the traditional oracle (which returns a single labeled instance) into two parts. PageOracle generates example pages, and LabelOracle produces correct labels for these instances. PageOracle is specific to a particular information resource, while LabelOracle is composed from heuristics that are reusable across domains.

**PAC analysis** is used to terminate the learning process, so the system takes as input accuracy ($\epsilon$) and

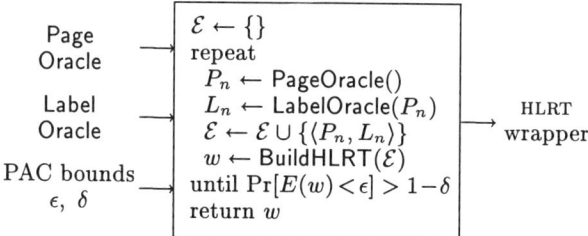

Figure 3: *The wrapper induction algorithm.*

confidence ($\delta$) parameters.

With this framework in place, we now present the wrapper induction algorithm; see Fig. 3. Wrapper induction proceeds by accumulating a set $\mathcal{E}$ of labeled example pages. On each iteration, BuildHLRT is called with $\mathcal{E}$, which returns wrapper $w$. Learning stops when $w$ satisfies the PAC bound. In Secs. 4–6, we describe the algorithm's main components:

BuildHLRT constructs an HLRT wrapper from a set of labeled example pages; see Sec. 4.

$\Pr[E(w) < \epsilon] > 1 - \delta$ is a PAC-theoretic termination condition, testing whether enough examples have been seen to be confident that a satisfactory wrapper has been learned; see Sec. 5.

LabelOracle is a function from a page to a label. In Sec. 6 we describe how to compose a correct labeling oracle from (possibly imperfect) heuristic knowledge.

## 4 Building HLRT wrappers

BuildHLRT takes as input a set of labeled pages, and returns an HLRT wrapper that is *consistent* with each labeled page. A wrapper is consistent with a labeled page if it generates the label for the page. Fig. 4 shows the BuildHLRT algorithm.

BuildHLRT reasons about the conditions that must hold if wrapper $\langle h, t, \ell_1, r_1, \ldots, \ell_K, r_K \rangle$ is to be consistent. For example, in Fig. 1(b), the string <I> is a valid value for $\ell_2$, because <I> actually precedes every instance of the second attribute. Such constraints apply to each $r_k$, and to each $\ell_k$ for $k > 1$.

BuildHLRT is complicated by the fact that $\ell_1$, $t$, and $h$ interact. For example, to determine whether <B> is acceptable as $\ell_1$ (even though the head and tail contain <B>), BuildHLRT must find an $h$ and $t$ such that <B> does in fact reliably mark the start of the first attribute. In this case, $h = $ <P> and $t = $ <HR> are satisfactory. Lines (a–d) in Fig. 4 capture the constraints that $\ell_1$, $t$, and $h$ must satisfy. BuildHLRT examines all possible combinations of $\ell_1$, $h$, and $t$, stopping when it finds values that jointly satisfy these constraints.

To summarize, BuildHLRT iterates over all choices for the $2K + 2$ delimiters, stopping when a consistent wrapper is encountered. BuildHLRT's search is made more efficient by decomposing the constraint satisfaction prob-

BuildHLRT(labeled pages $\mathcal{E} = \{\ldots, \langle P_n, L_n \rangle, \ldots\}$)
 *Note that each label $L_n$ partitions page $P_n$ into its attributes, separated by the strings between tuples and between the $K$ attributes within a tuple.*
 for $k = 1$ to $K$
  $r_k \leftarrow$ any common prefix of the strings following each
   (but not contained in any) attribute $k$
 for $k = 2$ to $K$
  $\ell_k \leftarrow$ any common suffix of the strings preceding each
   attribute $k$
 for each common suffix $\ell_1$ of the pages' heads
  for each common substring $h$ of the pages' heads
   for each common substring $t$ of the pages' tails
    if (a) $h$ precedes $\ell_1$ in each of the pages' heads; and
     (b) $t$ precedes $\ell_1$ in each of the pages' tails; and
     (c) $t$ occurs between $h$ and $\ell_1$ in no page's head; and
     (d) $\ell_1$ doesn't follow $t$ in any inter-tuple separator
    then return $\langle h, t, \ell_1, r_1, \ldots, \ell_K, r_K \rangle$

Figure 4: *The BuildHLRT algorithm.*

lem into three independant subproblems: finding values for (1) the $r_k$; (2) the $\ell_k$ ($k > 1$); and (3) $h$, $t$, and $\ell_1$.

In [Kushmerick, 1997], we prove that: *(1)* BuildHLRT is *sound* (if BuildHLRT returns a wrapper, then it is consistent) and *complete* (if a consistent wrapper exists, BuildHLRT finds it); and *(2)* under reasonable assumptions, BuildHLRT runs in time $O(KNMS^3)$, where each tuple has $K$ attributes, the shortest of the $N$ example pages has length $S$, and $M$ is maximum number of tuples in any single example.

Appendix A formally describes the conditions under which an HLRT wrapper is consistent with a labeled page.

## 5 PAC analysis

PAC analysis answers the question, 'How many examples must a learner see to be confident that its hypothesis is good enough—*i.e.*, to be *probably approximately correct*?'; see [Kearns & Vazirani, 1994] for an introduction. A PAC model defines an error metric over hypothesis: $E(w)$ is the probability that hypothesis $w$ will incorrectly label the next instance. The learning task is then analyzed in order to bound the number of examples which ensure that $\Pr[E(w) > \epsilon] < \delta$, for any given *accuracy* parameter $\epsilon$ and *confidence* parameter $\delta$. In [Kushmerick, 1997], we prove the following theorem.

**Theorem 1 (HLRT sample complexity)** *Suppose BuildHLRT($\mathcal{E}$) returns wrapper $w$, where $\mathcal{E}$ contains collectively $T$ tuples, each with $K$ attributes. If*

$$\left(1 - 2\left(1 - \frac{\epsilon}{2}\right)^T\right)^{2K} \left(1 - 2\left(1 - \frac{\epsilon}{2}\right)^{|\mathcal{E}|}\right)^2 > 1 - \delta,$$

*then $\Pr[E(w) > \epsilon] < \delta$, for any $0 < \epsilon < 1$ and $0 < \delta < 1$.*

For example, with $\epsilon = \delta = 0.1$, $K = 4$, and an average of 5 tuples/page, BuildHLRT must examine at least 72 examples to satisfy the PAC criteria.

This bound is relatively tight compared to typical PAC results. For example, the number of possible HLRT wrappers is infinite, but our bound does not depend on the number of wrappers. Thus clearly the stated bound is tighter than obtainable under simple PAC models (*e.g.*, [Valiant, 1984; Blumer *et al.*, 1987]), in which sample complexity grows with the number of hypotheses. The bound is also tighter than obtainable using Vapnik-Chervónenkis analysis [Haussler, 1988]. To understand these results, recall that BuildHLRT is essentially computing common prefixes and suffixes of sets of strings, which are highly constrained after relatively few examples; see [Kushmerick, 1997] for a detailed discussion.

## 6 Composing oracles

A key to induction is an oracle that labels examples. So far, we have assumed that LabelOracle is provided as input. We now describe how to compose LabelOracle from modular heuristic knowledge, which we call *recognizers*. A recognizer finds *instances* of a particular attribute on a page. For example, a country name recognizer would identify the four countries contained in Fig. 1(b)'s HTML. These recognized instances are then *corroborated* to label the entire page. For example, given a recognizer for countries and another for country codes, corroboration produces an oracle that labels pages containing pairs of these attributes.

Corroboration is trivial if each recognizer is perfect.[3] But an important feature of our approach is that it handles imperfect recognizers. Recognizers are either *perfect* (accept all positive instances and reject all negative instances of their target attribute), *incomplete* (reject all negative instances but reject some positive instances), *unsound* (accept all positive instances but accept some negative instances), or *unreliable* (reject some positive instances and accept some negative instances).

We require that each recognizer be annotated with the kind of error it makes. We expect this annotation to be natural for many kinds of recognizers. For example, a company name recognizer based on Fortune-500 data is incomplete, while a country code recognizer accepting any digit sequence is unsound. The intent is that recognizers are reusable across domains; a company name recognizer, for example, can be used with any information resource displaying companies.

Recall that LabelOracle is a function from a page to a label for the page. A label is an array, where rows correspond to tuples, and columns are attributes. A recognizer is a function from a page to a set of *instances* (subsequences of the page). The set of recognized instances is a column of the overall label array. The *corroboration problem*, then, is to build the entire label array from the individual columns. Note that the attributes' ordering within tuples is not part of the input; the corroboration algorithm must determine this ordering.

---

[3]Wrappers are needed even with perfect recognizers, because recognizers might be slow, while wrappers must be fast.

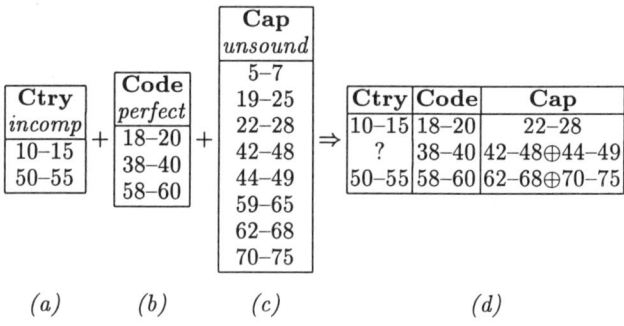

Figure 5: *Corroborating (a–c) yields (d).*

The basic idea of corroboration is that the location of some instances greatly constrains the possible location of others. Suppose recognizer $A$ is unsound and identifies an instance at position 10–20, while perfect recognizer $B$ finds an instance at 14–16. Since attributes never overlap, the $A$ at 10–20 must be a false positive (FP) and thus must be ignored, while the $B$ at 14–16 is a true positive (TP).

In the remainder of this section, we describe corroboration by walking through an example, present the Corrob corroboration algorithm, and describe how our work is extended to handle imperfect recognizers.

**Example.** Fig. 5 extends the country/code example to include an additional attribute, the country's capital.[4] Corroboration begins by noting the type of error made by each recognizer: in this simple example, assume the recognizer for the codes is perfect, but the country recognizer is incomplete and the capital recognizer is unsound.

Next, note that since the Code recognizer is perfect, all Code instances are TPs. Thus Code can be simply copied to the label array. The incomplete Ctry column is almost as easy: it is copied verbatim, but Code is used to align each Ctry instance. This leaves a "hole" in the Ctry column. Next, the corroborator processes the unsound Cap column. 5–7 must be a FP, because were it a TP, Code would have included an additional instance prior to 5. Next, since 19–25 overlaps with Code's 18–20, 19–25 must be a FP. Since 22–28 is the only remaining possibility for the first tuple's Cap instance, 22–28 must be a TP. However, for the second tuple, there is no way to choose between 42–48 and 44–49: one must be a TP and the other a FP, but there is no way to decide. Corroboration thus uses 42–48⊕44–49, indicating that *exactly one* of the two instances is a TP. Finally, corroboration rejects 59–65 because it overlaps with 58–60.

**The Corrob algorithm.** Fig. 6 shows Corrob, an algorithm for corroborating imperfect recognizers, for the case when at least one recognizer is perfect.[5] As indicated in the example, Corrob builds the label array by

---

[4]The indices in this example do *not* match Fig. 1(b).

[5]People can be used as perfect recognizers, though we seek to automate wrapper construction as much as possible.

Corrob(recognizers $\{\ldots, R_k, \ldots\}$, instances $\{\ldots, I_j^k, \ldots\}$)
  Notation: $I_j^k$ is the $j^{\text{th}}$ instance recognized by $R_k$.
  $A \leftarrow$ BuildArray($\{I_j^k : R_k$ is perfect or incomplete$\}$)
  for each $I_j^k$ such that $R_k$ is unsound or unreliable
    if $I_j^k$ is possibly a TP (based on the TPs in $A$), then
      $m \leftarrow$ RowOf($I_j^k, A$)
      $A_{m,k} \leftarrow A_{m,k} \oplus I_j^k$
  return $A$
BuildArray(necessarily TP instances $\{\ldots, I_j^k, \ldots\}$)
  Build array with each $I_j^k$ installed in the correct cell.
RowOf(possibly TP instance $I_j^k$, array $A$)
  Determine the row $m$ of $A$ to which $I_j^k$ belongs, which is always determined when at least one $R_k$ is perfect.

Figure 6: *The Corrob algorithm.*

first installing instances recognized by perfect or incomplete recognizers. These TPs are then used to categorize the remaining instances as either *necessarily* FPs (meaning they can be ignored), or *possibly* TPs (meaning they are inserted using '$\oplus$'). In [Kushmerick, 1997], we describe Corrob in more detail and prove that it is correct.

**Handling mistakes.** Note that Corrob's output might contain attributes that are *missing* ('?' indicates attributes falsely rejected by their recognizers) or *ambiguous* ('$\oplus$' indicates under-constrained attributes). But BuildHLRT assumes that LabelOracle produces a perfect label. We now describe how to extend our work to handle this discrepancy.

Missing attributes require only minor changes to BuildHLRT: the algorithm must simply generalize from fewer examples. For example, recall the original country/code resource in Fig. 1. Suppose that corroboration yields a label that is correct except that the country Congo is missing. In this case, when learning $\ell_1$, the algorithm generalizes from just the three occurrences of </I><BR>⇓<B>[6] that precede the recognized instances of the first attribute, and the algorithm may well fail to generate the correct wrapper. Only if the first country name is correctly recognized on a subsequent example page will BuildHLRT realize that $\ell_1$ must be a suffix of <HTML><TITLE>Some Country···Codes</B><P>⇓<B> as well as of </I><BR>⇓<B>.

The PAC model must also be extended to accommodate missing attributes. To do so, we generalize Thm. 1 so that, instead of assuming exactly $N$ head and tail and $T$ left and right delimiter examples, the model counts the actual numbers, based on the non-missing attributes. So, in the previous example, there are four examples for $l_2$, $r_2$ and $t$, but only three for $l_1$ and $r_1$, and zero for $h$.

Ambiguity requires more substantial changes. There are two kinds of ambiguity. First, as described earlier, Corrob uses '$\oplus$' to indicate that more than one recognized instance is consistent with a particular cell in the label array. Second, recall that Corrob must determine the ordering of attributes within tuples. But the recog-

---
[6]'⇓' indicates a carriage return character.

nized instances may be consistent with more than one ordering. For example, in Fig. 5, a valid label exists for the ordering (**Cap**, **Ctry**, **Code**) as well as for (**Ctry**, **Code**, **Cap**). (We previously ignored ordering ambiguity to simplify the presentation of Corrob.)

We extend BuildHLRT to handle both types of ambiguity as follows. An ambiguous label actually corresponds to a set of unambiguous labels, one for each way to resolve each ambiguity. Exactly one such label is correct; the rest either contain FPs or corresponds to an incorrect attribute ordering. Faced with ambiguity, BuildHLRT iterates over each possibility, stopping when a wrapper can be induced.

Clearly the number of such possibilities grows exponentially in the number of ambiguities. In practice this growth is tolerable for both unsound and incomplete recognizers, even with very high error rates; see Sec. 7. However, Corrob is impractical for unreliable recognizers, because the number of ambiguities grows quickly with an unreliable recognizer's error rate.

We extend the PAC model to handle ambiguous labels by accounting for situations in which BuildHLRT considers an incorrect way to resolve a label's ambiguity, and yet a consistent wrapper exists anyway. For example, in Fig. 5(d), if the second tuple's **Cap** attribute is actually 42–48 rather than 44–49, but BuildHLRT tries 44–49 first and successfully finds a wrapper, then BuildHLRT has probably made a mistake. Similarly, if BuildHLRT considers the ordering (**Cap**, **Ctry**, **Code**) before (**Ctry**, **Code**, **Cap**), then BuildHLRT is probably wrong if it finds a consistent wrapper.

We model this effect by assuming that such a situation happens with probability at most $\mu$ per opportunity, and thus the left-hand side of the bound in Thm. 1 is multiplied by $(1-\mu)^R$, where $R$ is the number of opportunities that a mistake of this type could have occurred as BuildHLRT was enumerating the possible labels. In practice, we find that $\mu$ is extremely close to zero and $R$ is relatively small, and thus ambiguity has a negligible effect on the PAC results. In [Kushmerick, 1997], we compare this noise model to others in the PAC literature.

## 7 Empirical evaluation

In this section, we present preliminary evidence demonstrating the feasibility of HLRT learning. Our Lisp implementation requires between 4 and 40 SGI Indy CPU seconds per example page, depending on the domain. Normalizing for the number of attributes ($K$) and the size of the example pages, our system requires about 0.21 CPU sec. per attribute per KB of example data.

Our first experiment verifies the utility of the HLRT bias. Learnability aside, can a significant fraction of interesting information resources be wrapped by HLRT? We surveyed 100 Internet resources selected randomly from an independent organization's index (search.com), and found that 48% can be wrapped by HLRT. We take this result to be evidence that HLRT is genuinely useful.

Our second experiment measures the robustness of the

system to the recognizers' error rates. We tested our system on (i) the OKRA email service, okra.ucr.edu/okra; and (ii) the BIGBOOK telephone directory, bigbook.com.

By hand, we constructed perfect recognizers for each attribute; OKRA has four attributes and BIGBOOK has six. As a baseline, we ran our system with these perfect recognizers. We then increased the error rates up to 40% (creating both incomplete and unsound recognizers for each attribute) and increased the number of imperfect recognizers from zero until all but one were imperfect.[7] We tested our system using two termination conditions: (a) we ran the system until the PAC criteria was satisfied (for $\epsilon = \delta = 0.1$); and (b) we required that the learned wrapper be 100% correct on a suite of test pages.

Fig. 7 shows the number of pages needed to induce a wrapper, as a function of the error rate, for each termination condition, and for each domain. Each curve within a graph represents a different number of imperfect recognizers. For example, the points marked "perfect" represent trials in which all recognizers are perfect, while the points marked "30% error rate of each recognizer" on the "2 imperfect recognizers" curves indicate trials in which two of the recognizers are imperfect (yielding either 30% FPs or 30% FNs) while the remaining recognizers are perfect. Thus in each graph, increasing the abscissa or examining curves with additional imperfect recognizers corresponds to trials in which the recognizers make more mistakes.

Figs. 7(i–ii.b) indicate that, from a practical perspective, relatively few examples are needed before the system learns the correct wrapper; across all conditions, about 4.9 examples suffice for OKRA and 29 for BIGBOOK. We conclude that the number of examples required is small enough that HLRT wrapper induction is practical, even for extremely high recognizer error rates.

Figs. 7(i–ii.a) show that the PAC bound is relatively loose. Across all conditions, about 105 examples are needed required to satisfy the PAC criteria. Thus the PAC bound is too loose by about an order of magnitude. We conclude that the current PAC model is too weak to tightly constrain the induction process. Nevertheless, since wrapper construction is intended to be an off-line process, the bound is not so loose as to be useless.

Finally, we have developed WIEN (pronounced "Vienna"), a wrapper induction environment. Using a Web browser, a user shows WIEN an example information resource page, and then uses the mouse to label the page. WIEN then tries to learn a wrapper for the resource. When the user shows WIEN a second example, it uses the learned wrapper to automatically label the new example. The user then corrects any mistakes, and WIEN generalizes from both examples. This process repeats until the user is satisfied. WIEN provides a complete implementation of BuildHLRT, though the user is assumed to label pages perfectly, so WIEN implements neither attribute recognition nor corroboration.

---

[7]Recall that Corrob is impractical for unreliable recognizers and requires at least one perfect recognizer.

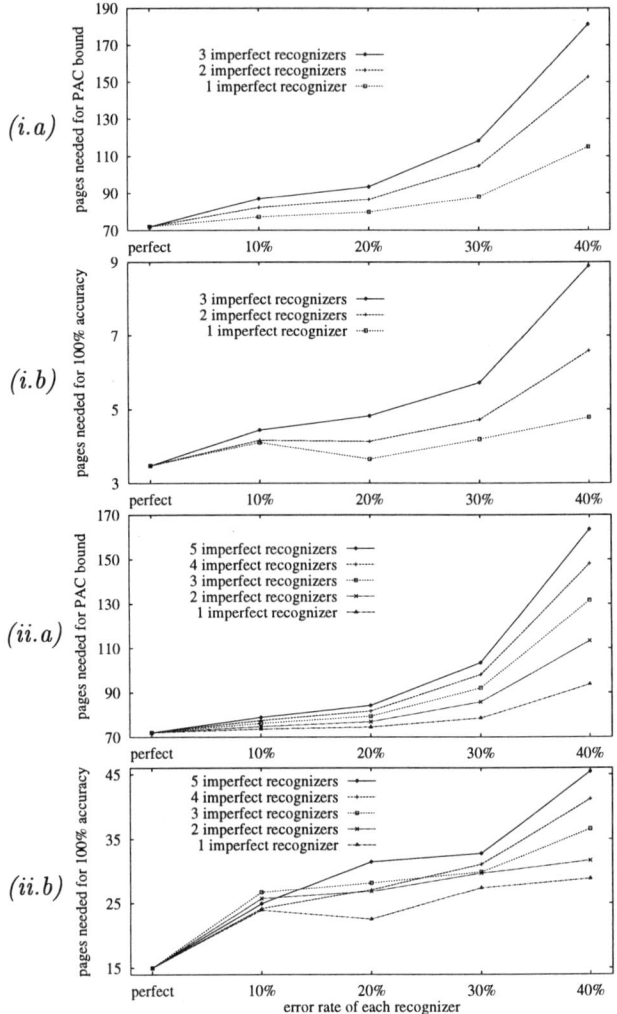

Figure 7: *Effect on learning curve of recognizer error, in the (i)* OKRA *and (ii)* BIGBOOK *domains, for the (a) PAC and (b) 100% accurate termination conditions.*

WIEN and the experimental data are available at www.cs.washington.edu/homes/nick/research/wrappers.

## 8 Related work

As suggested at the outset, wrapper construction is motivated by the software engineering issues involved with deploying software systems that rely on external information resources; examples include [Chawathe et al., 1994; Etzioni & Weld, 1994; Arens et al., 1996; Kirk et al., 1995]. While data interchange protocols (e.g. KQML [Finin et al., 1994]) have been proposed to address these issues, they require cooperation on the part of information providers, and such cooperation is rare.

From a formal perspective, in Sec. 1 we discussed the relationship between HLRT and FSA induction.

From an application perspective, our work is similar to [Ashish & Knoblock, 1997]. Their system learns a more expressive wrapper class than HLRT, but relies on many heuristics that are specific to HTML. In contrast, our systems treats HTML tags just as ordinary text. Moreover, their system requires human intervention to correct its mistakes, while our corroboration process is intended to correct mistakes automatically. A second related application is SHOPBOT [Doorenbos et al., 1997]. Though in many respects SHOPBOT is more ambitious, its wrapper language is less expressive than HLRT.

Finally, our recognition knowledge is similar to work on semantically labeling natural text, such as the MUC-6 "Named Entity" task [DARPA, 1995], though relatively little work has been done on corroborating multiple such knowledge sources.

## 9 Conclusions

Wrapper induction is a new technique for automatically constructing wrappers. We have made three contributions. First, we have formalized the wrapper construction problem as induction. Second, we have defined the HLRT bias, which is efficiently learnable in this framework. Third, we have shown how to use heuristic knowledge to compose the algorithm's oracle. Though our work has involved primarily Internet information resources, we expecte that our results are applicable to similar information-extraction tasks in other domains.

We intend to extend our framework in several ways. In addition to the biases shown in Fig. 2, we want to design wrappers that can handle non-tabular pages, such as pages organized hierarchically. The research issues involve exploring the tradeoff between expressiveness and learnability. We also hope to tighten the PAC model so it is more useful in practice as well as more predictive of observed learning curves.

**Acknowledgments.** This research is funded in part by ONR Grant N00014-94-1-0060, by NSF Grant IRI-9303461, by ARPA / Rome Labs grant F30602-95-1-0024, and by a gift from Rockwell International Palo Alto Research. Thanks to the UW CS&E AI group, Steve Minton and David Smith for helpful discussion, and to Boris Bak for help with the `search.com` survey.

## A HLRT consistency conditions

In this Appendix, we list the conditions under which HLRT wrapper $w = \{h, t, \ell_1, r_1, \ldots, \ell_K, r_K\}$ outputs label $L = \{\langle\langle b_{1,1}, e_{1,1}\rangle, \ldots, \langle b_{1,K}, e_{1,K}\rangle\rangle, \ldots, \langle\langle b_{M,1}, e_{M,1}\rangle, \ldots, \langle b_{M,K}, e_{M,K}\rangle\rangle\}$ for page $P$. This notation indicates that that $P$ contains $M$ tuples having $K$ attributes each, where the $k^{\text{th}}$ attribute of the $m^{\text{th}}$ tuple begins at index $b_{m,k}$ of $P$ and ends at $e_{m,k}$. Note that $L$ partitions $P$ as follows ('·' indicates concatenation): $P = S_{0,K} \cdot A_{1,1} \cdot S_{1,1} \cdot A_{1,2} \cdots S_{1,K-1} \cdot A_{1,K} \cdot S_{1,K} \cdots A_{M,1} \cdot S_{M,1} \cdot A_{M,2} \cdots S_{M,K-1} \cdot A_{M,K} \cdot S_{M,K}$. The $A_{m,k}$ are the attribute values: $A_{m,k} = P[b_{m,k}, e_{m,k}]$. The $S_{m,k}$ separate the tuples, and attributes within a tuple: $S_{m,k} = P[e_{m,k}, b_{m,k+1}]$ (except that $S_{0,K} = P[0, b_{1,1}]$, $S_{M,K} = P[e_{M,K}, |P|]$, and $S_{m,K} \equiv S_{m+1,0} = P[e_{m,K}, b_{m+1,1}]$).

Under this notation, $w$ is consistent with $P$ and $L$ iff:

1. the $\ell_k$ immediately precede their attributes
 $(S_{0,K}/h)/\ell_1 \neq \sharp \ \land \ \forall_{m,k>1} |S_{m,k-1}/\ell_k| = 0$
2. the $r_k$ follow (but don't occur within) their attributes
 $\forall_{m,k} \ r_k \cdot ((A_{m,k} \cdot S_{m,k})/r_k) = S_{m,k}$
3. $h$ occurs in the head and $t$ occurs in the tail
 $S_{0,K}/h \neq \sharp \ \land \ S_{M,K}/t \neq \sharp$
4. $t$ never precedes $\ell_1$ in an inter-tuple separator
 $\forall_{m<M} \ S_{m,K}/t \neq \sharp \ \Rightarrow \ |\ell_1| > |t \cdot (S_{m,K}/t)|$
5. $t$ doesn't occur between $h$ and $\ell_1$ in the head
 $(S_{0,K}/h)/t \neq \sharp \ \Rightarrow \ |\ell_1| > |t \cdot ((S_{0,K}/h)/t)|$
6. $t$ precedes $\ell_1$ in the tail
 $S_{M,K}/\ell_1 \neq \sharp \ \Rightarrow \ |t \cdot (S_{M,K}/t)| > |\ell_1 \cdot (S_{M,K}/\ell_1)|$

(where $s/s'$ is the substring of $s$ after the first occurrence of $s'$, with $s/s' = \sharp$ indicating that $s$ doesn't contain $s'$).

## References

[Angluin, 1982] D. Angluin. Inference of reversible languages. *J. ACM*, 29(3):741–65, 1982.

[Arens et al., 1996] Y. Arens, C. Knoblock, C. Chee, & C. Hsu. SIMS: Single interface to multiple sources. TR RL-TR-96-118, USC Rome Labs, 1996.

[Ashish & Knoblock, 1997] N. Ashish & C. Knoblock. Semi-automatic wrapper generation for Internet information sources. In *Proc. Cooperative Information Systems*, 1997.

[Blumer et al., 1987] A. Blumer, A. Ehrenfeucht, D. Haussler, & M. Warmuth. Occam's razor. *Information Processing*, 24(6):377–80, 1987.

[Chawathe et al., 1994] S. Chawathe, H. Garcia-Molina, J. Hammer, K. Ireland, Y. Papakonstantinou, J. Ullman, & J. Widom. The TSIMMIS project: Integration of heterogeneous information sources. In *Proc. IPSJ Conf*, 1994.

[Cowie & Lehnert, 1996] J. Cowie & W. Lehnert. Information extraction. *C. ACM*, 39(1):80–101, 1996.

[DARPA, 1995] DARPA. *Proc. 6th Message Understanding Conference*. Morgan Kaufmann, San Francisco, 1995.

[Doorenbos et al., 1997] R. Doorenbos, O. Etzioni, & D. Weld. A scalable comparison-shopping agent for the World-Wide Web. In *Proc. Autonomous Agents*, 1997.

[Etzioni & Weld, 1994] O. Etzioni & D. Weld. A softbot-based interface to the Internet. *C. ACM*, 37(7):72–6, 1994.

[Etzioni, 1996] O. Etzioni. The World Wide Web: quagmire or gold mine? *C. ACM*, 37(7):65–8, 1996.

[Finin et al., 1994] T. Finin, R. Fritzson, D. McKay, & R. McEntire. KQML: A language and protocol for knowledge and information exchange. In *Knowledge Building and Knowledge Sharing*. Ohmsha and IOS Press, 1994.

[Haussler, 1988] D. Haussler. Quantifying inductive bias. *Artificial Intelligence*, 36(2):177–221, 1988.

[Kearns & Vazirani, 1994] M. Kearns & U. Vazirani. *An introduction to computational learning theory*. MIT, 1994.

[Kirk et al., 1995] T. Kirk, A. Levy, Y. Sagiv, & D. Srivastava. The Information Manifold. In *AAAI Spring Symposium: Information Gathering from Heterogeneous, Distributed Environments*, pp. 85–91, 1995.

[Kushmerick, 1997] N. Kushmerick. *Wrapper Construction for Information Extraction*. PhD thesis, Univ. of Washington, 1997. In preparation.

[Valiant, 1984] L. Valiant. A theory of the learnable. *C. ACM*, 27(11):1134–42, 1984.

# INFORMATION RETRIEVAL

Information Retrieval 2

# COSPEX: A System for Constructing Private Digital Libraries

Masanori Sugimoto*  Norio Katayama  Atsuhiro Takasu

Research and Development Department
National Center for Science Information Systems
3-29-1, Otsuka, Bunkyo-ku, Tokyo, 112, Japan

## Abstract

In this paper we describe a system called COSPEX (COnceptual SPace EXplorer) which aims at assisting users in constructing their own private digital libraries. Information servers on the Internet have their own purposes. Each of the servers is constructed distributedly and managed autonomously, and its semantics is different from each other. On the other hand, users' information needs are different from each other. Now that almost infinite information resources are accessible through the Internet, a new method for promoting the mediation between the semantics of information servers and users' information needs is inevitable for users to fully utilize these information resources. COSPEX is a system which can enhance such mediation. When users of COSPEX access a certain information server, its database schema is loaded and visualized on the user interface. Users can interactively construct their own queries by viewing the visualized database schema. COSPEX can also assist users in their information discovery. By visualizing the relations among objects and attributes in a database, it assists users in finding their necessary information effectively.

Users of COSPEX can create their own collection from large information resources, that is, a private digital library, which is useful for their intelligent activities.

## 1 Introduction

In this paper we describe a system called COSPEX (COnceptual SPace EXplorer) which aims at assisting users in constructing their own private digital libraries.

Information servers on the Internet, for instance digital libraries [ACM, 1996; Schatz and Chen, 1996] have their own purposes. Each of the servers is constructed distributedly and managed autonomously, and its semantics is different from each other. On the other hand, users' information needs are different from each other. Now that almost infinite information resources are accessible through the Internet, a new method for promoting the mediation between the semantics of information servers and users' information needs is inevitable for users to fully utilize these information resources. COSPEX is a system which can enhance such mediation.

COSPEX is composed of three modules, Visual Query Editor (VQE), Conceptual Space Visualizer (CSV) and Private Digital Library Manager (PDLM). VQE assists users in their query formulation. It first loads and visualizes a schema of a database accessed by a user. By viewing it, he/she can easily understand the semantics of the database and how to construct his/her query. CSV assists users in their information discovery. It can configure a metric space to visualize the relations among objects and attributes in a database. By looking at it, users can know the semantic relations among objects and find necessary information effectively. PDLM is a manager of a user's private digital library. It stores and manages object sets which a user has retrieved through VQE and CSV. Users of COSPEX can organize and evolve their own collection, that is, a private digital library from large information resources.

In Section 2 the configuration of COSPEX is described. In Section 3 experiments and evaluations are shown. Section 4 describes related works. Section 5 gives conclusions and future works.

## 2 Configuration of COSPEX

### 2.1 Overview

Figure 1 shows an overview of COSPEX. COSPEX is programmed by Java language and is executable on a WWW (World Wide Web) browser like Netscape Navigator. In order to realize the accessibility to multiple information servers and visualize their database schema and object-attribute relations, each information server is augmented with the following information and components for COSPEX:

- Schema information

---

*Currently a visiting researcher in Department of Computer Science, University of Colorado at Boulder, CO, USA.

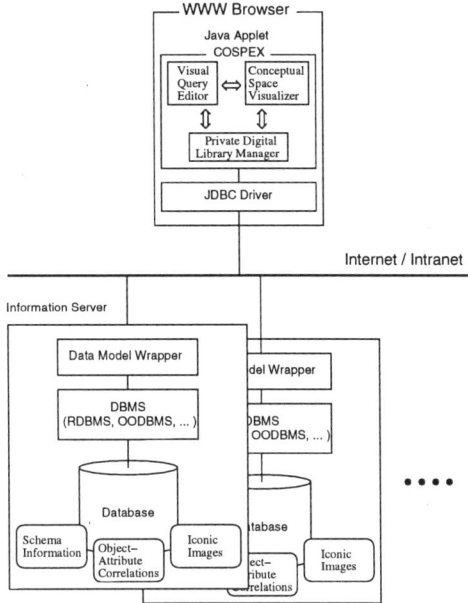

Figure 1: Overview of COSPEX

Each information server provides its own schema information, i.e., the description of its database schema, which is expressed based on the entity-relationship model and consists of the list of classes, predicates, and domain constraints.

- Object-attribute correlations
  Each information server provides object-attribute correlations to visualize relations among database contents. Object-attribute correlations are provided in the form of ternary relation, $(o, a, c)$, where $o$ is an object, $a$ is an attribute, and $c$ is the correlation between $o$ and $a$. Typical examples of object-attribute correlations are term frequency of documents, color histograms of images, etc. These correlations are either statically stored in databases or dynamically computed on demand.

- Iconic images
  Each information server has iconic images to visualize its database schema and to be used for a visual query.

- JDBC drivers
  Each information server has its own driver of Java Database Connectivity (JDBC) [Sun Microsystems, 1996]. JDBC is one realization of the concept of Open Database Connectivity (ODBC) and allows Java applets to contact with remote database servers.

- Data model wrappers
  COSPEX employs the entity-relationship model as the global data model, i.e., requests to and replies from databases are expressed based on the entity relationship model. Therefore, if an information server uses a DBMS (database management system) of other data models, e.g. RDBMS (relational DBMS) or OODBMS (object-oriented DBMS), a data model wrapper is introduced to translate the global model to the local model and vice versa.

In this paper COSPEX is used as a query interface for a digital library. However, it is an universal query interface because it is applicable to every information server which satisfies these conditions.

## 2.2 Visual Query Editor (VQE)

The role of VQE [Katayama et al., 1996] is to assist users in their query formulation. Figure 2 shows an example of a query expression on VQE. It consists of two parts: a palette and a canvas. The palette is located above and contains icons of classes. The contents of the palette is determined by the schema information supplied by each server. The canvas below is a workbench of query expressions. When users access an information server, VQE first loads and visualizes a schema of its database on its palette. They construct their query expressions in the form of a network structure. Retrieved object sets by VQE are stored and managed by PDLM.

For example, the query expression in Figure 2 is formulated as follows:

(1) A user selects what he/she wants to retrieve or what he/she is interested in by dragging and dropping a class icon from the palette to the canvas. In Figure 2, an article icon is dropped on the canvas.

(2) By double clicking an icon on the canvas, a user can see a list of predicates applicable to the class. The list is constructed from the loaded schema information and reflects the domain constraints of predicates. By selecting a predicate from the list, a template of a link is laid on the canvas.

(3) By selecting the class and attached it to the link end, a link between two classes is accomplished. A user repeats this sequence of operations until the expression represents what he/she wants to retrieve.

(4) A user can assign a constant value to an icon via a dialog box. The dialog box pops up by double clicking an icon with pressing a shift key. In Figure 2 a character string, "John Smith," is assigned to the name of an author.

(5) Finally, a user specifies what to retrieve. By clicking an icon, he/she can select the query target. The selected icon is marked with a round rectangle. For example, an article icon is selected for the query target in Figure 2. This expression is ready to be issued and means that there exists an article, it is written by an author, and the author has a name, "John Smith."

A formulated expression is converted into a logic-based query expression and issued to the information server. For example, the expression in Figure 2 is converted into the following expression:

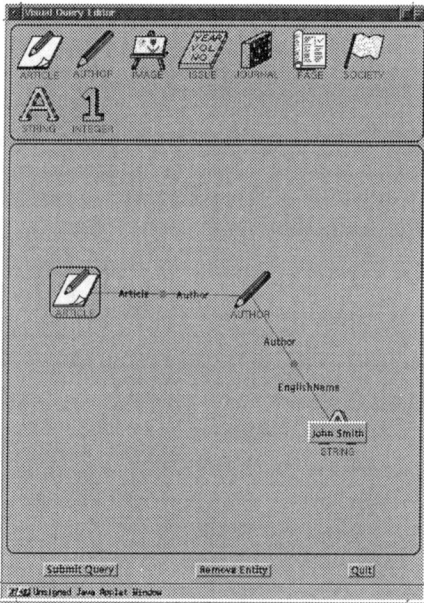

Figure 2: Visual Query Editor

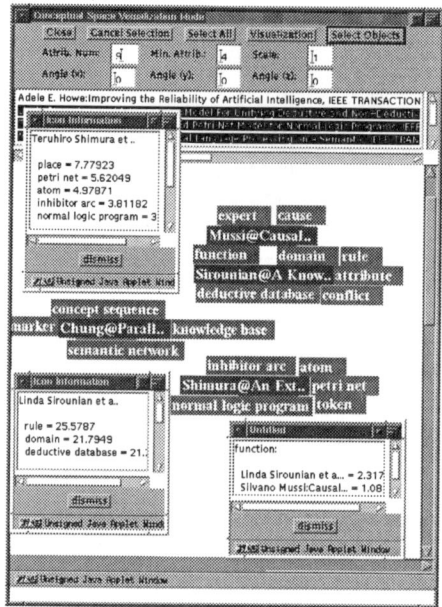

Figure 3: Conceptual Space Visualizer

```
[ Article=$article, Author=$author ],
[ Author=$author, Name="John Smith" ],
[ Class=/class/article, Instance=$article ],
[ Class=/class/author, Instance=$author ]
```

The first predicate corresponds to the link between an article and its author. The second corresponds to the link between the author and the author's name. The third and fourth specifies domain constraints for articles and authors. The correspondence between the entity relationship model and predicate logic is very simple. A class icon with no assignment corresponds to a variable, an icon with an assignment to a constant, and a link to an atom.

The major feature of VQE is that it can guide users to formulate their query expressions interactively by visualizing a database schema information loaded from the server. By viewing it, users can interpret its semantics and know how to construct syntactically valid query expressions which retrieve what they need.

### 2.3 Conceptual Space Visualizer (CSV)

The role of CSV is to help users to select and discover objects by visualizing object-attribute relations [Sugimoto et al., 1995a]. CSV starts with receiving an object set stored in PDLM or retrieved by VQE. Figure 3 shows an example of CSV which shows the relations among objects (in this case, articles) and their attributes (in this case, terms) in a two-dimensional metric space. By viewing it, users can know their semantic relations. Through the interaction between users and CSV, they can find new objects or new relations among objects, and reflect them to their query expressions on VQE. Retrieved objects by using CSV are stored and managed by PDLM.

CSV is composed of two submodules, an automatic indexing submodule and a statistical analysis submodule. By loading the corresponding object-attribute correlations from the server, CSV starts with an automatic indexing submodule. It configures an attribute vector for each object by using an automatic indexing algorithm developed by Salton [Salton, 1975] with our own extension. The weight of $attribute\ j$ in $object\ i$ among $n$ objects ($w_{ij}$) is calculated as follows:

(1) Calculate the densities of $attribute\ j$ in $object\ i$ ($d_{ij}$) and in $n$ objects ($\overline{d_j}$)

$$d_{ij} = \frac{\text{frequency of } attribute\ j \text{ in } object\ i}{\text{frequency of all the attributes in } object\ i}$$

$$\overline{d_j} = \frac{\text{frequency of } attribute\ j \text{ in } n \text{ objects}}{\text{frequency of all the attributes in } n \text{ objects}}$$

(2) Calculate the distribution of $attribute\ j$ ($v_j$)

$$v_j = \frac{\sum_{i=1}^{n}(d_{ij} - \overline{d_j})^2}{n-1}$$

(3) Calculate the weight of $attribute\ j$ in $object\ i$ ($w_{ij}$)

$$w_{ij} = \frac{d_{ij} \times v_j}{\overline{d_j}^2}$$

CSV successively starts a statistical analysis submodule. This submodule implements the dual scaling method (See [Nishisato, 1980] for details) and analyzes attribute vectors in order to visualize the relations among objects. Most of statistical analyses clarify the relations only among objects (in this case, articles) in data. The main feature of the dual scaling method is that it clarifies the relations not only among objects but also among attributes (in this case, terms) and among objects and attributes by giving them scores.

Through these calculations, the submodule creates a score table in which the scores of the objects and their attributes are described. CSV then configures a space which visualizes object-attribute relations. Figure 3 shows the relations among five articles. From this space users can interpret the semantic relations, for example, the articles '*Mussi@Causal..*' and '*Sirounian@A Know..*' have a relation with each other by the term '*function*,' and that the terms '*semantic network*' and '*knowledge base*' are key words for '*Chung@Parall...*'

By looking at the space, users can find articles which they would like to read. When users cannot make their information needs clear at the beginning of their search processes [Taylor, 1968; Belkin *et al.*, 1982], they can articulate their needs by discovering new terms.

CSV enables users their interactive operations. When they discover interesting objects or attributes on a space which CSV configured, they can select a region around them. Then a different space is configured where new attributes and detailed relations among the selected objects are shown. Such interactions between users and CSV can enhance their information retrieval process because relevance feedback [Salton, 1989] is effectively realized.

CSV can visualize the ternary relation in the form of $(o, a, c)$ where $o$ is an object, $a$ is an attribute, and $c$ is the correlation between $o$ and $a$. In Figure 3 CSV applies its automatic indexing submodule and statistical analysis submodule to the ternary relation $(o, a, c)$ where $o$ is an article, $a$ is a term, and $c$ is the correlation between an article and a term. The choice of object-attribute relations and definition of their correlations, i.e., what attributes to be used and what algorithm to be used for calculating correlations, are unrestricted and at the disposal of database administrators. For example, CSV can also visualize journal-term relations (based on the term frequency of each journal), academic society-author relations (based on the number of papers of each author carried in the publications of each society), and so on.

### 2.4 Private Digital Library Manager (PDLM)

PDLM is a manager of a user's own private digital library. Figure 4 shows an example of PDLM. When a user selects one of the accessible information servers, its contents are shown above and the contents of his/her private digital library are shown below on PDLM. Then a user can start VQE in order to construct query expressions or start CSV in order to visualize object-attribute relations. A workspace shown in the middle of PDLM is used to store objects retrieved through VQE and CSV.

PDLM has two major functions as a manager of a private digital library. The first function is the management of object sets retrieved by users using VQE and CSV. Object sets can be freely added to or removed from their private digital libraries to compose their useful collection. This function captures the fundamental function of the classical, i.e., non-digital, private libraries.

In order to make a desirable object set, a user can also carry out operations such as intersection, union, and difference on object sets which are retrieved from the same server. The second function which is currently in the planning and implementing stage is that it can autonomously collect object sets by accessing information servers on the Internet. PDLM stores not only object sets but also rules which has been used to retrieve them. Rules stored in PDLM are query expressions constructed by a user on VQE and classification rules generated by a decision tree algorithm [Quinlan, 1993] on CSV. By using these features PDLM can assist users in organizing and evolving their own private digital libraries.

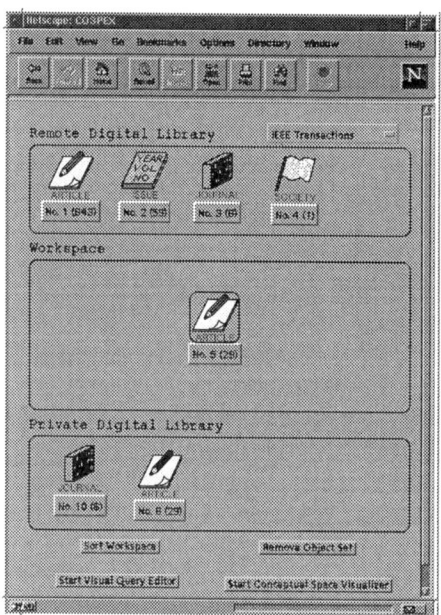

Figure 4: Private Digital Library Manager

### 2.5 Usage of COSPEX

Figure 5 illustrates an example of an usage of COSPEX. Users interact with VQE, CSV and PDLM alternately. In this example, a user first selects one of the accessible information servers from his/her URL (Universal Resource Locator) list.

(a) Query formulation.

Suppose that the user thinks of a keyword "agent." Then, he/she starts VQE and formulates a query to retrieve articles with a term "agent."

(b) Target alteration.

If the query result is too large to be displayed or visualized, COSPEX allows users to alter the query target for seeking more suitable search conditions. In this example, the target is altered from articles to journals to obtain a smaller result. This operation corresponds to altering the abstraction degree of the query target.

(c) Attribute discovery.

Once the query result is small enough to be visualized, CSV enables users to visualize relations among

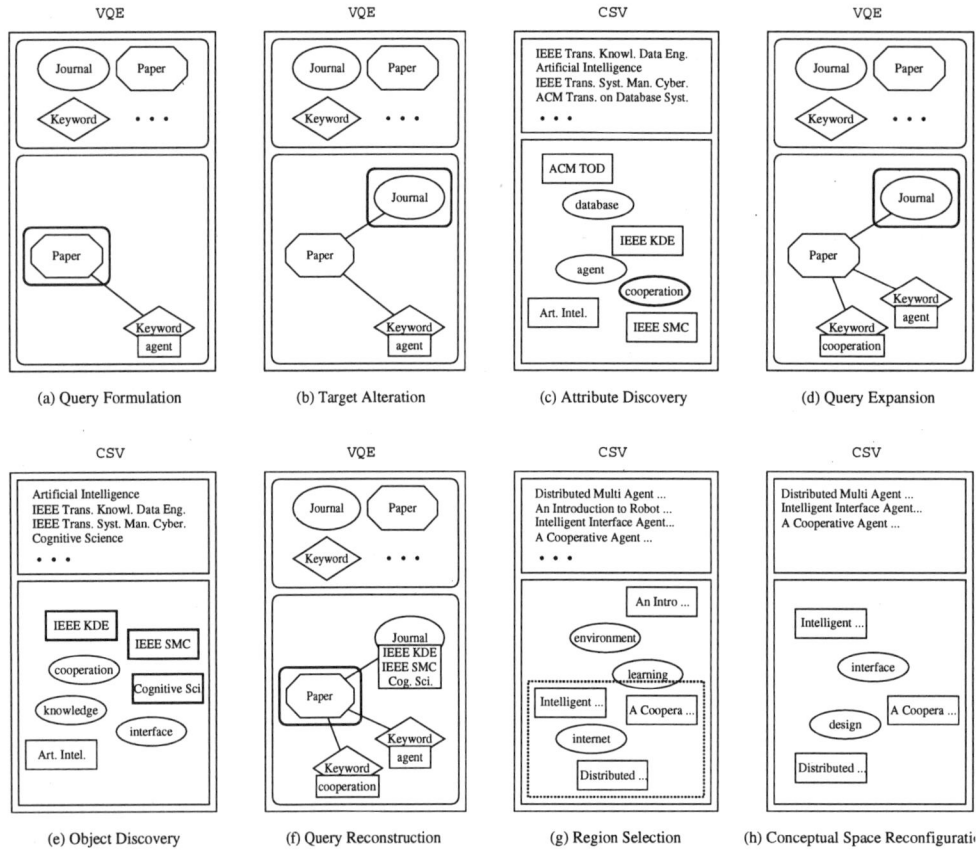

Figure 5: An example of an usage of COSPEX

objects and their attributes. In this example, relations among journals and their key terms are visualized. In this example, rectangles and ovals represent journals and terms, respectively.

One of the distinctive features of CSV is the discovery of attribute values. Because CSV places not only objects but also attributes within a single conceptual space, users can discover new attribute values relative to the objects. Discovered values can be used for query expansion in the next step.

(d) Query expansion.

Suppose that the user discovered a new term "cooperation" in step (c). Then, he/she can expand the query by adding the new term to the expression. Now, the expression means to retrieve journals which include articles with terms, "agent" and "cooperation."

(e) Object discovery.

In this step, the user visualizes relations among journals which are obtained by the query in step (d). The configured conceptual space is different from the one in step (c) because retrieved journals are reduced by the query expansion in step (d).

This step illustrates another feature of CSV, i.e., the discovery of objects. CSV enables users to discover objects which attract their interest. In this example, the user discovers that the journals, "IEEE KDE," "IEEE SMC," and "Cognitive Sci.," are relevant to his/her interest and that the others are irrelevant. Then, he/she can pick up relevant objects and feed them to the query expression of VQE.

(f) Query reconstruction.

Objects selected in CSV are fed into the query expression and disjunctive search condition is formulated in VQE. This expression means to retrieve articles which have terms, "agent" and "cooperation," and which are included in "IEEE KDE," "IEEE SMC," or "Cognitive Sci."

(g) Region Selection.

In this step, CSV visualizes articles obtained in the previous step. This step illustrates the interactive operation to a space configured by CSV. When a user feels interested in specific terms or articles, he/she can select a region around them. Then a different space is configured where new terms and detailed relations among selected objects are shown.

(h) Conceptual Space Reconfiguration.

This step illustrates the reconfigured conceptual space. It shows newly weighted key terms and detailed relations. This interactive operation to CSV enables users to reduce the result set by analyzing relations among them.

Users finally find their necessary objects and store them to PDLM as new components of their own private digital libraries. As described above, COSPEX assists users in their information retrieval by enhancing the mediation between the semantics of information servers and their information needs.

## 3 Experiments and Evaluations

Many experiments with our system have been carried out so far [Sugimoto et al., 1995b]. Here we describe the experiments about the visualization effects of CSV. Nine subjects consisting of graduate students and staff of Tokyo University used CSV in the following three different modes of visualization capability:

- (V0): No visualization (Boolean search only)
- (V1): Visualization of object relations
- (V2): Visualization of object-attribute relations

They rated each mode (very bad = 1; bad = 2; average = 3; good = 4; very good = 5) by considering the following question: could you discover objects or attributes effectively?

The hypothesis test[1] shows that users can discover objects and attributes more effectively in (V1) than (V0) at the 0.005 level of significance and in (V2) than (V1) at the 0.025 level of significance. This shows that (V2) is more effective than (V0) and (V1) in helping users discover objects and attributes. From the post-experiment interviews users told that the visualization of object-attribute relations is useful for the articulation of their information needs.

Table 1 shows the retrieval effectiveness on (V0) and (V2). Subjects are asked to search articles which they need. The fact that (V2) is higher than (V0) in recall and precision ratio explains that CSV can assist users in finding their necessary articles effectively.

## 4 Related Works

A large number of visual interfaces have been proposed in the database literature [Goldman et al., 1985; Czejdo et al., 1990; Vadaparty et al., 1993; Mohan and Kashyap, 1993; Miura and Moriya, 1992; Miura, 1994; Merz and King, 1994; Balkir et al., 1996]. Compared with these predecessors, the current implementation of VQE has limited expressive power in terms of query formulation. We need to extend its capability to permit more complex queries. However, VQE is superior to them in terms of database schema visualization because its interactive visualization technique is unique to VQE and helps users to capture database schemata and guides them to formulate syntactically valid query expressions.

---

[1] We carried out one-sided and paired t-tests, because the samples are paired and small.

|  | recall | precision |
|---|---|---|
| (V0) mode of CSV | 50% | 63% |
| (V2) mode of CSV | 58% | 78% |

Table 1: A summary of the retrieval effectiveness

Many techniques for visualization and interactive operations have been proposed in the studies of information retrieval systems. $I^3R$ [Thompson and Croft, 1989] is a distributed expert system based on a blackboard model. It configures a space by calculating the similarity among documents. BEAD [Chalmers and Chitson, 1992] also configures a document space by calculation based on the analogy of potential fields in physics. These two systems don't visualize object-attribute relations (e.g. document-term relations) simultaneously but object relations. InfoCrystal [Spoerri, 1993] assists users by visualizing their vector-space searches and Boolean searches. It differs from CSV because it only shows the search terms which users input and doesn't show the other weighted terms and relations among them. Scatter/Gather [Cutting et al., 1992], which classifies a large collection of documents into a small number of group, differs from CSV because it doesn't visualize users' search processes. In LyberWorld [Hemmje et al., 1994] users navigate along the links between documents and terms in a space during their search processes. CSV, on the other hand, configures a space where all the objects are visualized and users can select a specific region from it.

In the current configuration of COSPEX, we employ a wrapper mechanism to achieve the interoperability of multiple digital libraries. However, more sophisticated mechanisms are being studied in the digital library literatures [Atkins et al., 1996; Paepcke et al., 1996]. For example, the UMDL (University of Michigan Digital Library) project proposes a distributed agent architecture which permits multiple specialized agents to be dynamically coordinated and work together in order to accomplish a given task [Atkins et al., 1996]. This architecture promotes the modularity, flexibility and incrementality of digital library services. This kind of architecture is not incompatible with the architecture of COSPEX. It can be applicable to COSPEX and seems to be effective for promoting the modularity of its implementation.

## 5 Conclusions and Future Works

In this paper we proposed a system called COSPEX. By enhancing the mediation between the semantics of information servers and users' information needs, COSPEX assists users in constructing their own collection, that is, a private digital library from multiple information servers on the Internet.

We have several plans to extend the functions of COSPEX. One of them is to apply it for a collaboration testbed. When object sets and search rules are stored and shared by several users, COSPEX can assist users in their knowledge sharing and facilitate their collaboration. Class identification and object identification are also critical in managing a private digital library. A user collects objects from multiple information servers whose semantics are different from each other. In order to integrate the object sets from multiple information resources, a new class may need to be defined by merging existing classes through the interaction with a user. We will implement these extensions to make COSPEX more useful.

# References

[ACM, 1996] ACM. *First ACM International Conference on Digital Libraries*, 1996.

[Atkins et al., 1996] D. Atkins, W. Birmingham, E. Durfee, , et al. Toward Inquiry-Based Education Through Interacting Software Agents. *IEEE Computer*, 29(5):69 – 76, 1996.

[Balkir et al., 1996] N. H. Balkir, E. Sukan, G. Ozsoyoglu, et al. Visual: A graphical icon-based query language. In *Proc. of 12th Int. Conf. on Data Eng.*, pages 524 – 533, Feb 1996.

[Belkin et al., 1982] N. J. Belkin, R. N. Oddy, and H. M. Brooks. Ask for information retrieval: part I. background and theory. *Journal of Documentation*, 38(2):61 – 71, June 1982.

[Chalmers and Chitson, 1992] M. Chalmers and P. Chitson. BEAD: Explorations in Information Visualization. In *Proc. of SIGIR'92*, pages 160 – 169, 1992.

[Cutting et al., 1992] D. R. Cutting, D. R. Karger, J. O Pedersen, and J. W Tukey. Scatter/gather: A cluster-based approach to browsing large document collections. In *Proc. of SIGIR'92*, pages 318 – 329, 1992.

[Czejdo et al., 1990] B. Czejdo, R. Elmasri, M. Rusinkiewicz, et al. A graphical data manipulation language for an extended entity-relationship model. *IEEE Computer*, 23(3):26 – 36, Mar 1990.

[Goldman et al., 1985] K. J. Goldman, S. A. Goldman, P. C. Kanellakis, et al. Isis: Interface for a semantic information system. In *Proc. of SIGMOD'85*, pages 328 – 342, May 1985.

[Hemmje et al., 1994] M. Hemmje, C. Kunkel, and A. Willett. LyberWorld - A Visualization User Interface Supporting Fulltext Retrieval. In *Proc. of SIGIR'94*, pages 249 – 259, 1994.

[Katayama et al., 1996] N. Katayama, M. Sugimoto, and J. Adachi. A universal query interface for heterogeneous distributed digital libraries. In *Proc. of the 7th Int. Workshop on Database and Expert System Applications (DEXA'96)*, pages 332 – 339, Sep 1996.

[Merz and King, 1994] U. Merz and R. King. Direct: A query facility for multiple databases. *ACM Trans. Inf. Syst.*, 12(4):339 – 359, Oct 1994.

[Miura and Moriya, 1992] T. Miura and K. Moriya. On the completeness of visual operations for a semantic data model. *Data & Knowl. Eng.*, 9(1):19 – 44, 1992.

[Miura, 1994] T. Miura. Nesting quantification in a visual data manipulation language. *Data & Knowl. Eng.*, 12(2):167 – 196, Mar 1994.

[Mohan and Kashyap, 1993] L. Mohan and R. L. Kashyap. A visual query language for graphical interaction with schema-intensive databases. *IEEE Trans. Knowl. and Data Eng.*, 5(5):843 – 858, Oct 1993.

[Nishisato, 1980] S. Nishisato. *Analysis of Categorical Data: Dual Scaling and Its Application*. Univ. of Toronto Press, 1980.

[Paepcke et al., 1996] A. Paepcke, S. B. Cousins, H. Garcia-Molina, et al. Using Distributed Objects for Digital Library Interoperability. *IEEE Computer*, 29(5):61 – 68, 1996.

[Quinlan, 1993] J. R. Quinlan. *C4.5: Programs for Machine Learning*. Morgan-Kaufmann, 1993.

[Salton, 1975] G. Salton. *Dynamic Information and Library Processing*. Prentice-Hall, 1975.

[Salton, 1989] G. Salton. *Automatic Text Processing - The Transformation, Analysis, and Retrieval of Information by Computer*. Addison-Wesley Publishing Co., 1989.

[Schatz and Chen, 1996] B. Schatz and H. Chen. Digital Library Initiatives. *IEEE Computer*, 29(5), 1996.

[Spoerri, 1993] A. Spoerri. InfoCrystal: A visual tool for information retrieval & management. In *Proc. of the 2nd Int. Conf. on Information Knowledge Management*, pages 11 – 20, 1993.

[Sugimoto et al., 1995a] M. Sugimoto, K. Hori, and S. Ohsuga. A Document Retrieval System for Assisting Creative Research. In *Proc of the 3rd Int. Conf. on Document Analysis and Recognition (ICDAR'95)*, pages 167 – 170, Aug 1995.

[Sugimoto et al., 1995b] M. Sugimoto, K. Hori, and S. Ohsuga. A System for Assisting Creative Research Activity. In *Proc. of 6th Int. Conf. on Human-Computer Interaction*, pages 685 – 690, Jul 1995.

[Sun Microsystems, 1996] Sun Microsystems. *JDBC: A Java SQL API*, 1996. http://splash.javasoft.com/jdbc/index.html.

[Taylor, 1968] R.S. Taylor. Question-Negotiation and Information Seeking in Libraries. *Colleges and Research Libraries*, 29:178 – 194, 1968.

[Thompson and Croft, 1989] R. H. Thompson and W. B. Croft. Support for browsing in an intelligent text retrieval system. *Int. Journal of Man-Machine Studies*, 30:639 – 668, 1989.

[Vadaparty et al., 1993] K. Vadaparty, Y. A. Aslandogan, and G. Ozsoyoglu. Towards a unified visual database access. *SIGMOD Record*, 22(2):357 – 366, Jun 1993.

# Using a Bayesian Network Induction Approach for Text Categorization

**Wai Lam** and **Kon Fan Low** and **Chao Yang Ho**
Department of Systems Engineering and Engineering Management
The Chinese University of Hong Kong
Shatin
Hong Kong

## Abstract

We investigate Bayesian methods for automatic document categorization and develop a new approach to this problem. Our new approach is based on a Bayesian network induction which does not rely on some major assumptions found in a previous method using the Bayesian independence classifier approach. The design of the new approach as well as its justification are presented. Experiments were conducted using a large-scale document collection from Reuters news articles. The results show that our approach outperformed the Bayesian independence classifier as measured by a metric that combines precision and recall measures.

## 1 Introduction

Document categorization has become an important area in information retrieval and gathering. The objective of document categorization is to assign categories to a document. The categories are drawn from a set of pre-specified categories. It is possible to assign multiple categories to a document if they are appropriate. A document here refers to a piece of text. Traditionally this categorization task is performed manually by domain experts. Each incoming document has to be read and comprehended by an expert who assigns its categories. Obviously a large amount of human resources are required to carry out such categorization tasks. For instance, the OHSUMED document collection, composed of medical journal articles, requires a large amount of resources for manual categorization [Hersh et al., 1994].

A promising way to solve this problem is to automatically categorize documents based on their content. This can be achieved by learning a categorization scheme from a set of training examples. This problem can be viewed as a machine learning problem where an element may be classified into multiple category labels. Recently, some work has been done on this problem [Apte et al., 1994; Lewis et al., 1996; Yang, 1994; Larkey and Croft, 1996]. We present a new approach based on a Bayesian network induction. We show that it relaxes some major assumptions found in a previous approach based on the Bayesian independence classifiers [Lewis, 1992]. In this paper, the basic idea of the Bayesian independence classifier approach is first reviewed. We then introduce our approach for inducing Bayesian networks. We also contrast its relationship to the Bayesian independence classifier approach. Experiments were conducted using a large-scale document collection concerning with economic and financial news articles form Reuters. The results show that our approach outperformed the Bayesian independence classifier as measured by a metric that combines precision and recall measures.

## 2 Bayesian Independence Classifier Approach

In order to handle multiple category assignments, a separate classifier is constructed for each category. Each classifier is then used to determine whether a given document should be assigned to this category. This yields a set of categories for each document.

Lewis [Lewis, 1992] used a Bayesian approach for text categorization problems. Suppose we need to build a classifier for a particular category. Let the random variable $C_j$ denote the membership of a document in the $j$-th category and it can take on binary values 0 and 1. Let $d$ be an incoming document to be categorized. According to Bayes theorem,

$$P(C_j = 1 \mid d) = \frac{P(d \mid C_j = 1)P(C_j = 1)}{P(d)} \quad (1)$$

To represent document $d$, we use a vector $\vec{T_j}$ of chosen predictor terms $(T_{j_1}, \ldots, T_{j_p})$. Each term usually represents a particular word. Note that we may use different terms to represent the same document under different categories. Each predictor term can take on binary values 0 and 1 denoting absence or presence of the corresponding word in a document. Hence, Equation 1 becomes:

$$P(C_j = 1 \mid \vec{T_j}) = \frac{P(\vec{T_j} \mid C_j = 1)P(C_j = 1)}{P(\vec{T_j})} \quad (2)$$

Suppose the value of each term in vector $\vec{T_j}$ is $(t_{j_1}, \ldots, t_{j_p})$. Equation 2 can be expressed as:

$$P(C_j = 1 \mid \vec{T_j})$$
$$= \frac{P(T_{j_1} = t_{j_1}, \ldots, T_{j_p} = t_{j_p} \mid C_j = 1)P(C_j = 1)}{P(T_{j_1} = t_{j_1}, \ldots, T_{j_p} = t_{j_p})} \quad (3)$$

This formula requires estimates for the joint probabilities $P(T_{j_1} = t_{j_1}, \ldots, T_{j_p} = t_{j_p} \mid C_j = 1)$ and $P(T_{j_1} = t_{j_1}, \ldots, T_{j_p} = t_{j_p})$. Typically these probabilities are difficult to estimate from the training examples.

To overcome this problem, the following two assumptions are usually made regarding the marginal and conditional independence of predictor terms as follows:

$$P(\vec{T_j} \mid C_j = 1)$$
$$= P(T_{j_1} = t_{j_1} \mid C_j = 1) \cdots P(T_{j_p} = t_{j_p} \mid C_j = 1) \quad (4)$$

and

$$P(\vec{T_j}) = P(T_{j_1} = t_{j_1}) \cdots P(T_{j_p} = t_{j_p}) \quad (5)$$

Substituting Equations 4 and 5 into Equation 3, we get:

$$P(C_j = 1 \mid \vec{T_j}) = P(C_j = 1) \prod_i \frac{P(T_{j_i} = t_{j_i} \mid C_j = 1)}{P(T_{j_i} = t_{j_i})} \quad (6)$$

Ideally, we would like a feature term which is strongly related to the ideas that is conveyed by the document. Sometimes the word appears in the document and we can of course assign it as a predictor feature. However, it is also possible that the word does not appear in a document and the term may still relate to the ideas conveyed by the document. We are therefore not sure whether an index term should be assigned to a document. To incorporate this kind of uncertainty, the probability $P(\vec{T_j} \mid d)$ is introduced to indicate how close $\vec{T_j}$ is to the ideal indexing of the document. Each element $T_{j_i}$ in $\vec{T_j}$ can have binary values 0 or 1. Instead of directly estimating $P(C_j = 1 \mid d)$ as in Equation 1, we need to consider $P(\vec{T_j} \mid d)$. Hence the formula for $P(C_j = 1 \mid d)$ is modified as follows:

$$P(C_j = 1 \mid d) = \sum_{\vec{T_j} \in \{0,1\}^p} P(C_j = 1 \mid \vec{T_j})P(\vec{T_j} \mid d)$$

Applying Bayes' Theorem, we get

$$P(C_j = 1 \mid d)$$
$$= \sum_{\vec{T_j} \in \{0,1\}^p} P(C_j = 1) \frac{P(\vec{T_j} \mid C_j = 1)}{P(\vec{T_j})} P(\vec{T_j} \mid d) \quad (7)$$

In order to compute the above expression, the independence assumption in Equations 4 and 5 are still required. In addition, it requires the following assumption:

$$P(\vec{T_j} \mid d) = P(T_{j_1} = t_{j_1} \mid d) \cdots P(T_{j_p} = t_{j_p} \mid d) \quad (8)$$

Substituting Equations 4, 5 and 8 into 7, we get:

$$P(C_j = 1 \mid d)$$
$$= P(C_j = 1)$$
$$\sum_{t_{j_i} \in \{0,1\}} \frac{P(T_{j_1} = t_{j_1} \mid C_j = 1) \cdots P(T_{j_p} = t_{j_p} \mid C_j = 1)}{P(T_{j_1} = t_{j_1}) \cdots P(T_{j_p} = t_{j_p})}$$
$$P(T_{j_1} = t_{j_1} \mid d) \cdots P(T_{j_p} = t_{j_p} \mid d)$$
$$= P(C_j = 1)$$
$$\sum_{t_{j_i} \in \{0,1\}} \prod_i \frac{P(T_{j_i} = t_{j_i} \mid C_j = 1)P(T_{j_i} = t_{j_i} \mid d)}{P(T_{j_i} = t_{j_i})} \quad (9)$$

To evaluate the expression in Equation 9, Lewis [Lewis, 1992] developed an approximation as follows:

$$P(C_j = 1)$$
$$\prod_i [\frac{P(T_{j_i} = 1 \mid C_j = 1)P(T_{j_i} = 1 \mid d)}{P(T_{j_i} = 1)} +$$
$$\frac{P(T_{j_i} = 0 \mid C_j = 1)P(T_{j_i} = 0 \mid d)}{P(T_{j_i} = 0)}] \quad (10)$$

## 3 Bayesian Network Approach

A major assumption in the Bayesian independence classifiers is that the features are marginally independent as well as conditionally independent to $C_j$ or $d$. This assumption is used to estimate those high order probabilities such as $P(\vec{T_j})$, $P(\vec{T_j} \mid C_j)$, and $P(\vec{T_j} \mid C_j)$ as shown in Equations 4, 5, and 8. The accuracy of these estimations directly affect the performance of the classifier. Poor estimation will probably not result in good classification performance. In our text categorization problem, it is likely that the predictor terms are not independent. We develop a new approach based on a Bayesian network induction that eliminates this assumption.

A Bayesian network consists of nodes and arcs connected together to form a directed acyclic graph. Each node represents a variable that can take on a discrete set of domain-specific states. Each (directed) arc represents a probabilistic dependency between the source and destination variables.

The idea is to learn a Bayesian network for the set of variables representing the predictor terms $\vec{T_j}$ as well as the category $C_j$. The network provides a means for us to exploit the inherent dependency among these variables. Instead of relying on estimating high order probabilities as described above, we can compute $P(C_j = 1 \mid \vec{T_j})$ directly from the Bayesian network which already encodes the dependencies among the terms in $\vec{T}_j$ as well as the category $C_j$.

Apart from an acyclic graph structure, a Bayesian network consists of a set of conditional probability parameters associated with each node in the network. Suppose $X$ denotes a node representing one of the domain variables in a Bayesian network. Let $\Upsilon_X$ be the set of parents of node $X$ in the network structure (i.e., there is a directed arc from each node in $\Upsilon_X$ pointing to $X$). There is a conditional probability distribution $P(X \mid \Upsilon_X)$ associated with $X$. If $X$ does not have parents in the network structure, there is a prior probability distribution $P(X)$ associated with $X$.

After the network is constructed, it can be used for conducting reasoning. A common and useful kind of reasoning is to perform probabilistic inferences. In a typical scenario, the states of some variables are observed. We are interested in calculating the posterior probability distribution of some other variables given the instantiation of the observed variables.

To use Bayesian networks as classifiers in our text categorization problem, we first construct a Bayesian network for variables in $\{C_j\} \cup \vec{T}_j$ (i.e., $\{C_j, T_{j_1}, \ldots, T_{j_p}\}$). Machine learning techniques can be used to automatically construct a network from training documents. Once this network is constructed, it can be used to categorize documents, using probabilistic inferences on the network. More specifically, the feature variables representing the document (i.e., variables in $\vec{T}_j$) are instantiated appropriately. The posterior probability $P(C_j = 1 \mid \vec{T}_j)$ is then computed. The document will be determined as being assigned to the category if this probability value exceeds a certain threshold. The main advantage of this approach over the Bayesian independence classifier is that it relaxes those assumptions regarding feature independence. The network structure can capture the inherent dependency during the induction process.

The next issue is how to learn a Bayesian network from the training documents. There has been some research on learning Bayesian networks from data [Heckerman et al., 1995; Spirtes and Meek, 1995; Lam and Bacchus, 1994]. It is well known that learning arbitrary networks is quite computationally expensive. Since our purpose is to make use of the network for classification, we restrict our consideration to a family of network structures known as *classification-based network structures*. This can reduce the learning complexity considerably and at the same time achieve our objective for performing classification. A classification-based network structure is basically an arbitrary Bayesian network structure with the restriction that $C_j$ has no parents (i.e., $\Upsilon_{C_j} = \emptyset$). Figure 1 shows an example of a classification-based network structure. The rationale behind these structures is that the states of the feature variables may depend on the membership of the category, but not the other way around, if such dependencies exist. Bayesian networks have been adopted to build other classifiers [Friedman and Goldszmidt, 1996]. We focus on exploring Bayesian networks to solve large-scale document categorization problems and propose classification-based network structures to solve these tasks.

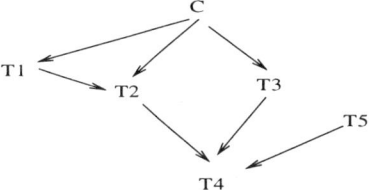

Figure 1: An Example of a Classification-Based Network Structure

## 4 Automatic Construction of Bayesian Networks

To learn a Bayesian network from training examples, we make use of a technique based on the *Minimum Description Length* (MDL) principle derived from [Lam and Bacchus, 1994]. Central to this approach is a *score metric* for a candidate network structure, where the score metric is a function representing the *total description length*, $L(B)$, of a candidate network structure $B$.

The total description length is decomposed into a function $L_{\text{total}}$ associated with each individual node in the network as follows:

$$L(B) = \sum_X L_{\text{total}}(X, \Upsilon_X)$$

where $X$ is a node in the network. $L_{\text{total}}$ of a node $X$ depends only on its parent set. Hence, $L_{\text{total}}$ is a constant if the parent set is fixed. The domain variables in our current categorization problem are $\{C_j\} \cup \vec{T}_j$ (i.e., $\{C_j, T_{j_1}, \ldots, T_{j_p}\}$). The network structures we consider are classification-based network structures as described in Section 3. Thus, node $C_j$ has a fixed set of parents, i.e., empty parent set. As a result, the score metric of a candidate network becomes:

$$\sum_{T_{j_i} \in \vec{T}_j} L_{\text{total}}(T_{j_i}, \Upsilon_{T_{j_i}})$$

The function $L_{\text{total}}$ is composed of two components:

$$L_{\text{total}}(T_{j_i}, \Upsilon_{T_{j_i}}) = L_{\text{network}}(T_{j_i}, \Upsilon_{T_{j_i}}) + L_{\text{data}}(T_{j_i}, \Upsilon_{T_{j_i}})$$

where $L_{\text{network}}$ and $L_{\text{data}}$ are called *network description length* and *data description length* respectively.

As described in [Lam and Bacchus, 1994], the network description length for a variable $X_i$ is given as follows:

$$L_{\text{network}}(X_i, \Upsilon_{X_i}) = |\Upsilon_{X_i}| \log_2(n) + Q(s_i - 1) \prod_{j \in \Upsilon_{X_i}} s_j$$

where $n$ is the number of variables. $s_j$ is the number of possible states of variable $X_j$, and $Q$ represents the number of bits required to store a numerical value.

This network description length has the property that the higher the topological complexity of the network the longer is its description length. It is widely recognized that conducting inference on highly-connected networks is likely to be intractable [Cooper, 1990], thus they are not very useful in practice. By minimizing this length function, we favor more useful networks, that is, those having simple topology.

For our categorization problem, there are $p + 1$ variables and each variable can take on binary values. Thus the network description length should be refined to:

$$L_{\text{network}}(T_{j_i}, \Upsilon_{T_{j_i}}) = |\Upsilon_{T_{j_i}}| \log_2(p+1) + Q 2^{|\Upsilon_{T_{j_i}}|}$$

Apart from the simplicity issue, we also need to consider the accuracy issue which refers to how accurate a network structure represents the training data. The accuracy issue is captured by the data description length as follows:

$$L_{\text{data}}(T_{j_i}, \Upsilon_{T_{j_i}}) = \sum_{T_{j_i}, \Upsilon_{T_{j_i}}} M(T_{j_i}, \Upsilon_{T_{j_i}}) \log_2 \frac{M(\Upsilon_{T_{j_i}})}{M(T_{j_i}, \Upsilon_{T_{j_i}})}$$

where the summation is taken over all possible instantiations of that variable and its parents. $M(.)$ is the number of cases that match a particular instantiation in the training data. The log function will be 0 if $M(T_{j_i}) = 0$. It can be shown that this data description length has the property that the more accurate the network structure the shorter is the description length. By minimizing this length function, we favor accurate networks.

Simplicity and accuracy cannot usually be maximized simultaneously. Given a particular network structure, it will become more accurate if we increase its topological complexity appropriately. As a result, we are faced with a tradeoff between simplicity and accuracy. It is desirable to learn an accurate network, and at the same time the structure should be as simple as possible. The score metric $L_{\text{total}}$, based on the MDL measure, offers a principled means to perform such tradeoff by considering the sum of the network description length $L_{\text{network}}$ and the data description length $L_{\text{data}}$ (note that both $L_{\text{network}}$ and $L_{\text{data}}$ are non-negative).

Within the framework, a shorter length $L_{\text{total}}$ corresponds to a better network. In this text categorization problem, we would like to find a classification-based network which has the lowest $L_{\text{total}}$. In situations where an optimal solution cannot be obtained due to limited computing resources, we wish to find a network with $L_{\text{total}}$ as low as possible. To find a good structure, we employ a best-first search strategy guided by the MDL metric. The heuristic function of a candidate network is basically composed of the MDL metric of the network structure and the arc to be added to the network. When we generate a candidate network structure, we impose a restriction that an arc cannot direct from a node representing a term to the node representing the category. Thus classification-based network structure topology can be maintained. This constraint increases the search efficiency because non-classification-based structures will be ignored during the search process.

## 5 Experimental Results

We have conducted experiments to demonstrate the performance of our approach. We implemented our approach as well as the Bayesian independence classifier approach. Thus we can directly compare the performance of both approaches. The document collection used in this experiment are economic and financial news articles from Reuters. There are 22,173 full text documents in the corpus. We used 21,450 documents between Feb. 26 1987 and Oct. 20 1987 in our experiment. These texts vary in length from single line bulletins to multi-page articles. Each document is assigned to one or more categories of economic interest. We partitioned the corpus into training and testing collection in the same way as the experiment done in [Lewis, 1992]. The texts before Apr. 7 1987 belongs to the training collection while those after this date belongs to the testing collection. There are 14,704 and 6,746 documents in the training and testing collection respectively. There are 112 categories in the training collection and the performance evaluation is based on these categories.

For each category, we select features based on the mutual information gain of the candidate feature and the category. The mutual information of a term $T_i$ and category $C_j$ is defined as:

$$\sum_{t \in \{0,1\}} \sum_{c \in \{0,1\}} P(T_i = t, C_j = c) \log_2 \frac{P(T_i = t, C_j = c)}{P(T_i = t) P(C_j = c)}$$

After mutual information of each term is calculated, the terms are ranked in descending order of mutual information and the top $p$ terms are chosen as the predictor features for the category.

For the evaluation of our categorization system, the performance of each category is measured. The overall effectiveness is then computed by the mean performance across each category. Consider a particular category,

| category | Bayesian indep. approach | Bayesian network approach | category | Bayesian indep. approach | Bayesian network approach | category | Bayesian indep. approach | Bayesian network approach |
|---|---|---|---|---|---|---|---|---|
| fishmeal | 1.000 | 1.000 | jobs | 0.636 | 0.714 | aust. dlr | 0.000 | 1.000 |
| mpiah | 1.000 | 1.000 | rice | 0.409 | 0.794 | coffee | 0.651 | 0.889 |
| lit | 1.000 | 1.000 | trade | 0.559 | 0.378 | wool | 1.000 | 1.000 |
| castor seed | 1.000 | 1.000 | lin meal | 1.000 | 1.000 | grain | 0.760 | 0.788 |
| ringgit | 1.000 | 1.000 | sun meal | 0.000 | 0.000 | nickel | 0.800 | 1.000 |
| pork belly | 0.000 | 1.000 | fuel | 0.000 | 0.353 | wheat | 0.652 | 0.879 |
| cornglutenfeed | 1.000 | 1.000 | nat gas | 0.328 | 0.333 | soy oil | 0.316 | 0.846 |
| tapioca | 0.000 | 1.000 | crude oil | 0.677 | 0.673 | iron steel | 0.171 | 0.400 |
| rand | 0.000 | 0.000 | carcass | 0.476 | 0.642 | soy meal | 0.588 | 0.696 |
| nzdlr | 0.000 | 0.000 | groundnut | 0.000 | 0.000 | ship | 0.769 | 0.726 |
| jet | 0.000 | 0.000 | linseed | 1.000 | 1.000 | pet. chem | 0.455 | 0.400 |
| peseta | 1.000 | 1.000 | yen | 0.364 | 0.517 | cocoa | 0.591 | 0.900 |
| cpu | 0.000 | 0.000 | sorghum | 0.522 | 0.774 | orange | 0.778 | 0.769 |
| inventories | 0.000 | 0.000 | strategic metal | 0.111 | 0.105 | tin | 0.167 | 0.914 |
| corn oil | 1.000 | 1.000 | dkr | 1.000 | 1.000 | lei | 0.750 | 0.800 |
| castor oil | 0.000 | 0.000 | nkr | 0.000 | 0.000 | acq. | 0.691 | 0.610 |
| naphtha | 0.000 | 0.400 | skr | 1.000 | 1.000 | money supply | 0.300 | 0.330 |
| dfl | 0.000 | 0.000 | stg | 0.000 | 0.000 | veg. oil | 0.810 | 0.684 |
| copra cake | 0.000 | 0.000 | cotton | 0.513 | 0.898 | reserves | 0.512 | 0.449 |
| red bean | 1.000 | 1.000 | sun seed | 0.545 | 0.667 | oil seed | 0.514 | 0.696 |
| rye | 0.000 | 0.000 | copper | 0.488 | 0.760 | tea | 0.250 | 0.667 |
| housing | 0.364 | 0.571 | dmk | 0.000 | 0.000 | palm kernel | 0.000 | 0.000 |
| cotton oil | 0.000 | 0.000 | soybean | 0.472 | 0.613 | zinc | 0.455 | 0.722 |
| oat | 0.500 | 0.667 | lead | 0.381 | 0.462 | palm oil | 0.960 | 0.615 |
| potato | 0.400 | 0.857 | livestock | 0.508 | 0.519 | alum | 0.615 | 0.686 |
| sun oil | 0.000 | 0.000 | hog | 0.286 | 0.316 | gnp | 0.612 | 0.634 |
| rape oil | 0.400 | 0.000 | cpi | 0.393 | 0.273 | platinum | 0.200 | 0.000 |
| groundnut oil | 0.000 | 0.000 | rape seed | 0.500 | 0.762 | interest | 0.472 | 0.289 |
| lin oil | 0.000 | 0.000 | corn | 0.611 | 0.894 | ipi | 0.516 | 0.345 |
| palladium | 0.000 | 0.000 | barley | 0.500 | 0.941 | rubber | 0.516 | 0.638 |
| l. cattle | 0.000 | 0.000 | bop | 0.746 | 0.513 | retail | 0.333 | 0.333 |
| lumber | 0.600 | 0.727 | gas | 0.400 | 0.571 | money fx. | 0.446 | 0.336 |
| can | 0.000 | 1.000 | heat | 0.615 | 0.533 | income | 0.167 | 0.222 |
| cruzade | 1.000 | 1.000 | meal feed | 0.667 | 0.750 | dlr | 0.526 | 0.231 |
| instal debt | 0.667 | 0.000 | silver | 0.118 | 0.125 | wpi | 0.222 | 0.462 |
| propane | 0.500 | 0.667 | coconut | 0.667 | 0.571 | earn | 0.814 | 0.740 |
| coconut oil | 0.000 | 0.400 | gold | 0.579 | 0.500 | | | |
| plywood | 0.000 | 1.000 | sugar | 0.740 | 0.764 | | | |

Table 1: The performance of each category measured by $F_1$ measure in the Bayesian independence classifier approach and our Bayesian network approach

the effectiveness of the categorization can be shown in a contingency table as follows:

|   | Expert Says Yes | Expert Says No |   |
|---|---|---|---|
| System Says Yes | $q$ | $r$ | $q + r$ |
| System Says No | $s$ | $t$ | $s + t$ |
|   | $q + s$ | $r + t$ | $q + r + s + t$ |

Table 2: A Contingency Table Showing Categorization Decisions For a Category

where $q$ is the number of documents belonging to the category and assigned to the category; $r$ is the number of documents not belonging to the category and assigned to the category; $s$ is the number of documents belonging to the category but not assigned to the category; $t$ is the number of documents not belonging to the category and not assigned to the category.

Some common effectiveness measures can then be defined in terms of these values:

$$(\text{recall})R = q/(q+s)$$
$$(\text{precision})S = q/(q+r)$$

Recall is the proportion of documents belonging to the category that the system successfully assigns to the category. Precision is the proportion of documents assigned to the category by the system that really belong to the category. An ideal categorization system would have recall and precision equal to 1. However, perfect recall can be achieved by a system that puts every document in the category, while perfect precision can be achieved by a system that puts no documents in the category, so just using either recall or precision does not provide a fair evaluation to a system.

Therefore, an effectiveness measure called $F$ measure, proposed by [Van Rijsbergen, 1979; Lewis et al., 1996], combines the recall and precision into a single score as follows:

$$F_\beta = \frac{(\beta^2 + 1)SR}{\beta^2 S + R}$$

$\beta$ ranges from 0 to infinity. When $\beta$ is 0, $F_\beta$ is the same as the precision. When $\beta$ equals to infinity, $F_\beta$ is the same as recall. The variation of $\beta$ between 0 and infinity corresponds to the variation of the weight to recall and precision. In our evaluation, $\beta$ is set to 1 (i.e., equal weight on recall and precision). After we compute all $F_1$ values for each category, the mean $F_1$ value across categories is computed so as to compute the overall evaluation of our categorization system.

Table 1 shows the performance of the Bayesian independence classifier approach and our Bayesian network approach measured using $F_1$ for each category in the experiment. The mean $F_1$ measure of the Bayesian independence classifier is 0.443 whereas the mean $F_1$ measure of our Bayesian network induction approach is 0.542. Therefore, there is roughly a 22.3% improvement for our Bayesian network approach. There are 88 out of 112 categories where the performance of the Bayesian network approach is the same or better than the Bayesian independence classifier approach.

We also investigate the sensitivity of the probability threshold on the performance of the Bayesian network approach. Figure 2 shows the performance of the Bayesian network approach when we varied the probability threshold.

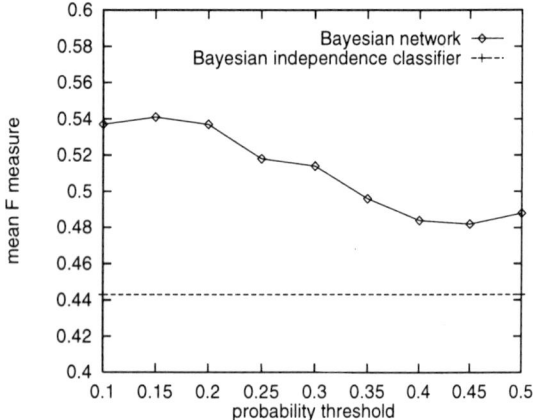

Figure 2: Sensitivity of Probability Threshold on the Bayesian Network Approach

## 6 Conclusion

Bayesian methods for automatic document categorization are investigated in this paper. We first reviewed an existing technique based on the Bayesian independence classifier approach. After discussing its drawbacks, we introduced a new approach based on Bayesian networks. The theoretical justifications of our new approach is also described. Experiments were conducted using a large document collection from Reuters news articles. The results show that our approach outperformed the Bayesian independence classifier as measured by a metric that combines precision and recall measures.

## References

[Apte et al., 1994] C. Apte, F. Damerau, and Weiss S.M. Automated learning of decision rules for text categorization. *ACM Transactions on Information Systems*, 12(3):233–251, 1994.

[Cooper, 1990] G. F. Cooper. The computational complexity of probabilistic inference using Bayesian belief networks. *Artificial Intelligence*, 42:393–405, 1990.

[Friedman and Goldszmidt, 1996] N. Friedman and M. Goldszmidt. Building classifiers using Bayesian networks. In *Proceedings of the Thirteenth National Conference on Artificial Intelligence*, pages 1277–1284, 1996.

[Heckerman et al., 1995] D. Heckerman, D. Geiger, and D. M. Chickering. Learning Bayesian networks: The combination of knowledge and statistical data. *Machine Learning*, 20(3):197–243, 1995.

[Hersh et al., 1994] W. Hersh, C. Buckley, T.J. Leone, and D. Hickman. OHSUMED: An interactive retrieval evaluation and new large test collection for research. In *Proceedings of the Seventeenth International ACM SIGIR Conference on Research and Development in Information Retrieval*, pages 192–201, 1994.

[Lam and Bacchus, 1994] W. Lam and F. Bacchus. Learning Bayesian belief networks: An approach based on the MDL principle. *Computational Intelligence*, 10(3):269–293, 1994.

[Larkey and Croft, 1996] L. S. Larkey and W. B. Croft. Combining classifiers in text categorization. In *Proceedings of the Nineteenth International ACM SIGIR Conference on Research and Development in Information Retrieval*, pages 289–297, 1996.

[Lewis et al., 1996] D. D. Lewis, R. E. Schapore, J. P. Callan, and R. Papka. Training algorithms for linear text classifiers. In *Proceedings of the Nineteenth International ACM SIGIR Conference on Research and Development in Information Retrieval*, pages 298–306, 1996.

[Lewis, 1992] D. D. Lewis. Feature selection and feature extraction for text categorization. In *Proceedings of Speech and Natural Language Workshop*, 1992.

[Spirtes and Meek, 1995] P. Spirtes and C. Meek. Learning Bayesian networks with discrete variables from data. In *Proceedings of the First International Conference on Knowledge Discovery and Data Mining*, pages 294–299, 1995.

[Van Rijsbergen, 1979] C. J. Van Rijsbergen. *Information Retrieval*. Butterworths, London, 1979.

[Yang, 1994] Y. Yang. Expert network: Effective and efficient learning from human decisions in text categorization and retrieval. In *Proceedings of the Seventeenth International ACM SIGIR Conference on Research and Development in Information Retrieval*, pages 13–22, 1994.

# Toward Structured Retrieval in Semi-structured Information Spaces

**Scott B. Huffman** and **Catherine Baudin**
Price Waterhouse Technology Centre
68 Willow Road, Menlo Park, CA 94025-3669
{huffman, baudin}@tc.pw.com

## Abstract

A semi-structured information space consists of multiple collections of textual documents containing fielded or tagged sections. The space can be highly heterogeneous, because each collection has its own schema, and there are no enforced keys or formats for data items across collections. Thus, structured methods like SQL cannot be easily employed, and users often must make do with only full-text search. In this paper, we describe an intermediate approach that provides structured querying for particular types of *entities,* such as companies, people, and skills. Entity-based retrieval is enabled by *normalizing* entity references in a heuristic, type-dependent manner. To filter results further, entities are categorized as playing particular *roles* (e.g., company as *client*, as *vendor*, etc.) in particular *collection types* (directories, client engagement records, etc.). The approach requires minimal meta-information – each collection's type and a list of fields containing entities in various roles – and much of this is derived automatically. On a set of typical user queries in a large corporate information space, the approach produces a dramatic improvement in retrieval quality over knowledge-free methods like full-text search.

## 1 Introduction

Decentralized information sharing architectures like the World Wide Web and Lotus Notes make it easy for individuals to add information, but as the space grows, retrieval becomes more and more difficult. *Semi-structured* information sharing systems, including Lotus Notes and a variety of meta-tagging schemes being developed for the World Wide Web (e.g. Apple's Meta-Content Format), address part of this problem by providing the ability to structure local parts of the information space. In a semi-structured information space, documents are sectioned into weakly-typed fields according to user specifications, and documents with the same field structure can be grouped into collections. Within a collection, field values can be used as indexes for easier retrieval.

Unfortunately, semi-structuring document collections does not solve the problem of retrieving information across a large information space. Even if individual collections are well designed for retrieval, users can be overloaded with the sheer number of collections. Retrieval across the entire space is difficult because it is highly heterogeneous. Each collection has its own local schema, and there are no enforced keys or formats for data items within or across collections.

In this paper, we address the problem of finding information across collections in a large semi-structured information space. Our goal is to provide querying that is more powerful and precise than full-text search, but without requiring the collections to be strongly typed, data normalized, and fully mapped to a global schema, as methods like multidatabase SQL require. In addition, to maintain the advantages of decentralized information sharing, we do not want to impose constraints like integrity checks when users create new documents.

Our approach is to provide high quality retrieval of information related to important *entities* in the information space. In our organization (a large professional services firm), important types of entities include people, companies, and consulting skills. A review of our largest collections revealed that nearly always, references to important entities are fielded rather than buried in free-running text. Because the same entity can be referred to in many different ways across a heterogeneous information space, our entity retrieval system *normalizes* references to entities in a heuristic, type-dependent manner. For instance, the person names "Mr. Bob Smith", "Smith, Robert", and "R. J. Smith" are normalized such that a query for any one (or a number of other possible forms) will retrieve documents containing any of them.

In addition to normalization, there are two entity-related *retrieval filters* provided by our retrieval system: *entity roles* and *collection types*. Entities often play identifiable *roles* within collections. For instance, a person can play the role of a partner on an engagement, a manager, a contact at a client, etc.; a company can play the role of a client, a vendor, a newsmaker, etc. These can be useful in organizing or filtering retrieved information ("find mentions of IBM as a client").

Based on entities and roles they contain, collections as a whole can be classified into a small number of

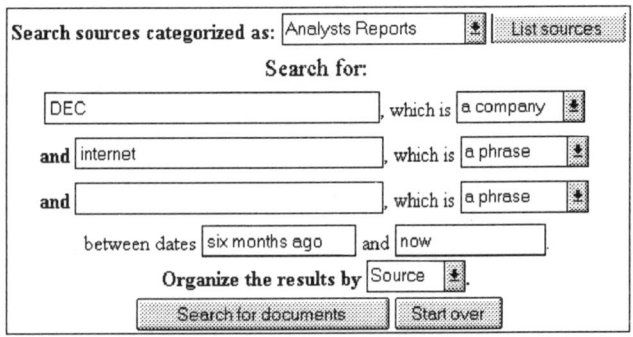

Figure 1. An NX search form

*collection types*. Examples include *client-engagements* collections, which contain clients, partners, and managers; *directories*, which contain staff names and phone numbers, etc. Like roles, collection types provide a useful way to organize or filter retrieved documents.

We have implemented an entity-based retrieval system called *PW Notes Explorer* (NX) that operates over a large semi-structured information space. The space includes corporate Lotus Notes collections and a small set of web collections. NX provides full-text search, entity-based search for people, companies, and skills, and role and collection-type organization and filtering of results. In addition, it uses meta-information about entity types such as phone numbers, office sites, and dollar amounts to allow queries for specific properties of entities (e.g. "find Bob Smith's phone number").

A key hypothesis behind this work is that *a relatively small amount of meta-information* – much less than that required to normalize and map collections to a global schema – *can give a large gain* in query power and precision over knowledge-free methods like full-text search. NX is one illustration of this hypothesis. It requires minimal meta-information about each collection – its collection type and an indication of fields containing entities in various roles – and uses it to produce a dramatic improvement in retrieval quality for entity-related queries. Much of the required meta-information is inferred automatically based on field names and data within the collections, using a simple heuristic classifier (see section 3.1).

In what follows, we first motivate entity-based retrieval with real-world example. Next, we describe the main components of our retrieval system, and present an empirical comparison between entity-based retrieval and full-text search for a set of typical queries. We conclude by discussing related and future work.

## 2 Entity-based retrieval

In a corporate setting, information in different documents is frequently linked through references to entities with business importance, such as people and companies. Often, users search for information about *particular* entities (e.g., "What is Bob Smith's phone number?" or "Who's the manager for the XYZ Co. account?") as opposed to ungrounded, aggregate queries across sets of entities (e.g. "Show me all managers with more than five clients over $5 million in sales"). We designed NX to support this type of search.

Consider a typical example from our organization. A staff member is writing a proposal to XYZ Company for some consulting work on XYZ's new customer tracking system. She needs answers to questions like:

(a) How large is XYZ Company? E.g., what are their assets, revenues, etc.?
(b) Does our organization have a prior relationship with XYZ? Have we done other consulting work for them in the past?
(c) Who did that work, and how can they be contacted?
(d) Do we have staff with expertise on customer tracking systems? How can they be contacted?

Each question refers to entities of various types – XYZ Company, staff members, phone numbers, skills, etc. – and these entities may be referred to differently in different documents. The entities play particular roles; for instance, XYZ as a client, but not as a vendor or a newsmaker. Some questions involve information that may be found in many collections of the same type – e.g., information about prior work for XYZ (b) might be found in numerous collections containing client engagements. Others involve linking information about XYZ with information about another entity -- e.g., question (c) requires finding staff names in documents that list XYZ engagements, and then finding contact information for those staff names.

To find answers using the source collections directly would require that the user:
- know the relevant collections and their locations.
- construct searches that accounts for different forms of an entity (e.g. "XYZ Corporation" vs. "X Y Z Inc.").
- discard hits where entities play irrelevant roles (e.g. XYZ as a vendor in an engagement, not as the client), and hits in irrelevant types of collections (e.g. XYZ mentioned in news rather than client engagements).

NX overcomes these requirements. The next section examines its key components.

## 3 PW Notes Explorer

NX operates over a semi-structured information space of hundreds of collections, currently containing about 300,000 documents. It is delivered over an intranet using HTML.

Figure 1 shows one of the system's search forms. The pick-box with "Analysts Reports" chosen allows a user to choose among collection types. The pick-boxes along the right side, such as the one with "a company" chosen, allow a user to choose an entity type for each search string; "a phrase" indicates that a full-text search for the string should be performed. Figure 2 shows part of the set of results retrieved by the search in Figure 1.

NX includes three main components:
1. **Semi-automatic field classification**. To build an index of entity references of different types, we must

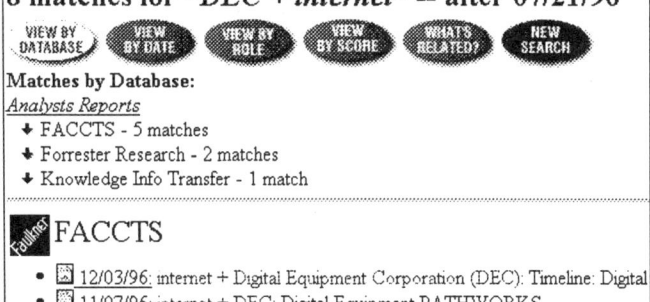

Figure 2. NX search results

identify where those types occur within collections. NX's field classifier uses field names and sample values from a collection to classify fields as containing entity types (people's names, company names, phone numbers, dollar amounts, etc.) and roles. As classification is not 100% accurate or complete, an interface is provided to alter the entity and role types for each collection's fields.

2. **Entity normalization.** When collections are indexed, entity references are extracted from entity-typed fields and *normalized* in a heuristic manner, using formatting knowledge and synonym tables specific to each entity type. At retrieval time, entity search terms are normalized in the same way and used to find matches in the index.

3. **Entity-related result organization and filtering.** Once a set of documents is retrieved, it can be viewed or filtered based on entity roles, collection types, and frequently co-occurring entities.

Next, we give a brief description of each component.

### 3.1 Semi-automatic field classification

Field classification attempts to identify an entity type and role for each collection field. The current version recognizes person names, company names, telephone numbers, geographic locations, office names, and dollar amounts. As in [Li & Clifton, 1994], the attributes used by the classifier are the *field name* and *sample values* from the documents in the collection.

The field classifier proceeds in three steps:

1. *Field name analysis*: Field names are tokenized according to capitalization and other separators. The tokens are then analyzed using a domain-dependent dictionary of entity types and roles. Tokens are matched against the entries in the dictionary using a set of matching heuristics that recognize different types of abbreviations. For instance, the tokens "Tel" and "Phone" match the entity type telephone, the token "Ptr" matches the entity role partner, and the token "Name" is indicative of a name of any type: person, company, or office. In addition, the field name analyzer uses a set of simple patterns. For instance, the token "By" in the last position of a field name can indicate an action (e.g.,

"ServedBy"), suggesting that the field contains person or company names, which are the primary actors in our domain. Field name analysis produces a weighted set of potential entity types.

2. *Field value analysis*: Sample data values drawn from documents in the collection are analyzed in a type-specific manner for each potential entity type produced by field name analysis. For each entity type, the sample values are analyzed to determine how confidently they can be considered that type of entity. These confidence levels are combined for each potential type. Three criteria are used to analyze values:

a) *Value lookup:* the system looks up field values in tables of common values for different entity types. These include peoples' first names, large company names, and office sites. The confidence level for value lookup is high.

b) *Common keywords:* the system looks for common words that appear in entity values. For instance, company names often contain "Co." or "Inc.". Person names can contain titles like "Mr." or "Ms.". The confidence level for common keywords is medium.

c) *Formatting:* type-dependent routines attempt to match common formats for values of each entity type. For instance, person names have a small number of capitalized words, possibly initials, etc. The confidence level for formatting criteria is low.

If the confidence level returned by field name analysis is low for a potential entity type, the system increases the number of sample values, in order to collect enough evidence to make a classification. An entity type is assigned if confidence exceeds a threshold.

3. *Role identification:* If an entity type is assigned, the classifier attempts to assign a role to the field based on its type and a second set of field name patterns. For example, a field of type *person name* with tokens "editor", "author", or "owner" indicates the role *author*.

In a small test, the field classifier analyzed 670 fields in 26 collections. Of these, 159 were of recognizable types. The field classifier correctly classified 154 of the 159 (97%). Most failures were due to failing to recognize tokens in the field name. Of the 154 classified fields, 61 were of the roles client, manager, partner, author, reviewer or vendor. The system correctly recognized 58 of these (95%).

More work is needed to extend the classifier's generality. It relies on a hand-built dictionary of entity type and roles, but it should be possible to use standard induction techniques to learn such a dictionary from examples. The field classifier may also be extended to recognize other types, such as skill descriptions, vendor products, and technical terminology.

### 3.2 Entity Normalization

In a standard relational database, tuples from different tables that contain information about the same entity each contain a *key* for that entity allowing the tuples to be joined. In a semi-structured document space, however, there are rarely unique keys shared by collections. Rather, entities are referred to within text strings in a variety of formats, with a variety of synonyms and abbreviations.

Therefore, to allow search over entities, entity references must be normalized and matched. For maximum retrieval speed, NX normalizes entity references at indexing time. Its entity index stores both the original form and a normalized form of each entity reference. At retrieval time, a normalized form of the user's search string is created and used to retrieve matches from the normalized entity index. In some cases, values are only partially normalized, and the original forms of retrieved matches and the search string are compared to verify the match.

Normalization and matching of strings that refer to entities has two important properties. First, it is *heuristic*. Generally, it is impossible to know with certainty whether two strings refer to the same entity. "Bob Smith" and "Robert A. Smith" may or may not be the same person; the retrieval system must make a reasonable guess.

Second, it is *type dependent*. Normalizing different types of entities requires different processing: e.g., normalizing person names is different than normalizing company names. Different variations in format are allowed for each type; "Smith, Bob" is a variation of "Bob Smith", but "Computer, Apple" is not a variation for "Apple Computer". In addition, different synonyms, abbreviations, and stop words apply. For company names, for instance, including or omitting words like "Corp." is only a small variation; for person names, "Mr." is a small variation unless the other string contains a female designator like "Ms." NX employs different normalization routines for different types, including small tables of synonyms, abbreviations, and stop words.

Due to space limitations, we will only briefly describe NX's normalization routines here.

For company names, the string containing the entity reference is first processed to produce uniform capitalization, combine leading initials (e.g., "I B M" becomes "IBM"), and expand common abbreviations. Next, we look up the leading words of the string in a table of company name synonyms. If a match is found, a designated synonym is used as the normalized form of the string. For example, the normalized form of "IBM" is "International Business Machines This could result in the same normalized form for different names, the differences are accounted for at retrieval time as described below.

At retrieval time, the search string for a company is normalized in the same way. Matches for the normalized form are retrieved from the entity index table. If the search string was found in the synonym table, all matches from the index are returned as retrieved matches, on the basis that appearing in the same synonym set implies a fairly strong match. For instance, all matches normalized to "International Business Machines" would be returned by a search for "IBM". In cases where no synonym is found, entity strings with the same first word are retrieved from the index and scored against the search string on a word-by-word basis. Words that appear in one but not the other reduce the match score by an amount proportional to the total number of words; common company words like "Corporation" are penalized less. Each entry with a match score over a threshold is returned as retrieved match. The search string "Foobaz Circuit Supply", for instance, matches "Foobaz Circuit Corp." but not "Foobaz Lawn and Garden Products".

The algorithm for person names is slightly more complex, since it must account for more possible variations in word-order, title words ("Dr.", etc.), generation designators ("Jr.", "III", etc.) and so on.

In addition, pre-processing is required to find the portions of the input string containing entity references. Often, a field will contain multiple entity values in a single string, with spurious information interspersed. For example, a typical person name field value might be "Bob J. Smith Jr. – partner; Sue Jones, 415-555-1212, Palo Alto." NX's normalization routines extract "Bob J. Smith Jr." and "Sue Jones" out of this field value.

### 3.3 Entity-related result filtering

Documents that match a query can be filtered and sorted based on the collection types they are drawn from, the roles played by entities in the documents, and other entities that co-occur in the documents. In addition, standard filters/sorts such as date ranges and relevance scores are provided.

In Figure 2, results are displayed sorted by the collection type and collections they are drawn from. Here, the search has been filtered using the "Analysts Reports" collection type. In cases where documents are retrieved from many collection types, the display provides a useful breakdown of the major kinds of information retrieved.

The buttons along the top of the page indicate alternative sortings. 'View by role' sorts the documents according to the roles of entities that matched the query within each document.

The 'What's related?' function retrieves the most frequently occurring other entities (people, companies, and skills) in the set of retrieved documents. Because they have been normalized, the co-occurring entities can be properly grouped independent of how they were referred to in the source documents. In essence, this uses the normalized entity index as a simple data warehouse, enabling an aggregation over entities in document sets. Entity co-occurrence often indicates an ongoing relationship, such as partners with clients, staff with skills, companies as industry peers, etc.

The "What's related?" function is a good example of entity-based retrieval's use of a small amount of meta-information to go beyond knowledge-free methods like

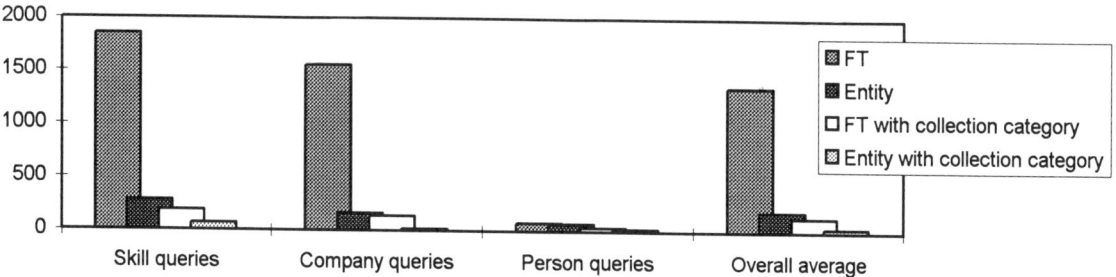

Figure 3. Average number of results for various query types

full-text search. A knowledge-free method cannot perform this type of function as meaningfully because it does not take advantage of the structure present in semi-structured collections. "What's related?" is a step toward structured retrieval. It does not retrieve structured relationships between data items, but approximates them using co-occurrence.

## 4 Results

To evaluate NX's entity-based retrieval and filtering functionality, we used a typical set of user questions from our organization's staff (drawn from NX logs, consultant training material, and online bulletin boards). For each question, we tested four conditions:
1. *Naïve full-text search:* Our users are not sophisticated full-text users. Thus, we used simple full-text searches for the entities in the question.
2. *Full-text search with collection-type filtering:* To (1), we added that only results from the appropriate collection type be returned.
3. *Entity search:* Instead of full-text search, we used NX's entity search for the entities in the question.
4. *Entity search with collection-type filtering:* To (3), we added that only results from the appropriate collection type be returned.

We tested these conditions for a set of 24 questions. The results are shown in Figure 3. Collection-type filtering produces a much smaller result set than full-text search alone, and collection-type plus entity search reduces the result set further. The average naïve full-text result set was 1358 documents from 15 collections; the average entity-search with collection-type filtering result set was 37 documents from 2 collections.

Without pre-scored document collections, it is difficult to obtain strong measures of recall and precision for large information spaces (too large to search by hand). In the absence of scored answer keys, we have resorted to approximate measures. One way to approximate recall is by comparing to a method known to produce high recall. In this case, the results returned by entity + filtering searches are basically equivalent to the results that would be found by a user submitting a query against each collection's local schema, if the user knew the proper collections, field names and synonymous field values for each query. Thus, entity + filtering search produces a comparable level of recall.

Assuming comparable recall levels, we can obtain a rough measure of precision by examining the total number of documents returned by each retrieval method. Our test questions called for fairly focused answers, as opposed to "deep background" questions, like "find all documents that mention the internet", and thus small result sets are appropriate. In practice, large result sets containing hundreds or thousands of documents are much less useful than small sets, because they require so much hand filtering. In this sense, we consider the dramatically smaller result sets produced by entity + collection filtered search to be a large improvement over naïve full-text search, for entity-related queries.

Because of the synonyming capability of entity search, there were often documents retrieved by entity searches that were not retrieved by naïve full-text searches. For the first ten questions, on average, 14% of the documents returned by entity searches were *not* returned by naïve full-text search. In some cases – in particular, searches for contact information for people – missed documents contained the crucial information to answer the question at hand.

## 5 Discussion and Future Work

The goal of our work is to provide better information retrieval across a large semi-structured space than full-text search, while avoiding excessive meta-information overhead. Our approach is based on observing that in an information space used by a particular organization, important entity types link information together and can be used as a central retrieval cue. This data-driven approach can be contrasted with schema-driven approaches used by multidatabase systems (e.g., [Arens et al., 1993]), and similar systems attempting to integrate structured world-wide web sources [Levy et al., 1996; Farquhar et al., 1995]. In schema-driven approaches, each local schema is mapped to a central global schema, and mapping rules are used to translate between data formats used by different sources (e.g. [Collet et al., 1991]). These approaches are appropriate for relatively small numbers of tables where the data within each table is well-specified; however, semi-structured information spaces can include hundreds of sources, and data even within single sources can have multiple formats. A schema integration phase would be burdensome in such a large space [Goh et al. 1994].

Instead, NX relies on heuristics to categorize fields into a small number of entity and role types, and normalizes entity values for retrieval (as in [Huffman and Steier, 1995]). The resulting retrieval system is more broadly applicable but less general than multidatabase systems, because queries must refer to an entity.

One area for future work is classification and normalization of a larger variety of entity types. New types we are considering include locations, product names, and technical terminology appearing in the body of a document. Field classification could be improved by using non-local information, such as data values' appearances in other collections. Indexing could be improved by extracting entity values from free-running text, in addition to tagged fields. This would require text processing tools like name finders [Rau, 1991; Hayes, 1994], technical-term extractors [Justeson and Katz, 1995; Chen and Chen, 1994], parsers that find characteristic phrases using fonts and other cues [Krulwich and Burkey, 1995], or shallow linguistic approaches such as those used by information extraction systems [MUC 93].

A related idea we are experimenting with is using text classification and thesaurus induction methods (e.g. [Grefenstette, 1994]) to attach organization-wide categories to documents as a whole. These categories can be used as retrieval cues or filters, similar to roles.

Finally, we are investigating ways to move further towards structured retrieval by allowing users to retrieve specific types of information related to entities, rather than simply documents. For instance, since our meta-information indicates which collections are directories and which fields within those collections contain people and phone numbers, it should be possible (with a small increment of additional meta-information) to allow users to retrieve a value for "Bob Smith's phone number" instead of just documents containing Bob Smith. Like the "What's related?" functionality, retrieval of specific related information makes use of the entity index as a data warehouse.

## 6 Conclusion

Semi-structured systems are an intermediate point between unstructured collections of textual documents (e.g., untagged Web pages) and fully structured tuples of typed data (e.g., relational databases). Our goal is to provide stronger retrieval than full-text search, but without requiring the strong typing, schema and data mapping needed for SQL-style querying. Based on observing how information is typically retrieved and used within our organization, we have developed an entity-based retrieval system over a large semi-structured information space. The system incorporates semi-automatic classification of fields, normalization of field values, and filtered retrieval using entity roles and collection types. For typical queries containing entities, the system provides much more focused and normalized retrieval than full-text search.

## References

[Arens et al., 1994] Arens, Y.; Chee, C. Y.; Hsu, Chun-Nan; and Knoblock, C. Retrieving and integrating data from multiple information sources. *Int'l J. on Intelligent and Cooperative Information Systems.*

[Chen and Chen, 1994] Chen, Kuang-Hua, and Hsin-Hsi Chen, Extracting Noun Phrases from Large-Scale Texts: A Hybrid Approach and its Automatic Evaluation. *In Proceedings of the 32nd Annual Meeting of the ACLV.*

[Collet et al., 1991] Collet, C, Huhns, M, and Shen, W. Resource integration using a large knowledge base in Carnot. *IEEE Computer*, pages 55-62.

[Farquhar et al., 1995] Farquhar, A., Dappert, A., Fikes, R., and Pratt, W. Integrating information sources using context logic. In *AAAI Spring Symposium on Information Gathering from Heterogeneous Distributed Environments.*

[Goh et al., 1994] Goh, Cheng Hian; Madnick, Stuart E.; and Siegel, Michael D. Context Interchange: Overcoming the challenges of large-scale interoperable database systems. *In Proceedings of the 3rd Int'l Conference on Information and Knowledge Management.*

[Grefenstette, 1994] Grefenstette, Gregory. *Explorations in automatic thesaurus discovery.* Kluwer Academic Press.

[Hayes, 1994] Hayes, P. NameFinder: Software that finds names in text. Carnegie Group Inc. technical report.

[Huffman and Steier, 1995] Huffman, Scott; Steier, David. Heuristic joins to integrate structured heterogeneous data. *In AAAI Spring Symposium on Information Gathering in Heterogeneous Distributed Environments.* AAAI.

[Justeson and Katz, 1995] Justeson, John, and Slava Katz, Technical Terminology: Some Linguistic Properties and an Algorithm for Identification in Text. *In Natural Language Engineering V 1.1.*, 1995.

[Krulwich and Burkey, 1995] Krulwich, Bruce; Chad, Burkey.. ContactFinder: Extracting indications of expertise and answering questions with referrals. *AAAI Spring Symposium on Knowledge Navigation.* AAAI, 1995.

[Levy et al., 1996] Levy, A., Rajaraman, A., and Ordille, J. Query-answering algorithms for information agents. In *AAAI-96*, pp. 40-47, 1996.

[Li & Clifton, 1994] Li, W. and Clifton, C. Semantic integration in heterogeneous databases using neural networks. *In 20th Int'l Conf. on Very Large Databases.*

[MUC, 1993], Proceedings of the Fifth Message Understanding Conference (MUC-5). Morgan Kaufmann.

[Rau, 1991] Rau, L. Extracting company names from text. *IEEE Conference on AI Applications.*

# INFORMATION RETRIEVAL

Information Retrieval 3

# The Self-Organizing Desk

Daniela Rus    Peter de Santis

Department of Computer Science
Dartmouth College
Hanover, NH 03755

{rus,gwitto}@cs.dartmouth.edu

## Abstract

*The self-organizing desk is a system that enhances a physical desk-top with electronic information. It can remember, organize, update, and manipulate the information contained in the documents on a desk. The system consists of a simple robot eye that can survey the desk, a module for smart extraction of information from the images taken by the robot, a module for representing this information in multiple views, and a module that allows a user to interact with this information.*

## 1  Introduction

We wish to create smart physical worlds, that can augment reality with electronic information. Such spaces will keep track of their own contents, indexing and organizing their objects in electronic views. We hope to achieve this vision by using sensors to extract information about the physical world. The self-organizing desk, a system that can keep track of its contents autonomously, is an example of such systems. Consider the flow of paper that arrives for processing at someone's desk. Many paper documents potentially relevant to planning and scheduling arrive every day. These documents are filtered and filed in filing cabinets, or in random piles. In this space one has to manage queries such as "where is the letter from John Hopcroft?", "where is the paper I received last week that has a red table in the upper right corner?", "where is all the information I need to complete this report?", *etc*. Answers to these queries should be of the form "the top, right pile, of the desk, about halfway down". Such query processing would be much easier if all the data on the desk-top were available electronically. The self-organizing desk described in this paper implements these ideas.

Systems capable of self-organization present their users with accurate summaries at various levels of detail, even when the data changes dynamically. Self-organization is implemented by using sensors to observe and extract information about the objects in a physical space. The sensori-data is captured, indexed, and organized in electronic views. The views are aggregated as a database and stamped with spatial and temporal information. In selecting electronic views, our premise is that textual cues, visual cues, and topic-subtopic relationships are equally important for locating and manipulating documents.

In this paper we describe a system that implements the self-organization metaphor on the documents on a desk-top. The self-organizing desk uses a robot eye (a steerable, 3 degree of freedom computer control camera) to survey the documents on a desk-top, to capture and index their attributes (words, color, images, tables, location, time, *etc.*). The system interacts with users through a GUI to help them locate items on the piles on the desk top, or visualize conceptually the content of the desk. It supports operations to add a document to a random location on the desk, to remove a document from the desk top, and to shift a stack of documents from one desk-top location to another.

The paper is organized as follows. First, we present the system that implements the self-organization metaphor for a desk. Second, we describe our experiments with the desk. Finally, we discuss related work and future extensions.

## 2  The System Description

The architecture of the self-organizing desk system is shown in Figure 1. The system uses a camera that surveys a desk top by sweeping above its surface continuously, looking for changes. We assume that the only objects on the desk are standard size papers[1]

---

[1] The presence of other objects for example coffee cups, does not affect the performance of the system. At the moment such objects are simply not recognized or processed by the system.

and that these papers undergo changes one at a time. The change can be one of the following: *add a new paper, remove a paper, or shift a stack of papers from one location to another.* Change is detected by the camera via the segmentation module. Each time the system notices a change, the following set of operations are executed. Suppose, for simplicity, that the system has noticed a new paper that was added to the desk. The approximate coordinates of the paper are computed (by the segmentation module) and the camera is automatically positioned (by the camera control module) so as to capture a picture of maximum detail of the new document. The resulting image is passed through a variety of filters including OCR, color, and tables and the filtered data is indexed in a database. In addition to layout, the database also contains space, time, and history information for each document. History captures the order in which the documents arrive and is used for defining stacks (or piles) of documents. This database is indexed, updated, organized and summarized on-line (as its contents changes dynamically) by the information access module. The user can query this module by specifying full text, visual attributes, and/or requests for content summaries, content organization, and visualization. Queries are answered in the form of a GUI that points the user to the location of the relevant documents.

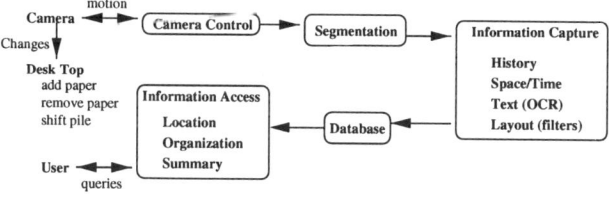

Figure 1: The system components of the self-organizing desk. The camera surveys a desk top for changes. The images taken by the camera are segmented and filtered for multiple representations in a database. The database is indexed, organized, and available for queries.

The following sections describe these modules in greater detail.

## 2.1 Surveying the Desk

The desk-top surveillance system consists of a Canon VC-C1 camera with zooming capabilities that is mounted on a pan-tilt computer steerable platform. The camera is connected to a Silicon Graphics Indy

Figure 2: The left image shows a desk-top picture taken with an Indy camera. The right image shows the result of running a Canny edge detector of the left image. Indy cameras are very low-resolution and inexpensive. Yet, they provide enough detail to be useful for surveying certain features.

(SGI) that controls its operation. Our initial experiments established that the manufacturer's motion control for this camera is inadequate for two reasons: it is inaccurate and inflexible. Instead of using the commercially supplied controller we developed an expressive interface and accurate control. The following operations are currently supported: (1) pan or tilt along the entire sweeping range, (2) pan or tilt to a specific location, (3) zoom to a specific location, and (4) capture the image.

## 2.2 Extracting Documents from Images

Currently, it is near-impossible to extract conceptual information about arbitrary three dimensional objects that are subject to noise and occlusion. However, since we restrict the objects in our application to be standard paper documents, we can develop effective segmentation algorithms. Consider Figure 2. It represents the image of a desk-top taken with a very low-resolution camera and the corresponding edge-detected image. The edge-detected image has significant noise but enough detail that can be detected and parsed automatically.

One important feature of a segmentation algorithm for this application is robustness in the presence of noise. Noise arises naturally here in two ways. Every time a human touches a paper on the desk, its position may be shifted by a small amount. The segmentation algorithm should tolerate this. Second, when papers are stacked on top of one another as in Figure 2(right) the edges of the stack are not perfectly aligned. The segmentation algorithm should find one set of edges only to approximate the location of the stack.

We have developed a statistical segmentation algorithm that has noise tolerance built-in. This algorithm

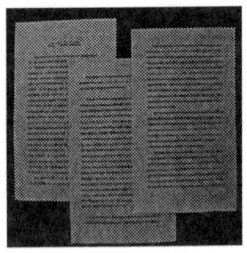 

Figure 3: The left image shows a desk-top image taken with our Cannon VC-C1 camera in our experimental setup. The right image shows the result of running our statistical border-finder on the left image.

has four steps. Step 1 compares the base image taken when the last event was detected against the current image. When a change is detected, the area of change is extracted from the image. Step 2 generates a new image of maximal detail that contains the entire area identified in Step 1. This is accomplished by aiming and zooming the camera. The correct camera motion (pan/tilt/zoom parameters) is computed by using the absolute camera and desk space coordinates, and the relative coordinates of the area of interest in the space. Step 3 identifies the enclosing border of the area of interest. This algorithm relies on object features and uses statistics on pixels to recognize and enclose the page boundaries. The basic idea is to do a walk along pixels that identifies a simple polygon, called the border polygon. The greatest challenge of this algorithm is noise: for the case of a desk-top, any form of object manipulation is likely to displace the paper by a small amount. Our algorithm uses statistics to address the slop problem by fitting a line through the pixels in an incremental fashion. Figure 3 shows a snapshot from the execution of this algorithm. Finally, the fourth step of the algorithm takes the border polygon and parses it to identify the pages. For the self-organizing desk application, this reduces to finding a cover of this polygon by rectangles, that is consistent with the document history. In our current implementation we assume that the documents on the desk top can only be translated relative to each other. We use this assumption to determine the vertices of a new document when the difference between the current image and the previous image returns 1, 2, or 3 vertices only. This situation arises when the new document overlaps existing documents on the desk.

## 2.3 Capturing Electronic Information about the Desk

The output of the border detection and document extraction algorithm consists of the relative coordinates of the page within the physical space. These coordinates are used to compute a camera configuration that can capture a new image of the object of interest, of maximal detail. This new image is ready for information extraction. A multitude of features is useful for defining attributes users can query on: arrival time of the document, physical location of the document, textual content, figures, color, tabular data, and layout content (for instance, is the document a letter?). Visual and layout cues like tables, color, and figures, complement the textual content of documents and they play an equally important role in information access.

The self-organizing desk supports queries that combine textual with layout information through multiple representations, each representation corresponding to an attribute detectable by a filter. This information extraction is carried by a library of smart filters. Our current library consists of three filters: OCR, color, and tables, but the architecture is expandable and we will add new filters in the future.

**OCR.** We have investigated the use of text filters (OCR systems) on camera images. We found that commercial OCR systems do not work well on images because the resolution is not high enough. Our approach was to enhance the resolution by dividing the image into n subimages (n is determined by the desired OCR accuracy), doing a linear interpolation for each pixel in the resulting images, and feeding the resulting images to the OCR system. This method improved the performance of the OCR system from 60% to 95% in character recognition accuracy.

**Color.** The color filter works by building a color histogram annotated with layout information for each object. The filter determines the 24 most prevalent colors occuring in the document and the location of each color. Location is a layout attribute determined by placing a $3 \times 3$ grid on the paper.

**Space, time, and history.** Each paper gets assigned a location on the desk by using the coordinates computed by the camera. Each paper is also time stamped. In addition, each paper is associated with a list of papers above it and papers below it, to capture the history of the desk-top. This information is necessary to implement the stack-shift operation. It is also necessary to estimate the location of a document on the desk if the document is covered by other documents.

## 2.4 Searching, Organizing, and Visualizing the Desk

The filters defined in Section 2.3 generate a web of representations for each document. We compile this multiplicity of representations in a database. In this database we also capture temporal information for each document (its arrival time on the desk), spatial information (its location on the physical desk), and history information (if the paper is in a pile, which documents are above it, and which documents are below it.) This database supports the following desk operations: adding a paper to a random location on the desk, removing a paper from the desk, and re-locating a paper or a stack on papers on the desk. The information in this database changes dynamically, as driven by these operations. In response to each event, the database is updated automatically.

The database comprises a collection of inverted indices, one for each attribute. An inverted index associates each attribute instance with a list of documents containing it. The advantage of this representation scheme is a speed-up in search: given a specific attribute (an word, a color, etc.), the list of documents containing this attribute is available in constant time.

We implemented several methods for searching, organizing, and visualizing the contents of this database.

**Search.** The self-organizing desk can be queried with keywords, with an entire document (full text), with color and layout information, with table information, and any boolean combination of these attributes. We use an augmented version of the *Smart System* [Sal91], which is a sophisticated text-retrieval system. We augmented Smart to also support color, layout, and table indices. Smart copes well with partially corrupted (by OCR) text. Its basic premise is that two documents are similar if they use the same words. Documents and queries are modeled as points in a *vector space* defined by the important words occurring in the corpus. When all texts and text queries are represented as weighted vectors, a similarity measure can be computed between pairs of vectors that captures the text similarity. We use this similarity measure as the basis for computing hyperlinks between documents that are similar to each other in this statistical framework.

**Organization.** The textual information contained in the documents on the desk is organized by topic using our own clustering algorithm (called the star algorithm) on the vectors in the document space. Our implementation uses a modification of the Smart system and the underlying cosine metric. The star algorithm gives a hierarchical organization of a collection into clusters. Each level in the hierarchy is determined by a threshold for the minimum similarity between pairs of documents within a cluster at that particular level in the hierarchy. This method conveys the topic-subtopic structure of the corpus according to the similarity measure used. The star algorithm is accurate in that it produces dense clusters that approximate cliques with provable guarantees on the pairwise similarity between cluster documents, yet are computable in $O(E)$, where $E$ is the number of edges above threshold in the document collection[2]. The documents in each cluster are tightly inter-related and a minimum similarity distance between all the document pairs in the cluster is guaranteed. This resulting structure reflects the underlying topic structure of the data.

**Summaries.** A topic summary for each cluster computed by the star algorithm is provided by the center of the underlying star for the cluster.

**Visualization.** We developed a visualization method for organized data that presents users with three views of the data: a Euclidean projection of the documents on the planar representation of the desk, a graph that shows the similarity relationship between all the documents, and a graph that shows the topic structure of the desk. The views allow users to select objects with a mouse and they are connected in that when an object is selected in one view, it gets highlighted in the other views. For instance, the user may select a cluster, which will highlight the areas of the desk where the documents in the cluster are located.

## 3 Experiments

We implemented all the modules described in Section 2 and constructed the following experiment. The VC-C1 camera is mounted on the tripod and connected to an SGI. The OCR package we use is EasyRead (which is a version of ScanWorks ported to SGIs.) The desk-top is set up vertically, on a wall. Documents are added/removed by pasting/removing them on the wall. Users interact with this system by physically moving papers, by entering queries in the GUI (see Figure 4), and by observing the visualization of the results returned by the search engine.

---

[2] $E$ is at most $N^2$, where N is the number of documents.

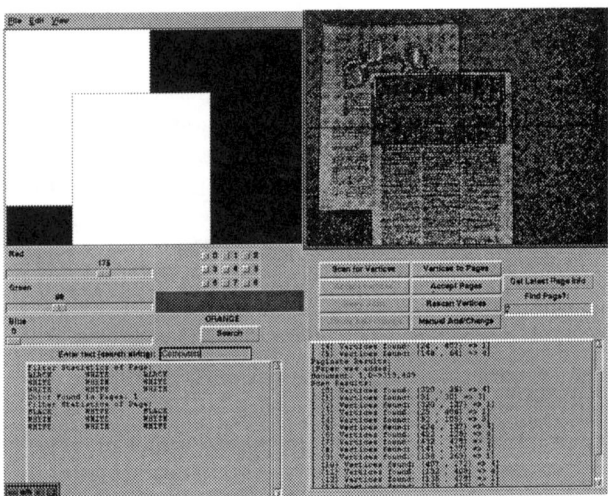

Figure 4: This figure shows the GUI to the desk. The top right section contains the actual image captured by the camera. The top left section contains the representation of the desk-top computed by the system. The bottom half portion of the GUI contains buttons for specifying attributes and queries and for displaying results.

We have repeated the following two experiments over fifty times recently. In the first experiment, three papers were added sequentially to the desk. The second paper was placed to overlap the first paper and the third paper was placed to overlap the previous two. In the second experiment we created a desk top by adding three papers sequentially. The first paper was placed randomly. The second paper was placed as in the first experiment. The third paper was placed approximately above the first paper so as to make up a stack. The addition of each paper was noticed by the camera, which triggered the main loop through the self-organization system. The location of each paper was computed by the segmentation module, the camera was then zoomed to capture the document location, the extracted images was passed through the OCR filter and the color filter. The results of these operations were entered in the database along with the space, time, and history of the document and the database was updated with this information. The database was then organized by content and summarized using the topic clustering and summarization algorithms. The experimental data for the segmentation and OCR aspects of the system is presented in Figure 5. Our human user input the following queries: "orange in upper right corner and distributed mobile robotics", "green

| Task | Tries | Success | Reliability |
|---|---|---|---|
| Segmentation | 50 | 45 | 90 % |
| OCR | 50 | 50 | 90 % avg. |

Figure 5: This table contains reliability data for the self-organizing desk operations. The first line shows accuracy of the segmentation algorithm whose task is to identify the papers. OCR is measured as average character recognition rate per run.

anywhere", and "manipulation". For all these queries the relevant documents were identified and the user was given pointers to their physical location of the desk. The only query failures we observed are for keyword searches on words corrupted by OCR.

The system takes approximately 30 seconds to determine a change on the desk. The enhanced OCR scheme (that processes and then merges 9 individual images per page) takes approximately 15 minutes, the initialization and update of the database takes negligible time (less than 1 second), and the queries take negligible time (less than 1 second.)

The failures of the system are due to the camera and the OCR. The current camera resolution and the current implementation of the segmentation algorithm make it impossible to distinguish corners that are closer than 1cm to each other. OCR is too slow. The character recognition is poor even with our enhancements. This problem could be solved by using a higher-resolution camera and a better OCR package.

## 4 Related Work

Efforts to enhance physical environments with electronic information include The Intelligent Room project at the AI Lab at MIT and the ALIVE project at the Media Lab at MIT. The goal of the Intelligent Room project is to create a room surveyed by cameras that can recognize and understand physical gestures. Progress on this project has been reported in [Tor95]. The ALIVE project allows users to interact with animated electronic characters and has been descibed in [Mae95].Other related projects include efforts from Euro Xerox and Hitachi to create interactive desks, where the user can write with a stylus pen on the desk top. The desk-top consists of a display that can capture the user's input. A camera mounted on the desk top is used to project on the desk top rather than extract information [AM+95].

The self-organizing desk draws from progress made in several areas: self-organizing systems [Koh90, CKP93], information retrieval and organization [Sal91, RA95], robotics and vision [MRR96,

HKR93], automated document structuring [TA92, RS95b, NSV92], and user interfaces [CAC93].

## 5 Discussion

We have described a system that implements the self-organization metaphor to enhance a physical space with electronic information. This system uses a number of key technologies in robotics, computer vision, OCR, information retrieval, filtering, and organization, and integrates them into a system that solves a new application. Our experiments demonstrate the feasibility and power of enhancing a desk-top with the electronic information extracted from the objects on the desk top. Applications of the technology demonstrated by this work reach beyond the self-organizing desk, into the area of using smart sensors for information processing in augmented reality. Candidates are dynamically changing physical space whose objects are simple enough that segmentation and feature extraction is possible. The smart sensors will support indexing, searching, and organizing the physical space. One immediate extension is a self-organizing bookcase.

The biggest computation sink in this system is the image segmentation module and the OCR. OCR performs poorly on camera-extracted images. We bypassed this problem by scanning images. We are currently building a self-organizing filing cabinet where documents are actively scanned in before being filed away. Our preliminary tests show that this system is fast and more reliable. In parallel with this effort, we are also developing better image segmentation tools for desk tops, smarter filters, and better GUIs.

## Acknowledgements

The camera control package was implemented by Scott Silver. The first version of the image segmentation package was implemented by Michael Taylor. The color filter and GUI was implemented by Hans Kieserman and the information organization system was implemented by Katya Pelekhov. We thank Carlo Tomasi for the initial discussion on image segmentation in the presence of slop.

## References

[AM+95] T. Arai, K. Machii, S. Kuzunuki, and H. Shojima, InteractiveDESK: A computer augmented desk which responds to operations on real objects, in *Proceedings of Computer Human Interactions*, 1994.

[CKP93] D. Cutting, D. Karger, and J. Pedersen. Constant interaction-time scatter/gather browsing of very large document collections. In *Proceedings of the $16^{th}$ SIGIR*, 1993.

[HKR93] D. Huttenlocher, G. Klanderman, and W. Rucklidge, Comparing images using the Hausdorff distance, in *IEEE Transactions on Pattern Analysis and Machine Intelligence*, 15(9):850–863, 1993.

[Koh90] T. Kohonen, The self organizing map, in *Proceedings of the IEEE*, 78(9):1464–1480, 1990.

[Mae95] P. Maes, Artificial Life meets Entertainment: Interacting with Lifelike Autonomous Agents, Special Issue on New Horizons of Commercial and Industrial AI, Vol. 38, No. 11, pp. 108-114, *Communications of the ACM*, ACM Press, November 1995.

[MRR96] R. Manmatha, S. Ravela, and E. M. Riseman, Retrieval from Image Databases Using Scale Space Matching, in *Proceedings of the ECCV '96*.

[NSV92] G. Nagy, S. Seth, and M. Vishwanathan. A prototype document image analysis system for technical journals. *Computer*, 25(7), 1992.

[RA95] D. Rus and J. Allan. Structural queries in electronic corpora. In *Proceedings of DAGS95: Electronic Publishing and the Information Superhighway*, May 1995.

[RS95b] D. Rus and K. Summers. Using whitespace for automated document structuring. *Advances in digital libraries*, eds. N. Adam, B. Bhargava, and Y. Yesha, Springer-Verlag, LNCS 916, 1995.

[Sal91] G. Salton. The Smart document retrieval project. In *Proceedings of the Fourteenth Annual International ACM/SIGIR Conference on Research and Development in Information Retrieval*, pages 356-358.

[Tor95] Mark C. Torrance, Advances in Human-Computer Interaction: The Intelligent Room, in *Working Notes of the CHI 95 Research Symposium*, 1995.

[TA92] S. Tsujimoto and H. Asada. Major components of a complete text reading system. In *Proceedings of the IEEE*, vol. 80, no. 7, 1992.

[Tur90] H. Turtle. Inference networks for document retrieval. PhD thesis. University of Massachusetts, Amherst, 1990.

[CAC93] Special issue on visualization. *Communications of the ACM*, 36(4), 1993.

# A Learning System for Selective Dissemination of Information

**Gianni Amati**
Fondazione Ugo Bordoni
via B. Castiglione, 59
I-00142 Rome, Italy

**Fabio Crestani**[*]
Dept. of Computing Science
University of Glasgow
Glasgow G12 8QQ, Scotland

**Flavio Ubaldini**
Fondazione Ugo Bordoni
B. Castiglione, 59
I-00142 Rome, Italy

## Abstract

New methods and new systems are needed to filter or to selectively distribute the increasing volume of electronic information being produced nowadays. An effective information filtering system is one that provides the exact information that fulfills a user's interest with the minimum effort by the user to describe it. Such a system will have to be adaptive to the user changing interest. In this paper we present a learning system for information filtering and selective information dissemination. The learning algorithm is described and the effectiveness of the system is evaluated in a true information filtering style.

## 1 Introduction

Information overload is an increasing problem in many domains. New information services (e.g. news services, electronic mail, libraries and databanks) deliver to the user an increasing volume of digital information. Often the information delivered to the user does not match the user interest and it ends up overloading the user, who will have to manually select the interesting information from the "noise". It is true that the user could switch off such automatic delivery of information, but in that case she will have to go after the information herself. In such a dynamic environment it is also difficult for the user, equipped with only conventional search capabilities, to keep up with the fast pace of information generation. Instead of making the user go after the information or having to go through the large amount of incoming information, information should more selectively flow to the interested user. Traditionally, libraries and databanks, provide such kind of service to user. A user is asked to provide a description of the classes of document in which she is interested. Such descriptions are then used as static queries that are submitted to a system that send to a passive user the documents matching the queries. Such systems are becoming increasingly important and will form an indispensable tool for global information systems.

In this paper we present an system called ProFile that learns by interacting with the user what is the user interest. ProFile constructs and adaptively tunes a representation of the classes of user's interests using a learning algorithm that is derived from the generalised probabilistic model of IR presented in [Amati and van Rijsbergen, 1995]. ProFile selectively distributes documents from a continuous stream to multiple users with several classes of interests. The fast learning and adapting capabilities of ProFile enables it to effectively perform information filtering and dissemination.

## 2 Selective Dissemination of Information

*Document filtering*, also known as *selective dissemination of information*, has a long history, most of it based on the unranked Boolean retrieval model of information retrieval (IR) [van Rijsbergen, 1979]. Most of the recent research on document filtering is based on the assumption that effective IR techniques are also effective document filtering techniques. The TREC conference (see [Harman, 1996] for the last TREC conference) is a good example of this practice.

Recently, the term *information filtering* (IF) has started being used in place of the old style document filtering, to emphasise the possibility of selectively distributing multimedia information. In the context of this paper we will use this term too, since the technique here presented can also be used to perform multimedia document filtering and dissemination.

In IF there is the twofold problem of determining what information is relevant to any user and how this decision can be automatically taken by the information system. Parametric and qualitative descriptions of what information is of her interest must be generated. These factors constitute on the whole what is generally referred to as the *user profile*, but what is needed for expressing them may be difficult to circumscribe. The user profile may consist of a set of keywords, that represent the topics of user's interest, similar to the one generated by the in-

---

[*]Previously at Dipartimento di Elettronica e Informatica, Universitá di Padova, Padova, Italy.

dexing process in IR. However this representation may result insufficient because the user could also need to know some additional modalities to which information is related, such as its novelty, urgency or purpose. Therefore it is not clear what factors are important for predicting the relevance of information to user's interests. However some limiting hypotheses on the type of user and/or data sources should be done to present a workable model of information filtering. Users may range from casual to specific ones and data types may vary from unstructured (images or "textual" data) to structured ones (such as relational tables). In the following we assume that the system is able to process the data and to provide them with precise (quantitative) descriptors of their contents, that is with a derivative structured information. Such data are called of *semi-structured type* [Belkin and Croft, 1992]. This is what typically happens with the indexing of textual data in IR: by using different techniques it is possible to assign sets of keywords to the data items. These sets are then organised into a matrix made out of vectors of weighted descriptors: weights are computed on the basis of a statistical analysis of the occurrences of words.

A second assumption upon which our work is based is that the user is *casual*, that is her profile is not predefined by some specific keywords chosen by the user in a *controlled language*, that is among a set of possible terms. On the contrary, we assume that the system does not necessarily know any initial definition of the user's profile. Rather, this is long-term defined. The underlying idea is that the system is trained by the user herself and that her profile "converges" to a stable description as soon as the number of the interactions with the information sources has become "large enough". This requirement for the user model makes a non-trivial difference with IR where the system is concerned with a single session at a time. Differently from IF, IR does not depend on time, though user's relevance feedback introduces an implicit temporal factor. Indeed, by adapting a variant of van Rijsbergen's model of IR to IF, as described in [Amati and van Rijsbergen, 1995], we are able to define a new model of information filtering. In particular we will show that the profile of a casual user consists of a vector whose descriptors are in the set of *uncontrolled* terms of the sources data.

## 3 Some Considerations about Related Work

Before considering the crux of the IF, we would like to make some remarks on how statistics may influence an IF model. Useful theoretical tools, that range from elementary probability theory to decision theory and statistical methods, are generally used to draw inferences for processing language and analysing linguistic structures. An example of stochastic model for indexing in IR is that based on the *expected mutual information measure* (EMIM) of van Rijsbergen's [van Rijsbergen, 1979], derived from Shannon's theory of information. This model was used to find out the word associations in the English language for the last edition of the Collins' dictionary [Church and Hanks, 1989]. Here a word is interpreted as a "message" which carries a measure of uncertainty (entropy) defined in accord to the probability laws. This entropy is maximum when the message is unclear while is null when the message is "free" from any noise. An alternative approach is *vector space model* (VSM) by Salton [Salton and McGill, 1983] based on the Zipf's rank-frequencies law. In [Amati and van Rijsbergen, 1995] we have explored the connection of our model with both the EMIM and the VSM. In this paper we will outline the connections of our IF model with IR models rather than arguing how statistics can be better applied to IF. For the purpose of this paper, which is merely concerned with the modelling user's behaviour in IF, we will not focus on whether information theory or rank-frequencies is the best representation of the document content. Indeed, we assume the system to have the "best" statistical method for drawing linguistic knowledge from a sample of data. Our model is predictive of the behaviour of the user rather than explicative of the content of documents in the sample.

A large number of systems have been currently developed to filter information, and special efforts are devoted to filter the Internet messages. Despite of this exploition, results concerning the evaluation of these systems are rarely available. The evaluation is generally left to the end users. As an example the NewsWeeder [Lang, 1995] has been evaluated by monitoring for a year two users which used a very large amounts of training data. Only two precision values w.r.t. the set of selected documents are given (44% and 59%) and the relative recall values are not available. Hence, very little knowledge is available on whether and how fast the end users experience the filter.

## 4 The ProFile System

The *ProFile* (PRObabilistic FILtEring) system has been developed at Fondazione Ugo Bordoni in Rome (Italy) in 1996 and has been in used since then by many researchers of that institution for filtering the Usenet News [Amati et al., 1995]. Despite being born with the purpose of filtering news, ProFile can be adapted to filter any incoming stream of information, like email, newswires, or newspaper articles.

### 4.1 The ProFile Architecture

In ProFile each user may define a number of conceptual classes to classify the filtered documents: each class has its own profile. IF systems have two ways for assigning a document to a conceptual class. The first one consists in ranking documents according to a similarity values between the profiles of conceptual classes. A document is then assigned to the conceptual class with the highest level of similarity. This technique is appropriate when conceptual classes cover the set of all possible documents. Differently, another technique consists in defining a relation to be satisfied by each couple class–document. If the document satisfy the relation, then it is classified into that class, otherwise it is discarded. If a document satisfy relations with more than one class, then it is either classify into all classes or one is chosed (an arbitrary one or the one with the strongest relation, if that can be quantified). The model used by ProFile follows this second approach by exploiting semantic information theory [Bar-Hillel and Carnap, 1953; Hintikka, 1970] and decision theory [Jeffrey, 1965]. ProFile operates according to the following procedure:

*Definition of the conceptual classes.* The user defines a set of conceptual classes in which he wants to filter and classify the incoming stream of documents. ProFile requires from the user a set of keywords for an approximate description of each conceptual class.

*Training phase.* The initial description of the user interests is utilised as a query by the FIFT service (Fub Information Filtering Tool) [Amati et al., 1995], a customised version of SIFT, a filtering system developed at Stanford (see Section 3). FIFT filters out of the document collection a set of documents that will be used as the "training set". The user go through the documents of the training set and assigns them relevance values with respect to each conceptual class. The relevance values are chosen from a scale of eleven values of interests (from 0 to 10). The user does not need to go through all the documents retrieved. The number of documents used in the training phase constitutes the *training data*. ProFile's relevance feedback process uses the probabilistic learning model that will be describe in detail in Section 4.2. The pre-filtering phase can go on as long as the user requires, with as many retrieval runs (performed by FIFT) and user relevance feedback as the user chooses.

*Filtering phase.* The user decides to activate the filtering phase when he believes that the definition of the conceptual classes built by FIFT using relevance feedback are accurate enough. The filtering phase is made up of two sub-phases:

1. *Filtering.* ProFile filters the documents and delivers to the appropriate user's conceptual class. The user can see the filtered documents classified into his personal conceptual classes.

2. *Tuning.* The user can either accept the result of the filtering and let ProFile keep working with the current profiles or otherwise he can modify the profiles providing additional information. This can be achieved by giving relevance values to the filtered documents in the same way it is done in the training phase. The additional information enables ProFile to tune to the user perception of relevance and adapt the profiles of the conceptual classes. This phase can be repeated as many times as the user wants.

It should be noticed that the initial training phase is very important for the effectiveness of the filter. Indeed, in the limit case of no relevant document in the training set (i.e. no document has been marked as relevant by the user before starting the filtering phase) the system will not retrieve any document and the user will not have any chance for correcting his profile with the tuning phase. We observed that the best training set is obtained by balancing the number of relevant and non-relevant documents present in the training set (see Section 5). Our way of training the system can be assimilated to the uncertainty sampling. In [Lewis and Gale, 1994], Lewis and Gale observed a better performance in using uncertainty sampling instead of relevance sampling [Ghosh, 1991] when the sample size is small in comparison with the number of positive examples in the set of non-evaluated data. This is an important feature of ProFile, because the first set of evaluated document in the training set is very small. Typically, a user wants to activate the filtering phase after only 20 or 30 documents have been examined.

### 4.2 The Information Filtering Model

In this section we describe in detail our probabilistic learning model. The model is derived from the generalised probabilistic model of IR presented in [Amati and van Rijsbergen, 1995].

**Learning theory**

At the abstract level IF can be seen as a process dealing with a repetitive event: a document is delivered to the user or not according to his current profile. A profile is a description of what the user is interested at. We assume that the document is represented by a set of terms (phrases, indexes, words or lexical units). The semantic relations between terms in the set $\mathcal{T}$ are implicitly explained by means of the set $\Omega(\tau)$ of documents which have been examined by the filter up to the current instant of time $\tau$. In statistics this set can be considered as a *sample* of the *population*. Relations between terms are often expressed using frequency values. The user relevance assessments also provide a way of expressing semantic relations between terms.

A learning theory for IF is a triple $<\Omega, \mathcal{A}, \mathcal{P}>$. $\Omega$ depends on a temporal parameter $\tau$, $\Omega(\tau)$ being the set of all documents processed before the time $\tau$. Here we assume that $\Omega$ is the set of documents which have constituted the data stream up to the current moment, so that $\tau$ can be omitted. $\mathcal{A}$ is the power set of $\Omega$, namely the set of all subsets of $\Omega$. $\mathcal{P}$ is defined by the user starting from the mutually exclusive elementary events, that is the elements $d$ of $\Omega$. This function is lifted from the elementary events to all the events $e_i$ of the space $\mathcal{A}$ by using the additivity axiom.

In a finite space, a probability can be then obtained by conditioning. The *conditioning* of $\mathcal{P}$ is defined, provided $\mathcal{P}(e_2) > 0$ as:

$$\mathcal{P}(e_1|e_2) = \frac{\mathcal{P}(e_1 \wedge e_2)}{\mathcal{P}(e_2)}$$

Functions defined from $\Omega$ to the set of real numbers are called *random variables*. In our model a random variable is associated to each term $t \in \mathcal{T}$. With a little abuse of language we denote this random variable with $t$ itself. Given a document $d \in \Omega$, the value $t(d)$ of the random variable $t$ is the statistics on the term $t$ in the document $d$, for example the $tf$ weighting (the relative frequency of $t$ in $d$), or the $idf$ weighting (defined as $idf(t) = -log(n/N)$, where $n$ is the number of documents in which $t$ occurs and $N$ is the number of documents in the collection) [Salton and McGill, 1983]. In ProFile we use $tf$ since $idf$ needs a complete information on the set of incoming data which is unrealistic in filtering or require a high expensive processing.

In other words if we denote by $\langle a_t^d \rangle_{d \in \Omega, t \in \mathcal{T}}$ the matrix $\langle t(d) \rangle_{d \in \Omega, t \in \mathcal{T}}$, then a row associated to $d$ is the vector $\langle t(d) \rangle_{t(d) \in \mathcal{T}}$ made out of the statistics of the set of terms in the document $d$, while the random variables $t \in \mathcal{T}$ are obtained by the columns of the matrix. In IR the matrix $\langle t(d) \rangle_{d \in \Omega, t \in \mathcal{T}}$ is called the *inverted file* of the collection $\Omega$. We can define the *conditioning expectation* of a discrete random variable $t$ with respect to the measure $\mathcal{P}$ as:

$$E_{\mathcal{P}}(t) = \frac{\sum_{d \in \Omega} t(d) \mathcal{P}(d)}{\mathcal{P}(\Omega)} \quad (1)$$

Note that if $0 \leq t(d) \leq 1$ then $0 \leq E_{\mathcal{P}}(t) \leq 1$.

In [Amati and van Rijsbergen, 1995], an IR model is introduced as follows. $\mathcal{P}$ corresponds to a subjective measure $R$ of relevance on the event space $\Omega$, its form is a scale of relevance weights $R(d)$, with $0 \leq R(d) \leq 1$, arbitrarily generated by the user. In ProFile, for example, we used a scale of 11 degree of relevance that are naturally mapped to the real numbers in the interval $[0, 1]$, but the whole continuous interval could be used. $\langle R(d) \rangle_{d \in \Omega}$ may be defined as a subjectively held vector and can be seen as a person's belief at the current instant of time. The dual measure of non-relevance, $\neg R(d) = 1 - R(d)$, can be also defined. $\langle \neg R(d) \rangle_{d \in \Omega}$ can be seen as a person's disbelief on $\Omega$.

As already pointed out, a random variable $t$ takes the values $t(d)$ by means of statistics. Since $t(d)$ is related to frequencies we may suppose that $0 \leq t \leq 1$. $E_R(t)$ can be considered as a relevance\frequency weight of the term $t$, while $E_{\neg R}(t)$ as a non-relevance\frequency weight of the term $t$.

When the system must decide whether a term is relevant or not on the basis of the expected measures of relevance and non-relevance of documents, an error can occur and then a loss is produced. To make this decision the system computes the *expected monetary value* of decision theory [Amati and van Rijsbergen, 1995], that is:

$$EMV(t) = \lambda_1 E_R(t) + \lambda_2 E_{\neg R}(t) \quad (2)$$

where $\lambda_1$ is the "gain" when $t$ is relevant to the user, while $\lambda_2$ is the "loss" when $t$ is not relevant to the user. The event "$t$ is relevant" produces a benefit whenever $EMV(t) > 0$.

$EMV$ can be equivalently given by the formula:

$$EMV1(t) = log \frac{\lambda_1 * E_R(t)}{\lambda_2 * E_{\neg R}(t)} \quad (3)$$

## Decision theory and semantic information

Let us assume that the user has to decide whether to use the term $t$ or not. $t$ has the "a priori" relevance value $E_R(t)$. Suppose also that $t$ is relevant to the information need of the user. $\lambda_1$ would be then the "award" if he takes $t$ while $\lambda_2$ would be the "cost" if he discards $t$ (with a priori probability $E_{\neg R}(t)$). If "$t$ is relevant", then the user will gain the amount of information of non-relevance of $t$: let us denote it by $Inf_{\neg R}(t)$. On the other hand, the loss $\lambda_2$ can be quantified by the amount of information of relevance of $t$, that is $Inf_R(t)$. In both information theories (semantic and frequency–based) the amount of information is taken to be inversely proportional to probability, that is $Inf_{\mathcal{P}}(e) = -log\, \mathcal{P}(e)$ or by the similar entropy expression. They share the principle that a sentence is more informative if it excludes more alternatives, that is, if it has a low probability (in particular tautologies are not informative at all because no alternatives can be excluded). Hintikka [Hintikka, 1970] suggests to use as a measure of information of a sentence the relative number of alternatives that the sentence excluded, more generally this can be formalised as $inf(e) = 1 - \mathcal{P}(e)$. In our case we have to assign the amount of information to random variables instead to sentences, that is, we may define the amount of information as $Inf_{\mathcal{P}}(t) =_{def} 1 - E_{\mathcal{P}}(t)$. Let us define $\neg t = 1 - t$, then: $Inf_{\neg R}(t) = 1 - E_{\neg R}(t) = E_{\neg R}(\neg t)$ and $Inf_R(t) = 1 - E_R(t) = E_R(\neg t)$ Substituting the values of the $\lambda$'s into (3), we have the *absolute relevance of the term*, which must satisfy the constraint:

$$w(t_i) = log \frac{E_R(t_i) * E_{\neg R}(\neg t_i)}{E_R(\neg t_i) * E_{\neg R}(t_i)} > 0 \quad (4)$$

The above model derives the probabilistic model of $IR$ [Robertson and Sparck Jones, 1976; van Rijsbergen, 1979]

$$w(t_i) = log \frac{\frac{r^i}{n_R - r^i}}{\frac{n^i - r^i}{N - n_R - n^i + r^i}} > 0 \quad (5)$$

under the hypothesis that: (a) $R$ is the counting measure for the relevance of documents i.e. $R$ takes a value $R(d) = 0$ or $R(d) = 1$ for every document according to the user relevance feedback; (b) $a_i^d$ is the *counting document-term matrix*, that is: $a_i^d = 1$, if the i-th term occurs in $d$, and $a_i^d = 0$ otherwise.

In the formula $n_R$ denotes the cardinality of the relevant set of documents, $N$ the cardinality of $\Omega$, $r^i$ the cardinality of the set of relevant documents in which the term $t_i$ occurs, $n_{\neg R}^i$ the cardinality of the set of non relevant documents in which the term $t_i$ occurs, and finally $n^i$ the cardinality of the set of documents in which the term $t_i$ occurs.

More generally, $w(t)$ can be used as a weight of relevance of the term $t$ for the user: greater is the value of $w_t$, higher is the degree of relevance of $t$. The vector $\langle w_t \rangle_{t \in \mathcal{T}}$ in ProFile can be thus considered as a weighted description of the user's profile.

Let us now define ProFile's learning model.

The expected probability of relevance for IR can be easily adapted to define a filtering function. Let us assume that $n$ *conceptual classes* $C_1, C_2, \ldots C_n$ are associated to a single user. These conceptual classes can possibly be reduced to two: the user's class of relevant documents and the set of uncertain documents. Let us examine one document $x = \langle x_t \rangle_{t \in \mathcal{T}}$, on the set $\mathcal{T}$ of terms, at a time from a stream of documents. Then the probabilistic model $< \Omega, \mathcal{A}, R_C >$, as described above, can be applied to each class by using the weights:

$$w_C(t) = \log \frac{E_{R_C}(t) E_{\neg R_C}(\neg t)}{E_{R_C}(\neg t) E_{\neg R_C}(t)} \qquad (6)$$

To summarise, ProFile works in the following way:

1. For each incoming document and for each conceptual class $C$ the user provides a relevance measure $R_C$, $0 \leq R_C \leq 1$.

2. By applying the decision theory we are able to provide a term $t$ with a weighting formula $w_C(t)$ (see equation (6)).

3. When a new document $y = \langle y_t \rangle_{t \in \mathcal{T}}$ is evaluated, the weighting formulas $w_C(t)$ are easily updated.

4. Finally, the vector space similarity function $SIM$ is applied to the vectors $x = \langle x_t \rangle_{t \in \mathcal{T}}$ and $w_C = \langle w_C(t) \rangle_{t \in \mathcal{T}}$ to compute a real number value for the membership of $x$ to $C$. The conceptual classes containing the document $x$ are such that: $SIM(x, w_{C_j}) > s_C$ where $s_C$ is a *threshold* value. From a theoretical point of view $s_C$ must be equal to 0. However, this threshold is experimentally greater than 0. Note also that if the user always gives the maximum uncertain value $\frac{1}{2}$ to each document in the stream of documents then $w^C$ is the null vector.

## 5  Evaluation Framework and Results

In the context of the work reported in this paper we intended to evaluate the performance of our IF learning model, in particular when little training data is provided. The collection we used is the *TREC-5 B* [Harman, 1996] a subset of the collection used in the experiments done in 1996 in the context of the TREC 5 initiative. The collection is made of 3 years (1990-92) of selected full text articles of the Wall Street Journal. The total number of documents (articles) in the collection is about 75.000. Each document is, on average, about 550 words in length. The size of the collection is about $260 Mbyte$. This is quite a

| Recall | 0.1 | 0.3 | 0.5 | 0.7 | 0.9 |
|---|---|---|---|---|---|
| Precision | 0.54 | 0.31 | 0.20 | 0.10 | 0.3 |

Table 1: Performance of ProFile for the base run.

large collection for IF and IR standards. We also used a set of 50 already prepared queries (or topics, as they are called in TREC) with the corresponding set of relevant documents that were used for the training and for the evaluation. The topics are complex and long and they can be regarded as almost complete description of the information need of a user. We regarded these topics as examples of relevant documents.

The evaluation was performed in true IR style, since this is the current practice for IF systems (see the evaluation methodology used in the various TREC conferences). The main retrieval effectiveness measures used in IR are Recall and Precision. *Recall* (R) is the proportion of all documents in the collections that are relevant to a query and that are actually retrieved. *Precision* (P) is the proportion of the retrieved set of documents that is also relevant to the query. Experimentally these measures have proved to be related in such a way that high precision brings low recall and viceversa. In order to give a measure of the learning performance of the filtering algorithm, R and P have been evaluated with different dimension of the set of training examples.

At each run we trained the system with only very few documents. The training data of each run was a subset of up to 32 relevant and 32 non relevant documents, randomly chosen. The filtering runs shown in Tables 1, 2, 3 and 4 are thus incremental. We did not exploit the *idf* weighting function which would have required the processing of the whole collection in advance. Moreover, we only used a stop list without the stemming. We therefore used a minimum amount of information about the text collection at each run. This is the normal situation in which many filters work, e.g. filtering systems for the net news. We made the hypothesis that the system cannot process in advance the incoming data.

We divided the possible users into three categories: user A demands a high precision performance from the system and is happy with low recall performance (a recall value 0.3, that is 30% of the total number of relevant documents in the collection), the user B requires average levels of recall and precision (a recall of 0.5), and the user C who wants to retrieve most of the relevant information stored in the collection (a recall value of at least 0.7).

Table 2 and Table 3 show that the learning must be restricted to only high frequent terms in the training data. They also shows that if the information need of an end user is stable in the long-term, learning is in general no faster using only positive documents compared with a balanced training set; negative examples are neither harmful nor useless when combined with positive information (notice the better behavior respectively of the runs 8R-8N and 16R-16N-AT with respect to the runs 8R and 16R, which have the same number of relevant

| User | 8R | 16R | 32R-AT | 32R-TT | 32R-HF |
|---|---|---|---|---|---|
| A | +6.3 | +9.9 | +10.6 | +13.8 | +15.5 |
| B | +5.4 | +9.6 | +3.2 | +13.5 | +14.3 |
| C | +17.7 | +18.3 | +22.5 | +26.9 | +27.3 |

Table 2: Precision increment in percentage w.r.t. the base run by using only relevant documents (R) as training. AT = all terms of the training data and of the topic in the profile, TT = only terms in the topic, HF = only the high frequency terms and terms in the topic.

| User | 4R-4N | 8R-8N | 16R-16N AT | 16R-16N TT | 16P-16N HF |
|---|---|---|---|---|---|
| A | +6.7 | +8.6 | +11.0 | +8.9 | +13 |
| B | +7.5 | +8.8 | +12.5 | +13.2 | +16 |
| C | +18.9 | +19.3 | +24.6 | +24 | +23 |

Table 3: Precision increment in percentage w.r.t. the base run with a balanced set of relevant (R) and non relevant (N) documents.

documents).

Table 4 shows that a training set made up of only negative examples do not contribute significantly in the tuning phase. The initial improvement with respect the base run does not growth by increasing the size of the training data. Notice that the performance of the run which does not add extra term to the base run profile (denoted by 32N-TT), is equal to the performance of the run (32N-AT) in which also the terms occurring in the training are weighted. This result shows that negative counterexamples contribute to eliminate the noise brought about by non-discriminant terms of the topic, but new negative terms do not help to discriminate the relevant documents.

Even though the topics were long and complex, the results show that few training documents improve substantially the performance of the system, hence a short tuning phase is indeed necessary especially when the document sources are different and not known in advance. Nevertheless, results shown that a relatively small subset of the relevance judgments works quite well with respect to the full set.

## 6 Conclusions and Future Work

We presented a learning algorithm to perform effective information dissemination and filtering. Our future work will be devoted to further experimentation in order to determine the best possible learning strategy among the many that can be performed using the proposed model.

| User | 8N | 16N | 32N-AT | 32N-TT | 32N-HF |
|---|---|---|---|---|---|
| A | +6.1 | +5.9 | +6.2 | +6.1 | +6.1 |
| B | +8.8 | +9.1 | +9.3 | +9.5 | +9.5 |
| C | +20.9 | +20.9 | +21.1 | +21.2 | +21.2 |

Table 4: Precision increment in percentage w.r.t. the base run using only non relevant documents

## References

[Amati and van Rijsbergen, 1995] G. Amati and C.J. van Rijsbergen. Probability, information and information retrieval. In *Proceedings of the First International Workshop on Information Retrieval, Uncertainty and Logic*, Glasgow, Scotland, UK, September 1995.

[Amati et al., 1995] G. Amati, D. D'aloisi, and V. Giannini. A framework for dealing with email and news messages. In *Proceedings of AICA 95*, pages 27–29, Cagliari, Italy, September 1995.

[Bar-Hillel and Carnap, 1953] Y. Bar-Hillel and R. Carnap. Semantic information. *British Journal of the Philosophy of Science*, 4:147–157, 1953.

[Belkin and Croft, 1992] N.J. Belkin and W.B. Croft. Information Filtering and Information Retrieval: two sides of the same coin? *Communication of the ACM*, 35(12):29–38, 1992.

[Church and Hanks, 1989] K.W. Church and P. Hanks. Word association norms, mutual information and lexicography. In *Proccedings of ACL 27*, pages 76–83, Vancouver, Canada, 1989.

[Ghosh, 1991] G. Ghosh. *A brief history of sequential analisys*. Marcel Dekker, New York, USA, 1991.

[Harman, 1996] D. Harman. Overview of the fifth text retrieval conference (TREC-5). In *Proceeding of the TREC Conference*, Gaithersburg, MD, USA, November 1996.

[Hintikka, 1970] J. Hintikka. On semantic information. In *Information and inference*. Synthese Library, Reidel, Dordrecht, The Netherlands, 1970.

[Jeffrey, 1965] R.C. Jeffrey. *The logic of decision*. McGraw-Hill, New York, USA, 1965.

[Lang, 1995] K. Lang. NewsWeeder: learning to filter netnews. In *Proceedings of ML 95*, pages 331–339, 1995.

[Lewis and Gale, 1994] D.D. Lewis and W.A. Gale. A sequential algorithm for training classifiers. In *Proceedings of ACM SIGIR*, pages 3–11, Dublin, Ireland, July 1994.

[Robertson and Sparck Jones, 1976] S.E. Robertson and K. Sparck Jones. Relevance weighting of search terms. *Journal of the American Society for Information Science*, 27:129–146, May 1976.

[Salton and McGill, 1983] G. Salton and M.J. McGill. *Introduction to modern Information Retrieval*. McGraw-Hill, New York, 1983.

[van Rijsbergen, 1979] C.J. van Rijsbergen. *Information Retrieval*. Butterworths, London, second edition, 1979.

# WebWatcher: A Tour Guide for the World Wide Web

**Thorsten Joachims**
Universität Dortmund
Informatik-LS8
Baroper Str. 301
44221 Dortmund, Germany

**Dayne Freitag**
Carnegie Mellon University
School of Computer Science
5000 Forbes Avenue
Pittsburgh, PA 15213, USA

**Tom Mitchell**
Carnegie Mellon University
School of Computer Science
5000 Forbes Avenue
Pittsburgh, PA 15213, USA

## Abstract

We explore the notion of a tour guide software agent for assisting users browsing the World Wide Web. A Web tour guide agent provides assistance similar to that provided by a human tour guide in a museum – it guides the user along an appropriate path through the collection, based on its knowledge of the user's interests, of the location and relevance of various items in the collection, and of the way in which others have interacted with the collection in the past. This paper describes a simple but operational tour guide, called WebWatcher, which has given over 5000 tours to people browsing CMU's School of Computer Science Web pages. WebWatcher accompanies users from page to page, suggests appropriate hyperlinks, and learns from experience to improve its advice-giving skills. We describe the learning algorithms used by WebWatcher, experimental results showing their effectiveness, and lessons learned from this case study in Web tour guide agents.

## 1 Introduction

Browsing the World Wide Web is much like visiting a museum. In a museum the visitor has general areas of interest and wants to see relevant artifacts. But visitors find it difficult to locate relevant material given that they do not initially know the contents of the museum. In many cases their initial interests are poorly defined, becoming clear only after they begin to explore. In a museum the user might rely on a tour guide who is familiar with the museum and how people interact with it. The visitor could describe his or her initial interests to the tour guide, who could then accompany the user, point out items of interest, and suggest which directions to turn next. During the tour the visitor could communicate with the guide, express interest in certain artifacts, ask and answer questions as they explore, and refine their interests.

People browsing collections of Web pages often behave like museum goers. For example, a visitor to CMU's Computer Science Web home page might have a general interest in "experimental research on intelligent agents." However, with no specific knowledge about the contents of the collection, the user may find it difficult to locate relevant information. Until they become aware that CMU conducts significant research on robotics, for example, they may not think to mention "learning robots" when describing their general interest in intelligent agents.

Here, we report research into software agents that act as tour guides for the Web. In its most general form, the metaphor of Web agent as tour guide is very broad, suggesting systems that carry on general natural language dialogs with their users, possess detailed knowledge about the semantic content of the Web pages they cover, and learn with experience. Here we describe a first experiment with a more restricted, but operational, tour guide. In particular, we describe WebWatcher [Armstrong et al., 1995], a system that accompanies the user as he or she browses the Web. Like a museum tour guide, WebWatcher interactively suggests where to go next. The user can communicate with the system and give feedback. WebWatcher acts as a *learning apprentice* [Mitchell et al., 1994], observing and learning from its users' actions. Over time WebWatcher learns to acquire greater expertise for the parts of the World Wide Web that it has visited in the past, and for the types of topics in which previous visitors have had an interest.

WebWatcher differs in several key respects from keyword-based search engines such as Lycos and Altavista. First, such search engines require that the user describe their interest in terms of specific words that match those in the target Web page. In contrast, WebWatcher can learn that a term such as "machine learning" matches a hyperlink such as "neural networks" or "Avrim Blum," even though these phrases share no words in common. Furthermore, current search engines do not take into account that documents are designed as hypertext. In many cases only a sequence of pages and the knowledge about how they relate to each other can satisfy the user's information need.

In the following we present the design of WebWatcher, as well as experimental results obtained from over 5000 tours given by WebWatcher to various visitors on the

Figure 1: The Front-Door page with WebWatcher's additions.

Web. We describe how WebWatcher learns to improve the quality of its advice, and present experimental results comparing different learning methods. Finally, we summarize the lessons and perspectives gained from these first experiments with a tour guide agent for the Web.

## 2 Trace of WebWatcher

WebWatcher was in operation from August, 1995, to February, 1997. A typical WebWatcher tour proceeded as follows. We enter the CMU School of Computer Science Web at the front door page. On this page an instance of WebWatcher can be invoked by clicking on the hyperlink "The WebWatcher Tour Guide". Clicking on this hyperlink leads us to a page where we are asked for a short description of our current interest. In this example, we enter the phrase "intelligent agents" as our interest. WebWatcher now returns us to the initial page, prepared to guide our tour (figure 1). We are no longer browsing alone, but will be accompanied by WebWatcher on any sequence of hyperlinks we follow from this point forward.

We can tell that WebWatcher accompanies us from the additions it makes to the original page. A page augmented with WebWatcher's additions is shown in figure 1. Among others they include:

- A list of *WebWatcher Commands* is inserted above the original page. The user can invoke these commands to communicate with WebWatcher as he browses from page to page.

- Selected hyperlinks from the original page have now been highlighted by WebWatcher, to suggest directions relevant to our browsing interests. WebWatcher highlights these hyperlinks by inserting eyeball icons ($\widetilde{oo}$) around the hyperlink, as shown in the "projects" link in figure 1 (recall in this example the user expressed an interest in "intelligent agents"). The advice WebWatcher provides in this fashion is based on knowledge learned from previous tours.

If we follow WebWatcher's suggestion in figure 1, we reach a page containing a long list of research project pages. WebWatcher again inserts the command list on top of the page and suggests three hyperlinks it judges relevant to our interest in "intelligent agents."

In general, the user may click on any hyperlink, recommended or not. Each time the user selects a hyperlink, WebWatcher accompanies the user to the next page, and logs this hyperlink selection as a training example for learning to improve future advice.

In addition to highlighting hyperlinks WebWatcher also provides other forms of assistance. In particular, it provides a keyword search using a variant of the Lycos search engine applied to the set of pages previously visited by WebWatcher. It also provides a user command "Show me similar pages" in the command list, which causes WebWatcher to display a list of pages which are similar based on a metric derived from hypertext structure [Joachims et al., 1995]. Clicking on the command "how many followed each link?" asks WebWatcher to display for each hyperlink the number of previous visitors who took that link. The command "Email me if this page changes" tells WebWatcher to periodically monitor the current page and send the user email if it changes.

WebWatcher accompanies the user along any hyperlink anywhere on the World Wide Web. To end the tour, the user clicks on one of two options in the command list: "Exit: Goal reached" or "Exit: Goal not found." This exit provides the user with a way of giving final feedback to WebWatcher.

## 3 Accompanying the User

WebWatcher is implemented as a server on a separate workstation on the network and acts much like a proxy. Before returning a page to the user it makes three modifications:

1. The WebWatcher command list is added to the top of the page.

2. Each hyperlink URL in the original page is replaced by a new URL that points back to the WebWatcher server.

3. If WebWatcher finds that any of the hyperlinks on this page are strongly recommended by its search control knowledge, then it highlights the most promising links in order to suggest them to the user.

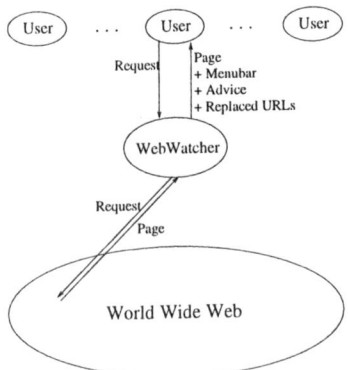

Figure 2: WebWatcher is an interface agent between the user and the World Wide Web.

While it waits for the user's next step, it prefetches any Web pages it has just recommended to the user to minimize network delays. Figure 2 depicts one cycle of user interaction. When the user clicks on a new hyperlink, WebWatcher updates the log for this search, retrieves the page (unless it has already been prefetched), performs similar substitutions, and returns the copy to the user. This process continues until the user elects to dismiss the agent.

## 4 Learning in WebWatcher

What is the form of the knowledge required by WebWatcher? In general, its task is to suggest an appropriate link given an interest and Web page. In other words, it requires knowledge of the following target function:

$$LinkQuality : Page \times Interest \times Link \rightarrow [0, 1]$$

The value of *LinkQuality* is interpreted as the probability that a user will select *Link* given the current *Page* and *Interest*. In the following we present three approaches to learning this target function from experience. The first approach uses previously given tours as a source of information to augment the internal representation of each selected hyperlink. The second approach is based on reinforcement learning. The idea is to find tours through the Web so that the amount of relevant information encountered over the trajectory is maximized. The third approach is the combined method that includes both of the first two approaches.

### 4.1 Learning from Previous Tours

In the first approach, learning is accomplished by annotating each hyperlink with the interests of the users who took this hyperlink on previous tours. Thus, whenever a user follows a hyperlink the description of this hyperlink is augmented by adding the keywords the user typed in at the beginning of the tour. The initial description of a hyperlink is the underlined text. Figure 3 illustrates the keywords accumulated by hyperlinks, and the way in which these are used to influence subsequent advice.

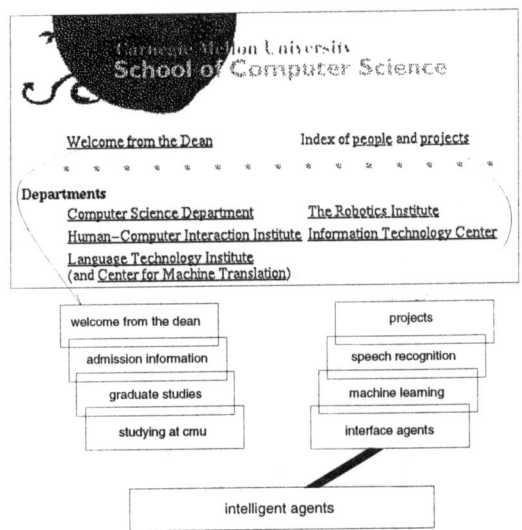

Figure 3: Here the interest of a new user in "intelligent agents" matches the "projects" hyperlink better than the "Welcome from the Dean" hyperlink, because of the keywords accumulated by this hyperlink during previous tours.

To suggest hyperlinks during a tour WebWatcher compares the current user's interest with the descriptions of all hyperlinks on the current page. WebWatcher suggests those hyperlinks which have a description sufficiently similar to the user's interest.

The metric used to compute similarity between a user's stated interest and a hyperlink description is based on a technique from information retrieval [Salton, 1991]. Interests and hyperlink descriptions are represented by very high-dimensional feature vectors, each dimension representing a particular word in the English language. The elements (called word-weights) of a vector are calculated using the TFIDF heuristic [Salton, 1991]. Based on this vector representation similarity is calculated as the cosine between vectors.

The algorithm WebWatcher uses to suggest hyperlinks considers all hyperlinks on the current page. For each hyperlink, the list of associated keywords (including the original underlined words) is used to calculate its similarity to the current user's interest. The value of *LinkQuality* for each hyperlink is estimated to be the average similarity of the $k$ (usually 5) highest ranked keyword sets associated with the hyperlink. A hyperlink is suggested if its value for *LinkQuality* is above a threshold. The maximum number of hyperlinks suggested on a page is three.

### 4.2 Learning from Hypertext Structure

The previous section describes a learning method that augments a given hyperlink with the stated interests of earlier users who selected it. In this section we describe a second learning method that augments a given hyperlink using words encountered in pages downstream of it. This approach is based on reinforcement learning. The

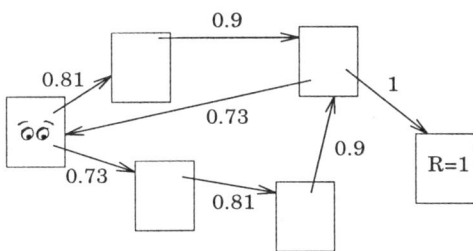

Figure 4: Example state space.

objective is to find paths through the Web which maximize the amount of relevant information encountered.

**Reinforcement Learning**

Reinforcement learning allows agents to learn control strategies that select optimal actions in certain settings. Consider an agent navigating from state to state by performing actions. At each state $s$ the agent receives a certain reward $R(s)$. The goodness of an action $a$ can be expressed in terms of an evaluation function $Q(s, a)$, defined for all possible state-action pairs. The value of $Q(s, a)$ is the discounted sum of future rewards that will be obtained if the agent performs action $a$ in state $s$ and subsequently chooses optimal actions. If the agent can learn this function, then it will know how to act in any state. More precisely,

$$Q(s_t, a) = \sum_{i=0}^{\infty} \gamma^i \cdot R(s_{t+1+i})$$

where $s_t$ is the state the agent is in at time $t$, and where $\gamma$ is a discount factor $0 \leq \gamma < 1$ that determines how severely to discount the value of rewards received further into the future. Under certain conditions, the Q function can be iteratively approximated by updating the estimate for $Q(s, a)$ repeatedly as follows (see [Watkins, 1989]):

$$Q_{n+1}(s, a) = R(s') + \gamma \max_{a' \in actions\_in\_s'} [Q_n(s', a')]$$

$s'$ is the state resulting from performing action $a$ in state $s$. Once $Q(s, a)$ is known, the optimal control strategy for the agent is to repeatedly pick the action $a$ that maximizes $Q(s, a)$ for its current state $s$.

Figure 4 gives an example. Boxes represent possible states of the agent. The edges represent actions that bring the agent from one state to another. The edges are annotated with values of the function $Q(s, a)$. In the rightmost state the agent receives a reward of 1. The reward is 0 in all other states. If the agent always follows the action with the highest Q value, it will get to the reward state in the smallest number of steps and thus maximize the discounted reward it receives.

**Reinforcement Learning and Hypertext**

Imagine a Web agent looking for pages on which the word "intelligent" occurs. For this agent, states correspond to Web pages, and actions correspond to hyperlinks. In this case, we define the reward $R_{intelligent}(s)$ for a particular page $s$ to be the TFIDF value of "intelligent" for $s$. The agent will then learn a $Q_{intelligent}(s, a)$ function whose value for page $s$ and hyperlink $a$ is the sum of discounted TFIDF values of "intelligent" over the optimal tour beginning with $a$. It can then use this $Q$ function to choose the best link at each step in the tour.

WebWatcher uses a separate reward function $R_w(s)$ and learns a distinct $Q_w(s, a)$ function for every word $w$. At runtime the system recommends hyperlinks for which the sum of the $Q_w(s, a)$ for the words in the user's interest description is highest. An additional reward is given if the interest matches the underlined text of the hyperlink.

Because WebWatcher cannot expect that users will always stick to pages it has already seen, a core question in implementing this approach is how to learn a general approximation for each of the Q-functions $Q_w(s, a)$ that applies even to unseen states (pages) and actions (hyperlinks). We chose a distance-weighted 3-nearest neighbor function approximator [Mitchell, 1997] for this purpose, because of the many features needed to describe pages and hyperlinks, and because of certain theoretical advantages [Gordon, 1995]. Each hyperlink $a$ is described by the TFIDF vector representation of the underlined anchor text, each page $s$ analogously by its title. We define the similarity between the hyperlink $a_1$ on page $s_1$ and the hyperlink $a_2$ on page $s_2$ to be the distance between $a_1$ and $a_2$, plus (heuristically) twice the distance between $s_1$ and $s_2$. The distance between two vectors is defined to be the cosine of the angle between the vectors, in keeping with the standard similarity measure used in information retrieval.

## 4.3 Experimental Setup

The experiments presented in this section were conducted using 1777 of the 5822 traces starting at the SCS-FrontDoor page WebWatcher collected between August 2, 1995, and March 4, 1996. We used only those tours where the user typed in an interest and where the tour was at least four steps long. Five different test-training splits were used with the pages and hyperlinks from 70% of the traces used for training.

In addition to the reinforcement learning approach RL and the learning method ANNOTATE from section 4.1, the results for four other methods are presented. RANDOM suggests hyperlinks at random from the current page. POPULARITY suggests those hyperlinks which have been followed most frequently in the past, ignoring information about the current user's interest. MATCH suggests hyperlinks according to the TFIDF-cosine similarity between their underlined text and the user's interest. Finally, the COMBINE method combines the predictions of RL, POPULARITY, ANNOTATE, and

|  | Accuracy |
|---|---|
| Random | 31.3% (.9) |
| Popularity | 41.9% (.4) |
| Match | 40.5% (.6) |
| Annotate | 42.2% (.7) |
| RL | 44.6% (.5) |
| Combine | 48.9% (.3) |

Figure 5: Results averaged over five different test training splits (with one standard error).

MATCH using logistic regression. 10% of the available training data is used for regression.

### 4.4 Experimental Results

The results in figure 5 provide a comparison of the learning methods. For each example in the test set each learner was allowed to choose three hyperlinks from the corresponding page. Accuracy measures the percentage of examples for which the user followed one of these chosen hyperlinks. We required each learning algorithm to make a prediction regardless of how confident it was.

The method WebWatcher used while accessible to the public was Annotate. The accuracy of 42.2% in the offline experiment here approximately matches the fraction of times (43.9%) users followed WebWatcher's advice during actual use of the system.

Reinforcement learning performs significantly better than the other basic methods. Nevertheless the combination of all methods achieves the highest accuracy, outperforming each of the individual methods. Our conjecture is that this is due to the diversity of the methods and sources of information that are combined. Popularity uses frequency information derived from user behavior on a particular page, Match uses the underlined text in hyperlinks, Annotate uses the interest descriptions from previous user traces, and RL makes use of hypertext structure that is downstream of the hyperlink in question.

### 4.5 Comparison with Human Experts

To get a better feel for the nature of the task of suggesting hyperlinks and to evaluate WebWatcher's performance, we conducted an experiment in which we asked humans to perform WebWatcher's task. To make the experiment more tractable we focused on one page, namely an earlier version of the SCS-Front-Door page shown in figure 1. From the 408 training examples available for this page we took 8 random subsets of 15 examples and presented these to eight people who were already well-informed about this page and its locale. For each example, they were given the stated interest of the user, then asked to suggest the 3 hyperlinks the user was most likely to follow out of the 18 hyperlinks on this page.

As figure 6 summarizes, humans achieved an accuracy of 47.5% in predicting which hyperlink the user would follow. The learning method Annotate achieved an accuracy of 42.9% under the same conditions. The 22.4%

|  | Accuracy |
|---|---|
| Random | 22.4% |
| Annotate | 42.9% |
| Human | 47.5% |

Figure 6: A comparison to human performance on the SCS-Front-Door page.

for random prediction again provides a baseline. These results suggest that predicting which hyperlink the user is going to follow is a fairly difficult task in a general setting. In comparison to human performance the learning approach does reasonably well.

## 5 Related Work

*Letizia* [Lieberman, 1995] is similar to WebWatcher in the sense that the system accompanies the user while browsing. One difference is that Letizia is located on a single user's machine and learns only his or her current interest. By doing lookahead search Letizia can recommend nearby pages.

*Syskill and Webert* [Pazzani et al., 1996] offers a more restricted way of browsing than WebWatcher and Letizia. Starting from a manually constructed index page for a particular topic, the user can rate hyperlinks off this page. The system uses the ratings to learn a user specific topic profile that can be used to suggest unexplored hyperlinks on the page. Syskill and Webert can also use search engines like LYCOS to retrieve pages by turning the topic profile into a query.

*Lira* [Balabanovic and Shoham, 1995] works in an offline setting. A general model of one user's interest is learned by asking the user to rate pages. Lira uses the model to browse the Web offline and returns a set of pages that match the user's interest.

## 6 Summary and Future Research

WebWatcher provides one case study of a tour guide agent for the World Wide Web. By "tour guide agent" we mean any agent that accompanies users from page to page, providing assistance based on a partial understanding of that user's interests and of the content of the Web pages. During its 18 months of operation, WebWatcher served thousands of people browsing CMU's School of Computer Science Web pages. The system has the following properties:

- WebWatcher provides several types of assistance, but most importantly highlights interesting hyperlinks as it accompanies the user.

- WebWatcher learns from experience. We found a multi-strategy approach to be the most effective learning method.

- WebWatcher runs as a centralized server so that it can assist any Web user running any type of Web browser as well as combine training data from thousands of different users.

Our experience with WebWatcher led us to believe that self-improving tour guide agents will play an important role on the Web in the future. WebWatcher demonstrates that it is possible for such an agent to provide helpful advice to many users, and that it is possible to automatically learn from the thousands of users with whom it interacts. Given that popular Web sites are typically visited by many thousands of users, and that the content of the Web changes frequently, it appears that machine learning will play a crucial role in future tour guide agents.

Our experience with WebWatcher has also shown that despite its ability to help some users, its highlighted hyperlinks match those followed by users in only 48% of all cases. Interestingly, when we assigned expert humans the same task, they could do no better. Examining many specific tours given by WebWatcher, we find the following partial explanation for this level of accuracy. Users tend to have a fairly short attention span and are distracted from their stated interest. Furthermore there is great diversity in the interests of users browsing CMU's SCS Front Door so that even after thousands of tours it is not uncommon for the next user to express an interest that WebWatcher has not yet encountered. Together, these two factors suggest that WebWatcher might achieve higher accuracy if the Web locale it had to cover had a narrower, more focused scope.

More generally, our experience with WebWatcher suggests a number of topics for future research:

- Personalized WebWatcher. Whereas WebWatcher learns to specialize to a specific Web locale, one could instead develop tour guide agents that learn to specialize to a particular user. Such an agent could learn a model of the longer-term interests of users by observing which pages they do and do not visit. We have recently begun experiments with such a personalized WebWatcher [Mladenic, 1996].

- Combining user-specific and Web locale-specific learning. At the same time WebWatcher learns a model of one particular user, it could retain its capability to annotate hyperlinks based on tours given to many users. In this context, one could explore a variety of methods for combining the benefits of single-user modeling and learning simultaneously from/about multiple users.

- Richer dialogs with users. One major shortcoming of the current WebWatcher is that it allows the user only to express a few keywords, and only at the beginning of the tour. A more flexible approach would involve an ongoing dialog with the user, much more like that a museum visitor might have with a human guide.

- New machine learning algorithms for classifying hyperlinks. The learning methods used by WebWatcher succeed in improving its performance over time. However, a large space of possible learning methods for this problem remains unexplored, i.e. the combination of linguistic analysis with the statistical techniques described here.

- Intelligent distributed hyperlinks. WebWatcher learns by associating new information with hyperlinks based on its experience. Note this learning could be performed in a much more distributed fashion, with each hyperlink separately building up its own model of itself and making recommendations to the user.

## Acknowledgements

This research is supported by a Rotary International fellowship grant, an NSF graduate fellowship, and by Arpa under grant number F33615-93-1-1330.

## References

[Armstrong et al., 1995] R. Armstrong, D. Freitag, T. Joachims, and T. Mitchell. Webwatcher: A learning apprentice for the world wide web. In *AAAI Spring Symposium on Information Gathering from Heterogeneous, Distributed Environments*, March 1995.

[Balabanovic and Shoham, 1995] M. Balabanovic and Y. Shoham. Learning information retrieval agents: Experiments with automated web browsing. In *AAAI Spring Symposium on Information Gathering from Heterogeneous, Distributed Environments*, 1995.

[Gordon, 1995] G. Gordon. Stable function approximation in dynamic programming. In *International Conference on Machine Learning*, 1995.

[Joachims et al., 1995] T. Joachims, T. Mitchell, D. Freitag, and R. Armstrong. Webwatcher: Machine learning and hypertext. In K. Morik and J. Herrmann, editors, *GI Fachgruppentreffen Maschinelles Lernen*. University of Dortmund, August 1995.

[Lieberman, 1995] H. Lieberman. Letizia: An agent that assists web browsing. In *International Joint Conference on Artificial Intelligence, Montreal*, August 1995.

[Mitchell et al., 1994] T. Mitchell, R. Caruana, D. Freitag, J. McDermott, and D. Zabowski. Experience with a learning personal assistant. *Communications of the ACM*, 37(7):81–91, July 1994.

[Mitchell, 1997] T. Mitchell. *Machine Learning*. McGraw-Hill, 1997.

[Mladenic, 1996] D. Mladenic. Personal webwatcher: Implementation and design. Technical report, J. Stefan Institute, Ljubljana, Slovenia, 1996.

[Pazzani et al., 1996] M. Pazzani, J. Muramatsu, and D. Billsus. Syskill & webert: Identifying interesting web sites. In *AAAI Conference, Portland*, 1996.

[Salton, 1991] G. Salton. Developments in automatic text retrieval. *Science*, 253:974–979, 1991.

[Watkins, 1989] C. Watkins. Learning from delayed rewards. Technical report, King's College, Cambridge, England, 1989.

# INFORMATION RETRIEVAL

Information Retrieval 4

# Recursive Plans for Information Gathering

**Oliver M. Duschka**
Department of Computer Science
Stanford University
Stanford, California 94305
U.S.A.

**Alon Y. Levy**
AT&T Laboratories
600 Mountain Avenue
Murray Hill, New Jersey 07974
U.S.A.

## Abstract

Generating query-answering plans for information gathering agents requires to translate a user query, formulated in terms of a set of virtual relations, to a query that uses relations that are actually stored in information sources. Previous solutions to the translation problem produced sets of *conjunctive plans*, and were therefore limited in their ability to handle information sources with binding-pattern limitations, and to exploit functional dependencies in the domain model. As a result, these plans were incomplete w.r.t. sources encountered in practice (i.e., produced only a subset of the possible answers). We describe the novel class of *recursive* information gathering plans, which enables us to settle two open problems. First, we describe an algorithm for finding a query plan that produces the maximal set of answers from the sources in the presence of functional dependencies. Second, we describe an analogous algorithm in the presence of binding-pattern restrictions in the sources, which was not possible without recursive plans.

## 1 Introduction

The problem of information integration (a.k.a. information gathering agents) has recently received considerable attention due to the growing number of structured information sources available online. Information integration systems (e.g., the Internet Softbot [Etzioni and Weld, 1994], SIMS [Arens et al., 1996], TSIMMIS [Chawathe et al., 1994], the Information Manifold [Levy et al., 1996], Occam [Kwok and Weld, 1996], Infomaster [Duschka and Genesereth, 1997b]) provide a *uniform* query interface to the multiple information sources, thereby freeing the user from having to locate the relevant sources, query each one in isolation, and combine manually the information from the different sources.

Information integration systems are based on the following general architecture. The user interacts with a uniform interface in the form of a set of *virtual* relation names that are used in formulating queries. The actual data is stored in external sources (called the *source relations*). In order for the system to be able to answer queries, we must specify a *mapping* between the virtual relations and the source relations. The most common method to specify these mappings (employed in [Levy et al., 1996; Kwok and Weld, 1996; Duschka and Genesereth, 1997b]) is to describe each source relation as the result of a *conjunctive query* (i.e., a single Horn rule) over the virtual relations. For example, an information source containing papers authored by AI researchers would be described as follows:

$$db_1(P, A) :- paper(P), author(P, A), ai(A).$$

The relations *paper*, *author* and *ai* are virtual relations that can be used in formulating queries, and $db_1$ is a source relation.

Given a query from the user, formulated in terms of the virtual relations, the system must translate it to a query that mentions *only* the source relations, because only those relations are actually available. That is, the system needs to find a query expression, (or, a set of conjunctive queries), that mentions only the source relations, and is equivalent to the original query. The new query is called a *query plan* (or a *query rewriting*), and the translation problem is known as the problem of *rewriting queries using views*. The views (a term used in database systems to refer to predefined queries) are the relations in the sources. The rewriting problem has also been investigated in the database literature because of its importance for query optimization and data warehousing [Yang and Larson, 1987; Chaudhuri et al., 1995; Levy et al., 1995; Rajaraman et al., 1995; Qian, 1996; Duschka and Genesereth, 1997a].

Previous results on the rewriting problem are of limited practical use for several reasons. First, they have concentrated on showing a bound on the size of the resulting query plan [Levy et al., 1995; Rajaraman et al., 1995]. These results establish the complexity of the rewriting problem, but yield only non-deterministic algorithms for its solution. Second, they considered only the problem of finding an *equivalent* rewriting of the query using the source relations. In practice, the collection of available sources may not contain *all* the information needed to answer a query, and therefore, we need to resort to *maximally-contained* rewritings. A maximally-contained rewriting provides all the answers that are possible to obtain from the sources, but the expression describing the rewriting may not be equivalent to the original query. For example, if we only have the $db_1$ source available, and our query asks for all papers by Computer Science researchers, then the following is a maximally-contained rewriting:

$q(P, Y, A) :- db_1(P, Y, A)$.

The third problem is that many sources encountered in practice have limitations on the binding patterns they support, or may satisfy certain functional dependencies. As an example of binding-pattern restrictions, a name server of an institution, holding the addresses of its employees, will not provide the list of all employees and their addresses. Instead, it will provide the address for a *given* name. In the case of equivalent rewritings, Rajaraman et al. [Rajaraman *et al.*, 1995] describe a bound on the size of the query plans that need to be considered. However, Kwok and Weld [Kwok and Weld, 1996] show that if we restrict our plans to be sets of conjunctive queries, then there may *not be* a finite maximally-contained rewriting. As an example of a functional dependency, the year of a conference functionally determines its location. The presence of functional dependencies further complicates the rewriting problem because it allows rewritings that are not valid otherwise. By ignoring the functional dependencies, we may miss answers to the query.

In this paper we introduce the new class of *recursive* query plans for information gathering. Instead of plans being only sets of conjunctive queries, they can now be recursive sets of function-free Horn rules. Using recursive plans, we are able to settle two open problems. First, we describe an algorithm for finding the maximally-contained rewriting in the presence of functional dependencies. Second, we describe an algorithm for finding the maximally-contained rewriting in the presence of binding-pattern restrictions, which was not possible without recursive plans.

Another significant advantage of our method is that it is *generative*, rather than *descriptive*. Our algorithms generate the rewriting in time that is polynomial in the size of the query. In contrast, previous methods [Levy *et al.*, 1995; Rajaraman *et al.*, 1995] describe the space of possible candidate rewritings, and propose heuristics for searching this space [Kwok and Weld, 1996; Levy *et al.*, 1996].[1] These methods combine the process of finding a rewriting with the process of checking whether it is equivalent to the original query (which is NP-hard). In contrast, our method isolates the process of generating the maximally-contained rewriting, which can be done much more efficiently.

## 2 Preliminaries

**Relations and queries:** Our representation of the domain and of the information sources includes a set of relations. For every relation, we associate an *attribute name* to each of its arguments. For example, the attribute names of the binary relation *author* may be *Paper* and *Person*. For a tuple $t$ of a relation $R$, we denote by $t.A$ the value of the attribute $A$ in $t$.

A function-free Horn rule is an expression of the form

$$p(\bar{X}) :- p_1(\bar{X}_1), \ldots, p_n(\bar{X}_n),$$

where $p$, and $p_1, \ldots, p_n$ are relation names, and $\bar{X}, \bar{X}_1, \ldots, \bar{X}_n$ are tuples of variables and constants such that any variable appearing in $\bar{X}$ appears also in $\bar{X}_1 \cup \ldots \cup \bar{X}_n$. The *head* of the rule is $p(\bar{X})$, and its *body* is $p_1(\bar{X}_1), \ldots, p_n(\bar{X}_n)$. The *base relations* of a set of Horn rules are the relations that appear *only* in the bodies of the rules and not in the heads (and therefore their extension includes only explicitly stored tuples). Given a set of rules, we can define a dependency graph, whose nodes are the relations appearing in the rules. There is an arc from the node of relation $p_i$ to the node of predicate $p$ if $p_i$ appears in the body of a rule whose head relation is $p$. The rules are said to be *recursive* if there is a cycle in the dependency graph.

A *query* is a set of function-free Horn rules. A *conjunctive query* is a single non-recursive Horn rule. One of the relations in the query is distinguished as the *query relation*, and its extension is the answer to the query.

**Query containment:** In our discussion we will need to compare between different queries. We use the condition of containment to compare queries. Let us denote by $q(D)$ the result of evaluating query $q$ on database $D$. Given two queries $q_1$ and $q_2$, we say that $q_1$ is *contained* in $q_2$ if for every database $D$, $q_1(D) \subseteq q_2(D)$. The fundamental results on containment of conjunctive queries and function-free Horn rules are described in [Chandra and Merlin, 1977; Sagiv and Yannakakis, 1980; Shmueli, 1993]

---

[1] The algorithm in [Levy *et al.*, 1996] checks whether the plans can be executed given the binding-pattern restrictions, but is not guaranteed to produce the maximally-contained rewriting when these restrictions are present. The algorithm in [Kwok and Weld, 1996] produces only conjunctive plans that are guaranteed to adhere to the limitations on binding patterns, but is not guaranteed to compute the maximally-contained plan.

**Functional dependencies:** An instance of a relation $p$ satisfies the *functional dependency* $A_1, \ldots, A_n \to B$ if for every two tuples $t$ and $u$ in $p$ with $t.A_i = u.A_i$ for $i = 1, \ldots, n$, also $t.B = u.B$. We will abbreviate a set of attributes $A_1, \ldots, A_n$ by $\bar{A}$.

When the relations satisfy a set of functional dependencies $\Sigma$, we refine our notion of containment to *relative containment*: Query $q_1$ is *contained* in query $q_2$ relative to $\Sigma$, denoted $q_1 \subseteq_\Sigma q_2$, if for each database $D$ satisfying the functional dependencies in $\Sigma$, $q_1(D) \subseteq q_2(D)$.

In order to decide containment of conjunctive queries in the presence of functional dependencies, Aho et al. [Aho *et al.*, 1979] show that it suffices to precede the containment algorithm by applying the *chase* algorithm to the contained query. The chase algorithm applies the following transformation to the body of a conjunctive query $q$ until no changes can be made. If the functional dependency $\bar{A} \to B$ holds for a relation $p$, and a conjunctive query $q$ has two subgoals of $p$, $g_1$ and $g_2$, with the same variables or values for the attributes $\bar{A}$, and $g_1$ has a variable $X$ for attribute $B$, then we replace the occurrences of $X$ in $q$ by the value or variable for $B$ in $g_2$.

**Modeling information sources and query plans:** The domain model of an information agent is a set of *virtual* relations. The relations are virtual because they are only meant to provide the user a uniform interface to a multitude of information sources, and the agent does not actually store the extensions of these relations. In our discussion, we assume that a user query is a conjunctive query over the virtual relations.[2]

The agent models the contents of the external information sources by a set of *source relations*, that are disjoint from the virtual relations. To answer user queries, the agent must also have a mapping between the virtual and source relations. The mappings, called *source descriptions*, are specified by a set of conjunctive queries, whose bodies contain only virtual relations and their heads are source relations. The meaning of such a mapping is that all the tuples that are found in the information source satisfy the query over the virtual relations.[3] Given a query $q$ from the user, the agent needs to formulate a *query plan*, which is a query that bottoms out in the source relations and produces answers to $q$. A query plan is a set of Horn rules whose base predicates include *only* the source relations.

---

[2] Our results apply also to the case in which user queries are recursive [Duschka and Genesereth, 1997a].

[3] Several authors have distinguished the case in which the source contains *all* the tuples that satisfy the query from the case in which some tuples may be missing from the source [Etzioni *et al.*, 1994; Levy, 1996]. For our discussion this distinction does not matter.

**Example 2.1:** Consider a domain model where *parent*, *male* and *female* are virtual relations. The mappings below say that the source relations $v_1$ and $v_2$ store the father and mother relation, respectively.

$v_1(X, Y) :- parent(X, Y), male(X)$
$v_2(X, Y) :- parent(X, Y), female(X)$

The following query plan determines all grandparents of *ann* from the available sources:

$answer(X) \quad :- parent(X, Z), parent(Z, ann)$
$parent(X, Y) :- v_1(X, Y)$
$parent(X, Y) :- v_2(X, Y)$

The *expansion* of a query plan $\mathcal{P}$, denoted $\mathcal{P}^{exp}$, is obtained from $\mathcal{P}$ by replacing all source-relation literals by their definitions. Existentially quantified variables in a source description are replaced by fresh variables in the expansion.

A query plan $\mathcal{P}$ is *maximally-contained* in a query $q$, relative to a set of functional dependencies $\Sigma$, if $\mathcal{P}^{exp} \subseteq_\Sigma q$, and for every query plan $\mathcal{P}_1$, if $\mathcal{P}_1^{exp} \subseteq_\Sigma q$ then $\mathcal{P}_1^{exp} \subseteq_\Sigma \mathcal{P}^{exp}$.

## 3 Functional Dependencies

We use the following example throughout this section to illustrate the difficulties introduced by functional dependencies and to present our algorithm. Suppose we have the virtual relations

$conference(Paper, Conference)$,
$year(Paper, Year)$,
$location(Conference, Year, Location)$

describing the conference at which a paper was presented, the publication year of a paper, and the location a conference was held at in a given year. A paper is only presented at one conference and published in one year. Also, in a given year a conference is held at a specific location. Therefore we have three functional dependencies:

$conference$: $Paper \to Conference$
$year$: $\quad\quad Paper \to Year$
$location$: $\quad Conference, Year \to Location$

We have the following information sources:

$v_1(P, C, Y) \quad :- conference(P, C), year(P, Y)$
$v_2(P, L) \quad\quad :- conference(P, C), year(P, Y),$
$\quad\quad\quad\quad\quad\quad location(C, Y, L)$

$v_1$ tells us in which conference and year a paper was presented, and $v_2$ stores the location of the presentation of a paper directly with the paper. Assume a user wants to know where IJCAI '91 was held:

$q(L) :- location(ijcai, 1991, L)$

The following plan would answer the query:

$answer(L) :- v_1(P, ijcai, 1991), v_2(P, L)$

The query plan finds *some* paper presented at IJCAI '91 using $v_1$, and then finds the location of the conference this paper was presented at using $v_2$. This plan is correct *only* because every paper is presented at one conference and in one year. In fact, if these dependencies would not hold, there would be no way of answering this query using the sources. It is also important to note that view $v_1$ is needed in the query plan even though the predicates in $v_1$, *conference* and *year*, don't appear in the query $q$ at all. Without functional dependencies, only views that contain predicates appearing in the user query need to be considered [Levy *et al.*, 1995].

In the following we are going to give a construction of query plans that is guaranteed to be maximally-contained in the given queries, even in the presence of functional dependencies. The key to the construction is a set of *inverse rules*, whose purpose is to *recover* tuples of the virtual relations from the source relations. In the following definition we use a set of function symbols; for every source relation $v$ with variables $X_1, \ldots, X_n$ in the body but not in the head of the source description, we have a function symbol $f_{v,i}$. These function symbols can later be removed from the query plan[4] [Duschka and Genesereth, 1997a].

**Definition 3.1:** *(inverse rules) Let $v$ be a source description*

$$v(\bar{X}) :- p_1(\bar{X}_1), \ldots, p_n(\bar{X}_n).$$

*Then for $j = 1, \ldots, n$,*

$$p_j(\bar{X}'_j) :- v(\bar{X})$$

*is an inverse rule of $v$. We modify $\bar{X}_j$ to obtain the tuple $\bar{X}'_j$ as follows: if $X$ is a constant or is a variable in $\bar{X}$, then $X$ is unchanged in $X'_j$. Otherwise, $X$ is one of the variables $X_i$ appearing in the body of $v$ but not in $\bar{X}$, and $X$ is replaced by $f_{v,i}(\bar{X})$ in $\bar{X}'_j$.*

We denote the set of inverse rules of the views in $\mathcal{V}$ by $\mathcal{V}^{-1}$.

Continuing with our example, the inverse rules for $v_1$ and $v_2$ are:

$r_1: conference(P,C) :- v_1(P,C,Y)$
$r_2: year(P,Y) :- v_1(P,C,Y)$
$r_3: conference(P, f_1(P,L)) :- v_2(P,L)$
$r_4: year(P, f_2(P,L)) :- v_2(P,L)$
$r_5: location(f_1(P,L), f_2(P,L), L) :- v_2(P,L)$

For example, rule $r_5$ extracts from $v_2$ that *some* conference in *some* year was held in location $L$. Suppose

---

[4] It is not necessary to remove function symbols. Function symbols are *only* introduced in inverse rules, and can never become nested. Therefore, termination of bottom-up evaluation is guaranteed even in the presence of function symbols. Only tuples not containing function symbols are considered answers to a query plan.

that $v_1$ stores the information that the paper "Fuzzy Dynamic Systems" was presented at IJCAI '91, and $v_2$ stores the information that "Fuzzy Dynamic Systems" was presented in Sydney. The inverse rules derive the following facts:

*conference*
$\langle fuzzy, ijcai \rangle$ (with $r_1$)
$\langle fuzzy, f_1(fuzzy, sydney) \rangle$ ($r_3$)

*year*
$\langle fuzzy, 1991 \rangle$ ($r_2$)
$\langle fuzzy, f_2(fuzzy, sydney) \rangle$ ($r_4$)

*location*
$\langle f_1(fuzzy, sydney), f_2(fuzzy, sydney), sydney \rangle$ ($r_5$)

The inverse rules don't take into account the presence of the functional dependencies. For example, because of the functional dependency in relation *conference*, $Paper \rightarrow Conference$, it is possible to conclude that the function term $f_1(fuzzy, sydney)$ must actually be the same as the constant $ijcai$. We model this inference by introducing a new binary relation $e$. The intended meaning of $e$ is that $e(c_1, c_2)$ holds if and only if $c_1$ and $c_2$ must be equal under the given functional dependencies. Hence, the extension of $e$ includes the extension of $=$ (i.e., for every $X$, $e(X,X)$), and the tuples that can be derived by the following chase rules ($e(\bar{A}, \bar{A}')$ is a shorthand for $e(A_1, A'_1), \ldots, e(A_n, A'_n)$):[5]

**Definition 3.2:** *(chase rules) Let $\bar{A} \rightarrow B$ be a functional dependency satisfied by a virtual relation $p$. Let $\bar{C}$ be the attributes of $p$ that are not in $\bar{A}, B$. The chase rule corresponding to $\bar{A} \rightarrow B$, denoted $chase(\bar{A} \rightarrow B)$, is the following rule:*

$$e(B, B') :- p(\bar{A}, B, \bar{C}), p(\bar{A}', B', \bar{C}'), e(\bar{A}, \bar{A}').$$

We denote by $chase(\Sigma)$ the set of chase rules corresponding to the functional dependencies in $\Sigma$.

In our example, the chase rules are

$e(C,C') :- conference(P,C), conference(P',C'),$
$\qquad\qquad e(P,P')$
$e(Y,Y') :- year(P,Y), year(P',Y'), e(P,P')$
$e(L,L') :- location(C,Y,L), location(C',Y',L'),$
$\qquad\qquad e(C,C'), e(Y,Y')$

The chase rules allow us to derive the following facts in relation $e$:

*e*
$\langle f_1(fuzzy, sydney), ijcai \rangle$
$\langle f_2(fuzzy, sydney), 1991 \rangle$

---

[5] We only require relation $e$ to be reflexive for ease of exposition. For every rule $r$ having a subgoal $e(X,Y)$ in its body, we could add a modified version of rule $r$ with subgoal $e(X,Y)$ removed and $X$ replaced by $Y$. The resulting set of rules wouldn't require $e$ to be reflexive.

The extension of $e$ is reflexive by construction, and is symmetric because of the symmetry in the chase rules. To guarantee that $e$ is an equivalence relation, it is still needed to enforce transitivity of $e$. The following rule, denoted by $\mathcal{T}$, is sufficient for guaranteeing transitivity of relation $e$:

$$e(X,Y) :- e(X,Z), e(Z,Y)$$

The final step in the construction is to rewrite query $q$ in a way that it can use the equivalences derived in relation $e$.

We define the query $\bar{q}$ by modifying $q$ iteratively as follows. If $c$ is a constant in one of the subgoals of $q$, we replace it by a new variable $Z$, and add the subgoal $e(Z,c)$. If $X$ is a variable in the head of $q$, we replace $X$ in the body of $q$ by a new variable $X'$, and add the subgoal $e(X',X)$. If a variable $Y$ that is not in the head of $q$ appears in two subgoals of $q$, we replace one of its occurrences by $Y'$, and add the subgoal $e(Y',Y)$. We continue until we cannot apply this rule anymore. Our example query would be rewritten to

$$\bar{q}(L) :- location(C,Y,L'),$$
$$e(C, ijcai), e(Y, 1991), e(L', L)$$

Note that evaluating query $\bar{q}$ on the reconstructed virtual relations and the derived equivalence relation $e$ yields the desired result: IJCAI '91 was held in Sydney.

Given a query $q$, a set of source descriptions $\mathcal{V}$, and a set of functional dependencies $\Sigma$, the constructed query plan includes $\bar{q}$, the inverse rules $\mathcal{V}^{-1}$, the chase rules $chase(\Sigma)$ and the transitivity rule $\mathcal{T}$. The following theorem shows that this query plan is maximally-contained in $q$ relative to $\Sigma$.

**Theorem 3.1:** *Let $\Sigma$ be a set of functional dependencies, $\mathcal{V}$ a set of conjunctive source descriptions, and let $q$ be a conjunctive query over the virtual relations. Let $\mathcal{R}$ denote the set of rules $\mathcal{V}^{-1} \cup chase(\Sigma) \cup \mathcal{T}$. Then, $\bar{q} \cup \mathcal{R}$ is maximally-contained in $q$ relative to $\Sigma$. Furthermore, $\bar{q} \cup \mathcal{R}$ can be constructed in time polynomial in the size of $q$, $\mathcal{V}$, and $\Sigma$.* □

**Proof:** The key to the proof is to show that for every conjunctive query plan $\mathcal{P} \subseteq_\Sigma q$, $\mathcal{P}^{exp} \subseteq_\Sigma (\bar{q} \cup \mathcal{R})^{exp}$. Because recursive query plans can be seen as an encoding of the union of infinitely many conjunctive query plans, it suffices to prove the claim for all conjunctive query plans. We prove the following statement by induction on $k$: if $q$ is a query, $\mathcal{P}$ is a conjunctive query plan, and $e_1,\ldots,e_k$ is a sequence of queries with $e_1 = \mathcal{P}^{exp}$, $e_k \subseteq q$, and $e_{i+1}$ results from $e_i$ by applying a chase step, then $\mathcal{P}^{exp} \subseteq_\Sigma (\bar{q} \cup \mathcal{R})^{exp}$. This would prove that $\bar{q} \cup \mathcal{R}$ is maximally-contained in $q$ relative to $\Sigma$.

For $k=1$, $\mathcal{P}^{exp}$ is contained in $q$. As shown in [Duschka and Genesereth, 1997a], this implies that $\mathcal{P}^{exp}$ is contained in $(q \cup \mathcal{V}^{-1})^{exp}$. It follows that $\mathcal{P}^{exp}$ is contained in $(\bar{q} \cup \mathcal{R})^{exp}$ relative to $\Sigma$.

For the induction step, let $k > 1$ and assume $e_{k-1} \not\subseteq q$. Let $\bar{A} \to B$ be the functional dependency that holds for relation $p$ and that is applied from $e_{k-1}$ to $e_k$. Then $e_{k-1}$ contains two subgoals of $p$, $g_1$ and $g_2$, with the same values/variables for the attributes in $\bar{A}$, and $g_1$ contains a variable $X$ for attribute $B$ that is replaced by some value/variable in $e_k$. Let $h$ be the containment mapping [Chandra and Merlin, 1977] that shows that $q$ contains $e_k$. Replace every value/variable $X_i$ in an argument position in $q$ that is mapped by $h$ to an argument position in $e_k$ that used to be variable $X$ in $e_{k-1}$ by a new variable $X'_i$. For each of the new variables $X'_i$, add two subgoals of $p$ to $q$ with the identical new variables for the corresponding attributes $\bar{A}$, $X_i$ and $X'_i$ for attribute $B$ respectively, and new variables for the remaining attributes. We can now find a containment mapping from query $q'$ to query $e_{k-1}$. This shows that $e_{k-1}$ is contained in $q'$. Therefore, $\mathcal{P}^{exp} \equiv e_1,\ldots,e_{k-1}$ is a chase sequence with $e_{k-1} \subseteq q'$. By the induction hypothesis we have that $\mathcal{P}^{exp} \subseteq (\bar{q}' \cup \mathcal{R})^{exp}$. Using the chase rule $chase(\bar{A} \to B)$, the transitivity rule, and the reflexivity of relation $e$, we can show that $\bar{q}' \cup \mathcal{R} \subseteq \bar{q} \cup \mathcal{R}$. It follows that $\mathcal{P}^{exp} \subseteq_\Sigma (\bar{q} \cup \mathcal{R})^{exp}$.

Query $\bar{q}$ contains all subgoals in $q$, and at most as many additional subgoals of $e$ as the sum of all arities of the subgoals in $q$. Also, there are as many inverse rules as there are subgoals in all source descriptions in $\mathcal{V}$ together. Finally, there are exactly as many chase rules as there are functional dependencies in $\Sigma$. We can conclude that $\bar{q} \cup \mathcal{R}$ can be constructed in time polynomial in the size of $q$, $\mathcal{V}$ and $\Sigma$. □

## 4 Limitations on binding patterns

Some information sources cannot answer arbitrary atomic queries on the relation they contain. To model source capabilities, we attach to each source relation an *adornment* (see [Ullman, 1989], Chap. 12), specifying which binding patterns the source supports.[6] An adornment of a source relation $v$ is a string of $b$'s and $f$'s of length $n$, where $n$ is the arity of $v$. The meaning of the adornment is that the source only supports queries in which the arguments with $b$ adornments are bound. The other arguments may be either bound or free. For example, the adornment $v^{bf}$ means that the first argument must be bound in queries on $v$. We define an *executable* Horn rule as follows.

**Definition 4.1:** *(executable Horn rule) Let $\mathcal{V}$ be a set of relations with binding adornment, and let $r$ be the following Horn rule whose body relations are in $\mathcal{V}$:*

---

[6] For simplicity of exposition, we assume that each source relation has a single adornment.

$$q(\bar{X}) :- v_1(\bar{X}_1), \ldots, v_n(\bar{X}_n)$$

The rule $r$ is executable *if the following holds for* $i = 1, \ldots, n$: *let* $j$ *be an argument position of* $v_i$ *that has a* $b$ *adornment, and let* $\alpha$ *be the* $j$*'th element in* $\bar{X}_i$. *Then, either* $\alpha$ *is a constant, or* $\alpha$ *appears in* $\bar{X}_1 \cup \ldots \cup \bar{X}_{i-1}$.

A query plan includes source relations and other relations, which we model as having the all-free adornment (i.e., $f^n$, where $n$ is the relation's arity). A query plan $\mathcal{P}$ is executable if for every rule $r \in \mathcal{P}$, $r$ is executable.

When sources have limitations on binding patterns, it turns out that there may *not* be a finite maximally-contained plan, if we restrict ourselves to plans without recursion. The following example, adapted from [Kwok and Weld, 1996], illustrates the point.

**Example 4.1:** Consider the following sources:
$$v_1^f(X) :- ijcaiPapers(X)$$
$$v_2^{bf}(X,Y) :- cites(X,Y)$$
$$v_3^b(X) :- awardPaper(X).$$

The first source stores IJCAI papers, the second is a citation database, but only accepts queries where the first argument is bound, and the third source will tell us whether a *given* paper won an award. Suppose our query is to find all the award papers:
$$q(X) :- awardPaper(X)$$

For each $n$, the following is an executable conjunctive query plan that is contained in $q$:
$$q_n(Z_n) :- v_1(Z_0), v_2(Z_0, Z_1), \ldots, v_2(Z_{n-1}, Z_n),$$
$$v_3(Z_n).$$

Furthermore, for each $n$, $q_n$ may produce answers that are not obtained by any other $q_i$, for any $i$. Intuitively, a paper will be in the answer to $q_i$ if the number of links that need to be followed from an IJCAI paper is $i$. Therefore, there is no bound on the size of the conjunctive queries in the maximally-contained plan.

We now show that by allowing recursive plans we *can* produce a maximally-contained plan. On our example, the construction will yield the following query plan. The construction is based on inventing a new recursively-defined relation, *papers*, whose extension will be the set of all papers that can be reached from the papers in $v_1$.

$$papers(X) :- v_1^f(X)$$
$$papers(X) :- papers(Y), v_2^{bf}(Y,X)$$
$$q(X) :- papers(X), v_3^b(X).$$

We now describe the construction for a given set of adorned source relations $\mathcal{V}$ and a query $q$. The recursive plan includes a unary relation $dom$ whose intended extension is the set of all constants that appear in the query or in the source descriptions, or that can be obtained by iteratively querying the sources. The rules involving $dom$ are the following.

**Definition 4.2:** *(domain rules) Let* $v \in \mathcal{V}$ *be a source relation of arity* $n$. *Suppose the adornment of* $v$ *says that the arguments in positions* $1, \ldots, l$ *need to be bound, and the arguments* $l+1, \ldots, n$ *can be free. Then for* $i = l+1, \ldots, n$, *the following rule is a domain rule:*
$$dom(X_i) :- dom(X_1), \ldots, dom(X_l),$$
$$v(X_1, \ldots, X_n).$$

*Also, if* $c$ *is a constant appearing in the source descriptions in* $\mathcal{V}$ *or in query* $q$, *then the fact* $dom(c)$ *is a domain rule.*

We denote by $domain(\mathcal{V}, q)$ the set of rules described above for defining the predicate $dom$. Notice that all domain rules are executable, and that relation $dom$ has adornment $f$. Every query plan $\mathcal{P}$ can be transformed to an executable query plan by inserting the literal $dom(X)$ before subgoals $g$ in $\mathcal{P}$ that have a variable $X$ in an argument position that is required to be bound, and $X$ does not appear in the subgoals to the left of $g$ in the body. The resulting query plan, denoted by $\mathcal{P}^{exec}$, is executable. Moreover, we can show that $\mathcal{P}^{exec}$ is equivalent to $\mathcal{P}$. Combining this result with the one of the previous section, we can conclude with the following theorem:

**Theorem 4.1:** *Let* $\Sigma$ *be a set of functional dependencies,* $\mathcal{V}$ *a set of conjunctive source descriptions with binding adornments, and let* $q$ *be a conjunctive query over the virtual relations. Then* $\bar{q} \cup chase(\Sigma) \cup \mathcal{T} \cup domain(\mathcal{V}, q) \cup (\mathcal{V}^{-1})^{exec}$ *is maximally-contained in* $q$ *relative to* $\Sigma$. □

Finally, we note that the query plan can be constructed in time polynomial in the size of $q$, $\mathcal{V}$ and $\Sigma$.

## 5 Conclusions

We introduced a novel approach to creating information gathering plans, that allows for recursive plans. We have shown that recursive plans enable us to solve two open problems. We described algorithms for obtaining a maximally-contained query plan in the presence of functional dependencies and in the presence of limitations on binding patterns. Our results are also of practical importance because functional dependencies and limitations on binding patterns occur very frequently in information sources in practice (e.g., the WWW).

Recursive information gathering plans have another important methodological advantage. Query plans can be constructed by *describing* a set of inferences that the information agent needs to make in order to obtain data from its sources. We are currently extending our algorithms to deal with order predicates (e.g., $\leq, <, \neq$) and with local completeness information about the sources [Etzioni *et al.*, 1994; Duschka, 1997].

## Acknowledgements

We thank Harish Devarajan and Dan Weld for discussions and comments on earlier versions of the paper.

# References

[Aho et al., 1979] Alfred V. Aho, Yehoshua Sagiv, and Jeffrey D. Ullman. Equivalences among relational expressions. *SIAM Journal on Computing*, 8(3):218–246, May 1979.

[Arens et al., 1996] Yigal Arens, Craig A. Knoblock, and Wei-Min Shen. Query reformulation for dynamic information integration. *International Journal on Intelligent and Cooperative Information Systems*, 6(2/3):99–130, June 1996.

[Chandra and Merlin, 1977] Ashok K. Chandra and Philip M. Merlin. Optimal implementation of conjunctive queries in relational data bases. In *Proceedings of the Ninth Annual ACM Symposium on the Theory of Computing*, pages 77–90, 1977.

[Chaudhuri et al., 1995] Surajit Chaudhuri, Ravi Krishnamurthy, Spyros Potamianos, and Kyuseak Shim. Optimizing queries with materialized views. In *Proceedings of the Eleventh International Conference on Data Engineering*, IEEE Comput. Soc. Press, pages 190–200, Los Alamitos, CA, 1995.

[Chawathe et al., 1994] Sudarshan Chawathe, Hector Garcia-Molina, Joachim Hammer, Kelly Ireland, Yannis Papakonstantinou, Jeffrey Ullman, and Jennifer Widom. The TSIMMIS project: Integration of heterogeneous information sources. In *Proceedings of the 100th Anniversary Meeting*, pages 7–18, Tokyo, Japan, October 1994. Information Processing Society of Japan.

[Duschka and Genesereth, 1997a] Oliver M. Duschka and Michael R. Genesereth. Answering recursive queries using views. In *Proceedings of the Sixteenth ACM SIGACT-SIGMOD-SIGART Symposium on Principles of Database Systems*, Tucson, AZ, May 1997.

[Duschka and Genesereth, 1997b] Oliver M. Duschka and Michael R. Genesereth. Query planning in Infomaster. In *Proceedings of the 1997 ACM Symposium on Applied Computing*, San Jose, CA, February 1997.

[Duschka, 1997] Oliver M. Duschka. Query optimization using local completeness. In *Proceedings of the Fourteenth National Conference on Artificial Intelligence*, Providence, RI, July 1997.

[Etzioni and Weld, 1994] Oren Etzioni and Daniel S. Weld. A softbot-based interface to the internet. *Communications of the ACM*, 37(7):72–76, 1994.

[Etzioni et al., 1994] Oren Etzioni, Keith Golden, and Daniel Weld. Tractable closed world reasoning with updates. In *Proceedings of the 4th International Conference on Principles of Knowledge Representation and Reasoning*, pages 178–189, San Francisco, CA, June 1994.

[Kwok and Weld, 1996] Chung T. Kwok and Daniel S. Weld. Planning to gather information. In *Proceedings of the AAAI Thirteenth National Conference on Artificial Intelligence*, 1996.

[Levy et al., 1995] Alon Y. Levy, Alberto O. Mendelzon, Divesh Srivastava, and Yehoshua Sagiv. Answering queries using views. In *Proceedings of the 14th ACM SIGACT-SIGMOD-SIGART Symposium on Principles of Database Systems*, San Jose, CA, May 1995.

[Levy et al., 1996] Alon Y. Levy, Anand Rajaraman, and Joann J. Ordille. Query-answering algorithms for information agents. In *Proceedings of the Thirteenth National Conference on Artificial Intelligence*, Portland, OR, August 1996.

[Levy, 1996] Alon Y. Levy. Obtaining complete answers from incomplete databases. In *Proceedings of the 22nd International Conference on Very Large Databases*, pages 402–412, Bombay, India, 1996.

[Qian, 1996] Xiaolei Qian. Query folding. In *Proceedings of the 12th International Conference on Data Engineering*, pages 48–55, New Orleans, LA, February 1996.

[Rajaraman et al., 1995] Anand Rajaraman, Yehoshua Sagiv, and Jeffrey D. Ullman. Answering queries using templates with binding patterns. In *Proceedings of the 14th ACM SIGACT-SIGMOD-SIGART Symposium on Principles of Database Systems*, 1995.

[Sagiv and Yannakakis, 1980] Yehoshua Sagiv and Mihalis Yannakakis. Equivalence among relational expressions with the union and difference operators. *J. ACM*, 27(4):633–655, 1980.

[Shmueli, 1993] Oded Shmueli. Equivalence of datalog queries is undecidable. *Journal of Logic Programming*, 15:231–241, 1993.

[Ullman, 1989] Jeffrey D. Ullman. *Principles of Database and Knowledge-Base Systems*, volume 2. Computer Science Press, 1989.

[Yang and Larson, 1987] H. Z. Yang and P.-Å. Larson. Query transformation for PSJ-queries. In *Proceedings of the Thirteenth International Conference on Very Large Data Bases*, pages 245–254, Los Altos, CA, 1987.

# Efficiently Executing Information-Gathering Plans

Marc Friedman    Daniel S. Weld

Department of Computer Science and Engineering
University of Washington, Box 352350  Seattle WA  98195-2350  USA
{friedman, weld}@cs.washington.edu

## Abstract

We describe Razor, a planning-based information-gathering agent that assists users by automatically determining which Internet information sites are relevant to their query, accessing those sites in parallel, and integrating the results. Razor uses a disjunctive graph-based plan representation. It then uses a novel and powerful form of local completeness reasoning in order to transform those plans into contingent plans of high quality. These contingent plans can be efficiently executed, obtaining more answers at less cost than the original plans. We focus in this paper on the algorithms underlying the plan transformation process.

## 1  Introduction

Thanks to the Internet, thousands of structured information sources are available for querying, and the number and variety of these sites is growing rapidly. While a wide range of questions can be answered via the Internet, the morass of sources means that users cannot easily get the information they need. Humans face three problems when trying to gather information. First, they must determine which of the myriad sites has information relevant to their question. Second, they must learn to navigate the sites' idiosyncratic interfaces. Third, for many queries they must integrate the data returned by several different sites.

By automating this process, a software agent can greatly simplify the task of gathering information. For example, a user could ask for reviews of all movies starring Marlon Brando playing in Boston. To gather the desired information, the agent must reason about the contents and capabilities of different information sources. In this case, no single information source can answer the query, and there are several choices of how to do it. The agent might first go to the Internet Movie Database to get a list of movies starring Marlon Brando, then go to MovieLink to see which of these movies is showing in Boston, finally to Ebert to get reviews of each of the relevant movies. Because most information sources are incomplete, it is often necessary to execute more than one such plan. For example, since Ebert contains only a fraction of the movie reviews on the web, the agent can return more information by also going to Cinemachine.

### 1.1  Execution Policies

Often we find two sites with overlapping information content. In the simplest case, this occurs when mirror sites are created for a popular (and hence heavily loaded) resource. For example, the Internet Movie Database has three mirror sites (US, UK, and Italy). This overlap creates a dilemma for agent designers — on the one hand we wish to tell the agent about the existence of all three sites (what if the US and UK sites are down?), but on the other hand we don't wish the agent to visit all sites in parallel since this wastes resources and undermines the very purpose of creating mirror sites.

Furthermore, mirror sites are just the simplest case of overlap. Some sites aggregate information that may be found from other sources. For example, the SABRE airline system provides information about all United flights, but doesn't report on the flights of Southwest Airlines. We say that SABRE *subsumes* United (defined in Section 3), but not Southwest. Given that some sequence of actions A subsumes some other sequence B, there are at least three execution policies:

- *Brute force* – Just ignore subsumption, and execute everything greedily.

- *Aggressive* – Execute both alternatives in parallel, but cancels all communication with B once A has successfully returned.

- *Frugal* – Initially, run only A. When A fails (perhaps because SABRE is overloaded), begin B. If subsumption is reflexive (as is the case with mirror sites) then the source with fastest expected response is executed at first.

While an aggressive policy is more thrifty with valuable resources than a brute force one, ultimately only something like a frugal policy will scale. There are good reasons to use one of these resource-conscious strategies rather than a brute force one. Unnecessary accesses to the internet waste memory for network connections and buffering. They waste time copying data, parsing text, and doing database operations. In the future, information sources may charge a micropayment for each access as well. If enough rapacious information agents populate the internet, then servers and the internet itself will degrade. Worst of all, on a slow network (*e.g.* modem or wireless link), excess accesses may flood the connection and slow the agent.

## 1.2 Contributions

This paper describes Razor, an information-gathering agent that automatically plans how to gather information that answers a query, reasons about the value of accessing the different sources, and then executes the actions in a manner that reduces wasteful accesses to overlapping sites. Specifically, we make the following contributions:

- We define an expressive form of *local completeness* statements (*e.g.*, conditions under which an information source will return *all* relevant information, or strictly more than another source) and provide algorithms for reasoning with them. Our methods improve on previous formulations [Etzioni *et al.*, 1994; Levy, 1996] by allowing a richer set of constraints scoping these statements.

- We present an algorithm that reasons about local completeness to determine when one portion of an information-gathering plan subsumes another, in the sense of returning at least as much information. We compare portions of the graph that return sets of intermediate values, and look at how those sets overlap.

- We describe a labeling algorithm that uses subplan subsumption to generate a contingent plan which implements the aggressive (or frugal) policy.

In Section 2 we review the basic information-gathering problem. In Section 3 we look at what happens when we introduce local completeness. Section 4 describes how to use local completeness to determine subsumption relationships within a plan. Section 5 explains how to use subsumption to implement a policy, such as aggressive or frugal, in a contingent plan. The paper concludes with a discussion of Razor's implementation status, related work, and conclusions.

## 2 Background: Sites, Queries, & Plans

Razor represents knowledge about the Internet using a logical theory consisting of two types of relations: world relations (written in `courier font`), which are a part of the user's *world ontology*, and site relations (written in SMALL CAPS). World relations express relationships of interest to a user: site relations merely label the collections of data accessible on the internet.

For example, the world ontology might represent information about movies with four relations: `actor-in`($Movie, Role, Actor$), `review-of`($Movie, Review$), `year-of`($Movie, Year$), and `shows-in`($Movie, City, Theater$). When posing queries, the user need only mention these world relations; Razor shields the user from the details of the available information sources by reasoning about site descriptions.

### 2.1 Site Descriptions

We model an information source with a *site description*. The site description defines a new site predicate in terms of a conjunction of relations in the world ontology. For example, given the world ontology shown above, the source description in Equation 1 says that the Internet Movie Database takes one input (the name of a movie) and returns tuples of the form $\langle M,A \rangle$ such that the actor $A$ plays some role $Role$ in movie $M$. The inputs are returned implicitly.[1]

$$\text{IMDBCAST}(\$M, A) \Rightarrow \texttt{actor-in}(M, Role, A) \quad (1)$$

The description's *head* (*i.e.*, its left hand side) shows the type signature of the tuples returned, while its *body* (RHS) lists the world relations satisfied by the tuple elements. Site descriptions can encode the fact that an information source supports only a subset of full relational algebra. Specifically, by annotating head arguments with $, we specify that the site *requires* values for those variables as input [Rajaraman *et al.*, 1995]. Thus a call to IMDBCAST can only return the data in the table for specified movie $M$.

Formally, site descriptions have a simple semantics; each one is syntactic sugar for a first-order sentence forming part of the agent's logical theory $\Delta$. For example, the IMDBCAST definition translates into:

$$\forall M, A \;\; \text{IMDBCAST}(M, A) \Rightarrow \quad\quad\quad (2)$$
$$\exists\, Role \mid \texttt{actor-in}(M, Role, A)$$

This makes it clear that IMDBCAST doesn't report *what* role the actor plays, just that *there exists* a role. Furthermore, the implication sign $\Rightarrow$ in Equation 2, rather than $\Leftrightarrow$, indicates that most sites contain *correct* but possibly *incomplete* information. This means that querying IMDBCAST with $M$ bound to the movie The Godfather returns some subset of the tuples that satisfy the world relation `actor-in(The Godfather`, $Role, A)$; the site is *not* guaranteed to return all such actors.

Figure 1 defines the sites used throughout the paper.

### 2.2 Queries

A user queries Razor with a conjunctive, function-free (*i.e.* datalog) query. For instance, we can encode the information-gathering problem from this paper's introduction as:[2]

`query1`($M, Review$) $\Leftarrow$ `actor-in`($M, Role,$ `Brando`) $\wedge$
    `shows-in`($M,$ `Boston`, $Thtr$) $\wedge$
    `review-of`($M, Review$).

The query head indicates what information is desired (tuples of $\langle M, Review\rangle$ ). The query body is a conjunction of equality constraints that define the inputs to the query (the start state), and world relations to be satisfied by all the tuples returned (the goal). In planning terms, the inputs map to the start state and the world relations to be satisfied are the goal. The agent satisfies the query by returning as many tuples of $\langle M, Review\rangle$ as it can find that satisfy this start state and goal.

The query can be viewed as syntactic sugar for a logical sentence defining a new relation `query1`:

---

[1] Each site also has a corresponding wrapper. The *wrapper* defines interface code to connect to the information source, feed it some inputs, and parse the response into tuples. The wrapper allows Razor to treat each source as a database table.

[2] The symbol $\Leftarrow$ can be read as 'if'.

IMDBACTOR($Actor, Movie) ⇒ actor-in(Movie, Role, Actor)
EBERT($Movie, Review) ⇒ review-of(Movie, Review)
CINEMACHINE($Movie, Review) ⇒ review-of(Movie, Review)
MOVIELINK($City, Movie, Theater) ⇒ shows-in(Movie, City, Theater)
METROCINEMA($City, Movie, Theater) ⇒ shows-in(Movie, City, Theater)
IMDBCAST($Movie, Actor) ⇒ actor-in(Movie, Role, Actor)
EBERTACTOR($Actor, Movie, Year) ⇒ actor-in(Movie, Role, Actor) ∧ year-of(Movie, Year)

Figure 1: Site descriptions of a few movie sites; dollar signs denote arguments that must be bound as input.

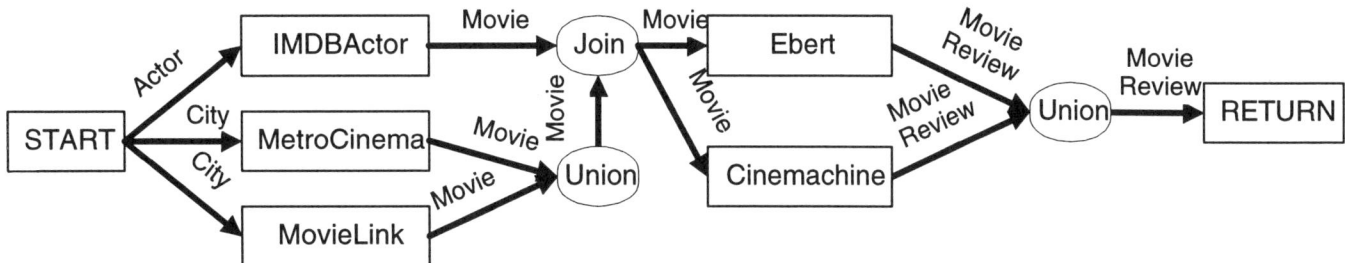

Figure 2: A solution plan to query1 using the first five sites.

$\forall M, Review$ query1$(M, Review) \Leftrightarrow$ (3)
$\exists Thtr, Role \mid$ actor-in$(M, Role, \text{Brando}) \wedge$
shows-in$(M, \text{Boston}, Thtr) \wedge$
review-of$(M, Review)$

## 2.3 Plans

A plan is a directed graph of information-gathering actions representing data flow dependencies. Specifically, a plan has three types of nodes: information-gathering operators (i.e. instances of site descriptions), database operators (JOIN, SELECT, and UNION), or the special nodes START and RETURN. The START node provides the input assignments, while RETURN node collects the answers. Edges of the plan graph are labelled with the attributes whose values propagate; tuples of those attributes flow along each directed edge, and each edge potentially performs a database projection operation on its inputs. Figure 2 shows a plan for query1.

Plans have both operational and declarative semantics. In operation, the plan is a data flow graph of queries to network information sources. Multiple tuples of data flow along edges. Whatever tuples flow into the RETURN node are collected and returned to the user as answers. In this example, METROCINEMA and MOVIELINK perform the same function, returning movies playing in a city. Since any movie returned by either is correct, we take the union. Because the same movie $M$ needs to play in the city *and* star the actor, we join the results of the UNION and IMDBACTOR, equating the movie columns. We then use the resulting values of $M$ as inputs to EBERT and CINEMACHINE. Declaratively, this graph is equivalent to the following datalog program defining predicate plan1:

plan1$(M, Review) \Leftarrow$ IMDBACTOR(Brando, $M$) $\wedge$
tempm(Boston, $M$) $\wedge$
tempr$(M, Review)$.

tempm$(City, M) \Leftarrow$ MOVIELINK$(City, M, Thtr)$.
tempm$(City, M) \Leftarrow$ METROCINEMA$(City, M, Thtr)$.
tempr$(M, Review) \Leftarrow$ EBERT$(M, Review)$.
tempr$(M, Review) \Leftarrow$ CINEMACHINE$(M, Review)$.

A datalog program captures the graph's logical content, so we will freely convert between the two forms. Note that cycles are indeed possible in the plan; this corresponds to recursion in the datalog form. A *solution plan* is just one that is executable, and semantically correct. A plan is executable if the input variables required by each step are connected to variables of the correct type along each in-edge, and those variables are available from the predecessor node. A graph is semantically correct if it returns only answers to the query (see also [Kwok and Weld, 1996; Levy, 1996]).

There are several ways to construct plans. Our first implementation used a partial-order version of the Occam planner [Kwok and Weld, 1996] to produce a (complete yet potentially infinite) stream of conjunctive plans; this stream was truncated and merged into a graph. The Razor system currently under development uses the recursive plan compilation procedure described in [Duschka and Levy, 1997] augmented with additional type inference and modified to produce acyclic graphs.

## 3 Representing Local Completeness

A single plan may contain many alternative ways to get an answer. Since most information sources are incomplete, Razor has to execute *all* of these alternatives to guarantee the maximal set of answers. However, in some cases a source may be complete, and by reasoning about this completeness, Razor can avoid redundant work. For instance, the SABRE online flight reservation system lists all United and American Airlines flights. If the user wanted a United flight and had already contacted SABRE, then there is no reason to go to

United's web page. But if the user were flexible with respect to airlines, she might try Southwest as well.

We represent these relationships *between* sites with *local completeness* axioms. This section describes our local completeness representation, and explains how they simplify and generalize those of [Etzioni *et al.*, 1994; 1997] and [Levy, 1996].

Intuitively, a *local completeness* (LC) *rule* states the conditions under which a site contains exhaustive information, *i.e.* when the site contains *all* information in the corresponding part of the world ontology. At the syntactic level, an LC rule takes the form *head* $\Leftarrow$ *body* where the head specifies a site relation and the body is a conjunctive constraint of world relations, site relations, and built-in relations like inequality. Moveover, the head is an information site, and the body must be subsumed by the body of the site description, using only the theory $\Delta$ of site descriptions (more on subsumption in section 4). To make this concrete, we present some examples of local completeness statements, in order of increasing complexity.

Suppose MOVIELINK lists all the movies playing anywhere, and their theaters. We encode this with LC as follows:

$$\text{MOVIELINK}(City, Movie, Theater) \Leftarrow \qquad (4)$$
$$\texttt{shows-in}(Movie, City, Theater).$$

This is the simplest kind of LC statement, equivalent to the local closed world declarations of [Etzioni *et al.*, 1994] — a site is complete over the entire relation in the body of its site description. Assuming the site is available, one would never need to go anywhere else but to MOVIELINK to find movies playing in some locale.

As a second example, suppose that it takes some time for a movie's statistics to be entered in IMDBCAST, but we know it is complete up until last year. We write this as:

$$\text{IMDBCAST}(M, A) \Leftarrow \texttt{actor-in}(M, R, A) \land \qquad (5)$$
$$\texttt{year-of}(M, Y) \land$$
$$(Y \leq 1996).$$

IMDBCAST's completeness is local, because it is limited to the subspace constrained by $\texttt{year-of}(M, Y) \land (Y \leq 1996)$. Note that this case can be expressed in the formulation of [Levy, 1996], but is not solvable by Levy's polynomial time algorithm because there is a world relation ($\texttt{year-of}$) in the constraint.

As a third example, note that the content of some information sources is highly *correlated*. [Levy, 1996] does not allow the expression of relationships between two sites, but we encode them as follows:

$$\text{MOVIELINK}(City, Movie, Theater) \Leftarrow \qquad (6)$$
$$\text{METROCINEMA}(City, Movie, Theater).$$

Here we said that MovieLink lists all of MetroCinema's reviews. Correlations are especially common because Razor treats each kind of information request as a separate information source, yet some requests rely on the same underlying database. For a non-movie example, note that the SABRE airline reservation system contains all flights shown on the American Airlines web page, but SABRE does not contain *all* flights, so an agent should consider Southwest and other information sources.

A more interesting case occurs when the sites include different types of data. For instance, whenever Ebert reviews a movie, he also lists the movie's cast:

$$\text{EBERTACTOR}(A, M, Y) \Leftarrow \text{EBERT}(M, Review) \land \qquad (7)$$
$$\texttt{actor-in}(M, Role, A) \land$$
$$\texttt{year-of}(M, Y).$$

In our final example, IMDBCAST has local completeness over a part of itself. We are confident that if a movie title is listed there, then all of its cast is listed, since casts of movies do not change. However, we cannot say whether a movie will be listed there at all, other than by actually checking. It is easy to encode this dependency:

$$\text{IMDBCAST}(M, A) \Leftarrow \texttt{actor-in}(M, R, A) \land \qquad (8)$$
$$\text{IMDBCAST}(M, A_2).$$

We add to our theory $\Delta$ all the local completeness axioms.

## 4 Subsumption

Consider the entire set of data that flows along each edge into a UNION node throughout plan execution. Ideally, we would like to infer set cover relationships among the edges leading into each UNION. Then we would know what actions we could defer without jeopardizing completeness.

The reasoning tool at our disposal is *subsumption*, which is essentially set inclusion. Razor considers the edges into a UNION node pairwise, and tries to discover pairwise set cover relationships (*i.e.*, subsumption relationships) deducible from the local completeness statements in $\Delta$. The rest of this section elaborates this idea, which is given as an algorithm in Figure 3.

### 4.1 Alternatives

Razor tests for subsumption on the portions of the graph leading into the same UNION node. More specifically, they should produce the same type of outputs from the same inputs. We call such subplans *alternatives*, and we characterize their graph properties as follows. Consider UNION node $U$, with at least two predecessor nodes called $M$ and $N$. Subgraphs $G_M$ and $G_N$ are alternatives whenever:

- $M$ is the sink node of $G_M$ and $N$ is the sink node of $G_N$.
- All nodes in $G_M$ are connected to $M$ by a path entirely within $G_M$. Same for $G_N$.
- Predecessors($G_M$) = Predecessors($G_N$), where the predecessors of $G$ are the predecessor nodes of the nodes in $G$ which are not themselves in $G$.

These graph properties roughly realize the natural notion of alternativeness. Function FindSubsumingAlternatives looks at all pairs of disjoint alternatives in $O(|P|^3)$ operations and $O(|P|^2)$ calls to Subsumes*.

### 4.2 Subsumption over Alternatives

Function Subsumes* is a subgraph subsumption reasoning algorithm based on local completeness.[3] We use the

---
[3] An infinite depth limit makes it complete in the absence of built-in functions.

```
FindSubsumingAlternatives(acyclic plan P)
  Each node has one marker per possible sink node
  Reset all markers on all nodes
  Foreach union node U
    Foreach immediate predecessor node R of U
      Traverse the links backwards from R
      marking all nodes reached with tag R.
      (All other nodes have implicit tag ¬R)
    Foreach pair ⟨M,N⟩ of predecessors of U
      A_M = { nodes marked (M ∧ ¬ N)}
      A_N = { nodes marked (N ∧ ¬ M)}
      If Subsumes*(A_M,A_N) Then Found(A_M,A_N,U)
      If Subsumes*(A_N,A_M) Then Found(A_N,A_M,U)
Subsumes*(subgraph G, subgraph G')
  If G contains a union node
    Construct G_1 ... G_n |
      G_i has the ith alternative in place of the union
    Return ⋁_i Subsumes*(G_i,G')
  Else If G' contains a union node
    Construct G'_1 ... G'_n |
      G'_i has the ith alternative in place of the union
    Return ⋀_i Subsumes*(G,G'_i)
  Else Return Subsumes(G,G')
Subsumes(conjunctive subgraphs G, G')
  Return Contains*(G,G' ∧ Expansion(G'))
Contains*(conjunction C, conjunction C')
  If Contains(C, C') then return true
  Else Choose a site relation T in C
    Choose an LC-rule R with head matching T and mgu σ
    Let E = RemoveConjunct(C,T) ∧ Body(R)|_σ
    If Contains*(E,C') return true
```

Figure 3: Pseudocode for subsumption detection. Expansion takes a conjunction of actions and replaces each action with the instantiated rule body from its site description. Contains searches for a containment mapping from $C$ to $C'$[Ullman, 1989, p 881]; if found this proves that the extention of $C$ contains that of $C'$. Choose marks a nondeterministic choice; backtracking must be used to consider all possibilities for completeness.

fact here that acyclic graphs are equivalent to logical recursive rules with database operations replaced by conjunction and disjunction. We say that subgraph $G$ *subsumes* subgraph $G'$ if from the site definitions and LC statements, $\Delta$, one can deduce that $G'$ is *contained* in $G$. At the heart of the algorithm is a containment mapping routine for conjunctions [Ullman, 1989, p 881]. If $G$ contains $G'$, executing $G$ will gather at least the tuples gathered by $G'$.[4]

Looking again at our running example, we would compare the two subgraphs (call them $G$ and $G'$) consisting of MOVIELINK and METROCINEMA respectively. The two graphs have the same input $City$, and their other variables are renamed away:

$$G = \text{MovieLink}(City, M, Theater) \qquad (9)$$
$$G' = \text{MetroCinema}(City, M', Theater') \qquad (10)$$

---
[4]In terms of formal logic, Subsumes* is just searching for a proof that for some substitution $\sigma$, $\Delta \models G'|_\sigma \Rightarrow G|_\sigma$. Contains* does backward-chaining search through $\Delta$. Contains does unification.

Since MOVIELINK contains all possible movies and theaters (Equation 4), Subsumes* is able to determine that $G$ subsumes $G'$. Let's walk through the process. Since neither graph contains a union node, Subsumes* quickly calls Contains*$(G, G')$. This causes a call to Contains with $C$ set to the conjunction of $G'$ and the expansion of its single conjunct.

$$\begin{aligned}C' = &\text{MetroCinema}(City, M', Theater') \land \\ &\texttt{shows-in}(M', City, Theater')\end{aligned} \qquad (11)$$

$C$ is just $G$. There is no containment mapping from $C$ to $C'$ so Contains returns false. At this point Contains chooses a conjunct and an LC-rule from $\Delta$ that will let it prove the implication. Let $T$ perforce be MOVIELINK$(City, M, Theater)$, and suppose then that $R$ is Equation 4. $E$ becomes

$$E = \texttt{shows-in}(M', City, Theater') \qquad (12)$$

The next (recursive) call to Contains* has $C'$ unchanged (Equation 11), but $C$ is now set to the previous value of $E$ (Equation 12). Now there *is* a containment mapping: setting $M = M'$ and $Theater = Theater'$ makes $C$ match the last conjunct of $C'$. Contains* returns true.

This algorithm works because containment is transitive. We build a chain connecting $G$ to $G'$. $C'$ is $G' \land \text{Expansion}(G')$, which contains $G'$ by construction. Similarly, $C$ contains $E$ on each recursive call. So whenever $E$ contains $C'$, the chain is complete.

In the worst case Subsumes* takes time exponential in three quantities:

- the number of UNION nodes in series in $G$ and $G'$ (breaking loops),
- the depth limit on recursion, and
- the number of repeated predicates in the containment mapping.

However in practice, these quantities are all very small integers. Furthermore, preliminary experiments suggest that subsumption reasoning actually takes much less time than actually execution of the subsumed information-gathering actions, even given a fast network connection.

## 5 Contingent Planning

Razor uses the subsumption relationships it finds to turn the original plan into a more resource-conscious one.

Instead of executing all possible alternative subgraphs in a plan, Razor determines which ones can be skipped without losing tuples. However, we cannot simply prune away subsumed subgraphs of the plan, since that would eliminate the possibility of recovering from unavailable source failures. For example, if the SABRE site is down, Razor should contact United airlines, but this would be impossible if we have removed all alternatives to SABRE from the graph. Our approach is to generate a *contingent plan* by annotating the graph with *guards*. This allows Razor to execute either an *aggressive* or *frugal* policy.

The contingent plan differs from the original plan in that a node may fire only when its guard is true. Guards are logical formulae that refer to the *status variables* of other nodes. There is one status variable per node, with the possible values SLEEPING, RUNNING, FAILED, and DONE. Guards make the execution of a node contingent on the execution status of other parts of the graph. Whatever policy is chosen, the executor's task is then simply to:

- fire RUNNING nodes when their inputs are available,
- complete RUNNING nodes when their data arrives,
- update the status variables accordingly, and
- awaken or put to sleep nodes whose guards changed value.

## 5.1 Generating the Guards

In our example, using the aggressive policy, the only change between the original plan and the contingent plan is one guard on the METROCINEMA node, causing it to sleep if the MOVIELINK call is DONE. Here is a close-up of that part of the contingent plan:

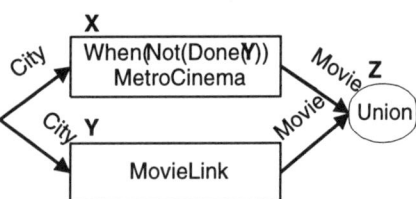

In fact, the aggressive execution policy only tests when sink nodes are DONE. No guards start out false; only when a subsuming alternative becomes DONE do some computations get put to sleep—those completely subsumed by alternatives that are DONE. This is conservative, in the sense that it will not lose any answers that the brute force method will find, and its use of internet resources is at least as good. The algorithm in Figure 4 implements the aggressive policy. It labels the nodes in subsumed alternatives $A_N$, from the list of subsumption relationships found by FindSubsumingAlternatives. Note that the guard of a node $P$ is always *true* unless $P$ is subsumed by some DONE alternative for every UNION node that uses $P$. Otherwise, the guard is the negated conjunction of the conditions on each UNION making $P$ redundant. The guard says, in effect, that if at least one of the subsuming alternatives on every path from $P$ is DONE, then $P$ is unnecessary.

The frugal execution policy (not detailed here) has a similar structure, with some extra machinery. The guard on node X will be simply `failed(Y)` rather than `not(done(Y))`. In addition the frugal policy needs an extra module to prevent deadlock. This module has to detect and break cycles in the subsumption relation between nodes, by using estimates of utility (*e.g.*, expected speed or micropayment cost) to prevent alternatives of higher utility from waiting for alternatives of lower utility.

## 6  Status & Future Work

We have fully implemented one information-gathering agent and are now constructing its successor Razor. The first, "Razor96" is written in Common Lisp and uses the Occam planner [Kwok and Weld, 1996] as its planning

```
GenerateGuards()
  Declare clause[all nodes,union nodes]
  Initialize all clauses to false
  Foreach ⟨A_M, A_N, U⟩ we found
    Foreach node P in A_N
      Add status(M, DONE) to clause[P,U]
  Foreach node P
    S_P = { union nodes reachable from P without passing
            through any other union nodes }
    P.guard = ¬ ⋀_{U ∈ S_P} (clause[P,U])
```

Figure 4: Pseudocode for aggressive guard generation.

module. Since Occam produces a stream of conjunctive plans, the subsumption module just compares pairs of conjunctive plans. Subsumes* extends and improves this algorithm. Furthermore, Razor96 does not use guards; rather, it prunes any subsumed conjunctive plans. It is therefore not insured against information source failures. This pruning algorithm prunes over 99% of the plans Occam produces on various movie-related queries. We found that most plans were pruned on their own (lack of) merit; either because they contain unnecessary actions (nonminimality), or are found redundant without using the recursive clause of Contains*. Using equation 4 further reduced the number of plans by a factor of 2.67. The entire subsumption and pruning process increased planning time only 53%.

The frailty of our original Lisp implementation, combined with an improved understanding of the ideal system architecture, led us to design the Razor system described in this paper. Our new C++ implementation uses recursive plans [Duschka and Levy, 1997] to construct the plan graph. Recursive plans are already far more compact than Occam plans, so we anticipate that the improvement in graph *size* from labelling will be far less drastic than in our previous implementation, however we still expect dramatic improvement over brute force execution. We are currently porting the subsumption and execution systems, and coding guard-generation. Razor will learn utility estimates for sites, and implement the frugal policy. Further down the road we will look at *optimal* strategies for maximizing the answers returned by a deadline.

## 7  Related Work

Razor draws together work in databases and AI planning. Several implemented systems (*e.g.*, [Chawathe *et al.*, 1994; Adali *et al.*, 1996; Daruwala *et al.*, 1995]) integrate multiple database systems using hand-tailored query plans. They focus on orthogonal problems, such as expressing ontologies and converting between them. The Internet Softbot [Etzioni and Weld, 1994] applies AI planning techniques to the problem of accessing online information as well as update. The SIMS database integration system [Arens *et al.*, 1996] applies AI planning to the problem of relational database integration with a rich hierarchical type system. The Information Manifold (IM) [Levy *et al.*, 1996a] is similar to SIMS and Razor. Like SIMS, IM handles object type hierarchies. Like Razor, IM is geared toward users of the web. This incarnation of the IM system is incapable of handling information sources other than full relational databases. [Levy et al., 1996b] describes an apparently later version of the IM which allows bound variables in site descriptions. This version uses an incomplete algorithm, however, for generating bounded-length information-gathering plans. For example, the IM can't find plans that require more steps than there are goal conjuncts; see [Rajaraman et al., 1995] for examples.

[Etzioni et al., 1994] introduced the notion of local closed world and presented fast algorithms for inference and update. [Levy, 1996] generalizes Etzioni et al.'s local completeness framework and presents a reduction of the answer-completeness problem to the query-independence problem. We further extend Levy's representation and solve the extended problem directly. Although there is a progression in expressiveness from the Internet Softbot to the IM and Razor only the Internet Softbot handles the case of update since its actions can affect the information sources.

Several planners in the AI literature represent contingencies within the plan [Peot and Smith, 1992; Draper et al., 1994; Pryor and Collins, 1996]. However, the absence of causal (state-changing) actions in an information-gathering context leads to significant differences. In a sense, Razor's non-contingent plan corresponds to a *universal plan* [Schoppers, 1987] and we introduce contingencies only for efficiency's sake.

## 8 Conclusions

Razor is an information-gathering agent that plans how to answer user queries, reasons about the redundancy of accessing the different sources, and then executes the actions in a manner that attempts to get the maximum information to the user without useless work. Razor uses a library of expressive *local completeness* declarations in order to determine which alternative subplans obtain similar information. Razor then constructs a *contingent plan* that encodes an efficient execution policy which is resilient in the face of information-source failure.

## 9 Acknowledgements

This research would not have been possible without the efforts of Chung Kwok, whose previous research laid the foundations for ours. He also contributed his time and code. We also thank Bob Doorenbos, Oren Etzioni, Keith Golden, Nick Kushmerick, Marc Langheinrich, Neal Lesh, Alon Levy, Kelly Murray, Dan Shiovitz, and David Smith, each of whom helped in some invaluable way. This research was funded by Office of Naval Research Grant N00014-94-1-0060, by National Science Foundation Grant IRI-9303461, by ARPA / Rome Labs grant F30602-95-1-0024, by a National Science Foundation Graduate Fellowship, and by a gift from Rockwell International Palo Alto Research Lab.

## References

[Adali et al., 1996] S. Adali, K. Candan, Y. Papakonstantinou, and V. Subrahmanian. Query caching and optimization in distributed mediator systems. In *Procceedings of SIGMOD-96*, 1996.

[Arens et al., 1996] Y. Arens, C. Knoblock, C. Chee, and C. Hsu. SIMS: Single interface to multiple sources. TR RL-TR-96-118, USC Rome Labs, 1996.

[Chawathe et al., 1994] S. Chawathe, H. Garcia-Molina, J. Hammer, K. Ireland, Y. Papakonstantinou, J. Ullman, and J. Widom. The TSIMMIS project: Integration of heterogeneous information sources. In *Proc. IPSJ Conf*, 1994.

[Daruwala et al., 1995] A. Daruwala, C. H. Goh, S. Hofmeister, K. Hussein, S. Madnick, and M. Siegel. The context interchange network. In *IFIP WG2.6 Sixth Working Conference on Database Semantics*, 1995.

[Draper et al., 1994] D. Draper, S. Hanks, and D. Weld. Probabilistic planning with information gathering and contingent execution. In *Proc. 2nd Intl. Conf. on AI Planning Systems*, June 1994.

[Duschka and Levy, 1997] O. Duschka and A. Levy. Recursive plans for information gathering. In *Proc. 15th Int. Joint Conf. on AI*, 1997.

[Etzioni and Weld, 1994] O. Etzioni and D. Weld. A softbot-based interface to the Internet. *C. ACM*, 37(7):72–6, 1994.

[Etzioni et al., 1994] Oren Etzioni, Keith Golden, and Dan Weld. Tractable closed-world reasoning with updates. In *Proc. 4th Int. Conf. on Principles of Knowledge Representation and Reasoning*, pages 178–189, 1994.

[Etzioni et al., 1997] O. Etzioni, K. Golden, and D. Weld. Sound and efficient closed-world reasoning for planning. *Artificial Intelligence*, 89(1–2):113–148, January 1997.

[Kwok and Weld, 1996] C. Kwok and D. Weld. Planning to gather information. In *Proc. 13th Nat. Conf. on AI*, 1996.

[Levy et al., 1996a] A. Levy, A. Rajaraman, and J. Ordille. Query-answering algorithms for information agents. In *Proc. 13th Nat. Conf. on AI*, 1996.

[Levy et al., 1996b] A. Levy, A. Rajaraman, and J. Ordille. Querying heterogeneous information sources using source descriptions. In *Proceedings of the 22nd VLDB Conference*, 1996.

[Levy, 1996] A. Levy. Obtaining complete answers from incomplete databases. In *Proceedings of the 22nd VLDB Conference*, 1996.

[Peot and Smith, 1992] M. Peot and D. Smith. Conditional Nonlinear Planning. In *Proc. 1st Intl. Conf. on AI Planning Systems*, pages 189–197, June 1992.

[Pryor and Collins, 1996] L. Pryor and G. Collins. Planning for contingencies: A decision-based approach. *J. Artificial Intelligence Research*, 1996.

[Rajaraman et al., 1995] A. Rajaraman, Y. Sagiv, and J. Ullman. Answering queries using templates with binding patterns. In *Proc. ACM Symp on Principles of Database Systems*, 1995.

[Schoppers, 1987] M. Schoppers. Universal plans for reactive robots in unpredictable environments. In *Proc. 10th Int. Joint Conf. on AI*, pages 1039–1046, August 1987.

[Ullman, 1989] J. Ullman. Database and knowledge-base systems. In *Database and knowledge-base systems*. Computer Science Press, 1989.

# Author Index

Adamson, Kenneth 226, 252
Agostini, Alessandro 1539
Akaho, Shotaro 880
Aleven, Vincent 335
Amati, Gianni 764
Anbulagan 366
Arima, Jun 210
Armstrong, Aaron 620
Asada, Minoru 24
Ashley, Kevin D. 335
Asker, Lars 860
Asoh, Hideki 880
Atkins, Ella M. 1267

Badea, Liviu 142
Bajcsy, Ruzena 1360
Balbiani, Philippe 190
Baral, Chitta 1441
Barette, Olivier 405
Baudin, Catherine 751
Baumgartner, Peter 460
Bell, David A. 226, 252
Benjamins, V. Richard 1246
Besnard, Philippe 150
Bessière, Christian 398, 412
Beveridge, J. Ross 1476
Bhatta, Sambasiva R. 294
Bhuiyan, Md. Shoaib 1496
Bibel, W. 1549
Binder, John 1292
Bistarelli, Stefano 352
Biswas, Gautam 568
Bittencourt, Guilherme 310
Biundo, Susanne 1240
Bjäreland, Marcus 1420, 1447
Blanchon, Hervé 1042
Blum, Christian 1072
Bobrow, Daniel G. 1511
Boden, Margaret A. 1563
Boufkhad, Yacine 122
Boutilier, Craig 1156
Boyd, Sarah 1540
Brafman, Ronen I. 1156
Braspenning, Peter J. 1372
Bratko, Ivan 914
Bresina, John L. 1260
Bretier, P. 1030
Brown, Ralf D. 708
Bundy, Alan 528
Burgard, Wolfram 1346

Callaway, Charles B. 952
Carbonell, Jaime G. 708
Carmel, David 606
Castelfranchi, Cristiano 1567
Castell, Thierry 372
Cayrol, Michel 372
Chen, Johnny 832
Chen, Shifu 89, 95
Chen, Zhaoqian 89
Choe, Yoonsuck 1120
Choi, Key-Sun 972
Christensen, Heneik I. 1360
Chung, Jae-Moon 1482
Collier, Nigel 1106
Coradeschi, Silvia 24
Costello, Tom 1426
Crestani, Fabio 764
Cristani, Matteo 1460
Crnogorac, Lobel 647

Dai, Honghua 1304
Dale, Michael 434
Davidsson, Paul 840
Dayan, Peter 1114
Dean, Thomas 1163
Debruyne, Romuald 412
Dechter, Rina 1297
Decker, Keith 578
Défourneaux, Gilles 216
De Giacomo, Guiseppe 1221
de Jong, H. 486
Delgrande, James P. 168
del Val, Alvaro 508
Demri, Stéphane 190
Denecke, Matthias 1036
Denzinger, Jörg 102
de Santis, Peter 758
Deville, Yves 405
Dickmanns, Ernst D. 1577
Dimopoulos, Yannis 900
d'Inverno, Mark 600
Doi, Shinichi 977
Donini, Francesco M. 136
Doorenbos, Robert 729
Drakengren, Thomas 1447, 1466
Druschel, Peter 832
Dubitzky, Werner 226, 252
Dubois, Didier 62
Dudek, Gregory 1340

Durfee, Edmund H. 612, 620, 1267
Duschka, Olivier M. 778
Dymetman, Marc 1002
Džeroski, Saso 900

Ernst, Michael D. 1169
Estlin, Tara A. 1227
Etzioni, Oren 16

Fürnkranz, Johannes 852
Faltings, Boi V. 392, 418
Fan Low, Kon 745
Fariñas del Cerro, Luis 62
Farquhar, Adam 494
Fehrer, Detlef 959
Feiner, Steven K. 1056
Ferret, Olivier 280
Fiedler, Armin 965
Fikes, Richard 494
Finin, Tim 584
Fisher, Michael 316
Foo, Norman Y. 82, 549
Fox, Dieter 1346
Fröhlich, Peter 460, 466
Frank, Jeremy 384
Frasconi, Paolo 1066
Frederking, Robert E. 708
Freed, Michael 322
Freitag, Dayne 770
Friedman, Marc 785
Friedman, Nir 10
Fuchs, Marc 102
Fuchs, Matthias 102
Furbach, Ulrich 460
Furuse, Osamu 983

Ganascia, Jean-Gabriel 698
Gardent, Claire 1016
Gat, Erann 1234
Geffner, Hector 555
Geib, Christopher 1156
Gelle, Esther 392
Geller, James 55
Geng, Yibing 708
Gent, Ian P. 1396
Gerevini, Alfonso 1460
Givan, Robert 1163
Goel, Ashok K. 294, 562
Goh, Han Chong 942
Goldszmidt, Moises 10

Gori, Marco 1066
Grégoire, Éric 122
Grau, Brigitte 280
Graves, Christopher R. 1476
Green, Ian 528
Griffith, Anna 668
Grudic, Gregory Z. 804
Grupen, Roderic A. 1366

Hahn, Peter 1072
Hahn, Udo 996, 1010
Hara, Isao 880
Haruno, Masahiko 894
Hawkins, Dale K. 1353
Hayamizu, Satoru 880
Hayes, Caroline Clarke 288
Heckerman, David 10
Heizmann, J. 1525
Hendler, James 1246
Herbrich, Ralf 798
Herzig, Andreas 62
Higuchi, Tetsuya 1146
Hiraki, Kazuo 301
Ho, Chao Yang 745
Ho, Tu Bao 258
Horacek, Helmut 959
Howard, Michael D. 1186
Hsu, Wynne 930
Huang, Timothy 1276
Huang, Xiaorong 965
Huber, Manfred 1366
Huffman, Scott B. 751
Hughes, John G. 226, 252
Huhns, Michael N. 1511
Hustadt, Ullrich 202

Iida, Hitoshi 983
Ikeda, Mitsuru 342, 501
Inaba, Masayuki 1593
Inoue, Katsumi 176
Ishida, Yoshiteru 1084
Ishiguro, Hiroshi 36
Ishii, Naohiro 1496
Ito, Takayuki 592
Itoh, Hidenori 1050
Iwahori, Yuji 1496
Iwasaki, Yumi 494

Jamnik, Mateja 528, 1541
Joachims, Thorsten 770
Jonsson, Peter 1466

Joshi, Aravind K. 1651
Junghanns, Andreas 692
Junker, Ulrich 162

Kaelbling, Leslie Pack 920
Kaindl, Hermann 662
Kajitani, Isamu 1146
Kakas, Antonis 900
Kambhampati, Subbarao 44
Kamei, Shin-ichiro 977
Kameya, Yoshitaka 1330
Kanade, Takeo 1488
Karlsson, Lars 1420, 1434
Kashino, Kunio 1126
Kato, Wataru 1496
Katayama, Norio 738
Kautz, Henry 50
Kawabata, Takeshi 30
Keane, Mark 232
Keesing, Ron 1234
Kindo, Toshiki 716
King, Margaret 1511
King, R. D. 4
Kingston, John 668
Kinley, Andrew 246
Kitamura, Yoshinobu 501
Kitano, Hiroaki 24, 1511
Kobayashi, Masashi 268
Kohlhase, Michael 1016
Kohri, Takayuki 820
Kojima, Takuya 684
Koller, Daphne 1316
Korb, Kevin 1304
Kramer, Stefan 662
Ku, Liang-Ping 930
Kudoh, Yoshimitsu 176
Kurumatani, Koichi 542
Kushmerick, Nicholas 729

Labrou, Yannis 584
Lam, Wai 745
Lamperti, Gianfranco 446
Large, Edward W. 1360
Larrosa, Javier 424
Lawrence, Peter D. 804
Leake, David B. 246
Leckie, Christopher 434
Lee, Danny 708
Lee, Ho Soo 1402
Lelouche, Ruddy 329
Lesh, Neal 1208
Lespérance, Yves 1221
Lesser, Victor R. 640
Lester, James C. 952

Levesque, Hector J. 1221
Levy, Alon Y. 778
Li, Chu Min 366
Li, Hongbing 95
Li, Jun-Jie 972
Liberatore, Paolo 68
Lin, Fangzhen 1412
Ling, Charles X. 874
Liu, Bing 930
Lobo, Jorge 1441
Loe, Kia-Fock 1134
Loiez, Etienne 474
Long, Derek 1022
Lopes, Philippe 252
Lowe, Will 1092
Luck, Michael 600
Lukaszewicz, Witold 1215
Lydiard, Terri 668
Luong, Chi Mai 258

Maclin, Richard 860
Madalińska-Bugaj, Ewa 1215
Marin, Roque 1454
Markert, Katja 1010
Markovitch, Shaul 606
Marquis, Pierre 122
Matsubara, Hitoshi 24
Matsubayashi, Kei 820
Matsui, Toshihiro 880
Maynard-Reid II, Pedrito 508
Mazure, Bertrand 122
McAllester, David 50
McCarragher, Brenan J. 514
McCreath, Eric 1310
McKeown, Kathleen R. 1607
McVey, Charles B. 1267
Meseguer, Pedro 1382
Meyer, Christophe 698
Miikkulainen, Risto 1120
Milios, Evangelos E. 1340
Millstein, Todd D. 1169
Mima, Hideki 983
Mitchell, Tom 770
Miura, Jun 1194
Mizoguchi, Riichiro 342, 501
Montanari, A. 196
Montanari, Ugo 352
Mooney, Raymond J. 1227
Morgenstern, Leora 655, 1613
Morimoto, Tetsuro 716

Morin, Jean-François 329
Morris, Robert A. 1260
Mosterman, Pieter J. 568
Motoda, Hiroshi 810, 1622
Motomura, Yoichi 880
Mouaddib, Abdel-Illah 1201
Muggleton, S. H. 4
Munos, Rémi 826
Murakawa, Masahiro 1146
Muraki, Kazunori 977
Murase, Hiroshi 1126
Murphy, Kevin 1292
Murphy, Robin R. 1353
Muscettola, Nicola 1234

Nagano, Saburo 684
Nagendra Prasad, M. V. 640
Nakamura, Mari 542
Nakamura, Tsuyoshi 1050
Nakamura, Yuichi 1488
Nakano, Ryohei 1078
Nakatani, Tomohiro 30
Nardi, Daniele 136
Navarrete, Isabel 1454
Nayak, Abhaya 549
Nayak, P. Pandurang 1178
Nebel, Bernhard 522
Nejdl, Wolfgang 460, 466
Neveu, Bertrand 358
Noda, Itsuki 24
Nomura, Tatsuya 936
Numao, Masayuki 268, 274
Nunes de Barros, Leliane 1246

Ohnishi, Noboru 1482
Ohsawa, Yukio 722
Okamoto, Seishi 238
Okuno, Hiroshi G. 30
Osawa, Eiichi 24

Padgham, Lin 1531
Pagnucco, Maurice 549
Panaget, F. 1030
Parzen, Michael I. 288
Pell, Barney 1234
Peltier, Nicolas 216
Penders, Jacques 1372
Peppas, Pavlos 549
Perkowitz, Mike 16
Pfeffer, Avi 1316
Phillips, Steven 301
Plaisted, David A. 108
Pogliano, Paolo 446

Policriti, A. 196
Poole, David 1284
Postma, Eric 1140
Prade, Henri 62

Régin, Jean-Charles 398
Rahmel, Jürgen 1072
Ramamohanarao, Kotagiri 647
Rao, Anand S. 647
Reed, Chris 1022
Reiter, Ray 1511
Rekleitis, Ioannis M. 1340
Remington, Roger 322
Renz, Jochen 522
Ricaud, Patrick 678
Rice, James 494
Rodgers, Stuart M. 1260
Rosati, Riccardo 136
Rossi, Francesca 352
Rouveirol, Céline 888
Rus, Daniela 758
Russell, Stuart 10, 1276, 1292

Saïs, Lakhdar 122
Sadek, M. D. 1030
Saito, Kazumi 1078
Sakaniwa, Katsuyuki 268
Sammut, C. 908
Sandholm, Tuomas 632
Sasaki, Yutaka 894
Sashima, Akio 301
Sato, Taisuke 1330
Satoh, Ken 182
Satoh, Kenji 977
Satoh, Shin'ichi 1488
Schaeffer, Jonathan 692
Schaerf, Andrea 1254
Schaub, Torsten H. 150, 168
Scheffer, Tobias 798
Schlee, Stefan 662
Schmidt, Renate A. 202
Schoppers, Marcel J. 1353
Schuster, Alfons 226, 252
Sebag, Michèle 888
Seki, Hirohisa 1050
Selman, Bart 50
Seta, Kazuhisa 342
Sgouros, M. Nikitas 990
Shang, Yi 378
Sharma, Arun 1310
Shatkay, Hagit 920
Shimura, Masamichi 274

Shin, Kang G. 1267
Shintani, Toramatsu 592
Shirai, Yoshiaki 1194
Shiraz, G. M. 908
Shoham, Yoav 508, 626, 1324
Sikka, Pavan 514
Singh, Moninder 655
Singh, Munindar P. 1511
Sison, Raymond 274, 1542
Smith, Ben 1234
So, Young-pa 612
Song, Hongjun 1543
Sperduti, Alessandro 1066
Srinivasan, A. 4
Staab, Steffen 996
Steels, Luc 1632
Steinborn, Jim 1476
Stephan, Werner 1240
Sternberg, M. J. E. 4
Stone, Peter 24
Straccia, Umberto 128
Stroulia, Eleni 562
Struss, Peter 480
Stumptner, Markus 440
Subramanian, Devika 832
Šuc, Dorian 914

Sugimoto, Masanori 738
Sycara, Katia 578

Taillibert, Patrick 474
Takasu, Atsuhiro 738
Tambe, Milind 24
Tamura, Shinko 977
Tanaka, Katsumi 626
Taylor, Guy 1531
Teller, Astro 1544
ten Teije, Annette 454
Teow, Loo-Nin 1134
Thrun, Sebastian 1346
Ting, Kai Ming 866
Tokoro, Mario 820
Trombettoni, Gilles 358
Trumbo, Mark 1402
Tsukimoto, Hiroshi 1098

Ubaldini, Flavio 764
Ueda, Kazuhiro 684

van den Bosch, Antal 1140
van den Herik, H. Jaap 140
van Harmelen, Frank 454
Van Hentenryck, Pascal 405, 1642
van Raalte, F. 486

Veale, Tony 232
Veloso, Manuela 24
Virili, Francesco 1545
Voronkov, Andrei 114

Wah, Benjamin W. 378
Wakaki, Toshiko 182
Wallace, Chris 1304
Walsh, Toby 1388, 1396
Wang, Handong 874
Wang, Hongxue 1546
Wang, Ke 942
Wang, Lisong 1050
Wang, Minkai 1050
Wang, Xianchang 156
Washio, Takashi 810
Watanabe, Taisuke 716
Webb, Geoffrey I. 846
Weigel, Rainer 418
Weijters, Ton 1140
Weld, Daniel S. 729, 785, 1169
White, John A. 252
Williams, Brian C. 1178
Williams, Mary-Anne 74
Williamson, Mike 578
Wilson, David 246
Witten, Ian H. 866

Woodham, Robert J. 1496
Wooldridge, Michael 600
Wotawa, Franz 440
Wu, Xindong 1304

Xu, Gang 1502

Yachida, Masahiko 722
Yamabana, Kiyoshi 977
Yang, Yiming 708
Ygge, Fredrik 632
Yip, Kenneth 534
Yoshida, Hideyuki 716
Yoshida, Kenichi 1622
Yoshizawa, Shuji 1146
You, Jia-Huai 156
Yuan, Li Yan 156
Yugami, Nobuhiro 238

Zelinksy, A. 1525
Zemel, Richard S. 1114
Zhang, Dongmo 89, 95
Zhang, Yan 82, 549
Zhang, Zhengyou 1502
Zhou, Michelle X. 1056
Zhu, Wujia 89, 95
Zhu, Yunshan 108
Zilberstein, Shlomo 1201
Zucker, Jean-Daniel 698